COLLOIDAL SILICA

Fundamentals and Applications

SURFACTANT SCIENCE SERIES

1. Nonionic Surfactants, *edited by Martin J. Schick* (see also Volumes 19, 23, and 60)
2. Solvent Properties of Surfactant Solutions, *edited by Kozo Shinoda* (see Volume 55)
3. Surfactant Biodegradation, *R. D. Swisher* (see Volume 18)
4. Cationic Surfactants, *edited by Eric Jungermann* (see also Volumes 34, 37, and 53)
5. Detergency: Theory and Test Methods (in three parts), *edited by W. G. Cutler and R. C. Davis* (see also Volume 20)
6. Emulsions and Emulsion Technology (in three parts), *edited by Kenneth J. Lissant*
7. Anionic Surfactants (in two parts), *edited by Warner M. Linfield* (see Volume 56)
8. Anionic Surfactants: Chemical Analysis, *edited by John Cross*
9. Stabilization of Colloidal Dispersions by Polymer Adsorption, *Tatsuo Sato and Richard Ruch*
10. Anionic Surfactants: Biochemistry, Toxicology, Dermatology, *edited by Christian Gloxhuber* (see Volume 43)
11. Anionic Surfactants: Physical Chemistry of Surfactant Action, *edited by E. H. Lucassen-Reynders*
12. Amphoteric Surfactants, *edited by B. R. Bluestein and Clifford L. Hilton* (see Volume 59)
13. Demulsification: Industrial Applications, *Kenneth J. Lissant*
14. Surfactants in Textile Processing, *Arved Datyner*
15. Electrical Phenomena at Interfaces: Fundamentals, Measurements, and Applications, *edited by Ayao Kitahara and Akira Watanabe*
16. Surfactants in Cosmetics, edited by Martin M. Rieger (see Volume 68)
17. Interfacial Phenomena: Equilibrium and Dynamic Effects, *Clarence A. Miller and P. Neogi*
18. Surfactant Biodegradation: Second Edition, Revised and Expanded, *R. D. Swisher*
19. Nonionic Surfactants: Chemical Analysis, *edited by John Cross*
20. Detergency: Theory and Technology, *edited by W. Gale Cutler and Erik Kissa*
21. Interfacial Phenomena in Apolar Media, *edited by Hans-Friedrich Eicke and Geoffrey D. Parfitt*
22. Surfactant Solutions: New Methods of Investigation, *edited by Raoul Zana*
23. Nonionic Surfactants: Physical Chemistry, *edited by Martin J. Schick*
24. Microemulsion Systems, *edited by Henri L. Rosano and Marc Clausse*
25. Biosurfactants and Biotechnology, *edited by Naim Kosaric, W. L. Cairns, and Neil C. C. Gray*
26. Surfactants in Emerging Technologies, *edited by Milton J. Rosen*
27. Reagents in Mineral Technology, *edited by P. Somasundaran and Brij M. Moudgil*
28. Surfactants in Chemical/Process Engineering, *edited by Darsh T. Wasan, Martin E. Ginn, and Dinesh O. Shah*
29. Thin Liquid Films, *edited by I. B. Ivanov*
30. Microemulsions and Related Systems: Formulation, Solvency, and Physical Properties, *edited by Maurice Bourrel and Robert S. Schechter*
31. Crystallization and Polymorphism of Fats and Fatty Acids, *edited by Nissim Garti and Kiyotaka Sato*
32. Interfacial Phenomena in Coal Technology, *edited by Gregory D. Botsaris and Yuli M. Glazman*
33. Surfactant-Based Separation Processes, *edited by John F. Scamehorn and Jeffrey H. Harwell*
34. Cationic Surfactants: Organic Chemistry, *edited by James M. Richmond*

35. Alkylene Oxides and Their Polymers, *F. E. Bailey, Jr., and Joseph V. Koleske*
36. Interfacial Phenomena in Petroleum Recovery, *edited by Norman R. Morrow*
37. Cationic Surfactants: Physical Chemistry, *edited by Donn N. Rubingh and Paul M. Holland*
38. Kinetics and Catalysis in Microheterogeneous Systems, *edited by M. Grätzel and K. Kalyanasundaram*
39. Interfacial Phenomena in Biological Systems, *edited by Max Bender*
40. Analysis of Surfactants, *Thomas M. Schmitt* (see Volume 96)
41. Light Scattering by Liquid Surfaces and Complementary Techniques, *edited by Dominique Langevin*
42. Polymeric Surfactants, *Irja Piirma*
43. Anionic Surfactants: Biochemistry, Toxicology, Dermatology. Second Edition, Revised and Expanded, *edited by Christian Gloxhuber and Klaus Künstler*
44. Organized Solutions: Surfactants in Science and Technology, *edited by Stig E. Friberg and Björn Lindman*
45. Defoaming: Theory and Industrial Applications, *edited by P. R. Garrett*
46. Mixed Surfactant Systems, *edited by Keizo Ogino and Masahiko Abe*
47. Coagulation and Flocculation: Theory and Applications, *edited by Bohuslav Dobiás*
48. Biosurfactants: Production Properties Applications, edited by Naim Kosaric
49. Wettability, *edited by John C. Berg*
50. Fluorinated Surfactants: Synthesis Properties Applications, *Erik Kissa*
51. Surface and Colloid Chemistry in Advanced Ceramics Processing, *edited by Robert J. Pugh and Lennart Bergström*
52. Technological Applications of Dispersions, *edited by Robert B. McKay*
53. Cationic Surfactants: Analytical and Biological Evaluation, *edited by John Cross and Edward J. Singer*
54. Surfactants in Agrochemicals, *Tharwat F. Tadros*
55. Solubilization in Surfactant Aggregates, *edited by Sherril D. Christian and John F. Scamehorn*
56. Anionic Surfactants: Organic Chemistry, *edited by Helmut W. Stache*
57. Foams: Theory, Measurements, and Applications, *edited by Robert K. Prud'homme and Saad A. Khan*
58. The Preparation of Dispersions in Liquids, *H. N. Stein*
59. Amphoteric Surfactants: Second Edition, *edited by Eric G. Lomax*
60. Nonionic Surfactants: Polyoxyalkylene Block Copolymers, *edited by Vaughn M. Nace*
61. Emulsions and Emulsion Stability, *edited by Johan Sjöblom*
62. Vesicles, *edited by Morton Rosoff*
63. Applied Surface Thermodynamics, *edited by A. W. Neumann and Jan K. Spelt*
64. Surfactants in Solution, *edited by Arun K. Chattopadhyay and K. L. Mittal*
65. Detergents in the Environment, *edited by Milan Johann Schwuger*
66. Industrial Applications of Microemulsions, *edited by Conxita Solans and Hironobu Kunieda*
67. Liquid Detergents, *edited by Kuo-Yann Lai*
68. Surfactants in Cosmetics: Second Edition, Revised and Expanded, *edited by Martin M. Rieger and Linda D. Rhein*
69. Enzymes in Detergency, *edited by Jan H. van Ee, Onno Misset, and Erik J. Baas*
70. Structure-Performance Relationships in Surfactants, *edited by Kunio Esumi and Minoru Ueno*

71. Powdered Detergents, *edited by Michael S. Showell*
72. Nonionic Surfactants: Organic Chemistry, *edited by Nico M. van Os*
73. Anionic Surfactants: Analytical Chemistry, Second Edition, Revised and Expanded, *edited by John Cross*
74. Novel Surfactants: Preparation, Applications, and Biodegradability, *edited by Krister Holmberg*
75. Biopolymers at Interfaces, *edited by Martin Malmsten*
76. Electrical Phenomena at Interfaces: Fundamentals, Measurements, and Applications, Second Edition, Revised and Expanded, *edited by Hiroyuki Ohshima and Kunio Furusawa*
77. Polymer-Surfactant Systems, *edited by Jan C. T. Kwak*
78. Surfaces of Nanoparticles and Porous Materials, *edited by James A. Schwarz and Cristian I. Contescu*
79. Surface Chemistry and Electrochemistry of Membranes, *edited by Torben Smith Sørensen*
80. Interfacial Phenomena in Chromatography, *edited by Emile Pefferkorn*
81. Solid–Liquid Dispersions, *Bohuslav Dobiás, Xueping Qiu, and Wolfgang von Rybinski*
82. Handbook of Detergents, *editor in chief: Uri Zoller* Part A: Properties, *edited by Guy Broze*
83. Modern Characterization Methods of Surfactant Systems, *edited by Bernard P. Binks*
84. Dispersions: Characterization, Testing, and Measurement, *Erik Kissa*
85. Interfacial Forces and Fields: Theory and Applications, *edited by Jyh-Ping Hsu*
86. Silicone Surfactants, *edited by Randal M. Hill*
87. Surface Characterization Methods: Principles, Techniques, and Applications, *edited by Andrew J. Milling*
88. Interfacial Dynamics, *edited by Nikola Kallay*
89. Computational Methods in Surface and Colloid Science, *edited by Malgorzata Borówko*
90. Adsorption on Silica Surfaces, *edited by Eugène Papirer*
91. Nonionic Surfactants: Alkyl Polyglucosides, *edited by Dieter Balzer and Harald Lüders*
92. Fine Particles: Synthesis, Characterization, and Mechanisms of Growth, *edited by Tadao Sugimoto*
93. Thermal Behavior of Dispersed Systems, *edited by Nissim Garti*
94. Surface Characteristics of Fibers and Textiles, *edited by Christopher M. Pastore and Paul Kiekens*
95. Liquid Interfaces in Chemical, Biological, and Pharmaceutical Applications, *edited by Alexander G. Volkov*
96. Analysis of Surfactants: Second Edition, Revised and Expanded, *Thomas M. Schmitt*
97. Fluorinated Surfactants and Repellents: Second Edition, Revised and Expanded, *Erik Kissa*
98. Detergency of Specialty Surfactants, *edited by Floyd E. Friedli*
99. Physical Chemistry of Polyelectrolytes, *edited by Tsetska Radeva*
100. Reactions and Synthesis in Surfactant Systems, *edited by John Texter*
101. Protein-Based Surfactants: Synthesis, Physicochemical Properties, and Applications, *edited by Ifendu A. Nnanna and Jiding Xia*
102. Chemical Properties of Material Surfaces, *Marek Kosmulski*

103. Oxide Surfaces, *edited by James A. Wingrave*
104. Polymers in Particulate Systems: Properties and Applications, *edited by Vincent A. Hackley, P. Somasundaran, and Jennifer A. Lewis*
105. Colloid and Surface Properties of Clays and Related Minerals, *Rossman F. Giese and Carel J. van Oss*
106. Interfacial Electrokinetics and Electrophoresis, *edited by Ángel V. Delgado*
107. Adsorption: Theory, Modeling, and Analysis, *edited by József Tóth*
108. Interfacial Applications in Environmental Engineering, *edited by Mark A. Keane*
109. Adsorption and Aggregation of Surfactants in Solution, *edited by K. L. Mittal and Dinesh O. Shah*
110. Biopolymers at Interfaces: Second Edition, Revised and Expanded, *edited by Martin Malmsten*
111. Biomolecular Films: Design, Function, and Applications, *edited by James F. Rusling*
112. Structure–Performance Relationships in Surfactants: Second Edition, Revised and Expanded, *edited by Kunio Esumi and Minoru Ueno*
113. Liquid Interfacial Systems: Oscillations and Instability, *Rudolph V. Birikh, Vladimir A. Briskman, Manuel G. Velarde, and Jean-Claude Legros*
114. Novel Surfactants: Preparation, Applications, and Biodegradability: Second Edition, Revised and Expanded, *edited by Krister Holmberg*
115. Colloidal Polymers: Synthesis and Characterization, *edited by Abdelhamid Elaissari*
116. Colloidal Biomolecules, Biomaterials, and Biomedical Applications, *edited by Abdelhamid Elaissari*
117. Gemini Surfactants: Synthesis, Interfacial and Solution-Phase Behavior, and Applications, *edited by Raoul Zana and Jiding Xia*
118. Colloidal Science of Flotation, *Anh V. Nguyen and Hans Joachim Schulze*
119. Surface and Interfacial Tension: Measurement, Theory, and Applications, *edited by Stanley Hartland*
120. Microporous Media: Synthesis, Properties, and Modeling, *Freddy Romm*
121. Handbook of Detergents, *editor in chief: Uri Zoller* Part B: Environmental Impact, *edited by Uri Zoller*
122. Luminous Chemical Vapor Deposition and Interface Engineering, *HirotsuguYasuda*
123. Handbook of Detergents, editor in chief: Uri Zoller Part C: Analysis, *edited by Heinrich Waldhoff and Rüdiger Spilker*
124. Mixed Surfactant Systems: Second Edition, Revised and Expanded, *edited by Masahiko Abe and John F. Scamehorn*
125. Dynamics of Surfactant Self-Assemblies: Micelles, Microemulsions, Vesicles and Lyotropic Phases, *edited by Raoul Zana*
126. Coagulation and Flocculation: Second Edition, *edited by Hansjoachim Stechemesser and Bohulav Dobiás*
127. Bicontinuous Liquid Crystals, *edited by Matthew L. Lynch and Patrick T. Spicer*
128. Handbook of Detergents, editor in chief: Uri Zoller Part D: Formulation, *edited by Michael S. Showell*
129. Liquid Detergents: Second Edition, *edited by Kuo-Yann Lai*
130. Finely Dispersed Particles: Micro-, Nano-, and Atto-Engineering, *edited by Aleksandar M. Spasic and Jyh-Ping Hsu*
131. Colloidal Silica: Fundamentals and Applications, *edited by Horacio E. Bergna and William O. Roberts*
132. Emulsions and Emulsion Stability, Second Edition, *edited by Johan Sjöblom*

COLLOIDAL SILICA

Fundamentals and Applications

Edited by

Horacio E. Bergna
H. E. Bergna Consultants
Wilmington, Delaware

William O. Roberts
Wilmington, Delaware

CRC Press is an imprint of the
Taylor & Francis Group, an **informa** business
A TAYLOR & FRANCIS BOOK

Published in 2006 by
CRC Press
Taylor & Francis Group
6000 Broken Sound Parkway NW, Suite 300
Boca Raton, FL 33487-2742

First issued in paperback 2020

ISBN 13: 978-0-367-57793-3 (pbk)
ISBN 13: 978-0-8247-0967-9 (hbk)

Visit the Taylor & Francis Web site at
http://www.taylorandfrancis.com

and the CRC Press Web site at
http://www.crcpress.com

Library of Congress Cataloging-in-Publication Data

Colloidal silica : fundamentals and applications / edited by Horacio E. Bergna, William O. Roberts.
p. cm. -- (Surfactant science series ; v. 131)
Includes bibliographical references and index.
ISBN 0-8247-0967-5
1. Silica. 2. Colloids. I. Bergna, Horacio E., 1924- II. Roberts, William O., 1936- III. Series.

QD181.S6C65 2005
546'.6832--dc22 2005050208

Preface

Colloidal phenomena have an essential importance in our lives. Living tissues are colloidal systems and therefore the diverse processes involved in our metabolism are for the major part of colloidal nature.

Industrial use of colloidal silicas is growing steadily in both traditional areas and ever-increasing numbers of novel areas. Colloidal silicas are found in fields as diverse as catalysis, metallurgy, electronics, glass, ceramics, paper and pulp technology, optics, elastomers, food, health care and industrial chromatography. However in spite of the apparent simplicity of silica's composition and structure, fundamental questions remain about the formation, constitution, and behavior of colloidal silica systems. As a result, a broad and fascinating area of study is open to scientists interested in fundamental aspects of silica chemistry and physics and to technologists looking for new uses of silica and for answers to practical problems.

This book is dedicated to the field of colloidal silica and although it is aimed at technical people not familiar with colloid science and silica chemistry, we have introduced new information on colloid science related to silica chemistry found in the current literature. In this respect the reader is encouraged to review the selected books listed in the reference list at the end of each chapter.

The work presented here includes both theoretical and experimental aspects of some of the most significant areas of colloidal silica science and technology. This book constitutes an update of the science and technology of colloidal silica since Ralph K. Iler, the distinguished silica scientist, published the definitive book on silica chemistry in 1979 and the American Chemical Society published *Colloid Chemistry of Silica* in 1994.

The book will increase the reader's understanding of the most important problems in this area of science. It is written by some of the most outstanding silica scientists of Argentina, Australia, Canada, China, Japan, Europe, New Zealand, Russia, Ukraine, the United Kingdom, and the United States.

In sum we believe that this book is useful not only to technical people unfamiliar with the subject but also to colloid and silica chemists.

Editors

Horacio E. Bergna is chairman of H.E. Bergna Consultants. Bergna received his licenciado and doctorate in chemistry at the National University of La Plata in Argentina and worked on his doctoral thesis (honors) on clay electrokinetics under Marcos Tschapek at the National Institute of Soils in Buenos Aires, Argentina. Bergna taught at the School of Chemistry in La Plata and worked at the Laboratory of Testing Materials and Technological Research of the Province of Buenos Aires. He did post-doctoral work at the Sorbonne in Paris and the Massachusetts Institute of Technology, where he worked with E.A. Hauser as a guest of the Department of Chemical Engineering and as a research staff member of the Division of Industrial Cooperation with J. Th. Overbeek, A. Gaudin and P. de Bruyn. He studied humanities at the City of London College and Columbia University.

Bergna is the author of 30 papers and holds 31 U.S. patents and more than 200 foreign corresponding patents in subjects such as colloidal silica syntheses, silica, alumina, aluminosilicates, zeolites, vanadyl phosphate catalysts, submicron grained products for metallurgy, and binders for foundry sands. Bergna received 18 Oscar Awards for the patents received by Dupont.

In 1990, Bergna organized and chaired the Ralph K. Iler International Symposium on the Colloid Chemistry of Silica held at the 200th ACS National Meeting. Bergna co-authored the colloidal silica section of the 1993 edition of the *Ullman Encyclopedia of Chemistry* and guest edited two special issues of the Elsevier international journal *Colloids and Surfaces*. He edited and co-authored *Colloid Chemistry of Silica*, published by the American Chemical Society, 1994.

In 1997 Bergna received the Pedro J. Carriquiriborde Prize from the Argentine Chemical Society in Buenos Aires.

William O. Roberts earned his bachelor's degree in chemistry at the Massachusetts Institute of Technology and his Ph.D. at Syracuse University. In 1963, he began working under Ralph Iler at the DuPont Experimental Station in Wilmington, Delaware. There he helped to develop micrograin cutting tools, and this evolved into a production venture that eventually moved to the DuPont plant in Newport, Delaware. In 1972, when the Newport facility closed, Roberts returned to Wilmington to work at the Chestnut Run technical facility. He remained there until his retirement in 1999. Various assignments there involved catalyst work and colored pigments, but the bulk of his last 27 years were spent on Ludox® colloidal silica. The technical service laboratory at Chestnut Run was the main source of new product development for Ludox®, and Roberts developed two new colloidal silica-based products that were patented.

Because of DuPont's involvement in the colloidal silica industry, Roberts became their representative to the Investment Casting Institute (ICI), for which he served on the Ceramics Committee, and eventually became chairman. He was elected to the ICI Board of Directors and served for 13 years until his retirement.

ASSOCIATE EDITORS

Contributors

Cheryl A. Armstrong
Department of Chemistry
Colorado State University
Pueblo, Colorado

F.J. Arriagada
Department of Materials Science and Engineering and the Particulate Materials Center
Pennsylvania State University
University Park, Pennsylvania

Michael R. Baloga
DuPont Company
New Johnsonville, Tennessee

Bhajendra N. Barman
FFFractionation, Inc.
Salt Lake City, Utah

Jonathan L. Bass
The PQ Corporation
Conshohocken, Pennsylvania

Theo P.M. Beelen
Schuit Institute of Catalysis
Eindhoven University of Technology
Eindhoven, The Netherlands

Horacio E. Bergna
DuPont Experimental Station
Wilmington, Delaware

J.D. Birchall
Department of Chemistry
Keele University
Keele, Staffordshire, UK

G.H. Bogush
Department of Chemical Engineering
University of Illinois
Urbana, Illinois

E.J. Bottani
Research Institute of Theoretical and Applied Physical Chemistry (INIFTA)
La Plata, Argentina

Harald Böttner
Fraunhofer Institute for Physical Measurement Techniques
Freiburg, Germany

C. Jeffrey Brinker
Sandia National Laboratories and Center for Micro-Engineered Ceramics
University of New Mexico
Albuquerque, New Mexico

U. Brinkmann
Degussa AG
Hanau-Wolfgang
Düsseldorf, Germany

A. Burneau
Laboratory of Physical Chemistry and Microbiology for the Environment
Villers-lès-Nancy, France

C. Carteret
Laboratory of Physical Chemistry and Microbiology for the Environment
Villers-lès-Nancy, France

I-Ssuer Chuang
Department of Chemistry
Colorado State University
Fort Collins, Colorado

A.A. Chuiko
Institute of Surface Chemistry
National Academy of Sciences of Ukraine
Kiev, Ukraine

Bradley K. Coltrain
Eastman Kodak Company
Corporate Research Laboratories
Rochester, New York

J.B. d'Espinose de la Caillerie
Quantum Physics Laboratory
The City of Paris Industrial Physics and Chemistry Higher Educational Institution
Paris, France

L.E. Cascarini de Torre
Research Institute of Theoretical and Applied Physical Chemistry (INIFTA)
La Plata, Argentina

F. Dumont
Free University of Brussels
Brussels, Belgium

M. Ettlinger
Degussa AG
Hanau-Wolfgang, Germany

James S. Falcone, Jr.
Department of Chemistry
West Chester University
West Chester, Pennsylvania

Horst K. Ferch
Degussa AG
Department of Applied Research and Technical Services
Silicas and Pigments
Degussa AG
Frankfurt, Germany

Lawrence E. Firment
DuPont Company
Wilmington, Delaware

D. Neil Furlong
Division of Chemicals and Polymers
Common wealth Scientific and Industrial Research Organization
Clayton, Australia

J.P. Gallas
Department of Material Sciences and Radiation
University of Caen
Caen, France

Miguel Garcia
Department of Chemistry
Colorado State University
Pueblo, Colorado

J. Calvin Giddings
Field-Flow Fractionation Research Center
Department of Chemistry
University of Utah
Salt Lake City, Utah

Dhanesh G.C. Goberdhan
Atomic Energy Authority
Harwell Laboratory
Oxford, UK

Chad P. Gonzales
Department of Chemistry
Colorado State University
Pueblo, Colorado

A.G. Grebenyuk
Institute of Surface Chemistry
National Academy of Sciences of Ukraine
Kiev, Ukraine

Peter Greenwood
Eka Chemicals (Akzo Nobel)
Bohus, Sweden

Vladimir M. Gun'ko
Institute of Surface Chemistry
National Academy of Sciences of Ukraine
Kiev, Ukraine

Michael L. Hair
Xerox Research Centre of Canada
Mississauga, Ontario, Canada

Marcia E. Hansen
FFFractionation, Inc.
Salt Lake City, Utah

Thomas W. Healy
School of Chemistry
University of Melbourne
Parkville, Victoria, Australia

H. Hommel
Quantum Physics Laboratory
The City of Paris Industrial Physics and Chemistry Higher Educational Institution
Paris, France

B. Humbert
Laboratory of Physical Chemistry and Microbiology for the Environment
Villers-lès-Nancy, France

Alan J. Hurd
Ceramic Processing Science Department
Sandia National Laboratories
Albuquerque, New Mexico

Bruce A. Keiser
Nalco Chemical Company
Naperville, Illinois

Larry W. Kelts
Corporate Research Laboratories
Eastman Kodak Company
Rochester, New York

Martyn B. Kenny
Department of Chemistry
Brunel University
Uxbridge, Middlesex, UK

D. Kerner
Degussa AG
Hanau-Wolfgang, Germany

J.J. Kirkland
DuPont Experimental Station
Central Research and Development Department
Wilmington, Delaware

T. Kobayashi
Kyushu Institute of Technology, Tobata
Fukuoka, Japan

Hiromitsu Kozuka
Institute for Chemical Research
Kyoto University
Uji, Kyoto-Fu, Japan

R. Krasnansky
Department of Chemistry and Biochemistry
University of Notre Dame
Notre Dame, Indiana

A.P. Legrand
Quantum Physics Laboratory
The City of Paris Industrial Physics and Chemistry Higher Educational Institution
Paris, France

Donald E. Leyden
Department of Chemistry (retired)
Condensed Matter Sciences Laboratory
Colorado State University
Fort Collins, Colorado

Guangyue Liu
Field-Flow Fractionation Research Center
Department of Chemistry
University of Utah
Salt Lake City, Utah

Luis M. Liz-Marzan
Department of Chemistry
University of Vigo
Vigo, Spain

V.V. Lobanov
Institute of Surface Chemistry
National Academy of Sciences of Ukraine
Kiev, Ukraine

J.-L. Look
Department of Chemical Engineering
University of Illinois
Urbana, Illinois

Gary E. Maciel
Department of Chemistry
Colorado State University
Fort Collins, Colorado

Sally J. Markway
Department of Chemistry
Colorado State University
Pueblo, Colorado

Egon Matijević
Center for Advanced Materials Processing
Clarkson University
Potsdam, New York

Akihiko Matsumoto
Atomic Energy Authority
Harwell Laboratory
Oxford, UK

A.J. McFarlan
Department of Chemistry
University of Ottawa
Ottawa, Ontario, Canada

A.R. Minihan
Unilever Research Port Sunlight
Bebington, UK

I.F. Mironyuk
Institute of Surface Chemistry
National Academy of Sciences of Ukraine
Kiev, Ukraine

David T. Molapo
Department of Chemistry
University of Ottawa
Ottawa, Ontario, Canada

Myeong Hee Moon
Field-Flow Fractionation
Research Center
Department of Chemistry
University of Utah
Salt Lake City, Utah

Barry A. Morrow
Department of Chemistry
University of Ottawa
Ottawa, Ontario, Canada

Paul Mulvaney
School of Chemistry
Nanotechnology Laboratory
University of Melbourne
Victoria, Australia

K. Osseo-Asare
Department of Materials, Science, and Engineering
and the Particulate Materials Center
Pennsylvania State University
University Park, Pennsylvania

Jan-Erik Otterstedt
Emeritus of Engineering Chemistry
Chalmers University of Technology
Gothenburg, Sweden

Eugène Papirer
Research Center for Physicochemistry
National Center for Scientific Research (CNRS)
Mulhouse, France

Robert E. Patterson
Research and Development Center
The PQ Corporation
Conshohocken, Pennsylvania

Charles C. Payne
Nalco Chemical Company
Naperville, Illinois

A.A. Pentyuk
Institute of Surface Chemistry
National Academy of Sciences of Ukraine
Kiev, Ukraine

V.K. Pogorelyi
Institute of Surface Chemistry
National Academy of Sciences of Ukraine
Kiev, Ukraine

Kristina G. Proctor
Department of Chemistry
Condensed Matter Sciences Laboratory
Colorado State University
Fort Collins, Colorado

John D.F. Ramsay
Atomic Energy Authority
Harwell Laboratory
Oxford, UK

S. Kim Ratanathanawongs
Field-Flow Fractionation Research Center
Department of Chemistry
University of Utah
Salt Lake City, Utah

William O. Roberts
DuPont Company (retired)
Wilmington, Delaware

Sumio Sakka
Institute for Chemical Research
Kyoto University
Uji, Kyoto-Fu, Japan

George W. Scherer
Princeton University
Princeton, New Jersey

R. Schmoll
Degussa AG
Hanau-Wolfgang, Germany

Helmut Schmidt
Institute for New Materials
Saarland University
Saarbrücken, Germany

J. Shimada
Kyushu Institute of Technology, Tobata
Fukuoka, Japan

Kenneth S.W. Sing
Brunel University
Department of Chemistry
Uxbridge, Middlesex, UK

P. Somasundaran
Langmuir Center for Colloids
and Interfaces
Columbia University
New York, New York

Stephen W. Swanton
Atomic Energy Authority
Harwell Laboratory
Oxford, UK

Dennis G. Swartzfager
DuPont Company
Wilmington, Delaware

J.K. Thomas
Department of Chemistry and Biochemistry
University of Notre Dame
Notre Dame, Indiana

V.V. Turov
Institute of Surface Chemistry
National Academy of Sciences of Ukraine
Kiev, Ukraine

Brenda L. Tjelta
Department of Chemistry
University of Utah
Field-Flow Fractionation Research Center
Salt Lake City, Utah

K.K. Unger
Department of Inorganic and
Analytical Chemistry
Johannes Gutenberg University
Mainz, Germany

Alfons van Blaaderen
Soft Condensed Matter Group
Debye Institute
Utrecht University
Utrecht, The Netherlands

Rutger A. van Santen
Schuit Institute of Catalysis
Eindhoven University of Technology
Eindhoven, The Netherlands

Alain M. Vidal
Research Center for Physicochemistry
National Center for Scientific
Research (CNRS)
Mulhouse, France

E.F. Voronin
Institute of Surface Chemistry
National Academy of Sciences of Ukraine
Kiev, Ukraine

A. Vrij
Van't Hoff Laboratory
University of Utrecht
Utrecht, The Netherlands

D.R. Ward
Unilever Research Port Sunlight
Bebington, UK

William A. Welsh
W.R. Grace & Company
Columbia, Maryland

W. Whitby
Unilever Research Port Sunlight
Bebington, UK

Peter W.J.G. Wijnen
Schuit Institute of Catalysis
Eindhoven University of Technology
Eindhoven, The Netherlands

Paul C. Yates
DuPont Company (retired)
Wilmington, Delaware

Akitoshi Yoshida
Central Research Institute
Nissan Chemical Industries, Ltd.
Chiba, Japan

K. Yoshinaga
Kyushu Institute of Technology, Kobata
Fukuoka, Japan

V.I. Zarko
Institute of Surface Chemistry
National Academy of Sciences of Ukraine
Kiev, Ukraine

L. Zhang
Langmuir Center for Colloids
and Interfaces
Columbia University
New York, New York

A.N. Zhukov
Department of Colloid Chemistry
St. Petersburg State University
St. Petersburg, Russia

Yu.L. Zub
Institute of Surface Chemistry
National Academy of Sciences of Ukraine
Kiev, Ukraine

L.T. Zhuravlev
Institute of Physical Chemistry
Russian Academy of Sciences
Moscow, Russia

C.F. Zukoski
University of Illinois
Department of Chemical Engineering
Urbana, Illinois

Table of Contents

Dedication

This book is dedicated to Ralph K. Iler who devoted his career to exploratory and industrial research in the chemistry of colloidal material. He is recognized worldwide for his unique contributions to a unified understanding of the colloidal chemistry of silica and silicates.

Ralph K. Iler not only made outstanding contributions to science and industry, but he was an individual sensitive to the beauty of nature and the works of humanity.

Dr. Iler's biography and portrait appear in *Colloid Chemistry of Silica* (American Chemical Society, Washington, D.C., in 1994). His book *The Chemistry of Silica*, published in 1979, is the definitive book on silica chemistry and a primary source of reference.

The ACS book mentioned above and the current volume, *Colloidal Silica*: Fundamentals and Applications, constitute an updating of Dr. Iler's book.

1 Colloid Science

Horacio E. Bergna
DuPont Experimental Station

CONTENTS

DEFINITION

Colloid science is generally understood to be the study of systems containing kinetic units which are large in comparison with atomic dimensions [1]. Such systems may be those in which the particles are free to move in all directions, or they may be derived systems, as a coagulum or a gel (discussed subsequently), in which the particles have lost their mobility either partially or entirely, but have maintained their individuality.

The colloidal state of subdivision comprises particles with a size sufficiently small ($\leq 1\ \mu m$) not to be affected by gravitational forces but sufficiently large (> 1 nm) to show marked deviations from the properties of true solutions.

SUBDIVISION OF PARTICLES AND THE COLLOIDAL STATE

Matter can be subdivided into progressively smaller parts, fragments or particles, until the dimensions of molecules, ions and atoms are reached. The subdivision process produces matter of progressively smaller particle size. In the size range much larger than the atomic range but much smaller than particles observable with the naked eye, the physical and chemical properties of the surface of these particles assume a preponderant role in the behavior of the system.

This range, where units are made up of a few hundred to perhaps a few billion atoms is said to be in the colloidal state. The range of colloidal dimensions does not have rigorous boundaries, since the threshold of increment varies for different properties in tens or hundreds of angstrom units. However, it can be stated that the colloidal range comprises particles with size between around 10 and 10,000 Å units.

In other words, colloidal particles are those with a size or with one dimension between 10 and 10,000 Å units (1 nm and 1 μm). Considering the size of the constituent atoms, this means that colloidal particles are made of associations or colonies of approximately 10^3 to 10^9 atoms. These atoms can be arranged in a crystalline or in an amorphous structure. It is pertinent to remark that the colloidal particles may either be crystalline or amorphous in nature (Figure 1.1). There is no antinomy between the terms "colloidal" or "crystalline state."

Figure 1.2 helps to situate the colloidal range in a linear scale in comparison with the wavelength of common radiations and expands the comparison to mesh openings of common sieves and microscope ranges.

The characteristic properties of the colloidal range vary gradually toward both ends of the dimensional boundaries and they tend to undergo a sudden increment near one or both ends of the range (Figure 1.3).

The process of subdivision of matter implies creation of new surfaces. Subdivision of matter increases its surface/volume ratio. A cube of 1 m of edge, for instance, has a volume of 1 m^3 and a surface area of 6 m^2. The mass of the same cube when the cube is subdivided in one thousand smaller cubes of 10 cm of edge occupies, of course, the same volume of the mass of the original cube, but the total surface area of all the new cubes is now 60 m^2. The surface: volume ratio increases ten times.

In the colloidal range, the surface: volume ratio is extremely high. All properties related to surfaces are therefore accentuated in this range. The limit or boundary between two homogeneous phases, the interface, shows characteristic properties. These properties of the interface, the "surface properties," play a predominant role in colloidal systems. This is why it is sometimes said that

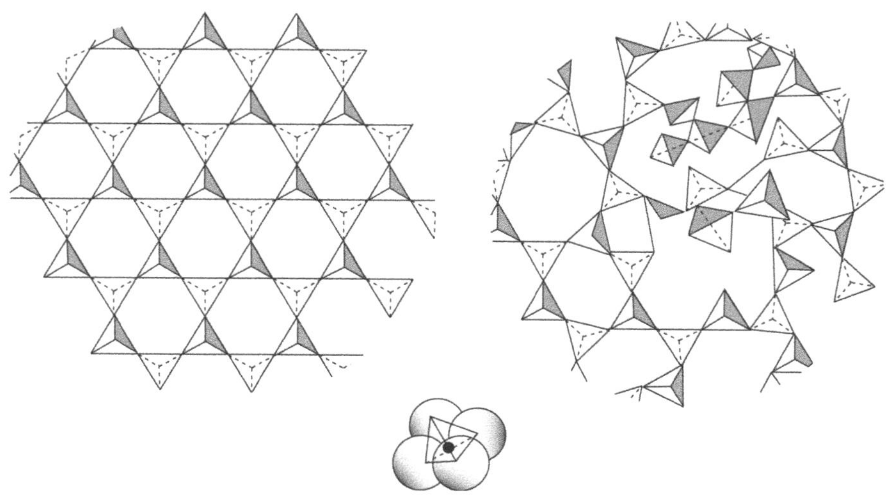

FIGURE 1.1 Crystalline and amorphous structures.

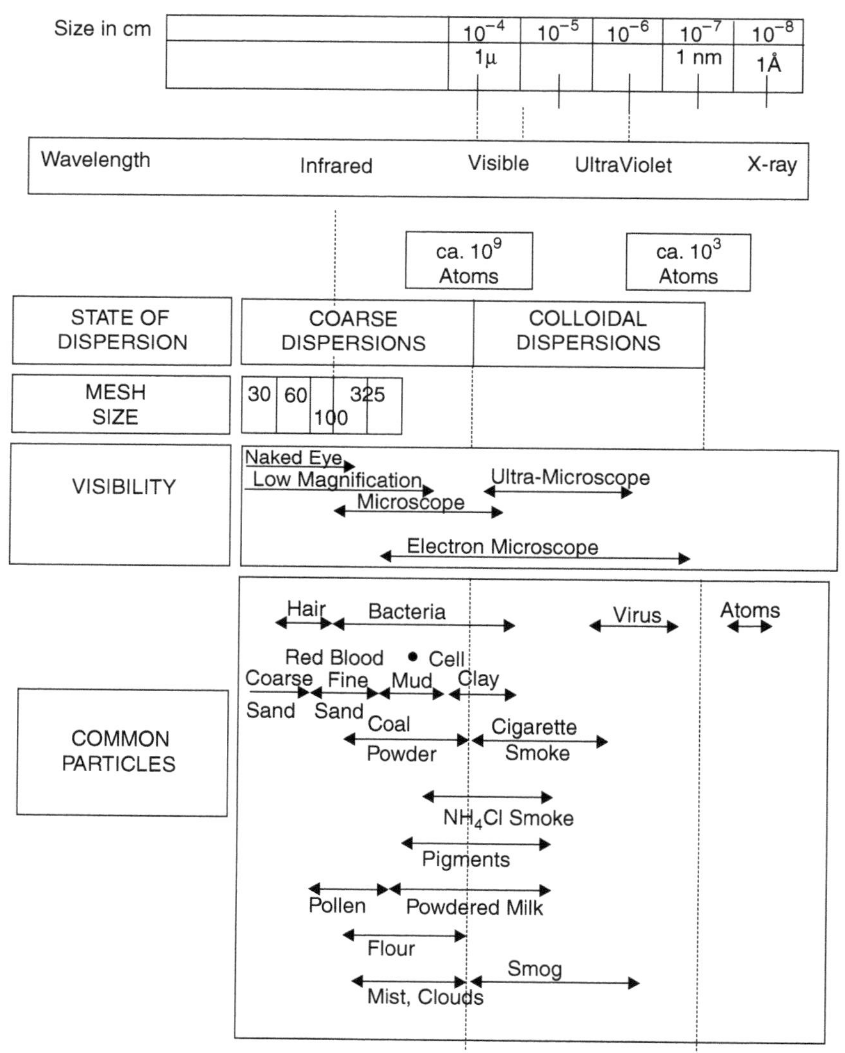

FIGURE 1.2 Particle sizes of dispersed systems.

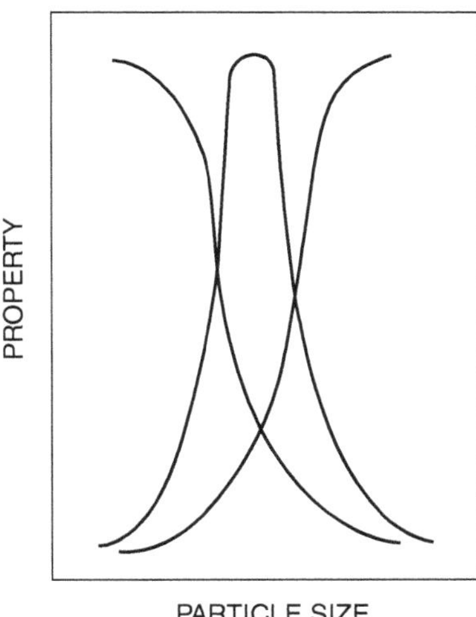

FIGURE 1.3 Plot of property vs. particle size.

colloidal properties are those of a large surface concentrated in a small volume.

Colloid Systems

Colloid systems are mostly based on very small particles dispersed in a solution. This is why it is sometimes said that colloidal properties are those of a large surface concentrated in a small volume.

Colloidal particles are commonly found distributed as a separate phase, the disperse phase, into another substance or substances, the dispersant or continuous phase. In this sense, colloidal systems are heterogeneous systems.

Either of the two phase can be in any of the states of matter: solid, liquid, or gas. A very common case is the dispersion in a liquid of colloidal particles of a solid.

All three dimensions need not be in the colloidal range: in fibers or needle-shaped particles only two dimensions are in this range and in thin films or disk-shaped particles only one dimension is in colloidal range. Nor must the units of a colloidal system be discrete: continuous-network structures, the basic units of which are of colloidal dimensions, also fall in this class, for example, porous solids and foams in addition to gels.

Today the science that they helped create is common knowledge to scientists and even many non scientists. This prompted Prof. Robert B. Dean to write that nearly everything we see is colloidal. The common molecules of inorganic chemistry and the common small molecules of organic chemistry are molecules of substances which are rarely encountered in a pure state in everyday life. When you get up in the morning you wash with colloidal soap, pout on your colloidal clothes, read a colloidal newspaper while eating a colloidal breakfast. The house you live in and the pavement you walk on are both colloidal; even you, yourself consist entirely of colloidal materials. The mineral kingdom is partly colloidal, the vegetable and animal kingdoms wholly so. Colloid Science is the link joining chemistry to all the biological sciences. It is also the most frequently encountered branch of applied chemistry in industrial practice [2].

History

Colloid science, as defined at the beginning of this chapter, is a discipline which determines and attempts to explain and predict the properties of substances based on certain dimensions. The term "colloid science" was created by Wolfgang Ostwald in 1929. A. Buzagh and E.A. Hauser joined Ostwald in pointing out that the term "colloid chemistry" was outdated and should be supplanted by the words "colloid science" since this is a field which cannot be considered as merely an appendix to physical chemistry [3].

As early as 1747, Pott made a "semisolution of silica," and as early as 1820, a reference is made to the preparation of a sol of "hydrated silica" [4].

Selmi (1843) was the first to investigate colloids systematically. He prepared colloidal solutions of sulphur, Prussian blue and casein, performing numerous experiments. He came to the conclusion that these were not true solutions but suspensions of small particles in water [5].

Graham (1861) is usually regarded as the founder of classical experimental colloid science. He classified all substances into two groups: "crystalloids" and "colloids." According to him the former could be easily crystallised, but not the latter. Colloids can be dissolved or dispersed and exposed to a semipermeable membrane the so called crystalloids pass through the membrane easily, but the colloids do not. This procedure is called dialysis. By 1864, silica were being prepared not only by the dialysis of gels but also by hydrolysis of silicate esters.

The name colloid was proposed by Graham (1862), because he considered all colloids to be more or less like glue and for this reason he gave them the Greek name ΚΟΛΛΟ. For colloids in a liquid suspension he used the name "sol." When the sols transformed into solid jellies under suitable conditions he called them "gels."

The work of Graham was of fundamental importance but his classification of all substances into crystalloids and colloids is not always right; many colloids, like some proteins, can be crystallized. On the other hand, almost all so-called crystalloids can be prepared in the colloidal state.

Faraday (1857) was another scientist who made interesting discoveries about colloids. He observed that a narrowly defined beam of light passing through a gold "sol" (colloidal suspension) appears as a whitish path.

The phenomenon was further studied by Tyndall and now bears his name, the "Tyndall effect" [7].

Schulze (1883) investigated the stability of colloidal solutions (sols) working mainly with inorganic colloids. He investigated thoroughly the phenomenon of flocculation or coagulation, to find out the flocculating power of different reagents.

Freundlich investigated adsorption phenomena and enunciates his law of adsorption in 1903. Siedentopf and Zigmondy (1903) invented the ultramicroscope based on the previously mentioned observation, of Faraday and Tyndall. The ultramicroscope was of great utility to study colloids until the invention of the electron microscope.

Important contributions toward the solution of the problem of particle size as well as sedimentation, movement and coagulation of particular were made by Smoluchowski (1906), Svedberg (1906), Perrin (1908), and Einstein (1908).

P.O. von Weimarn (1879–1935), James W. McBain (1887–1953), Harry N. Holmes, Harry B. Weiser (1887–1950), and Lloyd H. Reyerson also made important contributions to the development of modern Colloid Science.

REFERENCES

1. Verwey, E.J.W.; Overbeck, J.Th.G. Theory of the Stability of Lyophobic Colloids. Elservier, 1948.
2. Dean, Robert B. Modern Colloids, D. Van Nostrand Company, Inc. New York, 1948.
3. Hauser, E.A. Silicic Science; Van Nostrand: Princeton, NJ, 1955; p. 54.
4. Fremy, E. Am. Chem. Phys. 1853 (3), Bd 38, S 312–344.
5. Jirgensons, B.; Straumanis, M.E. A Short Textbook of Colloid Chemistry. Pergamon Press Ltd., London, 1954.
6. Graham, T. Am. Chem., 1862, Bd 123, S 860–861.
7. Jirgensons, B.; Straumanis, M.E. A Short Textbook of Colloid Chemistry. Pergamon Press Ltd., London.

2 The Language of Colloid Science and Silica Chemistry

Horacio E. Bergna
DuPont Experimental Station

CONTENTS

This section provides brief explanations for the most important terms that may be encountered in a study of the fundamental principles, experimental investigations and industrial applications of colloid science and silica chemistry.

The definition of some important terms has been given in the Colloid Science section of this book, Chapter 1. Others are given subsequently.

SOLS, GELS, AND POWDERS

A stable dispersion of solid colloidal particles in a liquid is called a *sol.* Stable in this case means that the solid particles do not settle or agglomerate at a significant rate. If the liquid is water, the dispersion is known as an *aquasol* or *hydrosol.* If the liquid is an organic solvent, the dispersion is called an *organosol.* The term *gel* is applied to systems made of a continuous solid skeleton made of colloidal particles or polymers enclosing a continuous solid skeleton made of colloidal particles or polymers enclosing a continuous liquid phase. Drying a gel by evaporation under normal conditions results in a dried gel called a *xerogel.* Xerogels obtained in this manner are often reduced in volume by a factor of 5 to 10 compared to the original wet gel as a result of stresses exerted by capillary tension in the liquid.[1]

An *aerogel* is a special type of xerogel from which the liquid has been removed in such a way as to prevent any collapse or change in the structure as liquid is removed [1]. This is done by drying a wet gel in an autoclave above the critical point of the liquid so that there is no capillary pressure and therefore relatively little shrinkage. The product is mostly air, having volume fractions of solid as low as about 0.1% [2], hence the term aerogel.[1]

An *aerosol* is a colloidal dispersion of particles in gas. *Fumed* or *pyrogenic* oxides, also known in the case of silica as *aerosols*, are powders made by condensing a precursor from a vapor phase at elevated temperatures [3]. (Usage has converted Aerosil, the trademark of Degussa's pyrogenic silica, into a generic term that includes other pyrogenic silicas, such as the Cabot Corporation's Cab-O-Sil.) Dried gels obtained by dispersing aerosols in water and then drying are called by some authors *aerosilogels.* Powders obtained by freeze-drying a sol are known as *cryogels.*[1]

Commercial colloidal silicas are commonly available in the form of sols or powders. The powders can be xerogels, dry precipitates, aerogels, aerosols, or dried and calcined coacervates. The ultimate unit for all of them is a *silica particle*, the size of which determines the specific surface area of the product.

The formation of silica sols, gels, and powders — a genealogical tree of colloidal silicas can be seen represented in Figure 4.4 of Chapter 4.

[1] *The Colloid Chemistry of Silica*, edited by Horacio E. Bergna, American Chemical Society, Washington, DC, 1994.

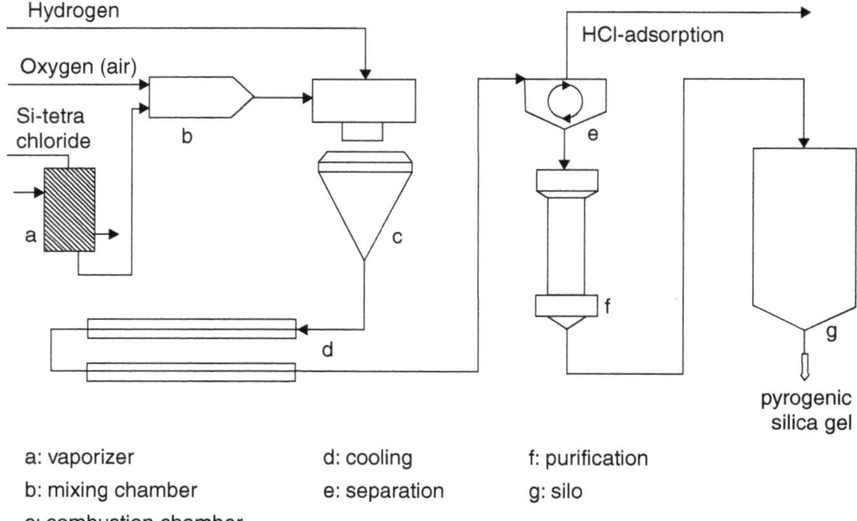

FIGURE 2.1 Flow — chart of production process.

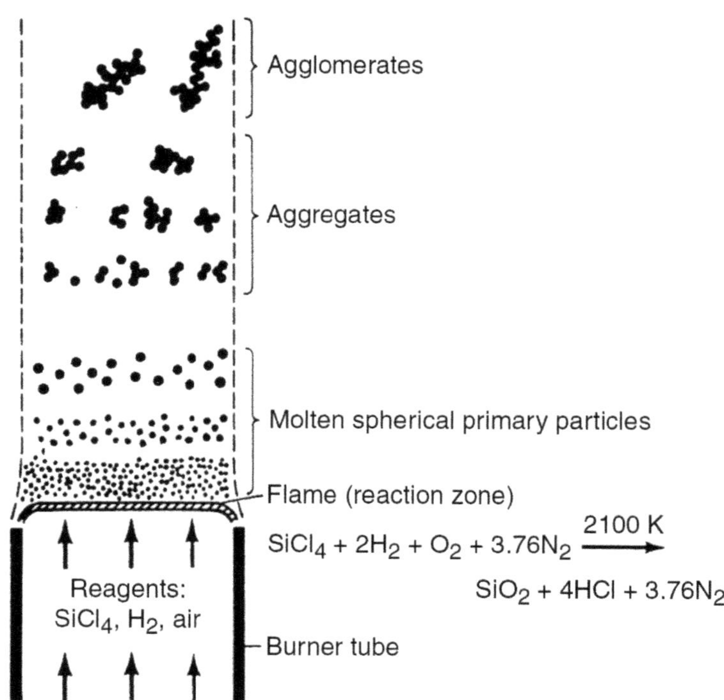

Fumed silica is formed when silicon tetrachloride reacts in a hydrogen flame to form single spherical droplets of silicon dioxide. These grow through collision and coalescence to form larger droplets. As the droplets cool and begin to freeze, but continue to collide, they stick but do not coalesce, forming solid aggregates, which in turn continue to collide to form clusters known as agglomerates

FIGURE 2.2 Collisions of flame-formed particles form larger aggregates and agglomerates.

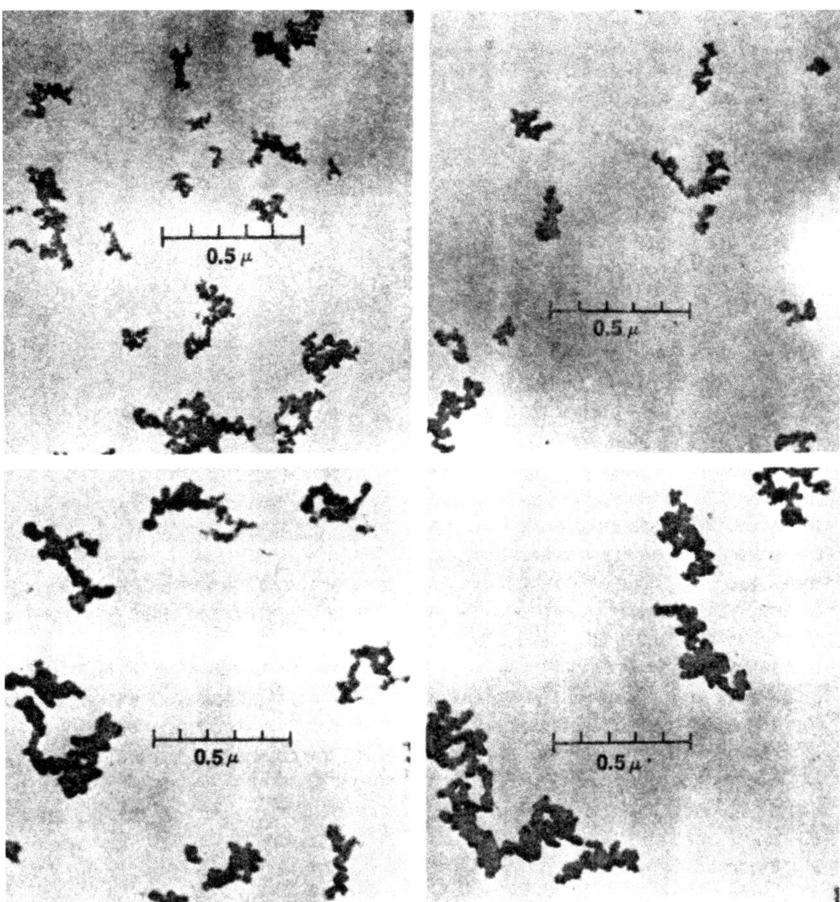

FIGURE 2.3 Growth in size of fumed silica particles as they are carried further from the flame is shown by these four electron micrographs. All samples were taken from the same flame but at different distances from the flame front: upper left, 8 ms residence time, specific surface 360 m^2/g; upper right, 13 ms, 350 m^2/g; lower left, 86 ms, 200 m^2/g; lower right, 137 ms, 150 m^2/g.

3 Colloid Chemistry of Silica: An Overview

Horacio E. Bergna
DuPont Experimental Station

CONTENTS

Silicon dioxide is the main component of the crust of the earth. Combined with the oxides of magnesium, aluminum, calcium, and iron, it forms the silicate minerals in our rocks and soil.

Over millions of years silicon dioxide, or silica, has been separated from the original silicate rocks by the action of water to appear as quartz. In a few places it was deposited in the amorphous form as opal.

Our English word silica has a very broad connotation: it includes silicon dioxide in all its crystalline, amorphous, soluble, or chemically combined forms in which the silicon atom is surrounded by four or six oxygen atoms. This definitely excludes all the organosilicon compounds made by man in which carbon atoms have been linked directly to silicon atoms — commonly referred to as "silicones," which do not occur in nature. Silica is soluble enough in water to play important roles in many forms of life. It forms the skeletons of diatoms, the earliest form of life that absorbed sunlight and began to release oxygen into the atmosphere. Many plants use silica to stiffen stems and form needles on the surface for protection.

As animals developed, the role of silica became less obvious. But each one of us contains about half a gram of silica, without which our bones could not have been formed, and probably also not our brains.

Silica has played a key role since the beginning of civilization, first in flint for tools and weapons and in clay and sand for pottery. The high strength and durability of Roman cement 2000 yr ago is now known to be due to the use of a special volcanic ash that is an almost pure form of amorphous colloidal silica. Today there is active research on the use of the somewhat similar silica fume from electric furnaces to make a super-strong Portland cement.

Our present technology would be very different without the silica for the catalysts of our oil refineries, for the molds for casting the superalloys in our jet engines, for modern glass and ceramics, electronic micro-circuits, quartz crystals, and fiber optics.

Ralph K. Iler, Alexander Memorial Lecture, Australia, 1989. Reprinted with permission from *Chemistry in Australia*, October 1986, p. 355.

Silicon dioxide, silica, can be natural or synthetic, crystalline or amorphous. This book is concerned mostly with synthetic amorphous silica in the colloidal state.

The building block of silica and the silicate structures is the SiO_4 tetrahedron, four oxygen atoms at the corners of a regular tetrahedron with a silicon ion at the center cavity or centroid (Figure 3.1). The oxygen ion is so much larger than the Si^{4+} ion that the four oxygens of a SiO_4 unit are in mutual contact and the silicon ion is said to be in a *tetrahedral hole* [1]. Natural silicas can be crystalline, as in quartz, cristobalite, tridymite, coesite, and stishovite, or amorphous, as in opal. Crystalline silica polymorphs are divided according to their framework density (SiO_2 groups per 1000 $Å^3$) into pyknosils and porosils, and the latter are further divided into clathrasils and zeosils depending on whether the pores are closed or open, that is, accessible to adsorption (see Chapter 23).

Familiarity with the structure of crystalline silica is helpful in understanding the bulk and surface structure of amorphous silica. All forms of silica contain the Si—O bond, which is the most stable of all Si—X element bonds. The Si—O bond length is about 0.162 nm, which is considerably smaller than the sum of the covalent radii of silicon and oxygen atoms (0.191 nm) [2]. The short bond length largely accounts for the partial ionic character of the single bond and is responsible for the relatively high stability of the siloxane bond. Although in most silicas and silicates the silicon atom is surrounded by four oxygen atoms, forming the tetrahedral unit $[SiO_4]^{4-}$, a sixfold octahedral coordination of the silicon atom has also been observed in stishovite and coesite [3]. The arrangements of $[SiO_4]^{4-}$ and $[SiO_6]^{8-}$ and the tendency of these units to form a three-dimensional framework structure are fundamental to silica crystal chemistry.

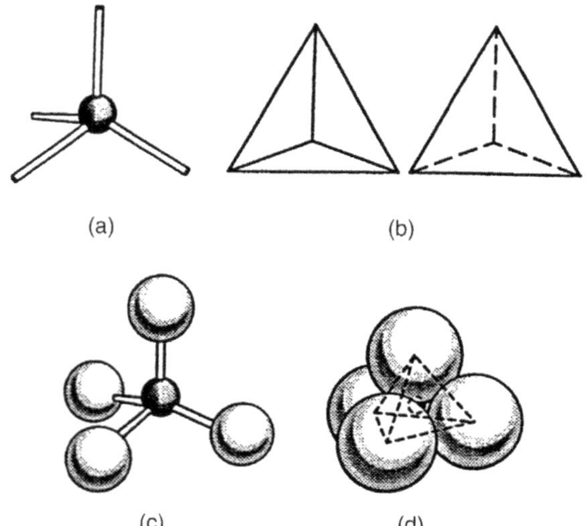

FIGURE 3.1 Methods of representing the tetrahedral coordination of oxygen ions with silicon: (a) ball and stick model, (b) solid tetrahedron, (c) skeletal tetrahedron, and (d) space-filling model based on packed spheres. (Reproduced with permission from reference 95. Copyright 1974.)

The silicates are built up in a manner analogous to that of the polyborates and the polyphosphates by sharing of oxygen atoms. In practice, two different SiO_4 groups may share only one oxygen atom, but any or all of the four of the oxygen atoms on a SiO_4 group may be shared with adjacent groups. Sharing of two oxygen atoms per unit yields a *chain*, three oxygen atoms a *sheet*, and four oxygen atoms a *three-dimensional network* [1]. The crystalline silicas quartz, tridymite, and cristobalite are in truth network silicates, each silicon being bound to four oxygens and each oxygen being bound to two silicons. Quartz is the stable form of crystalline silica below 870°C, tridymite below 1470°C, and cristobalite below 1710°C, but either of the two high-temperature forms can exist for long periods of time at room temperature and atmospheric pressure without turning to quartz [2].

The polymorphism of silicas is based on different linkages of the tetrahedral $[SiO_4]^{4-}$ units [2]. Quartz has the densest structure, and tridymite and cristobalite have a much more open structure. All three forms exist in α- and β-forms, which correspond to low- and high-temperature modifications, respectively. The α- and β-modifications differ only slightly in the relative positions of the tetrahedral arrangements. This similarity is evident from the fact that the conversion $\alpha \leftrightharpoons \beta$ is a rapid displacing transformation that occurs at relatively low temperatures. Quartz is the most stable modification at room temperature; all others forms are considered to be metastable at this temperature [2].

In amorphous silica the bulk structure is determined, as opposed to the crystalline silicas, by a random packing of $[SiO_4]^{4-}$ units, which results in a nonperiodic structure (Figure 3.2). As a result of the structural differences the various silica forms have different densities (Table 3.1).

The structure, Si—O bond length, and Si—O—Si bond angle in crystalline and amorphous silicas have been studied by x-ray, electron, and neutron diffraction and by infrared spectroscopy. Three strong absorption bands at 800, 1100, and 1250 cm^{-1} measured by infrared transmission techniques are attributed to fundamental Si—O vibrations and do not differ greatly in the various silica modifications, whereas in the high-frequency region (2800–4000 cm^{-1}) certain distinct differences are observed [2]. Figure 3.3 is a schematic representation of adjacent SiO_4 tetrahedra that shows the Si—O—Si bond angle [4]. Diffraction measurements have shown a difference between the Si—O—Si bond angle of quartz (142°), cristobalite (150°), and fused quartz (143°).

Silicate glasses are conventionally regarded as silicate frameworks in which cations are distributed at random. However, Gaskell et al. [5], using neutron scattering with

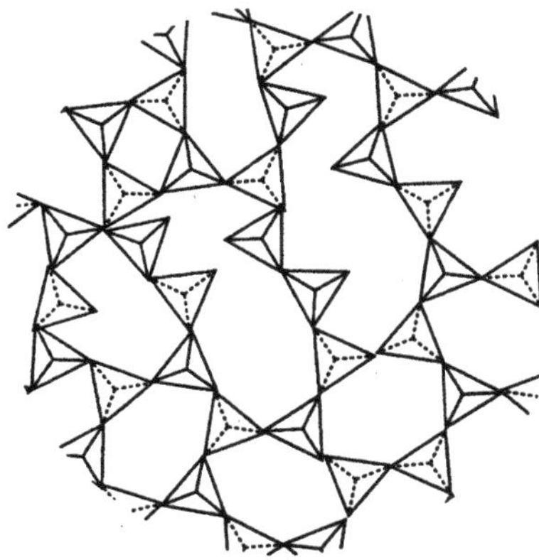

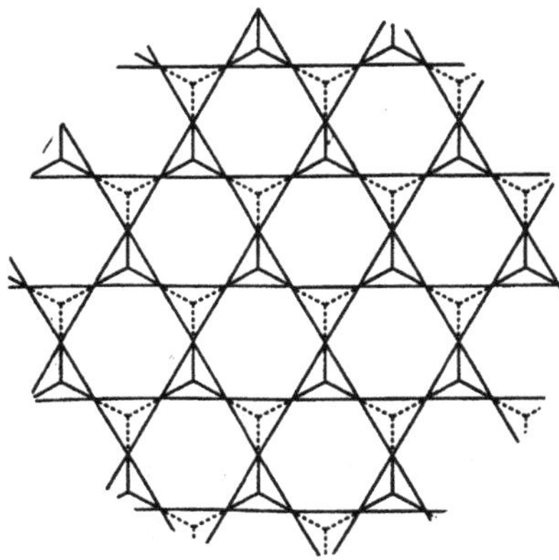

FIGURE 3.2 Two-dimensional representation of random versus regular packing of $(Si{-}O_4)^{4-}$ tetrahedra: amorphous (top) and crystalline silica. (Crystalline diagram reproduced with permission from reference 96. Copyright 1960.)

isotopic substitutions of Ca in a calcium silicate glass, revealed a high degree of ordering in the immediate environment of Ca over distances approaching 1 nm. The technique was later extended to obtain a direct measurement of the Ca—Ca distribution and provided what they

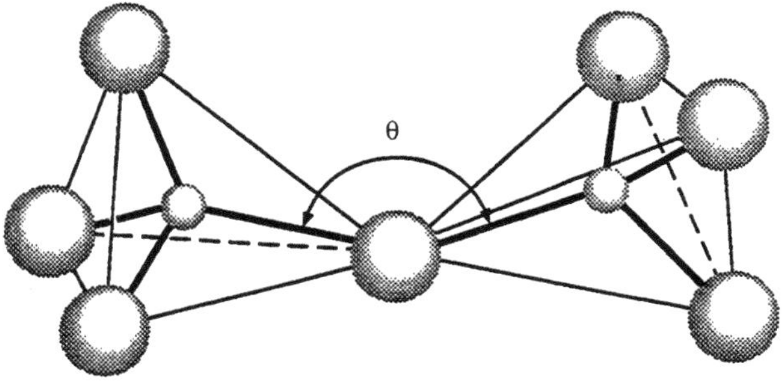

FIGURE 3.3 Schematic representation of adjacent SiO_4 tetrahedra that shows the Si—O—Si bond angle. Small circle, Si; large circle, O. (Reproduced with permission from reference 97. Copyright 1976 John Wiley & Sons, Inc.)

considered strong evidence that such glasses and possibly other amorphous oxides are more extensively ordered than previously seemed possible. These findings on silicate glasses made some researchers review the largely discredited notion, originally based on the observation of broad x-ray diffraction peaks centered in the range of the crystalline silica's strong peaks, that amorphous silica may also have limited domains with a high degree of ordering.

TABLE 3.1
Density (*d*) of Crystalline and Amorphous Silicas

Silica	Density (g/ml at 273 K)
Coesite	3.01
α-Quartz	2.65
β-Quartz	2.53
β-Tridymite	2.26
β-Cristobalite	2.21
Amorphous silica	2.20

Source: Reproduced with permission from reference 2. Copyright 1979 Elsevier Science Publishing Co., Inc.

COLLOIDAL DISPERSIONS AND COLLOID SCIENCE

As previously pointed out, this book deals mostly with *colloidal silicas*, that is, disperse systems in which the disperse phase is silica in the colloidal state of subdivision. The colloidal state of subdivision comprises particles with a size sufficiently small ($\leq$1 μm) not to be affected by gravitational forces but sufficiently large ($>$1 nm) to show marked deviations from the properties of true solutions. In this particle size range, 1 nm (10 Å) to 1 μm (1000 nm), the interactions are dominated by short-range forces, such as van der Waals attraction and surface forces. On this basis the International Union of Pure and Applied Chemistry (IUPAC) suggested that a colloidal dispersion should be defined as a system in which particles of colloidal size (1–1000 nm) of any nature (solid, liquid, or gas) are dispersed in a continuous phase of a different composition or state [6]. If the particles are solid they may be crystalline or amorphous. The disperse phase may also be small droplets of liquids, as in the case of emulsions, or gases, as for example in foams.

By way of comparison, the diameters of atoms and molecules of classical chemistry are below 0.5 nm. On the other end of the colloidal range, at about 1000 nm, the region of suspensions begins. Thus, colloid science, concerned with the intermediate range, is generally understood to be the study of systems containing kinetic units that are large in comparison with atomic dimensions [7]. Such systems may be those in which the particles are free to move in all directions, or they may be derived systems, as a coagulum or a gel (discussed subsequently), in which the particles have lost their mobility either partially or entirely, but have maintained their individuality.

All three dimensions need not be in the colloidal range: fibers or needle-shaped particles in which only two dimensions are in this range and thin films or disk-shaped particles in which only one dimension is in this range may also be treated as colloidal [7]. Nor must the units of a colloidal system be discrete: continuous-network structures, the basic units of which are of colloidal dimensions, also fall in this class, for example, porous solids and foams in addition to gels.

A more modern approach to colloidal dispersions is based on fractal geometry. The fractal approach, as explained later, provides a new basis for the definition and characterization of colloidal systems.

COMMERCIAL COLLOIDAL SILICAS

Commercial colloidal silicas are produced by many companies both in the Americas, in Europe, and in Japan as dispersions in water or organic solvents in different particle sizes. Current types are listed in Tables 3.2–3.6 and 3.8.

SOLS, GELS, AND POWDERS

A stable dispersion of solid colloidal particles in a liquid is called a *sol*. Stable in this case means that the solid particles do not settle or agglomerate at a significant rate. If the liquid is water, the dispersion is known as an *aquasol* or *hydrosol*. If the liquid is an organic solvent, the dispersion is called an *organosol*. The term *gel* is applied to systems made of a continuous solid skeleton made of colloidal particles or polymers enclosing a continuous liquid phase. Drying a gel by evaporation under normal conditions results in a dried gel called a *xerogel*. Xerogels obtained in this manner are often reduced in volume by a factor of 5 to 10 compared to the original wet gel as a result of stresses exerted by capillary tension in the liquid.

An *aerogel* is a special type of xerogel from which the liquid has been removed in such a way as to prevent any collapse or change in the structure as liquid is removed [8]. This is done by drying a wet gel in an autoclave above the critical point of the liquid so that there is no capillary pressure and therefore relatively little shrinkage. The product is mostly air, having volume fractions of solid as low as about 0.1% [8], hence the term aerogel.

An aerosol is a colloidal dispersion of particles in gas. *Fumed* or *pyrogenic* oxides, also known in the case of silica as aerosils, are powders made by condensing a

TABLE 3.2
Properties of Commercial Silica Sols Listed by Manufacturer

Sol (Manufacturer)	Grade	SiO_2 (%)	Stabilizer Type	Stabilizer (%)	Ratio SiO_2:Na_2O	pH	Particle Diameter (nm)	Specific Surface ($m^2\ g^{-1}$)	Technical Bulletin
W.R. Grace & Company Columbia, MD	HS-40	40	Na_2O	0.41	95	9.7	12	230	E10260 (1976)
	HS-30	30	Na_2O	0.32	95	9.8	12	230	E10260 (1976)
	TM	50	Na_2O	0.21	240	9.0	21	130	E10260 (1976)
	SM	30	Na_2O	0.56	54	9.9	7	360	E10260 (1976)
	AS[a]	40	NH_3	—	—	9.0	21	130	E10260 (1976)
	LS	30	Na_2O	0.10	300	8.2	12	130	E10260 (1976)
	WP[b]	35	Na_2O	0.62	130	11.0	21	130	E08913 (1976)
	(AS)[c]	30	NH_3	—	—	9.6	13–14	210–230	A82273 (1974)
	AM[d]	30	Na_2O	0.13	230	9.0	15	210	A21163
			Positively charged sols — Al_2O_3 coating						
Ondeo Nalco Naperville, Il.	1115	15	Na_2O	0.8	19	10.4	4	750	CTG-1115
	2326	14.5	NH_3	0.01	—	9.0	5	600	CTG-2326
	1130	30	Na_2O	0.65	46	10.2	8	375	CTG-1130
	1030	30	Na_2O	0.40	75	10.2	13	230	CTG-1030
	1140	40	Na_2O	0.40	100	9.7	15	200	CTG-1140
	1050	50	Na_2O	0.35	143	9.0	21	143	CTG-1050
	1034A	34	H		—	3.0 Max	19	158	CTG-1034
	2327	40	NH_3	0.10	—	9.3	23	130	CTG-2327

TABLE 3.3

Grade	15/500	30/360	30/220	30/80	305	40/220	40/130	50/80	F 45	30 NH_3/220	CAT80
SiO_2 wt %	15	30	30	30	30	40	40	50	45	30	40
Surface area m^2/g	500	360	220	80	220	220	130	80	80	220	—
Particle size nm	6	9	15	40	15	15	25	40	40	15	40
Na_2O wt %	0.40	0.55	0.30	0.13	0.30	0.40	0.18	0.22	0.20	<0.10	—
pH	10.0	10.0	9.7	9.6	9.5	9.7	9.0	9.3	9.5	9.0	4.0
Density g/cm^3	1.1	1.2	1.2	1.2	1.2	1.3	1.3	1.4	1.36	1.2	1.32
Viscosity mPas	<5	<8	<7	<6	<10	<25	<10	<15	<15	<10	<15

TABLE 3.4

Grade	215	830	1430	1440	2040	2050	9950	2040 NH_4	2034DI
SiO_2 wt %	15	30	30	40	40	50	50	40	34
Particle size nm	4	10	14	14	20	20	100	20	20
Na_2O wt %	0.83	0.55	0.40	0.50	0.38	0.47	0.12	—	—
pH	11.0	10.5	10.3	10.4	10.0	10.0	9.0	9.0	3.0
Density g/cm^3	1.10	1.22	1.21	1.30	1.30	1.40	1.40	1.30	1.23
Viscosity mPas	5	8	7	16	13	50	15	15	7

TABLE 3.5a
Silica — Typical Values

Product	Metal Oxide	Wt. % Metal Oxide	Media	% H_2O (Karl Fischer)	Specific Gravity *SG of SiO_2	pH *50/50 wt. in water +5 wt.% in aqueous slurry	Mean Particle Diameter nm	Mean Particle Diameter μm	Particle Charge
DP5480	Silica	30	EG	1.0	1.3	3.0*	50		Negative
DP5540	Silica	30	EG	1.0	1.3	3.0*	100		Negative
DP5820	Silica	30	EG	1.0	1.3	3.0*	20		Negative
Nyasil 5	Silica	92	Powder	N/A	2.2*	4.0 +		1.8	N/A
Nyasil 20	Silica	95	Powder	N/A	2.2*	4.0 +		1.4	N/A
Nyasil 6200	Silica	96	Powder	N/A	2.2*	4.0 +		1.7	N/A

Note: N/A, not applicable

precursor from a vapor phase at elevated temperatures. (Usage has converted Aerosil, the trademark of Degussa's pyrogenic silica, into a generic term that includes other pyrogenic silicas, such as the Cabot Corporation's Cab-O-Sil.) Dried gels obtained by dispersing aerosils in water and then drying are called by some authors aerosilogels. Powders obtained by freeze-drying a sol are known as *cryogels*.

Commercial colloidal silicas are commonly available in the form of sols or powders. The powders can be xerogels, dry precipitates, aerogels, aerosils, or dried and calcined coacervates. The ultimate unit for all of them is a silica particle, the size of which determines the specific surface area of the product.

Figure 3.4 shows the formation of silica sols, gels, and powders — a genealogical tree of colloidal silicas.

COLLOIDAL SILICA — STABILITY AND AGGREGATION

Figure 3.5 is a diagram representing a particle of silica, the unit of all colloidal silicas. In a restricted sense the term

TABLE 3.5b
Silica — Typical Values

Product	Bulk Density g/cc (lb/ft^3) Untamped	Tamped	Oil Absorption ASTM D1483-84	% Loss on Ignition (loss after 2 hrs., 1000°C)	% Free Moisture (loss after 2 hrs., 105°C)	Viscosity	Appearance
DP5480	N/A	N/A	N/A	N/A	N/A	60 cP	clear liquid
DP5540	N/A	N/A	N/A	N/A	N/A	60 cP	clear liquid
DP5820	N/A	N/A	N/A	N/A	N/A	85 cP	clear liquid
Nyasil 5	0.15 (9.36)	0.18 (11.23)	140 g of oil/ 100 g of powder	8	2.5	N/A	white powder
Nyasil 20	0.29 (18.10)	0.32 (20.00)	80 g of oil/ 100 g of powder	5	1.0	N/A	white powder
Nyasil 6200	0.29 (18.10)	0.31 (19.35)	77 g of oil/ 100 g of powder	4	0.7	N/A	white powder

Note: N/A, not applicable.

TABLE 3.6
Nyasil Silica Powders — Typical Analysis

Product	Particle Size Malvern Mastersizers D(v,0.1)	D(v,0.5)	D(v,0.9)	Surface Area, m^2/g	β Analysis Total Pore Vol., ml/g	Avg. Pore Radius, Å	Impurity Analysis ppm Sodium as Na	Iron as fe	Aluminum as Al	Color Analysis L	a	b
Nyasil 5	0.7	1.8	3.2	279	0.29	20.5	<1000	46	178	93.1	−0.41	0.36
Nyasil 20	0.5	1.4	2.7	167	0.33	39.1	<1000	37	60	93.5	−0.38	0.27
Nyasil 6200	0.6	1.7	3.8	64	0.24	74.4	<1000	23	33	93.4	−0.38	0.33

Note: N/A, not applicable.

TABLE 3.7
Silica Applications

Product	Abrasion resistance	Anti-block	Deep-dyed fiber	Coatings
DP5480	■	■	■	
DP5540	■	■	■	
DP5820	■	■	■	
Nyasil 5		■	■	■
Nyasil 20		■	■	■
Nyasil 6200		■	■	■

"colloidal silica" is often used to refer to concentrated stable dispersions or "sols" made of discrete, dense particles of amorphous silicas of uniform particle size from about 5 to 1500 nm and especially from about 5 to 100 nm. In a broad sense, silica gels and powders are also colloidal silicas because they consist of silica particles of colloidal size that range from 1 to 1000 nm in diameter. Because colloidal silica is composed of discrete particles, the difference with silica glass (also referred to as fused silica or, inappropriately, quartz glass) is that although both are amorphous, the structure of the glass is macroscopically continuous.

TABLE 3.8

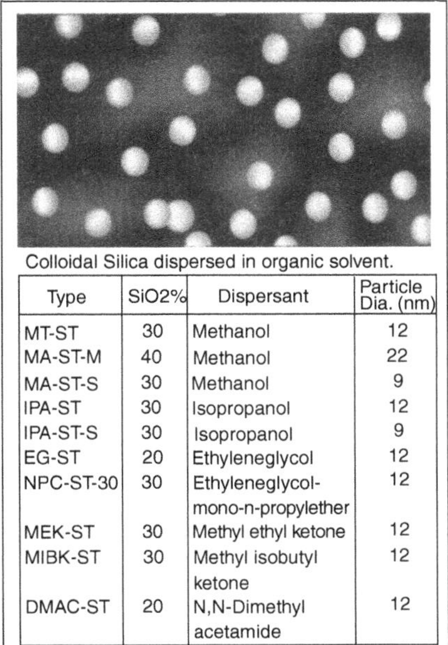

Colloidal Silica dispersed in organic solvent.

Type	SiO2%	Dispersant	Particle Dia. (nm)
MT-ST	30	Methanol	12
MA-ST-M	40	Methanol	22
MA-ST-S	30	Methanol	9
IPA-ST	30	Isopropanol	12
IPA-ST-S	30	Isopropanol	9
EG-ST	20	Ethyleneglycol	12
NPC-ST-30	30	Ethyleneglycol-mono-n-propylether	12
MEK-ST	30	Methyl ethyl ketone	12
MIBK-ST	30	Methyl isobutyl ketone	12
DMAC-ST	20	N,N-Dimethyl acetamide	12

Many more products are available

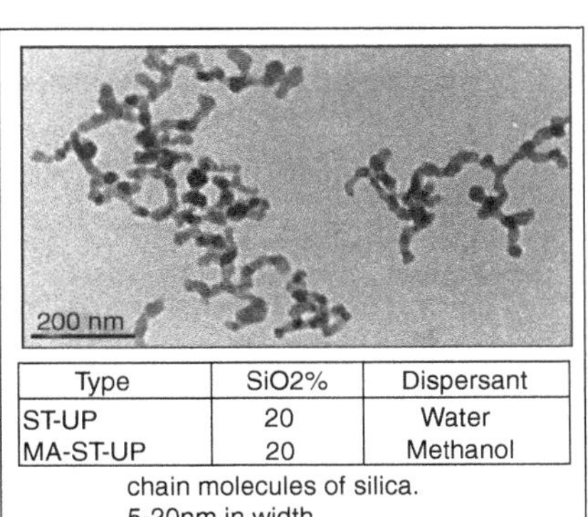

Type	SiO2%	Dispersant
ST-UP	20	Water
MA-ST-UP	20	Methanol

chain molecules of silica.
5-20nm in width
40-300nm in length

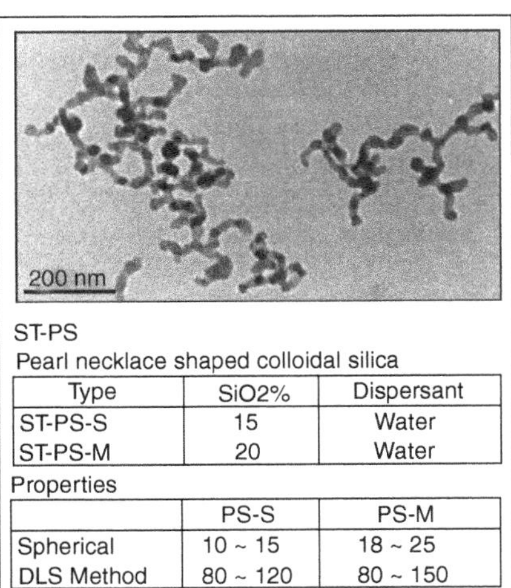

ST-PS
Pearl necklace shaped colloidal silica

Type	SiO2%	Dispersant
ST-PS-S	15	Water
ST-PS-M	20	Water

Properties

	PS-S	PS-M
Spherical	10 ~ 15	18 ~ 25
DLS Method	80 ~ 120	80 ~ 150

Japanese Company

Silica fibers show some colloidal properties and are often included in the studies of colloidal silica. Figure 3.6 is an electron micrograph of synthetic silica fibers with a specific surface area of 400 m^2/g [9].

Silica aquasols with particle size in the 5-nm to about the 60–100-nm range may remain for very prolonged periods of time without significant settling or loss of stability. *Stability* in colloid science is used not only in the thermodynamic sense but also in a strictly colloidal sense. "Colloidally stable" means that the colloidal particles do not settle and do not aggregate at a significant rate [6]. An *aggregate* in colloid science is a group of particles held together in any possible way. The term aggregate is used to describe the structure formed by the cohesion of colloidal particles.

Silica sols lose their stability by aggregation of the colloidal particles. Colloidal silica particles can be linked together or aggregate by gelation, coagulation or flocculation, or coacervation.

GELATION, COAGULATION, FLOCCULATION, AND COACERVATION

There is a basic difference between gelling or gelation and coagulation or flocculation. Both involve colloidal particles or polymers linking together and forming three-dimensional networks. But when a sol is gelled, it first becomes viscous and then develops rigidity and fills the volume originally occupied by the sol. On the other hand, when a sol is coagulated or flocculated, a precipitate is formed. In a concentrated sol the precipitate may be too voluminous to separate and will remain as a thixotropic mass, but in a dilute sol the precipitate will settle out. The difference between a sol, gel, and precipitate is illustrated in Figure 3.7 [3].

The terms coagulation (from the Latin "to drive together") and flocculation are commonly used interchangeably, but some authors prefer to introduce a distinction between coagulation — implying the formation of compact aggregates leading to the macroscopic separation of a coagulum — and flocculation — implying the formation of a loose or open network, a floc, that may or may not separate macroscopically [10].

Iler [3] distinguished the way in which colloidal particles aggregate or link together in the following manner:

1. Gelling, where the particles are linked together in branched chains that fill the whole volume of sol so that there is no increase in the concentration of silica in any macroscopic region in the medium. Instead, the overall medium becomes viscous and then is solidified by a coherent network of particles that, by capillary action, retains the liquid.

TABLE 3.9
The Grades Which Put the Final Touch

Product	Unit	50/50%	100/45%	200/30%	300/30%
Concentration	%	50	45	30	30
Na_2O content	%	0.09	0.23	0.15	0.35
Density	g/cm^3	1.387	1.343	1.206	1.209
Viscosity	mPas	max. 30	max. 15	max. 5	max. 7
Specific surface	m^2/g	50	100	200	300
Particle size	nm	approx. 50	30	15	9
Ionicity		anionic	anionic	anionic	anionic
Appearance		milky	milky	opalescent	slightly opalescent
Odor		odorless	odorless	odorles	odorless
pH		9	10	9	10
Additives		—	—	—	—
Shelf life	months	18	18	18	18
EU safety data sheet		319000	003429	003429	003429

2. Coagulation, where the particles come together into relatively close-packed clumps in which the silica is more concentrated than in the original sol so the coagulum settles as a relatively dense precipitate. Industrial precipitated silicas are powders formed when the ultimate silica particles are coagulated as loose aggregates in the aqueous medium, recovered, washed, and dried. A simple way to differentiate between a precipitate and a gel is that a precipitate encloses only part of the liquid in which it is formed.
3. Flocculation, where the particles are linked together by bridges of the flocculating agent that are sufficiently long so that the aggregated

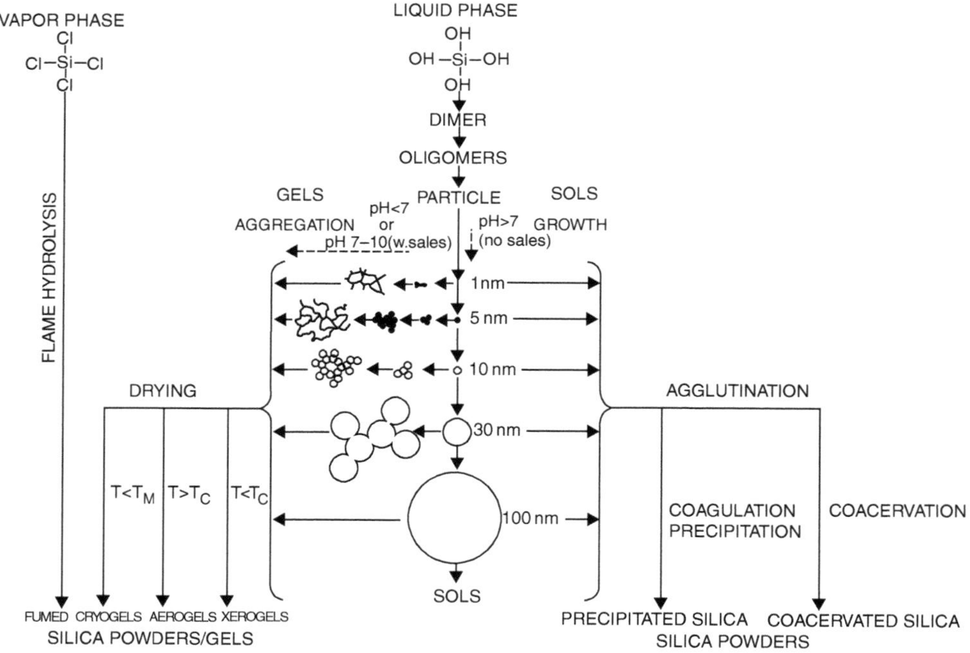

FIGURE 3.4 Formation of silica sols, gels, and powders by silica monomer condensation–polymerization followed by aggregation or agglutination and drying. Growth of nascent colloidal particles with a decrease in numbers occurs in basic solutions in the absence of salts. In acid solutions or in the presence of flocculating sols the colloidal silica particles form gels by aggregation into three-dimensional networks. The diagram is an expansion of Iler's classic Figure 3.1 [3] and constitutes a genealogical tree of colloidal silicas.

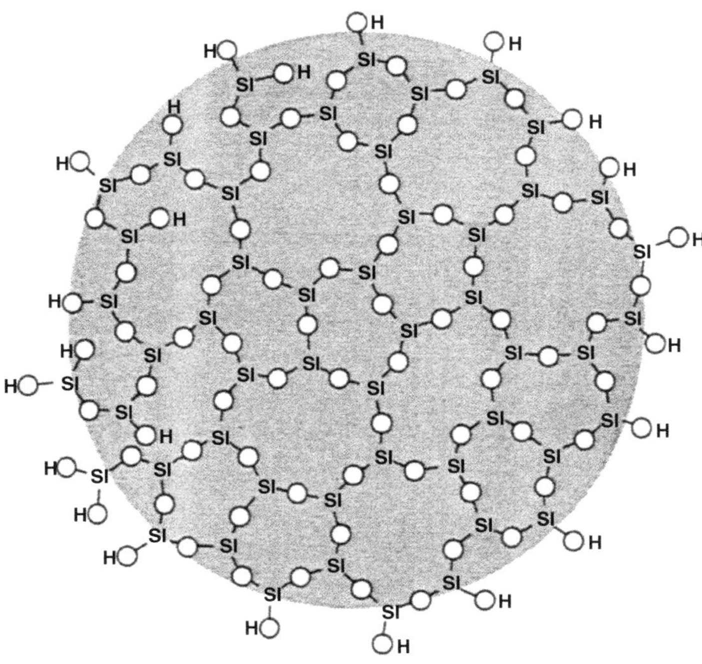

FIGURE 3.5 Schematic representation in two dimensions of a dehydrated but fully hydroxylated colloidal silica particle. The fourth oxygen coordinated with Si is above or below the plane of the paper. The figure is only a diagram, not a model. In an amorphous silica model the Si—O—Si bond angle may vary, but the Si—O distances are constant; each oxygen ion is linked to not more than two cations; the coordination number of oxygen ions about the control cation is 4 or less; oxygen tetrahedra share corners, not edges or faces; and at least two corners of each tetrahedron are shared.

structure remains open and voluminous. It is apparent that these differences will be noted mainly in dilute sols containing only a few percent of silica. In concentrated mixtures one can distinguish a gel, which is rigid, but not between a coagulate and a flocculate.

4. Coacervation, a fourth type of aggregation, in which the silica particles are surrounded by an adsorbed layer of material that makes the particles less hydrophilic, but does not form bridges between particles. The particles aggregate as a concentrated liquid phase immiscible with the aqueous phase.

100 μm

FIGURE 3.6 Porous silica fiber synthesized by W. Mahler. Details of preparation are given in reference 9. Scanning electron micrograph by M. L. Van Kavelaar. (Reproduced from reference 14. Copyright 1989.)

FRACTAL APPROACH TO COLLOID SYSTEMS

One of the most exciting developments of the past decade in the study of colloidal silica is the application of the fractal approach to the study of sols and gels. Fractals are disordered systems for which disorder can be described in terms of nonintegral dimension. The concept of fractal geometry, developed by Mandelbrot [11] in the early 1980s, provides a means of quantitatively describing the average structure of certain random objects. The fractal dimension of an object of mass M and radius r is defined by the relation

$$M \propto r^{D_f}$$

where D_f is named the mass fractal dimension of the object. For Euclidean (nonfractal) objects, D_f equals the dimension of space D. In three dimensions the mass of a sphere scales as the radius cubed, that is, $D \propto r^3$. Fractals

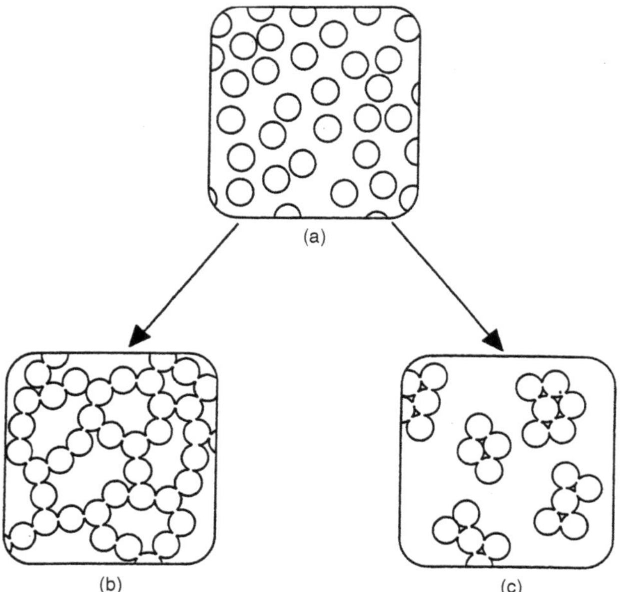

FIGURE 3.7 Silica gel versus precipitate: (a) sol, (b) gel, and (c) flocculation and precipitation. (Reproduced with permission from reference 3. Copyright 1979.)

are objects for which D_f is less than the dimension of space: $D_f < 3$. Because D_f has many of the properties of a dimension but is often fractional, it is called fractal dimension [12].

Fractal geometry also describes surface fractals, in which the surface area S is related to the radius by a fractional power:

$$S \propto r^{D_s}$$

where D_s is called the surface fractal dimension and is smaller than D but larger than $D - 1$.

In sum, objects with $D_f < D$ are referred to as mass fractals, and objects with $D_f = D$ and $D > D_s > D - 1$ have very rough surfaces and are referred to as surface fractals [12]. Brinker and Scherer [8] give a tree as an example of a mass fractal, because its branches become wispier as they move away from the trunk, so the mass of the tree increases more slowly than the cube of its height. On the other hand, they think of a piece of paper crumpled into a ball as a surface fractal because the area increases as the radius of the ball cubed ($D_s = 3$), but it is not a mass fractal because its mass also increases as r^3. Rarity [12] notes that a two-dimensional projection of a tree's root system looks very much like an overhead view of a river delta. Both are random fractals, and the similarities arise from similar large-scale rules of growth that seem not to be influenced by their detailed constituents.

For Rarity the attraction of fractal theories of nature is that the application of large-scale geometrical rules can lead to simple universal properties in many important and complex random-growth processes such as colloidal aggregation [12]. The fractal dimensions of objects with radii on the order of 10^2–10^4 Å may be measured in a small-angle x-ray scattering (SAXS) or static light-scattering experiment [13].

Application of scaling concepts has resulted in a much deeper understanding of the structure of colloidal aggregates and the kinetics of their formation [14]. Two distinct, limiting regimes of irreversible colloid aggregation have been identified: diffusion-limited (DL) and reaction-limited (RL) cluster aggregation. Lin et al. [14] compared electron micrographs of aggregates of colloidal silica with colloidal gold and polystyrene and showed compelling visual evidence for universalities in their reactions (Figure 3.8). Scaling analysis of light-scattering data

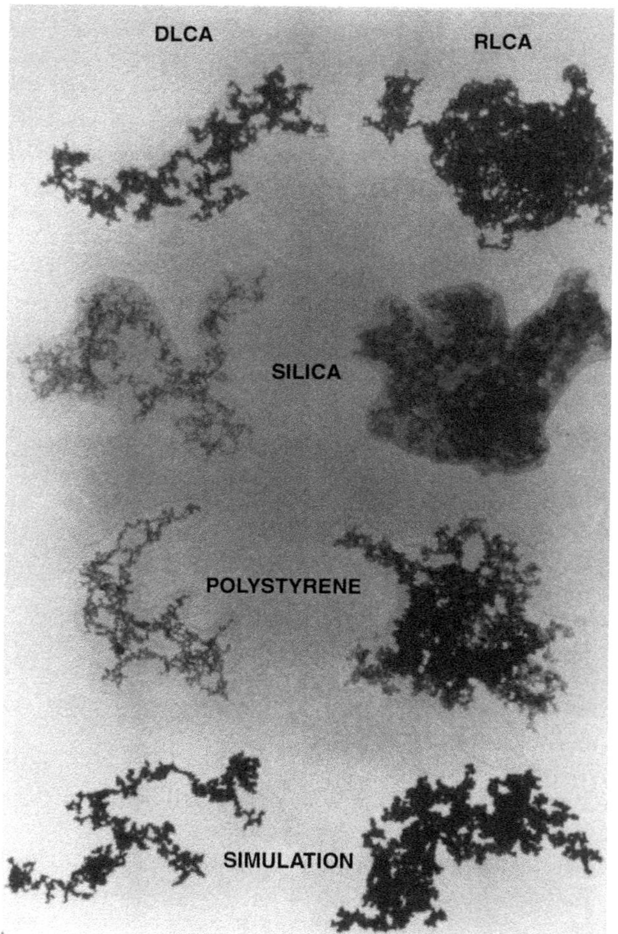

FIGURE 3.8 Transmission electron micrographs of typical clusters of gold, silica, and polystyrene colloids, prepared by both diffusion-limited and reaction-limited cluster aggregation and by computer simulation. There is a striking similarity in the structure of the clusters of different colloids in each regime. (Reproduced with permission from reference 14. Copyright 1989.)

used to compare the behavior of the three systems under both DL and RL aggregation conditions provided convincing experimental evidence to the authors that the two regimes of aggregation are indeed universal. They used static light scattering to measure the fractal structure of the aggregates and quasielastic light scattering (QELS) to measure the aggregation kinetics and to probe the shape of the cluster mass distributions.

On the basis of experimental observations of aggregation, including electron microscopy and QELS, Matsushita [15] concluded that at least some colloidal aggregates can be well described in terms of fractal geometry and that the size distribution of aggregates and the kinetics of their aggregation can also be related to their fractal structure.

Aubert and Cannell [16] showed that silica spheres about 22 nm in diameter in an aquasol (Ludox AS-40) can form aggregates through addition of salt to the sol, that is, salting out, with fractal dimensions of 1.75 ± 0.05 and 2.08 ± 0.05. Slow aggregation (RL) always yielded clusters with $D = 2.08 \pm 0.05$, whereas fast aggregation (DL) produced clusters with either $D = 1.75 \pm 0.05$ or $D = 2.08 \pm 0.05$, depending on the pH and silica concentration. Moreover, aggregates with $D = 1.75$ were observed to restructure to those with $D = 2.08$.

Legrand et al. [17] applied the concept of fractal dimensionality to describe texture and porosity, giving a better description of the process underlying the tiling of the surface of silica powders. Taking fractality into consideration, they were able to interpret consistently different experimental results (isotherms and SAXS) and to reveal clearly significant differences between pyrogenic silica and samples of silicas prepared in an aqueous medium. They found a difference in fractal dimensionality between pyrogenic and precipitated silicas of about 0.3; the pyrogenic silica was characterized by a smaller dimensionality. Suggested reading about the fractal approach to colloid science includes the letter to *Nature* by Mandelbrot and Evertsz [18] in addition to all the references cited in this section.

SILICA NUCLEATION, POLYMERIZATION, AND GROWTH: PREPARATION OF MONODISPERSE SILICA SOLS

The possibility of producing uniform and reproducible ceramic micro-structures through sol–gel techniques requires a better understanding of the processes of silica nucleation, growth, and polymerization. In other areas of technology, monodisperse silica sols of large and uniform particle size are required, and the field of colloidal silica synthesis is being broadened and extended from the classic aquasols of 5- to about 100-nm particle diameter to sizes up to 1 and 2 μm in the form of aquasols, organosols, and powders.

The classic silica aquasols 5–100 nm in particle diameter are prepared by nucleation, polymerization, and growth in aqueous systems. The particle size range can be extended to at least 300 nm by autoclaving. Silica organosols can be obtained by transferring the aquasols to an organic solvent.

Monodisperse silica sols of particle size up to about 2–3 μm in diameter were first obtained by Stober in an alcohol–ammonia system with enough water to hydrolyze a silane precursor [22].

The general theory of nucleation and polymerization in aqueous systems, in which silica shows some solubility, is discussed in detail in Iler's book [3]. However, very little was known at the time the book was published (1979) about the polymerization of silica when $Si(OH)_4$ is formed in nonaqueous systems. Progress made up to 1990 in the understanding of the hydrolysis and condensation of silicon alkoxides that leads to silica gels or to silica sols of large particle diameter are lucidly discussed by Brinker and Scherer [8]. Brinker's chapter in this book includes a clear explanation of the difference between hydrolysis and condensation of aqueous silicates and silicon alkoxides.

B.A. Keiser's contribution to this book (the introduction to the section "Preparation and Stability of Sols") constitutes an excellent introduction to silica nucleation, polymerization, and growth in both aqueous and alcoholic systems for the preparation of silica sols. Yoshida's chapter (Chapter 6) focuses on industrial development in the preparation of monodisperse sols from sodium silicate and predicts further progress in the development of silica sols that have shapes other than spherical, such as elongated, fibrous, and platelet. Colloidal silica particles with these shapes show novel properties and open the possibility of new industrial applications.

Although they are not discussed in Yoshida's chapter, new silica aquasols, organosols, and powders with particle diameters from 100 to about 2000 or 3000 nm (2–3 μm) are already being produced on industrial or semi-industrial scale in Japan for special uses.

Van Blaaderen and Vrij's chapter (Chapter 8) constitutes an excellent contribution to the understanding of the mechanisms of nucleation and growth of silica spheres in the alcohol–ammonia–water system to achieve particle sizes much larger than those of the classic silica sols synthesized in water. Kozuka and Sakka (Chapter 10) provide detailed conditions and the mechanism of formation of micrometer-sized particles of gels synthesized in highly acidic solutions of tetramethoxysilane (TMOS).

Using ultracentrifugation and a series of advanced experimental techniques such as small-angle neutron scattering (SANS), photon correlation spectroscopy (PCS), and ^{29}Si NMR, Ramsay et al. (Chapter 7) were able to study in detail the oxide–water interface of silica aquasols ~7 to 30 nm in particle diameter and found that all the sols investigated contained a significant proportion of oligomeric silicate species that could be associated with the core particle surface. This fact may not only explain the high charge density and exceptional stability of silica sols, but it may also affect the surface and pore structure of gels obtained by dehydrating these sols. Legrand et al. [17] also found evidence of polysilicic acid on the surface of silica aquasols as residues of the synthesis process that may be responsible in some cases for high counts of surface silanols.

In addition to these chapters, a picture of the state of the art as of 1991 in the synthesis of silica by hydrolysis and condensation of alkoxides is given in this book by Coltrain and Kelts (Chapter 48) and by Schmidt and Bottner (Chapter 49) and in the open literature by Minihan and Messing [23] and Bailey and Mecartney [24].

In sum, using as a landmark the 1979 book by Iler, in which he states that little is known about the polymerization of silica when $Si(OH)_4$ is formed in nonaqueous systems, one can say today that not only the knowledge of the theory of nucleation and polymerization of silica in aqueous systems is being fairly well consolidated, but also that significant progress is being made in the understanding of the hydrolysis and condensation of silicon alkoxides leading to silica gels and large-particle-size silica sols.

STABILITY OF SILICA SOLS

Derjaguin [25] distinguished three types of stability of colloidal systems.

1. *Phase stability*, analogous to the phase stability of ordinary solutions.
2. *Stability of disperse composition*, that is, stability with respect to change in dispersity (particle size distribution).
3. *Aggregative stability*, the most characteristic for colloidal systems. *Colloidally stable* means that the particles do not aggregate at a significant rate [6]. As explained earlier, *aggregate* is used to describe the structure formed by the cohesion of colloidal particles.

On the basis of work done in the years just before World War II, Derjaguin and Landau [26] were able to explain in 1941 many of the complex phenomena involved in aggregative stability on the basis of forces of interaction between colloidal particles, namely the van der Waals–London forces of attraction and the electrostatic forces of repulsion. In the meantime, as a result of theoretical investigations and calculations performed in the years 1940–1944 and without the benefit of much of the literature that appeared during the war years, Verwey and Overbeek [7] formulated a theory of stability of lyophobic colloids and published it as a book in 1948. Because their ideas were virtually identical to those of Derjaguin and Landau, the theory became known as the DLVO theory, and it appeared to allow the science of colloids to enter a new stage, less empirical, in which the experimental study of better defined objects could be guided by more quantitative ideas rather than by qualitative rules or working hypotheses. For example, the DLVO theory provided an explanation and a quantitative refinement of the empirical, qualitative Schulze–Hardy rule. It also predicted correctly the small influence of the co-ion for electrolytes of the 1–2 and 2–1 types and provided a basis for the calculation of the potential energy–distance relationship of colloidal particles.

The enunciation of the theory of stability of colloids stimulated an enormous volume of literature, most of which involved use of the theory in experimental and applied work. On this basis, Derjaguin wrote in 1989 a new monograph setting forth the present state of the theory of colloid stability [25].

The advent of concentrated monodisperse silica sols in the 1950s appeared to offer an ideal model to test the DLVO theory: a stable system of solid spheres with a particle diameter that could be varied in a broad range from about 5 to 100 or 300 nm. However, it soon became quite evident to many researchers, both in Iler's laboratories and elsewhere, that silica sols do not conform to the DLVO theory as originally formulated [27–32]. As an example, Figure 3.9 illustrates the problem, showing an area in the stability–pH curve of experimentally proven relative stability (metastability) of silica sols at around the zero point of charge where the theory predicts minimum stability. In addition, the plot of experimental values shows an unexpected minimum in what ought to be, according to the theory, a gradual rise from pH 2 to maximum stability at higher pH values. Another example is constituted by Horn's measurements [33]. Horn made direct studies of the forces between amorphous silica surfaces and found them to be repulsive in nature down to a few angstroms of separation. Horn did not observe in the case of silica what Dumont considers the "undeniable existence of overwhelming positive forces of attraction," a key factor in the DLVO theory (see the introduction to the section "Surface Chemistry of Silica").

Two possibilities are open: amend the DLVO theory or design a new theory based on typical properties of colloidal silica. Lyklema [34] believes the first approach is promising if features such as solvation layers, hydrophobic bonding, and surface charge fluctuations are incorporated into the DLVO theory.

Healy (Chapter 20) and Dumont also prefer the first approach. Healy sets down a model based on the control of coagulation by surface steric barriers of polysilicate plus bound cations. Healy's electrosteric barrier model is designed to stimulate new experimental initiatives in the study of silica sol particles and their surface structure.

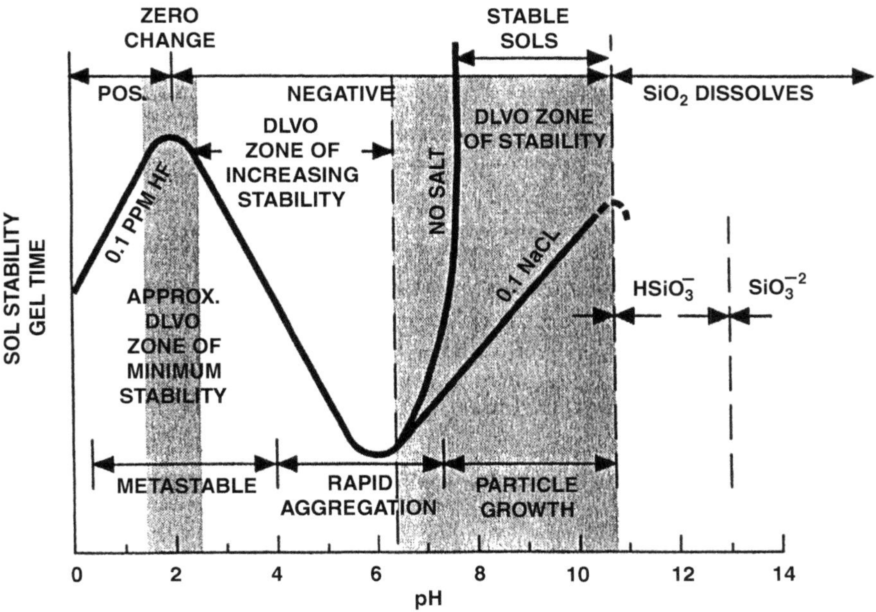

FIGURE 3.9 Effect of pH on the stability (gel time) of the colloidal silica–water system. Thick solid lines represent experimental results [3]. Shaded areas and white area in between are approximate zones corresponding to behavior predicted by the DLVO theory [7,26], some in contrast with experimental results: minimum stability predicted at pH around 2–3, increasing stability predicted at pH between 3 and 6–8, and maximum stability predicted at pH higher than 8. (Reproduced with permission from reference 3. Copyright 1979 John Wiley & Sons, Inc.)

Dumont believes that many particular aspects of the stability of silica hydrosols could be explained not only by the low value of the Hamaker constant but also by the relative importance of the static term of the Hamaker equation.

Yates [35] pointed out that if an approach such as the DLVO theory is to be generally applicable, it should be extended at least to include solvent effects and adsorption. It should also allow for substantial chemical activation energies for the formation of surface chemical bonds in irreversible coagulations and particle growth. However, Yates believes that a better approach might be to attempt a thermodynamic rather than a kinetic analysis of changes in the free energy of the system as a whole, attendant upon coagulation and separation into a more dilute suspension and a coagulated phase. Yates pointed out that Langmuir [36] suggested such an approach many years ago that was further extended by Iler [3], and although the mathematical treatment Langmuir attempted may not be applicable, the basic idea retains merit.

What Derjaguin considers the central issue of colloidal solutions remains largely unresolved for silica sols. This book mentions the ideas of the proponents of both the kinetic and the thermodynamic approach to the problem of stability of silica sols and is intended to stimulate the continuation of the healthy controversy started at the R.K. Iler Memorial Symposium. In this manner a consensus should eventually be reached that will allow the establishment of common quantitative parameters in the treatment of stability of silica sols and other disperse phase materials composed of polyvalent atoms linked by strong covalent bonds and the explanation of their experimentally observed behavior.

SILICA SURFACE

Many of the adsorption, adhesion, chemical, and catalytic properties of silicas depend on the chemistry and geometry of their surfaces. Because of the importance of these properties in determining the practical applications of silica, its surface chemistry is a subject of intensive studies. Considerable progress was made in the 1960s and 1970s in determining the silica surface structure by the combination of thermogravimetry and IR spectroscopy with chemical surface reactions and deuterium-exchange methods. The state of the art in that general period is covered by classic textbooks and monographs by Hair [37], Kiselev and Lygin [38], and Little [39]. Information on the silica surface structure obtained specifically by transmission IR spectroscopy before 1980 has been summarized by Hair [40].

More demanding needs for materials for the new technologies of the 1980s required further understanding of the silica surface. Significant advances in computer instrumentation and the development of new techniques for surface analysis are now allowing significant progress in the elucidation of the structure of silica surfaces.

Discussions of old and new techniques used to study the surface of silica, including diffuse reflectance Fourier transform (DRIFT) infrared spectroscopy, ^{29}Si cross polarization magic-angle nuclear magnetic resonance (CP MAS NMR) spectroscopy, as pioneered by Maciel and Sindorf [41] and others, and deuterium-exchange methods, may be found in review essays by Unger [2] and in articles by Legrand et al. [17] and Kohler et al. [42]. The use of computational chemistry and computational vibration spectroscopy vis à vis experimental spectra of inelastic neutron scattering is discussed in monographs by Chuiko et al. [43], and Khavryutchenko et al. [44].

Brinker and Scherer [8] pointed out that the area of a surface is defined largely by the method of surface area measurement. Many of the measurements of surface areas in work reported before the 1980s were based on the method of determining monolayer capacity of an adsorbent molecule of known cross-sectional area. In the Brunauer–Emmett–Teller (BET) method [45] the apparent surface area is determined from nitrogen adsorption. However, because the nitrogen molecule surface area is 16.2 Å^2, this definition of the surface excludes microporosity that is accessible, for example, to water molecules.

For materials such as porous gels, the monolayer capacity of water should be a more accurate definition of the surface. Belyakova et al. [46] reviewed the adsorption behavior of water on porous silicates as revealed by water adsorption isotherms. Another method for surface area measurement that uses water as a structural probe is based on proton nuclear magnetic resonance techniques. Application of the NMR method to porous gels is discussed by Gallegos et al. [47].

The most modern méthods of dealing with irregular, rough surfaces are based on fractal geometry. Examples of the application of the fractal approach to the analysis of silica surfaces are included in the Legrand et al. monograph [17] and in Brinker and Scherer's textbook [8].

In 1934 Hofman [48] postulated the existence of silanol groups (≡Si—OH) on the silica surface. Various analytical techniques allowed silica scientists to confirm and expand the view of the silica surface in terms of silanol groups, siloxane bridges, and hydrogen-bonded water. It is now generally accepted that surface silicon atoms tend to have a complete tetrahedral configuration and that in an aqueous medium their free valence becomes saturated with hydroxyl groups, forming silanol groups. Silanol groups in turn may condense to form, under proper conditions, siloxane bridges: ≡Si—O— Si≡.

Silanol Groups, Siloxane Bridges, and Physically Adsorbed Water

Most of the following postulated groups involving Si—O bonds either as silanols or as siloxanes have been identified on the surface or in the internal structure of amorphous silica (Figures 3.10 and 3.11).

- single silanol groups, also known as free or isolated silanols
- silanediol groups, also called geminal silanols
- silanetriols; postulated but real existence not yet generally accepted
- hydrogen-bonded vicinal silanols (single or geminal), including terminal groups
- internal silanol groups involving OH groups, sometimes classified as structurally bound water
- strained and stable siloxane bridges and rings
- physically adsorbed H_2O hydrogen-bonded to all types of surface silanol groups

Silanol groups are formed on the silica surface in the course of its synthesis during the condensation–polymeization of $Si(OH)_4$ or as a result of rehydroxylation of thermally dehydroxylated silica when treated with water or aqueous solutions (Scheme I). The silanol groups on the silica surface may be classified according to their nature, multiplicity of sites, and type of association [17].

An isolated silanol includes an OH group located at a distance sufficiently far from neighboring hydroxyl groups to prevent hydrogen bonding. A silicon site of this kind is designated as Q^3 in NMR Q^n terminology, where n equals the number of bridging oxygens (SiO—) bonded to the central silicon. The isolated silanol shows a sharp band at around 3750 cm^{-1} in the infrared spectrum.

Germinal silanols are silanediols groups located in Q^2 silicon sites [1]. Their existence was postulated by Peri and Hensley [49] but not confirmed experimentally until the advent of solid state ^{29}Si CP MAS NMR spectroscopy. On the basis of this technique, Fyfe et al. [50] determined in the particular case of a commercial silica gel that the ratio of $Q^4/Q^3/Q^2$ silicon sites was 8.8/5.7/1.

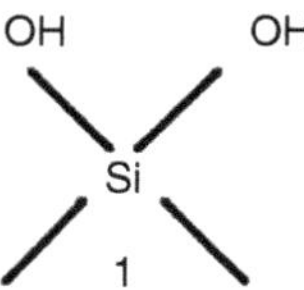

Vicinal or H-bonded or associated silanols are Si—OH groups located on neighboring Q^3 sites in which the OH to O distance is sufficiently small that hydrogen bonding occurs. Hydrogen bonding causes a reduction in the O—H stretching frequency, the magnitude of which depends on the strength of the hydrogen bond, and thus on the O—H to O distance [8]. The characteristic band of vicinal groups in the IR spectrum occurs at about 3660 cm^{-1} [40].

Germinal Q^2 silanol sites bonded to a neighboring Q^3 silicon through a single siloxane bridge also result in a hydrogen-bonded pair. The remaining OH experiences very weak hydrogen bonding. The identification of various configurations of surface silanol species by

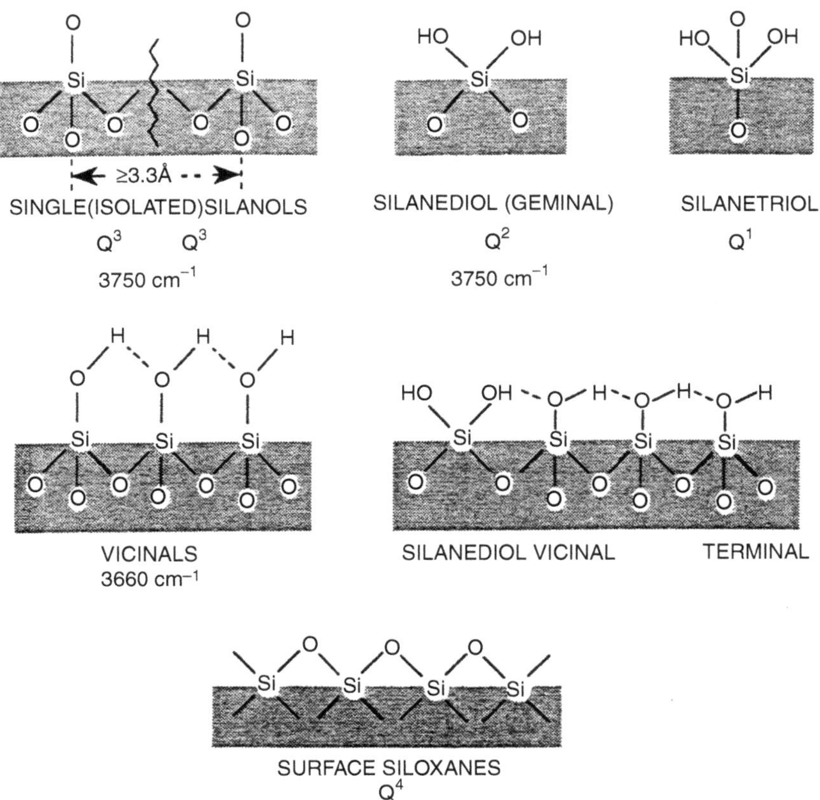

FIGURE 3.10 Silanol groups and siloxane bridges on the surface of colloidal silicas. Characteristic infrared bands at 3750 and 3660 cm^{-1} are shown for single and vicinal groups. Q^n terminology is used in NMR; n indicates the number of bridging oxygens (—O—Si) bonded to the central silicon ($n = 0–4$).

high-resolution IR spectroscopy is discussed by Hoffman and Knozinger [51].

Silanol groups may be found not only on the surface but also within the structure of the colloidal particles. These groups are designated internal silanols and in some cases are considered *structurally bound water*. Internal silanol groups are present within colloidal silica particles at various concentration levels depending on

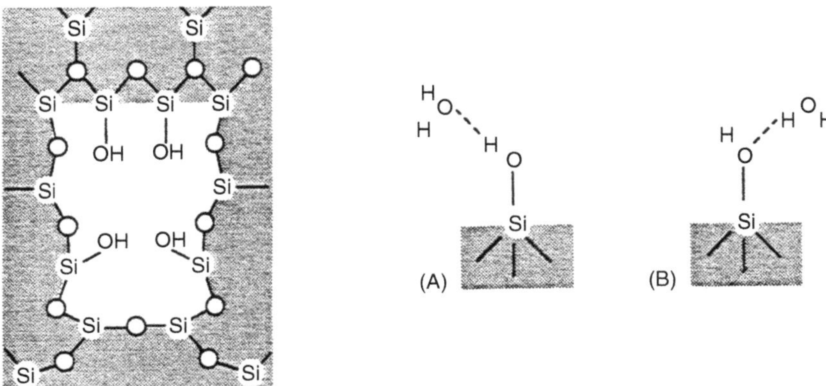

FIGURE 3.11 Left diagram: internal silanol groups. The fourth oxygen is above or below the plane of the paper. Right diagrams: two basic types of orientation of the water molecule with respect to the silanol group on the SiO_2 surface [53]. Type A, Oxygen atom in the H_2O molecule bonded to the hydrogen atom of the ≡Si—OH group. Type B, Hydrogen atom in the H_2O molecule bonded to the oxygen atom of the ≡Si—OH group.

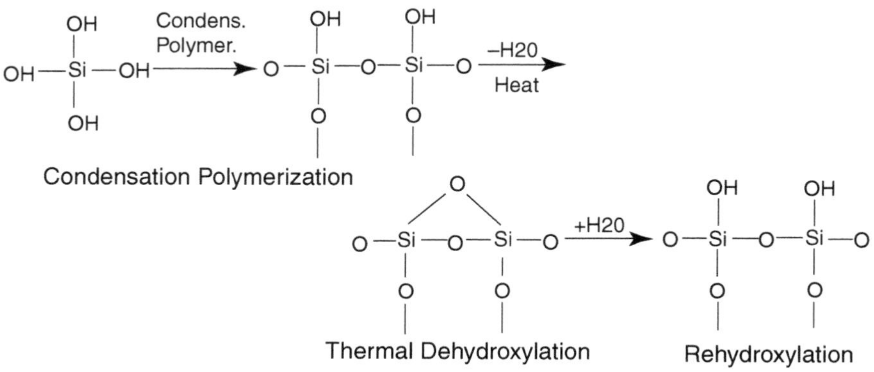

Condens. Polymer.
−H20 Heat
Condensation Polymerization
+H20
Thermal Dehydroxylation
Rehydroxylation

Scheme I.

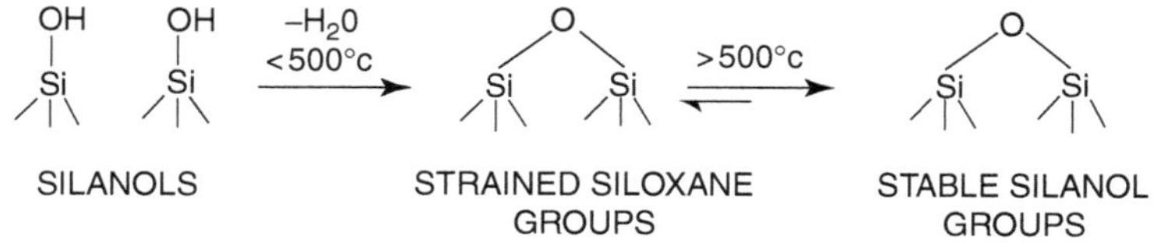

SILANOLS
$-H_2O$ <500°c
STRAINED SILOXANE GROUPS
>500°c
STABLE SILANOL GROUPS

Scheme II.

the synthesis temperature and other variables. According to Rupprecht [52], as much as 20% of the silanols present in hydrogels after surface dehydration may be internal. It is estimated that at about 600–800°C, and sometimes at lower temperatures, the internal silanols begin to condense. Complete evolution of internal silanols occurs at higher temperatures.

Surface and internal silanol groups may condense to form siloxane bridges. *Strained siloxane* bridges are formed on the hydroxylated silica surface by thermally induced condensation of hydroxyl groups up to about 500°C. At higher temperatures, the strained siloxane groups are converted into *stable siloxane* groups [2].

Strained siloxane bridges completely rehydroxylate upon exposure to water. Stable siloxane bridges also rehydroxylate but at a slower rate. For example, a wide-pore silica sample of 340 m^2/g calcined in air at 900°C took 5 years of contact with water at room temperature to rehydroxylate completely [53].

The silica surface OH groups are the main centers of adsorption of water molecules [53]. Water can be associated by hydrogen bonds to any type of surface silanols and sometimes to internal silanol groups. On the basis of thermograms obtained with various kinds of pure amorphous silicas, Zhuravlev [53] suggested that there may be two types of physically adsorbed water (physisorbed water) on the silica surface at a low degree of surface coverage, one with activation energies of desorption in the range of 6–8 kcal/mol and another one with activation energies in the range of 8–10 kcal/mol.

Impurities, such as Na, K, or Al, are sometimes picked up during the synthesis of aquasols in alkaline medium and may be occluded inside the colloidal particles, taking the place of the silanol protons (as with sodium or potassium) or forming isomorphic tetrahedra with an extra negative valence on the surface or inside the particles (as with aluminum) (Figure 3.12).

Surface silanol groups of silica aquasols stabilized in an alkaline medium exchange protons for the alkaline ion such as Na^+, K^+, or NH_4^+ (Figure 3.13). The surface

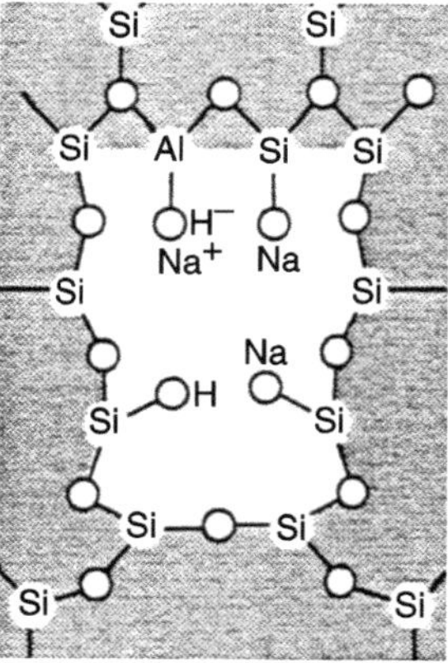

FIGURE 3.12 Occluded Na impurity and internal hydroxyl and Al impurities that form an isomorphic tetrahedron by substitution of Si. The fourth oxygen coordinated with Si is above or below the plane of the paper.

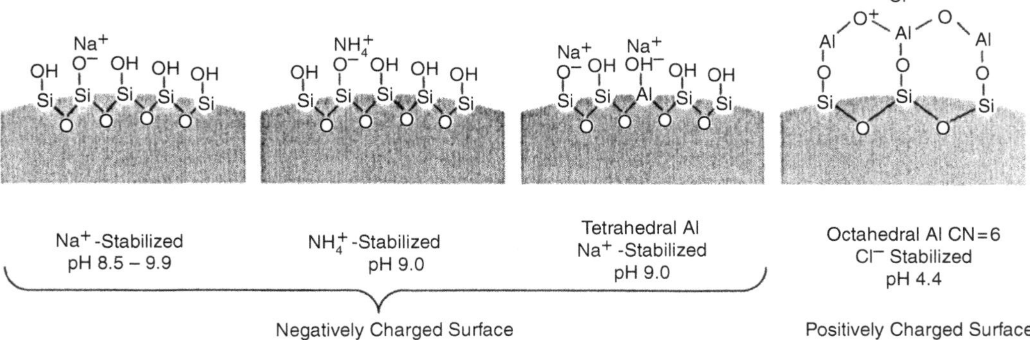

FIGURE 3.13 Sodium- and ammonium-stabilized and aluminum-modified surfaces of colloidal silicas.

silanol groups can be esterified, as in silica organosols, or silanized (silylated) (Figure 3.14). Derivatization of the silica surface is the basis for the use of silica in analytical and process chromatography.

Sindorf and Maciel [41,54–60] first used CP MAS NMR spectroscopy as a valuable surface-selective strategy to observe local silicon environments on the silica surface and to study the effects of dehydration and rehydration or derivatization on colloidal silicas in general. From a more recent work involving a combination of NMR techniques referred to as CRAMPS (combined rotation and multiple-pulse spectroscopy) emerges the picture of a possible silica surface structure of large hydrogen-bonded clusters of SiOH groups in complex arrays of hydrogen-bonding patterns and $Si(OH)_2$ islands isolated spatially from the $(SiOH)_n$ clusters. (See Figure 14.5 in Chapter 14.)

A description of surface heterogeneity dependent on the mode of preparation for both amorphous and crystalline silicas was attained by Vidal and Papirer (Chapter 31) by calculation of the distribution of the energy of adsorption of alkane probes on the solid surfaces. The experimental data were obtained after the changes of the silica surfaces upon chemical treatment or heat treatments by inverse gas chromatography (IGC).

The concentration of hydroxyl groups and the dehydration, dehydroxylation, and rehydroxylation of the silica surface, as well as the nature of the structurally bound water inside amorphous silica particles, have been studied in detail by the Russian school of Kiselev, Zhuravlev, and colleagues. Two sections that follow (Dehydroxylation of the Silica Surface and Hydroxylation of the Silica Surface) are annotated excerpts from an overview monograph presented by Zhuravlev at the R.K. Iler Memorial Symposium [53]. By comparing the views expressed here with the concepts advanced by authors such as Unger (Chapter 23), Legrand et al. [17], Maciel et al. (Chapter 34), Humbert et al. (Chapter 26), Morrow et al. (Chapter 24), Krasnansky and Thomas (Chapter 30), Vidal and Papirer (Chapter 31), Kohler et al. [42], Brinker et al. [61,62], and others whose work is more readily available in the West, and the views expressed in the work in Ukraine of Chuiko [63], Chuiko et al. [43], and Khavryutchenko et al. [44], the reader will have an understanding of the status of silica

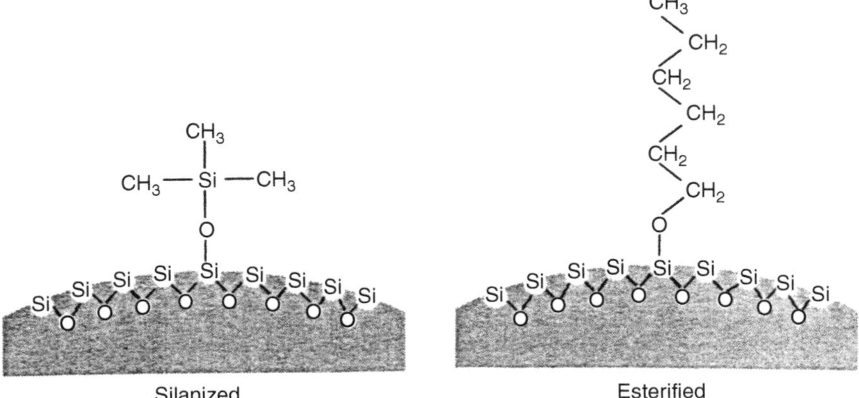

FIGURE 3.14 Silanized (silylated) and esterified surfaces of colloidal silica.

surface structure at the beginning of this decade. On this basis researchers can be prepared to follow the new insights that are developing in the 1990s, with the promising advent of relatively new and potentially very powerful techniques such as computational chemistry and the fractal approach, in addition to not yet developed novel methods for surface analysis.

Concentration of Hydroxyl Groups on the Silica Surface

The concentration of silanol groups on the silica surface expressed in number of OH groups per square nanometers, α_{OH}, is often called the *silanol number*. Zhdanov and Kiselev [64] first suggested that the silanol number of dehydrated but *fully hydroxylated* amorphous silica is a physicochemical constant independent of origin and structural characteristics of the silica. The numerical value Zhuravlev obtained for this constant is $\alpha_{OH} = 4.6$ OH groups per square nanometer (discussed later). This value is fairly close to the α_{OH} values obtained by other authors and is in good agreement with the surface concentration of silicon atoms on the octahedral face of β-cristobalite ($\alpha_{Si} = 4.55$ Si atoms per square nanometer) and on the rhombohedral face of β-tridymite represented in Figure 3.15 [65]. These two crystalline silicas are selected as references because they have densities and refractive indices close to those of amorphous silica.

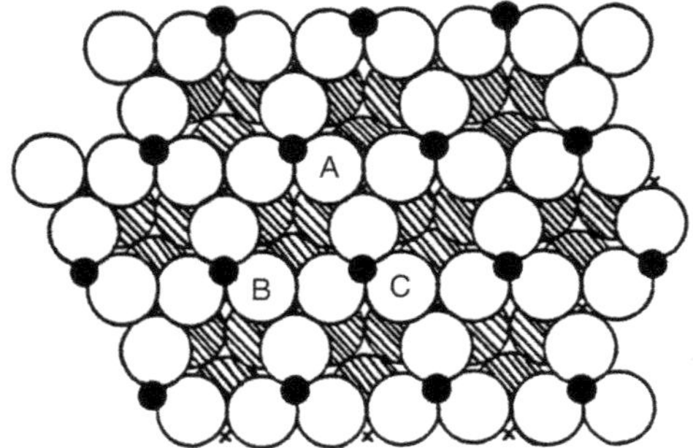

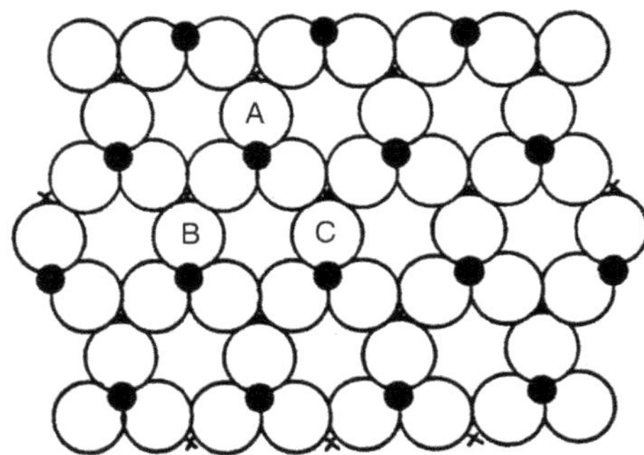

FIGURE 3.15 Top: Octahedral face of β-cristobalite. Bottom: Rhombohedral face of β-tridymite. The large open circles represent surface oxygen atoms. The small black circles represent surface silicon atoms, each of which carries one hydroxyl group when the surface is fully hydrated. The small crosses represent second-layer silicon atoms. The second-layer oxygen atoms are not shown, being directly under the second-layer silicon atoms. The partly visible third-layer oxygen atoms of β-cristobalite are shown as shaded circles. The third-layer oxygen atoms of β-tridymite are hidden because they are directly under the first-layer oxygen atoms. (Reproduced with permission from reference 65. Copyright 1961.)

Legrand et al. [17], on the basis of results obtained by chemical, thermogravimetric, and spectroscopic methods, showed how the structure of silica surfaces as *synthesized* (before rehydration) is strongly dependent on the method of preparation. For example, pyrogenic silicas as synthesized were found to have isolated OH statistically distributed in the order of 3 OH molecules per square nanometer, whereas precipitated silicas as synthesized showed a concentration of surface OH from 8 to 15 depending on the measuring technique. However, Legrand et al. attributed these high values to the presence of internal silanols and to a complex surface structure involving a very heterogeneous distribution of hydroxyls due to the presence of polysilicic acid.

Morrow and McFarlan (Chapter 24), using IR and vacuum microbalance techniques to study the accessibility of the silanol groups on silica surfaces to chemical and hydrogen–deuterium exchange probes, also found that the silanol number is larger in precipitated silicas ($\alpha_{OH} = 6.8$) than in pyrogenic silica as synthesized ($\alpha_{OH} = 3$), but the fraction of SiOH groups that exchange with a given probe is the same on both silicas, and this fraction decreases as the size of the exchange molecule increases. They observed that the reaction with H-bonded silanols on both silicas stopped well before that of the isolated silanols; this observation illustrates the blocking effect of the adjacent derivatized vicinal silanols, which prevents further reaction whatever the size of the reactant.

The Kiselev–Zhuravlev constant α_{OH} value of 4.6 was obtained with a deuterium-exchange method that distinguished between surface and bulk OH and with a mass spectrometric thermal analysis (MTA) method in conjunction with temperature-programmed desorption (TPD) [66].

One hundred fully hydroxylated silica samples prepared by different processes and having different structural characteristics such as specific surface area, type or size distribution of the pores, particle packing density, and structure of the underlying silica skeleton were analyzed two or more times each. Eighty-five percent of the 231 measurements were from samples with a specific surface area less than 400 m^2/g. The OH groups in ultramicropores, into which only water molecules can penetrate, were not classified as surface silanol groups but as structurally bound water.

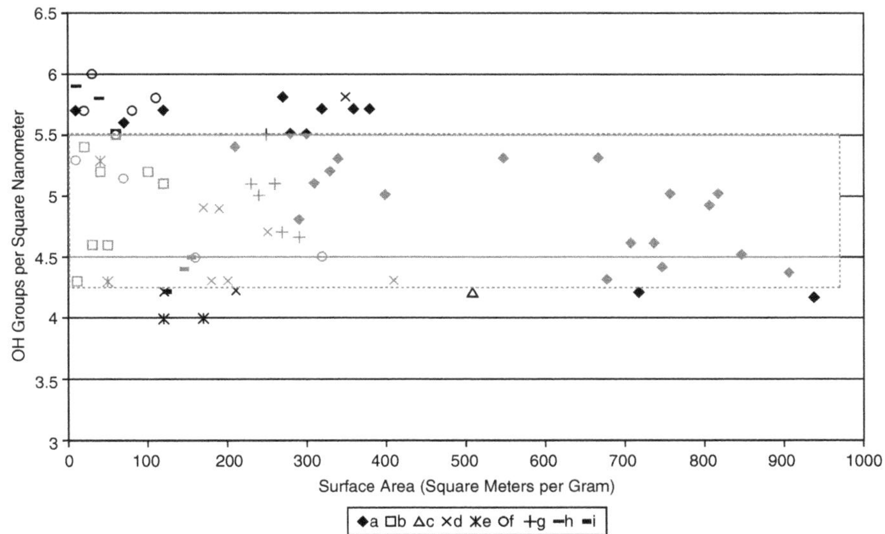

FIGURE 3.16 Concentration of the surface hydroxyl groups (the silanol number) α_{OH} for silicas having different specific surface areas S when the surface has been hydroxylated to a maximum degree. The letters a through i represent different types of amorphous silica (see text). The shaded area is a range of experimental data (100 samples of SiO_2 with different specific surface area, from 9.5 to $950\,m^2/g$). The average value of α_{OH} is 4.6 OH groups per square nanometer (least-squares method). (Reproduced with permission from reference 53. Copyright 1993.)

The silanol numbers obtained for the 100 samples are plotted versus specific surface area in Figure 3.16 [66]. The average silanol number of the 231 independent measurements is 4.9 OH groups per square nanometer. Calculations by the least-squares method gave 4.6 OH groups per square nanometer as an average silanol number.

Dehydration of the Silica Surface

Dehydration of the silica surface, that is, removal of physisorbed water, occurs at relatively low temperatures. The threshold temperature corresponding to the completion of dehydration and the beginning of dehydroxylation by condensation of surface OH groups is estimated to be $190 \pm 10°C$ [53].

In Zhuravlev's model [53] of adsorbed water the activation energy of water desorption increases from 6 to 10 kcal/mol as the fraction Θ of silica surface covered with physically adsorbed water decreases from $\Theta = 1$ to $\Theta = 0$.

Dehydroxylation of the Silica Surface

The concentration of OH groups on the surface decreases monotonically with increasing temperature when silicas are heated under vacuum. Most of the physisorbed water is removed at about 150°C. At 200°C all the water from the surface is gone so that the surface is made of single, geminal, vicinal, and terminal silanol groups and siloxane bridges. By about 450–500°C all the vicinal groups condense, yielding water vapor, and only single, geminal, and terminal silanol groups and strained siloxane bridges remain. The estimated ratio of single to geminal silanol groups on the surface is about 85/15 (a value close to the one reported by Fyfe et al. [50]) and is believed not to change with temperature, at least to about 800°C. Internal silanols begin to condense at about 600–800°C and in some cases at lower temperatures. At higher temperatures, up to about 1000–1100°C, only isolated (single) silanol groups remain on the silica surface.

At a sufficient surface concentration the OH groups make the silica surface hydrophilic. On the other hand, predominance of siloxane bridges on the silica surface makes the surface hydrophobic.[1]

In Figure 3.17 the concentration of surface hydroxyl groups α_{OH} is plotted versus the temperature of thermal treatment in vacuum up to 1100°C. The α_{OH} experimental values were obtained by analyzing, via the deuterium-exchange method, 16 different samples with specific surface areas ranging between 11 and $905\,m^2/g$. Porosity also varied within a wide range. For comparison, Figure 3.18 includes the values reported in 1966 by Kiselev and Lygin [38], obtained by analyzing 11 samples of aerogels, xerogels, and pyrogenic silica and a more recent measurement obtained via proton NMR

[1]Derjaguin (25) pointed out that the term "lyophilic colloids" is not really accurate, suggesting that it is better to speak of a "lyophilic state of colloids," because various types of lyophobic colloids, regardless of their material composition, can be brought into the lyophilic state by adsorption of surface-active substances or, in the case of silica, by rehydroxylation.

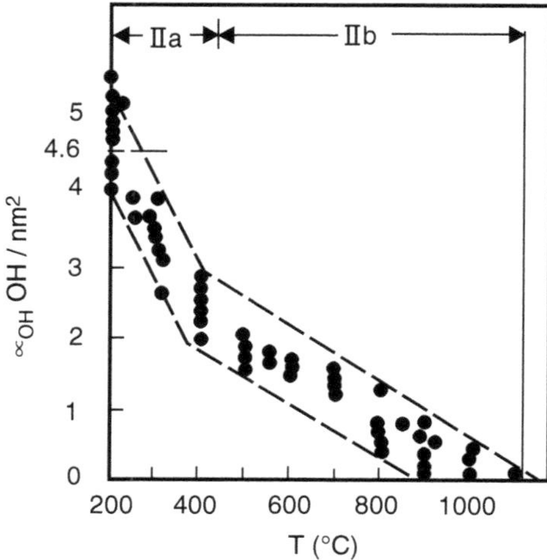

FIGURE 3.17 Silanol number as a function of the temperature of pretreatment in vacuo for different samples of SiO_2. The dashed lines indicate the area range of experimental data (16 samples with different specific surface areas from 11 to 905 m^2/g). Subregions of dehydroxylation: IIa, from 200 to ~450°C; and IIb, from 450 to ~1100°C. (Reproduced with permission from reference 53. Copyright 1993.)

TABLE 3.10
Temperature Dependence of α_{OH} and Θ_{OH}

Temperature of Vacuum Treatment (°C)	α_{OH}	Θ_{OH}
180–200	4.60	1.00
300	3.55	0.77
400	2.35	0.50
500	1.80	0.40
600	1.50	0.33
700	1.15	0.25
800	0.70	0.15
900	0.40	0.09
1000	0.25	0.05
1100	<0.15	<0.03

Note: Values for α_{OH} are reported as OH groups per square nanometer.

Source: Reproduced from reference 40. Copyright 1980 American Chemical Society.

analysis of Zorbax PSM 60 (a coacervated fully hydroxylated porous silica of about 400-m^2/g specific surface area) treated in vacuum at temperatures ranging between 200 and 900°C [67]. Despite the differences between samples, the value of α_{OH} at a given temperature of treatment and the descrease in α_{OH} as a function of temperature is similar in all cases. The most probable values of α_{OH} averaged from the plot of Figure 3.17 and the degree of surface coverage by hydroxyl groups Θ_{OH} are shown in Table 3.10 [53].

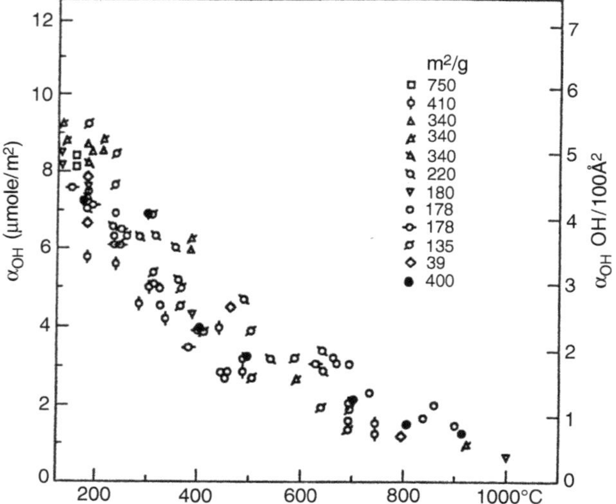

FIGURE 3.18 Concentration of surface hydroxyls on Zorbax PSM 60 by proton NMR [67] and on various samples of silica gel, aerogel, and pyrogenic silica by analytical methods other than NMR as a function of temperature. (Reproduced with permission from reference 98. Copyright 1966.)

Infrared analysis of the silica samples involved in this study showed no bound hydroxyls, only free hydroxyl groups, in the samples calcined in vacuum at temperatures 400°C and higher. On the basis of the infrared quantitative analysis of free hydroxyl groups, $\alpha_{OH\ free}$, and the total α_{OH} values obtained from Figure 3.17, the concentration of bound OH groups, $\alpha_{OH\ bound}$, throughout the temperature range 200–400°C and the concentration of siloxane bridges, $\alpha_{Si-O-Si}$, throughout the range 200–1100°C, can be calculated. In the lower part of Figure 3.19 the plots showing the monotonic decrease with temperature of silanol groups on the silica surface have been combined with plots of the corresponding increase of surface siloxane bridges with the estimated temperature limits for the presence of physisorbed water and geminal, vicinal, and isolated silanol groups on the silica surface.

Rehydroxylation of the Silica Surface

Rehydroxylation occurs and silanol groups are formed when the dehydroxylated silica surface reacts with water in a vapor or liquid state. Complete rehydroxylation was once thought possible only in samples that had been subjected to preliminary treatments at temperatures below 400°C. Rehydroxylation of silica samples activated at temperatures in the range of 400°C to 1000–1100°C is now known to be completely restored to the maximum hydroxylated state (α_{OH} = 4.6 OH groups per square nanometer) by treatment with water at room temperature. For samples subjected to extensive dehydroxylation, more

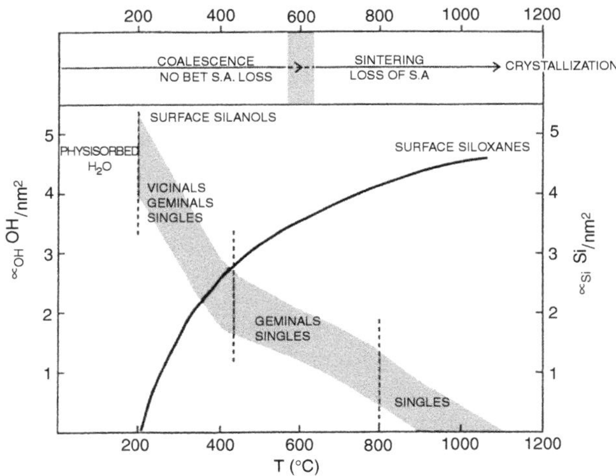

FIGURE 3.19 Different types of surface groups as a function of pretreatment of silica in vacuo (shaded area); average concentration of all OH groups is given. Broken vertical lines mark temperature limits for various types of silanol groups. Water may be physisorbed below 190°C. Hydroxyl groups bound through hydrogen bonds (vicinal silanols) may be present on the silica surface up to about 450°C, geminal silanols up to around 800°C, and free or isolated silanol groups up to around 1000°C. The curve represents the concentration of the surface Si atoms that are part of siloxane bridges. The upper part of the figure illustrates the effect of heat on silica powders obtained by drying silica aquasols synthesized in alkaline medium. (Reproduced with permission from reference 38. Copyright 1993.)

time is required for complete rehydroxylation. Thus, in the example given before, a starting wide-pore silica sample of specific surface area $S = 340\ m^2/g$ that had been calcined in air at 900°C ($\alpha_{OH} = 0.66$ OH groups per square nanometer), about 5 years was required, during which time the sample was in contact with water at room temperature, before the complete rehydroxylation of its surface was achieved ($\alpha_{OH} = 5.30$ OH groups per square nanometer). Accelerated rehydroxylation can be achieved by subjecting dehydroxylated silica samples to a hydrothermal treatment at 100°C. For example, an aerosilogel ($S = 168\ m^2/g$) was calcined in vacuo at 1100°C for 10 h. After this treatment its silanol number was $\alpha_{OH} = 0.06$ OH groups per square nanometer. The sample was then boiled in water for 60 h. The specific surface area was now 108 m^2/g and the silanol number $\alpha_{OH} = 4.60$ OH groups per square nanometer.

The rehydroxylation of a wide-pore silica gel sample calcined in air at 850°C and held in water at 100°C for periods covering 1 to 100 h was found to take 5–10 h for complete rehydroxylation. These and other results indicate that rehydroxylation of dehydroxylated silica (calcined at >400°C) in the presence of water requires considerable energy to activate the process of dissociative adsorption, E_d. Chemisorption of water appears to take place, resulting in the formation of hydroxyl groups bound through valence bonds to the SiO_2 surface. The reaction that takes place on the surface, which involves the breaking of the surface siloxane bonds, is the opposite of the condensation reaction.

Thus, the rehydroxylation of silica can be considered a process taking place in two subregions, below and above the temperature of the preliminary treatment at 400°C. This view agrees with the two subregions, IIa and IIb, observed in the hydroxylation process (Figure 3.20).

In subregion IIb the silica surface is occupied only by free hydroxyl groups and siloxane bridges. In subregion IIa, on the other hand, there are also OH groups bound together by hydrogen bonds (Figure 3.20).

For strongly dehydroxylated silica the surface concentration of siloxane groups is high (Figure 3.19). Because of the shift in the electronic density in going from O atom to Si atom, the formation of a hydrogen bond between the oxygen on the surface of the sample and water molecules is not favored. Thus, the siloxane surface is hydrophobic.

The hydroxylated surface with a predominance of silanol groups is hydrophilic. In the terminal ≡Si—O—H group, the electronic density is believed to become delocalized from the O—H bond to the neighboring Si—O bond. This delocalization permits the silanol groups to form strong hydrogen bonds with water molecules.

The data in Figures 3.19 and 3.20 appears to show that the high rate of rehydroxylation of silica subjected to preliminary activation at 200–400°C (subregion IIa) is apparently due to the fact that the concentration of the siloxane bridges is still low and each ≡Si—O—Si≡ bridge is surrounded by OH groups. Such a location of the surface groups weakens the Si—O bond in the siloxane bridge itself. During the rehydroxylation, additionally introduced

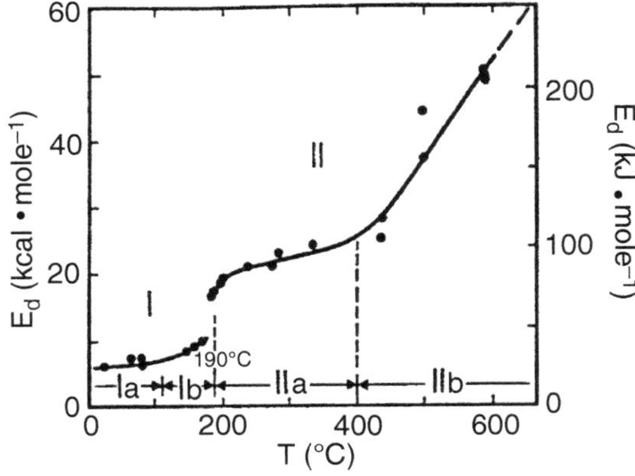

FIGURE 3.20 Activation energy of water desorption, E_d, as a function of temperature of the preliminary treatment of silica in vacuo. Ia, Ib — subregions of region I; IIa, IIb — subregions of region II. (Reproduced with permission from reference 53. Copyright 1993.)

water molecules first become adsorbed on silanol groups and have a direct effect on the neighboring weakened siloxane groups. This situation results in the splitting of the groups and the formation of new OH groups on the surface of silica.

Upon a preliminary activation of SiO_2 at >400°C (subregion IIb), the concentration of the siloxane bridges increases sharply. These bridges form whole hydrophobic regions on the surface, whereas the concentration of OH groups drops with an increase in temperature (Figure 3.19). But even under the condition of maximum activation (1000–1100°C) free OH groups still exist (Table 3.10), but at a large distance from one another. These OH groups act as the centers of adsorption when an additional amount of water is introduced, and rehydroxylation takes place first in the vicinity of the silanol groups. The localized areas of hydroxylated spots gradually expand. Rehydroxylation may be thought as proceeding along the boundary separating the hydrophilic and hydrophobic sections. Such rehydroxylation of the surface requires a considerable energy of activation of adsorption, E_a. Thus, rehydroxylation due to dissociative adsorption (chemisorption) of H_2O, with the splitting of siloxane bridges and the formation of new OH groups, proceeds according to a different mechanism, depending on the coverage of the surface with OH groups, Θ_{OH}.

At $1 \geq \Theta_{OH} > 0.5$ (subregion IIa, Figure 3.21) the activation energy of chemisorption, E_a, is close to zero. At $\Theta_{OH} < 0.5$ (subregion IIb, Figure 3.21) chemisorption proceeds very slowly at room temperature, but the rate of reaction increases sharply with an increase in the reaction temperature to 100°C. The physical adsorption of water on

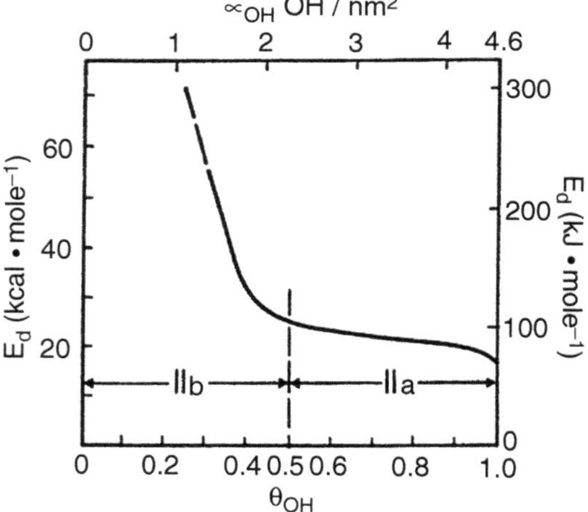

FIGURE 3.21 Activation energy of water desorption, E_d in region II, as a function of the surface concentration of OH groups, α_{OH}, or as a function of the surface coverage of SiO_2 with OH groups, Θ_{OH}. (Reproduced with permission from reference 53. Copyright 1993.)

silanol groups precedes chemisorption, which requires a much higher energy of activation: $E_a \gg O$.

A novel approach to the study of the hydroxylation and dehydroxylation of fumed and precipitated silicas is given in the study by Burneau et al. (Chapter 10), which was based on FTIR and Raman spectroscopy techniques.

Structurally Bound Water in Silica Particles

Deuterium-exchange measurements have shown that various types of amorphous dispersed silica contain not only surface hydroxyl groups but also structurally bound water inside the silica skeleton and fine ultramicropores. According to infrared spectral measurements [53], such bound water consists of silanol groups inside the silica sample (the adsorption band of stretching vibrations is about 3650 cm^{-1}). The distribution of OH groups between the surface and the bulk of the sample depends on a number of factors, but mainly on the method of preparation of the silica sample and its subsequent treatment.

Coalescence and Sintering

In the process of drying and heating, the particles of colloidal silica first coalesce extensively, but the area of contact between particles is not large enough to allow the loss of surface area to be detected by nitrogen sorption methods. The degree of coalescence is estimated by measuring the coalescence factor.[2] For silica powders made by drying an aquasol synthesized in an alkaline medium such as Ludox, the coalescence factor can be as high as 29 or 30, depending on particle size, even when the sol is freeze-dried or spray-dried at low temperatures. At this high range of coalescence the powders cannot be redispersed to the original level of discreteness by simply slurrying in water, although the diameter of the necks formed in between particles is not large enough to result in a loss of specific surface area as measured by nitrogen sorption. This range of coalescence is referred to as the range of incipient sintering. Beginning at about 600°C, again depending on particle size and degree of purity, coalescence may be so extensive that significant loss of BET surface area takes place. This is the onset of extensive sintering. Depending on the purity of the powder, especially Na content, pressed colloidal silica powders may sinter to theoretical density at about 1000 to 1100°C. Higher temperatures mark the onset of crystallization, generally with transition of amorphous silica to cristobalite. Figure 3.19 illustrates the overall effects of heat on the surface and bulk of colloidal silica.

[2]The coalescence factor is the present silica that has to be dissolved to restore the light transmission under standard conditions of a silica powder redispersed in water [68].

PARTICLE SIZE AND CHARACTERIZATION TECHNIQUES

The behavior of colloidal silica is based on its morphology and structure and on the chemistry of its surface. To fully characterize colloidal silicas one must measure not only the particle size, particle size distribution, surface area, and degree of aggregation, but also the structural and surface properties.

THE CONCEPT OF ZETA POTENTIAL

Colloidal silica particles have a negative charge on the surface as shown in the figures, surrounded by stationary positive charges. The zeta potential is the difference between the charge of this removable layer and that of the bulk of the suspending liquid. The negative change of the silica particles can be made positive by adsorption, for example, of octahedral aluminum ions of coordination number 6.

COLLOIDAL DISPERSIONS

ELECTROKINETIC EFFECTS AND THE CONCEPT OF ZETA POTENTIAL

The concept of the zeta potential. Most colloidal particles have a negative charge, as shown here, surrounded by stationary positive charges. These in turn are surrounded by a diffuse layer of negative charges. The zeta potential is the difference between the charge of this moveable layer and that of the bulk of the suspending liquid. For silica zeta potential is negative.

Methods of characterization utilized before the 1980s were discussed by Iler [3]. Significant advances in instrumentation have made possible the optimization of the classical techniques and the development of new ones.

One of the most important developments in the 1980s in the search for novel concepts to characterize colloidal systems such as silica sols and gels was the advent of the fractal approach. Another very important development in the past 10 yr was the application of ^{29}Si CP MAS NMR methods to the study of the silica surface. This technique made it possible, for example, to identify without ambiguity the presence of silanediol groups on the silica surface [55].

A survey of the physical and chemical techniques to characterize the surface structure of amorphous and crystalline silica is presented by Unger in this book (Chapter 23). Methods to measure particle size and particle size distribution and surface area are discussed by Kirkland (Chapter 40) and by Allen and Davies [69]. The use of some of these techniques by Morrow et al. (Chapter 24), Humbert et al. (Chapter 26), Vidal and Papier (Chapter 31), Kohler et al. [42], and Legrand et al. [17] to provide new insights into the silica surface structure was already mentioned in the section "Silica Surface" in this chapter.

Also discussed in that section was the information obtained by Zhuravlev, who used mainly classic methods, such as differential thermogravimetry combined with mass spectroscopy and deuterium-exchange. A novel and modern approach for the study of silica surfaces is based on the combined use of computational chemistry and inelastic neutron scattering spectroscopy [43,44].

As already mentioned in the discussion of fractals, small-angle X-ray scattering (SAXS) and static light-scattering techniques may be used to measure the fractal dimension of objects with radii on the order of 10^2-10^4 Å [12].

The use of one experimental method alone is usually not sufficient to obtain a complete picture to characterize the features of the various types of silica under study. The authors mentioned in the preceding text used complementary techniques to obtain a balanced view of the silica characteristics. The same approach was applied by Bergna, Firment, and Swartzfager (Chapter 53) to study amorphous silica coatings on the hydroxylated surface of a particulate alumina. In this case, transmission electron microscopy, BET nitrogen sorption, particle size measurements via Sedigraph and Coulter counter, zeta potential measurement, x-ray photoelectron spectroscopy (ESCA), high-resolution electron microscopy, diffuse reflectance infrared Fourier transform (DRIFT) spectroscopy, and secondary ion mass spectroscopy (SIMS) were used to characterize the silica coatings and to postulate a possible mechanism for the deposition of silica on hydroxylated surfaces.

The use of complementary techniques is well illustrated by Minihan et al. (Chapter 46) for measurements of pore size and pore size distribution of silica powders. Mercury intrusion and nitrogen sorption techniques are commonly used in the structural characterization of porous solids, often independently, despite the fact that very often the pore size distributions obtained by the two techniques fail to agree. For silicas Minihan et al. were able to demonstrate that mercury intrusion can lead to compression of silica structures and that this compression can account for differences in pore size distribution as measured by nitrogen sorption and may lead to misconceptions regarding the structure of the material under investigation.

SOL–GEL SCIENCE AND TECHNOLOGY

The growing and more sophisticated needs of modern technology have attracted the attention of scientists from diverse disciplines to the study of formation and properties of colloidal silica. A good example of the development and growth of an important area of the chemistry and physics of colloidal silica in the past decade by an interdisciplinary effort is provided by the extraordinary increase in the volume and the quality of published work on the so-called sol–gel science and technology. The term sol–gel science was coined in the 1950s to refer to the art of manufacturing materials by preparation of a sol, gelation of the sol, and removal of the solvent. In the field of ceramics and glass the term sol–gel applies not only to processes in which sols are gelled, but also to processes in which it is never clear whether discrete colloidal particles may at some time exist. If there is no evidence of a sol, Johnson [70] suggested that the use of the term sol–gel can better be thought of as an abbreviation of solution–gelation.

The sol–gel techniques have been practiced willingly and unwillingly since the first encounters with colloidal materials. One can imagine the frustration of the early Chinese and Egyptians laboring to find a suitable protective agent such as gelatin or gum arabic to manufacture ink before finally succeeding in avoiding gelation, or of the alchemists trying to prevent the gelation of their "aurum potabile," or of T. Bergman before finding in 1779 a way to stabilize what Hauser described as a "colloidal silicic acid" [71]. It is to the credit of contemporary material scientists to have converted a source of frustration into an art, then into a technology, and now into a science. At the time the term sol–gel became common, many colloid scientists felt like Molière's Monsieur Jourdan, who discovered at a very late age that, unknowingly, he had spoken prose all his life. In the same manner, some colloid scientists realized at one point that unknowingly and often unwillingly, they had been practitioners of sol–gel technologies.

What is now called sol–gel technology was originally developed by mineralogists for the preparations of homogeneous powders for use in studies of phase equilibria [72–74], by chemists for manufacturing nuclear fuel pellets [75,76], and by ceramists for the preparation of advanced ceramic materials [77,78].

The explosion of literature in the field of sol–gel science in the past decade made it difficult for newcomers to catch up to the state of the art and overwhelmed researchers already involved in the subject. The book by Brinker and Scherer [8] and the chapter by Brinker (Chapter 47) fill the need for a coherent account of the principles of sol–gel processing. Additional detailed discussions on this subject are included in the section on sol–gel technology in this book, in the collection of articles edited by Klein [72], and in the Proceedings of the Materials Research Society [79–81] and the Ultrastructure Processing meetings [82–84].

The development of sol–gel science and related colloidal techniques is contributing significantly to the development of the advanced materials required by the high technologies. For example, a crucial issue in the manufacture of ceramics for electronics is reliability. Improved processes are required to produce improved products, to minimize the number of rejects, and to achieve the high quality required for electronic parts. Product reliability (reproducible performance) comes with microstructure control, which requires process reproducibility, which in turn requires materials reproducibility. In this respect Turner [85] points out that the sol–gel process exhibits a number of advantages over conventional ceramic processing. The first advantage is increased chemical homogeneity in multicomponent systems. A second advantage of the sol–gel process is the high surface area of the gels or powders produced, which leads to relatively low sintering temperatures. A third advantage is that relatively high chemical purity can be maintained because of the absence of grinding and pressing steps. Finally, a fourth advantage of the sol–gel process according to Turner [85] is that a range of products in the form of fibers, powders, coatings, and spheres can be prepared with relative ease from simple solutions. Suggested reading on sol–gel science and its applications include the book by Brinker and Scherer [8] and references 72 through 85.

GELS AND POWDERS

Silica powders in the form of precipitated silica, pyrogenic or fumed silicas (aerosils), and silica gels are the largest volume specialty silicas. "Silica gel" is the term commonly used to refer to the xerogels obtained by drying silica hydrosols.

According to the Chemical Economics Handbook [86], the value of the precipitated silicas used annually in the United States is in the order of \$100 million. For fumed silicas and silica gels these values are about \$85 million and \$65 million, respectively. In comparison, the value for silica aquasols ("colloidal silica") is in the order of \$50 million.

The introduction to Part 5, "Silica Gels and Powders," by W. Welsh constitutes an introduction to the study of silica powders. Detailed accounts of the synthetic processes and applications of fumed silicas, silica gels, and precipitated silicas are given by Ferch (Chapter 14) and Patterson (Chapter 60). For scientists, silica powders are of special interest because they offer the opportunity of working with very pure systems with well-controlled ultimate particle size and specific surface area. One of the most important aspects of silica powders is their adsorptive properties. These properties are the subject of the work by Kenny and Sing (Chapter 44), which includes the crystalline zeolitic silica known as silicalite.

In contrast to organic gels, gels prepared from an aqueous precursor solution consist of a broad spectrum of structurally different silicate species. Chapter 45 (Wijnen et al.) describes the use of modern spectroscopic techniques in the study of the formation of aqueous silica gels. Silicon-29 NMR was used to study the oligomerization process of monomeric silicic acid, showing how the steady increase in monomer concentration induced dimerization at a critical monomer concentration followed by preferential formation of cyclic as opposed to linear trimeric silicate species at high pH values. SAXS was used to follow the aggregation of primary silica particles of molecular size (<1 nm), revealing reaction-limited aggregation as the process of gel formation at pH 4 in all systems studied. NMR and SAXS were combined with physical sorption techniques to follow the aging of the as-synthesized gel structures and thus allowed the proposal of a new model for the aging process. The Wijnen et al. model suggests that monomeric silicic acid is transported via the solution from the periphery of the aggregate into the core of the solution, as such contributing to a lower value for the observed fractal dimensionality of the aged silica gel.

Silica gels constitute one of the oldest silica commercial products in the colloidal state of dispersion, and yet much has to be learned about them. A deeper understanding of their structure, mode of formation, and properties should increase their relatively small rate of growth [now projected as about 1.5–2% for the next several years [86]] as a commercial product by consolidating current uses and expanding the range of applications.

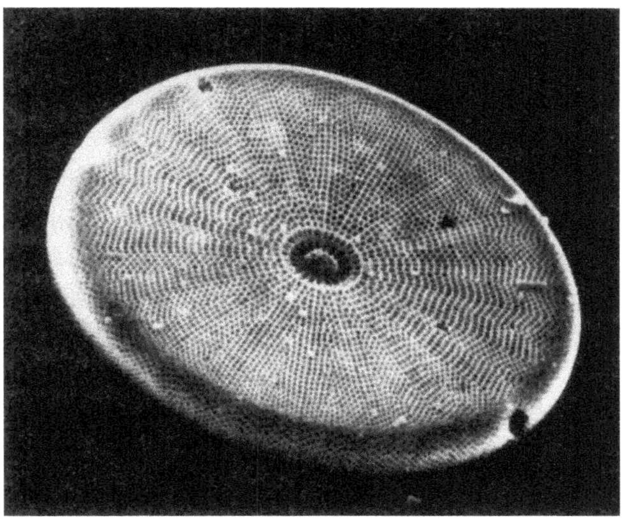

FIGURE 3.22 The diatom.

USES OF SILICA — FROM THE CAVES OF ALTAMIRA AND CRO-MAGNON TO SILICON VALLEY AND OUTER SPACE

When asked, "Where has colloidal silica been used?" Charles C. Payne answers that a more pertinent question is, "Where hasn't it been used?" (Chapter 54). This question applies also to silica in general. From the caves of Cro-Magnon and Altamira, where flint was used to make early tools and weapons, to Silicon Valley, where colloidal silica is used to polish sophisticated microcircuit parts, and to outer space, where silica has been used to help in the safe reentry of space vehicles, silica has played a role in the development of Western, and for that matter, world civilization.

In the modern world, colloidal silica powders and sols have different fields of application, with a few overlapping areas. As modern technology develops, new opportunities for silica proliferate, and new uses are constantly replacing the classic applications soon to become overshadowed by even newer uses.

Ferch, Payne, Patterson, and Falcone (Chapters 14, 54, 66, and 55, respectively) cover the history of applications of fumed silicas, silica gels and sols, and silicates. To this list the use of silica coacervates in analytical and process chromatography should be added. Furlong (Chapter 52) discusses the field of silica coatings; Ralph K. Iler was a pioneer in this area, and it is still dominated by his inventions. The starting point for uses of silica sols in modern technology is marked by the advent of monodisperse, concentrated, stable aquasols made of discrete particles [87–89]. For fumed and coacervated silicas, the starting point is constituted by the inventions of Kloepfer [90] and Iler [91], respectively, and by Iler again for silica-coated particulates [92].

At the risk of the list soon becoming obsolete, the current major applications of colloidal silica powders and sols can be summarized as follows [93]:

- colloidal silica: investment casting, silicon wafer polishing, and fibrous ceramics
- fumed silica: silicone rubber, unsaturated polyester, and specialty coatings
- precipitated silica: rubber, food, health care, battery separators, pesticides, and catalysts
- silica gel: food, health care, and industrial coatings

New challenges should open in the future for colloidal silicas and silicates. Opportunities exist, for example, in microelectronics, a $500 billion industry worldwide, spanning from consumer products such as compact disc players to supercomputers [94]. As Tummala said [94]:

> Microelectronics involve a broad spectrum of technologies, including semiconductors, packaging for processing information, magnetic and optical media for storing information, cathode ray tubes and liquid crystals with thin-film transistors for displaying information, electrophotography for displaying information, and optical fibers for transferring information. Computers in the 1990s are expected to perform functions not even dreamed in the

1970s. Microelectronic packaging is one of the key technologies that will make all the advances possible.

In fact, packaging has been identified recently as one of the 10 critical technologies for the advancement of mankind in the 1990s. Sol–gel technology — based on a better understanding of the formation and properties of sols and gels — and the coating of ceramic particles should represent the key for the invention of vastly improved ceramics and glass ceramics. Imagination is the only limit to the range of uses that scientists will find for old and new forms of colloidal silica to contribute to the quality of life of people around the world.

ACKNOWLEDGMENT

I would like to finish this overview by paying special homage to the creator of some of the most beautiful forms of silica, yesterday and today: the diatom (Figure 3.22). May the diatom continue oeuvres of art and science for many more millions of years without the impediment of our foolishness, but instead with the admiration of humanity.

REFERENCES

1. Gould, E. S. *Inorganic Reactions & Structures;* Henry Holt & Company: New York, 1957.
2. Unger, K. D. *Porous Silica;* J. Chromatogr. Library; Elsevier: Amsterdam, Netherlands, 1979, p. 16.
3. Iler, R. K. *The Chemistry of Silica;* Wiley: New York, 1979.
4. Kingery, W. D.; Bowen, H. K.; Uhlmann, D. R. *Introduction to Ceramics*, 2nd ed.; Wiley: New York, 1976.
5. Gaskell, P. H.; Eckersley, M. C.; Barnes, A. C.; Chieux, P. *Nature* **1991**, *350*, 675–677.
6. Everett, D. H. *Symbols and Terminology for Physicochemical Quantities and Units;* International Union of Pure and Applied Chemistry; Butterworths: London, 1971.
7. Verwey, E. J. W.; Overbeek, J. Th. *Theory of the Stability of Lyophobic Colloids;* Elsevier: New York, 1949.
8. Brinker, C. J.; Scherer, G. W. *Sol–Gel Science*; Academic: San Diego, CA, 1990.
9. Mahler, W.; Bechtold, M. F. *Nature* **1980**, *285*, 27–28.
10. La Mer, V. K.; Healy, T. H. *Pure Appl. Chem.* **1963**, *13*, 112–133.
11. Mandelbrot, H. H. In *A Fractional Approach to Heterogeneous Chemistry;* Avnir, D., Ed.; Wiley: New York, 1989.
12. Rarity, J.; *Nature* **1989**, *339*, 340–341.
13. Keefer, K. *Science of Ceramic Chemical Processing;* Hench, L. L.; Ulrich, D. R., Eds.; Wiley: New York, 1986.
14. Lin, M. Y.; Lindsay, H. M.; Weitz, D. A.; Ball, R. C.; Klein, R.; Meakin, P. *Nature* **1989**, *339*, 360–362.
15. Matsushita, M. In *The Fractal Approach to Heterogeneous Chemistry;* Avnir, D., Ed.; Wiley: Chichester, United Kingdom, 1989.
16. Aubert, C.; Cannel, D. S. *Phys. Rev. Lett.* **1986**, *56(7)*, 738–741.
17. Legrand, A. P. et al. *Adv. Colloid Interface Sci.* **1990**, *33*, 91–330.
18. Mandelbrot, B. B.; Evertsz, C. J. G. *Nature* **1990**, *348*, 143–145.
19. Barby, D. In *Characterization of Powder Surfaces;* Parfitt, G. D.; Sing, K. S. W., Eds.; Academic: New York, 1976.
20. *Colloids Surf.* **1992**, *63*(1/2); **1993**, *74*, 1.
21. Ulrich, D. D. *Chem. Eng. News* **1991**, *69*, 32.
22. Stober, W.; Fink, A. *J. Colloid Interface Sci.* **1968**, *26*, 62.
23. Minehan, W. T.; Messing, G. L. *Colloids Surfaces* **1992**, 63, 181–187.
24. Bailey, J. K.; Mecartney, M. L. *Colloids Surfaces* **1992**, *63*, 151–161.
25. Derjaguin, B. V. *Theory of Stability of Colloids and Thin Films;* Consultants Bureau: New York, 1989.
26. Derjaguin, B. V.; Landau, L. *Acta Physiochim.* **1941**, *14*, 633.
27. Allen, L. H.; Matijevic, E. *J. Colloid Interface Sci.* **1970**, *33*, 420.
28. Matijevic, E.; Allen, L. H. *Environ. Sci. Technol.* **1969**, *3*, 264.
29. Allen, L. H.; Matijevic, E. *J. Colloid Interface Sci.* **1969**, *31*, 287.
30. Allen, L. H.; Matijevic, E. *J. Colloid Interface Sci.* **1971**, *35*, 66.
31. Metijevic, E.; Allen, L. H. *J. Colloid Interface Sci.* **1973**, *43*, 217.
32. Milonjic, S. K. *Colloids Surfaces*, **1992**, *63*, 113–119.
33. Horn, R. G. *Abstracts of Papers, 200th ACS National Meeting;* American Chemical Society: Washington, DC, 1990; Division of Colloids and Surfaces, paper no. 41.
34. Lyklema, J. *Croatica Chemica Acta* **1977**, *50(1–4)*, 77–82.
35. Yates, P. C. *Abstracts of Papers, 200th ACS National Meeting;* American Chemical Society: Washington, DC, 1990; Division of Colloids and Surfaces, paper no. 235.
36. Langmuir, I. *J. Chem. Phys.* **1938**, *6*, 893.
37. Hair, M. L. *Infrared Spectroscopy in Surface Chemistry;* Dekker: New York, 1967.
38. Kiselev, A. V.; Lygin, V. I. In *Infrared Spectra of Surface Compounds;* Wiley: New York, 1975.
39. Little, L. H. *Infrared Spectra of Adsorbed Species;* Academic: New York, 1966.
40. Hair, M. L. In *Vibrational Spectroscopies for Adsorbed Species;* Bell, A. T.; Hair, M. L., Eds.; ACS Symposium Series 137; American Chemical Society: Washington, DC, 1980.
41. Maciel, G. E.; Sindorf, D. W. *J. Am. Chem. Soc.* **1980**, *102*, 7606–7607.
42. Kohler, J.; Chase, D. B.; Farlee, R. D.; Vega, A. J.; Kirkland, J. J. *J. Chromatogr.* **1986**, *352*, 275–305.
43. Chuiko, A. A.; Khavryutchenko, P.; Nechitajlov, P. *Abstracts of Papers, 200th ACS National Meeting;* American Chemical Society: Washington, DC, 1990; Division of Colloids and Surfaces, paper no. 87.

44. Khavryutchenko, V.; Ogenko, V.; Nechitajlov, P.; Sheha, E.; Musychka, A.; Natkaniec, I. *Abstracts of Papers, 200th ACS National Meeting;* American Chemical Society: Washington, DC, 1990; Division of Colloids and Surfaces, paper no. 100.
45. Brunauer, S.; Emmett, P. H.; Teller, E. *J. Am. Chem. Soc.* **1938**, *60*, 309.
46. Belyakova, L. D.; Dzhigit, O. M.; Kiselev, A. V.; Muttik, G. G.; Shcherbakova, K. D. *Russ. J. Phys. Chem.* (Engl. transl.) **1959**, *33*, 551.
47. Gallegos, D. P.; Smith, D. M.; Brinker, C. J. *J. Colloid Interface Sci.* **1988**, *124*, 186–198.
48. Hofman, V.; Endell, K.; Wilm, D. *Angewan. Chem.* **1934**, *30*, 539–558.
49. Peri, J. B.; Hensley, A. L., Jr. *J. Phys. Chem.* **1968**, *72:8*, 2926–2933.
50. Fyfe, C. A. et al. *J. Phys. Chem.* **1985**, *89*, 227–281.
51. Hoffman, P.; Knozinger, E. *Surf. Sci.* **1987**, *188*, 181.
52. Rupprecht, H. *Mitt. Dtsch. Pharm. Ges.* **1970**, *40*, 3–24.
53. Zhuravlev, L. T. *Colloids Surf.* **1993**, *74(1)*, 71.
54. Maciel, G. E. *Science* **1984**, *226*, 282–287.
55. Maciel, G. E.; Sindorf, D. W.; Batuska, V. J. *J. Chromatogr.* **1981**, *205*, 438–443.
56. Sindorf, D. W.; Maciel, G. E. *J. Am. Chem. Soc.* **1981**, *103*, 4263–4265.
57. Sindorf, D. W.; Maciel, G. E. *J. Phys. Chem.* **1983**, *87*, 5516–5521.
58. Sindorf, D. W.; Maciel, G. E. *J. Am. Chem. Soc.* **1982**, *86*, 5208–5219.
59. Sindorf, D. W.; Maciel, G. E. *J. Am. Chem. Soc.* **1983**, *105*, 1487–1493.
60. Sindorf, D. W.; Maciel, G. E. *J. Am. Chem. Soc.* **1983**, *105*, 3767–3776.
61. Brinker, C. J.; Kirkpatrick, R. J.; Tallant, D. R.; Bunker, B. C.; Montez, B. *J. Non-Crystalline Solids* **1988**, *99*, 418–428.
62. Brinker, C. J.; Brow, R. K.; Tallant, D. R. *J. Non-Crystalline Solids* **1990**, *120*, 26–33.
63. Chuiko, A. A. *Abstracts of Papers, 200th ACS National Meeting;* American Chemical Society: Washington, DC, 1990; Division of Colloids and Surfaces, paper no. 177.
64. Zhdanov, S. P.; Kiselev, A. W. *Zhur. Fiz. Khim.* **1957**, *31*, 2213.
65. Hockey, J. A.; Pethica, B. A. *Trans. Faraday Soc.* **1961**, *57*, 2247–2262.
66. Zhuravlev, L. T. *Langmuir* **1987**, *3(3)*, 316–318.
67. Bergna, H. E.; Letter to J. J. DeStefano, June 8, 1982.
68. Bergna, H. E.; Simko, F. A., Jr., U.S. Patent 3,301,635, 1967.
69. Allen, T.; Davies, R. *Abstracts of Papers, 200th ACS National Meeting;* American Chemical Society: Washington, DC, 1990; Division of Colloids and Surfaces, paper no. 37.
70. Johnson, D. W. *Ceramic Bulletin* **1985**, *64*, 1597.
71. Hauser, E. A. *Colloidal Phenomena*; McGraw-Hill: New York, 1939.
72. Roy, D. M.; Roy, R. *Am. Mineralogist* **1955**, *40*, 147.
73. Ewell, R. H.; Insley, H. *J. Res.* NBS **1935**, *15*, 173–186.
74. Barrer, R.; Hinds, L. *Nature* **1950**, *166*, 562.
75. Dell, R. M.; In *Reactivity of Solids;* Anderson, J. S.; Roberts, M. W.; Stone, F. S., Eds.; Chapman and Hall: New York, 1972; pp. 553–566.
76. Woodwead, J. L.; *Silic. Ind.* **1972**, *37*, 191–194.
77. Roy, R. *J. Am. Ceram.* **1956**, *39(4)*, 145–146.
78. Roy, R. *J. Am. Ceram.* **1969**, *52(6)*, 344.
79. *Better Ceramics Through Chemistry;* Brinker, C. J.; Clark, D. E.; Ulrich, D. R., Eds.; Materials Research Society: Pittsburgh, PA, 1984.
80. *Better Ceramics Through Chemistry;* Brinker, C. J.; Clark, D. E.; Ulrich, D. R., Eds.; Materials Research Society: Pittsburgh, PA, 1986.
81. *Better Ceramics Through Chemistry;* Brinker, C. J.; Clark, D. E.; Ulrich, D. R., Eds.; Materials Research Society: Pittsburgh, PA, 1988.
82. *Ultrastructure Processing of Ceramics, Glasses, and Composites;* Hench, L. L.; Ulrich, D. R., Eds.; Wiley: New York, 1984.
83. *Science of Ceramic Chemical Processing;* Hench, L. L.; Ulrich, D. R., Eds.; Wiley: New York, 1986.
84. *Ultrastructure Processing of Advanced Ceramics;* Mackenzie, J. D.; Ulrich, D. R., Eds.; Wiley: New York, 1980.
85. Turner, C. W. *Ceramic Bull.* **1991**, *70*, 1487–1490.
86. *Chemical Economics Handbook, Silicates and Silicas Report;* SRI International: Palo Alto, CA, 1990.
87. Bird, P. G., U.S. Patent 2,244,325, 1945.
88. Bechtold, M. F.; Snyder, O. E., U.S. Patent 2,574,902, 1951.
89. Rule, J. M., U.S. Patents 2, 577,484 and 2,577,485, 1951.
90. Kloepfer, H. German Patent 7,62,723, 1942.
91. Iler, R. K.; McQueston, H. J., U.S. Patent 3,855,172, 1971.
92. Iler, R. K., U.S. Patent 2,885,336, 1959.
93. *Opportunities in Special Silicas*; Kline & Co.: Fairfield, NJ, 1992.
94. Tummala, R. R. *J. Am. Ceram. Soc.* **1991**, *74*, 895–908.
95. Breck, D. W. *Zeolite Molecular Sieves;* Wiley: New York, 1974, p. 32.
96. Pauling, L. *The Nature of the Chemical Bond*, 3rd ed.; Cornell University Press: Ithaca, NY, 1960; p. 556.
97. Kingery, W. D.; Borren, H. K.; Uhlman, D. R. *Introduction to Ceramics*, 2nd ed.; Wiley: New York, 1976.
98. Kiselev, A. V.; Lygin, V. I. In *Infrared Spectra of Adsorbed Species;* Little, L. H., Ed.; Academic: London, 1966; p. 275.

4 Silicic Acids and Colloidal Silica

Horacio E. Bergna
DuPont Experimental Station

CONTENTS

Silicic acid can be in the form of a monomer or in the form of low molecular weight polymeric units. Monomeric silicic acid $Si(OH)_4$ has not been isolated or obtained in a concentrated solution without considerable polymerization, therefore is not a practical form to use in catalyst slurries to spray dry. Also not very practical is silicic acid formed in a way that does not separate it from the electrolyte products of the forming reaction. Residual electrolytes increase the ionic strength of the solution and result in destabilization followed by premature gelling of the silicic acid.

The preferred form of silicic acid to use in our preparations is polysilicic acid (PSA). The term "polysilicic acid" is used in the literature to distinguish certain products from monosilicic acid or "soluble silica," and from colloidal silica. For example, in [1], "The term polysilicic acid is generally reserved for those silicic acids that have been formed and partially polymerized in the pH range 1–4. They consist of ultimate silica particles generally smaller than 3–4 nm diameter" [1]. Monosilicic acid is represented by the formula $Si(OH)_4$. Polysilicic acid, that is oligomers of monosilicic acid, are polymers with molecular weights (as SiO_2) up to about 100,000, whether consisting of highly hydrated "active" silica or dense spherical particles less than about 3 nm in diameter and generally smaller than 304 nm diameter.

Polysilicic acid made of 1.5–4 nm diameter discrete particles may partially or completely polymerize into chains and three dimensional networks. Once a polysilicic acid solution has been exposed to alkaline conditions, it is rapidly converted to colloidal silica particles larger than 4–5 nm diameter.

Thereafter, silica assumes different characteristics and can be stabilized as colloidal silica sols in the pH range 8–10. Thus, colloidal silica as opposed to PSA is made of highly polymerized species or particles larger than about 5 nm.

The term polysilicic acid is justified, particularly by the fact that the silica has a very high specific surface area and contains a high proportion of SiOH groups per unit weight of silica. The terms may be somewhat misleading in the sense that in the low pH range where PSA is temporarily stable, the SiOH groups are essentially nonionized. Nevertheless, these silanol groups form silicon–oxygen–metal atom bonds with polybasic metal cations, as in the case of monosilicic acid. However, PSAs differ from the monomer in that they form addition complexes with certain classes of polar organic molecules through hydrogen-bond formation. Also, they can be isolated and esterified not only with alcohols under dehydrating conditions, but also with trimethylsilanol, even in aqueous solution.

The term "active" silica has sometimes been used in referring to PSA. For example, a distinction has been made by J. M. Rule [2] between "active" silica and other forms of polymeric or "colloidal" silica. "Active" silica is defined as "any silica in molecular or colloidal aqueous solution, in such a state of polymerization that when diluted with sodium hydroxide solution to a pH of 12, and concentration of about 0.02% SiO_2 at 30°C, the silica will be depolymerized substantially completely to monomer is not more than 100 min." The monomer is determined by the molybdic acid method. We prefer to use the term "polysilicic acid" rather than active silica.

PREPARATION OF PSA

Solutions of PSA can be prepared by adding a thin stream of sodium silicate solution with SiO_2:Na_2O ratio of 3.25:1.0 into the vortex of a violently stirred solution of H_2SO_4 kept at 0–5°C, and stopping the addition when the pH rises to about 1.7. Polysilicic acid solutions can be made continuously by bringing together solutions of sodium silicate and acid in a zone of intense turbulence and in such proportions that the mixture has a pH about 1.5–2.0.

Solutions free from the sodium salt can be obtained by hydrolyzing methyl or ethyl silicate in water at pH 2 with a strong acid as a catalyst for hydrolysis and temporary stabilizer for the silicic acid.

Our preferred method for the preparation of PSA is by deionization of a sodium silicate solution with an ion exchange resin at room temperature. In this way the PSA solution is substantially free of electrolytes and therefore more stable.

The pH is ca. 3.0 and the medium particle size is 20 A as calculated from the specific area obtained by Sears titation. Long exposure to fresh, strong ion exchange resin tends to lower the pH but the time required to reach pH values near 2 makes it impractical to use this procedure as a technique to stabilize the PSA. These facts are very important in view of the use of PSA in the preparation of catalyst/PSA products in an industrial scale.

After the first two Elan campaigns Rob Orlandi concluded that it would be necessary to improve the stability of the PSA made by ion exchange in order to obtain attrition resistant catalysts. Based on preliminary work done by John Orlan at the East Chicago Laboratory, we found that at pH 3 PSA solutions of 5 w/o SiO_2 could be stabilized by lowering the pH to pH 1.9–2.1. We can understand this behavior of PSA if we look at the stability curve of colloidal silica. See Figure 4.1.

Table 4.1 below shows the stability of PSA as estimated by Brookfield viscosity measurements at silica concentrations between 4 and 8 w/o, and 6, 21, and 35°C temperatures. Twenty-one degrees centigrade are considered "room temperature." Six degrees is an example of levels of stability obtained by refrigeration. Thirty-five degrees centigrade are presented as an example of a plant in the summer without air conditioning.

APPENDIX I: PREPARATION OF PSA 5% SiO_2

MATERIALS

(1) Deionized or distilled water
(2) DuPont JM grade sodium silicate: 29.6 w/o SiO_2, 9.1% Na_2O, 3.25 ratio
(3) Dowex grade HCR-W2-H ion exchange resin (1.8 equivalents/l)

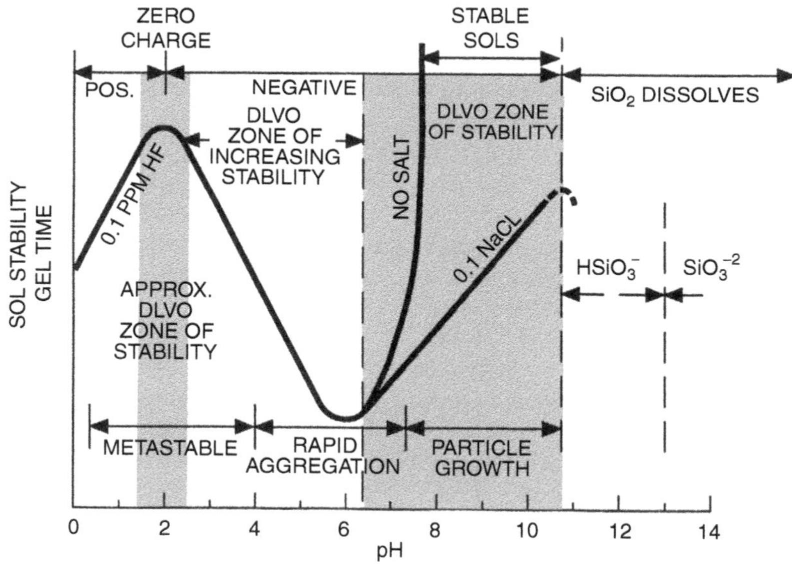

FIGURE 4.1 Effect of pH on the stability (gel time) of the colloidal silica–water system. Thick solid lines represent experimental results [3]. Shaded areas and white area in between are approximate zones corresponding to behavior predicted by the DLVO theory [7,26], some in contrast with experimental, results: minimum stability predicted at pH around 2–3, increasing stability predicted at pH between 3 and 6–8, and maximum stability predicted at pH higher than 8. (Reproduced with permission from reference 1.)

TABLE 4.1
Polysilicic Acid Viscosity (Brookfield LVDV-III Viscometer) — Temperature — Silica Concentration using pH = 3.0

			Viscosity at T = 6°C (cP)					Viscosity at T = 21°C (cP)					Viscosity at T = 35°C (cP)				
Elapsed time	Time, h	Day	4%	5%	6%	7%	8%	4%	5%	6%	7%	8%	4%	5%	6%	7%	8%
0	8:00	Mon	1	1	1	1	1	1	1	1	1	1	1	1	1	1	1
	9:00		1	1	1	1	1	1	1	1	1	1	1	1	1	1	1
	10:00		1	1	1	1	1	1	1	1	1	1	1	1	1	1	1
	11:00		1	1	1	1	1	1	1	1	1	1	1	1	1	1	143
	12:00		1	1	1	1	1	1	1	1	1	1	1	1	1	1	gel
	13:00		1	1	1	1	1	1	1	1	1	1	1	1	1	110	
	14:00		1	1	1	1	1	1	1	1	1	1	1	1	260	gel	
	15:00		1	1	1	1	1	1	1	1	1	20	1	1	gel		
8 h	16:00		1	1	1	1	1	1	1	1	1	gel	1	200			
24 h	8:00	Tue	1	1	1	1	250	1	gel	gel	gel		gel	gel			
	10:00		1	1	1	1	gel	1									
	12:00		1	1	1	1		1									
	14:00		1	1	1	1		26									
2 days	8:00	Wed	1	1	305	gel		gel									
	10:00		1	1	gel												
	12:00		1	1													
	14:00		1	1													
3 days	8:00	Thu.	1	45													
	10:00		1	60													
	12:00		1	150													
	14:00		1	gel													
1 week		Mon	gel														
		Tue															
		Wed															
		Thu															

Equipment

(1) 8 l SS beaker
(2) Air stirrer
(3) PH meter
(4) 4 l filter flask and 11 in. porcelain funnel
(5) Filter cloth and SLS 520 B $\frac{1}{2}$ folded filter paper
(6) Conductivity meter

Method

A 5 w/o SiO_2 PSA solution is prepared by diluting 1014 g of JM grade sodium silicate solution (300 g SiO_2) with 4985 g of distilled water in an 8 l SS beaker. The solution is stirred for a few minutes then filtered through SLS 520 B $\frac{1}{2}$ folded filter paper to give a clear water-like filtrate.

The clear sodium silicate filtrate pH 11.3 is stirred vigorously in an 8 l SS beaker at room temperature while Dow HCR-W2-H resin is added to reduce the pH to 3.0 ± 0.1. When the pH is around 6.8, excess resin is added to reduce the pH below 5.5 rapidly, thus avoiding microgel formation. The resin is filtered off and the clear filtrate is used within an hour.

The 5 w/o SiO_2 PSA solution has a pH of 3.0 ± 0.1 and a conductivity of 0.29 ± 1E uMho/cm @25°C. We refer to this solution as the 5% PSA solution.

To stabilize it, enough sulfuric acid is added to bring the pH to 1.9–2.1 (about 3 ml H_2SO_4 6.9% for 500 g of PSA 5 w/o SiO_2 solution).

REFERENCES

1. Iler, R. K., The Chemistry of Silica, Wiley, New York, 1979, p. 287.
2. Rule, J.M., U.S. Patent 2,577,484, DuPont, 1951.

Part 1

Preparation of Sols

Bruce A. Keiser
Nalco Chemical Company

Since the discovery of colloidal silica systems by Graham in the 1860s [1], investigators have been interested in understanding the formation of colloidal silica. In 1955 [2] and again in 1979 [3], Ralph K. Iler provided concise references regarding every aspect of silica chemistry, including the formation, preparation, and reactions of colloidal silica systems. Since his most recent publication, several investigators have added to our knowledge in the area. The information provided in this section is not intended to be an all-inclusive summary of work since the publication of Iler's book in 1979; rather, it will serve as an overview of recent findings and research directions.

The nucleation, polymerization and growth, and preparation of monodisperse sols is a focal point of research for many reasons. The nature of our understanding and the ease of formation make the silica system ideal for studying nucleation and growth of particles in a disperse state. Silica offers many diverse routes to similar products, allowing for customizing of product properties.

In recent years there has been renewed interest in silica systems. Some examples are taken from articles by Yamaguchi et al. [4], Righetto et al. [5], Chen et al. [6], and Philipse et al. [7]. These studies cover subjects such as kinetics of particle formation and growth, characterization, and application of colloidal silica. Work involving silica is not localized, but occurs in laboratories throughout the world. Reports issue routinely from countries such as the Netherlands, the United States, Great Britain, Japan, Australia, China, and Russia. Recent interest in silica formation and growth has focused on the area of sol–gel formation. Recently, a very thorough review was compiled by Brinker and Scherer [8].

This section contains chapters by Yoshida, Ramsay et al., van Blaaderen and Vrij, Arriagada and Osseo-Asare, Kozuka and Sakka, and Healy. These chapters provide an overview of work in silica nucleation, polymerization and growth, and preparation of monodisperse sols in a variety of systems. The work presented covers research areas currently under investigation throughout the world. The subjects include aqueous and nonaqueous systems, silicates and alkoxides, and acid- and base-catalyzed systems and can serve as a basis for understanding silica nucleation, polymerization, and growth of monodisperse sols.

Yoshida's chapter focuses on commercial methods of preparation that use silicates as raw materials and an aqueous continuous phase. The Ramsay et al. chapter elaborates on the impact that the method of preparation has on the physical characteristics and composition of silica particles prepared by methods discussed by Yoshida.

In contrast, the van Blaaderen and Vrij chapter is an excellent overview of the synthesis of silica from alkoxysilanes in aqueous systems. This overview includes both acid- and base-catalyzed systems. Further information regarding this area of interest is found in the section of this book entitled "Sol–Gel Technology."

The Arriagada and Osseo-Asare chapter presents findings from hydrolysis reactions carried out in mixed solvent systems. Here reverse micelles are used to orchestrate the hydrolysis of tetraethylorthosilicate.

The Kozuka and Sakka chapter demonstrates how nucleation and polymerization play a role in the formation of silica gels and subsequently silica-based glasses. The section concludes with Healy's chapter, which presents a model of the variations in silica sol coagulation effected by pH changes and addition of electrolytes. As a final note, a list of references [1–15] is included.

REFERENCES

1. Graham, T. *Ann. Chem.* **1862**, *121*, 36; **1865**, *135*, 65.
2. Iler, R. K. *The Colloidal Chemistry of Silica and Silicates*; Cornell University Press: Ithaca, NY, 1955.
3. Iler, R. K. *The Chemistry of Silica*; Wiley: New York, 1979, p. 287.
4. Yamaguchi, M.; Morikawa, H.; Soma, I. *Chem. Express* **1987**, *2(6)*, 333–336.
5. Righetto, L.; Polissi, A.; Comi, D.; Marcandalli, B.; Bellobono, I. R.; Ridoglio, G. *Ann. Chim.* (Rome) **1987**, *77(3–4)*, 437–455.
6. Chen, R.; Mao, Y. *Huangong Jinzhan* **1985**, *2*, 10–15.
7. Philipse, A. P.; Vrij, A. *J. Chem. Phys.* **1987**, *87(10)*, 5634–5643.
8. Brinker, C. J.; Scherer, G. W. *Sol–Gel Science: The Physics and Chemistry of Sol–Gel Processing*; Academic Press: Boston, MA, 1990.
9. Lasic, D. *Colloids Surf.* **1986**, *20(4)*, 265–275.
10. Yoshizawa, K.; Sugah, H.; Ochi, Y. *Sci. Ceram.* **1988**, *14*, 125–131.
11. Yoshida, A. *Kagaku Keizai* **1988**, *35(6)*, 22–28.
12. Coenen, S.; DeKruif, C. G. *J. Colloid Interface Sci.* **1988**, *124(1)*, 140–210.
13. Pnoomareva, E. I.; Abevova, T. A.; Dzhumagaziev, M. T.; *Kompleksn. Ispol's. Miner. Syr'ya* **1987**, *10*, 45–49.
14. Balboa, A.; Partch, R. E.; Matijevic, E. *Colloids Surf.* **1987**, *27(1–3)*, 121–131.
15. Rule, J. M., U.S. Patent 2,577,484, DuPont, 1951.

5 Science and Art of the Formation of Uniform Solid Particles

Egon Matijević
Clarkson University, Center for Advanced Materials Processing

Without science we should have no notion of equality, without art no notion of liberty.
W.H Auden

CONTENTS

INTRODUCTION

This presentation deals with an area of materials science, which is causing a great deal of excitement for academic and practical reasons, that is, with the so called monodispersed colloids. While the subject is an old one, dating back to Faraday's gold sols [1], it has become a topic of wide-spread interest only relatively recently.

There are some intriguing aspects of this field of science and technology which justify some comments. First, it is obvious from the title, that the size of the "solid particles" to be discussed is missing. Presently we have two groups of scientists interested in finely dispersed matter, some dealing in the nanometer and the others in the micrometer range. In many cases these systems are treated as separate worlds, yet it will be shown that they are closely connected, especially when these dispersions are prepared by chemical reactions in solutions.

Another interesting aspect is that the uniformity of particles has seldom been encountered in natural environments. Indeed, there are only a few examples of indigenous monodispersed systems, opal being probably the best known. Yet scientists have produced a large number of such solids over a broad range of modal sizes, of simple or mixed (internally or externally) compositions, of different structures, and in a variety of shapes. The reasons for the discrepancy between the natural occurrences and research achievements will be addressed in the presentation.

It will also be shown that, despite this progress, many questions remain to be resolved, which represent major challenges to the workers in the field. One of the problems is the actual mechanism (or mechanisms) by which uniform particles are formed. A recent development on the subject will be described. Another aspect is the predictability of shapes and structures of monodispersed colloids, to which there is still no answer.

Finally, the title of the lecture requires further explanation. The word "art" has two meanings in the English language: it is used to describe skilled craftsmanship or something of beauty. In dealing with well defined particles there is room for both of these interpretations.

EXAMPLES OF UNIFORM PARTICLES

There are many different techniques, which may yield particles of uniform size and shape, but none can be used to produce every kind of dispersions of desired properties. Due to the simplicity, versatility, and practicality, the precipitation from solutions is the method of choice. In the past this procedure yielded a number of monodispersed systems, although mostly by serendipity. About a quarter of a century ago systematic studies were initiated, which resulted in a multitude of well defined dispersions as reviewed in several articles [2–7]. In most instances the processes have involved either mixing reactants or decomposing complexes, normally under mild temperatures and in moderate concentrations.

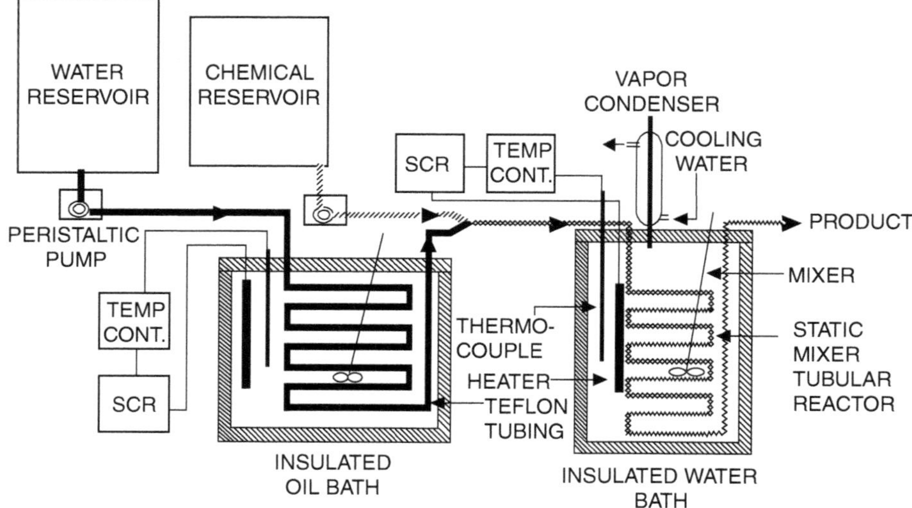

FIGURE 5.1 Schematic presentation of the apparatus for continuous flow precipitation.

Using the same technique, it is also possible to precipitate composite particulates. The latter can be homogeneous of exact stoichiometry, as exemplified by pure or doped barium titanates [9]. To achieve these conditions rapid mixing is required, such as by using the controlled double jet precipitation process. In contrast, slow precipitation results, as a rule, in internal inhomogeneity, that is, the composition changes from the center to the periphery, although the particles may still be perfectly spherical, as observed with mixed alumina/silica [10] or copper/lanthanum and copper/yttrium oxides [11].

Another major area of interest are coated particles. Again, it is possible to produce uniform surface layers of varying thickness on inorganic cores with either organic or inorganic coatings or organic cores with inorganic coatings [12,13].

It is essential to note that conditions needed to obtain a given material as a uniform dispersion are sensitive to a great degree to the experimental parameters, such as the temperature, concentration of reactants, pH, ionic strength, solvent composition, and so on. In some cases even a small change in these conditions can not only affect the particle uniformity, but it may yield solids of different chemical composition, structure, or morphology. This sensitivity explains the paucity of naturally occurring monodispersed particles. It should also be noted that the laboratory preparations require minutes or hours, while the processes in nature extend over geological times.

A final comment in this section refers to scaling up the production of well defined powders. Interestingly, the two materials first obtained in large quantities are the polymer latex and silica. The first is not a subject of this presentation, while silica will be treated extensively in this symposium.

One important aspect of scaling up precipitation processes, which yield uniform particles, is the necessity that any engineering design must consider the optimum conditions established in small batches. Much success has been achieved by using a plug-flow type of a reactor for continuous precipitation of a variety of uniform colloid dispersions, such as yttria, silica, aluminum hydroxide, and barium titanate [14]. The schematic presentation of this equipment is given in Figure 5.1 [9].

MECHANISM OF THE FORMATION OF MONODISPERSED PARTICLES

As one would expect, the understanding of the mechanism of the formation of monodispersed colloids by precipitation has been of major concern to workers in the field. For a long time the concept developed by LaMer was generally accepted; that is, such dispersions should be generated, if a short lived burst of nuclei in a supersaturated solution is followed by controlled diffusion of constituent solutes onto these nuclei, resulting in the final uniform particles. This mechanism is indeed operational in some, albeit limited, cases, and more often only at the initial stage of the precipitation process.

For a long time the writer of this article (abstract!) has been puzzled by some experimental observations. These perfectly spherical particles, obtained by precipitation in ionic solutions, exhibit X-ray characteristics of a known mineral which, for example, in case of ZnS is sphalerite. Obviously, it is not easily understood why would such homogeneously precipitated perfect spheres have crystalline characteristics. Importantly, low angle X-ray measurements showed these particles to be made up of essentiality identical nanosized subunits. Electron microscopy and other methods of evaluation demonstrated on numerous other dispersions, that particles of different morphologies and chemical compositions clearly exhibited particulate substructures [15,16].

Based on the illustrated sample and many others, it is now firmly established that the prevailing mechanism of the formation of monodispersed particles proceeds in several stages: (1) nucleation, (2) growth to nanosize particles, and (3) aggregation of nanosize particles to uniform final colloids. Extensive studies have indeed documented the existence of the aggregation stage [17]. For example, the electron micrographs, X-ray analysis, and independently prepared precursor particles all yielded the average diameter of crystallite subunits of 35 ± 5 nm in the gold spheres.

It was then necessary to derive a mechanism, which would account for the aggregation of a huge number of nanosized particles into identical larger colloids. Recently a kinetic model has been developed that explains this size selection mechanism [18]. The latter is based on the assumption that the nucleation is followed by a rapid formation of singlets, that is primary (nanosized) particles, which are sufficiently sparsely populated, and once formed are not further generated. Thus, their concentration decays by aggregation, when the conditions in the system eliminate repulsion between them. The latter can be due either to an increase in the ionic strength or a change in the pH in course of the process. It is also assumed that the diffusion constant of singlets is larger than that of aggregates. The dominance of the irreversible singlet capture in the growth process can, under certain set of conditions, result in the size selection (that is uniformity) of the final particles.

The calculated size distributions of the secondary (final) spheres for three reaction times, using the parameters for the precipitation of the mentioned gold sol, are given in Figure 5.2. The model describes reasonably well the experimental observations, considering the simplification used in the calculations.

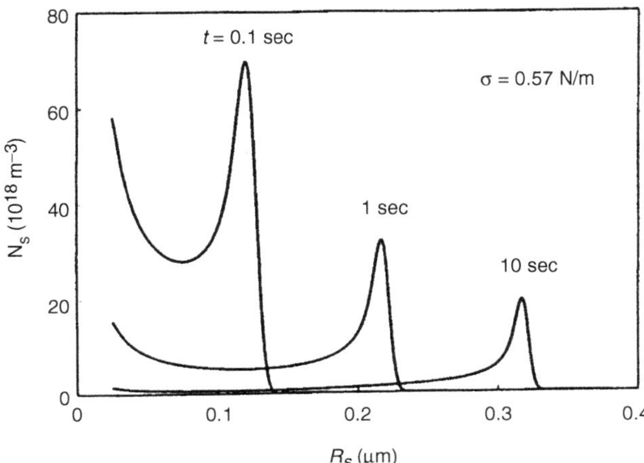

FIGURE 5.2 Distribution of secondary particles by their sizes at 0.1, 1, and 10 sec, calculated for the precipitation of spherical gold particles, using the model described in (18).

An interesting consequence of the described mechanism, is the fact that precipitation in solutions yields, as a rule, nanosized particles. If the process is arrested at this stage by additives, such as surfactants or microemulsions, one obtains stable nanosystems. However, in the absence of stabilizers, these particles aggregate (rather than to grow) into larger final products, which under appropriate conditions consist of monodispersed colloids. Thus, here we have established the "bridge" between the two "worlds" mentioned earlier!

"ART" AND SCIENCE

The new developments in the understanding of the mechanisms of formation of monodispersed colloids have greatly advanced the scientific aspects of this area of materials. Yet in actual preparations the art, that is skills, still play an essential role, especially since we do not know how to predict and control some properties, such as the shape or even the composition of the resulting particles.

The finely dispersed matter offers much in terms of the other aspect of art, that is the beauty. The latter can be affected by shapes or color or both. Examples of such artistic impressions can be seen in electron micrographs of monodispersed particles and their surfaces. Even more importantly, pigments, marbles, metals, and so on. are made of fine particles, without which we would have no paintings, sculptures, and other works of art, which so much embellish our lives.

REFERENCES

1. M. Faraday: Experimental Relations of Gold (and other Metals) to Light, *Phil. Trans. Roy. Soc.*, 147, 145–181 (1857).
2. E. Matijević: Controlled Colloid Formation. *Current Opinion in Colloid & Interface Science*, 1, 176–183 (1996).
3. E. Matijević: Formation of Monodisperse Inorganic Particulates. In *Controlled Particle, Droplet and Bubble Formation* (D. J. Wedlock, Ed.), Butterworth-Heinemann, London, 1994, pp. 39–59.
4. E. Matijević: Uniform Colloid Dispersions — Achievements and Challenges. *Langmuir*, 10, 8–16 (1994).
5. E. Matijević: Preparation and Properties of Uniform Size Colloids. *Chem. Mater.*, 5, 412–426 (1993).
6. T. Sugimoto: Preparation of Monodispersed Colloidal Particles. *Adv. Colloid Interface Sci.* 28, 65–108 (1987).
7. H. Haruta and B. Delmon: Preparation of Monodispersed Solids. *J. Chem. Phys.*, 83, 860–868 (1986).

8. Y.-S. Her, E. Matijević, and M.C. Chon: Preparation of Well Defined Colloidal Barium Titanate Crystals by the Controlled Double Jet Precipitation. *J. Mater. Res.*, 10, 3106–3114 (1995). Controlled Double-Jet Precipitation of Uniform Colloidal Crystalline Sr- and Zr-Doped Barium Titanates. *J. Mater. Res.*, 11, 3121–3127 (1996).
9. S. Nishikawa and E. Matijević: Preparation of Uniform Monodispersed Spherical Silica-Alumina Particles by Hydrolysis of Mixed Alkoxides. *J. Colloid Interface Sci.*, 165, 141–147 (1994).
10. F. Ribot, S. Kratohvil, and E. Matijević: Preparation and Properties of Uniform Mixed Colloidal Particles. VI. Copper(II)-Yttrium(III) and Copper(II)-Lanthanum(III) Compounds. *J. Mater. Res.*, 4, 1123–1131 (1989).
11. M. Ohmori and E. Matijević: Preparation and Proeprties of Uniform Coated Inorganic Colloidal Particles. VII. Silica on Hematite. *J. Colloid Interface Sci.*, 150, 594–598 (1992).
12. N. Kawahashi and E. Matijević: Preparation of Hollow Spherical particles of Yttrium Compounds. *J. Colloid Interface Sci.*, 143, 103–110 (1991).
13. Y.-S. Her, S.-H. Lee, and E. Matijević: Continuous Precipitation of Monodispersed Colloidal Particles. II. SiO_2, $Al(OH)_3$, and $BaTiO_3$. *J. Mater. Res.*, 11, 156–161 (1996).
14. W.P. Hsu, L. Rönnquist, and E. Matijević: Preparation and Properties of Monodispersed Colloidal Particles of Lanthanide Compounds. II. Cerium(IV). *Langmuir*, 4, 31–37 (1988).
15. D.V. Goia and E. Matijević: Tailoring the Size of Monodispersed Colloidal Gold. *Colloids Surf.*, 146, 139–152 (1999).
16. S.H. Lee, Y.-S. Her, and E. Matijević: Preparation and Growth Mechanism of Uniform Colloidal Copper Compounds by the Controlled Double-Jet Precipitation. *J. Colloid Interface Sci.*, 186, 193–202 (1997).
17. V. Privman, D.V. Goia, J. Park, and E. Matijević: Mechanism of Formation of Monodispersed Colloids by Aggregation of Nanosize Precursors. *J. Colloid Interface Sci.*, 213, 36–45 (1999).

See also: E Matijević: Monodisperse Colloids: Preparations and Interactions. *Progs. Colloid and Polymer Sci.*, 10, 38–44 (1966).

6 Silica Nucleation, Polymerization, and Growth Preparation of Monodispersed Sols

Akitoshi Yoshida
Nissan Chemical Industries, Ltd., Central Research Institute

CONTENTS

The industrial development of silica sol manufacturing methods is reviewed. Primary attention is focused on the preparation of monodispersed sols from water glass by the ion-exchange method. Details are given for variations of manufacturing process and for the characteristics of both the processes and sols obtained. Furthermore, the following surface modifications of particles are demonstrated: silica sols stabilized with ammonia, amine, and quaternary ammonium hydroxide; aluminum-modified or cation-coated silica sol; and lithium silicate. Finally, future trends in silica sol manufacturing are discussed from the viewpoint of not only raw materials and improvement of the procedures but also the function of the silica sols and their particle shape.

Monodispersed sol is commonly referred to as silica sol or colloidal silica. The term "colloidal silica" here refers to a stable dispersion of discrete, amorphous silica particles.

The industrial development of silica sols first began with the initial research by Graham [1] in 1861 involving the addition of hydrochloric acid to an aqueous solution of sodium silicate followed by dialysis to obtain dilute silica sol. In 1933, silica sol containing 10% SiO_2 was first marketed. This step was followed in 1941 by the announcement of an ion-exchange process [2], including procedures for stabilization and for concentration of the sol by heating, that used an ion-exchange resin to remove sodium ions from the dilute aqueous solution of sodium silicate. This sodium removal process is the most common today. In 1951, a process [3] for creating colloidal silica particles of uniform and controlled size was announced for the first time. In 1956, a method [4] was established for making stable sols consisting of microscopic particles having a diameter of only 8 nm yet containing >30% of silica. The history of the development of silica sols and a description of the fundamental and application research up to and throughout the 1950s was summarized in detail by Iler [5].

Various raw materials can be used in the manufacturing of monodispersed sol. Examples of these materials include silicon metals [6], silicon tetrachloride [7], ethyl silicate [8], water glass [2], and silica powder [9]. In this chapter, I focus attention on the preparation of monodispersed sols from water glass, a raw material that is presently used in large amounts industrially for the inexpensive production of silica sols.

The manufacturing processes of silica sols can be broadly divided into the following three steps:

1. formation of active silicic acid by removal of alkali ions from a dilute aqueous solution of water glass
2. formation of a dilute silica sol by nucleation from the active silicic acid and growth of discrete silica particles by polymerization
3. concentration of the dilute silica sol

In addition, I will discuss modification of the surfaces of silica particles.

MANUFACTURING METHODS OF SILICA SOLS AND THEIR CHARACTERISTICS

HISTORY OF SILICA SOL MANUFACTURING METHODS

Although numerous techniques have been proposed for manufacturing of silica sols, including dialysis [10], electrodialysis [11], peptization [12], acid neutralization [13], and ion exchange [3,14], the last three methods have come to be used most commonly. At present, the ion-exchange method is considered the most prominent technique.

Although the dialysis method, which involves the reaction of dilute water glass with an acid followed by dialytic removal of the sodium salt as the formed electrolyte, was proposed in 1861, it was never applied on an industrial basis. Furthermore, the electrodialysis method was attempted in various ways as a method for electrically removing sodium salt from water glass. But, unlike the last three methods mentioned, it did not reach the level of practical application.

Peptization Method

As shown in Figure 6.1, an acid such as sulfuric or hydrochloric is added to a dilute aqueous solution of water glass while stirring, or while heating as necessary, to neutralize and obtain a silica gel containing salt. Next, the crude silica gel is washed with water to remove the salt and obtain a silica wet gel. Then, water and an aqueous sodium hydroxide solution are added to the gel so that the pH of the resulting solution is 8.5–10 and a silica gel slurry is formed. The slurry is then heated for several hours in an autoclave at 120–150°C to allow the gel to peptize and form a sol. Thus, a silica sol is prepared. For example, the resulting silica sol contains 30 wt% SiO_2, has a pH of 10, has a molar ratio of SiO_2:Na_2O of about 100, and consists of irregular particles having a diameter of 10–20 nm. However, it is difficult to obtain silica sols having a desired particle size or high purity with this method.

Acid-Neutralization Method

As is indicated in Figure 6.2, an acid, such as sulfuric or hydrochloric, is added to a dilute aqueous solution of water glass while heating and stirring. This addition results in silica sol nucleation and particle growth to obtain a dilute silica sol containing salt. The salt is removed by either dialysis or electrodialysis, and the solution is concentrated to obtain a silica sol. In recent years, as the removal of salt has been made easier through the use of ultrafiltration membranes, this method

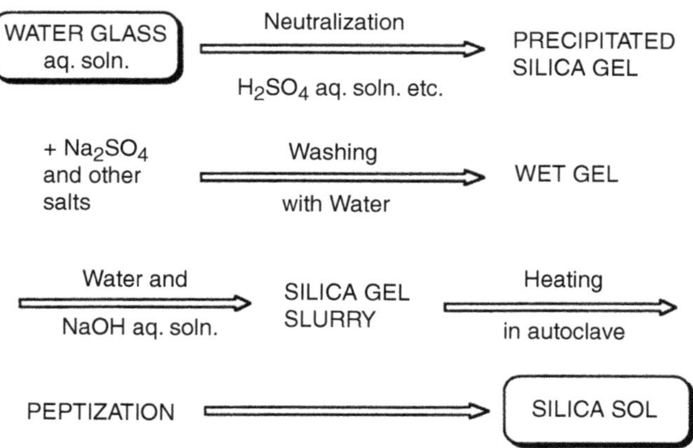

FIGURE 6.1 Flow chart of peptization method for manufacturing silica sols.

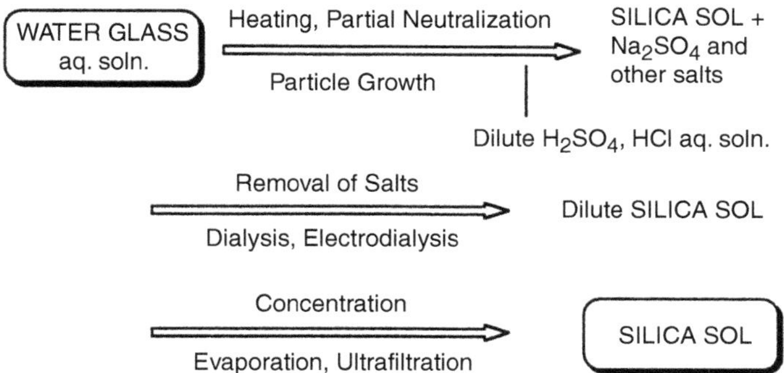

FIGURE 6.2 Flow chart of acid-neutralization method for manufacturing silica sols.

has come to be considered as having potential for the future.

Ion-Exchange Method

Sophisticated ion-exchange resins have been developed to efficiently remove sodium ions at the industrial application level. This method is currently the primary means of manufacturing silica sols. The following section provides a detailed description of this method.

Silica Sol Manufacturing Using the Ion-Exchange Method

The silica sol manufacturing method using ion-exchange is as shown in the basic flow chart of Figrue 6.3.

Ion Exchange

An aqueous water-glass solution as the raw material in this process can have a SiO_2 content of 30%, a Na_2O content of 10%, a molar ratio of SiO_2:Na_2O of 3.1, a specific gravity of 1.2, and a pH of 12–13. Water is added to this aqueous water-glass solution while stirring to obtain a dilute aqueous water-glass solution containing 2–6% SiO_2. In this state the silica takes the form of polysilicate anion. Next, this dilute aqueous water-glass solution is passed through a bed of cation-exchange resin in a column for which hydrogen ions have been regenerated in advance to allow the sodium ions to be adsorbed onto the resin bed and leave an aqueous solution of active silicic acid. This liquid is a microscopic colloidal solution with a pH of 2–4 containing 2–6% SiO_2 comprising particles with a diameter of 2 nm or less. The colloidal solution is also in an unstable state and easily gels upon standing because the degree of polymerization is several tens to several hundreds.

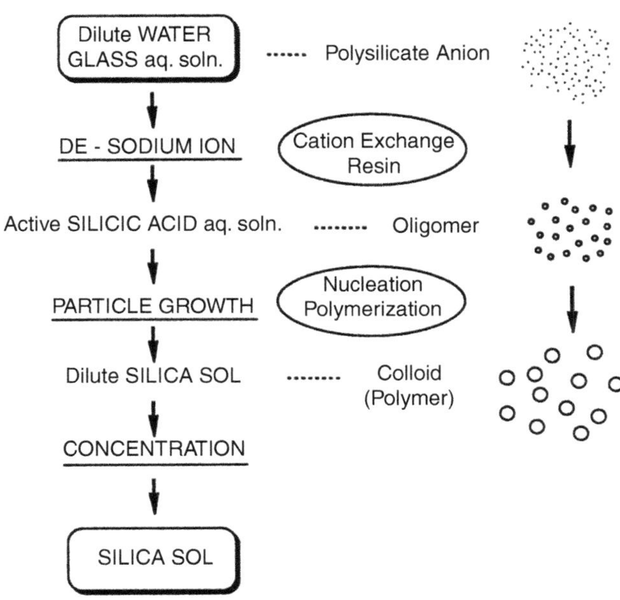

FIGURE 6.3 Flow chart of ion-exchange method for manufacturing silica sols.

Particle Growth

Nucleation [3], polymerization [15], and particle growth [16,17] are performed on this active silicic acid in the presence of alkali at a temperature of at least 60°C, a pH of 8–10.5, and a molar ratio of SiO_2:Na_2O of about 20–500, and thus a dilute silica sol is formed. The process of particle growth is discussed in more detail in the next section. The formed dilute silica sol generally consists of spherical particles having a diameter of 4–100 nm, a pH of 8–10.5 and a SiO_2 content of 2–6%. This silica sol is a colloidal liquid in a stable state, and therefore it causes no gelling. As illustrated in the flow sheet, the dilute aqueous water-glass solution is in the form of polysilicate anion. In contrast to this ionic state, the active silicic acid is in the form of an oligomer, that is, a colloid of microscopic particles. The dilute silica sol is a polymer in the colloidal state.

Concentration

Next, this dilute silica sol is concentrated so that the SiO_2 content is increased to 15–60%. The concentration is needed not only to ensure efficient transport of a small volume of sol but also for use in silica sol applications, for example, as a binder, a catalyst carrier, and a gelation agent. Because the concentration process of the dilute silica sol efficiently removes the dispersion medium of water, the most common way [3,14] of accomplishing this step in the past was by using steam under reduced pressure or normal pressure. However, more recently, use of the ultrafiltration membrane [18–20], originally intended for the separation of colloidal size particles, has been evaluated for ion removal; this approach is energy saving. It is actually employed in some applications.

Finished products are obtained after the concentration, pH, and conductivity of the concentrated sol are adjusted to maintain their stability. In actuality, several types of manufacturing methods are used, depending on the particular combination of particle growth and concentration.

Mechanism of Particle Growth

As shown in Scheme I, the silicic acid undergoes dehydrating condensation polymerization in the presence of alkali. The higher the temperature and the pH and the longer the reaction time, the faster the polymerization proceeds. As a

$$\equiv Si\text{-}OH \; (Na^+) + HO\text{-}Si\equiv \; (Na^+) \Longrightarrow \equiv Si\text{-}O\text{-}Si\equiv + Na^+ + OH^-$$

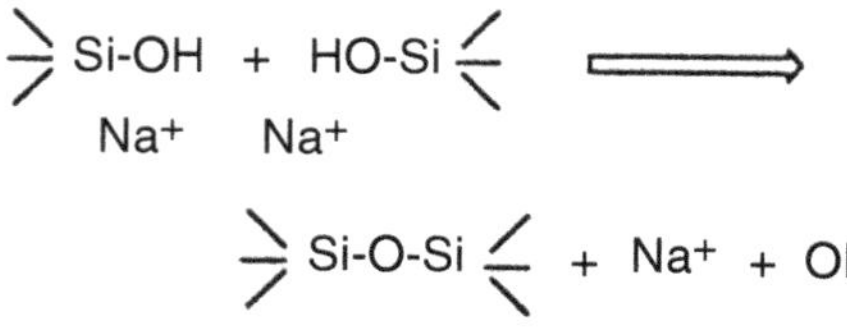

SCHEME I Polymerization of silicic acid by dehydrating condensation.

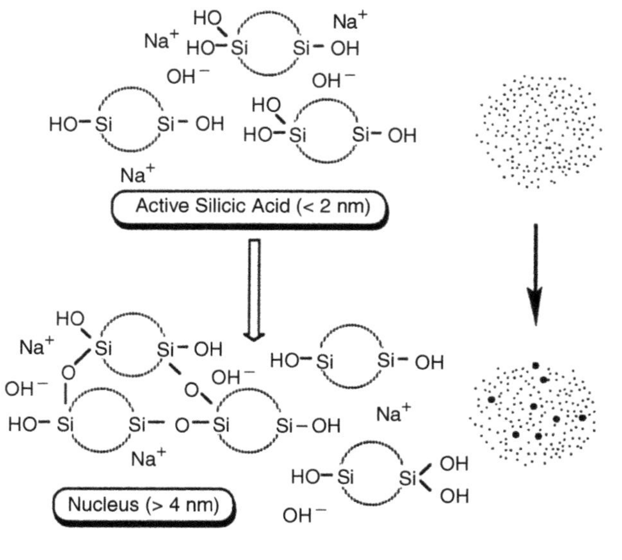

SCHEME II Polymerization of active silicic acid to form nuclei.

result of polymerization, Na^+OH^- that was adsorbed onto the active silicic acid is released and increases the pH.

As shown in Schemes II and III, active silicic acid is polymerized by heating in the presence of alkali and first forms nuclei. Active silicic acid then polymerizes around the nuclei, and thus particles are formed. This process is often referred to as "buildup." In actuality, nucleation and particle growth take place simultaneously. When the temperature is high (such as in autoclave heating) or when the concentration of silicic acid is high during particle growth, particles grow through bonding between nuclear particles or between larger particles, respectively. During nucleation and particle growth as well, Na^+ and OH^- ions are released, and thus the pH of the reaction system is increased. The higher the pH is, the faster the

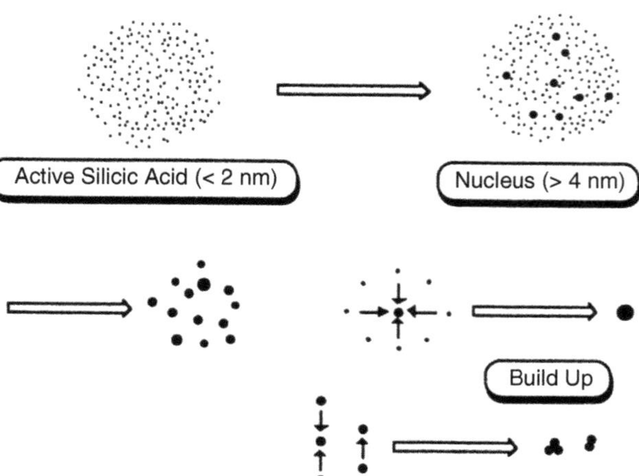

SCHEME III Polymerization of active silicic acid around nuclei to form particles.

particles grow. However, dissolution occurs if the pH is too high, so pH control is considered to be the most important factor.

In addition, when the pH of this liquid exceeds 10.5, the formed particles undergo dissolution and hydrolysis and then precipitate, and such dissolution and precipitation take part in the particle growth. The main factors that dominate particle growth are amount of alkali such as Na_2O, silica concentration, SiO_2:Na_2O molar ratio, temperature, time, and method of addition.

Variations of Silica Sol Manufacturing Process

As shown in Figure 6.4, the four methods, A-1, A-2, B-1, and B-2, use different technical combinations of nucleation, particle growth, and concentration. In any of these methods the raw material aqueous water-glass solution is diluted, and sodium is removed with a cation-exchange resin to obtain an active silicic acid. The characteristics of these four processes are shown in Table 6.1.

Methods A-1 and A-2

In Methods A-1 and A-2, a small amount of an aqueous NaOH solution is added, while stirring, to the acidic active silicic acid having a SiO_2:Na_2O molar ratio of 500–1000 until the SiO_2:Na_2O molar ratio is, for example, 80–100 and the solution is stable.

In Method A-1, a fixed amount of this liquid is first charged in an evaporator that is heated from the outside with steam. Simultaneous to the beginning of solution concentration under reduced or normal pressure, aqueous active silicic acid solution is continuously metered into the evaporator to carry out particle growth and concentration at the same time.

On the other hand, in Method A-2, particle growth and concentration are carried out separately. In the particle-growth step, a fixed amount of the stable aqueous active silicic acid solution is charged into a heating–aging tank and heated. Continuous charging of the aqueous silicic acid solution then follows to carry out buildup of particles. Alternatively, in another method the stable aqueous active silicic acid solution that was initially charged in the heating-aging tank is simply heated to carry out particle growth. A temperature of 60–150°C is commonly used for the aging under heating. The dilute silica sol prepared in this manner has a SiO_2 content of 2–6%, a pH of 9–10.5, and a particle size of 4–20 nm. Next, this sol is passed on to the concentration step. Because the silica particles have already grown and are in a stable state, concentration can be performed in an arbitrary manner, for example, through evaporation or using an ultrafiltration

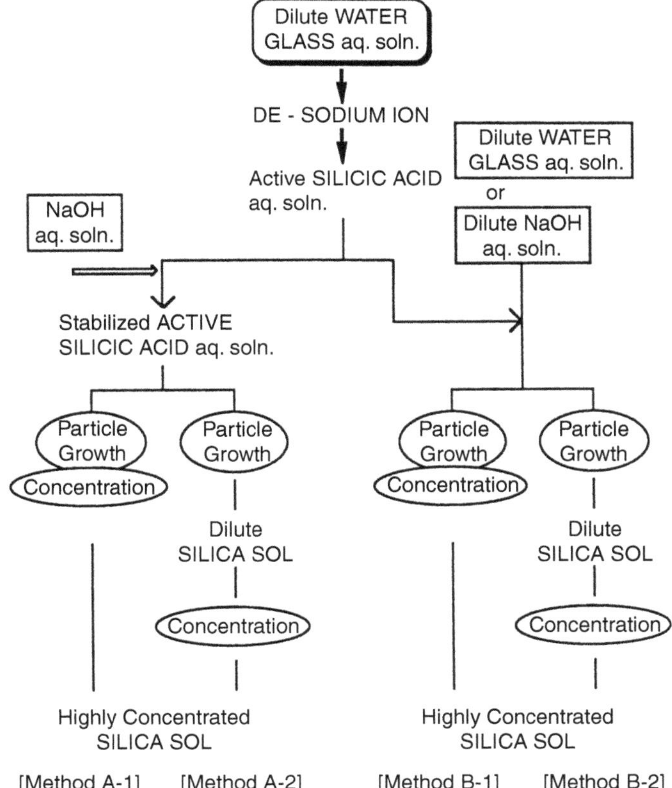

FIGURE 6.4 Four methods of silica sol manufacturing.

TABLE 6.1
Characteristics of Silica Sol Manufacturing Methods

Characteristics	Method A-1	Method A-2	Method B-1	Method B-2
Intermediate material	dilute silica sol		silicic acid and water glass	
Particle growth and concentration	simultaneous	separate	simultaneous	separate
Change in pH during particle growth	$7 \rightarrow 9$–10.5		$11.5 \rightarrow 9$–10.5	
Particle growth rate	slow		fast	
Size of nuclei	small		small–large	
Nucleation mechanism	polymerization of particles		initial: hydrolysis of water glass; intermediate to final: polymerization of particle	
Manufacturing of large-particle sols	difficult (normal pressure) possible (with pressurization)		possible	
Ease of buildup	more difficult than in Process B		easy	
Control of particle size	possible for microscopic- to medium-sized particles		possible for microscopic- to large-sized particles	
Control of particle size distribution	impossible or difficult	possible	impossible or difficult	possible
Manufacturing time	long (longer than in B-1)	short (longer than in B-2)	long	short

membrane. The pH, concentration, and conductivity of the highly concentrated sol are then adjusted to obtain a finished sol product.

Methods B-1 and B-2

Characteristically in Methods B-1 and B-2, the aqueous solution of active silicic acid is continuously charged into either a dilute aqueous water-glass solution or dilute aqueous NaOH solution that has preliminarily been heated, and stirred if necessary, to effect particle growth. In Method B-1, concentration and particle growth are carried out simultaneously under conditions similar to those of Method A-1. In Method B-2, particle growth and concentration are carried out separately, and in this regard it is essentially the same as Method A-2. However, in Method B-1 or B-2, larger diameter silica sol particles, 4–100 nm, can be formed. On the other hand, Methods A-1 or A-2 allows production only up to medium-sized silica particles having a diameter of 4–20 nm.

Manufacturing of Large-Particle Silica Sols

A method for manufacturing silica sols consisting of large-diameter particles of 50–200 nm is as follows. The buildup method is used because the growth of particles is difficult in Method A-1 or A-2 even when medium-diameter silica sols are manufactured.

As indicated in Scheme IV, stable dilute sol is nucleated to obtain sols having a particle size of, for example, 20 nm, which is used as a heel. Then, in the next step, an active silicic acid or stable dilute sol is continuously charged onto the heel to allow the silica particles to grow to a diameter of 50 nm. Thus, the process consists of two separate steps. In either Method B nucleation is carried out in the initial stage of the process, and buildup occurs from the intermediate to the final stages. However, this method is not clearly separated into two steps.

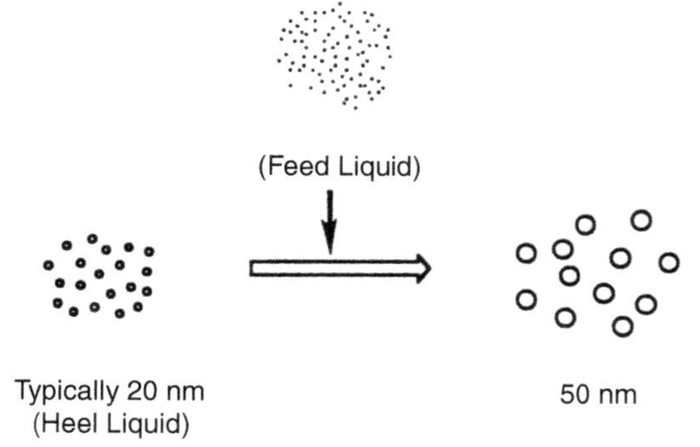

SCHEME IV Preparation of silica particles from stable dilute sol using the buildup method.

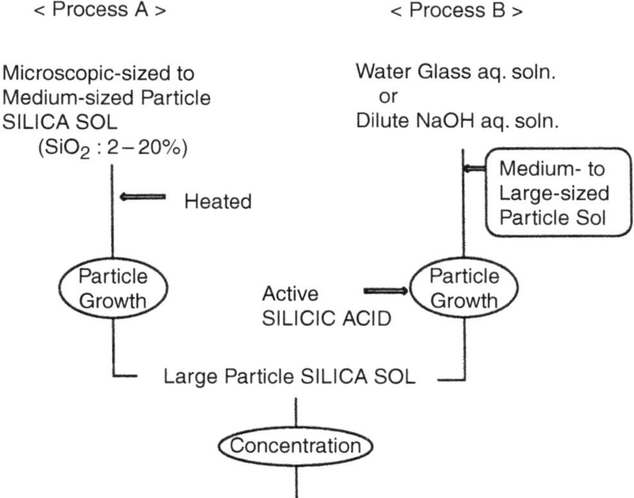

FIGURE 6.5 Flow chart of production of large-particle silica sols.

Figure 6.5 illustrates the production of large-particle silica sols. In Process A, silica sol consisting of microscopic- to medium-sized particles is heated in an autoclave at 120–130°C for several hours, and the hydrothermal treated silica sol is concentrated to obtain a highly concentrated large-particle silica sol. In Process B, medium- to large-particle-size sols, which serve as nuclei, are added to an aqueous water-glass solution or a dilute aqueous NaOH solution. The mixture is then heated to carry out aging while stirring, and in this state active silicic acid is charged to allow buildup of particles. Factors that dominate particle growth are the size and concentration of the silica particles that serve as the nuclei, the amount of active silicic acid to be added, the aging temperature, aging time, pH behavior, and others. Method B characteristically facilitates formation of sols with large particle sizes showing a uniform particle size distribution ranging up to about 200 nm.

Characteristics of Silica Sols Produced by the Four Methods

The characteristics of the sols manufactured according to Methods A and B are shown in Table 6.2.

MANUFACTURING METHODS OF SURFACE-MODIFIED SILICA SOLS

The silica sols described in the manufacturing methods are all generally stabilized with NaOH, and the majority of the silanol groups on the surface of the silica sol are covered with Na^+ ions, as indicated in Scheme V. Na^+ ions are strongly adsorbed onto the silanol group, and this condition sometimes is expressed as Si–ONa.

For manufacturing a surface-modified silica sol, an acidic silica sol [20] from which these Na^+ ions have been removed is used in most cases. As shown in

TABLE 6.2
Characteristics of Sols Formed by the Different Methods

Characteristics	Method A-1	Method A-2	Method B-1	Method B-2
Particle size distribution	broad	narrow	broad	narrow
Particle shape	irregular	spherical	spherical	spherical
Degree of silica particle distribution	low	high	high	high
Illustrations of particle shape				

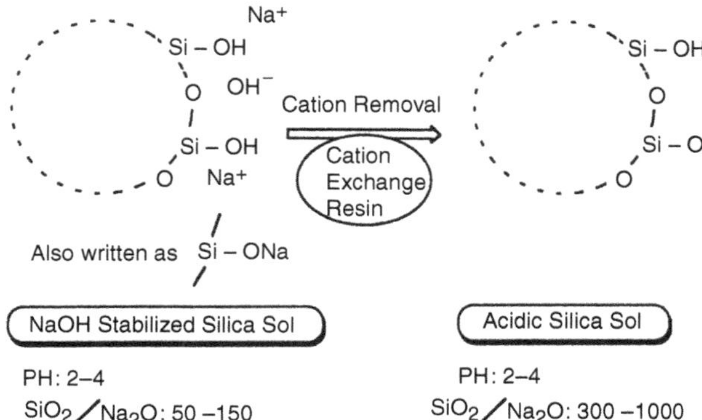

SCHEME V Removal of cations from stable silica sol to form acidic silica sol.

Scheme V, silica particles are brought into contact with a cation-exchange resin, and if necessary, with an anion-exchange resin, to obtain an acidic silica sols of pH 2–4. This acidic sol is stable because it is negatively charged even at pH 2–4, according to zeta-potential measurements. Starting from such acidic silica sol, surface-modified silica sols are manufactured.

Silica Sols Stabilized with Ammonia, Amine, and Quaternary Ammonium Hydroxide

As shown in Figure 6.6, ammonia [21], amine [22], or quaternary ammonium hydroxide [22] is added to the acidic silica sol while stirring, and if necessary the mixture is aged with heating to obtain a correspondingly stabilized silica sol. Such sols have a SiO_2 content of 20–50% and a pH of 8–12. This type of sol fits new applications because the sols remain mixed for a fixed period with strongly alkaline water glass, CaO, and MgO, with which ordinary silica sols can hardly be mixed because of gelling.

Aluminum-Modified Silica Sol

As shown in Figure 6.6, an aluminum-modified silica sol [23] can also be formed by adding an aqueous sodium aluminate solution to the acidic silica sol while stirring at room temperature so that the SiO_2:Al_2O_3 molar ratio is approximately 350 and aging the mixture with heating at 100°C. This modified silica sol has a SiO_2 content of 20–40%, and a pH of 8–10, and the surfaces of the silica particles are negatively charged. A major characteristic of this type of sol is that it does not gel and is stable in the neutral pH region. Ordinary silica sols cannot be used in this neutral region because they undergo gelation but the aluminum-modified sol has enabled various new applications. The aluminum-modified silica sol is the first surface-modified silica sol that has considerable industrial value.

Cation-Coated Silica Sol

As shown in Figure 6.7, an aqueous solution of basic aluminum chloride is added to an acidic or alkaline silica sol while stirring, and then a dilute aqueous NaOH solution is added until the pH of the mixture is 4–6. The mixture is then aged by heating to 80–100°C to form a cation-coated silica sol [24]. This silica sol is used in applications where the pH is in the acidic region and where it is used as a mixture with a cationic aqueous solution. Moreover, the cation-coated silica sol can be mixed easily with water-soluble organic solvents. Besides aluminum salt, other basic salts such as of Zr and Ti and cationic surface-active agents [25] can also be used for cation coating.

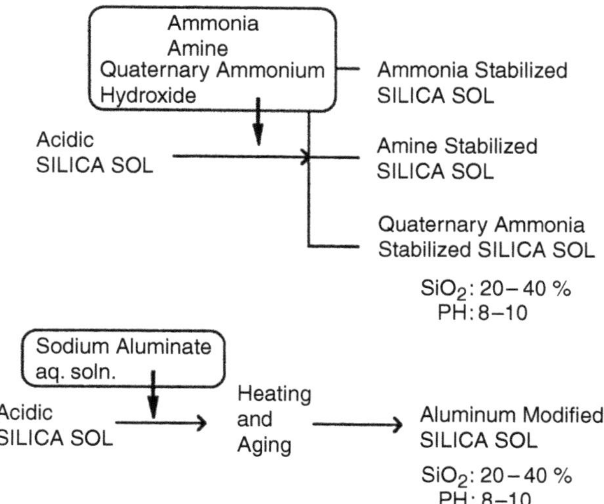

FIGURE 6.6 Flow chart of production of silica sols stabilized with ammonia, amine, or quaternary ammonium hydroxide.

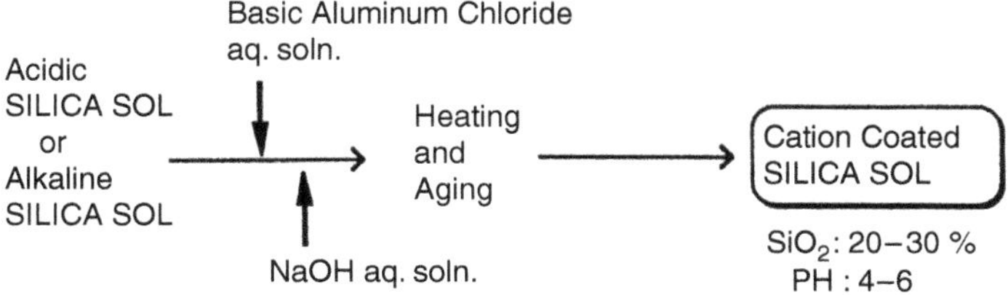

FIGURE 6.7 Flow chart of production of cation-coated silica sol.

LITHIUM SILICATE

In this chapter, the term lithium silicate means very small silica sol particles whose surfaces are coated with lithium hydroxide. As shown in Figure 6.8, this lithium silicate [26] can be prepared by adding an aqueous solution of lithium hydroxide to the acidic silica sol while stirring, and then heating the mixture to a temperature of 60°C or less for an extended period of time, for example, a half day to several days. This lithium silicate typically has a SiO_2 content of 10–25%, a SiO_2:Li_2O molar ratio of 3–9, and a pH of about 11. Lithium silicate is a kind of water glass like sodium silicate and potassium silicate, but it is used as a binder because it has particularly high levels of water resistance and heat resistance.

FUTURE TREND IN SILICA SOL MANUFACTURING

Silica sols will be required to have more unique and sophisticated characteristics in the future to meet the needs of various application fields [27]. Therefore, raw materials will no longer be limited to the aqueous water-glass solutions, but a variety of raw materials will be used corresponding to the specific properties required for the desired silica sol. For example, large- to very large-sized silica sols that are of high purity and as truly spherical as possible are required in electronics applications. Ethyl silicate will most likely be used as a raw material for this purpose. In addition, organosilica sols dispersed in organic solvents, such as methanol silica sol [28] and dioctyl phthalate sol [29], are beginning to be used in the field of plastics modification.

Furthermore, in an effort to discover unknown characteristics of silica sols, further progress will occur in the development of silica sols [30] having shapes other than spherical, such as elongated, fibrous, and platelet. We are currently working on the development of elongated silica sols. Compared with the spherical silica sols, the elongated silica sols have high film-forming properties and demonstrate unique properties as a binder or coating agent for the formation or surface treatment of inorganic fibers. As shown in the transmission electron microscope photographs of Figure 6.9, this new silica sol is elongated. For example, the particles have a diameter of about 10 nm and a length of 50–100 nm. To prepare such elongated silica sols, a calcium salt and an aqueous sodium hydroxide solution are added to an active silicic acid or an acidic silica sol, and the mixture is heated in an autoclave at 100–150°C for several hours. Thus, particle polymerization takes place in nonuniform directions, and an elongated silica sol [31] can be formed.

Besides varying the shapes of silica sols, the functions of silica sols will be varied in the future. Examples include a sol that gives a film with a high refractive index made up of a composite of silica and titanium oxide; a conductive silica sol whose surface is covered with an electrically conductive oxide or metal; and a porous silica sol that is made from ethyl silicate as the raw material.

On the other hand, in terms of manufacturing processes, the procedures employed will have to be improved continually, to improve the efficiency of the cation-exchange procedure, to manufacture silica sols continuously, and to enhance the efficiency of the concentration processes.

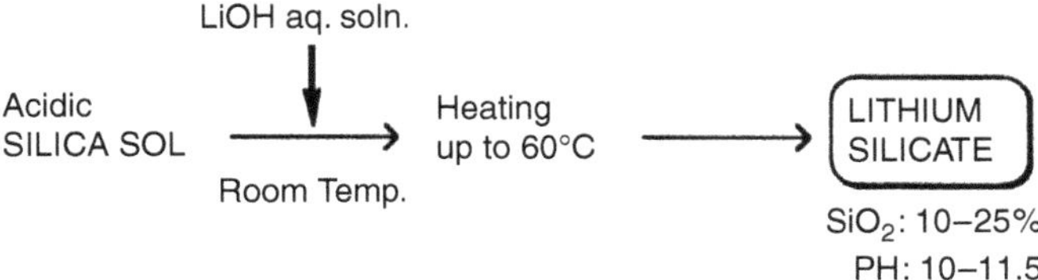

FIGURE 6.8 Flow chart of production of lithium silicate.

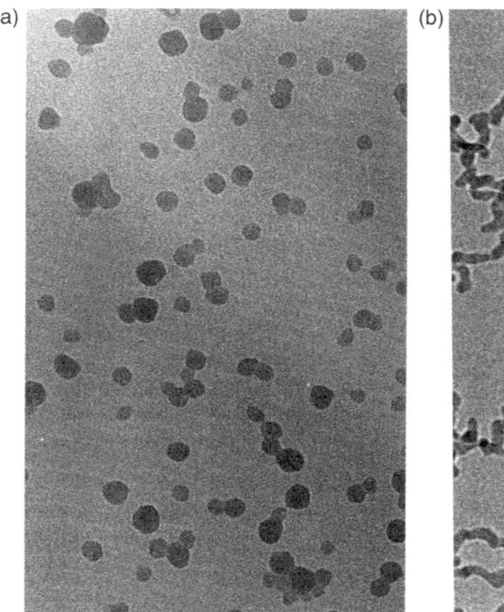

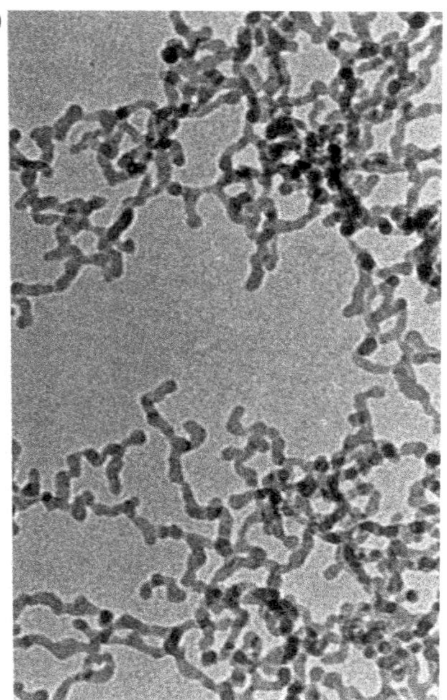

FIGURE 6.9 Silica sols with spherical (a) and elongated (b) particles.

CONCLUDING REMARKS

From the 1940s to the 1960s, the fundamental technology for manufacturing silica sols was fully established, and wide variety of silica sol products was developed, mainly owing to the achievements of Iler. From the 1970s to the 1980s, mass production processes were developed to achieve cost reduction to meet large-scale applications. And now, the manufacturing of silica sols is entering a new era. Even in this new era, we will surely continue to formulate our ideas and thoughts on the basis of the valuable research results achieved by Ralph Iler.

REFERENCES

1. Griessbach, R. *Chem. Ztg.* **1993**, *57*, 253–274.
2. Bird, P. G. U.S. Patent 2,244,325, 1941.
3. Bechtold, M. F.; Snyder, O. E. U.S. Patent 2,574,902, 1951.
4. Alexander, G. B. U.S. Patent 2,750,345, 1956.
5. Iler, R. K. *Colloid Chemistry of Silica and Silicates*; Cornell University Press: Ithaca, NY, 1955.
6. Balthis, J. H. U.S. Patent 2,614,994–5, 1952.
7. Radczewski, O. E.; Richter, H. *Kolloid-Z.* **1941**, *96*, 1.
8. Stöber, W.; Fink, A. *J. Colloid Interface Sci.* **1968**, *26*, 62.
9. Loftman, K. A.; Thereault, J. R. U.S. Patent 2,984,629, 1961.
10. Graham, T. *J. Chem. Soc. London* **1864**, *17*, 318.
11. Sanchez, M. G. Canadian Patent 586,261, 1959.
12. White, J. F. U.S. Patent 2,375, 1945.
13. Alexander, G. B.; Iler, R. K. U.S. Patent 2,601,235, 1952.
14. Bird, P. G. U.S. Patent 2,244,325, 1941.
15. Alexander, G. B. *J. Am. Chem. Soc.* **1954**, *76*, 2094.
16. Alexander, G. B.; Mcwhorter, J. R. U.S. Patent 2,833,724, 1958.
17. Broge, E. C.; Iler, R. K. U.S. Patent 2,680,721, 1954.
18. Chiton, H. T. J. British Patent 1,148,950, 1969.
19. Iler, R. K. U.S. Patent 3,969,266, 1976.
20. Mindick, M.; Reven, L. E. U.S. Patent 3,342,747, 1967.
21. Akabayashi, H.; Syoji, H. Japanese Patent 288,231, 1961.
22. Iler, R. K. U.S. Patent 2,692,863, 1954.
23. Alexander, G. B. U.S. Patent 2,892,797, 1959.
24. Alexander, G. B.; Bolt, G. H. U.S. Patent 3,007,878, 1961.
25. Akabayashi, H.; Yoshida, A. Japanese Patent 447,161, 1965.
26. Iler, R. K. U.S. Patent 2,668,149, 1952.
27. Yoshida, A. *Chem. Econ. (Jpn)* **1988**, *6*, 22.
28. Akabayashi, H.; Yoshida, A. *Kogyo Kagaku Zasshi* **1966**, *68*, 429.
29. Akabayashi, H.; Yoshida, A. *Kogyo Kagaku Zasshi* **1966**, *69*, 1832.
30. Broge, E. C. U.S. Patent 2,680,721, 1954.
31. Watanabe, Y.; Ando, M. European Patent Appl. Publication No. A2,0335195, 1989.

7 The Formation and Interfacial Structure of Silica Sols

John D.F. Ramsay, Stephen W. Swanton, Akihiko Matsumoto, and Dhanesh G.C. Goberdhan
Atomic Energy Authority, Harwell Laboratory

CONTENTS

Several techniques, including small-angle neutron scattering (SANS), ultracentrifugation, photon correlation spectroscopy, and ^{29}Si NMR spectroscopy, were used to investigate the nature of the oxide–water interface of silica sols and its significance in the formation and growth of colloidal particles in aqueous solution. These studies were performed with a range of commercial silica sols of different diameters in the range ≈7–30 nm. When the diameter is small the sols contain a significant proportion of oligomeric silicate species that may be associated at the surface of the particles. For sols of the largest diameter, the relative proportion of oligomers is much smaller. In all the sols the core of the particles has a highly condensed Si–O–Si structure.

The classic description of the structure and mechanisms of formation of silica sols by the hydrolysis and condensation of silicates in aqueous media was given by Iler in 1979 [1]. According to Iler, polymerization may occur in essentially three stages: [1] the polymerization of monomers to oligomers and then to primary particles, [2] growth of particles, and [3] particle aggregation to form networks that eventually give rise to a gel structure extending throughout the liquid medium. Stages 2 and 3 depend on the pH and salt concentration, as discussed by Iler. These general processes have subsequently been confirmed by many workers using a range of microscopic techniques [2] such as ^{29}Si NMR and IR spectroscopy, light scattering, small-angle X-ray scattering (SAXS), and small-angle neutron scattering (SANS), although less attention has been given to stage 2.

Considerable advances in understanding stage 1 have been obtained from ^{29}Si NMR investigations [3–5] and stage 3 from scattering measurements (light, X-rays, and neutrons) [6,7]. Stage 1 has been shown to give rise to a range of complex polysilicic acid structures. In general extensive condensation takes place in which monomers form ring structures that associate and condense further, eventually to form the core of silica particles containing few silanol groups. In stage 3 the present understanding of the double-layer interaction between particles and the mechanisms of aggregation to give gel formation is extensive and has been advanced by descriptions based on fractal theory [8].

In this chapter we will describe some recent investigations of the formation and interfacial structure of a series of commercially produced silica sols (Nalco, Dupont, and Ludox) with different diameters in the range ≈7–30 nm. This type of sol, which has wide industrial applications, has been used as a model system in numerous studies of colloidal silica, although in general, the nature of the silica–water interface has received little attention. The interfacial structure may well explain some of the unusual properties of silica sols, such as the high surface charge and exceptional colloidal stability together with the enhanced capacity for sorption and complexation of ionic species in solution. Indeed in the past these features have been ascribed to a surface "gel layer" resulting from extensive hydroxylation within a few ångstroms at the surface [9], although definitive evidence of this gel layer is still lacking. Here we have applied several techniques (photon correlation spectroscopy (PCS), SANS, analytical ultracentrifugation, and ^{29}Si NMR spectroscopy) to explore the properties of these silica sols. Particular interest attaches to the sol of smallest diameter because here we anticipate that any effects due to incomplete condensation of oligomeric components will be relatively more marked, especially in the diffusion behavior and effective diameter of the spherical particles.

EXPERIMENTAL DETAILS

Silica Sols. Concentrated (ca. 15 to ca. 40% w/w) silica sols of different particles size (S1 to S4) were obtained commercially from designated sample batches, viz., S1 is Nalcoag 1115 from Nalco Chemical; and S2, S3, and S4 are Ludox SM, HS, and TM respectively, from DuPont. These sols are prepared by hydrothermal treatment of sodium silicate solutions [1] and contain Na^+ as the counterion. These sols have already been studied extensively [6,10,11]. The mean diameters of the sol particles, as previously determined from transmission electron microscopy [10] together with the effective particle radius as derived from the specific surface areas, S_{BET}, of the outgassed gels [10] are given with other properties in Table 7.1. Measurements were made with sols of different concentration by diluting the stock samples with demineralized water. The stock sol, S1, was also dialyzed repeatedly against water (denoted $S1^D$) and used in further studies.

Small-Angle Neutron Scattering. Measurements were made as described previously (10,11) using a multidetector instrument installed in the PLUTO reactor at Harwell and also with the D11 instrument at the Institut Laue-Langevin, Grenoble, France.

Photon Correlation Spectroscopy. Measurements were made with a commercial 96-channel photon correlator (Malvern K7023) using a helium–cadmium laser and with a 256-channel correlator (Malvern K7032) using a helium–neon laser with a standard spectrometer system (Malvern 4700), adopting procedures as described previously [12].

Analytical Ultracentrifugation. Sedimentation coefficients were determined with an ultracentrifuge (Beckman L8–70M) fitted with a schlieren analytical attachment. Photographic images of schlieren (refractive index gradient) profiles were analyzed with a profile projector (Nikon, model V10) using standard procedures [13].

^{29}Si NMR Spectroscopy. ^{29}Si NMR spectra of the stock sols contained in 1-ml plastic vials were recorded at a field strength of 39 MHz using a commercial instrument (Bruker-Physics CXP200) at a 10-s pulse repetition rate.

RESULTS AND DISCUSSION

SMALL-ANGLE NEUTRON SCATTERING

The intensity of small-angle scattering, $\boldsymbol{I}(\mathbf{Q})$, for a concentrated colloidal dispersion of identical particles is given by [14]

$$\boldsymbol{I}(\mathbf{Q}) = K(\rho_p - \rho_s)^2 V_p^2 n_p \boldsymbol{P}(\mathbf{Q}) S(\mathbf{Q}) \quad (7.1)$$

where $\mathbf{Q}$ is the scattering vector, defined as

$$|\mathbf{Q}| = 4\pi \sin\Theta/\lambda$$

for a scattering angle 2Θ and wavelength λ; ρ_p and ρ_s are, respectively, the mean scattering length densities of the particles and solvent; V_p is the volume of each particle; n_p is the particle number density; K is an experimental constant; and $\boldsymbol{P}(\mathbf{Q})$ is the particle form factor, which for

TABLE 7.1
Properties of Silica Sols

Sol	Stock concentration (w/v)	Particle diameter (nm)[a]	Surface area of gel, S_{BET} (m^2/g)	Particle radius (nm)[b]	Na^+ content (mg/ml)	pH
S1	16.1	8	410	3.3	5.6	10.5
S2	32.8	12	260	5.2	6.0	10.1
S3	40.3	16	210	6.5	4.9	9.8
S4	47.2	30	130	10.5	3.2	9.0

[a]From electron microscopy.
[b]From S_{BET}, assuming silica density of 2.2 g/ml.

spheres of radius R is given by

$$P(\mathbf{Q}) = \left[\frac{3[\sin(\mathbf{Q}R) - \mathbf{Q}R\cos(\mathbf{Q}R)]}{\mathbf{Q}^3R^3}\right]^2 \quad (7.2)$$

The structure factor, $S(\mathbf{Q})$, is determined by the nature of the particle interaction potential [7]; for noninteracting systems $S(\mathbf{Q}) = 1$ (viz., for systems in the limit of very low concentration where electrical double-layer interactions are negligible).

Extensive SANS investigations have been performed previously with silica sols of the present type [6,10,11]. The scattering behavior of the sols can be closely described on the basis of the foregoing theoretical treatment for monodispersed spherical particles. Furthermore, detailed analysis of the form of the scattering curves [relative intensity $I(\mathbf{Q})$ vs. $\mathbf{Q}$] for sols of different concentration has provided information on the size of particles and the nature of the interaction potential. Typical values of R, derived from experimental fits of extensive scattering data [11] to Equation (7.1), are ~5 nm for S1, ~7 nm for S2, ~9 nm for S3, and ~15 nm for S4. These values confirm that R, although slightly larger, is in satisfactory accord with that determined from transmission electron microscopy (TEM). There is more discrepancy with the sol of smallest particle size. The possible explanation for this discrepancy will be discussed subsequently.

The structure and composition of the particles themselves are determined during the formation process. Information on the nature and composition of the particles can be obtained from contrast variation studies, as described previously [7,14]. Thus for the simple case of a uniform and homogeneous particle, the scattering length density, ρ_p, can be derived from variations of the solvent scattering length, ρ_s. This derivation is readily achieved with water using H_2O–D_2O mixtures (Table 7.2). ρ_p is derived from the solvent composition of zero contrast (i.e., $\rho_p = \rho_s$). This fact is illustrated by the scattering of sol S2 in different H_2O–D_2O mixtures (Figure 7.1), in which scattering

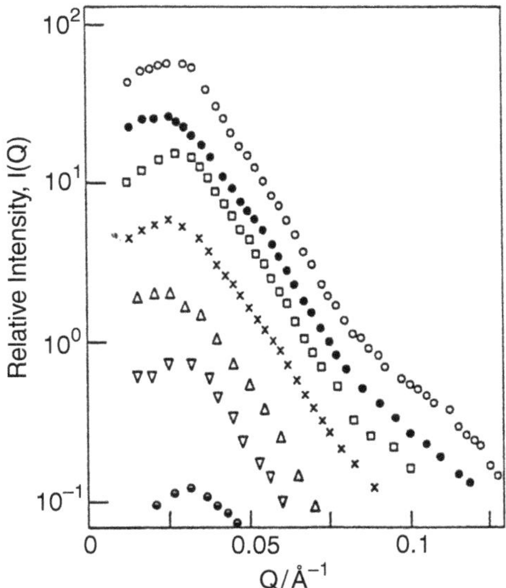

FIGURE 7.1 SANS results for silica sol S2 (0.15 g/cm^3) in different D_2O–H_2O mixtures. Key (% v/v D_2O): ○, 0; □, 30; △, 50; ▽, 55; ◒, 65; ×, 80; and ●, 100. (Reproduced with permission from reference 18. Copyright 1991 Academic press.)

is negligible for a composition of ~65% (v/v) D_2O. A precise determination of ρ_p is obtained from the linear relationship of $I(\mathbf{Q})^{1/2}$ vs. %v/v D_2O (cf. Equation 7.1) as illustrated in Figure 7.2. The experimentally determined value of ρ_p (3.6×10^{-10} cm^{-2}) is in close accord with that

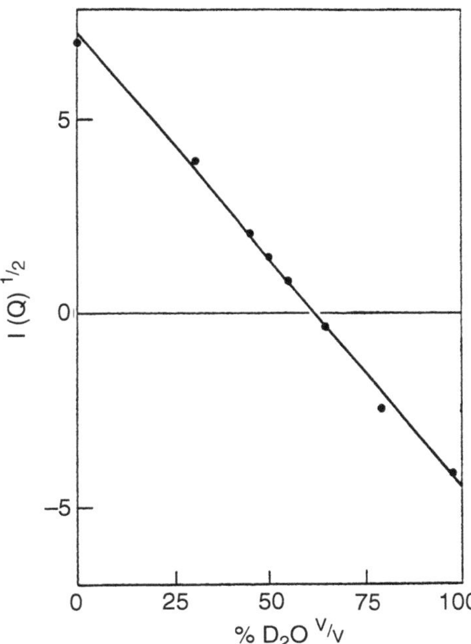

FIGURE 7.2 SANS contrast variation results for silica sol S2 in D_2O–H_2O mixtures. Intensity corresponds to $\mathbf{Q}$ (Å^{-1}) of 2.5×10^{-2}.

TABLE 7.2
Molecular Scattering Lengths, $\sum_i b_i$, and Corresponding Coherent Scattering-Length Densities, ρ, for Neutrons for Silica and Water of Mass Densities, δ

Compound	$\sum_i b_i$ (10^{-12}/cm)	δ (g/cm^2)	ρ (10^{10}/cm^{-2})
H_2O	−0.168	1.00	−0.56
D_2O	1.914	1.10	6.36
SiO_2	1.575	2.20	3.47
$SiO^a_{2\,exp}$			3.6

[a]Experimentally determined value.

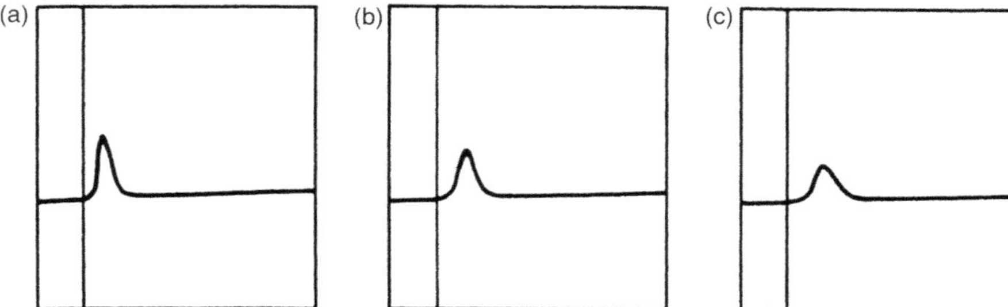

FIGURE 7.3 Schlieren photographs of silica sol S4 taken during ultracentrifugation at 1100 rpm after (a) 3.5 min, (b) 5.5 min, and (c) 7.5 min. The vertical lines toward the left side of the photographs indicate the air–solution meniscus; sedimentation is from left to right.

calculated for amorphous silica (3.47×10^{10} cm^{-2}) assuming a framework density of 2.2 g/mL (Table 7.2). This agreement indicates that the sols have a dense silica core and that any surface-hydroxylated species are readily exchangeable with D_2O. The form of the scattering curves in Figure 7.1, which show a maximum at $\mathbf{Q} \approx 3 \times 10^{-2}$ and a decrease in intensity at lower **Q**, is due to the maximum in the structure factor $S(\mathbf{Q})$ (cf. Equation 1). This feature becomes progressively pronounced as the sol concentration is increased and indicates that appreciable interaction occurs between the sol particles at the concentration (0.15 g/cm^{-3}) here.

Ultracentrifugation Studies

On ultracentrifugation the silica dispersions show a single solute boundary separating from the meniscus, which is observed as a peak in the schlieren (refractive index gradient) profile of the cell. Schlieren photographs recorded for a dilute dispersion of S4 are shown in Figure 7.3 and are typical of those observed for all the silicas. With increasing sedimentation time the boundary broadens as a result of diffusion and the centrifugal separation of particles of slightly different size. The observed boundary shapes indicate that the dispersions are reasonably monodisperse but with some increase in polydispersity for the smaller sized sols S1 and S2.

The effect of sol concentration on the sedimentation coefficients, s^{25}, of all four sols is illustrated in Figure 7.4. The values of s^{25} decrease as the size of the sol particle becomes smaller, as would be expected, and furthermore s^{25} is relatively insensitive to concentration in the range studied ($\sim 2 \times 10^{-3}$ to 6×10^{-2} g/mL). From the values of s_0^{25}, in Table 7.3 an effective radius, r_s can be derived by applying the Svedberg equation

$$M = \frac{RTs_0}{D(1 - \nu\delta_w)} \tag{7.3}$$

where M is the particle "molecular weight", R is the gas constant, T is absolute temperature, D is the diffusion coefficient, ν is the volume per unit mass ($= \delta_s^{-1}$) of the particles, and δ_w is the density of water. For anhydrous, spherical particles

$$r_s = \left(\frac{9}{2}\frac{\eta s_0 \nu}{1 - \nu\delta_w}\right)^{1/2} \tag{7.4}$$

where η is the viscosity of water.

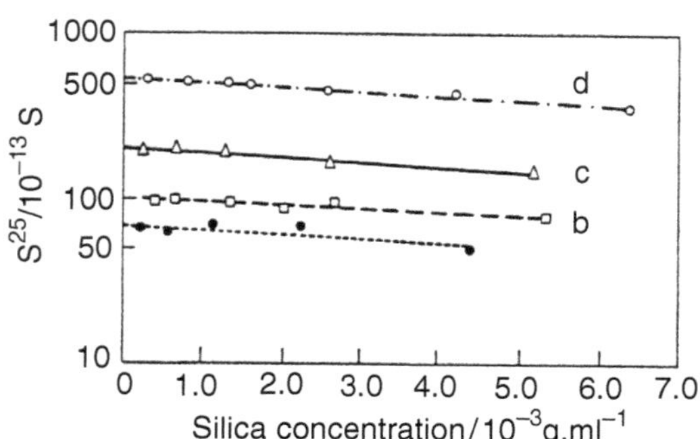

FIGURE 7.4 Variation of the sedimentation coefficient, s^{25}, with concentration for diluted silica sols as determined by analytical ultracentrifugation for (a) S1, (b) S2, (c) S3, and (d) S4.

TABLE 7.3
Sedimentation Coefficients, s_0^{25} of Silica Sols

Sol	$10^{13}s_0^{25}$ (s)	Effective radius r_s (nm)[a]
S1	68	4.8
S2	98	5.7
S3	198	8.1
S4	518	13.1
S1[D]	52	4.1

[a]Corresponds to sol radius assuming particle density, δ_s, of 2.2 g/ml.

The values of r_s for sols S2, S3, and S4 are all in reasonable accord with the sizes derived from SANS and TEM. That for S1 is considerably larger than that determined by TEM (~3.5 nm). However on extensive dialysis of the S1 sol a significant reduction occurs (Table 7.3). Furthermore, the change in the shape of the schlieren peaks that occurs after dialysis is marked as illustrated in Figure 7.5. The significance of these changes will be discussed in more detail later. After dialysis there appears to be a much smaller component that sediments considerably more slowly than the original sol particle. The areas under the two schlieren peaks after corresponding times of sedimentation are almost identical (Figure 7.5), a result showing that the mass of silica species is virtually unchanged. However, the skewing of the peak shape close to the meniscus can be ascribed, on semiquantitative examination of the peaks, to a smaller species that contributes up to ~20% of the total silica in the system. Furthermore, after dialysis the major component (measured in terms of the maximum ordinate of the peak) sediments more slowly, an observation indicating that it is reduced in effective size. Such effects can be tentatively ascribed to a change in the structure of the sol after dialysis. Thus in the initial sol the particles may be composed of a dense core surrounded by adsorbed oligomeric species. On dialysis and reduction of the ionic strength, a large proportion of these oligomeric species may be released into solution as a freely sedimenting fraction. Further support for such a process will be given later.

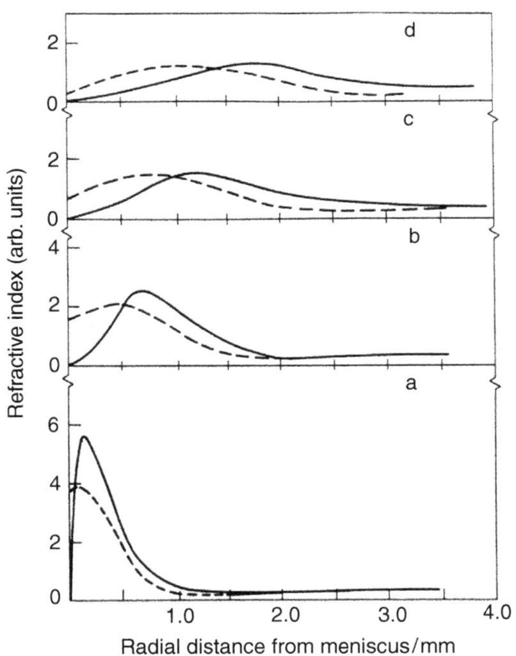

FIGURE 7.5 A comparison of the profiles of schlieren images for dilute (2.2×10^{-3} g/cm^3) samples of the dialyzed (- - -) and undialyzed (—) silica sol S1 after equivalent sedimentation times at 22,000 rpm. Time (min): a, 2; b, 6; c, 10; and d, 14.

PCS Studies

Translational diffusion coefficients, D_T, as a function of sol concentration were determined from PCS measurements. Measurements were made at different scattering angles ($2\Theta = 45°$, $90°$, and $135°$), and D_T was determined from initial slopes of the autocorrelation functions.

Typical autocorrelation functions, $g(\tau) - 1$, for dilute (<5 mg/mL) S3 and S4 sols (Figure 7.6) showed good fits to a single exponential decay, over a wide range of $\mathbf{Q}^2\tau$, a result indicating little polydispersity and particle interaction. ($\mathbf{Q} = 4\pi \sin \Theta\tilde{n}/\lambda$, where $\tilde{n}$ is the refractive index of the medium and τ corresponds to a delay time). For sols S1 and S2 single exponential fits were less satisfactory, a condition suggesting greater polydispersity. On the basis of a more detailed cumulants analysis of the form of g(τ), polydispersity indices for S3 and S4 were shown to be in the range ~0.13–0.2, whereas for S1 and S2 the index was considerably greater (0.25–0.4). Also, both light scattering and SANS give a Z-average size, and in consequence this size will be larger than that derived from TEM for polydispersed samples.

The variation of D_T with concentration for diluted stock sols S1, S2, S3, and S4 is shown in Figure 7.7 and Figure 7.8. In the range of concentration here a slight increase in D_T occurs with concentration. Although it is not appropriate to discuss in detail here, this effect can be ascribed [15] to the effects of electrical double-layer interactions. Values of D_T obtained on extrapolation to zero concentration and the corresponding hydrodynamic radii, R_H, are given in Table 7.4. For S4 R_H is slightly greater, although in reasonable accord with the radius derived from scattering and ultracentrifuge measurements. This difference becomes more significant for S3. However, for sols S2 and S1, R_H is considerably greater.

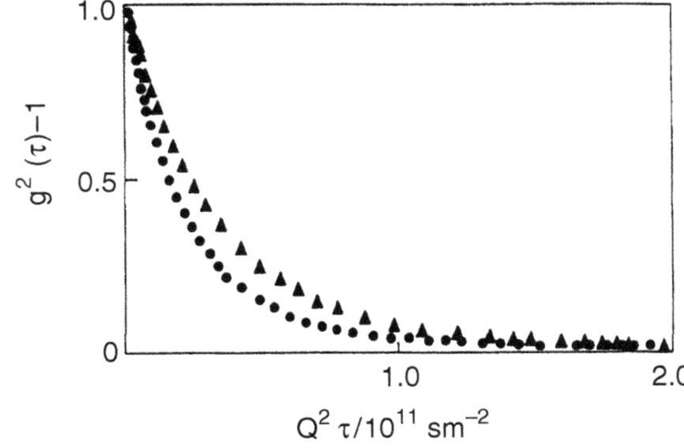

FIGURE 7.6 A comparison of normalized autocorrelation functions determined by PCS for diluted samples (1×10^{-3} g/cm^3) of silica S3 (●) and S4 (▲).

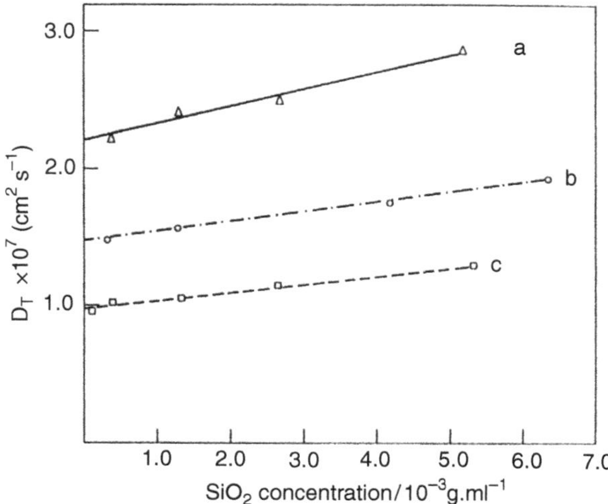

FIGURE 7.7 Translational diffusion coefficients determined by PCS as a function of concentration for dilute sols S3 (a), S4 (b), and S2 (c).

A possible explanation may arise from an extension of the shear plane from the particle surface due to adsorbed or associated oligomeric silicate species, as discussed earlier. Such an explanation is consistent with the changes in D_T that occur on dialyzing the S1 sol (cf. Figure 7.8): Dialysis results in a marked increase in D_T and a corresponding reduction in R_H (Table 7.5), presumably due to the release of oligomeric species to the aqueous solution.

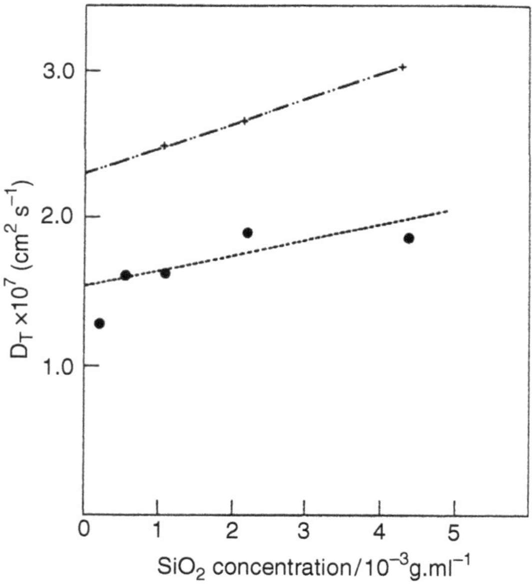

FIGURE 7.8 The variation of translational diffusion coefficients, D_T, measured by PCS with different sol concentrations for dilute samples of the dialyzed (+) and undialyzed (●) silica sol S1.

TABLE 7.4
Translational Diffusion Coefficients, D_T, and Hydrodynamic Radii, R_H, of Silica Sols at Infinite Dilution

Sol	$10^7 D_T$ (cm^2/s)	R_H (nm)
S1	1.5	16.5
S2	0.96	25.5
S3	2.2	11
S4	1.5	16.5
$S1^D$	2.3	9

A reduction in scattered light intensity (by ~65%) for sols of the same SiO_2 concentration after dialysis is also consistent with a reduction in the effective size of the particles. Any free oligomeric species will not make any significant contribution to the total scattered intensity because of their much smaller size. In this respect light scattering differs from ultracentrifugation in being sensitive only to the larger colloidal component. Measurements of D_T on sols of higher concentration show the effects of hydrodynamic interactions that lead to marked reductions in D_T (Figure 7.9). However, the onset of interaction sets in at considerably higher concentration for sol S4 (~0.15 g/ml) compared with sols of smaller particle size (~0.05 g/ml for S2). This feature may reflect a more extended range of hydrodynamic interactions in the sols of smaller particle size because of oligomers associated with the sol particles.

^{29}Si NMR Spectroscopy of Colloidal Silica

^{29}Si NMR spectroscopy has been used extensively to study aqueous solutions of silicates and has provided detailed information on the types of polymeric species in solution [3–5]. However, despite the power of the technique it has not been exploited previously to investigate colloidal silica. The usefulness of the technique arises from the

TABLE 7.5
Proportions of Q^4, Q^3, and Q^2 Species in Silica Sols

Sol	Concentration (% w/w)	Q^4	Q^3	Q^2
S1	15	68	23	9
S2	15	76	20	4
S3	40	78	17	5
S4	50	86	13	1

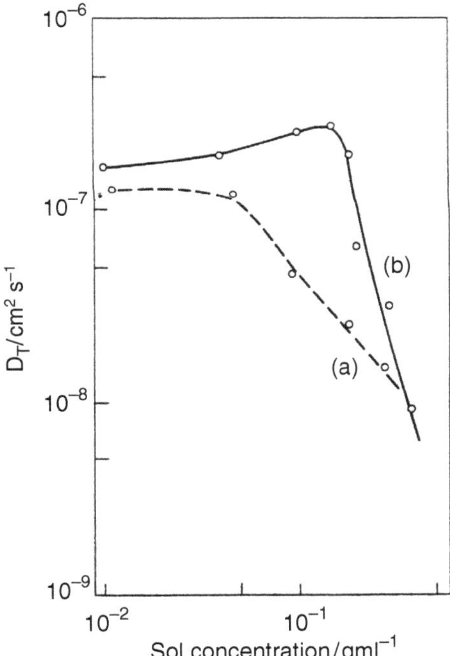

FIGURE 7.9 The effect of sol concentration on translational diffusion coefficients, D_T, determined by PCS for silica sols S2 (a) and S4 (b).

successive shift of the ^{29}Si resonance to high field on replacement of Si–OH bonds in silicate species with siloxy bonds, Si–O–Si. Thus monomeric species are observed furthest downfield ($\mathbf{Q}^\circ$) and other resonances are observed at intervals of ~10 ppm for end units ($\mathbf{Q}^1$), middle units ($\mathbf{Q}^2$), branching units ($\mathbf{Q}^3$), and tetrafunctional units ($\mathbf{Q}^4$). (**Q** refers to four possible coordination bonds; the superscript refers to the actual number.) From the relative intensities of the different resonances the relative proportions of the various structural entities can be derived.

^{29}Si NMR spectra of the four sols of different particle size (Figure 7.10) show three resonance peaks. The main intense resonance at −107 ppm can be assigned [16] to $\mathbf{Q}^4$ units corresponding to four siloxy bridges. Those at −96 and −85 ppm are assigned to $\mathbf{Q}^3$ and $\mathbf{Q}^4$ (corresponding to silicon with respectively three and two siloxy bridges) units, respectively. As the particle size decreases, the intensities of the two lower field peaks (−85 and −96 ppm) increase relative to the more prominent upfield resonance (−107 ppm). The relative proportions of these species can be derived form Gaussian fits to the peak profiles, and the results of such an analysis are given in Table 7.5. More detailed interpretation and further details of these investigations will be reported elsewhere [17]. However, the $\mathbf{Q}^4$ resonance corresponds to the silica core of the sol particles, and the $\mathbf{Q}^3$ and $\mathbf{Q}^2$ components to polyhydroxy units, either bound to the silica surface or as free oligomers in solution. For the smallest particle size sol, S1, the high proportion (32%) of these hydroxylated species is far in excess of that corresponding to a simple hydroxylated plane surface. Thus these results are consistent with the association of oligomers at the sol surface, as indicated from PCS and ultracentrifuge investigations.

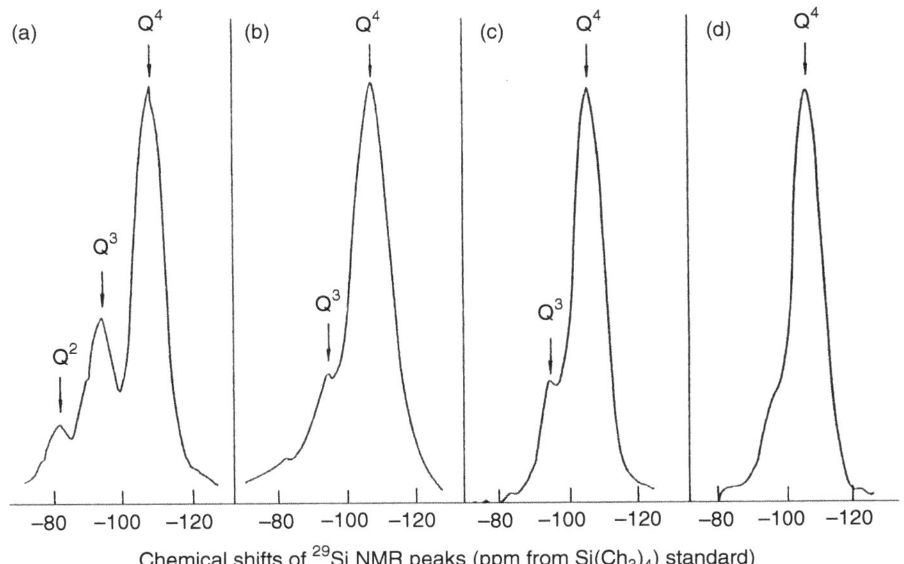

FIGURE 7.10 High-resolution ^{29}Si NMR spectra of silica colloids in water: (a) S1, (b) S2, (c) S3, (d) S4. The different chemical shifts (measured as parts per million with respect to $Si(CH_3)_4$) correspond to Si atoms in different environments: $\mathbf{Q}^4$ to SiO_4^{4-} tetrahedra in a three-dimensional structure; $\mathbf{Q}^3$ and $\mathbf{Q}^2$ to Si atoms that are coordinated to one and two hydroxyl groups, respectively.

CONCLUSION

The experimental investigations reported here provided further insight into the mechanisms of formation and growth of silica sols in aqueous solution. For sols of very small particle size (~7 nm), colloidal particles coexist with a high proportion (>20%) of oligomeric species. These species are predominantly associated with the colloidal particle surface. After dialysis, however, these species may be partially released into solution. This process probably results from the reduction of ionic strength and the consequent increase in electrostatic repulsion between the negatively charged surface and the anionic oligomers. The relative proportion of oligomeric silica decreases as the size of the sol particles is increased. Particle growth may thus arise from the condensation of monomers and oligomers at the particle surface in accord with the mechanism proposed by Iler [1].

Furthermore, the particle core has a density consistent with amorphous silica. The adsorption of oligomers around the core of the particles has implications that may explain the high charge density and exceptional stability of silica sols [9]. Enhanced stability may also arise from solvation forces. Such an interfacial structure has important consequences in both determining the mechanism and enhancing the capacity for sorption of other ionic species from solution.

Finally, the surface and pore structure of gels obtained after dehydrating sols may be affected by the presence of oligomeric silica. This effect will be particularly evident with sols of small particle size in which partial particle coalescence may occur, together with the generation of small micropores (<2 nm), as has been previously observed [18].

ACKNOWLEDGMENTS

The work described was undertaken as part of the Underlying and Corporate Research Programme of the UKAEA. We acknowledge the access to neutron scattering facilities, provided at the ILL, Grenoble, France.

REFERENCES

1. Iler, R. K. *The Chemistry of Silica*; Wiley: New York, 1979.
2. See, e.g., Brinker, C. J.; Scherer, G. W. *Sol–Gel Science*; Academic Press: New York, 1990.
3. Harris, R. K.; Knight, C. T. G.; Smith, D. N. *J. Chem. Soc. Chem. Commun.* **1980**, 726.
4. Arkakai, I.; Bradley, M.; Zerda, T. W.; Jones, J. *J. Phys. Chem.* **1985**, *89*, 4399.
5. Oriel, G.; Hench, L. L. *J. Non-Cryst. Sol.* **1986**, *79*, 177.
6. Ramsay, J. D. F.; Avery, R. G.; Benest, L. *Faraday Discuss. Chem. Soc.* **1983**, *76*, 53.
7. Ramsay, J. D. F. *Chem. Soc. Rev.* **1986**, *15*, 335.
8. Schaefer, D. W.; Martin, J. E.; Wiltzius, P.; Cannell, D. S. *Phys. Rev. Lett.* **1984**, *52*, 2371.
9. Lyklema, J. *Croat. Chem. Acta* **1971**, *43*, 249.
10. Ramsay, J. D. F.; Booth, B. O. *J. Chem. Soc. Faraday Trans. I* **1983**, *79*, 173.
11. Penfold, J.; Ramsay, J. D. F. *J. Chem. Soc. Faraday Trans. I* **1985**, *81*, 117; Bunce, J.; Ramsay, J. D. F.; Penfold, J. *J. Chem. Soc. Faraday Trans. I* **1985**, *81*, 2845.
12. Avery, R. G.; Ramsay, J. D. F. *J. Colloid Interface Sci.* **1986**, *109*, 448.
13. Swanton, S. W., Ph.D. Thesis, Bristol University, Bristol, England, 1989.
14. Jacrot, B. *Rep. Prog. Phys.* **1976**, 39, 911.
15. Klein, R.; Hess, W. *Faraday Discuss. Chem. Soc.* **1983**, *76*, 137.
16. Sindorf, D. W.; Maciel, G. E. *J. Am. Chem. Soc.* **1980**, *102*, 7606.
17. Dobson, C. J.; Goberdhan, G. C.; Ramsay, J. D. F. unpublished.
18. Ramsay, J. D. F.; Wing, G. *J. Colloid Interface Sci.* **1991**, *141*, 475.

8 Synthesis and Characterization of Colloidal Model Particles Made from Organoalkoxysilanes

A. van Blaaderen and A. Vrij
University of Utrecht, Van't Hoff Laboratory

CONTENTS

Monodisperse colloidal silica spheres were prepared by hydrolysis and condensation of tetraethoxysilane (TES) in a mixture of water, ammonia, and a lower alcohol. These silica spheres were coated with the coupling agent 3-aminopropyltriethoxysilane (APS) in the reaction medium. A new colloidal model system that was prepared consisted of a stable dispersion of monodisperse, hybrid, organic–inorganic "silica" spheres. Particles were characterized by ^{13}C and ^{29}Si NMR spectroscopy, elemental analysis, transmission electron microscopy, and static and dynamic light scattering. The particles made from TES alone (with radii between 15 and 100 nm) consisted of partially condensed siloxane structures, which were approximately the same for all particle sizes. Thus, the siloxane structure did not reflect differences in the particle shape and surface roughness. The percentage of silicon atoms not bonded to four other silicons, but bonded to one hydroxyl or ethoxy group, was close to 30%. A few percent of silicons had only two siloxane bonds, and at least a few percent of the ethoxy groups were present in the core of the particles, having never been hydrolyzed. Thus, hydrolysis probably is the rate-determining step in base-catalyzed particle growth. Further, the particle radius could be altered only by addition of $LiNO_3$, if the salt was added to the reaction mixture at the start of the reaction. Thus, the ionic strength provides a new parameter to study the reaction mechanism. Its influence confirms that the particles form through a controlled aggregation process of homogeneously precipitating nuclei early in the reaction. Monodispersity and a smooth particle surface are achieved through subsequent monomer addition. The new particles prepared from a mixture of APS and TES contained, on a silicon basis, as much as 26% APS distributed through the particle interior.

Monodisperse colloidal silica spheres with various surface coatings are of considerable interest to many fields, such as ceramics, colloids, catalysis, chromatography, and glass preparation. We are interested in these particles for their use as a model colloid. Previous work [1–9] has shown that by using scattering techniques, much insight can be gained in the interparticle structure and dynamics of concentrated dispersions. To avoid, for example, multiple scattering, it is important to determine and change the chemical microstructure of the (single) colloidal particles. The surface coating determines the solvents in which the colloids can be dispersed and the kind of interactions between colloidal spheres. An understanding of the mechanisms responsible for particle formation and the coating of its surface is very important to tailor the scattering properties and particle interactions.

Through the hydrolysis and condensation of tetraalkoxysilanes in mixtures of ammonia, water, and alcohols, (uncoated) silica particles are easily prepared by using a method developed by Stöber et al. [1]. The experimental procedure is very simple, but the mechanisms responsible for the formation and growth of the charge-stabilized silica spheres are not. A generally accepted scheme has not yet been presented. A tentative mechanism for the nucleation and growth is described in this chapter, based on an analysis of the final particle morphology and chemical microstructure. A description is also given of the use of the silane coupling agent 3-aminopropyltriethoxysilane (APS) to prepare model colloids.

APS is known for its stability in basic solutions [10]. The amine functionality facilitates the coupling of other desired molecules to a particle. For these reasons, this alkoxide was chosen to coat the surface of the Stöber silica spheres. A simple procedure was followed, first described by Philipse and Vrij [4], who used other coupling agents. Badley et al. [11] also described the coupling of APS and other organosilanes to silica.

APS and tetraethoxysilane (TES) were also used in the synthesis of a new kind of monodisperse colloidal system. Spherical particles were made by starting from a mixture of the organoalkoxysilanes TES and APS. This procedure will probably be applicable to other mixtures of organoalkoxysilanes as well.

The microstructure and particle morphology (size, shape, and surface roughness) of the colloids were studied with ^{13}C and ^{29}Si NMR spectroscopy, elemental analysis, transmission electron microscopy, and static and dynamic light scattering.

To facilitate the interpretation and to place the proposed mechanism into perspective, a literature survey is presented. Relevant details concerning the base-catalyzed chemistry and characterization of alkoxysilanes by NMR spectroscopy and some of the proposed particle formation and growth theories are given.

THEORY AND EARLIER WORK

Reactions and Chemical Mechanisms

Alkoxysilanes in a mixture of water, ammonia, and a lower alcohol may undergo many different reactions. Reactions between the different silane intermediates can be represented as follows:

$$-\overset{|}{\underset{|}{Si}}OR + HOR' \rightleftarrows -\overset{|}{\underset{|}{Si}}OR' + HOR \quad (8.1)$$

$$-\overset{|}{\underset{|}{Si}}OR + HO\overset{|}{\underset{|}{Si}} \rightleftarrows -\overset{|}{\underset{|}{Si}}O\overset{|}{\underset{|}{Si}}- + HOR \quad (8.2)$$

Here R and R′ stand for a hydrogen atom or an alkoxy group (methoxy, ethoxy, or propoxy). The other atoms bonded to silicon that are not depicted can either be carbon (in the organoalkoxides), or oxygen (belonging to a silanol and alkoxy group or a siloxane bond). The breaking of a C–Si bond and direct ester exchange between the alkoxides are not considered possible [12]. We will discuss only base-catalyzed reactions.

The reactions in Equation (8.1) constitute ester exchange, hydrolysis, and their reversals. In the base-catalyzed case these reactions proceed through a nucleophilic attack on the silicon atom, resulting in a pentacoordinate transition state [12–15].

We first consider the hydrolysis and the influence on its rate by some reaction conditions. OH^- is acting as the nucleophile; an increase in the concentration of this catalyst will increase the reaction rate. OH^- is produced by reaction of NH_3 with H_2O. The rate increases almost linearly with $[NH_3]$ [16]. As expected, the hydrolysis appears to be first order in TES [16,17]. The dependence of the rate constant on $[H_2O]$ is even higher, according to Harris et al. [18], than the dependence on $[NH_3]:[H_2O]^{1.5}$. The effect on the reaction rate of changing the solvent alcohol is quite complex, and further experimental work and a theoretical explanation are needed [15–18].

The influence of a substituent on the silicon atom can be derived from its ability to withdraw electrons or its steric hindrance to an attacking nucleophile. For these reasons, the lower alkoxides are hydrolyzed faster than the higher and branched alcohol derivatives. Hydrolysis of a given alkoxide proceeds in successive steps. Because the alkoxy group is less electron-withdrawing and more bulky than a hydroxy group, each successive loss of an alkoxy group is accompanied by an increase in the rate of loss of the next.

The base-catalyzed condensation reactions in Equation (8.2) also take place through a base-catalyzed nucleophilic attack on silicon resulting in a pentavalent transition complex [12,15,19,20]. The influence of $[H_2O]$ and $[NH_3]$ on the reaction rate and the type of alcohol used is almost the same as for the hydrolysis [12,18].

An important question to be explained by the chemical mechanism is why base catalysis often leads to condensed structures. In 1950, Aelion et al. [13] pointed out that the condensation reaction in base-catalyzed systems was faster than with acid catalysis and that the microstructure of the final product was different. Many of these facts are due to the opposite effects of, for example, substituents, on silicon on the stabilization of the transition state in base- and acid-catalyzed reactions [15]. For the base-catalyzed condensation reaction to take place, a silicon atom has to be attacked by a deprotonated silanol oxygen: the nucleophile. The acidity of the silanol proton increases as the basicity of the other groups bonded to the silicon decreases. This feature implies that polysilicic acid is a stronger acid than $Si(OH)_4$ [20]. Therefore, monomers react preferentially with higher polymerized species.

Under certain conditions (for instance, high [TES] or low water concentrations), the reaction between not fully hydrolyzed species must also be considered. In this case the condensation and hydrolysis reactions can occur in one step: After attack of a silanolate ion, the transition state can subsequently lose an alkoxy group. Alkoxy groups that end up on a highly condensed unit are then much more difficult to hydrolyze because of steric constraints.

The replacement of an alkoxy group causes an inductive effect. Binding of an organofunctional group through a C–Si bond decreases both the hydrolysis rate and the condensation rate through the same mechanism. The less electronegative carbon makes a nucleophilic attack on silicon more difficult.

This reasoning also applies to APS: Compared to TES, base-catalyzed hydrolysis and condensation are slower for APS. However, in some respects this molecule does not behave like other silane coupling agents. Solutions of APS in water are quite stable at basic pH values, whereas solutions of other coupling agents rapidly form insoluble precipitates under the same conditions [21]. This difference in behavior is explained by Plueddemann [21] by assuming that the hydrolyzed APS forms a six- or five-membered chelate ring. The six-membered ring is thought to form by binding of the amine with a silanolate ion and the pentacoordinate complex by binding between nitrogen and silicon. No direct evidence has been found for one of these structures. Most likely, one of them exists, because compounds with one CH_2 group more or less in the aliphatic chain do not display the increased stability of APS and rapidly form precipitates in water. Moreover, these precipitates remained insoluble after they were acidified, a result indicating that the insolubility is not caused by the amine portion [15].

Nucleation and Growth: Results from the Literature

Before discussing different mechanisms of nucleation and growth that have appeared in the literature, we first present some general observations about the hundreds of silica sols that have been described [1,2,4–7,16–18,22–28] and synthesized according to Stöber.

Most of the work was done with the alkoxide TES and the alcohol ethanol as solvent. Lowering the temperature always results in larger particles. Sometimes the radius can become 4 times larger by lowering the temperature 30°C [26]. In the range of TES concentrations in which particles seem to be stable, larger particles generally tend to be more monodisperse. Standard deviations in particle size of only 2% have been observed.

Mostly, particles are larger when synthesized from higher concentrations of TES, although the increase in radius is quite small, and sometimes the opposite trend is described [2]. After the reactants are mixed, the turbidity of the solution suddenly rises after a certain induction time and then slowly increases to its final value. The induction time is closely related to the total time of the particle growth. If the particles reach their final radius in a short time, the induction period is also short.

Increasing $[H_2O]$ and $[NH_3]$ results in faster reactions and initially, at low concentrations, in larger radii. However, both H_2O and NH_3 show a concentration for which a maximum radius is achieved. Increasing the concentration any further still gives shorter reaction times, but the radius decreases. Near these maxima the sols sometimes flocculate and are not so monodisperse. This instability is also found in the limits of high TES or high NH_3 concentrations. Under some conditions, particles with a grainy surface are observed.

Increasing the length of the solvent *n*-alcohol results in larger particles [1,16–18,22]. Densities of the resulting homogeneous particles were reported in the range 1.9–2.1 g/mL, and the refractive index was found close to 1.45 [2,3,26].

Matsoukas and Gulari [17] convincingly showed that the particle mass grows exponentially with the same time constant that describes the first-order hydrolysis of TES. They used the plasma lines in a Raman spectroscopy experiment to observe not only the intensities of the Si–O–Et and Et–OH bands, but at the same time used the light scattering of the particles to follow their growth. The induction period found in the scattering experiments was not found for the hydrolysis reaction.

Subsequently, Matsoukas and Gulari [22] worked out a model describing the nucleation and growth in more detail. In a purely kinetic theory, they formulated a monomer addition model in the presence of a rate-determining first-order initiation step and investigated the effect of different particle growth models upon the final particle size and polydispersity. In their model the dynamic competition between nucleation and growth is controlled by the hydrolysis that releases the active monomer. This monomer is produced in a slow hydrolysis reaction that inhibits nucleation and, therefore, promotes growth of large particles.

In the model, two limits are considered: a reaction-limited growth, which is characterized by a strong size dependence of the growth rate, and diffusion-limited growth, with much weaker size dependence. These limits refer only to the bonding between a particle and a monomer; the overall rate is still governed by the hydrolysis.

The actual rate equations are based on a succession of irreversible steps. Hydrolysis is modeled as first order in TES and produces the active monomer. These monomers can react with another monomer forming the (second-order) nucleation step, or can react with an already higher condensed species. All the subsequent addition steps are considered elementary and are described with different rate constants k_i (with i the number of monomers that formed the particle). Generally, k_i depends on the radius (R) of the particle. The limit that k_i is independent of the particle size was assumed to represent a reaction-limited growth (the correct dependence is $k_i \sim R$), and k_i proportional to the particle volume (R^3) was assumed to represent diffusion-limited growth (the correct dependence is $k_i \sim R^2$). Under these assumptions, Matsoukas and Gulari [22] derived for the final particle radius (R) the following expressions:

$$R \sim (k_p c/k_h)^{1/9} \quad \text{diffusion-limited growth} \qquad (8.3)$$

$$R \sim (k_p c/k_h)^{1/6} \quad \text{reaction-limited growth} \qquad (8.4)$$

where k_p and k_h are the polymerization and hydrolysis rate constants, respectively, and c is the starting TES concentration.

Using the correct dependencies of the rate constants k_i on the particle size, they also obtained expressions relating the final polydispersity to the final mean particle radius $\langle R \rangle$:

$$\sigma^2 \sim \langle R \rangle^{-3} \quad \text{diffusion-limited growth} \qquad (8.5)$$

$$\sigma^2 \sim \langle R \rangle^{-2} \quad \text{reaction-limited growth} \qquad (8.6)$$

In the experimental part the relations between Equation (8.3) and Equation (8.6) were tested; the data were best described with a reaction-limited growth (in the presence of a rate-limiting hydrolysis of the monomers). The increase in particle size with increasing ammonia concentration was rationalized by stating that, although this effect resulted in an increase in k_h, k_p would have to increase even further.

Bogush and co-workers [24,26,27] investigated the Stöber synthesis using electron microscopy, conductivity measurements, and the (small) change in reaction medium volume. They concluded that all the TES hydrolyzes completely in the first few minutes and that the Stöber silica particles are formed through a size-dependent, controlled, coagulative nucleation and growth mechanism. They argued that nucleation from the extremely supersaturated solutions continued almost to the point at which the particles reached their final size. In the beginning of the reaction, small unstable particles aggregate until a critical size is reached at which the probability of two particles of equal size sticking together becomes negligible. Homogeneous nucleation still continues, and the freshly formed small particles are taken up by the large stable particles that now remain constant in number. Classical nucleation expressions were used to estimate nucleation rates per unit volume [29,30]. Their proposed size-dependent nucleation–aggregation mechanism is analogous to a similar model developed to explain the emulsion polymerization of styrene [31].

Philipse [5] also assumed that fast hydrolysis created an active monomer bulk. He studied the growth of silica nuclei, already synthesized, after extra addition of different amounts of TES with static light scattering. To explain his growth curves (radius versus time), he used a diffusion-controlled particle growth in a finite bulk of monomers or subparticles. The model contained equations from classical flocculation theories. It takes into account the exhaustion of the monomer bulk and the retarding influence of an (unscreened) electrostatic repulsion between growing spheres and monomers.

Harris and co-workers [16,18] concluded that the Stöber system obeyed La Mer's homogeneous nucleation and growth model [32]. In this scheme the slow hydrolysis of TES builds up the critical concentration of silicic acid required to form nuclei. The short nucleation phase then lowers the concentration of silicic acid below its critical value, followed by growth of monodisperse silica particles. Monodispersity is always achieved if the nucleation phase is short enough, whether the growth process is diffusion- or reaction-controlled [33]. They found that under conditions where the amount of silicic acid remained relatively high (that is, a few times the equilibrium value) during an important part of the reaction, the resulting sol was polydisperse.

Schaefer and Keefer [14,19,34,35] used a model developed by Eden to describe the growth of cell colonies to explain their X-ray scattering experiments. They used this simple model of nucleation and chemical limited growth to mimic the growth of silica structures in TES solutions (1 *M*) with a relatively low water concentration (1–4 *M*) and with very little catalyst, 0.01 *M* NH_3.

In their version of the Eden model, one starts with a seed, that is, one site on a two-dimensional lattice, and randomly chooses and occupies one of the four neighboring sites. The next growth site is picked from the resulting six neighboring sites, etc. This method results in spherical particles with uniform interiors and smooth surfaces relative to their radii.

The fact that the interior becomes uniform seems an artifact, because growth is even allowed in free places inside a particle. The formation of such regions in three-dimensional growth is very unlikely, so allowing this kind of growth inside a particle in the two-dimensional model makes comparison with three-dimensional studies more realistic.

The effects on the particle shape of particle formation from partially hydrolyzed monomers were also investigated. By choosing certain mixtures of monomers, Schaefer and Keefer generated all kinds of particle morphologies, from porous clusters to surface and mass fractals.

Silicon Environments and Solid-State NMR Spectroscopy

Magic-angle spinning (MAS) and high-power proton decoupling (HPD) make it possible in solid-state NMR spectroscopy to obtain high-resolution ^{13}C and ^{29}Si spectra. These techniques average the chemical-shift anisotropy and dipolar interactions that otherwise would cause very broad lines. In a liquid this averaging is accomplished by the rapid thermal motions of the molecules in the magnetic field.

The NMR solid-state technique of cross-polarization (CP) consists of a transfer of magnetization from ^{1}H nuclei to ^{13}C or ^{29}Si nuclei through dipolar interactions. In this way, selective detection of only those ^{13}C and ^{29}Si nuclei that are near protons becomes possible. For silica consisting mainly of SiO_2, this technique entails a selective detection of only those nuclei close to the surface where OH groups and other proton-containing species are present. In addition to this selectivity, CP enhances the sensitivity of the NMR experiment and shortens the necessary delay between acquisitions of successive transitions to a few seconds. For the ^{29}Si nucleus that can have spin-lattice relaxation times (T_1) of minutes, this enhanced sensitivity can be important.

Through siloxane bonds, a silicon atom can be bonded to a maximum of four other silicon nuclei. If a ^{29}Si nucleus is bonded to four other silicon atoms, the chemical shift of this so-called $\mathbf{Q}^4$ species lies around −110 ppm. The $\mathbf{Q}^4$ representation comes from the old siliconen-chemistry [12]; the **Q** stands for quaternary (that is, having the possibility of forming four siloxane bonds). The number 4 gives the actual number of siloxane bonds in which the nucleus is participating (Figure 8.1). For every siloxane bond less, a shift of about 10 ppm occurs. Therefore, $\mathbf{Q}^2$ has a chemical shift around $-110 + 2 \times 10 = -90$ ppm.

The exchange of one Si–O bond with a Si–C bond makes a significant difference in chemical shift. For APS this exchange results in a shift +40 ppm. The organosilane is now designated T^0 to T^3 (T from ternary).

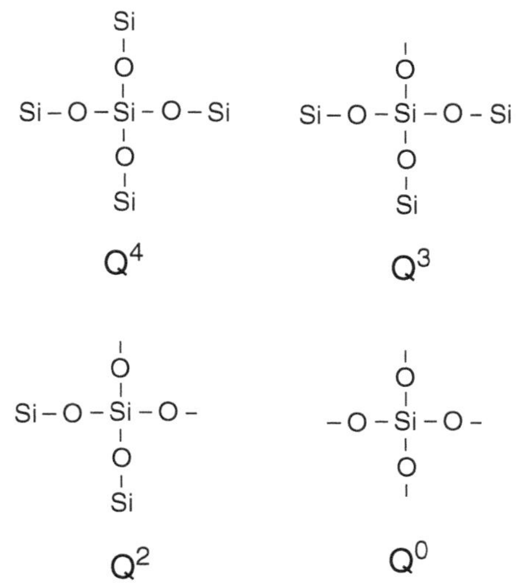

Si
|
O
|
Si – O – Si – O – Si
|
O
|
Si

Q^4

|
O
|
Si – O – Si – O – Si
|
O
|
Si

Q^3

|
O
|
Si – O – Si – O –
|
O
|
Si

Q^2

|
O
|
– O – Si – O –
|
O
|

Q^0

FIGURE 8.1 Silicon environments. Groups not depicted are alkoxy or silanol.

Clearly these features make silicon NMR spectroscopy a valuable technique in examining the microstructure of silica and its derivatives [36–40]. However, almost no difference in chemical shift is seen between a silicon atom that has an alkoxy and one that has a hydroxy group bonded to it. Carbon solid-state NMR spectroscopy is useful to determine the presence of alkoxy groups and to detect the carbon atoms of a coupling agent if present.

EXPERIMENTAL DETAILS

Materials and Particle Preparation

Methanol (Baker), ethanol (Merck), and 1-propanol (Baker) were of analytic reagent quality. Absolute technical grade ethanol (Nedalco) was used only for the large-scale (9 L) synthesis of Al. Solvents, tetraethoxysilane (Fluka, purum grade) and γ-aminopropyltriethoxysilane (Janssen) were freshly distilled before each synthesis. Ammonium hydroxide (Merck, 25%) was of analytical reagent quality and contained 14.0 mol/L NH_3 as indicated by titration. The silicas Ludox AS40 and Compol were kindly provided by DuPont and Fujimi.

Alcosols, silica particles dispersed in the reaction medium, were synthesized at 20°C according to the method of Stöber et al. [1] and the detailed description given by Van Helden et al. [2]. Glassware was cleaned with 6% hydrogen fluoride and rinsed with deionized water and absolute ethanol. Ammonium hydroxide, ethanol, and, if necessary, deionized water were mixed in a reaction vessel and placed in a constant-temperature bath. After allowing the temperature to come to equilibrium, the TES was added under vigorous stirring. After a few minutes the reaction was continued using a slow

TABLE 8.1
Reactant Concentrations

System	[TES] (*M*)	[APS] (*M*)	[H_2O] (*M*)	[NH_3] (*M*)	Solvent
A1	0.155		0.855	0.318	ethanol
A1APS[a]		0.477	0.855	0.318	ethanol
A2	0.0119		5.51	0.682	ethanol
A3	0.17		1.65	0.68	ethanol
A4	0.167		1.58	0.512	ethanol
A5[b]	0.159		3.01	1.12	ethanol
A5APS[a]		0.205	3.01	1.12	ethanol
A6	0.160		2.65	0.986	ethanol
A7[c]	0.160		2.65	0.986	eth–$LiNO_3$
A8[c]	0.160		2.65	0.986	eth–$LiNO_3$
Mix1	0.0809	0.0774	2.34	0.870	ethanol
M1	0.161		2.50	0.928	methanol
P1	0.178		1.03	0.383	propanol

[a]Alcosol taken as reaction medium for the coating reaction.

[b]Reaction temperature was 25°C; all other reactions were at 20°C.

[c][$LiNO_3$] = 1.0 m*M*. For A8 the $LiNO_3$ was added before TES, and A7 was added 15 min after TES.

stirring speed. The total amount of alcosol varied between 0.4 and 9 L. Concentrations of reactants used are given in Table 8.1; in their calculation, volume contraction was assumed to be absent.

For the coating reactions with APS (Table 8.1), a procedure similar to that described by Philipse and Vrij [4] was applied. Philipse and Vrij used the coupling agent 3-methacryloxypropyltrimethoxysilane to coat colloidal silica. Because an alcosol was used, the actual concentrations of water as listed in Table 8.1 have to be corrected for the (not exactly known) amount consumed in the hydrolysis — condensation reactions. After addition of the APS to 400 mL of alcosol, the solution was stirred slowly for an hour. After 2 h of refluxing, 250 mL was distilled slowly for an hour. The remaining unreacted APS was removed in four centrifugation redispersion steps. The ultracentrifuge used was from Beckmann (L5-50B), and the sediment was redispersed each time in absolute ethanol.

Silica spheres with APS distributed in the particle interior were prepared by adding APS (Table 8.1) first to the reaction mixture. Subsequently an equal volume of TES was added within 1 min. Free APS was again separated from the particles by centrifugation and redispersing in ethanol. Particle densities were measured by drying, under dry nitrogen for 24 h at 100°C, a known volume of a concentrated dispersion in absolute ethanol and weighing the residue.

Alcosols are referred to in this chapter as A*x*, with *x* being a numeral. The A is replaced by a P or a M if the alcohol used was not ethanol but 1-propanol or methanol, respectively (Table 8.1). Coated particles are designated A*x*APS and A*x*S; the last code stands for a coating with stearyl alcohol according to the procedure developed by Van Helden et al. [2]. *Mix* denotes particles prepared from a mixture of the alkoxides TES and APS. Some of these systems have been used in earlier work with different names, so some translation is necessary. A3S is designated in several investigations as SJ9 [9]. SM2, known here as A1, is described by Penders and Vrij [41]. The Ludox and Compol systems are described elsewhere by Duits et al. [42].

Light Scattering

Light-scattering measurements were made at 25 ± 0.1°C on dust-free, very dilute dispersions in ethanol. The dispersions were made by adding one drop of alcosol to 25 mL of ethanol followed by filtration through Millipore filters (the pore was typically several particle diameters). Cuvets with a diameter of 2 cm were cleaned by continuously rinsing with freshly distilled acetone.

Static light scattering (SLS) was performed with a Fica-50 photometer using vertically polarized light ($\lambda = 436$ and 546 nm). A correction was made for scattering of the solvent. The Rayleigh–Gans–Debye approximation could be used, and the particles were assumed to be spherical and to have a homogeneous refractive index. Under these assumptions, the particle radius (R_o) was obtained from a fit of the calculated form factor to the scattered intensity as a function of the scattering angle θ ($20^\circ \leq \theta \leq 150^\circ$). Intensities at low angles are sensitive to dust or clustered particles [2]; no irregularities were found, a result indicating monodisperse, nonclustered sols.

Dynamic light scattering (DLS) results were obtained with an argon ion laser (Spectra Physics Series 2000) operating at 488.0 and 514.5 nm. Autocorrelation functions were measured with a Malvern Multibit K7025 128-point correlator. Diffusion coefficients were obtained from a second-order cumulant fit [43,44] by using autocorrelation functions obtained from six scattering angles between 35° and 145°. From the diffusion coefficient, a hydrodynamic radius R_h was calculated by using the Stokes–Einstein relation. Except for dispersion Al, the normalized second cumulant yielded values smaller than ~0.05; the value for Al was around 0.12. More information can be found elsewhere about the procedures and equipment of SLS [3,6] and DLS [45,46].

Elemental Analysis

Elemental analysis was carried out by Elemental Microanalysis Limited (Devon, U.K). Prior to shipping, the samples were dried for 24 h at 100°C under nitrogen; before the measurements were made, they were dried again for 3 h under the same conditions.

Electron Microscopy

Transmission electron micrographs were made by dipping copper 400-mesh carrier grids in a dilute dispersion. The grids were covered with carbon-coated Formvar films, and the photographs were made of particles on the film. Philips EM301 and Philips CM10 transmission electron microscopes were used, with the magnification calibrated with a diffraction grating.

Some of the particles were studied after supercritical point drying. The particles were dispersed in acetone by several centrifugation steps (45,000 × g). In a critical point dryer (Balzers Union, CPD 020), the carbon-coated grids and the dispersion in acetone were placed. Under pressure, the acetone was exchanged against liquid CO_2; after increasing the temperature and passing the critical point, the particles adsorbed on the film were supercritically dried.

Particle radii of 500–2000 particles were measured with an interactive image analysis system (IBAS). From these data, a number-averaged particle radius $\langle R \rangle$ and a standard deviation σ defined by:

$$\sigma = \left[\frac{\langle R^2 \rangle - \langle R \rangle^2}{\langle R \rangle^2} \right]^{1/2} \tag{8.7}$$

were determined.

NMR Spectroscopy

High-resolution solid-state NMR spectra were measured at room temperature on a Bruker AM 500 spectrometer (silicon frequency 99.4 MHz and carbon, 125.7 MHz) equipped with a Bruker solid-state accessory. Spectra were obtained with a broad-band probehead with a 7-mm double air-bearing magic-angle spinning (MAS) assembly. Spinning speeds around 4000 Hz were employed. The 90° pulse lengths for the nuclei ^{13}C, ^{29}Si, and ^{1}H were around 5.5 μs. The spectra that were obtained with cross-polarization had contact times between 0.5 and 2 ms for carbon and between 0.5 and 6 ms for silicon; pulse sequences were repeated after 4 s. The number of accumulated free induction decays (FIDs) per spectrum ranged between 200 and 4000, depending on the system being investigated.

Quantitative silicon spectra were obtained by using 90° ^{29}Si pulses; it was not necessary to use high-power proton decoupling. Even at the high magnetic field used, the relatively small chemical-shift anisotropy and dipolar coupling with protons could be removed completely by MAS alone. Because, for some samples, the $\mathbf{Q}^4$ silicon atoms had spin–lattice relaxation times of about 90 s, the time delay between 90° pulses was chosen as long as 400 s. Typically, 200 FIDs were accumulated. Deconvolution of the spectra was performed by manually adjusting the height, width, and frequency of the Gaussian line shapes to obtain the best visual fit.

The samples used for NMR spectroscopy were freeze-dried to prevent any reaction during the drying process. Adamantane and the trimethylsilyl ester of double four-ring octameric silicate, $\mathbf{Q}^8\mathbf{M}^8$ were used to optimize experimental parameters and as external secondary (relative to TMS) chemical-shift references for ^{13}C and ^{29}Si, respectively. Both the T_1 measurements and a discussion on the use of ^{29}Si NMR spectroscopy for quantitative measurements will be described elsewhere [47].

RESULTS AND DISCUSSION

In this section, experimental results obtained with several different experimental techniques and measured on several different colloidal silica dispersions are described and discussed. The ultimate goal of the investigations is the ability to synthesize stable, monodisperse colloidal model spheres from organoalkoxysilanes with different chemical compositions and surface properties [47–52].

The first step was the synthesis of several colloidal systems with different radii from TES alone (A1–A6), and comparison of these particles with commercially available systems that were prepared in a different way, such as Ludox and Compol [20]. The stabilities of the sol and the radii were probed with static and dynamic light scattering (SLS and DLS, respectively) because these techniques are sensitive to particle clustering. The polydispersity in the particle size, particle shape, and surface roughness (all these particle properties are further referred to as the particle morphology) were investigated with transmission electron microscopy (TEM). The chemical microstructure of the particles or, more precisely, the siloxane structure of the final particles and the presence of alkoxy groups were investigated with solid-state silicon and carbon NMR, combined with elemental analysis. To obtain more information on the particle formation mechanism, two colloidal systems were made by using a different alcohol as solvent (M1 and P1), and two systems were made after addition of $LiNO_3$ to change the ionic strength (A7 and A8).

The second step consisted of the coating of several "Stöber" silica particles with APS and the synthesis of a new kind of particles from a mixture of APS and TES. The colloidal dispersions obtained were also characterized by the techniques just mentioned.

TABLE 8.2
Radii According to Transmission Electron Microscopy and Static and Dynamic Light Scattering

System	TEM[a] (nm)	SLS (nm)	DLS (nm)
A1	9.01 (20%)		9 ± 2
A1APS	9.15 (20%)		14 ± 3
A2	50.2 (150%)		
A3S	30.9 (13%)	39.6 ± 0.5	38.5 ± 0.5
A4	31.2 (13%)	40.2 ± 0.5	36.6 ± 1.6
A5	62.4 (11%)	71.5 ± 0.4	72.1 ± 1.2
A5APS	66.1 (11%)	73.6 ± 0.5	80 ± 3
A6	89.4 (6%)	107.2 ± 0.3	108.1 ± 0.7
A7	93.2 (8%)	109.4 ± 0.4	
A8	154.3 (8%)	179.3 ± 0.4	
Mix1	92.2 (11%)	128 ± 2	118 ± 3
M1	18.2 (16%)	26.1 ± 0.9	30 ± 2
P1	24.2 (12%)	33.1 ± 0.5	35.1 ± 1.3
Ludox	11.1 (20%)	17.5 ± 0.3	20.8 ± 0.5
Compol	50.5 (16%)		

[a]The number in parentheses is the relative standard deviation.

Particle Morphology (SLS, DLS, and TEM)

In this section the results of the "Stöber" silica particles are presented, followed by the Ludox and Compol particles, and finally the particles made with APS. Then, the results relevant to a tentative formation and growth model are discussed.

All the synthesized alcosols except M1 remained stable for months, as indicated by light scattering. M1 flocculated 2 weeks after the synthesis but did not show any sign of instability directly after the reaction.

The general trends on the dependence of the final particle radius on the concentrations of NH_3 and water, on the alcohol used as solvent, and on the temperature can be found in Table 8.1 and Table 8.2. Increasing the concentrations of NH_3 and water results in larger radii (compare for example, A1, A4, and A5). This effect is no longer true for very high concentrations of water (A2). Increasing the temperature decreased the radius (compare A5 and A6). Going from methanol to propanol increases the radius (compare M1, A1, and P1).

An interesting new result is the strong dependence of the radius on the ionic strength if this is increased *before* the addition of TES (A6, A7, and A8, Table 8.1 and Table 8.2). $LiNO_3$ was chosen to change the ionic strength because it dissociates completely in this concentration range [53]. The flocculation concentration of the sol A6 was 5 mM of $LiNO_3$.

The radii R_o and R_h as obtained with SLS and DLS, respectively, show reasonable-to-good agreement (Table 8.2). The small differences and the small corrections that can be made if the polydispersity is taken into account are discussed in more detail elsewhere [9,49]. The differences between the light-scattering radii and the radii as obtained with TEM are not negligible and cannot be explained by the differences in the averages that are determined with the different experimental techniques [2,9]. As was suggested by Van Helden et al. [2] we attribute these differences to radiation damage that causes the particles to shrink somewhat. (The drying process does not influence the radius determined with TEM, because no differences were observed between normally and supercritically dried samples.)

Despite the fact that the absolute radii are probably too small, the polydispersities determined with TEM are correct. As is known from the literature [1,2,26], the polydispersities decrease as the mean radius increases. The only exception is A2, but this system was made from a very low concentration of TES (Table 8.1).

The appearance of "Stöber" silica particles may depend strongly on the conditions under which they were made [2,4–7,24–27]. Particles have been described as being irregular, roughly spherical, spherical, and large, flocculated, and fused structures [28]. All the different types of morphology can be seen in the electron micrographs of the various alcosols shown in Figure 8.2, except for the flocculated structures. The irregularly shaped

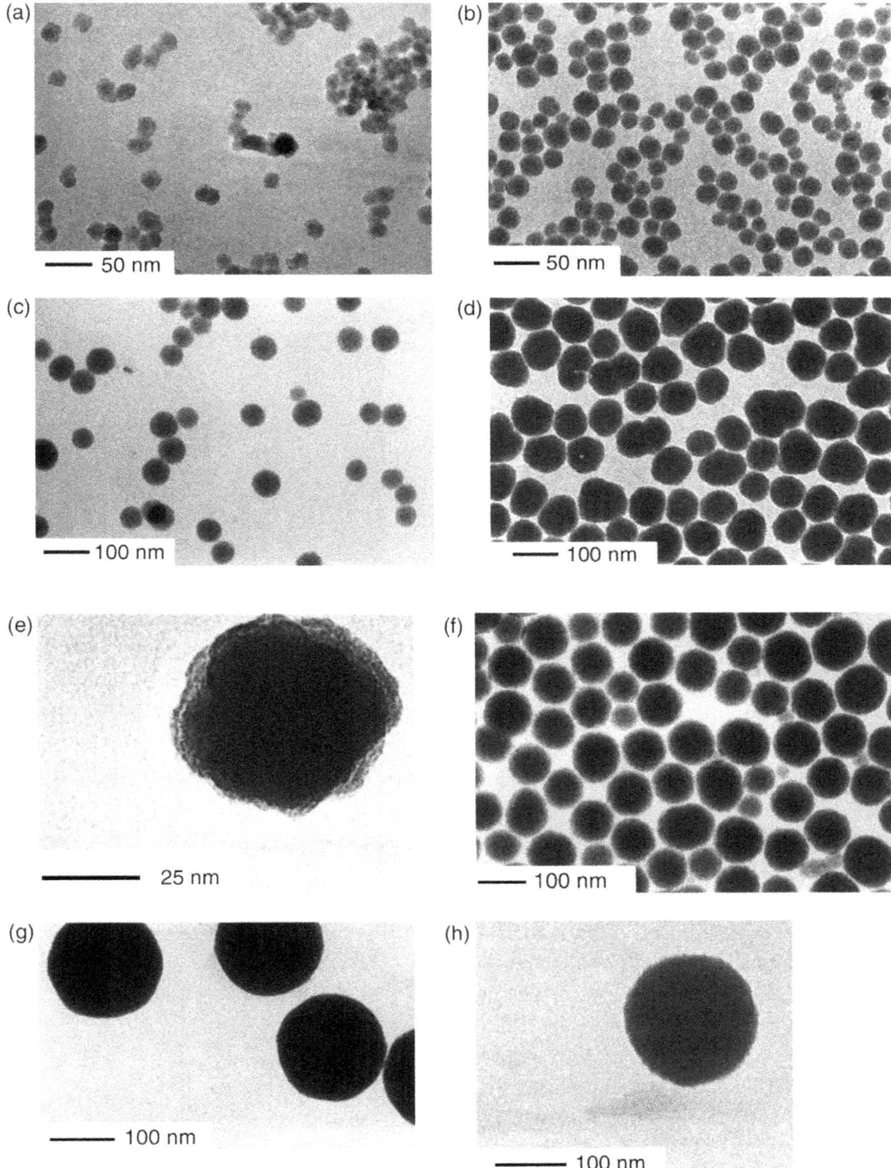

FIGURE 8.2 Transmission electron micrographs. (a) A1, $\langle R \rangle = 9.01$ nm (20%), supercritically dried. (b) Ludox, $\langle R \rangle = 11.1$ nm (20%). (c) A2, $\langle R \rangle = 50.2$ nm (150%). (d) A3S, $\langle R \rangle = 30.9$ nm (13%). (e) $\langle R \rangle = 25$ nm, (f) $\langle R \rangle = 50$ nm, (g) A6, $\langle R \rangle = 89.4$ nm (7%). (h) Mix 1, $\langle R \rangle = 92.2$ nm (11%).

particles were not deformed in any way by capillary forces [20], because some samples (A1 and A4) were supercritically dried, a condition that eliminates the effects of the surface tension. Probably because of the high speed of centrifugation needed to settle these small particles, a fraction of them appeared to be clustered after redispersing in acetone. However, the amount of separately laying particles (Figure 8.2a) was sufficient to clearly demonstrate that both the form and radius were the same when compared with a sample of A1 that was not supercritically dried. Figure 8.2e shows another highly magnified supercritically dried particle (A4). The particle shown in Figure 8.2e had the characteristics of particles that are smaller than ~70-nm radius and are synthesized from concentrations of TES around 0.17 *M*. These smaller "Stöber" particles are only roughly spherical and possess a grainy irregular surface (Figure 8.2d). However, the particles A2 prepared from a low concentration of TES and a high concentration of water (Table 8.1) are small, but almost perfect spheres with very smooth surfaces. Only the polydispersity is very high.

In Figure 8.2g a typical larger Stöber particle (A6) is shown. These particles are smooth and almost perfect spheres. Comparing the shape of the Ludox particles

(Figure 8.2b) and the Compol particles (Figure 8.2f) with the Stöber particles of about the same size (Figure 8.2a and 8.2d) shows only small differences. The particles not synthesized from TES are slightly more spherical, and the surfaces are somewhat less rough.

The micrographs of the APS-coated particles were not different from the uncoated spheres and are therefore not shown. In Figure 8.2h the particles Mix1 are shown. Considering the amount of APS that is part of these particles, it is amazing how spherical and monodisperse these particles are. Clearly, all the organic groups (disrupting the siloxane structure) inside the particles do not influence the final particle shape. When compared with A6, they appear only slightly rougher. The difference in density between these two systems is considerable and is visible in the micrographs. The density of Mix1 is 1.51 g/mL, and that of A6 is 2.08 g/mL [50].

Particle Microstructure (NMR Spectroscopy and Elemental Analysis)

The ^{29}Si CP NMR spectra were much easier to obtain than the direct excitation spectra, because of the long spin–lattice relaxation time of the silicon nuclei. Nevertheless, direct excitation was used to obtain the quantitative results of the siloxane structure, because not all $\mathbf{Q}^4$ nuclei were detected with CP (Figure 8.3 and Table 8.3) [47].

Surprisingly, all the quantitative spectra of the alcosol particles looked similar. As is also demonstrated in Table 8.3, the siloxane structure of all the Stöber silica spheres is more or less constant and independent of the particle size and thus the reaction conditions. Even the coated stearylsilica particle A3S is no exception, despite the 3 h at 200°C necessary for the coating. Thus, the differences mentioned in the previous section in the particle morphology are not correlated with the siloxane structure.

The colloidal particles Ludox and Compol consist of a more condensed silica structure (Table 8.3). As mentioned, CP detects only those nuclei that have dipolar interactions with protons. Because of the condensed structure, a significant number of $\mathbf{Q}^4$ silicon nuclei is too far away from protons and is not detected in a CP experiment (contrary to $\mathbf{Q}^4$ nuclei in the alcosol particles). An example of this effect can be seen in Figure 8.3a and 8.3c of the Compol particles. The spectrum in Figure 8.3a was obtained by using direct polarization through 90° pulses, MAS, and repetition times of 400 s. CP–MAS was used for Figure 8.3c. The CP spectrum, Figure 8.3c, contains more $\mathbf{Q}^3$ signal than $\mathbf{Q}^4$, whereas the quantitative spectrum, Figure 8.3a, shows no detectable amount of $\mathbf{Q}^2$, and only 12% of $\mathbf{Q}^3$ against 88% $\mathbf{Q}^4$.

The Ludox silica structure can be calculated to be more or less fully condensed. The surface and volume of a Ludox sphere are calculated from the radius (e.g., 11 nm, Table 8.2). Assuming 4.5 **OH** groups per square nanometer [20], the expected number of silanol groups as a percentage of the number of silicon atoms per particle can be determined. The number of silicon atoms in the particle can be obtained from the density (2.2 g/mL [20]) and the molecular weight, which is, taking 15% $\mathbf{Q}^3$ into account, 61.43 g/mol. These calculations result in a percentage of silanol groups of 5.7%. This amount is small

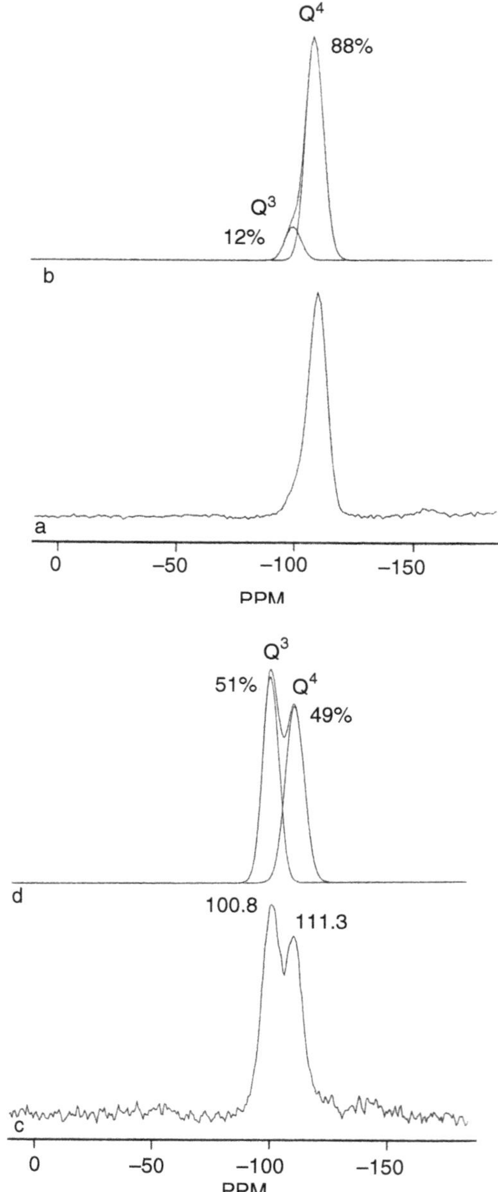

FIGURE 8.3 ^{29}Si NMR spectra of Compol. (a) Direct polarization MAS, 300 accumulations, and repetition time of 400 s. (b) Deconvoluted spectrum using Gaussian line shape. ^{29}Si NMR spectra of Compol. (c) CP–MAS, 2000 accumulations, repetition time of 4 s, and contact time of 4 ms. (d) Deconvoluted spectrum using Gaussian line shape.

TABLE 8.3
Relative Intensities of the Different Types of Silicon Environments

System	Q^4	Q^3	Q^2	T^3	T^2
A1[a]	73[b]	26	1.7		
A1APS	67	21	—	7.2	4.4
A2	69	32	—		
A3S	65	31	4.1		
A5[a]	64	31	5.5		
A5APS	65	30	4.0	—	—
A6	66	30	4.2		
Mix1	40	31	3.8	19	7.2
M1	64	30	6.5		
P1	66	28	5.6		
Ludox	85	15	—		
Compol	88	12	—		

Note: Data are reported as percentages. Blank cells indicate that the nuclei could not be present in the sample. Dashes indicate that the numbers were too small to be determined.

[a]The same reaction procedures as for the coating were followed, except for the addition of APS.

[b]Estimated errors for $\mathbf{Q}^4$ and $\mathbf{Q}^3$ were $\pm 2\%$; for the other species, $\pm 1\%$.

compared to the experimental number of 15% as obtained by NMR spectroscopy. It may be clear, however, that some surface roughness, as was observed by TEM, will increase the surface area. Because the siloxane structure of the Compol particles is close to that of Ludox, and the radius is about 5 times larger, it is also clear that even the Compol particles are not fully condensed SiO_2 with only silanol groups at the particle surface.

^{13}C CP–MAS NMR spectroscopy showed the presence of ethoxy groups for all the alcosols prepared in ethanol, in accordance with elemental analysis (Table 8.4). The carbon next to oxygen was found close to 60 ppm, and the methyl group was at around 17 ppm (Figure 8.4a). The stearyl-coated A3S was no exception: an important part of the carbon signals originated from ethoxy groups. The carbon content of the larger particles, A5 and A6, is even higher than that of the particles A1. This result seems to indicate that for the larger particles most of the ethoxy groups are trapped inside the particles. This hypothesis is corroborated by a washing procedure of A6 (Table 8.4), which did not remove much of the carbon content. For the particles with a (much) higher surface area, the effect of washing with water (P1 in Table 8.4) or the refluxing of the particles A1 was visible in a significant change in carbon content. The almost complete removal of the carbon content for P1 after washing seems to suggest that the particles are microporous.

The chemical shifts of a carbon next to an OH group or next to O–Si are very close together. If only one O–C carbon is found, it is difficult to distinguish between ethoxy groups chemically bound and ethanol molecules

TABLE 8.4
Elemental Analysis (Weight Percentages)

System	C (%)	H (%)	N (%)
A1[a]	1.68 ± 0.01[b]	0.97 ± 0.08	
A1	1.11 ± 0.02	0.99 ± 0.08	
A1APS	7.29 ± 0.01	2.20 ± 0.03	2.18 ± 0.02
A2	0.59 ± 0.01	0.85 ± 0.02	
A3S	9.7	2.7	
A5[a]	5.07 ± 0.12	1.66 ± 0.07	
A5APS	6.03 ± 0.05	2.05 ± 0.04	0.32 ± 0.01
A6	3.95 ± 0.04	1.55 ± 0.08	
A6[c]	3.54 ± 0.03	1.45 ± 0.08	
Mix1	12.95 ± 0.02	3.97 ± 0.07	4.49 ± 0.05
P1	3.84 ± 0.02	1.55 ± 0.04	
P1[c]	0.27 ± 0.04	1.34 ± 0.10	
Compol	0.1 ± 0.01	0.23 ± 0.02	

[a]Samples underwent the coating conditions.

[b]Errors indicate differences in duplicate results.

[c]Samples were washed for 3 h with water before drying.

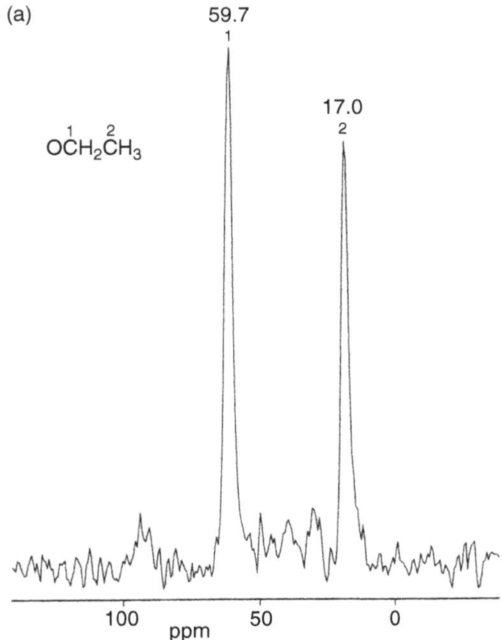

FIGURE 8.4 ^{13}C CP–MAS NMR spectra of alkoxy groups in silica; contact time of 2 ms. (a) A1, 1000 accumulations. (b) P1, 500 accumulations. (c) M1, 3000 accumulations.

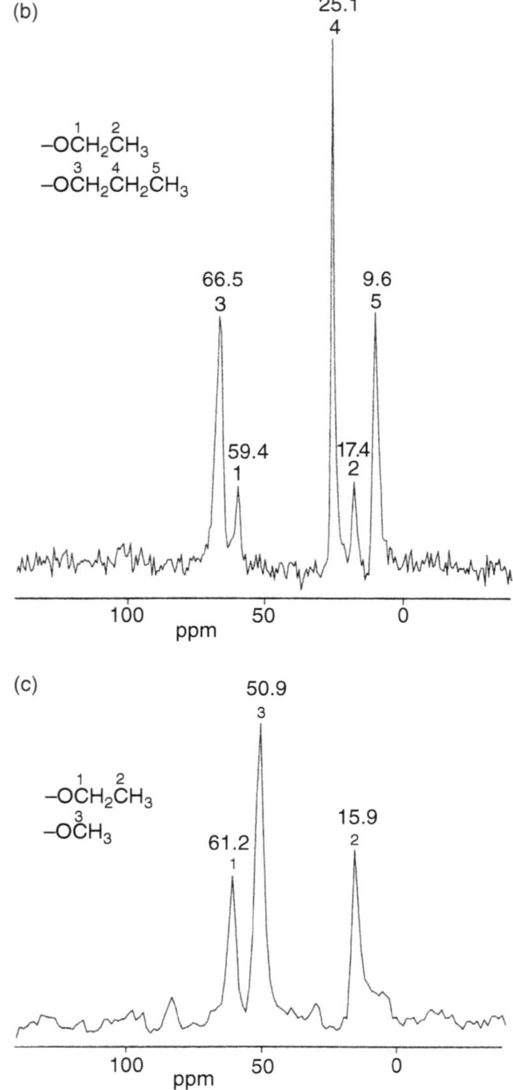

FIGURE 8.4 *Continued.*

bonded through hydrogen bonds with silanol groups. Another possibility for the origin of the ethoxy signals in the ^{13}C alcosol spectra is reesterification of hydrolyzed silanol groups.

To resolve the questions raised in the previous paragraph, particles were synthesized in methanol and 1-propanol. Complete hydrolysis and reesterification in these solvents would result in the appearance of only methoxy and propoxy groups. The ^{13}C CP–MAS spectra of A1, P1, and M1 are shown in Figure 8.4. Ethoxy groups can still be seen in both particles prepared in methanol and propanol. No reesterification of ethoxy groups is possible in these solvents; thus the detected groups must never have left the TES molecule. Although care must be taken in using CP spectra quantitatively, it is probably safe to conclude from Figure 8.4 and Table 8.4 that between 10% and 30% of the detected carbon in the elemental analysis belongs to ethoxy groups from TES. The amount of silicon atoms with ethoxy groups bonded to them that have never left the original TES molecule is at least a few percent, whereas some reesterification or ester exchange seems to take place as well.

If the carbon and proton weight percentages are subtracted from 1 g of particles (Table 8.4), the remainder of the weight can be equalized to $n\mathrm{SiO}_x$, where x can be determined from the relative intensities of **Q** species as determined by NMR spectroscopy (Table 8.3). The carbon content can be used to calculate the number of ethoxy groups. The number of silicon atoms with bonded ethoxy groups ranges from almost zero (for the particles A2 made in a relatively high concentration of water) to 14% for sample A5.

For the sample A3S, the amount of silicon was determined: 39.0% [elemental analysis described in Ref. (4)]. Subtracting C and H and using the siloxane structures from NMR as before, yields 38.94% Si in perfect agreement with the measured value. Apparently, the amount of physically absorbed water can be neglected in this case. This is not so strange, considering the heating to 190°C for 3 h during esterification. Jansen et al. [8] stated that 5% H_2O was present inside the particles, but this error was made because it was assumed that only $\mathbf{Q}^4$ was present.

Figure 8.5 shows all the possible silicon signals that can be obtained from reactions between TES and APS. Carbon assignments will be described in more detail elsewhere [50]. The chemical shifts compare well with literature data [36–40,54–56]. This test sample was made by

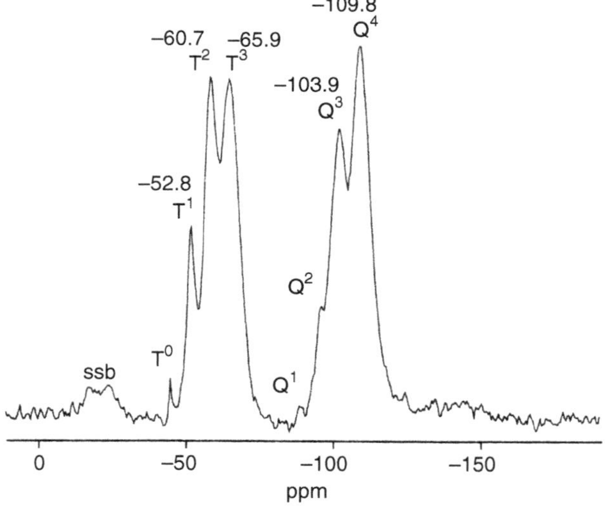

FIGURE 8.5 ^{29}Si CP–MAS NMR spectra of a mixture of TES–APS hydrolyzed by atmospheric water; 2000 accumulations and contact time of 4 ms.

mixing equal amounts of TES and APS and letting the mixture hydrolyze by atmospheric water. The white powder thus obtained (no colloidal spheres) was crushed and placed in a spinner. The quantitative silicon spectra obtained from A1APS and Mix1 did not contain signals from T^1, but were similar.

For the APS-coated A1APS, the silicon nuclei of APS could be determined quantitatively with NMR spectroscopy (Table 8.3). For 1 g of A1APS, as explained before, one calculates 1.68 mmol of APS. Using the N percentages gives 1.56 mmol of APS. The difference, 0.12 mmol, can be explained by assuming that a few percent physically adsorbed water was also present. The number of $\mathbf{Q}^3$ and $\mathbf{Q}^2$ becoming $\mathbf{Q}^4$ because of the coating reaction is clearly lower than the number of T species formed (Table 8.3). This result shows that APS did not form all its siloxane bonds with **Q** species.

Unfortunately, the coating of A5 contained too few APS to be determined with NMR spectroscopy. Comparing the surface-to-volume ratio of A1 and A5 using the TEM radius gives a value of 6.9; the ratio of the N percentages is 6.8. Probably, the amount of coating scales with the surface available and the reaction stops if the silanols belonging to **Q** species are not available anymore. For A5 this means that only 1.1% of the available APS actually ended up on the particle surface. Without the silica particles, the APS does not form any structure that scatters light [50].

These surface coverages are somewhat less than the coating with methacryloxypropyltrimethoxysilane as described by Philipse and Vrij [4]. He calculates ~10 groups per square nanometer. Using the values of A5 one finds ~2 APS molecules per square nanometer. Badley et al. [11] found for particles of radius 58.1 nm with elemental analysis 0.90% N. The difference with this work was that the concentration of APS was approximately 10 times smaller and no refluxing or distillation was used, but the reaction was continued for 20 h. The methoxy derivative of APS also was used, probably resulting in faster hydrolysis. Apparently, APS reacts very slowly with silica.

The results obtained for Mix1 are amazing: 36.3% of the APS used ended up inside the particles (Table 8.4). This value contrasts sharply with the small amount found in the coating reaction, even though the concentrations used there were higher. The APS did not end up as a surface coating, as is indicated by the very low amount of $\mathbf{Q}^4$. This result can only be explained if the APS is distributed through a large portion of the colloid.

As before, calculating the amount of APS per gram using SiO_x gives 3.28 mmol and, from the percent N one gets 3.19 mmol. The small difference is again probably due to a few percent H_2O. The amount of carbon also indicates that 5% of the Si atoms can still be bonded to an ethoxy group.

Nucleation and Growth Mechanism

In the following sections, we will test some of our findings with predictions and assumptions of the proposed mechanisms in the literature.

The finding that a few percent of ethoxy groups never leave the TES molecule is in accordance with the findings of Harris and co-workers [16,18] and Matsoukas and Gulari [17,22] that the growth rate of the silica particles is limited by the hydrolysis of TES. Under these conditions, some monomers may polymerize before the hydrolysis of all the groups on the silicon atom is complete. Were the hydrolysis complete in the first few minutes, it would indeed seem strange to find ethoxy groups in the methanosol and propanosol. This kind of reasoning can also be used to explain the large amount of APS that was build inside the Mix1 Particles. Here, hydrolyzed TES molecules could have reacted with the slower hydrolyzing APS molecules.

We find a contradiction with the assumptions made by Bogush and co-workers [24–27] that in the first few minutes all TES hydrolyzes and that during the whole particle growth the nucleation continues. If growth continued through aggregation of subparticles during the whole process, the surfaces of the spheres that can be seen in Figure 8.2c and 8.2g could not have been as smooth.

In our work and from the literature [1,2,24–27], it can be seen that particles with a radius of 15 nm to about 400 nm can be made with a single concentration of TES by changing only the amount of ammonia. This is a 25-fold increase in radius. Using Equation (8.3) and Equation (8.4) derived by Matsoukas and Gulari in their kinetic model yields for the ratio of hydrolysis and polymerization rate constants 2.4×10^8 and 3.8×10^{12} for the diffusion and reaction-limited cases, respectively. Both the hydrolysis and polymerization are caused by a nucleophilic reaction mechanism; hence, such an increase of one over the other is very unlikely. The factor 2.4×10^8 is also very far from the differences in reaction rates that have been determined experimentally by Harris and co-workers [16,18]. Therefore, although the model of Matsoukas and Gulari [22] seems to explain the polydispersity and growth of a particle distribution, it fails to predict the final size. This deficiency is because their kinetic model is also used to describe the particle nucleation process.

However, if we use the expressions for the classical nucleation rate, we also do not find the large increase in radius that is possible. Furthermore, Weres et al. [57] state that addition of NaCl to a homogeneous nucleating solution of silicic acid in water has only a moderate effect if the concentrations are below 1 *M*. Addition of salt to the reaction mixture of the Stöber process has, however, very large effects. Clearly, classical nucleation cannot explain all the features of the Stöber synthesis either.

One possible solution is to assume that the number of nuclei that have been formed through homogeneous nucleation decreases early in the reaction by a controlled aggregation process. After, and perhaps during, the short aggregation period, monodispersity is reached by the addition of monomers of small oligomers. The assumption that small silica particles are not stable against aggregation can be made plausible as follows: The silica particles in an alcosol derive their stability from the negative surface charge originating from dissociated silanol groups. Preliminary conductivity and electrophoretic measurements on the A6 particles in the reaction mixture indicate a surface potential of about 50 mV, a low, reasonable value. This value has also been reported by Bogush et al. [24] but it is very difficult to make an accurate interpretation of the measurements. Many of the important physical properties of the alcosol mixture are not exactly known. A better indication of the stability of the particles comes from the observation of the flocculation concentration of 5 mM $LiNO_3$. At a concentration of 1 mM, the sol remains stable, but the final particle radius is almost doubled (A6–A8, Table 8.1 and Table 8.2) only if the salt is added before the first particles are visible! When a concentration of 2 mM was added, the sol flocculated during its formation.

The Derjaguin–Landau–Verwey–Overbeek (DLVO) theory of charged colloids [58] predicts a substantial decrease in stability against flocculation with decreasing particle radius. Most likely the newly formed nuclei are not yet stabilized, and stability sets in only after a certain radius is obtained. After this size is reached, particles grow through monomer addition either reaction- or diffusion-limited, but with an overall rate still depending on the hydrolysis.

The finding that particles increase in size by using more ammonia (Table 8.1), in spite of the fact that this increases the surface charge, probably involves an increase in ionic strength that compresses the double layer. As reported by Bogush [27], during the first minutes of the reaction, there is a sharp increase in solution conductivity that decreases slowly during the remainder of the reaction. The production of silicic acid, partly dissociated by NH_3, causes this increase. When the particles grow, the number of available silanol groups slowly decreases, and so does the conductivity. This sudden increase in ionic strength is probably higher for higher concentrations of ammonia and results in larger particles. Clearly, these predictions have to be verified experimentally [49].

Eventually, the effect on the surface charge of the increase in $[NH_3]$ and $[H_2O]$ overcomes the increase in ionic strength, and the decrease in the number of nuclei is less, so that the particle radius decreases.

The model as proposed can also explain why solvents with a higher dielectric constant produce smaller particles and why, under conditions where the largest particles are formed, the sols sometimes flocculate.

Finally, the particle morphology must be explained in terms of the nucleation or growth mechanism. We tentatively assume that the irregular shape of some particles (for instance, A3 and A1) is caused by the proposed controlled aggregation mechanism early in the formation of the particles. For these systems the monomer growth was not yet able to create a smooth surface, as present on the larger spheres like A6. A2 also has very spherical and smooth particles, but here the low TES and high water concentration stabilized the aggregating particles at a small radius before most of the TES was used and a smooth particle surface could be achieved.

CONCLUSIONS

We conclude that hydrolysis is the rate-limiting step in growth because a few percent of the ethoxy groups of TES are not hydrolyzed fully before they polymerize and these groups are also included in the interior of the larger particles. This conclusion can also be drawn from more direct measurements presented in the literature [16–18,22].

We propose that silica particles synthesized according to the method of Stöber are formed through a controlled aggregation of nuclei formed through homogeneous nucleation, followed by growth through monomer addition. The ionic strength is an important parameter in the determination of the final particle size, and it is easily changed by addition of salt [49].

Silica particles can be coated by the silane coupling agent APS in the reaction medium to obtain stable colloids. These systems can be used to further bind relevant groups to the particle surface [48,51,52]. Quantitative ^{29}Si NMR spectroscopy appears to be a valuable tool in characterizing a new kind of stable colloidal particles synthesized from a mixture of the alkoxides APS and TES. These monodisperse, spherical particles can already be prepared with a low polydispersity and contain APS molecules distributed through an important part of the bulk of the particle. It would be interesting to see whether this method is also applicable to other mixtures of alkoxides. More work concerning the characterization of these new systems has to be done [50].

ACKNOWLEDGMENTS

We thank G. Nachtegaal and Dr. A.P.M. Kentgens for their help with the NMR measurements at the SON-NWO HF-NMR facility (Nijmegen, the Netherlands). We also thank J. Suurmond and J. Pieters for performing the electron microscopy, and A. Philipse for his comments.

This work was supported by the Netherlands Foundation for Chemical Research (SON) with financial aid from the Netherlands Organization for Scientific Research (NWO).

REFERENCES

1. Stöber, W.; Fink, A.; Bohn, E. *J. Colloid Interface Sci.* **1968**, *26*(1), 62–69.
2. Van Helden, A. K.; Jansen, J. W.; Vrij, A. *Colloid Interface Sci.* **1981**, *81*(2), 354–368.
3. Van Helden, A. K.; Vrij, A. *J. Colloid Interface Sci.* **1980**, *78*(2), 312–329.
4. Philipse, A. P.; Vrij, A. *J. Colloid Interface Sci.* **1989**, *128*(1), 121–136.
5. Philipse, A. P. *Colloid Polym. Sci.* **1988**, *266*(12), 1174–1180.
6. Philipse, A. P.; Smits, C.; Vrij, A. *J. Colloid Interface Sci.* **1989**, *129*(2), 335–352.
7. Philipse, A. P.; Vrij, A. *J. Chem. Phys.* **1987**, *87*(10), 5634–5643.
8. Jansen, J. W.; De Kruif, C. G.; Vrij, A. *J. Colloid Interface Sci.* **1986**, *114*(2), 481–491.
9. Moonen, J.; Pathmamanoharan, C.; Vrij, A. *J. Colloid Interface Sci.* **1989**, *131*(2), 349–365.
10. Plueddemann, E. P. *Silane Coupling Agents*; Plenum: New York, 1982.
11. Badley, R. D.; Ford, W. T.; McEnroe, F. J.; Assink, R. A. *Langmuir* **1990**, *6*, 792–801.
12. Noll, W. *Chemie und Technologie der Silicone*; 2nd ed.; Verlag Chemie GmbH: Weinheim, Germany 1968.
13. Aelion, B. R.; Loebel, A.; Eirich, F. *J. Am. Chem. Soc.* **1950**, *72*, 5705–5712.
14. Keefer K. D. *Mater. Res. Soc. Symp. Proc.* **1984**, *32*, 15–24.
15. Brinker, C. J.; Scherer, G. W. *Sol–Gel Science*; 1st ed.; Academic Press: Boston, MA, 1990.
16. Byers, C. H.; Harris, M. T.; Williams, D. F. *Ind. Eng. Chem. Res.* **1987**, *26*(9), 1916–1923.
17. Matsoukas, T.; Gulari, E. *J. Colloid Interface Sci.* **1988**, *124*(1), 252–261.
18. Harris, M. T.; Brunson, R. R.; Byers, C. H. *J. Non-Cryst. Solids* **1990**, *121*, 307–403.
19. Schaefer, D. W.; Keefer, K. D. *Phys. Rev. Lett.* **1984**, *53*(14), 1383–1386.
20. Iler, R. K. *The Chemistry of Silica*; Wiley & Sons: New York, 1979.
21. Plueddemann, E. P. In *Silylated Surfaces*; Leyden, D. E.; Collins, W. T., Eds.; Gordon & Breach: New York, 1980; pp 31–53.
22. Matsoukas, T.; Gulari, E. *J. Colloid Interface Sci.* **1989**, *132*(1), 13–21.
23. Coenen, S.; De Kruif, C. G. *J. Colloid Interface Sci.* **1988**, *124*(1), 104–110.
24. Bogush, G.; Zukoski, C. *The Colloid Chemistry of Graving Silica Spheres*; Ceramic Microstructures 1986; Plenum Press: New York, 1987; pp 475–483.
25. Brinker, C. J.; Scherer, G. W. *Sol–Gel Science*; 1st ed.; Academic Press: Boston, MA, 1990; pp 199–203.
26. Bogush, G. H.; Tracy, M. A.; Zukoski, C. F., IV *J. Non-Cryst. Solids* **1988**, *104*, 95–106.
27. Bogush, G. H.; Dickstein, G. L.; Lee, P.; Zukoski, C. F., IV *Mater. Res. Soc. Symp. Proc.* **1988**, *121*, 57.
28. Adams, J.; Baird, T.; Braterman, P. S.; Cairns, J. A.; Segal, D. L. *Mater. Res. Soc. Symp. Proc.* **1988**, *121*, 361–371.
29. Makrides, A. C.; Turner, M.; Slaughter, J. *J. Colloid Interface Sci.* **1973**, *73*(2), 345–367.
30. Nielsen, A. E. *Kinetics of Precipitation*; Pergamon Press: Oxford, England, 1964.
31. Feeney, P. J.; Napper, D. H.; Gilbert, R. G. *Macromolecules* **1984**, *17*, 2520–2529.
32. La Mer, V. K.; Dinegar, R. H. *J. Am. Chem. Soc.* **1950**, *72*, 4847.
33. Overbeek, J. Th. G. *Adv. Colloid Interface Sci.* **1982**, *15*, 251–277.
34. Keefer, K. D. In *A Model for the Growth of Fractal Silica Polymers*; Hench, L. L.; Ulrich, D. R., Eds.; Science of Ceramic Chemical Processing; Wiley & Sons: New York, 1986; pp 131–139.
35. Schaefer, D. W.; Keefer, K. D. *Mater. Res. Soc. Symp. Proc.* **1986**, *72*, 277–287.
36. Sindorf, D. W.; Maciel, G. E. *J. Am. Chem. Soc.* **1981**, *103*, 4263–4265.
37. Sindorf, D. W.; Maciel, G. E. *J. Am. Chem. Soc.* **1983**, *105*(12), 3767–3776.
38. Sindorf, D. W.; Maciel, G. E. *J. Phys. Chem.* **1982**, *86*(26), 5208–5219.
39. Fyfe, C. A.; Gobbi, G. C.; Kennedy, G. J. *J. Phys. Chem.* **1985**, *89*(2), 277–281.
40. Fyfe, C. *Solid State NMR for Chemists*; CFC Press: Ontario, Canada, 1983.
41. Penders, M. H. G. M.; Vrij, A. *Colloid Polym Sci.* **1991**, *286*, 823–831.
42. Duits, M. H. G.; May, R. P.; Vrij, A.; De Kruif, C. G. *J. Chem. Phys.* **1991**, *94*, 4521–4531.
43. Koppel, D. E. *J. Chem. Phys.* **1972**, *57*(11), 4814.
44. *Dynamic Light Scattering: Applications of Photon Correlation Spectroscopy*; Pecora, R., Ed.; Plenum: New York, 1985.
45. Kops-Werkhoven, M. M.; Fijnaut, H. M. *J. Chem. Phys.* **1981**, *74*(3), 1618–1625.
46. Van Veluwen, A.; Lekkerkerker, H. N. W.; De Kruif, C. G.; Vrij, A. *J. Chem. Phys.* **1988**, *89*(5), 2810–2815.
47. Van Blaaderen, A.; Van Geest, J.; Vrij, A. *J. Colloid Interface Sci.* **1992**, *154*(2), 481–501.
48. Van Blaaderen, A.; Vrij, A. *Langmuir* **1992**, *8*, 2921–2931.
49. Van Blaaderen, A.; Kentgens, A. P. M. *J. Non-Cryst. Solids* **1992**, *149*, 161–178.
50. Van Blaaderen, A. Vrij, A. *J. Colloid Interface Sci.* **1993**, *156*, 1–18.
51. Van Blaaderen, A.; Peetermans, J; Maret, G.; Dhont, J. K. G. *J. Chem. Phys.* **1992**, *96*(6), 4591–4603.

52. Van Blaaderen, A. *Adv. Mater.* **1993**, *5*(1), 52–54.
53. De Rooy, N. Ph.D. Dissertation, University of Utrecht, Utrecht, Netherlands, 1979.
54. Chiang, C. H.; Liu, N. I.; Koenig, J. L. *J. Colloid Interface Sci.* **1982**, *86*(1), 26–34.
55. Vankan, J. M. J.; Ponjee, J. J.; De Haan, J. W.; Van de Ven, L. J. M. *J. Colloid Interface Sci.* **1988**, *126*(2), 604–609.
56. De Haan, J. W.; Van den Bogaert, H. M.; Ponjee, J. J.; Van de Ven, L. J. M. *J. Colloid Interface Sci.* **1986**, *110*(2), 591–600.
57. Weres, O.; Yee, A.; Tsao, L. *J. Colloid Interface Sci.* **1980**, *84*(2), 379–402.
58. Verwey, E. J. W.; Overbeek, J. Th. G. *Theory of the Stability of Lyophobic Colloids*; Elsevier: New York, 1948.

9 Synthesis of Nanometer-Sized Silica by Controlled Hydrolysis in Reverse Micellar Systems

F.J. Arriagada and K. Osseo-Asare
Department of Materials Science and Engineering and the Particulate Materials Center, Pennsylvania State University

CONTENTS

The synthesis of nanometer-sized silica particles by the base-catalyzed, controlled hydrolysis of tetraethoxysilane (TEOS) in a nonionic reverse micellar system is described. Spectrofluorometric techniques were used to characterize the reverse micellar solutions. Particle characterization was conducted by transmission electron microscopy. The effect of the water-to-surfactant molar ratio (R) on particle size and size distribution was investigated over a wide range of R values (0.50 to 3.54). Stable dispersions of amorphous silica with mean particle diameters in the range of 46 to 68 nm were produced. Small (46 nm) and extremely monodisperse particles (polydispersity below 4%) were obtained at intermediate R values (1.4), whereas both particle size and polydispersity increased at lower and higher R values. The effects of R on particle size and size distribution are discussed in terms of water "reactivity" (i.e., proportion of bound to free water), concentration of reverse surfactant aggregates, distribution of hydrolyzed TEOS molecules among aggregates, and dynamics of intermicellar matter exchange. A mechanistic model for particle nucleation and growth in these systems is proposed.

The synthesis of uncoated and surface-modified silica particles via the hydrolysis of tetraethoxysilane (TEOS) in a homogeneous alcoholic solution of water and ammonia is well documented in the literature [1–7]. TEOS and other metal alkoxides (including those of titanium and iron) can be hydrolyzed in reverse microemulsion systems for the production of gels or metal oxide particles [8–16]. Systems incorporating surfactant aggregates have also been exploited in the synthesis of a number of other ultrafine particulate materials [13,15,17–20].

Yanagi et al. [9] reported that silica particles could be produced by the base-catalyzed hydrolysis of TEOS in a nonionic reverse micellar system. Yamauchi et al. [12], on the other hand, focused on the characterization of the product particles synthesized in an anionic system. Nitrogen adsorption and IR spectral data showed that the silica nanoparticles were highly porous; this result suggested a mechanism of formation involving primary particle aggregation [12]. The importance of microemulsion structural and dynamic factors in the nucleation and growth of silica particles in reverse micellar systems was recently discussed [14,15]. It was demonstrated [14] that the nanometer-sized particles produced had an extremely narrow size distribution that had not been achieved before in the conventional technique of TEOS hydrolysis in homogeneous ethanolic media. Emphasis was also given to the connection between microemulsion properties and silica particle size in an attempt to further understand the detailed mechanisms of particle formation. In the limited range of water-to-surfactant molar ratio (*R*) that was investigated (0.55 to 1.7), it was found that both particle size and size distribution decreased as *R* increased. Furthermore, particle formation statistics (calculated on the basis of micellar concentration and terminal particle size) indicated that the solute contents of about 10^5 aggregates were involved in the formation of each particle. Under the experimental conditions used, each aggregate was considered to contain enough solute (hydrolyzed TEOS) to form a stable nucleus. Thus, it was suggested that the nuclei initially formed aggregated during intermicellar collisions; this aggregation dramatically reduced the particle number densities to those found in the final stages of growth [14].

In this chapter, the results obtained in the synthesis of silica over a wide range of *R* are presented. As illustrated by these new results, particle size and size distribution are complex functions of the initial microemulsion formulation (essentially defined by *R* and the initial TEOS content).

EXPERIMENTAL DETAILS

Materials

The nonionic surfactant Igepal CO-520 (GAF Chemicals Corporation) is a polydisperse preparation of polyoxyethylene nonylphenyl ether with an average of five oxyethylene groups per molecule (hereafter NP-5). The material as received contained ca. 0.2% water, as determined by Karl Fisher titration. Water and low-boiling impurities (such as free alcohols) were removed by treatment at 60–70°C under vacuum for 26 h. Reagent-grade cyclohexane (Aldrich) was used after storage under molecular sieves. Ammonium hydroxide (Baker, 29.5 wt% NH_3), TEOS (Alfa, 99 wt%), and the fluorescent probe rutheniumtris(bipyridyl) chloride (Aldrich) [hereafter $Ru(Bpy)_3^{2+}$] were used without further purification.

Preparation and Characterization of Particles

The synthesis experiments were conducted at 23°C. As in previous work [14], *R* was the main variable studied. Different *R* values in the range of 0.5 to 3.5 were obtained by changing the concentration of the surfactant (in the range of 0.056 to 0.277 *M*) in the micellar system. The overall TEOS concentration and the water-to-TEOS molar ratio (*h*) were kept constant at 0.023 *M* and 7.8, respectively. A microemulsion plus second reactant synthesis route [15] was used; that is, TEOS was added to the NP-5–cyclohexane–ammonium hydroxide microemulsion to initiate the synthesis. Dispersion samples were extracted at different reaction times and deposited on transmission electron microscope (TEM) grids. Number-average and weight-average particle diameters and size distributions were determined on enlarged TEM micrographs with a ZIDAS image analysis system (Zeiss). Several hundred particles were measured for each sample.

Microemulsion Characterization

Fluorescence spectra were measured on a Shimadzu RF-5000 spectrofluorometer. Samples with *R* values in the range of 0.50 to 4.80 were prepared in a manner similar to that used for the samples in the synthesis experiments, that is, by using a fixed amount of aqueous phase and different surfactant concentrations. The probe $Ru(Bpy)_3^{2+}$ concentration was 9×10^{-6} *M*. The excitation wavelength was 460 nm.

RESULTS

Particle Characteristics

Under the experimental conditions investigated, particle growth was essentially completed after 160 h. The dispersions exhibited a bluish opalescence and remained stable for weeks after preparation. The TEMS in Figure 9.1 show that the resulting silica particles are essentially monodisperse and spherical in shape. The particles obtained after 21 and 160 h of reaction time at low (0.50), intermediate (1.35), and high (3.54) *R* values are shown in Figures 9.1a to 9.1c and 9.1d to 9.1f, respectively. The presence of the surfactant allows the formation of an ordered particle arrangement upon volatilization of cyclohexane. Similar filmlike aggregates formed in the presence of surfactant molecules were described by Alexander and Iler [21] and by Aksay [22].

The number-average particle diameter (d_n) at different *R* values obtained after 21 and 160 h of reaction time are shown in Figure 9.2. As *R* increases in this range, the terminal particle size decreases sharply, reaches a minimum at *R* values of about 1.4, and then increases again. A similar trend is observed for the particle size at earlier reaction times. Thus, at up to 21 h of reaction time, particle growth is much faster at the lower *R* range. Particles produced at low *R* values (below about 1.3) have achieved about 80% of their final size in this time, whereas the percentage decreases to 75, 70, and 65% for *R* values of 1.73, 2.43, and 3.54, respectively.

The size distribution of the particles, expressed as the standard deviation over the number-average diameter

FIGURE 9.1 TEM micrographs of silica particles obtained by hydrolysis of TEOS in reverse NP-5–cyclohexane–NH_4OH microemulsions. Temperature was 23°C with $t = 2$ h for $R =$ (a) 0.50, (b) 1.35, and (c) 3.54; and $t = 160$ h for $R =$ (d) 0.50, (e) 1.35, and (f) 3.54.

(polydispersity parameter in percentage), is shown in Figure 9.3. The size distribution is narrower at intermediate R values, and furthermore it decreases as time proceeds. This latter effect seems more pronounced in the higher R range. In the complete R range, however, the size distributions are significantly narrower than those typically obtained in the alcoholic hydrolysis of TEOS [3,5] for particles in this size range. For example, the polydispersity parameter for 46.3-nm (d_n) particles obtained at R of 1.73 is only 3.8% (Figure 9.3). The calculated weight-average particle diameter (d_w) for this sample is 46.5 nm, so the uniformity factor (d_w/d_n) is only 1.005. These results compare favorably with the higher uniformity factor of 1.033 for 57.7-nm (d_n) particles reported by Badley et al. [3] and underscore the potential of reverse microemulsion systems to produce colloidal particles of extremely narrow size distribution.

Microemulsion Characterization

The effects of R on the aggregation number N (i.e., the number of surfactant molecules per micelle) and on the nature of the solubilized water (i.e., bound or free) for this microemulsion system were reported [14]. As R increases, dipole–dipole interactions between hydrated oxyethylene groups of the surfactant molecule provide the driving force for aggregation [23]. Within the R range of interest, the aggregation number N of NP-5 molecules increases linearly from 23 at R of 0.5 to 135 at R of 3.54 [14,24]. By combining these aggregation numbers with the concentration of NP-5 used in each sample, the number of aggregates (N_a) in the total microemulsion volume (5.06 mL) can be calculated. Furthermore, the average number of TEOS molecules per aggregate (N_t) can be calculated from the total TEOS content and the

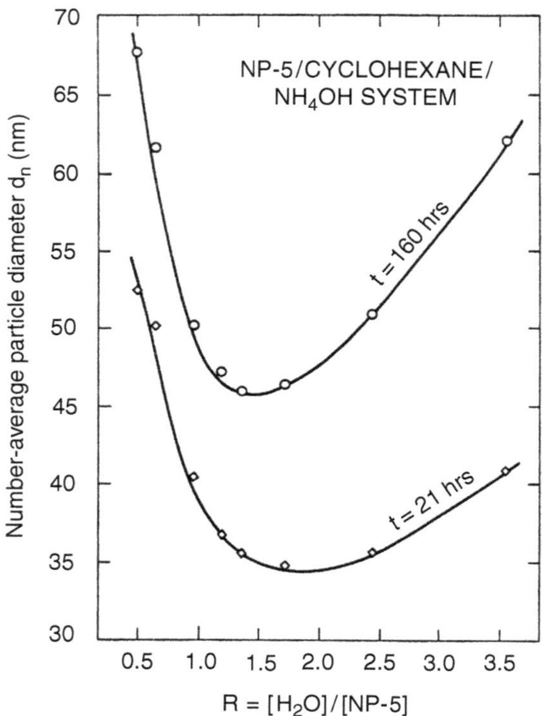

FIGURE 9.2 Effect of R on d_n for silica particles at 21 and 160 h; $T = 23°C$, $h = 7.8$, and [TEOS] = 0.023 M.

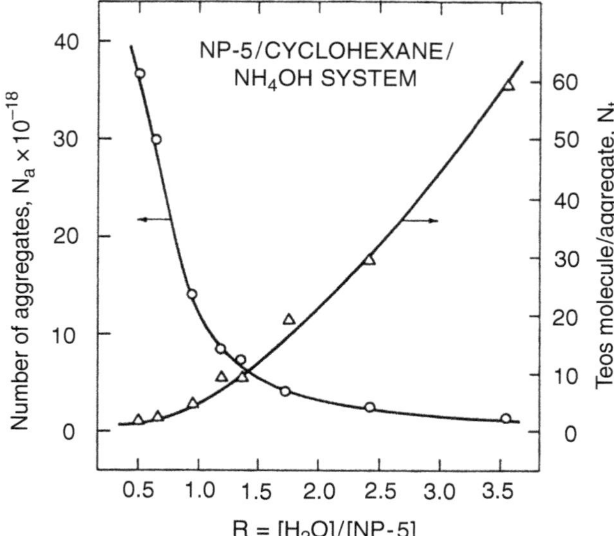

FIGURE 9.4 Effect of R on N_a and N_t; $T = 23°C$, $h = 7.8$, and [TEOS] = 0.023 M.

value of N_a at each R value. The results are presented in Figure 9.4. The number of aggregates decreases about 30-fold when R increases from 0.5 to 3.54. Thus, as R increases, fewer (but larger) surfactant aggregates are present in the system. As a result, N_t increases from 2 to about 60 as R increases.

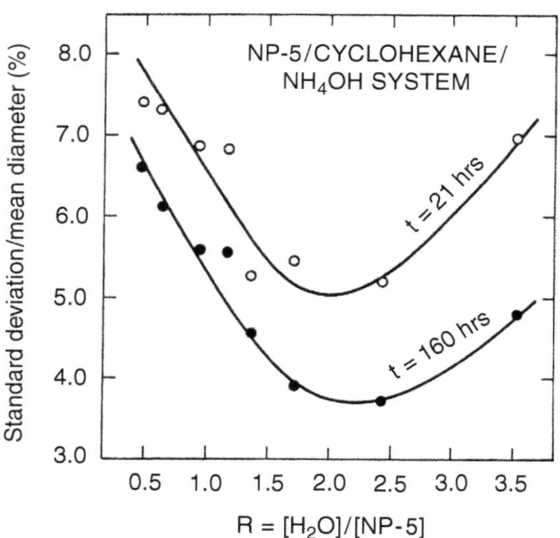

FIGURE 9.3 Effect of R on the size distribution of silica particles at 21 and 160 h; $T = 23°C$ and, $h = 7.8$. [TEOS] = 0.023 M.

Not only is aggregation favored as R increases, but there is also a change in the nature of the solubilized water. Initially, water is tightly hydrogen-bonded to the oxyethylene groups of the nonionic surfactant. Further water addition induces aggregation (dipole–dipole interactions), and a point is reached where unbound or "free" water molecules are present in the hydrophilic (polar) domain. The state of water in the polar domain is relevant to the formation of particles, because the initial hydrolysis of TEOS in the reverse micelle is expected to be favored as more "free" water molecules are available. Such an effect is expected because the hydrolysis of titanium alkoxides (naturally more reactive than TEOS toward water) is strongly inhibited in reverse microemulsions formulated with nonionic or anionic surfactants at low R values [10,16].

The fluorescence emission spectra of hydrophilic probes solubilized in the polar domains of reverse micelles provide a convenient means of ascertaining the nature of the solubilized water [25] and thus the expected reactivity of TEOS with these water molecules. For example, the emission spectra of $Ru(Bpy)_3^{2+}$ is blue-shifted with respect to water in media of reduced polarity [26]. Figure 9.5 shows the spectra of $Ru(Bpy)_3^{2+}$ obtained at different R values. At low R values (below about 0.9) the emission spectra have a maximum intensity at about 575 nm with a red shoulder at 615 nm. As the water content increases, the intensity at 615 nm increases and the emission at 575 nm decreases. For R values above 1, the spectra resemble that of the probe in water [26]. These results agree very well with previous results [14]

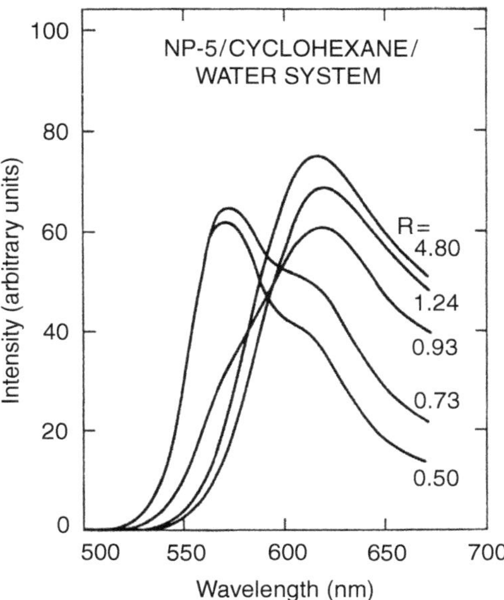

FIGURE 9.5 Fluorescence spectra of $Ru(Bpy)_3$ solubilized in the NP-5–cyclohexane–water reverse micellar system at different R values; $\lambda_{ex} = 460$ nm, $[Ru(Bpy)_3] = 9 \times 10^{-6}$ M, and $T = 23°C$.

for the fluorescence of $Ru(Bpy)_3^{2+}$ in NP-5–cyclohexane microemulsion samples of constant surfactant concentration. In addition, replacement of water by ammonium hydroxide results in spectral changes very similar to those of Figure 9.5 [14]. As suggested by Handa et al. [26], the probe is located close the oxyethylene groups at low R values. As R increases, $Ru(Bpy)_3^{2+}$ is partitioned to the hydrophilic "pool" and behaves as in bulk water.

DISCUSSION

TEOS–Microemulsion System

In these experiments the overall concentrations of water, ammonia, and TEOS were kept constant. The water-to-surfactant molar ratio R was varied by changing the total surfactant concentration. Thus, the observed effect of R on particle size and size distribution is due to the presence of surfactant aggregates, which result in the localization of reagents in well-defined polar (hydrophilic) and nonpolar (hydrophobic) domains. The NP-5–cyclohexane–NH_4OH system is visualized as consisting of small reverse micelles before TEOS addition. As R increases within the range investigated, larger but fewer aggregates are present [14,24], and a larger number of TEOS molecules will on average interact with each micelle (Figure 9.4). In addition, as R increases, a larger proportion of the water molecules is free within the micellar core (Figure 9.5). To interpret the effect of R on particle size and size distribution shown previously, the following basic assumptions have been made.

Amphiphilic Nature of Hydrolyzed TEOS

The TEOS molecules, readily solubilized in the external oil phase, interact with the water molecules present within the aggregates to produce hydrolyzed species. It is assumed that the products of hydrolysis, once formed, remain bound to the micellar aggregates (i.e., solubilized within the surfactant film or the aqueous core or both) because of their enhanced amphiphilic character (brought about by the formation of silanol groups). The exact locale for the solubilization of hydrolyzed TEOS molecules in a reverse micelle is expected to be dependent on the degree of hydrolysis; for example, the most polar species $[Si(OH)_4]$ would be expected to reside within the aqueous core. Once the hydrolyzed TEOS becomes solubilized, all further reactions (further hydrolysis and condensation) are restricted to the locale of the surfactant aggregates.

Locale of Evolving Solid Particles

The adsorption of polyoxyethylene nonylphenyl ethers (such as NP-5) on silica surfaces is dependent on pH. In relatively alkaline medium, the silica surface is deprotonated and highly negatively charged, and adsorption of the oxyethylene groups is inhibited [27–30]. Accordingly, because alkaline conditions prevail in these synthesis experiments, stabilization of the particles in cyclohexane (the external microemulsion phase) by direct attachment of NP-5 molecules to the particles is unlikely. The hydrophilic silica surface is thus suggested to be surrounded by a thin film of water molecules as well as polar reaction products (such as ethanol); this film is then stabilized by NP-5 molecules. Under these conditions, the micellar system can be viewed as containing two different populations of aggregates, that is, particle-filled aggregates and empty micelles. This situation has been well documented in studies of the solubilization of enzymes and proteins in reverse microemulsion systems [31–33] by "segregation" or "water shell" models.

Distribution of Surfactant Molecules

As a first approximation, the reverse micellar aggregates present in the solution are assumed not to be significantly affected by the addition of TEOS molecules nor by the subsequent reactions, and in particular the aggregation numbers (hence the micellar concentration) are assumed to remain unchanged. Consideration of the total number of particles produced (N_p) indicates that only a minor fraction of the total surfactant present would be associated with the growing particles.

Model of Particle Formation

In the reverse micellar system, the nucleation process will in principle be governed by the ease with which [1] TEOS

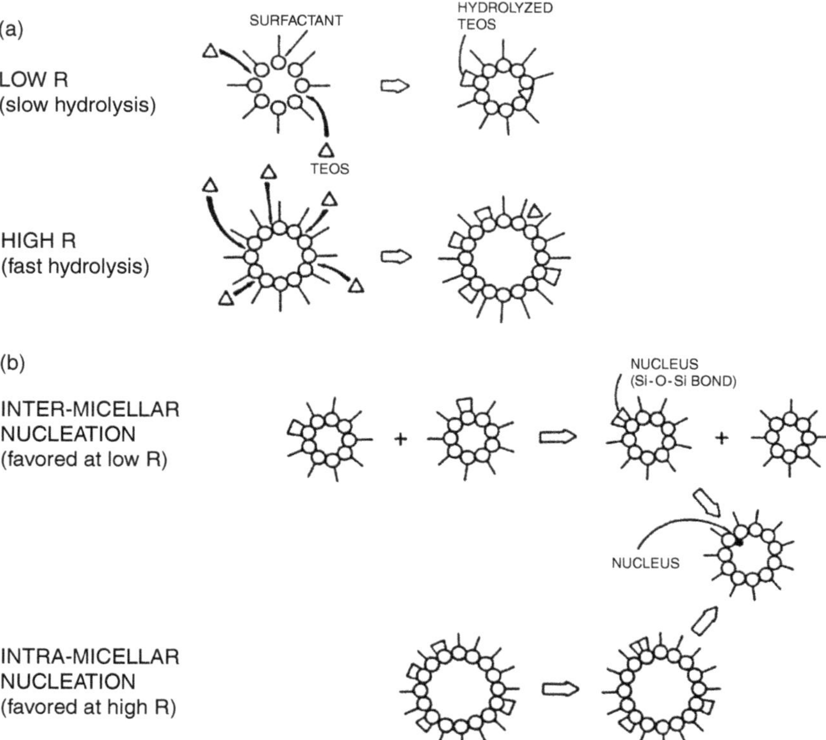

FIGURE 9.6 Formation of silica nuclei by hydrolysis of TEOS in reverse micellar systems: (a) partial hydrolysis of TEOS and association to reverse micelles; and (b) inter- and intramicellar nucleation.

molecules are hydrolyzed to form silanol groups and [2] silanol groups interact to form Si–O–Si bonds. Interaction of TEOS molecules with the micellar aggregates, hydrolysis with formation of silanol groups, and attachment of the partially hydrolyzed molecules to the micelles are illustrated in Figure 9.6a for low and high R values. Consideration of the effect of R on the proportion of free ("reactive") water in the micellar aggregates (Figure 9.5) indicates that hydrolysis of TEOS to form silanol groups is not favored at low R values, because water is mostly bound to the surfactant molecules. Furthermore, there are few silanol groups per aggregate (given the low values of N_t), so nucleation through intermicellar interactions is expected to predominate. In addition, the concentration of aggregates under these conditions is higher, and therefore the rate of intermicellar collisions is higher [34]. At high R values, on the other hand, the number of silanol groups per micelle is higher (based on the larger value of N_t at high R). Under these conditions, the probability of interactions between neighboring silanol groups in a given micelle to form Si–O–Si bonds (a nucleus) is higher, and therefore intramicellar nucleation would be favored. The essential features of the nucleation process at both low and high R values are shown in Figure 9.6b.

According to these ideas, the formation of a high number of nuclei is favored at high R values. Thus, the final particle size should decrease continuously as R increases, but as shown in Figure 9.2 this theory does not conform with the experimental data. To explain the minimum particle size observed at intermediate R values, the effect of R on both the number of aggregates and the distribution of hydrolyzed TEOS molecules among aggregates needs to be considered. On the one hand, the number of nuclei formed is expected to be proportional to the number of aggregates in the microemulsion volume, which as shown in Figure 9.4 decreases significantly as R increases. On the other hand, only a fraction of the total number of aggregates may contain sufficient hydrolyzed TEOS molecules to form a stable nucleus. This fraction can be quantified as a function of R if the average number of TEOS molecules per micelle (N_t), the number of hydrolyzed TEOS molecules required to form a stable nucleus (i_c), and the probability distribution of TEOS molecules per micelle are known. A Poisson statistical law can be assumed; this law has been shown [35,36] to satisfactorily describe the distribution of solute species among micelles in reverse microemulsions.

The probability (P_k) of having k hydrolyzed TEOS molecules per aggregate provided that the average value is N_t is therefore given by

$$P_k = \frac{N_t^k e^{-N_t}}{k!} \tag{9.1}$$

If the minimum number of hydrolyzed TEOS molecules per micelle required to form a stable nucleus is i_c, then the number of nuclei (N_n) formed at each R value can be calculated from

$$N_n = N_a \sum_{k=i_c}^{\infty} P_k \tag{9.2}$$

where N_a is the number of aggregates present at the given R value and ΣP_k ($k = i_c$ to infinity) is the probability that an aggregate contains i_c or more hydrolyzed TEOS molecules. The value of N_n was evaluated as function of R for different i_c values, which were chosen to be in a reasonable range [37]. The results obtained are shown in Figure 9.7. The number of nuclei formed displays a maximum at intermediate R values; this maximum furthermore decreases and shifts toward higher R values as the value of i_c increases. Although no quantitative assessment on a unique value for i_c can be offered at this time, the trends shown by these results may explain the minimum in particle size obtained at intermediate R values; that is, the minimum in particle size is the result of a maximum in the production of nuclei.

Particle Growth

A comparison of the relative number of nuclei formed and particles produced at each R value suggests a significant role of nuclei aggregation in the particle formation process. If a single (limited time) nucleation event occurred upon addition of TEOS to the micellar system, and if particle growth occurs only by addition of monomers (hydrolyzed TEOS) to condensed species, then the number of particles produced should correspond to the number of nuclei formed, which may be assumed to be of the order of magnitude given in Figure 9.7. Calculation of the population of particles (on the basis of their terminal diameter and assuming complete TEOS conversion), however, indicates that there are about 5 orders of magnitude fewer particles than nuclei. Forming only one stable nucleus out of 10^5 aggregates that are potentially capable of nucleation seems too low a probability. Therefore, it is proposed that a large number of nuclei is initially formed and that aggregation of nuclei during intermicellar collisions reduces dramatically the particle number density. Similar processes of particle growth via aggregative mechanisms have been suggested by Towey et al. [38] for CdS colloids growing in an anionic reverse microemulsion system and by Bogush and Zukoski [39] in the alcoholic hydrolysis of TEOS.

This particle aggregation process may occur only during a limited period of time, after which a constant particle number density of the growing particles is achieved. In a manner similar to the nucleation process, the aggregation may be governed by the number of micelles and the probabilistic fraction of aggregates containing a nucleus. In addition, it will strongly depend on the rate of intermicellar exchanges. After this process is completed, particle growth occurs by addition of monomers to existing particles, which then involves the interaction between particle-filled aggregates and empty micelles (containing hydrolyzed TEOS). Such a mechanism of growth is believed to be predominant at 21 h and longer times, as suggested by the narrow size distributions obtained (Figure 9.3).

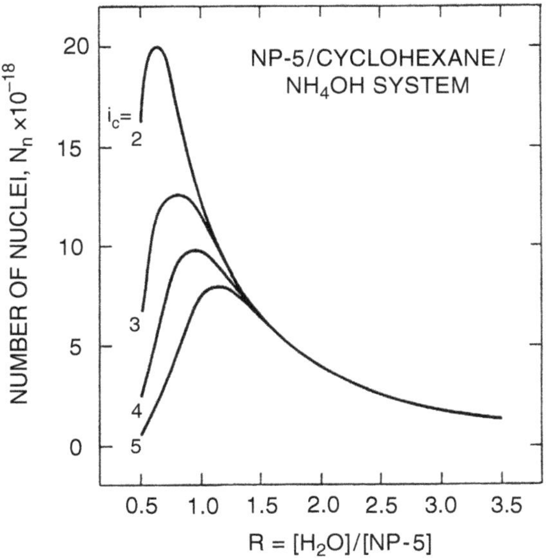

FIGURE 9.7 Effect of R on N_n at different i_c values; [TEOS] = 0.023 M.

Assessment of Particle Formation Mechanisms

The mechanisms of particle nucleation and growth in the nonionic micellar system discussed may in principle explain the observed particle size data. However, this is most likely a simplified model of the underlying processes that may be operative during the course of the synthesis reactions. The aggregation number of the surfactant molecules in the bulk oil phase (i.e., the population of empty micelles) may be affected by the presence of guest molecules, such as hydrolyzed TEOS, ethanol product, or both. The amphiphilic partially hydrolyzed TEOS (silanol groups as polar head) may also act as a cosurfactant, affecting the rigidity of the surfactant-stabilized oil–water interface and therefore the dynamics of intermicellar interactions [34,35]. On the other hand, the ethanol molecules produced are expected to remain within the polar domain because of their hydrophilic nature. Under these conditions, the aqueous phase is being enriched in ethanol as time proceeds (in fact, it may contain a significant fraction of ethanol under the conditions of these experiments). At the same time, water is being progressively consumed. Such significant changes in the nature of the solubilized aqueous phase (including an effective decrease in R) may dramatically affect the aggregation behavior.

Thus, the situation is much more complex if the characteristics of the reverse micellar system (e.g., aggregation, micellar concentration, and dynamic behavior) are functions of time. In light of the available results, no definitive explanation can be offered at this time to account for the remarkable monodispersity obtained at intermediate R values, at which maximum number of nuclei is formed. Quantitative evaluation of the main features of the proposed model must await further experimental data. Of particular interest in this connection are the development of a quantitative description of the nuclei-aggregation process and the determination of surfactant aggregation numbers with water–ethanol mixtures as aqueous solubilizate.

CONCLUSIONS

Further work is clearly needed to unravel the detailed mechanisms of the hydrolysis and polymerization reactions in these micellar systems. Of particular interest is a better description of the initial stages of growth, which are believed to be responsible for the narrow monodispersity of the particles obtained within the entire R range. Other issues of interest are the location of hydrolysis products and other intermediates, changes in the aggregation number of surfactant molecules due to changes in the nature of the solubilized aqueous phase, and a quantitative description of the particle-filled and empty reverse micelle populations.

The potential capabilities of microemulsion synthesis of colloidal silica and related materials are yet to be realized. For example, the compartmentalization of reagents at the molecular level in these media may allow close control of chemical homogeneity in the synthesis of glass (e.g., Si–Al) particles. In situ synthesis of silica-supported catalysts and synthesis protocols involving seeding techniques, coating techniques, or infiltration are also possible.

ACKNOWLEDGMENTS

This work was supported in part by the The Ben Franklin Partnership Program, Advanced Technology Center of Central and Northern Pennsylvania, Inc., and the Cooperative Program in Metals Science and Engineering, The Pennsylvania State University.

REFERENCES

1. Stober, W.; Fink, A.; Bohn, E. *J. Colloid Interface Sci.* **1968**, *26*, 62.
2. Van Helden, A. K.; Jansen, J. W.; Vrij, A. *J. Colloid Interface Sci.* **1981**, *81*, 354.
3. Badley, R. D.; Ford, W. T.; McEnroe, F. J.; Assink, R. A. *Langmuir* **1990**, *6*, 792.
4. Matsoukas, T.; Gulari, E. *J. Colloid Interface Sci.* **1989**, *132*, 13.
5. Bogush, G. H.; Zukoski IV, C. F. In *Ultrastructure Processing of Advanced Ceramics*; Mackenzie, J. D.; Ulrich, D. R., Eds.; Wiley: New York, 1988; p 477.
6. Byers, C. H.; Harris, M. T.; Williams, D. F. *Ind. Eng. Chem. Res.* **1987**, *26*, 1916.
7. Brinker, C. J.; Scherer, G. W. *J. Non-Cryst. Solids* **1985**, *70*, 301.
8. Tricot, Y. M.; Rafaeloff, R.; Emeren, A.; Fendler, J. H. In *Organic Phototransformations in Nonhomogeneous Media*; Fox, M. A., Ed.; ACS Symposium Series 278; American Chemical Society: Washington, DC, 1985; p 99.
9. Yanagi, M.; Asano, Y.; Kandori, K.; Kon-no, K. *Abs. 39th Symp. Div. Colloid Interface Chem.*; Chemical Society of Japan: Tokyo, Japan, 1986; p 386.
10. Guizard, C.; Stitou, M.; Larbot, A.; Cot, L.; Rouviere, J. In *Better Ceramics Through Chemistry III*; Brinker, C. J.; Clark, D. E.; Ulrich, D. R., Eds.; *Mat. Res. Soc. Symp. Proc.* Vol. 121; Materials Research Society: Pittsburgh, PA, 1988; p 115.
11. Friberg, S. E.; Yang, C. C. In *Innovations in Materials Processing Using Aqueous, Colloid and Surface Chemistry*; Doyle, F. M.; Raghavan, S.; Somasundaran, P.; Warren, G. W., Eds.; Minerals, Metals, and Materials Society: Warrendale, PA, 1988; p 181.
12. Yamauchi, H.; Ishikawa, T.; Kondo, S. *Colloids Surf.* **1989**, *37*, 71.
13. Ward, A. J. I.; Friberg, S. E. *MRS Bull.* **1989** (December), 41.
14. Osseo-Asare, K.; Arriagada, F. J. *Colloids Surf.* **1990**, *50*, 321.
15. Osseo-Asare, K.; Arriagada, F. J. In *Ceramic Powder Science III*; Messing, G. L.; Hirano, S.; Hausner, H., Eds.; American Ceramic Society: Westerville, OH, 1990; p 3.
16. Arriagada, F. J.; Osseo-Asare, K. In *Refractory Metals: Extraction, Processing and Applications*; Liddell, K. C.; Sadoway, D. R.; Bautista, R. G., Eds.; Minerals, Metals and Materials Society: Warrendale, PA, 1991; pp 259–269.
17. Fendler, J. H. *Chem. Rev.* **1987**, *87*, 877.
18. Sugimoto, T. *Adv. Colloid Interface Sci.* **1987**, *28*, 65.
19. Leung, R.; Hou, M. J.; Shah, D. O. In *Surfactants in Chemical/Process Engineering*; Wasan, D. T.; Ginn, M. E.; Shah, D. O., Eds.; Surfactant Science Series Vol. 28; Dekker: New York, 1988; p 315.
20. Robinson, B. H.; Khan-Lodhi, A. N.; Towey, T. In *Structure and Reactivity in Reverse Micelles*; Pileni, M. P., Ed.; Studies in Physical and Theoretical Chemistry Vol. 65; Elsevier: Amsterdam, Netherlands, 1989; p 198.

21. Alexander, G. B.; Iler, R. K., U.S. Patent 2,801,902, 1957.
22. Aksay, I. L. In *Ceramic Powder Science R*; Messing, G. L.; Fuller, E. R.; Hausner, H., Eds.; American Ceramic Society: Westerville, OH, 1988; Vol. 1, p 663.
23. Ravey, J. C.; Buzier, M.; Picot, C. *J. Colloid Interface Sci.* **1984**, *97*, 9.
24. Kitahara, A. *J. Phys. Chem.* **1965**, *69*, 2788.
25. *Surfactant Solutions: New Methods of Investigation*; Zana, R., Ed.; Surfactant Science Series Vol. 22; Dekker: New York, 1987.
26. Handa, T.; Sakai, M.; Nakagaki, M. *J. Phys. Chem.* **1986**, *90*, 3377.
27. Iler, R. K. *The Chemistry of Silica*; Wiley: New York, 1979.
28. Partyka, S.; Zaini, S.; Lindheimer, M.; Brun, B. *Colloids Surf.* **1984**, *12*, 255.
29. Van den Boomgaard, Th.; Tadros, Th. F.; Lyklema, J. *J. Colloid Interface Sci.* **1987**, *116*, 8.
30. Travalloni Louvisse, A. M.; Gonzalez, G. In *Surfactant-Based Mobility Control*; Smith, D. H., Ed.; ACS Symposium Series 373; American Chemical Society: Washington, DC, 1988; p 220.
31. Luisi, P. L. *Angew. Chem. Int. Ed. Engl.* **1985**, *24*, 439.
32. Brochette, P.; Petit, C.; Pileni, M. P. *J. Phys. Chem.* **1988**, *92*, 3504.
33. *Structure and Reactivity in Reverse Micelles*; Pileni, M. P., Ed.; Studies in Physical and Theoretical Chemistry Vol. 65; Elsevier: Amsterdam, Netherlands, 1989.
34. Fletcher, P. D.; Howe, A. M.; Robinson, B. H. *J. Chem. Soc. Faraday Trans. I* **1987**, *83*, 985.
35. Atik, S. S.; Thomas, J. K. *J. Am. Chem. Soc.* **1981**, *103*, 3543.
36. Nagy, J. B. *Colloids Surf.* **1989**, *35*, 201.
37. Nielsen, A. E. *Kinetics of Precipitation*; International Series of Monographs on Analytical Chemistry Vol. 18; Pergamon: New York, 1964.
38. Towey, T. F.; Khan-Lodhi, A.; Robinson, B. H. *J. Chem. Soc. Faraday Trans.* **1990**, *86*, 3757.
39. Bogush, G. H.; Zukoski, C. F. *J. Colloid Interface Sci.* **1991**, *142*, 19.

10 Formation of Silica Gels Composed of Micrometer-Sized Particles by the Sol–Gel Method

Hiromitsu Kozuka and Sumio Sakka
Kyoto University, Institute for Chemical Research

CONTENTS

Formation of silica gels composed of micrometer-sized particles from highly acidic solutions of tetramethoxysilane (TMOS) is described. In this method a limited amount of H_2O and a large amount of HCl are used in hydrolyzing TMOS. Detailed conditions and the mechanism of the formation of micrometer-sized particles in the solutions are discussed. Applications of the gels are also described.

Silica gels with continuous large pores are ideal as precursors for bulk silica glasses. Low-temperature synthesis of bulk silica glasses by the sol–gel method consists of hydrolysis and polycondensation of silicon alkoxide in a solution, gelation of the alkoxide solution, drying of the wet gel to remove residual liquid components, and heat treatment of the dried gel to remove the organic substances and for sintering. Because large, continuous pores reduce the capillary pressure on the gel framework during drying, synthesis of silica gel monoliths with large pores is desirable for the sol–gel process for the production of bulk glasses [1–5].

Acid or base is used as a catalyst for the hydrolysis of silicon alkoxide. Acid or base with a mole ratio to alkoxide as low as 0.01 or less is usually enough for the catalytic effect. On the other hand, water with a mole ratio to alkoxide as high as 10 or more is used for the complete hydrolysis of alkoxide. We have found, however, that monolithic opaque gels composed of micrometer-sized silica particles can be prepared from tetramethoxysilane (TMOS) when a limited amount of water and a large amount of hydrochloric acid are used for the hydrolysis [6,7]. Because of the large particle size and the accordingly large pore size, the gels can be dried without cracking.

The Stöber route [8] is a well-known method for providing submicrometer- or micrometer-sized silica particles by hydrolysis and condensation of silicon alkoxide; an excess of base and water is used in the reaction. Compared with this method, ours has quite different reaction conditions, namely, the use of a limited amount of water and a large amount of acid. In contrast to the reaction of silicon alkoxide with a large amount of water in basic conditions, Sakka and Kamiya [9] noticed from the measurement of the intrinsic viscosity of silica sols that linear particles or polymers, not round particles, are formed with acidic conditions and the addition of a small amount of water. Therefore, the reaction conditions for this method for producing round micrometer-sized particles is new, and the mechanism of formation of round particles is of interest.

In this chapter, detailed conditions for the formation of silica gels composed of micrometer-sized particles from highly acidic TMOS solutions are described, and the mechanism of the sol–gel reaction is discussed. Application of the gels is also discussed.

CONDITIONS FOR THE FORMATION OF MICROMETER-SIZED PARTICLES

PROCESS FOR MAKING GELS

Figure 10.1 shows the flow diagram for the sol–gel process for making gels. TMOS, methanol, 36% hydrochloric acid, and ion-exchanged water are used as the starting materials. Silicon alkoxide solutions of the desired mole ratios are prepared by mixing the reagents under vigorous stirring at room temperature. Fifty milliliters of the alkoxide solution is kept at 40°C until gelation. The quantity of acid reported refers to hydrogen chloride (HCl), and that of water includes water from the hydrochloric acid solution.

FIGURE 10.2 SEMs of dried gels formed from solutions with mole ratios $TMOS:H_2O:HCl:CH_3OH$ of 1 : 1.53 or 2.00 : 0.01–0.40 : 2. Mole ratios were as follows: (a) HCl/TMOS = 0.01 and H_2O/TMOS = 1.53, (b) HCl/TMOS = 0.25 and H_2O/TMOS = 1.53, (c) HCl/TMOS = 0.40 and H_2O/TMOS = 1.53, and (d) HCl/TMOS = 0.40 and H_2O/TMOS = 2.00.

EFFECT OF H_2O AND HCL CONTENT OF THE STARTING SOLUTIONS

In the studies of gelation of TMOS solutions having mole ratios $TMOS:H_2O:HCl:CH_3OH$ of 1 : 1.44–2.00 : 0.01–0.40 : 2, gels composed of larger particles were formed from the solutions with higher HCl content and lower H_2O content [6]. Figure 10.2 shows the scanning electron micrographs (SEMs) of the fracture surface of the dried gels. In the series of gels formed from the solutions with the same H_2O content at H_2O/TMOS = 1.53, larger particles in the gel skeleton are visible in the gels from the solutions with greater HCl concentrations. In the gel formed from the solution with HCl/TMOS = 0.40, particles 5 μm in diameter and connected at the neck are visible. Dependence of the particle size on the HCl content of the starting solution is shown in Figure 10.3. As the particle size increases, the transparency of the resultant dried gel is lost, and as shown in Figure 10.3, the bulk density of the gels decreases. Greater concentrations of H_2O, however, decrease the particle size, as shown in Figure 10.2d.

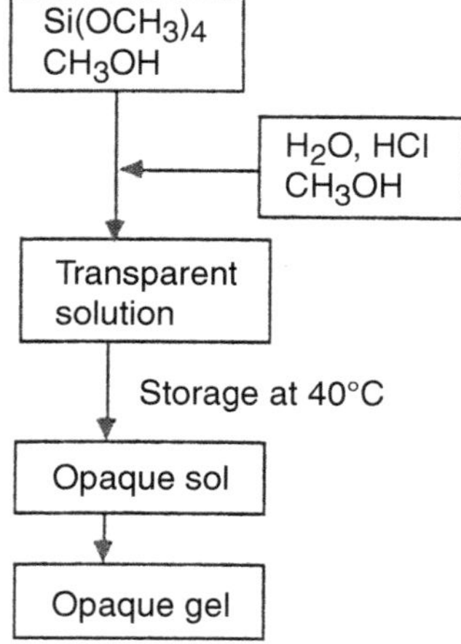

FIGURE 10.1 Flow diagram of the sol–gel process for making silica gel monoliths composed of micrometer-sized particles.

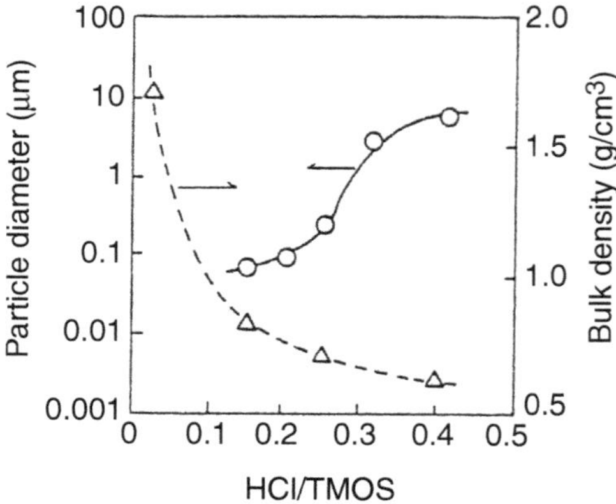

FIGURE 10.3 Dependence of particle size and gel bulk density on HCl/TMOS mole ratios. $TMOS:H_2O:HCl:CH_3OH$ is 1 : 1.53 : 0.01–0.40 : 2.

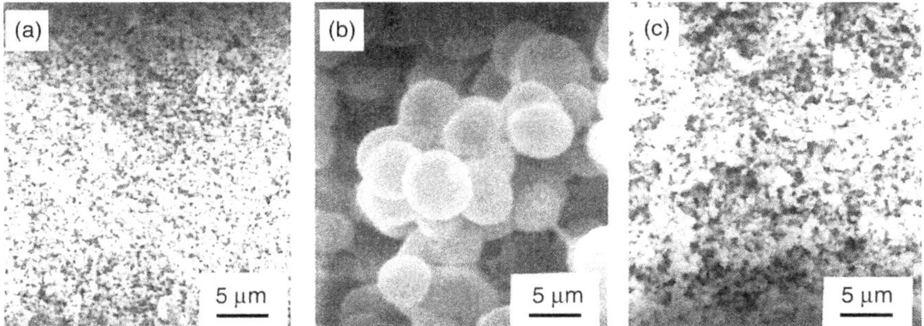

FIGURE 10.4 SEMs of the dried gels prepared from the solutions with mole ratios TMOS : H_2O : HCl : CH_3OH of 1 : 1.53 : 0.40 : x, where x = 0.5 (a), 3 (b), and 5 (c).

Effect of the Concentration of the Solutions

The structure of the gel also depends on the concentration of the solutions, that is, the amount of alcohol in the solutions. Figure 10.4 shows the SEMs of the fracture surface of the dried gels derived from various solutions. Micrometer-sized particles are visible in the gel formed from the solution with $CH_3OH/TMOS = 3$ (Figure 10.4b) as well as that from the solution with $CH_3OH/TMOS = 2$ (Figure 10.2c). Solutions with smaller ($CH_3OH/TMOS = 0.5$) and larger ($CH_3OH/TMOS = 5$) CH_3OH concentrations, however, resulted in gels composed of fine particles (Figure 10.4a and 10.4c).

Effect of the Kind of Alkoxides, Acids, and Alcohols

In the experiments on the gelation of solutions of tetramethoxysilane, tetraethoxysilane, tetraisopropoxysilane, and tetra-*n*-butoxysilane, in which solutions with H_2O/alkoxide = 1.5, HCl/alkoxide = 0.40, and [alkoxide] = 2 mol/l with CH_3OH as solvent were kept at 40°C in air-sealed flasks, an opaque gel formed only from the TMOS solution, whereas no gelation took place in the solutions of the other alkoxides [7].

Substitution of methanol by isopropyl or *n*-butyl alcohol also prevented formation of opaque monolithic gels composed of micrometer-sized particles [7]. TMOS solutions with mole ratios TMOS : H_2O : HCl : alcohol of 1 : 1.53 : 0.40 : 2 converted to opaque gel monoliths when methanol was used, whereas slightly opalescent gel fragments formed when isopropyl or *n*-butyl alcohol was used.

MECHANISM OF THE FORMATION OF MICROMETER-SIZED PARTICLES

^{29}Si NMR Spectra of the Gels

Molecular structure of the polymerized species is compared for three kinds of materials: the opaque gel composed of micrometer-sized particles formed from a highly acidic TMOS solution with mole ratios TMOS : H_2O : HCl : CH_3OH of 1 : 1.53 : 0.40 : 2, the transparent gel derived from a weakly acidic TMOS solution with mole ratios TMOS : H_2O : HCl : CH_3OH of 1 : 1.53 : 0.01 : 2, and the silica particles formed from a solution with mole ratios TMOS : H_2O : NH_3 : CH_3OH of 1 : 30 : 5 : 171. The first and second gels correspond to those shown in Figure 10.2c and 10.2a, respectively. The third one is made through a method similar to that of Stöber.

Figure 10.5 shows the ^{29}Si NMR spectra of these materials. There is less cross-linking in the gel from the

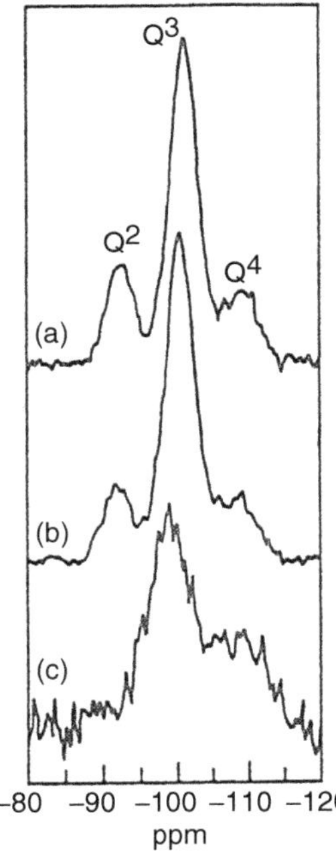

FIGURE 10.5 Silicon-29 NMR spectra of (a) a silica gel derived from a solution with molar composition TMOS : H_2O : HCl : CH_3OH of 1 : 1.53 : 0.4 : 2, (b) a silica gel derived from a solution with molar composition TMOS : H_2O : HCl : CH_3OH of 1 : 1.53 : 0.01 : 2, and (c) silica particles derived from the solution with molar composition TMOS : H_2O : NH_3 : CH_3OH of 1 : 30 : 5 : 171.

highly acidic solution with a limited amount of water than in the silica particles from the highly basic solution with an excess of water; fewer Q^4 species, Si atoms having four bridging oxygens, and more Q^2 species, Si atoms having two bridging oxygens, are present in the gels from the acidic solutions. The extent of cross-linking of the polymerized species is totally different, although gels from both highly acidic and highly basic solutions have micrometer- or submicrometer-sized particles that can be seen microscopically.

In contrast, the gels from the highly acidic and weakly acidic solutions have similar NMR spectra. This similarity indicates that the extent of cross-linking of the silica polymers from the highly acidic solution is as low as that from the weakly acidic solution. It would be possible to assume that the round particles formed from the highly acidic solution consist of linear polymers or particles, which have been proposed to be polymerized species in the weakly acidic solutions with a limited amount of water [9]. The sol from the weakly acidic solution has spinnability, that is, gel fibers can be drawn from the sol, and the resultant gel shows no particulate microstructure, whereas the sol from the highly acidic solution has no spinnability [6] and has particulate structure. It is plausible to think that these differences in the microstructure of the gels and the rheological properties of the sols do not arise from different polymer structures but from different aggregation states of the polymerized species.

The gels derived from the solutions with various CH_3OH concentrations, which have different microstructures, as mentioned in the section "Effect of the Kind of Alkoxides, Acids, and Alcohols", have almost the same ^{29}Si NMR spectra as those in Figure 10.6. This similarity indicates that the different microstructures observed for the gels from the solutions with different CH_3OH concentrations do not result from different structures of the polymers but rather from differences in the aggregation states of the polymers.

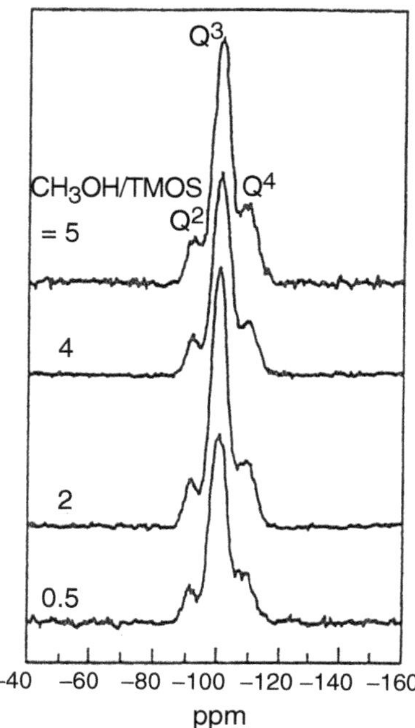

FIGURE 10.6 Silicon-29 NMR spectra of gels prepared from solutions with mole ratios $TMOS:H_2O:HCl:CH_3OH$ of $1:1.53:0.40:x$, where $x = 0.5$, 2, 4, and 5.

Instability of Particles under Centrifugation

The sol particles, the source of the translucence of the sols, are not stable against centrifugation. Figure 10.8 shows the dried gel obtained by centrifuging a sol with mole ratios $TMOS:H_2O:HCl:CH_3OH$ of $1:1.53:0.40:2$ before the occurrence of opalescence. Round, closed pores are visible instead of particulate structure; this observation suggests that the micrometer-sized, round particles formed in the reaction are not stable against mechanical forces and may consist of weakly cross-linked, flexible polymers.

Instability of Particles in Organic Liquids

Sol particles formed in the TMOS solutions with excess HCl and a limited amount of H_2O are not stable in nonpolar organic solvents; a translucent sol derived from a solution with mole ratios $TMOS:H_2O:HCl:CH_3OH$ of $1:1.53:0.40:2$ becomes transparent when benzene is added. Thus, the particles, which are the source of the translucence of the sol, are composed of polymers or primary particles soluble in nonpolar organic solvents. Figure 10.7 shows SEMs of dried gels derived from the sols in which nonpolar benzene was added and polar methanol was added after the occurrence of translucence. Much finer microstructure is evident in the gel from the benzene-added sol, whereas micrometer-sized particles are seen in the gel from the methanol-added sol.

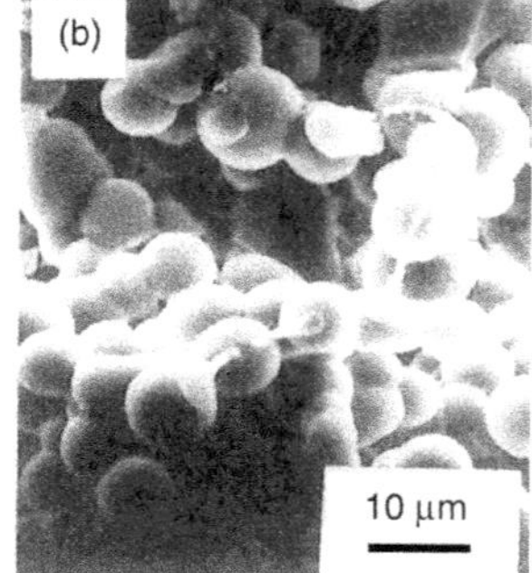

FIGURE 10.7 SEMS of the fracture surface of dried gels derived from solutions to which benzene was added (a) and methanol was added (b) after the occurrence of opalescence.

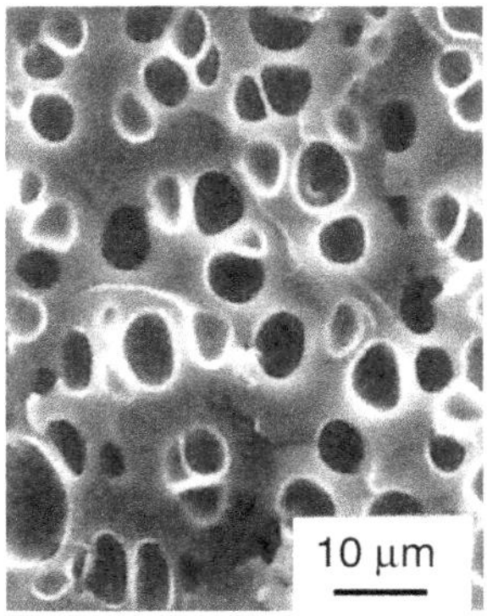

FIGURE 10.8 SEM of dried gel obtained with centrifugation.

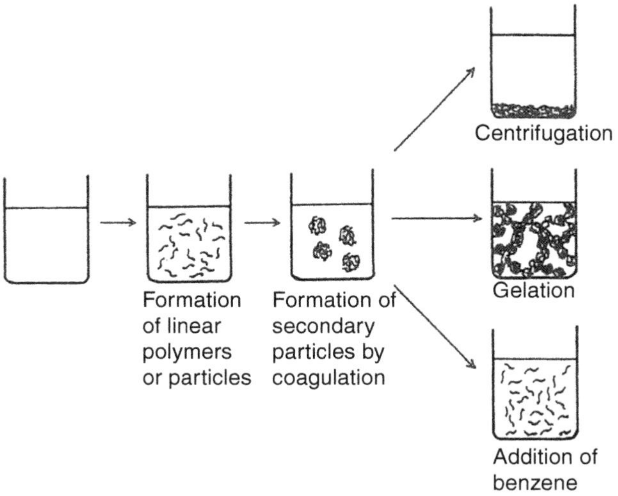

Centrifugation

Formation of linear polymers or particles

Formation of secondary particles by coagulation

Gelation

Addition of benzene

FIGURE 10.9 Illustration of the changes in the gel structure caused by the addition of organic compounds and by centrifugation.

Mechanism of the Formation of Round Particles

Because of the small amount of water used in the hydrolysis reaction in the starting solutions that give large particles, a large number of unhydrolyzed alkoxy groups are left on the polymerized species. These alkoxy groups give rise to the lipophilic nature of the polymerized species. A larger portion of methanol may be protonated and the ionic nature of the solvent may increase when a larger amount of HCl or an acid with a larger dissociation constant is used; these conditions result in a decrease in the solubility of the polymerized species in the solutions.

In the starting solutions with a large amount of water, the polymerized species would be soluble in the solvent because of the larger number of hydroxyl groups. Acids with lower dissociation constants were assumed to decrease the number of protonated alcohols, which reduce the ionic nature of the solvent. Alkoxides with longer hydrocarbon chains than methanol — which will release the alkoxy groups in the hydrolysis reaction to generate alcohols with longer chains than methanol — and alcohols other than methanol would increase the solubility of the polymerized species in the solvent.

The amount of methanol in the solution will change the state of the aggregation of the polymerized species. The aggregation of the lipophilic polymerized species discussed here is also regarded as phase separation of the sol, in which oil-like polymerized species are separated from the solutions. The change of the ratio of the solvent to the polymers is expected to change the degree of phase separation.

Because of the lipophilic nature of the primary particles, translucent sols become transparent as a result of dissolution of the primary particles in the solvent when nonpolar organic solvents are added. Flexibility of the primary particles results in instability of the secondary particle, as seen in the centrifugation experiment. Changes in the gel structure by addition of organic compounds and by centrifugation and the formation of round particles are illustrated in Figure 10.9.

Surface Changes in the Ambient Atmosphere

Large numbers of unhydrolyzed methoxy groups in the particles cause instability of the surface structure of the gel particles. Figure 10.10 shows SEMs of gels obtained from a TMOS solution and kept at 40°C in the ambient

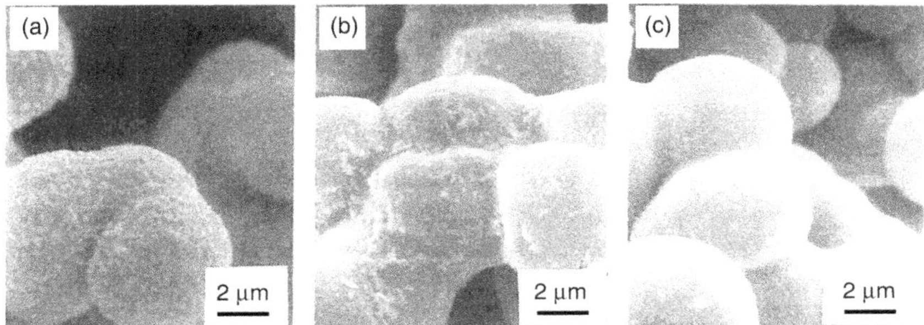

FIGURE 10.10 SEMS of dried gels prepared from solutions with mole ratios TMOS : H_2O : HCl : CH_3OH of 1 : 1.53 : 0.40 : 2 kept at 40°C in the ambient atmosphere for (a) 2 days, (b) 5 days, and (c) 13 days after gelation.

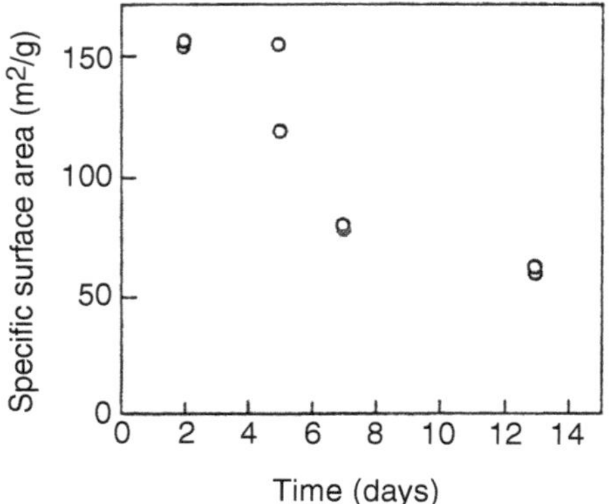

FIGURE 10.11 Change of the specific surface area of a gel with time after gelation. The gel was obtained from a solution with molar ratios TMOS : H_2O : HCl : CH_3OH of 1 : 1.53 : 0.40 : 2.

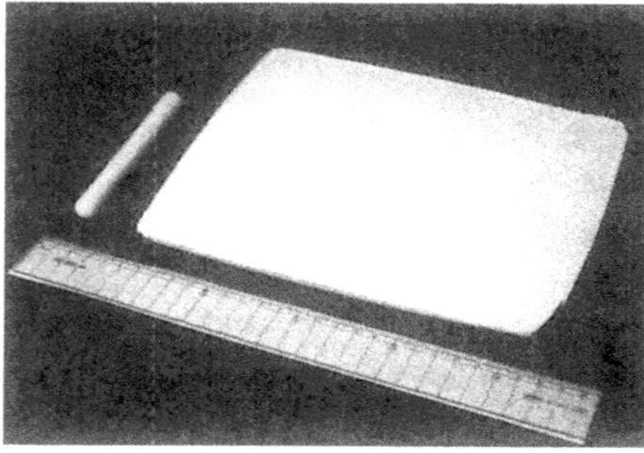

FIGURE 10.12 A gel plate and a rod obtained from a solution with mole ratios TMOS : H_2O : HCl : CH_3OH of 1 : 1.53 : 0.25 : 2 and dried at 40°C for about 1 month.

atmosphere for various times after gelation. Because the round particles are secondary particles, gels kept at this temperature for 2 days have rough surfaces. The surfaces of the particles, however, become more smooth with time. Reaction of the unhydrolyzed methoxy groups with the vapor in the atmosphere takes place after gelation, and polycondensation at the hydrolyzed sites may lead to the smoothing of the particle surface.

Figure 10.11 shows the change of the specific surface area of the gel with time after gelation. Specific surface area is measured by the Brunauer–Emmet–Teller (BET) method by using N_2 gas as the adsorptive agent. The surface area decreases with time; this observation corresponds to Figure 10.10, in which a rough particle surface turned smooth.

APPLICATIONS

As mentioned in the introduction, the most serious problem encountered in sol–gel process for making bulk silica glasses is the fracture of gels during drying. At the interface between the residual liquids and the pore walls, large capillary forces are generated, which cause cracking of the gel body. Preparation of gels having large pores is effective in preventing fracture, because the capillary force decreases as the pore radius increases. The method described in this chapter can produce such gels easily, as shown in Figure 10.12, where a crack-free gel plate of 21 cm × 17 cm × 0.9 cm prepared from a solution with mole ratios TMOS : H_2O : HCl : CH_3OH of 1 : 1.53 : 0.25 : 2 and dried at 40°C for about 1 month is shown.

These gels can be used as the precursor for porous silica glass bodies as well as for pore-free bulk silica glasses. Conventionally, porous silica glasses have been made by preparing sodium borosilicate glasses by the melt–quench method, heating the resultant glasses for phase separation, and leaching the Na_2O- and B_2O_3-rich phase out of the phase-separated glasses [10]. High-purity porous silica glasses should be produced by a simple method, however, if silica gels are used as the precursor. Figure 10.13 shows a bulk porous silica glass made by the sol–gel method and an SEM of the fracture surface. The gel was prepared from a solution with mole ratios TMOS : H_2O : HCl : CH_3OH of 1 : 1.53 : 0.40 : 2 and dried at 40°C for a few months. The dried gel was then heated at a rate of 0.5°C/min to 1300°C and kept at that temperature for 3 h. A crack-free porous glass was successfully obtained,

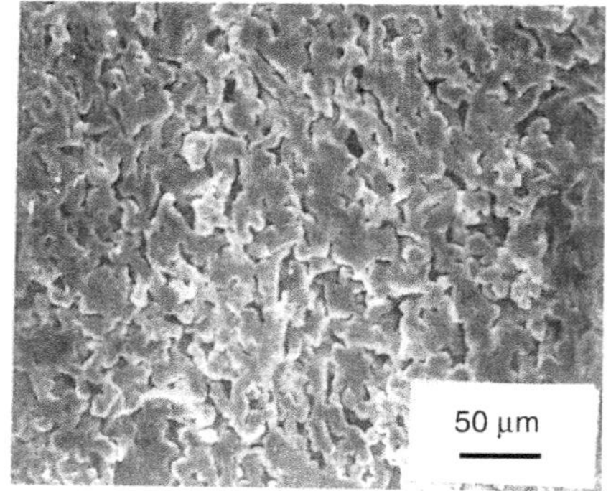

FIGURE 10.13 Appearance (left) and SEM of the fracture surface (right) of a porous silica glass prepared by the sol–gel method.

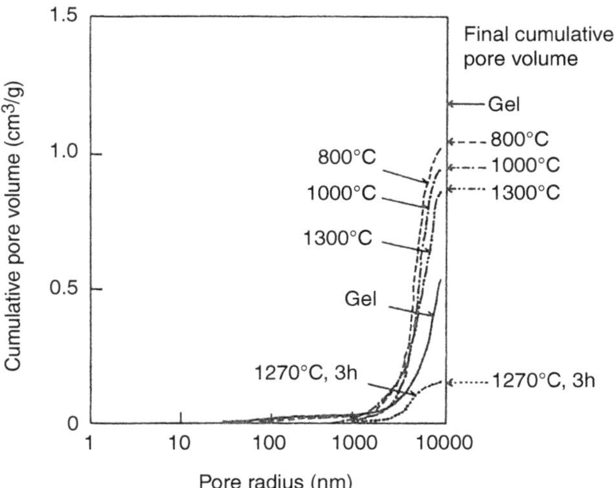

FIGURE 10.14 Pore-size distribution curves of silica gel and gel-derived porous silica glasses. The gel was obtained from a solution with mole ratios TMOS : H_2O : HCl : CH_3OH of 1 : 1.53 : 0.40 : 2. The gel was dried at 40°C for 7 days and heated at a rate of 0.5°C/min to various temperatures.

and continuous pores more than 10 μm in diameter are visible.

The pore characteristics can be controlled not only by changing the composition of the starting solution but also by changing the heat treatment conditions of the gels. Figrue 10.14 shows pore-size distributions of a gel and its derivatives made with differing heat treatments. Because measurement was possible only at pressures higher than 1 atm (101 kPa), measurement of the pore-size distribution could not be made for pores larger than 7.5 μm in radius. However, pores larger than 7.5 μm in radius are present in the gel, and the pore volume decreases and the pore-size distribution shifts to smaller radius when the upper heat treatment temperature is increased.

Such porous silica glasses of micrometer-sized pores can serve as filters and enzyme and microbe supports. If the small pores on the particle surface or the roughness of the particle surface are designed to be retained after heat treatment, the resultant porous glasses can be used as filters that function as catalyst supports.

REFERENCES

1. Rabinovich, E. M.; Johnson, D. W., Jr.; MacChesney, J. B.; Vogel, E. M. *J. Am. Ceram. Soc.* **1983**, *66*, 683.
2. Rabinovich, E. M. *J. Mater. Sci.* **1985**, *20*, 4295.
3. Scherer, G. W.; Luong, J. C. *J. Non-Cryst. Solids* **1984**, *63*, 163.
4. Prassas, M.; Phalippou, J.; Zarzycki, J. *J. Mater. Sci.* **1984**, *19*, 1656.
5. Toki, M.; Miyashita, S.; Takeuchi, T.; Kanbe, S. *J. Non-Cryst. Solids* **1988**, *100*, 479.
6. Kozuka, H.; Sakka, S. *Chem. Mater.* **1989**, *1*, 398.
7. Kozuka, H.; Yamaguchi, J.; Sakka, S. In *Proceedings of the XVth International Congress on Glass*; Mazurin, O. V., Ed.; Nauka: Leningrad, U.S.S.R., 1989; Vol. 2a, p 32.
8. Stöber, W.; Fink, A.; Bohm, E. *J. Colloid Interface Sci.* **1968**, *26*, 62.
9. Sakka S.; Kamiya, K. *J. Non-Cryst. Solids* **1982**, *48*, 31.
10. Nordberg, M. E. *J. Am. Ceram. Soc.* **1944**, *27*, 299.

11 Silica Aquasol Process to Prepare Small Particle Size Colloidal Silica by Electrodialysis

Horacio E. Bergna
DuPont Experimental Station

CONTENTS

Concentrated aqueous silica sols of small silica particles about 10 nm or less in diameter and uniformly sized are prepared by a process for maintaining a constant number of silica nuclei particles prior to the deposition process. The process for maintaining a constant number of silica nuclei particles in the sol consists of forming the particles at a pH 8–10.5 and a low temperature then heating the sol through a transition temperature to a deposition temperature maintaining the pH, removing alkali metal ions and simultaneously adding alkali metal silicate from the system at a rate such that the number of silica particles in the sol remains constant while the temperature is being raised.

INTRODUCTION

This chapter comprises work on electrodialysis of sodium silicate solution to produce relatively concentrated silica aquasols of very small particle size, equal or less than 7 nm. United States Patent number 4,410,405 issued on October 18, 1983 covers this process. This new process is in principle a relatively pollution free way to make silica aquasols. A 1974 economic evaluation of the patented process versus the ion-exchange resin process indicated cost savings, based on installation of a grass root plant.

Full potential process was demonstrated in a small laboratory batch unit (0.3 ft 2 area, but no basic design data was obtained. This chapter concerns work done with a one square foot, twin cell commercial unit to develop the basic data needed for a firm economic evaluation.

OBJECTIVES

- To demonstrate feasibility of electrodialysis process to make concentrated silica sols on a commercial unit
- To work out process conditions for small particle size of present silica aquasols

SUMMARY AND CONCLUSIONS

The technical feasibility of the electrodialysis process to make relatively concentrated equal or less than 7 nm particle size silica sols was demonstrated on a commercial electrodialysis unit and samples of sols that meet silica aquasols manufacture specifications were prepared.

- A new variable temperature method to make concentrated silica sols and specially suited to make very small particle size (4–7 nm) silica sols was demonstrated.
- Cost estimate to compare the electrodialysis process with the ion exchange resin process were made and showed that an electrodialysis plant would cost less than a completely new ion exchange resin plant.

Guidelines for the construction of viable electrodialysis cells for silica sols process were prepared. This was needed because the commercial one square foot electrodialysis twin cell used in this project develops eternal and internal leaks under the temperature and flow rate conditions specified by our colloidal silica process.

The following table summarizes the main economic features of the electrodialysis process to make silica sols in comparison with a plant's ion exchange resin process.

Savings over consol*
No ion exchange resin expense
Less steam for evaporation
Caustic by-product
No effluent lime treatment
Reduced sulfuric consumption
Lower solids treatment burden
Expense over ion exchange resin
Higher electrical cost
Higher maintenance cost
Higher overhead, depreciation, taxes, etc.
Initial resin and expense

*Consol: conventional process to prepare colloidal silica. See Figure 11.1.

On the basis of this work we conclude that

- The electrodialysis process to make silica aquasols commercial electrodialysis units is technically feasible. The new variable temperature method allows to make very small particle size sols (4–6 nm) and concentrated small particle sols (7 nm) at higher, more efficient temperatures.
- Electrodialysis remains an attractive technical option for new plant: an electrodialysis plant would cost less than a completely new ion exchange resin plant. Electrodialysis is an energy intensive process but it is essentially pollution-free and recovers substantial NaOH values for credit. Location of a plant in a low-cost power area could be attractive. Electrodialysis would also be a technical option in case environmental pressures require further reduction of sodium sulfate level in the silica sol waste of the ion exchange resin process.

PREPARATION OF SILICA AQUASOLS BY ELECTRODIALYSIS

Electrodialysis is a preferred process to prepare silica aquasols in areas where electric power is not expensive. The process consists in the electrolysis of an alkali metal silicate solution to continuously remove alkali metal ions until a silica aquasol of the desired particle size is obtained.

Figure 11.1 and Figure 11.2 show the design of an electrodialysis cell. Figure 11.1 shows the places occupied by the membranes that separate the feeds, anolyte and catholyte. Figure 11.2 is a simplified diagram of the silica aquasol electrodialysis process apparatus.

Dilute sodium silicate flows into the electrodialysis cell and is converted into an aquasol of small particle

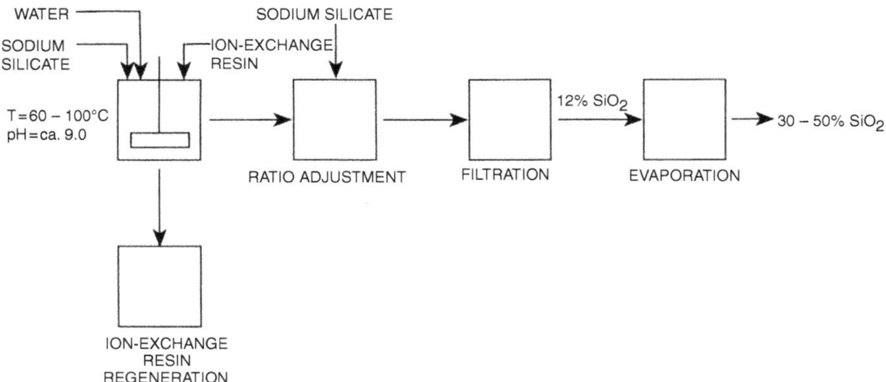

FIGURE 11.1 The design of an electrodialysis cell.

size. This sol is re-fed into the cell at the proper temperature until the desired particle size is achieved. A 25% silica sol of 15 nm particles can be prepared directly by this process. The electrolyte is then removed by ion exchange, the pH adjusted for optimum sol stability, and the sol concentrated to 30–50% silica [1].

There is essentially no consumption of acid except the small amount needed at the start of each batch to neutralize a dilute solution of sodium silicate (0.5% SiO_2) to pH 9 at 60–90°C to form silica nuclei to start the process. Narrow uniform spacing between the membranes is required to minimize power cost and avoid silica deposition. Water is added to the anode compartment since it is slowly transported to the cathode compartment, from which sodium hydroxide solution is constantly withdrawn. Anolyte and catholyte are circulated from the corresponding electrode compartments to separators for the removal of oxygen and hydrogen gases [1].

What follows is a description of a new method to make concentrated silica aquasols showing its advantages over the ion exchange processes in areas where electric power is not too expensive.

SILICA SOL PROCESS

Abstract

Concentrated aqueous silica sols of small silica particles about 10 nm or less in diameter and uniformly sized are prepared by a process for maintaining a constant number of silica nuclei particles prior to the deposition process. The process for maintaining a constant number of silica nuclei

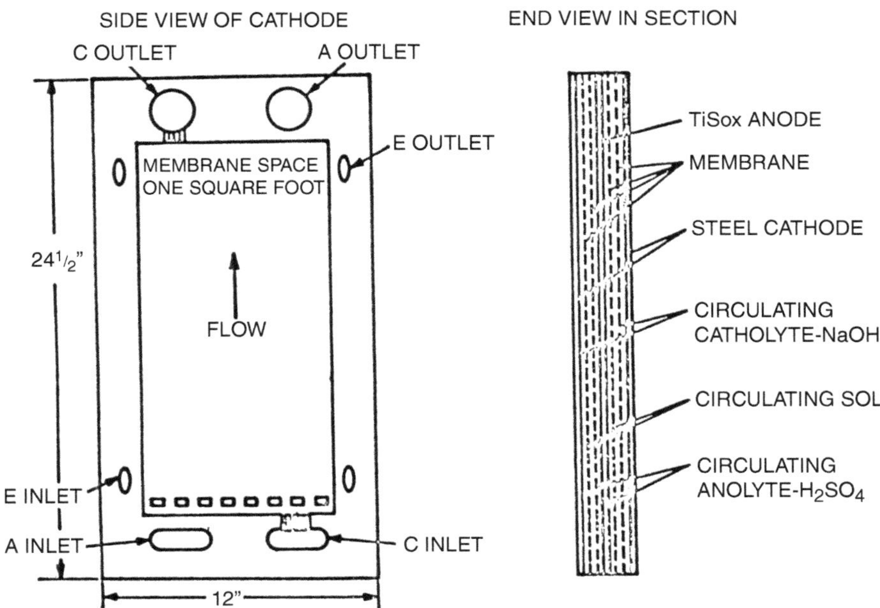

FIGURE 11.2 Simplified diagram of silica aquasol electrodialysis process apparatus.

particles in the sol consists of forming the particles at a pH 8–10.5 and a low temperature then heating the sol through a transition temperature to a deposition temperature, maintaining the pH, removing alkali metal ions and simultaneously adding alkali metal silicate from the system at a rate such that the number of silica particles in the sol remains constant while the temperature is being raised.

Description

Technical Field

This invention relates to a process for the preparation of concentrated aqueous silica sols of small and uniformly mixed silica particles of about 10 or less nanometers (nm) in diameter.

Background Art

Silica sols are made by adding sodium or potassium silicate to a dilute aqueous dispersion of small silica particles or nuclei and removing the sodium or potassium ions from the dispersion. This results in silicic acid being deposited upon the nuclei which grow to larger size and thus increase the silica concentration of the sol. In all processes of the prior art for making concentrated sols of silica particles of about 10 nm or less in diameter, it has been necessary to grow the particles in a dilute solution and then concentrate them by the evaporation of water which requires energy. In the following prior art the final sols contain particles larger than 10 nm in diameter.

There are at least three quite feasible processes by which silica sols can be made economically. These can be classified by the method of removing the cations from the dilute aqueous dispersion: (1) with ion exchange resins; (2) nonelectrolytically through membranes; and (3) by electrodialysis.

The use of ion exchange resins for making silica sols is well known and is described in U.S. Pat. Nos. 2,244,325 to Bird and U.S. Pat. No. 2,631,134 to Iler et al. U.S. Pat. No. 3,789,009 to Irani describes the use of ion exchange resins at high temperatures (60–150°C) to obtain dilute (10–16% SiO_2) "large particle" silica sols having particle number average diameter greater than about 15 nm which can be concentrated by evaporation. U.S. Pat. No. 2,601,235 to Alexander describes the processes in which nuclei of high molecular weight silica preferably made by heating a heel of silica sol above 60°C, are mixed with active silica and heated above 60°C to obtain dilute built-up silica sols. Within the temperature range disclosed in the art (60–150°C) the initial number of silica nuclei decreases and the larger particles increase in average size at the expense of the smaller particles which dissolve at high temperatures. Thus, the number of nuclei during the heat-up period is not maintained constant in number and the size of the nuclei is not kept small so that the final particles are larger than 10 nm (10 μm) in diameter, for example, 15–130 nm (Column 6, line 1).

Nonelectrolytic removal of the cations through membranes is described in U.S. Pat. No. 3,756,958 to Iler.

U.S. Pat. No. 3,668,088 to Iler describes the preparation by electrodialysis at high temperature of aqueous sodium or potassium silicate, of dilute silica sols of large particle size and very dilute sols of small particle size larger than 10 nm which can be concentrated by evaporation. The patent recites that at lower temperatures only extremely small particles of colloidal silica are obtained and the sols can therefore not be concentrated without gelling (Column 2, lines 27–29). Again in this case, the number of nuclei was not kept constant during the heat-up period so that in all examples the final particle size is larger than 10 nm.

The above-referred growth of silica particles by further accretion of silica as taught by the prior art is carried out in "hot solution," that is, 50–100°C, for a variety of reasons:

(1) In the electrodialysis process, to reduce the electrical resistance of the electrolyte and thus reduce the power required;
(2) In the ion exchange process when using a weak acid type of cation-exchange resin, to accelerate the rate of removal of sodium from solution; and
(3) To increase the rate at which soluble or active silica, released from soluble alkali metal silicate by removal of alkali metal ion, is deposited upon silica particles present in the solution.

Basically, nuclei particles for the earlier methods described of making silica sols are formed whenever a dilute aqueous solution of soluble silicate, which has a pH of over 11, is reduced in pH to below about 10.5. The lower the temperature, the smaller are the resultant nuclei.

Heretofore, it has been the practice to take such nuclei dispersions, heat them to "hot solution" temperature and then to enlarge the particles by silica accretion as described earlier.

However, when nuclei made at a lower temperature are heated to the higher temperatures (50–100°C.) at which particle growth is to be carried out, the nuclei are not stable especially when the particle diameter is less than about 5 nm. Smaller particles are dissolved and the remaining particles increase in average size.

The art processes effect the release of active silica under conditions of pH, temperature and rates of addition of active silica which do not result in the formation of additional silica nuclei, but in a decrease of the total number of silica nuclei. Therefore, the total number of nuclei present in the system is not kept constant during the process. When soluble silicate is added to the sol to grow the particles and to increase silica concentration, the final particle size depends on the number of surviving nuclei.

If too few nuclei survive the heating, the added active silica will be deposited in a smaller number of particles and therefore there will be faster growth and the final particles may be too large for certain applications, for example, as a binder for ceramic bodies and refractory fibrous insulation. For those reasons, there has been a need for a way to make dispersions of nuclei particles of about 10 nm or less in diameter and then to prevent their being dissolved at the higher temperature needed for efficient particle growth. This has proved to be a serious problem especially when concentrated sols of very small particle size (e.g., 15 or 20% SiO_2, particles smaller than about 10 nm in diameter) are to be made by electrodialysis. However, it is also difficult to produce similar concentrated sols of small particles directly by ion exchange. Dilute sols of small particles can be made by these processes and such sols can be concentrated by evaporation of water, but this requires the expenditure of extra energy. Uniformity of particle size is important in the case of product sols of particles smaller than about 6 or 7 nm in diameter in order to avoid spontaneous growth in size during storage. If there is a spread in size, the particles increase in average size but decrease in number, thereby changing in properties.

This invention teaches a procedure to maintain constant the original number of nuclei particle in the sol while the particles are grown so that no nuclei are lost in the growth process and therefore more concentrated products of small and quite uniformly sized silica particles are obtained.

Theory of the Invention

The solubility of very small particles is greater than that of larger particles. This is not ordinarily encountered in suspensions of solids where particles are large enough to be seen under a microscope. But in the case of silica particles of colloidal size, 5 nm particles, for example, have a solubility of 140 ppm expressed as SiO_2 in water, while 10 nm particles have a solubility of 110 ppm [1, p. 55, Figure 1.10a, Line A]. When silica nuclei having diameters in the range of about 3 to 5 nm, as obtained when a dilute solution of sodium silicate is neutralized at 25°C, is heated to 95°C for further processing and particle growth, the number of particles is greatly diminished. The smaller ones dissolve and the silica is deposited upon the larger ones. When any given particle starts to dissolve it becomes smaller and progressively still more soluble so that it quickly disappears.

This invention discloses how to heat a sol of very small nuclei to a higher temperature without having any of the nuclei disappear.

In none of the processes disclosed in the prior art has the problem been solved. It is possible to make dilute sols containing less than 15% by weight of silica and particle diameter less than 10 nm by prior art processes and these sols can be concentrated by evaporation of water. For still smaller particles which are needed as mentioned earlier, it has not been possible to obtain sols more concentrated than about 15% silica with particles smaller than about 10 nm, while growing the particles at 80–100°C by adding concentrated sodium silicate to the system and simultaneously removing sodium because the nuclei at the beginning of the growth process at 90°C are too large. By the time silicate has been added to obtain a concentrated sol, the particles are too large.

The objective of maintaining a constant number of nuclei particles according to this invention is obtained simply by maintaining a concentration of soluble silica in solution that is greater than the solubility of the smallest of the nuclei particles during the time the solution is being heated to the higher temperature. It is true that under these conditions, the aqueous phase is supersaturated with silica with respect to the solubility of all the nuclei including the smallest. Thus all nuclei grow somewhat in size during the heating period but no nuclei dissolve and the total number of nuclei particles remains unchanged. When more soluble silica is added to the system in the form of sodium silicate and the concentration of silica is increased, the final particle size is much smaller since the total number of particles is larger than if a certain number of the nuclei had disappeared during the heat-up period.

Summary of the Invention

The invention is directed to an improvement in the manufacture of silica sols by which more concentrated, more stable silica sols of small particle size (particle diameter less than 10 nm) can be obtained as a result of the more uniform number and size of the silica particles therein. This effect is achieved by making a certain number of small nuclei and keeping this number of nuclei substantially constant (decreasing no more than 25%) while the solution is being heated from the temperature at which the nuclei are formed, through a transition temperature to a higher temperature at which particle growth is to be carried out.

More specifically, a process has been found for maintaining a substantially constant number of silica nuclei particles in a sol at pH 8–10.5 while the sol is being heated from the maximum temperature at which the particles have been formed in the range of 10–50°C, to a temperature at least 10°C higher in the range of 50–100°C, at which the particles are to be grown, at a rate of about 1–4°C per minute, preventing change in the number of nuclei by adding a solution of alkali metal silicate and removing alkali metal ions from the system simultaneously to maintain the pH in the range, the rate of addition of silicate being such as to introduce silica at a rate of 0.5–6 preferably 1–4 g/h/1000 m^2 of surface of the silica nuclei in the system when the growth temperature is less than 70°C., and at a rate of 1–10 preferably 2–4 g/h/1000 m^2 when the growth temperature is in the range of 70–100°C. The product has a silica concentration from 15–30 g of silica per 100 ml of sol depending on the particle size. The larger the particle size of the product, the more concentrated it can be made. For this reason the designation of "concentrated" or "dilute" silica aquasols varies with particle size. A silica sol of 7 nm particle and 16 g SiO_2 per 100 ml of sol is considered to be concentrated but a silica sol of the same concentration with 14 mm particle size is considered to be dilute. In the present invention a silica sol of 7 mm particle size can be made by electrodialysis with a concentration of at least 27.5 g SiO_2 per 100 ml of sol. The concentration of the sols as made by electrodialysis can be increased further by evaporation. However, direct preparation of a concentrated sol, as disclosed in the present invention, is of practical importance since in this way much less water has to be evaporated to reach the final desired concentration thus gaining substantial savings in the energy consumed for production of the sols.

In particular, the invention is directed to a method for the preparation of concentration aqueous silica sols of small silica particles of about 10 nm or less in diameter comprising the sequential steps of

(a) forming an aqueous solution of alkali metal silicate containing 0.2–3.0% weight alkali metal silicate, basis SiO_2 in water, in which the molar ratio of SiO_2 to alkali metal oxide is from 2.5:1–3.9:1;
(b) at a temperature of 10–50°C, reducing the solution pH to 8–10.5, thereby spontaneously forming silica nuclei particles which are dispersed in the solution; and
(c) removing alkali metal ions from the nuclei-containing alkali metal silicate solution while (1) maintaining the solution at pH 8–10.5 by addition of alkali metal silicate at a rate sufficient to maintain a constant number of particles and (2) simultaneously raising the temperature of the solution by at least 10°C to 50–100°C

Stable concentrated silica sols are obtained from the solutions by the processes described earlier by continuing the removal of alkali metal ions from the solution at the temperature to which it was raised in step (c) above and maintaining the pH at 8–10.5 by addition of alkali metal silicate at a rate such that the number of particles remains substantially constant for a time sufficient to enlarge the silica particles to a preselected larger average particle size.

Detailed Description of the Invention

Definitions

The terms "silica sol" and "colloidal silica" refer to a dispersion of colloidal size silica particles in a liquid medium. As used herein, the liquid medium is water. Colloidal size particles include particles of about 10 to no more than 1000 Å (1–100 nm).

The term "percent weight" as applied to aqueous solutions of alkali metal silicates refers to the percent by weight of silicate, measured as SiO_2, in the total solution.

"Average particle size" as used herein is calculated from specific surface area as estimated by measuring the amount of alkali adsorbed from solution as pH of the sol is raised from 4–9 according to a modification of the method of Sears, [6]. The method is set out in the Analytical Procedures infra.

Frequently, the word "silicate" is used herein as a more convenient means for expression of the term "alkali metal silicate" or sodium silicate.

As used herein with reference to the generation of nuclei, the term "spontaneous" refers to the fact that the nuclei are generated without the addition of either seed materials or of preformed silica nuclei from another source.

Nucleation

The silica nuclei used in the process of the invention are made by conventional techniques. Starting material for the nucleation step is an aqueous solution of 0.2–3% weight alkali metal silicate in which the weight ratio of silica to metal oxide is from about 2.5 to 3.9. For purpose of obtaining controlled nucleation, it is preferred to start with a silicate solution having a pH above 10.2 at which nucleation occurs. A pH of 10.2 results from a silicate concentration of about 0.1%, measured as silica. Therefore, it is preferred to use a silica concentration above 0.1%, especially 0.2% and higher. A pH of about 10.3 results from a silicate concentration of about 0.2%, basis silica. The upper allowable concentration of silica is determined by one or two factors depending on what method of cation removal is used. In all cases, if the silica concentration exceeds about 3.0% weight, the nuclei tend to aggregate and thus produce particles of uneven size. Furthermore, in cases where cation removal is accomplished by electrodialysis, it is preferred that the silica concentration not exceed about 1.0%. Above that amount, the excess of silicate ions in the solution tend to migrate toward and clog up the dialysis membrane.

It is desirable from an energy utilization viewpoint to minimize the quantity of alkali metal ions which must be removed from the solution. For this reason, a practical minimum weight ratio of silica to alkali metal oxide in the nucleating solution is about 2.5:1. On the other hand, if the ratio of silica to alkali metal oxide (M_2O) is too high, the solution is likely to contain nuclei of uncontrolled size. It is preferred that the molar ratio of silica to alkali metal oxide (SiO_2/M_2O) be from about 3.0 to 3.9. An SiO_2/M_2O ratio of 3.25 is particularly preferred because of its ready availability in commercial quantities a low cost.

While alkali metal silicates in general can be used in the process of the invention, sodium and potassium silicates are preferred. Because of its greater availability and economy, sodium silicate is particularly preferred and normally be used.

The temperature of the nucleation step should be from about 10 to 50°C. Within this range, quite uniformly small nuclei are produced spontaneously within a short period. In general, the higher the nucleation temperature, the bigger are the nuclei. A temperature of 20–50°C and especially 30–50°C is preferred to obtain appropriately sized nuclei, for example, particles smaller than about 5 nm in diameter.

Spontaneous nucleation of the silicate solution is accomplished by reducing the pH of the solution to within the range of 8–10.5 by either of two general ways: (1) by partial neutralization of the silicate solution with acid; or (2) by removal of alkali metal ions from the solution.

In the case of the former method of reducing pH, several acids might be used. However, to simplify the problem of disposing of the resultant salt, it is preferred to use sulfuric acid. As has been noted by Iler et al., it is essential that mixing of the acid with the silicate be accomplished with extreme rapidity in order to prevent the formation of silicic acid gels which occurs if even local regions of pH 5–6 exist for an appreciable time [7]. In the case of the latter general method, the alkali metal ions are removed by adsorption upon an ion exchange resin or by dialysis through a cation-selective membrane. It should, however, be noted that the rate of cation removal is much too slow by simple dialysis to be economical. Thus, it is preferred to employ electrodialysis when dialytic separation is employed.

At pH less than 7 or 8, silica sols are not stable. Agglomerates of colloidal particles form and the silica sol can gel. This effect grows more pronounced as the concentration of the sol increases. In very dilute sols, low pH can be tolerated but in general should be avoided.

A solution pH of as high as 10.5 may be used when ion exchange resins are used for nucleation. However, the pH of the alkali metal silicate solution should not exceed about 9.5 when nucleation is effected by electrodialysis or by acid neutralization. At pH higher than 9.5, silicate ions are present which tend to migrate to and deposit on the membrane thus restricting the free flow of cations therethrough. Therefore, while a pH above 9.5 can be tolerated for a short time, inversely related to the pH level, it is preferred to limit operation of the process at a pH above 9.5 to short periods of time, for example, several minutes. Furthermore, the nuclei formed by either electrodialysis or by acid neutralization tend to become unstable and poorly dispersed in the presence of an electrolyte in the solution when pH of the solution exceeds about 9.5. This maximum pH will, however, fluctuate somewhat depending on size of the nuclei.

In some instances, it will be desired to provide larger particles for the preparation of concentrated sols. A preferred way of doing this by which uniformity of particle size can be maintained is to enlarge the nuclei obtained in the manner described earlier for a short time at the nucleation temperature. This is done by removing alkali metal ions from the nucleated solution while simultaneously maintaining the pH of the solution within the range of 8–10.5 by the addition of silicate at a rate such that the number of nuclei remains constant until the nuclei have become enlarged to the desired size, which will ordinarily be less than about 5 nm, for example 2–5 nm.

Transition

The process of the invention is a transition step by which the nucleated solution is prepared for rapid particle growth to a high concentration of silica. In this step, the nucleated solution is raised by at least 10°C to the temperature at which more rapid particle growth and concentration are to be carried out. However, in order to maintain uniform particle size of the enlarged particles and thus in many instances to be able to achieve higher silica concentrations without agglomeration, it is essential that this transition be carried out within the above-referred pH limits of 8–10.5 in such manner that the number of particles in the solution is substantially constant, that is, the number of silica particles in the system at the end of the transition step does not differ more than about 25% and preferably no more than 10% from the number of particles present at the beginning of the transition step.

It has been found that this can be done by maintaining a concentration of soluble silica in the solution greater than the solubility of the smallest particles so that they cannot dissolve. This is done by (1) adding alkali metal silicate to the solution and (2) removing alkali metal ions from the system simultaneously with raising the temperature of the system.

In order to raise the solution to an adequate particle growth temperature without loss of nuclei due to their being dissolved therein, it is essential to add alkali metal silicate to the solution at a carefully controlled rate. To avoid the formation of new nuclei, it is essential that the rate of adding soluble silica be no faster than the rate at which the soluble silica is deposited on the particles already present. In practice this required balance between the introduction and removal of soluble silica can be achieved by varying the rate of alkali metal silicate addition in response to variations in the relationship between the specific surface area and the weight of silica in the solution.

Quantitatively, it has been found that at solution transition temperatures up to about 70°C, an addition rate of 0.5–6 preferably 1–4 g/h/1000 m^2 of active silica will prevent the disappearance of particles. However, if the solution transition temperature exceeds 70°C, the rate should be about 1–10 preferably 2–4 g/h/1000 m^2. At slower rates of addition of active silica for each given temperature, some particles tend to disappear by dissolution because not enough active silica is added during the interval, therefore the total number of particles in the system would not remain constant. The particles would decrease in number and increase in size. On the other hand at faster rates the total number of particles in the system would not remain constant because more nuclei would be created.

In general, a preferred range of operation of the invention is to add active silica at a rate of 1–4 g/h/1000 m^2 at the beginning of the heating-up (transition) interval. Nevertheless, it is better to make actual determinations of the silica surface area and from there determine whether a given rate is suitable or not.

It is possible to determine whether new nuclei are being formed by measuring change in specific surface area of the silica in the sol during step (c). Let S be the measured specific surface area of silica in the sol being processed which, at a given time, contains a total of W grams of silica, and let S′ be the calculated specific area of silica in the sol which at a later time contains a total of W′ grams of silica and Sm is actual measured area at the later time. Then if no new nuclei are formed:

$$\frac{S'}{S} = \left(\frac{W}{W'}\right)^{\frac{1}{2}}$$

If the measured area (Sm) at the later time is greater than the calculated area (S′) at that same later time, then new nuclei are being formed. Conversely, if the measured area (Sm) at the later time is less than the calculated area (S′) at the same time, than the number of nuclei are becoming less. An alternative way of expressing the relationship is in terms of average particle diameter D, where D in nm equals 2750/S where S is in m^2/g. Therefore

$$\frac{D}{D'} = \left(\frac{W}{W'}\right)^{\frac{1}{2}}$$

The average number of particles can be determined from the silica concentration and average particle diameter using the relation $n = C/1.216\ D^3$, where C is the concentration of SiO_2 in gms/100 ml of solution and D is average particle diameter in centimeters ($1\ nm = 1 \times 10^{-7}$ cm).

When the removal of alkali metal ions is carried out by electrodialysis in any step of the process, it is essential that the sol-electrolyte contain an alkali metal salt of a nonsiliceous anion to function as a supplementary elctrolyte, as is disclosed in U.S. Pat. No. 3,668,088 referred to herein above. The purpose of the supplementary electrolyte is to lower the electrical resistance of the sol-electrolyte by providing anions other than silicate and hydroxyl ions to carry the current toward the anode. In the absence of supplementary electrolyte, the current is carried by hydroxyl ions and silicate ions which are present only at very low concentrations in the pH range 8–9.5. Furthermore, when the current is carried by silicate ions, these ions migrate towards the anode and deposit silica upon the membrane, thus further increasing the resistance and eventually terminating the process. There is an upper limit to the amount of supplementary electrolyte that can be present. If more than about 0.1 normal sodium salt such as sodium sulfate is present, the colloidal silica particles in the mixture tend to aggregate, even at the preferred pH of 9.

Even in the presence of supplementary electrolyte, silicate still migrates toward the anode if the pH of the sol-electrolyte is appreciably higher than about 9.5, but little migration occurs when the pH is maintained between 8 and 9.5. In this pH range the concentration of silicate ions in solution is very low relative to the non-siliceous anion.

The supplementary electrolyte may be added in the form of the sodium salt, or a source of non-siliceous anion may be added which will form the supplementary electrolyte in situ. Preferred nonsiliceous anions are those derived from a strong acid having a dissociation constant in water of a least 1×10^{-2} of which sulfuric acid is especially preferred.

In a preferred aspect of the process of this invention, the supplementary electrolyte of sodium sulfate is employed at a concentration of about 0.03–0.08 normal. Within this concentration range, the sol electrolyte has a sufficiently low electrical resistivity to permit the use of a practically low voltage on the cell, while at the same time the colloidal silica concentration can be greatly increased in the course of the process by the continued addition of concentrated sodium silicate solution, so that a silica sol containing as much as 35 g or more of SiO_2 per 100 ml, can be obtained without appreciable deposition of silica on the ion exchange membranes or formation of silica aggregates or gel.

As has been discussed hereinabove, addition of sodium silicate introduces active silica to the sol and increases the pH of the sol. As the sol is circulated past the ion exchange membrane the sodium ions in the sol are replaced by hydrogen ions and the pH falls. The rate of sodium silicate addition and rate of circulation of the sol must be coordinated with the rate of ion exchange through the membrane so that the sol entering the ion exchange apparatus will have a pH not higher than 9.5 and sol leaving the ion exchanger will have a pH not lower than about 8.

The rate at which the nucleated solution is heated to the particle growth temperature is to a great extent a matter of choice which is in recognition of the following factors:

(1) the more rapidly the heating is carried out, the smaller the particles will be when the growth temperature is reached
(2) the more rapidly the heating is carried out, the faster must be the rate of silicate addition.

As a general rule, it will be preferred to heat the nucleated solution to growth temperature as rapidly as possible. Though heating periods in the laboratory of only a few minutes are possible, practical heating periods in commercial scale equipment will generally be from about $1\frac{1}{2}$–2 h.

Particle Growth

Upon completion of the temperature transition of the sol, the particles will in some instances already be of the desired size in which case further particle growth is unnecessary, for example, particles 5 nm in diameter. However, when still larger particles and concentration are needed, they are produced by conventional techniques, for example, depositing further amounts of silica on the particles at a temperature from about 50–100°C. The lower limit is governed chiefly by the reaction rate desired while the upper limit is determined by the boiling point of water in the electrolyte. However, in order to obtain the advantages of the invention as to uniformity of larger particle size for high concentrations, it is necessary that the temperature during the growth step exceed the nucleation temperature by at least about 10°C and preferably 20°C.

To permit the growth of silica particles by accretion of silica from alkali metal silicate to a size sufficiently larger that the sol can be subsequently concentrated to a still higher silica content and remain stable, the growth step, like the transition step, is carried out in such manner as to avoid the formation of any substantial number of new particles as described hereinabove.

Upon completion of particle growth to the desired level, ions therein which tend to destabilize the sol, can be removed, for example, by precipitation, ion exchange or ultra filtration. The sol can then be restabilized by addition of a small amount of alkali and is thereafter very stable under ordinary storage conditions. Alternatively, if a dilute sol is produced, the supplementary electrolyte can be removed and the sol can be concentrated, for example to within the range of 10–40% by weight of silica (depending upon particle size) and stabilized with suitable amounts of alkali.

Membranes which may be used in the removal of alkali metal ions by electrodialysis are those which are impermeable to anions, but which allow the flow therethrough of cations. Such cation-selective membranes should, of course, possess chemical durability, high resistance to oxidation and low electrical resistance in addition to their ion-exchange properties. Homogeneous-type polymeric membranes are preferred, for example, network polymers such as phenol, phenosulfonic acid, formaldehyde condensation polymers and linear polymers such as sulfonated fluoropolymers and copolymers of styrene, vinyl pyridine and divinylbenzene. Such membranes are well known in the art and their selection for use in the method of the invention is well within the skill of the art.

When the process of the invention is carried out by electrodialysis, it is ordinarily done in equipment of the general type described in U.S. Pat. No. 3,668,088. As noted therein, it is preferred that the electrolyte be agitated strongly at least near the surfaces of the membranes in order to dislodge gas bubbles, to prevent deposition of silica and to prevent the development of low pH regions. Similarly, the sol should be agitated sufficiently so that high pH and high sodium concentrations are avoided at the point of sodium silicate addition.

When the removal of alkali metal ions in the process of the invention is done by adsorption upon ion exchange resins, the resins used will ordinarily be synthetic resins of the weak acid type which contain carboxylic groups as the functional sites. Such resins are usually based upon crosslinked copolymers of styrene and divinylbenzene in particulate form. These resins can be used in either the H^+ or NH_4^+ form. The solutions involved in the process can be contacted with the ion exchange resins either by passing them through a fixed bed of resin or by admixing solution and resin and then separating the resin from the solution by filtration, settling, centrifugation and the like.

The invention will be better understood by reference to the examples given subsequently.

Analytical Procedures

Specific Surface Area of Silica Particles in Aqueous Sols

A sample of sol containing 1.50 g SiO_2 is diluted to 135 ml with distilled water. To the diluted sample is added enough 4 *N* HCl to lower the pH to between 3.0 and 4.0, after which 30 ± 1 g NaCl are added and pH of the sample is then adjusted to 4.0 by addition of 0.1 *N* NaOH. The sample is titrated to pH 9.0 with 0.1 *N* NaOH and the required volume in milliliters of NaOH solution (V_t) is noted. Titration of a sample blank is also required. This is done by dissolving 30 ± 1 g NaCl in 135 ml distilled water, adjusting to pH 4.0, titrating the blank to pH 9.0 in the same manner as the sol sample and recording the required volume of NaOH (V_b). The specific surface area (S) in m^2/g is then determined by the relationship $S = 26.5\ (V_t - V_b)$. Any ammonia contained in the initial sol must be removed prior to initiating the above procedure. This is done by treating the ammonia-containing sample with a strong acid ion exchange resin, filtering out and rinsing the resin with water and combining the rinse water with the sol sample.

Soluble Silica in Presence of Colloidal Silica: Silicamolybdate-yellow method for SiO_2 based on E. B. Alexander, *JACS*, 75, 5655 (1953).

Total Silica: Fluoride method for determining silica, S. M. Thomsen, *Anal. Chem.*, 23, 973 (1951).

Experimental Apparatus (See Figures 11.2–11.4)

In the examples following in which the invention is demonstrated in whole or in part by electrodialysis, the electrodialysis unit is a "cell pair" comprised of a single coated titanium base anode common to two cells in the following sequence:

Blank End Plate
Nickel Plated Steel Cathode
Cation-Selective Membrane
Spacer
Anion-Selective Membrane
Coated Titanium Anode
Anion-Selective Membrane
Spacer
Cation-Selective Membrane
Nickel Plated Steel Cathode
Blank End Plate.

Thus, each side of the anode is divided into three compartments separated by the ion-selective membranes. Tie bolts keep the stack tightly sealed. The membrane separating the catholyte from the electrolyte is an Ionics CR-6170 alkali-resistant membrane. The membrane separating the anolyte from the electrolyte is an Ionics CR-61-CZL-183 acid-resistant membrane (Ionics, Inc., Watertown, Mass). The two membranes are separated by a $\frac{1}{4}$ in. thick plastic spacer. Effective area of the membranes is one square foot.

Both membranes are suitably framed to separate them from the cathode and the anode. Plastic electrode screens prevent the membranes from touching the electrodes. Flows through the cell compartments are from the bottom of the cell to the top, which provide good

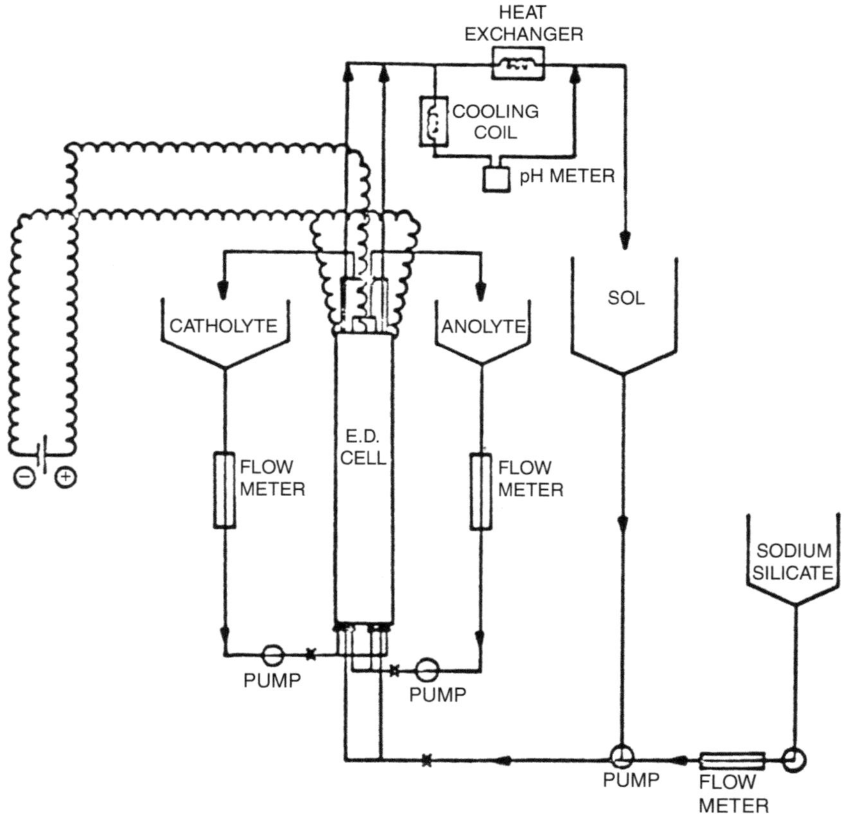

FIGURE 11.3 Simplified diagram of silica sol electrodialysis process apparatus.

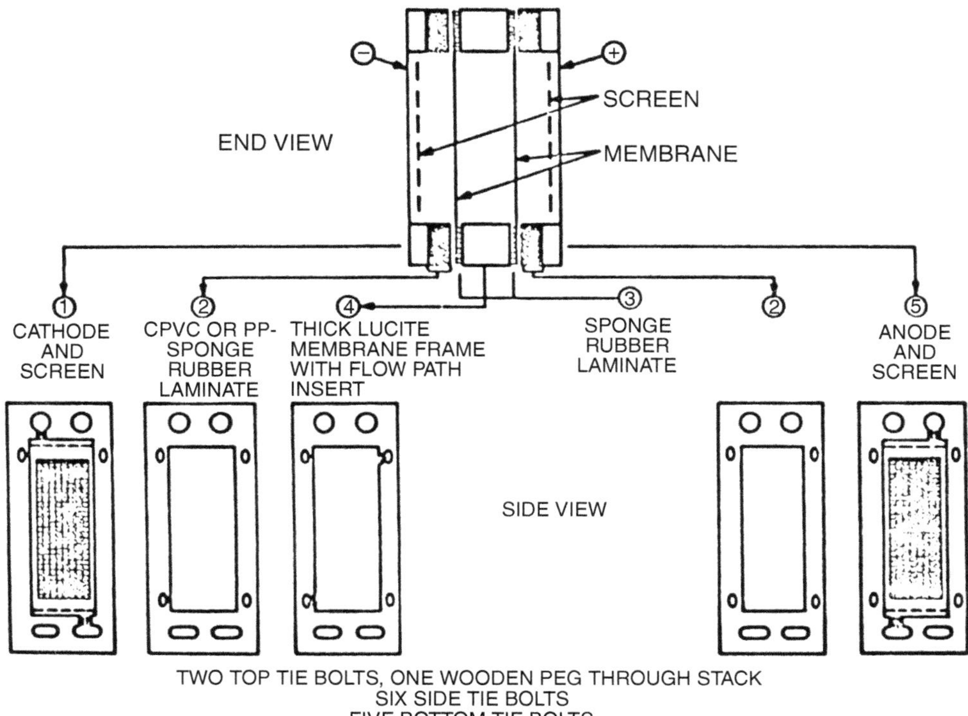

FIGURE 11.4 Electrodialysis cell construction.

flow configuration and easier exit of gases formed in the cell. Anolyte and catholyte enter via channels in the bottom of their compartments. Flow is diagonally across the cell.

Electrolyte enters from the side near the bottom of two rigid, plastic end blocks supporting the stack and exists from the diagonal top sides. The two electrolyte compartments are fed independently. When running the cell the center (electrolyte) compartment is overpressured by one psi, thus forcing the membrane to lie against the electrode screening.

The electrolyte is circulated at a rate of about 3 gallons per minute from the cell to a storage tank from which it is fed to the circulating pump. After leaving the pump, it passes through a steam-jacketed heat exchanger where it is heated according to the temperature desired in the cell and returned to the inlet of the electrolyte compartment. A sidestream of electrolyte is cooled and circulated through a cooler to a cell containing electrodes for measuring pH. Unless expressly set out otherwise, all pH measurements are measured at 25°C.

The two electrodes are connected to a source of direct current capable of supplying a total of up to 180 amperes to the cell. The current is adjustable by varying the voltage of the power supply.

Out of a total of 2736 g of silica introduced as sodium silicate, 2400 g or 87.7% of the original was recovered as silica sol. The specific surface area of the silica recovered was 262 m^2/g. Concentration of the sol was 25.4 g of silica per 100 ml. The volume of the sol was about 9450 ml.

Out of the total 27.36 equivalents of Na_2O introduced as sodium silicate, 24.56 equivalents were recovered as sodium hydroxide in the catholyte compartment which corresponds to 89.6% efficiency.

The silica sol obtained was treated with a mixture of a cationic and anionic exchange resins to remove ions from the medium.

The pH of the double deionized sol was 3.4 and was then adjusted to 9.1 by adding enough 3 *N* sodium hydroxide solution while stirring vigorously in a blender. The sol was concentrated to 30 wt% silica by evaporation. A small sample was saved for characterization and the rest of the sol was further concentrated to 40 wt% silica. Both the 30% wt silica sol sample and the 40 wt% silica sol were filtered through Celite filter aid. Even the 40% silica sol was perfectly stable under normal storage conditions. The 30% sol had a specific surface area (S.A.) of 230 m^2/g and when diluted to 9 wt% SiO_2 had a viscosity of 5.1 cps at 25°C.

TABLE 11.1
Preparation of Silica Sol Having 10 nm Particle Size by Electrodialysis

Feed solution	36 g SiO_2/100 ml
Electrolyte outflow pH	9
Temperature — Nucleation (T_N)	50°C
Temperature — Particle growth (T_G)	70°C
Time — Nucleation	70 minutes
Time — Transition	8 minutes
Time — Particle growth	317 minutes
Silica feed rate — at T_N	5.01 g SiO_2/min
Silica feed rate — at T_G	7.35 g SiO_2/min
Silica conc. — at end of T_N	0.2 g SiO_2/100 ml
Silica conc. — at beginning of T_G	3.4 g SiO_2/100 ml
Current — at T_N	100 amps
Current — at T_G	144 amps
Voltage	11.5 volts
Total volume of product	9450 ml

Time (min)	Temp. (°C)	Silica conc. (g SiO_2/100 ml)	S.A. (m^2/g)	Part. diam. (nm)	Part. per 100 ml × $^{10-19}$
20	49	1.0	790	3.4	2.0
39	52	1.9	668	4.0	2.3
70	51	3.4	520	5.2	2.0
97	71	4.9	462	5.8	2.0
138	71	7.2	405	6.7	2.0
185	71	10.5	358	7.5	2.0
232	71	13.6	344	7.8	2.3
261	71	15.6	321	8.4	2.1
303	72	18.5	302	8.9	2.0
337	74	21.3	284	9.5	2.0
380	73	23.4	275	9.8	2.0
407	71	23.5	262	10.3	1.9

Details of the above-described operation are given in Table 11.1.

EXAMPLE I

In this example, the invention was carried out by electrodialysis to make a silica sol in which the average particle diameter was about 10 nm.

Using the above-described "cell pair," a starting electrolyte solution was prepared by dissolving 43.2 g sodium sulfate in water to a total volume of 10 l to obtain a 0.06 Na_2SO_4 solution. An anolyte solution was made by dissolving 312.5 g of 96% sulfuric acid in water and adding water to a total volume of 6 l to obtain a 5% H_2SO_4 solution. Catholyte solution was prepared by dissolving 600 g of sodium hydroxide in water and adding water to a total volume of 6 l to obtain a 10% or 2.5 *N* NaOH solution.

A sodium silicate solution was prepared by mixing 900 ml of "F" grade commercial sodium silicate solution (SiO_2:Na_2O mole ratio of 3.25) with 100 ml of water and filtering the product through grade 615 Eaton-Dikeman Lab Filter Paper. The sodium silicate solution thus prepared contained 360 g of silica per liter.

The sodium sulfate electrolyte solution was circulated through the cell and heated to 50°C which also heated the cell to this temperature. Sixty cubic centimeters of the sodium silicate solution was fed into the electrolyte and the electrolyte-sodium silicate mixture circulated to obtain a homogeneous solution. Silica concentration of the homogeneous solution was 0.21 g SiO_2 per 100 ml. The pH of the solution was 10.35. The total volume of the solution was 10,285 ml.

Power was applied and the voltage raised quickly to 11.2 volts. Current reading was 160 amps. Operation was continued for 3 min by which time the pH value dropped to 9.5. At this time the voltage was 11.7 and the current was 95 amps. As operation was continued, the pH dropped to 9.0 and sodium silicate solution was fed at a rate of about 845 ml/h to adjust and keep the pH constant at 9.0.

The operation was continued and small samples of both the electrolyte containing the silica being formed and the catholyte were periodically removed and analyzed.

When the silica concentration in the electrolyte solution reached 3.5 g SiO_2 per 100 ml the solution was heated to 70°C in about 8 min while continuing the electrodialysis operation. The operation was continued at 70°C and pH 9 by adjusting the feeding rate of the sodium silicate solution in the sol.

From the foregoing results, it can be seen that the number of nuclei were kept substantially constant within quite narrow limits (plus 15%, minus 5%) during both the transition and growth steps therefore yielding a very stable sol of uniform particle diameter.

Therefore, silica feed rate during the first step at T_N (temperature at which the nuclei were formed, 50°C) was 5.01 g SiO_2/min which based on the silica concentration during this step (1 to 3.4 g SiO_2/100 ml), the total volume of the electrolyte solution (ca. 10 l), and the specific surface area of the silica particles measured during this period (790–520 m^2/g) as shown in the earlier table corresponds to values varying between 1.8 and 6.4 g SiO_2/h/1000 m^2.

Based on similar measurements, silica feed rate during the period of transition of temperature between 50°C and 70°C was 1.5–2 g SiO_2/h/1000 m^2, and during the period at temperature T_G (temperature at which silica particles are further accreted for particle growth, 70°C) the silica feed rate varied between 1.75 and 0.75 g SiO_2/h/1000 m^2. During this T_G period the silica concentration increased from 7.35 to 25.4 g SiO_2/100 ml, the specific surface area of the silica particles decreased from 405 to 262 m^2/g, and the total volume of the elecrolyte solution was kept constant at around 10 l.

EXAMPLE II

In this example, using same equipment and general procedures of Example I, a sol was prepared by electrodialysis having an average measured particle diameter of about 7.5 nm. The sol was nucleated at 30°C from a solution containing 0.2 g SiO_2/100 ml, grown at the nucleation temperature for about 2 h and then heated to 70°C for the growing step. Total volume of the electrolyte solution was kept constant at about 10 l throughout the run. Detailed operating conditions and the properties of the resultant sol are given in Table 11.2.

From the loss of nuclei after nucleation, during the step at T_N 30°C it is apparent that the silica feed rate was inadequate. However, by proper adjustment

TABLE 11.2
Preparation of Silica Sol Having 7.5 nm Particle Size by Electrodialysis

Feed solution	36 g SiO_2/min avg
Electrolyte outflow pH	9
Temperature — Nucleation (T_N)	30°C
Temperature — Particle growth (T_G)	70°C
Time — Nucleation	132 minutes
Time — Transition	11 minutes
Time — Particle growth	119 minutes
Silica feed rate — at T_N	4.2/7.6 g SiO_2/min avg
Silica feed rate — at T_G	7.6 g SiO_2/min avg
Silica conc. — at end of T_N	0.2 g SiO_2/100 ml
Silica conc. — at beginning of T_G	5 g SiO_2/100 ml
Current — at T_N	85 amps
Current — at T_G	150 amps
Voltage	11.6 volts
Total volume of product	about 10 l

Time (min)	Temp. (°C)	Silica conc. (g SiO_2/100 ml)	S.A. (m^2/g)	Part. diam. (nm)	Part. per 100 ml $\times 10^{-19}$
28	30	1.16	863	3.1	3.2
52	30	2.16	583	4.6	1.8
86	30	3.66	575	4.7	2.9
122	30	5.22	550	4.9	3.6
163	70	7.5	453	6.1	2.7
201	70	10.6	430	6.2	3.6
236	70	13.9	390	6.9	3.5
257	70	16.0	363	7.5	3.2

of the silica feed rate, the original number of nuclei were restored and maintained during the transition to 70°C.

The measurement results given in the table correspond to addition rates of silica varying during the first step a T_N 30°C between 0.9 and 3 g $SiO_2/h/1000\ m^2$; during the transition step between 30°C and 70°C, around 1.3 g $SiO_2/h/1000\ m^2$; and during the growth period at T_G 70°C, varying between 0.9 and 1.5 g $SiO_2/h/1000\ m^2$.

During the 6.8 h operation a total of 7600 of sodium silicate feed solution were added continuously and power applied with constant voltage while the current gradually increased from 95 to 156 amps and current density from 47.5 ASF (amps per square foot) to 78 ASF.

The volume of the electrolyte solution was kept at 10 l ± 6% throughout the run when necessary by direct addition of hot water to the electrolyte compartment.

At the end of the run the current was turned off and the silica sol product removed from the cell and allowed to cool down to room temperature.

EXAMPLE III

In this example, again using the same equipment and general procedures of Example I, a sol was prepared having an average measured particle diameter of about 10 nm. The sol was nucleated at 30°C from a solution containing 0.6 g SiO_2/ml, grown at the nucleation temperature for 20 min and then heated to 80°C for the growing step. Detailed operating conditions and the properties of resultant sol are given in Table 11.3.

The above data shows that the silica feed rate during the transition step was sufficient to maintain the number of particles at the end of the step within about 15% of the starting number (3.9–3.4). However, the silica feed rate during the growth step at 80°C was inadequate to maintain the number of particles at the original level. The resultant sol, while having good uniformity as compared to the prior art practice, would have been improved still further had the number of particles been maintained substantially constant during the growth step.

Based on the data given in the table, the silica feed rate during the transition heating step between 30 and 80°C

TABLE 11.3
Preparation of Silica Sol Having 10 nm Particle Size by Electrodialysis

Feed solution	36 g SiO_2/100 ml
Electrolyte outflow pH	9
Temperature — Nucleation (T_N)	30°C
Temperature — Particle growth (T_G)	80°C
Time — Nucleation	20 minutes
Time — Transition	38 minutes
Time — Particle growth	213 minutes
Silica feed rate — at T_N	5.5 g SiO_2/min.
Silica feed rate — at T_G	18.5 g SiO_2/min
Silica conc. — at end of T_N	0.6 g SiO_2/100 ml
Silica conc. — at beginning of T_G	1 g SiO_2/100 ml
Current — at T_N	75 amps
Current — at T_G	156 amps
Voltage	11.5 volts
Total volume of product	about 10 l

Time (min)	Temp. (°C)	Silica conc. (g SiO_2/100 ml)	S.A. (m^2/g)	Part. diam. (nm)	Part. per 100 ml $\times 10^{-19}$
20	30	0.97	990	2.7	3.9
38	62	1.98	805	3.3	4.5
58	80	3.82	595	4.5	3.4
79	80	6.3	430	6.3	1.8
107	80	7.5	405	7.5	1.5
145	80	9.9	368	7.35	2.0
195	80	14.1	332	8.15	2.15
232	80	15.0	300	9.0	1.7
255	80	19.5	284	9.5	1.9
271	80	22.5	265	10.0	1.9

started at 3.9 g SiO_2/h/1000 m^2 and varied between 3.4 and 4.5 g SiO_2/h/1000 m^2. During the growth period at T_G 80°C the silica feed rate varied between about 1.5 and 3.4 g SiO_2/h/1000 m^2.

EXAMPLE IV

In this example, a silica sol was prepared in accordance with the invention by a process in which a weak acid resin was fed simultaneously with a sodium silicate solution into a heel of dilute sodium silicate. Silica nuclei formation was performed at 35°C with a transition to 70°C.

The dilute sodium silicate heel (0.65 g SiO_2/100 ml) was prepared in the following manner: 45 ml of "F" grade sodium silicate (SiO_2:Na_2O mole ratio of 3.25) diluted to 36 g SiO_2/100 ml and filtered were added to 500 ml of tap water at 35°C in a 2 l graduated cylinder. The total volume in the graduated cylinder was brought to 2 l with 35°C tap water and added to a 4 l beaker. Five hundred milliliters of 35°C tap water were used to rinse the graduated cylinder to make the total volume of the solution in the beaker 2500 ml. The pH of this solution was 10.7. Thereupon 60 g of a weak acid type wet, drained resin were added at a uniform rate while stirring gently in a period of 15 min to bring the pH down to 9.0. The resin was Rohm and Haas IRC-84 Special.

The slurry was then filtered and the filtrate was analyzed: specific area of the silica was 785 m^2/g, corresponding to an average particle diameter of 3.45 nm and the estimated number of silica particles was 1.35×10^{19} per 100 ml of filtrate. One liter of the filtrate was used for this example. The filtrate had a pH 8.9 and the temperature was 35°C. Simultaneously, but separately, samples of wet, drained IRC-84 resin and sodium silicate were added to the filtrate. The wet, drained resin contains about 50% solids. About 50 ml of resin weighing about 30–40 g were added every 5–10 min to keep the pH between 9 and 9.5. The sodium silicate added was prepared by diluting with water 900 ml of "F" grade sodium silicate to a total of 1000 ml and then filtering. This sodium silicate solution feed with a concentration of 36 g/100 ml was fed into the reaction vessel (an open beaker) at a rate of about 11 ml/min. During this process pH was held to between 9 and 9.5 and the temperature at 35°C.

The additions were continued until 160 ml of the sodium silicate feed had been added and the concentration of the solution in the reaction vessel was 5 g SiO_2/100 ml. At this time a sample taken from the reaction vessel upon analysis showed that the specific surface area of the silica was 790 m^2/g, which corresponds to an average particle diameter of 3.4 nm, and that the number of silica particles per 100 ml of solution was about 11×10^{19}.

After the above-referred sample was extracted for analysis, simultaneous addition of resin and sodium silicate was continued while (1) the reaction vessel was heated at a uniform rate to 70°C in 15 min and (2) the solution pH was kept between 9 and 9.5. The additions were continued over a period covering a total time of 45 min during which time 505 ml of sodium silicate feed solution was added. At the end of the run the slurry was filtered and the filtered product was analyzed and found to contain 13.3 g of SiO_2 per 100 ml of solution. The amount of silicate added had been calculated to give 1505 ml of product with a concentration of 12 g SiO_2/100 ml. The difference found was due to evaporation of water during the process.

Specific surface area of the sol obtained was 609 m^2/g which corresponds to an average silica particle diameter of 4.45 nm. The calculated number of nuclei per 100 ml of solution corrected for evaporation effect was 12×10^{19}, showing that particle growth has taken place between the time when temperature was 35°C and silica sol concentration 5% and the time when temperature was 70°C, and silica sol concentration more than 12% without substantial change in the total number of nuclei.

Further evidence that no new nuclei were formed was that the ratio

$$\frac{S'}{S} \text{ was approximately equal to} \left(\frac{W}{W'}\right)^{1/2} \text{ where}$$

$$S = 790\,\mathrm{m^2/g},$$

$$S' = 610\,\mathrm{m^2/g},\ W = 50\,\mathrm{g\ of\ SiO_2\ and}$$

$$W' = 120\,\mathrm{g\ of\ SiO_2},$$

then

$$\frac{S'}{S} = \frac{610}{790} = 0.77 \text{ and}$$

$$\left(\frac{W}{W'}\right)^{1/2} = \left(\frac{50}{120}\right)^{1/2} = (0.4166)^{1/2} = 0.74$$

For storage the silica sol product was concentrated by evaporation to 15% by weight and the SiO_2/Na_2O weight ratio adjusted to 25 by addition of the necessary amount of 3 N sodium hydroxide. The resultant sol was quite stable.

EXPERIMENTAL

APPARATUS

The diagrams of Figures 11.2–11.4 illustrate the electrodialysis apparatus used which consisted of several parts:

- Modified Ionics Incorporated electrodialysis cell where reaction takes place and both silica and subproducts are formed.

- A rectifier as a source of DC current for the electrodialysis cell.
- Feed tanks for the electrolyte-sol mixture and the catholyte and anolyte solutions.
- A heat exchanger to control the temperature of the electrolyte-sol mixture throughout the operation.
- A pH monitoring system with pH electrodes to measure pH of electrolyte-sol at room temperature.
- Pumps, values and flow meters to control the flow rate of each of the solutions through the electrodialysis cell.

The electrodialysis cell is described in the following section and the rest of the components of the unit in Appendix C.

Electrodialysis Unit

The modified Ionics Incorporated electrodialysis unit was a "cell pair" consisting of a single-coated titanium base ($TiSO_x$) anode common to two cells in the following sequence:

1. Steel end plate.
2. Large CPVC end block.
3. Nickel plated steel cathode.
4. Electrode plastic screen.
5. CPVC-sponge rubber membrane frame with sponge against membrane.
6. Ionics CR 6170 alkali-resistant anion-selective membrane of size equal to stack cross-section.
7. Thick (1/4″) Lucitee spacer with flow path inserts and sponge rubber facing on both sides.
8. Ionics CR 61-CZL-183 acid resistant, cation-selective membrane of size equal to stack cross-section.
9. CPVC-sponge rubber membrane frame with sponge against membrane.
10. Electrode plastic screen.
11. Coated titanium $TiSO_x$ anode.

The reverse sequence was used on the other side of the anode to complete a symmetrical "cell pair." Thus, each side of the anode was divided into three compartments separated by the ion-selective membranes. Thirteen tie bolts kept the stack tightly sealed. One wooden peg through the stack top kept the stack components from sliding and disaligning the assembly. Effective area of the membranes was one square foot. The membrane frames separated the membranes from the electrodes. The plastic electrode screen prevented the membranes from touching the electrodes. The cell construction or stack assembly is illustrated in Figure 11.2. Figure 11.1 also shows a side view of the cathode.

The Ionics membranes must be kept wet at all times or they dry out and crack in a matter of an hour or less. The CR 6170 membranes were wet with a weak NaOH solution and the CR 61-CZL-183 in water

Flow of Liquids within the Cell and Electrical Connections of Electrodes

Flows through the cell compartments were from the bottom of the cell to the top, which provided good flow configuration and easier exit of gases formed in the cell. Anolyte and catholyte entered via channels in the bottom of their compartments. Flow was diagonally across the cell.

The electrolyte entered from the side near the bottom of the two rigid, plastic end blocks supporting the stack and exited from the diagonal top sides. The two electrolyte compartments were fed independently. When running the cell, the center (electrolyte) compartment was overpressured by one psi, thus forcing the membrane to lie against the electrode screening.

The electrolyte was circulated at a rate of about 3 gal/min from the cell to the storage tank from which it was fed to the circulating pump. After leaving the pump, it passed through the steam-jacketed heat exchanger where it was heated (or cooled with tap water) according to the temperature desired in the cell and returned to the inlet of the electrolyte compartment. A sidestream of electrolyte was cooled and circulated through a cooler to the pH cell. Thus, all pH measurements were measured at about 25°C.

The three electrodes were connected in parallel to the rectifier capable of supplying a total of up to 180 amps to the cell. The current was adjustable by varying the voltage of the power supply.

Membranes

CR-61 membranes supplied by Ionics Incorporated, (Watertown, MA), were the only kind used in our electrodialysis experiments. Nafione membranes were mounted on electrodialysis cell frames and in some cases tested for leaks, but we did not have the opportunity to use them in an actual electrodialysis run.

Ionics' CR-61 membranes are cation-selective membranes comprising sulfonated copolymers of vinyl compounds. The membranes are homogeneous films cast in a sheet form on a synthetic cloth backing. Thickness of these membranes is 20–25 mils. Ionics' Bulletin No. CR-61.0C gives the properties and characteristics of all Ionics' CR-61 membranes.

We used Ionics' CR-6170 and CR-61 CZL-183 membranes as cation permeable membranes to exclude anions from the cathode and anode compartments, respectively.

CR-6170 is a polypropylene-backed, cation-transfer membrane use in electrodialysis and electrolysis operations as a means for transporting cations while excluding anions. It is chemically resistant to both acids and bases and due to the presence of both strong and weak acid groups, it is said to give good conductivity at any pH value.

The properties and characteristics of CR-6170 given in the manufacturer's Bulletin CR-6170 0-C are reproduced in Table 11.3. In addition, water content and cation exchange capacity are 38.2% and 2.24 meq/dry gram resin, respectively.

Ionics' CR-61 CZL-183 is a Dynel-backed, cation-transfer membrane. Properties and characteristics of this membrane given in Ionics' Bulletin No. CR-61.4-C are reproduced in Table 11.4.

Du Pont's Nafione membranes are perfluorinated ion exchange resins fabricated into reinforced and unreinforced membranes. Nafione Membrane 425, which we used in some of our tests, is a homogeneous film 5 mils thick of 1200 equivalent weight perfluorosulfonic acid resin laminated with T-12 fabric of Teflone TFE resin to make a 50 mil membrane. Properties and characteristics of Nafione membranes are given in reference 2.

All membranes should remain wet and, in the case of Ionics' CR-6170, the liquid should be alkaline. The Ionics membranes will crack upon drying and the Nafione will shrink and break away from its frame.

Materials of Construction

Parts and materials of construction of the electrodialysis apparatus are given in Appendix C.

Solutions

The following solutions were used:

- Electrolyte: 0.06 *N* Na_2SO_4 aqueous solution (tap water) — Lower Na_2SO_4 concentrations were used in Runs No. 10 (0.04 *N*) and No. 11 (0.001 *N*) to study the effect of electrolyte concentration. No electrolyte was used in Run No. 14.
- Anolyte: 10% NaOH (ca. 2.5 *N*) aqueous solution.
- Catholyte: 5% H_2SO_4 (ca. 1 *N* H_2SO_4) aqueous solution.
- Feed: dilute Du Pont Sodium Silicate No. 20: 360 g SiO_2/l, 110 g Na_2O/l (3.6 *N* NaOH) prepared by diluting 990 ml of sodium silicate No. 20 (40 g SiO_2/100 ml or 28.4 percent weight SiO_2) to 1000 ml with H_2O. Specific gravity of this solution is 1.386 g/ml.

Measurements

The following measurements were made and recorded during the electrodialysis operation:

- Time
- Temperature of electrolyte-sol mixture
- Electrolyte-sol, catholyte and anolyte tank volumes
- Flow rate and pressure of electrolyte-sol, catholyte and anolyte entering and leaving the cell
- Feed rate of sodium silicate
- pH of sol-electrolyte
- Voltage and amperage of the electrodialysis cell and calculated electrical resistance of the electrolyte and current density in the electrolyte cell compartment
- Silica concentration of the electrolyte-sol
- Specific surface area of the silica sol
- NaOH concentration of the catholyte

Analytical Procedures

- Silica Concentration. The fluoride method [5] was used to determine silica in the electrolyte-sol.
- Specific Surface Area. Sears' titration method [6] was used to measure specific surface area of the silica in the electrolyte-sol.

TABLE 11.4
Comparison of Electrodialysis Silica Sol (Run No. 6) and Ludox® HS

	Ludox® HS-30 Consol East Chicago	Silica Sol Run No. 6 Electrodialysis Exp. Sta.
Specific Surface Area, m^2/g	198–258	230
Viscosity, cps at 25°C	3.7–6.4	5.1
%S (Theoretical Maximum 93)	86–88	90
Electron Micrographs	Discrete Spheres of Uniform Diameter	Discrete Spheres of Uniform Diameter

- Soluble Silicate. The molybdate method was used to measure soluble silicates [3].
- Silica percent in dispersed phase, %S, was measured using ICD Standard Method L750.500.
- Turbidity was measured using ICD Standard Method L750.800.
- Occluded sodium was measured using the method of reference [3].

Calculations

In some cases where soluble silicates could introduce an important error in the measurement of specific surface area, the results were corrected by determining soluble silicates and calculating the corrected specific surface area.

Equivalent average diameter of particles was calculated using the formula:

$$D(\text{nm}) = \frac{6000}{(\text{SiO}_2 \text{ dens.})\,(\text{SiO}_2 \text{ Specific Surface Area, m}^2/\text{g})}$$

and this value was used in turn to calculate the mass of the particles.

Number of silica nuclei or silica particles was estimated by dividing the silica concentration by the mass of a particle.

Operation

In a typical run operation started by circulating the sodium sulfate electrolyte solution, the sulfuric acid anolyte solution and the sodium hydroxide catholyte solution through their respective compartments in the cell. The electrolyte solution was then heated to the temperature specified for the run. Sixty milliliters of the sodium silicate solution were fed into the electrolyte and the electrolyte-sodium silicate mixture circulated to obtain a homogeneous solution. Silica concentration of the homogeneous solution was 0.21 g SiO_2 per 100 ml. Silica concentration at this point of the operation is referred to as "sodium silicate concentration at the time of nuclei formation" or SS_N. The pH of the solution was 10.35.

Power was applied and the voltage raised quickly to 11.2 V. Current reading was 160 amps. Operation was continued for 3 min by which time the pH value dropped to 9.5. At this time the voltage was 11.7 and the current was 95 amps. As operation was continued, the pH dropped to 9.0 and sodium silicate solution was fed at a rate of about 1350 ml/h to adjust and keep the pH constant at 9.0.

The operation was continued and small samples of both the electrolyte containing the silica being formed and the catholyte were periodically removed and analyzed.

When the silica concentration in the electrolyte solution reached a prespecified value, usually between 3.5 and 5 g SiO_2 per 100 ml, the solution was heated to the specified temperature in a period of time between 5 and 15 min which varied according to the temperature to be reached, while continuing the electrodialysis operation. The operation was continued at the specified temperature and pH 9 by adjusting the feed rate of the sodium silicate solution in the sol.

In a typical operation at 70°C during the 6.8 h operation, a total of 7600 ml of sodium silicate feed solution were added continuously and power applied with constant voltage while the current gradually increased from 95 to 156 amps and current density from 47.5 ASF (amps per square foot) to 78 ASF.

At the end of the run the current was turned off and the silica sol product was removed from the cell and allowed to cool down to room temperature.

Efficiency of the operation was determined by measuring the amount of silica and Na_2O recovered as percentage of silica and Na_2O introduced in the system as sodium silicate.

Samples were obtained at this time to measure silica concentration and specific surface area.

The silica sol obtained was treated with a mixture of a cationic and an ionic exchange resin to remove ions from the medium.

The pH of the double deionized sol was 3.4 and was then adjusted to 9.1 by adding enough 3 *N* sodium hydroxide solution while stirring vigorously in a blender. The sol was concentrated to the desired weight percent silica by evaporation and filtered through Celite filter aid. A sample was saved for characterization.

Table 11.5 is an example of the records kept for each of the electrodialysis runs. Table 11.6 summarizes equipment characteristics and conditions common to all electrodialysis runs included in this report.

Table 11.7 shows the variable conditions for the electrodialysis runs conducted in the course of this project and the specific surface area and concentration of the products.

RESULTS

Nucleation and buildup or growth of the silica particles during the electrodialysis process were followed by measuring silica concentration and specific surface area of the silica sol as a function of time from the moment electric current begins to flow through the sol-electrolyte solution. Average equivalent particle diameter of the silica particles and estimated number of silica particles

TABLE 11.5
Comparison of Electrodialysis Silica Sol (Run No. 7) and LUDOX® SM

	Ludox® SM		Electrodialysis silica sol (Run No. 7)			
	Intermediate	Product	Intermediate		Concentrated product	
			Sample 56	Sample 57	Sample 56	Sample 57
Specific Surface Area, m^2/g	380–420	320–400	430	390	323	295
Silica, wt%	11–12	29–31	10.6%	14%	30%	30%

at given moment were calculated on the basis of silica concentration, amorphous silica density (2.2 g/cc) and specific surface area. In some cases the relative percentage of soluble silica was substantial (very low silica concentration and/or very high pH). An appropriate correction in the calculations was made.

TABLE 11.6
Properties and Characteristics of Ionics Inc.'s Type CR 6170 Cation Transfer Membranes

Backing:	Polypropylene
Chemical Stability:	Unattacked by acids, bases, and stable to nonoxidizing solutions below pH 11. It may be used up to temperatures up to 55°C and sometimes up to 95°C depending on conditions and the nature of the media which it is contacted.
Physical Stability:	If kept wet, physical stability is excellent. Upon drying, the membrane shrinks and cracking may occur.
Backing:	Polypropylene (various types; see below)
Membrane Thickness:	0.025 cm to 0.065 cm depending on type of PP backing. 0.035 333 type PP nonwoven 0.040 372 type PP backing — woven 0.025 371 type PP backing — woven 0.065 374 type PP backing — woven
Burst Strengths:	250–500 psi
Water Content:	25–30% of wet resin only
Capacity:	Weak acid 2.60–2.70 meg/dgr Strong acid 1.50–1.90 meg/dgr
Area Specific Resistivity: 1000 Hz ohm-cm^2	0.5 N NaOH 8–10 0.01 *N* NaCl 15–20

Note: The property data are typical values only and no warranty as to such properties is given.

Results of the measurements and calculations for each electrodialysis are included in the data sheets and in the silica concentration versus particle size, specific surface area and number of particle plots kept in research notebooks E7073 and E7186.

Characterization of the products obtained in runs number 6 and 7 are given in Table 11.1 and Table 11.2 in comparison with Ludox® HS and SM, respectively.

DISCUSSION OF RESULTS

Influence of Process Variables

There are several factors that should be considered in designing specifications for the variable temperature process to obtain sols of desired silica concentration, particle size and stability.

Effect of temperature, temperature change and silica concentration of the heel are discussed subsequently. This project did not include a comprehensive study of variables. However, results obtained in the runs completed allow us to speculate on the effect of some of the other variables.

Effect of Temperature

The plots of Figures 11.5, 11.6, 11.7 show the effect of temperature on the growth of the silica particles by electrodialysis as the concentration of silica is increased by gradual addition of sodium silicate into the electrodialysis cell. In Figure 11.5 the specific surface area of the silica in m^2/g, as measured by the Sears titration method is plotted versus silica concentration in g/100 ml. In Figure 11.6 the average silica particle diameter in nm, as calculated from the specific surface area value assuming that the sol is made of dense spheres of amorphous silica with a density of 2.2 g/cc is plotted, versus silica concentration.

Higher temperatures produce silica sols of larger particle diameter. In the conditions of our experiments, running the electordialysis operation at 50°C, the product obtained when SiO_2 reached a concentration of 20 g/100 ml had a particle diameter corresponding to

TABLE 11.7
Properties and Characteristics of Ionics Inc.'s Type 61CZL183 Cation-Transfer Membranes

Ionics' Dynel-backed cation-transfer membrane 61CZL183 has a tight matrix. Therefore, it is suggested for use in electrodialysis as a means for transporting cations while excluding anions in applications where loss of solvent and of non-ionized product must be reduced below that obtainable with membrane 61AZL183.

Backing-Type:	Dynel		Specific Weight:	14 mg/cm^2
Weight:	4 oz/yd^2		Content:	34 wt% (dry)
Membrane Thickness:	24 mils (0.6 mm)			
Burst Strength (Mulen):	115 psi (8 kg/cm^2)			
Water Content:	40% of wet resin only			
Capacity:	2.7 meg/dry gram resin			
	0.01 *N* NaCl	0.1 *N* NaCl	1.0 *N* NaCl	3.0 *N* NaCl
Area Specific Resist. (ohm-cm^2)	13	11	8	5
Spec. Conductance (moh/cm)	5×10^{-3}	6×10^{-3}	8×10^{-3}	12×10^{-3}
Sucrose Transport % (g/Faraday)	16 ma/cm^2, 30% sucrose in 0.2 *N* KCl/0.02 *N* KCl			

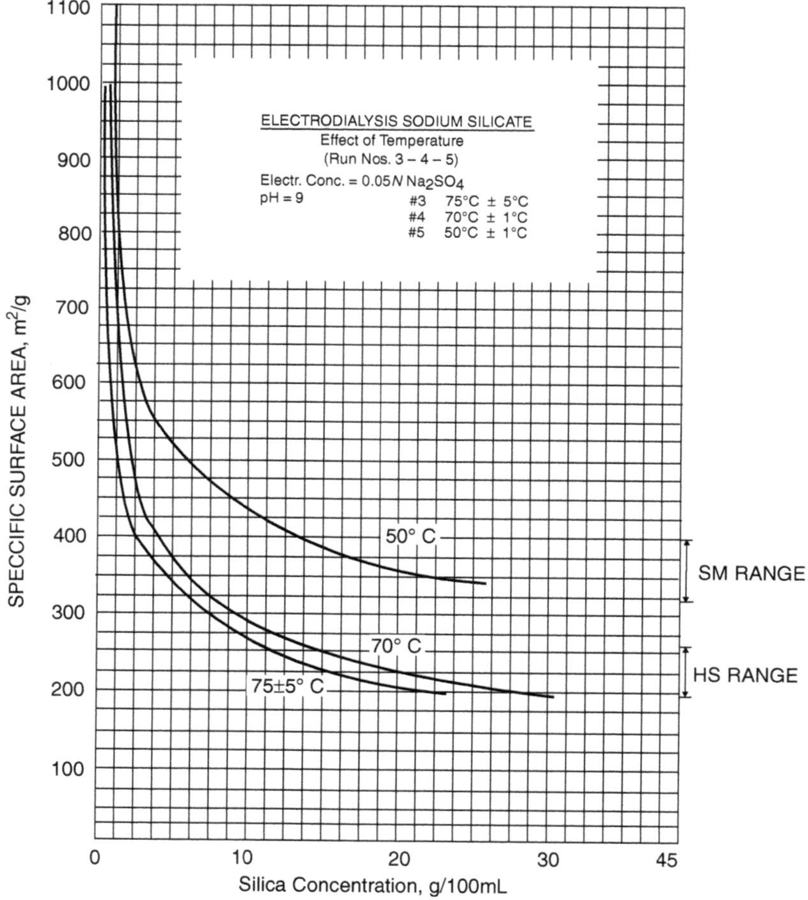

FIGURE 11.5 Specific surface area of silica in m^2/g versus silica concentration in g/100 ms as measured by sears titration method.

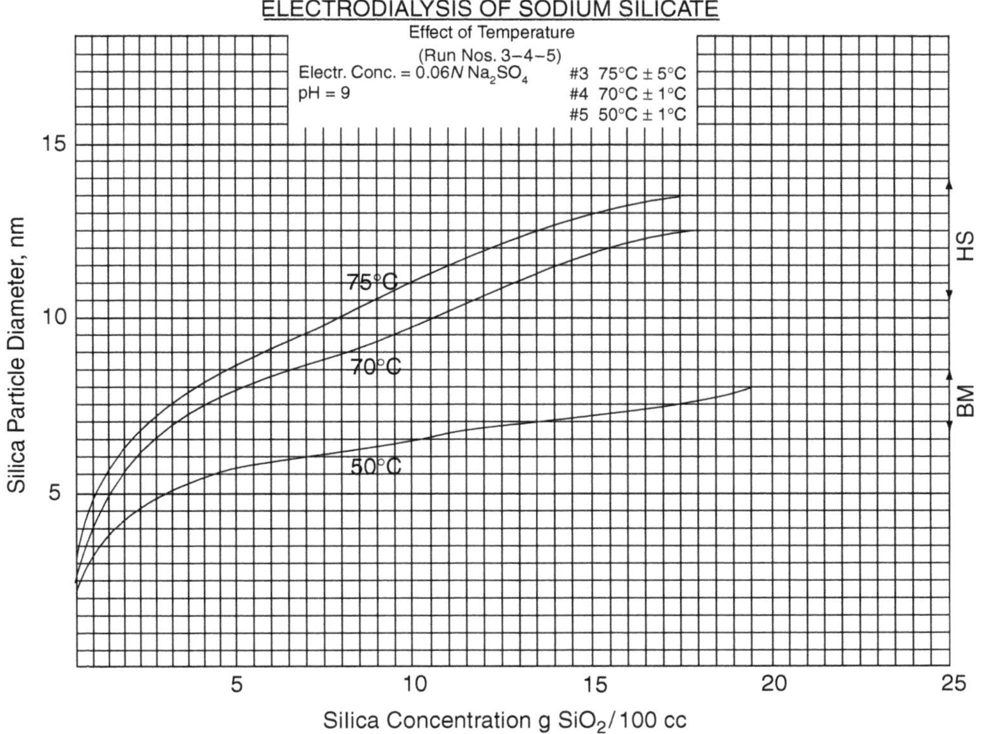

FIGURE 11.6 Average silica particle diameter versus silica concentration.

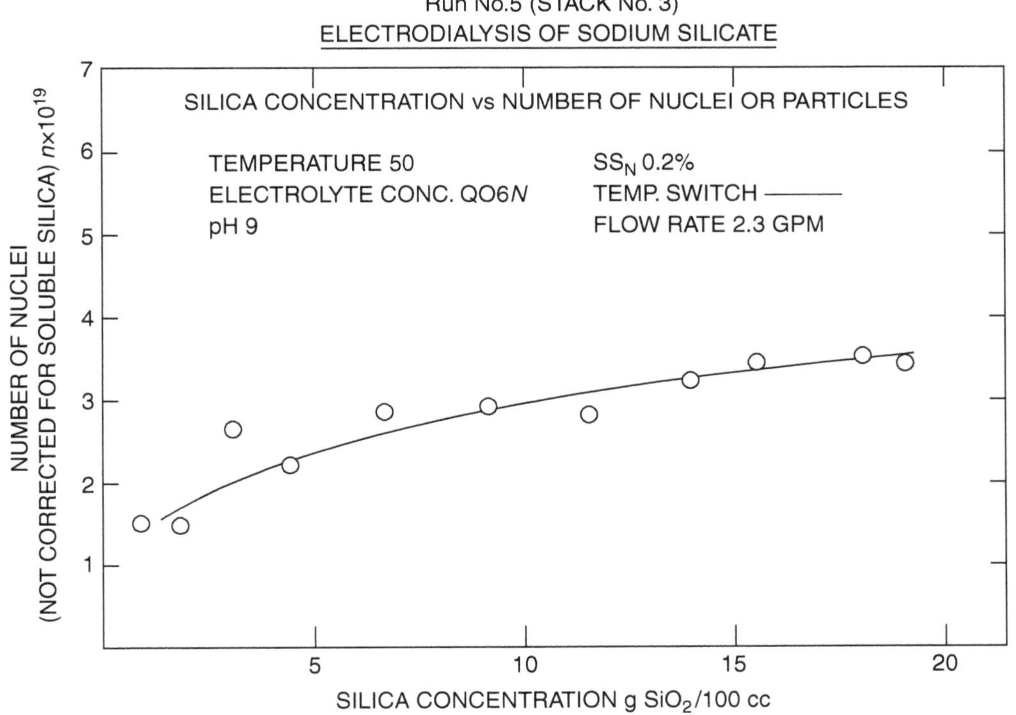

FIGURE 11.7 Calculation of the number of particles as a function of $\%SiO_2$ in run no. 5 at 50°C.

TABLE 11.8
Example of Electrodialysis Run Data Sheet Preparation of Silica Sol by Electrodialysis of Sodium Silicate Silica (grade: 12 nm particle size)

Run No. 6
Code E7168–10

Nucleation Temperature: 50°C

Growth temperature: 70°C

SS_N: 0.21

Silica concentration at temperature step-up: 3.3%

Electrolyte concentration: 0.06 *N* sodium sulfate
Electrolyte-sol volume: 10 l
Cell outflow pH: 9

Concentration of feed solution: 36 g SiO_2/100 ml

Total silica made: 2400 g
Total NaOH made: 24.52 equivalents
Total time: 6.78 h

Sample no.	Time, min.	Temp. °C	A amp	V volts	W watts	S w/v	S.A. m^2/g	D nm	Calculated number of silica particles per 100 ml of solution $\times 10^{19}$
1	20	49	95	11.7	1111	1.0	790	3.4	2.0
2	39	52	105	11.7	1228	1.89	668	4.05	2.3
3	70	51	107	11.7	1252	3.37	520	5.2	1.97
4	97	71	136	11.5	1115	4.85	462	5.85	1.99
5	138	71	141	11.5	1587	7.2	405	6.66	2.0
6	185	71	146	11.4	1664	10.5	358	7.5	2.04
7	232	71	146	11.4	1664	13.6	344	7.85	2.31
8	261	71	147	11.4	1676	15.6	321	8.45	2.11
9	303	72	147	11.4	1676	18.5	302	8.95	2.12
10	337	74	146	11.4	1664	21.3	284	9.5	2.04
11	380	73	146	11.4	1664	23.4	275	9.85	2.01
12	407	71	156	11.5	1794	25.4	262	10.3	1.86

Silica Yield = 87.7%
NaOH Yield = 89.6%

that of Ludox® SM. At 70°C when the SiO_2 reached the same concentration the product had a particle diameter corresponding to that of Ludox® HS.

According to calculations based on measured specific surface area, SiO_2 concentration, and extrapolation to the SS_N value (concentration of silica in the sol when silica nuclei begin to form), a larger number of smaller nuclei form at the lower temperatures. Since the incoming silica has to deposit on the surface of a large number of smaller nuclei, the addition of the same amount of silica produces smaller particles at lower temperatures.

The surface and plots show a steep drop below 2% SiO_2 followed by a more gradual decrease as higher SiO_2 concentrations are reached. The shapes of the curves indicate that continuing the electrodialysis to 25–30% SiO_2 sol concentration at 70°C or higher would give products with larger particle size than Ludox® HS. On the other hand to obtain Ludox® HS at 50°C, electrodialysis would have to stop at about 20% SiO_2.

Calculations of the number of particles as a function of % SiO_2 in run No. 5 at 50°C show that new nuclei are being formed as the process progresses (see Figure 11.7).

The shape of the surface-derived curves suggests that larger particle size products would be obtaind by nucleating at low temerpatures such as 30–50°C (therefore forming a large number of very small nuclei) and gradually switching to higher temperatures (70–90°C) to accelerate the growth of the particles. The approximate point of % SiO_2 at which temperature should be switched could be determined on the basis of the slope of the SiO_2/% SiO_2 curves at the temperature selected for the growth of the particle. Shifting the SA — % SiO_2 curve in the plot of Figure 11.9 (Run No. 4 at 70°C) to make the end point fall in the Ludox® HS surface dried range at 30% SiO_2 suggests that if the sol was nucleated at 50°C, a shift to 70°C at 3 to 5% SiO_2 should give a sol of S.A. corresponding to Ludox® HS at 30% SiO_2.

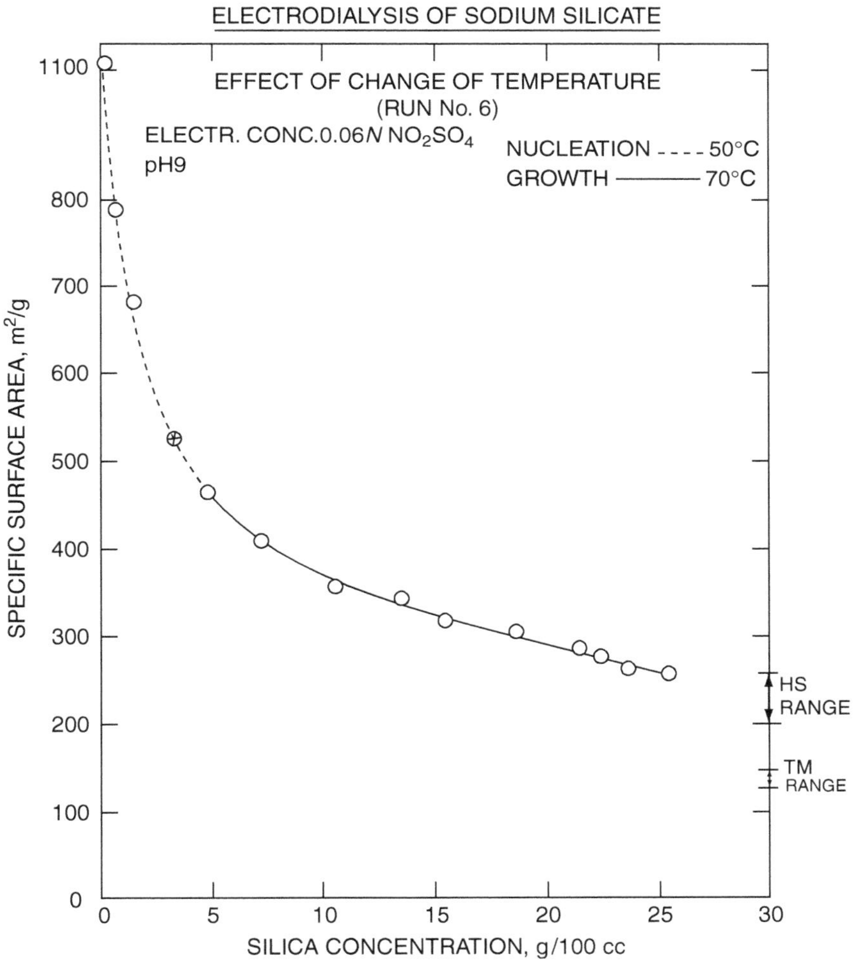

FIGURE 11.8 Specific surface area versus %SiO_2.

Effect of Sodium Silicate Concentration in the Nucleation of Silica Sols

The diameter of the silica nuclei or particles formed at the initial nucleation stage appears to be independent of sodium silicate concentration in the range of 0.6 (perhaps 0.3) to 4% SiO_2. This conclusion is based on nucleation results obtained by deionizing aqueous sodium silicate with Amberlite IR-84S to pH 9 at 20 and 35°C.

The results are given in Table 11.8. The table gives silica particle diameter calculated from the specific surface area obtained by the Sears titration method after soluble silica correction. Values obtained at 0.3% SiO_2 are subject to more error due to higher amount of silica monomer present relative to colloidal silica. Results obtained by electrodialysis nucleation and by acidification with H_2SO_4 are included for comparison.

These results should guide in determining the optimum concentration of sodium silicate to use in the first nucleation step of the variable temperature process.

Effect of Change of Temperature During the Process

Figure 11.8 (specific surface area vs. %SiO_2 plot) shows results obtained by nucleating silica by electrolysis at 0.21% SiO_2 and 50°C, continuing the electrodialysis until a 3.5% SiO_2 concentration is reached and then heating the cell to 70°C while continuing the electrodialysis (Run No. 6).

A comparison of plots of runs 4, 5, and 6 (Figure 11.10 and Figure 11.9) illustrates the advantage of this temperature switch which makes it possible to obtain 30% SiO_2 sols of particle size in the range of Ludox® HS.

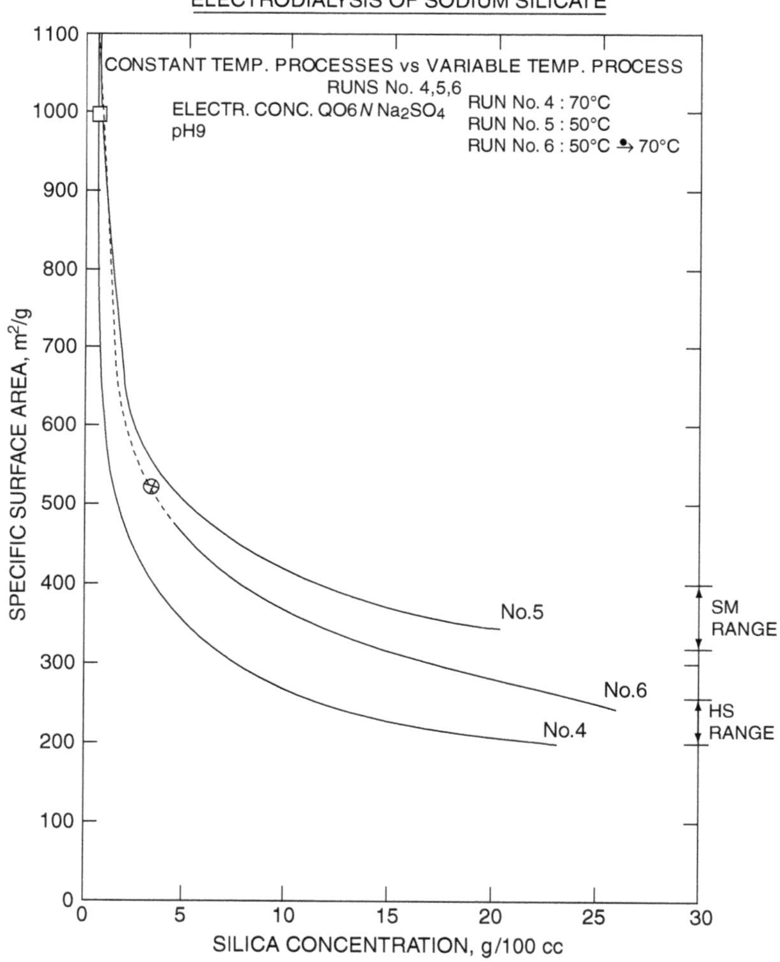

FIGURE 11.9 Constant temperature versus variable temperature process – a comparison of plots of runs 4, 5 and 6 in terms of specific surface area.

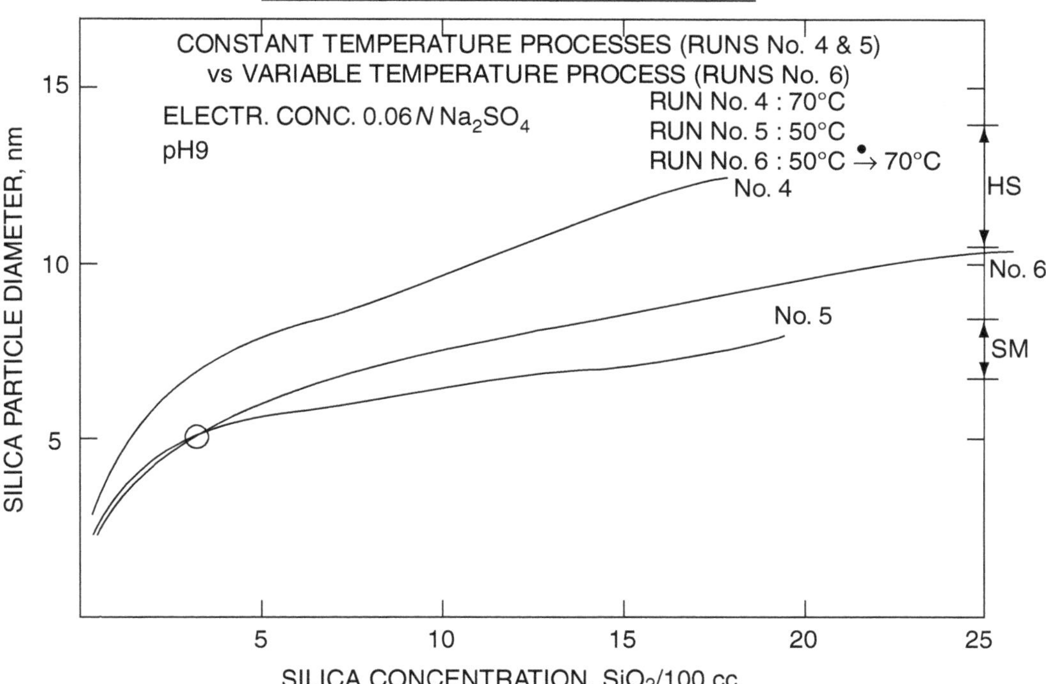

FIGURE 11.10 Constant temperature vs. variable temperature process — a comparison of plots of runs 4, 5 and 6 in terms of silica particle diameter.

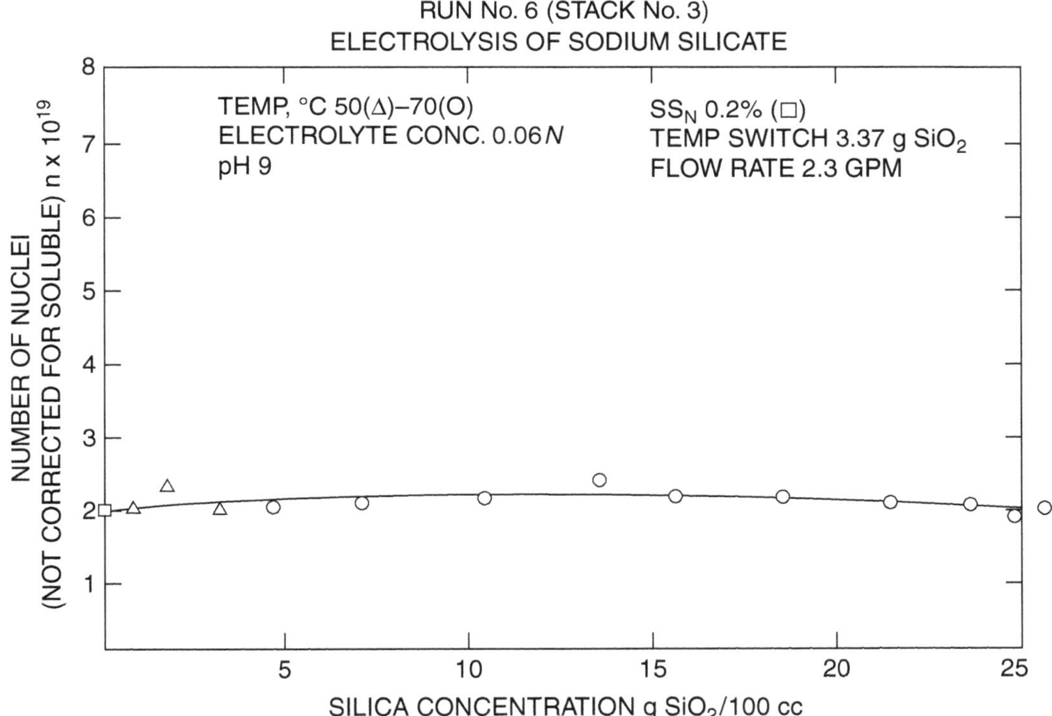

FIGURE 11.11 Electrolysis of sodium silicate (Run no. 6).

Calculations of the number of particles in the sol based on measured specific surface area and silica concentration show that contrary to the case of Run 5 at constant temperature 50°C, in the case of Run 6 at variable temperature the number of particles remain constant throughout the electrodialysis up to a concentration of 25% SiO_2 (see Figure 11.9). The fact that the number of particles remains constant after the initial formation of the nuclei suggests that silica added during the process contributes predominantly to the growth of the original nuclei without new nuclei being formed during the operation. Sols formed under these conditions are expected to have more uniform particle size distribution than those made with continued formation of new nuclei during the process.

Concentration of Sodium Silicate at the Time of Initial Nucleation, SS_N, and at the Point of Temperature Increase

Assuming that for a given temperature the size of the nuclei is independent of sodium silicate concentration, as suggested by the deionization experiments described above, SS_N will determine the number of nuclei formed at the start of the process. Other conditions such as rate of addition of sodium silicate to the heel being equal, the growth rate would be faster for lower SS_N. This is because the same amount of added silica will be distributed in fewer nuclei.

When temperature is increased growth rate increases. Therefore the decision of when to increase temperature based on colloidal silica concentration should be made by previously analyzing the effect of SS_N on the process at the selected temperature.

APPENDIX A

STACK TYPE ELECTRODIALYSIS CELL DESIGN

The electrodialysis cell used in this project was supplied by Ionics, Inc., Watertown, Mass., and modified in our laboratory. The original cell was designed by Ionics on the basis of specifications supplied by R.K. Iler

TABLE 11.9
Equipment and Conditions Common to Electrodialysis Runs Included in This Report

Electrodialysis Cell	
Type:	Ionics, Inc. "Cell pair"
Size:	One square foot
Cathode:	Ni plated steel
Anode:	$TiSO_X$ coated Al alloy
Cathode Membrane:	Ionics CR 6170
Anode Membrane:	Ionics CR 61–CZL–183
Membrane Spacing:	1/4″
Experimental Conditions Materials	
Heel:	0.18 to 2.0 SiO_2 (depending on the run) in 0.06 *N* Na_2SO_4 aqueous solution (more dilute solutions in three special runs)
Intake (Feed):	Sodium silicate 3.25 ratio, 36 g/100 g SiO_2
Sol-Electrolyte:	
Anolyte:	10% NaOH (ca. 2.5 *N*) aqueous solution
Catholyte:	5% H_2SO_4 (ca. 1 *N* H_2SO_4) aqueous solution
Operation	
Circulation Rates:	Sol-electrolyte ca. 3 gl/min
	Intake
	Anolyte
	Catholyte
Temperature (Sol-electrolyte):	Varied for each run
pH (Sol-electrolyte):	9
Current: Voltage:	ca. 11.5 V
Intensity:	ca. 85 Amp at the start, ca. 155 Amp at the end of most runs with 0.06 *N* Na_2SO_4 as electrolyte
Density:	ca. 42 SF at the start, ca. 75 ASF at the end of most runs with 0.06 *N* Na_2SO_4 as electrolyte

TABLE 11.10
Effect of Sodium Silicate Concentration On Nuclei Diameter of Silica Made by Deionization with Ion Exchange Resins, Electrodialysis and Acidification

SS_N sodium silicate concentration, % SiO_2	Nucleation Process			
	Ion exchange resin	Electrodialysis	Acidification	
	Nucleation temperature			
	35°C	35°C	30°C	30°C
	Nuclei Diameter, nm			
0.3	2.8	N.A.	N.A.	N.A.
0.6–0.7	3.7	3.15– 3.4– 3.15	N.A.	3.2
1	4.2	N.A.	2.5	N.A.
2	3.9	N.A.	2.6	N.A.
4	3.8	–	–	–

in 1973–1974 (see Appendix A). The cell as received in November 1974 was not operational under the specifications of Iler's colloidal silica process due to the type of materials and design used by the manufacturer.

Since making a new cell would have taken longer than the time assigned to this project, we only modified the Ionics cell so that it could operate for period of time long enough to provide useful data and to prove the feasibility of the colloidal silica electrodialysis process under specific conditions. A description of the cell used in our experiments is given in the Experimental section of this report (see Apparatus). Specific problems encountered in the original Ionics cell are described in detail in Appendix F. We discuss subsequently possible alternatives in cell design based on the basic problem of the Ionics design.

The stack type Ionics electrodialysis cell was put together according to the manufacturer's instructions. Preliminary tests with hot water developed external leaks which were corrected. Further tests with hot sodium silicate and in actual electrodialysis operation developed both internal leaks and new external leaks. The nature of some of the leaks was inherent to the design of the cell and could not be corrected without changing the design of the cell.

The problem with the cell design is the concept of circulating the three liquids (electrolyte-sol, catholyte and anolyte) through both sides of the membranes. To work at our specified temperature and flow rate, a cell of this design would require a high torque on the stack, rigid, solid spacers and frames, and strong membranes. The Ionics cell had only six tie bolts and therefore under our working conditions could not provide sufficient torque to prevent internal and external leaks through the laminated spacers and frames with which it was equipped. In addition, the membrane in the cathode compartment was not sufficiently rigid and under the hydraulic pressure of the circulating fluids tended to pull the sponge rubber frame into the cell compartments, thus developing internal leaks. Since the outside perimeter of the membranes was smaller than the outside perimeter of the stack, and not much larger than the perimeter of the cell compartments, little sliding of the membrane was needed before the edge reached the border of the liquid compartment and opened a passage to the continuous compartment.

AMJ has a completely new design comprising the use of tubular cells. A one-membrane prototype unit was built by AMJ and successfully tested as a nickel recovery system by Special Products. It appears that the tubular design could be adopted to make two-membrane cells for our electrodialysis process. Details are given in Appendix B.

TUBULAR CELL DESIGN FOR ELECTRODIALYSIS BY AMJ CHEMICAL, INC.

AMJ, located in Great Neck, NY, sells industrial equipment for electrodialysis and Donnan dialysis. Its special products division has tested successfully a prototype unit designed and built by AMJ for nickel recovery systems. At the time we were working in electrodialysis (1974), AMJ was engaged in further development work with Special Products under a secrecy agreement.

AMJ Electro Cell System

AMJ designs and sells the AMJ Electro Cell System for plastic preplate etch. This is an electrodialysis operation that requires the reoxidation of the reduced chromium in solutions that contain pure chromic acid.

The AMJ Electro Cell System is an electrodialysis unit with a tubular design claimed to be free of leaks and more efficient than the conventional rectangular stack design. This is a two-compartment system which according to AMJ can be adapted to make the three-compartment cell required by our sodium silicate electrodialysis process to make colloidal silica.

It appears that cells with the AMJ tubular design would be simpler to put together and to service, more rugged to operate and would occupy less space than the rectangular stack cells.

AMJ claims that temperature of operation, flow rate of fluids, thickness of electrodyalizate compartment, materials for electrodes, membrane supports and separators for our process should not offer any problems.

AMJ offered to build a 15 square foot unit guaranteed to perform as they claim which we could test for 30 to 60 days with their help. The cost of the electrodialysis (ED) cell as such would be around $18,000. Fifty percent would be paid on delivery and the balance would be paid only if the cell performs as guaranteed.

The cell as designed would be the first unit of a series of identical cells that could be linked to the system in case the operation is expanded.

The sketch below illustrates the AMJ design for an electrodialysis cell.

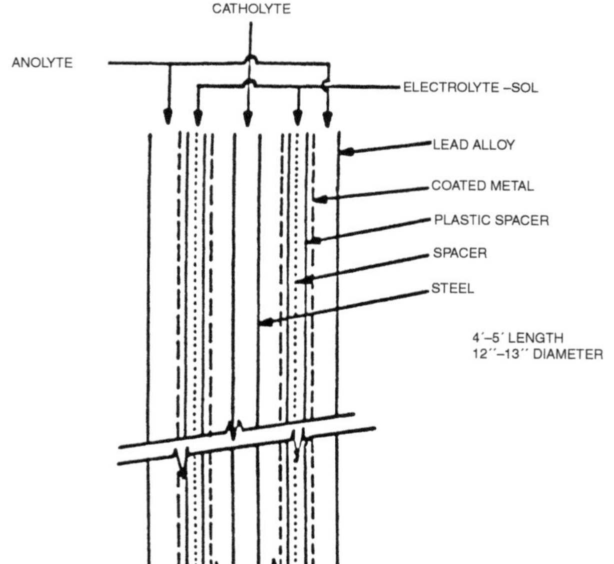

APPENDIX B

ELECTRODIALYSIS COMPANIES

USA

- Ionics, Inc., Watertown, Mass. This is the company that furnished our 2 square foot lab cell.
- Aqua-Chem, Waukesha, Wisc., a division of Coca-Cola. This is the company that will demonstrate for us in their labs a unit operating at 80°C and 5 g/min of about the size of ours.

Great Britain

- William Boby and Co., Hertfordshire, England, WD 31 HP. Mr. Lacey said he had not much success in getting any experimental data from this company.

Japan

- Mitsubishi Shoji Kaisha, Ltd.
  20 Marunouchi-2-Chome
  Chiyoda-Ku, Tokyo, Japan
- Asahi Chemical Industry Co.
  Kaisha, Japan
- Tokuyamo Soda Co., Ltd.
  Yamaguchiken, Japan
- Asahi Glass Co., Tokyo, Japan

APPENDIX C

COMPONENTS OF ELECTRODIALYSIS APPARATUS–MATERIALS OF CONSTRUCTION

H.E. Bergna–H.J. McQueston

The following list includes all components of the ED apparatus and gives the materials of construction used to fabricate them.

Tanks

1. One 5 gallon hold tank 304 SS with glass sight tube. E
2. Two 2 gallon polyethylene tanks. One for anolyte and one for catholyte.

PUMPS

1. Eastern two-stage centrifugal pump, model 2J, grade 316 SS to circulate electrolyte. E
2. Masterflex tubing pump variable speed drive with solid state speed controller, cat. No. 7545, to pump the sodium silicate into the system. SS
3. Two magnetic drive pumps, polypropylene housing, The Little Giant Pump Co., model F33 HXEHE-2027, one to pump the catholyte, the other one to pump the anolyte. A–C

HEAT EXCHANGERS

1. American Standard SSCF, 4.3 sq ft, grade 216 SS. E
2. 2–12″ (glass NLL7HN condensers. A–C
3. 304 SS 1/4″ (tubing coil, 7 rings 2″ × 3/4″. 7H

ROTOMETERS

1. 2 — Brooks Type 1110-09, grade 316 SS E
2. 2 — Brooks Type 1110-08, grade 316 SS A–C
3. 2 — Brooks Type 1122-1355, grade 316 SS SS

GAUGES

1. 4 — 0–100 psi Ashcroft Co., 2 1/2 in. (dial 316 SS E
2. 2 — 0–30 psi Ashcroft Co., 3 in. (dial chemical seal-Teflon® diaphragm C
3. 2 — 0–30 psi Ashcroft Co., 3 in. (dial chemical seal-Hastelloy C diaphragm A

pH METER

1. Orion model 401
2. (a) Electrodes: Fisher Calomel Reference Cat. No. 13-639-52
   Beckman High Sodium Cat. No. 39301

RECORDER

1. Honeywell Brown Electronik 0–1 mv over 10 inches (to monitor pH).

POWER SUPPLY

1. W.A. Reynolds Corporation (Phila., Pa.) type 300.12 Reycor silicon rectifier with self-contained powerstat control providing 300 amps at 12 V DC output with a 230 V, 3 phase, 60 cycle AC input. Unit had maximum of 5% ripple and was fan cooled. Cabinet size 22 in. *W* × 17 in. D × 241/2 in. H.
2. Phelps DO4GF 4/0 ang. cable wire to electrodialysis cell.

ELECTRODIALYSIS CELL

1. Ionics, Inc., electrodialysis cell, two-three compartment cells, approx. 2 sq ft cathode area. 2 Nickel coated steel cathodes — one common center $TiSO_X$ coated Al alloy anode.
   Membranes: 2 — Ionics CR-6170 (cathode side)
   2 — Ionics CR-61-CZL-183 (anode side)
   Cell construction material CPVC.
1. Electrolyte Circulating System
   (a) 1/2″ 304 SS pipe tubing and fittings with 1/2″ Apollo 316 SS ball valves (used in heat-up portion.)
   (b) 1/2″ polyethylene tubing with fast and tight polypropylene fittings and 1/2″ Apollo 316 SS ball valvee (from heat exchanger to electrodialysis cell.)
   (c) 1/2″304 SS pipe + 3/4″ID × 3/16″ wall black rubber tubing from electrodialysis cell to hold tank.
2. Anolyte and Catholyte Circulating System 3/8″ polyethylene tubing with fast and tight polypropylene fittings.
3. pH Monitor System
   (a) Gum rubber tubing 3/16″ ID, 1/16″ wall.
   (b) 1/4″ 304 SS tubing coil.
   (c) 1/4″ 304 SS ball valve.

APPENDIX D

OPERATION OF ELECTRODIALYSIS CELL H.E. BERGNA–H.J. MCQUESTON

This Appendix provides a checklist for the operation of the electrodialysis cell described in the Experimental section of this report. It is most important to check the cell before operation to make sure that it has been assembled properly and to keep an even torque on the cell tie bolts (250 in.-lbs) during operation to prevent external leaking.

Water Cleanup of Cell Compartments

1. The electrolyte, anolyte and catholyte hold tanks and cell compartments are drained.
2. Rubber tubing is connected to the exit lines of the cell compartments and run down a sewer drain.
3. Tap water is charged to all hold tanks.
4. The electrolyte circulating pump is turned on and the inlet valve to the cell is opened just enough to move the rotometer float. Pumping is continued in this fashion until the cell compartment is full and water is running down the sewer drain.
5. The anolyte and catholyte pumps are turned on in a like manner.
6. When all the cell compartments are full and water is running down the drain the circulation rate is increased to 0.1–0.2 gal/min in the anolyte and catholyte compartments and 2–2.25 gal/min in the electrolyte compartment.

   These circulation rates give about a one psi difference between the electrolyte compartment and the anolyte and catholyte compartments. The difference is required to keep the membranes apart and against the membrane screens, thus maintaining a constant distance between membranes.
7. Water is pumped through the system until the effluents are clear water white and neutral to pH paper. Then the pumps are turned off and the system is drained.
8. The return lines to the hold tanks are reconnected.

Alkali Cleanup of Electrolyte Compartment

System startup same as described in water cleanup of cell compartments.

1. The electrolyte system is drained and rinsed free of silica sols by pumping water through system until the effluent is clear.
2. Charge 15–18 l of 2% NaOH solution to the electrolyte hold tank. Start circulating.
3. Anolyte and catholyte circulation is started.
4. When all systems are circulating the steam valve to the electrolyte heat exchanger is opened to 3 psi inlet pressure, 0 psi outlet pressure.

   The alkali solution is heated to 75–80°C and circulated through the system for 1–1½ h. Then the steam is turned off and the system is washed free of alkali with water.

Start-up of Drained System

1. 6 l of 5% H_2SO_4 (1.02 N) solution and 6 l of 10% NaOH solution are charged to respected anolyte and catholyte hold tanks.
2. 10 l of H_2O containing sufficient Na_2SO_4 to give the desired electrolyte normality are added to the electrolyte hold tank.
3. The electrolyte pumps are turned on using same method described under water cleanup.
4. Sufficient dilute "JM" grade sodium silicate (36 g SiO_2/100 ml) is added to the circulating electrolyte to give the desired nucleating silica concentration.
5. Circulation rates and psi adjusted to same values as in Step 6 water cleanup, that is 0.1–0.2 gal/min anolyte, catholyte and 2–2.5 gal/min electrolyte with a one psi pressure difference between the membranes.

Under these operating conditions the volume in the hold tanks for anolyte and catholyte will be about 5.2 and 7 l for the electrolyte.

Should any sudden change in volume occur and there are no visible external leaks, check to see if there has been a change in psi between cell compartments. This allows the membranes to flex, thus causing a change in volume. If no change in psi has occurred check for internal leaking.

We observed deformation of the CPVC flow path inserts when we ran the cell with water at 70°–80°C. This may mean that CPVC would not be an adequate material to use in cells under our specified conditions of maximum temperature (80°C). Our changed design involved the use of Lucite® flow path inserts, but we did not establish what effect prolonged use of the cell at 80°C would have on the CPVC frames coated with sponge rubber and on the CPVC end block.

All the troubles developed when we were testing the equipment with only cold water or cold or hot salts, acid or base solutions.

REFERENCES

1. Iler, R.K., The chemistry of silica, John Wiley & Sons, 1979.
2. Grot, W.G.F. et al., Perfluorinated Ion Exchange Membranes, 141st National Meeting of the Electrochemical Society, Houston, May 1972.
3. Silicomolybdate-Yellow Method for Souble Silica Based on G.B. Alexander, J.A.C.S. 75, 5655 (1953).

4. Iler, R.K., Method of Producing Colloidal Silica by Electrodialysis of a Silicate, U.S. 3,668,088, June 6, 1972.
5. Iler, R.K., Electrodialysis Process for Making Ludox® Colloidal Silica, Dec. 1, 1973.
6. Sears, G.W., Surface Area by NaOH Titration of Ludox® Colloidal Silica *Anal Chem.*, 28, 1981 (1956).
7. Matijević, E., ed. Surface and Colloid Science, Wiley & Sons, New York, 1973.

12 Manufacturing and Applications of Water-Borne Colloidal Silica

William O. Roberts
DuPont Company (retired)

CONTENTS

INTRODUCTION

Silica is all around us. We are most used to seeing it as sand at the beach and in the form of glass, or as a component of the stones and concrete that surrounds us. In these forms silica appears to be quite inert. The aqueous colloidal silicas are by contrast quite reactive. Jet airplanes, computers, photographic film, steam irons, food wrappers, the space shuttle, cardboard boxes, the wine you had with dinner, and even the knee replacement you might need someday all exist or are a little better because of colloidal silica.

Many will define "colloidal" as just a size range without regard how the silica came to be in that size range. For example some will define "fumed" silica, which is produced by burning $SiCl_4$ in oxygen to give very finely divided silica, as being "colloidal silica." However, I take a much narrower view and this chapter deals exclusively with water-borne colloidal silicas that have been generated by polymerizing the particles from monomer and small silica units in aqueous medium. They are delivered to the user in water. If they are to be dried in order to accomplish their purpose, the user is the one who does it.

Once they are dried, they become much more like the inert forms of silica we spoke of above. Once dried, the particles can not be re-generated except by dissolving them to form sodium silicate and starting the production process all over again. While they may retain a fraction of the chemically active surface they had as monodisperse particles they are not the same.

The numbers of applications which colloidal silica is involved in is surprising. I listed a few above that we will touch on in the Applications section of this article. We will not exhaust the applications, but will try to cover a representative variety. Over the years the history of commercial colloidal silicas has been that for every use that disappears, two more are generated.

MANUFACTURING

Most commercially viable manufacturing processes of aqueous colloidal silica use an alkali silicate, such as sodium or potassium silicate as the source of silica. Most, or all, of the alkali is removed from the solution by ion exchange or electro-dialysis and is replaced by hydrogen ions thus forming silicic acid. The silicic acid is then polymerized to give particles of colloidal silica. This is not to imply that other processes cannot make colloidal silica, many of which are described in Iler [1], but these have not been found to be useful on a commercial scale for the most part.

Because of its lower cost, compared to the other alkali silicates available commercially, sodium silicate is the preferred starting material. Sodium silicates are not stoichiometric compounds but are low melting glasses which vary in their ratio of silica:sodium oxide, (SiO_2:Na_2O) content from about 1:2 to about 3.25:1. Generally, the higher (>3) ratio silicates are generally preferred since a major cost in the conversion is the removal of sodium from the silicate. The use of the ratio of silica:sodium oxide is a common characterization of colloidal silica sols and is undoubtedly a carry over from silicate technology. The ratio is generally given as a weight ratio, but since the molecular weights of SiO_2 and Na_2O are nearly identical (about 60 and 62 respectively) the weight ratio is very close to the molar ratio. The mole ratio is usually a more useful number to think about for a chemist.

One might think that simple acidification of sodium silicate could produce the desired silicic acid and lead to a route to colloidal silica products. But such an approach leads to a solution containing not only the desired silica units, but also a very large amount of the sodium salt of the acid used for the neutralization This high level of salt makes it virtually impossible to produce stable colloidal sols. In fact, one of the easiest ways to destabilize colloidal suspensions is to add significant amounts of electrolytes. Hence all highly successful approaches involve methods whereby sodium is removed from the silicate and held apart from the silicic acid units. Most commonly, this is done using ion exchange resins, but routes using electro-dialysis have also been developed.

In general terms, the processes used to make colloidal silica involve the following steps:

- The removal, by ion exchange, dialysis, or Electro-dialysis, of most or all of the sodium hydroxide from a solution of sodium silicate converting it to silicic acid and small oligomers of silicic acid.
- Adjusting the alkali content to give a silica:alkali (SiO_2:Na_2O) ratio appropriate to the desired finished particle size.
- Polymerization under controlled conditions to give colloidal silica particles.
- Concentration to give normal shipping concentrations.

These basic principles apply to virtually all the processes devised to produce colloidal silica commercially. However, a wide variety of variations on these principles have been devised over the years to produce enhanced effects such as larger and smaller particles, and broad or narrow particle size distributions, depending on the properties needed for a particular end use. The large variety of end uses that have been developed for colloidal silica is one of the most interesting aspects of working with these products. But we'll leave that discussion for later.

SODIUM SILICATES

Heating silica sand and sodium carbonate together in a furnace to about 1000–1200°F produces Sodium Silicate glasses. During the heating, carbon dioxide is given off and the resulting sodium oxide and silica react to give a low melting glass. The higher the alkali content the lower the melting point of the glass. Also, the higher the alkali content the more readily water-soluble is the glassy product. Because of it's glassy appearance and its solubility in water, sodium silicate has historically been known as "water glass."

Generally, a premix of silica and sodium carbonate is fed into one end of a furnace. Baffles and the high viscosity of the molten glass prevent rapid mixing of the incoming materials with the molten glass. The material after traversing the furnace overflows through an opening in the far end of the furnace and is typically caused to fall onto a water-cooled conveyer belt. The glass is rapidly cooled to a solid and the thermal shock of the rapid cooling causes the material to be broken into pieces which can be easily conveyed and fed to the dissolver.

Potassium silicate glasses are soluble in boiling water, but high (>3) ratio sodium silicate glasses cannot be dissolved in boiling water at atmospheric pressures. However, under autoclave conditions the solution rate is practical and hence sodium silicate dissolvers are typically pressure vessels. High-pressure steam provides both water and heat to dissolve the glass. Silicate glasses of 3.25 ratio can be dissolved in water under autoclave conditions to give solutions containing up to about 30% SiO_2.

PRINCIPLES OF THE MANUFACTURING PROCESS

Silicic acid can be considered to be H_4SiO_4, a tetrahedral molecule such as shown in Figure 12.1. As implied by the term *acid* in the name, a few of the hydrogens can be ionized off as Hydrogen ions. Silicic acid is a weak acid having a pH of about 3. Silicic acid polymerizes by the raction of the OH groups to split out a water molecule and form a Si-O-Si bridge between corners of two tetrahedra. This process is illustrated in Figure 12.2. This process can be continued at any corner and since there is free rotation of the tetrahedra with respect to each other the particles can grow in three dimensions. This results in the formation of roughly spherical particles.

In the stoichiometric sodium salt of silicic acid, Na_4SiO_4, the silica:soda ratio is 1:2. In silicate solutions having a silica:soda ratio of 3.25, much less sodium hydroxide is present to ionize off the weak hydrogen

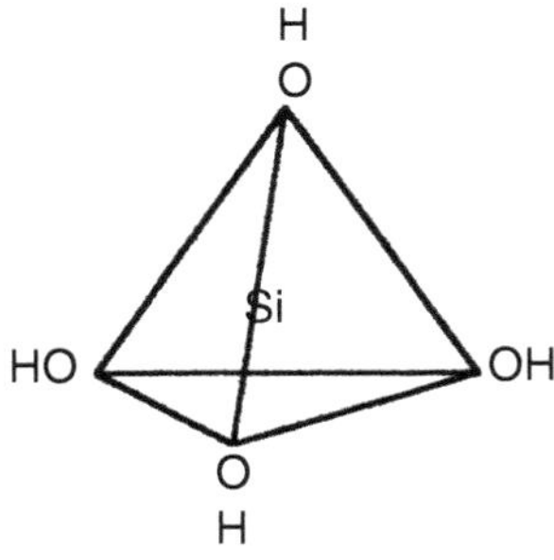

FIGURE 12.1 Silicic acid.

ions and only about 15% of the oxygen sites can be charged sites. It is the presence of these charged sites those causes the small oligomers to repel each other and avoid coagulation.

In high ratio silicate solutions the silica units have been found to be present in a variety of small polymeric forms as well as some monomer. One way of looking at the system is that it responds to changes in alkali content, by increasing or decreasing it's level of polymerization inversely with the alkali level. In essence the silica present reduces its surface area (the number of Si-OH units at the surface) to accommodate having less alkali present to form ionized sites. As Sodium ions are replaced by Hydrogen ions and the $—O^-$ groups present the neutralized (re-protonated), the charged sites also disappear. Condensation of the oligomers to small particles with a lower total surface area affords a way to maintain sufficient surface charge density to avoid gellation.

Several types of oligomers must exist for any particular ratio since they must of necessity be made up of whole numbers of silica units. The numbers of silica units in each must vary and the equilibrium number of each must vary to accommodate the variations in the overall ratio. Thus every particular ratio of sodium silicate solution has it's own equilibrium distribution of oligomers at a given Silica concentration and temperature. The distribution of oligomers tends to define the viscosity of the solution. The stability of these units would appear to be fairly substantial. It is observed that if the ratio is adjusted it may take some hours or days before a constant viscosity is reached.

If one mixes two silicate solutions of different ratio to give a solution having an intermediate ratio and known silica content and compares its viscosity with a silicate solution of identical ratio and silica content but which was produced weeks earlier, we find that the two solutions have different viscosities. Immediately after mixing the high and low ratio silicates, the viscosity of the mixture is far different from the aged sample. However, on standing for several hours or sometimes a few days, the viscosities of the two solutions are observed to be identical. This would indicate that the molecular rearrangements leading to the equilibrium distribution of oligomers occurs slowly. And this implies that at least some of the oligomers are fairly stable geometric forms and that transitions from one to the other are somewhat difficult.

This could have some ramifications in the manufacture of colloidal silica. Typically, sodium silicate is diluted prior to being deionized and polymerized. It is likely that the equilibrium distribution of oligomers will be different depending on whether the solution is freshly made or has been allowed to stand diluted for some time. While it is not clear that this variation in silica sources could have a significant effect on the properties of the colloidal silica made from them, it is a variable

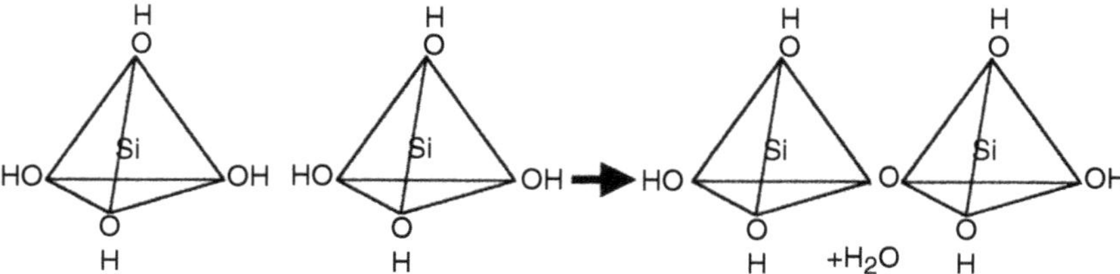

FIGURE 12.2 Polymerization of silicic acid.

which may or may not need to be controlled, depending on the details of the process steps.

Ion Exchange Resins

Cation ion exchange resins are long chain polymers, insoluble in water, which contain acid groups along the length of the chain. These acid groups can react with cations from the solution to form the salt form of the acid groups along its chain. In so doing it gives up hydrogen ions to the solution. Similarly anion exchange resins exist which can exchange hydroxyl ions for anions in solution. Used together, a mixed bed of these resins affords a way to purify water by replacing cations with hydrogen ions and anions with hydroxyls. The combination of hydroxyl and hydrogen ions gives water. The ion exchange resins are converted to their salt forms.

Cation exchange resins are used in the manufacture of colloidal silica to remove sodium ions and replace them with hydrogen ions, converting sodium silicate to silicic acid, which is then polymerized to colloidal silica particles. These resins are often made in the form of small porous beads about a millimeter in diameter to aid in handling. The beads have high surface area due to the open porous structure that maximizes the number of acid groups available to contact the solution of silicate. One feature that makes these resins so useful is that they can be "re-generated" or converted back and forth between the acid and salt form. The acid groups can be either in the acid form or in a salt form. When the acid form of the resin is contacted with a source of cations, like a solution of sodium silicate, the resin exchanges Hydrogen ions for sodium ions and is converted to the sodium form of the resin. The silicate ions are converted to the acid form, silicic acid. Conversely, treating the salt form of the resin with excess acid will convert it back into the acid form. The ability to be re-generated into the acid form allows these resins to be used over and over again.

When many of us took our first general chemistry course we were told, as a rule of thumb, that there were no insoluble sodium salts. However, the sodium form of an ion exchange resin can be thought of as an insoluble salt of sodium, since the sodium is removed from solution and is bound to the polymer. When washed with a strong acid, such as sulfuric acid, the sodium is released as the sodium salt of the strong acid (sodium sulfate) and the ion exchange resin is converted to the acid form. It can be rinsed free of excess acid and essentially becomes an insoluble acid. This reversibility of the form in which the resin can be found allows it to be re-used for many cycles.

Basic Types of Manufacturing Processes

Many of the processes for colloidal silica production are based on the following basic steps: (the following discussion is based on sodium silicate but would be applicable to other alkali silicates)

- Complete removal of sodium form a dilute solution of sodium silicate.
- Readjustment of the silica:soda ratio using sodium hydroxide or sodium silicate.
- Heating to produce particles.
- Continued addition of silica values to grow the particles (this step may or may not be part of the process).
- Concentrating the dilute sol.

Ion Exchange Processing

In fixed bed ion exchange the resin is held stationary in a column and the dilute silicate is passed up or down through the acid form resin. This process is illustrated in Figure 12.3. As it passes through the column, sodium is removed, and transferred to the ion exchange resin. The hydrogen ions from the resin are transferred to the silica

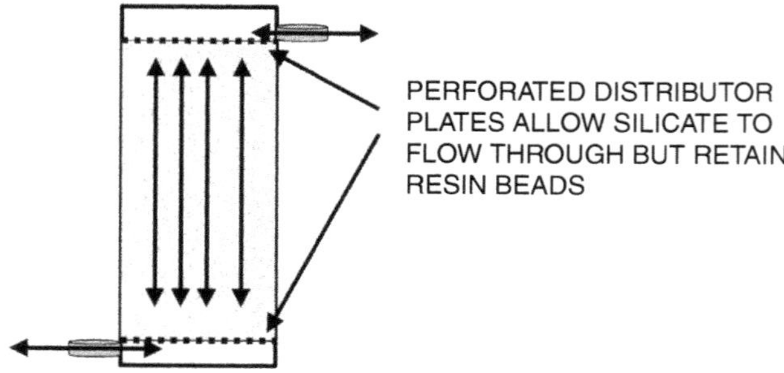

FIGURE 12.3 Fixed bed ion exchange process dilute silicate can flow either upwards or downwards through the resin.

units and the charged Si—O$^-$ sites are converted to the Si—OH form. Typically the dilute silicic acid has a pH of around 3. This eliminates virtually all of the charged sites on the silica oligomers and the tendency towards agglomeration increases significantly. However, it should be noted that silica is metastable in the pH range around 3 and hence the working time is longer than one might expect. Because of this increased tendency for the silica to form a gel it is necessary to keep the silica concentration low in order to prevent gellation in the column before particles can be grown. Typically, 3–4% silica solutions are used, but by pre-cooling the silicate solution one can operate at slightly higher concentrations.

In a stirred ion exchange bed, the ion exchange resin is literally stirred as the silicate is added to afford more efficient deionization of the silicate. Somewhat higher deionized silicate concentrations can be realized, but the system is more mechanically difficult to engineer and operate, and the higher attrition of the ion exchange resin caused by the agitation is a cost factor that must be considered.

In both of these processes there is always excess resin present and hence the deionization of the silicate is complete, with virtually all the sodium ions being removed and replaced with hydrogen ions. The pH is typically about 3–4. After the silicate solution is deionized some alkali (usually about 1% of that contained in the original sodium silicate), is normally added back to give a Silica: Soda ratio approximating the desired range of the finished sol. This provides sufficient charge sites to stabilize the finished product after particle growth. It should be noted that if the deionized solution were allowed to stand at room temperature, it would eventually thicken and form a continuous gel. However, by heating the dilute silicate the rate of agglomeration is increased, but so is the densification of the agglomerates into particles. Thus a combination of heating and agitation leads not to continuous gels, but discrete particles.

One of the simplest processes could be thought of as having the following steps:

- Deionizing the dilute silicate.
- Adding back sodium hydroxide or sodium silicate to adjust the silica:soda ratio.
- Heating to form particles.
- Evaporating the dilute product to a higher concentration.

This process is illustrated in Figure 12.4. This process tends to form fairly small particles some of which then agglomerate during the evaporation step to give a broad particle size range. Many of the particles will be made up of multiple particles that have agglomerated and then grown together during the evaporation process. This type of result is illustrated in Figure 12.5.

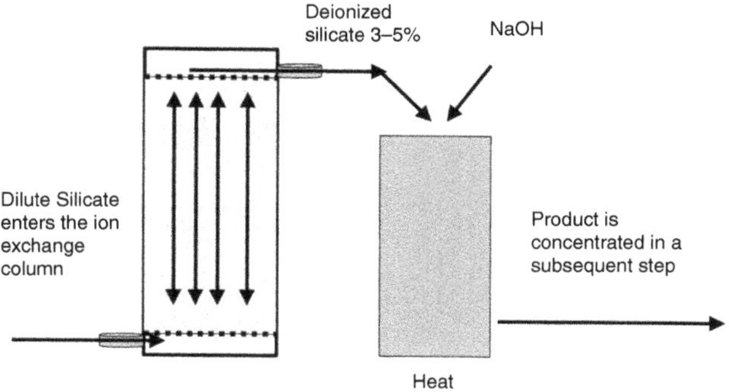

FIGURE 12.4 One of the simplest processes produces a dilute product in the reactor.

A variation on this process is to grow the particles and evaporate water at the same time by adding dilute deionized silicate to replace the water being removed using a constant volume evaporator.

- Complete removal of sodium from a dilute solution of sodium silicate.
- Readjustment of the silica:soda ratio using sodium hydroxide or sodium silicate.
- Heating to produce particles and evaporate water.
- Continued addition of dilute deionized silicate values to maintain constant volume and grow the particles to larger size and simultaneously concentrating the sol.

This process is shown in Figure 12.6. The evaporator is filled with deionized silicate and heating started. Particles begin to grow and as water is eliminated, it is replaced with dilute deionized silicate. Properly controlled, most of the added silica values deposit on already formed particles producing larger particles. The silica percentage in the evaporator grows as water is removed. The concentration of particles per volume remains constant. The product can be evaporated to its normal shipping strength, (typically 30–50%) depending on particle size.

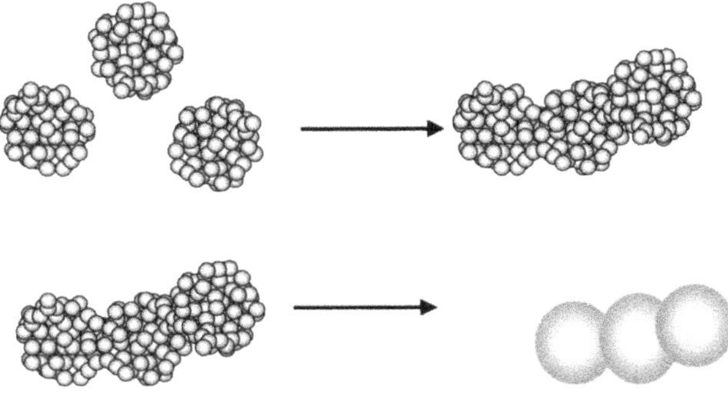

FIGURE 12.5 Formation of irregular particles.

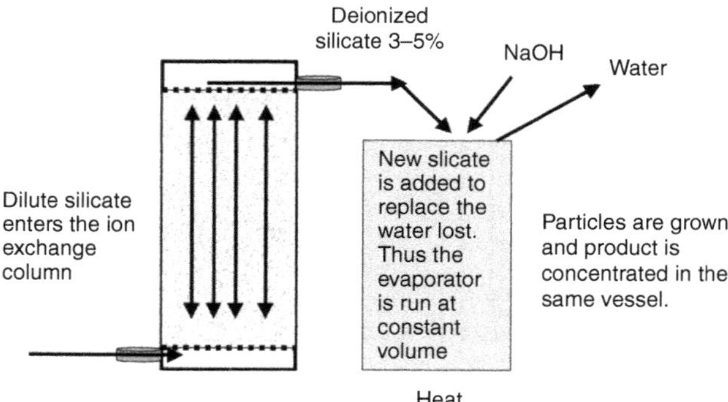

FIGURE 12.6 The reactor and evaporator become the same vessel allowing production of larger particles.

Another interesting variation on this process was developed by Monsanto and used to produce Syton® HT-50. HT-50 has a very broad particle size range and very large particles (ca. 100+ nm). The process uses a reactor in which the dilute deionized sol is heated to grow particles prior to concentration as described in the first example. However, a "heel" of product is always left in the reactor after each particle growth cycle. This process is illustrated in Figure 12.7. Hence, some particles in the reactor will have been through many growth cycles, some for only a few cycles as well as small particles that have been freshly formed. The particle size distribution tends towards a very broad "steady state" distribution. The resulting product contains particles from about 10 to 100+ nm.

A process developed and used at DuPont to make Ludox® colloidal silica does not completely deionize the silicate that enters the reactor. Rather, both sodium silicate and ion exchange resin are fed to a heel of water or silica sol. The relative rates are controlled in order to remove most, but not all the alkali from the system. This avoids the need to add back alkali to establish the correct pH for particle growth. The system is run at an appropriate elevated temperature to produce particles in the reactor. The high degree of control over the flow rates and the fact that the silicate is deionized more or less homogeneously throughout the reactor leads to the production of more uniform particle distributions than is routinely found in the other processes described above. This type of reactor is illustrated schematically in Figure 12.8. Since the silicate is supplied to the reactor with silica content greater than 25%, the final product is typically in the 12–15% range prior to concentration. Because the product was produced in a more concentrated state the process became referred to as the "consol" process, short for "concentrated sol." The fact that less water needs be evaporated is one advantage of this process, but the main advantage is the greater uniformity of particle size, which results from the deionized silicate being generated fairly homogeneously throughout the reactor.

Of course, the spent ion exchange resin must be separated from the product as it comes from the reactor and recycled for regeneration, while the product is moved to the concentration step. The mechanical handling of the ion exchange resin as it enters the reactor require extra equipment and controls not needed in the previously described processes.

Electrodialysis Route to Colloidal Silica

The use of electrodialysis as a method of producing colloidal silica products is potentially a very desirable process in that there is very little acid consumption, virtually no salts are produced which must be disposed of, and the by-products, hydrogen, oxygen, and sodium hydroxide are themselves useful.

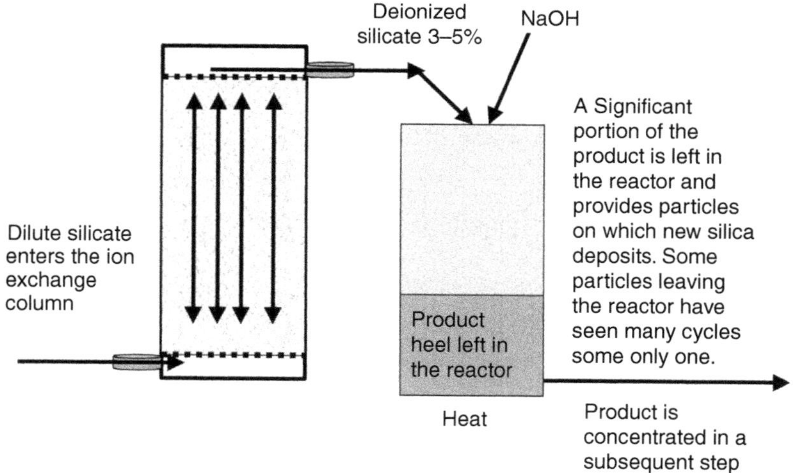

FIGURE 12.7 A variation allows production of a broad particle size distribution.

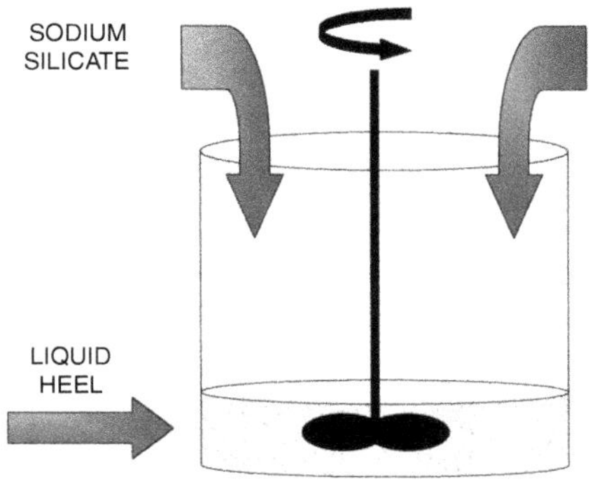

FIGURE 12.8 Consol type reactor.

Iler patented a process using electrodialysis [2] (see Figure 12.9 and Figure 12.10) which consists of three chambers. The outer chambers are anode and cathode. These border the inner chamber, which is made of two parallel and closely spaced porous membranes. Dilute sulfuric acid solution at the anode where oxygen gas is produced is circulated to a separator to remove the gas and is then recirculated to the anode. The sodium hydroxide solution, which is produced at the cathode along with hydrogen gas, is also circulated to a gas separator and the sodium hydroxide continuously removed. A process stream consisting of colloidal silica or colloidal silica nuclei, a small amount of sodium sulfate which acts as a conductor or supporting electrolyte to which sodium silicate is added to adjust the pH to about 9.5 is circulated rapidly between the membranes at 60–90°C. Current flow and flow rates are adjusted to maintain a pH of at least 8 as the stream leaves the cell. The silicate liberated deposits on the particles or nuclei present allowing them to grow to the desired size. A 15 nm sol containing 25% silica can be prepared directly by this method. The auxiliary electrolytes can be removed by ion exchange, the ratio adjusted, and the sol concentrated to 30–50% depending on particle size.

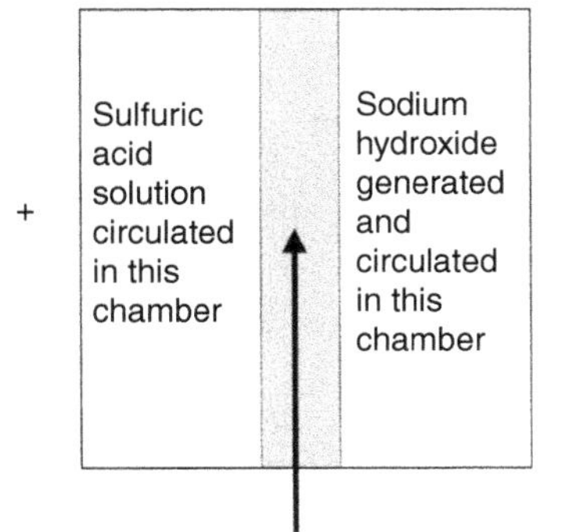

FIGURE 12.9 Electrodialysis route to colloidal silica described by R.K. Iler — U.S. Patent 3,668,088.

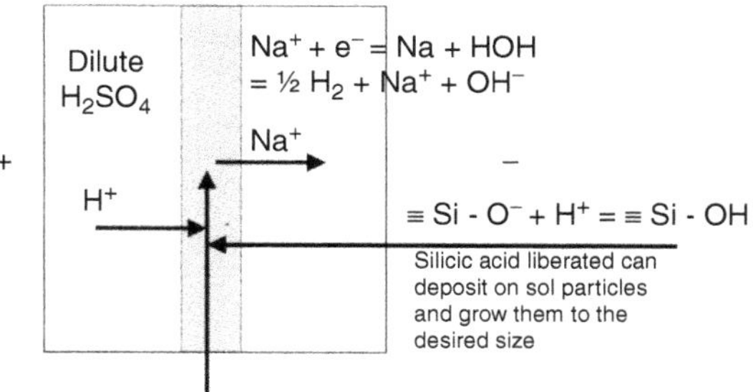

FIGURE 12.10 Electrodialysis route to colloidal silica described by R.K. Iler — U.S Patent 3,668,088.

Particle Growth vs. Nucleation

The solubility of amorphous silica is dependent on the particle size. See Iler [1], page 241. For very small, 1–2 nm particles, the solubility is well in excess of 300–400 ppm and falls rapidly as the particles increase in size. Near 100C there is a very significant drive for small particles to dissolve and the silica values to deposit on slightly larger particles. However, once the particles are larger, ~12–15 nm the difference in their solubility and that of larger particles is very small. Hence the driving force for the 12-nm particles to dissolve and re-deposit on larger particles also becomes small. What does this mean in a process attempting to grow particles much larger than 12–15 nm?

Freshly deionized silica tends to agglomerate quite rapidly to small (1–2 nm) particles. If the silica concentration is high and the temperature is maintained low, the particles do not grow, and their relatively high concentration causes them to form a continuous gel. If the silica concentration is maintained low and heat is applied, the particles can grow to larger size and in so doing decrease the number of particles. See Figure 12.11 and Figure 12.12.

This seems obvious now, but consider the brave researcher who decided to heat a solution and speed up

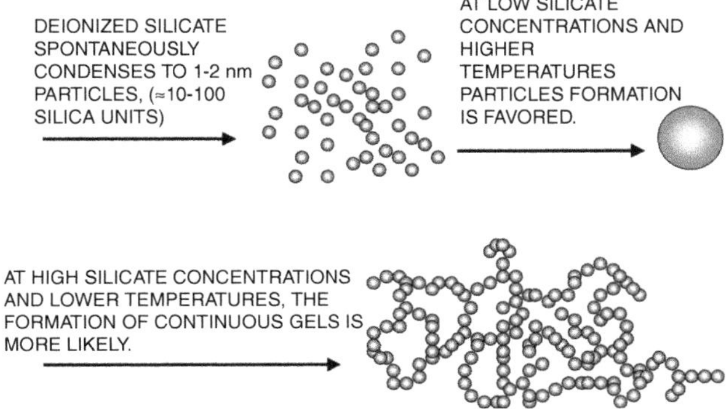

FIGURE 12.11 Particles or silica gel?

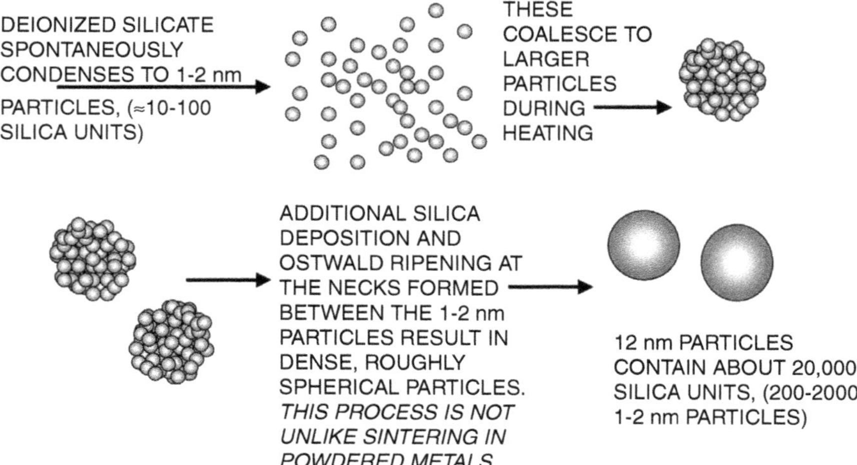

FIGURE 12.12 Particle growth.

the gellation process rather than slow it down by maintaining the system at a low temperature.

If the small units encounter a larger particle the very large difference in their solubility will cause the small silica units to be deposited on and increase the size of the larger particle. This ideal situation is illustrated in Figure 12.13. However, if the deionized silica units only run into similar size particles they will coalesce into new small particles which can then act as nuclei for a new smaller family of particles and the resultant sol will have a broad particle size range. This could happen if the silica is added faster than the larger particles can absorb it or if the concentration of larger particles becomes so small as to make it improbable that the incoming silica will encounter one.

Consider a reaction where 5% deionized silicate is heated to form small nuclei and then additional silica values are added in the form of 5% deionized silicate in order to grow the particles to a larger size. If all of the incoming silica is deposited on the previously formed nuclei then the number of particles in the system will remain constant. However, their concentration will diminish rapidly as the total volume of the system increases. At some point it will be more probable that incoming silica will coalesce to form new small particles or nuclei before they encounter a "growing" original particle, because the particles are few and far between. See Figure 12.14.

In order to prevent this effect processes have been developed where the active silica is added to a heel of nuclei while water is evaporated in order to maintain a constant volume in the system. This causes the concentration of the growing particles to remain constant. By maintaining a sufficiently slow addition of silica to the system the particles can be grown to large size without the formation of new nuclei. This type of process was discussed above and illustrated in Figure 12.6.

As described above in the production of Syton® HT-50, the development of larger particles depends on the probability of the "nascent" silica units running into a growing particle. As the particles grow larger the amount of silica needed to grow them further increases as the cube of the diameter. Hence, even if the nascent silica encounters a very large particle it will produce

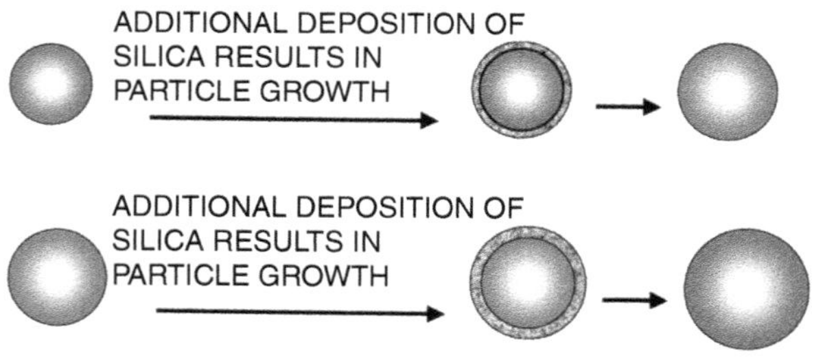

ADDITIONAL DEPOSITION OF SILICA RESULTS IN PARTICLE GROWTH

ADDITIONAL DEPOSITION OF SILICA RESULTS IN PARTICLE GROWTH

FIGURE 12.13 Continued silica deposition to give larger particles.

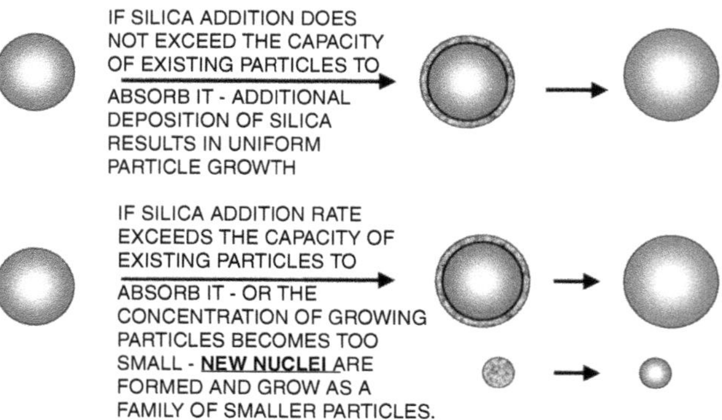

FIGURE 12.14 The key to narrow particle size distribution.

very little change in the diameter of that particle. Unless all of the nascent silica is deposited on large particles the excess will form a large number of small particles. These small particles will then compete with the larger particles for the incoming "nascent silica." The result is a sol having a broad particle distribution.

COLLOIDAL SILICA STABILITY

Colloidal silica as it is manufactured is almost always stabilized with sodium hydroxide. The purpose of the alkali is to react with a fraction of the weak acid —OH groups on the particle surface and form water and leave behind a charge site:

$$OH^- + \text{—Si—OH} = HOH + \text{—Si—O}^-$$

This process is shown in Figure 12.15. With enough of these charge sites present the particles tend to repel each other and cannot come close enough to form inter-particle bonds. Up to about pH 10–10.5, the higher the alkali content, the higher the charge density and therefore the higher the particle to particle repulsion. However, above that pH the particles can begin to dissolve to form sodium silicate (depolymerize) and become less stable. The presence of the sodium silicate formed acts in the same fashion as added electrolyte to further destabilize the system. Thus the system gels at least locally. This localized effect can be demonstrated by dropping some 5–10% NaOH solution into colloidal silica without high shear stirring. As the drop enters the solution, the localized effect is to cause a small gel structure that tends to remain stable in spite of continued low shear stirring.

Removing hydroxyl ions decreases the pH, and the charge density decreases as the number of surface silanol groups which are ionized, decreases. The gel rate increases with the decrease in charge density. With the decrease in charge density, there is less repulsion between particles and more opportunity for inelastic collisions. In considering the effect of pH changes one needs to keep in mind that the pH scale is a logarithmic scale. Each change of 1 pH unit corresponds to an order of magnitude change in the actual number of hydroxyl ions present. This is illustrated in Figure 12.16. It follows that the number of charge sites also decreases by about an order of magnitude when the pH drops a unit. Because of this, minor pH fluctuations can have a serious impact on the stability of colloidal silica.

Hydroxide ions have reacted with three acid sites producing water and leaving behind three negative sites which are counter-balanced by the three sodium ions originally associated with the hydroxide ions

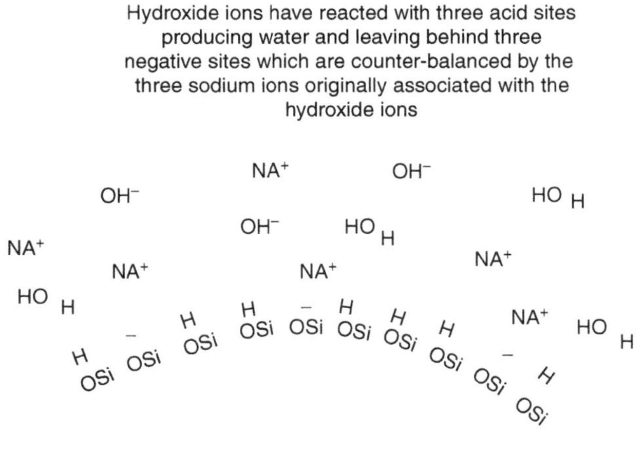

FIGURE 12.15 "Sodium" stabilization of colloidal silica.

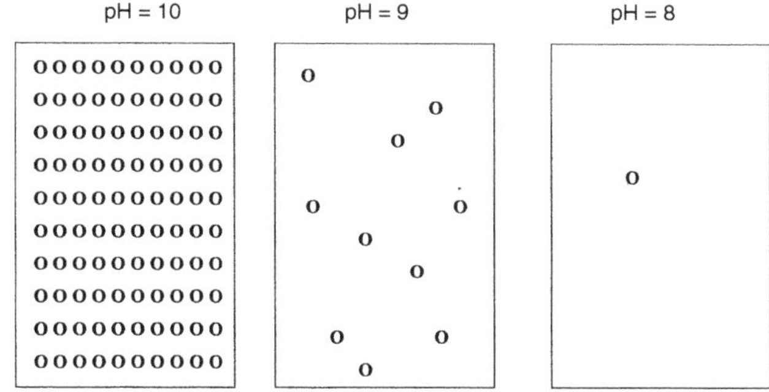

FIGURE 12.16 Representation of the pH scale.

Figure 12.17 is a combination of Figure 3.2 and Figure 4.13 found in Iler's *The Chemistry of Silica*. It illustrates the effect of pH coupled with the effects of salts contained in the sol. The effect of salt is more pronounced as its concentration is increased. Polyvalent metal ions such as calcium or magnesium have a much more drastic effect on stability because of their higher charge density.

DENSITY OF COLLOIDAL SILICA PARTICLES

All manufacturers of water-borne colloidal silica publish charts of specific gravity versus silica content. Both manufacturers and users have long used the specific gravity of colloidal silica to determine its concentration. Hence, the correlation between the two has been well established. There is very little difference in the published values of this correlation from manufacturer to manufacturer, indicating that the density of the particles they produce is very similar. The following is a calculation based on such data for Ludox® colloidal silica to calculate the density of the silica particles. They are considered to be representative of most of the commercially available silica sols. The one assumption is that the water molecules in the immediate vicinity of the silica particles still retain the density of normal water. Considering the high pressures needed to increase the density of water this would appear to be reasonable.

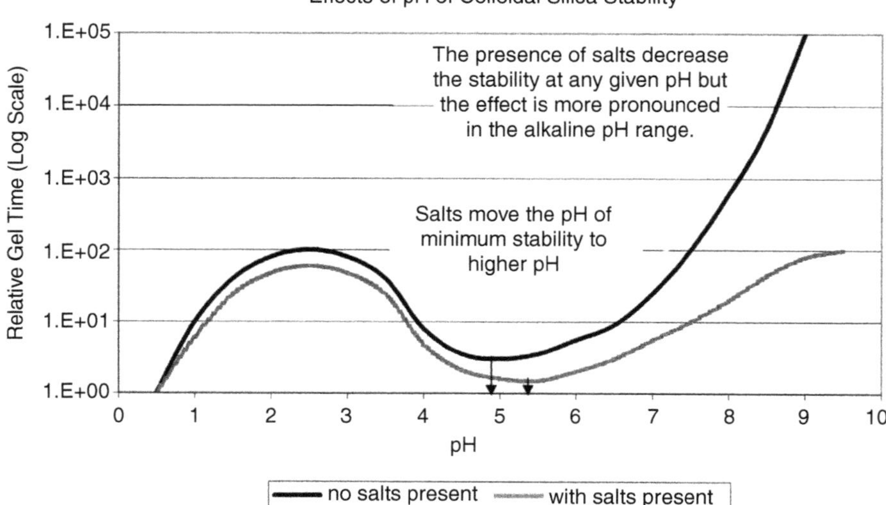

FIGURE 12.17 Effects of pH on colloidal silica stability.

It should be pointed out that the data is based on deionized silica sols to eliminate the effects of the alkali content, which vary with the particle size in commercial sols.

The following example illustrates the calculation for a 10% silica sol. By weight the sol is 90% water and 10% silica.

The specific gravity at 25°C is measured as 1.058.
The volume of 100 grams of the sol is then = 100/1.058 = 94.52 cc.
The volume of 90 g of water at 25°C where the density is 0.9997 = 90.03 cc.
The volume of the 10 g of silica = 94.52 − 90.03 = 4.49 cc
The density of the silica then = 10/4.49 = 2.23 g/cc.

Table 12.1 lists similar calculations over a broad range of silica concentrations.

CALCULATING SILICA UNITS PER PARTICLE

Much of the work done with colloidal silica depends on the chemistry of the surface layer of silanol groups, as they are the only part of the particle that is active chemically. Hence it is desirable to have a way to calculate the numbers of active silanols in a sol having a particular size and concentration. The approach shown below is based strictly on a simple geometrical approach.

The size of the average repeating unit is taken as the volume of a mole of silica (60 g) divided by its density 2.2 g/cm^3, divided by Avogadro's number. Expressed in cubic angstroms this volume = 45.3 Å^3.

TABLE 12.1
Calculated Particle Density

% Water	% Silica	Sp.G. 25°C	Wt. silica	Total Volume	Vol. silica (difference)	Calculated density
98	2	1.011	2	98.91	0.88	2.27
95	5	1.028	5	97.32	2.30	2.18
90	10	1.058	10	95.52	4.49	2.23
80	20	1.125	20	88.89	8.86	2.26
70	30	1.202	30	83.19	13.17	2.28
60	40	1.290	40	77.52	17.5	2.29
50	50	1.388	50	72.05	22.03	2.27

The value for the average density is 2.25 g/cc.

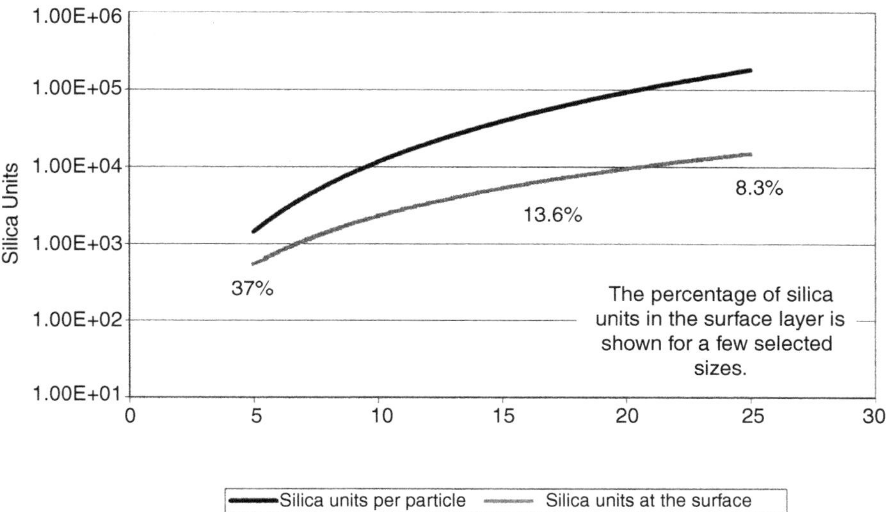

FIGURE 12.18 Silica units vs. particle size.

To simplify things the repeating unit was assumed to have a cubic shape, and the length of a side then is equal to the cube root of 45.3 Å = 3.57 Å

The calculation of the particle volume given its diameter (d) is straightforward, Volume $= \frac{4}{3}\pi \mathbf{r}^3$. Dividing the particle volume by the volume of one silica unit gives the total number of silica units. Calculating the number of silica units in a particle smaller in diameter by 3.57 Å (d-3.57), and substracting the number of silica units in that particle from the previous result gives the number of surface silica units in the particle of diameter (d).

Example: The following illustrates the calculations for 12 nm particles.

1 mole of silica weighs 60 grams.
The volume of 1 mole of silica = 60/2.2 = 27.27 cubic centimeters.
1 cubic centimeter = 10^{24} cubic Angstroms.
The volume of 1 silica unit = $27.27 \times 10^{24}/6.023 \times 10^{23}$ = 45.3 cubic Angstroms
Assuming the silica units to be cubic – the length of a side of the cube = $45.3^{1/3}$ = 3.57 Å.
The volume of a particle having a diameter of 12 nm:
Radius r = 6 nm = 60 Å.
Volume of the particle $= \frac{4}{3}\pi \times (60)^3 = 904320$ cubic Angstroms
Number of total silica units in the particle = 904320/45.3 = 19963
The volume of a particle having a radius one silica unit less:
Radius r' = 60 − 3.57 = 56.43 Å
Volume of the particle $= \frac{4}{3}\pi \times (56.43)^3 = 752313$ cubic Angstroms
Number of total silica units in the particle = 752313/45.3 = 16607
Therefore the number of silica units in the outer layer of the 12-nm particle is: 19963 − 16607 = 3356 silica units.
The percentage of silica in the outer layer (particle surface) = 3356 × 100/19963 = 16.8%

Figure 12.18 is a plot of the number of silica units in both particles and at the surface of the particles.

SILANOL GROUPS PER SQUARE NANOMETER ON THE COLLOIDAL SILICA SURFACE

If we take the number of surface silanols calculated and divide by the area of the particle surface we can calculate the surface density of silanol groups. Such calculations are plotted in Figure 12.19 for several particle sizes in the range of 5 to 100 nm. For larger particles the calculated values asymptotically approach 7.9. In the normal range of colloidal silica sizes sold (5–25 nm) the calculated values range from about 7–7.6.

Many determinations of the numbers of Silanol groups have been done in the past and have generally determined that there are about 4–5 silanol groups per square nanometer. L.T. Zhuravlev [3] has summarized much of this work. From this work the most likely value is 4.6 silanol units per nm^2 was determined. However, all of this work was done on silica dried at temperatures above 100°C. Most of the work was done on samples dried above

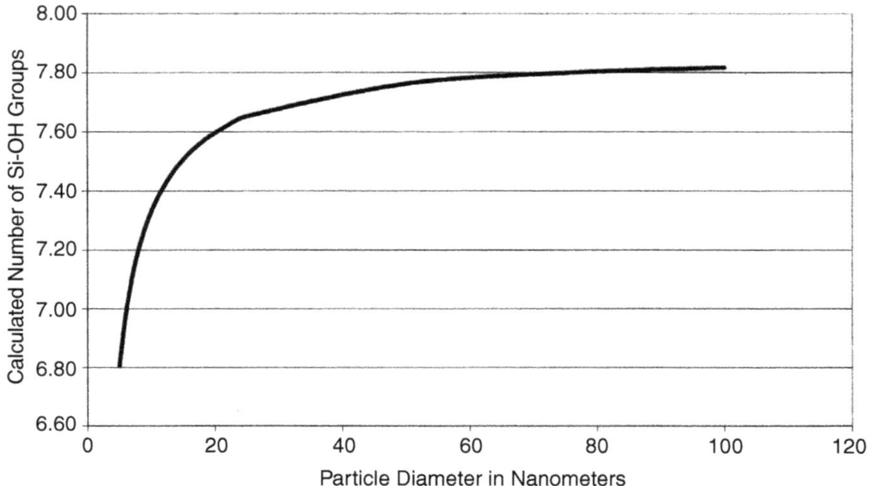

FIGURE 12.19 Calculated number of surface silanols.

200°C. The number of surface silanols was then determined by rehydrating the surfaces thus formed.

It is my belief and observation that a large percentage of the surface silanols are reacted during gellation and that these sites are not capable of rehydration. Drying even at ambient or lower temperature produces inter-particle bonding that is irreversible. Many of the end uses of colloidal silica as a binder depend on this property. Therefore determinations based on silanol re-hydration are consistently low in their estimates of the numbers of silanols which were on the original colloidal silica source. The coalescence factor for colloidal silica stabilized with alkali metals is often as high as 29–30 even if the sol is spray dried or freeze-dried at low temperatures. This means that 29–30% of the silica would have to be dissolved to restore the original light transmission to the powder slurried in water. Silica gels can be formed in aqueous media and never dried. A few processes have been described for peptizing gels to colloidal silica but these involve after treatments while maintaining the gels in the "wet" stage.

It is clear from the data shown in Figure 12.20, reproduced from Zhuravlev, that the higher the heating temperature, the lower the number of silanols found by rehydration. Therefore it seems unlikely to me that work done on dried colloidal silica will yield a genuine picture of the surface as it existed before drying. The most probable value for each temperature is given in Table 12.2 that is also reproduced from Zhuravlev.

If we plot this data as in Figure 12.21 and extrapolate backward to ambient temperatures the result is a silanol number of 7–7.5. I do not wish to imply here that any of the work referenced is in any way incorrect. I just would like the reader to be impressed with the fact that once

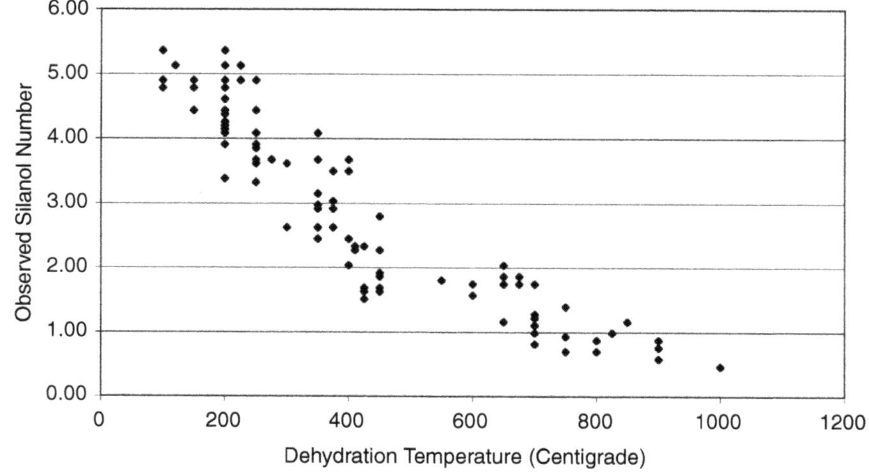

FIGURE 12.20 Silanol per nm^2 determined by several methods vs. dehydration temperature.

TABLE 12.2
Most Probable Value of Silanol Number vs. Temperature of Dehydration

Temperature of vacuum treatment (centigrade)	Silanol number
180–200	4.6
300	3.55
400	2.35
500	1.80
600	1.50
700	1.15
800	0.70
900	0.40
1000	0.25
1100	<0.15

colloidal silica particles have been allowed to coalesce they are very different from their condition when monodispersed.

AFTER TREATMENTS

The manufacturing methods described above can only control the particle size and particle size distribution of the resulting silica particles. If sodium silicate has been used, as we have assumed, the sols will all be stabilized with sodium hydroxide.* For many uses this is no problem, but for many others, such as very high temperature use where sodium lowers the refractoriness of the silica, or catalysis where sodium can act as a poison, it would be desirable to have an alternate counter ion. Ammonium hydroxide is an excellent choice since the ammonia is lost along with the evaporating water when the silica is dried in its final form.

Ion exchange is again the technology used to remove sodium ions from the sol and replace them with hydrogen ions. The sol is contacted with the acid form of a strong acid ion exchange resin and sodium ions are absorbed by the resin and replaced by hydrogen ions from the resin. Where very high purity is required, an anion exchange resin may also be used. An anion exchange resin is usually supplied in the basic (hydroxyl) form. It absorbs anions such as sulfate from the sol and replaces them with hydroxyl ions. The combined effect is to replace the salts present with water. It is not uncommon to use a mixed ion exchange column where both types of resins are blended together and the sol can contact both types almost simultaneously. This type of system can only work if there is a difference in density of the two resin types. In this case they can be separated into separate layers in the column by passing water up through the column. The operation of a mixed ion exchange bed is illustrated in Figure 12.23–Figure 12.25.

After ion exchange, ammonium hydroxide can be added to bring the pH up to the stable range ~9–10. In principle there is no reason that another base, either organic or inorganic could not be used, but commercially, the most often used alternative is ammonium hydroxide.

While high purity is the goal, total deionization of all anions from concentrated sols produces a problem. The sol becomes extremely viscous, and if this actually happens in

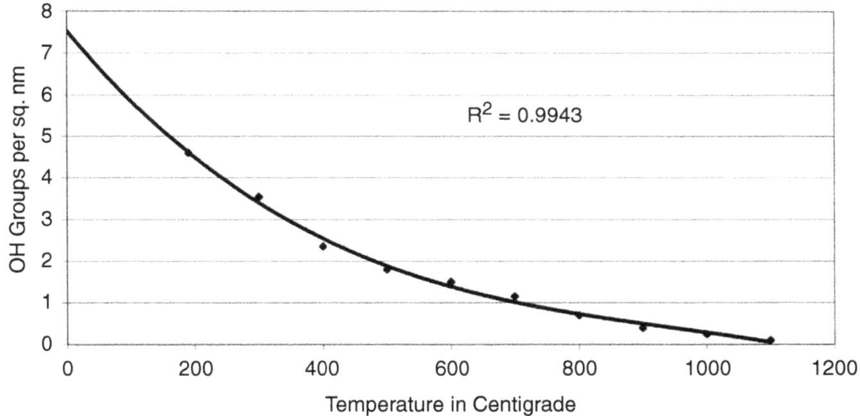

FIGURE 12.21 Temperature dependence of silanol number data taken from Table 12.2 extrapolated to ambient temperatures.

*Note: Colloidal silica is often referred to as being "sodium stabilized, or ammonium stabilized," but this is a misnomer. Colloidal silica is typically "hydroxyl ion stabilized" as it is the presence of OH^- ions which remove protons, forming water and leaving behind the negatively charged sites which repel other particles and thus prevent coagulation. See Figure 12.22. The sodium or ammonium is only there because they came along with the hydroxyl ions. That said, we will use the terminology sodium stabilized, ammonia stabilized etc. to indicate a sol stabilized with sodium hydroxide, ammonium hydroxide, and so on.

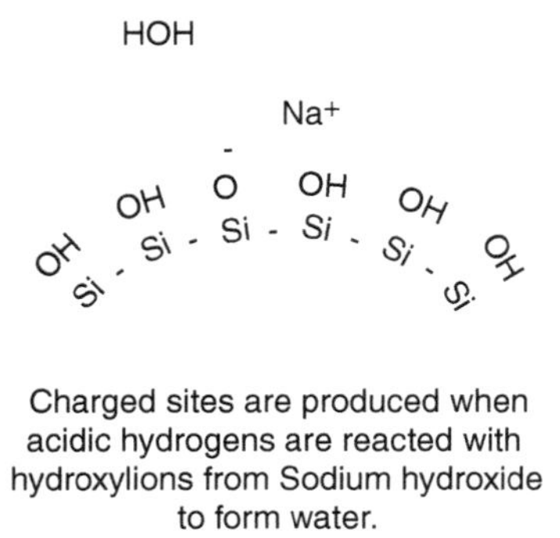

HOH

Na^+

Charged sites are produced when acidic hydrogens are reacted with hydroxylions from Sodium hydroxide to form water.

FIGURE 12.22 Surface charge sites produced by reaction with alkali.

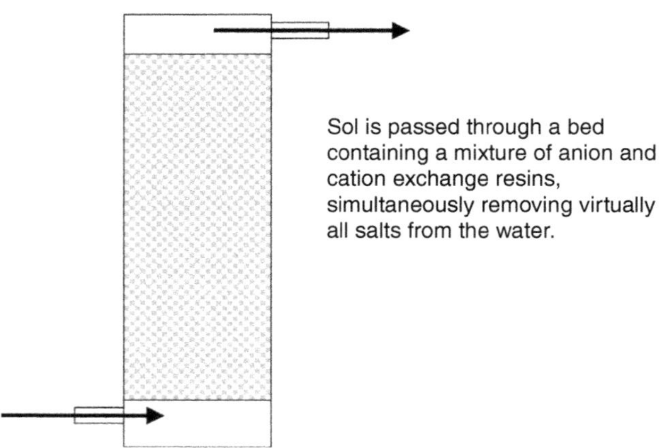

FIGURE 12.23 Mixed bed ion exchange.

the column, the column may be blocked and flow stopped. The explanation of this phenomenon illustrates some very fundamental and interesting aspects of the interaction of the colloidal particles and the water surrounding them.

ELECTROLYTE EFFECTS ON VISCOSITY AND STABILITY

It is well known, but not completely understood, that increased levels of electrolytes increase the rate of coagulation of colloidal silica particles. However, it is also seen that where fairly concentrated sols are virtually completely deionized using both cation and anion exchange resins the sols undergo a huge rise in viscosity, and can have a viscosity similar to "Vaseline." At first glance one would assume that the sol has gelled and been irreversibly ruined. However, if a small amount (a few hundred ppm) of an electrolyte such as sodium sulfate is added, the sol thins out and assumes a normal viscosity. It would appear that it is possible to do too good a job of purifying sols.

It is known that water-borne colloids have a surrounding "cloud" of water and electrolytes. This is often called the "double-layer" because it has been described as made up of two layers. Closest to the particle surface is a layer of tightly bound counter ions, held there by charge attraction called the Stern layer. Beyond that is a layer of diffuse cations and anions, which have a somewhat higher concentration than the bulk liquid. It was described by Guoy. It is estimated that colloidal silica particles have about a 2 nm thick "double layer" of water and electrolyte surrounding them. This thickness corresponds to a layer roughly 6–8 water molecules thick. Strong bonding to the silica surface and hydrogen bonding within this layer makes it tend to be part of the particle and move with it. Thus we might consider that there are two types of water present in a colloidal silica sol: "bound" water, and "free" water. Figure 12.26 is an illustration of this effect. Since the surface of the particle is charged negatively there exists counter-balancing charged ions (sodium ions in this illustration) which tend to be fairly close to the surface of the particle. They are held there by charge attraction. Other than the charges on the silica particle hydroxyl ions and

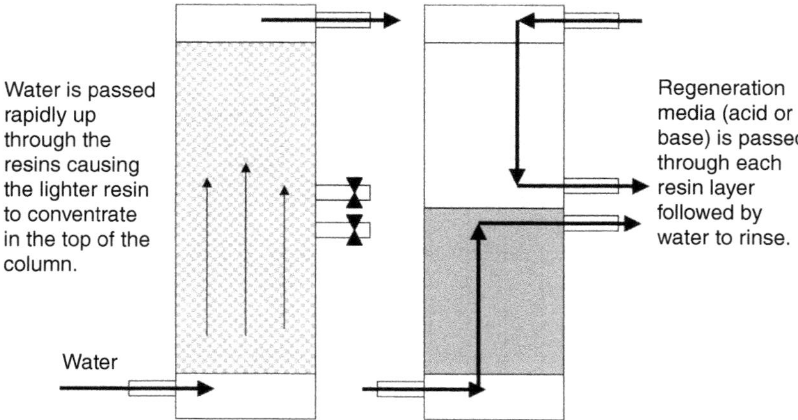

FIGURE 12.24 Mixed bed ion exchange — first step in regeneration.

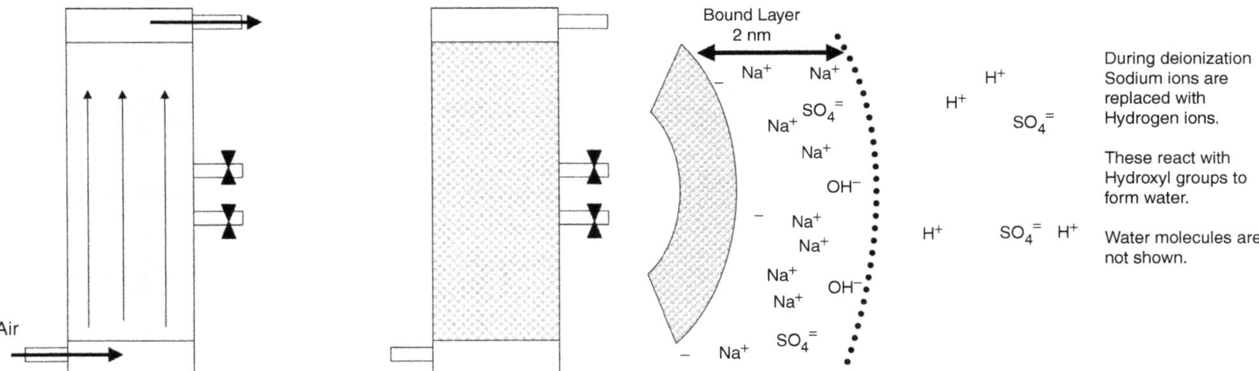

FIGURE 12.25 Regenerated resins are re-mixed using air pressure to induce turbulence.

FIGURE 12.27 Electrolyte distribution around surface after deionization.

some of the impurity anions present such as the sulfate ions shown offset the charge of the attracted sodium ions. The effect of the electrolyte found near the surface is to also hold water molecules by hydrogen bonding. Thus near the particle surface there is a water layer that is strongly associated with, or "bound" to the particle. The coagulation caused by high electrolyte levels has been interpreted as a diminishing of the thickness of the "bound" layer thus allowing closer approaches of particles. It is reasonable to assume then that as electrolyte is removed the "bound" layer expands and becomes larger.

The charged surface of the particles attracts around it a slightly higher concentration of ions than is found in the surrounding "free" water. This picture is a homespun version of the Stern and Guoy double layer interpretation of the surface condition. These have been examined extensively, and much more rigorously than we will attempt here.

Figure 12.27 illustrates the situation if a cationic ion exchange resin is used to remove the sodium ions from the sol. The water molecules are not shown. The sodium ions are replaced with hydrogen ions and these interact with the hydroxyls present to form water. The sulfate levels are left in tact. The surrounding water layer is reduced to the equivalent of a dilute sulfuric acid solution.

The ions in the bound layer are not significantly affected by the deionization, but soon after some diffuse into the surrounding water and establish a new equilibrium level. It is typical to see the pH of a freshly deionized sol to rebound to higher pH because of this re-distribution of electrolytes. Further deionization steps can further purify the sol. See Figure 12.28.

If instead of just a cation exchange resin a mixed bed or an additional anion exchange resin is used to also remove the sulfate anions then virtually all of the electrolytes may be removed. If the electrolyte level drops dramatically, the bound layer may increase in diameter as shown in Figure 12.29. If the sol is sufficiently concentrated, the bound layers may overlap as shown in Figure 12.30 with the result that the particles will be bonded to one another through hydrogen bonding. This can be seen as the cause of the viscosity rise.

The addition of small amounts of electrolyte "thin" the material immediately, indicating that the sols have not

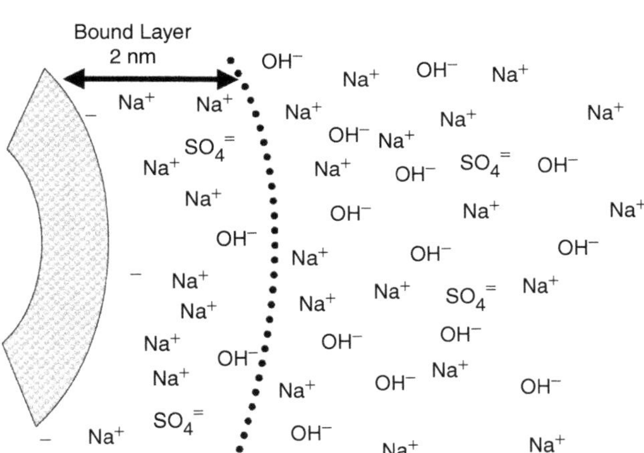

FIGURE 12.26 Electrolyte distribution around surface.

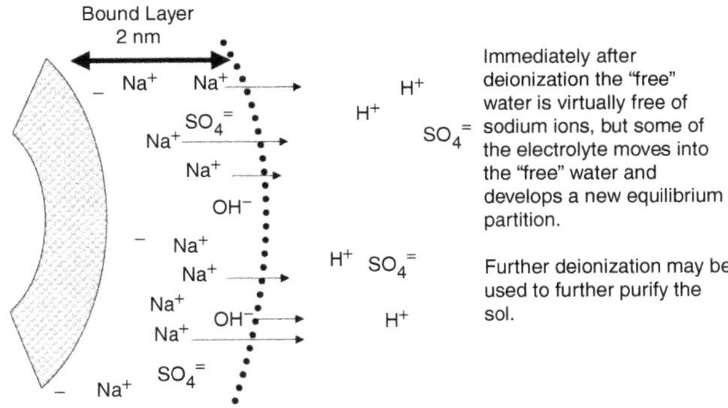

FIGURE 12.28 Redistribution of electrolyte in bound layer after deionization.

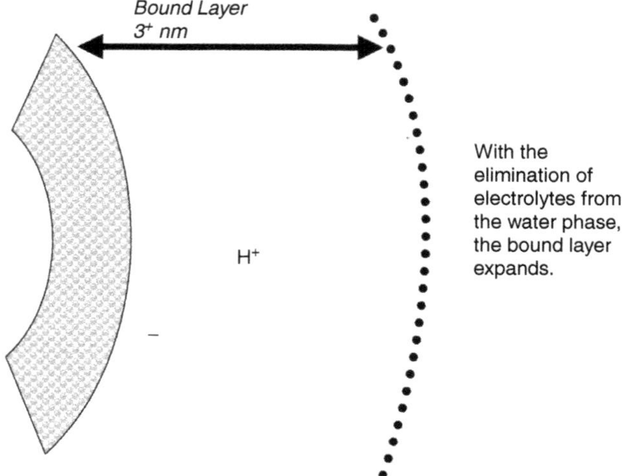

FIGURE 12.29 Electrolyte distribution around surface.

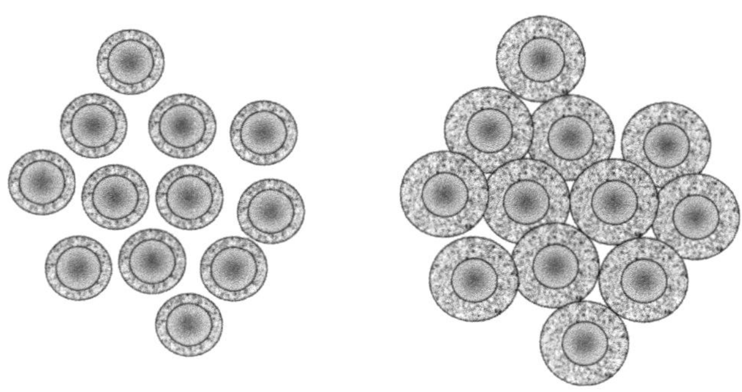

FIGURE 12.30 Expansion of the "bound" water layer causes increased viscosity.

actually formed Si–O–Si bonds between particles. To put this in some perspective, we can calculate that 250 ppm of sodium sulfate based on the weight of silica, in 12-nm silica sol represents 1–2 sodium sulfate units per particle.

One might question how a small shift in the thickness of this "bound" layer might have such a profound effect, but if we calculate the actual amount of water in the bound layer it is not surprising. Using the calculations above applied to a 40% 12 nm sol, we can calculate the number of particles in a given weight of silica and the volume of "bound" water can then be determined after calculating the volume for one particle. If we consider the "bound" water as part of the "solid" or particle phase, we see that even at 2 nm the volume of free water is less than the volume of the particles. Even with slight expansion of this "bound" layer, the free water rapidly becomes too small a fraction to avoid significant overlap of the "bound" layers. These calculations are illustrated in Figure 12.31.

One of the most interesting after treatments developed is the surface treatment with sodium aluminate. Aluminate ions, hydrated in water are isomorphic with silicic acid, but carry a negative charge because of the lower valence of the aluminum. This is illustrated in Figure 12.32. Because of the similarity aluminum in the form of aluminate can be substituted into the surface of the colloidal silica particles as illustrated in Figure 12.33. Wherever an aluminate becomes part of the surface it brings with it a negative charge independent of pH. If sufficient numbers of aluminate sites are inserted the sol can be acidified to a mildly acid pH and remain stable for hundreds or perhaps thousands of times longer than non aluminate treated sols. Replacement of only 0.5–1% of the surface silica units with aluminate is all that is required. However, acidifying the sol does produce salts, which do cause the eventual gellation of the acidified sol.

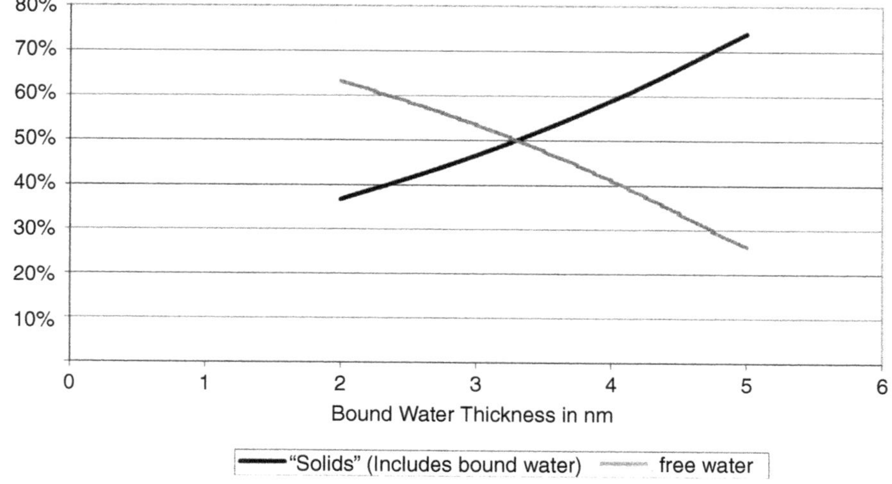

FIGURE 12.31 Volume % of both "Solids" and free water.

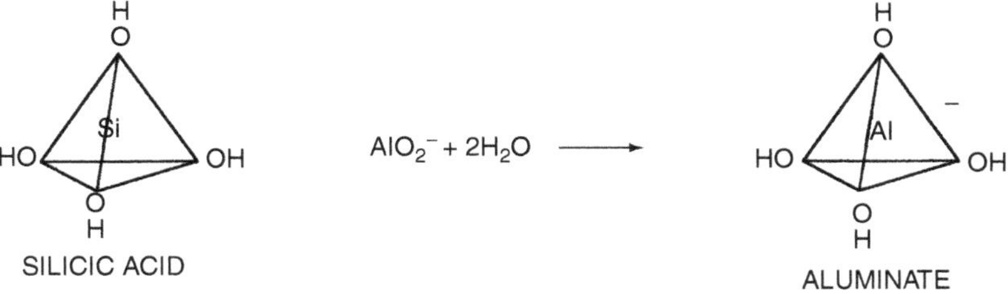

FIGURE 12.32 Silicic acid and hydrated aluminate ion.

However, if the sol is deionized into the acid range, thus avoiding the inclusion of salts, it can remain perfectly stable for years. The alumination of the surface allows us to make stable products over a pH range of 3–10 without concern for the exact pH. Thus one can select the pH based on the other components in a system and not have to be concerned that the pH chosen will de-stabilize the colloidal silica.

Commercially, aluminum is used to produce these products even though other amphoteric oxides such as tin, zinc, and lead are named in the original patent (4). It is clear from comparative studies using the others that aluminum is superior. This is perhaps not surprising since the others do not exist in a +3 valence and therefore would not necessarily be isomorphous with silicate. Gallium, while very expensive, is more similar to aluminum than the others, having a similar valence. Roberts found that surface treatment with small amounts of sodium gallate provided similar to superior stability for colloidal silica compared with aluminum. The product was produced by replacing about 0.3–0.5% of the surface silica with gallate ion. A portion of the surface modified sample was deionized and this product and the nondeionized material remained stable in a 60°C oven for over a year before it was discarded. It is considered that a year in the oven at this temperature is equivalent to about 10 yr at ambient temperature.

This type of colloidal silica carries a negative charge just as the alkali-stabilized product does. However, it is also possible to produce an acid sol that is positively charged. By coating the entire particle surface with a basic aluminum salt, the particle charge can be converted from negative to positive. Essentially, the particle appears from the outside to be colloidal alumina. A sketch of the surface is shown in Figure 12.34.

Typically, the colloidal silica is acidified to pH 7–8 and added to the basic aluminum salt solution using a high shear mixer to promote very rapid mixing of the two oppositely charged products. The rule of thumb is to put the material being coated into the coating material. This assures that there is always an excess of coating material present. The closer to pH 7 the pH of the colloidal silica the less sediment will be produced during the preparation. However, if the pH is taken much lower than 7, say 6.5 the mixture first appears quite clear and then gels rapidly. It is assumed that there is some critical level of surface charge that must be left on the colloidal silica in order to attract the colloidal alumina to coat it. If the surface charge is lowered below this threshold, by moving the pH lower than 7, then the basic aluminum chloride acts as a polyvalent metal salt and rapidly gels the system. Of course, if the pH is allowed to remain too high, much above 8, there is so much alkali present that the basic aluminum salt is gelled as aluminum hydroxide upon mixing.

One of the interesting aspects of positive sols is that they appear to exist a long time in a stable, but not completely reacted state at their surface. This first came to our attention when a user found that when he combined

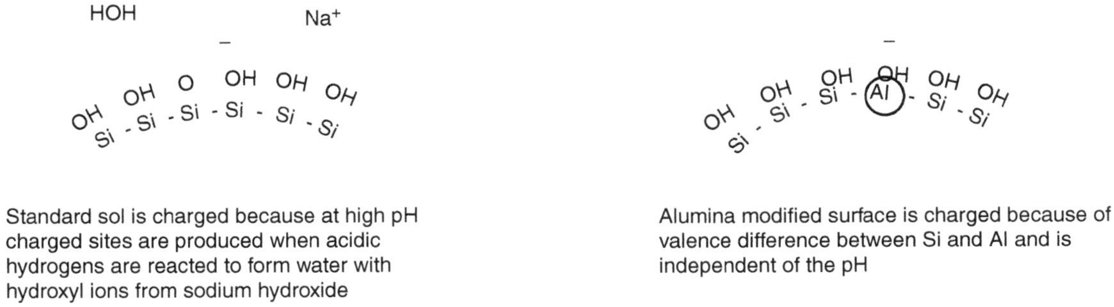

HOH Na+

Standard sol is charged because at high pH charged sites are produced when acidic hydrogens are reacted to form water with hydroxyl ions from sodium hydroxide

Alumina modified surface is charged because of valence difference between Si and Al and is independent of the pH

FIGURE 12.33 Aluminum in the surface produces a charged site independent of pH.

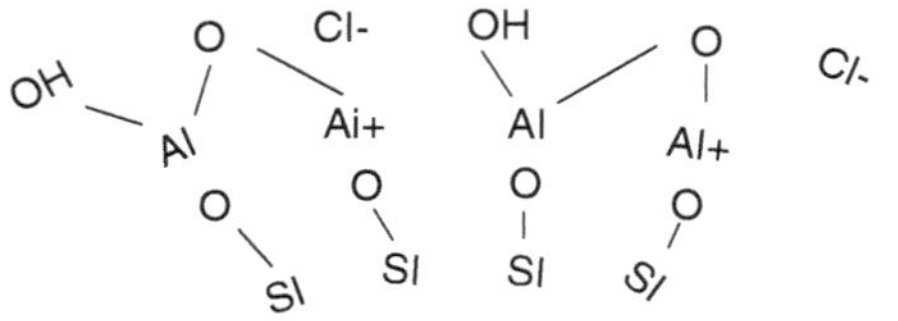

FIGURE 12.34 Alumina coated silica (positive sol).

freshly made (2–3 weeks old) with a resin system that heavy turbidity resulted. This conflicted with his original lab tests in which he produced a crystal clear mixture with a sample provided. The only answer we were left with after going down several blind alleys appeared to be the difference in age of the two products. The lab sample was about a year old when he used it but was made in normal production. The material he used in production was only a few weeks old.

Heat aging freshly made materials for practical times like 24 h did not appear to be the answer nor did making changes in the stoichiometry of the mix of silica and alumina salt.

To summarize a lengthy investigation, we developed a picture of the system, which implied that there was a fairly immediate adsorption of the basic aluminum salt around the silica particles. This produced a product that was quite stable with respect to aggregation. However, at this stage the product appeared to retain the ability to coagulate some materials sensitive to negative charge sites. This implied that the alumina was somewhat labile, came and went, revealing the negatively charged silica beneath it. Perhaps it was being displaced by something having a higher positive charge. After a few months of aging, however, the firm attachment of the alumina layer is complete or at least sufficient to avoid this effect.

Another possible explanation of this phenomena is that basic aluminum chloride (the most commonly used aluminum source) is thought to be itself very small 1–2 nm colloidal particles. If this is so, then the lengthy time for final reaction may be due to the time it takes to depolymerize the basic salt back to molecules so that full surface coverage can take place.

A final confirmation of the lengthy reaction time came from the customer who found that the same material that made bad product when fresh made excellent material after it had aged for 6 months. Since then this "aging" phenomenon has been observed in a few additional applications but for the most part is not a problem in most applications of the positive sols.

APPLICATIONS

When I interviewed with DuPont so many years ago I had the privilege of meeting Ralph Iler. The group was working on other colloidal systems than silica, but one couldn't spend any time with Ralph without hearing about Ludox® colloidal silica. I remember thinking at the time, "What could anyone find to do with *sand and water*?" After all, every chemist knew that silica was virtually inert. At least I was smart enough not to express those thoughts in front of Ralph. I have thought many times over the years how wrong I was. One of the most interesting aspects of dealing with colloidal silica is the wide variety of end uses that have been found for it. We will most certainly leave some out of this discussion because the many details are held as trade secrets and so are not known to us. But we will try to cover as broad a spectrum as we can.

In some instances where the process is large and/or complicated and the technology known to us the discussions are presented in the form of a short process summary followed by a section in which the details are discussed more extensively. The figures in each section are meant to tell the story of the application and you may wish to look at them even if you only read the summary.

The following categories will be discussed below:

- High Temperature Bonding
  - investment casting
  - fiber bonding
  - Space Shuttle tiles
  - stool coating
  - hot top boards
- Surface Modification
  - frictionizing
  - plastic films
    - anti-blocking
    - food wrap application
  - anti-soil
- Flocculation
  - paper making
  - wine and juice fining
  - water purification
- Silicon wafer polishing
- Miscellaneous
  - steam promoter
  - chromatography
  - catalysis
  - photography

HIGH TEMPERATURE BONDING

If one were to list all the bonding agents or glues in the world in order of their bonding strength at room temperature, colloidal silica would be near the bottom of the list. Near the bottom but not at the bottom. Colloidal silica when used as a binder for ceramic powders has a "green" (green here means as dried at room temperature) strength on the order of several hundred psi. Good glue might have a green strength of 10–100 times that.

However, if one repeats the rating at about 1000°C colloidal silica would be at or very near the top of the list. All the organic resins and adhesives that superceded it on the room temperature list would have burned away and many of the inorganics like sodium silicate which form stronger bonds at room temperature would be molten at this temperature. The bond strength of colloidal silica typically increases with firing and will be of the order double its "green" strength. Equally important is that it does not lose strength at intermediate temperatures while being heated to the high temperature. When cooled back to room temperature colloidal silica retains much of its strength. In some uses this is a benefit, in others a problem.

Investment Casting

Investment casting is the single largest use for colloidal silica in the area of high temperature bonding. The process is summarized below:

Summary

- Wax or plastic patterns (replicas) of the metal parts needed are produced by plastic injection molding techniques.
- Slurry of colloidal silica and refractory powder (silica, alumina zircon, etc.) is prepared. Slurries typically have the consistency of a thin pancake or crepe batter.
- The patterns are immersed in the slurry, drained, coated with dry refractory, and allowed to dry.
- The dipping is repeated as above until a ceramic shell of the required thickness (typically 0.2–0.5 in.) is built up on the patterns.
- The wax is melted from the ceramic shell.
- The shell is fired to burn off any residual wax and to gain strength.
- Molten metal is poured into the shell and allowed to solidify.
- The shell is broken away and the metal replica of the original wax is recovered.

The process is used to make parts of complex geometry such as jet engine blades and vanes as well as a myriad of industrial parts. The only investment in complex or expensive machining is in the production of the die used to make the patterns. Only one is needed and can be used for the manufacture of thousands of identical parts. These parts can be produced very close to or at finished tolerances, minimizing or eliminating much final machining.

Details

Probably the largest end use for colloidal silica as a binder is in the investment casting industry. Investment casting is also known as the "lost wax" process. Some of the objects from the Egyptian tombs were cast this way. In ancient times an artist molded a wax original probably in beeswax and molded soft clay around it. When the clay dried, the assembly was heated to melt the wax and harden the clay. The wax was "lost" in the process, hence the name "lost wax." The interior of the clay pot formed in this way was the negative image of the original wax. The clay container was filled with molten metal and allowed to cool. The clay was broken away leaving behind a copy of the original wax in metal. This process is illustrated in Figure 12.35.

Solid Mold Casting

This process evolved to be used for relatively large statuary where smaller parts were molded and then welded together. For much smaller parts like jewelry items a solid molding process was evolved. Essentially, the wax pattern was immersed in a fluid plaster-like material

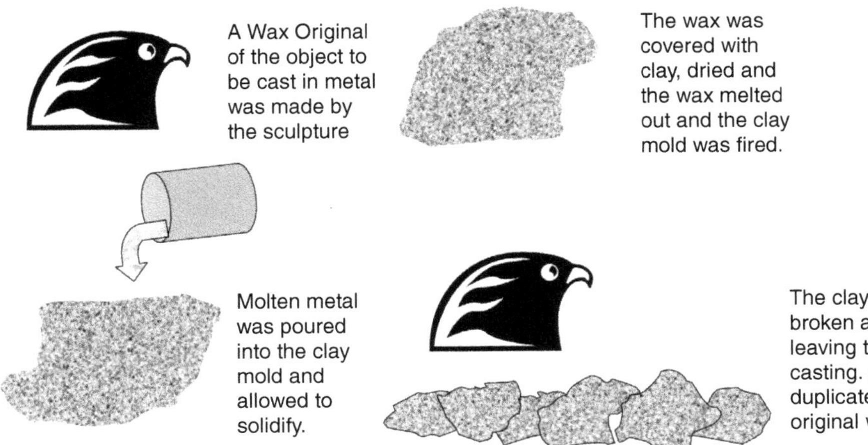

FIGURE 12.35 Lost wax casting as it may have been practiced.

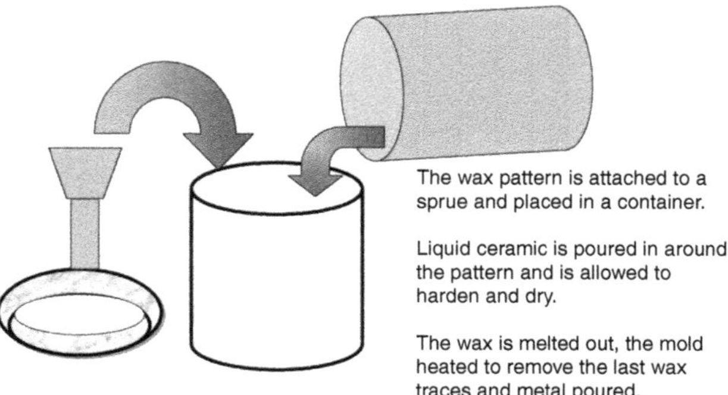

FIGURE 12.36 Solid mold investment casting.

such as "plaster of Paris" or a similar material that set chemically to a solid. The process is illustrated in Figure 12.36. The volume of ceramic material used in this process is quite large compared to the volume of the parts cast. Hence, the process is limited to articles of significant value where the high cost of the process is not a big issue. The process is still used today in dental labs and in the manufacture of fine jewelry items. However, the process is also limited in the types of alloys that can be cast this way. The high temperature alloys in use in many of the products we take for granted like the jet engine could not be cast in this type of mold system.

In the early 1940s several developments met with a need and resulted in the beginning of the modern investment casting industry.

The Need

The development of the jet engine required the manufacture of multiple blades to be made of very high temperature alloys. These blades looked very similar to miniature propellers. The complex airfoil geometry of the blades was very difficult to machine with the technology that existed at the time. Remember this was many years before the numerically controlled tools we take for granted today. This produced a need for a way to manufacture such blades and vanes at virtually finished tolerances to minimize the machining needed.

The Developments

- At about this time the development of plastic injection molding provided a way to produce multiple identical wax patterns.
- Colloidal silica binders became available commercially just after World War II, providing a material sufficiently strong and refractory to bond the molds together.

The process as it is practiced now actually produces a ceramic shell of modest thickness (about 0.2–0.5 in. thick) around the wax pattern. The relatively thin shell when compared to solid mold casting not only saves material but also allows for more uniform and rapid cooling of the metal parts. This is an important factor in maintaining fine grain size that results in stronger parts.

Shell Investment Casting

The conversion of plastic injection molding machines to run at the lower temperatures and pressures required for making parts of wax instead of plastic was a natural technological jump. A great deal of machining expertise goes into making a mold for the plastic injection molder, but only one is needed for each type of part. Using the master mold in conjunction with the automation of the injection molding machine patterns can be produced rapidly. Cycle time for the machine is usually 10–30 sec for small parts. As long as pressures, temperatures, and times are maintained fairly consistent the parts themselves will be consistent. A molding operator then can use the master mold to produce thousands of wax patterns in a relatively automated uniform process.

When the parts are small they are often affixed to a central wax bar or sprue. This array is often referred to as a "tree," since it resembles a tree trunk with branches attached. The process is illustrated in Figure 12.37–Figure 12.41.

The Function of Colloidal Silica

Colloidal silica serves two functions in the process. Since it is a low viscosity water-based liquid it is used as the vehicle to slurry the ceramic powders which make up the ceramic shell. The viscosities of slurries used in the process are usually similar to a thin crepe or pancake batter. They must be thin enough to flow into the surface details of the wax patterns such as the narrow grooves in a golf club head, but thick enough to leave a continuous

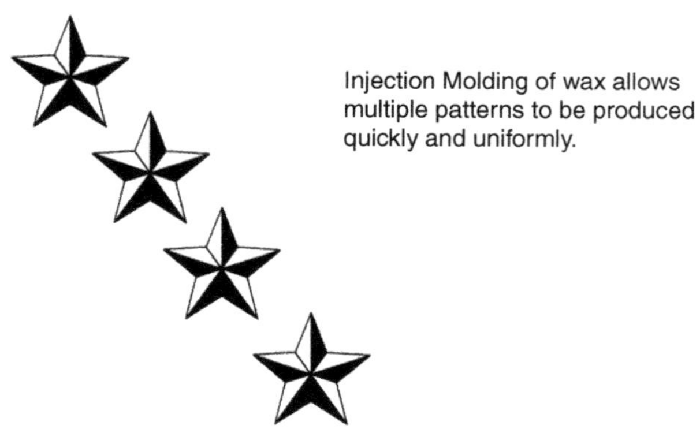

FIGURE 12.37 Modern shell investment casting pattern production.

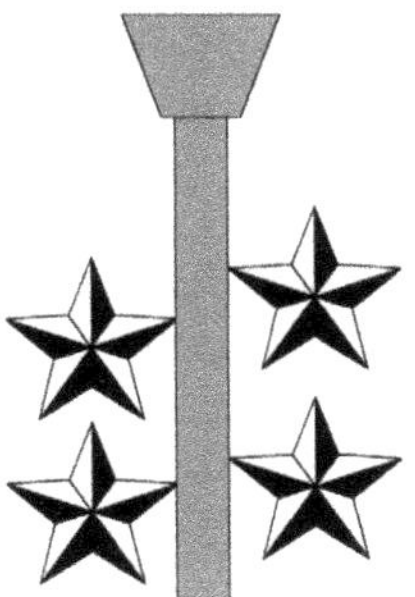

FIGURE 12.38 Wax patterns mounted to a central wax sprue.

coating of smooth ceramic. When the water is allowed to dry, the colloidal silica particles act as the glue that bonds the ceramic particles together.

FIGURE 12.40 The pattern tree is drained of excess slurry and stucco sand is added, the shell is dried and additional coats are added for strength.

Face Coat Slurries

Typically at least two slurries are used. The first of these is called the "face coat" or "prime coat" slurry. It is made up of very fine ceramic flour usually 200 mesh or finer. This slurry is the first slurry the wax patterns are dipped into and the very fine texture is required to faithfully replicate the fine details of the pattern. It is also the surface that the molten metal comes in contact with, when it is poured into the mold. Because of this, it is very important that the ceramic selected for this first or "prime" layer be inert chemically to the molten metal being poured. If not, the molten metal will react with the ceramic and lose its desirable metallic character.

Another characteristic the first dip must have is that it has to "wet" the wax surface and form a uniform layer. Since waxes are seldom hydrophilic, wetting agents are added to the face coat slurry to enable it to wet the surface. In addition, the wax patterns are typically cleaned in order to remove any traces of the mold release (usually a silicone) which was used during the pattern molding process. In some cases the wax patterns after cleaning, are dipped into a surfactant solution, allowed to dry and then dipped into the face coat slurry, which may or may not contain additional surfactant.

It would be even more desirable if the face coat actually bonded to the wax pattern, but this does not happen. The wetting agents form a layer on the surface of the wax that inhibits actual contact with the colloidal silica particles. DuPont developed and patented a product called Ludox® IWS [5]. IWS is a mixture of colloidal silica and zirconia, which actually appears to bond the organic surfaces at least strongly enough to resist it has being washed away by repetitious rinsing.

If a hydrophobic wax pattern is dipped in a surfactant-water mixture uniform wetting of the surface can be observed. However, if the pattern is rinsed in water, the surfactant is rinsed away and the surface again becomes

FIGURE 12.39 Pattern tree is dipped in ceramic slurry.

FIGURE 12.41 The wax is melted from the completed shell the shell is fired to remove the last wax traces molten metal is poured into the shell and allowed to solidify the shell is removed and the metal parts removed from the sprue.

hydrophobic. Water beads on the surface. If the same pattern is dipped in IWS the same uniform wetting of the surface is observed, but is not eliminated with rinsing. This would appear to indicate that at least a monolayer of colloidal oxide is retained on the surface of the wax pattern, rendering it hydrophilic. IWS acts in a similar way on many polymer surfaces in addition to wax. It is interesting to note that neither of the individual colloidal oxides alone act to hydrophilize the wax surface, only the mixture of the two. Wax patterns treated with IWS do not require the use of surfactants in the face coat slurry to achieve uniform wetting.

After the pattern is dipped in the prime slurry, the excess slurry is drained away in such a way as to leave a uniform coating over the entire surface of the pattern. Then a thin, uniform layer of dry, fine sand typically about 100 mesh is applied to the wet slurry coating. This layer of dry material is referred to as "stucco." One of its purposes is to absorb moisture from the slurry. This raises the viscosity of the slurry and slows or stops its flow off the pattern. It also serves as a rough surface that improves the adhesion of the next slurry layer.

The prime coat of refractory on the wax is usually only about 0.010–0.025 in. thick and is not strong enough by itself to support the load of the molten metal being poured. It is dried and then additional layers of slurry are added, stuccoed, and dried. Usually no more than two prime coats of very fine refractory are used. The rest of the shell is made up of "back-up" coats.

Back-Up Slurries

Back-up slurries are typically thinner in viscosity and made up of larger refractory grains 200 mesh or larger. Similarly, the stucco sizes used in the back-up coats are larger than those used in the prime coats. Even though the viscosity is thinner, the larger stucco sizes cause the back-up coats to be thicker than prime coats typically. It is a fact that the inherent strength of these back-up slurries is less than that of the prime slurries. One might ask then: Why not make the shells entirely of finely divided flours and stuccoes? There are a couple of good reasons. If we did make the entire shell in this way it would be much thinner after an equal number of coats than one made with the normal back-up slurries. This is because the larger refractories, especially the large stucco grains build the thickness of the shell more rapidly. The strength of the fine refractory measured as its modulus of rupture will be higher. However, the load bearing capacity of the shell increases as the square of the thickness, but only linearly with the MOR.

$$L = C \times MOR \times W \times T^2$$

where C = a constant; MOR = the modulus of rupture; W = the width of the specimen; T = the thickness of the specimen.

Think about a toothpick and a 2 × 4. They will have the same MOR assuming they are made of the same kind of wood. However, you can not walk across a toothpick, it is too small in cross-section to support the load. The extra width and especially the added thickness of the 2×4 allow it to support a much greater load. The same is true of the thicker ceramic shell. As an example let us consider two ceramic systems one (A) with a modulus of 600 psi and (B) with a modulus of 400 psi. If 5 coats of shell system A have a thickness of 0.25 in. and 5 coats of shell system B have a thickness of 0.5 in. and we assume they both have the same width then the load bearing capacity will be in proportion the square of their thickness as follows:

$$L(A) = C \times W \times 600 \times (0.25)^2 \quad \text{and}$$

$$L(B) = C \times W \times 400 \times (0.5)^2$$

$$L(A):L(B) = (600 \times 0.0625)/(400 \times 0.25) = 37.5:100$$

The weaker (lower MOR), but thicker shell can support more than 2.5 times the load of the stronger, but thinner shell system.

There is another important reason why shells are constructed in this manner. The very fine refractories used in the prime or face-coat slurries make fairly dense ceramics that are not very permeable to air movement through the shell. It is important that air be able to move through the shell walls as the metal flows in and displaces it. Unlike sand castings that have riser tubes to allow the air to escape, shell investment castings depend on the air being able to escape through the shell wall. If the mold cavity cannot fill completely with metal the cast part will be incomplete and useless. The larger refractories used in the back up coats have larger pores between them and therefore are more permeable to air.

Optimizing Shell Strength

Colloidal silica is not a good film former. If it was, one might think that using colloidal silica at maximum would result in the highest strength in the ceramic shell. However, this is not the case.

As water evaporates from colloidal silica, two things happen:

1. The colloidal silica concentration increases
2. The volume of the system decreases

All colloidal silicas have a concentration at which they will gel. The smaller the particle size, the lower the gelling concentration. Typically, this gel concentration will be about 35% silica solids for a 7-nm sol, 45% for a 12-nm sol and 55% for a 22-nm sol. Since the silica is more than twice as dense as water, the % volume solids of such gels will be between about 15 and 25%, and they will contain 75–85% water by volume. The bonds formed between particles are generally not very elastic and the continuous networks of particles formed through the water are brittle. As water continues to evaporate, it exerts very high capillary pressures on the silica structure and causes it to fracture into small pieces.

If however, the colloidal silica is used as a thin layer between refractory particles and the refractory particles were in close contact so that, only the thickness of a few particles were needed to bridge the gap between them a better bond should result.

If you consider how you glue two boards together, the same principles apply:

- Use a thin layer of glue; cover all the contact surfaces. (No one would consider leaving a quarter inch layer of glue between the boards.)
- Clamp securely to further thin the glue line and push the two board surfaces together. (In the case of slurries, the surface tension of the liquid water does the clamping for us — pulling the particles into close contact.)

Of course, if very little or no colloidal silica particles are present in the shell will have little or no strength. Remember the sandcastles we build on the beach. They barely hold together while wet, and fall apart when dry. The strength builds as the binder silica content increases. At the point where the binder solids gel just as the refractory comes in closest contact, the strength reaches it is maximum. Past this point where the binder silica concentration causes it to gel before closest contact of the refractory is achieved the strength falls off.

These three conditions and the typical form of the strength curve are illustrated in Figure 12.42–Figure 12.47. Sears first quantified this relationship [6]. His definition of the optimum colloidal silica content was based on knowing the surface area of both the binder colloidal silica and the refractory powders being used. While colloidal

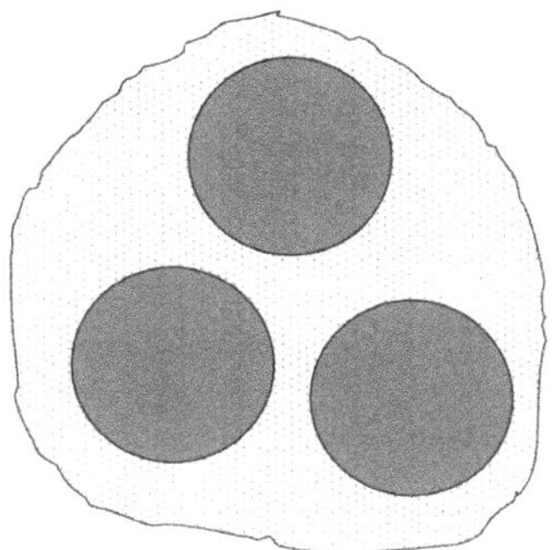

FIGURE 12.42 Optimizing bond strength of colloidal silica in investment casting slurries.

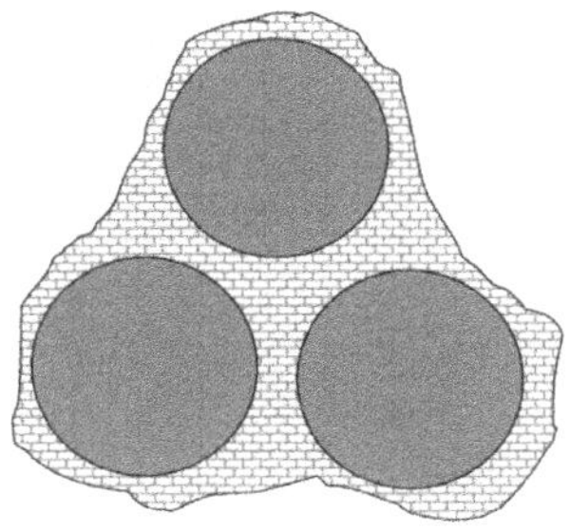

FIGURE 12.43 Slurry shrinks during water loss pushing the refractory particles closer and concentrating the colloidal silica.

silica manufacturers routinely determined the surface area of the binders (using the method Sears invented); refractories manufacturers do not commonly measure the surface area of their products.

The application of a log-log equation as part of the everyday formulation work in smaller foundries was difficult. While it is solution gave the appropriate binder silica needed for optimum strength, the amount of water needed to achieve the best viscosity characteristics had to be determined after the fact.

Because of this Roberts [7] took a different approach and determined the strength of slurries made at the same viscosity with binders containing different levels of colloidal silica. The binders were pre-diluted to the desired level and refractories added to make slurries. The strength of each was measured, plotted, and the optimum binder silica content determined from the plot. In all cases, the slurries were weak at low binder silica levels, went through a maximum and then became weaker when binder having higher silica contents were used.

The general form of the curve is followed for virtually all particle size sols although the actual value of the binder solids maximum varies with particle size. The strength maximum usually occurs at about 70–80% of the maximum possible binder strength for that particle size. This is indicated in Table 12.3.

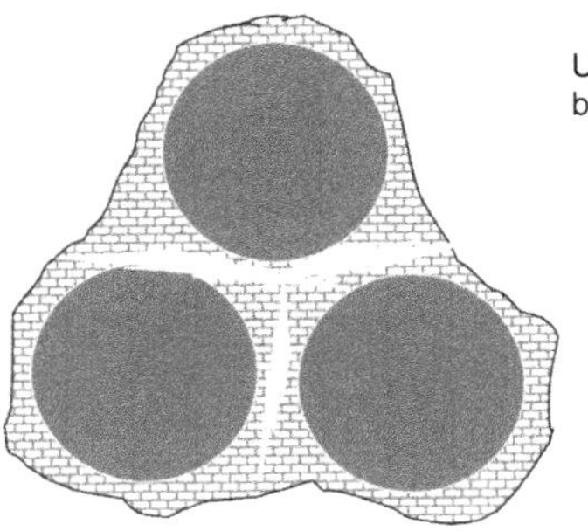

FIGURE 12.44 Colloidal silica concentration too high.

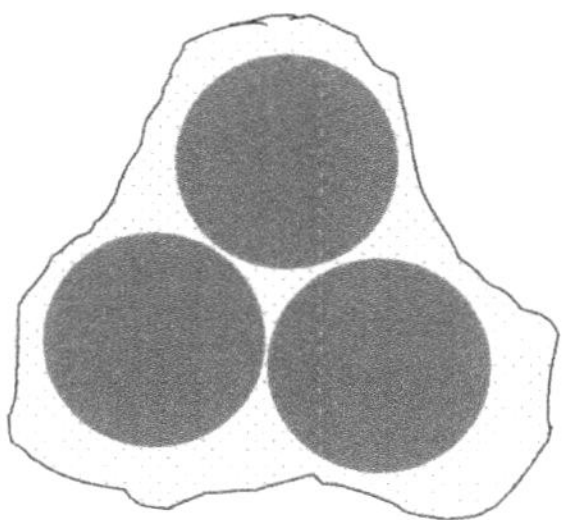

FIGURE 12.45 Colloidal silica concentration too small.

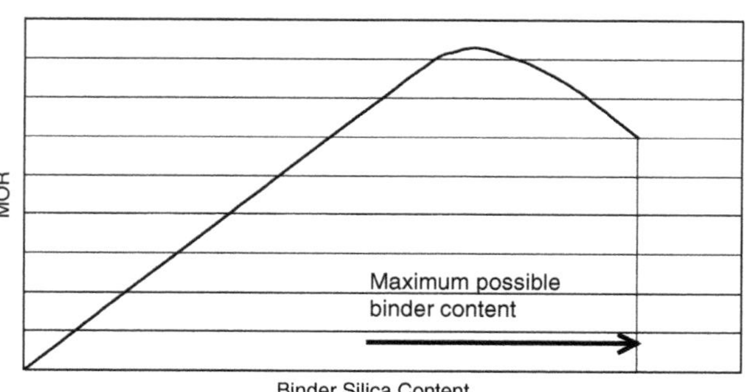

FIGURE 12.47 Typical MOR strength response of a slurry vs. binder silica content.

It is interesting to note that 12–14 nm sols were the first products to become commercially available and at that time only 30% product was available. This product was used as delivered for most of the early shell investment casting work. The smaller particle size sols were not available then and the ability to produce the 12–14 nm sols at 40%, which we take for granted now, had not been developed. Thus, fortuitously, the first product available was produced at the optimum concentration for use as an investment casting binder.

Maintaining Slurries

Investment casting slurries are made to a fairly precise formula utilizing the desired binder silica in the slurry to give maximum strength and the amount of refractory to give the slurry the proper rheology. The slurries must be continuously agitated in order to maintain the refractory in suspension. This is occasionally done with an agitator projecting down into the tank, but the presence of an agitator limits the space available for dipping patterns and more often the tank itself is rotated and a fixed L-shaped provides the needed turbulence to keep the slurry in uniform suspension. This type of setup is illustrated in Figure 12.48. The bar, being off to the side and at the tank bottom makes the bulk of the tank surface and volume available for dipping parts.

Since dipping parts is a routine part of the process, performed frequently, the slurry tanks in most foundries are left open to the atmosphere, at least during the time when production is taking place. Hence, evaporation of water is continuous and is probably increased by the need to continuously agitate the tanks. Water must be added frequently to account for this continuous evaporation. The purity of the water used to replace the evaporated water is important since the impurities do not evaporate and are concentrated in the slurry over time.

Other impurities can enter the slurry by being leached from the refractories themselves. Calcium and Magnesium are impurities that can have very deleterious effects on the stability of colloidal silica. Some refractories like Zircon are acidic in their reaction with colloidal silica and tend to reduce the pH over time. The stability of alkali stabilized colloidal silicas depend both on minimizing the presence of impurities (especially polyvalent metal ions) as

TABLE 12.3

Particle size	Maximum binder silica	Strength maximum
7–8 nm	30%	21–24% Silica
12–14 nm	40%	28–31% Silica
22 nm	50%	35–40% Silica

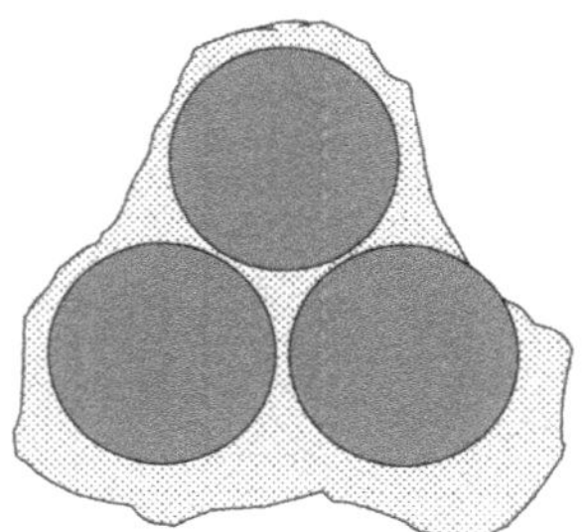

FIGURE 12.46 Collloidal silica at optimum concentration.

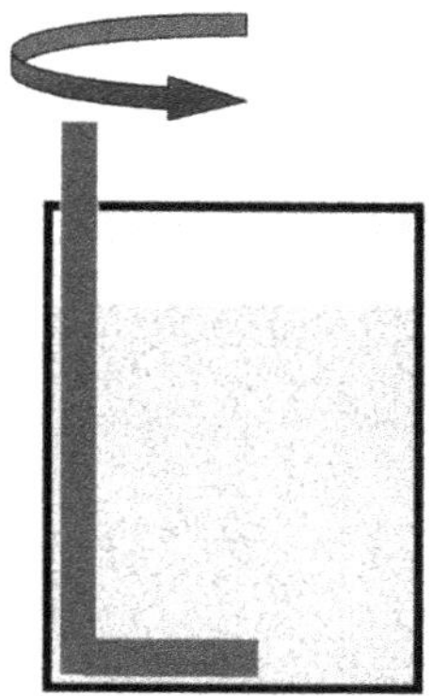

FIGURE 12.48 Rotating slurry tank for investment casting.

well as maintaining pH. Slurry operators then typically use distilled water to replace water lost and monitor the pH of the slurries. If the pH is drifting into an unstable range corrective efforts to maintain pH in the desired range are usually taken.

Adding sodium hydroxide to the system could be done to maintain pH; although adding NaOH at a higher concentration than about 5% usually results in some irreversible coagulation of the colloidal silica. Such a coagulum would result in a weak spot in the shell should it become part of the structure. In addition to the problem of coagulation, adding sodium hydroxide decreases the SiO_2:Na_2O ratio. This has the effect of decreasing the refractory character of the colloidal silica binder and decreasing its useful maximum temperature.

For those reasons, pH maintenance is more often done with the addition of ammonium hydroxide. Although a weaker base of sodium hydroxide can be used to maintain the pH at 9.5+. The good news is, that it evaporates with the water when the shell is dried and does not change the chemistry of the dried refractory. The bad news is, it is lost to the atmosphere and must be replaced frequently. If excess amounts are used then the smell of ammonia fills the air and is unpleasant at best. One of the most satisfactory ways to add ammonia to the slurry by routine water addition needed to compensate for evaporation. In some plants this addition is done in the form of a continuous drip into the slurry. Spot checks and experience allow the operators to adjust the drip rate to keep up with the evaporation rate. This is made easier since the slurry rooms are kept at constant temperature are humidity.

Binder Not Dependent on pH

A binder developed at DuPont called Ludox® SK [8,10] avoids the need for pH adjustments in order to maintain slurry stability. The product is based on aluminate modified colloidal silica. The aluminated product is deionized and because the alumina sites provide sufficient charge density the product is stable. Deionization to remove sodium hydroxide results in a pH which is acidic. As long as the pH remains above about 3, there is little danger of the alumina being leached from the product surface.

This deionized colloidal silica is not a strong binder. The removal of the sodium hydroxide, which appears to act as a binder catalyst, and/or the presence of the alumina on the surface, weakens the overall bond strength developed. However, this acidic version of colloidal silica does form stable mixtures with several water soluble polymers, the most important of which is polyvinyl alcohol (PVA). PVA is itself a good film former and imparts this property of much improved film formation to colloidal silica even when present as only about 3–5% of the total solids. The mixture of PVA and acid stabilized colloidal silica has been patented and commercialized as Ludox® SK. Slurries made with Ludox® SK have excellent slurry life with only water addition to correct for evaporative losses needed to maintain the slurries.

Refractory Fiber Bonding

Summary

- Refractory fibers, alumino-silicate typically, are suspended in colloidal silica to form thin slurry.
- The colloidal silica can be present as a major component of the liquid or as a minor component when coupled with a green strength binder such as starch.
- The function of the colloidal silica then is to help flocculate the starch and fibers.
- The slurry is drawn, usually with vacuum through a screen. The screen may have been shaped to form the desired part geometry.
- The liquid flows through the screen and the fibers remain behind on the screen.
- When sufficient fiber thickness has been developed, the form is removed from the slurry, the excess liquid is vacuumed off, and the part separated from the form.
- It may be dried and used as is, or
- The part may be re-saturated with higher strength colloidal silica to harden the finished part, and then dried, or
- It may be shipped wet to the customer to be dried at his facility, or
- The part may be dried, post-impregnated with colloidal silica and dried again. This can be repeated in order to build a dense, ceramic-like body.

Ceramic shapes made in this way have excellent insulating characteristics and relatively low weight when compared to firebrick types of insulation. Fiber bonded refractories also have very good thermal shock characteristics.

Details

Refractory fibers are generally alumino-silicate compositions although even more refractory fibers such as pure alumina are available. These fibers have much higher melting points than the glass wool fibers which most of us use as insulation in our homes. For this reason they find utility in high temperature insulation material such as might be required in kilns and high temperature ovens and furnaces found in laboratories and the manufacture of ceramics, china and porcelain.

While kilns made of firebrick have done the job for centuries, fiber insulation has an advantage in weight. The same level of insulation can be achieved with much less weight of refractory fiber insulation than firebricks.

The larger thermal mass of the firebricks requires that a kiln made with bricks must be pre-heated much longer (requires more energy input) to come to the same temperature compared to the kiln insulated with fiber insulation. Also, fiber insulation is less prone to thermal shock during heating or cooling than firebricks.

The manufacture of refractory fiber bonded shapes is very similar to paper making. In papermaking the slurry of wood fibers in water flows onto a screen. The water goes through the screen and the wood fibers remain on the screen and form a sheet of paper. In preparing refractory fiber items the process is similar except that a vacuum is used to pull the slurry through the screen allowing the water to pass and collecting the fibers on the screen. The screen can be shaped and hence articles having simple shapes can be made directly. This process is illustrated in Figure 12.49–Figure 12.51.

Colloidal silica can serve a number of uses in this process. Clearly, after the parts have been fired, the colloidal silica that is contained in the product is the only binder that remains. However, during the forming process a green strength binder may be used such as starch. If only cationic starch is added to the water along with the fibers, some of the starch would be attracted to the fibers that are somewhat anionic in charge. However most of the starch remains in solution. During the vacuum forming, much of it would go through the screen with the water and be lost.

However, if a small amount of colloidal silica is added, the negatively charged silica particles flocculate the excess starch as well as the starch coated fibers. The large flocs of fibers are readily and efficiently filtered from the water. In addition, the flocs of excess starch are trapped within the fiber matrix where they can act as binder. This increased efficiency in filtration has the effect of leaving the water phase much less contaminated with fibers and starch which presents much less of a disposal problem. This is illustrated in Figure 12.52 and Figure 12.53.

In some cases the colloidal silica is used at fairly high concentration in the water phase without the addition of starch or other green strength additive. In this case the amount of colloidal silica left to bind the fibers together is proportional to the concentration in the water phase. This is perhaps the simplest fiber bonding technique although it gives rise to a couple of complications.

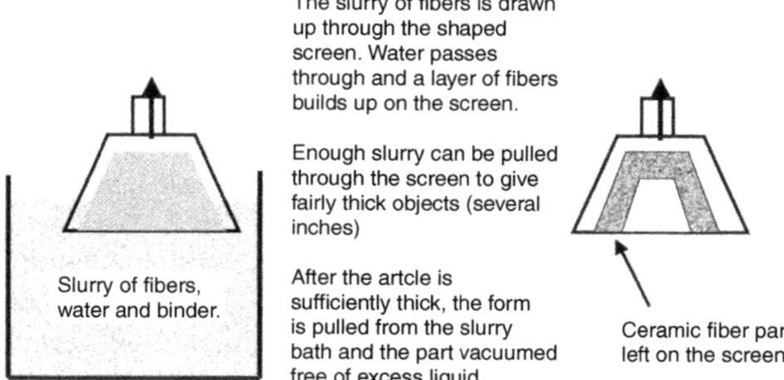

FIGURE 12.50 Vacuum is used to draw the slurry through the screen form.

First, the water must be recycled in order to make efficient use of the colloidal silica. Since typically the fiber loading in the slurry is only 1–3% by weight, only a very small amount of colloidal silica can be retained on the fibers as they are filtered out on the screen. In recycling the "white water" as it is called, small amounts of broken fibers and impurities leached from the fibers can compromise the utility of the remaining colloidal silica.

In addition, without a flocculating agent present or some other additive which prevents migration of the colloidal silica particles, drying the product, especially if it is a thick cross-section can lead to migration of the colloidal silica particles to the evaporating surface. Water has a natural tendency to be carried to the surface of fiber articles along the small orifices that form between fibers. The colloidal silica products remain entrained in the water phase as it is carried by capillary action to the surface. As the particles arrive at the surface and water continues to evaporate the particles are left behind and form a hard layer at the surface. However, at the interior of the part, where most of the fibers are virtually free of colloidal silica, the material may be very soft. This phenomenon is illustrated in Figure 12.54.

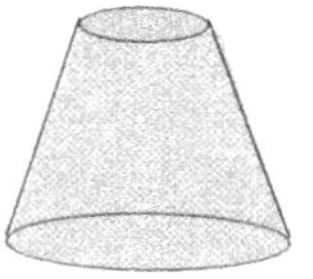

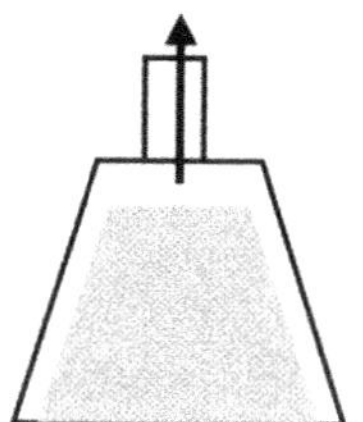

FIGURE 12.49 A screen form is used to allow the passage of the liquid but not the fibers in the slurry.

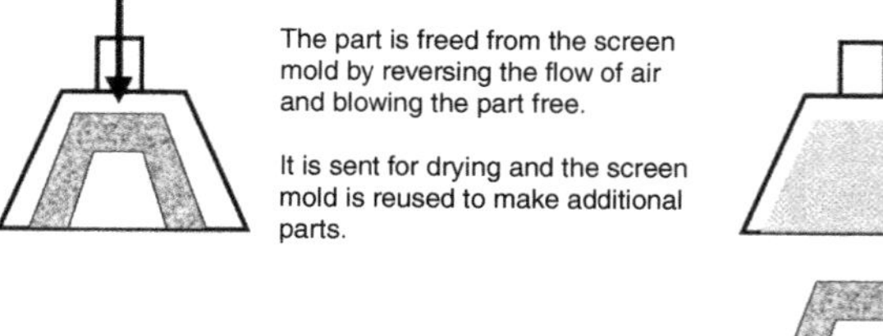

FIGURE 12.51 Formed parts are removed from the screen and dried. The screen is reused to produce more parts.

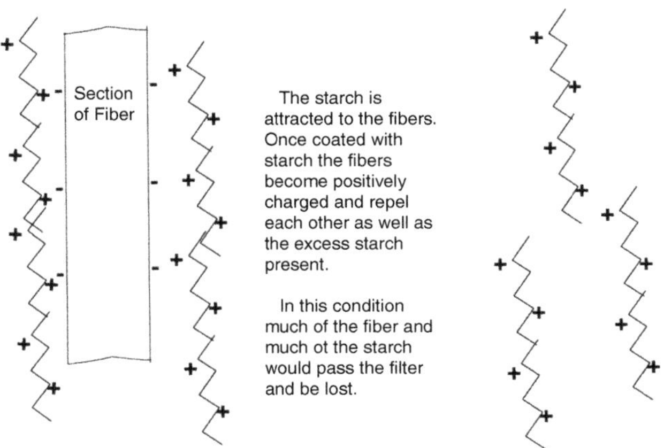

FIGURE 12.52 Silica as a flocculent.

In some products it is desirable to have a hard surface and a less dense (better insulating) interior. Parts whose surface might be exposed to the erosive effects of high velocity hot gasses would be examples of parts where a harder surface would prevent premature breakdown. However, if starch, or some other cationic agent is present, it captures much of the colloidal silica by charge attraction and prevents the migration from being as pronounced. But it can only hold back an amount of colloidal silica proportional to its charge density. Charge sites already occupied by colloidal silica particles will allow both water and silica particles to flow by.

Post Impregnation

It is often desired that parts be "hardened" by the addition of more colloidal silica to the already formed parts. One such product is a ceramic fireplace log. It is formed from fibers and post impregnated to give relatively high densities. Fillers (ceramic powders) may be incorporated in the fiber slurry from which the logs are formed giving a higher density than if the fibers alone were formed into the shapes. In some cases more than one impregnation may be needed in order to achieve the desired density.

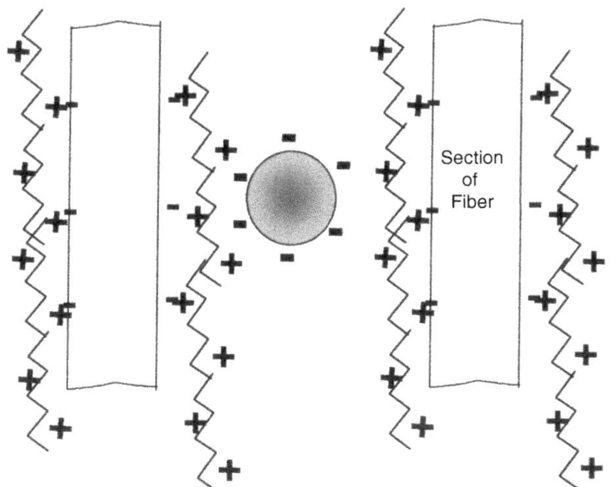

FIGURE 12.53 Colloidal silica flocculates fibers and starch by charge attraction.

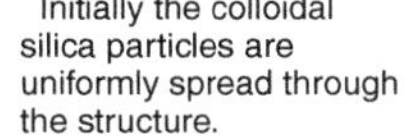

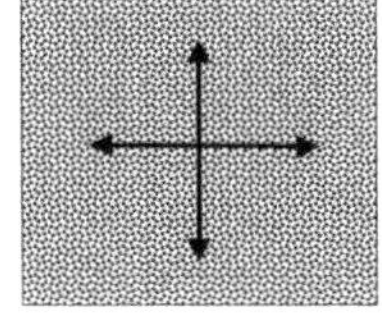

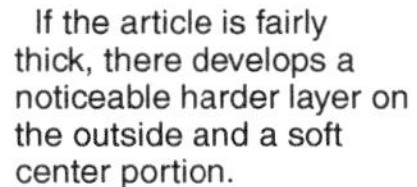

FIGURE 12.54 Migration of colloidal silica in fiber bonded parts.

Wet Blankets

One of the interesting ways fiber bonded materials are incorporated into furnaces is with the use of blankets of fibers wet with colloidal silica from the manufacturing process. The fibers, colloidal silica and additives are formed into the simple shape of a blanket of fibers. While still wet, the blankets are put in plastic bags to prevent water evaporation. These are shipped to the user and the wet blanket is removed from the plastic bag just prior to its being incorporated into the furnace wall. The water is then free to evaporate and the colloidal silica particles bond the structure together into the given shape.

Ammonia to Replace Sodium

Sodium plays numerous roles in stabilizing and affecting the properties of colloidal silica. The sodium-stabilized products are routinely used in applications up to about 2300–2400°F (circa 1300°C). Above these temperatures, even the small amount of sodium present acts to rapidly catalyze the amorphous colloidal silica products into a crystalline form of silica, cristobalite. Amorphous silica has a low coefficient of expansion up to its melting point. Cristobalite has a fairly high expansion rate and changes structure on cooling. This disrupts the structure of the bonds, weakening them. A furnace insulated with such a lining could not hold up in a condition where the temperature was cycled on and off at intervals. During the heated cycle conversion of the amorphous silica to cristobalite would occur. When the furnace was shut down the cooling would trigger the structure changes and the bonding would be weakened significantly. With several cycles it is possible that the insulation would begin to disintegrate.

One way to minimize the problem is to remove as much of the sodium hydroxide as possible and replace it with ammonium hydroxide to maintain the desired pH. When the product is dried the ammonia goes off with the water and leaves a purer form of amorphous silica behind. With less sodium content to catalyze the conversion to cristobalite the useful temperature range of the insulation can be increased, perhaps as much as 200 degrees. The use of mixed bed ion exchange resins in the elimination of sodium content and the manufacture of the ammonia-stabilized grades is discussed in the Manufacturing section above.

Occluded Alkali

Colloidal silica deionized and re-stabilized with ammonia still retains some of its sodium content. Even with several deionizations, it is difficult to get the sodium content below about 100 ppm. To put this in perspective this is equivalent to 5 sodium atoms being left in each particle, assuming 12 nm particles. Considering the small size of sodium ions and their fairly high concentration during the particle growth cycle of the manufacturing, it is not unlikely that these difficult to remove impurities were entrapped in the particles during particle growth. Once the particles have been formed and have achieved their finished density, removal of these sodium ions must depend on their diffusing to the particle surface where they can be removed by deionization. Solid state diffusion would be anticipated to be quite slow and so seven with aging and heating in between deionizations the sodium content is very difficult to get below 100 ppm.

In spite of the difficulties this technique of multiple deionizations and stabilization with ammonia is the technique used to generate the high purity silica needed for the insulating tiles that protect the space shuttle from burning up on re-entry. We will discuss the use in the space shuttle tiles below.

The presence of alumina in the surface of colloidal silica particles not only provides for charge sites independent of pH but also inhibits the conversion to cristobalite. Hence it is possible to add alumina to the surface of colloidal silica by reacting with sodium aluminate. The manufacture of such products is discussd above in the first part of this chapter. After alumina is added to the particle surface, the particles are deionized to remove the sodium. This provides a product that is both stable, low in sodium, and has a pH of 3–5. Adjusting the pH with ammonia is not necessary. The product is stable at the lower pH that results from being deionized because the alumina provides sufficient charge density to avoid coagulation. If mixed bed deionization is used then the absence of salts further enhances the stability of such a product.

It should be pointed out however, that the "green" bonds of such a product are not as strong as sodium stabilized products. The sodium catalyzes the formation of bonds between particles and other oxide surfaces and without in bond strength is cut by roughly 40–50%.

New fibers are being developed which dissolve in the slightly alkaline pH of the lung. These fibers may eventually replace the more common alumino-silicate fibers especially in use where human contact with breathable fibers is likely. Alumino-silicate fibers do not have the same small size as asbestos fibers, but the mere fact that they are fibers, could become airborne, and the body has no rejection mechanism, has prompted some countries to view them with suspicion. Having fibers which provide the same thermal properties but which would dissolved in the lung would preempt the possible problem. However, fibers which dissolve in alkaline conditions cannot be used with alkali stabilized colloidal silicas. They need a binder which is stable but acidic in pH as described above. Such products were first developed at DuPont and are now available through W.R. Grace who acquired the Ludox® colloidal silica business from DuPont.

Space Shuttle Tiles

If you ever visited the Kennedy Space Center in Florida you may have seen a surprising demonstration. A small block of the silica material that makes up the tiles used on the shuttle is heated to well above red heat. The demonstrator turns off the flame and almost immediately picks up the block with bare hands to demonstrate how fast the material can dissipate heat. Some of the silica that goes into this remarkable material is multiply deionized colloidal silica re-stabilized with ammonia. The material is aged in between deionizations to allow sodium to leave the particles and enter the water phase where they can be removed. The product is shown to be suitable for use when it can be heated to a designated temperature and avoid significant conversion to cristobalite.

Should tiles that could convert to cristobalite be used on the shuttle, multiple uses of the vehicle would not be possible. The amorphous tiles would convert to crystalline material from the heat of re-entry, but would likely remain intact until the temperature fell below the conversion temperature for cristobalite to convert to another form at about 700°C. This conversion and thermal shock could cause sufficient weakening of the material to cause the tiles to spall allowing the shuttle to overheat.

If the tiles survived the first reentry it would be very unlikely to survive a second one. The tiles converted to crystalline silica on the first mission would go through the destructive forces of changing crystal structure both in the heating and cooling during reentry on the next mission and would almost certainly disintegrate.

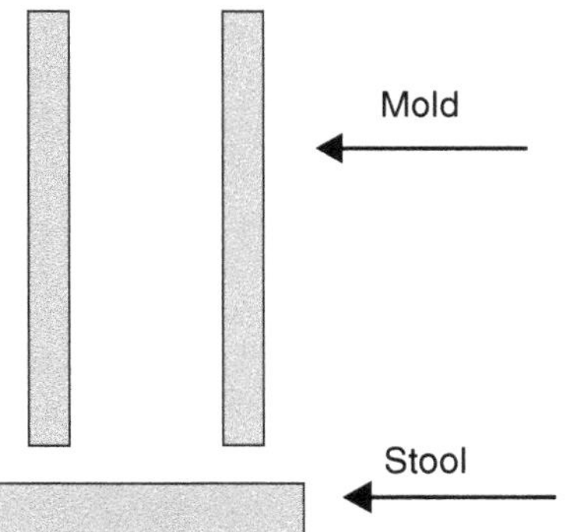

FIGURE 12.55 Ingot mold in cross-section.

Stool Coating

Large steel ingots are cast using an open-ended mold that is set on a large metal plate that forms the bottom of the mold. See Figure 12.55. This bottom plate is called the "stool". If molten metal were poured directly onto the stool, it would likely weld itself to the stool and it would become impossible to remove the solidified ingot. In order to prevent this from happening, the stool is coated with a ceramic material that can use colloidal silica as a bonding agent. Stool coating formulations are sprayed onto the hot stool (hot from previous use in the casting process). The water must vaporize off rapidly; leaving behind a coherent refractory layer which can withstand the erosive effects of the molten metal and act as a release agent for the finished ingot.

Hot Top Boards

When large ingots of molten steel are coast there is a tendency for the outside and top of the ingot to cool rapidly and solidify first from contact with the relatively cold mold and exposure to the air. The remaining molten metal tends to be colder near the walls and bottoms of the mold. The center stays somewhat warmer. The metal cools and shrinks in volume as the material nearest the wall solidifies and the remaining molten material cools to the melting point. Because of the continuous shrinkage of the remaining molten metal a significant depression in the top of the ingot may develop. In extreme cases, the depression may go deep into the interior of the ingot. This process is illustrated in Figure 12.56. If such an ingot were used for rolling into sheet stock the interior "pipe" as it is called would result in severely flawed metal sheet.

By placing insulating material near the top of the mold to prevent the upper portion of the melt from losing heat rapidly, the upper portion of the ingot can be kept from starting to solidify immediately. This leaves a reservoir of molten metal at the top of the ingot and eliminates piping. The insulating materials used are called "Hot Top boards" since they are designed to keep the top of the ingot hot. Typically the board is made from relatively low temperature insulating material such as the mineral wool material used in some suspended ceiling tiles. By itself this material would collapse and lose its insulating value almost immediately after being exposed to the heat of the molten metal. However, if the board is impregnated with colloidal silica and dried it becomes tough enough to hold its structure at least through a single use.

Surface Modifications

Frictionizing

Summary

For those of you old enough to remember when floors were routinely coated with waxes, you may remember your

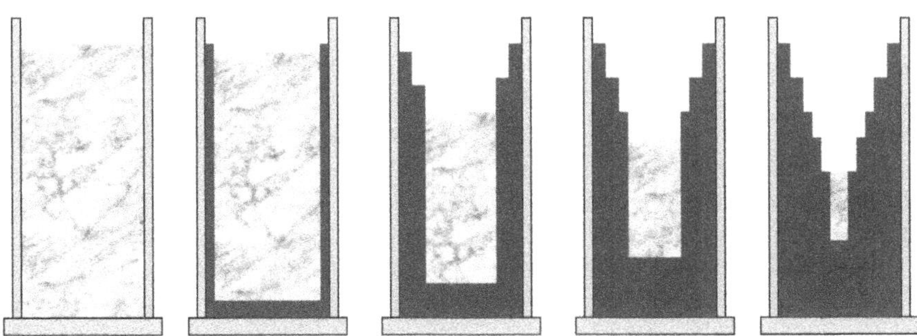

Molten metal poured above its melting point is chilled at the walls and bottom of the ingot mold and begins to solidify. The molten metal in the center cools more slowly and contracts as it cools. As the metal solidifies at the exterior of the molten zone and contraction continues a depression or in its worst condition a "pipe" develops in the center of the ingot.

FIGURE 12.56 "Piping" in metal ingots.

mother warning you to be careful because the floor was very slippery when freshly waxed. After it picked up some dirt and grit it became less slippery. Of course, it looked dirty and did not shine, so it had to be cleaned and re-waxed. Solving this problem of slippery waxed floors provided the first major application of colloidal silica. Waxes were supplied as water-borne emulsions. By adding colloidal silica to these waxes, it was possible to produce a waxed surface in which the surface was impregnated with colloidal silica particles (transparent dirt and grit) which improved the traction of the floor but did not affect the gloss of the floor.

While the frictionizing of floor polishes is not a big outlet for colloidal silica at this time, the frictionizing of paper and boxes is a significant use area. Slippery paper bags and boxes are a problem. They cannot be stacked easily without falling and they cannot be transported up steep conveyer lines. Paper bags and boxes that have high coefficients of friction can be stacked higher, conveyed up steeper conveyer belts, and transported with forklifts with less danger of spillage.

With the advent of recycling, paper has become more slippery than when it was primarily made from virgin fibers. In some mills almost all the paper is 100% recycled pulp and it is necessary to frictionize the paper during manufacture to make it possible to handle and transport without loss. Colloidal silica applied to the surface of paper can be thought of as converting it to extremely fine sandpaper and increases its coefficient of friction significantly.

One problem encountered in the manufacture of garments is that seams can slip during sewing, producing a pucker in the seam and therefore a reject. By treating the fabric with colloidal silica this tendency toward seam slippage can be reduced. Small amounts are used so the "hand" (the feel) of the fabric will not become stiff.

The use of colloidal silica in coatings applied to plastic films is fairly common and serves a variety of purposes. The most closely related to those above is its use as a frictionizing agent in order to make the conveying of the film through various coating stations easier. Freshly produced plastic film has a tendency to adhere to itself. This phenomenon is called "blocking." If such a film is rolled up, and adheres to itself so that it cannot be unrolled it is useless. By coating the film with a thin layer of colloidal silica particles, the tendency to "block" is eliminated. Once dried, colloidal silica particles no longer will bond to each other and their presence makes it impossible for adjacent polymer surfaces to come into direct contact.

Plastic films make up a great deal of the food packaging we are familiar with. Potato chips, pretzels and candy bars are just a few of the items whose principal packaging is plastic film. These films are coated with colloidal silica formulations to improve their friction and avoid blocking as above but they also have improved heat seal strength in the area where the two hot surfaces are forced into one another. This is presumably because the colloidal silica particles act as a reinforcing agent in the melt area. In addition better release characteristics for the area against the heating surface is observed probably because the colloidal silica particles prevent a large contact area of softened plastic from coming in contact with the heated surface. The coatings are also said to improve the printability of the film. This is important since food packaging is usually heavily printed.

Details

Floor Wax

At one time water based emulsions of waxes were routinely used on floors in the home and the work place. They imparted a wonderful shine to the floor but the coatings were very slippery underfoot. Slips and falls were a significant hazard. As dirt was tracked onto the floors the sheen and the slippery character disappeared. What was needed was an invisible "dirt" which would reduce the slip and not affect the shine.

By mixing colloidal silica in with the wax emulsion and applying it to the floor the resulting film has a high concentration of colloidal silica at the surface. In fact it accumulates in the surface, where it acts as transparent dirt and girt raising the coefficient of friction without affecting he gloss. In discussing binder migration in the section on fiber bonding was talked about how the colloidal silica particles tend to follow the water to the evaporating surface as long as liquid water is being transported through fine pores by capillary action. The same mechanism is operating here. The finished wax layer is diagrammed in Figure 12.57.

The action of the colloidal silica here is, in may ways, the same as the dirt tracked in onto the floor with time. However, the presence of the colloidal silica is not obvious to the eye and so it does not affect the look or gloss of the wax layer. In addition, the colloidal silica tends to harden the surface of the wax layer making it harder for dirt and grit to become imbedded in it. In this sense it acts like an anti-soil as well as a frictionizing material. The concept of the colloidal silica acting as invisible or transparent dirt is implicit in a number of applications.

For many of the applications the standard alkali stabilized colloidal silicas available were fine, but since the colloidal silica depended on the maintenance of fairly high pH in order to remain stable it was not compatible with lower pH wax emulsions. While the initial mixes might look quite stable the lower pH would eventually lead to slow coagulation of the silica resulting in poor shelf life for these products. The evolution of the aluminate modified materials which had much less dependence on pH for stability solved many of these problems and provided

Colloidal silica particles are uniformly dispersed in the wax emulsion when applied to the floor. As the water evaporates, the colloidal silica particles accumulate at the surface of the coating, providing a surface which is much less slippery than the wax alone.

Because the particles are so small, they do not affect the gloss of the wax.

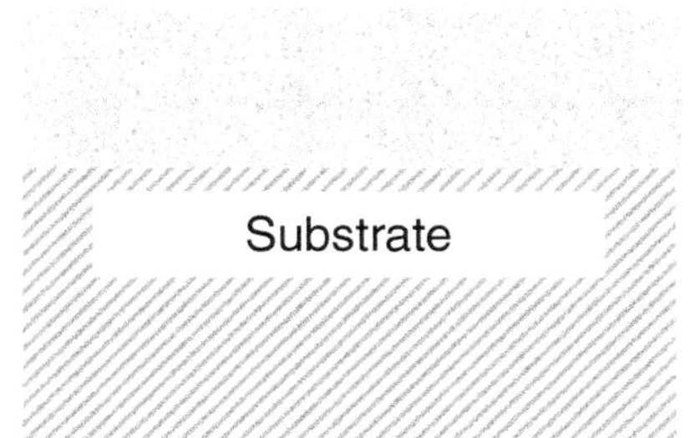

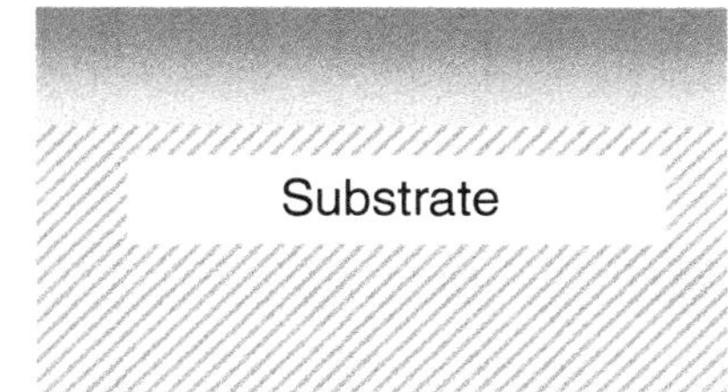

FIGURE 12.57 Nonslip floor wax.

a grade suitable for use in a broader selection of wax emulsions. Ludox® AM colloidal silica was the first such product made available.

Paper Frictionizing

One of the first industries interested in avoiding boxes slipping over one another was the glass bottle industry. It was the practice for bottles to be sent to the purchaser in cardboard boxes, which after the bottles were filled, provided the shipping container for the finished product. If at any stage a box of bottles fell from the stack, one could reasonably expect that the contents would be lost. Increasing the coefficient of friction without changing the appearance or the printability of the paper was an important concern. Colloidal silica when applied at only about 0.1–0.2 pounds per thousand square feet of paper provides a significant change in the coefficient of friction without changing the appearance of the paper surface. In fact it is so close to being invisible that an indicator for its presence has to be used.

Paper mills today make much of their products from 100%-recycled paper. In the recycling process previous coatings become contaminants and fibers become shorter, with the result that the paper becomes both weaker and slipperier. One might ask what difference a loss of friction could make to paper rolled up at the end of a papermaking machine? If you have ever rolled up a sheet of paper into a tube and then pulled it out into a "telescope" you have discovered the potential problem. Rolls of paper are removed from the machine by a clamping type arrangement which circles the roll like fingers and applies hydraulic pressure. The paper rolls are much wider than the aisles in the paper mills so they must be turned vertically to allow them to be moved up and down the aisles. If the clamping pressure is not sufficient to compensate for the slippery character of the paper, the roll of paper will "telescope," making it useless, except to be recycled again. See Figure 12.58.

Application Methods

Where colloidal silica is to be coated onto paper during manufacture, it is usually done using a spray system. A dilute suspension of colloidal silica is sprayed on the paper in between the last dryer stage and the wind-up area. The paper as it comes from the dryers is warm and the water is rapidly evaporated off leaving the colloidal silica in place.

Paper is wound in the horizontal position (left) on the paper machine. However, to transport it, it must be turned into the vertical position (center) to pass down the aisles.

If the paper is too slippery, the roll will telescope (right) and cannot be used.

Applying frictionizing to improve the coefficient of friction during manufacture is often the solution.

FIGURE 12.58 "Telescoping" paper rolls.

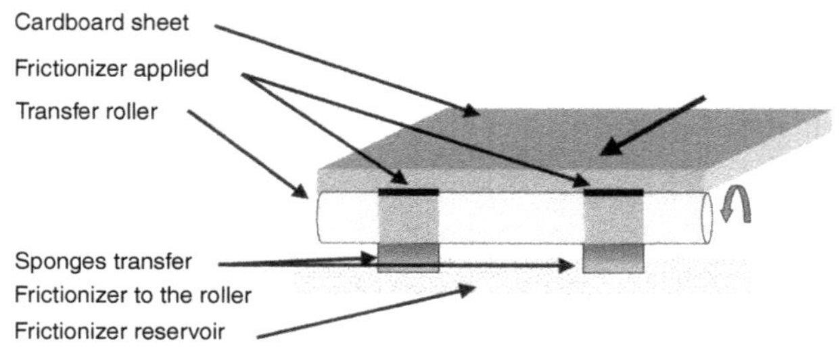

FIGURE 12.59 Frictionizing cardboard at specific areas which will become the top and bottom flaps.

When frictionizing cardboard boxes usually only the areas of primary contact (tops and bottoms) are treated with frictionizing. The material may be applied through the use of sponges that transfer the frictionizing material to a roller that then coats it onto the paper. By placing the sponges strategically at the points of the sheet that will become the bottom and top flaps of the folded box, it is possible to accomplish the maximum improvement with the minimum material. Such a system is sketched in Figure 12.59.

Typically in the paper industry the frictional characteristics are defined by determining the slide angle at which two pieces of paper or boxes will slide over one another. A typical lab device for determining the slide angle of a paper sample is shown in Figure 12.60.

A typical test procedure is outlined below:

- A paper or boxboard sample is cut into two appropriate size pieces.
- One is attached to the equipment by means of a clip.
- The other is attached to the sled (usually with double sticky tape)
- The two frictionized surfaces are placed together and the sled is put into its starting position which closes some type of contact electrical switch.
- This switch allows the motor that raises the incline to run.
- The incline of the machine increases until the sled slips from its position, breaking the electrical connection and stopping the movement of the incline.
- The slide angle is read from a protractor attached to the equipment.

Typically several measurements, (perhaps as many as ten) are taken to ensure that the coefficient of friction remains improved and does not decay significantly with movement of the two surfaces over one another. One normally finds some decay over the first one or two measurements but that the slide angle remains fairly constant after that.

One other mechanism appears to operate when fractioning paper and boxboard. The application of water to the surface will raise the coefficient of friction by lifting fibers out of the surface of the paper. If only water is applied, these fibers tend to go back into the surface if another box or paper bag is passed over its surface. However, if these fibers are coated with colloidal silica particles from the application, they are stiffened and strengthened and tend to remain upright creating a rougher surface.

Typical application rates are usually of the order of 0.1 pounds of colloidal silica solids per 1000 square feet of paper. The actual concentration of colloidal silica needed depends on how much liquid is being applied to the paper. Sponges and rollers that are fed directly from the colloidal silica reservoir usually apply a fairly "wet" application and in these cases the concentration used might be only 2–3% colloidal silica. Paper bags are often coated with very little liquid and hence require a higher (5–7%) concentration to get the right level. A typical response of the slide angle or coefficient of friction to application rate is shown in Figure 12.61.

Particle Size Effects

Up to a point, the larger the particle size of the colloidal silica used, the greater the improvement in the coefficient

The sled in the upper position closes a switch and allows the incline to move up untill the sled and paper sample slip.

The connection is broken, the incline stops moving and remains in position so the slide angle can be determined.

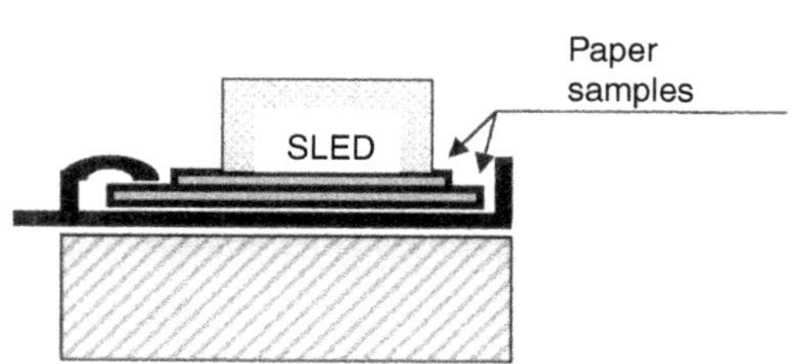

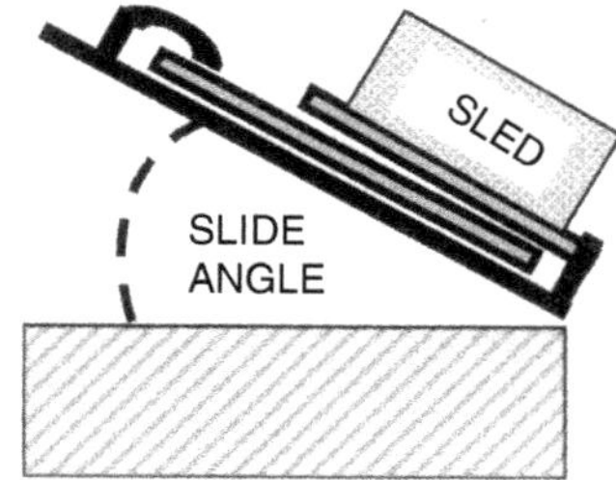

FIGURE 12.60 Slide angle measurement.

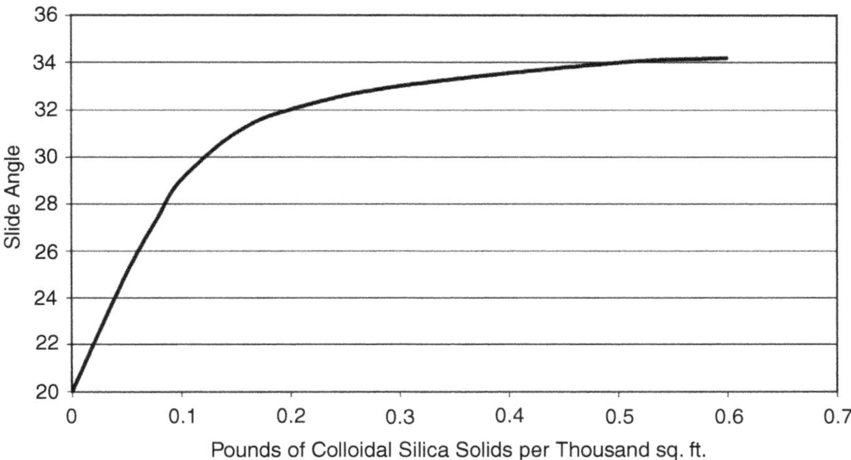

FIGURE 12.61 Typical slide angle response.

of friction. However, as the particle size increases the number of particles per unit area will decrease unless the concentration of the suspension being applied is increased. Consequently, there is relatively little increase as the particle size exceeds 30–40 nm.

Particle Charge Effects

The charge of the colloidal silica used can make a significant difference. If one remembers that paper fibers are negatively charged we can assume there is no charge attraction between the fibers and negatively charged colloidal silica. The colloidal silica particles tend to flow into the paper surface with the water and then some are retained in the interior of the paper where they do not improve friction. Many of the particles do return to the surface, with the evaporating water, but if starch or other cationic additives are used in the paper fabrication, these will tend to trap some of the particles that get to the interior.

However, if the particles are positively charged, the negative fibers tend to strip the treating solution of particles by charge attraction. Thus, the surface of the paper is most likely to be saturated with colloidal silica particles where they have the maximum effect on improving the coefficient of friction.

Plastic Films

Freshly produced plastic films can have a tendency to adhere to themselves (called "blocking") when being rolled up. If this happens the utility of the film is lost since it can't be unrolled. By applying a thin layer of colloidal silica this tendency to "block" is avoided. Colloidal silica is best applied from a water-based bath or spray. If the film is not hydrophilic then some formulation of the colloidal silica with wetting agents and/or co-binders is usually done. The application is usually such a thin layer of material that the general appearance of the film is not altered. However, much of the film is heavily printed making this less of a concern. It is an important feature of these coatings that the printability is typically improved.

Films are made into food packaging typically by heat sealing one or more seams in the film together. The colloidal silica acts to improve the strength of the heat-sealed seam, as well as making the film less likely to stick to the heated surface of the sealing device. We have tried to illustrate some of these features in Figure 12.62 and Figure 12.63.

Anti-Soil

Many surfaces have pores and cracks that are microscopic in nature. Soil can get into these cracks and be very difficult to remove. If the soil is dark colored, the surface affected appears darker and "dirty." By intentionally putting transparent "dirt" in these soil receptors regular soil is prevented from occupying the sites. Colloidal silica particles fit the bill as being transparent dirt.

Long before the current fluorocarbon-based carpet treatments were developed, colloidal silica was used as

Freshly formed polymer films can adhere to each other. The application of a thin layer of colloidal silica prevents the two polymer surface from meeting.

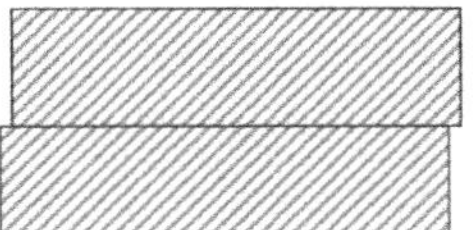

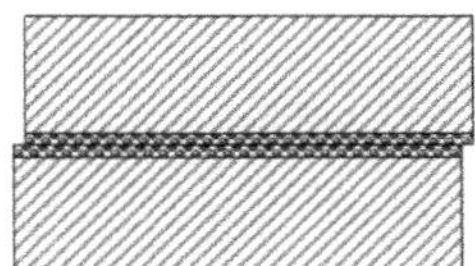

FIGURE 12.62 Colloidal silica coating prevents "blocking" (self-adhesion) of plastic films.

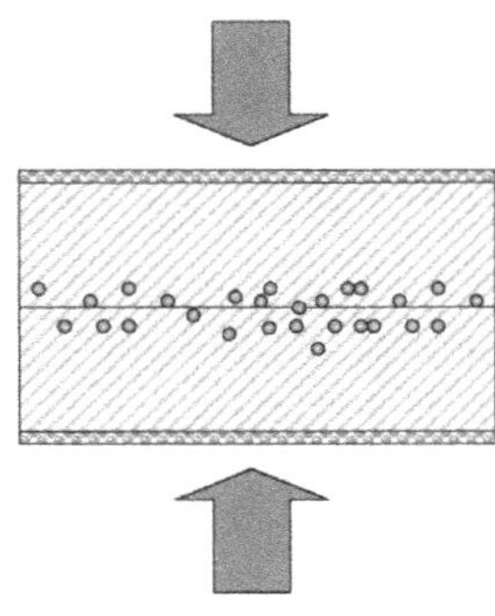

FIGURE 12.63 Colloidal silica coatings improve heat sealing of plastic films.

an anti-soil on carpets and was formulated into carpet cleaning compositions. The idea was to clean the carpet and leave behind some colloidal silica in the soil receptors to provide anti-soil character to keep the carpet cleaner longer. Using colloidal silica as an anti-soil on fabrics or carpets has one very significant disadvantage. The feel or "hand" of the garment or carpet is made harsher. Thus the soft carpet that might otherwise feel so good to bare feet might not feel so good when treated will colloidal silica. Just as the fibers of the paper are coated and stiffened when treated with colloidal silica, the same thing happens to the carpet fibers. In addition, while a colloidal silica treatment is quite effective against dry dirt, greasy stains are not repelled. Here the fluorocarbon treatments work very well.

Colloidal silica was also used commercially as an anti-soil on large-scale outdoor tanks such as chemical or gasoline storage tanks. In one demonstration, after cleaning, half the tank was coated with a dilute suspension of colloidal silica. The coated half of the tank retained its clean appearance and repelled dirt much longer.

One of the most interesting applications of this property has come in the last ten years or so. The Air Force developed and tested a colloidal silica based anti-soil treatment for fighter aircraft to help keep them from becoming coated with fumes and particulate from jet engine exhaust as well as normal dust and dirt. Even a modest accumulation of dust and dirt on the surfaces of high-speed jets has a significant effect on their performance. Because of this, a great deal of time and effort is put into cleaning aircraft. During cleaning, the planes are obviously not available to fly. Minimizing this downtime is the function of the anti-soil treatment.

Flocculation

Summary

Flocculation can be thought of as a process whereby colloidal or near colloidal particulate which are very difficult to filter or settle, are attracted to each other usually with the help of an additive of opposite surface charge, and form large agglomerates. These agglomerates are more readily filtered or settled. Flocculation is a fundamental process involved in municipal, industrial and food manufacturing processes. Perhaps one of the oldest uses where this process is most important is in the art of making wine.

Years ago I visited the Krug winery in California. Our tour guide was the master vintner himself. I have remembered his description of the start of the wine making process, vividly: "Everything, grapes, vines, a few leaves, and spiders go into the process." Particularly, the spiders have stuck in my mind.

Obviously, finding spider parts in your wine would ruin the finest meal, so wine manufacturers expend a great deal of effort to clarify the wine by precipitating every bit of particulate they can from the wine before they bottle it. Wine, at the end of fermentation, is cloudy and carries some undesirable tastes associated with fragments of the grapes that are very slow to settle if left to themselves. The fining process not only improves the color and the clarity, it also improves the flavor of the wine. Historically, egg white, clays, and gelatin have all been used to clarify or "fine" wine. These materials tend to have positive charges and therefore attract the negatively charged debris from the grapes. In modern times colloidal silica has been used. Since colloidal silica is also negatively charged, one might ask how it could possibly be helpful in coagulating other negative colloids. In fact by itself it is not very useful. But used in conjunction with positive agents such as gelatin, the combination is extremely effective. We will discuss this in more detail later.

In white wine manufacture, the juice is pressed from the grapes and fermented free from contact with the wineskins that would introduce a color, (red wines are fermented in contact with the skins). Usually, several pressings of the grapes are made. The best wine is usually thought of as coming from the first pressing: the least contaminated with skin or stem solids. However, a second and third pressing are often made in order to increase the overall yield. The third pressing is typically heavily laden with plant solids. Some wineries have found that clarifying this third pressing juice before it is fermented results in a much better quality wine. Fruit juices other than grape juice are sometimes clarified using similar techniques before they are marketed.

Water purification is a process that many of us take for granted until something goes wrong and the water runs turbid from our taps. We depend on our municipalities to clarify the water for us as well as remove any harmful contaminants. Turbid water is treated with alum (aluminum sulfate) or other agents that flocculate the offending particulates and cause them to settle out or become more easily filtered. In the past, some water treatment engineers found it useful to use some dilute silicate along with a reduced pH to provide what was termed "activated silica." The treatment was not always reproducible even by the same operators, but when it worked it was very effective as a clarification agent. In recent years methods and equipment to reproducibly produce such activated silica have been developed and patented. The product has limited stability, but is very effective as an aid to flocculation when used within a few days or weeks.

In manufacturing paper, finely dispersed cellulose fibers dispersed in water flow onto a moving screen. Having a flocculating agent present that will rapidly cause the paper fibers to floc into easily filtered agglomerates makes the separation of the fibers and water much more efficient. Such agglomeration also helps in the retention of any particulate filler or pigments which might be added to the paper. The use of cationic starch and high surface area colloidal silica has been patented to provide faster more efficient flocculation.

Details

Very often the use of colloidal silica in the flocculation process is that of a secondary flocculating agent. Most surfaces in the world are negatively charged, as is colloidal silica. Hence, putting colloidal silica in with other negatively charged colloids merely gives a mixture of negatively charged material and does not give rise to the rapid coagulation that mixing oppositely charged colloids does. However, if a positively charged colloid such as gelatin in the case of wine, or starch in the case of paper manufacture is added to the negative materials to be settled, they tend to attract each other into larger agglomerates which tend to settle more rapidly. Even so, the settling may still take hours or days to accomplish. By adding colloidal silica to those agglomerates which now have a good deal of positive charge imparted from the gelatin or starch, the agglomerates are agglomerated by charge attraction to the colloidal silica particles, and very large flocs develop and rapidly settle. We have tried to illustrate this in Figure 12.64 and Figure 12.65.

Remember "Goldilocks and the Three Bears?" One bowl of porridge was too hot, another too cold and the third "just right." Performing a flocculation is a little like that. You need the amount of flocculating agent to be "just right" to do the best job. Flocculating charged species with other charged particles or molecules can be thought of as similar to the titration of acids and bases. There is an end point where just the right amount of flocculating agent has been added that results in the largest flocs possible and the fastest settling time. Not adding this amount and flocculation and settling will be less effective. Perhaps surprisingly, an excess of flocculating agent can also result in very little flocculation and not improve the settling rate at all.

Negatively charged dispersed material is treated with a positively charged species like gelatin or cationic starch.

The two combine to give small flocs which will settle in time, but it is still slower than desired.

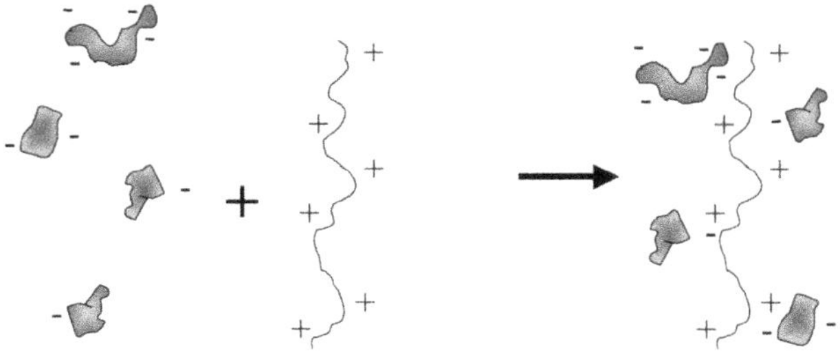

FIGURE 12.64 Simple flocculation using a single flocculating agent.

We interpret this fact as indicating that the surfaces of the original negatively charged particulates have been completely covered with the positively charged species. These coated particles are positively charged, repel each other, and hence there is no coagulation by charge attraction. These principles are illustrated in Figure 12.66.

If one examines the effects of colloidal silica on the optimum flocculation point, one finds that the amount required is always related to the surface area of the colloidal silica used. If it takes 5 grams of colloidal silica with a surface area of 400 m^2/gm to optimize the flocculation, then it will take 10 grams of colloidal silica with a surface area of 200 m^2/gm and 20 grams if the surface area is 100 m^2/gm. One must supply the same total surface area to achieve optimal results. However, there is a major difference in the activity of the colloidal silica

Colloidal silica added to the small positively charged flocs results in larger flocs which settle more rapidly.

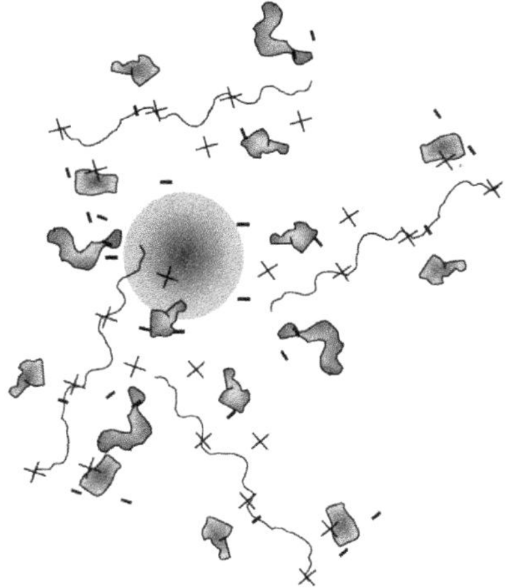

FIGURE 12.65 Using colloidal silica as a secondary flocculating agent.

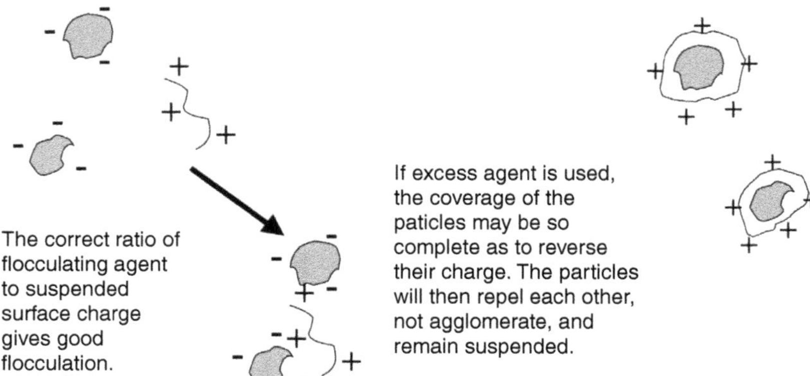

FIGURE 12.66 The amount of flocculating agent needs to be "just right."

used that is dependent on particle size. The speed of flocculation or re-flocculation is faster with smaller particles. If one subjects the system to high shear, breaks up the agglomerates, eliminates the shear, and then measures the rate at which the flocs reform, one observes that the rate at which the floc reforms is faster, the smaller the particles. This seems reasonable since the higher the specific surface area (smaller particles) the more particles it takes to make up a gram of material. We can illustrate this by cutting a cube into smaller cubes and seeing what happens to the number of cubes versus the area generated. See Figure 12.67. The number of particles increases as the cube of the reduction in size whereas the area increases linearly with the reduction in size. Therefore in our example above, the 400 m^2/gm sample has 64 times more particles per gram than the 100 m^2/gm sample. But it takes 4 times as much of the latter material to flocculate the system. Even taking that into account, the small particle material will have 16 times more particles at the optimum flocculation point. Having more particles present increases the probability of a meaningful encounter with a debris particle. Since the overall speed of the reaction depends on the probability of interactions the small particle sol re-flocs the system faster than the larger particles. The speed of the re-flocculation is very important in the paper making process. The slurry of paper fibers and other additives is delivered to the papermaking machine under conditions of intense mixing right up to the point where it is distributed across the moving screen. This distributor is called the "head box". As the slurry flows onto the moving screen, the faster the pulp flocculates the faster and more efficient will be the separation of the solids and the water. The efficiency at which filler particles such as pigments or extenders are trapped in the paper structure is also dependent on the flocculation of the mixture. If the flocculation is very fast, resulting in rapid filtration of fibers and fillers from the white water, then the papermaking machine can be run faster and produce more paper. Also if the separation of fibers and water is very efficient the recycling of the "white water" is less problematic.

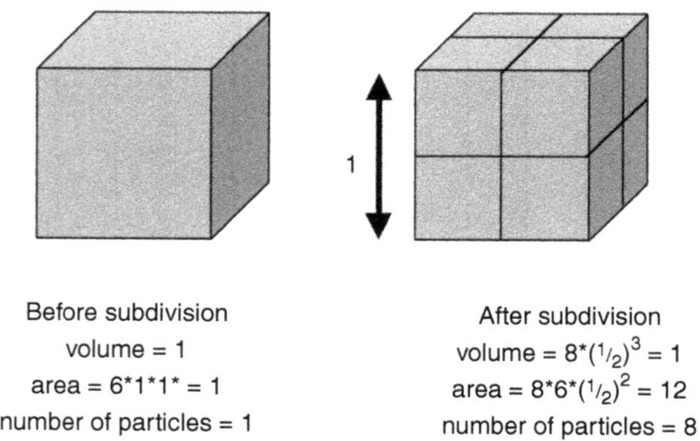

FIGURE 12.67 Subdivision to smaller particles increases surface area but not the volume.

The use of alkali stabilized, high surface area colloidal silicas works well in alkaline papermaking. The higher pH of the system provides for significant charge to be on the colloidal silica. However, in acid conditions, alkali stabilized colloidal silicas lose most of their charge. To provide products suitable in these conditions, alumina modified colloids are used. In the chapter on Manufacturing, we discussed these modified products. Aluminate sites in the silica surface have permanent negative character that is independent of the pH. Thus these products can provide adequate surface charge density to perform well even in acid.

Silicon Wafer Polishing

Summary

In the manufacture of the amazing computer 'chips' which have become an integral part of our lives, the first step is to grow large, pure, single crystals of silicon. These single crystals are cut into discs and then polished to a virtually flawless surface. The circuitry laid down on the silicon surface is so miniaturized that even minute scratches are chasms by comparison. The first steps in the polishing process use somewhat standard abrasive polishing methods to smooth the surface and chemical etchants to remove impurities. However, to achieve the final result an abrasive of very small size is needed. Colloidal silica provides such an abrasive. A typical polishing machine is shown in Figure 12.68.

Although it is not shown in the sketch, the silicon wafer is mounted in a carrier and is pushed down on a rotating platen. For those of you old enough to remember phonograph records, the set-up will look slightly familiar. The platen is covered with a polishing pad that is somewhat the consistency of felt and is porous enough to hold some liquid polishing medium. These pads are often made of polyurethane and are resistant to abrasive wear. The polishing medium or slurry flows onto the pad and is distributed by centrifugal force. It flows off the side of the pad and is either recycled or discarded. The concentration of colloidal silica in the polishing slurry is typically about 3–5%.

It should be noted that the polishing mechanism is not just simply abrasive polishing but includes a chemical reaction component also. Alkali stabilized colloidal silica is used and it can be shown that the polishing rate increases with the pH of the sol. This is thought to indicate that the alkali reacts with the silicon to form an alkaline silicate at the surface and that one function of the colloidal silica is to remove the silicate layer so that fresh surface can be exposed.

$$4NaOH + Si \longrightarrow 2H_2 + Na_4SiO_4$$

We have run experiments where only high pH water is supplied in the polishing machine and shown that without the presence of some colloidal silica, the removal rate goes to nil. This would imply that the alkali silicate salts from a somewhat protective skin over the silicon surface. Likewise, the lower the pH, the lower the removal rate. One further proof that a chemical reaction is involved is the

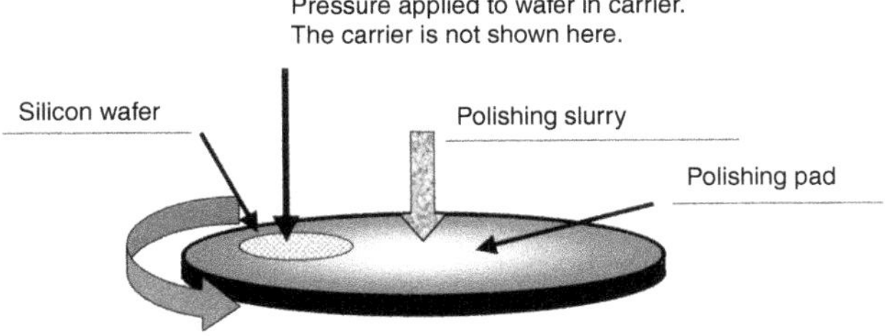

FIGURE 12.68 Typical silicon wafer polishing.

fact that the polishing rate increases significantly with the temperature. If only abrasion was involved, we would not expect any significant effect of temperature.

In the first phase of the polishing an aggressive chemical-mechanical approach is used to remove surface silicon at a rapid rate. This stage of the polishing is called the "stock removal" phase. Normally, about 0.001 inches (25 μm) are removed during the stock removal phase. A second polishing using much less aggressive conditions provides the final finish, much the way rough and fine abrasives are used to achieve a final finish on other materials.

Details

Effect of pH

Most colloidal silicas are stabilized in pH range of 9–10, but usually more alkali is added to the polishing slurry to maximize the alkali content. However, the ability to add alkali is very limited. As discussed earlier, the effect of excess alkali is to tend to dissolve the colloidal silica particles and convert them back to sodium silicate. If enough sodium silicate is present it acts like any other salt and causes the colloidal silica particles to agglomerate and/or gel. In any case, the increase in pH is short lived as the pH falls as the particles begin to dissolve and form sodium silicate. One might consider the particles as acting as a slow-release silicic acid.

Alumino-silicates are known to be less soluble in alkali than pure silica and Sears [9] used this fact to produce and patent a product having a high degree of alumina surface modification which resisted alkaline attack and therefore would retain a high pH for a longer period. This allowed it to be a more effective polishing agent.

In general, the presence of a very small ion like Na^+ is not desirable in polishing slurry as the ion is thought to be able to be retained in or on the silicon where it can cause malfunctions. Hence, adding large excesses of alkali is not considered desirable. It has been found that amines generally have a greater effect on the removal rate during polishing than alkali hydroxides. This may be because they can produce higher pH without acting to dissolve the silica particles or it may have to do with a tendency to form species more readily released from the silicon surface. The inference of some type of complex ion formation is found from the fact that in general di-amines are more effective at improving removal rate than monoamines.

Effect of Pressure, Temperature and Flow Rate

Pressure, temperature, and flow rate are considered together here as they are all interconnected in their effect on most polishing machines. When pressure is increased, the temperature, unless separately controlled, goes up. The effect of the temperature rise is to increase the chemical attack component of the polishing mechanism. Under conditions of constant pressure and slurry flow, the temperature will achieve a relatively constant value and, in turn, the removal rate will be constant. If the flow rate is increased, the additional slurry carries away more heat and the temperature drops. If less slurry is supplied, more heat remains in the pad and the temperature increases.

Using lots of slurry, particularly if the slurry is not recycled, but discarded, is not only expensive, but the lower temperature reduces the rate at which silicon is removed from the wafer. However, using a very low slurry flow to increase the temperature and therefore the chemical attack rate runs the risk of developing so much heat that the slurry dries, and the silica particles aggregate into scratch-producing sediments. Therefore, as with "Goldilocks and the Three Bears" one flow is too hot, one is too cold and one is "just right'.

In fact, the rate of chemical attack and the rate of abrasive polishing of the colloidal silica need to be well matched. if the chemical attack is so fast compared to the abrasive character that the abrasive action cannot level the high spots left on the etched surface then pits will be left in the polished surface. If these cannot be smoothed out in the final polish, then they will represent defects in the finished wafer surface.

Final polishing may be done with the same colloidal silica products without added chemical additives, or it may be done with colloidal silica designed for the job. Generally, the final polishing slurries will be designed to be less chemically aggressive and depend on abrasive character to remove any damage left behind by the aggressive chemical attack of the stock removal slurry. The goal is a surface that is extremely flat and free of defects. Circuitry is laid down in part by imaging the circuit on the chip surface. Because of the miniature dimensions involved, any deviations from flatness can cause the image to be out of focus and therefore be defective. As circuitry has become smaller and smaller, to achieve greater speed and higher circuit density the need for flatter wafers, even fewer defects, and even less contamination has become the norm.

Planarization

Early in the evolution of electronic chip manufacture, the circuits laid down on the silicon surface were one layer deep. Gradually, as the need to pack more and more circuitry on a single chip has evolved, the circuit components have become smaller, requiring better surface quality of the polished substrate. In addition, as an approach to greater device density, several layers of circuitry are placed one over the other. An insulating layer separates the circuitry layers. Here and there conductors that join the different levels of the circuit penetrate the insulating layer. We have tried to illustrate this in Figure 12.69. As different parts of each circuit are laid down and then covered with an insulating layer, the surface tends to become bumpy and is no longer flat enough to allow for good focus of the optical equipment used to image the next layer of circuitry. In order to overcome this, a polishing process called "planarization" is used which essentially polishes the bumpy insulating layer to a flat surface. Good planarization agents must be capable of polishing not only the insulating layer, but also polish the conductors at close to the same rate. If the slurry over-polishes the conductors they will end up below the level of the insulating layer and not make the proper connection between layers. If they are left above the insulating layer, they may also cause bad connections, short circuits or interfere with the imaging of the next layer of devices. These conditions are illustrated in Figure 12.70.

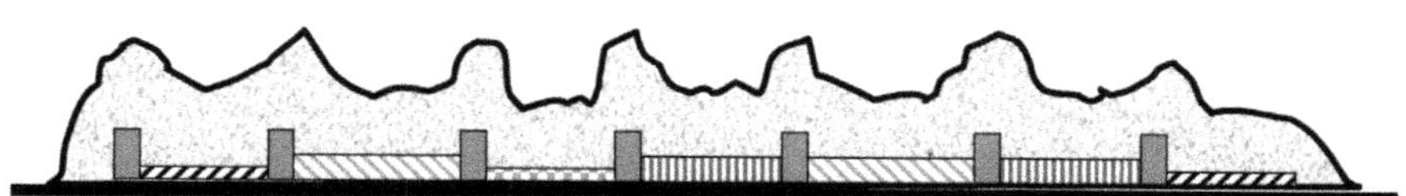

FIGURE 12.69 Planarization.

Above: The conductors polish faster than the insulating layer leaving them below the surface of the new circuitry.

Below: The conductors are polished slower than the insulating layer and protrude above the plane of the next circuitry layer.

FIGURE 12.70 Planarization must be equal for both insulation and conductors.

In these small chips, where the circuitry is so miniaturized the effect of even very low levels of contamination with stray ions can be very significant. For this reason the purity requirements for the colloidal silicas used have been extremely high. The thought being that metal contaminants in the colloidal silica could diffuse from the particles to the surface of the chips and contaminate the circuitry. To my mind, this fear is overstated from the standpoint that:

- Most of the contaminants are interior to the particles.
- Metal silicates are some of the most stable and insoluble materials in nature making easy solubility and transfer unlikely.
- Since the temperatures are usually less than 60°C solid state diffusion is very slow.
- The particles used are nearly 100% dense so there are no pores through which migration could rapidly take place.
- The average contact time of any particle with the surface is likely to be quite brief.

However, one can argue that there are significant particles even at this low level that could transfer material to the surface of the wafer as below:

- The typical contamination levels of the really bad actors is usually less than 100 ppb
- The particle size is typically about 50 nm for polishing
- With this particle size and 100 ppm contaminant there would be about 1 atom of contaminant for each 3 particles.
- The surface layer is about 5% of the particle.
- Therefore we would expect to find a contaminant ion on the surface of 1–2% of the particles.

After polishing the surface of silicon wafers are acid treated to remove metallic impurities but this is not always possible in the case of planarization of circuitry. Hence the concern for contamination in multilayer devices is potentially much more valid.

Miscellaneous Uses

Steam Promoter

Steam irons are intended to produce steam smoothly and efficiently by having drops of water fall on a small hot area of the metal sole plate called the steam chamber. The steam generated exits through holes in the bottom of the sole plate that is in contact with the fabric. However, when the iron is new, the clean metal surface of the steam chamber is likely to act like a metal griddle. The water droplet will dance around the surface on a thin layer of steam. The contact with the metal surface is minimal and so is the rate of steam production.

As the iron is used the condition will typically correct itself because minerals dissolved in the water will be deposited on the steam chamber surface. The resulting layer of deposits act as an inorganic sponge, absorbing the water, allowing it to spread out over a much larger surface and be heated to steam uniformly and efficiently. However, people expect a new steam iron to work well right out of the package. A multi-week break-in period would be unacceptable. So, to produce steam irons that work well right out of the box, most manufacturers apply a thin layer or layers of colloidal silica to the surface of the steam chamber during manufacturing in order to provide the needed porous layer of heat resistant material.

Two criteria must be met in order to get an adherent layer of colloidal silica on the steam chamber:

- The surface must be hydrophilic in order to get good adhesion of the colloidal silica to the metal.
- The colloidal silica must be applied in several thin layers that dry quickly before the next layer is applied.

In order to accomplish these criteria, the sole plates are cleaned before the material is applied. This can be done by chemically cleaning the surface with an alkaline metal cleaner, or by heating the surface above 600°F. Irons, which are being coated with a product like Teflon®, are typically heated above this temperature to cure the Teflon® and so a secondary heating can be avoided if the colloidal silica is applied before the plates have cooled. The colloidal silica is usually applied by spraying a 5% sol on the hot sole plates with a second or two to allow for drying between spray applications. Typically 5 coats are applied, which leave a thin, porous layer of silica adhering to the steam chamber surface. The presence of this silica layer causes the water to spread over and through it and be turned to steam efficiently. The irons produced this way are not dependent on the accumulation of mineral deposits and in hard water areas it is sometimes recommended that distilled water be used instead of tap water.

Chromatography

R.K. Iler developed a way of producing small porous microspheres of silica by combining the tendency for urea and formaldehyde to form very small spherical polymer agglomerates with the tendency for colloidal silica to gel under controlled acid conditions. By initiating the gelation and at the appropriate time simultaneously add the urea and formaldehyde to form polymers. The urea-formaldehyde engulfed the colloidal silica and both were incorporated in the polymer particles. The product was washed, dried, and fired to burn off the organic portion, leaving the porous silica behind. This product was marketed under the name Zorbax®. It was primarily used as the absorbing medium in chromatography. Figure 12.71 illustrates the process.

Another method of preparing interesting media for chromatography is to spray dry colloidal silica. During spray drying, water evaporates from the droplet surface and colloidal silica particles become more and more concentrated as the droplets become smaller in volume. At some point, the colloidal silica particles are concentrated enough to begin to gel. When this happens, if the gel forms a barrier to the evolution of the remaining water, the droplet explodes leaving virtually all the material as small hollow spheres with a single hole through which the remaining water vapor escaped. This process is illustrated in Figure 12.72. Proprietary solutions to this problem were developed at DuPont and perhaps other manufacturers or users. Small spheres without "blow-out" holes can now be made by spray drying. These particles can mimic the material made by the Zorbax® process.

Catalysis

In many catalyzed processes, the reactant gases are passed up through the catalyst bed that is made up of beads of catalytically active material. The particles actually are lifted up by the gas flow and are essentially fluidized by the gas flow. Such a system is called a "fluidized bed". The particles of catalyst bump into each other and tend to wear away the surface. This has the benefit that new surface is presented to the reaction mass continuously, but eventually the particles become so small as to be entrained in the gas stream and are swept out of the reaction zone. The harder and tougher the particles the longer they last in the reactor. Some catalysts are produced by either spray drying a mixture of catalyst material and colloidal silica or by extruding a mixture of colloidal silica and catalyst material as the interaction of the materials cause the colloidal silica to gel and bond the materials together into small catalyst pellets. The primary function of the silica is to bond the catalyst together into an attrition resistant particle that will last a long time in the reaction zone.

Since sodium often acts as a catalyst poison, ammonia stabilized versions of colloidal silica are most often used in the preparation of catalysts. In some cases freshly deionized silicate is used as the silica source. This form of active silica is much higher in surface area and is therefore a more efficient binder. However, its use causes the

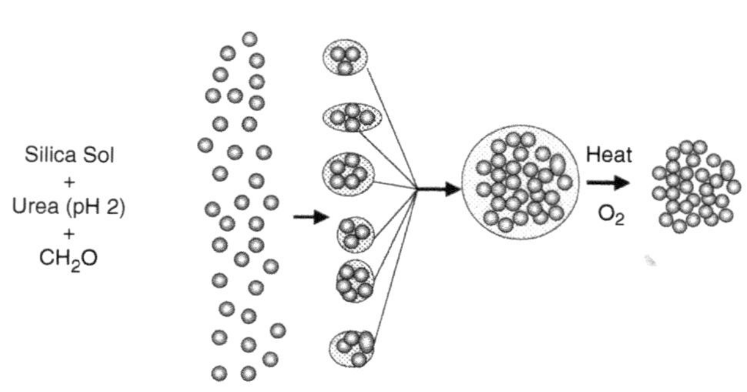

FIGURE 12.71 Formation of Zorbax® porous silica microspheres.

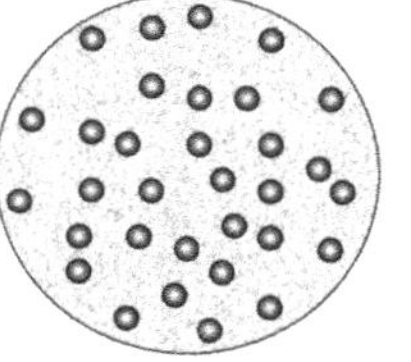
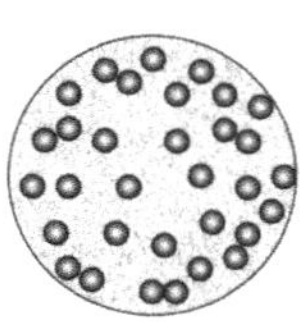
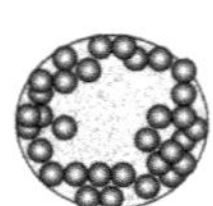

Colloidal silica particles become more and more concentrated as water is lost and volume decreases. Finally, the particles get and the last water evolved blows a hole in the gel surface in order to release the pressure.

The resulting aggregate shape shown at the right is typical. The colloidal silica is all at the surface and the interior is essentially empty

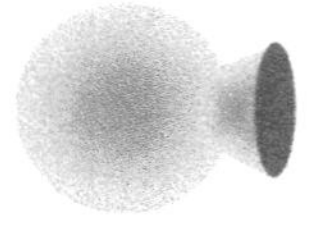

FIGURE 12.72 Progression of spray dried colloidal silica.

catalyst manufacture to be a harder process to control since the deionized silica changes very rapidly. Small variations in concentration, temperature and pH can make significant changes in the results.

Photography

Perhaps the simplest use of colloidal silica in this area is its use to frictionize and coat the film base material to both make it easier to be conveyed through the process and also to prevent it from "blocking" or adhering to itself when rolled up. The use of colloidal silica as an anti-blocking agent is discussed above under the "Surface Modification" of plastic films.

In black and white photography, the use of colloidal silica can help control the particle size of the silver deposits in the film or the prints. Increasing the number of silver crystals formed in the process that in turn limits their individual size increases the contrast of the film and probably the resolution. The black in a "black and white" negative or print is the result of light being diffused or adsorbed at the surface of the small silver particles. If the silver particles grow large enough they begin to act as small silver mirrors and reflect some of the light giving a brown rather than a black coloration to the film or print. Those of you old enough will remember that the first Polaroid prints had such a brownish tone, the result of such overgrowth of the silver crystals during the print development. The use of colloidal silica helped dramatically improve the contrast of the prints when it was introduced to the process.

Biology

In biological work on small, perhaps single-celled animals or plants, a separation of species or forms in necessary. In some cases the species being studied is only found surrounded by a myriad of other life forms and inert debris. It is necessary to effect a separation without killing or changing the species to be studied. One way this can be accomplished is to use the unique specific gravity of the individual to separate it from the others. The difference in density of the desired species from the others present is likely to be very small, so a fairly precise method is needed.

By using a fluid of known density to suspend the mixture of creatures in one will find that those of greater density will sink and those with lighter density will rise to the surface. Centrifuging the mixture can increase the rate of "settling". The real problem is what to use for the fluid. If one chooses to use salt solutions to increase the density of water then osmosis takes place as the salt tries to flow across the cell membrane of the creatures. The increase in salt content may kill the creature and they most assuredly increase their density making the separation difficult. By using colloidal silica suspensions diluted appropriately, one can achieve any density desired from 1.0 to about 1.4 (50% silica). The colloidal particles do not cross the cell membrane and are non-toxic to the system. However, most colloidal silicas are treated with a biocide as manufactured and so samples free of biocide are desirable for this application. Typically, the ammonia-stabilized versions do not contain additional biocides (the ammonia acts as a biocide). In some cases workers have acquired samples of these materials and heated them to drive off the ammonia. However, these materials must be used fairly quickly, as the loss of the ammonia results in a loss of stability and eventually the samples will gel. Some workers choose to use the aluminated sols since they can be pH adjusted to any desirable pH with less chance of becoming unstable. They typically find a way to remove the biocide or request samples without biocide.

I should point out that these research type samples are generally provided by the manufacturers as a courtesy to the scientific community as the quantities involved have no commercial significance and special requests such as leaving out biocides, etc. almost always cost the manufacturer money.

Secondary Oil Production

Before I became familiar with the oil recovery process, my ideas were all based on Hollywood's version of the "gusher" which sprayed oil into the air. I knew the pressure fell and eventually the oil had to be pumped out of the underground "pool" of oil. I pictured this as a large underground cavern filled with oil. Perhaps none of you were so naïve. I was surprised when someone from the oil industry showed me a rock, seemingly completely solid and told me that it was slightly porous and that much of the earth's oil supply was held in the pores of such rocks. Oil is recovered from such rocks by pumping water through the strata and collecting it at various points around the perimeter. This is illustrated in Figure 12.73. While only two recovery wells are shown it should be noted that many such wells are used around the perimeter of the anticipated water flow through the strata.

At some point an area will develop where oil is preferentially removed perhaps because the porosity of the rock is greater, the oil less viscous, etc. In any case, the water flowing through this area will increase in volume and velocity further increasing the removal rate. Eventually, this area consumes much of the total water flow. Some recovery wells see much decreased water flow and therefore much less oil production. The area has become the "path of least resistance" to the water flow and because it "steals" most of the water flow, it is known in the industry as a "thief zone." The development of such a "thief zone" is illustrated in Figure 12.74. Colloidal silica can be used to seal off such a zone and help restore higher production

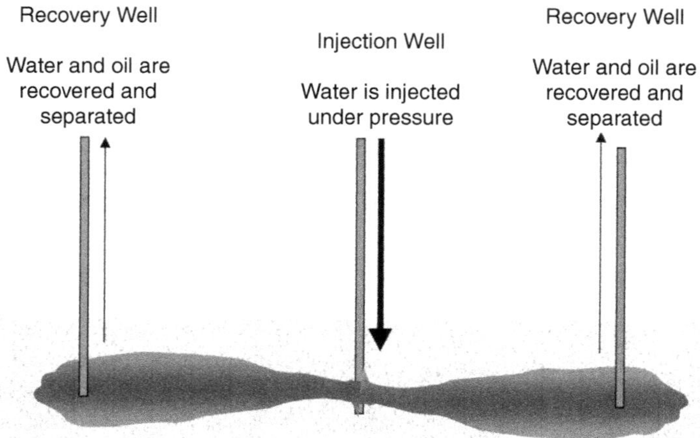

FIGURE 12.73 Secondary oil recovery.

levels. Figure 12.75 illustrates the injection process to seal the thief zone. Figure 12.76 illustrates the production improvement resolting from sealing the theif zone.

By careful adjustment of the pH, particle size and salt content the gel time of colloidal silica can be accurately controlled in the range of several minutes to several days. In particularly large fields the travel time of the water from the injection site to the recovery well might be very long (several days). It is possible to change the formulation with time to shorter and shorter gel times so that when pumping is concluded the entire thief zone gels nearly simultaneously. Using the shortest gel times possible is important. We observe silica gels to form as flow ceases as the bonding of particles form a continuous network. When the flow ceases we often refer to this as the gel time. However, the gel continues to cross-link and strengthen. If one measures the gel strength one finds that the maximum strength is reached at about 10× the original observed gel time. If one were to wait for maximum strength to develop in a gel which had a two-day gel time it would take almost three weeks. Fortunately, the strength develops rapidly at first and then asymptotically approaches the maximum. Because of this about 80% of the maximum develops at only about 2-3× the initial gel time.

The gel time can be adjusted by varying

- Silica concentration
  Rate increases with the concentration. The ultimate gel strength also increases with the silica content, but because of the cost of

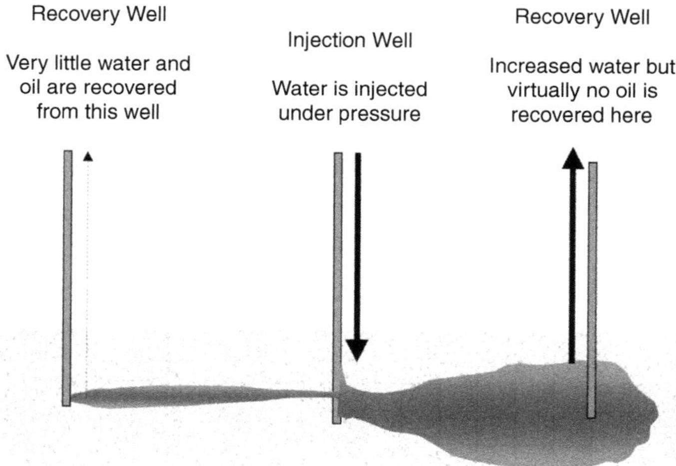

FIGURE 12.74 Thief zone development.

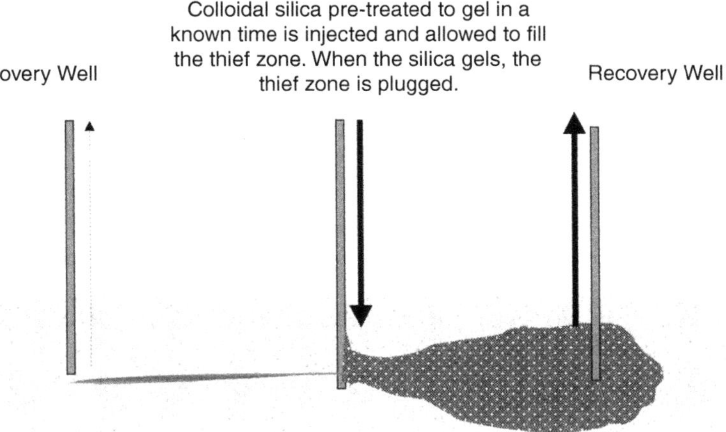

FIGURE 12.75 Sealing the thief zone.

colloidal silica the lowest concentrations where continuous gels are formed are usually chosen. The smaller the particle size the lower this minimum concentration becomes. At 7–8 nm the silica concentration can be as low as about 5% and still form continuous, strong gels. 12–14 nm sol might have to be 10% silica to form continuous strong gels.

- pH

  As shown in Figure 12.17 above, the pH should be adjusted into the 5–6.5 range for maximum effect.

- Salt content

  The higher the salt content the faster the gel rate up to a point. At very high salt contents continuous gels do not form, but the sol is destabilized as a flocculent precipitate. Polyvalent metal salts like magnesium or calcium salts are much more effective at increasing the gel rate than monovalent salts like sodium chloride. However, it was discovered that some soils can readily ion exchange magnesium and calcium and leave the solution depleted of electrolyte.

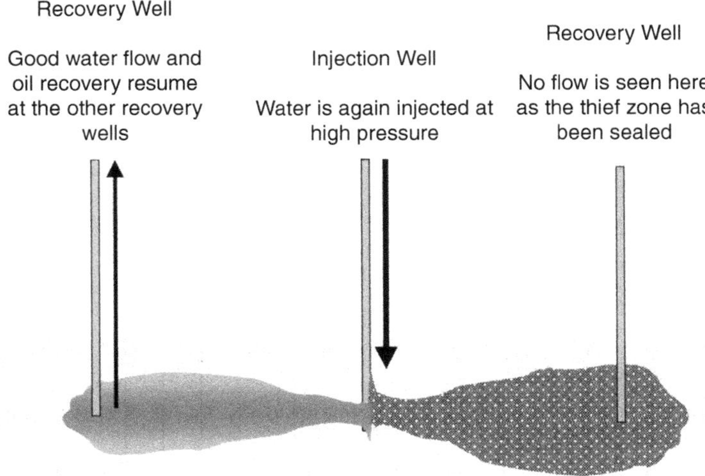

FIGURE 12.76 Oil production resumes.

This increases the gel time far beyond that observed in the lab when the soil is not present. It is important; to have soil samples present when these gel time determinations are done to avoid this problem. Predominantly sandy soils are not likely to give this problem, but clay soils which themselves may be calcium, magnesium alumino-silicates may readily ion exchange.

In order to change the gel time over a significant range both the acid and salt levels need to be changed simultaneously. Another method, developed in our laboratory uses a constant electrolyte-acid mixture and modifies the silica by adding controlled amounts of aluminate to it to make it more or less susceptible to gelation. We found that gel times from an hour to a week could be accomplished with this method. The principle here is the observation that small amounts of aluminate (about 3–4% surface coverage with aluminate) such as is found in Ludox® AM afford significant stability to acidic pH. This is illustrated in Figure 12.77.

The data is taken from a Ludox® Colloidal Silica bulletin: It compares Ludox® AM and Ludox® HS which are both 12 nm sols containing 30% silica. In fact Ludox® AM is made using HS as the colloidal silica to which sodium aluminate is added to introduce alumina sites into the particle surface. Ludox® AM contains only about 0.2% alumina, which means that about 3–4% of the surface silanols are substituted with aluminate ions. The effect of this modest surface change on the stability is quite remarkable. It has been my observation that nature seldom produces changes in a completely discontinuous fashion. It seemed logical that lesser amounts of aluminate additions would produce lesser improvements in stability as illustrated in Figure 12.78. Therefore using a constant acid-salt mixture and varying the level of alumina on the particle surface could alter the gel time.

Changing the alumination level in the field might seem more difficult than changing the salt-acid mix in the field, but in fact it may be simpler. In the work in the oil fields, using small particle size sols that form good gels at only 5% silica concentration is the most cost effective. Using these sols, we showed that adding the aluminate immediately prior to the addition of the salt-acid solution produced the same gel times as waiting several hours for the aluminate to react. This indicated to us that the reaction with the aluminate was complete virtually instantaneously. This work may become significant in oil production at some time in the future, but for now, oil prices are low enough so that the added cost of this technology will limit its use.

One may question why sodium silicate, which can also be caused to gel at known rates, is not a candidate for use in this area. It is certainly much less expensive than colloidal silica and forms strong gels. The time span

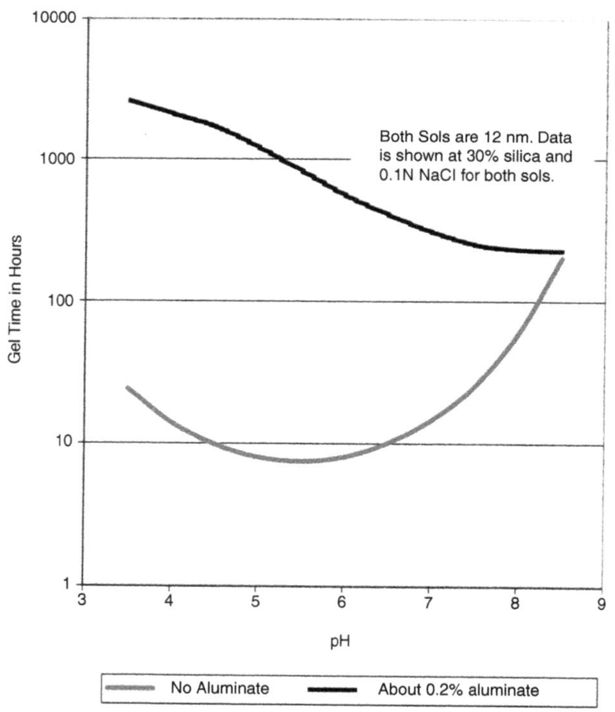

FIGURE 12.77 Effect of aluminate on gel time.

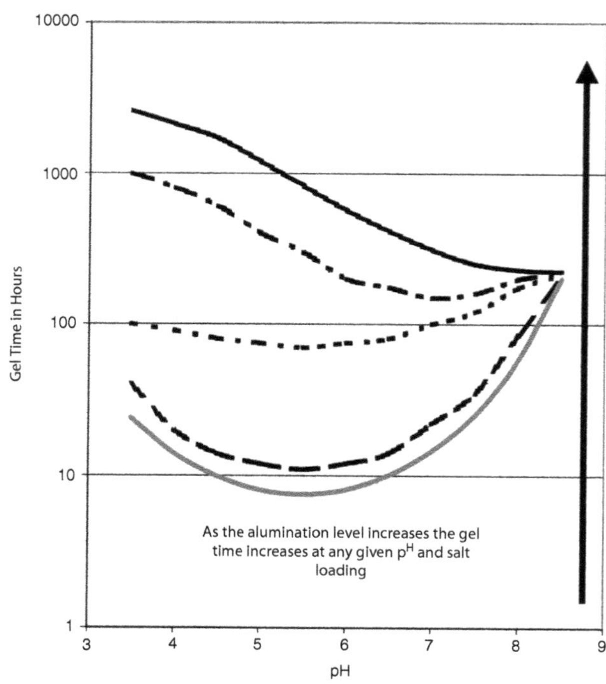

FIGURE 12.78 Effect of varying levels of aluminate.

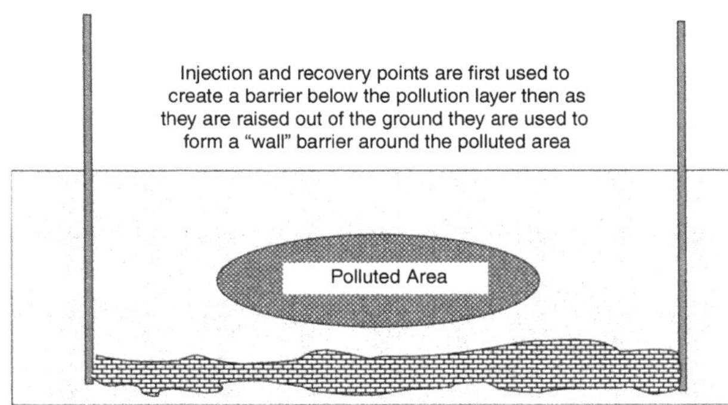

FIGURE 12.79 Colloidal silica used to isolate underground pollution.

over which one can control sodium silicate gelation is much more limited than with colloidal silica. Achieving gel times of several days would require using fairly low concentrations of silicate and the resulting gels would shrink in volume during the "strengthening" period. This process of gels decreasing in volume, similar to metal powders "sintering" to more dense solids, is called syneresis. It is accompanied by expulsion of liquid from the gel. Thus a gel of silicate would become fragmented in the strata and would be much less effective at blocking water flow. In addition, silicate gels, subjected to the high temperatures of deep wells would hydrothermally convert to quartz or other crystalline species and completely lose the volume-filling gel structure desired. Colloidal silica gels are much more stable to temperature.

Environmental

Using similar technology, it is possible to isolate underground pollution from spreading into the water supply. In this case, injection wells and recovery wells could be used as shown in Figure 12.79 to form a barrier below the polluted layer, followed by the creation of a "wall" type barrier around the area. If shown to be effective, such treatments would be much cheaper and less disruptive than digging up and disposing of the contaminated soil.

CONCLUSION

I have tried to cover at least a representative variety of the end uses for colloidal silica products. I have by no means exhausted the list. I feel sure that new uses will continue to be developed especially in an environmentally conscious climate of experimentation. After all what could be more environmentally friendly than "*sand and water*"?

REFERENCES

1. Iler, R.K. *The Chemistry of Silica*, John Wiley and Sons 1979.
2. Iler, R.K. U.S. Patent 3,668, 088 (DuPont) 1974.
3. Zhuravlev, L.T. *Colloids Surf.* 1993 74(1), 71.
4. Alexander, G.B. and Bolt, G.H. U.S. Patent 3,007,878 (DuPont) 1961.
5. Roberts, W.O. U.S. Patent 5,723,181 (DuPont) 1998.
6. Rusher, R.L. "Strength Factors of Ceramic Shell Molds", *Cast Metals Research Journal*, Vol. 10, No. 4 December 1974 and Vol. 11, No. 1 March 1975.
7. Roberts, W.O. "Factors Affecting Shell Strength" Investment Casting Institute Proceedings 25th Annual Meeting.
8. Roberts, W.O. U.S. Patent 5,118,727 (DuPont) 1992.
9. Sears, G.W. U.S. Patent 3,922,393 (DuPont) 1975.
10. DuPont. "Ludox Colloidal Silica: Properties, Uses, Storage, and Handling." E.I. DuPont de Nemours and Co., Industrial Chemicals Dept. Product Information Bulletin H-47771-2, Wilmington, DE.

13 Enterosorbent Silics: Properties and Clinical Application

A.A. Chuiko, A.A. Pentyuk, and V.K. Pogorelyi
National Academy of Sciences of Ukraine, Institute of Surface Chemistry

CONTENTS

A new enterosorbent silica has been developed and introduced into medicinal practice. This synthetic, a highly disperse silica with an extended specific surface is characterized by its chemical purity, stability, and physiological innocuousness. The regular structure of its surface as well as the presence of a large number of surface reactive sites insure a high adsorptive capacity of Silics with respect to water, protein molecules, toxins, pathogenic microorganisms and viruses. At present new technologies are being described for application of Silics as an individual medicinal preparation of sorptive action and as an active basis for a novel generation of composite drugs for multimodality therapy.

In recent years the efferent methods of treatment which are based on the removal of toxic and ballast substances of exo- or endogenous origin from an organism have developed into an independent trend of modern pharmacotherapy. Medicinal sorbents differ considerably in their chemical nature and production process and include various modifications of activated charcoal, ion-exchange resins, silicas, polymers, and other natural and synthetic substances.

Scientific workers of the Institute of Surface Chemistry of the NAS of Ukraine and the Vinnytsya State Medical University have designed a new enterosorbent Silics and introduced it into medicinal practice. The enterosorbent is a synthetic amorphous highly disperse silica (HDS). Owing to salient features of its surface, HDS is employed not only as a sorbent with biocorrecting properties but also as a matrix used in the capacity of a carrier of composite medicinal agents [1].

From the chemical standpoint a nucleus of an HDS particle is a three-dimensional polymer whose structural units are silicon-oxygen tetrahedra bonded by disiloxane bridges Si—O—Si. On the surface of HDS particles there are groups O—H chemically bonded with silicon atoms (silanol groups SiOH). The hydroxylic cover of HDS gives rise to a high hydrophilicity of its surface and, correspondingly, to its ability to sorb polar molecules, especially water molecules. The surface of HDS has weak proton-donating properties. Its isoelectric point is attained at pH = 2. The surface of HDS enters into an adsorption interaction with charged molecules and substances capable of forming hydrogen bonds O—H···O. The mechanism of adsorption is to a great extent dependent on pH of medium [2].

The main salient feature of HDS which is brought about by its physicochemical properties is a high protein-sorbing (proteinonektonic) ability that forms the basis of application of HDS for removal of exotoxins, endotoxins, pathogenic immunocomplexes, products of degradation of necrotic tissues, and other harmful substances of protein origin from an organism as well as for fixation of microorganisms [3].

MECHANISM OF BIOLOGICAL ACTIVITY OF HIGHLY DISPERSE SILICA

The biological activity of Silics is based on the following HDS properties which are brought about by the chemical nature of its surface:

1. High hydrophilicity of the surface of HDS
2. High biosorbing activity

3. Fixation of large amounts of microorganisms and microbial toxins
4. Adsorption of low-molecular substances.

Let us consider these properties in detail.

1. High hydrophilicity of HDS is due to the presence of a hydroxylic cover and electron-accepting silicon atoms of silanol groups. The surface of HDS is able to sorb polar molecules. Water wets the surface of HDS well, which leads to formation of thin slurries (suspensions) or gels of various consistencies depending on the ingredient ratio.

The hydrophilic properties of HDS have found application to eliminate edemas and to decrease exudation in the case of local treatment of wounds at a stage of inflammation, to bind and structurise water in the intestine of a patient suffering from diarrhea. Besides, HDS is widely used as a drying agent in dermatological practice.

2. High protein-sorbing activity of HDS is effected through all reactive sites of silica (hydroxyl groups, silicon atoms of silanol groups, and electron-donating oxygen atoms of siloxane bonds), and, probably, through co-ordinatively bound water. There are three types of interaction between protein molecules and silica surface, namely electrostatic interactions, hydrogen bonding, and hydrophobic interactions. Adsorption of protein molecules on HDS can be interpreted applying a theoretical model based on interaction of a polyelectrolyte with a charged surface of HDS. According to this model, the affinity of a protein to the surface is determined mainly by electrostatic forces. This inference is corroborated by the dependence of adsorption value for the protein on pH of medium, with the dependence graph displaying a maximum at the isoelectric point (IEP) of the protein [3,4].

Thus, in the situation with serum albumin (IEP is equal to about 5.0) the mutual interaction is observed over the pH interval from the zero-charge point of the surface (pH 2.0) and to the protein IEP when interacting agents bear different charges. At pH values that do not fall within this interval the affinity will be determined by other factors, namely by hydrogen bonds and hydrophobic interactions. The adsorption value maximum at the albumin IEP is attributed to the fact that uncharged globules of the protein have minimal size and a unit surface may accommodate a larger number of them.

A study has been made of adsorption of proteins on HDS within the scope of the model of interaction between the silica sorbent and biologic fluids of an organism (wound exudate, blood serum, etc.). The model proteins have been bovine serum albumin (BSA), egg albumin, dried human blood plasma, horse hemoglobin, gelatin. The kinetic tests have indicated that the main mass of the protein (not less than 90%) is adsorbed on HDS from solution within the first 10 min of the contact, with the adsorption rate being independent of type of protein. A similar high adsorption rate may be attributed to the fact that HDS has a nonporous structure. As regards the influence of acidity of medium, the researches conducted have shown that the maximum adsorption of proteins on HDS is observed at pH values close to the corresponding IEP of these protein. For instance, in the case of the bovine serum albumin the maximum adsorption value is attained at pH 5.0 (IEP = 4.8).

The results presented in Table 13.1 give evidence for the fact that the protein-adsorbing capacity of HDS substantially exceeds the corresponding capacity of other familiar sorbents applied in the medicinal practice. With increasing ionic strength of solutions the adsorption value increases.

TABLE 13.1
Degree of Removal (in %) of Proteins from an Aqueous Medium by HDS and Other Medical Sorbents

Sorbent	Protein preparation	Degree of removal (%)		
		Distilled water, pH 6.5	0.9% NaCl solution, pH 6.5	0.1 M phosphate buffer, pH 5.7
HDS	BSA	27	60	62
	Plasma	64	90	95
SUGS	BSA	1.8	2.4	3.0
	Plasma	0	5.4	4.6
SKN	BSA	4.0	0	0
	Plasma	6.0	6.0	5.3
AUVM Dnipro-MN	BSA	9.2	5.2	6.4
	Plasma	10	12	14
Debrisan	BSA	0	0	0
	Plasma	0	1.4	1.5

Besides, the adsorption value also increases as pH of solutions approaches the IEP of the albumin.

Of interest is also the following adsorption potentiality of HDS. In the case of polymer-containing disperse systems with a certain ratio of a solid phase and adsorbate one can observe flocculation that involves merging of disperse particles through formation of bridges with polymer molecules. The flocculation process results in large aggregates that quickly precipitate. The process is especially characteristic of protein-containing systems, which is due to polyvalence of protein macromolecules.

A study has been made of the aggregative stability of colloidal solutions of HDS as a function of concentration of a protein in the solution. It has been found that the presence of a protein at concentrations lower than 4 mg mL does not affect the stability of a colloidal solution of HDS because such a low amount of protein molecules is not sufficient for coalescence of silica particles into aggregates. At a high concentration of a protein (5 mg mL and more) the colloidal solution is stable owing to the colloid protection effect. The lower and upper limits of the dispersion stability make up an interval of concentrations (equivalence zone) suitable for flocculation of a colloidal solution of HDS to take place. In this case a colloidal solution of HDS possesses a higher proteinonektonic ability in comparison with an ordinary dispersion [5].

The results achieved when studying the specificity of interaction between HDS and human blood plasma *in vitro* give evidence for an increased affinity of disperse silica to lipoproteins in comparison with proteins that do not contain lipids (Table 13.2).

It can be seen that degrees of fixation of lipid components of blood serum by the sorbent amount to 90% and more (cholesterol: 77–78%) while those of the total protein amount to only about 26%. The increased sorption of lipid-containing proteins on the silica surface is attributed to fixation of a whole lipoprotein particle (micelle) whose lipid phase content is predominant in comparison with the content of other phases and whose external surface is covered by proteins and phospholipids that possess an increased affinity to HDS. The data collected give grounds to infer that it will be possible to apply HDS for extracorporal hemosorption in the case of hyperlipidemia [6].

On the basis of the above-outlined observations one can, to some extent, explain hypolipidemic activity of HDS in the case of its peroral application because lipids in a gastrointestinal tract are in the form of micelles consisting of proteins, phospholipids, and bile acids. The sorbent provides fixation of food fats and lipid components that appear in an intestine as a result of hepatoenteric recirculation, which interferes with their absorption. It has been shown that ingestion of a medicinal agent developed on the basis of HDS by patients who suffer from disturbances of metabolism of lipids already within 3–4 weeks brings about a substantial decrease in the level of cholesterol in blood serum, and this decrease is the greater, the higher was the initial level of cholesterol at the beginning of the treatment.

As is known, sorbents of medicinal specification must meet certain requirements, with one of the most important being their physicochemical stability, especially their inertness with respect to the internal medium of an organism. Such a sorbent should not be dissolved when in contact with biologic fluids (exudate of a wound, content of a digestive tract). From the results achieved during experiments with test animals (rats and rabbits) it follows that

TABLE 13.2
Content of Proteins, Lipoproteins, and Cholesterol in Blood Serum before and after Treatment with HDS and Corresponding Adsorption Values (Percentage of Adsorbed Substance)

Serum components		Time of contact with the sorbent (min)			
		0	15	30	60
Total protein	$g\ L^{-1}$	74.00	58.60	56.1	54.8
	wt.%		20.80	24.2	25.9
Low-density lipoproteids	$g\ L^{-1}$	6.50	0.38	0.13	0.10
	wt.%		94.20	98.00	98.50
Total lipids	$g\ L^{-1}$	9.40	2.20	1.50	1.30
	wt.%		76.60	84.00	86.20
Phospholipids	$g\ L^{-1}$	2.81	0.20	0.13	0.08
	wt.%		92.90	95.4	97.20
Triglycerides	$g\ L^{-1}$	1.93	0.38	0.28	0.18
	wt.%		80.30	85.50	90.60
Total cholesterol	$mmol\ L^{-1}$	11.0	2.30	2.50	2.40
	wt.%		78.00	77.0	78.00

after a single intragastric administration of even a large dose of HDS (1 g kg^{-1}) any statistically reliable variations in concentration of silicon in blood are not revealed. It has been found that rats which were given the above-said dose of HDS for 30 days display a tendency towards increasing in excretion of silicon. However, in this case any reliability of disagreements has not been verified either. One of the causes of the absence of any marked absorption of HDS in a digestive tract may be its low solubility over the pH interval from 2 to 8. It is also possible that there is a specific physiologic mechanism which prevents penetration of silica microparticles through intestine walls. Thus, it has been shown that HDS satisfies one of the most important requirements to sorbents of medicinal specification, namely it possesses a high physicochemical stability and is not dissolved in the internal medium of an organism [6].

The high protein-sorbing ability displayed by HDS makes the basis for its application for fixation and removal of bacterial toxins, pathogenic immunocomplexes, products of decomposition of necrotic tissues, and other injurious substances of protein origin from an organism.

3. The protein-sorbing properties characteristic of HDS impart preparations on its basis the ability to fix microorganisms. The interaction between silica and microorganisms is not distinguished for any specific nature. It is attributed to the affinity of silica particles to glycoproteid structures and to phospholipids in membranes of microbe cells.

A research has also been made into interaction of HDS with enteropathogenic *Escherichia coli, Staphylococcus aureus, Proteus vulgaris, Bacillus pynocyaneus* [7]. The procedure is as follows. A certain amount of HDS is added to 3 mL of a diurnal culture of microorganisms, the mixture is stirred for 2–3 min and filtrated, following which the filtrate is quantitatively sowed on a nutrient medium. After incubation for 24 h at 37 °C a count is taken of the colonies grown, with their number being equal to the number of bacteria that were not fixed by the sorbent. The results of the researches show that even at low concentrations of HDS (0.33–1.33 wt.%) practically all the microorganisms which were in the solution (up to 3.5 milliard of microbial bodies per gramme of the sorbent) are fixed, with the fixation value being virtually independent of type of microorganisms. The interaction of the microorganisms with HDS is distinguished for some particularities. Firstly, sizes of HDS particles (4–40 nm) are considerably smaller than those of microorganisms (1–10 μm), so that HDS seems to bring about the effect of agglutination, which substantially increases its adsorptive capacity with reference to microorganisms. Thus, since HDS particles are much smaller than microbial cells, it is the sorbent particles that are sorbed on microbial cells and not vice versa. Even at a low concentration, HDS particles are able to agglutinate microorganisms, i.e. to act in the capacity of a glue that unites microorganisms into a conglomerate. Evidently, it is this phenomenon of agglutination of microorganisms by particles of HDS which explains its unique ability to bind enormous amounts of microorganisms in comparison with other sorbents. Secondly, it has been found that after a contact of microorganisms with HDS they vary some of their properties. For example, they become more sensitive to antibiotics, especially to erythromycin, gentamycin, and streptomycin (from 40–60% to 100%). Thirdly, after a contact with HDS the microorganisms become sensitive to the action of proteolytic enzymes and cationic and anionic SAS, such as bile acids and phospholipids, i.e. to natural components of intestinal and gastric juices. The high affinity of HDS to microorganisms and its influence on processes of vital activity of microorganisms provide an explanation for mechanism of its curative effect because possibility of appearance of infectious diseases and gravity of their progress are directly dependent on values of contagious dose, number of bacteria that are accumulated in an intestine in the course of colonisation.

4. Adsorption of low-molecular substances. Of interest also are researches into regularities of adsorption of medicinal substances of HDS because the results of such researches form a scientific basis for development of preparations with a modulated pharmakinetics. A study has been made of adsorption of orthophen, quinidine, scopolamine, amphotericin, and some vitamins on HDS from aqueous solutions [8].

By way of example, it has been shown that in the case of slightly soluble antibiotic amphotericin B it is possible to increase the rate of absorption of the curative substance by simultaneous introduction of the substance and HDS into an intestine. When amphotericin B (at a dose of 3–7 g) is administered perorally, its concentration does not exceed 0.3 mg mL^{-1} and the biologic availability makes up only 3%. The experiment was carried out on 56 rats subdivided into 6 groups. The first group included animals that were administered 2 mL of 0.2% solution of the preparation (4 mg). The rats of groups II–VI were administered 2 mL of a mixture which contained 4 mg of amphotericin B and different amounts of HDS. The control group was given 2 mL of distilled water. For the half of the animals the experiment was terminated in 4 h (for the rest of them in 24 h). The results achieved are presented in Table 13.3.

From the data presented it is clear that introduction of the antibiotic simultaneously with an aqueous suspension of HDS leads to an increase in the maximum concentration of the preparation in blood from 2 to 21 μg mL^{-1}. The sharp increase in the absorption of the curative substance provides an increase in its biologic availability. Analogous effects were observed for substances of other classes, such as alkaloids (quinidine), carbohydrate (xylose), organic substances (voltarene). Thus, the studies of absorption showed that maximum concentration of quinidine in

TABLE 13.3
Variations in Concentration of Amphotericin B in Blood of Rats in the Case of Its Simultaneous Introduction together with HDS

Groups of animals and doses	Concentration (μg mL^{-1})	
	4 h	24 h
I. Amphotericin (4 mg) + water (2 mL)	2.0 ± 0.1	0.9 ± 0.4
II. Amphotericin (4 mg) + HDS (60 mg kg^{-1})	21.2 ± 0.5	5.4 ± 1.9
III. Amphotericin (4 mg) + HDS (50 mg kg^{-1})	19.1 ± 0.3	6.0 ± 2.5
IV. Amphotericin (4 mg) + HDS (30 mg kg^{-1})	14.7 ± 0.2	4.9 ± 1.2
V. Amphotericin (4 mg) + HDS (20 mg kg^{-1})	9.6 ± 0.3	4.1 ± 1.3
VI. Amphotericin (4 mg) + HDS (0.2 mg kg^{-1})	2.1 ± 0.1	0.9 ± 0.3

blood increased from 2.6 to 4.6 μg mL^{-1} and that of voltarene (anti-inflammatory agent) from 16 to 26 μg mL^{-1}.

Further, a study has been conducted of the release of quinidine (common antiarrhythmic preparation) from various medicinal forms produced by immobilization of quinidine and its complexes with surface active substances or protein on HDS. On the basis of the results of the comparative analysis of the data of biopharmaceutic and pharmacokinetic researches it was inferred that the requirements to medicinal agents of prolonged action are most fully met by the preparation produced by coprecipitation of complexes of quinidine with molecules of serum albumin on the surface of HDS. Administration of this preparation does not give rise to any sharp peak of concentration of quinidine in blood and a slow decrease in concentration during a long term is observed. Besides, administration of this medicinal form makes it possible to provide the maximum bioavailability of quinidine.

The results of the comprehensive and thorough researches into physicochemical and biological properties of HDS conducted at the Vinnytsya State Medical University and at other medical institutions of Ukraine gave grounds to infer that HDS is an active medicinal substance which on its own can function as a therapeutic agent. Sorptive detoxication with the aid of HDS brings about a profound effect in the case of acute intestinal infections, diarrheas of various genesis, viral hepatitis, as well as in the situation of local treatment for pyoinflammatory diseases and purulent wounds. The pharmacotoxicological trials of HDS have substantiated innocuousness of its enteral and applicative usage at doses more high than those permitted for its application as an ancillary substance. Thus, the maximum dose of HDS (10 g per kg of a test animal body weight) which was introduced into the stomach of an animal proved to be nontoxic. Single and repeated peroral administration of HDS at a dose of 100 and 300 mg kg^{-1} for two species of animals (rabbits and rats) did not lead to any deviation of biologic, immunologic, pharmacologic or morphologic indices [1,4].

CLINICAL APPLICATION OF SILICS

The data about fields of application of HDS Silics at polyclinics for treatment for infectious diseases are presented in Table 13.4.

It is seen that the sphere of application of Silics is rather large and covers both intestinal infections and toxicoses which victimise infants, as well as viral hepatitises, botulism.

By way of illustration some concrete data are presented in Table 13.5. From the data it is clear that inclusion of Silics into complex treatment of patients suffering from salmonellosis, dysentery, and intestinal toxicosis accelerates normalization of clinic manifestations of these diseases by a factor of two and more. In the case of botulism the normalization of symptoms characteristic of lesions of nervous system is shortened by almost 4 days. If intestinal infections are not severe, Silics can be recommended as a single therapeutic agent. In the case of a considerable diarrheal syndrome it is more expedient to use it together with rehydration substances.

Inclusion of Silics into a complex of therapeutic agents for patients suffering from viral hepatitises substantially accelerates recovery rates of patients so that their normal level of bilirubin and activity of alanine aminotranspherase are recovered within shorter periods of time.

The mechanism of the therapeutic effect of HDS on treatment for intestinal infections seems to involve the following major aspects [1].

INFLUENCE ON INTESTINAL MICROFLORA

Direct Influence

- fixation of bacteria and their removal from an organism with stools; sorption of microbial toxins of protein nature and other pathogenic

TABLE 13.4
Clinical application of Silics for Treatment for Infectious Diseases

Field of application	Pathologic syndrome	Pharmacologic effect	Particularities of application	Number of patients
Intestinal in-infections (toxi-infections, salmonellosis, shigellosis, cholera, etc.)	Diarrhea, intoxication, dyspepsia	Fixation of microorganisms and their toxins, normalization of absorption and secretion in GIT	Monotherapy Combination with antibacterial agents and agents for rehydration	54 224
Intestinal toxicosis in children	Dehydration, diarrhea, intoxication, dyspepsia	Sorption of toxins, normalization of absorption and secretion of GIT	Monotherapy Combination with antibacterial agents and agents for rehydration	30 144
Viral hepatitises	Cholestasis, cytolysis of hepatocytes, intoxication	Sorption of viruses, bile acids, and bilirubin	Monotherapy. Combination with hepatoprotectors	33 181
Botulism	Neurotoxicosis, dyspepsia	Sorption of toxins, enhancement of action of immunopreparations	Combination with specific serums	103

proteins (neuraminidase, hyaluronidase, contact hemolysins) that promote pathogenic action of microorganisms

- bactericidal effect is possible (in the presence of bile acids and proteolytic enzymes)

Indirect Influence

- creation of conditions unfavorable for vital activity of pathogenic microorganisms (concentration of microorganisms on the sorbent results

TABLE 13.5
Duration of Symptoms of Intestinal Infections in the Course of Treatment with the Aid of Silics (M ± m)

Duration of symptoms (days)	Traditional treatment	Traditional treatment + Silics
Salmonellosis		
Diarrhea	5.60 ± 0.31	2.00 ± 0.45
Normalization of coprograms	5.60 ± 1.06	2.60 ± 0.35
Dysentery		
Diarrhea	7.00 ± 1.50	2.30 ± 0.43
Normalization of coprograms	7.20 ± 1.21	4.00 ± 1.25
Intestinal toxicoses in children		
Intoxication	2.70 ± 0.13	1.80 ± 0.12
Dehydration	3.70 ± 0.18	2.10 ± 0.12
Disturbances of microcirculation	3.80 ± 0.21	2.10 ± 0.35
Botulism (severe case)		
Neurotoxicosis	13.80 ± 1.39	9.90 ± 0.78

Dynamics of biochemical indices in the course of treatment of patients suffering from hepatitis (M ± m)

Indices	Traditional treatment		Traditional treatment + silics	
	Before treatment	In 7–10 days	Before treatment	In 7–10 days
Bilirubin (μmol L^{-1})	264 ± 9.7	192 ± 15.9	244 ± 28.7	151 ± 22.8
Alanine amino-transferase (μmol L^{-1})	3.20 ± 0.65	2.20 ± 0.19	3.00 ± 0.73	1.50 ± 0.28

in a local deficit of nutrients that are necessary for them, fixation of hemoglobin sets a limit on iron that is necessary for microorganisms, etc.)

- adjuvant action (concentration of microbial cells and their toxins on the sorbent enhances the antigenic action and immune response of an organism).

Interaction of HDS with Intestinal Walls and Intestine Content

- blocking of receptors of the mucous membrane of the stomach which are responsible for adhesion of microorganisms and fixation of toxins; intensification of transport of water, electrolytes, and other substances from the intestine into internal medium; modeling of baroreceptors and chemoreceptors of intestinal walls which are responsible for motility
- clearance of intestinal juice from toxic substances (products of vital activity of microorganisms and of microbial putrefaction of proteins), toxic metabolites of endogenous origin (bilirubin, bile acids, micelle complexes, medium-molecular peptides, etc.)
- fixation of cholesterol and other nonpolar lipids being members of complexes with proteins and phospholipids
- the sorbent particles perform the role of sites of concentration and transport of ingredients of the intestine content so that the sorbent acts as a con-enzyme thereby favoring the interaction between metabolities and accelerating the natural course of the process of their transformation, which leads to a decrease in the amount of intermediate products with toxic properties
- the enterosorbents present in the gastrointestinal tract induce immobilization of digestive enzymes and intensifies digestion (in particular hydrolysis of proteins), which reduces irritation of immune system and activates reactions of oxidation, decomposition of peroxide compounds, transamination, etc.

In the course of the above-mentioned reactions HDS remains unchanged and, correspondingly, preserves its activity within all the time of its residence in the intestine.

Besides, Silics may have much promise and potentiality in clinical treatment for internal diseases. The major lines of researches into application of this preparation in the relevant therapy are presented in Table 13.6.

Of significance is the ability of Silics to lower levels of cholesterol and triglycerides as well as to retard aggregation of thrombocytes. Thus, with the help of Silics it

TABLE 13.6
Application of Silics for Clinical Treatment for Infectious Diseases

Fields of application	Pathologic state	Pharmacologic effect	Particularities of application	Number of patients
Disturbances of lipoid metabolism	Hypercholesterolemia	Reduction of absorption and synthesis of cholesterol	Monotherapy. Combination with statines	29 42
	Hypertriglyceridemia	Reduction of absorption and synthesis of triglycerides	Combination with nicotinic acid and fibrates	43 33
Thrombocytarno-coagulative homeostasis	Hypercoagulation	Retardation of aggregation of thrombocytes	Monotherapy. Combination with aspirin	29 118
Allergology	Bronchial asthma, food allergy, psoriasis, eczema	Detoxication, sorption of immune complexes	Monotherapy. Combination with antihistaminic preparations, steroids	138 97 49 250 160
Correction for effect of toxicity of medicinal agents	Insufficient efficiency, side effects	Acceleration of absorption, decrease in dosage, detoxication	Amiodarone, nicotinate, symvastatine, orthophen, quinidine	43 32 25

becomes possible to correct main pathogenic factors of atherosclerosis, namely hyperlipidemia and hypercoagulation. In the case of profound disturbances of lipoid metabolism it proved advantageous to employ Silics together with other hypolipodemic agents. The use of the complexes Silics—symvastatine (synthetic inhibitor of biosynthesis of cholesterol) makes it possible to decrease the dose of the latter by a factor of two without lowering of the intensity of hypolipodemic action. Positive results have also been achieved with reference to reduction of allergic complications induced by symvastatine and of its hepatotoxic action.

Silics has also been successfully employed for treatment of patients suffering from allergic diseases (such as bronchial asthma), chronic obstructive lesions of lungs, food allergies, psoriasis, eczema. The profound detoxication effect of Silics and its high affinity to proteins and medium-molecular peptides form the foundation for its application for treatment of patients of this profile. The remote results of such a treatment have shown that the use of Silics for treatment of patients suffering from chronic obstructive lesions of lungs makes it possible to produce a more complete and more prolonged curative effect in comparison with the traditional treatment.

Important fields of application of Silics are surgery, dentistry, oncology, obstetrics, and gynecology (Table 13.7). As far as these fields are concerned, there are two directions of application of Silics, namely as an enterosorption agent and as an agent for local (apllicative) use.

Local applicative use of Silics may have much promise in treatment for purulent wounds, destructive pancreatitis, peritonitis, purulent pleurisy, odontogenous phlegmons. The mechanism of the curative action resides in sorption of pathogenic microorganisms and microbial toxins, in dehydration of wound tissues. The local applicative use of Silics reveals one more salient feature of the preparation, namely its hemostatic effect. It has been proved that HDS activates the first phase of coagulation of blood, so that it can be employed in the case of moderate external or internal hemorrhages. Depending on localization of a hemorrhage, Silics is used for applications or for drainage and insufflation. Hemostasis is usually attained after 1–2 treatment procedures involving the sorbent.

TABLE 13.7
Application of Silics in Surgery, Oncology, Dentistry, Obstetrics, and Gynecology

Fields of application	Pathologic state	Pharmacologic effect	Particularities of application	Number of patients
Operative surgery	Hemorrhages	Hemostasis	Applications	66
Purulent wounds, destructive peritonitis, ileus, purulent pleurisy	Localized aerobic and anaerobic infections	Sorption of microorganisms, endo- and exotoxins, dehydration	Applications, drainage of cavities	91 56 36 25
Oncology	Intoxication syndrome	Sorption of endo- and exotoxins, chemopreparations	Enterosorption	312
Dentistry	Pyoinflammatory processes	Sorption of microorganisms, endo- and exotoxins, dehydration, abrasive action	Applications, drainage of cavities	156 286
Gynecology	Endometritis, vaginitis, pelvioperitonitis	Sorption of microorganisms, endo- and exotoxins	Applications, lavage	83 56
Operative obstetrics	Hemorrhages, pyoinflammatory processes	Hemostasis, sorption of microorganisms, endo- and exotoxins, localized dehydration	Applications, lavage, combination with antibacterial agents	62
Pregnancy	Gestoses of pregnancy	Detoxication	Monotherapy. Combination with antioxidants	45 72

When Silics is used for treatment of purulent wounds or odontogenous phlegmons, it exerts a marked effect on all the phases of a disease process, which manifests itself in acceleration of healing of wounds, with the time interval necessary for restoration of the function of an organ that was injured being shortened by 3–4 days.

It is also possible to employ Silics for treatment in the case of puerperal infections occurring after cesarean sections. Lavage and drainage of the cavity of the uterus shortened the duration of fever, which, in its turn, decreased the time span of antibacterial therapy. Silics in the capacity of applications or as an ingredient for drainage is also successfully employed for treatment for gynecologic diseases proper (such as vaginitis, endometritis, pelvioperitonitis).

The general detoxicating action of Silics manifests itself in the case of its use of treatment of oncologic patients. Silics seems to decrease levels of endogeneous intoxication that was a side effect of application of chemotherapeutic preparations and radiation therapy, with the efficiency of enterosorption being comparable to that of plasmapheresis.

REFERENCES

1. Chuiko, A.A.; Pentyuk, A.A. In *Proc. of the Scientific Session of the Department of Chemistry of the NAS of Ukraine Dedicated to the 80th Anniversary of the NAS of Ukraine, 9–10 June, 1998*; Osnova: Kharkiv, 1998, pp 36–51 (in Ukrainian).
2. Chuiko, A.A.; Gorlov, Yu.I. *Surface Chemistry of Silica: Surface Structure, Active Sites, Sorption Mechanisms*; Naukova Dumka: Kiev, 1992 (in Russian).
3. Gerashchenko, I.I. Thesis, Khar'kov, Ukraine, 1997 (in Russian).
4. Chuiko, A.A., Ed., *Silicas in Medicine and Biology*; Kiev—Stavropol, 1993 (in Russian).
5. Chuiko, A.A., Ed., *Medicinal Chemistry and Clinical Application of Silicon Dioxide*; Naukova Dumka: Kiev, 2003 (in Russian).
6. Polesya, T.I. Thesis, Moscow, Russia, 1992 (in Russian).
7. Shtat'ko, E.I. Thesis, Institute of Surface Chemistry of the NAS of Ukraine, Kiev, Ukraine, 1993 (in Russian).
8. Il'chenko, A.V. Thesis, Institute of Surface Chemistry of the NAS of Ukraine, Kiev, Ukraine, 1992 (in Russian).

14 Industrial Synthetic Silicas in Powder Form

Horst K. Ferch
Degussa AG, Department of Applied Research and Technical Services, Silicas and Pigments

CONTENTS

Pure synthetic silicon dioxide in powdered form is discussed. After a brief history, the significance of this product group is shown by the total production quantity in the Western hemisphere. A clear classification of synthetic silicas is given, and the principal differences between thermal and wet-process products are illustrated. After-treated silicas are also discussed. Various applications of synthetic silicas are described in detail. Questions about useful handling methods, registration, approval, and toxicology are addressed.

The development of synthetic silicas in powder form took place in the 20th century. Nearly all of the manufacturing processes can be used on an industrial basis, although the extent to which they have been applied varies considerably. The ideas originated in North America and Germany (Table 14.1) [1].

Mention was made in 1887 of electric arc silicas [2], and a detailed account by Potter has been available since 1907 [3]. Industrial use of this process, which involves electricity costs, became possible only after further developments by the B.F. Goodrich Company [4].

Silica gels were first described in 1914 in the Göttingen thesis of Patrick [5], who then developed patents for methods of production in the United States [6] that were first implemented in 1920 [7]. The first silica aerogels were made by Kistler in 1931 [8], and production was started in 1942 [9]. Stöwener was an important early figure in silica gel production in Germany [10].

The first reinforcing silica for rubber, which was in the form of a calcium silicate and known as Silene, was introduced to the U.S. market as early as 1939 [11]. The first reinforcing fillers for rubber articles available in West Germany after World War II were also calcium and aluminum silicates. The production of Hi-Sil, a silicate with a high silica content, started in 1946 [12]. The first "pure" precipitated silica was brought to the European market in 1953 and was called Ultrasil VN 3 [13].

Although the patent was not issued until 1942, Aerosil was successfully produced for the first time by the original flame hydrolysis process in 1941 [14]. The inventor, H. Kloepfer, told me that the idea behind this high-temperature hydrolysis had taken shape in 1934. The details of this process were published in 1959 and later [15,16]. In 1955, Flemmert [17] succeeded in exchanging the $SiCl_4$ used in the aerosil process with SiF_4. This Fluosil process was used in Sweden for about 15 years, and a

TABLE 14.1
Overview of SAS

Process	Raw materials or kind of after-treatment	Inventor	Commercial name
Pyrogenic or thermal silicas			
Flame hydrolysis	$SiCl_4 + H_2 + O_2$	Kloepfer, 1941	Aerosil, Cab-O-Sil
Electric arc	quartz + coke	Potter, 1907	Fransil EL, TK 900
Silicas from wet-process			
Precipitation	water glass + acid	1940	Hi-sil, Ultrasil VN 3, Vulcasil, Zeolex
Spray drying	water glass + acid	approx. 1964	Ketjensil, Sipernat, Zeosil
Gels	acid + water glass	Patrick, 1914; Stöwener, 1924	Gasil, Sorbsil, Syloid
Overcritical conditions	acid + water glass	Kistler, 1931	Santocel
Hydrothermal process	sand + lime	1957	HK 125
After-treated silicas			
Physical after-treatment	thermal		Syloid 73
Chemical after-treatment	coating		OK 412
Chemical after-treatment	reaction on the surface		Aerosil R 972, Sipernat D 17

Note: The term SAS here does not include any salts or synthetic sodium aluminosilicate (SSAS), which are not discussed in this chapter.

recently built factory belonging to Grace commenced production in Belgium with this method in 1990.

Synthetic silicas in powder form are used in a wide range of applications. A breakthrough in certain fields was made possible only by the existence of these synthetic silicas, for example, the silicone rubber that was developed in the United States.

The existing market in 1990 in the Western hemisphere was estimated to be about 1 million metric tons per annum. Table 14.2 gives the approximate distribution of these products in powder form according to the different types. The capacity has doubled since 1974 [1]. If the Eastern hemisphere is included, the value of 1 million metric tons per year increases by approximately 20%.

The large quantities of "fly ashes" that originate from the generation of energy in power stations and from the "silica fume" byproduct of metallurgical processes should also be mentioned. Especially in toxicological literature, the terms "silica fume" and "fumed silica" are used synonymously; this usage is misleading, because the coarse particles of "silica fume," for instance, could cause irreversible changes in the lungs of mammals.

The production of "silica fume" for 1985 was 1.2 million metric tons per year [18]. If an annual increase rate of 4% is calculated, this figure should have risen to 1.46 million metric tons per year by 1990. Fly ashes and silica fume were released into the atmosphere on a virtually uncontrolled basis from about 175 production plants in 1985 [19]. In addition, plasma processes have been used since 1968 to produce synthetic silicas in powder form [20,21], although these are of no technical relevance.

TABLE 14.2
Estimated Distribution of Synthetic Silicas in Powder Form for 1990 in the Western Hemisphere

Silica type	Amount (1000 tons)	Percent of total amount	Percent of total value
Precipitated	810	80	51
Fumed	100	10	33
Gels	90	9	15
Arc	<1	<1	<1
Total	1000	100	100

NOMENCLATURE AND CLASSIFICATION OF TYPES

Until 1985 silicas were usually divided into two groups: natural and synthetic. Most of the naturally occurring SiO_2 modifications are crystalline, but this group also includes amorphous or mostly amorphous products, such as diatomaceous earth or kieselguhr. However, when the observations made at the end of the introduction are considered, this classification no longer proves sufficient, as fly ashes and silica fume are also synthetically produced, although not deliberately. These "airborne dusts" are not harmless [22].

For this reason, I submitted a proposal in 1985 [23] that a differentiation be made between "specifically produced" products and materials that necessarily accumulate.

TABLE 14.3
Classification of Some SiO_2 Terms (Montreal 1987)

Term	Origin	BET surface area (m^2/g)
Fly ashes	power stations	~3
Silica fume	metallurgical processes	~15
Synthetic amorphous silicas	specifically produced	>50

"Specifically produced" silicas are the desired main products, the quality of which is constantly monitored and controlled. This is not the case with the products that necessarily accumulate and are produced on an "uncontrolled" basis.

The products belonging to this uncontrolled group tend to be classified as amorphous [22,24,25]. However, it was shown with the aid of x-ray diffractometer curves that dusts of this nature display the first signs of a short-range order, unlike the specifically produced products [26].

In the old classification system, the groupings were

- natural silicas (mostly crystalline)
- synthetic silicas (amorphous)

Under the new system, the classifications would be

- natural silicas
- specifically produced silicas
- uncontrolled dusts

Table 14.3 shows the categories used to classify synthetic silicas during the Silica Fume in Concrete conference held in 1987 at Montreal.

The term "synthetic amorphous silicas" (SAS) dates to 1985, when a number of people from the American Society for Testing Materials (ASTM) formed a working group called SAS. When discussing amorphous silicas, researchers differentiate between

- pyrogenic or thermal silicas and
- wet-process silicas.

For many years there was considerable confusion regarding the nomenclature of SAS [27,28]. The categories used here for the production processes of SAS were first published in 1968 [29] and were also maintained in reference 1. Further information about the historical details surrounding this chosen system of classification is given in reference 1. The basic categories used in classifying SAS are

1. pyrogenic or thermal silicas (also called fumed silicas)
2. wet-process silicas
3. after-treated silicas, which include
   - physical,
   - chemically fixed, and
   - coated treatment

Further details are given in Table 14.4.

TABLE 14.4
Principal Differences between Characteristic Properties of Pyrogenic and Wet-Process Silicas

Property	Pyrogenic	Wet-process
SiOH density (per square nanometer)	2–3	~6
Drying loss (%)	2–3	~6
Ignition loss (%)	2	5
Aggregate particle size (μm)	—[a]	2–10
Impurities	slight	higher
Density (g/cm^3)	2.2	2.0
Pore volume	0	—[b]
Price	high	lower

[a]Not obtainable.

[b]Always present.

DIFFERENCES AMONG THE TYPES

A standard practice is to characterize SAS by physicochemical data [1]. Some difficulties are encountered when the data given by different manufacturers are compared: first, because test methods have not been generally adopted [dibutyl phthalate (DBP) adsorption, pore distribution, and methanol wettability], and second, because the correlations with the application-related properties, which are of interest to the user, are not widely known.

Nonetheless, it is possible to demonstrate several clear differences between pyrogenic and wet-process SAS (Table 14.4). The principal difference lies in the silanol group density given in the top line of this table, which in turn determines both the drying loss (generally expressed as moisture) and the ignition loss. Because pyrogenic silicas do not generally undergo a grinding process directly after production, it is not possible to state the agglomerate particle size of this group of products. Each method of analysis would provide a different result, depending on how the sample was handled prior to testing. The difference in the interstitial or pore volume is considerable: pyrogenic silicas are compact; that is, they have only an interparticle volume. Wet-process silicas always have a clearly definable share of internal pore volume.

The size of the specific surface area, which is often determined according to the Brunauer–Emmett–Teller (BET) method [30], is easily reproducible, but gives no indication as to the existence of an internal or external

TABLE 14.5
Comparison of SiO_2 Content and Total Impurities after 2 h at 1000°C

SiO_2 source	SiO_2 (%)	Impurities (ppb)
Wet-process method	98–99	<2,000,000,000
Pyrogenic method	>99.8	200,000
Aerosil in its purest form	>99.9999	<100

surface. The submicroscopic (and microscopic) pore volume is determined, for instance, according to the de Boer "*t*-curve method" [31,32].

Commercial pyrogenic silicas usually tend to be purer than wet-process silicas (Table 14.5). Data of a very high-grade silica (silica in its purest form), which is only available in pilot-plant amounts at present, are included in Table 14.5.

The differences between precipitated silicas and silica gels should be discussed. If the water-vapor adsorption behavior of two products with practically the same BET surface area is compared, distinct differences are apparent (Figure 14.1) [33].

The isotherms of silica gels can be recognized as being in the form of a standing "S," and those of precipitated silicas are shown as a lying "S." During the water-vapor adsorption, the silica gels approach the saturation point, whereas this point is not reached by the precipitated silicas.

The difference in the pore volumes is apparent if measured according to reference 31. Figure 14.2 gives pore distribution curves of the products shown in Figure 14.1. Figure 14.2 illustrates how precipitated silicas have relatively wide pores, and in particular a wide range of pore diameters, whereas silicas gels also have relatively narrow pores and display well-defined maximum peaks in the pore distribution curve [33]. By means of physical after-treatment, it is possible to subsequently enlarge the pores of silica gels, so that these products assume adsorptive properties normally restricted to precipitated silicas [34]. A further distinct difference becomes apparent [35] if the specific surface area according to Carman [36] is measured as a function of the porosity ε, which plays an important role in the Carman equation.

Synthetic amorphous silicas were characterized by Von Buzagh [37] as early as 1937. In 1953, Iler [38] wrote about the convenience of differentiating between precipitated silicas and silica gels. Stauff elaborated on this subject in 1960 [39], and this work was endorsed by Iler in 1973 [40]. The details are compiled in reference 1.

Electron microscopy is useful in studying silicas. Both transmission and scanning electron microscopy have greatly contributed to the understanding of the primary particle size and their distribution functions. These

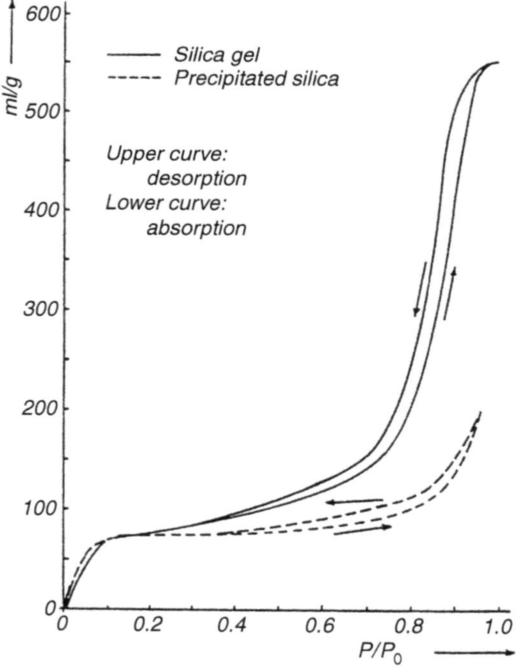

FIGURE 14.1 Water adsorption isotherms of a silica gel and a precipitated silica at room temperature. The existing pressure is *p*, and the pressure in the saturated state is P_0.

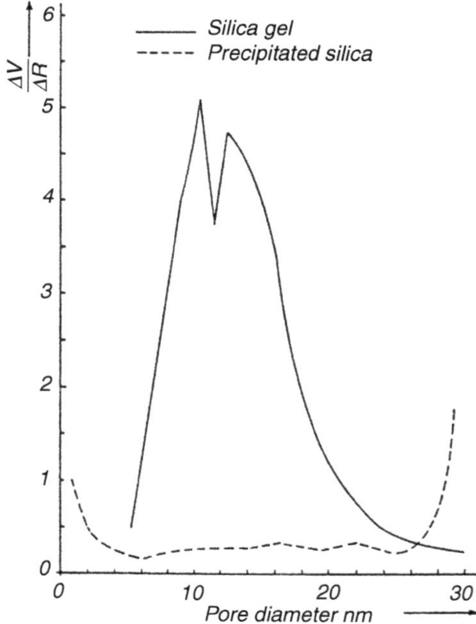

FIGURE 14.2 Comparison of the pore diameter distribution curves of a silica gel and a precipitated silica (33). *V* and *R* denote volume and radius, respectively.

microscopies have also led to a considerable increase in knowledge regarding the points of adhesion that result in the formation of aggregates. However, electron microscopy cannot be used to differentiate between SAS in every case. This subject is dealt with in reference 41.

AFTER-TREATED SYNTHETIC AMORPHOUS SILICAS

All forms of SAS may be after-treated. There are several kinds of possible after-treatment. Physical after-treatment — which does not include grinding in SAS — plays virtually no role here, although some such types were available on the market (e.g., Syloid 73).

The methods of after-treatment, together with the classifications used in the literature until 1973, were compiled in an article in 1976 that includes nearly 100 references [1]. The increase in patent literature on this theme has been particularly apparent. The materials that have been used for the chemical after-treatment of SAS and the date that their use was suggested are as follows [1]:

- 1944, compounds containing chlorine
- 1951, silicone polymers and other polymers
- 1954, compounds containing oxygen
- 1968, compounds containing nitrogen

Kistler was the first person (1944) to suggest the use of chemical after-treatment [42], and the first industrially produced after-treated SAS was brought onto the market in 1962 [43]. This product now bears the trade name Aerosil R 972.

The suggestions are of course given only in a rough and simplified form in the preceding list. For example, after-treatment with fluorinated compounds was also suggested [44,45]. A silica gel that has been after-treated with hydrofluoric acid develops hydrophobic characteristics to the same extent as Aerosil R 972, which is after-treated with dichlorodimethylsilane. Figure 14.3 [46] demonstrates the different behavior of Aerosil 130 (the untreated synthetic amorphous silica) and Aerosil R 972 for water adsorption.

Defining "hydrophobicity" is difficult. A comparison of the different test methods [47] shows that varying results are obtained. The level of the carbon content, in particular, is not a measure for the degree of hydrophobicity. Until a few years ago, the terms "after-treatment" and "hydrophobicity" were synonymous when used in connection with SAS. However, because of the constant increase in the number of new after-treated products, new applications are being discovered for which hydrophobic properties are not required.

The use of chemically aftertreated SAS to improve special properties is discussed in a later section. Coating procedures, which do not involve firmly bonded groups or molecules on the surface of the synthetic amorphous silica as a result of a chemical reaction, are also important, particularly for SAS that are used as flattening agents in the coatings industry. Both silica gels [48] and precipitated silicas [49] may be coated with wax. This coating improves the suspension behavior and the redispersibility in liquid coatings and improves the scratch resistance or "nail hardness" of the dry coating film. During this coating process, the waxes are attached to the surface by means of adsorption [33].

APPLICATIONS

As mentioned in the introduction, SAS are used in many fields. Only the most significant applications are discussed here. The extensive patent literature is not considered here.

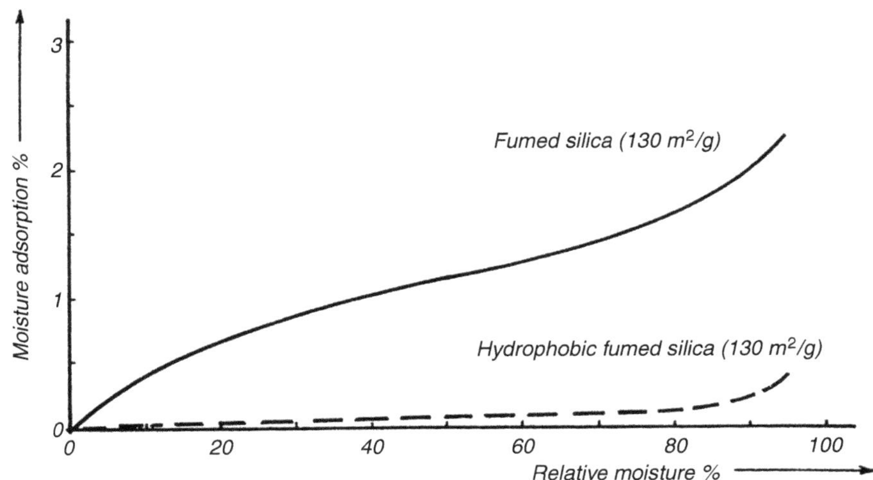

FIGURE 14.3 Water vapor adsorption isotherms at room temperature of the basic hydrophilic silica (top) and the hydrophobic grade produced with this material (bottom) (46).

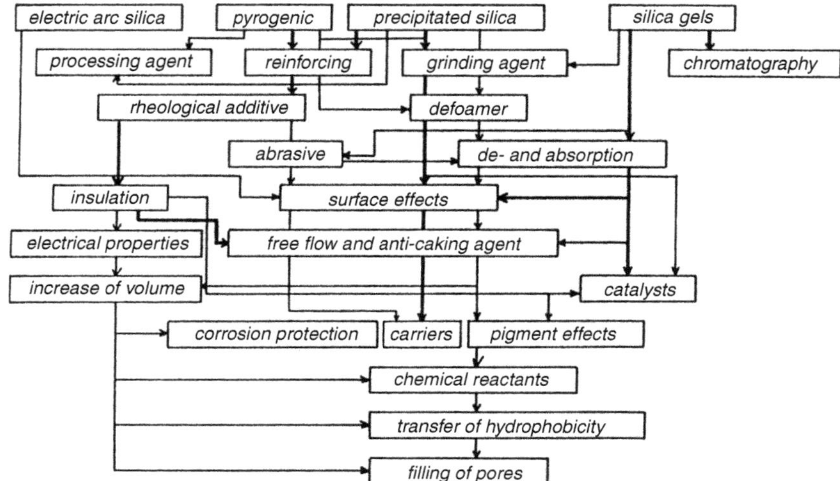

FIGURE 14.4 Overview showing the application of SAS. (Reproduced with permission from reference 50. Copyright 1982.)

Figure 14.4 [50], published in 1982, gives an overview of the applications together with the types of SAS generally used in each case.

Improvement of Mechanical Properties

The most important application of SAS, and one of the oldest, is the control of the mechanical properties of rubber. SAS are important additives for both styrene–butadiene rubber (SBR) and natural rubber (NR), second in importance only to carbon black [51,52]. Figure 14.5 demonstrates the increase in tensile strength at room temperature for silicone rubber with various reinforcing fillers and kieselguhr. An improvement is also brought about in the mechanical strength of fluoroelastomers and other special kinds of rubber [51]. Table 14.6 summarizes the improvements that may be achieved in other fields.

Rheological Additives

SAS serve as excellent rheological additives in a large number of liquid, pastelike, or thermoplastic systems and in solids. Between 0.1 and 2% of the additive is used for applications of this kind, whereas between 3 and 33% is used for the carrier applications discussed in the following section. Table 14.7 lists the applications in which SAS are used as rheological additives.

Carrier Applications

These applications are understood to mean fields of use in which the applied amounts of SAS exceed 2%. By far the most important application is choline chloride in a 60 or 75% aqueous solution that is transformed into a pressure-stable, free-flowing powder with the aid of spray-dried precipitated silicas. Another significant use is as plant protectants and pesticides; SAS are being used in a wide range of applications in this field. An overview of these applications is given in Table 14.8 [53].

SAS are also used in pharmaceutical applications, for example, in the conversion of plant extracts (such as extract valerianae) into free-flowing, pressure-stable powders that are then used to form tablets [54].

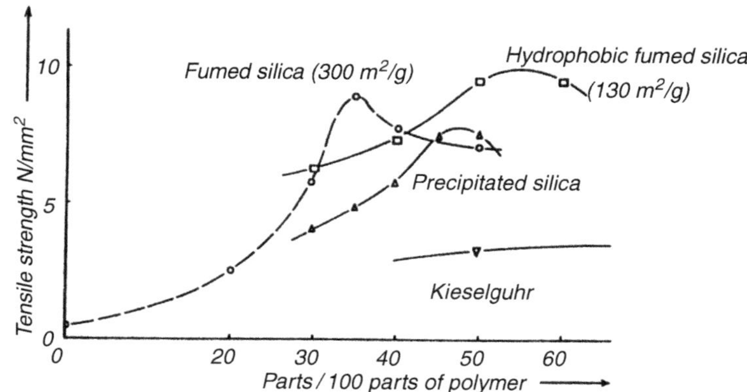

FIGURE 14.5 Increase in the tensile strength at room temperature of silicone rubber containing various fillers.

TABLE 14.6
Improvement of Mechanical Properties by SAS

Application	Improvement
Thermoplastic materials	increase of hardness improvement of the form stability by increased temperatures
Adhesives	heat stability
Decorative paints	sanding of household paints
Toothpastes	tailor-made abrasiveness
Wettable powders (pesticides) and dusts	grinding additive

Source: Reprinted with permission from reference 50. Copyright 1982.

Surface Effects

The surface properties of coatings and printing inks can be changed, as can the properties of plastic films and the surface of paper. The coatings industry uses the largest quantity of SAS. An addition of SAS allows a controlled roughening of the coating surface that produces a flattened or matte effect [55] (Figure 14.6). This effect is of particular interest for furniture coatings, coil coatings, and clear finishes for artificial leather. The desired degree of flattening can be regulated according to the particle size and effective particle volume of the SAS chosen and by the amount of silica added. All types of synthetic amorphous silica are used in this particular field of application.

Two problems can occur during the processing and application of plastic films; these are known as "blocking" and "slip." Blocking describes the state that occurs when two films lying on top of each other are difficult to separate because of unwanted adhesion. It is particularly important to prevent this effect with video tapes and other such films.

Slip describes the gliding motion of thick plastic films used for packaging; heavy goods piled on top of one another are in danger of sliding if the surface of these packaging films is too slippery. SAS are also used during the manufacture of cellulose films to reduce surface slip [33]. Another field of application is in the coating of high-grade papers that are used, for example, in the ink-jet or photocopying processes. An improvement in the adsorptive capacity of the paper results in a high uniformity of the surface [57].

As a rule, 3–7% of synthetic amorphous silica is required for applications in the coatings sector, whereas 0.2–2% is needed in the other fields of application. The market shares for the coatings, plastics, and paper sectors in 1990 were in the ratio 10:5:1, respectively.

Pigment Effects

These effects are achieved in practice only with precipitated silicates. This effect is of special significance in emulsion or latex paints and in paper coatings. These applications play a more minor role in the United States than they do in Europe and Asia. Part of the titanium dioxide content in the paint and paper-coating applications is substituted with precipitated silicates without causing a

TABLE 14.7
Fields in Which SASs Are Used to Improve Rheological Behavior

Field	Type[a]	Improvement
Liquid systems		
Adhesives	pp, ps, sg	brushability, no cobwebbing
Coatings and inks	ps, (sg, pp)	thixotropy, formation of yield points, sharper points
Effect coatings	ps	reinforcing the effects
Gel coats and casting resins	ps, (sg)	thicker layers, no sagging, prevention of sedimentation
Toothpastes	pp, ps, (sg)	workability
Powders		
Coffee, milk and fruit powders	ps, pp, sg	free flow, prevention of caking
Fire extinguishers	ps, pp	free flow, prevention of caking
Garlic, tomato and other powders	ps, sg	free flow, prevention of caking
Inorganic salts, fertilizers	pp, ps	free flow, prevention of caking
Powder coatings, fluid bed systems, toner powders	ps, (pp)	free flow, prevention of caking
Spices, vitamins, amino acids	ps, (pp, sg)	free flow, prevention of caking
Tablets	ps, (pp, sg)	free flow, prevention of caking
Wettable powders	pp, (ps)	free flow, prevention of caking
Thermoplastic Systems		
Plasticized PVC	pp, (sg)	prevention of plate-out

[a] Abbreviations are ps, pyrogenic silica; pp, precipitated silica; and sg, silica gel. Abbreviations in parentheses refer to less-significant applications.

TABLE 14.8
Most Important Uses of Silica in the Formulation of Plant Protectants and Pesticides

Form applied	Active ingredient content	Particle size	Typical use	Typical SiO_2 concentration (%)
Dust	<2%	<45 μm	free-flow agent	<3
			grinding additive	<10
Dust concentrates	40–90%	<45 μm	free-flow agent,	<3
			carrier and grinding additive	<40
Wettable powders	30–80%	<45 μm	carrier,	15–35
			grinding additive,	<10
			free-flow agent	<3
Granules	<5%	<3 mm	anticaking agent	<3
Water-dispersible granules	5–80%	<3 mm	carrier,	15–35
			grinding additive,	<10
			free-flow agent	<3
Suspension concentrates and flowables	30–60%	<5 μm	suspension stabilizer,	approx. 0.5
			control of crystalline growth	approx. 0.1

deterioration in the brightness or opacity. The amount of silicate added must not exceed 5% of the total paint because of the high oil adsorption of these silicates [56,57]. This application is gaining in importance as a result of the current shortage of TiO_2 on the market and the constantly rising price of TiO_2. Precipitated silicates and silicas are also being used more and more in the paper pulp itself. Because of differing demands, a clear regional splitting can be observed. The main area of use is Japan, followed by Europe. Only recently has the use increased in the United States and Canada. For price reasons, the amounts added in these applications are 1–2%. The main improvement is in the printability; the increased brightness is of less significance.

Electrical Effects

By means of adsorption, pyrogenic SAS reduce the number of charge carriers that contribute to the conductivity in poly(vinyl chloride) (PVC) high-voltage cables. As Figure 14.7 shows, this change brings about a reduction in the dielectric losses; that is, the cables can be used at higher temperatures [33]. When used in high-temperature vulcanization (HTV) silicone rubber, these pyrogenic SAS affect the electrical properties as a whole far less than precipitated silicas do; thus, these silicas are not used in cable applications [58].

An increasing use is also being made of pyrogenic silicas in the prevention of electrostatic charges in

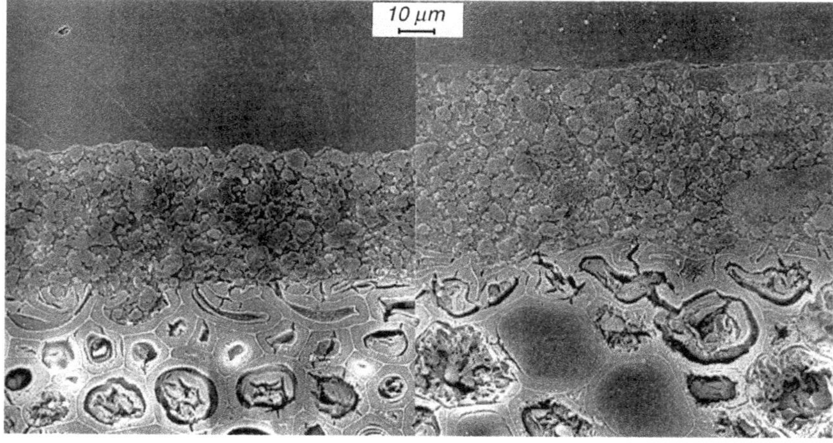

FIGURE 14.6 Scanning electron micrographs depicting microtome sections of wood glazes on absorbent wood: left, control paint; right, paint containing 1.5% fumed silica related to the total coating mass. The presence of this material prevents the coarser SiO_2 particles of the flattening agent from settling.

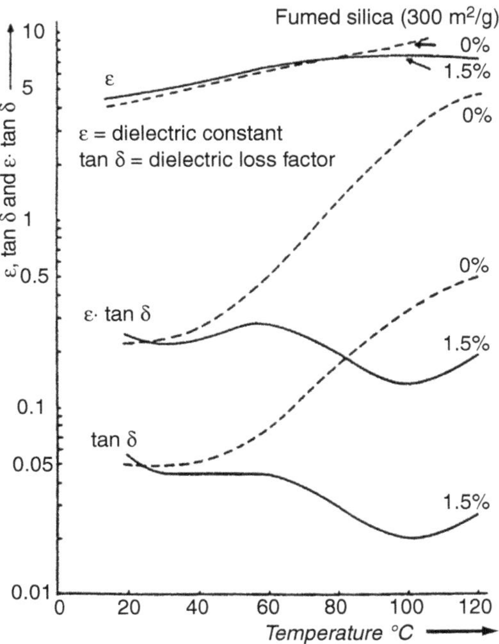

FIGURE 14.7 Dielectric properties of plasticized PVC with and without a fumed silica (300 m^2/g) as a function of temperature.

powdered substances [33]. These silicas play an important role, for instance, in the pharmaceutical industry, where they are used to maintain an undisturbed process of making tablets, or in the grinding of sulfur for pesticides.

After-treated pyrogenic SAS have been used lately to an increasing extent to regulate the triboelectric properties of toner powders. The extent of the effects that can be attained is shown in Table 14.9.

Adsorption

The adsorption of water vapor from mixtures of gas is the main application in industrial practice. This field is the domain of silica gels (Figure 14.1). The economic advantages afforded by the use of silica gels are further increased by the ability to regenerate the wet silica gels. silica gels are not only used in dynamic applications of this nature, but are also used for static applications, such as in the protection of chemicals or in the production of insulating glass for double glazing.

A less significant but nevertheless interesting application is the use of SAS to adsorb certain substances in beer to prevent the beer from becoming cloudy as a result of lengthy periods of storage or excessive cooling. Both silica gels and precipitated silicas are used for this application. Approximately 50–100 g/hl are required. This field of application is also undergoing changes. Europe has been the main area of use, but now the United States, Japan, and Australia are gaining in significance.

TABLE 14.9
Charge-to-Mass Ratios of Aerosil 200 and Some After-Treated Silicas Based on This Material

Sample	Ratio (μC/g)	Relative[a]
VT 501	–560	311
VT 502	–530	294
VT 500	–400	222
R 974	–330	183
R 811	–230	128
Aerosil 200	–180	100
X	–150	83
Y	–50	28
VT 330	–9	9

[a]Percent of Aerosil 200 value.

Similar use is made of pyrogenic SAS in the adsorption of blood components (γ-globulins) from human blood to increase the storage stability of the specially produced blood products. As in the application of SAS in breweries, the SAS do not remain in the beer or in the blood, but are removed almost without trace by filtration and centrifugation.

Catalysis

The production of catalyst supports for a wide range of applications will remain an important field for silica gels in the 1990s. Specially purified silica gels are shaped, dried, and coated. Because of their inferior deformability and stability, molded tablets of other SAS play a subordinate role in the manufacture of catalyst supports (e.g., for the polymerization of olefins). However, the use of pyrogenic SAS for this application is also growing, as the large number of patent applications proves [59]. As with adsorption processes, the size of the pore volume remaining after the deformation is of utmost importance. The systems SiO_2–Ni, SiO_2–Pt, and SiO_2–Ag are of particular interest. Although hydrogenation and reduction are the primary reactions connected with SAS, work has also been carried out on oxidation and polymerization reactions [60]. The so-called "spillover effect" on Pt-coated Aerosil is also worthy of note: ethylene is converted to dideuteroethane with the aid of D_2 [60].

Use of After-Treated SAS for the Improvement of Properties

After-treated, mainly hydrophobic SAS can be used in a large number of applications to produce special effects that cannot be achieved at all (or not as effectively) with the hydrophilic source silicas. An example is the

FIGURE 14.8 Zinc dust paints based on an epoxy ester after a 168-h salt spray test in accordance with DIN 50 021 and ASTM B 117-64: left, control paint; right, paint containing 2% Aerosil R 972 related to zinc dust mass.

improvement of the corrosion protection of primers for the car industry [61]. Figure 14.8 shows a typical effect. A well-defined hydrophobic effect brought about by after-treated SAS is used to regulate the water balance of offset printing inks [62].

Special effects have also been established during the improvement of the free-flow properties of products in powder form. Unfortunately, it is not yet possible to predict under which conditions a hydrophobic synthetic amorphous silica will produce better results than a hydrophilic (i.e., untreated) one. A similar situation exists in the area involving the thickening of technically important resins (e.g., special unsaturated polyester or epoxy resins [63]).

Other Uses

Not all of the possible applications of SAS have been discussed here. A number of other applications are listed in alphabetical order in Table 14.10. In the future, precipitated silicas may be used to a considerable extent for low-temperature insulations (e.g., for refrigerators). A moisture-resistant film with added silica has outstanding heat resistance and does not need any fluorinated hydrocarbons. SAS will likely be used as raw materials for pure silica glass for applications with rigorous requirements. The purity and uniformity of such SAS will have a great influence on the final properties. New applications will undoubtedly follow.

Summary

Table 14.11 is an estimation of the consumption figures for the different fields of application. These values were based on the 1 million metric tons per year for 1990 (Table 14.2). At the same time, an attempt was made to weight these values in such a way as to express the importance of the respective SAS types.

TABLE 14.10
Other Uses for SAS

Application	Type[a]
Battery separators	pp
Chewing gum	ps, pp
Chromatography	sg
Cigarette filters	sg, ps
Control of crystalline growth	pp, ps
Defoamer	pp, ps, sg
Fish Feed	pp, (ps)
Foundary additives	pp, ps
High-temperature insulation	ps
Instant soups	pp, (ps)
Matting of bulbs and fluorescent lamps	pp, ps
Phlegmatization of organic peroxides and explosives	pp, ps
Silica greases	ps
Special detergents	pp, ps
Stain remover	pp, ps
Suspension aid	ps
Thickening in car batteries	ps

[a]Abbreviations are defined in Table 14.7.

HANDLING OF SAS

The need to observe the respective values for the maximum concentration of a substance at the place of work has meant that questions involving the handling of SAS have assumed an important role. The current value

TABLE 14.11
Estimated Consumption of SAS in the Western Hemisphere for 1990

Application	Consumption (metric tons/year)	Percent of total	Type[a]
Reinforcement	730,000	73.0	pp (ps)
Rheology	8,000	0.8	ps
Carrier	75,000	7.5	pp
Surface effects	25,000	2.5	ps, pp, sg
Pigment effects	50,000	5.0	pp
Electrical effects	7,000	0.7	ps
Adsorption	25,000	2.5	sg
Catalysis	30,000	3.0	sg
Hydrophobic grades	5,000	0.5	
Various	45,000	4.5	
Total	1,000,000	100.0	

[a]Abbreviations are defined in Table 14.7.

is set at 4 mg/m^3 of total dust, both in the United States and West Germany.

Handling is also gaining significance for ecological reasons. In Europe, about 25% of SAS is still being delivered in paper sacks. Difficulties arise with the polyethylene-coated paper sacks, which cannot be recycled at present. As a result of a yearly increase of 8% (1988) in total waste produced in West Germany, SAS consumers are encountering increasing difficulties and costs to dispose of the empty paper sacks. For this reason, it is of major importance that the pneumatic conveyance of SAS functions well. Distances of up to 300 m and differences in height of up to 25 m pose no great technical problems. Moreover, the weighing and measuring of SAS needs to be carried out both continuously and discontinuously with sufficient accuracy.

Mechanical conveyance — for example, with angle-type bucket elevators — is only recommended for granulated products, such as those used in the rubber industry.

QUESTIONS REGARDING REGISTRATION AND APPROVAL

In the past, non-after-treated SAS bore the Chemical Abstracts Service (CAS) registry number 7631–86–9, aluminum silicates bore the number 58425–86–8, and calcium silicates the number 1344–95–2. SAS that were after-treated with dichlorodimethylsilanes were given the number 60842–32–2 (formerly 68611–44–9), and SAS after-treated with polysiloxanes the number 67762–90–7. Thanks to efforts by SASSI (Synthetic Amorphous Silica and Silicates Industrial Association), new numbers have been assigned since June 7, 1989, which are designed to allow a better differentiation between the amorphous and crystalline substances. The following numbers are now valid for the different types of SAS:

- for fumed amorphous silicas, 11.2945–52–5
- for precipitated silicas, 11.2926–00–8
- for silica gels, 11.2926–00–8

As these numbers are not yet listed in the legal inventories of toxic chemicals in the United States, Europe, Australia, and Canada, it is also necessary to give the old CAS registry numbers. This situation causes certain problems — at least in countries outside the United States.

Synthetic silicas are permitted for use in many countries as anticaking agents in foodstuffs in powder form. National regulations must be observed here. In the United States, silicon dioxide is used in an amount not to exceed 2% by weight of the food and is a direct "food additive" [64]. In West Germany, a direct allowance is given only for salt, chewing gum, and some other foodstuffs. Silica is registered in the European Community under E551 as a food additive. The peroral harmlessness of SAS for human beings is established by the World Health Organization–Food and Agriculture Organization with its classification "unlimited daily intake."

"Silica" is also mentioned in the United States as an indirect food additive without any limitation in reference 64, parts 175–178, for adhesives, components of coatings, paper and paperboard, organic polymers, production aids, auxiliaries, and sanitizers. In this respect, the guidelines in Germany are very similar.

Each subdivision of SAS (fumed, gels, and precipitated) is listed separately in several national pharmacopoeias, for example, in the USPXXI/NFXVI and in the pharmacopoeia of the European Community.

The testing of SAS according to the National Institute of Occupational Safety and Health (NIOSH) test number 7601, for example, can never serve as an exact procedure for the differentiation of crystalline or amorphous silica. The test may be useful (to a certain extent) for determining the content of respirable or total dust in airborne samples in order to ascertain the approximate degree of risk for the human lung.

TOXICOLOGY

SAS can affect the health of a human being by skin contact, inhalation, or peroral intake.

Skin Contact

Employees coming into contact with SAS sacks or dusts occasionally complain of itchiness or reddening of the skin. The individual sensitivity of a worker certainly plays a role here. I have not heard of any permanent complaints or damage during my 32 working years. The results obtained for Aerosil, for example, are compiled in reference 26, which includes numerous bibliographic references. In view of the possible occurrence of scleroderma, in 1989 the Industrial Injuries Insurance Institute for the Ceramics and Glass Industries said that no such case has been registered for more than 20 yr [65].

Inhalation

In 1962, reports issued by a production plant for precipitated silicas in the United States [66] stated that employees who had worked for more than 16.5 years at a dust concentration of 0.35–204 mg of SiO_2 per cubic meter showed no signs of silicosis or any other dust-related lung disease. This statement was confirmed for precipitated silicas in West Germany after regular medical examinations were carried out on 131 employees who had worked for a maximum of 38 yr [26]. Similar findings were confirmed for arc silicas in 1982 [67]. After exposure to the pyrogenic aerosil for 14 yr [68] and for more than 30 yr [26], no silicosis had occurred. Furthermore, no signs of toxic

TABLE 14.12
Acute Toxicity of Some SAS on Rats

Commercial name	Oral LD_{50} for rats (mg/kg)	Institution
Hydrophilic Aerosil types	>10,000	CIVO, TNO, NL (1979)
Hydrophobic Aerosil R 972	>5,000	Leuschner (1977)
Ultrasil VN 3 Calsil, Extrusil, and Pasilex (all hydrophilic)	>10,000	CIVO, TNO, NL (1979)
Coated OK 412	>1,000	Leuschner (1979)

Source: Reprinted with permission from reference 65. Copyright 1987.

effects were observed [26]. On the basis of experience gained to date, no harmful side effects are to be expected with silica gels either.

It is interesting to note that in West Germany the "increased occurrence of bronchial carcinomas in patients with silicosis cannot be based on statistical records" [65]. In other words, even if silicosis develops as a result of quartz dust, the silicosis cannot be associated with a cancerous disease. According to Mayer [69], the findings from animal experiments that showed that quartz dust also caused an increase in carcinomas cannot be applied to human beings.

Peroral Intake

The effect of orally ingested SAS plays a subordinate role in industrial toxicology [70]. This fact is also taken into account by the WHO, which says the daily intake of SiO_2 is unlimited. None of the silicas are resorbed following oral administration. The LD_{50} for rats is therefore very large, as Table 14.12 demonstrates.

REFERENCES

1. Ferch, H., *Chem. Ing. Techn.* **1976**, *48*, 922.
2. Anonymous *Am. Chem. J.* **1887**, *9*, 14.
3. Potter, H.N. U.S. Patent 875,674, 1907; 875,675, 1907.
4. BF Goodrich Co., German Patent 1,034,601, 1955; U.S. Patent 2,863,738, 1958.
5. Patrick, W. A. Thesis, Göttingen University, Germany, 1914.
6. Silica Gel Corp. U.S. Patent 1,279,724, 1918.
7. Patrick, W. A.; McGaunack, J. *J. Am. Chem. Soc.* **1920**, *42*, 946.
8. Kistler, S. S. *Nature (London)* **1931**, *127*, 741; *J. Phys. Chem.* **1932**, *36*, 52.
9. White, J. F. *Chem. Ind.* **1942**, *51*, 66.
10. IG-Farbenindustrie, German Patent 428,041, 1924.
11. Pittsburgh Plate Glass Co. U.S. Patent 2,287,700, 1942.
12. Boss, A. E. *Chem. Ing. News* **1949**, *27*, 677.
13. *75 Jahre Chemische Fabrik Wesseling AG*; Arch. f. Wirtschaftskunde GmbH: Darmstadt, Germany, 1955.
14. Degussa A. G. German Patent 762,723, 1942.
15. White, L. J.; Duffy, G. J. *Ind. Eng. Chem.* **1959**, *51*, 232.
16. Wagner, E.; Brünner, H. *Angew. Chem.* **1960**, *72*, 744.
17. NYNÄS Petroleum, German Patent 1,208,741, 1955.
18. Anonymous *Minaeraçao Metalurgia* **1985** July, *49*, 466.
19. Anonymous *Metal Bulletin Monthly* **1985** July, 53.
20. Monsanto German Patent 1,940,832, 1968.
21. Lonza A. G. German Patent 2,337,495, 1972.
22. Regidor, M. Thesis, Grenoble, France, 1981.
23. Ferch, H. *Staub, Reinh. Luft* **1985**, *45*, 237.
24. Vitums, V. C.; Edwards, M. J.; Niles, N. R.; Borman, J. O.; Lowry, R. D. *Arch. Environ. Health* **1977**, *32*, 62.
25. Brambilla, C.; Brambilla, E.; Regaud, D.; Perdrix, A.; Paramelle, B.; Fourcy, A. *Rev. Fr. Mal. Resp.* **1980**, *8*, 383.
26. Ferch, H.; Gerofke, H.; Itzel, H.; Klebe, H. *Arbeitsmed., Sozialmed., Präventivmed.* **1987**, *22*, 6, 23.
27. American Cyanamid, U.S. Patent 2,114,123, 1938.
28. Burak, N. Chem. *Process UK* **1967**, *8*, 10.
29. Ferch, H. IX. *Fatipec-Kongreßbuch* **1968**, 144.
30. Brunauer, S.; Emmett, P. H.; Teller, E. J. *J. Am. Chem. Soc.* **1938**, *60*, 309.
31. De Boer, H. J. *J. Catalysis* **1965**, *4*, 646.
32. De Boer, H. J.; Lippens, B. C.; Linsen, B. G.; Brockhoff, J. C. P.; Van Den Henvel, A.; Osinga, Th. J. *J. Coll. Interf. Sci.* **1966**, *21*, 405.
33. Ferch, H. In *Lehrbuch der Lacke und Beschichtungen*; Kittel, H., Ed.; W. A. Colomb Verlag: D-Oberschwandorf, 1974; Vol. 2.
34. W. R. Grace German Patent 1,036,220, 1955.
35. Kindervater, F. *XII. Fatipec-Kongreßbuch* **1974**, 399.
36. Carman, P. C. *J. Soc. Chem. Ind. London* **1938**, *57*, 825; **1950**, *69*, 134.
37. Von Buzagh, A. In *Colloidal Systems*; Technic. Press Ltd.: London, 1937; p. 149.
38. Iler, R. K. *Colloid Chemistry of Silica and Silicates*; Cornell University Press: Ithaca, NY, 1953.
39. Stauff, J. *Kolloidchemie*; Springer-Verlag: Berlin-Heidelberg, 1960.
40. Iler, R. K. In *Surface and Colloidal Science*; Matijevic, E., Ed.; Wiley: New York, 1973.
41. Ferch, H.; Seibold, K. *Farbe Lack* **1984**, *90*, 88.
42. Kistler, S. U.S. Patent 2,589,705, 1944.
43. Brünner, H.; Schutte, D. *Chem. Zeitg.* **1965**, *13*, 437.
44. Davison Chem. Co. U.S. Patent 2,625,492, 1950.
45. Wilska, S. *Suom. Kem.*; *Finland* **1959**, *32B*, 89.
46. Ferch, H. *Farbe Lack* **1979**, *85*, 651.
47. Ferch, H. *13th Eurocoat Congress; Nice, France*, **1987**, 128.
48. W. R. Grace German Patent 1,006,100, 1957.
49. Degussa A. G. U.S. Patent 3,607,337, 1968.
50. Ferch, H. *Progr. Org. Coat; Switzerland*, **1982**, *10*, 91.

51. Parkinson, H. *Reinforcement of Rubbers*; Lakemann: London, 1957.
52. Bachmann, H. J.; Sellers, J. W.; Wagner, M. P.; Wolf, R. F. *Hi-Sil*; Columb Southern Chemical Corp.: Barberton, Ohio, 1960.
53. Ferch, H.; Reisert, A.; Bode, R. *Kautschuk Gummi, Kunststoffe* **1986**, *39*, 1084.
54. Ferch, H.; Müller, K. H.; Oelmüller, R. In *Technical Bulletin Pigments No. 1*, 5th ed.; Degussa AG: Frankfurt, Germany, 1989.
55. Ferch, H. *Pharm. Ind.* Germany **1970**, *32*, 478.
56. Ferch, H.; Flach, V. *Wochenblatt f. Papierfabrikation* **1987**, *115*, 10.
57. Ritter, H. S. *Paint Varn. Prod.* **1963**, *53*, 31.
58. Ferch, H. *Verfkroniek; Netherlands*, **1965**, *38*, 9.
59. Bode, R.; Ferch, H.; Fratzscher, H. *Kautschuk Gummi, Kunststoffe* **1967**, *20*, 699.
60. Ferch, H.; Koth, D. *Chem. Ing. Techn.* **1980**, *52*, 628.
61. Lenz, D. H.; Conner, W. C. *J. Catalysis* **1987**, *104*, 288.
62. Ferch, H. *Farbe Lack* **1979**, *85*, 651.
63. Schumacher, W.; Gräf, H. *Farbe Lack* **1971**, *77*, 237.
64. *Code of Federal Regulations*, Title 21, Part 172, Section 480.
65. *Technical Bulletin Pigments No. 27*, 4th ed.; Degussa AG: Frankfurt/M., Germany, 1987.
66. Mayer, P., personal communication.
67. Plunkett, E. R.; De Witt, B. J. *Arch. Environ. Health* **1962**, *5*, 469.
68. Ferch, H.; Habersang, S. *Seifen-Öle-Fette-Wachse* **1982**, *108*, 487.
69. *Technical Bulletin Pigments No. 64*, 3rd ed.; Degussa AG: Frankfurt/M., Germany, 1987.
70. Volk, H. *Arch. Environ. Health* **1960**, *1*, 125.

15 High Ratio Silicate Foundry Sand Binders

Horacio E. Bergna
Dupont Experimental Station

CONTENTS

Binder solution for preparing sand cores of initial high strength but with essentially no strength after casting metals above 700°C, said binder solution comprising an aqueous solution of sodium, potassium or lithium silicate having an overall molar ratio of SiO_2/alkali metal oxide from 3.5:1 to 10:1, and containing sufficient amorphous silica so that the fraction of the total silica in the binder solution which is present as amorphous silica is from 2 to 75%, the amorphous silica having a particle size in the range from about 2 to 500 nm.

BACKGROUND

In the metal casting industry molten metal is cast into molds containing sand cores made from foundry sand and binders. These sand cores are conventionally bonded with organic resins which, during curing and casting of the metal, decompose and evolve byproducts which are odoriferous, offensive fumes which are not only skin irritants but in most cases toxic. The molds themselves are made from foundry sand bonded with oils, clays and organic resins. Thus, during their use, similar problems can occur.

A great percentage of the sand binders used by the foundry industry are made of phenol- and urea-formaldehyde resins, phenolic- and oil-isocyanate resins, and furan resins. Almost all of these binders and their decomposition products such as ketones, aldehydes, and ammonia are toxic. The principal effect on man is dermatitis which occurs not so much from completely polymerized resins, but rather from the excess of free phenol, free formaldehyde, alcohol or hexamethylenetetramine used as a catalyst. Formaldehyde has an irritating effect on the eyes, mucous membrane and skin. It has a pungent and suffocating odor and numerous cases of dermatitis have been reported among workers handling it. Phenol is a well-known poison and is not only a skin irritant but is a local anesthetic as well, so that burns may not be felt until serious damage has been done. Besides being capable of causing dermatitis it can do organic damage to the body. Furfuryl alcohol defats the skin and contact with it has to be avoided. Hexamethylenetetramine is a primary skin irritant which can cause dermatitis by direct action on the skin at the site of contact. Urea decomposes to carbon dioxide and ammonia, the latter of which is intolerable in toxic concentrations. In addition to the binders, some processes use flammable gases such as triethylamine as a curing agent. Capturing or destroying gases, smoke and objectionable odors are only temporary, stop-gap expensive solutions. New binders are needed that completely eliminate the sources of offensive odors and toxic gases.

Many of the organic binders are hot setting and therefore require heating to cure. Hot molds not only add hazards and complicate pollution control problems but add economical problems related to increased use of energy and increased equipment, maintenance and operation costs.

An alternative is to use inorganic cold setting binders, such as sodium silicate, which set at room temperature without producing objectionable gases or vapors. The use of silicates, however, results in the silicate bond remaining too strong after casting, so that the core is still coherent, and has to be removed by use of violent mechanical agitation or by dissolving the silicate bond with a strong, hot aqueous alkali. The problem may be lessened to a degree by using sodium silicate solutions admixed with organic materials such as sugar, but even in this case the core is still coherent after casting and requires extreme measures for removal such as violent mechanical agitation.

Thus, there is a need to create a binder for sand in making cores and molds for casting metals such as aluminum, bronze, or iron, that will have satisfactory high strength before the metal is cast, retain sufficient hot strength and dimensional stability during the hot metal pouring, but which will have such strength after the metal has been cast and cooled, that the sand can be readily shaken out of the cavities formed by the cores; the binder also should be one that will not evolve unreasonable amounts of objectional fumes when the sand cores and molds are subjected to molten metal.

SUMMARY

I have discovered that molds and sand cores of initial high strength but with essentially no strength after casting metals above 700°C can be made by bonding foundry sand with an aqueous solution of sodium, potassium, or lithium silicate or their mixtures and amorphous colloidal silica the amounts of silicate and amorphous colloidal silica being such that the overall molar ratio of SiO_2/alkali metal oxide (M_2O) is from 3.5:1 to 10:1, preferably 4:1 to 6:1, the fraction of the total silica present as amorphous colloidal silica is from 2 to 75% by weight, preferably 2 to 50%, and most preferably 10 to 50%, the amorphous colloidal silica having a particle size in the range from about 2 to 500 nm, and the 98 to 25% balance of the total silica being in the form of silicate ions. The percent solids of the aqueous binder solution being 20–55% by weight. The amorphous colloidal silica in the binder comprises both the amorphous colloidal silica component of the mixture and the amorphous colloidal silica fraction inherently present in aqueous solutions of alkali metal silicates of ratio more than about 2.5.

In alkali metal aqueous solutions containing more than 2.5 mols of SiO_2 per mole of M_2O, it is found by ultrafiltration, according to a procedure referred to herein as the Gore Procedure, that at the concentrations used in this invention part of the silica in solution is ionic and part of it is colloidal, the colloidal fraction being retained by the ultrafilter while the ionic silicate passes through. In the case of sodium silicate for example, concentrated commercial silicate solutions are available having a SiO_2/Na_2O ratio as high as 3.8/1.0 and these concentrated solutions therefore contain a substantial proportion of the silica present in the colloidal state. The colloidal fraction consists of a range of sizes less than 5 nm diameter and down to near 1 nm, with a substantial amount of 2 or 3 nm diameter. These units are so small that solubility equilibrium is rapidly established so that if the solution is diluted with water the units pass into solution forming lower molecular weight ionic species.

The higher the ratio of concentrated aqueous solutions of alkali metal silicates the higher the colloidal silica content, but for each ratio the colloidal silica content decreases with dilution of the solution.

To prepare a binder having SiO_2/M_2O ratio of 3.5 to 3.8 it is therefore not necessary to add any colloidal silica if an alkali metal silicate solution is used already in the ratio range. On the other hand, if an alkali metal silica solution with ratio lower than 3.5 is used, it is necessary to add at least some colloidal silica in the form of a sol to prepare out binder.

Silica aquasols (water dispersions of colloidal amorphous silica) containing only a small amount of alkali as a stabilizer are commercially available and are described in the preferred aspects of this invention.

In summary, binder compositions of our invention comprise (1) aqueous solutions of alkali metal oxide silicates with or without amorphous silica present therein and (2) amorphous colloidal silica, if the silicate does not have any amorphous silica present therein or if the level of amorphous silica in the silicate is not sufficient.

The core and mold compositions of the invention have the additional advantage in that they can be made cold-setting, that is, heating to set the binder system is not necessary. Thus, they can be set with CO_2 or a suitable acid releasing curing agent.

Preferred for use in the compositions of the invention are binder wherein the alkali metal silicate is sodium silicate and at least 10% of the amorphous silica is obtained from a silica sol.

In preferred embodiments of the composition of the invention carbonaceous materials and/or film forming resin adhesives are employed. These materials can add desirable properties with respect to shake-out and storage life. The employment of these optional, but preferred, materials is described in greater detail in the following paragraphs.

Thus, I have found sand core or mold compositions of foundry sand and binder wherein the composition consists essentially of 85–97 parts by weight of foundry sand and 3–15 parts by weight of an aqueous binder comprising an aqueous sodium, potassium or lithium silicate solution or mixtures thereof with 20–55% solids content and amorphous silica, the amorphous silica in the silicate solution determined by the Gore test procedure, the binder characterized by (1) a molar ratio of silica to alkali metal oxide of from 3.5:1 to 10:1; (2) a weight fraction of the total silica present as amorphous silica is from 2 to 75%; and (3) a weight fraction of the total silica present as silicate ions is from 98 to 25% and the amorphous silica has a particle size of from 2 to 500 nm and the sand core or mold possesses a compressive strength sufficiently low to permit easy crushing after said core or mold is used in preparing a metal casting.

Accordingly, the present invention also includes a method for making a sand core or a sand mold useful in the casting of molten metal which comprises mixing 85–97 parts by weight of foundry sand with 3–15 parts by weight of a binder which comprises an aqueous sodium, potassium or lithium silicate solution or mixtures thereof with 20–55% solids content with amorphous silica having a particle size of from 2 to 500 nm, the amount of silicate and amorphous silica being adjusted to form a binder with (1) a molar ratio of silica to alkali metal oxide ranging from 3.5:1 to 10:1; (2) the weight fraction of total silica present as amorphous silica of from 2 to 75%; and (3) a weight fraction of the total silica present as silicate ions of from 98 to 25%; the amorphous silica present in the silica solution is determined by the Gore test procedure, forming the sand and binder mixtures into the desired shape and setting the formed mixture.

DESCRIPTION

Foundry Sand

The compositions of the invention will contain between 85 and 97 parts by weight of foundry sand, preferably between 90 and 96 parts by weight. The amount of binder used is related to sand type and particle size in that with small sand particles and more angular surfaces, more binder mixture will be necessary.

The type of foundry sand used is not critical and the useful foundry sands include all of the ones conventionally used in the metal casting industry. Thus, these sands can be zircon sands (zirconium silicates), silica sands, e.g., quartz, aluminum silicate, chromite, olivine, staurolite, and their mixtures.

The particle size of the foundry sand again is not critical and American Foundrymen's Society (AFS) particle sizes of 25–275 GFN can be employed. GFN stands for grain fineness number and is approximately the number of meshes per inch of that sieve which would just pass the sample if its grains were of uniform size, that is, the average of the sizes of grains in the sample. It is approximately proportional to the surface area per unit weight of sand exclusive of clay.

The useful sands can be washed sands or they can be unwashed sands and contain small amount of impurities, that is, clay. If recycle sands are used, an adjustment may have to be made to the binder mixture to take into account any silicate present in such sands.

Various minerals can be used as sand additives to optimize mold or core performance. For instance, alumina or clay powders can be used to improve the high-temperature strength and shake-out characteristics of the sand cores.

Conventional refractory grain alumina powders, kaolin, and Western bentonite can be used. Kaolin is preferred in amounts between 0.5 and 10% by weight of the sand. An example of a kaolin grade useful for this purpose is Freeport Kaolin Co.'s "Nusheen" unpulverized kaolin material which consists of kaolinite particles with a specific surface area of about 16 m^2/g.

Binder System

The compositions of the invention contain 3–15 parts, per 100 parts of sand binder mixture by weight, of a binder system comprising a water soluble alkali metal silicate, amorphous colloidal silica and water. The key is to have very finely divided amorphous silica particles of colloidal size dispersed within the alkali metal silicate bond. It is inherent in the nature of water soluble alkali metal silicates having a molar ratio SiO_2/alkali metal oxide (M_2O) above about 2.5, that colloidal silica is present. In the case of silicates having a ratio higher than 3.5, the colloidal silica content is such that they may be employed without adding more colloidal silica, but in the case of alkali metal silicates of lower silica/alkali metal oxide ratio there is little or no amorphous colloidal silica present so that amorphous colloidal silica must be added in order to produce the cores and molds of the present invention.

In order for the foundry core or mold to become weak after heating and cooling, it is helpful to have crystalline silica such as cristobalite formed throughout the binder mass by spontaneous nucleation at high temperatures. Such nucleation apparently occurs at the surface of particles of amorphous colloidal silica. Hence, the larger the area of such surface, the weaker the resulting core after heating and cooling. If enough amorphous silica is colloidally subdivided and dispersed within the silicate, then within one gram of such silicate binder there can exist dozens of square meters of amorphous silica surface. The smaller the particles, the more rapid the loss of core strength after heating at 700°C, and cooling.

The useful water soluble silicate component of the mixture includes the commercially available sodium, potassium or lithium silicate, or their mixtures. Sodium silicate is preferred. These silicates are usually used as solutions; however, their hydrates can be used provided that water is mixed into the binder, either prior to or during application to the sand. The useful sodium silicate aqueous solutions have a weight ratio of silica to sodium oxide ranging from 1.9:1 to 3.75:1 and a concentration of silica and sodium oxide of about 30–50% by weight. As stated above, a fraction of the silica in the useful water soluble sodium silicate of SiO_2/M_2O ratio higher than 2.5 is in the form of very small particle size amorphous colloidal silicate. Alkali metal silicates with SiO_2/alkali metal oxide ratio higher than about 3.5:1 are referred to as high ratio alkali silicates or alkali polysilicates although they contain in fact a certain proportion of colloidal silica. In essence high ratio alkali metal silicate aqueous solutions can be conceived as mixtures of alkali metal ions, silicate ions and colloidal silica. High ratio alkali metal silicate solutions contain varying amounts of monomeric silicate ions, polysilicate ions and colloidal silica micelles or particles. The type, size of the ions and micelles or particles, and distribution depend for each alkali metal on ratio and concentration. Aqueous solutions of moderate concentration of the metasilicate ratio, namely SiO_2/alkali metal oxide 1:1, or more contain mainly the monomeric silicate ions. In disilicate aqueous solutions of moderate concentration, with SiO_2/M_2O of 2/1 only the simple metasilicate and disilicate ions are present. Aqueous solutions of silicates with greater ratios contain monomeric silicate ions, dimeric silicate ions, and polymeric silicate ions (trimers, tetramers, pentamers, etc.)

The degree of polymerization of the silica in silicate solutions may be expressed as the number of silicate groups formed in the average molecule of silicic or polysilicic acid corresponding to the alkali metal silicate. The degree of polymerization increases with the ratio of the silicate. Whereas for example a sodium silicate solution of ratio 0.5:1 may have an average silica molecular weight of 60 corresponding to one molecule of SiO_2, sodium silicate solutions of ratio 1, 2, 3.5, and 4.0 are formed to have average molecular weights of about 70, 150, 325, and 400, respectively. This is the reason why as mentioned above high ratio silicates containing a large proportion of polymeric ions are also known as "polysilicates."

Silicate polymer ions with a corresponding silica molecular weight above about 600 are sufficiently large to be considered as very small silica particles and will hereinafter be referred to as colloidal silica or colloidal SiO_2. Colloidal particles are generally defined as particles with a particle size between about 1 and 500–1000 nm. This particle size range constitutes the colloidal range and is not limited by a sharply defined boundary.

Alkali metal silicates with an "average" silica molecular weight higher than around 200–300 have a fraction of their silicate ions present as polysilicate ions in the colloidal range. The higher the average molecular weight the higher the fraction of polysilicate ions in the colloidal range and the higher the molecular weight or particle size of polymer ions or particles in the colloidal range. For example, a sodium silicate solution ratio 3.35:1 may contain more than 2 or 3 and as much as 15 percent by weight of the total silicate or silica in the form of colloidal silica. Sodium silicate solutions ratios 3.75:1 and 5:9 may contain more than 8 or 10 and as much as 33% by weight of the total silica, respectively in the form of polysilicate ions or colloidal silicate. Higher ratio sodium silicate solutions of various ratios eventually reach a state of equilibrium in which the colloidal silica fraction has a certain particle size distribution. In the case of sodium silicate aqueous solutions ratio 3.25 to 4 at equilibrium the colloidal silica fraction has a particle size smaller than 5 nm.

High ratio sodium silicate solutions may be prepared by simply adding dilute silica aquasols (colloidal dispersions of silica in water) to dilute low ratio sodium silicate solutions. In this case and until equilibrium is reached, average particle size of the colloidal silica fraction will be determined by time and silica particle size distribution of the original sol and the original silicate solution.

Increase in the ratio of alkali metal silicate solutions containing a constant concentration of silica causes an increase is viscosity even to the point of gelling or solidification. For this reason the maximum practical concentrations for alkali metal silicate solutions decrease with increasing ratio. Maximum practical concentration is the maximum concentration of SiO_2 plus Na_2O in solution at which the silicate solution flows like a fluid by gravity and is stable to gelation for long periods of time. The following table illustrates as an example the case of sodium silicate aqueous solutions.

Approximate SiO_2/Na_2O molar ratio	Approximate maximum practical concentration, % Wt.
1.95	55
2.40	47
2.90	43
3.25	39
3.75	32
5.0	<20

Above a certain concentration which decreases with increasing silica-soda ratio as explained above, sodium silicate aqueous solutions become very viscous and are stable for only a limited period of time. Stability in this case means resistance to gelling. More stable solutions can be made at lower sodium silicate concentrations but this may become impractical in a foundry binder. The high water content of very high ratio (more than 4 to 5) sodium silicate solutions at practical viscosities prevent their extended use as a foundry binder in the present invention. Excessively high water content in a foundry binder means unacceptably weak sand molds or cores and detrimental quantities of steam evolving when the molten metal is poured into the sand mold-core assembly.

Thus, the compositions of this invention involve percent solids in the aqueous binder of from 20 to 55%. Based on 3–15 parts by weight of binder in the composition of this invention this translates to 1.35–12% by weight water in the binder based on the final composition.

I have discovered ways of using high ratio alkali metal silicates as foundry sand binders without introducing excessive amounts of water into the sand and without employing unstable commodities.

A practical way of using high ratio silicate as binders for foundry sands is to mix concentrated silica aquasols and concentrated sodium silicate aqueous solutions in situ, that is on the surface of the sand grains, thus forming the high ratio silicate on the sand surface.

Concentrated sodium silicate aqueous solutions cannot be mixed with concentrated silica aquasols without almost immediate gelling. It would be very impractical or simply impossible to mix gels formed in this manner with sand using the means available today in common foundry practice.

However, I have discovered that effective mixing and binding effect is obtained with sand if the concentrated silica sol is mixed first with the sand to form a uniform and continuous film on the surface of the sand grains. The concentrated sodium silicate solution is then added to the sand mass in a second, separate step and the sodium silicate then mixed with the colloidal silica film on the surface of the sand, gelling in situ to form an intimately and uniformly mixed binder within the sand mass. The sand mix thus formed in the mixer can be molded by any of the various processes available in foundry practice and hardened to form strong molds or cores.

When sand molds or cores made with low ratio (less than about 3.5) silicate binders get dry either by exposure to a dry atmosphere or by heating, they become harder. On the other hand, when sand molds or cores made with very high ratio silicate as binders get dry either by exposure to a dry atmosphere or by heating they tend to become weak and friable. This is because the overall strength of the mold or core is primarily dependent on the mechanical properties of the solid film formed by the silicate adhesive when it sets. The separation of adhesive bonds is rarely the breaking away of the solid–liquid interface but more generally a rupture either within the adhesive film or within the body of the material to which the adhesive was applied. Cracks or other faults within the adhesive film are more likely to account for low bond strength than rupture at the interface.

The formation of crystalline silica within the mass of the binder contributes to weaken the bond between sand grains after heating and cooling the molds and/or cores, therefore, providing easier core shake-out and separation of the metal from the mold. Conventional sodium silicate binders form a glass on the surface of the sand grains when the molds or cores are heated to high temperatures. When the mold or core cools down to room temperature the glass becomes very rigid forming a very strong bond, therefore, hardening the mold or core. For this reason a core made with such a binder is very difficult to break up and remove from the cavity of a cast metal during the foundry operation known as shake-out.

When colloidal silica is embedded in a matrix of sodium silicate it tends to crystallize and form cristobalite at the temperatures to cores reach when metals are cast. Due to the difference in thermal expansion coefficient, the expansions and contractions of the cristobalite crystals embedded in the glass matrix tend to crack the binder film surrounding the sand grains therefore weakening the mold or core. This weakening effect has to be added to the already mentioned weakening effect due to the cracking of high ratio silicate films on dehydration. Due to these weakening mechanisms a core made with the high ratio silicates covered by this invention is very easy to break up and remove or separate from the cast metal during the shake-out operation.

Thus the difference in behavior between low and high ratio silicate binders for sand molds and cores can be understood by observing films formed on silica glass plates by slow evaporation of for example aqueous solutions of sodium silicate of various ratios.

The low silicate/soda ratio (2.0) sodium silicate solution dries in air at room temperature very slowly forming a very viscous, smooth and clear film. At higher ratio (2.4) drying is faster and the silicate film obtained shows some cracks. At very high ratios (3.25 and 4.0) sodium silicate solutions include substantial amounts of very small particle size colloidal silica and drying is even faster: cracking is even more extensive and the film tends to lose integrity. A silica sol of particle size 14 nm and SiO_2/Na_2O ratio 90 does not form a continuous film under the same drying conditions.

Low ratio silicate binders thus form on the sand surface viscous, smooth films which do not form cracks on drying. On the other hand, the films formed on the sand surface by high ratio silicate binders, crack on drying thus weakening the sand core or mold. For these reasons cores made with low ratio silicate binders outside the scope of the present invention become stronger when they are heated at high temperatures by molten metals in the pouring operation of the casting process. On the other hand, cores made with high ratio silicate binders within the present invention are reasonably strong when just made, but become weak and friable during the casting operation.

In the practice of this invention a compromise has to be made when choosing a binder composition by selecting one with a SiO_2/Na_2O ratio not so high that the sand molds or cores will weaken to unacceptable levels by merely drying at room temperature when exposed to the atmosphere, and not so low that the sand molds or cores will form a cohesive, solid glass bond when the core or mold is heated in the casting operation so that the core or mold becomes very strong when cooled down to room temperature and cannot be separated easily from the metal casting. The room temperature, as-made strength of sand molds or cores obtained with high ratio silicate binders of this invention may be upgraded by the addition to the silicate bonded sand mix of a fugitive film-forming resin adhesive in the form of a water solution or water dispersion. In this case, as explained below in more detail, the molds or cores become stronger by drying at room temperature. However, when heated to high temperatures during the casting process the resin adhesive decomposes evolving harmless vapors and the weakened core and mold can be easily separated from the cast metal.

If a preformed sodium polysilicate having a molar ratio of silica to alkali metal oxide in the range of 3.5–10 is employed before it gels, the same effects as with the amorphous silica sodium silicate system will be obtained. An aqueous sodium polysilicate containing 10–30% by weight silica and sodium oxide and having a silica to sodium oxide weight ratio of 4.2:1 to 6.0:1 can be produced as described in U.S. Pat. No. 3,492,137.

Similarly, the high ratio lithium silicates of Iler U.S. Pat. No. 2,668,149 or the potassium polysilicates of Woltersdorp, application Ser. No. 728,926, filed May 14, 1968, now Defensive Publication 728,926, dated Jan. 7, 1969, can be employed as the binder provided the requirements as to molar ratio, particle size and amount of amorphous silica are followed.

Furthermore, alkali metal polysilicates stabilized by quaternary ammonium compounds or guanidine and its salts can also be employed. Some stabilized polysilicates of this type are described in U.S. Pat. No. 3,625,722. This method, however, has the disadvantage of producing unpleasant odors on casting due to the thermal decomposition of the organic molecule.

Complexed metal ion stabilized alkali metal polysilicates can also be used, such as copper ethylenediamine hydroxide stabilized sodium polysilicate made by mixing copper ethylenediamine with colloidal silica and then the silicate, or the stabilized polysilicates of U.S. Pat. No. 3,715,224.

The useful amorphous silica are those having a particle size in the range from about 2 to 500 nm. In addition to the amorphous silica already present in aqueous solutions of high ratio alkali metal silicates, such silicas can be obtained from silica sol (colloidal dispersions of silica in liquids), colloidal silica powders, or submicron particles of silica. The silica sols and colloidal silica powders, particularly the sols, are preferred in view of the shake-out properties of the binders made from them.

Gore Procedure

The amount of colloidal silica present in an aqueous solution of high ratio alkali metal silicate can be determined for example by ultrafiltration. Ultrafiltration refers to the efficient selective retention of solutes by solvent flow through an anisotropic "skinned" membrane such as the Amicon "Diaflo" ultrafiltration membranes made by the Amicon Corporation of Lexington, Mass. In ultrafiltration solutes, colloids or particles of dimensions larger than the specified membrane "cut-off" are quantitatively retained in solution, while solutes smaller than the uniform minute skin pores pass unhindered with solvent through the supportive membrane substructure.

Amicon "Diaflo" ultrafiltration membrane offer a selection of macrosolute retentions ranging from 500 to 300,000 molecular weight as calibrated with globular macrosolutes. These values correspond to pore sizes between about 1 and 15 nm. Each membrane is characterized by its nominal out-off, that is, its ability to retain molecules larger than those of a given size.

For effective ultrafiltration, equipment must be optimized to promote the highest transmembrane flow and selectivity. A major problem which must be overcome is concentration polarization, the accumulation of a gradient of retained macrosolute above the membrane. The extent of polarization is determined by the macrosolute concentration and diffusivity, temperature effects on solution viscosity and system geometry. If left undisturbed, concentration polarization restricts solvent and solute transport through the membrane and can even alter membrane selectivity by forming a gel layer on the membrane surface—in effect, a secondary membrane — increasing rejection of normally permeating species.

An effective way of providing polarization control is the use of stirred cells. Magnetic stirring provides high ultrafiltration rates.

A recommended procedure is to use a Amicon ultrafilter Model 202, with a pressure cell of 200 ml capacity and a 62 mm diameter ultrafilter membrane operated at 25°C with magnetic stirring with air pressure at around 50 psi.

In the case of sodium silicate for example, an aqueous solution diluted with water, is placed in the cell. An Amicon PM-10 membrane, 1.8 nm diameter pores, is used. Pressure is applied and filtrate collected. In some cases, water is fed in to replace the volume passing through the filter into the filtrate. The solution in the filter cell is concentrated until the filtration rate is only a few ml per hour.

The filtrate is collected in progressive fractions, and they and the final concentrated solution from the cell are examined: volumes are noted and SiO_2 and Na_2O concentrations in grams per ml are determined by chemical analysis.

In some cases, the concentrated solution on the filter is further washed by adding water under pressure, as fast as filtrate is removed. In these cases there is further depolymerization or dissolution of the colloid fraction.

The percentage of colloidal silica, based on total silica, is indicated by the amount of residual silica that does not pass through the filter. These represent maximum values for the amount of colloid present, since some ionic soluble silica is still present. In further examples the residual soluble silica is subtracted and the composition of the colloid is calculated.

It is not necessary to isolate the pure colloid, but only to measure the concentration of SiO_2 and Na_2O as ultrafiltration proceeds. Since the concentration of "soluble" sodium silicate in the filtrate is about the same as in the solution in the cell if this colloid is present only at low concentration, the amount and composition of colloid can be calculated by difference.

Allowance should be made in interpreting results obtained with this method for the fact that evey time water is added to the system some depolymerization of colloid or polysilicate ions probably occurs.

The colloidal amorphous silicas useful in preparing the compositions of the invention have a specific surface area greater than $5\,m^2/g$ and generally in the range of 50–800 m^2/g and preferably in the range of 50–250 m^2/g. The specific surface area is determined by nitrogen adsorption according to the BET method. The ultimate particle size of the silica used is in the colloidal range, and is generally in the range of 20–500 nm, preferably 12–60 nm. Thus, the silica sols of the desired particle size range described by M.F. Bechtold and O.E. Snyder in U.S. Pat. No. 2,574,902; J.M. Rule in U.S. Pat. No. 2,577,484; or G.B. Alexander in U.S. Pat. No. 2,750,345 can be used.

Positive silica sols and alumina modified silica sols wherein the ultimate silica particles have been modified and/or made electrically positive by partially or completely coating the particle surface with aluminum compounds can also be used in the present invention as a source of amorphous silica. Such sols are described for example by G. B. Alexander and G. H. Bolt in U.S. Pat. No. 3,007,878 and by G.B. Alexander and R.K. Iler in U.S. Pat. No. 2,892,797. The advantage of these sols is that in some cases they form more stable mixtures with sodium silicate aqueous solutions than the unmodified silica sols.

Certain very finely divided colloidal silica powders such as those made by the "fume process" by burning a mixture of silicon tetrachloride and methane, have a sufficiently discrete, particulate structure that such powders can be dispersed in water by colloid milling to give a sol useful in this invention. It is also obvious that such a powder can also be colloid milled directly into a solution of silicate.

Very finely divided colloidal silica powders can also be obtained by treating certain silicate minerals such as clay or calcium silicate with acid, followed by suitable heat treatment in an alkaline medium. Similarly, finely divided colloidal silicas can be produced by precipitating silica from a solution of sodium silicate with carbon dioxide. Such precipitated silicas are commonly used as reinforcing fillers, for elastomers because they are extremely finely divided, and the ultimate particles are easily broken apart. Finely divided aerogels of silicas may be employed, such as those described by Kistler in U.S. Pat. Nos. 2,093,454 and 2,249,767.

The finely divided colloidal silica powders useful in the composition of the invention are characterized by having specific surface areas as determined by nitrogen adsorption according to the BET method, of from 5 to 800 m^2/g and preferably 50 to 250 m^2/g, and being further characterized by the fact that the aggregates of ultimate silica particles are generally less than 10 μ in diameter.

The amounts and types of amorphous silica that can be dispersed within the soluble silicate depends to a considerable extent on the amount of grinding or mixing that is done to disintegrate and disperse particles of amorphous silica in the silicate bond. Thus, for example, it is possible to start with fused silica glass and grind it to the point where a substantial amount is present as particles smaller than a micron. The inclusion of a high concentration of this type of material can provide sufficient surface for nucleation of cristobalite or tridymite within the alkali metal silicate glass bond when the sand core or mold reaches high temperature during the metal casting operation. Also, finely divided natural forms of silica such as volcanic glasses which, in the presence of alkali silicates, can be devitrified, may be used, providing they are sufficiently finely divided and well dispersed in the sodium, potassium or lithium silicate solution used as the binder.

The compositions of the invention will have 2–75% of the total silica present in the binder present as amorphous silica, preferably 10–50% the balance of the total silica being in the form of silicate ions. As the specific surface area of the amorphous silica increases, lesser amounts of it will be required in the binder mixture.

There is a practical maximum concentration of amorphous silica that can be dispersed in the aqueous silicate solution. It is often desirable to incorporate as high a concentration of amorphous silica as possible, yet still have a workable fluid binder to apply to the sand. If the proportion of amorphous silica to soluble silicate is too low, than the shake-out will be adversely affected. On the other hand, if the ratio of amorphous silica to soluble silicate is too high, the mixture will be too viscous and must be thinned with water. Also, there will not be enough binder to fill the spaces between the amorphous silica particles in the bond, and it will be weak. In generally, the higher the content of amorphous silica relative to sodium or potassium silicate, the weaker the initial bond as set by carbon dioxide. Conversely, the more silicate in the binder, the higher will be the initial and retained strengths.

The binder system should have a molar ratio of silica to alkali metal oxide which ranges from 3.5 to 10, preferably 3.5 to 7. This ratio is significant because the ratios of soluble potassium, lithium or sodium silicates commercially available as solutions lie within a relatively narrow range. Most of sodium silicates are within the range of SiO_2/Na_2O of about 2:1 to 3.75:1. Thus, overall ratios of binder compositions obtained by admixing colloidal silica, such as ratios of 4:1, 5:1, 7:1 are mainly an indication of what proportions of colloidal silica and soluble silicates were mixed since the amount of amorphous silica in the soluble silicate at ratios of 2:1 to 3.75:1 are small.

However, in the ratio range of about 3.5:1 to 4.0:1, compositions of a specified ratio are not necessarily equivalent. Thus, a potassium silicate having an $SiO_2/\text{-}K_2O$ ratio of 3.9:1, in which there is a distribution of polysilicate ions, but relatively small amount of colloidal silica, differs considerably from a mixture made by mixing a potassium silicate solution of SiO_2/K_2O of 2.0:1 with colloidal silica having a particle size of, for example, 14 nm. In the latter case, the colloidal particles will remain as such in solution over a considerable period of time. Such a composition has two advantages over the more homogeneous one in that the low ratio of silicate has a higher binding power giving greater initial strength, while the higher content of colloidal particles results in a major reduction in the strength in the core after casting the metal.

OPTIONAL ADDITIVES

In the casting of some metals, for example, iron or steel, very high casting temperatures are involved, that is, 2500–2900°F. If the mass of the core is small relative to the mass of the cast metal during such high temperature casting, there may be some vitrification of the silicate thus creating shake-out problems. To alleviate this situation a carbonaceous material can be added to the core composition. These carbonaceous materials assist the binder of the invention in providing excellent shake-out, particularly after the core has been subjected to very high temperatures.

The useful carbonaceous materials should have the following characteristics:

(a) It should not interfere with the binder system.
(b) It should have a particle size or primary aggregate equivalent diameter sufficiently large to leave discontinuities in the glass formed by the binder at very high temperatures, as it burns off partially or completely. It should also have particle size which is not large enough to weaken the sand core as fabricated, and specially not larger than the particle size of the sand itself. Thus the particle size or primary aggregate equivalent diameter should range between 0.1 μ and 75 μ, preferably between 5 μ and 50 μ. When the ultimate particle size of the carbonaceous material is smaller than 0.1 μ it is generally coalesced or it tends to coalesce in the sand mix into primary aggregates larger than 0.1 μ.
(c) It should not be too avid for water, otherwise it would subtract from the binder system, drying up the sand and making it impossible or difficult to mold.

Preferred for use are pitch, tar, coal-tar pitch, pitch compounds, asphaltenes, carbon black, and sea coal, and most preferred are pitch and carbon black.

Pitch is a by-product from coke making and oil refining and is distilled off at around 350°F. It has a melting range of from 285 to 315°F, is highly volatile, high in carbon and extremely low in ash. Following is a typical analysis of coal-tar pitch in weight percent:

Volatile	47.37%
Fixed Carbon	52.43
Ash	0.2
Sulfur	0.5

Pitch is a material resistant to moisture absorption and is often used as a binder or as an additive for foundry sand cores and molds.

Sea coal is a common name used to describe any ground coal employed as an additive to foundry sands. Sea coal is used in foundry sands primarily to prevent wetting of the sand grains by the molten metal, thus preventing burn-on and improving the surface finish of castings. It is also used as a stabilizer and to promote chilling of the metal.

Following is a typical analysis of sea coal given on a dry basis:

	Weight Percent
Ash	5.10%
Sulfur	0.51
Volatile carbonaceous material	40.00
Fixed Carbon	53.80
Ultimate analysis	
Hydrogen	5.20%
Carbon	81.29
Nitrogen	1.50
Oxygen	6.40
Sulfur	0.51
Ash	5.00

Tar is generally defined as a thick, heavy, dark brown or black liquid obtained by the distillation of wood, coal, peat, petroleum, and other organic materials. The chemical composition of a tar varies with the temperature at which it is recovered and raw material from which it is obtained.

Carbon blacks are a family of industrial carbons, essentially elemental carbon, produced either by partial combustion or thermal decomposition of liquid or gaseous hydrocarbons. They differ from commercial carbons such as cokes and charcoals by the fact that carbon blacks are particulate and are composed of spherical particles, quasigraphitic in structure and of colloidal dimensions. Many grades and types of carbon black are produced commercially ranging in ultimate particle size from less than 10–400 nm. In most grades ultimate particles are coalesced or fused into primary aggregates, which are the smallest dispersible unit of carbon black. The number of ultimate particles making up the primary aggregate gives rise to "structure"—the greater the number of particles per aggregate, the higher the structure of the carbon black.

When mixed with sand fine particle size carbon blacks are coalesced into aggregates in the sand mix, therefore they leave discontinuities in the binder phase when burned off during the high temperature casting operation.

An example of a commercial carbon black is Regal 660, sold by the Cabot Corporation of Boston, MA., which has the following characteristics:

Nigrometer Index	83
Nitrogen Surface Area	112 m^2/g
Oil (DBP) Absorption	62 cc/100 g
Fixed Carbon	99%

The carbonaceous material should be present in the core composition in the amount of 0.5 to 4 weight percent based on the foundry sand, preferably 1 to 2 weight percent.

The amount of carbonaceous material e.g., pitch, needed depends, to some degree, on the refractoriness of the binder used which is in turn a function of the silica/alkali molar ratio, and on the temperature to which the core will be subjected during casting. When a SiO_2/Na_2O ratio of 5:1 sodium polysilicate is used as a binder, no pitch is needed if the core is used for nonferrous metal castings since in these cases the core temperature will not exceed about 1200°C. If the same binder is used for small cores in massive iron castings, 2% of pitch is useful to help break up the silicate glass formed.

In the event it is desirable to make cores and store them for extended periods of time prior to use, I have discovered that the addition of a film-forming resin adhesive in the form of a water solution or water dispersion, drastically extends the storage life of foundry sand-cores made with the binder of the invention. Thus the use of these materials enable the formed cores to retain sufficient strength and hardness during storage.

Useful film-forming resin adhesives include polyvinyl esters and ethers and their copolymers and interpolymers with ethylene and vinyl monomers, acrylic resins and their copolymers, polyvinyl alcohol, water dispersion of polyolefin resins, polystyrene copolymers such as polystyrene butadiene, polyamide resins, natural rubber dispersions, and natural and modified carbohydrates (starch or carboxycellulose). Particularly preferred for use are aqueous dispersions of polyvinyl acetate and vinyl acetate-ethylene copolymers.

The polymer resin should be in a state of subdivision suitable for uniform distribution on the sand grains to form an adhesive film and hold the sand grains strongly together. It is preferred that resin dispersions be between 40 and 60% by weight solids. The higher the concentration of solids, the better, as less water will have to be removed, however, with concentrations above 60% by weight it can be difficult to mix the dispersion into the sand. With resin solutions, e.g., solutions of polyvinyl alcohol, concentrations of 4–20% solids are preferred.

Useful polyvinyl acetate dispersions are milk-white, high-solids dispersions of vinyl acetate homopolymer in water. Such dispersions have excellent mechanical and chemical stability. Typical properties of a preferred polyvinyl acetate dispersion are given in the following table. Commercially available dispersions with similar characteristics are Monsanto's S-55L, Borden's "Polyco" 11755, Air Products' "Vynac" XX-210, and Seydel Wooley's "Seycorez" C-79.

Typical Properties of a Preferred Polyvinyl Acetate Homopolymer Aqueous Dispersion

Solids, %	55
Brookfield viscosity, P*	8.5–10
pH	4–6
Molecular weight (number average)	30,000–60,000 (mostly crosslinked)
Average particle size, microns	1–2 (range from 0.1 to 4)
Density (25°C), approx. lb./gal.	9.2
Surface tension (25°C), approx. dynes/cm.	55
Min. film formation temperature**	
°C	17
°F	63
Residual monomer as vinyl acetate, % max.	1.0
Particle charge	Essentially nonionic

*Brookfield model LVF, No. 2 spindle at 6 rpm or No. 3
**ASTM D2354.

The useful vinyl acetate-ethylene copolymers are milk-white dispersions of 55 w/o solids in water with a viscosity between 12 and 45 poises. Du Pont's "Elvace" is a commercially available dispersion with these characteristics.

The useful polyvinyl alcohol (PVA) is a water soluble synthetic resin 85–99.8% hydrolyzed. Du Pont's "Elvanol" resins and Goshenol GL-05, 85% hydrolyzed, low viscosity PVA are examples of suitable commercially available materials. "Elvanol" grades give 4% water solutions with a viscosity ranging from 3.5 to 65 Cp at 20°C as measured by the Hoeppler falling ball method. Water solutions of PVA at low concentrations (up to about 10–15 weight percent) or concentrated aqueous colloidal dispersions of the water insoluble polymer resins mix uniformly with sand and provide good adhesion. Very concentrated water solutions of PVA (higher than 20 weight percent) are too viscous and do not mix well enough with sand.

To obtain optimum adhesion, the film forming resin dispersion or solution should be added such that it does not gel or coagulate either the silica or the sodium silicate before adding them to the sand. For instance, the polymer resin dispersions can be mixed with the silica before adding to the sand because both are compatible and do not gel when mixed together. The mixtures can be added to sand and they will form an adhesive film on the surface of the sand grains. After the silica and the polymer resin dispersion have been mixed with the sand, the sodium silicate solution can be added to the sand and although it will thicken in contact with the silica and the polymer resin dispersion, it will do so in situ, that is, fairly uniformly distributed on a preformed film of silica and polymer resin.

If before adding to the sand the sodium silicate is mixed with the concentrated polymer dispersion and the silica, it thickens and gels and it cannot subsequently be mixed adequately with the sand. Instead of distributing fairly uniformly on the surface of the sand grains, it would tend to form lumps and distribute unevenly in the sand.

Alcoholic solutions of the polymer resins may be used but are not recommended as additives to the silica-sodium silicate binder because they get very thick in contact with the binder and tend to gel faster than the aqueous dispersions and therefore do not distribute as uniformly on the sand grains. however, dilute alcoholic solutions of polymer resins can be used as such or mixed with commercial zircon core washes to coat the surface of the cores and give improved hardness and storage life to the cores. In this case the gel forms on the surface of the sand core already set, and it air dries fairly fast or it is dried almost instantaneously by lighting the alcohol to extinction of the flame, therefore preventing the possible diffusion of the alcohol into the core.

The use of a water solution or water dispersion of a polymer resin produces sand cores with the silica-sodium silicate binder having as gassed mechanical strength somewhat lower than that of sand cores made with silica-sodium silicate binder without the polymer resin solution or dispersion. This may be due to the weakening of the sodium silicate bond caused by the dilution produced by the water of the polymer resin solution or dispersion. However, drying of the core on storage, more than overcomes this effect and after very few days the cores show a much higher mechanical strength than the one obtained immediately after gassing with CO_2.

Two mechanisms may contribute to the hardening and strengthening on storage provided by the polymer resin. One is the thickening in situ of the adhesive film of silica-polymer resin-sodium silicate on the sand grains due to the "salting-out" effect caused by electrolyte formation on gassing with CO_2. More important is the thickening and solidification of the film caused initially by the CO_2 blown through the sand grains and specially the subsequent evaporation of the water from the sand core on storage.

Under these conditions the polymer resin macromolecules and/or colloidal particles are expected to coalesce and form an effective adhesive bond between the sand grains and reinforce the sodium polysilicate binder.

In the case of the polyvinyl esters the alkaline hydrolysis caused by the mixing with the sodium silicate will tend to form in the already formed uniform film, polyvinyl alcohol, perhaps an even better adhesive than the ester itself.

The colloidal silica-resin, e.g., polyvinyl acetate components of the binder can be used in the form of a stable liquid mixture, the carbonaceous material being optionally present. Thus uniform mixtures containing colloidal silica and polyvinyl acetate within the relative amounts specified in this invention, such as 1.94 parts by weight of 40% aqueous colloidal silica and 2 parts by weight of 55% polyvinyl acetate aqueous dispersion, can be made by mixing the two components in a beaker. The mixture is stable and uniform and can be used within the working day. Overnight the mixture tends to separate in two layers and can be stirred up to make it uniform.

One method of providing a stable, pourable mixture of colloidal silica-polyvinyl acetate with or without the carbonaceous material, e.g., pitch, is to make the liquid phase slightly thixotropic but not viscous. In other words, to make it so that it sets to a weak gel structure at once when undisturbed (to maintain all particles in uniform suspension) but when stirred, or even tilted to pour, the yield point is so weak as to permit ready transfer of the material and easy blending with the sand.

Thixotropic suspensions with the characteristics described above can be prepared using a three component suspending agent system disclosed in U.S. Pat. No. 3,852,085, issued Dec. 3, 1974. This system consists of (a) carboxymethyl cellulose and (b) carboxyvinyl polymer in a total amount of about 36–65 weight percent with the relative amount of (a) to (b) varying from a weight percent ratio of about 1:4 to 4:1 and (c) magnesium montmorillonite clay in a concentration of about 35–64 weight percent.

The useful compositions will contain between 95 and 99.5% by weight of the binder components and between 0.5 and 5% by weight of the suspending agent system. In a composition containing only the colloidal silica and resin, 15–35% of the binder will be silica solids and 15–35% of the binder will be resin solids. In a three component binder, 5–20% will be silica solids, 5–20% resin solids and 5–40% will be carbonaceous matter.

This suspension system can be used with dispersion containing a maximum solid content of 55% by weight of polymer resin and colloidal silica or polymer resin, colloidal silica and carbonaceous material such as pitch. The minimum solid content is only limited by the amount of water that is practical to add to the sand mix to obtain practical cores.

For example, to prepare a colloidal silica-polyvinyl acetate-pitch suspension 0.67 parts by weight of "Benaqua" (magnesium montmorillonite sold by the National Lead Co.) can be dispersed in 235 parts by weigh of water with low shear mixing; 0.67 parts by weight of CMC-7H (carboxymethyl cellulose) and 0.67 parts by weight of Carbopol 941 (water soluble carboxyvinyl polymer) can be added and dissolved using low shear mixing; 0.15 parts by weight of a 1% solution of GE-60 (silicone-based emulsion) can be added as an antifoam agent; 194 parts by weight of "Ludox" HS-40 (aqueous colloidal silica dispersion sold by E. I. du Pont de Nemours & Co.) can be added and mixed with moderate shear mixing; 200 parts by weight of Gelva S-55L (polyvinyl acetate aqueous dispersion sold by the Monsanto Company) can be added and mixed with moderate shear mixing; then 200 parts by weight of "O" Pitch sold by the Ashland Chemical Company can be added and mixed with moderate shear mixing. A fluid suspension containing colloidal silica-polyvinyl acetate and pitch is obtained at a suitable ratio to be used as a component of the silicate binder system of the invention.

Alternatively, 58 parts by weight of water can be used instead of 235 parts by weight of water and in this case a uniform, stable suspension is obtained which is more viscous than the previously described, but still pourable and mixes well with sand.

Alternatively, pitch can be omitted from the preparation, and fluid suspensions containing colloidal silica-polyvinyl acetate are obtained at a suitable ratio to be used as components of the silicate binder system of the invention.

Application of the Binder

The binder mixture of the invention can be applied to the sand in various ways. Thus, if the binder mixture has sufficient shelf life, it can be formulated, stored, and applied to the sand when needed. The silicate and amorphous silica can be stored separately and then mixed together when needed and applied. Furthermore, they can be applied separately to the sand. If this latter procedure is used, it is preferred to first apply the amorphous silica, mix it into the sand, then apply the silicate and mix again. However, the silicate can be applied first.

Uniform sand mixes can be prepared by adding the binder to the sand in conventional foundry mixer, muller, or mix-mixers, or laboratory or kitchen mixers, and mixing for sufficient time to obtain a good admixture of the sand and binder, e.g., for several minutes. When added separately, it is desirable to mix each component for less than 2 min to avoid undue drying.

If an alkali metal polysilicate solution is used as a binder, it should be mixed directly with the sand. If on the other hand colloidal silica and sodium silicate solution are added separately to the sand, it is preferable to add the silica sol first and to mix it thoroughly with the sand before adding the sodium silicate. Once the sodium silicate is added, the mix should not be kept too long in the mixer. A period of 2 min stirring is generally optimum for the sodium silicate.

Dry colloidal silicas such as pyrogenic amorphous silica do not mix well with the sand and in addition they tend to absorb water from the sand-binder system. Therefore, dry colloidal silica powders should be added to the sand in the form of a paste made with water or water should be added to the sand to help mix the dry silica powder. The amount of water made to use the paste should be enough to assure good mixing of the silica powder and yet not too much to affect the strength of the core or mold when it is hardened. Generally the amount of water needed in this case is no more than around 3% by weight of sand.

When the film forming resin or pitch are incorporated into the core composition, if the components are added separately to the sand, the resin should be added to the sand before the silicate. The resin can be added to the sand before or after the colloidal silica. The order in which the pitch is added is not critical with respect to either the silica or the silicate.

When materials such as clays or oxides are used as additives besides the binder, they should be mixed thoroughly with the sand in the sand mixer before adding the binder.

In some cases it is found convenient to use a release agent mixed with the sand to prevent the core or mold from sticking to the core box or pattern after setting. In these cases a conventional core or mold release such as kerosene or Mabco Release Agent "G" supplied by the M.A. Bell Company of St. Louis, MO, should be added to the sand mix in the last 20 sec of the 2 min period of mixing the sodium silicate.

If the sand mix is not going to be used immediately, it should not be allowed to dry or react with atmospheric CO_2. The mix should therefore be stored in a tightly closed container or plastic bag from where the air has been squeezed out before sealing until it is ready to be used. If a slightly hard layer forms on the top surface of the sand due to air left inside the container, the hard layer should be discarded before using the sand to make cores or molds.

A practical way of checking uniformity of the sand mix and observe changes in the sand mix, such as reaction with the atmospheric CO_2, is to add a few grams of an indicator such as phenolphthalein at the beginning of the mixing operation. The phenolphthalein can be added in the form of a fine powder before adding the sodium silicate or dissolved in the sodium silicate or in the silica sol. Usually 160 mg of phenolphthalein per kilogram of sand is sufficient to develop a deep pink color in the sand mix.

Conventional foundry practice can be followed to form and set the sand core or mold. The sand can be compacted by being rammed, squeezed or pressed into the core box either by hand or automatically, or can be blown into the core box with air under pressure.

The formed sand mix can be hardened very fast at room temperature by gassing the sand with CO_2 for a few seconds. Optimum gassing time can be determined either by measuring the hardness or the strength of the core or by observing the change of color of the sand mix when an indicator such as phenolphthalein has been previously added to the sand.

Thermal hardening can be used for cores made with the binder compositions of the invention instead of CO_2 hardening. For instance, high strength cores can be obtained in a very short time by forming the sand mix in a hot box at temperatures between 100 and 300°C. In general, the higher the temperature the shorter the time required to achieve a certain strength level. On the other hand at a fixed temperature in general, the core strength increases with time of heating. However, thermal hardening is not a preferred setting process for the compositions of the invention because cores made in this way do not have as good shake-out characteristics as those made by CO_2 hardening.

Another fast hardening process that can be used is CO_2 gassing in a warm box (about 60–80°C) or gassing with heated CO_2.

When fast hardening is not required, cores with the binders of the invention can be set with other common curing agents used for the systems known in the art as silicate no-bakes. These curing agents are organic materials which are latent acids such as ethyl acetate, formamide, and acetins. Most of these agents contain glycerol mono-, di-, or tri-acetates or any other material which can release or decompose into an acid substance which in turn produces hardening of the alkali metal silicate. Furthermore, such a hardening process can produce cores having long shelf life without the need for a film-forming resin adhesive, that is, polyvinyl acetate.

Conventional water based or alcohol based core washes can be used to treat the surface of the cores. This type of treatment is in some cases to improve the surface of the metal casting or the hardness and shelf life of the core. Shelf life is the period of time after making for which the sand core is useful.

Polyvinyl acetate homopolymers and copolymers can be used as core washes for sand cores as aqueous dispersions, in organic solvent solutions or mixed with zircon or graphite in aqueous or alcoholic suspensions. Polyvinyl alcohol or partially hydrolyzed polyvinyl alcohol can be used in aqueous solutions, organic solvent dispersions or mixed with zircon or graphite.

Polyvinyl alcohol or hydrolyzed polyvinyl acetate: 5% by weight to 20% by weight in water solutions or 5% by weight to 40% by weight in alcoholic solutions. More concentrated solutions are too thick to obtain uniform coating of the cores, more dilute solutions are too thin to provide satisfactory protective coating on the core surface.

Polymer resin aqueous dispersions and alcoholic solutions: 5% by weight to 40% by weight of polymer resin such as polyvinyl acetate homopolymer or copolymer in water solutions or 5% by weight to 25% by weight of polymer resin such as polyvinyl acetate homopolymer or copolymer in alcoholic solutions.

Polymer resin-zircon or graphite mixtures: In water based core washes: 15–25% by weight of polymer resin such as polyvinyl acetate homopolymer or copolymer and 30–50% by weight of zircon (25–50% by weight of water).

In alcohol based core washes: 5–10% by weight of polymer resin such as polyvinyl acetate homopolymer or copolymer and 30–50% by weight of zircon or graphite (40–60% alcohol).

The alcohols useful in the above core washes include methanol and ethanol.

Satisfactory polymer resin-zircon core washes are made for example by slurrying 1 part by weight of a commercial zircon core wash (as shipped by the supplier in the form of a wet powder) in 1 part by weight of 55% polyvinyl acetate aqueous dispersion if the core wash is intended to be used shortly after preparation. More dilute slurries are preferred for core wash compositions intended to be stored for some time before using. In this case the 1 part by weight of the zircon wet powder should be slurried in 1 part by weight of water before mixing with 1 part by weight of 55% polyvinyl acetate aqueous dispersion.

Aqueous polyvinyl acetate or zircon-polyvinyl acetate or graphite-polyvinyl acetate core washes are applied on the core surface by common foundry practices such as dipping, spraying, brushing, and so on, and allowing the core to air dry before using.

Sand cores coated with alcohol base polyvinyl acetate or zircon-polyvinyl acetate are lighted immediately after one wash application as in common foundry practice with alcohol base zircon core washes.

Concentration of polyvinyl alcohol aqueous solutions to give satisfactory core washes with adequate viscosity depends on molecular weight of the polymer. Polyvinyl alcohol solutions can also be used as a mixture with zircon or graphite core wash.

CASTING METALS

Sand molds and cores made with the binder compositions of the invention can be used to cast most metals, such as gray, ductile and malleable iron, steel, aluminum, copper-based alloys such as brass or bronze. Steel is usually cast at around 2900°F, iron at about 2650°F, brass and bronze at around 2100°F and aluminum at about 1300°F.

With the molds or cores of the invention it is desirable that the core have an initial strength such that it can be handled without undue care and that it will stand up during the casting of the molten metal, that is, will not wash away or distort. In standard American Foundrymen's Society lab tests this means that the core should have a compressive strength of at least 100 psi and preferably over 150 psi.

It is desirable that the hardness of freshly made cores exceed 5, preferably 10. The greater the hardness, the better, particularly at the time of metal pouring when it should exceed 10 and preferably 20.

Scratch, hardness of cured cores can be measured with commercial hardness tester No. 674 available from Harry W. Dietert Co., 9330 Roselawn Avenue, Detroit, MI. This is a practical, pocket-sized instrument for measuring the surface and sub-surface hardness of baked cores and dry sand molds.

The tester has three abrading points which are loaded by a calibrated spring which exerts a constant pressure. These abrading points are rotated in a circle 3/8 in. in diameter. To obtain the hardness values, the lower end of the instrument is held against the sand surface and the abrading points are rotated three revolutions. The hardness values are actually obtained by measuring the depth to which the abrading points penetrate. The maximum hardness value indicated by this tester is 100 for zero penetration. When the abrading points move down a distance of 0.250 in., the hardness of the core is 2000. Intermediate values are read from the instrument dials.

The core should, after the metal has been cast and cooled, have a retained strength such that it can be shaken out without the use of undue energy. This corresponds to a compressive strength in lab tests of, preferably, less than 50 psi.

The following examples are offered to illustrate various embodiments of the invention. All parts and percentages are by weight unless otherwise indicated.

EXAMPLE 1

This is an example of the use of guanidine stabilized sodium polysilicate (SiO_2/Na_2O ratio 5:1) prepared according to Example 1 of patent application Ser. No. 287,037, filed Sept. 7, 1972, as a binder for foundry sand cores. These sand cores were used to make aluminum castings in a nonferrous metal foundry.

The binder sample was made with 1890 g of sodium silicate Du Pont Grade No. 20 (SiO_2/Na_2O molar ratio 3.25:1, 28.4% SiO_2, 8.7% Na_2O), 56 g of water, 539 g of 1.3 M guanidine hydroxide and 1015 g of Ludox® HS, a commercial colloidal silica sol containing 30% SiO_2 of particle size of about 14 nm.

The sand mix was prepared in the following way: 90 g of kaolin and 2 g of phenolphthalein were added to 10 lb of sand while stirring in a 10-lb capacity Clearfield mixer. 0.5 lb of binder solution were also added to the sand while stirring and the sand was mixed for a total of 2 min.

The sand used was a mixture of 50 parts of Houston's subangular bank sand AFS No. 40–45 and 50 parts of

No. 1 Millcreek, OK, AFS 99 ground sand. The sand when used was at room temperature (75°F) Humidity of the room was about 80%. The binder mixed readily with the sand showing excellent mixing characteristics. Flowability of the mix was also excellent.

The sand max was placed in a polyethylene bag and sealed. The sand mix was used the following day to make sand cores. Three to four pound sand cores were made by filling wooden core boxes with the sand mix, compacting the sand by hand and gassing it for about 15–25 sec with CO_2 gas at an estimated pressure of 20–30 lb.

The color of the sand is deep pink due to the phenolphthalein added. After gassing the cores had the natural color of the original sand. Good release of the core was observed when the core box was opened to remove the core. The cores were immersed in a conventional alcohol zircon core wash and flamed before using. This is a common practice with core washes for sodium silicate sand cores.

The cores were assembled in a sand mold and used to make an aluminum casting. Aluminum was poured at a temperature of about 1375°F. When pouring was completed the casting was allowed to cool for about 15 min inside the sand mold assembly. The aluminum casting was removed from the mold when still hot and the sand core was observed before shake-out. Shake-out was very easy; the core broke up and flowed like un-bonded sand upon touching. No offensive odors were noticed during the casting and cooling.

The aluminum castings had very good surface finish and were used in normal production.

EXAMPLE 2

This is an example of the use of sand cores made with the binder solution of Example 1, to make gray iron castings.

Two 10 lb sand mix batches were made by adding 0.5 lb of the binder solution and 0.7 g of phenolphthalein to 10 lb of Houston subangular bank sand AFS 45–50, while stirring in a 10-lb capacity Clearfield mixer and mixing for 2 min. The binder mixed very well with the sand and gave a uniform sand mix containing 5% of binder by weight of sand. The sand mix showed excellent flowability. The sand mix was kept in a closed polyethylene bag for 4 h before using.

Two more 10 lb sand mix batches were made by adding 23 g of "Nusheen" kaolin powder furnished by the Freeport Kaolin Co., 0.5 lb of the binder solution and 0.7 g of phenolphthalein to 10 lb of the same Houston sand AFS 45–50, while stirring in a 10 lb Clearfield mixer, and mixing for 2 min. The kaolin powder and the binder mixed readily with the sand and a uniform sand mix with excellent flowability containing 5% of binder and 0.5% of kaolin by weight of sand was obtained in this manner. The sand mix was kept in a closed polyethylene bag for about 4 h before using.

Sand cores were made by placing the sand mixes into a half-bottle shaped aluminum core box with no parting agent, placing iron rods longitudinally in the mix, tapping the sand, and gassing the core with CO_2 until the core surface developed enough hardness but the sand still had a light pink color. The gassing was accomplished by placing a CO_2 probe for 5–10 in. in different parts of the sand core until it was uniformly hardened.

Six core halves with the shape of half-bottles were obtained in this manner and all were dried at 450°F for 1 min. No core wash was applied to the surface of the cores. Two half-bottle shaped parts made with sand mix containing no kaolin were assembled and glued together with a conventional silicate core paste furnished by the M. A. Bell Co. of St. Louis, MO, under the trade name of "Fast-Dry," to form a bottle-shaped sand core.

Two half-bottle shaped parts made with sand mix containing 0.5% of kaolin by weight of sand were also assembled and glued together with the same core paste to form a second bottle-shaped sand core.

A third bottle-shaped core was made by assembling and pasting together one half-bottle shaped core part prepared with sand containing 0.5% by weight of kaolin and one half-bottle shaped core part prepared with sand containing no kaolin.

Three full bottle-shaped sand cores were obtained in this manner and they were assembled inside a sand mold. Gray iron at about 2650°F was poured into the mold and allowed to cool for about 1 h before removing from the mold. Shake-out of all three cores, with and without kaolin, was very easy. The sand core broke up and flowed out when tapped with an iron bar.

EXAMPLE 3

This is an example of the use of a lithium polysilicate solution as a binder for foundry sand cores. The sand cores made with this binder were used to cast brass metal parts.

The lithium polysilicate solution contained 20 weight percent of silica and 2.1 weight percent of lithium oxide, therefore the SiO_2/Li_2O ratio was 4.8:1 Density of the solution in 9.8 lb/gal (specific gravity 1.17 g/cc); viscosity 10 cp; pH 11.

The sand mix was prepared by adding 0.1 lb of "Nusheen" kaolin powder, 2 g of phenolphthalein powder, and 1 lb of lithium polysilicate binder solution to 10 lb of a sand mixture (50 weight percent Houston sand AFS 50 and 50 weight percent #1 Millcreek, OK, sand AFS 90) in a 10 lb Clearfield sand mixer while stirring. The mix was stirred for 1 min and a half and 30 g of a conventional release agent commercially available from the M. A. Bell Co. of St. Louis, MO, under the trade name of Mabco Release Agent "G," was added while stirring. The mix was stirred for a total time of 2 min.

During the operation it was observed that the binder mixed readily with the sand. The sand mix obtained had very good flowability and it was kept overnight in a closed polyethylene bag before using to make sand cores.

Cores were made by ramming the sand mix with a tamper in a wood core box painted with aluminum paint. CO_2 gassing was applied for 5–10 sec from each end of the U shaped cores or through a center hole in the case of cylindrical type cores. When the core boxes were opened, the hard, strong sand cores released without difficulty. The cores were immersed in a conventional zircon-alcohol core wash and flamed before using.

The cores were assembled into sand molds and molten brass was poured at about 2100°F. The metal was allowed to cool to about room temperature. The sand core broke up very easily and flowed from inside the casting without difficulty.

EXAMPLE 4

This is an example of the use of the guanidine stabilized sodium polysilicate (SiO_2/Na_2O ratio 5:1) of Example 1 to make sand cores and test them according to American Foundrymen's Society standard methods.

The sand mix was prepared by adding 30 g of the binder solution and 100 mg of phenolphthalein powder to 570 g Portage 515 sand. Portage 515 is a sand from Portage, WI, with an AFS (American Foundrymen's Society) Grain Fineness Number as defined in pages 5–8 of the seventh edition (1963) of the AFS Foundry Sand Handbook, of 67–71. In this example the AFS number was 68. Phenolphthalein is added only as an indicator for optimum gassing time with CO_2.

The addition of the sodium silicate to the sand was made gradually while the sand was stirred at speed setting 2 in a "Kitchenaid" mixer Hobart K45. The sand was mixed for a total of 10 min.

AFS standard and specimens for foundry sand mixtures were used for making tests. The specimens are cylindrically shaped and exactly 2 in. $\pm$ 0.001 in. (508 cm) diameter and 2 in. $\pm$ 1/32 in. (508 cm) height prepared in a standard sand rammer. The standard sand rammer and the standard procedure to make test specimens are described in Sections 4–5 and 4–9, respectively of the above-mentioned Foundry Sand Handbook. In this example 170 g of the sand mixed were used to fall within AFS specimen height specifications after ramming.

AFS standard specimens prepared in this manner were strong enough to be handled and in this case they had a pink color due to the phenolphthalein indicator added to the alkaline mix.

A Dietert CO_2 gassing fixture set No. 655 supplied by the Harry W. Dietert Co. of Detroit, MI was used to harden the sand specimens by making CO_2 gas flow through them at a controlled rate for an optimum period of time. The CO_2 setting equipment consists of a pressure reducer and flow meter, and gassing fixtures for the standard 2 in. diameter precision specimen tube where the sand specimen in rammed.

The flow meter is calibrated in terms of gas flow at atmospheric pressure from 0 to 15 l/min. A constant gas flow of 3 l/min was used and the optimum gassing time of each sand mix was determined by testing a number of cores made at different gassing times. The change of color of the phenolphthalein in the sand during gassing indicated the degree of neutralization reached by the alkaline silicate and could be used as a preliminary guidance to try to estimate the hardening of the sample.

After gassing the compressive strength of the standard sand specimen was measured in a motor driven Dietert No. 400 Universal Sand Strength Machine equipped with a No. 410 high dry strength accessory to increase the range of compression strength to 280 psi.

Evaluation of the shake-out characteristics of the sand cores made with the binder compositions was made with the AFS nonstandard Retained Strength test. The standard, hardened-by-gassing. 2 × 2 in. sand specimens were soaked in an electric muffle furnace at 850°C. for 12 min in their own atmosphere, then removed from the furnace and allowed to cool to just above room temperature, and tested in the Universal Sand Strength Machine.

Some specimens made with commercial silicates as a comparison sometimes gave strength values higher than 280 psi and were therefore tested in an Instron Machine.

Gassing times and strength values obtained with guanidine stabilized sodium polysilicate bonded AFS 68 Portage 515 sand are given in the Table 15.1.

Employing the method of preparation of the sand mix, forming and hardening the sand core specimen, and testing compression strength given in this Example 4, different binder compositions of the invention were used to make

TABLE 15.1

Binder	Compressive strength psi	
	As gased	Retained after 850°C—12 min
guanidine stabilized sodium polysilicate	160	10
Example A	160	30
Example B	190	30
Example C	185	30
Example D	160	10
Example E	200	25
Example F	180	10
Example G	100	<10

and test and cores. The binder composition used are described below. Testing results obtained are included in the Table 15.1.

EXAMPLES

A. Kaolin (2% by weight) mixed with the sand before adding the 5% guanidine stabilized sodium polysilicate of this Example 4 and mixing for 2 min.
B. 5% Tetramehylammonium hydroxide (TMAH) stabilized sodium polysilicate made according to teachings of U.S. Pat. No. 3,625,722.
C. Kaolin (0.5% by weight) mixed with the sand before adding the T.M.A.H. stabilized sodium polysilicate of Sample B and mixing for 2 min.
D. 5% Of sodium polysilicate SiO_2/Na_2O molar ratio 3.75:1 made by dissolving fine colloidal silica powder (HiSil 233) in sodium silicate SiO_2/Na_2O molar ratio 3.25:1.
E. 5% Of sodium polysilicate SiO_2/Na_2O molar ratio 6.5:1 stabilized with copper ethylene-diamine hydroxide.
F. 10% of lithium polysilicate SiO_2/Li_2O molar ratio 4.8:1 made according to the teachings of U.S. Pat. No. 2,668,149.
G. 10% Of potassium polysilicate SiO_2/K_2O molar ratio of 5:1.

EXAMPLE 5

Amorphous silica-sodium silicate binder composition of SiO_2/Na_2O ratio 5:1 can be formed directly on the sand by addition of colloidal silica sol of uniform particle diameter about 14 nm to the sand, mixing, and then adding sodium silicate SiO_2/Na_2O molar ratio 3.25:1 and mixing for 2 min.

14.96 g of Du Pont Ludox® HS-40 (40 w/o SiO_2) is poured into 745 g of Portage 515 sand in a Hobart K-45 mixer while stirring at speed setting 2. Then 40 g of Du Pont sodium silicate grade No. 20 (SiO_2/Na_2O molar ratio 3.25:1) is added and mixed for 2 more min.

Standard AFS 2 × 2 in. samples made by ramming, then gassing for 30 sec with CO_2 at a flow rate of 3 l/min have a compressive strength of 200 psi and a retained compressive strength at room temperature after soaking in a furnace at 850°C for 12 min and cooling at 20 psi.

EXAMPLE 6

Amorphous silica-sodium silicate binder composition of SiO_2/Na_2O ratio 5:1 formed directly on the sand as in Example 5 but using a colloidal silica sol of uniform particle diameter about 25 nm instead of 14 nm, with the same sodium silicate.

12 g of Du Pont Ludox® TM-50 (50 w/o SiO_2)
40 g of Du Pont sodium silicate No. 20
748 g of Portage 515 sand
CO_2 gassing time = 30 sec
Compressive strength = 230 psi
Retained strength (850°C — 12 min) = 15 psi
Retained strength (1375°C — 12 min) = 35 psi

EXAMPLE 7

Amorphous silica-sodium silicate binder composition of SiO_2/Na_2O ratio 5:1 formed directly on the sand as in Example 5 but using a colloidal silica sol of uniform particle diameter about 25 nm instead of 14 nm, and using sodium silicate SiO_2/Na_2O molar ratio 3.75:1 instead of 3.25:1.

6.76 g of Du Pont Ludox® TM-50 (50 w/o SiO_2)
40 g of Phila. Quartz Co. sodium silicate grade S 35
753.24 g of Portage 515 sand
CO_2 gassing time = 30 sec
Compressive strength = 180 psi
Retained strength (850°C — 12 min) = 15 psi

EXAMPLE 8

Amorphous silica-sodium silicate binder composition made with the same components and using the same forming method directly on the sand as used in Example 5, except that relative amounts of silica sol and sodium silicate are calculated to give a final SiO_2/Na_2O molar ratio 8:1 in the mixture.

32 g of Du pont Ludox® TM-50
40 g of Du Pont sodium silicate No. 20
728 g of Portage 515 sand
CO_2 gassing time = 30 sec
Compressive strength = 210 psi
Retained strength (850°C — 12 min) = 20 psi

EXAMPLE 9

Amorphous silica-sodium silicate binder compositions made by first mixing the colloidal amorphous silica as a paste with the sand, then adding the sodium silicate and mixing for 2 min.

3.61 g of Ca-O-Sil M-5 pyrogenic silica powder mixed with 14.4 g of water made a thick paste which was mixed with 475 g of Portage 515 sand in a Hobart K-45 mixer.

To the uniform sand-silica mixture, 25 g of Du Pont sodium silicate No. 20 added and mixed for 2 min.

Standard AFS 2×2 in. samples made by ramming, then gassing for 30 sec with CO_2 at a flow rate of 3 l/min. Compressive strength measured: 210 psi. Retained strength (850°C — 12 min): 15 psi.

EXAMPLE 10

This is an example of the use of an amorphous silica-sodium silicate composition of SiO_2/Na_2O ratio 5:1 as a binder for foundry sand cores, a polyvinyl acetate aqueous dispersion as a co-binder and additive for durability, and pitch as an aid to improve shake-out and casting surface finish.

An amorphous silica-sodium silicate binder composition of SiO_2/Na_2O ratio 5:1 is formed directly on the sand by addition of colloidal silica sol of uniform particle diameter about 15 nm to the sand, mixing, and then adding sodium silicate SiO_2/Na_2O molar ratio 3.25:1 and mixing for an additional period of time.

The sand mix is prepared in the following way: 16 grams of "O" Pitch sold by Ashland Chemical Company of Columbus, OH are added to 800 g of Portage 515 sand supplied by Martin Marietta Aggregates of Rukton, IL, in a "Kitchen-Aid" Hobart K-45 mixer while stirring at speed setting 2 and mixed thoroughly with the sand.

14.70 g of "Ludox" HS-40 colloidal silica sold by E. I. du Pont de Nemours and Company, and 16 g of Gelva S-55L polyvinyl acetate aqueous dispersion, sold by Monsanto Chemical Company, are mixed in a plastic beaker, added to the sand-pitch mix and mixed in the Hobart mixer for 2 min.

Finally, 40 grams of Du Pont sodium silicate grade No. 9 (SiO_2/Na_2O molar ratio 3.25:1) are added and mixed for 2 more minutes.

AFS (American Foundrymen's Society) standard specimens for foundry sand mixtures are made immediately after the mixing is completed, as described in Example 4. The specimens are set by gassing with carbon dioxide using the equipment and procedure of Example 4. Optimum gassing time for the composition of this example is 20 sec.

Gassed cores are separated in two groups: one group of cores is left untreated, the second group of cores is coated by immersion in various core wash compositions given in Table 15.2. Cores coated with water based and methanol based core washes are allowed to air dry, whereas cores coated with ethanol based core washes standard, hardened-by-gassing, 2 × 2 in sand specimens are soaked in an electric muffle furnace at 850 or 1375°C for 12 min in their own atmosphere, then removed from the furnance and allowed to cool to just above room temperature, and tested in the Universal Sand Strength Machine. For all cores prepared as described in this example, both 850 and 1375°C retained strength values were less than 25 psi.

Some specimens made with commercial silicates as a comparison sometimes give retained strength values higher than 280 psi and are therefore tested on an Instron machine.

EXAMPLE 11

This example describes the preparation of sand cores bonded with amorphous silica, sodium silicate and polyvinyl acetate and their use in casting 2.5 in grey iron and brass pipe tees. The sand mix is prepared in a Carver "S" mixer by adding to 400 lb of sand (Whitehead Brothers "E" sand with an AFS number 92.2), a mixture consisting of 10.5 lb Du Pont "Ludox" HS-40 and 9 lb Monsanto Gelva S-55L polyvinyl acetate aqueous dispersion, and 9 lb of pitch (Ashland Chemicals "O" Grade). After 5 min 27 lb of Du Pont No. 9 sodium silicate are added and mixing is continued for a further 5 min. The free flowing, uniformly brown mix is then discharged to a storage bin.

The cores are formed by air blowing the mix into a steel pattern comprising twin 2.5 tees and gassing with carbondioxide at 65 psi for 3.5 sec. The cores are immediately removed from the pattern and placed on storage trays. 150 cores are made in 18 min, each weighing about 2.5 lb. No fumes or odors are detected during the mixing or core preparation and the cores have adequate strength for normal handling in the foundry. They have a very smooth surface with an AFS hardness number of about 20. The cores are positioned in oil bonded sand molds, enclosed by steel boxes and grey iron is poured at about 2700°F. Ninety cores are used within a few hours of preparation and 58 are stored for three days at relative humidity of about 25% at about 18°C. The cores which are stored for 3 days are both stronger and harder than when first made.

After pouring the iron the cores are cooled almost to room temperature. No offensive odors are detected during metal pouring or cooling. The cores are then very weak and shake-out readily with excellent surface peel from the iron. After final cleanup by wet drum tumbling and shot blasting, the pipe tees have a much smoother internal surface than those made in normal production using a commercial, proprietary silicate binder. In addition to having a rougher surface some of the tees made using cores with the commercial binder still had sand adhering to the internal surface after cleanup.

Two of the cores prepared as described above are coated by brushing on a slurry consisting of 50% zircon and 20% polyvinyl acetate methanolic dispersion (Monsanto Gelva V7-M50) and 30% methanol. The alcohol is allowed to air dry leaving a hard coating of zircon bonded with polyvinyl acetate. The hardness is measured

TABLE 15.2
Compressive Strength and Core Hardness Versus Elapsed Time on Storage at 73°F ± 2°F and 50% Relative Humidity of Portage 515 Sand Cores made with 5% Sodium Silicate –1.94% Silica sol – 2% Polyvinyl Acetate Aqueous Dispersion –2% Pitch Uncoated and Coated with Various Core Washes (Grades of Binder Components in Table 15.1)

Core wash		Core properties as made	Time elapsed on storage				
			1 day	2 days	3 days	1 week	1 month
No core wash	C.S.**	165	180	200	260	260	275
	Hardness**	30	40	45	45	45	45
Polyvinyl acetate	Compr. Str.	150		280			
water based core wash (75% Monsanto Gelva S-55L in water)	Hardness	90	90	90	90	90	90
Commercial zircon	C.S.	170			> 280		335
core wash (50% "Lite-Off" A ethanol dispersion)***	Hardness	60		65			60
Commercial graphite	C.S.	170			>280		
core wash (50% Pyrokote**** ethanol dispersion)	Hardness	55		75			75
Polyvinyl acetate	C.S.	170			> 280	> 335	
alcohol based core wash (75% Monsanto Gelva V7-M50 in methanol)	Hardness	> 100			> 100		> 100
Polyvinyl acetate-zircon	C.S.	130				280	
water based core wash (1 part Gelva S-55L, 1 part Lite-Off A, 1 part water)	Hardness	30				100	

*C.S. Compressive Strength, psi. American Foundrymen's Society Standard Method for Bulked Cores.
**Hardness. Core (Scratch) Hardness.
***Lite-Off A is a product of M. A. Bell Co., St. Louis, Mo.
****Pyrokote Supreme 14-5X supplied by Penna. Foundry Supply and Sand Co., Philadelphia, Pennsylvania.

TABLE 15.3
Mechanical Properties Versus Elapsed Time on Storage at 73°F ± 2°F and 50% Relative Humidity of Portage 515 Sand Cores Made with 5% Sodium Silicate* – 1.94% Silica sol – 2% Polyvinyl Acetate Aqueous Dispersion*** – 2% Pitch******

Mechanical properties	Core properties as made	Elapsed time since making core				
		1 day	2 days	3 days	1 week	1 month
Compressive Strength, psi	165	180	200	260	260	275
Core (Scratch) Hardness	30	40	45	45	45	45
Tensile Strength, psi	25	25	40	45	45	45

*Du Pont Sodium Silicate No. 9: 29 w/o SiO_2: 8.9 w/o Na_2O.
**Du Pont Ludox® HS-40:40 w/o SiO_2.
***Monsanto Gelva S:55L: 55 w/o polyvinyl acetate.
****Ashland Chemical "O" Pitch powder.

as 90 AFS and shows no change after storing for 3 days at about 25% relative humidity and about 18°C.

The cores are positioned in molds, and brass is poured at 2120°F. After cooling to room temperature the cores collapse readily and shake-out is easily accomplished with excellent peel from the metal surface. No offensive fumes are detected during metal pouring and cooling. The internal surface of the brass tees is very clean and smooth.

EXAMPLE 12

The procedure of Example 11 is repeated using Houston, subangular bank sand, AFS number 45, and omitting the pitch. Half of the cores are coated by immersing them in an agitated slurry containing 50% graphite (Pyrokote), 10% Monsanto Gelva S-55L polyvinyl acetate, and 40% alcohol, allowing them to drain and igniting the alcohol to burn off completely. The other half are similarly treated with an aqueous slurry containing 75% Monsanto (Gelva S-55L) polyvinyl acetate dispersion, allowing them to drain and air dry.

After storing for 2 weeks at about 80% relative humidity and 30°C all the cores are strong and hard (AFS hardness number 80–90). The cores are positioned in the molds and brass is poured at 2150°F and allowed to cool room temperature. No offensive fumes are detected during metal pouring and cooling. Core breakdown is very easy in all cases and the shake-out sand is granular and free flowing. Surface peel and internal surface finish are excellent in the case of tees made from cores treated with the graphite polyvinyl acetate wash and very good for cores coated with polyvinyl acetate alone. No sand residues are observed on the internal surfaces of tees cast from any of the cores.

EXAMPLE 13

This example describes the preparation of sand cores bonded with colloidal silica powder, sodium silicate and polyvinyl acetate ethylene copolymer and their use in the production of cast iron end plates for boilers.

Two thousand pounds of Portage No. 515 sand, AFS number 68 are charged to a batch muller. Forty pounds of pitch (Ashland Chemical Co. "O" grade) are thoroughly mixed with the sand over a period of 3 min. Twenty pounds of Cab-O-Sil M-5 pyrogenic silica powder, as a thick paste with 80 lb of water, and 40 lb of Du Pont's "Elvace" 1873, a 55% aqueous dispersion of polyvinyl acetate/ethylene copolymer (13% ethylene) are then added to the mulled mixture over a period of two minutes. 106 lb of Du Pont No. 20 sodium silicate are then added and the mixing continued for an additional 2 min. Half a minute from the end of the mixing period, 1.5 lb of M.A. Bell's "G" grade flow agent are added. The free flowing mix is discharged into a bin. Cores are made by hand ramming the mix into the two halves of a split core box. The two halves are clamped together and the core is gassed with carbon dioxide at 30 psi for a period of 30 sec. No fumes or odors are detected during mixing and core preparation. The core is then stripped from the pattern and after storing for several days at about 50% humidity and 25°C it is assembled in the mold. Iron is poured at 2650°F and after cooling to about 1500°F the molds are broken away. Examination of the cores shows them to be quite friable and they collapse immediately on a vibrator table and shake-out as granular lump free sand. The boiler end plates are free from defects, dimensionally accurate and have excellent surface finish.

EXAMPLE 14

This is an example of the use of esters as setting agents for the high ratio silicate binders of this invention.

The sand mix is prepared by mixing 14.7 g of "Ludox" HS-40 and 2 g of Triacetin (glycerol triacetate sold by Eastman Kodak), with 760 g of Portage 515 sand using a "Kitchen-Aid" mixer, Hobart K45. The sand is mixed for a total of 2 min and 40 g of sodium silicate ratio 3.25 (Du Pont No. 9) are then added. After an additional 2 min mixing the free flowing sand mix is used immediately to prepare standard 2 in. diameter cylinders as described in Example 4. Cores are similarly made using 2 g of ethyl acetate (ACS grade sold by Fisher Scientific Co.) in place of Triacetin. Cores are stored at 73°F and 50% relative humidity. The compressive strength, hardness and shake-out characteristics are evaluated as described in Example 4 and the results are tabulated in Table 15.4.

In addition to very good initial strength and hardness, both strength and hardness increase on storage and the loss of strength after heating the cores for 12 min at 850°C is indicative of good shake-out.

A 400 lb sand mix is made in a Carver "S" mixer as described in Example 11 adding 8 lb of pitch in addition to "Ludox" GS-40, sodium silicate No. 9 and Triacetin at the same levels on the sand as described above. Cores for 2.5 pipe tees are made as described in Example 11 except that the cores are not gassed with CO_2. After allowing them to harden in the pattern for 5 min the pattern is stripped and the cores are stored for 3 days before being assembled in the molds. Ductile iron is poured at about 2700°C and the castings are allowed to cool for about 2 h inside the mold assembly. After removing the castings from the molds the cores collapse readily in a vibrator and the recovered sand is granular and free from lumps. No odors are produced during the entire

TABLE 15.4
Mechanical Properties Versus Elapsed Time on Storage at 73°F. and 50% Relative Humidity of Portage 515 Sand Cores Made with 5.3% Sodium Silicate* −1.93% Silica Sol and Either 0.26% Triacetin*** or 0.26% Ethyl Acetate******

Mechanical properties		Elapsed time since making core				After heating
		1 day	3 days	5 days	one week	850°C. for 12 min.
Compressive	Triacetin		570	685		90
Strength, psi	Ethyl acetate	260			685	100
Core (Scratch)	Triacetin		80	85		
Hardness	Ethyl acetate	98			90	

*Du Pont Sodium Silicate No. 9; 29 w/o SiO_2; 8.9 w/o Na_2O.
**Du Pont Ludox® HS-40: 40 w/o SiO_2.
***Eastman Kodak glycerol triacetate.
****Fisher Scientific Co. ACS grade ethyl acetate.

operation and the castings have very good interior surface finish.

EXAMPLE 15

This is an example of heat setting the high ratio sodium silicate binder of this invention.

A sand mix is prepared in a Hobart K45 mixer by adding 12 g of "Ludox" TM-50, 16 g of polyvinyl acetate dispersion (Monsanto Gelva S-55L) and 40 grams of sodium silicate ratio 3.25 Du Pont No. 9) to 750 g of Portage 515 sand. The mixing time is 10 min and the free flowing mix is used to prepare standard 2 in diameter cylinders as described in Example 4. The cores are carefully removed from the compacting cylinder and heated for 1 h in an air oven at 100°C. The strength and hardness of the cured cores are as follows:

Compressive strength = 1200 psi
AFS Hardness = 95
Retained strength (850 + C — 12 min) = 150 psi

EXAMPLE 16

This is an example of the use of dextrin with high ratio sodium silicate binder to produce cores which retain excellent strength and hardness when stored for several weeks.

A sand mix is prepared as described in Example 14 by mixing 14.7 g "Ludox" HS-40, 16 grams of 50% aqueous solution of dextrin (sold by Industrial Products Chemicals, Pikesville, MD) previously mixed with 40 g of sodium silicate ratio 3.25 (Du Pont No. 9), 760 g of Portage 515 sand. Standard cores are prepared and set by gassing with carbon dioxide as described in Example 4. Compressive strength and hardness measurements when freshly made and after storing for 1 week at about 50% relative humidity and 23°C show these cores to have excellent storage life. Loss of strength after heating for 12 minutes at 850°C and 1375°C in indicative of good shake-out.

EXAMPLE 17

Amorphous silica-sodium silicate-polyvinyl acetate of SiO_2/Na_2O molar ratio 5:1 formed directly on the sand by addition of a uniform, stable mixture of aqueous silica sol of uniform particle diameter about 12 nm to the sand, mixing and then adding sodium silicate SiO_2/Na_2O molar ratio 3.25:1 and mixing for 2 min.

14.96 grams of Du Pont "Ludox" HS-40 (40 w/o SiO_2) are mixed in a beaker with 16 g of Monsanto Gelva S-55L polyvinyl acetate aqueous dispersion (55 w/o polyvinyl acetate) and poured into 740 g of Portage 515 sand in a Hobart "Kitchen-Aid" K45 mixer,

TABLE 15.5

	Initial	After 1 Week
Compressive Strength, psi	135	150
Core (Scratch) Hardness	30	50
Retained Strength (850°C — 12 min), psi	20	
Retained Strength (1375°C — 12 min), psi	50	

TABLE 15.6
Mechanical Properties Versus Elapsed Time on Storage at 73°F ± 2°F and 50% Relative Humidity of Portage 515 Sand Cores Made with 5% Sodium Silicate –1.94% Silica Sol – 2% Polyvinyl Acetate Aqueous Dispersion Uncoated and Coated with Various Core Washes

					Elapsed time on storage					
Core wash				Mechanical properties as made	1 day	2 days	3 days	1 week	2 weeks	1 month
	Strength, psi									
	Tensile	30	40	50	55	60				
No core wash	Strength, psi									
	Core	35	50	50	50	50				
	Scratch									
	Hardness									
Polyvinyl acetate-zircon	Compressive	150			290		440			
water-based core wash	Strength									
(1 part water)										
Monsanto Gelva S-55L;	Hardness	25	95			95		95		
1 part "Lite-Off" A; 1 part water)										
Polyvinyl acetate-zircon	C.S.	170			425			450		
alcohol-based core wash	Hardness				100			100		
(1.0 parts Gelva V7-50										
diluted with methanol to 20%										
polyvinyl acetate; 1 part										
"Lite-Off" A)										
Polyvinyl acetate water-based core wash	C.S.	170			275			315		
(75% Gelva S-55L in H_2O)	Tensile	25				60		140		
	Hardness	100				100	100			
Polyvinyl acetate alcohol-based	C.S				320			395		
core wash (75% Gelva V7-M50	Hardness				100			100		
in methanol)										

stirred at speed setting 2 for 2 min. Then adding 40 g of Du Pont sodium silicate grade No. 9 (29 w/o SiO_2, 8.9 w/o Na_2O) and mixing for 2 more minutes.

Standard AFS 2 × 2 in samples made by ramming. then gassing for 20 sec with CO_2 at a flow rate of 5 l/min, are allowed to age at about 23°C and 50% humidity, others are immersed in polyvinyl acetate or polyvinyl acetate-zircon water-based core washes and allowed to air dry. Samples treated with core washes are allowed to age under the same conditions as the untreated specimens. Compressive strength and core scratch hardness, and in some cases tensile strength is determined the day of making the cores and after several periods of time.

Results obtained are shown in Table 15.5.

16 Spray Dried Silica for Chromatography

Horacio E. Bergna
Dupont Experimental Station

CONTENTS

There is disclosed a process for preparing superficially porous macroparticles comprising spray drying a specified well-mixed slurry of core macroparticles, colloidal inorganic microparticles and a liquid, and sintering the resulting product to cause a 5–30% decrease in surface area.

BACKGROUND OF THE PROCESS

This invention relates to a process for preparing superficially porous macroparticles for use in chromatography and as catalysts or catalyst supports.

In chromatography it is customary to pass a mixture of the components to be resolved in a carrier fluid through a separative zone in a chromatographic apparatus. The separating or resolving zone generally consists of a material which is chromatographically sorptively active. Chromatographic apparatus generally employs packed columns of granular material. For analytical application the columns usually are of small internal diameter, while for preparative chromatography, larger diameter columns are employed. Support materials commonly employed for chromatography are granules having sorptively active surfaces or surfaces which have been coated with a substance which is sorptively active. Passing the mixture to be separated through the column results in repeated chemical interactions between the different components of the sample and the chromatographically active surfaces. Different compounds migrate at different speeds through the column because of these repeated, selective interactions. The separated components in the column effluent are generally passed through an analyzer or detector, for example a flame ionization detector in gas chromatography or an ultraviolet absorption detector in liquid chromatography, to determine when the resolved components emerge from the column and to permit the identification and quantitative measurement of each component.

Particles for use as catalysts or for catalyst supports should have overall size and porosity to permit ready access of reacting species to catalytically active sites within the particles.

It has long been recognized that superior chromatographic supports for liquid chromatography would consist of a plurality of discrete particles of regular shape, preferably spheres, having surfaces with a large population of superficial, shallow pores and no deep pores. The support granules should be regular and their surface characteristics readily controllable and reproducible. The same ready-access characteristics that make particles superior for chromatography also are desirable for catalysts and catalyst supports. Such particles have been very difficult to realize in practice with a result that the cost of superficially porous column chromatographic packings and particularly catalyst and catalyst supports has inhibited their use.

British Patent No. 1,016,635 discloses a chromatographic support made by coating a particulate refractory solid on an impermeable core. The coating is accomplished by dispersing the coating material in a suitable liquid in a slurry. The cores are then coated with the slurry, withdrawn, and dried to remove the liquid. The result is a rather loosely held, mechanical coating of non-uniform, disoriented particles. These coated cores may be used as chromatographic supports.

Kirkland (Kirkland, J.J., *Gas Chromatography 1964*, A. Goldup, editor, The Institute of Petroleum, London, pp. 285–300, 1965) has described the preparation of a chromatographic support by bonding successive layers of silica microparticles to glass beads by means of very thin fibrillar boehmite films. These coated cores may be employed as chromatographic or catalysts or catalyst supports.

Coated glass beads consisting of a single layer of finely divided diatomaceous earth particles bonded to the glass beads with fibrillar boehmite have also been described as a chromatographic support (Kirkland reference as above; Kirkland, J. J., *Anal. Chem.*, 37, 1458–1461, 1965).

A method of preparing superficially porous particles by depositing colloidal inorganic particles of a given size and ionic charge from aqueous dispersion onto the surface of a solid, a single monolayer of microparticles at a time, and by repeating the process, to coat the surface with any desired number of monolayers, is described in Canadian Patent No. 729,581.

U.S. Patent No. 3,505,785 issued to Kirkland on April 14, 1970, discloses a method of preparing superficially porous particles by first forming a coating consisting of alternate layers of colloidal inorganic microparticles and of an organic colloid, and then removing the alternate monolayers of organic matter so as to obtain a residual coating of layers of colloidal inorganic particles in which all the microparticles are alike.

SUMMARY

There is disclosed an improved process for preparing superficially porous macroparticles comprising:

1. Forming a well-mixed slurry of core macroparticles, colloidal inorganic microparticles to coat the macroparticles, and a liquid, said core macroparticles being (1) impervious and stable, (2) regularly shaped, and (3) about 5–200 μm in size and said microparticles being substantially uniform in size and shape and having a size of about 4–1000 nm; the ratio of the weight of the microparticles to weight of liquid being from about 0.05 to 0.5 and the ratio of volume of coating microparticles to total volume of coating microparticles and core macroparticles being from about 0.003 to 0.7;
2. Atomizing the slurry to form a fine spray using spraying conditions suitable to produce droplets of a size greater than the size of the core macroparticles;
3. Contacting the spray with a drying medium at a temperature of from about 130 to 400°C to evaporate the liquid or cooling the spray below the freezing point of the liquid and drying the resulting frozen particles without thawing them;
4. Heating the resulting dried, coated macroparticles at a temperature and for a time sufficient to cause sintering resulting in about 5 to 30% decrease in surface area of said dried, coated macroparticles.

BRIEF DESCRIPTION OF THE DRAWING

The drawing is a cross-sectional representation of a superficially porous particle made by the process of this invention.

Detailed Description of the Invention

The present invention relates to a process for preparing a powder of discrete, superficially porous macroparticles such as that shown diagrammatically in the drawing. The superficially porous macroparticle 1 comprises a core 2 which functions as a substrate for the remaining portion of the macroparticle. Adherred to the surface of the core is a porous coating 3, or crust composed of layers 4 of microparticles 5 which are characteristically in an almost regular close-packed configuration.

Any impervious material suitable for use in chromatography or as a catalyst or catalyst support may be used as

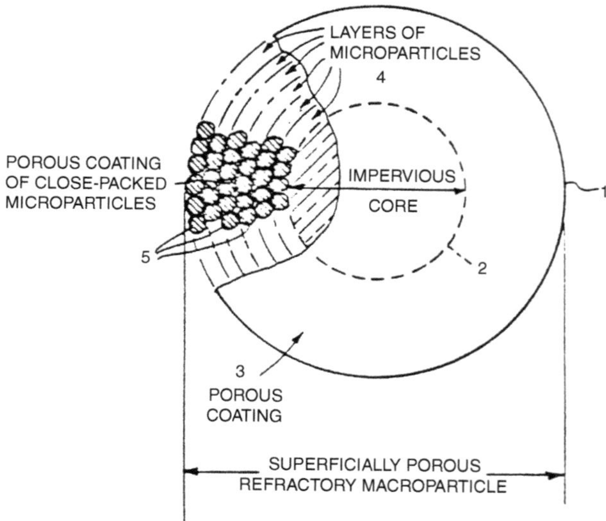

FIGURE 16.1 The drawing is a cross-sectional representation of a superficially porous particle made by the process of this invention.

the core or macroparticle support. By impervious material is meant a material having a surface sufficiently free from pores so that when employed as the substrate in a chromatographic process or as a catalyst or catalyst support, the substance passing through a zone of these particles will not enter the interior of the core. For most purposes the core should be impervious to nitrogen gas. The shape of the macroparticle cores used by the process of the present invention is generally regular, which is preferred in most chromatographic and catalyst applications. Any macroparticle shape suitable for use in chromatography or as catalysts can be employed, such as saddles, polyhedra, rings, rods, and cylinders. However, spheres are preferred because of their regular and reproducible packing characteristics, ease and convenience of handling, and their general resistance to mechanical degradation.

U.S. Patent No. 4,131,542, issued to Bergna et al. on December 26, 1978, discloses a process for preparing low-cost silica packing for chromatography comprising (a) spray drying with flowing air at a temperature from 130 to 400°C a silica sol containing from 5–60 wt% essentially nonaggregated spherical silica particles of uniform size wherein at least 75% of the particles have a diameter of from 0.5 to 2 times the weight average diameter and (b) sintering the resulting porous micrograins to reduce the surface area thereof from 5 to 20%.

Disadvantages of prior art products include coatings which are subject to easy removal as by chipping and flaking, lack of control of variables such as thickness and uniformity of the coating, chemically inhomogeneous surfaces, surface components which are deleterious as catalysts as to certain types of selective adsorption, inability to prepare structures with a uniform surface and with a certain predetermined porosity, and the requirement of many depositions of a single layer at a time. A method which eliminates or minimizes some of these disadvantages is desirable. Moreover it would be advantageous to have the coated materials irreversibly bonded to the core.

The composition of the core macroparticle is not critical except that it should be stable to the conditions necessary to prepare the coating and suitable for use in chromatography or as a catalyst. The cores can be, for example, glasses, sand, ceramics, metals, or oxides. In addition to truly impervious cores such as these, other types such as alumina silicate molecular-sieve crystals or small-pore porous oxide microspheres such as those described in U.S. Patent No. 3,855,172, issued to Iler et al. on December 13, 1974, can be used. In general, materials which have some structural rigidity are preferred. Glass beads are preferred macroparticles because of their uniformity of surface characteristics, predictability of packing characteristics, strength, and low cost.

The size of the cores is in general not critical. The upper size limit of the macroparticle is determined by the spray-drying equipment and its ability to spray the slurry of colloidal material. In general, the maximum size of the macroparticle will be about 500 μm. The important factors are that (a) the macroparticles must be maintained in a slurried state and (b) they must be able to pass through the atomizer to be formed as droplets. For spheres or similarly shaped bodies, a diameter or size of 5–200 μm prior to coating is preferred.

The microparticles that make up the coating can be of any desired substance compositionwise which can be reduced to a colloidal state of subdivision; however, they must be dispersible in a medium as a colloidal dispersion. Water is the best medium for dispersions of particles of varying ionic charges. Examples of suitable aqueous sols are amorphous silica, iron oxide, alumina, thoria, titania, zirconia, zircon, and alumina silicates, including colloidal clays such as montmorillonite, colloidal kaolin, attapulgite, and hectorite. Silica is preferred material because of its low order of chemical activity, its ready dispersibility, and the easy availability of aqueous sols of various concentrations.

The exterior and interior surface of the superficially porous refractory macroparticles to be used in chromatographic columns may be further modified by various treatments, such as reaction with organosilane alcohols, depending on the type of chromatographic separation required. For example, the superficially porous macroparticles can be modified with a variety of organosilane groups using the procedure described in U.S. Patent No. 3,795,313, issued on March 5, 1974, to Kirkland et al.

Drying of Spray by Moisture Removal

Separation of Dried Product from the Air

These stages are discussed in U.S. Patent No. 4,131,542, the relevant sections of which are incorporated herein by reference. Discussion herein is limited to factors which are affected somewhat differently by the difference in feed.

Ultimate Particle Size of the Feed

The feed is a slurry made of a well blended mixture of macroparticles and microparticles in a liquid, preferably water. The size of the macroparticles in essence determines the particle size of the dry product since the thickness of the coating produced by the microparticles is only a fraction of the diameter of the macroparticle. The size of the microparticle, on the other hand, determines the pore size and influences the pore size distribution of the coating. Preferably both macroparticles and microparticles are spherical.

Concentration of Solids in the Feed

Increase of the microparticle concentration in the slurry and increase of the number of microparticles relative to the number of macroparticles increase the thickness of the coating on the macroparticle. Obviously the higher the number of microparticles in the droplets of the spray, the larger the number of microparticles that will dry on the surface of the macroparticles. An increase in the thickness of the coating will, of course, increase the size of the particle of the product relative to the size of the original macroparticle.

The total concentration of the solids, including both macroparticles and microparticles, is limited by the viscosity of the slurry. Too high a concentration of solids will increase the slurry viscosity to the point that will make it unsuitable for atomizing using conventional equipment.

Concentration of microparticles in the feed also has an influence on the shape of the grains constituting the powder product. Depending on the concentration of microparticles relative to the concentration of macroparticles, a fraction of the spray droplets may have only microparticles and no macroparticles. Such droplets will dry producing amphora-shaped porous micrograins of the kind described in U.S. Patent No. 4,131,542. Therefore the dry product can be a mixture of coated macroparticles and hollow porous micrograins. Such a mixture can be screened or sedimented to separate the lighter, porous grains from the denser superficially porous, macroparticles.

Viscosity and Temperature of the Feed

The higher the viscosity of the feed, the coarser the spray at constant atomizing conditions will be. This factor has to be considered in order to control the droplet size relative to the size of the macroparticles. Viscosity is influenced by feed concentration and, in some cases, by temperature.

Other than the influence on viscosity, the effect of feed temperature is negligible. The possible increase in feed heat content is small compared to heat requirements for evaporation.

Feed Rate

Increase of feed rate at constant atomizing and drying conditions increases the size of the spray droplets. Therefore feed rate is an important factor to control the droplet size relative to the size of the macroparticles. Decrease of feed rate decreases, in general, the outlet temperature in the spray dryer and, therefore, there is less tendency to disrupt the microparticle layer coating the macroparticles. Changes in feed rate normally do not produce substantial changes in the deviation of the particle size distribution of the product.

Besides the feed (slurry) properties and feed rate, the following variables in dryer design and operation affect the characteristics of the dried product.

Atomization of Feed into a Spray

The characteristic features of spray drying are the formation of a spray, commonly referred to as "atomization", and the contacting of the spray with air. The atomization step must create a spray for optimum evaporation conditions leading to a dried product of required characteristics. Therefore, the selection and the manner of operation of the atomizer are of decisive importance in determining the kind of product obtained. Centrifugal, (rotary) pressure, kinetic energy, sonic and vibratory atomization can be used but centrifugal is preferred.

In all atomizer types, increased amounts of energy available for liquid atomization result in sprays having smaller droplet sizes. Higher energy of atomization means more break up of the liquid giving a finer mist and therefore smaller product particles. The size of the core macroparticle core will not be affected, but, on the other hand, excessive atomization energy may produce droplets of diameter smaller than the core macroparticle. In this case the microparticles dispersed in the droplet will dry separate from the glass bead forming a smaller porous micrograin (PMG) of the kind described in U.S. Patent No. 4,131,542. Atomization energy therefore should be kept at a level suitable for the production of droplets of a size larger than the macroparticles of the slurry.

Spray-Air Contact

Fast heating of droplets containing microparticles produces a dry skin of microparticles when the slurry does not contain macroparticles. In this case the dry skin of microparticles traps water inside the hollow sphere. Evaporation and evolution of the trapped water tends to produce a hole through the spherical grains obtained as a product. However, in the present process using a mixture of microparticles and macroparticles, the microparticles form a dry skin on the surface of the macroparticles. If the dry skin is too thick, fast heating may create craters on the otherwise uniform coating of the macroparticle.

There are dryer designs that incorporate both "co-current" and "counter-current" layouts, that is, mixed flow dryers. This type of design can also be used in the process of the invention.

Drying of Spray

When the droplets of the spray come into contact with the drying air, evaporation takes place from the droplet surface. Increase in the inlet temperature increases the dryer evaporative capacity at constant air rate. Higher inlet temperatures generally mean a more economic dryer operation. Increased temperature also may produce increased coating thickness due to a more randomized packing of the microparticles and therefore reduction in density of the coating. Generally the following air is at a temperature of from 130 to 400°C with from 150 to 300°C being the preferred range.

Fast evaporation could cause hollowness and even fracture of the coating. For this reason factors affecting drying rate have to be controlled to prevent or moderate lack of uniformity of the coating depending on the characteristics desired for each product.

The superficially porous macroparticles formed by spray of freeze drying are sintered and depending on the ultimate use of the particles, it may be desired to carry out acid washing treatments of the macroparticles to improve their properties for chromatography and as catalysts or catalyst supports. The sequence of these steps may be interchanged, depending on the needs of a particular system, but sintering is generally the preferred initial treatment.

17 Preparation of Monodisperse Ultrafine Hybrid Silica Particles by Polymer Modification

K. Yoshinaga, J. Shimada, and T. Kobayashi
Kyushu Institute of Technology

CONTENTS

Preparation of ultrafine monodisperse hybride particles, dispersible in low polar organic solvent, from monodisperse colloidal silica by two-step polymer modification was studied. Bindings of the secondary polymer to monodisperse poly(maleic anhydride-styrene)-modified colloidal silica particles (120 nm) have made the composites in low polar solvent dispersible. The dispersion of the particles in good solvent for the secondary polymer is due to the steric repulsion of solvated polymer chains. The dispersibility of the hybrid particles in poor solvent-rich solution was controlled by delicate balance between nonpolar–nonpolar interaction and electrostatic repusion among the particles.

INTRODUCTION

Fabrication of physical or chemical properties on ultrafine particles is promising to lead to development of new functional materials. Polymer modification enables to disperse inorganic colloidal particles in low-polar organic solvent due to steric repulsion of polymer chains in inter-particles. If it could control the surface hydrophobicity of surface potential of colloidal particles by polymer modification, it would be possible to construct a particle-arrayed composite, based on hydrophobic interaction among the particles. So far, we have reported that a modification of monodisperse colloidal silica with trimethoxysilyl-terminated poly(maleic anhydride-styrene)(P(MA-ST)-Si$(OMe)_3$) let to prevention of aggregation and improving dispersibility in organic solvent [1,2]. Moreover, a successive reaction of the composite with diisocyanates brought about an efficient crosslinking of the surface polymer layer, and a simultaneous formation of carboxyl or amino group on the surface [3]. In this work, we have investigated the preparation of ultafine polymer-silica particles, being dispersible in low-polar solvent, from monodisperse colloidal silica by two-step modification using P(MA-ST)-Si$(OMe)_3$ and the amino-terminated secondary polymer, poly(methyl acrylate) (PMA), poly (methyl methacrylate (PMMA) or poly(styrene) (PSt), as shown in Scheme I. Also, the effects of secondary

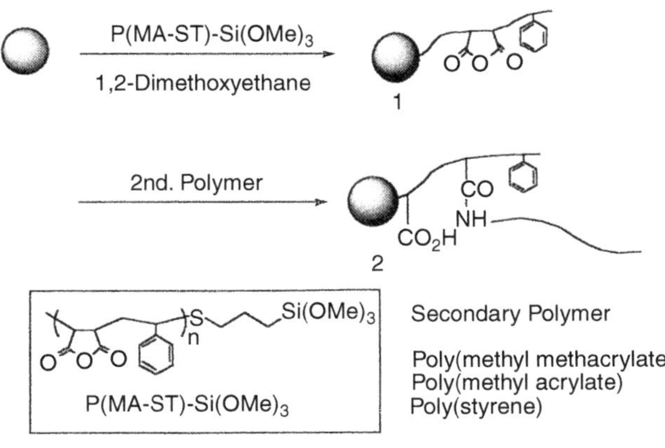

SCHEME I Two-step polymer modification of colloidal silica.

polymer on the dispersibility of the hybrid particles in organic solvent were studied by ESR study and ζ-potential of particles of the composite.

EXPERIMENTAL

Monodisperse colloidal silica (120 nm) suspended in ethanol was used. Trimethoxysilyl-terminated poly (maleic anhydride-styrene) (P(MA-ST)-Si(OMe)$_3$) was prepared by copolymerization of maleic anhydride and styrene in the presence of AIBN and 3-mercaptopropyl trimethoxysilane. Amount of binding polymer on silica particles was determined by a thermal gravimetric analysis. The dispersion-aggregation behavior of the composite in the cosolvent of ethyl acetate (AcOEt)-methanol (MeOH) was monitored by the absorbance at 500 nm. Particle size distributions in cosolvent were measured by a dynamic light scattering method. ζ-Potential in cosolvent was measured by a light scattering electrophoresis. The spin-labeled PMMA and PSt were synthesized by the reaction of respective polymer having carboxyl group with 4-amino-2,2,6,6-tetramethyl-piperidine-1-oxyl (4-amino-TEMPO) using *N,N*-dicyclohexylcarbodiimide.

RESULTS AND DISCUSSION

Binding of Secondary Polymer to Composite 1

The reactions of composite **1** with secondary polymer of amino-terminated poly(methyl acrylate), poly(methyl methacrylate) or polystyrene were carried out in acetone or tetrahydrofuran under refluxing for 24 h. The amount of the binding polymer decreased with increasing of the molecular weight, as shown in Table 17.1.

Dispersibility of Composite 2 in Cosolvent

Absorbance of suspension, containing the composite particles binding PMA or PMMA as a secondary polymer, were measured in an AcOEt-MeOH cosolvent at 500 nm. In Figure 17.1, dependencies of the absorbance of respective suspension containing PMA- and PMMA-P(MA-ST)SiO$_2$ on MeOH content in the cosolvent are shown. In general, the assembling of fine particles in a diluted suspension makes the absorbance increase, because of an increase of the light scattering. The suspension of PMA(2.5 k)-P(MA-ST)/SiO$_2$ showed a marked decrease of their absorbance at 60 vol% MeOH content to give zero absorbance, but the absorbance increased in a solvent containing more than 70

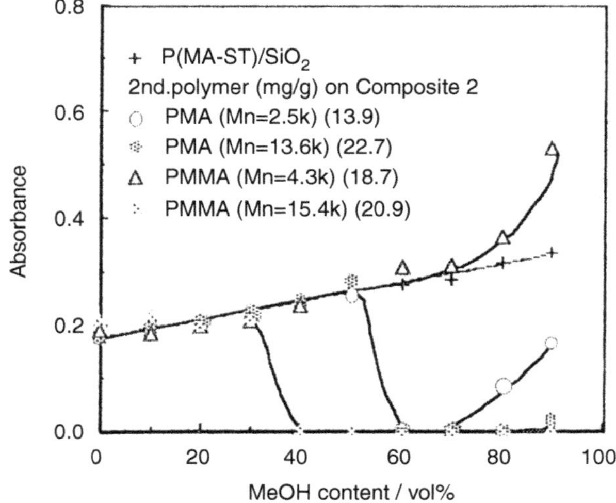

FIGURE 17.1 Absorbance changes of PMA- and PMMA-modified Composite 2 suspension with MeOH content in AcOEt/MeOH.

TABLE 17.1
Binding of Secondary Polymer on P(MA-ST$_2$)/SiO

	2nd-polymer			P(MA-ST)/SiO$_2$		Bound 2nd polym.	
	10^{-3} Mn	Xn[a]	g	mg/g[b]	g	mg/g	μmol/g
PMA	2.5	29	0.2	61.2	0.2	13.9	5.56
	13.6	168	0.6	61.2	0.15	22.7	1.67
PMMA	4.3	43	0.4	65.9	0.2	18.7	4.35
	15.4	154	0.2	65.9	0.15	20.9	1.36
PSt	4.5	43	0.05	57.9	0.2	32.8	7.29
	4.5	43	0.01	57.9	0.2	21.1	4.69
	11.0	106	0.6	57.9	0.2	33.2	3.02
	11.0	106	0.05	57.9	0.2	7.6	0.69

[a]Polymerization degree.
[b]Amounts of poly(maleic anhydride-styrene) on composite **1**.

vol% MeOH. It is considered that the nonpolar–nonpolar interaction among the polymer chains on the particles cause aggregation, but in higher-MeOH content, particles disperse base on electrostatic repulsion of silanol and carboxyl group produced by binding of PMA chain. Figure 17.2 shows changes in the absorbance of suspensions containing PSt-P(MA-ST)/SiO_2 with the MeOH content in the AcOEt-MeOH cosolvent. The suspension of PSt(11.0 k)-P(MA-ST)/SiO_2 showed an increase of the absorbance in higher-MeOH content due to weak aggregation. However, all particles bound PSt were observed not to be sedimented sedimentation in the cosolvent, and stabilized in the wide range of solvent polarity from ether to alcohol.

ESR Study of Spin-Labeled Secondary Polymer

In order to inquire into conformational change of secondary polymer in AcOEt-Hexane cosolvent, order parameter, S, was determined by the line-width on a ESR spectrum. The parameter S usually increases with increasing anisotropy of the radical species on polymer chains. When the species is in solid, the S becomes 1. As shown in Figure 17.3, S values of all composite particles increased with hexane content and decreased with increasing of the molecular weight of PMMA and PSt. These results suggest that molecular motion of polymer chains on the particle was restrained, because of shrinking due to the desolvation. Therefore, it was suggested that dispersibility of composite **2** in good solvent-rich solution was attributable to steric repulsion of polymer chain.

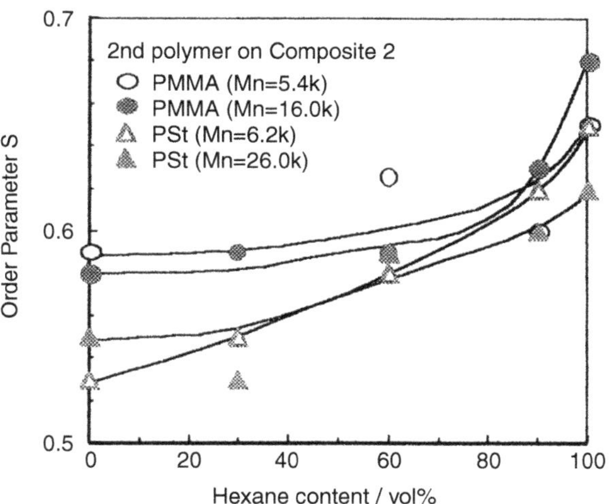

FIGURE 17.3 Dependence of order parameter S on hexane content in AcOEt/hexane.

ζ-Potential of Composite 2

In Figure 17.4, ζ-potential of composite **2** in AcOEt-MeOH cosolvent are shown. ζ-Potential of PMMA(4.3 k)-P(MA-ST)/SiO_2 was -100 mV in AcOEt, and increased with increasing of MeOH content. PMA(2.5 k)-P(MA-ST)/SiO_2 showed ζ-potential of -127 mV in AcOEt, and the maximum value (-23 mV) in about 60 vol% MeOH content. These results show that dispersibility of composite particles in this cosolvent is also related to ζ-potential. In Figure 17.4, ζ-potential of the PSt-modified particles in AcOEt-MeOH cosolvent are also shown. Unexpectedly, the composite showed highly negative

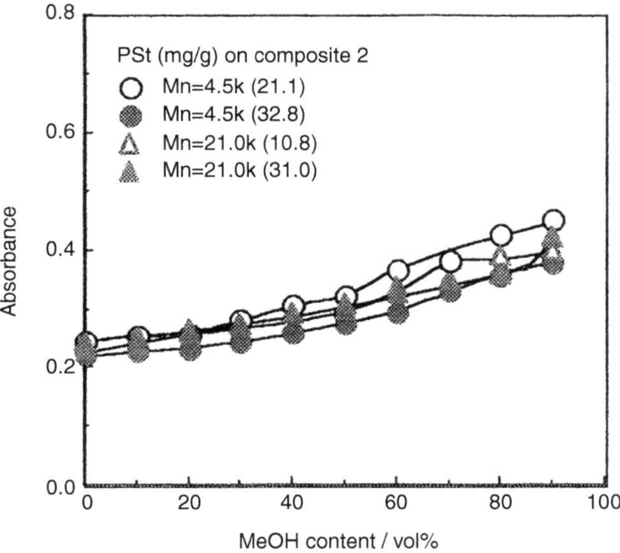

FIGURE 17.2 Absorbance changes of PSt-modified Composite **2** suspension with MeOH content in AcOEt/MeOH.

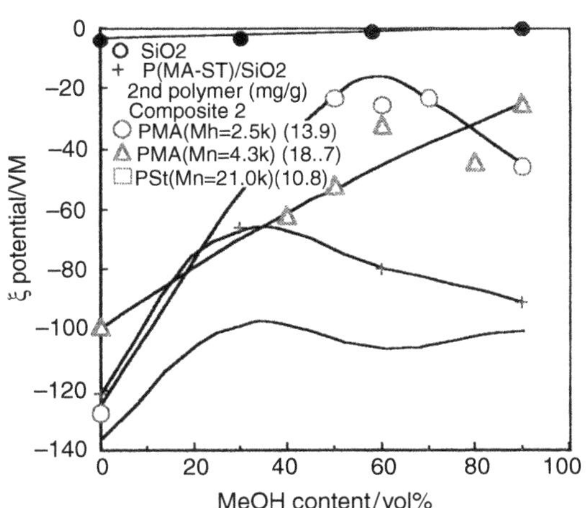

FIGURE 17.4 Dependence of ζ-potential on MeOH content in AcOEt/MeOH.

potential over −60 mV even in high MeOH content solution.

Therefore, dispersion-aggregation behavior of the composite **2** in poor solvent-rich solution was probably controlled by delicate balance between the electrostatic repulsion due to negatively high ζ-potential and non-polar–nonpolar interaction of shrunk secondary polymer chains in interparticles.

REFERENCES

1. K. Yoshinaga and K. Nakanishi, *Composite Interfaces*, **2**, 95 (1994).
2. K. Yoshinaga, K. Nakanishi, Y. Hidaka, and H. Karakawa, *Composite Interfaces*, **3**, 231 (1995).
3. K. Yoshinaga and M. Teramoto, *Bull. Chem. Soc. Jpn.*, **69**, 2667 (1996).

18 Monodisperse Core-Shell Silica Colloids from Alkoxysilanes

Alfons van Blaaderen
Utrecht University, Debye Institute, Soft Condensed Matter Group

A general seeded-growth procedure to grow extremely monodisperse core-shell silica spheres covering almost the whole colloidal size range has been developed. The core-shell seeds were grown in a water/ammonia in cyclohexane microemulsion made with a non-ionic sufactant: polyoxyethylene nonylphenyl ether (NP5). The core-shell seeds were obtained with radii (R) around 20 nm and polydispersities δ (relative standard deviations in size) as low as 3%. Synthesis [1] and characterization [2] of the very monodisperse and spherical seed particles as grown in the non-ionic microemulsion have been described before. Also, the use of the microemulsion procedure to make nano-crystalline CdS-core silica-shell colloids by mixing multiple microemulsions has been demonstrated [3]. We have found a way of extracting these seeds out of the microemulsion without losing colloidal stability by vacuum distillation and redispersion of the particles and non-ionic surfactant in a mixture of water, ethanol and ammonia in which the spheres can be grown larger using alkoxysilanes. This further growth of the seeds reduces the polydispersity as $1/R$ [4] resulting in extremely monodisperse spheres. The synthetic procedure can also be used to make more monodisperse fluorescent core-shell particles [5], which are very useful in quantitative 3D confocal microscopy studies [6]. Further, we have slightly modified the method developed by Liz-Marzan [7] to prepare metal-core/silica-shell systems. The surface properties of the core-shell particles can be modified in many ways, for instance, with short alkane chains (octadecyl chains), longer polymer brushes (poly-isobutyne) or silane coupling agents in order to change the interaction potential between the spheres and make the particles dispersable in a range of solvents [see e.g., 8 and Refs. cited]. The core-shell nature of the model particles gives additional control over, for example, the optical properties of the colloids. Some examples of particles synthesized from alkoxysilanes are shown in the following [9]: polar molecules is energetically more favourable in comparison with hydrogen-bonded structures. By experiment it has been established that after thermal vacuum treatment of silica samples at 470 K an adsorption complex contains two molecules of H_2O per silanol group and after that at 670 K it has one molecule of water [2,3]. The formation of the most stable coordination complexes is related to penetration of adsorbate molecules into a surface layer with the subsequent coordination in *trans*-positions to (Si)OH groups. The channels for such a penetration are hexagonal structural cavities typical of the (111) face of β-crystobalite.

Substitution of trimethylsilyl groups for hydroxyls does not eliminate sites of strong adsorption of small polar molecules. In the case of low surface coverages, molecules of hydrogen fluoride that is distinguished for the maximum affinity to silicon compounds do not react with silanes and form strong complexes capable of acting as intermediates of the fluorination reaction. The field desorption mass-spectrometry makes it possible to register three characteristic regions of the adsorbate removal with desorption maxima, and the number of these regions is in agreement with the number of different types of adsorption complexes of H_2O and CH_3OH (hydrogen-bonded complexes and coordination complexes of *cis*- and *trans*-structure). Besides, it has been also shown that during reactions with water-adsorbing reagents one can perform a separate introduction of different forms of adsorbed water. The x-ray photoelectron spectroscopy detects an increase in the binding energy for core $2s$- and $2p$-electrons of silicon atoms as SiO_2 undergoes dehydration, and this increase is most sharp in the range of the high-temperature desorption maxima. This points to molecules of coordinatively bonded water as the source of variation of the electron density on Si atoms of a surface layer.

The application of various functional capabilities of disperse silicas calls for thorough knowledge about the role of the state of their hydrated cover since it is of major importance for the completeness of proceeding of chemical modification reactions as well as for detailed structure of surface compounds. The systematic experimental and quantum-chemical researches [1–3] resulted in the consistent insights into the structure of the hydrated cover of SiO_2, mechanisms of its formation and destruction. The concepts formulated take into account the data on feasibility to draw a distinction between sorbed and chemically bonded water. It made it possible to give concrete expression to such fundamental notions as dehydration and dehydroxylation of surface. Dehydration is a process of desorption of water evolved upon decomposition of aquacomplexes having various structures. The temperature interval of silica dehydration in vacuum extends to 900 K (H-bonded species decompose at 300–370 K, coordination complexes of *cis*- and *trans*-structure decompose at 420–500 and 550–850 K respectively). Dehydroxylation is a chemical reaction of breakdown of a hydroxyl cover. Condensation of neighbouring silanediol groups on zones with the structure of (100) face of β-crystobalite proceeds simultaneously with dehydration of SiO_2 at 500–650 K, which is corroborated by the ^{29}Si NMR technique. Isolated silanol groups are removed at temperatures over 900 K. The mechanisms of the processes of thermal dehydroxylation of silica were considered in Refs. [2,3].

Between processes of adsorption and chemisorption there is a close relationship, since in the process of evolution of adsorbed species one can observe formation of transient complexes that are products of reactions proceeding by the Langmuir–Hinshelwood mechanism. According to the accepted classification of reactions in a surface layer of SiO_2 they are subdivided into two main classes, namely reactions with substitution of structural hydroxyl protons (S_Ei processes) and reactions with substitution of OH groups at silicon atoms (S_Ni processes). Reactions of heterolytic decomposition of siloxane bonds proceeding by the $Ad_{N,E}$ mechanism form a separate class.

By now we have studied experimentally numerous S_Ei reactions of the electrophilic substitution involving the replacement of protons of $\equiv$SiOH groups by positively charged fragments of molecules of hydro- and methylchlorosilanes, silazanes, alkoxysilanes, organosiloxanes, alkyl borates and phosphates, chlorides and oxochlorides of a number of elements, and so on. A detailed study has been also conducted on the nucleophilic substitution of functional groups at silicon atoms (reactions with halogen hydrides, alcohols, phenols, processes of hydrolysis of $\equiv$SiX groups of various types, etc.). Besides, we have considered a special class of reactions whose subsequent steps proceed by different mechanisms. In particular, such compounds as $SOCl_2$, PCl_5, WCl_6, $WOCl_4$, $MoOCl_4$, and so on interact with silanol groups by the S_Ei mechanism. These interactions lead to the formation of unstable intermediate compounds that in certain conditions undergo intramolecular rearrangements. As a result, $\equiv$SiCl groups are formed, which conforms to the substitution (by the S_Ni mechanism) of chlorine atoms for OH groups at silicon atoms, and molecules of a corresponding oxochloride split off [3].

The action of the mechanisms of all the above-mentioned processes can be given a proper rational explanation by quantum-chemical calculations of cross-sections of potential energy surfaces for the interaction of reacting systems [2,3], but these calculations are very tiresome and time-consuming. However, the goal can be achieved through a much more simple approach that lays emphasis on the stereochemistry of transient states and takes into account a predominant contribution of effects of reagent deformation to the reaction activation energy. Within the scope of the developed deformation model [2,3] the activation barrier of S_Ei reactions is determined by the structure deformation of only electrophilic reagent molecules. In the case of S_Ni reactions the main contribution to this barrier is made by the deformation of the tetrahedral environment of a silicon atom with a functional group. This approach enables consideration of processes proceeding by the $Ad_{N,E}$ mechanism as well. The insights into the deformational nature of activation barriers of chemical transformations have made the foundation for the uniform description, analysis, and prediction of the reactivity of modifying reagents and surface functional groups, and, consequently, of paths of reactions in a SiO_2 surface layer.

The results of the comprehensive studies into the structure of the pyrogenic silica surface, its hydration sheath, regularities and mechanisms of adsorption processes and chemical transformations have provided the scientific foundation of searches for the most promising routes of chemical modification of SiO_2 and various goal-directed syntheses in a surface layer as well as for the development of many novel materials with a wide range of physical and chemical properties.

The fundamental results achieved in the field of surface chemistry of silica formed the basis for the evolution of a novel direction of researches — chemical science of nanostructural materials. Within the scope of this scientific direction over 200 new materials with designed properties have been developed which find much use in care of public health, agriculture, ecology, civil engineering, instrument making, and other branches of the national economy. The materials developed include new medicinal preparations of sorptive action with a controlled pharmacodynamics; drug delivery systems; photosetting composites for orthopedic stomatology; enterosorbents and aerosolic vaccines for cattle breeding; preparations for clarification and stabilization of wines, beers, fruit juices, and other beverages; protective and stimulative compositions for increasing yield of agricultural crops; reagents for induced rain precipitation; sorbents for removal of petroleum and its products from surface of water; thermostable general-duty lubricants; water-repellent coatings; fillers for varnishes, paints, and enamels; polishing compositions for finishing treatment of electronic devices; heat- and sound-insulation materials; fireproof covers; fire-extinguishing powders, and many others.

REFERENCES

1. Chuiko, A.A., *React. Kinet. Catal. Lett.*, 1993, V. 50, No 1–2, p. 1–13.
2. Chuiko, A.A., Gorlov, Yu.I., *Surface Chemistry of Silica: Surface Structure, Active Sites, Sorption Mechanisms*, Kiev, Nauk. Dumka, 1992 (in Russian).
3. Chuiko, A.A., *Izv. Akad. Nauk SSSR*, 1990, No. 10, p. 2393–2406.
4. Ogenko, V.M., Rosenbaum, V.M., Chuiko, A.A., *A Theory of Vibrations and Reorientations of Surface Atomic Groups*, Kiev, Nauk. Dumka, 1991 (in Russian).

19 Preparation of Silica Solid Microspheres by Hydrolysis of Tetraethyl Ortho Silicate (TEOS) and Silica Porous Microspheres by Spray Drying Aggregated Colloidal Silica

Horacio E. Bergna
DuPont Experimental Station

Silica porous microspheres can be prepared by hydrolysis of tetraethyl ortho silicate (TEOS) see Figure 19.1. The reproducibility of the method is demonstrated in Figure 19.2. The product of a scale batch is shown in Figure 19.3. The effect of impurities in TEOS, in Figure 19.4.

The build up process with TEOS at 70°C from original solution with particles of 445 nm in diameter to 1 micrometer can be observed in Figure 19.5 and Figure 19.6.

The effect of drying silica microspheres from water and from ethanol slurries then redispersing in ethylene glycol is seen in Figure 19.7. The effect synthesis temperature, 30°C versus 60°C, is shown in Figure 19.8.

Finally, Figure 19.9 shows the silica porous microspheres product obtained in a 100 gallon kettle scale up process.

Table 19.1 shows the calculated and measured particle diameter of 445 mm silica TEOS slurry gradually built up at 75°C.

With reference to the silica porous microspheres obtained by spray drying aggregated colloidal silica (framed a pyrogene) Figure 19.10 and Figure 19.11 are micrographs at various magnifications, 300 to 3000×, of silica porous microspheres with a rotary atomizer at 45 m rpm.

Figure 19.12 shows silica micropheres made with a Bowen 1425 two fluid hozzle, 60 psi.

Figure 19.13 shown the mesopore size distribution obtained nitrogen absorption–desorption of an aggregated colloidal silica slurry.

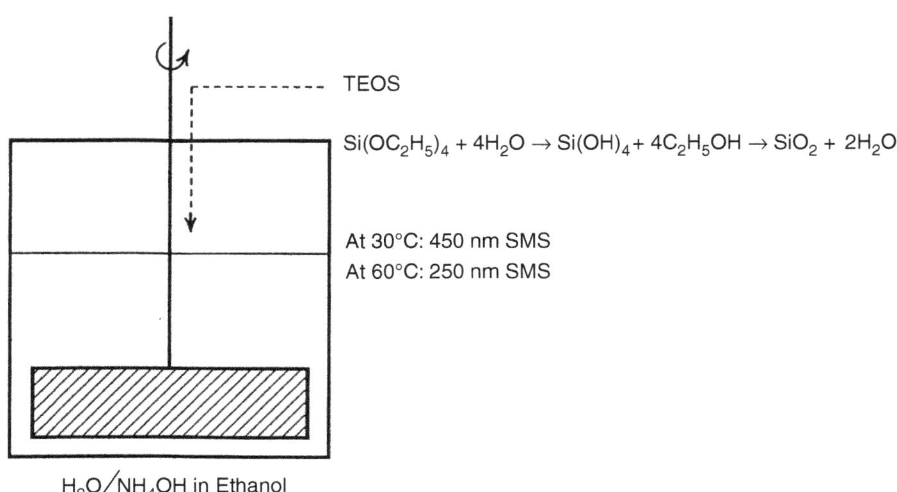

FIGURE 19.1 Preparation of silica solid microspheres by hydrolysis of TEOS.

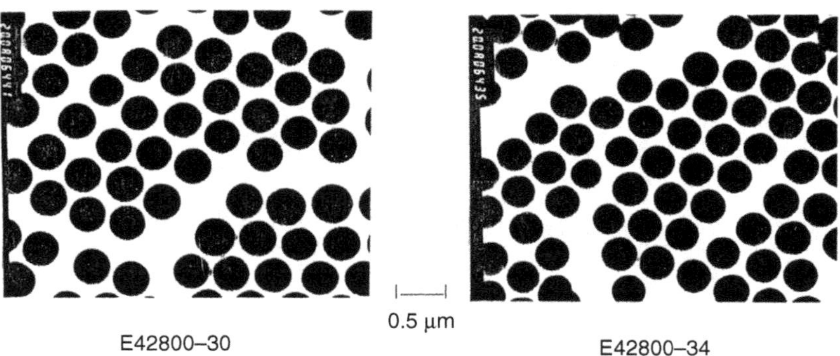

FIGURE 19.2 Tem (Van Kavelaar) silica solid microspheres.

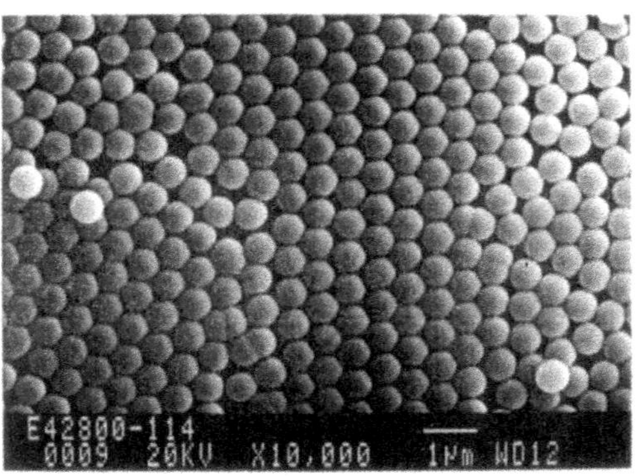

FIGURE 19.3 Silica solid microspheres scale up batch SEM.

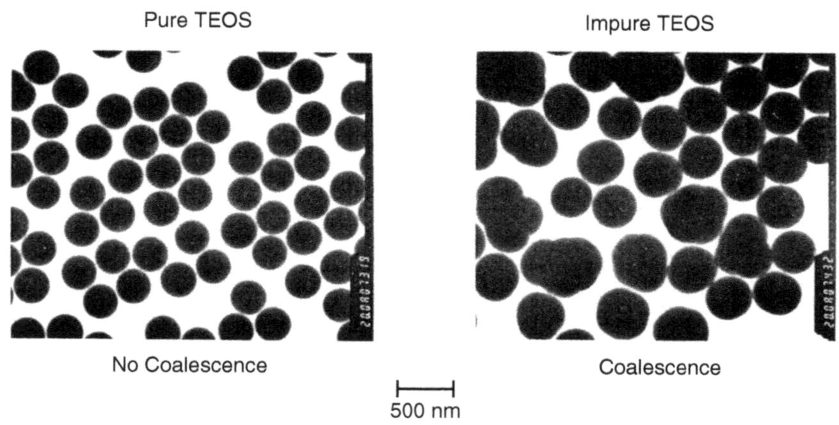

FIGURE 19.4 Silica solid microspheres TEOS hydrolysis process effect of impurities in TEOS.

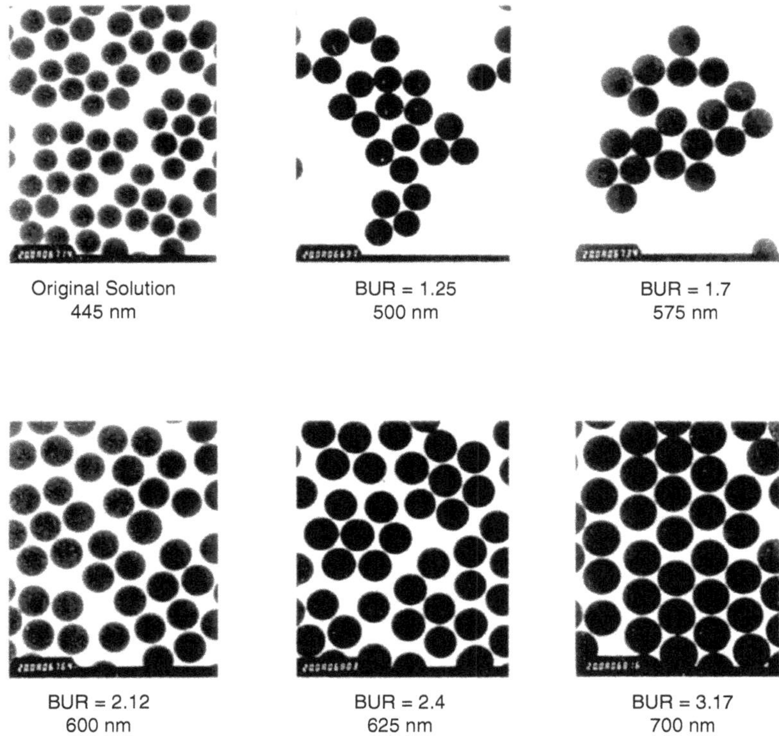

FIGURE 19.5 Silica solid microspheres build up process with TEOS-70°C.

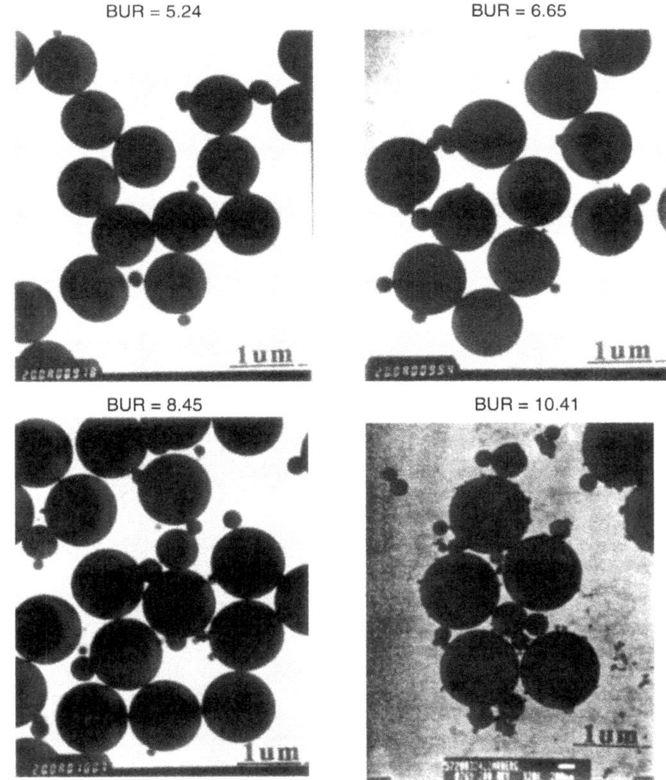

FIGURE 19.6 Build up of silica SMS by hydrolysis of TEOS.

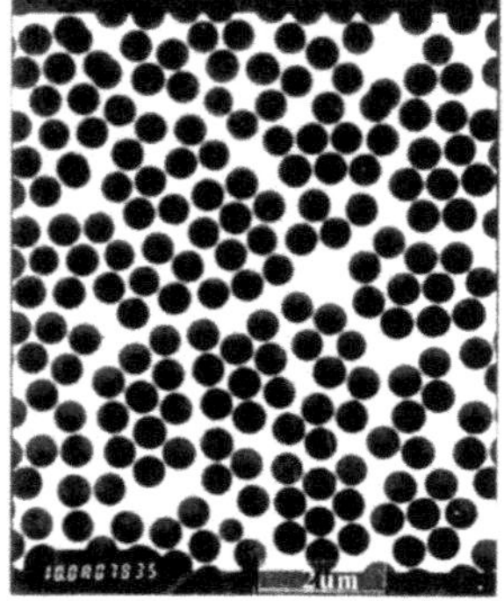

Dried From Water Dried From Ethanol

FIGURE 19.7 Silica solid microspheres dried from water and from ethanol slurries, then redispersed in ethylene glycol.

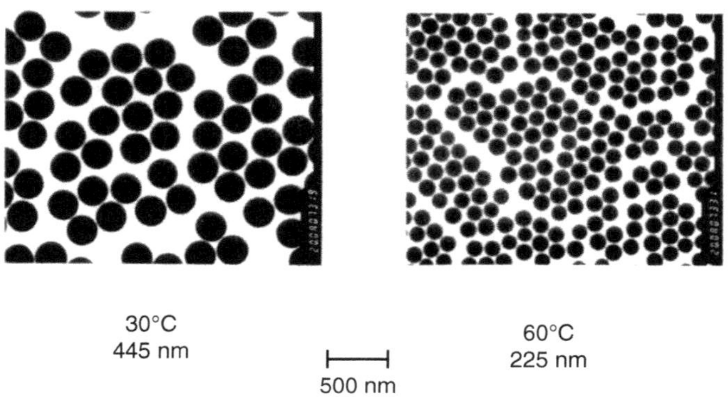

FIGURE 19.8 Silica solid microspheres TEOS hydrolysis process effect of temperature.

FIGURE 19.9 TEOS hydrolysis process 100 gal kettle scale up.

TABLE 19.1
Build Up Ratio, Calculated and Measured Particle Diameter of 445 mm Silica SMS Built Up at 75°C with TEOS (E57200–32)

				Quantimet 970 analysis			Dispersion		
Code	BUR	Calcul. D, nm	TEM D, nm	D, nm	S.D.	% Coal.	MONO/BIM.	Sec. Part. nm	Coalesc.
1	1.0	445	445						
19	1.23	475	475	495	22	1.2	M		Negligible
72	1.98	560	600	600	22	0.4	Molly M	Fur 200–250	
96	2.37	595	Ch. 600	635	28	1.4	M		Some
120	2.79	625	625	624	23	1.6	M		None
144	3.16	650	675	687	28	1.3	M		Some
179	3.63	685	700	700	72	14.4	B: Some 75–200 col. w. 700		
202	4.08	710	Ch. 725	706	32	0.2	B: Some 100–350		
226	4.55	735	Ch. 740	760	44	12.2	B: Some 70–250		
250	5.24	775	775				B: Some 100–250		
274	5.90	805	800	Not yet completed (5/2/89)			B: Some 100–200 + finer apprxy. debris		
301	6.65	835	850				B: Some 100–200 + finer apprxy. debris		
321	7.17	860	850				B: Some 75–200		
344	7.80	880	900				B: None abundant 75–500		
368	8.45	905	900				B: None abundant 75–500		
392	9.33	935	925–950				B: ditto + some fine aggr. debris		
422	10.41	970	1000				B: ditto		

BUR = Build up Ratio, Calcul. = Calculated DIAM from BUR, TEM: Estimated D from TEM, S.D. = Standard Deviation, MONO/BIM = Monomodal or Bimodal Dispersion, SEC Part = Secondary Particle Approximate D, COAL. = Coalescence.

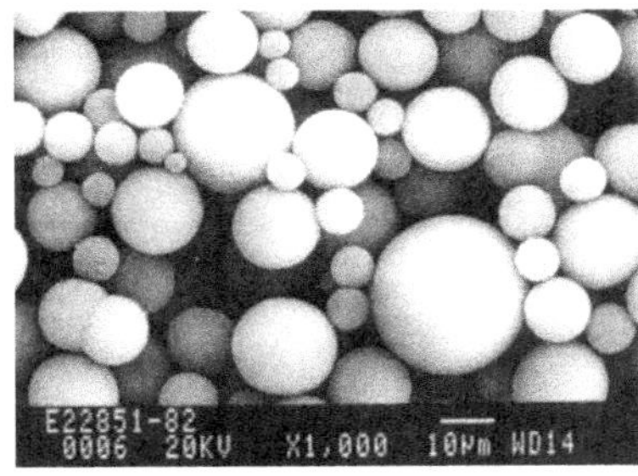

FIGURE 19.10 Silica porous microspheres spray dried 5 W/O aqueous slurry of Cab-O-Sil EH—5 rotary atomizer 45M RPM.

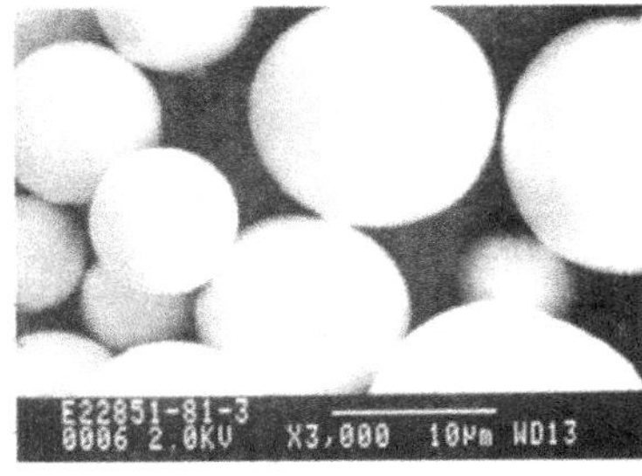

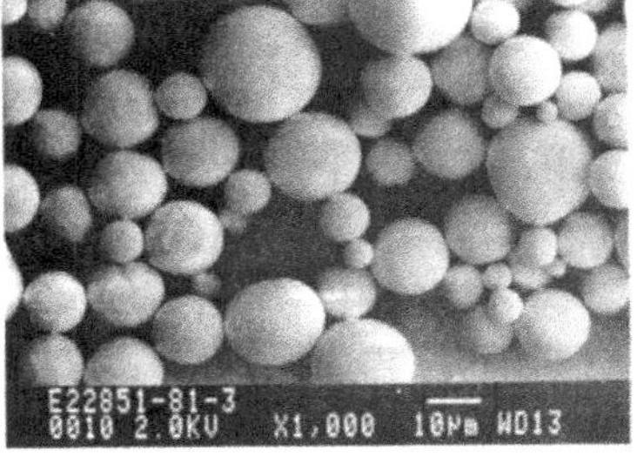

FIGURE 19.11 Silica porous microspheres spray dried 5 W/O aqueous slurry of Cab-O-Sil M-5 rotary atomizer 45M RPM.

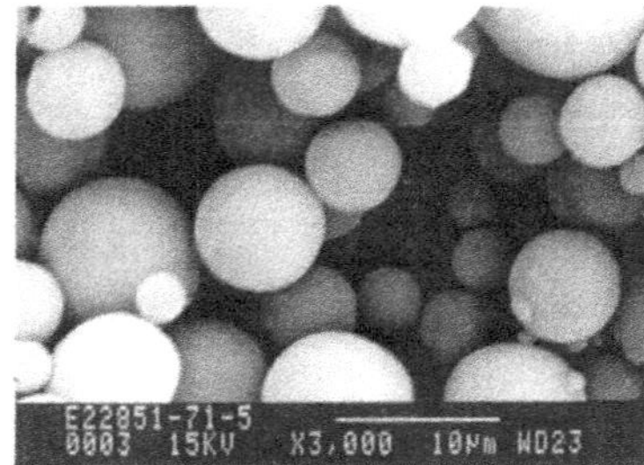

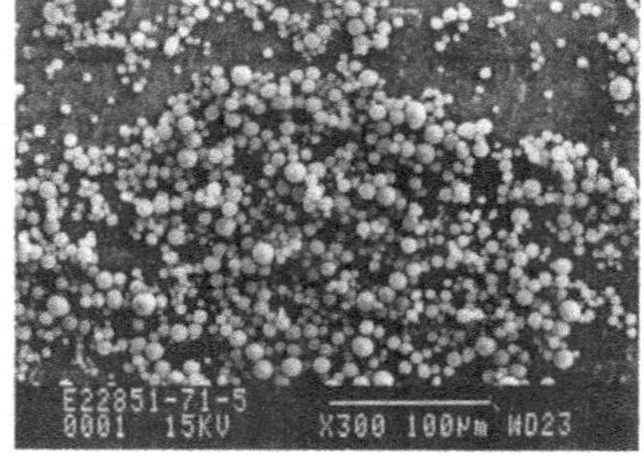

FIGURE 19.12 Silica porous microspheres spray dried 5 W/O aqueous slurry of Cab-O-Sil EH-5 bowen 1425 – two fluid nozzle, 60 PSI.

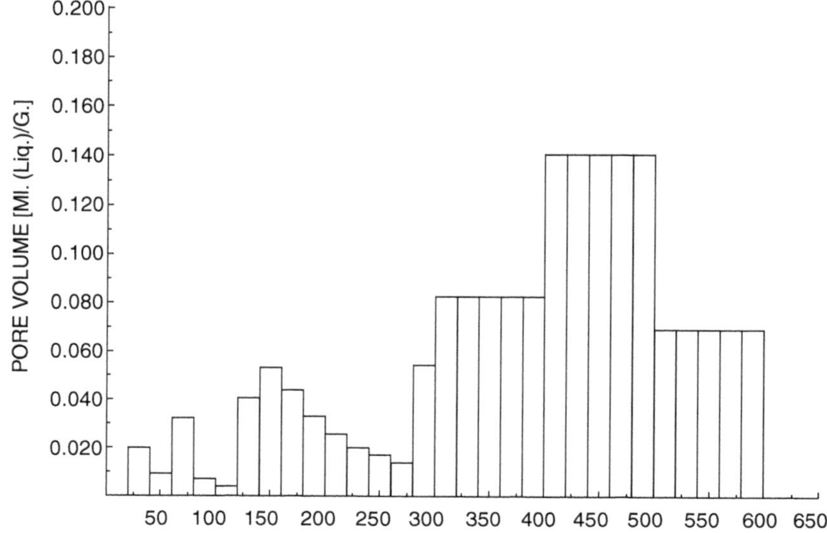

FIGURE 19.13 Silica porous microspheres: spray dried Cab-O-Sil M-5 mesopore size distribution by nitroghen adsorption–desorption.

Part 2

Stability of Sols

Do the Silica Hydrosols Obey the DLVO Theory?

F. Dumont
Free University of Brussels

Amorphous silica has many special properties that are reported in the outstanding book of Ralph Iler [1] and in this book. This often unexpected behavior raises difficult questions for those working in the field.

For example, it is well known that the silica hydrosols are stable at their point of zero charge (pzc) and that they also coagulate in alkaline solutions, in which their electrical surface charge is high and should therefore increase their stability. Such behavior is very unusual indeed, and this question arises immediately: Why does the Derjaguin–Landau–Verwey–Overbeek (DLVO) theory seem to be unable to cope with the silica hydrosols while it explains satisfactorily, at least to the best of our knowledge, the behavior of all other colloidal systems?

This question may be answered in two manners, as represented by communications to the Ralph K. Iler Memorial Symposium. On the one hand, Yates proposes a thermodynamic approach to replace the failing DLVO theory. This approach can explain several experimental facts, but suffers from poor generality. According to Healy, on the other hand, the DLVO theory should give a coherent description of the hydrosol behavior on the condition that all the forces that play a role in the interaction are introduced in the model. This correction leads to a good description of the observed properties and to a better description of the interface structure.

These two approaches appear quite conflicting; however, some kind of reconciliation between them could be initiated by a short review of the theory foundations. According to DLVO theory [2], at least two kinds of forces exist between colloidal particles: electrostatic forces, which result from the particle electrical charge, and the universal London–van der Waals forces. Between identical particles, the electrostatic forces are repulsive, whereas the London–van der Waals forces are attractive. Their resultant force determines the stability of the hydrosol, and their existence is unquestionable, as are the foundations of the theory. However, the equations used could be more reliable. The expressions of the electrical forces are all derived from the Poisson–Boltzmann equation, which is based on two major approximations: the solvent is considered a structureless dielectric continuum, and the exact field generated by the ions is approximated by a mean field. Moreover, the London–van der Waals forces calculation was based, at least until recently, on the microscopic theory of the dispersion forces, which suffers strongly from the very questionable additivity hypothesis. The difficulty in assessing a priori the magnitude of the errors introduced by all these assumptions opens the door to criticism.

Fortunately, important progress has been made in the past decade; Monte Carlo simulations [3] have demonstrated that the Poisson–Boltzmann equation is quantitatively correct when the counterions are monovalent, assuring the correctness of the electrical forces calculation.

TABLE 1
Hamaker Constant of Silica and Titanium Dioxide in Water

Element	ε^+	n^a	$A_{st}(10^{-20}$ J)	A_{disp} $(10^{-20}$ J)	A $(10^{-20}$ J)	A_{st}/Δ_{tot} 100
H_2O	78.36	1.33	—	—	—	—
TiO_2	86–170	2.50	±0.07	23.2[a]	23.27[b]	0.3
SiO_2	3.8	1.45	0.267	0.384	0.651	41

[a]These values are taken from the *Handbook of Chemistry and Physics*, CRC Press, 1976.
[b]Due to the fact that the absorption frequencies of TiO_2 and water are very different, this calculated value of TiO_2 must be considered as approximate.

Moreover, the decisive theoretical work of an Australian group in Canberra [4] has significantly improved the Lifshitz exact macroscopic theory of the dispersion forces, leading not only to the correct calculation of the London–van der Waals forces but also to the finding that the previous microscopic theory gives the right order of magnitude of the interaction energy. At the same time, the Canberra group discovered experimentally the existence of new short-range forces that play an important role in the interaction process and must therefore be added to those already accounted for by the original DLVO theory. These are known as the structural forces [5]. For a better understanding of the particular behavior of the silica hydrosols, it is necessary to come back in some detail to the London–van der Waals interaction energy. The attraction energy between two identical spherical particles (medium 1) dispersed in a medium 3 is given by the equation

$$V_{\text{att}} = -\frac{A_{13}}{12}H(x,y) \tag{1}$$

where A_{13} is the Hamaker constant and $H(x, y)$ the Hamaker function, which depends on the interparticle distance and on the particle radius [2]. The Hamaker constant A_{13} can be calculated with an error less than 5% by an equation directly deduced from macroscopic theory [4]. This calculation requires knowledge of the spectral properties of the media 1 and 3. Nevertheless, Israelachvili (5) gave an approximate expression for A_{13}, which is sufficient for this discussion:

$$A_{13} = \frac{3}{4}kT\left[\frac{\varepsilon_1 - \varepsilon_3}{\varepsilon_1 + \varepsilon_3}\right]^2 + \frac{3h\nu_e}{16\sqrt{2}} \cdot \frac{(n_1^2 - n_3^2)^2}{(n_1^2 + n_3^2)^{3/2}} \tag{2}$$

where ε_1 and ε_3 are the static dielectric constant of the media, n_1 and n_3 are their refractive indexes in the visible range, ν_e is the absorption frequency of the media assumed to be the same for both of them, k is the Boltzmann constant, T is temperature, and h is Planck's constant.

The first term of the right-hand side of Equation (2) describes the static contribution the interaction. Its value cannot be greater than $3kT/4$, and its most important property is that it is screened by the solution ions at a distance amounting to the Debye length [4]. The second term represents the spectral or dispersive part of the energy and is insensitive to the solution ionic strength but subject to the propagation retardation effect, which is not included in the approximate Equation (2).

Table I reports the values of the static (A_{st}) and dispersive (A_{disp}) parts of the Hamaker constant of silica in water, calculated from Equation (2). The corresponding values for TiO_2, a typical electrocratic colloidal oxide, are also included for comparison, which is probably the key to an explanation of the special behavior of the silica hydrosols. These data show that the Hamaker constant of SiO_2 is approximatively 35 times smaller than that of TiO_2. Thus, the attraction energy between two silica particles is 35 times smaller than that between two TiO_2 particles of the same size. This weakness of the attraction energy enhances the role of the afore-mentioned structural forces, which are strongly dominated by the London–van der Waals attraction in the case of TiO_2.

Moreover, the SiO_2 Hamaker constant itself could also be at the origin of some unexpected properties. As seen in Table I, the static term accounts for approximately 40% of the attraction energy, and as the attraction is screened by the solution ions, an increase of the electrolyte concentration will cause a substantial decrease in attraction. Because the refractive index n_3 of the solution increases with the electrolyte concentration, the dispersive term of the Hamaker constant will also slightly decrease, contributing also to a weakening of the overall attraction energy: such a stabilizing action by an increase in ionic strength is not common in colloidal chemistry. This effect is expected to be of no importance for TiO_2

because its static term accounts for less than 0.3% of the Hamaker constant and because the high value of the TiO_2 refractive index renders the dispersive term insensitive to slight variations of the solution refractive index.

In conclusion, it seems that many particular aspects of the stability of silica hydrosols could be explained not only by the low value of the Hamaker constant but also by the relative importance of the static term. Thus, SiO_2 is a valuable system for the experimental study of the structural forces. It is likely that the unique chemical properties of silica and silicic acid might also give rise to surprising results.

REFERENCES

1. Iler, R. K. *The Chemistry of Silica*; Wiley: New York, 1979.
2. Verwey, E. J. W.; Overbeek, J. Th. G. *Theory of the Stability of Lyophobic Colloids*; Elsevier: Amsterdam, Holland, 1948.
3. Johnson, B.; Wennerstrom, H.; Halie, B. *J. Phys. Chem.* **1980**, *84*, 2179.
4. Mahanty, J.; Ninham, D. W. *Dispersion Forces*; Academic Press: New York, 1979.
5. Israelachvili, J. N. *Intermolecular and Surface Forces*; Academic Press: San Diego, CA, 1987.
6. *CRC Handbook of Chemistry and Physics*, 56th ed., 1975–1976; CRC Press: Boca Raton, FL.

20 Stability of Aqueous Silica Sols

Thomas W. Healy
University of Melbourne, School of Chemistry

CONTENTS

The control of silica sol coagulation by pH and by addition of simple electrolytes is said to be "anomalous" in that it is not simply predicted by conventional Derjaguin–Landau–Verwey–Overbeck (DLVO) theory. This chapter describes a model based on the control of coagulation by surface steric barriers of polysilicate and bound cations. The model suggests new experimental directions.

The coagulation–dispersion behavior of aqueous silica sols is central to almost all processes requiring their unique adsorption, dispersion, gelation, and sol–gel properties. Aqueous silica sols are of particular interest in colloid science because their coagulation–dispersion behavior is said to be "anomalous," that is, their stability in terms of electrolyte–pH control does not follow the pattern followed by almost all other oxide and latex colloidal materials. This chapter examines aqueous silica sol coagulation effects in light of studies of macroscopic silica–water interfaces and in particular the electrical double layer at such interfaces.

It is difficult to define precisely the term "aqueous silica sols" and thereby contrast them with other forms of silica (colloidal silica, colloidal quartz, pyrogenic silica, etc. and so forth). Bulk chemical distinctions are not very useful. The definition chosen here follows Iler's terminology [1]. Aqueous silica sols are characteristically composed of spherical particles nucleated and grown by alkaline hydrolysis of sodium silicate solutions. They are often monodisperse systems and have particle diameters in the range 1–100 nm (density, $\sim$2.2 g/cm^3) that lead to sols that vary from optically transparent to opalescent.

The colloidal behavior referred to in the preceding text as "anomalous" helps to further define the term "aqueous silica sols." Iler [2] noted, in a second major monograph, that such silica sols were colloidally stable at their isoelectric point pH (pH_{iep}). Iler also noted that their remarkable stability in high salt concentration at near-neutral pH.

The first definitive study of the "anomalous" coagulation behavior of silica sols was by Allen and Matijévic [3] in 1969. They focused on the measurement of the "critical coagulation concentration" of Ludox AM & HS sols 15 nm in diameter [210 m^2/g Brunauer–Emmett–Teller (BET) surface area], as determined by standard Rayleigh ratio light-scattering measurements. Figure 20.1 is a summary of the key results of Allen and Matijévic [3]. In particular, when Na^+ and Li^+ were the coagulating cations, the critical coagulation concentration decreased with increasing pH. Further, the silica sols were colloidally stable below the pH limit of Figure 20.1 and stable at the observed pH_{iep} of about pH 2–3. For K^+ and Cs^+ the silica sols were stable from pH 2–3 to pH 6, and these ions showed an irregular series effect in the pH 6–11 range, as shown schematically in Figure 20.1.

Depasse and Watillon [4] in a similar study in 1970 observed the same "anomalous" coagulation trends, again using standard light-scattering techniques with 50-nm-diameter silica sol particles prepared in a way similar to the technique of Allen and Matijévic [3]. Figure 20.2 is a summary of the Na^+ and Li^+ coagulation data of Depasse and Watillon [4] for an addition of electrolyte of 1.5 *M*.

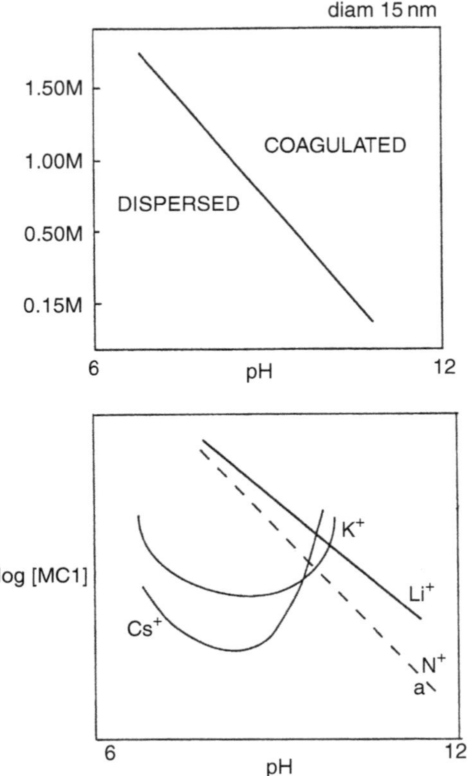

FIGURE 20.1 Schematic view of the coagulation domains of Ludox silica sols observed by Allen and Matijévic. (Adapted from Ref. [3].)

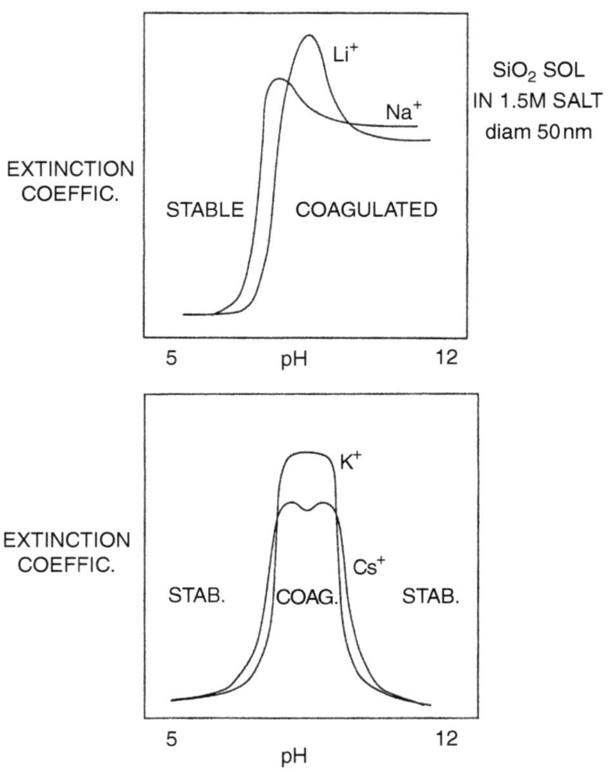

FIGURE 20.2 Schematic view of the coagulation domains of SiO_2 sols as determined by Depasse and Watillon. (Adapted from Ref. [4].)

The simplest way of focusing on the "anomalous" character of these results is to present the variation of the electrokinetic or zeta potential of the silica sols as a function of pH and added salt. The general form of the electrokinetic results obtained for a vitreous silica [5] and for all silica sols is shown schematically in Figure 20.3.

The silica sols, and indeed all oxide sols, show an increasing negative zeta potential with increasing pH as the pH is raised above the pH_{iep}. The magnitude of the zeta potential decreases uniformly at each pH as the salt concentration is increased. There are subtle effects in the electrokinetics as the counterion is varied from Li^+ to Cs^+, but these effects are minor compared with the general reduction in zeta potential as the pH is moved toward the isoelectric point or as 1:1 electrolyte is added at any pH.

The usual result of such variation in the zeta potential with pH and 1:1 electrolyte concentration is that the critical coagulation concentration normally varies with pH as shown schematically by the theory line in Figure 20.4. The critical coagulation concentration increases as the pH is increased above the isoelectric point and peaks at high values as shown. Many examples of these trends are confirmed for oxide and latex colloids for which H^+

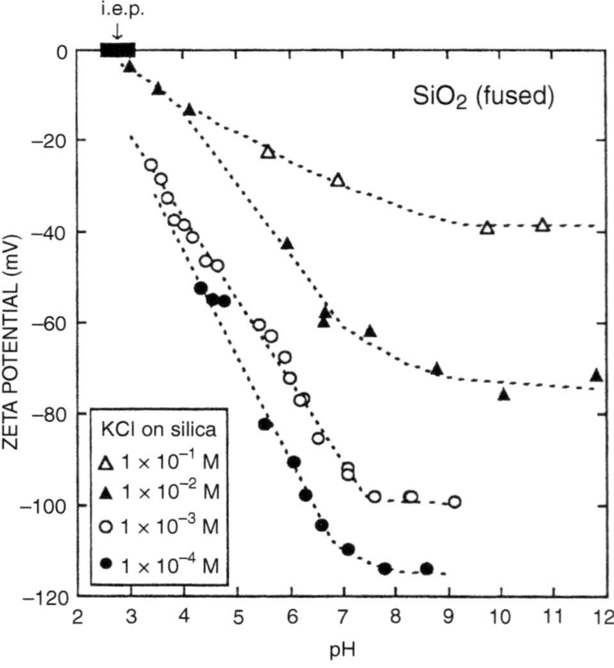

FIGURE 20.3 Plot of the zeta potential of a fused silica capillary versus pH in aqueous solutions of KCl as determined by the flat-plate streaming potential technique. (Adapted from Ref. [5].)

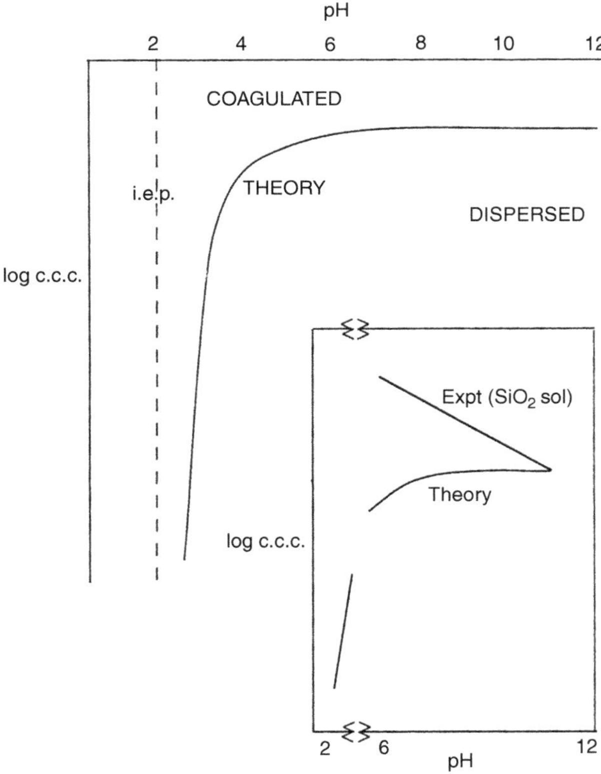

FIGURE 20.4 Schematic variation of the DLVO theoretical stability domain for a pH_{iep} colloid of pH 2 (critical coagulation concentration, c.c.c.; isoelectric point, i.e.p.). The insert shows this theoretical prediction compared to that observed for silica sols.

and OH^- are potential-determining ions [6,7]. The inset of Figure 20.4 illustrates this normal or theoretical variation with pH expected of the critical coagulation concentration, together with the form of the observed results for silica sols. In this inset the high-pH critical coagulation concentration values, for both theory and experiment, are shown coinciding for reasons to be highlighted subsequently. The challenge for researchers is to seek an understanding in terms of classic DLVO theory, and more recent formulations of stability theory of the increased stability with decreasing pH and the observed stability at the pH_{iep}.

ORIGINS OF COLLOID STABILITY

The intention in this section is to define key concepts and terms; detailed descriptions of the forces between colloidal particles are in standards texts [8,9].

ELECTROSTATIC REPULSION

The electrostatic energy of repulsion (V_R) between like charge surfaces arises from the overlap of the electrical double layers on the two particles. The range of interaction is of the order of the Debye–Hückel reciprocal length (κ^{-1}). For 1:1 electrolytes the value of κ^{-1} is approximately 10, 3, and 1.0 nm for concentrations of 10^{-3}, 10^{-2}, and 10^{-1} M, respectively. The magnitude of V_R at any given value of separation between two particle surfaces at given values of particle size, electrolyte concentration, and temperature is proportional to the surface potential (ψ_o) on each particle. A measure of ψ_o and hence the repulsive energy of interaction can be obtained from the experimentally accessible, measured electrokinetic or zeta potential (ζ) by equating ζ to ψ_d, the diffuse-layer potential.

The link between the surface charge and the potentials ψ_o, ψ_d, and ζ is not trivial in the case of the silica–water interface [10]. Unlike other oxides, the SiO_2 pH–charge relationship shows a characteristically small charge at small values of ΔpH for pH values exceeding pH_{iep} [11,12]. According to Healy and White [13], this relationship means that ψ_d ($\equiv \zeta$), within 2–3 units of the isoelectric point, will be expected to be smaller than that of oxides such as TiO_2, Al_2O_3, and Fe_2O_3. Such a relatively slow increase in potential is seen in some studies of the ζ–pH relationship of silica sols [3] but is less evident in studies of the zeta potential of macroscopic vitreous silica or quartz surfaces [5,14].

Again, silica, unlike other oxides, will not regulate charge during the approach of two surfaces [15] and may demonstrate a subtly different V_R–separation curve. Recent computations [16] suggest such charge-regulation effects will be small, but will be seen most especially around pH_{iep}, at which point the SiO_2 surface charge is low.

Consideration of the stability of silica sols, that is, the so-called "anomalous" stability, will, in later sections of this chapter, focus on 1:1 electrolyte concentrations of 0.1 *M* and greater. At this salt concentration, the range of the repulsive electrostatic forces between particles is small, and any subtle differences in the electrostatics between silica and other oxide sols in itself cannot provide the necessary repulsion to stabilize silica sols at such high electrolyte levels.

VAN DER WAALS ATTRACTION

The attractive van der Waals energy of interaction (V_A) for spheres in the 10- to 100-nm size range for silica sols discussed here varies as the inverse of the separation distance, and at any separation V_A is directly proportional to particle size. The Hamaker constant (A), which controls the magnitude of the variation of van der Waals attraction with particle radius (a) and separation (H_o) between surfaces, is for silica–water–silica not a large number. Further, the known hydration–polysilicic acid formation at silica–water interfaces will further reduce the overall Hamaker constant in the silica sol–water–silica sol system.

The simple attractive van der Waals term (V_A) and the simple electrostatic (double-layer) repulsive interaction energy (V_R) are now summed. Silica sol particles in the 10–100 nm radius size range, for salt concentrations of 0.1 *M* and greater and for pH values of 5 or more above the pH_{iep} of silica are now considered. The isoelectric point is taken as pH 2–3. Thus for 25-nm-radius particles the variation of total energy V_T ($=V_A + V_R$) with particle separation is attractive at all separations, and the sols are therefore expected to be unstable over this entire pH range >5 as shown by the theory line in Figure 20.4 and in the inset in Figure 20.4.

It is important to dispel any hope that adjustment of the electrostatic or van der Waals term will bring theory into line with experiment for silica sols. At about 0.1 *M* 1:1 electrolyte, reductions in the Hamaker constant to values near that of water itself or potentials of near −100 mV at, for example, pH 9–10 would be required. The experimental evidence [17] clearly points to a value more like −20 mV. Furthermore, several successful fits of double-layer theory to the observed charge and potential behavior of silicas are illustrated by the results (points) and theory (solid lines) in Figure 20.5, taken from the work of James [17] for a well-studied pyrogenic silica. James [17] invoked site dissociation, that is,

$$-\text{SiOH} \longrightarrow -\text{SiO}^- + \text{H}^+$$

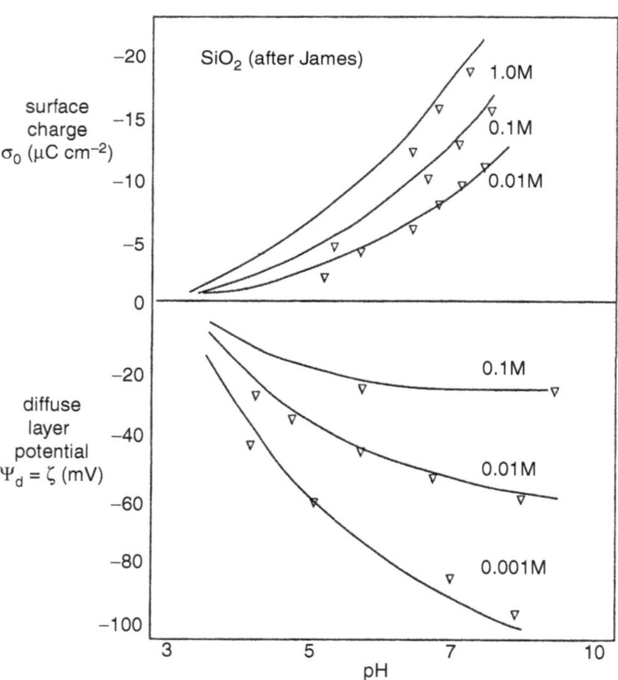

FIGURE 20.5 Comparison of theory and experiment in the variation of surface charge and electrokinetic potential for a pyrogenic silica. (Adapted from Ref. [17].)

and counterion binding, for example,

$$-\text{SiO}^- + \text{Na}^+ \longrightarrow -\text{SiO}^-\text{Na}^+$$

and fitted simultaneously (as shown in Figure 20.5) the observed charge and potential data. Similar fits to observed charge and potential data for many forms of silica, including silica sols, have been made, and only minor adjustment of the (sodium) ion binding constants are required.

Steric Stabilization and "Anomalous" Stability

An explanation of the "anomalous" stability of Iler's silica sols in terms of steric stabilization effects requires that oligomeric or polymeric silicate species are present at the silica–water interface and that steric repulsion results during overlap of such layers. This mechanism is appealing in that soluble silicates, usually sodium silicates, are universal dispersants of many electrostatic colloids. Again, well-hydrated silicas [2] and other colloids exposed to aqueous silicate [18] acquire high adsorption densities of aqueous silica.

To construct a model for "anomalous" stability, observations of silica sols at pH 11 are used. Allen and Matijévic [19] observed a critical coagulation concentration of ~0.15 *M* for Na^+ at pH 11.0. If they take this pH 11 sol to, for example, pH 8.0, it remains stable in salt concentrations much >0.15 *M*, and the critical coagulation concentration increases even further as the pH is reduced from pH 8 to pH 6.

At pH 11.0, if the silica sol is treated as a "normal" 30-nm-diameter silica (without adsorbed silicate layers), DLVO calculation generates a coagulation condition ($V_{Tmax} = 0$) at just under 2-nm surface separation. The electrostatic repulsion is inside this separation and cannot overcome the van der Waals attraction.

A key element of this postulate is that the silica sol is "normal" at pH 11.0; that is, it has no protective (steric) layer of silicate. Iler [2] and more recently Furlong et al. [18] noted the desorption (or lack of adsorption) of oligomeric and polymeric aqueous silica at pH values >10.5. Conversely, these workers observed adsorption of oligomeric and polymeric aqueous silica below pH 10.5.

Thus, as the pH is changed from 11 to 8 in 0.15 *M* salt, silicate steric layers are generated. For these layers to stabilize the sol under these conditions, they must extend farther than the $V_{Tmax} = 0$ condition of 2 nm. The hypothesis is therefore that a 2-nm steric silicate oligomer–polymer barrier protects a silica sol at pH 8.0 in 0.15 *M* salt.

To explore this very simple picture further, the ion-exchange behavior of silica sols must be considered. In their 1970 paper Allen and Matijévic [19] linked the observed coagulation behavior to the observed ion exchange. In this elegant study they introduced the concept of a "critical exchange curve," namely, the increase in exchange capacity with increasing pH at the critical coagulation concentration condition at each pH. These ion-exchange results can be used to consider what happens as salt is added to the pH 8, 0.15 *M* salt case that we have proposed is stabilized by a steric silicate polymer layer.

The stabilizing adsorbed polysilicate layer is the actual exchange volume or layer, and addition of salt above 0.15 *M* must lead to further binding of Na^+ to the stabilizing layer. It is therefore important to view the barrier at the surface as a M^+–polysilicic acid coating, which will increase in thickness as the pH is decreased because of decreased –SiOH ionization; conversely, it will decrease in thickness as the amount of bound cation increases (i.e., it will "exchange") in the language of Allen and Matijévic.

Indeed, the presence of increasing amounts of bound Na^+ must switch off the electrosteric contribution of the adsorbed polysilicate layer, and, for the pH 8.0 sol at its critical coagulation concentration, which is greater than 0.15 *M*, a secondary minimum well of $>2kT$ must open up as a result of the compression of the thickness and screening length of the adsorbed layer (*k*, Boltzmann constant; *T*, temperature).

Similar effects will operate at pH 6.0, initially at 0.15 *M* salt, where again the sol is electrosterically stabilized. As salt is added at pH 6, again the thickness and range of the electrosteric coating will eventually decrease to yield a $>2kT$ secondary minimum at the critical coagulation concentration, which is very much >0.15 *M*. These effects are summarized schematically in energy–distance curves in Figure 20.6.

The sequence is as follows:

1. pH 11; 0.15 *M* NaCl—unstable "normal" sol behavior
2. pH 8.0; 0.15 *M* NaCl—stable because of electrosteric coating
3. pH 8.0; >0.15 *M* NaCl—secondary minimum coagulation
4. pH 6.0; 0.15 *M* NaCl—stable; electrosteric barrier
5. pH 6.0; $\gg 0.15$ *M* NaCl—secondary minimum coagulation

The proposed mechanism has several interlocking postulates that will need experimental testing. They are as follows:

- At pH 11, oligomeric–polymeric silicate is absent from the sol particle surface
- At pH < 10.5, oligomeric–polymeric silicate coatings appear
- These oligomeric–polymeric coatings are electrosteric stabilization barriers

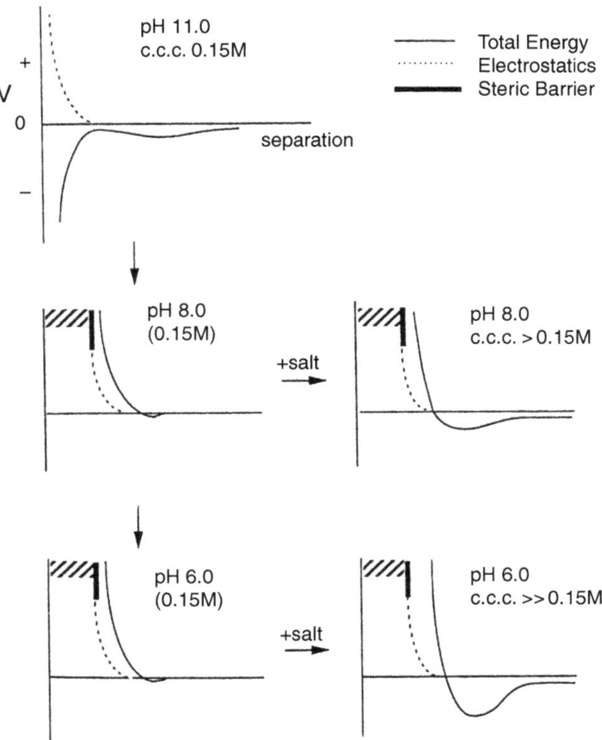

FIGURE 20.6 Total energy–distance curves for silica sols initially at pH 11 and 0.15 *M* 1:1 electrolyte, together with an illustration of the electrosteric barrier effect at pH values of 8.0 and 6.0 and electrolyte concentrations in excess of 0.15 *M*.

- The range of steric thickness plus electrostatic screening increases as the pH is reduced from 11 to 6 and decreases as the layers bind more Na^+ as the total coagulating salt concentration is increased above 0.15 *M*.

The final application of the hypothesis concerns the behavior of silica sols at, say, 0.15 *M* salt as the pH is reduced from 11 to 2. The energy–distance curves for such cases are shown schematically in Figure 20.7. The top starting curve illustrates the case of a sol initially at pH 11 but now at pH 2. The $V_{Tmax} = 0$ condition is maintained, and at pH 2 such a sol should be coagulated. Coating by oligomeric–polymeric silica generates a thick (low degree of ionization) coating at pH 2 that gives significant steric, rather than electrosteric, protection. Addition of salt fails to diminish this coating thickness at the low cation-exchange state or low cation-exchange capacity state of pH 2. Secondary minimum effects would be pushed out to unrealistic values of greater than 5 nm.

Hydration Effects

In this discussion of silica sols at pH 11, 8, 6, and 2, the cation effects or critical coagulation concentration effects are

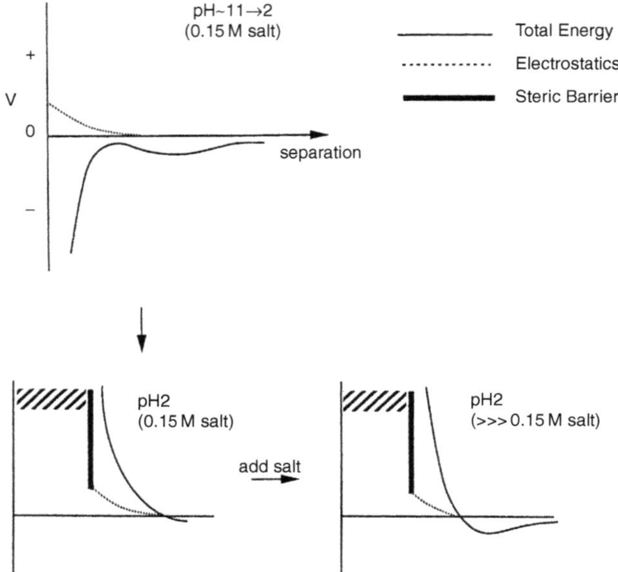

FIGURE 20.7 Total energy–distance curves for silica sols initially at pH 11 and 0.15 *M* 1:1 electrolyte, together with an illustration of the steric barrier effect at pH 2 and electrolyte concentrations very much in excess of 0.15 *M*.

focused on Na^+. As indicated in Figure 20.1 and Figure 20.2, the silica sol coagulation behavior, with respect to the sequence Cs^+ through Li^+, is not simple. Thus, in general terms, Li^+ behaves in a manner similar to Na^+ in that the critical coagulation concentration increases as the pH moves from 11 to 6; Li^+ contains sols more "anomalously" stable than those with Na^+ as counterion. The behavior with Cs^+ is "normal"; that is, the critical coagulation concentration decreases slightly as the pH moves from 11 to 8.5, but again increases as the pH moves from 8.5 to 6.0.

Viewed in terms of the electrosteric mechanisms outlined in Figure 20.6 and Figure 20.7, the 1:1 electrolyte cation effects indicate that relative to Na^+, Li^+ enhances any stability imposed by the barrier of polysilicate plus bound cation, whereas Cs^+ diminishes such effects. This observation is consistent with a "hydration stabilization" (6) mechanisms. Thus, the more strongly hydrated the cations in the polysilicate exchange layer,

- The lower the binding energy,
- The lower the extent of exchange,
- The thicker the layer at any given pH and salt molarity, and
- The more effective the electrosteric barrier.

CONCLUSIONS

The origin of the coagulation behavior of Iler's "silica sols" is far from understood. The electrosteric barrier model herein proposed is designed to stimulate new experimental initiatives in the study of colloidal silica sol particles and their surface structure. The adsorbed steric layer, impregnated with bound (exchanged) cations, at the surface of 1–100-nm-diameter silica sol particles has the general properties needed to understand the "anomalous" coagulation behavior. The details await experimental and theoretical input.

ACKNOWLEDGMENTS

This work was supported by an Australian Research Council Program grant. Participation in the ACS R. K. Iler Memorial Symposium in September 1990 was made possible by a generous grant from the Corporate Colloid Group of ICI (United Kingdom).

REFERENCES

1. Iler, R. K.; *The Colloid Chemistry of Silica and Silicates;* Cornell University Press: Ithaca, NY, 1955.
2. Iler, R. K.; *The Chemistry of Silica;* Wiley: New York, 1979.
3. Allen, L. H.; Matijévic, E. *J. Colloid Interface Sci.* **1969**, *31*, 287.
4. Depasse, J.; Watillon, A. *J. Colloid Interface Sci.* **1970**, *33*, 430.
5. Scales, P. J.; Grieser, F.; Healy, T. W.; White, T. W.; Chan, D. Y. C. *Langmuir*, in press.
6. Healy, T. W.; Homola, A.; James, R. O.; Hunter, R. J. *Faraday Discuss. Chem. Soc.* **1978**, *65*, 156.
7. Wiese, G. R.; Healy, T. W. *J. Colloid Interface Sci.* **1975**, *51*, 427.
8. Hunter, R. J. *Foundations of Colloid Science;* Oxford University Press: New York, 1989; Vol. 1.
9. Russel, W. B.; Saville, D. A.; Schowalter, W. R. *Colloidal Dispersions;* Cambridge University Press: Cambridge, UK, 1990.
10. Yates, D. E.; Levine, S.; Healy, T. W. *J. Chem. Soc. Faraday Tarns. 1* **1974**, *70*, 1807.
11. Tadros, Th. F.; Lyklema, J. *J. Electroanal. Chem. Interfacial Electrochem.* **1968**, *17*, 267.
12. Yates, D. E.; Healy, T. W. *J. Colloid Interface Sci.* **1976**, *55*, 9.
13. Healy, T. W.; White, L. R. *Adv. Colloid Interface Sci.* **1978**, *9*, 303.
14. Wiese, G. R.; James, R. O.; Healy, T. W. *Faraday Discuss. Chem. Soc.* **1971**, *52*, 302.
15. Healy, T. W.; Chan, D. Y. C.; White, L. R. *Pure Appl. Chem.* **1980**, *52*, 1207.
16. Metcalfe, I. M.; Healy, T. W. *Faraday Discuss. Chem. Soc.*, in press.
17. James, R. O. *Adv. Ceram.* **1987**, *21*, 349.
18. Furlong, D. N.; Freeman, P. A.; Lau, A. C. M. *J. Colloid Interface Sci.* **1981**, *80*, 21.
19. Allen, L. H.; Matijévic, E. *J. Colloid Interface Sci.* **1970**, *33*, 420.

21 Stabilization Against Particle Growth

Paul C. Yates
DuPont Company (retired)

Polysilicates or sols of very small particles are stabilized by having sufficient alkali in the system. Yates (41) proposed that alkali-stabilized sols are stable not only against gelling, but also in the thermodynamic sense. He pointed out that there are thermodynamic factors that prevent the spontaneous growth of particles or their aggregation and stabilize the high interfacial solid–liquid interface in the silica–water system. The main factor that counterbalances the free energy change involved in loss of surface area in the silica–water system is the strong adsorption of the continuous liquid phase or of stabilizing counterions or of other adsorbed species at the surface of the dispersed phase, thus changing the free energy of the interface.

The effect can most easily be understood by considering the changes in free energy that occur during the following hypothetical, reversible, isothermal steps in the three-component system SiO_2–H_2O–NaOH starting with an alkali-stabilized sol:

1. Desorb NaOH from the surface of the colloidal particles: ΔF_1.
2. Desorb water from the surface of the colloid; that is, the free energy of wetting the surface: ΔF_2.
3. Decrease the surface area of the colloid to the minimum possible geometrical area allowed by its density; that is, the free energy of the solid colloid: ΔF_3.
4. Return solvent to the surface of the collapsed colloid of minimum area: ΔF_4.
5. Return the NaOH to the system: ΔF_5.

By definition,

$$\Delta F_1 + \Delta F_5 = \text{free energy of adsorbed NaOH on colloid}$$

$$\Delta F_2 + \Delta F_3 + \Delta F_4 = \text{interfacial free energy of colloid in a binary } SiO_2\text{–}H_2O \text{ system}$$

If

$$\Delta F_1 + \Delta F_5 = \Delta F_2 + \Delta F_3 + \Delta F_4 = 0$$

then the initial colloid system of three components is at thermodynamic equilibrium. In other words, if the free energy of adsorption of sodium hydroxide on the surface of the colloid equals the free surface energy of the system in the absence of NaOH, then the sol is thermodynamically stable.

Heretofore all colloid systems such as sols have been considered to be thermodynamically unstable.

In a silica sol the interfacial free energy of amorphous silica versus water is of the order of 50 ergs cm^{-2}. The decrease in free surface energy accompanying the loss of surface, assuming about 8 silicon atoms nm^{-2}, amounts to 900 cal g-$mole^{-1}$ of surface silicon atoms. If the colloidal particles are stabilized thermodynamically, the counterbalancing free energy term must be of about this magnitude. This stabilizing energy comes from the adsorption of OH^- and the counter Na^+ ions on the colloid surface.

Yates calculated the amount of alkali necessary to stabilize silica particles against growth as follows.

The free energy change connected with the ionization of surface groups depends on the acid dissociation constant which, in turn, varies with the salt concentration and the degree to which the ionization of surface groups has already occurred. The free energy change under constant conditions will be

$$\Delta F = \alpha(-RT \ln K') \tag{1}$$

where α is the fraction of surface groups ionized and K' is the instantaneous value of the dissociation constant at that particular value of α, other conditions being fixed, and assuming a surface containing one mole of silanol groups.

As will be discussed in connection with the charge on particles, Yates derived a formula for the value of K' as a function of pH, sodium ion concentration in solution, and α, the degree of dissociation of the surface.

$$\text{pH} = \text{p}K - n \log \frac{1-\alpha}{\alpha} - 0.74 \log(A\,\text{Na}^+) \tag{2}$$

where K is the overall dissociation constant and (A Na^+) is the activity of the sodium ion times the sodium ion normality.

From titration data at a fixed concentration of Na^+ ion, a plot of $-\log[(1-\alpha)/\alpha]$ against pH gives a line of slope n and intercept pK. The value of n from the experimental data is 3.47.

The corresponding expression involving K' is

$$\mathrm{pH} = \mathrm{p}K' - \log\frac{1-\alpha}{\alpha} \tag{3}$$

where

$$\mathrm{p}K' = \mathrm{p}K - (n-1)\log\frac{1-\alpha}{\alpha} - 0.74\log(A\,\mathrm{Na}^+) \tag{4}$$

From equations 1 and 4,

$$\Delta F = 2.3\alpha RT\left[-\log K - (n-1)\log\frac{1-\alpha}{\alpha} - 0.74\log(A\mathrm{Na}^+)\right] \tag{5}$$

It can be shown that,

$$\alpha = \frac{2430}{rS} \tag{6}$$

where r is the weight ratio of SiO_2:Na_2O in the system and S is the specific surface area of the silica in square meters per gram; Yates assumed the density of silica to be 2.3 g cm^{-3}.

Substituting the numerical values for K, α and n gives

$$\Delta F = \frac{3.38\times10^6}{rS}\left[12.08 - 2.47\log\frac{rS-2430}{2430} - 0.74\log(A\,\mathrm{Na}^+)\right] \tag{7}$$

which expresses the free energy change associated with moles of surface groups at any level of ionization or salt content.

In order for a silica sol to lower its specific surface free energy by particle growth or aggregation to form a sol of lower surface area, it is necessary to reverse the ionization and return the adsorbed ions to the intermicellar liquid. According to Yates the free energy change will be γS, where γ is the value of the specific interfacial free energy between the silica surface and water. The latter is a function of particle size and thus of S.

Let the free energy change occurring when particles of diameter D lose ΔS cm^2 of surface be

$$\Delta F_D = -\gamma_D\,\Delta S \tag{8}$$

A similar expression for a flat surface is

$$\Delta F_0 = -\gamma_0\,\Delta S \tag{9}$$

$$\Delta F_D - \Delta F_0 = (\gamma_0 - \gamma_D)\Delta S \tag{10}$$

This is the difference in free energy between a colloid of particle diameter D and the same colloid if it had the free energy of a flat surface. This would be $-RT\ln(S_D/S_0)$, where S_D and S_0 are the corresponding solubilities of the curved and flat surfaces of silica.

Yates used the equation published by Iler in 1955 (8):

$$\ln\frac{S_D}{S_0} = \frac{3.8}{D} \tag{11}$$

where D is in millimicrons. This was based on an interfacial energy of 80 mg cm^{-2}. However, as shown in Chapter 1 (see also Figures 1.10*a* and 3.32), the value is more likely to be about 50 mg cm^{-2} so that the equation should probably be

$$\ln\frac{S_D}{S_0} = \frac{2.4}{D}$$

Then

$$\Delta F_D - \Delta F_0 = -RT\frac{2.4}{D} \tag{12}$$

From equations 8 and 10,

$$(\gamma_0 - \gamma_D)\Delta S = -2.4\frac{RT}{D} \tag{13}$$

$$\gamma_D = \gamma_0 + 3.8\frac{RT}{D\Delta S} \tag{14}$$

$$\Delta F_D = -\gamma_0^1\Delta S + 2.4\frac{RT}{D} \tag{15}$$

However, D can be related to S by taking into account the density of silica, assumed by Yates to be 2.3 g cm^{-3}:

$$D = \frac{2554}{S} \tag{16}$$

Then

$$-\Delta F_D = \gamma_0\Delta S + \frac{2.4\,RT}{2554}S \tag{17}$$

Since at 25°C, $T = 298$°K, $R = 1.987$ cal deg^{-1} $mole^{-1}$ and

$$-\Delta F_D = \gamma_0\Delta S + 0.56S \tag{18}$$

Assuming $\gamma_0 = 50$ mg cm^{-2}, ΔS equals the area of 6×10^{23} surface silicon atoms. Assuming 8 silicon atoms per square nanometer, the value of S is 7.5×10^8 cm^2. Changing ergs to calories gives

$$-\Delta F_D = 50(2.39 \times 10^{-8})(7.5 \times 10^8) + 0.56S$$

$$-\Delta F_D = 896 + 0.56S \text{ cal} \tag{19}$$

Then $-\Delta F_D$ from equation 19 must equal ΔF from equation 7:

$$896 + 0.56S = \frac{3.38 \times 10^6}{rS} 12.08 - 2.47 \log \frac{rS - 2430}{2430} - 0.74 \log(A\,\text{Na}^+) \tag{20}$$

Rearranging gives

$$r = \frac{3.38 \times 10^6 \{12.08 - 2.47 \log[(rS - 2430)/2430] - 0.74 \log(A\,\text{Na}^+)\}}{S(896 + 0.56S)} \tag{21}$$

In the absence of foreign salts, the concentration of Na^+ can be calculated simply from the concentration of silica, C_s, and the ratio r since

$$\frac{[SiO_2]}{[Na_2O]} = r$$

$$[Na_2O] = \frac{[SiO_2]}{r}$$

Since the sodium concentration in stable sols seldom exceeds 0.1 N, we may assume the activity is about unity:

$$\text{Na}^+ = \frac{2C}{r}$$

where C is the molar concentration of SiO_2 in the system:

$$r = \frac{3.38 \times 10^6 \{12.08 - 2.47 \log[(rS - 2430)/2430] - 0.74 \log(2C/r)\}}{S(896 + 0.56S)} \tag{22}$$

For specific examples, assuming values for S and C, the following values of r were calculated from the equation.

For a 15% sol of 4.5 nm particles (600 m^2 g^{-1}) the calculated SiO_2:Na_2O ratio is 47:1. For a 39% sol of 13.4 nm particles (200 m^2 g^{-1}) the calculated ratio is 170:1. (Yates's earlier calculations gave 32:1 and 115:1.)

In commercial sols corresponding ratios of about 25:1 and 100:1 are actually used for these particle sizes. As shown in Chapter 2, particles of about 16 nm diameter stabilized at 100:1 ratio grow very little in 20 yr. However, particles 4–5 nm in diameter in some cases undergo a slow 10–20% increase in diameter in a year or so even though stabilized at 25:1 ratio. However, lot-to-lot variations can occur, probably because of variations in the initial particle size distribution.

The above formulas probably do not apply to particles much under 7–8 nm in diameter since the increasing solubility of the particles involves the presence of appreciable concentrations of silicate ions which must be taken into account in the stabilized systems. This would probably require more stabilizing alkali than the present equations indicate, that is, lower values of r.

Part 3

Surface Chemistry of Silica

Michael L. Hair
Xerox Research Centre of Canada

The history of investigation of the molecular surface chemistry of silica can be divided into three time periods. The period between 1960 and 1970 saw the introduction of IR spectroscopy and the use of chemical reactivity to follow reactions on a silica surface at a molecular level. The period between 1970 and 1980 was less active, but the advances in computer instrumentation that occurred in that decade (in particular Fourier transform IR and NMR spectroscopy) have led to the current revitalization of the subject. The chapters that follow will question some old interpretations and will reinterpret some old ideas.

In the early studies of the silica surface, one of the major enthusiasms for the work was not the silica surface itself but the fact that it was the means to an end [1]. For the first time it became possible to "view" the molecular interactions that were occurring at a surface by using readily available instrumentation. Provided that the surface area of the material was sufficiently high, silica and its surface derivatives could be directly observed by using transmission IR spectroscopy. Rigid materials were easily prepared from porous glasses, and both porous silicas and fumed silicas could be pressed into thin, self-supporting discs that could be treated in a vacuum while in the path of an IR beam. The discs were transparent from 4000 cm^{-1} to around 1300 cm^{-1}, and bands were observed that could be identified as being due to hydroxyl groups, which were an integral part of the surface. Both chemical and physical interactions could be followed by monitoring the disappearance of these hydroxyl groups and the appearance of bands due to the new surface structures. It became clear that the surface chemistry of silica was dominated by these surface hydroxyl groups, and investigations of the structure and reactivity of these groups dominated silica research for more than 30 yr. The basic results are shown in Figure 1. Several groups can be identified in the OH stretching region: molecular water, inaccessible SiOH groups, isolated and geminal SiOH groups (which are indistinguishable in the IR, but were first identified by chemical reactivity and confirmed by the more recent NMR studies). Also now identified is the highly active SiOH group at 3720 cm^{-1} — a result of modern FTIR spectroscopy. The number and distribution of these surface hydroxyl species is dependent upon the pretreatment of the silica and, therefore, so is the surface chemistry.

In much of the definitive IR work on the silica surface researchers have chosen to work with fumed silica. This choice was mainly for experimental reasons (the ease of preparing the self-supporting disk), but also because it minimizes another important issue — the nature of porous silica surface. A major advance in the past decade has been in the controlled synthesis of many silica polymorphs with variable pore size. Accordingly, the past decade has seen a renewed enthusiasm for the study of porous silicas, their reaction with chemical probes, and H_2–D_2 exchange reactions. An increasing body of evidence indicates that the basic silica structure is similar in both cases, but that accessibility and derivatization of the porous silicas can sterically alter the process and the kinetics of the reactions.

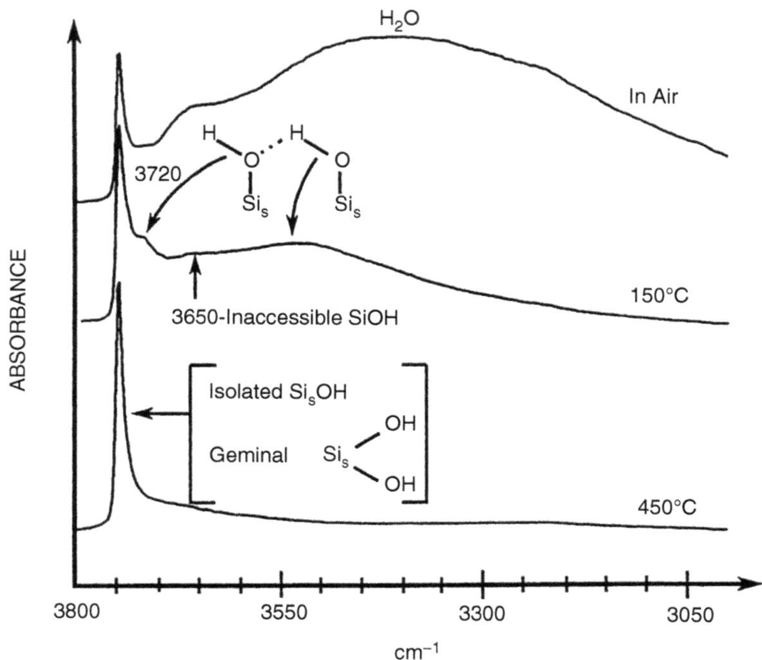

FIGURE 1. Transmission IR spectra of silica in air and at 150 and 450°C. (Adapted from reference 2.)

The most important development in the past decade has undoubtably been the application of NMR spectroscopy to the study of the surface silanol groups. The freely vibrating Si_3—OH groups occur at 3747 cm^{-1} in the IR spectrum, but chemical reactions have shown that this band is due to both single and geminal OH groups (i.e., two hydroxyls attached to the same surface silicon atom). Magic-angle spinning techniques, particularly when coupled with cross-polarization, enable distinction between the isolated and geminal groups. About 15% of the total number of the surface silanol groups are geminal — a number that is much lower than predicted by the early literature. The new technique is particularly useful when used in conjunction with IR measurements. Experimental application of the NMR technique at variable temperatures and under evacuated conditions is difficult, but the progress is excellent. One of the significant pictures arising from both NMR and IR studies is possible reinterpretation of the nature of the silica surface. Earlier pictures considered uniform distributions of hydroxyl groups over the total surface.

Derivatization of the silica surface is of importance particularly in the preparation of materials for chromatographic supports and for the support of biologically active molecules such as proteins and enzymes. The active groups are usually attached to the silica surface by reaction with the surface hydroxyl groups via an intermediate coupling agent, usually a silane. The reactions of these coupling agents with the surface hydroxyl groups have consequently been studied for many years, and much of the early work defined the stoichiometry and kinetics of such reactions. Despite this large body of information, it is still often unclear whether the reaction has occurred with the surface hydroxyl group itself (and is therefore genuinely chemically bonded), or whether water that is physically absorbed on the surface has caused polymerization of the silane (i.e., the polymer then participates in the subsequent reaction with the biological materials). The role of water is elusive and needs investigation, particularly in the nonaqueous colloidal systems where such reactions are normally carried out. Traces of water in such systems affect not only the chemical reactivity of the surface and the nature of the chemical reactions, but also the zeta potential and presumably the stability of these particles in suspensions.

Other "impurities" such as Al, B, and P strongly affect the acidity and reactivity of the hydroxyl groups. Clearly this in turn grossly affects adsorption, reactivity, and chromatographic performance, but few controlled studies have been carried out. This area will become increasingly important in bridging the gaps between our knowledge of silica, alumina, and the very important alumino silicates.

Although research emphasis has clearly been on the surface hydroxyl groups, it would be remiss not to mention the active sites that are produced by high-temperature treatments of silica or its organic derivatives. Although small in number, their high reactivity (especially under evaluated conditions) must be considered [3,4].

In addition to the instrumental developments in IR spectroscopy, Morrow and his co-workers have now shwon that it is possible to prepare ultra-thin films of silica [5]. This step enables observation of the IR spectrum of silica and its reactions throughout the IR region from 4000 to 400 cm^{-1}. This region contained many bending modes that previously could not be observed, and, for instance, the experimentalist is now able to view the formation of $Si_s—O—Si_s$ bridges on the surface. A very recent application of this technique is seen in the reaction of chlorosilanes with the subsequent interaction with water [6].

REFERENCES

1. Hair, M.L. *Infrared Spectroscopy in Surface Chemistry*; Dekker: New York, 1967.
2. Morrow, B.A. In *Spectroscopic Analysis of Heterogeneous Catalysts, Part A: Methods of Surface Analysis*; Fierro, J.L.G., Ed.; Elsevier: Amsterdam, Netherlands, 1990; Chapter 3.
3. Morrow, B.A.; Cody, I.A. *J. Phys. Chem.* **1976**, *80*, 1995, 1998.
4. Low, M.J.D.; McNelis, E. *J. Catal.* **1986**, *100(2)*, 328.
5. Morrow, B.A.; Tripp, C.P.; McFarlane, R. A. *J. Chem. Soc. Chem. Commun.* **1984**, 1292.
6. Tripp, C.P.; Hair, M.L. *Langmuir* **1991**, *7*, 923.

22 The Surface Chemistry of Silica – The Zhuravlev Model

L.T. Zhuravlev
Russian Academy of Sciences, Institute of Physical Chemistry

CONTENTS

The study of the silica-water system is important both from the theoretical point of view and for practical applications [1]. In this connection, an investigation of the so-called structurally bound water [2] in dispersed amorphous silica is of interest. This term describes OH groups that are bound via the valence bond to Si atoms on the silica surface (hydroxyl coverage) and, in some cases, to Si atoms inside the particle of silica itself. Numerous spectral and chemical data unambiguously confirm the presence of silanol (SiOH) groups on silica surface. The past decades saw a rapid growth in the science and technology domains that deal with the production and utilization of various colloid and microheterogeneous forms of silica developing high surface areas: sols, gels and powders. The properties of such silicas are determined in the first place by: (i) the chemical activity of the surface; this activity depends on the concentration, the distribution and the type of hydroxyl groups, and on the presence of siloxane SiOSi bridges; and (ii) the porous structure of the silica. Various problems related to silica surface characteristics are encountered in different areas of science and technology: physics, chemistry and physical chemistry, agriculture, soil science, biology and medicine, electrical energetics, oil processing industry, metallurgical and mining industries, some fields of geology, etc. Different types of silica are widely used as selective adsorbents, active phase carriers in catalysis, polymer fillers, thickening agents for dispersed media, binding agents in molding, adsorbents and supports in gas and liquid-chromatography, etc. Chemical modification of dispersed silica makes it possible to regulate its adsorption properties and the technological characteristics of composite materials. Presently, the use of SiO_2 is increasing for the manufacturing of high quality materials (in microelectronics, optics, fiber optics).

Since 1950, many reviews have appeared on surface chemistry of amorphous silica. Accordingly, is there a need for yet another review in this field? The author of the present paper has carried out numerous experimental studies on the subject in question, systematized such important characteristics as the concentration and the distribution of different types of silanol groups, established the energetic heterogeneity of the surface in a wide temperature range of pretreatment and investigated the characteristics of bound water inside the SiO_2 particles. Besides, he made a careful study of the structural characteristics of many different silica samples. On the basis of these researches, the author was able to construct an original physico-chemical model [3–5], describing the surface properties of amorphous silica (referred in literature as the Zhuravlev model). Therefore, it seems pertinent to compare the results obtained by the author with those reported in literature.

In this model of an amorphous silica surface, the determining factor is the presence of silanol groups. The concentration of these groups depends on the conditions of thermal pretreatment in vacuum (or on other types of pretreatment). The decrease in the surface density of OH groups is accompanied by a loss of hydrophilicity. When increase of the concentration of SiOSi bridges results in an increased of the hydrophobicity of silica. Also it is necessary to take into account possible changes occurring simultaneously in the structure of the surface and of the skeleton of the silica matrix. These factors determine the starting conditions that are necessary for working out the model for the amorphous silica surface. To avoid the introduction of complicating factors, such as the possible effect of impurities on the silica surface properties, different active sites, and so on, we shall not consider them at this stage.

The various methods of deuterium exchange (MDE), developed by the author [3–5], were used for a quantitative analysis of the OH groups on the silica surface. The advantage of MDE is that, under certain conditions, the isotopic exchange with D_2O is limited to the surface and does not involve structurally bound water inside silica. The processes of the dehydration (the removal of physically adsorbed water from the surface) and the dehydroxylation (the removal of OH groups from the surface) have been investigated using the mass spectrometric thermal analysis method (MTA) in conjunction with a temperature programmed desorption technique (TPD). The MTA-TPD method, worked out by the author [3–5], makes it possible to obtain the kinetic parameters of the processes. About 150 samples of purified amorphous silica samples of different types — silica gels, aerosils, aerosilogels and porous glasses — were investigated. The silanol number, that is, the number of OH groups per unit surface area, α_{OH}, was determined for the hydroxylated state of the surface. Detailed structural characterizations of the silica samples, before and after thermal treatments, were also performed.

In order to investigate the water removal from the silica surface, in the absence of undesirable effects (diffusive retardation in fine pores, readsorption, evolution of bound water from inside the sample at higher temperatures, effect of impurities and unstable structure of SiO_2 skeleton), we prepared a standard S-79 sample [4,5]. This SiO_2 sample met the following requirements: (i) high purity (very low concentration of impurities on the surface and inside the sample); (ii) total absence of bound water in the bulk of the sample; (iii) total absence of fine pores (micro- and ultramicropores), with only wide pores (51 nm in diameter); (iv) stabilized structure of the SiO_2 skeleton obtained by multiple heating and cooling of the sample; (v) presence of silanol groups only at the surface of the sample, hydroxylated to a maximum degree on the initial sample.

The comparison of the results obtained by the author [3–5] with those reported in the literature covers the following topics: (i) limiting temperature for removing physically adsorbed water from the hydroxylated surface; (ii) completely hydroxylated state of the surface; (iii) dehydroxylation of the surface; (iv) rehydroxylation of the surface; (v) structurally bound water inside silica particles. Owing to the lack of space in this abstract, references are made only to those results which closely correspond to our data. A complete comparison is given in our review article.

LIMITING TEMPERATURE, T_l, FOR REMOVING PHYSICALLY ADSORBED WATER FROM THE HYDROXYLATED SURFACE OF AMORPHOUS SILICA

To elucidate the nature of the hydroxyl coverage and to quantitatively determine the concentration of OH groups on the silica surface, such groups must be differentiated from the molecularly adsorbed H_2O and from bound water inside the silica particles. The separation of the two processes — dehydration and dehydroxylation — is a difficult task.

Although many investigations have been carried out, using various methods, for the determination of the value of T_l, there is no general agreement between the values obtained so far. T_l values were obtained in three temperature intervals: (i) pretreatment of SiO_2 in vacuum at temperatures close to room temperature; (ii) pretreatment from 100 to 200°C; (iii) there are statements [6,7], that the presence of coordinated molecular water on the SiO_2 surface is observed at higher temperatures from 400 up to ~650°C. We carried T_l measurements by MTA-TPD for the S-79 sample that has been subjected to pretreatment in vacuum under different conditions, at temperatures $\leq$ 200°C. A detailed analysis of the differential kinetic curves passing through a maximum showed [4,5] that: (i) the major amount of the physically adsorbed water, including polymolecular adsorption, is removed from the hydroxylated surface at room temperature; (ii) a small amount of molecularly adsorbed water, within the limits of a monolayer, remains on the hydroxylated surface up to ~200°C; (iii) the exact magnitude of T_l that corresponds to complete dehydration and the beginning of dehydroxylation is $T_l = 190 \pm 10$°C; (iv) at 190°C, the kinetic parameters calculated from the experimental data show an abrupt increase: the activation energy of water desorption, E_D, changed from ~10 to ~17 kcal/mole, and the kinetic order of the thermal desorption reaction changed from 1 to 2. Our value of $T_l = 190$°C is close to that 180°C obtained by Young and Bursh [8,9] on the basis of a differential thermal analysis and of changes in the heat of wetting of the surface.

The sharp increase we observed for the kinetic parameters, E_D and ν, at $T_l = 190$°C was predicted by Rebinder [10]. He said that there would be an inevitable abrupt increase in the isotherm free energy, $-\Delta G$, which characterizes, as does E_D, the binding strength of water on the surface of dispersed materials. This increase should occur in the process of drying of such materials during the transition from the completion of dehydration to the beginning of the removal of chemically bound water (i.e., OH groups on the SiO_2 surface).

COMPLETELY HYDROXYLATED STATE OF THE SILICA SURFACE

For the development of a model describing the silica surface, it is first of all necessary to have reliable quantitative data on the concentration of OH groups, as a function of the preliminary thermal treatment in vacuum of SiO_2 samples. This is particularly important for the fully hydroxylated state of the silica surface. In 1957, De Boer and Vleeskens compared silanol numbers for completely hydroxylated surfaces of amorphous silica and crystalline modifications of silica (β-cristobalite, β-tridymite) whose density is close to that of amorphous SiO_2 [11]. They pointed out that since β-cristobalite crystallizes in octahedra, the silanol number should be calculated from the (1.1.1) plane of the octahedral face. The calculated concentration of OH groups was $\alpha_{OH} = 4.55\ OH/nm^2$. The theoretical (crystallographic) value of silanol numbers for crystalline modifications of silica was found to be $\alpha_{OH} = 4.6–4.9\ OH/nm^2$. In their thermogravimetric investigations of wide-pore amorphous SiO_2, subjected to a pretreatment at ~600°C and rehydroxylation in an autoclave or heat treatment and repeated rewetting, followed by the removal at 120°C of the physically adsorbed water, the authors found that $\alpha_{OH} = 4.5–5.0\ OH/nm^2$ [11]. This experimental value is practically the same as the theoretical one. So, De Boer and Vleeskens were the first to obtain a reliable value for α_{OH}. However, owing to the insufficiently sensitive thermogravimetric method they used, the authors reached the erroneous conclusion that, for the initial hydrated samples of amorphous silica which were not subjected to pretreatment at ~600°C and then rehydroxylated, $\alpha_{OH} = 6–8\ OH/nm^2$. Such a conclusion fails to take into account the possible presence of structural water inside the silica particles and attributes the total weight loss of the sample, after high-temperature annealing, only to the loss of OH groups on the silica surface. In 1958, Zhuravlev and Kiselev obtained, for the first time by the MDE method, the silanol number $\alpha_{OH} = 5.2\ OH/nm^2$ for the initial hydrated SiO_2 sample (the wide-pore silica gel) which contained OH groups on the surface as well as structural water inside the SiO_2 particles. Since, Zhuravlev and co-workers determined the values of α_{OH} of about 150 hydroxylated SiO_2 samples which were prepared by different methods and had different structural characteristics [3–5]. When measuring α_{OH}, one of the most important factor is the specific surface area, S_{kr}, as determined by the BET method, using low temperature adsorption of krypton [12]. This inert adsorbent is chosen because the adsorption of Kr atoms on the silica surface is nonspecific, that is, insensitive to changes in the concentration of OH groups on the silica surface. Thus, the silanol number α_{OH} was determined on the surface of pores that are accessible to the Kr atoms. In those cases, where the biporous samples of SiO_2 contained very narrow pores (ultramicropores) in addition to wide pores (mesopores), the samples were considered to be widepore ones, and the very narrow pores were excluded for the determination of α_{OH}. Ultramicropores are comparable in diameter with water molecules, and therefore only water molecules can penetrate them. The OH groups in these very narrow pores were classified, not as surface silanol groups, but as bound water molecules inside the silica particles.

The values of the silanol number, α_{OH}, of 100 silica samples, with a completely hydroxylated surface, were established [3–5]. The average silanol number (arithmetical mean) was found to be $\alpha_{OH,av} = 4.9\ OH/nm^2$. Calculations by the least-squares method yielded $\alpha_{OH,av} = 4.6\ OH/nm^2$. These values are in agreement with those reported by De Boer and Vleeskens [11] as well as with results reported by other researchers. To sum up, the magnitude of the silanol number, which is independent of the origin and structural characteristics of amorphous silicas is considered to be a physicochemical constant. The results fully confirmed the idea predicted earlier by Kiselev and co-workers [13,14] on the constancy of the silanol number for a completely hydroxylated silica surface. This constant now has a numerical value: $\alpha_{OH,av} = 4.6 \pm 0.5\ OH/nm^2$ [3–5] and is known in literature as the Kiselev-Zhuravlev constant.

DEHYDROXYLATION OF THE SILICA SURFACE

We used the MDE method to determine the average value of the silanol number α_{OH} of samples heat treated in vacuum (up to 1000–1100°C), starting from the maximum state of hydroxylation of the surface. Experimental results obtained on 16 samples of amorphous silica are shown in Figure 6 of Ref. 5. The samples differed from one another in the method of their synthesis and in their structural characteristics (the specific surface area of the samples varied from 11 to 905 m^2/g, and their porosity also varied within a wide range). Despite all these differences, the value of α_{OH} at a given treatment temperature is similar for all the samples, and the decrease in the value of α_{OH} under similar heating conditions also follows approximately the same pattern. The average values of α_{OH} (and the corresponding degree of surface coverage with OH groups, θ_{OH}) are presented in Table 22.1.

As seen from these data, the values of α_{OH} decrease considerably in the range from 200 to 400–450°C. Between 400 and 1100°C, this decrease becomes notably smaller. Correspondingly, the value θ_{OH} in the first steep section of the plot decreases from 1 to about 0.5, and in the second, more flat section, it drops from 0.5 to very low values approaching zero. From these results, it

TABLE 22.1
Silanol Types and Degree of Coverage with Total OH Groups, as a Function of Treatment Temperature (the Zhuravlev Model)

Temp. OH* (°C)	Total OH (OH/nm^2)	Degree of coverage, θ_{OH}	Isolated OH (OH/nm^2)	Vicinal OH (OH/nm^2)	Geminal (OH/nm^2)
180–200	4.60	1.00	1.15	2.85	0.60
300	3.55	0.77	1.65	1.40	0.50
400	2.35	0.50	2.05	0	0.30
500	1.80	0.40	1.55	0	0.25
600	1.50	0.33	1.30	0	0.20
700	1.15	0.25	0.90	0	0.25
800	0.70	0.15	0.60	0	0.10

*Data in column 6 — the correction for geminal OH groups [16,17].

becomes possible to assess the most probable values of α_{OH} and θ_{OH} within a wide temperature range. Many researches use in their studies our values of $\alpha_{OH} = f(T°C)$ and $\theta_{OH} = f(T°C)$ as a set of physicochemical constants (at fixed temperatures), and these data are widely cited in literature. The types of silanol numbers — the silanol number of free, isolated OH groups $\alpha_{OH,free}$ and the silanol number of OH groups bound through the H-bond, $\alpha_{OH,bound}$, on the silica surface were determined by MDE and IR spectroscopic measurements, depending on the temperature of the pretreatment in vacuum [4,5,15]. The correlation between the silanol number (determined on 400°C treated samples) and the intensity of the IR absorption band of OH groups is based on the fact that, for the sample calcined in vacuum at 400°C, there are practically only free OH groups left (the SiO_2 starting sample, aerosilogel, $S = 330\ m^2/g$, with a fully hydroxylated surface, had no ultra narrow pores) [15]. This means that the total concentration of OH groups on the silica surface, at 400°C, corresponds to the concentration of free OH groups α_{OH}. The concentration of bound (vicinal) OH groups, $\alpha_{OH\ bound}$, and the concentration of siloxane SiOSi bridges, α_{SiOSi}, throughout the range of temperatures was determined from the total concentration of OH groups and the concentration of free OH groups (Figure 22.1 and Table 22.1). Based on data reported in literature, Vansant et al. compared seven different models describing the distribution of free OH groups as a function of pretreatment temperature [16]. If the free silanol groups distribution proposed by Zhuravlev (Figure 22.1) is used as a reference (with an error of $\pm 0.5\ OH/nm^2$), then practically almost all reported distribution data obtained by different methods are within this interval of errors. This indicates a satisfactory agreement between the various data and confirms the existence of a real temperature dependent distribution of free silanol groups with a maximum at ~400°C.

Vansant et al. [16] also analyzed the distribution of all types of OH groups, according to the Zhuravlev model (Figure 22.1) and correcting by taking into account the results of Maciel et al. [17] using ^{29}Si NMR for the evaluation of the geminal groups (Table 22.1). Practically, this means that the free hydroxyl curve (Zhuravlevís data) is decomposed in single (free) and geminal (free) hydroxyls. Differential thermokinetic curves obtained by MTA-TPD, for a series of S-79 samples that had been subjected to pretreatment at ≥200°C, show the sole presence of OH groups on the surface, at different degrees of coverage. These OH groups are removed from the surface via the condensation reaction leading to the formation of SiOSi

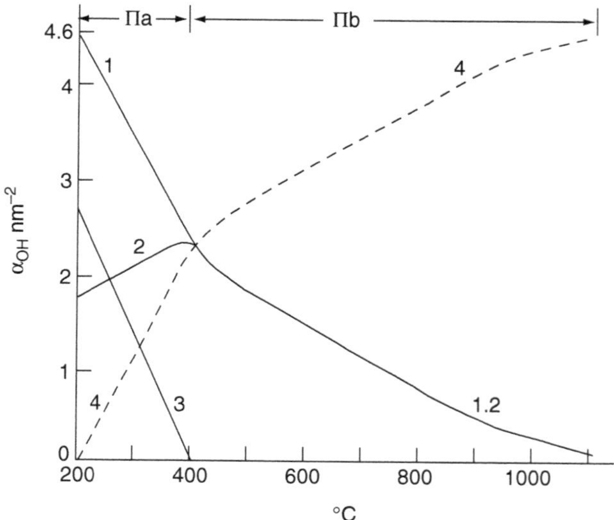

FIGURE 22.1 Different types of surface groups as a function of pretreatment of silica in vacuum [5]: Curve 1, average content of all OH groups; Curve 2, free OH groups content; Curve 3, content of the vicinal OH groups; curve 4, content of surface Si atoms in siloxane bridges.

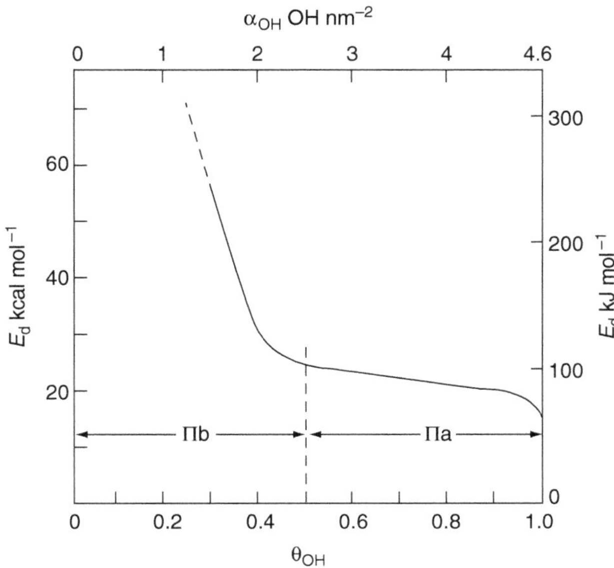

FIGURE 22.2 Activation energy of the associative desorption of water, E_D, as a function of the surface content of OH groups, α_{OH}, or as a function of the surface coverage of SiO_2 with OH groups, θ_{OH} [5].

bridges and molecular water:

$$(\equiv Si—OH)+(\equiv Si—OH) \longrightarrow (\equiv Si—O—Si\equiv)+H_2O \quad (22.1)$$

From the kinetic curves, the energy E_D was computed as a function of temperature (see Figure 10, in Ref. 5). The results show that $E_D = f(T°C)$ is characterized by two approximately rectilinear sections: in the range from 200 to ~400°C and above 400°C, with a notable change in the slope as one goes from the first to the second section. From the known temperature dependence of the activation energy and the silanol number, we obtained the dependence of the activation energy of dehydroxylation E_D from the concentration of OH groups, α_{OH}, or the dependence of E_D on the degree of coverage with OH groups, θ_{OH} (Figure 22.2). The functionalities, $E_D = f(\alpha_{OH})$ and $E_D = f(\theta_{OH})$, are of great importance for this model: they relate the energetic characteristics of the SiO_2 surface with the degree of the surface hydroxylation. We discern two linear sections. In the case of the high coverage ($1 \geq \theta_{OH} > 0.5$), the main process is the dehydroxylation of vicinal (bridged) silanols. The desorption energy E_D is almost independent of the silanol concentration and is mainly determined by a set of perturbations due to H-bonded OH groups. These perturbations cease with the disappearance of the H-bonded OH groups at ~400°C (at $\theta_{OH} \cong 0.5$). Thus, the subregion IIa is characterized by the presence of lateral interactions (H-bonds) between neighboring OH groups. This, in turn, exerts an effect on the SiOSi bridges which are under stress in this subregion IIa. In the low coverage subregion IIb ($\theta_{OH} < 0.5$), the main role is played by free OH groups and SiOSi bridges. For this subregion, E_D is strongly dependent on the concentration of OH groups. When there are only free OH groups, surrounded by SiOSi bridges, the latter groups can acquire a relatively large area following a high temperature treatment of silica. Under these conditions, the main mechanism describing the transfer of OH groups corresponds to condensation [Equation (22.1)] via disordered migration of protons on the surface (a process of the activated diffusion of OH groups). At the final stage, water is evolved, owing to the interaction of two OH groups that accidentally approach each other to a distance of ~0.3 nm. A probable mechanism for such migration of protons at elevated temperatures involves the interaction of protons with O atoms of the neighboring SiOSi bridges, resulting in the formation of new surface OH groups, which are displaced relative to their initial position.

REHYDROXYLATION OF THE SURFACE (CHEMISORPTION OF WATER)

When the dehydroxylated SiO_2 surface reacts with water, silanol groups are formed. Or, a rehydroxylation takes place owing to dissociative adsorption (chemisorption) of H_2O, according to the opposite process of Equation (22.1). As has been noted by many researchers and confirmed by our investigations [1,5,8,9,11,18,19], a complete hydroxylation of the surface readily occurs for those SiO_2 samples which were subjected to pretreatment at temperatures ≤400°C. After calcination at higher temperature, only partial rehydroxylation takes place. However, as has been shown in our work [5, 18], the strongly dehydroxylated surface can be restored to the fully hydroxylated state by treatment with water at room temperature, but this process takes much time (up to several years). To accelerate the rehydroxylation process, we subjected such silica samples to hydrothermal treatment at 100°C. From these data [5,18], it follows that the rehydroxylation proceeds by different reaction mechanisms in subregions IIa and IIb (Figure 22.1 and Figure 22.2), that is, below and above the threshold temperature of ~400°C. The probable mechanisms of the restoration of the hydroxylated coverage in these subregions are given elsewhere [5]. It is necessary to emphasize that the rehydroxylation process, which is due to the dissociative adsorption of water involving the splitting of siloxane bridges and the formation of new OH groups, takes place via: (i) the rapid, non activated (or weakly activated) chemisorption (subregion IIa), and (ii) the slow, strongly activated chemisorption (subregion IIb).

STRUCTURALLY BOUND WATER INSIDE THE PARTICLES OF AMORPHOUS SILICA

Measurements carried out by MDE and the combined thermogravimetric methods showed that different types of amorphous dispersed silica contain not only surface OH groups, but also structurally bound water inside the silica skeleton and inside the very fine ultramicro-pores [4,5,20,21]. IR spectral investigations show [22,23] that such bound water consists of silanol groups inside the silica samples (the absorption band of stretching vibrations is about $3650\ cm^{-1}$). The distribution of OH groups between the surface and the volume of the sample depends on the method of preparation of the silica sample and its subsequent treatment. This question has been studied in detail [21] on the basis of our data and of other data reported in literature. It should be emphasized that any sample of amorphous silica, in a state of complete hydroxylation of the surface, has a constant silanol number (the physicochemical constant, see above), which is $\alpha_{OH,av} = 4.6\ OH/nm^2$. Higher values for α_{OH} are due to the presence of bound water inside the silica skeleton. The removal of such internal OH groups from the silica particles, according to our investigations, takes place when the sample is heated in vacuum at 600–700°C.

The results of our experimental and theoretical investigations, together with the data published in the literature, permit us to construct a model for the surface of amorphous silica. The silanol groups on the silica surface are of basic importance in the proposed model. For a more detailed analysis of the chemical properties of the silica surface, it is necessary to take into account the presence on the silica surface of: impurities, structural defects, other functional groups, active sites, and so on. The findings established for an amorphous silica surface in the proposed model can be applied to other, more complex silica containing systems and related materials (silicates) as well as to various solid oxide substances containing OH groups on their surface.

REFERENCES

1. R.K. Iler, *The Chemistry of Silica*. Wiley-Interscience, New York, 1979.
2. S. Mattson, *Soil Science*, *33* (1932) 301.
3. L.T. Zhuravlev, *Langmuir*, *3* (1987) 316.
4. L.T. Zhuravlev, *Pure Appl. Chem.*, *61* (1989) 1969.
5. L.T. Zhuravlev, *Colloids Surfaces* A, *74* (1993) 71.
6. L.A. Ignat'eva, V.I. Kvlividze, V.F. Kiselev, In: *Bound Water in Dispersed Systems*. V.F. Kiselev, V.I. Kvlividze, Eds., MGU Press, Moscow, 1970, p. 56.
7. Yu.I. Gorlov et al., *Theoret. Experim. Khimiya*, *16* (1980) 202.
8. G.J. Young, *J. Colloid Sci.*, *13* (1958) 67.
9. G.J. Young, T.P. Bursh, *J. Colloid Sci.*, *15* (1960) 361.
10. P.A. Rebinder, In: *All-Union Sci. Conf. on the Intensification of the Drying Processes*. Profizdat, Moscow, 1958, p. 20.
11. J.H. De Boer, J.M. Vleeskens, Proc. Koninkl. Ned. Akad. Wetenschap., Ser. B, *60* (1957) 23, 45, 54; *61* (1958) 2, 85.
12. A.V. Kiselev et al., *Kolloid Zh.*, *22* (1960) 671.
13. S.P. Zhdanov, A.V. Kiselev, *Zh. Fiz. Khim.*, *31* (1957) 2213.
14. L.D. Belyakova, A.V. Kiselev et al., *Zh. Fiz. Khim.*, *33* (1959) 2624.
15. A.A. Agzamkhodzhaev, G.A. Galkin, L.T. Zhuravlev, In: *Main Problems of Physical Adsorption Theory*. M.M. Dubinin, Ed. Nauka, Moscow, 1970, p. 168.
16. E.F. Vansant, P. Van Der Voort, K.C. Vrancken, *Characterization and Chemical Modification of the Silica Surface*. Elsevier, Amsterdam, 1995.
17. G.E. Maciel, D.W. Sindorf, *J. Phys. Chem.*, *87* (1983) 5516.
18. A.A. Agzamkhodzhaev, L.T. Zhuravlev et al., *Kolloidn. Zh.*, *36* (1974) 1145.
19. V.V. Strelko, In: *Adsorption and Adsorbents*, No.2, Naukova Dumka, Kiev, 1974, p.65.
20. L.T. Zhuravlev, A.V. Kiselev, *Kolloidn. Zh.*, *24* (1962) 22.
21. L.T. Zhuravlev, In: *Main Problems of Physical Adsorption Theory*. M.M. Dubinin, Ed., Nauka, Moscow, 1970, p. 309.

23 Surface Structure of Amorphous and Crystalline Porous Silicas: Status and Prospects

K.K. Unger
Johannes Gutenberg University, Department of Inorganic and Analytical Chemistry

CONTENTS

Substantial progress in the elucidation of the surface structure of crystalline and amorphous silicas has been achieved by means of high-resolution spectroscopic techniques, for example, ^{29}Si cross-polarization magic-angle spinning NMR spectroscopy and Fourier transform IR spectroscopy. The results lead to a better understanding of the acidity, dehydration properties, and adsorption behavior of the surface. These properties are key features in the design of novel advanced silica materials. The current methods of characterization are briefly reviewed and summarized.

The surface chemistry of silica was a subject of intensive study in the period between 1960 and 1970 as a consequence of the widespread industrial use of colloidal, pyrogenic, and precipitated silicas, as well as silica hydrogels and xerogels. Chemical surface reactions and IR spectroscopy were the most-applied methods in surface structure elucidation. Significant contributions to the understanding of the silica surface were made by Fripiat [1], Kiselev and co-workers [2], Hair [3], Little [4], Peri [5], and others. In contrast to this active period, little progress was since made until about 1980. Advances in surface and materials science caused a search for novel materials with controllable properties, and the surface structure of silica regained considerable interest. Three major developments were important: first, the experience gained in silicate chemistry [6], particularly in the area of synthetic zeolites [7]; second, the increasing use of surface-modified silicas as packings in high-performance liquid chromatography (HPLC) [8]; and third, the progress made in high-resolution spectroscopic techniques applied to surface characterization. This chapter summarizes the most important achievements made in this field and puts the problems involved into perspective.

STRUCTURAL ASPECTS IN SILICA CHEMISTRY

The term silica, as applied to solid forms having the stoichiometric composition of SiO_2, has many meanings. Thus, it is useful to attempt to classify the different solid forms and modifications of silica according to distinct structural characteristics [9].

The bulk structures of silicas are classified as crystalline and amorphous polymorphs. More than 35 well-defined crystalline silicas are known, which are well-characterized by the Si—O length, the Si—O—Si bond angle, and the Si—O bond topology and coordination [10]. Some of the crystalline polymorphs are collected in Table 23.1. Because of the lack of sufficiently precise methods to assess the long-range structural order, amorphous silicas remain poorly characterized. They can be

TABLE 23.1
Crystalline Phases That Have Topologically Distinct SiO_2 Frameworks

Name	d_f	Formula; unit cell content
Stishovite	43.0	SiO_2
$SiO_2(Fe_2N\text{-type})$	42.8	SiO_2
Coesite	29.3	SiO_2
Quartz	26.6	SiO_2
Moganite	26.3	SiO_2
Keatite	25.1	SiO_2
Cristobalite	23.2	SiO_2
Tridymite	22.9	SiO_2
Nonasils	19.2	$88SiO_2 8M^8 8M^9 4M^{20}$
Melanophlogites	19.0	$46SiO_2 2M^{12} 6M^{14}$
Dodecasils 3C (Silica ZSM-39)	18.6	$136SiO_2 16M^{15} 8M^{16}$
Dodecasil 1H	18.5	$34SiO_2 3m^{12} 2M^{12} 1M^{20}$
SIGMA-2	17.8	$64SiO_2 8M^9 4M^{20}$
Silica sodalites	17.4	$12SiO_2 2M^{14}$
Decadodecasils 3R	17.6	$120SiO_2 6M^{10} 9M^{12} 6M^{19}$
Decadodecasils 3H	17.6	$120SiO_2 6M^{10} 9M^{12} 1M^{15} 4M^{19}\ 1M^{23}$
Silica ZSM-23	20.0	$24SiO_2(CH_3)_2N(CH_2)_7N\ (CH_3)_2$
Silica ZSM-48	19.9	$48SiO_2H_2N(CH_2)_8NH_2$
Silica ZSM-22	19.7	$24SiO_2HN(C_2H_5)_2$
Silica ferrierite	19.3	$36SiO_2 2H_2N(CH_2)_2NH_2$
Silica ZSM-12	18.5	$28SiO_2N(C_2H_5)_3$
Silica ZSM-50	18.2	$112SiO_2 n\text{-}[(CH_3)_3N(CH_2)\ _6N(CH_3)_3](OH)_2$
Silicalite II (Silica ZSM-11)	17.9	$96SiO_2 n\text{-}[N(C_4H_9)_4]OH$
Silicalite I (Silica ZSM-5)	17.8	$96SiO_2 4[N(C_3H_7)_4]F$
Fibrous silica	19.6	SiO_2

Note: Abbreviations are as follows: d_f, framework density in number of SiO_2 groups per 1000 $Å^3$; M, guest molecule located in a cage that has $f = m_j$ faces.
Source: Reproduced with permission from reference [10]. Copyright 1988.

loosely discriminated according to their dispersity, bulk density, and type of pore structure.

The first classifying quantity in the characterization of crystalline silica polymorphs is the framework density d_f, which is expressed as the number of SiO_2 groups per 1000 $Å^3$ [10]. Values of d_f range from 17 to 43 SiO_2 groups per 1000 $Å^3$ (Table 23.1).

According to the value of d_f, crystalline silicas were divided by Liebau [10] into pyknosils ($d_f > 21$ SiO_2 groups per 1000 $Å^3$) and porosils ($d_f < 21$ SiO_2 groups per 1000 $Å^3$). Pyknosils are defined as "polymorphs with frameworks too dense to enclose guest molecules that are larger than helium and neon" [10]. Pyknosils are nonporous.

The second quantity is the type of porosity and pore structure. Phases with silica frameworks that have pores wide enough to accommodate larger guest molecules are called porosils, irrespective of whether their pores are filled or empty [10]. Porosils have a micropore system of pores with widths between 0.4 and 0.8 nm. In contrast to the well-defined microporous crystalline silicas, porous amorphous silicas lack any long-range structural order. The average pore diameter of amorphous silica materials covers the range between a few to several thousand nanometers. Porosils are further divided into clathrasils and zeosils, depending on whether the pores are closed or open, that is, accessible to adsorption. Clathrasils form cagelike pores that are schematically described by polyhedra. A typical representative of the clathrasil family is Dodecasil 1H, the crystals of which are depicted in Figure 23.1 [11]. Zeosils can be considered as aluminum-free and zeolites, which are microporous crystal-line-network aluminosilicates [10]. The end-member of the pentasil ZSM-5 family, for instance, is silicalite-I, with an Si:Al ratio of >1000 [12]. Silicalite-I exhibits an open bidirectional pore system of straight and zigzag channels about 0.5–0.6 nm in width [12]. Silicalite-I can be synthesized with a defined phase composition, with high purity, and as large crystals with a narrow particle size distribution [13]. Because of these properties and its

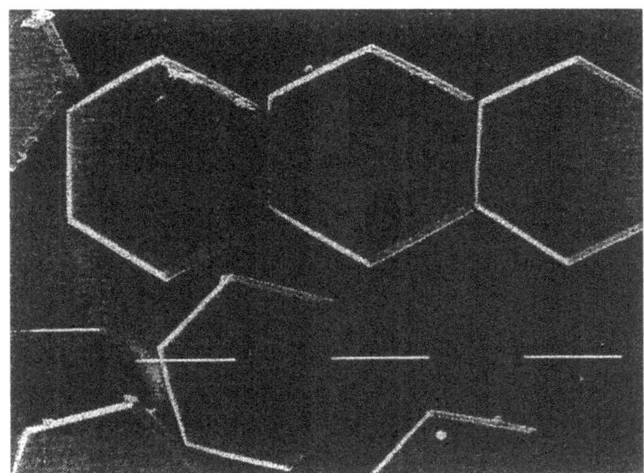

FIGURE 23.1 Crystals of Dodecasil 1H, synthesized with 1-amino adamantan. (Reproduced with permission from reference 11. Copyright 1989.)

high thermal stability up to 1200 K, silicalite-I is well-suited as a reference material for crystalline and amorphous porous silicas.

A third classifying quantity relates to the surface structure of silicas, which is characterized by the coordination of surface silicon atoms; the resulting functional groups; their density, topology, and distribution; the degree of hydroxylation; the hydration–dehydration behavior; the acidic and basic properties of surface functional groups; and their adsorption behavior and chemical reactivity. The pattern of the surface structure in terms of these properties is discussed in the section "Current View of the Silica Surface." The following section reviews the methods by which reliable information on these properties is obtained.

SURVEY OF METHODS FOR CHARACTERIZING THE SILICA SURFACE

According to the underlying principle, the methods are grouped into spectroscopic, thermal and calorimetric, adsorption and wetting, isotopic exchange, microscopic, scattering, and chemical reactions techniques (Table 23.2) [14–29]. This chapter is not a comprehensive treatment of all methods in depth. Results of spectroscopic and adsorption methods are discussed for aspects of content of information, validity, applicability, and limitations.

SPECTROSCOPIC METHODS

Fourier Transform Infrared (FTIR) Spectroscopy

FTIR is carried out either in the transmission mode or as diffuse reflectance FTIR [14] with pellets, self-supporting sample disks, or loosely packed powders. Spectra are commonly recorded in the frequency range between 400 and 4000 cm^{-1}. Figure 23.2 [30,31] shows the adsorption bands and band assignments of three different types of silicas: an amorphous highly disperse nonporous silica (Aerosil 200, Degussa), a crystalline nonporous silica (α-quartz), and a microporous crystalline silica (silicalite-I). The band at 3750 cm^{-1} is assigned to the stretching vibration of free hydroxyl groups that occur at the surface of amorphous silicas. This band usually overlaps with the adsorption bands originating from hydrogen-bonded hydroxyl groups and from adsorbed water. Adsorbed water on amorphous silica is removed under vacuum at 383 to 473 K. Above 473 K, hydrogen-bonded hydroxyl groups condense, and the corresponding absorption band diminishes. Free surface hydroxyl groups still exist after annealing the silica at 1273 K at a low content. Figure 23.3 [32] shows the IR spectra of a porous silica measured at different pretreatment temperatures. Thus, for IR spectroscopic measurements it is essential to control the pretreatment conditions (vacuum, temperature, and moisture) to achieve reproducible results.

There is still a discussion on the appearance of the absorption band due to geminal groups. Although Camara et al. [33] suggested that the band at 3710 cm^{-1} was specific for geminal groups, Morrow and Gay [34] claimed that they absorb at 3750 cm^{-1}. Discrimination between bulk and surface hydroxyl groups is effected by subjecting the silica to deuteration with D_2O and monitoring the absorption of the corresponding –OD bands [22]. Bulk hydroxyls and bulk water do not take part in the isotopic exchange, and thus their absorption bands remain unaffected. FTIR spectroscopy is a technique to identify surface hydroxyl groups and to follow their adsorption behavior toward selected probe substances. Quantitation is difficult to achieve because of the lack of absorption coefficients.

Several types of cells have been constructed and are commercially available that allow an *in situ* measurement or have a movable holder to evacuate and heat the sample. With these cells it is also possible to admit gas or vapor to subsequently measure the frequency shift upon adsorption. This technique can reveal additional information about the surface properties of the material. Carbon monoxide, for instance, has been applied as a probe to detect hydroxyl groups at the silica surface. Upon adsorption of CO at 77 K, the band due to free hydroxyl groups shifts to a lower frequency by 78 or 93 cm^{-1} [35–37].

Brønsted and Lewis acid sites are distinguished by the adsorption of pyridine monitored by means of IR spectroscopy [38]. Pyridine adsorbed on Brønsted sites gives rise to an absorption band at about 1550 cm^{-1}; pyridine adsorbed on Lewis sites generates a band between 1440 and 1460 cm^{-1}, depending on the type of Lewis acid sites.

It is interesting to compare the absorption pattern of the three types of silicas that differ in bulk and surface structure (Figure 23.2). α-Quartz shows two types of

TABLE 23.2
Physical and Chemical Methods for Characterizing the Surface Structure of Silica

Method	Information	References
Spectroscopic:		
FTIR	Types of hydroxyl groups, acidity	
Diffuse reflectance FTIR adsorption of probe molecules (pyridine, carbon monoxide, and so forth, monitored by FTIR)	Brönsted and Lewis acid sites, adsorption behavior, and chemical reactivity	[14]
^{29}Si CP–MAS[a] NMR spectroscopy	Short-range structural order of surface silicon atoms, bond	[15–18]
^{29}Si CP–MAS NMR spectroscopy, ^{1}H MAS NMR spectroscopy	Length, bond angles, acidic properties (protonic sites) of silica, and concentration of protons	[19]
Secondary ion mass spectrometry (SIMS)	Chemical composition of the surface as a function of the depth	
Extended x-ray absorption fine-structure spectroscopy	Local structural surrounding, nearest neighbor distances, coordination number, and bond lengths	
Electron spectroscopy for chemical analysis (ESCA)	Chemical compositions of the surface	
Photophysical studies of direct energy transfer between excited donor and ground-state acceptor molecules	Pore and surface morphology	[20]
Thermal and calorimetric:		
Microcalorimetry	Heat of adsorption, phase transitions of adsorbates	
Thermogravimetry	Weight loss as a function of temperature	
Differential thermogravimetry (DTG), differential scanning calorimetry (DSC), and thermoporometry (monitoring liquid–solid phase transitions of a pure liquid capillary condensate in a porous system)	Enthalpic (exothermic or endothermic) changes upon heating surface area, pore size distribution, and average pore diameter	[21]
Isotopic exchange:		
Heterogeneous isotopic exchange using deuterated (D_2, D_2O, CH_3OD, CF_3COOD, and so forth) and tritiated (HTO) substances, combined with mass spectrometry, IR spectroscopy and ^{1}H NMR spectroscopy	Total content of surface hydroxyl groups, content of physisorbed water	[22–24]
Adsorption and wetting:		
Adsorption of gases and vapors, combined with IR spectroscopy, microcalorimetry, and so forth	Heats of adsorption, specific surface area, pore size distribution, average pore diameter, and fractal dimension	
Temperature-programmed desorption of absorbed substances (ammonia, pyridine, and so forth, coupled with mass spectrometry and IR spectroscopy	Acidic functional groups, relative acid strength	
Adsorption of bases (*n*-butylamine) in aprotic solvents using Hammett or arylmethanol indicators	Total content of surface acid sites, acidity distribution	[25]
Wetting with liquids	Surface energy, surface wettability, and contact angle	
Microscopy:		
Transmission and scanning electron microscopy, scanning tunneling microscopy	Texture of silica, geometrical surface structure, pore shape, and pore homogeneity	[26]
Scattering:		
Small-angle x-ray and neutron scattering	Pore morphology, roughness of pore surface, and pore size distribution	[27,28]
Chemical Methods:		
Using chemically reactive substances such as chlorine, metal halides, grignard compounds, and reactive chloro- and alkoxysilanes	Stoichiometry of surface reactions, reactivity of surface hydroxyl groups	[29]

[a]CP–MAS denotes cross-polarization–magic-angle spinning.

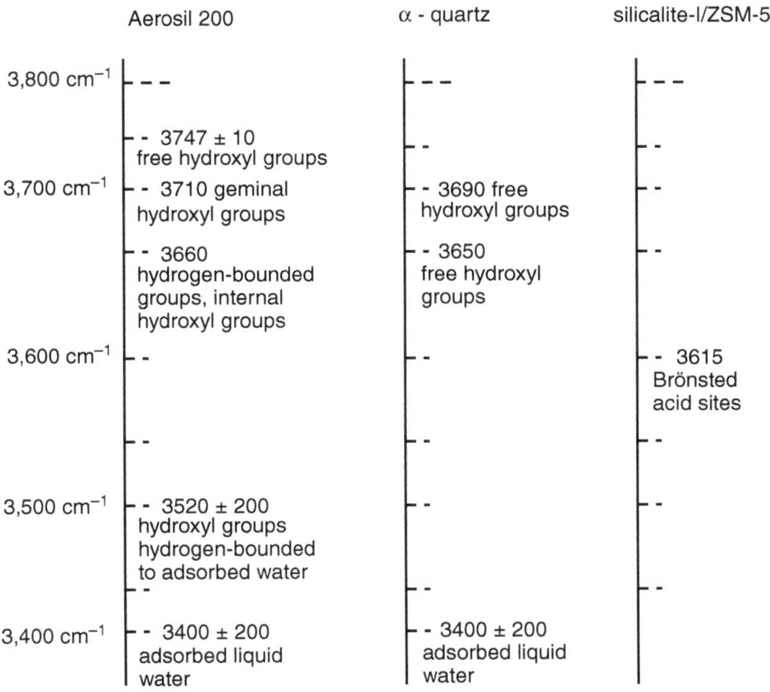

FIGURE 23.2 Band assignment of surface silica species in the high-frequency region (2000–4000 cm^{-1}). (Reproduced with permission from reference 30. Copyright 1979.)

free hydroxyl, with absorption bands at 3690 and 3620 cm^{-1} [32,39]. On silicalite-I, a band at 3750 cm^{-1} with low intensity is assigned to free hydroxyl groups at defect sites of the crystals. Increasing the aluminum content, that is, forming ZSM-5 zeolite, leads to an appearance of a band at 3615 cm^{-1}, which is assigned to a Brønsted acid of the following type [31]:

```
                    H
O      O      O      O      O
 \    /  \   /  \   /  \   /
   Si      Al      Si      Si
 /    \  /   \  /   \  /   \
O      O      O      O      O
```

^{29}Si CP-MAS NMR Spectroscopy

This method is very useful for characterizing silica [15–17]. The main information derived from an NMR spectrum is the chemical shift, the intensity, and the line width. The ^{29}Si chemical shift is determined by the number and type of tetrahedral framework atoms connected to tetrahedral silicon atoms. The spectrum thus allows the detection of the number of structurally inequivalent kinds of silicon atoms of various Si $(O-)_{4-n}(OSi)_n$ units in silicates and as Q^4 (*m* Al) for Si $(OSi)_{4-n}(OAl)_m$ units in framework alumosilicates (Figure 23.4 and Figure 23.5) [40].

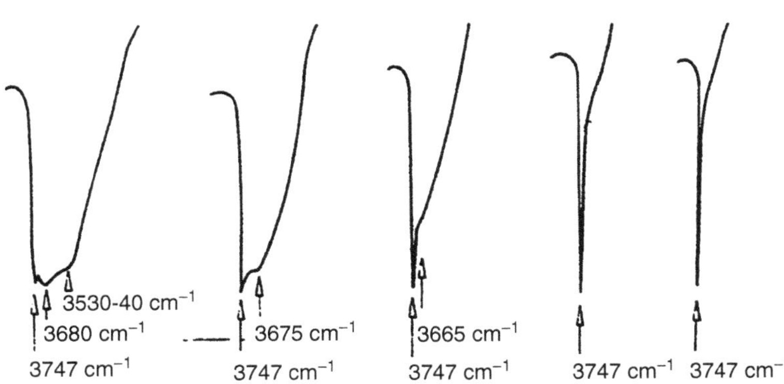

FIGURE 23.3 IR spectra of a porous silica (specific surface area $a_s = 475\ m^2/g$; average pore diameter p = 7 nm) obtained after evacuation at (left to right) 473, 673, 773, 873, and 973 K. (Reproduced with permission from reference 32. Copyright 1979.)

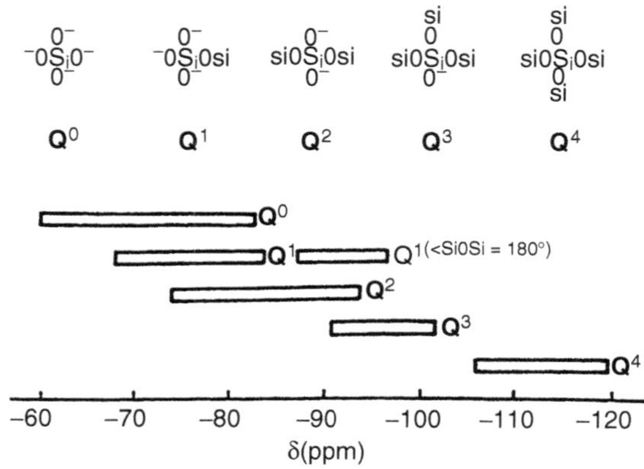

FIGURE 23.4 Typical ranges of ^{29}Si chemical shifts of $\mathbf{Q}^n$ units in silicates. (Reproduced with permission from reference 40. Copyright 1990.)

A ^{29}Si CP–MAS NMR spectrum of amorphous silica shows three signals, at −91, −100, and −109 ppm, that are assigned to geminal hydroxyl groups ($\mathbf{Q}_2$), free hydroxyl groups ($\mathbf{Q}_3$), and siloxane groups ($\mathbf{Q}_4$) (Figure 23.6). The amplitude of all three signals is a function of the contact time [18]. $\mathbf{Q}_2$ and $\mathbf{Q}_3$ show comparable maxima at a contact time of 6 to 8 ms, whereas a broad maximum is observed for $\mathbf{Q}_4$ between 8 and 25 ms [18]. With known relaxation times, the relative ratios of $\mathbf{Q}_2/\mathbf{Q}_3/\mathbf{Q}_4$ can be assessed [19]. Dynamic studies have been performed to detect changes in the surface structure after certain treatments. On acid-treated silicas, for example, two types of $\mathbf{Q}_4$ groups appeared with different T_{SiH} and $T_{\mathrm{i}\rho\mathrm{H}}$ values, where T_{SiH} is the cross-relaxation constant and $T_{\mathrm{i}\rho\mathrm{H}}$ the proton spinlattice relaxation time in the rotating frame [41]. The $\mathbf{Q}_4$ units with low T_{SiH} and $T_{\mathrm{i}\rho\mathrm{H}}$ values are considered polycrystalline domains at the surface of amorphous silica. By independent neutron diffraction measurements on acid-treated silica samples, a tridymite-like phase could be monitored [42]. The $\mathbf{Q}_4$ units with high T_{SiH} and $T_{\mathrm{i}\rho\mathrm{H}}$ values are believed to be of amorphous type [42].

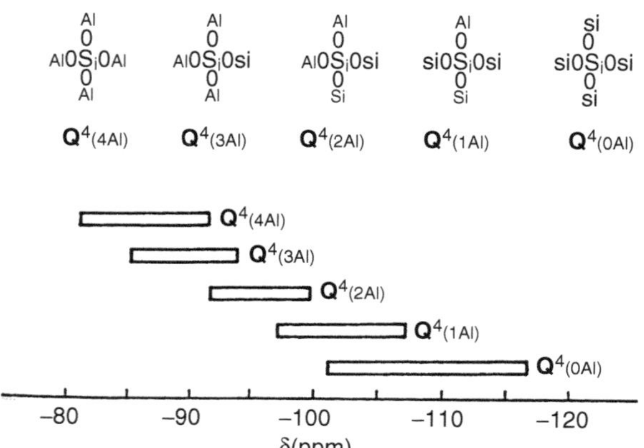

FIGURE 23.5 Typical ranges of ^{29}Si chemical shifts of $\mathbf{Q}^4$ (mAl) units in aluminosilicates. (Reproduced with permission from reference 40. Copyright 1990.)

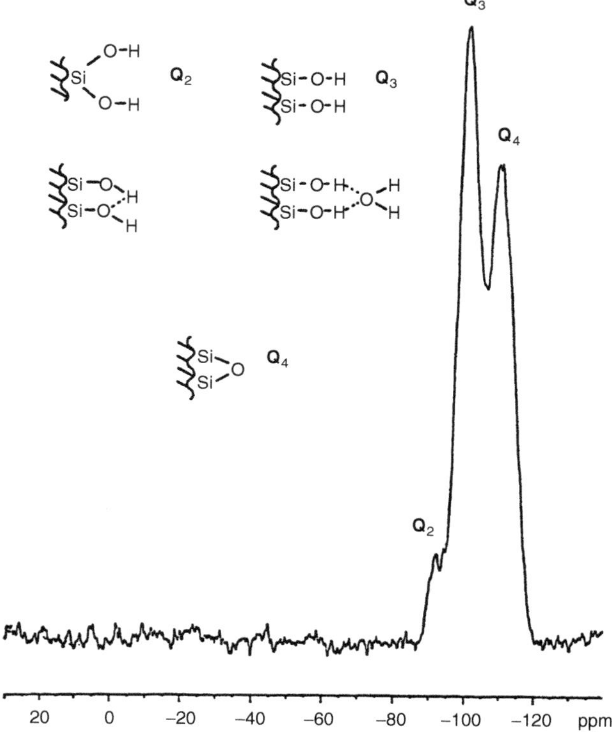

FIGURE 23.6 ^{29}Si CP–MAS NMR spectrum of an amorphous porous silica. (Reproduced with permission from reference 18. Copyright 1991.)

Adsorption Methods

Adsorption studies are a classical method of characterizing silicas. A new generation of high-resolution dynamic volumetric adsorption equipment now commercially available allows automatic measuring of several thousand data points of the isotherm down to a relative pressure ratio p/p_0 of about 10^{-5} with high precision [43]. Adsorption measurements are extremely useful for characterizing silicas, in particular microporous types. The progress achieved is demonstrated by the following examples of microporous crystalline silicas. Sorption experiments were performed with argon and nitrogen at 77 K on large and uniform crystals of silicalite-I (Figure 23.7) [44]. The results obtained drastically deviated from the expected data: instead of a Langmuir type of isotherm, an isotherm with distinct steps was obtained (Figure 23.8) [45]. These steps are interpreted as a successive filling of the microporous channels where the gas molecules are adsorbed at distinct adsorption sites. The nitrogen isotherm on silicalite-I shows a

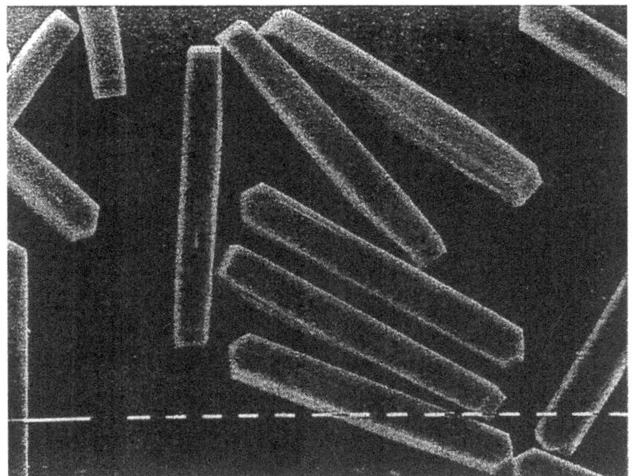

FIGURE 23.7 Large crystals of silicalite-I that were synthesized alkaline-free with tetrapropylammonium bromide as template. (Reproduced with permission from reference 46. Copyright 1990.)

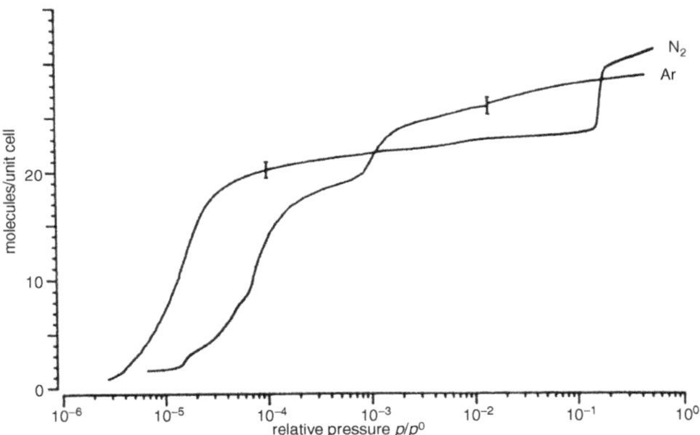

FIGURE 23.8 Adsorption isotherms of argon and nitrogen at 77 K on silicalite-I obtained with high-resolution dynamic volumetric sorption equipment. (Reproduced with permission from reference 45. Copyright 1989.)

second peculiar phenomenon: a pronounced hysteresis occurs between the adsorption and desorption branch at p/p_0 of about 0.1–0.15 (Figure 23.9) [46]. This additional uptake is probably due to a phase change of the liquid adsorbate into a solidlike one, which is evidenced by concurrent microcalorimetric and *in situ* neutron diffraction measurements [47].

CURRENT VIEW OF THE SILICA SURFACE

SURFACE CHARGE

For a number of applications the surface charge of silica is of eminent interest. The isoelectric point (point of zero charge) can be derived from electrophoretic mobility

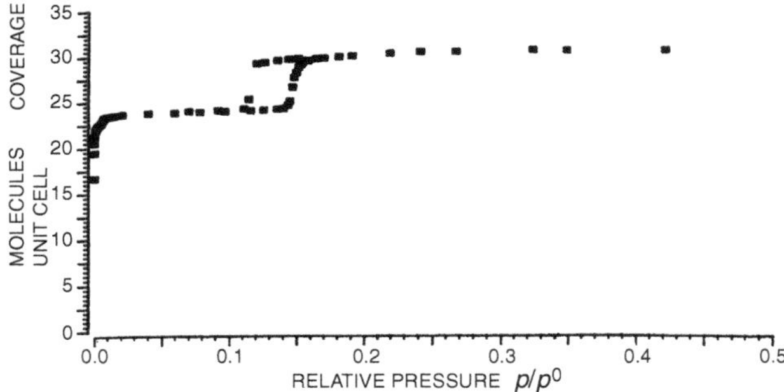

FIGURE 23.9 Adsorption isotherm of nitrogen at 77 K on silicalite-I obtained with conventional gravimetric sorption equipment. (Reproduced with permission from reference 46. Copyright 1990.)

(zeta potential) measurements on silica suspensions of varying pH. The isoelectric point was measured to be between pH 1 and 2 [48]. Above pH 2 the particles are negatively charged; below pH 1 the particles bear a positive charge. In one study the isoelectric point of silica was reported to be at about pH 5 [49]. Recently, a chromatographic-grade silica (Zorbax RX) became commercially available, and it exhibits distinct anion-exchange properties at about pH 5 [50]. The origin of this peculiar behavior is not yet clear.

SURFACE FUNCTIONALITY

Several groups contribute to Brønsted acidity, as free (isolated), paired (geminal), and hydrogen-bonded (vicinal) hydroxyl groups. Bridging hydroxyl groups in microporous amorphous silicas may also contribute [51]. They are identified by their absorption bands in the frequency region between 3000 and 4000 cm^{-1} and by ^{29}Si and ^{1}H CP–MAS NMR spectroscopy. Much attention has been paid to geminal hydroxyl groups, which were first postulated by Peri [5].

Geminal hydroxyl groups have been shown to exist at the surface of Aerosil 200 even at high thermal pretreatment of 1073 K [34]. Geminal hydroxyl groups are formed as a result of acid treatment of amorphous porous silicas [42]. Geminal hydroxyl groups have been found to be more reactive in chemical surface modification of amorphous porous silicas than free hydroxyl groups [8].

The relative proportion of the hydroxyl groups heavily depends on the method used for estimation. The relative content varies from silica to silica depending on the manufacturing process and after-treatment conditions.

The total concentration of surface hydroxyl groups at a fully hydroxylated surface has been assessed to be about 5 per square nanometer, independent of the type of silica [52]. The concentration decreases monotonically with increasing temperature when silicas are thermally pretreated under vacuum. Physisorbed water is first removed at temperatures between 380 and 420 K. Simultaneously, hydrogen-bonded hydroxyl groups condense to strained siloxane groups, which act as Lewis sites according to the following equation:

$$\underset{/|\backslash}{\overset{\text{OH}}{\underset{|}{\text{Si}}}} + \underset{/|\backslash}{\overset{\text{OH}}{\underset{|}{\text{Si}}}} \xrightarrow{-H_2O} \underset{/|\backslash}{\overset{+}{\text{Si}}}\;\underset{/|\backslash}{\overset{\text{O}^-}{\underset{|}{\text{Si}}}} \longrightarrow \underset{/|\backslash}{\text{Si}}\overset{\text{O}}{\diagup\;\diagdown}\underset{/|\backslash}{\text{Si}}$$

Strained siloxane groups are formed up to about 770 K and then convert into stable siloxane groups [53]. Strained siloxane groups completely rehydroxylate upon exposure to water, whereas stable siloxane groups rehydroxylate slowly. With the removal of hydroxyl groups and the formation of siloxane groups, the silica surface loses its hydrophilic character and becomes hydrophobic. The movement of the hydrophilic to hydrophobic character can be followed by measuring the heats of adsorption of water and other polar adsorptives.

The protonic sites at the surface of amorphous silica are weakly acidic compared to the acid centers in crystalline aluminosilicates (e.g., zeolites). On ZSM-5, Brønsted sites are formed by protons adjacent to aluminum atoms in the tetrahedral framework. The concentration of acid sites increases with the aluminum content. The total acidity as well as the acid strength distribution can be determined by using *n*-butylamine and Hammet or arylmethanol indicators [25]. Depending on the pK_a of the indicator, a relative scale of the strength distribution is obtained [54]. Results for a series of amorphous porous silicas of graduated pore size are shown in Figure 23.10. The acidity varies between pK_a -1 and $+9$ and is nearly the same for all silicas studied. The results are specifically valid for this method only and cannot be compared with those derived from other methods.

One of the major factors that determines the acidity values is the accessibility of surface sites, which is a function of the pore size of the silica, the surface morphology, and the size of the amine molecule used for the acidity assessment [54]. A great deal of work has been carried out to elucidate the chemical and geometrical heterogeneity of the silica surface. The geometrical heterogeneity can be described by the fractal dimension D, which varies between 2 and 3 [55]. The surface heterogeneity in terms of thermodynamic aspects can be visualized by the surface energy and the energy distribution [56]. Appropriate functions are derived from potential energy equations that aid adsorption measurements. Although an enormous literature exists on the aspects of surface heterogeneity, this approach has so far gained little importance in the evaluation of silicas in practice.

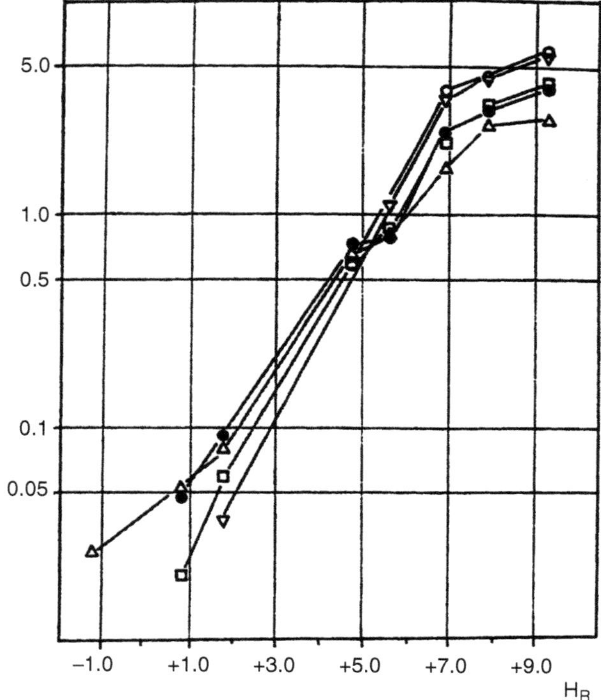

FIGURE 23.10 Cumulative acidity in micromoles per square meter on silica-20 (Δ), silica-40 (□), silica-60 (●), silica-100 (▽), and silica-200 (○) as a function of the acidity strength H_R. (Reproduced with permission from reference 54. Copyright 1980.)

CONCLUSIONS

The silica surface represents a dynamic system that undergoes slight to notable changes depending on the conditions of the environment. To assess the dynamic character of the surface, studies are required that enable monitoring of structural changes during silica formation, after-treatment, use, and storage. Spectroscopic methods are the most suitable in this respect. In this context, the use of selected probe molecules embedded, adsorbed, or chemically bonded opens a new dimension in gaining substantial information and is exemplified by photophysical studies. A particular focus will be on *in situ* measurements to directly monitor subtle structural properties of the surface. The combined application of sophisticated, *in situ*, high-resolution, physicochemical methods will shed much light on the understanding of the silica surface.

REFERENCES

1. Fripiat, J. J.; Uytterhoeven, J. *J. Phys. Chem.* **1962**, *66*, 800.
2. Kiselev, A. V.; Lygin, V. I. *Infrared Spectroscopy of Surface Compounds*; Wiley Interscience: New York, 1975.
3. Hair, M. L. *Infrared Spectroscopy in Surface Chemistry*; Dekker: New York, 1967.

4. Little, L. H. *Infrared Spectra of Adsorbed Species*; Academic Press: London, 1966.
5. Peri, J. B.; Hensley, A. L., Jr. *J. Phys. Chem.* **1968**, *72*, 2926.
6. Liebau, F. *Structural Chemistry of Silicates*; Springer Verlag: Heidelberg, Germany, 1985.
7. Van Bekkum, H.; Flanigen, E. M.; Jansen, J. C. *Introduction to Zeolite Science and Practice*; Elsevier: Amsterdam, Netherlands, 1991.
8. Unger, K. K. *Porous Silica: Its Properties and Use in Column Liquid Chromatography*; Elsevier: Amsterdam, Netherlands, 1979.
9. Iler, R. K. *The Chemistry of Silica and Silicates*; Wiley Interscience: New York, 1979.
10. Liebau, F. In *Silicone Chemistry*; Correy E. R.; Correy, J. Y.; Gasper, P. P., Eds.; Ellis Horwood: Chichester, UK, 1988, pp 309–323.
11. Müller, U.; Reich, A.; Unger, K. K. In *Recent Advances in Zeolite Science*; Klinowski, J.; Barrie, P. J., Eds.; Elsevier: Amsterdam, Netherlands, 1989; pp 241–252.
12. Jacobs, P. A.; Martens, P. A. *Synthesis of High-Silica Aluminosilicate Zeolites*; Elsevier: Amsterdam, Netherlands, 1987.
13. Müller, U.; Unger, K. K. *Zeolites* **1988**, *8*, 154–156.
14. Leyden, D. E.; Murthy, R. S. S. *Trends Anal. Chem.* **1988**, *7*, 164.
15. Maciel, G. E.; Sindorf, D. W. *J. Am. Chem. Soc.* **1980**, *102*, 7607.
16. Sindorf, D. W.; Maciel, G. E. *J. Am. Chem. Soc.* **1983**, *105*, 1487.
17. Engelhardt, G.; Michel, D. *High-Resolution Solid State NMR of Silicates and Zeolites*; Wiley Interscience: New York, 1987.
18. Albert, K.; Bayer, E. *J. Chromatogr.* **1991**, *544*, 345.
19. Köhler, J.; Chase, D. B.; Farlee, R. D.; Veega, A. J.; Kirkland, J. J. *J. Chromatogr.* **1986**, *352*, 275.
20. Zachariasse, K. A. In *Photochemistry on Solid Surfaces*; Matsumo, T.; Anpo, M., Eds.; Elsevier: Amsterdam, Netherlands, in press.
21. Eyraud, C.; Quinson, J. T.; Brun, M. In *Characterization of Porous Solids*; Unger, K. K.; Rouquerol, J.; Sing, K. S. W.; Kral, H., Eds.; Elsevier: Amsterdam, Netherlands, 1988; p. 295–316.
22. Zhuravlev, L. T.; Kiselev, A. V.; Nadina, V. P.; Polyakov, A. L. *Russ. J. Phys. Chem.* **1963**, *37*, 113, 1216.
23. Holik, M.; Matejkova, B. *J. Chromatogr.* **1981**, *213*, 33.
24. Unger, K. K. *Porous Silica: Its Properties and Use in Column Liquid Chromatography*; Elsevier: Amsterdam, Netherlands, 1979; pp 72–76.
25. Tanabe, K. *Solid Acids and Bases*; Academic Press: New York, 1970.
26. Kuk, Y., Silverman, P. J. *Rev. Sci. Instrum.* **1989**, *60*, 165.
27. Schmidt, P. W. In *Characterization of Porous Solids*; Unger, K. K.; Rouquerol, J.; Sing, K. S. W.; Kral, H., Eds.; Elsevier: Amsterdam, Netherlands, 1988; pp 35–48.
28. Ramsay, J. D. F. In *Characterization of Porous Solids*; Unger, K. K.; Rouquerol, J.; Sing, K. S. W.; Kral, H., Eds.; Elsevier: Amsterdam, Netherlands, 1988; pp 23–34.
29. Unger, K. K. *Porous Silica: Its Properties and Use in Column Liquid Chromatography*; Elsevier: Amsterdam, Netherlands, 1979; pp 64–68.
30. Unger, K. K. *Porous Silica: Its Properties and Use in Column Liquid Chromatography*; Elsevier: Amsterdam, Netherlands, 1979; p 10.
31. Tissler, A. Ph.D. Thesis, Johannes Gutenberg-Universität, Mainz, Germany, 1989.
32. Unger, K. K. *Porous Silica: Its Properties and Use in Column Liquid Chromatography*; Elsevier: Amsterdam, Netherlands, 1979; pp 69–70.
33. Camara, B.; Dunken, H.; Fink, P. *Z. Chem.* **1968**, *8*, 155.
34. Morrow, B. A.; Gay, I. D. *J. Phys. Chem.* **1988**, *92*, 5569.
35. Zaki, M. I.; Knoezinger, H. *Mater. Chem. Phys.* **1987**, *17*, 201.
36. Knözinger, H. In *Acid-Base Catalysis*; Tanabe, K.; Hattori, H.; Yamaguchi, T.; Tanaka, T., Eds.; Kodansha Ltd.: Tokyo, Japan, 1989; pp 147–167.
37. Baila, R. M.; Kantner, T. R. *J. Phys. Chem.* **1966**, *70*, 1681.
38. Ward, J. W. *J. Catal.* **1967**, *9*, 225.
39. Gallei, E. *Ber. Bunsenges. Phys. Chem.* **1973**, *77*, 81.
40. Engelhardt, G. *Trends Anal. Chem.* **1989**, *8*, 343–347.
41. Pfleiderer, B.; Albert, K.; Bayer, E.; Van De Ven, L.; De Haan, J.; Cramers, C. *J. Phys. Chem.* **1990**, *94*, 4189–4194.
42. Unger, K. K.; Lork, K. D.; Pfleiderer, B.; Albert, K.; Bayer, E. *J. Chromatogr.* **1991**, *556*, 395–406.
43. Reichert, H.; Unger, K. K. *Trends Anal. Chem.* **1990**, *100*, 44–48.
44. Müller, U.; Brenner, A.; Reich, A.; Unger, K. K. In *Zeolite Synthesis*; Occelli, M. L.; Robson, H. E., Eds.; ACS Symposium Series 398; American Chemical Society: Washington, DC, 1989; pp 346–359.
45. Müller, U.; Reichert, H.; Robens, E.; Unger, K. K.; Grillet, Y.; Rouquerol, F.; Rouquerol, J.; Dongfen Pan, A.; Mersmann, A.; Fresenius, Z. *Anal. Chem.* **1989**, *333*, 433.
46. Müller, U. Ph.D. Thesis, Johannes Gutenberg-Universität, Mainz, Germany, 1990.
47. Reichert, H.; Müller, U.; Unger, K. K.; Grillet, Y.; Rouquerol, J.; Coulomb, J. P. In *Characterization of Porous Solids*; Rodrigues-Reinoso, F.; Rouquerol, J.; Sing, K. S. W.; Unger, K. K., Eds.; Elsevier: Amsterdam, Netherlands, 1991; pp 535–542.
48. Unger, K. K. *Porous Silica: Its Properties and Use in Column Liquid Chromatography*; Elsevier: Amsterdam, Netherlands, 1979; p 138.
49. Denoyel, R.; Rouquerol, F.; Rouquerol, J. In *Fundamentals of Adsorption*; Liapis, A. I., Ed.; Engineering Foundation: New York, 1987; pp 199–210.
50. Kirkland, J. J. personal communication.
51. Kondo, S. personal communication.
52. Zhuravlev, L. T. *Langmuir* **1987**, *3*, 316.
53. Knözinger, H.; Stählin, W. *Progr. Colloid Polymer Sci.* **1980**, *67*, 33.
54. Kittlemann, U.; Unger, K. K. *Progr. Colloid Polymer Sci.* **1980**, *67*, 19.
55. Avnir, D.; Farin, D.; Pfeifer, P. *New J. Chem.* **1992**, *16*, 439.
56. Oscik, J. *Adsorption*; Ellis Horwood: Chichester, England, 1982.

24 Infrared Study of Chemical and H–D Exchange Probes for Silica Surfaces

B.A. Morrow and A.J. McFarlan
University of Ottawa, Department of Chemistry

CONTENTS

The accessibility of surface silanols toward hydrogen-sequestering agents with steric dimensions that increase in the order $ZnMe_2$, BCl_3, $TiCl_4$, $AlMe_3$, and $Me_3SiNHSiMe_3$ (HMDS) were compared spectroscopically and gravimetrically for a fumed and a precipitated silica. The H–D exchange reaction was similarly studied with D_2O, ND_3, and deuterated methanol, isopropyl alcohol, and tert-*butyl alcohol. The surface areas of the two silicas are similar, but the silanol density on the precipitated silica is about twice that on the fumed silica. Approximately equal fractions of the total number of OH groups per square nanometer undergo H–D exchange with a given probe, and this fraction decreases as the size of the molecule increases. The fraction of silanols that can be derivatized by chemisorption is about twofold greater on the fumed silica, but the surface densities of derivatized silanols on the two silicas are nearly equal for the same chemisorption probe. The degree of exchange is much greater than the degree of chemisorption of molecules of similar size because the chemisorbed product on the surface inhibits further reaction.*

The surface properties of amorphous silicas are largely influenced by the nature of the surface silanol (SiOH) groups [1–3]. Lewis acid–base sites are absent unless the silica has been activated at very high temperatures, Brønsted acidity at the gas–solid interface is low or nonexistent, and the siloxane bridges are relatively unreactive toward most molecules. This chapter discusses some methods that employ chemical modification and H–D exchange to probe the nature of the surface hydroxyl groups on silica. Infrared spectroscopy is the main technique used, and one objective has been to compare the silanol groups on a fumed silica with those on a precipitated silica.

EXPERIMENTAL DETAILS

This work was carried out with (1) a fumed silica (Cab-O-Sil HS5; designated A-x, where x is the temperature of activation for 1 h in vacuum) having a Brunauer–Emmett–Teller (BET) (N_2) surface area of $325 \pm 5\ m^2/g$ and (2) a high-purity (Na 60 ppm, Al <100 ppm, Fe 20 ppm, and Ti <20 ppm) nonporous precipitated silica (designated P-x) from Rhône-Poulenc (France) having a surface area of $285 \pm 5\ m^2/g$. The powder (50 mg total, $10\ mg/cm^2$) was compacted at 10^7 Pa into thin self-supporting disks for IR transmission studies. The IR cell [4] was constructed of quartz, had a volume of about

300 ml, and was connected via a lightly greased ball joint to a Pyrex vacuum line (300 ml in volume) capable of attaining a base pressure of about 10^{-7} torr (1.3×10^{-5} Pa). All reactants were transferred as gases from the main manifold to the reaction cells, and pressures (measured with a capacitance nanometer) or does are indicated in the text. Fourier transform infrared (FTIR) spectra were recorded with a Bomem DA3-02 instrument (mercury cadmium telluride, or MCT, detector) or a Bomen Michelson MB100 instrument at a resolution of 2 cm^{-1}, except for those shown in Figure 24.7, for which a resolution of 4 cm^{-1} was used. Vacuum microbalance experiments were carried out with a Sartorius model 4433 instrument having a sensitivity of 0.1 μg.

RESULTS

UNTREATED SILICA

Figure 24.1A through Figure 24.1C show infrared spectra of fumed silica in the 3800- to 3000-cm^{-1} spectral region after activation in vacuum at 150, 450, and 800°C, respectively. Figure 24.2A through Figure 24.2C show the corresponding spectra for the precipitated silica activated at the same temperatures. Both sets of spectra for all temperatures of activation exhibit the well-known [1–3,5–7] sharp intense band near 3747–3740-cm^{-1} due to isolated noninteracting SiOH groups, this peak being the sole feature after activation at 800°C.

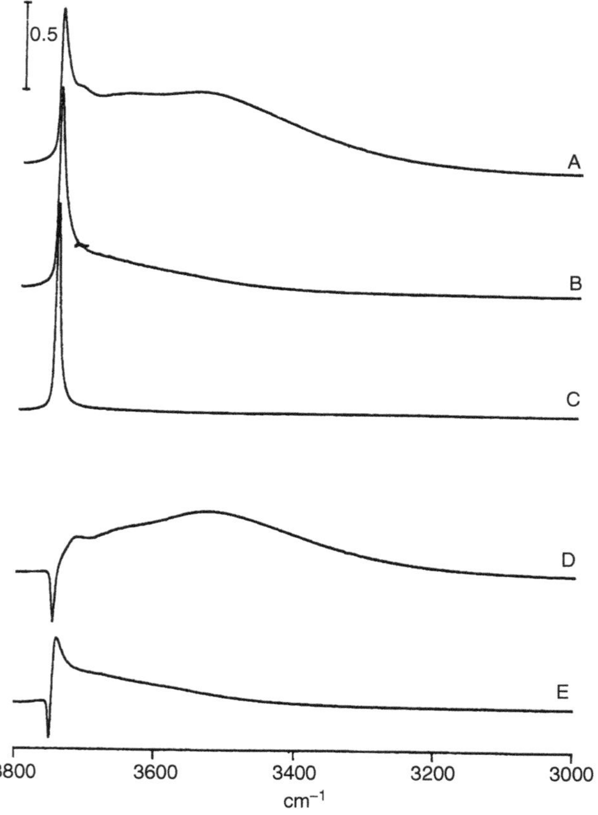

FIGURE 24.1 Infrared spectra of fumed silica in the SiOH stretching region after vacuum activation for 1 h at (A) 150, (B) 450, and (C) 800°C. Curves D and E are the difference spectra A–B and B–C, respectively.

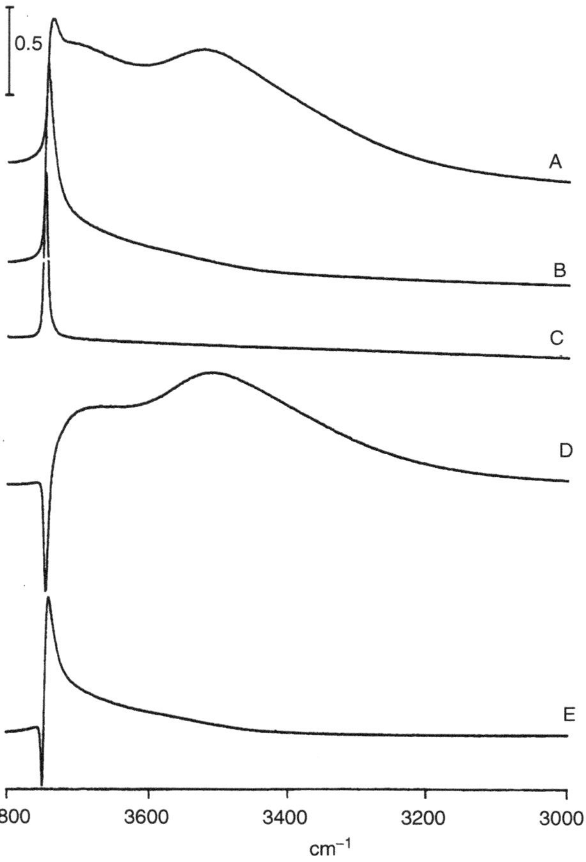

FIGURE 24.2 Infrared spectra of precipitated silica in the SiOH stretching region after vacuum activation for 1 h at (A) 150, (B) 450, and (C) 800°C. Curves D and E are the difference spectra A–B and B–C, respectively.

For 150°C activation, the peak due to isolated silanols is at 3747 cm^{-1} for A-150 and at 3738 cm^{-1} for P-150, and for both silicas there is a broad band centered near 3520 cm^{-1}. For A-150 (Figure 24.1A) there is a shoulder that extends to low wave number of the 3747-cm^{-1} peak near 3720–3715 cm^{-1}, and the 3520- and 3720-cm^{-1} bands have been assigned [6,7] to the H-bonded and free silanol groups of a pair or chain of silanols as follows:

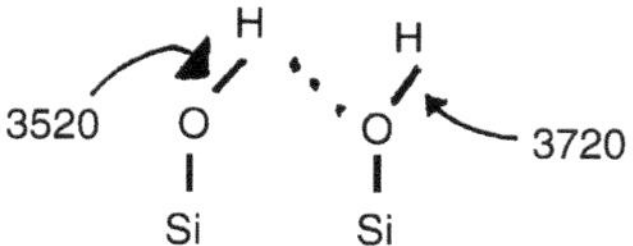

A slight inflection near 3660 cm^{-1} can just be discerned in Figure 24.1A. This band is due to perturbed or inaccessible internal silanols [8–10], the numbers of which increase as the pressure used to prepare self-supporting disks is increased. They are largely inaccessible to many reactant gases, and this topic is discussed in more detail later in the chapter. The 3720- and 3660-cm^{-1} features are not resolved in the spectrum of the precipitated silica (Figure 24.2A), which instead exhibits a broad feature near 3695 cm^{-1}.

Activation at 450°C eliminates most of the intensity due to the H-bonded and inaccessible silanols and leaves an asymmetric peak at 3747 cm^{-1} for A-450 and at 3745 cm^{-1} for P-450; this asymmetry largely disappears following activation at 800°C, at which point the frequency of both peaks is at 3748 cm^{-1}. The most noticeable difference between the fumed and precipitated silicas is that for the precipitated silica, and particularly for P-150, the isolated SiOH peak intensity is lower, and that due to H-bonded silanols is higher.

Figure 24.1D and Figure 24.2D show the difference spectra upon heating from 150 to 450°C, that is, curve 1A–1B and curve 2A–2B. Peaks going upward indicate bands that have decreased in intensity upon going from 150 to 450°C, and those going downward indicate bands that have increased in intensity. For both, there is a loss of intensity associated with H-bonded silanols and an increase in the isolated free SiOH intensity with increasing temperature. This loss can be attributed [6,7] to a condensation process whereby a chain of H-bonded silanols containing an odd number of silanols (three in the example) condense to liberate water, thereby creating a siloxane bridge site and an isolated silanol:

$$\text{Si–O–H}\cdots\text{O(H)–Si}\cdots\text{O(H)–Si} \dashrightarrow \text{Si–O–Si} + \text{Si–O–H} + H_2O \qquad (24.1)$$

Such effects have been observed for other silicas [5,11], and the changes are much more pronounced for the precipitated silica.

Figure 24.1E and Figure 24.2E show the corresponding spectral changes upon heating from 450 to 800°C. The changes here are mainly associated with the disappearance of the asymmetric tail to low wave number of the 3747–3745-cm^{-1} peaks, and the change is again more pronounced for the precipitated silica.

A comparison of the spectra of A-150 and P-150 after 150°C activation (Figure 24.1A and Figure 24.2A, respectively) shows that the band area of the precipitated silica is apparently greater than that of the fumed silica. In a series of spectra recorded with identical 2-cm^{-1} resolution, the integrated intensities (band areas) were 150 and 292 cm^{-1} for the same quantity (10 mg/cm^2) of fumed or precipitated silica (for convenience the unit cm^{-1} will be omitted in future discussions of integrated intensities). These integrated intensities have no quantitative meaning because the extinction coefficient of an H-bonded OH oscillator varies with wave number [12], but they suggest that there are more silanol groups on the precipitated silica under these conditions. As is discussed in more detail in the next section, the number of OH groups per square nanometer is about 3.1 and 6.8 for A-150 and P-150, respectively.

The strong peaks in the spectra of A-450 and P-450 are asymmetric to low wave number, and much of this asymmetry disappears after activation at 800°C. This asymmetry has been attributed in the past to isolated nonhydrogen-bonded geminal silanol groups [13], $Si(OH)_2$, which are expected to exhibit two OH stretching modes. However, theoretical calculations indicate [14] that the shift between these modes might be as small as 1 or 2 cm^{-1}. IR spectroscopy has not been able to demonstrate whether geminal species indeed exist on silica, although ^{29}Si NMR spectroscopy evidence has apparently demonstrated that these species can be distinguished, even on vacuum-activated fumed silica up to 1000°C [13,15]. The fraction of geminal silicon sites is always about 17 ± 3% of total silicon sites bearing hydroxyls, regardless of the method of preparation of the silica or the activation temperature under vacuum [13,15–18]. Thus, this 17% figure seems to be characteristic of all amorphous annealed silicas. In the work discussed here, no distinction is made between isolated and geminal silanols insofar as the 3747–3737-cm^{-1} peak can be assigned to either species.

Chemical and H–D Exchange Probes for 150°C Activated Silica

The infrared spectrum of fumed silica after vacuum activation at 150°C for 1 h (A-150) is identical to that observed after simple evacuation for 1 h at ambient temperatures [6,19]. Experiments with a vacuum microbalance have confirmed that there was insignificant mass loss upon going from 1-h evacuation at 22°C to further evacuation at 150°C for 1 h. Therefore, the spectrum shown in Figure 24.1A can be considered to be that of an "as received" hydroxylated fumed silica that contains no adsorbed water. The infrared spectrum of P-150 is almost identical to that observed after evacuation at ambient temperature for 1 h, except that there was about 5% decrease in the intensity of the broad 3520-cm^{-1} feature upon going to 150°C. The microbalance indicated a small decrease in mass of the sample upon going from 22°C vacuum treatment to 150°C (discussed later). After evacuation of either silica for 1 h at 22°C, there was no indication of an infrared band at 1620 cm^{-1} characteristic of the bending mode of H_2O; thus, the ambient temperature vacuum treatment was sufficient to remove all adsorbed water. Therefore, the loss of mass of P-22 upon heating to 150°C can probably be attributed to the condensation of pairs of H-bonded silanols, which results in the desorption of water.

One notable difference between A-150 and P-150 was that there was no mass loss with A-150 during prolonged evacuation at 150°C, whereas with P-150 there was a continuing very slow decrease in mass. Thus, for this precipitated silica the condensation process is a continuous slow process even at 150°C. This situation poses difficulties in specifying the number of SiOH groups that are present on P-150. Because most of the data described here for both silicas were obtained after a standard 1-h activation at 150°C, the initial SiOH density was measured under these conditions only, with the understanding that for the precipitated silica these numbers would be slightly lower if longer activation times were used. Before these results are given, the question of the accessibility of surface silanols is first addressed.

Figure 24.3 shows infrared spectra of A-150 and P-150 before and after complete reaction with excess $TiCl_4$. (A "complete reaction" is considered to have occurred when no further spectral changes are observed; this time was about 1 min for $TiCl_4$. The time evolution of these spectral changes is discussed later.) Curves C and C′ show the difference spectra, illustrating the spectral *change* as a result of the reaction. Curves B and B′ show that not all of the silanols react with $TiCl_4$, because there is a large residual intensity with a maximum near 3660 cm^{-1}. The band at 3660 cm^{-1} has been attributed to perturbed silanols that are at points of interparticle contact and are inaccessible to some reactants [8–10]. The number of these inaccessible silanols on fumed silica increases with the pressure used to prepare self-supporting disks.

In principle, $TiCl_4$ might react with single silanols, vicinal pairs, or geminal pairs of silanols on silica as shown in reaction 2, 3, and 4. Although these reactions are not shown, it is possible that only one or both of the OH groups of a vicinal pair or geminal pair of silanols might react as depicted for an "isolated" silanol.

$$\text{isolated}\quad \mathrm{SiOH} + \mathrm{TiCl_4} \longrightarrow \mathrm{SiOTiCl_3} + \mathrm{HCl} \tag{24.2}$$

$$\text{Vicinal pairs:}\quad \begin{matrix}\mathrm{SiO{-}H}\\ \mathrm{SiO{-}H}\end{matrix} + \mathrm{TiCl_4} \longrightarrow \begin{matrix}\mathrm{SiO}\\ \mathrm{SiO}\end{matrix}\!\!>\!\mathrm{TiCl_2} + 2\mathrm{HCl} \tag{24.3}$$

$$\text{Geminal}\quad \mathrm{Si}\!<\!\begin{matrix}\mathrm{OH}\\ \mathrm{OH}\end{matrix} + \mathrm{TiCl_4} \longrightarrow \mathrm{Si}\!<\!\begin{matrix}\mathrm{O}\\ \mathrm{O}\end{matrix}\!>\!\mathrm{TiCl_2} + 2\mathrm{HCl} \tag{24.4}$$

For other reactive hydrogen-sequestering (HS) agents, the extent of reaction would be expected to be different as the size of the reactant changed. Figure 24.4 shows a series of spectra observed after the complete reaction of A-150 with various HS agents (the exchange reaction with D_2O is included for comparison; in this reaction accessible SiOH is converted to SiOD), and Figure 24.5 shows the corresponding spectra for P-150. Inspection shows that there is a considerable difference in the numbers of residual or inaccessible silanols according to the nature of the reactant.

We have attempted to quantity the effect of the size of the reactant on the number of silanols that react with a given HS agent. Table 24.1 shows the cross-sectional area of each reactant, the integrated intensity of the residual silanol band profile after reaction with each HS agent (and D_2O), and the percentage decrease of the initial SiOH band intensity. The reactant area was calculated by using the same method used to calculate BET cross-sectional areas, that is, on the basis of values of the density of the liquid reactant and with the assumption of close-packed spheres. These values are at best approximate; for example, a molecule such as $ZnMe_2$ having a

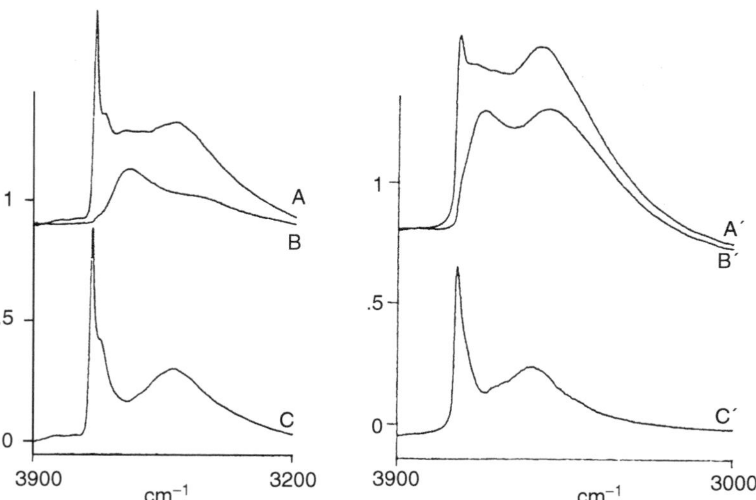

FIGURE 24.3 Infrared spectra of A-150 and P-150 before (A and A′, respectively) and after (B and B′) complete reaction at 22°C with $TiCl_4$. Curves C and C′ show the difference spectra (A–B and A′–B′).

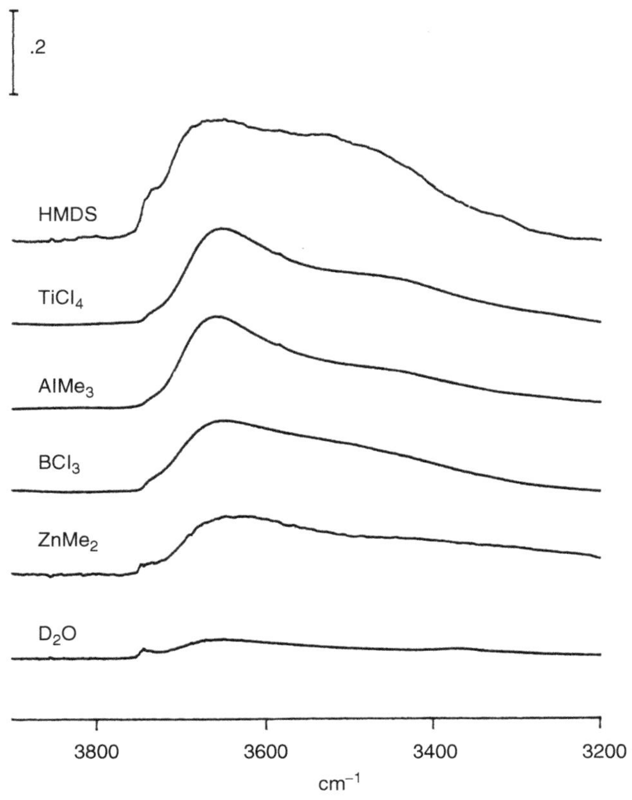

FIGURE 24.4 Infrared spectra of fumed silica A-150 after complete reaction at 22°C with the indicated hydrogen-sequestering agent or after H–D exchange with D_2O. The integrated intensities are shown in Table 24.1.

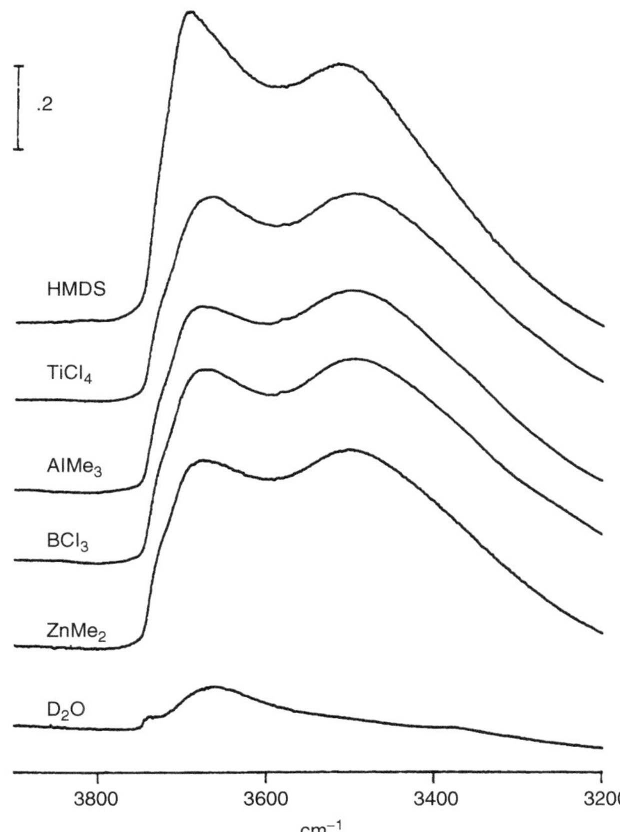

FIGURE 24.5 Infrared spectra of precipitated silica P-150 after complete reaction at 22°C with the indicated hydrogen-sequestering agent or after H–D exchange with D_2O. The integrated intensities are shown in Table 24.1.

linear C–Zn–C configuration is probably far from spherical in its reactive cross section. The data show that the area of the residual band after reaction with A-150 increases as the size of the reactant increases. However, for P-150 there is little change upon going from $TiCl_4$ to $ZnMe_2$, and then a very large change upon going from $ZnMe_2$ to exchange with D_2O.

The data shown in Table 24.1 were obtained with A-150 or P-150 sample disks that were pressed under identical conditions and that contained the same quantity of silica, 10 mg/cm^2. The integrated area of P-150 before reaction was about twice that of A-150; this observation suggests that there are a greater number of silanols initially on P-150. As stated earlier, the relative areas before reaction have no quantitative significance because the extinction coefficient of an H-bonded OH oscillator is expected to vary with the strength of the interaction and is generally larger the greater the shift to lower wave number [12]. Therefore, the silanol density (SiOH groups per square nanometer) was measured for both silicas after 1 h of activation. This attempt immediately poses the problem of whether the accessible silanol density or the total silanol density is wanted; for the accessible silanol density, the "size" of the reactant will obviously influence the results. Therefore, three procedures were used to obtain a measure of this parameter. By using the vacuum microbalance, the mass change was measured, first after exchange with either D_2O or ND_3, these being the "smallest" probe molecules used (as will be discussed, the molecules have similar degrees of exchange and similar cross-sectional areas), and then after complete reaction with HMDS (hexamethyldisilazane), this being the largest probe used. In the third method, the total SiOH density was determined by measuring the mass loss upon heating the silicas from 150 to 450°C (this heating eliminates most of the H-bonded and perturbed inaccessible silanols as water). Then the remaining free isolated silanols were totally exchanged with D_2O, and the mass change for the conversion of SiOH to SiOD was measured. This third method gives the total silanol density on these silicas. The results are summarized in Table 24.2.

A comment on these silanol densities is necessary. A value of about 4.5–5.0 OH/nm^2 is generally accepted for "fully hydroxylated" silicas, although there is considerable scatter in the literature data [1,2,20,21]. Fumed silicas that

TABLE 24.1
Chemisorption Data for Reactions on A-150 and P-150

Reactant	Reactant area (Å^2)	A-150 Intensity[a] I_A	A-150 % SiOH change[b]	P-150 Intensity[c] I_p	P-150 % SiOH change[b]
HMDS[b]	54.2	100	33	248	15
$TiCl_4$	35.2	54	64	200	32
$AlMe_3$	32.1	45	70	189	35
BCl_3	30.3	47	69	196	33
$ZnMe_2$	25.7	36	76	188	36
D_2O	10.5	12	92	22	92

[a]Integrated intensity I_A after reaction; initial intensity $I_{0A} = 150$.

[b]Percent decrease in the SiOH intensity after reaction, I/I_0 for A-150 and P-150.

[c]Integrated intensity I_p after reaction; initial intensity $I_{0P} = 292$.

have not been intentionally rehydrated generally have a density lower than 4.5, as has been found by others [22,23]. Precipitated silicas and silica gels can have silanol densities much greater than 5 if polymerization is not complete, or if there is a large number of geminal silanols. The values reported here are well within the ranges expected for both types of silica considered to be "as received" materials. Thus, the most important result of the data in Table 24.2 is that the total silanol density of P-150 is about 2.2 times greater than that on A-150, and this factor also is reflected in the number accessible to D_2O or ND_3 for exchange. However, *equal numbers of silanols* are capable of reacting with HMDS.

The exchange reaction itself depends on the reactant area. Studies of this dependence have been carried out by using exchange molecules of differing steric dimension (D_2O, ND_3, and OD-containing methanol, isopropyl alcohol, and *tert*-butyl alcohol). After an initial spectrum of A-150 or P-150 was recorded the sample was exposed at 22°C to 10 torr (1300 Pa) of a given molecule for 10 min, followed by evacuation for 10 min. This process was repeated a total of five times, and after the last exposure the sample temperature was raised to 150°C while the sample was evacuated, held at 150°C for 10 min, and cooled to 22°C before the final spectrum was recorded. There was very little change between the fourth and fifth exposure, and all final spectra for both silicas are shown in Figure 24.6. Table 24.3 shows the area of each exchange molecule, the integrated intensity after exchange, and the percent change in intensity as a result of the exchange. As expected, the degree of exchange increases as the size of the reactant decreases. However, for the same probe molecule, *the percent exchange is virtually identical for each silica.* A comparison of the data in Table 24.1 and Table 24.3 shows that for molecules of comparable size, those that exchange are accessible to many more silanol groups than those that chemisorb. For example, in a comparison of i-PrOD or *t*-BuOD (i-Pr is isopropyl; *t*-Bu is *tert*-butyl) with the series $ZnMe_2$, BCl_3, $AlMe_3$, and $TiCl_4$, the contrast is particularly dramatic for P-150, for which only about 33% of the silanols were accessible to the molecules that chemisorb on silica, whereas about 80% underwent exchange with i-PrOD or *t*-BuOD.

TABLE 24.2
Silanol Density on A-150 and P-150 Determined by Different Methods

Method	A-150	P-150
D_2O–ND_3 exchange	2.5 ± 0.1	5.6 ± 0.3
HMDS reaction	1.42 ± 0.02	1.48 ± 0.02
Total SiOH	3.1 ± 0.1	6.8 ± 0.3

Note: Densities are reported as SiOH groups per square nanometer.

Relative Rates of Reaction of Various Silanol Types

The question of whether there is a difference of reactivity at 22°C between the H-bonded accessible silanols (3520 cm^{-1}) and free isolated silanols (3748–3738 cm^{-1}) with the most reactive HS agents is now addressed. This question was studied by using the rapid scan facility of a DA3 FTIR spectrometer, in which a single 4-cm^{-1} resolution spectrum can be obtained in 0.25 sec. The procedure involves the addition of excess reactant and following the spectral changes as a function of time. The results for $AlMe_3$, BCl_3, and $TiCl_4$ on both silicas are reported here. These reactions are essentially complete after 12, 24, and 60 sec, respectively, after a reactant exposure of

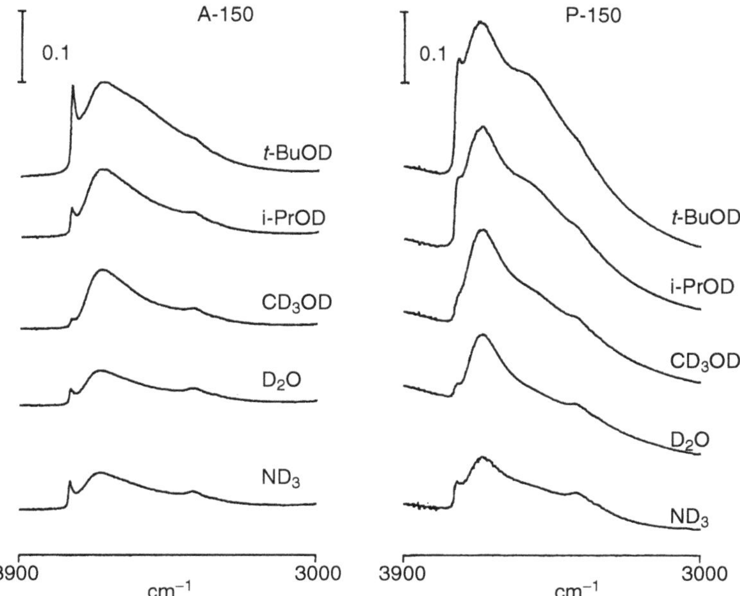

FIGURE 24.6 Infrared spectra of A-150 and P-150 after H–D exchange at 22°C with the indicated molecule. The integrated intensities are shown in Table 24.3.

4–8 mmol/g of silica (this exposure corresponds to about a three- or four-fold excess of reactant when compared with the number of SiOH groups that react).

The results are presented as difference spectra in Figure 24.7, which show the spectral *changes* that occurred in a given time interval. The number of scans and the interval between spectra are shown in Table 24.4. Only the first three intervals are shown because 50–70% of the silanols had reacted at the end of the third sequence, and the time of each interval was chosen to represent about the same degree of reaction for each reactant. The signal-to-noise level in these spectra is poor, particularly for BCl_3 and $AlMe_3$, for which a single scan was used. However, for the A-150 series an inspection, particularly of the first interval, shows that the broad 3520-cm^{-1} band due to H-bonded silanols has a greater intensity in the order $BCl_3 > TiCl_4 > AlMe_3$. This trend continues for the other intervals. A parameter "*R*", which is the ratio of the peak height of the isolated SiOH peak ($\approx$3745 cm^{-1}) to that at 3520 cm^{-1}, was determined. From the curves shown and other sets of experimental data, the *R* values were estimated to be 5.0 ± 0.3 for $AlMe_3$, 2.7 ± 0.2 for $TiCl_4$, and 1.9 ± 0.1 for BCl_3. A low value of *R* indicates that the H-bonded silanols are initially relatively more reactive than the isolated silanols in the order $BCl_3 > TiCl_4 > AlMe_3$.

TABLE 24.3
Exchange Data for Reactions on A-150 and P-150

Reactant	Reactant area (Å^2)	A-150 Intensity[a] I_A	A-150 % SiOH change[b]	P-150 Intensity[c] I_p	P-150 % SiOH change[b]
t-BuOD	31.7	36	76	74	75
i-PrOD	27.6	23	85	54	81
CD_3OD	18.0	17	89	36	88
ND_3	12.9	13	92	21	93
D_2O	10.5	12	92	22	92

[a]Integrated intensity I_A after exchange; initial intensity $I_{0A} = 150$.

[b]Percent decrease in SiOH intensity after exchange, I/I_0 for A-150 and P-150.

[c]Integrated intensity I_p after exchange: initial intensity $I_{0P} = 292$.

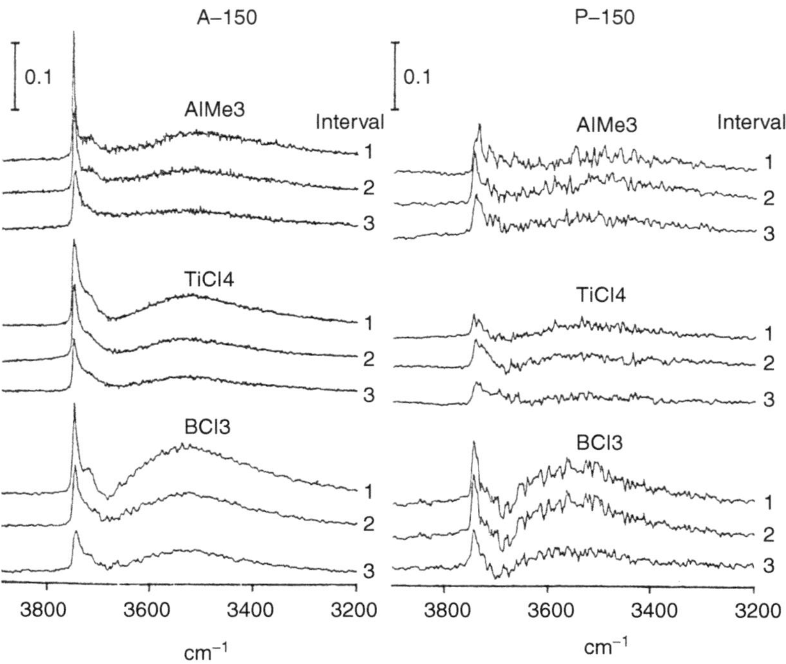

FIGURE 24.7 Difference spectra for A-150 and P-150 as a function of time after reaction with the indicated HS agents at 22°C. The "interval" corresponds to the recording of the first three sets of successive spectra with the time lapse indicated in Table 24.4 (seconds between spectra).

For the P-150 series, the intensity is low, and there is not such a clear difference in the extent of "disappearance" of the 3520- versus ≈3745-cm^{-1} peaks with any reactant. Qualitatively, the trend follows the same order as for A-150, but because of the very poor signal-to-noise level, characterization via the *R* factor was not attempted.

Although not shown in Figure 24.7, from about the fifth interval on for any reactant and for either silica, there was very little change in intensity due to H-bonded silanols, whereas the sharp peak at 3745 cm^{-1} continued to "disappear", albeit to a lesser extent than in the earlier intervals. Thus, the reaction with H-bonded silanols ceases before that with isolated silanols. This observation is consistent with the results obtained for the static and exchange reactions.

DISCUSSION

The purpose of this work has been to compare the chemical reactivity and the H–D exchange characteristics of the silanol groups on two hydroxylated "as received" silicas of different origin. The silicas have similar surface areas, and both are essentially nonporous. This discussion addresses only the properties of the A-150 and P-150 materials.

The total SiOH density on P-150 is about twice that on A-150 (Table 24.2; 6.8/3.1 = 2.2). This factor of about 2 occurs frequently. For example, the number of SiOH groups per square nanometer accessible to exchange with D_2O and ND_3 also differs by about two (Table 24.2; 5.6/2.5 = 2.2). For each of the exchange

TABLE 24.4
Time and Scan Data for Fast-Scan Spectra

	A-150		P-150	
Reactant	**Time between spectra (sec)**	**No. of scans[a]**	**Time between spectra (sec)**	**No. of scans[a]**
$AlMe_3$	1	1	2	1
BCl_3	2	1	2	1
$TiCl_4$	4	4	4	4

[a]A single scan requires 0.25 sec. For multiple scans, one 0.25-sec scan can be acquired every 0.5 sec.

reagents listed in Table 24.3, about the same percent of H–D exchange occurred on each silica, meaning that about 2 times the absolute number of silanols on P-150 were inaccessible to each reagent as on A-150. Finally, for the chemisorption reagents listed in Table 24.1 (ignoring the exchange data for D_2O), the "percent SiOH change" for A-150 is about twice that observed for P-150 (see columns 4 and 6). However, the absolute number of silanols that reacted with HMDS (Table 24.2) was about equal for both silicas.

The exchange data shows that the texture or porosity of both surfaces is probably similar, because the same fraction of silanols underwent H–D exchange on both silicas when a variety of exchange probes of differing size was used. Moreover, the chemisorption data for the various HS agents show that the reaction with H-bonded and inaccessible silanol groups ceases when about the same total number of silanols have been derivatized, regardless of the silica type. Given the factor of 2 difference in the silanol densities per square nanometer, the absolute number of silanols that "react" with a given HS agent is about the same on both silicas (and this fact was directly verified for HMDS; see Table 24.2). That is, when a similar density of chemisorbed product per square nanometer is formed on either silica, then further reaction is essentially inhibited. This point of inhibition presumably occurs when a shroud of chemisorbed product covers the accessible external surface, thereby preventing diffusion of the reactant to other silanols that are normally accessible to the exchange probes. Because the silicas have similar surface areas, this shroud is formed when the same absolute number of silanols have reacted. This observation again suggests that the surface texture of these silicas is similar.

The rate studies showed that there is a difference in the reactivity of H-bonded versus isolated silanols. About 98% of the spectral changes occurred within 12, 24, or 60 s for $AlMe_3$, BCl_3, and $TiCl_4$, respectively. These molecules have comparable sizes (see Table 24.1), so differences in the rates of reaction are probably chemical in origin, although diffusion might also play a small role. As has been discussed for A-150 (6), the greater ability of the chlorine-containing HS agents to react initially with the H-bonded silanols may be related to the ability of these agents to react bifunctionally with vicinal pairs of silanols (reaction 3). Further, the BET "areas" are probably not a realistic measure of the differences in the reactive cross-sectional areas of these molecules. A methyl group may occupy more space than a Cl atom on the basis of van der Waals dimensions so that, relative to BCl_3 or $TiCl_4$, $AlMe_3$ may have a greater effective cross-sectional area than is suggested by the numbers in Table 24.1. Whatever the interpretation of these results, the reaction with H-bonded silanols on *both* silicas stopped well before that of the isolated silanols; this observation again illustrates that the blocking effect of adjacent derivatized vicinal silanols (reaction 3) is effective in preventing further reaction, whatever the size of the reactant.

Finally, this work has demonstrated the utility of probing the silanol groups on silica by using both chemisorption and H–D exchange. The conclusions reached could not have been arrived at without this combined approach that used probes of differing steric dimensions. The chemisorption probes clearly inhibit reaction by blocking access to other silanols, whereas this inhibition does not occur with the H–D exchange probes. It would be very interesting to extend this study to compare nonporous silicas with meso- or micropores to assess the possible role of proton migration in the exchange reaction.

ACKNOWLEDGMENT

We are grateful to the Natural Sciences and Engineering Research Council of Canada for financial support.

REFERENCES

1. Kiselev, A. V.; Lygin V. I. *Infrared Spectra of Surface Compounds*; Wiley: New York, 1975.
2. Iler, R. K. *The Chemistry of Silica*; Wiley: New York, 1979.
3. Hair, M. L. *Infrared Spectroscopy in Surface Chemistry*; Dekker: New York, 1967.
4. Morrow, B. A.; Ramamurthy, P. *J. Phys. Chem.* **1973**, *77*, 3052.
5. Hoffmann, P.; Knozinger, E. *Surface Sci.* **1987**, *188*, 181.
6. Morrow, B. A.; McFarlan, A. J. *J. Non-Cryst. Solids* **1990**, *120*, 61.
7. Morrow, B. A. *Stud. Surf. Sci. Catal.* **1990**, *57A*, A161.
8. Tyler, A. J.; Hambleton, F. H.; Jockey, J. A. *J. Catal.* **1969**, *13*, 35.
9. Armistead, C. G.; Tyler, A. J.; Hambleton, F. H.; Mitchell, S. A.; Hockey, J. A. *J. Phys. Chem.* **1969**, *73*, 3947.
10. Hambleton, F. H.; Hockey, J. A.; Taylor, J. A. G. *Nature (London)* **1965**, *208*, 138, and *Trans. Faraday Soc.* **1966**, *62*, 801.
11. Burneau, A.; Barrès, O.; Gallas, J. P.; Lavalley, J. C. *Langmuir* **1990**, *6*, 1364.
12. Paterson, M. S. *Bull. Minéral.* **1982**, *105*, 20.
13. Morrow, B. A.; Gay, I. D. *J. Phys. Chem.* **1988**, *92*, 5569.
14. Sauer, J.; Schröder, K. P. *Z. Phys. Chem.* **1985**, *266*, 379.
15. Maciel, G. E.; Sindorf, D. W. *J. Am. Chem. Soc.* **1980**, *102*, 7606.
16. Sindorf, D. W.; Maciel, G. E. *J. Am. Chem. Soc.* **1983**, *105*, 1487.
17. Fyfe, C. A.; Gobbi, G. C.; Kennedy, G. J. *J. Phys. Chem.* **1985**, *89*, 277.

18. Legrand, A. P.; Hommel, H.; Tuel, A.; Vidal, A.; Balard, H.; Papirer, E.; Levitz, P.; Czernichowski, M.; Erre, R.; Van Damme, H.; Gallas, J. P.; Hemidy, J. F.; Lavalley, J. C.; Barrès, O.; Burneau, A.; Grillet, Y. *Adv. Colloid Interface Sci.* **1990**, *33*, 91.
19. Ghiotti, G.; Garrone, E.; Morterra, C.; Boccuzzi, F. *J. Phys. Chem.* **1979**, *83*, 2863.
20. Zhuravlev, L. T. *Langmuir* **1987**, *3*, 316.
21. Tanabe, K.; Misono, M.; Ono, Y.; Hattori, H. *Stud. Surf. Sci. Catal.* **1989**, *51*, 92.
22. Gay, I. D.; McFarlan, A. J.; Morrow, B. A. *J. Phys. Chem.* **1991**, *95*, 1360.
23. Mathias, J.; Wannemacher, G. *J. Coll. Interface Sci.* **1988**, *125*, 61.

25 Infrared Studies of Chemically Modified Silica

Barry A. Morrow and David T. Molapo
University of Ottawa, Department of Chemistry

CONTENTS

Vibrational spectroscopy has had a profound effect on our understanding of the surface chemistry of silica. Indeed, it has been the model system for the use of IR spectroscopy for probing the surface chemistry of oxides, the first studies having been carried out in late 1950. The books written by Hair [1] and Little [2] in the sixties have become 'classics' with respect to the early use of IR spectroscopy for studying the surface properties of silica and adsorbed species thereon.

The most common method which has been used to record the IR spectrum of silica has been to compact a high surface area silica (200 to 300 m^2/g) into a self-supporting disc which contains from 5 to 10 mg of SiO_2 per cm^2. By using this preparation technique the disc can then be conveniently mounted in a suitable evacuable chamber where subsequent vacuum activation can be carried out in order to remove adsorbed water or other contaminants, and adsorption experiments at the gas-solid interface can easily be carried out.

This paper will only be concerned with the surface chemistry of silica in vacuum and/or in the presence of an adsorbate. The IR spectrum of a typical self-supporting disc of a pyrogenic or fumed silica after heating under vacuum for 1 h at 150°C is shown in Figure 25.1A. Pyrogenic or fumed silicas [some trade names are Aerosil and Cab-O-Sil] are made by the flame hydrolysis of $SiCl_4$ at 1000°C. These non-porous silicas have a low bulk density and adsorbed water can be removed by evacuation at 20°C. However, the spectral properties are identical when evacuation is carried out at 150°C; evacuation at the latter temperature is preferred in order to remove any trace impurities which may be present. The spectrum is characterized by a sharp absorption band at 3747 cm^{-1} with a broad tail to low wavenumber having a maximum near 3550 cm^{-1}. The sharp peak is due to the OH stretching vibration of isolated non-hydrogen-bonded SiOH or $Si(OH)_2$ groups and the broad tail having a peak near 3550 cm^{-1} is due to these groups when they are H-bonded. The weak broad features between 2000 and 1300 cm^{-1} are due to overtones and combination modes of bulk SiO_2 vibrations and the region of total absorption between 1300–1000 cm^{-1} and from 850–750 cm^{-1} are due to absorption of IR radiation by bulk modes of SiO_2.

The IR spectrum in the OH stretching region is shown on an expanded wavenumber scale in Figure 25.2A. Additional weak features near 3720 and 3650 cm^{-1} can be discerned and we will comment on these later.

The hydrogen bonded hydroxyls start to condense to liberate water when silica is heated above 150°C [3], the reaction being:

$$\mathrm{Si{-}O{-}H\cdots O(H){-}Si} \longrightarrow \mathrm{Si{-}O{-}Si} + \mathrm{H_2O} \qquad (1)$$

The majority of the H-bonded silanols condense in this way when silica is heated at 450°C under vacuum, and a typical IR spectrum is shown in Figure 25.2B. There is an intensified peak at 3747 cm^{-1} due to isolated silanols and an asymmetric tail to low wavenumber.

The spectral changes are more clearly seen from 'difference' spectra, that is, subtracting the spectrum for the 450°C activated sample from the 150°C one, as shown in Figure 25.2C. Peaks going down represent new features which are created upon heating from 150 to 450°C and peaks going up represent spectral features

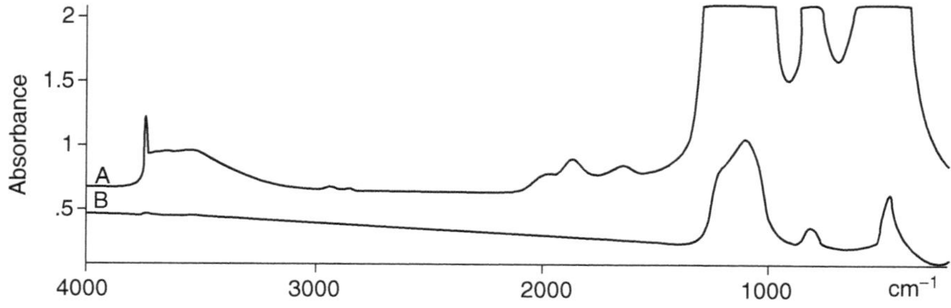

FIGURE 25.1 A, IR spectrum of a self-supporting disc and B, a thin film of SiO_2 after heating under vacuum at 150°C for 1 h.

lost during the heat treatment. Consider a chain of H-bonded silanols which contain an odd number of OH groups. In the example of three OH groups below, the terminal OH which is not H-bonded absorbs at 3720 cm^{-1} whereas those which are H-bonded absorb at about 3550 cm^{-1}. Heating results in the process:

$$\text{Si–O(H}\cdots\text{)–Si–O(H}\cdots\text{)–Si–OH} \longrightarrow \text{Si–O–Si} + \text{Si–OH} + H_2O \quad (2)$$

which creates one siloxane site, one water molecule, and an isolated SiOH group (3747 cm^{-1}). The net effect is an increase in the number of isolated SiOH groups and an increase in the intensity of the 3747 cm^{-1} IR band; this would not occur if the chain contained an even number of OH groups.

There is a weak spectral feature near 3650 cm^{-1} which can be seen in Figure 25.2A and 25.2C. This is known to be due to SiOH groups which are perturbed by interparticle contact; such OH groups are inaccessible to many reactants, depending its steric size [3].

Much of the earlier IR work which was related to the adsorption of molecules on silica was concerned with studying the disappearance of features associated with SiOH stretching vibrations, and, if applicable, looking at the appearance of new features due to adsorbed species. This has been relatively straight forward if the adsorbed moiety contains a 'light' functional groups, generally one which contains a hydrogen atom. Therefore, adsorbed species which contain CH_x or NH_x functionalities are easily detected because their CH or NH stretching and deformation vibrations lie above 1300 cm^{-1}. Below this, self-supporting discs of silica are opaque to IR radiation except for two regions of partial transparency between 1000 and 850 cm^{-1} and from about 750 to 550 cm^{-1}. [The situation is more extreme for discs of other oxides, for example Al_2O_3, TiO_2 or ZrO_2, which are totally absorbing below 1000 to 800 cm^{-1} without any windows of transparency [4].]

Accordingly, the traditional methods of IR transmission spectroscopy do not permit access to the low wavenumber spectral region, a region where most of the important functionalities other than hydrogen are expected to absorb, for example, most metal-oxygen modes, or Y=O modes (Y = a transition metal or S, P, and so on). Thus, when $TiCl_4$ adsorbs on silica to yield a $SiOTiCl_3$ surface species, the SiOTi modes are expected to lie near 1000 cm^{-1} and the $TiCl_3$ stretching modes are expected to lie near 500 to 400 cm^{-1} and both spectral regions are inaccessible to IR radiation when using the self-supporting disc IR transmission method.

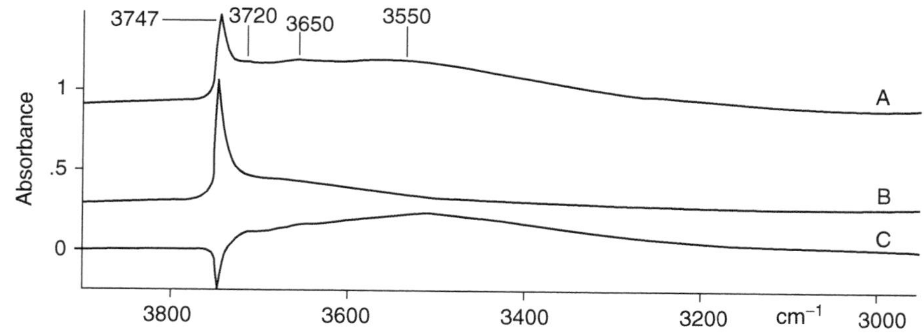

FIGURE 25.2 IR spectra of SiO_2 discs: A, after activation under vacuum at 150°C and B, 450°C; C, the difference spectrum, that is subtraction of curve B from A.

It can be argued that other methods permit access to the low wavenumber spectral region, such as Raman spectroscopy and diffuse reflectance spectroscopy (DRIFTS). It is beyond the scope of this paper to comment on the deficiencies of these methods; there are many when compared to the well understood transmission method, both from the experimental and theoretical points of view.

In 1984 our laboratory developed a new method for obtaining IR transmission spectral data for adsorbed species on silica [5]. With the then advent of FTIR methods and the ability to carry out spectral subtraction, we reasoned that if the quantity of SiO_2 could be reduced to the extent that even the strongest bulk SiOSi mode was not totally absorbing, then one should be able to detect new vibrational modes even in spectral regions where the silica is strongly absorbing. The reduction in the quantity of adsorbent was achieved by forming a thin film of silica on an optically transparent substrate. The latter was initially NaCl (low wavenumber limit about 600 cm^{-1}), but more recently this has included ZnSe (500 cm^{-1}), KBr (400 cm^{-1}), AgCl (300 cm^{-1}), CsI (200 cm^{-1}) and silicon (<200 cm^{-1}). Wide range mid-IR spectrometers have at best a CsI beam splitter, and the practical low wavenumber limit of transmission is 200 cm^{-1}. [Other beam splitters will permit lower wavenumber access, but at the expense of not being able to observe spectral features in the 4000 to 1000 cm^{-1} region, and for surface studies, it is essential to also be able to observe spectra in the latter spectral region.]

We will show a recent example of the utility of the method. The IR spectrum of a thin film of silica is shown in Figure 25.1B. In spite of there being strong infrared absorptions at 1100, 800 and 480 cm^{-1}, spectral subtraction of this spectrum from that which is observed following chemisorption of a reactant can yield spectral features which were formerly obscured when using a self-supporting disc. We will illustrate the power of this method from the adsorption of $P(CH_3)_2Cl$ on silica, a reaction which eventually yields a stable $SiOP{=}O(CH_3)_2$ surface species [6]. [In that study, oxygen-18 shifts and ^{31}P soild state NMR were used to confirm the identity of the surface species.]

Figure 3 (top) shows the spectra (1500–750 cm^{-1}) of silica before and after the reaction, respectively. Figure 25.3 (bottom) shows the difference spectrum at about ten times the original absorbance scale. All of the peaks are better resolved and the peaks at 1244 cm^{-1} (peak A) and 1045 cm^{-1} (peak B), which previously could not be clearly seen, are characteristic of P=O and SiOP vibrational modes respectively. These bands also underwent the expected ^{18}O shift when the reactant was adsorbed on an ^{18}O exchanged silica. The remaining modes, which are due to CH_3 deformation and rocking modes, did not exhibit any shift after ^{18}O exchange. The P=O and SiOP modes could never be detected using a self-supporting disc of SiO_2, and this example serves to illustrate the utility and methodology behind the thin film, or 'TF' technique.

The 'TF' method has been used extensively in our laboratory, and by Tripp and Hair [7–9]. Tripp was a graduate student in our group during the developmental stages, and he has independently applied this method with Hair to some very successful recent studies of the chemisorption of various silane coupling agents on silica. Although the early work was confined to silica as an adsorbent, we have recently demonstrated its utility for studying the low wavenumber spectra of adsorbed species on other oxides [4].

This paper is concerned with new applications of the TF method for studying adsorption processes on silica. In particularly, we have re-visited earlier work, the adsorption of $TiCl_4$ on SiO_2, a problem which has been under investigation using IR spectroscopy for about 30 yr, and we will show that some preconceived ideas about the reaction of this molecule on silica may have to be modified [10–14]. We will discuss past notions of this reaction later, but first we will discuss the experimental conditions.

EXPERIMENTAL

The silica used in this work was grade M5 Cab-O-Sil, having a BET (N_2) surface area of 205 m^2/g. The basic details have been described previously [3], but some important details of the method are worth repeating. The thin film of silica can usually be achieved by spreading a layer of fumed silica on an optically transparent window (CsI was used here) and wiping off the excess so that about 0.1 to 0.2 mg of SiO_2 per cm^2 remains. One wants to have an absorbance between 1.0 and 1.5 at 1100 cm^{-1}. The film is inevitably slightly inhomogeneous and it is essential that all spectral measurements be carried out *in situ* so that the sample position does not change, thus avoiding spectral artifacts which would arise after subtraction of the background spectrum. Spectra were recorded using a CsI equipped Bomem Michelson MB100 FTIR with a DTGS detector using 400 scans at a resolution of 4 cm^{-1}. Samples are designated Sx where x is the temperature of activation under vacuum for 1 hour.

RESULTS

Titanium tetrachloride is a very reactive molecules which will readily hydrolyze to yield titanium dioxide. Its reaction with the hydroxyl groups on silica has been studied extensively and it is generally assumed that it can react monofunctionally with single SiOH groups or bifunctionally with pairs of hydrogen bonded SiOH

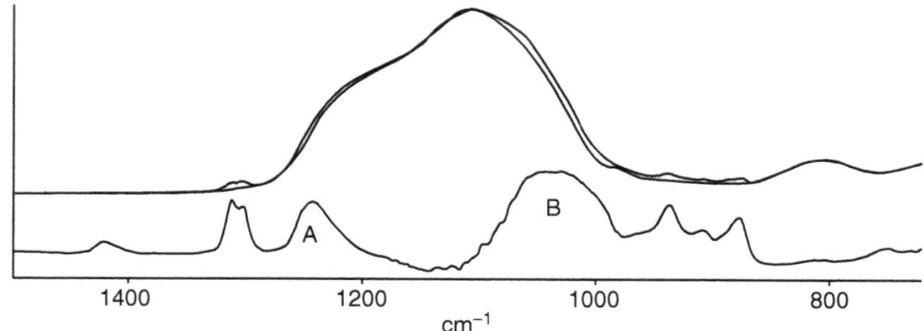

FIGURE 25.3 Top: thin film spectra of silica before (solid line) and after (dotted line) the chemisorption of $P(CH_3)_2Cl$. Bottom: difference spectrum (dotted minus solid) scaled about 10-fold relative to the top spectra.

groups as follows [10–14]:

$$SiOH + TiCl_4 \longrightarrow SiOTiCl_3 + HCl \quad (3)$$

$$(SiO-H)_2 + TiCl_4 \longrightarrow (SiO)_2TiCl_2 + 2\,HCl \quad (4)$$

The reaction at room temperature with the accessible SiOH groups on silica is over in less than 60 seconds when excess $TiCl_4$ is in contact with a self-supporting disc of SiO_2 [11].

Thin film infrared spectra of excess $TiCl_4$ in contact with S20 and S400 are shown in Figure 25.4A and Figure 25.4B respectively.

The very strong sharp band at 500 cm^{-1} is due to the antisymmeric TiCl stretching mode of $TiCl_4$ and the bands near 1000 to 900 cm^{-1} are due to SiOTi modes. For S400, which has fewer adjacent H-bonded silanols, the sharp band at 1028 cm^{-1} has been previously attributed to a $SiOTiCl_3$ species which reacts via reaction (3) [5,10]. For S20, which additionally has a larger number of H-bonded silanols, there is another strong band at 920 cm^{-1} accompanied by weaker bands from 800 to 700 cm^{-1}, and these have been attributed to a cyclic bifunctionally adsorbed species as would be created by reaction (4).

There are a number of weaker bands in both spectra which cannot be due $TiCl_4$; the symmetric $TiCl_4$ stretching mode at 388 cm^{-1} is forbidden in the IR and the TiCl deformation modes are below 200 cm^{-1} and cannot be observed [12]. On the other hand, the surface products of reactions (3) and (4) are expected to have TiCl stretching modes which lie between 500 and 388 cm^{-1} and which would be allowed in the infrared, and some of the weak bands observed in this spectral region are undoubtedly due to these modes.

The time evolution of the spectra are important. Figure 25.5 shows the spectral changes observed for the 400°C activated sample with increased time of contact with $TiCl_4$, and following evacuation of $TiCl_4$.

The most striking and predictable effect is the rapid decrease in the intensity of the peak at 500 cm^{-1} due to removal of unreacted $TiCl_4$. More subtle, is the slow decrease over 15 h evacuation of the peaks at 1028, 960 (shoulder) and the growth of a weak peak near 785 cm^{-1}. The lower wavenumber region will be discussed later. Figure 25.6 shows the spectral changes for the 20°C activated sample. Here, the spectral changes

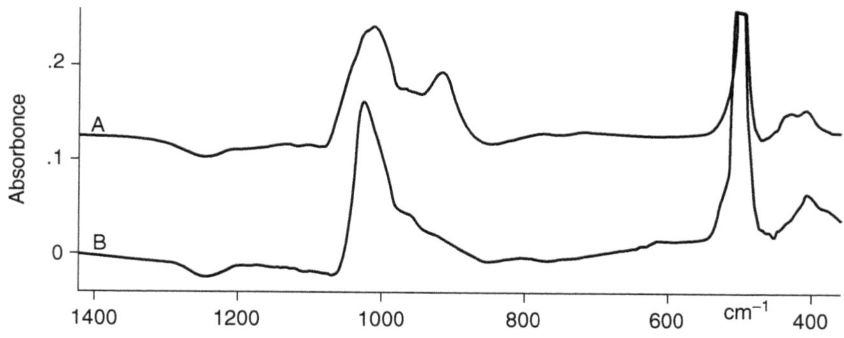

FIGURE 25.4 $TiCl_4$ adsorbed on SiO_2 activated at A 20°C and B 400°C.

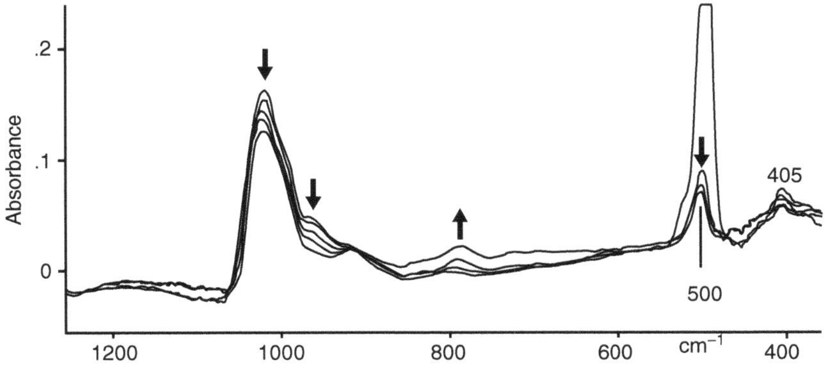

FIGURE 25.5 $TiCl_4$ adsorbed on 400°C activated SiO_2 for 5 and 30 min., and subsequently evacuated for 5 min., 30 min. and 15 h. The arrows indicate the direction of the spectral changes with increasing time.

are much more dramatic. Arrows indicate the direction of spectral change with time, but the most important feature is the dramatic increase in intensity in the 800–650 cm^{-1} spectral region. These changes appear to be correlated with a decrease in the intensity of the peak near 500 cm^{-1}, and the changes below 500 cm^{-1} are even more complex.

Figure 25.7 shows, on an expanded wavenumber scale, the spectral region from 550 to 350 cm^{-1} from Figure 25.6. Apart from a steady decrease in the intensity of the peak near 500 cm^{-1}, there is a complex change in the region from 450 to 400 cm^{-1}. For the first spectrum observed immediately after the addition of $TiCl_4$ there is a weak peak at 410 cm^{-1} with a shoulder to high wavenumber (these and other positions of peak maxima are shown on the Figure). With increasing time, the maximum appears at 436 cm^{-1} and by the last spectrum after 15 h evacuation there is a slight appearance of a secondary shoulder at 457 cm^{-1}.

The behaviour is suggestive of a process whereby chlorine atoms are sequentially eliminated from a $SiOTiCl_3$ fragment, giving … $TiCl_2$ and, finally … TiCl. In general the antisymmetric stretching mode is not expected to change in wavenumber significantly during this process and the average $TiCl_x$ frequency remains approximately constant. For example, for $TiCl_4$, the average of the triply degenerate stretching mode at 500 cm^{-1} and the nondegenerate symmetric mode at 388 cm^{-1} is 472 cm^{-1}. This average should not change significantly as the number of halogen atoms is reduced (similar correlations exist for M-carbon and M-chlorine stretching modes, e.g., in GeMeCl and SiMeCl compounds [12]). This being accepted, we can calculate the expected position of the symmetric $TiCl_x$ stretching mode for other fragments, it being assumed that the antisymmetric mode is at 500 cm^{-1}. For … $TiCl_3$ it is 416 cm^{-1}, for … $TiCl_2$ it is 444 cm^{-1}, and for … TiCl it is the average, 472 cm^{-1}.

Therefore, assuming that the adsorption of $TiCl_4$ initially gives $SiOTiCl_3$ we expect a band near 416 cm^{-1} (observed 410 cm^{-1}). A species $(SiO_2)TiCl_2$ would be expected to have its symmetric stretch at 444 cm^{-1} (there is a shoulder at 428 cm^{-1}). With time, the 428 cm^{-1} shoulder becomes a peak at 436 cm^{-1} which suggests conversion of the trichlorinated species to the

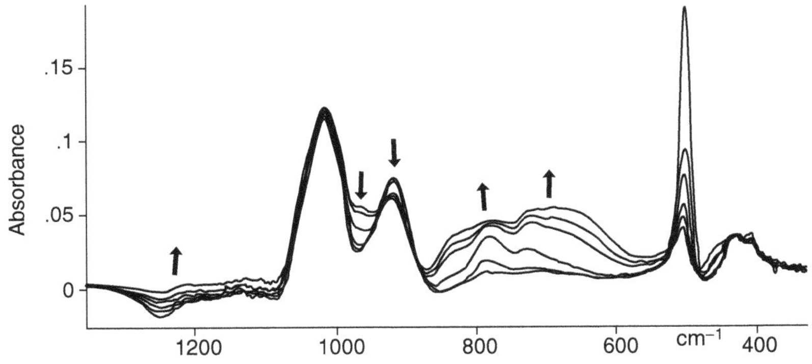

FIGURE 25.6 $TiCl_4$ adsorbed on 20°C activated SiO_2 and subsequently evacuated for 5 s, 15 min, 90 min, 6 h and 15 h. The arrows indicate the direction of the spectral changes with increasing time.

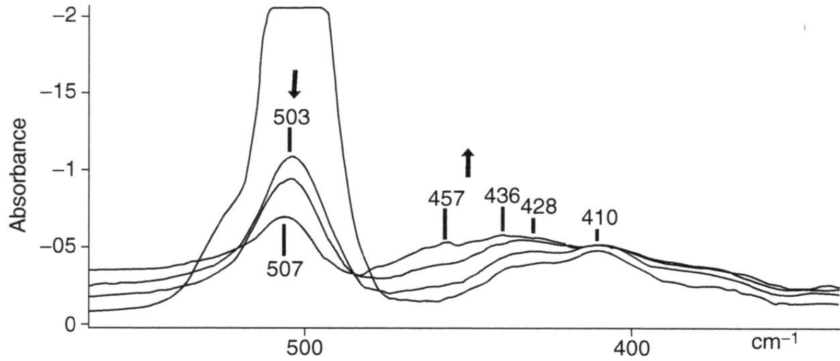

FIGURE 25.7 $TiCl_4$ adsorbed on 20°C activated SiO_2 and subsequently evacuated for 5 s, 15 min, 90 min, 6 h and 15 h (as in Figure 25.6, but on an expanded wavenumber scale). The arrows indicate the direction of the spectral changes with increasing time.

dichlorinated species (little apparent decrease at 410 cm^{-1} is due to overlap by the much broader 436 cm^{-1} feature). Finally, a higher wavenumber shoulder grows at 457 cm^{-1}; the calculated wavenumber for the single chlorinated species is 472 cm^{-1}). Although we stress that the calculated wavenumbers are not to be considered more than a guide of the expected shifts during this process (replacement of Cl by O would have more complicated effects on the exact wavenumber of the remaining TiCl vibrations), it is clear that such a successive substitution is probably responsible for the observations.

There is an overall decrease in the intensities of the $TiCl_x$ modes with time and a considerable increase in the intensities of bands in the 800–650 cm^{-1} spectral region. The overall effect suggests that as the number of chlorine atoms bound to Ti decreases, additional SiOTi bonds are formed. Schematically, we might have:

$$SiOTiCl_3 \rightarrow (SiO)_2TiCl_2 \rightarrow (SiO)_3TiCl \rightarrow (SiO)_4Ti \qquad (5)$$

the latter species corresponding to insertion of titanium into the silica lattice.

DISCUSSION

It is clear that the thin film method can be successfully used to probe low wavenumber vibrations of metal-chlorine containing species. This is the first time these vibrations have been observed for this molecule using IR spectroscopy, and to our knowledge, the only other IR example of the observation of MCl_x vibrations is from the work of Tripp and Hair for some methylchlorosilanes on silica [7–9]. Spaces does not permit us to discuss results we have obtained following the chemisorption of other reactive chloro species such as $VOCl_3$ and BCl_3 [4].

An unexpected result was the progressive apparent dechlorination of $SiOTiCl_3$. We have verified that this phenomenon was not related to the presence or absence of $TiCl_4$ either physically adsorbed or in the gas phase. We could also observe the growth of the same IR bands between 1000 and 600 cm^{-1} using a self-supporting disc. Therefore, the dechlorination of $TiCl_4$ on silica and the eventual incorporation of Ti as a random mixed metal surface oxide is probably entropy driven. Although the initial chemisorption follows reaction (3) and (4), further dechlorination probably results in the formation of SiCl surface species. The vibrations of this near 700 cm^{-1} would be impossible to detect with a thin film given the low extinction coefficient [15], and in any case, they would be masked by the much stronger SiOTi vibrations. Finally, the results have implications for mixed oxide catalysts which are prepared by chemical vapor deposition. Structural models which are based on the notion that only reactions like those depicted in schemes (3) and (4) occur are probably not valid.

REFERENCES

1. Hair, M. L. "*Infrared Spectroscopy in Surface Chemistry*", Marcel Dekker, N.Y., 1967.
2. Little, L. H. "*Infrared Spectra of Adsorbed Species*", Academic Press Inc., London, 1966.
3. Morrow, B. A. "*Studies in Surface Science and Catalysis*", **57A**, A161–A224 (1990).
4. Molapo, D. T. *Ph. D. Thesis*, University of Ottawa, 1998.
5. Morrow, B. A., Tripp, C. P. and McFarlane, R. A. *J. Chem. Soc. Chem. Commun.*, 1282 (1984).
6. Morrow, B. A. and Lang, S. J. *J. Phys. Chem.*, **98**, 13319 (1994).
7. Tripp, C. P. and Hair, M. ,L. *Langmuir*, **7**, 923 (1991).
8. Tripp, C. P. and Hair, M. L. *Langmuir*, **8**, 1961 (1992).
9. Tripp, C. P. and Hair, M. L. *J. Phys. Chem.*, **97**, 5693 (1993).

10. Morrow, B. A. and McFarlan, A. J. *J. Non-Cryst. Solids*, **120**, 61 (1990).
11. Morrow, B. A. and McFarlan, A. J. *Langmuir*, **7**, 1695 (1991).
12. Morrow, B. A. and Hardin, A. H. *J. Phys. Chem.*, ***83***, 3135 (1979).
13. Haukka, S., Lakomaa, E. L. and Root, A. *J. Phys. Chem.*, **97**, 5085 (1993).
14. Kytokivi, A. and Haukka, S. *J. Phys. Chem. B*, **101**, 10365 (1997).
15. Lang, S. J. and Morrow, B. A. *J. Phys. Chem.*, **98**, 13314 (1994).

26 Fourier Transform Infrared and Raman Spectroscopic Study of Silica Surfaces

B. Humbert, C. Carteret and A. Burneau
Laboratory of Physical Chemistry and Microbiology for the Environment

J.P. Gallas
University of Caen, Department of Material Sciences and Radiation

CONTENTS

Infrared (between 400 and 8000 cm^{-1}) and Raman (between 5 and 400 cm^{-1}) spectra of silica powders are interpreted to discuss the differences between fumed, gel and precipitated samples. Site models of the surfaces of fumed silica are proposed. These models are based on surfaces resembling faces {111} of β-cristobalite, edges and steps. The site $SiOH \cdots O_b(H)SiO_aH$, located at steps between planes {111}, is related to the absorptions at 3500, 3715, and 3742 cm^{-1}, for the hydroxyl groups from left to right, respectively. Because of the comparatively strong hydrogen bond involved in this kind of site, a weak heating of the sample dehydroxyls these silanols into SiOSiOH to give a siloxane bridge which is a part of two fivefold $(\text{-SiO-})_5$ rings. The models of single silanols are hydroxyl groups placed about 0.4 nm apart on faces {111}. Those silanols are weakly interacting and are condensed around 400°C into three-fold $(\text{-SiO-})_3$ rings the breathing mode of which gives rise to the D_2 Raman band at 607 cm^{-1}. Most of the latter rings need a high relative humidity to be rehydroxylated. Vicinal single silanols on edges could also condense above 500°C into very reactive $(\text{-SiO-})_2$ rings. A more disordered surface must be considered to explain the heterogeneous distribution of surface groups for gel and precipitated silica samples.

the surface reactivity of silica powders depends on the nature, distribution, and accessibility of the surface sites. The definition of such sites on the atomic scale is rather difficult, all the more because it is not independent of the underlying network structure. These sites involve one or more Si—OH groups. In the following, an Si—OH group at the surface of silica will be called silanol, although this group should not be confused with the silanol molecule H_3SiOH. A $Si(OH)_2$ will be called geminal silanol. The silanol groups either single or geminal, are of particular interest. The infrared transmission spectra of pressed, self-supporting silica disks have been used to study the νOH stretching mode of silanols in the 3200–3800 cm^{-1} region [1]. The principal results obtained before 1980 have been summarized by Hair [2a] and between 1980 and 1997 by Burneau and Gallas [2b]. The broad absorption around 3530 cm^{-1} is due to H-bonded silanols. The amount of bonding is much greater on a silica that is precipitated from solution than on a sample prepared by flame oxidation [1]. Another component, observed at about 3715 cm^{-1} on fumed silicas, has been assigned to terminal silanols, which are only proton acceptors and have their own proton free for an additional H-bond [3]. A narrow band at 3747 cm^{-1} remains alone at the end of any thermal treatment. It is assigned to isolated silanols because environment effects are minimized in this configuration. The distance of such an isolated OH group to a first neighbor (fn) has been estimated to be 0.44 nm [4]. Some silanols are not exchangeable with heavy water. They are called internal and absorb around 3660 cm^{-1}. The surface chemistry also depends on the configuration of the silicon and oxygen atoms building up the surface. These heavy atoms are involved in vibrations at wavenumbers much lower than νOH. Before 1980, some of these modes were studied with infrared spectra, despite a strong absorption of the bulk [5,6].

During the period 1980–1995, the main advances in the spectroscopic characterization of silica powders have come from ^{29}Si NMR, Raman diffusion, and Fourier transform infrared (FTIR) studies. NMR studies have given two types of results. First, single and geminal silanols have been quantitatively differentiated as a function of dehydroxylation by thermal treatment and subsequent rehydroxylation by liquid water [7]. The fraction of geminal silanols

$$f_g = \frac{[Si(OH)_2]}{[Si(OH)_2] + [SiOH]}$$

is between 0.15 and 0.20 on a fully hydroxylated pure silica surface [7]. Such a clear characterization is not possible from infrared spectra. Second, surface dehydroxylation induces a diminution of the average of the four Si-O-Si angles around a part of the silicon atoms. In conjunction with results from Raman spectra, this effect has suggested the formation of small planar rings on the surface [8,9].

The Raman characterization of silica gels is a by-product of researches on the sol-gel process to obtain silica glasses at low temperature from solutions of tetra-alkoxysilanes. These spectra have two puzzling features that are also observed in the Raman spectrum of fused, vitreous (v) silica. Two peaks, at 490 cm^{-1} (called D_1) and 604 cm^{-1} (called D_2), are superimposed on the broad band to about 440 cm^{-1}, which is the most intense signal in the spectrum of v-SiO_2. These two peaks are unusually sharp for a noncrystalline solid and have not been explained with a continuous random network model. Their intensities in v-SiO_2, particularly for the D_2 peak, increase with "fictive" temperature, that is, the temperature (up to 1500°C) from which the sample has been quenched to room temperature [10]. The D_2 peak also dramatically increases with neutron irradiation [11,12]. Numerous structural models have been proposed for D_1 and D_2, mainly defects such as a broken Si-O bond, or the breathing modes of planar rings of order $n = 4$ (four-fold) or 3 (three-fold), where n is the number of (SiO) groups in a ring [13].

The Raman spectrum of a gel before any heating displays the D_1 band, but not the D_2 band. The D_2 peak appears only after the gel has been preheated above 200°C. Its intensity is maximum after a pretreatment around 600°C. With a gel of high surface area, this intensity is then greater than that with untreated v-SiO_2 and similar to that observed with irradiated v-SiO_2. The D_2 peak decreases above 600°C and becomes similar to that observed for a fused silica after the gel has been consolidated to dense amorphous silica at 1100°C [14–16]. For a preheated gel, The D_2 intensity is reduced by exposure to water vapor [17]. Exchanges with $H_2^{18}O$ have shown that, in a gel, both the D_1 and D_2 modes involve surface sites [16]. After some controversy, the D_2 band has been convincingly assigned to three-fold rings [8,16], and a surface model has been proposed on this basis. Brinker et al. have suggested that the precursor structure could be the {111} faces of β-cristobalite. The condensation of two "isolated vicinal" silanols on such a surface would result in equal numbers of three-, five-, and six-membered rings [9]. The assignment of D_1 to four-fold planar rings is not so straightforward. A very intense band at 490 cm^{-1} has been found in the Raman spectrum of a gel before any drying. With the assumption of two quasi-degenerate features, this band has been called D_0 and assigned to symmetric stretching mode $\nu_sO_3Si(OH)$ [18]. Such an assignment of a Raman peak near 490 cm^{-1} to single silanols is also supported by NMR results (19). Since about 1985, the high sensitivity and the excellent signal/ratio of the FTIR spectrometers have allowed to:

1. access to extended spectral ranges, from far to near infrared (e.g., 20–24)
2. easier data processing, for example, for differences between successive spectra and quantitative analysis [2b,23–26]
3. sampling by methods other than transmission, for example, with a diffuse reflectance (DR) attachment [27–29].

As far as possible, three types of comparisons are made in this chapter:

1. between samples (a fumed silica and silica powders obtained from solutions)
2. between two kinds of infrared spectra (transmission in near and middle infrared ranges through self-supporting silica disks and also DR of a powder)
3. between infrared and Raman spectra

A summary is given for spectral variations as a function of dehydration in vacuum, controlled rehydration, dehydroxylation by thermal pretreatment up to 700°C, rehydroxylation, and grafting. On the basis of NMR, x-ray, and Raman results, the near and middle infrared spectra are further assigned in terms of surface sites related to the local structure of the underlying SiO network. Emphasis is given to a detailed structure analysis of a fumed silica. Additional features are necessary to describe other types of silica.

EXPERIMENTAL DETAILS

The experimental details have been described (22–24, 27–30). The silica samples are:

1. a fumed (or pyrogenic) silica called A (Aerosil 200, Degussa, N_2 specific surface area $A_s = 215\ m^2\ g^{-1}$, as provided by the supplier)
2. a precipitated silica called P (Zeosil 175 MP, Rhône Poulenc, $A_s = 175\ m^2\ g^{-1}$, as provided by the supplier)
3. a silica gel called G (Rhône Poulenc, $A_s = 320\ m^2\ g^{-1}$, as provided by the supplier)
4. a called chromatographic silica gel, called C (Merck K60, $A_s = 420\ m^2\ g^{-1}$).

The apparent surface densities of silanols deduced from thermogravimetric measurements were 4, 14, 12, and 7 OH groups per nm^2 for samples A, P, G, and C, respectively [23,28]. From ^{29}Si NMR measurements, the fractions of geminal sites were evaluated to be $f_g = 0.21, 0.18, 0.17,$ and 0.15 for samples A, P, G, and C, respectively [29–30]. These values correspond, respectively, to 0.35, 0.30, 0.29, and 0.26 for the fractions of geminal hydroxyl groups

$$x_g = \frac{2[Si(OH)_2]}{2[Si(OH)_2] + [SiOH]}$$

The preparation of grafted silica samples by reaction with methanol under high pressure at 200°C for 2 h has been described [31]. These samples are called AC_1, PC_1, and GC_1.

The middle infrared (MIR) spectra were obtained with FTIR spectrometers (Nicolet 60SX, resolution = 2 cm^{-1}; Perkin Elmer System 2000, resolution 2 cm^{-1}, both the spectrometers equipped with deuterated triglycine sulfate, DTGS, detector), either by transmission through self-supporting silica disks or from diffuse reflectance of silica powder. Silica disks pressed under 50 MPa could be evacuated with a residual pressure under 10^{-3} Pa. The spectra are shown in a normalized absorbance scale, according to

$$\text{Normalized absorbance} = \text{measured absorbance}\frac{s}{(mA_s)} \tag{26.1}$$

with m the mass in grams of the pellet of hydroxylated silica, and s the pellet area in square centimeters.

Powder DR data were collected with a Harrick DRA-2CI attachment and a HVC-DRP heatable cell, evacuable below 0.1 Pa. To collect spectra below 4000 cm^{-1}, the silica powder was generally mixed with potassium bromide, without any pressure, so that there was no close interaction between KBr and the silica surface. The measured temperature, called apparent temperature, is that of the stainless steel cup containing the sample; the difference from the real temperature of the sample surface analyzed by infrared radiations has been examined [27]. The temperature used for RD spectra is the temperature measured on the sample support.

The near infrared (NIR) spectra have been obtained with a FTIR Perkin Elmer 2000 spectrometer equipped with a DTGS detector, a tungsten-halogen source and a quartz beamsplitter. The transmission spectra, with a resolution of 2 cm^{-1}, are obtained through the pure silica plates placed in a chamber in which the samples may be evacuated (pressure about 10^{-5} Pa) and treated by heating until 750°C.

The inelastic scattering spectra between 100 and 4000 Raman shifted wavenumber (cm^{-1}) of self-supporting disks of silica A, P, and C were obtained with a spectrometer Coderg T800 and a spectral resolution of 7 cm^{-1}. The 514.5 nm emission line of an Ar^+ laser was used with a power of about 200 mW at the sample. The low frequency Raman between 2 and 30 cm^{-1} are collected with a high resolution of 0.3 cm^{-1}.

All the displayed spectra are recorded at room temperature.

RESULTS

HYDRATION AND DEHYDRATION AT ROOM TEMPERATURE

The ν_2 mode (bending or δ mode) of water molecule has been used in the past to determine the quantity of adsorbed water on silica sample [22]. However, the use of this absorption is made difficult by the absorptions of combination and overtone of network silica modes between 1500 and 1850 cm^{-1}. It is easier to use the characteristic combination $\nu_2 + (\nu_1$ or ν_3, respectively the symmetric and the asymmetric stretching modes) of water molecules as shown recently [28]. The spectral range 5000–5350 cm^{-1} characterizes these combinations of water molecules. Because this range is fairly isolated from the silica absorptions, it allows to study separately the physisorbed water molecules [2b]. Figure 26.1 shows clearly that a pure silica A pellet evacuated at 25°C under a residual of pressure of 10^{-4} Pa does not display any absorption between 4750 and 5500 cm^{-1}. Despite some controversy about this point [32], most of the silica samples are readily dehydrated by evacuation at room temperature

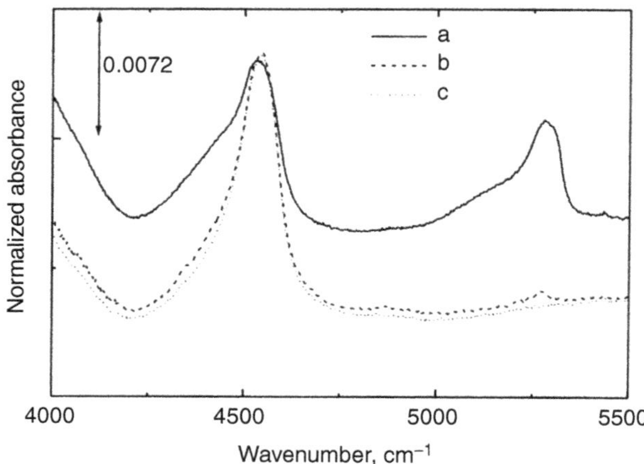

FIGURE 26.1 Evolution of the near infrared transmission spectra, in the range 4000–5500 cm^{-1}, of pure silica A (Aerosil 200) pellet by progressive evacuation: (a) sample at 20°C in equilibrium under a water relative humidity of 0.5; (b) evacuation 1 min at 25°C under 10^{-4} Pa; (c) evacuation 7 h at 100°C under 10^{-4} Pa. All spectra have been recorded at 25°C.

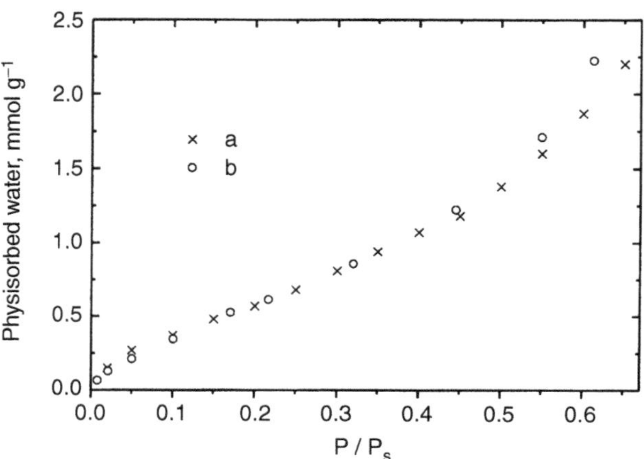

FIGURE 26.2 Water adsorption isotherms on AerosilA200 after evacuation 10 h at 100°C: a) gravimetric measurements using vacuum balance; b) spectroscopic measurements using the $H_2O(\nu + \delta)$ absorption. Physisorbed water is the adsorbed water in 10^{-3} mol per gram of dry silica and P/P_s is the relative pressure of water at equilibrium.

[5,22,25,33–35]. This behavior is not specific of pyrogenic silica samples, but it is more general, for instance, the DR spectra of pure powder G shows that the $\nu_2 + \nu_3$ water absorption disappears below 0.1 Pa (curve d of Figure 3 in reference 22). Moreover, the $\nu_2 + \nu_3$ band gives some insight into the states of adsorbed water. A shoulder at 5315 cm^{-1} corresponds to water molecules with at least one unbonded OH group on the outer surface [22]. This water species is the most readily desorbed from silica G. A supplementary advantage in using the near infrared absorption lies in the fact the corresponding molar absorption coefficient ε is constant with the amount of the adsorbed water [28]. This last point allows us to use the Beer law to quantify the amount of the adsorbed molecules. For example, Figure 26.2 compares two adsorption isotherms; the first is measured, as usually, by gravimetric measurements using vacuum quartz spring balance while the second is obtained by integration of the massif $\nu_2 + (\nu_1$ or $\nu_3)$. Both the adsorption isotherm curves are very close. The BET transform of those isotherms gives a specific area value for the water. As previously observed for fumed silica [30], the specific area given by water adsorption is lowest than the value measured by nitrogen adsorption. Thus, for the fumed silica A, the value obtained from the water adsorption is 57 m^2/g against 215 m^2/g for the nitrogen adsorption, if the molecular surface area of water is assumed to be 0.148 nm^2.

The νOH absorption of the liberated silanols appears at 3740 cm^{-1} for samples G and P (Figure 26.3, curve b-a) correlated to the decrease of the shoulder at 5315 cm^{-1}. A further dehydration eliminates water molecules H-bonded through their two OH groups (22) and causes an increase in silanol absorption between 3730 and 3675 cm^{-1} (Figure 26.2).

Silica A has an unusual spectrum. The νOH absorption of the silanols that become non-H-bonded to water increases first at 3746 cm^{-1}, which corresponds to isolated silanols, then between 3746 and 3700 cm^{-1}, with a shoulder near 3715 cm^{-1} (Figure 26.2). In a last stage, the absorption sharply decreases at 3748 cm^{-1}, whereas it goes on increasing around 3742 and 3715 cm^{-1} (curve d-c). Although puzzling, this last stage has been confirmed in various experiments and with different spectrometers, either by diffuse reflectance or by transmission (22). It was concluded that the absorbance diminution at

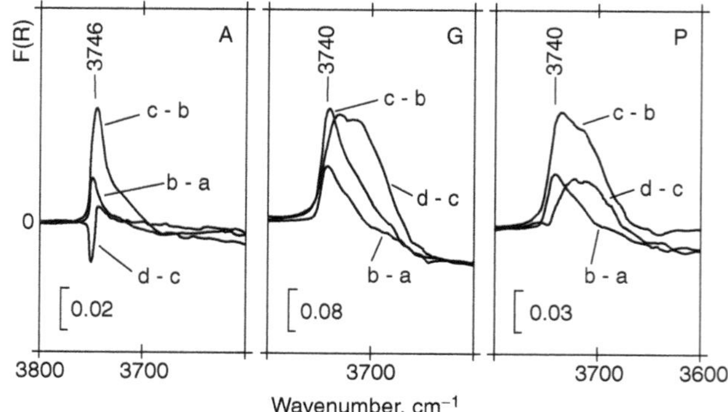

FIGURE 26.3 Successive variations of the diffuse reflectance spectra of silica, A, G, and P (mass fraction 0.12 in KBr) by progressive evacuation at 20°C; a: in air; b: 30 Pa; c: 3 Pa and d: 0.4 Pa. (Reproduced from reference 22. Copyright 1990 American Chemical Society.)

$3748\ cm^{-1}$ at the end of dehydration is not an artifact in spectrum subtraction. This feature is assigned to a real shift of the νOH band to low wavenumber when vapor pressure becomes smaller than about 10 Pa. Such a shift might itself be related to a temperature increase of the sample due to the strong absorption of infrared radiation below $1300\ cm^{-1}$.

Silanol Heterogeneity

The νOH absorbance profile of fully dehydrated silica A, G, P is shown in Figure 26.4. A narrow absorption at $3747\ cm^{-1}$ dominates the spectrum of silica A and characterizes isolated silanols on surfaces that are not completely hydroxylated. This last statement is also based on the Raman spectrum discussed later. About 9% of the silanols are internal, inaccessible to D_2O molecules at room temperature ($\nu OH = 3670\ cm^{-1}$; Figure 26.4). The weakness of the components at 3715 and $3530\ cm^{-1}$ is also characteristic of fumed silica samples [3].

The middle infrared spectra of samples G and P are very different from that of silica A. The absorptions of both the H-bonded and the weakly perturbed silanols are very strong. Because most of the silanols are exchangeable with heavy water (Figure 26.4), these absorptions cannot be assigned to internal silanols. A body of results suggests the presence of inner silanols on the silicas prepared from solutions before any thermal treatment [22]. These inner silanols are not taken into account by nitrogen or argon adsorption, although their protons are exchangeable [29]. They explain the anomalously large apparent surface densities of silanols on silicas G and P. The structure of such inner silanols is not straightforward. According to Yates

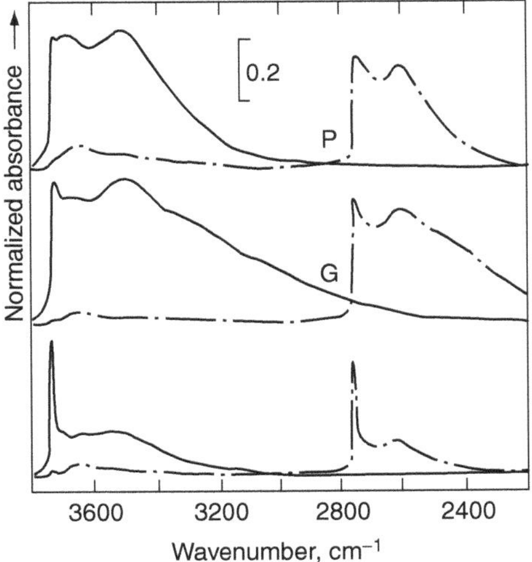

FIGURE 26.4 Normalized absorbances of self-supporting silica disks evacuated at room temperature, before (solid line) or after (broken line) D2O exchange at room temperature. (Reproduced from reference 22. Copyright 1990 American Chemical Society.)

and Healy [35], the surface of disperse SiO_2 obtained from solutions would be made of a gel layer of incompletely condensed polysilicic acid, which remains permeable to water molecules although more or less compacted. Another model is suggested by x-ray diffraction of silica gels [36]. Those silicas would be made of very small primary particles that associate to bigger clusters in the course of gelation or precipitation. Various types of inner silanols could result from this association [37]. For instance, the primary polyhedron postulated by Himmel et al. [36] can build up a surface resembling a {111} face of β-cristobalite, but with some Si atoms lacking. A hole of SiOH corresponds to three single silanols, each being exchangeable and H-bonded as in a cyclic trimer [29,37]. However, argon or nitrogen adsorption would probe nearly the same area as on a regular surface.

In order to try to quantify these different populations, this middle infrared absorbance profile is compared in Figure 26.5 with both the collected profiles in the NIR spectrum and in the inelastic Raman Stokes scattering spectrum, respectively for the vibrational transitions 2νOH et νOH. The advantage of this kind of comparison is to give a better representation of the distribution of the silanol populations [2b,23]. Indeed, it is well known that hydrogen bonding decreases the characteristic wavenumber of the stretching vOH but also it increases the intensity of the MIR transition corresponding. Thus the recorded absorbance, in the MIR range, may be assumed to be, in a first approximation, as the result of the "multiplication" of the distribution of silanols by a spectral function strongly decreasing with the wavenumber [23,38]. To reach the true distribution of the hydroxyl groups, one could divide the MIR spectrum by one function $B(\nu)$ representative of the variation with the hydrogen bonds of the square of the absorption transition moment. From a study of weak hydrogen bonds in the liquid phase, the derivative of the dipole moment could be related to the wavenumber by a linear relation (38 and reference 3 in 38). The function $B(\nu)$ may be then written as: $B(\nu) = (6462 - 1.687\nu)^2$. And the distribution is obtained by multiplying the MIR spectrum by the ratio $B(3747)/B(\nu)$. These approach may be applied to amorphous silica samples [38] and crystalline layered silicate samples [39]. However it would be better to have a straightforward experimental observation of the OH distribution. Both the NIR absorbance and the Raman scattering intensity are weakly dependent of the H-bond effects [2b,38]. Figure 26.5 shows that the distribution of silanol populations in silica A, characterized by a sharper asymmetric contribution with a maximum at $3747\ cm^{-1}$ on the Raman spectrum or $7328\ cm^{-1}$ in NIR, is different from that of sample C, with the maximum at about 3742 or 7320 NIR. The wavenumbers obtained for a silica sample preheated at 750°C in vacuum that bears then at its surface only isolated silanol groups are 7331 in NIR

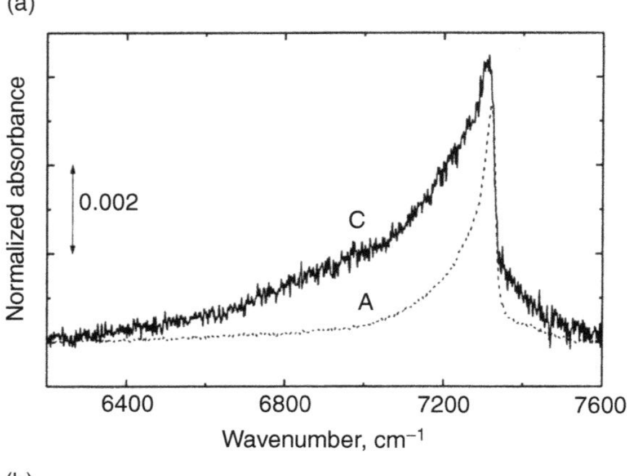

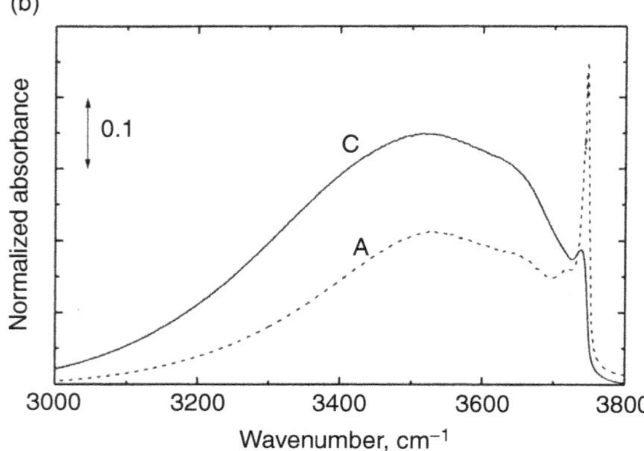

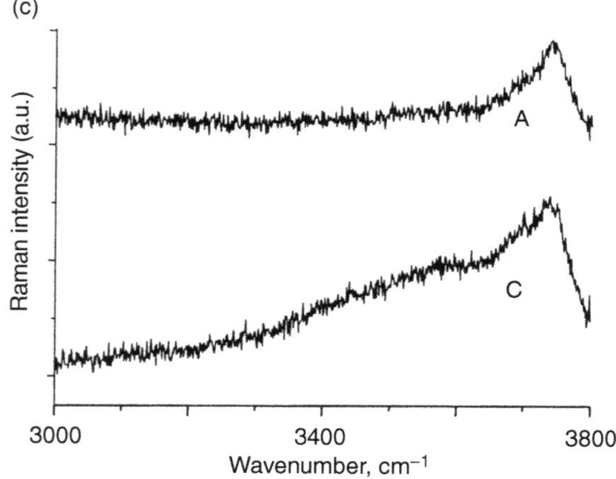

FIGURE 26.5 Spectral profiles in the OH stretching ranges: (a) NIR, 2nOH; (b) MIR, nOH; and (c) Raman, nOH. The spectra are collected under a residual pressure lower than 0.1 Pa.

and 3748 in MIR (24). Thus, the sample A is characterized by a narrow distribution of hydroxyl groups usually called isolated silanols because environment effects are minimized in this configuration. These isolated silanol groups dominate the population of superficial external hydroxyls of the sample A. The populations at about 3715 shifted Raman wavenumber (7265 in NIR) and at 3530 (6900 respectively) are then weakest. For the sample C, one can clearly distinguish (Figure 26.5) a narrow component centered at about 3742 (7320 in NIR), a second broader component at 3690 (7210 in NIR) with an equivalent integrated intensity and a third very broad population centered at 3500–3520 (about 6820–6900) with a higher value of the integrated intensity.

Thermal Treatments

Infrared Spectra

The successive variations of the νOH absorptions of self-supporting silica disks degassed at 160, 260, 460, 645, and 745°C have been studied at room temperature [2b,28]. For every sample, the two main features are as follows:

1. The higher the degassing temperature, the higher is the wavenumber of the absorbance diminution. This fact corresponds to a preferential dehydroxylation of the silanols that interact the most strongly with neighbors.
2. A concomitant creation of free silanols is observed in the range 3742–3747 cm^{-1}.

More accurately, three silanol distributions are successively evidenced by increasing the activation temperature, even though these distributions are not completely separated:

1. Up to 260°C, the eliminated silanols are strongly H-bonded. An important difference between the samples exists, however. For silica A, the νOH diminution around 3500 cm^{-1} is observed simultaneously with another component, weaker but well resolved, at 3715 cm^{-1}. Most of the H-bonded silanols on silica A are thus involved in pairs, in agreement with previous studies of fumed silicas (3). In contrast, no component is resolved at 3715 cm^{-1} for silicas G and P. Those spectra are explained by additional clusters of H-bonded silanols larger than pairs, for example, by the hydroxyl triplets mentioned, which correspond to "holes of SiOH."
2. Between 260 and 460°C, the preceding reaction is completed while a second group of weakly bonded silanols, absorbing between 3620 and 3700 cm^{-1}, are dehydroxylated. At this stage, silica A is again differentiated from other samples, because the absorbance diminution is rather localized at 3620 cm^{-1}.
3. Between 460 and 645°C, the absorption diminution peaks near 3735 cm^{-1} and is very asymmetrical toward low wavenumbers, with a sub-maximum near 3680 cm^{-1} only for silica A.

Another difference between the silicas concerns the wavenumber of the absorption augmentation. For silica A, the maximum of the variation is close to 3747 cm^{-1} whatever the pretreatment temperature. For the other silica samples, it shifts from 3742 to 3747 cm^{-1} when the pretreatment temperature is increased.

The same spectral evolutions were observed for the samples mixed with KBr, pretreated in the DR cell at high temperature, and then cooled down to room temperature. The successive variations of the νOH absorptions are shown in Figure 26.6 for samples G and P. An Apparent temperature of 500°C is necessary to eliminate the most strongly H-bonded silanols (curve a). In contrast, the distribution of weakly perturbed silanols is desorbed between 540 and 580°C (curve c), which corresponds to a rapidly increasing efficiency of the treatment above 540°C. After a treatment at the apparent temperature of 580°C, the νOH profiles are similar to those of self-supporting silica disks pretreated at the real temperature of 645°C. An intermediate distribution of desorbed silanols, with two broad peaks at about 3570 and 3675 cm^{-1}, is also shown in Figure 26.6 (curve b).

Raman Spectra between 150 and 1200 cm^{-1}

As provided by the supplier, the Raman spectrum of the sample A is quite different from that usually observed for a silica gel before a thermal pretreatment, because it displays a comparatively strong band D_2 at 607 cm^{-1} (Figure 26.7). Moreover the spectra of samples C and P display a more intense band at 975–980 cm^{-1}. The intensity of this last band which is assigned at the stretching Si-OH mode, reflects a relative highest amount of hydroxyl present in the samples C and P than in the sample A. Indeed, when the ratio of the integrated intensity of the band at 980 to the one of the massif at 800 cm^{-1} (assigned to the symmetric stretching O-Si-O modes) is assumed to be proportional to the density of hydroxyl, the Raman spectra give a density of hydroxyl for C 3.5 and 1.5 times higher than for A and P respectively. These relative values are in a good agreement with the other experimental data obtained for our samples. Thus the relative variations of the bands centered at 980 cm^{-1} allow to qualitatively follow the evolution of the hydroxyl density during the thermal treatments.

The fluorescence during the heating treatment of samples G and P prevented the recording of their Raman spectra at different level of dehydration. Only silica A and C were successfully studied, in spite of a poor signal-to-noise ratio (Figure 26.8). A preheating

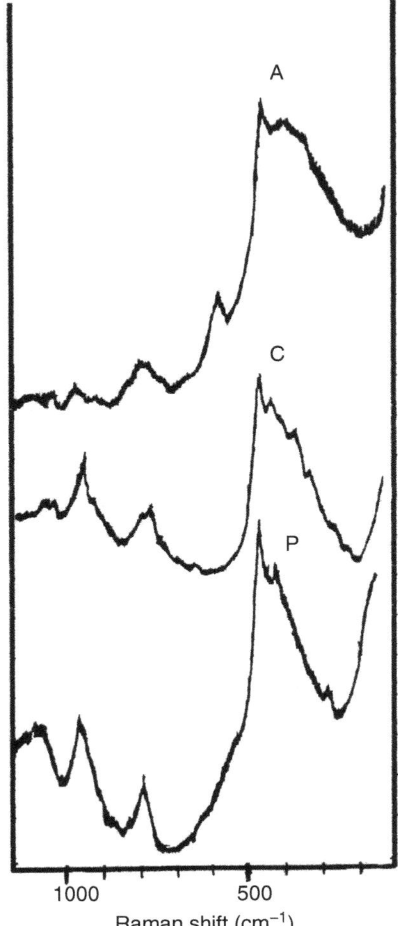

FIGURE 26.7 Raman spectra of samples A, C, and P between 150 and 1100 cm^{-1}. The scattered band at 607 cm^{-1} is displayed only on the spectra of the sample A. The 980 cm^{-1} component is large for the samples C and P but weak for sample A.

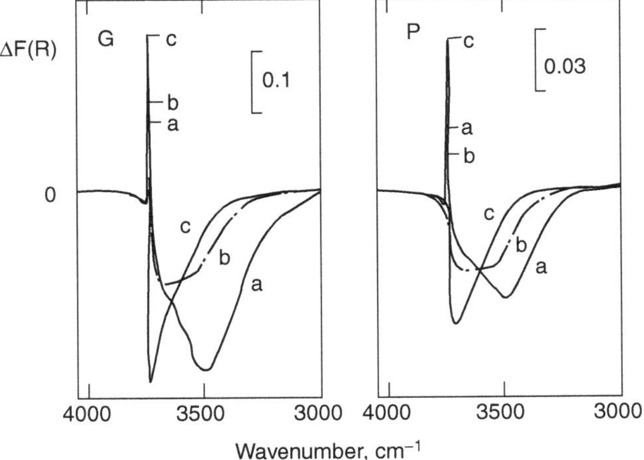

FIGURE 26.6 Successive variations of the diffuse reflectance spectra, at room temperature, of samples G and P (mass fraction 0.12 in KBr), induced by preheating the samples in vacuum below 0.1 Pa. Apparent temperatures of the pretreatment: a = 500–350; b = 540–500 and c = 580–540°C.

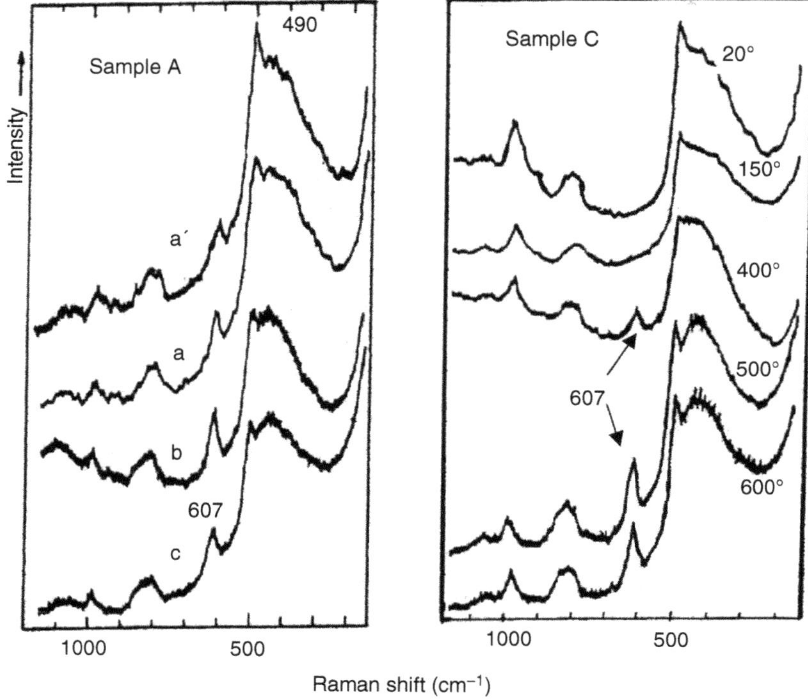

FIGURE 26.8 On the left: Raman spectra of sample A: (a) as received, (a′) exposed for 120 h to 100% relative humidity, and preheated in air to 400°C (b) and to 580°C (c). On the right: Raman spectra of the sample C as received and preheated to the indicated temperatures.

of sample A at 400°C in air induces an augmentation of the D_2 band by a factor 1.5. Simultaneously, the intensity of the D_0 or D_1 peak or both at 490 cm^{-1} is lowered by a factor of about 0.7 (curve 8b). A further preheating at 580°C does not increase the D_2 band any more. On the contrary, both the D_0-D_1 and the D_2 components decrease somewhat (curve 8c). The exposure of the initial silica A to 100% relative humidity (RH) induces a diminution of band D_2 by 50% and an obvious augmentation of the signal at 490 cm^{-1} (curve 8a′). The preheating of sample C induces the creation of D_2 peak only if the temperature is higher than 400°C. Indeed, the preheating at 150°C decreases the D_1-D_0 component and the 980 cm^{-1} intensity, without creating the D_2 peak. The D_2 intensity increase with a preheating at 500°C and is constant between the 500 and 600°C treatments. The heating in air of the sample C at 600°C has divided by about three the integrated intensity of the component at 980 cm^{-1}.

Three conclusions can be drawn:

1. The silica A surface is not fully hydroxylated. About two-thirds of the sites that are precursors of type D_2 siloxane bridges are already dehydroxylated in the initial sample. That observation is consistent with both the high manufacturing temperature of the fumed silica and its low hydroxyl surface density (≈3.60 OH groups per square nanometer; Table 26.1) but is not in agreement with Brinker's view that the hydrolysis of three-fold rings D_2 is easy [9]. In contrast, the primary hydration of silica A should take place on sites other than faces {111} of β-cristobalite [9], which is further discussed later in the chapter
2. the strong decrease of the Raman scattering at 490 cm^{-1} by heating up to 400°C is related to dehydroxylation rather than to the opening of four-fold rings. The spectra (8a′) to (8b) of the sample A, in Figure 26.8, although not normalized, suggest that the diffusion around 490 cm^{-1} involves vibrations of some silanol sites. Thus, the band at 490 cm^{-1} would be assigned to structure D_0 [18]
3. whereas species D_0 appears as a precursor of D_2 for the hydroxyl condensation observed below 400°C, it does not do so neither between 25 and 150°C for silica C nor between 400 and 580°C. Indeed, the signal at 490 cm^{-1} decreases without the creation of a corresponding amount of structure D_2. That observation shows that there are condensations into structures other than D_2. For temperature higher than 600°C, some sintering is also possible, which would

TABLE 26.1
Hydroxyl and Methoxyl (OMe) Surface Densities on Initial Sites of Silica A

Sample	Group	Total	$S_m + S_e$	S_p	G_p	G_e	D_2
A	OH	3.65–3.95	1.6	0.65	1.3–1.6	0.1	
AC_1	OH	1.35–1.65		0.65	0.65–0.95	0.05	
(Figure 26.9, curve a)	OMe	2.9	1.6		0.65	0.05	0.6–0.9
AC_1, 460°C	OH	0.7–1.0			0.65–0.95	0.05	
(Figure 26.9, curve b)	OMe	2.25	1.6			0.05	0.6–0.9
AC_1, 460°C, MeOH	OH	0.05				0.05	
(Figure 26.9, curve c)	OMe (eq. 5a)	4.9	1.6		0.65	0.05	2.6
	OMe (eq. 5b)	4.9	1.6	0.65	1.3	0.05	1.3

Note: Surface densities are per square nanometer. Abbreviations are as follows: S_m, single silanol surrounded by other single silanols, as in the middle of a {111} face of β-cristobalite (Figure 26.10a); S_e, single silanol on an edge (Figure 26.10b); S_p, single silanol perturbed by H-bonding with a neighboring hydroxyl of a site G_p at a step between {111}-like faces (Figure 26.10a); G_p, geminal silanol at a step; G_e, geminal silanol at an edge; and D_2, surface siloxane bridge belonging to an $(SiO\text{-})_3$ ring.

not be in contradiction with the constant value found for the specific surface area of silica A up to 650°C (30), because A_s should somewhat increase by hydroxyl condensation alone because of the mass diminution at nearly constant surface area

4. the density of the three-folds rings in initial sample A may be estimated from the increase of the integrated intensity of the component at 610 cm^{-1} between 200 and 400°C which corresponds to a decrease of 0.85 OH/nm^2 measured by thermogravimetry (40). Indeed, if we assume that to obtain one D2 ring two silanol groups condense, then the integrated intensity of D2 signal is correlated to a density of OH. Thus we deduce that the initial density of D2 is about $1/nm^2$. However one-third of these rings for sample A are certainly localized in the bulk, since a strong exposure at a saturated vapor of this sample keeps about one-third of the normalized signal of D2 (Figure 26.8a′, 40). Consequently a rehydroxylated sample A by opening of the D2 rings would display about {3.6 (initial) $+2 \times 0.66$} OH/nm^2, that is 5 OH/nm^2!

Methyl Grafts

Infrared spectra show that the reaction of surface silanols with methanol (MeOH) at 200°C for 2 h under high pressure is incomplete. As previously described (30), every silica grafted under these conditions (samples AC_1, GC_1, and PC_1) displays a resolved component between 3737 and 3743 cm^{-1}, in addition to internal and H-bonded silanols. For instance, curve a of Figure 26.9 shows the spectrum of AC_1 with a density of 2.9 methoxyl groups per square nanometer (Table 26.1) (31). Compared with the H-bond band, the component near 3740 cm^{-1} is the most intense for sample AC_1. Such an incomplete reaction is mainly assigned to geminal sites in the particular case of silica A. A site $SiOH\cdots O_b(H)SiO_aH$ (called S_p-G_p) is speculated: it is made of a single silanol (S_p) and two geminal hydroxyls (G_p). On the initial silica A, S_p would be H-bonded ($\nu OH \approx 3500$ cm^{-1}) to the terminal O_bH group of site $G_p(\nu O_bH \approx 3715$ $cm^{-1})$; O_bH would interact somewhat with the second geminal group O_aH ($\nu O_aH \approx 3742$ cm^{-1}). In this structure, the hydroxyl most reactive toward alcohols is thought to be O_bH, which is the most proton donor group because of the

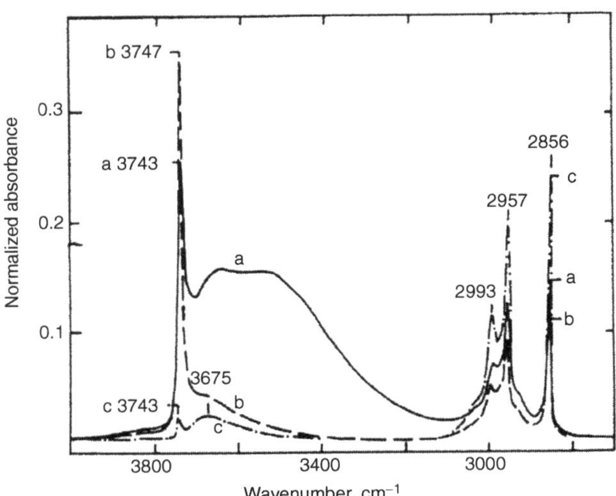

FIGURE 26.9 Normalized absorbances recorded at room temperature of a self-supporting disk of grafted silica AC1: a, evacuated at room temperature; b, predegassed at 460°C; and c, after 1 h of contact with 4700 Pa methanol at 460°C and evacuation.

cooperative effect in the OH···O_b(H) bond. The reaction of silica A with methanol displays an absorption decrease mainly at 3715 and 3747 cm^{-1}, the latter decrease corresponding to the methoxylation of isolated single silanols. The presence of unreacted hydroxyls absorbing at both 3743 and about 3500 cm^{-1} on AC_1 could be due to a partial methoxylation of site S_p–G_p according to:

$$\text{SiOH}\cdots\text{O(H)SiOH} + \text{MeOH} \longrightarrow \text{SiOH}\cdots\text{O(Me)SiOH} + \text{H}_2\text{O} \quad (26.2)$$

Curve b of Figure 26.9 shows the spectrum of AC_1 after degassing at 460°C. The methoxyl density, deduced from the variation of the CH_3 stretching bands, decreases to 2.25 per nm^2 (Table 26.1). The free silanol νOH wavenumber shifts from 3743 to 3747 cm^{-1}, whereas the H-bonded hydroxyls disappear and a few internal silanols remain at 3675 cm^{-1}. By using the value $\varepsilon = 350$ L mol^{-1} cm^{-1} [2b,30] for the molar decadic absorption coefficient of silanols that have been isolated by thermal pretreatment, the density of these OH group is found to be close to 0.7 per square nanometer. The value of 350 for the extinction coefficient ε has been discussed in the literature (e.g., 23,24 and references herein); if an extreme other value of 230 is used [24], then a density of about 1 OH/nm^2 is found. Consequently, the density of the spectrum (Figure 26.9, curve b) is between 0.65 and 1.0 OH/nm^2 (Table 26.1). Whatever the final state of the degassed molecules, the following reaction is in agreement with the spectral changes of AC_1 from curve a to b:

$$\text{SiOH}\cdots\text{O(Me)SiOH} \longrightarrow \text{SiOSiOH} + \text{MeOH} \quad (26.3)$$

Thus, the isolated single silanols on thermally treated sample AC_1 would be located on the ex-G_p sites (Table 26.1). The density of these hydroxyls gives an estimate of the density of sites S_p-G_p on silica A to be about 0.65 (=2.9–2.25) per square nanometer. As a consequence, the OH density on initial sample AC_1 would be between 1.3 and 1.7 per square nanometer. The sum of this value and 2.9 (the methoxyl density) gives a total larger than the surface OH density on silica A. Thus when sample AC_1 is prepared, some methanol is added to siloxane bonds according to:

$$\text{SiOSi} + 2\,\text{MeOH} \longrightarrow 2\text{SiOMe} + \text{H}_2\text{O} \quad (26.4)$$

This conclusion is in agreement with the previously discussed Raman results showing that the surface of initial silica A is not fully functionalized: sites D_2 are the most likely location of this addition (Table 26.1).

A further exposure of 4700 Pa MeOH at 460°C for 1 h gives final densities of 4.9 OMe per square nanometer, 0.05 isolated OH per square nanometer, and a few internal hydroxyls (curve c of Figure 26.9 and Table 26.1). In this stage, the change of S_p-G_p sites can be schematized in two extreme ways: either the change only induces the methoxylation of the remaining 0.65 OH per square nanometer according to:

$$\text{SiOSiOH} + \text{MeOH} \longrightarrow \text{SiOSiOMe} + \text{H}_2\text{O} \quad (26.5a)$$

or every S_p-G_p site is completely methoxylated in spite of steric hindrance according to

$$\text{SiOSiOH} + 3\text{MeOH} \longrightarrow \text{SiOMe MeOSiOMe} + 2\,\text{H}_2\text{O} \quad (26.5b)$$

This methoxylation would not be observed at 200°C because of a high activation energy. In both hypotheses, the final density of 4.9 OMe per square nanometer involves the complementary opening of siloxane bridges (Table 26.1). This last value is very closed to the one discussed previously only on the base of the Raman spectra, for which an opening of the all accessible D2 rings by the water molecules could give a superficial density of 5 OH/nm^2.

DISCUSSION

A description of silica surfaces should not be made only in terms of hydroxyl groups without reference to the underlying atoms. Various experimental results suggest some local order on the surface, even though the bulk network is amorphous. The presence of typical sites on the atomic scale is all the more likely because constraints due to disorder should relax on a hydroxylated surface. Therefore, network and surface models based on crystalline structures are needed to define possible silanol sites and relate them to the spectral features discussed. References to crystalline silicas for studying divided samples have often been made [7,9,29,34,35,41–44]. Recently, Chuang and Maciel [44] have published an excellent general discussion on the use of β-cristobalite to explain the different published chemical-physic results obtained on the amorphous silica, hydration–dehydration, formation of D2 rings or yet dehydroxylation. The models that follow use a few structures on the scale of some SiO_4 tetrahedra (in a radius of about 0.5 nm) and disorder at longer distances, between 1 nm and the diameter of silica particles, which is about 10 nm for silicas A and P and 5 nm for silicas C and G [30]. A consequence of this view is that the whole surface can be described with a fractal model, an assumption supported by both small-angle x-ray [30] and Raman [29,42] scatterings.

NETWORK MODELS

With a value increasing from about 2.12 g cm^{-3} at 200°C up to 2.31 g cm^{-3} at 900°C, the skeletal (or true) mass per unit volume (or density) of silica gels is consistent with a network structure looking like either tridymite or cristobalite [45,46]. X-ray results have suggested for a long time that the structure of silica powders most closely resembles that of cristobalite. Recent results confirm that silica gels are built up of interconnected six-membered rings of $[SiO_4]$ tetrahedra. One polyhedron type, with four rings similar to those in cristobalite, has been found predominant [36]. By using small-angle x-ray scattering, displacing phase transitions have also been observed around 100°C for silica gels that were not preheated [36]. On that basis, adequate models of the structure of silica powders are α-cristobalite at low temperature and β-cristobalite at high temperature. The values of the true density between 200 and 900°C mentioned are consistent with that of β-cristobalite (2.175 g cm^{-3} at 290°C) [46], provided the surface is taken into account. The augmentation of density with temperature is indeed related to the condensation of the surface silanol layer into siloxane bridges. The effect can be estimated as follows. With a thickness of 0.16 nm and a surface density of 4.5 SiOH per square nanometer, the mass per unit volume of a silanol layer would be 4.5×10^{14} $(16+1+28/2)/(6 \times 10^{23} \times 0.16 \times 10^{-7}) = 1.45$ g cm^{-3}. In contrast, the mass per unit volume of this layer after complete condensation into a siloxane layer 0.06 nm thick would be 2.25×10^{14} $(16+28)/(6 \times 10^{23} \times 0.06 \times 10^{-7}) =$ 2.74 g cm^{-3}. These values show that, when hydroxyl condensation proceeds, the apparent density, corresponding to both the network and the surface layer, increases from a value smaller to a value larger than the network density. Whatever the temperature and the corresponding structure of cristobalite, the cubic unit cell of β-cristobalite is used to discuss surface models, because the α-structure can also be described with a similar pseudo-cube [46].

SURFACE MODELS

A surface corresponding to the {111} faces of β-cristobalite would bear single silanols on alternate Si atoms [29,42]. The O–O distance of two fn silanols is about 0.51 nm in an idealized structure of β-cristobalite with the orientation *Fd3m* [46]. The corresponding surface density is 4.5 OH per nm^2. The condensation of two fn silanols would generate a three-fold siloxane ring and a five-fold siloxane ring.

Faces {110} would show parallel rows of vicinal single silanols, with two SiO directions making ideally an angle of 70.5°. The smallest O–O distance between silanols, both inside a row and in two neighboring rows, would ideally be 0.44 nm. The O–O distance between second neighbor (sn) silanols in a row is 0.51 nm. The corresponding surface density is 5.5 OH per square nanometer. The condensation of fn silanols in two neighboring rows would generate non-planar four-fold siloxane rings [42]. The condensation of two vicinal silanols in a row would give a very strained two-fold ring. The condensation of an silanols inside a row would give a three-fold ring as on a plane {111}.

Faces {100} could display geminal silanols with a surface density of 7.8 OH per square nanometer. The $SiO_2(H_2)$ groups are located in parallel planes containing an axis ⟨110⟩ [42]. In this direction, an idealized O–O distance of 0.25 nm is found for both silanols of a given geminal site and of two fn sites. The distance between rows is 0.51 nm. The most likely condensation is between two strongly H-bonded hydroxyl groups of two neighboring (nonvicinal) sites inside a row. This condensation would generate a couple of nonplanar five-fold siloxane rings.

The adequacy of faces {110} in explaining the experimental results has not been fully investigated, but faces {111} appear, up to now, to be the best model of a funiform surface for silica. With a density of about 135 siloxane bridges per cubic nanometer, a hydroxylated {111}-like surface is indeed likely to be very stable. In opposition to this view, it has been claimed [7] that reducing the hydroxyl density below 4.5 per nm^2 appears rather infeasible, because all OH groups would be separated by at least 0.5 nm. Such a description of faces {111} comes in fact from an over-idealised cristobalite, in the orientation *Fd3m*. In contrast, the condensation of fn silanols inot three-fold siloxane rings appears quite feasible by reference to the real structures of α- or β-cristobalite (orientations $P4_12_1$ or $P2_13$, respectively) [46]. Furthermore, the formation of such rings is supported by NMR and Raman spectra [8,9]. However, the extension of the faces should not be overestimated, even though *n*-hexadecyl grafts 2 nm long can lie in an ordered structure on such faces [27]. Edges and steps must be taken into account. Four sites are further defined:

1. S_m is a single silanol, somewhere on the middle of a surface {111} (Figure 26.10a). The idealized O–O distance with a fn silanol, which is 0.51 nm, only applies to one-fourth of the silanols in a more realistic model of β-cristobalite in the orientation $P2_13$ [46]. The three other silanols, which are related by a three-fold symmetry axis, are brought closer together, to 0.41 nm (they are displayed by the triplet i-j-k in Figure 26.11). With the α-cristobalite model in the orientation $P4_12_1$ [46], the hydroxyl groups would be rotated towards each other as street-lights on opposite sides of a street. Across this "avenue," the fn O–O distance is 0.40 nm for

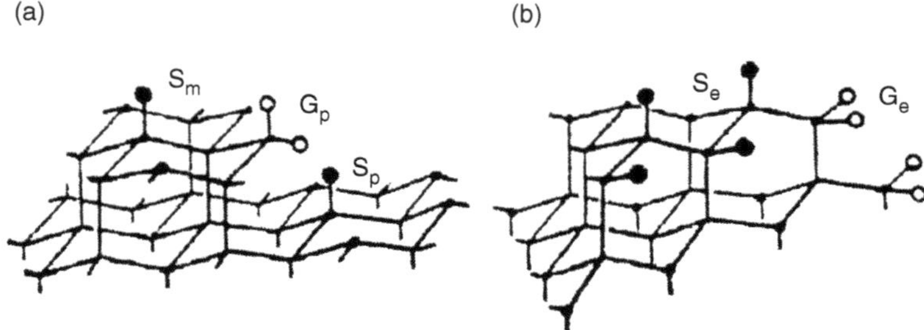

FIGURE 26.10 Definition of some silanol sites at steps (a) or at edges (b) between planes {111} of β-cristobalilte. The silicon atoms bonded to hydroxyl groups are shown by small dots. Only a few silanols are shown for sake of clarity. Full and open circles denote single and geminal hydroxyls, respectively.

all silanols (Figure 26.12). The sn O–O distance is 0.60 nm outside this avenue. With a fn O–O distance of about 0.4 nm, silanols S_m should not be strongly H-bonded, although a further distortion, compared to the cristobalite models, could allow a significant perturbation.

2. S_e is a single silanol on the edge between two faces {111} making a dihedral angle of 109.5° (Figure 26.10b). The edge is a zig-zag chain of staggered silanols and has the characteristics of a silanol row on a plane {110}. For two vicinal silanols of this row, the O–O distance is calculated between 0.37 and 0.50 nm on the basis of the real structures of α- and β-cristobalite, near the idealized value of 0.44 nm.

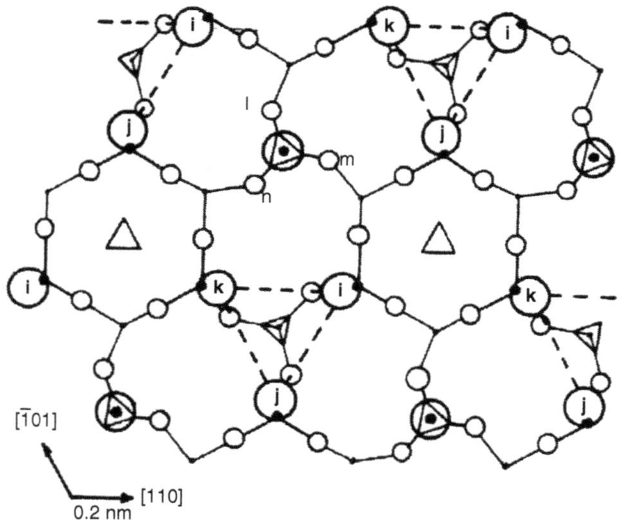

FIGURE 26.11 Projection along $[1\bar{1}1]$ of the surface atoms of face $(1\bar{1}1)$ of β-cristobalite in the orientation $P2_13$ (46). Small dot, Si atom at height ≈ 0; big dot, Si atom at height ≈ 0.1 nm; small open circle, O atom; large open circle, OH group at height ≈ 0.26 nm; triangle, three-fold symmetry axis and dashed line, possible weak H-bond between nearest neighbor OH groups.

3. G_e is a geminal site on the edge between two faces {111} making an angle of 70.5°. The $SiO_2(H_2)$ plane is perpendicular to the edge. The fn O–O distances between the silanols of the edge are similar to those with neighboring silanols S_m, in the range of 0.4–0.5 nm.
4. Geminal sites are also involved in steps between parallel surfaces {111}. Such steps between two successive planes can be built either with one site G_e (Figure 26.10b) or with one geminal site G_p and one single silanol S_p (Figure 26.10a). The O(H) atoms of G_e are parallel to the step, whereas they are perpendicular in a site S_p-G_p.

A step of more than one $[SiO_4]$ tetrahedron would involve a portion of {111} plane, with more than one H-bonded G_p site. Sindorf et al. explained their NMR

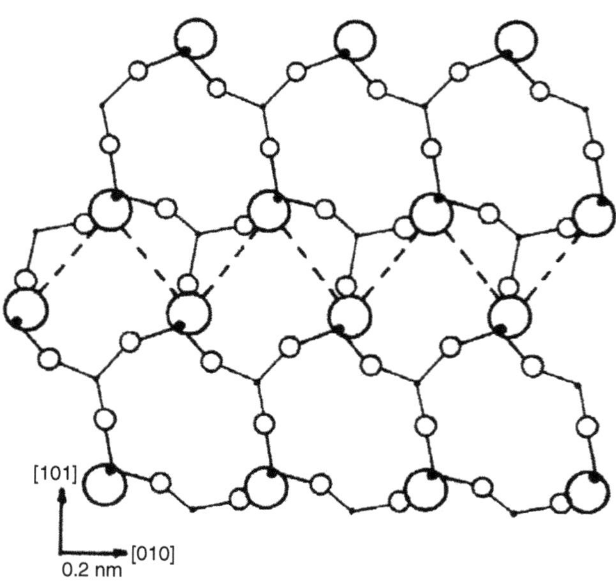

FIGURE 26.12 Projection along $[\bar{1}01]$ of the surface atoms of face $(\bar{1}01)$ of α-cristobalite in the orientation $P4_12_1$ (46).

results with a model made of both {111} and {100} faces (see Figure 12 of reference 7, 44). However, two facts suggest that faces {100} do not participate much on the silica surfaces and that geminal silanols are rather located at short steps between faces {111} on a consolidated silica surface. First, the fractional of geminal sites on a hydroxylated silica is rather low, with a value $f_g \approx 0.15$ [7,27]. Second, no augmentation of a single silanol population was observed [7] by heating around 200°C, which causes the density of geminal sites to decrease by more than 10%. This result does not agree with the expectation that two neighboring G_p sites should be transformed into two single silanols at the very beginning of dehydroxylation, according to

$$\cdots O(H)SiOH \cdots O(H)SiOH \longrightarrow \cdots O(H)SiOSiOH + H_2O \quad (26.6)$$

In contrast, the reaction

$$SiOH \cdots O_b(H)SiO_aH \longrightarrow SiOSiO_aH + H_2O_b \quad (26.7)$$

leaves unchanged the single silanol density, because S_p is replaced by O_aH.

Indeed, the possible arrangements of silanols on a surface are quasi-infinite, even with the hypothesis of a cristobalite-like local structure of the network, and we do not claim to completely describe the surface with these sites. However, strongly H-bonded hydroxyls are not located on the silicon atoms ending one siloxane bridge, according to the structure HO-Si-O-Si-OH: the condensation of these hydroxyls into a two-fold ring would be unlikely at low temperature, in contrast with experiment. The H-bonded hydroxyls are rather bonded to silicon atoms that are brought close together through a siloxane chain [HO-(SiO-)$_4$Si-OH] and condense easily into unstrained (-SiO)$_5$ rings. For silica A, H-bonded hydroxyls are hereafter described as pairs involving geminal silanols of faces {100}, the extension of which is probably limited to steps. Other sites are involved in the larger clusters of H-bonded hydroxyls evidenced on samples G, C, and P through thermal treatment.

Surface of a Fumed Silica

Several results obtained for silica A, as received and after some contact with air, can be rationalized in the following way. The low silanol surface density (about 3.65 OH per square nanometer, internal silanols excluded), the comparatively high fraction of geminal sites ($f_g = 0.21$), and the presence of a rather strong D_2 band in the Raman spectrum indicate an only *partial* and *selective* hydrolysis of the surface after the manufacturing of silica A at high temperature.

From silanol densities measured between 4.2 and 5.7 OH per nm^2, Zhuravlev [47] found a mean value of 4.9 OH per nm^2 at 180–200°C under vacuum for 100 samples of different, fully hydroxylated, amorphous silicas. Thermogravimetry of silica A showed that the silanol density that must be compared to these values, in the same conditions, is about 3.3 OH per square nanometer. As received, silica A is thus not completely hydroxylated. Other results corroborate this conclusion: the D_2 Raman band decreases under a further exposure to water vapor (Figure 26.8, curve a′, 40), and the methoxyl density of 4.9 per nm^2 observed after reaction with methanol at 460°C is larger than the initial OH density (Table 26.1).

According to the NMR study of Sindorf and Maciel [7], the f_g value measured for the initial silica A is comparatively high and suggests a preferential hydrolysis of sites that are precursors of geminal silanols. Conversely, the thermal treatment at 260°C mainly condenses geminal silanols [7]. Thus, the absorbance diminutions centered at 3500 and 3715 cm^{-1} and the augmentation at 3747 cm^{-1} mainly correspond to reaction 7, even though other contributions are not excluded, particularly contributions of aggregates larger than pairs. However, the Figure 26.5 clearly displays, with Raman and NIR spectra, that the population at 3500 cm^{-1} is about as large as the population at 3715 cm^{-1}, that is about one S_p group for one O_b(H). Thus we can conclude for the sample A that the thermal treatment at 260°C is governed by the reaction 7. This assignment agrees with etherification by methanol leading to the structure $SiOH \cdots O_b(CH_3)$-SiO_aH on sample AC_1. It gives additional support to the structure $SiOH \cdots O_b(H)SiO_aH$ located at steps between planes {111}. Because the first hydration stage of silica A induces the decrease of two bands near 3715 and 3742 cm^{-1}, the most stable adsorption of water should take place at steps S_p-G_p, from which water clusters are growing toward the middle of faces {111}. The formation of water clusters has been shown [22,48]. This mechanism is in agreement with the kinetic of deuteration by D_2O [49] and with the hydration process previously described [50], provided each pair of so-called "vicinal single silanols" is changed into a site S_p-G_p, the silicon atoms of which are not vicinal (*vicinal* meaning bridged by an oxygen atom).

A further pretreatment of silica A at 460°C eliminates weakly H-bonded groups (νOH $\approx$ 3620–3680 cm^{-1}). This variation is assigned not only to internal silanols (28), but also to silanols S_m (and perhaps S_e) with fn O—O distances around or smaller than 0.4 nm. The simultaneous increase of the Raman band D_2 supports the formation of three-fold rings on cristobalite faces {111}, as proposed by Brinker et al. [9]. However, the regular condensation model suggested by Brinker, which could be possible on α-cristobalite, is unlikely because the network rather resembles β-cristobalite at such a temperature. We suggest that (-SiO-)$_3$ rings are first made from triplets i-j-k of nearly equivalent hydroxyls on faces {111} (Figure 26.11), according to

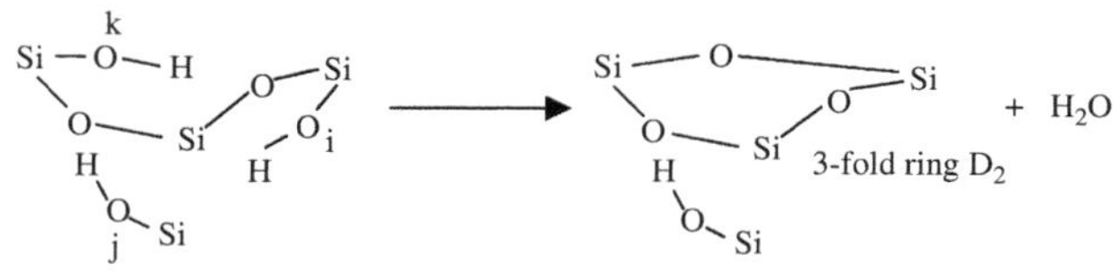

$$\longrightarrow \text{3-fold ring } D_2 + H_2O \qquad (26.8)$$

The following dehydroxylation stage, between 460 and 645°C, can be interpreted by the condensation of about 0.3 OH per nm^2 absorbing around 3680 cm^{-1} (full width at half height (fwhh) $\approx$180 cm^{-1}), internal silanols excluded, and of nearly the same number with νOH $\approx$ 3736 cm^{-1} (fwhh $\approx$20 cm^{-1}) [28]. Although not firmly proved, a condensation as pairs of weakly proton donor and acceptor hydroxyls is suggested:

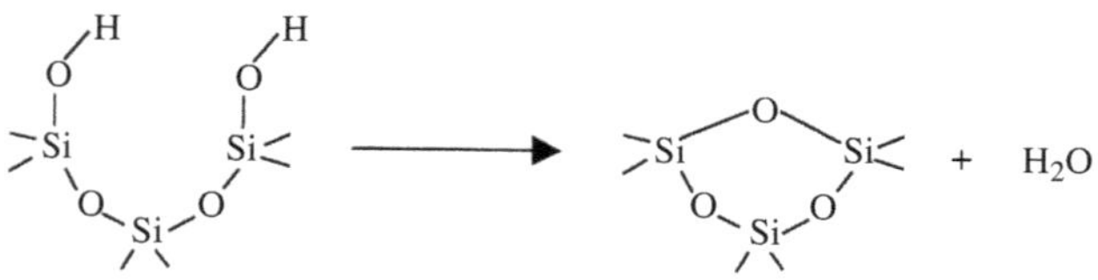

$$\longrightarrow + H_2O \qquad (26.9)$$

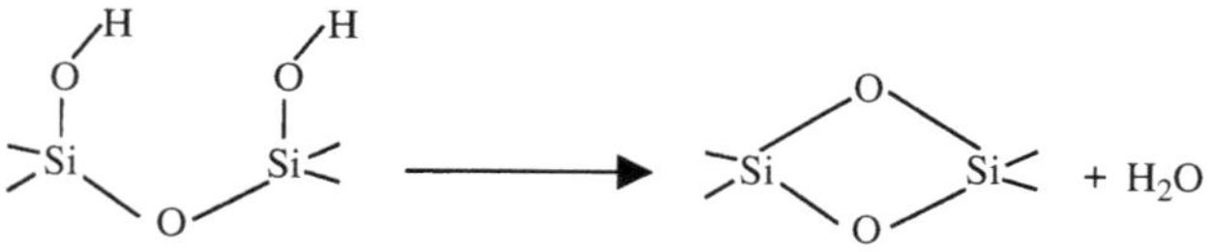

$$\longrightarrow + H_2O \qquad (26.10)$$

Reaction (26.10) would characterize vicinal silanols S_e. It is in agreement both with the Raman spectrum, because band D_2 is not increased and with the infrared components simultaneously observed at 908 and 888 cm^{-1}. The resulting two-fold ring, which would be located at edges, is known to be very reactive [3].

A tentative assignment of the main features of the infrared spectrum of silica A is summarized as follows:

- 3500 cm^{-1}, mainly S_p
- 3620 cm^{-1}; closest S_m in triplet sites i-j-k
- 3670 cm^{-1}, internal silanols
- 3680 cm^{-1}, weak proton donors S_m, S_e, and G_e in pairs
- 3715 cm^{-1}; terminal G_p; 3736–3742 cm^{-1}, weakly perturbed G_p, S_m, S_e, and G_e
- 3747 cm^{-1}, isolated S_m, S_e, and single silanols ex-G_p and ex-G_e.

This assignment gives also a possible explanation of the Raman components appearing at about 3685 and 3615 cm^{-1} by rehydroxylation of a silica gel pretreated at 600°C [14].

Sites S_p-G_p achieve a somewhat concave step, accommodating what has been called inner silanols [22]. Our assignment gives an improved picture of the surface and of the hydration mechanism of a fumed silica. However, the infrared spectrum is intricate and its assignment is still partly speculative. Whereas the comparison of the infrared and Raman spectra supports the conclusion that the formation of three-fold rings is mainly due to the condensation of weakly perturbed silanols (that absorb above about 3600 cm^{-1}), the reciprocity is not warranted. For instance, the contribution of isolated silanols (νOH = 3750 cm^{-1}), postulated by Brinker et al. [16], is excluded.

A rough picture of a fumed silica surface that comes out of the spectroscopic measurements on both initial and grafted silica A is summarized in Table 26.1. The hydroxyl and methoxyl populations refer to the sites of the initial silica. For instance, the 0.6 OMe per nm^2 located on ex-D_2 sites of sample AC, come from reaction 4. This evaluation is based on the assumption that only three-fold rings D_2, mainly precursor of S_m, are not hydroxylated. In fact, some five-fold rings, precursor of S_p-G_p, are not excluded, because the initial sample has been exposed only to low relative humidity. According to equation 5b given for sample AC_1 methoxylated at 460°C, a completely hydroxylated silica A would accommodate 4.95 OH per nm^2 (3.65 OH per nm^2 + 0.65 Siloxane bridge per nm^2 opened on the initial sample), with a fraction of geminal sites $f_g = 0.16$, internal silanols excluded. This evaluation is in agreement with the results of literature for equivalent samples [7,47], which suggests that steric hindrance at the steps S_p-G_p does not preclude species SiOMeO(Me)SiOMe. A density of about 1.95 OH per nm^2, including the free O_aH group, would lie at these steps. This value is rather high: it corresponds, in the hypothesis corresponding to equation 5b in Table 26.1, to 40% of the hydroxyls on the fully hydroxylated sample. Such steps are related to the fractal dimensions $D_s = 2.1 \pm 0.05$ in the framework of a fractal model of the surface (30). As for the surface reactivity, these sites are involved in intricate mechanisms, because they accommodate the strongest proton donor ObH, although they show some steric hindrance.

Surface of a Gel or a Precipitated Silica

Samples G and P display a very large apparent silanol density (12–14 OH per nm^2 with Ar or N_2 specific surface area). Most of these hydroxyls can be deuterated and many are engaged in H-bonded aggregates larger than pairs. Thus, for these samples the sites S_p-G_p

provide a minor part of the H-bonded silanols: the surface picture of silica A cannot be extensively applied. Without describing here detailed aggregation mechanisms of these samples, [e.g., described in 29,37], we focus on the site called the "hole of SiOH" earlier. Suppose that on a face {111} of β-cristobalite the missing SiOH is located on a "three-fold axis": the three neighboring O atoms, *l-m-n* in Figure 26.11, become a triplet of strongly H-bonded hydroxyls on account of their short distance. The first dehydroxylation stage could mainly condense triplets 1-m-n of strongly H-bonded silanols according to:

5-fold ring
5-fold ring
$+ H_2O$

(26.11)

The new siloxane bridge is part of two five-fold siloxane rings and is not involved in the Raman band at 607 cm^{-1} that characterizes $(\text{-SiO-})_3$. This model is consistent with the fact that the D_2 band only appears for a gel preheated above 200°C (Figure 26.8, spectrum 150°C of the sample C) and mainly around 600°C [14–16]. Such holes on gel or precipitated silica samples add to the steps mentioned to define the roughness of the surface, which is modeled with higher fractal dimensions ($D_s = 2.4–2.5$) than for a fumed sample [30].

Unified Interpretation of the D_0-D_1 and D_2 Raman Bands of Disperse Silicas, Monolithic Gels, and Fused Silicas

There is an apparent discrepancy about the occurrence of these bands. On the one hand, the D_2 peak is maximum in a gel preheated at about 600°C, whereas the Raman signal is strongly decreased in the range 480–500 cm^{-1} compared to the initial, hydroxylated state. Further heating decreases both signals to the typical intensities of v-silica [15,16]. On the other hand, these features are enhanced with the fictive temperature of v-silica. The hydroxyl content of v-silica has been reported [10] not to change the defect concentration, but only the dynamic of the structural relaxation. The opposite temperature dependences of the gel and v-silica Raman spectra obviously correspond to different mechanisms. However, the similarity of the spectral signatures suggests a similarity of the involved structures.

We assign the 490 cm^{-1} peak to the unstrained sites $(\text{-SiO})_3SiO'$, where O′ is an oxygen atom fully bonded to only one silicon atom. The type of motion involved could be similar to that of the network, giving rise to the strong and broad Raman signal around 440 cm^{-1}, that is, O atom displacements bisecting the SiOSi angles. A particular configuration of such an anchored O_3SiO' tetrahedron about the network would determine a combination of these transverse oxygen displacements that induces a strong polarizability change [51]. Thus, the 490 cm^{-1} band of a disperse silica would be assigned to some O_aSiO' tetrahedra bearing single silanols (structure D_0), as previously speculated [18]. The population of the single silanol sites involved in band D_0 and that giving rise to structure D_2 probably are not fully coincident, even for a fumed sample. For instance, silanols S_p can contribute to D_0, whereas they condense with G_p into five-fold rings. Similarly, vicinal silanols S_e could condense into two-fold rings.

We postulate that in v-silica a relaxed O_3SiO' structure ensures the particular oxygen arrangement that generates the D_0 band and that D_0 can lead to three-fold rings giving rise to band D_2. This interpretation is supported by the tremendous augmentation of D_2 by neutron irradiation, whereas D_0 is not enhanced much by such irradiation [12]. This fact suggests that D_0 is an intermediate species between the initial network and three-fold rings D_2. Similarly, D_2 is more favored than D_0 above 1300°C [10]. In a first stage, high temperature or neutron irradiation can induce a relaxed bond $O_3SiO\text{-} + SiO_3$ in a highly strained region of the network. Because six-fold rings are dominant, the most likely mechanism is then

O^- +

(26.12)

which gives two three-fold rings from one relaxed bond. This reaction is catalyzed by OH groups because a proton can attack a strained siloxane bridge and, upon completion of reaction 12, it becomes available to diffuse toward another strained site. According to the experimental results of reference 10, the standard enthalpy for reaction 12 is 1.3 eV. The main objection to this interpretation for v-silica is that the band $\nu SiO'$ is not observed in the range 950–1050 cm^{-1} [12]. A possible explanation is that the scattering activity of pure SiO stretching modes is very weak in the silica network because of a polarizability compensation in the SiO′Si structure [52]. This effect is not involved when a SiO^- group lies in front of a counter-ion, for instance, in mixtures of SiO_2 and Na_2O [43]. Only a weak signal around 850 cm^{-1} can be related to the increase of D_1 in Figure 26.3 of reference 10. This band could correspond to the breathing mode $\nu_sO_3(SiO')$ previously shown in disperse silica samples [38].

REFERENCES

1. McDonald, R. S. *J. Phys. Chem.* **1958**, *62*, 1168.
2. (a) Hair, M. L. In *Vibrational Spectroscopies for Adsorbed Species*; Bell, A. T.; Hair, M. L., Eds.; ACS Symposium Series 137; American Chemical Society: Washington, DC, 1980; pp 1–11. (b) Burneau A.; Gallas J. P. In *The Surface Properties of Silicas*. Chapter 3A *Vibrational Spectroscopies*; Legrand, A. P. Ed., 1998, p 147.
3. Morrow, B. A.; Cody, I. A.; Lee, L. S. M. *J. Phys. Chem.* **1976**, *80*, 2761.
4. Davydov, V. Ya.; Zhuravlev, L. T.; Kiselev, A. V. *Russ. J. Phys. Chem.* **1964**, *38*, 1108.
5. Hino, M.; Sato, T. *Bull. Chem. Soc. Jpn.* **1971**, *44*, 33.
6. Boccuzzi, F.; Coluccia, S.; Ghiotti, G.; Morterra, C.; Zecchina, A. *J. Phys. Chem.* **1978**, *82*, 1298.
7. Sindorf, D. W.; Maciel, C. E. *J. Am. Chem. Soc.* **1983**, *105*, 1487. (Ajouter d'autres réf RMN de Si pour 0.15!!).
8. Brinker, C. J.; Kirkpatrick, R. J.; Tallant, D. R.; Bunker, B. C.; Montez, B. *J. Non-Cryst. Solids* **1988**, *99*, 418.
9. Brinker, C. J.; Brow, R. K.; Tallant, D. R.; Kirkpatrick, R. J. *J. Non-Cryst. Solids* **1990**, *120*, 26.
10. Galeener, F. L. *J. Non-Cryst. Solids* **1985**, *71*, 373.
11. Stolen, R. H.; Krause, J. T.; Kurkjian, C. R. *Discuss. Faraday Soc.* **1970**, *50*, 103.
12. Bates, J. B.; Hendricks, R. W.; Shaffer, L. B. *J. Chem. Phys.* **1974**, *61*, 4163.
13. Galeener, F. L. *Solid State Commun.* **1982**, *44*, 1037.
14. Gottardi, V.; Guglielmi, M.; Bertoluzza, A.; Fagnano, C.; Morelli, M. A. *J. Non-Cryst. Solids* **1984**, *63*, 71.
15. Bertoluzza, A.; Fagnano, C.; Morelli, M.A.; Gottardi, V.; Guglielmi, M. *J. Non-Cryst. Solids* **1986**, *82*, 127.
16. Brinker, C. J.; Tallant, D. II.; Roth, E. P.; Ashley, C. S. *J. Non-Cryst. Solids* **1986**, 82, 117.
17. Krol, D. M.; Van Lierop, J. C. *J. Non-Cryst. Solids* **1984**, *68*, 163.
18. Mulder, C. A. M.; Damen, A. A. J. M. *J. Non-Cryst. Solids* **1987**, *93*, 387.
19. Lippert, J. L.; Melpolder, S. B.; Kelts, L. M. *J. Non-Cryst. Solids* **1988**, *104*, 139.
20. Hoffmann, P.; Knözinger, E. *Surf. Sci.* **1987**, *188*, 181.
21. Yamauchi, H.; Kondo, S. *Colloid Polym. Sci.* **1988**, *266*, 855.
22. Burneau, A.; Barrès, O.; Gallas, J. P.; Lavalley, J. C. *Langmuir* **1990**, *6*, 1364.
23. Carteret, C. Thesis, Université UHP, Nancy 1, **1998.**
24. Burneau, A., Carteret, C. *Phys. Chem. Chem. Phys.* **2000**, *2*, 3217.
25. Zhdanov, S. P.; Kosheleva, L. S.; Titova, T. I. *Langmuir* **1987**, *3*, 960.
26. Baumgarten, E.; Wagner, R.; Lentes-Wagner, C. *Fresenius Z. Anal. Chem.* **1989**, *335*, 375.
27. Burneau, A.; Barrès, O.; Vidal, A.; Balard, H.; Ligner, G.; Papirer, E. *Langmuir* **1990**, *6*, 1389.
28. Gallas, J. P.; Lavalley, J. C.; Burneau, A.; Barrès, O. *Langmuir* **1991**, *7*, 1235.
29. Humbert, B. Thesis, Université de Nancy I, Nancy, **1991.**
30. Legrand, A. P.; Hommel, H.; Tuel, A.; Vidal, A.; Balard, H.; Papirer, E.; Levitz, P.; Czernichowski, M.; Erre, R.; Van Damme, H.; Gallas, J. P.; Hemidy, J. F.; Lavalley, J. C.; Barres, O.; Burneau, A.; Grillet, Y. *Adv. Colloid Interface Sci.* **1990**, *33* (2–4), 91–330.
31. Zaborski, M.; Vidal, A.; Ligner, G.; Balard, H.; Papirer, E.; Burneau, A. *Langmuir* **1989**, *5*, 447.
32. Fripiat. In *Soluble Silicate*; Falcone, J. S., Jr., Ed.; ACS Symposium Series 194; American Chemical Society: Washington, DC, **1982**; p 165.
33. Young, G. J. *J. Colloid Sci.* **1958**, *13*, 67.
34. Hockey, J. A. *Chem. Ind. (London)* **1965**, 57.
35. Yates, D. E.; Healy, T. W. *J. Colloid Interface Sci.* **1976**, *55*, 9.
36. Himmel, B.; Gerber, Th.; Bürger, H. *J. Non-Cryst. Solids* **1987**, *91*, 122.
37. Burneau, A.; Humbert, B. *Colloids and Surfaces* **1993**, *A75*, 111.
38. Burneau, A.; Barres, O.; Gallas, J. P.; Lavalley, C. *Proceedings of the International Workshop on FTIR Spectroscopy*; Vansant, E. F.; Department of Chemistry, University of Antwerp, Belgium, **1990**; p 108.
39. Eypert-Blaison, C., Michot, L., Humbert, B., Pelletier, M., Villièras, F. d'Espinose de la Caillerie, J. B. *J. Phys. Chem.* **2002** (in press).
40. Humbert, B. *J. Non-Cryst. Solids* **1995**, *191*, 29.
41. Peri, J. B.; Hensley, A. L., Jr. *J. Phys. Chem.* **1968**, *72*, 2926.
42. Humbert, B.; Burneau, A.; Gallas, J. P.; Lavalley, J. C. *J. Non-Cryst. Solids* **1992**, *143*, 75.
43. Chuang, I.S.; Maciel, G.; *J. Am. Chem. Soc.* **1996**, *118*, 401.
44. Chuang, I.S.; Maciel, G.; *J. Phys. Chem. B* **1997**, *101*, 3052.
45. Vasconcelos, W. L.; De Hoff, R. T.; Hench, L. L. *J. Non-Cryst. Solids* **1990**, *121*, 124.
46. Wyckoff, R. W. G. *Crystal Structures*, 2nd ed.; Interscience: New York, **1963**; Vol. 1, pp 312–319.
47. Zhuravlev, L. T. *Langmuir* **1987**, *3*, 316.
48. Klier, K.; Shen, J. H.; Zettlemoyer, A. C. *J. Phys. Chem.* **1973**, *77*, 1458.
49. Van Roosmalen, A. J.; Mol, J. C. *J. Phys. Chem.* **1978**, 82, 2748.
50. Van Roosmalen, A. J.; Mol, J. C. *J. Phys. Chem.* **1979**, *83*, 2485.
51. Furukawa, T.; Fox, K. E.; White, W. B. *J. Chem. Phys.* **1981**, *75*, 3226.
52. Stolen, R. H.; Walrafen, G. E. *J. Chem. Phys.* **1976**, *64*, 2623.

27 Adsorption on Silica and Related Materials

*L.E. Cascarini de Torre and E.J. Bottani**

Research Institute of Theoretical and Applied Physical Chemistry (INIFTA)

CONTENTS

INTRODUCTION

Since the publication in 1979 of Iler's book *The Chemistry of Silica* [1] the interest on silica has not decreased. Scientists working either on basic or applied research continue to publish papers concerning the chemistry and properties of silica. During the 1990s a new family of solid silica derivates (mesoporous silica) has been synthetized expanding both technological applications and basic research opportunities of silica-based materials. Applications in the areas of separation, catalysis, nanotechnology, molecular assembly, molecular recognition, and so on [2–7] are studied by a large number of scientists.

The chemistry and surface properties of porous and non-porous silica were extensively reviewed [8,9]. We have organized this review in several sections starting with a very brief account of mesoporous solids synthesis and their main structural characteristics (see the section on "Mesoporous Silica"). This section also comments on the preparation of chemically modified solids and their properties as well as how the mesoporous structure can be controlled. We also present some theoretical results obtained through computer simulations. Those studies model the solid structure of silica that will be employed in other simulations. Section III is devoted to the characterization of the surface using different techniques. Adsorption from the gas phase is reviewed. Probably the solid–liquid interface is the most intensively studied. In the section, Adsorption from Solution, we present a series of experimental results, followed by some results obtained employing different forms of chromatography. A few paragraphs are devoted to adsorption kinetics and we conclude with some examples of general studies with a more theoretical character.

PREPARATION AND SOLID STRUCTURE

MESOPOROUS SILICA

Methods for preparation of silica samples exhibiting or not porosity are known since a long time [10]. Before the discovery of mesoporous silica [11] zeolites were the most important materials with micropores forming a regular array of channels with uniform size [12–14]. Mesoporous materials were mainly represented by amorphous silica [1], pillared clays, silicates [15,16] and certain forms of alumina. A common characteristic of those materials is the irregular spatial distribution of pores and a wide spectrum of pore sizes [17]. In 1992 Beck et al. announced the synthesis and characterization of new mesoporous silica-based materials [11,18]. These mesoporous molecular sieves received the general

*Deceased.

denomination of M41S. They can be prepared from different silica sources and in different forms (silicate or aluminosilicate). The synthesis, generally speaking, includes mixing a siliceous compound with a surfactant at certain pH. The resulting gel is then heated to eliminate water and finally calcined in nitrogen atmosphere. The final material obtained presents characteristics strongly dependent on the size and concentration of the surfactant.

Surfactant/silicon ratio and reaction conditions are critical in determining the characteristics of the final product [11]. For example, the solid obtained using cationic cetyltrimethylammonium as surfactant produces, after calcination, a hexagonally ordered porous solid with pore sizes between 2 and 3 nm [19,20]. The most studied material of this type that is known as MCM-41 has a hexagonal array of uniform channels. Beck et al. [11] using quaternary ammonium surfactants ($C_nH_{2n+1}(CH_3)_3N^+$) determined the effect of the carbon chain length (n) on the pore diameter. Their results show that changing the chain length from $n = 8$ to $n = 16$ increases the pore size from 1.8 nm up to 3.7 nm. The specific surface area determined by the BET method is approximately 700–1000 m^2/g.

The success achieved in the preparation of ordered mesoporous materials using the so-called supramolecular templating method was very poor. Only a few attempts among several have produced useful materials [21–23].

Beck et al. [11] proposed two general mechanisms that can lead to the formation of mesoporous silicates. According to the liquid crystal templating (LCT) mechanism [18] the structure is determined by the assembly of surfactant molecules into micellar liquid crystals that serve as templates for the formation of the mesoporous ordered solid. The second mechanism elaborated by Beck et al. indicates that the liquid crystalline phase that is formed during the synthesis is trapped by silicate condensation. Later on, Monnier et al. [24] elaborated a more detailed mechanism to explain the formation and morphologies of mesoporous solids. This mechanism identifies three processes starting by the binding of silicate oligomers (multidentate) to the cationic surfactant. The second process consists of the polymerization of silicate in the interface region. The last one is the charge density matching between the surfactant and the silicate. In Figure 27.1 a schematic representation of the proposed mechanism is shown. This pictorial representation shows the transition from a lamellar to a hexagonal mesophase. The transition occurs through the corrugation of the surfactant-silicate sheets.

The structural details of MCM materials were determined by Edler et al. [25,26]. Based on x-ray diffraction data [25] a two-layer wall structure was proposed (Figure 27.2) for MCM-41. The framework of the solid usually called the "wall" is made of silica with a density ca. 0.99 g/cm^3. The walls are coated by a layer of lower density silica (0.87 g/cm^3) called the "lining". The central region constitutes the "central hole". In Figure 27.2 all these regions and their dimensions are indicated. There are two basic questions concerning the roughness of the central hole walls and the position of the channel centroid relative to the hexagonal lattice along the channel. Edler et al. [25,26] concluded that the walls are smooth and that the channels exhibits a certain degree of curvature. They support these conclusions with data obtained from several sources [26].

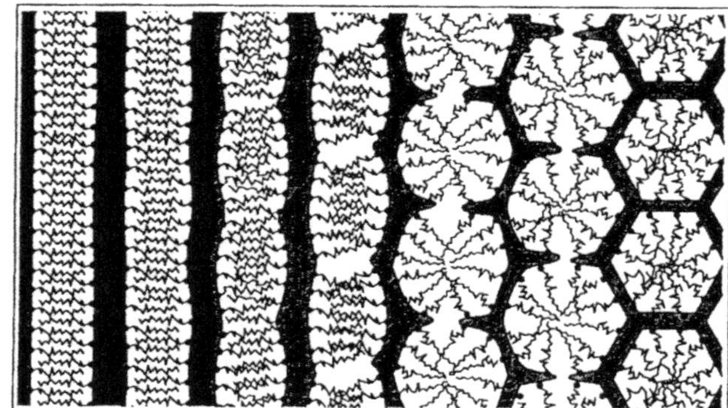

FIGURE 27.1 Pictorial representation of the mechanism for the transformation of surfactant–silicate mixture from the lamellar structure to the hexagonal mesophase. (Reprinted from reference 24. With permission.)

Another member of the M41S family is the MCM-48 that can be synthetized by the classical hydrothermal route [24] or using a room temperature technique [27,28]. The main difference between both methods is that the last

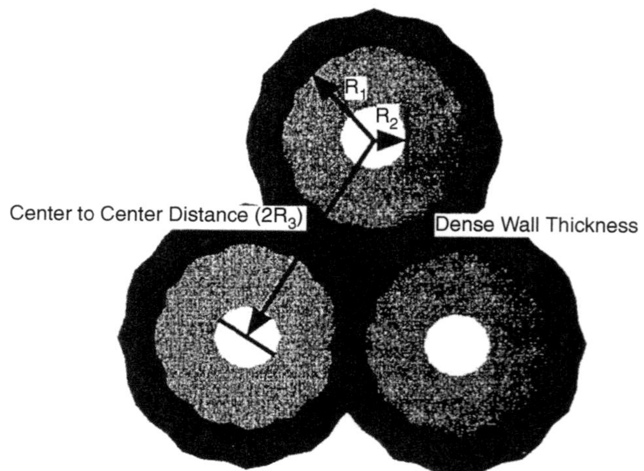

FIGURE 27.2 Schematic representation of the proposed structure for MCM-41 silica. R_2 is the radius of the hole = 0.7 nm; R_1 is the radius the more dense projection = 2.1 nm and $2R_3$ is the lattice unit cell = 4.8 nm. (Reprinted from reference 26. With permission.)

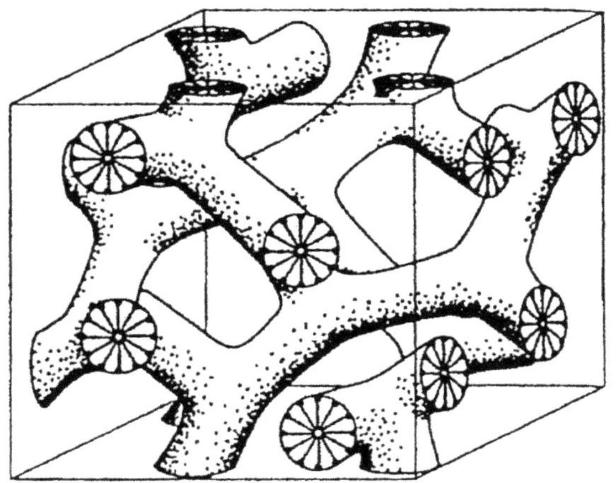

FIGURE 27.3 Pictorial representation of the proposed structure of MCM-48. (Reprinted from reference 28. With permission.)

one produces nonagglomerated uniform spheres [see Figure 27.6 in reference 28]. MCM-48 presents a characteristic cubic-arranged pore system (Figure 27.3). Schumacher et al. [28] reported that this material presents walls with average thickness ca. 0.9 nm (this value is obtained by averaging over a set of samples). The method employed to estimate the wall thickness proposed by Ravikowitch et al. [29] combines x-ray diffraction data with gas physisorption results.

The sol-gel process based on the self-organization between surfactant and silicate was successfully employed to prepare silica films, plates, and fibers which can be optically transparent [30].

Recently it became possible to prepare materials with larger macroscopic ordering length scales by using amphiphilic block copolymers as templating agents. Silica monolithic mesophase was obtained using block copolymers like Pluronic F127 [31]. Two-dimensional x-ray diffraction data indicates that the obtained solids present orientationally ordered domains as large as $0.5\ \mathrm{cm}^2 \times 0.1$ cm. Another characteristic reported [31] is the presence of cubic domains along the edges of the monolith. Recently it was reported a synthesis leading to transparent silica monoliths with pore size controlled by surfactant or cosurfactant, cosurfactant/surfactant ratio, and hydrothermal treatment before calcination [32].

In 1998 Zhao et al. [33] reported the use of triblock copolymer to prepare mesoporous silica (SBA-15) with a hexagonal orientational order and tunable pore sizes between 5 and 30 nm. The synthesis is performed at moderate temperatures (300–355 K) and after calcination the products, mainly depending on the copolymer, present BET surface areas ranging from $600\ \mathrm{m}^2/\mathrm{g}$ up to $1100\ \mathrm{m}^2/\mathrm{g}$. Pore sizes go from 4.5 nm up to 10 nm and wall thickness is ca. 4 nm. According to Zhao et al. [33] pore size can be expanded upto 30 nm by changing the sizes of the blocks that constitute copolymer. Miyazawa and Inagaki [34] reported that SBA-15 microporosity, which is originated in the walls, is determined by the synthesis temperature and the Si/surfactant ratio in the starting reaction mixture.

Fractal Silica

The use of fractals to describe materials with irregular structures has provided to be a promising formalism [35]. In 1998 Aikawa et al. [36] reported the synthesis of a new porous silica that exhibits self-similarity and a wide-range pore size distribution. The templating agent was the bicontinuous structure of surfactant molecules. A mixture of polyoxyethylenedodecylether, water and isooctane has been employed to prepare the bicontinuous layer. Nitrogen adsorption, determined at 77 K, showed a hysteresis of type H3 [37] indicating the presence of slit-shaped mesopores. The adsorption isotherm also shows a significant adsorption at low pressures, which is an indication of microporosity. Avnir and Pfeifer [38] demonstrated that the pore size distribution is related to the fractal dimension through:

$$-\frac{dV_p}{dr} \propto r^{2-D} \tag{27.1}$$

where V_p is the pore volume, r is the pore radius, and D is the fractal dimension. The sample obtained by Aikawa et al. [36] has a fractal dimension $D = 3.2$ which means that the porous structure is very well 3D developed. Aikawa et al. [39] described in a more detailed way the behavior, structure, morphology, and preparation of fractal silica. They reported the preparation of mesoporous silica by hydrolysis and polycondensation of tetraethylorthosilicate in cyclohexane–nonylphenylether–water mixtures. Their main conclusion is that poly-condensation in bicontinuous microemulsions produces silica particles with layered structures and broad pore size distribution (micro to macropores). The macropore size distribution follows the fractal rule with dimension 1.7.

Modified Materials

Silica is a very versatile material that can be easily modified either by chemical or thermal treatments [40,41]. The most relevant changes concern the alternations of: silanol number, pore size distribution and specific surface area.

The synthesis of self-assembled mesoporous silica-based materials with ordered structures has created a renewed interest in silica chemistry [see reference 42 and references therein]. To achieve those modifications there are two basic techniques: modification of a material after its synthesis and the so-called one-pot synthesis. Both methods cannot guarantee a full preservation of the ordered structure. Antochshuk et al. [3] studied in detail the introduction of functional groups into the structure of FSM-16 mesoporous silica. They concluded that the modification process occurs in several steps. It begins with the interaction of silane molecules with the surface, after other reactions the final step consists in conformational rearrangements of the attached species. NMR results obtained by the same authors show that only half of the available silanol groups are isolated thus not all of them can be functionalized. Another interesting finding concerns the increase in thermal stability with the order degree of the alkyl phase attached.

Antochshuk et al. [42] proposed a new method, template displacement with organosilanes, where the functionalization is achieved without a calcination or surfactant extraction. It is shown that this procedure not only introduces functional groups but it also contributes to preserve and to stabilize the structure [43]. The authors also studied several MCM-41 samples substituted with Ce. Their results indicate that the incorporation of Ce into the framework of silica produces an increase in BET surface area as well as in pore volume. They also found some improvement in thermal and hydrothermal stability of Ce-substituted samples. Nevertheless a decrease in long-term stability has been detected [43].

The incorporation of metal atoms in mesoporous silica structure is of high interest in catalysis. Ravikovitch et al. [44] reported the hydrothermal synthesis and characterization of V/MCM-41 catalysts. These catalysts are known to be very active in hydrocarbon oxidation with hydrogen peroxide in liquid phase. Van der Voort et al. [45] reported the preparation of catalysts with VO_x species on the surface of MCM-48 silica. Vanadylacetylacetonate is sublimed in a vacuum reactor where it reacts with a heated substrate. After the reaction is completed the catalyst precursor is calcined at 773 K in air, the final product has ca. 8.7% in weight of vanadium. The reaction proceeds through either a ligand-exchange mechanism or a hydrogen-bonding interaction. UV-vis diffuse reflectance spectrum clearly shows that tetrahedrally coordinated VO_x species are on the surface of the MCM-48 sample.

Araujo et al. [46] reported the preparation of MCM-41 like mesoporous silica with lanthanum and cerium incorporated. The final materials obtained have similar structural properties as MCM-41, the main difference is in the higher acidity of the La and Ce-doped silicas as tested with n-butylamine.

Recently Cabrera et al. [47] developed a new synthesis of ordered mesoporous solids involving the use of atrane complexes as precursors. This technique was tested to obtain silica, Al and Ti mesoporous oxides and simply and doubly doped oxides with a large list of metals [see Table 3 in reference 47]. Other dopants or chemical groups were added to MCM type materials. For example, Ni and B were used as dopants and the catalytic activity tested [48a]; other authors reported use of Ti and Ti-Al as dopants [48b]; and CuCl [48c]. Sulfonic acid was used to functionalize the surface of MCM [48d]; β-cyclodextrin [48e] was also employed as modifier as well as phenil groups were employed to graft MCM surface [48f].

Computer Simulations

Computer simulation proved to be an excellent tool to study a large variety of problems. Structure of solids and interfacial processes were largely studied by numerical simulations of the corresponding real systems [49–52]. The specific cases of bulk silica and silicate glasses were reviewed quite recently [53]. Amorphous silica was the material chosen to study by computer simulation in several cases [41,54–56].

Bakaev and Steele performed computer simulations employing a random net model to obtain a bulk of amorphous silica [54]. The validity of this model is confirmed by diffraction data collected from different sources [57]. The structure inferred is mainly composed of almost regular tetrahedra in which O-Si-O angles are very close to the theoretical one. On the other hand Si-O-Si angles exhibit a wide distribution (generally between 140–150 degrees). To obtain the bulk silica they first built an amorphous solid imposing periodic boundary conditions in the three directions. This task is done with a Monte Carlo Canonical Ensemble algorithm [50]. Once the equilibrium configuration is achieved the structure is annealed using a molecular dynamics algorithm [50,54,55]. The obtained structure is employed to study physical adsorption of water. Bakaev and Steele [54] demonstrated that hydrophilicity of silica surface is due to an increase in the surface electric field, which in turn is due to the existence of simple and triple coordinated oxygen atoms as well as very distorted SiO_4 tetrahedra. They also demonstrated that a completely hydrophobic surface contains only bridging oxygen atoms.

Horbach et al. [56] have employed the model developed by van Beest et al. [58]. This model gives a quite reasonable description of several properties of silica glass [59,60]. Horbach et al. [56] performed molecular dynamics simulations within the harmonic approximation to calculate the specific heat of amorphous silica. Their results are qualitatively in agreement with experimental values obtained by other authors [56]. They claim that the discrepancy with experimental results is in part due to size effects of the simulation box employed. A simpler model has been employed by Cascarini de Torre et al. [41]. To simulate the adsorption properties of amorphous silica the authors described the structure of the solid as a collection of randomly packed spheres, Bernal's model [61], representing the oxygen anions.

Grand Canonical Ensemble Monte Carlo simultations of nitrogen physisorption performed with this model solid reproduced the experimental isotherms. Moreover different silanol numbers were simulated by randomly changing surface oxides anions by less adsorbing atomic groups of the same size. This model neither can predict any mechanical property of the solid nor its chemical reactivity since it does not take into account the chemical structure of the real solid. We shall later discuss the results obtained with this model.

Hammond et al. [62] developed a periodic wall model to calculate the structure factor of MCM-41 mesoporous silica. This model is able to explain changes observed in the Bragg x-ray scattering when the surfactant is removed from the pores of the solid. Moreover the model is sensitive to pore packing order/disorder thus it could be useful to study other related materials like MSU-X, MCM-48, and so on.

SURFACE CHARACTERIZATION

Any dispersed material exhibits specific properties that are a consequence of its surface structure. As surface structure we mean topography and chemical composition. Both characteristics contribute to determine the behavior as adsorbent of the material thus affecting its catalytic performance and its chemical reactivity.

There are several techniques that can be employed to characterize the surface of a dispersed solid. We will mention several methods to finally focus our attention on adsorption as a technique that provides a characterization of the solid surface through the behavior of the adsorbed phase.

Despas et al. [63] described the use of high-frequency impedance measurements to study the reactivity of silica gel and Stöber silica with respect to molecular and ionic bases of various sizes and strengths in aqueous solution. The experimental setup has been previously described [64]. With this technique it is possible to distinguish between silanol and silanolate groups. Stöber silica shows a different behavior compared to silica gel due to the microporosity present in the former.

Glinka et al. [65] described a new method to determine the adsorption energy distribution function for water molecules adsorbed on dispersed silica. The method involves the reaction between hydrated uranyl species and active sites on the surface. Since uranyl groups are luminescent under selective laser excitations it is possible to study how these groups are attached to the surface. The nonuniform broadening of the emission spectra is attributed to variations in the degree of perturbation of uranyl group due to its attachment to the surface active site. This technique was employed in combination with nitrogen adsorption and high-resolution thermogravimetry results.

To characterize the strongest adsorption sites on the surface of modified fused silica, Wirth et al. employed fluorescence microscopy [66]. They conclude from their experimental results that fused silica presents very active adsorption sites having a topographical origin. This conclusion is also supported by the fact that homogenously redeposited silica gel surface has a lower silanol activity. White et al. [67] reported the results obtained from thin film infrared spectra. They studied hydrogen-bonded and chemisorbed methoxymethylsilanes on fumed silica through IR absorption bands below 1300 cm^{-1}. A complete description of the experimental setup has been reported in a previous paper [68]. They conclude that at room temperature all methoxy groups interact with surface hydroxyl groups through hydrogen bonds. At higher temperatures (ca. 423 K) silanes are chemisorbed forming a Si-O-Si bond, the other methoxy group of each molecule interacts with surface hydroxyl groups. At temperatures higher than 673 K the reaction mechanism is complicated by methanol reaction with the surface.

The so-called gas-phase titration is a chromatographic method proposed by Nawrocki [69] to study the strongest adsorption sites. This method that is based on the use of adequate probes, gives a good approximation to the surface concentration and adsorption energy of those sites. Recently Bilínski [70] concluded from gas-phase titration data that surface silanols condensation eliminates only the specific interaction features leaving the nonpolar ones intact.

A combination of x-ray diffraction data with nitrogen adsorption and electron microscopy has been employed by Clerc et al. [71] to describe the structure of SBA-15 mesoporous silica. The model they propose is similar to the one presented earlier by Edler et al. [25]. The presence of a lining, according to Edler et al., or a corona, according to Clerc et al., of lower density than the walls and the existence of occluded molecules in this region is enough to explain the origin of microporosity when those molecules are removed. SBA-15 structure will be further discussed in the next section.

Matsumoto et al. [72] employed adsorption calorimetry to study water adsorption on FSM-16 mesoporous silica. Their experiments clearly show a transition from a hydrophobic to a hydrophilic state of the surface when water adsorption–desorption cycles are performed. The hydrophobicity attained cannot be reverted even by heating the solid up to 823 K. They concluded that the hydrolysis of siloxane bridges produces silanol groups accounting for the hydrophilic character of the surface.

ADSORPTION FROM GAS PHASE

Adsorption of gases is the most often chosen tool to characterize the surface of a solid [73,74]. Gas physisorption provides information on surface characteristics (energetic and topographic) through the behavior of the adsorbed molecules. In principle, the most important features of a solid that are relevant in several applications are: the pore size distribution and the adsorption energy distribution function. Both of them can be determined from adsorption–desorption isotherms. In this section we present some results obtained on silica and glass surfaces. A second group of results concerns the determination of the adsorption energy distribution function. Finally we summarize several recent papers dealing with the pore size distribution of mesoporous silica. The discussion of the structure of SBA-15 silica will be retaken at the end of this section.

Adsorption on Silica and Glass

One of the advantages of physical adsorption as a test is that it is a nondestructive one. It is possible in consequence to employ different molecules as probes on the same sample. The adsorbate is selected according to two main characteristics. In first place is the molecular size that limits topographic features which can be detected with a given probe. The second relevant property is the electronic structure of the adsorbate. If the molecule is nonpolar and has no quadrupole moment [75] it will not show specific interactions with surface species. Oxygen is one example of this kind of adsorbate when it only interacts physically with the surface. Nitrogen is very similar in size to oxygen and both have the same shape, the difference between them is the large quadrupole moment exhibited by nitrogen which enables the molecule to experience specific interactions. Argon is a monoatomic adsorbate smaller than N_2 and O_2. It cannot show specific interactions even though it has a large polarizability. Several characteristics of the surface can be inferred from a comparative study of a set of adsorbates chosen according to the information needed [76]. Cascarini de Torre et al. [41] modified silica samples by chemical and thermal treatments and detected the modifications using nitrogen adsorption. Figure 27.4 shows N_2, Ar and O_2 adsorption isotherms on modified silica. Only nitrogen is capable of detecting differences between the samples. This fact indicates that specific interactions are operating through nitrogen quadrupole moment. The authors have interpreted the experimental results with the aid of computer simulations, which will be discussed in the next section.

Bakaev et al. [77] employed a similar approach to study CO_2 adsorption on glass surfaces. Their results confirm that CO_2 molecule is sensitive to the structure of the surface while Ar does not show this characteristic. In another paper Bakeva et al. [78] studied CO_2 adsorption on glass fibers in a wide temperature range. Employing the classical volumetric technique they have determined the adsorption isotherms and also calculated the isosteric heats of adsorption. They have also analyzed their data in terms of the independent adsorption sites model [74,79,80]. The most important equation of this model is the generalized adsorption isotherm:

$$N(p) = N_m \int_{U_{\min}}^{U_{\max}} \theta^L(p, U) f(U)\, \mathrm{d}U \tag{27.2}$$

where $N(p)$ is the overall experimental isotherm, N_m is the monolayer capacity, U is the adsorption energy, $[U_{\min}, U_{\max}]$ is the energy interval, $\theta^L(p, U)$ is the local isotherm, and $f(U)$ is the adsorption energy distribution function. This equation has been known since a long time ago and several problems arise while trying to use to calculate $f(U)$ from an experimental isotherm [74,80,81]. Several approaches proposed to solve the problem need the definition of the local adsorption isotherm and the distribution function. In their paper Bakaeva et al. [78] employed Langmuir equation as the local isotherm. The authors also expressed the adsorbed quantity as a function of the chemical potential thus they could obtain the adsorption free energy distribution function that can be related to the entropy of molecules adsorbed on different sites. From the isosteric heat of adsorption profiles it is concuded that the surface of glass fibers is heterogeneous.

Quiñones et al. [82] calculated the adsorption energy distribution function for a series of chlorinated hydrocarbons adsorbed on silical gel. To solve Equation (27.2) they took the Jovanovic isotherm as the local one and assumed that the Jovanovic–Freundlich model [83] describes the overall isotherm. The obtained distributions are single peaked for all the substrates studied. Glinka et al. [65] analyzed N_2 adsorption on dispersed silica. They deconvoluted the adsorption energy distribution functions using gaussian components and assign the peak located at 13.1 kJ/mol to nitrogen adsorption on water molecules linked to the surface. The intensity of this peak decreases with thermal treatment of the solid, thus at ca. 1100 K it reaches the lowest intensity due to silanol condensation and their removal of the surface. They have also identified two other peaks located at 5.6 kJ/mol and 8.9 kJ/mol, but they do not assign them. It is interesting to note that the same peaks are observed with alumina as adsorbent. Based on the results obtained by Cascarini de Torre et al. [41], it is possible to confirm Glinka's et al.'s [65] conclusions about the origin of the peak located at ca. 13 kJ/mol. According to Cascarini de Torre et al., the peak ca. 9 kJ/mol could be assigned to adsorption on a pure amorphous silica surface. Nevertheless these authors did not find a peak at lower energies the shape of the $f(U)$ curves are very similar to the ones obtained by Glinka et al.

Adsorption on Porous Silica

Much work has been done where the adsorbent is a mesoporous silica. Those studies aim to a characterization

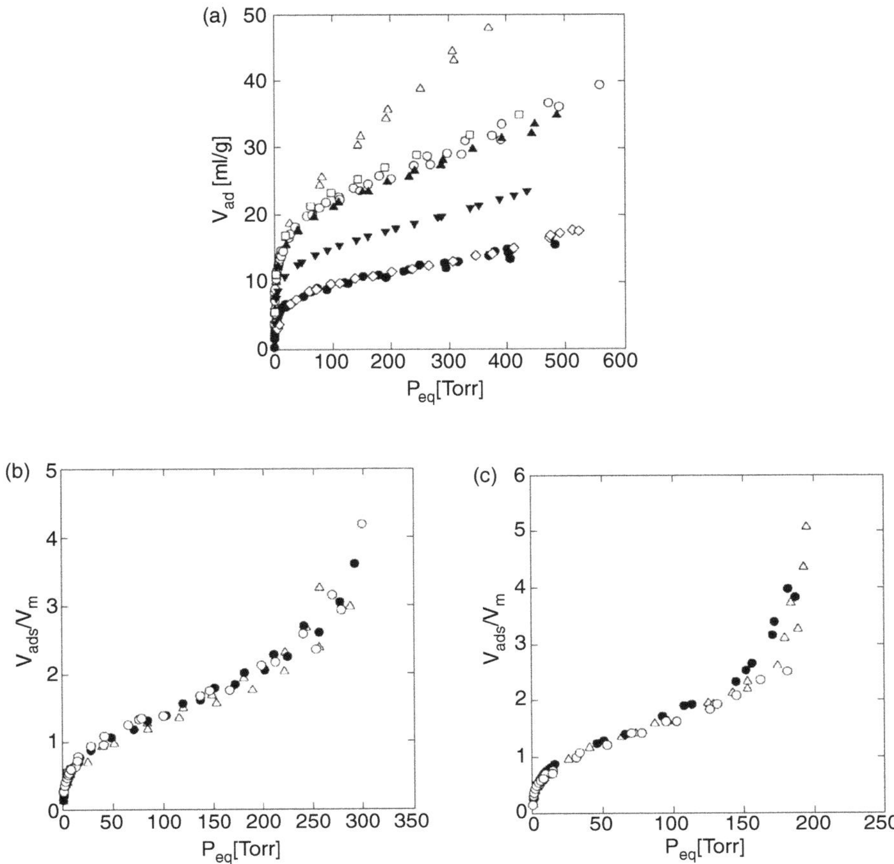

FIGURE 27.4 (a) Nitrogen adsorption isotherms corresponding to all the samples studied at 80.2 K. (Δ) SO_4H_2; (▲) ClH; (○) original; (•) high temperature; (◇) PO_4H_3; (▼) ClK; (□) ClH-2. (b) Argon adsorption isotherms obtained at 80.2 K. (Δ) SO_4H_2; (○) original; (•) high temperature. (c) Oxygen adsorption isotherms obtained at 80.2 K. (Δ) SO_4H_2; (○) original; (•) high temperature. (Reprinted from reference 41. With permission.)

of the pore structure and to increase our knowledge on hysteresis and the mechanism of adsorption on mesoporous materials. Edler et al. [84] studied hydrogen adsorption on MCM-41 and they found that the adsorbed hydrogen state is more like a solid than a liquid. They also found that MCM-41 surface is heterogeneous as revealed by the isosteric heat of adsorption. Branton et al. [85] analyzed nitrogen and carbon tetrachloride adsorption on MCM-41 silica. The mean pore diameter is 3.4 nm for their sample as determined from N_2 adsorption. Nitrogen isotherm is also fully reversible as could be expected from the pore diameter. CCl_4 adsorption was investigated in the temperature range between 273.2 and 323 K. The first interesting characteristic of CCl_4 isotherms is that they do not show a point B thus it is not possible to detect the point of monolayer completion. The second characteristic is the existence of hysteresis in the isotherms at temperatures below 303 K. Finally the isotherms at 303 and 323 K show a linear portion at low relative pressures, <0.2. This fact that is usually associated with Henry's law, is difficult to explain based on usual arguments. Moreover the isosteric heat of adsorption decreases in the same surface coverage region by ca. 2 kJ/mol.

Even though Kelvin's equation is of questionable validity for narrow mesopores it was employed and the results are unexpectedly in good agreement with nitrogen results.

The density functional theory was employed by Neimark et al. [44,86,87] to study the adsorption of N_2 and Ar on MCM-41 mesoporous silica and V catalysts supported on MCM-41. In another group of papers, Neimark et al. employed the density functional theory to study N_2 and Ar adsorption MCM-48 mesoporous silica [28,29]. Their main goal was to characterize the pore size distribution of those mesoporous solids. According to the density functional approach it possible to calculate the adsorption isotherm on an individual pore. If a cylindrical symmetry is assumed for the pore, the adsorption isotherm is given by:

$$N_p(x) = \frac{8}{D^2}\int_0^{D_2} r(\rho(r) - \rho_o)\,\mathrm{d}r \qquad (27.3)$$

where $N_p(x)$ is the adsorbed quantity per unit of pore volume at relative pressure x, D is the internal pore diameter, ρ_o is the bulk density of the adsorbate and $\rho(r)$ is the local density of the fluid. Evidently the amount adsorbed calculated with Equation (27.3) is a function of the pore diameter thus it should be better indicated as $N_p(x, D)$. An equivalent expression to the general adsorption isotherm, Equation (27.2), can be obtained which uses the adsorption isotherm on individual pores, Equation (27.3), as kernel. The final expression is:

$$N_{\mathrm{exp}}(x) = \int_{D_{\mathrm{min}}}^{D_{\mathrm{max}}} N_{\mathrm{p}}(x, D) f(D)\,\mathrm{d}D \qquad (27.4)$$

where $N_{\mathrm{exp}}(x)$ is the experimental adsorbed quantity at relative equilibrium pressure x, $f(D)$ is the pore size distribution function and $N_p(x, D)$ is the adsorption isotherm, Equation (27.3), on an individual pore of diameter D. There are several standard mathematical procedures to invert Equation (27.4) that give the pore size distribution. The density functional theory has been tested in many cases from which it can inferred that any Kelvin-based approach to calculate the pore size distribution underestimate the average pore size by ~1 nm [44].

Lukens et al. [88] studied N_2 adsorption on a large series of MCM-41, SBA-15 and spherical pore mesocellular foams. They proposed the use of a simplified method with respect to Sing's α_s one. The quantity α_s is defined by:

$$\alpha_s = \frac{v(x)}{v_{\mathrm{ref}}(0.4)} \qquad (27.5)$$

where $v(x)$ is the adsorbed quantity at relative pressure x, and $v_{\mathrm{ref}}(0.4)$ is the adsorbed quantity on a reference solid at relative pressure $x = 0.4$. The simplification developed by Lukens et al. consists in replacing $v_{\mathrm{ref}}(0.4)$ with the statistical thickness of the adsorbed layer defined with Frenkel–Halsey–Hill theory [89]. The resulting expression is a simple one from the mathematical point of view and can be employed in the relative pressure range comprised between 0.05 and 0.995. Another advantage of this method, known as β_s method, is that it does not require the adsorption isotherm on a reference material. The major disadvantage resides in the fact that the Frenkel–Halsey–Hill equation exponent must be known [90]. In the same paper, Lukens et al. [88] proposed a modification to the Broekhoff-de Boer method. The modification consists in replacing the original definition of the adsorbed layer thickness with the approximation due to Hill [89]. The results obtained by the authors with the modified Broekhoff-de Boer method are in good agreement with transmission electron microscopy data.

Nguyen et al. [91] developed a new model to describe adsorption on porous media. In the cited reference they extended the model to the case of cylindrical mesopores and employed it with nitrogen and benzene data on MCM-41 samples. Their starting point is the definition of what they called pore-enhanced pressure. Adsorption forces within a pore could be up to 4 times the force experienced by the adsorbate on the open surface. According to Nguyen et al. [91] interpretation of this effect implies that adsorptive molecules are "attracted" to the interior of the pore and "compressed" in a liquid like state. They proposed the following expression to calculate the pore-enhanced pressure, $p_p(r)$:

$$p_p(r) = p_o \exp\left[-\frac{E_p(r)}{\mathrm{RT}}\right] \qquad (27.6)$$

where p_o is the bulk gas pressure and $\langle E_p(r)\rangle$ is the average potential energy of a gas-phase molecule within the pore. Finally, they employ the Kelvin equation in which the adsorbed layer thickness is written as a function of the pore-enhanced pressure and the pore radius. The corresponding equation is:

$$r - t(p_p, r) = \frac{2\gamma v_m \cos\theta}{\mathrm{RT}\ln\left(p_p / p_s\right)} \qquad (27.7)$$

where γ, v_m and θ are the surface tension, the molar volume and the contact angle respectively. The thickness of the adsorbed layer can be calculated from the thickness on a flat surface, $t_f(p)$, using the expression suggested by Kiselev et al. [92]:

$$t(p, r)^2 - 2rt(p, r) - 2rt_f(p) = 0 \qquad (27.8)$$

where r is the empty pore radius. With Equations (27.6)–(27.8), it is possible to calculate the adsorption isotherm for a given pore. If a set of pore radii is selected to represent the pore size distribution it is possible to calculate the overall adsorption isotherm. Changing the pore size spectrum it is possible to reproduce the experimental isotherm with the calculated one. Using the mathematical procedure described elsewhere [93] the pore size distribution can be calculated. In the model it is necessary to calculate the potential energy of adsorbed molecules within the pores to use Equation (27.6) thus the interaction parameters must be known. The problem concerning the interaction parameters could be solved by adopting a reference surface according to Nguyen and Do [91]. In our opinion the problem can be overcome just by adjusting the interaction parameters to reproduce the experimental isotherm in a computer simulation [94].

Gas adsorption has been successfully employed to elucidate the structure of SBA-15 mesoporous silica. Ryoo et al. [95] have corroborated the existence of micropores and a certain degree of connectivity in this material. The method employed combines adsorption data with selective pore blocking via chemical bonding of organosilanes. To analyze the connectivity a method involving inverse platinum replicas of the mesoporous structure is described [95]. SBA-15 structure presents not only large, uniform, and ordered channels but also smaller complementary pores which can be removed by calcinations at ca. 1300 K. To study the location of the complementary porosity, platinum is deposited in the pores and then the silica framework is dissolved. Then the Pt particles are observed with transmission electron microscopy (TEM). When a Pt loading of ca. 30% in weight is deposited, TEM images show the formation of parallel Pt wire bundles. These bundles are coherent with the SBA-15 geometrical structure details. The same technique applied to MCM-41 does not show bundle formation but isolated Pt nanowires [95,104] because there is no connection between the pores. According to the authors these experiments undoubtfully demonstrate the existence of connectivity in SBA-15 mesoporous silica.

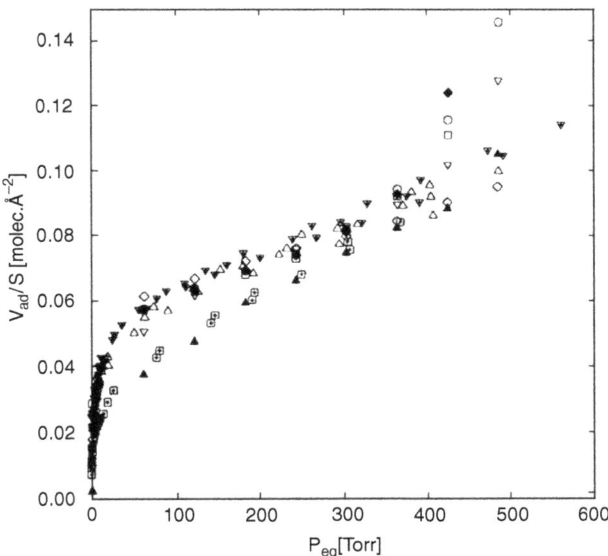

FIGURE 27.5 Nitrogen simulated and experimental adsorption isotherms at 80.2 K. BET specific surface areas have been calculated from each isotherm to express the adsorbed quantity as molecules/Å^2. (⊡) SO_4H_2; (□) N = 10; (○) N = 20; (◆) N = 30; (▽) N = 60; (▲) N = 90; (v) original; (◇) N = 0; (△) high temperature. (Reprinted from reference 41. With permission.)

Computer Simulations

Computer simulation of the adsorbed phase can be employed in several ways. it is possible to determine the adsorption isotherm and to calculate almost any property of the system; to study the effect of system variables upon adsorption and to analyze the behavior of the adsorbed phase. Computer simulation needs a model describing the adsorbent. Moreover numerical simulations of its adsorption characteristics can be employed as criterion to validate the proposed model for the solid [41,77,96].

We found what we called a local heterogeneity effect [41] due to the presence of surface atomic groups different from oxygen and silicon and is related to nitrogen specific interactions. The model employed to describe the solid exhibits the same characteristics as adsorbent of real amorphous silica. It is possible to reproduce the adsorption isotherms of different gases on pure silica and on thermally or chemically modified silica. Figure 27.5 shows nitrogen adsorption isotherms as an example. The model also provides the adsorption energy distribution functions, which are shown in Figure 27.6. These distribution functions are in perfect agreement with the results obtained by Glinka et al. [65] that have been discussed in the section, "Adsorption on Silica and Glass." It must be pointed out that the only difference between both results is that our distributions do not show a peak at very low adsorption energies.

Bakaev et al. [77] simulated CO_2 adsorption on silica surfaces to test the model they have elaborated to describe this kind of solid. Their main conclusion is that highly annealed model silica surfaces are able to reproduce the experimental characteristics.

Some time ago, Vuong et al. [97] studied the effect of surface roughness on model heterogeneous porous silica.

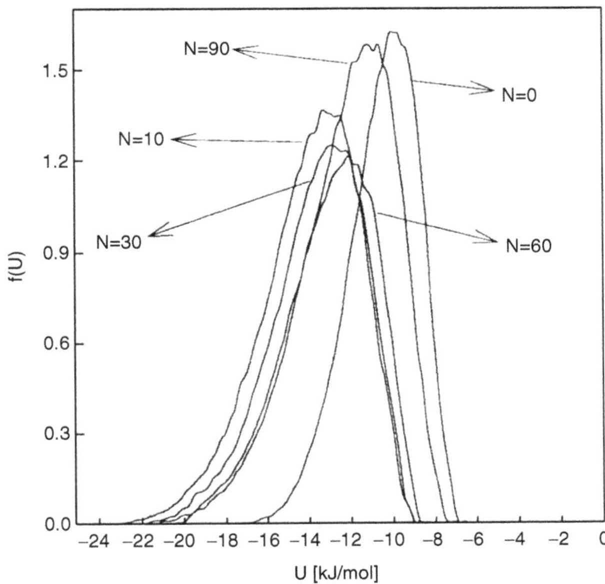

FIGURE 27.6 Adsorption energy distribution functions obtained from the simulated isotherms. N is the number of surface atoms replaced. (Reprinted from reference 41. With permission.)

The basic model solid was developed by Kaminsky et al. [98] and consists of microspheres with the interaction sites uniformly distributed on their surfaces. Computer simulations performed with their model solid reproduce quite well the adsorption of methane on silica gel at low surface coverage. The agreement between simulation and experiment is not good enough at higher densities of the adsorbed phase. The observed differences could be due to several reasons. In first place, it must be mentioned imperfections in modeling interactions between adsorbed molecules [94]. According to Vuong et al. [97] surface roughness is the main cause of discrepancy between simulations and experiments. They tested some other model solids and concluded that surface roughness alters the maximum density of the adsorbate attainable on the surface. As could be expected the effect on the isosteric heat is more marked at low densities of the adsorbed phase. This is due to the broadening of the adsorption energy distribution function. Vuong et al. also conclude that surface roughness is not the only cause of disagreement between calculated and experimental data studied in the cited paper. This is why is our opinion gas–gas interaction potentials should be improved before drawing any conclusion.

Computer simulation on model solids is also employed to test theoretical models. One of the most interesting characteristics of adsorption on porous materials is the existence of hysteresis. The subject has deserved a lot of work and usually assigned to vapor condensation in the pores at lower pressures than in the bulk. No matter all the effort devoted to this problem it has not been possible to obtain an accurate relationship between hysteresis loops and capillary condensation. One criticism to the use of Monte Carlo simulations to study hysteresis is that the mechanism of adsorption and desorption implicit in the algorithm is not equal to the real one [99]. This question has been discussed by Sarkisov et al. [100] who concluded that Grand Canonical Monte Carlo (GCMC) simulations can describe very well the observed hysteresis. In an effort to add more evidence Sarkisov et al. developed a new kind of simulation, which is assumed to better mimic the physical process. This method, GCMD (Grand Canonical Molecular Dynamics), consists in having the system divided into two cells. One cell contains the adsorbent and the other the gas. Both cells are subjected to periodic boundary conditions and gas molecules can diffuse from one cell to the other by two mechanisms: Monte Carlo or Molecular Dynamics random walk. A few months before Sarkisov et al. [100] paper was published, Neimark et al. [101] submitted another paper in which they described a similar simulation technique that they call gauge cell method. The basic idea underneath both methods is the same. In Neimark's paper [101] a more careful thermodynamic description is presented. The method is based on thermodynamic integration. In theory it is possible to join vapor-like and liquid-like limits of stability (spinodals) with a continous path of equilibrium states and to determine the phase coexistence with the aid of Maxwell's equal areas rule. The proposed method assumes that the phase diagram of the gauge cell is known. Neimark et al. [101] also calculate the energy barriers between metastable and stable states. The calculated barriers together with the Arrhenius factors for the transition between metastable and stable states determine the conditions of spontaneous condensation or evaporation. According to the authors this method can provide new physical insight into the vapor–liquid coexistence and the origin of the isotherm hysteresis observed in nanoporous solids. The simulated isotherm of Ar on MCM-41 reproduces the vertical steps observed in the experiments. The other parts of the isotherm are qualitatively reproduced [see Figure 27.9 in reference 101].

Neimark et al. [102] also studied the problem of adsorption–desorption hysteresis with the nonlocal density functional theory (NLFT). They compared NLFT results with Monte Carlo (MC) simulations. Their main conclusion is that both methods, NLFT and MC, can quantitatively predict the adsorption and desorption branches of the isotherm provided that the fluid–fluid and fluid–solid interaction parameters are adequate.

In a recent communication Bartkowiak et al. [103] reported the results obtained with dielectric relaxation spectroscopy and computer simulation of the melting-freezing behavior of a confined fluid. The solids studied were porous glass, Vycor glass and MCM-41. With dielectric relaxation spectroscopy data it is possible to determine melting points and orientational relaxation times. The obtained results are compared with Monte Carlo computer simulations. If σ is the diameter of the fluid molecule it is found that the fluid freezes into a single crystalline structure in pores of diameter greater than 20σ. For pores with a diameter smaller than 15σ even a partial crystallization does not occur. In the case of pores with diameters between 15σ and 20σ a partial crystallization with the formation of amorphous regions is observed.

ADSORPTION FROM SOLUTION

In this section we present a brief account of the most recently published studies and techniques concerning the solid–liquid interface involving silica and related materials. Interactions between solutions and solid surfaces play an important role in several processes such as ore flotation, colloidal stabilization, oil recovery, soil pollution and so on [105].

Gomez et al. [106] studied the behavior of nonwetting liquids in porous hydrophobic silica. They proposed a new method, liquid intrusion microcalorimetry, to determine the contact angle of a fluid in a mesoporous material. The obtained results are in agreement with data from other sources.

NMR and IR spectroscopy have been employed by Boiadjiev et al. [107] to study the series of chemical reactions occurring on the surface of a high-area-silica highly hydroxylated with dimethylznic and *n*-alkanethiols of different chain lengths. The wicking method was employed [108] to study the adsorption of several *n*-alkanes, water, formamide, and diiodomethane on silica gel. The obtained data allowed the calculation of several components of the surface free energy. Bremmell et al. [109] have analyzed the problem of flotation efficiency of silica particles employing ionic surfactants (sodium dodecylbenzenesulfonate, SDBS, and cetyltrimethylammonium bromide, CPB). The experiments on CPB adsorption do not show any unusual feature up to concentrations corresponding to the zero charge point of the solid particles. If more surfactant, CPB, is added a change in orientation of some surfactant ions in the interface is observed. In fact the reorientation implies that the charged head-groups are pointing to the solution. This effect has been previously described as hemi-micellization or bilayer formation or reverse orientation model [109]. The authors also have determined the direct interaction force between a silica sphere and a mica surface using atomic force microscopy (AFM).

AFM was also employed by Kanda et al. [110] to study the behavior of silica surfaces in contact with water–alcohol mixtures. They measured the interaction and adhesive forces between a mica plate and silica particles. According to their data, alcohols heavier than 1-propanol are adsorbed in a vertical position forming a structured monolayer.

One experimental problem that arises in the study of the state of adsorbed species from solution is that the same species is also present in the solution in large quantities. This means that any technique that directly monitors those species must be able to subtract the contribution of nonadsorbed molecules or ions. To overcome this problem Ström et al. [111] have adapted the time-resolved fluorescence quenching technique to study the state of cationic surfactants on silica surfaces. According to this technique fluorescence decay is recorded following the single-photon counting method. The experimental setup has been described elsewhere [112]. One of the surfactants studied was dodecyl-1,3-propylene-pentamethyl-bis (ammonium chloride). AFM images clearly show that this surfactant forms a regular hexagonal packing of spherical micelles on the cleavage plane of mica.

The adsorption of liquid mixtures leads to the study of competition for the adsorption sites. Goworek et al. [113] studied the adsorption of mixtures of hydrocarbons on a series of silica gels with different porosity. They found evidences that most of the systems included in their work tend to form mixed surface phases. They also conclude that excess adsorption and the observed selectivity depend on the pore size.

Calorimetry is a technique very often employed to study the solid–liquid interface. Calorimetric experiments have contributed to establish that the adsorption of nonionic surfactants on both hydrophilic and hydrophobic surfaces progresses in two steps. At the beginning a low-affinity adsorption is observed that mainly involves monomer exothermic adsorption. This leads to a low surface coverage. In the next step that begins at surfactant concentrations close to the critical micelle concentration, a highly cooperative process is observed. The measured enthalpies are of the same magnitude as the micelle formation in the bulk [114]. Király et al. [115] reported a comparative study of the adsorption of nonionic surfactants (N,N-dimethyldecylamine-N-oxide and *n*-octyl-β-D-monoglucoside) on silica glass and graphite. Their calormetric determinations confirm the two-stage process described above. During the second step globular surface aggregates are endothermically formed. They concluded that the driving forces operating in both steps are hydrophobic interactions.

Zajac [116] studied the interfacial aggregation of quaternary ammonium surfactants on dispersed silica employing microcalorimetry and adsorption experiments. His main conclusion is that interfacial aggregation on dispersed silica substrates is driven by the same molecular interactions that govern micelle formation in the bulk. Dawidowicz et al. [117] studied the adsorption of a series of alcohols on polymer-coated porous silica glasses. The technique employed in their work was flow-microcalorimetry and the solvent used was *n*-hexane. They proposed a two-step process to explain the experimental data. A rather quick first one in which the alcohol molecules are attached to the polymer film outer region followed by a slower process by which the incoming molecules start to penetrate deeper in the polymer film. They also studied the desorption process by replacing the flowing solution with pure solvent, *n*-hexane. Since nonpolar hexane molecules show a low affinity for the polymer it tends to minimize its volume thus favoring the removal of the adsorbed alcohol. This model can explain the difference between the heats of adsorption and desorption that amount ca. 2 J/g being the desorption heat greater than the adsorption one.

Another example of the usefulness of microcalorimetry to study the solid–liquid interface has been reported by Draoui et al. [118]. They analyzed the adsorption of Paraquat on different minerals, which are part of the soil, to study the retention process of this pesticide by soils. In the particular case of silica they concluded that the main driving force for adsorption is electrostatic. They found a linear decrease of the adsorption enthalpy in the range of −25 to −20 kJ/mol. Nevertheless, the Paraquat molecule has two charges; the heat evolved during adsorption is of the same order as in the case of other single-charged molecules. This fact is explained assuming that the distance between two charges on silica surface does not fit with the corresponding ones in the Paraquat molecule.

Infrared spectroscopy is another technique employed to study the adsorption from solution of different species. White et al. [119] studied the adsorption of methoxymethylsilanes with silica catalyzed by amines using a thin film IR technique which has been described elsewhere [68]. The results obtained by White et al. [119] are in agreement with an independent work reported by Ahmad et al. [20] who studied the adsorption of acetophenones on silica. In both cases hydrogen-bond interaction is one of the most important factors in the adsorption process. In both studies [119,120] the structure of the adsorbed species are described based on IR absorption bands.

Using in-situ IR spectroscopy Combes et al. [121] studied the reaction of organosilanes with silica. The originality in their work resides in the fact of using supercritical CO_2 as solvent for the organosilanes. It is possible to analyze the progress of the reaction by simply venting the CO_2. This technique has the advantage that eliminates the solvent and makes easier to obtain the IR spectra and its interpretation. Combes et al. [121] demonstrated that the physisorption of octadecyltrichlorosilane (OTS) occurs via a weak interaction with surface hydroxyl groups. Another advantage of using supercritical CO_2 as solvent is that it avoids the tedious drying process necessary to eliminate water from silica and the reagents employed. It is known that some water is needed to hydrolyze and adsorb OTS but it also promotes polimerization of silane products on the surface.

Fourier-transform IR spectroscopy has been employed by Bose et al. [122] to study the adsorption of functionalized siloxane polymers with mesoporous silica. The interest of these polymers resides in the fact that they are used to coat silica to prepare a range of gas chromatographic columns with moderate polar sites. The main goal of the work reported by Bose et al. [122] is to study the effect of thermal cycling on the structure of mesoporous silica coated with cyanopropylmethyl-phenyl-methyl-siloxane. To accomplish this task nitrile stretching bands located between 2150 and 2350 cm^{-1} are analyzed. Thermal treatments seem to produce a redistribution of the polymer on the surface of silica thus reducing the degree of hydrogen bonding in the highly coated samples.

Another variant of IR spectroscopic technique is known as Fourier Transform IR Attenuated Total Reflection Spectroscopy (ATR FT-IR) has been described by Poston et al. [123] to study adsorbates at the silica–solution interface. The authors reported the preparation of ZnSe internal-reflection elements coated with a porous silica layer of ca. 700 nm thick. They studied the adsorption of ethylacetate from *n*-heptane solutions. This technique allows the determination of IR spectra in-situ and its dependence on the solution concentration. They found a nonlinear adsorption isotherm of ethylacetate on silica which could be explained assuming surface heterogeneity. A multidimensional least-squares fit with an appropriate isotherm can be employed to resolve the spectra into components due to each interfacial species. In a recent paper Rivera et al. [124] pursued the previous study. In this work they confirmed the presence of surface heterogeneity effects. They could also confirm that isolated silanols and surface water adsorbed on vicinal silanols constitute the main adsorption sites. One weak point in their model is the use of a two-site Langmuir equation to describe the adsorption.

Other *in-situ* spectroscopies have been employed to study different aspects of adsorbed species. Among them it is possible to mention fluorescence spectroscopy [125], diffuse reflectance of solvatochromic probes [126], Raman spectroscopy [127], and surface-enhanced Raman spectroscopy [128].

CHROMATOGRAPHIC STUDIES

One problem that arises in gas chromatography is originated in the poor hydrolytical and thermal stability of the stationary phase. The existence of residual silanol groups on the surface is the cause of surface heterogeneity and asymmetry in the chromatographic peaks of polar species. Fluorinated stationary phases show high selectivity towards halogenated compounds, as is the case of silica having polyfluoroalkyl groups on its surface.

Gas chromatography studies of surface modified silicas provided new insight concerning the adsorption properties of new stationary phases [129]. Inverse gas chromatography (IGC) can be a useful tool to determine the adsorption properties of low surface area solids. The work done by Bakaeva et al. [130] is a good example of IGC use to study low-surface-area-silica glasses. In the cited paper Bakaeva et al. employed IGC to analyze the adsorption of butanol and hexane on E-glass fiber.

Gas chromatography allows the determination of the adsorption isotherm by integration of the retention curve that shows the dependence of the retention volume on the equilibrium vapor pressure of the adsorbate [131]:

$$N_{\mathrm{ad}} = \frac{1}{\mathrm{RT}} \int_0^p V_R \, \mathrm{d}p \qquad (27.9)$$

where N_{ad} is the amount adsorbed, V_R is the retention volume and p the equilibrium pressure. Bakaeva et al. [130] demonstrated that the surface of E-glass is homogeneous with respect to hexane and heterogeneous for butanol. This conclusion is in agreement with previous results obtained with computer simulation [54,77].

Gas phase titration method that has been mentioned in the section, "Surface Characterization" [69,70] is based on gas chromatography. The method consists of preadsorbing a given molecule, called blocking agent, on the most active sites and to analyze how the adsorption of a probe is altered. Another way to obtain information concerning the adsorption sites is to determine the adsorption energy distribution function, $f(U)$ (Equation 27.2). From gas chromatographic data it is possible to calculate $f(U)$ from:

$$f(U) = -\frac{j}{a_m}\left(\frac{p}{\mathrm{RT}}\right)^2 \frac{\partial V_R}{\partial p} \qquad (27.10)$$

where j is the James-Martin compressibility factor and a_m is the monolayer capacity. To deduce Equation (27.10) it is necessary to use the asymptotically corrected condensation approximation [74 and references therein]. Biliński [132] demonstrated that gas chromatographic results using Equation (27.10) are in good agreement with the gas phase titration method. He confirmed that the strength of surface–adsorbate interaction decreases in the sequence benzene, chloroform, n-octane as a result from the strong donor–acceptor interaction for benzene. The effect of thermal treatment of silica gel was also studied by gas chromatography [133] and the results compared with the gas phase titration method [134]. From gas chromatography [133] it is concluded that thermal treatment of silica gel produces important changes in its adsorption properties. Increasing the heating temperature eliminates surface silanol groups resulting in a decreasing magnitude of specific interactions. The $f(U)$ calculated with Equation (27.10) shows large differences between the treated samples in the case of octane adsorption. For adsorbates that interact via specific forces there is a large difference when the treatment is performed at ca. 1100 K. Biliński [134] concluded that gas phase titration method is useful to analyze the small population of the strongest interaction sites of silica gel.

IGC has also been employed by Donnet et al. [135] to characterize silica xerogels. From chromatographic data the authors estimated the BET equation C parameter which is a measure of the gas–solid interaction energy. They found that increasing the silylation of the surface decreases the value of C in agreement with the results obtained by Cascarini de Torre et al. [41] with gas adsorption and computer simulations.

Liquid chromatography is also employed to investigate the liquid–solid interface. High efficiency adsorption-based separation procedures like batch elution chromatography and simulated moving bed chromatography are today important purification techniques, particularly in the pharmaceutical industry. Lanin et al. [136] elaborated a model to determine monomolecular adsorption constants from binary solutions chromatography. The experimental system studied is anisole in n-hexane solution using hydroxylated silica as adsorbent. The authors claim to obtain good agreement between calculated isotherms with their method and the one proposed by Glueckauf [137]. It must be pointed out that the model is very simple and its main postulates are:

1. the surface of the adsorbent is both chemically and geometrically homogeneous
2. there are no molecular associations in the adsorbed layer
3. both the solution bulk and the solution in direct contact with the surface exhibit ideal behavior

It is evident that the first postulate is far from being followed by any real system. Moreover a lot of work has been done demonstrating severe effects due to surface heterogeneity. With respect to the ideal behavior of the solution it has been demonstrated that the ideal solution theory is a good approximation [138].

Elution on a plateau is chromatographic technique related to frontal analysis that has been developed a long time ago [139] but scarcely used. Rheinländer et al. [140] employed this method to study the adsorption of surfactants on silica gel. They compared this method with other ones and found a good agreement. Their adsorption isotherms show two steps. The first plateau is at least two orders of magnitude lower than the second one. The surface excess concentration corresponding to the first plateau decreases with increasing the temperature. This behavior could be related to the fact that in the first plateau the concentration of surfactant molecules on the surface is very low leaving enough free space to allow molecules to lie flat on the surface. When a smaller homologue of the surfactant is adsorbed, the same temperature behavior is observed for both plateaus [140]. The adsorption isotherms can be described quite well with the general equation obtained by Zhu and Gu [141]:

$$n_\mathrm{ad} = \frac{n_\mathrm{max} K_1 x\left[1/n + K_2 x^{\mathrm{n}-1}\right]}{1 + K_1 x(1 + K_2 x^{\mathrm{n}-1})}; \quad x = \frac{c}{c_o} \qquad (27.11)$$

where n_ad is the surface excess, n_max is the maximum surface excess, c is the concentration and $c_o = 1\ \mathrm{mol \cdot m^{-3}}$, K_1 and K_2 are equilibrium constants for monolayer adsorption in the first layer and hemi-micelle formation respectively. Finally n is the aggregation number of hemi-micelles. If $K_2 \to 0$ and $n \to 1$ the isotherm Equation (27.11) reduces to Langmuir equation.

ADSORPTION KINETICS

We were able to detect a reduced number of published papers in the last few years concerning the study of adsorption kinetics involving silica and related materials. In our best knowledge there are only five papers published in the period 1998–2000. For this reason we shall limit the discussion of this topic to a brief description of those papers. Gas adsorption kinetics has been reviewed a long time ago by Kreuzer et al. [142].

The kinetics and mechanism of cationic surfactant adsorption on silica was studied by Pagac et al. [143] using scanning angle reflectometry. They also expanded the scope of their work to include coadsorption with cationic polyelectrolytes always in aqueous solutions. They were able to identify two regimes below and above the critical micelle concentration (cmc). Below cmc the surfactant is adsorbed as monomer, when cmc is reached micelle adsorption occurs. The adsorbed micelles form a close-packed structure. The comparison of these conclusions with the ones derived for nonionic surfactants [113–116] suggests that the way adsorption progresses is very similar in both cases at least in their gross details. The close-packed structure of micelles suggested by Pagac et al. [143] have been observed by Ström [111] using AFM in the case of a different cationic surfactant adsorbed on mica.

Hansen et al. [144] studied the adsorption kinetic of small molecules on silica using total internal reflection fluorescence correlation spectroscopy. They detected several unusual effects that could not be explained in the moment. Adsorption and desorption rates obtained with this technique were slower than previously accepted values. According to the authors the explanation could be related to surface heterogeneity.

The third paper in this subject that we were able to retrieve is due to Biswas et al. [145]. In their introduction to the paper they said that dynamic and mechanistic aspects of adsorption of surfactants at the solid–liquid interface, particularly silica surface, were rare and quoted six papers. The most recent among them was due to Tiberg [146] in 1996. Adsorption kinetics was studied by Biswas et al. [145] using classical batch experiments. They found that the adsorption follows a two-step first-order rate equation. From the calculated rate constants they obtained the activation energies and entropies concluding that both processes are entropy controlled.

The next paper we will comment on in this section is a letter by Nakatami et al. [147] in which they describe a microscale technique to study the dynamics of adsorption. This technique, the single-microparticle injection, is basically an optical method that uses Lambert-Beer law to follow the concentration of methylene blue on the surface of a silica gel microparticle. Their main conclusions are that equilibrium is attained within 20 min, Langmuir equation describes the experimental adsorption isotherm, methylene blue molecules penetrate into the pores and the whole process is controlled by adsorbate diffusion in water.

A gravimetric technique to study the adsorption from gas phase was described by Lobanov et al. [148]. They employed a piezoelectric crystal as measuring device. It is necessary to adequately coat the crystal surface to attain the desired selectivity. With this experiment they were able to determine the adsorption and desorption kinetics. Monoethanolamine adsorbed on silica was the system chosen for this work. Their conclusion, supported by quantum mechanical calculations, is that adsorption at low temperatures occurs by hydrogen bonding on silica surface. At higher temperatures conformational changes allow the adsorbate to also interact with oxygen atoms of vicinal terminal hydroxyl groups of the surface.

MISCELLANEOUS STUDIES

In this section we review several papers dealing with different aspects of adsorption on silica and related materials from a more theoretical point of view.

Adsorption of nonionic surfactants on porous solids has been studied by Huinink et al. in a series of papers [149,150]. They elaborated a thermodynamic approach that accounts for the major features of experimental adsorption isotherms. It is a very well known fact that during the adsorption of nonionic surfactants there is a sharp step in the isotherm. This step is interpreted as a change from monomer adsorption to a regime where micelle adsorption takes place. Different surfactants produce the step in a different concentration range. The step is more or less vertical depending on the adsorbate. The thermodynamic analysis made by Huinink et al. is based on the assumption that the step could be treated as a pseudo first order transition. Their final equation is a Kelvin-like one, which shows that the change in chemical potential of the phase transition is proportional to the curvature constant (Helmholtz curvature energy of the surface).

In the second paper of the series [150] the authors extend their treatment with a mean field lattice theory. They found that the adsorbed amount in the pores is smaller and the step in the isotherm shifts to lower chemical potential than in a flat surface in the same conditions. They also established that the influence of the curvature on the phase transtition increases with the length of the headgroup. The shift of the phase transition increases with the adsorption energy.

Under certain circumstances it is possible to obtain adsorption isotherms that correspond to a layer-by-layer mechanisms. This is a very well known fact in gas adsorption on homogeneous flat surfaces [151] and is known as layering. At the solid–liquid interface stepwise isotherms are also observed, nevertheless one step and sometimes two are found [114]. Sellami et al. [152] studied the adsorption of 2,5-dimethylpyridine (DMP) on silica from aqueous solutions. The experimental results [153] clearly show the stepwise character of the isotherms, which could either mean multilayer adsorption or layering. In their paper [152] they addressed the question of using the adsorption isotherm to discriminate between both phenomena. Their experimental data has been obtained using a batch method on a wide temperature range (298–329 K). From the analysis of the isotherms they concluded that this is the first case of pure layering observed at the liquid–solid interface.

The interfacial behavior of block copolymers is of interest in several fields like stabilization of emulsions, foams, and wetting control [154]. Gerdes et al. [155] studied the wetting behavior of aqueous solutions of triblock copolymers on silica. The experimental approach was based on the use of a Wilhelmy force balance and direct images of contact angle. Their results show that the three-phase contact line advances in jumps over the surface when it is immersed at constant speed into the copolymer solution. Apparently the stick-slip spreading mechanism is the same as has been proposed for short chain cationic surfactants.

The prediction of multicomponent equilibria based on the information derived from the analysis of single component adsorption data is an important issue particularly in the domain of liquid chromatography. To solve the general adsorption isotherm, Equation (27.2), Quiñones et al. [156] have proposed an extension of the Jovanovic-Freundlich isotherm for each component of the mixture as local adsorption isotherms. They tested the model with experimental data on the system 2-phenylethanol and 3-phenylpropanol mixtures adsorbed on silica. The experimental data was published elsewhere [157]. The local isotherm employed to solve Equation (27.2) includes lateral interactions, which means a step forward with respect to, that is, Langmuir equation. The results obtained account better for competitive data. One drawback of the model concerns the computational time needed to invert Equation (27.2) nevertheless the authors proposed a method to minimize it. The success of this model compared to other resides in that it takes into account the two main sources of nonideal behavior: surface heterogeneity and adsorbate–adsorbate interactions. The authors pointed out that there is some degree of thermodynamic inconsistency in this and other models based on similar - assumptions. These inconsistencies could arise from the simplifications included in their derivation and the main one is related to the monolayer capacity of each component [156].

As adsorbents all solids are characterized by several properties: adsorption energy, distribution function, roughness, and porous structure. The existence of a porous structure can be detected by several features of the adsorption isotherm of simple gases. The most often employed as it has been recommended by IUPAC is nitrogen. Micropores (pore width less than 2 nm) make themselves evident through a large adsorption at very low pressures; mesopores (pore width between 2 nm and 50 nm) are recognized as the origin of hysteresis in the adsorption–desorption branchs of the isotherm. There is a clear dependency of hysteresis on pore width, at least in nitrogen case that has been shown, among others, by Lewellyn et al. [158]. The hysteresis is usually analyzed with the Kelvin equation [89]. All the theories or methods employed to determine the pore size distribution that are based on Kelvin equation have known drawbacks but they are still in use due to the lack of something better. To complicate the situation it has been shown that there is a case in which a mesoporous solid will not exhibit hysteresis in nitrogen adsorption–desorption isotherms. This situation is attained when the pore width is equal to or less than 4 nm. Inoue et al. [159] addressed this problem and developed a way to predict hysteresis vanishing as a function of the pore width. In this aspect mesoporous silica with ordered array of regular pores like MCM-41 and FSM are unique materials that play an important role in experimental and theoretical studies. In fact they are as close as possible to ideal solids thus computer simulations and theoretical models results can be directly tested against experimental data.

Kelvin equation is not appropriate to describe adsorption in narrow mesopores due to problems concerning the contact angle included in the equation. Cole et al. [160] developed a long time ago a model in which they included explicitly fluid–solid interactions. Based on this model Inoue et al. [159] have analyzed nitrogen adsorption on FSM-16 silica and found that the critical pore width is ca. 3.6–3.8 nm. Narrower pores than 4 nm will not produce hysteresis. This method is general enough to be applied to other materials and adsorbates. The same material has been studied by Xie et al. [161] using a combination of x-ray diffraction and ^{1}H-NMR data.

The next paper we will review concerns the synthesis of hexagonally ordered nanoporous carbonaceous materials [162]. Siliceous materials are related to these new ones because the formers are employed as templates. The mesoporous silica is treated with an organic compound in solution, that is, sucrose, and dried several times before the material is heated in vacuum at ca. 1173 K. The final step consists of dissolving the siliceous template with hydrofluoric acid at room temperature. The carbonaceous product is almost Si-free and shows a very similar structure to MCM-41 except that the carbonaceous material shows random interconnection of the tubes through micropores located on the walls.

Silica is also employed to prepare microporous inorganic membranes suitable for gas separation. De Vos et al. [163] reported the preparation of silica membranes with a very low defect concentration. They employed a sol–gel synthesis starting from tetraethylorhosilicate. These membranes consist of a microporous layer on top of a supported mesoporous γ-Al_2O_3 membrane. The support layer provides mechanical strength to the selective silica top layer. The prepared membranes have a thick silica layer (ca. 30 nm) on the support. Increasing the sintering temperature from 673 to 873 K results in a much denser membrane with smaller pores.

Chevrot et al. [164] presented a new experimental technique to study low temperature constant rate thermodesorption of different adsorbates. In the cited paper they analyze water thermodesorption from mesoporous silica MCM-41 and the results are compared with data obtained with adsorption gravimetry. According to the authors this technique has better sensitivity and resolution than the traditional temperature programmed desorption technique. Other advantages of this technique are that a small quantity of sample is needed, solids with lower surface area can be studied, and the lower temperature at which the equipment can operate (163 K).

CONCLUSIONS

Several general conclusions could be stated. MCM-41 walls according to Edler results [25,26] are smooth. Preparation of mesoporous silica is almost a standard technique. In FSM-16 only half of the available silanol groups are isolated thus can be functionalized. Thermal stability increases with the order degree of the attached alkyl phase. Metallic dopants increase thermal and hydrothermal stability but a decrease in the long-term stability is detected. SBA-15 has micropores that exhibit certain degree of connectivity. Nonionic surfactant adsorption on both hydrophobic and hydrophilic surfaces progresses in two stages. At low surface coverage exothermic adsorption of monomers is the main process. Then close to cmc a highly cooperative process takes place. The use of supercritical fluids probed to be useful in IR spectroscopy studies. Gas titration method and adsorption energy distribution function produce results that are in good agreement. Surfactant adsorption progresses according to a two-stage mechanism independently of the ionic character of the surfactant. Mesoporous silica with ordered array of regular pores is as close as possible to an ideal solid thus computer simulations and theoretical models can be tested against experimental data. Simulating an adsorption isotherm and comparing with experiments can test solid structure models.

ACKNOWLEDGMENTS

The authors acknowledge the collaboration given by E.S. Flores and E.A. Fertitta during the preparation of this review. Both authors are researchers of the Comisión de Investigaciones. Científicas de la Provincia de Buenos Aires (CIC), EJB is Visiting Associate Professor of the National University of El Litoral.

REFERENCES

1. Iler, R.K.; "The Chemistry of Silica"; John Wiley (**1979**).
2. Aikawa, K.; Kaneko, K.; Fujitsu, M.; Tamura, T. and Ohbu, K.; *Langmuir*; *14*, 3041 (**1998**).
3. Antochshuk, V. and Jaroniec, M.; *J. Phys. Chem. B*; *103*; 6252 (**1999**).
4. Kaneko, K.; *J. Membr. Sci.*; *96*; 59 (**1994**).
5. Pinnavaia, T.J.; Thorpe, M.F.; *Access in Nanoporous Materials.* Plenum Press (**1995**).
6. Tolbert, S.H.; Sieger, P.; Stucky, G.D.; Ankin, S.M.J.; Chi-Cheng, W. and Hendrickson, D.N.; *J. Am. Chem. Soc.*; *119*, 8652 (**1997**).
7. Schacht, S.; Huo, Q.; Voigt-Martin, I.G.; Stucky, G.D. and Schüth, F.; *Science*; *273*; 768 (**1996**).
8. Bergna, H.E.; *The Colloid Chemistry of Silica.* Advances in Chemistry Series 234. ACS (**1994**).
9. Papirer, E.; *Adsorption on Silica Surfaces.* Marcel Dekker (**2000**).
10. Okkerse, C.; in *Physical and Chemical Aspects of Adsorbents and Catalysts.* Linsen, B.G. Editor. Academic Press (**1970**), Chap. 5..
11. Beck, J.S. et al.; *J. Am. Chem. Soc.*; *114*, 10834 (**1992**).
12. Jacobs, P.A. and van Santen, R.A. editors; *Zeolites: Facts, Figures, Future.* Elsevier, Amsterdam (**1989**).
13. Öhlmann, G.; Pfeifer, H. and Fricke, R. editors; *Catalysis and Adsorption by Zeolites.* Elsevier, Amsterdam (**1991**).
14. Murakami, Y.; Iijima, A. and Ward, J.W. editors; *New Developments in Zeolite Science and Technology.* Elsevier, Amsterdam (**1987**).
15. Pinnavaia, T.J.; *Science*; *220*, 365 (**1983**).
16. Tindwa, R.M.; Ellis, D.K.; Peng, G.Z. and Clearfield, A.; *J. Chem. Soc. Faraday Trans. I*; *81*, 545 (**1985**).
17. Interrante, L.V.; Hampden-Smith (Editors) *Chemistry of Advanced Materials.* Wiley-VCH (**1998**).
18. Kresge, C.T.; Leonowicz, M.E.; Roth, W.J.; Vartuli, J.C. and Beck, J.S.; *Nature*; *359*, 710 (**1992**).
19. Sayari, A.; *Chem. Mater.*; *8*, 1840 (**1996**).
20. Huo, Q.; Margolese, D.I. and Stucky, G.D.; *Chem. Mater.*; *8*, 1147 (**1996**).
21. Sayari, A. and Lin, P.; *Microporous Mater.*; *12*, 149 (**1997**).
22. Ciesla, U. and Schüth, F.; *Microporous Mater.*; *27*, 131 (**1999**).
23. Behrens, P.; *Angew. Chem. Int. Ed. Engl.*; *35*, 515 (**1996**).
24. Monnier, A.; Schüth, F.; Huo, Q.; Kumar, D.; Margolese, D.; Maxwell, R.S.; Stucky, G.D.; Krishnamurty, M.; Petroff, P.; Firouzi, A.; Janicke, M. and Chmelka, B.F.; *Science; 261*, 1299 (**1993**).
25. Edler, K.J.; Reynolds, P.A.; White, J.W. and Cookson, D.; *J. Chem. Soc. Faraday Trans.*; *93*, 199 (**1997**).
26. Edler, K.J.; Reynolds, P.A. and White, J.W.; *J. Phys. Chem. B*; *102*, 3676 (**1998**).
27. Schumacher, K.; Grüin, M. and Unger, K.K.; *Microporous Mater.*; *27*, 201 (**1999**).
28. Schumacher, K.; Ravikovitch, P.I.; Du Chesue, A.; Neimark, A. V. and Unger, K.K.; *Langmuir*; *16*, 4648 (**2000**).

29. (a) Ravikovitch, P.I. and Neimark, A.V.; *Langmuir*; *16*, 2419 (**2000**); (b) Ravikovitch, P.I. and Neimark, A.V.; *Studies in Surf. Sci. and Catal.* Vol 129, Sayari, A. et al. (Editors). Elsevier, p. 597 (**2000**).
30. Ryoo, R.; Ko, C.H.; Cho, S.J. and Kim, J.M.; *J. Phys. Chem. B*; *101*, 10610 (**1997**).
31. Melosh, N.A.; Davidson, P. and Chmelka, B.F.; *J. Am. Chem. Soc.*; *122*, 823 (**2000**).
32. Feng, P.; Bu, X. and Pine, D.J.; *Langmuir*; *16*, 5304 (**2000**).
33. Zhao, D.; Feng, J.; Huo, Q.; Melosh, N.; Fredrickson, G.H.; Chmelka, B.F. and Stucky, G.D.; *Science*; *279*, 548 (**1998**).
34. Miyazawa, K. and Inagaki, S.; *Chem. Commun.*; 2121 (**2000**).
35. Avnir, D. Editor, *The Fractal Approach to Heterogeneous Chemistry*. J. Wiley and Sons (**1989**).
36. Aikawa, K.; Kaneko, K.; Fujitsu, M.; Tamura, T. and Ohbu, K.; *Langmuir*; *14*, 3041 (**1998**).
37. Sing, K.S.W. et al.; *Pure Appl. Chem.*; *57*, 603 (**1985**).
38. Pfeifer, P. and Avnir, D.J.; *J. Chem. Phys.*; *79*, 3566 (**1983**).
39. Aikawa, K.; Kaneko, K.; Tamura, T.; Fujitsu, M. and Ohbu, K.; *Colloids and Surfaces A*; *150*, 95 (**1999**).
40. Goworek, J.; Borówka, A. and Kusak, R.; *Colloids and Surfaces A*; *157*, 127 (**1999**).
41. Cascarini de Torre, L.E.; Flores, E.S. and Bottani, E.J.; *Langmuir*; *16*, 1896 (**2000**).
42. Antochshuk, V. and Jaroniec, M.; *Chem. Mater.*; *12*, 2496 (**2000**).
43. Antochshuk, V.; Araujo, A.S. and Jaroniec, M.; *J. Phys. Chem. B*; *104*, 9713 (**2000**).
44. Ravikovitch, P.I.; Wei, D.; Chuch, W.T.; Haller, G.L. and Neimark, A.V.; *J. Phys. Chem. B*; *101*, 3671 (**1997**).
45. Van der Voort, P.; Morey, M.; Stucky, G.D.; Mathieu, M. and Vansant, E.F.; *J. Phys. Chem. B*; *102*, 585 (**1998**).
46. Araujo, A.S. and Jaroniec, M.; *J. Coll. Interf. Sci.*; *218*, 462 (**1999**).
47. Cabrera, S.; El Haskouri, J.; Guillén, C.; Latorre, J.; Beltrán-Porter, A.; Beltrán-Porter, D.; Marcos, M.D. and Amorós, P.; *Solid State Sciences*; *2*, 405 (**2000**).
48. (a) Wong, S.T.; Lee, J.F.; Chen, J.M. and Mon, C.Y.; *J. Molec. Catal. A*; *165*, 159 (**2001**); (b) Davies, L.J.; McMorn, P.; Bethell, D.; Page, P.C.B.; King, F.; Hancok, F.E. and Hutchings, G.J.; *J. Molec. Catal. A*; *165*, 243 (**2001**); (c) Li, Z.; Xie, K.C. and Slade, R.C.T.; *Appl. Catal. A*; *205*, 85 (**2001**); (d) Díaz, I.; Mohino, F.; Pérez-Pariente, J. and Sastre, E.; *Appl. Catal. A*; *205*, 19 (**2001**); (e) Phan, T.N.T.; Bacquet, M. and Morcellet, M.; *J. Incl. Phen. and Macrocy. Chem.*; *38*, 345 (**2000**); (f) Roshchina, T.M.; Shoniya, N.K.; Kitaev, L.E.C.; Gurevich, K.B. and Kazmina, A.A.; *Russian J. of Phys. Chem.*; *74*, 2026 (**2000**).
49. Ciccotti, G.; Frenkel, D. and McDonald, I.R.; Eds. *Simulation of Liquids and Solids.* North-Holland Personal Library (**1990**).
50. Allen, M.P. and Tildesley, D.J. *Computer Simulation of Liquids.* Oxford University Press (**1990**).
51. Nicholson, D. and Parsonage, D. *Computer Simulation and the Statistical Mechanics of Adsorption.* Academic Press (**1982**).
52. Haile, J.M. *Molecular Dynamics Simulation.* John Wiley and Sons (**1992**).
53. Poole, P.H.; McMillan, P.F. and Wolf, G.H.; *Structure, Dynamics, and Properties of Silicate Melts.* Edited by Stebbins, J.F.; McMillan, P.F. and Dingwell, D.B. Reviews in Mineralogy. Vol. 32 (Mineralogical Soc. Amer. Washington, D.C.) (**1995**).
54. Bakaev, V.A. and Steele, W.A.; *J. Chem. Phys.*; *111*, 9803 (**1999**).
55. Bakaev, V.A; *Phys. Rev. B*; *60*, 10723 (**1999**).
56. Horbach, J.; Kob, W. and Binder, K.; *J. Phys. Chem. B*; *103*, 4104 (**1999**).
57. Wright, A.C.; *J. Non-Cryst. Solids*; *179*, 84 (**1994**).
58. Van Beest, B.H.; Kramer, G.J. and van Santen, R.A.; *Phys. Rev. Lett.*; *64*, 1955 (**1990**).
59. Vollmays, K.; Kob, W. and Binder, K.; *Phys. Rev. B*; *54*, 15808 (**1996**).
60. Horbach, J.; Kob, E. and Binder, K.; *J. Non-Cryst. Solids*; *235*, 320 (**1998**).
61. Bernal, J.D.; *Proc. R. Soc. London*; *A284*, 299 (**1964**).
62. Hammond, W.; Prouzet, E.; Mahanti, S.D. and Pinnavaia, T.J.; *Microporous and Mesoporous Mat.*; *27*, 19 (**1999**).
63. Despas, C.; Walcarius, A. and Bessière, J.; *Langmuir*; *15*, 3186 (**1999**).
64. Bessière, J.; Chlihi, K. and Thiebaut, J.M.; *Electrochemica Acta*; *31*, 63 (**1986**).
65. Glinka, Y.D.; Jaroniec, C.P. and Jaroniec, M.; *J. Coll. and Interf. Sci.*; *201*, 210 (**1998**).
66. Wirth, M.J.; Ludes, M.D. and Swinton, D.J.; *Anal. Chem.*; *71*, 3911 (**1999**).
67. White, L.D. and Tripp, C.P.; *J. Coll. and Interf. Sci.*; *224*, 417 (**2000**).
68. Tripp, C.P. and Hair, M.L.; *Langmuir*; *7*, 923 (**1991**).
69. Nawrocki, J.; *Chromatographia*; *31*, 193 (**1991**).
70. Biliński, B.; *J. Coll. and Interf. Sci.*; *228*, 182 (**2000**).
71. Clerc, I.M.; Davidson, P. and Davidson, A.; *J. Am. Chem. Soc.*; *122*, 11925 (**2000**).
72. Matsumoto, A.; Sasaki, T.; Nishimiya, N. and Tsutsumi, K.; *Langmuir*; *17*, 47 (**2001**).
73. Suzuki, M. Editor. Studies in Surface Science and Catalysis. Vol. 80; *Fundamentals of Adsorption.* Elsevier (**1993**).
74. Rudzinski, W. and Everett, D.H.; *Adsorption of Gases on Heterogeneous Surfaces.* Academic Press (**1992**).
75. Hirschfelder, J.O.; Curtis, C.F. and Bird, R.B.; *Molecular Theory of Gases and Liquids.* John Wiley (**1961**).
76. Martínez-Alonso, A.; Tascón, J.M.D. and Bottani, E.J.; *J. Phys. Chem. B*; *105*, 135 (**2001**).
77. Bakaev, V.A.; Steele, W.A.; Bakaeva, T.I. and Pantano, C.G.; *J. Chem. Phys.*; *111*, 9813 (**1999**).
78. Bakaeva, T.I.; Bakaev, V.A. and Pantano, C.G.; *Langmuir*; *16*, 5712 (**2000**).
79. Mikhail, R.Sh. and Robens, E. *Microstructure and Thermal Analysis of Solid Surfaces.* John Wiley (**1983**).

80. Jaroniec, M. and Madey, R. *Physical Adsorption on Heterogeneous Solids.* Elsevier (**1988**).
81. Cascarini de Torre, L.E. and Bottani, E. J.; *Colloids and Surf. A*; *116*, 285 (**1996**).
82. Quiñones, I.; Stanley, B. and Guiochon, G.; *J. of Chromatography A.*; *849*, 45 (**1999**).
83. Quiñones, I. and Guiochon, G.; *J. Coll. and Interf. Sci.*; *183*, 57 (**1996**).
84. Edler, K.J.; Reynolds, P.A.; Branton, P.J.; Trouw, F.R. and White, J.W.; *J. Chem. Soc. Faraday Trans.*; *93*, 1667 (**1997**).
85. Branton, P.J.; Sing, K.S.W. and White, J.W.; *J. Chem. Soc. Faraday Trans.*; *93*, 2337 (**1997**).
86. Ravikovitch, P.I.; Domhail, S.C.O.; Neimark, A.V.; Schüth, F. and Unger, K.K.; *Langmuir.*; *11*, 4765 (**1995**).
87. Neimark, A.V.; Ravikovitch, P.I.; Grün, M.; Schüth, F. and Unger, K.K.; *J. Coll. and Interf. Sci. 207*, 159 (**1998**).
88. Lukens, W.W.; Schmidt-Winkel, P.; Zhao, D. Feng, J. and Stucky, G.; *Langmuir*; *15*, 5403 (**1999**).
89. Gregg, S.J. and Sing, K.S.W. *Adsorption, Surface Area and Porosity.* 2nd Ed. Academic Press (**1995**).
90. Carrott, P.J.M. and Sing, K.S.W.; *Pure Appl. Chem.*; *61*, 1835 (**1989**).
91. Nguyen, C. and Do, D.D.; *J. Phys. Chem. B; 104*, 11435 (**2000**).
92. Karnaukhov, A.P. and Kiselev, A.V.Z.; *Fiz. Khim.*; *34*, 1019 (**1960**).
93. Nguyen, C. and Do, D.D.; *Langmuir*; *16*, 1319 (**2000**).
94. Bottani, E.J. and Bakaev, V.A.; *Langmuir*; *10*, 1550 (**1994**).
95. Ryoo, R.; Ko, C.H.; Kruk, M.; Antochshuk, V. and Jaroniec, M.; *J. Phys. Chem. B*; *104*, 11465 (**2000**).
96. Cascarini de Torre, L.E. and Bottani, E.J.; *Langmuir*; *11*, 221 (**1995**).
97. Vuong, T. and Monson, P.A.; *Langmuir*; *14*, 4880 (**1998**).
98. Kaminsky, R.D. and Monson, P.A.; *J. Chem. Phys.*; *95*, 2936 (**1991**).
99. Schoen, M.; Rhykerd, C.L.; Cushman, J.H. and Diestler, D.J.; *Mol. Phys.*; *66*, 1171 (**1989**).
100. Sarkisov, L. and Monson, P.A.; *Langmuir*; *16*, 9857 (**2000**).
101. Neimark, A.V. and Vishnyakov, A.; *Phys. Rev. E*; *62*, 4611 (**2000**).
102. Neimark, A.V.; Ravikovitch, P.I. and Vishnyakov, A.; *Phys. Rev. E*; *62*, R1493 (**2000**).
103. Bartkowiak, M.S.; Dudziak, G.; Sikorski, R.; Gras, R.; Radhkrishnan, R. and Gubbins, K.E.; *J. Chem. Phys.*; *114*, 950 (**2001**).
104. Coleman, N.R.B.; Morris, M.A.; Spalding, T.R. and Holmes, J.D.; *J. Am. Chem. Soc*; *123*, 187 (**2001**).
105. Manne, S. and Gaub, E.H.; *Science*; *270*, 1480 (**1995**).
106. Gomez, F.; Denoyel, R. and Rouquerol, J.; *Langmuir*; *16*, 4374 (**2000**).
107. Boiadjiev, V.; Blumenfeld, A.; Gutow, J. and Tysoe, W.T.; *Chem. Mater.*; *12*, 2604 (**2000**).
108. Holysz, L.; *Coll. and Surf. A*; *134*, 321 (**1998**).
109. Bremmell, K.E.; Jameson, G.J. and Biggs, S.; *Coll. and Surf. A*; *146*, 75 (**1999**).
110. Kanda, Y.; Iwasaki, S. and Higashitani, K.; *J. Coll. and Interf. Sci.*; *216*, 394 (**1999**).
111. Ström, C.; Hansson, P.; Jönsson, B. and Söderman, O.; *Langmuir*; *16*, 2469 (**2000**).
112. Almgren, M.; Hansson, P.; Mukhtar, E. and van Stam, J.; *Langmuir*; *8*, 2405 (**1992**).
113. Goworek, J.; Derylo-Marczewska, A. and Borówka, A.; *Langmuir*; *15*, 6103 (**1999**).
114. Ottewill, R.H.; Rochester, C.H. and Smith, A.L. (Eds.) *Adsorption from Solution.* Academic Press (**1983**).
115. Király, Z. and Findenegg, G.H.; *Langmuir*; *16*, 8842 (**2000**).
116. Zajac, J.; *Colloids and Surfaces*; *167*, 3 (**2000**).
117. Dawidowicz, A.L.; Patrykiejew, A. and Wianowska, D.; *J. Coll. and Interf. Sci.*; *214*, 362 (**1999**).
118. Draoui, K.; Denoyel, R.; Chgoura, M. and Rouquerol, J.; *J. Therm. Anal. and Cal.*; *58*, 597 (**1999**).
119. White, L.D. and Tripp, C.P.; *J. Coll. and Interf. Sci.*; *227*, 237 (**2000**).
120. Ahmad, I.; Dines, T.J.; Anderson, J.A. and Rochester, C.H.; *J. Coll. and Interf. Sci.*; *195*, 216 (**1997**).
121. Combes, J.R.; White, L.D. and Tripp, C.P.; *Langmuir*; *15*, 7870 (**1999**).
122. Bose, A.; Gilpin, R.K. and Jaroniec, M.; *J. Coll. and Interf. Sci.*; *226*, 131 (**2000**).
123. Poston, P.E.; Rivera, D.; Uibel, R. and Harris, J.M.; *Appl. Spect.*; *52*, 1391 (**1998**).
124. Rivera, D.; Poston, P.E.; Uibel, R.H. and Harris, J.M.; *Anal. Chem.*; *72*, 1543 (**2000**).
125. Carr, J.W. and Harris, J.M.; *Anal. Chem.*; *59*, 2546 (**1987**).
126. Jones, J.L. and Putan, S.C.; *Anal. Chem.*; *63*, 1318 (**1991**).
127. (a) Matzner, R.A.; Bales, R.C. and Pemberton, J.E.; *Appl. Spectrosc.*; *48*, 1043 (**1994**); (b) Ho, M.; Cai, M. and Pemberton, J.E.; *Anal. Chem.*; *69*, 2613 (**1997**).
128. Thompson, W.R. and Pemberton, J.E.; *Chem. Mater.*; *7*, 130 (**1995**).
129. (a) Roschina, T.M.; Gurevich, K.B.; Fadeev, A.Y.; Astakhov, A.L. and Lisichkin, G.V.; *J. Chromatogr. A*; *844*, 225 (**1999**); (b) Kimura, M.; Kataoka, S. and Tsutsumi, K.; *Coll. Polym. Sci.*; *278*, 848 (**2000**).
130. Bakaeva, T.I.; Pantano, C.G.; Loope, C.E. and Bakaev, V.A.; *J. Phys. Chem. B.*; *104*, 8518 (**2000**).
131. Dorris, G.M. and Gray, D.G.; *J. Coll. and Interf. Sci.*; *71*, 93 (**1979**).
132. Biliński, B.; *J. Coll. and Interf. Sci.*; *225*, 105 (**2000**).
133. Biliński, B.; *J. Coll. and Interf. Sci.*; *201*, 180 (**1998**).
134. Biliński, B.; *J. Coll. and Interf. Sci.*; *201*, 186 (**1998**).
135. Donnet, J.B.; Wang, T.K.; Li, Y.J.; Balard, H. and Burns, G.T.; *Rubber Chem. and Technol.*; *73*, 634 (**2000**).
136. Lanin, S.N.; Ledenkova, M.Yu. and Nikitin, Yu.S.; *J. Chromatogr. A*; *797*, 3 (**1998**).
137. Glueckauf, E.; *J. Chem. Soc.*; 1302 (**1947**).
138. (a) Quiñones, I.; Cavazzini, A. and Guiochon, G.; *J. Chromatogr. A*; *877*, 1 (**2000**); (b) Bakaev, A.V. and Steele, W.A.; *Langmuir*; *13*, 1054 (**1997**).

139. Reilley, C.N.; Hildebrand, G.P. and Ashley, J.W.; *Anal. Chem.*; *34*, 1097 (**1962**).
140. Rheinländer, T.; Klumpp, E.; Schlimper, H. and Schwuger, M.J.; *Langmuir*; *16*, 8952 (**2000**).
141. Zhu, B.Y. and Gu, T.; *J. Chem. Soc. Faraday Trans. I*; *85*, 3813 (**1989**).
142. Kreuzer, H.J. and Gortel, Z.W.; *Physisorption Kinetics.* Springer Verlag (**1986**).
143. Pagac, E.S.; Prieue, D.C. and Tilton, R.D.; *Langmuir*; *14*, 2333 (**1998**).
144. Hansen, R.L. and Harris, J.M.; *Anal. Chem.*; *70*, 4247 (**1998**).
145. Biswas, S.C. and Chattoraj, D.K.; *J. Coll. and Interf. Sci.*; *205*, 12 (**1998**).
146. Tiberg, F.; *J. Chem. Soc. Faraday Trans.*; *92*, 531 (**1996**).
147. Nakatami, K. and Sekine, T.; *J. Coll. and Interf. Sci.*; *225*, 251 (**2000**).
148. Lobanov, V.V.; Chuiko, A.A.; Burlaenko, N.A.; Klymenco, V.E.; Tertykh, V.A.; Yanishpolskii, V.V. and Teretz, M.I.; *J. Therm. Anal. Cal.*; *62*, 381 (**2000**).
149. Huinink, H.P.; de Keizer A.; Leermakers, F.A.M. and Lyklema, J.; *Langmuir*; *13*, 6452 (**1997**).
150. Huinink, H.P.; de Keizer A.; Leermakers, F.A.M. and Lyklema, J.; *Langmuir*; *13*, 6618 (**1997**).
151. (a) Dash, J.G. *Films on Solid Surfaces.* Academic Press (**1975**); (b) Thomy, A. and Duval, X.; *Surf. Sci.*; 299, 415 (**1994**); (c) Suzanne, J.; Coulomb, J.P. and Bienfait, M.; *Surf. Sci.*; *40*, 414 (**1973**) and *44*, 141 (**1974**); (d) Steele, W.A.; *Langmuir*; *12*, 145 (**1996**).
152. Sellami, H.; Hamraoui, A.; Privat, M. and Olier, R.; *Langmuir*; *14*, 2402 (**1998**).
153. Hamraoui, A.; Privat, M. and Sellami, H.; *J. Chem. Phys.*; *106*, 222 (**1997**).
154. Adamson, A.W.; *Physical Chemistry of Surfaces*, 4th Ed. John Wiley and Sons (**1982**).
155. Gerdes, S. and Tiberg, F.; *Langmuir*; *15*, 4916 (**1999**).
156. Quiñones, I. and Guiochon, G.; *J. Chromatogr. A.*; *796*, 15 (**1998**).
157. Zhu, J.; Katti, A.M. and Guiochon, G.; *J. Chromatogr.*; *552*, 71 (**1991**).
158. Lewellyn, P.L.; Grillet, Y.; Schüthe, F.; Reichert, H. and Unger, K.K.; *Microporous Mater*; *3*, 345 (**1994**).
159. Inoue, S.; Hanzawa, Y. and Kaneko, K.; *Langmuir*; *14*, 3079 (**1998**).
160. Cole, M.W. and Saam, W.F.; *Phys. Rev. Lett.*; *32*, 985 (**1974**).
161. Xie, X.; Satozawa, M.; Kunimori, K. and Hayashi, S.; *Microporous and Mesoporous Materials*; *39*, 25 (**2000**).
162. Jun, S.; Joo, S.H.; Ryoo, R.; Kruk, M.; Jaroniec, M.; Liu, Z.; Ohsuna, T. and Terasaki, O.; *J. Am. Chem. Soc.*; *122*, 10712 (**2000**).
163. de Vos, R.M. and Verweij, H.; *J. Membr. Sci.*; *143*, 37 (**1998**).
164. Chevrot, V.; Llewellyn, P.L.; Rouquerol, F.; Godlewski, J. and Rouquerol, J.; *Thermochimica Acta*; *360*, 77 (**2000**)

28 Structure of Disperse Silica Surface and Electrostatic Aspects of Adsorption

A.A. Chuiko, V.V. Lobanov, and A.G. Grebenyuk
National Academy of Sciences of Ukraine, Institute of Surface Chemistry

CONTENTS

The structure has been examined and a description has been given of the active sites of silica surface as well as of the principal macroscopic and microscopic parameters generally describing its structure. The influence has been studied of the electric field of SiO_2-surface layer on the properties of adsorption complexes and its role has been investigated in the run of physico-chemical processes at the interface of solid–gas. The distribution of near-surface electrostatic field has been shown to allow predicting the structure of primary adsorption complexes, to estimate the energetics of molecule transformation on solid surface, and to define the properties of surface compounds formed. Within the frameworks of notions on distribution of molecular electrostatic potential within the subsurface layer of modified silicas, their hydrophobic and hydrophilic properties have been studied. It has been discovered that there are small compact areas of negative potential value within subsurface layer of even completely trimethylsilylated silica controlling its residual hydrophility.

Last three or four decades are characterized by a violent development of biochemical investigations and, in more wide sense, of the whole complex of the sciences concerning the life — its origin, evolution, and chemical and physico-chemical procedures of separation, purification, and analysis of the substances of living nature on the base of classic biology, bioorganic, and biochemistry, new scientific trends arise: molecular biology, bioorganic and bioinorganic chemistry, biophysical chemistry, physical biochemistry, and others — more narrow and specialized ones. Practically all the trends enumerated contain the sections devoted to the interaction of biologically active molecules with solid surfaces, the surface of silicon dioxide (in various crystalline and amorphous modifications) occupying an important position among them. Such a unique silica position is conditioned by the fact that about 87% of lithosphere mass — the cradle of life — consists of silica and silicates. The atomic part of silicon in the lithosphere is equal to 20% [1] whereas that of carbon — the main element of living nature — only to 0.15%. Moreover, it should be taken into account that, according to the substantiated estimation [2], about 10 billion tons of silicon is involved into the cycle of living matter functioning.

Silicon permanently enters into living organisms with food and water; the solubility of amorphous silica in water attains 100 mg/l [3]. Silicon also enters into respiratory organs with dust and aerosol particles that are always present in air. All this results in the migration of silicon compound within an organism, in their accumulation in the most tissues, and in metabolism [4]. Silicon and its compounds are not only present in all living beings [5–9], but also are necessary elements for them [8,9]. Despite the life is built by carbon compounds, silicon compounds and, in particular, silica and silicates played an important role in the process of its origin. Thus, according to A.I. Oparin's ideas [10,11], living matter was formed within mineral nature by abiogenic way in primary ocean. Nevertheless, the concentrations of initial substances were too low to form the first complex organic molecules in marine water. The data on adsorption capabilities of natural minerals [12] turned to the idea that the necessary concentration of the reagents could be created at the surface of mineral particles, firstly at that of silicates and silica which adsorbed organic substances from dilute water solution and simultaneously catalysed their sequential transformations. In particular, arbitrary process of chromatographic separation of the molecules forming living matter could occur on clay minerals. An important role in life origin could be played by silica acid gel [13].

Thus, the permanent contact of bioobjects from bioactive molecules to living organisms with silica and its derivatives at different stages of the evolution development, and its immediate active participation in vital processes actualized the studies on SiO_2 use in biology, agriculture, and medicine.

THE CONCEPT OF AMORPHOUS SILICA "SURFACE"

An investigation of various forms of amorphous SiO_2 as carriers for biopreparations and medicines of different chemical nature assumes knowledge, first of all, on its structure and surface peculiarities as well as on the structure of adsorption sites. It is also important to have at one's disposal methods of modifying all surface and, in particular, its adsorption sites that could, from one hand, elongate their action. For understanding nature of interaction of the SiO_2 surface with biopreparations as well as tissues of inner and outer organs of alive being, it is necessary to take into account the structure of silica surface layer and the status of its hydrate and hydroxylic cover. Then, the data on surface dynamic properties (adsorption and chemisorption processes) as well as on profundity of its relaxation and reconstruction changes under effect of such important external factors as temperature, pressure, electromagnetic ray treatment, and so on also play on essential role in the purposeful change of therapeutic properties of the medicines supported on SiO_2 surface.

As all the following expositions will suppose a microscopic, that is, atomic-molecular approach to investigation of surface properties, it is expedient to give a definition of this notion. The term "surface" has a distinct synonymous by defined sense only at examination of the limit of nonporous solid phase in cases of macroscopic phenomena. According to Iler's ideas [14], at the examination of adsorption with porous solids the "surface" means a limit of nonpermeability for nitrogen molecules. Nevertheless it should be stressed that the choice of nitrogen is conditioned only by the fact that it is the most widely used adsorbate for measurement of specific area. When other types of probe molecules are used, the limit of permeability should be naturally shifted either inside adsorbent bulk phase, or to liquid or gaseous phase contacting with solid.

Such a definition is unfit for silica with high value of specific surface (starting from the smallest colloid particles up to macroporous samples of silica gels) and its use can result in definite type misunderstandings obscuring the essence of the phenomena under consideration. It is obvious that in the case given some distinctive peculiarities of SiO_2 structure should be used. To our mind, such a peculiarity can be a three-dimensional siloxane net and its appearance (at conventional movement from the phase contacting with solid into adsorbent bulk) should be considered as a surface separating adsorbent bulk phase from gaseous or liquid one containing adsorbed substance. We shall consider the part of massive adsorbent that directly adjoins the surface to be an undersurface layer, and the layer of gaseous or liquid phase contacting with surface — to be subsurface one. It is assumed at such a definition of surface that functional groups are disposed on it. Of course, in some experiments, especially when silica surface is chemically modified with silicon-containing compounds or their vapors are deposited onto SiO_2 samples, even such a notion of surface would be not quite accurate.

CHARACTER OF SURFACE BONDING

After surface formation as a result cutting off some part of unlimited solid, atoms of the first atomic layer become bound not so strongly as before in the bulk or as equivalent atoms are bound of internal layers after surface forming [15–17]. There is a lack of at least one nearest neighbouring atom at every surface one. In case of covalent bond that is typical of many oxides and, in particular, of silica, this means that one or some bonds are ruptured and unsaturated or broken bonds appear. There are two mechanisms of restoration of the saturated character of surface atomic bonds, the first one consisting in the addition of gaseous phase atoms (adatoms), that is, chemisorption, and the second — in the change of surface structure. It should be noted that the classification is used in this work of adsorption processes proposed by Langmuir [18]. According to Langmuir, chemisorption should be comprehended as an interaction of adatoms with surface atoms having free valences. Chemisorption takes place in any case, when fresh-formed surface contacts with gaseous or liquid phase. Due to chemisorption, structural changes of the surface appear to be less strong than those without it, and they often do not occur at all, because the ideal surface structure is stabilized due to chemisorption of adatoms. Speaking about ideal surface, we shall imply that atoms on surface and directly under it are in the same points of the elementary cell in unlimited crystal. If chemisorption is excluded, the only probable way of surface stabilization consists in the change of its structure. This change is directed to restoration of the bond strength of atoms of the first surface layer without excessive weakening atomic bonds of the next layers. The goal mentioned can be reached at the expense of the creation of formerly absent bonds of surface atoms or due to strengthening already existing bonds between themselves or with the second layer atoms. It can be assumed that the more effective mechanism of surface stabilizing should be the origin of new bonds of surface atoms. In other words, the origin of surface from unlimited crystal is accompanied by atomic shift of some subsurface layers as compared with their positions within initial solid. These shifts can be divided into two classes depending on their influence on translation symmetry. If it keeps unchanged, they say that the changes evoke a surface relaxation. In this case equivalent atoms in different elementary cells are shifted in the same way.

A concrete kind and scopes of relaxation and reconstruction changes depend on type of chemical bonds of surface atoms. First of all within the framework of a generalized approach it is necessary to take into account four bond types: van-der-Waals one, ionic, covalent, and metallic ones, not depending on the predominant binding in bulk solid. It is possible that there would be covalent-bonded surface atoms in the crystal with presumably ionic bond type or, vice versa, an ionic bond can take place at the surface of covalent crystal.

As the matter under our consideration — silica — belongs to compounds having a covalent bond with a great part of ionicity, let us dwell on these bond types and analyze them together.

It is known from the theory of chemical bond [19] that covalent bond is characterised by a strong angular dependence what has to result in relatively large shifts of surface atoms. The most effective covalent bond is realised via σ-type bonds with participation of hybridized sp^3-orbitals. In bulk silica any silicon atom forms such bonds with the nearest oxygen atoms stationed at knots of regular tetrahedron. For surface silicon atom, one σ-bond is ruptured, that is, hybridized sp^3-orbital is not closed any more by sp^3-hybrid orbitals as a rule. Surface silicon atoms are expected to leave the plane they occupied in the SiO_2 crystal. Such a structural change can be conditioned by partial weakening orbital hybridization of surface silicon atoms. The sp^3-hybridization is known to be closely connected with tetrahedral coordination of silicon atoms. A similar fourfold coordination of surface atom does not take place any more. A question arises, if the sp^3-hybridization is profitable for surface silicon atoms having less than four, namely three nearest neighbors. For an isolated silicon atom having no neighbours, the sp^3-hybridization is energetically unprofitable because it is connected with excitation of s-electrons from lower energy into more high as for energy p-state. In bulk crystal, this disadvantage in energy, as compared with a situation without hybridization, is not only compensated, but is even exceeded by a considerable advantage in covalent bond energy of four bonding hybridized orbitals per each silicon atom. For silicon atoms on an ideal surface with a number of covalent bonds less than four, an energy advantage due to covalent bond formation is not just the same as in the bulk. That is why sp^3-hybridization can appear to be energetically disadvantageous for the orbitals of surface silicon atoms. A partial weakening can occur of hybridization accompanied by the formation of one p-type orbital and sp^2-ones. Such a transformation of the orbitals directly forming chemical bonds obtained a name of dehybridization. Dehybridized orbitals have other directions in the space than hybridized ones what results in involving surface silicon atoms into bulk phase.

Now let us take into account partially ionic character of Si—O bonds in silica. Unlike covalent bond, relatively large forces of ionic one act not only between nearest neighbors, but also between distant atoms because of long range character of Coulomb interaction. Taking into account indirectness of ionic bond, it can be expected that, if the surface stability is secured by ionic bond forces, relatively small structural changes are preferable and these changes should be the weakest whenever the bulk bond character is also ionic. Thus, accounting for partial bond ionicity in silica gives a reason to believe that involving surface silicon atoms into bulk should be not so considerable as it could be expected basing on an idea on total covalence of Si—O bonds.

PHYSICAL PARAMETERS DESCRIBING SURFACE STATE OF AMORPHOUS SILICA

When simulating bulk phase of amorphous silica, one has already to meet both considerable difficulties in computing and complication of representing data on its structure within the frameworks of customary notions of the theory of reactivity of the solid crystalline surfaces [20,21]. When turning to the surface of amorphous silica, analogous problems arise, their grade increasing because of absence of the structural characteristics generally describing its structure. Nevertheless one can make an attempt to imagine amorphous silica surface as a partially disordered conglomeration of sections, any of them being described with different crystallographic faces of the most wide-spread crystalline silica modifications. The necessity of such an approximation arises any time when one wants to give a self-consistent explanation for numerical values of some characteristics that are commonly used to describe surface and bulk phase properties of amorphous silica. Usually such characteristics are as follows:

1. Minimal temperature (T_m) of total removing physically adsorbed water from hydroxylated surface
2. Concentration of hydroxyl groups on completely hydroxylated silica surface
3. Dependence of hydroxyl group concentration on the temperature

Before turning to a description of the amorphous silica surface at atomic level, let us examine in detail the characteristics listed because they form a basis allowing to obtain a self-consistent notion about this structure.

Minimal Temperature of Total Removing Physically Adsorbed Water from Hydroxylated Surface

In order to clarify the nature of hydroxylic cover and to determine the hydroxyl group concentration on silica surface in a quantitative way, usually it is necessary first of all to distinguish those groups from molecularly sorbed water and from that contained inside silica particles, that is, to make a differentiation of the processes of dehydroxylation and dehydration. Despite many investigators determined the T_m value by use of various methods, as a whole, there is no synonymous agreement between the values obtained up to now. In this connection, the results obtained in [22,23] by methods of thermal programming desorption for this standard sample (named as S-79 by the author of the papers mentioned) are the most reliable. The samples undergone to a preliminary preparation *in vacuo* under different conditions at the temperature of 25 to 200°C. A detailed analysis of various kinetic curves showed that: the main quantity of physically adsorbed water including that adsorbed in polymolecular way is removed from the hydroxylated surface at the temperature of 29 to 50°C;

1. A small quantity of adsorbed water within monolayer is kept on hydroxylic surface up to 200°C
2. the T_m value that accounts for total dehydration and start of dehydroxylation is of 190 ± 10°C
3. kinetic parameters calculated from the experimental data for this temperature interval sharply increase. In particular, the activation energy of water desorption, E_d, increases from 10 up to 17 kcal/mol, and the reaction itself from monomolecular becomes bimolecular

The value $T_m = 190$°C well agrees with the results of works [3,14] on differential thermal analysis and moistening heats of silica surface. The probable jump of kinetic parameters within some temperature interval was already noted in the paper [24] where it was stressed that, when disperse materials were dried, a transition was observed from complete dehydration to the start of removing chemically bound water, a sharp increase in isotermal free energy being a characteristic if this transition.

Concentration of Hydroxyl Groups on Hydroxylated Silica Surface

For detail development of the model for amorphous silica surface, first of all it is necessary to have data on OH-group concentration (C_{OH}) dependent on the temperature of preliminary thermal treatment of SiO_2 samples *in vacuo*. Already in 1957 the values of C_{OH} were obtained for completely hydroxylated surfaces of amorphous silica and crystalline SiO_2 modifications (β-crystobalite and β-tridimite), their densities being close to that of amorphous silica [25]. The authors of this paper found the C_{OH} to be of 2.55 groups OH/nm^2 for (111) face of β-crystobalite. Theoretically calculated values of silanol number for different crystalline faces of β-crystobalite and β-tridimite appeared to be within 4.6 ÷ 4.9 groups OH/nm^2. According to the experimental data [25] for amorphous silica, $C_{OH} = 6.0 \div 6.9$ groups OH/nm^2, those should be considered to be wrong due to lack of sensitivity of the gravimetric method used by the authors.

Up to how, a great experimental and theoretical material has been accumulated on the content of hydroxyl groups on the surfaces of disperse silicas. Among chemical methods of C_{OH} measurement, the most quantity is composed by the works using high reactivity of surface hydroxyl groups in the reactions with diborane [26–30], lithium aluminum hydride [31,32], D_2O [33–35], potassium vapors [36,37], thionil chloride [38–40], organometallic compounds [38,41–44], and halogenides [24,45–47]. The data on chemical analysis allow us to make suitable calibrations for quick determination of the concentration of silanol groups on SiO_2 surface with use of IR-spectroscopy. It was shown in [48] that the content of silanol groups of pyrogenic silica surface changes antibately to its specific area (C_{OH} value increased from 1.7 to 9.5 $mcmol/m^2$ when specific area changed from 360 to 59 m^2/g).

The analysis of data mentioned shows that the concentration of surface hydroxyls on various SiO_2 samples fluctuates from 1.7 to 12.0 $mcmol/m^2$ what corresponds to 1.0–7.2 groups of OH per 1 nm^2. The same interval includes the C_{OH} value for silica gel obtained in [48] by

NMR ^{1}H method (7.0 mcmol/m^2). The great divergence of the values of concentrations of structural hydroxyl groups apparently should be referred to concrete experimental conditions, sample biography, and completeness of the removal of sorbed water.

A monotonous decrease of C_{OH} with increasing the temperature of the preliminary preparation of SiO_2 samples can be indirect evidence in favor of the latter circumstance.

A more narrow range is pointed out in [49,50] of C_{OH} (7–9.5 mcmol/m^2) independently on the conditions of preliminary preparation of silica samples and their specific area. It allowed one to consider the concentration of structural hydroxyl groups of extremely hydroxylated samples to be some reproducible physicochemical constant. The dependence observed of C_{OH} on the value of specific area [14] was explained in [49] by the presence of hydroxyl groups in internal cavities of silica particles, the contents of cavities are maximal provided their are synthesized by a pyrogenic way. It should be noted, that this idea is not believed by all the investigators.

In spite of importance of reliable results about the concentration of hydroxyl groups on the surface, a necessity is obvious of authentic information on the way of their distribution within the surface silica layer. According to [49], hydroxyl groups are randomly distributed through the surface dependent on the type of surface hydroxylated silicon atom (see Figure 28.1). The scheme presented in that figure supposes that there are some types of structural hydroxyls on the surface: isolated single one (1), double or geminal in silanediol groups (2), neighboring or vicinal ones (3). A considerable content of near-stationed structural hydroxyls with hydrogen bonding between them is a distinctive feature of the structural scheme of silica hydroxyl cover proposed in [49] on the base of analysis of SiO_2 IR-spectra. Such an extremely hydroxylated state of silica, as for [49], is a characteristic of only narrow temperature interval. Even relatively moderate (about up to 400–500 K) heating samples can result in their partial dehydroxylation.

An analogous structural scheme of silica hydroxylic cover is performed in paper [14] with the only except

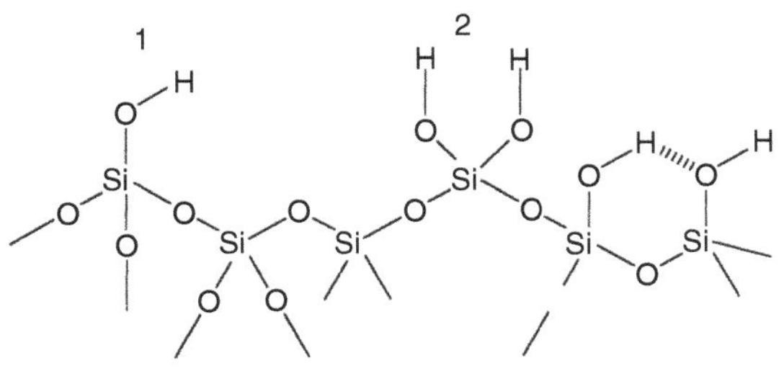

FIGURE 28.1 Types of OH-groups forming amorphous silica hydroxylic cover.

that an existence is supposed of metastable silanetriol groups besides silanol and silanediol ones on the fresh-made from water solution SiO_2 surface. When heating hydro gels, as a result of condensation of the part of structural hydroxyls, there kept only silandiol groups randomly stationed on the surface.

The high content supposed of vicinal hydroxyls on the surface [49] and allegedly connected with this easiness of its dehydroxylation at low temperatures of thermal treatment of silica samples needs a consideration on one more kind of surface structural element of strained disiloxane bridges.

Such cyclic structures should have a high reactivity, nevertheless, probability of their realising on pyrogenic silica surface is very small. At the same time, it is assumed in [51] that disiloxane bridges are typical of the surface of disperse silicas dehydrated within wide temperature interval, and they can play a noticeable role in chemical transformation within surface layer, in particular in the reactions with some element halogenides, element organic compounds, in the processes of esterification, rehydroxylation, and so on.

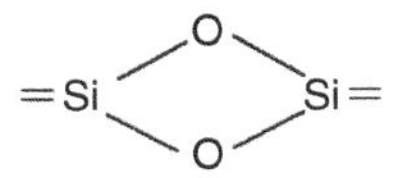

Dependence of Hydroxyl Group Concentration on Temperature

The average value of silanol number was determined in [35] for the samples thermoevacuated at different temperatures up 1000–1100°C with use of heavy water. The initial state was of a completely hydroxylated surface. The experimental results obtained for 16 samples of amorphous silica are given in Figure 28.2. The initial samples differed one from another in preparation method and structural characteristic (specific area of the samples varied within 11 to 905 m^2/g).

Despite all these distinctions, the average value of C_{OH} at definite temperature was the same. The dependence of silanol group concentration on the temperature for various samples also obeyed approximately the same rule for all the samples. The averaged values of C_{OH} obtained within temperature interval of 180–800°C are performed in Table 28.1.

It is seen from the data given that the C_{OH} value decreases considerably within the temperature interval from 200 to 400–500°C. At more high temperatures (400–1100°C) this decrease is considerably lesser. Relatively to this degree of surface occupation by OH groups (θ_{OH}) decreases from 1 to 0.5 within the first temperature interval and then slowly decreases from 0.5 drawing near zero. An opportunity appears from the results considered to estimate the most probable values of C_{OH} and θ_{OH}

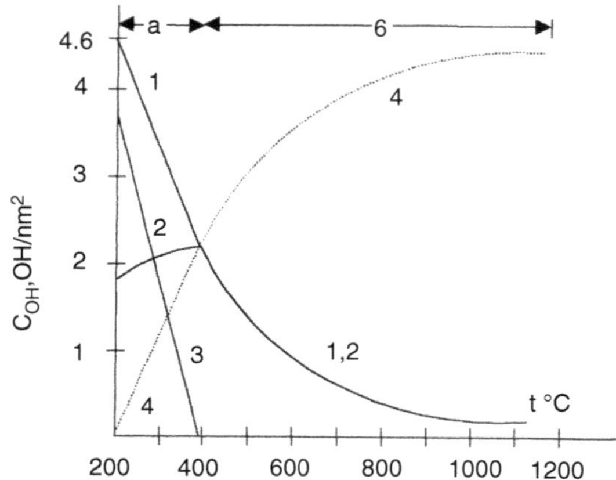

FIGURE 28.2 A plot of the concentration of various types of hydroxyl groups on the temperature of preliminary treatment of the silica samples in vacuum [46,47] (curves 1–3): 1, total concentration of all the types of OH-groups; 2, concentration of alone isolated OH-groups; 3, concentration of vicinal OH-groups. The curve 4 gives a dependence of the content of silicon atoms in siloxane bridges on the temperature.

within rather wide temperature range (200–1100°C). That is why the dependence $\theta_{OH} = f(T°C)$ and $C_{OH} = f(T°C)$ can be used as reproducible physicochemical constants along with the value of hydroxyl group concentration at normal temperature (25°C).

The free silanol groups and surface hydroxyls bound with hydrogen bonds were also differentiated in [5,34,52]. With this purpose some investigations were carried out with simultaneous application of IR-spectroscopy and reactions of isotope exchange of silanol group hydrogen with deuterium. A correlation between value of silanol number and intensiveness of relative bands in IR-spectra is conditioned by the fact that under thermoevacuation of silica gel samples up to the temperature of 400°C, the surface is left by only alone isolated hydroxyl groups. This gives an opportunity to calculate the concentration in free hydroxyl groups (C_{OH}, free) and that of hydroxyls bound with hydrogen bound (C_{OH}, bond). Basing on the data [34,35,52], a comparison was made in work [53] of seven different models describing the distribution of free surface hydroxyls as dependent on the temperature of preliminary preparation of silica samples. The author jumped to the conclusion on reproducibility of the function $C_{OH} = f(T°C)$ within the models examined, what proves the introduction of the notion on concentration of hydroxyl groups on amorphous silica surface as a constant characterizing its state to be substantiated. The distributions were also analyzed of all the types of OH groups, accordingly to data given in Table 28.1. and in Figure 28.2 as well as with taking into account the results of work [54] where concentrations of geminal hydroxyls were calculated. This gave an opportunity to estimate the contributions of alone isolated OH groups (free) and geminal hydroxyls (free). The calling mass spectrometry analysis data combined with thermoprogrammed desorption [55,56] for the samples undergone preliminary thermoevacuation at the temperature of 200°C gave an opportunity to estimate the value of C_{OH} at different degrees of occupation. A mechanism was proposed in [55,56] for elimination of such groups from silica surface *via* reaction of condensation resulting in formation of SiOSiO bridges and water molecules

$$(\equiv Si{-}OH) + (\equiv Si{-}OH) \rightarrow \equiv Si{-}O{-}Si\equiv + H_2O$$

The analysis of kinetic curves obtained in [22,23] gave an opportunity to calculate the activation energy (E_a) of the reaction examined. The curve of dependence $E_a = f\,(T°C)$ can be performed as two approximately linear sections: one of them belongs to the interval of 200 to 400°C, the other — to temperatures above 400°C. Using these data, a dependence was obtained in [35] of the activation energy of dehydroxylation (E_{dh}) on C_{OH} or on the occupation degree of the surface with hydroxyl groups (see Figure 28.3). The dependence of $E_{dh} = f(C_{OH})$ and of $E_{dh} = f(\theta_{OH})$ connects energy characteristics of hydroxylated surface

TABLE 28.1
Concentration of Silanol Groups of Various Types as Dependent on the Temperature of Preliminary Treatment

Temperature, (°C)	Total concentration of OH-groups, (OH/HM^2)	Occupation degree, Θ_{OH}	Concentration of isolated OH-groups, (OH/HM^2)	Concentration of vicinal OH-groups, (OH/HM^2)	Concentration of geminal OH-groups, (OH/HM^2) [16,17]
180–200	4.60	1.00	1.15	2.85	0.60
300	3.55	0.77	1.65	1.40	0.50
400	2.35	0.50	1.05	0	0.30
500	1.80	0.40	1.55	0	0.25
600	1.50	0.33	1.30	0	0.20
700	1.15	0.25	0.90	0	0.25
800	0.70	0.15	0.60	0	0.10

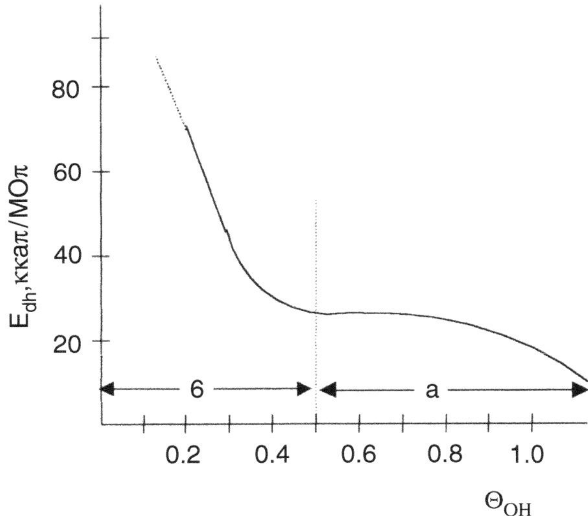

FIGURE 28.3 A plot of the energy of water associative desorption on the occupation extent, Θ_{OH}, of silica surface by hydroxyl groups.

with density of hydroxylic cover on it. Two linear sections are clearly distinguished in Figure 28.3. At high degree of surface occupation ($1 \geq \theta \geq 0.5$) a dehydroxylation occurs mainly at the expense of vicinal silanols. The dehydroxylation energy almost does not depend on silanol group concentration and is determined by distortions caused by hydrogen bond formation between them. These distortions vanish at the temperature of 400°C ($\theta_{OH} \cong 0.5$). Thus, a hydrogen bonding is a characteristic of the region (a) (see Figure 28.3) between neighboring hydroxyl groups. For low degrees of occupation ($\theta_{OH} < 0.5$), free hydroxyl groups and Si—O—Si bridges play the main role. In this region (b) E_{dh} strongly depends on the concentration of surface hydroxyl groups. When only isolated hydroxyl groups surrounded with Si—O—Si bridges are present, the latter can occupy rather big areas that are peculiar to silica samples obtained as a result of high temperature treatment. Under these conditions the elimination of hydroxyl groups from the surface is realized due to a condensation mechanism as a result of proton migration through the surface. At the final stage, a separation of water takes place due to interaction of two neighboring hydroxyls that can be at the distances up to 0.3 nm. The probable mechanism of proton migration at elevated temperatures includes its interaction with the bridge oxygen atoms of neighboring Si—O—Si groups as an intermedium stage that results in the formation of new surface OH groups shifted relative to their initial positions.

TOPOGRAPHY OF HYDROXYL AND HYDRATE COVER OF SILICA SURFACE

In the hydroxyl cover of silica surface is a definite part of surface active sites, so the hydrate cover structure should be to some extent a derivative from silanol group distribution on the surface.

Among physical methods of investigation of silica surface, IR-spectral measurements give a direct information about the presence of chemically definite groups. Kiselev and Sidorov [49,57] are considered to be pioneers in the region of investigation of infrared absorption spectra of organic compounds adsorbed on microporous glass (dehydrated and partially dehydrated amorphous silica). They were the first to show, that in this region of the spectra of amorphous silicas (independent on the way of their obtaining) a narrow absorption band was observed with a maximum at 3749 cm^{-1}, that was assigned by the majority of investigators to stretching vibrations of isolated surface hydroxyl groups [49,58–60]. It was postulated in the work [24] without sufficient grounds that isolated silanediol groups $=Si(OH)_2$ (a formation of hydrogen bonds between two hydroxyl being improbable for them) could absorb at the same frequency as alone surface silanol groups. The attempts were unsuccessful to differentiate the absorption band at 3750 cm^{-1} into separate components relative to the vibrations of silanediol groups on the samples prepared at the temperatures up to 750°C. According to theoretical calculations [61], the frequencies of stretching vibrations of isolated single and geminal (double) OH-groups differ for not more than 1–2 cm^{-1}. So it is out of success to determine such differences in a valid way.

Nowadays serious arguments have been obtained supporting existence of silanediol groups on silica surface by ^{29}Si and ^{1}H high performance NMR spectroscopy within solid with use of cross-polarisation and sample rotation under magic angle [62–68]. The ^{29}Si NMR spectra of silica samples show the signals (chemical shifts of −109.3, −99.8 and −90.6 m. p.) that are assigned respectively to silicon atoms without OH-groups; silicon atoms bound with one or two hydroxyls [64,65]. Nevertheless, the values of silanol number calculated on the base of these data are considerably different.

In order to solve the problem whether isolated silanol groups are stationed presumably homogeneously or as isolated islands what can be connected with disorder in surface silicon-oxygen tetrahedra in amorphous silicas, a chemical method was used in works [69–74]. It consisted in consequent inserting into surface silica layer firstly groups of $\equiv Si—O—SiCl_3$ by action on the latter with $SiCl_4$, and then relative alkoxysilylic groups due to treatment with alcohols of chlorinated at the first stage surface. It was assumed that, when varying the length of aliphatic alcohol n-radial in the reaction with grafted trichlorosilylic groups, one can estimate an average distance between silicon atoms of silanol groups on aerosil surface. It was shown that only in case of methanol, $—SiCl_3$ groups could be totally involved into chemical reaction. In case of n-butanol, only 2/3 of surface Si—Cl bonds could take part of the reaction.

The data obtained allowed to estimate the distance between isolated groups Si—OH of calcined at 400–600°C aerosil to be about 0.6–0.7 nm what justifies the predominance of distance structural hydroxyls that are relatively homogeneously stationed on the surface. It means practically that only every second silicon atom of pyrogenic silica surface bears a hydroxyl group [75]. It should be expected that on such a surface section the chemisorption of polyfunctional reagent molecules (for example, $SiCl_4$, BBr_3, CH_3SiCl_3) can be realised only with participation of single silanol group. Then, some authors [76–80] believe the participation of two different groups Si—OH (or of single $=Si(OH)_2$ one) in the reaction to be probable. All this indicates probable considerable fluctuations in distribution density of silanol groups without mutual perturbation of OH-vibrations, or there is a sufficient concentration of groups $=Si(OH)_2$ on the surface that was estimated by authors of [77–79] to be 40–60% of total quantity of hydroxyl groups, the surface being pretreated at 600–800°C. Such a conclusion (as it is mentioned in [81]) can be a consequence of the overestimation of silanol group concentration on initial silica.

The investigations carried out in [82] of the reaction of pyrogenic silica pretreated at 600°C with BBr_3 vapors did not detect a considerable concentration of silanol groups on the surface. The results obtained showed the ratio of B:Br in surface chemical compounds to be of 1/2, and the quantity of grafted $\equiv SiOBBr_2$ groups (2.3 mcmol/m^2) was close to the concentration of silanol groups on initial aerosil (2.5 mcmol/m^2) what (as for authors) testified a homogeneous distribution of isolated silanol groups on the surface of dehydrated silica.

A narrow absorption band of the stretching vibrations of hydroxyl groups in both overtone and ground regions of the spectrum is observed only at rather high temperatures of silica heating where sorbed water and part of hydroxyl groups are eliminated from the surface. If adsorbed water is available on SiO_2 surface, a wide intensive band is observed with absorption maximum within 3400–3450 cm^{-1} in low-frequency region of the spectrum stretching vibrations of OH-groups, that is assigned by all the investigators to the vibrations of the OH-groups of water molecules bound one with another by hydrogen bonds. The intensiveness of this band is sharply increased under thermoevacuating silica and vanishes completely under heating at 150–200°C.

When water is eliminated due to rise of the temperature of amorphous silica heating, the band of stretching vibrations of OH-groups can be decomposed into three bands with maxima $\nu'_{OH} = 3750\,cm^{-1}$, $\nu''_{OH} = 3680\,cm^{-1}$, and $\nu_k = 3540\,cm^{-1}$ [83–85]. The narrow ν'_{OH} bond is related to isolated silanol groups, the average distance between them being of ≥ 0.3 nm. The wide bands with diffuse maxima (ν'', ν_k) the majority of investigators assign to the hydroxyls distant for ≤ 0.3 nm and bound with hydrogen bonds:

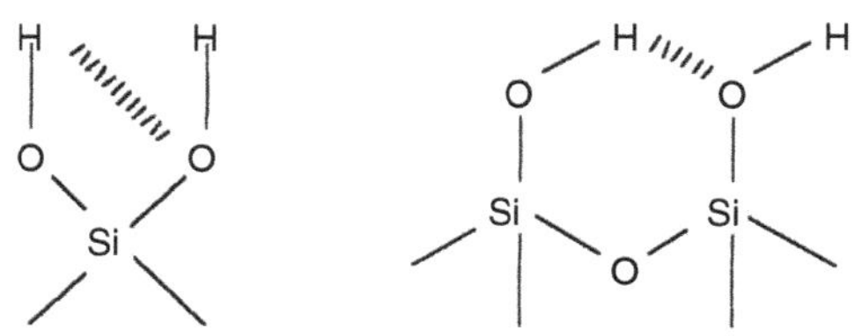

The bond ν_k vanishes from the spectrum under heating silica sample *in vacuo* up to 400°C, and when the temperature increases up to 600°C, the bond $\nu'_{OH} = 3680\,cm^{-1}$ also vanishes. Within long-wave region of the IR-spectrum of hydrated amorphous silica a band $\nu_d = 1610$–$1640\,cm^{-1}$ is present that is usually related to bending vibrations of molecularly sorbed water. This band always appears along with that in the region of OH-group stretching vibrations ($\nu_k = 3540\,cm^{-1}$). All these bands vanish in the spectra of dehydrated samples. This fact allowed to assign the ν_k band to strongly adsorbed water molecules [86,87].

In many cases the presence of some types of the vibrations of surface OH-groups or the asymmetry of the ν_{OH} band within the region of low frequencies can be explained by the disordered surface structure (including subsurface layer of solid) without involving the hypothesis on hydrogen bridges [88,89]. The deformation of surface tetrahedra caused by valence angle and chemical bond length variations affects the force constants and so shifts the frequencies of OH-groups being in their peaks. For example, if the maximal frequency $\nu'_{OH} = 3750\,cm^{-1}$ relates to the OH-groups of undistorted tetrahedra, then low-frequency maximum of 3660 cm^{-1} is typical of the frequencies of the OH groups of deformed ones. There are some different ideas on the band near 3680 cm^{-1}.

It was supposed on the base of spectral studies on silica–D_2O interaction that the band $\nu''_{OH} = 3680\,cm^{-1}$ increasing in intensity for the samples hydrothermally pretreated relates to interglobular water or, more exactly, to hydroxyl groups inside globulae perturbed by weak hydrogen bonds [90]. A hypothesis on interglobular hydroxyl groups related to the ν'_{OH} absorption band is accepted by many researchers. An analysis of literature data [81] showed the absorption near 3680 cm^{-1} to be the less sensitive to the reagents forcing hydrogen. This band appears in aerosil spectrum after conducting a chemical reaction with chlorosilanes, boron halogenides, silicon tetrachloride, and thionyl chloride. Meanwhile at high temperatures of the contact with small-size molecules (methyl alcohol, hydrogen fluoride) the ν'_{OH} band does not appear in the spectra. This gives a reason to suppose that the absorbance near 3680 cm^{-1} is caused be the interaction of hydroxyl groups at hard-accessible for large molecules sites of the contacts of different silica globulae [91,92]. An opinion was expressed in [93] that the vibrations of OH-groups forming a weak hydrogen bond and stationed in the pores of molecular dimensions — ultrapores (cavities) of silica frame are responsible for the ν'_{OH} absorption [58].

The adsorption of water vapours on preliminary dehydrated aerosil surface results in the appearing in IR-spectrum firstly a band of 3550 cm^{-1}, then after continuous contact with water vapors at room temperature an absorption is observed near 3660–3680 cm^{-1} [84, 85]. When explaining this experimental fact, the authors of work [94] introduced a notion on "resistance against rehydration" and assumed that water molecules diffused into the sample bulk and then interacted with siloxane bonds and became low-movable. The windows can exist on initial silica surface formed by six tetrahedra combination where a penetration is possible of water molecules [93,95,96]. Nevertheless, as it is noted in [81], the questions remain unclear about the nature of the forces pressing H_2O molecules to diffuse into solid as well as stabilized forms of these molecules.

It is generally accepted that hydroxyl groups of dehydrated silica surface are the main sites of adsorption of the molecules capable to form hydrogen bonds with these sites (water, alcohol, amines). This conclusion was made on the base of investigation of the shift of vibration bands of free Si—OH groups to low frequency region [49,50]. The following models are proposed, for example, for the adsorption complexes of water on silica surface:

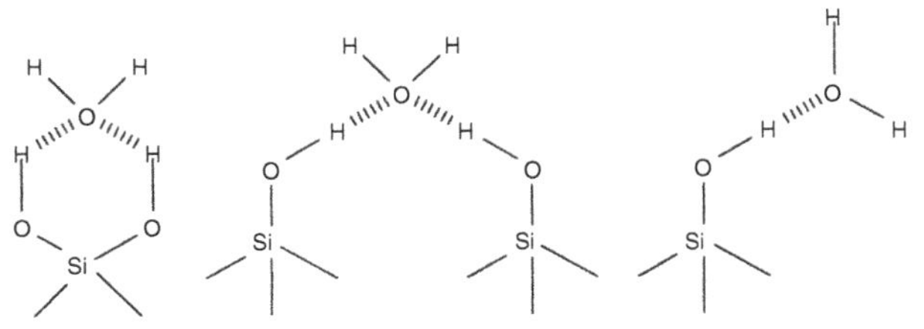

The supposition on the presence on SiO_2 surface of another (not hydroxylic) sites of strong molecular adsorption appeared for the explanation of the absence of position shift and change in intensiveness of absorption band of surface hydroxyl groups. Then, the presence was noted of adsorbed molecules on the surface that were kept by it even after evacuation of the sample under weak heating.

Firstly this phenomenon was noticed in [97,98] when the adsorption was examined of H_2O on porous glass pretreated at 400°C. Analogous results were obtained also for methanol adsorption, what served as a base for the assumption on the presence of another sites (along with hydroxyl groups) on the surface that were called by the authors "second kind sites." When the studies were carried out on thin plates of porous glass, a decrease was noticed of the intensiveness of absorption band of surface hydroxyl groups only under large relative pressures of water vapors. The ammonia adsorption resulted in near-total vanish of the absorption bond of Si—OH groups that was reduced to initial intensiveness as a result of evacuation at room temperature. After that the absorption bands remained in the spectrum of stretching vibrations of NH-groups [97]. All this allowed to make a conclusion that such sites of "second type" could be "valence-unsaturated silicon atoms of the porous glass surface" [97–99].

Kiselev and Lygin in the monograph [49] explain spectral manifestation of adsorption and the nature of another (not hydroxyl) sites (sites of "second kind") by strong specific adsorption on admixture atoms (for example, those of boron or aluminium) usually possessing strong electron acceptor properties. They believe it to be proved in case of porous glass that can include up to 5% of B_2O_3 [100]. The absence of the intensiveness decrease of the absorption band of the free hydroxyl groups of SiO_2 surface (ν'_{OH}) due to adsorption of water molecules is explained by the influence, firstly, of water chemisorption occurring at siloxane sections of the surface of partially dehydroxylated samples, secondly, by heating silica with infrared radiation as this heating decreases the water adsorption.

The studies of the IR-spectra on very pure silica gel samples [86] and an aerosil [83–85] showed that when small portions of water were adsorbed, the intensiveness of the band of free hydroxyls did not change noticeably. The adsorption band ν'_{OH} changed very little with adsorption increase up to occupation degree of $\Theta = 0.5$ whereas the intensiveness of ν_d and ν_k bonds simultaneously increased. All this testified nondissociative adsorption at the sites that were more active than OH-groups [89].

The studies on adsorbed water phase by NMR methods in the region of low temperature showed that water adsorption in the quantity of $\cong 0.3\alpha_m (\alpha_m = 6.2\,\text{mcmol/m}^2$ for BET) did not result in the noticeable signal of hydroxyl groups. Thus, the NMR data also indicated that isolated OH-groups could not be primary adsorption sites for polar molecules [89,101]. No signals were observed of Si—O$^\bullet$ or —Si$^\bullet$ radicals as well as of paramagnetic admixtures in the ESR spectra of the silica samples used for IR- and NMR studies and heated up to 800°C [89]. The spectral performance of the water molecules adsorption on silica surface in the work [102] was interpreted without use of the notions on existence of nonhydroxylic adsorption sites. It was assumed that water molecules, when bound with silanol groups as small quantities, firstly reacted as proton donor ones (I), and only under large occupations they behaved as proton acceptors forming hydrogen bonds with surface OH-groups *via* unshared electron pairs of oxygen atoms (II):

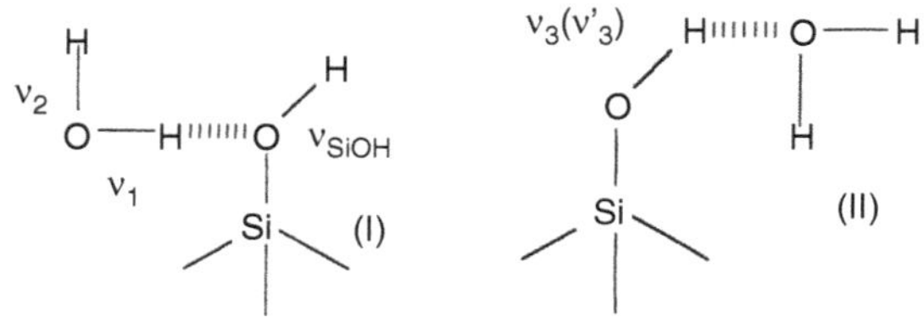

The assignment of the frequencies was made as follows: $\nu_{SiOH} = 3749\,cm^{-1}$, $\nu_1 = 3440\,cm^{-1}$, $\nu_2 = 3665\,cm^{-1}$, $\nu_3 = 3050\,cm^{-1}$, and $\nu_3 = 3100\,cm^{-1}$. Nevertheless the cause remained with no explanation of the appearance of two bands discovered by them (ν_3 and ν_3) due to water adsorption on hydroxylated silica surface. The high initial heats of water adsorption (17–20 kcal/mol) on SiO_2 surface [103–105] that cannot be explained *via* hydrogen bonding between H_2O molecules and hydroxyl groups testify the presence of the sections within silica surface layer with greater activity than that of hydroxyl groups.

Kiselev and co-workers obtained IR-spectral, adsorption, and radiospectroscopic data testifying the presence of strongly bound molecular water on the silica surface (coordinated to hydroxylated silicon atoms of SiO_2 surface). The absorption near $3550\,cm^{-1}$ was interpreted as stretching vibrations of the OH-groups of strongly bound water molecules. The concentration of coordinately bound water is of $\cong 10^{-2}$ ($\cong 1\%$) of the quantity of OH-groups on SiO_2 surface evacuated at 200–300°C [89]. Meanwhile the studies carried out of water adsorption on pyrogenic silica [83] showed that at the same dehydration temperatures the amount of strongly bound water corresponded to the number of free silanol surface groups. This fact gave the authors an opportunity to assume that hydrated SiO_2 surface had free OH-groups and equal quantity of water molecules coordinately bound with silicon atoms of silanol groups.

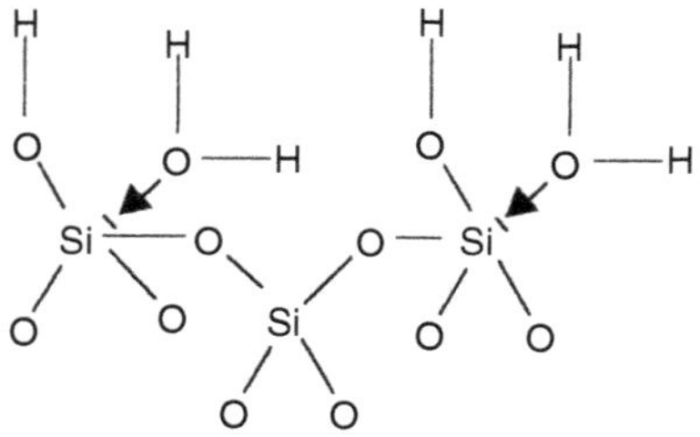

The adsorption of small water amounts on dehydrated silicas results in the formation of acidic proton sites that arise due to coordination binding H_2O molecules of silanol groups and are more active adsorption sites than OH-groups [86]. When the occupation degree of surface with adsorbed molecules increases, coordinately bound water molecules should be sites of swarm (or cluster) formation, the latter including 10–12 H_2O molecules [89].

The carrying out of combined spectral-gravimetric experiment in the study of SiO_2 surface dehydration as well as water and methanol adsorption [83,106] suggested the presence of active adsorption sites on dehydrated *in vacuo* silica surface, these sites can be coordinately unsaturated silicon atoms of silanol groups. it was shown that the amount of separated water molecules due to calcination of pyrogenic silica *in vacuo* at 200 to 400°C corresponds to the concentration of hydroxyl groups [106]. As the desorption of water molecules proceeds similarly to dehydration of initial surface, one can assume that the latter also contains free Si—OH groups and related amount of water molecules coordinately bound with silicon atoms of silanol groups. The study of chemisorption processes on aerosil surfaces with such sensible to H_2O presence reagents as trimethylchlorosilane [107], $SiCl_4$ [71], and dimethylchlorosilane [108] also indicated the presence of more strong nucleophilic reagents than isolated silanol group on the surface of silica calcined at 200°C.

The IR-spectroscopy investigations of thionyl chloride interaction with aerosil surface [109] gave, at bottom, direct evidences that adsorption near $3550\,cm^{-1}$ was caused by strongly adsorbed water. Indeed, the inserting aerosil sample pretreated at 250°C (hydrated surface) as well as that pretreated at 400°C (according to [59,110], the surface loses about half of Si—OH groups as a result of the condensation of near-stationed hydroxyls; band near $3550\,cm^{-1}$ is absent) with $SOCl_2$ resulted in the same adsorption values. The presence of strongly sorbed water on the surface was proved by heating chlorinated aerosil (pretreatment of the sample at 400°C in closed volume) what resulted already at 520°C in band vanishing with maximum at $3680\,cm^{-1}$ and reduction of SiOH-group band. Basing on the results mentioned, the authors of [109] suggested the occurrence of some forms of molecular water on the aerosil surface according to the scheme (at small occupation degrees):

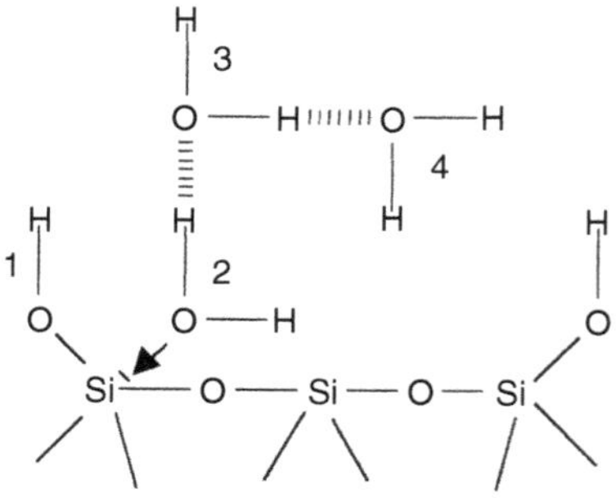

where 1, isolated structural silanol groups ($3750\,cm^{-1}$); 2, water molecule connected with acidic water proton of type 2 by strong hydrogen bonds [86,100] ($3550\,cm^{-1}$); 4, physically adsorbed water molecule (3450 and $1640\,cm^{-1}$).

An opportunity of the formation of coordinate bonds between silicon atoms of silanol groups and electron donor molecules found a convincing confirmation, when adsorption was studied of small electron donor molecules such as water, methanol, and hydrogen fluoride [83,111–114].

The analysis of ^{1}H NMR spectra of high resolution in solid phase as well as the results of mass-spectrometry investigation of water and methanol desorption from trimethylsilylated aerosil surface with no OH-groups [64,115] testify a nonhydroxylic nature of the sites of adsorption of small water amounts. A correspondence has been fixed of three characteristic regions of the sorbed water and methanol elimination with desorption maxima related to probable destruction of hydrogen-bonded (low-temperature maximum) and coordinate adsorption complexes of various structure (high-temperature maximum) by the help of desorption-field mass-spectrometry [116,117].

A concentration was estimated in the work [84] of strongly bound water under different conditions of vacuum sample preparation. As for these data, at the temperature $\approx$60°C (a sample within the ray of IR-spectrometer) on hydrated aerosil surface there can be approximately six or seven adsorbed water molecules per silanol group. It is quite possible that adsorbed molecules form hexa-coordinated adsorption complexes (hexameric structures similar to those described in [118] (see Figure 28.4). After evacuation at 200°C, there are kept two water molecules per silanol group on aerosil surface, and after calcination at 400°C — only one H_2O molecule per group ≡Si—OH [129].

A probability has been grounded by methods of quantum chemistry of the formation of two forms of coordination-bonded water molecules with silicon atoms of silanol groups [72,121,122]. At the initial stage of adsorption the complexes are formed with transposition of water molecules relatively to silanol group:

An adsorption band of 3680 cm^{-1} corresponds to such a complex. The hexagonal cavities ("hexagons") with concentration comparable with content of structural hydroxyl groups can serve as channels for an-out-of activation penetration of H_2O molecules into the subsurface layer. The occupation of "hexagon" with water molecule creates a potential barrier at the month of surface six-link cycles for the penetration of the next water molecules into three nearest "hexagons" [123]. Thus, the concentration of primary water adsorption sites is at least ten times less that the content of structural hydroxyl groups. The formation of a potential barrier can be connected with deformation of the hexagonal cavities of surface layer, when dimensions of particles decrease due to elimination of some amount of strongly bound water [124]. The presence of potential barrier explains a relative passiveness in the reactions with reagent sensitive to water as well as abnormally low velocity of rehydration of the SiO_2 samples calcined at high temperature [84].

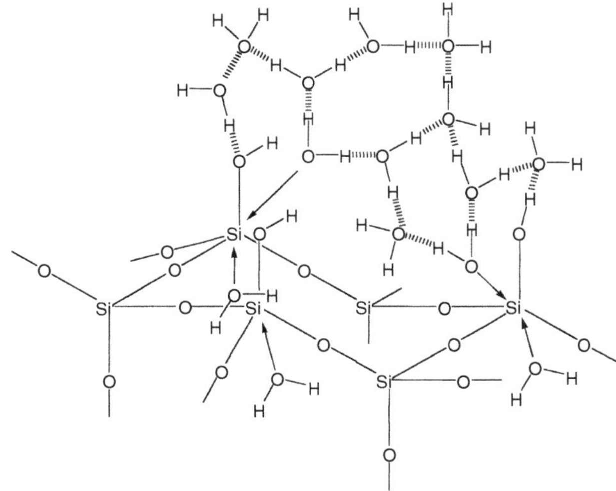

FIGURE 28.4 A schematic image of the structure of primary adsorption complexes of water molecules on SiO_2 surface.

At the next stage of adsorption coordinate complexes are formed with cis-position of water molecule relatively to hydroxyl group, those are characterized by the adsorption near 3550 cm$^{-1'}$.

The addition of new water portions occurs *via* coordinately-bound H_2O molecules that are centres of cluster formation in accordion to scheme [123] (at small occupation degrees):

At the consequent stages of water adsorption structural hydroxyl groups are involved into the process of cluster formation. The dehydration of SiO_2 samples is characterised by essentially reverse sequence of corresponding process.

ELECTROSTATIC ASPECTS OF ADSORPTION ON SILICA SURFACE

An a priori analysis on the reactivity and peculiarities of chemical behavior of molecules is a rather difficult but quite solvable task of theoretical chemistry. If molecules interact with a solid surface, the complexity of its solution increases repeatedly. This is cause by the circumstances as follows: firstly, an interaction occurs between two systems of different nature — molecule and surface that can be considered to be endless at the scale of partner; secondly, it is difficult to simulate a surface adequately that is a macrodefect of the crystal periodic structure. Moreover, a definite grade of amorphization of surface layer is a characteristic of even typical crystal [125]. Taking into account probable relaxation and reconstruction of real surface as compared with ideal one, obtaining valid structural information on surface and subsurface layer of solids seems to be rather problematic. A cluster model of solid and its surface that is natural for chemists operating terms of local chemical bonds (despite that it is not quite suitable for the systems with covalent bond) may be considered to be fit for the objects with ionic bonds that are objects of our investigation.

In the latter case we have a system of alternating charge densities what presses to take into account long-range electrostatic interactions caused by charges at dots of the crystalline lattice of support as well as by those of the atoms limiting its surface. Such a counting of electrostatic effects is not commonly used for description of the reactivity and chemical transformations of admolecules within subsurface layer that affects the correctness of the results obtained.

From the concept aspect, a microscopic description of the reactions at solid surface with ionic bond type should include the following moments:

1. An analysis of electronic and nuclear subsystems of the molecule-reagent
2. An examination of electronic and nuclear subsystems of solid what is possible provided subsurface layer is simulated at appropriate level
3. A calculation of the interaction between them at the account of surface electrostatic field (orientation and polarization of admolecule, deformation of admolecule and local surrounding of reactive site of the surface
4. The final stage is an act of chemical transformation at solid surface what assumes knowledge on multidimensional surface of potential energy.

This program can be realized, in principle, at two levels: the nonempirical one foresees calculations of admolecule and small cluster, that is, badly simulates the surface itself, and the semiempirical one allows to examine, besides admolecule, a large clusters that performs the structure of the most characteristic surface parts in a proper way. Nevertheless, the approach described can not be realized practically in the pure form in case of the systems interesting for chemistry, and so a necessity arises of the adequate taking into account of electrostatic field and potential of solid object. A problem of expanding procedure of the calculation of molecular electrostatic potential (MESP) accepted in the theory of molecule reactivity [126] to the interaction of admolecules with solid surface is rather complicated and laborconsuming but we were a success in its solution and in developing effective scheme of calculation for the case of ionic crystals. Taking silica surface as an example, let us demonstrate the importance and fruitfulness of counting electrostatic field of solid support.

A fragment of hydroxylated SiO_2 surface is performed in the Figure 28.5 that has a structure of β-crystobalite (111) face, and the electrostatic potential distribution itself within subsurface layer is given in the Figure 28.6 [127]. It is seen from this figure that there is a local minimum ρ within the plane of silanol group at the distance of about 1 Å from its oxygen atom, and continuous regions of positive potential are characteristic of hydrogen and silicon atom surroundings. It allows us to assume that small polar molecules (e.g., H_2O, HF) are capable of penetration into coordination sphere of hydroxylated silicon atom. A region of high positive ρ value in the hexagonal cavity of surface layer and near entrance into it forming the most considerable adsorption potential for binding anions and small electron donor molecules is of particular interest. A region with little negative potential value stationed above entrance into the cavity seems to determine relatively low potential barrier for penetration of anions and small electron donor molecules into the cavity.

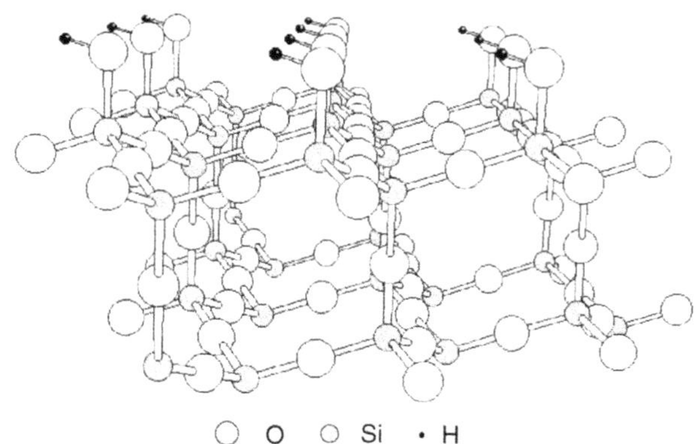

FIGURE 28.5 A schematic representation of the structure of silica surface similar to the β-crystobalite face (111).

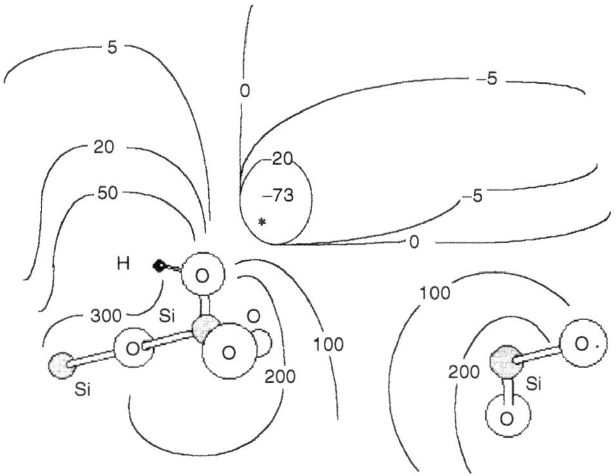

FIGURE 28.6 The profile of electrostatic potential (in J/mol) in the surroundings of β-crystobalite face (111).

It should be noted that the qualitative conclusions on the peculiarities of the ρ spatial distribution obtained for model systems are applicable also to disperse silicas (silica gels and aerosil) with the surface resembling structurally the crystallographic faces of β-crystobalite.

Chemical modification is known to be capable to change substantially sorption properties of silica surface [72]. Due to identity of support for a hydroxylated and chemically modified silica changes in sorption properties of different samples are to be connected naturally with corresponding peculiarities of ρ within the surrounding of functional groups. Thus in case of chlorinated silica (Figure 28.7) it should be noticed that ρ has substantially lesser heterogeneity above surface near a functional group as compared with that of hydroxylated SiO_2 [128]. The smaller negative value of potential near chlorine atom as compared with that the oxygen atom of silanol group as well as negative values above hexagonal cavity indicate a considerable rise of potential barrier for embedding small ligands into coordination sphere of the chlorosilyl group silicon atom. Probably just this fact can explain an initial hydrophobic properties and reduced speed of the primary stages of hydrolysis of the chlorinated SiO_2 surface. An analysis of the electrostatic potential profiles near the chlorinated, aminated (Figure 28.8), and hydrogenated (Figure 28.9) silica surface discovers a distinct dependence on the nature of functional groups and allows to describe clearly a comparative activity of these groups in the processes of polar compounds adsorption [129].

An analysis of the peculiarities of electrostatic potential spatial distribution within subsurface layer can give an important information on initial stages of silica surface hydration and on peculiarities of cation sorption without carrying out labor-consuming computations on potential energy surface [130]. It is known that besides hydroxyl groups almost always there are water molecules on silica surface bound to it in various ways. According to the literature, the main features of hydrate cover in the regions of isolated hydroxyl groups can be described within frameworks of three models. A formation of hydrogen bonds between surface silanol group and H_2O molecule is a characteristic of the models I and II (Figure 28.10 and Figure 28.11). The first model has a water molecule as proton donor, whereas the second one has a silanol group. Within model I (Figure 28.10) there

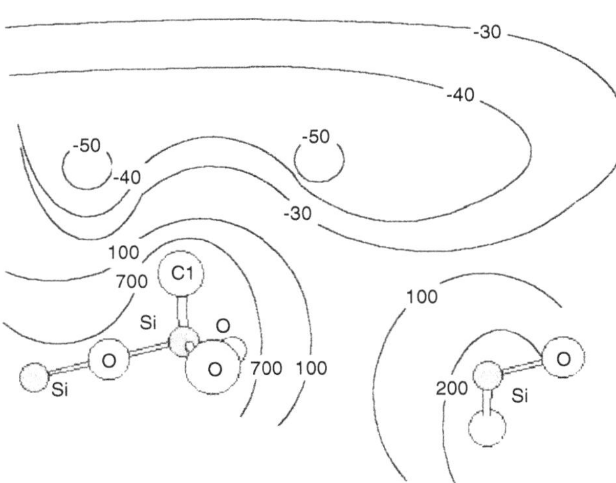

FIGURE 28.7 The profile of electrostatic potential (in J/mol) in the surroundings of totally chlorinated β-crystobalite surface face (111).

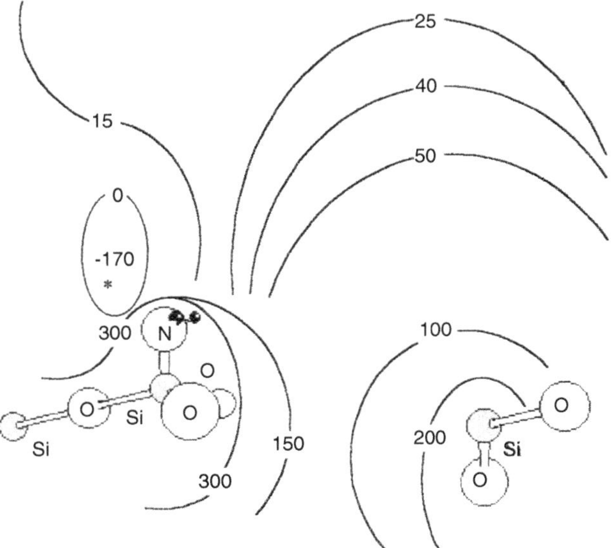

FIGURE 28.8 The profile of electrostatic potential (in J/mol) in the surroundings of totally aminated β-crystobalite surface face (111).

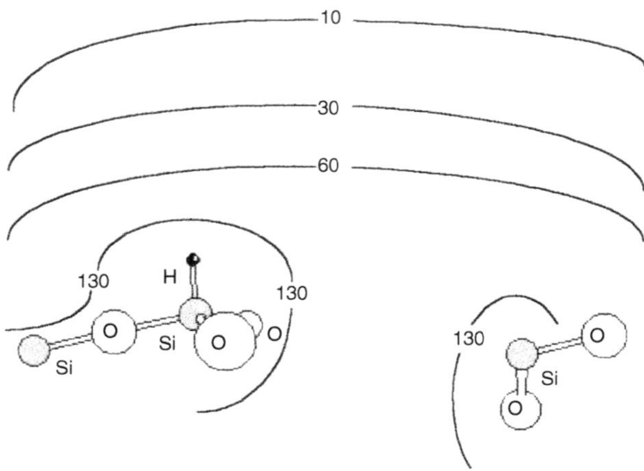

FIGURE 28.9 The profile of electrostatic potential (in kJ/mol) in the surroundings of totally hydridesilylic β-crystobalite surface face (111).

are no regions of the ρ negative values near oxygen atom of silanol group. Such a region is present near oxygen atom of admolecule, the minimum being rather considerable and equating to -260 kJ/mol. The access of cations into these regions from solution or vacuum is out of barrier, that is, their trajectories completely pass through the regions of negative potential values. It is follows from this figure that when sorbing, the cations firstly are localized within the regions of minimal ρ values near the water molecule oxygen atom followed by exchange with the nearest proton that does not take part of hydrogen bond formation. Analogous conclusions are true also for the model II (Figure 28.11).

The model III (Figure 28.12) assumes a coordinate bond between the H_2O molecule oxygen atom and surface hydroxylated silicon atom and is common used for an explanation of the firm keeping of water with disperse silica surfaces. The presence of one compact region of the negative potential values is a characteristic of it. Similar to the models I and II the ρ minima are

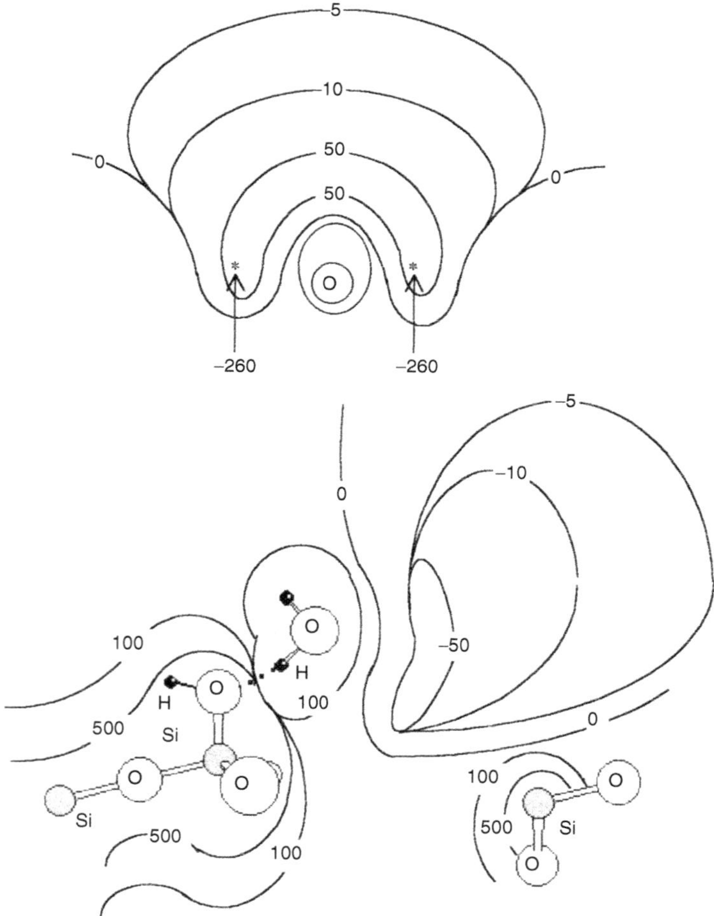

FIGURE 28.10 The electrostatic potential distribution for the model I of hydrated state of silica surface. The ρ distribution within the plane perpendicular to the water molecule plane is given in upper left part of the picture. Isopotential lines correspond to ρ values in kJ/mol.

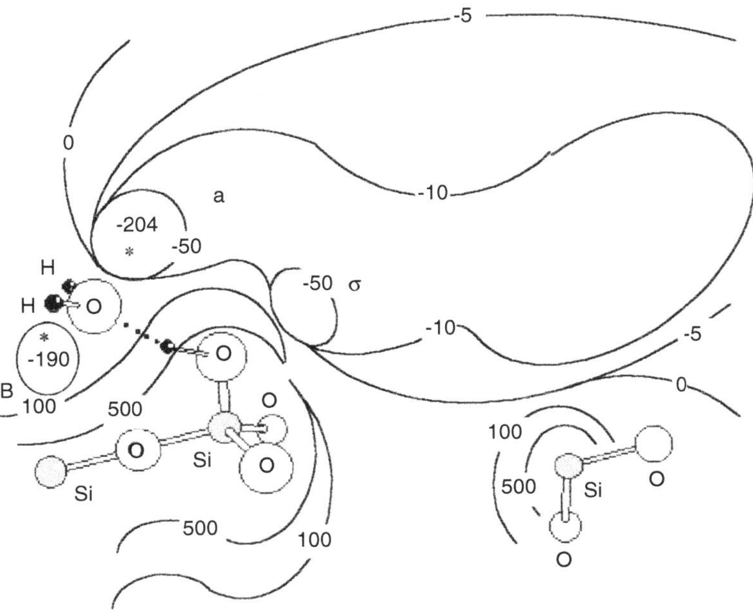

FIGURE 28.11 The distribution of electrostatic potential for the model II of silica surface hydrated state. Isopotential lines correspond to the ρ values in kJ/mol.

near the water molecule oxygen atoms and corresponds to −240 mJ/mol. Nevertheless in this case the difference in the potential values near oxygen atoms between silanol group and coordinately bonded water is not such considerable as in the models examined above but does not exclude an opportunity of the initial localization of the cations sorbed in these minima.

Thus the data on the ρ distribution confirm the initial stage of cation sorption with silica surface to occur accompanied by the molecules of sorbed water rather than by structural silanol groups [131].

Use of idea on the role of electrostatic potential in oxide surface chemistry allows also to simulate the change in surface properties in clear way when various

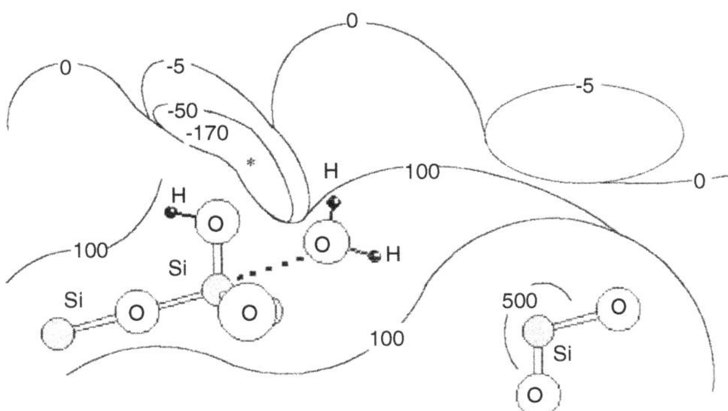

FIGURE 28.12 The distribution of electrostatic potential for the model III of silica surface hydrated state. Isopotential lines correspond to the ρ values in kJ/mol.

admixture ions are introduced into oxide matrix [132]. For example, a modification of silica is probable via isomorphous substitution of aluminum atoms for some quantity of silicon atoms. A rather considerable part of disperse silica surfaces, in particular those of aerosil and silica gels, being characterized by the structural elements that are typical of the β-crystobalite (111) face, it is naturally to assume the model with structural parameters of this face that is slightly distorted by introducing aluminum atoms into silicate matrix to be suitable for aluminosilicates. To describe the properties of aluminosilica surface more completely, two models should be examined that simulate the structure of its most important sections. In the model IV (Figure 28.13) an aluminum atom substitutes a hydroxylated silicon atom in the initial model of the β-crystobalite (111), and in the model V (Figure 28.14) an aluminum atom is inserted in place of subsurface silicon atom bound with four siloxane bounds to neighboring Si atoms. The topology of the ρ distribution within subsurface region of aluminosilica is more complicated than that of initial hydroxylated silica what is connected with the greater variety of active sites (three-coordinated silicon atoms, bridge hydrogen atoms, and silanol groups).

The Figrue 28.15 performs the ρ distribution near 3-coordinated aluminum atom. The presence of the region of negative potential values adjoining the bridge oxygen atom is characteristic of it what testify its proton acceptor properties. Such regions are absent for the oxygen atom binding two neighboring silicon atoms in β-crystobalite. The compact region of the ρ positive values neighboring with the aluminum atom is surrounded with continuous

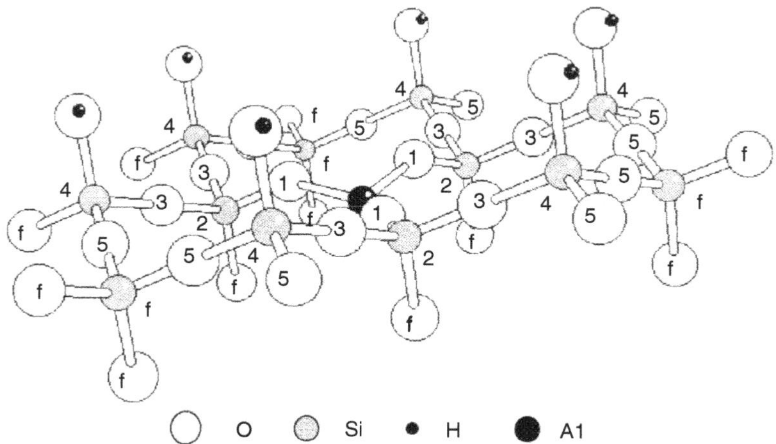

FIGURE 28.13 A scheme of the cluster IV modeling a part of aluminosilica surface realized due to the change of hydroxylated silicon atom by aluminum one within surface layer of β-crystobalite. The atoms numbering is given in dependence on their distances from aluminum one. The atoms are labeled with "f" letter that kept their positions under optimization unchanged and the same as those in β-crystobalite.

region of negative potential values what creates a rather perceptible potential barrier for the penetration of electron donor molecules to the Lewis acidic site that is 3-coordinated aluminum atom in the case given. For the cluster V the potential near 4-atomic fragment of ≡Al—OH—Si≡ has only positive values (Figure 28.16), that is, the access of proton acceptor molecules or anions to the Brönsted acidic site is practically out of barrier [133].

The examination of the local state densities (LSD) of the systems under consideration can serve as one more example of obtaining useful information on the properties of concrete atoms in surface structures [134]. The Figure 28.17 performs these state densities of silicon and aluminum atoms calculated for the cluster IV. Thus, the one of the maxima of LSD of aluminum atom is stationed within the energy region from -1 to $+1$ eV, and the energy of the highest occupied orbital is of -8.4 eV.

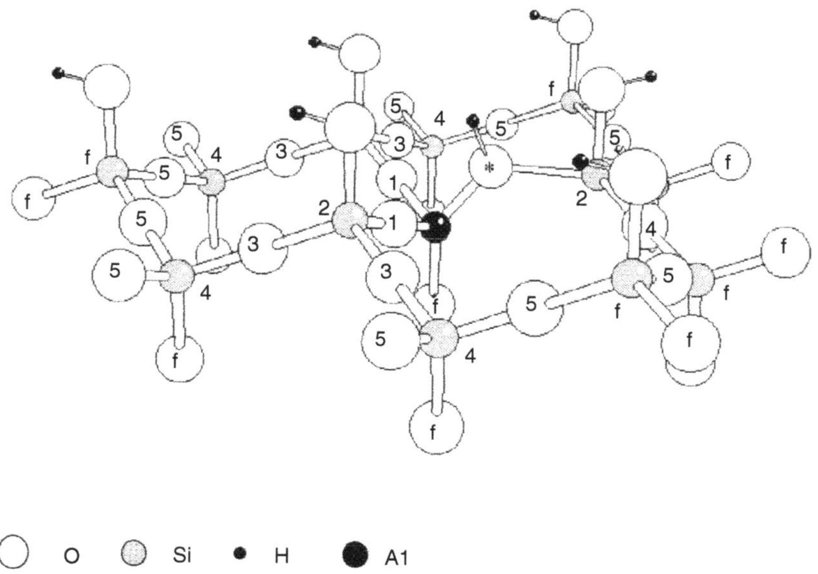

FIGURE 28.14 A scheme of the cluster V modeling a part of aluminosilica surface realized due to the change of bulk subsurface silicon atom by aluminum one within surface layer of β-crystobalite. The atoms numbering is given in dependence on their distances from aluminum one. The atoms are labeled with "f" letter that kept their positions under optimization unchanged and the same as those in β-crystobalite.

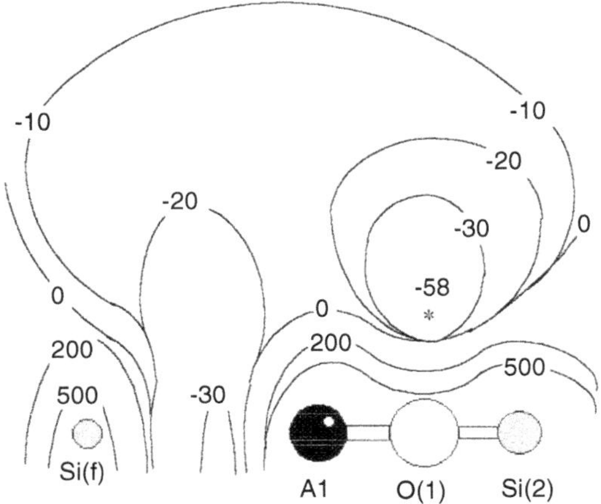

FIGURE 28.15 The distribution of electrostatic potential for the cluster IV within the plane perpendicular to the plane of the $Al[O(1)Si(2)]_3$ fragment and crossing the Si(2), O(1), Al, and Si(f) atoms. The potential values are given in kJ/mol.

The $3p_z$ aluminum atom orbital pays the main contribution to this maximum what gives grounds to assume a localization of the surface electronic state at this atom displaying as an electron donor site. The principal maximum of a hydroxylated surface silicon atom is localized within energy region corresponding to the highest occupied orbitals, and there is a maximum within -1 to $+1$ eV indicating the electron acceptor properties of this atom. A coordinate bond of electron donor molecules with hydroxylated surface silicon atom seems to be realized due to the presence of the LSD maximum within the region of vacant levels. For a subsurface silicon atom, all the maxima are within the region of lower energies. Analogous features of the mutual stationing the LSD maxima of various type silicon atoms are characteristics of the β-crystobalite.

Basing on ideas on the importance of taking the electrostatic potential within subsurface layer of oxides into

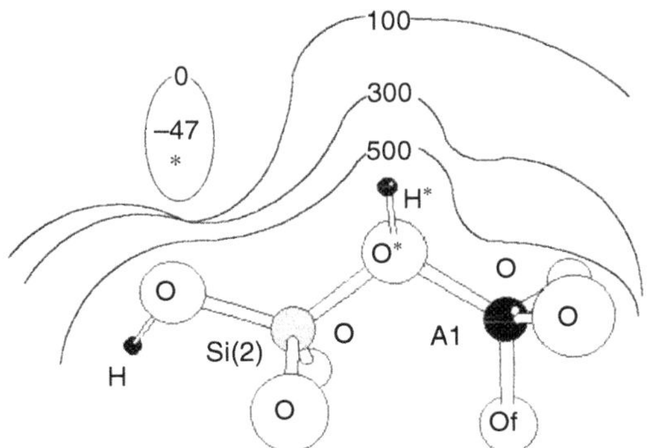

FIGURE 28.16 The distribution of electrostatic potential for the cluster V within the plane of (HO)Si(2)O(*)AlO(f). The potential values are given in kJ/mol.

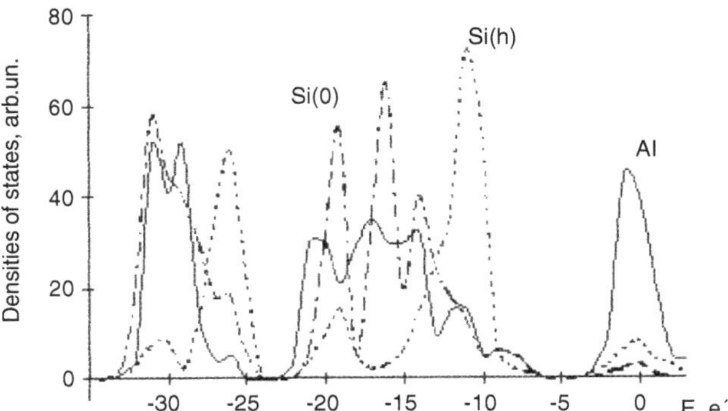

FIGURE 28.17 The local state densities of the Si(o), Si(h) and Al atoms calculated for the cluster IV. The values of state densities are given in relative units.

account, let us examine the peculiarities of the elementary acts of the chemical transformation at the surface. Both a modifying molecule and the surface reactive site with neighboring atoms undergo to deformation changes in the processes of the chemical modifying of surface. It was shown in the literature [72] that the main contribution to the activation energy of the reaction of electrophilic substitution of the silanol group proton

$$\equiv\text{Si—OH} + \text{XY} \rightarrow \equiv\text{Si—OX} + \text{HY}$$

is paid by the deformation process in admolecule and its characteristics near surface should be another than those of isolated state (Figure 28.18). To estimate this difference, generally speaking, two approaches can be used. According to the first one the results should be compared of quantum chemical calculations of isolated molecule with analogous data for the same molecule near surface, the configuration of transition state being applied. The second approach makes it possible to change the study of molecule near surface with its examination within some effective electrostatic field [135,136]. Of course the total calculation on the effect of the field of complicated distribution is practically impossible, but general conformities to natural laws of the influence of surface on admolecule can be elucidated within the approximation of homogenous electric field.

The second approach being applied, let us examine what is the role of subsurface electrostatic field in the reactions of electrophilic substitution of the proton of isolated silanol group with trimethylsilylic radicals $—SiR_1R_2R_3$. These reactions run via four-centered transition states (see Figure 28.18). A deformation is essential for such transition states of the admolecule silicon atom tetrahedral surrounding up to planarity of the XR_1R_2 atomic group as well as a trigonal-bipyramidal configuration of the system: modifying molecule + silanol group oxygen atom.

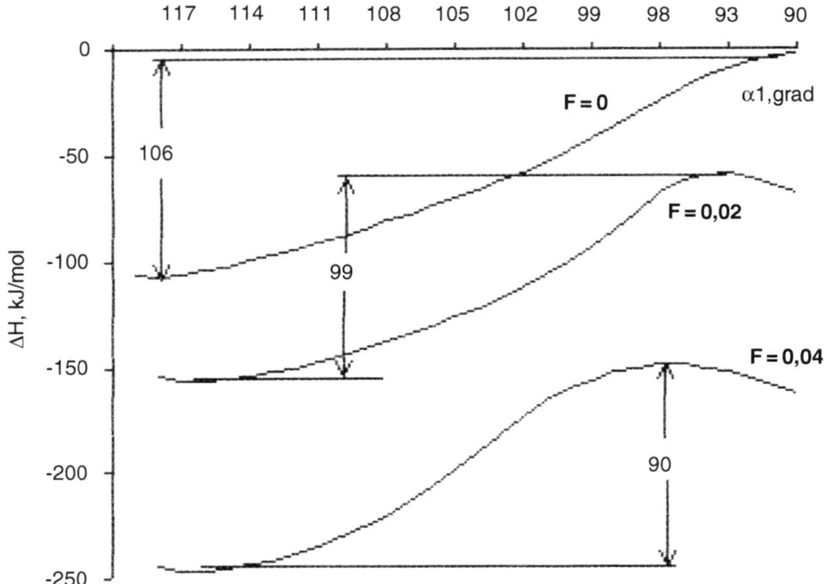

FIGURE 28.18 The deformation energies of trimethylaminosilane molecule in zero field and those in the fields of F = 0.02 and 0.04 a.u.

A degree of distortion of the central silicon atom tetrahedral surroundings is controlled by angles α_1 and α_2 that are equal to 90° for a planar conformation of the group mentioned.

The plots of the heat of formation on angle α_1 are given in the Figure 28.18 calculated for the isolated $NH_2Si(CH_3)_3$ molecule at two values of outer field and without it. One can make two important conclusions from this figure. First, taking the outer field into account created by lattice and surface atoms in the region where the modifying molecule is stationed for the transition state structure results in decrease of deformation energy E_{def} with rise of field. Second, if the E_{def} maximum is achieved at $\alpha_1 = 90°$ for zero field then at $F = -0.02$ a.u. this maximum corresponds to $\alpha_1 = 92°$ and at $F = 0.04$ a.u. the following rise of α_1 to 95° occurs. Calculating subsurface field leads to the situation that the transition state of exothermic reactions considered is achieved at the α_1 angle values (α_1 controls moving on along reaction coordinate) that precede the planar structure of $NH_2Si(CH_3)_3$ fragment, that is, the transition state structure at zero field is nearer to that of initial reagents as compared with the situation for zero field. The values of E_{def} and of angle α_1 corresponding to its maximum for the molecules of some electrophilic reagents are given in the table. It is seen from it first of all that the maximum of E_{def} value is achieved at the α_1 angle values $<90°$ for the endothermic reactions of trimethylpseudohalogensilanes $(CH_3)_3SiN_3$, $(CH_3)_3SiNCS$, and $(CH_3)_3SiNCO$ interaction with silanol groups. This testifies a considerable distortion of molecule structures in transition states.

Thus, applying this idea on subsurface electrostatic field allows us to describe more correctly these reactions within the framework of the deformation model for activation barriers of the reactions of electrophilic substitution of silanol group protons and to elucidate factors determining a degree of distortion of the spatial structure of modifying molecules near an adsorbent surface.

The advantages of this approach consist in simplicity (there is no necessity to examine interaction of admolecules with solid surface that appears only as effective electrostatic field created by it and a lattice), in refusal from standard supermolecular description within the frameworks of usual approaches of zero differential overlap that in principle fail in description of potential barriers. Also an opportunity arises to compare reactivity of molecules in the reactions of proton electrophilic substitution not only in the row of related compounds (for example methylchlorosilanes) but of those containing various functional groups.

In conclusion let us consider problems on the formation of hydroxylic cover on fresh-made surfaces. The electrostatic fields created by lattice ions near surface limiting crystal play the most important and determining role in its formation. Depending on oxide chemical composition, definite factors are promoted to the foreground, but general conformities to natural laws of hydroxylic cover formation can be observed using some simplest models. Thus, the Figure 28.19 demonstrates a fresh-made surface of β-crystobalite obtained by rupture of crystalline lattice according to the (111) incline plane. Three-coordinated silicon atoms are stationed in the

TABLE 28.2
The Deformation Energies, E_{def} (kJ/mol), Heat Effects, Q (kJ/mol), and Angles α_1^{max} (Degrees) of the Reaction of Silylation

	F = 0.00			F = 0.02		F = 0.04	
Molecule	E_{def}	−Q	α_1^{max}	E_{def}	α_1^{max}	E_{def}	α_1^{max}
$(CH_3)_3$ SiN_3	151	98	90	141	88	130	85
$(CH_3)_3$ SiNCS	175	120	90	163	86	148	83
$(CH_3)_3$ SiNCO	172	168	90	164	85	153	80
$CH_3)_3$ SiCl	162	—	90	151	88	148	86
$(CH_3)_3$ SiBr	140	—	90	128	98	114	102
$(CH_3)_3$ SiJ	126	—	90	108	100	86	105

plane of this face, rows of single-coordinated oxygen atoms arising above them. As the valences of surface silicon and oxygen atoms are totally unsaturated, strongly heterogeneous electrostatic fields of high stress are created within their surroundings. Gradients of these fields can be judged from Figure 28.20. When hitting hollows at surface of the face discussed (Figure 28.21), a water molecule (there is an excess of them always and everywhere) takes its bearings such that its oxygen atom is stationed near a silicon one and its hydrogen atoms are stationed near oxygen ones. Calculations indicate that an electrostatic field of about 0.02 a.u. is created [135] within the region of the water molecule oxygen atom localization. When in such fields, water molecule electronic shells are so deformed that its ionization takes place following by a molecular ion $H_2O^{+\bullet}$ [136] formation (Figure 28.22). Effect of field on this ion results in its consequent decomposition into $OH^{\bullet}$ radical and H^+. A hydroxyl radical is bound with three-coordinated silicon atom and proton migrates into nearest minimum of negative potential near an oxygen atom. As a result two surface hydroxyl groups are formed due to decomposition of one molecular ion.

Some conclusion can be made from the data presented, in particular:

1. There is a clear and synonymous concretizing sites of polar molecule primary adsorption.
2. A heterolytic character of chemical reactions on SiO_2 surface follows from electrostatic potential distribution near silica surface with polar functional groups. An existence of potential regions with considerable values of opposite sign near these groups at the distances of about chemical bond length indicated strong polarization of the reactive site of the reacting molecule as well as increase in bond polarity
3. A change in the nature of functional groups, hydration of silanol groups, inserting admixture atoms into silica matrix results in essential redistribution of electrostatic field and in respective varying adsorbent surface properties

It should be noted that the information examined on active sites nature was obtained without carrying out complicated quantum chemical calculations within a supermolecular approach. It is important that a clear analysis is probable on this basis of the chemical consequences of any change in composition and surface and bulk structure of oxides and their compositions (from ionic implantation to destruction of the system simulated) what allows to recognize calculations with real counting electrostatic

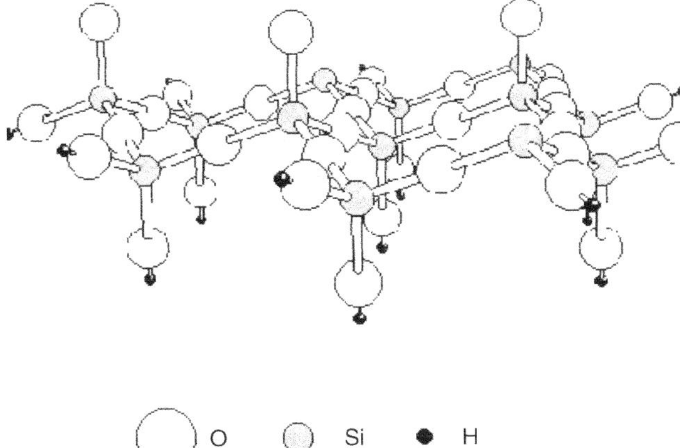

FIGURE 28.19 A fresh-made surface of crystalline β-crystobalite.

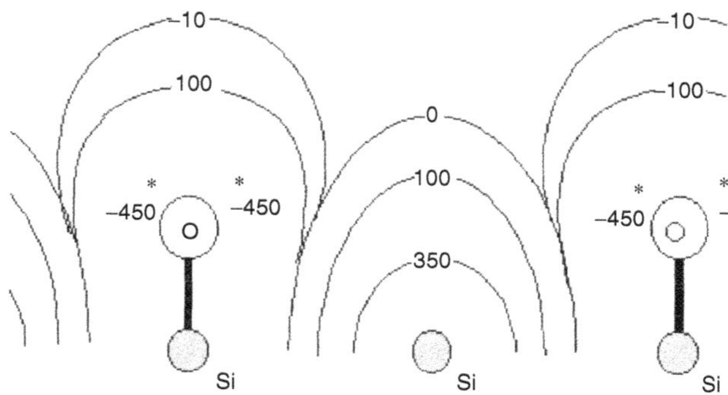

FIGURE 28.20 The distribution of electrostatic potential near fresh-made surface of β-crystobalite.

field to be an effective tool of computing experiment and theoretical design of solid systems with the properties assigned.

HYDROPHOBIC AND HYDROPHILIC PROPERTIES OF TRIMETHYLSILYLATED SILICA SURFACE

In the previous section the adsorption concept was entered of adsorption molecular electrostatic potential and it was remarked that the first and rather acceptable approximation could be obtained by its help for constructing hydrate cover of any type of solid surface. However, within the framework of notions on MESP it is also possible to predict some other important properties of adsorption complexes of polar molecules on silica surface. Thus, it was shown in [137] that hydrophility of the hydroxylated surface of silicion dioxide could be explained by availability of local minima of the ρ potential within the subsurface region created by the atoms of surface functional groups and solid substrate. Here in water this approach was applied to explain adsorption properties of completely chlorinated, aminated, and hydrogenated surfaces [128].

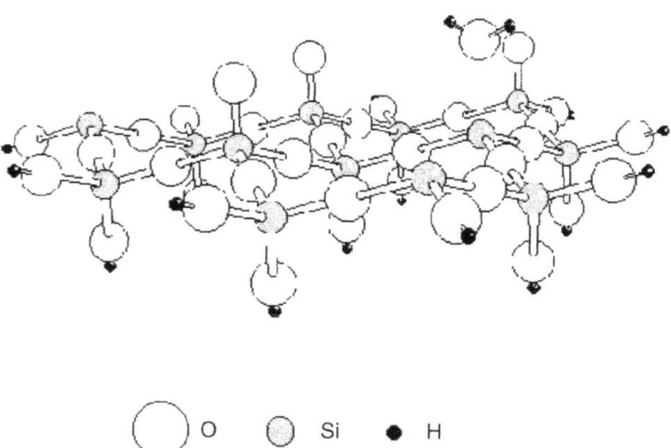

FIGURE 28.21 Water molecule near fresh-made surface of β-crystobalite.

However in this work the problem was not studied at all on correlation of the subsurface distribution of ρ with adsorption properties of grafted layer containing functional groups of various nature. The importance of its consideration is stipulated by the fact that adsorption layer grafted on any solid surface in a covalent way is always multifunctional because it usually contains initial functional groups besides grafted ones. The deriving of the monofunctional stratum as a result of the silanization reaction is practically impossible because of originating steric troubles, energy heterogeneity of surface active sites, topography of grafted groups, mutual influence of grafted molecules, their interaction with the substrate, and also due to other adsorption factors of various degrees of importance [137].

In the section given the electronic structure has been considered by the semiempirical methods of quantum chemistry (PM3 [138]) of a row of clusters simulated for the most typical section of the surface of hydroxylated, completely trimethylsilylated silica, and also of silicas with the variable amount of hydroxyl and trimethylsilyl (TMS) groups. The initial unmodified surface of hydroxylated silica was performed with adsorption cluster of $Si_{21}O_{56}(OH)_{12}(Si^*)_{24}$ (cluster 0) with the structure relative to face (111) of β-crystobalite. The following structural parameters were used: R(Si—OH) = 1.746 Å, R(SiO—Si(CH$_3$)$_3$) = 1.789 Å, R(Si—C) = 1.873 Å, R(C—OH) = 1.08 Å, ∠Si—O—H = 108.6°, ∠Si—O—Si(CH$_3$)$_3$ = 118.6°. They were obtained *via* optimization of the space structure of the cluster (1) with gross-formula of $Si_{24}O_{56}(OH)_{11}(Si^*)_{24}[Si(CH_3)_3]$ (Figure 28.23) with one TMS group in position 4 (Figure 28.24). The optimization was carried out in the supposition of "frozen" silicon-oxygen substrate, for which the experimental data were used on the space structure of β-crystobalite (R (Si—O) = 1.65 Å, ∠Si—O—Si = 180° [139].

Schematically the reaction of sequential grafting TMS group to model cluster (0) can be presented by the equation

$$Si_{24}O_{56}(OH)_{12-n+1}(Si^*)_{24}[Si(CH_3)_3]_{n-1} + ClSi(CH_3)_3 \rightarrow Si_{24}O_{56}(OH)_{12-n}(Si^*)_{24}[Si(CH_3)_3]_n + HCl,$$

where n = 1 ÷ 12.

The enthalpy of this reaction at each stage (except n = 0) should depend not only on n, but also on that, what position relatively earlier grafted groups is entered by the next TMS group.

The schematic view of cluster (0) from above is presented in Figure 28.23, the figures designating adsorption sequence of the substitution of TMS groups for the hydrogen atoms of surface silanols, and the dependence of reaction enthalpies (ΔH) on the amount of these groups is given in Figure 28.24. ΔH values were calculated

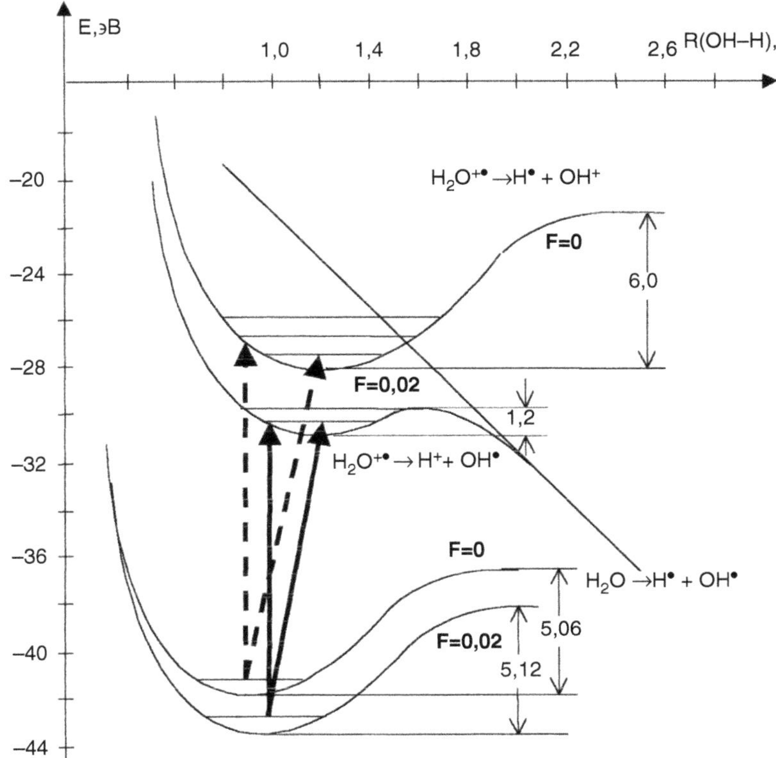

FIGURE 28.22 A schematic performance of the processes of ionizing and decomposition of water molecule as well as of its molecular ion in zero field and in the field of 0.02 a.u.

as differences between the sums of total energies for reaction products and for initial reagents. The total energy of cluster (0) was accepted to be a conventional zero.

The reaction enthalpy at n = 1 (ΔH_1), that is, for substitution of one TMS group for silanol in position 1 is of −82 kJ/mol. The introduction of the second such group into position 6 gives $\Delta H_2 = -130$ kJ/mol. The absolute value of ΔH_2 is noticeably lower than the double value of $|\Delta H_1|$. It testifies that a rather noticeable repulsive interaction takes place between two grafted to silica surface trimethylsilyl groups divided by a distance at 14.3 Å which can be estimated approximately as $|2\Delta H_1 - \Delta H_2| = 54$ kJ/mol. The filling of site 12, equidistant from site 1 and 2, gives $\Delta H_3 = -175$ kJ/mol.

Hereinafter within the framework at the model developed, the positions 2, 9, and 10 were occupied what

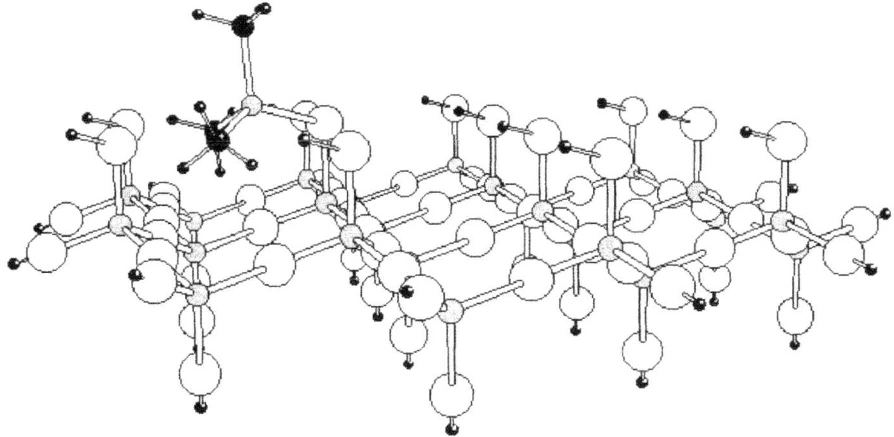

FIGURE 28.23 Cluster 1 simulated for silica surface with alone TMS group in position 4 surrounded with 11 silanols.

resulted in appearance of three isolated pairs of TMS groups with minimal distances between them. A minimum was achieved for the plot of ΔH on n at n = 7, that is, for such a cover where each TMS group had two analogous neighbors. The ΔH value at minimum is equal to −262 kJ/mol (that related to one TMS group gives a value of ≈37 kJ/mol, the latter being rather close to the experimentally measured reaction heat of the silylation of silica surface with trimethylchlorosilane that is of 42 kJ/mol (according to the data of work [75]). Further filling (n = 8 ÷ 12, sites 5, 3, 11, 4, 8, and 7 respectively) keeps the values $\Delta H_8 - \Delta H_{12}$ negative but they decrease as absolute values. This decrease is especially noticeable at n = 10 (the number of the nearest TMS groups is equal to 4), n = 11 (5 neighboring TMS groups), and n = 12 (all six nearest sites are already occupied with TMS groups).

The model presented of step-by-step silylation of silica surface alongside with obviousness and consistence of the approach is not deprived of certain defects. Their removal is connected with sharp magnification of the range of calculations, which result in an insignificant correction of numerical values of ΔH_{11}. The improvement of the model implies a possibility of optimization of the space structure of the clusters with a rising number of TMS groups. This should result in some lowering of the bottom of the curve indicated in Figure 28.24 without shifting the position of its minimum at n = 7. A probable insignificant distortion of the silicon-oxygen skeleton can also be taken into account that occurs due to substitution of trimethylsilyl groups for the hydrogen atoms of surface silanols. It is possible that use of the cluster of greater sizes should give a more detailed information on the dependence of the silylation reaction heat on the sequence of grafting TMS groups to silica surface.

When comparing a calculated theoretically heat of any reaction occurring on solid surface, obviously, it is impossible to limit a consideration by an interaction with only one site. This is connected with the fact that, unlike gas phase reactions where an interaction occurs between molecules of gas phase or solution react with solid surface, each following molecule, strictly speaking, interacts already with another substance differing from that having reacted with previous molecules. It is seen well from the equation mentioned above of the interaction of trimethylchlorosilane with model cluster (0). Of course, when a finite-size molecule interacts with real solid surface that can be believed to be boundless in scale of the partner, such a differentiation should become noticeable at the occupation degree approximating to one. To our opinion, in such cases every time one has to build dependencies of ΔH on n which not always look like the curve in Figure 28.24. According to the accounts, for substitutes occupying an area on solid oxide surface is comparable with that related to one silanol group, the curve of ΔH_{12} (4) becomes a horizontal line already for n = 5–8. This indicates that the interactions between molecules and between surface silanol groups are of the same order.

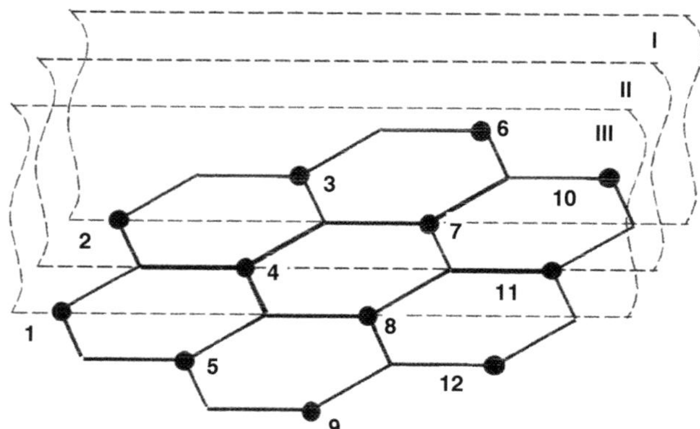

FIGURE 28.24 Numbering of the sites of functional group stationing at silica surface and image of the planes used for construction of electrostatic potential.

A cluster reproducing completely trimethylsilylated surface (TMS-SiO_2) of silica was derived from the cluster (0) due to a substitution of —$Si(CH_3)_3$ groups for the hydrogen atoms of all silanols. The surfaces realized due to partial substitution of these hydrogen atoms with TMS groups were presented by clusters of $Si_{24}O_{56}(OH)_{12-n}(Si^*)_{24}[Si(CH_3)_3]_n$, where $n = 1 \div 11$. The influence of a discarded part of crystalline lattice was taken into account by introduction of pseudoatoms (Si*) at the periphery of clusters adjoining the bulk phase of substrate.

The calculation on electrostatic potential distribution was carried out in accordance with the procedure described in [132] in two-centred approximation without a preliminary orthogonalization of atomic basis.

The cluster 1 is presented in Figure 28.22 that simulates the silica surface formed at initial stages of the silylation reaction or due to lack of modifying silane within reaction volume. The presence of a common region of negative values of ρ localized near atoms of O(4) and of carbon of grafted group is a characteristic of the potential distribution (Figure 28.25) with the plane II (Figure 28.23) crossing as a perpendicular to the plane of hydroxylated silicon atoms *via* the chain of bonds ≡Si(cryst.)—O—Si—C of TMS group. The formation of this region is conditioned by negative charges of TMS group carbon atoms, atom O(4) and also by contribution from surrounding silanol groups. A minimum ρ near atom O(4) reaches the value of −113 kJ/mol what is noticeably lower than that near the oxygen atom of silanol group in cluster 1 (right upper part of the picture) or in cluster 0 (Figure 28.26).

The date on electrostatic potential indicates that, when compounds with small polar molecules are adsorbed, there are the most probable such localizing site and admolecule

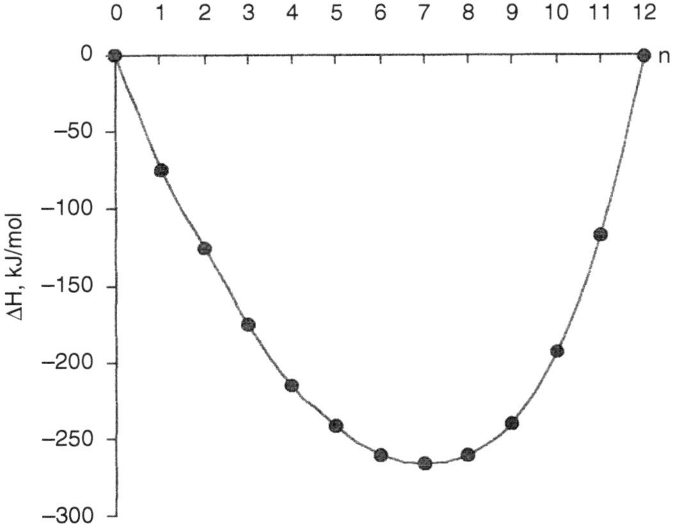

FIGURE 28.25 A plot of the reaction enthalpy (ΔH) of consequent silylation of silica surface on the amount (n) of grafted TMS groups.

orientation that its positively charged part is stationed within the region of negative values of potential, whereas a negatively charged one — within the region of positive values of potential. A penetration of electron donor group or atom into the coordination sphere of silicon atom bearing a TMS group is likely, and the potential near it has high positive values, whereas the proton of admolecule is stationed near atom O(4) in a region with negative values of ρ. The comparison of potential distributions near atoms O(4) and O(11) gives a reason to believe that the formation energy of the complexes with participation of trimethylsilylated silicon atom should be higher than that for hydroxylated one. It is important to notice that a grafted TMS group cannot bind water with hydrogen bond where a surface functional group acts as a proton donor. Probably this can explain the well-known experimental fact [139,140] that, in spite of increase of the formation energy of the adsorption complexes of water molecules with partially trimethylsilylated silica surface as compared with hydrosilylated one, total amount of the water kept under such a modification decreases.

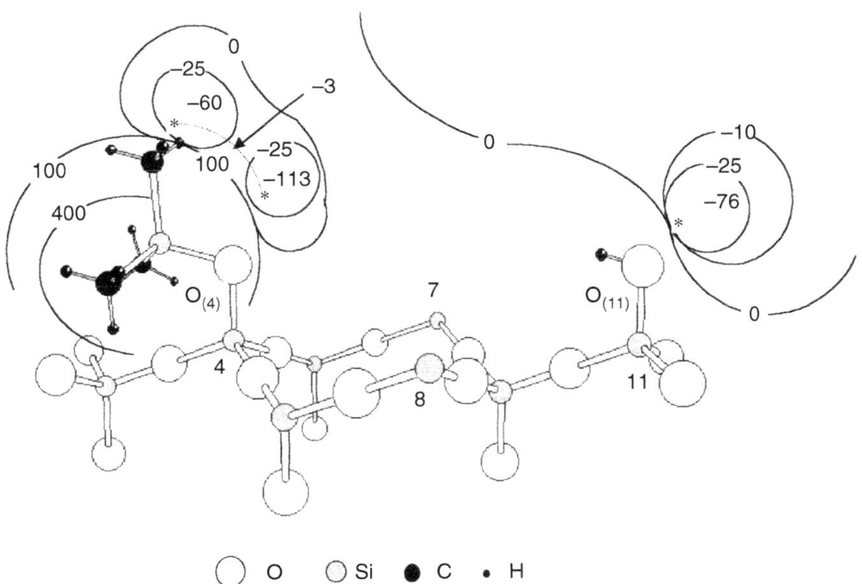

FIGURE 28.26 Profiles of the electrostatic potential (kJ/mol) within the plane II for subsurface silica layer simulated with cluster 1.

Let us examine the ρ distribution within the plane I which is parallel to the plane II and distant from the latter for 2.69 A (Figure 28.27). It crosses over two hydroxyl groups (sites 2 and 7) that are at the distance of 5.38 A from a grafted TMS group. The effect of a TMS group is especially felt on the potential distribution near silanol in the position 2. Thus the region size of ρ negative values near atom O(2) is much less than that in case of hydroxylated silica, the potential minimum being only of -38 kJ/mol. There is also a compact region of positive potential values appearing due to close position of the methyl group of TMS radical in site 4. The distribution of potential near silanol in position 7 depends also on the presence of TMS group in position 4, though not so essentially as for the hydroxyl in position 2. Common features of the ρ distribution in the neighborhood of atom O(7) are similar to those for hydroxylated silica. This testifies a probability of the formation of adsorption complexes of small electron donor and proton donor molecules with some surface silanols of partially trimethylsilylated surface of the same type as for hydroxylated silica. However, another type of hydroxyls appears with properties sharply distinguishing from those of silanols of the initial unmodified silica.

When considering the peculiarities of further silica silylation, it should examine (at least briefly) the models for surface occupation available in the literature. The main models are follows:

1. Island-like or cluster model [141,142] suggesting that grafting of one modifier molecule to any surface active site promotes grafting of consequent molecules in the neighborhood of the initial one. This model, as a rule, describes the surface reactions with participation of modifying molecules having two or more anchor groups and an electrophilic functional group
2. Casual or statistical model [143] that supposes that the site of grafting each consequent molecule does not depend on the position of earlier grafted molecules
3. Uniform of lattice models [144] assume that the consequent molecule reacts with the surface, being placed onto it under a certain rule

The last two models are characteristic of silanes with one anchor group and a chemically inactive functional group. In our case, obviously, one should prefer the last two models. However, if the data on reaction energetics of sequential silica silylation are taken into account, it is possible to approve in great probability that each sequential TMS group is stationed as far as possible from that already grafted on the surface.

Within the framework of the finite cluster model of the surface we examine (12 sites) two TMS groups should be placed, for example, in positions 1 and 10, and three ones — in sites 1, 6, and 12.

The potential distribution in cluster 2 simulated for a surface section with two TMS groups, as a whole, differs from that examined above. Thus, it is seen from Figure 28.28 that there is no united region of negative ρ values in the neighborhood of TMS group grated into position 1, and the minima near atoms of carbon and of O(1)

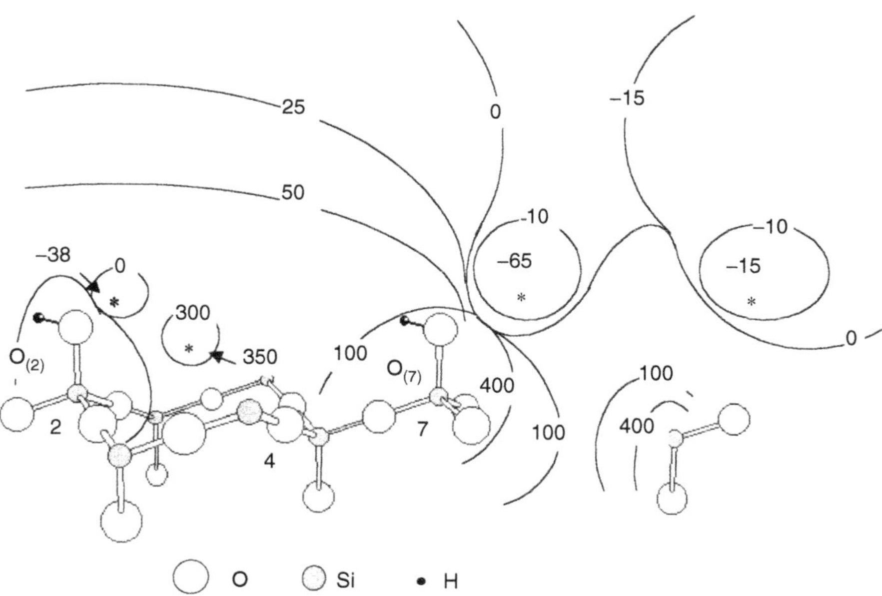

FIGURE 28.27 Profiles of the electrostatic potential (kJ/mol) within the plane I for subsurface silica layer simulated with cluster 1.

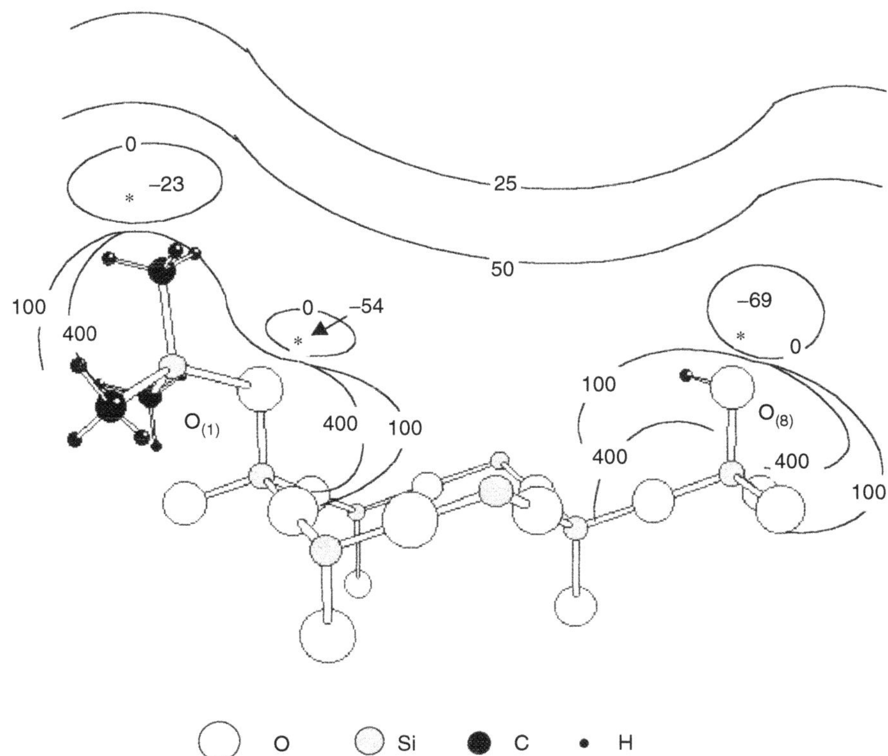

FIGURE 28.28 Profiles of the electrostatic potential (kJ/mol) within plane III for subsurface silica layer simulated with cluster 2.

are not so deep as for cluster 1. All this is a result of the influence of the TMS group grafted into position 10. The latter also influences on the potential distribution near hydroxyl groups. This influence is the most left near silanol in position 7 that is in the neighborhood of TMS group (Figure 28.29). It is expressed as a decrease in size of the region of negative potential values and as an appearance of the potential maximum localized not near atomic nuclei but at a distance of 3.5 Å from hydroxylated silicon atom, whereas continuous regions of negative potential values with a gradual rise of the absolute value while drawing near the minimum point are characteristic of the surfaces simulated with clusters 0 and 1. The presence of compact, relatively small regions of negative ρ values with large gradient of potential is typical of the surface represented by cluster 2. The same tendency is also followed up for the clusters with $n = 4-8$.

The cluster with nine TMS groups (silanol groups remain only in sites 4, 7, and 8) simulated for the surface section with three hydroxyls surrounding with TMS groups has a small region of negative potential near atom O(7) and no such region near atom O(2) (Figure 28.23). A region of small negative potentials is formed near carbon atom of trimethylsilylated group what is a characteristic of completely trimethylsilylated silica surface. Such a distribution of ρ testifies a weak manifestation of the properties typical of silanols surrounded with TMS groups.

A total hydrophobization of SiO_2 surface sharply decreases its capability to bind polar molecules independently on their stereochemistry and electron donor properties. At the same time, it is follows from the analysis of the results on investigation of the regularities of formation and destruction of adsorption structures of small polar compounds (e.g., H_2O, CH_3OH) [72] within surface layer of pyrogenic silica, where practically all the terminal hydroxyls are substituted with trimethylsilyl groups (TMS-SiO_2) that a definite amount of adsorbate is strongly kept by the surface. It is necessary to note that a hydrophobic trimethylsilylic cover hinders from strong adsorption of small polar molecules only at the initial stages of the process and its dynamics are defined mostly by relative vapor pressure rather than by duration of the contact adsorbent-adsorbate [72,145]. Water and methanol are eliminated from the surfaces of both hydroxylated and modified silicas practically within the same temperature interval [84].

The analysis of the peculiarities of adsorption of the compounds mentioned on TMS-SiO_2 surface typical of activation process gave a reason to believe that the adsorption structures within ultramicrocavities of surface layer were responsible for strong binding adsorbate molecules localized in the coordination spheres of silicon atoms bearing functional groups (as in the case of hydroxylated surface) [72]. This may be attributed to the incapability

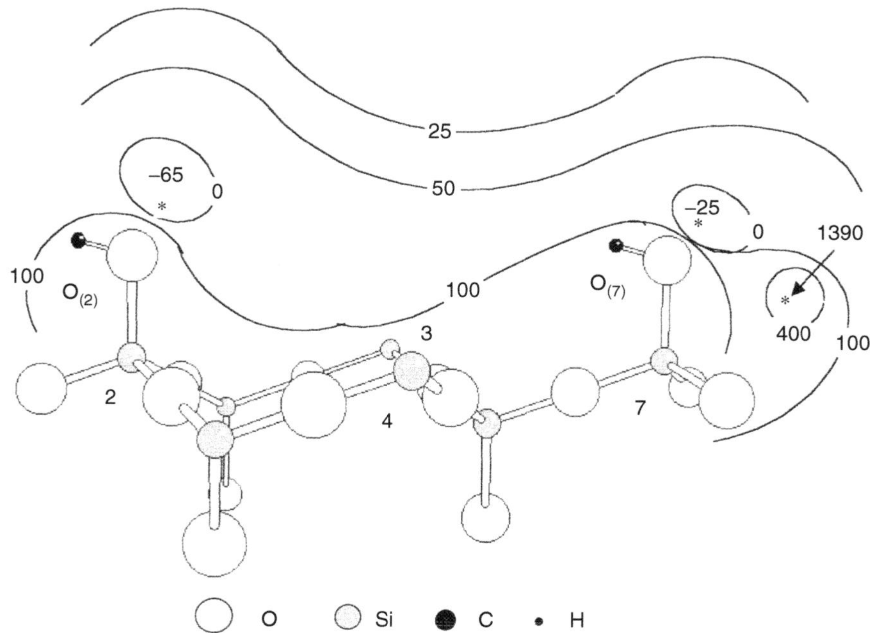

FIGURE 28.29 Profiles of the electrostatic potential (kJ/mol) within plane I for subsurface silica layer simulated with cluster 2.

of rather dense trimethylsilylic cover to block the matrix from the penetration of small molecules of polar compounds into structural microcavities. Collective effects within the grafted layer were considered as the main causes [146–148] probably resulting in the appearance of cavity-like spots within a dense cover [143] providing its penetrability; nevertheless, this possibility does not concretize the nature of the sites of strong adsorption of electron donor molecules. Let us note that, despite the notions on coordination of small electron donor molecules within the microcavities of SiO_2 surface layer are rather substantiated [72], it is still incomprehensible why the

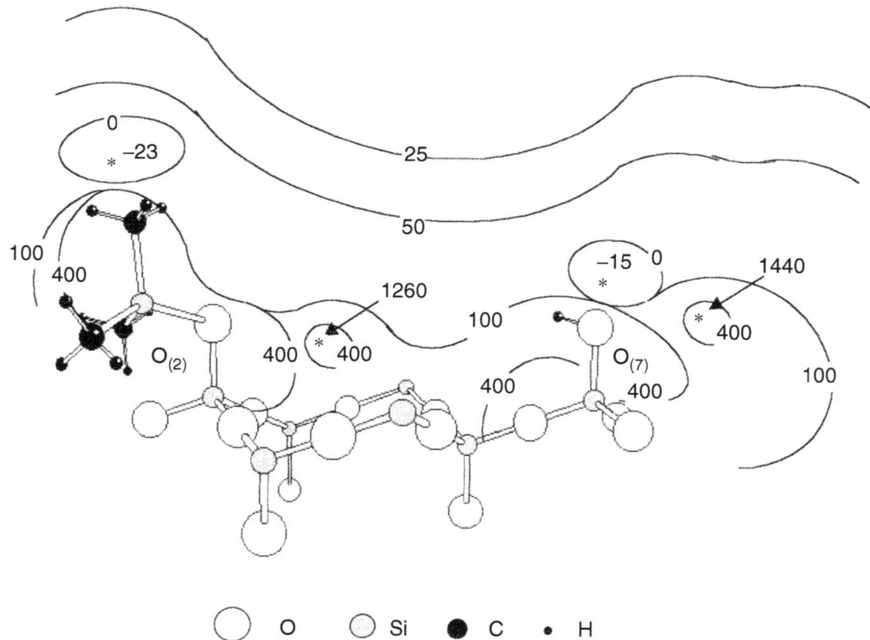

FIGURE 28.30 Profiles of the electrostatic potential (kJ/mol) within plane I for subsurface silica layer simulated with cluster with nine TMS groups.

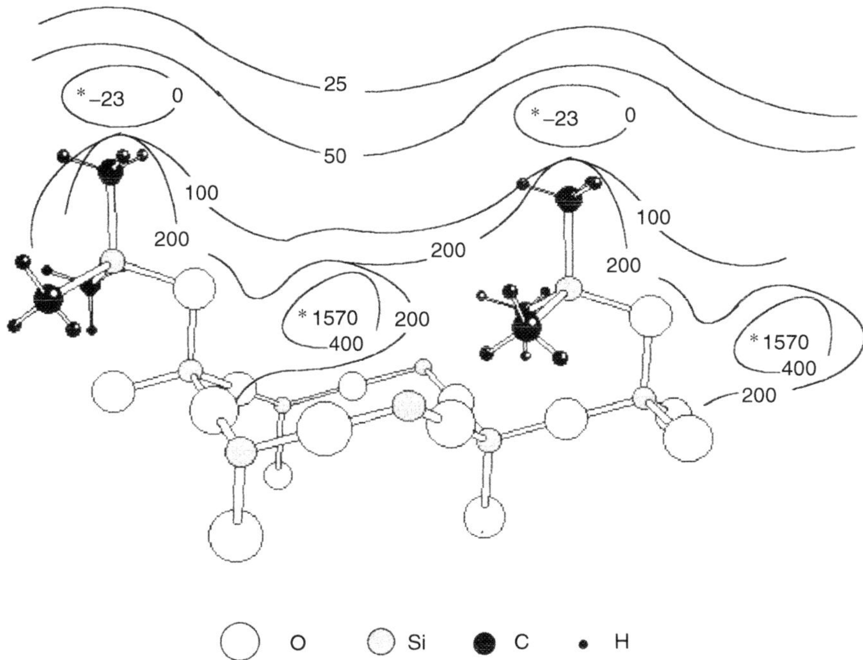

FIGURE 28.31 A distribution of the electrostatic potential (kJ/mol) within the neighborhood of completely trimethylsilylated silica surface.

stability of adsorption complexes is practically the same for such compounds on the surface of both hydroxylated silica and TMS-SiO_2, the certainly layer negativity of SiOH group as compared with that of $SiOSi(CH_3)_3$ one being taken into account. As the functional cover and naturally electrostatic potential are changes on the transition from hydroxylated SiO_2 to TMS-SiO_2, it seems to be purposeful to compare the peculiarities of space distribution of ρ near the surface of initial unmodified silica (Figure 28.26) and those of hydrophobized TMS-SiO_2 (Figure 28.31). A presence is seen from this picture of local compact areas near the entrance into structural microcavities with abnormally high positive potential values at rather large distances from silicon atoms what excludes a coordination mechanism of binding electron donor molecules (or anions) within the surface layer of TMS-SiO_2. However, in this case a high adsorption potential of electron acceptor nature is also formed within microcavities.

As any polar molecule contains fragments with alternating charge density, its movement within the regions with increasing values of positive potential occurs under conditions of increasing disbalance between attracting atoms with enlarged values of electron density and repulsing atoms with decreased electron density, adsorption of small electron donor molecules is characterized by reduced kinetics in perfect argument with experimental [72]. At the same time near potential maxima the rate of its increase is large and a localization of the molecules with heterogeneous electron distribution with extreme areas becomes energetically profitable, and thus explains the experimentally discovered considerable stability of adsorption water and methanol complexes on the TMS-SiO_2 surface [72,84].

The comparison of the peculiarities of spatial ρ distribution within the surface layer of both hydroxylated and trimethylsilylated silica surface (Figure 28.26 and Figure 28.31) allows us to explain naturally the close stability of the strong adsorption complexes of small electron donor molecules despite their different structure. Though in the case of hydroxylated surface there is no site with such high positive values, this is compensated by the bipolarity of distribution displayed in the existence of areas with rather large potential values of opposite sign at the distances comparable with lengths of chemical bonds that promotes a stabilization of adsorption complexes.

Thus, the permeability of dense hydrophobicity over the SiO_2 surface with grafted trimethylsilyl groups concerning small polar molecules can be explained on the base of the analysis of the peculiarities of space distribution of the electrostatic potential near surface and without calling notions on collective effects in the ensembles of surface functional groups. It can be believed that the strong adsorption complexes formed within surface layer at TMS-SiO_2 have an electrostatic nature and are stabilized in the neighborhood of structural microcavities in positions with high positive potential values. The grafted layer of partially trimethylsilylated silica surface can be assigned to a bimodal phase including simultaneously functional groups of two types: hydrophobic (TMS) and hydrophilic (silanes) ones. At all the stages of TMS group (from n = 1 to up n = 12) grafting on the surface there are always compact, relatively small areas of negative ρ values controlling its hydrophobic properties. They do not vanish even at large concentration of TMS groups. Moreover, when n increases, areas of negative potential values (-26 kJ/mol) are formed near the carbon atoms of TMS groups. This gives a reason to believe that even completely trimethylsilylated silica surface can be thought to be conventionally hydrophobic. It should be also noticed that at initial stages of trimethylsilylation of silica surface its hydrophobic properties even increase what is testified by the presence of more deep minimum of negative potential value as compared with those of completely hydroxylated silica.

REFERENCES

1. Vinogradov, A.P. *Regularities of Chemical Element Distribution within Terrestrial Crust*; Moscow State University: Moscow, 1955. (In Russian).
2. Vernadsky, V.I. *Sketches on Geochemistry*; Nauka: Moscow, 1983. (In Russian).
3. Iler, R.K. *Colloid Chemistry of Silica and Silicates*; Gosstroyizdat: Moscow, 1959. (In Russian).
4. Voronkov, M.G.; Zelgan, G.I.; and Lukevits, E.Ya. *Silicon and Life*; Zinatne: Riga, 1978. (In Russian).
5. Vernadsky, V.I. *Biosphere*; Mysl: Moscow, 1967. (In Russian).
6. Vernadsky, V.I. *Chemical Structure of Biosphere of The Earth and Its Environment*; Nauka: Moscow, 1965.
7. Scheresmann, E.H. *Chem. Rundschau* **1964**, *17*, 281.
8. Wiessner, J.H.; Mandel, N.S.; Sohnle, P.G.; and Mandel, G.S. *Exp. Lung Res.* **1989**, *15*, 801.
9. Vallyathan, V.; Mega, J.F.; Shi, X.; Dalal, N.S. *Am. J. Respir. Cell. Molec. Biol.* **1992**, 6, 404.
10. Oparin, A.I. *Origin and Initial Development of Life*; Medicina: Moscow, 1966. (In Russian).
11. Oparin, A.I. *Origin of Life on The Earth*; AN USSR: Moscow, 1957. (In Russian).
12. Bernal, J. *Origin of Life*; Mir: Moscow, 1969. (In Russian).
13. Vysotsky, Z.Z.; Danilov, V.I.; Strelko, V.V. *Uspekhi Sovremennoy Biologii.* **1967**, *63*, 362. (In Russian).
14. Iler, R.K. *The Chemistry of Silica*; Wiley: New York, 1979.
15. Neterenko, B.A.; Snitko, O.V.; *Physical Properties of Atomically Pure Semiconductor Surfaces*; Naukova Dumka: Kiev, 1983. (In Russian).
16. Harrison, W.A. *Electronic Structure and Properties of Solids: The Physics of The Chemical Bond*; W.H. Freeman and Company: San Francisco, 1980.
17. Harrison, W.A. *Surf. Sci.* **1976**, *31*, 165.
18. Gerasimov, Ya.M. *A Course of Physical Chemistry*; Khimiya: Moscow, 1973; Vol 2. (in Russian).
19. Pauling, L. *The Nature of Chemical Bond*; Goskhimizdat: Moscow, 1947. (In Russian).
20. Belashchenko, D.K. *Fiz. Met. & Metalloved* **1985**, *60*, 1076. (In Russian).
21. Nakano, A.; Kalia, R.N.; Vashishta, P. *J. Non-Cryst. Solids.* **1994**, *171*, 157.
22. Zhuravlev, L.T. *Langmuir* **1987**, *3*, 316.
23. Zhuravlev, L.T. *Colloids. Surf. A.* **1993**, *74*, 71.
24. Peri, J.B.; Hensley, A.L. *J. Phys. Chem.* **1968**, *72*, 2926.
25. De Boer, J.H.; Vleeskers, J.M. *Proc. Koninnl. Ned. Acad. Wetenschap., Ser. B.* **1957**, *60*, 54.
26. Shapiro, I.; Weiss H.G. *J. Phys. Chem.* **1953**, *57*, 219.
27. Naccache, C., Imelik, B. *C.R. Acad. Sci.* **1960**, *250*, 2019.
28. Naccache, C.; Imelik, B. *Bull. Soc. Chim. France* **1961**, 553.
29. Mathieu, M.V.; Imelik, B. *J. Chim. Phys. et Phys. Chim. Biol.* **1962**, *59*, 1189.
30. Tongelen, M.U.; Uytterhaeven, J.; Fripiat, J.J. *Bull. Soc. Chim. France.* **1965**, 2318.
31. Deuel, H.; Wartmann, J.; Hutschenker, K. et al. *Helv. Chem. Acta.* **1959**, *42*, 1160.
32. Tchertov, V.M.; Dzhambayaev, D.B.; Platchinda, A.S.; Neimark, I.Ye. *Dokl. AN SSSR.* **1965**, *161*, 1149. (in Russian).
33. Lisitchkin, G.B., ed. *Modified Silicas in Sorption, Catalysis, and Chromatography*; Khimiya: Moscow, 1986. (In Russian).
34. Zaitsev, V.N. *Complex Forming Silicas. Synthesis, Structure of Grafted Layer, and Surface Chemistry*; Folio: Kharkov, 1997. (In Russian).
35. Zhuravlev, L.T. In *An International Conference on Silica Science and Technology, From S (Synthesis) to A (Applications), Mulhouse, France* **1998**, “Silica 98” 293.
36. Bliznakov, T.M.; Bakyrdzhiev, I.V.; Mazhdraganova, M.B. In *Adsorbents, Their Preparation, Properties, and Application*, Nauka: Lenigrad, 1971. p. 109. (In Russian).
37. Blisnakov, T.M.; Bakyrdchyev; I.V.; Machraganova, M.B. *J. Catal.* **1969**, *19*, 135.
38. Unger, K.; Gallei, K. *Kolloid. Z. und Z. Polym.* **1970**, 3584.
39. Tertykh, V.A.; Pavlov, V.V.; Mashchenko, V.M.; Chuiko, A.A. *Dokl. AN USSR.* **1971**, *201*, 913.
40. Chuiko, A.A. *Surface Chemistry of SiO_2, Nature and Role of Active Sites in Adsorption and Chemisorption Processes*: Thesis Diss. Doct. Chem. Kiev, 1971. (In Russian).
41. Van Der Voort, P.; Possemiers, K.; Vansant, E.F. *J. Chem. Soc., Faraday Frans.* **1996**, *92*, 834.
42. Hank, W. *Z. anorg. und allg. Chem.* **1973**, *395*, 191.
43. Sato, M.; Kabayashi, T.; Shima, J. *J. Catal.* **1967**, *7*, 342.
44. Fripiat, J.J.; Uytterhoeven, J. *J. Phys. Chem.* **1962**, *66*, 800.
45. Hambleton, F.H.; Hockey, J. *Trans. Faraday Soc.* **1966**, *62*, 1694.

46. Kohlscütter, H.; Bogel, U. *Verhantungshen. Kolloid. Ges.* **1971**, *24*, 29.
47. Armistead, C.G.; Tyler, A.I.; Hambleton, F.H. *J. Phys. Chem.* **1966**, *70*, 3947.
48. Sobolev, V.A. *Quantitative Study on High Disperse Silica-Aerosil Surfaces by Method of IR-Sepctroscopy:* Thesis Diss. Candidate. Chem. Kiev, 1971. (In Russian).
49. Kiselev, A.V.; Lygin, V.I. *Infrared Spectra of Surface Compounds and Adsorbed Substances*; Nauka: Moscow, 1972. (In Russian).
50. Little, L.H. *Infrared Spectra of Adsorbed Species*; Academic Press: London, New York, 1966.
51. Laskorin, B.N.; Strelko, V.V.; Strazhesko, D.N.; Denisov, V.I. *Silica Based Sorbents in Radiochemistry*; Atomizdat: Moscow, 1977. (In Russian).
52. Galkin, G.A.; Kiselev, A.V.; Lygin, V.I. *Zhurn. Fiz. Khimii.* **1967**, *41*, 40. (In Russian).
53. Varsant, E.F.; Vandervoort, K.C. *Characterization and Chemical Modification of the Silica Surface*; Elsevier: Amsterdam, 1995.
54. Maciel, G.E.; Sindorf, D.W. *J. Phys.* **1983**, *87*, 5516.
55. Zhuravlev, L.T.; Kiselev, A.V. Kolloidn. Zh. **1962**, 22. (In Russian).
56. Zhuravlev, L.T. In *Ground Problems of Theory of Physical Adsorption*/Ed. M.M. Dubinin: Nauka: Moscow, 1970. P. 309. (In Russian).
57. Sidorov, A.N. *Dokl. AN USSR.* **1954**, *95*, 1235. (In Russian).
58. McDonald, R.S. *J. Phys. Chem.* **1958**, *62*, 168.
59. Kiselev, A.V.; Lygin, V.I. Kolloidn. Zh. **1959**, *21*, 581.
60. Peri, J.B. *J. Phys. Chem.* **1966**, *70*, 2937.
61. Sauer, J.; Schroder, K.-P. *Z. Phys. Chem. (Leipzig).* **1985**, 379.
62. Morrow, B.A.; Gay, I.D. *J. Phys. Chem.* **1988**, *9*, 5569.
63. Fyfe, C.A.; Gobbi, G.C.; Kennedy, G.J. *J. Phys. Chem.* **1985**, *89*, 277.
64. Lippmaa, E.T.; Samsonov, A.V.; Brey, V.V.; Gorlov, Yu.I. *Dokl. AN USSR.* **1981**, *259*, 403.
65. Leonardelli, S.; Facchini, L.; Fretigny, C. *et al. J. Amer. Chem. Soc.* **1992**, *114*, 6412.
66. Daniels, M.W.; Sefcik, J.; Francis, L.F.; Mc Cormick, A.V. *Coll. Interf. Science.* **1999**, *219*, 351.
67. Schenk, U.; Hunger, M.; Weitkamp, *J. Magnet. Reson. Chem.* **1999**, *37*, 75.
68. Landay, M.V.; Varvey, S.P.; Herskowtiz, M.; Regev, O.; Pevzner, S.; Set. T.; Luz, Z. *Micropor. Mesopor. Mat.* **1999**, *31*, 149.
69. Sobolev, V.A.; Tertykh, V.A.; Bobryshev, A.I.; Chuiko, A.A. *Zhurn. Prikladn. Spektroskop.* **1970**, *13*, 863.
70. Tertykh, V.A.; Mashchenko, V.M.; Chuiko, A.A. *Dokl. AN USSR.* **1971**, *5*, 863. (In Russian).
71. Tertykh, V.A.; Mashchenko, V.M.; Chuiko, A.A.; Pavlov, V.V. *Fiz.-Khim. Mekhan. & Liofil. Disp. Sist.* **1973**, *4*, 37. (In Russian).
72. Chuiko, A.A.; Gorlov, Yu.I. *Silica Surface Chemistry: Surface Structure, Active Sites, Sorption Mechanisms*; Nauk. Dumka: 1992. (In Russian).
73. Curthoys, G.; Davydov, V.Ya.; Kiselev, A.V.; *J. Colloid. & Interface Sci.* **1974**, *48*, 58.
74. Tertykh, V.A.; Pavlov, V.V.; Tkatchenko, K.I.; Chuiko, A.A. *Teor. & Eksperim. Khimiya.* 1975. *11*, 415. (In Russian).
75. Sobolev, V.A.; Chuiko, A.A.; Tertykh, V.A.; Mashchenko, V.M. In *Bound Water in Disperse Systems*; Moscow State University: Moscow, 1974; Issue 3, P. 621. (In Russian).
76. Hair, M.L.; Hertl, W. *J. Phys. Chem.* **1969**, *73*, 2372.
77. Evans, B.; White, T.E. *J. Catal.* **1968**, *11*, 336.
78. Hair, M.L.; Hertl, W. *Ibid.* **1969**, *15*, 307.
79. Kiselev, V.I.; Lygin, V. I.; Shchepalin, K.L. *Zhurn. Fiz. Khimii.* **1985**, *59*, 1521. (In Russian).
80. Brey, V.V.; Gorlov, Yu.I.; Korol, E.N. *Teor. & Eksperim. Khimiya.* **1982**, *18*, 122. (In Russian).
81. Tertykh, V.A.; Belyakova, L.A. *Chemical Reactions with Participation of Silica Surface*; Naukova Dumka: Kiev, 1991. (In Russian).
82. Tertykh, V.A.; Varvarin, A.M.; Simurov, A.V.; Belyakova, L.A. *Teor. & Eksperim. Khimiya.* **1989**, *25*, 374. (In Russian).
83. Chuiko, A.A.; Sobolev, V.A.; Tertykh, V.A. *Ukr. Khim. Zhurn.* **1972**, *38*, 774.
84. Chuiko, A.A.; Ogenko, V.M.; Tertykh, V.A.; Sobolev, V.A. *Adsorption and Adsorbents* **1975**, 69.
85. Lygin, V.I. *Zhurn. Fiz. Khim.* **1989**, *63*, 289. (In Russian).
86. Ignatyeva, L.A.; Kiselev, V.F.; Tchukin, G.D. *Dokl. AN USSR.* **1968**, *181*, 914. (In Russian).
87. Kvlividze, V.I.; Iyevskaya, N.M.; Egorova, T.S. *et al. Kinetika & Kataliz* **1962**, *3*, 91. (In Russian).
88. Zarifyants, Yu.A.; Kiselev, V.F.; Khrustaleva, S.V. In *Bound Water in Disperse Systems*; Moscow State University: Moscow, 1974. Issue 3. P.74. (In Russian).
89. Kiselev, V.F.; Krylov, O.V. Adsorption Processes on Surfaces of Semiconductors and Dielectrics; Nauka: Moscow, 1978. (In Russian).
90. Davydov, V. Ya; Kiselev, A.V. *Zhurn. Fiz. Khimii.* **1963**, *37*, 2593. (In Russian).
91. Hambleton, F.H.; Hockey, J.A.; Taylor, J.A. *J. Trans. Faraday Soc.* **1966**, *62*, 801.
92. Tyler, A.J.; Hambleton, F.H.; Hockey, J.A. *J. Catal.* **1969**, *13*, 35.
93. Strelko, V.V. *Adsorption & Adsorbents* **1974**, 65 (In Russian).
94. Doremus, R.H. *J. Phys. Chem.* **1971**, *75*, 3147.
95. Strelko, V.V.; Kartel, N.T.; Burushkina, T.N. *Dokl. AN USSR.* **1974**, *216*, 360. (In Russian).
96. Strelko, V.V.; Burushkina, T.N.; Kartel, N.T. *Adsorption & Adsorbents* **1976**, 38. (In Russian).
97. Sidorov, A.N. *Zhurn. Fiz. Khimii.* **1956**, *30*, 995.
98. Nikitin, A.N.; Sidorov, A.N.; Karyakin, A.V. *Zhurn. Fiz. Khimii.* **1956**, *30*, 117. (In Russian).
99. Sidorov, A.N. *Opt. & Spektroscop.* **1960**, *8*, 806. (In Russian).
100. Chapman, I.D.; Hair, M.L. *Trans. Faraday Soc.* **1965**, *61*, 1507.
101. Kvilividze, V.I. *Dokl. AN USSR.* **1964**, *157*, 158. (In Russian).

102. Karyakin, A.V.; Muradova, G.A.; Maisuradze, G.V. *Zhurn. Prikladn. Spectroskop.* **1970**, *12* 903. (In Russian).
103. Egorova, T.S.; Kiselev, V.F.; Krasilnikov, G.K. *Dokl. AN USSR.* **1958**, *123*, 1060. (In Russian).
104. Egorova, T.S.; Zarifyants, Yu. A.; Kiselev, V.F.; Krasilnikov, G.K.; Murina, V.V. *Zhurn. Fiz. Khimii.* **1962**, *36*, 1458. (In Russian).
105. Kiselev, A.V.; Yashin, Ya. I. *Gas Adsorption Chromatography*; Nauka: Moscow, 1967. (In Russian).
106. Tertykh, V.A.; Chuiko, A.A.; Sobolev, V.A.; Bobryshev, A.I. *Ukr. Khim. Zhurn.* **1971**, *37*, 1242. (In Russian).
107. Tertykh, V.A.; Chuiko, A.A.; Mashchenko, V.M.; Pavlov, V.V. *Zhurn. Fiz. Khimii.* **1973**, *47*, 158. (In Russian).
108. Chuiko, A.A.; Mashchenkov, V.M.; Khaber, N.B. *et al. Fiz.-Khim. Mekhan. & Liofil. Disp. Sist.* **1973**, 43. (In Russian).
109. Tertykh, V.A.; Pavlov, V.V.; Mashchenko, V.M.; Chuiko, A.A. *Dokl. AN USSR.* **1971**, *201*, 913. (In Russian).
110. Tchukin, G.D.; Ignatyeva, L.A. *Zhurn. Prikladn. Spektroskop.* **1968**, *8*, 872. (In Russian).
111. Tertykh, V.A.; Chuiko, A.A.; Pavlov, V.V.; Ogenko, V.V. *Dokl. AN USSR.* **1972**, *206*, 893. (In Russian).
112. Tertykh, V.A.; Ogenko, V.M.; Voronin, E.F.; Chuiko, A.A. *Zhurn. Prikladn. Spektriskop.* **1975**, *23*, 464. (In Russian).
113. Voronin, E.F.; Tertykh, V.A.; Ogenko, V.M.; Chuiko, A.A. *Teor. & Eksperim. Khimiya.* **1978**, *14*, 638. (In Russian).
114. Gavrilyuk, K.V.; Gorlov, Yu.I.; Konoplya, M.M. *et al. Ibid.* **1979**, *15*, 212. (In Russian).
115. Gavrilyuk, K.V.; Gorlov, Yu.I.; Nazarenko, V.A. *et al. Ibid.* **1983**, *19*, 364. (In Russian).
116. Gorlov, Yu.I.; Golovaty, V.G.; Konoplya, M.M.; Chuiko, A.A. *Ibid.* **1980**, *16*, 202.
117. Gorlov, Yu.I.; Golovaty, V.G.; Korol, E.N. *et al. Ibid.* **1982**, *18*, 90. (In Russian).
118. Hertl, W.; Hair, M.L. *Nature* **1969**, *223*, 1150.
119. Nazarenko, V.A.; Furman, V.I.; Guzikevich, A.G.; Gorlov, Yu.I. *Teor. & Ekserim. Khimiya.* **1985**, *21*, 66. (In Russian).
120. Low, M.J.D.; Ramasubramanian, N. *J. Phys. Chem.* **1967**, *21*, 730.
121. Chuiko, A.A.; Gorlov, Yu.I. *Visnyk AN Ukr. SSR.* **1982**, 39. (In Ukrainian).
122. Gorlov, Yu.I. *Zh. Fiz. Khimii.* **1985**, *59*, 1213. (In Russian).
123. Gorlov, Yu.I.; Konoplya, M.M.; Chuiko, A.A. *Teor & Eksperim. Khimiya.* **1980**, *16*, 333. (In Russian).
124. Ogenko, V.M.; Ivashechkin, V.G.; Chuiko, A.A.; Mironyuk, I.F. *Ibid.* **1985**, *21*, 745. (In Russian).
125. Bechstedt, F.; Enderlein, R. *Semiconductor Surfaces and Interfaces. Their Atomic and Electronic Structures*; Academie-Verlag: Berlin, 1988.
126. Politzer, P. In *Homoatomic Rings, Chains, and Macromolecules of Maingroup Elements*; Rheingold, A.L., Ed; Elsevier: Amsterdam, 1977. p. 95.
127. Lobanov, V.V.; Gorlov, Yu.I. *Khimiya, Fizika i Tekhnologiya Poverkhnosti [Russian Journal of Surface Chemistry, Physics, and Technology].* **1992**, *1*, 3. (In Russian).
128. Lobanov, V.V.; Gorlov, Yu.I. *Zhurn. Fiz. Khimii.* **1995**, *69*, 407. (In Russian).
129. Lobanov, V.V.; Gorlov, Yu.I. *Zhurn. Fiz. Khimii.* **1995**, *69*, 652. (In Russian).
130. Lobanov, V.V. The Sorption of Cations on Hydrated SiO_2 Surface and the Character of Its Electrostatic Potential; Kyiv, 1997. (Preprint of NTUU "KPI"). (In Russian).
131. Strelko, V.V.; Strazhesko, D.N.; Soloshenko, N.I.; Rubanic, S.K.; Beran, A.A. *Dokl. AN USSR.* **1969**, *86*, 1362. (In Russian).
132. Lobanov, V.V.; Gorlov, Yu.I.; Chuiko, A.A.; Pinchuk, V.M.; Yu.I. Sinekop, Yu.I.; Yakimenko, Yu.I. *The Role of Electrostatic Interactions in Adsorption on Solid Oxide Surfaces* Vec+ Ldt.:Kyiv, 1999. (In Russian).
133. Lobanov, V.V.; Terez, M.I.; Gorlov, Yu.I. *Structure of Alumina-Silica Surface and Properties of Its Active Sites*; Kyiv, 1997. (Preprint NTUU "KPI"). (In Russian).
134. Salahub, D.R.; Messmer, R.P. *Phys. Rev. B.* **1977**, *16*, 2526.
135. Lobanov, V.V.; Aleksankin, M.M. *Teor. & Eksperim. Khimia.* **1980**, *6*, 477. (In Russian).
136. Korol, E.N.; Lobanov, V.V.; Nazarenko, V.A.; Pokrovsky, V.A. *The Physical Foundations of the Field Mass-Spectrometry*; Korol, E.N., Ed.; Naukova Dumka: Kiev, 1978. (In Russian).
137. Mironyuk, I.F.; Lobanov, V.V.; Ogenko, V.M. *Teor. & Eksperim. Khimiya.* **2000**, *36*, 287. (In Russian).
138. Stewart, J.J.P. *J. Comput.-Aided Mol. Design.* **1990**, *4*, 1.
139. Liebau, F. *Structural Chemistry of Silicates*; Springer-Verlag: Berlin, Heidelberg, 1985.
140. Balard, H.; Papirer, E.; Khalfi, A.; Barthel, H.; Weis, J. In *An International Conference on Silica Science and Technology, From S (Synthesis) to A (Applications), Mulhouse, France (1–4 Sept. 1998) "Silica 98"*. p. 403.
141. Kudryavtsev, G.V.; Staroverov, S.M. *Zhurn. Vsesoyuzn. Khim. Obshchestva.* **1989**, *34*, 308. (In Russian).
142. Becker, O.M.; Ben-Shaul, A. *Phys. Rev. Lett.* **1988**, *61*, 2869.
143. Wirth, M.J.; Fatunbi, H.O. *Anal. Chem.* **1993**, *65*, 822.
144. Gerasimowicz, W.V.; Garrovay, A.N.; Miller, J.B. *J. Phys. Chem.* **1992**, *96*, 3658.
145. Fadeev, A.Y.; Eroshenko, V.A. In *An International Conference on Silica Science and Technology, From S (Synthesis) to A (Applications), Mulhouse, France (1–4 Sept. 1998) "Silica 98"*. P. 99.
146. Foti, G.; Kovats, E.S. *Langmuir.* **1989**, *5*, 232.
147. Golubev, V.B.; Matveyev, V.V.; Startoverov, S.M.; Lisitchkin, G.V. *Teor. & Eksperim. Khimiya.* **1988**, *24*, 701. (In Russian).
148. Foti, G.; Martinez, C.; Kovats, E. *J. Chromatograph.* **1989**, *461*, 233.

29 Variable-Temperature Diffuse Reflectance Fourier Transform Infrared Spectroscopic Studies of Amine Desorption from a Siliceous Surface

Donald E. Leyden[1] *and Kristina G. Proctor*[2]
Colorado State University, Department of Chemistry, Condensed Matter Sciences Laboratory

CONTENTS

Variable-temperature diffuse reflectance infrared Fourier transform spectroscopy was used in conjunction with pyridine desorption studies to assess the acidity of a siliceous surface. An amorphous, porous silica substrate was investigated. The results contribute to an understanding of the acidic strength and the distribution of acidic sites on this material. A hydrogen-bonding interaction was observed between pyridine and the surface. Isothermal rate constants and an activation energy for the desorption process are reported and can be used as direct measures of surface site acidity.

Siliceous substrates are used extensively in industry and applied research. These materials can undergo a variety of chemical modifications that make them useful for applications such as catalysis and chromatography. An increased understanding of the nature of surface reactions and reaction products would facilitate current and future applications of these substrates. Acquisition of such information requires innovative surface characterization methods.

One characteristic of interest is the surface propensity toward adsorption, hydrogen bonding, and acid–base interactions with chemical reagents. For example, surface silylation by alkoxysilanes may involve hydrolysis of the alkoxy groups by surface-adsorbed water. The overall rate of reaction (from dry, aprotic solvents such as toluene) may be limited by the rate of hydrolysis, which is likely dependent on the fraction of alkoxysilane associated with the substrate surface. A convenient measure of the distribution of a solute between the solvent and substrate is the chromatographic capacity factor (k') [1]. Figure 29.1 shows a plot of the relative rate of hydrolysis for several alkoxysilanes (*n*-octyltriethoxysilane, 3-mercaptopropyltriethoxysilane, 3-cyanopropyltriethoxysilane, and 3-aminopropyltriethoxysilane) on the surface of controlled-pore glass versus the amount of silane adsorbed. The hydrolysis rate is followed by measuring the amount of ethanol produced in toluene solution. The amount of adsorbed silane is represented by the quantity $S[k'/(1 + k')]$, where S is the total amount of alkoxysilane in solution [2]. The rate of hydrolysis of the silane alkoxy

[1]Current address: Retired.
[2]Current address: Department of Chemistry, University of Southern Colorado, Pueblo, CO 81001.

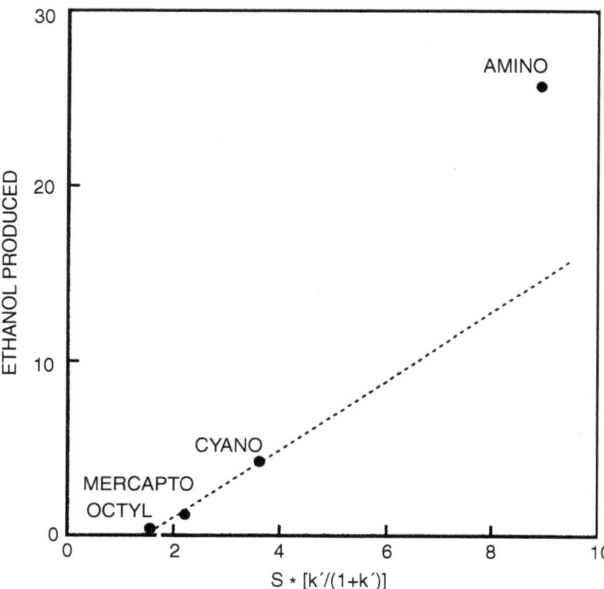

FIGURE 29.1 Relative amount of ethanol produced versus amount of silane adsorbed on the substrate, calculated from the chromatographic capacity factor *k*′.

groups is linearly related to the fraction of the reagent adsorbed to the substrate for three of the compounds. However, in 3-aminopropyltriethoxysilane the rate of reaction is much faster than that predicted by the simple adsorption model. Knowledge of such differences is of considerable importance to a better understanding of the nature of surface substrate reactions.

Another area of interest is the acidity of siliceous substrates. Current methods of surface characterization provide a variety of information about the acidity of a surface. Visible indicators covering a range of pK_a values may be used to estimate the acidity as defined by the Hammett acidity function [3,4]. This method is vague in interpretation and can only provide a measure of relative acidic strength. Other methods involving the adsorption and desorption of gaseous bases can also assess relative acidic strengths. However, assignments to specific surface sites are subject to ambiguity of interpretation. These methods include differential thermal analysis (DTA) [5], thermal gravimetric analysis (TGA) [6], and titration calorimetry [7]. The catalytic efficiency of a material can also be used to assess a general measure of surface acidity [8,9].

Investigations of the acidity of specific surface sites may be accomplished by studies coordinated with spectroscopic methods, such as infrared (IR) spectroscopy, nuclear magnetic resonance (NMR) spectroscopy, or mass spectrometry (MS). Surface characterization with Fourier transform infrared (FTIR) spectroscopy can provide quantitative results with experimental methods that are easily performed. However, the transmission sampling techniques traditionally employed for infrared studies may introduce experimental artifacts on the analyzed surface [10,11]. To minimize this problem, self-supporting pellets are usually prepared, and the minimum pressure required to form a mechanically stable pellet is used. Although much information related to surface acidity has been obtained with transmission techniques, there remains some question regarding the extent of surface alteration due to sampling procedures.

Diffuse reflectance FTIR (DRIFT) spectroscopy provides an alternative to transmission infrared spectroscopy with respect to sampling procedures. DRIFT spectroscopy requires dispersion of the sample in a finely ground, nonabsorbing matrix such as KCl or KBr. The integrity of the sample surface is ensured because no pressure is used in the preparation. Variable-temperature DRIFT (VT-DRIFT) spectroscopy can be performed with commercially available, heatable–evacuable sample cells that can be interfaced for computer temperature control [12]. Gaseous base adsorption and desorption processes can be followed directly; thus, specific surface sites can be identified and quantified and their acidic strength can be assessed. Previously, such results could be obtained only by combined methods such as IR–TGA [13] or IR–TPD–MS (TPD, temperature-programmed desorption) [14]. Such combinations require variable-temperature experiments of independently prepared samples or elaborate instrumental design for measurements taken from the same sample. VT-DRIFT spectroscopy provides a direct and independent means for the characterization of acidic surfaces in their native form.

This chapter describes the results of the acidity characterization of a selected silica surface with VT-DRIFT spectroscopy. Examples of the capabilities of the method are demonstrated by the qualitative determination of the adsorption and thermal desorption characteristics of pyridine on amorphous, porous silica gel. Procedures for the determination of isothermal desorption rate constants and activation energy of desorption are presented and discussed as a means of assessing acid site strength.

EXPERIMENTAL DETAILS

MATERIALS

Amorphous silica gel was used (J.T. Baker), 290-m^2/g surface area, 60–200 mesh, and 126-Å mean pore diameter. Pyridine (Baker, reagent grade) was used as received.

INSTRUMENTATION

Spectra were acquired with a Nicolet 60SX FTIR spectrometer, continuously purged with dry air and equipped with a liquid-nitrogen-cooled, wideband mercury–cadmium telluride detector. Coaddition of 100 interferometer scans at 8-cm^{-1} resolution was employed. The location of absorption maxima was confirmed by spectra taken at 1-cm^{-1} resolution. All spectra were converted into Kubelka–Munk units prior to use. Integration of peak areas was accomplished by using software available on the Nicolet 60SX. All peak areas were normalized to the 1870-cm^{-1} Si–O–Si combination band [15].

The diffuse reflectance accessory (model DRA-2CN, Harrick Scientific) was modified with a three-dimensional translational stage to optimally position the sample for maximum radiation throughoutput [15]. The sample cell (model HVC-DRP, Harrick Scientific) was heated by a resistive heater contained within a post that housed a sample cup. The base of the sample cell also contained an external connector for evacuation and a second port, which was sealed with a septum and used for the introduction of pyridine by microsyringe. The sample cell cover contained a channel and connectors for water cooling and ZnSe windows (12 mm in diameter) for the IR radiation. The base and cover of the cell were sealed vacuum-right with an O-ring.

Temperature of the sample cell was monitored and controlled by an interface to an Apple II^{+} computer. The temperature of the sample cell was sensed by an internal Fe–constantan thermocouple connected to a digital thermometer (Omega Engineering, model 199AJC-D). The digital value of the temperature was available as a binary decimal and was read by an input–output card (John Bell Engineering, model 79-295) in an expansion slot of the computer. The cell temperature was controlled to 1°C by using software written in Applesoft and assembly languages and a signal to an exterior switch card that controlled the on–off status of a variable-voltage transformer.

Procedures

Prior to use, silica samples were calcined at 500°C for 5 h in an open-ended tube furnace and stored in a desiccator after rehydroxylation. Samples were dispersed (15% w/w) in finely ground KCl by mixing in a Wig-L-Bug capsule without the grinding ball (Crescent Dental Manufacturing). Dispersions containing 4–6 mg of sample were spread over a bed of KCl in the sample cup and flattened by light compression with a smooth object. Samples were heated to 200°C under slight vacuum (100–150 mmHg, or 13–20 kPa) to remove physisorbed water then cooled to room temperature for the adsorption of pyridine. Liquid pyridine was introduced into the evacuated cell containing the sample at 25°C by injection with a Hamilton microsyringe (0.5 μL). A volume of 3 μL resulted in saturation of the sample surface sites. Equilibrium occurred within 10 min under these conditions. The isothermal desorption of pyridine from silica gel was followed after adsorption of the base at room temperature. The sample cell was evacuated at room temperature to remove excess pyridine, and the temperature was then quickly ramped (30 s) and maintained at the desired temperature for the desorption. Spectra were recorded at intervals of 1, 5, 10, 20, and 30 min during the progression of 5-h desorption studies conducted at 50, 60, 70, 80, and 90°C.

The possibility of interaction of pyridine with the dispersion matrix material was investigated by collecting spectra of pure KCl and of pyridine adsorbed on pure KCl before and after evacuation. All traces of pyridine were removed from the KCl spectrum following a few minutes of evacuation.

RESULTS AND DISCUSSION

The surface of amorphous silica gel was qualitatively characterized by the adsorption of pyridine. Because pyridine is a weak base (pK_a 5.25), it selectively interacts with the more acidic surface sites. This interaction is relevant because metal oxide promoted catalysis is believed to occur by mechanisms involving acid-induced intermediate species (e.g., carbonium ions). Ring vibrations of pyridine give rise to infrared absorptions in the 1400–1700-cm^{-1} region. The absorption bands of pyridinium ion, covalently bonded pyridine, and hydrogen-bonded pyridine were assigned by Parry [16]. These species are formed upon interaction with Brönsted acid, Lewis acid, and hydrogen-bonding sites, respectively. The absorption bands of these interactions are distinguishable and can serve as a tool for the qualitative identification of surface acidic sites.

The VT-DRIFT spectrum of silica gel in the silanol absorption region before the adsorption of pyridine is shown in Figure 29.2. The absorption band at 3740 cm^{-1}

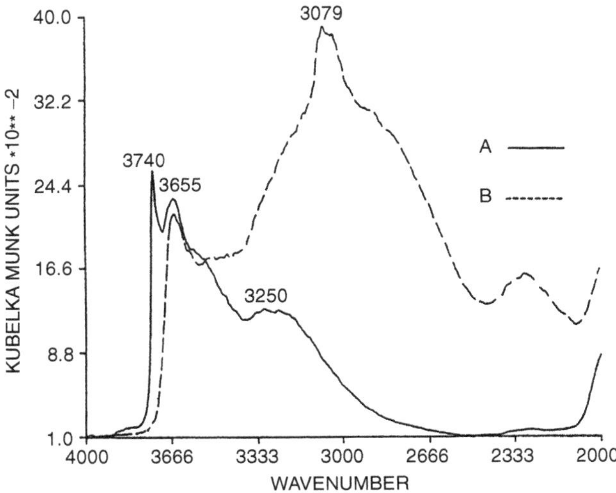

FIGURE 29.2 DRIFT spectrum of silica gel in the silanol absorption region at 25°C (A) after removal of physisorbed water and (B) after adsorption of pyridine.

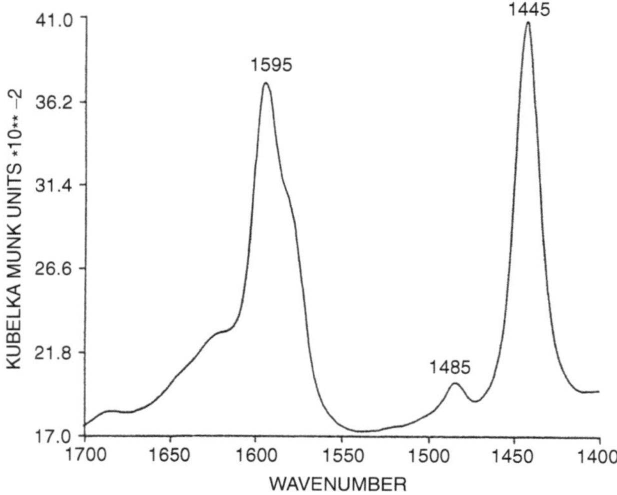

FIGURE 29.3 DRIFT spectrum of pyridine adsorbed on silica gel.

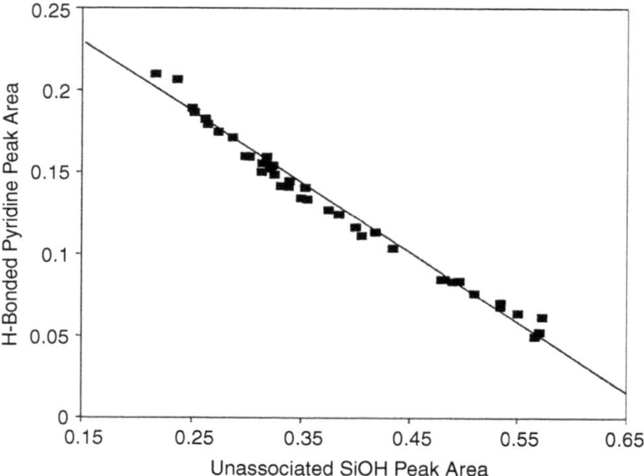

FIGURE 29.4 Appearance of unassociated silanol sites with desorption of hydrogen-bonded pyridine.

was assigned to silanols that do not hydrogen-bond to other silanols or to surface-adsorbed water [17]. These sites are referred to as unassociated silanol sites. The absorption at $3740\ cm^{-1}$ is slightly downshifted from that assigned to a freely vibrating silanol at $3750\ cm^{-1}$ [17]. The other absorption maxima observed at $3655\ cm^{-1}$ and $3450\ cm^{-1}$ arise from surface silanols that hydrogen-bond to each other and are hydrogen-bonded by surface-adsorbed water, respectively [17]. These two groups of silanols are collectively referred to as associated silanols.

The VT-DRIFT spectrum of pyridine adsorbed on unmodified silica gel is shown in Figure 29.3, Absorption maxima are observed at 1595, 1485, and $1445\ cm^{-1}$. According to the assignments made by Parry [16], these bands indicate hydrogen-bond formation. The absorptions result from in-plane C–C stretching modes 8a, 19a, and 19b, respectively [18]. The acidic strength of the surface sites is insufficient to generate pyridinium ion. This observation is consistent with previous studies of silica gel that used transmission infrared spectroscopic techniques [19].

The silanol absorption region changes as pyridine is desorbed from the silica gel surface. If pyridine is involved in reversible hydrogen-bonding with silanols, adsorbed pyridine should shift at least part of the unassociated silanol absorption to the associated silanol band. Conversely, desorption should shift the respective absorption band areas in the opposite manner. Figure 29.4 and Figure 29.5 show results obtained upon desorbing pyridine from the silica gel surface. In Figure 29.4, the peak area of the unassociated silanol band increases in a linear relationship with pyridine desorption, as measured by the peak area of hydrogen-bonded pyridine. Conversely, the peak area for the associated silanol band decreases linearly with pyridine desorption in Figure 29.5. These data provide strong evidence that the unassociated silanols are the preferred sites of pyridine adsorption because they have greater acidic strength than the associated silanols.

The nature of the interaction between pyridine and surface silanols was investigated further by using VT-DRIFT. Hydrogen-bonded pyridine was desorbed from the silica gel surface isothermally, and the process was followed by using the integrated absorption of the band at $1445\ cm^{-1}$. According to the definition of the Kubelka–Munk function, this peak area is directly proportional to the surface concentration of hydrogen-bonded pyridine. The data were fit to integrated rate law equations representing zero-, first-, and second-order kinetics. As shown

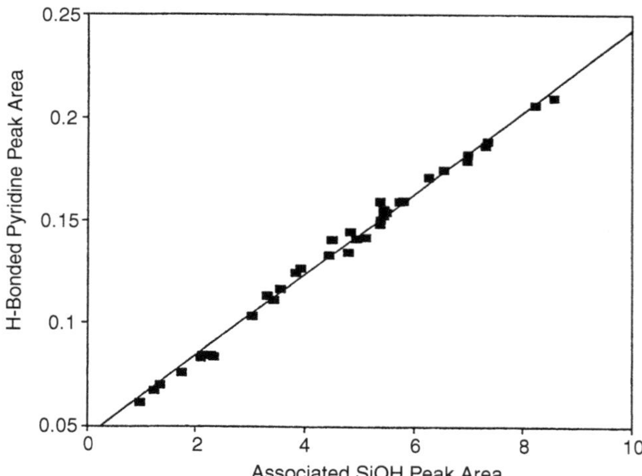

FIGURE 29.5 Disappearance of associated silanol sites with desorption of hydrogen-bonded pyridine.

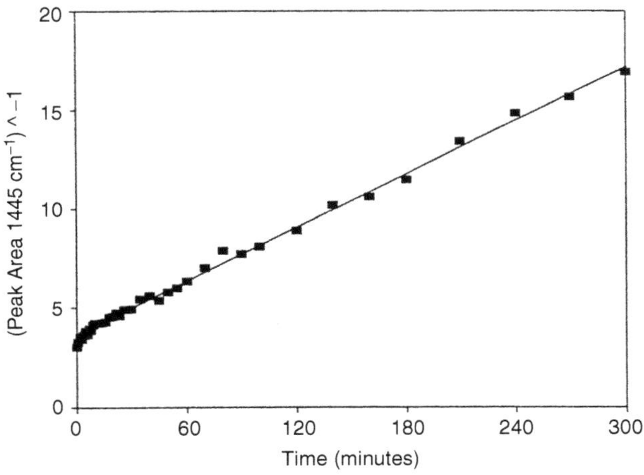

FIGURE 29.6 Reciprocal integrated absorption at 1445 cm^{-1} versus time for the desorption of pyridine from silica gel at 80°C.

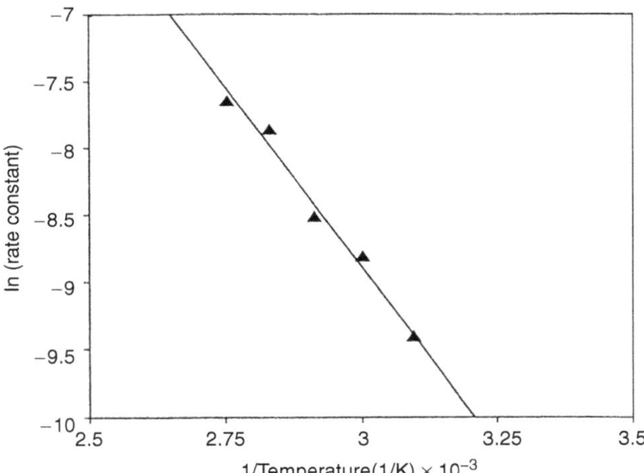

FIGURE 29.7 Arrhenius plot for the desorption of pyridine from silica gel.

in Figure 29.6, an excellent fit (linear correlation coefficient $r = 0.998$) was obtained when a reciprocal equation of the form $1/y = mx + b$ was used. The second-order rate law for this process can be expressed as

$$\frac{d[HPYD]}{dt} = -k_2[HPYD]^2 t$$

where [HPYD] is the concentration of H-bonded pyridine, k_2 is the second-order rate constant, and t is time. Integration of this rate law yields the integrated rate equation

$$\frac{1}{[HPYD]} = -k_2 t + c$$

in which a linear relationship between reciprocal concentration and time is predicted (Figure 29.6). The slope in Figure 29.6 represents the negative value of the second-order rate constant for the isothermal desorption process. This constant is a direct measure of the strength of the pyridine–silanol interaction and thus represents the degree of acidity of the surface acidic sites on amorphous silica gel.

The observed second-order desorption of pyridine is believed to be evidence, that the majority of surface silanols are paired on the surface in either a vicinal or geminal configuration. Further evidence for the pairing of the majority of silanols on the silica surface was presented elsewhere [20] and is in direct agreement with studies conducted by Peri and Hensley [21].

Isothermal desorption rate constants were determined from pyridine desorption studies conducted in the range of 50–90°C. An Arrhenius plot of the natural logarithm of the second-order rate constants versus the reciprocal of the desorption temperature is shown in Figure 29.7. From these data, a value of 43.6 ± 4.2 kJ/mol is obtained for the activation energy of desorption. This value indicates a strong hydrogen bond between pyridine and surface silanols [22]. The desorption activation energy may further be compared with the heat of adsorption of pyridine on silica, 45.2 ± 3.5 kJ/mol, obtained by Hertl and Hair [23]. The close agreement between these values indicates that the activation energy for the adsorption of pyridine on silica is near zero.

CONCLUSIONS

Variable-temperature diffuse reflectance Fourier transform infrared spectroscopy provided qualitative and quantitative information concerning the acidity of surface sites on silica. Although the use of pyridine as a probe has been well known for many years, the ability of the technique to permit easily accomplished investigations in a short period of time without disturbing the integrity of the sample is significant. Pyridine selectively adsorbs to the unassociated silanol sites that are more acidic than the sites hydrogen-bonded to surface water or other silanol sites. The activation energy for the desorption of pyridine is indicative of a strong hydrogen bond and is nearly equal to the heat of adsorption. Thus, the activation energy for the adsorption of pyridine is near zero.

ACKNOWLEDGMENTS

This research was supported in part by Grants CHE–8513247 and CHE-8712457 from the National Science Foundation and funds from Dow Corning Corporation. The Nicolet 60SX FTIR spectrometer was purchased in part from a grant from the National Science Foundation (CHE–8317079).

REFERENCES

1. Karger, B. L.; Snyder, L. R.; Horvath, C. *An Introduction to Separation Science*; Wiley: New York, 1973; pp 30–31.
2. Morrall, S. W. Ph.D. Thesis, Colorado State University, 1984, p 162.
3. Mishima, S.; Nakajima, T. *J. Chem. Soc. Faraday Trans. 1* **1986**, *82*, 1307.
4. Take, J.; Tsuruya, T.; Sato, T.; Yoneda, Y. *Bull. Chem. Soc. Jpn.* **1972**, *45*, 3409.
5. Stone, R.; Rase, H. *Anal. Chem.* **1957**, *29*, 1273.
6. Clark, A.; Holm, V. C. F.; Blackburn, D.M.; *J. Catal.* **1962**, *1*, 244.
7. Airoldi, C.; Santos, Jr., L.; *Thermochim. Acta* **1986**, *104*, 11.
8. Holm, V. C. F.; Clark, A. *J. Catal.* **1963**, *2*, 16.
9. Hattori, H.; Takahashi, O.; Takagi, M.; Tanabe, K. *J. Catal.* **1981**, *68*, 132.
10. Hambleton, F. H.; Hockey, J. A.; Taylor, J. A. G. *Nature (London)* **1965**, *208*, 138.
11. Blitz, J. P.; Murthy, R. S. S.; Leyden, D. E. *Appl. Spectroscopy* **1986**, *40*, 829.
12. Murthy, R. S. S.; Blitz, J. P.; Leyden, D. E. *Anal. Chem.* **1986**, *58*, 3167.
13. Ballivet, D.; Barthomeuf, D.; Pichat, P. *J. Chem. Soc. Faraday Trans. 1* **1972**, *68*, 1712.
14. Schwarz, J. A.; Russel, B. G.; Harnsberger, H. F. *J. Catal.* **1978**, *54*, 303.
15. Murthy, R. S. S.; Leyden, D. E. *Anal. Chem.* **1986**, *58*, 28.
16. Parry, E. P. *J. Catal.* **1963**, *2*, 371.
17. Hair, M. L. *Infrared Spectroscopy in Surface Chemistry*; Dekker: New York, 1967; pp 79–139.
18. Basila, M. R.; Kantner, T. R.; Rhee, K. H. *J. Phys. Chem.* **1964**, *68(11)*, 3197–3207.
19. Scokart, P. O.; Declerck, F. D.; Sempels, R. E.; Rouxhet, P. G. *J. Chem. Soc. Faraday Trans. 1* **1976**, *75*, 359–371.
20. Proctor, K. G. Ph.D. Thesis, Colorado State University, 1989, pp 40–87.
21. Peri, L. B.; Hensley, Jr., A. L. *J. Phys. Chem.* **1968**, *72*, 2926.
22. Cotton, F. A.; Wilkinson, *G. Advanced Inorganic Chemistry*; Wiley-Interscience: New York, 1980; p 21.
23. Hertl, W.; Hair, M. L. *J. Phys. Chem.* **1968**, *72*, 4676.

30 Surveying the Silica Gel Surface with Excited States

R. Krasnansky[1] and J.K. Thomas[2]
University of Notre Dame, Department of Chemistry and Biochemistry

CONTENTS

Time-resolved and steady-state fluorescence probing was used to study gas–solid and liquid–solid silica gel interfaces. The molecular surveying probes pyrene, 1-aminopyrene, pyrenecarboxylic acid, and 9,10-diphenylanthracene adsorbed at the silica gel surface give information about probe environment, mobility, and accessibility. Reactions of surface-bound arene singlet excited states with molecular oxygen were monitored over a range of temperatures; the reactions were assigned to a unique Langmuir–Hinshelwood mechanism. The kinetics reflect the details of oxygen adsorption and movement on the silica gel surface.

Surfaces are of interest because of their media-thickening tendency [1], chromatographic separation ability [2], and catalytic activity [3]. Although surfaces have been long exploited, understanding the nature of surfaces has not always kept pace. Key questions about surface phenomena revolve around the understanding of the chemical environment present at, the mobility found on, and the molecular accessibility allowed at a surface.

Answers to the key questions about surface phenomena have been obtained by use of luminescence probing techniques. In general, a lumophore is isolated at an interface. On the basis of steady-state and time-resolved spectral properties of the lumophore in neat solution or in simple organized media [4], the energy, spectral shape, and time-resolved decay profile of a probe's emission are interpreted to address the key questions of local environment, mobility, and accessibility. Specific modifications of the lumophore probe impinge unique discriminating abilities on the probe. By virtue of their photophysical properties, pyrene and anthracene derivatives have gained a significant place among luminescence probes.

Because of its surface functionality, thermal stability, and at times porous intraparticle structure, silica gel presents an attractive surface for photophysical and photochemical study. This chapter deals with both surface functionality identification and with molecular mobility and reactivity occurring at the silica gel surface.

Pyrene has a nonallowed $0 \rightarrow 0$ transition (ground state to excited state) that is markedly medium-dependent. Media effects are readily monitored through pyrene's fluorescence vibrational structure [5]. The compounds 1-aminopyrene (1-AP) and 1-pyrenecarboxylic acid (PCA) each possess two chromophores that interact in the excited state; the extent of the interaction depends on the media. The lone pair of the amine chromophore of 1-AP acts as an internal switch for photophysics. Previous work [6] showed that the photophysical behavior of 1-AP is dominated by protonation, or "blocking," of the lone pair (Figure 30.1). Protonation of the amino group leads to a complex, 1-APH$^+$, that exhibits photophysical properties similar to those of pyrene; the free 1-AP system

[1]Current address: Rohm and Haas Company, Research Laboratories, 727 Norristown Road, Spring House, PA 19454
[2]Corresponding author

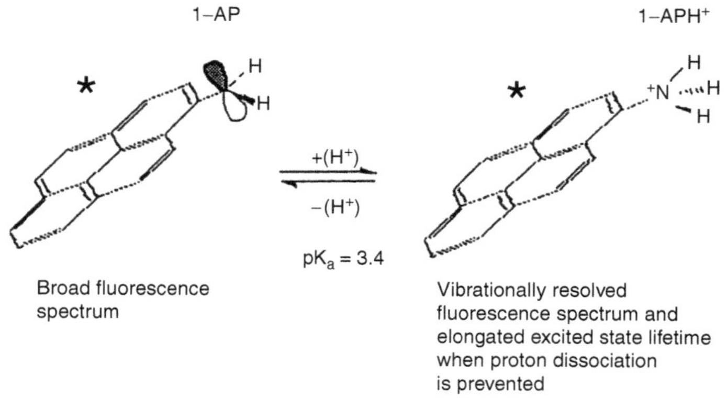

1-AP

1-APH+

+(H+)

−(H+)

$pK_a = 3.4$

Broad fluorescence spectrum

Vibrationally resolved fluorescence spectrum and elongated excited state lifetime when proton dissociation is prevented

FIGURE 30.1 Acid–base photophysical behavior of 1-AP.

exhibits different photophysics. In similar work [7], the photophysical properties of PCA have been correlated to its acid–base properties and media effects. The compounds 1-AP and PCA have been used to distinguish the relative population of geminal and nongeminal silanol functionality at the silica gel surface.

The nonporous silica gel Cab-O-Sil is a convenient system to study molecular geometry during the fluorescence quenching of ^{1}arenes* by oxygen at the gas–solid interface without the complication of partitioning the probe between an external and an internal porous surface. When quenching molecules are present in both the gaseous and the adsorbed states, either a Langmuir–Rideal [8] reaction between a gas-phase molecule and a surface-bound molecule or a Langmuir–Hinshelwood [9] reaction between two surface-bound molecules may occur.

The fluorophores incorporated for the oxygen-quenching study were pyrene, monitoring the $^{1}B_{3u} \rightarrow {}^{1}A_{1g}$ transition [10], and 9,10-diphenylanthracene (DPA), monitoring the $^{1}B_{2u} \rightarrow {}^{1}A_{1g}$ transition [10,11]. The adsorption of fluorophores onto the Cab-O-Sil surface is not homogeneous, but rather can be characterized by a distribution of adsorption sites; each adsorption site presents a microenvironment that is reflected by the unimolecular decay rate of a fluorophore residing at the site. The distribution of fluorophore unimolecular decays is modeled by a Gaussian distribution in natural logarithmic space about a mean unimolecular decay rate. The observable excited-state decay rate in the presence of quencher also has a Gaussian distribution. The oxygen quenching of surface-bound excited-state fluorophore is considered predominantly Langmuir–Hinshelwood.

EXPERIMENTAL DETAILS

INSTRUMENTATION

Steady-state fluorescence spectra were obtained on an SLM–Aminco SPF-500 spectrofluorometer equipped with an LX300 UV illuminator, a 1200-grooves per mm grating, and a Hamamatsu R 928P photomultiplier tube in conjunction with a Zenith Z-368 computer. A neutral density filter, optical density 1.0, was placed in the excitation line to prevent photodecomposition of surface-bound fluorophore.

Time-resolved fluorescence decay profiles were obtained with a PRA Nitromite nitrogen flow laser, model LN-100, with a 0.12-ns full width at half-maximum, 70-μJ, 337.1-nm pulse. Emitted light was collected at 90° to the excitation line; a Kopp 4-96 band-pass filter removed collected scattered light. The monitoring wavelength, λ_{ob}, was selected with a Bausch and Lomb 33-86-02 monochromator equipped with a 1350-grooves per millimeter grating and detected by a Hamamatsu R-1644 microchannel plate with a response time of 0.2 ns. The signal was digitized via a Tektronix 7912HB programmable digitizer equipped with a 7B10 time base and either a 7A16A amplifier (response time 1.6 ns) or a 7A23 amplifier (response time 0.7 ns). Decay profile simulations were performed on a Zenith Z-368 computer by a nonlinear least-squares fitting method.

SAMPLE PREPARATION

Samples were prepared by exposing the silica gel, previously dried at 150°C for 24 h, to either cyclohexane or pentane solutions containing selected amounts of the fluorophore. The solvent was carefully removed under vacuum when required. Complete probe adsorption of liquid–solid samples was verified with absorption spectroscopy. Less than 0.07% of the silica surface was typically covered by the probe. Immediately before data collection, dry samples were evacuated under vacuum at 125–130°C for 30 min. The total dehydration procedure was sufficient to remove the physisorbed water while leaving the surface silanol functionality intact [12]. Selected amounts of oxygen were introduced into a constant-position sample call by a series of stopcock manipulations and a vacuum line. Liquid–solid samples were deoxygenated ($[O_2]_{final} < 10^{-6}$ M) by bubbling the samples with solvent-saturated nitrogen for 30 min.

Various temperatures were achieved by incorporation of a quartz Dewar flask and a stream of chilled nitrogen gas. Nitrogen was cooled by passing the gas through a coiled copper tube submerged in liquid nitrogen. The sample temperature was monitored via a thermocouple attached to the sample cell wall and varied by regulating the nitrogen gas flow. Exterior Dewar fogging was prevented with a second, room-temperature stream of nitrogen.

Oxygen adsorption isotherms were obtained with a differential pressure analysis apparatus. The apparatus consisted of equal-volume spheres connected by a meter-high U-tube, filled halfway with distilled and degassed dimethylpolysiloxane, and a series of three-way stopcocks. Each sphere possessed a cell port equipped with a two-way stopcock. The total volumes were calibrated such that the volume of the left sphere equalled that of the right. A sample of known weight was placed in one of the cells, and the whole system was evacuated. With the two-way stopcocks closed, the system was equilibrated with a given amount of oxygen; the left and the right spheres were then isolated, and the two-way stopcocks were opened. The sample and the empty reference cells were equivalently submerged into various chilling baths. The amount of gas adsorbed was determined from the difference in the heights of the dimethylpolysiloxane columns and a calibration curve. The bulk pressure was measured with a Hastings vacuum gauge equipped with a DV-300 Raydist gauge tube.

HS-5 Cab-O-Sil having a Brunauer–Emmett–Teller surface area (S_{BET}) of $325 \pm 25\ m^2/g$, an accessible pyrene surface area (S_{pyrene}) of $348 \pm 40\ m^2/g$, and a particle diameter of 0.008 mm was donated by the Cabot Corporation. Mathenson Coleman and Bell silica gel (MCB) possesses an S_{BET} of $672 \pm 70\ m^2/g$, an S_{pyrene} of $250 \pm 50\ m^2/g$, and a mesh of 28–200. Each silica has a silanol concentration of 4.9 ± 0.9 silanols per square nanometer [13]. Pyrene was purchased from Aldrich and passed three times down an activated silica gel–cyclohexane column. A sample of 1-aminopyrene (97%) was purchased from Aldrich, recrystallized from ethanol, and passed down an activated silica gel–benzene column. The 9,10-diphenylanthracene (99%), high-performance liquid chromatography grade cyclohexane, and gold label pentane were used as received from Aldrich. Oxygen was used as received from Mittler. Dimethylpolysiloxane was purchased from Sigma. A vacuum of 10^{-4} torr (10^{-2} Pa) was achieved with a Duo-seal model 1400 vacuum pump.

RESULTS AND DISCUSSION

Photophysics of Pyrene on Various Silica Gel Surfaces

Pyrene gives information about its environment via changes in its fluorescence fine structure [14]. Typically, five vibronic bands are identified; the ratio of the III band at 392 nm to the I band at 372 nm, the III/I ratio, increases in noninteracting (nonpolar) media with a concomitant increase in fluorescence lifetime [15]. The versatility of the pyrene probe arises from the "forbiddenness" of the $S_0 \rightarrow S_1$ transition; any intensity for the transition comes from vibronic coupling with higher excited states [10]. Interactive or polar solvents, via their interaction with the arene, increase the intensity of the $0 \rightarrow 0$ transition in both the adsorption and the emission. This increase in intensity is reflected in the increase in peak I with respect to the change in peak III. Pyrene adsorbed on a silica gel surface exhibits a typical III/I ratio of 0.58 [15]. This value is between the values observed for pyrene in water and in methanol and reflects the interaction of the arene probe with the surface silanol functionality of the silica gel surface. Similar data are obtained on solid Al_2O_3.

Photophysics of 1-Aminopyrene on Various Silica Gel Surfaces

Figure 30.2 presents the fluorescence spectra of 1-AP adsorbed on MCB and FS-662 silica gel in cyclohexane. The fluorescence spectrum of MCB-bound 1-AP is quite typical of 1-APH^+, whereas the fluorescence spectrum of FS-662-bound 1-AP is quite typical of 1-AP. The surface-bound 1-APH^{+*} decays with an inherent unimolecular lifetime of 135 ± 2 ns, and the surface-bound 1-AP* decays with an inherent unimolecular lifetime of 4.9 ± 0.1 ns. The different photophysical behavior of the 1-AP indicates that the adsorption sites for 1-AP are different on these silica gel samples at the given probe loadings.

Moderate heat treatment of the MCB silica gel at 450°C for 24 h, a temperature below the sintering temperature, alters the surface such that the fluorescence spectra of 1-AP adsorbed on this surface have characteristics of both 1-AP and 1-APH^+; a heating temperature of 650°C yielded an adsorbed 1-AP spectrum identical to that of 1-AP adsorbed on FS-662. A several-hour concentrated nitric acid treatment of the FS-662 silica, followed by washing with water until the wash maintained a neutral pH and drying, alters the surface such that the fluorescence spectrum of 1-AP adsorbed to this surface resembles that obtained from the MCB surface.

The thermal [16] and the chemical [17] treatments of the silica gels alter the degree of clustering of the silanol functionality on the silica surface. The available surface silanol functionality is shown in Figure 30.3. Hair et al. [17,18] showed that the geminal silanol configuration gives rise to an adsorption site that, when occupied by aniline, yields a protonated adsorbed form of aniline; a similar correlation would account for the observed photophysical behavior of 1-AP on the silica gel surfaces. Milosavljevic and Thomas [7] used PCA to probe the

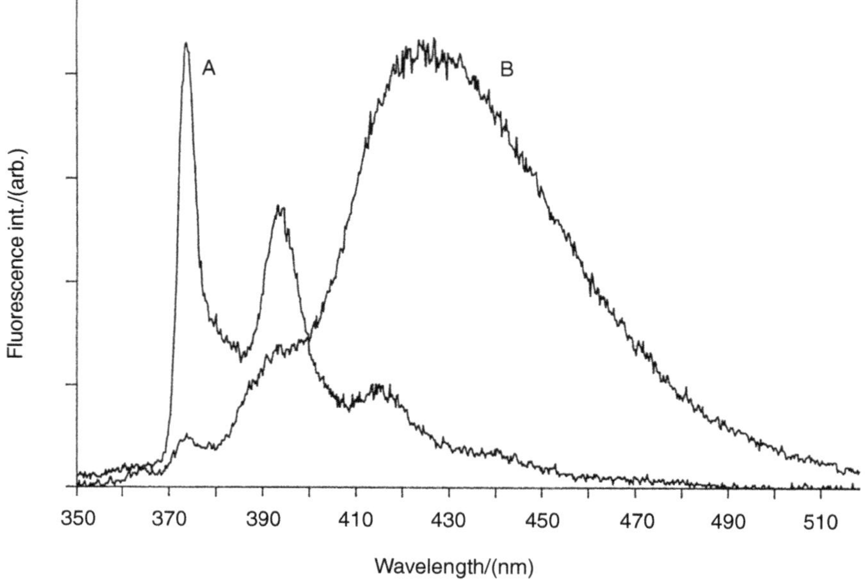

FIGURE 30.2 Fluorescence spectra: 1.9×10^{-7} mol/g 1-AP on (A) MCB and (B) FS-662 silica gel in deoxygenated cyclohexane at an excitation wavelength (λ_{ex}) of 337 nm.

microacidity of the silanol functionality. Scheme I shows the photophysics of PCA. In particular, the observed fluorescence decay rate constant was diagnostic for the determination of the apparent pH of the microenvironment. MCB-bound PCA demonstrated a neutral form, singly protonated carboxylic acid group fluorescence, whereas FS-662-bound PCA demonstrated a mixed anionic–neutral fluorescence. The observed fluorescence decay rate constants of $1.9 \times 10^8\ s^{-1}$ and $3.8 \times 10^7\ s^{-1}$ correspond to an apparent pH of 1.6 and 4.1 for the microenvironments of the geminal and nongeminal silanols, respectively.

The ^{29}Si NMR [19] cross-polarization magic-angle spinning (CP MAS) and ^1H NMR magic-angle spinning with multiple-pulsed line narrowing (CRAMPS for combined rotation and multiple-pulse spectroscopy) techniques of Maciel et al. [20,21] indicate that FS-662

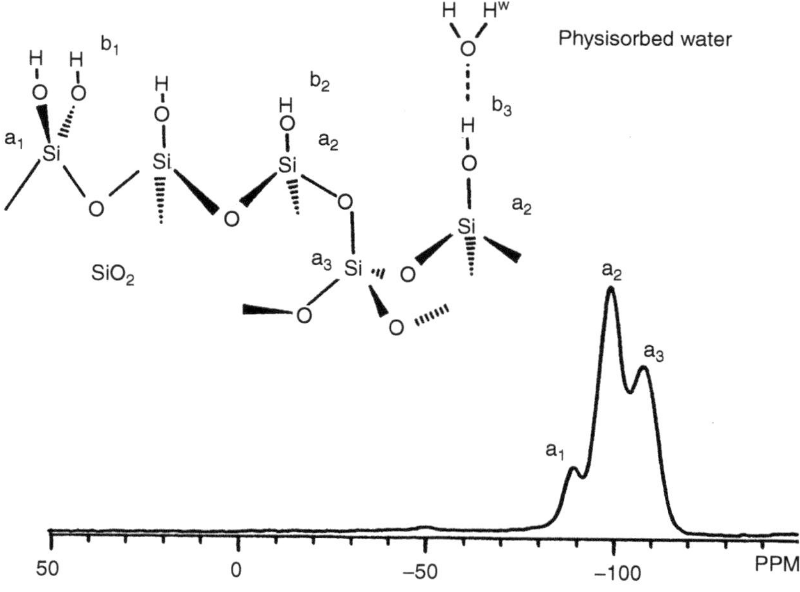

FIGURE 30.3 An 11.88-MHz ^{29}Si CP MAS NMR spectrum of MCB silica gel evacuated at 100°C with a pictorial view of the silica gel surface.

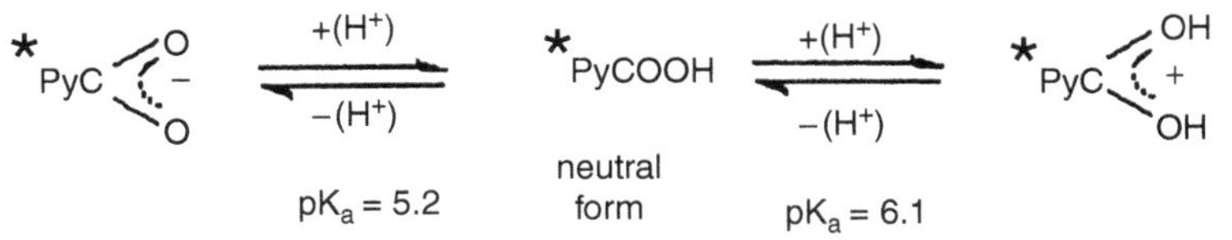

PyC
O
O
+(H+)
−(H+)
pKa = 5.2
PyCOOH
neutral
form
+(H+)
−(H+)
pKa = 6.1
PyC
OH
OH

SCHEME I Acid–base photophysical behavior of PCA.

silica gel has 9% of its silanol functionality in the geminal configuration. Figure 30.3 presents the ^{29}CP MAS NMR spectrum of the MCB silica gel. The percentage of MCB surface silanol functionality in the geminal configuration was determined to be 23%. Figure 30.4 presents the ^{1}H CRAMPS NMR spectra of the MCB silica gel sample dried under vacuum at both 100 and 500°C as well as signal deconvolution. The deconvolution clearly shows the resolution of a fourth set of transitions that are present with the MCB silica at 396.6 Hz but not with

A
15
10
5
0
−5
−10
ppm
B
15
10
5
0
−5
−10
ppm
A
D
b2
b3
b1
w
C

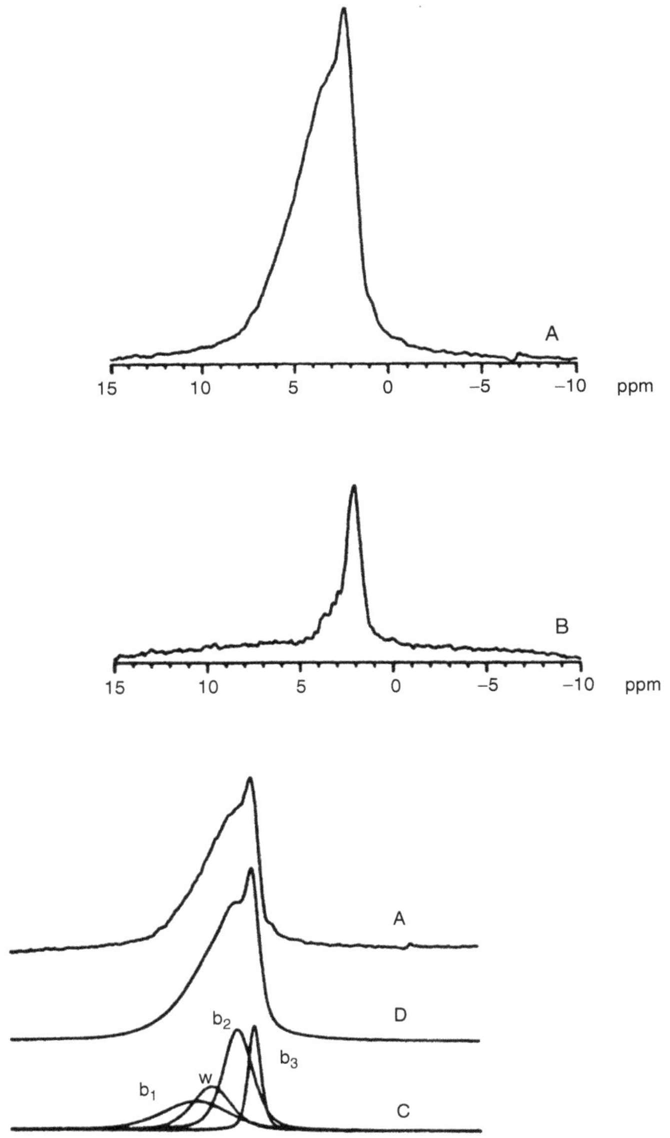

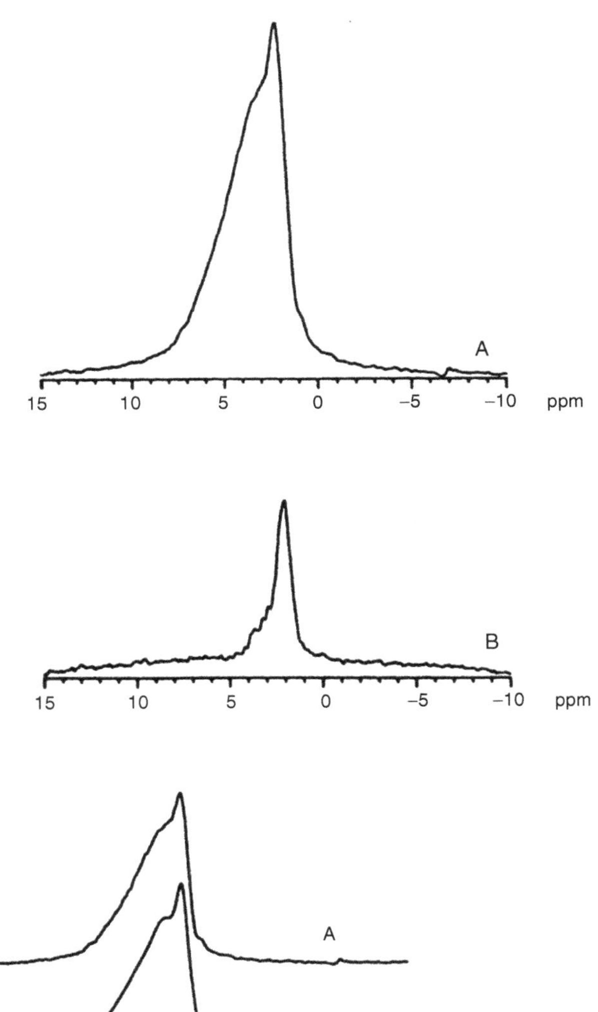

FIGURE 30.4 187-MHz ^{1}H CRAMPS spectra of MCB silica gel (A) evacuated at 100°C and (B) evacuated at 500°C. Plot C is a deconvolution of spectrum A, and plot D is a computer simulation based on C.

FS-662. The 396.6-Hz ^{1}H transition is interpreted as originating from the geminal silanol configuration. The geminal configuration is completely removed, and the hydrogen-bonded vicinal configuration is partially removed, by heating of the MCB silica at 500°C. The redistribution of surface silanol functionality as observed by NMR spectroscopy supports the redistribution observed by photophysical probing technique. The ^{1}H CRAMPS NMR analysis of the MCB silica gel yields a surface silanol functionality distribution of 30% geminal, 48% vicinal, and 22% isolated.

The higher acidity geminal silanol configuration is seen to mandate the photophysical behavior of 1-AP. Increasing the surface 1-AP concentration progressively increases the contribution of the neutral form to the fluorescence spectrum of MCB-bound 1-AP. The redistribution of spectral character equates to a titration of the geminal configuration. The percentage of silanols that exist in the geminal configuration is

$$\%\text{geminal} = 100\left(\frac{2[1-\text{AP}]_{\text{titration}}N(1\times10^{-18})}{S_{\text{probe}}[\text{silanol}_{\text{total}}]_{\text{surface}}}\right) \tag{30.1}$$

where S_{probe} is the probe-accessible surface area in square meters per gram, $[1\text{-AP}]_{\text{titration}}$ is the titration point in moles per gram, N is Avogadro's number, and the surface silanol concentration ($[\text{silanol}_{\text{total}}]_{\text{surface}}$) is taken as 4.9 silanols per square nanometer; the factor 2 accounts for two silanols constituting each geminal configuration. Deconvolution of the room-temperature fluorescence spectra at various probe loadings and use of Equation (30.1) initially indicated that only 0.1% of the MCB silanols were in the geminal configuration, a conclusion in disagreement with the NMR spectroscopy study. As in solution, dropping the temperature increases the proportion of the fluorescence spectrum attributable to the 1-APH^{+*} species (Figure 30.5). The temperature redistribution of the fluorescence spectrum indicates that a deprotonation of the 1-APH^{+*} state competes with fluorescence of 1-APH^{+*}. The deprotonation accounts for the dramatic undercounting of geminal configurations. The deprotonation is apparently frozen out at 77 K; unfortunately, the high 1-AP loading required to reach the true titration point introduces self-absorption problems to the fluorescence study. The absorption process, however, occurs on a time scale at which deprotonation does not occur. Figure 30.6 presents the ratio of observed protonated form to neutral form of 1-AP on FS-662 as a function of probe loading and as monitored by deconvolution of the absorption spectra. Figure 30.6 and Equation (30.1) indicate that 3% of the FS-662 silanol configurations are geminal, a result in agreement with the NMR spectroscopy study. The lower counting of the absorption technique reflects the blocking of a proportion of the silanol functionality by the pyrene.

Gaussian Distribution Model

As reported earlier [22], neither pyrene nor DPA fluorophore yields monoexponential unimolecular decays once adsorbed onto the silica gel surface. The low surface coverage and the absence of pyrene excimers indicate that the surface probe exists as an isolated species. Previously, deviations from the first-order decay kinetics were treated by invoking a multiexponential model [23]. The biexponential luminescence decay function is

$$I(t) = I(0)\{A\exp(-k_1t) + [1-\text{AP}]\exp(-k_2t)\} \tag{30.2}$$

where $I(0)$ is the initial fluorescence intensity and $I(t)$ is the fluorescence intensity at time t. The physical interpretation of the given biexponential model mandates that the probe molecule is partitioned by a percent factor A between two uniquely identifiable environments characterized by the rate constants k_1 and k_2. The biexponential model is weakened by its intrinsic assumption of two possible environments. Each additional exponential term added to the model also brings an additional partitioning factor. Because, in the absence of two chemically unique adsorption sites, true representation of multiple adsorption sites on a surface would require the summation of each adsorption site, the number of variables in the decay assimilation quickly increases to the point where an interpretation of those values becomes questionable. Albery et al. [24] approached the problem of mathematically assimilating heterogeneous systems by expanding the work of Scott et al. [25] and laid the foundation for the development of the Gaussian distribution model. Heterogeneous systems can be characterized through a mean rate constant, $\bar{k}$, and a distribution parameter, γ. The dispersion in the first-order rate constants for $-\infty \le x \le \infty$ becomes

$$\ln k = \ln\bar{k} + \gamma x \tag{30.3}$$

The observed decay profile is composed of the summation of the contributions from each microscopic species. Integration over the distribution $\exp(-x^2)$, when the fluorescence intensity is proportional to the probe's excited state concentration, yields

$$\frac{I(t)}{I(0)} = \frac{\int_{-\infty}^{\infty}\exp(-x^2)\exp[-\bar{k}t\exp(\gamma x)]\mathrm{d}x}{\int_{-\infty}^{\infty}\exp(-x^2)\mathrm{d}x} \tag{30.4}$$

where

$$\int_{-\infty}^{\infty}\exp(-x^2)dx = \sqrt{\pi}$$

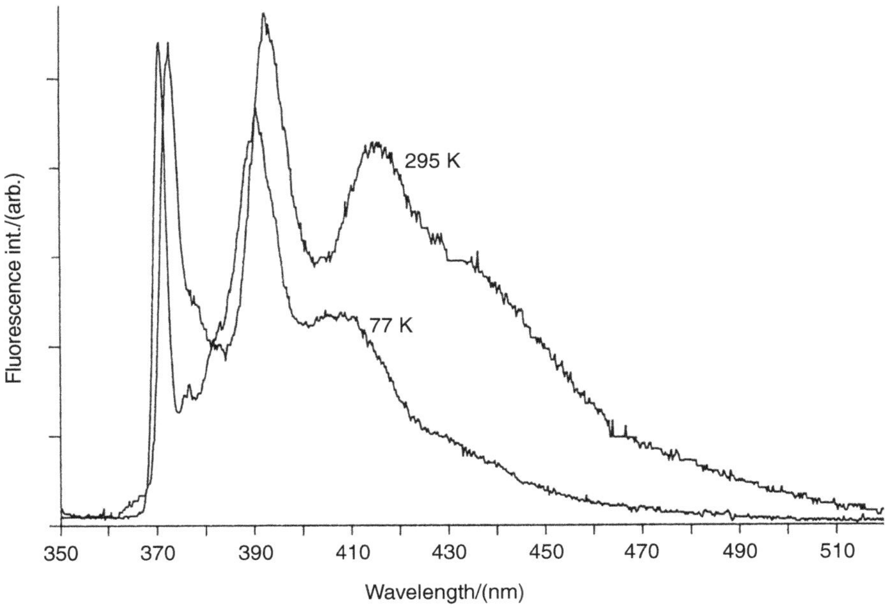

FIGURE 30.5 Fluorescence spectra: 9.82×10^{-7} mol/g 1-AP on dry MCB silica gel under vacuum; excitation wavelength was 337 nm at 295 and 77 K.

and x is the integration range. After transformation of the variable, $x = \ln(\lambda)$ for $x < 0$ and $x = -\ln(\lambda)$ for $x > 0$, the integration of the numerator of Equation (30.4) can be carried out by using the extended Simpson's rule [26]; division of this integration by the denominator yields Equation (30.5):

$$\frac{I(t)}{I(0)} = \frac{1}{\sqrt{\pi}}\int_0^1 g(\lambda)\,d\lambda$$

$$g(\lambda) = \lambda^{-1}\exp\{-[\ln\lambda]^2\}\{\exp(-\bar{k}t\lambda^{\gamma}) + \exp(-\bar{k}t\lambda^{-\gamma})\}$$

$$\frac{I(t)}{I(0)} = \frac{0.2}{3\sqrt{\pi}}\{2[g(0.1) + g(0.3) + g(0.5) + g(0.7) + g(0.9)] + g(0.2) + g(0.4) + g(0.6) + g(0.8) + \exp(-\bar{k}t)\} \quad (30.5)$$

A representation of the Gaussian model is shown in Figure 30.7.

The Gaussian model was found to describe the photophysical behavior of both pyrene and 9,10-diphenylanthracene adsorbed on silica gel well. The physical interpretation of the Gaussian model in these systems would be that each probe experiences slightly different geometric perturbations due to surface irregularities on the scale of the probe's cross-sectional area or various surface-bound strengths. The Gaussian model permits the removal of the constraint of a very limited number of types of adsorption sites found in the biexponential model and leaves only two variable parameters, $\bar{k}_{obs}$ and γ, in the mathematical simulation of the decay profile.

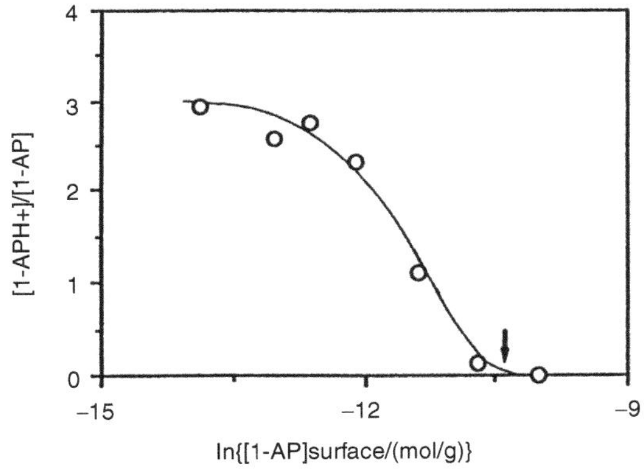

FIGURE 30.6 Plot $[1\text{-APH}^+]/[1\text{-AP}]$ versus ln [1-AP], for 1-AP adsorbed on FS-662 silica gel in deoxygenated cyclohexane monitored by deconvolution of absorption spectra.

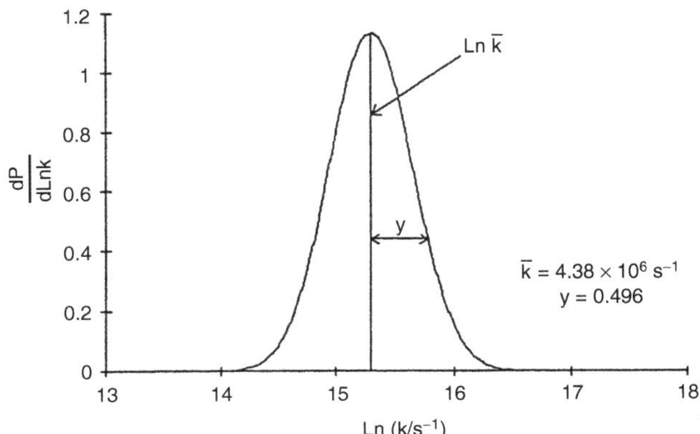

FIGURE 30.7 Representation of the Gaussian distribution model.

Bimolecular quenching reactions of excited states are typically represented by the pseudo-first-order reaction kinetics, Equation (30.6a). When a distribution of probe environments exists, the difference between the macroscopic fluorescence decay profile observed and microscopic fluorescence decay rates present must be recognized. The composite distribution of both unimolecular decay rates and the respective bimolecular reaction rates are combined in the distribution of observable decay rates. Hence, the parameters in Equation (30.6a) have to be redefined in the Gaussian distribution terminology to yield Equation (30.6b).

$$k_{obs} = k_0 + k_q \text{ [quencher]} \quad (30.6a)$$

$$\bar{k}_{obs} = \bar{k}_0 + k'_q \text{ [quencher]} \quad (30.6b)$$

Whereas the observed decay profile no longer is characterized by a single decay rate, the steady-state fluorescence intensity becomes dependent on both γ_{obs} and $\bar{k}_{obs}$. The typical Stern–Volmer plot is no longer represented by Equation (30.7a), but rather by Equation (30.7b), where $\bar{k}_{obs}$ is defined by Equation (30.6b), k'_q is the bimolecular quenching rate constant, $\bar{k}_0$ is the probe's mean excited-state unimolecular decay rate constant, $\bar{k}_{obs}$ is the mean observed decay rate constant, γ_0 is the distribution parameter of the Gaussian for the unimolecular decay, and γ_{obs} is the distribution parameter for the observed unimolecular decay rate.

$$\frac{I_0}{I} = 1 + \frac{k_q}{k_0} \text{ [quencher]} \quad (30.7a)$$

$$\frac{I_0}{I} = \exp\left(\frac{\gamma_0^2 - \gamma_{obs}^2}{4}\right)\frac{\bar{k}_{obs}}{\bar{k}_0} \quad (30.7b)$$

Reference 27 formally addresses the mathematical derivations associated with the Gaussian distribution model as applicable to luminescence probing.

Oxygen Quenching of Singlet Excited States of Pyrene and DPA [27]

Because of the dramatic difference in unimolecular decay rates of pyrene and DPA, a large concentration range of oxygen could be examined and compared. Steady-state and time-resolved oxygen-quenching studies were performed at various temperatures and correlated as a function of bulk gaseous oxygen concentration. Oxygen pressures were converted to concentration units through the ideal gas equation to account for density changes at the various temperatures.

The temperature dependence on the steady-state quenching is typified by DPA (Figure 30.8). There was a dramatic increase in the degree of quenching at a given concentration of oxygen with a decrease in temperature when oxygen gas-phase concentrations were used as the independent variable.

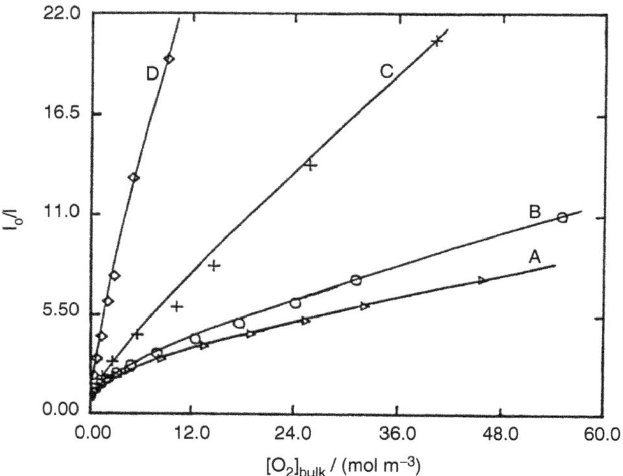

FIGURE 30.8 Steady-state oxygen quenching of 7.76×10^{-7} mol/g DPA on Cab-O-Sil ($\lambda_{ex} = 365$ nm and $\lambda_{ob} = 390$–520 nm) at (A) 23, (B) −27, (C) −50, and (D) −90°C.

The Gaussian distribution model was applied to the decay profiles at various concentrations of oxygen. Time-resolved fluorescence decay profiles for the oxygen quenching of the DPA–Cab-O-Sil system are shown in Figure 30.9. Initial intensities of the fluorescence decay profiles are fixed at the value obtained in the absence of quencher to indicate that the quenching is dynamic in nature; short time intervals in the decay profiles are lost in the response time of the system. When the quenching is plotted in terms of bulk gaseous oxygen concentrations, as would be appropriate for a Langmuir–Rideal scheme, there is an increase in quenching efficiency with a decrease in temperature (Figure 30.10). Simulation of the fluorescence decay becomes difficult at high degrees of fluorophore quenching because an uncertainty in initial fluorescence intensity arises from an instantaneous component that enters the quenching scheme.

Recently, Turro and co-workers [28] derived the expressions for the bimolecular rate constant for a Langmuir–Rideal type quenching scheme both for a smooth surface, Equation (30.8a), and for a porous solid in the Knudsen regime, where the pore diameter is much less than the gas-phase mean free path, Equation (30.8b).

$$-\frac{d[\text{probe}^*]}{dt} = \frac{1}{4}\frac{\alpha\langle v\rangle\sigma_{probe}}{kT}P_q[\text{probe}^*] + \frac{[\text{probe}^*]}{\tau_0} \quad (30.8a)$$

$$-\frac{d[\text{probe}^*]}{dt} = \frac{4}{6}\frac{\pi\alpha r_{ab} g R_p\langle v\rangle}{kT}P_q[\text{probe}^*] + \frac{[\text{probe}^*]}{\tau_0} \quad (30.8b)$$

where

$$\langle v\rangle = \left(\frac{8kT}{\pi m_w}\right)^{1/2}$$

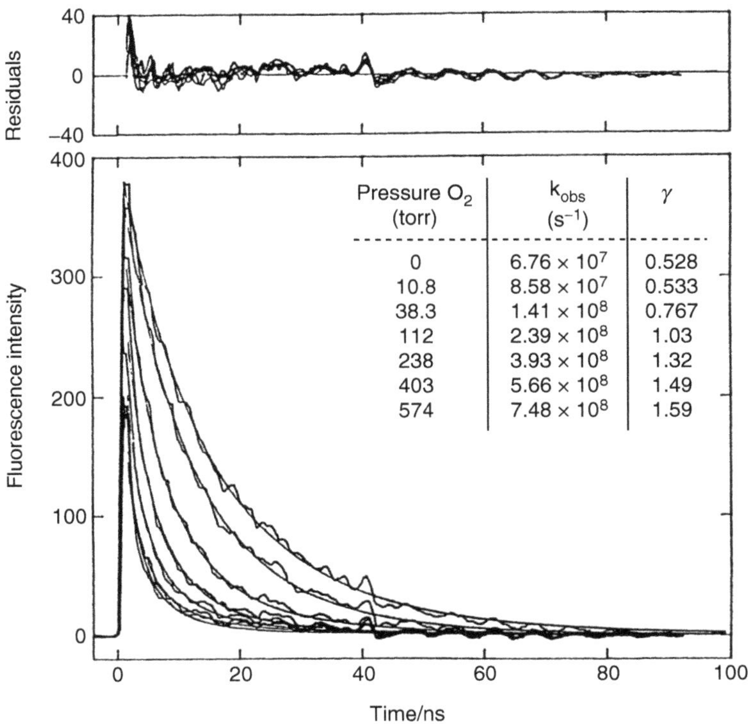

FIGURE 30.9 Transient fluorescence decay profile of 4.76 × 10^{-7} mol/g DPA on Cab-O-Sil, with Gaussian model simulated decay profiles at 18°C and various pressures of oxygen (λ_{ex} = 337 nm and λ_{ob} = 430 nm).

α is the efficiency of the bombardment reaction, σ_{probe} is the cross-sectional area of the probe, r_{ab} is the interaction radius for the reactants, g is a geometric factor ($g > 0$), R_p is the pore radius, τ_0 is the lifetime of the probe in the absence of quencher, P_q is the bulk gaseous quenching pressure, m_w is the molecular mass in kilograms per molecule of the gas, T is the absolute temperature, and k is the Boltzmann constant. Converting pressures to concentration units through the ideal gas law and substituting for the mean velocity $<v>$ gives Equations (30.9a) and (30.9b) (R is the ideal gas constant) from Equations (30.8a) and (30.8b), respectively.

$$k_q = R\,\sigma\,\alpha\,(2\pi m_w k)^{-1/2}\sqrt{T} \qquad (30.9a)$$

$$k_q = \frac{4}{3} R\,R_p\,r_{ab}\,\alpha\left(\frac{2\pi}{m_w k}\right)^{1/2}\sqrt{T} \qquad (30.9b)$$

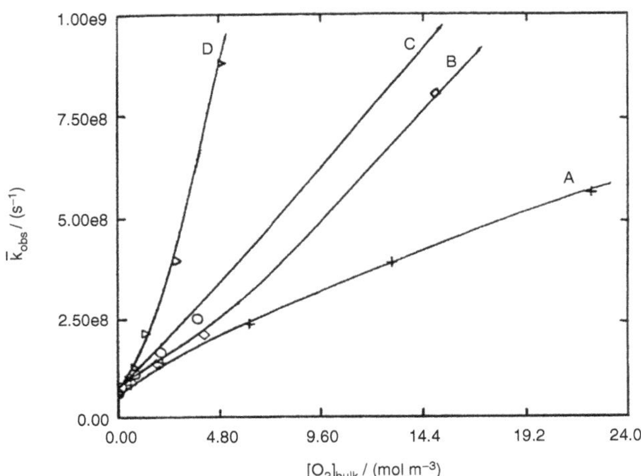

FIGURE 30.10 Time-resolved oxygen quenching in 4.76 × 10^{-7} mol/g DPA (λ_{ex} = 337 nm and λ_{ob} = 430 nm) showing the response of $\bar{k}_{obs}$ to bulk gaseous oxygen concentration at (A) 18, (B) −28, (C) −50, and (D) −83°C.

Equations (30.9a) and (30.9b) show that a gas-phase bombardment quenching mechanism predicts that the quenching efficiency will increase with the square root of temperature; both probe systems show decreasing oxygen quenching efficiencies with increasing temperatures.

Oxygen Adsorption Isotherms

Oxygen adsorption isotherms on the Cab-O-Sil surface at various temperatures up to −47°C were measured. The isotherms obtained followed Langmuir adsorption at the temperatures examined. Theoretical treatment of the oxygen adsorption isotherm data [27] yields a heat of adsorption of 3.54 kcal/mol (14.8 kJ/mol) and the assimilation times of 23.2, 44.1, 99.5, and 238 ps at −47, −64, −80, and −98°C, respectively.

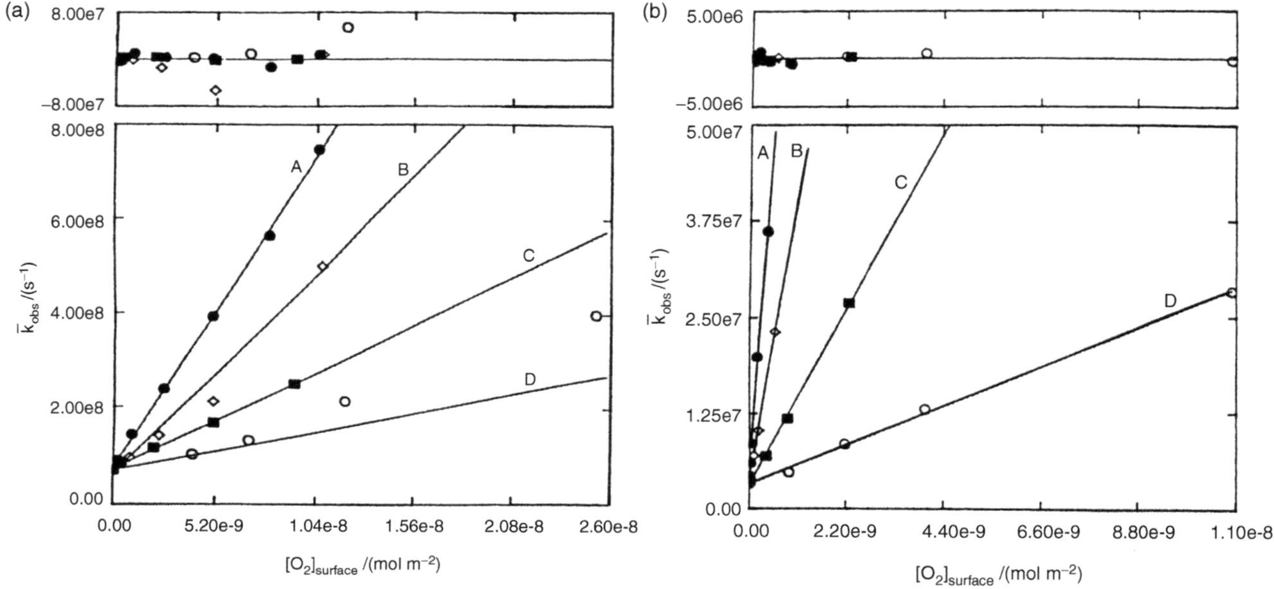

FIGURE 30.11 (a) Time-resolved oxygen quenching ($\lambda_{ex} = 337$ nm) with a Gaussian fit of 4.76×10^{-7} mol/g DPA ($\lambda_{ob} = 430$ nm) at (A) 18, (B) −28, (C) −50, and (D) −83°C. (b) Time-resolved oxygen quenching with Gaussian fit of 4.74×10^{-7} mol/g pyrene ($\lambda_{ob} = 397$ nm) at (A) 19, (B) −26, (C) −66, and (D) −92°C on Cab-O-Sil.

Langmuir–Hinshelwood Approach to Oxygen Quenching

Surface oxygen concentrations for the various gaseous bulk oxygen concentrations were determined with the adsorption isotherm parameters. From a steady-state point of view, the amount of oxygen adsorbed on the silica surface in the range where quenching occurs is quite low; at ambient temperatures, the surface quencher concentration even approaches the surface probe concentration. The amount of time a given oxygen resides on the surface is also quite short. Such short assimilation times would not be expected to yield the observed dynamic quenching if the oxygen initially present on the surface at the moment of probe excitation was completely responsible for the quenching of the probe's excited state. The probe's excited-state lifetime is immensely greater than the assimilation time of the oxygen. The number of times an adsorbed oxygen is replaced, at some other arbitrary surface location, by an oxygen from the gas phase is great during the lifetime of the probe. The random placement of oxygen on the surface would be crucial in describing the apparent surface migration distance incurred during the lifetime of the probe's excited state. A large surface oxygen exchange rate also upholds pseudo-first-order conditions.

Figure 30.11 shows that the observed decay rate constants of both DPA and pyrene vary linearly with surface oxygen concentration when surface oxygen concentration is used as the independent variable. The bimolecular quenching rate constants obtained from the slopes of these plots decrease with a decrease in temperature, as would be expected for a Langmuir–Hinshelwood mechanism. Under higher probe surface coverage conditions, it has been shown through the formation of static, as opposed to dynamic, excimers of pyrene on the silica gel surface that the fluorophore does not migrate on the time scale of the excited state [29]; all motion, therefore, arises from the motion of the oxygen. An Arrhenius treatment of the observed quenching rate constants yields activation energies for the bimolecular Langmuir–Hinshelwood type quenching reaction of 3.31 and 2.31 kcal/mol (13.8 and 9.67 kJ/mol) for the pyrene–Cab-O-Sil/oxygen and the DPA–Cab-O-Sil/oxygen systems, respectively. A representation of the dynamics of the oxygen quenching is presented in Figure 30.12.

The silica surface plays a role in the quenching scheme analogous to the solvent cage in solution. The interaction time of the probe and quencher is increased, and excess translational, vibrational, and rotational energy of the oxygen is dissipated by the adsorption process. The adsorption process, thus, increases the quenching probability.

CONCLUSIONS

Luminescence probing techniques have demonstrated their versatility in characterizing the environment found on, the mobility and accessibility allowed at, and the reactivity occurring at gas–solid and liquid–solid interfaces. Pyrene can be used to monitor the interactive nature of the silica surface, and 1-AP and PCA faithfully distinguish the silanol functionality present at the silica gel surface. MCB silica gel possesses a higher relative geminal

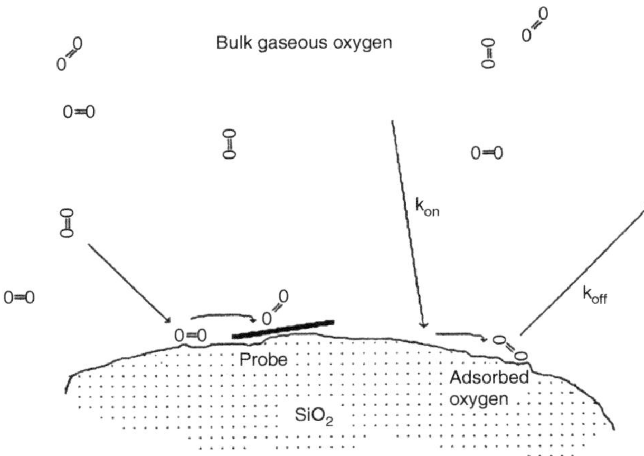

FIGURE 30.12 Oxygen quenching of Cab-O-Sil bound fluorophore.

silanol concentration compared to the FS-662 silica gel; this greater concentration gives rise to strong hydrogen bonding to the weak 1-AP base. Protonation of the amine's lone pair stops the mixing of the bichromophoric excited states and causes photophysical behavior typical of an isolated, or more correctly an inertly substituted, pyrene moiety.

The heterogeneous adsorption and the bimolecular quenching of the Cab-O-Sil bound pyrene and DPA by molecular oxygen were computer-simulated with a Gaussian distribution model. The bimolecular quenching rate constants followed the opposite trend with temperature when the gas-phase concentrations of oxygen were used as the independent variable as mandated by a Langmuir–Rideal mechanism. The bimolecular quenching rate constant followed the expected trend with temperature when surface concentrations of oxygen were used as the independent variable as mandated by a Langmuir–Hinshelwood mechanism. The quenching observed could not be accounted for by the oxygen present on the surface at the time of excitation; the exchange of surface-bound oxygens with oxygens from the bulk gas reservoir during the lifetime of the excited state had to be recognized. We assign the predominant quenching of the singlet excited states of pyrene and DPA located on the Cab-O-Sil surface by oxygen to a modified Langmuir–Hinshelwood scenario.

ACKNOWLEDGMENTS

We are grateful to the NSF (Grant CHE-8911906) and the EPA for financial support of this work and to the Cabot Corporation for donation of Cab-O-Sil. We thank Bratoljub H. Milosavljevic from the Boris Kidrich Institute for Nuclear Science in Belgrade, Yugoslavia, and Kasuhide Koike from the National Research Institute for Pollution and Resources in Onogawa, Tsukuba, Japan, for their stimulating discussions. Finally, we thank the Colorado State University NMR facility for acquiring the NMR spectra.

REFERENCES

1. Wightman, J.D.; Chessick, J. J. *J. Phys. Chem.* **1962**, *13*, 1217. Sirianni, A. F.; Meadus, F. W.; Puddington, I. E. *Can. J. Chem.* **1964**, *42(12)*, 2916.
2. Kirkland, J. J. *Anal. Chem.* **1965**, *37*, 1458. Kirkland, J. J. *Chromatogr. Sci.* **1969**, *7*, 7. Kirkland, J. J. *Chromatogr. Sci.* **1972**, *10*, 593. Nikitin, Y. S. *Anal. Chem.* **1964**, *36(8)*, 1526. Kirkland, J. J. *Chromatogr. Sci.* **1970**, *49*, 84.
3. Stiles, A. B. *Catalyst Supports and Supported Catalyst, Theoretical and Applied Concepts*; Butterworths: Boston, MA, 1987.
4. Organized media is defined as a chemical system that possesses a certain degree of self-assembly.
5. Thomas, J. K. *The Chemistry of Excitation and Interfaces*; ACS Monograph Series 181; American Chemical Society: Washington, DC, 1984.
6. Hite, P.; Krasnansky, R.; Thomas, J. K. *J. Phys. Chem.* **1986**, *90*, 5795.
7. Milosavljevic, B. H.; Thomas, J. K. *J. Phys. Chem.* **1988**, *92*, 2997.
8. Rideal, E. K. *Proc. Cambridge Philos. Soc.* **1939**, *35*, 130.
9. Hinshelwood, C. N. *Kinetics of Chemical Changes*; Clarendoni: Oxford, England, 1940; p 187.
10. Pariser, R. *J. Chem. Phys.* **1955**, *24(2)*, 250.
11. Birks, J. B. *Photophysics of Aromatic Molecules*; Wiley-Interscience: New York, 1970; p 71.
12. Iler, R. K. *The Chemistry of Silica Gel: Solubility, Polymerization, Colloid and Surface Properties, and Biochemistry*; Wiley: New York, 1979; p 630. Sindorf, D. W.; Maciel, G. E. *J. Am. Chem. Soc.* **1983**, *105*, 1487.
13. Zhuravlev, L. T. *Langmuir* **1987**, *3*, 316.
14. Kalyanasundaram, K.; Thomas, J. K. *J. Am. Chem. Soc.* **1977**, *99*, 2039.
15. Thomas, J. K. *J. Phys. Chem.* **1987**, *94*, 553; Beck, G.; Thomas, J. K. *Chem. Phys. Chem.* **1985**, *94*, 553; Bauer, R. K.; de Mayo, P.; Okada, K.; Ware, W. R.; Wu, K. C. *J. Phys. Chem.* **1983**, *87*, 460. Bauer, R. K.; de Mayo, P.; Natarajan, L. V.; Ware, W. R. *Can. J. Chem.* **1984**, *62*, 1279.
16. Young, G. T. *J. Colloid Sci.* **1958**, *13*, 67.
17. Anderson, J. H.; Lambard, J.; Hair, M. L. *J. Colloid Interface Sci.* **1975**, *50(3)*, 519. de Mayo, P. *Pure Appl. Chem.* **1982**, *54(9)*, 1623. Iler, R. K. *The Chemistry of Silica*; Wiley: New York, 1979; pp 109–211.
18. Hair, M. L.; Hertle, W. J. *J. Phys. Chem.* **1969**, *73*, 4269.
19. Sindorf, D. W.; Maciel, G. E. *J. Am. Chem. Soc.* **1983**, *105*, 1487.
20. Bronnimann, C. E.; Zeigler, R. C.; Maciel, G. E. *J. Am. Chem. Soc.* **1988**, *110(7)*, 2023.

21. Sindorf, D. W.; Maciel, G. E. *J. Phys. Chem.* **1982**, *86*, 5208.
22. Bauer, R. K.; de Mayo, P.; Okada, K.; Ware, W. R.; Wu, K. C. *J. Phys. Chem.* **1983**, *87*, 460.
23. Lochmuller, C. H.; Colborn, A. S.; Hunnicatt, M. L.; Harris, J. M. *J. Am. Chem. Soc.* **1984**, *106*, 4077.
24. Albery, W. J.; Bartlett, P. N.; Wilde, C. P.; Darwent, J. R. *J. Am. Chem. Soc.* **1985**, *107*, 1854.
25. Scott, K. F. *J. Chem. Soc. Faraday Trans. I* **1980**, *76*, 2065.
26. Riddle, D. F. *Calculus and Analytic Geometry*, 3rd ed.; Wadsworth Publishing: Belmont, CA, 1979; p 219.
27. Krasnansky, R.; Koike, K.; Thomas, J. K. *J. Phys. Chem.* **1990**, *94*, 4521.
28. Drake, J. M.; Levitz, P.; Turro, N. J.; Nitsche, K. S.; Cassidy, K. F. *J. Phys. Chem.* **1988**, *92*, 4680.
29. Hara, K.; de Mayo, P.; Ware, W. R.; Weedon, A. C.; Wong, G. S. K.; Wu, K. C. *Chem. Phys. Lett.* **1980**, *69(1)*, 105. Krasnansky, R., Ph.D. Thesis, University of Notre Dame, 1990.

31 Surface Chemistry and Surface Energy of Silicas

Alain M. Vidal and Eugène Papirer
National Center for Scientific Research (CNRS), Research Center for Physicochemistry

CONTENTS

The establishment of relationships between the surface chemistry and the surface free energy of silicas is important for practical applications of these materials. Inverse gas chromatography, either at infinite dilution or finite concentration, appears to be an effective method for the detection of changes of surface properties induced by chemical or thermal treatments. Silicas of various origins (amorphous or crystalline) with surface chemistries modified by chemical (esterification) or heat treatment were compared. The consequences of these modifications on surface energetic heterogeneities were assessed.

Silica exists in a broad variety of forms, in spite of its simple chemical formula. This diversity is particularly true for divided silicas, each form of which is characterized by a particular structure (crystalline or amorphous) and specific physicochemical surface properties. The variety results in a broad set of applications, such as chromatography, dehydration, polymer reinforcement, gelification of liquids, thermal isolation, liquid-crystal posting, fluidification of powders, and catalysts. The properties of these materials can of course be expected to be related to their surface chemistry and hence to their surface free energy and energetic homogeneity as well. This chapter examines the evolution of these different characteristics as a function not only of the nature of the silica (i.e., amorphous or crystalline), but also as a function of its mode of synthesis; their evolution upon modification of the surface chemistry of the solids by chemical or heat treatment is also followed.

Only two kinds of functional groups can be found on silica surfaces: siloxane bridges and hydroxyls (silanols). However, among the hydroxyls, different types can be identified, namely single free silanols, geminal hydroxyls, hydroxyl pairs associated through hydrogen bonding (either vicinal or brought together, for example, at points of contact between particles or in micropores), inner hydroxyl groups, and adsorbed water [1–3].

Various methods, either chemical [4–8] or physical [9–15], can be used for the determination of these surface groups, and their number and type can be easily modified by chemical (e.g., esterification upon reaction with alcohols) or heat treatment. However, for heat treatment, as shown by Fripiat [16], the modification of the surface chemical properties is much more complex than would be expected when only considering the curves relating weight loss to temperature. Thus it should be of interest to relate the evolution of surface silanol groups to the surface free energy of silica samples.

The surface free energy of a solid (γ_s) can be expressed as a sum of two components: γ_s^d (the dispersive component), describing London-type interactions, and γ_s^{sp} (the specific component), including all other interactions (H-bonding, polar, and so forth).

$$\gamma_s = \gamma_s^d + \gamma_s^{sp}$$

Two methods can be used for the assessment of the γ_s of divided solids: contact-angle measurements and adsorption processes. The drawbacks of the contact-angle measurements are associated with surface roughness of the samples. As for the adsorption process, determination of the components of the surface free energy of the solid is based on interpretation of adsorption isotherms, either complete (calculation from spreading pressures) or only from the first linear part of the isotherm. In this respect, inverse gas chromatography (IGC), which appears to be the technique of choice [17], was extensively used in this study.

EXPERIMENTAL DETAILS

MATERIALS

Eight silicas from different synthesis processes were studied:

1. Five amorphous silicas:
   - A, a fumed silica, Aerosil 200 (Degussa, 200 m^2/g)
   - P, a precipitated silica, Zeosil 175 MP (Rhône-Poulenc, 175 m^2/g)
   - G, a gel of silica, RP1 (Rhône-Poulenc, 230 m^2/g)
   - C, a colloidal silica, FDR (Rhône-Poulenc, 10 m^2/g)
   - F, a fibrillar silica [18], FATM 220/1 (180 m^2/g)
2. Three crystalline silicas:
   - L_1, synthetic [19] H_2SiO_5 (84 m^2/g)
   - L_2, synthetic H-Magadiite (48 m^2/g)
   - L_3, synthetic H-Kenyaite (18 m^2/g)

A and F were high-purity silicas. All samples but A were slightly porous.

Alcohol-Modified Silicas

The silicas were modified by reaction with alcohols, either methanol (C_1) or hexadecanol (C_{16}), according to a method previously described [8]. Degree of esterification was determined by elemental analysis, microgravimetry (weight loss associated with the pyrolysis of modified silicas), or radiochemistry (use of ^{14}C-labeled alcohols). The corresponding modified silicas were identified as XC_1 and XC_{16}.

Inverse Gas Chromatography and Heat Treatment of Silicas

Silica particles of adequate size (0.25–0.5 mm in diameter) obtained by compression and sieving were used to fill chromatographic columns (stainless steel, 30 cm long, 3 mm in diameter) connected to a gas chromatograph fitted with a flame ionization detector. Helium was used as carrier gas at a flow rate of 20 cm^3/min [the flow rate conditions were selected so as to have the best efficiency for the chromatographic columns (obtained at the minimum of the Van Deemter curve)]. Measurements were made at a column temperature of 60°C. Thermal treatment of silica was done under helium flow by heating the column to the desired temperature. After being heated for 30 min, the column was cooled to the analysis temperature. With alkane probes, symmetrical retention peaks were observed. For polar probes, skewed peaks were usually recorded; for such peaks an integrator was used to determine the peak first-order moment. Calculation of γ_s^d and specific interaction parameters from chromatographic data were described elsewhere [20] and are only briefly described here.

When minute amounts of solute are adsorbed, the adsorption process can be described by the initial part of the adsorption isotherm, which is practically linear. Under these conditions Henry's law applies. It is then possible to relate the thermodynamic parameters of adsorption, such as the variation in free energy upon adsorption of the solute at zero coverage (ΔG^0), to the retention volume of the probe (V_N):

$$\Delta G^0 = -RT \ln (C'V_N)$$

where R is the gas constant, T is the temperature, and C' is a constant depending on the reference state for the adsorbed molecule.

It is known from earlier studies that ΔG^0 varies linearly with the number of carbon atoms of a homologous alkane series. It is thus possible from the preceding equation to calculate the adsorption free energy increment associated with one CH_2 group:

$$\Delta G_{(CH_2)} = -RT \ln[V_{N(n)}/V_{N(n+1)}]$$

where $V_{N(n)}$ and $V_{N(n+1)}$ are the retention volumes of n-alkanes with n and $n+1$ carbon atoms, respectively. The quantity $\Delta G_{(CH_2)}$ can provide an estimation of the dispersive interactions between one CH_2 group and an adsorbent and is related to the γ_s^d of the solid by

$$\Delta G_{(CH_2)} = 6.023 \times 10^{23} \times a_{(CH_2)} \times 2[\gamma_s^d \gamma_{(CH_2)}]^{1/2}$$

where $a_{(CH_2)}$ is the surface area of a CH_2 group (0.06 nm^2) and $\gamma_{(CH_2)}$ is the surface tension of a surface made of CH_2 groups only, for example, polyethylene (35.6 mJ/m^2 at 20°C). For interactions of polar probes with polar surfaces, the free energy of adsorption can be expressed by

$$\Delta G = \Delta G^d + \Delta G^{sp}$$

where ΔG^d and ΔG^{sp} are the contributions to the free energy of adsorption of the dispersive and specific interactions, respectively. The generally accepted way of separating London dispersion effects from specific effects, in infinite dilution chromatography, is to compare the chosen solute probe with an *n*-alkane of approximately the same geometry and polarizability. As a consequence, the specific interaction parameter I_{sp} is obtained by subtracting ΔG^d, corresponding to nonspecific interactions, from ΔG, measured by inverse gas chromatography:

$$I_{sp} = \Delta G^{sp} = \Delta G - \Delta G^d$$

The accuracy of the measurements was equal to ± 0.001 min for the retention time, $\pm 0.1°C$ for the column temperature, ± 20 Pa for the atmospheric pressure, and ± 100 Pa for the pressure drop. Therefore, net retention volumes were known with a precision of about 5%. Consequently, the absolute error for free energy of adsorption and for the specific interaction parameter are estimated to be ± 0.1 and ± 0.2 kJ/mol, respectively.

RESULTS AND DISCUSSION

London Component of the Surface Free Energy of Heat-Treated Silicas

Figure 31.1 shows the evolution of γ_s^d for the different types of silicas versus heat-treatment temperature. The origin of the sample as well as the thermal treatments applied are important in determining γ_s^d.

Amorphous silicas A, P, and G exhibit similar behaviors, corresponding to an increase of γ_s^d, from about 60–70 mJ/m^2 at 60°C to 100 mJ/m^2 at 500°C, followed by a decrease at temperatures up to 700°C. These complex variations must be associated with complex chemical changes occurring on the surface of silicas upon heat treatment. Colloidal (C) and fibrous (F) silicas have similar general trends characterized by a continuous increase of γ_s^d up to 600°C (e.g., sample C; at 600°C γ_s^d is 160 mJ/m^2).

The evolution of γ_s^d for crystalline silicas is completely different from that of amorphous silicas. Starting from a very high value at low temperatures, it levels off between 200 and 400°C, then increases again for higher heat-treatment temperatures. The extremely high values obtained, particularly for the L_2 sample (γ_s^d is in the 120–470-mJ/m^2 range), raise a fundamental question as they do not have a physical meaning. These results can be associated with the intercalation of the alkane probes between the sheetlike structure of crystalline silicas, a process that was demonstrated by use of branched alkanes.

The behavior of sample L_1 is a mix of that of crystalline and amorphous silicas, because it behaves as a crystalline silica at low temperature (60–200°C) and as an amorphous silica above 200°C (this behavior parallels the evolution of its crystalline organization, which is unstable above 200°C).

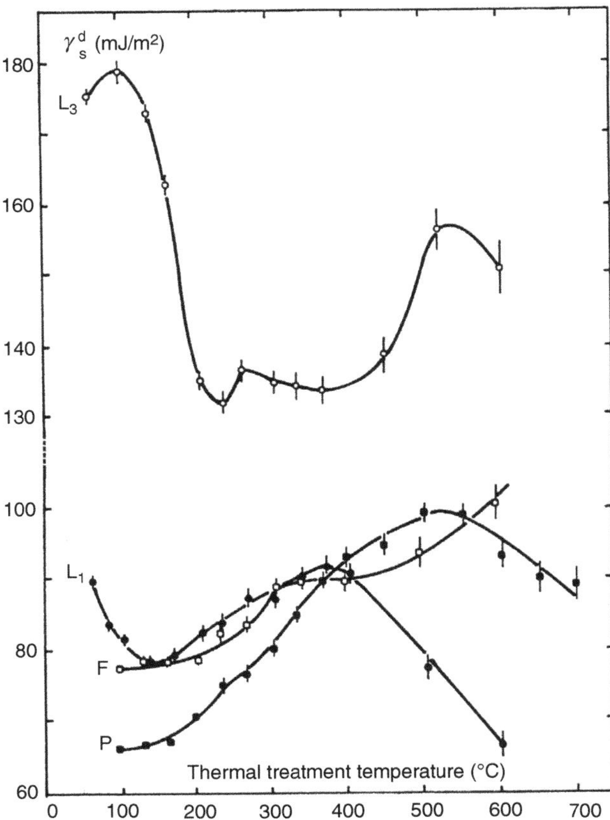

FIGURE 31.1 Evolution of the dispersive component of the surface free energy of silicas versus heat treatment temperature.

These curves point not only to the differences but also to the similarities exhibited by amorphous and crystalline samples. The main differences appear essentially at low treatment temperatures, at which the surface chemical properties of both types of silica are very much different. Interpretation of the evolution of γ_s^d versus temperature in amorphous silicas would suggest that the probes interact with siloxane bridges preferentially through dispersive interactions. This explanation was supported by Brinker et al. [12,13], who identified by ^{29}Si NMR and Raman spectroscopies two types of silicon–oxygen species on the surface of silicas: tetra (unstrained) and trisiloxane (strained) rings. The trisiloxane rings, nonexistent at low temperature, form at intermediate temperatures and become prevalent in the 350–650°C range. At higher temperatures trisiloxane cycles rearrange to yield less-strained cyclotetrasiloxane units.

The evolution of the physicochemical characteristics of the surface of amorphous silicas upon thermal treatment can be envisioned as follows. At room temperature the surface of the solids is covered by a multilayer of water, the external layers of which are eliminated at 30°C under vacuum. At higher temperatures, only a monomolecular layer of water interacts with the surface (strongly with silanol groups, weakly with siloxane bridges). At about 100°C, part of the water is evacuated, and thus the exposed surface will be available for interactions with alkane probes, a process associated with an increase of γ_s^d. At 250°C all of the physically adsorbed water, but not molecules trapped in pores, has been eliminated. The phenomena observed between 250 and 500°C can probably be attributed to the condensation of vicinal silanols [8], which yield trisiloxane cyclic compounds. At 500°C, geminal silanols begin to disappear [21] and trisiloxane rings rearrange to yield tetrasiloxane cycles. Thus, above 500°C the γ_s^d plot may be taken to represent the condensation of geminal and isolated silanols, a process that may be the reason, or the consequence, of a surface rearrangement that occurs at temperatures up to 800°C [22].

Specific Component of the Surface Free Energy of Heat-Treated Silicas

Specific interaction capacities of heat-treated silicas, that is, their ability to interact with polar molecules, were examined with chloroform (Lewis acid probe) and toluene and benzene (amphoteric molecules). Figure 31.2 provides examples of the evolution of the specific interaction parameter I_{sp} of the different silicas with chloroform as a probe.

All amorphous silicas except C showed an I_{sp} that decreased with temperature of treatment, with a more or less pronounced step in the 200–400°C temperature range. This evolution, which parallels the silanol content of the solids [23], suggests that I_{sp} reflects the interaction

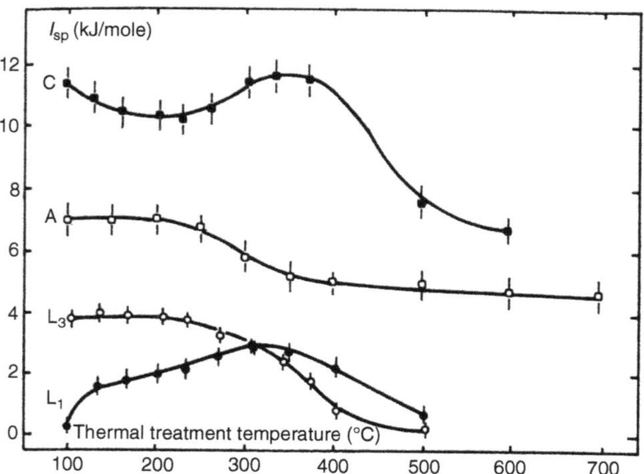

FIGURE 31.2 Evolution of the specific interaction parameter of silicas with chloroform versus heat treatment temperature.

of the probe with silanol groups. The I_{sp} of the silicas treated at the highest temperature also appear to be quite different; thus, a nonequivalent surface chemical state of the various amorphous samples is implied.

In crystalline silicas the variation of I_{sp} versus temperature is much more complex. L_1 behaves as an amorphous sample, in agreement with the evolution of its crystalline structure with temperature, whereas L_2 and L_3 show maxima at 450 and 350°C, respectively. Such a result means that interactions with chloroform are increasing while the total number of surface hydroxyls, which are theoretically responsible for these interactions, is decreasing. Thus, two antagonistic mechanisms have to be envisioned, one involving intercalation of the probe within the lamellar layers of the crystalline silicas (up to 350°C; the interlamellar distance is decreased, and walls covered with silanol groups are thus closer — as a consequence their influence on inserted chloroform molecules will increase, yielding higher interaction parameters), the other (above 350°C) associated with the condensation of hydroxyls, resulting in a loss of active sites and thus in a decrease of chloroform interactions [a process confirmed by ^{29}Si cross-polarization–magic-angle spinning (CP–MAS) NMR spectroscopy [23]].

Deactivation of silica surfaces by grafting of alkyl chains [esterification with short-chain (C_1) or long-chain (C_{16}) alcohols] was reported [24] to be associated with strong decreases in γ_s^d as well as I_{sp}, which for C_{16}-modified samples are very close to those exhibited by polyethylene (known to be a surface of very low energy). Thus, the behavior of esterified silicas versus temperature was of interest.

Evolution of the Surface Free Energy of Esterified Silicas with Heat-Treatment Temperature

Figure 31.3 shows the evolution versus temperature of γ_s^d of initial and esterified *G*. The modifications are associated with a decrease of the dispersive component of the surface free energy, a process very much dependent on the number of carbon atoms of the grafted alkyl chain. Thus, the silica reacted with hexadecanol can be considered completely coated by a hydrocarbon layer, whereas the part of the surface that reacted with methanol, which is unhindered by the methyl grafts, remains available for further interactions. The curves also show the thermal stability of the grafted alkyl chains. Starting at about 250°C, a steep increase of γ_s^d (particularly with C_{16}-modified silicas) is evident and is probably related to the pyrolysis of the grafted chains. This process seems to be completed by 500°C, because the curves corresponding to the esterified samples merge with that of initial silica at this temperature. The increase in γ_s^d of the modified silicas can thus be considered to result from a combination of phenomena

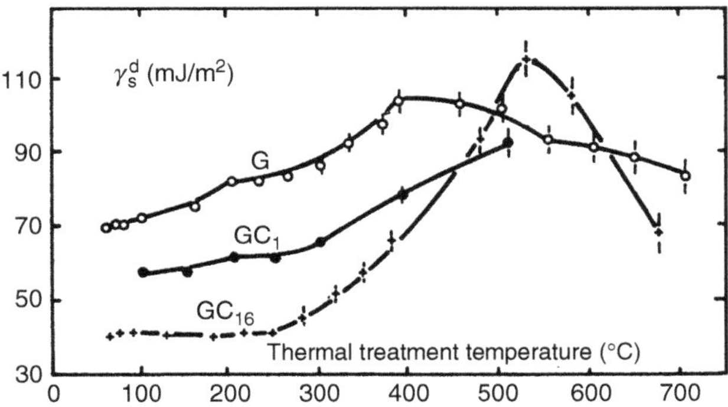

FIGURE 31.3 Evolution of the dispersive component of the surface free energy of esterified silicas versus heat treatment temperature. (Reproduced with permission from reference 26. Copyright 1990.)

corresponding to the behavior of ungrafted silicas and to the degradation of the grafted alkyl chains.

Specific interaction parameters are strongly reduced upon surface modification (hexadecylated silicas have I_{sp} values close to zero). In methanol-reacted silicas, I_{sp} values decrease when heat-treatment temperature increases. This result suggests that after grafting with methanol, some free hydroxyls are still available for further interactions. This scenario is also likely for C_{16}-grafted silicas, but for these silicas the residual silanol groups are completely shielded by the grafted chains and are thus inaccessible to the probes.

The results obtained with the different silicas point to the importance of the mode of preparation on their surface characteristics and thus on their surface heterogeneity. Understanding of the surface heterogeneity can be attained by calculation of the distribution function of the energy of adsorption of alkane molecules on the surface of the solids.

Distribution Function of Adsorption Energy

From chromatographic data it is possible to relate the amount of solute adsorbed on a solid to the equilibrium pressure and thus to plot its adsorption isotherm. For a heterogeneous surface, the experimentally measured adsorption isotherm can be described as a sum of local isotherms corresponding to different surface-active sites. The isotherm can then be represented by the following integral equation:

$$V(p) = \int_0^\infty \theta(p,\epsilon)\phi(\epsilon)\mathrm{d}\epsilon$$

where $\theta(p,\epsilon)$ is the local adsorption isotherm on sites corresponding to an adsorption energy ϵ, $\theta(\epsilon)$ is the distribution function of adsorption energy, and p is pressure [25,26]. Thus, by knowing the complete isotherm and using an approximation of local isotherm, it is possible to calculate $\theta(\epsilon)$. Because these are physical adsorption processes, only alkane probes were considered. Figure 31.4 shows the distributions of energy measured on silicas A, G, and L_3 with hexane as a solute. The distributions are bimodal and point to the existence of two different types of adsorption site on the surface. Moreover, the distributions appear to be dependent on the nature of the silica. The distribution calculated for silica L_3 is much narrower than those of silicas A and G. This result is in agreement with the surface topology of hydroxyls, which are known to be homogeneously distributed on the surface of L_3, whereas A exhibits a flat but chemically more heterogeneous surface, and G, which has a higher silanol content, is very heterogeneous [23]. However, it is difficult to assign the peak corresponding to a given energy of interaction to a particular type of surface functional group. Nevertheless, the distribution energy curves yielded by initial silica G were compared to those yielded by the same silica grafted with methanol (GC_1) and hexadecanol (GC_{16}). Esterification would be expected to be associated with a complex modification of the silica surface. It appears (Figure 31.5) that grafting is indeed followed both by a shifting of the distribution curve toward smaller energies (the longer the chains, the larger the shift) and by a decrease of the height of the peak corresponding to the smaller energy. These shifts can be related to a decrease of the accessibility of the surface due to the steric hindrance of the grafted alkyl chains. Moreover, because the area under the curves is proportional to the number of adsorption sites, the decrease of peak height could be related to the decrease of the number of accessible silanols. As a consequence, the low energy peak could tentatively be linked with the energy of interaction between the alkane probe and surface hydroxyls.

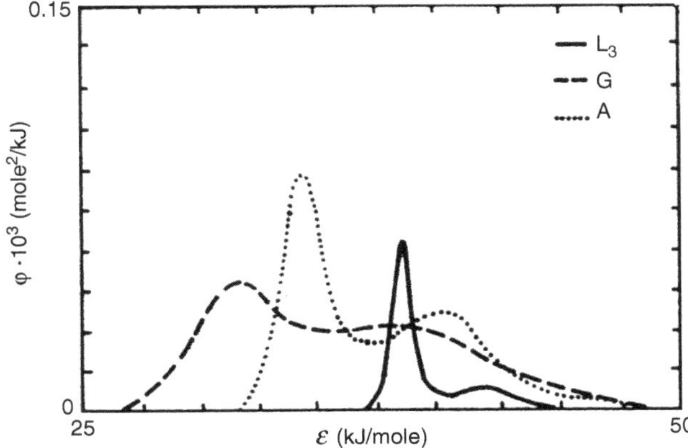

FIGURE 31.4 Distribution of the energy of adsorption of hexane for different silicas. (Reproduced with permission from reference 27. Copyright 1990.)

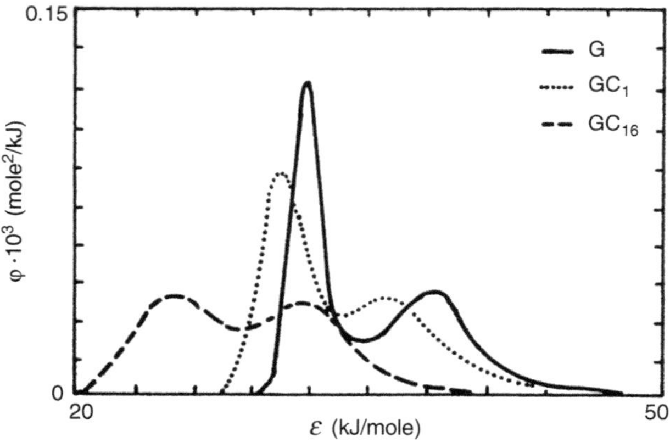

FIGURE 31.5 Distribution of the energy of adsorption of hexane for initial and esterified silicas. (Reproduded with permission from reference 27. Copyright 1990.)

CONCLUSIONS

The evolution of the surface free energy components of the different samples showed that the physicochemical surface characteristics of silicas and their surface heterogeneity are dependent on the mode of preparation. An approximation of surface heterogeneity was attained by calculation of the distribution function of the energy of adsorption of alkane probes on the solid surfaces.

REFERENCES

1. Iler, R. K. *The Chemistry of Silica*; Wiley Interscience: New York, 1979.
2. Kondo, S. *Nippon Kagaku Kaishi* **1985**, *6*, 1106.
3. Wagner, M. P. *Rubber Chem. Technol.* **1976**, *49*, 703.
4. Hertl, W. *J. Phys. Chem.* **1968**, *72*, 1248.
5. Armistead, C. G.; Tyler, A. J.; Hambleton, F. H.; Mitchell, S. A.; Hockey, J. A. *J. Phys. Chem.* **1969**, *73*, 3947.
6. Evans, B; White, T. E. *J. Catal.* **1968**, *11*, 336.
7. Vidal, A.; Papirer, E.; Donnet, J. B. *J. Chim. Phys.* **1974**, *71*, 445.
8. Zaborski, M.; Vidal, A.; Ligner, G.; Balard, H.; Papirer, E.; Burneau, A. *Langmuir* **1989**, *5*, 447.
9. Morrow, B. A.; Cody, I. A. *J. Phys. Chem.* **1972**, *77*, 1465.
10. Hair, M. L. *J. Non-Cryst. Solids* **1975**, *19*, 199.
11. Sander, L.; Callis, J. B.; Field, L. R. *Anal. Chem.* **1983**, *55*, 1068.
12. Brinker, C. J.; Kirkpatrick, R. J.; Tallant, D. R.; Bunker, B. C.; Montez, B. *J. Non-Cryst. Solids* **1988**, *99*, 418.
13. Brinker, C. J.; Tallant, D. R.; Roth, E. P.; Ashley, C. S. *J. Non-Cryst. Solids* **1986**, *82*, 117.
14. Maciel, G. E.; Sindorf, D. W. *J. Am. Chem. Soc.* **1980**, *102*, 7606.
15. Miller, M. L.; Linton, R. W.; Maciel, G. E.; Hawkins, B. L. *J. Chromatogr.* **1985**, *319*, 9.
16. Fripiat, J. In *Soluble Silicates*; Falcone, J. S., Ed.; ACS Symposium Series 194, p 165.
17. Conder, J. R.; Young, C. L. *Physicochemical Measurements by Gas Chromatography*; Wiley Interscience: New York, 1979.
18. Aulich, H. A.; Eisenrith, K. H.; Urbach, H. P. *J. Mater. Sci.* **1984**, *19*, 1710.
19. Le Bihan, M. T.; Kalt, A.; Wey, R. *Bull. Soc. Fr. Minéral. Cristallogr.* **1971**, *94*, 15.
20. Papirer, E.; Balard, H.; Vidal, A. *Eur. Polym. J.* **1988**, *24*, 783.
21. Sindorf, D. W.; Maciel, G. E. *J. Am. Chem. Soc.* **1983**, *105*, 1487.
22. Feltl, L.; Lutovsky, P.; Sosnova, L.; Smolkova, E. *J. Chromatogr.* **1974**, *91*, 321.
23. Ligner, G.; Vidal, A.; Balard, H.; Papirer, E. *J. Colloid Interface Sci.* **1990**, *134*, 486.
24. Vidal, A.; Papirer, E.; Wang, M. J.; Donnet, J. B. *Chromatographia* **1986**, *23*, 227.
25. Rudzinski, W.; Jagiello, J.; Grillet, Y. *J. Colloid Interface Sci.* **1982**, *48*, 478.
26. Jagiello, J.; Ligner, G.; Papirer, E. *J. Colloid Interface Sci.* **1990**, *137*, 128.
27. Legrand, A. P. et al. *Adv. Colloid Interface Sci.* **1990**, *33*, 91.

32 Diffuse Reflectance FTIR Spectroscopic Study of Base Desorption from Thermally Treated Silica

Kristina G. Proctor, Sally J. Markway, Miguel Garcia, Cheryl A. Armstrong, and Chad P. Gonzales
Colorado State University, Department of Chemistry

CONTENTS

Diffuse reflectance Fourier transform infrared spectroscopy (DRIFTS) is a convenient and useful tool to decipher information regarding silanol (SiOH) site populations on porous silica surfaces. Studies that employ probe base species that interact with accessible silanols, provide spectral evidence of distinct silanol populations having differing affinity for, and strength of interaction with the base. Pyridine is a weak base that was selected for use due to its ability to exhibit variable strengths of hydrogen bonding interaction with silanols. Desorption processes have been monitored in real-time via DRIFTS using the unassociated silanol site band near 3740 cm^{-1}and a ring vibration of hydrogen bonded pyridine at 1445 cm^{-1}. The resultant kinetic data provides insight into the organization of silanols on the surface by virtue of the kinetic order of the desorption process with respect to the base concentration. Pyridine desorption studies have been applied to two porous silica gels having differing properties and intended applications; a general purpose type A gel, and a high purity, chromatographic grade, type B silica. Both silicas were thermally treated at different temperatures in the 200–400°C range and desorption studies conducted on the resultant surfaces. Major findings for the study include second order desorption kinetics with respect to pyridine concentration, evidence for three individual rate processes, increasing desorption rate constant thus diminished acidity with increasing pretreatment temperature, and a greater general acidity for silanol sites on type A versus type B silica. Data and results of the pyridine desorption studies followed by DRIFTS are presented and discussed.

Amorphous silica is a versatile material that has been employed as a substrate in a number of important applications including catalysis and chromatography [1]. The popularity in use of porous silica is based upon several important properties including abundance in nature, thus cost effectiveness in production, large surface area, rigidity, and surface reactivity that allows for selective modification to impart desirable properties. The wide-spread utility of silica is demonstrated by a corresponding, voluminous amount of research aimed at elucidating the structure and chemical behavior of surface constituents. The research has spanned over 60 years and many useful summaries of the literature have been written [2–3].

Silanol sites (SiOH) impart much of the surface behavior of silica and serve as the point of chemical attachment via derivitization processes [4]. Basic agreement has been achieved with respect to the existence of three primary configurations of silanol sites. The orientations are depicted in Figure 32.1 and include isolated or unassociated silanols, vicinal sites that are engaged in

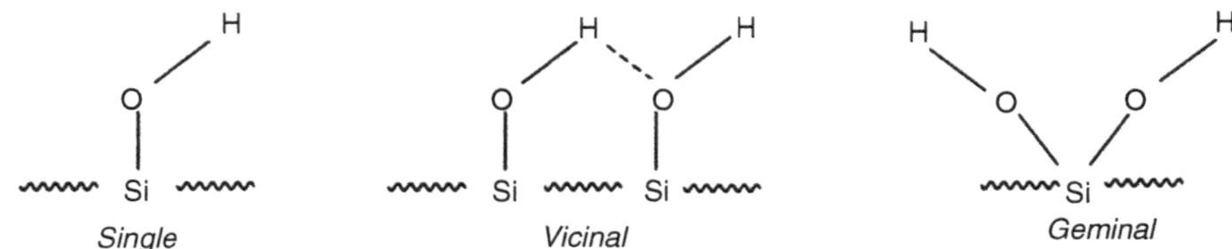

FIGURE 32.1 Configuration of silanol sites on the silica surface.

hydrogen bonding with neighboring sites, and geminal sites that have a diol structure [1–2]. Although agreement has been found with the types of sites that exist, many inconsistencies are reported in regard to the exact concentration, surface distribution, and degree of chemical reactivity and acidity associated with the different sites. Discrepancies in the literature can be attributed to multiple sources including the surface selectivity of the analytical technique employed and variability of the production procedures and specific type of silica surface studied (e.g., aerosils, fumed silica, precipitated silica, porous gels, etc.). In addition, many factors influence the number, type, orientation and chemical environment of the silanol sites. The list includes degree of surface hydration and contamination as influenced by pretreatment procedures used to cleanse the substrate prior to analysis, surface curvature, mean-pore diameter, particle morphology and diameter.

Numerous established and novel surface characterization methods and approaches have been brought to bear in an effort to advance the current level of understanding of the silica surface. Recent contributions to the literature have included chemometrics [5], atomic force microscopy [6], FT-Raman spectroscopy [7] and theoretical approaches [8]. The instrumental techniques of Fourier transform infrared (FTIR) spectroscopy and solid-state nuclear magnetic resonance spectroscopy (NMR) have been used extensively due to the information that can be deciphered regarding the chemical nature, orientation, number and interaction of silanol sites [9]. The techniques have several key and complementary advantages in the type of information that can be derived. Solid-state ^{29}Si NMR spectroscopy is powerful in that distinguishable spectral peaks are generated and can be quantified for the single and geminal silanol site populations [10–12]. Alternatively, infrared spectra of silica allow for distinction between silanol sites that experience differing degrees of perturbation through hydrogen bonding with neighboring sites or an external adsorbate [13–14]. A variety of surface sampling techniques of FTIR spectroscopy have been developed [15] and several summaries have been written [16–17]. The method of diffuse reflectance Fourier transform infrared spectroscopy (DRIFTS) is advantageous in maintaining the structural integrity of the surface in contrast to traditional transmission analysis using pressed pellets which has been shown to induce physical and chemical surface alterations [18–20]. DRIFTS is the most selective of the surface sampling options in FTIR analysis [16]. DRIFTS is advantageous in that instrumentation is relatively inexpensive, a spectrum can be acquired in a fraction of a second allowing for signal averaging thus excellent signal to noise ratio, and commercially available chambers allow for flexible experimental control of the sample environment at the time of spectral acquisition [17]. DRIFTS can be employed to follow real time kinetic processes of silica surfaces including those involving a selected molecular probe. Careful acquisition and interpretation of DRIFT spectra allows for quantification [17,21] of spectral peaks including those arising from surface silanols and the surface modifier or probe species.

Provided in Figure 32.2 are DRIFT spectra of two porous silica gel surfaces after the removal of physisorbed water in the OH stretching region by mild evacuation at 100°C. The dry spectrum of silica demonstrates the presence of three primary bands. The sharp band centered near 3740 cm^{-1} has been assigned to silanols that are completely unassociated, or very weakly associated sites. Truly isolated silanols that are entirely free of perturbation are known to exhibit an absorption at 3750 cm^{-1} [13]. The feature near 3665 cm^{-1} is due to vicinal silanols that undergo mutual hydrogen bonding. The broad band centered near 3500 cm^{-1} has been assigned to sites that interact with residual physisorbed water. Variations in the peak maxima and the shape of the bands for the two DRIFT spectra, demonstrate the differences in properties of the two surfaces and their respective surface sites. Silica has been categorized into two types of substrates, A or B, based upon the position of the unassociated silanol band, the level of hydroxylation of the surface, and the relative acidity of silanol sites [4]. Type A silicas are believed to demonstrate a silanol band at or above 3740 cm^{-1} due to a non fully hydroxylated surface which results in silanol sites having greater acidity than corresponding type B silicas that demonstrate a silanol band position below 3740 cm^{-1} [22–23]. Zorbax Rx-SIL silica is a chromatographic grade, type B silica that has been exhaustively purified to remove metal ion contamination. Davisil is a more general-purpose porous silica gel having irregular particle morphology, a

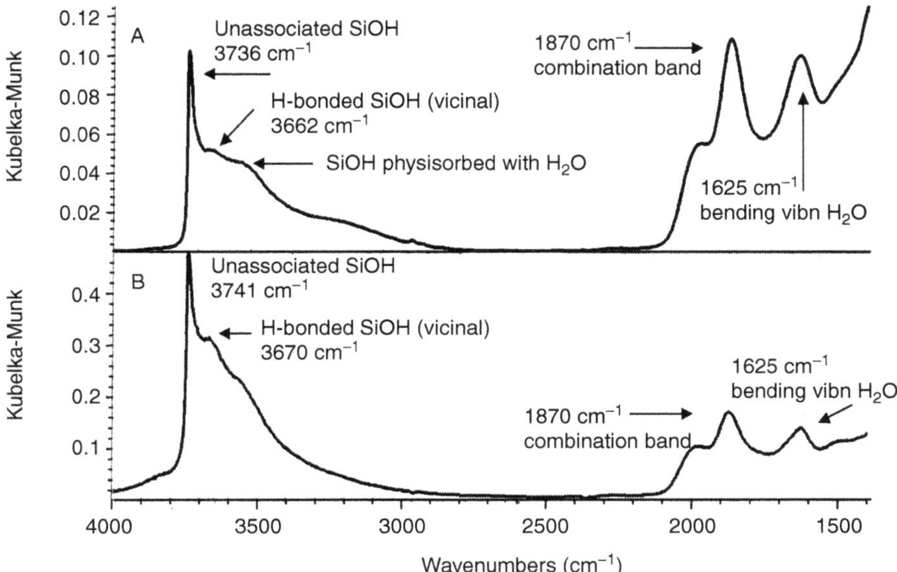

FIGURE 32.2 DRIFT Spectra of porous silica; (A) type B, Zorbax Rx-Sil, and (B) type A, Davisil, after the removal of physisorbed water from the surface by mild evacuation at 100°C.

relatively large and irregular particle diameter, and lower purity. The presence of geminal silanols is not directly distinguishable in either DRIFT spectrum in Figure 32.2. Previous IR studies conducted at high resolution have found evidence that the feature near 3740 cm^{-1} has contributions from not only single isolated silanols but also geminal sites [24–25]. Although unique IR bands are not present in the DRIFT spectrum of silica for all silanol species, evidence of different populations can be obtained through kinetic studies involving interaction of silanols with probe molecules. Quantitation of individual silanol populations can be determined through indirect procedures such as a surface titration with a probe base species that is monitored spectroscopically [26].

Pyridine is a weak base that serves as a useful probe for silanols due to a variable affinity for, and selectivity in, interaction with sites having differing acidity. Studies that follow surface interactions with pyridine benefit from data that provide information regarding silanols that are physically accessible by the base. Silanol sites that reside below the surface do not participate in the chemical equilibria associated with separation or catalytic processes, thus, are of little relevance from an applications standpoint. Parry [27] provided the original infrared band assignments for pyridine interaction with siliceous-based oxides and found distinguishable ring vibrations for different types of interactions. Pyridine that hydrogen bonds to weakly acidic silanols exhibits characteristic absorptions at 1445 and 1595 cm^{-1}. Interaction with very acidic Brönsted acid silanols, results in formation of the pyrdinium ion, giving rise to a distinctive band at 1540 cm^{-1}. Lewis acid sites, such as those found in silica-alumina (SiO_2-Al_2O_3), result in formation of a coordinate covalent bond with pyridine which is known to give rise to distinctive bands at 1450, 1490, and 1617 cm^{-1}. Presented in Figure 32.3 is a DRIFT spectrum of pyridine adsorbed on a dry and unmodified silica gel surface in the 1700–1400 cm^{-1} region. The bands that are present at 1445 and 1595 cm^{-1} indicate a hydrogen bonding interaction between silanols and pyridine thus the sites are not acidic enough to protonate the base.

Studies of pyridine interaction with siliceous surfaces are numerous and have been completed using a variety of techniques including temperature programmed desorption [28], DRIFTS [29] and many others. The original contribution made by this author in the first edition of

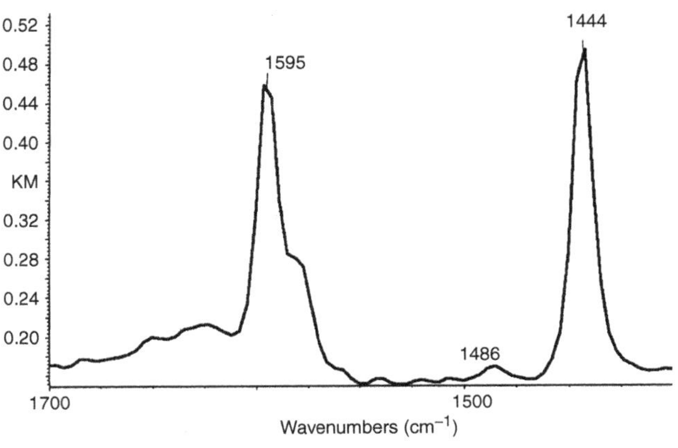

FIGURE 32.3 DRIFT spectrum of pyridine hydrogen bonded on a dried silica gel surface in the ring vibration region for pyridine.

The Colloid Chemistry of Silica included results for a study of the desorption kinetics of pyridine as it was removed from the surface of an amorphous, porous silica gel [30]. Included in this chapter are data, results, and discussion regarding pyridine desorption experiments that have been carried out on two different silica surface types which were exposed to thermal pretreatment in the 200–400°C range. Pyridine interactions have been studied to achieve an enhanced understanding of the types, configuration, distribution and relative acidity and concentrations of silanol sites present. The research completed includes studies of a general-purpose, type A, porous silica gel, and a type B, low acidity, chromatographic grade silica.

EXPERIMENTAL DETAILS

MATERIALS

Two amorphous silica gels were utilized in the investigations. Zorbax Rx-SIL (DuPont Inc., acquired from Agilent Technologies Newport) is a chromatographic grade, type B, low acidity silica that is exhaustively purified to remove metal ion contamination. Properties of the Zorbax silica are as follows; purity (99.999%), particle size (5 μm), mean-pore diameter (80 Å), surface area (182 m^2/g) and spherical particle shape. Davisil (Grace Davison Inc.) is a general-purpose silica with irregular particle shape, particle diameter (127–254 μm), purity (99 + %) purity, mean-pore diameter (150 Å), and surface area (300 m^2/g). SEM photos of the two silicas are provided in Figure 32.4 and Figure 32.5. KCl (Aldrich) was used for sample dispersions and was ground to a particle diameter (≤5 μm) with a Wig-L-Bug grinder (Crescent Dental Manufacturing) using the grinding ball for 2 min of operation. Pyridine (spectrophotometric grade, Mallinckrodt) was used for the desorption experiments.

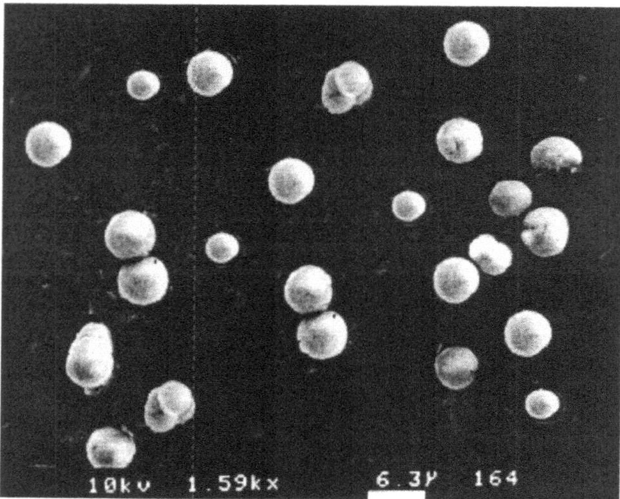

FIGURE 32.4 SEM photo of Zorbax Rx-Sil, type B silica using a scale of 6.3 μm.

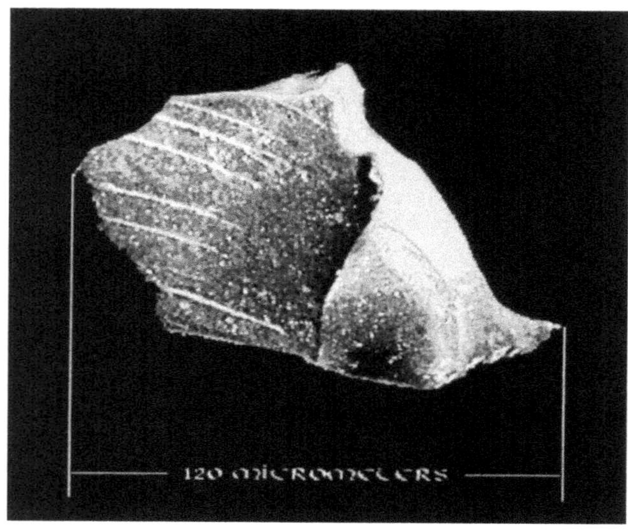

FIGURE 32.5 SEM photo of Davisil, type A silica using a scale of 120 μm.

INSTRUMENTAL ANALYSIS

DRIFT spectra were acquired using a Thermo-Nicolet 5SXC FTIR, single beam spectrometer equipped with a mercury-cadmium-telluride, type B, broad-band detector. A commercial DRIFT accessory (Spectra-Tech Collector^{TM}) and an environmental chamber (SpectraTech, high temperature/vacuum) were used. Unless otherwise noted, spectra were acquired by signal averaging 100 scans at 4 cm^{-1} resolution. Single beam spectra of silica dispersions were ratioed to a previously recorded background spectrum of dry and finely ground KCl. Spectra were collected and saved as percent reflectance and converted to Kubelka-Munk units prior to peak analysis. Peak positions and peak areas were identified using standard OMNIC software utilities on the FTIR. The areas of all peaks of interest were recorded and interpreted as a ratio of the peak area of the 1870 cm^{-1} SiOSi combination band, which has been shown to provide normalization to an internal standard [17]. A series of spectra were converted to both Kubelka-Munk and log(1/R) units and peak areas analyzed and compared. There was no appreciable difference in relative peak area trend data thus Kubelka-Munk units were used throughout the investigations.

PROCEDURES

Silica was thermally treated by heating in porcelain boats in an open-ended tube furnace (Thermolyne Model 1100) for 24 hours at the desired temperature (200–400°C).

Samples were allowed to cool within the tube furnace for 1 hour and immediately transferred to a glass vial and stored in a dessicator. Prior to spectral acquisition, KCl was placed in the sample cup of a DRIFT environmental chamber and heated under mild vacuum at 100°C to remove physisorbed water. The chamber was cooled to 25°C and a DRIFT spectrum recorded and stored for the background file. Dispersions of silica (15% w/w) were carefully prepared in finely ground KCl and stored in a 100°C drying oven overnight prior to spectral analysis. An aliquot of the dispersion having known %w/w, was accurately weighed and transferred to a bed of KCl previously placed in the sample cup of the DRIFT environmental chamber. The dispersion was mixed with the top layer of the KCl and compacted lightly with a smooth object to obtain a horizontally uniform surface and constant sample depth and compaction. A dry spectrum of the silica dispersion at room temperature was acquired after removing physisorbed water by heating at 100°C under mild vacuum for 30 min.

Desorption experiments were carried out at 25°C by introducing an excess amount of pyridine (4 μl) into the sealed and evacuated DRIFT environmental chamber containing a dry silica dispersion (4 mg silica) and allowing the system to equilibrate for 30 min. An initial DRIFT spectrum was recorded and the stopcock to a standard vacuum line was opened. Subsequent DRIFT spectra were recorded at various time intervals for 5 hours. Pertinent peak areas of the unassociated silanol band (3740 cm^{-1}), hydrogen-bonded pyridine (1445 cm^{-1}) band and SiOSi combination band (1870 cm^{-1}) were recorded, ratios calculated, and kinetic plots generated for the silanol and pyridine peak area ratios versus time.

DATA, RESULTS AND DISCUSSION

DESORPTION KINETICS

Data collected from the desorption experiments are presented in Table 32.1. A plot of the relative change in peak area of the hydrogen bonded pyridine band (1445 cm^{-1}) and the unassociated silanol band (near 3740 cm^{-1}) is provided as a function of time in Figure 32.6 for the desorption experiment conducted with the Davisil sample pretreated at 200°C. Similar plots result from data collected for the other samples analyzed in the study. Removal of the excess pyridine was found to be complete within the first 1–2 min of the desorption procedure after which there was a concomitant inverse change in the areas of the pyridine and unassociated silanol bands. As anticipated, removal of pyridine with time results in an increased band area of the unassociated silanols.

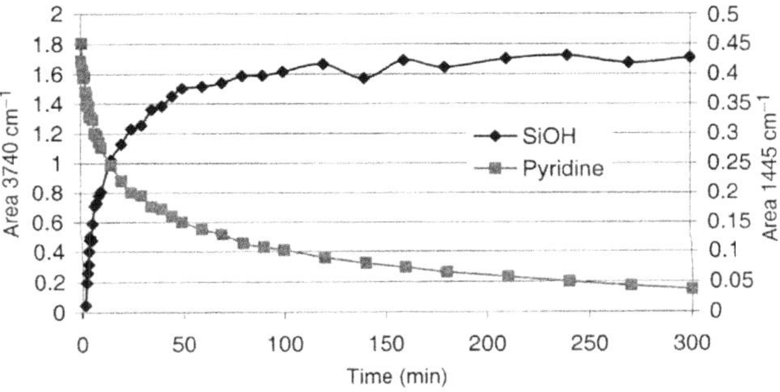

FIGURE 32.6 Change in the unassociated silanol and pyridine peak areas as a function of time of desorption for a Davisil silica sample pretreated at 200°C.

Included in Figure 32.7 is a graph demonstrating the relative rate of change of the 1445 and 3740 cm^{-1} band areas for the 200°C Davisil sample. Three important inferences can be made regarding the data. First, there are two different relative rates of change with respect to the removal of hydrogen bonded pyridine and the appearance of the unassociated silanol band. Second, the initial process occurs within the first 50 min of desorption and the latter between 50–300 min. Third, the first process occurs at a rate that is 3 times greater than the latter process. Taken together, the data indicates the existence

TABLE 32.1
Kinetic Data for Pyridine Desorption from Thermally Treated Silica

Silica	Pretreatment temp (°C)	2nd Order rate equation: Slope × 10^{-2}	y-int	R^2	Unassoc. SiOH (cm^{-1})	Pyridine remaining after 5 h: 1445 cm^{-1} area	%
Davisil	200	7.42	2.68	0.993	3740	0.037	9.1
	300	9.48	2.69	0.994	3743	0.032	8.4
	400	13.7	2.31	0.991	3745	0.013	3.6
Zorbax	200	15.9	3.23	0.995	3738	0.019	5.1
	300	30.8	3.79	0.984	3742	0.010	4.2
	400	54.6	4.13	0.984	3744	0.002	1.1

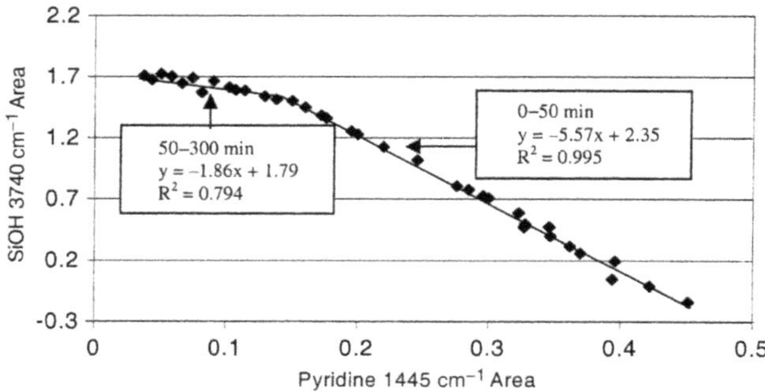

FIGURE 32.7 Relative change in the unassociated silanol and hydrogen bonded pyridine band areas for the Davisil silica sample thermally treated at 200°C.

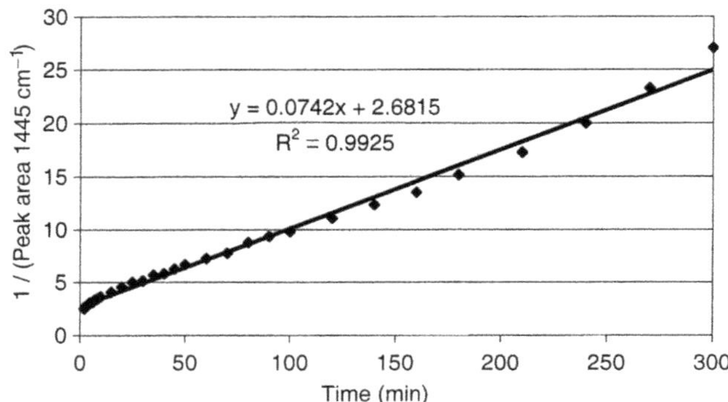

FIGURE 32.9 Second order kinetic plot of the 1445 cm^{-1} pyridine peak area data for the Davisil silica sample thermally treated at 200°C.

of at least two unassociated silanol populations, one in which pyridine is most easily removed during the initial 50 min, and a second that retains pyridine longer and releases pyridine at a much slower rate. A logical interpretation supported by previous research [24–25] would be that these populations are single and geminal silanols. It is likely that the silanol sites that initially release pyridine most quickly are single sites and that geminal silanols retain the base for a longer period of time. This is supported by previous titration experiments of these silicas in which it was determined that geminal silanols comprise 33.1% and 31.4% of the total silanol population for Zorbax and Davisil respectively, and that pyridine preferentially seeks out these sites during the initial titration process [31].

The kinetic order of the desorption process provides useful information as to the manner in which probe species interact and are removed from silanols on the surface. Provided in Figure 32.8 and Figure 32.9 are respective first and second order plots for the desorption data of the 1445 cm^{-1} peak area of pyridine as it was removed from the surface of the Davisil sample pretreated

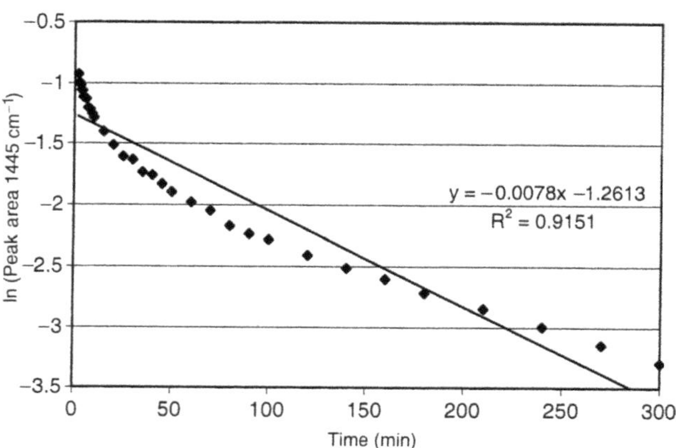

FIGURE 32.8 First order kinetic plot of the 1445 cm^{-1} pyridine peak area data for the Davisil silica sample thermally treated at 200°C.

at 200°C. Similar plots exist for all Davisil and Zorbax samples analyzed using this procedure. The graphs were created by calculating and plotting data using the first (32.1) and second (32.2) order integrated rate equations shown below. In the equations, t is time, [H-Pyd] represents the 1445 cm^{-1} peak area thus relative concentration of the hydrogen bonded pyridine band at time = 0 or time = t, and k is the rate constant for the desorption process.

$$\ln[\text{H-Pyd}]_{t=t} = -kt + \ln[\text{H-Pyd}]_{t=0} \qquad (32.1)$$

$$[\text{H-Pyd}]_{t=t}^{-1} = kt + [\text{H-Pyd}]_{t=0}^{-1} \qquad (32.2)$$

If the data for the entire desorption experiment is to be described by a singular kinetic process, the second order plot is overwhelmingly a more accurate model for the removal of pyridine from the surface silanols. The corresponding correlation coefficients for the first and second order plots are 0.901 and 0.993 respectively, demonstrating a much greater linear agreement with the second order plot. A similar result was observed in the research presented by Leyden and Proctor in the first edition of *The Colloid Chemistry of Silica* [30]. The overall observation implies that the vast majority of surface silanols are configured and behave as "paired" sites due to their proximity to one another. A reasonable explanation is that disruption of the hydrogen bond between pyridine and one silanol of the pair is followed in close succession by disruption of a second silanol-pyridine hydrogen bond. This would occur if the two silanol sites were close enough in proximity that perturbations to the electropositivity of the first hydrogen of the pair induces changes to the second silanol as the first pyridine molecule is removed. The explanation is consistent with the wide-spread acceptance of the existence of vicinal silanols that are close enough to experience mutual hydrogen bonds and geminal silanols that are diol species. The proposed mechanism is also consistent with evidence that silanol sites are clustered in patches on the surface [31] of the Zorbax silica, thus configured in a manner whereby the vast majority of silanols are close to neighboring sites.

Although most pyridine was removed from the sample surfaces during the 5 hour experiments, some silanols were able to retain the base. Experimental data for the Davisil sample treated at 200°C, indicates that 9.1% of the pyridine is retained on the surface after 5 hours of desorption. The silanols that strongly hydrogen bond pyridine exhibit a greater Brönsted acidity toward base species and obviously comprise a small relative population of silanols on the surface. The structural nature of these strongly acidic silanols cannot be determined from the data; however, others have speculated that these species could be single, truly isolated sites, or terminal (end) vicinal Si–OH, which do not hydrogen bond to a neighbor. The likelihood of the former will be addressed in the discussion of the data from the samples pretreated at increasing temperatures that follows.

Effects of the Thermal Pretreatment Temperature

Comparative data for the samples pretreated at 200, 300, and 400°C are presented in Table 32.1 and in Figure 32.10–Figure 32.14. DRIFT spectra of the Davisil and Zorbax samples treated at the various temperatures are included in Figure 32.10 and Figure 32.11. Inspection of the DRIFT spectra in the SiOH stretching region verifies a positive shift in the position of the unassociated silanol band with increasing treatment temperature for both silicas studied. The data presented in Figure 32.12 illustrates a 6 cm^{-1} shift for the Zorbax silica, and a 5 cm^{-1} shift for Davisil as the pretreatment temperature increases from 200–400°C. In all respective cases, the band position for the unassociated silanol band on the type A silica is higher than that of the type B silica. The band shift in the positive direction suggests that the silanols are becoming more isolated on the surface and are experiencing reduced perturbation from neighboring sites. Truly isolated SiOH are known to exhibit an absorption at 3750 cm^{-1}.

Provided in Figure 32.13 and Figure 32.14 are the second order kinetic plots for data collected from the 200, 300 and 400° C treated samples of Davisil and Zorbax, respectively. The corresponding data for the linear regression equations and the pyridine band areas remaining on the surface after 5 hours are included in Table 32.1. Several key conclusions can be drawn from the data. First, correlation coefficients for the linear fit to the second order integrated rate equation are large, positive values for all experiments conducted. Thus, a second order model accurately describes the desorption experiments for all samples if a singular process is assumed to occur. Second, the desorption rate constant increases with increasing pretreatment temperature for both Davisil and Zorbax substrates. If rate constants are compared for the 400°C samples relative to the 200°C samples, pyridine is removed 1.8 and 2.6 times more quickly on the 400°C samples of Davisil and Zorbax, respectively.

The rate constant data indicates that the rate, thus ease, of pyridine removal from surface silanol sites increases with increasing pretreatment temperature for both silicas. If diminished acidity of silanol sites results in a weaker hydrogen bond to pyridine, then the data indicates that increasing treatment temperature results in not only fewer silanols, but also silanols with reduced acidic strength. An important correlation is that thermal treatment of silica in the 200–400°C range is known to result in the condensation of silanols that experience mutual hydrogen bonds. Most commonly these are vicinal sites

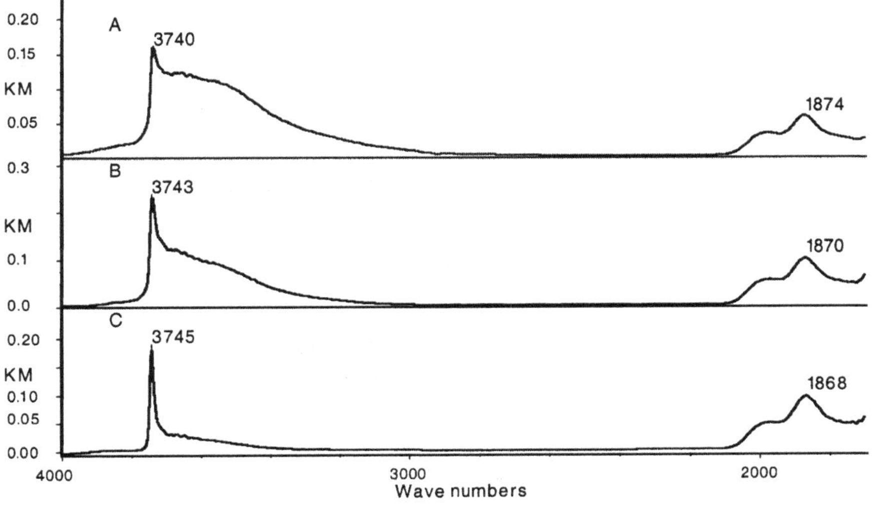

FIGURE 32.10 DRIFT spectra of Davisil silica after thermal treatment at (A) 200°C, (B) 300°C, and (C) 400°C.

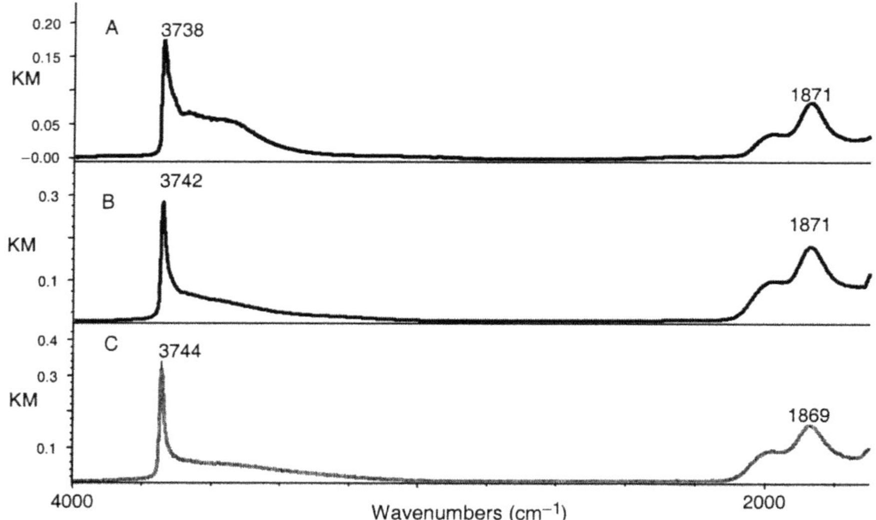

FIGURE 32.11 DRIFT spectra of Zorbax silica after thermal treatment at (A) 200°C, (B) 300°C, and (C) 400°C.

but could also be geminal sites that are able to hydrogen bond to sites on one or both sides of the diol species. A widely accepted outcome of thermal treatment and condensation of silanols, supported by NMR, IR and other evidence, is the creation of a greater number of isolated silanol species. Taken together, the results imply that as vicinal sites are reduced and more isolated species are generated by treatment at higher temperatures, the comparative acidity of silanol sites decreases, for both substrates studied.

The relative acidity of the silanol sites on the two types of silica surfaces studied can be compared with respect to the rate constants of the desorption process, and the amount of pyridine retained on the surface at the end of the desorption study. For the samples prepared at 200°C, the second order desorption rate constant was 7.42×10^{-2} for Davisil and 15.9×10^{-2} for Zorbax. Pyridine was removed more quickly from silanol sites on the low acidity, type B silica by a factor of 2.1, as anticipated. For the corresponding samples prepared at 300 and 400°C, the desorption rate constant for Zorbax was found to be 3.2 and 4.0 times greater than that for the Davisil samples, respectively. Therefore, increasing pretreatment temperature has a more profound effect on reducing silanol site acidity for the type B silica. This could be due to a greater relative population of vicinal silanols on the type B silica which are known to be eliminated with increasing pretreatment temperature.

The relative amount of pyridine retained on the surface after the 5 h desorption procedure may be taken as an indicator of the presence of strongly acidic silanols, and is measured by the peak area of the 1445 cm^{-1} band for the DRIFT spectra recorded at 300 min. A percentage of pyridine remaining can also be calculated from the ratio of the 1445 cm^{-1} peak area at time 300 min relative to that at 1.0 minute, before which excess pyridine is

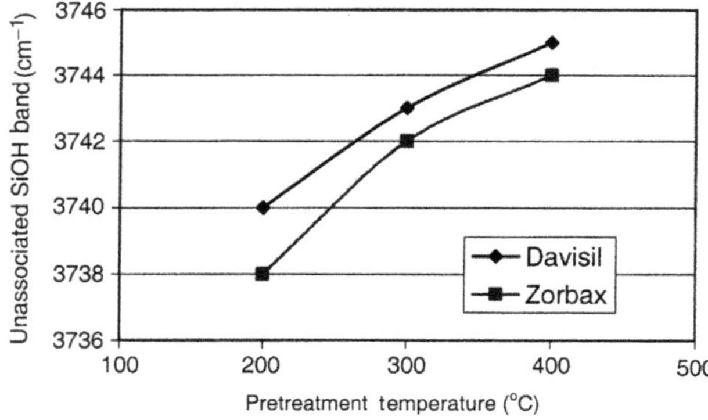

FIGURE 32.12 Change in the peak position of the unassociated silanol band for Davisil and Zorbax silicas thermally treated at 200–400°C.

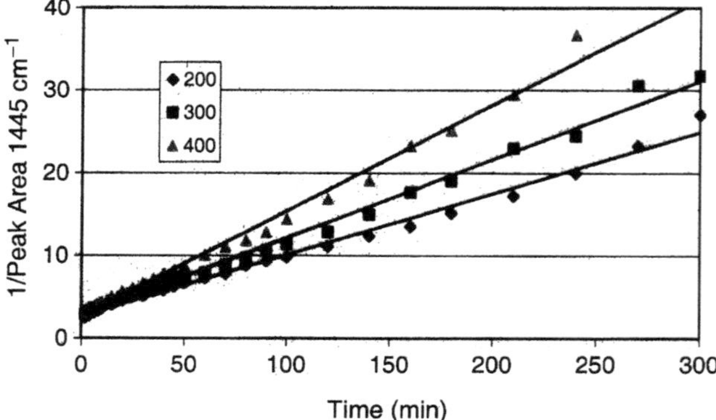

FIGURE 32.13 Second order kinetic plot for the desorption of pyridine from Davisil silica thermally pretreated at different temperatures.

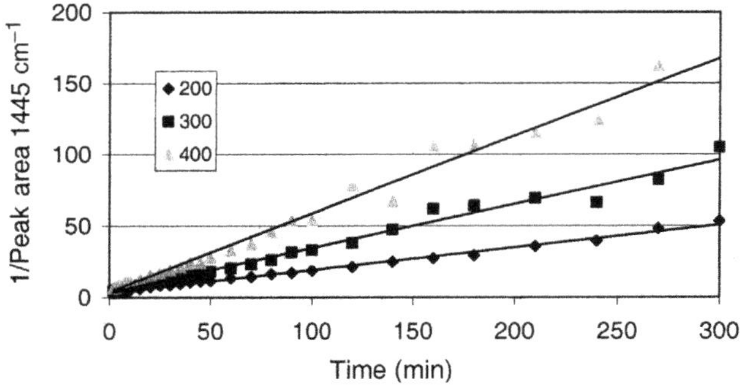

FIGURE 32.14 Second order kinetic plot for the desorption of pyridine from Zorbax silica thermally pretreated at different temperatures.

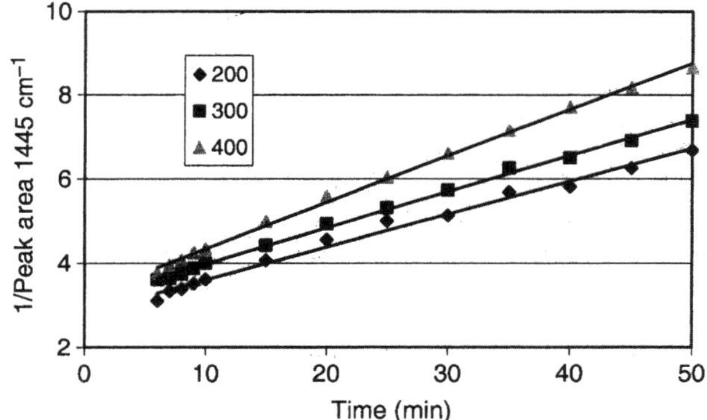

FIGURE 32.16 Second order kinetic plot for the desorption of pyridine from Davisil silica pretreated at different temperatures in the 6–50 minute region.

removed. The corresponding data is presented in Table 32.1. For Davisil samples treated at 200, 300 and 400°C, the percentage of pyridine retained on the surface was found to be 9.1, 8.4, and 3.6% respectively. The percentage pyridine retained data is in agreement with the increasing values for the desorption rate constants for the same samples, which indicate a general decrease in the relative acidity of silanol sites on the surface. Similar data is observed for the Zorbax samples where the percent pyridine remaining is 5.1, 4.2, and 1.1% for samples treated at 200, 300, and 400°C, respectively. An interesting and expected comparison is that the low acidity type B silica, demonstrates a significantly lower concentration and relative percentage of strongly acidic silanols, than does the type A silica, for respective samples prepared at the same pretreatment temperature.

Although the kinetic data derived in these experiments can be well represented by a singular second order desorption model, closer inspection of the data presents evidence for three different, and successive second order processes that are indicative of desorption from distinct types of silanol populations (e.g., vicinal, geminal, single isolated, etc.). Presented in Figure 32.15–Figure 32.17 are second order kinetic plots for the Davisil samples in the 2–6 min, 6–50 min, and 50–300 min time periods, respectively. Similar plots are provided in Figure 32.18–Figure 32.20 for the Zorbax samples. The data provide compelling evidence for three sequential desorption processes for all samples studied.

Included in Table 32.2 are numerical rate equation data for the linear regression analysis of the second order plots shown in Figure 32.15–Figure 32.20. Several inferences can be drawn from the data. First, the desorption rate constant associated with the process occurring in the 2–6 min period is larger than the rate constants for the 6–50 min and 50–300 min intervals, for all samples. This implies the presence of a group of unassociated silanols, contributing to the infrared band near 3740 cm^{-1}, which are the least acidic and most readily release pyridine. Second, rate constants for the 6–50 min

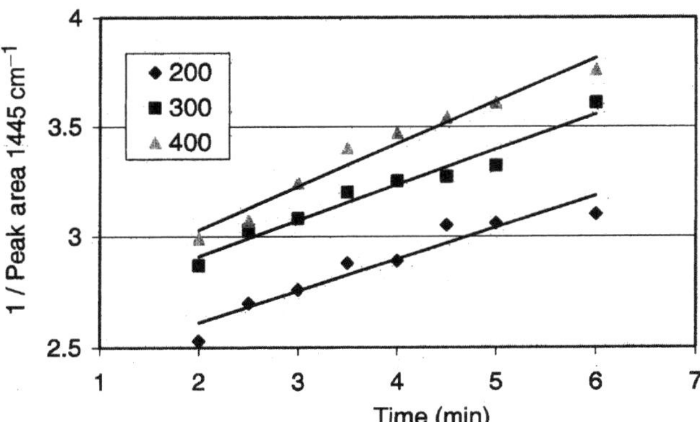

FIGURE 32.15 Second order kinetic plot for the desorption of pyridine from Davisil silica pretreated at different temperatures in the 2–6 min region.

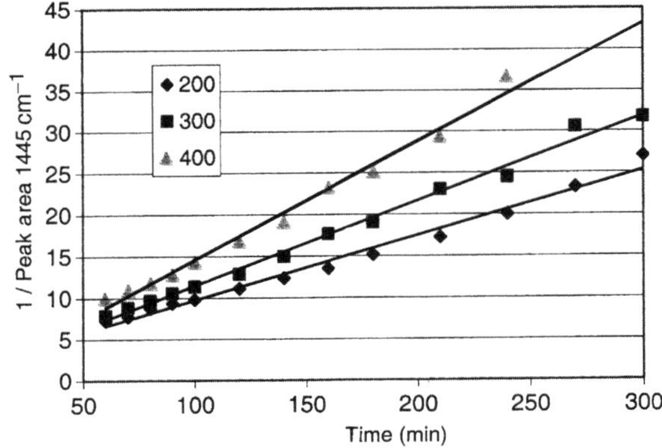

FIGURE 32.17 Second order kinetic plot for the desorption of pyridine from Davisil silica pretreated at different temperatures in the 50–300 min region.

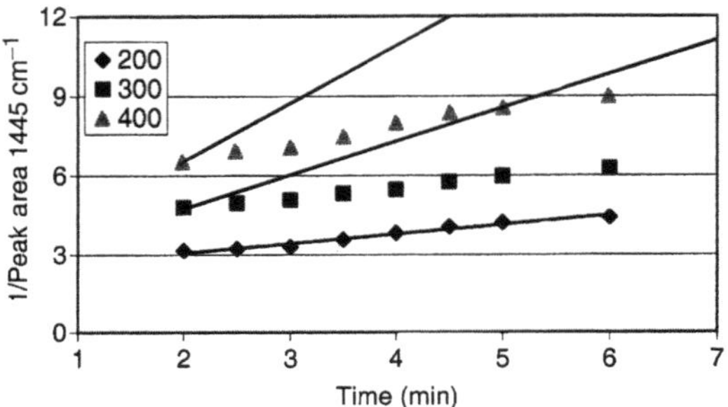

FIGURE 32.18 Second order kinetic plot for the desorption of pyridine from Zorbax silica pretreated at different temperatures in the 2–6 minute region.

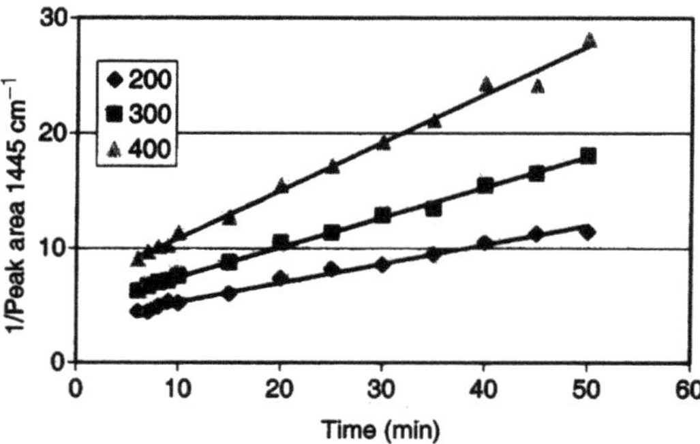

FIGURE 32.19 Second order kinetic plot for the desorption of pyridine from Zorbax silica pretreated at different temperatures in the 6–50 minute region.

period are smaller than those for the 2–6 min period, but are generally greater than the rate constant for the process occurring in the 50–300 min interval. The results suggest the presence of two additional and distinguishable groups of silanols. When pyridine is removed from silanols in the 6–50 min time interval, the unassociated silanol band grows at a similar rate as when the base is removed from the first, least acidic, population.

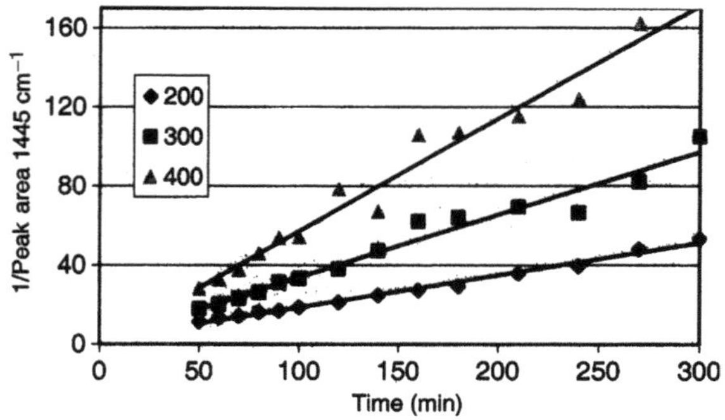

FIGURE 32.20 Second order kinetic plot for the desorption of pyridine from Zorbax silica pretreated at different temperatures in the 50–300 minute region.

TABLE 32.2
Kinetic Data for Individual Desorption Processes

		Zorbax			Davisil		
		2nd order rate equation			2nd order rate equation		
Time (min)	Pretreat. temp (°C)	Slope $\times 10^{-2}$	y-int	R^2	Slope $\times 10^{-2}$	y-int	R^2
2–6	200	37.5	2.36	0.970	14.3	2.33	0.914
	300	38.2	3.99	0.991	16.2	2.59	0.953
	400	65.3	5.25	0.981	19.7	2.64	0.967
6–50	200	16.8	3.61	0.989	7.80	2.82	0.990
	300	26.2	4.89	0.997	8.60	3.12	0.998
	400	41.7	6.72	0.993	11.1	3.24	0.998
50–300	200	16.2	2.53	0.990	7.80	1.96	0.984
	300	31.6	2.25	0.961	10.2	1.28	0.992
	400	57.0	0.053	0.963	14.3	0.281	0.987

A logical inference is that pyridine is removed from single, unassociated silanols during the initial desorption process (2–6 min) followed by removal from geminal silanols in the 6–50 min period. The disappearance of pyridine in the 50–300 min interval is most likely attributable to removal from vicinal silanols. This can be substantiated by the greatly reduced rate of increasing peak area change of the 3740 cm^{-1} band relative to change in the pyridine 1445 cm^{-1} band area, beyond 50 minutes (see Figure 32.7).

Consideration of the linear regression data included in Table 32.2, provides for two additional observations and subsequent conclusions. First, increasing pretreatment temperature of respective silica samples, results in increasing rate constant values within each of the successive desorption processes (2–6, 6–50 and 50–300 min). The observation implies that condensation of hydrogen bonding silanols to form a greater number of single, more isolated species, results in reduced acidity of all types of silanols on the surface. And finally, the rate constant values for removal of pyridine from all types of silanols on the type B silica, Zorbax, are consistently greater than the corresponding values on the type A silica, Davisil. Thus, this provides further evidence for the reduced capacity of silanols on the type B silica, having presumed lower acidity and capacity to hydrogen bond weak bases such as pyridine.

CONCLUSIONS

Diffuse reflectance FTIR spectroscopy has been shown to be a useful tool with which to study base desorption from siliceous surfaces *in situ*. Related experiments have yielded insight regarding the types and relative acidity of silanol sites. Kinetic order for the desorption process provides information regarding the configuration of silanol sites on the surface. In this study, all processes were observed to follow second order kinetics with respect to the base, strongly suggesting the great majority of silanol sites are paired in some capacity across the entire silica surface. Desorption rate constants provide a measure of strength of interaction between the acidic silanols and the probe base species. Rate constant data from this investigation suggest overwhelmingly that increasing pretreatment temperature results in silanol site populations having diminished acidity. Results from this study also demonstrate that all silanols on a type B silica have lower acidity than those on a type A silica. Three different silanol populations were found to exist on all of the respective silica samples pretreated in the range of 200–400°C. DRIFT spectral evidence suggests that the relative acidity of silanol sites is greatest for vicinal silanols, and that geminal silanols are more acidic than single, unassociated sites.

ACKNOWLEDGMENTS

The authors would like to respectfully thank the following agencies and individuals for financial support or other contributions to the research; the National Institutes of Health Minority Biomedical Research Support Program (grant 2 S06 GM0187 16-19), the Colorado Commission on Higher Education — Programs of Excellence (grant CRS 23-1-118), and Dr. J.J. Kirkland of Agilent Technologies-Newport for contribution of Zorbax Rx-Sil silica.

REFERENCES

1. K.K. Unger, ed., *Packings and Stationary Phases in Chromatographic Techniques*, Marcel-Dekker, New York, 1990.
2. R.K. Iler, *The Chemistry of Silica. Solubility, Polymerization, Colloid and Surface Properties, and Biochemistry*, Wiley-Interscience, New York, 1979.
3. K.K. Unger, *Porous Silica, Its Properties and Use as a Support in Column Liquid Chromatography, J. Chromatography Library*, Vol 16, Elsevier, Amsterdam, 1979.
4. J. Nawrocki, *The Silanol Group and Its Role in Liquid Chromatography, J. Chromtograph. A*, **779**, pp. 29–71, 1997.
5. M. Peussa, S. Härkönen, J. Puputti, and L. Niinistö, Application of PLS Multivariate Calibration for the Determination of the Hydroxyl Group Content in Calcined Silica by DRIFTS, *J. Chemometrics*, **14**, pp. 501–512, 2000.
6. M.J. Wirth, M.D. Ludes, and D.J. Swinton, Spectroscopic Observation of Adsorption to Active Silanols, *Anal. Chem.*, **71**, pp. 3911–3917, 1999.
7. S. Ashtekar, J.J. Hatings, P.J. Barrie, and L.F. Gladden, Quantification of the Number of Silanol Groups in Silicalite and Mesoporous MCM-41: Use of FT-Raman spectroscopy, *Spectroscopy Lett.*, **33(4)**, pp. 569–584, 2000.
8. J. Casanovas, F. Illas, and G. Pacchioni, *Ab initio* Calculations of ^{29}Si Solid State NMR Chemical Shifts of Silane and Silanol Groups in Silica, *Chem. Phys. Lett.*, 326, pp. 523–529, 2000.
9. B.A. Morrow and I.D. Gay, Infrared and NMR Characterization of the Silica Surface, *Adsorption on Silica Surfaces* (Series: Surfactant Science Series 90), E. Papirer, ed., Marcel Dekker Inc., New York, pp. 9–33, 2000.
10. G.E. Maciel and P.D. Ellis, NMR Characterization of Silica and Alumina Surfaces, NMR Techniques in Catalysis, *Chemical Industries*, **55**, pp. 231–309, 1994.
11. I.S. Chuang, and G.E. Maciel, Probing Hydrogen Bonding and the Local Environment of Silanols on Silica Surfaces via Nuclear Spin Cross Polarization Dynamics, *J. Am. Chem. Soc.*, **118**, pp. 401–406, 1996.
12. G.E. Maciel, C.E. Bronnimann, R.C. Zeigler, I-Ssuer Chuang, D.R. Kinney, and E.A. Keiter, Multinuclear NMR Spectroscopy Studies of Silica Surface, in *The Colloid Chemistry of Silica*, ACS Advances in Chemistry Series no. 239, pp. 269–282, 1994.
13. M.L. Hair, *Infrared Spectroscopy in Surface Chemistry*, Marcel Dekker, New York, 1967.
14. M.L. Hair and W. Hertl, Adsorption on Hydroxylated Silica Surfaces, *J. Phys. Chem.*, **73(12)**, pp.4269–4276, 1969.
15. P.R. Griffiths, *Chemical Infrared Fourier Transform Infrared Spectroscopy*, Vol. 3, Academic Press, New York, 1982.
16. J.P. Blitz, R.S.S. Murthy, and D.E. Leyden, Comparison of Transmission and Diffuse Reflectance Sampling Techniques for FT-IR Spectrometry of Silane Modified Silica Gel, *Appl. Spectrosc.*, **40(6)**, pp. 829–831, 1986.
17. J.P. Blitz, Diffuse Reflectance Spectroscopy, in *Modern Techniques in Applied Molecular Spectroscopy*, Francis M. Mirabella (ed.), Wiley-Interscience, New York, pp. 185–219, 1998.
18. F.H. Hambleton, J.A. Hockey, and J.A.G. Taylor, An Infra-red Investigation of the Effect of Pressure on Silica Powders, as Revealed by Deuterium Oxide Exchange, *Nature*, **208**, pp. 138–139, 1965.
19. P.W. Yang and H.L. Casal, *In-Situ* Monitoring of Solid-State Photochemical Reactions by Diffuse Reflectance Infrared Spectroscopy, *Appl. Spectrosc.*, **40(7)**, pp. 1070–1073, 1986.
20. V.A. Bell, V.R. Citro, and G.D. Hodge, Effect of Pellet Pressing on the Infrared Spectrum of Kaolinite, *Clays and Clay Minerals*, **39(3)**, pp. 290–292, 1991.
21. R.S.S. Murthy and D.E. Leyden, Quantitative Determination of 3-Aminopropyl Triethoxysilane on Silica Gel Surface Using Diffuse Reflectance Infrared Fourier Transform Spectrometry, *Anal. Chem.*, **58**, pp. 1228–1234, 1986.
22. J. Kohler, J.J. Kirkland, *J. Chromatogr.*, **385**, p. 125, 1987.

23. J. Kohler, D.B. Chase, R.D. Farlee, A.J. Vega, J.J. Kirkland, *J. Chromatogr.*, **352**, p. 275, 1986.
24. P. Hoffmann and E. Knözinger, Novel Aspects of Mid and Far IR Fourier Spectroscopy Applied to Surface and Adsorption Studies on SiO_2, *Surface Science*, **188**, pp. 181–198, 1988.
25. A.J. McFarlan and B.A. Morrow, Infrared Evidence for Two Isolated Silanol Species on Activated Silicas, *J. Phys. Chem.*, **95**, pp. 5388–5390, 1991.
26. K.G. Proctor, S.K. Ramirez, K.L. McWilliams, J.L. Huerta, and J.J. Kirkland, *The Progressive Effect of Surface Silylation on the Silanol Population of Silica*, Chemically Modified Surfaces, J.J. Pesek, M.T. Matyska and R.R. Abuelafiya, Eds., Royal Society of Chemistry, Cambridge, UK, pp. 45–60, 1996.
27. E.P. Parry, An Infrared Study of Pyridine Adsorbed on Acidic Solids. Characterization of Surface Acidity, *J. Catal.*, **2**, pp. 371–379, 1963.
28. I. Gillis-D'Hamers, I. Cornelissens, K.C. Vrancken, P. Van Der Voort, and E.F. Vansant, Modeling of the Hydroxyl Group Population Using an Energetic Analysis of the Temperature-Programmed Desorption of Pyridine from Silica, *J. Chem. Soc. Faraday Trans.*, **88(5)**, pp. 723–727, 1992.
29. K.G. Proctor and D.E. Leyden, Surface Acidity Characterization of Siliceous Materials by Variable Temperature Diffuse Reflectance FTIR, *Chemically Modified Oxide Surfaces*, Vol. 3, D.E. Leyden and W.T. Collins Eds., Gordon and Breach Science Publishers, New York, pp. 137–149, 1990.
30. D.E. Leyden and K.G. Proctor, Variable-Temperature Diffuse Reflectance Fourier Transform Infrared Spectroscopic Studies of Amine Desorption from a Siliceous Surface, *The Colloid Chemistry of Silica*, ACS Advances in Chemistry Series No. 234, H. Bergna Ed., pp. 257–267, 1994.
31. K.G. Proctor, K.M. Murphy, C.B. France, S.K. Ramirez, and A.R. Gennuso, DRIFTS Study of Silica Selection in Preparation of Substrates for HPLC, 7th International Conference on Chemically Modified Surfaces, Evanston, IL, June 1999.
32. H.A.M. Velhurst, L.J.M. van de Ven, J.W. deHaan, H.A. Claessens, F. Eisenbeiss, and C.A. Cramers, J. Chromatogr. A., 687, p. 213, 1994.

33 Salient Features of Synthesis and Structure of Surface of Functionalized Polysiloxane Xerogels

Yu.L. Zub and A.A. Chuiko
National Academy of Sciences of Ukraine, Institute of Surface Chemistry

CONTENTS

On the basis of the authors' works a consideration is given to salient features of synthesis of a new class of sorbents and supports (namely polysiloxane xerogels with nitrogen-, oxygen-, and sulfur-containing functional groups) which are produced by the sol–gel method. The structure of the synthesized xerogels is determined and ascertained by a number of physical techniques (SEM, TEM, IR and Raman spectroscopy, solid-state NMR spectroscopy, ESR spectroscopy, thermal analysis). A consideration is also given to structure-adsorption characteristics of the xerogels and to dependence of the characteristics on various factors.

When responding to challenges of practice (adsorption technologies, chromatography, analytical chemistry, sensor technology, etc.), over the past two decades the chemistry of modified silicas made a substantial progress [1–5]. Up to now, the grafting of carbonfunctional organic silicon or organic compounds to silica continues to be one of the promising methods for synthesizing materials of new types. It should be noted, however, that modified silicas produced in this way have a number of drawbacks. The first of them is a relatively low content of functional groups. The second drawback is bound up with the necessity to employ multistage syntheses (a method of surface assembly) for creating a surface layer with a complex structure. It gives rise to a further decrease in the content of surface functional groups, with the decrease being sometimes quite essential. Besides, when following the modification procedure, creation of surface layers with high hydrolytic and thermal stabilities is often problematic. The third limitation is due to the fact that this method does not enable one to exert control over numbers of grafted groups. The grafting procedure itself calls for a preliminary treatment of starting silicas and solvents (e.g., their thorough dehydration in order to avoid formation of secondary polycondensation products).

At the same time, there is a simple and convenient route for production of organic silicon materials whose surfaces contain diverse functional groups [6]. The route is based on the reaction of hydrolytic polycondensation of alkoxy- or chlorosilanes [Scheme (1)]. This route is referred to as a sol–gel method [a use is also made of such terms as a sol–gel procedure or sol–gel technology [7]. In essence, the method is based on the fact that during the course of addition of water (usually in the presence of catalysts) there proceeds hydrolysis of alkoxysilanes with formation of silanol groups $\equiv$Si—OH. The latter readily interact with each other (or with alkoxysilyl groups RO—Si$\equiv$) with formation of siloxane bonds $\equiv$Si—O—Si$\equiv$, which brings about emergence of oligomers. The subsequent condensation of these oligomers results in formation of polymers of various structures. Growth of polymer chains leads to appearance of colloidal particles (appearance of a sol) whose size enlargement gives rise to formation of aggregates and, in the long run, to a sol–gel transition. The subsequent treatment of the gel (aging, washing, drying, etc.) gives a xerogel.

$$Si(OR)_4 + (RO)_3SiR' \xrightarrow[-ROH]{+H_2O} (SiO_2)_x \cdot (O_{3/2}SiR')_y \qquad (33.1)$$

When producing functionalized polysiloxane xerogels, a use is most often made of two-component systems [in terms of alkoxysilanes; Scheme (1)], where one of the components is tetraalkoxysilane [most often tetraethoxysilane $Si(OC_2H_5)_4$] which performs the role of a structure-forming agent, while the second component is trialkoxysilane whose fourth substituent at the silicon atom becomes an introduced functional group R′. Instead of alkoxysilanes in the capacity of starting reagents it is possible to use chlorinated derivatives of silicon as well. However, preference is usually given to alkoxysilanes that are more convenient in the preparative aspect.

The employment of the sol–gel method for synthesizing organic silicon materials offers a number of advantages in comparison to modification of silicas. The major advantage consists in the possibility of using multiconstituent systems (in terms of alkoxysilanes). In principle, one can vary both the nature and ratio of structure-forming agents [$E(OR)_n$, E = Si, Al, etc.] and composition and ratio of alkoxysilanes of the $(RO)_3SiR'$ type. In the latter case an appropriate one-stage process enables one to produce polysiloxane xerogels with a complex surface layer that contains several functional groups R′ varying in their nature. Of note here is the fact that the total content of such groups may reach a value (of 3.5–4.0 mmol/g) that is higher (by a factor of about 4–5 and more) than that in the situation with modified silicas. Moreover, if a use is made of monocomponent systems, the total content of accessible functional groups turns to be even higher (almost by an order of magnitude) in comparison to modified silicas [8,9], which is of essential importance for application of such materials in adsorption technology. It should also be noted that substituents R′ in trialkoxysilanes $(RO)_3SiR'$ with an end electron-donating (or accepting) group [of the —$(CH_2)_3NH_2$, =$[(CH_2)_3NH]_2C$=S, —CH=CH_2 type] are able to enter into various chemical reactions including reactions that lead to formation of metal complexes [Scheme (2)]. The formed coordination compounds with peripheral alkoxysilyl groups can enter into a subsequent reaction of hydrolytic polycondensation. In the presence of structure-forming agent $Si(OR)_4$ such a reaction can give gels whose appropriate treatment yields xerogels with incorporated metal complexes. Such systems are of an indubitable interest both as specific sorbents and as catalysts [10]. Generally speaking, employment of the sol–gel method offers ample scope both for designing novel materials and for modifying their surfaces.

$$(RO)_3SiR' + ML_m \longrightarrow (RO)_3SiR' \cdot ML_m \qquad (33.2)$$

It should also be noted that the possibility for a wide choice of hydrolytic polycondensation reaction parameters (nature, composition, and ratio of reacting components; nature of a nonaqueous solvent; amount of a hydrolyzing agent; pH and temperature of a reaction medium; nature of intermicellar liquid; conditions of aging and drying of a gel; etc.) allows one to exercise some control over properties of final products (in the first place over structure-adsorption characteristics of xerogels).

With allowance for the fact that in terms of numbers of components (including those containing various functional groups) the systems under consideration are practically unrestricted, we may infer that their use will have much potentiality for designing novel materials with unique properties. From this standpoint it is not surprising that this field of chemistry made a rapid progress, especially during the past decade. At the same time, it should be noted that, as far as we know, the first works intended to produce sorbents with functional groups of basic [$\equiv Si(CH_2)_3NH_2$] and acidic [$\equiv Si(CH)_2COOH$] character were published as early as at the beginning of the 1960s [11,12]. However, later on the available papers were concerned mainly with researches into organic silicon adsorbents whose surfaces contained alkyl and aryl radicals [13]. The physicochemical foundation of synthesis of nonspecific adsorbents by the sol–gel method was formed by the cycle of works performed by Slinyakova's team. In 1977 this team together with Voronkov's team published a paper [14] which provided a description of synthesis of xerogels using a reaction of hydrolytic polycondensation of mercaptomethyltrimethoxysilane. It was shown that on the surface of this xerogel there were sulfhydryl groups that made the substance an effective sorbent of a number of heavy and noble metals. This publication gave rise to a series of papers concerned with sulfur-containing organic silicon sorptive materials [8,9]. Publication of papers by Parish and his collaborators at the end of the 1980s [15,16] provided a new stimulus to designing methods for synthesizing such materials with other functional groups and to researches into properties and application of these materials. It is evident that presence of diverse groups during the course of the sol–gel synthesis will exert an influence on the composition and structure of xerogels in general and on their surface in particular. The present brief survey of our relevant works is an attempt to give a consideration to salient features of such systems.

SYNTHESIS OF FUNCTIONALIZED POLYSILOXANE XEROGELS

SYNTHESIS OF XEROGELS WITH AMINO GROUPS IN A SURFACE LAYER

As it was mentioned above, as far back as 1966 a method was advanced for producing "aminoorganosilica gel" by hydrolysis of 3-aminopropyltriethoxysilane (APTES) [or of its mixtures with tetraethoxysilane (TEOS)] in an alkali medium [11]. The procedure for producing poly (3-aminopropyl)siloxane (PAPS) xerogel involved interaction of a mixture TEOS/APTES (42:15 cm^3) in 50 cm^3 of methanol in the presence of 10 cm^3 of a 0.5 *M* solution of NaOH. The obtained gel was washed for 7 days up to the neutral reaction of washing waters and dried at 120°C up to constant sample weight. The adsorptive capacity of the sorbent with respect to methacrylic acid was equal to 1.6 cm^3/g.

In 1989 Parish and co-workers gave an independent description of somewhat different procedure for synthesizing PAPS and of some properties of this material [15]. During several years that followed this xerogel was repeatedly prepared by numerous groups of researchers (see Table 33.1) and studied in more detail in comparison to other xerogels. The aim of practically all the works concerned with synthesis of PAPS was to establish such synthetic procedure which would yield a sorbent with as much as possible extended surface area that would be distinguished for the maximal content of amine groups. To our mind, however, in the description of the procedures for synthesizing PAPS there are some discrepancies. Thus, referring to reference [15] Yang and co-authors [17] described a synthesis of PAPS in methanol in the presence of a 0.20 *M* solution of HCl, with molar ratio TEOS/APTES being equal to 2:1. However, the authors of reference [15] described the method for synthesizing PAPS according to which no use was made of any solvent. In addition, the synthesis described in reference [15] did not involve a catalyst either since in this case APTES itself acted as an "inner" catalytic agent. Besides, in accord with the procedure suggested in reference [15], at first water should be added to TEOS (it is not out of place to mention here that TEOS is not mixed with water) and only then it is necessary to add an amine-containing silane, which does not follow from the procedure advanced in reference [17]. In this connection, we made several attempts [18–20] to identify all the factors that exert an influence on structure-adsorption characteristics of xerogel PAPS and, in this way, to work out an optimal procedure for its production. It was proved that this procedure should involve a reaction of joint hydrolytic polycondensation of APTES and TEOS [Scheme (3)] in an alkaline medium whose creation in the system after addition of water was due to the fact that APTES possessed aminopropyl groups. With this aim in view we studied such relevant conditions and parameters as a ratio of reactive components, amount of water for hydrolysis, nonaqueous solvent nature, temperature of synthesis, duration of aging at the gel stage, washing/drying regime.

$$Si(OC_2H_5)_4 + (C_2H_5O)_3Si(CH_2)_2NH_2 \xrightarrow[-C_2H_5OH]{+H_2O}$$

$$(SiO_2)_x[O_{3/2}Si(CH_2)_3NH_2]_y(H_2O)_z \tag{33.3}$$

Let us consider some of the factors in more detail. First of all, it should be noted that in accord with our procedure of the synthesis [18–20] the 2:1 ratio of TEOS/APTES was maintained (with the exception of some experiments), but the preliminary hydrolysis of TEOS [15] was not carried out: the two alkoxysilanes were mixed directly (with a solvent if required), which was expected to lead to a uniform distribution of 3-aminopropyl groups on the xerogel surface. However, it was necessary to cool the starting mixture on an ice bath because otherwise we failed to introduce the required amount of water before a gel appeared. The amount of water used was equal to the half of the amount necessary for complete hydrolysis of all ethoxy groups to occur. In such a system a gel was formed within several minutes. In 24 h the gel was comminuted and dried in vacuum at 105°C for 6 h. After washing with 1000 cm^3 of water the gel was dried again under the same conditions. These conditions were also maintained if a use was made of a group of nonaqueous solvents. However, in this case the time period necessary for a gel to form was, as a rule, somewhat longer, and, therefore, the reaction mixture was not cooled. These conditions enabled us to synthesize xerogels with reproducible composition and main characteristics. The content of amine groups calculated from the data of the elemental analysis for nitrogen was close to that determined from the uptaking proton number. For most of the samples synthesized it ranged between 3.3 and 3.9 mmol/g (see Table 33.1, samples *8–11*, *13–18*, *21*) and only for four samples it fell outside the limits, namely for sample *7* that was not washed with water, for sample *12* whose synthesis involved hydrolysis of alkoxysilanes with the amount of water which made a quarter of the one necessary for complete hydrolysis of all the ethoxy groups, and for samples *19* and *20* which were prepared using DMF

TABLE 33.1
Some Characteristics of Functionalized Xerogels

Sample	Functional group, R′	Component ratio[a]	Solvent	$C_{R'}^{s}$ [b] (mmol/g)	$C_{R'}^{s}$ [c] (mmol/g)	Structure-adsorption characteristics S_{sp} (m^2/g)	V_s (cm^3/g)	d_{eff} (nm)	References
1	$-(CH_2)_3NH_2$	2:1	—	3.28	3.06	—	—	—	[15]
2	$-(CH_2)_3NH_2$	2:1	—	3.7	3.6	—	—	—	[30]
3	$-(CH_2)_3NH_2$	1:1	—	4.3	4.1	—	—	—	[30]
4	$-(CH_2)_3NH_2$	2:1	MeOH[d]	3.59	—	—	—	—	[17]
5	$-(CH_2)_3NH_2$	2.6:1	EtOH	3.1	3.3	191	1.08	—	[31]
6	$-(CH_2)_3NH_2$	5.2:1	EtOH	1.5	1.0	107	0.97	—	[31]
7	$-(CH_2)_3NH_2$	2:1	—	4.2	4.0	264	1.382	17.7	[20]
8	$-(CH_2)_3NH_2$	2:1	—	3.5	3.4	315	0.712	8.2	[20]
9	$-(CH_2)_3NH_2$	2:1	—	3.7	3.3	155	0.806	16.5	[20]
10	$-(CH_2)_3NH_2$	2:1	—	3.9	3.8	205	0.612	9.0	[20]
11	$-(CH_2)_3NH_2$	2:1	—	4.0	3.9	140	0.679	14.7	[20]
12	$-(CH_2)_3NH_2$	2:1	—	3.7	2.8	92	0.121	4.7	[20]
13	$-(CH_2)_3NH_2$	2:1	MeOH	3.5	3.7	171	0.649	12.2	[20]
14	$-(CH_2)_3NH_2$	2:1	EtOH	3.7	3.6	146	0.617	14.8	[20]
15	$-(CH_2)_3NH_2$	2:1	EtOH	3.3	3.6	150	0.799	17.3	[20]
16	$-(CH_2)_3NH_2$	2:1	*n*-PrOH	3.6	3.3	142	0.705	16.1	[20]
17	$-(CH_2)_3NH_2$	2:1	*n*-BuOH	3.5	3.3	132	0.655	17.2	[20]
18	$-(CH_2)_3NH_2$	2:1	CH_3CN	3.7	3.4	146	0.605	12.9	[20]
19	$-(CH_2)_3NH_2$	2:1	DMF	3.8	2.8	146	0.260	5.1	[20]
20	$-(CH_2)_3NH_2$	2:1	DMF	3.8	2.8	222	0.541	7.2	[20]
21	$-(CH_2)_3NH_2$	2:1	Et_2O	3.5	3.3	40	0.140	12.2	[20]
22	$-(CH_2)_3NH_2$	1.5:1.5	—	—	3.0	16	0.025	3.8	[20]
23	$-(CH_2)_3NH_2$	1:2	—	—	4.6	5	0.003	4.1	[20]
24	$-(CH_2)_3NH(CH_2)_2NH_2$	2:1	—	5.07	4.95	—	—	—	[15]
25	$-(CH_2)_3NH(CH_2)_2NH_2$	2:1	—	6.2	5.8	—	—	—	[21]
26	$-(CH_2)_3NH(CH_2)_2NH_2$	1:1	—	7.8	7.0	—	—	—	[21]
28	$-(CH_2)_3NH(CH_2)_2NH_2$	2:1	—	6.0	5.5	—	—	—	[30]
29	$-(CH_2)_3NH(CH_2)_2NH_2$	1:1	—	6.7	6.1	—	—	—	[30]
30	$-(CH_2)_3NH(CH_2)_2NH_2$	2:1	EtOH	6.0	6.2	4	0.005	4.2	[32]
31	$=[(CH_2)_3]_2NH$	2:1	EtOH	3.7	3.7	280	0.378	3.7	[32]
32	$-(CH_2)_3NC_3H_5N$	2:1	EtOH	2.7	2.7	22	0.058	6.8	[32]
33	$-(CH_2)_3NHCH_3$	2:1	EtOH	3.4	—	5	0.004	3.9	[32]
34	$-(CH_2)_3NH_2/-CH_3$	1:1:1	EtOH	3.7	3.7	106	0.186	7.3	[33]
35	$-(CH_2)_3NH_2/-CH_3$	1:1:1	DMF	—	—	170	0.633	9.7	[33]
36	$-(CH_2)_3NH_2/-C_6H_5$	1:1:1	EtOH	3.0	3.3	35	0.078	13.5	[33]

[a] TEOS: the second component containing functional group R′ (: the third one).
[b] In terms of elemental analysis for nitrogen.
[c] In terms of uptake of protons.
[d] The catalyst was hydrochloric acid.

(Table 33.1). It is evident that washing of a xerogel leads to removal of a greater portion of uncondensed amine-containing oligomers from its surface. However, washing with an ample amount of water (e.g., with 5 l of water) resulted in a decreased content of SiO_2 (sample *9*, Table 33.1), which was probably caused by hydrolysis of bonds Si—O—Si in the alkaline medium [21,22]. It is also evident that in the situation with DMF (distinguished for a high donor number) some DMF molecules were entrained by the matrix during the course of the sol–gel process (which is evidenced for by the IR spectroscopy data), with the entrainment blocking access of protons to amine groups and inhibiting formation of hydrogen bonds.

The PAPS samples prepared in conformity to the above-described procedure had a composition which could approximately be represented by the formula $(SiO)_{2.6}(O_{3/2}Si(CH_2)_3NH_2) \cdot H_2O$ and possessed a high thermal stability (see later). Micrographs of sample *8* made with the help of SEM and TEM are shown in Figure 33.1 [19]. The first two of them (Figures 33.1a and 33.1b) provide evidence for the fact that the material consists of large irregularly shaped particles on whose surface there are numerous highly disperse particles. From the TEM micrograph (Figure 33.1c) it is easily seen that the xerogel has a 3-dimensional skeleton formed by condensed globules of spherical form. The diameter of these globules is close to 20 nm.

Our preliminary experiments showed that the products obtained could be dried at 200°C in vacuum without affecting their structure-adsorption characteristics; all the data listed in Table 33.1 were gathered after such a treatment. At this temperature it is possible to remove water without affecting the aminopropyl functional groups. As is evident from Table 33.1, all the xerogels produced are classified as mesoporous adsorbents. By way of example, Figure 33.2 shows an adsorption isotherm for nitrogen in the case of sample *7*. The isotherm displays a capillary condensation hysteresis with a wide loop, which provides evidence for the presence of large pores. It was to be expected since the major factor governing the value of the specific surface area is the size of primary particles (globules) [22,23]. These globules are products of hydrolysis of ethoxysilanes and of condensation of mono- and oligomers formed during the course of the reaction. Allowing for the fact that in a medium with pH >7 the processes of (poly)condensation proceed rather rapidly [24], formation of relatively large globules is an inevitable outcome.

When analyzing the influence of one or another factor on structure-adsorption characteristics of PAPS, it is necessary to bear in mind that any variation of the

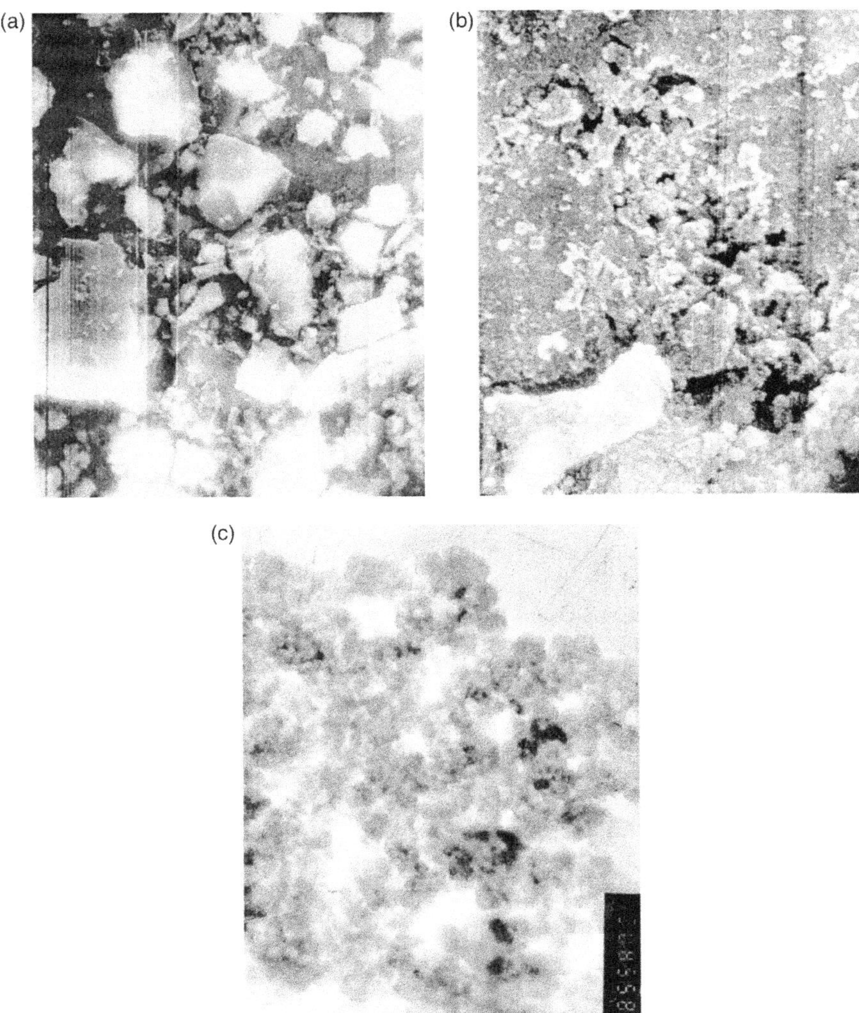

FIGURE 33.1 Micrographs of sample *8*, made with the help of SEM (a: 200×; b: 5000×) and TEM (c: the scale bar (14.7 mm) is 50 nm).

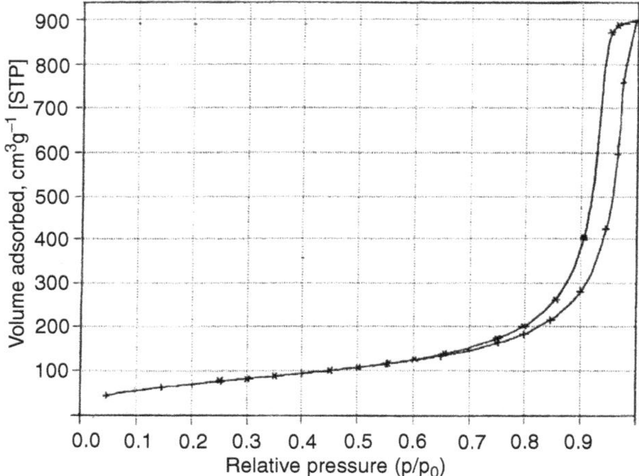

FIGURE 33.2 Isotherms of nitrogen adsorption(+)–desorption(*) for sample *7*.

properties can be due to both structural (geometrical) parameters (sizes of globules, their packing density, etc.) and to chemical phenomena (e.g., hydrolysis of Si—O—Si bonds in an alkaline medium resulting in the passage of some portion of the surface layer of the adsorbent in the form of oligomers of various compositions into solution). With due regard for the above-mentioned factors, let us consider the influence of washing of the PAPS samples with water after their drying in vacuum on their texture. On the one hand, the washing should be accompanied by the xerogel particle hydrophilization that is brought about, in the first place, by hydrogen bonds formed between water molecules and aminopropyl and silanol groups. In this case the xerogel skeleton should become more elastic and more amenable to subsequent shrinkage upon reheating [23,25]. Such a sample would possess a more fine-pored structure, which is actually observed when comparing the characteristics of sample *8* to those of sample *7* (Table 33.1). Besides, some contribution to hydrophilization during the course of washing may also be made by unlocking of a certain portion of fine pores as a result of removal of aminosiloxane oligomers (in going from sample *7* to sample *8* the amine group content decreases from 4.0 to 3.5 mmol/g). On the other hand, washing of the PAPS xerogels with distilled water can, in fact, be regarded as their washing using a solution with pH > 7 (pH of the filtrate formed upon washing of PAPS is equal to ca. 9.6). An alkaline medium, especially in the situation of a prolonged washing (as it was the case with sample *9*), promotes the reactions involving interactions between xerogel globules. Formation of chemical bonds at the points of contact of xerogel particles should lead to their aggregation, which facilitates subsequent strengthening of the xerogel skeleton. Under this condition the skeleton is subject to a less deformation in the process of drying, which will lead to formation of a large-porous structure. Indeed, sample *9* washed with 5,000 cm^3 of water has higher values of V_s and d_{eff} and lower values of S_{sp} in comparison with the corresponding values for sample *8* washed with 1,000 cm^3 of water (see Table 33.1). It is seen, as one would expect, that the hysteresis loop on the adsorption isotherm for sample *9* is shifted towards the region of higher relative pressures as against the position of the loop for sample *8* (see Figure 33.3). Certainly, hydration of xerogel takes place during the course of its washing both with 1,000 cm^3 of water and with 5,000 cm^3 of water. However, in both cases the contribution made by the hydration seems to be the same, while in the second case the contribution made by the aggregation of xerogel globules is predominant.

The porosity variation observed in going from sample *7* to sample *8* is difficult to explain, for example, by a decrease in the size of globules as a result of the removal of some portion of the surface layer of particles during washing. In such a case, any increase in the

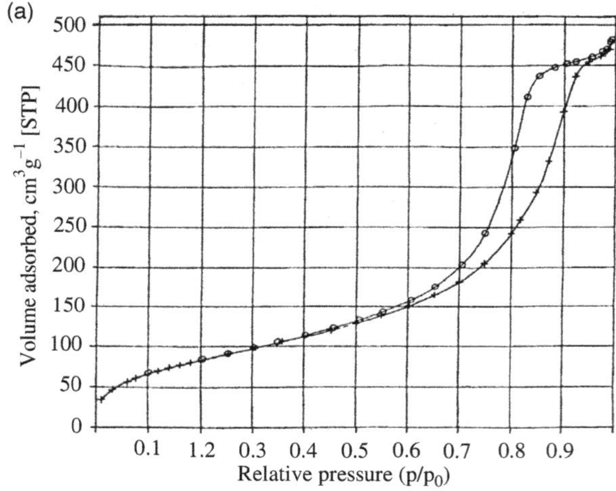

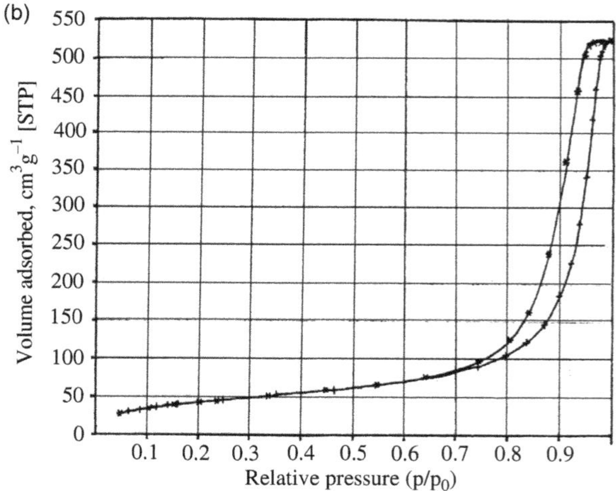

FIGURE 33.3 Isotherms of nitrogen adsorption(+)–desorption(*) for sample *8* (a) and sample *9* (b).

amount of washing water, as in the situation with sample *9*, should lead to enhancement of this effect, which is not observed by experiment.

It is known that an increase in the time of setting of silicic acid hydrogel leads to an increase in the volume and diameter of pores in the formed xerogel at the constant or decreased value of the specific surface [22,25]. However, the results of the comparison between the characteristics of the porous structures for sample *8* (ageing time: 24 h) and for sample *10* (ageing time: 168 h) provides evidence for an anomalous change of V_s, namely: upon the aging of the PAPS gel it does not increase, as it is to some extent the case with d_{eff}, but decreases (Table 33.1). In view of this fact it is possible to assume that the aging of the PAPS gel is not accompanied with any noticeable reaction involving condensation of surface OH groups of primary particles, since in this case a strengthening of the gel skeleton would cause an increase in the V_s value. In addition, it may be noted that the reaction of the hydrolytic polycondensation of 0.096 mol of TEOS and 0.048 mol of APTES should yield 0.528 mol of ethanol. In terms of 9.5 cm^3 of water the yield should amount to 40.5 cm^3 if it is assumed that the reaction involved condensation of all the silanol groups (the mentioned quantity of water includes 0.264 mol introduced at the initial reaction stage). Thus, during the course of maturation of a PAPS gel the intermicellar liquid is a mixture of water and ethanol, with the content of the latter being predominant (23.5 and 76.5 vol.% respectively). The presence of organic solvents (including ethanol) in a system retards, as a rule, the coalescence of primary particles leading to formation of particles with larger sizes [25]. Therefore, the growth of globules upon the ageing of sample *10* resulting in the observed decrease of S_{sp} is due also to other reasons. It is quite possible that there is some growth of primary particles at the expense of unreacted aminosiloxane oligomers remained in the gel. This explanation seems to be plausible and is indirectly evidenced for by the increase in the content of amine groups in sample *10* in comparison with sample *8* (Table 33.1). To all appearance, the precipitation of such oligomers leads not only to an increase in the globule size but also to 'healing' of fine pores. It is the last phenomenon that brings about some decrease in V_s. In this case, one may anticipate that the d_{eff} quantity for samples *10* and *8* will take on closely spaced values, which is observed by experiment (Table 33.1).

In our opinion, the changes in the structure of sample *11* dried at the atmosphere pressure (Table 33.1) are caused by similar reasons. The relatively slow removal of the intermicellular liquid in the situation with this sample in an alkaline medium at 105°C can facilitate both the continuation of the growth of large particles at the expense of small ones and the diffusion of aminosiloxane oligomers. As a result, in the texture of this sample there are more marked changes in comparison to sample *10*. The above-mentioned molecular processes should not vary the xerogel skeleton as a whole, that is why the values of V_s for samples *11* and *8* are very close (see Table 33.1).

The influence of temperature on the formation of the PAPS porous structure has been considered, by way of example, for two samples synthesized in the presence of ethanol and DMF. It is known that any increase in the temperature of the silica gel precipitation leads to an increase in the sorptive capacity and effective diameter of pores, that is promotes formation of adsorbents with a large-porous structure [26]. This effect is usually related to variations in the mobility of particles and, as a consequence, in their aggregative stability. A similar pattern of increasing in values of V_s and d_{eff} is also observed for samples *14* and *19* in comparison to samples *15* and *20* prepared at a low temperature and at room temperature respectively (see Table 33.1). This effect is especially pronounced in the situation when the non-water solvent is DMF. However, in this case, with increasing temperature, in the S_{sp} value there is not a decrease but an increase from 146 to 222 m^2/g. Here, of some importance seems to be the nature of DMF as a solvent that readily forms strong hydrogen bonds. This results in that some portion of DMF molecules is not removed upon vacuum drying of samples *19* and *20*. In other words, samples *19* and *20* somewhat differ in their composition from other samples of PAPS characterized in Table 33.1, and, therefore, the comparative study of their structural and adsorptive properties would not be correct.

Let us dwell on the influence of the nature of a nonaqueous solvent upon the texture of PAPS xerogels. To begin with, it should be noted that the synthesis procedure followed by us differs considerably from those applied when preparing polyorgano- or polymetal-organosiloxanes [13]. Firstly, the amount of water used for the synthesis is half the stoichiometric quantity necessary for hydrolysis of all the ethoxy groups of reacting alkoxysilanes. Therefore, water should be regarded as one of the reactants and not only as a solvent. Thus, in the case of a system consisting only of ethoxysilanes and added water, the formation of siloxane bonds during the course of synthesis should be caused by two parallel processes, namely by the condensation of silanol groups ($\equiv$Si—OH + HO—Si$\equiv$ $\rightarrow$ $\equiv$Si—O—Si$\equiv$ + H_2O) and by the condensation of silanol and ethoxysilane groups ($\equiv$Si—OH + $C_2H_5OSi\equiv$ $\rightarrow$ $\equiv$Si—O—Si$\equiv$ + C_2H_5OH). Secondly, the nature of a nonaqueous solvent introduced before the onset of the hydrolytic polycondensation reaction should exert a direct effect at its first step because at its last steps the quantity of the released ethanol and water becomes comparable (in terms of volume) to the quantity of the introduced organic additive (see earlier). As a consequence, the intermicelllar liquid in the gel before its drying is, in essence, a solution consisting of water, ethanol, and the above-mentioned additive. Then, if it is assumed that in all the cases the volume of the formed ethanol and water is the same, one can, in principle, evaluate the effect of the organic additive nature. Thus, judging by the data listed in Table 33.2, in all the cases the introduction of a nonaqueous solvent brings about a decrease in S_{sp} and an increase in d_{eff}, as is easily seen from Table 33.1, suffice it to compare samples *13–18* and *21* with sample *8*. However the V_s values varied slightly and fluctuated about the V_s value for sample *8*. Especially marked was the effect exerted by diethyl ether (see the characteristics of sample *21*). As a matter of fact, its introduction caused an about eightfold decrease in the specific surface as against the specific surface for sample *8*.

Therefore, in the system under study nonaqueous solvents exert an essential effect on proceeding of colloid-chemical processes which result in structurization of a porous material. The outcome of such an effect is, as a rule, a decrease (sometimes rather substantial) in the specific surface value as well as a decrease in the sorptive volume of pores and increase in their size. Analogous inferences were also made by the authors of reference [19] who considered the isotherms of adsorption of *n*-hexane by samples *13–19*. All the curves (with the exception of the isotherm for sample *19*) were S-shaped and were distinguished for a steep rise of hysteresis loop at $P/P_0 > 0.5$. This is known to be characteristic of mesoporous adsorbents [27]. The isotherm for sample *19* prepared in the presence of DMF was more flat and differed markedly from the isotherms of samples *13–18*. The observation is in complete agreement with the porous structure parameters of this sample (see Table 33.1).

It is clear that in all the cases the introduction of an organic additive reduces the degree of hydration, which can increase probability of interaction of globules with each other. The process of aggregation of globules must proceed the more vigorously the lower is the affinity between water and an introduced nonaqueous solvent. Certainly, the presence of aminopropyl groups provides additional favourable conditions for hydration. However, any substantial decrease in the solvent amount causes a rapid aggregation of globules accompanied with a considerable decrease in the xerogel porosity, as it was the case for diethyl ether. According to the above-mentioned, with increasing molecular weight of alcohols in the homologous series, their introduction should shift the sorbent porosity towards macroporous structures [26]. As a whole, this trend manifests itself in our experiments (Table 33.1, samples *13*, *15–17*), but neither in our case nor in other cases there is no complete parallelism between one or another physical parameter of the used nonaqueous additives and structure-adsorptive properties of the synthesized xerogels. Thus, for example, in going from methanol to *n*-butanol S_{sp} decreases consistently, while the increases in d_{eff} lack any systematic character. Moreover, as one would expect, in going from methanol to ethanol the V_s value increases, however, subsequently, in going from ethanol to *n*-butanol this value decreases. It cannot be ruled out that this fact is bound up with some changes in the nature of mass transfer which, in its turn, is related to the elevated solubility of the products of hydrolytic polycondensation of TEOS and APTES (oligomers) and their stabilization in the solved state as a result of the action of the mixed solvent [28]. Of course, some lyophilization of the system as a consequence of the addition of organic solvents exerts its effect on the action of capillary forces in the course of formation of the porous structure during the lyogel → xerogel transition (drying). In other words, the texture of the synthesized sorbents is the overall result of the action of a number of factors. Therefore, in such a complex system it is feasible to ascertain dependences only of a general character.

In conclusion, it should be noted that in the course of the syntheses the TEOS/APTES molar ratio was maintained equal to 2:1. The fact is that as this ratio reduces towards 1:1, the sample porosity decreases substantially, and at the ratio of 1:2 the xerogel becomes practically nonporous (Table 33.1, samples *22* and *23* respectively). Besides, it may be noted that when the TEOS/APTES ratio in solution was equal to 2:1, the $[SiO_2]/[O_{3/2}Si(CH_2)_3NH_2]$ ratio in the xerogel amounted to 2.6:1 (Table 33.1, sample *8*). At the same time, when the TEOS/APTES ratio in solution was equal to 1:2, the corresponding ratio in the xerogel made up only 1.6:1 (Table 33.1, sample *23*). Or else, if in the first case in the space unit there were 13 $SiO_{4/2}$ tetrahedra and 5 $O_{3/2}Si(CH_2)_3NH_2$ tetrahedra, then in the second case the ratio of the corresponding tetrahedra was equal only to 8:5. It becomes evident that the reduction of the TEOS/APTES ratio during the course of synthesis leads to formation of nonporous polymers that seems to be similar to those described earlier by Andrianov [29]. It cannot be ruled out that the prevalence of formation of

TABLE 33.2
Value of Adsorption by Amine-Containing Xerogels

Sample	Value of adsorption, $a \times 10^{-2}$ cm^3/g, at $P/P_0 = 0.2$		
	n-Hexane	Acetonitrile	Acetic acid
8	5.8	5.6	18.0
13	6.0	5.6	14.8
14	5.6	4.2	12.2
31	9.0	5.5	24.0
34	6.4	4.0	21.4
36	2.8	5.6	10.0

such polymers in the situation with a decreased amount of a hydrolyzing agent (water) results in the synthesis of a small porosity sample. Thus, PAPS *12* whose synthesis was effected using half the quantity of water substantially differs in its structure-adsorption properties from PAPS *8* and approaches samples *22* and *23* (see Table 33.1).

The aminoxerogels synthesized are stable up to about 200°C and retain their surface structure up to this temperature. As would be expected, strong heating in air destroys them. For example, PAPS *8* loses 26.8% of its mass upon heating to 650°C in air. The DTA data provide evidence for endothermicity with a maximum at 110°C [20] accompanied by a loss in mass of 6.9%, associated predominantly with loss of water. Loss of propylamine groups begins above 270°C, and there are endotherms with maxima at 288, 378, 438, 522, and 620°C, with mass losses totalling 19.0%. On the basis of these observations, formula $(SiO)_{2.6}[O_{3/2}Si(CH_2)_3NH_2] \cdot H_2O$ may be derived, which corresponds well with the observed compositions, especially for the samples prepared without a solvent [20]. In the other cases, the observed carbon content and C/N ratio are slightly higher, indicating the formation of a trace of the ammonium hydrogen carbonate (about 1%). The possibility of incomplete hydrolysis of Si—OEt groups is excluded by the NMR data (see subsequently).

Thus, on the basis of the above-drawn inferences about directional effects of diverse factors during the course of synthesis it is possible to produce PAPS xerogels with programmed (within certain limits, of course) structure-adsorption characteristics and desired contents of functional groups on their surface. It is assumed that the described method for controlling porous structure is applicable to the whole class of amine-containing xerogels.

The above-considered approach involving ethanol to prevent formation of a phase interface during maturation of a gel was employed by us to prepare xerogels with such functional groups as ethylenediamine groups (Table 33.1, [(*30*)], secondary amine groups (*31*), imidazolinyl groups (*32*), methylamine groups (*33*). Sample *31* arouses a special interest because in terms of a number of its characteristics (with the exception of the value of sorptive volume of pores) it is similar to sample *8*. Besides, the average size of its pores is lower by a factor of about two. One cannot rule out that these differences are due to the fact that on the surface of sample *31* there appear so called arched structures ($\equiv Si(CH_2)_3$ $—NH—(CH_2)_3Si\equiv$). The synthesis of xerogels with ethylenediamine groups was previously described by the authors of Refs. [15,21,30] who did not use any solvent. According to these authors the ratio $\equiv$EOS/ $(CH_3O)_3Si(CH_2)_3NH(CH_2)_2NH_2$ ranged between 2:1 and 1:1. However, since any structure-adsorption characteristics of the sorbents synthesized were not presented, their comparison with our samples is impossible. Such a comparison would be of some significance because we have found (32) that even at a ratio of 2:1 the corresponding process in ethanol leads to formation of a practically nonporous matrix (Table 33.1, sample *30*). The inference about existence of this tendency (i.e., decreasing of the specific surface area and sorptive volume of pores with increasing size of functional groups) is also corroborated by the structure-adsorption characteristics of samples *32* and *33* (Table 33.1). A similar effect was also observed in the case of xerogels with alkyl and aryl functional groups [13]. Increasing sizes of functional groups retards, as a rule, a gel formation process. It is assumed that they hinder spatial cross-linking of macromolecules so that as a result of their growth their sizes may become very large.

The authors of reference [32] made an attempt to prepare amine-containing xerogels with a bifunctional surface layer, with the functional species being aminopropyl and methyl (or phenyl) groups [see Scheme (4)].

$$
\begin{aligned}
&Si(OC_2H_5)_4 + (C_2H_5O)_3Si(CH_2)_3NH_2 + (C_2H_5O)_3 \\
&\quad \times SiR' \xrightarrow[-C_2H_5OH]{+H_2O} (SiO_2)_x[O_{3/2}Si(CH_2)_3NH_2]_y \\
&\quad \times (O_{3/2}SiR')_z(H_2O)_k, \qquad (33.4)
\end{aligned}
$$

where R′ is CH_3 (MTES) or C_6H_5 (PTES).

The molar ratio of reacting silanes TEOS: APTES:MTES (or PTES) in a starting mixture was equal to 1:1:1. The syntheses were effected in the presence of ethanol since our preliminary experiments showed that absence of a solvent led to appearance of an interface between phases during gel formation. Elemental analysis of these phases (in the case of MTES) showed that in terms of their composition they almost do not differ from each other. Thus, the upper phase contained 18.5 wt.% of C, 5.2 wt.% of N, and 4.7 wt.% of H while the content of these elements in the lower phase made up 17.0, 4.8, and 4.4 wt.% respectively. The causes for such a separation into layers are not known but the fact observed suggests an idea that it is necessary to exercise a more close control over gel formation processes especially in those systems where such a process is accompanied by a strong opalescence. The systems under study are just systems of this kind.

The procedure for synthesizing xerogels with a bifunctional surface layer did not differ from that for xerogels with a monofunctional surface layer [18–20]. The sorbents prepared are white powders which readily form air suspensions due to the presence of a large number of finely divided fractions. This is seen from micrographs for a sample that contains aminopropyl and phenyl groups (Figure 33.4). The micrographs made with the help of a scanning electron microscope give evidence for

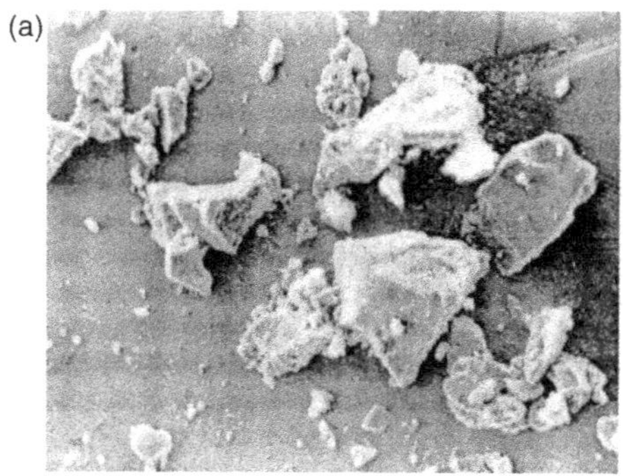

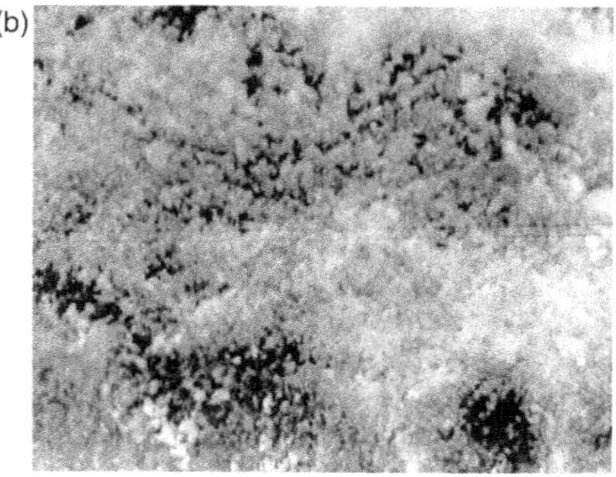

FIGURE 33.4 Scanning electron micrographs of sample *36* [(a) 200×, (b) 3500×].

the fact that the xerogels obtained are aggregates of particles of an irregular form, which is typical of xerogels. On the basis of the elemental analysis data the composition of the prepared adsorbents can be represented by the following formulae: $[(SiO_2)_{1,7}(O_{3/2}Si(CH_2)_3NH_2)(O_{3/2}SiCH_3)(H_2O)]$ and $[(SiO_2)_2(O_{3/2}Si(CH_2)_3NH_2)(O_{3/2}SiC_6H_5)(H_2O)_{0,6}]$. They agree with the acid–base titration results presented in Table 33.1 (samples *34* and *36* respectively). As was mentioned, this table also displays structure-adsorption characteristics of the synthesized xerogels. It is seen that they possess a lower specific surface area in comparison to sample *14* with a monofunctional surface layer synthesized under the same conditions (including the presence of ethanol). Especially small is the sorptive volume of pores so that it can be assumed that the appearance of hydrophobic alkyl or aryl groups on the surface leads to formation of large-sized globules and loosely packed structures. The validity of the assumption is corroborated by the data obtained for sample *36* with the aid of a transmission electron microscope (33). From the micrograph shown in Figure 33.5, it is seen that the average globule size makes up about 50 nm. This value agrees with the estimate of the average globule size made for this sample by the familiar formula, namely $d = 6000/\rho S$ (33).

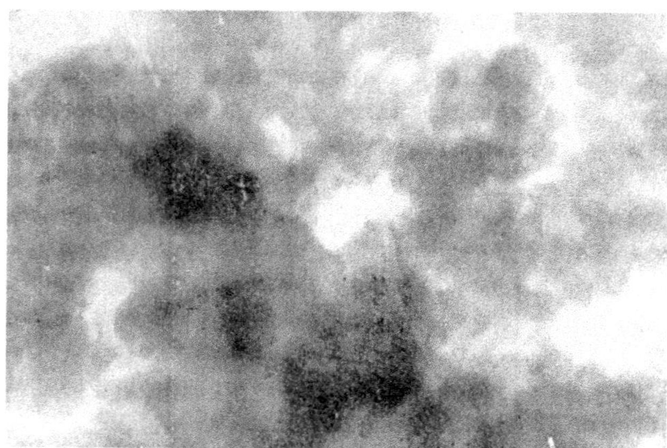

FIGURE 33.5 Transmission electron micrograph of sample *36*. The scale bar (15 mm) is 100 nm.

Thus, it can be inferred that a xerogel on whose surface there are aminopropyl and phenyl groups is formed by globules with sizes that are larger by a factor of about 2.5 than those of globules of a xerogel on whose surface there are aminopropyl groups (Figure 33.1c). Therefore, the nature of the second type of functional groups has a substantial significance for creating a bifunctional surface layer. It is known to have an influence on the rate of hydrolysis of a starting silane, on the degree of hydration of primary particles and, as a consequence, on the rate of their growth. Steric and electronic effects of groups of the second type may also exert an influence on the way of packing of globules that are formed.

In conclusion, let us compare structure-adsorption characteristics of sample *35* with those of sample *34* (the method for synthesizing the former differed from that for synthesizing the latter only in the solvent, namely instead of ethanol a use was made of DMF) (see Table 33.1). As is seen, the specific surface area of sample *35* and especially the sorptive volume of its pores turn to be markedly larger while the effective diameter of the pores turns to be only slightly smaller. To all appearance, the introduction of DMF (which creates strong hydrogen bonds and possesses a higher viscocity in comparison to ethanol) impedes growth of globules and brings about essential changes in the structure of a formed xerogel. Thus, we get one more support for the thesis that by varying one or another condition during the course of a sol–gel synthesis it is possible to exert a purposeful influence on structure-adsorption characteristics of functionalized polysiloxane xerogels.

Further, it was of interest to compare adsorptive properties of samples with mono- and bifunctional surface layer with respect to sorbates of various nature [33]. Figure 33.6 shows isotherms of adsorption of *n*-hexane, acetonitrile, and acetic acid by samples *14* and *36* in the relative pressure interval 0.0–0.5 (it should be noted here that the isotherms for samples *8*, *13*, *31*, and *34* are analogous to those for sample *14*). It is known that

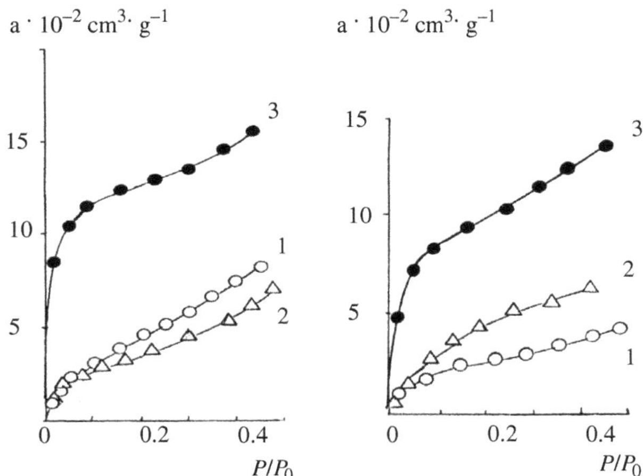

FIGURE 33.6 Comparison of initial portions of adsorption isotherms of n-hexane (1), acetonitrile (2), and acetic acid (3) for samples *14* (a) and *36* (b) at 20°C.

molecules of n-hexane can participate in adsorption interactions only at the expense of dispersion forces. Acetonitrile is distinguished for the fact that its nitrogen atom exhibits basic properties while CH_3COOH shows proton-donating properties. As is seen from Figure 33.6, in the case of the both samples the curve for adsorption of acetic acid is situated substantially higher than that of n-hexane and acetonitrile. It provides evidence for a relatively strong interaction of protons of carboxyl groups of the acid with amino-containing groups present on the surface of the xerogels. As far as the isotherm of adsorption of acetonitrile for sample *14* is concerned, in the initial range of coverage it coincides with that of adsorption of n-hexane (Figure 33.6a). However, with increasing P/P_0 it is situated lower than the adsorption curve for n-hexane. This experimental result may be due to an enhancement of the electrostatic repulsion between surface amino groups and CH_3CN molecules. The effects observed are supported by the values of adsorption (a) for the above-said adsorbates. The adsorption values determined for a number of xerogels are listed in Table 33.2. From the table it is seen that in the case of sample *36* the values of adsorption of acetonitrile are higher than those in the case of sample *14*. This result is graphically illustrated in Figure 33.6b, where for the sample with aminopropyl and phenyl groups the isotherm of adsorption of acetonitrile is positioned higher than that of n-hexane. In other words, in the situation with this sample we observed an inversion of the positions of adsorption for these sorbates in comparison with sample *14*. The cause of this phenomenon seems to be the fact that on the surface of sample *36* there appeared phenyl groups (in addition to aminopropyl groups). On the other hand, xerogel *34* (which was prepared under conditions identical to those employed for the synthesis of xerogel *36* but whose surface layer contained methyl radicals instead of phenyl radicals) in terms of its adsorptive properties was much more close to samples *8*, *13*, *14*, and *31* than sample *36* (see Table 33.2).

This inference can also be made by comparing the curves of distribution of pores in their size (Figure 33.7). One of the few properties common to samples *34* and *36* is hydrophobicity of their surface layers. Thus, by varying the composition of a surface layer of xerogels it is possible to exert an effect on selectivity of their sorption.

The analysis of the data in Table 33.2 gives grounds to draw one more inference concerning sample *31* and consisting in that in its sorptive properties this sample differs markedly from the rest of the xerogels irrespective of the nature of a sorbate. Such a difference is also noted when comparing its adsorption values with those for sample *8* whose specific surface area value is close to that for sample *31* although its sorptive pore volume is higher by a factor of about 2 (see Table 33.1). Nevertheless, at $P/P_0 = 0.2$ the sorptive capacity of sample *31* with respect to n-hexane and acetic acid is greater by a factor of about 1.6 and 1.3 respectively in comparison with sample *8*. The differences seem to be related to salient features of the surface of sample *31* in general and to formation of the above-mentioned arched structures on its surface in particular.

Synthesis of Xerogels with Urea and Thiourea Groups in a Surface Layer

The sorptive properties of sulfur-containing xerogels with respect, in the first instance, to ions of heavy and noble metals has been studied in detail in references [11,12]. Synthesis of such xerogels is performed in a one-pot route using hydrolysis of appropriate trialkoxysysilanes, that is a use is made of monocomponent systems. The medium (water or water/dioxane) was, as a rule, basic. This, in principle, can lead to a partial hydrolysis or oxidation of sulfur-containing groups. Besides, specific surface areas of such sorbents are relatively low. With allowance for this fact and with a view to expand possibilities for designing surfaces of sulfur-containing xerogels we made an attempt to work out a somewhat different approach to their synthesis. As an example, a choice was made of groupings [—NH—C(S)—NH—]; for the sake of comparison we also performed synthesis of a number of xerogels with urea groups [—NH—C(O)—NH—] [34].

At first, we prepared some starting trialkoxysilanes with oxygen- and sulfur-containing functional groups. Their synthesis involved the well-known reaction of primary amines with iso(thio)cyanates [see Scheme (5)]. Usually, this reaction does not lead to formation of by-products. It can often proceed at room temperature and allows one to obtain a

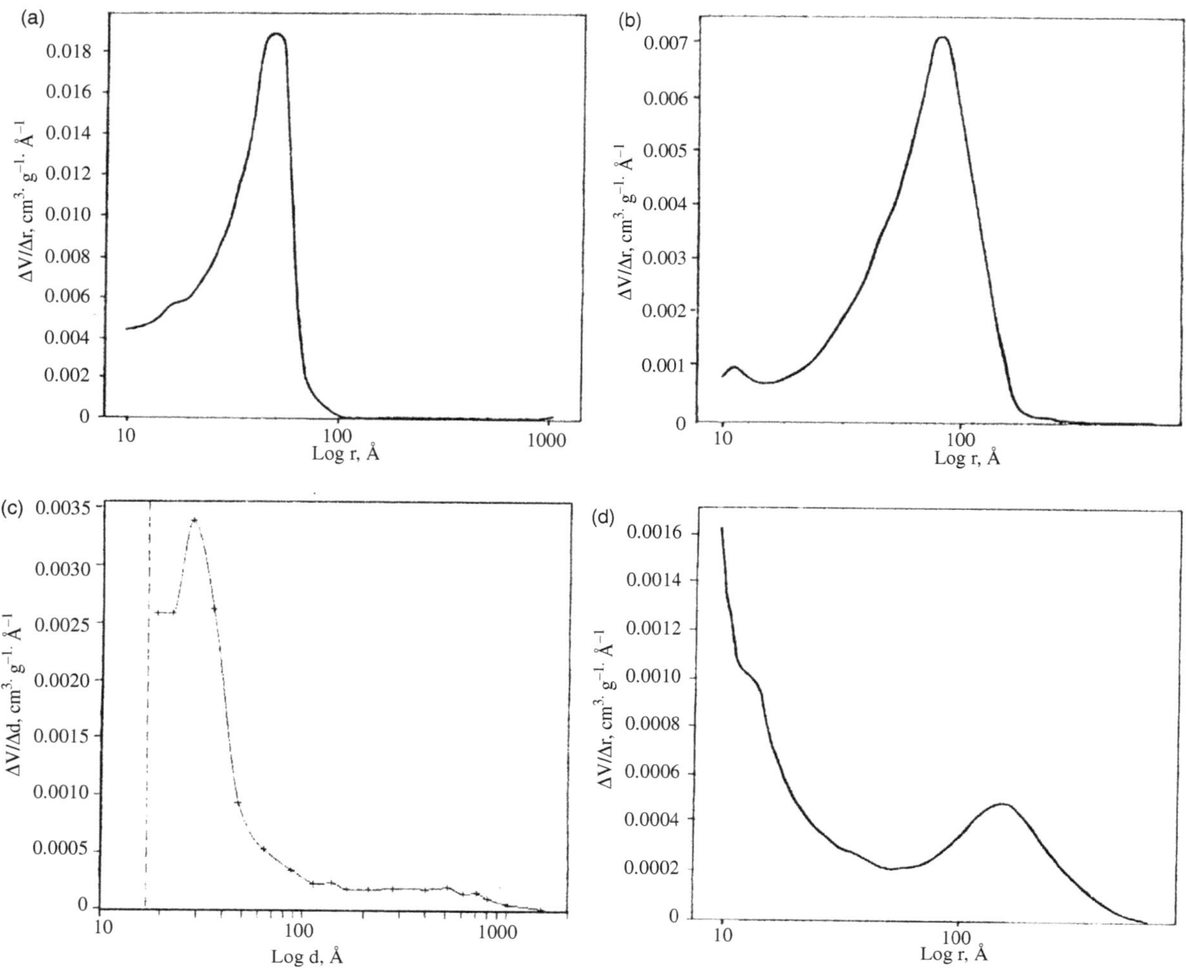

FIGURE 33.7 Pore size distributions for samples *8* (a), *14* (b), *34* (c), and *36* (d).

broad gamut of trifunctional silanes with a high yield.

$$(RO)_3Si(CH_2)_3NH_2 + O(S)CN - R'' \xrightarrow{\text{Solv.}} (RO)_3Si(CH_2)_3NHC(O(S))NHR'' \quad (33.5)$$

Thus, we have prepared triethoxysilanes which contain radicals R'' [such as n-C_3H_7 (O I, S II), C_6H_5 (O III, S IV), $(CH_2)_3Si(OC_2H_5)_3$ (O V, S VI), $CNSC_6H_4$ (O VII, benzothiazole radical)]. It should be noted that V and VI were synthesized earlier [35–38] using the transamination reaction. In this case V was obtained as an oil with a yield of 12.2%. Irrespective of solvents (n-hexane or benzene), reaction [5] results in formation of a substance which crystallizes as a white solid (melting point 61–63°C) in quantitative yield. The spectral characteristics of this product are consistent with those described in [38], except the NH position in ^{1}H NMR spectra, with this signal for the crystalline product being shifted to strong fields [an analogous situation is observed for substance IV, whose ^{1}H NMR spectrum was described in [39]. A single strong band in the range 3000–4000 cm^{-1} was identified in the IR spectrum of V (at 3336 cm^{-1}) and attributed to ν(NH). After drying in vacuum this substance is hardly dissolved well in alcohols, benzene, and so on. The product is air- and moisture-resistant. Earlier, substance IV was described both as an oil [39] and a solid product [40]. However, our aim was to prepare triethoxysilane precursors for the sol–gel synthesis which, as a rule, does not call for any special purification. According to the spectral data gathered for the precursors their purity was quite satisfactory for a further use. At the same time, it should be said that most triethoxysilanes can form oils, because they can keep traces of solvents, for example ethanol which can be seen in their ^{1}H NMR spectra. In the situation with our precursors, however, that was not a problem because they were prepared in ethanol which is the solvent used for the subsequent synthesis of gels.

All the precursors obtained have been characterized by IR spectroscopy on the basis of their intense absorption bands characteristic of their typical functional groups. Groups NHC(O)NHR″ show a strong narrow ν(NH) adsorbtion band in a range of 3325–3350 cm^{-1}. In the case of the NHC(S)NHR″ groups the ν(NH) band is broader and less intense and is usually shifted into the region of lower frequencies (3265–3275 cm^{-1}). Besides, in a range of 1500–1700 cm^{-1} we observe strong and narrow absorption bands [δ(CNH) + ν(CN)] and ν(CO) for urea-containing silanes [41]. In the case of S-containing silanes, the ν_{as}(NCN) is a single broad band (also in the case of different groups on nitrogen atoms) [42].

All the precursors prepared were used in the synthesis of gels [34]. By the reaction of hydrolytic co-condensation we synthesized xerogels using two- and three-component systems, where by the term 'component' we mean a silane-containing precursor [see Scheme (6)].

$$\begin{aligned} &Si(OC_2H_5)_4 + (C_2H_5O)_3Si(CH_2)_3NHC(O\text{ or }S)NHR'' \\ &+ (C_2H_5O)_3Si(CH_2)_3R''' \xrightarrow[-C_2H_5OH]{+H_2O/F^-} SiO_2/SiO_{3/2} \\ &\times (CH_2)_3NHC(O\text{ or }S)NHR''/SiO_{3/2}(CH_2)_3R''', \quad (33.6) \end{aligned}$$

where $R'' = n\text{-}C_3H_7$, C_6H_5, $(CH_2)_3Si(OC_2H_5)_3$, $CNSC_6H_4$; $R''' = —NH_2$ or $—NH(CH_2)_2NH_2$.

The solvent and catalyst used were ethanol and NH_4F respectively. The ratio F^-/Si was usually 1:100 or smaller. The amount of water was half of the one needed for complete hydrolysis of all the ethoxy groups. The gels synthesized were opalescent, which is characteristic of the sol–gel process in a basic or nucleophilic medium. After aging (usually for 24 h) the gels were crushed, dried in vacuum, crushed again, washed with water and dried again in vacuum. The drying conditions for sulfur-containing gels were milder than in the rest of the cases. The elemental analysis (C, H, N, S) data for the obtained xerogels are presented in Table 33.3.

Depending on the composition of their surface layer, the xerogels synthesized can be subdivided into two groups, namely the xerogels with a monofunctionalized (*37*, *38*, *40*, *41*, *44–46*, *49*, *50–53*) and bifunctionalized (*39*, *42*, *43*, *47*, *48*, *54*) surface layer (Table 33.3). The first group xerogels are hydrophobic and posses a small value of specific surface area (S_{sp}) if the TEOS/trifunctional silane ratio is 2:1. If the TEOS/trifunctional silane ratio is 4:1, the xerogels prepared have hydrophilic properties and porous structure. With increasing amount of TEOS in the system (e.g., up to a ratio of 8:1), the xerogels obtained preserve their hydrophilic properties, but their specific surface area becomes markedly larger (Table 33.3). The hydrophilic character of their surface was also displayed by xerogels with a bifunctional surface layer. This seems to be due to appearance of amino groups in this layer. Besides, as distinct from the xerogels with a monofunctional (thiourea) surface layer, these xerogels are porous even at a starting component ratio of [2:(0.5 + 0.5)].

The C:N:S ratio calculated using the elemental analysis data (Table 33.3) indicates that in most cases the ratio between introduced functional groups remains close to that used in the starting mixture of alkoxysilanes. It is appropriate to mention here that the IR spectroscopy gives evidence for the absence of any changes in the composition and structure of introduced functional groups (see subsequently).

As one would expect, the xerogels whose synthesis involved precursors with O-containing functional groups display a higher thermal stability than their S-containing counterparts. For instance, thermal decomposition of the surface layer of xerogel *38* begins to take place at 250°C, while the corresponding temperature for its sulfur-containing analogue is equal to 225°C [34].

Thus, if a use is made of trialkoxysilanes with voluminous (thio)urea groupings, by varying ratios TEOS/trialkoxysilane it is possible to synthesize xerogels whose surfaces show different degrees of hydrophilicity. Noteworthy is the fact that decreasing amount of trialkoxysilane in a starting mixture results in xerogels with a more extended surface.

Synthesis of Xerogels with Thiol Groups in a Surface Layer

When preparing polysiloxane xerogels with such a functional group as $\equiv Si(CH_2)_3SH$, one may encounter some difficulties. Firstly, in this case in order to effect an appropriate hydrolytic polycondensation reaction it is necessary to introduce a catalyst. Most often a use is made of such catalysts as $(n\text{-}Bu)_2Sn(CH_3COO)_2$ and hydrochloric acid (as a rule, of low concentration) but, nevertheless, formation of gels and their maturation takes much time. Moreover, according to the ^{13}C and ^{29}Si CP MAS NMR spectroscopy [17] the xerogels prepared in this way contain substantial amounts of unreacted alkoxy groups and fragments of the $HS(CH_2)_3Si(OSi\equiv)_2(OR)$ and $HS(CH_2)_3Si(OSi\equiv)(OR)_2$ type. In other words, polysiloxane skeletons of these xerogels do not have strong effective cross-links. Secondly, application of such a catalyst as $(n\text{-}Bu)_2Sn(CH_3COO)_2$ leads usually to xerogels with very low values of S_{sp}, that is, to formation of practically nonporous matrices (Table 33.4, sample *57*) [43]. Besides, by the Mössbauer spectroscopy it was shown that the polymeric matrices contained tin [16]. Therefore, in our work the catalyst was fluoride [43]. Thirdly, in the situation

TABLE 33.3
Elemental Analysis Data and Specific Surface Areas for Xerogels with Urea and Thiourea Functional Groups

Xero-gel	Composition of reacting mixtures and ratio of components	Elemental analysis data				C/N/S ratio	C^a (mmol/g)	S_{sp} (m^2/g)
		C	H	N	S			
37	TEOS/I (2:1)	24.7	5.0	7.5	—	7.6:2:0	2.9	<1
38	TEOS/I (4:1)	18.5	3.4	5.2	—	8.4:2:0	1.9	
39	TEOS/I/APTES (4:1:1)	14.9	3.3	5.3	—	9.8:3:0	1.3	
40	TEOS/II (2:1)	24.6	4.7	7.5	9.2	7.1:1.9:1	3.2	<1
41	TEOS/II (4:1)	20.6	3.8	5.9	9.4	5.8:1.4:1	3.3	189
42	TEOS/II/APTES (4:1:1)	19.0	4.0	6.3	2.7	19.0:5.4:1	0.9	109
43	TEOS/I/II (4:1:1)	25.5	5.1	7.6	4.4	15.5:4.0:1	1.4	
44	TEOS/IV (2:1)	31.0	3.8	6.8	8.3	10.1:1.9:1	2.9	<1
45	TEOS/IV (4:1)	24.6	3.2	5.2	6.5	10.1:1.8:1	2.2	175
46	TEOS/IV (8:1)	17.2	2.2	3.4	2.0	22.9:3.9:1	0.6	276
47	TEOS/IV/APTES (4:1:1)	25.2	3.5	6.5	6.1	11.0:2.4:1	2.0	70
48	TEOS/III/IV (4:1:1)	32.8	4.2	6.9	5.1	17.1:3.1:1	1.7	
49	TEOS/V (2:1)	26.7	5.4	5.8	—	10.7:2:0	2.2	<1
50	TEOS/V (4:1)	14.4	3.4	4.5	—	7.4:2:0	1.7	
51	TEOS/VI (4:1)	15.7	3.6	4.4	7.0	6.0:1.4:1	2.4	
52	TEOS/VII (2:1)	32.0	3.4	9.0	7.2	11.9:2.9:1	2.4	
53	TEOS/VII (4:1)	25.1	2.8	6.9	7.4	9.1:2.2:1	2.5	
54	TEOS/VII/TMPED (4:1:1)	27.6	4.4	9.6	5.6	13.2:3.9:1	1.9	

[a]Functional group concentration calculated on the basis of the elemental analysis data.

with systems containing 3-mercaptopropyltrimethoxysilane (MPTMS) one can ordinarily observe formation of a phase interface (in the presence or in the absence of methanol). However, if in a given case the two formed phases are mixed in 24 h [43], in the situation with other groups the phenomenon may not take place (see earlier). When a use is made of the procedure suggested in reference [43], the synthesis of xerogel *58* proceeds very smoothly, with a transparent elastic gel being formed within 1 min; then, a slight heating up is observed, and this exothermic effect gives evidence for continuation of the polycondensation process, with the gel becoming more and more opalescent. In 24 h such a gel is readily amenable to processing designed to obtain a xerogel. The HS-containing xerogel

TABLE 33.4
Some Characteristics of Xerogels with Thiol Groups

Xero-gel	Molar ratio of components TEOS/MPTMS/APTES in a starting mixture	Solvent	C^s_{SH} (mmol/g)	Structure-adsorption characteristics			References
				S_{sp} (m^2/g)	V_s (cm^3/g)	d_{eff} (nm)	
55	2:1:0	MeOH[a]	3.8	—	—	—	[*15*]
56	2:1:0	MeOH[a]	3.4	—	—	—	[*17*]
57	2:1:0	MeOH[a]	4.0	<1			[*43*]
58	2:1:0	MeOH[b]	4.5	216	0.152	2.8	[*43*]
59	2:0.75:0.25	MeOH	3.3 (1.0)[c]	214	0.358	3.6; 6.4	[*43*]
60	2:0.50:0.50	MeOH	2.1 (1.8)[c]	217	0.302	4.0; 4.7	[*43*]
61	2:0.25:0.75	MeOH	1.0 (2.8)[c]	234	0.675	11.5	[*43*]

[a]The catalyst was a tin compound.
[b]The catalyst was fluoride.
[c]The figure between the brackets is the content of amino groups.

synthesized following this procedure is distinguished for its extended surface (Table 33.4, sample *58*).

Further, we described a procedure [43] for producing three more xerogels containing a bifunctional surface layer of the SH/NH_2 type [following the procedure advanced in reference [18] but without using any catalyst] (Table 33.4, samples *59*, *60*, and *61*). Ratio MPTMS/APTES in a starting mixture was varied from 3:1 to 1:3. The functional group contents presented in Table IV were calculated from the results of the elemental analysis and acid–base titration. It can be seen that the SH/NH_2 functional group ratios in the final products are close to those specified during the course of the synthesis. Thus, in the case of sample *59* it is equal to 3.5 (3.00 in the starting mixture); sample *60*: 1.1 (1.00); sample *61*: 0.4 (0.3). It is also seen that the mercaptopropyl functional groups are less susceptible to washing procedures (more strongly bound) in comparison to aminopropyl groups.

All the synthesized xerogels (samples *57–61*) are white powdery substances (sometimes they are tinged with yellow). Their micrographs made with the help of a scanning electron microscope are shown in Figure 33.8. These micrographs display clearly discernible trends. Xerogel *57* consists of irregular particles, frequently of triangular shape with sharp edges. The particle size is 150–200 μm. In going from sample *59* to sample *61* we notice that the connectivity between the particles is increasing so that sizes of agglomerates range from 50–100 μm (*59*) up to 400 μm and more (*61*). Also, the content of aminopropyl groups grows in going from *59* to *61*. It gives us grounds to assume that during the course of the synthesis the increasing number of amine groups and, therefore, the increasing pH value lead to a higher solubility of the silica matrix, which results in formation of smoother particles that are bound together.

The thermoanalytical curves for the xerogels synthesized are shown in Figure 33.9 and Figure 33.10. The small weight losses observed for samples at about 100°C (sample *57*, 2.6%; *58*, 1.1%; *59*, 3.1%; *60*, 4.4%; *61*, 5.3%) are assigned to desorption of water whose presence in the samples is also evidenced for by their IR spectra. In general, the thermograms of xerogels *57* and *58* are similar

FIGURE 33.8 Scanning electron micrographs of xerogels: (a) sample *57* (200×); (b) sample *59* (300×); (c) sample *60* (500×); (d) sample *60* (1000×); (e) sample 61 (200×); (f) sample *61* (1000×).

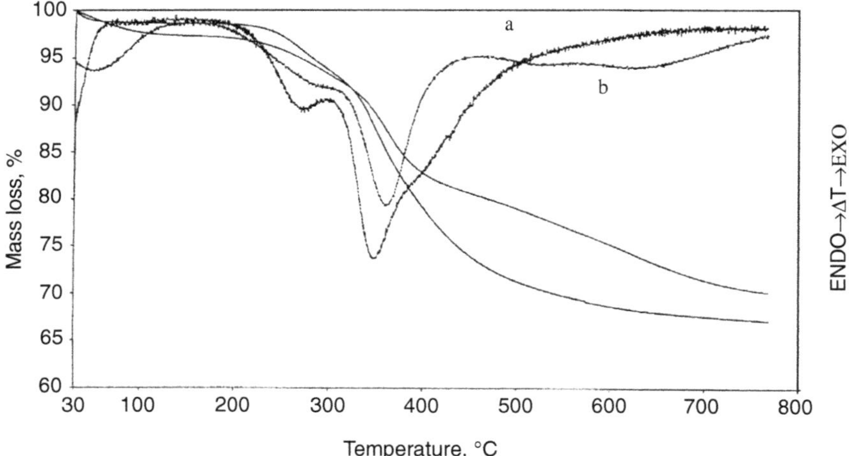

FIGURE 33.9 Thermoanalytical curves for xerogels *57* (a) and *58* (b).

but there are some slight differences. In the case of xerogel *58* the weak exothermal band at 280°C is more pronounced than for xerogel *57* (Figure 33.9). The intensive burning out of a surface layer for sample *58* begins at 250°C, while for sample *57* at 265°C. And though the total weight loss values for both the samples are very close (27.2% for *57* and 26.5% for *58*), sample *57* differs by showing two exothermal effects at temperatures above 500°C. The possible cause of these differences seems to be a higher porosity of sample *58*, which results in a larger availability of functional groups. Besides, the synthesis of xerogel *57* was catalyzed, with $(n\text{-}C_4H_9)_2$-$Sn(CH_3COO)_2$ being used as a catalyst. The Mössbauer spectroscopy (16) has shown that this molecule remains incorporated into/adsorbed on the final gel. The thermogram of sample *61* which contains the least amount of mercapto groups (Figure 33.10e) turns to resemble those considered above. The thermograms of samples *59* and *60* (Figure 33.9c and d) display only one strong exothermal effect at 350°C. Thus, it is possible to infer that the surface layer of the xerogels with a functional group ratio SH/NH_2 equal to 3:1 and 2:2 possess the highest thermal stability. Poly(3-aminopropyl)siloxane xerogel synthesized at a TEOS/APTES ratio of 2:1 has its first exothermal band at 270°C [20]. Therefore, it may be hypothesized that in the situation with samples *59* and *60* there is an effect of synergism.

It should be noted that xerogel *58* whose synthesis involved fluoride as a catalyst has a specific surface area of 200 m^2/g and more (Table 33.4). Xerogel *57* whose synthesis was catalized by a tin compound is entirely non-porous. Nonporous (or low-porous) materials seem to be the poly(3-mercaptopropyl)siloxane xerogels synthesized by the authors of references [15,17,44] who used the same tin derivatives. Polymercaptomethylsiloxane synthesized in an alkaline medium (single-component system) has a surface area of 80 m^2/g (from *n*-hexane adsorption) and 162 m^2/g (from argon adsorption) [45]. Xerogels *59*, *60*, and *61* containing a bifunctional surface layer are distinguished for very high values of

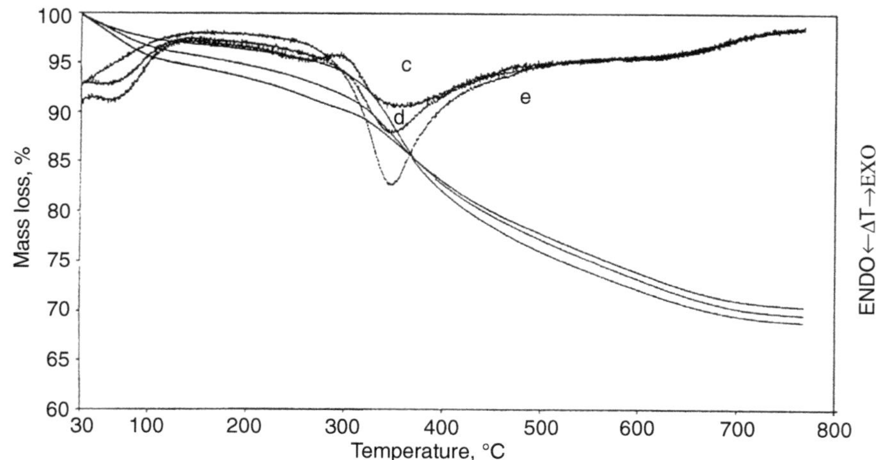

FIGURE 33.10 Thermoanalytical curves for xerogels *59* (c), *60* (d), and *61* (e).

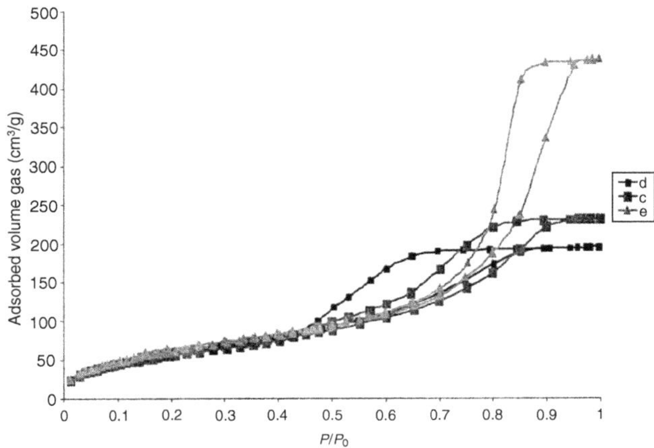

FIGURE 33.11 Nitrogen isotherms for samples *59* (c), *60* (d), and *61* (e).

specific surface areas and pore volumes (see Table 33.4). Besides, in the case of all the samples with a bifunctional surface layer one can observe a capillary condensation hysteresis pointing to the presence of mesopores (Figure 33.11).

Worthy of note is also the fact that in the situation with xerogel *61* containing the smallest amount of amine groups in comparison with xerogels *59* and *60* the hysteresis loop turns to be broader and is shifted to larger relative pressures (see Figure 33.11). These observations are in agreement with a high sorptive pore volume value which is characteristic of this sample. Really, it is by a factor of about 2 higher than the corresponding values for samples *59* and *60* (Table 33.4) and approaches pore volumes characteristic of polyaminosiloxane xerogels produced under analogous conditions [20]. The specific surface area of these samples varies within a small interval. However, with increasing content of amine groups, that is, in going from sample *59* to sample *61*, it increases gradually (Table 33.4). At the same time, as is seen from this table, there is not such a correlation between sorptive volumes of pores and their effective diameter. In this connection, it is interesting to consider curves of distribution of pores in their effective diameter shown for samples *58–61* in Figure 33.12. For instance, in the case of sample *58* with a monofunctional surface layer this curve gives evidence for a narrow distribution, and the effective diameter value is close to d_{eff} characteristic of the boundary between meso- and micropores (see Table 33.4). When in the surface layer there appear aminopropyl groups (sample *59*), one can observe formation of a biporous structure (Figure 33.12c), with the both types of pores being mesopores (Table 33.4). With increasing number

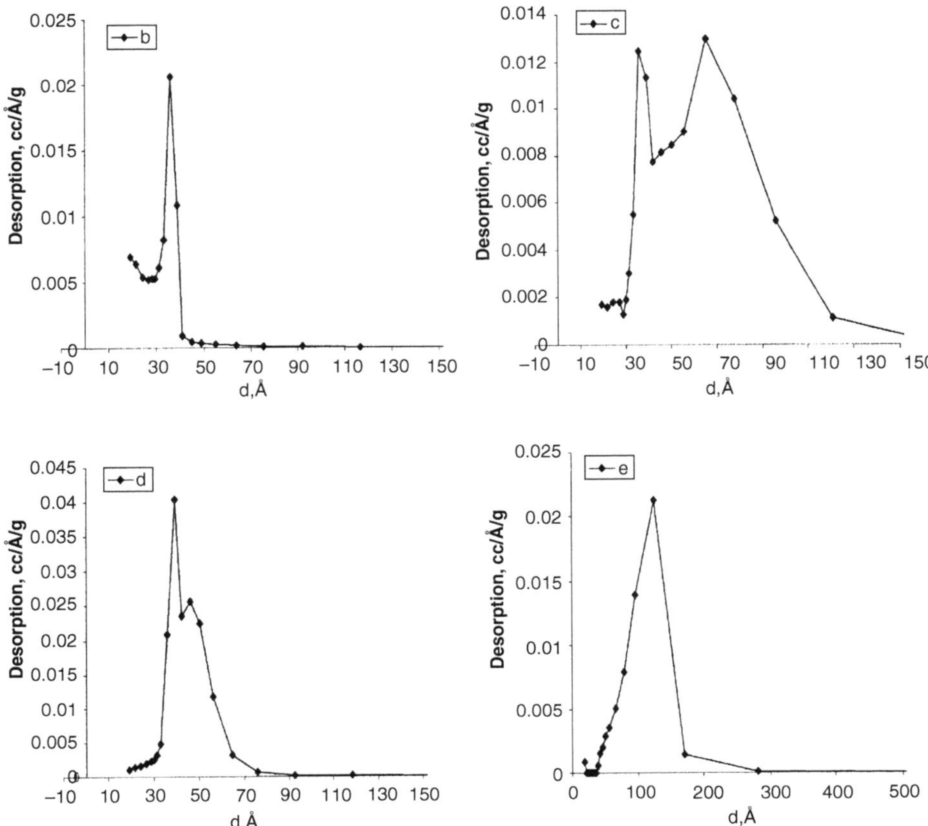

FIGURE 33.12 Pore size distributions for samples *58* (b), *59* (c), *60* (d), and *61* (e).

of amine groups (Figure 33.12d), the biporous structure is still to some extent preserved. However, in the case of the sample with the maximal content of amino groups in the surface layer (Figure 33.12d) it is possible to draw an unambiguous inference about its absence. Xerogel *61* is a monoporous sample with large mesopores (Table 33.4). It is interesting to note that in samples *59* and *60* with a bifunctional surface layer the effective diameter of the first system of pores is close to the diameter for sample *58* which contains a layer of only thiol groups. It becomes clear that in the given case there is not any simple dependence between the composition of starting mixtures of silanes and structure-adsorptive characteristics of xerogels synthesized on their basis.

From the above-stated it follows that synthesis of polyorganosiloxane xerogels with a bifunctional surface layer is possible when a use is made of functional groups of various nature. In this case the composition of a surface layer can be varied over a wide range (in our experiments the SH/NH_2 ratio was equal to 4:0, 3:1, 2:2, or 1:3).

Synthesis of Xerogels with Carboxyl Groups in a Surface Layer

There were several attempts made with a view to synthesize such xerogels. One of them was designed to employ hydrolytic polycondensation of such organosilicon monomers as trichlorosilylbuturic acid chloride or silyl ester of triethoxysilylbuturic acid [46]. However, the insoluble cross-linked polymers formed were liable to swelling. In addition, they were thermally stable only at temperatures below 200°C. Thus, they practically did not have any advantages over common organic cationites. This may also be said about the product of co-polymerization of acrylic acid and allyl triethoxysilane [47]. Besides, an attempt was made to effect hydrolytic polycondensation of $(CH_3O)_3Si(CH_2)_2SCH_2COOSi(CH_3)_3$ in an acidic or alkaline medium [48]. In neither of the cases, however, desired polymers with silsesquioxane structure were synthesized, with the only outcome being formation of a mixture of oligo- and polysiloxanes. It was assumed that the result was due to the fact that the trimethylsilanol released during the course of hydrolysis participated in polycondensation processes. Therefore, with the view of synthesizing silico-organic polymers the $(CH_3O)_3Si(CH_2)_2SCH_2COOH$ adduct obtained by the authors of reference [48] was used without its further purification. In this case, one could observe formation of the corresponding polyorganilsilsesquioxanes with a yield of 69 and 21% in an acidic and alkaline medium respectively. It became evident that the presence of sulfur in alkyl radicals lowered the stability of such polymers. From this standpoint, of interest is the patent [12] concerned with the sol–gel method for synthesizing xerogels with group $\equiv Si(CH_2)_2COOH$. The essence of the suggested approach consists in an acid hydrolysis of cyanoethyltriethoxysilane $(C_2H_5O)_3Si(CH_2)_2CN$ (CETES) in the presence of TEOS. After drying at 100°C in vacuum for 6 h the white product proved to be a good sorbent of $NHEt_2$ and Py. Later on, this approach was considered in more detail by the authors of references [49,50]. For the sake of comparison a consideration was also given to synthesizing xerogels with carboxyl-containing groups by a variant of the method whose first step led to formation of matrices with nitrile groups on their surface. The subsequent saponifying of these groups should give carboxyl groups. However, the substance with nitrile groups synthesized on the basis of a two-component system (TEOS/CETES) in the presence of the catalyst $(n\text{-}C_4H_9)_2Sn(CH_3COO)_2$ turned to be nonporous so that any subsequent hydrolysis of a sample of this substance would hardly result in a sorbent with a developed porous structure. Then, an attempt was made to synthesize xerogels with carboxyl-containing groups by a method which provided proceeding of two concurrent processes, namely of a reaction of hydrolytic polycondensation and saponifying of nitrile groups. In this case, a use was made not of a base catalyst but of an acid catalyst which was a mixture of two acids (acetic and sulfuric) in water at a component ratio of 1:1:3. Unfortunately the saponifying of nitrile group was not complete. Because of the fact that in the system there proceeded two processes (hydrolytic polycondensation and saponifying), it was rather difficult to choose conditions required for a complete saponifying of nitrile groups before formation of a gel. Therefore, with due account of the method described in reference [12] the authors of Refs. [49,50] advanced a new synthetic procedure. According to this procedure the sol–gel process took place only after saponifying of nitrile groups [see Scheme (7)]. At first, it is necessary to perform conversion of nitrile into an imino ester, which is attained by passing of gaseous hydrogen chloride through the system and subsequent introduction of 96% ethanol. Then, there proceeds hydrolysis of the ester and hydrolytic polycondensation reaction (in an acidic medium in the presence of TEOS), with the outcome being a carboxyl-containing gel. Washing and drying of the gel gives a xerogel.

$$(C_2H_5O)_3Si(CH_2)_{2\text{ or }3}CN \xrightarrow[+C_2H_5OH]{+HCl}$$

$$\times\, (C_2H_5O)_3Si(CH_2)_{2\text{ or }3}$$

$$\times\, C(OOC_2H_5)NH \cdot HCl \xrightarrow[-C_2H_5OH,\ -NH_4Cl]{+H_2O,\ +HCl,\ +TEOC}$$

$$(SiO_2)_x(O_{3/2}Si(CH_2)_{2\text{ or }3}COOH)_y. \qquad (33.7)$$

From the elemental analysis data and from the IR spectroscopy (see subsequently), it follows that the powdery

TABLE 33.5
Composition and Some Characteristics of Xerogels with Residues of Carboxylic Acids on Their Surfaces

Xero-gel	Molar ratio of components TEOS/ functional silane in a starting mixture	Thermal analysis data[d]		Structure-adsorption characteristics		
		H_2O	Organic part	S_{sp} (m^2/g)	V_s (cm^3/g)	d_{eff} [e] (nm)
62[a]	TEOS/CETES = 4:1	17.1 (17.9)	14.3 (14.5)	675	0.37	2.2(1.7)
63[b]	TEOS/CETES = 4:1	22.9 (23.4)	12.1 (12.6)	680	0.36	2.1(1.8)
64[a]	TEOS/CPTES = 4:1[c]	24.0 (22.5)	15.5 (16.1)	520	0.25	1.9(1.5)
65[a]	TEOS/CETES = 2:1	—	—	370	0.23	2.5(1.7)

[a]Drying was performed in vacuum.
[b]Drying was performed at the atmospheric pressure.
[c]CPTES - $(C_2H_5O)_3Si(CH_2)_3CN$.
[d]The values in brackets were calculated on the basis of the elemental analysis data.
[e]Pore size in brackets is the maximum of the pore size distribution (w_m) obtained according to [51].

white xerogels prepared in this way do not contain nitrile groups. From Table 33.5 it is evident that different conditions of drying of samples (at vacuum and under normal pressure) exert only a slight influence on their water content. The analytical data are in good agreement with the results of the thermoanalytical studies of the synthesized sorbents. Firstly, over a temperature range between 20 and 140°C all the samples display an endothermal effect related to a loss of water (by way of example, Figure 33.13 shows a thermogram for sample *64*).

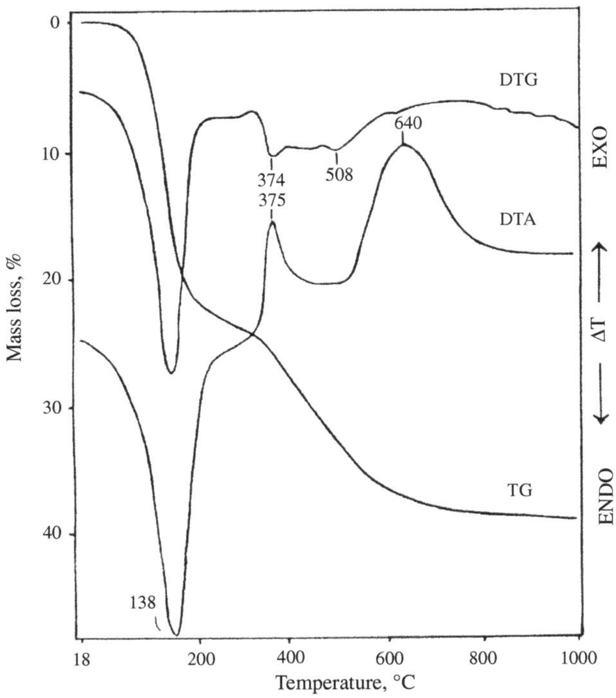

FIGURE 33.13 Curves of DTA and thermogravimetry of carboxyl-containing sample *64*.

Secondly, in air all the samples undergo decarboxylation. Their mass losses correspond to the complete oxidation of their organic components (Table 33.5). In the case of samples *62* and *63* the oxidation takes place at T_{max} = 438°C. The DTA curve for sample *64* shows two exothermal effects (at 375 and 640°C).

Table 33.5 presents also some data on the structue-adsorption characteristics of the synthesized sorbents. It is seen that the synthesized substances have an extended porous structure (their specific surface area ranges from 370 to 680 m^2/g). The average diameter of pores gives grounds to classify them as mesopores, but, at the same time, its value is rather close to the corresponding value characteristic of the boundary between micro- and mesopores, which manifests itself in the isotherm shape (Figure 33.14).

One of the salient features of the xerogels obtained is the fact that according to the results of the potentiometric titration which was performed in order to determine contents of carboxyl groups on their surfaces the number of

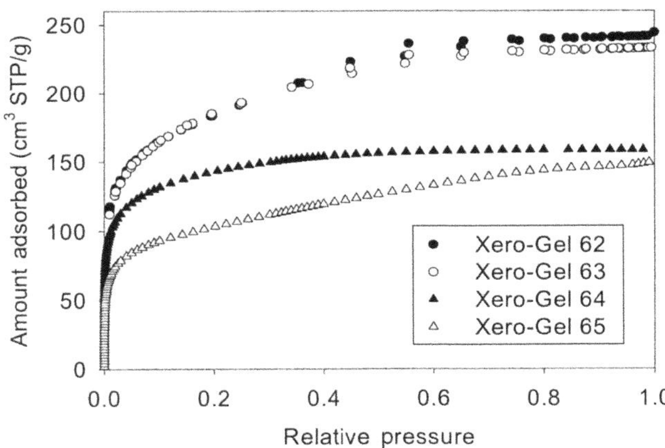

FIGURE 33.14 Nitrogen isotherms for samples *62*, *63*, *64*, and *65*.

these groups is always lower than that calculated from the elemental analysis data. Possible causes of this effect are considered in the section concerned with the IR spectroscopy of the samples.

Thus, it may be inferred that the sol–gel method allows one to synthesize diverse xerogels including xerogels on whose surface there are residues of carboxylic acids (for instance, of propionic and butyric acids).

STRUCTURE OF FUNCTIONALIZED POLYSILOXANE XEROGELS

IDENTIFICATION OF XEROGELS BY THE VIBRATIONAL SPECTROSCOPY

The vibrational spectroscopy (first of all the IR spectroscopy) is a traditional technique that makes it possible to relatively quickly and in general easily identify presence of one or another type of functional groups in xerogels. However, when applying this technique, one may also encounter some difficulties. They are as follows. Firstly, practically all xerogels contain marked amounts of water so that in a region above 3000 cm^{-1} their IR spectra have intense and very broad band $\nu(OH)$ (Figure 33.15). As a rule, this band masks absorption bands of other groups (e.g., of amine groups). Besides, this role is also often performed by deformation vibration band $\delta(H_2O)$ at about 1630 cm^{-1}. One of the possible ways to overcome the difficulty is a thermal vacuum treatment of polysiloxane xerogels since such a treatment enables one to remove water molecules from their surface. However, it should be kept in mind that in this case there may proceed one or another rearrangement of a surface layer formed by functional groups, which entails changes in IR spectra of samples. Besides, a number of functional groups may turn to be of a low thermal stability. Secondly, since it is not feasible to increase the content of functional groups in a synthesized sample, sometimes we cannot observe in its IR spectrum the absorption bands that are of a low intensity in the IR spectrum of a starting silane [recall, for instance, absorption band $\nu(SH)$]. Nevertheless, the data gathered with the aid of the IR spectroscopy of the xerogels are rather informative and useful both for determining of their composition and for elucidation of their surface layer structure [52]. Let us consider specific examples.

Figure 33.15 shows the IR spectra of xerogels on whose surface there are amine groups of different nature. In a region of 1030–1190 cm^{-1} all of them contain the most intense absorption band which usually has a shoulder

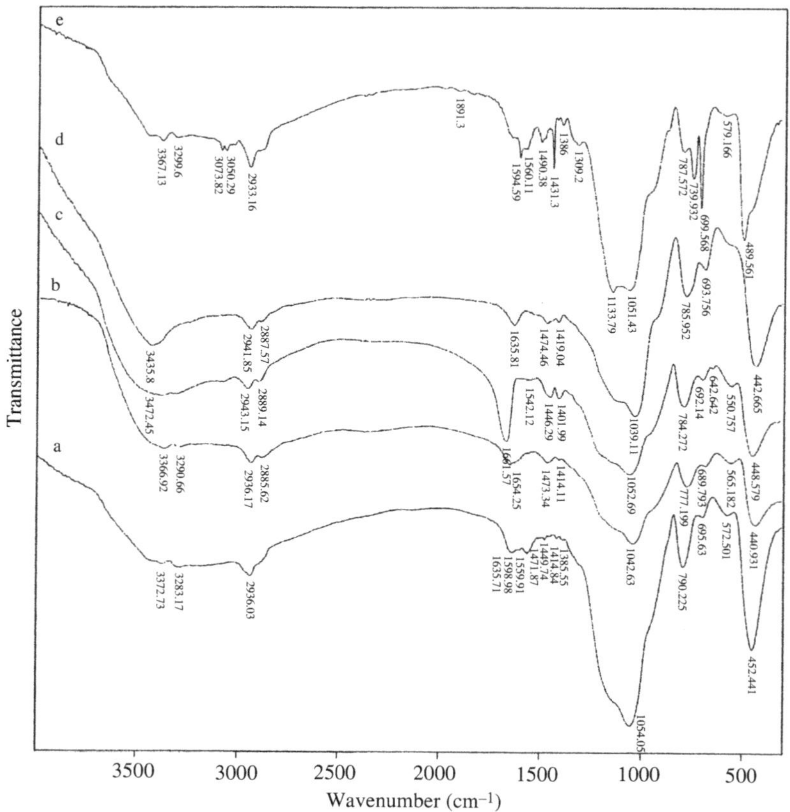

FIGURE 33.15 IR spectra of xerogels *8* (a), *30* (b), *32* (c), *31* (d), and *36* (e).

on the side of higher frequencies. Appearance of this band is typical for the three-dimensional ≡Si—O—Si≡ framework [53]. It should be noted that the xerogel whose surface layer contains phenyl groups (sample *36*) is distinguished for a specific spectral curve in this region, namely there are two peaks of a practically identical intensity at 1051 and 1134 cm^{-1} which can be used for its identification. In the region 2800–3000 cm^{-1} there appear sharp absorption bands that are characteristic of valence vibrations of bonds C—H of propyl chains. In the region 1250–1500 cm^{-1} these radicals have a number of low-intensity bands related to vibrations of methylene groups. It is anticipated that in these two regions (2800–3000 and 1250–1750 cm^{-1}) there should appear additional absorption bands if a use is made of more complex functional groups (of the 3-(2-imidazolin-1-yl)propyl type) or if several functional groups (of the APTES/MTES or APTES/PTES type) are concurrently introduced into a surface layer. Indeed, in the first case we can observe an intense absorption band at 1662 cm^{-1} assigned to double bond C=N of the imidazolinyl functional group (Figure 33.15, sample *32*). In the second case (for the xerogel whose synthesis involved APTES/PTES) in the IR spectra there is a number of sharp absorption bands of a medium intensity at 1431 and 1595 cm^{-1} (700 and 740 cm^{-1}) typical for the IR spectra of aryls. Besides, at 3050 and 3074 cm^{-1} one can clearly discern two weak absorption bands characteristic of valence vibrations ν(CH) of phenyl rings (Figure 33.15, sample *36*). When performing our synthesis of amine-containing xerogels, we used DMF and we proceeded on the assumption that some portion of the solvent was entrained during the course of the sol–gel process and was not removed during their drying in vacuum. Indeed, in the IR spectra of samples of such xerogels there is a band of a low intensity at 1660 cm^{-1} attributed to ν(CO) of DMF [52].

Thus, it is clear that the IR spectroscopy technique can help, firstly, corroborate the inference that the surface of the absorptive materials prepared contain one or another type of functional groups which were introduced during the course of the synthesis involving alkoxysilanes; secondly, detect the presence (or reveal the absence) of molecules of solvents which were used during the course of the synthesis; thirdly, detect the appearance of hydrogen bonds and even new groupings. As far as such new formations are concerned, amino- and thiol-containing xerogels are worthy of a more detailed consideration. The IR spectra of xerogels *57* and *58* which contain only 3-mercaptopropyl functional groups (Figure 33.16) almost do not differ from the IR spectra of xerogels with 3-aminopropyl groupings (Figure 33.15, sample *8*). One of the slight differences is the appearance of a weak absorption band at 2565 cm^{-1} that is assigned to ν(SH). With reducing fraction of thiol groups on the xerogel surface, its intensity decreases (samples *59* and *60*) and

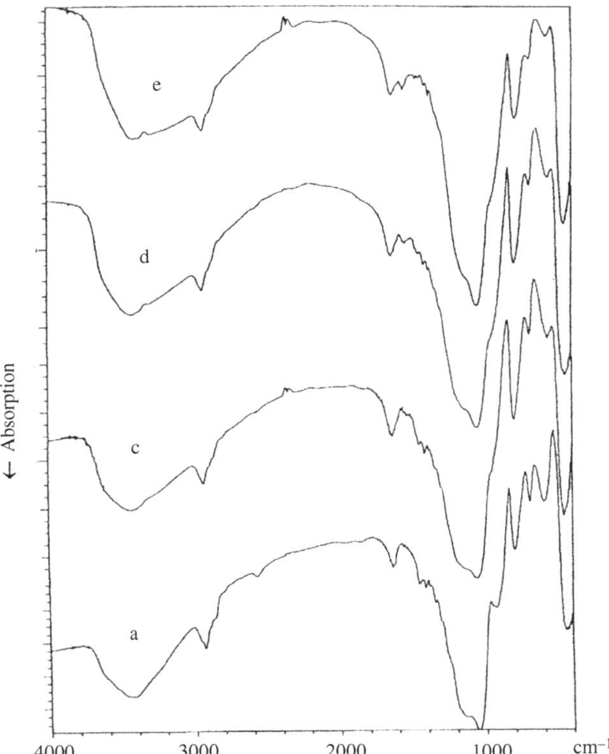

FIGURE 33.16 IR spectra of xerogels *57* (a), *59* (c), *60* (d), and *61* (e).

in the IR spectrum of xerogel *61* it becomes practically invisible (Figure 33.16e). At the same time, in the IR spectra of samples *60* and *61* at 1560 cm^{-1} there appears a weak absorption band whose intensity increases in going from sample *60* to sample *61*, that is with increasing number of amine groups in surface layers. The appearance of this band can be related to deformation vibrations $\delta(RNH_2)$. It may be assumed that this band is formed as a result of a contact between thiol and amino groups like $—SH\cdots NH_2—$. However, in the IR spectra of sample *8* that has a surface layer consisting of only 3-aminopropyl groups we can also observe a weak absorption band at 1560 cm^{-1} (Figure 33.15), which gives evidence that aminopropyl chains on the surface of xerogel *8* are also involved in formation of strong hydrogen bonds. It may be hypothesized that the source of protons are silanol groups because thiol groups are absent in these samples. With allowance for the fact that line ν(SH) in Raman spectra is much more intense, we recorded such spectra for samples *51–61* (see Figure 33.17). From Figure 33.17d it is seen that in the case when the $—(CH_2)_3SH/—(CH_2)_3NH_2$ ratio on the surface of xerogel *60* was equal to 1.1:1, at 2576 cm^{-1} there was a very intense line attributed to valence vibrations of bond S—H. Moreover, that line was also present in the Raman spectrum of xerogels *59* and *61* where the ratio of the

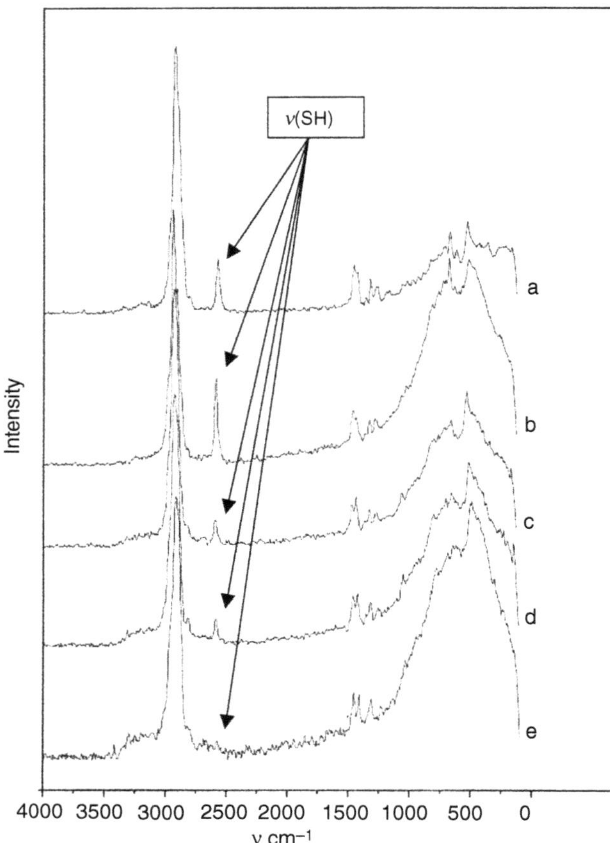

FIGURE 33.17 Raman spectra of xerogels *57* (a), *58* (b), *59* (c), *60* (d), and *61* (e).

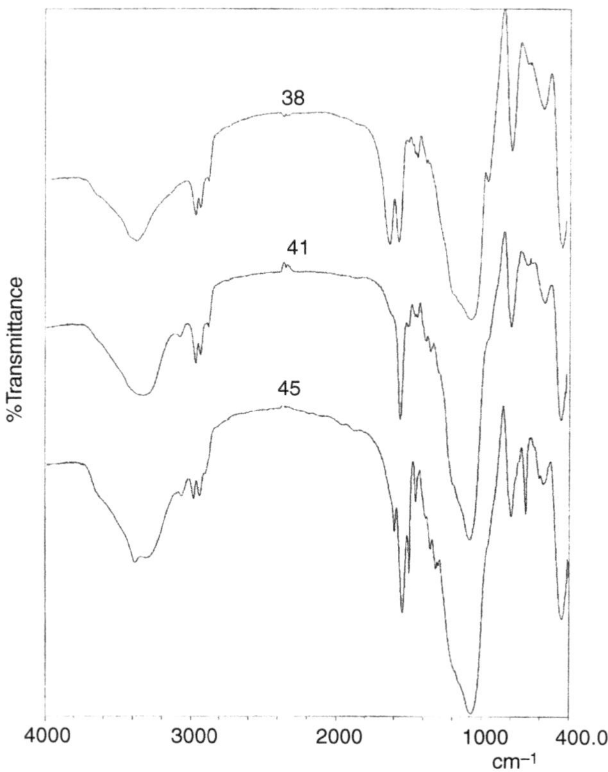

FIGURE 33.18 IR spectra of xerogels whose surfaces contain urea groups (sample *38*) and thiourea groups (samples *41*, *45*).

above-indicated groups was equal to 3.3:1 and 0.4:1 respectively (Figure 33.17 c and e). Therefore, the source of protons in xerogels *59–61* with a bifunctional surface layer is not thiol groups. Then, such a proton source may be silanol groups despite a marked difference between their pK_a. It should also be noted that in the IR spectra of all the amine-containing xerogels (Figures 33.15 and 33.16) against a background of band ν(OH) of sorbed water (3100–3600 cm^{-1}) there are two additional weak absorption bands (at ~3300 and ~3370 cm^{-1}) which are ascribed to $\nu_{s,as}$(NH) of amine groups involved in hydrogen bonds. These observations provide evidence for a complex structure of surface layers in these xerogels.

If in such a surface layer of xerogels there appear functional groups that are more complex in terms of their composition, the IR spectra of the xerogels become more complex as well, which is observed for xerogels containing (thio)urea groups (Figure 33.18).

However, these functional groupings give a number of intense absorption bands, with the absorption bands in the region 1500–1700 cm^{-1} being especially informative. Table 33.6 lists main vibration frequencies and their assignment for a number of xerogels containing a mono- or bifunctional surface layer. In most cases it is possible to make unambiguous assignments and, as a consequence, identifications of xerogels of this class.

In conclusion, let us consider the inferences which were made when analyzing the IR spectra of carboxyl-containing xerogels (Figure 33.19). First of all, it can be noted that their pattern is typical for IR spectra of xerogels functionalized by a direct method. Thus, all of them contain an intense absorption band at ~1065 cm^{-1} (with a shoulder at ~1130 cm^{-1}) characteristic of samples with a carbon-containing polysiloxane skeleton. Of importance is also the fact that in the IR spectra of these xerogels in the region 2200–2300 cm^{-1} there are no absorption bands characteristic of vibrations ν(C≡N). Hence, the advanced method for synthesizing xerogels provides complete saponifying of nitrile groups. At the same time, in these spectra in a region of 1720–1730 cm^{-1} there appears an absorption band of a medium intensity characteristic of compounds with hydrogen-bonded groups COOH [41]. When the ratio of reacting components (TEOS and nitrile-containing trialkoxysilane) in the starting mixture is 2:1, the corresponding spectrum displays a new weak shoulder at 1645 cm^{-1}, that is nearby the above-mentioned band (Figure 33.19). The shoulder may be attributed to deformation vibrations of water molecules present in the samples. However, when the ratio of

TABLE 33.6
Main Absorption Bands (cm^{-1}) in the IR Spectra of Xerogels with (Thio)urea Functional Groups and Their Assignment

Xerogel	ν(NH) + ν(OH)	ν(CH)	ν(CO)	[δ(CNH) + ν(CN)] or ν_{as}(NCN)	ν(SiO)
37	3373	2870, 2936, 2969	1636	1574	1072, 1134
38	3380, 3440	2876, 2936, 2976	1639	1574	1077, 1164
39	3303, 3362	2880, 2939, 2965	1639	1574	1059, 1144
40	3283	2874, 2933, 2965	—	1559	1074, 1141
41	3327	2880, 2936, 2968	—	1558	1079, 1162
42	3289, 3357	2877, 2939, 2965	—	1559	1064, 1141
43	3323	2876, 2933, 2970	1653	1535	1070, 1125
45	3301, 3383	2894, 2936, 2980	—	1543	1077, 1164
46	3385	2896, 2941, 2992	—	1550	1075, 1160
48	3316, 3385	2885, 2943, 2992	1665	1546	1051, 1133
49	3371	2975, 2933, 2891	1652	1570	1074, 1102
50	3415	2895, 2933, 2985	1660	1580	1068
51	3384	2896, 2925, 2978	—	1567	1072
54	3290, 3368	2880, 2940, 2980	1706	1548	1048, 1133

the above-indicated components is 4:1, the IR spectrum shows a sharp absorption band at 1637 cm^{-1} instead of the shoulder. It is not ruled out that the appearance of this line gives evidence for formation of bonds between carboxyl groups and ethanol typical for esters.

Characterization of Functionalized Polysiloxane Xerogels by the NMR Method

Solid-state ^{29}Si and ^{13}C NMR spectra can furnish not only an independent corroboration of the inferences about composition and structure of the surface of xerogels drawn on the basis of the IR spectroscopy data but also an additional useful information. Figure 33.20 shows the ^{13}C CP MAS NMR spectra of some PAPS xerogels. The spectrum of sample *8* contains three strong signals [at 10.2 ppm ($SiCH_2$), 22.1 ppm (CCH_2C), and 43.0 (CH_2NH_2)] which are related to different methylene groups of propyl chains (Figure 33.20a). A note should be made that in this spectrum there is no signal from carbon atoms and groups $SiOC_2H_5$. Therefore, this xerogel (as well as others) does not contain alkoxy

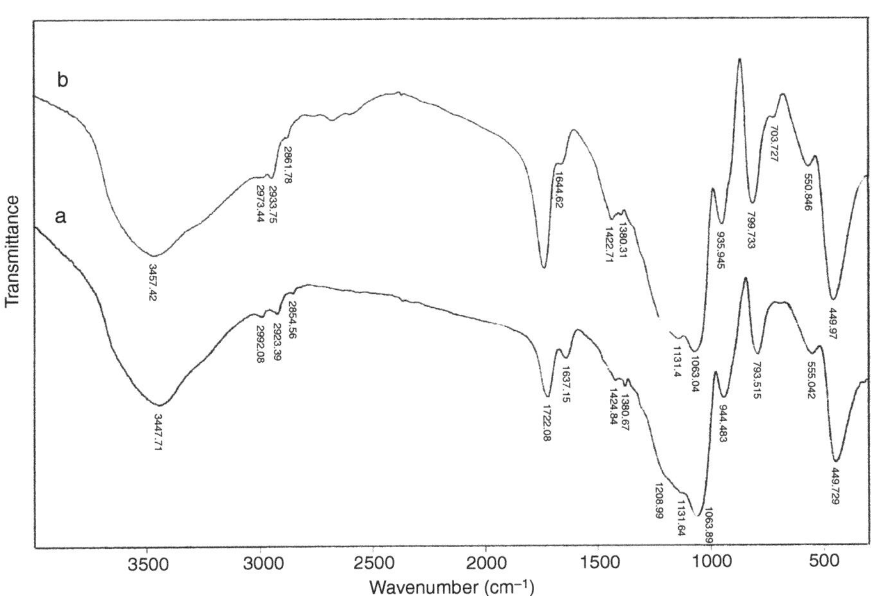

FIGURE 33.19 IR spectra of xerogels with residues of propionic acid. The reacting component ratio in a starting solution was 4:1 (a, sample *62*) and 2:1 (b, sample *65*).

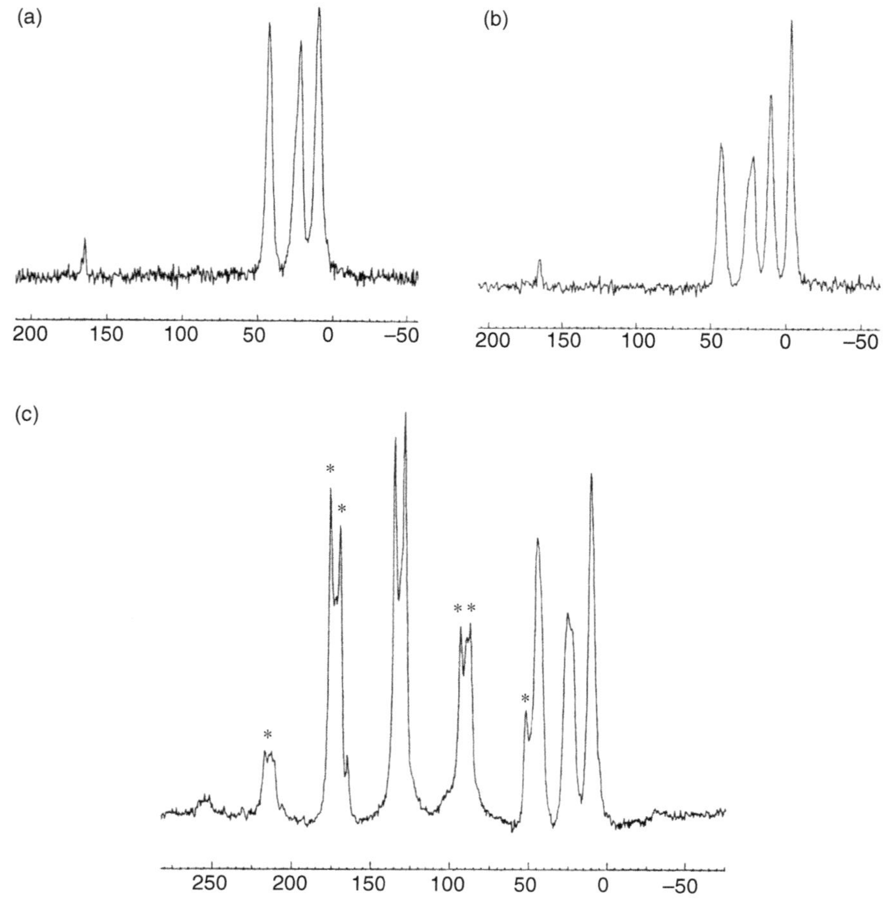

FIGURE 33.20 ^{13}C CP MAS NMR spectra of samples *8* (a), *34* (b), and *36* (c) (* is the symbol for sidebands).

groups. It is known that the chemical shift value of the central carbon atom of propyl radical C[$\mathbf{CH_2}$]C points to the status of amino groups. In the spectrum for 'free' APTES this resonance line is at 26–29 ppm while in the case of protonation of amine groups it is shifted into the strong field region and is seen at 21–22 ppm [54]. Thus, aminopropyl chains on the surface of xerogel *8* should be protonated. This inference is in agreement with the results of the works by Chiang et al. [55], Caravajal et al. [54], and Maciel et al. (56), but it is not in agreement with the IR spectroscopy data (see earlier). It is not ruled out that aminopropyl chains on the surface of xerogel *8* are involved in formation of strong hydrogen bonds like $\equiv$SiOH···NH_2—.

The ^{13}C CP MAS NMR spectra of samples with a bifunctional surface layer prove to be more complex. Thus, the sample with aminopropyl and phenyl groups (xerogel *36*) has signals at 10.6 ppm (Si$\mathbf{CH_2}$); 22.2 ppm (shoulder) and 25.6 ppm (C$\mathbf{CH_2}$C); 44.6 ppm ($\mathbf{CH_2}NH_2$); 128.0 ppm and 134.2 ppm (carbon atoms of phenyl radicals) (Figure 33.20c). An analogues spectrum of sample *34* which was synthesized using system TEOS/APTES/MTES (Figure 33.20b) bears a strong resemblance to the above-described spectra. In this spectrum there are three signals attributed to carbon atoms of methylene groups of propyl chains [10.8; 22.3, and 25.0 (shoulder); 43.7 ppm]. The forth signal at −2.6 ppm is related to carbon atom of a methylic group that is bonded directly with silicon atom. With allowance for the position of the central atom of a propyl radical in these samples it is possible to make an inference that only some part of amine groups are involved in formation of strong hydrogen bonds. The other part of these groups are close in terms of their nature to amino groups of "free" APTES.

The most likely assignment for the small peak at about 165 ppm is hydrogen carbonate anions formed during the air-storage of this sample. In the presence of water vapour, carbon dioxide would be absorbed to form the ammonium hydrogen carbonate [57] [Scheme (8)]. Such salts would decompose during the course of heating at above 100°C. Similar observations have been reported for silica modified with APTES [58].

$$RNH_2 + H_2O + CO_2 \longrightarrow RNH_3^+HCO_3^- \qquad (33.8)$$

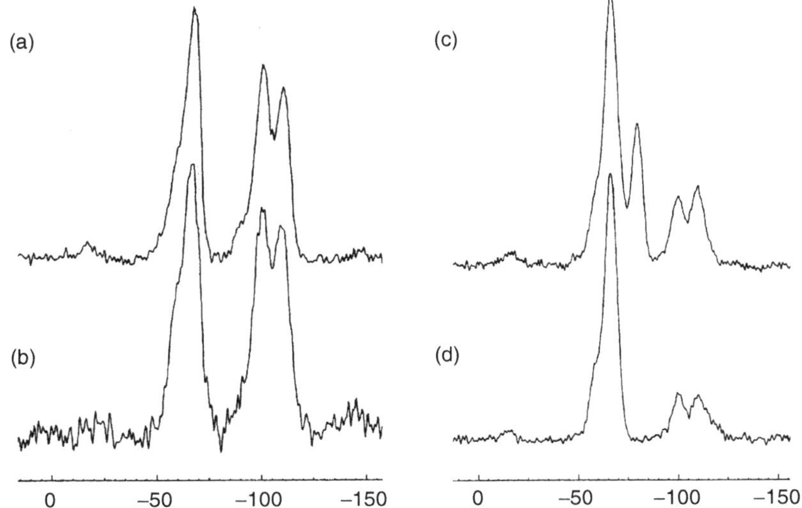

FIGURE 33.21 ^{29}Si CP MAS NMR spectra of samples *8* (a), *21* (b), *36* (c), and *34* (d).

The ^{29}Si CP MAS NMR spectra recorded for amine-containing xerogels show two or three major sets of resonances from −110 to −50 ppm related to various structural units (Figure 33.21). The first region contains, as a rule, two intense peaks and one very weak peak (at −109.8, −99.9, and ca. −90 ppm) which are assigned to $(SiO)_4Si$, $(SiO)_3SiOH$ and $(SiO)_2Si(OH)_2$ [59]. The presence of signals in the second region is characteristic of samples with phenyl groups. As is seen from Figure 33.21c, there is one signal at −80.3 ppm with a shoulder at −78 ppm, which gives evidence for existence of structural units of the $(SiO)_3SiC_6H_5$ and $(SiO)_2Si(OH)C_6H_5$ type. The third region also contains one signal at −66 ppm with a shoulder at −57 ppm, which is in agreement with the structural units of the previous type, that is with $(SiO)_3Si(CH_2)_3NH_2$ and $(SiO)_2Si(OH)(CH_2)_3NH_2$. The fact that in the case of sample *34* in this region there is no signal from carbon atom bonded with a methyl group can be explained by overlapping of this signal with a signal from silicon atom bonded with a methylene group of aminopropyl radical. It should also be noted that the ^{29}Si CP MAS NMR spectra of samples *8* and *21* are practically identical, although the xerogels differ substantially on their structure-adsorption characteristics (Figure 33.21).

Thus, irrespective of the functionality a surface layer, nature of functional groups, and structure-adsorption characteristics, the polyaminosiloxane xerogels always contain silicon atoms of three types, with their environment consisting of only oxygen atoms. Besides, there are two more types of silicon atoms that are bonded (through alkyl chains) with functional groups of the $\equiv Si(CH_2)_3NH_2$ and $=Si(OH)(CH_2)_3NH_2$ type which under normal conditions are involved in formation of strong hydrogen bonds. When surface layers contain hydrophobic groups as well, only some part of amine groups turn to be bonded in such way; the other part of these groups are close in terms of their nature to amino groups of 'free' APTES.

Application of Metal Microprobe Method for Characterizing the Structure of a Xerogel Surface

It was of interest to find out in what way the structural units whose existence was ascertained with the help of the NMR spectroscopy are arranged on surfaces of the synthesized xerogels. With a view to answer the question we employed the metal microprobe method which allowed us to study sorption of copper(II) ions by amine-containing xerogels from acetonitrile solutions [52,60,61]. It was established that irrespective of surface coverage by the metal on the surface there proceeded formation of the same complexes. As an example, Figure 33.22 shows EPR spectra of copper(II) ions sorbed on the surface of a poly(3-aminopropyl)-siloxane xerogel at different initial ratios $Cu(NO_3)_2$/sorbent. The composition of these complexes can be represented as $[CuO_2N_2]$, that is in the equatorial plane of copper(II) coordination polyhedron there are two nitrogen atoms. It is unambiguously corroborated by the EPR spectra of copper(II) complexes formed on the xerogel surface with groups $NH[(CH_2)_3Si\equiv]_2$ at a metal/ligand ratio of 1:10. In this case in region $g_\perp$ there appears a hyperfine structure (Figure 33.23). The presence of five components of the hyperfine-structure splitting provides unambiguous evidence for the fact that ion Cu^{2+} is in environment of two equivalent nitrogen atoms. The comparative analysis of parameters of the EPR spectra recorded for copper(II) complexes which are formed on the surface of polyaminosiloxane xerogels (Table 33.7) gives

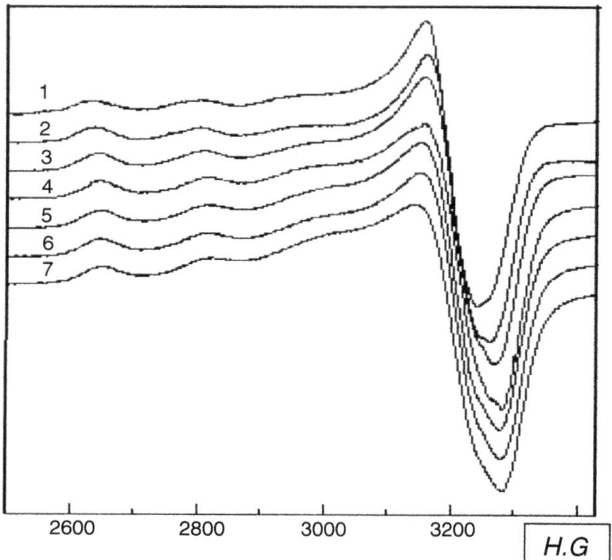

FIGURE 33.22 EPR spectra of ions Cu^{2+} sorbed by sample *8* at various values of ratio C^S_{Cu}/C^S (NH_2) (1, 0.63; 2, 0.57; 3, 0.49; 5, 0.25; 6, 0.20; 7, 0.10).

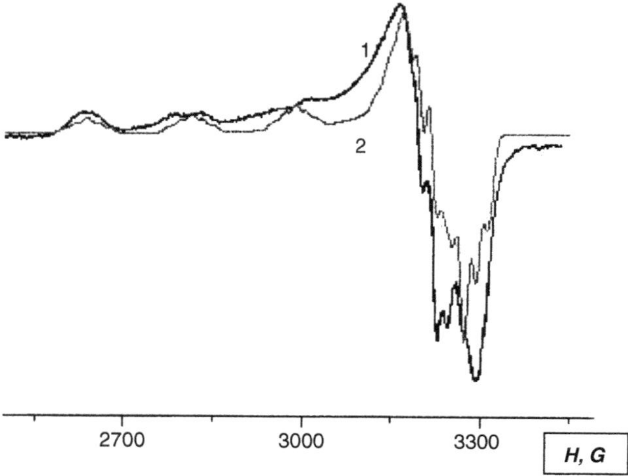

FIGURE 33.23 EPR spectra of copper(II) complexes on the surface of sorbent *31* containing groups $NH[(CH_2)_3Si{\equiv}]_2$ (1, experimental spectrum at the C^S_{Cu}/C^S_R ratio of 1:10; 2, model spectrum).

grounds for the inference that irrespective of nature of amine groupings and structure-adsorption characteristics of sorbents the coordination sphere of a copper(II) ion always contain two nitrogen atoms. This may provide evidence for the fact, firstly, that, such polyaminosiloxane xerogels have similar structures of their surfaces and, secondly, that on their surfaces there may exist oligomers with the above-mentioned aminopropyl groupings formed during the hydrolytic polycondensation reaction (at least, there may exist dimers whose composition can, for example, be represented by the formula $=O_2Si[(CH_2)_3NH_2]{-}O{-}Si(OH)[(CH_2)_3NH_2]{-}O{-}$).

TABLE 33.7
Parameters of the EPR Spectra Recorded for Copper(II) Complexes Formed on Surfaces of Various Amino-Containing Supporters

Composition of the coordination sphere of a Copper(II) ion	$g_\perp$	$g_\parallel$	$A^{Cu}_\parallel$ (10^4 cm^{-1})	References
CuN_2O_2[a]	2.070	2.280	136	[15]
CuN_2O_2[a]	2.062	2.297	176	[60]
CuN_2O_2[b]	2.055	2.247	184	[61]
CuN_2O_2[c]	2.060	2.260	190	[61]
	2.060	2.240	167	
CuN_2O_2[d]	2.008	2.240	195	[52]
$[Cu(pic)_2(OSi{\equiv})_n]$[e]	2.061	2.240	179	[62]
	2.066[f]	2.243[f]	179.4[f]	
$[Cu(RNHR')_2(H_2O)_4]^{2+}$	2.060	2.290	—	[63]
$C^S_{Cu}/C^S_{NH_2}$ = 1:11.2	2.063	2.243	170	[64]
$C^S_{Cu}/C^S_{NH_2}$ = 1:5.6	2.064	2.238	170	[64]
$Cu[NH_2(CH_2)_3Si{\equiv}]_2(H_2O)_2$	2.055	2.255	166	[65]
$Cu[NH_2(CH_2)_3Si{\equiv}]_2(H_2O)_2$	2.055	2.321	151	[66]

[a]Xerogel with groups $\equiv Si(CH_2)_3NH_2$.
[b]Xerogel with groups $[\equiv Si(CH_2)_3]_2NH$.
[c]Xerogel with groups 3-(2-imidazolin-yl).
[d]Xerogel with groups $\equiv Si(CH_2)_3NH_2/CH_3$.
[e]$OSi{\equiv}$ is for silica surface; pic is for α-picolinic acid.
[f]Another batch of silica.

CONCLUSIONS

From the above-presented experimental material it is possible to draw the following inferences. In most cases the advanced procedures enable us to synthesize readily polysiloxane xerogels with reproducible characteristics and surface layers that contain such functional species as amine, thiol, carboxyl, and (thio)urea groups. The structure-adsorptive properties of the produced xerogels with a monofunctional surface layer are strongly affected by character of functional groups and conditions of synthesis. Further, we designed simple synthetic procedures which allow one to produce polysiloxane xerogels with a bifunctional surface layer that can contain not only amine and thiol or amine and thiourea groups but also amine and alkyl (aryl) groups. In this case, the ratio of reacting alkoxysilanes and nature of an "additional" functional group exert a profound effect both on degree of hydrophobicity of the surface of such materials and on their porosity; at certain ratios of reacting components it is possible to produce mesoporous materials with a biporous structure. The data collected with the help of a number of physical methods give evidence for the fact that, on the one hand, on the surface of functionalized polysiloxane xerogels there are present structural units of an identical nature and, on the other hand, the surface structures of these materials are similar in their character.

ACKNOWLEDGMENTS

The authors would like to thank Prof. M. Jaroniec for his help and fruitful discussion. This research was partially sponsored by NATO's Scientific Affairs Division in the framework of the Science for Peace Programme (Grant SfP 978006).

REFERENCES

1. *Chemistry of Silica Surface*, Vol. 2; Chuiko, A.A., Ed.; Institute of Surface Chemistry: Kyiv, 2001 (in Russ.).
2. Zaitsev, V.N. *Complexing Silicas: Synthesis, Structure of Bonded Layer and Surface Chemistry;* Folio: Kar'kov, 1997 (in Russ.). Kholin Yu.V.; Zaitsev, V.N. *Complexes on a Surface of Chemically Modified Silicas;* Folio: Kharkov, 1997 (in Russ.).
3. Vansant, E.F.; Voort, Van Der P.; Vrancken, K.C., *Characterization and Chemical Modification of the Silica Surface;* Elsevier: Amsterdam, 1995.
4. Tertykh, V.A.; Belyakova, L.A. *Chemical Reactions with Participation of Silica Surface;* Naukova Dumka: Kyiv, 1991 (in Russ.).
5. Lisichkin, G.V.; Kudryavtsev, G.V.; Serdan, A.A. *Modified Silicas in Sorption, Analysis and Chromatography;* Khimiya: Moscow, 1986 (in Russ.).
6. Zub, Yu.L.; Parish, R.V. *Stud. Surf. Sci. Catal.* **1996**, *99*, 285.
7. Brinker, C.J.; Scherer, G.W., *Sol–Gel Science: The Physics and Chemistry of Sol–Gel Processing;* Academic Press: San Diego, 1990.
8. Voronkov, M.G.; Vlasova, N.N.; Pozhidaev, Yu.N. *Zh. Prikl. Khim.* **1996**, *69*, 705 (in Russ.).
9. Voronkov, M.G.; Vlasova, N.N.; Pozhidaev, Yu.N. *Appl. Organometal. Chem.* **2000**, *14*, 287.
10. Yakubovich, T.N.; Zub, Yu.L.; Chuiko, A.A. In *Chemistry of Silica Surface*, Vol. 1; Chuiko A.A., Ed.; Institute of Surface Chemistry: Kyiv, 2001, p. 54–97 (in Russ.). Yakubovich, T.N.; Teslenko, V.V.; Veisov, B.K.; Zub, Yu.L.; Parish, R.V. *Koord. Khimiya.* **2001**, *27*, 597 (in Russ.). Yakubovich, T.N.; Teslenko, V.V.; Kolesnikova, K.A.; Zub, Yu.L.; Leboda, R. *Stud. Surf. Sci. Catal.* **1995**, *91*, 597.
11. Chuiko, A.A.; Pavlik, G.Ye.; Budkevich, G.B.; Neimark, I.Ye., USSR Patent 182,719, 1966.
12. Chuiko, A.A.; Pavlik, G.Ye.; Neimark, I.Ye., USSR Patent 164,680, 1964.
13. Slinyakova, I.B.; Denisova, T.I. *Organo-silicon Adsorbents: Production, Properties, Application;* Naukova Dumka: Kyiv, 1988 (in Russ.).
14. Finn, L.P.; Slinyakova, I.B.; Voronkov, M.G.; Vlasova, N.N.; Kletsko, F.P.; Kirillov, A.I.; Shklyar, A.V. *Dokl. AN SSSR.* **1977**, *235*, 1426 (in Russ.).
15. Khatib, I.S.; Parish, R.V. *J. Organomet. Chem.* **1989**, *369*, 9.
16. Parish, R.V.; Habibi, D.; Mahammadi, V. *J. Organomet. Chem.* **1989**, *369*, 17.
17. Yang, J.J.; El-Nahhal, I.M.; Maciel, G.E. *J. Non-Crystal. Solids.* **1996**, *204*, 105.
18. Zub, Yu.L.; Kovaleva, L.S.; Zhmud', B.V.; Orlik, S.N.; Uzunov, I.; Simeonov, D.; Klisurski, D.; Teocharov, L. *Proc. 7th Int. Symp. Heterog. Catal., Bourgas (Bulgaria)*, 1991; Vol. 1, pp. 565–571. Zub, Yu.L.; Gorochovatskkkaya, M.Ya.; Chuiko, A.A.; Nesterenko, A.M. *Ext. Abstr. Fourth Int. Conf. Fundamentals of Adsorption, Kyoto (Japan)*, 1992; pp. 461–463.
19. Matkovskii, O.K.; Yurchenko, G.R.; Stechenko, O.V.; Zub, Yu.L. *Scientific Notes of Ternopil' State Training Univ. Series 'Khimiya'.* **2000**, 40 (in Ukrain.).
20. Zub, Yu.L.; Parish, R.V.; Stechenko, O.V.; Chuiko, A.A. *J. Non-Crystal. Solids.* **2003** (to be submitted).
21. El-Nahhal, I.M.; Parish, R.V. *J. Organometal. Chem.* **1993**, *452*, 19.
22. Iler, R.K. *The Chemistry of Silica;* Wiley: New York, 1979.
23. Neimark, I.Ye.; Sheifain, R.Yu. *Silica Gel, Its Production, Properties, and Application;* Naukova Dumka: Kyiv, 1973 (in Russ.).
24. Okkerse, C. In *Physical and Chemical Aspects of Absorbents and Catalysts;* Linsen, B.G., Ed.; Academic Press: London and New York, 1970; p. 233.
25. Neimark, I.Ye. *Synthetic and Mineral Adsorbents and Catalyst Supports;* Naukova Dumka: Kyiv, 1982 (in Russ.).

26. Komarov, V.S.; Dubnitskaya, I.B. *Physico-chemical Principles for Control over Porous Structure of Adsorbents and Catalysts;* Nauka i Tekhnika: Minsk, 1981 (in Russ.).
27. Gregg, S.J.; Sing, K.S.W. *Adsorption, Surface Area and Porosity;* Academic Press: London, 1982.
28. Gordon, J.E. *The Organic Chemistry of Electrolyte Solutions;* Wiley: New York, 1975.
29. Andrianov, K.A. *Organo-silicon Compounds*; Goskhimizdat: Moscow, 1955 (in Russ.). Andrianov, K.A. *Organo-silicon Polymer Compounds;* Gosenergoizdat: Moscow, 1946 (in Russ.).
30. Yang, J.J.; El-Nahhal, I.M.; Chuang, I-S.; Maciel. G.E. *J. Non-Crystal. Solids.* **1997**, *209*, 19.
31. Zhmud, B.V.; Sonnefeld, J. *J. Non-Crystal. Solids.* **1996**, *195*, 16.
32. Zub, Yu.L.; Chuiko, A.A.; Stechenko, O.V. *Reports of NAS of Ukraine.* **2002**, 150 (in Ukrain.). Stechenko, O.V.; Zub, Yu.L.; Parish, R.V. *Proc. 3rd Int. Sypm. "Effects of Surf. Heterogeeneity in Adsorp. and Catal. on Solids", Torun (Poland)*, 1998; pp. 231–232. Zub, Yu.L.; Drozd, L.S.; Chuiko, A.A. *Abstr. IUPAC Symp. on the Characterization of Porous Solids, Marseille (France)*, 1993; p. 95.
33. Stechenko, O.V.; Yurchenko, G.R.; Matkovskii, O.K.; Zub, Yu.L. *Scientific Bulletin of Uzhgorod Univ. Series 'Khimiya'* **2000**, 107 (in Ukrain.).
34. Zub, Yu.L.; Melnyk, I.V.; Chuiko, A.A.; Cauzzi, D.; Predieri, G. *Chemistry, Physics and Technology of Surface.* **2002**, No 7, 35. Zub, Yu.L.; Seredyuk, I.V.; Chuiko, A.A.; Cauzzi, D.; Predieri, G. *Abstr. 3dr Int. Conf. "Chemistry of High-organized Substances and Scientific Fundamentals of Nanotechnology," St. Petersburg (Russia)*, 2001; pp. 261–263.
35. Gilkey, J.U.; Kraenke, R.E. U.S. Patent 3,208,971, 1965.
36. Vlasova, N.N.; Pestunovich, A.E.; Voronkov, M.G. *Izv. AN SSSR. Ser. Khim.* **1979**, 2105 (in Russ.).
37. Voronkov, M.G.; Pestunovich, A.E.; Kositsyna, T.I.; Sterenberg, B.Z.; Pusechkina, T.A.; Vlasova, N.N. *Z. Chem.* **1983**, *23*, 248.
38. Voronkov, M.G.; Pestunovich, A.E.; Kositzina, E.I.; Shterenberg, B.Z.; Pushechkina, T.A.; Vlasova, N.N. *Zh. Obschei Khimii.* **1984**, *54*, 1098 (in Russ.).
39. Ferrari, C.; Predieri, G.; Tiripicchio, A.; Costa, M. *Chem. Mater.* **1992**, *4*, 243.
40. Baigozhin, A. *Zh. Obschei Khimii.* **1972**, *43*, 1408 (in Russ.).
41. Lin-Vien, D.; Colthup, N.B.; Fateley, W.G.; Grasselly, J.G. *The Handbook of Infrared and Raman Characteristic Frequencies of Organic Molecules;* Academic Press: San Diego, CA, 1991.
42. Jensen, K.A.; Nielsen, P.H. *Acta Chem. Scand.* **1966**, *20*, 597.
43. Melnyk (Seredyuk), I.V.; Zub, Yu.L.; Chuiko, A.A.; Voort, Van Der P. *Chemistry, Physics and Technology of Surface.* **2002**, No 8, 125.
44. El-Nahhal, I.M.; Yang, J.J.; Chuang, I.C.; Maciel, G.E. *J. Non-Cryst. Solids.* **1996**, *208*, 105.
45. Voronkov, M.G.; Vlasova, N.N.; Kirillov, A.I.; Zemlyanushnova, O.V.; Rybakova, M.M.; Kletsko, F.P.; Pozhidaev, Yu.N. *Dokl. AN SSSR.* **1984**, *275*, 1095 (in Russ.).
46. Sobolevskii, M.V.; Grinevich, K.P.; Demchenko, M.I.; Vasyukov, S.Ye.; Fedotov, N.S.; Rybalka, M.G. *Plasticheskiye Massy.* **1977**, 68 (in Russ.).
47. Belinskaya, Z.I.; Baskevich, D.N.; Petukhov, G.G.; Konovalov P.G.; Subbotina, A.I., USSR Patent 209,748, 1968.
48. Pozhidaev, Yu.N.; Zhyla, G.Yu.; Yarosh, O.G.; Kirillov, A.I.; Vlasova, N.N.; Voronkov, M.G. *Zh. Obschei Khimii.* **1997**, *67*, 1097 (in Russ.).
49. Yashina, N.I.; Zub, Yu.L.; Chuiko, A.A. *Abstr. 3rd Polish-Ukr. Symp. "Theor. Experim. Stud. Interfacial Phenomena and Their Technol. Application", Lviv (Ukraine)*, 1998; p. 77. Prybora, N.A.; Dzyubenko, L.S.; Zub, Yu.L.; Jaroniec, M. *Chemistry Sciences. Collected Sci. Papers of Nat. Training M.P.Dragomanov's Univ.* **1999**, 41 (in Ukrain.).
50. Prybora, N.A.; Zub, Yu.L.; Chuiko, A.A.; Jaroniec, M. *Abstr. 2nd Int. Conf. on Silica Science and Technol., Mulhouse (France)*, 2001; p. 171.
51. Kruk, M., Jaroniec, M., Sayari, A. *Langmuir.* **1997**, *13*, 6267.
52. Zub, Yu. L. Dr. Sc. Thesis, Institute of Surface Chemistry, NAS of Ukraine, 2002.
53. Finn, L.P.; Slinyakova, I.B. *Colloid. Zh.* **1975**, *37*, 723 (in Russ.).
54. Caravajal, G.S.; Leyden, D.E.; Quinting, G.R.; Maciel, G.E. *Anal. Chem.* **1988**, *60*, 1776.
55. Chiang, C.H.; Liu, N.I.; Koenig, J. L. *J. Colloid Interf. Sci.* **1982**, *86*, 26.
56. Maciel, G.E.; Bronnimann, C.E.; Zeigler, R.C.; Chuang, I-S.; Kinney, D.R.; Keiter, E.A. In *The Colloid Chemistry of Silicas;* Bergna, H.E., Ed.; American Chemical Society: Washington, DC, 1994; p. 269.
57. Culler, S.R.; Naviroj, S.; Ishida, H.; Koenig, J.L. *J. Colloid. Interf. Sci.* **1983**, *96*, 69.
58. Zaper, A.M.; Koenig, J.L. *Polymer Compos.* **1985**, *6*, 156.
59. Engelhardt, G.; Michel, D. *High-resolution Solid-state NMR of Silicates and Zeolites;* Wiley: Chichister, 1987.
60. Yakubovich, T.N.; Teslenko, V.V.; Zub, Yu.L.; Chuiko, A.A. *Chemistry, Physics and Technology of Surface.* **1997**, No 2, 62 (in Russ.).
61. Stechenko, Ye.V.; Yakubovich, T.N.; Teslenko, V.V.; Veisov, B.K.; Zub, Yu.L.; Chuiko, A.A. *Chemistry, Physics and Technology of Surface.* **1999**, No 3, 46 (in Russ.).
62. Kokorin, A.N.; Vlasova, N.N.; Pridantsev, A.A.; Davidenko, N.A. *Izv. KN SSSR. Ser. Khim.* **1997**, 1765 (in Russ.).
63. Kholin, Yu.V.; Zaitsev, V.N.; Mernyi, S.A.; Varzatskii, N.K. *Zh. Neorg. Khimii.* **1995**, *40*, 1325 (in Russ.).
64. Filippov, A.P.; Zyatkovskii, V.M.; Karpenko, G.A. *A. Teor. Eksp. Khim.* **1981**, *17*, 363 (in Russ.).
65. Solozhenkin, P.M., Semikopnyi, A.I.; Sharf, V.Z.; Lisichkin, G.V. *Zh. Fiz. Khim.* **1988**, *62*, 477 (in Russ.).
66. Golubev, V.B.; Kudryavtsev, G.V; Lisichkin, G.V.; Mil'chenko, D.V. *Zh. Fiz. Khim.* **1985**, *59*, 2804 (in Russ.).

34 Multinuclear NMR Studies of Silica Surfaces

Gary E. Maciel and I-Ssuer Chuang
Colorado State University, Department of Chemistry

CONTENTS

The solid-state NMR revolution of the past quarter century has had an enormous impact of the state of knowledge of silica and modified silicas. $^{1}H \rightarrow ^{29}Si$ cross polarization (CP) with magic-angle spinning (MAS) is an invaluable approach for observing local silicon environments on the silica surface. ^{1}H NMR approaches distinguish clustered (hydrogen-bonded) and isolated surface silanols. Correlations between ^{29}Si and ^{1}H NMR behaviors in silicas have led to detailed structural models of the silica surface based on intersections of OH-terminated 100 and 111 faces of β-cristobalite. Other nuclides (e.g., ^{1}H, ^{2}H, ^{13}C, and ^{15}N), along with ^{29}Si, provide valuable approaches for the characterization of local structure and motion in derivatized (e.g., silylated) silicas. ^{17}O NMR shows great promise for structure elucidation in modified silicas in which some of the Si sites of silica are replaced by other metal centers.

INTRODUCTION

The "high-resolution" NMR study of silica and derivatized silicas is a subject first explored in about 1979 by then graduate student, Dean Sindorf, when we realized that $^{1}H \rightarrow ^{27}Si$ cross polarization (CP) would yield ^{29}Si MAS (magic-angle spinning) spectra that would dramatically emphasize silicon nuclei at or very near the surface (i.e., near hydrogen nuclei) [1,2] Since that time, ^{29}Si CP-MAS experiments on silica-based samples, and subsequent experiments with other nuclides (e.g., ^{1}H, ^{13}C, ^{17}O), have become very popular, and publications reporting this kind of work have become almost too numerous to follow, certainly too numerous to review comprehensively in this paper. However, although there has been a large volume of such work that has been reported, most of it has involved largely the same approaches introduced in Sindorf's Ph.D. thesis or in subsequent important extensions based on time-domain $^{1}H-^{29}Si$ CP-MAS, and other, techniques by Research Associate, Dr. I-Ssuer Chuang [3], and on ^{1}H NMR studies by then postdoc, Charles Bronnimann [4], and in specific applications by other members of this research group [5], some carried out in collaboration with other research groups. Much of this work has been reviewed elsewhere [6,7], some of it in the previous volume of this series [6a]. Accordingly, in preparing this paper, we have strongly emphasized the work about which we could be most authoritative, namely from our own research group, and we have drawn heavily from material summarized in our previous reviews. This material has been supplemented by material from our research group that has not been reviewed earlier and by noteworthy results from other groups. Because of this arbitrary choice of material to include, we have omitted some excellent work by other groups, for example, the extensive work by

Legrand and co-workers [8] and the important ^{29}Si NMR work by Brinker and co-workers on the formation of silica in the sol-gel process [9]. We assume that much of this is covered elsewhere in this volume. In addition, the huge subject of template-synthesized mesoporous silicas, for which there is a very large recent NMR literature, is omitted from this paper, which focuses on what is commonly referred to as *amorphous* silica, e.g., silica gels, precipitated silicas and fumed silicas.

Much of the recent and current research interest in silica and silica-based materials focuses on surface properties [10]. With any NMR (or other spectroscopic) experiment on a solid, a technique that provides no major detection advantage to nuclei at the surface will generate spectra that are dominated by peaks due to nuclei that are in positions in the interior (bulk) of a particle, because the number of nuclei that constitute the bulk of a particle will typically be much larger than the number of corresponding nuclei in analogous structural sites at the surface (unless the surface area is very large, say $\gg 100\ m^2\ g^{-1}$). This intensity dominance by peaks due to bulk sites can be overcome if (1) the nuclei being observed are located only (or largely) at the surface or (2) the method of generating the polarization to be observed discriminates strongly in favor of nuclei at the surface. The former situation often obtains for protons in a typical silica, because most of the protons in such systems exist at the surface as covalently attached -OH groups (*vide infra*), as physisorbed H_2O, or as covalently attached structures that result from derivatization processes.

To date, the most popular and generally successful surface-selective polarization strategies have been based on $^1H \rightarrow X$ CP [11], where X is a nucleus present at the surface (and presumably also within the bulk).[2a,d] These strategies are often based on the assumption that essentially all (or, at least, most) of the protons in the system are present at the surface and on the fact that cross polarization depends upon a static component of the ^1H–X dipolar interaction, which has an inverse cube dependence on the ^1H–X internuclear distance (in many cases the cross polarization rate appears to have an essentially r^{-6} dependence). Cross polarization was developed during the early 1970s by Pines, Gibby, and Waugh [11]. Its early impact was primarily the dramatic increase in ^{13}C signal-to-noise ratio in NMR experiments on organic solids, which rendered such experiments especially attractive when carried out with MAS, as demonstrated first by Schaefer and Stejskal [12]. From the point of view of surface applications, at least as important as the effective increase in sensitivity is the above-mentioned dependence of the cross-polarization rate on internuclear distances. Figure 34.1 displays the essence of a surface-selective CP strategy in which only those X nuclei that are close enough to the surface (within, say, 5–6 Å, as represented by the "cross-hatched" area) can be cross polarized by surface protons. The more remote X nuclei in the "interior" or "bulk" of the material are not cross polarized efficiently. Hence, the efficiency, or dynamics, of cross polarization can be used to discriminate in favor of the surface nuclei.

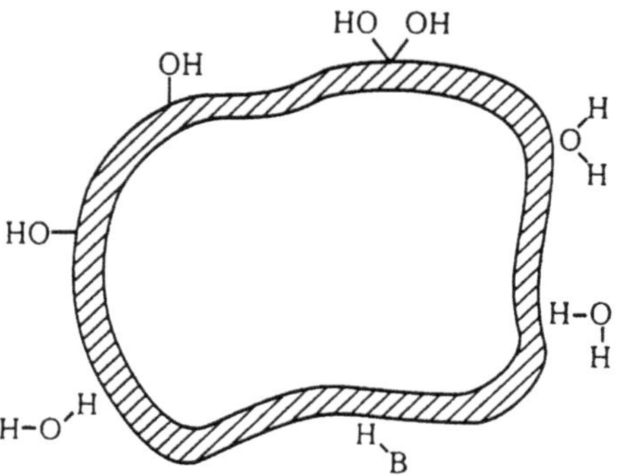

FIGURE 34.1 Cross polarization as a surface-selective strategy, showing only protons near the (hatched) surface region, as covalently attached hydroxyls, physisorbed water, or physisorbed acids (B–H) of some other type. From reference [6c]. With permission.

Perhaps the most obvious strategy for preferentially or selectively polarizing surface nuclei in the presence of an overwhelmingly larger number of nuclei in analogous structural sites in the interior would be to use a relaxation reagent that can, at least briefly, interact with the surface and thereby relax the surface nuclei. Although surprisingly little effort seems to have been expended in this direction, one must note the elegant ultra-low-temperature (~10 mK) NMR studies of Waugh and co-workers [13], who used the relaxation effect of ^3He impinging on a surface for the selective relaxation of nuclei at the surface. Other surface-selective (or preferential) relaxation mechanisms would seem possible via the dipolar mechanism of impinging species with large nuclear magnetic moments (say, ^1H or ^{19}F) or even the electron spin magnetic moments of paramagnetic relaxation agents. In the case of paramagnetic surface relaxants, the possibility of dynamic nuclear polarization (DNP) [14] of surface nuclei from adsorbed paramagnetic species seems attractive, possibly with the Overhauser mechanism operating if the adsorption is rapidly reversible, or the solid state mechanism if the adsorption/desorption process is very slow. Possibilities would also appear to exist for surface-selective relaxation mechanisms based on quadrupolar relaxation of a nuclide with $I > 1/2$ (e.g., ^2H or ^{17}O) due to rapid, reversible adsorption/desorption causing a modulation of the local electric field gradient. Recent progress in the surface transfer of ^{129}Xe polarization that has been dramatically enhanced via electron → nuclear polarization transfer (e.g., from optically pumped rubidium) is promising for not only providing a surface-selective NMR strategy, but also for providing enhancements in the effective sensitivity of NMR detection at surfaces [15].

The most popular line-narrowing technique in modern solid-state NMR spectroscopy is MAS, in which the sample is mechanically rotated rapidly (thousands of revolutions per second) about an axis that makes an angle of 54.7° relative to the direction of the static magnetic field [16]. Sufficiently rapid MAS brings about the coherent averaging of inhomogeneous line-broadening effects, such as the chemical-shift anisotropy (CSA) and inhomogeneous magnetic dipole–dipole interactions. This effect is a coherent, mechanical analog of the incoherent motional averaging that is accomplished naturally in liquids by Brownian motion.

In 1980, Sindorf and Maciel[2a,d] published the first high-resolution example of the use of $^{1}H \rightarrow X$ CP for surface-selective X detection in a demonstration of $^{1}H-^{29}Si$ CP in silica gel. Since that time, $^{1}H \rightarrow {}^{29}Si$ CP has remained the most popular application of this strategy, although there has been significant success with applications to other types of systems. Figure 34.2 shows typical ^{29}Si spectra of silica gel and related samples obtained by CP-MAS and direct polarization (DP) MAS (based on ^{29}Si spin-lattice relaxation, not CP).

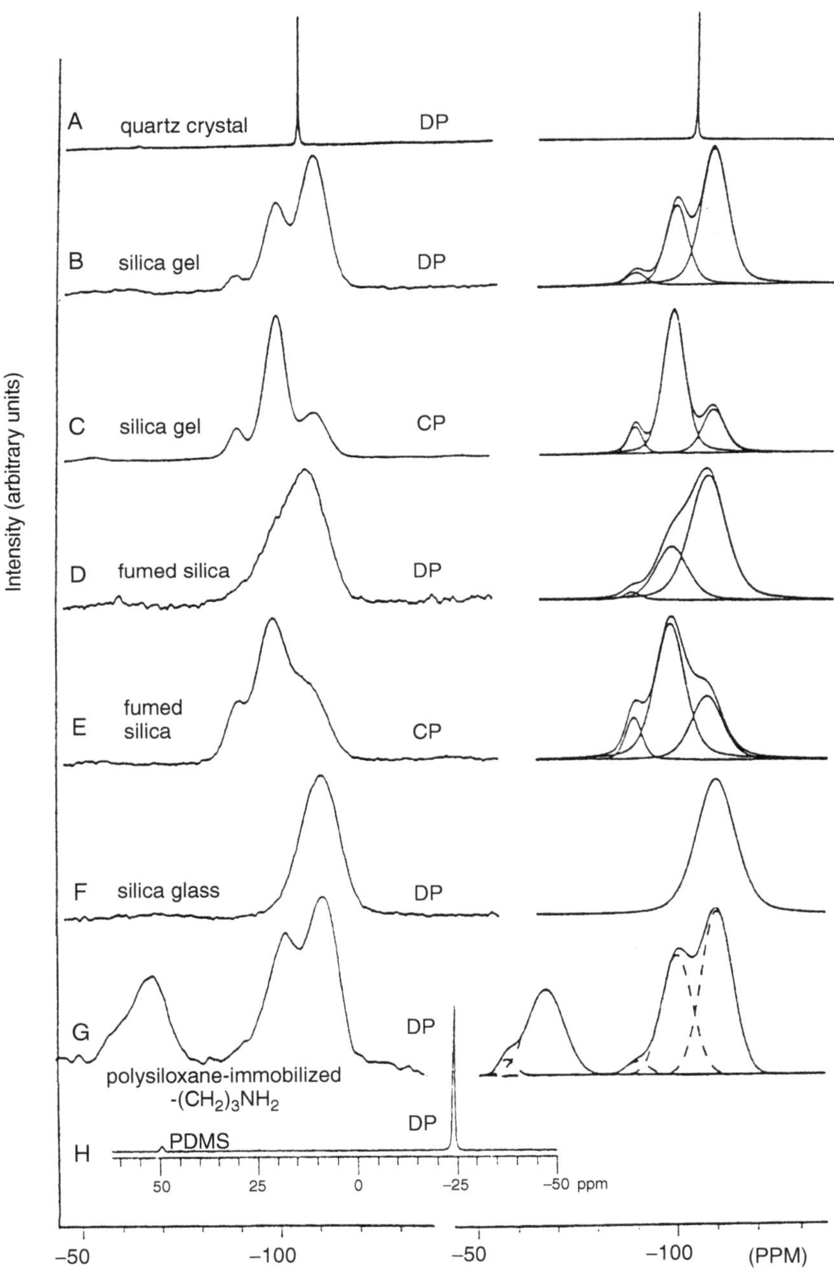

FIGURE 34.2 ^{29}Si MAS spectra of polysiloxanes, based on DP or CP. A) Quartz crystal, B) and C) silica gel, D) and E) fumed silica, F) silica glass, G) polysiloxane-immobilized $-CH_3CH_2CH_2NH_2$, H) polydimethylsiloxane (PDMS) rubber. Left side, experimental spectra. Right side, computer-deconvoluted spectra. From reference [6e]. With permission.

The DP-MAS spectrum of an undried silica gel (Figure 34.2B) is dominated by the peak due to ($\geqslant$SiO)$_4$**Si** (siloxane, Q_4) sites, which represent the bulk (nonsurface regions) of silica particles. The ^{29}Si CP-MAS spectrum of an undried silica gel (Figure 34.2C) selects primarily surface sites and shows the following three peaks: a peak at −89 ppm (relative to liquid TMS) due to ($\geqslant$SiO)$_2$**Si**(OH)$_2$ (geminal, Q_2) sites; a peak at −99 ppm arising from ($\geqslant$SiO)$_3$**Si**OH (single silanol, Q_3) sites; and a peak at −109 ppm from the Q_4 (siloxane) sites near the surface. These peak assignments can be made on the basis of the usual kinds of empirical chemical shift correlations with structure from liquid-sample data on silicic acid solutions. However, the dynamics of the ^{1}H–^{29}Si CP process can also be used to make these assignments.

Figure 34.3 shows the results of a variable contact time CP experiment, in which the ^{1}H → ^{29}Si CP contact period (t_{cp}) is varied in order to elucidate the ^{1}H → ^{29}Si CP (relaxation) time constant (T_{HSi}) for each ^{29}Si peak. The early (small t_{cp}) part of such curves is typically dominated by the rate of CP transfer, as characterized by the rate constant T_{HSi}^{-1}, and the latter part of such curves is usually determined by the rate constant of the rotating frame spin-lattice relaxation of the protons responsible for polarization transfer to the observed silicons, as characterized by the time constant, $T_{1\rho}^{H}$ (assuming $T_{1\rho}^{H} > T_{HSi}$). These curves can be analyzed mathematically in terms of well-known equations [17].

This analysis shows that T_{HSi}^{-1} for the −89 ppm peak is roughly twice that of the −99 ppm peak, which in turn is an order of magnitude larger than the T_{HSi}^{-1} value of the −109 ppm peak. In terms of the number and distances of nearby protons, the ^{29}Si chemical shift assignments given above in terms of Q_2, Q_3, and Q_4 sites are entirely consistent with these T_{HSi} determinations.

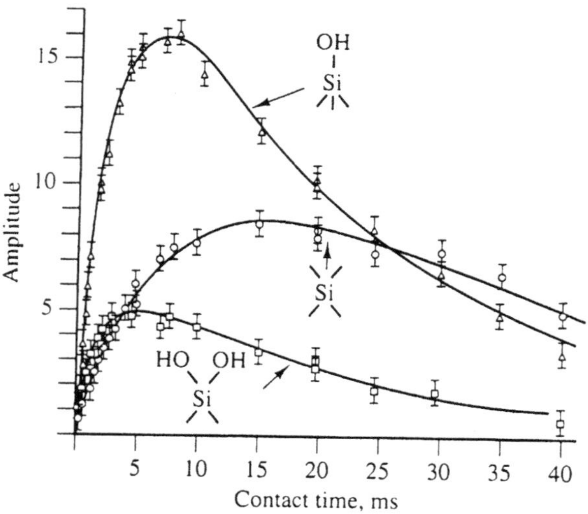

FIGURE 34.3 Variable contact time ^{29}Si CP-MAS plots for silica. From reference [2a]. With permission.

UNDERIVATIZED SILICA

^{29}Si NMR Patterns

Linewidth

Although the major peak of each ^{29}Si DP-MAS spectrum shown in Figure 34.2 is centered in the −100 to −110 ppm region characteristic of **Si**(—O–Si$\leqslant$)$_4$ local structure, there are substantial differences among the spectra [3e]. The four spectra obtained via DP show a range of linewidths spanning from about 0.3 ppm for crystalline quartz to about 13 ppm for silica glass. This range of linewidths represents a corresponding range or order, or crystallinity, as has been reported previously [18] and as discussed in more detail below. The occurrence of shoulders at about −90 ppm and about −100 ppm in the spectra of silica gel and fumed silica (Figuress 34.2B–E) is due to the fact that the surface areas of these two samples are large, so the populations of ($\geqslant$Si—O—)$_3$**Si**OH sites and ($\geqslant$Si—O—)$_2$**Si**(OH)$_2$ sites are accordingly substantial in relation to the dominant ($\geqslant$Si—O—)$_4$**Si** sites. The dramatically increased relative intensities of the Q_2 and Q_3 peaks, relative to the Q_4 peak, in the CP-MAS spectra of silica gel (Figure 34.2C) and fumed silica (Figure 34.2E), in comparison to relative peak intensities in the corresponding DP-MASS spectra (Figures 34.2B and D), reflect the selection of surface sites by the ^{1}H → ^{29}Si CP process [2a].

As indicated above, the varying linewidths seen in the ^{29}Si MAS NMR spectra of the various types of silica showing in Figure 34.2 reflect varying degrees of structural order, or crystallinity, in the samples. In our view, a "pure crystal" is at one end of the "order spectrum," having perfect short-range and perfect long-range order. A "partially crystalline" solid is usually considered to be a material with a high degree of short-range order (the identities and geometrical arrangements of nearest and next-nearest neighbor atoms or moieties), but the structural order is attenuated rapidly with distance. An amorphous material is one with a substantial degree of short-range order, but essentially no long-range order. A glassy material typically has only limited short-range order (with some degree of uniformity in only the *identities* of nearest and next-nearest neighbor atoms or moieties), with substantial variations in local geometry (e.g., bond angles and lengths with respect to nearest and next-nearest neighbor atoms). Recently, some investigations have started to reveal that even silica glass and vitreous silica glass possess medium-range order resembling cristobalite and/or tridymite [19]. This entire spectrum of structural order is spanned by silica, as is amply reflected in the ^{29}Si MAS NMR spectra presented in

Figure 34.2. One sees the very narrow ^{29}Si NMR peaks (linewidth <1 ppm) observed for the Q_4 silicon sites seen in Figure 34.2A for quartz (or for β-cristobalite) [18a–e], reflecting the high degree of order, that is a homogeneous structural situation, as is characteristic of true crystals. The other end of the "order spectrum" in silicas is represented by the very broad ^{29}Si Q_4 peak of a glassy SiO_2, as seen in Figure 34.2F (linewidth ~12 ppm). One sees from the intermediate Q_4 linewidths in Figure 34.2 for silica gel (~6 ppm) and fumed silica (~9 ppm) that these materials are intermediate between a highly amorphous (glassy) and a completely crystalline material. The fact that medium-range order is more extensive in silica gel than in fumed silica (with its wider distribution of bond angels and distances resulting from its formation at much higher temperatures) is manifested in the fact that the Q_4 peak of silica gel is appreciably narrower than that of fumed silica. Indeed, the relatively rapid formation processes for silica gels and fumed silicas (especially the latter), may preclude the dominance of certain thermodynamic factors that can otherwise determine the characteristics of crystal surfaces. We believe that the amorphous materials, silica gel and even fumed silica, are properly represented as having glasses and crystalline materials. In particular, we believe that the *surfaces* of these amorphous high-surface-area materials can be modeled in terms of crystal faces of β-cristobalite (*vide infra*) [3e]. No doubt some other characteristics may be modeled in terms of glass silica, but we do not focus on such properties in this instance. Burneau et al. have indicated that various experimental results suggest some local order on the silica surface even with an amorphous bulk network [20].

The spectrum (Figure 34.2G) of a functionalized polysiloxane in Figure 34.2 represents a polysiloxane in which pendant —$CH_2CH_2CH_2NH_2$ groups are attached to the polysiloxane framework, replacing one siloxane linkage on some of the silicon atoms. The Q_4 peak in this spectrum has a linewidth of about 8 ppm, comparable to those in the amorphous silicas (Figures 34.2B–E), and substantially less than that of silica glass. This implies that the distribution of local bond angles and bond lengths at silicon atoms is narrower for the functionalized polysiloxanes (of which the example with pendant $-CH_2CH_2CH_2NH_2$ groups is representative) than for the silica glass, in which one expects a substantial amount of strain that is "locked-in" as the fused silica is cooled to form a glass.

Related to the issues of crystallinity, order or randomness in silica samples, as reflected in ^{29}Si MAS linewidths, are the matters of "homogeneous" and "inhomogeneous" contributions to the linewidth. The latter results from a superposition/overlap of numerous closely spaced peaks, each representing a specific structural environment; that is, inhomogeneous linewidth contributions arise from a degree of randomness in the structure — structural dispersion. A highly crystalline sample should have a very small inhomogeneous contribution of linewidth. The homogeneous contribution can be determined by a measurement of the transverse (spin-spin) relaxation time, T_2: $(\Delta\nu_{1/2}(\text{homog}) = (\pi T_2)^{-1}$. Such measurements on silica gel samples corresponding to the ^{29}Si MAS spectra of Figures 34.2B–C, obtained by the rotor-synchronized Hahn echo method,[21] yielded results of 70 ± 10 ms for the largest two peaks [3f]. This corresponds to a homogeneous linewidth contribution of about 4.5 Hz, or about 1 ppm on the 200 MHz (^1H) spectrometer employed. Hence, most of the ~6 ppm ^{29}Si MAS linewidths observed for silica gel arises from structural inhomogeneity.

Dehydration of Silica Gel

The ^{29}Si CP-MAS spectra collected in Figure 34.4 represent two important classes of chemical transformations of silica surfaces that were studied via ^{29}Si CP-MAS spectra by Sindorf and Maciel.[2] Figure 4(c) shows a spectrum obtained on a silica gel sample that has been dehydrated under vacuum at 209°C. One can see in this spectrum the subtle redistribution of peak intensity and the dramatic line broadening relative to the spectrum observed on an "air hydrated" silica gel sample. These changes, especially the line broadening, are presumably due to a redistribution of hydrogen bonding patterns, and perhaps bond angles and bond lengths (and strains) introduced with the removal of the (predominantly physically) adsorbed water of the "hydrated" and essentially "annealed" silica surface represented in Figure 34.4(b). Such experiments carried out over a wide range of dehydration temperatures, and corresponding rehydration experiments, have revealed valuable information on the effects of dehydration and rehydration on the silica gel surface. The ^{29}Si CP-MAS spectrum shown in Figure 34.4(d) represents a sample prepared by the silylation of a silica gel by $(CH_3)_3SiCl$; silylated products will be discussed in greater detail in a later section.

An interesting set of ^{29}Si CP-MAS experiments on silica samples that have been "equilibrated" with deuterated water show that a small fraction of the silanols are not readily accessible to D_2O exchange, even in boiling samples; such "trapped" silanols have been characterized via time-domain ^{29}Si CP-MAS studies to be primarily nonhydrogen-bonded Q_3 silanols [3c,5q]. Figure 34.5 shows ^{29}Si CP-MAS spectra of a silica gel that had been equilibrated with D_2O at 25°C (Figure 34.5a) and then exposed to air for various periods of time, until eventually (essentially 90 min) the original, unexchanged sample is recovered (Figure 34.5f). One sees that the Q_2 (geminal) silanol peak is essentially gone with the D_2O-exchanged sample, although a Q_3 signal is certainly present; after

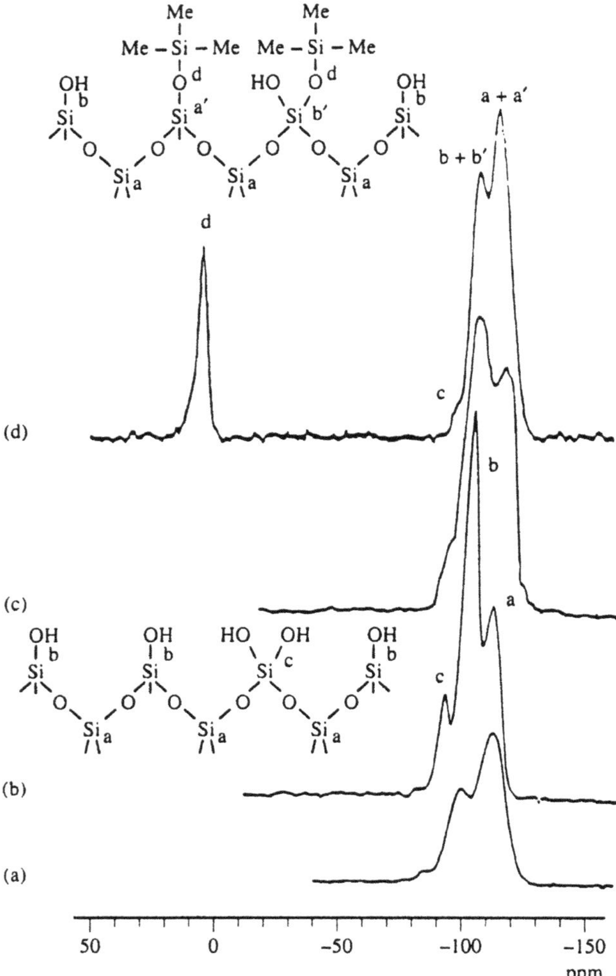

FIGURE 34.4 ^{29}Si MAS spectra of silica gel samples. (a) DP-MAS spectrum of an undried sample. (b) CP-MAS spectrum of the same sample as in (a). (c) CP-MAS spectrum of a sample of dehydrated under vacuum at 209°C. (d) Sample derivatized with $(CH_3)_3SiCl$. Taken from reference 6c. With permission.

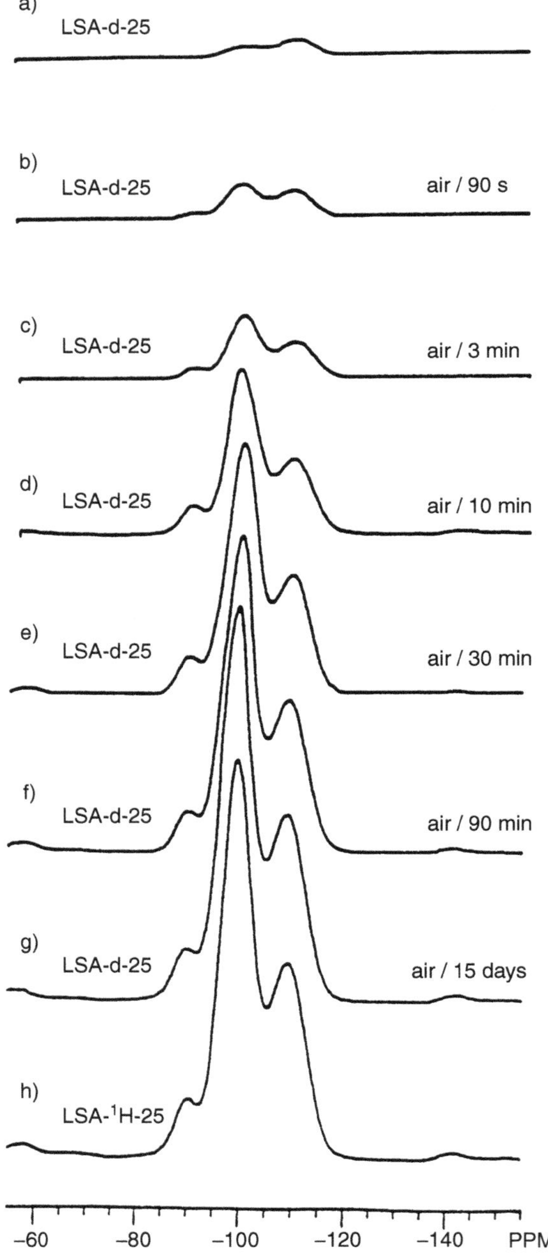

FIGURE 34.5 ^{29}Si CP-MAS spectra of a silica gel sample that has been D_2O- exchanged at 25°C (a) and then exposed to air for the indicated periods of time (b–g), spectrum of the unexchanged (initial) silica (h). Taken from reference 3c. With permission.

only 90 s of exposure to air, both the Q_2 and Q_3 signals are seen to be growing in.

A related set of experiments is shown in Figure 34.6, which includes results for a silica gel sample that was D_2O exchanged at 100°C. From the combination of such studies [3c], it has been determined that, for a typical silica gel, 91–97% (less for a fumed silica) [5r] of the silanols are exchangeable with D_2O; the exact percentage depends on the time and temperature of exchange. The geminal silanol signal is completely depleted by D_2O exchange and Figure 34.6 shows that the geminal silanol signal is "immediately" restored to its equilibrium intensity shortly after a D_2O-exchanged silica is exposed to moisture in air. Hence, none of the inaccessible or subsurface silanols are of the Q_2 type. Comparing the relative intensities of Q_3 and Q_4 type silicons in Figure 34.6 for H_2O-treated and D_2O-treated silica gels also indicates that the inaccessible silanols are mainly "interior" in nature. Inside silica gel's internal cavities, all single silanols are totally surrounded by siloxane (Q_4) bridges; therefore, protons of these trapped single silanols are able to cross polarize more ^{29}Si nuclei in Q_4 structures (per ^1H) than would be the case with surface silanols.

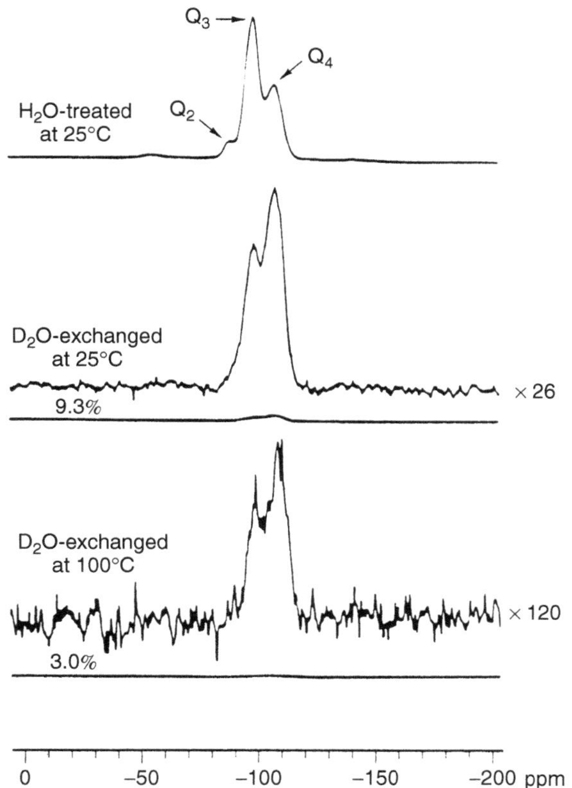

FIGURE 34.6 ^{29}Si CP-MAS spectra of silica gels stirred in H_2O or D_2O at 25°C or 100°C, as indicated. Taken from reference 3c. With permission.

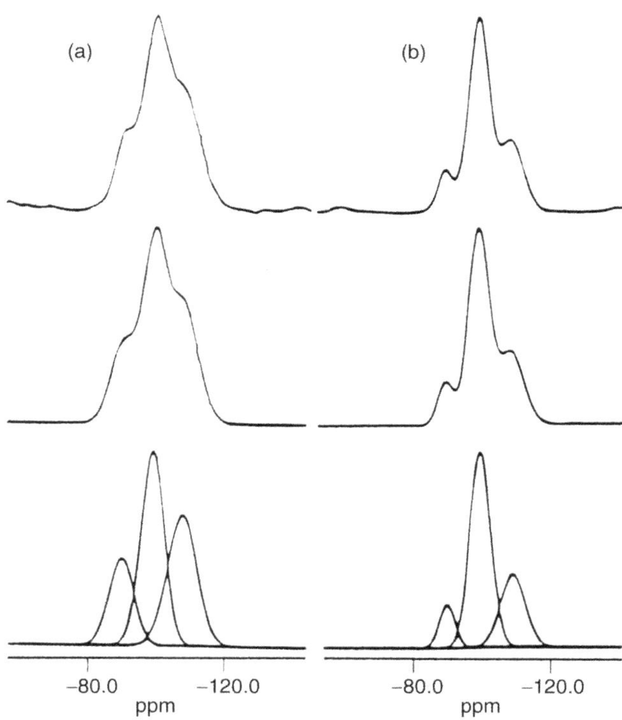

FIGURE 34.7 ^{29}Si CP-MAS spectra of (a) Cab-O-Sil fumed silica and (b) silica gel. Top: experimental spectra. Bottom: individual Q_2, Q_3 and Q_4 peaks by deconvolution. Middle: computer sum of the contributions shown at the bottom. Taken from reference 5r. With permission.

Extensive studies of spin dynamics in D_2O-exchanged samples have revealed a substantial amount of detail about the local environment of motional dynamics of "internal" silanols in silica gel [3c]. The rate constants measured for $^{1}H \rightarrow {}^{29}Si$ cross polarization indicate that the hydroxyl groups of "internal" silanols rotate freely about the Si—O axis in a manner similar to the rotation of nonhydrogen-bonded silanols on a dehydrated silica surface. Analogous studies have also been carried out on a fumed silica Cab-O-Sil [5r]. While the D_2O-exchange behavior and various features of ^{1}H and ^{29}Si spin dynamics in Cab-O-Sil are qualitatively similar to what is discussed above for silica gel, some significant and possibly important differences are observed [5r], perhaps because of interparticle (particle-bridging) silanols that have been suggested by some authors for fumed silicas.

Fumed Silica

Other types of silicas (and derivatized silicas), besides those based on silica gels, have also been studied by ^{29}Si CP-MAS (and DP-MAS) experiments, for example, by Brinker and co-workers [9], by Legrand and co-workers [8c,d] and by Liu and Maciel [5q,r]. Figure 34.7 shows a comparison of ^{29}Si CP-MAS spectra of a fumed silica (a Cab-O-Sil, formed by the vapor-phase combustion of $SiCl_4$) and a silica gel equilibrated to about the same H_2O vapor pressure [5r]. The same peaks are present, but they are broader in the case of the fumed silica, and the percentage of surface silica sites that are single silanols is seen to be smaller for the fumed silica than for silica gel. The greater linewidth presumably relates to the greater dispersion of local surface geometries (and chemical shifts) in the Cab-O-Sil structure, which is formed at a higher temperature, and possibly to the potential effects of so-called "interparticle sites" at the junctures of the primary particles that contribute to the overall topography of a fumed silica [5q]. Detailed studies of ^{29}Si CP-MAS spin dynamics of fumed silicas, especially when viewed in relationship to analogous silica gel results, appear promising for identifying the main similarities and differences between these two types of silica surfaces [5r].

^{1}H NMR Studies

High-resolution ^{1}H NMR spectroscopy has proved to be highly useful in studying the surfaces of silicas and a variety of other solids. In order to obtain high-resolution ^{1}H NMR spectra of solids (including their surfaces), it is necessary to average not only the chemical shift anisotropy (easily done by MAS), but also $^1H—^1H$ dipolar interactions [4d,f]. The latter can be very large (tens of kHZ). Magnetic dipole-dipole interactions manifest an inverse cube dependence on internuclear distance. Therefore, such interactions, and the $^1H—^1H$ spin-spin flip-flops that they generate, are especially strong if the protons are situated in close proximity to each other, for example in a typical organic solid, but also presumably in hydrogen-bonded clusters of hydroxy groups on a surface. Hence, *a priori* one can feel confident that moderate-speed MAS experiments (say, <20 kHZ) are adequate for studying silica surfaces only for substantially dehydrated samples. For nondehydrated or nondeuterated samples, in which the *local* surface density of protons can be substantial, either multiple-pulse techniques (*vide infra*) or ultrafast MAS (>20 KHz) may be required for eliminating the line-broadening effects of $^1H—^1H$ dipolar interactions. Alternatively, a clever strategy employed by Vega and co-workers is to ensure that the ^1H concentration is small by exchanging protons at the surface with deuterons [22].

In 1968, Waugh and co-workers [23] introduced a *multiple-pulse* approach for averaging strong homonuclear dipolar interactions. The strategy of this kind of approach is that, over the *entire* period of each individual multiple pulse cycle (four 90° pulses in the original work), the average Hamiltonian that governs the evolution of the spins over the entire cycle does *not* include the homonuclear dipolar interaction. A nonvanishing chemical shift effect, albeit scaled down, in present in the average Hamiltonian. Hence, if one acquires one data point stroboscopically between each adjacent pair of multiple pulse cycles in a long string of such cycles, the resulting time-dependent signal (analogous to a free induction decay) is modulated by chemical shift effects, but not by homonuclear dipolar effects.

The original homonuclear line-narrowing pulse sequence (WAHUHA) [23] was a four-pulse sequence; later elaborations involve more pulses in the total cycle and offer compensation for pulse imperfections and/or higher order averaging of the homonuclear dipolar interaction [24]. Currently, the most popular multiple pulse sequences are the eight-pulse MREV-8 sequence [24a] and the 24-pulse BR-24 sequence [24b].

For powdered or amorphous samples, in order to avoid line broadening due to chemical shift anisotropy, MAS is employed (which also averages various heteronuclear dipolar couplings). Gerstein and co-workers [25] were the first to combine a multiple pulse homonuclear line-narrowing technique with MAS, and introduced the acronym CRAMPS for Combined Rotation And Multiple-Pulse Spectroscopy [4,25].

Figure 34.8 shows the difference in ^1H line-narrowing capabilities between CRAMPS and MAS-only (no multiple-pulse sequence) with a modestly high MAS speed (10–11 kHz) [4e]. For the relatively proton-dilute (~2.5 atom%, mostly situated on the surface) silica gel system, the ^1H NMR experiment yields a modestly narrowed ^1H spectrum even on a static sample, and 4.7 kHz MAS yields a spectrum that is superficially similar to that obtained by CRAMPS. However, close inspection reveals important details that differentiate the ^1H spectra obtained by these two techniques (Figures 34.8(d), (e)). All of the remaining ^1H spectra discussed in this article were obtained by the CRAMPS technique.

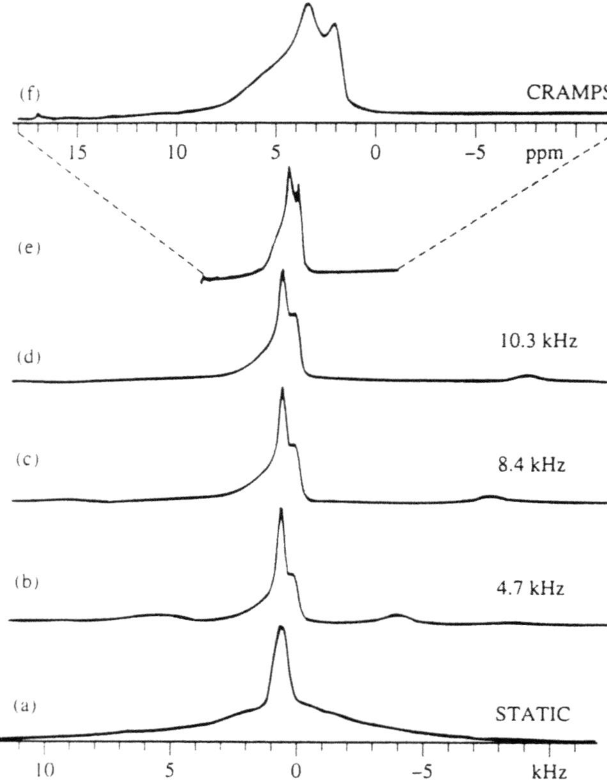

FIGURE 34.8 187 MHz ^1H NMR spectra of partially dried (in a dry box for 2 hr.). silica gel. (a) Static sample, single pulse. (b), (c), (d) MAS-only with indicated MAS speed. (e) CRAMPS. (f) CRAMPS on an expanded scale. Taken from reference 4e. With permission.

One should note that the CRAMPS technique is not necessarily a panacea for ^1H NMR studies on surfaces, or for any other type of sample. Dynamics associated with motion and/or chemical reactions with a time constant that is comparable to the cycle time of the multiple-pulse sequence interferes with a multiple pulse averaging of $^1H-^1H$ dipolar interactions [3b,4f]. Of course, an analogous problem can also arise with MAS, for which the critical period is ν_{MAS}^{-1}, where ν_{MAS} is the MAS frequency [26]. To quench this kind of dynamic interference with the line-narrowing efficiency of any cyclic technique, like MAS and/or a multiple-pulse sequence, one would have to

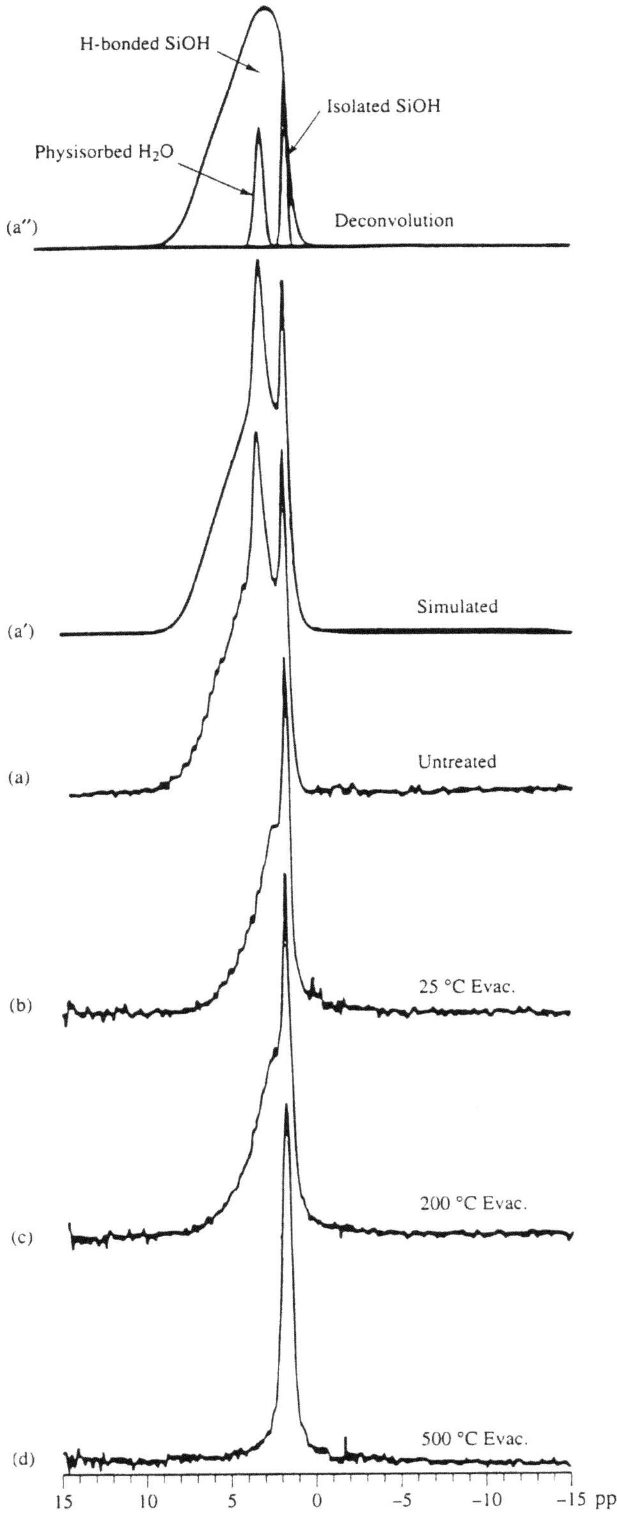

FIGURE 34.9 187-MHz ^{1}H CRAMPS spectra of Fisher S-679 silica gel. (a) Partially dried (in a dry box for 2 hr). (b) Evacuated at 200°C. (c) Evacuated at 25°C. (d) Evacuated at 500°C. (a″) Deconvolution of spectrum (a). (a′) Computer simulation based on (a″). Taken from reference 4c. With permission.

change the timescale of molecular motion, for example, by altering the sample temperature.

As an example of the application of ^{1}H CRAMPS technique to the study of silica surfaces [4c], Figure 34.9 shows the ^{1}H CRAMPS spectra obtained on partially dried (Figures 34.9(a), (a′), (a″)) and dried (Figures 34.9(b), (c), (d)) silica gel samples. Comparison of spectra (a) and (b) show that the relatively sharp peak at about 3 ppm in the spectrum of a partially dried silica is removed by moderate dehydration; this peak is therefore identified as physisorbed water. The broad nonsymmetrical peak that is largely to the left of the sharp 1.9 ppm peak is unaffected by sample evacuation at 200°C, but removed by evacuation at 500°C; this behavior suggests that this peak can be identified tentatively with clustered (or hydrogen-bonded) silanols, which are close enough together that suitable dehydration pathways are available. The sharp peak at 1.9 ppm remains after evacuation at 500°C; this behavior implies that the silanol groups represented by this peak are sufficiently far from each other to make dehydration difficult, so it is assigned to isolated (nonhydrogen-bonded) silanols. These assignments are consistent with 35 years of literature results on liquid-solution samples that show that hydrogen bonding typically reduces proton shielding, and with the reasonable point of view that a wide range of hydrogen-bonding structures exist on a silica surface (chemical shift dispersion: broad peak). These assignments are also consistent with the results of proton dipolar-dephasing experiments (*vide infra*). Analogous ^{1}H CRAMPS results have been obtained on the hydration/dehydration state of a fumed silica, as shown in Figure 34.10 (with a reference peak present at 0 ppm).

In addition to ^{1}H NMR applications based almost entirely on the chemical shift, as in the cases described above, there have also been productive ^{1}H NMR studies of silicas based primarily on time-domain approaches. One of the simplest and most useful of these approaches is a ^{1}H CRAMPS version [4c] of the popular "dipolar-dephasing" experiment used routinely in ^{13}C CP-MAS NMR. In this ^{1}H CRAMPS version, a "dephasing period" 2τ is inserted between the initial $\pi/2$ pulse that generates transverse magnetization and the multiple-pulse line-narrowing pulse train that detects the transverse magnetization. During the dephasing period, those proton magnetic moments that experience strong dipolar interactions undergo a corresponding degree of dephasing, and the resulting detected signal will be accordingly attenuated. Figures 34.11(a) and (b) show the results of applying the dipolar-dephasing CRAMPS experiment to partially dried silica gel (Figure 34.11(a)) and to a silica gel sample evacuated at 200°C (Figure 34.11(b)); in both cases, one sees that the broad (3.0) ppm peak is attenuated by a dephasing period (2τ) of 160 μs, whereas the sharp (1.7 ppm) peak is largely unattenuated. From the

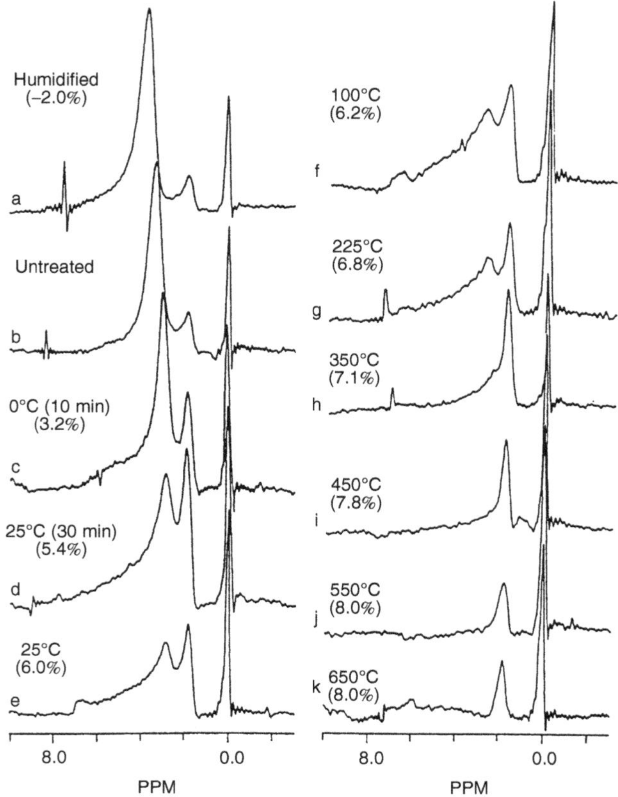

FIGURE 34.10 360 MHz ^{1}H CRAMPS spectra of Cab–O–Sil fumed silica as a function of hydration/dehydration state (weight loss due to drying shown in parentheses). Sharp peak at 0.05 ppm due to a PDMS reference. Taken from reference 5q. With permission.

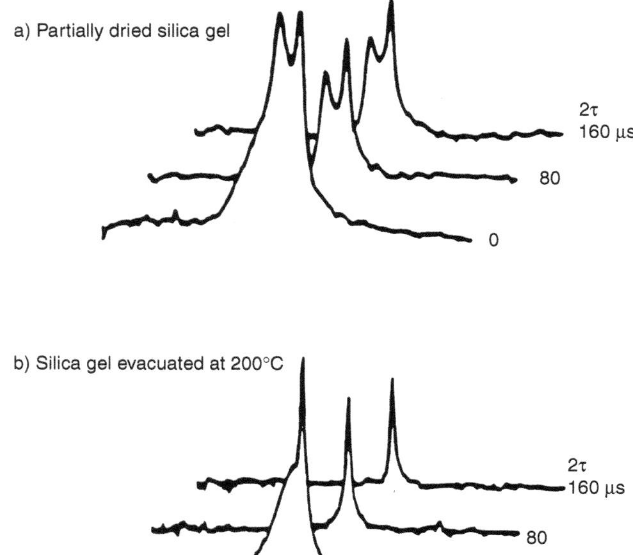

FIGURE 34.11 ^{1}H CRAMPS dipolar-dephasing experiment. (a) Results for partially dried (in dry box for 2 h) silica gel (showing τ values). (b) Results for sample evacuated at 200°C (showing τ values). Taken from reference 4c. With permission.

relationship between hydrogen bonding and proximity (between atoms that participate in a hydrogen bond) and the r^{-3} dependence of dipolar interactions, this dipolar-dephasing behavior provides strong support for identifying the 1.7 ppm peak as arising from isolated (nonhydrogen-bonded) silanols and the broad 3.0 ppm peak as clustered (hydrogen bonded) silanols. This is the same conclusion regarding proximity (and hydrogen bonding) that one would reach from the dehydration behavior shown in Figure 34.9 and Figure 34.10. It would be difficult to rationalize the elimination of water via the "dehydration" of silanols that are *isolated* from each other. Since the physisorbed water peak (3.5 ppm) is only slightly attenuated by the 160 μs dephasing period (Figure 34.11(a)), one can conclude that water protons experience only weak resultant dipolar interactions, the result of efficient motional averaging.

Other CRAMPS-based ^{1}H time domain approaches have also been useful in studying silica surfaces. Figure 34.12 summarizes a CRAMPS-detected T_1^H determination on silica gel samples, based on an alternating, two-sequence scheme of the Freeman-Hill type [27]. For each of the two silica samples, partially dried and 25°C evacuated, all components of proton magnetization relax essentially homogeneously, which suggests that proton spin diffusion among the different spin isochromats is much more efficient than spin-lattice relaxation, which is characterized overall by T_1^H values of 0.7–4 s. Furthermore, the marked increase in T_1^H that results from sample drying indicates that the physisorbed water in the sample of Figure 34.12(a)) provides an important relaxation source, and is presumably in a rapid state of motion at room temperature [3b].

Having the ability to establish, via dipolar dephasing, different spin polarizations for hydrogen bonded and isolated silanols, one can explore spin exchange between these two spin sets by an experiment of the type represented in Figure 34.13, in which, after permitting the proton magnetization to evolve in a dipolar-dephasing period (say, 80 μs, to strongly attenuate the magnetization from protons experiencing strong dipole-dipole interactions — that is hydrogen bonded), a "mixing period" is introduced before multiple-pulse detection. During that mixing period, depending on its duration and depending on relevant internuclear distances, a proton spin set that has been depleted during the dipolar-dephasing period can have magnetization restored via spin-spin flip-flop transfer from a spin set that had not been depleted.

One sees for the partially dried silica (Figure 34.13a) that in a mixing time of a few ms after eliminating ^{1}H magnetization due to hydrogen-bonded silanols, there is a substantial transfer of magnetization (spin exchange)

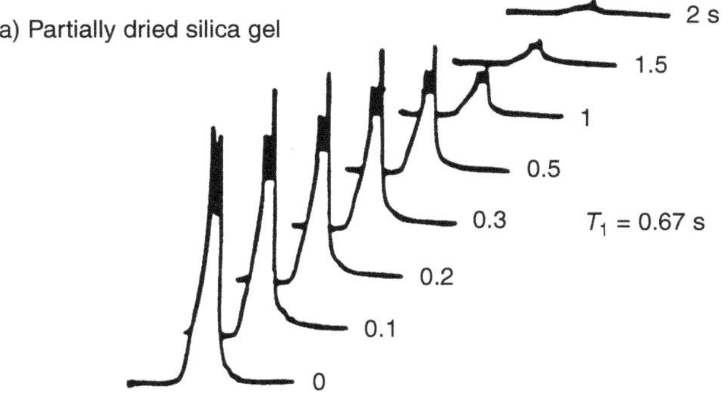

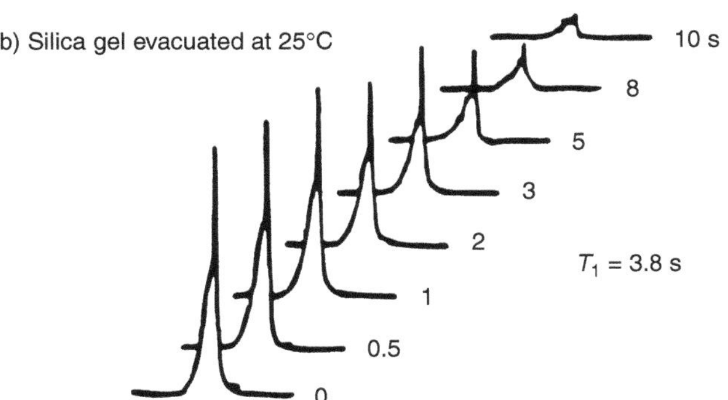

FIGURE 34.12 ^{1}H CRAMPS T_1 determination. (a) Results for partially dried (in dry box for 2 h) silica gel (showing τ values). (b) Results for sample evacuated at 25°C. Taken from reference 4c. With permission.

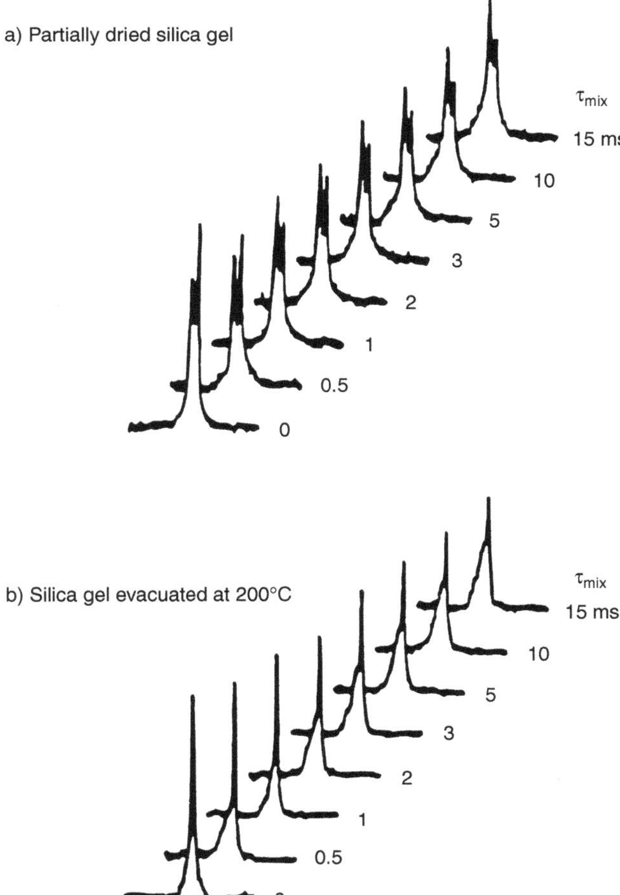

FIGURE 34.13 ^{1}H CRAMPS spin exchange experiment. (a) Results for untreated silica gel (showing τ_{mix} values). (b) Results for sample evacuated at 200°C (showing τ_{mix} values). Taken from rererence 4c. With permission.

from the isolated silanol reservoir to the hydrogen-bonded silanol protons. In any attempt to interpret or model the structure of the silica surface, the rate of this spin exchange places constraints on the spatial proximities of these two proton spin sets in terms of the dipole-dipole interactions that can be responsible for ^{1}H—^{1}H flip-flops, and/or chemical (H^+) exchange.

A variety of CRAMPS-based ^{1}H NMR experiments of the types represented in Figures 34.8–34.13 have been carried out on silica samples in which there have been systematic variations of physisorbed water content or dehydration, and often temperature. Figure 34.14 shows ^{1}H CRAMPS spectra of a hydrated (exposed to air) silica gel over a temperature range from 138 K to 298 K [3b]. An effective freezing point depression of about 45 K can be noted from these kinds of measurements; this result is consistent with previous reports based on other types of experiments [28].

Figure 34.15 shows ^{1}H CRAMPS dipolar-dephasing results obtained at room temperature on three different silica gels [3b]. The behavior shown is consistent with the ideas described above.

While in the past the CRAMPS technique was so demanding technically that very few laboratories were willing to invest the required the effort, modern commercial solid-state NMR spectrometers have been designed so that multiple-pulse experiments of many types are within the reach of any experimentalist who wishes to examine or utilize the potential for CRAMPS in his/her systems. In addition, commercially available MAS probes with speeds in the 30 kHz range render the MAS-only approach relatively routine for ^{1}H NMR studies on silica-type samples.

^{17}O NMR Experiments

In addition to cross polarization to ^{29}Si in silica or to ^{13}C (or ^{15}N, ^{31}P, etc.) in derivatized silicas, cross polarization to ^{17}O in isotopically-enriched silicas has also been illuminating. Oldfield and co-workers have, during the past several years, made significant progress in the application of solid state ^{17}O NMR techniques for the characterization of inorganic materials. Walter, Turner, and Oldfield[29] have demonstrated that $^1H \rightarrow {}^{17}O$ CP experiments are not only

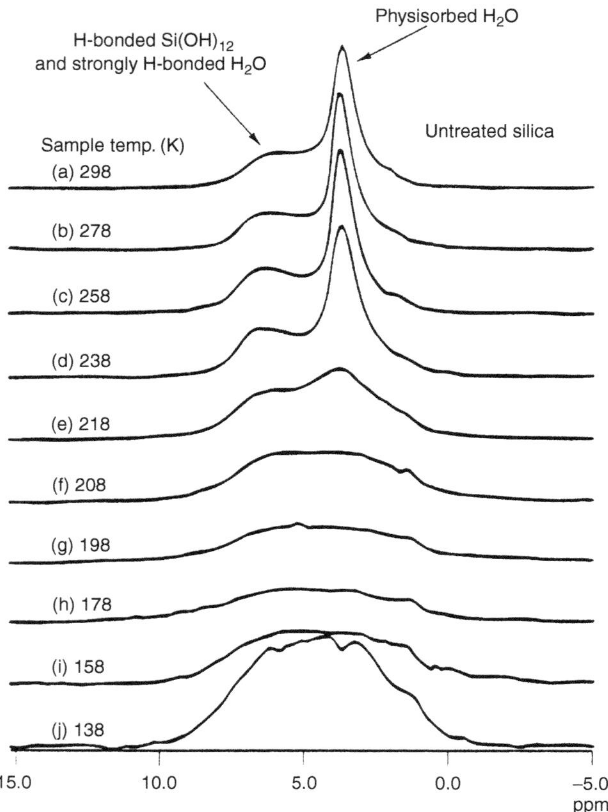

FIGURE 34.14 ^{1}H CRAMPS spectrum of an untreated silica gel as a function of temperature as shown. Taken from reference 3b. With permission.

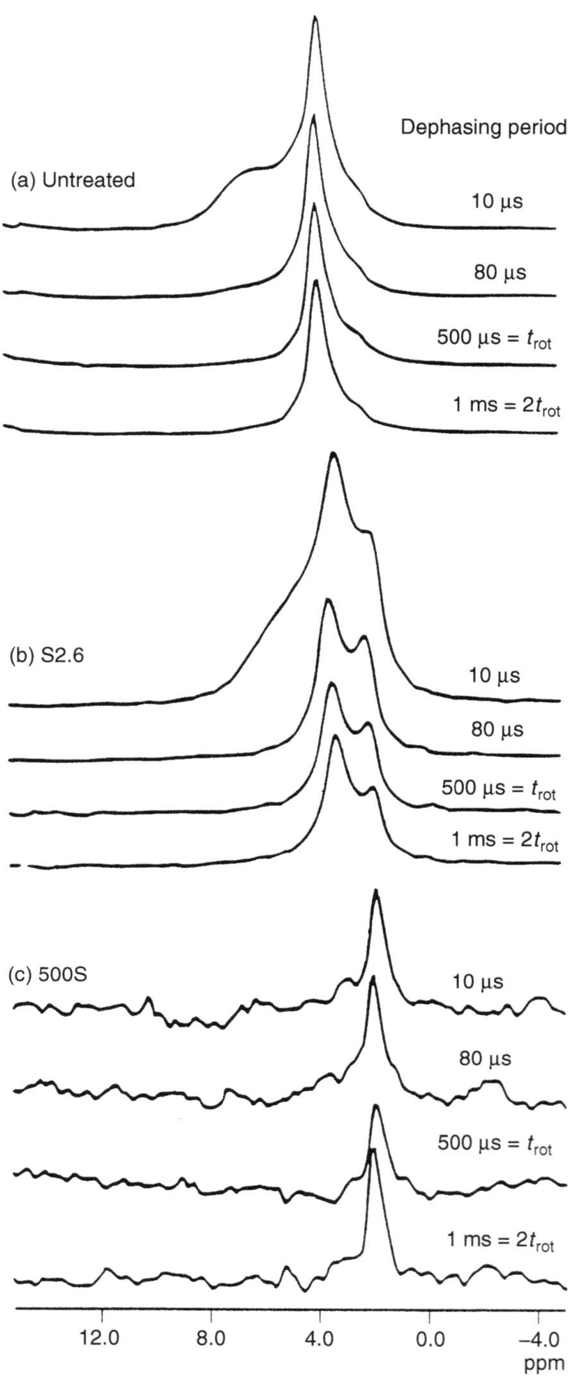

FIGURE 34.15 ^{1}H CRAMPS dipolar-dephasing spectra of three different silica gels. (a) Untreated. (b) Modestly dehydrated (2.6% weight loss). (c) Drastically dehydrated (at 500°C). Taken from reference 3b. With permission.

feasible, but they are also very informative from the point of view of *editing* ^{17}O spectra, by discriminating against ^{17}O signals from oxygen sites with no directly bonded hydrogen. Figure 34.16 shows $^1H \rightarrow {}^{17}O$ CP spectra of amorphous silica and a model SiOH system. From a comparison of the spectra obtained from static samples and by MAS, with and without CP, it was possible to assign the ^{17}O signal due to SiOH groups at the surface.

With the advent of higher-field spectrometers that substantially reduce line-broadening due to the second-order quadrupole effect [30] and with the use of multiple-quantum techniques for eliminating such line-broadening effects [31], one can expect that ^{17}O NMR techniques will play an increasingly important role in the study of silica and modified silica systems. Of course, the very low natural abundance of this nuclide remains a substantial constraint on such applications.

The Interplay between ^{1}H and ^{29}Si Spins in Silica Systems

As we have seen above, ^{29}Si CP-MAS experiments can distinguish via chemical shift among single silanols (Q_3), geminal silanols (Q_2) and siloxane moieties (Q_4), and ^{1}H CRAMPS (or fast MAS) can distinguish, in terms of chemical shift and dipolar-dephasing behavior, between hydrogen-bonded and nonhydrogen-bonded (isolated) OH groups.

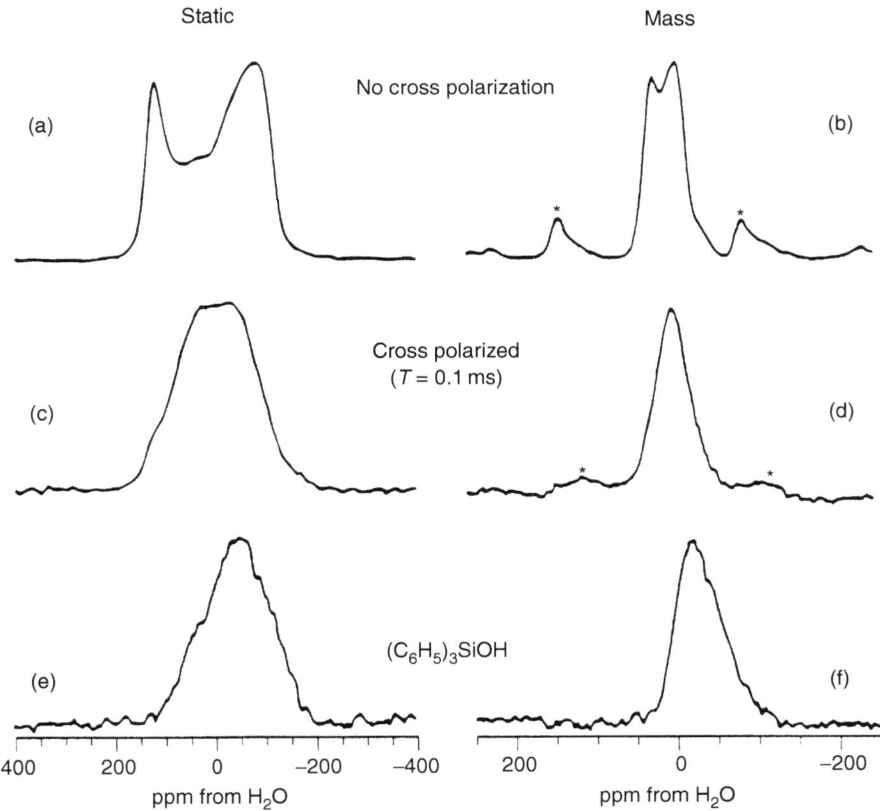

FIGURE 34.16 Static and MAS ^{17}O spectra of amorphous SiO_2 and polycrystalline $(C_6H_5)_3SiOH$ obtained at 67.8 MHz. (a) ^{1}H-decoupled static spectrum of SiO_2 without CP: 108 scans. (b) ^{1}H-decoupled MAS spectrum (of SiO_2): 100 scans, 7.6 kHz spinning speed (indicates spinning sidebands). (c) $^{1}H \rightarrow {}^{17}O$ static spectrum of SiO_2: 200 scans, 0.1 ms contact time. (d) $^{1}H \rightarrow {}^{17}O$ MAS spectrum of SiO_2: 200 scans, 0.1 ms contact time. (e) ^{1}H-decoupled static spectrum of $(C_6H_5)_3SiOH$ without CP: 500 scans. (f) ^{1}H-decoupled MAS spectrum $(C_6H_5)_3SiOH$: 800 scans, 4.0 kHz spinning speed. All spectra were obtained using a 2 s recycle time. Taken from reference 29. With permission.

It would be highly desirable to correlated these two types of information. Ideally, one would base such a correlation on two-dimensional (2D) $^{1}H—^{29}Si$ heteronuclear chemical shift correlation (HETCOR) experiments of the general type that are employed routinely for $^{1}H—^{13}C$ correlation in liquids and more recently in solids [32,33]. Indeed, Vega has reported such experiments based on $^{1}H \rightarrow {}^{29}Si$ CP for the polarization transfer step.[34] However, as is seen below, rotating-frame ^{1}H spin diffusion among the protons in a typical silica gel during the spin-lock state in a $^{1}H \rightarrow {}^{29}Si$ CP experiment can substantially scramble what one would hope are discrete $^{1}H \leftrightarrow {}^{29}Si$ CP correlations in the time frame (>200 μs) required for relatively efficient CP transfer. Furthermore, the multiple-pulse approaches used in $^{1}H—^{13}C$ HETCOR experiments on solids are not very efficient in these $^{1}H—O—^{29}Si$ systems, although successes reported in $^{1}H—^{31}P$ 2D HETCOR experiments on P—O—H systems [35] indicate that $^{1}H—^{29}Si$ HETCOR in silica may yet be attractive, albeit inefficient and extremely difficult to quantify. In any case, as an alternative to the 2D HETCOR approach, a variety of ^{29}Si-detected $^{1}H—^{29}Si$ CP experiments have been carried out in which the behavior of protons is monitored by ^{29}Si, establishing the correlation.

The simplest of such experiments is a ^{29}Si CP-MAS experiment in which ^{29}Si detection is carried out *without* proton decoupling. MAS should still average the $^{1}H—^{29}Si$ dipolar interaction during detection, yielding a corresponding sideband pattern to the extent that this interaction behaves inhomogeneously, that is to the extent that the $^{1}H—^{29}Si$ dipolar interaction is not altered (by chemical reaction, motion or $^{1}H—^{1}H$ flip-flops) during a MAS rotor period [3a]. Figure 34.17 shows a comparison of spectra obtained with and without ^{1}H decoupling; it is clear that the geminal silanol peak suffers most dramatically from the absence of high-power ^{1}H decoupling, implying that ^{1}H spin exchange (presumably associated with hydrogen bonding) is most efficient in the protons of geminal silanols. The ^{1}H spin exchange rate decreases with increasing MAS speed.

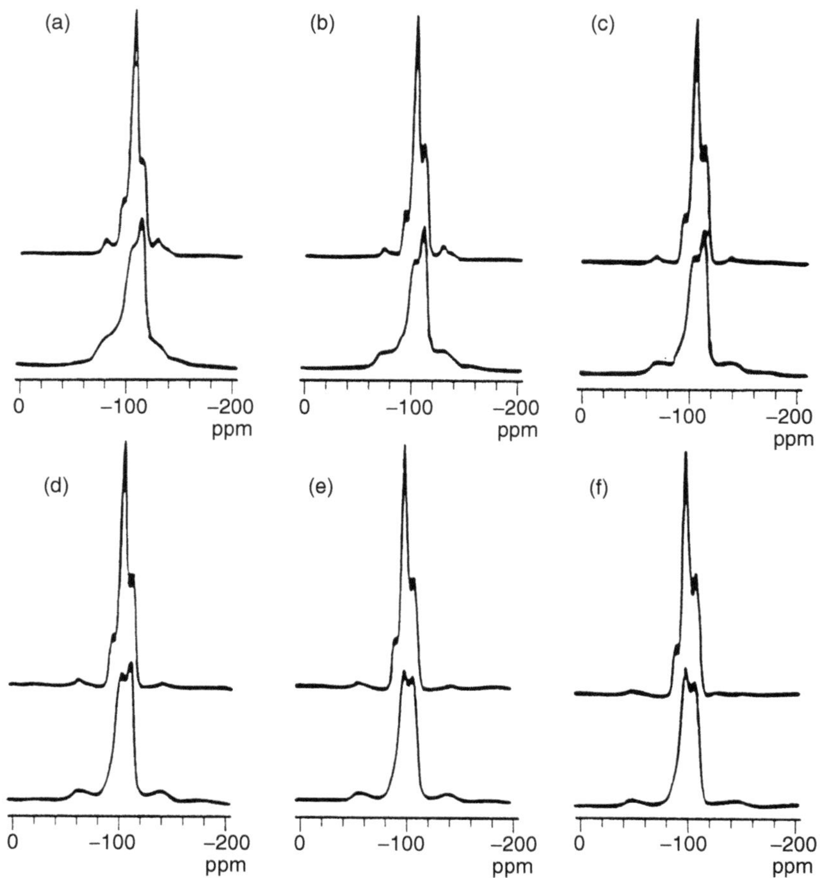

FIGURE 34.17 Proton-decoupled (top spectrum of each set) and proton-coupled (bottom spectrum of each set) 39.75 MHz ^{29}Si CP-MAS NMR spectra of Fisher S-679 silica gel at six different MAS speeds. CP time, 5 ms. (a) 1.0 kHz, 1096 accumulations. (b) 1.1 kHz, 3000 accumulations. (c) 1.4 kHz, 720 accumulations. (d) 1.6 kHz, 2000 accumulations. (e) 1.8 kHz, 2000 accumulations. (f) 2.0 kHz, 2000 accumulations. Taken from referebce 3a. With permission.

An analogous set of results, given with computer-deconvoluted peak contributions, is shown in Figure 34.18 for a fumed silica. The interpretations are the same as for the silica gel.

Another ^{29}Si CP-MAS experiment useful for correlating ^{1}H and ^{29}Si spin behaviors is the ^{1}H—^{29}Si analog of the common ^{1}H—^{13}C dipolar-dephasing experiment. In this technique, rotational and Hahn echo formation occur for a dephasing period (2τ) corresponding to two MAS rotor periods ($2\tau_{rot}$) for the isotropic ^{29}Si chemical shift, the ^{29}Si chemical shift anisotropy and the ^{1}H—^{29}Si dipolar interaction (to the extent that it behaves inhomogeneously) [3a]. Figure 19 shows results of the ^{1}H—^{29}Si CP-MAS experiment with the dephasing period ranging over more than $4t_{rot}$. Focusing on the points at $2\tau = 0$, $2t_{rot}$ and $4t_{rot}$, one sees very little dephasing decay of the Q_4 (siloxane) signal, and more efficient decay for the Q_2 (Geminal silanol) signal than for the Q_3 (single silanol) signal. This again suggests more efficient ^{1}H—^{1}H spin diffusion among the $\geqq$Si(OH_2) protons than among $\geqq$SiOH protons.

A very direct correlation of ^{1}H CRAMPS dipolar-dephasing behavior with ^{29}Si CP-MAS signals is obtained in an experiment in which there is a 2τ ^{1}H—^{1}H depolar-dephasing period *before* CP transfer to ^{29}Si [3a]. Taking account of the rotational echo behavior of ^{1}H magnetization, for $2\tau = 2nt_{rot}$ essentially all relevant proton interactions refocus except the ^{1}H—^{1}H dipolar interaction. Hence, the magnetization of those protons involved in the strongest (shortest, least mobile) hydrogen bonds is most effectively dephased during $2\tau = 2nt_{rot}$ and unavailable for CP transfer to ^{29}Si. Figure 34.20 shows the results obtained on a silica gel sample.

For a CP contact time (t_{cp}) that is small enough (100 μs) to avoid the rotating-frame spin diffusion that scrambles the desired ^{1}H—^{29}Si correlation (*vide supra*), the geminal silanol peak at −89 ppm is the one that suffers most from ^{1}H—^{1}H dipolar dephasing for $2t_{rot}$ period. The effect of rotating frame proton spin diffusion is also clear from the spectra in Figure 34.20; if a long CP contact period (e.g., 5 ms) is employed, essentially

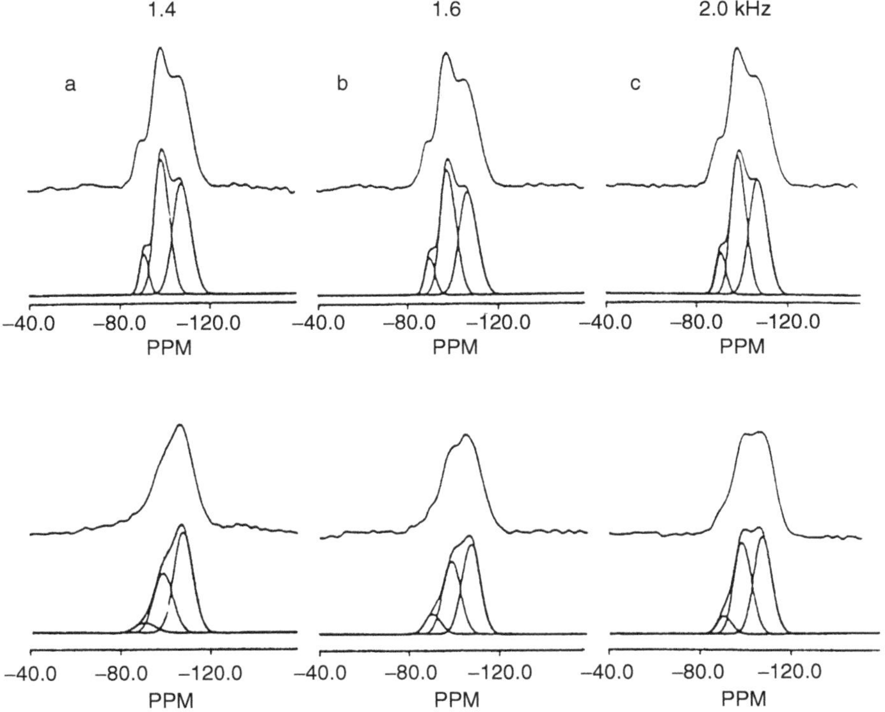

FIGURE 34.18 ^{1}H-decoupled (top spectrum of each set) and ^{1}H-coupled (bottom spectrum of each set) ^{29}Si CP-MAS spectra of untreated Cab–O–Sil fumed silica at three different MAS speeds, as indicated. Computer-simulated spectrum and individual deconvoluted contributions are shown below each experimental spectrum. Taken from refrecne 5r. With permission.

the same relative peak intensities are obtained whether or not the $2t_{rot}$ ^{1}H—^{1}H dipolar-dephasing period is included in the experimental sequence.

Analogous ^{29}Si-detected ^{1}H—^{1}H dipolar-dephasing results for a fumed silica are shown in Figure 34.21. Again, the interpretation mirrors that given for silica gel.

Structural Models

One overriding theme emerges from a large body of results of the type described above: the protons that are primarily responsible for cross polarization to geminal silanol silicons are much more extensively involved in proton spin diffusion (facilitated geometrically by hydrogen bonding) than are the protons primarily responsible for CP to single silanol silicons. This theme is consistent with structural models of the silica surface (or, at least fragments of it) that correspond to specific faces of a β-cristobalite crystal [3b,c], as suggested by Sindorf and Maciel.

Figure 34.22 shows views looking "into" the 111 and 100 faces, which contain single silanols (Q_3) and geminal silanols (Q_2), respectively. From the O—O distances between hydroxyl oxygen of these surfaces, we see that one should expect hydrogen bonding between adjacent geminal silanols, but not between adjacent single silanols; this is in agreement with the NMR results summarized above. Of course, the silica surface is not a homogeneous one, and may be describable as a composite of these two types of surfaces, with suitable interfaces. The geometrical relationship between silanols that are adjacent *across* these interfaces is important in the overall hydrogen-bonding patterns at silica surfaces. Furthermore, the presence of water on the surface dramatically changes the pattern of hydrogen bonding, including the establishment of hydrogen-bonding networks among the single silanols.

Chuang and Maciel have built upon the general ideas embodied in the modes of Figure 34.22, taking full advantage of the extensive spin dynamics results obtained in this laboratory and the extensive vibrational spectroscopy data obtained elsewhere [3e]. They concluded that about 46% of geminal silanols and 53% of single silanols on a Fisher silica gel surface evacuated at 25°C are not hydrogen bonded, and the rest of the silanols are hydrogen bonded; that is both Q_3 and Q_2 silanols can be hydrogen bonded or not hydrogen bonded, depending on circumstances. In order to explain all these findings, as well as non-NMR data in the literature, they generated more detailed models in which the clear distinction between hydrogen-bonded Q_2 silanols and nonhydrogen-bonded Q_3 silanols implied in Figures 34.22(a) and (b) are altered in predictable ways at intersections between 100

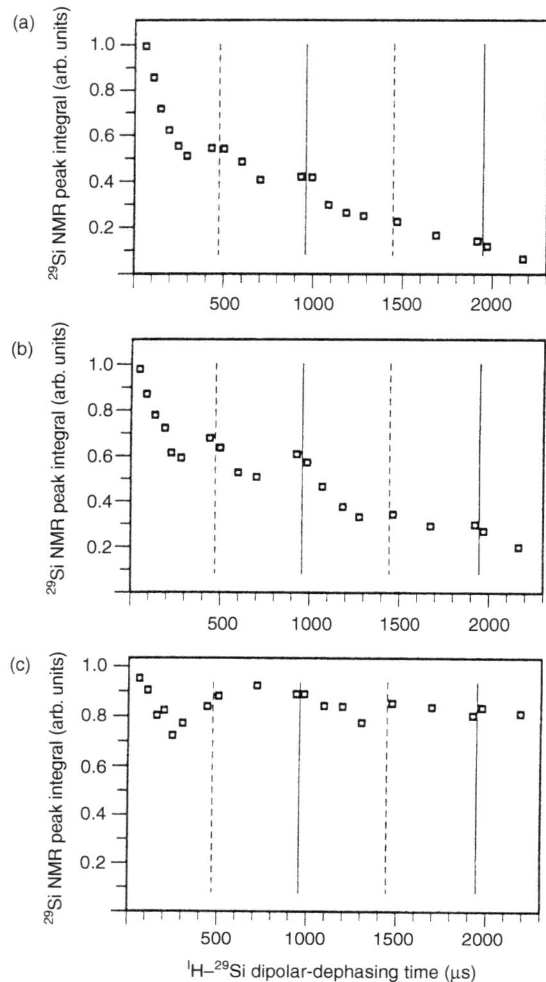

FIGURE 34.19 Plots of deconvoluted peak integrals of the 39.75-MHz ^{29}Si CP-MAS NMR spectra of Fisher S-679 silica gel versus ^{1}H—^{29}Si dipolar-dephasing time up to four rotor periods. CP contact time, 5 ms; MAS speed, 2.0 kHz. Vertical dashed lines show odd numbers of rotor periods and vertical solid lines show even numbers of rotor periods. (a) −89 ppm peak (geminal silanols). (b) −99 ppm peak (single silanols). (c) −109 ppm peak (siloxane silicons). Taken from reference 3a. With permission.

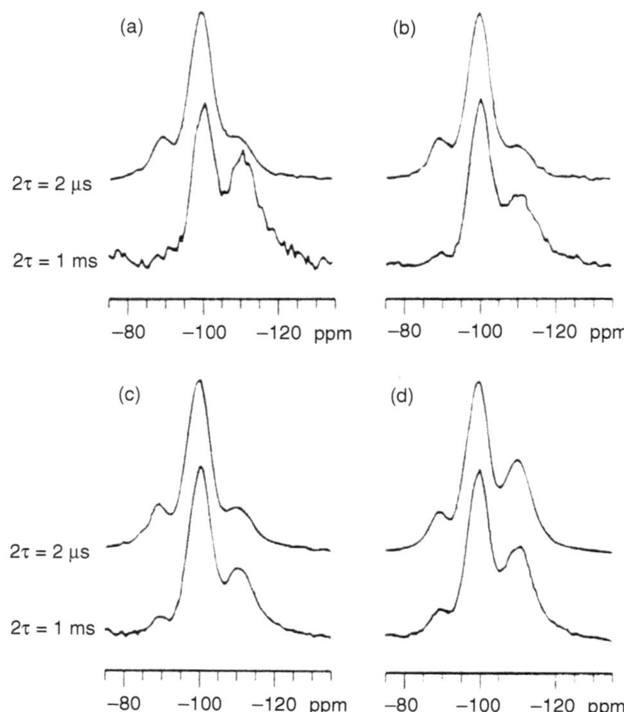

FIGURE 34.20 ^{29}Si CP-MAS NMR spectra of Fisher S-679 silica gel obtained with 2 μs (top spectrum of each set) and two rotor periods (1.04 ms; bottom of each set) of ^{1}H–^{1}H dipolar-dephasing prior to four different ^{1}H–^{29}Si CP contact times. MAS speed, 1.9 kHz, (a) $t_{cp} = 100$ μs (top spectrum, 7376 accumulations; bottom spectrum, 82,504 accumulations). (b) $t_{cp} = 300$ μs (top spectrum, 2400 accumulations; bottom spectrum, 46,200 accumulations). (c) $t_{cp} = 1$ ms (top spectrum, 432 accumulations; bottom spectrum, 21,232 accumulations). (d) $t_{cp} = 5$ ms (top spectrum, 600 accumulations; bottom spectrum, 8320 accumulations). Taken from reference 3a. With permission.

and 111 faces. Invoking such intersections also makes it possible to generate models of pores [3e]. Figure 34.23 shows representative models of surface segments generated by intersections between 100 and 111 faces. Experiments are underway to test/quantify the existence of such structures in silica.

Derivatized Silicas

Silylation

The ^{29}Si CP-MAS spectrum in Figure 34.4(d) represents a sample prepared by the silylation of a silica gel by $(CH_3)_3SiCl$. This silylation reaction is a member of an important class of derivatization of the silica surface, represented by the following chemical equation:

$$\text{R—Si—X} + \text{H—O—}\langle\text{Si}\rangle \longrightarrow \text{R—Si—O—}\langle\text{Si}\rangle + \text{HX} \quad (34.1)$$

where X represents a labile leaving group (e.g., Cl or OCH_2CH_3) and ⟨Si⟩ represents a silanol site on the reactant silica surface or a corresponding derivatized silica site in the reacted sample. Such reactions are important, or potentially important, technologically — for the preparation of stationary phases for chromatographic separations, as a means of immobilizing reactive chemical centers (e.g., catalytic sites), in coupling agents for composite materials, and for a wide range of other applications in which it is desired to anchor or "immobilize" a chemically important

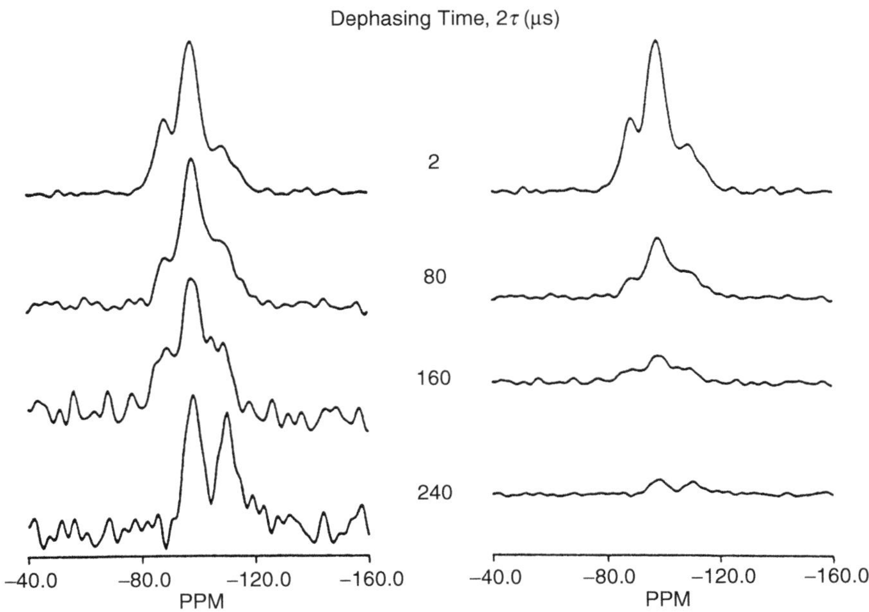

FIGURE 34.21 ^{29}Si CP-MAS spectra of Cab—O—Sil fumed silica obtained with four different $^{1}H—^{1}H$ dipolar-dephasing times (as indicated) prior to 100 μs CP period. Spectra on the left scaled to a common height of the −99 ppm peak; and spectra on the right plotted on a common absolute intensity scale. Taken from reference 5r. With permission.

moiety. Peak d in the spectrum of Figure 34.4(d), at ca. 10 ppm, is assigned to the trimethylsilyl group covalently attached to the silica surface. Clearly the silylation process has brought about a change of intensities of peaks a, b and c in Figure 34.4(b), corresponding to the Q_2 sites, the Q_3 sites and the Q_4 sites, at the underivatized surface. At the relatively low level of structural detail represented by Q_2, Q_3 Q_4 notation, the silylation process transforms Q_2 sites into Q_3 sites and the Q_3 sites into Q_4 sites, and these changes are seen by comparing the spectra in Figures 34.4(b) and (d). By examining such intensity changes systematically, it has been possible to elucidate important reactivity patterns in these systems [1,2]. Indeed, ^{29}Si CP-MAS NMR, along with supporting data from ^{13}C CP-MAS experiments, can serve as an analytical technique for monitoring chemical reactivity patterns, for example, the relative reactivities of the various types of silanols on a silica surface. Figure 34.24 shows the stepwise silylation of a precipitated silica by the silylation agent, $CH_3(CH_2)_{17}Si(CH_3)_2Cl$, forming a version of the important chromatographic agent, sometimes referred to as "C18".

The complex chemistry that can occur on a silica surface after silylation by a reagent with more than one leaving group (X), for example, $RR'SiX_2$ or $RSiX_3$, is also amenable to study by CP-MAS NMR [2f]. In the ^{29}Si CP-MAS spectra shown in Figure 34.25, one sees that the product formed initially from the reaction of silica with $(CH_3)_2Si(OCH_2CH_3)_2$ depends on the reaction conditions and predrying of the silica, and can be converted by the moisture in air (Figures 34.25(c,d)) to products in which $\geqslant$—Si—OCH_2CH_3 moieties are replaced by $\geqslant$Si—OH and ultimately $\geqslant$Si—OSi$\leqslant$ moieties. In this case, ^{13}C CP-MAS spectra are also useful, because they detect the presence and amount of residual $\geqslant$Si—OCH_2CH_3 moieties.

The ^{29}Si CP-MAS spectra shown in Figure 34.26 represent an even more complex derivatized silica system, a series of samples prepared by the silylation of silica with 3-amino-propyltriethoxysilane (APTS) under a variety of conditions (pretreatment temperature = 200°C for the 200 series, or 110°C for the 110 series, 25°C for the RT series, or with silylation carried out in an aqueous slurry, AQ series) [5m]. APTS-derivatization of the silica surface is of interest for such diverse applications as variety of composite materials (in which APTS serves as a coupling agent between a silica-like component and, usually, an organic polymer) and for metal complexation agents. One sees from the spectra of Figure 34.26 that increasing the amount of water in the silylation process or increasing the post-silylation "curing" temperature brings about changes in attached silane populations from species with one Si—OSi attachment (−49 ppm) to species with two such attachment (−58 ppm) to three such attachments (−66 ppm). Although ^{13}C spectra are primarily useful for monitoring the residual $\geqslant$Si—OCH_2CH_3 moieties in this system. A careful analysis of the ^{13}C CP-MAS spectra (Figure 34.27) of samples corresponding to those represented in Figure 34.26 reveals that the ^{13}C chemical shift of the central carbon

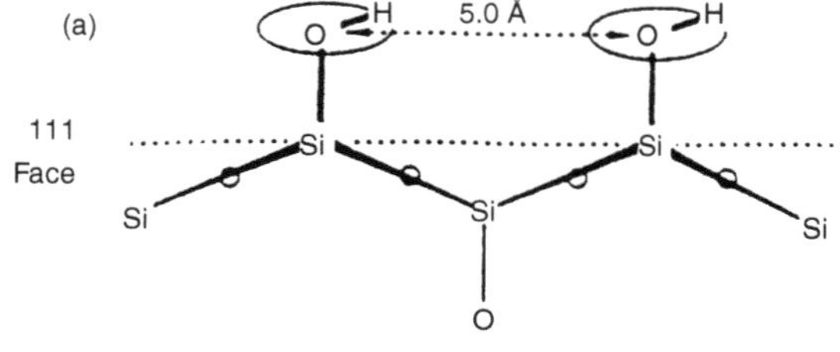

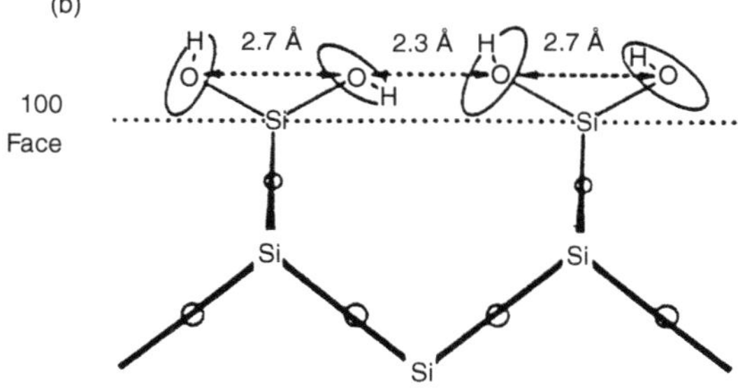

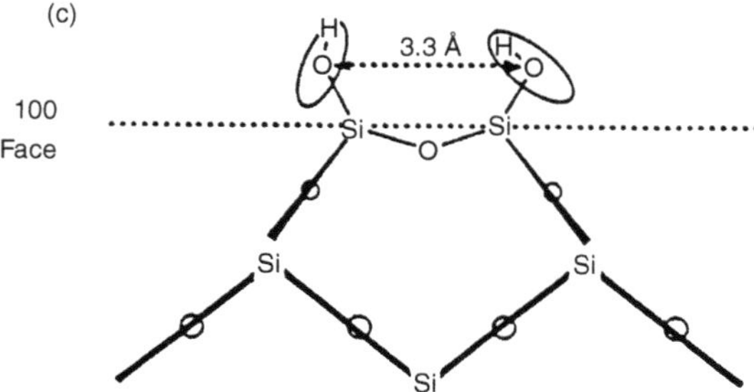

FIGURE 34.22 Side views of specific silicon planes (dashed line representing an edge of such a plane) of β-cristobalite. Drawing approximately to scale. (a) 111 face. (b) 100 face. (c) vicinal sites from dehydration of the 100 face. Taken from reference 3a. With permission.

of the pendent —$(CH_2)_3$— group originating from APTS is sensitive to protonation or hydrogen bonding of the amino group. However, ^{13}C chemical shifts are typically useful primarily just for identification of residual alkoxy leaving groups or of pendent groups attached to the silica surface. ^1H CRAMP spectra (*vida infra*) and ^{15}N CP-MAS spectra (Figure 34.28) can also be useful for studying this important issue.

Figure 34.29 and Figure 34.30 show ^1H CRAMPS spectra of silica surfaces that have been derivatized with $(CH_3)_3SiCl$ and $(CH_3CH_2O)_3SiCH_2CH_2CH_2NH_2$ (APTS), respectively. These spectra show that this technique can provide a rich level of structural detail on derivatized silica samples. The difference between the spectra of unprotonated and protonated APTS-derivatized samples (Figure 34.30) is the absence and presence, respectively, of a clearly distinct peak corresponding to the amino group of APTS [4f]. This difference reflects, at least in part, the interference of ^{14}N quadrupole effects of the $-NH_2$ group with MAS averaging of $^1H—^{14}N$ dipolar interactions. Calculations on $\gt$N—H groups show that, because of the small internuclear N–H distances for directly bonded pairs, very high NMR fields (>1000 MHz) would be required to reduce this broadening to an acceptable level by the "brute force" approach of simply going to a higher field [36].

Alkylsilane-Modified Silica

The silylation process represented in Figure 34.24, as well as analogous processes employing reagents of the type $CH_3(CH_2)_{17}CH_2SiX_3$ (X = Cl or OCH_3, etc.), yields a type of derivatized silica that is extremely popular as a solid support for chromatography. Such systems have been studied by a variety of techniques, including ^{13}C NMR and ^2H NMR. Figure 34.31 shows ^{13}C MAS spectra of a C_{18}-derivatized silica obtained by direct polarization (DP, single pulse, no CP) cross polarization (CP). Examples are included in which the C_{18}-silica has been "wetted" with representative chromatographic solvents. This system has also been studied as a function of surface loading of the C_{18} chain.

One can see in Figure 34.31 that some of the carbon sites (e.g., C3, C16, C18, Cl′) are individually discernable, while others have resonances that overlap with the resonances of one or more other carbon sites (the large central peak being due to carbons C4–C15). The ability of CP to select static components of a surface derivative relative to more mobile components, which would be emphasized by non-CP (DP) techniques that rely on direct ^{13}C spin-lattice relaxation, can be seen in Figure 34.31, which provides a comparison between the DP (single-pulse) and cross-polarization ^{13}C MAS NMR results on four C_{18}-silica samples. Dramatic differences in lineshapes and intensities, which can be interpreted at least qualitatively in terms of local motion within the C_{18} chain, can be noted readily in these comparisons.

More detailed information on motion within the C_{18} chains, at least the timescales that are relevant, are obtained by ^{13}C NMR relaxation studies [5p]. Several ^{13}C NMR studies of C_{18}-derivatized silicas and other *n*-alkyl analogs have been reported [5n,p,37] Even more detail on the motion can be obtained by wide line ^2H NMR spectroscopy (a technique that does not have high resolution) on C_{18}-silica samples in which deuterium has been selectively substituted for protons [50]. In this approach, the line-narrowing effects of motion on the broad, quadrupole-based ^2H NMR linewidth of a mechanically static sample is modeled theoretically for specific trial motions to elucidate the detailed nature of the motion. As an example, Figure 34.32 shows experimental ^2H spectra, and the corresponding theoretical simulation for a dry sample of (1-d_2)-C_{18}-silica as a function of temperature. From the theoretical simulations, based on trial

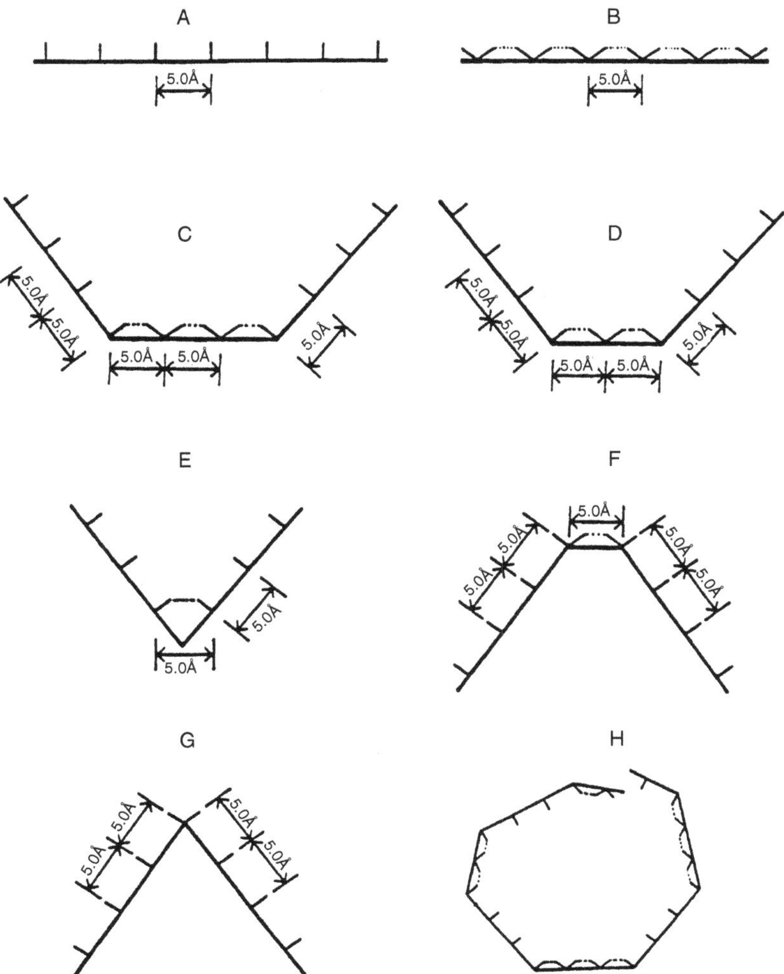

FIGURE 34.23 Parts A and B show shorthand notation for single (111)-type and (100)-type β-cristobalite faces, respectively, with short lines representing OH groups and long lines representing (111)-type and (100)-type faces, respectively. Specific configurations are made possible by intersections of two (111)-type β-cristobalite faces with each other (E and G) or with a (100)-type face (C, D and F), as represented by shorthand notation. (H) A hypothetical model of a defect structure or pore structure of silica, as represented by shorthand notation. Dotted lines represent hydrogen bonds. Taken from reference 3e. With permission.

motional models, one can obtain exquisite detail on both the types and timescales (correlation times) for the chain motions, as well as activation energies when variable-temperature strategies are used.

Other Surface Modifications

Materials prepared by other types of silica-surface modification have also been studied by NMR. In a combination ^{13}C and ^{29}Si (and, to a lesser extent, ^{27}Al) NMR study, Tao and Maciel [5t] have examined the reactivity of the silica surface to reaction with the "simple" organometallic methylating agents, CH_3Li, CH_3MgBr, $(CH_3)_2Zn$ and $(CH_3)_3Al$, as well as with the chlorination agent $SOCl_2$ and with $SiCl_4$. The formation of Si—CH_3 and Si—Cl bonding was examined. CH_3Li was found to be very destructive to the silica framework, while no obvious evidence was found that $(CH_3)_2Zn$ reacted at all to form Si–CH_3 bonds with dry silica or with $SOCl_2$-modified or $SiCl_4$-modified silica.

Figure 34.33 shows ^{29}Si (left column) and ^{13}C (right column) CP-MAS spectra of the reaction products of $SOCl_2$-chlorinated silica (parts A and a) with the four methyl-metal reagents mentioned above. One can see in Figure 34.33A that the chlorinated silica prepared by the $SOCl_2$ treatment shows a decrease of the Q_3/Q_4 intensity ratio (compared with Figures 34.2C, 34.4B, 34.24A) due to the replacement ⟨Si⟩—OH by ⟨Si⟩—Cl on the silica surface. Fewer protons in the sample of Figure 34.33A are available to contribute to ^{1}H–^{29}Si CP and fewer silanol signals are available to contribute to the Q_3 signal.

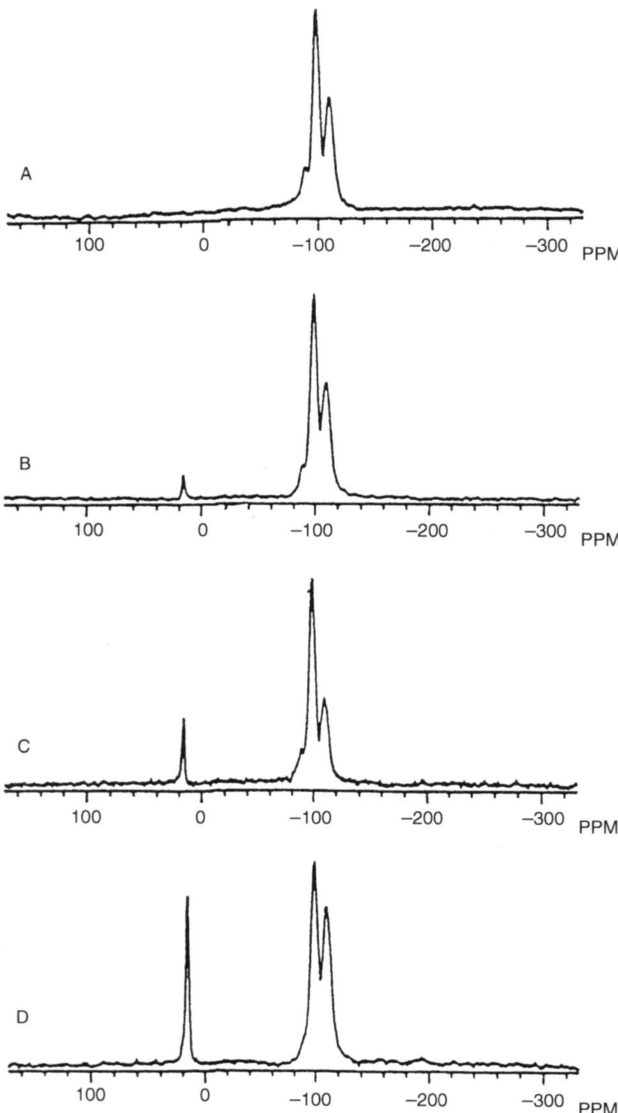

FIGURE 34.24 ^{29}Si CP-MAS spectra of silica (Whatman Partisil) silylated with $CH_3(CH_2)_{17}Si(CH_3)_2Cl$. A) Unsilylated. B)–D) increasing degree of silylation. Taken from reference 6e. With permission.

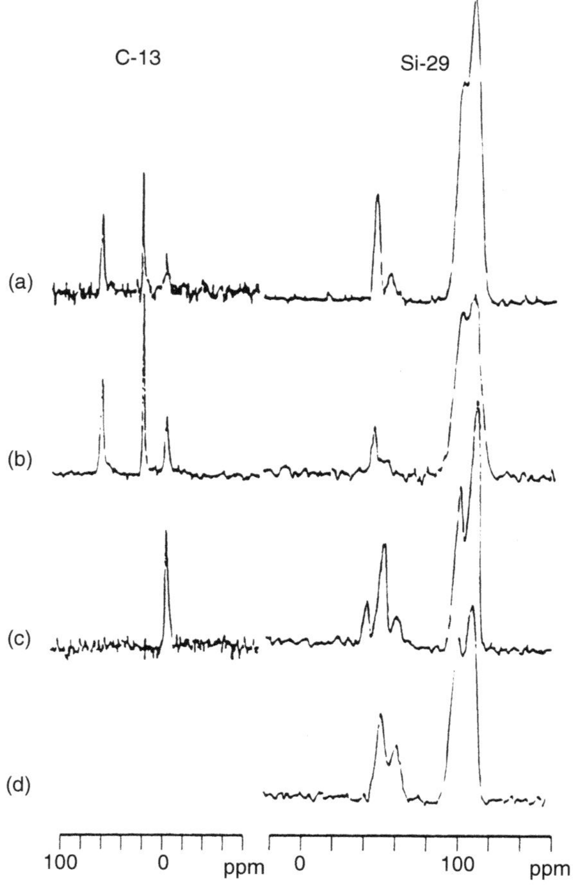

FIGURE 34.25 ^{29}Si (right) and ^{13}C (left) CP-MAS spectra of silica gel derivatized with $(CH_3)_2Si(OCH_2CH_3)_2$. (a) Product of reaction with predried silica at 138°C. (b) Product of reaction with predried silica at 240°C. (c) Product of reaction with undried silica at 115°C. (d) Sample (c) heated in air at 150°C. Taken from reference 2f. With permission.

Upon examining Figure 34.33B, representing treatment of $SOCl_2$-treated silica with CH_3Li, it is not straightforward to explain why high ^{29}Si NMR intensities of *both* ⋛SiO**Si** $(CH_3)_3$ (13 ppm) and (⋛SiO)$_2$**Si**$(CH_3)_2$ (−15 ppm) are seen, since there is little ⋛SiO**Si**$(OH)_3$ on an unmodified silica surface (none seen in Figures 34.2C, 4b, 24A); it therefore seems unlikely that the $SOCl_2$ chlorination approach could produce a substantial amount of (⋛SiO)**Si**$(Cl)_3$, which ostensibly would subsequently react with methyllithium to form (⋛SiO)**Si**$(CH_3)_3$. Therefore, it appears that methyllithium plays a much more complicated ("destructive") role, breaking down a portion of the silica framework, and ultimately creating a complex array of surface functionalities, including (⋛SiO)**Si**$(CH_3)_3$ and (⋛SiO)$_2$**Si**$(CH_3)_2$. In the ^{29}Si CP-MAS spectrum of the sample resulting from reaction of chlorinated silica with methylmagnesium bromide (Figure 34.33C), one sees, in addition to peaks corresponding to the silica framework, additional major peaks corresponding to the (⋛SiO)$_3$ **Si**CH_3 functionality (peak at −63 ppm) and the (⋛SiO)$_2$ **Si**(HO)CH_3 functionality (peak at −52 ppm), with a much weaker peak at −15 ppm, indicating the presence of (⋛SiO)$_2$**Si**$(CH_3)_2$ and almost no (⋛SiO)**Si**$(CH_3)_3$ (13 ppm). The two stronger peaks at −63 and −52 ppm could be due to species generated from the conversion of the original silanol functionalities, ⟨**Si**⟩ $(OH)_n$, on the silica surface (n = 1 or 2), without requiring any "damage" of the silica framework. The (⋛SiO)$_2$(HO) **Si**CH_3 peak at −52 ppm might result from the partial conversion of geminal silanols, ⟨**Si**⟩ $(OH)_2$, or from partial degradation of single silanol moieties, ⟨**Si**⟩OH. Analogous,

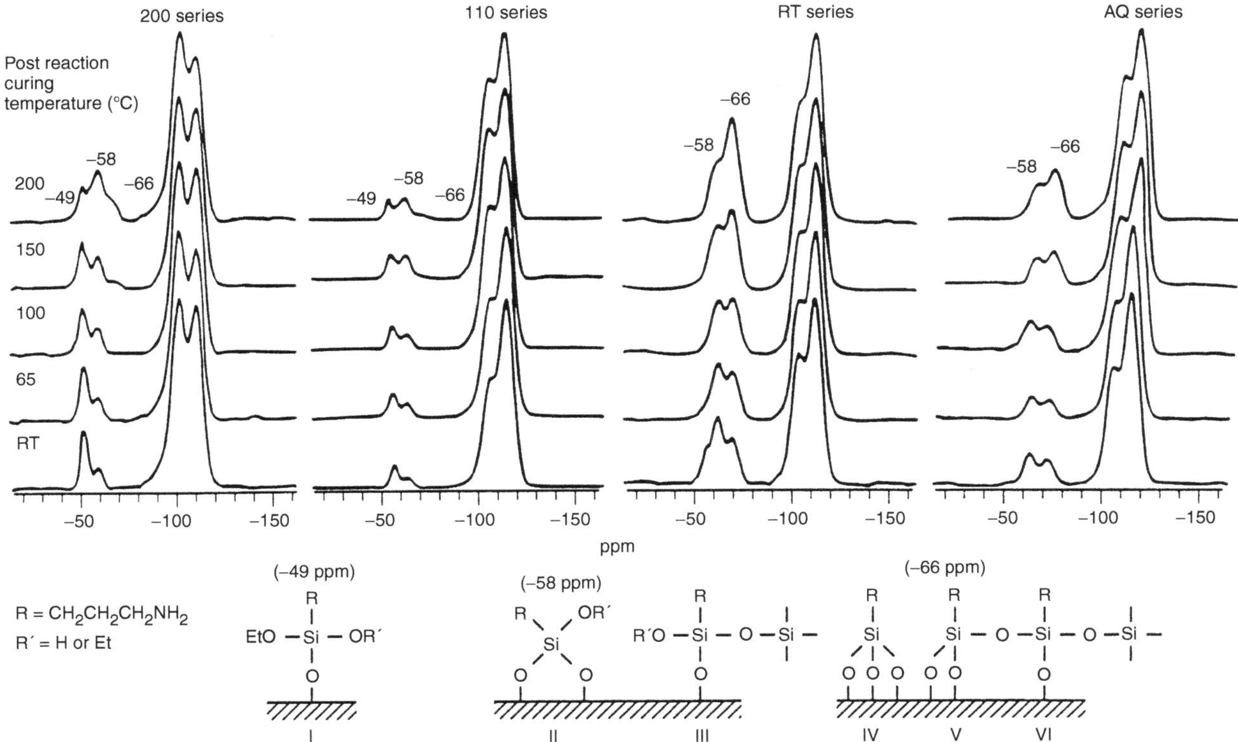

FIGURE 34.26 ^{29}Si CP-MAS spectra of APTS-modified silica gels. Each column of spectra corresponds to the drying temperature of silica gel (°C; RT = room temperature) under vacuum prior to reaction in dry toluene, or aqueous reaction conditions (AQ). Post-reaction treatment (curing) temperature shown on the left. Structural assignments given at the bottom. Taken from reference 5m. With permission.

but quantitatively different, results on $SOCl_2$-treated and CH_3Li-treated silica gel, as well as n-butyllithium-treated silica gel, were reported earlier by Bush and Jorgenson in an extensive near infrared study supported by less detailed NMR results [38]. Yamamoto and Tatsumi have used ^{13}C and ^{29}Si NMR to demonstrate the attachment of methyl groups directly to silicon atoms of silica by reaction of methyl Grignard reagent with a *n*-butoxy-modified silica [39a].

The NMR results obtained in the Tao-Maciel work provide no obvious evidence that dimethylzinc reacts with the chlorinated silica *to form direct silicon–carbon bonds.* Figure 34.33D shows ^{29}Si CP-MAS spectrum of a dimethylzinc-treated chlorinated-silica sample; this spectrum appears to be essentially identical with the spectrum of the original silica *before chlorination.* The fact that the spectrum in Figure 34.33D does not look like that of chlorinated silica in Figure 34.33A is due to the use of water to decompose excess dimethylzinc in the workup procedure used to prepared the sample of Figure 34.33D; in this procedure the ⟨**Si**⟩—Cl functionalities have been hydrolyzed to ⟨**Si**⟩—OH. The ^{29}Si CP-MAS spectrum of trimethylaluminum-treated chlorinated silica

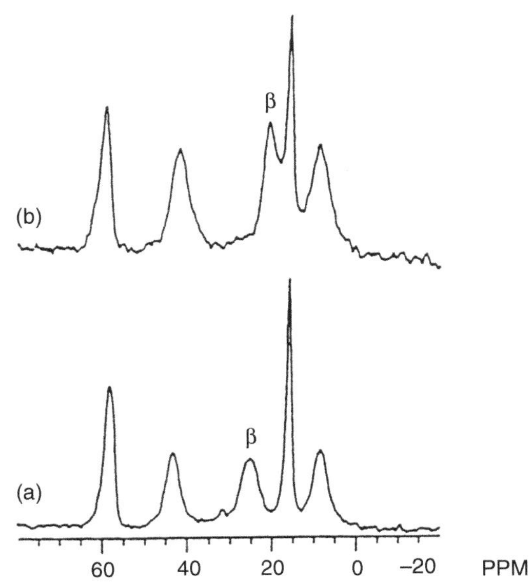

FIGURE 34.27 ^{13}C CP-MAS spectra of APTS-modified silica gels (a) and HCl-treated sample (b). Taken from reference 6a. With permission.

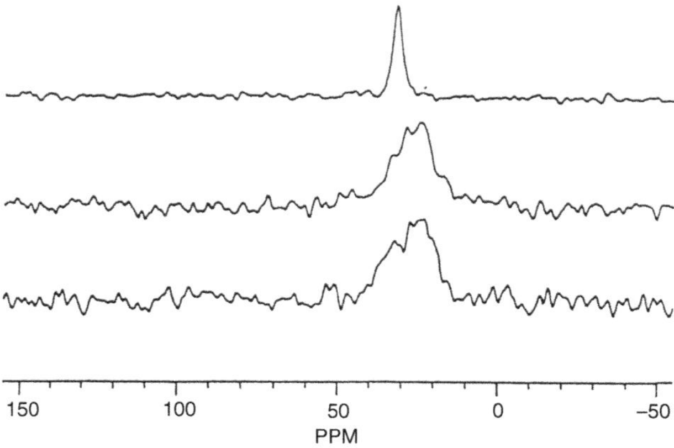

FIGURE 34.28 ^{15}N CP-MAS spectra (natural abundance) of APTS-modified silica gels; (top) untreated, (middle) after treatment with H_2SO_4 and (bottom) after treatment with 0.1 M NaOH. Taken from reference 6a. With permission.

(Figure 34.33E) shows very little intensity of ($\geqslant$SiO)$_3$-**Si**CH_3 at −60 ppm. These "negative" results from the dimethylzinc treatment of chlorinated silica and the trimethylaluminum treatment of chlorinated silica, as well as analogous results (not shown here) for $SiCl_4$-modified silica, in contrast to the cases of CH_3Li or CH_3MgBr reagents, are somewhat surprising in terms of the reactive natures typical of these CH_3M reagents. Boiadjiev and co-workers [39b] used solid-state NMR to examine the reaction products between n-alkanethiols and Me_2Zn-modified silica, but showed no NMR results on samples that had not been treated with a thiol.

Solution-state ^1H NMR analyses (not shown here) performed on the decanted solutions present after treatments of chlorinated silica with $(CH_3)_2Zn$ and $(CH_3)_3Al$

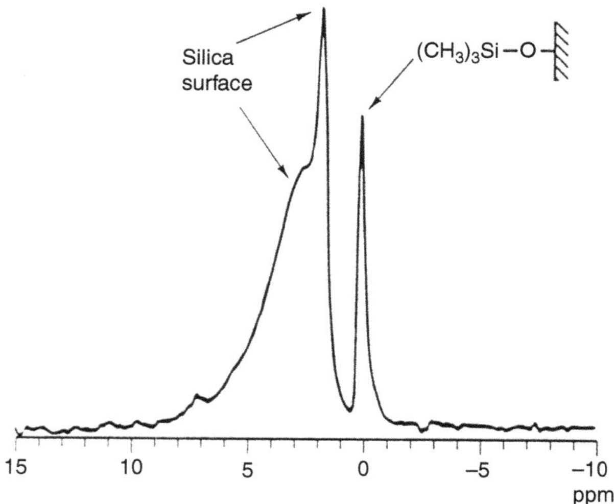

FIGURE 34.29 ^1H CRAMPS spectrum of trimethylsilyl-derivatized silica gel. Taken from reference 4f. With permission.

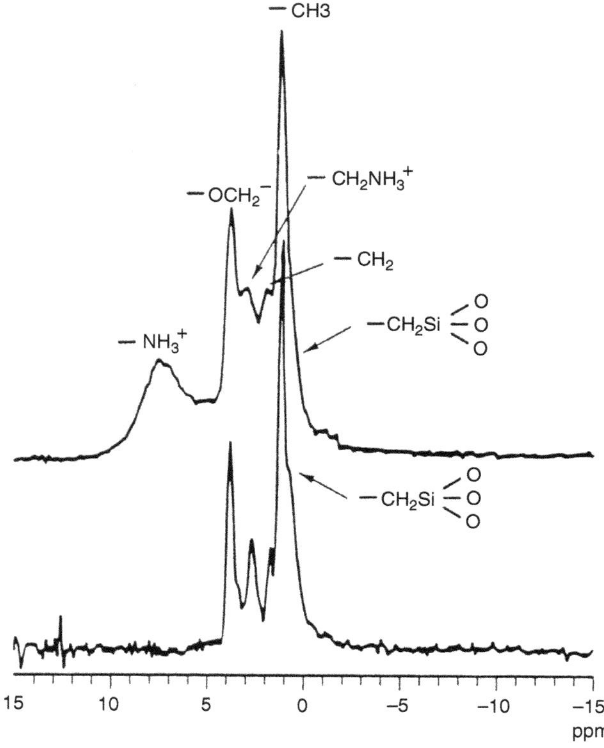

FIGURE 34.30 ^1H CRAMPS spectra of APTS-modified silica gel. Lower, untreated. Upper, treated with HCl. Taken from reference 4f. With permission.

for 24 h show that the majority of the dimethylzinc ($\delta_H = -0.7$ ppm) and trimethylaluminum ($\delta_H = -0.5$ ppm) introduced in excess at the beginning of the treatments remain unreacted [5t]. The amount of initially added $(CH_3)_2Zn$ or $(CH_3)_3Al$ that is not accounted for in the supernatant liquids of the final reaction mixtures is about 1.8 mmol/g of SiO_2 or 2.0 mmol/g of SiO_2, respectively; these numbers can be compared with the 1.2 mmol of Cl/g of SiO_2 in the chlorinated silica starting material (by elemental analysis). Thus, it appears that substantial amounts of $(CH_3)_2Zn$ and $(CH_3)_3Al$ react, or at least strongly interact, with the chlorinated silica surface, but not to form Si—CH_3 bonds.

The ^{13}C NMR signals between 1 and −5 ppm in Figure 34.33, which correspond to CH_3 groups attached to silicon, are seen in the ^{13}C CP-MAS spectra of CH_3Li-treated $SOCl_2$-chlorinated silica (Figure 34.33b) and of CH_3MgBr-treated $SOCl_2$-chlorinated silica (Figure 34.33c), while no ^{13}C signals were found in the CP-MAS spectrum of the $(CH_3)_2Zn$-treated $SOCl_2$-chlorinated silica sample (Figure 34.33d), and only a very small ^{13}C signal was found for $SOCl_2$-chlorinated silica that was treated with trimethylaluminum (Figure 34.32e). By analyzing the ^{13}C NMR lineshapes, one can discern structural differences among the samples represented in Figure 34.33. The ^{13}C CP-MAS spectrum in

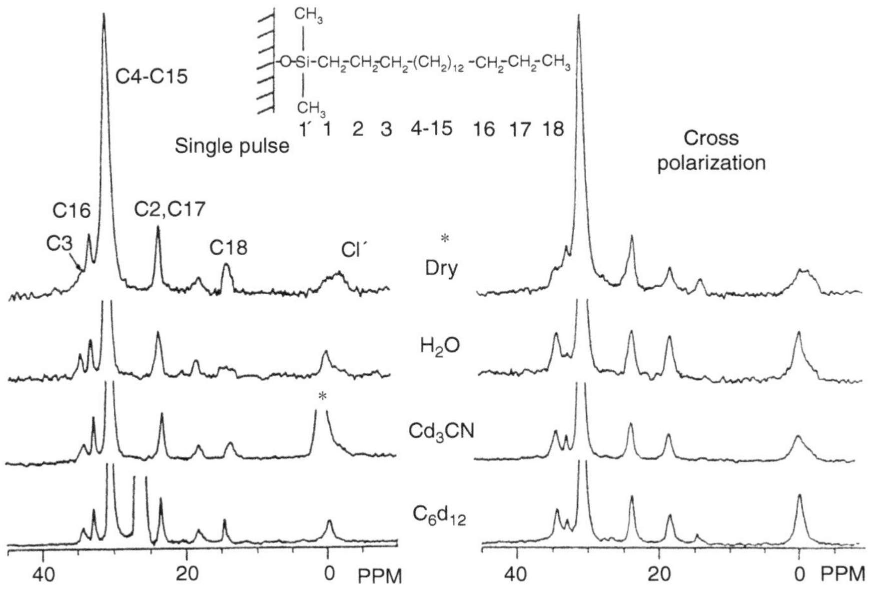

FIGURE 34.31 ^{13}C MAS spectra of C_{18}-silica samples, treated with the indicated liquids. Taken from reference 5p. With permission.

Figure 34.33b has a dominant peak at about 1 ppm, which corresponds to the resonance from the ($\geqslant$SiO)**Si**($CH_3)_3$ structural unit. Some intensity between -1 and -5 ppm in Figure 34.33b is due to ^{13}C resonances from ($\geqslant$SiO)$_2$-**Si**($CH_3)_2$ and ($\geqslant$SiO)$_3$**Si**CH_3 functionalities [36], although they were not resolved as distinct peaks. The CH_3MgBr-treated sample in Figure 34.33c shows another type of surface modification, compared to the sample of Figure 34.33b. The highest peak in Figure 34.33c, at -5 ppm, indicates a high population of ($\geqslant$SiO)$_3$**Si**CH_3

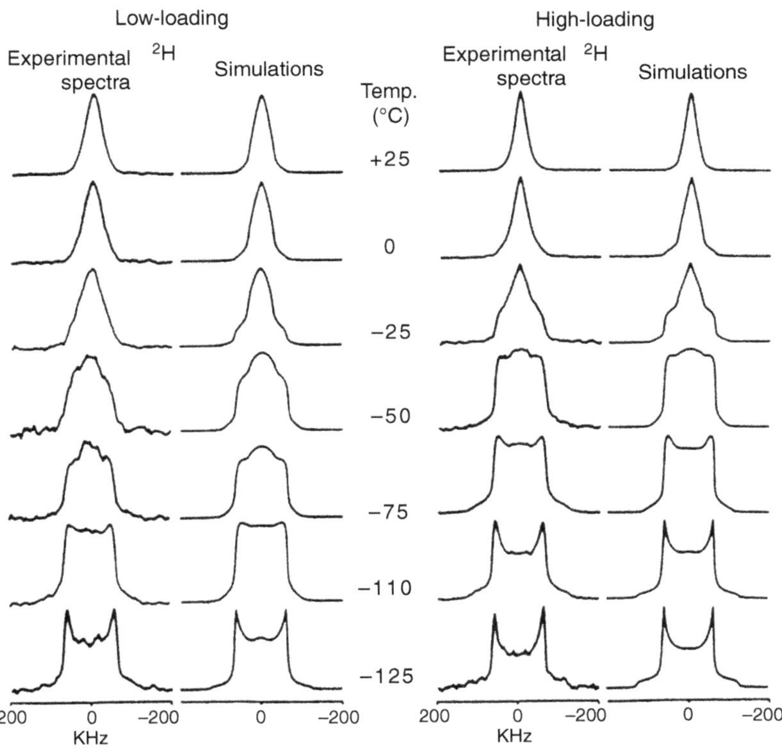

FIGURE 34.32 Comparison of simulated ^2H NMR spectra with the experimental spectra ($\tau = 50$ μs) of dry, low-loading and dry, high-loading [1,1-d_2]-C_{18}-silica samples taken over a range of temperatures. Taken from reference 5o. With permission.

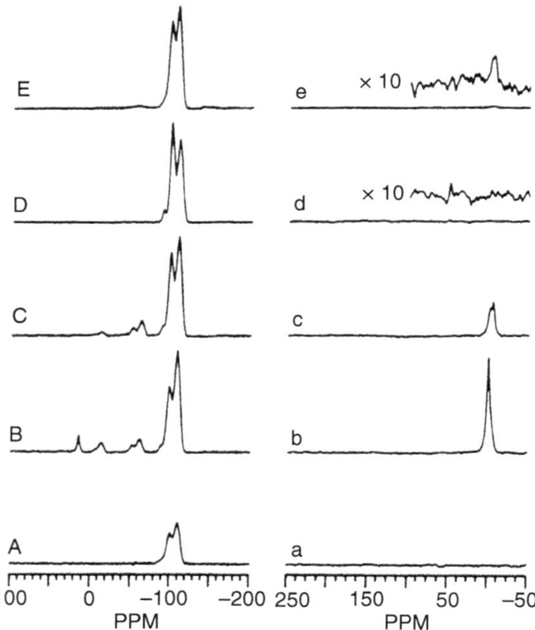

FIGURE 34.33 Solid state ^{29}Si (left column) and ^{13}C (right column) CP-MAS NMR spectra of the reaction products of $SOCl_2$-chlorinated silica (A and a) with methyllithium (B and b), methylmagnesium bromide (C and c), dimethylzinc (D and d), and trimethylaluminum (E and e). Each ^{29}Si spectrum was obtained with the following conditions: ^{1}H 90° pulse, 6.0 μs; CP contact time, 10 ms; ^{1}H decoupling, 43 kHz; 1 s repetition delay; and 7200 scans. Each ^{13}C spectrum was obtained with the following conditions: ^{1}H 90° pulse, 6.0 μs; CP contact time, 5 ms; ^{1}H decoupling, 43 kHz; 2 s repetition delay; and 7200 scans. Taken from reference 5t. With permission.

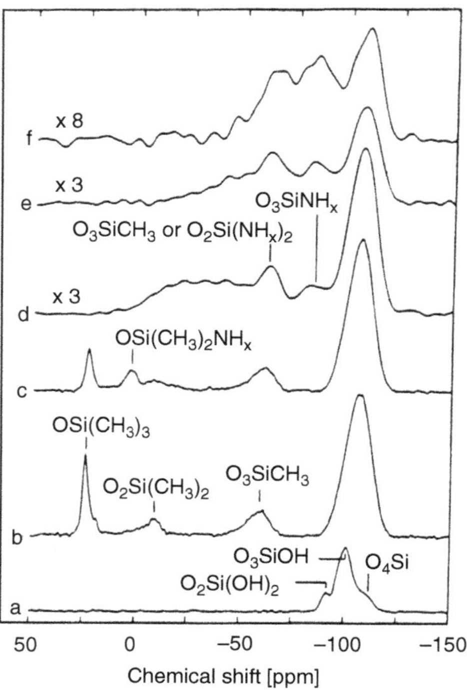

FIGURE 34.34 ^{29}Si CP-MAS NMR spectra of (a) silica dehydroxylated at 1023 K, (b) silica reacted with TMA at 423 K and (c) TMA-modified silica reacted with ammonia at 423 K, (d) 623 K, (4) 723 K and (f) 823 K. Taken from reference 40b. With permission.

species, while some intensity appears as a shoulder at − 1 ppm, which is the contribution from $(\geqslant SiO)_2\mathbf{Si}(CH_3)_2$ species. There is no distinct signal at 1 ppm in Figure 34.33c for $(SiO)Si(CH)_3)_3$ species. The thrust of these ^{29}Si and ^{13}C CP-MAS spectra of Figure 34.33 is that methylmagnesium bromide is by far the most suitable of these four CH_3M reagents to synthesize methyl-modified silica from $SOCl_2$-chlorinated silica, while largely maintaining the integrity of the silica framework. Further studies are underway to determine the nature of the species present at various stages of these CH_3 metal/silica treatments.

There have been numerous reports of studies of the reaction of silica with Me_3Al (TMA) *vapor* by optical spectroscopy techniques, and at least a few solid-state NMR studies have been reported [40]. These studies have led to chemical conclusions that involve structures that are similar to what are discussed above, although in different amounts, reflecting the very different reaction conditions employed, for example, the pretreatment of the silica and vapor phase-vs-solution. In addition, Anwander and co-workers reported a detailed study, including solid-state ^{1}H, ^{13}C, and ^{27}Al NMR results, on the reaction of the *mesoporous* material, MCM-41, with a solution of $AlMe_3$ in *n-hexane* [41]. In the vapor deposition studies, the main emphasis has often been on the subsequent reaction of Me_3Al-treated silica with another reagent, for example, NH_3 (for preparing silica-supported AlN films), H_2O (for preparing Al_2O_3 films) or a catalytic center. For example, Figure 34.34 shows ^{29}Si CP-MAS spectra obtained by Puurunen et al. [40b] on silica reacted with Me_3Al vapor (Figure 34.34b) and then with ammonia at four different temperatures (Figures 34.34c–f). Corresponding ^{13}C CP-MAS spectra are shown in Figure 34.35. The ^{29}Si NMR spectra (Figure 34.34) show clearly the formation of Si—C and Si—N bonds on the silica surface and the ^{13}C NMR spectra (Figure 34.35) show the presence of Al—C and Si—C bonds in product species.

Other solid-state NMR studies of silica-surface reactions have included studies with aqueous AlC_3 [42], and with $AlCl_3$ vapor, gaseous phosphorous compounds (e.g., PCl_3, $POCl_3$) and ammonia vapor (*vide infra*). Of course, in such cases there are additional NMR nuclides from which potentially useful data can be obtained, for example, ^{27}Al and ^{31}P. An example is given in Figure 34.36, which shows ^{27}Al MAS NMR spectra reported by Sato and Maciel [5v] of catalyst material

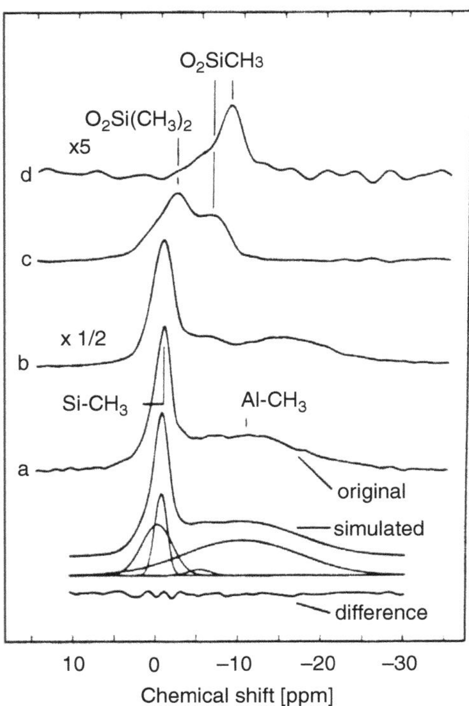

FIGURE 34.35 ^{13}C CP-MAS NMR spectra of (a) TMA-modified silica (original and simulated) and (b) TMA-modified silica reacted with ammonia at 423 K, (c) 623 K and (d) 823 K. Taken from reference 40b. With permission.

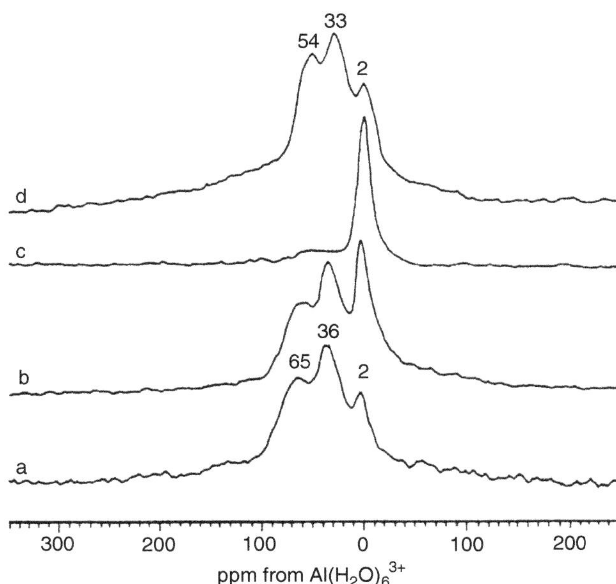

FIGURE 34.36 ^{27}Al MAS NMR spectra of silica-grafted aluminum chloride prepared at 300°C. a: the sample was sealed with sealing tape. b: exposed to air without sealing tape for 4 h. c: for 48 h. d: sample "a" hydrolyzed by contacting water vapor (180 mmol · g-sample^{-1} at 500°C for 1 h. Taken from reference 5v. With permission.

prepared by the reaction of $AlCl_3$ vapor with silica at 300°C. The spectrum of the unadulterated product (Figure 34.36a) has three main peaks, at 2, 36, and 65 ppm. When the sample was exposed to air for 4 h, intensities of the peaks at 36 and 65 ppm decreased, and the intensity of the peak at 2 ppm increased (Figure 34.36b). The peak at 2 ppm is due to 6-coordinate aluminum species. The peaks at 36 and 65 ppm changed into the peak at 2 ppm after 48 h of exposure to air (Figure 34.36c). This indicates that water molecules are readily chemisorbed to 4- or 5-coordinate aluminum species and that all aluminum species are thereby changed into 6-coordinate aluminum.

Figure 34.36d exhibits the ^{27}Al MAS-NMR spectrum of the silica-grafted aluminum chloride sample after hydrolysis. The hydrolyzed sample displays three chemical shifts, around 2, 33, and 54 ppm. In aluminosilicates, 4-coordinate aluminum atoms have ^{27}Al chemical shifts at around 60 ppm, while 6-coordinate species appear around 0 ppm [7,43]. In addition, 5-coordinated aluminum has been found at about 35 ppm in several alkoxides [44a] and minerals [44b–d]. The three peaks at 2, 33 and 54 ppm of the hydrolyzed sample (Figure 34.36d) are assigned to aluminum species coordinated with 6, 5 and 4 oxygen atoms, respectively. The 4-coordinate and possibly the 5-coordinate aluminum peaks are different from those of the samples before hydrolysis: this treatment moves the chemical shifts to lower frequency (higher shielding) when chlorine ligands are exchanged with hydroxyl groups. This indicates that the coordination numbers of aluminum atoms grafted onto silica are retained after hydrolysis, and that the aluminum species are dispersed atomically on the surface. Thus, the peak at 2 ppm in Figures 34.36(a) and (b) is due to aluminum species coordinated with six chlorine and oxygen atoms, including water molecules, and the peaks at 65 and 36 ppm are 4- and 5-coordinate species, respectively. The peak at 65 ppm is considered to be due to aluminum species coordinated either with two chlorines and two oxygens or with three chlorines and one oxygen atom. Since one hydrogen chloride molecule is produced during the reaction of each silanol with aluminum trichloride, the elemental analysis results (not shown here) indicate that the peak at 65 ppm is due to aluminum species coordinated with two chlorines and two oxygen atoms. The peak at 36 ppm is due to aluminum species coordinated with two chlorines and three oxygen atoms. Similar studies examined the effect of the $AlCl_3$-silica reaction temperatures and the dependence on silica pretreatment temperature and correlated the results with catalytic activity.

Not surprisingly, since ^{31}P ($I = 1/2$, 100% natural abundance) is such a convenient/easy nuclide for NMR, there is a substantial literature based primarily on ^{31}P NMR studies of materials prepared by the treatment of silica with PCl_3 and $POCl_3$ and related species, for example, H_3PO_3, $P(OMe)_3$, H_3PO_4 and $MePCl_2$ [45].

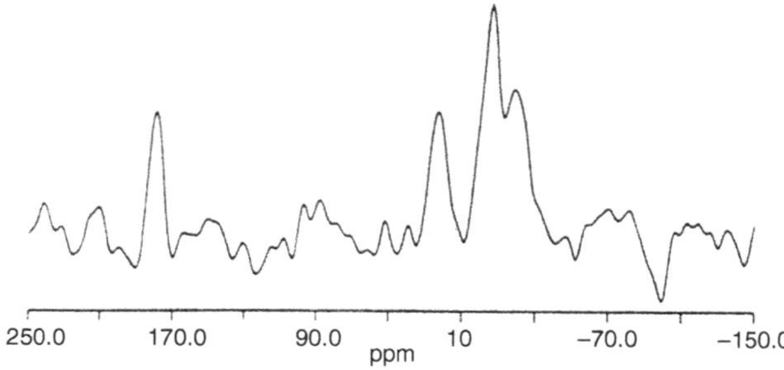

FIGURE 34.37 Cross-polarization ^{31}P MAS NMR spectrum of pure PCl_3 after 18 h of adsorption on silica. Taken from reference 45c. With permission.

Although these studies have provided useful data, it is perhaps unfortunate that the published ^{31}P NMR results have, in most cases, not been accompanied by corresponding ^{29}Si NMR data. Figure 34.37 shows the ^{31}P MAS NMR spectrum reported by Morrow, Lang and Gay on a sample of PCl_3 adsorbed on silica (after 18 hours) [45c]. Although only a peak at about 219 ppm due to physisorbed PCl_3 is observed initially at room temperature, after 18 hours at least four new peaks were observed: 184 ppm (assigned to ⟨Si⟩ $OPCl_2$), 24 ppm (not assigned), −5 ppm (tentatively assigned to ⟨**Si**⟩—O—P(H)(OH)=O and −16 (tentatively assigned to (⟨**Si**⟩—O—$)_2$P(H)=O. The role of CP and dephasing due to dipolar interaction between ^{31}P and ^{1}H are helpful in making such assignments. Gay, Morrow and Lang used ^{31}P NMR to study the physisorption, chemisorption and silica-catalyzed isomerizations that occur when trimethyl phosphite, $(CH_3O)_3P$, is adsorbed on silica [45b].

An interesting observation made by Bernstein and co-workers in a study of the reactions of both gaseous and liquid PCl_3 with silica gel is the appearance of a peak at −207 ppm in the ^{29}Si CP-MAS spectrum of the product obtained from the liquid PCl_3 case [45a]. This peak was interpreted as being due to hexacoordinate silicon sites formed by a reaction between PCl_3 and silica.

In their work on Me_3Al-treated silica and its interaction with gaseous ammonia Puurunen and coworkers studies the reaction of silica with ammonia via ^{1}H MAS and ^{29}Si CP-MAS NMR [40b]. The spectra, represented in Figure 34.38, show the formation of Si–N attachments at the silica surface.

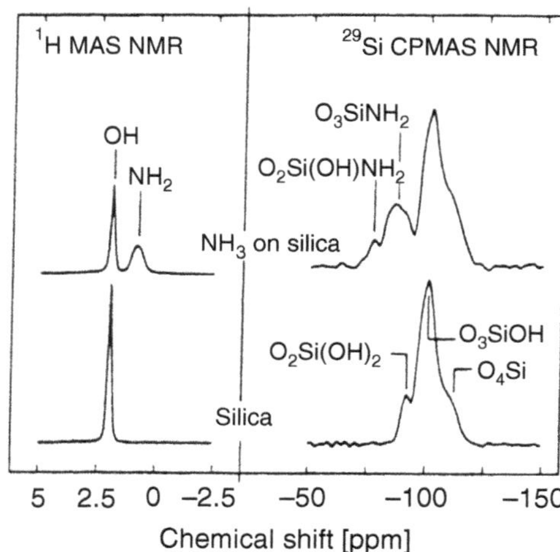

FIGURE 34.38 ^{1}H MAS NMR and ^{29}Si CP-MAS NMR spectra of silica dehydroxylated at 1023 K (below) and the same silica reacted with ammonia at 823 K (above). Taken from reference 40b. With permission.

Substitutionally-Modified Silicas

At least in part because of the greater versatility and flexibility of modern sol-gel synthesis methods [46], there has been a major effort worldwide in the preparation and characterization of materials that are closely related to silica, but with one (or more) of the five atoms in the basic SiO_4 unit partially replaced by another atom (possibly with a change in coordination number). One of the most common types of such systems is the polysiloxane class, in which one of the leaving groups (X) of a portion of the initial SiX_4 reagent is replaced by some desired pendent group. There is a huge literature on polysiloxanes, including several papers from this research group [47], describing the preparation and NMR characterization of numerous polysiloxane systems. The NMR characteristics of these materials are in most respects similar to those of the corresponding surface-derivatized silicas, and polysiloxanes will not be covered in this paper.

Another important class of materials that have been receiving a substantial amount of attention devoted to their preparation and characterization over the past dozen years is the class of silica-like materials in which some of the Si sites of a normal silica are occupied by other elements. The most frequently studied of these types of systems are the molecular sieves, or zeolite-type, materials, especially with Al-for-Si substitution. However, these materials, which are the subject of an enormous literature, including a very large NMR literature, typically are substantially or highly crystalline and are not covered in this chapter, which deals with essentially amorphous silica materials.

The classic example of amorphous materials of this class is the silica-aluminas [48]. The lack of crystallinity in typical silica-aluminas precludes the occurrence of the relatively well separated ^{29}Si MAS NMR peaks found for zeolites for different n values in **Si**$(OAl)_n(OSi)_{4-n}$ structural units. Nevertheless, similar ^{29}Si

chemical shift ranges are seen in MAS spectra of silica-aluminas, as shown in Figure 34.39, for a set of silica-alumina samples synthesized, with the assistance of ultrasonics, by Xiong and co-workers.[48a] These authors attributed the deconvoluted ^{29}Si MAS contributions at −90, −100 and −110 ppm to **Si**$(OSi)_2(OAl)_2$, **Si**$(OSi)_3$(OAl) and **Si**$(OSi)_4$, respectively, although similar spectra are obtained with highly amorphous silicas (see Figure 34.2), and these peak positions also correspond to the Q_2, Q_3 and Q_4 sites of pure silicas. ^{27}Al MAS spectra of silica-aluminas usually display a broad tetra-coordinate-aluminum peak in the 60 ppm region, associated with the structural framework, and a peak in the 0 ppm region, often somewhat sharper, due to AlO_6, often non-framework, sites. To date, ^{27}Al spectra have not provided easy access to additional structural information, although detailed and elegant studies have been reported [48d]. Several ^{1}H NMR studies [4a,48b,49] and NMR base/probe studies [5b,c,k,l] on the surface acidities of silica-aluminas have been reported.

A great deal of interest, including solid-state NMR study, has been focused in the last ten years on materials of the SiO_2—TiO_2 type [50]. Unfortunately, although ^{29}Si NMR does show clear sensitivity to relevant structural issues (e.g., Si—O—Ti vs. Si—O—Si linkages, the coordination number of Ti in Si—O—Ti moieties) in *crystalline* samples, it has not proved to be highly informative regarding the existence, amounts and detailed nature of Si—O—Ti linkages in *amorphous* samples. ^{29}Si chemical shifts of Si—O—Ti moieties overlap the region found for Q_2, Q_3 and Q_4 peaks of silica itself. More promising for these systems is ^{17}O NMR [50d,e]. Figure 34.40 shows

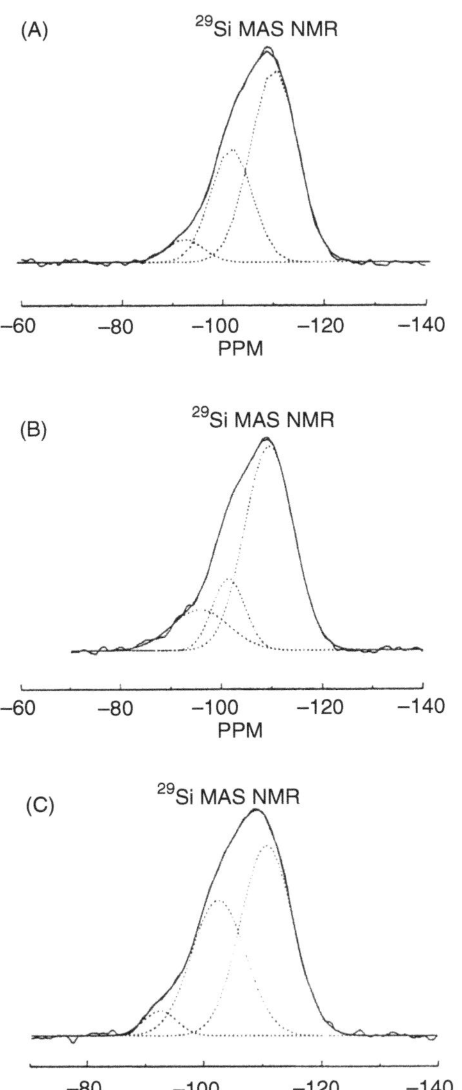

FIGURE 34.39 ^{29}Si MAS NMR spectra of different solid samples: (A) UMSA1; (B) UMSA2; (C) UMSA3 (silica-aluminas). Taken from reference 48a. With permission.

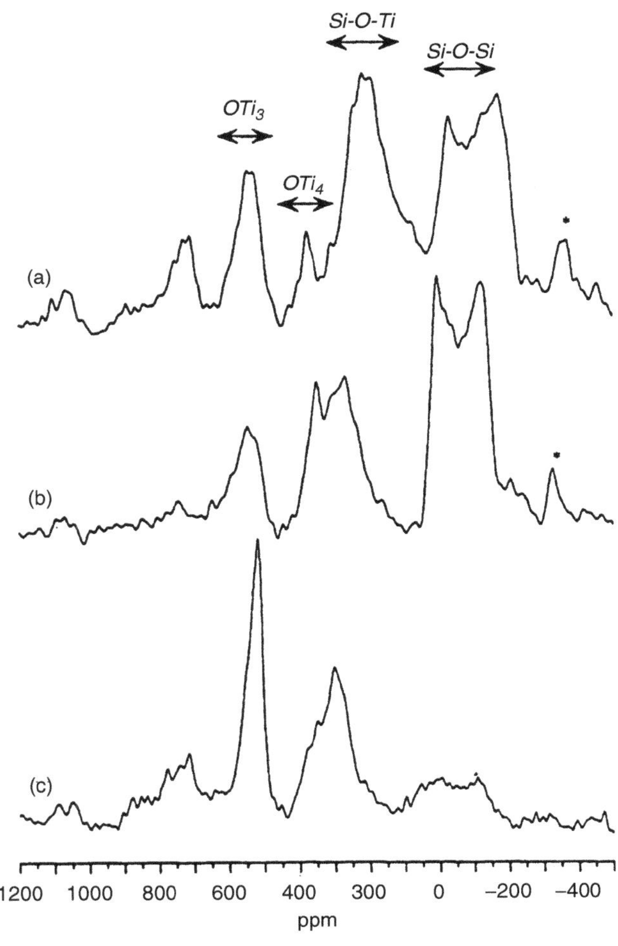

FIGURE 34.40 ^{17}O MAS NMR spectra recorded at 5.6 T of (a) QTi, (b) Tti, and (c) DTi samples (*spinning sidebands). Taken from reference 50d. With permission.

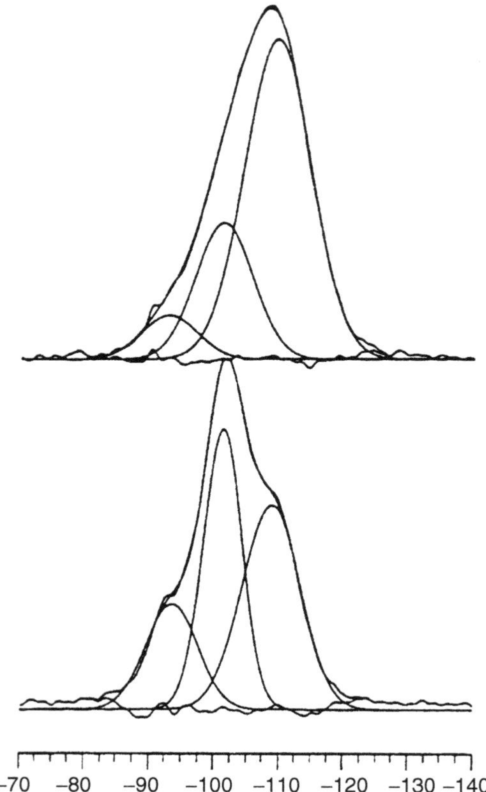

FIGURE 34.41 ^{29}Si MAS NMR spectra showing deconvolution by Gaussian fitting for a $(Ta_2O_3)_{0.05}(SiO_2)_{0.95}$ xerogel in the unheated form (bottom) and after heat treatment to 750°C. Taken from reference 51. With permission.

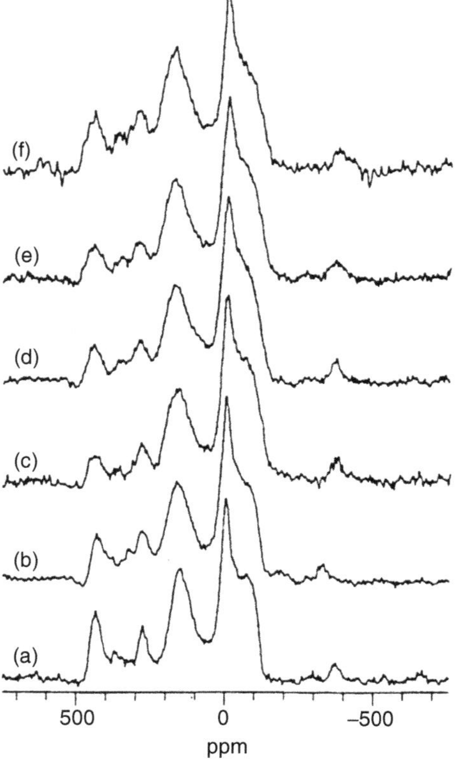

FIGURE 34.42 ^{17}O MAS NMR spectra of a $(Ta_2O_5)_{0.25}(SiO_2)_{0.75}$ xerogel after various heat treatments: (a) no heat treatment; (b) 125°C; (c) 250°C; (d) 350°C; (e) 500°C and (f) 750°C (peaks at 360 and −360 ppm are spinning sidebands). Taken from reference 51. With permission.

^{17}O MAS spectra (obtained at 5.6 T) on three SiO_2—TiO_2 gels studied by Gervais, Babonneau and Smith [50d]. These spectra show distinct patterns due to SiOSi linkages, SiOTi linkages and various titania-like moieties (OT_3, OT_4). By obtaining spectra of three magnetic field strengths, they were able to determine the ^{17}O isotropic chemical shifts and nuclear electric quadrupole parameters.

A combination of ^{29}Si and ^{17}O MAS measurements on a $(Ta_2O_5)_x(SiO_2)_{1-x}$ series of xerogels of various compositions and preparation histories has been reported by Pickup and co-workers [51]. This is apparently another case in which solid-state ^{29}Si NMR results are not as informative in structural detail as are ^{17}O NMR results. The former spectra display overlapping peaks (Figure 34.41) that over the same ranges as for pure silica. In contrast, the ^{17}O MAS spectra (Figure 34.42) show peaks/patterns identified with primarily Si–O–Si linkages (~0 ppm), Ta—O—Ta linkages (~285 and ~445 ppm for OTa_3 and OTa_2 sites, respectively) and Ta—O—Si linkages (~160 ppm). Figure 34.42 shows the effect of heat treatment on the ^{17}O MAS spectra of a $(Ta_2O_5)_{0.25}(SiO_2)_{0.75}$ sample, again displaying substantial second-order quadrupole effects.

The SiO_2–ZrO_2 system is apparently yet another material in which the broad ^{29}Si peaks observed in the solid-state spectra are useful, albeit not definitive [50c,52]. Sato and co-workers have shown ^{29}Si MAS spectra of SiO_2—ZrO_2 samples prepared from the reaction of quartz glass beads and $ZrO(OH)_2$ in an aqueous ammonia medium [52]. They observed broad ^{29}Si peaks, the shapes and positions of which depend on the sample preparation details, and inferred that silicon exists only in tetracoordinate sites linked to a zirconia framework via Si—O—Zr bonds in this system.

The vanadia/silica system [53] is another one in which ^{29}Si NMR data, at least for the sol-gel derived glasses on which such data have been reported, provide only rough qualitative support for conclusions based primarily on other types of data [53f]. In the V_2O_3/SiO_2 system, ^{51}V NMR results have been more useful to date [53a,b]. Figure 34.43 shows ^{51}V NMR spectra, both MAS and static-sample results, on a 10% vanadia/silica system under ambient (wet) conditions

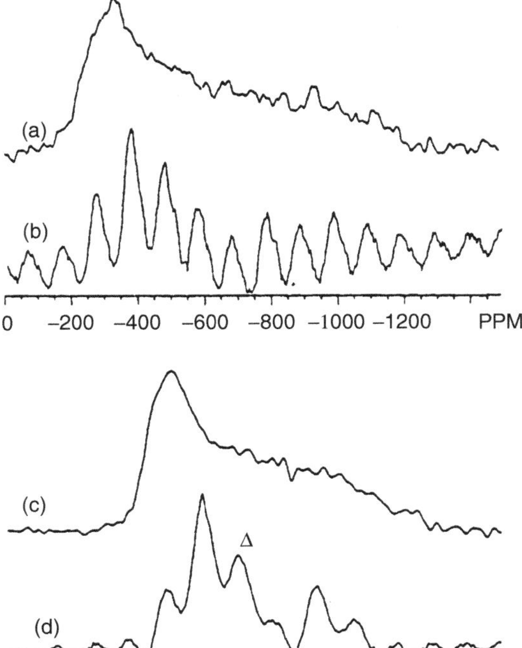

FIGURE 34.43 The 79.0-MHz solid-state ^{51}V NMR spectra of vanadia supported on silica (10 wt% V_2O_5): (a) wide-line NMR spectrum of the hydrated material present under ambient conditions, (b) MAS-NMR spectrum (at 8-kHz spinning speed) of this sample, (c) wide-line NMR spectrum following dehydration at 700°C and (d) MAS-NMR spectrum (at 8-kHz spinning speed) of the dehydrated sample. The central resonances of the MAS-NMR spectra are indicated by the symbol Δ.

(a, b) and after dehydration (c, d). Determination of the principle elements of the ^{51}V chemical shift tensors obtained by a spin-echo technique, from both MAS and static sample experiments, permitted comparisons with analogous results from model compounds. These comparisons led to the conclusion that the vanadium deposited on the silica surface in an ambient (wet) sample is in a V_2O_5-like 5-coordinate structure, whereas in dehydrated samples, the vanadium is in a 4-coordinate structures of the $(\geqq Si{-}O)_3V = O$ type.

SUMMARY AND CONCLUSIONS

Solid-state NMR, especially experiments based on $^1H \rightarrow {}^{29}Si$ CP-MAS experiments, have proven to be enormously informative in the study of silicas and modified silicas. Important structural details that would be difficult or impossible to determine by other methods have been derived from NMR results.

Although there has been a very large volume of interesting NMR-based research on silicas during the past twenty years, the *nature* of the knowledge secured from this work has not progressed dramatically since the pioneering work of Dean Sindorf in his Ph.D. thesis. The one element that has been missing in most of the reported NMR studies is capitalizing on one of the potential strong points of NMR — its inherent quantitativeness. Although NMR intensities quantitatively reflect sample populations only if proper account is taken of spin dynamics (e.g., relaxation or cross-polarization rates), in most silica studies little or no attention has been paid to these issues, so one can view the intensities in only a qualitative manner. After nearly 25 years of "preliminary" (qualitative) results in solid-state NMR studies of silica systems, it is time to take the trouble to quantitate a substantial fraction of experiments.

ACKNOWLEDGMENTS

The authors gratefully acknowledge that much of the Colorado State University part of this research was supported by National Science Foundation grant CHE-9021003 and Department of Energy grant DE-FG-03-95ER14558.

REFERENCES

1. D. W. Sindorf, "Si-29 and C-13 CP/MAS NMR Studies of Silica Gel and Bonded Silane Phases," *Ph.D. Thesis*, **1982**, Colorado State University.
2. a. G. E. Maciel and D. W. Sindorf, *J. Am. Chem. Soc.* **1980**, *102*, 7606.
   b. D. W. Sindorf and G. E. Maciel, *J. Am. Chem. Soc.* **1981**, *103*, 4263.
   c. G. E. Maciel, D. W. Sindorf and V. J. Bartuska, *J. Chromatogr.* **1981**, *205*, 438.
   d. D. W. Sindorf and G. E. Maciel, *J. Phys. Chem.* **1982**, *86*, 5208.
   e. D. W. Sindorf and G. E. Maciel, *J. Am. Chem. Soc.* **1983**, *105*, 1487.
   f. D. W. Sindorf and G. E. Maciel, *J. Am. Chem. Soc.* **1983**, *105*, 3767.
   g. D. W. Sindorf and G. E. Maciel, *J. Phys. Chem.* **1983**, *87*, 5516.
   h. D. W. Sindorf and G. E. Maciel, *J. Am. Chem. Soc.* **1983**, *105*, 1848.
3. a. I-S. Chuang, D. R. Kinney, C. E. Bronnimann, R. C. Zeigler and G. E. Maciel, *J. Phys. Chem.* **1992**, *96*, 4027.
   b. D. R. Kinney, I-S. Chuang and G. E. Maciel, *J. Am. Chem. Soc.* **1993**, *115*, 6786.
   c. I-S. Chuang, D. R. Kinney and G. E. Maciel, *J. Am. Chem. Soc.* **1993**, *115*, 8695.
   d. I-S. Chuang and G. E. Maciel, *J. Am. Chem. Soc.*

1996, *118*, 401.
e. I-S. Chuang and G. E. Maciel, *J. Phys. Chem. B* **1997**, *101*, 3052.
f. I-S. Chuang and G. E. Maciel, unpublished results.

4. a. C. E. Bronnimann, I-S. Chuang, B. L. Hawkins and G. E. Maciel, *J. Am. Chem. Soc.* **1987**, *109*, 1562.
b. C. E. Bronnimann, R. C. Zeigler and G. E. Maciel, "Proton NMR Studies of Silica Surfaces and Modified Silica Surfaces," in *Chemically Modified Surfaces. Volume 2: Chemically Modified Surfaces in Science and Industry*, D. E. Leyden and W. T. Collins (eds.), Gordon and Breach Science Publishers, New York, pp. 305–318 (1988).
c. C. E. Bronnimann, R. C. Zeigler and G. E. Maciel, *J. Am. Chem. Soc.*, **1988**, *110*, 2023.
d. C. E. Bronnimann, B. L. Hawkins, M. Zhang and G. E. Maciel, *Anal. Chem.* **1988**, *60*, 1743.
e. S. F. Dec, C. E. Bronnimann, R. A. Wind and G. E. Maciel, *J. Magn. Reson.* **1989**, *82*, 454.
f. G. E. Maciel, C. E. Bronnimann and B. L. Hawkins, "High-Resolution ^{1}H Nuclear Magnetic Resonance in Solids via CRAMPS," in *Advances in Magnetic Resonance: The Waugh Symposium, Vol. 14*, W. S. Warren (ed.), Academic Press, Inc., San Diego, CA, pp. 125–"150 (1990).

5. a. A. B. Fischer, J. A. Bruce, D. R. McKay, G. E. Maciel and M. S. Wrighton, *Inorg. Chem.* **1982**, *21*, 1766.
b. G. E. Maciel, J. F. Haw, I-S. Chuang, B. L. Hawkins, T. A. Early, D. R. McKay and L. Petrakis, *J. Am. Chem. Soc.* **1983**, *105*, 5529.
c. J. F. Haw, I-S. Chuang, B. L. Hawkins and G. E. Maciel, *J. Am. Chem. Soc.* **1983**, *105*, 7206.
d. D. K. Liu, M. S. Wrighton, D. R. McKay and G. E. Maciel, *Inorg. Chem.* **1984**, *23*, 212.
e. R. W. Linton, M. L. Miller, G. E. Maciel and B. L. Hawkins, *Surface and Interface Anal.* **1985**, *7*, 196.
f. M. L. Miller, R. W. Linton, G. E. Maciel and B. L. Hawkins, *J. Chromatogr.* **1985**, *319*, 9.
g. W. E. Rudzinski, T. L. Montgomery, J. E. Frye, B. L. Hawkins and G. E. Maciel, *J. Chromatogr.* **1985**, *323*, 281.
h. W. E. Rudzinski, T. L. Montgomery, J. E. Frye, B. L. Hawkins and G. E. Maciel, *J. Catalysis* **1986**, *98*, 444.
i. G. S. Caravajal, D. E. Leyden and G. E. Maciel, "Solid-State NMR Studies of Aminopropylsilane Modified Silica," *in Chemically Modified Surfaces. Volume 1: Silanes, Surfaces and Interfaces*, D. E. Leyden (ed.), Gordon and Breach Science Publishers, New York, pp. 283–303 (1986).
j. G. E. Maciel, R. C. Zeigler and R. K. Taft, "NMR Studies of C_{18}-Derivatized Silica Systems," in *Chemically Modified Surfaces. Volume 1: Silanes, Surfaces and Interfaces*, D. E. Leyden (ed.), Gordon and Breach Science Publishers, New York, pp. 413–429 (1986).
k. L. Baltusis, J. S. Frye and G. E. Maciel, *J. Am. Chem. Soc.* **1986**, *108*, 7119.
l. L. Baltusis, J. S. Frye and G. E. Maciel, *J. Am. Chem. Soc.* **1987**, *109*, 40.
m. G. S. Caravajal, D. E. Leyden, G. R. Quinting and G. E. Maciel, *Anal. Chem.* **1988**, *60*, 1776.
n. R. C. Zeigler and G. E. Maciel, "A Study of Chain Motion and Configuration in Dimethylocta-decylsilane-Modified Silica Surfaces," in *Chemically Modified Surfaces. Volume 2: Chemically Modified Surfaces in Science and Industry*, D. E. Leyden and W. T. Collins (eds.), Gordon and Breach Science Publishers, New York, pp. 319–336 (1998).
o. R. C. Zeigler and G. E. Maciel, *J. Am. Chem. Soc.* **1991**, *113*, 6349.
p. R. C. Zeigler and G. E. Maciel, *J. Phys. Chem.* **1991**, *95*, 7345.
q. C. Liu and G. E. Maciel, *Anal. Chem.* **1996**, *68*, 1401.
r. C. Liu and G. E. Maciel, *J. Am. Chem. Soc.* **1996**, *118*, 5103.
s. V. H. Pan, T. Tao, J-W. Zhou and G. E. Maciel, *J. Phys. Chem. B* **1999**, *103*, 6930.
t. T. Tao and G. E. Maciel, *J. Am. Chem. Soc.* **2000**, *122*, 3118.
u. T. Tao, V. H. Pan, J-W. Zhou and G. E. Maciel, *Solid State Nucl. Magn. Reson.* **2000**, *17*, 52.
v. S. Sato and G. E. Maciel, *J. Molec. Catal. A* **1995**, *101*, 153.

6. a. G. E. Maciel, C. E. Bronnimann, R. C. Zeigler, I-S. Chuang, D. R. Kinney and E. A. Keiter, "Multinuclear NMR Studies of Silica Surfaces," in *The Colloid Chemistry of Silica, Advances in Chemistry*, **234**, H. Bergna (ed.), American Chemical Society, Washington, D.C., pp. 269–282 (1994).
b. G. E. Maciel and P. D. Ellis, "NMR Characterization of Silica and Alumina Surfaces," in *NMR Techniques in Catalysis*, A. Pines and A. Bell (eds.), Marcel Dekker, Inc., New York, pp. 231–310 (1994).
c. G. E. Maciel, "Silica Surfaces: Characterization," in *Encyclopedia of Nuclear Magnetic Resonance*, D. M. Grant and R. K. Harris (eds.), John Wiley & Sons, Ltd., Sussex, England, Vol. 7, pp. 4370–4386 (1996).
d. G. E. Maciel, "NMR in the Study of Surfaces. ^{1}H CRAMPS and ^{29}Si CP-MAS Studies of Silica," in *Nuclear Magnetic Resonance in Modern Technology*, G. E. Maciel (ed.), Kluwer, Dordrecht, Netherlands, pp. 401–446 (1994).
e. G. E. Maciel, "Solids Based on the Siloxane Bridge, from Silica to Silicones," in *Solid State NMR of Inorganic Materials*, J. J. Fitzgerald (ed.), *ACS Symposium Series 717*, Washington, D.C., 1999, pp. 326–356.
f. G. E. Maciel, "Solid-State NMR Studies of Heterogeneous Catalysts," in *Heterogeneous Catalysis, Proc. of the Second Symposium of the Industry-University Cooperative Chemistry Program of the Department of Chemistry, Texas A&M University*, B. L. Shapiro (ed.), Texas A&M University Press, College Station, TX, pp. 349–381 (1984).

7. G. Englehardt and D. Michel, *High Resolution Solid-State NMR of Silicates and Zeolites*, Wiley, New York (1987).

8. a. A. P. Legrand, H. Hommel, A. Tuel, A. Vidal, H. Balard, E. Papirer, P. Levitz, M. Czernichowksi, R. Erre, H. Van Damme, J. P. Gallas, J. F. Hemidy, J. C. Lavalley, O, Barres, A. Burneau and Y. Grillet, *Adv. Colloid Interface Sci.* **1990**, *33*, 91.
b. A. Tuel, H. Hommel, A. P. Legrand and E. Kovats, *Langmuir*, **1990**, *6*, 770.
c. A. Tuel, H. Hommel, A. P. Legrand, Y. Chevallier and J. C. Morawski, *Colloids Surf.* **1990**, *45*, 413.
d. S. Leonardelli, L. Facchini, C. Fretigny, P. Tougne and A. P. Legrand, *J. Am. Chem. Soc.* **1992**, *114*, 6412.
e. A. P. Legrand, H. Taibi, H. Hommel, P. Tougne and S. Leonardelli, *J. Non-Cryst. Solids* **1993**, *155*, 122.
9. a. C. J. Brinker, R. J. Kirkpatrick, D. R. Tallant, B. C. Bunker and B. Montez, *J. Non-Cryst. Solids* **1988**, *99*, 418.
b. C. J. Brinker, R. K. Brow, D. R. Tallant and R. J. Kirkpatrick, *J. Non-Cryst. Solids* **1990**, *120*, 26.
10. a. R. K. Iler, *The Chemistry of Silica. Solubility, Polymerization, Colloid and Surface Properties and Biochemistry*, Wiley-Interscience, New York (1979).
b. *The Colloid Chemistry of Silica*, H. E. Bergna (ed.), *Advances in Chemistry Series 234*, American Chemical Society, Washington, D.C. (1994).
11. A. Pines, W. G. Gibby and J. S. Waugh, *J. Chem. Phys.* **1973**, *59*, 569.
12. J. Schaefer and E. O. Stejskal, *J. Am. Chem. Soc.* **1976**, *98*, 1031.
13. P. C. Hammel, P. L. Kuhns, O. Gonen and J. S. Waugh, *Phys. Rev. B* **1986**, *34*, 6543.
14. R. A. Wind, M. J. Duijvestijn, C. Van Der Lugt, A. Manenschijn and J. Vriend, *Prog. NMR Spectrosc.* **1985**, *17*, 33.
15. D. Raftery, H. Long, T. Meersmann, P. J. Grandinetti, L. Reven and A. Pines, *Phys. Rev. Lett.* **1991**, *66*, 584.
16. a. E. R. Andrew, *Philos. Trans. R. Soc. (London)* **1981**, *A299*, 505.
b. E. R. Andrew, *Prog. NMR Spectrosc.* **1971**, *8*, 1.
c. I. J. Lowe, *Phys. Rev. Lett.* **1959**, *2*, 285.
d. H. Kessemeier, R. E. Norberg, *Phys. Rev.* **1967**, *155*, 321.
17. M. Mehring, *High Resolution NMR in Solids*, Springer, Berlin, pp. 135 (1983).
18. a. F. Devreux, J. P. Boilot, F. Chaput and B. Sapoval, *Phys. Rev. Lett.* **1990**, *65*, 614.
b. J. V. Smith and C. S. Blackwell, *Nature* **1983**, *303*, 223.
c. G. Engelhardt and R. Radeglia, *Chem. Phys. Lett.* **1984**, *108*, 271.
d. B. L. Phillps, J. C. Thomson, Y. Xiao and R. J. Kirkpatrick, *Phys. Chem. Minerals* **1993**, *20*, 341.
e. D. R. Spearing, I. Farnan and J. F. Stebbins, *Phys. Chem. Minerals* **1992**, *19*, 307.
f. R. F. Pettifer, R. Dupree, I. Farnan and U. Stronberg, *J. Non-Cryst. Solids*, **1988**, *106*, 408.
g. Y. Xiao, R. J. Kirkpatrick and Y. J. Kim, *Phys. Chem. Minerals* **1995**, *22*, 30.
19. a. G. S. Henderson, M. E. Fleet and G. M. Bancroft, *J. Non-Cryst. Solids* **1984**, *68*, 333.
b. T. Gerber and B. Himmel, *J. Non-Cryst. Solids* **1986**, *3*, 324.
c. B. Himmel, T. Gerber, W. Heyer and W. Blau, *J. Mater. Sci.* **1987**, *22*, 1374.
20. A. Burneau, B. Humbert, O. Barres, J. P. Gallas and J. C. Lavalley, "Fourier Transform Infrared and Raman Spectroscopic Study of Silica Surfaces," in *The Colloid Chemistry of Silica*, H. E. Bergna (ed.), *Advances in Chemistry Series 234*, American Chemical Society, Washington, D.C., pp. 199–222 (1994).
21. W. L. Earl and D. L. Vander Hart, *Macromol.* **1979**, *12*, 672.
22. Z. Luz and A. J. Vega, *J. Phys. Chem.* **1987**, *91*, 374.
23. J. S. Waugh, L. M. Huber and U. Haeberlen, *Phys. Rev. Lett.* **1968**, *20*, 180.
24. a. W.-K. Rhim, D. D. Elleman and R. W. Vaughan, *J. Chem. Phys.* **1973**, *58*, 1772.
b. D. P. Burum and W. K. Rhim, *J. Chem. Phys.* **1979**, *71*, 944.
25. B. C. Gerstein, R. G. Pembleton, R. C. Wilson and L. M. Ryan, *J. Chem. Phys.* **1977**, *66*, 361.
26. D. Suwelack, W. P. Rothwell and J. S. Waugh, *J. Chem. Phys.* **1980**, *73*, 2559.
27. R. Freeman and H. D. Hill, *J. Chem. Phys.* **1971**, *54*, 3367.
28. R. J. Wittebort, M. G. Usha, D. J. Ruben, D. E. Wemmer and A. Pines, *J. Am. Chem. Soc.* **1988**, *110*, 5668.
29. T. H. Walter, G. L. Turner and E. Oldfield, *J. Magn. Reson.* **1988**, *76*, 106.
30. A. J. Vega, "Quadrupolar Nuclei in Solids," in *Encyclopedia of Nuclear Magnetic Resonance*, D. M. Grant and R. K. Harris (eds.), Wiley, New York, pp. 3869–3889 (1996).
31. A. Medek, J. S. Harwood and L. Frydman, *J. Am. Chem. Soc.* **1995**, *117*, 12779.
32. D. P. Burum and A. Bielecki, *J. Magn. Reson.* **1991**, *96*, 645.
33. C. E. Bronnimann, C. Ridenour, D. R. Kinney and G. E. Maciel, *J. Magn. Reson.* **1992**, *97*, 522.
34. A. J. Vega, *J. Am. Chem. Soc.* **1988**, *110*, 1049.
35. R. A. Santos, R. A. Wind and C. E. Bronnimann, *J. Magn. Reson. B* **1994**, *105*, 183.
36. R. A. Lewis, "Characterization of Intermediates Produced by the Pyrolysis of Hydridopolysilazane," Ph.D. Dissertation, Colorado State University (1994).
37. a. M. Pursch, D. L. Vanderhart L. C. Sander, X. Gu, T. Nguyen, S. A. Wise and D. A. Gajewski, *J. Am. Chem. Soc.* **2000**, *122*, 6997.
b. M. Pursch, L. C. Sander and K. Albert, *Anal. Chem.* **1999**, *71*, 733A.
c. M. Pursch, L. C. Sander, H.-J. Egelhaaf, M. Raitza, S. A. Wise, D. Oelkrug and K. Albert, *J. Am. Chem. Soc.* **1999**, *121*, 3201.
d. S. Strohschein, M. Pursch, D. Lubda and K. Albert, *Anal. Chem.* **1998**, *70*, 13.
e. M. Raitza, M. Herold, A. Ellwanger, G. Gauglitz and K. Albert, *Macromol. Chem. Phys.* **2000**, *201*, 825.

f. B. Buszewski, R. M. Gadzata-Kopciuch, M. Markuszewski and R. Kaliszan, *Anal. Chem.* **1997**, *69*, 3277.
g. M. Pursch, L. Sander and K. Albeit, *Anal. Chem.* **1996**, *68*, 4107.
h. M. Pursch, S. Strohschein, H. Handel and K. Albert, *Anal. Chem.* **1996**, *68*, 386.

38. S. G. Bush and J. W. Jorgenson, *J. Chromatog.* **1990**, *503*, 69.
39. a. K. Yamamoto and T. Tatsumi, *Chem. Lett.* **2000**, 624.
b. V. Boiadjiev, A. Blumenfeld, J. Gutow and W. T. Tysoe, *Chem. Mater.* **2000**, 2640.
40. a. E.-L. Lakomaa, A. Root and T. Suntola, *Appl. Surface Sci.* **1996**, *107*, 107.
b. R. L. Puurunen, A. Root, S. Haukka, E. I. Iiskola, M. Lindblad and A. O. Krause, *J. Phys. Chem. B* **2000** 104, 6599.
c. R. L. Puurunen, A. Root, P. Sarv, S. Haukka, E. I. Iiskola, M. Lindblad and A. O. I. Krause, *Appl. Surface Sci.* **2000**, *165*, 193.
d. A.-M. Uusitalo, T. T. Pakkanen, M. Kröger-Laukkanen, L. Nünistö, K. Hakala, S. Paavola and B. Löfgren, *J. Mol. Catal. A: Chemical* **2000**, *160*, 343.
41. R. Anwander, C. Palm O. Groeger and G. Engelhardt, *Organometallics* **1998**, *17*, 2027.
42. S. Jun and R. Ryoo, *J. Catal.* **2000**, *195*, 237.
43. J. M. Thomas and J. Klinowski, *Adv. Catal.* **1985**, *33*, 199.
44. a. O. Kriz, B. Casensky, A. Lycka, J. Fusek and S. Hermanek, *J. Magn. Reson.* **1984**, *60*, 375.
b. M. C. Cruickshank, L. S. D. Glasser, S. A. I. Barri and I. J. F. Poplett, *J. Chem. Soc., Chem. Commun.* **1986**, 23.
c. L. B. Alemany and G. W. Kirker, *J. Am. Chem. Soc.* **1986**, *108*, 6158.
d. L. F. Nazar, G. Fu and A. D. Bain, *J. Chem. Soc., Chem. Commun.* **1992**, 251.
45. a. T. Bernstein, P. Fink, F. M. Mastikhin and A. A. Shubin, *J. Chem. Soc. Faraday Trans.* **1986**, *82*, 1879.
b. I. D. Gay, A. J. McFarlan and B. A. Morrow, *J. Phys. Chem.* **1991**, *95*, 1360.
c. B. A. Morrow, S. J. Lang and I. D. Gay, *Langmuir* **1994**, *10*, 756.
d. S. J. Lang, I. D. Gay and B. A. Morrow, *Langmuir* **1995**, *11*, 2534.
e. P. Kohli and G. J. Blanchard, *Langmuir* **2000**, *16*, 695.
46. J. J. Brinker and G. W. Scherer, *Sol-gel Science: The Physics and Chemistry of Sol-Gel Processing*, Academic Press, Boston (1990).
47. G. E. Maciel, "NMR Characterization of Functionalized Polysiloxanes," in *Solid State NMR of Polymers*, I. Ando (ed.), Elsevier, Tokyo, Ch. 25 (1998).
48. a. N. Yao, G. Xiong, K. L. Yeung, S. Sheng, M. He, W. Yang, X. Liu and X. Bao, *Langmuir* **2002**, *18*, 4111.
b. C. Dorémieux-Morin, C. Martin, J.-M. Bregeault and J. Fraissard, *Appl. Catalysis* **1991**, *77*, 149.
c. E. Oldfield, J. Haase, K. D. Schmitt and S. E. Schramm, *Zeolites* **1994**, *14*, 101.
d. A. D. Irwin, J. S. Holmgren and J. Jonas, *J. Mater. Sci.* **1988**, 2908.
49. a. M. Hunger, D. Freude, H. Pfeifer, H. Bremer, M. Jank and K. P. Wendlandt, *Chem. Phys. Lett.* **1983**, *100*, 29.
b. M. Hunger, D. Freude and H. Pfeifer, *J. Chem. Soc. Faraday Trans.* **1991**, *87*, 657.
c. P. Batamack, C. Doremieux-Morin, J. Fraissard and D. Freude, *J. Phys. Chem.* **1991**, *95*, 3790.
d. L. Heeribout, R. Vincent, P. Batamack, C. Dorémieux-Morin and J. Fraissard, *Catal. Lett.* **1998**, *53*, 23.
50. a. J. P. Rainho, J. Rocha, L. D. Carlos and R. M. Almeida, *J. Mater. Res.* **2001**, *16*, 2369.
b. Y. Liu, H. Du, F.-S. Xiao, G. Zhu and W. Pang, *Chem. Mater.* **2000**, *12*, 665.
c. S. Bachmann, L. F. C. Melo, R. B. Silva, T. A. Anazawa, I. C. S. F. Jardim, K. E. Collins, C. H. Collins and K. Albert, *Chem. Mater.* **2001**, *13*, 1874.
d. C. Gervais, F. Babonneau and M. E. Smith, *J. Phys. Chem. B* **2001**, *105*, 1971.
e. M. A. Holland, D. M. Pickup, G. Mountjoy, E. S. C. Tsang, G. W. Wallidge, R. J. Newport and M. E. Smith, *J. Mater. Chem.* **2000**, *10*, 2485.
51. D. M. Pickup, G. Mountjoy, M. A. Holland, G. W. Wallidge, R. J. Newport and M. E. Smith, *J. Mater. Chem.* **2000**, *10*, 1887.
52. S. Sato, R. Takahashi, T. Sodesawa, S. Tanaka, K. Aguma and K. Ogura, *J. Catal.* **2000**, *196*, 190.
53. a. A. E. Stiegman, H. Eckert, G. Plett, S. S. Kim, M. Anderson and A. Yavrouian, *Chem. Mater.* **1993**, *5*, 1591.
b. N. Das, H. Eckert, H. Hu, I. E. Wachs, J. F. Walzer and F. J. Feher, *J. Phys. Chem.* **1993 97**, 8240.
c. B. Taouk, M. Guelton, J. P. Grimblot and J. P. Bonnelle, *J. Phys. Chem.* **1988**, *92*, 6700–6705.
d. O. B. Lapina, V. M. Mastikhin, A. V. Nosov, T. Beutel and H. Knozinger, *Catal. Lett.* **1992**, *13*, 203.
e. J. P. Solar, P. Basu and M. P. Shatlock, *Catal. Today* **1992**, *14*, 211.
f. M. D. Curran, D. D. Pooré and A. E. Stiegman, *Chem. Mater.* **1998**, *10*, 3156.

35 Modified Silicas: Synthesis and Applications

A.A. Chuiko
National Academy of Sciences of Ukraine, Institute of Surface Chemistry

Chemical modification of the disperse silica surface offers ample scope for development of materials with preassigned properties for various duties. In order to optimize the modification processes it is necessary to know their mechanisms and have at one's disposal the reliable data about such characteristics as the structure of the SiO_2 surface layer, topography of its hydroxyl coating, state of its hydration sheath depending on the conditions of synthesis and pretreatment of silicas. Substantial differences in the degree of purity, completeness of hydration, and pore structure quite often impede establishment of correlations between the aforementioned data for numerous varieties of disperse SiO_2. Quite detailed information on the structure of surface layers has been collected in the case of the pyrogenic silica that is distinguished by its high purity, absence of pores and labile structural elements. By modern physicochemical methods it has been found that the pyrogenic SiO_2 surface is formed from sections with the structure of (111) face and, to a lesser extent, of (100) face of β-crystobalite with terminal silanol and silanediol groups spaced 0.6–0.7 nm apart. The bulk phase of the pyrogenic SiO_2 is represented to approximately the same extent by motifs of quartz and crystobalite [1,2].

The SiO_2 hydroxyl coating formed by groups with a substantial dipole moment has local and collective properties. The local characteristics of structural hydroxyls define their sorptive activity and reaction capacity. The collective properties manifest themselves in such phenomena as variation of the spectral line shape in the IR range, orientation phase transitions in the lattice of OH groups. The theory of rotation mobility of such groups developed earlier [3] reveals a significant role of their collective properties in adsorption and chemisorption processes.

We have achieved comprehensive experimental results on the nature of active sites of surface, regularities of proceeding of sorption processes [2,4]. The techniques of chemical probing, IR spectroscopy, ^{29}Si and ^{1}H NMR, desorption mass-spectrometry, x-ray photoelectron spectroscopy have provided strong evidence for the nonhydroxylic origin of primary sites of adsorption of small polar molecules (H_2O, HF, CH_3OH, etc.). Our analysis of profiles of electrostatic potential (ESP) and electron density in a surface layer of SiO_2 has given results that are indicative of high positive values of the potential (low values of the density) at silicon atoms with polar terminal groups. It points to a coordinative unsaturation of such atoms. The quantum-chemical calculations of various levels of approximation with allowance for the ESP values of the SiO_2 lattice showed that the formation of coordination complexes of small polar molecules is energetically more favourable in comparison with hydrogen-bonded structures. By experiment it has been established that after thermal vacuum treatment of silica samples at 470 K an adsorption complex contains two molecules of H_2O per silanol group and after that at 670 K it has one molecule of water [2,4]. The formation of the most stable coordination complexes is related to penetration of adsorbate molecules into a surface layer with the subsequent coordination in *trans*-positions to (Si)OH groups. The channels for such a penetration are hexagonal structural cavities typical of the (111) face of β-crystobalite.

Substitution of trimethylsilyl groups for hydroxyls does not eliminate sites of strong adsorption of small polar molecules. In the case of low surface coverages, molecules of hydrogen fluoride that is distinguished for the maximum affinity to silicon compounds do not react with silanes and form strong complexes capable of acting as intermediates of the fluorination reaction. The field desorption mass-spectrometry makes it possible to register three characteristic regions of the adsorbate removal with desorption maxima, and the number of these regions is in agreement with the number of different types of adsorption complexes of H_2O and CH_3OH (hydrogen-bonded complexes and coordination complexes of *cis*- and *trans*-structure). Besides, it has been also shown that during reactions with water-absorbing reagents one can perform a separate introduction of different forms of adsorbed water. The x-ray photoelectron spectroscopy detects an increase in the binding energy for core 2*s*- and 2*p*-electrons of silicon atoms as SiO_2 undergoes dehydration, and this increase is most sharp in the range of the high-temperature desorption maxima. This points to molecules of coordinatively bonded water as the source of variation of the electron density on Si atoms of a surface layer.

The application of various functional capabilities of disperse silicas calls for thorough knowledge about the role of the state of their hydrated cover since it is of major importance for the completeness of proceeding of chemical modification reactions as well as for detailed structure of surface compounds. The systematic experimental and quantum-chemical researches [1,2,4] resulted in the consistent insights into the structure of the hydrated cover of SiO_2, mechanisms of its formation and destruction. The concepts formulated take into account the data on feasibility to draw a distinction between sorbed and chemically bonded water. It made it possible to give concrete expression to such fundamental notions as dehydration and dehydroxylation of surface. Dehydration is a process of desorption of water evolved upon decomposition of aquacomplexes having various structures. The temperature interval of silica dehydration in vacuum extends to 900 K (H-bonded species decompose at 300–370 K, coordination complexes of *cis*- and *trans*-structure decompose at 420–500 and 550–850 K respectively). Dehydroxylation is a chemical reaction of breakdown of a hydroxyl cover. Condensation of neighboring silanediol groups on zones with the structure of (100) face of β-crystobalite proceeds simultaneously with dehydration of SiO_2 at 500–650 K, which is corroborated by the ^{29}Si NMR technique. Isolated silanol groups are removed at temperatures over 900 K. The mechanisms of the processes of thermal dehydroxylation of silica were considered in Refs. [2,4].

Between processes of adsorption and chemisorption there is a close relationship, since in the process of evolution of adsorbed species one can observe formation of transient complexes that are products of reactions proceeding by the Langmuir-Hinshelwood mechanism. According to the accepted classification of reactions in a surface layer of SiO_2 they are subdivided into two main classes, namely reactions with substitution of structural hydroxyl protons (S_Ei processes) and reactions with substitution of OH groups at silicon atoms (S_Ni processes). Reactions of heterolytic decomposition of siloxane bonds proceeding by the $Ad_{N,E}$ mechanism form a separate class.

By now we have studied experimentally numerous S_Ei reactions of the electrophilic substitution involving the replacement of protons of $\equiv$SiOH groups by positively charged fragments of molecules of hydro- and methylchlorosilanes, silazanes, alkoxysilanes, organosiloxanes, alkyl borates and phosphates, chlorides and oxochlorides of a number of elements, etc. A detailed study has been also conducted on the nucleophilic substitution of functional groups at silicon atoms (reactions with halogen hydrides, alcohols, phenols, processes of hydrolysis of $\equiv$SiX groups of various types, etc). Besides, we have considered a special class of reactions whose subsequent steps proceed by different mechanisms. In particular, such compounds as $SOCl_2$, PCl_5, WCl_6, $WOCl_4$, $MoOCl_4$, etc. interact with silanol groups by the S_Ei mechanism. These interactions lead to the formation of unstable intermediate compounds that in certain conditions undergo intramolecular rearrangements. As a result, $\equiv$SiCl groups are formed, which conforms to the substitution (by the S_Ni mechanism) of chlorine atoms for OH groups at silicon atoms, and molecules of a corresponding oxochloride split off [2].

The action of the mechanisms of all the above-mentioned processes can be given a proper rational explanation by quantum-chemical calculations of cross-sections of potential energy surfaces for the interaction of reacting systems [2,4], but these calculations are very tiresome and time-consuming. However, the goal can be achieved through a much more simple approach that lays emphasis on the stereochemistry of transient states and takes into account a predominant contribution of effects of reagent deformation to the reaction activation energy. Within the scope of the developed deformation model [2,4] the activation barrier of S_Ei reactions is determined by the structure deformation of only electrophilic reagent molecules. In the case of S_Ni reactions the main contribution to this barrier is made by the deformation of the tetrahedral environment of a silicon atom with a functional group. This approach enables consideration of processes proceeding by the $Ad_{N,E}$ mechanism as well. The insights into the deformational nature of activation barriers of chemical transformations have made the foundation for the uniform description, analysis, and prediction of the reactivity of modifying reagents and surface functional groups, and, consequently, of paths of reactions in a SiO_2 surface layer.

The results of the comprehensive studies into the structure of the pyrogenic silica surface, its hydration sheath, regularities and mechanisms of adsorption processes and chemical transformations have provided the scientific foundation of searches for the most promising routes of chemical modification of SiO_2 and various goal-directed syntheses in a surface layer as well as for the development of many novel materials with a wide range of physical and chemical properties.

The fundamental results achieved in the field of surface chemistry of silica formed the basis for the evolution of a novel direction of researches — chemical science of nanostructural materials. Within the scope of this scientific direction over 200 new materials with designed properties have been developed which find much use in care of public health, agriculture, ecology, civil engineering, instrument making, and other branches of the national economy. The materials developed include new medicinal preparations of sorptive action with a controlled pharmacodynamics; drug delivery systems; photosetting composites for orthopedic stomatology; enterosorbents and aerosolic vaccines for cattle breeding; preparations for clarification and stabilization of wines, beers, fruit juices, and other beverages; protective and stimulative compositions for increasing yield of agricultural crops; reagents for induced rain precipitation; sorbents for removal of petroleum and its products from surface of water; thermostable general-duty lubricants; water-repellent coatings; fillers for varnishes, paints, and enamels; polishing compositions for finishing treatment of electronic devices; heat- and sound-insulation materials; fireproof covers; fire-extinguishing powders, and many others.

REFERENCES

1. Chuiko, A.A., *React. Kinet. Catal. Lett.*, 1993, V. 50, No 1–2, p. 1–13.
2. Chuiko, A.A., Gorlov, Yu.I., *Surface Chemistry of Silica: Surface Structure, Active Sites, Sorption Mechanisms*, Kiev, Nauk. Dumka, 1992 (in Russian).
3. Chuiko, A.A., *Izv. Akad. Nauk SSSR*, 1990, No 10, pp. 2393–2406.
4. Ogenko, V.M., Rosenbaum, V.M., Chuiko, A.A., *A Theory of Vibrations and Reorientations of Surface Atomic Groups*, Kiev, Nauk. Dumka, 1991 (in Russian).

36 Electric Surface Properties of Silica in Nonaqueous Electrolyte Solutions

A.N. Zhukov
St. Petersburg State University, Department of Colloid Chemistry

The origin of the surface charge and electric surface properties of solid inorganic oxides in aqueous electrolyte solutions are fairly well known. Main mechanisms of the generation of the surface charge in such systems are either the loss (-SO^-) or the gain (SOH_2^+) of proton by surface hydroxilic groups (-SOH) due to their reactions of acid–base equilibria with water and the adsorption of the background electrolyte ions (K^+ and A^-) on ionized hydroxilic groups (-SO^-–K^+ and -SOH_2^+–A^-). But the mechanisms operating in nonaqueous solutions are not clearly understood.

The main objective of this work was to study and to compare the electric surface properties (electrokinetic potential ζ and charge σ; adsorption of ions Γ; surface conductivity K_s) of powdery silica (quartz and aerosil) in solutions $C = 10^{-5}$–10^{-2} M of LiBr, KBr and NaBr in five solvents differing from water: two amphoteric solvents (ethanol (EtOH), 1-butanol (BuOH)); three aprotic solvents (dimethylsulfoxide (DMSO)), dimethylformamide (DMFA) and acetone (Ac) and in ethanol–water mixture (E + W).

The various electrokinetic effects (electrical conductance, streaming current, streaming potential, electrophoretic mobility) were measured for the plugs and the suspensions of powdery silica in the mentioned solutions. A pair of Ag–AgBr electrodes was used to contain the plugs and to mesure both the streaming current or potential and the conductivity of plugs with A.C. Winston bridge. Using a conductometric cell with a pair of platinum electrodes, the conductivities of diluted suspensions were measured at the different volume fractions of silica. These conductometric data and the various theoretical expressions were used for the calculations of surface conductivity of silica (Figure 36.1). The dependencies of K_s on log C show that in all cases the surface conductivity has abnormally high values several orders higher than

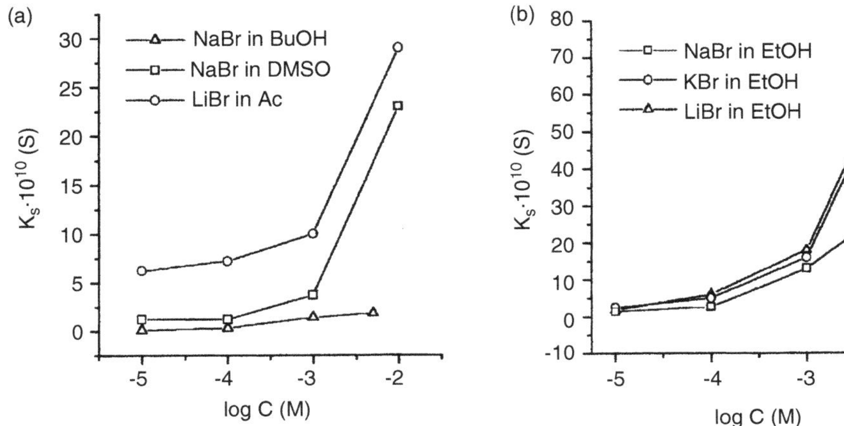

FIGURE 36.1 Plots of surface conductivity of quartz versus log C of the alkaline metal bromides in various solvents.

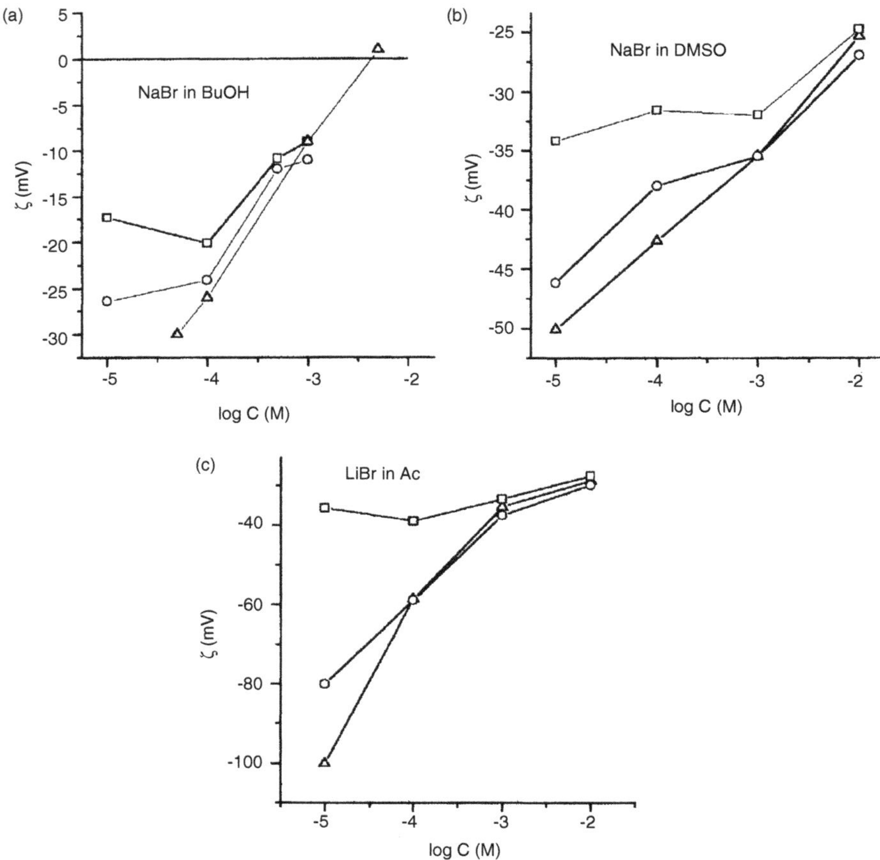

FIGURE 36.2 Dependencies of the zeta-potential of quartz on the log C calculated using electrokinetic data and expressions of Smoluchowsky (□) and Henry (○) for electrophoresis and an expression for the streaming potential (Δ) modified by taking into account the experimental values of K_s.

the values calculated according to Bikerman theory, continually increasing with the growth of C and having non-zero values at the isoelectric points.

Electrokinetic data were converted in ζ-potential values (Figure 36.2) using the expressions of Smoluchowsky and Henry for electrophoresis and the Smoluchowsky's expression for streaming potential modified by taking into account the experimental values of K_s. It is obvious that the incorporation of the K_s in the theories of electrokinetic phenomena brings to the best compatibility of the dependencies ζ (log C) obtained by different methods and especialy in the region of the low electrolyte concentration.

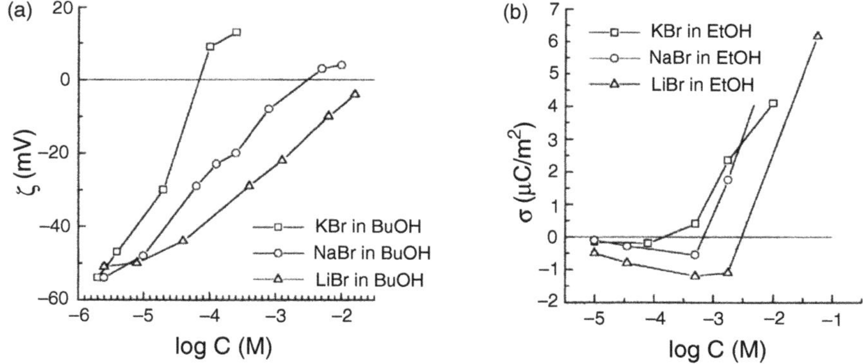

FIGURE 36.3 Plots of electrokinetic potential (a) and electrokinetic charge (b) of quartz versus log C of the alkaline metal bromides in BuOH and EtOH.

The general trends represented by examples of the dependencies of ζ-potential and electrokinetic charge of quartz on the concentrations of alkali metal bromides in BuOH and EtOH (Figure 36.3) are the following: values of ζ and σ in the diluted solutions are negative and they become more positive with increasing values of C. The increase of electrolyte concentration brings to the more positive values of ζ and σ while depending on the nature of electrolyte LiBr < NaBr < KBr and gives the inversion of their sign. The solubility of these electrolytes and solvatation ratios of their cations decrease in the same order.

The isotherms of ion adsorption onto silica surfaces were determined by the depletion method and calculated from the difference between initial ion concentration and the one measured in the supernatant after 24 hours of equilibration. The isotherms of cation and anion adsorbtion $\Gamma(C)$ (Figure 36.4) show their nonequivalent character (adsorption of cations is always greater than of anions and increases in the order Li < Na < K).

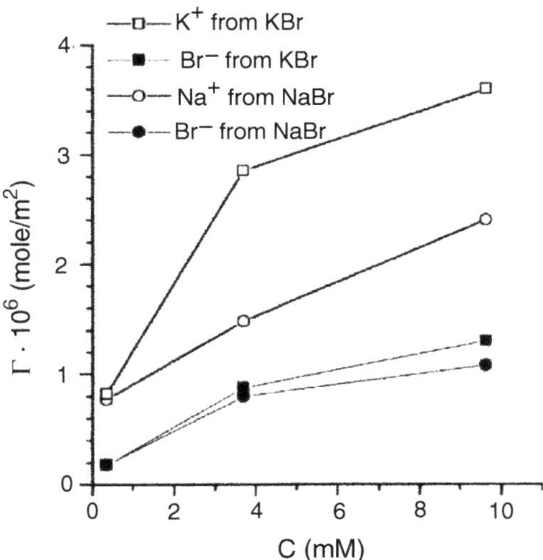

FIGURE 36.4 Isotherms of ion adsorption onto silica surface for KBr and NaBr solutions in EtOH.

The analysis of the results brings to the following view on the mechanism of the charge formation on oxides in the electrolyte solutions in the studied solvents. The origin of the surface charge in such protic liquids as alcohols is due to the acid–base interactions similar to that in water and involving proton as a potential-determining ion. Value of this charge in the diluted solutions is defined by the relative acidity of both the liquids (EtOH > BuOH) and the surface OH-groups of SiO_2. The surface charge on the silica in aprotic liquids is determinated by the donor–acceptor interactions. In the case of DMSO and DMFA with their high donicity, this charge is formed by the proton dissociation of the OH-groups. The adsorption of background electrolyte ions also influences not only the value but the sign of the electrokinetic charge. The increase of the negative charge of σ with the growth of electrolyte concentration and then its sign inversion may be explained by the fact that these ions are adsorbed not only on the ionized OH-groups (-SiO^-, -$SiOH_2^+$) but also on nonionized ones (-SiOH). That means that their specific interactions with the surface groups are mainly determined by the solvatation ratio of ions.

The surface conductivity in the studied systems is due to the mobility of all ionic species in diffuse and Stern layers. The experimental values of K_S are in good correlation with the theoretical ones calculated using expression $K_s = F\,(u^+\Gamma^+ - u^-\Gamma^-)$ and the experimental values of Γ (Figure 36.4) if the surface mobilities of adsorbed ions (u^+ and u^-) have the 0.7–0.8 of their bulk values.

Author acknowledges a grant 96-03-34313a from Russian Foundation for Fundamental Researches.

37 Chemical Reactions at Fumed Silica Surfaces

Vladimir M. Gun'ko and Alexey A. Chuiko
National Academy of Sciences of Ukraine, Institute of Surface Chemistry

CONTENTS

Unmodified and modified fumed silicas are widely used as fillers, additives, thickeners, adsorbents, carriers, and so on [1–11]. Chemical modification of the fumed silica surfaces with respect to silanol and siloxane bonds by different reactants (organic, hetero-organic, and inorganic compounds) as well as preparation of nonstandard fumed silicas allows us to extend applications of these materials in medicine, biotechnology, environmental and human protection, and industry [11], since the surface properties of modified silicas differ substantially from those of pristine oxide materials that was clearly shown by Iler and other authors [1–11]. The classification of chemical reactions occurring at the silica surfaces has a long history incipient from the classifications of reactions of both organic and organosilicon compounds but revised with respect to interface processes and justified on the basis of investigations performed by experimental (such as reaction kinetics, TPD MS, spectroscopy, etc.) and theoretical (quantum chemistry, MC modeling, etc.) methods [1–11]. Therefore a major portion of the reaction mechanisms characteristic for the silica surfaces is listed below with a brief description.

The lion's share of reactions at the fumed silica surfaces occurs with the participation of surface silanol groups, since they possess significantly greater reactivity than siloxane bonds [1–22]

$$\equiv\text{Si-OH} + \text{HX} \rightarrow \equiv\text{Si-O(H)}\cdots\text{HX} \rightarrow \equiv\text{Si-X} + \text{H}_2\text{O} \quad (37.1)$$

$$\equiv\text{SiO-H} + \text{XY} \rightarrow \equiv\text{SiO-H}\cdots\text{XY} \rightarrow \equiv\text{Si-OY} + \text{HX} \quad (37.2)$$

The first kind of reactions can be assigned to nucleophilic substitution mechanism $S_Ni(Si)$, since nucleophilic X^-group with electron-donor α-atom (X = F, OR, etc.) forms donor-acceptor ($Si \leftarrow X_\alpha$) bond substituting OH^- in a quasi-cyclic transition state (TS) (typically four-centered with active atoms Si, O, H, and X_α). The second type of reactions is related to an electrophilic substitution mechanism S_Ei, as H^+ is substituted by an electrophilic Y^+ group in the quasi-cyclic TS [four-centered with O, H, X, and Y; for example, X = Hal, $Y = MHal_{n-1}$ (M is a metal, P, etc.; Hal is a halogen); $X = NR'$, $Y = R''$, etc.]. These reactions can be assigned to heterolytic ones akin to intramolecular processes, since back attack to ≡SiOH is practically impossible but frontal or side attack can occur [10,11,18]. Formation of the cyclic TS cannot provide high energetic barrier of the reactions (therefore certain processes can fast occur even at room temperature), as total separation of the negative (X^-, OH^-, SiO^-) and positive (H^+, Si^+, Y^+) charges does not take place in the prereaction complexes and the TS. However, some enhancement of the atom charges on the reactive bonds is observed in the TS comparing with those in the prereaction complexes [14].

Other types of surface reactions such as $Ad_{N,E}$ related, for example, to cleavage of siloxane bonds

$$\equiv\text{Si-O-Si}\equiv + \text{XY} \rightarrow \equiv\text{Si-OY} + \text{XSi}\equiv \quad (37.3)$$

nucleophilic addition (Ad_N), for example, of unsaturated organics to Si-H

$$\equiv\text{Si-H} + \text{CH}_2\text{=CHR} \rightarrow \equiv\text{Si(CH)}_2\text{R} \quad (37.4)$$

electrophilic addition (Ad_E), for example, of ethylenimine to Si-O

$$\equiv\text{SiO-H} + \text{CH}_3\text{CHNH} \rightarrow \equiv\text{SiO(CH}_2)_2\text{NH}_2 \quad (37.5)$$

elimination (E) from surface functionalities

$$\equiv\text{Si-X} \rightarrow \equiv\text{Si-Y} + \text{Z} \quad (37.6)$$

condensation (E_2)

$$\equiv\text{Si-OH} + \text{H-OSi}\equiv \rightarrow \equiv\text{Si-O-Si}\equiv + \text{H}_2\text{O} \quad (37.7)$$

reactions with the participation of different defects

$$=\text{Si=O} + \text{H}_2\text{O}{=}\text{Si(OH)}_2 \quad (37.8)$$

$$\equiv\text{Si-O}^- + \text{XY} \rightarrow \text{SiOY} + \text{X}^- \quad (37.9)$$

$$\equiv\text{Si-O}^\bullet + \text{XY} \rightarrow \text{SiOY} + \text{X}^\bullet \quad (37.10)$$

$$\equiv\text{Si}^+ + \text{XY} \rightarrow \text{SiX} + \text{Y}^+ \quad (37.11)$$

are worthy of notice as well as multistage reactions occurring through different mechanisms [1–11]. There are many factors affecting the reaction mechanism (i.e., reaction rate, activation energy, reaction constants, structures of prereaction complexes and TS, lifetime of these complexes, reaction time and depth, reaction dynamics and kinetics as a whole, etc.) such as the reaction and pretreatment temperatures, reactant pressure, chemical, spatial and electronic structures of reactants, structural features of silica (e.g., fumed silicas with different specific surface areas and primary particle size distribution, porous silica gels, etc.), availability of other components at the interfaces (e.g., adsorbed water, amines, etc.), or liquid media (water, organics, etc.) promoting or inhibiting reaction as well as external forces (mechanical or electromagnetic), and so on. On studying surface reactions, one should consider the difference in the reactivity of the same functional groups in reacting molecules and surface functionalities, since they differ not only in their mobility and vibrational spectra but also in the electronic structure, and they are differently affected by the surface electrostatic field or compounds (e.g., water) presented at the air/oxide interfaces [11,14]. Thus, an overall picture of the chemical reactions at the silica surfaces can be very complex.

There are several models to describe and understand the reaction mechanisms at the silica surfaces. One of the models is very simple and based on the strength of bonds breaking and forming during reactions. Clearly, that so rough a model cannot be used for appropriate detailed and quantitative description of the surface reactions, since electronic (polarization, charge transfer, dispersion interaction, electron correlation and exchange, etc.), geometric (deformation of reactants in the TS), diffusive (reagent diffusion at the interfaces) and other factors playing an important role in the reaction kinetics and dynamics are ignored [1–11]. Application of a model based only on deformation of molecular reactants and active surface groups in the TS can give useful information on comparison of the reactivity of similar compounds of a given series [11]. However, the deformation model of the activation barrier can give inappropriate values of the activation energy as changes in the electronic structure of reactants in the TS are ignored. Consequently, it cannot be successfully used if reactants have geometrically close groups (e.g., OH and NH) possessing substantially different electronic properties, which result in different reactivity; for example, the reactivity of the surface groups such as ≡SiOH and $\equiv SiNH_2$ differs too much strongly to be described using this deformation model. Thus, a variety of the factors (electronic and geometric) should be considered on the investigations of the mechanism of surface reactions that allows us to be sure of the accuracy of obtained results. Despite numerous investigations of chemical reactions at the silica surfaces [1–10] comprehensive experimental and theoretical description of these factors is not completed. In this brief review, we analyze the corresponding works performed mainly at the Institute of Surface Chemistry (Kiev) and devoted to fumed silicas and related oxides as the main objects of the synthesis and investigations carried out by Chuiko and dozens of his coworkers during the last decade and described in detail in several monographs and article collections [11]. Main directions of these investigations are:

(1) Synthesis of standard and nonstandard (using, e.g., nonstochiometric ratio between reagents differently distributed in the flame with varied temperature gradient, different flow velocities, and turbulence, etc.) fumed oxides based on silica and characterized by the specific surface area from 30 (mixed oxides, 50 for pure silica) to 500 m^2/g, different morphology and dispersivity of secondary particles (aggregates of primary particles, agglomerates of aggregates, flocks)
(2) development of new hybrid materials (organo-oxide, carbon-mineral, etc.)
(3) chemical modification of fumed silicas by organic, organosilicon, organometallic, and inorganic compounds using one or several reaction cycles
(4) exploration of the physicochemical characteristics of modified silicas in different media using a variety of experimental and theoretical methods
(5) investigations of kinetics, dynamics, and mechanism of reactions at the oxide surfaces and development of new models, ideas or conceptions related to the surface chemistry of metal oxides
(6) search of new applications of unmodified and modified fumed oxides and related materials in medicine, biotechnology, environmental and human protection, and industry.

Clearly, based on the physicochemical properties of silica and related materials, their most important characteristics depend on the presence of water in the form of intact molecules or different (single, twin, vicinal, localized in dense islands of OH groups, free and hydrogen bonded, placed in different pores or adjacent primary particles) hydroxyl groups [1–11]. Therefore the first subject of this review is the surface and interface waters.

SURFACE WATER

Intact water molecules and silanols (as a product of dissociative adsorption of water molecules) present at the silica surfaces play a very important role in the chemical modification of silicas, since ≡SiOH groups are the main adsorption and reactive sites, and water molecules can influence H^+ transferring (or H^- or $H^\bullet$ transferring; however, electron transferring can precede later, that is, H^+ transferring can remain as limiting), which is a limiting stage of many reactions listed earlier. Every type of adsorbed water has a corresponding temperature interval over which desorption occurs (e.g., the temperature onset for dehydration of thermally treated silica in vacuum lies at ca. 200°C, and entire dehydroxylation is observed at ca. 1100°C) [23]. These intervals can overlap; therefore, experimental separation of desorption processes is difficult and interpretation of desorption data is complicated. Additionally, sample origin, morphology of the particles, chemical composition of the surfaces, and pretreatment conditions influence features of adsorption–desorption of water and these processes at crystalline (e.g., quartz, α-alumina, rutile, etc.) or amorphous and porous (silica gel) or nonporous (fumed silica, alumina, and mixed oxides) highly disperse oxides are numerous due to the variations in both structural and energetic characteristics of their surfaces.

Despite the large body literature related to water desorption from silica and related oxides [1–4,11,23,24] a consensus regarding the origin of water desorbed from these materials at various temperatures has not been reached and the relationship between the structure of the surface sites with distinct hydroxyl groups and their propensity to desorption remains somewhat unclear. Therefore, it was of interest to study water desorption from fumed silica comparing with other oxides such as titania, silica/titania (ST), and silica/alumina (SA) using an improved TPD technique, for example, one-pass temperature-programmed desorption (OPTPD) controlled by the time-of-flight mass spectrometry (TOFMS) [25]. According to the literature, the values of the activation energy ($E^{\neq}$) of associative desorption of water from silicas vary from 80 to 280 kJ/mol (experimental) or from 75 to 400 kJ/mol (theoretical), but for titania and alumina the $E^{\neq}$ values are lower (120–200 kJ/mol by experiment). Furthermore, theoretical estimations of $E^{\neq}$ obtained by different methods are significantly distinct.

The value of the activation energy $E^{\neq}$ of desorption strongly depends on a surface coverage (Θ_{OH}) and increases with decreasing Θ_{OH}. The corresponding desorption equation can be written as follows

$$d\Theta/dt = -k_o\Theta^x \exp(-E^{\neq}(\Theta)/RT) \qquad (37.12)$$

where $x = 1 + (1/f)$, f is the fractal dimension; $k = k_o\exp(-E^{\neq}/RT)$ is the rate constant. If diffusion can limit the reaction then k and $E^{\neq}$ are given by

$$k = k_D k_r/(k_D + k_r) \qquad (37.13)$$

$$E^{\neq} = (k_D E_r + k_r E_D)/(k_D + k_r) \qquad (37.14)$$

where the D and r subscripts correspond to diffusion and reaction, respectively. Readsorption of molecules after their associative desorption gives $x \neq 2$ in Equation (37.12). Consideration of these effects and the influence of $H^{\bullet(+)}$, $OH^{\bullet(-)}$, and H_2O diffusion extends the possible range of k_o to 10^{-7}–10^4 cm^2 s^{-1} which is wider than the 10^{-4}–10^4 cm^2 s^{-1} range presented in the literature for associative desorption [26].

Fumed silica (A-300, specific surface area $S_{BET} \approx$ 300 m^2 g^{-1} estimated from nitrogen adsorption at 77.4 K); fumed titania (50 m^2 g^{-1}); fumed silica/titania (250 m^2 g^{-1}) at $C_{TiO_2} = 22$ wt.% (ST_{22}) (such C_{TiO_2} was chosen as water adsorption is close to a maximum at $C_{TiO_2} \approx 20$ wt.%); silica/alumina (170 m^2 g^{-1}) with $C_{Al_2O_3} = 23$ wt.% (SA_{23}) (Pilot Plant at the Institute of Surface Chemistry, Kalush, Ukraine); and fumed silica/CVD-titania (210 m^2 g^{-1}) (ST_{21}(CVD) obtained by the chemical vapor deposition (CVD) technique by hydrolysis of $TiCl_4$ chemisorbed on fumed silica A-300 during several reaction cycles; the CVD-synthesis of ST was described in details elsewhere [27,28]) were studied [25]. Rehydrated (for 0.5 h) samples are labeled "R".

An advantage of OPTPD-TOFMS is detection of desorbed products in their one pass from a sample to a mass-spectrometer detector through the ionization region with no possibility for other particles (desorbed but not detected at once) to return here. A regime of desorption and observation was realized in the installation of a chamber of "black" type ("black" is due to specific conditions allowing to detect practically only particles desorbed from sample in the one-pass regime, that is, other processes do not distort the signal) [25].

Unimolecular (desorption of intact molecules $\equiv MOH \cdots OH_2 \rightarrow \equiv MOH + H_2O$) or associative ($\equiv M(OH){-}O(H){-}M\equiv \rightarrow \equiv M\text{-}O\text{-}M\equiv + H_2O$ or $\equiv M(OH){-}O{-}(HO)M\equiv \rightarrow \equiv M\langle^{O}{}_{O}\rangle M\equiv + H_2O$) desorption of water molecules can be described by the rate with Equation (37.12) of first or second order, respectively. The relationship between the ion current measured and the reaction rate constants was described elsewhere [29]. The fumed oxide surfaces are heterogeneous and every type of surface sites can influence the corresponding desorption peak and the corresponding center (E_o) in a desorption energy distribution for this peak. If the TPD spectrum is convoluted into several peaks without any restriction that the activation energy calculated over the total temperature ranges for each of the peaks can be underestimated due to the overestimation of the peak width. Therefore, we used some modification of the $E^{\neq}$ calculations described in detail elsewhere [25,29].

Studied samples contained 5 wt.% (silica) to 13 wt.% (ST_{20}) adsorbed water at room temperature (Table 37.1). The lion's share of this water is molecularly adsorbed (the corresponding first TPD-DTG peak is the highest and contributes more than 70% of the total intensity) and it can be desorbed on heating to 450–500 K. In the case of ST, the amount of adsorbed water increases as the titania concentration grows and it is maximum for $C_{TiO_2} \approx$ 20 wt.% [25]; therefore, we chose the ST_{22} sample for the study by OPTPD-TOFMS. When C_{TiO_2} increases, the first DTG peak (integral intensity about 80%) splits into two peaks due to the differences in the energy of formation (E) and sizes of the hydrogen-bonded clusters of water molecules on the silica and titania phases and their interfaces (this difference in E can reach to 20 kJ/mol) [25].

TABLE 37.1
DTG Peak Temperature (T_{max}), Amount of Desorbed Water (a_i) Corresponding to Every Approximating *i*-Curve and Total Concentration (a_Σ) of Desorbed Water

Sample	T_{max} (K)	a_i (wt.%)	a_Σ (wt.%)
A-300	346	2.79	4.91
	596	1.28	
	926	0.84	
TiO_2	337	3.76	9.06
	405	2.54	
	628	2.76	
ST_{20}	343	4.17	12.91
	397	4.83	
	414	2.02	
	789	1.89	
$ST_{21,CVD}$	385	4.99	8.98
	603	3.31	
	806	0.68	
SA_{23}	371	4.99	7.37
	535	0.81	
	858	1.57	

According to the literature [2,23], we suggest that the first OPTPD (for $m/z = 18$) or DTG peak for all studied oxides is caused by unimolecular desorption of intact H_2O molecules; that is, the reaction equation is the first order ($x = 1$ in Equation (37.12)). All other peaks at higher temperatures were assigned to associative desorption of water molecules in a second order reaction (as bimolecular process), that is, ($x = 2$ in Equation (37.12)). Maybe some part of water observed at T above 550–600 K can be due to water desorbed through the unimolecular process (e.g., water desorption from micropores, defects, and primary particle volume), but this contribution is not dominant [23]. Besides, the amount of intact water is minor in the OPTPD measurements performed at 10^{-9} Torr as the amount of adsorbed molecules depends linearly on the pressure.

The DTG data (Table 37.1) correspond to desorption of ca. 7 H_2O/nm^2 for silica (less than the number of molecules from one molecular layer as the area for one adsorbed H_2O is about 0.1 nm^2), near to 40 H_2O/nm^2 for TiO_2 (i.e., the titania surface is covered by several molecular layers), and about 60 H_2O/nm^2 for ST_{20}. Consequently, OPTPD (Table 37.2, N_Σ) gives only a minor portion of this value: 33% for silica, 13% for titania, and 3% for ST.

The OPTPD spectrum of water desorbed from fumed silica has three well-defined maxima (Figure 37.1a), but our calculation gives four peaks (relatively narrow peaks) [25]; a shoulder at 580 K (as the second peak) is discernible. The lower limits of the $E^{\neq}$ and k_o values for associative desorption (Table 37.2, the first values) are close to those for amorphous silica at the same S_{BET} value. The number of water molecules desorbed per one nm^2 (Table 37.2) corresponds to four OH groups (plus 0.25 of molecules of molecular-adsorbed H_2O per nm^2). The amount of OH groups (α_{OH}) is consistent to the data from reference [23]. In addition, the α_{OH} value estimation for fumed silica heated to 400–600°C gives 1.5–1.7 OH/nm^2, which corresponds to the N_{H_2O} ($N_{H_2O} = 2\alpha_{OH}$) value for the last peak (Table 37.2). We calculated $E^{\neq}$ and k_o (Table 37.2, the first values for $x = 2$ and $l = 2–4$)

TABLE 37.2
Parameters for the OPTPD Spectra: l is the Number of Peak, x is the Reaction Order, T_{max} is the Peak Temperature, k_o is the Pre-exponential Factor, n_{mol} and N_{H_2O} are the Number of Water Molecules per mg and per nm^2, Respectively, N_Σ is the Sum of N_{H_2O}

Sample	T_{max} (K)	$E^{\neq}$ (kJ/mol)	k_o (s^{-1})	$n_{mol} \times 10^{17}$	N_{H_2O} (nm^{-2})	N_Σ, (nm^{-2})
SiO_2	451	66	1.6×10^6	0.75	0.25	2.34
	580–596	87–106	2.8×10^6–9.5×10^7	1.10	0.37	
	725–747	114–171	2.2×10^6–4.9×10^{10}	2.76	0.92	
	879–883	187–252	9.4×10^{10}–5.2×10^{14}	2.40	0.80	
SiO_2 (R)	474	69	4.3×10^6	0.49	0.16	1.03
	607–631	91–112	1.3×10^7–1.2×10^8	0.81	0.27	
	707–709	109–160	5.1×10^8–4.8×10^{10}	1.26	0.42	
	812–818	174–209	8.8×10^9–2.7×10^{12}	0.55	0.18	
TiO_2	480	71	3.6×10^6	0.85	1.70	5.06
	549–566	85–96	9.8×10^6–4.0×10^7	1.26	2.52	
	638–645	100–138	1.5×10^7–1.9×10^{10}	0.43	0.86	
TiO_2 (R)	438	63	2.3×10^6	1.25	2.50	3.38
	536–544	84–87	9.6×10^6–1.2×10^7	0.44	0.88	
TS_{22}	438	63	2.7×10^6	1.81	0.72	1.94
	511–576	84–88	1.4×10^7	0.34	0.14	
	678–694	108–154	1.3×10^7–3.1×10^{10}	1.51	0.60	
	834–835	179–221	1.1×10^{10}–1.1×10^{13}	1.21	0.48	
TS_{22} (R)	398	55	1.9×10^6	0.77	0.31	1.23
	537–545	84–91	1.7×10^7–3.2×10^7	1.03	0.41	
	657–679	100–144	1.8×10^6–1.2×10^{10}	0.52	0.21	
	944–950	195–278	1.7×10^{10}–1.5×10^{15}	0.74	0.30	
$TS_{21,CVD}$	430	60	1.3×10^6	1.84	0.88	2.51
	524–549	83–92	1.4×10^7–2.9×10^7	0.91	0.43	
	664–676	100–145	2.4×10^6–1.5×10^{10}	1.72	0.82	
	787	166–196	3.6×10^9–5.7×10^{11}	0.79	0.38	
$TS_{21,CVD}$ (R)	411	61	2.3×10^7	0.50	0.24	0.63
	472–485	72–75	3.1×10^7–9.5×10^6	0.36	0.17	
	606–615	95–109	1.6×10^7–1.3×10^8	0.22	0.22	
AS_{23}	458	67	3.0×10^6	2.09	1.23	4.18
	557–566	86–102	3.6×10^6–6.7×10^7	1.18	0.69	
	703–718	108–160	1.8×10^6–3.7×10^{10}	2.77	1.63	
	858–867	179–232	5.3×10^9–3.2×10^{13}	1.06	0.62	
AS_{23} (R)	429	61	2.0×10^6	1.44	0.85	2.49
	532–551	83–91	7.3×10^6–2.3×10^7	0.96	0.56	
	669–693	102–152	1.8×10^6–1.5×10^{10}	0.98	0.58	
	892	191–254	5.3×10^{10}–5.7×10^{14}	0.85	0.50	

Note: The intervals of the T_{max}, $E^{\neq}$, and k_o values are given for the reactions of associative desorption of water molecules.

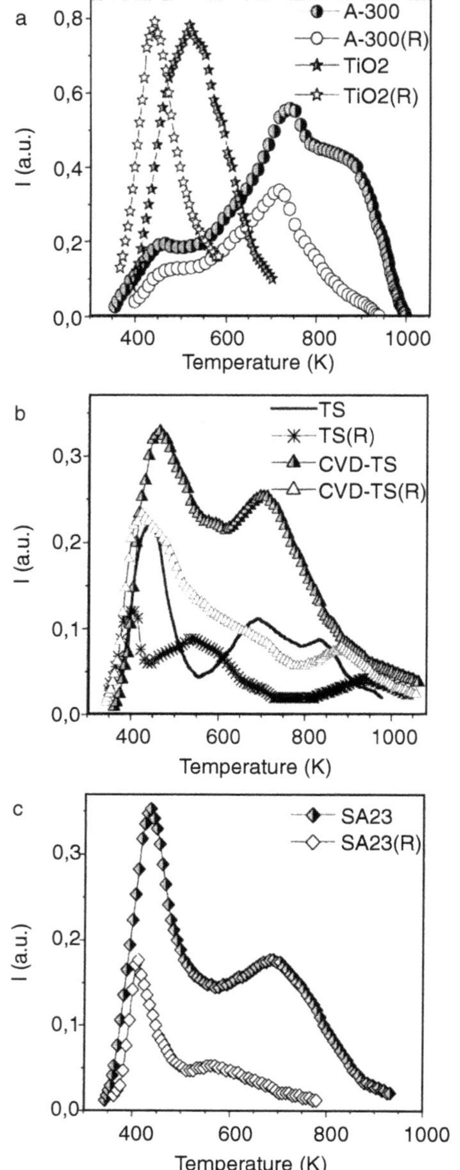

FIGURE 37.1 OPTPD spectra of water desorbed from (a) fumed silica A-300 and fumed titania; (b) fumed silica/titania and CVD-TiO_2/fumed SiO_2; and (c) fumed silica/alumina; re-hydration (R) time is 0.5 h.

according to the literature [23], but obtained $E^{\neq}$ values are lower than experimental heats of dissociative adsorption of water ($Q = 100-280$ kJ/mol) on different oxides, therefore, we recalculated $E^{\neq}$ and k_o values for bimolecular processes (Table 37.2, the second values for $x = 2$ and $l = 2-4$). An increase of $E^{\neq}$ leads to a growth of the k_o values, which corresponds better to the defined range for k_o. According to the literature [30,31], dense islands of several (3–10) OH groups can be formed on the silica surface. In such islands, the distances between some OH groups are relatively short and this can provide relatively easy dehydroxylation so that the islands disappear after evacuation at T about 770 K [31].

Therefore, we can assume that the second OPTPD peak (Figure 37.1a) corresponds to water desorbed from such islands, which contain not only vicinal OH groups but also twin $>Si(OH)_2$ groups. Such islands can be formed near contacts between primary particles in their aggregates where adsorption of intact water molecules is preferred [14].

The third peak (Figure 37.1a) can be linked to water desorbed due to the interaction of vicinal or twin OH groups but not from the dense clusters of OH groups. Dissociative adsorption of water occurs easily for those sites, at which each metal (Si, Al, Ti) atom already bonds one OH group or more [14,25,27]; that is, the products of such adsorption correspond to the structures as $M_1(OH)_2$-O(H)-M_2(OH) or $M_1(OH)_2$-O-$M_2(OH)_2$. Therefore, the $E^{\neq}$ value for the reaction

$$\begin{aligned}&\text{—Si(OH)}_2\text{—O—Si(OH)}_2\text{—} \longrightarrow \\ &\text{—(HO)Si}\langle_{\text{O}}{}^{\text{O}}\rangle\text{Si(OH)}(\leftarrow \text{OH}_2)\text{—} \longrightarrow \\ &\text{—(HO)Si}\langle_{\text{O}}{}^{\text{O}}\rangle\text{Si(OH)—} + \text{H}_2\text{O}\end{aligned} \tag{37.15}$$

is lower than that for $>Si(OH)\text{-}O\text{-}Si(OH)<\rightarrow> Si\langle_O{}^O\rangle Si < +H_2O$ as bond strain is lower for equation (37.15) due to the lesser number of the lattice Si-O bonds in the active sites.

The fourth OPTPD peak can be caused by the reaction between isolated OH groups spaced at a larger distance, when reaction is limited by $H^{\bullet}$ or $OH^{\bullet}$ diffusion from one OH group to another or by strong deformation of the surface, for example, near contact between the primary particles in aggregates and agglomerates at small Θ_{OH}.

The OPTPD spectrum of water desorbed from silica after rehydration of the heated sample displays strong changes relative to initial spectrum (Figure 37.1a). The maximal intensity decreases by half, and the last peak is reduced by 80 percent, but the relative contribution of molecularly adsorbed water increases. The third and fourth peaks shift toward lower temperatures, and the $E^{\neq}$ values decrease (Table 37.2). These effects are caused by close localization of hydroxyl groups formed on dissociative adsorption of water during rehydration; that is, the amount of isolated OH groups spaced at a large distance is smaller than for the initial oxide, as such isolated groups make the main contribution to the highest-temperature peak. The intensity of the second peak decreases slightly after rehydration, which suggests our assumption that the second peak is due to water desorbed from the dense islands of OH groups with the participation in the reaction of twin groups. Additionally, after heating and rehydration, the $E^{\neq}$ and T_{max} values for the first and the second peaks do not decrease

(Table 37.2). This is due to smaller amounts of readsorbed water giving a higher enthalpy of adsorption than for original samples and to the occurrence of the associative reactions between OH groups in the dense islands (for the second peak). The third peak can be connected with the condensation of OH groups linked by hydrogen bonds (vicinal hydroxyls) or single but closely placed OH groups (not involved in dense clusters of OH groups). In this case the contribution of the deformation energy to $E^{\neq}$ increases and k_o decreases as the number of active neighboring sites (i.e., possible pathways of associative reactions) reduces. The highest-temperature peak strongly decreases after rehydration, as the number of isolated groups is small (Table 37.2, N_{H_2O}) due to the relatively short time (0.5 h) of rehydration. It should be noted that for low Θ_{OH} (i.e., for high-temperature peaks) the k and $E^{\neq}$ can strongly depend on diffusion of $H^{\bullet}$ that, according to Equations (37.13) and (37.14), leads to a reduction of k_o, but increasing reaction temperature can compensate this effect (Table 37.2).

The number of hydroxyl groups and adsorbed water molecules per nm^2 for fumed titania is higher than that for silica (Table 37.2) but the OPTPD spectrum shape is more simple for titania (Figure 37.1a). Water desorption from titania occurs under milder conditions than for silica due to a lower value of $E^{\neq}$ for associative desorption of water from neighboring terminal (TiOH) and bridging (TiO(H)Ti) groups than that for two terminal SiOH groups (the energy in the corresponding process $Ti^{VI} \rightarrow Ti^{V}$ is lower than for $Si^{IV} \rightarrow Si^{III}$) [27]. Additionally, α_{OH} (or N_{Σ}) for titania is higher than that for silica (Table 37.2), that is, the number of dense islands of OH groups can be higher for titania and this leads to lowering $E^{\neq}$ and the desorption temperature.

After heating and rehydration the high-temperature two peaks disappear (Figure 37.1a). This effect can be explained from the absence of isolated ≡TiOH or ≡TiO(H)Ti≡ groups; that is, all dissociatively adsorbed water molecules form only adjacent ≡TiOH and ≡TiO(H)Ti≡ groups (dense clusters) as H or OH can diffuse relatively slow at 300 K (due to a high energy of the bond cleavage of ≡M-OH or ≡MO-H) [14,27,28]. Water desorption with the participation of such pairs makes a contribution only to the second OPTPD peak which is lower after rehydration. Consequently, the third low peak for original titania is caused by water desorption with the participation of residual hydroxyl groups (spaced at a larger distance) in a bimolecular process. The first peak shifts to lower temperatures and grows, that is, the number of water molecules, which do not dissociate on the surface, is relatively high.

The amount of water desorbed from ST in OPTPD measurements is lower than that for silica or titania (Table 37.2), but under standard conditions ST contains a greater amount of water than silica or titania (Table 37.1). A major portion of molecularly adsorbed water on ST can locate at the interfaces of TiO_2-SiO_2 in the form of large clusters [11,14,25,27,28]. Because of this and the fast desorption of physisorbed water upon pumping to high vacuum, the amount of residual water on ST after pumping is lower than that on silica or titania as the heterogeneity of the individual oxides is lower and water molecules distribution on them is more uniform than on ST. This leads to stronger interactions between these molecules and the surfaces of individual oxides, as molecules near the surface have a lower free energy [28,32–36]. These suppositions are supported by a maximum intensity of the first peak (Figure 37.1). In addition for all peaks, T_{max} is lower for original ST_{22} than that for original silica (for ST, two high-temperature peaks lie lower than that for silica, but their shape is close to those for silica) or titania. A high-temperature shoulder above 900 K (Figure 37.1b) can be due not only to isolated ≡SiOH groups but also to bulk hydroxyl groups at the phase boundary TiO_2-SiO_2, as these phases strongly differ in the crystalline structures determining the appearance of the strained bonds relaxed via hydrolysis upon dissociative adsorption of water. However, for pure silica, this region of the TPD spectrum is linked only with the interaction between isolated OH groups.

After heating and rehydration of ST_{22}, the OPTPD spectrum changes greatly especially in the region of the peaks 2 and 3 (Figure 37.1b), and a decrease in the $E^{\neq}$ and T_{max} values is observed for all except the fourth peak (Table 37.2). However, the high-temperature shoulder is more intensive for ST_{22}(R) than for ST_{22}. These changes in the OPTPD spectrum can be caused by the alterations in the phase boundary TiO_2-SiO_2 induced by heating-cooling and dehydration–hydration cycles. In addition, according to the intensity of the corresponding peaks, the rehydration of the silica phase is lower (desorption at 700–900 K) than that for the interfaces and the titania phase (desorption at 400–600 K). The interfaces of ST in the bulk can influence these processes as well as in the case of ST(CVD). However, heating and rehydration do not give such dramatic changes in the OPTPD spectrum in the region corresponding to desorption from the titania phase or ST(CVD) phase boundary at the surface. At the same time rehydration of the silica phase is not observed as the peaks corresponding to water desorbed from silica are absent, that is, a part of CVD-titania is deposed on the silica particles as small clusters (other part of titania presents as large individual titania particles [27,28]) which are the centers of water adsorption and desorption from ST(CVD). However, for ST(CVD), molecularly adsorbed water has a desorption peak close to that for fumed ST (Figure 37.1b, the first peak) but water desorption from the last oxide occurs at lower temperatures.

In the case of fumed SA molecularly adsorbed water desorbs at higher temperatures than from ST

(Figure 37.1c; Table 37.2) due to higher acidic properties of SA leading to stronger interaction of water molecules with the surface. Additionally, the similarity between alumina and silica units is higher than that for titania and silica (i.e., Al isomorphously substitutes Si in the silica lattice more easily than Ti), therefore, a separated alumina phase can be formed in fumed SA only for $C_{Al_2O_3}$ ≈20 wt.% (for fumed ST the corresponding C_{TiO_2} ≈9 wt.% and ≈3 wt.% for ST[CVD]) and the distribution of alumina in the silica matrix is more homogeneous than that for titania in ST [27,28,32–35]. It should be noted that the changes in the free energy of adsorbed water are greater for Al_2O_3 and Al_2O_3/SiO_2 than for silica [28,32–36]. Because of these facts, the SA surface interacts more strongly with water molecules, and the OPTPD spectra have only two separated maxima (Figure 37.1c). The total amount of water adsorbed on SA after rehydration is higher than that on ST (Table 37.2, N_Σ) as well as the intensity of the first peak. However, the spectrum shapes for SA and ST are similar at T above 550 K.

Thus, the OPTPD-TOFMS technique allowed the direct measurement of the amounts of associatively desorbed water per each type of the sites and per nm^2 and enabled to elucidate the details of dehydration and dehydroxylation of fumed silica, titania, ST, SA, and ST(CVD). Quantitative results were obtained regarding water desorption through various mechanisms with the participation of different hydroxyl groups contributing to different TPD peaks. The maximum activation energy of water desorption corresponds to the silica phase in ST and SA. The initial amounts of water adsorbed on ST or SA are higher than those for individual oxides, but the main body of adsorbed water is localized at the interfaces in the large clusters of water molecules, which can easily be desorbed by pumping to high vacuum or heating to approximately 400 K. Therefore, only a minor portion of initial amounts of adsorbed water (observed in TPD-DTG measurements) remains on the oxide surfaces in OPTPD measurements at the pressure of 10^{-9} Torr. Notice that results obtained using the OPTPD-TOFMS method are in agreement (at least qualitatively) with the data obtained by more simple TPD MS methods [11,23]; however, so detailed quantitative estimations of different forms of adsorbed water on fumed and CVD-oxides were obtained for the first time [25].

It should be noted that there are several points of view on the structure of adsorption states of water on the silica surfaces based on following initial assumptions: (i) intact water molecules form only hydrogen bonds with silanols [2–4,23]; (ii) the strongest adsorption complexes of water form due to donor-acceptor bonds such as ≡Si(OH) ← OH_2 or ≡Si(OH) ← $2OH_2$ and ≡SiOH...OH_2 or ≡SiO(H)...HOH bonds are absent [11]; and (iii) formation of water clusters and droplets (up to appearance of capillary condensation of water in "mesopores" as gaps between neighboring primary particles in their aggregates) on fumed silica occurs in the zone of contacts of primary particles (with possible different hydrogen bonds such as ≡SiOH...OH_2, ≡SiO(H)...HOH, ≡SiOH...OHH...O(Si≡)$_2$, ≡SiO(H)...HOH...O(Si≡)$_2$, etc.) and a low amount of water molcules can form complexes in the form of practically dissociated molecules such as (Si(OH)(←OH^-)-O(H^+)—Si≡ [14]. These assumptions are based on experimental and theoretical investigations; however, direct observations and identifying evidence of only given complexes are absent. However, there are indirect evidences of the availability of different complexes with participation of water molecules (or other compounds, for example, HF). For instance, the surface conductivity of strongly hydrated silica is higher than that for bulk water by several orders; that is, the proton mobility (as this conductivity is not an electron one and it is provided mainly by mobile protons) at the silica surfaces is high at the presence of adsorbed water. Secondly, a portion of active surface sites observed on hydrated silica can be stronger than silanol groups or molecularly adsorbed water. Thirdly, free silanol groups (IR band at 3750 cm^{-1}) are observed independently on the concentration of adsorbed water even in the amounts larger than that of monolayer coverage by order or more; additionally, according to the data obtained by dielectric relaxation spectroscopy, adsorbed water does not form a continuous layer even at the concentration of 30 wt.%. Besides, even low amounts of adsorbed water affect the chemical reactions at the silica surfaces [1–14]. The third model of adsorbed water in the form of clusters and droplets with the participation of intimate ion pairs in =Si(OH)(←OH^-)—O(H^+)–Si≡ can explain many observed phenomena related to interfacial water on fumed silica surfaces not only with respect to adsorbed water per se but also to such effects as reduction of reactivity of the hydrogen-bonded silanols (as they are localized in narrow places between adjacent primary particles) and others [14].

From the practical point of view with respect to the silica modification to obtain different functionalities, the electrophilic substitution reactions are very important; additionally, they are interesting from the theoretical point of view due to a variety of possible reactants affecting the TS structure and third components (catalysts) promoting H^+ transferring [1,2,11].

S_Ei REACTIONS

Reactions of organosilicon compounds (OSC) such as $SiHal_nR_{4-n}$ (Hal = Cl, Br, or I; R = H, $(CH_2)_nCH_3$, $(CH_2)_nNH_2$, etc.) with the silica surfaces occurring through the S_Ei mechanism [Equation (37.2)] fill a large place in practical modification of silicas [1–11].

Sophisticated analysis of different factors affecting the reactivity of OSCs and surface silanols [10–14,16–18,20,22,37,38] (based on the experimental data obtained by different authors used a variety of contradictory models, ideas, and conceptions to explain these data) reveals the tendencies between changes in quantum chemical indexes of organic molecules (their maximum charges and energies of the frontier orbitals) and their empirical donor–acceptor parameters as well as heats of the hydrogen-bonding complex formation with OH groups on silica surface and the chemisorption activation energies on this surface. The electron donor ability of surface sites and electron acceptor ability of gas reactant play a leading part in the rate of electrophilic substitution of H^+ in the surface OH groups, whereas the opposite properties of these reactants affect the stability of hydrogen-bonded precursors in these reactions. The overall activation reaction barrier reduces with increasing preliminary adsorption complex stability and overall reaction rate is growing. The reaction selectivity depends on the mechanism of surface reaction. As follows from the quantum chemical data [14,37,38], the OH groups on the oxide surfaces are characterized by different heats of bond dissociation and charges of the OH group atoms, that is, by chemical heterogeneity. In this case the adsorption equilibria and chemisorption kinetics can be described by means of distribution functions on the quantum chemical indexes, or on empirical acid–base parameters of the active sites, or the deformation model of the activation barriers with consideration of the impact of the surface electrostatic fields [11]. The linear dependence between hexamethylsiloxane chemisorption activation energies and average donor component of organic molecules adsorption energy on the individual and mixed Si, Ti, and Al oxides has been obtained. Features of the chemisorption kinetics of organic compounds onto silica and some other metal oxides were also analyzed in detail for several mechanisms (S_Ei, S_Ni, $Ad_{N,E}$, etc.) with consideration for the nonuniformity and the fractality of the surfaces [11,16].

It should be noted that the electronic states of atoms (charges, local density of the electron states) significantly change on both thermal vibrations and motion along the reaction pathway shown in Figure 37.2 for interaction between the bridging ≡AlO(H)Si≡ group and CH_3OH or H_2O. This circumstance provides adjustment of the parameters of interacting reactants to reduce the activation energy of reactions with both charge and orbital control, since the interaction energy is composed from several components (electrostatic, polarization, charge transfer, electron exchange, etc.) differently dependent on atomic charges, intermolecular bond length, and localization and energy of molecular orbitals (MOs). These contributions to the energy of different hydrogen bonds (Table 37.3) are estimated by the Kitaura–Morokuma method.

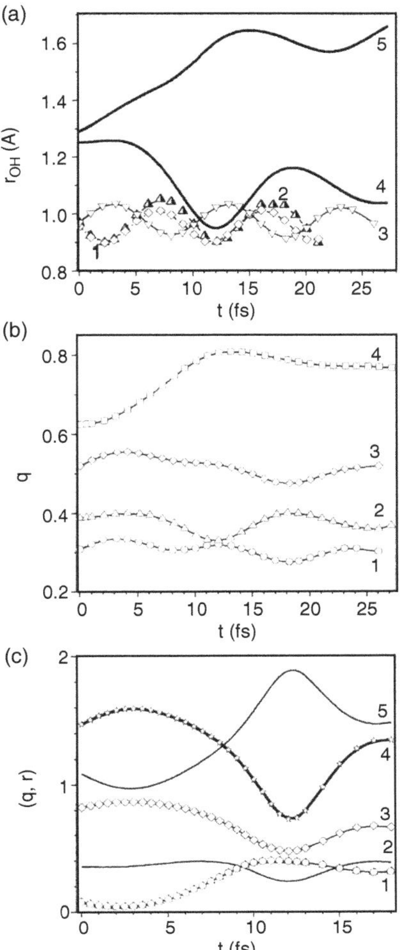

FIGURE 37.2 (a) Changes in the bond O-H length as a function of time for: (1) ≡AlO-H, (2) (Al(O-H)Si≡, (3) (≡Si,Al≡) OH...O(H)CH_3, TS of CH_3OH reaction with ≡AlO(H)Si≡ at O-H from (4) (≡Si,Al≡)OH...O(H)CH_3 and (5) ≡AlO(H)Si≡; (b) changes in atomic charges of (1, 2) H and O (3, 4) from ≡AlO(H)Si≡ on formation of (1, 3) the hydrogen bond and (2, 4) in the S_Ei reaction; and (c) changes in charges (1) $-q_{O,H_2O}$ (2) q_H, (3) $-q_{O,AlOSi}$ and the bond length (4) $r_{(Al,Si)O...H}$ and (5) $r_{H...O,H_2O}$ on H^+ transferring from H_3O^+ to the O atom in ≡Si(O^-)Al≡.

The n value in $SiHal_nR_{4-n}$ affects the reaction energy (Table 37.4, ΔE_t) and the activation energy $E^{\neq}$ of OSC reaction with ≡SiO-H. The heat effect ($Q \approx -\Delta E_t$) increases but the activation energy decreases with increasing n (as well as with the atomic number of Hal).[11,37,38]

A detailed survey of the reaction between hexamethyldisilazane (HMDS) and silanol groups to form trimethyl silyl groups

$$2{\equiv}SiOH + (CH_3)_3\text{-}Si\text{-}NH\text{-}Si\text{-}(CH_3)_3 \longrightarrow 2{\equiv}Si\text{-}O\text{-}Si\text{-}(CH_3)_3 + NH_3 \quad (37.16)$$

TABLE 37.3
Different Contributions to Hydrogen Bonding Energy According to Analysis by Kitaura–Morokuma Method

Complex	$-\Delta E_t$	$E_{electrostatic}$	E_{exchan}	$E_{polarization}$	$E_{charge\ transfer}$	q_H
[a] $H_2O...HOH$	23.4	−31.0	17.2	−2.3	−7.1	0.277
[a] $CH_3OH...OH_2$	23.8	−30.9	17.7	−2.6	−7.3	0.268
[a] $H_3N...HOH$	26.8	−39.1	23.9	−3.8	−7.7	0.282
[a] $\equiv SiOH...OH_2$	29.9	−39.4	21.9	−3.8	−8.4	0.333
[b] $\equiv SiOH...OH_2$	29.2	−39.0	21.9	−3.3	−8.4	0.395
[b] $\equiv SiOH...NH_3$	36.5	−54.8	33.9	−5.3	−10.1	0.419
[a] $\equiv SiO(H)...HOCH_3$	17.9	−26.1	16.3	−2.8	−5.3	0.283
[a] $\equiv SiO(H)...HF$	36.7	−53.5	33.3	−7.2	−10.0	0.375
[b] $SiOH...O(H)R_{Gly}$	19.3	−26.7	19.2	−2.4	−8.9	0.381
[b] $\equiv SiOH...NH_2R_{Gly}$	32.4	−49.4	34.8	−5.3	−12.4	0.413

Note: [a] 6-311G(d,p) and [b] 6-31G(d,p). Gly is glycine; $\equiv SiOH$ is $Si(OH)_4$; energy in kJ/mol.

was performed based on experimental and theoretical data and analyzing the impact of the reaction media [9,39–41]. It was shown that the activation energy of reaction (37.16) can be well estimated using ab initio quantum chemical method at the MP2/6-31G(d,p) level (notice that B3LYP/6-31G(d,p) gives slightly worse results) to search the TS. Ignorance of the electron-correlation effects, for example, using the 6-31G(d,p) basis set, leads to a significant overestimation (by a factor of 2.5) of the $E^{\neq}$ value. Amino silane R_3SiNH_2 formed on reaction of HMDS with $\equiv SiOH$ interacts with silanol at substantially lower activation energy (by 20 kJ/mol); that is, limitation of reaction (37.16) is linked to the first stage, namely reaction between HMDS and silanol. Analysis of components of the interaction energy in the prereaction complex (hydrogen bonding) and the TS reveals that the main contribution is caused by the electron exchange effect (Table 37.5), which is in agreement with a strong influence of the electron-correlation and exchange effects on the $E^{\neq}$ value.

Analysis of the liquid media influence on reaction (37.16) based on the experimental and theoretical studies shows that polar aprotic solvents (acetone, acetonitrile, etc.) can inhibit this process and the effective $E^{\neq}$ value increases by 20–30 kJ/mol (Table 37.6) [39].

Investigations of the S_Ei reactions of such amino silanes as $[(C_2H_5)_2N]_nSi(CH_3)_{4-n}$ with fumed silica showed that their reactivity decreases with increasing n value and an optimal reaction temperature increases from 20°C ($n = 1$), 50 (2), 120 (3), to 150°C ($n = 4$) [17].

Reactions of different cyclosilazanes (CSA) $[-(CH_2)_2SiNH-]_n$ with silica allowed Chuiko and coworkers

TABLE 37.4
Energy (ΔE_t) of Reactions $\equiv SiO\text{-}H + SiCl_nR_1'R_m'' \rightarrow \equiv SiOSiCl_{n-1}R_1'R_m'' + HCl$ ($R' = H$, $R'' = CH_3$, $n + 1 + m = 4$) Calculated in a Cluster Approach by Ab Initio Method (6-31G(d,p))

OSC	ΔE_t (kJ/mol)
$SiCl_4$	−31
$SiCl_3H$	−27
$SiCl_2H_2$	−16
$SiClH_3$	−4
$SiCl_3(CH_3)$	−26
$SiCl_2(CH_3)_2$	−18
$SiCl(CH_3)_3$	−7
$SiCl_2H(CH_3)$	−18
$SiClH_2(CH_3)$	−5
$SiClH(CH_3)_2$	−5

TABLE 37.5
The Energy Interaction Components (kJ/mol) for Hydrogen Bond $\equiv SiOH...NH_2SiH_3$ and the TS of H_3SiNH_2 Reaction with $\equiv SiOH$ by Kitaura–Morokuma Method (6-31G(d, p))

Energy components	H bond	TS
Electrostatic	−41	−233
Exchange	28	407
Polarization	−5	−110
Charge transfer	−9	−248
High order coupling	−0.3	71
ΔE	−27	−112
ΔE_{BSSE}	−22	−100
Deformation		13 (amine) 198 (silanol)

TABLE 37.6
Changes in $E^{\#}$ (kJ/mol) ($\Delta E^{\#}$ ($E^{\#}_{sol} - E^{\#}_{gas}$) of Silylation Due to Solvation (6-31G(d,p)) for Reaction of HMDS or H_3SiNH_2 with ≡SiOH

Solvent	HMDS[a]	H_3SiNH_2[b]
Cyclohexane	22	−3
Benzene	21	−6
Toluene	22	−5
Acetone	24	
Acetonitrile	33	−10
Water	−16	

Note: The geometry was calculated using [a]MP2/6-31G(d,p) and [b]6-31G(d,p).

to obtain the surface functionalities including reactive Si-N bonds in arched structures. For example, the reaction of CSA at $n = 3$ with fumed silica at 150°C leads to full substitution of surface OH groups through the S_Ei mechanism, and an intensive IR band at 3375 cm^{-1} depicts the availability of N-H bonds at the surface [17].

As it was mentioned above, electron-donor molecules (e.g., amines such as triethylamine, TEA) can strongly promote the S_Ei reactions. A similar effect can be also caused by adsorbed water, as elevating temperature of fumed silica dehydration reduces the loading of OSC (Figure 37.3a) as well as the depth of the reaction (Figure 37.3b and 37.3c) [11,12]. Similar effects were observed for many reactions between OSC or organics with fumed oxides [10–13].

OSCs $(CH_3)_3Si$-X (X = N_3, NCO, NCS, $NCNSi(CH_3)_3$) demonstrate a high reactivity in reactions with silica even at room temperature; however, these reactions were assigned by Tertykh as the $Ad_{N,E}$ cleavage of the siloxane bonds [18]. Elevating of the reaction temperature leads to change the reaction mechanism with increasing contribution of the S_Ei process with the participation of ≡SiOH groups [11,18].

The S_Ei reactions of chlorides PCl_3, PCl_5, WCl_5, $MoCl_5$, $TiCl_4$, etc. or oxichlorides $SOCl_2$, $POCl_3$, $MoOCl_4$, CrO_2Cl_2, as so on were used to modify the fumed silica surfaces [11,18,20,42–49]. Some of these reactions were characterized as the elimination (E) processes with formation of several intermediates including reactive Si-Cl bonds, which can be used in subsequent modification of the silica surfaces, for example

$$\equiv SiO\text{-}H + WCl_5 \rightarrow \equiv SiOWCl_4 + HCl \quad (S_Ei) \quad (37.17)$$

$$\equiv SiOWCl_4 \rightarrow \equiv SiCl + WOCl_3 \quad (E) \quad (37.18)$$

where reaction (37.17) corresponds to S_Ei, and reaction (37.18) is the elimination reaction with intra-functionality regrouping akin to S_Ni. Applications of models of irreversible and equilibrium polycondensations on CVD of chromium oxide allowed one to analyze the size distribution of the Cr_2O_3 clusters depending on the used technique [20]. Notice that Plyuto and coworkers published dozens of papers related to the investigations of chemical reactions between different metal chlorides and silica surfaces with formation of new phases of different morphology [20].

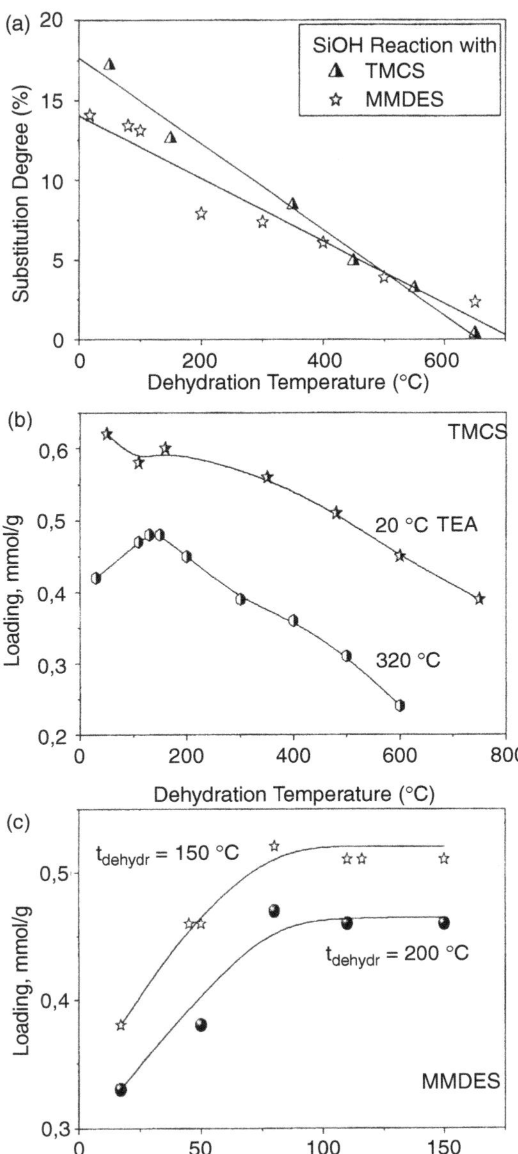

FIGURE 37.3 Dehydration temperature influence on reactions of trimethylchlorosilane (TMCS) and methacryloyloxymethylenemethyl diethoxysilane (MMDES) with silica at (a) room temperature; and (b) TMCS at 320°C and 20°C (with the presence of TEA); and (c) the reaction temperature influence on MMDES loading at different dehydration temperatures.

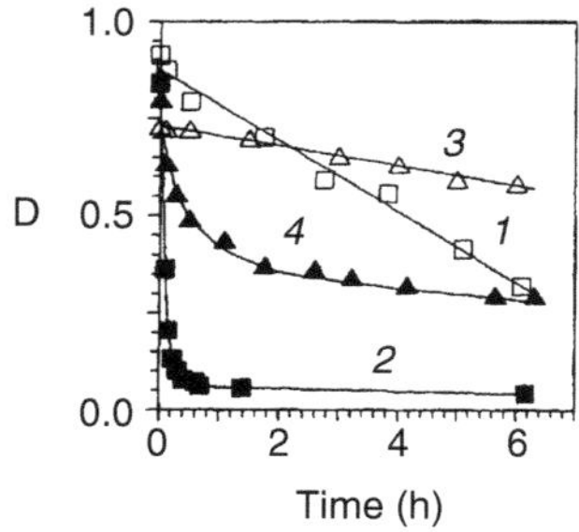

FIGURE 37.4 Kinetics of reduction of the optical density ($D = lg(I_0/I)$) of the IR band of free silanols at 3750 cm^{-1} due to reactions of PCl_3 (pressure 0.57 p/p_0) with fumed silica (1) degassed at 200°C for 1 h and (2) containing 6.9 wt.% of water; and reactions of $POCl_3$ (pressure 0.6 p/p_0) with silica (3) degassed at 200°C for 1 h and (4) containing 8.1 wt.% of water.

It should be noted that adsorbed water strongly affects of the reactions between PCl_3 or $POCl_3$ and fumed silica enhances their loading (Figure 37.4) [11,22], as well as in the case of other compounds reacting through S_Ei, S_Ni, and other mechanisms.

Analysis of the influence of solvents on the PCl_5 reactions occurring similar to reactions (37.17) and (37.18) reveals that the reaction rate increases in series $CH_3CN > C_6H_6 > CCl_4$ [20]. It was also noted that in the case of the PCl_5 reaction with silica from the gas phase, such compounds as CH_3CN, C_6H_6, CCl_4, CCl_3H, and C_6H_{14} promote the reaction, which does not occur without them under the same conditions [20]. Chemisorption of chlorides and oxichlorides of metals, subsequent hydrolysis and heating can result in formation of surface clusters with metal oxides, which can be used as catalysts for some organic reactions or to change important properties of disperse oxides used as fillers, additives, pigments, and so on [11].

The size of silica clusters used in ab initio quantum chemical calculations of the activation energy plays an important role, for example, the use of a cluster with four tetrahedron SiO_4 on interaction of $SiCl_4$ with the SiOH group through the S_Ei mechanism gives $E^{\neq} = 134$ kJ/mol at the 6-31G(d,p) basis set (experimental estimation gives 80–90 kJ/mol), that is, it is lower than that for the HMDS reaction (161 kJ/mol, experimental estimation gives 77 kJ/mol) at the same basis set but for a cluster with $Si(OH)_4$. However, the reactivity of HMDS is higher than that of $SiCl_4$ (as experimental $E^{\neq}$ is lower by 10–15 kJ/mol) [39]. More appropriate the $E^{\neq}$ values could be obtained with consideration for electron correlation and exchange effects using the DFT or MP perturbation theory [14,39].

According to Chuiko [11], Tertykh [10,18], Belyakova [19], Voronin [12,13], Bogillo [16], as a whole the reactivity of Si-X groups in reactions with ≡SiOH groups at the fumed silica surfaces under the S_Ei mechanism decreases at different X in following series: I > Br > NR > Cl > OR > OSi > H > CR [11,42]. According to Gun'ko [42], theoretical estimations give changes in the $E^{\neq}$ values for this series by approximately order. Main factors determining the reactivity of these Si-X bonds in the S_Ei reactions are their polarity and polarizability and electron-donor properties of α atom in X [11,42]. Electron-donor compounds (e.g., amines TEA, TMA, etc.) can significantly reduce the activation energy of the S_Ei substitution of H^+ in SiOH due to changes in H^+ transferring, which is the limitation stage (giving 60–85% of the activation energy) of the reaction [11,42–44].

The search for correlations between the electronic structure and the activation energy of the OSC reactions with silica [4,10,11,37,38] shows that there is no linear correlation for a wide range of such compounds (Figure 37.5, the dark points, correlation coefficient is less than 0.2), but for narrow series (e.g., $(CH_3)_{4-n}SiCl_n$ at $n = 1–4$) correlations are observed with the orbital (the energies of the frontier MOs) and charge characteristics (Figure 37.5, correlation coefficient is greater than 0.96).

On the basis of these correlations it can be supposed that the reactions of $(CH_3)_{4-n}SiCl_n$ with silica may be under mixed orbital-charge control. It should be noted that for this narrow series of OSCs practically all parameters of their electronic and spatial structure (as well as the activation energies estimated by using the deformation model of the activation barrier [11]) correlate to the changes in the activation energy, since changes in the number of Cl atoms give very concordant changes in other parameters of the molecules. Therefore, it should be distinct parameters which really affecting the OSC reactivity and parameters which correlate with changes in the activation energy of given reactions but don't affect the activation energy.

The appreciable increase of the rate of the OSC reactions with silica in the presence of bases [10,11,37,38, 42–44] which mostly affect the proton transfers, underlines the importance of the contribution of these processes to the activation energy, which follows from a quantum-mechanical dynamic examination of the reactions [50,51]. Analysis of the trajectory of the movement of H^+ from the silanol group (S_Ei mechanism) to the Cl atom (the noncatalytic reaction) or to N in NR_3 and then to Cl (the catalytic reaction) during the reaction of R_3SiCl (R = H, CH_3) with silica in the presence of NR_3 (R = CH_2CH_3, CH_3, H) in relation to the noncatalytic transformation indicates a significant change in the trajectory right up to the formation of R_3NH^+. However, the use of the usual gradient method of search for the transition state without taking account of the dynamics of the movement of the atoms with various masses leads to high values of $E^{\neq}$ as a result of involvement in movement of the polar molecule of the catalyst along the generalized coordinate immediately after H^+ [42,44,50,51]. This movement is

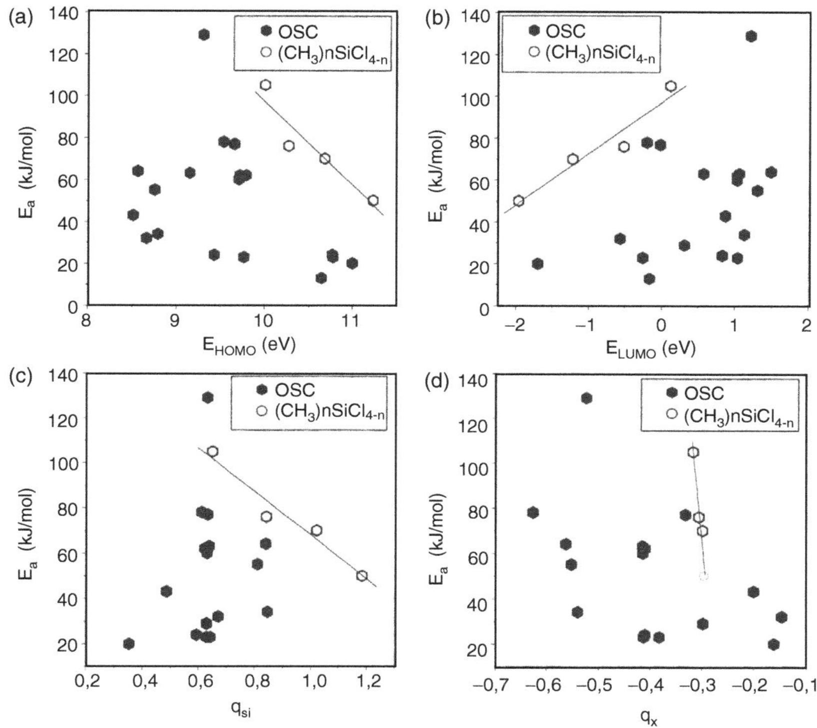

FIGURE 37.5 Relationships between the activation energy (E_a) of OSC reactions with silica and (a) energy of the highest occupied MO (E_{HOMO}) of OSCs; (b) energy of the lowest unoccupied MO (E_{LUMO}) of OSCs; (c) atomic charge of Si in OSCs; and (d) charge of α atom in X group bound to Si in OSCs [37,38].

hindered by the nonuniform electrostatic field of the surface of the oxide [11], which gives rise to an increase in the calculated value of $E^{\neq}$. Only in terms of the method of a dynamic reaction coordinate (DRC) or with the use of mass-weighted changes of the coordinates during the search for the transition state (with the inclusion of electron correlation) is it possible to obtain a decrease of $E^{\neq}$ in the presence of NR_3. The movement of the system along the generalized coordinate is calculated on the basis of the potential energy gradient and the kinetic energy of the atoms. The potential energy is determined by a quantum-chemical method. By varying the rate of dissipation of the kinetic energy it is possible to model the effects due to scattering of the energy of the activated functional group at the phonons of the solid or in reaction with molecules from the gas phase [50,51]. Notice that according to a general rule, enhancement of the exothermicity of reactions corresponds to a greater similarity in the geometrical structures of the prereaction complexes and the TS at the fumed silica surfaces, as well as enhancement of the endothermicity of surface reactions leads to increase in the difference in the geometry of the prereaction states and the TS [11,42]. Nevertheless a maximal contribution (up to 85%) to the activation energy of these processes is caused by H^+, H^-, or $H^\bullet$ transferring, that is, by elongation of X-H bonds [42], but this contribution decreases (to 50–65%) for endothermic reactions, as contribution of deformation of other bonds increases. The later was investigated for decomposition (E_2 reactions) of surface functionalities [25,29,42].

In the processes with proton transfer, the tunneling effects can probably play an important role. Thus, the critical temperature for the tunneling of H^+ at $E^{\neq}_{H+} =$ 100 kJ/mol and with transfer of H^+ along the hydrogen bond amounts to 500–520 K. Therefore, the theoretical modeling of the surface reactions, particularly the catalytic reactions taking place at room temperature or reactions involving charged or defect structures $\equiv SiO^-$, $=Si=O$, $=SiOOSi=$, and others taking place at lower temperatures right down to 77 K, during transfers of H^+, H^-, or $H^\bullet$ can give a high activation energy (compared with the experiment), unless the contribution from tunnel processes is taken into account [50,51].

Reactions of MR_3 or $M(OR)_3$ (M = B, Al, Zn, etc.) with the silica surfaces are interesting from the theoretical and practical points of view [11]. Investigations performed by Brey and coworkers [21] revealed that some of these reactions can occur at room temperature through the S_Ei mechanism since the surface functionalities always contain Si-O-MR groups in dependent on the kind of reactants (MR_3 or $M(OR)_3$). Additionally, such atoms as Al can easily increase their coordination number from three to four, therefore the donor–acceptor bond ≡SiO(H) → MR_3 or ≡SiO(H) → $M(OR)_3$ can be easier formed than similar bonds with the participation of the Si atoms. Reactivity of $B(CH_3)_3$ depended on the temperature nonlinearly (maximal at 160°C) [21] which was considered in the terms of the efficiency of the reactant flow to the reactive centers under the Langmuir-Hinshelwood or Eley-Rideal mechanisms [21]. However, there is another explanation of these effects depending on the lifetime of the prereaction complexes decreasing with elevating temperature, since formation of the prereaction complexes is requirement for reaction occurring according to the Langmuir-Hinshelwood mechanism and the Eley-Rideal mechanism is less effective than the first mechanism for similar reactions [42]. The surface functionalities after modification of the fumed silica surfaces by n-C_4H_9-$B(OCH_3)_2$ and $B(OCH_3)_3$ were analyzed using the NMR and TPD MS methods and it was found that their structures (as well as their moisture stability) strongly depends on the electron-acceptor properties of the B atoms (S_Ei mechanism) changing due to the hydrophobic group with n-C_4H_9 [21]. These boron-silicas modified by propylene oxide (through the Ad_E mechanism at room temperature) demonstrated enhanced stability against hydrolysis even at 300°C [21].

The influence of molecule diffusion on their interaction (e.g., in reactions occurring through the Langmuir-Hinshelwood mechanism) can be clearly shown using quenching of pyrene by DMA at the fumed silica surfaces. This effect depends not only on the observation temperature but also on the temperature of silica treatment (Figure 37.6) [11].

The interaction of alcohols, phenols, and organic acids with the surface of silica [1–11,44,52–54] usually takes place by the $S_Ni(Si)$ mechanism, since the TS with the donor–acceptor bond ≡Si(OH) ← O(H)R [i.e., $S_Ni(Si)$] is lower (by 50–120 kJ/mol or even more) on the potential energy hypersurface than the transition state

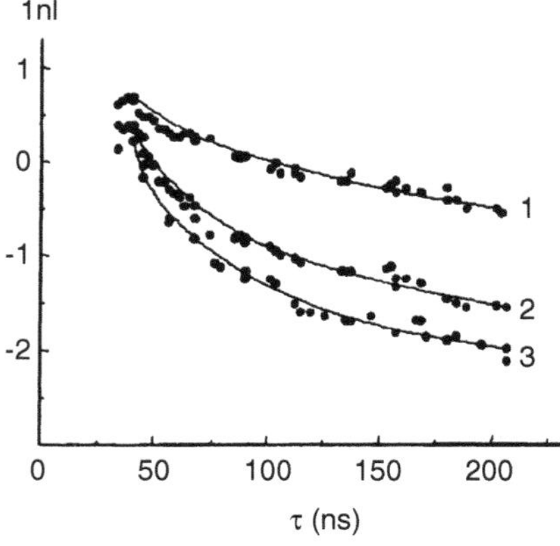

FIGURE 37.6 Kinetics of fluorescence decay of pyrene ($C = 5$ μmol/g) adsorbed on fumed silica A-300 with the presence of a fluorescence killer DMA (N,N-dimethylaniline) at C_{DMA} = (1) 2 μmol/g; (2) 20 μmol/g; and (3) 200 μmol/g.

with the ≡SiO(H) → CO(H)R bond (S_Ei mechanism), since the electron-acceptor characteristics of the Si atom (in ≡SiOH) are higher than for the C atom in CO(H)R. The S_Ei mechanism on interaction of methanol with ≡SiOH and the $Ad_{N,E}$ mechanism on its interaction with ≡Si—O—Si≡ are also possible, but nevertheless with lower probability (S_Ei) or at higher temperature ($Ad_{N,E}$) than $S_Ni(Si)$; additionally, the methanol reaction with silica is characterized by a low heat effect [10,11,42]. An important role is played by the amount of residual water, which can promote the substitution of the surface OH groups particularly in the case of difficulty reacting phenols [52].

Ad_N REACTIONS

The effectiveness of modified silicas for applied purposes may depend on the hydrolytic and thermal stability of the functional coating. One method of increasing the stability of the surface layer is to create ≡Si—O—Si—CR bonds, hydrolytically more stable than ≡Si—O—Si—O—CR, by using ≡Si—O—Si—H groups in reactions with olefins or other unsaturated compounds [10,11]. It is possible to produce Si—H bonds in the surface layer of silica by various methods, using the reactions of ≡SiCl groups with various hydrides, the thermolysis of ≡$SiOCH_3$ groups, or the reaction of silica with silanes already containing Si-H bonds [10,11,19]. The reactivity of the Si—H bonds during the Ad_N reaction with unsaturated organic compounds can depend not only on the structure of the latter but also on the environment of the Si—H bond at the reaction center. For example, according to Tertykh [10,18] and Belyakova [19], the reactivity decreases in series ≡$SiOSi(Cl)_2H$ > ≡Si—H > $SiOSi(Cl)(CH)_3H$. However, the availability of the Si-Cl bonds, which can be easily hydrolyzed by residual water, can lead to an appearance of more complex arched surface functionalities for which reactivity differs from that for bound chlorosilanes. It can be supposed that the reactivity of the surface Si—H groups in reaction with olefins will differ from the reactivity of the same groups in grafted silanes. However, the multifactorial nature of the effect of the environment prevents detailed analysis of these differences on the basis of purely experimental data [10,11].

For a deeper understanding of the effects it is possible to use theoretical investigations of the stereochemical and electronic structure of the functional coating and model the reactions of the Si—H group with unsaturated compounds in various environments. As an example of such a reaction, we will examine the reaction of 1-hexene and ethylene with Si—H groups in various environments, using the Gaussian 94 [55] and GAMESS [56] program packages with the 6-31G(d,p), LANLDZ, or AMl [57] basis sets. The electron correlation and exchange effects were considered in terms of MP4 or DFT/B3LYP theory [55].

Comparison of the experimental data [10,11] and theoretical calculations shows that the reactivity of the Si-H bonds in reaction with olefins increases with decrease in their polarity, while the exothermic effect of the reaction increases. In addition, in the transition state of the reaction of hexene (AMl) or ethylene (B3LYP/6-31G(d,p) with the Si-H group (in a cluster of four polyhedra (AMl) and SiH_4 in B3LYP/6-31G(d,p) calculations) the charge at the H atom being transferred decreases in relation to the reagents $\Delta q_H = -0.2$ (AMl) or -0.05 (B3LYP/6-31G(d,p) with $q_H = -0.03$ in the transition state). The DRC/AMl calculations of the movement of the system along the generalized coordinate from the transition state toward the reagents show that the transition of the system to the local minimum closest to the transition state comes after 0.9×10^{-14} s at change in energy amounts to about 80% of the activation energy of the reaction. This is due mainly to the rapid movement of the H atom toward the Si atom, since the position of the remaining active atoms (Si, α-C, β-C) remains almost unchanged for such a short time. Consequently, the main contribution to the activation energy can come from H transfer, while the transition state is close to a biradical state. It should be noted that the AMl method overestimates the activation energy [$E^{\neq} \approx 300$ kJ/mol for $\equiv$Si—H + C_6H_{12} → $\equiv$Si$(CH_2)_5CH_3$, inclusion of limited configuration interaction (3 × 3) reduces $E^{\neq}$ by only 6 kJ/mol], as also for the ab initio calculations at the 6-31G(d,p) basis set ($E^{\neq} = 308$ kJ/mol for H_3-Si—H + $H_2C = CH_2$ → $H_3SiCH_2CH_3$ with allowance for the change in the energy of the zero vibrations in the transition state but without electron correlation). The calculations with electron correlation give more acceptable values (starting from the low temperature of about 200°C for such reactions), and $E^{\neq} = -206$ kJ/mol (MP4/6-31G(d,p) the geometry of the transition state was calculated with the inclusion of electron correlation in the MP4/6-31G(d,p) basis set) or 217 kJ/mol (B3LYP/6-31G(d,p)//6-31G(d,p), that is, the geometry was calculated in the HF/6-31G(d,p) basis set without electron correlation while the energy was calculated with electron correlation). Thus, calculations of the activation energy of the reactions with a biradical transition state ($H^{\bullet}$ transferring) must necessarily include those made with the essential inclusion of electron correlation. This requirement is not so rigid for the reactions taking place by the *$S_N i$(Si)* and *$S_E i$* mechanisms, for which the controlling stage involves proton transfer (i.e., the localization of the electron density in the transition state can increase), and the contribution from the ground electronic configuration is predominant [25,33,37,38,42,47] in contrast to the biradical transition state with limiting $H^{\bullet}$ transfer.

An important aspect of investigations into the reaction mechanisms is the search for the structure of the transition state. There are several approaches: synchronous movement to the transition state along the generalized reaction coordinate from the points corresponding to the reagents and the products on the potential energy hypersurface – the quasi-Newton method QST2 [55]. The effectiveness of this method is increased if the approximate geometry of the transition state is taken into account (the QST3 method). When the geometry of the system is close to the transition state, the algorithm operating with the Hessian eigenvalues (EF) can be used for its refinement: in the true transition state there can only be one negative eigenvalue. Various versions of the Berni algorithm, the Newton-Raphson method, abnd the EF/G1 or EF/G2 methods can also be used. In our searches for the structure three methods were usually employed in succession: QST2 → QST3 → EF.

$Ad_{N,E}$ REACTIONS

Cleavage of the siloxane bonds through the *$Ad_{N,E}$* mechanism can occur by different compounds such as $(CH_3)_3$Si-X (X = N_3, NCO, NCS, NCNSi$(CH_3)_3$), HF, H_2O, strong bases, etc., and the reaction temperature depends strongly on the type or amounts of reactants and promoters [1–11]. In the case of reaction with the water molecule alone, the activation energy of reaction (37.3) is approximately 200 kJ/mol [33]. The $E^{\neq}$ energy decreases if the Si atoms in the Si—O—Si bond already have OH groups, as well as if several molecules of water or OH^- take part in the reaction (as beginning of silica dissolution). Additionally, the activation energy decreases on cleavage of asymmetrical bridges such as Ti—O—Si or Al—O—Si (up to 75 kJ/mol or lower for reactions in the aqueous media) [11,14,33,42]. Therefore, the Ti-O-Si bonds are not observed in CVD-TiO_2/SiO_2 after contact with water vapor, for example, a characteristic IR band at 950–940 cm^{-1} related to valence Si-O vibrations in the Si-O-Ti bridges disappears for samples in air; however, it is observed in vacuum [13,28]. It should be noted that reactions of the silanone =Si=O groups with electron-donor molecules can occur at the temperature significantly lower than 273 K (even at 77 K) due to substantial contribution of H^+ tunneling at effective activation energy close to zero [42]. For simplicity, the *$Ad_{N,E}$* and *Ad_N* reactions can be considered as *$S_N i$(Si)*, as interaction of adsorbate molecules (or ions) with the Si atoms plays the main role in the cleavage of the siloxane bonds or other related reactions [42].

$S_N i$ REACTIONS

The *$S_N i$(Si)* reactions (37.1) as well as *$S_E i$* can be limited by H^+ transferring to O atom in $\equiv$SiOH groups [11,14,42]. Since Si atom in $\equiv$SiOH groups possesses relatively low electron-acceptor properties (see Figure 37.7, DMAAB adsorbed onto dried silica forms

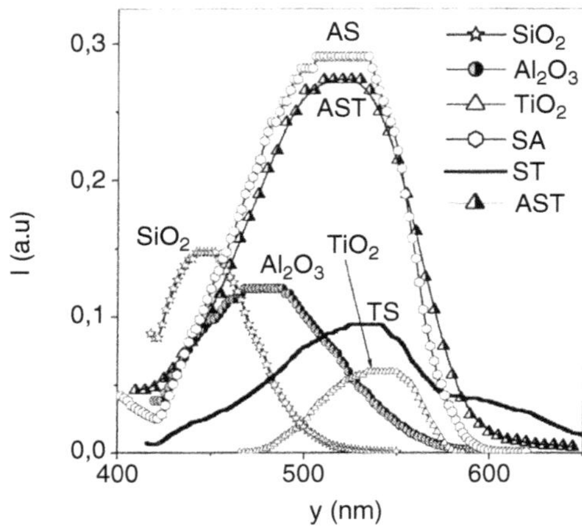

FIGURE 37.7 Optical spectra of (dimethylamino)azobenzene adsorbed on fumed silica (treated at 373 K for 1 h, A-300), alumina (473 K), titania (373 K), SiO_2/TiO_2 (373 K, 37 wt% TiO_2), Al_2O_3/SiO_2 (473 K, 30 wt% Al_2O_3) and $Al_2O_3/SiO_2/TiO_2$ (373 K).

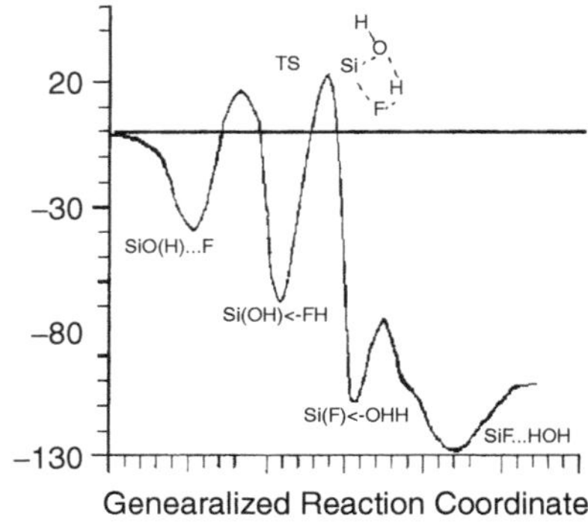

FIGURE 37.8 Changes in the potential energy (kJ/mol) along the generalized reaction coordinate for HF interaction with SiOH group according to quantum chemical calculations [58,61].

only the hydrogen bonds with no H^+ transfers in contrast to mixed oxides ST, SA, and AST, or even alumina and titania) and hardly transforms its surrounding from four-fold to five-fold that only strong electron-donor atoms (such as F^- in adsorbed and strongly polarized HF or O in OH^-) with an appropriate size can relatively easy substitute OH^- through the S_Ni mechanism. In other cases (e.g., for reaction with HCl) the reaction temperature should be high or catalysts (e.g., amines TEA, TMA, etc.) should be used [11,42]. Reaction of HF with silanol groups is a striking example of the S_Ni mechanism, and this reaction was explored in detail by means of experimental and theoretical methods [42,43,58–61]. The behavior of HF molecules adsorbed on the silica surfaces depends on the sample origin, pretreatment conditions, temperature of adsorption, the presence of water or other electron-donor compounds, (such as amines). If $T < 300$ K and water is absent that HF forms complexes which give the IR band at 905 cm^{-1}. Similar complexes can be observed on modified silica with ≡Si—NR, ≡Si—OCR, and ≡Si—Hal (Hal = Cl, Br, I) groups; however, in the case of ≡Si—H, ≡Si—F, and ≡Si—O—$Si(CH_3)_3$ functionalities, similar complexes are not observed in the IR spectra, as the electron-donor properties of the α atom and the electron-acceptor properties of the Si atom, as well as the polarizability of this bond in the surface groups are too weak [11,58,59].

Theoretical estimations of the activation energy of reaction (37.1) at X = F give the values from 70 to 100 kJ/mol for alone HF molecules [42]. The profile of the potential energy of the HF reaction with SiOH group calculated by MNDO/H method and shown in Figure 37.8 is relatively complex. The ab initio calculations give different picture of this profile with less deep the second minimum corresponding to formation of the donor-acceptor bond ≡Si(OH) ← FH and higher second peak (120–250 kJ/mol dependent on the use basis set and consideration for electron correlation and exchange effects). The presence of water or amines sharply reduces the activation energy up to 25–30 kJ/mol for the aqueous medium (dissolution of silica in hydrofluoric acid) and the reaction easily occurs at room temperature. The reaction depth depends on the concentration of adsorbed HF and the IR intensity of Si-F band increases with C_{HF} (Figure 37.9). Semiempirical DRC/PM3 calculations (modeled conditions correspond to room temperature) show a relative stability of the ≡Si(OH) ← FH bond (Figure 37.10). It should be noted that interaction of HF with fumed silica unmodified and modified containing different functionalities capable (e.g., containing N or O atoms) or incapable (C, H, etc.) to form strong hydrogen bonds up to H^+ transferring from HF is characterized by the IR band at 905 cm^{-1}, which depicts (with simultaneous appearance of the IR

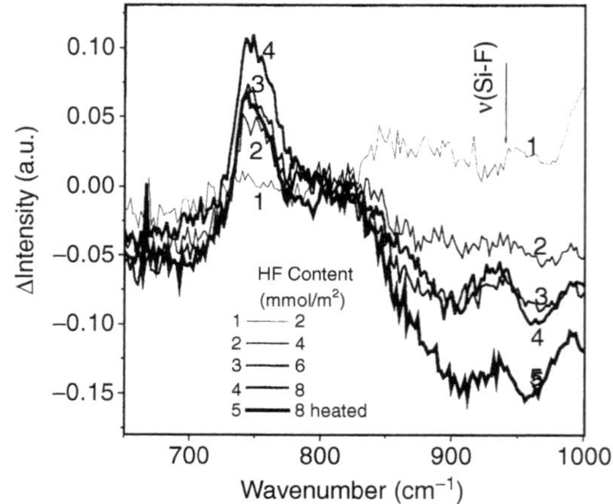

FIGURE 37.9 IR spectra of silica after reaction with different amounts of HF.

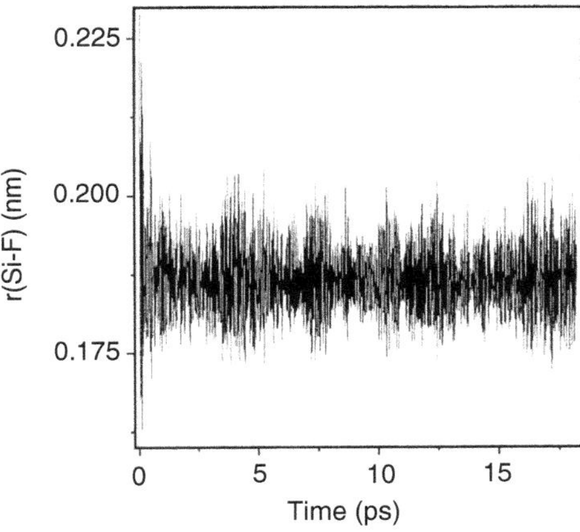

FIGURE 37.10 Dynamics of thermal vibrations of donor–acceptor bond Si ← F in prereaction complex calculated by DRC/PM3 method.

bands characteristic for H-NR_3^+ structures) formation of such complexes as ≡Si(X) ← F^-...H^+NR-Si ≡ (HF interacting, e.g., with the fumed silica surfaces modified by cyclosilazanes) [11,12,42].

Another typical example of the $S_Ni(Si)$ reactions taking place on the silica surfaces is the hydrolysis of the functional groups ≡Si—X (X = OR, CR, H, Hal, etc.) with a cyclic structure of the transition state and with H^+ transfer as a limitation stage [14,25,42,44,62–67]. According to experimental and theoretical investigations, the hydrolytic instability of the Si-X bonds decreases in series I > Br ≈ NR > Cl > OCR > F > H > C with increasing activation energy by approximately order. Even in terms of the semiempirical methods (AM1, MNDO/H) it is possible to obtain an activation energy close to experimental values [42], for example, the error in the theoretical determination of $E^{\neq}$ for the hydrolysis of the ≡$SiOCH_3$ group is not greater than 10%, which is comparable with the experimental errors. The $E^{\neq}$ value in such reactions depends on the polarizability of the Si—X bond, the electron-acceptor characteristics of Si (which depend on the structure of the group X in ≡Si—X), and the number of water molecules taking part in the reaction at one center. These factors are more important than the dimensions of the α-atom in X (the steric effect may be more significant in processes not limited by H^+ transfers), for example, the ≡Si—H or ≡Si—F groups are hydrolytically more stable than ≡Si—Br or ≡Si—I. The polarizability of the Si-Hal bond increases with increase in the atomic number of the halogen, while the hydrolytic stability decreases. Increase in the size of the water cluster near the group being hydrolyzed leads to greater stabilization of the transition state than of the prereaction complex, since the charges of the active atoms in the transition state (particularly q_H) are greater than the corresponding charges in the prereaction complex [14,25,42,44,62–67]. In addition, further channels for H^+ transfers involving water molecules appear (i.e., the pre-exponential of the reaction rate constant increases), and the contribution from tunneling increases (the effective activation energy decreases).

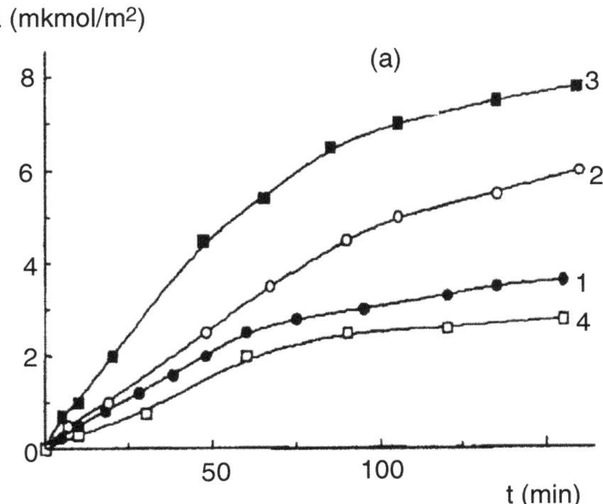

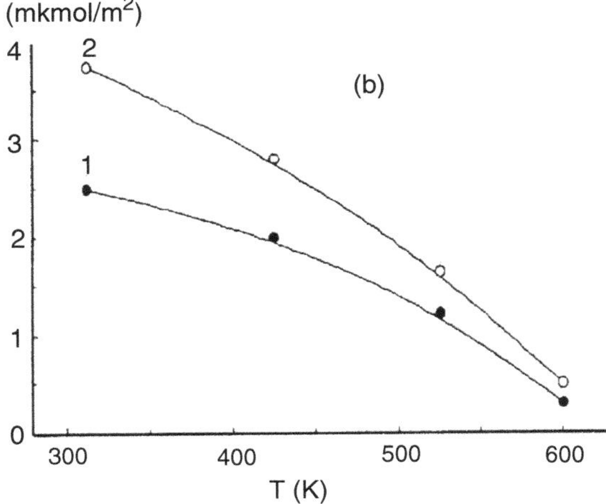

FIGURE 37.11 (a) Kinetics of sorption of monoethanol amine (MEA) onto fumed silica degassed in vacuum at (1) 298 K, (2) 673, (3) 923 K, and (4) onto sylilated (by TMCS) silica; (b) desorption of MEA from silica previously treated at (1) 673 K and (2) 923 K.

Chemisorption of monoethanol amine onto fumed silica reveals dependence of the MEA loading on the state of the surfaces, as it grows with increasing dehydration level (Figure 37.11) [11].

DECOMPOSITION OF SURFACE FUNCTIONALITIES

An important characteristic of the functional coating is its thermal stability. The thermal destruction of the functional groups ≡SiOCR usually takes place under milder conditions than the destruction of the organosilicon functionalities containing Si-CR bonds, since the activation energy of transfer of H^+ to the O atom in $\equiv$SiOCR is lower than the $E^{\neq}$ value for the destruction of the Si-CR group [10,11,14,29,42,44,68–73]. Investigations of the mechanism of dissociation of the $\equiv$SiOR groups [R = CH_3, CHO, COOH, CH_2COOH, $(CH_2)_3CH_3$, $(CH_2)_6CH_3$, C_6H_5, $CH_2C_6H_5$, $(CH_2)_2C_6H_5$, etc.] in terms of the transition state, RRKM, and DRC theories using various quantum chemical methods [14,29,42, 44,68–73] show that proton transfer occurs during the formation of the $\equiv$SiOH groups (the product from the dissociation of $\equiv$SiOR), and in the case of decomposition $\equiv SiOCH_3 \rightarrow \equiv Si{-}H + CH_2O$, which takes place at higher temperatures, the transfer of H^- or $H^{\bullet}$ is possible.

Analysis of the TPD curves allowed Pokrovskiy and coworkers [15,29,69] to determine the activation energy of desorption (E_d) and the pre-exponential factor (k_o) for the desorption rate constant in desorption equation, which can be represented as follows [11,15,29,73]

$$d\Theta/dt = -k\Theta^n \tag{37.19}$$

where Θ is the surface coverage, n is the reaction order (assumed to be known from experimental data) and the desorption rate constant is

$$k = k_o \exp(-E_d/RT) \tag{37.20}$$

where R is the gas constant. At the initial condition $\Theta_{t=0} = 1$, we have the following solutions of Equation (37.19):

$$\begin{aligned} \Theta(t) &= \exp[-\Phi(t)], && n = 1 \\ \Theta(t) &= 1/[1 + \Phi(t)], && n = 2 \\ \Theta(t) &= 1/[1 + 2\Phi(t)]^{1/2}, && n = 3 \end{aligned} \tag{37.21}$$

where

$$\Phi(t) = \int_0^t k\,dt \tag{37.22}$$

According to Equations (37.21), the desorption rate, which is directly proportional to the ion current, can be obtained as

$$\begin{aligned} d\Theta(t)/dt &= -k\exp[-\Phi(t)], && n = 1 \\ d\Theta(t)/dt &= -k\{1/[1 + \Phi(t)\}^2, && n = 2 \\ d\Theta(t)/dt &= -k\{1/[1 + \Phi(t)\}^{3/2}, && n = 3 \end{aligned} \tag{37.23}$$

Equations (37.23) illustrate the typical temperature dependence of the ion current, and the second-order desorption curve being nearly symmetrical, the first-order curve decreasing faster on the high-temperature side and the converse being true for the third-order process. Taking into account the direct proportionality of $d\Theta/dt$ to the ion current, a simple procedure of treating the experimental data $I(t)$ may be proposed [15,29,73]. The $\Theta(t)$ dependence on I(t) may be written as

$$\Theta(t) = \Psi(t)/S \text{ or } d\Theta/dt = -I/S \tag{37.24}$$

where

$$\Psi(t) = \int_t^{\infty} I\,dt; \quad S = \int_0^{\infty} I\,dt$$

Using Equations (37.23) and (37.24), we obtain

$$\begin{aligned} \ln k &= \ln[I/\Psi(t)], && n = 1 \\ \ln k &= \ln[IS/\Psi^2(t)], && n = 2 \\ \ln k &= \ln[IS^2/\Psi^3(t)], && n = 3 \end{aligned} \tag{37.25}$$

If all the assumptions are allowable and the reaction order n is properly chosen, then the function $\ln k = f(1/T)$ expressed according to Equations (37.25) is linear over the full temperature range. An advantage of the procedure involving Equations (37.25) is the utilization of all the data obtained experimentally, including a high-temperature portion of the thermograms, which are most important in estimating the adequacy of the model and, in particular, the reaction order n. Applications of some of presented formulae to TPD MS experimental data were described in detail elsewhere [11,29,69,73].

If we assume that the ionic current intensity is proportional to the desorption rate, then it is possible to calculate the rate constant from the temperature dependence of the ionic current. The analysis of the TPD curves indicates that the reaction order can be estimated from the curve shape. At $n = 2$, a curve has nearly symmetrical shape. Faster changes in the right part of a curve are observed at $n = 1$ and in the left part at $n = 3$. It is difficult to assume the occurrence of desorption at $n = 3$, as the desorption occurs typically through uni- or bimolecular processes. Silica modified by pyrolysis of a mixture of heptanol and phenylmethanol at 773 K in an autoclave was used as example for the TPD MS investigations. Treatment of TPD curves were performed using Equation (37.25) at $n = 1$ and 2 (Table 37.7). The differences in the activation energies calculated at $n = 1$ and 2 (using different portions of the TPD curves) are relatively small except for several cases. To calculate the k_o and E_d values at $n = 1$, only the portion of the TPD curves to the left of the maximum was used and mainly the right portion of the TPD curves was used at $n = 2$ (Table 37.7).

As follows from the data (Table 37.7), with increasing temperature the first to desorb from the surface are

TABLE 37.7
The Kinetic Parameters Obtained on the Basis of the Data from TPD Mass Spectrometry during Thermal Destruction of Modified Silica

Product	M/z	k_o (n = 1)[a] (s^{-1})	[a]$E^{\#}$(kJ/mol)	k_o (n = 2) (s^{-1})	$E^{\#}$(kJ/mol)	T_{max}(K)
Heptanol	31	1.25×10^{12}	95	1.47×10^{9}	104[c]	413
	45	1.28×10^{9}	73	2.30×10^{12}	121[c]	411
	59	7.50×10^{10}	86	1.13×10^{12}	112[c]	410
Heptene	98	7.62×10^{13}	117	5.59×10^{9}	114[c]	453
Hexane	85	3.39×10^{10}	91	1.74×10^{7}	87[b]	448
				5.10×10^{7}	99[c]	
	86			2.79×10^{11}	101[c]	383
Phenylmethanol	56	9.06×10^{9}	73	5.92×10^{8}	95[c]	383[d]
	77	6.62×10^{11}	98	2.37×10^{7}	90[c]	448[d]
	91	3.67×10^{12}	101	1.70×10^{10}	102[a]	413
				3.90×10^{9}	99[b]	
				2.10×10^{10}	101[c]	
Methylbenzene	92	5.11×10^{12}	101	5.53×10^{9}	101[b]	413
				3.21×10^{11}	117[c]	
Dimethylbenzene	105	3.49×10^{13}	113	7.07×10^{10}	113[c]	423
Benzene	78	2.66×10^{12}	93	2.68×10^{10}	108[c]	421
		1.22×10^{8}	121	5.98×10^{13}	228[c]	763
Diphenylene	152	4.96×10^{12}	123	3.39×10^{11}	143[c]	513
Biphenyl	154	4.42×10^{10}	103	8.06×10^{8}	107[b]	498[e]
Naphthalene	128	1.89×10^{13}	118	1.86×10^{13}	146[c]	463
		5.43×10^{12}	170	7.09×10^{14}	216[c]	703
Methylnaphthalene	142	3.85×10^{13}	126	1.62×10^{12}	139[c]	476

Note: Calculations for [a]the initial section of the TPD spectrum; [b]the full curve; [c]the final section; [d]shoulder at 413 K; [e]shoulder at 423 K; n is the reaction order.

molecules of phenyl-methanol, heptanol, heptene, hexane, benzene, methyl-benzene, and dimethyl-benzene. The E_d values for these compounds are over the 73–120 kJ/mol range. Consequently, compounds larger than adsorbed precursors (e.g., dimethyl-benzene, naphthalene, etc.) and desorbed at T about 373–473 K can be formed only during pyrolysis of the alcohols 773 K for 6 h, as E_d is relatively low. At slightly higher temperature and greater E_d values, naphthalene, methyl-naphthalene, and diphenyl desorb (Table 37.7). Moreover, in the case of benzene and naphthalene, marked high-temperature TPD maxima are observed.

Temperatures, at which these maxima appear, are close to the pyrolysis temperature (773 K). In this case, the activation energy of desorption is 220–230 kJ/mol. Benzene and naphthalene can be involved in the process of thermal decomposition taking place in the carbon phase at such a temperature. Due to the fact that silica surface carbonization was carried out under mild conditions (773 K), this process did not run until the end, which could correspond to formation of practically pure carbon phase. When the temperature of the mass spectrometer reactor reaches the value close to the pyrolysis temperature, further radical bond scission can take place resulting in volatile organic compounds. The fact that the carbonization of silica gel is not completed is indicated by marked amounts of different organics desorbed at high temperatures. Therefore on utilization of carbosils prepared at relatively low pyrolysis temperatures of 673–773 K, it should be taken into account the availability of adsorbed hydrocarbons, which can influence the carbosil surface properties.

Application of the RRKM theory (used for to reactions in the gas phase) to surface reactions required detailed analysis of contributions from vibrations that belong to the matrix to the statistical sums of the prereaction state and transition state [29,68]. Inclusion of the low-frequency phonons of the solid ($<200\ cm^{-1}$) leads to appreciable dependence of the preexponential of the rate constant (k_o) on the temperature. It should be noted that the k_o values calculated on the basis of the experimental thermagrams (TPD-MS) at various temperature sections on the assumption of the nondependence of k_o on T can differ by three orders of magnitude (Table 37.7).

During the application of the RRKM theory to the destruction of the surface groups [29,42,68,73] we assumed that:

(1) The vibrational states of the surface groups and lattice atoms correspond to the normal distribution, and effective exchange of energy occurs between them.
(2) The vibrations of the atoms of the surface groups and the nearest fragment of the surface are active, and the remainder of the solid is regarded as an adiabatic subsystem, that is, it does not introduce changes in the ratio of the statistical sums of the transition state and the prereaction state $Q^{\neq}/Q$ in the reaction process.
(3) The rotational degrees of freedom of the surface group are active, if they belong to the bonds undergoing changes during the transformation while the others belong to the adiabatic subsystem.
(4) The rate of deactivation of the activated groups is proportional to the pressure of the gaseous products, but as a result of the execution of the experiments under vacuum (10^{-3}–10^{-4} Torr) the effectiveness of deactivation is low (<0.1).
(5) The average distance between surface groups on fumed silica is 0.65–0.7 nm [10,11], and lateral interactions can be disregarded for small groups at $\Theta < 0.3$.
(6) The dissociation reactions are direct and without intermediates, their time corresponds to 10^{-12}–10^{-13} s, and the process takes place more quickly than the exchange of energy between the active and adiabatic subsystems.
(7) The electronic and translational statistical sums are unchanged in the dissociation process [29,68,73].

Poor agreement is observed between the experimental and theoretical k_o values [calculated by a combined method, that is, quantum-chemical calculation of the activation energy (Table 37.7 and Table 37.8) and the frequencies of the vibrations of the bonds in the prereaction complex and in the transition state and calculation of the rate constants on the basis of RRKM theory] for certain processes where the effects of electron correlation and the contributions of the excited electronic configurations are not predominant [68–73].

Many of homolytical reactions can occur upon pyrolysis of bound heptanol and phenyl-methanol, but only a portion of these processes was modeled using PM3 method with CI (3 × 3) (Table 37.8). The obtained reaction energies (difference between the total energies of the reactants and the products) show that homolytical breaking of Si–OR, SiO–CR, or SiOC–CR bonds can

TABLE 37.8
The Activation Energy (kJ/mol) of Some Decomposition Reactions

N	Reaction	$E^{\neq}$
1	$\equiv SiO(CH_2)_2C_6H_5 + H_2O \rightarrow \equiv SiOH + C_6H_5(CH_2)_2OH$	85
2	$\equiv SiO(CH_2)_2C_6H_5 \rightarrow \equiv SiOH + C_6H_5CH{=}CH_2OH$	260
3	$2(\equiv SiO(CH_2)_2C_6H_5 \rightarrow 2(\equiv SiO(CH_2)_2) + C_6H_5\text{-}C_6H_5$	399
4	$\equiv SiO(CH_2)_2C_6H_5 \rightarrow \equiv SiOCH{=}H_2 + C_6H_6$	291
4[a]	$\equiv SiO(CH_2)_2C_6H_5 \rightarrow \equiv SiOCH{=}H_2 + C_6H_6$	287
5	$\equiv SiO(CH_2)_2C_6H_5 \rightarrow \equiv SiOCH_2^{\cdot} + {}^{\cdot}CH_2C_6H_5$	440
5[a]	$\equiv SiO(CH_2)_2C_6H_5 \rightarrow \equiv SiOCH_2^{\cdot} + {}^{\cdot}CH_2C_6H_5$	316
6	$C_6H_5(CH_2)_2OH \rightarrow C_6H_6 + CH_2{=}CHOH$	271
6[a]	$C_6H_5(CH_2)_2OH \rightarrow C_6H_6 + CH_2{=}CHOH$	253
6[b]	$C_6H_5(CH_2)_2OH \rightarrow C_6H_6 + CH_2{=}CHOH$	360
6[c]	$C_6H_5(CH_2)_2OH \rightarrow C_6H_6 + CH_2{=}CHOH$	207

Note: [a]CI for six MOs and 400 configurations; [b]3-21 G(d); [c]MP2/3-21 G(d).

occur more easily in the triplet states. Typically, during a significant elongation of a valence bond (e.g., on heating or electron excitation on the TPD MS measurements), a triplet state contribution increases with a decreasing singlet state. For bound heptanol, the most probability is removal of hexane fragment (Table 37.8) that corresponds to the availability of hexane and its fragments in the TPD mass-spectra. The energy of elimination of a benzene fragment ($C_6H_5^{\bullet}$) is greater that that of $C_6H_5CH_2^{\bullet}$, which corresponds to the lower T_{max}, for methyl-benzene desorption in comparison with benzene (Table 37.7). Typically, the energy of homolytical breaking SiOC—C bond is lower than that of Si—OCC, SiO—CC bonds (Table 37.9). Therefore, the corresponding products can appear at lower temperatures in the TPD mass-spectra. Also, the energies of disproportionation reactions can be relatively low that can cause the formation of dimethyl-benzene at low temperatures. The energy of hydrolysis of Si—OR bonds is low and the activation energy of such a process is 80–90 kJ/mol; that can provide the appearance of molecules of n-heptanol and benzyl alcohol at temperatures of 400–550 K. The energies of formation of large molecules C_{10}–C_{12} are relatively low (Table 37.10). The activation energies of unimolecular decomposition of the corresponding bound hydrocarbon groups can be two or three times larger (200–300 kJ/mol). However, the E_d values calculated from the TPD data correspond to this range for only benzene and naphthalene (for the high-temperature maximum). This effect can be explained from desorption of a marked portion of observed compounds bound in carbon globules and pores of silica gel blocked by pyrocarbon in contrast to unimolecular desorption of hydrocarbons chemically bonded to the silica surface

TABLE 37.9
Energy of Bond Breaking Calculated Using Different Methods

Broken bond	Energy (kJ/mol)	Spin change	Basis set
$\equiv$**Si**—OCH_2CH_3	329	singlet → triplet	6-31 G(d,p)
	429		B3LYP/6-311 G(d,p)
	296		PM3
$\equiv$SiO—CH_2CH_3	255	singlet → triplet	6-31 G(d,p)
	379		B3LYP/6-311 G(d,p)
	289		PM3
$\equiv SiOCH_2{-}CH_3$	258	singlet → triplet	6-31 G(d,p)
	368		B3LYP/6-311 G(d,p)
	254		PM3

without its previous carbonization when the carbon phase is totally absent. It should be noted that carbonization of phenyl-ethanol at the silica gel surface at 673 K (i.e., 100°C below than that in this work) gives the greater activation energies of desorption of the corresponding products as well as the theoretical E_d values; however, the energy of desorption of bound phenyl-methanol and phenyl-ethanol and their fragments are close.

Thus, pyrolysis of the *n*-heptanol and benzyl alcohol mixture at the silica surfaces results in the formation of not only pyrocarbon deposit with carbon globules and tiny particles, but also short (e.g., different hydrocarbons C_6 such as benzene, hexane, and hexene) and larger (C_n at $n = 8$ and higher) compounds, which can desorb at relative low temperatures (350–500 K) and activation energies of 73–120 kJ/mol. Such a structure of partially carbonized oxide surfaces can be considered as a model of deactivation of solid oxide catalysts or catalyst supports. An increase in the temperature of pyrolysis of alcohols at the silica surface from 673 to 773 K leads to decrease of the activation energy and temperatures of desorption of residual hydrocarbons that can be caused by changes in packing of these compounds in the surface layer. Theoretical modeling shows the possibility of the formation of compounds observed in the TPD mass spectra through different mechanisms.

TABLE 37.10
Energy of Unimolecular and Bimolecular Reactions of Bound Phenyl-Methanol and Heptanol on Silica Cluster (PM3 Method)

N	Reaction	Energy (kJ/mol)
1	$\equiv SiOCH_2C_6H_5 \rightarrow \equiv SiOCH{=}C_6H_6$	194
2	$2(\equiv SiOCH_2C_6H_5) \rightarrow \equiv SiOC_6H_5 + \equiv SiOCH_2C_6H_5CH_3$	−27
3	$2(\equiv SiOCH_2C_6H_5) \rightarrow \equiv SiOCH_2\text{-}CH_2OSi\equiv + C_6H_5\text{-}C_6H_5$	69
4	$\equiv SiO(CH_2)_6CH_3 \rightarrow \equiv SiOCH + C_6H_{14}$	203
5	$\equiv SiO(CH_2)_6CH_3 \rightarrow \equiv SiOH + C_7H_{14}$	62
6	$\equiv SiO(CH_2)_6CH_3 \rightarrow \equiv SiOCH_3 + C_6H_{12}$	95
7	$\equiv Si(O(CH_2)_6CH_3){-}O{-}(C_6H_5CH_2O)Si< \rightarrow \equiv SiO(CH_2)_8OSi\equiv + C_6H_6$	38
8	$\equiv Si(O(CH_2)_6CH_3){-}O{-}(C_{10}H_7CH_2O)Si< \rightarrow \equiv SiO(CH_2)_8OSi\equiv + C_{10}H_8$	44
9	$\equiv Si(O(CH_2)_6CH_3){-}O{-}(C_{10}H_7CH_2O)Si< \rightarrow \equiv SiO(CH_2)_8OSi\equiv + C_{10}H_8CH_3$	28
10	$\equiv SiOCH_2C_6H_5 + H_2O \rightarrow \equiv SiOH + C_6H_5CH_2OH$	12
11	$\equiv SiO(CH_2)_6CH_3 + H_2O \rightarrow \equiv SiOH + C_7H_{15}OH$	−9

Thermal decomposition of small functional groups such as $\equiv SiOCH_3$, $\equiv SiOCHO$, and $\equiv SiOCOCH_3$ on fumed silica surfaces was reasonably well described by theories of absolute reaction rates and RRKM as unimolecular destruction with a main heat flux from the solid surfaces to functionalities [68,71,72]. Data of the quantum chemical methods were in accord with the experiment as estimates of the probability of various decomposition channels, but activation energy values are overestimated. It should be noted that if R in $\equiv SiOR$ is larger than CH_3 that the main channel of the decomposition leads to formation of $\equiv SiOH$ groups. Increase in the size of surface functionalities (e.g., $R \geq C_6$) results in enhancement of the probability of lateral interaction of them during decomposition and some products have the molecular weight greater than that of surface functionalities (e.g., decomposition of $\equiv SiOC_6H_5$ gives a low yield of biphenyl) [73].

Modification of the silica surfaces by different oxides M_xO_y at M = Al, Cr, Ti, P (Figure 37.12) and others (M = V, Zn, Sn) typically reduces the decomposition temperature [22]. The MS thermogram shape depends on the kind of CVD-oxides (C < 5 wt.%), that is, its catalytic capability in the decomposition of PDMS; however,

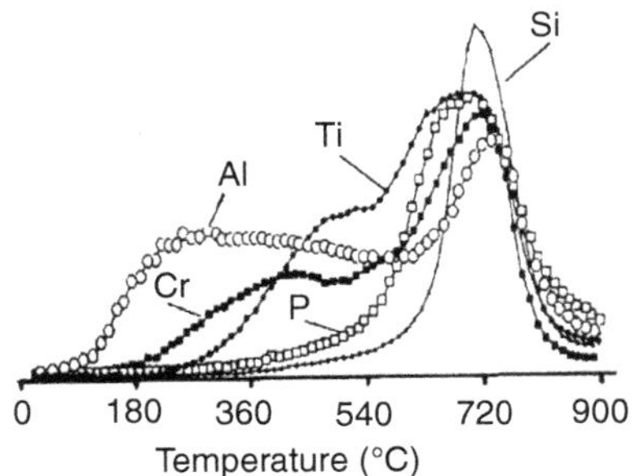

FIGURE 37.12 MS thermograms of methane (16 m/z) on destruction of polydimethylsiloxane (PDMS) $M_xO_y/SiO_2/$PDMS (10 wt.%) at M = Al, Cr, Ti, and P.

the peak connected with reactions at the silica surfaces remains (Figure 37.12).

As it was mentioned above, chemical modification of fumed silicas and related oxides extends possible applications of these materials, which are briefly analyzed below.

APPLICATIONS OF MODIFIED SILICAS

Dozens of reactions of free radicals ($SiCl_3^\bullet$, $SiCl_3O^\bullet$, $Cl^\bullet$, $OCl^\bullet$, $H^\bullet$, $OH^\bullet$, $HO_2^\bullet$, $\equiv Si^\bullet$, $\equiv SiO^\bullet$, etc.), charged particles, ion-radicals, O atoms, molecules and proto-particles in the flame with $SiCl_4$, O_2, H_2 and related compounds on synthesis of fumed silica are worthy of special attention, since variations in synthesis conditions allow one to prepare materials strongly different in their morphological (primary and secondary particle size distributions, type of contacts between adjacent particles) and surface (concentrations of silanols and intact water) properties. These problems as well as structural and adsorptive characteristics of different silicas were analyzed in details by Mironyuk and coworkers in a series of publications [36].

Modification of the oxide surfaces can lead to changes in their structural characteristics, which were studied with respect to different silicas [11,36,41,74]. Notice that systematic investigations of different structural, adsorptive, spectroscopic, electrophysical, electropheretic, kinetic and other properties using SVD-regularization applied to appropriate integral equations allow us to obtain a variety of the distribution functions with respect to the mentioned characteristics giving relatively general picture for unmodified and modified silicas [14,16,32–41,49,65,74–77].

Fumed silica Cab-O-Sil HS-5, silica gels Davisil 633 and 643 with grafted 3-aminopropyl dimethylsilyl (APDMS), butyl dimethylsilyl (BDMS), octadecyl dimethylsilyl (ODDMS), and trimethylsilyl (TMS) groups of different concentrations were synthesized by Blitz and coworkers and investigated using the nitrogen adsorption method [41]. Changes in the textural and energetic characteristics of modified silicas depend on features of the oxide matrices and grafted OSC. Results of the changes in structural parameters of narrow pore silica gel (Davisil 633, Table 37.11), and wide pore silica gel (Davisil 643, Table 37.12) were shown [41]. For example, a reduction of both the BET surface area and pore volume is seen with an increasing extent of surface modification for a given organosilane [41].

Observations can be made concerning the pore size distributions (PSD) data shown in Figure 37.13. The only dramatic changes in $f_V(R_p)$, pore volume as a function of pore radius, occurred for modified fumed silica at pore radius >10 nm. These changes result from rearrangement of primary particle swarms which are also reflected in pore volume data (Table 37.13). Marked changes in $f_V(R_p)$ for modified silica gels are seen as a displacement of the main peak from mesopores toward smaller pore radius. This is readily explained by a partial filling of mesopores with grafted organosilane.

TABLE 37.11
Structural Parameters of Unmodified and Modified Silica Gel Davisil 633

Modified surface structure	Organosilane loading a (mmol/g)	S_{BET} (m^2/g)	Pore volume V_p (cm^3/g)	Pore radius, R_p (nm)
Unmodified D633	0	451	0.892	3.96
Trimethylsilyl D633T(0.6)	0.6	407	0.827	4.1
Trimethylsilyl D633T(0.32)	0.32	399	0.810	4.1
Aminopropyl-dimethylsilyl D633A(0.6)	0.60	353	0.769	4.4
Aminopropyl-dimethylsilyl D633A(1.1)	1.1	320	0.634	4.0
Butyldimethylsilyl D633B(0.4)	0.40	373	0.730	3.9
Butyldimethylsilyl D633B(1.2)	1.2	356	0.630	3.5
Octadecyldimethyl-silyl D633O(0.6)	0.6	288	0.626	4.3
Octadecyldimethyl-silyl D633O(0.52)	0.52	244	0.524	4.3

TABLE 37.12
Structural Parameters of Unmodified and Modified Silica Gel Davisil 643

Modified surface structure	A (mmol/g)	S_{BET} (m^2/g)	V_p (cm^3/g)	R_p (nm)
Unmodified D643	0	344	1.144	6.7
Trimethylsilyl D643T(0.4)	0.4	303	1.038	6.9
Trimethylsilyl D643T(0.39)	0.39	279	0.993	7.1
Aminopropyl-dimethylsilyl D643A(0.4)	0.40	280	1.021	7.3
Aminopropyl-dimethylsilyl D643A(0.84)	0.84	259	0.919	7.1
Butyldimethylsilyl D643B(0.4)	0.40	305	1.029	6.7
Butyldimethylsilyl D643B(0.8)	0.80	277	0.897	6.5
Octadecyldimethyl-silyl D643O(0.4)	0.40	253	0.933	7.4
Octadecyldimethyl-silyl D643O(0.6)	0.60	192	0.646	6.7

These data have shown that the larger the functional groups the lower the residual microporosity. The surface becomes "smoother" with increasing length of the grafted functionalities with respect to adsorbed nitrogen molecules, since long CH groups can "lie" on the silica surface blocking micropores and reducing surface roughness. This may result from changes in the topological structure of the modified surfaces and/or differences in adsorption energy and adsorption potential.

The investigations of the textural and energetic characteristics of silica gels and fumed silica modified by different OSC having a group with various length and distinct chemical structure show their marked dependences not only on the morphology of the initial silicas but also on the characteristics and concentration of grafted modifiers. All the deposited OSC reduce the specific surface area of all the silicas, the pore volume of silica gels, and the adsorption energy differently. Minimal changes are observed for silica/TMS possessing the smallest size among the studied OSC and, perhaps, remaining free "windows" in the modifier layer to the silica surfaces with Si-O-Si bonds accessible for nitrogen molecules, since the TMS groups do not "lie" on the silica surface. For other modifiers, such windows are practically absent, especially in the case of relatively long octadecyl dimethylsilyl groups, which can "lie" on the surface, that results in decrease in the high-energy peak of the adsorption energy corresponding to direct interaction of nitrogen molecules with the silica surfaces. The mesoporous character of all the studied samples does not change dramatically upon surface modification. The low contribution of micropores for the pristine silicas becomes smaller for modified samples due to blocking by "lying" long groups except for sort TMS [41].

Disperse oxides unmodified or modified by organics (OC) or OSC are used as fillers, adsorbents, or additives [1–11]. OSCs are used as promotors of adhesion, inhibitors of corrosion, for the stabilization of monodisperse oxides and the formation of the nanoscaled particles. Oxide modification by alcohols or other OC is of interest for synthesis of polymer fillers, as such modification leads to plasticization and reinforcement of the filled coating, but in this case a question arises about hydrolyzability of the $\equiv$M—O—C bonds between oxide surface and alkoxy groups, as those are less stable than $\equiv$M—O—M$\equiv$ formed, for example, upon the silica modification by silanes or siloxanes. The high dispersity, high specific surface area, and high adsorption ability of fumed oxides have an influence on their efficiency as fillers of polymer systems.

This efficiency can be improved by the modification of their surface. For modification of oxide surfaces, such OSCs as $XSiR_3$, X_2SiR_2, and $XSi(R_2)R'Y$ (where X = Cl, NCO, NCS, N_3, CN, OR, etc., R is a hydrocarbon group) or more complex OSCs were widely used [1–11]. The characteristics of pure OSCs or these compounds bound to the surface may be different; for example, the properties of the bound groups can depend on the degree of surface coverage (Θ) due to lateral interactions, on the nature of the surface (e.g., unmodified silica and

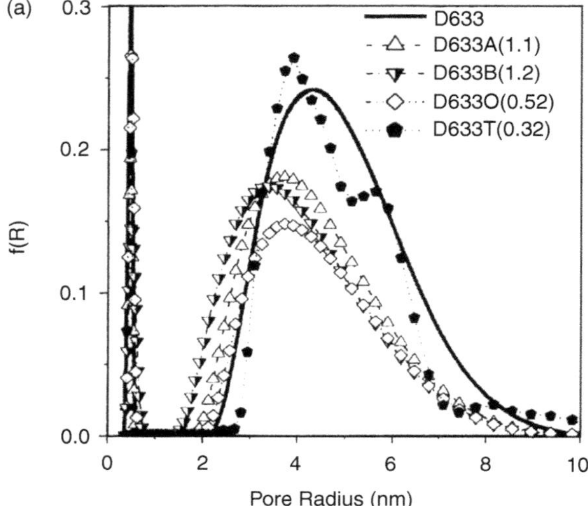

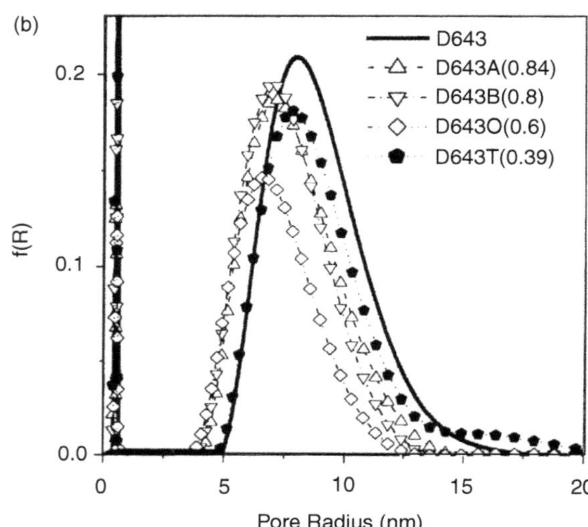

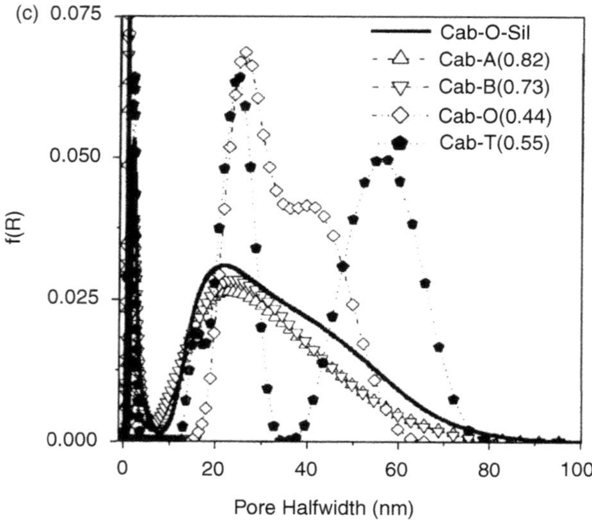

FIGURE 37.13 Pore size distributions for pristine and modified (a) D633, (b) D643, and (c) Cab-O-Sil at maximal OSC loading.

titania or mixed oxides such as TiO_2/SiO_2 have different OH groups $\equiv$MOH, $\equiv$MO(H)M$\equiv$, etc., and other surface sites with different acidic and basic properties and reactivities). Additionally, unmodified titania particles, for example, Degussa P25, can coagulate in aqueous suspensions at pH $\geq$ 5 and, for pH = 7–10, the effective diameter D_{ef} = 1.2–2.3 μm for primary particle swarms of this oxide. Modification of titania by inorganics or organics can allow one to stabilize the particle size distribution at D_{ef} = 0.2–0.5 μm (close to the optimal value for pigments) at pH = 3 − 11.

Adhesion interactions at the solids/polymers interfaces are first and foremost adsorption interactions between the solid surface and polymer molecules [1–11]. After polymerization there is a low molecular-weight fraction of coupling agents, which can decrease the cohesion and adhesion of the polymer film. If the molecules from this fraction interact with the filler particles preferentially (which can be reached due to the filler surface modification) instead of with the material surface covered, then the boundary layer of the film can be free from this fraction and adhesion increases as strengthening the boundary layer of the coating leads to stronger adhesion of the coating to the covered surfaces [46].

Commercial titania (anatase, S_{BET} = 20 m^2/g), fumed titania (50 m^2/g), fumed titania/silica, labeled as TS^P, (9, 14, 20, 29, and 36 wt.% of TiO_2 at S_{BET} = 215, 137, 70, 60, and 90 m^2/g respectively), titania/silica obtained by the CVD method using a fumed silica substrate, labeled as TS^H, (9, 20, and 30 wt.% of TiO_2 at S_{BET} = 273, 187, and 154 m^2/g respectively); commercial titania, rutile (labeled as TiO_2(rt)) and rutile covered by a layer of SiO_2 and Al_2O_3 (labeled as TiO_2(rt-M)) (TiO_2(rt) and TiO_2(rt-M)), also, fumed silicas A-300 and A-380 have been used in comparative investigations. 3-Aminopropyltriethoxysilane (APTES or AP), hexamethyldisilazane (HMDS or DS), and methacryloyloxymethylenemethyl diethoxysilane (MMDES, MM) (Table 37.14); additionally, a mixture of hexamethyltricyclosiloxane (15 wt.%), octamethyltetracyclosiloxane (80 wt.%), and 5 wt.% of the corresponding penta- and hexacyclosiloxanes (MD); di(ethylene glycol) (DEG), glycerol, butanol, methanol (spectrophotometric grade) were used for oxide modification [46].

The pretreatment temperature for titania or TS was 373 K for evacuation to 10^{-4} Torr for several hours to remove the adsorbed water. The ampoules with the heated samples were opened at room temperature and the required amount of OC or OSC was used for their modification. After adsorption, the system was evacuated to 10^{-2} Torr. Reaction temperature was 373 K and the reaction time was 3–4 h. To remove the gas-phase products and excessive content of reagents the samples were evacuated for 1 h at the reaction temperature. Modification of silica samples was described elsewhere [46].

TABLE 37.13
Structural Parameters of Unmodified and Modified Fumed Silica Cab-O-Sil

Modified surface structure	A (mmol/g)	S_{BET} (m^2/g)	V_p (cm^3/g)	R_p (nm)
Unmodified Cab-O-Sil	0	326	1.310	8.0
Trimethylsilyl Cab-T(0.4)	0.40	263	1.610	12.2
Trimethylsilyl Cab-T(0.55)	0.55	251	1.758	14
Aminopropyl-dimethylsilyl Cab-A(0.6)	0.60	208	1.315	12.6
Aminopropyl-dimethylsilyl Cab-A(0.82)	0.82	191	0.995	10.4
Butyldimethylsilyl Cab-B(0.4)	0.40	227	1.251	11.0
Butyldimethylsilyl Cab-B(0.73)	0.73	232	1.050	9.1
Octadecyldimethyl-silyl Cab-O(0.4)	0.40	177	1.210	13.7
Octadecyldimethyl-silyl Cab-O(0.44)	0.44	151	1.524	20.2

ADHESION OF FILLED COATINGS

The adhesion property was determined using a simple test of removal of the coatings [partitioned to small bands (2 mm)] from the steel plates (60 × 120 mm) by sticky tape. The estimation of adhesion was made from the number of the bands that remained on the steel plate after peeling off the tape. Every test was repeated 2–4 times and the relative average estimations on the influence of original or modified oxide additives on the adhesion properties of the filled polymer films relative to those for unmodified film are summarized in Table 37.15.

The addition of A-380 of 0.5–2.5 wt.% reduces the adhesion (Table 37.15) and this may be caused by weak interaction between the hydrophilic silica particles and the hydrophobic CH groups of ethyl cellulose. This assumption is supported by the increase in adhesion after the addition of modified silica. When TS^P are used as an additive, the adhesion increases for $C<1$ wt.%, but for

TABLE 37.14
Titania and TiO_2/SiO_2 (Unmodified and Modified) Used as Fillers to Ethyl Cellulose and Nitrocellulose on the Tests of Adhesion, UV Stability, Corrosion Resistance, and Water Absorption of Filled Films

Oxide	S (m^2/g)	C_X (wt%)	Modifier	$T_p = T_r$ (K)	C_m (mmol/g)
TiO_2	20		APTES; MMDES; HMDS;	373	0.2
	50		MMDES + APTES;		0.1 + 0.1
			MMDES + HMDS;		0.1 + 0.1
			APTES + HMDS		0.1 + 0.1
SiO_2/TiO_2	215	9	APTES	373	0.4
SiO_2/TiO_2	137	14			
SiO_2/TiO_2	70	20	APTES;	373	0.3
			MMDES + APTES		0.15 + 0.15
SiO_2/TiO_2	60	29			
SiO_2/TiO_2	90	36	APTES	373	0.3
SiO_2/CVD TiO_2	273	9			
SiO_2/CVD TiO_2	187	20			
SiO_2/CVD TiO_2	154	30			

TABLE 37.15
Influence of Oxide Additives on the Relative Adhesion Properties of the Ethyl Cellulose Coatings to a Steel Plate in Comparison with Nonfilled Ethyl Cellulose Coatings

Oxide	0.5 wt.%	1 wt.%	1.5 wt.%	2 wt.%	2.5 wt.%
TS_9^H	0.74	0.62	0.69	0.80	0.55
TS_{20}^H	0.80	0.95	0.42	0.54	0.55
TS_{30}^H	0.95	1.15	0.78	0.74	0.40
TS_9^P	0.69	0.46	0.46	0.31	0.23
TS_{14}^P	0.88	0.80	0.85	1.38	0.80
TS_{20}^P	1.23	0.92	0.38	0.46	0.38
TS_{29}^P	1.15	0.46	0.23	0.34	0.42
TS_{36}^P	0.77	0.95	0.31	0.19	0.42
TS_9^P-APTES	0.20	0.62	0.15	1.04	1.15
TS_{20}^P-APTES	1.14	1.18	0.95	0.99	1.13
TS_{36}^P-APTES	1.15	1.09	0.98	1.13	1.04
TS_{20}^P-AP/MM		1.03	0.79	1.01	0.99
A-380	0.78	0.77	0.65	0.58	0.42
A-380-MMDES		1.03	0.96	0.90	0.99
A-380-DEG		1.91	1.27	2.17	3.21

Note: AP = APTES, MM = MMDES.

$C > 1.5$ wt.%, TS gives the same effect as unmodified silica. The modification of TS or silica by OSCs or alcohols increases adhesion (Table 37.15). Analysis of the data obtained shows that good results can be obtained using TS^P (20–30 wt.% of titania) modified by OSCs with the active polar groups. Analogous modification of silica improves the efficiency of the additive.

WATER ABSORPTION BY FILLED FILMS

Protection properties of the coatings depend not only on adhesion characteristics but also on the absorption of water as a redox agent and a carrier of oxygen, which can promote the coating degradation. This effect depends on the structures of the coupling agents and surface layers at the pigment and filler particles, which can have active polar groups (O-H, N-H, etc.) or hydrophobic nonpolar groups (CH). Water absorption by unfilled and filled films (weight 0.3–0.5 g, thickness 0.03–0.04 mm) was studied (Table 37.16). The water absorption study was carried out in a box with $p/p_0 = 0.95$ at 290 K for 72 h. The films were weighed before and after absorption of water with instrumental errors of ± 0.1 mg. For the most part, the use of oxide fillers leads to a decrease in water absorption by the ethyl cellulose coatings (Table 37.16), which is a positive effect in improving the protection properties of the coatings. However, the addition of hydrophobic oxides sometimes gives an increase in the water absorption by the filled films, for example samples TiO_2 + HMDS, A_{BUT} (A-300 + butanol), and so on.

Thus, water absorption by the filled films depends not only on the hydrophilic/hydrophobic properties of the individual components, but also on their interaction in and specific spatial structure of the films; for example, the coupling/filler interaction has an influence on the packing density of molecules and particles in the films and on their adsorption properties. The amount of water absorbed by the nitrocellulose films filled by silica modified by alcohols is higher than that for pure nitrocellulose. This effect can be explained by the hydrophilic properties of the oxide fillers such as A_{DEG}, but A_{BUT} is hydrophobic; however, A_{BUT} can absorb some amount of water. The water absorption by the films filled by the modified oxides has a complicated nature. Such films have new adsorption sites at the oxide filler surface, which can form the hydrogen bonds with water molecules (energy 30–60 kJ/mol), but the hydrophobic groups such as $\equiv SiO(CH_2)_nCH_3$ or $-Si(CH_3)_3$ change entropy of adsorbed water (their cluster structure relative to that on pure oxides) due to hydrophobic interaction; in addition, the filler particles alter the polymer film structure. For example, hydrophobic CH groups on the surfaces of modified fillers interact with the hydrophobic fragment of polymers and alter the structure of the films and promote interaction between the water molecules and hydrophilic polar groups of polymers. In the case of hydrophilic groups on filler surfaces, the water molecules can interact with them, that is, the overall effect for differently modified fillers may be the same.

TABLE 37.16
Influence of Oxide Additives on the Relative Water Absorption by the Films with Ethyl Cellulose and Nitrocellulose(*) in Comparison with Nonfilled Films

Oxide	0.5 wt.%	1 wt.%	1.5 wt.%	2 wt.%	2.5 wt.%
TS_9^H	0.13	0.13	0.06	0.17	0.22
TS_{20}^H	0.30	0.48	0.80	0.53	0.46
TS_{30}^H	0.48	0.78	0.83	0.97	1.12
TS_9^P	0.18	0.23	0.48	0.47	0.58
TS_{14}^P	0.38	0.38	0.43	0.30	0.38
TS_{20}^P	0.35	0.46	0.27	0.36	0.38
TS_{29}^P	0.30	0.29	0.38	0.13	0.26
TS_{36}^P	0.015	0.26	0.29	0.37	0.22
TS_9^P-APTES	0.69	0.12	0.41	1.64	0.87
TS_{20}^P-APTES	1.00	1.04	0.92	0.83	0.77
TS_{36}^P-APTES	1.10	1.10	0.99	0.94	0.91
TS_{20}^P-AP/MM		1.75	1.17	1.17	1.08
TiO_2-APTES		1.12	0.87	1.12	1.21
TiO_2-HMDS(0.4)		0.98	1.10	1.13	1.13
TiO_2-HMDS(0.2)		0.70	0.74	0.70	0.91
A-380	0.31	0.14	0.15	0.14	0.24
A-380-MMDES		1.60	0.96	0.90	0.99
A_{BUT}^*		2.30	1.60	2.00	1.10
A_{DEG}^*		0.30	1.80	1.70	2.50

LIGHT RESISTANCE OF FILLED FILMS

Coating stability depends on the degradation rate connected with the probability of redox reactions with the participation of water and oxygen. These processes have an influence on photochemical reactions in the surface layer of the oxide pigments as coating degradation occurs as a result of photostimulated redox reactions with the participation of pigment (titania) surfaces, H_2O, O_2, different radicals and ions such as $HO^\bullet$, $HO_2^\bullet$, H^+, and HO^-. The light resistance of the titania pigments is usually improved by the thin Al_2O_3, ZnO, and SiO_2 layers or patches, which are deposited on the titania surface from liquids with pH = 6.5–7.1. The mechanism of photochemical stabilization of TiO_2 is complicated. The titania stabilization may also be improved by hydrophobizing of pigment surfaces by organosilicon compounds, as in this case the amount of water on the pigment particles is lower and the content of OH^-, $OH^\bullet$ and $HO_2^\bullet$ decreases. Basic groups in OSCs, for example, amine groups, can enhance this effect, due to neutralization (blocking) of the acidic surface sites, as those interact strongly with adsorbed water. In addition, deposition on surfaces of compounds (e.g., substituted phenols such as $C_6H_2(CCH_3)_3OH$), which can be the radical or electron traps, leads to inhibition of redox reactions [46].

We have studied the influence of chemical modification of titania by some OSCs on the light resistance of the filled coating with nitrocellulose (Table 37.17). These coatings were examined by accelerated tests for light resistance determination using climatic equipment with controlled intensity of UV light. The simulated times of sample exposure under UV light were equal to 2, 4, 7, 12, and 17 light years with normal moisture. The reflection spectra of the coatings were recorded with an SF-18 (LOMO) spectrophotometer. For simplicity, we assumed that the reflection coefficient for the samples without UV light action was equal to 1, that is, 100% of reflection or 0% of absorption. An estimation of the light resistance of the coatings was made on the basis of the difference in the peak intensity of the absorption band at 420–460 nm, which is sensitive to the coating degradation under UV light. It follows from results obtained (Table 37.17) that for the most part, for the polymer films filled by the modified pigments the light-resistance increases. Differences between titania modified by HMDS, APTES, and MMDES lead to variations of the coating degradation rates under UV light. The data presented in Table 37.17 show that the modification of anatase surface by OSC (degree of substitution of surface OH by a modifier, α, equals 1) increases the photochemical stability of the coating close to that for rutile.

The modification of the anatase surfaces by MMDES ($\alpha = 1$) and HMDS ($\alpha = 0.4$) is more efficient than that by APTES ($\alpha = 1$). More effective modification of TiO_2 surfaces is observed for HMDS at $\alpha = 0.4$. The light resistance of the samples with APTES is lower but as

TABLE 37.17
Relative UV Stability of the Coating Filled by Unmodified and Modified Anatase(*) or Rutile (rt) as a Function of Exposure (Years, t)

Filler	α	$t = 0$	$t = 2$	$t = 4$	$t = 7$	$t = 12$	$t = 17$
TiO_2*		1	0.93	0.80	0.67	0.66	0.64
TiO_2-MA*	0.5	1	0.88	0.70	0.63	0.56	0.55
TiO_2-MA*	1	1	1.00	0.93	0.63	0.63	0.63
TiO_2-AP*	0.5	1	0.90	0.85	0.74	0.66	0.65
TiO_2-AP*	1	1	1.00	0.87	0.80	0.78	0.78
TiO_2-AP/MM*	1	1	1.00	0.83	0.78	0.75	0.75
TiO_2-AP/DS*	1	1	0.93	0.80	0.72	0.65	0.64
TiO_2-MM/DS*	1	1	1.00	0.95	0.75	0.67	0.66
TiO_2-DS*	0.4	1	1.00	0.95	0.84	0.77	0.75
TiO_2-DS*	0.2	1	0.93	0.85	0.70	0.64	0.60
TiO_2(rt)		1	1.00	0.94	0.86	0.83	0.79
TiO_2(rt-M)		1	1.00	1.00	0.88	0.84	0.82
TiO_2(rt-M)-MM	0.5	1	1.00	0.88	0.78	0.73	0.71
TiO_2(rt-M)-AP	0.5	1	1.00	0.86	0.80	0.73	0.72
TiO_2(rt)-MM	0.5	1	1.00	1.00	0.90	0.84	0.80
TiO_2(rt)-AP	0.5	1	0.98	0.93	0.80	0.75	0.74
TiO_2(rt)-MM	1	1	0.95	0.90	0.82	0.78	0.76

the α value increases (as in the case of TiO_2-DS) the light resistance increases. We have studied the influence of the modification of titania by the OSC mixture on the light resistance of the coatings. Good results in the use of such mixture for the anatase modification were obtained for MMDES and HMDS (TiO_2-MM(DS)). TiO_2 was modified by APTES ($\alpha = 0.5$), then by MMDES (TiO_2-AP(MM)) or HMDS (TiO_2-AP(DS)) and the other sample was TiO_2 modified by MMDES ($\alpha = 0.5$) and HMDS ($\alpha = 0.5$)–TiO_2-MA(DS). Examination of these samples shows that additional modification of TiO_2-MMDES by HMDS improves the light resistance of the coatings, but TiO_2 modified by APTES and then by MMDES or HMDS does not give such an effect. It should be noted that the efficiency of the modification of titania by MMDES for the light fastness of the filled films decreases when the effective exposure of the coatings under UV light is above 7 yr, whereas the polymer film filled by titania with the bound methylsilyl groups (TiO_2-DS) is more stable. It may be expected that the use of such or analogous OSCs in combination with the inorganic oxide (silica, alumina, etc.) surface layer on the titania particles will be more efficient [11,46].

CORROSION RESISTANCE OF THE COATINGS WITH FILLED POLYMERS

The study of adhesion and water absorption gives important information about the protection properties of the coating but it is not direct information. Such information may be obtained via investigation of the corrosion resistance of the coatings on metal plates. Steel plates coated by films with filled ethyl cellulose were placed in a 3 wt.% of NaCl aqueous solution for 72 h and the protection properties of the coatings were estimated in accordance with the appearance and the intensity of corrosion using some scale. Estimation with magnitude (χ) of six on the six-scale corresponded to a coating without any corrosion; magnitude of five corresponded to a coating with point traces of corrosion; magnitude of four corresponded to small patches of corrosion; magnitude of three corresponded to corrosion of less than 5% of the coating; magnitude of two was to 10%; magnitude of 1 was above 10%. Diphenylamine (DPA) as a corrosion inhibitor was adsorbed in addition on oxide filler surfaces. It was found that the most stable coatings were those filled by $A_{MD,DPA}$ ($\chi = 5$–6), A_{DPA} ($\chi = 5$), TS^P_{20} ($\chi = 4$–5), TS^P_{29} ($\chi = 4$–5), TS^H_{20} ($\chi = 4$–5) (1–2 wt.%). Lower χ is observed for unmodified TS^P_{36} and silica A-380. It should be noted that some correlation between the corrosion resistance of the filled coatings and absorption of water and conductivity of them is observed for unmodified and modified TS or individual oxides. For example, specific conductivity of the coating filled by effective filler (e.g., modified silica with the addition of DPA) did not change after exposure in the electrolyte solution during 80 days, but in the case of poor filler (e.g., unmodified silica) this parameter increased by half [46].

Thus, the addition of small amount (0.5–2 wt.%) of highly disperse (fumed) modified titania, TiO_2/SiO_2 or silica allows us to improve the adhesion properties and light and corrosion resistance of filled coatings. Unmodified and modified fillers can have different influence on different protection properties of the filled polymer coatings. In other words, we can improve certain properties of the coatings by using fillers but other properties may be worse. The efficiency of the coating depends on the nature of the pigment, filler and modifier and the degree of substitution of the active surface sites of the solid particles. The modification of surfaces gives a stronger effect for pigments with weaker initial light resistance, such as anatase relative to rutile. The efficiency of chemical modification of pigment surfaces can be improved by the use of radical (electron)-trap compounds as a part of the modified surface layers of the pigment particles using OC or OSC mixture for modification and corrosion inhibitor for immobilization on the modified pigment surface. It was found that highly disperse titania/silica modified by APTES, MMDES, or fumed silica modified by butanol or di(ethylene glycol) with certain content and the optimal ratio of different functional groups at surfaces are effective fillers for improvement of the protection properties of filled coatings [11,46].

Marked differences in the electrophoretic potential ζ values of unmodified and modified (by MMDES) silicas are observed at pH of 2–4 and >7 (Figure 37.14a) [75]. For pH between 2 and 4, the D_{ef} values are large (>1 μm even at $C_{MMDES} = 0.2$ mmol/g); this can lead to a reduction of particle mobility as the diffusion coefficient depends on D_{ef}^{-1} (various for different swarms) but ζ depends on the particle curvature, which does not change. At pH > 7, ζ values of MMDES/silica are smaller due to a diminution of the number of ≡SiOH groups, which provide negative charging of the surface at pH > pH(IEP_{SiO2}) ≈ 2.2, and an increase in the C_{MMDES} value changes ζ(pH) due to a reduction of the silanol concentration (surface charges on particles are formed due to dissociation of the ≡SiO-H bonds) leading to a reduction of the surface acidity and an increase in the pH(IEP) value, however, this dependence is nonlinear. Notice that at pH > 5, the ζ potential curves for MMDES/silica lie above that of initial silica independently of a C_M value, but in the case of 3-methacryloyloxypropyltrimethoxysilane (MAPTMS)/silica, the opposite result is observed (Figure 37.14). This effect can be explained by the different amounts of hydroxyls, which form in the surface layer of modified silicas due to hydrolysis of residual ≡SiOR (R = CH_3 (MAPTMS), CH_2CH_3 (MMDES)) groups, as the possibility of the participation of three SiOR groups of a MAPTMS molecule in the reactions giving three ≡Si—O—Si≡ bonds is significantly lower than that for two ≡SiOR groups of MMDES. The modification of fumed silica by such OSC as MMDES, MAPTMS, APTES, HMDS, etc., can strongly change the particle size distribution in the aqueous suspensions and the dependence of zeta potential on pH. MMDES as a modifier gives the better results in comparison with its mixture with APTES or individual MAPTMS for reducing the particle (agglomerate) size of modified silica at the concentrations of the modifiers over the 0.05–0.2 mmol/g range. Grafted MMDES slightly reduces the hydrophilic properties of silica and diminishes the ζ potential, but bound MAPTMS markedly magnifies the negative ζ potential at pH > 5. However, samples of MAPTMS/silica at C_M > 0.12 mmol/g synthesised at 90°C and heated at 160°C demonstrate the marked hydrophobic properties and low dispersity in the aqueous suspensions [75,76].

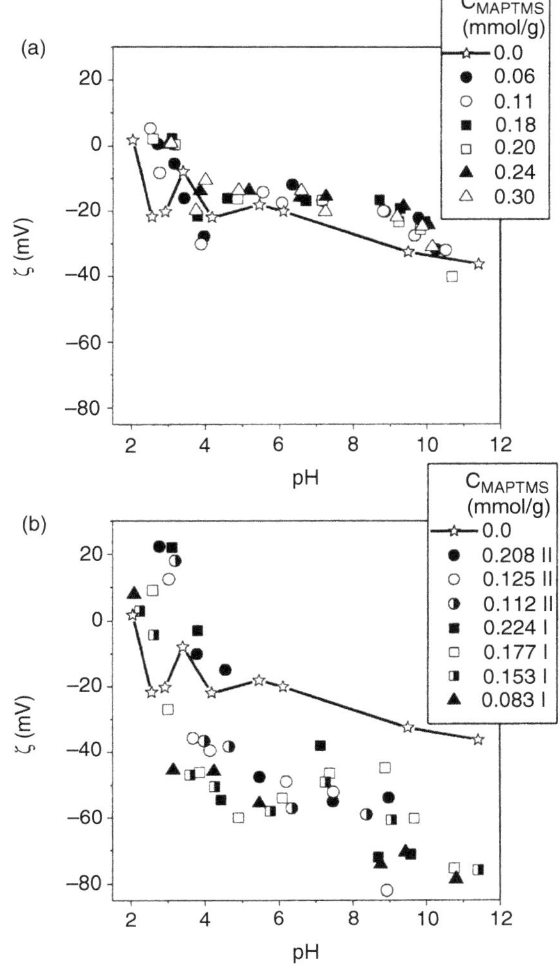

FIGURE 37.14 Zeta potential as a function pH at different concentrations of (a) MMDES, (b) MAPTMS (grafted at (I) 90°C, (II) then heated at 160°C) modifying the fumed silica surfaces.

Changes in the Gibbs free energy of the interfacial water in the aqueous suspensions of modified silica are maximal at a medium concentration of MAPTMS (0.153 mmol/g) grafted onto A-175, but at the maximal C_{MAPTMS} value, these changes are lower than those of the interfacial water in the suspension of pristine silica. Theoretical calculations of the solvation effects showed a reduction in the free energy due to hydrolysis of residual ≡SiOR groups or formed strained ≡Si-O-Si≡ bonds between silanes and the substrate, which is in agreement with electrophoresis investigations [75,76]. Thus, the physicochemical properties of modified silicas depend not only on the concentration and the chain length of grafted functionalities but also on the number of reactive groups in modifier molecules reacting with the silica surfaces or group-to-group.

CONCLUSIONS

Controlled changes in conditions of fumed oxide synthesis and subsequent reactions of surface modification selecting appropriate reactants (individual or mixed applied in mono- or multistage reactions) and medium allow us to obtain materials with a required degree of substitution in the ≡SiOH (or other surface hydroxyls) groups or cleavage of the siloxane bonds by organic, hetero-organic and inorganic compounds and characterized by various adsorptive (hydrophilicity, liophilic behavior, adsorption potential, etc.) and structural (surface nonuniformity, primary and secondary particle size and pore size distributions, etc.) properties and effective in different practical applications. Deeper understanding of the mechanisms of the surface reactions achieved on the basis of sophisticated analysis of experimental and theoretical results can be fruitful to predict optimal conditions for the surface modification and possible characteristics of modified materials appropriate for different purposes and applications in different media.

The states of intact and dissociatively adsorbed water, the impact of nonstandard synthesis conditions on the structural-adsorptive properties of fumed silicas and related oxides, developments of overall conceptions and ideas related to the kinetics, dynamics, and mechanisms of surface reactions, the classifications of these reactions, developments in synthesis of stable Si-C bonds, different CVD-oxides on the fumed silica surfaces, and other results obtained at the Institute of Surface Chemistry (Kiev, Ukraine) are noteworthy and show new possibilities in preparation of materials based on fumed silicas and related oxides for different applications in medicine, biotechnology, environmental, and human protection, as well as in industry.

ACKNOWLEDGMENT

Some presented investigations have been performed in collaboration with the colleagues from the Eastern Illinois University (Charleston, IL) and the Institute of Surface Chemistry (Kiev). The authors thank Prof. J.P. Blitz (Charleston), Dr. E.F. Voronin, and Dr. E.M. Pakhlov (Kiev) for fruitful cooperation.

REFERENCES

1. (a) Iler, R.K. *The Chemistry of Silica;* Wiley: Chichester, 1979; (b) Iler R. K. *Memorial Symposium on the Colloid Chemistry of Silica*, 200th ACS National Meeting, Division of Colloid and Surface Chemistry, American Chemical Society, Washington, D.C., August 26–30, 1990.
2. Legrand, A.P.; Ed.; *The Surface Properties of Silicas;* Wiley: New York, 1998.
3. (a) *Proceedings of International Conference on Silica Science and Technology "Silica 98"*, Sept. 1–4, Mulhouse (France), 1998; (b) *Proceedings of International Conference on Silica Science and Technology "Silica 2001"*, Sept. 3–6, Mulhouse (France), 2001.
4. Vansant, E.F.; Van Der Voort, P.; Vrancken, K.C. *Characterization and Chemical Modification of the Silica Surface*; Studies in Surface Science and Catalysis; Elsevier: Amsterdam, 1995; Vol. 93.
5. Leyden, D.E. *Silanes, Surfaces and Interfaces*, Gordon and Breach: New York, 1985.
6. Leyden, D.E.; Collins, W.T. *Chemically Modified Oxide Surfaces*, Gordon and Breach: New York, 1989.
7. Mottola, H.A.; Steinmetz, J.A. *Chemically Modified Surfaces*, Elsevier: Amsterdam, 1992.
8. Pesek, J.J.; Leigh, I.E. *Chemically Modified Surfaces*, Royal Society of Chemistry: Cambridge, UK, 1994.
9. Blitz, J.P.; Little, Ch.B.; Eds.; *Fundamental and Applied Aspects of Chemically Modified Surfaces;* Royal Society of Chemistry: Cambridge, 1999.
10. (a) Tertykh, V.A.; Belyakova, L.A. *Chemical Reaction Involving Silica Surface;* Naukova Dumka: Kiev, 1991; (b) Tertykh, V.A.; Belyakova, L.A., In: *Adsorption on New and Modified Inorganic Sorbents*, Dabrowski, A.; Tertykh, V.A.; Eds.; Amsterdam: Elsevier, 1996, pp 147–189; (c) Simurov, A.V.; Belyakova, L.A.; Tertykh, V.A. *Functional Materials* **1995**, *2*, 51; (d) Tertykh, V.A.; Tomachinsky, S.N. *Functional Materials* **1995**, *2*, 58; (e) Varvarin, A.M.; Belyakova, L.A.; Tertykh, V.A.; Leboda, R.; Charmas, B. *Colloids Surf.* A **1996**, *110*, 129; (f) Tertykh, V.A.; Yanishpolskii, V.V.; Bereza, L.V.; Pesek, J.J.; Matyska, M. *Therm. Anal. Cal.* **2000**, *62*, 539; (g) Tertykh, V.A.; Yanishpolskii, V.V. In: *Adsorption on Silicas*, Papirer, E.; Ed.; Marcel Dekker: New York, 2000, pp 523–564; (h) Tertykh, V.A.; Yanishpolskii, V.V.; Panova, O.Yu. *J. Therm. Anal. Cal.* **2000**, *62*, 545.
11. (a) Chuiko, A.A.; Ed.; *Chemistry of Silica Surface*, Institute of Surface Chemistry: Kiev, Vol. 1 and 2, 2001; (b) Chuiko, A.A.; Ed.; *Chemistry, Physics, and Technology of Surfaces*, Issues 4–6, KM Academia: Kiev, 2001; (c) Chuiko, A.A.; Ed.; *Chemistry, Physics, and Technology of Surfaces*, 7–8, KM Academia: Kiev, 2002; (d) Chuiko, A.A.; Gorlov, Yu.I. *Chemistry of Silica Surfaces*, Naukova Dumka: Kiev, 1992; (e) Chuiko, A.A.; Ed.; *Silicas in Medicine and Biology*, SMI: Stavropol, 1993.
12. (a) Voronin, E.F.; Chuiko, A.A. In *Chemistry of Silica Surface*, Chuiko, A.A.; Ed.; Institute of Surface Chemistry: Kiev, 2001, Vol. 1, pp 252–331;

(b) Pakhlov, E.M.; Voronin, E.F.; Bogillo, V.I.; Chuiko, A.A. *Dokl. AN UrSSR, Ser. B* **1989**, *No 8*, 50.

13. (a) Pakhlov, E.M.; Voronin, E.F. In *Chemistry of Silica Surface*, Chuiko, A.A.; Ed.; Institute of Surface Chemistry: Kiev, 2001, Vol. 1, pp 422–509; (b) Pakhlov, E.M.; Voronin, E.F.; Chuiko, A.A. *Dokl. AN USSR*, **1991**, *318*, 148; (c) Voronin, E.F.; Pakhlov, E.M.; Chuiko, A.A *Colloids Surf. A* **1995**, *101*, 123; (d) Pakhlov, E.M.; Voronin, E.F.; Borysenko, M.V.; Yurchenko, G.R. *J. Therm. Anal. Cal.* **2000**, *62*, 395.
14. Gun'ko, V.M. In *Chemistry of Silica Surface*, Chuiko, A.A.; Ed.; Institute of Surface Chemistry: Kiev, 2001, Vol. 2, pp 29–78.
15. (a) Pokrovskiy, V.A.; Chuiko, A.A. In *Chemistry of Silica Surface*, Chuiko, A.A.; Ed.; Institute of Surface Chemistry: Kiev, 2001, Vol. 2, pp 79–116; (b) Pokrovskiy, V.A. *Rapid Commun. Mass Spectrom.* **1995**, *9*, 588.
16. (a) Bogillo, V.I., In *Chemistry of Silica Surface*, Chuiko, A.A.; Ed.; Institute of Surface Chemistry: Kiev, 2001, Vol. 2, pp 117–216; (b) Bogillo, V.I.; Shkilev, V.P. *Langmuir* **1996**, *12*, 109; (c) Dabrowski, A.; Bogillo, V.I.; Shkilev, V.P. *Langmuir* **1997**, *13*, 936; (d) Bogillo, V.I.; Shkilev, V.P. *J. Thermal Anal. Calorimetr.* **1999**, *55*, 483; (e) Bogillo, V.I., In: *Adsorption on New and Modified Inorganic Sorbents*, Dabrowski, A.; Tertykh, V.A.; Eds.; Elsevier: Amsterdam, 1996, pp. 135–184; (f) Pokrovskiy, V.A.; Bogillo, V.I.; Dabrowski, A. In *Adsorption and its Application in Industry and Environmental Protection*, Dabrowski, A.; Ed.; Elsevier: Amsterdam, 1999, Vol. 2, pp. 571–634.
17. Makarov, O.A.; Pavlov, V.V.; Chuiko, A.A. In *Chemistry of Silica Surface*, Chuiko, A.A.; Ed.; Institute of Surface Chemistry: Kiev, 2001, Vol. 2, pp 217–226.
18. (a) Tertykh, V.A. In *Chemistry of Silica Surface*, Chuiko, A.A.; Ed.; Institute of Surface Chemistry: Kiev, 2001, Vol. 2, pp 271–301; (b) Tertykh, V.A. *Macromol. Symp.* **1996**, *108*, 55; (c) Tertykh, V.A., In: *Organosilicon Chemistry III. From Molecules to Materials*, Auner, N.; Weis, J.; Eds.; Weinheim: VCH, 1998, pp 670–681; (d) Sidorchuk, V.V.; Tertykh, V.A.; Leboda, R.; Hubicki, Z. *Adsorp. Sci. Technol.* **1995**, *12*, 231; 239.
19. (a) Belyakova, L.A.; Chuiko, A.A. In *Chemistry of Silica Surface*, Chuiko, A.A.; Ed.; Institute of Surface Chemistry: Kiev, 2001, Vol. 2, pp 302–326; (b) Belyakova, L.A.; Varvarin, A.M. *Adsorpt. Sci. Technol.* **2000**, *18*, 65; (c) Belyakova, L.A.; Varvarin, A.M. *Colloids Surf. A* **1999**, *154*, 285; (d) Belyakova, L.A.; Varvarin, A.M.; Linkov, V.M. *Colloids Surf. A.* **2000**, *168*, 45; (e) Belyakova, L.A. *Adsorpt. Sci. Technol.* **1996**, *69*, 885.
20. (a) Borisenko, N.V.; Gomenyuk, A.A.; Mutovkin, P.A.; Mikolaychuk, V.V.; Isarov, A.V.; Chuiko, A.A. In *Chemistry of Silica Surface*, Chuiko, A.A.; Ed.; Institute of Surface Chemistry: Kiev, 2001, Vol. 2, pp 327–368; (b) Borisenko, N.V.; Mutovkin, P.A.; Chuiko, A.A. *Kinet. Katal.* **1997**, *38*, 119; (c) Mutovkin, P.A.; Babich, I.V.; Plyuto, Yu.V.; Chuiko, A.A. *Dokl. Russ. AN* **1993**, *328*, 345; (d) Mutovkin, P.A.; Babich, I.V.; Plyuto, Yu.V.; Chuiko, A.A. *Ukr. Khim. Zh.* **1993**, *59*, 727; (e) Gomenyuk, A.A.; Babich, I.V.; Plyuto, Yu.V.; Chuiko, A.A. *Ukr. Khim. Zh.* **1993**, *59*, 269; (f) Plyuto, Yu.V.; Gomenyuk, A.A.; Babich, I.V.; Chuiko, A.A. *Kolloid. Zh.* **1993**, *55, No 6*, 85; (g) Gomenyuk, A.A.; Babich, I.V.; Plyuto, Yu.V.; Chuiko, A.A. *Zh. Fiz. Khim.* **1993**, *67*, 2455; (h) Gomenyuk, A.A.; Plyuto, Yu.V.; Babich, I.V.; Chuiko, A.A. *Zh. Fiz. Khim.* **1992**, *66*, 2903; (i) Plyuto, Yu.V.; Gomenyuk, A.A.; Babich, I.V.; Chuiko, A.A. *Dokl. Russ. AN* **1993**, *328*, 193; (j) Mikolaichuk, V.; Stoch, J.; Babich, I.; Isarov, A.; Plyuto, Yu.; Chuiko, A. *Surf. Int. Anal.* **1993**, *20*, 99; (k) Mikolaichuk, V.V.; Stoch, J.; Plyuto, Yu.V.; Chuiko, A.A. *Kinet. Katal.* **1993**, *34*, 527; (l) Mikolaichuk, V.V.; Isarov, A.V.; Plyuto, Yu.V.; Chuiko, A.A. *Dokl. Russ. AN* **1993**, *330*, 64; (m) Borisenko, N.V.; Plyuto, Yu.V.; Chuiko, A.A. *Ukr. Khim. Zh.* **1993**, *59*, 917; (n) Borisenko, N.V.; Plyuto, Yu.V.; Chuiko, A.A. *Ukr. Khim. Zh.* **1992**, *58*, 27; (o) Borisenko, N.V.; Plyuto, Yu.V.; Chuiko, A.A. *Ukr. Khim. Zh.* **1991**, *57*, 608; (p) Plyuto, Yu.V.; Babich, I.V.; Sheldon, R.A. *Appl. Surf. Sci.* **1999**, *140*, 176; (q) Plyuto, Yu.V.; Babich, I.V.; Sharanda, L.F.; de Wit, A.M. *Thermochim. Acta* **1999**, *335*, 87.; (r) Gomenyuk, A.A.; Babich, I.V.; Plyuto, Yu.V.; Chuiko, A.A. *Zh. Fiz. Khim.* **2000**, *74*, 904.
21. (a) Kaspersky, V.A.; Brey, V.V. In *Chemistry of Silica Surface*, Chuiko, A.A.; Ed.; Institute of Surface Chemistry: Kiev, 2001, Vol. 2, pp 369–399; (b) Brei, V.V.; Kaspersky, V.A.; Gulyaniskaya, N.E. *React. Kinet. Catal. Lett.* **1993**, *50*, 415; (c) Brei, V.V.; Kaspersky, V.A.; Khomenko, K.N.; Chuiko, A.A. *Adsorpt. Sci. Technol.* **1996**, *14*, 349; (d) Chernyavskaya, T.V.; Brei, V.V.; Grebenyuk, A.G.; Chuiko, A.A. *Polish J. Chem.* **1997**, *71*, 955.
22. (a) Bogatyrev, V.M.; Chuiko, A.A. In *Chemistry of Silica Surface*, Chuiko, A.A.; Ed.; Institute of Surface Chemistry: Kiev, 2001, Vol. 2, pp 447–486; (b) Bogatyrev, V.M. *React. Kinet. Catal. Lett.* **1999**, *66*, 177; (c) Bogatyrev, V.M.; Pokrovskiy, V.A. *Rapid Commun. Mass Spectrom.* **1995**, *9*, 580; (d) Bogatyrev, V.M.; Borysenko, M.V. *J. Therm. Anal. Cal.* **2000**, *62*, 335; (e) Borysenko, M.V.; Bogatyrev, V.M.; Dyachenko, A.G.; Pokrovskiy, V.A.; In *Chemistry, Physics, and Technology of Surfaces*, Chuiko, A.A.; Ed.; Issues 7–8, KM Academia: Kiev, 2002, pp. 11–18.
23. Zhuravlev, L.T. *Coloids Surf. A* **1993**, *74*, 71.
24. Glinka, Yu.D.; Krak, T.B.; Belyak, Yu.N.; Degoda, V.Ya.; Ogenko, V.M. *Colloids Surf. A.* **1995**, *104*, 17.
25. Gun'ko, V.M.; Zarko, V.I.; Chuikov, B.A.; Dudnik, V.V.; Ptushinskii, Yu.G.; Voronin, E.F.; Pakhlov, E.M.; Chuiko, A.A. *Int. J. Mass Spectrom. Ion Proces.* **1998**, *172*, 161.
26. Zhdanov, V.P. *Elementary Physicochemical Processes on Solid Surfaces*, Plenum Press: New York, 1991.
27. Gun'ko, V.M.; Zarko, V.I.; Chibowski, E.; Dudnik, V.V.; Leboda, R.; Zaets, V.A. *J. Colloid Interface Sci.* **1997**, *188*, 39.

28. Gun'ko, V.M.; Zarko, V.I.; Turov, V.V.; Leboda, R.; Chibowski, E.; Holysz, L.; Pakhlov, E.M.; Voronin, E.F.; Dudnik, V.V.; Gornikov, Yu. I. *J. Colloid Interface Sci.* **1998**, *198*, 141.
29. Gun'ko, V.M.; Pokrovsky, V.A. *Internation. J. Mass Spectr. Ion Proces.* **1995**, *148*, 45.
30. Hwang, S.J.; Uner, D.O.; King, T.S.; Pruski, M.; Gerstein, B.C. *J. Phys. Chem.* **1995**, *99*, 3697.
31. Bronnimann, C.E.; Zeigler, R.C.; Maciel, G.E. *J. Am. Chem. Soc.* **1988**, *110*, 2023.
32. Gun'ko, V.M.; Zarko, V.I.; Turov, V.V.; Leboda, R.; Chibowski, E.; Pakhlov, E.M.; Goncharuk, E.V.; Marciniak, M.; Voronin, E.F.; Chuiko, A.A. *J. Colloid. Interface Sci.* **1999**, *220*, 302.
33. Gun'ko, V.M.; Zarko, V.I.; Turov, V.V.; Voronin, E.F.; Tischenko, V.A.; Dudnik, V.V.; Pakhlov, E.M.; Chuiko, A.A. *Langmuir* **1997**, *13*, 1529.
34. Gun'ko, V.M.; Zarko, V.I.; Turov, V.V.; Leboda, R.; Chibowski, E. *Langmuir* **1999**, *15*, 5694.
35. Gun'ko, V.M.; Turov, V.V. *Langmuir* **1999**, *15*, 6405.
36. (a) Mironyuk, I.F.; Gun'ko, V.M.; Turov, V.V.; Zarko, V.I.; Leboda, R.; Skubiszewska-Zięba, J. *Colloid. Surf. A* **2001**, *180*, 87; (b) Mironyuk, I.F. Scientific *Basis of Controlled Synthesis of Fumed Silica and Its Physicochemical Properties*; Doctoral Thesis, Institute of Surface Chemistry: Kiev, 2001; (c) Gun'ko, V.M.; Mironyuk, I.F.; Zarko, V.I.; Voronin, E.F.; Turov, V.V.; Pakhlov, E.M.; Goncharuk, E.V.; Leboda, R.; Skubiszewska-Zięba, J.; Janusz, W.; Chibowski, S. *J. Colloid Interface Sci.* **2001**, *242*, 90.
37. Bogillo, V.I.; Gun'ko, V.M. *Langmuir* **1996**, *12*, 115.
38. Gun'ko, V.M.; Bogillo, V.I.; Chuiko, A.A. *ACH—Model. Chem.* **1994**, *131*, 561.
39. Gun'ko, V.M.; Vedamuthu, M.S.; Henderson, G.L.; Blitz, J.P. *J. Colloid Interface Sci.* **2000**, *228*, 157.
40. Blitz, J.P.; Gun'ko, V.M. In *Encyclopedia of Surface and Colloid Science*, Hubbard, A.T.; Ed.; Marcel Dekker, 2002, pp. 2939–2950.
41. Gun'ko, V.M.; Sheeran, D.J.; Augustine, S.M.; Blitz, J.P. *J. Colloid Interface Sci.* **2002**, *249*, 123.
42. Gun'ko, V.M. *Mechanisms of Chemical Reactions at Disperse Oxide Surfaces*, Doctoral Thesis, Institute of Surface Chemistry, Kiev, 1995.
43. Gun'ko, V.M. *Kinet. Katal.* **1993**, *34*, 716.
44. Gun'ko, V.M.; Voronin, E.F.; Pakhlov, E.M.; Chuiko, A.A. *Langmuir* **1993**, *9*, 716.
45. Gun'ko, V.M. *Zh. Fiz. Khim.* **1997**, *71*, 1268.
46. Gun'ko, V.M.; Voronin, E.F.; Zarko, V.I.; Pakhlov, E.M.; Chuiko, A.A. *J. Adhesion Sci. Thechnol.* **1997**, *11*, 627.
47. Gun'ko, V.M. In *Fundamental and Applied Aspects of Chemically Modified Surfaces*, Blitz, J.P.; Little, Ch.; Eds.; Royal Society of Chemistry, Cambridge, 1999, pp. 270–279.
48. Gun'ko, V.M.; Zarko, V.I.; Voronin, E.F.; Pakhlov, E.M.; Chuiko, A.A. In *Fundamental and Applied Aspects of Chemically Modified Surfaces*, Blitz, J.P.; Little, Ch.; Eds.; Royal Society of Chemistry, Cambridge, 1999, pp. 183–190.
49. Gun'ko, V.M. *Theoret. Experim. Chem.* **2000**, *36*, 1.
50. Gun'ko, V.M. *Colloid. Surf. A* **1995**, *101*, 279.
51. Gun'ko, V.M. *Kinet. Katal.* **1991**, *32*, 322.
52. Gun'ko, V.M.; Silchenko, S.S.; Bogomaz, V.I. *Teoret. Eksperim. Khim.* **1991**, *27*, 715.
53. Gun'ko, V.M.; Basyuk, V.A.; Chernyavskaya, T.V.; Chuiko, A.A. *Ukr. Khim. Zh.* **1990**, *56*, 571.
54. Gun'ko, V.M.; Turov, V.V.; Zarko, V.I.; Dudnik, V.V.; Tischenko, V.A.; Voronin, E.F.; Kazakova, O.A.; Silchenko, S.S.; Chuiko, A.A. *J. Colloid Interface Sci.* **1997**, *192*, 166.
55. *Gaussian 94, Revision E. 1*, Frisch, M.J.; Trucks, G.W.; Schlegel, H.B.; Gill, P.M.W.; Johnson, B.G.; Robb, M.A.; Cheeseman, J.R.; Keith, T.; Petersson, G.A.; Montgomery, J.A.; Raghavachari, K.; Al-Laham, M.A.; Zakrzewski, V.G.; Ortiz, J.V.; Foresman, J.B.; Cioslowski, J.; Stefanov, B.B.; Nanayakkara, A.; Challacombe, M.; Peng, C.Y.; Ayala, P.Y.; Chen, W.; Wong, M.W.; Andres, J.L.; Replogle, E.S.; Gomperts, R.; Martin, R.L.; Fox, D.J.; Binkley, J.S.; Defrees, D.J.; Baker, J.; Stewart, J.P.; Head-Gordon, M.; Gonzalez, C.; and Pople, J.A. Gaussian, Inc., Pittsburgh PA, 1995.
56. Schmidt, M.W.; Baldridge, K.K.; Boatz, J.A.; Elbert, S.T.; Gordon, M.S.; Jensen, J.J.; Koseki, S.; Matsunaga, N.; Nguyen, K.A.; Su, S.; Windus, T.L.; Dupuis, M.; Montgomery, J.A. *J. Comput. Chem.* **1993**, *14*, 1347.
57. Dewar, M.J.S.; Zoebisch, E.G.; Healy, E.E.; Stewart, J.J. *J. Am. Chem. Soc.* **1985**, *107*, 3902.
58. Gun'ko, V.M.; Voronin, E.F.; Chuiko A.A. *Zh. Obsch. Khim.* **1991**, *61*, 552.
59. Gun'ko, V.M.; Voronin, E.F.; Chuiko, A.A. *Zh. Neorg. Khim.* **1991**, *36*, 320.
60. Gun'ko, V.M. *Zh. Fiz. Khim.* **1991**, *65*, 398.
61. Gun'ko, V.M. *Kinet. Katal.* **1991**, *32*, 576.
62. Bredow, Th.; Jug, K. *Surf. Sci.* **1995**, *327*, 398.
63. Cordoba, A.; Luque, J.J. *Phys. Rev. B.* **1985**, *31*, 8111.
64. Kassab, E.; Seiti, K.; Allavena, M. *J. Phys. Chem.* **1988**, *92*, 6705.
65. Leboda, R.; Gun'ko, V.M.; Marciniak, M.; Malygin, A.A.; Malkov, A.A.; Grzegorczyk, W.; Trznadel, B.J.; Pakhlov, E.M., Voronin, E.F. *J. Colloid Interface Sci.* **1999**, *218*, 23.
66. Turov, V.V.; Gun'ko, V.M.; Zarko, V.I.; Bogatyr'ov, V.M.; Dudnik, V.V.; Chuiko, A.A. *Langmuir* **1996**, *12*, 3503.
67. Gun'ko, V.M. *Zh. Fiz. Khim.* **1997**, *71*, 1268.
68. Brei, V.V.; Gun'ko, V.M.; Dudnik, V.V.; Chuiko, A.A. *Langmuir* **1992**, *8*, 1968.
69. Pokrovskiy, V.A. *Adsorption Sci. Technol.* **1997**, *14*, 301.
70. Gun'ko, V.M.; Leboda, R.; Pokrovskiy, V.A. *Polish J. Chem.* **1999**, *73*, 1345.
71. Brey, V.V.; Gun'ko, V.M.; Khavryuchenko, V.D.; Chuiko, A.A. *Kinet. Katal.* **1990**, *31*, 1164.

72. Gun'ko, V.M.; Brey, V.V.; Chuiko, A.A. *Kinet. Katal.* **1991**, *32*, 103.
73. Gun'ko, V.M.; Leboda, R.; Pokrovskiy, V.A.; Charmas, B.; Turov, V.V.; Ryczkowski, J. *J. Analyt. Appl. Pyrolys.* **2001**, *60*, 233.
74. Gun'ko, V.M.; Zarko, V.I.; Leboda, R.; Marciniak, M.; Janusz, W.; Chibowski, S. *J. Colloid Interface Sci.* **2000**, *230*, 396.
75. Gun'ko, V.M.; Voronin, E.F.; Pakhlov, E.M.; Zarko, V.I.; Turov, V.V.; Guzenko, N.V.; Leboda, R.; Chibowski, E. *Colloid. Surf. A* **2000**, *166*, 187.
76. Gun'ko, V.M. Zarko, V.I.; Leboda, R.; Chibowski, E. *Adv. Colloid Interface Sci.* **2001**, *91*, 1.
77. Gun'ko, V.M.; Leboda, R.; In *Encyclopedia of Surface and Colloid Science*, Hubbard, A.T.; Ed.; Marcel Dekker, 2002, pp. 864–878.

38 Structural and Adsorptive Characteristics of Fumed Silicas in Different Media

Vladimir M. Gun'ko, V.I. Zarko, V.V. Turov, E.F. Voronin, I.F. Mironyuk, and A.A. Chuiko
National Academy of Sciences of Ukraine, Institute of Surface Chemistry

CONTENTS

Fumed silicas synthesized under varied conditions were studied by means of adsorption, ^{1}H NMR, photon correlation spectroscopy, and electrokinetic methods. Prepared silicas possess different specific surface area ($S_{Ar} = 85–512\ m^2/g$), structures of primary particles and their swarms, concentrations of silanols ($C_{OH} = 1.9–5.2\ \mu mol/m^2$), and weakly ($C_{w,105} = 0.4–2.4$ wt%) and strongly ($C_{w,900} = 0.4–2.2$ wt%) bound waters. There is correlation between the specific surface area (S) of fumed silica and the flow velocity v_f ($S \sim \ln v_f$ at $v_f < 25–30$ m/s). Decrease in the amounts of hydrogen/oxygen in the flame and elevating synthesis temperature or flame turbulence enhance the size of primary particles, which become slightly micropous on addition of hydrogen reacting at the flame periphery. Dividing of the flow in the burner to several smaller flows reducing the turbulence without changes in other synthesis conditions significantly enhances the specific surface area. The concentration of silanols increases with growing primary particle size, and the hydrophilicity ($C_{w,105} + C_{w,900}$) decreases at oxygen deficiency on the synthesis. The impact of polymers on the suspension and dried powder characteristics depends on the adsorption mechanism and conformation of polymer molecules (globular or unfolded) due to strong or weak intramolecular interactions. Globular proteins interacting with silica through the flocculation mechanism have a weaker effect on the textural characteristics of powders prepared by drying of the suspensions but strongly impact the aqueous suspension shifting the swarm size distribution toward larger sizes in comparison with poly(vinyl pyrrolidone), PVP, adsorbed in the unfolded state and giving nearly monomodal particle size distribution. Ionogenic surfactant 1,2-ethylene-bis-(N-dimethyl carbodecyloxymethyl) ammonium dichloride at $C_{Aet} < 0.01$ wt. % and ethanol at $C_{EtOH} = 10–50$ wt. % impact the swarm size distributions of silicas dependent nonlinearly on the concentrations and pH. Adsorbed metal ions have an influence on the surface charge density and the electrophoretic mobility of fumed silica particles increasing with concentration.

Fumed silicas synthesized by high-temperature hydrolysis of $SiCl_4$ in the $O_2(N_2)/H_2$ flame are amorphous and possess a large specific surface area (S up to $500\ m^2/g$) and a narrow primary particle size distribution [1]. Protoparticles (1–2 nm) [2], formed in the initial zone of the flame, collide, stick together, coalesce, and are covered by new silica layers to form primary particles (average diameter $d = 5–50$ nm, density $\approx$ 2.2 g/cm^3). Collision, sticking and fusing of primary particles bonded by ≡Si—O—Si≡ give primary aggregates. Subsequent attachment of primary particles to them enhances their sizes as well as collision and sticking together of primary aggregates (at lower T) that leads to formation of larger secondary aggregates of 100–500 nm through mainly hydrogen and electrostatic bonding [1–3]. For aggregates, mass fractal dimension $D_m \approx 2.5$ was calculated using equation $\rho_{aggregate} = \rho_{particle}(d_{aggregate}/d_{particle})^{D-3}$ at $\rho_{aggregate} \approx$ 0.7 g/cm^3 [3]. Clearly, bonding strength of primary particles in primary and secondary aggregates depends on their temperature and sizes, coordination numbers and type of bonding (chemical ≡Si—O—Si≡ or intermolecular such as direct ≡SiOH ... O(H)Si≡ or with the participation of water molecules ≡SiOH ... $(OH_2)_n$ ... O(H)Si≡, etc.) [1]. At lower temperatures and after hydration, aggregates form loose agglomerates (>1 μm, $D_m \approx 2.1–2.2$, bulk density ≈2–6% of the specific density) through hydrogen bonding (with the participation of water molecules) and electrostatic interactions [1–3]. It should be noted that estimation of the fractal dimension using small angle neutron (SANS) or x-ray (SAXS) scattering methods gives typically smaller values (for loosely packed samples) than those determined from the adsorption data [2,4]. For instance, from the SANS and light scattering data, $D \approx 1.8$ for such fumed silicas as Cab-O-Sil M5, HS-5, and EH-5; however, according to another method [2a], their surface fractal dimension in 2.0, 2.28 and 2.54, respectively, which agree with the values estimated from adsorption data. If $S \geq 100\ m^2/g$ that isolated primary particles are not observed separately without special treatment, while the smaller the particles, the stronger the bonding in the aggregates [1]. According to scanning electron microscope (SEM) findings, fumed silica does not change their morphology on heating at 1000°C for 7 days, but at 1200°C, it crosslinks to glass [1]. Clearly, changes in the flame temperature between 1000°C and 1700°C can strongly influence the characteristics of primary particles and their swarms [1,5].

Many physicochemical properties of fumed silica depend significantly not only on the primary and secondary particle size distributions but also on the concentration of adsorbed water (C_w) in the form of intact molecules and ≡SiOH groups (C_{OH}) [1–3,6,7]. For instance, marked amounts of adsorbed water can negatively affect the characteristics of fumed silica as a filler of liophilic media or polymers. There are several methods to change C_w and C_{OH} such as chemical modification (hydrophobization) of silica by organosilicon or organic compounds, heating at high temperature giving a tentative diminution of the hydrophilicity, etc. [1–3,6]. However, the first method enhances the material cost and changes the nature of the surfaces and the particle size distributions, which can be undesirable for utilization of dispersed silica. The use of the second method results in a decrease in S and changes in the structure of secondary particles. Therefore, production of fumed silica possessing initially a desirable surface hydrophilicity (C_w, C_{OH}) and an appropriate specific surface area [7,8] can be of interest from both theoretical and practical points of view. There are several noteworthy tendencies in preparation of fumed silicas: (i) formation of porous primary particles at lowering temperature of the synthesis in contrast to nonporous particles synthesized under standard conditions; (ii) variations in the concentration of surface hydroxyls and the hydrophilicity of powder as a whole; (iii) enhancement of the dispersity of primary particles due to changes in synthesis conditions and the use of external factors affecting the flame (e.g., strong external electrostatic field); (iv) utilization of different precursors (e.g. $SiCl_x(CH_3)_{4-x}$) and dopants to control important properties of fumed materials.

There is not a conventional model of primary particle bonding in aggregates of fumed silica, as some authors assume hydrogen and electrostatic bonds, but others believe that primary particle bonding occurs through ≡Si—O—Si≡ bridges [1–5]. From SEM or TEM images [1–3], it is difficult to determine the nature of this bonding, as some primary particles in aggregates seem tightly sintered (due to collision, sticking and fusing in the hot reaction zone of the flame possibly with formation of ≡Si—O—Si≡ bonds between adjacent particles), but others have small-area contacts, which can be formed at lower temperature at the flame periphery due to intermolecular, that is, nonsiloxane, binding. Obviously, a number of chemical and intermolecular bonds between adjacent primary particles in aggregates can be varied depending on their temperature on the collision and other conditions in the flame. Therefore one can assume that both types of primary particle bonding in aggregates are possible, which can differently show on silica interaction with polar environment (water, polymers, surfactants, solvents). Fumed silica composed of roughly spherical primary particles forming swarms can change the structural characteristics on suspending-drying, heating, adsorption of surfactants and polymers, etc., as this initial material does not exist in the form of individual primary particles and the characteristic size at each subsequent level of the structural hierarchy increases by ten times. Clearly, any pretreatment of fumed silica possessing structurally nonrigid swarms, whose stability typically decreases with their size, can alter many of the properties important for different applications [1–6]. The concentrated aqueous suspensions of fumed silica ($C_{SiO_2} \approx 5$ wt.%) possessing frequently multimodal size distributions of swarms (or secondary particles) of primary particles (SSD) [9] due to preservation of the mentioned swarms are relatively stable, and silica does not lose the adsorptive ability during long storage period [10]. Besides, at $C_{SiO_2} \approx 9-10$ wt.%, "jellification" of the silica dispersion is observed after its exposure during several hours due to formation of a continuous three-dimensional network of interparticle bonds. At the same time, the diluted suspensions ($C_{SiO_2} < 1$ wt.%) are less stable and characterized by more broadened SSDs [9].

Chemical modification of oxide surfaces, immobilization of polymers on them, addition of different surfactants or solvents (e.g., alcohols) to the aqueous suspensions can be responsible for alterations in the hydrogen bond network and the free energy of the interfacial layers, surface charge distribution, and so on, which can affect the dispersion stability and other characteristics of both suspensions and powders prepared from these suspensions [9–15]. The last with adsorbed molecules (e.g., drugs, PVP, proteins, cellulose, etc.) can be of interest for medicine, biotechnology, etc.

The aim of this work was to synthesize a variety of fumed silicas possessing different texture and hydrophilicity, to characterize the structural and adsorptive properties of these materials in air and aqueous media, and to study the impact of polymers (bovine serum albumin (BSA), egg albumin, gelatin, PVP), an ionogenic surfactant (drug Aethonium, 1,2-ethylene-bis(N-dimethyl carbodecyloxymethyl) ammonium dichloride) and polar solvent mixtures with water-ethanol, and aqueous metals (Cs(I), Sr(II), Pb(II)) on the characteristics of fumed silica in the aqueous suspensions and dried powders.

EXPERIMENTAL

MATERIALS

A variety of fumed silicas ($S_{Ar} = 85–512$ m^2/g) were synthesized using $SiCl_4$ burned (hydrolyzed/oxidized) in the oxygen/nitrogen-hydrogen flame ($O_2/N_2 \approx 0.25$) under controlled conditions (temperature, flow velocity (Figure 38.1) and turbulence, ratio of reagent amounts and their distribution in the flame, burner diameter $d_n = 36$, 42, 52 or 62 mm) to produce materials possessing varied hydrophilicity (C_w, C_{OH}), specific surface area and other structural characteristics (Table 38.1–Table 38.4). The flame temperature ($T_f = 1000–1320$°C) was measured using a Ranger II (Rayter) optical pyrometer. The modified technique of the fumed silica synthesis was described in details elsewhere [7,8].

To study adsorption of proteins, fumed silica (Pilot Plant of Institute of Surface Chemistry, Kalush, Ukraine; 99.8% purity, specific surface area $S_{Ar} = 102–411$ m^2 g^{-1}) was heated at 773 K for several hours to remove residual HCl and other adsorbed compounds. Samples A-200 ($S_{N2} \approx 190$ m^2 g^{-1}) and A-300 ($S_{N2} \approx 300–340$ m^2 g^{-1}) at the concentration $C_{SiO_2} = 1–6$ wt.% treated in a ball mill during several hours (ball-milled suspension, BMS), or strongly sonicated for 5 min by means of an ultrasonic disperser, or softly treated in an ultrasonic bath ($C_{SiO_2} = 0.1–6$ wt.%) were also used to explore the polymer/silica suspensions.

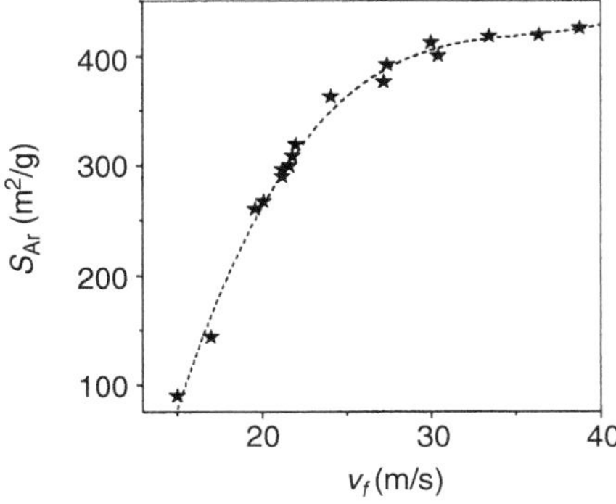

FIGURE 38.1 Relationship between the flow velocity v_f and the specific surface area (S_{Ar}) for the first-third series of fumed silicas.

ADSORPTION

Egg albumin (molecular weighting $\approx 4.4 \times 10^4$), bovine serum albumin, gelatin ($\approx 3.5 \times 10^5$), ethanol and Aethonium (pharmaceutical purity) were used as received. Adsorption of BSA and gelatin was studied without strong pretreatment of oxides and with no addition of the electrolyte buffer solution. Silica (40 mg of oxide per 5 cm^3 of protein solution or 1 g oxide per 125 cm^3 of the solution) was added to protein solution (0.6 wt. %), agitated for 0.5 h and adsorption was measured after exposure for 1 h (plateau adsorption), and the suspension was centrifuged at 6000 rpm for 15 min. The Biuret reactant (4 cm^3) was added to the supernatant (1 cm^3). After agitating, the solution was exposed for 0.5 h, then its optical density was measured at $\lambda = 540$ nm to calculate the adsorbed amount of proteins comparing with the initial solution. This method was described in detail elsewhere [16]. Notice that dependence of albumin adsorption onto fumed oxides on pH was studied previously [10c]. Protein/silica residual was dried at 313 K and degassed at 333 K for 2 h before nitrogen adsorption measurements.

Poly(vinyl pyrrolidone) (pharmaceutical purity) $(-CH_2\ CH(NC_4H_6O)-)_n$, $n \approx 100$, molecular weight of 12600 ± 2700, was used as received. The PVP solution was added to the BMS of silica, or fumed silica powder was added to the PVP solution then agitated (1000 rpm) for several hours, then sonicated for 5–6 min. The ratio between the concentrations $\gamma = C_{PVP}/C_{SiO_2}$ was between 0 and 1, and $\gamma \leq 0.1$ corresponds to practically irreversible adsorption of PVP, as it is not washed from silica. Estimated statistical monolayer ($\Theta = 1$) corresponds to $\gamma \approx 0.2$. Several experiments with PVP/silica were performed using the physiological buffer solution (PBS) with NaCl

TABLE 38.1
Characteristics of Fumed Silica (First Series) Possessing Low Hydrophilicity

Sample number	ρ_{ap} g/dm^3	S_{N2} m^2/g	S_α m^2/g	S_{Ar} m^2/g	V_p cm^3/g	R_p nm	d nm	D_{AJ} $p/p_0 < 0.85$	C_{w105} wt.%	C_{w900} wt.%
S1–1	37	144	145	118	0.261	3.7	18.9	2.618	1.0	1.1
S1–2	37	160	159	143	0.290	3.9	17.0	2.596	0.8	0.8
S1–3	49	206	219	216	0.411	4.0	13.2	2.588	1.0	0.9
S1–4	43	226	239	242	0.437	3.9	11.4	2.589	1.0	0.9
S1–5	46	337	340	328	0.608	3.6	8.3	2.608	1.2	1.0
S1–6	42	381	369	381	0.667	3.5	7.2	2.624	1.4	1.0

Note: $C_w = C_{w,105} + C_{w,900}$; $C_{w,105}$ is the amount of water desorbed on heating at $T < 105°C$, $C_{w,900}$ is the amount of water desorbed at $105 < T < 900°C$.

(5.5 g), KCl (0.42 g), $CaCl_2$ (0.5 g), $MgCl_2$ (0.005 g), and $NaHCO_3$ (0.23 g) per dm^3 of the distilled water. The PVP adsorption was determined spectrophotometrically at $\lambda = 420$ nm on the basis of the calibrated graph using the corresponding mixtures of the liquid residual of the centrifuged PVP/silica suspensions and the aqueous solutions of citric acid (38.4 g per liter of the distilled water) and iodine (0.81 g of I_2 plus 1.44 g of KI per dm^3 of the distilled water) forming characteristic complexes with PVP.

Nitrogen adsorption-desorption isotherms were recorded for the first series of silicas (possessing a low initial hydrophilicity) at 77.4 K using a Micromeritics ASAP 2010 (or 2405 N) adsorption analyzer. The specific surface area (Tables 38.1–38.4, S_{N2} and S_{Ar}) was calculated using the BET method [17]. The pore volume V_p was determined from the adsorption at relative pressure $p/p_0 \approx 0.98$–0.99. The S_{N2} and V_p were utilized to estimate average pore radius R_p. The specific surface area S_α (Table 38.1) was calculated using the α_S plot method [17] and silica gel Si-1000 as a reference material [18]. The BET surface area (S_{Ar}) for all the samples (Tables 38.1–38.4, Figure 38.1) was also evaluated using a Jemini 2360 (SVLAB) apparatus with argon adsorption just after the silica synthesis. Note that the nitrogen adsorption measurements were performed 2–3 months late; therefore, the differences between S_{Ar} and S_{N2} (Table 38.1) are rather random in the confidence BET range, while they are less than 18%. Calculation of the fractal dimension (D_{AJ}) [19] was performed on the basis of nitrogen adsorption at $p/p_0 \leq 0.85$. The average diameter of primary particles (Tables 38.1–38.4, d) was estimated as follows

$$d = 6000 \Big/ \left(S_{BET} \sum_{i=1}^{N} C_i \rho_i \right) \tag{38.1}$$

where C_i and ρ_i are the concentration and the specific density of i phase, respectively, and N is the number of phases ($N = 1$ for fumed silica and 2 or 3 for mixed oxides [9,10,12,13]). The pore size distribution $f(R_p)$ was calculated using the overall adsorption isotherm equation [20]

$$W = \int_{r_{\min}}^{r_k(p)} f(R_p)\mathrm{d}R_p + \int_{r_k(p)}^{r_{\max}} \frac{w}{R_p} t(p, R_p) f(R_p)\mathrm{d}R_p \tag{38.2}$$

where $r_{\min}$ and $r_{\max}$ are the minimal and maximal half-widths of pores, respectively; $w = 1$ for slitlike pores and 2 for cylindrical pores; $r_k(p)$ is determined by the modified Kelvin equation

$$r_k(p) = \frac{\sigma_s}{2} + t(p, R_p) + \frac{w\gamma v_m \cos\theta}{R_g T \ln(p_0/p)} \tag{38.3}$$

TABLE 38.2
Synthesis Conditions and Characteristics of Fumed Silicas (Second Series)

Sample number	d_n mm	$SiCl_4$ dm^3/h	γ	v_f m/s	R_e	T_f °C	S_{Ar} m^2/g	d nm
S2–1	36	65	1	27.2	51160	1092	376	7.2
S2–2	36	70	1	30.4	55292	1042	400	6.6
S2–3	36	80	1	33.4	62860	1002	416	6.5
S2–4	42	70	1	21.6	47290	1154	300	9.1
S2–5	52	120	1	24.1	67560	1202	362	7.5
S2–6	52	120	0.8	21.2	57440	1242	296	9.2

Note: $\gamma = 1$ corresponds to the stoichiometric ratio between H_2/O_2 and $SiCl_4$.

TABLE 38.3
Synthesis Conditions and Hydrophilicity of Fumed Silicas (Third Series)

Sample number	γ_{H_2}	γ_{O_2}	v_f m/s	S_{Ar} m²/g	d nm	C_{OH} µmol/m²	$C_{w,105}$ wt%	$C_{w,900}$ wt%	$a_{w,mono}$ µmol/m²
S3–1	1.0	1.0	21.6	300	9.1	3.32	1.8	1.7	8.3
S3–2	1.0[a]	1.0	20.1	267	10.2	3.67	1.6	1.4	8.2
S3–3	1.0[a]	0.8	21.2	290	9.4	3.48	1.0	0.8	9.0
S3–4	1.1[a]	0.65	19.6	260	10.5	3.81	0.6	0.3	4.9
S3–5	1.2[a]	0.8	17.0	144	18.9	5.00	0.4	0.4	2.6
S3–6	1.0[b]	0.8	21.8	308	8.9	3.30	1.3	1.5	4.4
S3–7	1.0[b]	1.0	21.6	299	9.1	3.32	1.8	2.0	8.2
S3–8	1.2[a]	0.65	22.0	319	8.5	3.30	0.5	0.6	3.0

Note: Flow velocity in the annular nozzle v_{f,H_2} = [a]2 m³/h or [b]8 m³/h; $\gamma_{H_2} = 1$ and $\gamma_{O_2} = 1$ correspond to the stoichiometric amounts of H_2 and O_2. The first sample was synthesized under standard conditions (stoichiometric ratio $H_2/O_2/SiCl_4$, laminar flow, etc.).

and $t(p, R_p)$ can be computed with the modified BET equation

$$t(p, R_p) = t_m \frac{cz}{(1-z)} \times \frac{[1 + (nb/2 - n/2)z^{n-1} - (nb+1)z^n + (nb/2 + n/2)z^{n+1}]}{[1 + (c-1)z + (cb/2 - c/2)z^n - (cb/2 + c/2)z^{n+1}]} \quad (38.4)$$

$t_m = V_m/S_{BET}$; $b = \exp(\Delta\varepsilon/R_gT)$; $\Delta\varepsilon$ is the excess of the evaporation heat due to the interference of the layering on the opposite wall of pores; $t(p, R_p)$ is the statistical thickness of adsorbed layer; V_m is the BET monolayer capacity; $c = c_s \exp((Q_p - Q_s)/R_gT)$; c_s is the BET coefficient for adsorption on flat surface $c_S = \gamma e^{E-Q_L/R_gT}$, Q_L is the liquefaction heat, E is the adsorption energy, γ is a constant; Q_s and Q_p are the adsorption heat on flat surface and in pores, respectively; $z = p/p_0$; n is the number (noninteger) of statistical monolayers of adsorbate molecules and its maximal value for a given r_k is equal to $(R_p - \sigma_s/2)/t_m$; σ_s is the collision diameter equal to the average size of nitrogen and surface atoms; and R_g is the gas constant. Desorption data were utilized to compute the $f(R_p)$ distributions with Equations (38.2)–(38.4) and the modified regularization procedure [27]

TABLE 38.4
Synthesis Conditions and Characteristics of Fumed Silicas (Fourth Series) with Different Specific Surface Area and Hydrophilicity

Sample number	d_n mm	$SiCl_4$ dm³/h	γ_{H_2}	γ_{O_2}	v_f m/s	R_e	T_f °C	S_{Ar} m²/g	d nm	C_{OH} µmol/m²	$C_{w,105}$ wt.%	$C_{w,900}$ wt.%
S4–1	36	97.5	1	1	32.1	107100	1082	368	7.4	2.69	1.96	1.78
S4–2	36	97.5	1	1.5	42.9	122000	1051	398	6.9	2.38	2.40	2.20
S4–3	42	82.5	1	1	20.1	77800	1100	325	8.4	3.48	1.88	1.80
S4–4	42	82.5	1.1	1	20.5	78190	1105	350	7.8	3.16	1.79	1.62
S4–5	42	82.5	1.3	1	21.5	79150	1100	376	7.3	2.82	1.32	1.26
S4–6	52	105	1	1	16.6	79920	1190	170	16.0	4.70	1.62	1.44
S4–7	52	127.5	1	1	20.2	97000	1202	255	10.7	3.67	1.64	1.52
S4–8	52	135	1	1	21.4	102800	1210	275	9.9	3.64	1.72	1.60
S4–9	52	135	1	0.7	19.2	98810	1320	320	8.5	3.28	0.66	0.72
S4–10	52	135	1	0.6	17.3	95370	1290	262	10.4	3.67	0.42	0.46
S4–11[a]	52	135	1	1	21.4	26690	1221	415	6.6	2.20	1.64	1.52
S4–12[a]	52	135	1	0.8	18.6	25350	1250	512	5.3	1.88	1.40	1.20
S4–13	62	150	1	1	16.7	95680	1264	85	32.1	5.20	0.80	0.90
S4–14	62	180	1	1	19.9	114600	1284	196	13.9	4.10	1.60	1.72
S4–15[a]	62	180	1	1	19.9	24350	1298	412	6.6	2.20	1.90	2.10

Note: [a]Flow in the burner is divided by smaller ones with the effective diameter of 13.5 mm.

under non-negativity condition ($f(R_p) \geq 0$ at any R_p) and at fixed regularization parameter $\alpha = 0.01$. For fumed oxides with spherical primary particles, Equation (38.3) should be replaced by [17]

$$\ln\frac{p_0}{p} = \frac{\gamma v_m}{R_g T}\left[\frac{1}{r_k(p)} - \frac{2}{\sqrt{(a+t'+r_k(p))^2 - a^2} - r_k(p) + a + t'}\right] \quad (38.5)$$

where a is the radius of primary particles, and $t' = t + \sigma_s/2$. Fumed oxides are characterized by the primary particle size distribution $f(a)$, which becomes broader with lowering specific surface area [1]. Using known $f(a)$ for fumed silica [1] with an appropriate average D value as an initial $f(a)$ function for mixed oxides, one can calculate an initial $f(R_p)$ distribution using Equation (38.2), Equation (38.4), and Equation (38.5). Then applying a self-consistent method with binary regularization (SCR) with respect to both $f_{SCR}(R_p)$ and $f(a)$ (by turns up to self-consistency for both distributions) one can obtain the corresponding matched solution of the integral equations for the distribution functions of the pore (gap) and primary particle sizes. For comparative investigations of the gap size distributions for fumed oxides, another simple model with cylindrical pores at $R_p > 2t_m$ and slitlike micropores at $R_p < 2t_m$ was also used with Equations (38.2)–(38.4). Application of Equations (38.2)–(38.4) to pores of different types (such as cylindrical, slitlike, gap between primary particles and mixed ones) within the scope of uni-regularization with respect only to $f(R_p)$ was described in detail elsewhere [10,12–14].

Water adsorption-desorption on silica samples S3-*i* (weighing 50–100 mg, pressed at approximately 10^4 Torr) was studied using an adsorption apparatus with a McBain–Bark quartz scale. After evacuation to 10^{-3} Torr for 1–2 h, samples were heated at 613 K for 3–4 h to a constant weight, then cooled to 293 ± 0.2 K, and adsorption of water vapor was studied at pressure (p) varied in the (0.06–0.999)p_0 range. The measurement accuracy was 1×10^{-3} mg with relative mean error ±5%. Water desorption in air (Tables 38.1–38.4, C_w) was studied by means of the Q-gravimetric method. The distributions of free energy changes (ΔG) upon water adsorption on the silica surfaces from air were calculated using the Langmuir equation [17] as the kernel in the overall adsorption isotherm in the form of Fredholm integral equation of the first kind [21,22].

Electrophoresis and Swarm Size Distribution (SSD)

Electrophoretic and swarm size distribution (SSD) investigations were performed using a Zetasizer 3000 (Malvern Instruments) apparatus based on the photon correlation spectroscopy (PCS) ($\lambda = 633$ nm, $\Theta = 90°$, software version 1.3). Deionized distilled water and 0.4 g of oxide per dm^3 of the water was utilized to prepare the suspensions, which were then ultrasonicated for 5 min using an ultrasonic disperser (Sonicator Misonix Inc.) (500 W, frequency 22 kHz). The pH values measured by a precision digital pH-meter were adjusted by addition of 0.1 M HCl or NaOH solutions, the salinity was constant (0.001 M NaCl). Electrophoretic behavior and SSD of fumed oxides in the aqueous suspensions studied by using PCS method was described in detail elsewhere [9].

According to the Smoluchowski theory [23], there is a linear relationship between electrophoretic mobility U_e and ζ potential $U_e = A\zeta$, where A is a constant for a thin electrical double layer (EDL) at $\kappa a \gg 1$ (where a is the particle radius, and κ is the Debye–Huckel parameter). For a thick EDL ($\kappa a < 1$), e.g., at pH close to the isoelectric point (IEP), the equation with the Henry correction factor is appropriate

$$U_e = 2\varepsilon\,\zeta/3\eta \quad (38.6)$$

where ε is the dielectric permittivity; and η is the viscosity. However, the relationship between the ζ potential and the mobility for aggregates of primary particles should be corrected taking into consideration the particle volume fraction (ϕ) and the shear-plane potential (Ψ_a) at the particle-fluid interfaces within the porous aggregate [24]

$$U_e = \frac{\varepsilon}{3\eta}\left(1 + \frac{\phi}{2}\right)(2\zeta + \Psi_a F) \quad (38.7)$$

where F is the electroosmotic flow factor;

$$\Psi_a = \frac{\zeta(1+\kappa a)}{B}; \quad (38.8)$$

$$B = \frac{(1-\kappa^2 ab)\sinh(\kappa b - \kappa a) - (\kappa b - \kappa a)\cosh(\kappa b - \kappa a)}{\sinh(\kappa b - \kappa a) - \kappa b\cosh(\kappa b - \kappa a)}; \quad (38.9)$$

b is the outer radius of a unit cell representative, in an average sense, of the aggregate interior defined by $\phi = a^3/b^3$ (e.g., for aggregates of fumed silicas $\phi \approx 0.3$–0.4 estimated from the pore volume of aggregates). Differences in the ζ potentials calculated using eqs (38.6) and (38.7) can reach up to 25% [24]. One can estimate the κ values according to Xu [25] and the average size of primary particles from the specific surface area using Equation (38.1) (or from direct observation of primary particles using microscopic methods) [1] or the primary particle size distribution $f(a)$

with minimal (a_{min}) and maximal (a_{max}) radii

$$U_e = \frac{\varepsilon}{3\eta}\int_{a_{min}}^{a_{max}} (1+\frac{\phi}{2})(2\zeta + \Psi_a F) f(a) da \qquad (38.10)$$

Potentiometric Titration

To evaluate the surface charge density, potentiometric titrations were performed using a Teflon thermostated vessel in nitrogen atmosphere free from CO_2 at 25 ± 0.2°C. The solution pH was measured using a PHM240 Research pH-meter (G202C and K401 electrodes) coupled with an REC-61 recorder. The surface charge density was calculated using the potentiometric titration data for a blank electrolyte solution and oxide suspensions ($C_{SiO_2} = 0.2$ wt.%), at a constant salinity of 10^{-3} *M* NaCl, from the difference of acid or base volume utilized to obtain the same pH value as that for the background electrolyte of the same ionic strength according to the following equation $\sigma_0 = \Delta VcF/mS_{BET}$, where $\Delta V = V_s - V_e$ is the difference between the base (acid) volume added to the electrolyte solution V_e and suspension V_s to achieve the same pH; *F* is the Faraday constant, *c* is the concentration of base (acid), *m* is the weight of the oxide.

Adsorption of Metal Ions

Adsorption of Pb(II) on the oxide surfaces ($C_{SiO_2} =$ 0.2 wt.%) was performed from the aqueous solution of $Pb(ClO_4)_2$ at the initial Pb(II) concentration of 10^{-5}, 10^{-4} or 10^{-3} *M* (concentration of radioactive species ^{210}Pb(II) was 10^{-6} *M*) with addition of a neutral electrolyte (10^{-3} *M* $NaClO_4$) using a Teflon cell (50 cm^3) temperature-controlled at T = 25 ± 0.2°C. The pH values were adjusted by addition of 0.1 *M* HCl or NaOH solutions. Determination of gamma radioactivity of the solution containing ^{210}Pb was performed using a Beckman Gamma 5500B counter. Adsorption of Sr(II) (with ^{90}Sr and 10^{-3} *M* NaCl as a neutral electrolyte) and Cs(I) (with radioisotope ^{137}Cs and without a neutral electrolyte) using a similar technique was performed at the metal ion concentrations of 10^{-5}, 10^{-4} or 10^{-3} *M*. In the case of the Sr(II) adsorption, the measurements were performed in two channels in order to eliminate ^{90}Y contribution.

To calculate the normalized adsorption constant (K_n) (relative binding strength) distributions $f(pK_n)$ of aqueous metals, the integral equation [26]

$$\theta(pH) = \int_0^\infty \frac{K_n z}{1+K_n z} f(pK_n) dK_n \qquad (38.11)$$

(where θ is the normalized cation uptake $C_{cat}/C_{cat,0}$; $C_{cat,0}$ is the initial concentration of aqueous metal; log z = −pH, $pK_n = -\log K_n$) was solved using the modified regularization procedure CONTIN with unfixed regularization parameter (automatically determined on the basis of F-test and confidence regions) [22] and nonnegativity condition for $f(K_n)$. To determine the equilibrium constants of cation adsorption (K_{Ct}), the triple layer model (TLM) was also applied using the GRIFIT program [27].

^{1}H NMR Spectroscopy

A high-resolution WP-100 SY (Bruker) NMR spectrometer with a bandwidth of 50 kHz was used for recording ^{1}H NMR spectra of water bound to the silica surfaces in the gas and liquid media. Relative mean errors were ±10% for signal intensity and ±1 K for temperature. The amount of interfacial unfrozen water (C_{uw}) in the aqueous suspensions of fumed silica frozen at 200 < T < 273 K was determined comparing an integral intensity (I_{uw}) of ^{1}H NMR signal of unfrozen water with that of water adsorbed on silica powder in air using a calibrated function $f(C^c)$ obtained on the basis of the measurements of the integral intensity for given amounts of water adsorbed onto the surface from the gas phase. The signals of surface hydroxyls and water molecules from ice were not detected due to features of the measurement technique and the short time (~10^{-6} s) of cross-relaxation of proton in solids. Gibbs free energy changes (ΔG) in the interfacial water were calculated (with relative mean error ±15%) using its known dependence for ice $\Delta G = 0.036(T-273)$. One can assume that water is frozen (T < 273 K) at the interfaces when $G = G_i$ and the value of $\Delta G = G - G_0$ equals $\Delta G_i = G_i(T) - G_i\,|_{T=273\,K}$ and corresponds to a decrease in *G* due to water interaction with the solid surfaces (G_0 denotes the Gibbs free energy of undisturbed bulk water). The amounts of silica in the aqueous suspensions were approximately constant (≈ 6 wt. %) for all the studied dispersions. To obtain the isotropic value of the chemical shift of ^{1}H ($\delta_{H,iso}$) (referenced to tetramethylsilane) at different temperatures, wet silica powders suspended in $CDCl_3$ (utilized to reduce noise signals) were used. This ^{1}H NMR technique with freezing-out of the bulk water was described in detail elsewhere [10a,10b,12a,12d,13d,28,29].

The adsorption potential distribution in respect to the interfacial water disturbed by silica or silica/PVP was computed as $f(A) = -dC_{uw}/dA$ (where $A = -\Delta G$ is the differential molar work). Additionally, ΔG as a function of the pore radius (R_p) or volume (V_p) filled by unfrozen water was calculated using the pore size distribution $f(R_p) = dV_p/dR_p$ assuming that water can be frozen in narrow pores at lower temperature than in larger ones. Since stable aggregates of primary particles exist in the aqueous suspension of fumed silica [9], one can assume that the channel structure (i.e., gaps between primary particles) in these aggregates does not strongly change in the aqueous suspensions; however, agglomerates can be rearranged significantly up to total decomposition. Since unfrozen water locates mainly in aggregates or on their outer surfaces, $\Delta G(C_{uw})$ can be transformed to $\Delta G(V_p)$ or $\Delta G(R_p)$ assuming that the unfrozen water density $\approx$ 1 g/cm^3 and this water fills the pores similarly to liquid nitrogen used to estimate $f(R_p)$ and V_p.

Infrared Spectroscopy

A-300 and PVP/A-300 prepared with the solid residual of the centrifuged dispersion dried at room temperature, then heated at 350K for 3 h and pressed (28 × 2 mm, ≈10–16 mg) or PVP (0.5–0.6 mg) stirred with KBr (≈60 mg) were used to record the IR spectra by means of a Specord M80 (Karl Zeiss, Jena) spectrophotometer. BSA/A-300 samples (22 × 5 mm, ≈12 mg) were prepared utilizing dried (at room temperature for 24 h), stirred and pressed solid residual of the centrifuged (6000 rpm for 30 min) BSA/silica dispersions (C_{SiO_2} = 5 wt. % and various C_{BSA}) exposed for 1 h.

RESULTS AND DISCUSSION

An increase in the reactant discharge rate at the stoichiometric ratio ($\gamma = 1$) between O_2/H_2 and $SiCl_4$ and at the same burner diameter d_n gives lowering flame temperature T_f, increase in the flow velocity v_f and the specific surface area S_{Ar} (decrease in the primary particle size d) (Table 38.2, samples 1, 2, and 3, Table 38.4, samples 6, 7, and 8) due to reduction of the reaction time in the flame. An enlargement of the burner diameter leads to elevating T_f (S2-2, S2-4, S2-5) and decreasing v_f and S_{Ar} (S2-2, S2-4). This effect can be connected to raising flow turbulence (Table 38.2 and Table 38.4, R_e) and more effective heat exchange with environment [8]. An expansion of the reaction zone at a larger burner diameter and the same injection reactant rate elevates the flame temperature, which can be also changed due to variation in the reactant ratios. For instance, a diminution of γ (for O_2/H_2 in respect to $SiCl_4$) to 0.8 (S2-6) gives elevating T_f (Table 38.2) due to possible changes in the reaction mechanism. On the H_2 deficiency, contribution of direct oxidizing of $SiCl_4$ to SiO_2 instead of hydrolysis of Si-Cl increases [8]. If $1.3 > \gamma_{H2} > 1$ that T_f can slightly decrease due to such endothermic reactions as $SiO_2 + H_2 \rightarrow SiO + H_2O$, $SiCl_4 + 2H_2 \rightarrow Si + 4HCl$, etc. At γ_{H2} between 1.3 and 1.4, T_f elevates (as well as v_f and S_{Ar}, see S4-3, S4-4, S4-5) and the surface hydrophilicity decreases (C_{OH}, C_w), but at $\gamma_{H2} > 1.4$, T_f decreases again [8]. At diminution of $\gamma_{O2} < 1$, S_{Ar} increases (S4-8, S4-9) then decreases (S4-10). This effect can be linked to formation of Si and Si–H, which then interact with oxygen drawn to the flame from the environment. If $\gamma_{O2} > 1$ (S4–2) (over-stoichiometrical oxygen provides direct oxidizing $SiCl_4$) that S_{Ar} increases (as well as C_w); however, S_{Ar} is maximal at $\gamma_{O_2} < 1$ (S4-12) and a low flame turbulence (Table 38.4, R_e). There is well-seen relationship between v_f and $S_{Ar} \sim ln v_f$ (at $v_f <$ 25–30 m/s, at $v_f >$ 30 m/s enhancement of S becomes significantly slower) or d estimated using Equation (38.1). At d_n = 42, 52, and 62 mm and $v_f >$ 30, 24, and 22 m/s (up to 100 m/s due to the automodel regime), respectively, the silica dispersity does not practically change [8]. Possible deviations from the graph shown in Figure 38.1 can be due to changes in the reaction mechanism at nonstoichiometric amounts of oxygen (e.g., S4-8, S4-9, S4-10) or the flame turbulence (S4-11, S4-12, S4-15).

Reynolds criterion $R_e = v_f d_n \rho_g / \eta$ (where ρ_g is reactant vapor density) corresponds to the turbulent flame (Tables 38.2 and 38.4, $R_e > 10^4$). Great turbulence promotes formation of larger particles with tight attachment on sticking in aggregates with subsequent layering of SiO_2 on their contacts. Dividing of the flow in the burner to several smaller flows (without changes of other synthesis conditions) reduces R_e and enhances S_{Ar} significantly (Table 38.4, samples 11, 12, and 15). In the case of the laminar flow (standard synthesis), primary particles are spherical, nonporous and their contacts in aggregates are less tight than those for nonstandard silica. Additional supply of a low amount (in comparison with the main flow) of hydrogen (on deficiency of O_2 in the flame) through the annular nozzle [8c] provides the hydrolysis of residual $SiCl_4$ and Si-Cl$_x$ at the flame periphery at 600–800°C that results in formation of porous surface layer (shallow micropores) on primary particles.

Changes in V_p and S (or primary particle diameter d, as $S \sim 1/d$) lead to marked alternations in nitrogen adsorption (Fig. 38.2a); however, the type of the isotherms is the same (II type) [17], as well as for other fumed oxides, due to their structural features [1–10,12,13]. The isotherms and the α_S plots (Figure 38.2) or R_p (Table 38.1) demonstrate that fumed silicas are rather mesoporous independently on average d. Since the gaps between primary particles in aggregates and between aggregates in loose agglomerates can be considered as mesopores of complicated shapes causing relatively narrow hysteresis loops (Figure 38.2a). Note that only the interior volume in aggregates is filled by nitrogen at $p/p_0 \rightarrow 1$ (external surfaces of aggregates can be also covered by several monolayers of nitrogen), while inter-aggregate volume (>15–20 cm^3/g) in agglomerates is significantly larger than V_p and is not filled by nitrogen (as well as by Ar or H_2O) at $p/p_0 \rightarrow 1$. The normalized α_S plots for fumed silicas do not practically deviate from the plot for Si-1000 ($S_{N2} \approx 26$ m^2/g) at $\alpha_S < 1.5$ (Figure 38.2b); consequently, micropores can give small contribution to the porosity of nonstandard fumed silicas of the first series (Table 38.1). This low contribution, which is provided by shallow micropores of

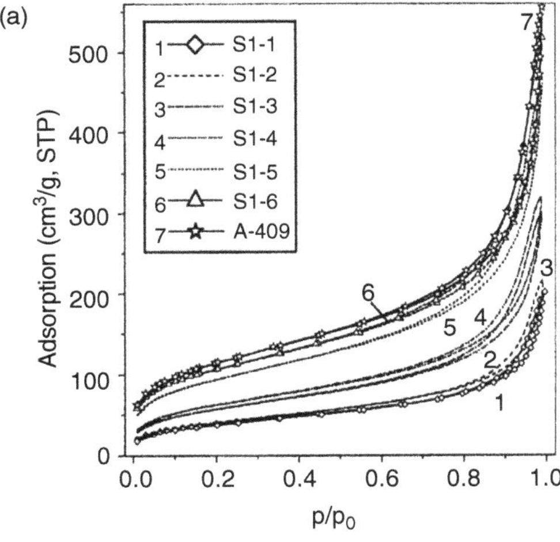

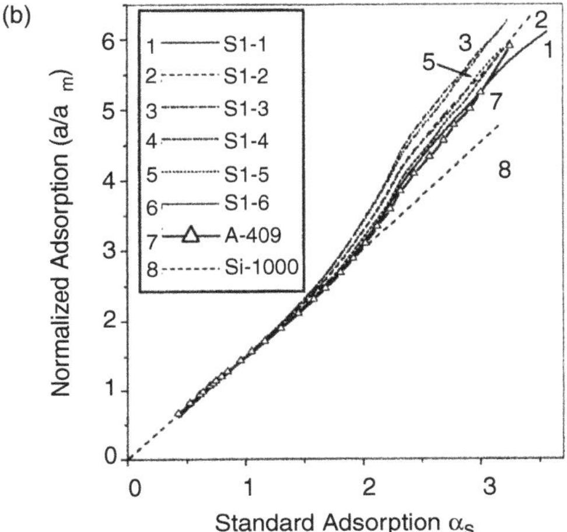

FIGURE 38.2 (a) Isotherms of nitrogen adsorption-desorption (77.4 K) on fumed silicas of the first series; and (b) the α_S plots for adsorption normalized by dividing by the BET monolayer capacity a_m.

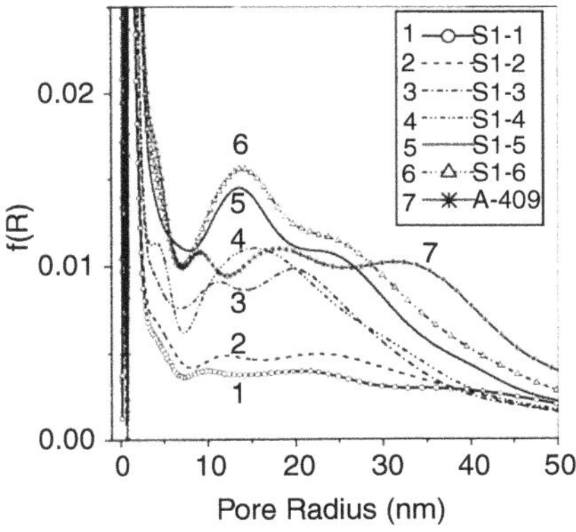

FIGURE 38.3 Pore size distributions for first series samples computed on the basis of the desorption data using the regularization procedure at fixed regularization parameter $\alpha = 0.01$ and a model of slitlike micropores at $R_p < 2t_m \approx 0.7$ nm and cylindrical mesopores at $R_p > 2t_m$.

(Figure 38.2 and Figure 38.3). The use of self-consistent binary regularization with the model of pores as gaps between spherical particles gives $f_{SCD}(R_p)$ (Figure 38.4) (as well as $f(a)$ (Figure 38.5)) different from $f(R_p)$ (Figure 38.3); however, common features are seen in these distributions related to gaps in aggregates at $R_p < 5$ nm and in agglomerates at $R_p > 10$ nm. The $f(a)$ distribution for A-409 reveals a broad distribution (Figure 38.5) due to formation of relatively large particles with shallow micropores synthesized under nonstandard conditions.

Changes in the apparent (bulk) density (Table 38.1, ρ_{ap}) do not correlate with S and V_p, which are mainly linked to the parameters of primary particles and aggregates, but the empty space in fumed silica ($V_{emp} \approx (1000 - (\rho_{ap}/\rho_{SiO_2}))/\rho_{ap}$) connected to ρ_{ap} (i.e., agglomerates and primary particles and contact zones between these particles, suggests that these contacts are relatively tight for non-standard fumed silica resulting also in its low initial hydrophilicity (Table 38.1, C_w) studied using fresh samples.

The similarity in the isotherm shape for silicas of the first series (Figure 38.2) corresponds to a similarity in the pore size distributions $f(R_p)$ (Figure 38.3) computed using the overall equation described in details elsewhere [7,10,12–14,20] with the regularization procedure. Large mesopores at $R_p > 10$ nm can correspond to inter-aggregate space in agglomerates, as channels in aggregates mainly correspond to $R_p < 5–8$ nm. There are three pairs of similar $f(R_p)$ corresponding to samples S1-1 and S1-2, S1-3 and S1-4, and S1-5 and S1-6, which (in pairs) have close values of V_p (notice that $f(R_p)$ is linked to dV_p/dR_p and normed to V_p) and S_{N2} (Table 38.1). A maximal contribution of micropores is observed for A-409 at $S_{N2} = 409$ m^2/g

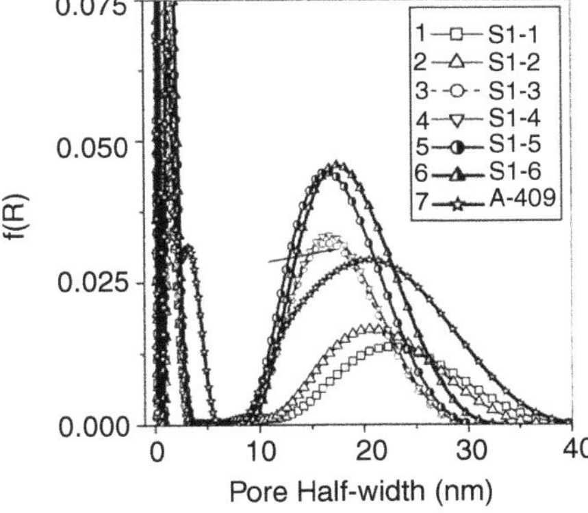

FIGURE 38.4 Pore size distributions for first series samples computed on the basis of the desorption data using the regularization procedure at fixed regularization parameter $\alpha = 0.01$ and a model of pores as gaps between spherical primary particles computed with simultaneous regularization with respect to $f(a)$.

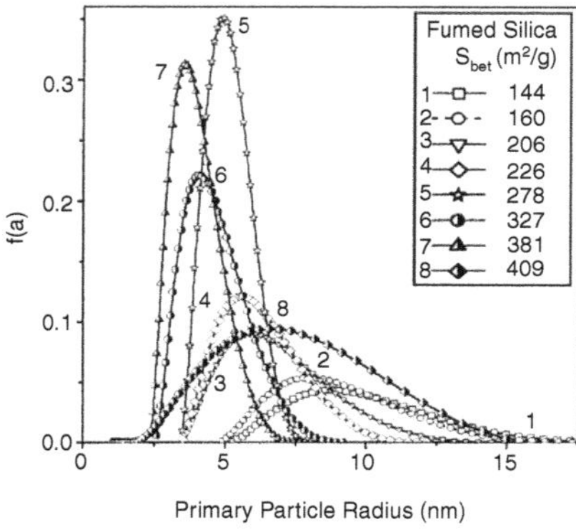

FIGURE 38.5 Primary particle size distributions for first series samples computed on the basis of the desorption data using the regularization procedure with parallel calculations of $f_{SCD}(R_p)$.

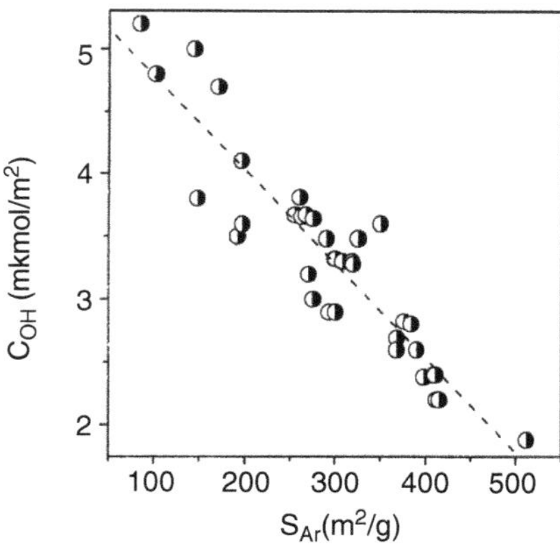

FIGURE 38.6 Relationship between the specific surface area S_{Ar} and the concentration of free silanols at $\nu_{OH} = 3750\ cm^{-1}$.

visible flocks) is significantly larger ($V_{emp} \approx 25\ cm^3/g$) than V_p ($<0.7\ cm^3/g$ close to the empty volume in aggregates). Thus, ρ_{ap} and S (and V_p) are related to different levels of the structural hierarchy of silica. However, an increase in ρ_{ap} (and R_p) correlate with a diminution of fractal dimension (D_{AJ}); that is, the increase in the apparent density of powders corresponds to smoother particle surfaces, and the enhancement of the dispersivity is accompanied by growing fractality. One can assume that standard fumed silica is not only mass fractal [2,3] but also pore (channel) and surface fractal (in respect to primary particles and aggregates). From the N_2 adsorption data for standard silica S3-1, $D_{AJ} = 2.62$ at $p/p_o < 0.85$ (which can characterize surface and pore fractality of aggregates). Nonstandard samples (prepared, e.g., at a greater flow of hydrogen through the annular nozzle) can possess larger surface fractality, e.g., for A−409, $D_{AJ} = 2.632$. At the same time, the flow turbulence resulting in formation of tighter contacts between primary particles can reduce the pore fractality (as their walls are smoothed, but primary particle surfaces have shallow micropores); therefore, D_{AJ} is higher for S3-1 (2.62) than for S1-5 (2.59), however, the last sample has larger S (or smaller d). We can assume that changes in synthesis conditions allow one to vary the parameters of fumed silica on all levels of its structural hierarchy [1−7]. Clearly, structural features of non-standard silicas and changes in the concentration of ≡SiOH groups (Tables 38.1−38.4, Figure 38.6) can reflect in the adsorption of polar H_2O to a greater extent than in the case of adsorption of nonpolar N_2 or Ar.

One can assume that the H_2O adsorption energy is greater on formation of water clusters near contacts between adjacent primary particles bonded one to another by the hydrogen bonds, but for tightly 'adnate' particles, formation of such clusters is less probably; therefore, the adsorbed water amounts are lower for nonstandard silicas (Table 38.1, 38.3 and 38.4). The concentration of strongly bound water desorbed between 105°C and 900°C ($C_{w,900}$) for less hydrophilic silicas is lower by several times (Tables 38.3 and 38.4). These structural and adsorptive features of fumed silicas cause a reduction not only of the amounts of weakly bound water ($C_{w,105}$) adsorbed from air and desorbed at $T < 105°C$ but also the monolayer capacity for water ($a_{w,mono}$ determined from adsorption data) adsorbed at controlled p/p_o at room temperature (Figure 38.7a).

For standard S3-1, adsorption-desorption of water does not give a large hysteresis loop (as well as the nitrogen isotherm has a narrow hysteresis loop similar to that shown in Fig. 38.2a), but for other samples from the third series, the water isotherms have marked hysteresis loops (desorption is not shown in Figure 38.7a); consequently, these samples can have porous primary particles (or aggregates with a significantly altered texture) in contrast to standard fumed silica. An increase in $\gamma_{H2} > 1$ reduces water adsorption (C_w) and monolayer capacity $a_{w,mono}$ (Table 38.3, samples 4, 5, and 8); that is, the hydrophilicity drops. Raising hydrogen flow through the annular nozzle to $8\ m^3/h$ impacts the adsorbed water amounts (compare C_w for S3-2 and S3-7 or S3-3 and S3-6). However, the concentration of free silanols C_{OH} changes slightly. Oxygen deficiency in the flame gives samples characterized by diminution of the amount of adsorbed water, but the monolayer capacity changes slightly (e.g., S3-3) or decreases (Table 38.3, samples 4, 5, 6, and 8).

Changes in the silica hydrophilicity result in marked differences not only in the water adsorption (Figure 38.7a) but also in alterations in the free energy distributions $f(\Delta G)$ (Figure 38.7b). For instance, reduction of water adsorption on nonstandard silicas (Figure 38.7a, Table 38.3, samples 4, 5, and 8) is accompanied by a

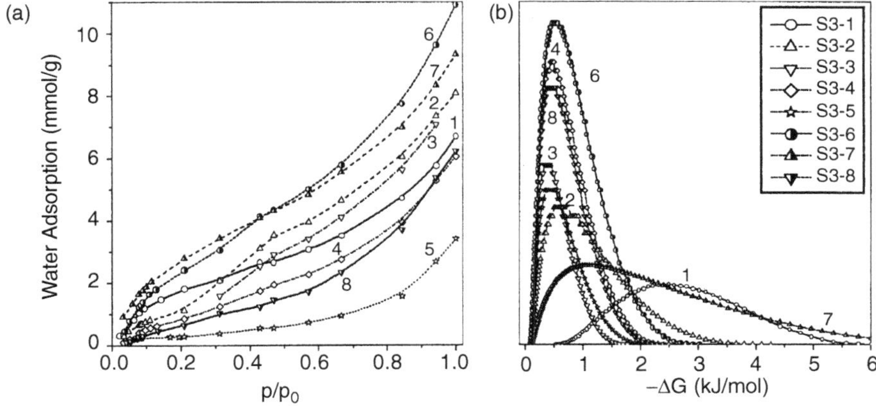

FIGURE 38.7 (a) Isotherms of water adsorption at 293 K on silicas of the third series and (b) the distributions of free energy changes due to water adsorption computed using the Langmuir equation.

decrease in $f(\Delta G)$ analyzed at coverage a less than the water monolayer $a_{w,\mathrm{mono}}$ estimated from the adsorption data using the integral Langmuir equation. Changes in the particle structure in more turbulent flame accompanied by reduction of surface hydrophilicity enhance the intensity of the $f(\Delta G)$ peak at 0.5 kJ/mol, which becomes narrower (Figure 38.7b).

Alterations in the structure of primary particles and their contacts influencing the water adsorption (Figure 38.7, Table 38.3, C_w, $a_{w,\mathrm{mono}}$) can be also analyzed using the IR spectra (Figure 38.8). For instance, sample S3−6 adsorbs water significantly larger (by two times) than standard fumed silica (Figure 38.7a, S3−1 and S3−6 at $p/p_0 \to 1$), which appears in great intensity of a broad band over 3000–3700 cm^{-1} (Figure 38.8, curves 1 and 4) linked to water adsorbed in different forms and disturbed surface silanols [1,2]. This effect is accompanied by a substantial but atypical reduction of the intensity of the band at 3750 cm^{-1} in comparison with the band intensity at 3400 cm^{-1} (Figure 38.8, curves 1 and 4), which is not observed for the standard silica characterized by a higher relative intensity of the band at 3750 cm^{-1} (Figure 38.8, curve 6). One can assume that this difference is connected to structural features of nonstandard silica, which can be more microporous (due to additional H_2 flow of 8 m^3/h through the annular nozzle) than standard fumed silica. Besides, the dependence $C_{\mathrm{OH}} \sim 1/S_{\mathrm{Ar}}$ (Figure 38.6) can be partially caused by the morphology of the secondary particles, while the larger the specific surface area, the denser the aggregates and the larger the coordination number of primary particles in them [1,2]. Therefore, a portion of disturbed OH groups (3550–3200 cm^{-1}) increases with S and aggregate density. Note that C_{OH} for samples, possessing large primary particles (e.g., S3-5, S4-6, S4−13, S4−14) and smaller aggregates, is close to $\alpha_{\mathrm{OH}} = 4.7$, which is typical value for many silicas [30,31].

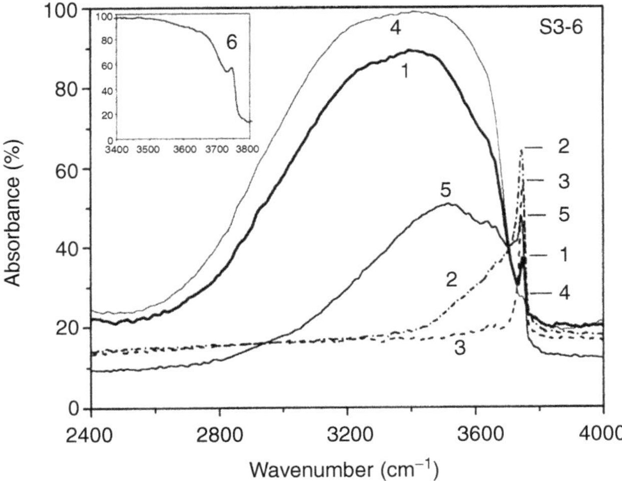

FIGURE 38.8 IR spectra of nonstandard fumed silica sample S3−6 (1) in air, after degassing at (2) 450°C and (3) 650°C, (4) letting in saturated water vapor, (5) degassing at room temperature; and (6) standard sample S3−1 in air.

Pretreated aqueous suspensions of fumed silica are characterized by multimodal SSDs and the first peak of the smallest particles (10–30 nm or slightly above, $S = 250$–350 m^2/g) corresponds to primary aggregates or even to the largest primary particles (smallest $d_{\mathrm{PCS}} \approx d + 2\kappa^{-1}$), while the size of secondary aggregates corresponds to 100–500 nm [3,9]. The second d_{PCS} peak (or two peaks) is often observed namely in this range. Additionally, one or two frequently observed peaks correspond to agglomerates at $d_{\mathrm{PCS}} > 1$ μm; however, in the concentrated (≈5 wt%) suspensions, large agglomerates are not observed. Diluted treated or nontreated suspensions can demonstrate instable agglomerates at d_{PCS} between 1 and 50 μm, which, however, can be easily rearranged to smaller or larger swarms even during short-time exposure or stirring for several minutes [9].

For bigger particles (Table 38.1, *d*), U_e is greater (Figure 38.9a), which is typical due to several reasons, e.g., greater charge density. Larger $D_{ef}(pH)$ is observed for a sample (UD) with minimal S_{N2} (Figure 38.9c). Opposite dependencies of D_{ef} and U_e (at pH = 3.4–4.4) versus S_{N2} are observed for silicas possessing low hydrophilicity (Fig. 38.9b), as U_e is a function of *d* and charge density, but D_{ef} depends not only on structural characteristics of primary and secondary particles and interaction forces between them but also on treatment of the suspension and it is not a stable characteristic (Fig. 38.9b, $D_{ef}(UD)$ and $D_{ef}(UB)$, and Ref. [9]). Relatively low amounts of ≡SiOH groups and structural features of non-standard samples can be responsible for a weak dependence of the SSDs on pH (Figs. 38.9c, 38.10 and 38.11).

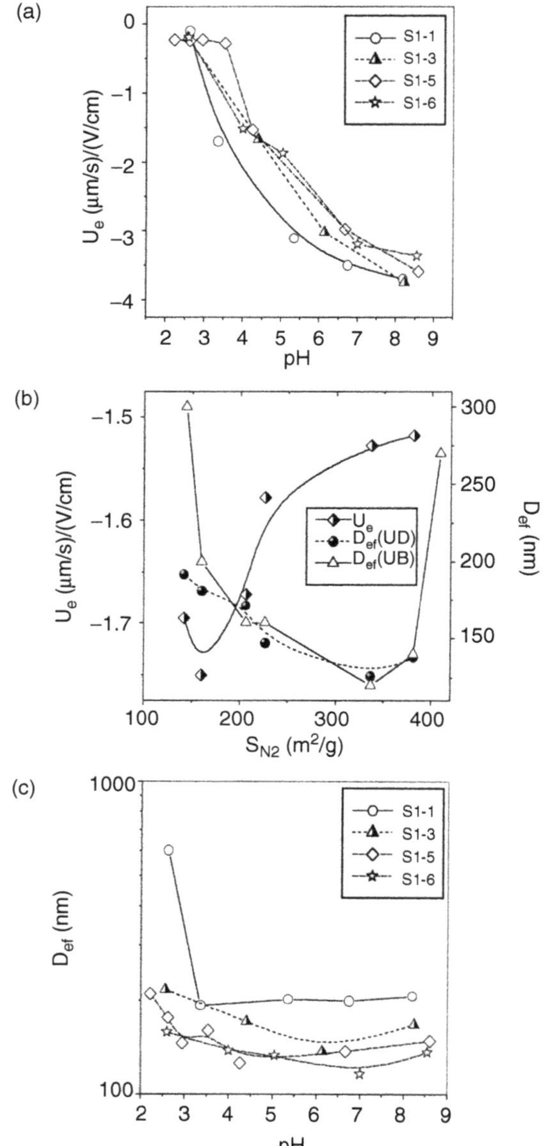

FIGURE 38.9 (a) Dependence of the electrophoretic mobility U_e on pH for samples of the first series; (b) relationship between S_{N2} and U_e or D_{ef} (treated using the ultrasonic disperser, UD, at pH = 3.4–4.4; or the ultrasonic bath, UB, at pH ≈5) without addition of the electrolyte; and (c) D_{ef} as a function of pH (UD).

A maximal light scattering intensity is observed at d_{PCS}- between 100 and 500 nm (secondary aggregates) (Figure 38.10), which are close to $\lambda_{PCS} = 633$ nm or $\lambda_{PCS}/2$. However, the distributions in respect to the number of particles *f(N)* have a maximal population at lower d_{PCS} values corresponding to primary aggregates or primary particles (Figure 38.11), which give lower light scattering (compare Figure 38.10 and Figure 38.11) as $d \ll \lambda_{PCS}$. The *f(N)* distributions are different for silicas (Figure 38.11) with different average primary particle size (Table 38.1, *d*), as a diminution of *d* leads to a larger contribution of smaller primary aggregates composed of several or dozens of primary particles. Secondary aggregates can involve thousands or even dozens of thousands of primary particles. In the case of standard fumed silica, a stronger dependence of the SSDs (multimodal with three or four peaks) on pH is observed [9]. Additionally, for less hydrophilic samples (Table 38.1), the D_{ef} values at pH close to pH(IEP_{SiO2}) ≈ 2.2 are smaller than those for standard fumed silicas. Because silica particles' capability to form a large number of the strong hydrogen bonds with water molecules from the first interfacial layer decreases, it leads to reduced D_{ef} and smaller changes in the free energy of adsorbed water (Figure 38.7b). The aqueous suspensions of fumed silica treated in an ultrasonic bath for 0.5 h are characterized by the SSDs [9,10] close to those measured at the larger concentration but stronger sonicated using the UD (Fig. 38.10). The plots of the relationship between D_{ef} and S_{N2} for these silica suspensions differently treated are close (Fig. 38.9b). However, the softer treated suspensions demonstrate a marked enhancement of the aggregate size (D_{ef}) at the smallest and largest S_{N2} values (Figure 38.9b). The last is in agreement with increasing strength of smaller primary particle bonding in aggregates [1–3]. Additionally, samples S1–1 (specific surface area of micropores $S_{mic} = 22$ m^2/g estimated using the Dubinin–Stoeckli method with correction on adsorption in mesopores) and S1–6 ($S_{mic} = 61$ m^2/g) are characterized by greater values of the fractal dimension (Table 38.1, D_{AJ}), as well as A–409 ($S_{mic} = 69$ m^2/g, see also *f(a)* in Figure 38.5) with $D_{AJ} = 2.632$; that is, the surface roughness of these samples is larger than that of others. Consequently, samples of the first series can be characterized by different contributions of the microporosity of non-standard primary particles and different strength of primary particle bonding in their aggregates. Another confirmation of contribution of the microporosity of the primary particles to the total porosity was obtained using the ^{1}H NMR spectra of adsorbed water.

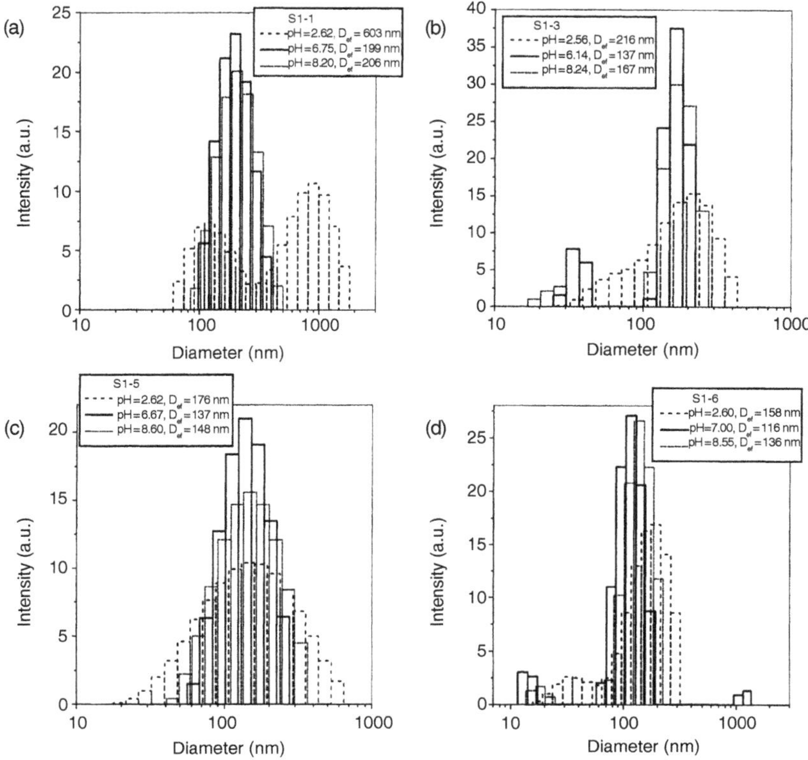

FIGURE 38.10 Relative PCS intensity for the sonicated (UD, 5 min) aqueous suspensions of samples (a) S1–1, (b) S1–3, (c) S1–5 and (d) S1–6 of the first series at different pH values shown in the legends (as well as D_{ef}), $C_{SiO_2} = 0.25$ wt%, and 10^{-3} M NaCl.

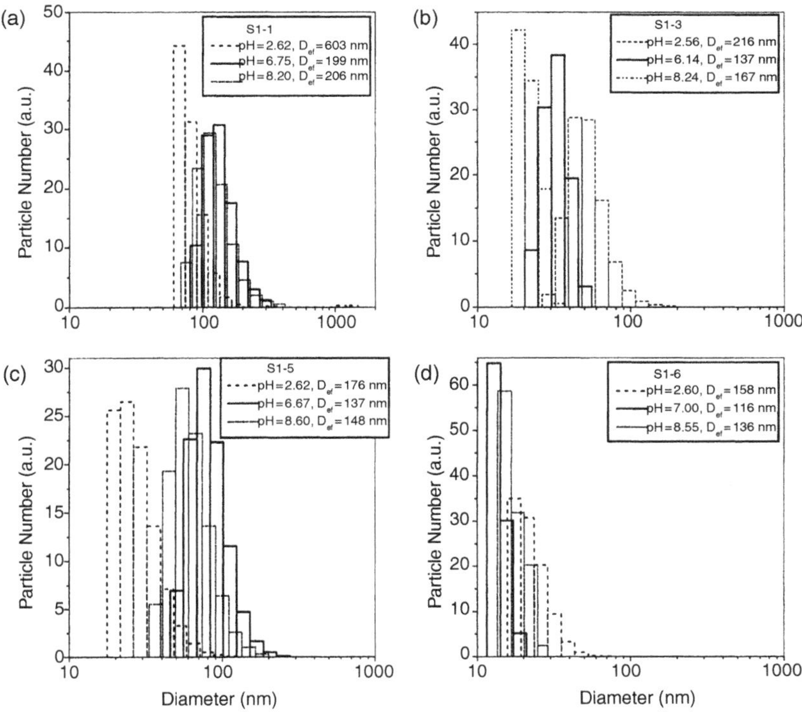

FIGURE 38.11 PCSs in respect to the particle number for the sonicated (UD) aqueous suspensions of samples (a) S1–1, (b) S1–3, (c) S1–5 and (d) S1–6 of the first series at different pH values shown in the legends, $C_{SiO_2} = 0.25$ wt%, and 10^{-3} M NaCl.

The chemical shift $\delta_H(T)$ of ^{1}H NMR of water adsorbed on silica samples of the first series changes over the 1.6–4.5 ppm range (Figure 38.12) and depends on T nonlinearly. These $\delta_H(T)$ graphs differ from that for standard silica S3–1 (Figure 38.12a). Observed complexity of $\delta_H(T)$ can be linked to structural features of primary particles (e.g., their atypical microporosity) or their aggregates and the possibility of formation of several types of water clusters in the confined space of micropores of primary particles or gaps between them in aggregates. To explain these effects one can assume that δ_H of a molecule, which does not take part in the hydrogen bonds as a proton-donor, is equal to that of a non-bonded molecule (dissolved in $CDCl_3$) $\delta_H = 1.5–1.7$ ppm. If a molecule forms four hydrogen bonds (similar to those in ice) that $\delta_H = 7$ ppm; that is, the participation of each H in the hydrogen bonds enhances average δ_H by 2.7 ppm.

Additionally, one can assume that the participation of a molecule in the hydrogen bonds through O (as electron-donor) does not change its average δ_H value. Several complexes can be drawn as follows with no consideration for H in ≡SiOH groups to estimate δ_H

$\delta_H = 1.7$ ≡Si-O-H...**OHH**; (I)

$\delta_H = 4.3$ ≡Si-(H)O...**H-OH**; (II)

$\delta_H = 4.3$ ≡Si-O-H...**(H)O-H**...O(H)-Si≡ (II')

$\delta_H = 7$ ≡Si-(H)O...**H-O-H**...O(H)-Si≡ (III)

$\delta_H = 5.7$ ≡Si-(H)O...**H-O-H**...O(H)Si≡ | **H-O-H** (IV) (38.12)

These complexes can transform one to another (due to the water molecule diffusion or the impact of the solvent $CDCl_3$) or change on addition of next water molecules. However, in the confined space of shallow micropores in non-standard primary particles, a possible number of molecules in the clusters (as well as the number of the hydrogen bonds per a molecule) is restricted and the molecular mobility differs from that in the larger gaps between primary particles in aggregates; therefore, the ^{1}H NMR spectra of water adsorbed on standard (with nonporous primary particles) and non-standard fumed silicas differ markedly (Fig. 38.12). The average δ_H values on transformation of these complexes correspond to

$$
\begin{aligned}
&(I) \leftrightarrow (II) - \delta_H = 3 \text{ ppm};\\
&(I) \leftrightarrow (III) - \delta_H = 4.3 \text{ ppm};\\
&(II) \leftrightarrow (III) - \delta_H = 5.1 \text{ ppm};\\
&(I) \leftrightarrow (II) \leftrightarrow (III) - \delta_H = 4.3 \text{ ppm};\\
&(III) \leftrightarrow (IV) - \delta_H = 6.3 \text{ ppm};\\
&(I) \leftrightarrow (IV) - \delta_H = 4.4 \text{ ppm}.
\end{aligned} \quad (38.13)
$$

The average chemical shift close to $\delta_H \approx 4.3$ ppm is observed for several types of the complexes in Equation (38.13). Besides, formation of larger complexes on the basis of initial I-IV (scheme 38.12) gives the average δ_H value close to 4.3 ppm, which also corresponds to the experimentally observed δ_H for different silicas at their high

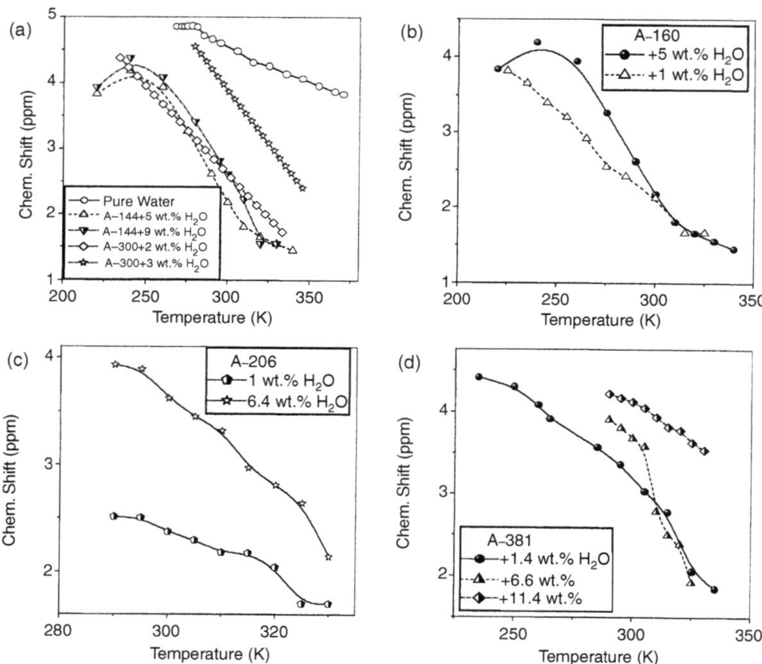

FIGURE 38.12 Chemical shift $\delta_{H,iso}$ as a function of temperature for samples of the first series (a–d) and (a) standard S3–1 at different hydration levels in air.

TABLE 38.5
Parameters of Bound Water Layer in Aqueous Suspensions of Fumed Silica (First Series)

Sample number	$-\Delta G^s$ kJ/mol	$-\Delta G^w$ kJ/mol	C^s_{uw} mg/g	C^w_{uw} mg/g	γ_S mJ/m^2
S1−1	3.8	0.8	200	900	304
S1−2	2.8	0.6	400	1200	315
S1−3	2.9	0.8	250	1150	245
S1−5	2.75	0.6	300	600	124
S1−6	2.75	0.7	230	270	67

hydration (Figure 38.12). Note that water adsorbed in mesopores of silica gels is characterized by greater δ_H values than that adsorbed on fumed silica over a larger temperature range due to the possibility of formation of larger water clusters in silica gel mesopores with a greater number of the hydrogen bonds per a molecule (stronger affected by the pore walls), and there even at T = 320–350 K δ_H = 3.6–4.6 ppm [28,29].

The total volume of weakly (C^w_{uw}) and strongly (C^s_{uw}) bound waters (unfrozen at T < 273 K) at the silica interfaces (Table 38.5) is markedly larger than V_p (but significantly lower than V_{emp}) with one exception for S1−6 (Table 1). With increasing specific surface area of the first series samples, there is tendency of reduction of the free surface energy (Table 5, γ_S) and the amount of weakly bound water (C^w_{uw}). Besides, the $\Delta G(C_{uw})$, $\Delta G(V_p)$ and $\Delta G(R_p)$ graphs for S1−6 lie over those for other silicas (Figure 38.13). These effects can be caused by contribution of unfrozen water localized in the interaggregate volume for samples with larger S and V_p, while the particle number in such aggregates is greater than that in the aggregates formed by primary particles with larger d values.

Changes in the free energy of strongly bound water (Table 38.5, ΔG^s), which is localized close to the silica surfaces and unfrozen at T significantly lower than 273 K, decrease with diminution not only of C_{OH} (Table 38.1) but also of a possible number of hydrogen bonds per a molecule in narrower pores (gaps) of aggregates of smaller primary particles for samples with increasing S (see $f(R_p)$ in Figs. 38.3 and 38.4). In contrast to marked dependences of the characteristics of strongly bound waters, alterations in ΔG^w (corresponding to weakly bound water disturbed due to long-range action of the silica surfaces) depend slightly on the silica characteristics (Tables 38.5 and 38.1). The relationships between ΔG and the volume V_p (Figure 38.13b) or the radius R_p (Figure 38.13c) of pores filled by unfrozen water have been computed assuming that water is frozen in narrower pores at lower temperatures. The $\Delta G(R_p)$ graphs (Figure 38.13c) show that changes in ΔG at R_p < 1 nm are relatively small but depend on the structural characteristics of silica as well as at R_p > 1 nm, and marked changes in $\Delta G(R_p)$ are observed at R_p between 1 and 10 nm (at R_p < 3 nm for S1−6). Obviously that $\Delta G(R_p)$

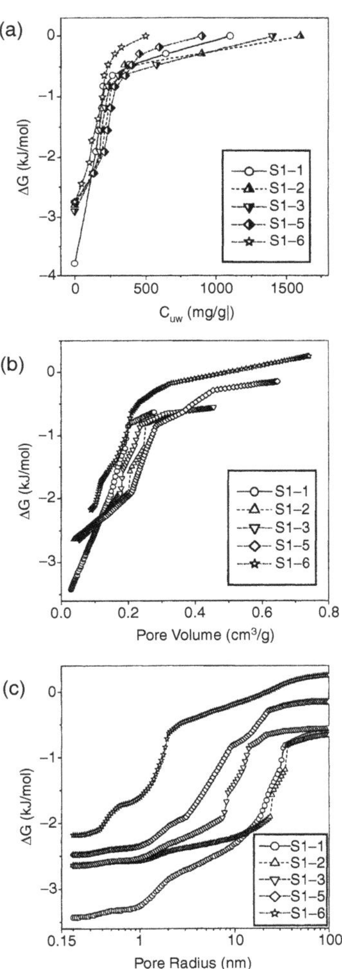

FIGURE 38.13 Relationships between changes in the Gibbs free energy of the interfacial water in the frozen aqueous suspension of silicas of the first series and (a) amounts of unfrozen water; (b) the volume and (c) the radius of pores filled by unfrozen water.

is more sensitive to these characteristics than $\Delta G(V_P)$ or $\Delta G(C_{uw})$ (Figure 38.13). On the basis of these plots, one can assume that the disturbance of water by the silica surfaces is observed on the distance up to 10–15 nm. The ΔG ranges determined using ^{1}H NMR (Fig. 38.13) and adsorption (Fig. 38.7) methods are in agreement in respect to changes in these values due to variation in the hydrophilicity of fumed silica samples of the first and third series, despite significant difference in the measurement techniques, that can be considered as the evidence of the reliability of obtained results. Thus, changes in the synthesis conditions allow one to produce fumed silicas characterized by a different ability to interact with such polar compounds as water due to structural features of primary particles and their swarms.

Polymer Adsorption

Increasing stability of particle swarms with decreasing primary particle size [1–3,9,10] can play an important role on polymer adsorption. Globular protein molecules can interact with silica mainly at the outer surfaces of undestroyed aggregates, as their channel radii (mainly R_p <5 nm) are less than the polymer molecule size. An increase in S_{Ar} (Table 38.6) enhances the plateau adsorption of BSA or gelatin in mg per gram of oxide studied at $C_{SiO_2} = 0.8$ wt.% and $C_{protein} = 0.6$ wt.%. Note that these silica samples synthesized under varied conditions are characterized by different hydrophilicity even at close S_{Ar} values (Table 38.1–Table 38.4) that can explain a scatter of the protein adsorption versus S_{Ar}. The adsorption in mg per m^2 of the surface area goes down with S_{Ar} due to diminution of the concentration of accessible silanols (C_{OH}), which are the main adsorption sites on silica. Besides, the adsorption rises with C_{OH} independently on the nature of proteins (Table 38.6). The C_{OH} values were determined from the integral intensity of the IR band at 3750 cm^{-1} according to a method described elsewhere [32].

PVP molecules simpler and smaller than protein molecules do not have strong intramolecular bonds in contrast to proteins, despite the availability of electron-donor N—C=O groups, as proton-donor groups are absent in PVP. Adsorbed PVP enhances pH of the isoelectric point (IEP) nearly linearly with C_{PVP}, and the ζ potential is reduced with increasing C_{PVP} at a constant C_{SiO_2} (Table 38.7). A large number (≈100 per a PVP molecule) of polar electron-donor N—C=O bonds at a short distance between the pyrrolidone rings can be responsible for practically irreversible adsorption of the polymer molecules on silica. PVP molecules are not washed from the silica surfaces at $C_{PVP}/C_{SiO_2} \leq 0.1$ due to bonding in multi-centered adsorption complexes and approximately 2/3 C=O groups at $\Theta < 1$ form the hydrogen bonds with silanols, as simultaneous breaking of all these bonds is unlikely. However, changes in the Gibbs free energy (ΔG) on adsorption of polar or ionogenic polymers on solid surfaces from the aqueous solution are relatively small ($\Delta G \approx 4$–$5\ k_BT$) [11]. Interaction between PVP and SiOH groups can cause buildup of the surface charge density on silica depending on C_{PVP} and the electrolyte concentration. However, strong adsorption of PVP enhances pH(IEP) and reduces ζ (Table 38.7) due to shielding of the oxide surfaces. Nevertheless D_{ef} of PVP/silica swarms (Table 38.7) decreases due to decomposition of agglomerates by PVP adsorbed in the unfolded state. Besides, adsorbed PVP can reduce the adsorption of albumin or gelatin with C_{PVP} (Table 38.8)

TABLE 38.6
Characteristics of Fumed Silicas and Plateau Adsorption of BSA and Gelatin

No.	S_{Ar},[a] m^2 g^{-1}	C_{OH}, μmol m^{-2}	C_{BSA}, mg g^{-1}	C_{BSA}, mg m^{-2}	$C_{gelatin}$, mg g^{-1}	$C_{gelatin}$, mg m^{-2}
1	102	4.8	200	1.96	170	1.67
2	148	3.8			243	1.62
3	197	3.6	210	1.06	295	1.49
4	192	3.5	230	1.19	200	1.04
5	270	3.2	310	1.14	275	1.01
6	293	2.9	280	0.95	320	1.09
7	300	2.9			346	1.15
8	275	3.0			315	1.14
9	308	3.0	320	1.03	307	1.0
10	308	3.0	350	1.14	380	1.23
11	368	2.6	380	1.03	250	0.68
12	384	2.8	450	1.17	420	1.09
13	390	2.6	410	1.05	470	1.21
14	410	2.4	430	1.05	450	1.10
15	411	2.4	380	0.92	200	0.48

TABLE 38.7
Parameters of Aqueous Suspensions of Fumed Silica, Pure PVP and PVP/Silica

Sample	C_{PVP}, wt.%	C_{A300}, wt.%	ζ, mV	pH	D_{ef}, μm
PVP	5	–	1.5	6.27	>50
A–300[a]	–	5	–15.1	5.21	0.78
A–300[b]	–	5	–12.9	5.43	0.76
A–300/PVP[a]	0.13	5	–8.0	5.93	0.50
A–300/PVP[a]	0.25	5	–15.6	6.16	0.47
A–300/PVP[a]	0.38	5	–12.2	6.16	0.37
A–300/PVP[a]	0.5	5	–14.1	6.07	0.34
A–300/PVP[c]	2.5	5	–7.1	5.72	0.31
A–300/PVP[c]	5.0	5	–6.2	5.75	0.36

Note: [a]BMS for 5 h and stored for 1 month; [b]fresh suspension prepared in the ultrasonic bath for 9 h; [c]PVP addition to the BMS of silica before (5 min) the measurements.

due to marked shielding of surface silanols responsible for strong intermolecular interaction with proteins.

The adsorption potential computed in respect to the disturbed interfacial water using the ^{1}H NMR spectra with entire freezing-out of the bulk water and layer-freezing-out of the interfacial water at 210 K < T < 273 K shows that the concentration of weakly bound water (Figure 38.14b, $A < 1$ kJ/mol, Table 38.9, C_{uw}^{w}) are lower for the pure silica suspension or the pure PVP solution. The decline in a linear portion of the $\Delta G(C_{uw})$ graph at small ΔG (Figure 38.14a) corresponding to weakly bound water is larger for pure silica suspension than that for PVP/silica; that is, the boundary between disturbed and undisturbed waters is more robust without PVP. However, in the case of the pure PVP solution, C_{uw}^{w} cannot be estimated, while nearly linear $\Delta G(C_{uw})$ allows one to calculate only C_{uw}^{s}, as the capability of noninogenic PVP to disturb the water to a large distance is low. This effect can be also caused by formation of PVP oligomers in the pure aqueous solution, as $D_{ef} > 50$ nm (Table 38.7), whose "outer surface area" is small. The *f(A)* intensity (Figure 38.14b) increases at $C_{PVP} = 5$ wt.% (as well as C_{uw}^{s} and C_{uw}^{w} (Table 38.9)), but at $C_{PVP} = 1$ wt.%, the opposite effect is observed at $A > 1$ kJ/mol, C_{uw}^{s} and the free surface energy γ_S in comparison with those for the individual silica suspension (or C_{uw}^{s} for PVP/PBS/silica). On the other hand, changes in ΔG versus V_p (Fig. 38.14c) or R_p (Fig. 38.14d) show that the silica impact on water in pores at R_p between 1 and 10 nm is significantly larger, as the free energy is lower, than that for the PVP/silica suspension ($C_{PVP} = 1$ wt.%, $\gamma = C_{PVP}/C_{SiO_2} \sim 0.17$). This effect is due to reduction in long-range components of surface forces for silica shielded by PVP. Clearly, in the case of large interparticle distances in agglomerates, unfrozen water can be located close to the silica particles, i.e., to the outer surfaces of aggregates. Therefore formation of a denser PVP layer at the silica surfaces at $\Theta < 1$ leads to decrease in C_{uw}^{s}. At large $C_{PVP} = 5$ wt.% ($\gamma \approx 0.83$), a major portion of PVP is in the nonimmobilized state or weakly interacts with silica but the amounts of waters weakly and strongly bound to PVP increase significantly (Table 38.9, Fig. 38.14a). Thus, at $\gamma < 0.2$ (i.e. $\Theta < 1$), polymer molecules adsorb strongly and the number of free pyrrolidone groups can

TABLE 38.8
Impact of PVP immobilized on A–300 on Protein Adsorption from the Aqueous Solution

C_{A30}, wt.%	C_{PVP}, wt.%	$C_{egg\ alb.}$, mg/g	$C_{gelatin}$, mg/g
3	0	170	380
3	0.15	80	210
3	0.3	30	116
4	0	180	224
4	0.2	102	190
4	0.4	40	100

TABLE 38.9
Parameters of Interfacial Water Layer for Pure Silica and Silica/PVP Suspensions ($C_{SiO_2} \approx 6$ wt.%)

System	$-\Delta G_s$, kJ/mol	$-\Delta G_w$, kJ/mol	C^s_{uw}, mg/g	C^w_{uw}, mg/g	γ_S, mJ/m^2
A–300	3.2	1.3	700	700	253
[a]A–300 + 0.3 wt.% PVP	2.5	1.0	500	900	186
A–200	3.0	1.4	730	680	279
1 wt.% PVP	1.6		300		
A–200 + 1 wt.% PVP	3.0	0.9	520	1400	220
A–200 + 5 wt.% PVP	3.0	1.0	1100	1700	403

Note: $S_{N2} \approx 300$ and 190 m^2 g^{-1} for A–300 and A–200, respectively; [a]in PBS; ΔG_s and ΔG_w are the changes in the Gibbs free energy of strongly and weakly bound waters, respectively; C^s_{uw} and C^w_{uw} are the concentrations of the unfrozen waters strongly and weakly bound to the surfaces; γ_S is the change in free surface energy.

be less than bonded ones; that is, the adsorbed PVP layer is relatively dense, and oxide particles or PVP molecules can disturb only a relatively thin interfacial water layer. At great C_{PVP}, a significant portion of PVP molecules has free tails or loops, which do not interact with silica but effectively interact with water molecules disturbing their hydrogen bond network. However, a decline in the ΔG dependence on C_{uw}, V_p or R_p (especially at $\Delta G > -0.7$ kJ/mol) is small for a relatively thick layer of weakly bound water (Fig. 38.14). Notice that in the case of the adsorption of protein hydrolysate, fermentatively hydrolyzed bull blood proteins with fractions from 4000 to 25000 Da, on fumed silica at $C_{SiO_2} = 3$ wt.% and $C_{protein} = 1$ wt.%, $C_{uw} = 300–1200$ mg/g [7a], i.e., small proteins can shield the silica surface stronger than larger globular proteins or PVP, as C_{uw} is lower.

The SSDs in the silica or polymer/silica suspensions (Figs. 38.15–38.19) can be multi- or monomodal

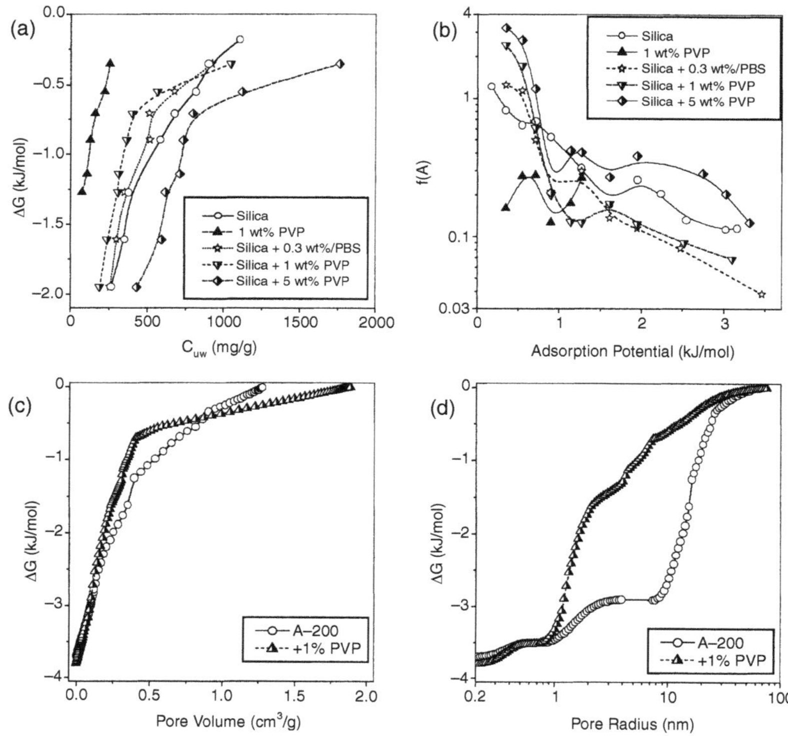

FIGURE 38.14 (a, c, d) Changes in the free energy and (b) water adsorption potential distributions for pure PVP solution and suspensions of silica, silica/PVP, and silica/PVP/PBS computed on the basis of ^{1}H NMR data; ΔG as a function of (c) filled pore volume and (d) pore radius for silica (6 wt.% A–200) and PVP/silica (1 wt.% PVP, 6 wt.% A–200) suspensions.

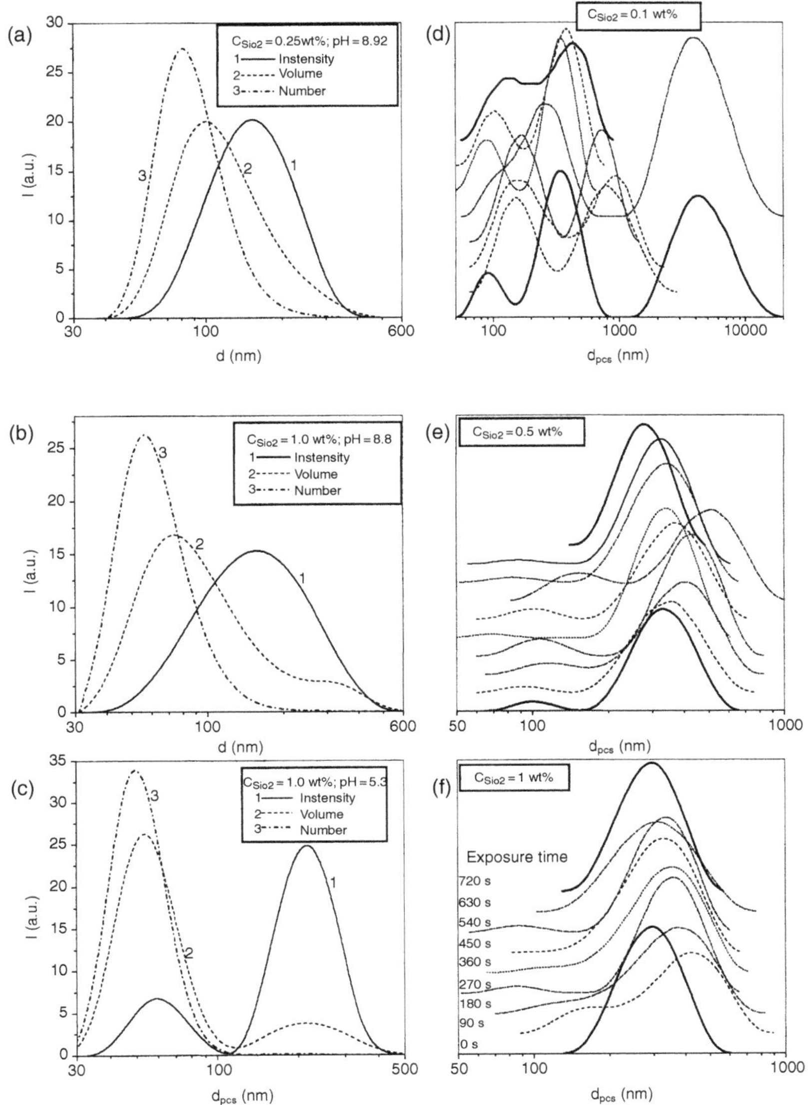

FIGURE 38.15 Particle size distributions in respect to the light scattering (solid lines), particle volume (dashed lines) and particle number (dot-dashed lines) for the aqueous suspensions of fumed silica A–300 at C_{SiO_2} = (d) 0.1 wt.%, (a) 0.25 wt.%, (e) 0.5 wt.% and (b, c, f) 1 wt.% and different pH values; suspensions (a–c) strongly sonicated by the disperser for 5 min and (d–f) softly sonicated (pH ≈ 5.3–5.6) in the ultrasonic bath for 30 min; (d–e) exposure time interval between each measurement was ≈ 90 s, the regularization parameter $\alpha = 0.01$.

depending on the component concentrations and pH [4]. However, small aggregates ($d_{PCS} < 100$ nm) give the main contributions to the distributions related to both light scattering (SSD_I) and particle volume (SSD_V) or number (SSD_N) for all pure silica suspensions (Figures 38.15a–38.15c), some samples with PVP (Figure 38.16 and Figure 38.17), silica suspended in ethanol/water [10b] or aqueous suspension of silica/Aethonium (Figure 38.19). This effect is caused by decomposition of agglomerates and large aggregates of primary silica particles by solvent or polymer molecules and due to electrostatic repulsion between negatively charged silica particles. Smaller aggregates give the main contribution not only to SSD_N but also to SSD_V (Figure 38.15 and Figure 38.16), but in comparison with SSD_N or SSD_V, SSD_I shifts toward larger d_{PCS}, as $\lambda_{PCS} = 633$ nm and the light scattering intensity is greater for particles with the size close to λ_{PCS}. The fumed silica dispersion stability decreases with lowering C_{SiO_2} [4] and even short-time exposure leads to marked growth of silica particle swarms especially at low $C_{SiO_2} = 0.1$ wt.% (Figures 38.15d–38.15f). However, at $C_{SiO_2} = 1$ wt.%, changes in the SSDs during several minutes are relatively small (Figure 38.15f, the first and last curves are very close).

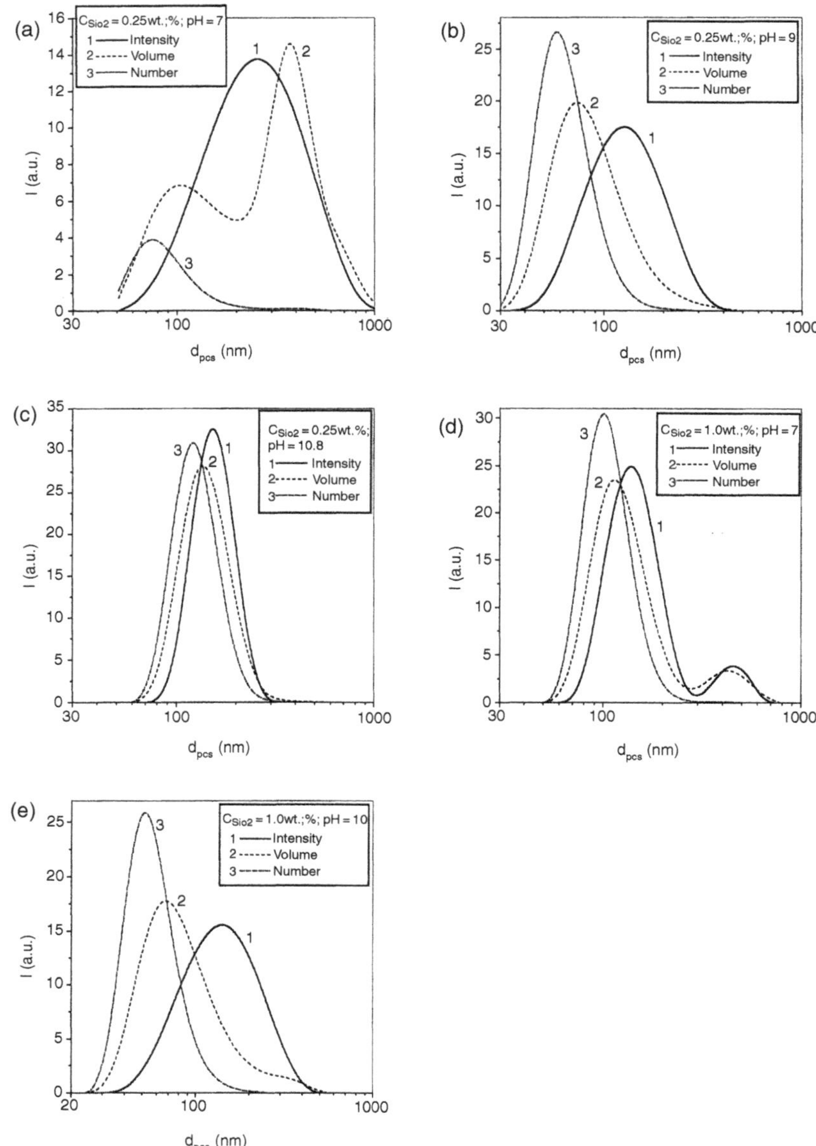

FIGURE 38.16 Particle size distributions in respect to the light scattering (solid lines), particle volume (dashed lines) and particle number (dot-dashed lines) for the aqueous suspensions of silica/PVP at C_{SiO_2} = (a–c) 0.25 wt.% and (d, e) 1 wt.% at $C_{PVP}/C_{SiO_2} = 0.04$.

Typically recording of the autocorrelation function in PCS is carried out at the angle of 90° corresponding to a maximal light scattering. The study of the suspension (C_{SiO_2} = 1 wt.%), which was sonicated for 30 min in the bath, aged for two months and agitated before measurements, at other scattering angles gives similar SSD_S [10b] but slightly shifted due to agglomeration on the suspension exposure of ≈90 s between each measurement starting from 90 to 15°. Additionally, two-month storage of the suspension results in growth of the particle size in comparison with the initial suspension [9,10]. Similar results related to the SSD displacement toward larger d_{PCS} due to long-time aging of the fumed silica suspensions were observed at different C_{SiO_2} values; however, increase in C_{SiO_2} typically reduces agglomeration [9,10].

Features of PVP/silica swarms reflect in interactions with living mobile flagellar microorganisms *Proteus mirabilis 187* ($C_{PM} \approx 10^6/cm^3$) depending on C_{PVP}. PVP/silica impacts the SSDs stronger than pure silica, as SSD markedly shifts toward larger d_{PCS} only for PM/PVP/silica (Fig. 38.17a). The proper mobility of flagellar PM decreases significantly (Fig. 38.17b) due to interaction with PVP/silica in contrast to PM/pure silica in the diluted suspensions (C_{SiO_2} = 0.1 wt.%). Consequently,

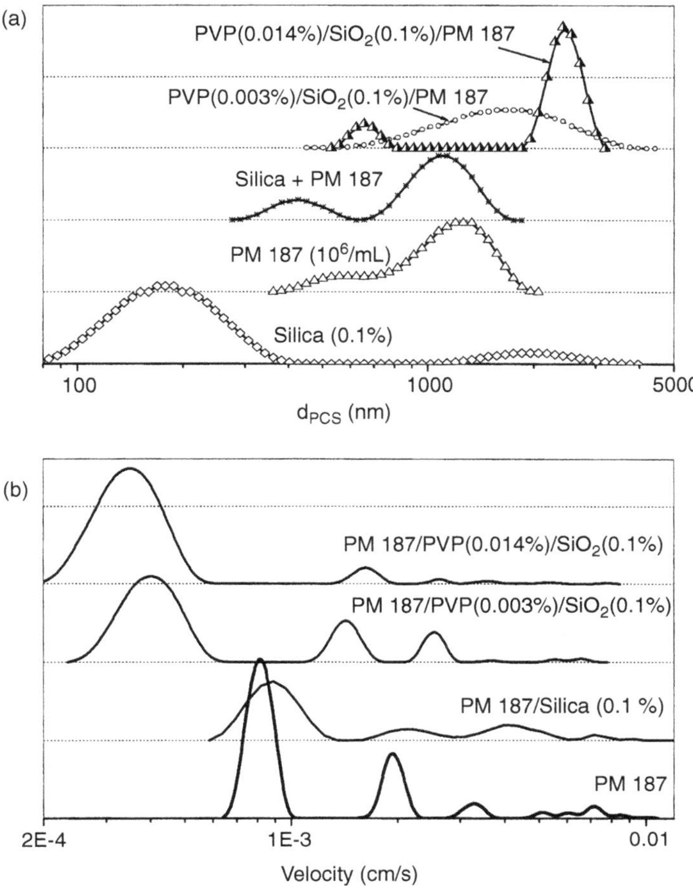

FIGURE 38.17 (a) SSDs for pure silica (BMS), living microorganisms *Proteus mirabilis 187* (PM 187), mixture of silica and PVP/SiO_2 with PM 187; and (b) velocities of PM 187 and their mixture with silica or PVP/silica computed at unfixed regularization parameter.

even at a low concentration, PVP/silica reduces the vital functions of PM significantly stronger than pure silica does due to high adhesion properties of PVP. Note that these in respect to the cell surfaces are used in some drug compositions.

For egg albumin/silica suspended-centrifuged-dried-suspended, the SSDs (Figure 38.18), which are shifted toward larger d_{PCS} in comparison with that for pure silica (Figure 38.15) or PVP/silica (Figure 38.16 and Figure 38.17), depend on the albumin concentration (C_{alb}), as the distributions become more broadened and complex with C_{alb}. At a maximal C_{alb} value, a large peak of SSD_V appears at $d_{PCS} > 1$ μm (Figure 38.18c) corresponding to swarms with the main contribution of protein molecules or their oligomers as SSD_I has a low intensity in this region. Clearly, the capability of dissolved proteins in the light scattering is lower than that of silica swarms. Small particles at $d_{PCS} > 30$ nm can correspond to both washed albumin molecules and small their oligomers or their swarms with silica particles, as SSD_I has a low intensity at d_{PCS} between 30 and 100 nm.

For all the protein/silica samples, the main SSD_V peaks lie to the left from SSD_I that corresponds to the availability of large agglomerates with protein molecules or oligomers at smaller contribution of silica particles. Therefore one can assume that interaction of egg albumin with the silica surface is weaker than intramolecular interactions in its molecules (i.e., globular structure of proteins is not decomposed and their portion can be washed from silica on secondary suspending [7a]) in contrast to PVP molecules, which interact with silica stronger than with other PVP molecules, as intramolecular interaction in PVP is weak. These features of polymer-silica interactions can impact differently the pore formation on drying of the solid residual of the centrifuged polymer/silica suspensions. One can assume that changes in the medium (e.g., water/ethanol frequently utilized to study the adsorption of compounds poorly water-soluble) for silica suspensions result in alterations in their SSDs and other characteristics.

Changes in the concentration of ethanol (C_{EtOH}) in the fumed silica suspensions ($C_{SiO_2} = 0.1-1.0$ wt.%, pH = 5.9–4.5) influence of the D_{ef} value only slightly

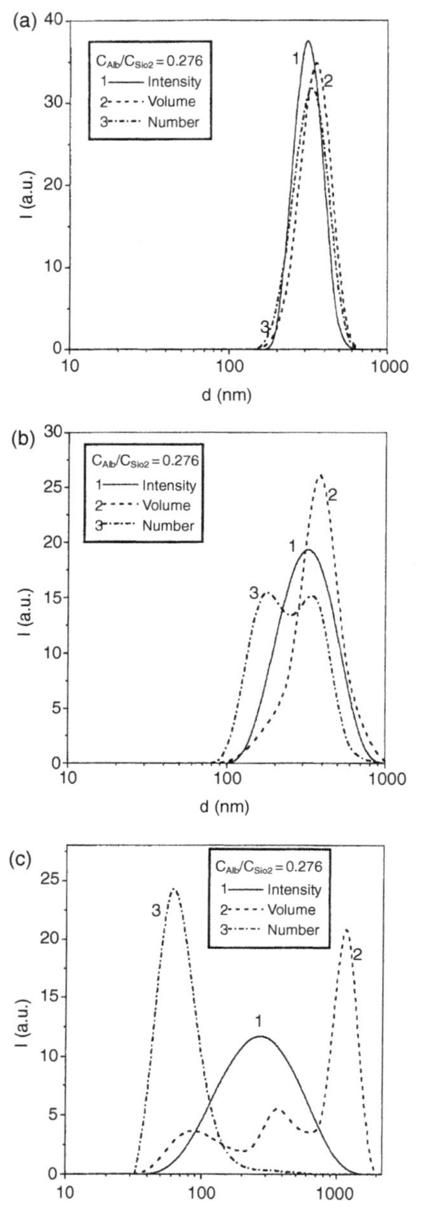

FIGURE 38.18 Swarm size distributions in respect to the light scattering (solid lines), particle volume (dashed lines) and particle number (dot-dashed lines) for the aqueous suspensions of silica/egg albumin at $C_{SiO_2} = 0.25$ wt.% and C_{alb}/C_{SiO_2} = (a) 0.172, (b) 0.23, and (c) 0.276.

but nonlinearly (Figure 38.20a) as well as the swarm size distributions [9,10]. Note that the pH value of similar suspensions decreases linearly with addition of ethanol as follows $pH \approx pH_0 - 0.028C_{EtOH}$ [21]. In contrast to ethanol, Aethonium significantly changes the SSDs (Figure 38.19 and Figure 38.20b) especially at pH not far from pH(IEP_{SiO_2}) when the negative surface charge density is low that prevents strong interaction with the positively charged surfactant ions. With the presence of NaCl (0.001 M) at $C_{SiO_2} = 1$ wt.% and low C_{Aet}, the multimodal distributions can be observed and characterized by the availability of small aggregates (15–30 nm) (Figure 38.19) close to the size of primary silica particles. Similar particles are frequently observed in the concentrated BMS of fumed silica [4]. In the case of lower $C_{SiO_2} = 0.1$ wt.%, similar small particles with the presence of Aethonium are observed too, as well as in the aqueous/ethanol medium at $C_{EtOH} = 10\%$ and $C_{SiO_2} = 1$wt.% [10b].

However, for other aqueous/ethanol suspensions at larger C_{EtOH}, $d_{PCS} > 50$ nm and both monomodal and multimodal SSDs are observed [9,10]. Addition of small amounts of Aethonium results in different changes in the size of the smallest particles (Figure 38.19); however, the D_{ef} value always increases on addition of Aethonium (Figure 38.20b). Thus, addition of different amounts of ethanol gives smaller changes in the SSDs than significantly lower amounts of Aethonium (Figure 38.19 and Figure 38.20) or polymers (Figure 38.17 and Figure 38.18). Note that a high concentration of Aethonium cations can cause formation of flocks with silica particles charged negatively at used pH values. These changes are nonlinear in respect to the additive concentration due to nonrigid swarm structures easily rearranged by surfactant ions, polar (PVP) or charged (protein) polymer molecules. However, these changes in the SSDs are nondramatic and solid sediment is not formed; consequently, the aqueous suspensions of fumed silica are relatively stable on changes in the polar medium composition that is important for fumed silica applications in different media. It is of interest to study the impact of this composition on the corresponding dried complex powders (suspended-centrifuged-dried), which can be utilized for different purposes.

Changes in the initial fumed silica properties or pretreatment conditions on drying of the solid residual of the centrifuged aqueous suspensions result in marked differences in the nitrogen adsorption-desorption isotherms (Fig. 38.21) and α_S plots (Fig. 38.22) reflecting alterations in the textural characteristics of solids (Table 38.10). The incline in the hysteresis loops for dried powders is larger than that for the initial silica; that is, their mesopore size distributions differ due to changes in large channels in aggregates and interaggregate space in agglomerates on suspending-centrifuging-drying (Fig. 38.23). A similar shape of the adsorption isotherms, which can be assigned to the II type with a narrow hysteresis loop [12], is observed for silica/PVP and silica/protein powders but characterized by decreased adsorption due to the reduced porosity in comparison with that of dried pure silica (Table 38.10, V_p) and reduced specific surface area (S_{N2}).

These changes are clearly observed in the α_S plots (Figure 38.22) (Si-1000 was used as a reference material

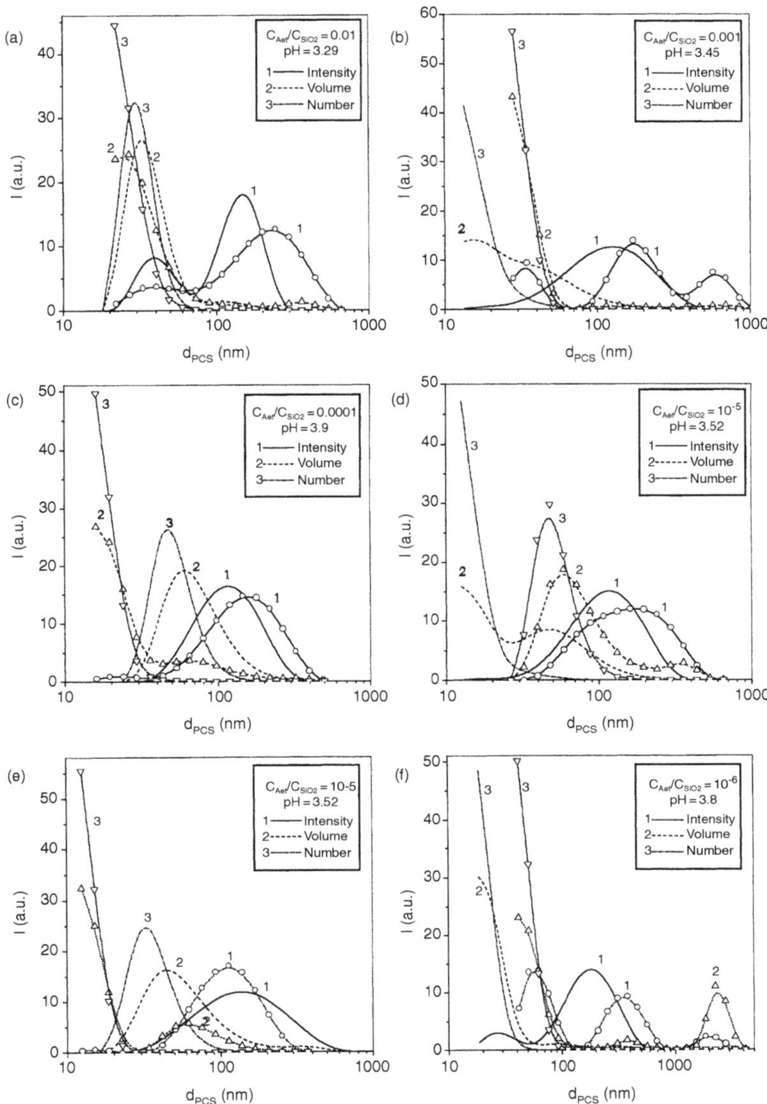

FIGURE 38.19 Particle size distributions in respect to the light scattering (SSD_I, solid lines), particle volume (SSD_V, dashed lines) and particle number (SSD_N, dot-dashed lines) for pure silica (lines) or water/Aethonium/silica (shaken for 0.5 min after addition of Aethonium) (lines + symbols).

[22]) showing a marked diminution of the adsorption in narrow pores and the opposite effect for large mesopores of the dried powders. Note that only the interior volume in aggregates is filled by nitrogen in $p/p_0 \rightarrow 1$ and external surfaces of aggregates can be also covered by several monolayers of nitrogen. Since interaggregate volume (>15–20 cm^3/g) in agglomerates is significantly larger than V_p and is not filled by nitrogen (as well as by Ar or H_2O) at $p/p_0 \rightarrow 1$. The availability of PVP gives a stronger structural effect in comparison with proteins, for example, the normalized α_S plots are nearly the same for all the samples with silica/albumin or silica/gelatin independent on their concentration and nature (Figure 38.22b) in contract to silica/PVP (Fig. 38.22d). Thus, structural changes caused by suspending-drying of silica and silica/polymers can be observed for both narrow and large pores of dried solid residuals. Structural features of silica particles suspended-dried are clear observed in SEM image of relatively large agglomerate >5 μm (Fig. 38.23). Secondary particles become denser in comparison with those of pristine fumed silica [1–3,9,10], however, they are less dense than silica gel particles, as transformation of fumed silica to silica gel requires more rigid reaction conditions. For example, HTT of A–380 in an autoclave at 150°C for 6 h reduces S_{N2} to 192 m^2/g but enhances V_p to 1.29 cm^3/g (initial $V_p = 0.67$ cm^3/g);

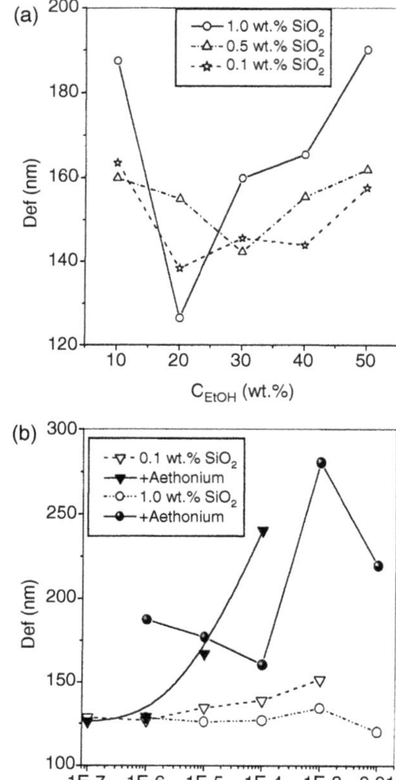

FIGURE 38.20 Dependence of effective diameter D_{ef} on (a) C_{EtOH} and (b) C_{Aet} (solid symbols, open symbols corresponds to pure silica suspensions before addition of Aethonium) at different C_{SiO_2} (a) without and (b) with NaCl (0.001 M).

that is, both changes are approximately twofold but opposite. At the same time these changes are close to those observed on suspending-drying of A–300 (Table 38.10) and linked to dissolution-condensation of the silica fragments [1,3]. Since direction of these processes is connected to dissolution of smaller particles and growth of larger particles with simultaneous compacting contacts between adjacent particles [3c], the specific surface area decreases. The V_p value determined using the nitrogen adsorption increases due to compacting of agglomerates (Figure 38.23), which enhances the capillary condensation of nitrogen, and its adsorbed volume increases at $p/p_0 \rightarrow 1$, i.e., estimated V_p rises.

Changes in the shapes of the isotherms (Fig. 38.21) and α_S plots (Figure 38.22) are in agreement with the adsorption potential distributions $f(A) = -da/dA$, where a denotes the adsorbed amount of nitrogen; $A = -\Delta G = R_gTln(p_0/p)$, also showing alterations related to the adsorption in narrow pores (at $A > 1$ kJ/mol) and large mesopores ($A < 0.1$ kJ/mol) in comparison with that for the pristine silica due to changes in the silica particle morphology on drying and filling of the gaps between them by polymer molecules. More detailed information about structural changes in dried silica and silica/polymer powders can be obtained on analysis of the pore size distributions $f(R_p)$.

Suspending-centrifuging-drying of polymer/silica changes the pore structure differently depending on a polymer kind and its concentration (Figure 38.24). The availability of egg albumin or gelatin at different concentrations causes a weak effect, related mainly to filling of pores by protein molecules or formation of their large oligomers, in contrast to PVP, markedly rearranging secondary silica particles and pores in them. Albumin and gelatin can adsorb on silica surfaces in the globular form due to strong intramolecular forces and hydrophobic interaction between nonpolar side groups of amino acids [4,7,19,23]. Typically albumin or gelatin adsorb on fumed silica in amounts of 300–600 mg per gram of oxide (through the flocculation mechanism) and their maximal adsorption is observed at pH close to IEP of proteins having the

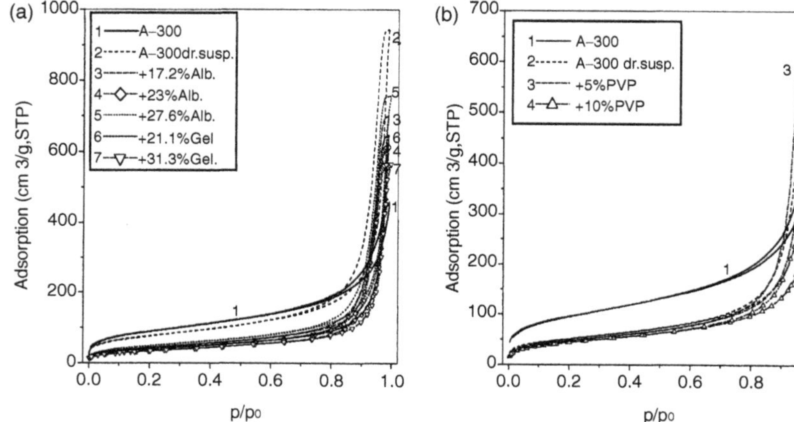

FIGURE 38.21 Nitrogen adsorption-desorption isotherms for pristine A–300 and dried silica and (a) silica/proteins and (b) silica/PVP prepared by drying of the corresponding aqueous suspensions.

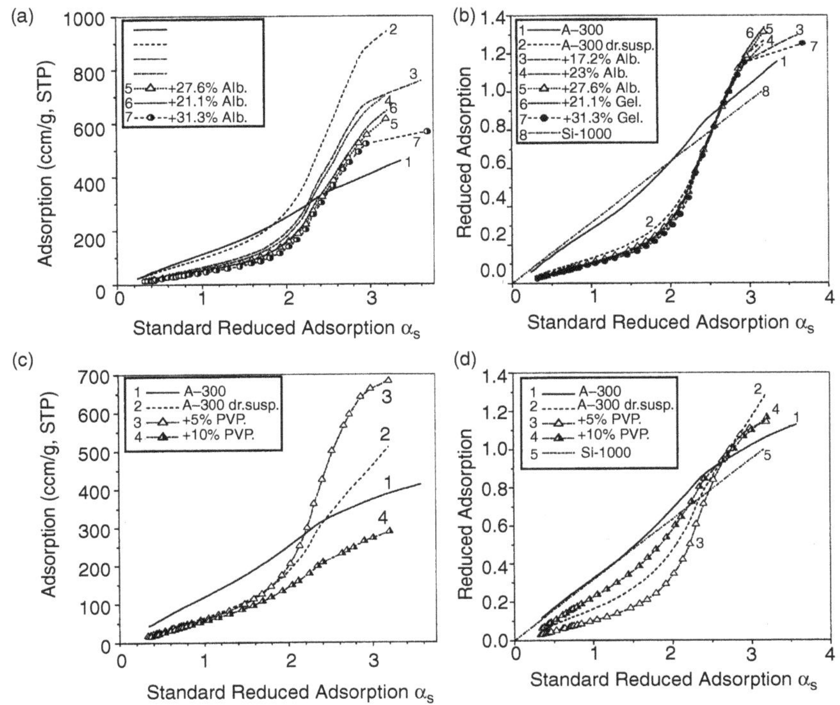

FIGURE 38.22 The α_S plots for (a, b) pristine and dried silica and silica/proteins and (c, d) pristine and dried silica and silica/PVP; (b, d) reduced adsorption V/V_p.

lowest cross-section on the surfaces at pH(IEP) due to their denser structures compacted by strong attractive intramolecular bonds with the absence of repulsive electrostatic forces between charged groups [4,9,19,23]. However, strong adsorption of PVP corresponds to lower amounts approximately 100–150 mg per gram of silica [4,19].

PVP in the amounts above 10 wt.% ($\gamma = 0.1$) provides disturbing of a major portion of accessible ≡SiOH groups, as the band at 3750 cm^{-1} practically disappears at $C_{PVP} = 17.5$ wt.% (Figure 38.25a). In the case of BSA/silica samples (Figure 38.26a), this band is else observed at significantly larger BSA concentrations, for example, $C_{BSA} = 30$ wt.%, than C_{PVP}. Besides, a baseline level in the IR spectra of BSA/silica samples elevates with C_{BSA} (Figure 38.26) in contrast to PVP/silica (Figure 38.25) due to scattering by larger protein/silica particles (also observed by the PCS method). These results confirm that PVP molecules adsorb in the unfolded form in contrast to proteins keeping the globular structure in the liquid media or adsorption state. Additionally, strong interaction between unfolded PVP molecules and primary particles of fumed silica leads to decomposition of a significant portion of particle swarms, as nearly linear PVP molecules can penetrate into the channels of aggregates and decompose them. Notice that in the atmosphere of the saturated water or ethanol vapors, PVP molecules can be well distributed on the surfaces of agitated fumed silica similarly to that observed in the aqueous suspension; that is, water or ethanol in the concentration of 10–30 wt.% can promote unfolding and transportation of PVP molecules decomposing a marked portion of secondary silica particles. Therefore, decreased silica-silica interaction

TABLE 38.10
Structural Parameters of Pristine Powders and Dried Suspensions of Fumed Silica, Silica/PVP, Silica/Egg Albumin and Silica/Gelatin

Powder	C_X, wt.% (X = PVP, Albumin)	S_{N2}, $m^2\ g^{-1}$	V_p, $cm^3\ g^{-1}$
[a]A-300	–	342	0.566
[b]A-300	–	182	0.612
[b]A-300/PVP	5	170	0.923
[b]A-300/PVP	10	157	0.383
[a]A-300*	–	322	0.613
[b]A-300*	–	275	1.147
[b]A-300*/Albumin	17.2	175	0.903
[b]A-300*/Albumin	23.0	157	0.868
[b]A-300*/Albumin	27.6	138	0.724
[b]A-300*/Gelatin	21.1	143	0.745
[b]A-300*/Gelatin	31.3	123	0.701

Note: [a]Pristine powder, [b]dried powders from solid residual of the suspensions.

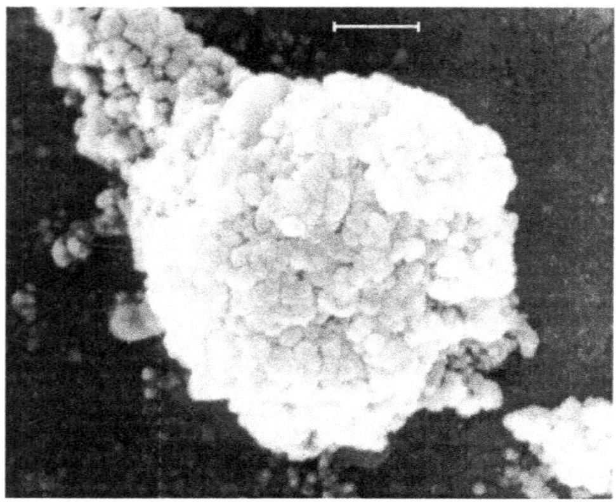

FIGURE 38.23 SEM image of aggregate of suspended-dried fumed silica A–300 (bar corresponds to 1 μm).

between particles covered by PVP ($\Theta \geq 1$) results in the textures characterized by markedly lower mesoporosity in comparison with that for dried pure silica or protein/silica powders (Figure 38.24).

For instance, at $C_{PVP} = 10$ wt.% ($\Theta < 1$), $f(R_p)$ has only a small peak at $R_p \approx 15$ nm shifted in comparison with that for pure dried silica, silica/PVP at $C_{PVP} =$ 5 wt.% or silica/proteins. At $C_{PVP} = 5$ wt.%, the pore volume increases (as well as for other dried powders with one exception of silica/PVP at $C_{PVP} = 10$ wt.%) (Table 38.10); however, the specific surface area decreases for all the dried samples. The distributions $f_S(R_p)$ computed in respect to dS_{N2}/dR_p show that narrow pores are responsible for the main contribution to the specific surface area (Figure 38.24c and Figure 38.24d), which decreases strongly for dried powders. In the aqueous suspensions, polymer molecules can penetrate into pores and block ≡SiOH groups, whose interaction is necessary to form the ≡Si–O–Si≡ bridges on drying. All these effects lead to various morphologies of the dried powders with silica or silica/polymer (Table 38.10, Figure 38.24).

Adsorption of ions depends on the surface properties of solids; on the other hand, the interfaces can be modified by adsorbed ions in respect to the value and the sign of the surface charge and the electrokinetic potential [33–40]. The density of metal ions adsorbed at the solid/liquid interfaces depends not only on the type of adsorbent and solution composition but also on kind and form of metal cations, especially for ions undergoing hydrolysis (such as Pb(II) and Sr(II)). Dramatic increase in the adsorption is observed over a narrow pH range, usually no more than two pH units. Specifically adsorbed ions (without bound OH groups) adsorbed at the inner Helmholtz plane close to the surface charge plane; therefore, they can affect surface charge density, shift PZC and influence the potential distribution within the electrical interfacial layer. Adsorption of metal ions onto fumed silica depends strongly on the type of metal, e.g., hydrolyzed aqueous metals such as Pb(II) and Sr(II) forming species

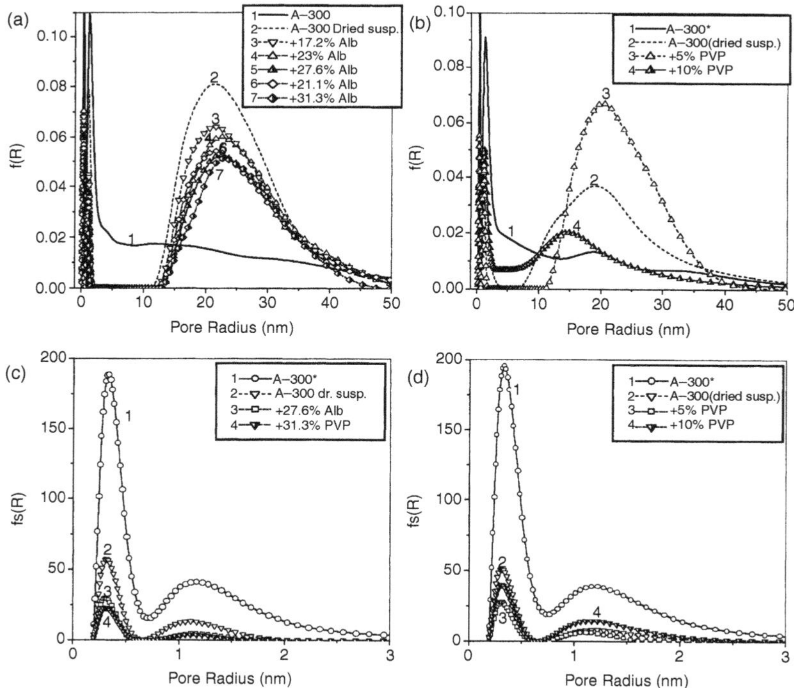

FIGURE 38.24 Pore size distributions in respect to (a, b) dV_p/dR_p and (c, d) dS/dR_p for pristine fumed silica A–300 and powders prepared by drying aqueous suspensions of (a, c) silica, egg albumin/silica, and gelatin/silica and (b, d) silica and PVP/silica.

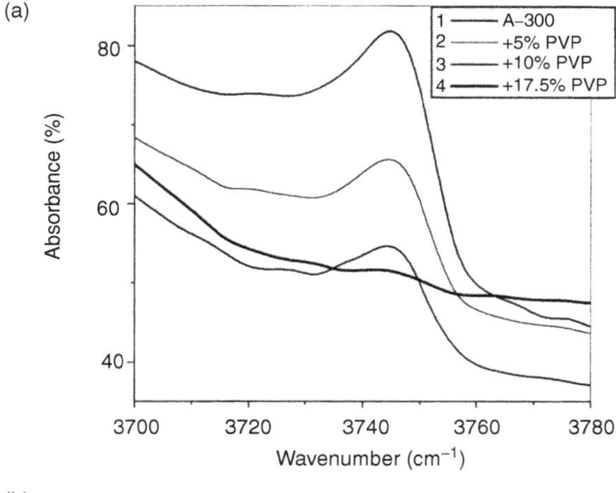

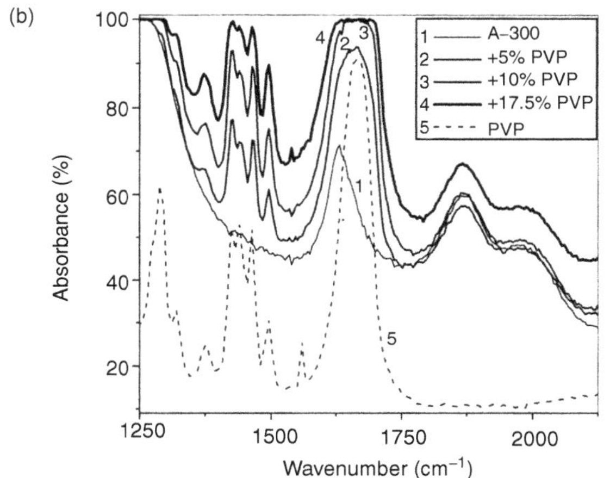

FIGURE 38.25 IR spectra of silica and PVP/silica at $C_{PVP} = 0$ (curve 1), 5 (2), 10 (3) and 17.5 (4) wt.% for (a) SiO-H band at 3750 cm^{-1} and (b) ν between 1250 and 2100 cm^{-1}; pure PVP (curve 5).

$M_x(OH)_y$ adsorb in significant amounts (up to 100% for Pb(II) at pH > 7) comparing with Cs(I) [14e]. Adsorption of Cs(I) on different oxides is approximately ten times as lower than that of Pb(II) or Sr(II) due to the difference in adsorption and solvation energies, as Cs(I) in contrast to Pb(II) and Sr(II) is in the form of only Cs^+[14e]. Adsorption of different metal ions on mixed oxides [14e] is significantly greater than that on silica A-300 (Figure 38.27) because of the availability of stronger acidic (≡Si–O(H)–Al≡ or ≡Si–O(H)–Ti≡) than ≡SiOH and basic (≡AlOH and ≡TiOH) sites reacting as follows

$$\equiv\text{Si}-\text{O(H)}-\text{Al}\equiv\text{Ct}^{n+}\rightarrow$$
$$\equiv\text{Si}-\text{O}(\text{Ct}^{(n+)-1})\text{Al}\equiv+\text{H}^+ \quad (38.14)$$
$$\equiv\text{AlOH}+\text{Ct}^{n+}\rightarrow\equiv\text{AlO(H)Ct}^{n+} \quad (38.15)$$

or with other sites (≡TiO(H)Ti≡, ≡TiOH, $(\equiv AlO)_2OH$, $(\equiv AlO)_3OH$, ≡TiO(H)Al≡, etc.). Complexation equilibrium constant for these reactions can be written as follows

$$K_{Ct}=\frac{[\equiv\text{SO}^-\text{Ct}^+][\text{H}^+]}{[\equiv\text{SOH}][\text{Ct}^+]}\frac{f_{\mu\text{Ct}}\gamma_\text{H}}{\gamma_0\gamma_\text{Ct}}\exp\left[\frac{e(\psi_\text{Ct}-\psi_\text{H})}{\text{kT}}\right] \quad (38.16)$$

where $f_{\mu Ct}$ is the surface activity coefficient, γ_i is the mean activity coefficient of i ion in solution, ψ_H and ψ_{Ct} is the mean potential at the planes of H and cation adsorption, respectively. For instance, on Cs(I) adsorption on anatase pK = 9.4 [33e] or silica gel pK = 5.5 (≡SiOCs) and −2.05 ($\equiv SiO(H)Cs^+$) with the presence of Na^+; [33f] Pb(II) on iron hydroxide it is in the 6.9–13 range; [36b] for Pb(II) on silica 7.75 or 5.09 on binding to one ≡SiOH group and 17.2 or 10.68 on binding to two ≡SiOH groups [34].

Calculations of the adsorption constant distributions (Figure 38.28) using Equation (38.11) show that bi- or tri-modal distributions $f(pK_n)$ are seen on adsorption of different cations [14e]. The $f(pK_n)$ peaks corresponding to strong interaction ($pK_n < 5$) appear due to significant adsorption of cations at low pH (Fig. 38.27). Sharp enhancement in adsorption up to 100% for Pb(II) gives a $f(pK_n)$ peak between 5 and 10 (which is in agreement with previous data [42] and results obtained using TLM (Table 38.11); however, in general, the constants K_n and K_{Ct} calculated using Equation (38.11) and Equation (38.16) differ as Equation (38.11) gives normalized ones). Broadened peaks at $pK_n > 10$ are caused by some changes in adsorption after the "edge" one and can be attributed to cation interaction with weaker sites or with two sites simultaneously (according to the literature [33]). In the case of Pb(II) uptake up to 100% by oxides with a high specific surface area [14e] but with relatively low density of adsorbed cations and a smooth curve at pH > 7 (Figure 38.27c), a similar broad $f(pK_n)$ peak is absent (Figure 38.28). This low-density adsorption can occur on strong active sites.

In the case of adsorption of Cs(I), the $f(pK_n)$ distributions (Figure 38.28) reveal relatively low adsorption constants since the maximal uptake of Cs(I) is not more than 10%. A maximal uptake of Sr(II) depends on its concentration and it can reach up to 100% at 10^{-5} M, but at 10^{-3} M, it is between 35–70% depending on the specific surface area of oxides and their kind [14e]. Additionally, adsorption of Sr(II) is characterized by relatively great initial values at pH < 6 and non-steep enhancement at higher pH values (Figure 38.27b). This results in $f(pK_n)$ with a large peak at $pK_n < 7$ and a broad peak at $pK_n > 10$ (Fig. 38.28). These differences in $f(pK_n)$ for Cs(I), Sr(II) and Pb(II) are linked to the differences in the form of cations (hydrated or not), composition of their shells with water molecules and the distributions of active surface sites most appropriate for adsorption of

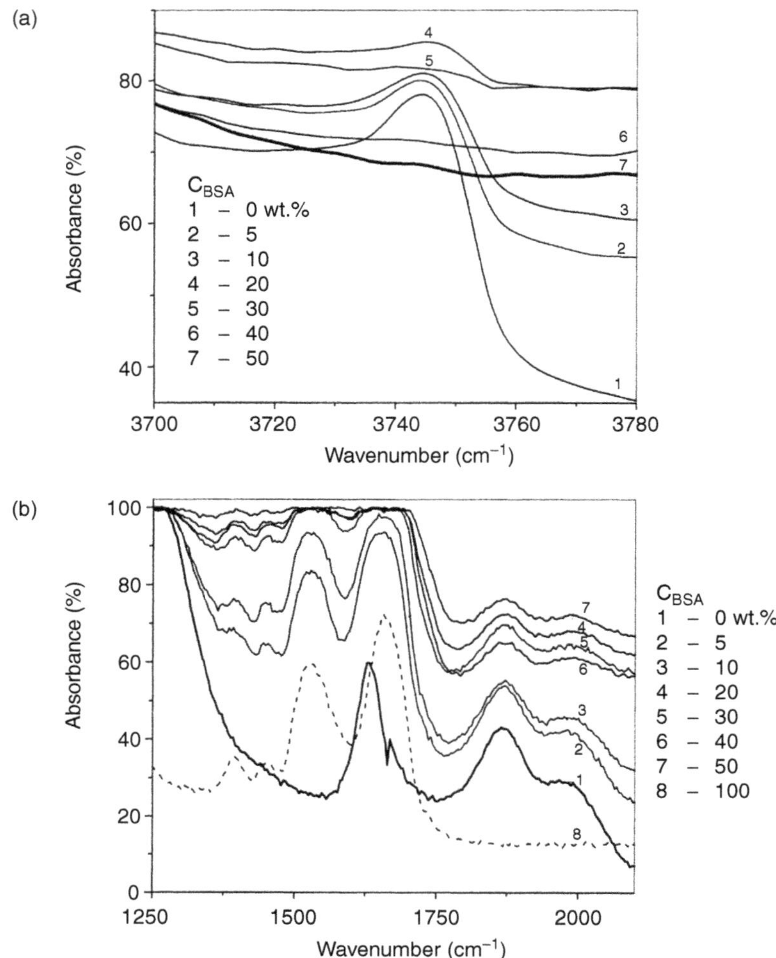

FIGURE 38.26 IR spectra of silica and BSA/silica at $C_{BSA} = 0$ (1), 5 (2), 10 (3), 20 (4), 30 (5), 40 (6), 50 (7) and 100 wt.% (8) for (a) SiO–H band at 3750 cm^{-1} and (b) ν between 1250 and 2100 cm^{-1}.

each species. As a whole, complexation equilibrium constants (Table 38.11) calculated within the scope of TLM and the $f(pK_n)$ distributions (Figure 38.28) are in agreement (observed differences are due to the differences in the used models and the corresponding equations). In the case of Pb(II)/SiO_2, the first two complexes in Table 38.3 give the main contribution that is in agreement with the main contribution of a $f(pK_n)$ peak at pK = 6 (Fig. 38.28). Similar pictures are seen for Sr(II)/SiO_2 (Table 38.11 and Figure 38.28). The difference between results for Cs(I)/SiO_2 is greater, as this cation adsorbs poorly that results in broadened $f(pK_n)$ peaks at pK > 8. In the case of mixed oxides, there is another reason of the differences between the constants computed with different models, as it is difficult in TLM to describe a lot of different surface sites interacting with aqueous metals.

Adsorption of metal cations impacts the surface charge density (Figure 38.29) and electrophoretic mobility (Figure 38.30) even up to changes in the sign of the electrokinetic potential. The nonuniformity and the heterogeneity of the surfaces of primary particles of binary and ternary oxides (distributions of individual phases on the surfaces, solid solution of one oxide in another with formation of surface asymmetrical bridges $\equiv M_1O(H)M_2\equiv$, etc.) comparing with fumed silica, as well as the availability of porous aggregates, result in large differences in the surface charge density σ_0(pH) for them, whose curves significantly differ from ζ(pH) [14e]. Concentration of adsorbed ions can play a significant role in changes in σ_0(pH) and ζ(pH) if it is higher than some critical value corresponding to dramatic changes in the surface charge density (Figure 38.29) and the electrophoretic mobility (Figure 38.30). The latter parameter was used instead of the electrophoretic potential, since for similar systems, the conductivity of the surface layer should be also considered on transform of U_e to ζ in addition to the porosity of aggregates that results in increase in the ζ magnitudes (i.e., this effect is opposite to that of the porosity) [39].

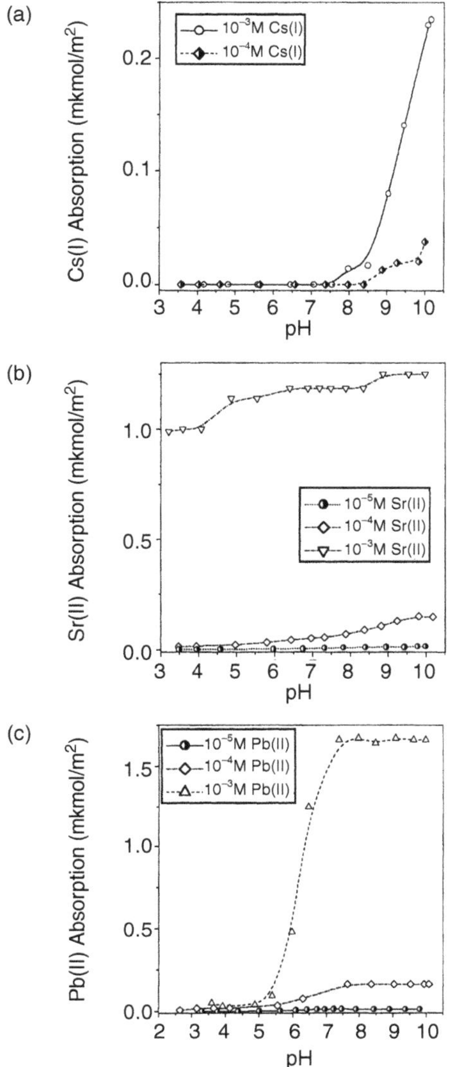

FIGURE 38.27 Adsorption of aqueous metals (a) Cs(I), (b) Sr(II), and (c) Pb(II) on fumed silica A–300.

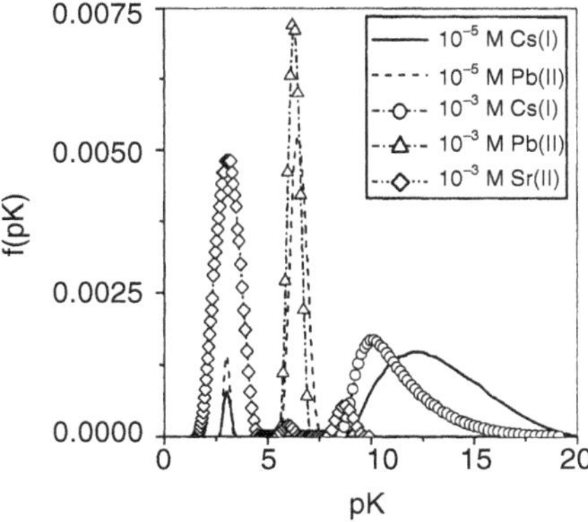

FIGURE 38.28 Adsorption constant distributions of aqueous metals Cs(I) at 10^{-3} M and 10^{-5} M; Pb(II) and Sr(II) (b) at 10^{-3} M on fumed silica at $C_{SiO_2} = 0.2$ wt.%.

However, corrections in ζ related to the solution conductivity and the porosity of aggregates of primary particles of fumed silica give close results even at a larger salinity (0.01 M NaCl) [9] than that used here. Therefore in our calculations, correction related to the conductivity was not considered (this correction considered in parallel to the porosity effect could result in a small reduction of the difference between the graphs obtained using Equations (38.6) and (38.7) or (38.10), and in the case of adsorption of Pb(II), the electrophoretic mobility directly measured was analyzed. Notice that upon adsorption of tetraalkylammonium or alkali ions, changes in the surface charge density on fumed silica Aerosil OX 50 with nonporous primary particles (average size of 40 nm) is significantly lower than that for porous silica (Stöber silica), since the porosity of the latter plays a substantial role [40].

This circumstance is in agreement with relatively weak effect of the conductivity on the ζ values of fumed silica [9]. It should be noted that the surface charge density increases for fumed silica with increasing mean size of primary particles (Figure 38.29) that is in agreement with enhancement of the amounts of free surface $\equiv$SiOH groups with growing d or decreasing S (Figure 38.6) as these deprotonated groups are responsible for the negative charge of the silica surface.

Character of changes in σ_0 and U_e due to adsorption of Pb(II) differs, since one can see enhancement of the negative value of σ_0(pH) (Figure 38.29) and appearance of positive U_e(pH) (Figure 38.30) at $C_{Pb} = 10^{-3}$ M (similar effects were described in details elsewhere [38]). This concentration corresponds to approximately 10 wt.% in respect to oxide and gives approximately one-two Pb(II) ions per nm^2 of A–300. There is another effect contributing change in the sign of U_e at $C_{Pb} = 10^{-3}$ M due to the difference in the mobility of oxide swarms and metal cations under applied external field, which results in assemblage of cations near or inside porous aggregates more than the equilibrium concentration of adsorbed ions (these cations can speed up oxide aggregates in the external electrostatic field). Clearly, this effect increases with increasing size and porosity of oxide aggregates and pH. The ion current near the oxide surfaces cannot compensate this effect entirely. However, the Pb(II) concentrations of 10^{-4} M or less provides too low surface coverage to change σ_0(pH) and U_e(pH) so substantially (as at $C_{\text{Pb}} = 10^{-3}$ M) due to both specific adsorption of cations and their assemblage near aggregates under the external field.

CONCLUSION

Variation of conditions of the fumed silica synthesis allows one to prepare materials characterized by various textures in respect to both primary particles (their porosity,

TABLE 38.11
Complexation Equilibrium Constants for Aqueous Metals Adsorbed on Different Oxides Calculated Using TLM (GRFIT)

Oxide	Cation	C_{Ct} (M)	Complexation reaction	$-\log K$
A-300	Pb(II)	10^{-4}	$\equiv SiOH + Pb^{2+} \rightarrow \equiv SiO^{-}Pb^{2+} + H^{+}$	5.6
			$\equiv SiOH + PbOH^{+} \rightarrow \equiv SiO^{-}Pb(OH)^{+} + H^{+}$	6.1
			$\equiv SiOH + Pb(OH)_2 \rightarrow \equiv SiO(H)Pb(OH)_2$	1.1
A-300	Sr(II)	10^{-4}	$\equiv SiOH + Sr^{2+} \rightarrow \equiv SiO^{-}Sr^{2+} + H^{+}$	14.3
			$\equiv SiOH + SrOH^{+} \rightarrow \equiv SiO(H)SrOH^{+}$	3.8
			$\equiv SiOH + SrOH^{+} \rightarrow \equiv SiO^{-}SrOH^{+} + H^{+}$	6.1
			$\equiv SiOH + Sr(OH)_2 \rightarrow \equiv SiOHSr(OH)_2$	1.3
A-300	Cs(I)	10^{-4}	$\equiv SiOH + Cs^{+} \rightarrow \equiv SiO^{-}Cs^{+} + H^{+}$	16.4
			$\equiv SiOH + Cs^{+} \rightarrow \equiv SiO(H)Cs^{+}$	3.3

amounts of water adsorbed in different forms, etc.) and swarms (their structure and size distribution, type of contacts in aggregates, etc.) and different levels of the hydrophilicity. Alterations in the texture of primary particles and their swarms allow one to control the adsorptive characteristics of fumed silicas in respect to nonpolar and polar compounds over wide ranges. According to the IR spectra, 20 wt.% of PVP and 40 wt.% of BSA adsorbed on the fumed silica surfaces disturb all the $\equiv$SiOH groups (appearing as the band at 3750 cm^{-1}) through direct interaction between polymer molecules and the silica surfaces or by means of water molecules. This effect can be explained by decomposition of agglomerates and large aggregates, as total surface area becomes accessible for both unfolded PVP and globular BSA molecules with the participation of water molecules. Consequently, one could assume that aggregates formed due to the siloxane bridges, which are not decomposed on polymer adsorption, should be relatively small, but larger aggregates are formed due to non-siloxane linkages; i.e., hydrogen and another intermolecular bonding. Ionogenic surfactant Aethonium ($C_{Aet} < 0.01$ wt.%) and ethanol (high concentrations 10–50 wt.%) have a lower effect on the particle size distributions than PVP or albumin. Globular proteins (albumin and gelatin) have a weaker influence on the silica structuring in suspending-drying in comparison with PVP due to the difference in their adsorption, as proteins are characterized by strong intramolecular bonds, but unfolded PVP molecules stronger rearrange silica particle swarms. Suspending-drying of silica and silica/polymers leads to reduction of the specific surface area but the effective pore volume accessible for adsorbed nitrogen increases in comparison with the pristine fumed silica. Polymers covering the surfaces of primary particles and their aggregates provide additional diminution in S but relative reduction in the pore volume in comparison with that for dried pure silica suspension is smaller, and addition of 5 wt.% PVP even enhances V_p. Covering of

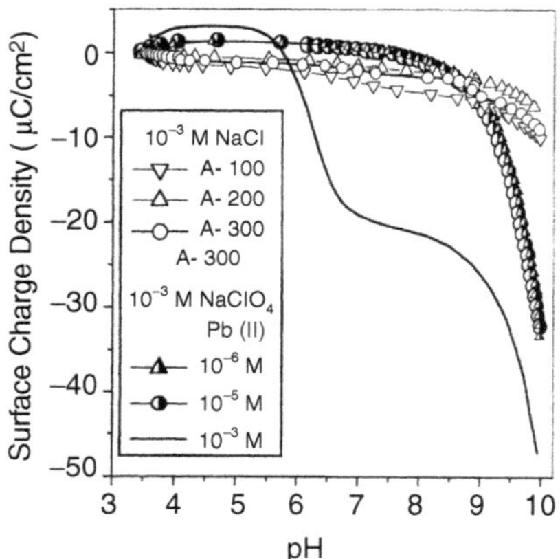

FIGURE 38.29 Surface charge density as a function of pH for different fumed silicas without and with the presence of Pb(II) at different concentrations.

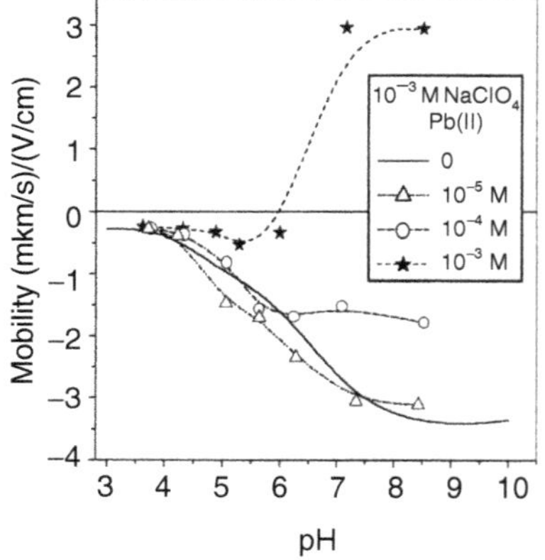

FIGURE 38.30 Electrophoretic mobility as a function of pH for the aqueous suspensions of fumed silica A-300 at different concentration of Pb(II).

the silica surfaces by polymers blocking surface silanols changes the adsorption potential distributions for adsorbed water and nitrogen adsorption energy distributions, since the adsorption potential is reduced at low coverage, as narrow pore contribution decreases, but enhanced at great coverage corresponding to secondary filling of large mesopores. Adsorbed metal cations can strongly change the surface charge density and elecrtokinetic mobility if their concentration is greater than a critical value.

ACKNOWLEDGMENT

Many of the presented investigations have been performed in collaboration with the colleagues from the Maria Curie-Sklodowska University (Lublin, Poland), Institute of Surface Chemistry (Kiev), Institute of Biochemistry (Kiev), and the authors thank Prof. R. Leboda, Prof. W. Janusz, Prof. S. Chibowski, Dr. J. Skubiszewska-Zięba (Lublin), Prof. I.I. Geraschenko, Prof. Yu.N. Levchuk, Dr. E.M. Pakhlov, Dr. N.N. Vlasova, E.V. Goncharuk, and N.V. Guzenko (Kiev) for fruitful cooperation.

REFERENCES

1. (a) *Basic Characteristics of Aerosil;* Technical Bulletin Pigments, N11; Degussa AG: Hanau, 1997; (b) Iler, R.K. *The Chemistry of Silica*; Wiley: Chichester, 1979.
2. (a) Legrand, A.P., Ed. *The Surface Properties of Silicas*; Wiley: New York, 1998; *(b) Proceedings of International Conference on Silica Science and Technology "Silica 98"*, Sept. 1–4, Mulhouse (France), 1998; (c) *Proceedings of International Conference on Silica Science and Technology "Silica 2001"*, Sept. 3–6, Mulhouse (France), 2001; (d) Kiselev, A.V.; Lygin, V.I. *IR Spectra of Surface Compounds and Adsorbed Substances;* Nauka: Moscow, 1972.
3. (a) Barthel, H. *Colloid. Surf. A* **1995**, *101*, 217; (b) Barthel, H.; Rosch, L.; Weis, J. in: *Organosilicon Chemistry II. From Molecules to Materials;* Auner, N.; Weis, J., Eds. VCH: Weinheim, pp. 761–778, 1996; (c) Barthel, H.; Heinemann, M.; Stintz, M.; Wessely, B. *Proceedings of International Conference on Silica Science and Technology "Silica 98"*, Sept. 1–4, Mulhouse (France), 1998, 323.
4. (a) Hurd, A.J. *Mater. Res. Soc. Symp. Proc.* **1990**, *172*, 3; (b) Hurd, A.J.; Flower, W.L. *J. Colloid Interface Sci.* **1988**, *122*, 178; (c) Schaefer, D.W.; Hurd, A.J. *Aerosol Sci. Technol.* **1990**, *12*, 876.
5. (a) Briesen, H.; Fuhrmann, A.; Pratsinis, S.E. *Chem. Engineer. Sci.* **1998**, 53, 4105; (b) Vemury, S.; Pratsinis, S.E. *J. Aerosol Sci.* **1996**, 27, 951; (c) Ehrman, S.H.; Friedlander, S.K.; Zachariah, M.R. *J. Aerosol Sci.* **1998**, 29, 687.
6. (a) Dabrowski, A.; Tertykh, V.A., Eds. *Adsorption on New and Modified Inorganic Sorbents;* Elsevier: Amsterdam, 1996; (b) Vansant, E.F.; Van Der Voort, P.; Vrancken, K.C. *Characterization and Chemical Modification of the Silica Surface;* Elsevier: Amsterdam, 1995; (c) Blitz, J.P.; Little, Ch.B., Eds. *Fundamental and Applied Aspects of Chemically Modified Surfaces;* Royal Society of Chemistry: Cambridge, 1999; (d) Gun'ko, V.M.; Vedamuthu, M.S.; Henderson, G.L.; Blitz, J.P. *J. Colloid Interface Sci.* **2000**, *228*, 157.
7. Mironyuk, I.F.; Gun'ko, V.M.; Turov, V.V.; Zarko, V.I.; Leboda, R.; Skubiszewska-Zięba, J. *Colloid. Surf. A* **2001**, *180*, 87.
8. (a) Mironyuk, I.F. *Fiz. Kondens. Vysokomolecul. System.* **1998**, *6*, 59; (b) Mironyuk, I.F.; Voronin, E.F.; Pakhlov, E.M.; Chuiko, A.A. *Ukr. Khim. Zh.* **2000**, *66*, No 10, 81; (c) Mironyuk, I.F. Scientific *Basics of Controlled Synthesis of Fumed Silica and Its Physicochemical Properties*; Doctoral Thesis, Institute of Surface Chemistry: Kiev, 2001.
9. Gun'ko, V.M.; Zarko, V.I.; Leboda, R.; Chibowski, E. *Adv. Colloid Interface Sci.* **2001**, *91*, 1.
10. (a) Gun'ko, V.M.; Turov, V.V.; Zarko, V.I.; Dudnik, V.V.; Tischenko, V.A.; Voronin, E.F.; Kazakova, O.A.; Silchenko, S.S.; Chuiko, A.A. *J. Colloid. Interface Sci.* **1997**, *192*, 166; (b) Gun'ko, V.M.; Zarko, V.I.; Voronin, E.F.; Turov, V.V.; Mironyuk, I.F.; Gerashchenko, I.I.; Goncharuk, E.V.; Pakhlov, E.M.; Guzenko, N.V.; Leboda, R.; Skubiszewska-Zięba, J.; Janusz, W.; Chibowski, S.; Levchuk, Yu.N.; Klyueva, A.V. *Langmuir* **2002**, *18*, 581; (c) Gun'ko, V.M.; Vlasova, N.N.; Golovkova, L.P.; Stukalina, N.G.; Gerashchenko, I.I.; Zarko, V.I.; Tischenko, V.A.; Goncharuk, E.V.; Chuiko, A.A. *Colloid. Surf. A* **2000**, *167*, 229; (d) Gerashchenko, B.I.; Gun'ko, V.M.; Gerashchenko, I.I.; Leboda, R.; Hosoya, H.; Mironyuk, I.F. *Cytometry*, **2002**, *49*(2), 56.
11. (a) Andrade, J.D. *Principles In Protein Adsorption*, In: *Surface and Interfacial Aspects of Biomedical Polymers*, vol. 2, Plenum Press: New York, 1985; (b) Landau, M.A. *Molecular Mechanism of Action of Physiologically Active Compounds*, Nauka: Moscow, 1981; (c) *Protein at Interfaces: Physicochemical and Biochemical Studies*, Brash, J.L.; Horbett, T.A., Eds., ASC Symp. Ser., vol. 342, Amer. Chem. Soc.: Washington, DC, 1987; (d) Parfitt, G.D.; Rochester, C.H., Eds.; *Adsorption from Solution at the Solid/Liquid Interface*, Academic Press: London, 1983; (e) Sato, T.; Ruch, R. *Stabilization of Colloidal Dispersions by Polymer Adsorption*, Marcel Dekker: New York, 1980.
12. (a) Gun'ko, V.M.; Mironyuk, I.F.; Zarko, V.I.; Voronin, E.F.; Turov, V.V.; Pakhlov, E.M.; Goncharuk, E.V.; Leboda, R.; Skubiszewska-Zięba, J.; Janusz, W.; Chibowski, S. *J. Colloid Interface Sci.* **2001**, *242*, 90; (b) Gun'ko, V.M.; Zarko, V.I.; Leboda, R.; Marciniak, M.; Janusz, W.; Chibowski, S. *J. Colloid Interface Sci.* **2000**, *230*, 396; (c) Gun'ko, V.M.; Zarko, V.I.; Turov, V.V.; Leboda, R.; Chibowski, E.; Gun'ko, V.V. *J. Colloid. Interface Sci.* **1998**, *205*, 106; (d) Gun'ko, V.M.; Zarko, V.I.; Turov, V.V.; Leboda, R.; Chibowski, E. *Langmuir* **1999**, *15*, 5694; (e) Gun'ko, V.M.; Zarko, V.I.; Turov, V.V.; Leboda, R.; Chibowski, E.; Pakhlov,

E.M.; Goncharuk, E.V.; Marciniak, M.; Voronin, E.F.; Chuiko, A.A. *J. Colloid. Interface Sci.* **1999**, *220*, 302.

13. (a) Gun'ko, V.M.; Skubiszewska-Zieba, J.; Leboda, R.; Zarko, V.I. *Langmuir* **2000**, *16*, 374; (b) Zarko, V.I.; Gun'ko, V.M. *Adsorption Sci. Technol.* **1996**, *14*, 331; (c) Zarko, V.I.; Gun'ko, V.M.; Chibowski, E.; Dudnik, V.V.; Leboda, R. *Colloids Surf. A* **1997**, *127*, 118; (d) Gun'ko, V.M.; Turov, V.V.; Zarko, V.I.; Voronin, E.F.; Tischenko, V.A.; Dudnik, V.V.; Pakhlov, E.M.; Chuiko, A.A. *Langmuir* **1997**, *13*, 1529; (e) Gun'ko, V.M.; Zarko, V.I.; Chibowski, E.; Dudnik, V.V.; Leboda, R.; Zaets, V.A. *J. Colloid. Interface Sci.* **1997**, *188*, 39; (f) Gun'ko, V.M.; Zarko, V.I.; Turov, V.V.; Leboda, R.; Chibowski, E.; Holysz, L. Pakhlov, E.M.; Voronin, E.F.; Dudnik, V.V.; Gornikov, Yu.I. *J. Colloid. Interface Sci.* **1998**, *198*, 141; (g) Gun'ko, V.M.; Zarko, V.I.; Leboda, R.; Voronin, E.F.; Chibowski, E. *Colloids Surf. A* **1998**, *132*, 241; (h) Gun'ko, V.M.; Voronin, E.F.; Pakhlov, E.M.; Zarko, V.I.; Turov, V.V.; Guzenko, N.V.; Leboda, R.; Chibowski, E. *Colloids Surf. A* **2000**, *166*, 187.
14. (a) Gun'ko, V.M.; Leboda, R.; Turov, V.V.; Villiéras, F.; Skubiszewska-Zięba, J.; Chodorowski, S.; Marciniak, M. *J. Colloid Interface Sci.* **2001**, *238*, 340; (b) Gun'ko, V.M.; Leboda, R.; Skubiszewska-Zięba, J.; Turov, V.V.; Kowalczyk, P. *Langmuir* **2001**, *17*, 3148; (c) Gun'ko, V.M.; Sheeran, D.J.; Augustine, S.M.; Blitz, J.P. *J. Colloid Interface Sci.*, **2002**, *249*, 123; (d) Gun'ko, V.M.; Zarko, V.I.; Sheeran, D.J.; Blitz, J.P.; Leboda, R.; Janusz, W.; Chibowski, S. *J. Colloid Interface Sci.*, **2002**, *252*, 109; (e) Gun'ko, V.M.; Zarko, V.I.; Mironyuk, I.F.; Goncharuk, E.V.; Borysenko, M.V.; Pakhlov, E.M.; Leboda, R.; Janusz, W.; Skubiszewska-Zięba, J.; Charmas, B.; Marciniak, M.; Matysek, M.; Chibowski, S. *Langmuir*, *Colloids Surf. A* **2004**, *240*, 9.
15. (a) Norde, W.; Lyklema, J. *Colloids Surf.* **1989**, 38, 1; (b) Norde, W.; Favier, P. *Colloids Surf.* **1992**, 64, 87; (c) Norde, W.; Anusiem, A.C.I. *Colloids Surf.* **1992**, 66, 73; (d) Urano, H.; Fukuzaki, S. *J. Fermentation Bioengineer.* **1997**, 83, 261.
16. (a) Menshikov, V.V. (Ed.), *Laboratory Investigation Methods in Clinic*, Medicine: Moscow, 1987; (b) Weichselbaum, T. *Amer. J. Clin. Pathol.* **1946**, *16*, 40; (c) Kochetov, G.A. *Practical Enzymology Manual*, Vysshaya Shkola: Moscow, 1980.
17. (a) Adamson, A.W.; Gast, A.P. *Physical Chemistry of Surface*, 6th ed., Wiley: New York, 1997; (b) Gregg, S.J.; Sing, K.S.W. *Adsorption, Surface Area and Porosity*, 2nd ed., Academic Press: London, 1982; (c) Toth, J., *Adv. Colloid Interface Sci.* **1995**, *55*, 1.
18. Jaroniec, M.; Kruk, M.; Olivier J. P. *Langmuir* **1999**, *15*, 5410.
19. Avnir, D.; Jaroniec, M. *Langmuir* **1989**, *5*, 1431.
20. (a) Nguyen, C.; Do, D.D. *Langmuir* **1999**, *15*, 3608; (b) Nguyen, C.; Do, D.D. *Langmuir* **2000**, *16*, 7218; (c) Gun'ko, V.M.; Do, D.D. *Colloids Surf. A* **2001**, *193*, 71.
21. Szombathely, M.V.; Brauer, P.; Jaroniec, M. *J. Comput. Chem.* **1992**, *13*, 17.
22. Provencher, S.W. *Comp. Phys. Comm.* **1982**, *27*, 213; 229.
23. (a) Hunter, R.J. *Zeta Potential in Colloid Sciences*, Academic Press: London, 1981; (b) Hunter, R.J. *Introduction to Modern Colloid Science*, Oxford University Press: London, 1993.
24. (a) Miller, N.P.; Berg, J.C. *J. Colloid Interface Sci.* **1993**, *159*, 253; (b) Miller, N.P.; Berg, J.C.; O'Brien, R.W. *J. Colloid Interface Sci.* **1992**, *153*, 237.
25. Xu, R. *Langmuir* **1998**, *14*, 2593.
26. Borkovec, M.; Rusch, U.; Cernik M.; Koper, G.J.M.; Westall, J.C. *Colloids Surf. A* **1996**, *107*, 285.
27. Ludwig, Chr. *GRIFIT, A Program for Solving Speciation Problems: Evaluation of Equilibrium Constants, Concentrations and other Physical Parameters, Internal Report*, University of Berne, 1992.
28. Turov, V.V.; Leboda, R. *Adv. Colloid Interface Sci.* **1999**, *79*, 173.
29. Gun'ko, V.M.; Turov, V.V. *Langmuir* **1999**, *15*, 6405.
30. Gun'ko, V.M.; Zarko, V.I.; Chuikov, B.A.; Dudnik, V.V.; Ptushinskii, Yu.G.; Voronin, E.F.; Pakhlov, E.M.; Chuiko, A.A. *Int. J. Mass Spectrom. Ion Processes* **1998**, *172*, 161.
31. Zhuravlev, L.T. *Colloids Surf. A* **1993**, *74*, 71.
32. Sobolev, V.A.; Tertykh, V.A.; Chuiko, A.A. *Zh. Prikladn. Spectros.* **1970**, *13*, 646.
33. (a) Janusz, W. *J. Colloid Interface Sci.* **1991**, *145*, 119; (b) Janusz, W.; Szczypa, J. *J. Dispersion Sci. Technol.* **1998**, *19*, 267; (c) Janusz, W.; Jablonski, J.; Szczypa, R. *J. Dispersion Sci. Technol.* **2000**, *21*, 739; (d) Janusz, W.; Staszczuk, W.; Sworska, A.; Szczypa, J. *J. Radioanal. Nuclear Chem.* **1993**, *174*, 83; (e) Janusz, W.; Kobal, I.; Sworska, A.; Szczypa, J. *J. Colloid Interface Sci.* **1997**, *187*, 381; (f) Marmier, N.; Delisee, A.; Fromage, F. *J. Colloid Interface Sci.* **1999**, *212*, 228.
34. Schindler, P.W.; Furst, B.; Dick, R.; Wolf, P.U. *J. Colloid Interface Sci.* **1976**, *55*, 469.
35. (a) Primet, M.; Pichat, P.; Mathieu, M.V. J. Phys. Chem. **1971**, *75*, 1216; (b) Boehm, H.P. Angew. Chem. **1966**, *78*, 617; (c) Jones, P.; Hockey, J.A. *J. Chem. Soc., Faraday Trans.* **1971**, *67*, 2679; **1972**, *68*, 907.
36. (a) Robertson, A.P.; Leckie, J.O. *J. Colloid Interface Sci.* **1997**, *188*, 444; (b) Farley, K.J.; Dzombak, D.A.; Morel, F. M.M. *J. Colloid Interface Sci.* **1985**, *106*, 226; (c) Vlasova, N. *J. Colloid Interface Sci.* **2001**, *233*, 227.
37. Cernik M.; Borkovec, M.; Westall, J.C. *Environ. Sci. Technol.* **1995**, *29*, 413.
38. (a) Kosmulski, M. *Colloids Sur. A* **1996**, *117*, 201; (b) Kosmulski, M.; Eriksson, P.; Gustafsson, J.; Rosenholm J.B. *J. Colloid Interface Sci.* **1999**, *220*, 128.
39. Sonnefeld, J.; Löbbus, M.; Vogelsberger, W. *Colloids Surf. A* **2001**, *195*, 215.
40. de Keizer, A.; van der Ent, E.M.; Koopal, L.K. *Colloid Surf. A.* **1998**, *142*, 303.

39 Adsorption of Surfactants and Polymers on Silica

P. Somasundaran and L. Zhang
Columbia University, Langmuir Center for Colloids and Interfaces

CONTENTS

INTRODUCTION

Silica is encountered in various industrial processes, and to control the performance of these processes, it is often necessary to modify its surface properties such as zeta-potential, suspension stability, hydrophobicity, and adsorption capacity. Surface properties of silica are a function of its state of hydrolysis as well as pretreatment. Adsorption of surfactants and polymers can also lead to marked changes in its interfacial properties and yield desired performance.

SILICA WATER CHEMISTRY

Silica particles in aqueous solution possess a surface charge due to preferential dissolution of surface species and interfacial ion-exchange. The magnitude and sign of the surface charge is dependent on the concentration of the potential determining ions and is a function of pH and ionic strength of the solution.

The dissolution process of silica in aqueous solution is illustrated in Figure 39.1. When fresh silica is contacted with water, hydrolysis of the surface species takes place giving rise to silanol groups at the surface (Figure 39.1a). When exposed to water for longer intervals, this surface tends to hydrolyze further to form $=Si(OH)_2$ (Figure 39.1b). Formation of more silanol groups is possible, particularly under low bulk concentration of alkali metals (Figure 39.1c). Further hydrolysis of the above surface can result in the dissolution of the surface species to form bulk silicic acid and expose new silicon atoms to the bulk water (Figure 39.1d). The silica surface will thus form successively/Si(OH), $=Si(OH)_2$, $-Si(OH)_3$ at the interface and $Si(OH)_4$ in solution. Surface hydrolysis followed by dissolution will take place till solid-solution equilibrium is reached [1].

The main cause for the surface charge generation is the dissociation of the silanol groups at the interface.

$$-Si-OH + H^+ \, \Omega \, -Si^+ + H_2O \text{ or } -Si-OH_2^+$$
$$-Si-OH + OH^- \, \Omega \, -Si-O^- + H_2O$$

Due to the acidic nature of the silanol surface, silica possesses a negative charge at neutral pH. The point of zero charge of silica is about pH 2. The number and type of silanol groups on the surface and the degree of hydration of the surface are the major factors that determine the surface properties of silica particles.

The rate of silica hydrolysis is a function of the solution pH, ionic strength, and temperature, and the surface properties are affected also by prolonged contact with solutions. It is to be noted that pretreatment of the silica by reagents such as HF, HNO_3 and NaOH, often employed in research, can produce drastic effects on its surface properties. These effects have been demonstrated by electrokinetic measurements. For example, quartz treated with hydrofluoric acid solution followed by warm sodium hydroxide solution can change the isoelectric point by as much as 4 pH units [2].

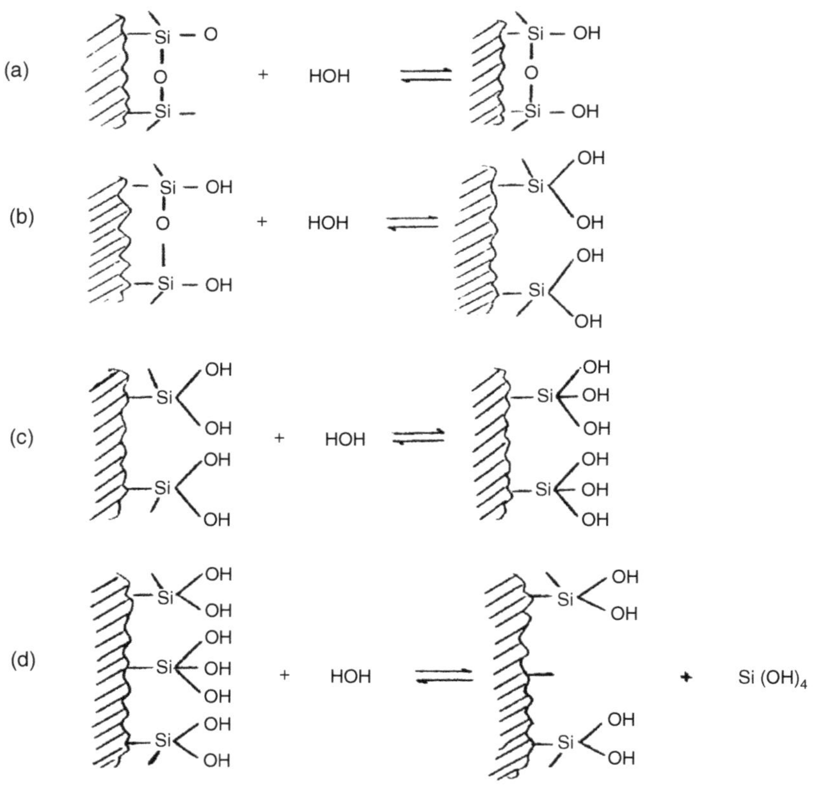

FIGURE 39.1 The hydrolysis of the surface species of silica [1].

MECHANISMS OF SURFACTANT AND POLYMER ADSORPTION ON SILICA

Surface properties can be adjusted by the adsorption of surfactants and polymers. Adsorption itself can essentially be considered to be preferential partitioning of the adsorbate into the into the interfacial region. It is the result of one or more contributing forces arising from electrostatic attraction, chemical reaction, hydrogen bonding, hydrophobic interactions, and solvation effects.

Cationic surfactants and polymers adsorb readily on silica due to electrostatic interaction since pH of most practical systems is above 2 and silica is negatively charged under these conditions. Anionic surfactants or polymers do not adsorb on silica at neutral pH due to the presence of similar charge on the solid and the adsorbate. However, anionic surfactants can adsorb on silica in the presence of multi-valent metal cations in the pH range in which metal ions hydrolyze to their first hydroxyl complex. This has been attributed to the chemisorption of the first hydroxyl species of metal ions on silica and modification of the quartz surface [3].

Nonionic ethoxylated surfactants and polyethylene oxide polymers have been found to adsorb on silica. The driving force for the adsorption of these reagents in considered to be hydrogen bonding between the ether oxygen of the surfactant/polymer and the surface hydroxyl groups of silica: $\text{—SiOH} \cdots \text{O(CH}_2\text{CH}_2)_2\text{=}$. Interestingly, another group of nonionic surfactants, the alkyl polyglucosides, is found to exhibit minimal adsorption on silica [4].

ADSORPTION OF SURFACTANTS AND POLYMERS

The relationship between adsorption and interfacial properties such as contact angle, zeta-potential and flotation recovery is illustrated in Figure 39.2 for cationic surfactant dodecylammonium acetate/quartz system (5). The increase in adsorption due to association of surfactants adsorbed at the solid-liquid interface into two dimensional aggregates called solloids (surface colloids) or hemimicelles occurs at about 10^{-4} M DAA. This marked increase in adsorption density is accompanied by concomitant sharp changes in contact angle, zeta-potential and flotation recovery. Thus these interfacial phenomena depend primarily on the adsorption of the surfactant at the solid-liquid interface. The surface phenomena that reflect the conditions at the solid-liquid interface (adsorption density and zeta-potential) can in many cases be correlated directly with the phenomena that reflect the

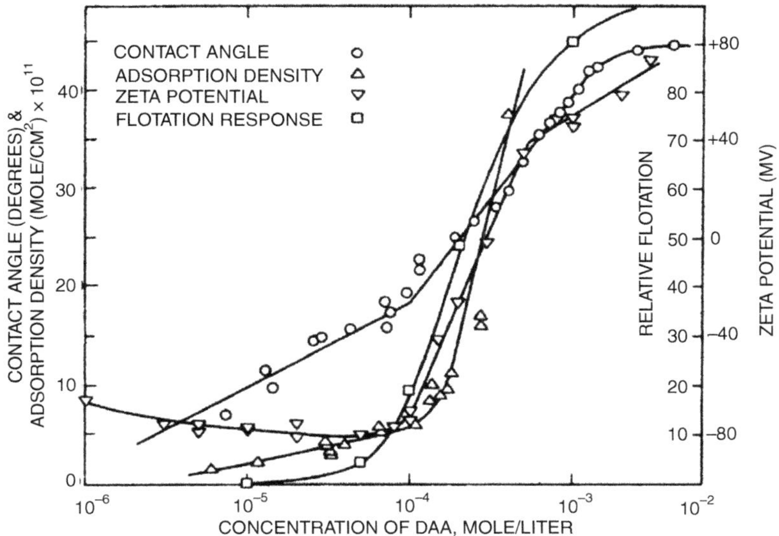

FIGURE 39.2 Correlation of adsorption density, contact angle, flotation response and zeta potential for quartz as a function of dodecylammonium acetate concentration at pH 6 to 7, 20 to 25°C [5].

conditions at the solid-liquid-gas three phase triple contact (contact angle and flotation).

Anionic surfactants adsorb on silica only in the presence of multivalent metal cations. This phenomenon has been utilized in the flotation of quartz where metal cations are added as activators. Significant adsorption of anionic surfactants on silica can also be induced by other surfactants that adsorb on silica. For example, adsorption of ethoxylated surfactant, octaethylene glycol mono-*n*-dodecyl ether($C_{12}EO_8$), on silica can induce adsorption of the anionic surfactant sodium *p*-octylbenzenesulfonate (C_8MS). The anionic surfactant adsorbs on silica either through interaction with the adsorbed ethoxyl chains of the non-ionic surfactant and/or through hydrocarbon chain-chain interactions with neighboring $C_{12}EO_8$ [6].

ADSORPTION OF SURFACTANT/POLYMER MIXTURES AND ITS ROLE IN FLOTATION

Polymers and surfactants are used together sometimes to obtain desirable effects. Polymer-surfactant interactions in solution and at the interfaces can change the interfacial properties of the solid directly or indirectly. It is shown that depending on the nature of the polymer and the surfactant, polymers can affect flotation of quartz by affecting the adsorption of the surfactant on it [7].

Quartz can be floated by cationic surfactant dodecylamine hydrochloride. Nonionic polymer, polyacrylamide (PAM), which does not adsorb on quartz and does not cause flotation by itself, increases the quartz flotation by amine slightly due to the uptake of water molecules by the polymer for hydration. The hydration of polymer causes an increase in the effective concentration of amine.

Anionic sulfonated polyacrylamide (PAMS) is also found to increase amine flotation of quartz. Although PAMS does not adsorb on the negatively charged quartz and cause no direct activation of amine adsorption, the polymer-surfactant electrostatic interaction can lead to the formation of complexes. This polymer-surfactant complex can reduce the armoring of bubbles and lead to flotation. The anionic polymer can also bridge the adsorbed amine to the amine on the bubble surface and enhance flotation under saturated adsorption conditions. The hydration effect of the polymer may also be responsible for the enhanced flotation in this case.

Flotation in solutions containing cationic polymers such as PAMD is depressed due to masking of the surface by the large polymer molecule. Competitive interactions between the polymer and surfactant at the adsorption sites on the quartz surface, if any, may also contribute to the decrease. Polymer itself does not float the quartz since it exposes a hydrophilic surface.

In the anionic flotation of quartz, activators such as multivalent metal ions are required to provide adsorption sites for surfactant. Cationic polymer can also adsorb on silica and offer adsorption site for surfactant. It is found that quartz can be activated by cationic PAMD for flotation by the anionic dodecylsulfonate [7].

SUMMARY

Silica surface properties are a function of its state of hydrolysis. The rate of the hydrolysis depends on pH, ionic strength, and previous treatment history of the sample. Surfactants and polymers can adsorb on silica through different mechanisms such as electrostatical interaction and hydrogen bonding. The adsorption of surfactant and polymer can drastically alter the interfacial physicochemical properties of the silica such as zeta-potential and hydrophobicity. Depending on the type of the surfactant and the polymer, the adsorption of their mixtures on silica shows either synergism or antagonism. These effects are significant for the modification of surface properties and have important implications in various industrial processes such as adhesion, flotation and flocculation/dispersion.

ACKNOWLEDGMENTS

The authors acknowledge financial support of the National Science Foundation (CTS-9622781, CTS-9632479 and EEC-9804618).

REFERENCES

1. R.D. Kulkarni and P. Somasundaran. in *Oxide-Electrolyte Interfaces*, (R.S. Alwitt ed.), Am. Electrochem. Soc., Princeton, NJ, 1973, p. 31.
2. R.D. Kulkarni and P. Somasundaran. *Int. J. Min. Proc.*, 4:89 (1977).
3. M.C. Fuerstenau, B.R. Palmer. in *Flotation A.M. Gaudin Memorial Vol. I*, (M. C. Fuerstenau, ed.), AIME, New York, 1976, p. 148–196.
4. L. Zhang, P. Somasundaran, and C. Maltesh. *J. Colloid Inter. Sci.*, 191:202 (1997).
5. D.W. Fuerstenau, T.W. Healy, and P. Somasundaran. *Trans AIME*, 229:131 (1964).
6. P. Somasundaran, E.D. Snell, Edward Fu, and Qun Xu. *Colloids & Surfaces*, 63:49 (1992).
7. P. Somasundaran and L.T. Lee. *Separation Sc. and Tech.*, 16:1475 (1981).

Part 4

Particle Size and Characterization Techniques

Jonathan L. Bass
The PQ Corporation

Average particle size and particle size distribution are very important properties of colloidal silica that influence the behavior of such materials in a broad number of applications. When colloidal silicas are aggregated in gels, precipitates, or fumed products, their texture can be described in terms of pore volume distributions. The texture of the aggregate also influences its behavior in applications. The chapters in this section describe methods used to determine particle size and pore volume distributions in silicas.

The most direct method to observe particle size distributions of colloidal silica, which lie in the submicrometer particle size range, is transmission electron microscopy. However, this method is not practical for most researchers and is useless as an analytical tool in the processing of colloidal silica for several reasons. To obtain useful micrographs the particles must be thoroughly dispersed on a microscope grid. Only an extremely small portion of a sample will be observed, so several replicate measurements must be made to have a meaningful statistical distribution of particles. The particles must be measured and counted according to size, which is a tedious procedure when done manually and not trivial when an image analysis program is used. The best use of transmission electron microscopy is as an occasional check of other methods.

Prior to the 1970s, methods such as light scattering and surface area measurements were used to obtain average particle size of colloidal silica. Both of these methods assume that the particles are spheres. Light scattering requires dilute samples, whereas surface area measurement requires careful drying so that the particles do not aggregate to the extent that a significant amount of surface area is lost. Another method, used by Sears [1], involves titration of the sodium ion used to stabilize the sol. It was found that empirical constants developed for this procedure had to be modified as more sols were analyzed.

The most significant advance in particle size distribution measurement in the submicrometer particle size region has been the development of field-flow fractionation methods [2] over the past 15 years. These methods have been applied successfully to colloidal silica. The methods separate the particles according to size so that an actual distribution is measured. The smallest particles are measured first. Dilution of commercial sols is required for these measurements. Separation of particles is also effected by the disk centrifuge method, but in this case detection of small particles is time consuming and difficult to quantify.

Particle size distributions have been reported with another new method, quasi-elastic light scattering. This method is most useful for the average particle size of monodisperse systems. Mathematical transforms are used to resolve polydisperse distributions, but no physical separation takes place. This method seems to work satisfactorily if the dispersion around each maximum in the polydisperse sol is narrow and the peaks are well separated in particle size.

Other well-established methods are used for particle size distributions of aggregates of colloidal silica when the particle size extends into the micrometer size range. These include optical and scanning electron microscopy, laser light scattering, sedimentation, conductivity changes through an orifice, and light block-age. Because each of these methods involves different principles of detection, agreement between methods is usually only seen for spherical particles.

The textural properties of colloidal silica aggregates may be determined by scanning and transmission electron microscopy, capillary condensation of gases, and mercury intrusion porosimetry. As mentioned earlier, a major drawback of electron microscopy is the question of whether the observed field is representative of the sample. Unless an aggregate is fractured or sectioned, scanning electron microscopy will only probe the aggregate's external surface. Because the aggregates are too thick for an electron beam to penetrate, the transmission electron microscopist can only examine edges of aggregates or must section them to 5 to 50 nm in thickness.

Nitrogen is typically used as the adsorbate gas in capillary condensation measurements, condensed at liquid nitrogen temperatures. It may be excluded, because of its larger cross-sectional area, from small micropores. Other gases such as oxygen or argon are better probes for small micropores. Water may be used as a qualitative probe of the hydrophilicity of the silica surface.

There have been recent reports on the use of thermoporosimetry as a method for determining pore volume distributions [3]. This method involves freezing and thawing of liquids in pores. The freezing point decreases as liquid is frozen in smaller pores. The enthalpy change at the transition is proportional to the pore volume at the pore diameter corresponding to the temperature differential.

Because the range of pore volume distribution determined by mercury intrusion overlaps the capillary condensation range to some extent, one might expect that the two methods could serve as a check on each other. In practice, there are often significant differences in the results between the two methods that can be usefully exploited for better understanding of textural properties. These differences can be attributed to factors such as aggregate strength and pore shape.

Future developments in field-flow fractionation methods will probably involve extending the detection range to encompass both larger and smaller particle sizes. Currently, field-flow fractionation methods are not able to detect particles below 10 nm, yet some commercial products are smaller than this. Thermoporosimetry may supplement capillary condensation as a pore volume distribution method because analyses can be completed in substantially shorter times.

REFERENCES

1. Sears, G.W., Jr. *Anal. Chem.* **1956**, *28*, 1981.
2. Giddings, J.C.; Yang, F.J.F.; Myers, M.N. *Anal. Chem.* **1974**, *46*, 1917.
3. Quinson, J.F.; Astier, M.; Brun, M. *Appl. Catal.* **1987**, *30*, 123.

40 New Separation Methods for Characterizing the Size of Silica Sols

J.J. Kirkland
DuPont Experimental Station, Central Research and Development Department

CONTENTS

Both instrumental and chemical methods traditionally are used for characterizing the size of silica sols. Instrumental methods include electron microscopy, light scattering, turbidity–absorbance, centrifugation, and low-angle x-ray scattering. Chemical procedures involve gas adsorption, titrations, preferential precipitation, and rate of particle dissolution. Recent development of methods based on high-resolution separations has greatly widened the scope, accuracy, and precision of silica sol characterizations. These new methods include sedimentation field-flow fractionation (FFF), flow FFF, size-exclusion chromatography, hydrodynamic chromatography, and capillary-zone electrophoresis. Certain techniques (e.g., sedimentation FFF) permit the accurate determination of particle-size distributions, providing information that is not available by established characterization methods. The advantages and limitations of these new separation methods for characterizing the size of silica sols and other inorganic colloids are highlighted.

The characterization of colloidal silica has been the subject of numerous studies involving both physical and chemical methods. Iler [1] summarized many of the available methods, with particular emphasis on chemical approaches. Other more recent reviews [2–4] featured instrumental methods that are useful for characterizing silica sols and other colloids. This chapter describes some of the relatively new separations methods for characterizing silica sols. The merits and limitations of these methods are summarized so that the potential user can critically evaluate the capability of each approach for a projected application.

SEDIMENTATION FIELD-FLOW FRACTIONATION

This method sedimentation field-flow fractionation (SdFFF) is one of a family of field-flow fractionation (FFF) methods that were originally devised by Giddings at the University of Utah [5]. Although the potential of SdFFF was predicted more than 20 years ago [6], only in the past few years has SdFFF become a useful laboratory tool. The emergence of commercial instrumentation has opened up the potential of this powerful tool for characterizing silica sols and a wide range of other colloidal particles, as well as soluble macromolecules [7–14].

SdFFF separations are carried out in very thin, open channels shaped like a belt or ribbon and suspended in a centrifuge. A schematic of a SdFFF apparatus is shown in Figure 40.1. Liquid mobile phase is precision-pumped through a sampling-valve loop containing the liquid mixture to be analyzed. The sample is swept from this sampling loop through a rotating seal into the channel within the centrifuge. Following separation in the channel, the individual sample components flow back through the rotating seal into a detector (typically a turbidimeter) that senses the particles. A computer controls the pump output, sample valve actuation, and rotor speed. It

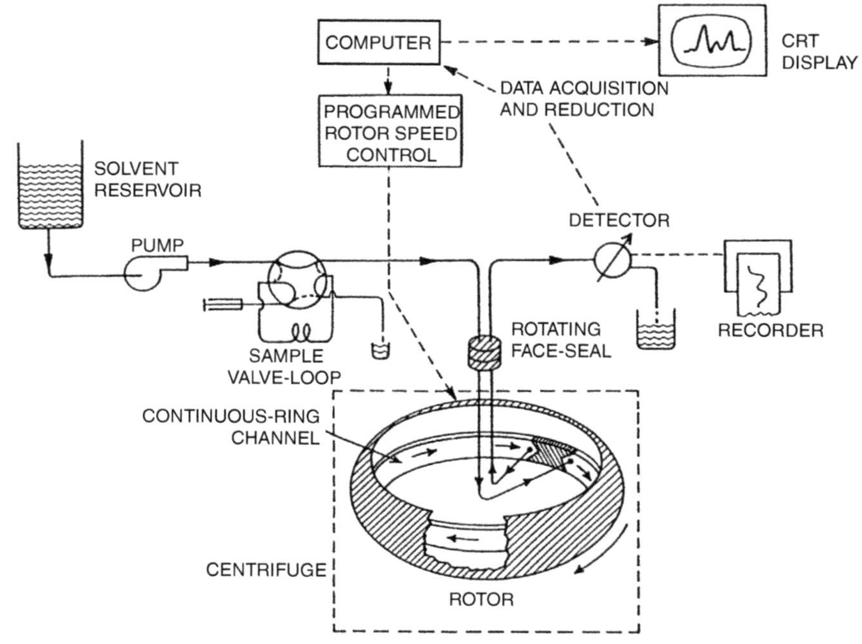

FIGURE 40.1 Schematic of SdFFF equipment. (Reproduced with permission from reference 36. Copyright 1982.)

also acquires data from the detector and transforms this output to a true concentration profile. In addition, calculations and graphics are provided for displaying data on particle size and particle-size distribution.

The development of a typical SdFFF separation is illustrated in Figure 40.2. With the liquid mobile phase stopped and the channel rotating at an appropriate speed, the sample mixture is injected into the channel (Figure 40.2a). The channel is rotated in this mode for a "relaxation" or pre-equilibration period that allows the particles to be forced toward the accumulation wall at approximately their sedimentation equilibrium position. Sample particles that have a density greater than the mobile phase are forced toward the outer wall. Diffusion opposite to that imposed by the centrifugal force field causes the particles to establish a specific mean-layer thickness near the accumulation wall as a function of particle mass. (Typically, the average distance of the particle "cloud" from the accumulation wall is about 5–20 μm.) Liquid mobile phase then is flowed continuously through

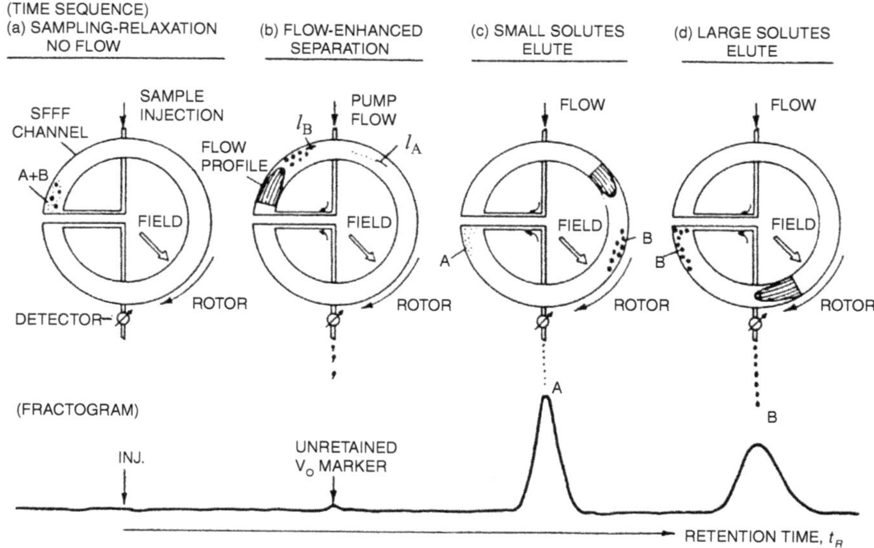

FIGURE 40.2 Development of a SdFFF separation. (Reproduced with permission from reference 36. Copyright 1982.)

the channel with a characteristic bullet-shaped laminar flow profile (Figure 40.2b). Solvent and particles that are unaffected by the centrifugal external force field are intercepted by the average of all flow streams and elute first from the channel in one channel volume ("dead" volume). Small particles are engaged by faster flow stream more quickly than larger particles and are eluted first (Figure 40.2c). Larger particles near the wall are intercepted by slower flow streams and are eluted later (Figure 40.2d). Thus, particles elute from the channel in order of increasing effective mass.

The resulting fractogram provides information on the masses of the sample components by means of quantitative relationships describing SdFFF retention [5]. Appropriate computer software permits the calculation of particle-size averages and particle-size distribution for the separated sample.

Force-field programming commonly is used in the SdFFF characterization of many colloidal particles to ensure that the entire particle-size distribution can be described in a convenient analysis time [10,15]. Constant force-field operation provides for the highest resolution of particles in the sample, with resulting highest precision. However, this mode of operation does not permit the rapid optimization of operating parameters for analyzing many samples. Also, characterization of samples with wide particle-size distributions is difficult with constant force-field operation. Force-field programming removes these limitations and provides a convenient and practical compromise for most applications.

A convenient and accurate method of force-field programming in SdFFF uses a time-delayed exponential decay (TDE–SFFF) procedure [9,10,15]. In this method, mobile-phase flow is initiated after sample injection, and the initial force field is held constant for a time equal to τ, the time constant of the subsequent exponential decay. After this delay, the force field is decayed exponentially with a time constant τ. In the TDE–SFFF mode, a simple log-linear relationship is obtained with both particle mass (and size) as a function of retention time. This simple relationship permits a convenient calculation of the quantitative information desired for the sample.

An example of the use of the TDE–SFFF method for characterizing a silica sol sample is shown in Figure 40.3. The upper plot show the response of the turbidimetric detector as a function of separation time. The middle plot is the differential curve. Because the response of the turbidimetric detector is particle-size dependent, the detector output signal is transformed with known, quantitative light-scattering relationships to produce a plot of the relative concentration as a linear function of particle diameter. To the left of the bottom plot are various particle-size averages that were calculated for this sample. The weight- and number-diameter averages of 21.5 and 20.0 nm, respectively, indicate that this silica sol sample had a narrow size distribution (polydispersity of 1.075).

Selecting the mobile-phase carrier is critical in characterizing silica sols by SdFFF. To prevent aggregation or particle flocculation, the pH should be in the range of 8–9, and the ionic strength should be low, generally <0.05. A convenient mobile phase is 1 m*M* ammonium hydroxide, which provides the proper pH ($\sim$8) and ionic strength. In addition, the positively charged ammonium ions adsorb to the negatively charged silica surface, so an environment that retards particle aggregation is created. Other satisfactory mobile phases include dilute organic buffers, for example, those containing ethanolamine or TRIS [tris (hydroxymethyl) aminomethane].

Advantages of SdFFF include accurate particle size distributions without the need for standards. Calculations require that the density of the colloid be known, although accurate and precise measurement of particle density can also be performed by SdFFF if required [16,17]. SdFFF is capable of characterizing silica sols in the 10–1000-nm range. Particle distributions of almost two orders of magnitude can be measured in a single analysis by using programmed force-field methods.

The high resolution of SdFFF permits excellent discrimination between particle sizes. Narrow particle-size distributions with a 10–15% difference in size produce bands that can be resolved to baseline. This high level of resolution produces accurate size measurements. Computer-controlled SdFFF instruments are capable of data with excellent reproducibility — results with $<5\%$ deviation (relative) are common. A typical analysis time for a silica sol is about 0.5 h. The SdFFF method can be applied to a wide range of colloidal particles, both inorganic and organic.

The SdFFF method is handicapped only by the somewhat high cost of equipment ($50,000–$100,000) and the need of the operator to acquire expertise that is not common to other more traditional instrumental methods. (Commercial equipment is available from Electronic Instruments and Technology, Inc., Sterling, VA, and FFFractionation, Inc., Salt Lake City, UT).

FLOW FIELD-FLOW FRACTIONATION

This method field-flow fractionation (FlFFF) is another member of the family, with measurement capabilities somewhat like those of SdFFF [18]. The basic separation process is retained in this method. Particles separate because they are intercepted by different flow stream velocities near the accumulation wall. However, in this FFF method, particles equilibrate at distances from the wall strictly as a function of their size (Stokes radius). The nearness to the wall is a balance of the cross flow in the channel

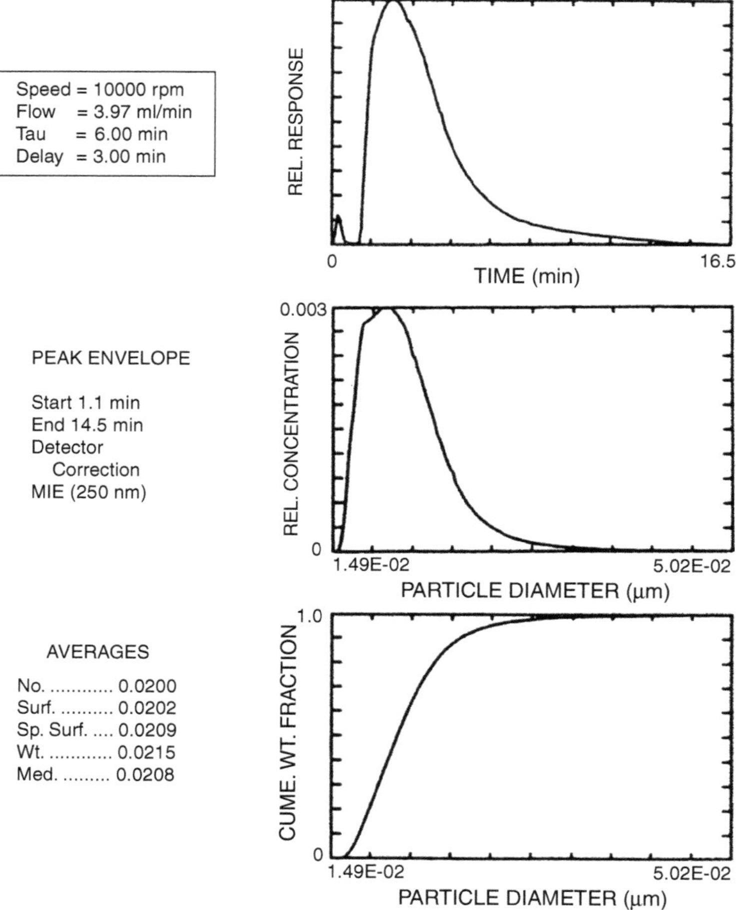

FIGURE 40.3 TDE–SFFF analysis of silica sol sample:mobile phase, 1 m*M* ammonium hydroxide; initial rotor speed, 10,000 rpm; mobile-phase flow rate, 3.97 ml/min; delay, 3.00 min; exponential decay time constant, 6.00 min; sample, 0.50 ml, 0.67%; and detection, UV, 220 nm.

pushing particles toward the wall and normal diffusion tending to move them away.

FlFFF separations are usually performed in thin channels constructed with flat beds. Figure 40.4 shows a schematic of a symmetrical channel [19]. In this design, the force field is created by flowing a liquid across the channel perpendicular to the normal flow down the channel. This flowing is accomplished by using semi-permeable membranes as both channel walls — solvent can flow through the membranes, but particles are retained within the channel for the separation. The remainder of the apparatus is quite similar to that previously described for SdFFF. Sample loading and detection is performed in the same manner.

A more recent and generally more advantageous design is an asymmetrical channel, shown schematically in Figure 40.5 [20]. In this design, a single semipermeable membrane serves as the accumulation wall. A flat glass plate rather than a membrane is used at the top of the channel. The force field is again created by flow through the bottom membrane, but here the flow velocity down the channel is asymmetrical. As a result of the mobile-phase carrier liquid being introduced at one end, the velocity of this liquid decreases as it proceeds down the channel. This change in mobile-phase carrier velocity somewhat complicates the retention theory for particles, but quantitative relationships for calculating particle size still are available [21].

A specific advantage of the asymmetrical channel design in FlFFF is that the injected sample can be prefocused into a very sharp band before separation. The result of this band sharpening is higher separation resolution and improved accuracy of particle-size measurement. This band prefocusing step is schematically shown in the upper plate of Figure 40.5. With no flow, the sample is injected through a port slightly downstream from the incoming mobile phase. With suitable valves, carrier liquid is then pumped into both ends of the channel at flow rates appropriate to focus the normally diffuse sample in a narrow band slightly below the

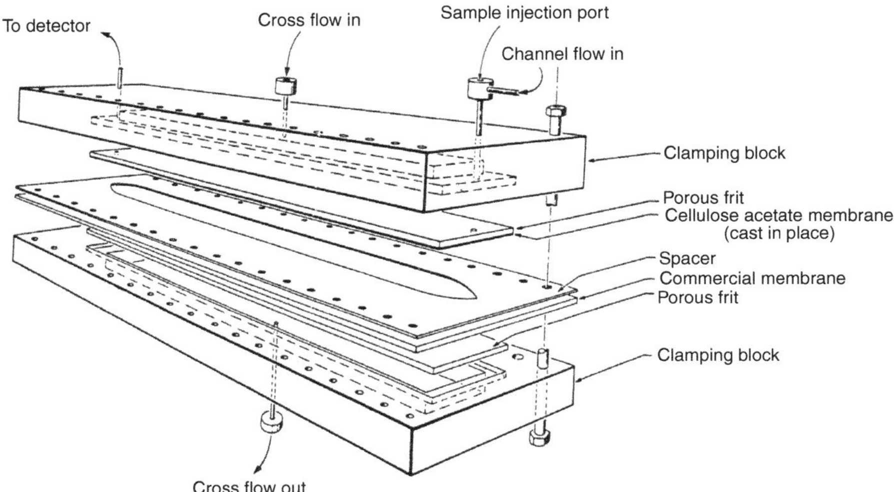

FIGURE 40.4 Symmetrical channel for FlFFF. (Reproduced with permission from reference 19. Copyright 1980.)

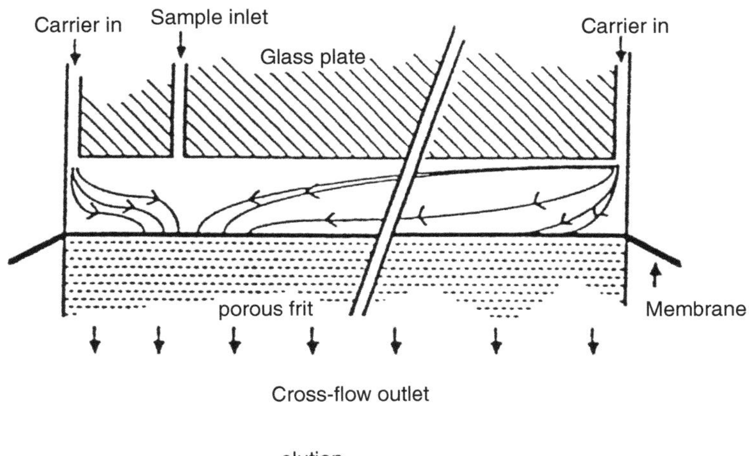

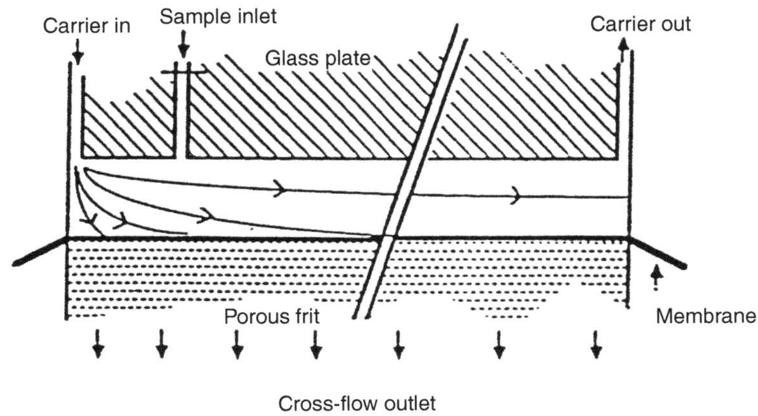

FIGURE 40.5 Asymmetrical channel for FlFFF. (Reproduced with permission from reference 20. Copyright 1989.)

injection port. Flow is then reinitiated only on the inlet of the channel to carry out the separation in the normal fashion, as shown in the bottom plate of Figure 40.5.

Particle-size calculations are performed with quantitative relationships involving observed particle retention as a function of known experimental parameters [21]. The fundamental calculated result of this retention is the diffusion coefficient distribution of the colloidal sample. Particle-size distribution can then be calculated; spheres are assumed with the relationship

$$d_p = \frac{2RTt_R}{(W^2 N\pi\eta)\ln\{[z/L - (V_c + V_{out})/V_c]/[1 - (V_c + V_{out}/V_c]\}} \quad (40.1)$$

where d_p is the particle diameter, R is the gas constant, T is absolute temperature, t_R is the retention time of the particle, W is the channel thickness, N is Avogadro's number, η is the viscosity of the mobile phase, z is the focusing distance from the channel inlet, L is the channel length, V_c is the mobile-phase flow rate across the channel, and V_{out} is the flow rate out of the channel.

Figure 40.6 [22] shows the separation of a silica sol mixture by a symmetrical-channel FlFFF with a programmed (exponentially decayed) force-field technique. The advantages of this method is that a wide range of particle sizes can be accessed in a practical working time in a single experiment. Resolved bands for the individual narrow size-distribution silica sols in Figure 40.6 were obtained for this mixture under the operating conditions used. Also given are differential and cumulative particle-size distribution plots for the middle-sized component of this mixture, with a weight average of 0.21 μm and a polydispersity of 1.22 for this individual silica sol.

The advantages and limitations of FlFFF for characterizing colloids are quite similar to those described for SdFFF. Accurate particle size and particle-size distributions can be obtained without the need for standards. A range of particle sizes of about 10–1000 nm can be characterized in 0.5–1 h. An optimized FlFFF separation provides resolution and particle discrimination similar to that of SdFFF, with good long-term reproducibility. FlFFF has a somewhat wider range of applicability than SdFFF. Soluble organic macromolecules down to about

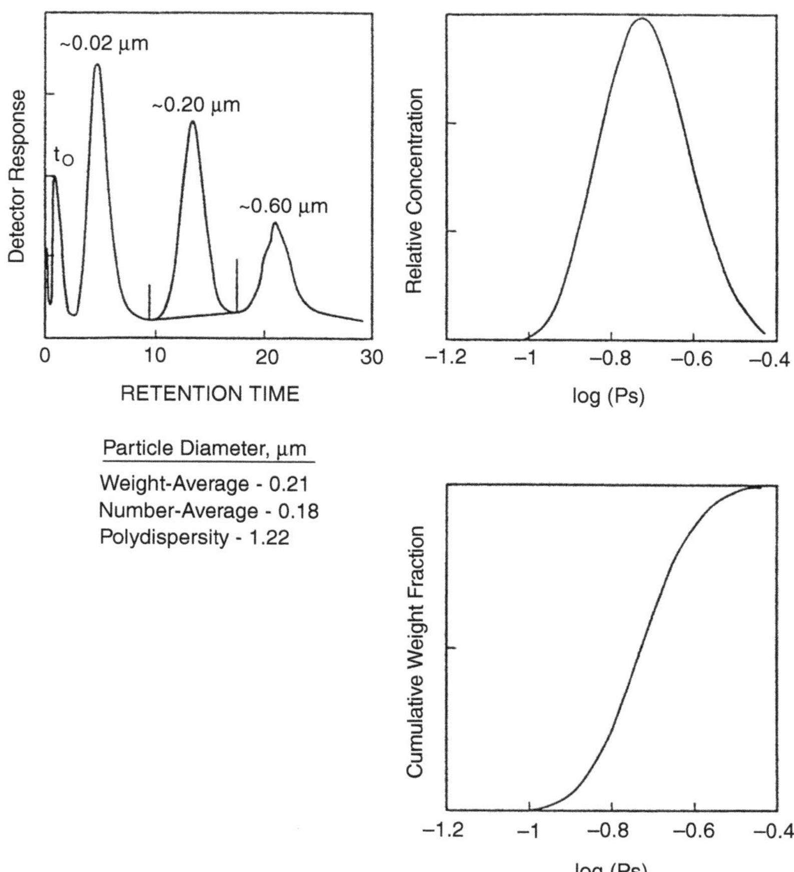

FIGURE 40.6 Separation of silica sol mixture by symmetrical-channel FFF with an exponentially decayed cross-flow force field. Retention time is given in minutes; Ps is the particle size. (Reproduced with permission from reference 22. Copyright 1992.)

5000 Da can be characterized by FlFFF, as well as particulates of many kinds.

Apparatus cost of FlFFF is somewhat less than that for SdFFF. Some special expertise is required for effective execution of the method. (Commercial equipment for FlFFF is now available from FFFractionation, Inc., Salt Lake City, UT.)

SIZE-EXCLUSION CHROMATOGRAPHY

This characterization method SEC is effective and relatively simple for silica sols and other colloids. This technique is commonly used for determining the molecular-weight distribution of a wide range of synthetic and natural macromolecules [23]. However, its application for characterizing colloids has not been widely practiced. SEC has been utilized to determine the particle-size distribution of various polymer lattices, but few applications of the method for characterizing silica sols have been reported [24,25].

The basic principle of SEC separations is illustrated in Figure 40.7. Separations are performed in a column packed with particles having pores mostly of a predetermined single size. A carrier mobile phase is passed through this bed so that liquid fills the pores inside and outside of the bed structure. As a mixture of colloidal particles passes through the packed bed, all colloids that are too large to enter the pores within the particles elute from the column first as a single band at the total exclusion volume. Intermediate-sized colloids (B in Figure 40.7) enter the pores and are retained according to the volume that can be accessed by the colloid — the smaller the colloid, the larger the volume that can be accessed within the pores and the more retained is the colloid (A in Figure 40.7). Solvent and very small colloids can access essentially all of the volume within the pores. These materials are most retained and elute last in the chromatogram as a single peak at the total permeation volume of the column. Thus, the fractionation range of the column is dictated by the size of the pores. All materials elute between the total exclusion and total permeation volumes, with fractionation occurring only in intermediate-volume selective-permeation range.

The apparatus used for SEC is relatively simple isocratic high-performance liquid chromatography (HPLC) equipment, as shown in Figure 40.8. The principal components are a precision high-pressure solvent delivery system, detector, and a column with particles having appropriate pore size. Detection is conveniently accomplished with a UV photometer or spectrophotometer acting as a turbidimetric device. Alternatively, other units such as a light-scattering detector, densimeter, or refractive index detector can be used for many applications. Porous particles for the columns can be semirigid polystyrene–divinylbenzene [24] gels or rigid, porous silicas [25].

Quantitation of silica sol sizes is accomplished by means of a calibration plot. Theory predicts and experiments confirm a linear plot of the log of the particle diameter versus the retention (or elution) time of the colloid. This straight-line relationship is seen in the data in Figure 40.9 for a series of silica sols in the 8–25-nm range. Silica sols characterized by SdFFF were used as "standards" for this calibration plot. The column used for these separations contained porous spherical silica particles that had surfaces covalently coated with "diol" silane groups.

To determine the size of unknown silica sols with relatively narrow distributions, the retention time of the band for an unknown sample is compared to a peak-position calibration plot such as that shown in Figure 40.9. Just as for the determination of molecular weighty for polymers, broad distributions of silica sols can be measured with appropriate software by using known calibration methods involving peak positions or by using known standards with broad distributions [26]. However, commercial

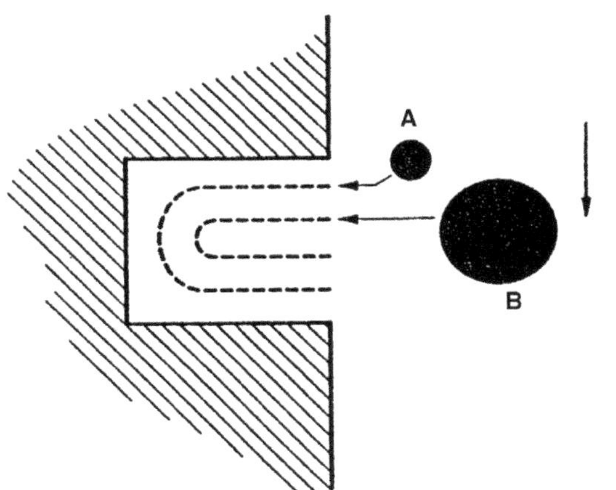

FIGURE 40.7 Size-exclusion fractionation within a single pore.

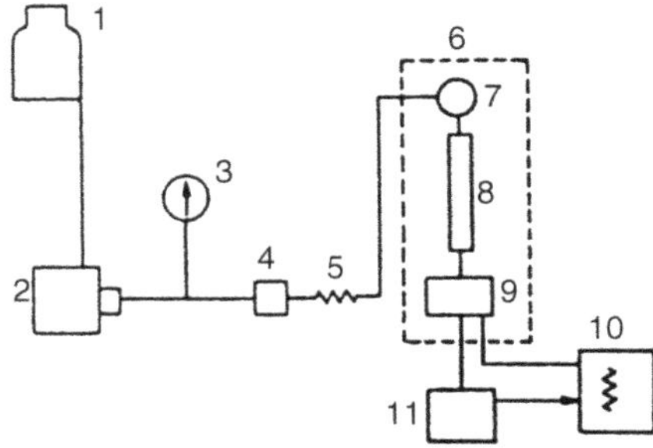

FIGURE 40.8 Schematic of typical apparatus for SEC: 1, mobile-phase reservoir; 2, solvent metering pump; 3, pressure gauge; 4 and 5, pulse damper; 6, thermostat; 7, sample injection valve; 8, SEC chromatographic columns; 9, detector; 10, recorder; and 11, computer.

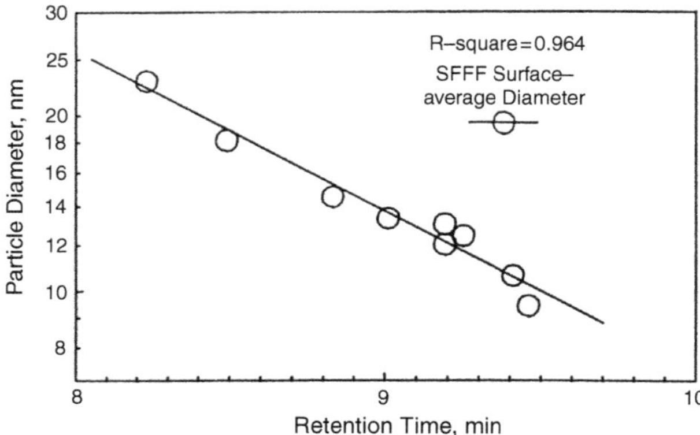

FIGURE 40.9 Size-exclusion calibration for silica sols: column, two 25 × 0.49 cm Synchropak-1000; mobile phase, 5 m*M* triethanolamine, pH 7.5; flow rate, 0.50 ml/min; sample, 0.020 ml, 0.6% in mobile phase; and UV detector, 240 nm.

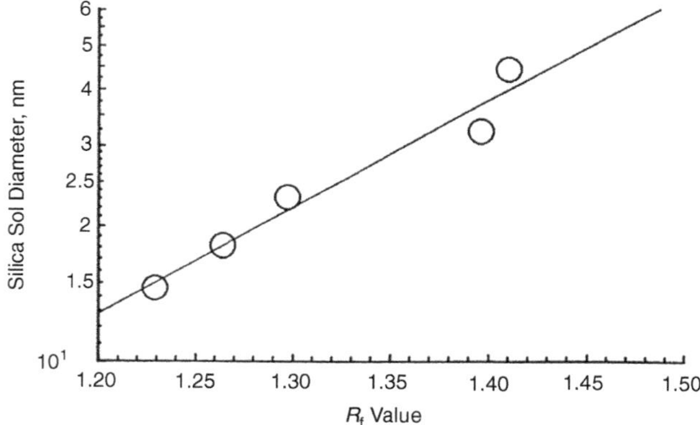

FIGURE 40.10 SEC calibration for silica sols, R_f method: column, two 25 × 0.62 cm Zorbax PSM-1000; mobile phase, 0.01 *M* TRIS, pH 7.0; flow rate, 1.0 ml/min; sample, 0.020 ml, 0.67% in mobile phase; and UV detector, 240 nm.

software for specifically characterizing silica sols and other colloids is apparently not yet available.

Selection of the mobile phase is critical in the characterization of silica sols by SEC. As with the other separation methods, pH should be slightly basic, and low ionic strength must be used to prevent particle aggregation. In addition, the mobile phase must interact with the surface of the packing-particle pores to neutralize undesirable charge effects. Negatively charged surfaces within the pore can result in ion-exclusion effects whereby negatively charged silica sols are unable to enter the pores for the desired size-exclusion process. Positively charged pore surfaces can result in irreversible retention of negatively charged colloids. For the data in Figure 40.9, 5 m*M* triethanolamine at pH 7.5 was found satisfactory. This mobile phase apparently levels out undesirable electrostatic effects so that the desired SEC mechanism can occur with this system.

Precision of the SEC measurement of silica sols can be improved by using an internal standard marker in the sample injected into the column for analysis. A totally permeating species such as potassium dichromate is convenient for this purpose. The marker serves to compensate for retention variations that can occur because of changes in flow rate, temperature, and other effects. With an internal retention marker, calibration and measurements are made with R_f values instead of retention times, as illustrated by the graph in Figure 40.10. R_f values are merely the ratios of the retention time of the marker to the retention time of the colloid peak, $t_{R,marker}/t_{R,colloid}$. For the data in Figure 40.10, a column of bare silica with 100-nm pores was used with a mobile phase of 0.01 *M* TRIS buffer, pH 7.0. A SEC chromatogram of a 32-nm silica sol (value from SdFFF) with the internal marker is shown in Figure 40.11. With this column and the R_f method, silica sols in the 10–50-nm range can be measured with reproducibilities of about ±2 nm (2σ).

A strong positive feature of SEC is that instrumentation is readily available in the form of HPLC apparatus. No special experience is needed for those acquainted with this widely practiced method. Relatively unskilled operators can quickly learn to perform the analysis satisfactorily. Average particle sizes are quickly measured by the peak-position method. However, it is also feasible to determine particle-size distributions if appropriate computer software is available. Separation times are predetermined, because all species elute between the total exclusion and total permeation volumes (provided the desired SEC process is the only retention). No special method development is required, other than ensuring that the proper mobile phase–stationary phase combination is

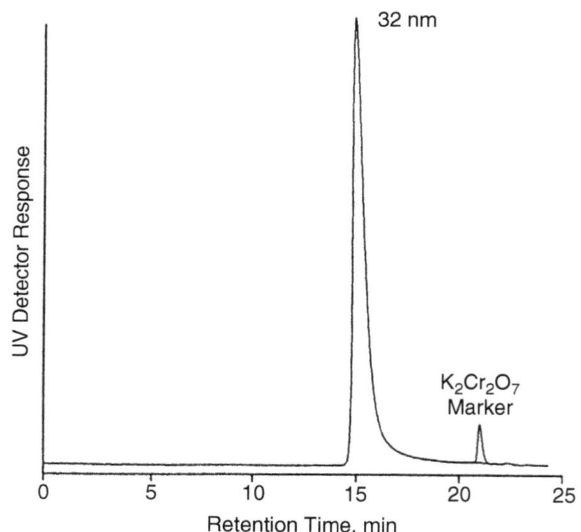

FIGURE 40.11 SEC calibration chromatogram of silica sol with marker; conditions as in Figure 40.10.

selected. Particle diameter is directly a function of retention or elution times.

A limitation of the SEC method is that standards are required. The method has lower resolution and particle discrimination than the FFF methods discussed earlier; poorer analysis precision results. A distinct limitation of the SEC method is that silica sol particles larger than about 60 nm cannot be analyzed by this approach. Silica sol particles of this size can enter larger-sized pores, but are partially or totally retained within the column. Presumably, particles of this size have such poor diffusion that attractive interaction with the pore walls (by van der Waals or other forces) cannot be overcome. Although silica particles smaller than about 2 nm can undergo the desired SEC process when columns of the correct pore size are used, such sols are difficult to detect by turbidimetry. Therefore, this level represents a practical lower limit of silica sol characterization by most separation methods. As noted earlier, for successful SEC the mobile phase must be carefully chosen, as for all of the characterization methods based on separations.

HYDRODYNAMIC CHROMATOGRAPHY

This separation method HDC was developed within the Dow Chemical Company in the early 1970s [27]. This method has been utilized for determining the particle size of many polymer lattices [27,28], but it also can be used for characterizing a wide range of silica sols. Separation by the hydrodynamic effect is illustrated in Figure 40.12 [29]. Colloids flowing between particles in a packed bed or within a capillary are subjected to different velocities as a result of the Poiseuille-like flow of the liquid mobile phase. Colloid particles are excluded from the wall interface, where the fluid velocity is lowest. Thus, larger colloid particles are intercepted by faster flow streams, have a higher mean velocity, and elute more quickly from the column compared to smaller colloids.

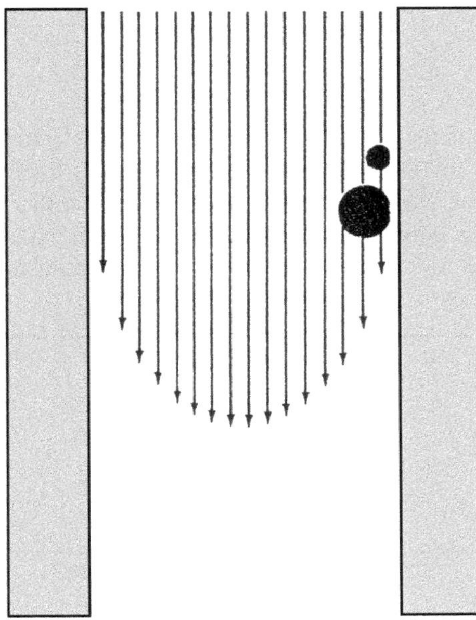

FIGURE 40.12 Hydrodynamic effect. (Reproduced from reference 29. Copyright 1982 American Chemical Soceity.)

Equipment and technique for HDC of silica sols are essentially the same as for SEC. The difference is that the packed bed of the separating column is composed of nonporous, rather than porous, particles. Typically, these particles are polystyrene-based beads, but glass or dense silica beads also are effective. Alternatively, a long, narrow capillary can be used as the separating medium [30].

Relatively small fractionating volumes are associated with HDC systems. Therefore, the R_f market method [27] typically is used to compensate for possible variations in the operating parameters during the separation. Figure 40.13 shows a HDC calibration plot that was obtained by the marker method. The arbitrary log sol diameter versus R_f plot produced a linear relationship for this series of 40–600-nm SdFFF-characterized sols as standards. With this particular HDC system, silica sols can be routinely measured with precisions of about $\pm 15\%$ (relative) with the peak-position calibration method. This precision level is a direct result of the relatively poor resolution of HDC separations.

Silica sol samples with wide particle-size distributions also can show a bias toward particle sizes larger than actual, largely because turbidimetric detectors respond much more strongly to larger sol sizes. As a result, the apex of the sol peak appears to be at a smaller retention time than actually is the case. This phenomenon causes the calculated size value to be somewhat higher than actual for sols that do not have a narrow size distribution.

With the HDC method, no special instrumentation or experience is needed for the simple peak-position method, and elution of the sol is directly correlated to particle size. Good repeatability for the peak-position method is

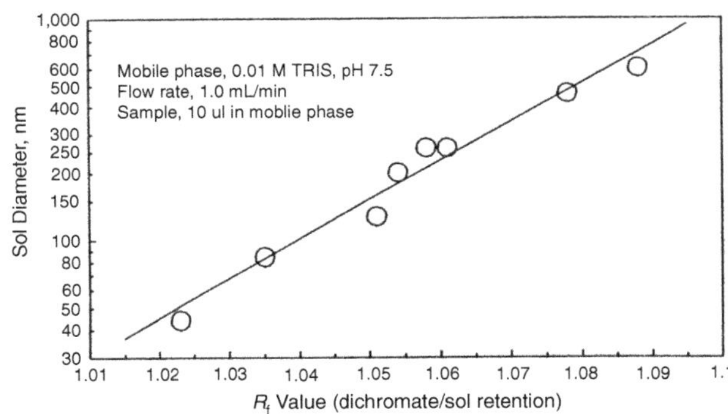

FIGURE 40.13 Hydrodynamic calibration for silica sols: column, two 25 × 0.94 cm columns of 20-μm glass beads; marker, 1 mM potassium dichromate; sample, 0.40% in mobile phase; and UV detector, 260 nm.

experienced if an internal marker is used. HDC can be used with silica sols in the range of about 6–600 nm. The upper limit of sol size has not been clearly defined. However, experiments have suggested that sols $\geq$600 nm do not completely elute from a packed bed of 20-μm glass beads.

The utility of HDC is somewhat limited by the relatively poor resolution and particle-size discrimination of the method, which restrict the precision of HDC in silica sol characterization. In principle, accurate particle-size distributions of silica sols also are possible with the HDC method. However, for such characterizations special software with corrections for the extensive band dispersion in HDC is required, along with a suitable band deconvolution method [28]. Commercial HDC apparatus with this sophisticated software package apparently is no longer available. Standards are generally required, although quantitative retention relationships have been reported for capillary HDC systems in characterizing polymers [37]. As with all of the other separation methods, careful selection of the mobile phase is required in HDC. Mobile phases generally are the same as those used for the FFF methods and SEC.

CAPILLARY-ZONE ELECTROPHORESIS

This new separations method CZE is rapidly gaining distinction for separating a wide range of ionizable species because of its very high resolution [32]. The usefulness of CZE for characterizing silica sols has been explored [33].

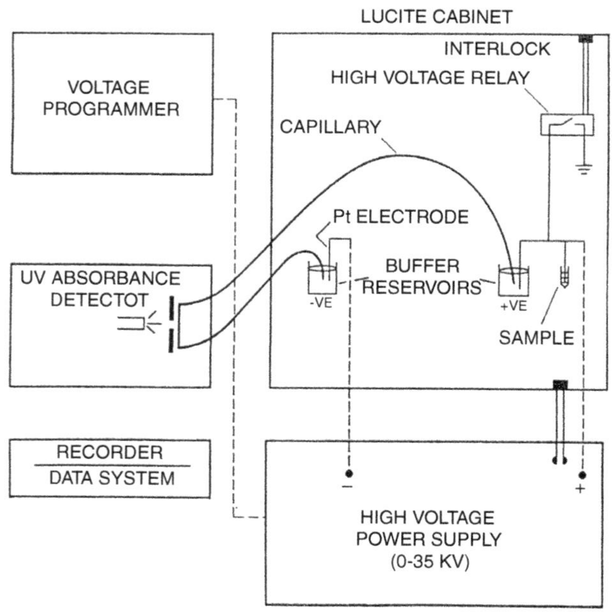

FIGURE 40.14 Schematic of CZE apparatus. (Reproduced from reference 34. Copyright 1988 American Chemical Society.)

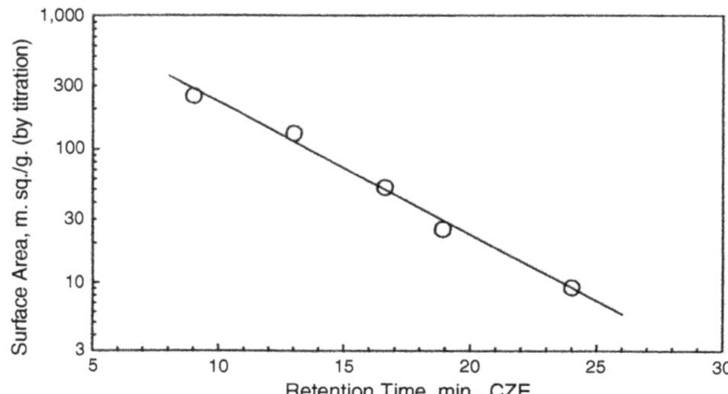

FIGURE 40.15 Correlation of silica sol surface area with CZE band retention: capillary, 60 cm × 50 μm; 30,000 V; mobile phase, 2.5 m*M* ammonium hydroxide and 4.7 m*M* ammonium chloride; and UV detection, 190 nm.

A schematic of a simple CZE apparatus is shown in Figure 40.14 [34]. In CZE, small fused silica or glass capillaries 50–100 μm in diameter are used as the separating medium. Capillary forces at the interface between the buffer and the capillary wall stabilize a liquid buffer in which the separation occurs, eliminating the need for a semirigid gel that is used in conventional slab-gel electrophoresis. The very narrow capillaries permit the use of high voltages (30 kV) with resultant small electrical currents. Heating is rapidly dissipated from the narrow capillary, so rapid separations are possible. In addition, because the process occurs in a UV-transparent capillary, a portion of the capillary can be used as an optical flow cell. Thus, on-line detection of separated bands can be accomplished by turbidimetry and other optical methods.

Migration of a charged species such as a silica sol down a capillary in CZE is described by the expression.

$$v = \mu V/L \tag{40.2}$$

where v is the migration velocity, μ is the electrophoretic mobility of the charged species, V is the total applied voltage across the capillary, and L is the capillary length [32]. The popularity of the CZE method is based on its potential to develop very large plate numbers, with

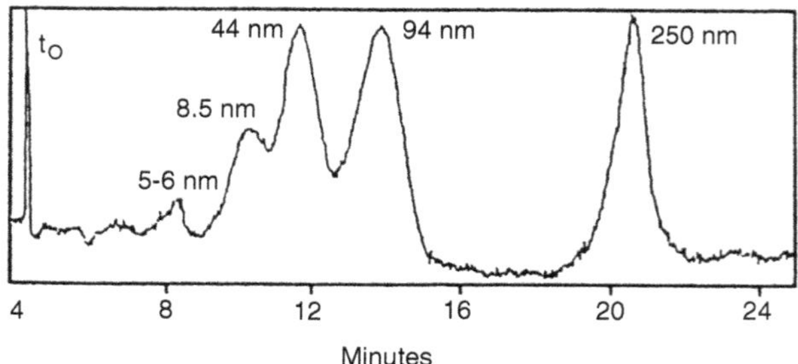

FIGURE 40.16 Separation of silica sols by CZE. Conditions as for Figure 40.15. (Courtesy of R. M. McCormick).

TABLE 40.1
Separations Methods for Characterizing Silica Sols

Feature	SdFFF	FlFFF	SEC	HDC	CZE
Range (nm)	10–1000	10–1000	5–60	10–600	5–600
Quantitation strength	Size, distribution	Size, distribution	Size	Size	Size
Equipment	Commercial	Commercial	Commercial	Commercial or research	Commercial
Experience needed	Moderate	High	Low	Low	High
Resolution and precision	High	High	Moderate	Poor to moderate	Very high

attendant high separation resolving power. The plate number N achievable in CZE can be described by

$$N = \mu V/2D \tag{40.3}$$

where D is the diffusion coefficient of the particle [32]. The length of the capillary does not directly affect separation efficiency. Thus, very high voltages are used across relatively short capillaries to achieve excellent separations in a short time (<0.5 h).

Silica sol standards are required for calibrating CZE retention. Figure 40.15 shows the correlation of CZE retention time with the surface area of several silica sols determined by a titration method [35]. In this case, the silica sol diameter is directly related to the surface area by the expression

$$d_p = 2720/S \tag{40.4}$$

where d_p is the sol diameter in nanometers for silica with a density of 2.2 g/cm^3 and S is the surface area in square meters per gram.

The excellent fractionating power of CZE for silica sols is shown by the electropherogram in Figure 40.16. Sols ranging from $\sim$5 to 250 nm were separated in a single experiment in about 23 min. The range of sols and the degree of separation can be controlled by selecting the mobile-phase ions and ionic strength, separating voltage, and other operating parameters [33].

The CZE method is capable of very high resolution and discrimination between silica sols — the highest of all of the methods discussed here. Resolving power is easily varied as needed by changing the voltage used in the separation or the ionic strength of the liquid within the capillary. Elution of the silica sols is correlated to the size (diffusion coefficient) of the silica sol. CZE will effectively separate over a wide particle-size range, at least <6 to >600 nm. Limiting particle sizes for silica sols have not yet been determined.

Limitations of CZE for characterizing colloids include the need for standards to establish calibrations. Moderately expensive equipment is required, but appropriate commercial apparatus is now widely available. At present, CZE is not well-suited for routine analyses, and some special experience is needed to operate the equipment. Quantitative techniques are relatively crude at this stage of CZE development. Current quantitation is limited to peak-position measurements that provide estimates of the average particle size for narrow distributions. In theory, the determination of particle-size distributions by CZE is feasible. However, appropriate software for this purpose has not yet been developed.

CONCLUSIONS

Table 40.1 summarizes the features of separations methods for measuring the size of silica sols. In theory, particle-size distribution measurements of silica sols could be devised for SEC, HDC, and CZE. However, these methods are currently restricted to estimating a particle-size average for relatively narrow distributions.

REFERENCES

1. Iler, R. K. *The Chemistry of Silica*; Wiley: New York, 1979; p 344.
2. *Modern Methods of Particle Size Analysis*; Barth, H. G., Ed.; Chemical Analysis Series, Vol. 73; Wiley: New York, 1984.
3. Allen, T. *Particle Size Measurement*, 3rd ed.; Chapman and Hall: London, 1981.
4. Kaye, B. H. *Direct Characterization of Fine Particles*; Wiley: New York, 1981.
5. Giddings, J. C.; Yang, F. J. F.; Myers, M. N. *Anal. Chem.* **1974,** *46*, 1917.
6. Giddings, J. C. *Separation Sci.* **1966**, *1*, 123.
7. Giddings, J. C.; Yang, F. J.; Myers, M. N. *Separation Sci.* **1975,** *10*, 133.
8. Yang, F. J.; Myers, M. N.; Giddings, J. C. *J. Colloid Interface Sci.* **1977,** *60*, 574.
9. Kirkland, J. J.; Yau, W. W.; Doerner, W. A.; Grant, J. W. *Anal. Chem.* **1980,** *52*. 1944.
10. Kirkland, J. J.; Rementer, S. W.; Yau, W. W. *Anal. Chem.* **1981,** *53*, 1730.
11. Kirkland, J. J.; Dilks, C. H., Jr.; Yau, W. W. *J. Chromatogr.* **1983,** *225*, 255.

12. Schallenger, L. E.; Yau, W. W.; Kirkland, J. J. *Science*, **1984,** *225*, 434.
13. Giddings, J. C.; Caldwell, K. D.; Jones, H. K. In *Particle Size Distribution: Assessment and Characterization*; Provder, T., Ed.; ACS Symposium Series 332; American Chemical Society: Washington, DC, 1987; pp 215–230.
14. Mercus, H. G.; Mori, Y.; Scarlett, B. *Colloid Polymer Sci.* **1989,** *267*, 1102.
15. Yang, F. J. F.; Myers, M. N.; Giddings, J. C. *Anal. Chem.* **1974,** *46*, 1924.
16. Giddings, J. C.; Karaiskskis, G.; Caldwell, K. D. *Separation Sci. Technol.* **1981,** *16*, 607.
17. Kirkland, J. J.; Yau, W. W. *Anal. Chem.* **1983**, *55*, 2165.
18. Giddings, J. C. *Anal. Chem.* **1981,** *53*, 1170A.
19. Giddings, J. C.; Myers, M. N.; Caldwell, K. D.; Fisher, S. R. In *Methods of Biochemical Analysis*, Vol. 26; Glick, D., Ed.; Wiley: New York, 1980; pp 79–136.
20. Wahlund, K. G.; Litzen, A. *J. Chromatogr.* **1989,** *461*, 73.
21. Wahlund, K. G.; Giddings, J. C. *Anal. Chem.* **1987,** *59*, 1332.
22. Kirkland, J. J.; Dilks, C. H. Jr.; Rementer, S. W.; Yau, W. W. *J. Chromatogr.* **1992,** *593*, 339.
23. Yau, W. W.; Kirkland, J. J.; Bly, D. D. *Modern Size-Exclusion Liquid Chromatography*; Wiley: New York, 1979; Chapters 12 and 13.
24. Singh, S.; Hamielec, A. E. *J. Liq. Chromatogr.* **1978**, *1*, 187.
25. Kirkland, J. J. *J. Chromatogr.* **1979,** *185*, 273.
26. Yau, W. W.; Kirkland, J. J.; Bly, D. D. *Modern Size-Exclusion Liquid Chromatography*; Wiley: New York, 1979; Chapter 9.
27. Small, H. *J. Colloid Interface Sci.* **1974,** *48*, 147.
28. McGowan, G. R.; Langhorst, M. A. *J. Colloid Interface Sci.* **1982,** *89*, 94.
29. Small, H. *Anal. Chem.* **1982,** *54*, 892A.
30. Silebi, C. A.; Dosramos, J. G. *J. Colloid Interface Sci.* **1989,** *130*, 14.
31. Tijssen, R.; Bos, J.; van Kreveld, M. E. *Anal. Chem.* **1986,** *58*, 3036.
32. Jorgenson, J. W.; Lukacs, K. D. *Anal. Chem.* **1981,** *53*, 1298.
33. McCormick, R. M. *J. Liq. Chromatogr.* **1991,** *14.* 939.
34. McCormick, R. M. *Anal. Chem.* **1988,** *60*, 2322.
35. Sears, G. W., Jr. *Anal. Chem.* **1956,** *28*, 1981.
36. Kirkland, J. J.; Yau, W. W. *Science* **1982,** *218*, 121.

41 Characterization of Colloidal and Particulate Silica by Field-Flow Fractionation

J. Calvin Giddings[1], S. Kim Ratanathanawongs[1], Bhajendra N. Barman[2,3], Myeong Hee Moon[1], Guangyue Liu[1], Brenda L. Tjelta[1], and Marcia E. Hansen[2]

[1]University of Utah, Department of Chemistry, Field Flow Fractionation Research Center
[2]FFFractionation, Inc.

CONTENTS

Particle-size and mass distribution curves, along with information on particle porosity, density, shape, and aggregation, can be obtained for submicrometer- and supramicrometer-size silica materials suspended in either aqueous or nonaqueous media by field-flow fractionation (FFF). Narrow fractions can readily be collected for confirmation or further characterization by microscopy and other means. Among the silicas examined were different types of colloidal microspheres, fumed silica, and various chromatographic supports. Size distribution curves for aqueous silica suspensions were obtained by both sedimentation FFF and flow FFF and for nonaqueous suspensions by thermal FFF. Populations of aggregates and oversized particles were isolated and identified in some samples. The capability of FFF to achieve the high-resolution fractionation of silica is confirmed by the collection of fractions and their examination by electron microscopy.

Particle-size distribution and other particle characteristics, including degree of aggregation, shape, and porosity, strongly influence the properties of numerous forms of colloidal and particulate silica [1]. The rapid and accurate measurement of these distributions and characteristics by conventional techniques (such as microscopy, light scattering, and sedimentation) has proven to be difficult or in many cases impossible; thus, alternate technologies should be considered. The capability of field-flow fractionation (FFF), a newer technology, to provide high-resolution particle-size distribution curves and particle densities is now well documented [2–9] and suggests that FFF might

[3]Present address: Texaco, Inc., P.O. Box 1608, Port Arthur, TX 77641

play a constructive role in silica characterization. The ability of FFF to separate out aggregates, impurities, and other complex subpopulations differing in particle mass, size, or density and to provide narrow fractions of these subpopulations for examination by microscopy and other means is entirely unique; this capability is important and may prove essential in meeting tightening demands for precise particle characterization. The extension of FFF methodology to the measurement of particle shapes and porosities, as discussed here, is illustrative of the potential versatility of FFF in providing multiple properties of the components of complex colloidal materials. The application of FFF to diverse silica materials, explored in this chapter, is an opportune combination of a challenging materials characterization problem and a relatively new and highly versatile particle characterization technology.

FFF is a family of chromatographic-like techniques that fractionate colloidal materials according to differences in physiochemical properties [2–6]. These properties include particle mass, size, and effective density; the specific combination of properties controlling the fractionation depends on the subtechnique of FFF. In an FFF run, the particles in the sample are separated according to the relevant property (e.g., particle mass) and are eluted from the FFF flow channel in a well-defined sequence (e.g., from low mass to high mass). The relative amounts of the sample eluting at different times are measured by a detector, and the resulting fractogram (detector signal versus time curve) is converted into a property distribution curve such as a mass or size distribution curve.

Figure 41.1 shows, for illustrative purposes, the fractograms for two different size ranges of polystyrene latex microspheres. (Polystyrene latex standards with narrow size distributions are ideal probes for testing the resolving power and accuracy of particle characterization methods. Those techniques unable to resolve close-lying latex standards are generally unable to provide detailed size distribution information on silica or other particulate matter). In the first fractogram, Figure 41.1a, submicrometer-size latex particles are separated according to size (or mass) — the smallest particles emerge first and the largest last — by sedimentation FFF with power programming [10]. (The spin rate (revolutions per minute) versus time curve for the power program is also shown; the initial spin rate is 2000, the flow rate is 1.6 mL/min, and the parameters t_1 (predecay time), t_{sf} (stop-flowtime), t_a (power programming time constant), and p (power parameter in power programming) are 5 min, 2 min, -40 min, and 8, respectively). Figure 41.1b shows the separation of supramicrometer-size microspheres at even higher resolution and speed with sedimentation–steric FFF. (Here the spin rate is 1100 rpm and the flow rate is 6.00 mL/min.) For particles in this size range, the elution sequence proceeds from large to small particles as shown in Figure 41.1b.

Although field-flow fractionation was first developed in the 1960s [11], the first major study of colloidal silica by FFF was not reported until 1978 [4]. At that time it was shown that the subtechnique of flow FFF could be used to fractionate colloidal silica down to a particle size of 0.01 μm (*see* Figure 41.2). The fractionation was verified by electron microscopy. Size distribution curves were obtained under different experimental conditions and shown to be consistent with one another. The effects of particle aggregation were examined.

Subsequent work on colloidal silica with sedimentation FFF was reported by Kirkland and co-workers [7,12] and by Yonker et al. [13], who used various organic liquids as the FFF carrier.

In other work carried out on two different occasions, both sedimentation FFF [14] and flow FFF [15] were used to characterize the size distribution of chromatographic silicas having particles ranging up to approximately

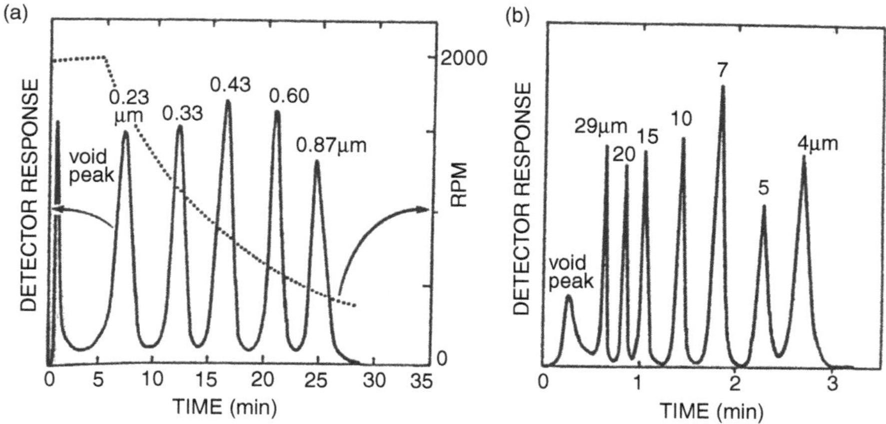

FIGURE 41.1 Illustration of the high resolving power of FFF for the separation of particles in different size ranges: (a) submicrometer polystyrene latex microspheres fractionated by power-programmed normal mode sedimentation FFF; and (b) supramicrometer-sized latex spheres fractionated by sedimentation–steric FFF.

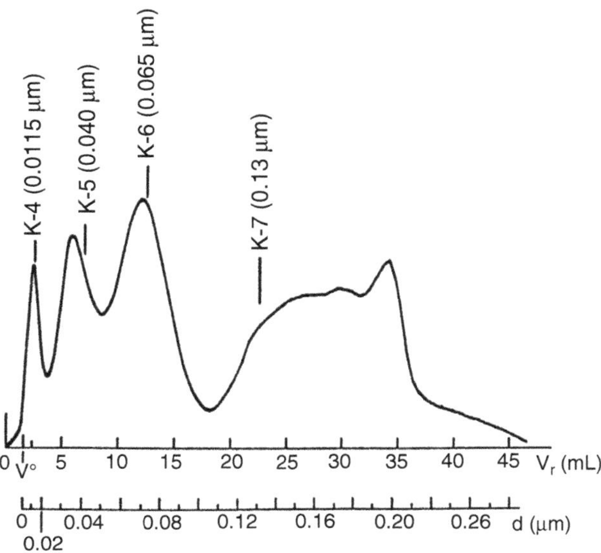

FIGURE 41.2 Early flow FFF fractograms of K-4, K-5, K-6, and K-7 colloidal silicas. The flow rates were $\dot{V} = 3.16$ mL/h and $\dot{V}_c = 11.1$ mL/h; the void volume was 1.62 mL. (Reprinted with permission from reference [4]. Copyright 1978 Academic Press.)

20 μm in diameter. In more recent work [16], sedimentation FFF was used to obtain both the size and apparent density (thus giving the porosity and pore volume) of silica support particles used in chromatography.

In this chapter, both colloidal silica and coarser silica materials such as chromatographic supports are examined. The particle sizes range from 0.01 to 20 μm. An arsenal of FFF subtechniques was used, including sedimentation FFF, flow FFF, and thermal FFF. Although most of these studies involve aqueous suspensions, thermal FFF is shown to be capable of fractionating and characterizing nonaqueous suspensions of silica as well.

THEORY

In FFF, particle size, mass, density, and so forth are determined by their relationship to particle retention time t_r, the time required for the passage of the particle through the FFF channel. The relationship between particle properties and t_r arises because a particle's position in the streamlines of a thin channel, and thus the particle's velocity, is determined by the force exerted on the particle by an external field directed at right angles to flow. The interactive force between the field and the particle can generally be expressed explicitly in terms of particle properties, thus a mathematical relationship between these properties and the retention time is provided. The mechanism by which particles of different properties are differentially eluted, and thus the form of the mathematical relationship between particle properties and t_r, depends on the particle size range. The so-called normal mechanism (or normal mode) of FFF, producing the separation illustrated in Figure 41.1a, is applicable to particles ranging from about 1 nm in size up to a somewhat flexible upper limit generally set in the 1–2-μm size range. The steric mechanism (or steric mode), utilized in Figure 41.1b, has an adjustable lower limit of about 0.3 μm and extends upward, if necessary, beyond 100 μm.

In either the normal or steric mode of FFF, a small sample of suspended particles is injected into a stream of carrier liquid flowing through a thin ribbonlike channel. The particles are carried down the channel by the flow of carrier. However, the velocity of any given particle at any instant is determined by the position of the particle along the transverse (or thin) dimension of the channel as a consequence of the parabolic flow between the two flat walls confining the channel (*see* Figure 41.3). FFF, in either the normal or steric mode, operates by applying an external field or gradient across these channel walls and thus directed transversely across the channel. The field must interact with the particles and drive them toward one wall. By driving particles to different equilibrium positions or distributions, different particle types are carried along in different portions of the parabolic flow stream, which imparts a differential flow velocity to different particle subpopulations and results in their separation.

NORMAL FFF

In normal mode FFF, used primarily for colloidal materials, the particles are driven toward one of the

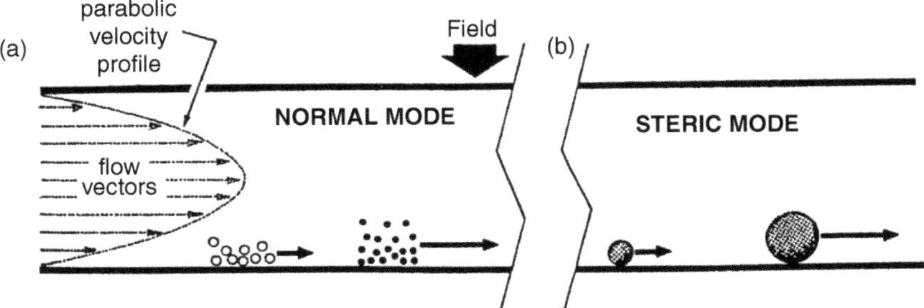

FIGURE 41.3 Diagram showing parabolic flow profile between FFF channel walls and the displacement by flow of particles according to (a) normal and (b) steric mechanisms.

channel walls (the *accumulation wall*) by the external field. As the particle concentration increases in the vicinity of the accumulation wall, diffusion begins to counteract the field-induced motion by driving particles away from the wall (Figure 41.3a). An equilibrium distribution is soon established in which the concentration c at any elevation x above the accumulation wall relative to the concentration c_0 at the wall is given by the exponential relationship [2–6]

$$\frac{c}{c_0} = e^{-x/\ell} \tag{41.1}$$

where the exponential ℓ equals the mean elevation of particles above the wall. The constant ℓ is related to the diffusion coefficient D, the field-induced velocity U, and the driving force F exerted by the field on a single particle according to the relationships

$$\ell = \frac{D}{U} = \frac{kT}{F} \tag{41.2}$$

where k is the Boltzmann constant and T is the absolute temperature. Because D, U, and F are related to the physiochemical properties of the particle, Equation (41.1) and Equation (41.2) establish a relationship between the distribution of particles along transverse axis x and these properties.

The velocity at which particles are swept through the channel, and thus their retention time t_r, also depends on the transverse distribution expressed by Equation (41.1) and Equation (41.2). More specifically, retention time t_r is given by

$$t_r = \frac{t^0}{6\lambda[\coth(1/2\lambda) - 2\lambda]} \tag{41.3}$$

which reduces to

$$t_r = \frac{t^0}{6\lambda} \tag{41.4}$$

when t_r is large compared to the void time t^0. Here λ is the dimensionless ratio of ℓ to channel thickness w; that is,

$$\lambda = \frac{\ell}{w} = \frac{kT}{Fw} \tag{41.5}$$

where the last equality (the relationship between λ and F) is based on Equation (41.2).

The retention parameter λ is the critical link between the experimental retention time given by Equation (41.3) [or Equation (41.4)] and the particle properties that determine the force F exerted on the particles by the field. The specific particle properties that are characterized by this relationship depend on the type of field applied. For a sedimentation field, F is related to the particle mass m by

$$F = m\left(\frac{\Delta\rho}{\rho_p}\right)G \tag{41.6}$$

where ρ_p is the particle density, $\Delta\rho$ is the difference between the particle and the carrier density, and G is the strength of the field measured as acceleration. For spherical particles of diameter d, F assumes the form

$$F = (\pi/6)\,d^3\,\Delta\rho\mathrm{G} \tag{41.7}$$

This equation applies to nonspherical particles as well, provided d is interpreted as the effective spherical diameter of the particle, equal to the diameter of a sphere having the same volume as the particle.

By combining either Equation (41.6) or Equation (41.7) with Equation (41.3) (or 41.4) and Equation (41.5), the experimental retention time t_r can be related to particle mass m or diameter d. To fully utilize this relationship, the particle density ρ_p (and the resulting difference $\Delta\rho$ between ρ_p and the carrier density) must be known or obtained. This density is usually known to a good approximation. If it is not known, FFF methods have been developed for determining particle density values [17]. However, uncertainties or complications introduced by the density dependence of sedimentation forces can be bypassed by using an alternate density-independent force in the FFF system, such as that utilized in flow FFF.

In flow FFF the particle is driven to the accumulation wall by the physical cross-flow of the carrier fluid. To implement this method, the walls of the channel must be permeable. A stream of carrier is then introduced through one wall and withdrawn from the opposite wall. The cross-flow fluid motion so generated is superimposed on the axial channel flow described earlier. This cross-flow displaces particles toward the accumulation wall as effectively as a sedimentation force. However, the driving force is density-independent; specifically,

$$F = 3\pi\eta d_s U \tag{41.8}$$

where U, defined earlier as the transverse displacement velocity induced by the field, is simply the cross-flow velocity of the fluid in flow FFF, and η is the viscosity. The particle property entering this equation is the Stokes diameter d_s, different (except for spheres) from the sedimentation FFF diameter d. (Although diameter d is defined by the volume of a particle, d_s is defined by the friction coefficient and Stokes law, $f = 3\pi\eta d_s$.) By virtue of Equation (41.8), Equation (41.5), and Equation (41.3), flow FFF can be used to calculate the diameter d_s for any particle eluting at time t_r independent of density.

A third and still different type of driving force that can be used in FFF is the effective force generated by a temperature gradient applied between the channel walls. The corresponding FFF subtechnique is known as thermal FFF. Although thermal FFF has been used primarily for the analysis of synthetic polymers, we have found that the method is applicable to particles suspended in organic liquids. The effective driving force for thermal FFF is given by

$$F = -D_T f\,dT/dx \tag{41.9}$$

where D_T is the thermal diffusion coefficient and dT/dx is the temperature gradient. Unfortunately, the dependence of D_T on particle properties is not fully understood, and thus empirical calibration has to be used to obtain size distribution and other data on particulate materials.

Steric FFF

In steric FFF, particles are driven almost to the point of contact with the accumulation wall (Figure 41.3b). The equilibrium position of these larger particles thus depends on their size. However, the particle position, and thus retention time, is complicated by hydrodynamic lift forces acting in opposition to the primary driving forces expressed by Equation (41.7), Equation (41.8), and Equation (41.9). These lift forces, which increase with the flow rate, drive the particles a short distance away from the channel wall and thus into higher positions within the parabolic flow profile, where the flow velocity is greater than it would be for particles immediately adjacent to the wall [18,19]. The major role played by these lift forces and the greatly diminished role of Brownian motion (which is usually negligible for the large particles analyzed by steric FFF) distinguish the steric mechanism from the normal mechanism of FFF. Because the hydrodynamic lift forces are not well understood [19], and are thus difficult to incorporate into theory, there are no theory-based equations for retention time analogous to Equation (41.3) and Equation (41.4) for normal FFF. Instead, a semiempirical approach is used with constants established by experimental calibration. In particular, the retention time t_r is best expressed in the logarithmic form [20]

$$\log t_r = -S_d \log d + \log t_{r1} \qquad (41.10)$$

where S_d is the diameter-based selectivity and t_{r1} is a constant representing the retention time of a particle of unit diameter. The value of S_d is somewhat less than unity when sedimentation is used as a driving force and somewhat greater than unity when cross-flow serves as the driving force [18]. Both S_d and t_{r1} are established by logarithmic plots of t_r versus d of well-characterized standards such as polystyrene latex beads. The value of t_{r1} depends on the field strength and, for sedimentation FFF, also on the particle density. It will be shown later that this dependence makes possible the determination of particle density and porosity values.

EXPERIMENTAL DETAILS

A battery of six different FFF systems was used to provide a comparison of results obtained not only from different systems of the same type but from systems of entirely different types (i.e., with different primary force fields). The characteristics of the six systems are summarized in Table 42.1. Included in the collection are two sedimentation FFF systems, three flow FFF systems, and one thermal FFF system. The characteristics and operation of these different categories of instruments are described in more detail in this section.

Sedimentation FFF

The two sedimentation FFF (or SdFFF) systems have similar specifications and design features that have evolved in two decades of work at the University of Utah. All have a horizontal rotation axis (radius 15.1 cm) that makes them suitable not only for the normal mode analysis of submicrometer-size particles, but also for the steric FFF of particles well above 1 μm in diameter (*see* Figure 41.1). System Sed I is a research instrument constructed at the University of Utah, and system Sed II is a commercial instrument (model S101) from FFFractionation. Although almost identical in

TABLE 41.1
Field-Flow Fractionation Systems Used

System designation	System type	System source	Channel dimensions (cm)			Void volume (mL)
			Thickness	Breadth	Length[a]	
Sed I	sedimentation FFF	FFFRC[b]	0.013	1.0	90.0	1.14
Sed II	sedimentation FFF	FFFractionation[c]	0.025	2.0	89.0	4.47
Flow I	flow FFF	FFFRC	0.018	2.1	38.5	1.35
Flow II	flow FFF	FFFRC	0.024	2.0	27.2	1.17
Flow III	flow FFF	FFFRC	0.013	2.0	27.2	0.63
Therm I	thermal FFF	FFFRC	0.022	2.0	46.5	2.02

[a]Tip-to-tip channel length.
[b]Field-Flow Fractionation Research Center, Department of Chemistry, University of Utah, Salt Lake City, UT 84112.
[c]FFFractionation Inc., P.O. Box 58718, Salt Lake City, UT 84158.

design and performance, Sed I has been fitted with a special channel of reduced dimensions (*see* Table 41.I). The reduced channel thickness ($w = 127$ μm) makes this instrument particularly effective for the rapid analysis (in the steric mode) of particles ranging from 0.5 to 40 μm in diameter (*see* Figure 41.1b). (Steric FFF in Sed II will fractionate particles up to ~80 μm.)

Conventional ancillary equipment was used with these sedimentation FFF systems. Sed I, for example, utilized a model 410 high-performance liquid chromatography (HPLC) pump from Kontron Electrolab (London) and a UV Spectroflow Monitor SF770 from Kratos Analytical Instruments (Westwood, NJ). Fractions of eluted silica were collected by means of a model FC-80K Microfractionator from Gilson Medical Electronics (Middleton, WI).

Flow FFF

The three flow FFF systems (Flow I, II, and III) are each constructed of two Lucite blocks with inset ceramic frits, a membrane, and a spacer. The frits provide a homogeneous distribution of cross-flow over the entire channel area. A membrane stretched over one frit surface serves as the accumulation wall and retains sample inside the channel. Systems Flow I and II were assembled with the YM-30 ultrafiltration membrane (Amicon, Danvers, MA) and Flow III with the Celgard 2400 polypropylene membrane (Hoechst-Celanese, Separations Products Division, Charlotte, NC). The frit of the second Lucite block defines the opposite (depletion) wall. The spacers, consisting of Teflon (Flow I) or Mylar (Flow II, Flow III), determine the channel thickness.

The flow FFF systems are characterized by the use of a second pump to drive carrier across the channel thickness; this setup provides the field that induces migration of sample toward the accumulation wall. The Flow I system was operated with an Isochrom LC pump (Spectra-Physics Inc., San Jose, CA) as the channel flow pump and a pulseless syringe pump (built in-house) as the cross-flow pump. Sample was injected via a Valco injector (Valco Instruments co., Houston, TX) with a 20-μL loop, and the eluted sample was detected at 254 nm with a UV–visible detector (UV-106, Linear Instruments, Reno, NV). The peripheral equipment employed in Flow II and III consisted of a Kontron model 410 channel flow pump, a syringe pump serving as the cross-flow pump, a Rheodyne (Cotati, CA) model 7010 pneumatic-actuated injection valve, and a model 757 Spectroflow UV–vis detector from Applied Biosystems (Ramsey, NJ) operated at 254 nm.

For some experiments, an evaporative light-scattering detector (ELSDII, Varex Corporation, Burtonsville, MD) was connected in series with the UV–vis detector of Flow I. The operating conditions were those recommended by the instrument manufacturer.

The main difference between the three flow FFF systems is the channel dimensions, given in Table 41.1. Although systems Flow I and II were constructed with 254-μm spacers, the protrusion of the compressible membrane material into the channel space reduced the thickness w below this value. The w values reported in Table 41.1 are back-calculated from void volume measurements and subsequently confirmed by polystyrene standards. Flow I, with an initial w of 181 μm, was reassembled at one point, after which w was found to be 152 μm. This variation in w due to membrane compressibility was not observed when the thin polypropylene (25 μm) membrane was used in Flow III. In this case, the channel thickness equals the spacer thickness. Thus the measured and geometric void volumes are in good agreement.

Thermal FFF

The thermal FFF system (Therm I) is a unit of conventional design in which the channel volume is cut and removed from a Mylar spacer and the spacer is sandwiched between two chrome- and nickel-plated copper bars [21]. The temperature gradient is instituted by heating one bar and cooling the other; temperature differences up to $\Delta T = 53$ K were utilized. Although the channel length and thickness (*see* Table 41.1) are fairly typical for thermal FFF, the channel thickness is 3 times that commonly used in our laboratories. Although this added thickness causes some loss of resolution and speed, the capability for fractionating particles should be adequate for the work.

The carrier for the thermal FFF system was reagent-grade acetonitrile from EM Industries (Gibbstown, NJ). The flow of this carrier was driven by a model M-6000A pump from Waters Associates. A model CSI UV detector from Cole Scientific (Calabasas, CA) operating at 254 nm was used to detect the eluting particles. The signals were collected on an Omniscribe chart recorder (Houston Instrument, Austin, TX). A pressure regulator was used to maintain the channel pressure at 100 psi (700 kPa) above atmospheric pressure to provide solvent and signal stability when the operating temperature of the hot wall approaches the solvent boiling point.

Standards and Samples

The polystyrene standards used were obtained from Duke Scientific (Palo Alto, CA) and Seradyn (Indianapolis, IN).

The silica samples examined by different FFF subtechniques and systems are shown in Table 41.2. These samples have particles that fall in three categories: nonporous colloidal microspheres, fumed silicas having a chainlike structure, and larger porous silica particles typical of

TABLE 41.2
Silica Samples Examined and FFF Systems Utilized

Sample identity	Source	Nominal size (μm)	FFF system
Colloidal silicas			
Ludox HS-30%	DuPont	0.012	Flow I, Flow II
TM		0.022	
SM		0.007	
Monospher 50	E. Merck	0.05	Therm I, Sed II
100		0.10	
150		0.15	
250		0.25	
500		0.50	
Nyacol 9950	PQ Corp.	0.1	Sed I, Sed II Flow III, Therm I
Fumed silicas (chainlike network)			
Cab-O-SiL L-90	Cabot Corp.	0.22	Sed II, Flow III
M-5		0.17	
EH-5		0.14	
Chromatographic silicas (porous spheres)			
A. Hypersil-5 (136 Å)	Shandon	5	Sed I
B. Hypersil-5 (120 Å)	Shandon	5	Sed I, Flow II
C. Spherisorb	Phase Separations	5	Sed I, Flow II
D. Nucleosil	Machery-Nagel	5	Sed I, Flow II
E. Hypersil-3	Shandon	3	Sed I, Flow II

chromatographic supports. The carriers and sample amounts injected for each type of silica analyzed by different FFF subtechniques are listed in Table 41.3.

Preparation of the Cab-O-Sil fumed silica, a multi-stepped procedure, was as follows. The fumed silica powder (25–125 mg) was weighed into a 10-mL vial and then suspended in 5 mL of a 0.0016 *M* NaOH solution. After an initial 10-s vortex stage, the sample vial and a sonic probe were both immersed in a small water bath. Sonication was carried out for 5 min at 200 W. (The recommended procedure called for 2-min sonication at 480 W with a microtip inserted directly into the sample solution. This procedure was modified because we did not have a microtip.) Immediately before each injection the sample was vortexed for 10 s.

The aqueous carriers were made from doubly distilled deionized water. Most of these carriers contained a surfactant (FL-70, Fisher Scientific, Fairlawn, NJ) and a bacteriocide (sodium azide, NaN_3).

Scanning Electron Microscopy

Electron micrographs were obtained by using a Hitachi S-450 scanning electron microscope (Hitachi Scientific Instruments, Tokyo, Japan). Specimens were prepared by filtering an aliquot of each collected fraction (of interest) onto a 13-mm Nuclepore membrane filter. The filter was mounted on an aluminum stub and subsequently coated with a thin gold and palladium layer. The magnification and acceleration voltages varied from 25,000× and 30 kV for the fumed silica to 750× and 20 kV for the chromatographic silica.

TABLE 41.3
Summary of Carrier and Amount of Sample Injected

Sample	System	Carrier	Amount injected (μg per run)
Ludox	Flow I	0.1% FL-70, 0.02% NaN_3; 0.001 *M* NH_4OH	60–120
	Flow II	0.01 *M* phosphate, pH 7.3	60–120
Monospher	Therm I	acetonitrile	40–70
	Sed II	0.1% FL-70, 0.02% NaN_3	40–70
Nyacol	Sed I	0.1% FL-70, 0.02% NaN_3	40
	Sed II	0.1% FL-70, 002% NaN_3	60
	Flow III	0.001 *M* NH_4OH	40
	Therm I	acetonitrile	70
Cab-O-Sil	Sed II	0.1% FL-70, 0.01% NaN_3; 0.02 *M* triethanolamine	560
	Flow III	0.001 *M* NH_4OH	500
Chromatographic silicas	Sed I	0.1% FL-70, 0.02% NaN_3	175–200
	Flow II	0.1% FL-70, 0.02% NaN_3	100

Transmission Electron Microscopy

A Philips model 201 (Arvada, CO) transmission electron microscope was used as an alternative method for determining the particle sizes of Ludox silicas. Specimens were prepared on formvar-coated, 200-mesh copper specimen grids. Typical acceleration voltages and magnifications were 80 kV and 65,000×.

Data Analysis

The computer program used for data analysis was developed at the Field-Flow Fractionation Research Center. The underlying theory is similar to that discussed by Giddings et al. [4]. For normal mode characterizations, the fractograms are converted to particle size distributions by using developed theory. However, for steric mode analyses, calibration curves are required [15,20].

RESULTS AND DISCUSSION

The application of FFF techniques to assorted silica samples including colloidal microspheres (Ludox, Monospher, and Nyacol), fumed silica, and chromatographic silica (*see* Table 41.2) are discussed in this section. In several cases, different FFF systems are applied to the same samples, and the results are compared. The acquisition of size distribution data is emphasized, but the possibilities for measuring densities and porosities for spherical particles and structural factors for nonspherical silica are also discussed (and in one case demonstrated).

Ludox Colloidal Silica

Ludox colloidal silicas (DuPont) were among the first silica materials studied by FFF. In the first paper describing the applicability of FFF to colloidal silica [4], Ludox and related silicas were fractionated by flow FFF; the fractionation was verified by transmission electron microscopy (TEM). The theory behind the fractionation and the acquisition of size distribution data was developed, and evidence of aggregation was examined. An example of the fractionation of four colloidal silicas from that study is shown in Figure 41.2. The primary drawback of this earlier work was the lengthy runs, in some cases requiring over 10 h. Most of the experimental runs on Ludox described in this section were completed in less than 10 min.

To examine data reliability, two of the flow FFF systems (Flow I and Flow II) described in Table 41.1 were used. As shown in Table 41.4, five different sets of conditions corresponding to different flow FFF systems, carriers, flow rates, and detectors were used. The detectors employed were a UV detector (254 nm) and an evaporative light-scattering (ELS) detector. Studies were performed to identify a range of sample amounts that could be injected without overloading the channel. The quantities injected were 600, 120, and 60 μg. The effect of overloading (decreased retention time) was evident in the first case but not in the latter two cases. Less than 120 μg of Ludox silicas were injected throughout the remaining flow FFF studies.

Figure 41.4a shows the fractograms obtained for the three different Ludox silicas from Flow I and the experimental conditions corresponding to set 1 of Table 41.4. The cross-flow and channel flow-rates are listed as $\dot{V}_c$ and $\dot{V}$, respectively. The run times, following a stop-flow time of 40 s for relaxation, are only 5–10 min. Similar fractograms of these samples were generated by using the other sets of conditions defined in Table 41.4.

TABLE 41.4
Comparison of Mean Particle Diameters for Three Ludox Colloidal Silicas Measured under Different Experimental Conditions

Experimental condition	Set 1	Set 2	Set 3	Set 4	Set 5
System	Flow I	Flow I	Flow I[a]	Flow I[a]	Flow II
Carrier	0.1% Fl-70	0.1% Fl-70	0.001 M	0.001 M	0.01 M
	0.02% NaN_3	0.02% NaN_3	NH_4OH	NH_4OH	phosphate buffer
$\dot{V}$ (mL/min)	4.04	3.11	0.41	0.41	2.69
$\dot{V}_c$ (mL/min)	3.24	3.14	1.00	1.00	1.70
Detector	UV	UV	UV	ELS	UV
Sample diameter (nm)					
TM	30	31	31	30	34
HS-30%	18	19	18	18	18
SM	11	10	13	13	14

[a]Reassembled channel.

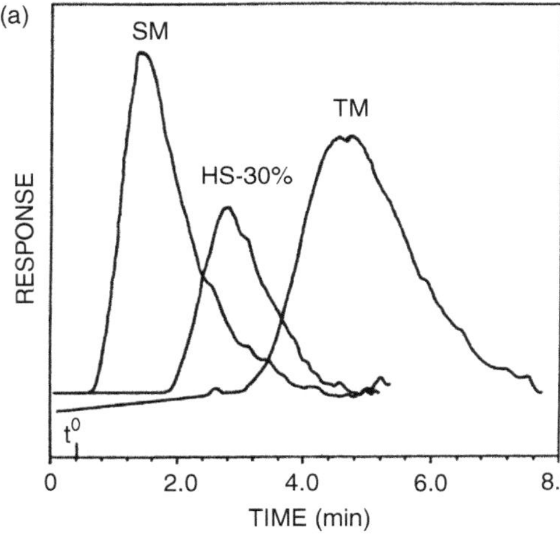

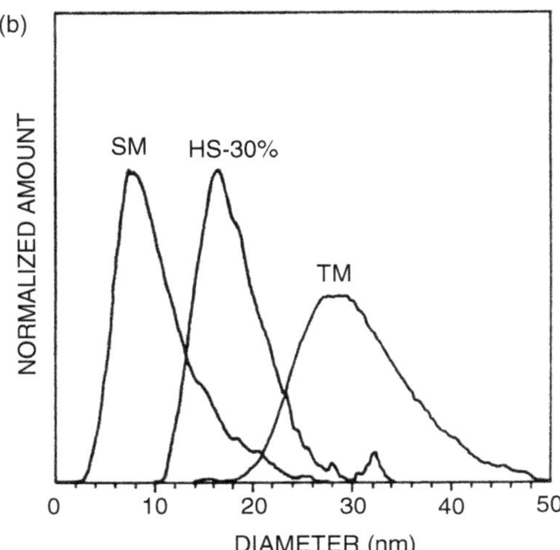

FIGURE 41.4 Flow FFF (system Flow I) characterization of three different Ludox colloidal silica suspensions: (a) fractograms; and (b) particle size distributions.

Standard data analysis software was used to convert the fractograms of Figure 41.4a into size distribution curves. The results are shown in Figure 41.4b. The mean diameters (corresponding to the first moments of the peaks) obtained from the different experimental sets are shown in the lower half of Table 41.4. These results are self-consistent and in satisfactory agreement with the TEM measurements of 30, 18, and 12 nm for TM, HS-30%, and SM silicas, respectively.

Not only are the mean diameters consistent among the different experimental sets, but the entire size distribution curve appears generally to be quite reproducible. Figure 41.5 illustrates this point by showing the size distributions obtained both for TM and HS-30% in the experimental conditions of sets 1, 2, and 5. These size distributions were not normalized so that the different curves can be differentiated from one another. Although the discrepancies appear to be small, experimental set 5

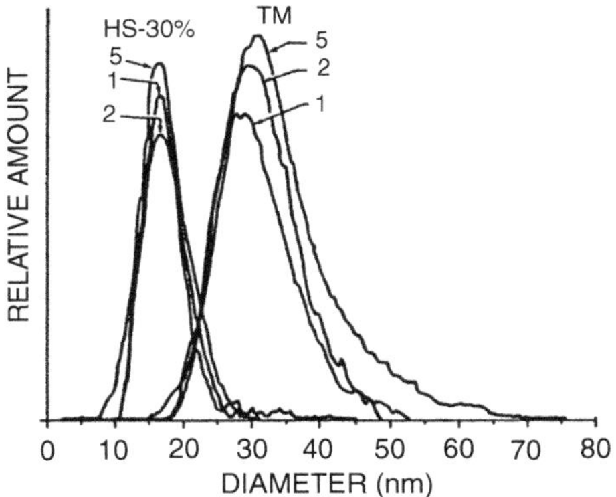

FIGURE 41.5 Comparison of the size distribution curves for two Ludox silicas obtained with two different flow FFF channels operated under three different sets of experimental conditions. The conditions are specified by matching the set numbers with those in Table 42.4.

provided broader distributions. However, theoretical calculations show that the system band broadening for set 5 conditions is excessive for the acquisition of accurate size distribution data. In particular, for TM the flow FFF system is calculated to generate only 69 theoretical plates, marginal for good fractionation. By contrast, set 1 and 2 conditions generate about 290 plates for TM, sufficient to produce accurate distributions. The observed increase in peak tailing for set 5 may also indicate a small degree of aggregation induced by the phosphate buffer.

The particle size distribution curves shown in Figure 41.4b and Figure 41.5 were obtained without correcting the UV detector signal for light scattering [6,22]. With a light-scattering correction, the population size was shifted to significantly lower values. Unfortunately, in this particle size range the light-scattering correction is remarkably sensitive to very low levels of light absorption, which cannot be assumed to be precisely zero.

To better evaluate the light-scattering correction, we compared the results obtained from the normal UV detector and the ELS detector; the latter, in theory, provides an absolute mass distribution curve. Figure 41.6 shows almost perfect agreement between the size distribution curves obtained from the UV and ELS detectors. Some, but certainly not all, of the (unexpected) agreement can be attributed to the use of nonoptimal flow rates, which provided less fractionation efficiency than the other experimental sets, as noted earlier. Clearly, the matter of applying light-scattering corrections to the signal from conventional UV detectors, particularly with colloidal silica, requires more study. When these corrections are made, the resulting diameters may or may not shift significantly relative to those reported here. These results show, nonetheless, that flow FFF is capable of producing meaningful and reproducible size distribution data for populations of colloidal silica spheres.

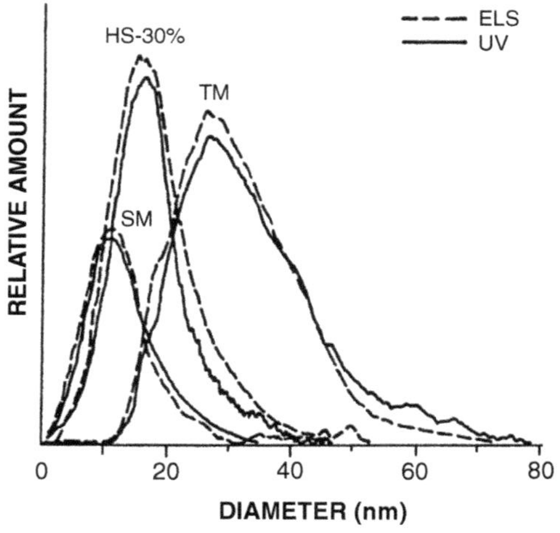

FIGURE 41.6 Comparison of size distribution curves of Ludox silicas obtained from Flow I with a conventional UV detector and an ELS detector.

Monospher Colloidal Silica

Superimposed thermal FFF fractograms (system Therm I) of various sizes of Monospher colloidal silicas (Merck) run in acetonitrile are shown in Figure 41.7. A ΔT of 53 K (cold wall temperature of 290 K) was used to obtain these data. The broadness and excessive tailing of the peaks suggests a high polydispersity for the particle populations. In addition, the proximity of elution of the Monospher 100 and 150 samples indicates that their mean sizes are not as different as suggested by the nominal diameters (0.10 and 0.15 μm) reported in Table 41.2. Although the thermal diffusion behavior of particles in liquids [reflected in the D_T term of Equation (41.9)] is not well understood, the systematic differential migration of different sizes of Monospher silicas illustrates the capability of thermal FFF for separating and characterizing particles in nonaqueous media [21].

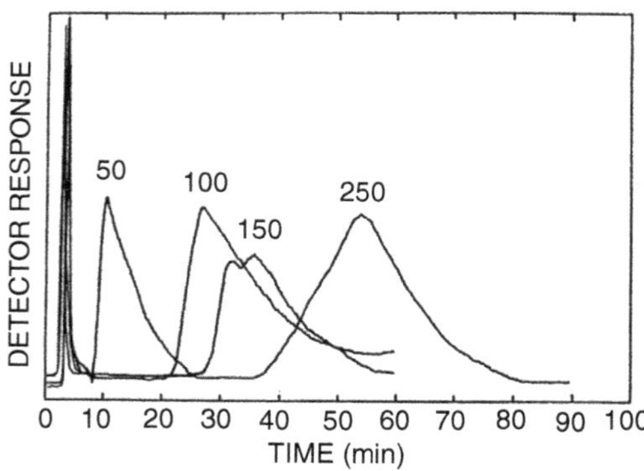

FIGURE 41.7 Fractograms of four different Monospher silicas suspended in acetonitrile by thermal FFF (system Therm I).

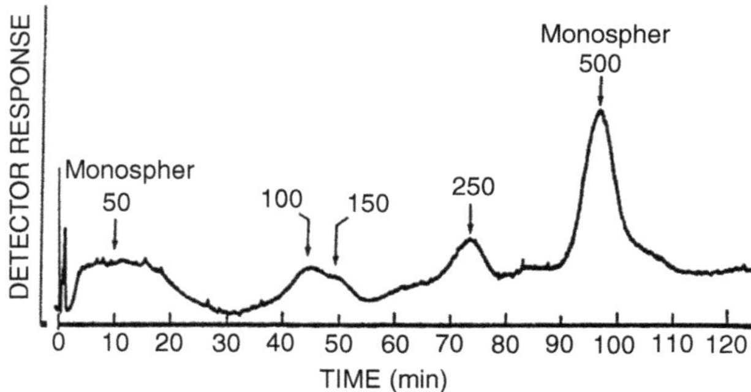

FIGURE 41.8 Separation of Monospher colloidal silicas with power-programmed sedimentation FFF (Sed II system).

The Monospher samples were further analyzed by sedimentation FFF. The fractogram shown in Figure 41.8 was obtained for a mixture of five different diameters of Monospher silica beads under the following power-programmed field conditions [10]: $G_0 = 973.2$ gravities (2400 rpm), $t_1 = 10$ min, $t_a = -80$ min, $t_{sf} = 6$ min, and $\dot{V} = 4.14$ mL/min. As in Figure 41.7, the Monospher 100 and 150 peaks are surprisingly close together. The broad and unusual slope of the peaks (including extra peaks) suggests that the population distributions for these samples are broad, possibly aggravated by aggregation.

The 250-nm sample was analyzed for possible aggregation. The fractogram shown in Figure 41.9 was obtained with a constant field of 15.2 gravities (300 rpm) and a flow rate of 4.0 mL/min. The micrographs in Figure 41.9 show the original sample and the fractions collected from a series of peaks as indicated on the fractogram. (In the absence of aggregation, a single peak would be expected.) The micrograph for the original sample shows uniform-sized beads, although many of them are in contact with one another. However, the clear presence of doublets and triplets in the second and third peaks, respectively, and of the assorted larger clusters in the peak obtained by turning off the field, indicates that the 250-nm sample is highly aggregated. This result illustrates the capability of FFF to probe colloidal populations, both through fractionation and through the collection of fractions for further analysis.

Nyacol Colloidal Silica

Of the variety of Nyacol (Nyacol Products, Inc.) colloidal silicas, only Nyacol 9950 was investigated. According to the product bulletin, this "silica is present as amorphous spheres 100 millimicrons in size" [23]. We undertook the further characterization of this material by using several FFF systems, including Sed I, Sed II, Flow III, and Therm I.

Power-programmed runs on Sed I and Sed II and constant-field runs on Therm I yielded broad and bimodal

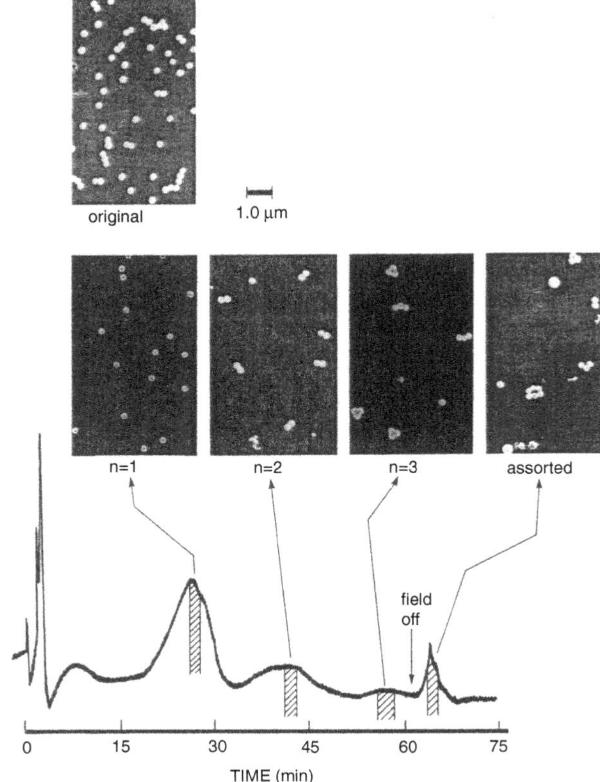

FIGURE 41.9 Sedimentation FFF (Sed II system) fractogram and electron micrographs of Monospher 250 aggregates.

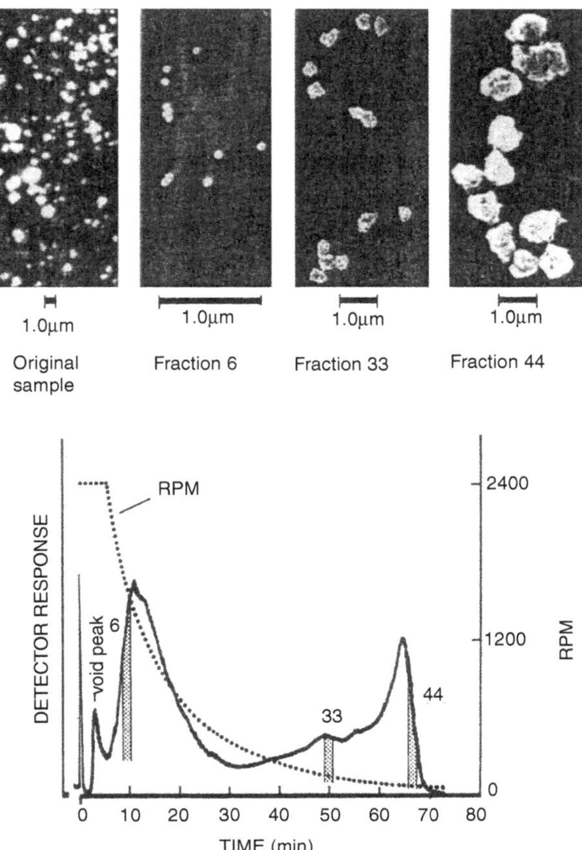

FIGURE 41.10 Fractograms and electron micrographs of fractions of Nyacol 9950 collected from a power-programmed sedimentation FFF (Sed I) run.

fractograms in all cases, a result that suggests that the colloidal material was distributed in two main population groups. The fractogram provided by the Sed I system [$G_0 = 973.2$ gravities (2400 rpm), $\dot{V} = 0.52$ mL/min, $p = 8$, $t_1 = 5$ min, $t_a = -40$ min, and $t_{sf} = 3$ min] is shown in Figure 41.10. This fractogram suggests that the size distribution is bimodal, with calculations showing that one mode is near 0.05 μm and another mode somewhat above 0.5 μm. To further elucidate this unexpected distribution, electron micrographs were taken of the original material and of three of the fractions collected at the positions indicated on Figure 41.10. These micrographs, shown at the top of Figure 41.10, verify the presence of both small and large particle populations. (The micrograph of the original sample was not properly focused to show the smaller particles.) Many of the particles in the larger population lie well above 0.5 μm in diameter, so significant steric effects may interfere with an accurate sizing of the second mode. Accordingly, a constant field run at 100 rpm and $\dot{V} = 0.2$ mL/min was instituted to reduce the excessive steric disturbances inherent in programmed runs made at high initial field strengths and high flow rates. This run yielded a mode at 0.63 μm, which is in reasonable agreement with the electron microscope results presented in Figure 41.10.

The Nyacol was also analyzed by the Sed II system, which was equipped with a thicker channel than Sed I (254 μm instead of 127 μm). Conditions similar to those described for the Sed I run were used, with the exception of $\dot{V} = 4.15$ mL/min. There is excellent agreement between Sed I and Sed II for the size distributions found in the vicinity of the first mode. However, the distributions of particles near the second mode are different, probably because of the variation in steric effects in the two runs.

If the Nyacol particles are assumed to be solid spheres, correction for light scattering suggests that the population of larger particles constitutes only a small mass fraction of the sample. However, if the dominant population is made up of the smaller particles, then the principal mode clearly lies nearer a diameter of 0.05 μm than the 0.1 μm indicated by the manufacturer. A variety of runs on different instruments verified the former mode value; specifically, we obtained 0.058 μm for Sed I, 0.058 μm for Sed II, 0.043 μm for Therm I, and 0.041 μm for Flow III. (Experimental conditions for Therm I were $\Delta T = 53$ K and $\dot{V} = 0.60$ mL/min, and for Flow III were $\dot{V} = 0.51$ mL/min and $\dot{V}_c = 0.41$ mL/min).

Cab-O-Sil Fumed Silica

Considerable work was done on the three fumed silica samples (Table 41.2) by using both sedimentation FFF and flow FFF. Figure 41.11 and Figure 41.12 demonstrate the effectiveness of both FFF systems in the fractionation of these chainlike particles. For both sedimentation and flow FFF, the L-90 sample was injected and fractions were collected at various elution time intervals for further analysis by scanning electron microscopy. The micrographs included in Figure 41.11 and Figure 41.12 show the expected elution order, from small to large particle size. The sedimentation FFF run shown in Figure 41.11 (Sed II system) utilized power field programming with the conditions $G_0 = 380.2$ gravities (1500 rpm), $\dot{V} = 2.05$ mL/min, $t_1 = 6$ min, $t_a = -48$ min, $t_{sf} = 5$ min, and $p = 8$. For the flow FFF analysis of Figure 41.12 (Flow III system), $\dot{V} = 0.50$ mL/min and $\dot{V}_c = 0.41$ mL/min.

Figure 41.13 shows the particle size distributions acquired for the three fumed silicas by using the Flow III system. (No light-scattering correction was attempted for these irregularly shaped particles.) Superimposed on these plots is the size distribution of L-90 from the fractogram shown in Figure 41.11, obtained by using the Sed II system. (For the sedimentation FFF calculations, a density of 2.4 g/mL was used.) A broader size distribution of larger average diameter for flow FFF is observed than that for SdFFF. Such a difference is expected because, as noted in the "Theory" section, the two techniques measure different diameters (discussed later). Some of the difference, however, may arise from hydrodynamic effects or from particle interactions with the flow FFF membrane. (The latter explanations are suggested by an anomaly in the flow FFF of fumed silica, in which we have observed some dependency of the calculated particle size distribution on the channel flow rate.)

A summary and comparison of mean particle diameters obtained from the size distribution curves of SdFFF and flow FFF (d_{SED} and d_{FLOW}, respectively)

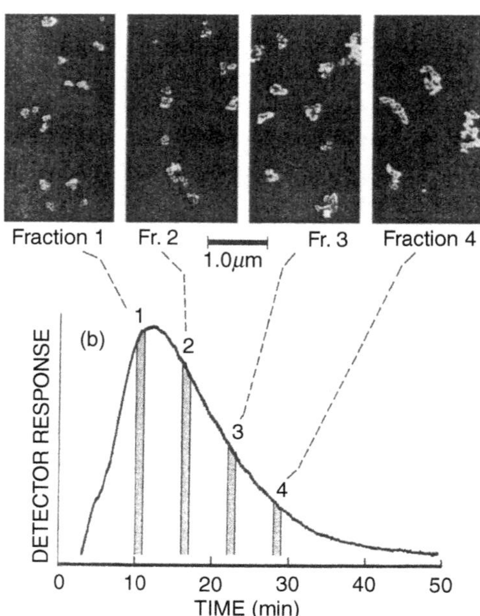

FIGURE 41.12 Fractograms and electron micrographs for collected fractions of Cab-O-Sil L-90 obtained with flow FFF (Flow III).

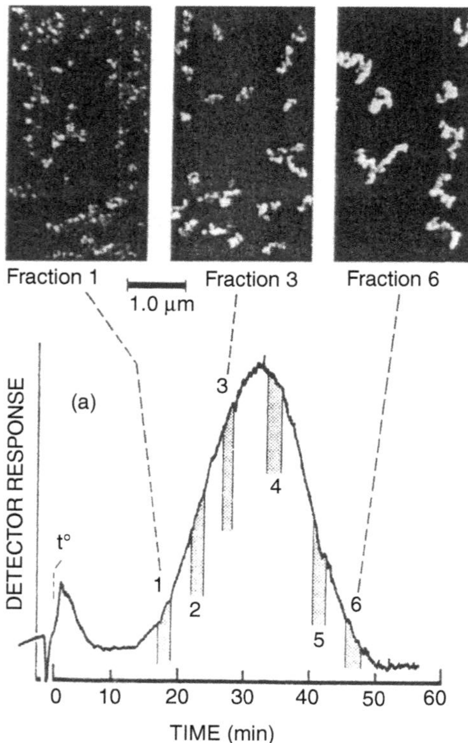

FIGURE 41.11 Fractograms and electron micrographs for collected fractions of Cab-O-Sil L-90 obtained by using sedimentation FFF (Sed II).

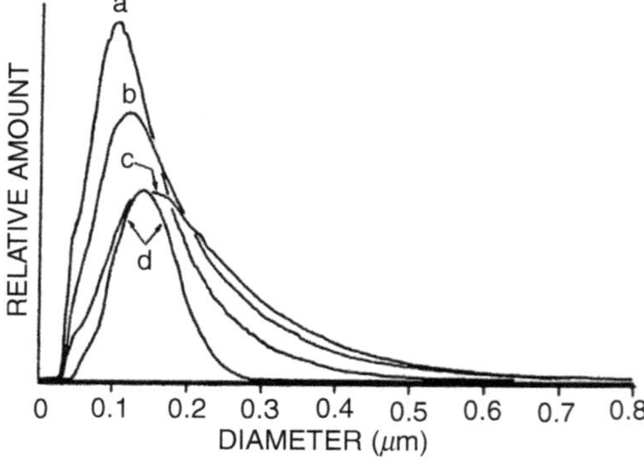

FIGURE 41.13 Particle size distributions of fumed silica samples obtained with flow FFF (a, EH-5; b, M-5; and c, L-90) and sedimentation FFF (d, L-90).

TABLE 41.5
Comparison of Diameter Values Measured by Flow FFF, Sedimentation FFF, and Image Analysis

Sample	d_{SED}	d_{FLOW}	d_{IA}
L-90	0.15	0.20	0.22
M-5	0.14[a]	0.17	0.17
EH-5	0.12[a]	0.14	0.14

Note: All values are given in micrometers.
[a]Carrier was 0.02 M triethanolamine.

with nominal values provided by Richard Geiger of Cabot Corporation based on image analysis d_{IA} is presented in Table 41.5. (The images were generated with a JEOL TEM system and analyzed on a Kontron image analysis system. The Kontron system was used to determine the surface area of the projected image and to convert the area into an equivalent diameter. Weight-averaged mean diameters were then determined.) All the results are in approximate agreement with one another. Some size discrepancies are expected because different "sizes" are measured by different techniques. This important point should eventually lead to more detailed shape and structure analysis, as explained later.

As noted in the "Theory" section, retention in sedimentation FFF is determined by particle mass [*see* Equation (41.6)]. Accordingly, the particle diameter emerging from sedimentation FFF calculations is the effective spherical diameter of the particle, which would be the diameter of a sphere occupied by the particle if it were fused into a spherical globule. Clearly, this diameter is smaller for these chainlike structures than that provided by most other methods, which to varying degrees measure the extension of the structure through space. Thus the average diameters reported in Table 41.2 based on the image area of the solid chains are larger than the modal diameter measured by sedimentation FFF. Flow FFF should yield a diameter [the Stokes diameter d_s; *see* Equation (41.8)] more nearly representative of the true hydrodynamic diameter of the particles and thus a diameter larger than that measured by sedimentation FFF. This is indeed the case, as shown by the results listed in Table 41.5. The ratio of the diameters measured by these alternative methods to the SdFFF diameter should provide a measure of the bulkiness (or fractal dimension) of the chainlike structures. More work is obviously needed to exploit the comparison of methods.

Chromatographic Silica

Silica particles used for liquid chromatographic supports are generally porous spheres in the diameter range 2–20 μm. Important properties of these particles bearing on chromatographic performance include mean size, size distribution, presence of aggregates and fines, and particle porosity. All these characteristics should be accessible to measurement by FFF; most are described later.

Chromatographic silica was studied by FFF on several previous occasions. An early study using sedimentation–steric FFF yielded size and size distribution information [14] but undoubtedly incurred errors because it preceded recent calibration techniques developed to account for hydrodynamic lift forces [20]. More recent work with flow FFF has provided size and size distribution information and has clearly signaled the presence of aggregates in one commercial sample [15]. In a study [16] using sedimentation–steric FFF combined with microscopy, we developed a strategy for obtaining not only size distributions, but also porosity and porosity distributions. Some of the data derived from this study are summarized in Table 41.6.

As suggested earlier, both sedimentation and flow FFF (in steric or hyperlayer modes) can be used to characterize particles in the size range of chromatographic silica. However, the two FFF approaches are more complementary than they are redundant. This is once again (as found also in the normal mode) a consequence of

TABLE 41.6
Application of Sedimentation-Steric FFF to Chromatographic Silica

Silica support	Apparent density (g/cm^3)	Porosity	Pore volume (mL/g)	Mode diameter (μm)	Diameter range[a] (μm)
A. Hypersil-5 (136 Å)	1.52	0.57	0.59	5.86	8.35–4.92
B. Hypersil-5 (120 Å)	1.49	0.59	0.66	6.13	7.59–5.18
C. Spherisorb	1.63	0.47	0.40	4.63	6.20–4.16
D. Nucleosil	1.45	0.62	0.75	4.71	5.92–4.04
E. Hypersil-3	1.49	0.59	0.66	3.43	4.95–2.88

[a]Range corresponds to 10–90% of cumulative mass distribution.
Source: Reproduced from reference [16]. Copyright 1991 American Chemical Society.

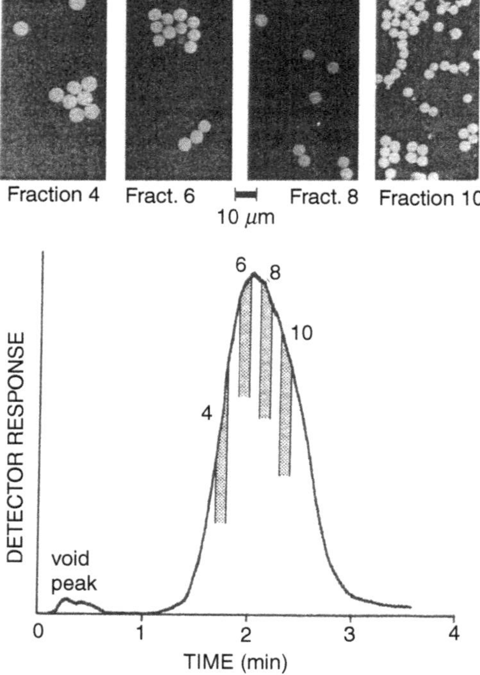

FIGURE 41.14 Fractogram from sedimentation FFF system Sed I of the chromatographic support Hypersil-5 (136-Å pore size) along with electron micrographs of particles collected in fractions 4, 6, 8, and 10.

the different force laws that control particle behavior in the two systems. The driving force in flow FFF depends only on the Stokes diameter d_s as shown in Equation (41.8). Thus for spherical silica particles, the diameter and diameter distribution, and nothing more, is characterized by flow FFF. However, the force acting on particles in a sedimentation FFF channel is a function of both particle diameter d and the density difference $\Delta\rho$ as shown by Equation (41.7). Measurement by sedimentation FFF alone thus yields a mix of diameter and density information. However, if d can be established independently by other means (such as flow FFF or microscopy), then $\Delta\rho$ can be obtained and the particle density (or density distribution) calculated. From the density values, porosities and pore volumes can be derived.

Figure 41.14 illustrates the fractionation of the chromatographic support Hypersil-5 (136-Å pore size) by sedimentation-steric FFF with system Sed I. The fractogram and the time intervals (6 sec) in which fractions were collected from the eluting stream are shown. The electron micrographs of four of the resulting fractions confirm the high-resolution size fractionation; the largest particles are seen to emerge first and the smallest last, as expected of the steric mechanism. (A similar result is found with flow-hyperlayer FFF). Fractograms like that illustrated in Figure 41.14 combined with particle diameter measurements from one or more of the micrographs can be used to calculate particle densities as explained in reference [16].

Figure 41.15 shows the fractograms obtained from high-speed runs with sedimentation FFF (system Sed I) and flow FFF (Flow II) on five different chromatographic support materials. The fractograms look fairly similar, but, as noted earlier, they bear different information on size and density parameters. As already noted, the fractograms for sedimentation FFF (Figure 41.15a), when used in conjunction with microscopy, yield density and porosity information, as summarized in Table 41.6. With the density and

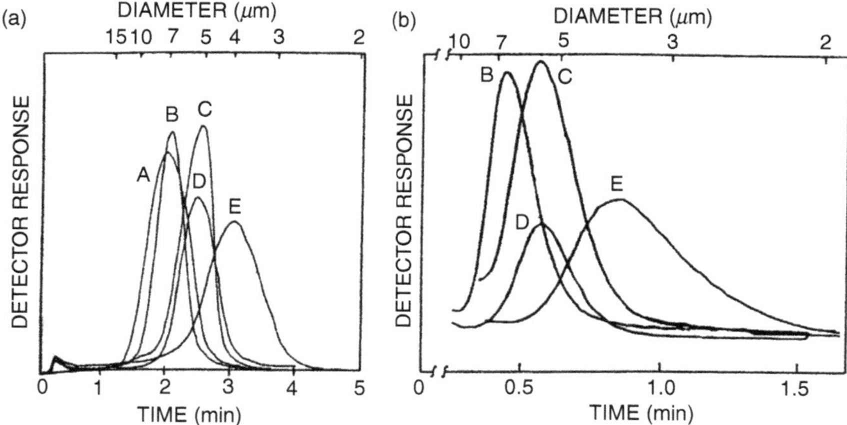

FIGURE 41.15 Fractograms of chromatographic silicas identified (by letter) in Table 42.2 obtained by using (a) sedimentation FFF system Sed I and (b) flow FFF system Flow II. The diameter scale at the top is obtained by using a calibration process based on Equation (42.10) and the measured retention times of polystyrene latex standards. For sedimentation FFF, density compensation is carried out by adjusting the spin rate for each support material in accordance with its density [20]. The corresponding spin rates utilized are A, 465; B, 479; C, 425; D, 500; and E, 475 rpm.

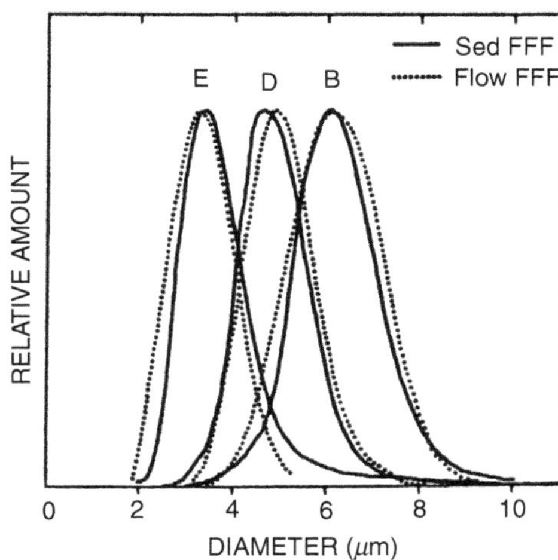

FIGURE 41.16 Comparison of size distribution curves of chromatographic silica obtained independently by sedimentation FFF and flow FFF. The three materials are B, Hypersil-5 (120-Å pore size); D, Nucleosil; and E, Hypersil-3.

thus $\Delta\rho$ known, the speed of the centrifuge can be adjusted such that the product $G\Delta\rho$ is the same for the porous silica particles as for various sizes of latex standards (such as those shown in Figure 41.lb) used for calibration. By this rotation speed adjustment (in which different rotation speeds are applied to different silica supports in Figure 41.15a to compensate for the various density differences), the particle diameter distribution can be obtained by using a calibration procedure based on Equation 41.10. (The procedures are detailed in reference [20].) The results are shown for three chromatographic supports in Figure 41.16.

The same supports are analyzed somewhat more conveniently (with a similar calibration method) by flow FFF because density effects need not be considered (nor can density values be obtained). Flow FFF thus yields independent size distribution curves; these are also shown in Figure 41.16. The agreement in the two sets of distribution curves is excellent; this agreement suggests that these two totally independent methods are capable of providing reliable and accurate size distribution information for particles with $d > 1$ μm. (The small discrepancy around 2 μm may result from the fact that the Sed I system was accurately calibrated only down to 4 μm.)

CONCLUSIONS

Although FFF clearly has many advantages in the characterization of colloidal and particulate silica, a number of experimental difficulties must still be resolved in certain areas. These results show, for example, that the quality of the experimental data, particularly for the Ludox and Cab-O-Sil silicas, would be greatly improved by finding detectors having higher sensitivity and by finding ways to remove the uncertainties of the light-scattering correction applied to conventional UV detectors. (Such uncertainties are common to many particle-sizing methods.) In addition, there are indications of occasional particle-wall interaction and adsorption problems found in the analysis of the colloidal materials, particularly with flow FFF.

Despite some present limitations, the data reported in this chapter demonstrate that FFF has some powerful and unique capabilities in silica characterization. The flexibility of FFF in analyzing particles at different size extremes is one asset. Perhaps the most outstanding advantage of FFF, however, stems from its utilization of high-resolution fractionation and the associated capability of collecting narrow fractions for further examination by other techniques, such as microscopy. The fruits of this capability were demonstrated in the isolation and identification of colloidal aggregates, in the acquisition of particle densities and porosities for chromatographic silica, and so on. The possibility of obtaining detailed structural information on fumed silica is very likely to be realized by combining sedimentation FFF and flow FFF; fractions would be shunted from one system to another to exploit the complementary information they provide.

The capability of FFF to produce high-resolution (and thus detailed) size distribution curves in the submicrometer-size range is particularly important. Submicrometer-size distributions extending down to 5-nm diameter, as obtained from flow FFF, are generally very difficult to obtain by other techniques. For such small particles, electron microscopy is a primary tool, but electron microscopy can be used even more beneficially in combination with FFF, particularly if aggregation or other morphological features of the sample materials must be examined.

A more recent development in FFF, demonstrated by the results presented, is found in its capability of producing both rapid (often in 2–3 min) and accurate size distribution data for supramicrometer-size particles. The extension of this capability to the measurement (with the aid of microscopy) of particle densities and porosities in this size range, as described here, is also significant.

ACKNOWLEDGMENTS

We thank Edward King of the Biology Department, University of Utah, for help and advice on electron microscopy; Richard Geiger of Cabot Corporation for providing the image analysis data for the Cab-O-Sil silicas; Andre Kumerrow of FFFractionation, Inc., for technical assistance; and the companies who contributed samples for this work. This work was supported by Grant CHE–9102321 from the National Science Foundation.

LIST OF SYMBOLS

c	particle concentration
c_0	particle concentration at accumulation wall
d	particle diameter
d_s	Stokes diameter
D	diffusion coefficient
D_T	thermal diffusion coefficient
F	driving force exerted by field on a particle
G	field strength measured as acceleration
G_0	initial field strength
k	Boltzmann constant
l	mean particle elevation
m	particle mass
p	power parameter in power programming
S_d	diameter-based selectivity
t_1	predecay time
t_a	power programming time constant
t^0	void time
t_r	particle retention time
t_{r1}	retention time of particle of unit diameter
t_{sf}	stop-flow time
T	absolute temperature
U	field-induced velocity
$\dot{V}$	channel flow rate
$\dot{V}_c$	cross-flow rate
x	distance above accumulation wall
w	channel thickness
ΔT	temperature difference between hot and cold walls
$\Delta\rho$	difference between particle and carrier density
η	viscosity
λ	retention parameter
ρ_p	particle density

REFERENCES

1. Iler, R. K. *The Chemistry of Silica*; Wiley: New York, 1979.
2. Giddings, J. C.; Yang, F. J. F.; Myers, M. N. *Sep. Sci.* **1975**, *10*, 133–149.
3. Giddings, J. C.; Myers, M. N.; Yang, F. J. F.; Smith, L. K. In *Colloid and Interface Science*, Vol. IV; Kerker, M., Ed.; Academic Press: New York, 1976; pp. 381–398.
4. Giddings, J. C.; Lin, G. C.; Myers, M. N. *J. Colloid Interface Sci.* **1978**, *65*, 67–78.
5. Giddings, J. C.; Karaiskakis, G.; Caldwell, K. D.; Myers, M. N. *J. Colloid Interface Sci.* **1983**, *92*, 66–80.
6. Yang, F.-S.; Caldwell, K. D.; Giddings, J. C. *J. Colloid Interface Sci.* **1983**, *92*, 81–91.
7. Kirkland, J. J.; Dilks, C. H., Jr.; Yau, W. W. *J. Chromatogr.* **1983**, *255*, 255–271.
8. Caldwell, K. D. *Anal. Chem.* **1988**, *60*, 959A–971A.
9. Giddings, J. C. *Chem. Eng. News* (October 10) **1988**, *66*, 34–45.
10. Williams, P. S.; Giddings, J. C. *Anal. Chem.* **1987**, *59*, 2038–2044.
11. Giddings, J. C. *Sep. Sci.* **1966**, *1*, 123–125.
12. Kirkland, J. J.; Yau, W. W.; Doerner, W. A.; Grant, J. W. *Anal. Chem.* **1980**, *52*, 1944–1954.
13. Yonker, C. R.; Jones, H. K.; Robertson, D. M. *Anal. Chem.* **1987**, *59*, 2573–2579.
14. Giddings, J. C.; Myers, M. N.; Caldwell, K. D.; Pav, J. W. *J. Chromatogr.* **1979**, *185*, 261–271.
15. Ratanathanawongs, S. K.; Giddings, J. C. *J. Chromatogr.* **1989**, *467*, 341–356.
16. Giddings, J. C.; Moon, M. H. *Anal. Chem.* **1991**, *63*, 2869–2877.
17. Giddings, J. C.; Karaiskakis, G.; Caldwell, K. D. *Sep. Sci. Technol.* **1981**, *16*, 607–618.
18. Giddings, J. C.; Chen, X.; Wahlund, K.-G.; Myers, M. N. *Anal. Chem.* **1987**, *59*, 1957–1962.
19. Williams, P. S.; Koch, T.; Giddings, J. C. *Chem. Eng. Commun.* **1992**, *111*, 121–147.
20. Giddings, J. C.; Moon, M. H.; Williams, P. S.; Myers, M. N. *Anal. Chem.*, 1991, *63*, 1366–1372.
21. Liu, G.; Giddings, J. C. *Anal. Chem.* **1991**, *63*, 296–299.
22. Kirkland, J. J.; Rementer, S. W.; Yau, W. W. *Anal. Chem.* **1981**, *53*, 1730–1736.
23. Nyacol 9950 colloidal silica product bulletin, Nyacol Products, Inc., Ashland, MA 01720.

42 Formation of Uniform Precipitates from Alkoxides

C.F. Zukoski, J.-L. Look, and G.H. Bogush
University of Illinois, Department of Chemical Engineering

CONTENTS

The mechanism of precipitation of uniform particles through the hydrolysis and condensation of metal alkoxides is discussed. Final particle sizes were found to be sensitive to the reaction medium ionic strength and the surface potential of the growing particles. In addition, the rate of loss for soluble metal-containing species was found to be independent of the surface area of the growing particles. These observations imply that the major growth pathway is through agglomeration of small gel particles produced by reactions between soluble species that proceed independent of the presence of particles. Calculations that are presented support this growth mechanism.

The length of the nucleation period in precipitation is commonly invoked as the primary control parameter for forming uniform particles in a reaction involving homogeneous nucleation and growth [1–3]. In the model originally proposed by La Mer and Dinegar [4] for the mechanism of formation of sulfur sols, uniform size distributions result if all the particles are formed in a short burst of nucleation, and then particle growth occurs by a mechanism in which large particles increase in diameter more slowly than smaller particles (as occurs when growth is limited by diffusion to the particle surface). Despite the influence this model has had on studies of inorganic particle precipitation chemistry, the model has seen little corroboration and indeed has been brought into question for the sulfur sol for which it was developed [5]. Although the La Mer mechanism would undoubtedly result in uniform particles, finding systems that meet the conditions required for the model to hold has been elusive.

Studies on the formation of polymer latex particles have provided an alternative mechanism whereby uniform particles can result from a homogeneous nucleation–precipitation reaction. In emulsion polymerization, an insoluble monomer is mixed with water and a water-soluble free radical initiator is added. Final particle size depends on reaction temperature, reagent concentration, and parameters controlling the colloidal stability of the growing particles (i.e., ionic strength and pH). The initial locus of the reaction is in the aqueous phase. Oligomers grow to a size at which they become insoluble and undergo a sol-to-gel transition. Because of their relatively low concentration, this transition involves only a few polymer molecules, and the gel phase grows by aggregation. The charge on the primary particles is small, but as the aggregation process proceeds, the charge per particle grows. As a consequence, aggregation rates of particles of equal size decrease, but the aggregation rate of particles of dramatically different size increases. The result is a bimodal particle (or gel-phase) size distribution, with one peak located at the primary particle size and the second peak representing particles that are stable to mutual coagulation and grow by scavenging smaller particles. Upon aggregation, the gel-phase particles coalesce, and thus the particles are able to retain a spherical shape. As the reaction proceeds, the growing particles reach a constant number density. These colloidally stable particles swell with monomer, and the locus of the reaction is transferred from the aqueous phase to the inside of the particle phase. Uniformity is achieved through control of the colloidal stability of the primary particles and aggregates of these particles [6–8].

A second organic analogy to the precipitation of inorganic particles occurs in dispersion polymerization. Here a solvent is chosen in which the monomer is soluble but the polymer is not. The reaction is initiated and proceeds through polymerization until the oligomers grow to a size at which they undergo a phase transition and precipitate to form polymer and solvent-rich phases. Uniform particles are achieved through the addition of steric stabilizers that control the aggregation of the growing gel phase [9].

The major distinction between the model of La Mer and that developed for uniform latex particles lies in the incorporation of colloidal stability of small particles. The La Mer model assumes that each nucleus is colloidally stable and survives at the end of the reaction at the center of a particle. The aggregation models argue that stabilizing primary small particles is difficult, but aggregation does not necessarily result in a broad particle-size distribution. When schemes for control of particle-size distribution are developed, the result of accepting the notion that colloidal stability can play an important role is that attention is focused away from the length of the nucleation period and towards the colloidal properties of the growing particles.

This chapter reviews our studies on the formation of uniform precipitates through the hydrolysis and condensation of titanium and silicon alkoxides. Our work has been aimed at elucidating whether uniformity is the result of the details of the chemistry or if particle-size distributions can be controlled by physicochemical means. In particular, the chapter focuses on experiments probing the length of the nucleation period in these systems and the effects of parameters that control particle interaction potentials. For the systems studied, the nucleation period appears to be a substantial fraction of the entire reaction period, and particle size is largely controlled by parameters related to particle interaction potentials. These results suggest that there are strong links between the physical chemistry underlying the formation of uniform latex particles and that controlling particle-size distributions of hydrous metal oxide precipitates [10–14].

UNIFORM SILICA PARTICLES

Stober et al. [15] developed a method of preparing remarkably uniform silica particles with sizes ranging from 50 nm to >1 μm in diameter. Their recipe involves hydrolyzing silicon alkoxides in aqueous alcoholic solutions containing ammonia. The resulting solids are amorphous and are 11–15% porous. We chose to use the hydrolysis and condensation of tetraethylorthosilicate, TEOS, in ethanol as a model precipitation reaction to study parameters leading to uniformity.

Our results [10–12] provide evidence that the La Mer mechanism is not at the heart of the uniformity of the resulting particles. Primarily, the rate of particle growth, the rate of loss of soluble silica (i.e., material that is able to pass through a 22-nm filter), and the rate of loss of TEOS (as determined from ^{29}Si NMR spectroscopy) are all well described by first-order processes with the same rate constant, k_1. These results are in keeping with those of Matsoukas and Gulari [16] and have been used to suggest that TEOS hydrolysis is the rate-limiting step in particle growth. The rate constant, k_1, was measured as a function of water and ammonia concentration. In addition, we found that the conductivity of the reaction medium first increases and then decreases. The rate of decrease in conductivity ($\sigma - \sigma_0$) is well fit by a first-order process with a rate constant with the same value as k_1. Here σ_0 is the initial conductivity of the reaction medium and σ is the suspension conductivity at reaction time t. The conductivity is found to be well modeled by [11,12]

$$\sigma - \sigma_0 = \{ \exp(-k_1 t) - \exp(-k_2 t) \}$$

where k_2 has also been characterized over a wide range of water and ammonia concentrations. For all of the reactions studied, $k_2 \geq 10k_1$. The conductivity increase is taken as an indication of the hydrolysis of TEOS, the production of silicic acid groups, and the subsequent deprotonation of these species by ammonia. Decreases in conductivity are associated with the loss of deprotonated silicic acid groups through condensation reactions. As a result of the strong link between the long time rate constant for the conductivity (i.e., k_1), and that measured for particle growth, as well as the rates of loss of soluble silica and TEOS, we feel conductivity is an accurate method of determining the relative amounts of hydrolyzed but uncondensed TEOS.

Soluble silica concentrations required for nucleation were determined by measuring the lag time for the increase of turbidity in a solution after the addition of TEOS. As the soluble silica concentration decreases toward the equilibrium value, C_{eq}, the lag time increases (but the final particle size is not greatly reduced). At an initial TEOS concentration of 5–10 times C_{eq}, the lag time becomes much longer than the time for the hydrolysis and condensation reactions to reach equilibrium. Thus, if the TEOS concentration exceeds 5–10 times C_{eq}, nucleation can proceed (i.e., 5–10 times C_{eq} represents the critical concentration that must be exceeded before appreciable nucleation is observed).

As the soluble silica concentration is considerably above this value for virtually the entire precipitation reaction (Figure 42.1), if nucleation does not occur throughout the reaction, the concentration of hydrolyzed TEOS must lie below the critical value. This situation may occur if the rate of hydrolysis is matched by the rate of deposition on particle surfaces. Two observations serve to discount this possibility. First, solution conductivities can be used to estimate the concentration of hydrolyzed TEOS.

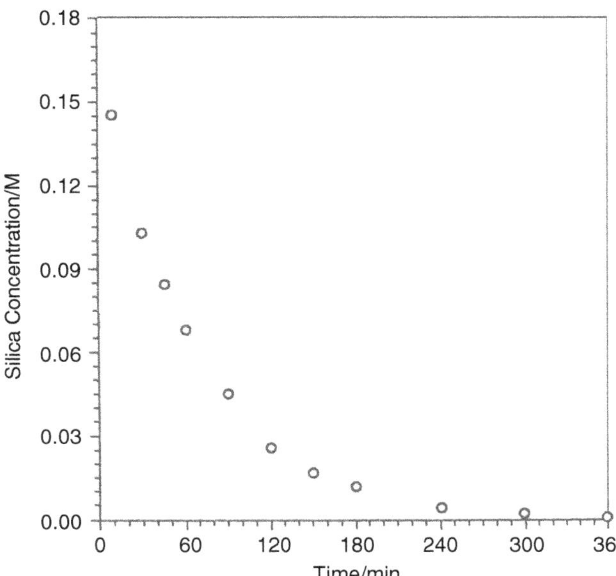

FIGURE 42.1 Soluble silica concentration as a function of time for a reaction at 25°C for 0.17 M TEOS, 1 M NH_3, and 3.8 M H_2O ($C_{eq} = 2.4 \times 10^{-4}$ M).

Using mobilities of KCl in water to characterize mobilities in the ions that give rise to the conductivity and assuming unit activity coefficients, we estimate that the concentration of hydrolyzed TEOS lies above the critical nucleation value for at least two-thirds of the time required for the soluble silica concentration to reach its final value (Figure 42.2). Second, if there is a balance between the rate of production of hydrolyzed TEOS and condensation of these species at particle surfaces, then absolute concentrations and rates of loss should depend on the surface area available for deposition. A series of reactions were performed in which particles were first precipitated and then diluted with their mother liquor such that their final particle density was decreased to a quarter of that in the as-precipitated suspension. TEOS was then added to these suspensions, and the rates k_1 and k_2 were determined. The absolute conductivity as well as the rate coefficients k_1 and k_2 were independent of particle density or even the presence of any seed particles [12]. These results suggest that hydrolyzed TEOS concentration is above the critical nucleation level for much of the reaction and that there is a decoupling of reactions between soluble species and growth of particles. Consequently, hydrolysis and condensation reactions proceed independently of the presence of particles. Thus, addition of soluble species to particle surfaces is not the major growth mechanism.

Studies by Klemperer and Ramamurthi [17] showed that under basic conditions silicon alkoxides react such that the molecular-weight distribution of soluble species remains peaked at the monomer level throughout the reaction. At early times, the reaction pathway is seen to follow the Flory–Stockmayer polymerization pathway developed for monomers with three to four reactive sites. The Flory–Stockmayer theory predicts the most probable molecular-weight distribution for monomers with various degrees of reactivity by assuming that all sites have equal reactivity and that there is no intramolecular cross-linking [18]. Although these assumptions begin to break down early in the reaction, the results of Klemperer and Ramamurthi indicate that under basic conditions, there will be a few high-molecular-weight species produced through a cascade-like series of reactions. The fate of these high-molecular-weight species is key to the nucleation and growth of the silica particles.

The Flory–Huggins theory of polymer solubility suggests that if a polymer is growing in a poor solvent, when it reaches a critical molecular weight, a phase separation will occur in which a polymer-rich or gel phase is formed, as is a polymer-poor, solvent-rich phase [18]. If the high-molecular-weight species are sufficiently dilute, this phase separation may involve only a few polymers, and thus the phase-separated regions may be small. Continued growth of the gel phase will then occur through molecular addition and aggregation. Because the number of reactive sites in the gel particle is greatly reduced once the soluble polymer collapses the rate of growth by molecular addition will be small. Instead, a more likely growth pathway of the gel phase may be by aggregation with continuing internal densification. The particle-size distribution of the gel phase will then be determined by kinetic parameters that control the size dependence of aggregation rates of gel particles.

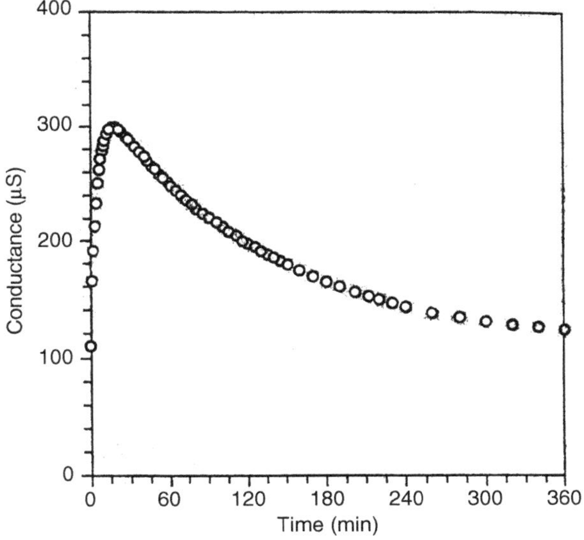

FIGURE 42.2 Conductivity of the reaction medium as a function of time after mixing (25°C, 0.17 M TEOS, 1.0 M NH_3, and 3.8 M H_2O).

This mechanism is hypothesized for the formation of uniform silica and titania particles, as it contains many of the elements observed in particle precipitation kinetics. Primarily, the continuous formation of high-molecular-weight species that undergo a phase transition at a critical size (as also suggested by the work of Bailey and Mecartney [19,20]) to form nuclei helps to explain the decoupling of reactions between soluble species and the total surface area available for deposition. In this model, growing polymers remain soluble and thus would rather be surrounded by solvent than by other polymer species. Thus, these oligomers are not lost to particle surfaces until they reach a critical size, undergo a phase transition, and become "sticky." Particle growth is then controlled by rates of aggregation between existing "gel" phase particles and freshly formed "gel" particles. Second, this mechanism suggests a pathway in which nucleation occurs for virtually the entire reaction period, in keeping with the experimental observations described earlier. Final particle-size distributions will be determined by parameters controlling the rates of aggregation of primary particles and the size at which aggregates become stable to mutual coagulation. Thus, without altering rates of reaction between soluble species, an increase in the ionic strength of the continuous phase would force particles to achieve a larger size before they become colloidally stable. As a consequence, final number densities of particles will be smaller and average diameters larger. To confirm these predictions, final particle sizes were measured as a function of ionic strength. Rate constants k_1 and k_2 were unchanged with [NaCl] up to 10^{-2} *M*. However, final particle diameters doubled [12–14].

SIZE-DEPENDENT AGGREGATION RATES

Models of particle growth based on aggregation require a minimum-size particle — in this case, the primary particle size. Because of a current lack of understanding of the thermodynamic properties of high-molecular-weight inorganic species, methods of determining primary particle sizes in an *a priori* manner have not been developed for the polymer collapse model discussed earlier. However, for a system that more closely follows classical nucleation theory, methods of estimating nucleus sizes and solid–liquid surface tensions have been developed on the basis of critical nucleation concentrations [21]. For TEOS and titanium alkoxides, critical nucleation concentrations occur in the range of 1–10 times $[C]_{eq}$, and classical nucleation theory suggests that nucleus surface tensions are on the order of 80–159 ergs/cm^2 and nuclei radii are 2–10 nm. As described later, these primary particle sizes are close to those required for an aggregation model to predict final particle diameters.

At the end of the precipitation reaction, the solid particles must be colloidally stable if a uniform particle-size distribution is to be observed. A question important to final uniformity is the particle size when this stability is achieved. The particles will always feel the long-range van der Waals attractive interactions. Interactions of an electrostatic or solvation origin can give rise to a repulsive barrier that can provide kinetic stabilization. At the end of the reaction, particles precipitated from TEOS and titanium alkoxides have final particle number densities, N_∞, of 10^{16}–10^{18} m^{-3}. These particles are suspended in a solvent with an ionic strength of approximately 10^{-4} *M* and have surface potentials of 10–35 mV. Our studies indicate that the particles also feel a short-range repulsive interaction that we have modeled as a solvation interaction with decay length of 1 nm and contact interaction potential of 1–1.5×10^{-3} J/m^2. With a Hamaker coefficient of 2.4 k*T*, classical Smoluchowski aggregation theory [22] can be used to determine the time required to halve the number density by aggregation,

$$t_{1/2} = 3\mu W_{ii}/4kTN_\infty \qquad (42.1)$$

where μ is the continuous-phase viscosity, W_{ii} is the stability ratio for aggregation of two particles of size *i*, *k* is the Boltzmann constant, and *T* is temperature. If the La Mer model is used to describe particle growth, then N_∞ particles will be produced during the nucleation period and will have a radius on the order of 2–10 nm. Classical aggregation theory can be used to determine that for particles of this size, the stability ratio is about unity; thus, half the nuclei at a density of N_∞ will be lost by aggregation in 0.2–100 s, a time much shorter than the reaction period. Thus, it appears that while nuclei are very difficult to stabilize, particles grow by some mechanism such that at the end of the reaction they are colloidally stable.

As mentioned earlier, nuclei can grow through the addition of soluble molecules or by aggregation. A model was developed to determine if aggregation and continued nucleation alone can give rise to the particle-size distributions observed in the titania and silica precipitation reactions [12]. A similar approach was taken by Harris et al. [23] for TEOS-derived silica and by Santacesaria et al. [24] for soluble-salt-derived titania. This model is based on the observed decoupling of reactions between soluble species (which are assumed to give rise to nuclei) and particle growth (which is assumed to occur by aggregation). Essentially, the model allows particles to nucleate and particles containing *i* and *j* primary particles to aggregate with a binary rate constant of $\beta(i,j)$. If $\beta(i,i)$ decreases as *i* increases but $\beta(1,i)$ increases with *i*, a uniform precipitate is predicted. In this model, primary particles of radius r_1 are formed at a rate $g_n(t)$, written

$$g_n(t) = g_s[\exp(-k_1 t) - \exp(-k_2 t)] \qquad (42.2)$$

where $g_s = 3v(C_o - C_{eq})/[4\pi r_1^3(1/k_1 - 1/k_2)]$. Here v is the molar volume of the solid phase and C_o is the initial alkoxide concentration. The overall particle-size distribution evolves in time through a coupled set of equations linking the number density of particles containing k primary particles, $n(k, t)$:

$$\mathrm{d}n(k, t)/\mathrm{d}t = \frac{1}{2}\sum_{i=1}^{k-i} \beta(i, k-i)\, n(i, t)\, n(k-i, t) - n(k, t)\sum_{i=1}^{\infty} \beta(k, i)\, n(i, t) + \delta_{k,1}\, g_n(t) \quad (42.3)$$

Here $\delta_{k,1} = 1$ if $k = 1$ and is otherwise zero.

Smoluchowski aggregation rate kernels modified to account for hetero-aggregation in the presence of repulsive barriers become

$$\beta(i, j) = (1 + r_i/r_j)2\beta_s/[(4W_{ij}/W_{11})r_i/r_j] \quad (42.4)$$

with $W_{ij}/W_{11} = (r_{11}/r_{ij})(r_i/r_j)\exp[Jr_{11}(r_{ij}/r_{11} - 1)]$ for $r_i < r_j$. Here W_{11} is the stability ratio of primary particles and $r_{ij} = 2r_ir_j/(r_i + r_j)$. The variable $\beta_s = 8kT/3\,\mu W_{ii}$ is the characteristic aggregation rate constant from $8kT/3\,\mu W_{11}$. The parameter J is found from the slope of a log (W_{ii}/W_{11}) versus $(r_{ii}/r_{11}) - 1$ curve. The stability ratio is found from particle interaction potentials, and J is found to increase with the maximum in the pair interaction energy.

Using measured surface potentials, estimates of Hamaker coefficients, and solvation parameters, the only unknown required for the aggregation model to make quantitative prediction of particle-size distributions is the size of the primary particle, r_1. Extensive calculations [12,25] indicate that Equation (42.3) predicts a particle-size distribution that rapidly evolves into two peaks. The larger particles grow to a size at which they become stable to mutual coagulation, and after sufficient time a constant number density of colloidally stable particles is established. For a wide range of Jr_{11}, with k_1t_s and k_2t_s much smaller than unity, the final number density depends in a simple manner on parameters governing the solution to Equation (42.3). Here $t_s = (g_s\beta_s)^{-1/2}$ is the characteristic time for nucleation and aggregation. For this range of conditions, the final number density of particles is well correlated by

$$N_\infty = 2.4 \times 10^{-2}(Jr_{11})^3 n_s \quad (42.5)$$

where $n_s = (g_s/\beta_s)^{1/2}$ is the characteristic number density scale. This model is similar to that developed by Feeney et al. [8] for emulsion polymerization of uniform latex particles. The breadth of the final particle size distribution is systematically narrower than that observed experimentally; this result emphasizes the conclusion that aggregation can result in uniformity.

As shown in Table 42.1, reasonable estimates of particle interaction parameters provide a narrow range of value of J. The remaining unknown parameter, r_1, is used to force a fit between the final measured average size and that predicted from Equation (42.5). Predicted values of r_1 fall in a physically meaningful range and are similar to those predicted by classical nucleation theory. Although the interaction potentials used in these calculations include estimated parameters, reasonable values have been used. These results indicate that particle growth by aggregation alone can reproduce experimentally observed final particle sizes. The model prediction of final average sizes rests on a prediction of final particle number density. Once N_∞ particles have been formed and are stable to mutual aggregation, the particles can grow by any mechanism that relieves the supersaturation. For example, in the formation of polymer lattices, the stable particles swell with monomer and nucleation of primary particles stops. In the precipitation of silica and titania particles, the high supersaturations and the decoupling of rates of loss of hydrolyzed alkoxides suggest that primary particles may be continuously formed throughout the reaction. Final particle porosity and granular appearance supports our suggestion that aggregation acts as a major growth mechanism throughout the entire reaction.

UNIFORM TITANIA PARTICLES

The formation of uniform particles through the hydrolysis and condensation of titanium alkoxides has proven more elusive than for systems involving silica. Reports of uniform particle formation through the hydrolysis of tetraethylorthotitanate (TEOT) in aqueous ethanol by Barringer and Bowen [26,27] have been difficult to reproduce. Jean and Ring [28,29] developed a technique involving use of hydroxypropylcellulose in which uniform particles that are formed are fluffy and densify poorly. Considerable effort to prepare uniform particles following the techniques of Barringer and Bowen proved futile until we found that the growing particles are very sensitive to shear-induced aggregation. Subsequently, we have been able to reproduce the results of Barringer and Bowen if the degree of agitation to which the precipitating suspensions are exposed is kept at a low level.

The titania particles precipitate under reaction conditions very similar to those of the silica systems discussed earlier. A critical nucleation concentration of 1.5–3 times $[C]_{eq}$ is measured. This low supersaturation level is not reached until very late in the precipitation reaction (Figure 42.3). The rate of loss of soluble titania is also independent of the presence of solid surface area. Finally, on the basis of measures of particle surface potentials, nuclei of sizes less than about 20 nm are expected to be unstable

TABLE 42.1
Particle Interaction Parameters

Sample	$[Me(OR)_4]$ (M)	$[H_2O]$ (M)	$[NH_3]$ (M)	[HCl] ($M \times 10^4$)	$[Si]_{eq}$ ($M \times 10^4$)	$[Ti]_{eq}$ ($M \times 10^4$)	k_1^a ($s^{-1} \times 10^4$)	k_2^b ($s^{-1} \times 10^3$)	$\langle D \rangle^c$ (nm)	$\delta_\infty/\langle D \rangle^d$ (%)	ψ_o^e (mV)	J^f (nm^{-1})	$W_{r_1}^g$	r_1^h (nm)
A	0.17[i]	15	0.5	0	9 ± 1	NA[j]	4.2 ± 0.4	8.3 ± 0.9	269	8	−13	0.35	1.6	3.0
B	0.17[i]	15	1.0	0	4 ± 1	NA	9.5 ± 1.0	17 ± 2	343	6	−13	0.35	1.7	3.1
C	0.17[i]	15	2.0	0	7 ± 1	NA	17 ± 1	22 ± 2	453	2	−13	0.35	1.3	1.7
D	0.17[i]	15	3.0	0	6 ± 1	NA	23 ± 2	32 ± 3	459	1	−13	0.35	1.3	1.6
E	0.05[k]	0.25	0	0	NA	130	0.25	2	1100	13	11	0.30	1.4	1.7
F	0.05[k]	0.25	0	0.5	NA	150	0.25	2	820	13	18	0.37	1.5	1.7

[a]Rate constant for loss of soluble metal-containing species.

[b]Rate constant for initial increase in hydrolyzed alkoxide concentration. For reactions containing TEOT, k_2 is large and not easily measured. For these reactions, k_2 is an estimate.

[c]Measured final particle diameter.

[d]Measured standard deviation in diameter distribution divided by the average particle diameter.

[e]Surface potential determined from particle electrophoretic mobilities.

[f]Parameter controlling rate of aggregation of particles of like size.

[g]Stability ratio for primary particles of radius r_1.

[h]Primary particle size required for aggregation model to predict $\langle D \rangle$.

[i]TEOS sample.

[j]NA, not applicable.

[k]TEOT sample.

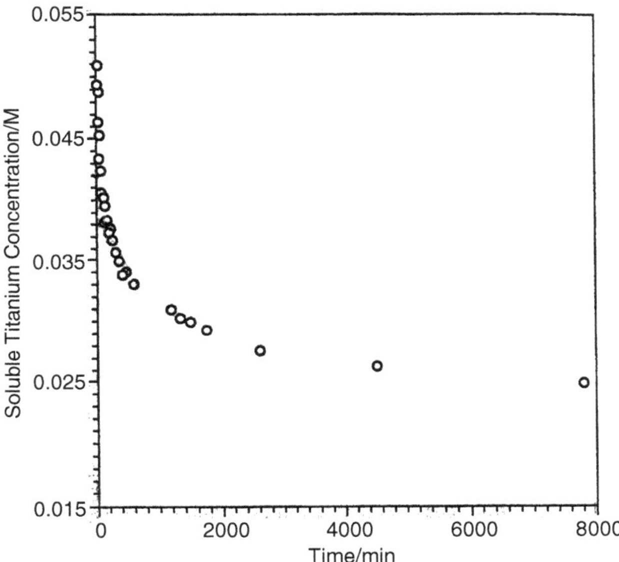

FIGURE 42.3 Soluble titanium concentration as a function of time for a reaction at 25°C; 0.05 *M* TEOT and 0.25 *M* H_2O ($C_{eq} = 0.025$ *M*).

and to rapidly aggregate. These results again indicate that during the precipitation of titania, nucleation may occur over much of the reaction period and final particle sizes may be determined by the aggregation of primary particles. These conclusions are supported by the transmission electron microscopy work of Diaz-Gomaz et al. [30].

The role of colloidal interactions is further supported by the observation that as the background concentration of HCl is raised to 10^{-3} *M*, the rate of loss of soluble titania is not altered, particle charge increases, and final particle size decreases [13,14]. Again, these observations demonstrate the importance of colloidal stability during the precipitation of uniform particles.

Shear-induced aggregation alters particle-size distributions after particles have reached a size at which they are stable to mutual aggregation by Brownian motion (i.e., aggregation is not observed in a quiescent fluid). In the absence of acid, the shear rate required to induce aggregation is $10\ s^{-1}$. Particles are observed to grow with a narrow size distribution until the average particle size is on the order of 600 nm. If the shear rate in the reactor is above $10\ s^{-1}$, large flocs are observed when particles exceed this size. At lower shear rates, the particles continue to grow, and at the end of the reaction, the particles have an average diameter of 1.3 μm with a standard deviation of approximately 15%.

CONCLUSIONS

Maintaining the stability of small colloidal particles has long been recognized to be difficult. Indeed, development of methods of stabilizing colloidal silica with a particle size under 10 nm was a major breakthrough. In the face of well-established models for colloidal interactions, researchers must demonstrate the stability of nuclei produced in precipitation reactions, rather than ask for evidence of their instability.

Silica particles precipitated from aqueous, basic alcoholic solutions to which silicon alkoxides have been added appear to grow as if the limiting reaction occurred at the particle surface [12,16]. However, the rate of loss of hydrolyzed TEOS is independent of particle surface area available for deposition. In addition, the concentration of soluble silica remains above the critical concentration required for nucleation until late in the precipitation reaction. These results and the observation that the final particles are uniform, with average size depending on ionic strength, lead us to conclude that the particles do not grow by the mechanism of addition of low-molecular-weight soluble species and that colloidal stability is important in the final particle-size distribution. Detailed calculations suggest that growth solely by aggregation of high-molecular-weight species (a mechanism that will result in changes in standard deviations in particle size distributions as observed in the seeded-growth studies [12,16]) can result in particles of narrow size distribution. This model for particle growth provides a mechanism for the formation of a constant number of growing particles early in the reaction — a necessary step in the formation of uniform precipitates [12,23]. In addition, this model incorporates the physical chemistry of particle interaction potentials and aggregation and provides an obvious explanation of the final particle porosity. The density and ability of the primary particles to coalesce will be determined by the solution thermodynamics of growing silicon- and oxygen-containing molecules in mixed basic solvents, a field that has seen little exploration.

Precipitation of uniform titania particles follows a similar pathway in that the particles appear to be the result of reactions between soluble titania species that culminate in a sol-to-gel phase transition of high-molecular-weight species and the resulting aggregation of these primary particles. Under typical conditions in which the particles are precipitated from TEOT in ethanol, particles of narrow size distributions are initially observed to grow to a dimeter near 600 nm. If the shear rate is kept below $10\ s^{-1}$, the particles continue to grow to >1 μm in size, with a relatively narrow final particle-size distribution. On the other hand, if the suspension is subjected to a larger shear rate, 600-nm particles aggregate and grow together, forming large open chains with well-formed interparticle necks. If particle surface charge is increased, final particle size is reduced, and a larger shear rate is required to induce aggregation. These results again provide graphic evidence that colloidal stability plays a major role in the formation of uniform precipitates.

The studies reviewed here show that precipitation of uniform particles requires two conditions to be met. First, at a point early in the reaction, a constant number of colloidally stable particles must be established. Second, these particles must remain stable to mutual coagulation throughout the subsequent reaction. If the primary particles formed are unstable with respect to larger particles, a short nucleation period is not a prerequisite of a narrow final particle-size distribution. Indeed, continuous slow nucleation with the correct aggregation rate kernels can produce very uniform particles. If the nuclei are unstable and aggregation is very fast, a broad particle-size distribution may result. For colloidally stable nuclei, a short nucleation period will be important in establishing uniform precipitates. However, nuclei with 1–15-nm radii are difficult to stabilize under typical precipitation conditions.

ACKNOWLEDGMENTS

This work was partially supported by a grant from the International Fine Particle Research Institute. We thank E. Lestan for help with the precipitation of titania. We also gratefully acknowledge stimulating conversations with W. Klemperer.

REFERENCES

1. La Mer, V. K. *Ind. Eng. Chem.* **1952**, *44*, 1270.
2. Matijevic, E. *Acc. Chem. Res.* **1981**, *14*, 22.
3. Matijevic, E. *Ann. Rev. Mater. Sci.* **1985**, *15*, 483.
4. La Mer, V. K.; Dinegar, R. H. *J. Am. Chem. Soc.* **1950**, *72*, 4842.
5. Kerker, M.; Daby, E.; Cohen, G. L.; Krab, J. P.; Matijevic, E. *J. Am. Chem. Soc.* **1963**, *67*, 2105.
6. Fitch, R. M.; Kamath, V. K. *J. Colloid Interface Sci.* **1976**, *54*, 6.
7. Lichti, G.; Gilbert, R.; Napper, D. H. *J. Polymer. Sci.* **1983**, *21*, 269.
8. Feeney, P. T.; Napper, D. H.; Gilbert, R. G. *Macromolecules* **1984**, *17*, 2570.
9. Croucher, M. D.; Winnik, M. A. In *An Introduction to Polymer Colloids*; Kluwer Academic: Dordrecht, Netherlands, 1990.
10. Bogush, G. H.; Zukoski, C. F. In *Ultrastructure Processing of Advanced Ceramics*; Mackenzie, J. D.; Ulrich, D. R., Eds.; Wiley: New York, 1988; p 477.
11. Bogush, G. H.; Dickstein, G. L.; Lee, K. C.-P.; Zukoski, C. F. *Mater. Res. Soc. Symp. Proc.* **1988**, *121*, 57.
12. Bogush, G. H.; Zukoski, C. F. *J. Colloid Interface Sci.* **1990**, *142*, 1; **1990**, *142*, 19.
13. Zukoski, C. F.; Chow, M. K.; Bogush, G. H.; Look, J.-L. In *Better Ceramics Through Chemistry IV*; Zelinsky, B. J. J.; Brinker, C. J.; Clark, D. E.; Ulrich, D. R., Eds.; Materials Research Soc.: Pittsburgh, PA, 1990.
14. Look, J.-L.; Bogush, G. H.; Zukoski, C. F. *Faraday Disc., Chem. Soc.* **1990**, *90*, 345.
15. Stober, W.; Fink, A.; Bohn, E. *J. Colloid Interface Sci.* **1968**, *26*, 62.
16. Matsoukas, T.; Gulari, E. *J. Colloid Interface Sci.* **1988**, *124*, 252; **1989**, *132*, 13.
17. Klemperer, W. G.; Ramamurthi, S. D. In *Better Ceramics Through Chemistry III*; Brinker, C. J.; Clark, D. E.; Ulrich, D. E., Eds.; Materials Research Society: Pittsburgh, PA, 1988.
18. Flory, P. J. *Principles of Polymer Chemistry*; Cornell University Press: Ithaca, NY, 1953.
19. Bailey, J. K. In *Proceedings of the 47th Annual Meeting of the Electron Microscopy Society of America*; Bailey, G. W., Ed.; San Francisco Press: San Francisco, CA, 1989; p 434.
20. Bailey, J. K.; Mecartney, M. L. *Mat. Res. Soc. Symp. Proc.* **1990**, *180*, 153.
21. Nielsen, A. E. *Kinetics of Precipitation*; Pergamon: New York, 1964.
22. Russel, W. B.; Saville, D. A.; Schowalter, W. R. *Colloidal Dispersions*; Oxford University Press: Oxford, United Kingdom, 1989.
23. Harris, M. T.; Branson, R. R.; Byers, C. H. *J. Non-Cryst. Solids* **1990**, *120*, 397.
24. Santacesaria, E.; Tonello, M.; Storti, G., Pace, R. C.; Carra, S. *J. Colloid Interface Sci.* **1986**, *111*, 44.
25. Bogush, G. H., Ph.D. Dissertation, University of Illinois, 1990.
26. Barringer, E. A.; Bowen, H. K. *J. Am. Ceram. Soc.* **1982**, *65*, C199.
27. Barringer, E. A.; Bowen, H. K. *Langmuir* **1985**, *1*, 414, **1985**, *1*, 428.
28. Jean, J. H.; Ring, T. A. *Langmuir* **1986**, *2*, 251.
29. Jean, J. H.; Ring, T. A. *Colloids Surf.* **1989**, *29*, 273.
30. Diaz-Gomaz, M. I.; Gonzalez Carreno, T.; Serna, C. J.; Palaios, J. M. *J. Mater. Sci. Lett.* **1988**, *7*, 671.

Part 5

Silica Gels and Powders

William A. Welsh
W.R. Grace & Company

Synthetic amorphous silicas are of wide interest not only to those who study silica syntheses and characterization, but also to the many users of silicas. An understanding of the chemistry and structure of these solids is necessary to better understand their functionality. Synthetic amorphous silicas have been found to have remarkable utility in a variety of applications. The diversity of end uses (adsorption, reinforcing, dentifrice abrasion, catalysis, catalyst transport, rheology modification, flatting of coatings, antiblocking films, insulation, anticaking, corrosion inhibition, and chromatography, to name several) is attributable to the microscopic and macroscopic structure of silica powders and the chemistry of the solid silica surface. It has thus been of great importance to understand the structure and chemistry of silicas to optimize performance in specific end uses. The ability to control the synthesis and formation of these solids has permitted this rather simple chemical composition to be available with a vast array of properties.

Central to the understanding of amorphous silicas are the structure and chemistry of the fundamental building block, the micelle, and the nature of the three-dimensional array of these building blocks. The three general classes of amorphous silicas — silica gels, precipitated silicas, and fumed silicas — all are composed ultimately of micelles, silicate tetrahedra polymerized in random geometries that form nearly spherical units of varying diameters. At the surface of the micelle, the chemistry can be complex; some tetrahedra may not be fully condensed, so hydroxyls or, in organic-derived silicas, —Si—OR species are exposed. Once these micelle units are formed, the interactions between micelles eventually create the three-dimensional structure of the solid. In fumed silicas this interaction is small and the micelle is the fumed silica particle. The interaction between fumed silica particles is, however, not insignificant, and in fact it leads to many unique properties and applications for fumed silicas. In silica gels and precipitated silicas, the micelle interaction proceeds through the condensation of —Si—OH groups between micelles and, depending upon the extent of this reaction and further intergrowth, leads to a porous network between the interconnected micelles. The pore structure will be determined by the micelle size distribution and the packing array of the micelles. The richness of the chemistry to be practiced to control this third level of structure is clear. Characterization of the porous structure is often more important than characterization of the solid because the porosity creates the functionality.

The final level of structure, the solid particle size and morphology, is first controlled by the geometry of the initial sol when gelation occurs. Thus, very large to micrometer-size silica gel particles are possible. Precipitated silica particles generally are grown into the 1–100-μm range. Fumed silica particles are typically 100–600 Å in diameter. All three solids can be further processed to form particles of varying sizes and shapes by mechanical or chemical means.

The chemistry of the silica surface is dependent upon the content of —OH or —OR groups. When the surface is dehydroxylated, the chemistry of the remaining siloxanes is markedly different than when fully hydroxylated.

Because silica's hydroxylated or dehydroxylated surface is relatively neutral in acid–base chemistry, and otherwise relatively inert, it can be strongly affected by other surface species such as salts, acids, or bases adsorbed or not washed out following solid formation. Moreover, any co-gelled oxides such as Al_2O_3 can dramatically alter the surface chemistry, leading to whole new families of properties. Surface chemical treatments can be applied to change the properties of the silica surface and strongly influence functionality.

Thus, a large technology base has arisen for the synthesis of amorphous silicas based on early colloidal chemistry observations. These materials have been used and developed for numerous applications, both as produced and modified with other components. The chemistry and technology of making amorphous silicas has been developed to tailor the desired material properties for end uses. The resultant solids can be characterized, and often more importantly their pore structure must be characterized to properly understand their properties and functionalities. Of course, the chemistry of these often high-surface-area solid surfaces must also be understood for synthesis and use. With an appreciation of these points, a basic understanding of amorphous silicas has begun.

43 Synthetic Amorphous Silicas

U. Brinkmann, M. Ettlinger, D. Kerner[1] and R. Schmoll
Degussa AG

CONTENTS

Pure silica (synthetic silicon dioxide) in powdered form is discussed. After a brief history, this product group is described from production to applications. A classification of the different synthetic silicas is given, and the principal differences between pyrogenic and wet-process products are illustrated. After-treated silicas are also discussed. Various applications of synthetic silicas are described in detail. Questions about useful handling methods as well as toxicology are addressed.

INTRODUCTION AND HISTORY

Synthetic amorphous silica (SAS) became industrially relevant in the second half of the 20th century. From all the developments mainly three different manufacturing processes are now used on an industrial basis. The ideas originated in North America and Germany (see Table 43.1). Ferch in [1] provides an overview and a classification of the different types of silica.

Electric arc silicas [2] were first described in 1887, and a detailed account by Potter has been available since 1907 [3]. Industrial use of this process involving rather high electricity costs, became possible only after further developments by the BF Goodrich Company [4] and production was performed by Degussa [5,6] from 1972 until 1995.

Silica gels were first mentioned in 1861 by Graham [7]. Commercial production began at the Silica Gel Corp. with the process invented by Patrick [8]. The first silica aerogels were made by Kistler in 1931 [9], and production was started in 1942, using supercritical drying conditions [10]. Recently, Cabot established an automated new production process using subcritical drying, based on an idea published in 1992 by Smith, Brinker and Desphande [11].

The production of Hi-Sil, a silicate with a high silica content, started in 1946 [12]. The first "pure" precipitated silica was brought onto the European market in 1951 and was called Ultrasil VN 3 [13].

Fumed (or pyrogenic) silica was successfully produced for the first time by the original flame hydrolysis process (Aerosil-process) in 1942 [14]. The details of this process were published in 1959 and later [15,16]. In 1955, Flemmert [17] succeeded in exchanging the $SiCl_4$ used in the Aerosil process with SiF_4. SiF_4 is available as a by-product from the phosphate production from Apatite. This Fluosil-process was used in Sweden for about 15 years, and a factory belonging to Grace commenced production in Belgium with this method in 1990 but closed down soon after.

[1]For communication use following address: Degussa AG, Aerosil & Silanes, Dr. Dieter Kerner, Rodenbacher Chaussee 4 D – 63457 Hanau-Wolfgang, Germany

TABLE 43.1
Historical Overview of SAS

Process	Raw materials or after-treatment	Inventor/Company	Year	Commercial product name
Thermal silicas				
Flame hydrolysis	$SiCl_4 + H_2 + O_2$	Kloepfer	1941	Aerosil, Cab-O-Sil
Electric arc[a]	Quartz + Coke	Potter	1907	Fransil EL, TK 900
Plasma[a]	Quartz + Water	Lonza	1972	Experimental Products
Silicas from a wet process				
Precipitation	Sodium silicate + Acid	Degussa, PPG	1940	Hi-Sil, Sipernat, Ultrasil, Zeosil
Gels	Sodium silicate + Acid	Grace, Ineos	1919	Gasil, Syloid
Aerogels	Sodium silicate + Acid	Kistler	1931	Santocel, Nanogel
After treated silicas				
Coating	Wax coating		1970	Acematt, Syloid
Chemical after-treatment	Surface reaction with silanes		1962	Aerosil, Cabosil, Sipernat

[a]Discontinued.

Fumed silica must not be confused with "silica fume," a by-product of power plants and metallurgical processes. Being a by-product, this silica fume has no well defined properties and therefore no valuable applications. In the past, fly ashes and silica fume were given off into the atmosphere on a virtually uncontrolled basis. With improved pollution control the amount released to the atmosphere could be distinctly reduced.

Synthetic silicas in powder form are used in a wide range of applications and are part of our day to day life (see Chapter 4).

Worldwide production capacities in 1990 were estimated to be about 1 million metric tons per annum. In 2000 they were at about 1.3 million metric tons.

PROPERTIES, CHARACTERIZATION, AND SURFACE CHEMISTRY

All synthetic amorphous silicas are white powders, chemically seen pure SiO_2. They are amorphous and no crystallinic part can be detected (detection limit 0.5%).

Depending on the production process they have tapped densities (DIN/ISO 787/XI) ranging from 50 g/l to 600 g/l. They all consist of primary particles, aggregates and agglomerates (DIN 53 206). During the production process, primary particles grow together to form chemically bonded aggregates. These aggregates then form agglomerates held together by van der Waals forces and hydrogen bonding. In powder form, only agglomerates are existing.

Primary particles can be identified as individual particles by means of suitable analytical methods (e.g., TEM, Figure 43.1).

Typically, these spherical primary particles have a diameter in the range of 2–20 nm [18]. Aggregates are assemblies of primary particles which are grown together in the form of chains or clusters. The aggregates are formed by the collision of particles, by particle growth and by the further deposition of silica onto these aggregates. In practice, the degree of aggregation defines the structure of the silica as aggregates can not be broken into primary particles by any dispersion process. On the other hand, when dispersed into a fluid agglomerates can be broken into aggregates. Typically, the mean particle sizes for agglomerates are between 2 and 250 μm for silicas in powder or microgranular form whereas granulated silicas have a mean particle size of up to 3000 μm.

One important method of characterization is the specific surface area, mainly measured by the BET-method [19] or modernized versions of this method [20]. The specific surface area of industrial precipitated silica varies widely from 30 m^2/g up to 900 m^2/g; for silica gels it is typically from 300 m^2/g to 1000 m^2/g and for fumed silicas it ranges from about 10 m^2/g to 400 m^2/g. BET measures both, the outer surface and the surface of accessible pores. The specific surface area is of importance especially in reinforcement, rheological behavior and adsorption applications.

The dibutyl phthalate (DBP) absorption (ASTM methods D 4222-83 and D 4642-87) and the dioctyl phthalate (DOP) absorption (ASTM methods D 2414) characterise the structure mainly of precipitated silicas and reflect the primary particle size and shape as well as aggregate size and shape. DBP/DOP absorption correlates with the structure of the silica. Common values are in the range of 175–320 g per 100 g silica.

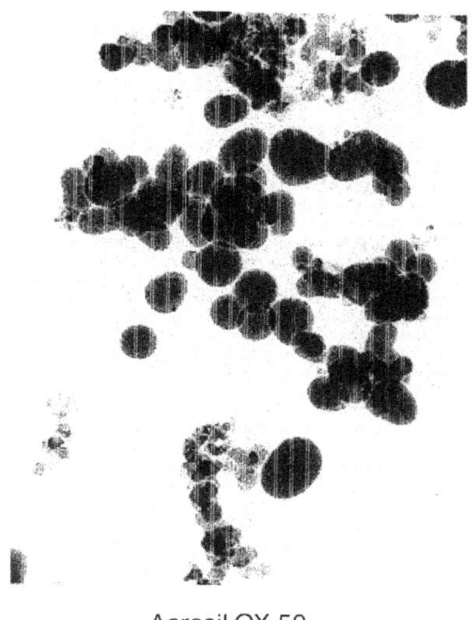

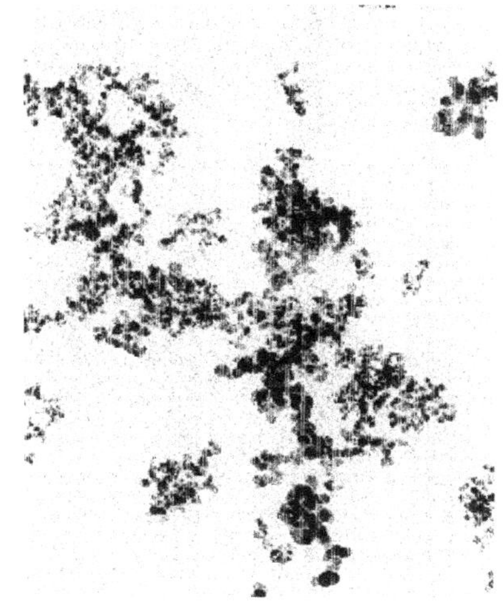

FIGURE 43.1 Particle size of fumed silica.

Another important parameter to describe precipitated silicas and silica gels is the porosity expressed by the pore size distribution since it describes the sorption characteristics of the silica. But it should be mentioned that the pore size distribution as well as the surface area are the results of an underlying parameter: the particle sizes of the aggregates. According to IUPAC [21] there are following classifications:

Micropores	Pore Diameter	0–2 nm
Mesopores	Pore Diameter	2–50 nm
Macropores	Pore Diameter	50–7500 nm

Precipitated silicas are typically macroporous materials where the mercury method is generally the best method available for the reliable determination of pore sizes above 30 nm. Washburn et al. [22] introduced the mercury intrusion method to measure the pore size of a porous silica.

Silica gels may largely be characterized by the same methods as applied for precipitated silicas, with the exception of porosity measurement. While precipitated silicas are macroporous and the Hg-intrusion method can be applied successfully, gels are mesoporous and N_2-adsorption is the method of choice. The BET-method is used to determine the overall surface area, BJH- and other algorithms are being applied to characterize the pore volume and pore size distribution characteristics. Relevant ASTM-methods are D 4222-83 and D 4682-87. The N_2-pore volume of typical gel products ranges from 0.3 to more than 2.0 ml/g with corresponding average pore diameters from below 2 nm to more than 30 nm.

Typically, precipitated silicas show a broad pore size distribution whereas silica gels exhibit a sharp maximum (Figure 43.2). Usually, fumed silicas have only outer surface without any pores. The outer surface is easily accessible in adsorption or desorption processes.

Hydrophilic precipitated silicas contain between 2–10 wt.% of physically adsorbed water, hydrophilic fumed silicas only up to 2%. After 2 h of drying in a drying cabinet at 105°C, the physically adsorbed water is removed. The weight difference before and after drying is equivalent to the mass of physically bonded water (method according to ISO 787/2, ASTM D 280) and is

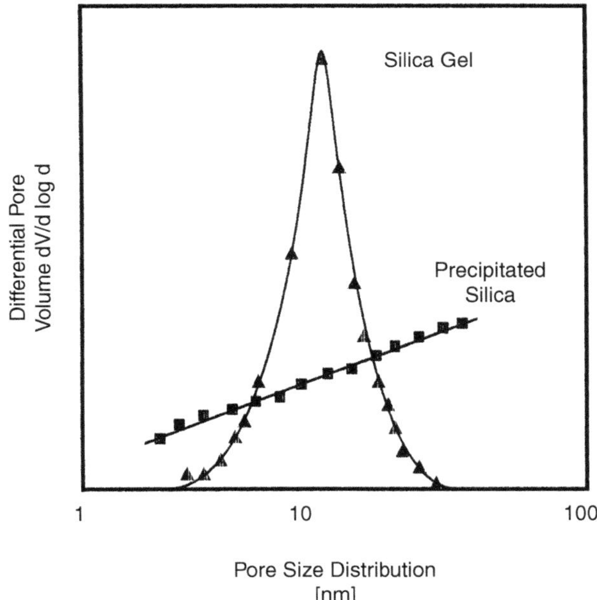

FIGURE 43.2 Pore size distribution of a precipitated silica and a silica gel.

called loss on drying. The amount of chemically bonded water is determined by heating a sample for 2 h at a temperature of 1000°C (loss on ignition, method according to ISO 3262/11, ASTM D 1208). The weight difference before and after heating at 1000°C is equivalent to the mass of physically bonded water (=loss on drying) plus the mass of chemically bonded water.

Generally, the pH value (ISO 787-9) of precipitated silicas is approximately 7, for hydrophilic fumed silica it is around pH 4.

The determination of the sieving residue (ISO 787/18) provides an indication of the difficult to disperse amount of the fractions in a precipitated silica.

In industrially produced precipitated silicas and silica gels besides SiO_2 as the main component, traces of other metal oxides like Na_2O, as well as sulfates and chlorides can be determined. In fumed silica, main trace is hydrochloric acid. The SiO_2 content is determined gravimetrically by fuming off with hydrofluoric acid. The analysis for metal oxides is then performed by means of atomic absorption spectroscopy (AAS) from the residue remaining from the fuming off process. Sulphate and chloride contents can be determined by potentiometric titration.

Typical values for the different silicas are given in Table 45.2.

SAS have two different functional groups on their surfaces: silanol (Si—OH) groups and siloxane (Si—O—Si) groups (Figure 43.3). These two groups substantially influence the surface properties and hence the application properties. The fully hydroxylated surface of precipitated silica has ~4.6 silanol groups per nm^2 [23–25], for silica gels the value is at ~5.5 [26] and for fumed silica ~2.5–3.5 [27] which results in the hydrophilic character of the products.

Whereas the siloxane groups are generally chemically inert, the reactivity of the silanol groups allows chemical surface modification [28,29]. Thus, the reaction with organosilanes [30–34], silicone fluids or chlorosilanes [35–38] leads to hydrophobic silicas.

Hydrophobization is carried out industrially both by wet processes (e.g., addition of organochlorosilanes to the precipitate suspension) and dry processes (e.g., the reaction of fumed silica with dimethyldichlorsilane [39]). A scheme of the reaction and change in surface from hydrophilic to hydrophobic shows Figure 43.4.

In coating processes, in contrast, no chemical reaction takes place; the coating agents are physically adsorbed on the silica surface. Wax coating with wax emulsions has become industrially important [40].

Since the early 1990s precipitated silica has been used as a high-performance reinforcing filler in rubber compounds for passenger car tires. To achieve a higher reinforcement silane coupling agents are needed. They provide a chemical link between precipitated silica and polymer [41–43] (see Figure 43.5). Most used silane

TABLE 43.2
Physicochemical Data of SAS

			Precipitated silica	Silica gel	Pyrogenic silica
CAS-Nr.			112926-00-8	112929-00-8	112945-52-5
Specific surface area	ISO 5794-1, Annex D	m^2/g	50–800	20–1000	50–400
Average primary particle size	TEM	nm	2–20	n.a.	7–40
Mean particle size[a]	—	μm	3–3000	0.1–5000	n.a.
Pore volume	IUPAC, App. 2,Pt. 1		Macroporous	Micro- and mesoporous	n.a.
Loss on drying	ISO 787-2	%	3–6	ca. 3	0.5–2
Loss on ignition	ISO 3362-11	%	3–12	5–6	0.5–2.5
pH-value	ISO 787-9		6–8	4–8	3–5
DBP-number	ASTM D 2414	g/100 g	50–350	n.a.	100–350
Tapped density	ISO 787-11	g/l	90–450	n.a.	50–150
SiO_2-content[b]	—	%	98–99	>99.5	>99.9
Al_2O_3[b]	—	%		<0.05	<0.05
TiO_2[b]	—	%			<0.03
Fe_2O_3[b]	—	%	<0.03		<0.003
Na_2O[b]		%	<1	<0.1	
HCl[b]	—	%	n.a.		<0.025
SO_3(sulfate)	—	%	<0.8		n.a.

[a]Various methods.
[b]Based on ignited substance.

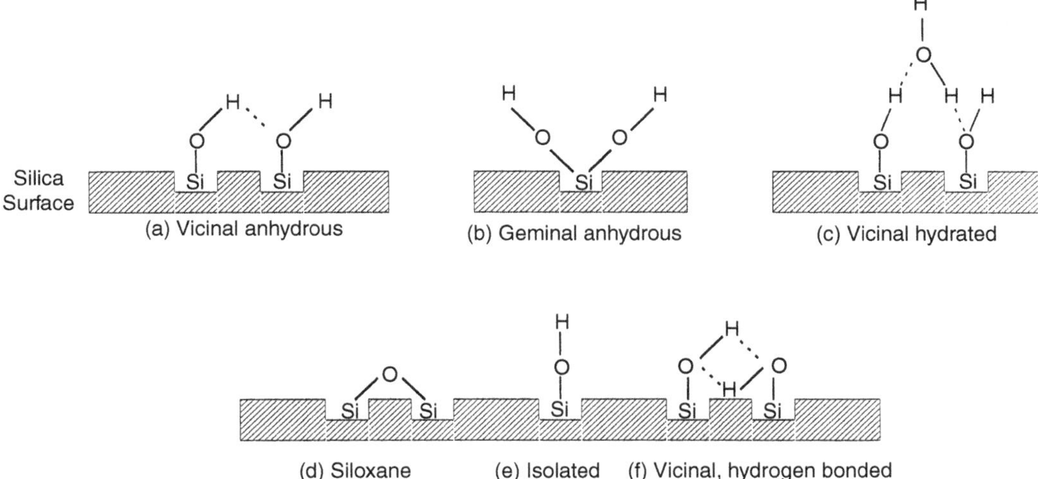

FIGURE 43.3 Different types of hydroxyl groups on silica surfaces.

for this application is bis(triethoxysilylpropyl)tetrasulfane (e.g., Si 69® from Degussa).

PRODUCTION PROCESSES

PRECIPITATED SILICA

Precipitated silicas have only been produced since the 1940s but they have grown to become the most important group of silica products on the basis of production tonnage. The worldwide production capacity of precipitated silica in 1999 was ca. 1,100,000 t, compared to ca. 400,000 t in 1970 [44]. The European production volume for precipitated silica for the year 2000 was 285,500 t [45].

The raw materials used for the industrial production of precipitated silicas are alkali silicate solutions, preferably sodium silicate, and acid. Silica is precipitated through a reaction of both components. Depending on the intended application of the silica precipitation parameters such as feed rates, stirring, shearing, temperature, pH, alkali content etc. can be adjusted allowing for a wide range of silica products with different characteristics, like for example, specific surface areas ranging from approximately 25–700 m^2/g. Typically, precipitated silicas are synthezised under alkaline conditions at pH values >7, while gels are typically produced under acidic conditions.

Industrially produced synthetic precipitated amorphous silicas are still made today through precipitation from diluted sodium- or potassium-silicate-solutions with the aid of mineral acids as sulfuric acid, hydrochloric acid or with the aid of carbon dioxide. In case of the use of sodium silicate and sulfuric acid the reaction follows Equation (43.1)

$$Na_2O \times 3.3\ SiO_2 + H_2SO_4 \longrightarrow 3.3\ SiO_2 + Na_2SO_4 + H_2O \qquad (43.1)$$

In general a module of 3.2–3.5 — that indicates the ratio between disodium monoxide and SiO_2-content in the sodium silicate — is used in order to reduce by-products. In all cases a corresponding by-product is produced that can be sodium sulfate, sodium chloride or sodium carbonate depending on the acid component used. In case of the use of potassium silicate the corresponding potassium salts develop. These by-products must be removed in the downstream process.

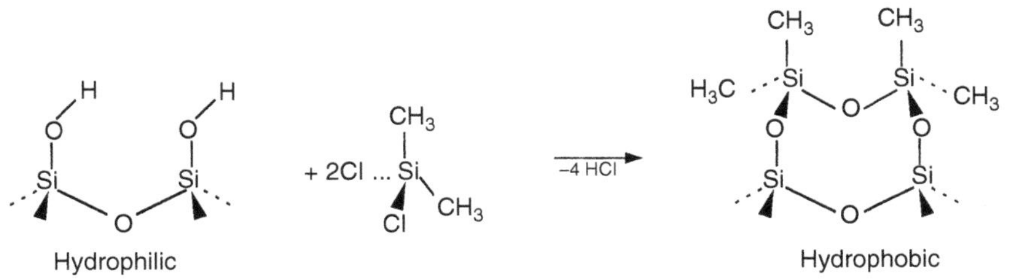

FIGURE 43.4 Reaction of the surface silanol groups with dimethyldichlorosilane.

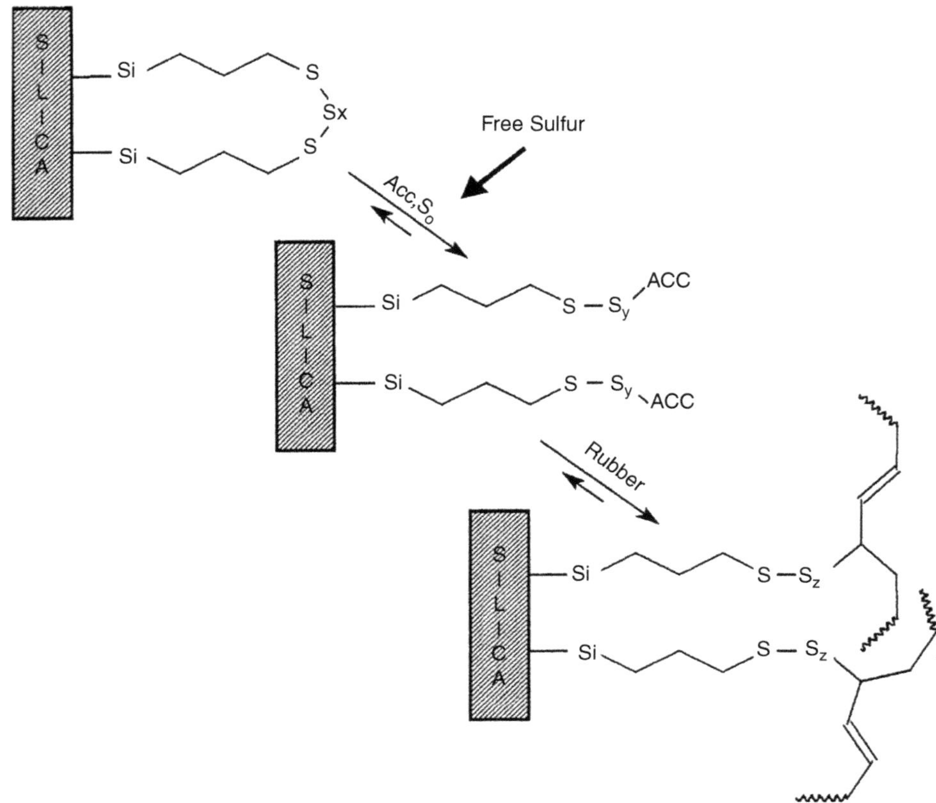

FIGURE 43.5 Simplified silica-rubber coupling reaction with Si 69.

Fundamentally, the polymerization reaction is according to Equation (43.2):

$$\text{Monomer} \longrightarrow \text{Dimer} \longrightarrow \text{Cyclic} \longrightarrow \text{Particle} \longrightarrow \text{Precipitated Silica} \tag{43.2}$$

The exact timing of the aggregation in the process as well as the duration of the particle growth can be controlled by the salt concentration, for example by adding sodium sulfate. The underlying mechanism of the process is described in detail in [46].

The production process consists of the following steps: precipitation, fitration, drying, grinding, and, in some cases, compacting and granulation.

So far only batch precipitation processes have attained economic importance [47,48], although continuous precipitations have also been reported [49]. In general, acids and alkali metal silicate solution are fed simultaneously into water in a stirred vessel building silica seeds. In the course of the precipitation, three-dimensional silica networks are formed, accompanied by an increase in viscosity. The networks are reinforced by further precipitation of oligomeric silica and grow further into discrete particles with a decrease in viscosity. The formation of a coherent system and thus a gel state is avoided by stirring and increasing the temperature.

The separation of the silica from the reaction mixture and the removal of the salts contained in the precipitate is done in filter aggregates such as rotary filters, belt filters, or chamber-, frame-, and membrane-filter presses.

Since the solid content of the filter cake products is only at 15–25%, about 300 to 600 kg of water must be evaporated for each 100 kg of final product. Depending on the desired properties of the end product, drying is carried out in spray, nozzle spray, plate, belt, or rotary dryers. Special product properties can be achieved by spray drying the filter cakes after redispersion in water or acid [50,51]. Figure 43.6 shows scanning electron micrographs of a nozzle spray-dried, unground silica and a conventionally dried, ground silica.

Various mills and, if required, classifiers [52] are used to control the desired fineness of the particles. Special degrees of fineness can be achieved by air-jet or mechanical grinding [53]. The silica is separated from the air in cyclones or filters.

Processes for the compression and granulation of precipitated silica have been developed [54,55] to reduce the volume for transportation and for certain uses and also to decrease the formation of dust on handling and processing.

Bags, big bags, containers, or transport silos are used for packaging and shipping. Information about the different methods of packaging and shipping is given in [56].

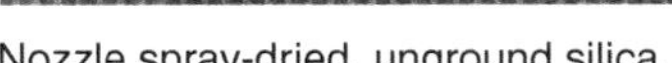
Nozzle spray-dried, unground silica

Conventionally dried, ground silica

FIGURE 43.6 SEM of different precipitated silicas.

In Germany the general dust limit of 4 mg inhalable dust per m^3 air must be adhered to when handling precipitated silicas. Similar regulations apply in other countries.

Silica Gel

The original process as developed by Patrick [57] is the basis of all modern silica gel processes. The main steps are synthesis (sol formation/gelation), washing/aging and drying followed by sieving, milling, or surface modification depending on the final product (Figure 43.7).

The reaction mechanism during gel formation differs from that of precipitated silicas and comprises sol formation followed by the gelation step, where the liquid precursor sol has the same chemical composition as the gel formed from it. This might be described by interaction of separate micelles in the sol first by hydrogen bonding, followed by condensation. More details may be taken from the overview in [58]. Removal of the excess salts by washing leaves a silica skeleton with a three-dimensional pore network. Process conditions during gelation, washing and subsequent washing define the size of the pore network. Typical silica gels exhibit pores in the mesopore regime (2–50 nm), which is comparable to the primary particle size of precipitated silicas.

Drying of the washed hydrogel leaves another option for structure modification [59]. Depending on the degree of surface tension of the solvent, the capillary pressure to be overcome during the drying step causes shrinking of the gel. Besides the nature of the liquid, the speed of drying is the main factor influencing the degree of shrinkage. As a rule of thumb slow drying causes more shrinkage and loss of pore volume than fast drying. If by proper choice of process conditions the openness of the structure is largely maintained, the gel is called an aerogel [60].

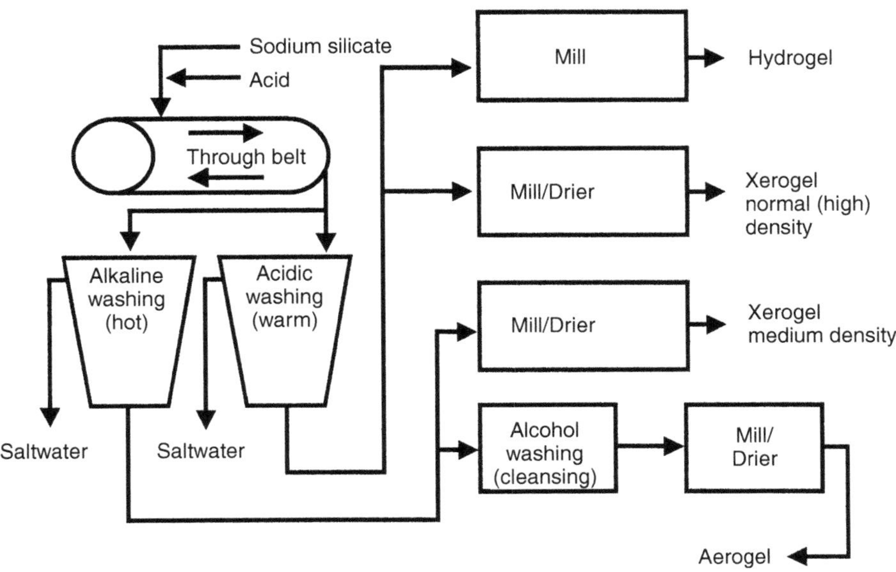

FIGURE 43.7 Industrial production process of silica gel.

Synthesis

The first synthesis step comprises the formation of a hydrosol, which is produced by the controlled mixing of water glass and (diluted or) strong acids, as sulphuric or hydrochloric acid.

The pH-value and SiO_2 content of this hydrosol are determined by the concentration of the raw materials and their mixing ratio. Typically acid excess is preferred, as under these conditions the intermediate sol is more stable and the process is less sensitive to feed fluctuations. During the sol-forming step an unstable intermediate-monomeric orthosilicic acid — is formed which then rapidly undergoes an acid-catalysed condensation reaction to form oligomers. When the molecular weight reaches ca. 6000, a sudden increase of both the viscosity and the modulus of elasticity is observed. This increase marks the transformation of the sol to a gel that will then further develop its internal structure.

In the hydrogel state, larger agglomerates are generated which are cross-linked to form an open, branched-chain structure. Choice of the gelation conditions can define the particle size and form of the hydrogel; industrial processes normally from either lumps or spherical beads.

Washing/Aging

During the subsequent washing process excess salts are removed in order to purify the gel, which in parallel also causes structural changes within the gel's framework. By choice of the washing conditions (e.g., pH, temperature, time) different specific surface areas can be achieved for the purified hydrogel. Washing can be performed in fixed beds or slurries, operating continuously or discontinuous.

The hydrogel formed has a continuous structure, giving a three-dimensional network of pores filled with water. The total volume of pores per mass-unit is called the pore volume and is a specific characteristic of the gel type. It is a main characteristic of silica hydrogels, that mechanical treatment like milling does not affect the pore structure. Even after ball-milling the hydrogel into submicron particles the structure is maintained [61].

Drying

Hydrogel can be used for a few applications but in most cases the gel must be dried. During the drying process, the surface tension of the solvent in the pores can act to shrink the hydrogel volume. During slow drying, as water is evaporated from a silica hydrogel, the structure collapses gradually due to the surface tension of water. Eventually, a point is reached where even though water is still evaporating the gel structure no longer shrinks. At this point the gel is called a xerogel. Fast drying can minimise the shrinkage — and the removal of water by solvent exchange followed by drying has the same effect.

Materials that are dried with negligible loss of pore volume are known as aerogel. First aerogels were made in the early 1930s by Kistler [62], who applied solvent exchange and supercritical drying techniques to remove pore fluid thus maintaining the pore structure. On larger scale silica aerogels were produced by Monsanto (Santocel®) in the 1950s, and BASF in the 1980s. Both companies discontinued aerogel production, mainly for cost reasons.

Several attempts were undertaken to replace the costly supercritical drying step by subcritical drying. Brinker and his co-workers [11] proposed a multistep hydrogel silylation and solvent exchange process that allows aerogel drying under subcritical conditions. This proposal led to intense process development work at Hoechst in Germany. The process know-how and pilot plant equipment were sold to Cabot, which announced the construction of a semiworks plant in Germany, that shall become operational by the end of 2002 [63].

A more comprehensive overview on Aerogels may be taken from the review articles [64] or [65].

PYROGENIC SILICA

Fumed silica is produced in a flame process using hydrogen, oxygen (air) and silicontetrachloride as raw materials according to Equation (3):

$$SiCl_4 + 2H_2 + O_2 \longrightarrow SiO_2 + 4HCl \qquad (43.3)$$

The "byproduct" HCl is recycled to produce silicontetrachloride and hydrogen from siliconmetal:

$$Si + 4HCl \longrightarrow SiCl_4 + 2H_2 \qquad (43.4)$$

Formally, in this closed loop SiO_2 (fumed silica) is produced by the oxidization of Siliconmetal:

$$Si + O_2 \longrightarrow SiO_2 \qquad (43.5)$$

As alternative raw materials, trichlorosilane and methyltrichlorosilane (from silicones' production) can be used. Schwarz [66] also used chlorine-free components like for example, hexamethyldisiloxane as raw material.

A scheme of the production process is shown in Figure 43.8.

Hydrogen, air and $SiCl_4$-vapor are fed to the burner forming fumed Silica and hydrochloric acid. The reaction products are cooled through heat exchangers before entering the solid/vapor separation which can be done by cyclones or filters. As there is still some HCl on the surface of the fumed silica, this part is removed in a deacidifier which can be for example, a rotary kiln or

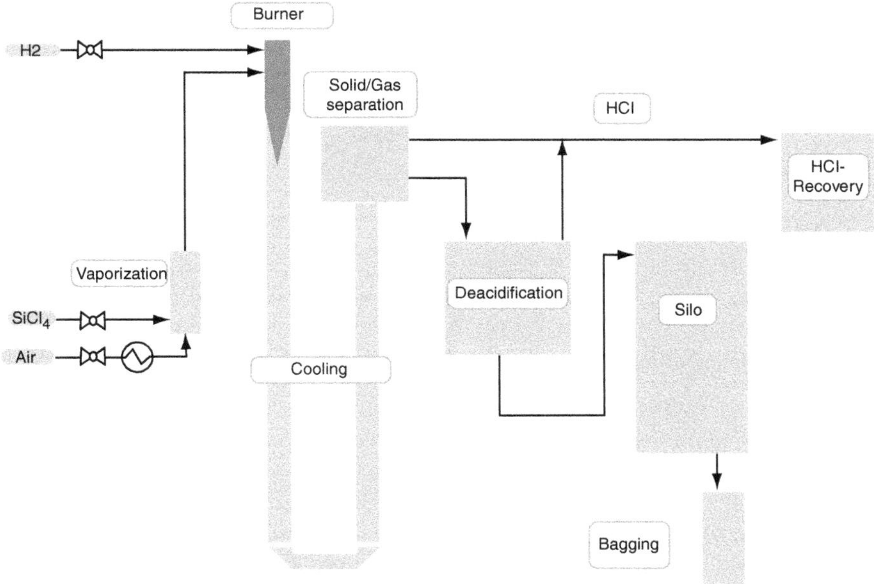

FIGURE 43.8 Industrial production process of fumed silica.

fluidized bed reactor. All the HCl is absorbed in the HCl-recovery unit. The fumed silica is stored in a Silo and — after densification — filled in bags, big bags, containers or silo trucks for shipment.

The properties of fumed silica can be determined during the production process by adjusting the feed rates of the three components — hydrogen, air and silane — to the burner. Basically, the specific surface area (particle size) is determined by the flame temperature.

APPLICATIONS

The most prominent applications for the different types of SAS are shown in Figures 43.9–Figure 43.11 [45]. Although these figures only represent the European market, there are no big deviations for the world-wide consumptions.

The most important field of use for precipitated silica is the reinforcement of elastomer products like tires, shoe soles and mechanical rubber goods (e.g., conveyor belts, mats, seals, etc.). In terms of tire performance, it is essential to use extremely good dispersible silica grades such as Ultrasil® 7000 GR (Degussa) or Zeosil® 1165 MP (Rhodia) together with a silane coupling agent (see Chapter 2). The use of these silicas in tread compounds leads to significant improvements in rolling resistance and wet traction of tires without compromising tread wear ("Magic Triangle"). Precipitated silica is now widely used in Europe, and demand in North America and Asia is growing owing to the need to further reduce rolling resistance to cut fuel consumption and CO_2 emissions. A typical passenger tire is shown in Figure 43.12.

Hydrophilic and hydrophobic fumed silicas are mainly used to improve the mechanical strength of RTV-, HTV- or liquid silicone rubber. Fumed silicas with higher surface area usually lead to higher transparency of the silicone rubber which is important for special applications

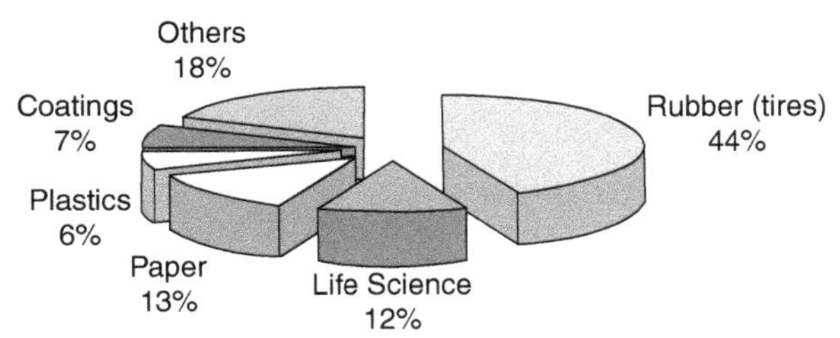

FIGURE 43.9 European consumption of precipitated silica in 2000.

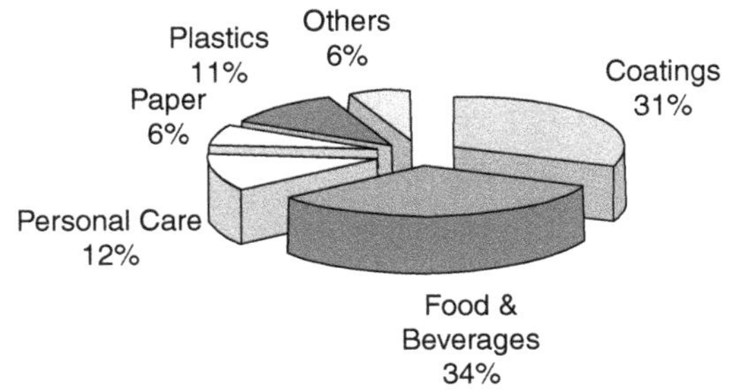

FIGURE 43.10 European consumption of silica gel in 2000.

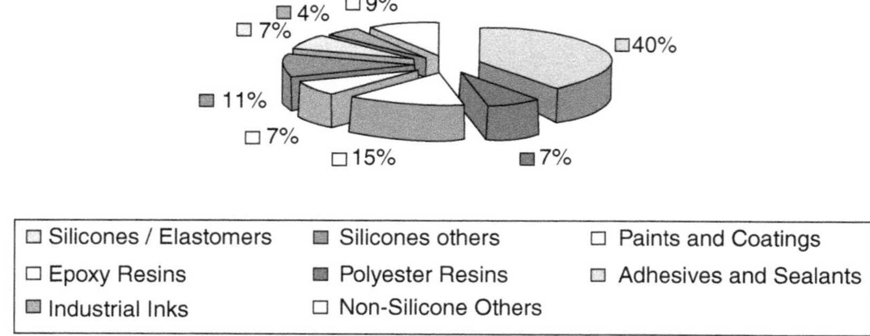

FIGURE 43.11 European consumption of fumed silica in 2000.

like medical tubes. In some applications with lower demand also precipitated silicas are used. The cross linkage between the polymer chains and the interaction of the silica with polymer matrix are both responsible for producing silicone rubber vulcanisates of mechanical strength. This reinforcing effect of silicas has been extensively investigated over the past years [67–69].

Fumed silicas are not only essential fillers in relatively new economically expanding polymers like special silicone rubber [70], they are also of importance in natural and synthetic rubber for applications, where outstanding properties regarding mechanical strength and temperature stability or reduced permeability for gases and liquids are required.

Precipitated silicas but also fumed silicas are widely used as carrier for liquids and semiliquids and as free flow agent for powder formulations, particularly of hygroscopic and caking substances. Hydrophobic pyrogenic silicas are used in technical products like fire extinguishing powders, pigments or powder coatings, but also in products for skin contact applications like cosmetic powders. Important parameters in these applications are the absorption capacity, the kinetics of the absorption, good flowability and a low dust content of the carrier silica (good mechanical stability). Well-performing precipitated silicas absorb liquids or solutions to provide powder concentrates that contain up to 70% of the liquid [71]. Liquid materials can be converted into a dry, free flowing product that can be mixed in any ratio with other dry substances. Typical materials converted into powder form are listed in Table 43.3 [72].

A special case is the use of hydrophobic pyrogenic silica in toners. There they not only prevent clogging, but also control or stabilize the triboelectric charge. Conventional toners charge negatively and therefore need a negatively charged hydrophobic fumed silica. The amount of remaining silanol groups after the hydrophobization treatment controls the negative charge. New toner generations have positive charges. For these toners,

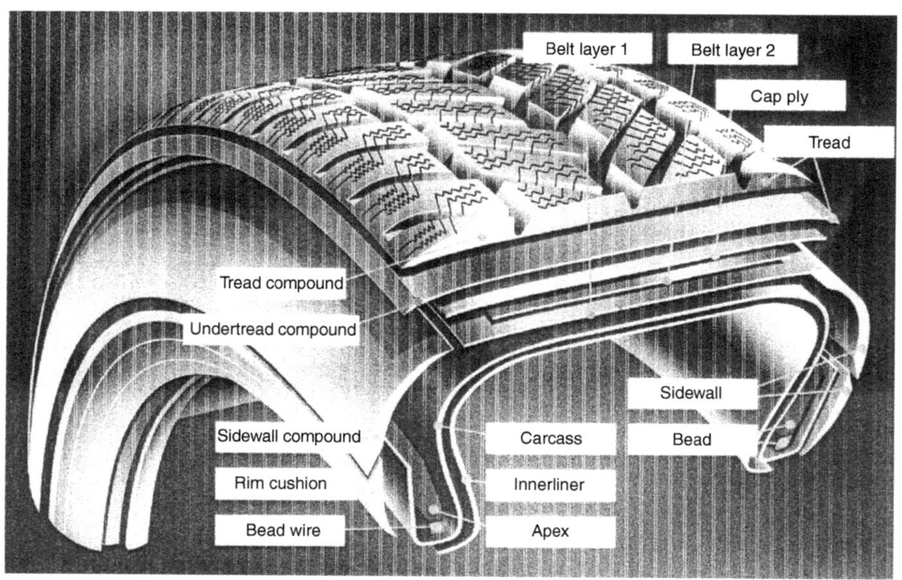

FIGURE 43.12 Typical passenger tire.

TABLE 43.3
Use of Precipitated Silica as Carrier for Other Products

Product	Amount in %
Formic acid	50
Cholinechloride 75%	33
Ethoxyquine	33
Hop extract	33
Peroxides (organic)	40
Plant protectives	5–35
Propionic acid	35
Vitamin E acetate	50

tailor-made hydrophobic fumed silicas with the required positive charges are made. Even small amounts between 0.5 and 1.5% hydrophobic fumed silica are sufficient to get the desired effects in toners.

In coatings all three types of SAS can be used. Special precipitated silicas and gels are used as matting agents for the matting of paints and varnishes. The matting effect results from the roughening of the surface on a microscopic scale, such leading to a diffuse light scattering. As already a small amount of silica is sufficient to get the wanted effect, it does not negatively affect the good properties of the coating system. Fumed silica is mainly used for rheology control purposes and as an anti-settling agent for pigments, because in many systems not only thixotropy but additionally a yield point is established [73].

Plastic films like polyethylene (PE) foils often tend to stick to each other which can be prevented by an addition of the appropriate silica as an anti-blocking agent. In PE battery separators, precipitated silicas are used to generate the porosity of the separator. The silica also leads to a more hydrophilic surface for better wetting of the PE separators. In lead acid batteries for cars it is essential to thicken the sulfuric acid with fumed silica: the gel-like acid will not get out of the battery in case of an accident. It has been found very early that hydrophilic fumed silicas can thicken liquids. The lower the polarity of the liquid, the stronger the thickening effect. In this case polarity means the tendency to form hydrogen bonds. In pure water hydrophilic fumed silicas have a relatively weak rheologic efficiency. In special cases for example, at pH values of below 0 or in the presence of electrolytes they can be extremely efficient. Therefore, they are used to thicken the sulfuric acid in lead batteries or the vanadium salt solutions in vanadium batteries.

Fumed silicas do not only increase the viscosity of a liquid, they also install a strong thixotropy [16]. For obtaining an optimal and reproduceable thickening effect an intensive dispersion is required, which can be done by a dissolver, a high speed rotor-stator-mixer, a pearl mill, a three roll mill or a kneader, depending on the system and wanted degree of dispersion. Fumed silicas are also used in polyester and epoxy resins for thixotropy control providing improved handling properties for these materials. For example in the production of boat bodies made of unsaturated polyester (UP) the clear coat contains usually approximately 0.8% and the gel coat approximately 2% of a hydrophilic fumed silica with a specific surface area of approximately 200 m^2/g. That prevents sagging and thicker layers can be applied in one working cycle [74].

Today the rheology of many liquid systems is controlled by hydrophilic or hydrophobic fumed silicas. An interesting example are 2P-epoxy resins. The resin can be thickened effectively with a hydrophilic fumed silica, but when the polyamine hardener is added, the viscosity raises and after a short time it drops tremendously and the application is impaired or even impossible. To overcome this problem surface treated silicas have been developed, which do not show detrimental interactions with the polyamine and which have a good rheological effect [75] (Figure 43.13).

Different theories exist about the thickening mechanism of fumed silicas. One of the first was the so-called chicken wire structure. That means the silica particles interact with each other via their silanol groups and form a three dimensional structure, which reduces the mobility of the liquid molecules. Under mechanical impact like shearing or shaking the structure is destroyed and the viscosity of the system decreases. After the end of the mechanical impact, the three dimensional network re-establishes itself and the viscosity increases again as a function of time. This mechanism may be valid in simple nonpolar liquids. In liquid mixtures or polymer solutions it is much more complicated and the adsorption pattern on the fumed silica surface seems to play an important role [76].

The past decade has already seen tremendous technological advances in non-impact printing (NIP) techniques. Most of the ink used in ink jet printing contains water soluble organic dyes and dispersed pigments with the majority of the typical ink formulation consisting of water. Once deposited on the substrate, the ink must dry quickly to avoid spreading through the paper fibers to the adjacent printing pixel. With accelerated print speed and reduced drop size, fast drying becomes crucial. A key to meeting these requirements is coating the surface of the substrate (e.g., paper) with silica based materials. These coatings allow for rapid ink absorption, thereby promoting sharp edge acuity, spheric and defined spreading of the ink drop, no shine-through or strike-through of the inks, and excellent image density. Due to the internal macroporic porosity of precipitated silicas

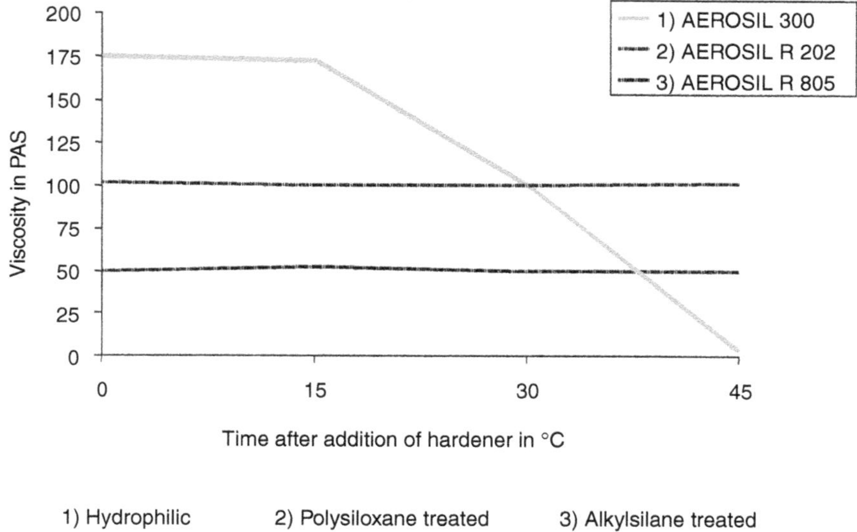

FIGURE 43.13 Viscosity of different fumed silicas in a 2P-epoxy resin.

these materials show perfect absorption behaviour. The available silanol groups on the surface contribute to hold the inks on the surface, keeping diffusion to a minimum.

Precipitated silicas and silica gels are used in toothpastes as abrasives and as rheology stabilizer. The hardness, the structure and particle size distribution of the silicas are the properties which control the abrasiveness and the rheology of the toothpaste. For transparent toothpastes also fumed silicas are in use.

Besides the above mentioned applications there are many other different uses of precipitated silicas for example, in plastics, construction, pharmaceuticals etc. Reference [77] provides an extensive overview about the variety of these different applications.

In applications like anti-slip, where mechanical stability of the micronized particle plays an important role, silica gel products are of advantage. As an example, silica gels withstand more easily the rather severe master-batch and extrusion conditions for PE- or PP-film production. As the particle size and structure is maintained, gel products are preferred versus precipitates.

Silica gels are applied for the stabilisation of beer, using the selective adsorption properties of gel products. Another adsorption application is the use of silica hydrogels in edible oil refining, where the gel acts as an adsorbent for phospholipids and color.

Another big group of applications that cannot be covered by other silicas is the usage of (coarse) granular gels for moisture adsorption, in chromatography or as catalyst carrier.

As fumed silica is a very finely divided powder with relatively soft particles, it is a smooth abrasive for extremely fine polishing. Dispersions of fumed silicas have been used for the final polishing of metal moulds or silicon wafers (CMP = chemical mechanical planarization). By this special process the interlayer dielectrics of silicon dioxide and the various metal layers for installing the electric conduction lines are planarized with a so-called CMP slurry to guarantee completely plane surfaces during the production of semiconductor wafers. The CMP slurry is in the most cases a dispersion of 10 to 15% fumed silicon dioxide in water, with amines as stabilizers and a pH value between 9 to 12 controlled by potassium hydroxide. For metal layers different CMP slurries are used which can contain acidic compounds, oxidizing and chelating agents. The more layers such a chip consists of, the more planar the individual layers must be and the more important the CMP process becomes to ensure the proper function [78].

In the field of high temperature thermal insulation advantage is taken of the special powder properties and the amorphous character of fumed silicas. In principle, mixtures of fumed silicas, an opacifier to reflect the heat radiation and a small quantity of mineral fibers for reinforcing are used for temperatures up to 1000°C. It is necessary to densify the mixture to approximately 200 g/l to obtain the minimum of the superposition of gas and the solid state thermal conduction. Important is also the absence of mineralizing ions, like sodium or potassium, to prevent sintering effects. Commercial thermal insulation systems based on this principle have thermal conductivities of approximately 25 mW/mK at a mean temperature of 200°C and of approximately 30 mW/mK at a mean temperature of 400°C. Therefore, they are used when a high insulation efficiency must be obtained within a small volume. The areas of application are storage heaters, electric cooker hobs, high temperature furnaces, insulations of pipes in ships and airplanes or aircraft turbines.

A very high purity is one of the outstanding features of fumed silicas. They can be used as raw material for the manufacturing of silica glasses for optical fibers or quartz crucibles. A problem is the quite low powder density of the pyrogenic silicas. Even densified products have a tapped density of only approximately 120 g/l, which is too low for a sinter process. For this reason highly concentrated dispersions are used, with pyrogenic silica alone or in combination with other silica sources like tetramethoxy silane and additives [79,80]. Fumed silicas can also be densified by wet processes. In this case it is possible to obtain powder densities, which are suitable for direct sinter processes [81]. Pyrogenic silicas can also be used for the synthesis of silicates, zeolites or special glasses.

HANDLING OF SAS

Silica dust is a nuisance at the working place. In most countries maximum concentrations are set by environmental regulations, for example, is the current value set at 4 mg/m^3 of inhalable dust, both in the United States and Germany. For this reasons, proper handling is of great importance. SAS are pneumatically conveyed in closed systems like from production to the silo where it is bagged or shipped in bulk. Dense or diluted phase conveying systems are most commonly used. The receiving vessel must be deaerated via filters.

Bagging is done in paper bags or plastic bags by using standard bagging machines. As none of these machines can avoid a slight powder loss during removal of the bag from the nozzle of the bagging machine, local suction at the bagging machine and high air ventilation are of importance.

TOXICOLOGY

In principle, SAS can affect the health of a human being by skin contact, inhalation, or oral intake. In working environments dermal and inhalation toxicity are of major importance. Therefore, numerous corresponding investigations have been performed and papers published [82–87] which will be discussed below. Also, reference is made to papers giving general overviews of the toxicological behavior of SAS [88,89].

Skin Contact

A great number of animal studies showed that the topical application of SAS did not lead to irritations. Also, eye contact did not cause irritation. Occupational physicians occasionally reported dryness of the skin which led in some cases to degenerative eczema. This must not be misinterpreted as a sign of sensitization or allergy as the reactions may be avoided by skin care. No signs of systemic toxicity were observed [88,89].

The results obtained for Aerosil, for example, are compiled in reference [90], which includes numerous bibliographic references.

Inhalation

In Germany regular medical examinations were carried out on 131 employees exposed to precipitated silica for a maximum of 38 yr [90]. No signs of silicosis or any other dust-related lung desease were observed. After exposure to the fumed silica Aerosil for 14 yr [91] and for more than 30 yr [90], no silicosis had occurred. Furthermore, no signs of toxic effects were observed. On the basis of experience gained to date, no harmful side effects are to be expected with silica gels either. In acute inhalation tests with rats no clinical signs of toxicity were observed. At necropsy, no macroscopic organ abnormalities were found. No mortality occurred during the exposure and observation period [84].

Oral Intake

The effect of orally ingested SAS plays a subordinate role in industrial toxicology. Based on the available data synthetic amorphous silica may be judged to be not toxic by the oral route [89]. This fact is also taken into account by the WHO, which says that the daily intake of SiO_2 is unlimited.

REFERENCES

1. H. Ferch, *Chem. Ing. Techn.* **48**, 922 (1976).
2. Anonymous, *Am. Chem. J.* **9**, 14 (1887).
3. US 875,674; US 875,675, **1907**, H. N. Potter,.
4. (a) DE 1,034,601, **1955**, BF Goodrich (b) USP 2,863,738, **1958**, BF Goodrich.
5. DE 1 180 723, **1963**, Degussa.
6. DE 1 933 291, **1969**, Degussa.
7. T. Graham, *J. Chem. Soc.* **17**, 318 (1864).
8. US 1,279,724, **1918**, Silica Gel Corp.
9. S.S. Kistler, Nature (London) **127**, 741 (1931); *J. Phys. Chem.* **36**, 52 (1932).
10. J.F. White, *Chem. Ind.* **51**, 66 (1942).
11. WO 94/25149, R. Desphande, D.M. Smith, C.J. Brinker.
12. A.E. Boss, *Chem. Ing. News* **27**, 677 (1949).
13. Degussa Archive, Degussa AG, Germany.
14. DE 762,723, **1942**, Degussa.
15. L.J. White, G.J. Duffy, *J. Ind. Eng. Chem.* **51**, 232 (1959).
16. E. Wagner, H. Brünner, *Angew. Chem.* **72**, 744 (1960).
17. DE 1,208,741, **1955**, NYNÄS Petroleum.
18. Kirk-Othmer, *Encyclopedia of Chemical Technology, Fourth Edition*, Vol. 21, Silica, p 998.

19. S. Brunauer, P.H. Emmet, E. Teller, *J. Am. Chem. Soc.* **59**, 2682 (1936).
20. J. Seifert, G. Emig, *Chem.-Ing. Techn.* **59**, 475 (1978).
21. IUPAC Manual of Symbols and Terminology, Appendix 2, Pt. 1, Colloid and Surface Chemistry, *Pure and Applied Chemistry*, **31**, 578 (1972).
22. E.W. Washburn, *Proc. Nat. Acad. Sci.*, USA **7**, 115 (1921).
23. L.T. Zhuralev, *Langmuir* **3**, 316 (1987).
24. L.T. Zhuralev, *Colloids and Surfaces A: Physicochemical and Engineering Aspects* **74**, 71 (1993).
25. E.F. Vansant, P. Van der Voort, K.C. Vrancken, Characterization and Chemical Modification of the Silica Surface, *Studies in Surface Science and Catalysis*, Vol. **93**, 88 (1995).
26. *Ullmann's Encyclopedia of Industrial Chemistry, 5th Edition* (1993), Wiley, New York, Vol. A 23, p. 631.
27. J. Mathias, G. Wannemacher, *J. Colloid Interface Sci.* **125**, 61 (1988).
28. A. Legrand: The Surface Properties of Silicas, J. Wiley & Sons, New York, **1998**, 286.
29. Eugene Papirer (Editor): Adsorption on Silica Surfaces, Marcel Dekker, New York, **2000**, ISBN 0-8247-0003-1.
30. S. Wolff, Kautsch. *Gummi Kunstst.* **41**, 675 (1998).
31. DE 1 172 245, **1963**, Degussa.
32. DE 2 729 244, **1977**, Degussa.
33. EP 0 466 958, Degussa.
34. EP 0 672 731, Degussa.
35. K. Albert, E. Bayer, B. Pfleiderer: *J. Chrom.* **506**, 343 (1997).
36. DE 1 074 559, **1959**, Degussa.
37. DE 2 628 975, **1976**, Degussa.
38. DE 19 757 210, Degussa.
39. EP 0 090 125, Degussa.
40. EP 0 341 383, **1989**, Degussa.
41. M.P. Wagner: The Consequences of Chemical Bonding in Filler Reinforcement of Elastomers, paper presented at the Colloques Internationaux, Sept. 24–26, **1973**, Le Bischenberg-Obernai, France.
42. F. Thurn, S. Wolff: Neue Organosilane für die Reifenindustrie, Kautsch. *Gummi Kunstst.* **28**, 733 (1975).
43. S. Wolff, Kautsch. *Gummi Kunstst.* **34**, 280 (1975).
44. *Ullmann's Encyclopedia of Industrial Chemistry, 5th Edition* (1993), Wiley, New York, Vol. A 23, Chapter 7.1.
45. CEFIC ASASP, BREF Working Group of Synthetic Amorphous Silica, **2002**, 10.
46. R.K. Iler: *The Chemistry of Silica*, J. Wiley & Sons, New York **1979**, p 174.
47. DE 1 467 019, **1963**, Degussa.
48. DE 1 283 207, **1961**, PPG.
49. DE 2 224 061, **1972**, Sifrance.
50. EP 0 018 866, **1980**, Rhone-Poulenc.
51. DE 2 505 191, **1975**, Degussa.
52. DE 1 467 437, **1965**, Degussa.
53. DE 1 293 138, **1965**, Degussa.
54. US 4 179 431, **1979**, Degussa.
55. DE 1 567 440, **1965**, Degussa.
56. Degussa, Technical Bulletin Pigments: The Handling of Synthetic Silicas and Silicates, Pig. 28-3-2-598, 3rd Edition, May **1998**.
57. US 1 297 724, **1918**, Patrick.
58. *Ullmann's Encyclopedia of Industrial Chemistry, 5th Edition* (1993), Wiley, New York, Vol. A 23, p. 629–635.
59. C.J. Brinker, G.W. Scherer, Sol-Gel Science, Academic Press, San Diego **1990**, 458–460.
60. IUPAC Compendium of Chemical Terminology, 2nd Edition **1997**.
61. WO 00/02814.
62. Kistler, *J. Phys. Chem.* **36**, 52 (1932).
63. Cabot Annual Report **2001**, p. 32.
64. J. Fricke, A, Emmerling, *J. Sol-Gel Sci. Tech.* **13**, 299–30 (1998).
65. N. Hüsing, U. Schubert, *Angew. Chem. Int. Ed.*, Vol **37**, 22–45 (1998).
66. DE 30 16 010, Degussa.
67. P. Vondracek, M. Schätz, *Kautschuk + Gummikunststoffe* **33**, 699 (1980).
68. D.M. Hoffmann, I.L. Chiu, *Appl. Polym. Sci. Proc.* **46**, 100 (1981).
69. S.J. Clarson, J.A. Semlyen: *Siloxane Polymers*, Prentice-Hall Inc., **1993**, ISBN 0-13-816315-4.
70. Burkus F.S., Amaresekera J., *Rubber World*, June **2000**, 26.
71. K.H. Müller, *Mühle Mischfuttertech.* **114**, 28 (1977).
72. Degussa, *Schriftenreihe Pigmente: Synthetische Kieselsäuren als Fließhilfsmittel und als Trägersubstanz*, Pig. 31-7-3-592, 5th Edition, May **1992**.
73. C.R. Hegedus, I.L. Kamel, *J. Coat. Tech.* **65**, 49 (1993).
74. Technical Bulletin Pigments No. 54, Company Brochure of Degussa AG, Frankfurt/Main, Germany **1999**.
75. Technical Bulletin Pigments No. 27, Company Brochure of Degussa AG, Frankfurt/Main, Germany **2000**.
76. K. Kobayashi, K. Araki, Y. Imamura, *Bull. Chem. Soc. Jpn.* **62**, 3421 (1989).
77. George Wypych: *Handbook of Fillers*, ChemTec Publishing, **1999**, ISBN 1-895198-19-4.
78. B.L. Mueller, J.S. Steckenrider, *Chemtech*, February **1998**, 38.
79. R. Clasen, *Glastechn. Ber.* **61**, 119 (1988).
80. WO 0204370, **2002**, Novara Technology.
81. E.M. Rabinovich et al., *Amer. Chem. Soc.* **66**, 683 (1983).
82. C.J. Johnston et al., *Toxicological Sciences* **56**, 405–413 (2000).
83. R. Merget. et al., *Occupational Hygiene* **5**, 231–251 (2001).
84. J. Lewinson, W. Mayr, H. Wagner, *Toxicology and Pharmacology* **20**, 37–57 (1994).
85. W. Klosterkötter, *Arch. Hyg. Bakteriol.* **149**, 577–598 (1965).
86. P. Reuzel et al., *Fd. Chem. Toxic.* **29**, 341–354 (1991).
87. Y. Sun et al., DAE 2001, Garmisch-Partenkirchen, September 6–7, **2001** (Abstract).
88. M. Maier, SILICA 2001, Mulhouse, September 3–6, **2001** (Abstract).
89. IUCLID Data Set of the European Commission: Silicon Dioxide (**2000**).
90. H. Ferch et al., *Arbeitsmed., Sozialmed., Präventivmed.* **22**, 6, 23 (1987).
91. H. Ferch, S. Habersang, *Seifen-Öle-Fette-Wachse* **108**, 487 (1982).

44 Adsorptive Properties of Porous Silicas

Martyn B. Kenny and Kenneth S.W. Sing[1]
Brunel University, Department of Chemistry

CONTENTS

Four main types of porous silica adsorbents have been identified: compacts of pyrogenic powders, precipitated silicas, silica gels, and zeolitic silicas. The importance of porosity relative to the adsorptive properties of each group is reviewed, with particular reference to the adsorption of nitrogen, argon, and water vapor. The differences in size and specificity of these adsorptive molecules may be exploited to explore the surface properties of each grade of silica. A notable feature of Silicalite I, which is the best known of the zeolitic silicas, is its remarkable hydrophobic character. Furthermore, the uniform tubular pore structure of this microporous silica is responsible for other highly distinctive properties.

Amorphous and crystalline forms of silica are now widely used as industrial adsorbents and catalyst supports. The preparation of a highly active and inexpensive silica adsorbent is not difficult, but the fine tuning of the adsorbent activity is somewhat more demanding. Hence, over the past 40 years the upgrading of the adsorptive properties of silicas has presented a challenge to many academic and industrial research workers.

The microstructure of various amorphous silicas was first discussed in a comprehensive manner by Iler in his early book *The Colloid Chemistry of Silica and Silicates* [1]. Iler drew attention inter alia to the importance of the dense silica particle size and particle packing density in controlling the surface and colloidal properties of sols, gels, and precipitates. In particular, he showed how a change in coordination number of these globular particles could affect the porosity and hence the adsorptive properties of the dried materials.

Since 1955, many others have extended these principles of particle packing, and now secondary and tertiary assemblages can be identified within the microstructures of certain silica gels and precipitates [2,3]. Iler had proposed [1] that the minimum size of the dense silica globule was about 1 nm. Later Barby [2], making use of transmission electron microscopy, came to the conclusion that in many amorphous silicas the primary particle size was indeed $1-1.5$ nm.

In their pioneering studies of silica sols, Alexander and Iler [4] employed low-temperature nitrogen adsorption to determine the surface areas of the colloidal particles after removal of the aqueous medium. The Brunauer–Emmett–Teller (BET) areas were found to be only slightly larger than the values obtained from the particle size distributions as determined by light scattering and electron microscopy. These remarkable measurements indicated little change in the particle size or shape after the stabilized silica sols were carefully dried.

Before the distinctive adsorptive properties of porous silica can be described, the different ranges of pore size that are of special importance to the mechanisms of physisorption must be identified. Micropores are the pores of the smallest width ($d < 2$ nm); mesopores are of intermediate

[1]Current address: Department of Chemistry, University of Exeter, Devon EX4 4QD, United Kingdom.

size ($d \sim 2$–50 nm); macropores are the widest pores ($d > 50$ nm) [5]. Amorphous silica gels tend to be mesoporous or microporous, whereas the crystalline zeolitic silicas possess intracrystalline microporosity. The precipitated silicas are macroporous and also, to a small extent, microporous. These and other aspects of the microstructures will be discussed in the following sections.

COMPACTS OF PYROGENIC POWDERS

From the standpoint of gas adsorption, the pyrogenic silicas can be regarded as essentially nonporous. Transmission electron microscopy has revealed that the high-temperature arc silicas, and to a lesser extent the flame-hydrolyzed "fume" silicas (e.g., Degussa Aerosils), consist of discrete spheroidal particles. According to Barby [2], these globules are in fact composed of primary particles of about 1 nm. The coordination number is so high that there is virtually no microporous structure; in their original, loosely packed state, these powders give physisorption isotherms (e.g., nitrogen at 77 K) of Type II in the Brunauer [6] and International Union of Pure and Applied Chemistry (IUPAC) [5] classification (Figure 44.1). This type of isotherm is associated with unrestricted monolayer–multilayer adsorption, the stage of monolayer completion being indicated by point B in Figure 44.1 [7].

Well-defined pore structures are developed as a result of compaction of the fume silicas [8,9]. If the compaction pressure is not too high [$\sim$10 tons in.$^{-2}$ (245 giganewtons m^{-2})] the nitrogen isotherm becomes similar to Type IV in Figure 44.1. The initial part of the isotherm is scarcely changed: The area of particle–particle contact is low, so there is little overall loss of surface area. However, at higher relative pressures capillary condensation occurs in the newly created mesopores and causes the isotherm to swing upwards away from the original (Type II) path until the pores are all filled and the isotherm reaches a plateau.

Very small particles of fume silica were compacted in the work of Avery and Ramsay [8], who found that high compaction pressures resulted in the conversion of the isotherm type from IV to I. A drastic loss of BET area accompanied this change (from 630 to 219 m^2/g), and Avery and Ramsay concluded that this change was associated with a marked increase in particle packing density. The shape and reversibility of this Type I isotherm was a clear indication that the effective pore width had been reduced to below 2 nm, that is, that the compact had become microporous.

Clearly, highly active mesoporous or microporous silicas cannot be produced by the compaction of nonporous powders, but Avery and Ramsay's [8] and other compaction studies [10–12] confirmed the importance of particle coordination in determining porosity and hence, adsorptive properties.

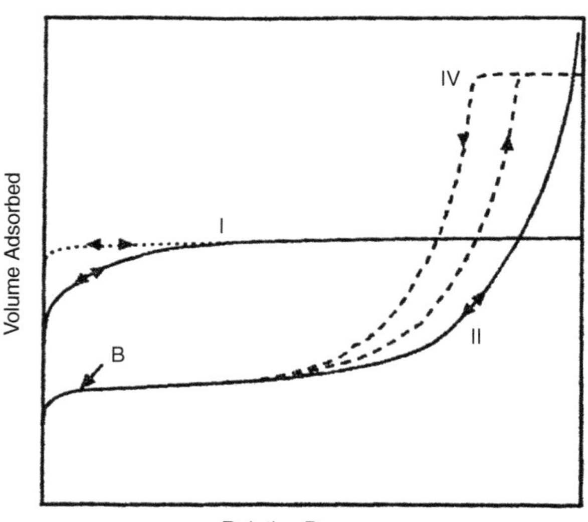

FIGURE 44.1 Types of physisorption isotherms given by porous and nonporous solids.

PRECIPITATED SILICAS

The surface properties of precipitated silicas have not been studied in as much detail as those of the fume silicas or gels. On the other hand, the extensive patent literature is an identification of the industrial importance of these materials [2,3].

The gas adsorption measurements by Zettlemoyer and co-workers [13,14] appeared to indicate that some precipitated silicas (e.g., HiSil 233 from Pittsburgh Plate Glass Company) behaved as nonporous adsorbents. Thus, reversible Type II isotherms of nitrogen and argon were obtained by Bassett et al. [14], who concluded that unrestricted monolayer–multilayer adsorption had occurred. More recent work [15] showed that this interpretation is probably an oversimplification of the physisorption process.

In contrast to the behavior of the pyrogenic silicas, the level of nitrogen physisorption by several precipitated silicas was especially sensitive to change of outgassing temperature. Analysis of the adsorption data by the conventional BET method indicated an apparent increase in the surface area of $\sim$26% over the range of outgassing temperature of 25–200°C (*see* Figure 44.2). A more rigorous interpretation of the nitrogen isotherms, by application of the: α_s-method [16], revealed that this change was misleading and that the increase in BET area was associated with the development of microporosity [15]. A_s is the external area obtained by affiliation of the α_s-method [16], and V_{mic} is the derived micropore volume.

The behavior of the precipitated silicas with respect to the adsorption of water vapor was even more anomalous [17,18]. Kiselev [11] and others [19] had demonstrated that in the fully hydroxylated form, a wide range of nonporous pyrogenic silicas gave rise to a common reduced water isotherm (i.e., adsorption per unit area versus relative

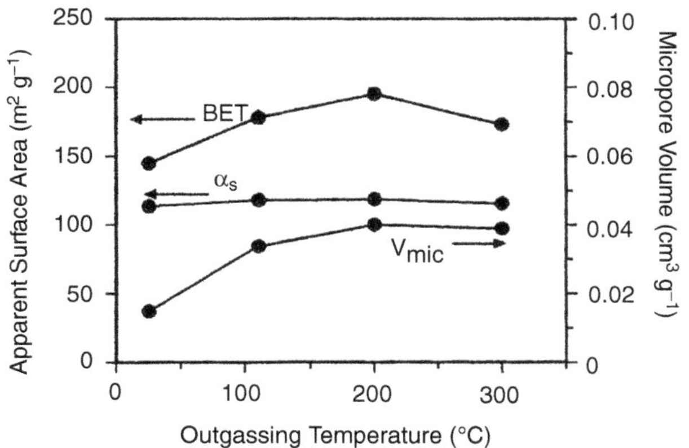

FIGURE 44.2 The effect of outgassing temperature on the BET area, α_s external area, and micropore volume (V_{mic}) for precipitated silica VN3.

pressure). However, in the precipitated silicas, the level of water adsorption was much higher than expected for monolayer adsorption. There seemed little doubt that water molecules were able to penetrate into very narrow pores that could not accommodate nitrogen or other molecules [17].

The abnormal behavior of the precipitated silicas appears to be due to the presence of trapped hydroxyl groups within the secondary particles. Thus, although the primary globules are densely packed within the secondary agglomerate, they apparently remain partially hydroxylated. The internal hydroxyls undergo hydrogen-bonding with water molecules, which are able to move in and out of the secondary particles. It is evident that the removal of these hydrogen-bonded water molecules also leads to the development of the small micropore volume.

SILICA GELS

A large number of gas adsorption studies [2,3,7,11,20] have been reported on high-area silica gels, but unfortunately much of this work was carried out on ill-defined samples of unknown origin. Silica hydrogels are usually prepared by reacting silicate and acid in an aqueous medium. The properties of the final product (normally a xerogel) are controlled by the conditions under which the condensation–polymerization reaction occurs and by the after-treatment (washing and removal of the liquid phase). Syneresis takes place when the wet hydrogel is allowed to stand, and considerable further shrinkage of the solid accompanies the hydrogel–xerogel conversion.

The dependence of the xerogel porosity on the conditions of gelation has been investigated [22]. In the early work of Sing and Madeley [21,22] a microporous product was obtained from the hydrogel prepared from sodium silicate and sulfuric acid at pH 3.5. Mesoporous structures developed when the reaction was carried out at higher pH (~6). On the other hand, changes in the silicic acid concentration over a wide range had very little effect, provided that the gelation was conducted in a buffered aqueous medium [21].

Acid washing appears to result in partial depolymerization of the hydrogel [23,24], which in turn leads to some enhancement in the adsorption activity. The effect is illustrated by the results in Figure 44.3. In this case, the original hydrogel was prepared at pH 5.4 and portions were then subjected to different forms of after-treatment [25]. Soaking in HCl (at pH 2.0 for 24 h) resulted in significant upward movement of the nitrogen isotherm, that is, increase in both the BET area (from 284 to 380 m^2/g) and the pore volume (from 0.44 to 0.55 cm^3(liquid)/g). However, the shape of the hysteresis loop remained almost unchanged, a result suggesting that the mesopore size distribution was not altered to any significant extent.

The most striking result illustrated in Figure 44.3 was obtained when the hydrogel was washed with ethyl alcohol. The vacuum-dried alcogel so obtained gave a very much larger uptake of nitrogen over the complete range of relative pressure. A twofold increase was evident in both the BET area and pore volume. Thus, by replacing water as the continuous liquid phase, it was possible to reduce the capillary forces that normally bring about a drastic shrinkage of the open hydrogel during normal drying.

The alcogel featured in Figure 44.3 had a BET area of 641 m^2/g and a pore volume of 0.93 cm^3(liquid)/g. An even larger pore volume can be obtained if the fluid phase is removed under supercritical conditions to give an aerogel, that is, a product having a very low particle coordination number. Such materials are macroporous and have a high surface area (Table 44.1), but they are

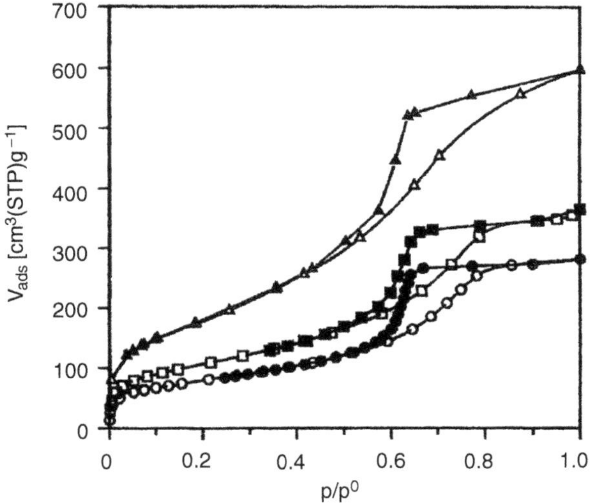

FIGURE 44.3 Nitrogen isotherms (77 K) for xerogel (○ and ●), acid-washed xerogel (HCl, pH 2.0, 24 h) (□ and ■), and alcogel (△ and ▲). Clear symbols denote adsorption; dark symbols denote desorption.

TABLE 44.1
Typical Surface Areas and Pore Volumes of the Silica Gels

Type	Porosity	BET-area (m^2/g)	Pore volume (cm^3/g)
Aerogel	macro	800	2.0
G-xerogel	meso	350	1.2
S-xerogel	meso	500	0.6
S-xerogel	micro	700	0.4

usually mechanically weak and unstable when exposed to water vapor. The upper limiting surface area of a silica composed of discrete primary particles would be ~2000 m^2/g, but so far it has not been possible to obtain areas approaching this magnitude.

Conventional silica gels (termed S-type by Barby [2]) are produced by the roasting (or oven-drying) of low-density hydrogels, which undergo drastic shrinkage with considerable loss of pore volume and surface area. A different type of adsorption is produced from the same initial hydrogel if it is subjected to hydrothermal aging prior to the final drying. In this case packing and fusion of the primary particles takes place [26], so that after drying the pore space is largely confined to the interstitial space between the secondary particles. The resulting G-xerogel has a somewhat lower surface area and larger and more uniform mesopore volume (Table 44.1 and Figure 44.4).

Low-temperature nitrogen adsorption is normally used for the determination of surface area and pore size distribution of porous materials. However, specific field-gradient quadrupole interactions play a significant role in the adsorption of nitrogen on hydroxylated silicas or other polar surfaces [7]. Accordingly, some authors [14,27] have proposed that a nonpolar adsorptive such as argon should be used instead of nitrogen for the determination of surface area.

The difference in shape of the nitrogen and argon isotherms, both determined at 77 K, on a mesoporous silica is illustrated in Figure 44.4a. In the middle range of relative pressure, the isotherms follow almost identical paths. The divergence at relative pressure $p/p^0 > 0.7$ is associated with the onset of capillary condensation of nitrogen in the mesopores and confirms that argon cannot be employed at 77 K for the assessment of the mesopore size distribution [7]. Of particular interest is the difference in shape of the isotherms at $p/p^0 < 0.2$, that is, in the monolayer region. The use of high-resolution adsorption (HRADS) [28] allowed the initial part of the isotherm to be explored in considerable detail (*see* Figure 44.4b): It reveals that the nitrogen isotherm is extremely steep at fractional coverages <0.1 and at $p/p^0 < 10^{-4}$. This steepness is probably due to localized adsorption on the highest energy sites of the surface. These results are consistent with calorimetric measurements of differential enthalpies of adsorption [27,29] and indicate that the surface was more heterogeneous with respect to the adsorption of nitrogen than argon. On the other hand, the lack of a well-defined point B suggests that argon is less reliable than nitrogen for the determination of surface area [7,30].

ZEOLITIC SILICAS

A number of microporous polymorphs of crystalline silica can now be prepared. One procedure is to attempt the dealumination of a readily available zeolite; another approach involves direct synthesis, for example, of ZSM-5, in the form of Silicalite I [31]. The considerable amount of recent interest shown in these Al-free zeolites (or porotectosilicates) has been stimulated by the uniformity of their channel structures and their hydrophobic nature.

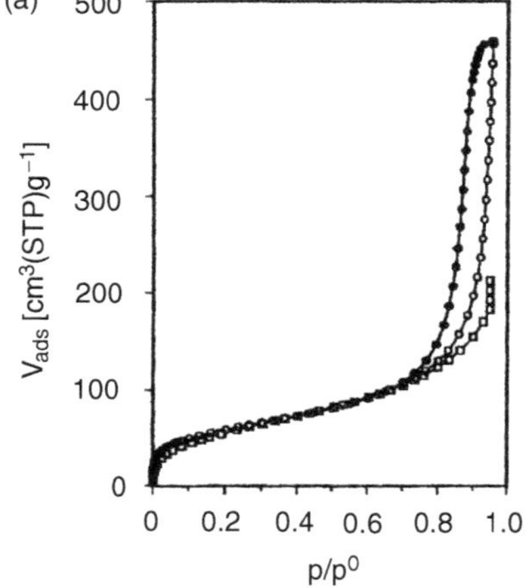

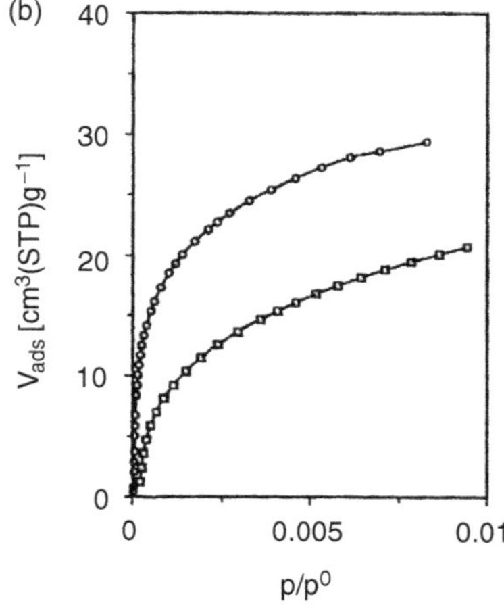

FIGURE 44.4 (a) Nitrogen (○ and ●) and argon (□) isotherms at 77 K for a mesoporous silica. (b) detailed low relative pressure data. Clear symbols denote adsorption; dark symbols denote desorption.

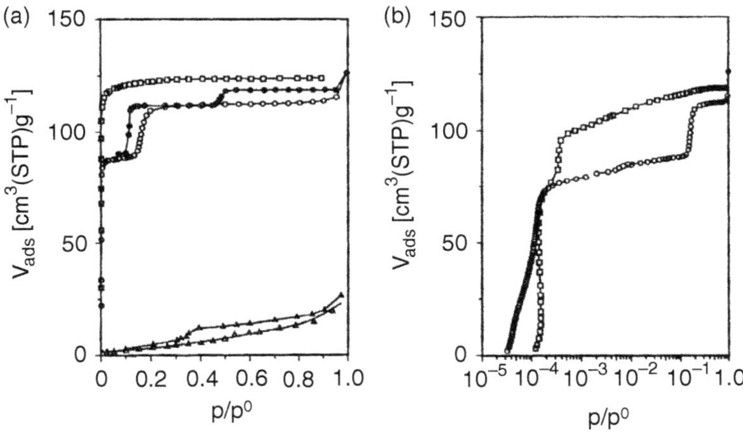

FIGURE 44.5 (a) Nitrogen (77 K) (○ and ●), argon (77 K) (□) and water vapor (298 K) (△ and ▲) isotherms for Silicalite I. (b) detailed low relative pressure data. Clear symbols denote adsorption; dark symbols denote desorption.

TABLE 44.2
Pore Volumes at $p/p^0 = 0.95$ of Molecular Sieve Silicas

Adsorbent	Nitrogen	Argon	Water vapor
Silicalite I	0.18	0.16	0.02
Microporous silica 3[a]	0.10	0.08	0.09
Precipitated silica VN3	0.12[b]	—	0.11[b]

Note: All values are given in cubic centimeters per gram.
[a]From reference 32.
[b]Data were recorded at $p/p^0 = 0.80$.

These features are illustrated by the results of argon, nitrogen, and water measurements on Silicalite I (Figure 44.5a and Table 44.2 [32]). The low-pressure steepness (high affinity) of the nitrogen and argon isotherms is indicative of the filling of very narrow micropores, but at higher relative pressure these isotherms are of very different shape. The hysteresis loop in the range $p/p^0 \sim 0.1-0.2$ is a characteristic feature of the nitrogen isotherm and is probably associated with a phase transition from a liquidlike to a solidlike structure [33,34]. The loop at higher relative pressure is the result of capillary condensation in the secondary pore structure outside the intracrystalline pores. Use of the HRADS technique reveals that further phase changes take place at very low relative pressures with both nitrogen and argon (Figure 44.5b) and also that the uptake of nitrogen occurs at lower relative pressures than argon, probably because of the stronger adsorbate–adsorbate interactions of the nitrogen quadrupoles and also the quadrupole interaction with any residual surface hydroxyl groups.

The uptake of water vapor by Silicalite I is remarkably low over the complete range of relative pressure. If we allow for the secondary pore filling, the amount of water adsorbed within the Silicate I crystals seems negligible (Table 44.2). An explanation for these anomalous results must take into account the pore geometry in relation to the hydrogen-bonded structure of water. Thus, the intracrystalline pores of Silicalite I are tubular (intersecting straight and sinusoidal channels) of diameter ~0.55 nm, and most likely, an array of water molecules cannot be adsorbed without considerable distortion of the directional hydrogen bonds. In our view [35], it is of particular significance that the uptake of water within the slit-shaped pores of 0.5–0.6-nm width in molecular sieve carbons is much greater and consistent with the thickness of an undistorted thin "slab" of water. Therefore, Silicalite I is more hydrophobic than any form of microporous carbon available at present. Although further progress can be made in the refinement of the pore size distribution of amorphous mesoporous silicas, the development of well-defined microporous silicas will depend on the preparation of new zeolitic structures and improvements in crystallinity of the existing zeolitic silicas.

ORDERED MESOPOROUS SILICAS

A new family of highly ordered mesoporous silicate structures was first described by Mobil scientists [36] in 1992 and has since attracted considerable attention. These materials were originally referred to as "mesoporous molecular sieves" and were generally designated M41S. The most thoroughly investigated members of this family are collectively known as MCM-41 (Mobil Catalytic Material, number 41), and comprise hexagonal arrays of non-intersecting tubular pores of controlled size–typically c. 4 nm diameter. In the original synthesis, the hydrothermal conversion of the silicate or aluminosilicate gel to the ordered structure was facilitated by the presence of a quaternary ammonium surfactant, which appeared to act as a template for the formation of a micellar or liquid-crystal type of intermediate structure. The washed and air-dried intermediate product was calcined at 450°C to remove residual organic material. As a result of the extensive investigations of Inagaki, Pinnavaia, Schuth, Stucky, Unger, White and others it is now evident that M41S structures can be produced from various precursors and under different experimental conditions.

Since 1992, physisorption isotherms have been reported [37] on many samples of MCM-41. Although all of these isotherms are essentially Type IV in the IUPAC classification, they are not all of exactly the same shape. The most striking feature, which was first noted in 1993 [38], is a sharp reversible pore filling/emptying step: this is located at a characteristic relative pressure, which is dependent on the adsorptive gas and the operational temperature. In the case of nitrogen adsorption at 77 K on samples of 4 nm MCM-41, this mesopore filling step starts at $p/p^0 \sim 0.42$ and ends at $p/p^0 \sim 0.46$. These findings are consistent with other evidence (hydraulic radius, electron micrographs, and X-ray diffraction patterns) which indicate that this form of MCM-41 has a narrow distribution of non-intersecting cylindrical pores. In an extended IUPAC classification [37], such reversible Type IV isotherms have been designated Type IVc isotherms.

Several investigations have been made of the dependence of isotherm shape on the pore size of MCM-41. It turns out that those samples with pores wider than ca. 4 nm all give nitrogen isotherms having well-defined hysteresis loops, whereas those samples with narrower pores (e.g., with pore diameters of 3.4 and 2.5 nm) give completely reversible nitrogen isotherms. However, in the latter case, the pore filling step tends to be less steep than that given by the 4 nm sample. We recall that $p/p^0 \sim 0.42$ is generally accepted [7] as the lower limit of capillary condensation hysteresis for nitrogen adsorption at 77 K and that a cooperative form of pore filling is believed to occur in the smaller 'supermicropores' at lower relative pressures [37]. Since the critical p/p^0 at which the capillary condensate becomes unstable is dependent on the adsorptive and temperature, it follows that certain samples of MCM-41 can give both reversible Type IV isotherms and hysteresis loops. Indeed, experimental and theoretical work on these model structures has already been of considerable value in providing an improved understanding of the mechanisms of adsorption hysteresis [37].

It is remarkable that it has been possible to prepare some MCM-41 silica samples with surface areas in excess of 1000 m^2g^{-1}. Although these materials are not microporous, it is evident that the pore walls must be quite thin and are probably puckered and it is not surprising to find that pronounced ageing occurs on prolonged exposure to water vapor [37]. Efforts are now being made to enhance the stability of these high-area products by surface modification. More work is also required to improve the uniformity of the pore structure and scale up the production of these important materials.

REFERENCES

1. Iler, R. K. *Colloid Chemistry of Silica and Silicates*; Cornell University Press: Ithaca, NY, 1955.
2. Barby, D. In *Characterisation of Powder Surfaces*; Parfitt, G. D.; Sing, K. S. W., Eds.; Academic Press: London, 1976; p 353.
3. Iler, R. K. *The Chemistry of Silica*; John Wiley & Sons: New York, 1979.
4. Alexander, G. B.; Iler, R. K. *J. Phys. Chem.* **1953**, *57*, 932.
5. Sing, K. S. W.; Everett, D. H.; Haul, R. A. W.; Moscou, L.; Pierotti, R. A.; Rouquérol, J.; Siemieniewska, T. *Pure Appl. Chem.* **1985**, *57*, 603.
6. Brunauer, S.; Deming, L. S.; Deming, W. S.; Teller, E. *J. Am. Chem. Soc.* **1940**, *62*, 1723.
7. Gregg, S. J.; Sing, K. S. W. *Adsorption, Surface Area and Porosity*; Academic Press: London, 1982.
8. Avery, R. J.; Ramsay, J. D. F. *J. Colloid Interface Sci.* **1973**, *42*, 597.
9. Gregg, S. J.; Langford, J. F. *J. Chem. Soc. Faraday Trans. 1* **1977**, *73*, 747.
10. Gregg, S. J. *Chem. Ind. (London)* **1968**, 611.
11. Kiselev, A. V. In *The Structure and Properties of Porous Materials*; Everett, D. H.; Stone, F. S., Eds.; Butterworths: London, 1958; p 68.
12. Gregg, S. J.; Sing, K. S. W. In *Surface and Colloid Science*; Matijevic, E., Ed.; Wiley: London, 1976; Vol. 9, p 231.
13. Zettlemoyer, A. C. *J. Colloid Interface Sci.* **1968**, *28*, 343.
14. Bassett, D. R.; Boucher, E. A.; Zettlemoyer, A. C. *J. Colloid Interface Sci.* **1968**, *27*, 649.
15. Carrott, P. J. M.; Sing, K. S. W. *Adsorption Sci. Technol.* **1984**, *1*, 31.
16. Sing, K. S. W. *Chem. Ind. (London)* **1967**, 829.
17. Carrott, P. J. M.; Sing, K. S. W. *Adsorption Sci. Technol.* **1986**, *21*, 9.
18. Carrott, P. J. M., Ph.D. Thesis, Brunel University, Uxbridge, Middlesex, England, 1980.
19. Baker, F. S.; Sing, K. S. W. *J. Colloid Interface Sci.* **1976**, *55*, 605.
20. Everett, D. H.; Parfitt, G. D.; Sing, K. S. W.; Wilson, R. *J. Appl. Chem. Biotechnol.* **1974**, *24*, 199.
21. Sing, K. S. W.; Madeley, J. D. *J. Appl. Chem.* **1953**, *3*, 549.
22. Sing, K. S. W.; Madeley, J. D. *J. Appl. Chem.* **1954**, *4*, 365.
23. Mitchell, S. A. *Chem. Ind. (London)* **1966**, 924.
24. Neimark, I. E.; Sheinfain, R. Yu.; Krugilkova, N. S.; Stas, O. P. *Kolloid Zhur.* **1964**, *26*, 595.
25. Wong, W.-K., Ph.D. Thesis, Brunel University, Uxbridge, Middlesex, England, 1982.

26. Kiselev, A. V. *Disc. Faraday Soc.* **1971**, *52*, 14.
27. Rouquerol, J.; Rouquerol, F.; Peres, C.; Grillet, Y.; Boudellal, M. In *Characterisation of Porous Solids*; Gregg, S. J.; Sing, K. S. W.; Stoeckli, H. F., Eds.; The Society of Chemistry and Industry: London, 1979; p 107.
28. Pieters, W. J. M.; Venero, A. F. In *Catalysis on the Energy Scene*; Kaliaguine, S.; Mahay, A., Eds.; Elsevier: Amsterdam, Netherlands, 1984; p 155.
29. Furlong, D. N.; Sing, K. S. W.; Parfitt, G. D. *Adsorption Sci. Technol.* **1986**, *3*, 25.
30. Carruthers, J. D.; Payne, D. A.; Sing, K. S. W.; Stryker, L. J. *J. Colloid Interface Sci.* **1971**, *36*, 205.
31. Flanigan, E. M.; Bennett, J. M.; Grose, R. W.; Cohen, J. P.; Patton, R. L.; Kirchner, R. M.; Smith, J. V. *Nature (London)* **1978**, *271*, 512.
32. Dollimore, D.; Heal, G. R. *Trans. Faraday Soc.* **1963**, *59*, 1.
33. Muller, U.; Unger, K. K. In *Characterisation of Porous Solids*; Unger K. K., Ed.; Elsevier: Amsterdam, Netherlands, 1988; p 101.
34. Muller, U.; Reichert, H.; Robens, E.; Unger, K. K.; Grillet, Y.; Rouquerol, F.; Rouquerol, J.; Pan, D.; Mersmann, A. *Fresenius Z. Anal. Chem.* **1989**, *333*, 433.
35. Kenney, M. B.; Sing, K. S. W. *Chem. Ind. (London)* **1990**, 39.
36. Kresge, C. T.; Leonowicz, M. E.; Roth, W. J.; Vartuli, J. C.; Beck, J. S. *Nature (London)* **1992**, *359*, 710.
37. Rouquerol, F.; Rouquerol, J.; Sing, K. *Adsorption by Powders and Porous Solids;* Academic Press, London, **1999**.
38. Branton, P.J.; Hall, P.G.; Sing, K.S.W. *J. Chem. Soc., Chem. Commun.* **1993**, 1257.

45 Silica Gels from Aqueous Silicate Solutions: Combined ^{29}Si NMR and Small-Angle X-Ray Scattering Spectroscopic Study

Peter W.J.G. Wijnen[1], *Theo P.M. Beelen, and Rutger A. van Santen**
Eindhoven University of Technology, Schuit Institute of Catalysis

CONTENTS

The use of modern spectroscopic techniques in the study of the formation of aqueous silica gels is described. The oligomerization process of monomeric silicic acid was studied by silicon-29 NMR spectroscopy; at high pH values cyclic trimeric silicate species were favored compared to the linear structure. Aggregation of primary silica particles of molecular size (<1 nm) was studied by analysis of small-angle x-ray scattering patterns. All systems studies (pH 4.0) indicate reaction-limited aggregation. Polyvalent cations influence the rate of aggregate formation in a negative way: aluminum at low concentrations (1 mol%) significantly inhibits aggregation. A new model for the aging process that proposes that monomeric silicic acid is transported (via solution) from the periphery of the aggregate into the core of the solution is given.

The formation of amorphous silica gels is an important process in modern industry because of their many applications. Most investigations in the field of silica gel preparation have contributed to the development of a physical chemical understanding of silica gel formation. As such, many procedures to synthesize silica gels tailor-made to the demands of specific applications have resulted [1].

Although much phenomenological and empirical knowledge is available, the polymerization process is still lacking in molecular chemical descriptions [1–3]. The preparation of silica has more or less become an art rather than a science based on fundamental knowledge of the preparation conditions. Insights into the underlying principles and molecular chemical aspects of silica gel formation are still limited. Understanding of the molecular chemical aspects of silica gel formation is required because small variations in preparation conditions and precursor solutions can result in different structural properties

[1]Current address: Experimental Station, DuPont (Nederland) B.V., P.O. Box 145, 3300 AC, Dordrecht, Netherlands.
*Corresponding author.

of the final product. Furthermore, the molecular aspects of silica gel formation also play a role in the synthesis of crystalline, microporous silicates and aluminosilicates (zeolites) [4], a process still poorly understood.

This chapter presents in situ spectroscopic investigations on the formation of silica gels from aqueous silicate solutions. The processes discussed deal with (1) the formation and growth of primary particles by polycondensation of small (monomeric) silicate anions, studied by silicon-29 NMR spectroscopy; (2) the aggregations of these primary particles into ramified silica aggregates and the role of cations in the aggregation process, studied by small-angle x-ray scattering (SAXS); and (3) the reconstruction of silica aggregates through chemical processes, studied with both silicon-29 NMR spectroscopy and SAXS. These three processes can be regarded as the key processes in formation of silica gels, and as such this chapter addresses the need for a more fundamental description of the process of silica gel formation.

SILICON-29 NMR SPECTROSCOPY

Most commercial production processes of silica gel make use of aqueous silicate solutions. In contrast to alcoholic precursor solutions previously used to study the hydrolysis and oligomerization processes of the monomeric precursor [2], aqueous silicate solutions (water glass) contain a broad spectrum of structurally different silicate species, all present as essentially dissolved silica. To probe the local atomic environment of silicon nuclei and investigate the degree of polymerization of silicate species, silicon-29 NMR spectroscopy has been applied since the early 1970s [5]. The large number of structurally different silicate species observed by ^{29}Si NMR spectroscopy contributes to the fact that the observed resonance lines of commercial waterglass solutions are rather broad: overlapping and small differences in resonance frequency of the different lines cause substantial line broadening. More alkaline solutions of aqueous silicate give rise to better resolved spectra [6]. Furthermore, the concentration of small oligomeric and monomeric silicate anions is increased with respect to the concentration of these anions in silicate solutions of lower alkalinity as a consequence of the higher alkalinity. Increasing pH values give rise to more depolymerized species.

Oligomerization of Aqueous Monomeric Silicic Acid

Because polymerization of the water-glass solution gives rise to a broad distribution of silicate anions and thus results in poorly resolved ^{29}Si NMR lines, the application of ^{29}Si NMR spectroscopy in this type of reaction provides minor information about the different oligomerization steps and the reaction mechanism of monomeric silicic acid. Therefore a different approach to the study of aqueous silicate solutions was applied. Because of the slow dissolution of amorphous silica gel in alkaline solutions, a gradual increase in the concentration of monomeric silicic acid can be expected. In Figure 45.1, typical NMR spectra of such a dissolution process are presented. In this example, amorphous, pyrogenic silica gel (Aerosil 200; Degussa) was dissolved in aqueous tetramethylammonium hydroxide (TMAOH; Janssen Chimica). TMAOH was chosen as the base because silicate species present in aqueous tetramethylammonium silicate solutions show a high degree of monodispersity: the number of different silicate species is rather small compared to alkali metal silicate solution [7], and this choice facilitates identification of oligomers and oligomerization pathways. Initially (Figure 45.1a) a single resonance line is observed, which is attributed to monomeric silicate anions. Because of the gradual dissolution of

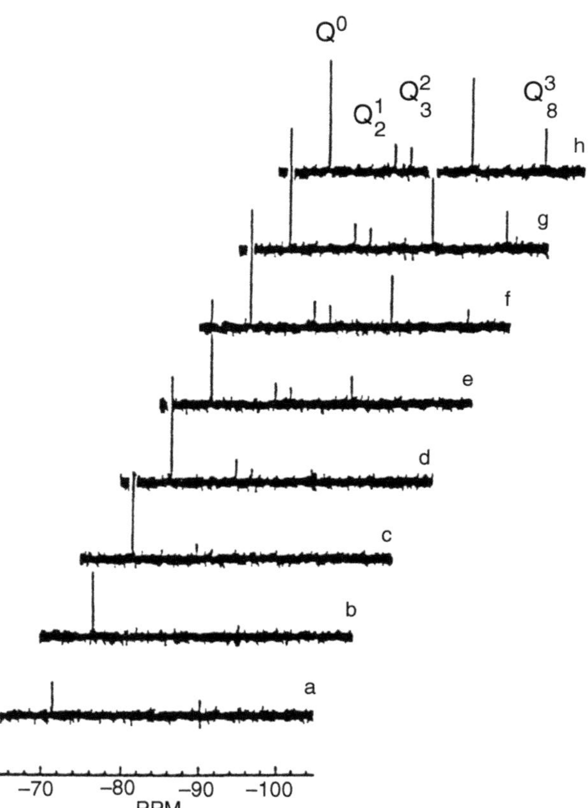

FIGURE 45.1 Dissolution of amorphous silica gel in aqueous tetramethylammonium hydroxide ($TMA_2O:SiO_2:36H_2O$) after (a) 83, (b) 250, (c) 833, (d) 1083, (e) 1250, (f) 1500, (g) 1666, and (h) 1833 min from initial mixing. The initially highly viscous suspension of tetramethylammonium silicate was introduced in zirconia magic-angle spinning NMR rotors and analyzed with a Bruker CXP-300 NMR (7.05) spectrometer. Five hundred free-induction decays were averaged that applied 45° pulses with a 10-s pulse delay. Magic-angle sample spinning was applied to average any chemical shift anisotropy arising from the highly viscous suspension.

the amorphous gel matrix, an increase in monomeric silica concentration is detected. Once a certain concentration of monomeric silicic acid has been reached, oligomerization of the monomers occurs and is reflected in the occurrence of a second resonance line. This line corresponds to dimeric silicate species. Further oligomerization takes place as a result of ongoing monomer dissolution.

From the development of resonance lines in the different spectra, a qualitative reaction mechanism can be extracted for the oligomerization process. In this mechanism, formation of cyclic structures is a predominant phenomenon. In the presence of TMA cations, no linear trimeric silicate anions are found in the silicate solutions, as can be deduced from Figures 45.1d through 45.1 h. The use of alkali metal hydroxides as bases shows, aside from the cyclic trimeric silicate anion, the linear structure; the relative concentration of each depends on the alkalinity and the alkali metal used. The formation of cyclic trimeric silicate anions occurs before formation of the linear trimeric species [6]. The formation of double-cyclic silicate anions can be observed in Figure 45.1e; two cyclic trimeric silicate anions combine to form the prismatic hexameric silicate anion (Q_6^3). This prismatic hexameric silicate anion forms, through addition of two monomers or one dimer, the well-known cubic octameric silicate anion. This cubic octameric silicate anion has been proposed to be predominantly stabilized by the tetramethylammonium cation. However, here it is shown that aside from stabilization of the Q_8^3 anion by TMA, other cyclic structures such as Q_6^3 and Q_3^2 are preferentially formed in aqueous silicate solutions containing tetramethylammonium cations. The observation that tetramethylammonium cations direct the oligomerization pathway may well be linked with the observation that during zeolite synthesis the zeolite structure needs to be built in a certain well-defined fashion. Cations in general will have an effect on the way oligomerization proceeds in aqueous silicate solutions.

Influence of Alkali Metal Hydroxides on Dissolution Rate

In a study [6] of the dissolution of amorphous silica gels in aqueous alkali metal hydroxides, the rate of dissolution was found to depend on the cation used in the dissolution reaction. A maximum in dissolution rate was found for potassium hydroxide solutions, whereas both intrinsically smaller and larger cations (lithium–sodium and rubidium–cesium) showed slower dissolution rates, as can be concluded from the concentration of dissolved silicate species (normalized peak areas) as a function of alkali metal cation (Figure 45.2). This result is contradictory to the expectation that a monotonic increase or decrease in dissolution rate is to be observed for the different cations used. One major effect that occurs at the high pH values of this study is that the majority of silanol groups (≡Si–OH) at the surface of the silica gel are ionized by hydroxyl anions present in the highly alkaline solution. This result implies that a large surface charge will be present on the silica. Because in aqueous solutions at high pH values contra-ions (cations) for the hydroxide (normally sodium) are present, the negative surface charges will be compensated by (alkali metal) cations.

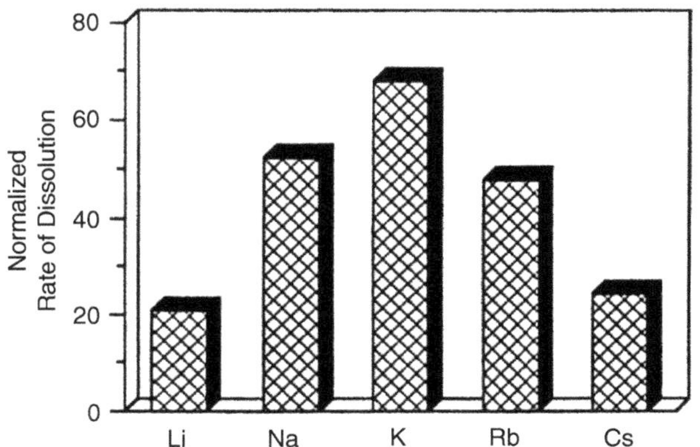

FIGURE 45.2 Normalized rate of dissolution of amorphous silica gel in alkali metal hydroxides as determined from the initial, integrated peak area of dissolved species (5 wt% silica suspensions; M_2O:$3SiO_2$:$180H_2O$). Other experimental details are given in the caption of Figure 47.1.

Because the only variable changed in this dissolution study was the type of alkali metal hydroxide, differences in dissolution rate must be attributed to differences in adsorption behavior of the alkali metal cations. The affinity for alkali metal cations to adsorb on silica is reported [8] to increase in a continuous way from Cs^+ to Li^+, so the discontinuous behavior of dissolution rate cannot simply be related to the adsorption behavior of the alkali metal cations. We ascribe the differences in dissolution rate to a promoting effect of the cations in the transport of hydroxyl anions toward the surface of the silica gel. Because differences in hydration properties of the cations contribute to differences in water bonding to the alkali metal cations, differences in local transport phenomena and water structure can be expected, especially when the silica surface is largely covered by cations. Lithium and sodium cations are known as water structure formers and thus have a large tendency to construct a coherent network of water molecules in which water molecules closest to the central cation are very strongly bonded; slow exchange (compared to normal water diffusion) will take place between water molecules in the nearest hydration shell and the bulk water molecules.

On the other hand, cesium, rubidium, and, to a lesser extent, potassium are known as water structure breakers; they cause the water molecules to be very disordered around the central cation. These larger cations have a small tendency to coordinate water molecules in a hydration shell, so lithium cations interact strongly with the (inner) hydration shell, whereas cesium cations barely have a hydration shell. As adsorption of (hydrated) cations will take place for aqueous silicate dispersions at high pH values, the state in which the adsorbed cations are present at the surface is crucial for promoting effects. Strong hydration forces (lithium) give rise to a slow exchange of water (hydroxyl) molecules, whereas no hydration (cesium) results in slow exchange as well. Potassium is the most favored alkali metal cation in transport reactions of hydroxyl anions towards the silica gel surface.

This phenomenon was confirmed by the introduction of symmetric tetraalkylammonium hydroxides in the dissolution of silica gel. In TMAOH the observed rate of dissolution was slow compared to the rate observed for cesium hydroxide dispersions, and cesium hydroxide has the lowest rate for the different alkali metal hydroxides. Results in Figure 45.3 clearly reveal an inhibition time between mixing of the silica gel with the aqueous TMAOH and the onset of dissolution. This observation is attributed to the strong interaction of the rather apolar TMA cation with the negatively charged silica gel surface. Because in this case no hydration shell is present, dissolution only occurs very slowly. The observed inhibition period of the dissolution reaction can be related to specific interactions of TMA cations with relatively large oligomeric species of the monomeric silicic acid. From the time that prismatic hexameric silicate anions (double three-membered ring, Q_6^3) are present in the solution (*see* Figure 45.1e; $\delta = -89.2$ ppm), TMA cations migrate from the silica surface to the solution; this migration leaves an almost noncovered silica gel surface, and dissolution occurs as fast as in alkali metal hydroxide solutions. A speculative model for the increased affinity of TMA cations for the silicate solution is provided by the clathrate structure of TMA silicate solutions. In this model, oligomers of silicic acid (cubic octameric silicate) are surrounded by tetramethylammonium cations in such a way that the geometric nature of the silicate oligomer matches the structure of the hole induced by the arrangement of TMA cations [9].

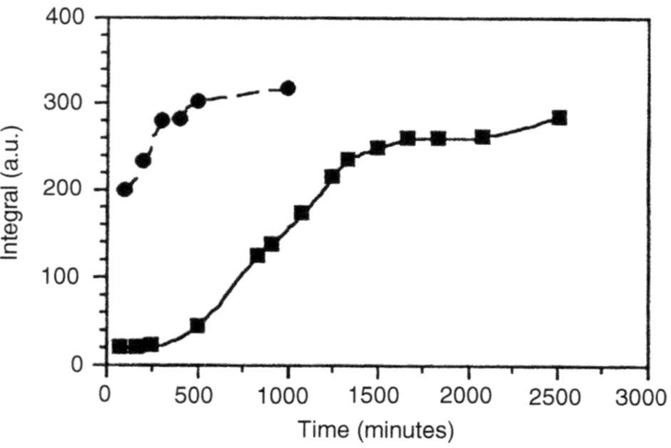

FIGURE 45.3 Total amounts of dissolved silicate anions (in terms of integrated peak area; a.u., arbitrary units) in solutions of cesium silicate (dashed line) and tetramethylammonium silicate as a function of reaction time (TMA_2O or Cs_2O:$3SiO_2$:$180H_2O$). The temperature was 25°C; other experimental details are given in the caption of Figure 47.1.

SMALL-ANGLE X-RAY SCATTERING

Silicon-29 NMR spectroscopy has thus been used to investigate the oligomerization of aqueous monosilicic acid. Prolonged polymerization of aqueous silicate solutions, however, yields a very broad distribution of polysilicic anions and colloidal structures [6,7]. This fact implies that with ^{29}Si NMR spectroscopy, only information with little detail on the ongoing polymerization process can be obtained, and so the polymerization process can be described only in a qualitative way in terms of relative changes in NMR line intensities and line widths. Therefore, a different spectroscopic technique was applied to study silica gel formation: Small-angle x-ray scattering SAXS [10]. As in x-ray diffraction, interference of scattered x-rays allows the identification of structural properties. From Bragg's relation ($n\lambda = d \sin 2\theta$; λ is the wavelength, d is the interatomic separation, 2θ is the angle between the incident and diffracted light, and n is an integer), it can be seen that large structural features should be investigated at relatively small scattering (diffraction) angles (typically $2\theta < 5°$). Systems containing colloidal or subcolloidal particles can be studied at small scattering angles. silica gels are constructed of a continuous network of particles of colloidal size (typically 3–50 nm in diameter). In the early 1980s, silica gels prepared from alcoholic precursor solutions (in contrast to the aqueous solutions discussed here) yielded scattering curves that were indicative of the formation of fractal structures [11].

FRACTAL BEHAVIOR IN SILICA GEL CHEMISTRY

In fractal theory a structure can be described in terms of its fractal or broken dimensionality [12]. This dimensionality, in contrast to the Euclidean dimensionality, which quantifies the space dimensionality embedding the structure, often has a noninteger value between 1 and 3. Fractal structures do not have a constant value for density; it gradually changes when traversing the system. For mass fractals, the way in which density varies is reflected in the fractal dimensionality D: density ρ varies with length scale r according to $\rho \propto r^{D-3}$. The fractal dimensionality can be used as a kind of fingerprint in the description of the process of aggregation of primary silica particles: different types of aggregation process result in different fractal dimensionalities [13].

Fractal behavior is reflected in a power-law relationship between scattered intensity I and scattering vector $\mathbf{Q}$ ($\mathbf{Q} = 2\pi/\lambda \sin 2\theta$). Therefore, in a log–log plot of scattered intensity versus scattering vector, fractality is observed as a linear region of the scattering curve; the

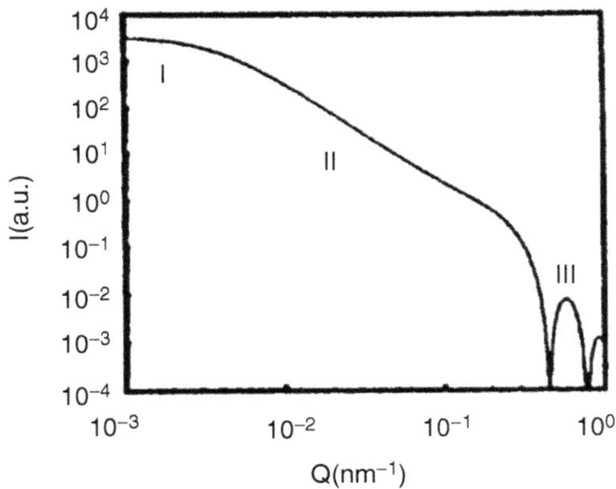

FIGURE 45.4 Simulated small-angle scattering curve from fractal geometry theorems (16). The small-angle scattering curve represents scattering from a fictitious fractal aggregate (R = 250 nm, D = 2.25, and the size of the primary building unit r_0 was 3 nm).

slope of this linear part is, in general, smaller than 3 if the material is fractal with respect to the mass (variation of density within the structure) and between 3 and 4 if the material is fractal with respect to the surface area [13–15]. In Figure 45.4, a computed small-angle scattering curve of a fractal system is presented [16]. From deviations from power-law scattering at small and large scattering vectors, information is extracted concerning the size of the fractal aggregates and the size of the primary particles constructing the aggregate, respectively. These two parameters determined from in situ small-angle scattering curves are the structural properties that are determined by aggregation kinetics imposed by the precursor composition. A continuous increase in size of the silica aggregates was observed during the formation of silica gel [17]. Moreover, after reaching the gelation point, the aggregate size still appears to increase as a function of time. At the gelation point, a continuous percolating network of silica particles exists in the solution, which has an infinite viscosity: twisting of the reaction vessel does not deform the meniscus.

According to classical theories of gelation, a continuous aggregate of infinite size should be present at the gelation point. However, from the SAXS curves a determinable, finite size of the aggregates is extracted. This observation suggests that the growth process of silica aggregates has to be anisotropic. The finite size of fractal silica aggregates in silica gels might indicate the presence of elongated, intermingled structures with a finite (mean) aggregate radius. On the other hand, according to Martin and Hurd [14], SAXS on gelated, nondilute silica systems gives no information about the size of the aggregates, and so no conclusion should be drawn from the development of aggregate size at relatively long reaction times. From a certain reaction time near the gelation point, the solutions can no longer be considered dilute, so determination of aggregate sizes becomes disputable. However, this chapter focuses on qualitative differences in the evolution of aggregate size at reaction times that are short compared to the gelation time.

The size of primary particles can be extracted from the deviation of fractal behavior at large scattering vectors **Q**. In silica gels freshly formed from aqueous silicate solutions, no deviation from this fractal power law could be observed for **Q** values as high as 2 nm^{-1}. This result implies that the primary particles of freshly prepared silica gels are smaller than 2 mm, maybe even of molecular size. Oligomers present in the aqueous silicate solutions used as precursor solutions are not subject to further growth in acidic solutions, but merely aggregate into continuous networks of silicate particles. At low pH values (pH $\approx$ 4) aggregation of oligomers is a fast process in relation to growth of the primary particles.

Influence of Total Silica Concentration on Aggregation Kinetics

Dilution of the silicate solution would be expected to give rise to a crossover in the type of aggregation behavior of the primary silicate particles. For relatively concentrated systems ($[SiO_2] \approx 5$ wt%), a reproducible fractal dimensionality of $D = 2.20 \pm 0.05$ was measured (Figure 45.5). This value of the fractal dimensionality is in fairly good agreement with fractal dimensionalities obtained from computer simulations of reaction-limited cluster–cluster aggregation [13]. In this limiting situation of reaction-limited aggregation, diffusion of particles or clusters of particles toward each other is faster than the kinetics of the chemical

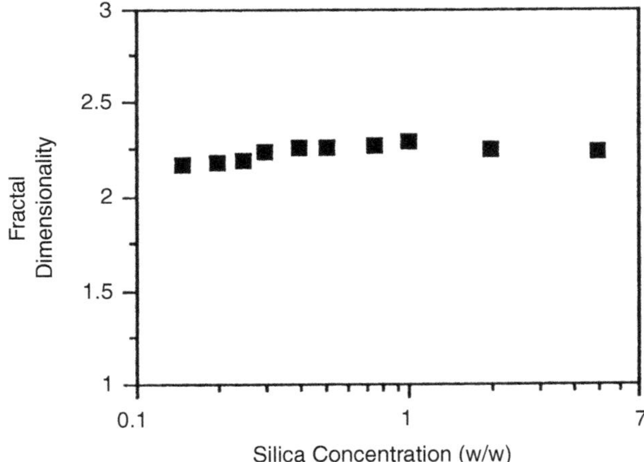

FIGURE 45.5 Fractal dimensionality of aqueous silica aggregates differing in total silica concentration obtained from SAXS spectra. The spectra were recorded after 5 days of reaction (pH 4.0 at 25°C) in sealed polyethylene bottles. SAXS spectra were recorded at the Synchrotron Radiation Source of Daresbury Laboratories, United Kingdom, on beam line 8.2.

reaction between the particles. Dilution of the silicate solution in aggregating particles (total silica concentration) should therefore result in an increasing importance of the diffusion rate of the aggregating particles. So *reaction*-limited aggregation should change into *diffusion*-limited aggregation, as was observed by Aubert and Cannell [18] for colloidal silica solutions. This transition in type of kinetic aggregation process is accompanied by a difference in fractal dimensionality. Experimentally, however, no influence of the total silica concentration ($0.1<[SiO_2]<$ 8 wt%) on the fractal dimensionality was observed at the length scales (1–30 nm), temperature, and time scales investigated (Figure 45.5). This result implies that diffusion of particles in the aqueous silicate solution or suspension must be a fast process compared to the reaction kinetics between two aggregating particles, or internal organizations must be fast compared to aggregation. Concentrations lower than 0.1 wt% do not lead to aggregation processes.

In contrast with these observations concerning the independence of fractal dimensionality from monomeric silicate concentration, the remarkable observation was made that the presence of oligomeric silicate anions increases the rate of aggregate formation considerably [17]. A decrease in the pH value of precursor solutions low in silica concentration (about 0.2 wt%), gives rise to oligomerization of the monomeric silicate anion. The oligomers formed appear to act as aggregation kernels for the silicate anions. The presence of these aggregation kernels induces rapid aggregation, as is reflected in the fractal dimensionality of the aggregates. Compared with usually observed fractal dimensionalities of aqueous silicate aggregates ($D = 2.2$), the fractal dimensionality of aggregates grown in solutions of low silica concentration and in the presence of oligomers is low ($D = 1.8$). This low fractal dimensionality points to a diffusion-limited cluster–cluster aggregation process. Reorganizations in the structural arrangement of the primary building units gradually cause an increase in fractal dimensionality to values observed for the reaction-limited cluster–cluster aggregation type [19,20].

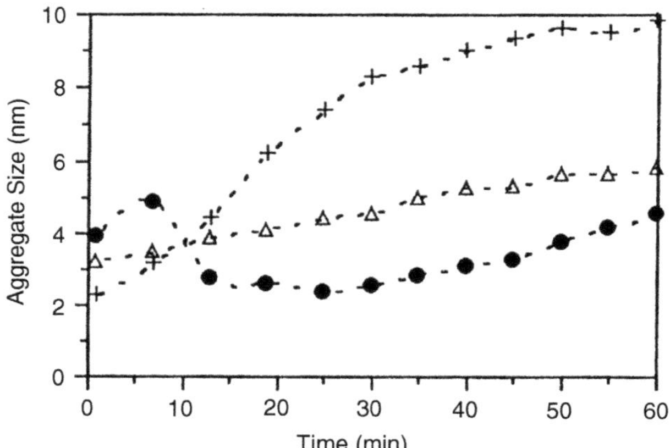

FIGURE 45.6 Size of aggregates as a function of reaction time for silicate solutions containing different cations: +, K only; Δ, Mg:Si = 0.01 mol/mol; •, Al:Si = 0.01 mol/mol. The total concentration of silica was 5 wt%. Silica gels were prepared at pH 4.0 at room temperature (25°C) by adding the silicate solution under vigorous stirring to a solution of hydrochloric acid (2 *N*) containing $MgCl_2$ or $Al_2(SO_4)_3$.

Influence of Polyvalent Cations on Aggregation Kinetics

This influence was investigated. No influence on the final fractal dimensionality of the gelating silicate solutions was observed for the cations investigated (Li^+, Na^+, K^+, Rb^+, Cs^+, TMA^+, Mg^{2+}, and Al^{3+}). The rate of aggregate growth, however, does strongly depend on the presence of polyvalent cations and hydrophobic monovalent cations (TMA) [21]. Aggregation is retarded by addition of aluminum cations (Al:Si = 0.01) and to some lesser extent by addition of magnesium cations (Figure 45.6). This difference is understood in terms of the valency of the cations. Investigation of the influence of different concentrations of aluminum cations revealed that increasing aluminum concentrations cause the observed inhibition with respect to aggregate growth to diminish (Figure 45.7). Fast reaction between aluminum cations and silicate oligomers causes the silica particles to become charge stabilized. Aggregation is slow because of the charges of the primary particles. The structure of the aggregates is such that in a model of screened aggregation, initially low

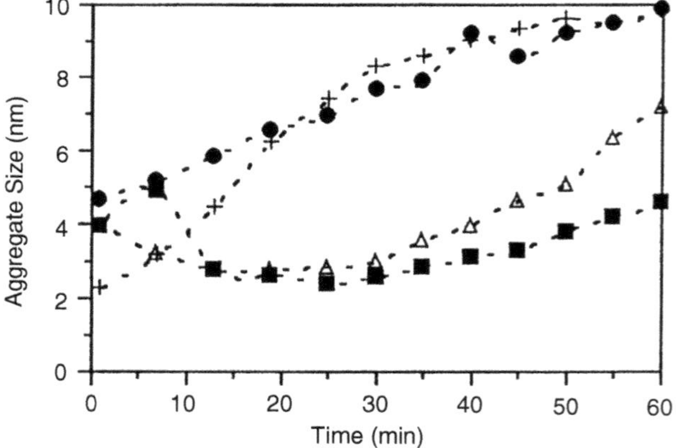

FIGURE 45.7 Size of aggregates as a function of reaction time for silicate solutions containing different concentrations of aluminum cations: +, K only, Al:Si = 0.0; ■, Al:Si = 0.01; Δ, Al:Si = 0.04; and •, Al:Si = 0.10. The pH was 4.0, and the temperature was 25°C. The total concentration of silica was 5 wt%. Silica gels were prepared at pH 4.0 at room temperature (25°C) by adding the silicate solution under vigorous stirring to a solution of hydrochloric acid (2 *N*) containing different concentrations of $Al_2(SO_4)_3$.

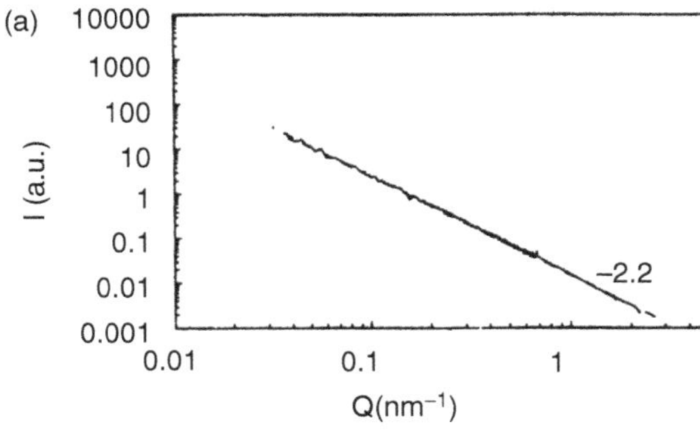

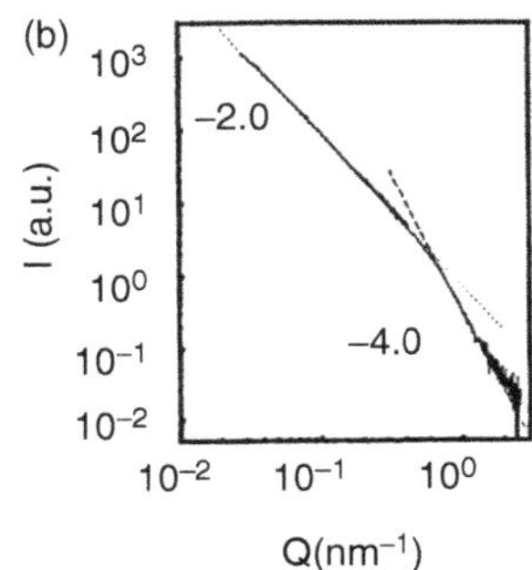

FIGURE 45.8 SAXS curves of a silica gel prepared at pH 3.9 from potassium water glass and hydrochloric acid: (a) freshly prepared silica gel recorded after 2 h of reaction; (b) same gel as in a, but after 1 year of aging at room temperature. The dotted line is a nonlinear least-squares fit of the fractal region in the curve and gives $D = 2.0$. The dashed line is from the Porod law ($I \propto \mathbf{Q}^{-4}$). In both a and b, two scattering curves measured at two different camera lengths (4.5 and 0.7 m) are combined to cover a broad scattering range (2 orders of magnitude in **Q** vector).

dimensionalities are observed. However, when the concentration of aluminum is increased, formation of homogeneous aluminosilicates causes charge stabilization to decrease. The aggregation kinetics resemble kinetics in solutions with no aluminum cations added.

Addition of tetramethylammonium cations results in an increase in aggregation rate, which can be attributed to the breaking of the local structure of water molecules surrounding the silicate particles. The activation energy of reaction between two primary particles is decreased.

AGING OF AQUEOUS SILICA GELS

So far, a discussion on the processes involved in the formation of silica gels has been presented. However before a silica gel, with its intrinsic properties of high specific surface area preserved, can be made, it must be aged in a way that prevents the framework of silica particles from collapsing upon drying. Until now, aging of aqueous silica gels was considered an Ostwald dissolution process of silicate units at surfaces with small radius of curvature (small particles) and subsequent deposition at surfaces with large or negative radius of curvature (large particles and necks between particles, respectively) (*1*, p 228). In Figure 45.8, small-angle scattering curves of a freshly prepared and an aged aqueous silica gel are presented; different scattering characteristics are evident. In the scattering profile in Figure 45.8b, the scattering of primary particles can be distinguished (part III in Figure 45.2) from the case of freshly prepared silica gels, for which no scattering of primary particles can be observed (Figure 45.8a). Thus, the mean size of the scattering primary particles increases during aging from molecular level (<0.5 nm) to colloidal level (ca. 3–5 nm). The fractal dimensionality on the other hand has changed to a value smaller than the initial fractal dimensionality, so the gradient in density within the aggregates has become larger. Here dissolution is presumed to preferentially occur at peripheral positions in the aggregates, where the density in silica is low compared to the more dense core of the aggregates. Subsequent deposition of the monomeric species in the core of the aggregates causes the difference in density between the core (high) and the peripherals (low) to increase; a decrease in fractal dimensionality results.

The driving force for structure transformation can be related to the entropy of the structures [22]. Very open structures are of highest entropy and thus will strive for a state that has less entropy. Because the structure of the silica aggregates is more or less rigid, the only way to decrease entropy is to gradually dissolve silica at places of highest entropy (i.e., the low-density peripherals of the fractal structures) and to deposit the dissolved monomers at places with higher density (i.e., lower entropy). An increase in solubility of the silica aggregates will contribute to a faster process of structure reorganization. As such, aging of aqueous silica gels can be accelerated by higher pH values, high temperatures, and the presence of fluorine anions [20]. The rate of dissolution of monomeric silicate anions is rate determining in the aging processes of aggregates formed by reaction-limited cluster–cluster aggregation. Interestingly, aging of aggregates formed by reaction-limited aggregation gives a decrease in fractal dimensionality with an increase in primary particle size. Aging of aggregates formed by diffusion-limited cluster–cluster aggregation results in an increase of fractal dimensionality at (nearly) constant primary particle size [20]. The time scales on which both transformations take place are very different. Aging of aggregates grown by diffusion-limited aggregation is fast compared to aging of reaction-limited aggregates. Both transformations correspond to a decrease in system entropy and potential energy.

REFERENCES

1. Iler, R. K. *The Chemistry of Silica*; Wiley: New York, 1979.
2. Brinker, C. J.; Scherer, G. W. *Sol Gel Science*; Academic: Boston, MA, 1990.
3. Hench, L. L.; West, J. K. *Chem. Rev.* **1990**, *90*, 33.
4. Barrer, R. M.; Coughlan, B. *Molecular Sieves;* Society of Chemical Industry London, 1968.
5. Engelhardt, G.; Jancke, H.; Mäge, M.; Pehk, T.; Lippmaa, E. *J. Organometallic Chem.* **1971**, *28*, 293.
6. Wijnen, P. W. J. G.; Beelen, T. P. M.; De Haan, J. W.; Rummens, C. P. J.; Van de Ven, L. J. M.; Van Santen, R. A. *J. Non-Cryst. Solids* **1989**, *109*, 85.
7. Wijnen, P. W. J. G.; Beelen, T. P. M.; De Haan, J. W.; Van de Ven, L. J. M. Van Santen, R. A. *Colloid Surf.* **1990**, *45*, 255.
8. Depasse, J.; Watillon, A. *J. Colloid Interface Sci.* **1970**, *33*, 430.
9. Keijsper, J. J.; Post, M. F. M. In *Zeolite Synthesis*; Occelli, M. L.; Robson, H. E., Eds.; Symposium Series 398; American Chemical Society: Washington DC, 1989; pp 28–48.
10. Guinier, A.; Fournet, G. *Small Angle Scattering of X-rays*; Wiley: New York, 1955.
11. Brinker, C. J.; Keefer, K. D.; Schaefer, D. W.; Ashley, C. S.; Assink, R. A.; Kay B. D. *J. Non-Cryst. Solids* **1982**, *48*, 47.
12. Mandelbrot, B. B. *The Fractal Geometry of Nature*; W. H. Freeman and Co. San Francisco, CA, 1982.
13. Meakin, P. *Adv. Colloid Interface Sci.* **1988**, *28*, 249.
14. Martin, J. E.; Hurd, A. J. *J. Appl. Cryst.* **1987**, *20*, 61.
15. Schmidt, P. W. In *The Fractal Approach to Heterogeneous Chemistry*; Avnin D., Ed.; Wiley: New York, 1989.
16. Teixeira, J. *J. Appl. Cryst.* **1988**, *21*, 781.
17. Wijnen, P. W. J. G. Ph.D. Thesis, Eindhoven University of Technology Eindhoven, the Netherlands, 1990.
18. Aubert, C.; Cannell, D. S. *Phys. Rev. Lett.* **1986**, *56*, 738.
19. Wijnen, P. W. J. G.; Beelen, T. P. M.; Rummens, C. P. J.; Saeijs, J. C. P. L.; Van Santen, R. A. *J. Appl. Cryst.* **1991**, *24*, 759.
20. Wijnen, P. W. J. G.; Beelen, T. P. M.; Rummens, C. P. J.; Saeijs, J. C. P. L.; Van Santen, R. A.; *J. Colloid Interface Sci.* **1991**, *145*, 17.
21. Beelen, T. P. M.; Wijnen, P. W. J. G.; Rummens, C. P. J.; Van Santen, R. A. In *Better Ceramics through Chemistry IV;* Zelinski, B. J. J.; Brinker, C. J.; Clark, D. E.; Ulrich, D. R., Eds.; *Mat. Res. Soc. Symp. Proc*; Materials Research Society: San Francisco, CA, 1990; Vol. 180; pp 273–276.
22. Kaufman, J. H.; Melroy, O. R.; Dimino, G. M. *Phys. Rev. A* **1989**, *39*, 1420.

46 Interpretation of the Differences between the Pore Size Distributions of Silica Measured by Mercury Intrusion and Nitrogen Adsorption

A.R. Minihan, D.R. Ward, and W. Whitby
Unilever Research Port Sunlight

CONTENTS

Measuring the nitrogen sorption isotherms of a number of silicas both before and after analysis by mercury intrusion demonstrates that mercury intrusion can lead to compression of silica structures and that this compression can account for differences in pore size distributions measured by the nitrogen sorption and mercury intrusion techniques. These techniques are widely employed in the structural characterization of porous solids, often independently, despite the fact that very often the pore size distributions obtained by the two techniques fail to agree. Compression effects must be recognized because use of incorrect information can lead to misconceptions regarding the structure of a material.

Mercury intrusion and nitrogen sorption are two common techniques used to analyze the structures of porous solids. However, they can give different pore size distributions or pore volumes for a given solid. Giles et al. [1] suggested that differences between pore size distributions as measured by mercury intrusion and by nitrogen sorption might be due to progressive rearrangement of the structure during mercury intrusion analysis followed by breakthrough into the voids between the globular particles when the particles reach a coordination number of 4. More recent work [2] on fume silicas and silica aerogels, which were examined by mercury intrusion after enclosure in an impermeable membrane, demonstrated that the bulk of the intrusion that takes place with such materials is associated entirely with compaction of the powder particles.

The effect of mercury intrusion analysis on structure was examined for a series of silica xerogels with different pore size distributions. This analysis was achieved by applying nitrogen sorption analysis to the silicas both before and after mercury intrusion analysis. The study required the development of a method for the removal of mercury from a sample after the initial intrusion measurement that does not damage the structure. The results show the potential for an elastic deformation of the structure during compression as well as irreversible compression during mercury intrusion.

EXPERIMENTAL DETAILS

MATERIALS

The silica samples were selected on the basis of high purity, narrow pore size distributions, and availability with a large particle size, such that the inter- and intraparticle porosity regions are clearly distinguishable in the mercury intrusion curves. This feature allows the examination of the internal pore structure without the confusion of overlapping interparticle porosity. An experimental sample of silica [similar to silica produced by Crosfield Chemicals, Warrington, United Kingdom, as a support for Phillips ethylene polymerization (EP) catalyst] and a series of silicas manufactured by Crosfield for chromatographic applications (Sorbsil C60, Sorbsil C200, and Sorbsil C500) were examined. The surface areas of these materials, determined from the nitrogen adsorption isotherms by the Brunauer–Emmett–Teller (BET) equation [3], are shown in Table 46.1. The effect of mercury intrusion on a sample of silica spheres (S980 G1.7, manufactured by Shell) was also examined. To eliminate any errors due to moisture sorption by the silicas, all samples were predried at 120°C for at least 2 h and stored in a desiccator until used.

NITROGEN GAS ADSORPTION ANALYSIS

The apparatus used was a Micromeritics ASAP 2400, a fully automatic nitrogen gas sorption apparatus that can be programmed to measure gas adsorption and desorption isotherms and calculate surface areas and pore volumes by using a number of widely accepted procedures. All samples were outgassed initially at room temperature until a pressure of less than 100 mtorr (13 Pa) was achieved. Outgassing was completed by heating the samples to 120°C and evacuating until a pressure of less than 5 mtorr (0.7 Pa) had been sustained for at least 2 h. The criteria for terminating the outgassing step was the attainment of a stable pressure less than 5 mtorr and was not based on the outgassing time. Typical outgassing times were between 12 and 18 h, and the samples were held under vacuum until the start of the analysis. After outgassing, the samples were cooled under vacuum and the tubes were back-filled with helium before being transferred to the analysis ports.

TABLE 46.1
Surface Areas of Silica Samples

Sample	BET surface area
EP silica	301
Sorbsil C60	511
Sorbsil C200	299
Sorbsil C500	114

Note: Data are reported as m^2/g.

The surface areas in Table 46.1 were obtained from the adsorption isotherms by using the BET method [3]. The relative pressure range used for BET analysis was selected to give the best linear correlation with the BET function, and the surface areas were calculated by assuming a molecular cross section for the nitrogen molecule of 0.162 nm^2.

The pore size distributions were calculated by using the desorption isotherm, following the method of Barrett, Joyner, and Halenda (BJH) [4]. In this procedure the Kelvin equation is used to calculate the radius r_p of the capillaries, which are assumed to be cylindrical:

$$\ln\left(\frac{p}{p_0}\right) = \frac{-nV_m\gamma}{r_pRT}\cos\Theta \tag{46.1}$$

Here V_m is the condensed molar volume (34.68 cm^3/mol for nitrogen); γ is the liquid–vapor surface tension (8.72×10^{-3} N/m for nitrogen); R is the gas constant; T is the temperature; p is the pressure of nitrogen above the sample; p_0 is the saturation vapor pressure of nitrogen at temperature T; and n is a unitless factor. The contact angle Θ is assumed to be zero, and the value of n is set to 2 for the desorption branch of the isotherm. The pore radius is then calculated from r_p by adding the thickness of the adsorbed layer present before capillary condensation takes place. This thickness (t) is calculated by using the Halsey [5] equation:

$$t = \sigma\sqrt[3]{\frac{5}{\ln(p_0/p)}} \tag{46.2}$$

A value of 0.354 nm is used for the average thickness σ of a single molecular layer of nitrogen. The algorithm used in the ASAP 2400 software is based on Faas's [6] implementation of the BJH method.

MERCURY INTRUSION ANALYSIS

Mercury intrusion measurements were carried out with a porosimeter (Micromeritics 9220) capable of intruding mercury with intrusion pressures (p) up to 414 MPa (60,000 psi). Pore size distributions were calculated from the intrusion curve by using the Washburn [7] equation:

$$p = \frac{-2\gamma\cos\Theta}{r_p} \tag{46.3}$$

A value of 140°C was used for the contact angle of mercury on the solid (Θ), and the surface tension of mercury (γ) was taken as 0.485 N/m. These values correspond to an effective working range for the instrument of 150 μm to 1.7 nm in pore radius. The samples were outgassed at room temperature to a pressure of 50 mtorr (7 Pa) immediately prior to analysis to facilitate filling the penetrometers with mercury. All data were fully corrected for mercury compression with calibrated penetrometers.

Experimental Procedure

Each sample was first characterized by both mercury intrusion and nitrogen sorption. Mercury intrusion measurements were replicated at least four times, and the solid residues from each analysis were collected and combined after the bulk of the mercury was decanted. These samples were washed free of mercury by using 50% nitric acid (25 mL per 0.5 g of solid) and then washed free of acid by filtering and reslurrying in demineralized water (six times with 50 mL per 0.5 g of solid). The washed samples were then rapidly cooled in liquid nitrogen and freeze-dried (Chemlab SB4). For comparison, samples of material that had not been analyzed with mercury intrusion were washed and dried in a similar manner to test for structural modification caused by the acid-washing technique.

After being dried, the samples were reexamined by nitrogen sorption and mercury intrusion, and a portion of the material was analyzed to determine the residual mercury levels. This analysis was achieved by acid digestion (10 mL of 50% aqua regia; sample sizes were approximately 0.2 g in all cases) in pressure-sealed poly(tetrafluoroethylene) (PTFE) tubes heated to 140°C for 10 min in a microwave oven (CEM). The solutions were analyzed after suitable dilution in distilled water with a graphite furnace atomic absorption spectrometer (Perkin Elmer 5100-PC). The detection limit for this method is estimated to be 6 ppm of mercury on the dry solid.

RESULTS AND DISCUSSION

The pore size distribution of the high-pore-volume silicas used for EP applications can be determined from the nitrogen adsorption isotherms by using the BJH method [3] described; a typical isotherm is shown in Figure 48.1. The isotherm has a "type A" hysteresis loop according to de Boer's classification [8], and this hysteresis indicates the pore structure has a uniform cylindrical form with no evidence of "ink bottle" pores. These structures, however, have a mercury intrusion curve similar to that shown in Figure 46.2. Three distinct regions of intrusion are usually observed; the first intrusion step at around 30 μm is associated with the voids between particles, and its exact position is dependent on the particle size distribution of the sample. This interparticle intrusion step is not of interest here, and the silicas selected for study were chosen because they have large particle sizes. Inter- and

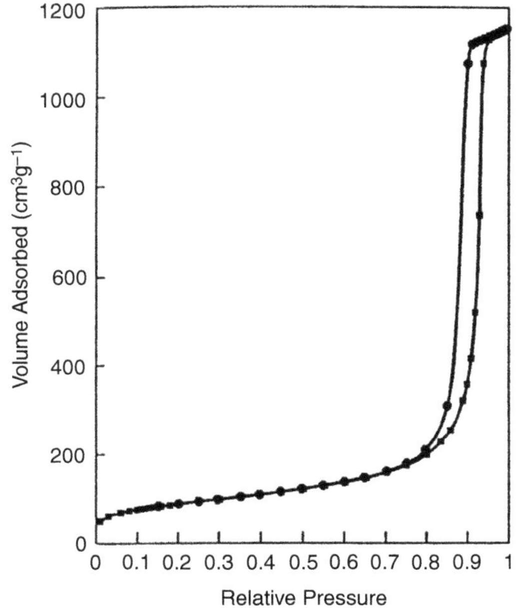

FIGURE 46.1 Nitrogen sorption isotherms of a typical high-pore-volume silica used in EP applications.

intraparticle pore size regions are thus easily resolved, and data interpretation in the remainder of this chapter concentrates solely on the internal porosity. The second and third intrusion steps are associated with the intraparticle porosity and should be directly comparable to the

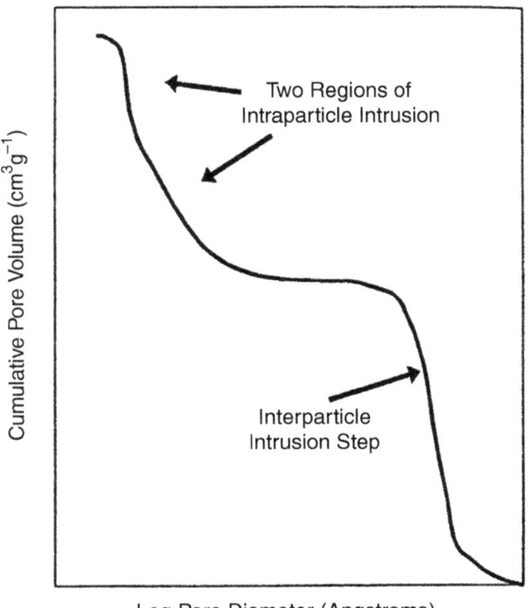

FIGURE 46.2 Mercury intrusion curve of a typical high-pore-volume silica used in EP applications.

nitrogen desorption pore size distributions. Figure 46.3 shows such a comparison for the model EP-type silica.

These curves show that the pore size distributions measured by the two techniques are different, although the total pore volumes are similar. This observation suggests that although the two techniques are measuring the same pore structure, either the models used in the interpretation of the data are inappropriate and do not adequately describe the pore structure or the structure is modified during analysis. Earlier workers [1,2] suggested that these differences are due to compression of the silica structure under the pressure applied in the mercury intrusion analysis. If this assessment is correct, such compression should be reflected in a change in the structure as determined by nitrogen sorption following the mercury intrusion experiment.

Effect of Mercury Removal Method on Silica Structure

To ensure that the washing technique used to remove the mercury from the samples after intrusion analysis does not alter the silica structures, some of the original samples of silica were treated in identical fashion and then reanalyzed by nitrogen sorption. Figure 46.4 and Figure 46.5 show the nitrogen desorption pore size distribution curves of the Sorbsil C60 and Sorbsil C200 samples before and after such treatment. These results demonstrate that no significant structural modification results from the acid-washing treatment.

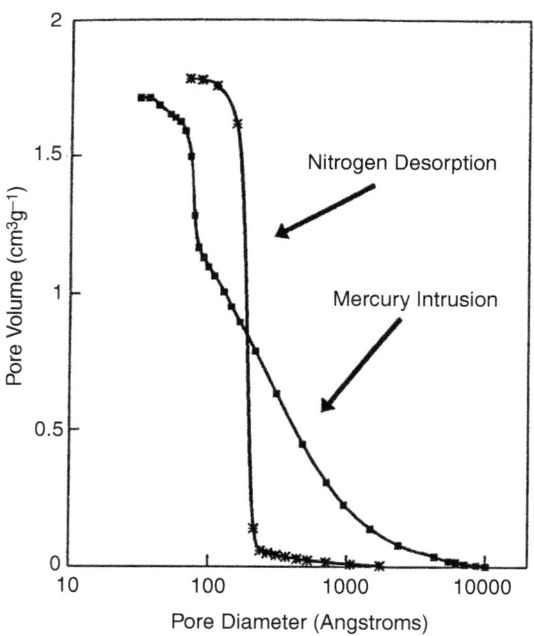

FIGURE 46.3 Comparison of the pore size distribution of a high-pore-volume silica measured by mercury intrusion and nitrogen desorption.

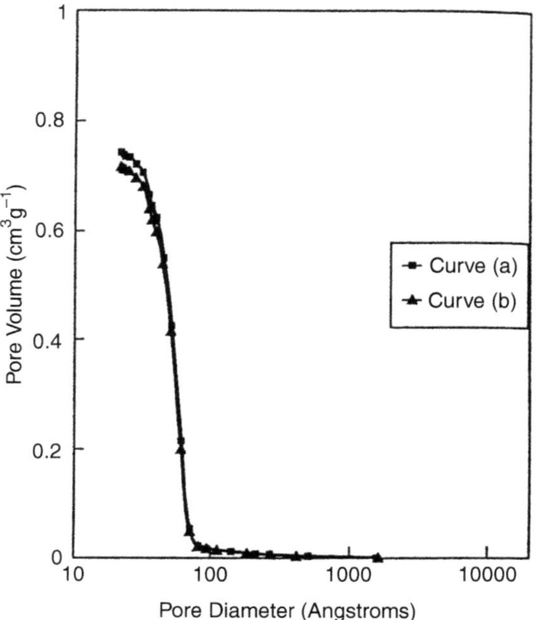

FIGURE 46.4 Effect of the mercury removal method on the pore structure of Sorbsil C60. Curve a is the material before treatment, and curve b is that after treatment.

Analysis for residual mercury with the digestion method described indicated that only trace amounts of mercury remain on the samples after the initial washing (Table 46.2). A detected level of 1000 ppm of residual mercury would result in an apparent loss in pore volume

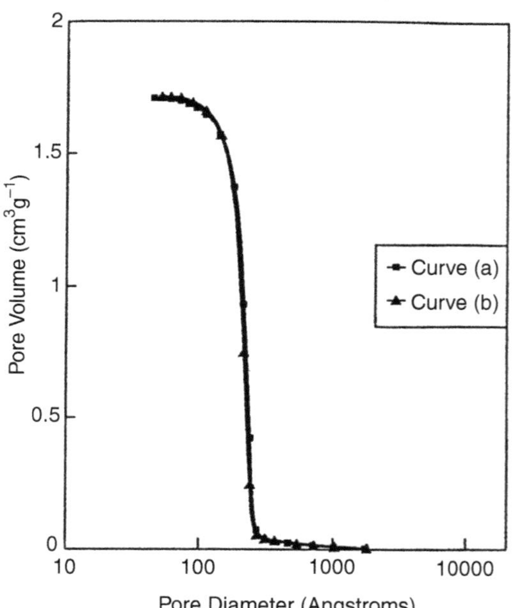

FIGURE 46.5 Effect of the mercury removal method on the pore structure of Sorbsil C200. Curve a is the material before treatment, and curve b is that after treatment.

TABLE 46.2
Mercury Levels after Intrusion and Washing

Sample	Mercury level detected
EP silica	320
Sorbsil C60	90
Sorbsil C200	1300
Sorbsil C500	60

Note: Data are reported as parts per million.

of 0.074 cm^3/g of silica. Thus the detected levels are always sufficiently low that any reduction in pore volume after intrusion cannot be explained in terms of residual mercury in the pore structure.

Effects of Mercury Intrusion

The facts that there is very little residual mercury in the solids after extraction and that the extraction method used does not cause any detectable structural modification have been established; now the earlier suggestion [1] that mercury intrusion causes compression of the pore structures can be examined.

Pore Structure of Sorbsil C500

Figure 46.6 depicts the internal cumulative pore volume as a function of pore diameter for Sorbsil C500. The total pore volumes as measured by the two techniques agree to within 0.05 cm^3/g; the pore volume measured by mercury porosimetry is slightly higher, possibly because this is an extremely wide-pored silica, with some pores too wide to be measured by nitrogen sorption.

As with the EP catalyst support, the nitrogen sorption technique (curve a) gives rise to a sharp, monotonic increase in pore volume over a narrow range of pore sizes, whereas the mercury intrusion technique yields a broad, almost linear increase in pore volume at large (>50 nm) pore sizes followed by a sharp increase in pore volume at pore diameters in the range 30–50 nm.

Whereas this broad portion of the intrusion curve represents the majority of the pore volume for the EP catalyst support (Figure 46.3), it corresponds only to approximately 20% of the total pore volume for the C500 silica. This fact is consistent with the concept of a compression step followed by an intrusion step, because the crossover between compression and intrusion would be expected to occur at a lower pressure for the more highly aged, wider pore silica.

Curve c in Figure 46.6 represents the pore size distribution as measured by nitrogen desorption after mercury intrusion analysis and subsequent mercury removal. The mercury intrusion clearly results in a significant loss in pore volume, an observation consistent with an irreversible

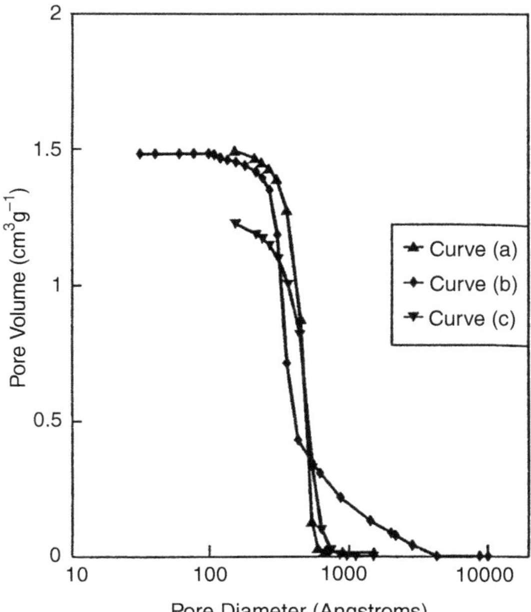

FIGURE 46.6 Effect of mercury intrusion on Sorbsil C500. The pore size distribution of the original material measured by nitrogen desorption is shown as curve a and that measured by mercury intrusion is shown as curve b. Curve c shows the nitrogen desorption pore size distribution after mercury intrusion and the removal of mercury.

compression of the silica during the intrusion process. This loss in pore volume as measured by nitrogen desorption is 0.32 cm^3/g, a value corresponding fairly closely to the 0.4 cm^3/g that represents the broad–diffuse portion of the intrusion curve. The reanalysis by nitrogen sorption thus provides strong evidence that the silica is irreversibly compressed during the mercury intrusion experiment. True intrusion into the pores only appears to occur when the work required to cause further compression is less than that required to force mercury into the pores. The intrusion pressure (and hence the apparent pore size) at which this situation occurs will be a function of the original pore size distribution and the strength of the silica structure.

Mercury Intrusion Experiments with Silica Spheres

These silica spheres (S980 G1.7 from Shell) were not examined in the same detail as were the other silica samples, but the photographs are included because they illustrate the effect of mercury intrusion on the integrity of the solid. These particular spheres have a typical pore volume of 1 cm^3/g and a pore diameter of 60 nm. The particles are also much larger than the Sorbsil materials (1.7 mm in diameter, compared to 40 to 60 μm for the Sorbsil materials).

Despite the fact that the pores in this material are quite large, the deformation caused by the compression effect is clearly demonstrated in Figure 46.7. The most noticeable feature in these pictures is the cracked and broken nature of the particle surfaces and some slight indications of concave surfaces between the cracks. The integrity of the spheres is otherwise substantially maintained. These spheres have pores slightly larger than those of the Sorbsil C500, and a similar effect might be expected from the compression of the C500. Rather than create large cracks in the surface, however, the smaller particle size material probably simply fractures into even smaller pieces. If this breakdown occurs, internal porosity is lost in favor of interparticle porosity.

Pore Structure of Sorbsil C200

The results obtained for the pore size distributions of Sorbsil C200 material are shown in Figure 46.8. Nitrogen adsorption (curve a) and mercury intrusion (curve b) again give different pore size distributions, the differences in the distributions being similar to those observed for the EP support (Figure 46.3). The internal pore volume as measured by mercury porosimetry is about 0.1 cm^3/g less than that obtained by nitrogen sorption. This difference may be due to the somewhat arbitrary choice of 1000 nm as the cutoff point between inter- and intraparticle porosity; some intrusion–compression may occur at larger pore sizes–lower pressures.

As was observed for the C500 silica, reanalysis of the pore structure by nitrogen sorption (Figure 46.8, curve c) following intrusion and removal of mercury indicates that the porosimetry experiment results in a permanent loss in pore volume and a shift to smaller pore sizes. In C200, however, this loss in pore volume no longer approximates that associated with the broad-diffuse area of the intrusion trace. Instead, the loss represents only about 40% of this region. Furthermore, a second intrusion experiment yields a trace (curve d) that confirms the loss in pore volume, but shows the same two regions, with an apparent compression phase followed by an intrusion step at virtually the same apparent pore size as in the initial intrusion experiment (curve b).

From this data the structure can be concluded to be compressed during the first mercury intrusion, but the compression region (i.e., the initial gradual slope) must contain two contributions, one associated with an irreversible collapse and a second associated with elastic compression. In Sorbsil C500, the irreversible collapse appears to account for virtually all of this region, whereas the two phenomena are of approximately equal magnitude in C200 silica.

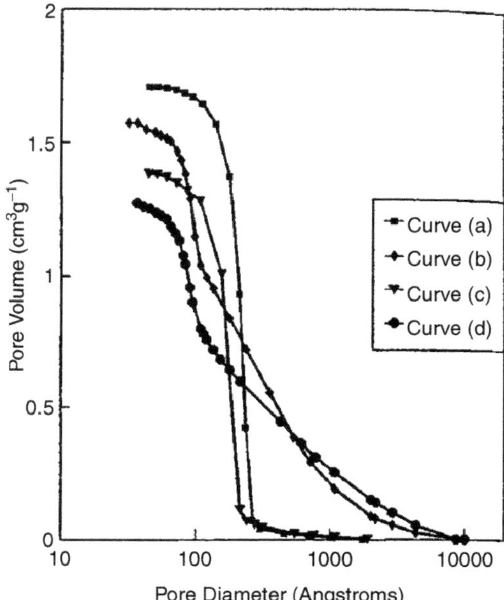

FIGURE 46.8 Effect of mercury intrusion on Sorbsil C200. The pore size distribution of the original material measured by nitrogen desorption is shown as curve a, and that measured by mercury intrusion is shown as curve b. The pore size distribution after removal of the mercury is shown as curve c (nitrogen desorption) and curve d (mercury intrusion).

(a)
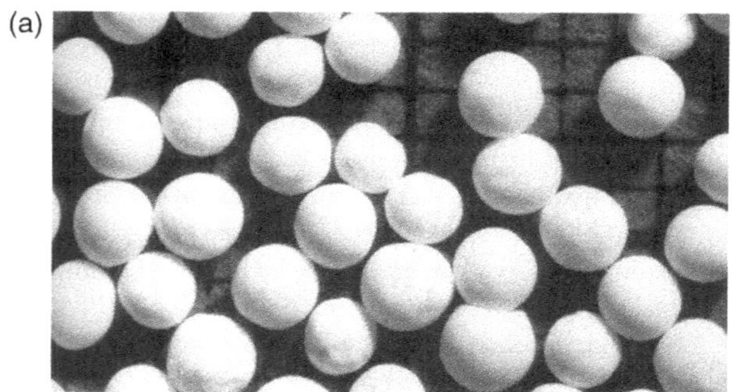
(b)

FIGURE 46.7 Effect of mercury intrusion on silica spheres (Shell). Top, starting material; and bottom, material after mercury intrusion (no attempt was made to remove the mercury).

Pore Structure of Sorbsil C60

Examination of a silica with a relatively small pore size, such as Sorbsil C60, produces a different picture. The results (Figure 46.9) indicate differences in both pore size distributions and pore volumes as measured by nitrogen

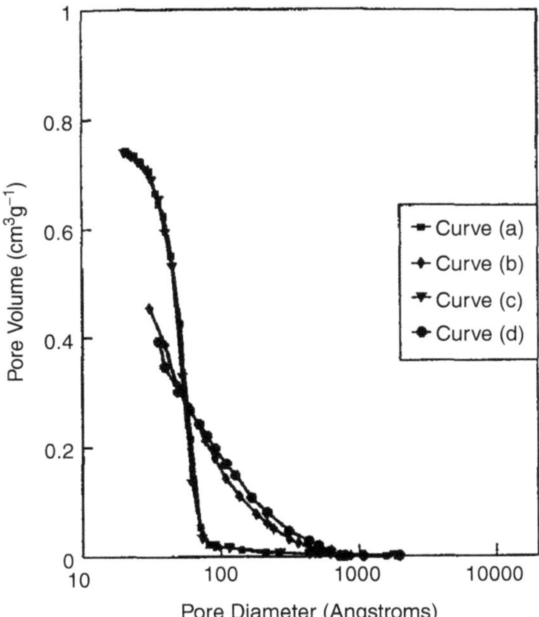

FIGURE 46.9 Effect of mercury intrusion on Sorbsil C60. The pore size distribution of the original material measured by nitrogen desorption is shown as curve a, and that measured by mercury intrusion is shown as curve b. The pore size distribution after removal of the mercury is shown as curve c (nitrogen desorption) and curve d (mercury intrusion).

desorption (curve a) and mercury intrusion (curve b). Pore volumes measured by mercury intrusion are now significantly lower than those measured by nitrogen sorption, and the intrusion trace does not show a sharp step as the intrusion pressure increases, but rather only the initial gradual slope, which extends throughout the measured range of the intrusion plot (40–1000 Å pore diameter). This observation suggests that the structure is still being compressed and that no intrusion has taken place when the upper pressure limit of the porosimeter has been reached.

Curves c and d in Figure 46.9 show nitrogen desorption and mercury intrusion traces, respectively, for pore size distributions after the initial intrusion experiment and subsequent mercury removal. In a surprising result, no permanent modification to the structure was caused by the initial intrusion experiment. In view of the results obtained with the silicas of larger pore size (Sorbsil C200 and C500), the results can be rationalized only in terms of a completely elastic compression of the structure during the intrusion experiment.

CONCLUSIONS

These results confirm the suggestions of earlier workers [1, 2] that the mercury intrusion method can lead to structural deformation of solids during analysis. For silicas, there appears to be both an elastic deformation and an irreversible compression effect that contribute to the differences in pore size distributions measured by nitrogen sorption and mercury intrusion. The irreversible compression appears to dominate in wide-pored silicas, which are presumably highly aged and therefore strong but brittle, whereas high-surface-area, low-pore-diameter silicas such as C60 undergo an almost completely elastic deformation.

With large-pore-size silicas (e.g., Sorbsil C500) the compressive effects of mercury intrusion are minimized because intrusion can take place into the pores at lower pressures. Pore volumes of such materials measured by mercury intrusion can be greater than those determined from nitrogen sorption because a significant fraction of pores lies outside the nitrogen sorption measurement range. Hence the pore size distributions measured by mercury intrusion can be more useful than the nitrogen sorption results.

Smaller pore size materials are subjected to greater compressive forces than wide-pore materials during mercury intrusion, because intrusion occurs at higher pressures and more marked structural changes occur before intrusion into the pores takes place. Some of this compression is an elastic deformation rather than an irreversible compaction. The pore sizes as measured by mercury intrusion for such materials are complex and depend not only on actual pore size distribution but on particle strength and deformability. As such, mercury intrusion is inappropriate for determining pore size distributions of these materials, although the technique can be used to determine total pore volumes as long as a sharp intrusion step is observed in the intrusion trace. Nitrogen sorption methods should therefore be used for the most realistic assessments of pore sizes for this type of material.

When the pore size of silicas is very small (as in C60), the compression of the structure caused by the mercury intrusion decreases the pore size to such an extent that it is outside the range for mercury intrusion analysis. The pore volumes measured represent only the compression region of the curve, and as a consequence the total pore volumes measured are lower than those measured for nitrogen. Nitrogen sorption methods are the only appropriate technique for this type of material.

The data presented show clearly the importance of using a combination of both mercury and nitrogen sorption methods if a complete understanding of porous solid structures is to be achieved. Even when mercury pore size distributions are capable of replication, they may not represent the true pore structure of the solid being examined.

ACKNOWLEDGMENT

We thank N. Whitehead for assistance in the development of the methodology for and determination of the residual mercury levels in the silica samples.

REFERENCES

1. Giles, C. H.; Havard, D. C.; McMillan, W.; Smith, T; Wilson, R. In *Characterisation of Porous Solids*; Gregg, S. J.; Sing, K. S. W.; Stoeckli, H. F., Eds.; Society of Chemical Industry: London, 1979; pp 267–284.
2. Smith, D. M.; Johnston, G. P.; Hurd, A. J. *J. Colloid Interface Sci.* **1990**, *135*, 227–237.
3. Brunauer, S.; Emmett, P. H.; Teller, E. *J. Am. Chem. Soc.* **1938**, *60*, 309.
4. Barrett, E. P.; Joyner, L. G.; Halenda, P. P. *J. Am. Chem. Soc.* **1951**, *73*, 373–380.
5. Halsey, G. D. *J. Chem. Phys.* **1948**, *16*, 931.
6. Faas, G. S., Masters Thesis, Georgia Institute of Technology, Atlanta, GA, 1981.
7. Washburn, E. W. *Phys. Rev.* **1921**, *17*, 273.
8. de Boer, J. H. *The Structure and Properties of Porous Materials*; Butterworth: London, 1958; p 68.

Part 6

Sol–Gel Technology

George W. Scherer
Princeton University

The term "sol–gel" processing is broadly applied to describe fabrication of inorganic materials by preparing a sol, inducing gelation, and then drying (and, usually, firing) the gel. The particles or polymers in the sol are usually grown by hydrolysis and condensation of metal halides or metal–organic compounds. The huge interest in this technology reflects its potential for making films, fibers, and powders of unusual quality; further, low-temperature processing offers the possibility of making novel materials by intimately combining organic and inorganic components. The following five chapters explore the science behind these applications. These discussions are unified by their emphasis on metal–organic precursors, which offer the greatest flexibility in control of gel structure.

An abundance of background information on sol–gel processing can be found in the proceedings of topical meetings, including the *International Workshop on Glasses and Ceramics from Gels* [1], the Materials Research Society symposia entitled *Better Ceramics Through Chemistry* [2], and the series of meetings entitled *Ultrastructure Processing* [3]. A coherent account of the science of sol–gel processing is available in a recent text [4], and the technology is discussed in a book edited by Klein [5]. A topic that is not covered in the following pages is the preparation of aerogels, which are made by supercritical extraction of the liquid from the pores of gels. This topic is discussed in the proceedings of biennial topical meetings [6].

The first chapter of this section, by Brinker (Sandia National Laboratories), provides an overview of the sol–gel process. Brinker illustrates how the chemistry of hydrolysis and condensation of alkoxides offers control over polymer growth, and how the behavior of alkoxides differs from that of the inorganic precursors used in aqueous systems. The relatively low solubility of silica in alcoholic solutions permits the growth of highly ramified fractal aggregates. Given such a sol, there are many processing pathways that can be followed. The aggregates may be allowed to grow and link together into a continuous gel network, which can be molded to make a monolithic ceramic object. Alternatively, by controlling the conditions of growth, researchers can make inorganic polymers that are suitable for growing particles, drawing of fibers, or deposition of films. The structure of the aggregates determines that of the gel, which has important implications for the subsequent processes of drying and sintering. Moreover, manipulation of the structure permits control of the optical and mechanical properties and chemical reactivity of films, as well as monolithic gels.

The chemistry of hydrolysis and condensation of silicon alkoxides is now understood in considerable detail, as indicated in the chapter by Coltrain and Kelts (Eastman Kodak Co.). Extensive use of nuclear magnetic resonance has revealed the influence of factors such as pH on the kinetics of the competing reactions. With this information it is possible to rationalize the structures of the aggregates, as revealed by studies of small-angle scattering of X-rays and neutrons. This level of understanding opens the possibility for deliberate control of gel structure and properties. Nonsilicate systems have received less attention, but there is a fair body of literature on aluminates, borates, and transition metals, which is summarized in reference 4; transitional metal oxide chemistry is also discussed in excellent reviews by Livage and co-workers [7,8].

One of the most exciting opportunities offered by sol–gel processing is the possibility of creating hybrid materials with organic and inorganic components mixed on the molecular level. Schmidt and Böttner (University of Saarlandes) discuss the preparation of porous hybrid materials. By appropriate choice of precursors and processing conditions, the porosity and — most important — the chemical nature of the interface can be controlled. The variety of organic ligands that can be incorporated into the gel permits a virtually infinite range of hydrophilic–hydrophobic or acid–base characteristics. In addition, unique optical and mechanical properties can be achieved in such hybrids. Methods of preparation and applications of organically modified gels are discussed in a review by Schmidt [9].

Arguably the most important application of sol–gel technology is the preparation of films to exploit their optical or electronic properties or their chemical or mechanical resistance. Hurd (Sandia National Laboratories) examines the physics of film deposition by dip coating, elucidating the factors that control the thickness and structure of the film. Direct observation of the deposition process reveals the kinetics of the drying process and the existence of flows driven by surface tension gradients when binary solvents are used. Extensive discussion of the preparation of, and applications for, such films are presented in references 4 and 5.

Most ceramics processing is done with powders, because of the convenience of shaping them (to minimize machining) and because fine powders permit sintering at relatively low temperatures. Considerable interest has developed for the use of sol–gel technology for the preparation of uniform oxide particles of small diameter for fabrication of high-quality ceramics. Following the pioneering work of Stöber, Fink, and Bohn [10], many workers have prepared a wide variety of monodisperse particles; however, the mechanism by which such uniform particles grow has not been understood. Finally, a clearer picture of the process is emerging, as indicated in the chapter by Zukoski et al. (University of Illinois). It seems that the inorganic polymers develop such a high degree of cross-linking that they undergo phase separation, providing nuclei for the growth of particles. Monodispersity results from the nature of colloidal forces, which make large particles stable against aggregation from one another yet permit smaller particles to continue to aggregate. Model calculations support this argument and indicate the most important parameters to control.

REFERENCES

1. *J. Non-Cryst. Solids*; Volumes 48, 63, 100, and 121.
2. *Materials Research Society Symposia Proceedings*; Materials Research Society: Pittsburgh, PA; Volumes 32, 73, 121, 180, and 271.
3. Proceedings published by Wiley (New York) under the titles *Ultrastructure Processing of Ceramics, Glasses, and Composites* (1984), *Science of Ceramics Chemical Processing* (1986), and *Ultrastructure Processing of Advanced Ceramics* (1988) and *Chemical Processing of Advanced Materials* (1992).
4. Brinker, C. J.; Scherer, G. W. *Sol–Gel Science*; Academic Press: New York, 1990.
5. *Sol–Gel Technology for Thin Films, Fibers, Preforms, Electronics, and Speciality Shapes*; Klein, L. C., Ed.; Noyes: Park Ridge, NJ, 1988.
6. *Aerogels*; Fricke, J., Ed.; Springer–Verlag: New York, 1986. Second International Symposium on Aerogels: *Rev. Phys. Appl.* **1989**, *24(C4)*.
7. Livage, J.; Henry, M.; Sanchez, C. *Progress Solid State Chem.* **1988**, *18*, 259–342.
8. Livage, J. In *Sol–Gel Science and Technology*; World Scientific: London, 1989; pp 103–152.
9. Schmidt, H. In *Sol–Gel Science and Technology*; World Scientific: London, 1989; pp 432–469.
10. Stöber, W.; Fink, A.; Bohn, E. *J. Colloid Interface Sci.* **1968**, *26*, 62–69.

47 Sol–Gel Processing of Silica

C. Jeffrey Brinker
Sandia National Laboratories, University of New Mexico and Center for Micro-Engineered Ceramics

CONTENTS

The sol–gel process for preparing silica and silicates from metal alkoxide precursors is reviewed and compared to the processing of aqueous silicates as described by Iler. Sol–gel processing combines control of composition and microstructure at the molecular level with the ability to shape material in bulk, powder, fiber, and thin-film form. In sol–gel processing of metal alkoxides, hydrolysis reactions replace an alkoxide group with a hydroxyl group. Subsequent condensation reactions involving the hydroxyl groups produce siloxane bonds. The structure of the evolving silicates is a consequence of the successive polymerization, gelation, aging, drying, and heating steps. Often the structures of polymers, gels, and dried gels (either xerogels or aerogels) may be characterized on the 1–20-nm length scale by a mass or surface fractal dimension. On longer length scales, dried gels are micro- or mesoporous, with surface areas often exceeding 800 m^2/g. During heating, these gels undergo continued polymerization, structural relaxation, and viscous sintering; dense amorphous silica essentially indistinguishable from its conventionally prepared counterpart ultimately results.

A sol is defined as a colloidal dispersion of particles in a liquid. A gel is a substance that contains a continuous solid skeleton enclosing a continuous liquid phase: the liquid prevents the solid from collapsing; the solid prevents (retards) the liquid from escaping. Thus the formal definition of sol–gel processing is the growth of colloidal particles and their linking together to form a gel. This definition has been expanded to include virtually all liquid-based processes for the preparation of ceramic materials. As illustrated in Figure 47.1, the formation of films, fibers, and unaggregated particles are all considered sol–gel processes, even though gelation may not occur (particle formation), and often the sol is composed of polymers rather than particles (fiber formation).

The excitement of sol–gel technology derives from the ability to control composition and microstructure at the molecular level combined with the ability to shape the material at room temperature, for example, by casting bulk gels in precision molds, spinning fibers, or dip coating thin films [1]. This chapter briefly reviews sol–gel technology for the preparation of silica and silicates (where silicate refers to any hydroxylated or alkoxylated forms of silica as well as multicomponent silicates, $M_xO \cdot SiO_2$). The discussion focuses on silicates prepared from metal alkoxide precursors. Where possible, comparisons are made with aqueous silicates as described by Iler in the classic monograph *The Chemistry of Silica* [2].

Additional information on sol–gel processing of ceramic materials in general includes review articles by Hench and West [3], Sakka [4], Masdiyasni [5], Roy [6], Zelinski and Uhlmann [7], and Zarzycki [8] as well as conference proceedings of the International Workshops on

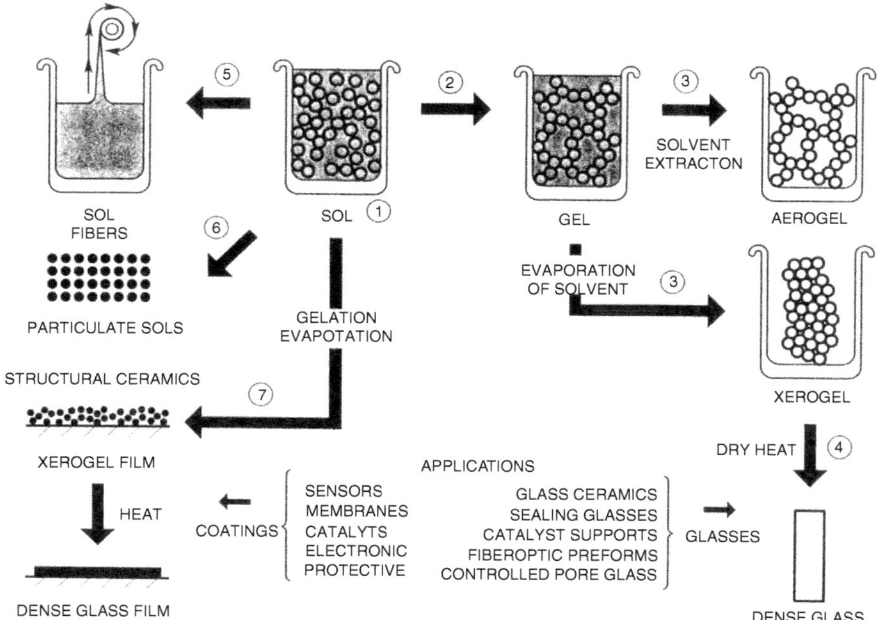

FIGURE 47.1 Illustration of the various stages of the sol–gel process. The numbers refer to the order in which these stages are presented in the text.

Gels [9–14], the International Conference on Ultrastructure Processing [15–19], and the Materials Research Society (MRS) Symposium on Better Ceramics Through Chemistry [20–23]. Various applications of sol–gel processing are described in *Sol–Gel Technology for Thin Films, Fibers, Performs, Electronics, and Specialty Shapes*, edited by Klein [1]. The underlying physics and chemistry are described in *Sol–Gel Science* by Brinker and Scherer [24].

HYDROLYSIS AND CONDENSATION OF AQUEOUS SILICATES

The most weakly hydrolyzed form of silica detectable in aqueous solution is orthosilicic acid, $Si(OH)_4$ [25], although it is generally believed that protonation of silanols to form cationic species $\equiv Si(OH_2)^+$ can occur below about pH 2. Above pH 7, further hydrolysis involves the deprotonation of a silanol group to form an anionic species [25]:

$$Si(OH)_4(aq) \longrightarrow Si(OH)_3O^- + H^+ \tag{47.1}$$

Because $Si(OH)_3O^-$ is a very weak acid, $Si(OH)_2O_2^{2-}$ is observed in appreciable quantities only above pH 12 [25].

By analogy to organic polymer systems, $Si(OH)_4$ may polymerize into siloxane chains that then branch and cross-link. However, Iler [2] states, "in fact, there is no relation or analogy between silicic acid polymerized in an aqueous system and condensation-type organic polymers." Iler recognizes three stages of polymerization: (1) polymerization of monomers to form particles; (2) growth of particles; and (3) linking of particles into branched chains, networks, and finally gels. Iler divides the polymerization process into three approximate pH domains: pH < 2, 2–7, and >7. A pH of 2 appears to be a boundary, because the point of zero charge (PZC), where the surface charge is zero, and the isoelectric point (IEP), where the electrical mobility of the silica particles is zero, both are in the range pH 1–3. A pH of 7 appears to be a boundary, because both the solubility and dissolution rates are maximized at or above pH 7, and because above pH 7 the silica particles are appreciably ionized (e.g., Equation 47.1) so that particle growth occurs without aggregation or gelation. For all pH ranges, the addition of salt promotes aggregation and gel formation [2] (*see* Figure 47.2).

Because gel times decrease steadily between pH 2 and 6, it is generally assumed that polymerization above the IEP occurs by a bimolecular nucleophilic condensation mechanism (S_N2-Si) involving the attack of hydrolyzed, anionic species on neutral species [2]:

$$\equiv SiO^- + \equiv Si{-}OH \longrightarrow \equiv Si{-}O{-}Si\equiv + OH^- \tag{47.2}$$

Because of inductive effects, the most acidic silanols and hence the most likely to be deprotonated, are the most highly condensed species [26]. Therefore condensation according to Equation (47.2) occurs preferentially between more condensed species and less condensed, neutral, species. As suggested in Figure 47.2, this situation leads to a typical condensation pathway: monomer, dimer, trimer, and tetramer. Tetramers tend to cyclize

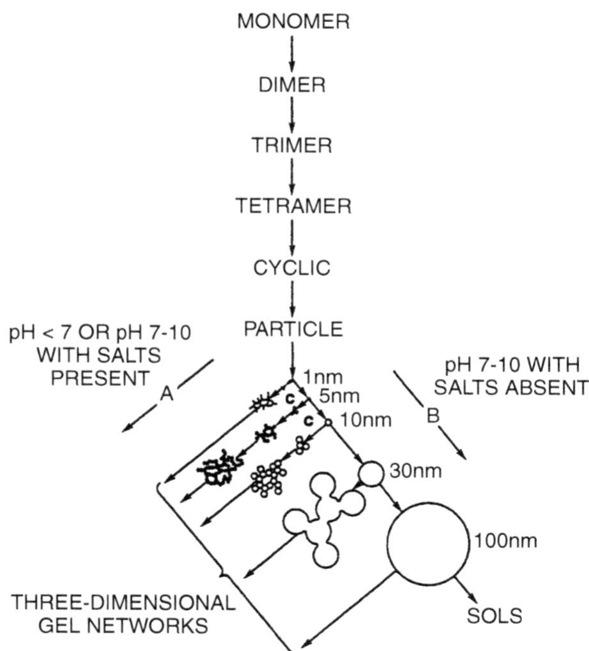

FIGURE 47.2 Polymerization pathway of aqueous silicates according to Iler [2]. Stages of growth recognized by Iler: polymerization of monomer to form particles, growth of particles, and linking of particles together into branched chains, networks, and, finally, gels. (Reproduced with permission from reference 2. Copyright 1978.)

because of the proximity of chain ends and the substantial depletion of the monomer population. Further growth occurs by addition of monomer and other low-molecular-weight species to cyclic species to create particles and by aggregation of particles to form chains and networks [2].

Growth above about pH 7 is distinguished from that below pH 7 by at least two factors: (1) Above pH 7, particle surfaces are appreciably charged, so particle aggregation is unlikely, whereas near the IEP there is no electrostatic particle repulsion, so the growth and aggregation processes occur together and may be indistinguishable. (2) Because of the greater solubility of silica and the greater size dependence of solubility above pH 7, growth of primary particles continues by Ostwald ripening, a process in which smaller, more soluble particles dissolve and reprecipitate on larger, less soluble particles. Growth ceases when the difference in solubility between the largest and smallest particles becomes negligible. Above pH 7, growth continues by Ostwald ripening at room temperature until the particles are 5–10 nm in diameter, whereas at lower pH growth stops after a size of only 2–4 nm is reached. Because of enhanced silica solubility at higher temperatures, growth continues to larger sizes, especially above pH 7 [2].

Because gel times decrease below the IEP, it is believed [2,4,27,28] that below about pH 2, condensation occurs by a bimolecular nucleophilic mechanism involving a protonated silanol:

$$\equiv SiOH_2^+ + HO{-}Si\equiv \rightarrow \equiv Si{-}O{-}Si\equiv + H^+ \quad (47.3)$$

[Unlike in carbon chemistry, there is no evidence for a siliconium ion $\equiv Si^+$ [29].]

HYDROLYSIS AND CONDENSATION OF SILICON ALKOXIDES

Tetramethoxysilane, $Si(OCH_3)_4$, abbreviated TMOS, and tetraethoxysilane, $Si(OCH_2CH_3)_4$, abbreviated TEOS, are the most commonly used metal alkoxide precursors in sol–gel processing of silicates [24]. Silicate gels are most often synthesized by hydrolyzing the alkoxides dissolved in their parent alcohols with a mineral acid or base catalyst. At the functional group level, three bimolecular nucleophilic reactions are generally used to describe the sol–gel process [24]:

$$\equiv Si{-}OR + H_2O \leftrightharpoons \equiv Si{-}OH + ROH \quad (47.4)$$

$$\equiv Si{-}OH + RO{-}Si\equiv \leftrightharpoons \equiv Si{-}O{-}Si\equiv + ROH \quad (47.5)$$

$$\equiv Si{-}OH + HO{-}Si\equiv \leftrightharpoons \equiv Si{-}O{-}Si\equiv + H_2O \quad (47.6)$$

The hydrolysis reaction [Equation (47.4)] replaces alkoxide groups with hydroxyl groups. Subsequent condensation reactions involving the silanol groups produce siloxane bonds plus the by-products alcohol [Equation (47.5)] or water [Equation (47.6)]. The reverse of hydrolysis is esterification, in which hydroxyl groups are replaced with alkoxides. The reverse of condensation is siloxane bond alcoholysis [Equation (47.5)] or hydrolysis [Equation (47.6)].

The roles of acid or base catalysts are illustrated schematically in Figure 47.3 [30]. The hydrolysis reaction appears to be specific acid or base catalyzed [27,29,31]. Acid catalysts protonate the alkoxide group [Equation (47.4)], making a better leaving group (ROH) and avoiding the requirement for proton transfer in the transition state [32]. Base catalysts dissociate water, producing a stronger nucleophile (OH^-) [32]. The condensation reaction depends on the acidity of the silicate reactants. Above about pH 2, acidic silanols are deprotonated; strong nucleophiles, $\equiv SiO^-$, are created [*see* Equation (47.2)]. Below about pH2, weakly acidic silanols or ethoxides are protonated, so good leaving groups (H_2O or ROH) are created and the requirement of charge transfer in the transition state [*see* Equation (47.3)] is avoided. The rate of siloxane bond hydrolysis increases above pH 4 and at very low pH [2]. Similar behavior is expected for siloxane bond alcoholysis reactions. The esterification of silanols was reported [33–35] to proceed much faster under acid-catalyzed conditions.

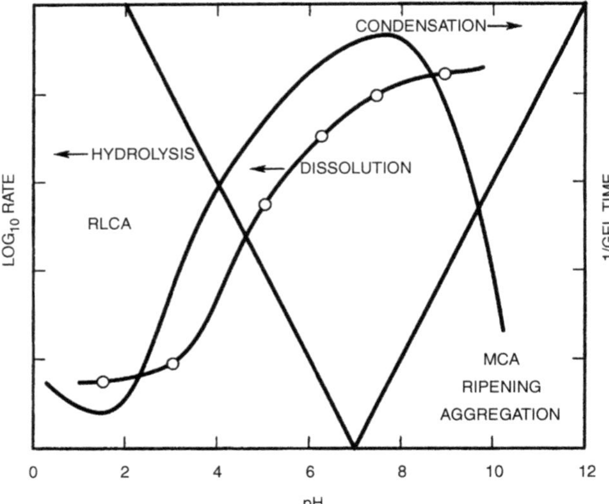

FIGURE 47.3 Illustration of the pH dependence of the hydrolysis, condensation, and dissolution rates of silica. Condensation rates are judged by the reciprocal of gel times. RLCA denotes reaction-limited cluster aggregation. (Reproduced with permission from reference 30. Copyright 1988.)

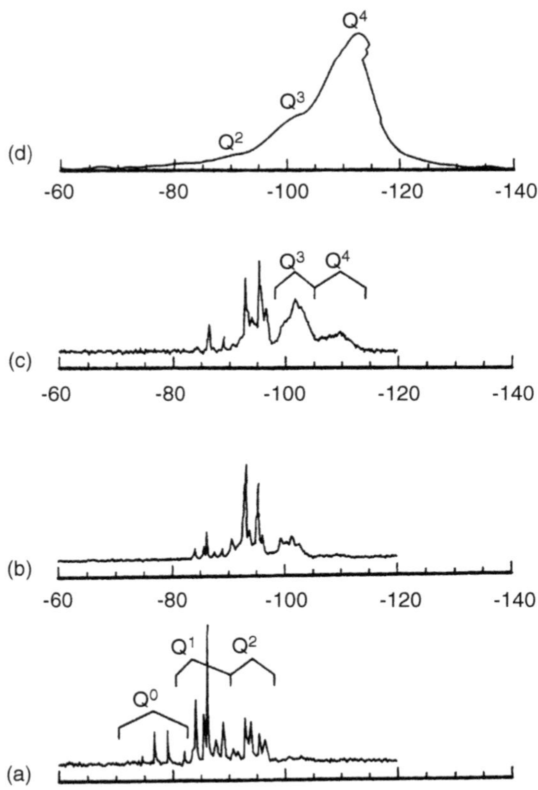

FIGURE 47.4 Comparison of ^{29}Si NMR spectra of an acid-catalyzed TEOS sol ($H_2O/Si = 2$) and a commercial aqueous silicate (Ludox HS40): (a) TEOS sol after 3 h; (b) TEOS sol after 3 days; (c) TEOS sol after 14 days of reaction; and (d) Ludox sol. Q notation refers to the number of bridging oxygens (−OSi) surrounding the central silicon atom (0–4). From Assink [41].

Although making clear that there is no analogy between the polymerization of aqueous silicates and condensation-type organic polymers, Iler [2] suggested that such (molecular) siloxane networks "might be obtained under conditions where depolymerization is least likely to occur, so the condensation is irreversible and siloxane bonds cannot be hydrolyzed once they are formed." The solubility of silica is reduced by a factor of 28 when water is replaced by a 90 wt% methanol and 10 wt% water mixture [2]; thus Iler's hypothesis might be realized under conditions in which silicon alkoxides are hydrolyzed with small amounts of water (generally, H_2O:Si ratios less than or equal to 4) especially below pH 7, at which the solubility and the dissolution rate are minimized (*see* Figure 47.3).

Evidence for molecular siloxane networks is abundant in numerous ^{29}Si NMR spectroscopy studies of alkoxides hydrolyzed under acidic conditions [35–40]. Figure 47.4 [41] shows a sequence of ^{29}Si NMR spectra of TEOS hydrolyzed with 2 mol of water under acidic conditions in ethanol and for comparison a ^{29}Si NMR spectrum of a commercial aqueous silicate (Ludox). With time the TEOS sol becomes more highly condensed, as evident from the disappearance of monomer ($\mathbf{Q}^0$) and the progressive formation of end groups and di-, tri-, and tetrasubstituted silicate species ($\mathbf{Q}^1$–$\mathbf{Q}^4$ species, respectively). However, even after 14 days, di- and trisubstituted species appear more prevalent than tetrasubstituted species. By comparison, ^{29}Si NMR spectra of aqueous silicates (Figure 47.4d) and base-catalyzed alkoxides (Figure 47.5) are dominated by monomer and tetrasubstituted species.

The prevalence of fully condensed $\mathbf{Q}^4$ species in aqueous silicate sols (Figure 47.4d) is consistent with Iler's view [2] that particle growth occurs in a manner that maximizes the extent of internal condensation. Iler [2] stated that "at the earliest stage of polymerization, condensation leads to ring structures followed by addition of monomer and linking together of the cyclic polymers to (form) larger three-dimensional molecules. These condense internally to the most compact state with SiOH groups remaining on the outside." The paucity of $\mathbf{Q}^1$–$\mathbf{Q}^3$ species in the NMR spectra of aqueous silicates and base-catalyzed alkoxides implies that under these conditions growth is dominated by the addition of monomer to highly condensed particles and that monomer addition is accompanied by extensive siloxane bond hydrolysis (ring opening) and reformation (so as to minimize $\mathbf{Q}^1$–$\mathbf{Q}^3$ species).

The distinction between particulate and polymeric silicate sols is also evident from small-angle scattering

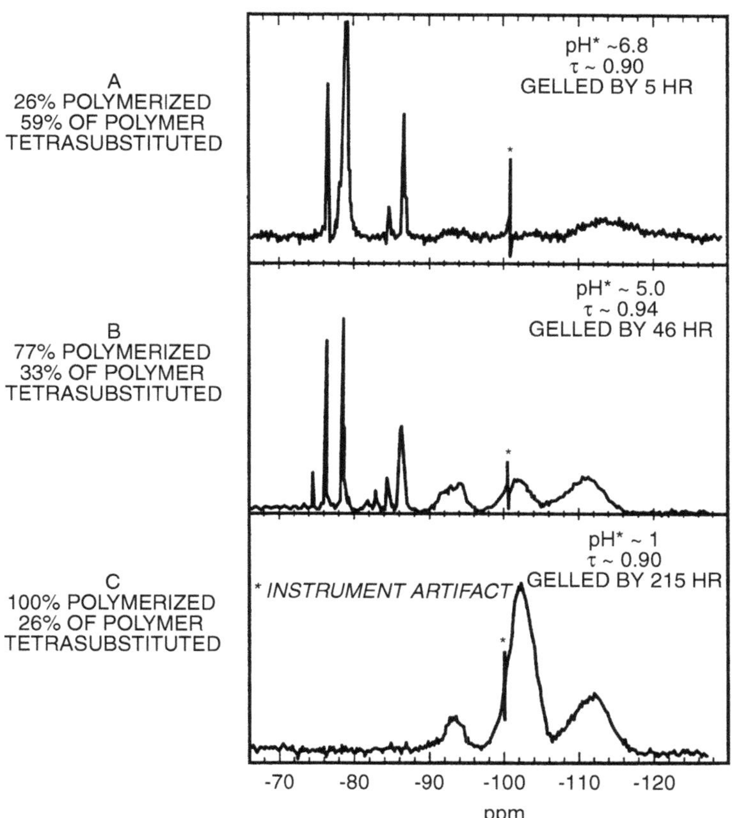

FIGURE 47.5 ^{29}Si NMR spectra of TEOS sols at $t/t_{gel} \geq 0.9$ for three different hydrolysis conditions. At pH* $\sim$1, all the monomer (~ -70 to -80 ppm) is depleted; further growth must occur by CCA, whereas under neutral or basic conditions monomer persists and growth is biased to MCA. (Reproduced with permission from reference 40. Copyright 1986.)

investigations (e.g., references 42–45) and from cryogenic transmission electron microscopy (cryo-TEM) investigations [46]. Figure 47.6 compares small-angle scattering data obtained from a commercial aqueous silicate to that obtained from a variety of silicate sols prepared from alkoxides [45]. The power–law relationships implied by the data can be interpreted on the basis of fractal geometry by the following expression [43]:

$$P = -2D + D_s \tag{47.7}$$

where P is the Porod slope, D is the mass fractal dimension, and D_s is the surface fractal dimension [47]. D relates an object's mass M to its radius r according to

$$M \sim r^D \tag{47.8}$$

Whereas for uniform (nonfractal) objects D would be the dimension of space 3, mass fractals are characterized by $D = D_s < 3$, so the density ρ of an object decreases with r as $\rho \sim 1/r^{(3-D)}$. D_s relates an object's area to its size. In three dimensions, D_s varies from 2 for a smooth (nonfractal) surface to 3 for a fractal surface that is so convoluted that it acquires the dimension of space. For nonfractal objects, $D = 3$ and $D_s = 2$, so $P = -4$, as shown by Porod and Kolloid [48].

Figure 47.6 shows that aqueous silicate sols are composed of uniform particles, whereas the various alkoxide-derived gels are either mass or surface fractals. These results have been rationalized by various reaction-limited kinetic growth models such as monomer–cluster aggregation (MCA) [49] and cluster–cluster aggregation (CCA) [50]. In MCA, growth occurs by the addition of monomers to higher molecular weight species (clusters). In CCA, growth occurs by the addition of clusters to both monomers and other clusters. Reaction-limited conditions imply that the condensation rate is sufficiently low with respect to the transport (diffusion) rate that the monomer or cluster can sample many potential growth sites before reacting at the most favorable one [49]. Reaction-limited conditions are obtained for most silicate synthesis schemes.

MCA requires a continual source of monomers, which, because of physical or chemical factors, condense preferentially with higher molecular weight species rather than themselves. Because growth occurs monomer-by-monomer under reaction-limited conditions, all potential growth sites are accessible; the result is compact, uniform (nonfractal) objects characterized by $P = -4$. ^{29}Si NMR

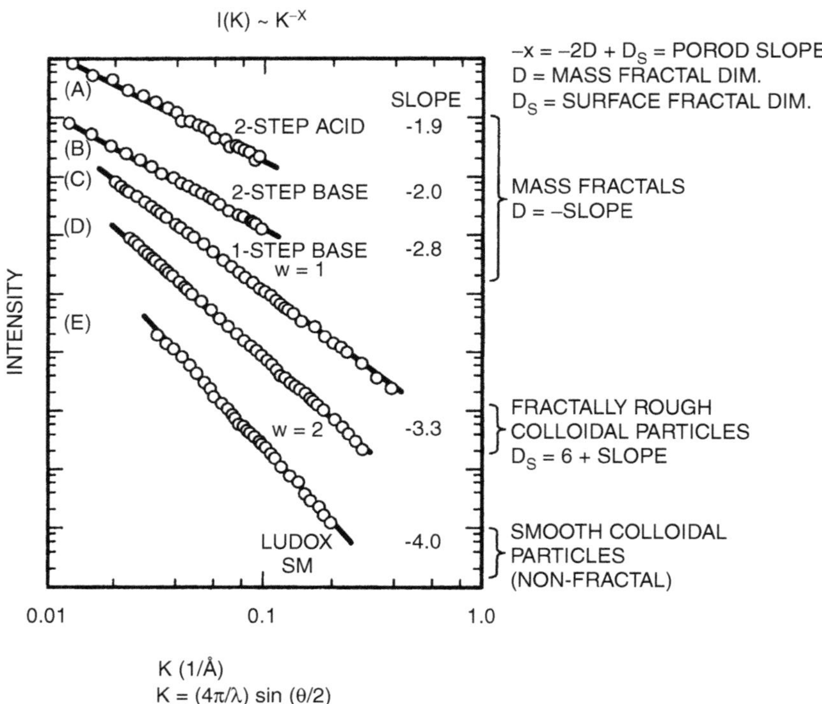

FIGURE 47.6 Log of scattered intensity versus log K obtained by small-angle x-ray scattering (SAXS) for alkoxide-derived gels prepared under different hydrolysis conditions and a commercial aqueous silicate (Ludox SM). (Reproduced with permission from reference 44. Copyright 1985.)

spectra (e.g., Figure 47.5) show that a major requirement for MCA, namely, a source of monomer, is met for alkoxide-derived sols prepared at neutral or basic pH. As discussed earlier, the base-catalyzed condensation mechanism [Equation (47.2)] favors the reaction of low- and high-molecular-weight species, so the growth is biased toward MCA.

In CCA, monomers are depleted at an early stage of the growth process, so further growth must occur exclusively between clusters. Strong mutual screening of cluster interiors leads to ramified objects characterized by a mass fractal dimension $D \approx 2$ (Porod slope $P \approx -2$). As shown in Figure 47.6 for the two-step acid-catalyzed sample, Porod slopes of ~ -2 are generally observed for silicate sols prepared from alkoxides under acid-catalyzed conditions. Corresponding ^{29}Si NMR spectroscopy data for the two-step acid-catalyzed sample indicated that monomer was essentially depleted before the onset of measurable growth. Similarly, ^{29}Si NMR spectroscopy data presented in Figure 47.5 for an acid-catalyzed silica sol show that for normalized gel time $t/t_{gel} \sim 0.9$, the sol is 100% polymerized, so any further growth must occur by CCA. (In fact, ^{29}Si NMR spectroscopy showed that for these conditions monomer was depleted at a very early stage of the polymerization process, $t/t_{gel} = 0.01$ [40].)

MCA and CCA are just two of many plausible growth models, and the predictions of $P = -4$ and $P = -2$ for these two aggregation processes are not unique. However, the qualitative predictions of these models, namely, compact, uniform structures when growth proceeds in the presence of monomer and ramified, fractal objects in the absence of monomer, are generally observed.

GELATION AND AGING

The gel point is defined as the time when an infinite, spanning polymer or aggregate first appears. For aqueous systems, Iler [2] observed the formation of three-dimensional gel networks below pH 7 or 7–10 with salts present (Figure 47.2) and attributed gelation to what is now known as ballistic cluster–cluster aggregation [51]. As discussed later, gelation in both aqueous and alkoxide-derived sols is consistent with a percolative process involving cluster–cluster aggregates. The physical and chemical changes that occur after gelation but before complete drying are referred to as aging [24].

The first theory that attempted to derive the divergences in cluster mass and average radius accompanying gelation is that of Flory [52] and Stockmayer [53]. In their model, bonds are formed at random between adjacent nodes on an infinite Cayley tree or Bethe lattice (*see* Figure 47.7). The Flory–Stockmayer (FS) model is qualitatively successful because it correctly describes the emergence of an infinite cluster at some critical extent of reaction and

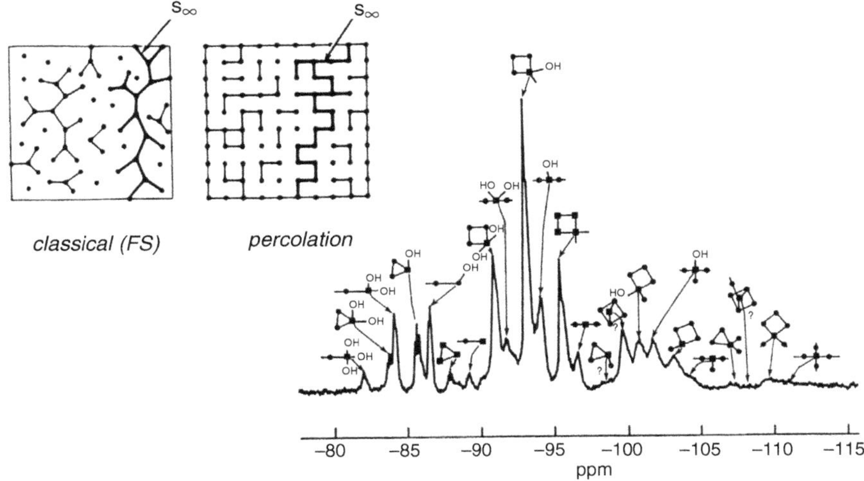

FIGURE 47.7 Illustrations of gelation according to the classical Flory–Stockmayer model and the percolation model. In the classical model, cyclic configurations are avoided, so the unphysical situation M ~ R^4 results. As illustrated by the ^{29}Si NMR spectrum of an acid-catalyzed TEOS sol [36], cyclic species are quite prominent sol components. (Reproduced with permission from reference 36. Copyright 1988.)

provides good predictions of the gel point. Although the polymerization of silicon alkoxides under acid-catalyzed conditions appears to produce molecular networks analogous to the organic polymers for which the FS model was developed, two inherent problems are encountered when adapting the FS approach to describe silicate polymers and gels. (1) Because of the nature of the Cayley tree, cyclic species are excluded, yet cyclic species are very prevalent in acid-catalyzed silicate sols [36] (*see* Figure 47.7). (2) Because cyclic configurations are avoided, the purely branched clusters formed on the Cayley tree are predicted to have a fractal dimension of 4 [54]; a divergent density for large clusters results ($M \sim r^4$).

Because of problems with the FS approach, Stauffer [55] and de Gennes [56] advanced bond percolation as a desription of polycondensation (*see* Figure 47.7). In the percolation model, bonds are formed at random between adjacent nodes on a regular or random *d*-dimensional lattice [57]. In this approach, cyclic molecules are allowed and excluded volume effects are directly accounted for.

Aggregation of either particles (*see* Figure 47.2) or polymers is generally believed to account for the growth of clusters far from the gel point. As the clusters grow, their density decreases as $\sim 1/r$, so their volume fraction increases. Eventually, the clusters overlap and become nearly immobile. Further condensation involves a percolative process in which bonds form at random between large ($\sim$1 μm) aggregates, which are equivalent to nodes located on a random three-dimensional lattice. Martin and co-workers [58,59] have shown that the percolation model accounts for the static structure factor of alkoxide-derived silicate sols on the 25–400-nm length scale near the gel point. The evolution of properties in the vicinity of the gel point is generally in agreement with critical behavior predicted by percolation theory and in contradiction to the classical theory [24].

As indicated in the percolation diagram in Figure 47.7, a relatively small fraction of the reactant species is part of the spanning network at the gel point. To incorporate more of the reactants in the gel network and thus impart strength and stiffness, gels are often aged prior to drying in either the mother liquor or other liquid. The aging process comprises [60–62] (1) continued polymerization; (2) syneresis; (3) coarsening; (4) phase separation; and in some cases (5) hydrolysis and esterification.

Polymerization continues both between monomers or polymers and the spanning network and within the network itself as reactive terminal groups diffuse into close proximity. This process appears to promote syneresis (sometimes called macrosyneresis) [63], the shrinkage of the gel network, resulting in expulsion of liquid from the pores. Scherer [64] states that the kinetics of syneresis depend on the driving force (polymerization), the mobility of the gel network, and the rate of fluid flow through the contracting network.

Iler's view of aging [2] is represented schematically in Figure 47.8a. The higher solubility of surfaces with positive curvatures causes dissolution there and reprecipitation on interparticle contacts that have negative curvatures and lower solubilities. This coarsening process, which is driven by a reduction in the solid–liquid interfacial energy, builds necks between particles that significantly strengthen the gel network. Figure 47.8b shows that aging an alkoxide-derived gel under basic conditions,

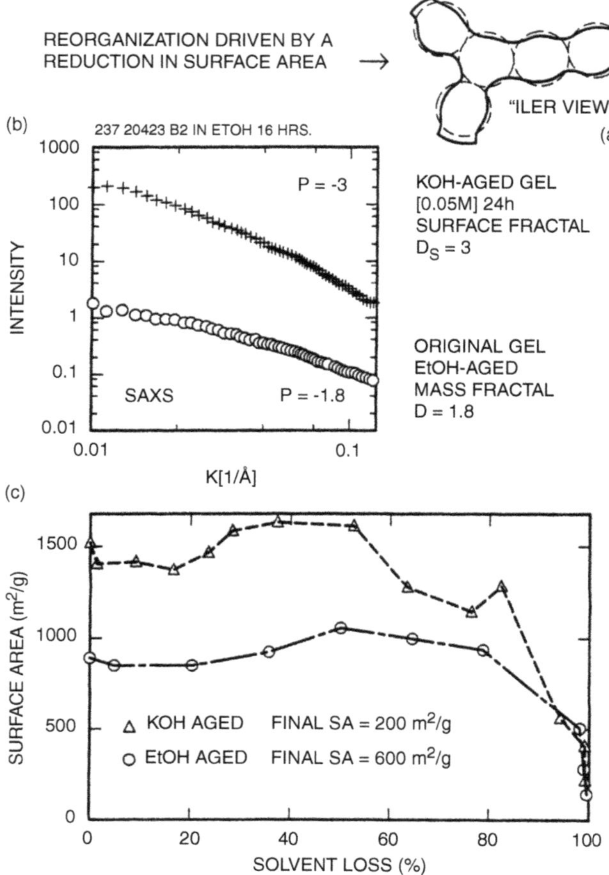

FIGURE 47.8 (a) coarsened structure that results from aging a network of particles under conditions in which there is partial solubility of the condensed phase. Material is removed from surfaces with positive curvatures and deposited at interparticle contacts that have negative curvatures; "neck" formation results. (Reproduced with permission from reference 2. Copyright 1978.) (b) Porod plots obtained by SAXS for an alkoxide-derived gel prepared at neutral pH in a 90% ethanol–10% water solvent (original gel EtOH-aged) and a similar gel aged for 24 h in 0.05 M KOH in ethanol (KOH-aged gel) [65]. (c) surface areas of EtOH- and KOH-aged gels as a function of percent solvent loss during drying; data obtained by proton spin relaxation methods. Final surface areas (SA) of fully dried gels were measured by using N_2 BET method. (Reproduced from reference 66. Copyright 1989. American Chemical Society.)

where coarsening is enhanced, causes reorganization of a mass fractal ($D = 1.8$) into a surface fractal ($D_s = 3$) [65] accompanied by an *increase* in the solid–liquid interfacial area from about 900 to 1500 m^2/g (Figure 47.8c), as measured in situ by proton spin-relaxation techniques [66]. Although coarsening occurs under these conditions, apparently the dissolution–reprecipitation process creates a microporous "skin" at the solid–liquid interface that accounts for the unexpected increase in surface area. As

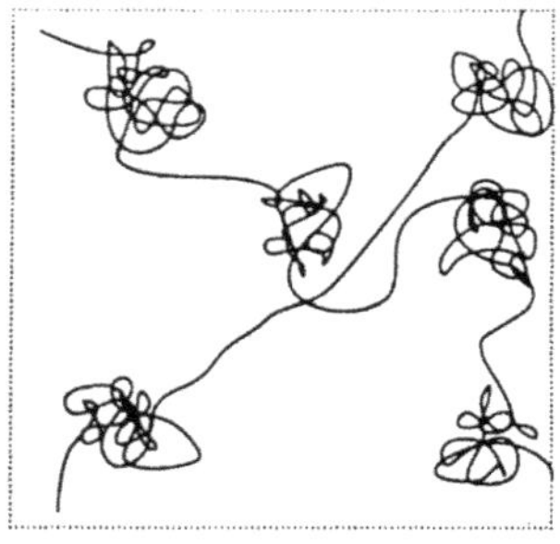

FIGURE 47.9 Representation of microsyneresis [24]. Immersion of a water-aged sample in ethanol is accompanied by partial esterification of the gel surface, network depolymerization, and solvation; smaller, more uniform pores result. The reverse process, immersion of an ethanol-aged sample in water, causes almost complete hydrolysis, network condensation, and apparent phase separation into water-rich and polymer-rich regions; larger pores with broader size distributions result. (Reproduced with permission from reference 67.)

discussed later, this microporous layer collapses during drying because of capillary forces, and the dry gel (xerogel) surface area is less for the base-aged sample than the original sample aged in mother liquor, consistent with Iler's original view (Figure 47.8a).

One type of phase separation process is microsyneresis, illustrated in Figure 47.9 [24,67]. Microsyneresis, which is common in organic gels, results from a greater affinity of the polymer for itself than for the liquid. This situation can arise when a gel is aged in a liquid other than the mother liquor. For example, Davis and coworkers [67] showed that sequential aging of alkoxide-derived gels in alcohol and then water causes hydrolysis, polymerization, and phase separation. The reverse sequence (water and then alcohol) causes esterification, depolymerization, and solvation (*see* Figure 47.9). Macroscopic phase separation occurs when a base-catalyzed TEOS-derived gel is immersed in water. Unreacted TEOS present at the gel point phase separates into sufficiently large droplets that light scattering is observed [61]. Artifacts of these droplets are observed by scanning electron microscopy (SEM) on fracture surfaces of the corresponding dried gels [61].

DRYING

Drying is generally accomplished by evaporation to form a xerogel (from the prefix xero, meaning dry). Scherer [68] divides the evaporative drying process into several stages (*see* Figure 47.10). In the first stage, or constant rate period, the body shrinks to accommodate the liquid lost by evaporation, and the liquid–vapor interface remains at the exterior surface. The second stage begins when the

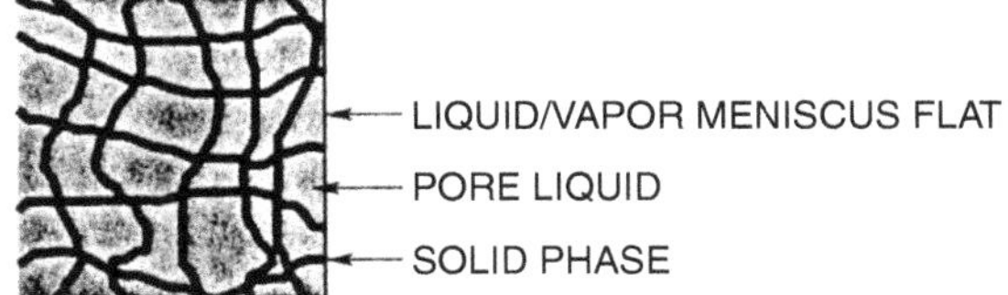

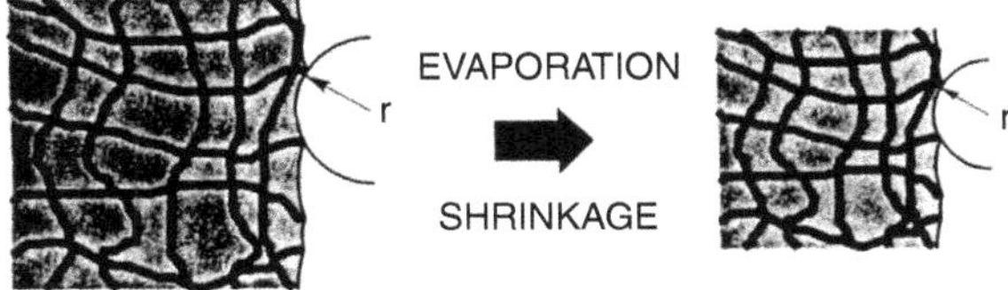

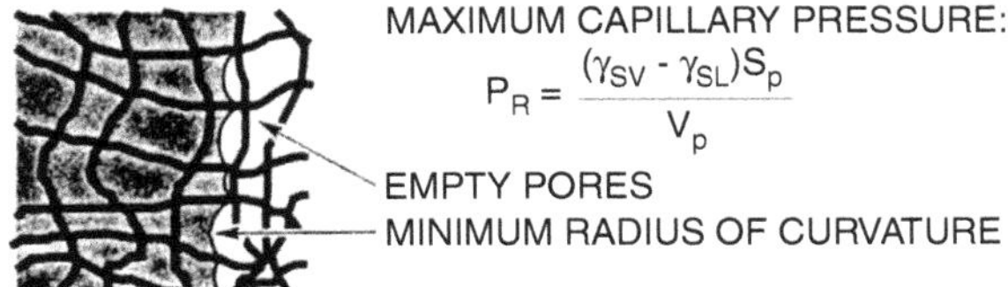

FIGURE 47.10 Illustration of drying process [24]. Capillary tension develops in liquid as it "stretches" to prevent exposure of the solid phase by evaporation, and the network is drawn back into liquid (a). The network is initially so compliant that little stress is needed to keep it submerged, so the tension in the liquid is low, and the radius of the meniscus (r_c) is large (b). As the network stiffens, the tension rises as r_c decreases. At the critical point, the radius of the meniscus becomes equal to the pore radius; the constant rate period ends and the liquid recedes into the gel (c). (Reproduced with permission from reference 24. Copyright 1990).

body is too stiff to shrink, so the liquid recedes into the gel interior. Initially, a continuous liquid film remains that supports flow to the exterior, where evaporation continues to occur (first falling rate period). Eventually the liquid becomes isolated into droplets, so evaporation of liquid in the gel and diffusion of vapor to the exterior (second falling rate period) is required.

In gels, the first stage of drying, where the liquid–vapor interface (meniscus) remains at the exterior surface, continues while the body shrinks to as little as one-tenth of its original volume. The most important pressure contributing to this shrinkage is the capillary pressure (P_c) that results from the radius of curvature (r_c) of the meniscus [68]:

$$P_c = -2\gamma_{LV}\cos(\theta)/r_c \qquad (47.9)$$

where γLV is the liquid–vapor interfacial energy (surface tension) and θ is the contact angle. Before evaporation begins, the meniscus is flat and $P_c = 0$. Capillary tension ($P_c > 0$) develops in the liquid at it "stretches" to prevent exposure of the solid phase by evaporation. At the initial stage of drying the gel is quite compliant and the network shrinks in response to this tension, so r_c remains large. However, shrinkage is accompanied by continued polymerization reactions within the network. As the network stiffens, r_c decreases and P_c increases. The maximum capillary tension (P_R) is attained when r_c is equal to the hydraulic radius of the pore ($2V_P/S_P$, where V_P is the pore volume and S_P is the surface area) [68]:

$$P_R = \gamma_{LV}\cos(\theta)S_P/V_P = (\gamma_{SV} - \gamma_{SL})S_P/V_P \qquad (47.10)$$

Any further drying causes the meniscus to recede into the gel interior.

Because the pore radius may approach molecular dimensions in some alkoxide-derived gels, P_R is an enormous tension, often exceeding several hundred megapascals. This tension is balanced by a compressive stress in the network that causes shrinkage. Despite the large magnitude of P_R, if P_R were uniform throughout the body, cracking would not be a major problem [68]. Unfortunately, because of the low permeability of the network, it is difficult to draw liquid from the gel interior, so a pressure gradient develops. As the pressure gradient increases, so does the variation in free strain rate, with the surface tending to contract faster than the interior. The spatial variation in strain (or strain rate) causes the stress that leads to fracture [68].

Several strategies reduce the tendency of gels to crack during drying [68]. Very slow drying reduces the gradient in strain by allowing the surface and interior to shrink at comparable rates. Aging strengthens the network and in some cases increases the pore radius, consequently reducing P_R. The use of surfactants or so-called drying control chemical additives (DCCA) [69] or, for example, the replacement of water with alcohol, reduces the pore fluid surface tension (or increases the contact angle), also reducing P_R. Supercritical drying [70] avoids liquid–vapor interfaces altogether and therefore capillary pressure.

Supercritical drying (also referred to as hypercritical drying) was first used by Prassas and Hench [71] to produce large silica monoliths without cracking. It involves the extraction of solvent above its critical point, where there is no distinction between liquid and vapor and, hence, no liquid–vapor interfacial energy ($\gamma_{LV} = 0$, so according to Equation (47.10), $P_R = 0$). In some cases drying is accomplished with no measurable shrinkage of the network. The resulting dry gels are called aerogels, because air may constitute over 99% of the volume [72]. Whereas Iler [2] referred to xerogels as "a contracted and distorted version of the gel originally formed in solution", aerogels should more closely represent the structure of the original gel. In support of this idea, Woignier et al. [73] found that the mechanical properties of aerogels are reasonably consistent with percolation theory; for example, the Young's modulus (E) varies with density (ρ) as follows:

$$E \propto \rho^{3.7 \pm 0.3} \qquad (47.11)$$

STRUCTURE AND CONSOLIDATION OF DRIED GELS

For both particulate and polymeric gels, the removal of pore liquid during drying exposes an interconnected porous network within the gel that surrounds the solid or skeletal phase. The average dimensions of the pores and the thickness of the skeleton depend on the structure that existed at the gel point and the extent of rearrangement and collapse of this structure that occurred during aging and drying. Quite often the dimensions of the pores and skeletal phase constituting the dried gel are sufficiently small that, despite high volume percent porosities, xerogels and aerogels are transparent or translucent [74,75]. In fact, primarily the small pore sizes and correspondingly the high surface areas of dried gels distinguish these materials from conventional porous ceramic green bodies or powder compacts. Table 47.1 [24] summarizes the porosities of several silicate xerogels and aerogels prepared from TEOS as determined from standard nitrogen adsorption–condensation measurements.

The structures of dried gels are determined on the molecular, mesoscopic, and macroscopic length scales by using a combination of solid-state magic-angle spinning (MAS) NMR, vibrational spectroscopy (IR and Raman), small-angle scattering spectroscopy, and nitrogen adsorption–condensation. Compared to wet gels, xerogels and aerogels are more highly condensed [76,77]. However, because the surface coverage of terminal OH or OR groups on the amorphous silica surface generally exceeds ~4 per square nanometer [2], the high surface areas of alkoxide-derived gels require that a significant number of $\mathbf{Q}^2$ and $\mathbf{Q}^3$ silicon sites remain. On longer length scales, small-angle scattering investigations have shown that fractal structures existing at the gel point are more or less preserved in the desiccated gel [78]. As discussed earlier, the absence of capillary pressure during supercritical drying causes fractal structures to be better preserved in aerogels than in xerogels [73].

When silica xerogels or aerogels are heated, they change size (ultimately shrink) and lose weight. Figure 47.11 compares the linear shrinkage and weight loss of two alkoxide-derived silica xerogels with that of a particulate xerogel prepared from a fumed silica [79]. The particulate xerogel differs from the other two in that it is composed of comparatively large, fully polymerized particles, similar in many respects to xerogels derived from aqueous silica sols.

The shrinkage curve can be divided into three approximate regions defined by the accompanying weight loss [80]. Below about 150°C, weight loss is attributed to desorption of physisorbed alcohol and water. There is little associated shrinkage; in fact, the most weakly condensed xerogels exhibit a measurable dilation because of their very high coefficients of linear thermal expansion (e.g., 470×10^{-7}/K for the acid-catalyzed sample in Figure 47.11 [24]). By comparison, the particulate xerogel (sample C), which is composed of a fully condensed silica skeleton, shows essentially no change in length below 150°C. At high temperatures (above

TABLE 47.1
Summary of Porosity of Silicate Xerogels and Aerogels

Sample	Pore volume (cm^3/g N_2 STP)	V_p^a	Surface area (m^2/g)	Pore diameter (Å) Adsorption	Pore diameter (Å) Desorption	Bulk density[b] (g/cm^3)
Two-step acid-catalyzed xerogel	345	0.54	740	10–50	18	1.54
Two-step acid-base-catalyzed xerogel	588	0.67	910	10–100	46	0.99
Particulate (one-step base-catalyzed xerogel)	686	0.70	515	10–200	125	~0.6
Two-step acid–base-catalyzed aerogel	1368	0.82[c]	858	10–500[c]	186[c]	0.30

[a]Volume fraction porosity is based on the theoretical SiO_2 skeletal density of 2.2 g/cm³.
[b]Measured at ~25% relative humidity.
[c]Because most of the adsorption occurs near P/P_0 near 1, pore volumes and pore size distributions may be inaccurate for aerogels.
Source: Reproduced with permission from reference 24. Copyright 1990.

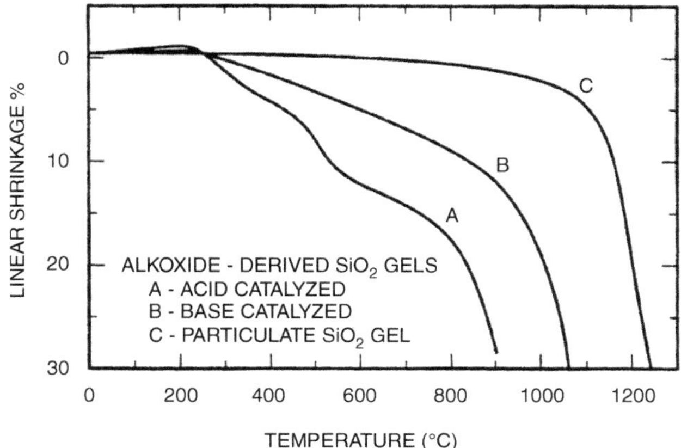

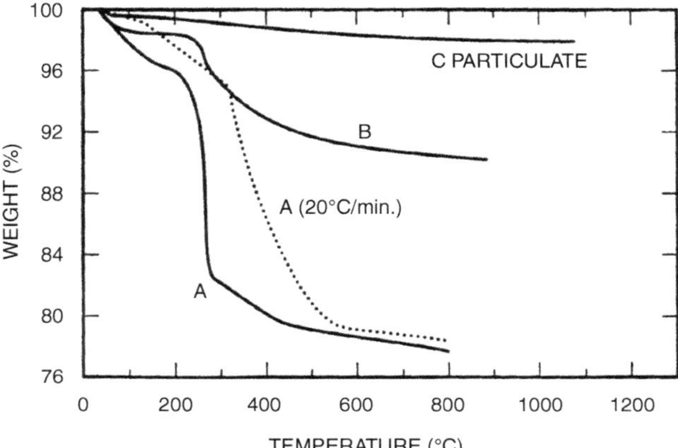

FIGURE 47.11 Linear shrinkage and weight loss versus temperature for three types of xerogels heated in air at 2°C/min. Sample A, two-step acid-catalyzed xerogel; sample B, two-step acid–base-catalyzed xerogel; and sample C, particulate gel prepared from fumed silica. Weight loss of sample A was also measured at 20°C/min. (Reproduced with permission from reference 79. Copyright 1984.)

800–1000°C, depending on the synthesis procedure), there is considerable shrinkage with little associated weight loss. In this region, shrinkage occurs primarily by viscous sintering [81]. At intermediate temperatures, both shrinkage and weight loss are substantial for the alkoxide-derived xerogels. Shrinkage at intermediate temperatures is attributed [80] to continued condensation reactions and structural relaxation [82]. Weight loss is attributed to the loss of water, the by-product of condensation, and pyrolysis of any residual organic compounds [80]. The particulate gel shows little change in length or weight at intermediate temperatures, because it is composed of essentially fully polymerized silica.

Condensation reactions and structural relaxation, that is, the approach of the network structure toward its equilibrium configuration [82], cause the skeletal density (the density of the solid phase that is inaccessible to helium) to increase toward that of silica glass. Figure 47.12 shows that this skeletal densification accounts for all of the shrinkage observed at intermediate temperatures for a multicomponent borosilicate gel [80]. The following data are skeletal densities of various xerogels after drying and the values approached (→) during heating:

- Commercial aqueous silica sols and gels [2], $2.20 \rightarrow 2.30\ g/cm^3$
- Base TMOS [58], $2.08 \rightarrow 2.20\ g/cm^3$
- Base NaBSi [58], $1.45 \rightarrow 2.40\ g/cm^3$
- Acid TEOS [59], $1.70 \rightarrow 2.20\ g/cm^3$
- NaAlBSi [55], $1.65 \rightarrow 2.27\ g/cm^3$

These data [2,80,83,84] indicate that skeletal densification at intermediate temperatures is observed in general for metal alkoxide derived silicate xerogels, whereas aqueous silicates have skeletal densities comparable to silica glass, and thus, like the fumed silica gel discussed earlier, exhibit little skeletal densification at intermediate temperatures.

Viscous sintering is a process of densification driven by interfacial energy [81]. Material moves by viscous flow in such a way as to eliminate porosity and thereby reduce the solid–vapor interfacial area. The rate of viscous sintering is proportional to the surface area and inversely proportional to viscosity and pore size. The differences in shrinkage observed in the high-temperature region in Figure 47.11 reflect differences in the rate of viscous sintering. The particulate gel has comparatively large pores, low surface area, and, because it is essentially fully polymerized silica, high viscosity. It begins to sinter substantially near 1100°C, whereas the acid-catalyzed sample, characterized by small pores, high surface area, and low viscosity, begins to sinter below 800°C.

The processes responsible for skeletal densification (polymerization and structural relaxation) also result in an increase in viscosity [81]. Although skeletal densification occurs primarily at intermediate temperatures, it may continue at higher temperatures, where it has a dramatic effect on the kinetics of viscous sintering. Figure 47.13 [79] plots the change in viscosity versus bulk density for an alkoxide-derived silica gel (sample A in Figure 47.11) that was heated to the indicated temperatures at either 2 or 20°C/min and held isothermally. These data, which were obtained by analysis of the shrinkage curves with a viscous sintering model developed by Scherer [85], illustrate several important trends commonly observed when sintering gels derived from alkoxides. (1) Unlike with well-annealed conventional glass, the viscosity is not a single-valued function of temperature. It increases isothermally by more than 3 orders of magnitude. (2) Because of the increasing viscosity, there is little associated densification. (3) Heating at a greater

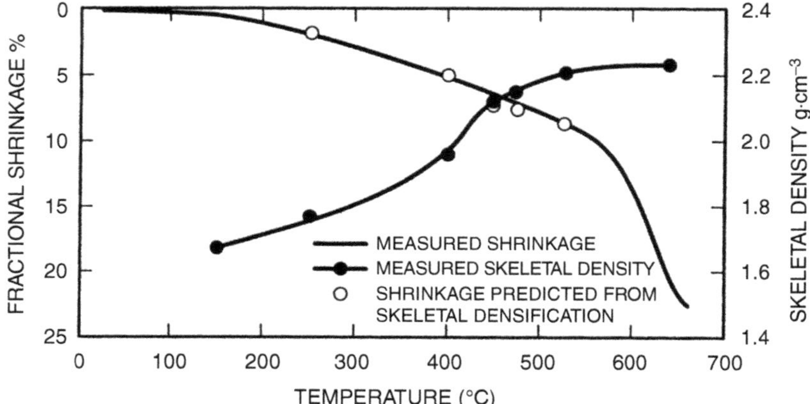

FIGURE 47.12 Linear shrinkage and skeletal density of a multicomponent borosilicate xerogel during heating in air at 2°C/min. Skeletal densification accounts for all the shrinkage observed between 250 and 550°C. (Reproduced with permission from reference 84. Copyright 1986.)

rate results in a lower viscosity and higher density at the beginning of the isothermal hold.

These trends reflect the competition between the rates of sintering, polymerization, and structural relaxation. Greater heating rates provide less time for viscous sintering, but also less time for polymerization and structural relaxation, processes that contribute to higher viscosity. Because the amount of viscous sintering is proportional to the integral of time divided by viscosity [81], the reduction in viscosity could more than compensate for the reduction in time, leading to the remarkable conclusion that faster heating promotes densification at lower temperatures [24]. From a practical standpoint, the trends in Figure 47.13 demonstrate that the more efficient densification scheme involves heating at a constant rate rather than employing an isothermal hold.

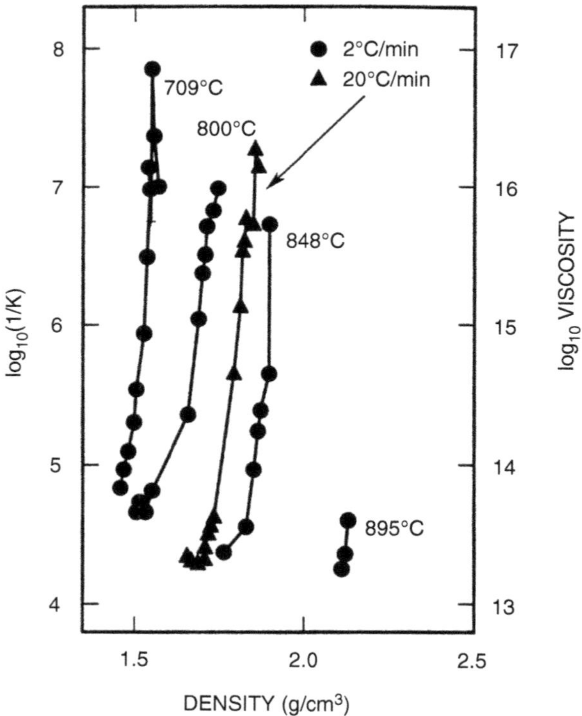

FIGURE 47.13 Sintering parameter $(1/K)$ and viscosity according to Scherer model [85] of viscous sintering for two-step acid-catalyzed xerogel. Samples were heated to indicated temperatures at 2 or 20°C/min and held isothermally. Corresponding bulk densities are plotted on the abscissa. (Reproduced with permission from reference 79. Copyright 1984.)

The low temperatures and correspondingly high viscosities involved in viscous sintering of gels (normally near the glass transition temperature T_g, defined by a viscosity of $10^{13.5}$ P) may permit the formation of metastable glasses [24]. For example, because phase separation or crystallization normally require diffusion over distances of several nanometers (at least), it is possible that viscous sintering could be complete before any measureable phase separation or crystallization occurred. For example, homogeneous glasses have been prepared below 1000°C in alkaline earth silicate systems that exhibit stable immiscibilities extending to over 2000°C [86–88].

The possibility of preparing metastable glasses from gels by low-temperature sintering has provoked the question of whether or not sintered gels generally differ from the corresponding melt-prepared glasses. On the basis of a detailed comparison of the relaxation kinetics of sintered borosilicate gels and melted borosilicate glasses, it was concluded [89] that, once the gel was processed in the vicinity of T_g, it was indistinguishable from the melted glass. Previous conclusions to the contrary (e.g., references 90 and 91) have in general been related to subtle differences in composition, so such comparisons require that the compositions be identical (including the OH content).

SURFACE STRUCTURE AND CHEMISTRY

Because of their high surface areas, Iler [2] believed that the properties of porous gels are dominated largely by the surface chemistry of the solid phase. The surfaces of silica xerogels dried at 100°C are terminated with hydroxyl groups (or in some cases, both hydroxyl and alkoxide groups). Zhuravlev [92] showed that the hydroxyl coverage of fully hydroxylated silica gels is 4.9 OH/nm^2, regardless of the surface area or manner of preparation. Although Iler [2] suggested that geminal silanols [$=Si(OH)_2$] do not exist on a dried surface, recent MAS ^{29}Si NMR spectroscopy results have shown that the hydroxylated silica surface is composed of $\mathbf{Q}^2$–$\mathbf{Q}^4$ silicon sites [for both aqueous [93] and alkoxide-derived [77] materials].

During the heat treatment procedures employed for consolidation, the surface is progressively dehydroxylated. Numerous Raman spectroscopy investigations (e.g., references 94–97 and Figure 47.14) have correlated surface dehydroxylation with intensification of a narrow Raman band at ~600 cm^{-1} labeled D2 (from its previous association with a defect). MAS ^{29}Si NMR spectroscopy experiments have in turn correlated the relative intensity of the D2 band with a reduction in the average Si–O–Si bond angle of $\mathbf{Q}^4$ silicons [77]. These results are consistent with the formation of cyclotrisiloxanes (three-membered rings) according to the following [77]:

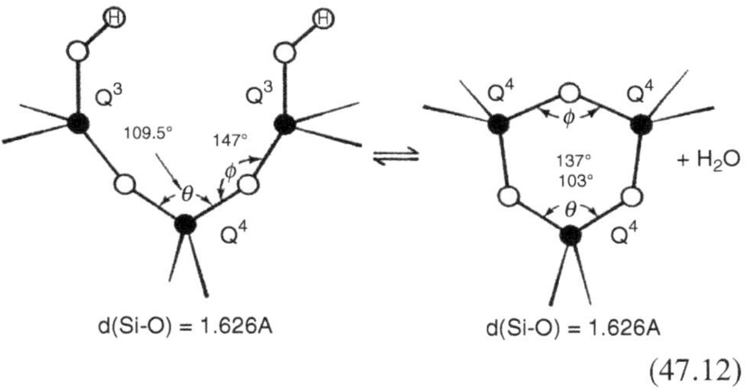

(47.12)

The concentration of the three-membered rings increases with surface area and the extent of dehydroxylation [98]. Although the rings are strained, as evident from the reduced Si–O–Si and O–Si–O bond angles and increased Si–O bond length (compare the left and right sides of Equation 47.12), such rings are apparently the preferred way to terminate the *dehydroxylated* silica surface [99], because they remain in high concentrations even in the vicinity of T_g (~1100°C), where the silicate network is able to reconstruct [100]. Because of their strain, the hydrolysis rate constant of cyclic trisiloxanes [reverse of Equation (47.12)] is about 75 times greater than the dissolution rate of conventional fused silica [101]. As such, cyclic trisiloxanes are preferred sites for surface rehydration [24].

Normally, sintering commences before dehydroxylation is complete [102] so that a substantial hydroxyl

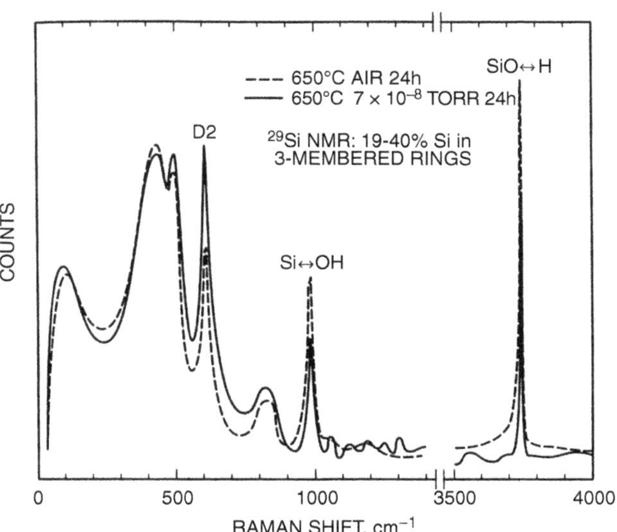

FIGURE 47.14 Raman spectra of silicate xerogel after a 24-h hold in air at 650°C or after 24 h in air and 24 h in vacuum at 650°C. Corresponding MAS ^{29}Si NMR spectra indicated that after the vacuum treatment up to 40% of the silicon atoms were contained in three-membered rings. (Reproduced with permission from reference 99. Copyright 1990.)

content is retained in the fully densified gel. This situation is problematic when it is necessary to reheat the densified gel in the vicinity of its softening temperature, for example, in fiber drawing or sealing operations, because the evolution of water causes bloating. In addition, strong absorption in the IR spectrum due to O–H stretching vibrations significantly degrades the transmission of optical fibers. To avoid this situation, halogen treatments have been employed [103,104]: reaction of the hydroxylated silica surface with sources of chlorine or fluorine above their dissociation temperatures results in quantitative replacement of OH with F or Cl. Although Si–Cl vibrations do not absorb at wavelengths of interest for optical communications, evolution of chlorine may also result in bloating. The chlorine-bloating problem is mitigated by a second heat treatment in dry oxygen below the sintering temperature (1000–1100°C). Because of the greater Si–F bond strength, significantly more fluorine can be incorporated in fully sintered gels without the bloating problems associated with chlorine.

By definition, the intrinsically high surface area of gels requires that gases are accessible to a substantial portion of the solid phase. For example, in a gel with a Brunauer–Emmett–Teller (BET) surface area of 850 m^2/g, 65% of the silicon atoms are on a surface. The accessibility of the surface makes gas- or liquid-phase reactions a viable means of "bulk" compositional modification. This situation has been exploited by numerous researchers who have reacted silicate xerogels with ammonia to form oxynitride glasses without melting [105–107].

FIBER FORMATION

Some highly viscous metal alkoxide derived sols can be drawn or spun into fibers useful for reinforcement or the production of refractory textiles. Such sols are called "spinnable". Spinnable sols are generally prepared from ethanolic solutions of TEOS hydrolyzed with 1–2 mol of water under acidic conditions and aged in open containers [108–110]. During aging the sols are concentrated by evaporation of alcohol. With time these sols show a progressive increase in the dependence of the reduced viscosity (η_{sp}/C) on the silica concentration (C) [110]. Such behavior is illustrated in Figure 47.15, where the concentration dependence of the reduced viscosity of an silica sol prepared with $H_2O/Si = 1$ is compared to that of a particulate silica sol (Ludox) and sodium metasilicate, a chainlike silicate. After shorter periods of aging ($t/t_{gel} < 0.5$), the sol exhibits little concentration dependence of η_{sp}/C, consistent with a system of noninteracting spherical particles (e.g., Ludox):

$$\eta_{sp}/C = k/\rho \tag{47.13}$$

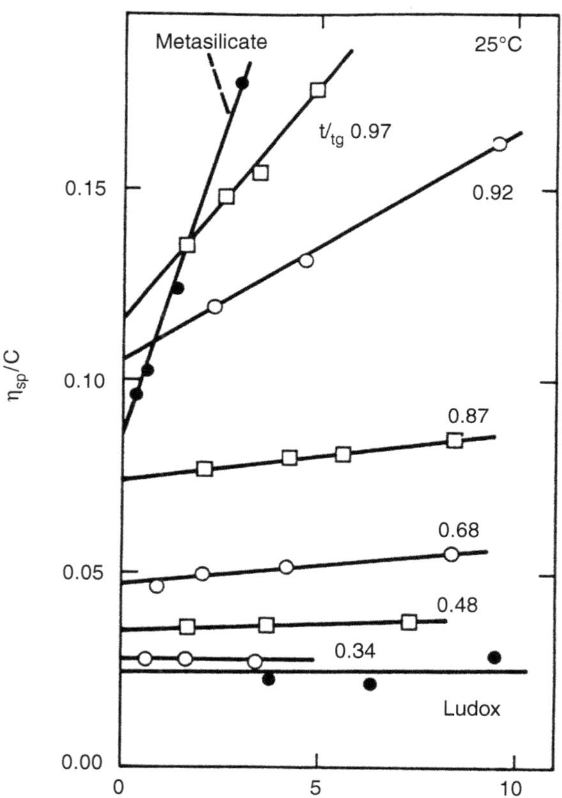

FIGURE 47.15 Concentration dependence of the reduced viscosity, η_{sp}/C, of an acid-catalyzed TEOS sol ($H_2O/Si = 1$) as a function of the normalized gel time t/t_{gel}. Data obtained for Ludox and sodium metasilicate are shown for comparison. (Reproduced with permission from reference 110. Copyright 1982.)

where k is a constant and ρ is the density of the particles. Further aging leads to extended, interacting polymers that exhibit a concentration dependence of the reduced viscosity described by the Huggins equation [111]:

$$\eta_{sp}/C = \eta + k'(\eta)^2 C \tag{47.14}$$

where k' is a constant and η is the intrinsic viscosity.

Compared to particulate sols, spinnable sols also show a greater molecular-weight dependence of the intrinsic viscosity:

$$\eta = k'' M_n \alpha \tag{47.15}$$

where k'' is a constant that depends on the kind of polymer, solvent, and temperature, α depends on the polymer structure, and M_n is the number-average molecular weight. According to theory [111], $\alpha = 0$ for rigid spherical particles, $\alpha = 0.5–1.0$ for flexible chainlike or linear polymers, and $\alpha = 1–2$ for rigid rodlike polymers. Sakka [108] found $\alpha = 0.64$ or 0.75 for spinnable silica sols prepared from TEOS with $H_2O:Si = 2$ or 1, respectively.

The value of α can be re-expressed in terms of a mass fractal dimension according to the following relationship [43]:

$$\eta \sim R_g^3/M_n \sim M_n^{(3/D)-1} \tag{47.16}$$

where R_g is the radius of gyration of the polymer. According to this expression, $\alpha = 0.64–0.75$ corresponds to $D = 1.83–1.71$, consistent with structures ranging from linear swollen polymers to swollen branched polymers. The ^{29}Si NMR spectrum of a spinnable silica sol (Figure 47.16) is composed of $\mathbf{Q}^1–\mathbf{Q}^4$ species [112] indicative of randomly branched polymers, rather than strictly linear ($\mathbf{Q}^2$) polymers or ladder ($\mathbf{Q}^3$) polymers.

On the basis of these observations, the high viscosities required for spinnability are a result of swollen (extended) randomly branched polymers that exhibit a strong concentration and molecular weight dependence of the reduced and intrinsic viscosities, respectively. During aging, solvent evaporation increases both the concentration of the sol and the molecular weight; these increases in turn cause [Equation (47.14) and Equation (47.15)] the viscosity to increase. However, premature gelation, which leads to elastic behavior and fiber fracture, is avoided because the combination of acid catalyst and low $H_2O:Si$ results in highly esterified polymers that exhibit low condensation rates on the fiber-drawing time scale.

PARTICLE FORMATION

As shown in Figure 47.2, polymerization of aqueous silicates at pH 7–10 in the absence of salt leads to the formation of unaggregated particles that may exceed 100 nm

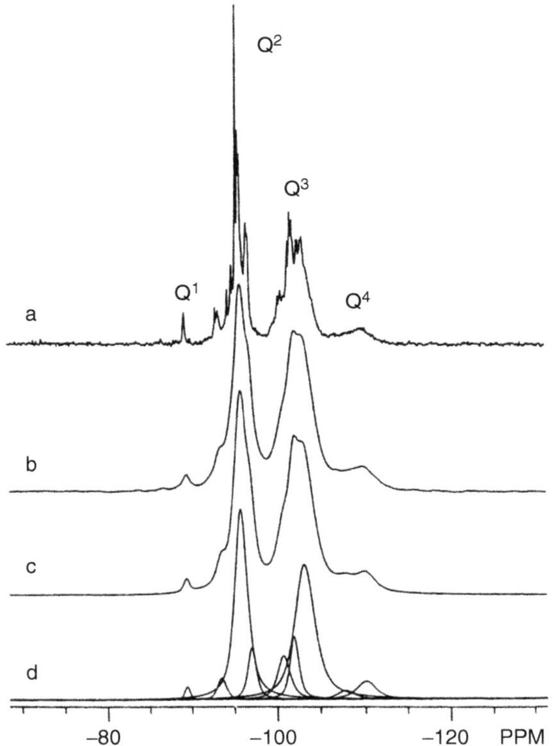

FIGURE 47.16 ^{29}Si NMR spectra of spinnable silica sol prepared from TEOS ($H_2O/Si = 1.5$). Spectrum a, experimental, 1.0-Hz exponential line broadening; spectrum b, experimental, 30-Hz exponential line broadening; spectrum c, computer simulation; and spectrum d, resonance components of computer simulation. (Reproduced with permission from reference 112. Copyright 1989.)

in diameter [2]. In this pH range, polymerization of monomer to oligomers and three-dimensional particles 1–2 nm in diameter occurs in only a few minutes. Once the concentration of monomer is substantially reduced, further growth occurs primarily by dissolution of smaller, more soluble particles and deposition of silica on larger, less soluble particles in a process called Ostwald ripening [2]. Aggregation is avoided because the larger particles are charged and thus repel each other.

In Ostwald ripening, the rate of growth depends on the particle size distribution. At a given temperature, growth stops when the size of the smallest particles is so large that the difference in solubility of the largest and smallest particles becomes negligible. At room temperature, this occurs when the particles exceed 4–5 nm in diameter [2]. Under hydrothermal conditions, where the solubility is enhanced, growth may continue to 150 nm [2]. According to Iler [2], the distribution of particle sizes is generally not known, but a Gaussian distribution is assumed.

Larger particles with narrower size distributions can be prepared from alkoxides by using a method developed by Stober, Fink, and Bohn (SFB) [113]. Dilute solutions of TEOS are hydrolyzed with large concentrations of water ($H_2O:Si = 7.5$ to >50) under quite basic conditions ($[NH_3] = 1-7$ M). The particles that form are spherical, and typically $<5\%$ of the particles differ by more than 8% from the mean size.

The growth of monodisperse particles is generally explained by the nucleation and growth model of La Mer and Dinegar [114]. According to this model, the supersaturation of hydrous oxide is increased, for example, by hydrolysis and condensation, above the critical concentration (C_N), where nucleation is extremely rapid. Nucleation causes the supersaturation to be reduced below C_N, where further nucleation is unlikely. Growth continues on the existing nuclei until the concentration is reduced to the equilibrium solubility. The single burst of nuclei confers monodispersity to the sol; if nuclei continue to form throughout the growth stage, a broad distribution of particle sizes is expected.

Bogush and Zukoski [115] performed a careful study to elucidate the operative growth mechanism accounting for monodispersity in the SFB process. By analysis of the number density of particles, their size, and the molar volume of silica, they determined that C_N was exceeded during the complete course of the process, so nucleation continued during the growth period. To account for monodispersity under conditions that did not meet the criteria of the classic nucleation and growth model, Bogush and Zukoski [115] proposed a nucleation and aggregation model. Although negatively charged, the initial primary particles (≤ 10 nm) are unstable because of their small size, so aggregation results. The larger, stable aggregates then sweep through the sol, picking up freshly formed primary particles and smaller aggregates. Monodispersity of the final particles is achieved through size-dependent aggregation rates.

Additional support for a nucleation and aggregation model is provided by cryo-TEM results of Bailey and Mecartney [116]. By fast-freezing thin liquid films of sol, they were able to follow the evolution of particle growth in situ without artifacts introduced by drying. They concluded that particle growth occurs by addition of small, low-density particles to larger, colloidally stable "seeds". A similar hypothesis was made by Iler [2] on the basis of some very early TEM work by Radczewski and Richter [117] on sols prepared by hydrolysis of $SiCl_4$.

FILM FORMATION

Prior to gelation, silicate sols can be deposited as thin films by such techniques as dipping or spinning. In dip coating, the substrate is normally withdrawn vertically from the coating bath at a speed U_0 [118] (*see* Figure 47.17). The moving substrate entrains the sol in a fluid mechanical boundary layer that splits in two above the bath surface,

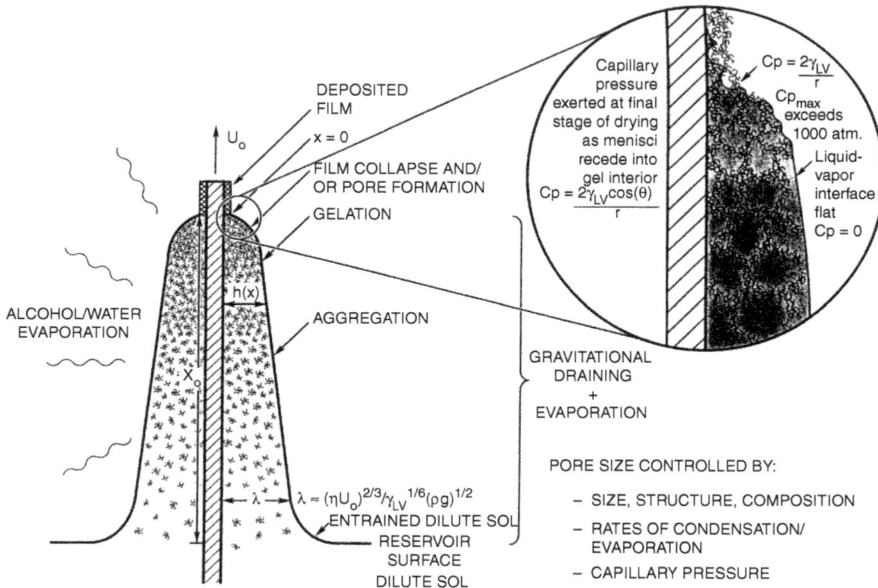

FIGURE 47.17 Illustration of sol–gel dip-coating showing sequential stages of deposition. Inset shows details of the final stage of drying.

returning the outer layer to the bath. The entrained sol is thinned and concentrated by gravitational draining and vigorous evaporation of solvent (normally an alcohol–water mixture). A steady-state drying profile terminated by a well-defined drying line ($x = 0$ in Figure 47.17) develops when the upward flux due to entrainment of the sol on the moving substrate is just balanced by the fluxes attributable to evaporation and draining [119]. The thickness (h) of the deposited film depends on U_0, the sol viscosity (η), and surface tension (γ_{LV}) according to the following relationship [120]:

$$h \sim (\eta U_0)^{2/3}/\gamma_{LV}^{1/6}(\rho g)^{1/2} \qquad (47.17)$$

where g is gravitational acceleration. Spin coating [121] differs in that the film thins primarily by centrifugation, which imposes a greater shear stress on the depositing sol and causes the evaporation rate to be greater because of forced convection caused by the spinning substrate.

The thin-film-forming process differs from the processes of gelation, aging, and drying described earlier for bulk gels (*see* Figure 47.1) in several fundamental ways [118]:

1. Whereas drying normally follows gelation for bulk gel processing, in dip or spin coating, drying overlaps the processes of polymer growth, gelation, and aging. This overlap of the deposition and drying stages establishes a very brief time scale for coating (several seconds) compared to bulk gel formation and drying (typically days or weeks).
2. As in bulk systems, the structure of the dried film (xerogel) depends on the competition between capillary forces that tend to compact the structure and aging processes that stiffen the structure and thus increase its resistance to compaction. However, the brief duration of the coating process provides little time for aging, so films remain compliant and are compacted to a greater extent by capillary forces.
3. Gravitational or centrifugal draining combined with possible flows driven by surface tension gradients may impose very high shear rates on the depositing sol. After gelation, continued shrinkage of the constrained film due to drying, skeletal densification, or sintering creates tensile stresses within the film. By comparison, bulk gels are generally not subjected to high shear rates and are free to shrink isotropically.

Polymer growth during the deposition stage probably occurs by CCA, with trajectories ranging from Brownian (dilute conditions) to ballistic in the final 5% of the deposition process, where strong convective forces may exist because of rapid concentration of the inorganic phase [119]. The gel point is ill defined but may be considered as the moment when the condensing network is sufficiently stiff to withstand flow due to gravity, yet still filled with solvent. From this point, further evaporation may collapse the film or create porosity within the film.

Compared to other thin-film-forming processes such as sputtering, evaporation, or chemical vapor deposition (CVD), sol–gel film formation has the advantage that both the composition and microstructure can be controlled on the molecular level. Thus, the films can be "tailored" for specific applications, for example, dense films for protective or optical applications and porous films for sensors or membranes.

The surface area, pore volume, and pore size of the deposited film depend on such factors as the size and structure (fractal dimension) of the entrained inorganic species, the relative rates of condensation and evaporation during deposition, and the magnitude of the capillary pressure [122]. The fractal dimension influences porosity through steric control. Mandelbrot [47] showed that if two objects of radius R are placed independently in the same region of space, the number of intersections ($M_{1,2}$) is expressed as

$$M_{1,2} \propto R^{D1+D2-d} \tag{47.18}$$

where $D1$ and $D2$ are the respective fractal (or Euclidian) dimensions and d is the dimension of space, 3. According to this expression, if each object has a fractal dimension less than 1.5, the probability of intersection decreases indefinitely with R. These structures are "mutually transparent" and, if deposited by dip or spin coating, should freely interpenetrate each other as they are concentrated on the substrate surface, so dense films results. If $D1$ and $D2$ are greater than 1.5, the probability of intersection increases algebraically with R. These structures are "mutually opaque". Strong screening of the cluster interiors prevents interpenetration, creating a porous structure much like an assemblage of "tumbleweeds".

Mandelbrot's relationship [Equation (47.18)] assumes that the objects are perfectly rigid and "stick" immediately and irreversibly at each point of intersection (chemically equivalent to an infinite condensation rate). In fact, fractal objects are more or less compliant, and the "sticking probability" is always $\ll 1$. These factors mitigate the criterion for mutual transparency; for example, if the condensation rate is reduced, screening is less effective, and objects with $D > 1.5$ can interpenetrate. Because the condensation rate of silica depends strongly on pH (*see* Figure 47.3), the "transparency or opacity" and ultimately the film porosity can be manipulated by the addition of acid or base catalyst [24].

For a particular withdrawal speed, U_o, the evaporation rate establishes the position of the drying line with respect to the reservoir surface (X_o in Figure 47.17) and thus the time scale. For reactive sols, a reduction in the evaporation rate provides more time for condensation reactions to occur, leading to stiffer networks that resist collapse by capillary forces. For colloidally stable sols (e.g., electrostatically stabilized SFB particles or highly esterified polymers), a reduction in the evaporation rate provides more time for ordering or interpenetration, leading to denser structures.

As in the case of drying bulk gels (illustrated in Figure 47.10), the maximum capillary pressure is exerted on the film network at the final stage of the deposition (drying) process, when liquid–vapor menisci recede into the film interior (*see* Figure 47.17). However, because of the short time scale of the coating process, film networks are more compliant than bulk gels, allowing more shrinkage to occur prior to the final stage. Thus, compared to bulk gels, the pore size is smaller and the maximum capillary pressure greater when the menisci finally recede into the film interior. It is conceivable that in some situations over 1000 atm (100,000 kPa) of pressure are exerted on the depositing film network. It is not surprising therefore that, under some conditions, film structures collapse to such an extent that no porosity is accessible to a nitrogen probe molecule (kinetic diameter ~0.4 nm) [118,123].

A particularly illustrative example of the dependence of film microstructure on the size, structure, and condensation rate of the depositing silicate phase is that of a borosilicate sol deposited at pH ~3 after various periods of aging at 50°C [122]. The borosilicate species are characterized by a mass fractal dimension $D \simeq 2.4$. Aging at 50°C

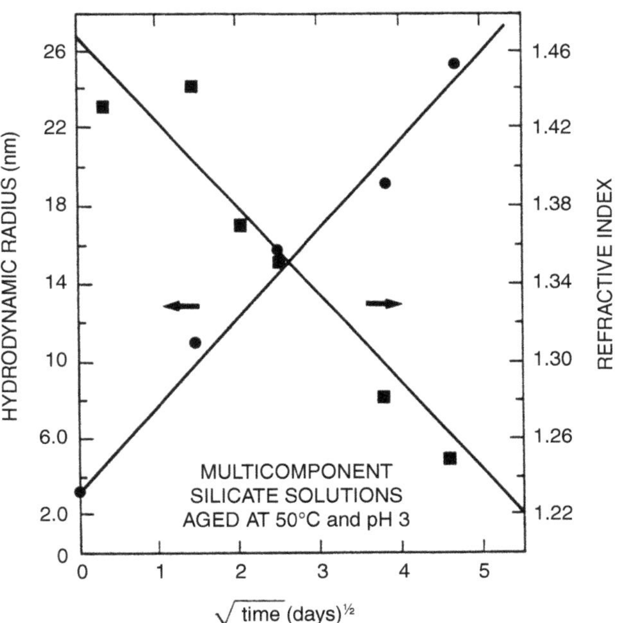

FIGURE 47.18 Reciprocal relationship between the hydrodynamic radius of multicomponent borosilicate species and the corresponding refractive index of films prepared from borosilicate sols as a function of root aging time at 50°C and pH ~ 3. (Reproduced with permission from reference 1. Copyright 1988.)

TABLE 47.2
Porosities of Films Prepared from Multicomponent Borosilicate Sols

Sample aging times[a]	Refractive index	%Porosity ads. N_2	Median pore radius (nm)	Surface area (m^2/g)	Applications
Unaged	1.45	0	<0.2	1.2–1.9	Dense, protective electronic and optical films
0–3 Days					Microporous films for sensors and membranes
3 Days	1.31	16	1.5	146	
1 Week	1.25	24	1.6	220	Mesoporous films for sensors, membranes, catalysts, optics
2 Weeks	1.21	33	1.9	263	
3 Weeks	1.18	52	3.0	245	

[a]Aging of dilute sol at 50°C and pH 3 prior to film deposition.
Source: Reproduced with permission from reference 118. Copyright 1990.

causes them to grow in size as indicated by the plot of hydrodynamic radius versus aging time in Figure 47.18. Corresponding refractive indices of films deposited from the aged sols are observed to decrease with increasing hydrodynamic radius. This behavior is consistent with a system of mutually opaque fractal clusters. Because the density of a mass fractal decreases with its size r as $\rho \sim 1/r^{(3-D)}$, the density (and refractive index) of an assemblage of noninterpenetrating (opaque) clusters would also decrease with r. For small r, the cluster density is greater and the effective screening (and opacity) is less, a situation that leads to much denser structures that have no accessible porosity [123]. Denser structures are also obtained by reducing the pH of the coating bath from about pH 3 toward pH 1 immediately prior to deposition [24]. The reduction in pH lowers the condensation rate. Thus branched structures are able to interpenetrate on the film deposition time scale, and compact films are produced.

Table 47.2 [118] shows that the surface area, pore volume, and pore size of the deposited films vary consistently with the aging times. Thus the film structures may be tailored for such applications as surface passivation, sensors, membranes, or catalysts by a simple aging process prior to film deposition. In addition, multiple deposition schemes involving different compositions or structures or both allow the formation of complex layered architectures potentially useful for optics, electronics, or sensors.

SUMMARY

The structures of silicate polymers formed from alkoxides by the sol–gel process vary from weakly branched "molecular networks" to highly condensed particles that are similar to commercial aqueous silicates described by Iler [2]. The sol–gel process combines this control of microstructure on molecular length scales with the ability to form specialized shapes such as fibers, films, monosized powders, and monoliths at room temperature. Silica gels may be exploited for their porosity or consolidated to dense glasses indistinguishable from their melt-prepared counterparts. Sintering occurs in the vicinity of T_g and permits the formation of novel metastable materials. Iler's view on many of these aspects of the sol–gel process have been largely confirmed by recent experiments.

ACKNOWLEDGMENTS

I have benefited from the continued support of the Department of Energy Basic Energy Sciences program "The Growth and Structure of Ceramic Precursors" and from associated collaborations with colleagues at Sandia National Laboratories (Roger Assink, Dale Schaefer, Keith Keefer, Joe Bailey, Carol Ashley, Alan Hurd, Scott Reed, and Dan Doughty) along with George Scherer at DuPont. This work was performed at Sandia National Laboratories and supported by the U.S. Department of Energy under Contract DE–AC04–76P00789.

REFERENCES

1. *Sol–Gel Technology for Thin Films, Fibers, Preforms, Electronics, and Specialty Shapes*; Klein, L. C., Ed.; Noyes: Park Ridge, NJ, 1988.
2. Iler, R. K. *The Chemistry of Silica*; Wiley: New York, 1978.
3. Hench, L. L.; West, J. K. *Chem. Rev.* **1990**, *90*, 33–72.
4. Sakka, S. *J. Non-Cryst. Solids* **1985**, *73*, 651–660.
5. Masdiyasni, K. S. *Sagamore Army Mat. Res. Conf. Proc.* **1985**, *30*, 285–336. Baes, E. F.; Mesmer, R. E. *The Hydrolysis of Cations*; Wiley: New York, 1976.
6. Roy, R. *Science (Washington, DC)* **1987**, *238*, 1664–1669.
7. Zelinski, B. J. J.; Uhlmann, D. R. *J. Phys. Chem. Solids* **1984**, *45*, 1069–1090.

8. Zarzycki, J. In *Glass Science and Technology*; Ulhmann, D. R.; Kriedl, N. J., Eds.; Academic: Boston, MA, 1984; Vol. 2, pp 209–49.
9. *J. Non-Cryst. Solids* **1982**, *48*, 1–230.
10. *J. Non-Cryst. Solids* **1984**, *63*, 1–300.
11. *J. Non-Cryst. Solids* **1986**, *82*, 1–436.
12. *J. Non-Cryst. Solids* **1988**, *100*, 1–554.
13. *J. Non-Cryst. Solids* **1990**, *121*, 1–492.
14. *J. Non-Cryst. Solids* **1992**, *147, 148*, 1–841.
15. *Ultrastructure Processing of Glasses, Ceramics, and Composites*; Hench, L. L.; Ulrich, D. R., Eds.; Wiley: New York, 1984.
16. *Science of Ceramic Chemical Processing*; Hench, L. L.; Ulrich, D. R., Eds.; Wiley: New York, 1986.
17. *Ultrastructure Processing of Advanced Ceramics*; Mackenzie, J. D.; Ulrich, D. R., Eds.; Wiley: New York, 1988.
18. *Ultrastructure Processing IV*; Uhlmann, D. R.; Ulrich, D. R., Eds.; Wiley: New York, 1992.
19. *Ultrastructure Processing V*; Hench, L. L.; West, J. K., Eds.; Wiley: New York, 1992.
20. *Better Ceramics Through Chemistry*; Brinker, C. J.; Clark, D. E.; Ulrich, D. R., Eds.; Elsevier: New York, 1984.
21. *Better Ceramics Through Chemistry II*; Brinker, C. J.; Clark, D. E.; Ulrich, D. R., Eds.; Materials Research Society: Pittsburgh, PA, 1986.
22. *Better Ceramics Through Chemistry III*; Brinker, C. J.; Clark, D. E.; Ulrich, D. R., Eds.; Materials Research Society: Pittsburgh, PA, 1988.
23. *Better Ceramics Through Chemistry IV*; Zelinski, B. J. J.; Brinker, C. J.; Clark, D. E.; Ulrich, D. R., Eds.; Materials Research Society: Pittsburgh, PA, 1990.
24. Brinker, C. J.; Scherer, G. W. *Sol–Gel Science*; Academic: San Diego, CA, 1990.
25. Baes, E. F.; Mesmer, R. E. *The Hydrolysis of Cations*; Wiley: New York, 1976.
26. Keefer, K. D. In *Better Ceramics Through Chemistry*; Brinker, C. J.; Clark, D. E.; Ulrich, D. R., Eds.; North-Holland: New York, 1984; pp 15–24.
27. McNeill, K. J.; DiCaprio, J. A.; Walsh, D. A.; Pratt, R. F. *J. Am. Chem. Soc.* **1980**, *102*, 1859.
28. Pohl, E. R.; Osterholtz, F. D. In *Molecular Characterization of Composite Interfaces*; Ishida, H; Kumar, G., Eds.; Plenum: New York, 1985; p 157.
29. Corriu, R. J. P.; Henner, M. *J. Organometallic Chem.* **1974**, *74*, 1–28.
30. Brinker, C. J. *J. Non-Cryst. Solids* **1988**, *100*, 30–51.
31. Aelion, R.; Loebel, A.; Eirich, F. *J. Am. Chem. Soc.* **1950**, *72*, 5705–5712.
32. Voronkov, M. G.; Mileshkevich, V. P.; Yuzhelevski, Y. A. *The Siloxane Bond*; Consultants Bureau: New York, 1978.
33. Brinker, C. J.; Keefer, K. D.; Schaefer, D. W.; Ashley, C. S. *J. Non-Cryst. Solids* **1982**, *48*, 47–64.
34. Brinker, C. J.; Keefer, K. D.; Schaefer, D. W.; Assink, R. A.; Kay, B. D.; Ashley, C. S. *J. Non-Cryst. Solids* **1984**, *63*, 45–59.
35. Pouxviel, J. C.; Boilot, J. P. *J. Non-Cryst. Solids* **1987**, *94*, 374–386.
36. Kelts, L. W.; Armstrong, N. J. In *Better Ceramics Through Chemistry III*; Brinker, C. J.; Clark, D. E.; Ulrich, D. R., Eds.; Materials Research Society: Pittsburgh, PA, 1988; p 519.
37. Lippert, J. L.; Melpolder, S. B.; Kelts, L. W. *J. Non-Cryst. Solids* **1988**, *104*, 139–147.
38. Artaki, I.; Bradley, M.; Zerda, T. W.; Jonas, J. *J. Phys. Chem.* **1985**, *89*, 4399–4404.
39. Klemperer, W. G.; Mainez, V. V.; Ramamurthi, S. D.; Rosenberg, F. S. In *Better Ceramics Through Chemistry III*; Brinker, C. J.; Clark, D. E.; Ulrich, D. R., Eds.; Materials Research Society: Pittsburgh, PA, 1988; pp 15–24.
40. Kelts, L. W.; Effinger, N. J.; Melpolder, S. M. *J. Non-Cryst. Solids* **1986**, *83*, 353–374.
41. Assink, R. A. personal communication.
42. Martin, J. E.; Hurd, A. J. *J. Appl. Cryst.* **1987**, *20*, 61–78.
43. Schaefer, D. W. *MRS Bull.* **1988**, *8*, 22–27.
44. Schaefer, D. W.; Martin, J. E.; Keefer, K. D. In *Physics of Finely Divided Matter*; Bocarra, N.; Daoud, M., Eds.; Springer–Verlag: Berlin, Germany, 1985; p 31.
45. Schaefer, D. W.; Keefer, K. D. In *Fractals in Physics*; Pietronero, L; Tosatti, E., Eds.; North-Holland, Amsterdam, Netherlands, 1986; pp 39–45.
46. Bailey, J. K.; Nagase, T.; Broberg, S. M.; Mecartney, M. L. *J. Non-Cryst. Solids* **1989**, *109*, 198.
47. Mandelbrot, B. B. *Fractals, Form, and Chance*; Freeman: San Francisco, CA, 1977.
48. Porod, G. *Kolloid Z.* **1951**, *124*, 83.
49. Keefer, K. D. In *Better Ceramics Through Chemistry III*; Brinker, C. J.; Clark, D. E.; Ulrich, D. R., Eds.; Materials Research Society: Pittsburgh, PA, 1986; pp 295–304.
50. Meakin, P. In *On Growth and Form*; Stanley, H. E.; Ostrowsky, N., Eds; Martinus-Nijhoff: Boston, MA, 1986; pp 111–135.
51. Sutherland, D. N. *J. Colloid Interface Sci.* **1967**, *25*, 373–380.
52. Flory, P. J. *J. Am. Chem. Soc.* **1941**, *63*, 3083; *J. Phys. Chem.* **1942**, *46*, 132.
53. Stockmeyer, W. H. *J. Chem. Phys.* **1943**, *11*, 45.
54. de Gennes, P. G. *Biopolymers* **1968**, *6*, 715–729.
55. Stauffer, D. *J. Chem. Soc. Faraday Trans. II* **1976**, *72*, 1354.
56. de Gennes, P. G. *Scaling Concepts In Polymer Physics*; Cornell Univ. Press: Ithaca, NY, 1979.
57. Zallen, R. *The Physics of Amorphous Solids*; Wiley: New York, 1983.
58. Martin, J. E.; Wilcoxon, J.; Adolf, D. *Phys. Rev. A* **1987**, *36*(4), 1803–1810.
59. Martin, J. E. In *Proceedings of Atomic and Molecular Processing of Electronic and Ceramic Materials: Preparation, Characterization, Properties*; Aksay, I. A.; McVay, G. L.; Stoebe, T. G.; Wager, J. F., Eds.; Materials Research Society: Pittsburgh, PA, 1987; pp 79–89.
60. Vega, A. J.; Scherer, G. W. *J. Non-Cryst. Solids* **1989**, *111*(2,3), 153–166.

61. Scherer, G. W. *J. Non-Cryst. Solids* **1989**, *109*, 183–190.
62. Yoldas, B. E. *J. Mater. Sci.* **1986**, *21*, 1087–1092.
63. Klimentova, Yu. P.; Kirichenko, L. F.; Vysotskii, Z. Z. *Ukr. Khim. Zh.* **1970**, *36*(1), 56–58 (Eng. trans.).
64. Scherer, G. W. In *Better Ceramics Through Chemistry III*; Brinker, C. J.; Clark, D. E.; Ulrich, D. R., Eds.; Materials Research Society: Pittsburgh, PA, 1988; pp 179–186.
65. Brinker, C. J.; Schaefer, D. W., unpublished work.
66. Glaves, C. L.; Brinker, C. J.; Smith, D. M.; Davis, P. J. *Chem. Mat.* **1989**, *1*, 34–40.
67. Davis, P. J.; Brinker, C. J.; Smith, D. M.; Assink, R. A.; Schaefer, D. W.; Tallant, D. R., submitted to *J. Non-Cryst. Solids.*
68. Scherer, G. W. *J. Am. Ceram. Soc.* **1990**, *73*, 3–14.
69. Hench, L. L. In *Science of Chemical Processing*; Hench, L. L.; Ulrich, D. R., Eds.; Wiley: New York, 1986; pp 52–64.
70. Kistler, S. S. *J. Phys. Chem.* **1932**, *36*, 52–64.
71. Prassas, M.; Hench, L. L. In *Ultrastructure Processing of Glasses, Ceramics, and Composites*; Hench, L. L., Ed.; Wiley: New York, 1984; pp 100–125.
72. Hrubesh, L. W.; Tillotson, T. M.; Poco, J. F. In *Better Ceramics Through Chemistry IV*; Zelinski, B. J. J.; Brinker, C. J.; Clark, D. E.; Ulrich, D. R., Eds.; Materials Research Society: Pittsburgh, PA, 1990; pp 215–319.
73. Woignier, T.; Phalippou, J.; Vacher, R. In *Better Ceramics Through Chemistry III*; Brinker, C. J.; Clark, D. E.; Ulrich, D. R., Eds.; Materials Research Society: Pittsburgh, PA, 1988; pp 697–702.
74. Wallace and Hench, L. L. In *Better Ceramics Through Chemistry*; Brinker, C. J.; Clark, D. E.; Ulrich, D. R., Eds.; Elsevier, North-Holland: New York, 1984; pp 47–52.
75. Russo, R. E.; Hunt, A. J. *J. Non-Cryst. Solids* **1986**, *86*, 219–230.
76. Klemperer, W. G.; Mainz, V. V.; Millar, D. M. In *Better Ceramics Through Chemistry II*; Brinker, C. J.; Clark, D. E.; Ulrich, D. R., Eds.; Materials Research Society: Pittsburgh, PA, 1986; pp 15–26.
77. Brinker, C. J.; Kirkpatrick, R. J.; Tallant, D. R.; Bunker, B. C.; Montez, B. *J. Non-Cryst. Solids* **1988**, *99*, 418–428.
78. Schaefer, D. W.; Keefer, K. D. In *Fractals in Physics*; Pietronero, L.; Tosatti, E., Eds.; North-Holland: Amsterdam, Netherlands, 1986; pp 39–45.
79. Brinker, C. J.; Drotning, W. D.; Scherer, G. W. In *Better Ceramics Through Chemistry*; Brinker, C. J.; Clark, D. E.; Ulrich, D. R., Eds.; Elsevier, North-Holland: New York, 1984; pp 25–32.
80. Brinker, C. J.; Scherer, G. W.; Roth, E. P. *J. Non-Cryst. Solids* **1985**, *72*, 345–368.
81. Scherer, G. W. In *Surface and Colloid Science*; Matijevic, E., Ed.; Plenum: New York, 1987; Vol. 14, pp 265–300.
82. Scherer, G. W. *Relation in Glasses and Composites*; Wiley: New York, 1986.
83. Tohge, N.; Moore, G. S.; Mackenzie, J. D. *J. Non-Cryst. Solids* **1984**, *63*, 95–104.
84. Brinker, C. J.; Roth, E. P.; Tallant, D. R.; Scherer, G. W. In *Science of Ceramic Chemical Processing*; Hench, L. L.; Ulrich, D. R., Eds.; Wiley: New York, 1986, 37–51.
85. Scherer, G. W. *J. Am. Ceram. Soc.* **1977**, *60*, 236–239.
86. Hayashi, T.; Saito, H. *J. Mater. Sci.* **1980**, *15*, 1971–1977.
87. Yamane, M.; Kojima, T. *J. Non-Cryst. Solids* **1981**, *44*, 181–190.
88. Bansal, N. P. *J. Am. Ceram. Soc.* **1988**, *71*(8), 666–672.
89. Scherer, G. W.; Brinker, C. J.; Roth, E. P. *J. Non-Cryst. Solids* **1986**, *82*, 191–197.
90. Weinberg, M. C.; Neilson, G. F. *J. Mater. Sci.* **1978**, *13*, 1206–1216.
91. Weinberg, M. C.; Neilson, G. F. *J. Am. Ceram. Soc.* **1983**, *66*(2), 132–134.
92. Zhuravlev, L. T. *Langmuir* **1987**, *3*, 316–318.
93. Fyfe, C. A.; Gobbi, G. C.; Kennedy, G. J. *J. Phys. Chem.* **1985**, *89*, 277–281.
94. Bertoluzza, A.; Fagnano, C.; Morelli, M. A.; Gottardi, V.; Guglielmi, M. *J. Non-Cryst. Solids* **1982**, *48*, 117–128.
95. Krol, D. M.; Van Lierop, J. G. *J. Non-Cryst. Solids* **1987**, *63*, 131–144.
96. Brinker, C. J.; Tallant, D. R.; Roth, E. P.; Ashley, C. S. In *Defects in Glasses*; Galeener, F. L.; Griscom, D. L.; Weber, M. J.; Eds.; Materials Research Society: Pittsburgh, PA, 1986; pp 387–411.
97. Brinker, C. J.; Tallant, D. R.; Roth, E. P.; Ashley, C. S. *J. Non-Cryst. Solids.* **1986**, *82*, 117–126.
98. Wallace, S. Ph.D. Thesis, University of Florida, Gainesville, FL, 1991.
99. Brinker, C. J.; Brow, R. K.; Tallant, D. R.; Kirkpatrick, R. J. *J. Non-Cryst. Solids* **1990**, *120*, 26–33.
100. Bartram, H. E.; Michalske, T. A.; Rodgers, J. W., Jr. *J. Phys. Chem.* **1991**, *95*, 4453–63.
101. Bunker, B. C.; Haaland, D. M.; Michalske, T. A.; Smith, W. L. *Surf. Sci.* **1989**, *222*, 95–118.
102. Gallo, T. A.; Brinker, C. J.; Kelin, L. C.; Scherer, G. W.; In *Better Ceramics Through Chemistry*; Brinker, C. J.; Clark, D. E.; Ulrich, D. R., Eds.; Elsevier: New York, 1984; pp 85–90.
103. Matsuyama, I.; Susa, K.; Satoh, S.; Suganuma, T. *Ceramic Bull.* **1984**, *63*, 1408–1411.
104. Rabinovich, E. M.; Wood, D. L.; Johnson, D. W., Jr.; Fleming, D. A.; Vincent, S. M.; MacChesney, J. B. *J. Non-Cryst. Solids* **1986**, *82*, 42–49.
105. Brinker, C. J.; Haaland, D. M. *J. Am. Ceram. Soc.* **1983**, *66*, 758–765.
106. Brow, R. K.; Pantano, C. G. *J. Am. Ceram. Soc.* **1987**, *70*, 9–14.
107. Kamiya, K.; Ohya, M.; Yoko, T. *J. Non-Cryst. Solids* **1986**, *83*, 209–222.
108. Sakka, S. In *Better Ceramics Through Chemistry*; Brinker, C. J.; Clark, D. E.; Ulrich, D. R., Eds.; Elsevier: New York, 1984; p 91.
109. Sakka, S.; Kamiya, K.; Makita, K.; Yamamoto, Y. *J. Non-Cryst. Solids* **1984**, *63*, 223–235.

110. Sakka, S.; Kamiya, K. *J. Non-Cryst. Solids* **1982**, *48*, 31–46.
111. Abe, Y.; Misono, T. *J. Poly. Sci. Chem.* **1983**, *21*, 41.
112. Brinker, C. J.; Assink, R. A. *J. Non-Cryst. Solids* **1989**, *111*, 48.
113. Stober, W.; Fink, A.; Bohn, E. *J. Colloid Interface Sci.* **1968**, *26*, 62–69.
114. La Mer, V. K.; Dinegar, R. H. *J. Am. Chem. Soc.* **1950**, *72*(11), 4847–4854.
115. Bogush, G. H.; Zukoski, C. F. In *Ultrastructure Processing of Advanced Ceramics*; Mackenzie, J. D.; Ulrich, D. R., Eds.; Wiley: New York, 1988; pp 477–486.
116. Bailey, J. K.; Mecartney, M. L. *Colloids and Surfaces*, submitted.
117. Radczewski, O. E.; Richter, H. *Kolloid Z.* **1941**, *96*, 1.
118. Brinker, C. J.; Hurd, A. J.; Frye, G. C.; Ward, K. J.; Ashley, C. S. *J. Non-Cryst. Solids* **1990**, *121*, 294–302.
119. Hurd, A. J.; Brinker, C. J. In *Better Ceramics Through Chemistry III*; Brinker, C. J.; Clark, D. E., Ulrich, D. R., Eds.; Materials Research Society: Pittsburgh, PA, 1988; pp 731–742.
120. Landau, L. D.; Levich, B. G. *Acta Physiochim, U.R.S.S.* **1942**, *17*, 42–54.
121. Bornside, D. E.; Macosko, C. W.; Scriven, L. E. *J. Imaging Technol.* **1987**, *13*, 122–129.
122. Brinker, C. J.; Hurd, A. J.; Ward, K. J. In *Ultrastructure Processing of Advanced Ceramics*; Mackenzie, J. D.; Ulrich, D. R., Eds.; Wiley: New York, 1988; p 223.
123. Frye, G. C.; Ricco, A. J.; Martin, S. J.; Brinker, C. J. In *Better Ceramics Through Chemistry III*; Brinker, C. J.; Clark, D. E.; Ulrich, D. R., Eds.; Materials Research Society: Pittsburgh, PA, 1988; pp 349–354.

48 The Chemistry of Hydrolysis and Condensation of Silica Sol–Gel Precursors

Bradley K. Coltrain and Larry W. Kelts
Eastman Kodak Company, Corporate Research Laboratories

CONTENTS

A brief overview of sol–gel hydrolysis and condensation reactions and the evolving structures is presented. Results from existing literature are summarized, but emphasis is placed on recent and unpublished work. A sound knowledge of sol–gel chemical reactions is necessary for a thorough understanding of silicate polymer growth and control of material properties. This understanding is complicated by concurrent hydrolysis and condensation reactions. Many workers are contributing to this knowledge base. This chapter represents a summary of the state-of-the art in this area.

The hydrolysis and condensation of silicon alkoxides is an area of intense interest. The sol–gel process uses high-purity monomers for low-temperature production of fibers, monoliths, coatings, and powders. Structures of the polymers produced in the sol ultimately dictate both gel and glass properties.

This chapter is a brief overview of silicon alkoxide hydrolysis and condensation and the resulting structures, emphasizing recent studies and unpublished work. Schmidt et al. [1] and more recently Brinker [2] have published excellent overviews of this chemistry; this chapter attempts to amplify and complement these reports.

A number of variables influece the structural evolution of silicate polymers: pH, solvent, water-to-silicon ratio (W), and monomer. Low-pH conditions with $W = 1-2$ are useful for fiber spinning; when $W = 4$ the sols are useful as coatings, whereas high W values or high pH produces colloidal suspensions or powders. Monoliths are frequently produced by addition of drying control chemical additives (DCCAs), which enhance drying rates while reducing cracking due to drying stresses. The following discussion elucidates how the chemistry is controlled to dictate sol–gel properties.

The two fundamental chemical reactions of the sol–gel process are (1) hydrolysis:

$$\equiv Si(OR) + H_2O \underset{k_E}{\overset{k_H}{\rightleftharpoons}} \equiv Si(OH) + ROH$$

and (2) condensation:

$$\equiv Si(OR) + (OH)Si\equiv \overset{k_A}{\rightleftharpoons} \equiv Si{-}O{-}Si\equiv + ROH \text{ (alcohol producing)}$$

$$\equiv Si(OH) + (OH)Si\equiv \overset{k_W}{\rightleftharpoons} \equiv Si{-}O{-}Si\equiv + H_2O \text{ (water producing)}$$

k_H, k_E, k_A, and k_W are rate constants for hydrolysis, esterification, alcohol-producing condensation, and water-producing condensation, respectively. Usually, hydrolysis and condensation reactions are concurrent. The following discussion separates the two in an attempt to conceptually simplify the chemistry. In some cases, because the two reactions are interdependent, they are discussed together, especially when gel times are considered.

HYDROLYSIS

Aelion et al. [3,4] showed that hydrolysis of tetraethoxysilane (TEOS) is acid or base catalyzed with a minimum rate at pH 7. Studies of alkyltrialkoxysilane hydrolysis in buffered aqueous solutions indicate that hydrolysis is both specific acid and specific base catalyzed [5,6].

Trivalent siliconium ion [1,7] and pentavalent silicon species [6] were proposed as condensation intermediates. Gas-phase mass spectrometric reactions of TEOS indicate a surprising stability of both trivalent and five-coordinate $[(OEt)_4SiOH]^+$ species [8], but the trivalent species was incapable of initiating condensation, at least under these conditions. Although the gas-phase studies may not directly correlate with solution behavior, they do suggest that three- and five-coordinate species are at least possible.

Most results to date support the presence of a five-coordinate intermediate. Hydrolysis rates decrease with increasing bulkiness of the alkoxide [5,9]. This feature and a large negative entropy of activation (-39 cal deg^{-1} mol^{-1}) [5] are consistent with associative mechanisms. Pohl and Osterholz [6] proposed intial protonation of an alkoxide under low-pH conditions followed by backside attack of a water molecule, producing a five-coordinate transition state with inversion at silicon. Flank-side attack without inversion was also proposed [10,11]. At high pH, hydroxyl anions are proposed to attack the silicon atom to produce either a five-coordinate transition state or stable intermediate. The low-pH mechanism involves protonation of an alkoxide, so silicon atoms with electron-donating groups should hydrolyze more readily. Because –OH and –OSi groups are more electron withdrawing than –OR, the reactions become more difficult with increasing hydrolysis and condensation. The opposite is expected for high-pH conditions.

However, NMR studies, coupled with statistical modeling, contradict these arguments. Pouxviel and co-workers [12,13] studied acid-catalyzed reactions of TEOS with a variety of W values. Simulated kinetic curves for temporal evolution of various silicon species observed by ^{29}Si NMR spectroscopy were consistent with relative hydrolysis rate constants for sequential hydrolysis of the four –OEt substituents of 1:5.3:20:36. These trends were confirmed by more recent ^1H NMR spectroscopy [14], which yielded values for the four successive hydrolysis steps for TEOS of 0.0143, 0.064, 0.29, and 1.3 min^{-1}, respectively. Clearly these results indicate that inductive effects cannot be solely used to explain the relative hydrolysis rates. Steric bulk of the alkoxy substituent relative to hydroxy may have a dominant effect on hydrolysis rates.

Pouxviel et al. [12] demonstrated that higher W values ($W = 10$) enhance hydrolysis rates and thereby make hydrolysis under acid conditions nearly complete prior to condensation. At intermediate W ($W = 4$), hydrolysis and condensation proceed concurrently, and residual —OR groups remain; alcohol-producing condensation reactions become important. At low W values ($W = 1-2$), —OH sites are limited; thus reactions are retarded because both alcohol- and water-producing condensations require —OH groups.

Sols with low W values can be spun into fibers [15,16]. Their spinnability was attributed to the presence of linear polymers [15]. However, some work shows that linear polymers are not produced under these conditions. Using ^{29}Si NMR and small-angle x-ray scattering (SAXS) spectroscopy. Brinker and Assink [17] showed that TEOS–ethanol sols with $W = 1.5-1.7$ react via a random growth process leading to branched polymers. The polymers are more extended, producing a high-viscosity sol, which is stable against gelation for extended periods and is conducive to spinnability. Sacks and Sheu [16] showed that spinnability is optimized when sols are highly shear thinning but not thixotropic.

Solvent effects are primarily due to polarity of the solvent and availability of labile protons [18]. The solvent's ability to solvate or hydrogen-bond with available anions or cations greatly alters rates of reaction and particle size. The most thoroughly studied solvent effect is the use of DCCAs for production of monolithic gels. Orcel and Hench [19] and Jonas [20] studied the effect of formamide on a tetramethoxysilane (TMOS)–methanol system and concluded that increased viscosity and strong hydrogen bonding under both acidic and basic conditions reduced the hydrolysis rate.

This conclusion is supported by low-temperature ^{29}Si NMR work of Boonstra and co-workers [21] on a two-step TEOS–ethanol sol–gel system. The results indicate that formamide acts as a base to reduce H^+ concentration and the hydrolysis rate, but more importantly it also reduces the dimerization rate. In a two-step process, less than stoichiometric acidic water is added for hydrolysis, and condensation remains relatively slow. Dimerization is retarded, and so the number of free silanols available for condensation in the second step (addition of basic water) is enhanced. Thus, addition of formamide decreases the gel time by increasing the number of silanols for condensation. Gels with larger particle sizes and larger mean pore size are produced, and drying stresses are reduced.

Probably the most widely studied and important reaction variables are pH and catalyst. Pope and Mackenzie [22] studied the effects of catalysts on gel times of TEOS reactions. Such studies combine information on catalytic effects for both hydrolysis and condensation. The gel time of HCl-catalyzed TEOS reactions plotted as a function of pH is a sigmoidal shape (Figure 48.1). This plot is consistent with Iler's results of polymerization of $Si(OH)_4$ [23]. Further, Pope and Mackenzie concluded

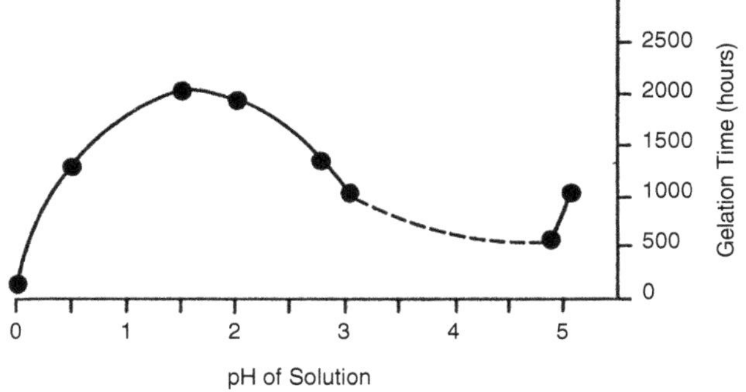

FIGURE 48.1 Gel time vs. pH for HCl-catalyzed TEOS sol (after Ref. 22).

that gel times depend on the catalytic mechanism of the catalyst as well as the pH, because sols catalyzed by HF and acetic acid have surprisingly short gel times (Table 48.1).

Flouride catalysis is well known [22,23] and is proposed as due to the strong affinity of Si for F^-, small size of F^-, and ability of F^- to attack Si and expand its coordination sphere to 5 or 6. Increasing Si coordination may facilitate the loss of —OR groups, generate more electrophilic species, and thereby accelerate polymerization [23,24].

Further, Jonas and coworkers [24,25] have shown that TMOS polymerization is greatly enhanced by addition of NaF. A comparison of TMOS-derived gels generated at pH 6.4 with and without addition of NaF shows that F^- catalyzed gels are more branched in the early reaction stages and have more monomers, dimers, and trimers present at gelation. However, scattering results show fractal dimensions of about 2.2 for both systems, a result suggesting a similar reaction-limited cluster–cluster growth model for both. The primary difference in structure was in the 100-Å range, with larger radii of gyration for F^- catalyzed systems.

Short gel times for acetic acid catalyzed sols is suggested [22] as due to displacement of —OR by an acetate ligand and subsequent attack by alcohol on the bound acetate, eliminating ethyl acetate and generating a silanol without hydrolysis:

$$Si(OR)_4 + CH_3COOH \xrightarrow{-ROH} (RO)_3SiO_2CCH_3$$
$$\xrightarrow{+ROH} (RO)_3SiOH + CH_3COOR$$

Thus, two pathways were proposed for silanol production: hydrolysis as previously described and direct elimination of ethyl acetate.

Our work [26] shows that carboxylic acids have no unusual catalytic effect on gel times when gel time versus pH* (pH* was obtained with a glass electrode in the reaction mixture) was plotted over the range of pH* = 0–7 for various catalysts (Figure 48.2). The reaction conditions of Mackenzie and coworkers [22,27] were used (TEOS:ethanol:water:acid = 1:4:4:var; monomer and alcohol were mixed 30 min prior to water addition). The discrepancy in the catalytic effect of carboxylic acids results from Mackenzie and coworkers' comparison of acids at similar concentrations, which is difficult with the wide range of pK_a values. Acetic acid is weaker than HCl and thus provides a higher solution pH at the same concentration. This property is accentuated in alcohol because acetic acid acts as a weaker acid in alcohol (versus water) whereas HCl maintains a more constant acid strength [28]. The sinusoidal curve obtained for gel time versus pH* plots for each of the acids is consistent with the results of Pope and Mackenzie [22] for HCl-catalyzed sols and with Iler for gelation of $Si(OH)_4$ [23]. Some differences in gel times were observed near pH 2,

TABLE 48.1
Gel Times and pH for TEOS Sols with Various Catalysts

Catalyst	Initial solution pH	Gel time (h)
HCl	0.05[a]	92
HNO_3	0.05[a]	100
H_2SO_4	0.05[a]	106
HF	1.90	12
HOAc	3.70	72
NH_4OH	9.95	107
No Catalyst	5.00	1000

[a]Between 0.01 and 0.05.
Note: TEOS:ethanol:water:acid = 1:4:4:0.05.
Source: Modified from Ref. 22.

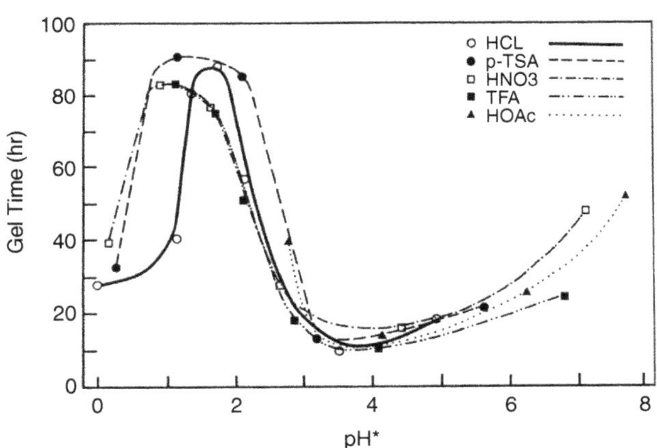

FIGURE 48.2 Gel time vs. pH* for TEOS sols catalyzed by HCl, trifluoroacetic acid (TFA), *p*-toluenesulfonic acid (*p*-TSA), HOAc, and HNO_3.

similar to results reported by Iler for silicic acid gelation using various catalysts [23]. However, the gel times for the sols catalyzed by carboxylic acid were not unusually short.

The silicon monomer chosen affects hydrolysis rates, gel times, and gel properties [2,3,27]. Increasing steric bulk of the alkoxy ligands decreases hydrolysis rates by about a factor of 6 going from ethoxy to hexyloxy [3]. Differences in volume shrinkage, bulk density, and surface area are also reported for gels derived from different monomers [27].

Although TMOS sols gel faster relative to TEOS (160 versus 321 h), this property is not entirely due to hydrolysis rates, but also to the structural evolution of the sol [29]. ^{29}Si NMR spectroscopic results show that TEOS-derived sols grow through intermediates containing more compact, cyclic structures than TMOS sols (Figure 48.3). Thus, TMOS polymers are more extended, chain-like structures that result in larger overlap of the polymers in solution and thus shorter gel times. These studies were done at low *W* values so residual —OR groups would be present. The increased steric bulk of the ethoxy groups may favor cyclization.

Substituting silicon tetraacetate, $Si(OAc)_4$, for TEOS in a typical sol–gel reaction (monomer:ethanol = 1:4, mixed 30 min prior to addition of 4 mol of 1.0 *M* HCl) results in a decrease in gel time from 242 to 1 h [27]. This decrease is attributed to the generation of SiO_2 by direct attack of alcohol on the bound acetates. However, recent work [30] shows that virtually no acetate ligands remain after reacting $Si(OAc)_4$–ethanol for 30 min (Figure 48.4). The acetates were replaced by ethanol via a rapid alcoholysis reaction. Also noteworthy is the lack of condensed silicon species in the ^{29}Si NMR spectrum and the absence of ethyl acetate in the ^{1}H NMR spectrum. Acetic acid was reported [31,32] to displace ethoxy groups on TEOS, but only upon refluxing. Acetate ligands are very stable on inorganic elements, such as Ti, that readily expand coordination to 5 or more [33]. Bidentate ligation of acetate blocks reaction sites in such cases. $Si(OAc)_4$, however, has monodentate ligation [34], the ligands are much more labile, and alcoholysis is facile. Attack is possible on silicon as well as the carbonyl of the acetate ligand.

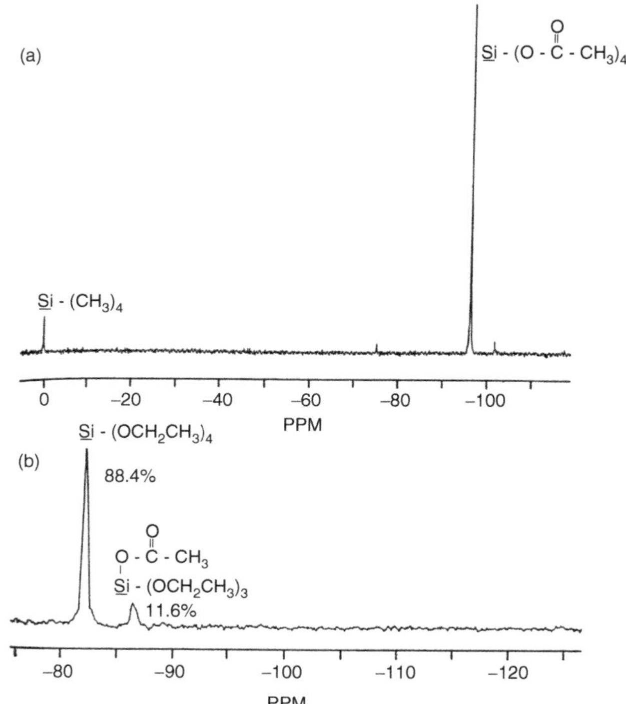

FIGURE 48.4 ^{29}Si NMR spectra of (a) $Si(OAc)_4$ in $CDCl_3$ and (b) $Si(OAc)_4$ in 4 mol of ethanol 30 min after mixing.

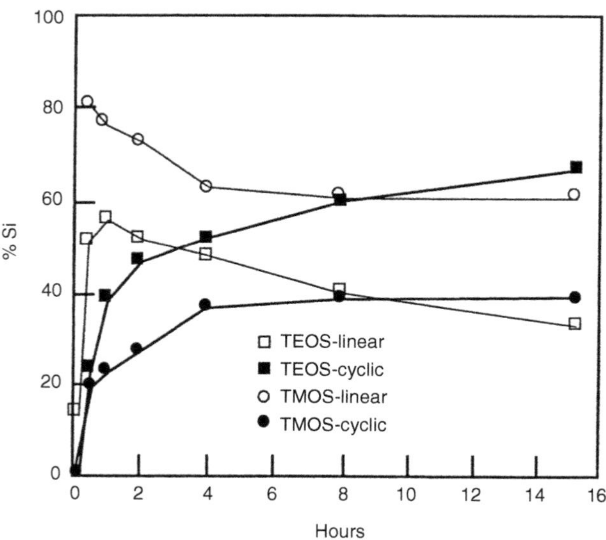

FIGURE 48.3 Linear vs. cyclic species calculated from ^{29}Si NMR spectra for TEOS–ethanol and TMOS–methanol sols (Si:water:acid = 1:2:0.02).

In this example, prior to water addition, alcoholysis causes the sol to consist primarily of TEOS in 4 equivalents of acetic acid. Thus the question becomes, why does TEOS gel much faster in acetic acid than in ethanol? Figure 48.5 shows ^{29}Si NMR spectra for the $Si(OAc)_4$–ethanol system just described and for TEOS in 4 mol of acetic acid, both spectra taken 45 min after addition of 4 mol of 1 *M* HCl. No evidence is seen in the NMR spectra for substitution of ethoxide ligands by acetate in the TEOS–acetic acid system. The two spectra are similar. (Additionally, the gel times for the two systems are nearly identical.) Condensation has proceeded rapidly to give a distribution of condensed species typical of acid-catalyzed sols. ^{1}H NMR spectra show rapid and nearly complete hydrolysis and the

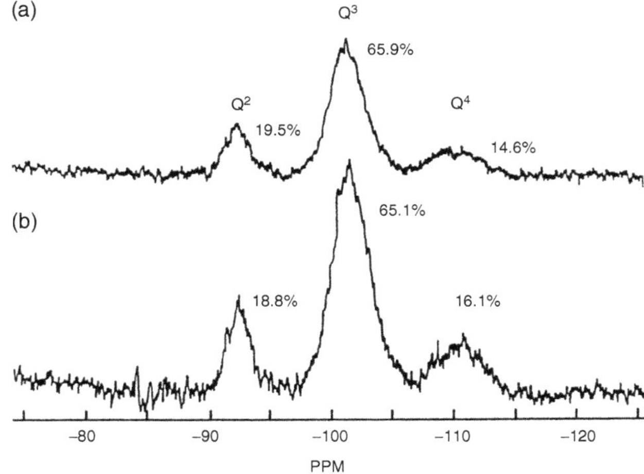

FIGURE 48.5 ^{29}Si NMR spectra of (a) TEOS–acetic acid sol and (b) $Si(OAc)_4$–ethanol sol, 45 min after addition of 4 equiv of 0.15 *M* HCl. **Q** refers to four possible condensation bonds; the exponent is the actual number of condensation bonds.

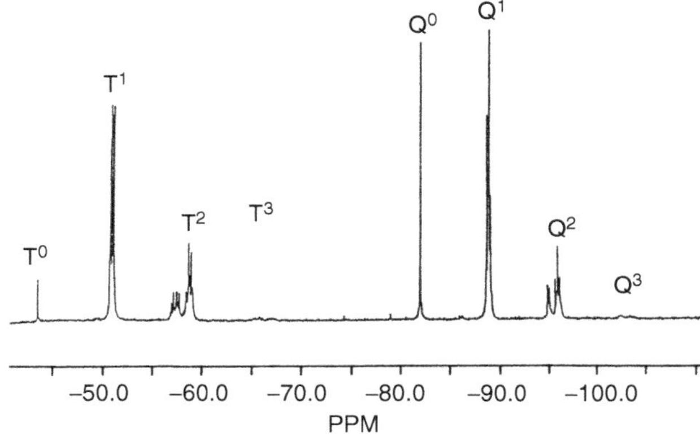

FIGURE 48.6 ^{29}Si NMR spectrum after 275 h of 1:1 MeTEOS–TEOS in ethanol with 0.5 equiv of 0.15 *M* DCl-D_2O. **T** and **Q** refer to the number of possible condensation bonds: **T** is three, **Q** is four; the exponent is the actual number of condensation bonds.

generation of ethyl acetate only after acidic water addition. Ethyl acetate arises from the acid-catalyzed reaction of free ethanol and acetic acid. The rapid gel time is not due to acetate ligation, but is apparently due to the very low pH of this system (acetic acid and HCl); below pH 2, condensation is catalyzed by H^+. Although water is a by-product in the generation of ethyl acetate, the rapid gel times could not be duplicated simply by increasing the water in TEOS–ethanol sols by the amount produced by the side reaction (determined by ^1H NMR spectroscopy). Additionally, solvent effects may play a role because the dielectric constant of the medium is reduced by acetic acid and ethyl acetate produced in the reactions.

Schmidt [35] demonstrated that alkyl substituents enhance ligand electron donation. Hydrolysis is enhanced under acidic conditions, but is retarded under basic conditions, by methyl substitution of ethoxysilane derivatives. This result is important for the generation of organic-inorganic hybrid materials. When TEOS is reacted with silicon monomers containing alkyl substituents (or organic polymers with trialkoxysilane groups), the relative reaction rates just described must be controlled to produce true hybrids and avoid "blocky" materials.

Recent work [36] shows the importance of this control. Methyltriethoxysilane (MeTEOS) and TEOS were mixed in equimolar amounts in ethanol followed by addition of 0.5 equiv of 0.15 *M* DCl. Such conditions limit hydrolysis and condensation. Figure 48.6 shows the ^{29}Si NMR spectrum of this mixture 275 h after water addition. MeTEOS has reacted further as evidenced by the smaller T^0 vs. Q^0 peaks. After only 0.5 h 69% of the MeTEOS monomers have homodimerized, whereas only 31% have reacted with TEOS. Forty-four percent of the MeTEOS silicons have become blocked with two other MeTEOS monomers. This outcome shows that mixing these two monomers together and adding sufficient water results in materials with MeTEOS- and TEOS-rich blocks. However, in our system (limited water) a point is reached where hydrolyzed MeTEOS reacts preferentially with hydrolyzed TEOS because late in the reaction the concentration of the TEOS is higher (most of the hydrolyzed MeTEOS has gone on to form oligomers). An understanding of the reaction variables affecting hydrolysis and condensation rates is needed to design and synthesize materials with new physical properties.

Reesterification, or the reverse of hydrolysis, is often important, particularly under acid conditions. Recent GC work on HCl-catalyzed TEOS sols at pH values ranging from 0.9 to 3.0 is shown in Table 48.2.

Both hydrolysis and reesterification rates increase sharply as pH is lowered. These data show that the

TABLE 48.2
Rate Constants from GC and Kinetic Model for HCl-Catalyzed Sols

pH	k_H	k_E	k_W	k_A
0.9	9.6	1.6	0.48	0.09
1.9	1.23	0.35	0.69	0.16
3.0	0.082	0.014	1.30	0.32

Note: TEOS:ethanol:water = 1:3.4:2. All values are in liters per mole hour.
Source: Modified from ref. 37.

esterification reaction is important under these conditions as hydrolysis is only 3.5–6.0 times faster. This reaction results in acid-catalyzed gels frequently containing residual OR groups even though hydrolysis is nearly complete [2,10].

CONDENSATION

Alcohol- or water-producing condensation generates a three-dimensional network. Pohl and Osterholtz [6] showed that condensation of alklysilane-triols is specific acid and base catalyzed. Above the isoelectric point of silica (about pH 2.5) condensation proceeds by nucleophilic attack of deprotonated silanols on neutral silicates [6,23]. Below the isoelectric point the reaction proceeds by protonation of silanols followed by electrophilic attack [2,6]. These reactions favor less highly condensed sites, because these are the most electron rich, and lead to more extended, ramified structures. Above pH 2.5, reactions favor more highly condensed sites.

One important parameter influencing the structural evolution of silicate polymers is the "molecular separation" of the reactive species [38,39]. Yoldas showed that TEOS concentration strongly influences the size of the silicate polymers. The effect is modified by the amount of water present; higher values of *W* result in higher degrees of polymerization. Concentrating the reacting solutions results in significant polymer growth if silanols are present, such as with high values of *W*.

The solubility of silica is also important. Figure 48.7 shows the relative dissolution rate for aqueous silicates as a function of pH [23]. The rate is slow at low pH so there is little bond redistribution. SAXS studies coupled with computer simulations have shown that under these conditions reaction-limited cluster–cluster aggregation is

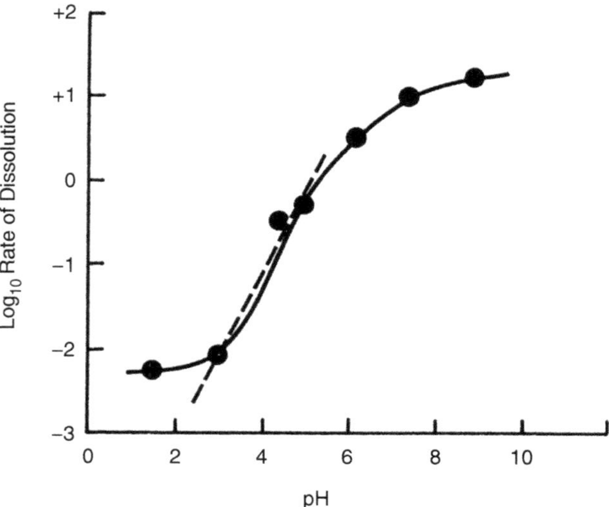

FIGURE 48.7 Silica dissolution rate as a function of pH (after Ref. 23).

favored and leads to branched structures with mass fractal dimensions (d_f) of about 2 [2,40,41]. Recent wide-angle x-ray scattering (WAXS) and SAXS results show that acidic conditions generate primary particles of about 1-nm size, which in turn aggregate to form clusters with d_f of 1.75–2.14, consistent with reaction-limited or diffusion-limited cluster–cluster aggregation [42]. These primary particles are proposed to consist of two silicate polyhedra made up of four six-membered rings [43].

At pH > 7, where the dissolution of silica is more favored, nucleation and growth is the predominant mechanism [2]. Dissolution ensures a constant supply of monomers with high mobility. This feature, coupled with condensation being favored at more highly condensed sites, leads to the generation of highly crossed-linked, large particles stabilized by electrostatic repulsions. This dissolution reaction as well as slow hydrolysis accounts for monomers frequently being observed in base-catalyzed systems even past the gel point [2,44]. ^{29}Si NMR spectroscopy shows that under basic conditions fully condensed or $\mathbf{Q}^4$ species and monomers are the predominant species present. Under low pH conditions, no $\mathbf{Q}^0$ or $\mathbf{Q}^1$ and a distribution of $\mathbf{Q}^2$, $\mathbf{Q}^3$, and $\mathbf{Q}^4$ species are present (if sufficient water is provided) [44].

Relative reaction rates of hydrolysis, condensation, reesterification, and dissolution must be understood and controlled to dictate structural evolution. However, accurate values for rate constants are difficult to obtain because of the enormous number of distinguishable reactions as next nearest neighbors are considered, and to the concurrency of these reactions. Assink and Kay [45] use a simplified statistical model assuming that the local silicon environment does not affect reaction rates, and the reactions for a particular silicon species are the product of a statistical factor and rate constant. These assumptions ignore steric and inductive effects. For example, this model predicts that the relative rate constants for the four sequential hydrolysis steps leading from TMOS to $Si(OH)_4$ would be 4:3:2:1. This model was applied to acid-catalyzed TMOS sols with *W* values ranging from 0.5 to 2.0. ^{29}Si NMR spectra on the temporal evolution of various silicon species show the model is in excellent agreement with experimental results. A lower limit for k_H was calculated as 0.2 L/mol-min. Values for k_W and k_A are 0.006 and 0.001 L/mol-min, respectively.

As mentioned earlier, Pouxviel and co-workers [12,13] studied acid-catalyzed reactions of TEOS with a variety of *W* values. The statistical model works better for TMOS than TEOS, possibly because of the increasing importance of steric factors with increasing bulk of the alkoxy substituent. Such effects are ignored in the statistical model. Although the relative hydrolysis rate coefficients for sequential hydrolysis of the four ethoxy groups was found by ^{29}Si NMR spectroscopy to be 1:5:12:5, reesterification reactions were ignored [12]. Simulated kinetics curves including the effects of k_E yielded the relative hydrolysis rate constants of 1:5.3:20:36, as noted earlier [13]. Reesterification rate constants increased steadily with increasing hydroxyl substitution. Condensation rate constants also increased with increasing hydroxyl substitution, but decreased with the degree of condensation. These relative trends are contrary to predictions based solely on inductive effects. These early applications of statistical models illustrate the complexity of sol–gel reactions and the fact that no variable can be readily ignored.

These trends were confirmed by more recent ^{1}H NMR work [14]. Gel permeation chromatographic (GPC) data by the same authors showed that acid-catalyzed polymerization of TEOS with $W = 4$ produced monomodal molecular weight distributions, whereas similar conditions with $W = 6$ produced bimodal distributions later in the reaction.

Kinetic measurements on the dimeric hexamethoxydisiloxane species generated values of $k_A = 0.0007$ l/mol-min and $k_W = 0.0011$ l/mol-min, versus 0.001 and 0.006 l/mol-min for the monomer [46]. The alcohol-producing condensation reaction is about the same for monomers and dimers, but the water-producing condensation is significantly lower for dimers. This difference is probably due to both increased steric crowding in the dimer and the inductive effect of an —OSi ligand retarding the reaction, as described earlier. Steric and inductive effects are important variables on reaction kinetics.

Klemperer and coworkers [47–50] investigated a molecular building block approach to silica sol–gels by hydrolyzing and condensing $[Si_2O](OCH_3)_6$, $[Si_3O_2](OCH_3)_8$, and $[Si_8O_{12}](OCH_3)_8$. With low acid or base concentrations and higher W values, polymerization is rapid relative to depolymerization, and the primary structures of the starting materials are maintained. With lower W values and higher acid or base concentrations, polymerization and depolymerization reactions are competitive, and some redistribution occurs.

Recent studies on the sol–gel transition for TMOS show no drastic change in structure at the gel point [51]. Percolation theory accurately accounts for gross structural features. Well past the gel point, motion remained at the molecular level and water freely diffused. Understanding and controlling structural evolution from the early reaction stages are important because these basic structures carry through to the gel state.

CONCLUSIONS

Structural control in sol–gel processes is complicated because many and diverse variables affect concurrent reactions differently. Inductive and steric factors contribute to the reaction rates. pH is probably the single most important variable in these reactions. It accounts for differences when DCCAs are used and for the rapid gel times in $Si(OAc)_4$ sols, as well as differences in the two predominant growth mechanisms: nucleation and growth and cluster–cluster aggregation.

When alkyl-substituted silicon alkoxides are used, other factors such as inductive effects on reaction kinetics become more important. The desire for new hybrid materials has made sol–gel science and technology one of the faster growing areas of chemistry. Although much has been learned, there remains a need for a better chemical understanding of the complex sequence of reactions.

REFERENCES

1. Schmidt, H.; Scholze, H.; Kaiser, A. *J. Non-Cryst. Solids* **1984**, *63*, 1.
2. Brinker, C. J. *J. Non-Cryst. Solids* **1988**, *31*, 100.
3. Aelion, R.; Loebel, A.; Eirich, F. *J. Am. Chem. Soc.* **1950**, *72*, 124.
4. Aelion, R.; Loebel, A.; Eirich, F. *Recueil* **1950**, *69*, 61.
5. McNeil, K. J.; DiCaprio, J. A.; Walsh, D. A.; Pratt, R. R. *J. Am. Chem. Soc.* **1980**, *102*, 1859.
6. Pohl, E. R.; Osterholtz, F. D. In *Molecular Characterization of Composite Interfaces*; Ishida, H.; Kumar, G., Eds.; Plenum: New York, 1985; p 157.
7. Swain, C. G.; Estere, W. M., Jr.; Jones, R. H. *J. Am. Chem. Soc.* **1949**, *71*, 965.
8. Campostrini, R.; Carturan, G.; Pelli, B.; Traldi, P. *J. Non-Cryst. Solids* **1989**, *108*, 143.
9. Aelion, R.; Loebel, A.; Eirich, F. *J. Am. Chem. Soc.* **1950**, *72*, 5705.
10. Keefer, K. D. In *Better Ceramics through Chemistry*; Brinker, C. J.; Clark, D. E.; Ulrich, D. R., Eds.; Elsevier North Holland: New York, 1984; p 15.
11. Uhlmann, D. R.; Zelinski, B. J.; Wnek, G. E., ibid., p 59.
12. Pouxviel, J. C.; Boilot, J. P.; Beloeil, J. C.; Lallemand, J. *J. Non-Cryst. Solids* **1987**, *89*, 345.
13. Pouxviel, J. C.; Boilot, J. P. *J. Non-Cryst. Solids* **1987**, *94*, 374.
14. Hui, Y.; Zishang, D.; Zhonghua, J.; Xiaoping, X. *J. Non-Cryst. Solids* **1989**, *112*, 449.
15. Sakka, S.; Kamiya, K. *J. Non-Cryst. Solids* **1982**, *31*, 48.
16. Sacks, M. D.; Sheu, R. S. *J. Non-Cryst. Solids* **1987**, *383*, 92.
17. Brinker, C. J.; Assink, R. A. *J. Non-Cryst. Solids* **1989**, *111*, 48.
18. Artaki, I.; Zerda, T. W.; Jonas, J. *J. Non-Cryst. Solids* **1986**, *81*, 381.
19. Orcel, G.; Hence, L. L. *J. Non-Cryst. Solids* **1986**, *79*, 177.
20. Jonas, J. In *Science of Ceramic Chemical Processing*; Hench, L. L.; Ulrich, D. R., Eds.; Wiley–Interscience: New York, 1986; p 65.
21. Boonstra, A. H.; Bernards, T. N. M.; Smits, J. J. T. *J. Non-Cryst. Solids* **1989**, *109*, 141.
22. Pope, E. J. A.; Mackenzie, J. D. *J. Non-Cryst. Solids* **1986**, *87*, 185.
23. Iler, R. K. *The Chemistry of Silica*; Wiley: New York, 1979.
24. Winter, R.; Chan, J. B.; Frattini, R.; Joans, J. *J. Non-Cryst. Solids* **1988**, *105*, 214.

25. Winter, R.; Hua, D. W.; Thiyagarajan, P.; Jonas, J. *J. Non-Cryst. Solids* **1989**, *108*, 137.
26. Coltrain, B. K.; Melpolder, S. M.; Salva, J. M. In *Ultrastructure Processing of Advanced Materials*; Uhlmann, D. R., Ulrich, D. R.; Eds.; Wiley: New York, 1992; p 69.
27. Chen, K. C.; Tsuchiya, T.; Mackenzie, J. D. *J. Non-Cryst. Solids* **1986**, *81*, 227.
28. Bates, R. G. *Determination of pH*; Wiley: New York, 1973.
29. Kelts, L. W.; Armstrong, N. J. *J. Mater. Res.* **1989**, *4*, 423.
30. Coltrain, B. K.; Kelts, L. W.; Armstrong, N. J.; Salva, J. M., unpublished results.
31. Sanchez, C.; Livage, J.; Henry, M.; Babonneau, R. *J. Non-Cryst. Solids* **1988**, *100*, 65.
32. Campero, A.; Arroyo, R.; Sanchez, C.; Livage, J. In *Ultrastructure Processing of advanced Ceramics*; Mackenzie, J. D.; Ulrich, D. R., Eds.; Wiley: New York, 1988; p 327.
33. Doeuff, S.; Henry, M.; Sanchez, C.; Livage, J. *J. Non-Cryst. Solids* **1987**, *89*, 206.
34. Mehrotra, R. C.; Bohra, R. *Metal Carboxylates*; Academic: London, 1983.
35. Schmidt, H. *J. Non-Cryst. Solids* **1985**, *73*, 681.
36. Kelts, L. W.; Coltrain, B. K., unpublished results.
37. Ro, J. C.; Chung, I. J. *J. Non-Cryst. Solids* **1989**, *110*, 38.
38. Yoldas, B. E. *J. Non-Cryst. Solids* **1986**, *82*, 11.
39. Yoldas, B. E. *J. Polym. Sci. A: Polym. Chem.* **1986**, *24*, 3425.
40. Keefer, K. D. In *Better Ceramics through Chemistry II*; Brinker, C. J.; Clark, D. E.; Ulrich, D. R., Eds.; Materials Research Society: Pittsburgh, PA, 1986; p 585.
41. Brinker, C. J.; Keefer, K. D.; Schaefer, D. W.; Assink, R. A.; Kay, B. D.; Ashley, C. S. *J. Non-Cryst. Solids* **1984**, *63*, 45.
42. Himmel, B.; Gerber, Th.; Burger, H. *J. Non-Cryst. Solids* **1990**, *119*, 1.
43. Himmel, B.; Gerber, Th.; Burger, H. *J. Non-Cryst. Solids* **1987**, *91*, 122.
44. Kelts, L. W.; Effinger, N. J.; Melpolder, S. M. *J. Non-Cryst. Solids* **1986**, *83*, 353.
45. Assink, R. A.; Kay, B. D. *J. Non-Cryst. Solids* **1988**, *107*, 35.
46. Doughty, D. H.; Assink, R. A.; Kay, B. D. In *Silica-Based Polymer Science: A Comprehensive Resource*; Advances in Chemistry Series 224; Ziegler, J. M.; Fearon, F. W. G., Eds.; American Chemical Society: Washington, DC, 1990; p 241.
47. Klemperer, W. G.; Mainz, V. V.; Millar, D. M. *Mater. Res. Soc. Symp. Proc.* **1986**, *73*, 3.
48. Klemperer, W. G.; Mainz, V. V.; Millar, D. M. *Mater. Res. Soc. Symp. Proc.* **1986**, *73*, 15.
49. Klemperer, W. G.; Ramamurthi, S. D. *Mater. Res. Soc. Symp. Proc.* **1988**, *121*, 1.
50. Klemperer, W. G.; Ramamurthi, S. D. *Polym. Prepr.* **1987**, *28*, 432.
51. Winter, R.; Hua, S. W.; Song, X.; Mantulin, W.; Jonas, J. *J. Phys. Chem.* **1990**, *94*, 2706.

49 Chemistry and Properties of Porous, Organically Modified Silica

Helmut Schmidt
Saarland University, Institute for New Materials

Harald Böttner
Fraunhofer Institute for Physical Measurement Techniques

CONTENTS

Sol–gel techniques were used to prepare porous, organically modified silica materials. The introduction of organic groupings was carried out with alkoxysilanes as precursors; methyl and propyl amino groups were used. The results show that high-porosity materials can be synthesized; the microstructure strongly depends on reaction conditions such as composition, solvent, catalyst type, and concentration. Microstructure tailoring affects mechanical as well as adsorption properties, and custom-made materials such as abrasives and adsorbents with special properties were synthesized.

Silica is an oxide with innumerable variations. On the basis of this structural variability, a variety of different types of materials have been prepared, such as compact fused silica, in the form of plates, fibers, or finely dispersed pellets, and porous materials. Because compact silica is mostly used for optical components or applications for which high chemical durability is required, porous and finely dispersed materials have completely different applications, for example, as fillers, adsorptive materials, or strengtheners in composites. Various aspects of the chemistry, physics, and applications of silica are described in detail in the book by Iler [1], with special consideration given to finely dispersed and porous materials.

The main preparation method for fused silica still involves melting technology, but a few other methods are used that involve sol–gel or related routes [2]. For fibers and coatings, sol–gel technology has become more and more important [3–5]. For dispersed materials, pyrolytic and precipitation processes are the dominating technologies for large-scale production [6–7]. To adapt these types of silica to a variety of special applications, various types of chemical surface modifications have been performed. A survey is given in reference 1. By surface modification (e.g., by organosilicon compounds) the hydrophobicity or hydrophilicity can be varied. For example, chromatographic materials (for reverse-phase chromatographic columns, for example) are prepared by reacting alkylsilanes that have well-defined alkyl chain-lengths (e.g., C_8) with the silica surface (Scheme I):

This type of surface modification, if adequate functional groups are used, can be useful in incorporating silica particles into polymeric matrices (e.g., into rubber for tires) or in increasing the hydrolytic stability of high-surface-area silica (e.g., that used for membranes). Surface modification of silica is a very important principle and is widely commercialized.

GENERAL CONSIDERATIONS

Another principle of modification, the modification of the network by organic groups, is not as important now compared to the surface modification principle. The exchange of an –O– group of a $[SiO_4]^{4-}$ tetrahedron by a CH_3 group drastically changes the properties of the modified

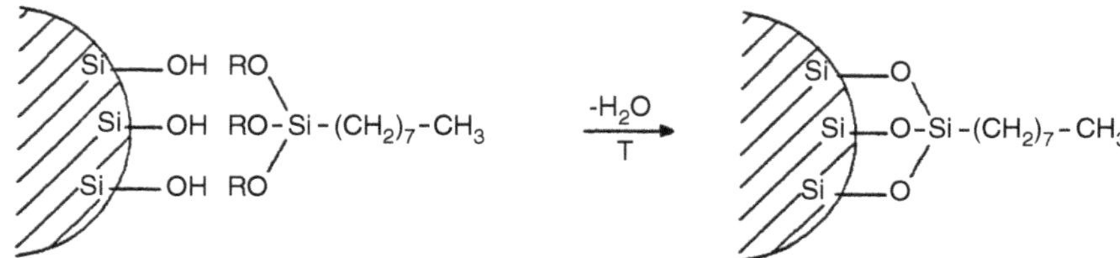

SCHEME I.

"glass" compared to fused silica. The density drops from about 2.2 (fused silica) to about 1.3 g/cm^3, and the thermal coefficient of expansion increases from 0.5×10^{-6} to 100×10^{-6} K^{-1}. Bulk materials modified by CH_3 are used as spin-on glasses for coating purposes in microelectronics. They can be oxidized to SiO_2 glassy films if desirable [8]. The partial substitution of ≡Si–O– bonds by ≡Si–CH_3 groups leads to densification temperatures far below the densification of inorganic SiO_2 sol–gel films. The examples indicate an interesting effect of organic modification of silica on properties as well as on structure. This chapter focuses on porous silica materials that have been bulk-modified by ≡Si–C– bonds and derived by sol–gel methods.

High porosity or high surface areas can be obtained by various preparation or special processing techniques, for example, by leaching of phase-separated sodium borosilicate glasses. Small dimensions of building units are required; small particles or small pores between units must be created (Figure 49.1). In materials with small pores between units, a high rigidity of material is required to avoid the collapse of the pores during the network synthesis.

The required network rigidity results from the three-dimensional cross-linking of the $[SiO_4]^{4-}$ tetrahedrons. Bulk modification by introduction of organic units through ≡Si–C– bonds leads to a change of network connectivity and should also affect porosity or surface area. On the other hand, in this case the modification becomes an intrinsic property that should not be affected by surface corrosion (Scheme II), because after removal of the surface layer by hydrolytic processes, the following layer exhibits the corresponding structure and properties. The properties to be developed determine whether a surface or a bulk modification is more appropriate.

PREPARATION OF POROUS MATERIALS

Reactivity of Precursors in Sol–Gel Processes

If multicomponent systems are prepared by hydrolysis and condensation of alkoxide precursors in methanol, the reaction rates of hydrolysis and condensation (including aggregation) of the different components become very important for the distribution of the different components. The simple question of how CH_3 groups might be distributed within a porous two-component system with $Si(OR)_4$ and $CH_3Si(OR)_3$ or $(CH_3)_2Si(OR)_2$ as precursors leads to serious mechanistic problems. A comparison of the hydrolysis rates with acid catalysis indicates an increase with increasing number of organic ligands [9–11] and the opposite with bases. The effect on specific surface areas of the resulting materials is shown in Figure 49.2. Similar results with $(CH_3)_2SiO$-containing systems for membrane formation were obtained with ethanol.

The surface areas were measured by multipoint Brunauer–Emmett–Teller (BET) techniques. In NH_3-catalyzed materials, the fraction of micropores is extremely low (<1 vol%). In acid catalysis, the micropore content increased with decreasing surface area. The CH_3 content was determined by IR spectroscopic analysis of CH_3 groups only in NH_3-catalyzed composites could a loss of CH_3-containing units be observed.

The experimental process for synthesizing the porous materials was standardized as follows. The silanes were mixed in the ratios as indicated in MeOH (Me, methyl) as solvent, and then at room temperature 1 mol of water per ≡SiOR group was added. The water contained the indicated concentration of catalysts. The reaction mixture was stirred for 5 min and then stored in a closed

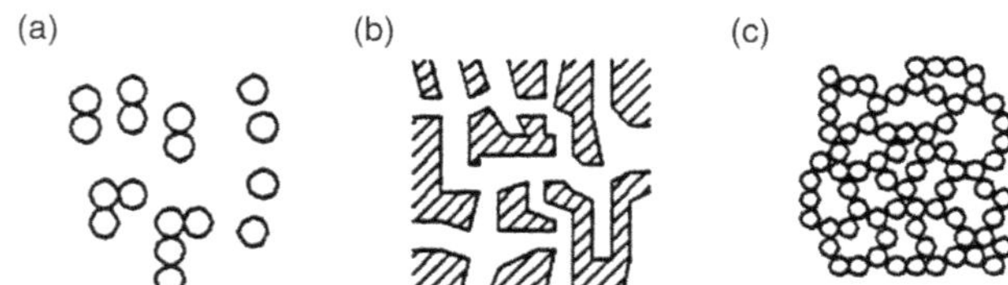

FIGURE 49.1 Model of high-surface-area materials: (a) small particles; (b) "reverse" system, small pores; and (c) porosity generated by aggregation of small particles (a).

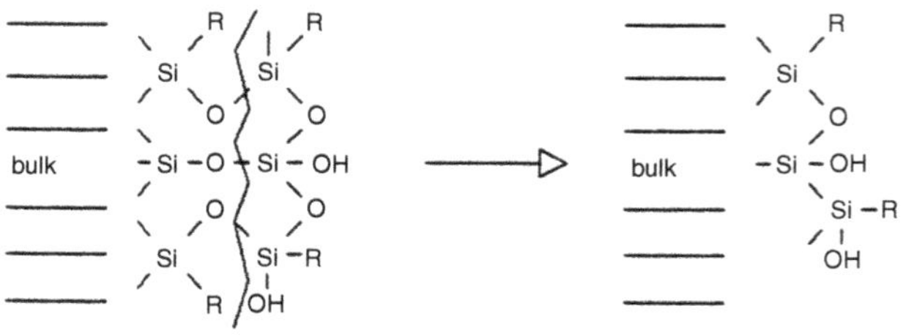

SCHEME II. Model of corrosion of organic modification of silica ("bulk" denotes the $-CH_3$-modified structure.)

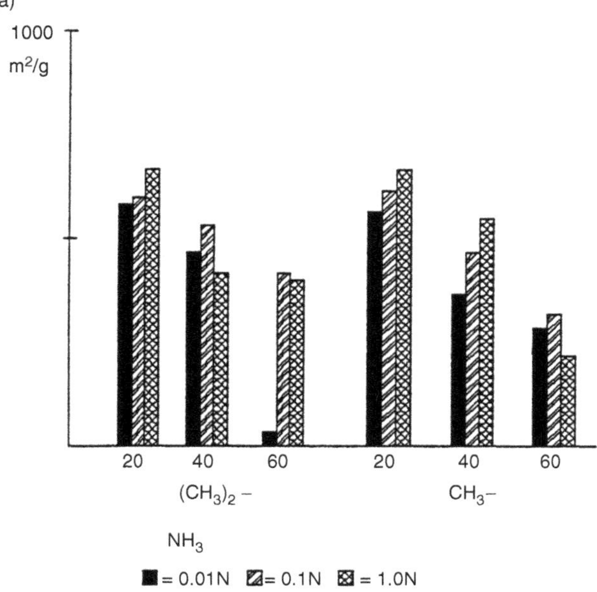

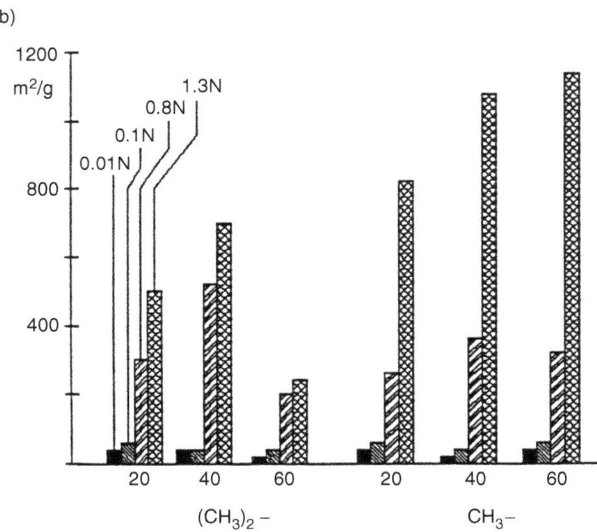

FIGURE 49.2 Effect of reaction conditions and composition on the specific surface area of $-CH_3$-modified silica [tetramethoxysilane (TMOS) as shown]. Part a: catalyst, NH_3; abscissa numbers, weight percent of $CH_3SiO_{3/2}$ or $(CH_3)_2SiO$. Part b: catalyst, HCl; abscissa numbers, same as in part (a).

flask. The two different catalysts show opposite tendencies with respect to increasing concentrations of CH_3 groups in the system. With NH_3, the surface areas decrease with increasing CH_3 group content in both systems (SiO_2–$CH_3SiO_{3/2}$ and SiO_2–$(CH_3)_2SiO$), but there is no significant difference between both systems.

In the acid-catalyzed systems, very high surface areas are observed in CH_3-containing composites. The effect of the HCl concentration is significant. In basic catalysis, the reaction conditions do not seem to influence structure and porosity to the same extent as in the acid case. In acid catalysis, the total surface area increases with decreasing gelling times and increases with increasing CH_3 concentrations. The gelling times are given in Table 49.1.

The increase of surface area with increasing CH_3 concentration (decreasing network connectivity) can be explained by the interaction of CH_3-covered pore walls with water; this interaction leads to reduced interfacial tensions. Synthesis yields are a function of catalyst and composition (Figure 49.3). This fact can be explained by a reaction rate consideration. As shown in reference 9, the hydrolysis rate of the CH_3-substituted silanes in NH_3 catalysis drops rapidly. This decrease does not necessarily lead to longer gelation times because the gel formation could be caused by the tetraorthosilicate network only. As shown in reference 11, the condensation rate increases with higher pH values, leading to shorter gelling times in the composite sol compared to the acid-catalyzed case. Because of the low hydrolysis rates in the basic case, the gelling occurs mainly from the $Si(OR)_4$ hydrolysis and condensation, and the unhydrolyzed CH_3-containing precursors can evaporate if the gels are dried shortly after gelling. IR spectra in the base-catalyzed case show a corresponding decrease of the CH_3 signal in basic gels, too. If the composites are stored in a closed system for several weeks without drying after gelation to allow further reaction, higher yields can be obtained from basic catalysis, too. This process was carried out for surface area and H_2O adsorption measurements.

To support the hypothesis of weaker pore wall interaction in the CH_3-modified case, H_2O adsorption experiments were carried out on several porous silica gels with different compositions (Figure 49.4). The CH_3-modified system shows almost no H_2O adsorption, a result indicating extremely hydrophobic pores. Increasing hydrophilicity (II $\rightarrow$ IV) changes the adsorption behavior gradually. During the condensation step of the CH_3-containing gels, a self-adjustment of the system seems to take place, and the CH_3 groups "turn" to the inner pore walls (Scheme III).

In summary, high-surface-area systems can be prepared from $CH_3{-}Si{\equiv}/SiO_2$ and $(CH_3)_2Si{=}/SiO_2$ composites even if the network stiffness is expected to be decreased substantially, perhaps because of the decrease of interaction between hydrophobic (self-arranged) pore walls and H_2O and the reduction of interfacial forces.

The influence of the catalyst does not only affect the physical properties; the morphology can be influenced, too. Figure 49.5 shows two gels with 60 wt% $(CH_3)_2SiO$ in a $(CH_3)_2SiO/SiO_2$ composite. Whereas the NH_3-catalyzed material is transparent, the HCl-catalyzed material is cloudy. Pore analysis shows a maximum at radius r of 4.2 and 500 nm. The larger radii can be of interest as "transport pores" for kinetics in liquid adsorption processes. This example shows how processing conditions

TABLE 49.1
Gelling Times of Various Composites

Catalyst type and concentration	80:20	60:40	40:60
1 N NH_3 (2 N HCl)	~1 (~3 × 10^5)	~1 (~3 × 10^5)	~1 (~3 × 10^5)
0.1 N NH_3 (0.1 N HCl)	300 (~23 × 10^5)	~840 (~11 × 10^5)	~2700 (~13 × 10^5)
0.01 N NH_3 (0.01 N HCl)	1200 (~47 × 10^5)	~10^5 (~80 × 10^5)	~3 × 10^5 (~60 × 10^5)

Note: Times are reported in seconds. In the composite ratios, first number is the weight percent of SiO_2 in SiO_2–$CH_3SiO_{3/2}$.

can be used for the generation of special microstructure properties.

Preparation of Materials with Special Functions

The porosity not only defines the adsorption parameters of a material, but also its mechanical stability. On the basis of this idea and the hydrolysis and condensation kinetics and its effect on microstructure, an abrasive powder was developed [12–13] with an abrasion-controlling mechanism for human skin. The investigation of the synthesis parameter shows a direct connection between the composition and the mechanical properties of the granular material (Figure 49.6). As expected from the previous experiments, HCl had to be used to provide reproducible CH_3 concentrations and well-defined, reproducible material properties.

The increase of the abrasion number (not representing higher abrasion) at higher $(CH_3)_2SiO$ contents is misleading and due to the increasing elasticity of the grains, which is not covered by the test. For the hydrolysis and condensation step, HCl had to be used because NH_3 did not lead to satisfying products for reasons pointed out in the section "Reactivity of Precursors in Sol–Gel Processes." The abrasive material was incorporated into a soap matrix and is used for the medical treatment of acne papulo pustulosa very successfully. It has been on the market for several years as a commercial product [14].

To customize the adsorption of functional molecules, the incorporation of organic groupings increasing the interaction between the adsorbent and the molecule to be adsorbed should be an adequate method. Several mechanisms can be taken into consideration, for example, an acid–base or a hydrophobic interaction mechanism. If larger molecules such as enzymes are considered, the pore size plays an important role, and the area of the accessible surface is an important parameter, too. To create basic sites, various types of silica with amino groups were synthesized by cohydrolyzing $(CH_3O)_3$ $Si(CH_2)_3NH_2$, $Si(OC_2H_5)_4$, and optionally $(CH_3)_2$ $Si(OC_2H_5)_2$ at room temperature with 50 vol% methanol as solvent.

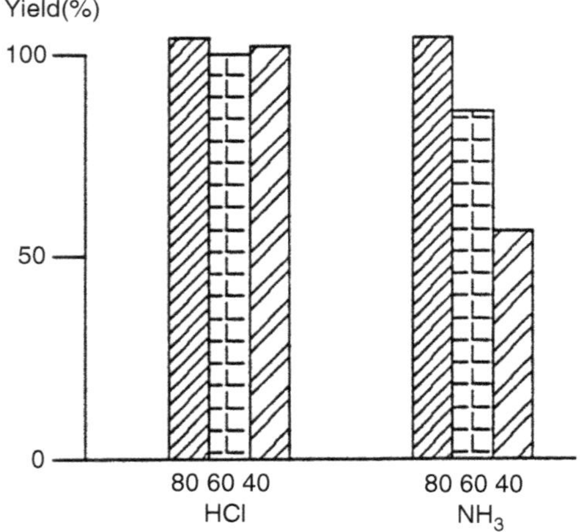

FIGURE 49.3 Percentage of product yield depending on catalyst and composition. The samples were dried after gelling to a constant weight at 110°C; yields above 100% are due to residual water. Catalyst concentrations were 1.0 *N* NH_3 and 1 *N* HCl; abscissa numbers denote wt% SiO_2 in SiO_2–$CH_3SiO_{3/2}$ composites.

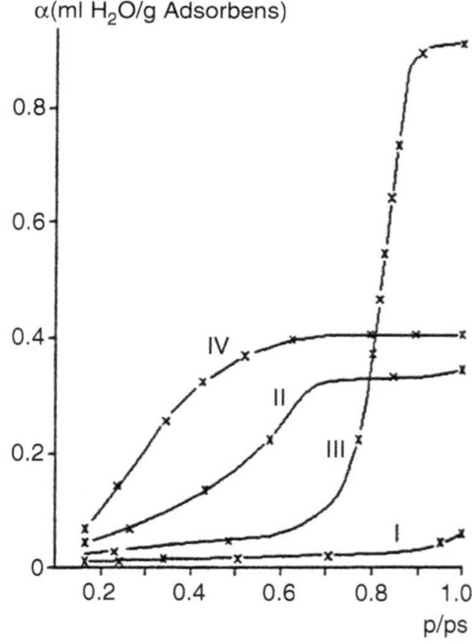

FIGURE 49.4 H_2O adsorption isotherms (partial pressure over saturation pressure) on four different silicas: I, CH_3-modified 300 m^2/g; II, $-CH_2-CH_2COOH-$, 800 m^2/g; III, $(CH_2)_3$ NH_2-, 310 m^2/g; and IV, unmodified, 450 m^2/g.

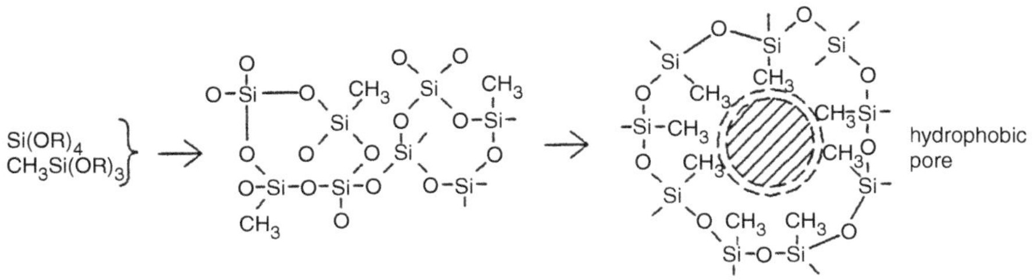

$Si(OR)_4$, $CH_3Si(OR)_3$ → → hydrophobic pore

SCHEME III. Self-arrangement of CH_3-modified systems.

Because amino groups act autocatalytically [15–17] in the presence of water, for acid catalysis an excess of HCl was used to overcompensate the formation of $-NH_3^+Cl^-$. In these cases, the gels were washed with methanol and water until no Cl^- could be detected in the filtrate. How far the incorporation of amino groups into silica could affect the adsorption of acid components was of interest. Lactic acid and a sulfonic acid (a commercially available dye named Telon Light Yellow) were chosen as test components [18]. In Figure 49.7 the adsorption isotherm of lactic acid is shown. Unmodified SiO_2 does not have remarkable adsorption in aqueous solution under these circumstances. The result shows the effect of the amino modification quite clearly, because the lactic acid load of the adsorbent is remarkable, and it is difficult to adsorb small water-soluble molecules in an aqueous environment.

The kinetics of the adsorption are shown in Figure 49.8, which demonstrates an almost ideal breakthrough curve that indicates fast kinetics. The breakthrough level represents almost the equilibrium value and does not depend on the flow rates used in the experiment. A glass tube 0.5 cm in diameter was used. Chemical analysis was carried out in steps of 5 mL to monitor the lactic acid concentration during the absorption experiment.

The sulfonic acid did not have remarkable adsorption under these circumstances. If, in addition to amino groups, a $(CH_3)_2Si{=}$ unit is incorporated, the adsorption rates increase remarkably. The best results were obtained with the composition SiO_2:$(CH_3)_2SiO$:amino = 40:60:2.5 with 9 *N* HCl as catalyst. The high concentration was found to lead to a maximum of load in a series of experiments with varying catalyst concentrations. The aromatic sulfonic acid could be loaded to 120 g per kilogram of adsorbent with an equilibrium concentration of 1.6 g/L and to 26 g/kg with 0.2 g/L (Figure 49.9). This fact can be attributed to a "double function" adsorption mechanism of a hydrophobic interaction of $-CH_3$ groups with the phenyl group of the sulfonic acid and the acid-based interaction. The kinetics of the flow experiments could be improved by microstructure tailoring (Figure 49.9) by variation of the catalyst concentration. Figure 49.9b

FIGURE 49.5 Images of gel surfaces: a, 0.1 *N* NH_3, 450 m^2; b, 6 *N* HCl, 270 m^2g. The bar is 4 μm.

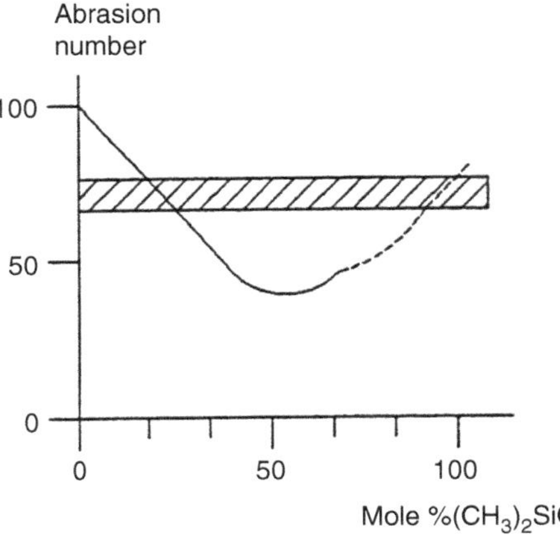

FIGURE 49.6 Abrasion numbers of a $(CH_3)_2SiO-SiO_2$ composite. The number represents the fraction of unaffected grains of an average diameter of 0.3 mm (in percentage) after a special abrasion test (15). The hatched area represents abrasive behavior recognized to be satisfying by test persons. The dashed line represents data that were not useful because of elasticity.

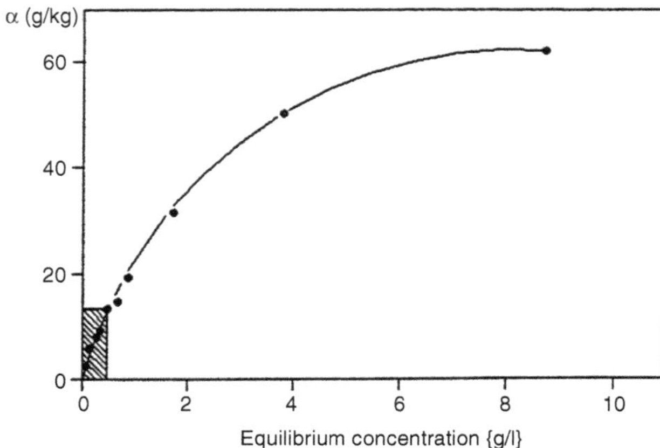

FIGURE 49.7 Adsorption isotherm of lactic acid on an amino-modified adsorbent. The composition was 5/95 (weight percent) in $NH_2(CH_2)_3SiO_{3/2}/SiO_2$. The mole surface was found to be 9.8×10^{-5}–NH_2/g; the method was that of reference 17. The BET surface area was 350 m^2/g; the hatched area indicates high loads even at low equilibrium concentrations.

shows the optimized microstructure with "transport pores" and a surface area about three times that of the adsorbent in Figure 49.9a.

Another interesting case is the adsorption of CO_2, which does not show dipole moment but is able to dissociate with water to an ionic compound according to $CO_2 + H_2O \rightleftharpoons H_2CO_3 \rightleftharpoons H^+ + HCO_3^-$. The question arises whether amino groups in combination with water adsorbed to the pore walls are able to act in a dissociative way for the adsorption of CO_2 according to $H^+CO_3^- + H_2N- \rightarrow HCO_3^- + H_3N^+-$. Therefore, a series of adsorbents with varying $-NH_2$ concentrations were prepared, and the adsorption isotherms were determined. The influence of H_2O on the adsorption isotherms was determined by thermogravimetric analysis (TGA) with a N_2 gas flow loaded with water vapor. The adsorbents loaded with CO_2 were flushed with the H_2O-containing N_2, and the H_2O content of the gels was determined after the experiment. The evaluation of the mass balance and the amount of residual water led to the conclusion that the H_2O completely replaces the adsorbed CO_2. On the other hand, as a function of the NH_2 content, remarkable amounts of CO_2 were able to be adsorbed on adsorbents dried at 150°C. These adsorbents show far lower OH peaks in the IR spectra than pure silica. This observation indicates that the ionic mechanism does not play any role, and a weak dipole interaction is postulated instead:

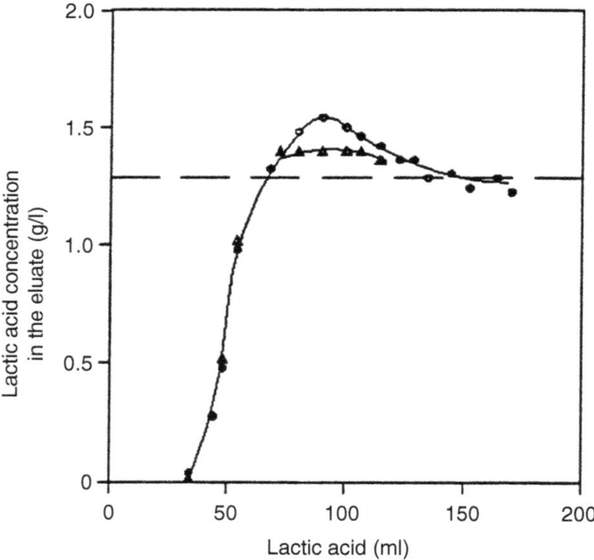

FIGURE 49.8 Breakthrough curves of a column-flow experiment. Flow rates were 2.0 (▲) and 0.83 (•) ml/min; 3 g of adsorbent was used. The dashed line indicates input lactic acid concentration.

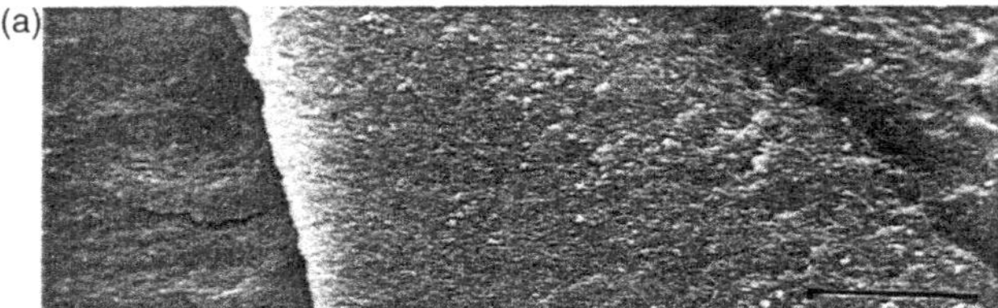

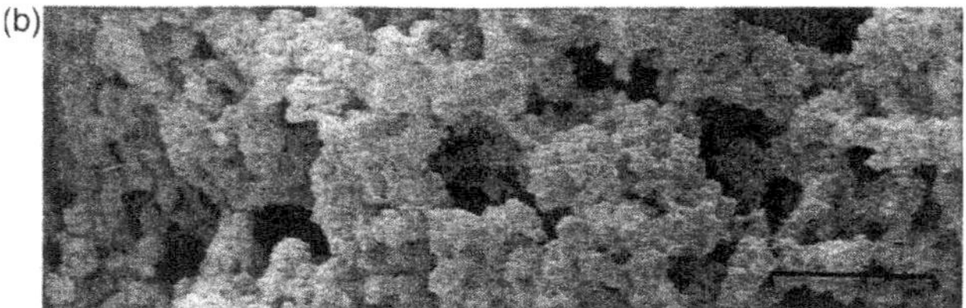

FIGURE 49.9 Amino- and CH_3-modified silica: a, 9 *N* HCl, 67 m^2/g; b, 12 *N* HCl, 188 m^2/g. The bars are 20 μm.

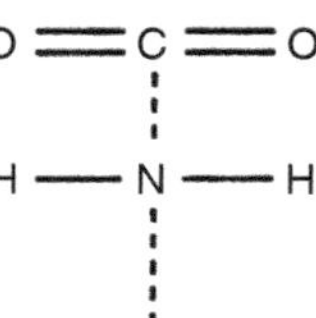

IR spectra, however, did not show a significant shift of the =C=O vibration. Figure 49.10 shows some of the most important results.

The NH_2 groups affect the amount of adsorbed CO_2 as well as the thermal behavior and illustrate the change of thermodynamics with composition. At higher pressures (not displayed in Figure 49.10), the amino-modified adsorbents show a saturation plateau around 5 bar CO_2 and a load of about 5–10 wt% CO_2, depending weakly on the surface area, whereas the CH_3-modified material

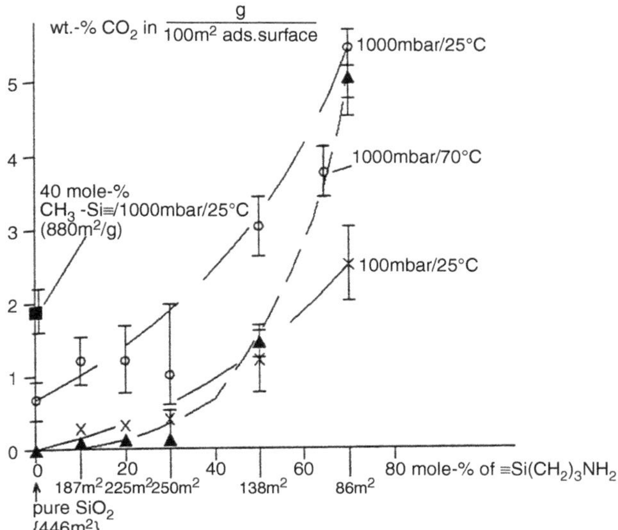

FIGURE 49.10 NH_2-containing silica with CO_2 at 100 and 1000 mbar and two different temperatures. The load is normalized on $100\text{-m}^2/\text{g}$ surface area. Data of pure silica and a $-CH_3$-modified silica are given for comparison; the numbers with m^2 as units at the abscissa display the surface areas of the investigated adsorbents.

($880\ m^2/g$) shows no significant plateau up to 15 bars (the end of the test) and a load of about 36 wt% at room temperature, which is an extremely high load. This observation leads to the question of whether these materials can be used for CO_2 storage or as propellants with CO_2. In reference 19, a heat pump is proposed; it uses two types of tailored adsorbents for CO_2 in a closed-system process.

The determination of the heat of adsorption shows higher values on adsorbents with amino contents above 10 mol% (e.g., SiO_2, 45 kJ/mol; 10 $-NH_2$, 43 kJ; 20 $-NH_2$, 58 kJ; 30 $-NH_2$, 65 kJ; experimental error, $\pm 7\%$) and a decay as a function of the CO_2 surface coverage (e.g., in an adsorbent with 20 mol % $-NH_2$ from 58 kJ/mol≡2 mL CO_2 per gram to 17 kJ/mol≡10 mL CO_2 per gram). Thus, the amino-modified materials interact rather strongly in the first step with the CO_2, and the interaction decreases with increasing load; this observation suggests a population of various adsorption sites.

CONCLUSIONS

The organic bulk modification of silica can be used for tailoring specific surface properties. One advantage of this reaction route is the possibility to use a one-step process, but the reaction must be controlled very carefully. Multifunctional materials can be synthesized, too.

ACKNOWLEDGMENTS

We thank the Bundesminister für Forschung und Technologie of the Federal Republic of Germany and Ruhrgas Company for financial support. We also thank J. Strutz for experimental work and helpful discussions.

REFERENCES

1. Iler, R. K. *The Chemistry of Silica*; Wiley: New York, 1979; pp 462–714.
2. Hench, L. L.; West, J. K.; Zhu, B. F. In *SPIE Proceedings of Sol–Gel Optics*; SPIE: San Diego, CA, 1990; Vol. 1328, pp 230–240.
3. Sakka, S. In *Sol–Gel Technology for Thin Films, Fibers, Preforms, Electronics and Speciality Shapes*; Klein, L. C., Ed.; Noyes Publications: Park Ridge, NJ, 1988; p 140.
4. Achtsnit, H.-D.; Wegerhoff, W. Oral presentation at German Glass Forum Meeting, May 1990.
5. Scherer, G. W.; Brinker, C. J. *The Physics and Chemistry of Sol–Gel Processing*; Academic Press: New York, 1990; pp 204–209.
6. *Aerosil* (Technical Bulletin); Degussa: Hanau, Germany, 1970.
7. *Lab-O-Sil, CGEN-7*, Cabot Corporation: Boston, 1970.
8. Bagley, B. G.; Quinn, W. E.; Khan, S. A.; Barboux, P.; Tarascon, J.-M. *J. Non-Cryst. Solids* **1990**, *121*, 454.
9. Schmidt, H.; Scholze, H.; Kaiser, A. *J. Non-Cryst. Solids* **1984**, *63*, 1.
10. Kaiser, A.; Schmidt, H. *J. Membr. Sci.* **1985**, *22*, 257.
11. Schmidt, H.; Kaiser, A.; Rudolph, M.; Lentz, A. In *Science of Ceramic Chemical Processing*; Hench, L. L.; Ulrich, D. R., Eds.; Wiley: New York, 1986; p 87.
12. Coltrain, B. K.; Melpolder, S. M.; Salva, J. M. In *Proc. IVth Intl. Conf. on Ultrastructure of Ceramics, Glasses and Composites*; Uhlmann, D. R.; Ulrich, D. R., Eds.; Wiley: New York, 1989; pp 69–76.
13. Schmidt, H.; Kaiser, A.; Patzelt, H.; Scholze, H. *J. Phys.* **1982**, *43 (C9, suppl. 12)*, 275.
14. Kompa, H. E.; Franz, H.; Wiedey, K. D.; Schmidt, H.; Kaiser, A.; Patzelt, H. *Ärztliche Kosmetologie* **1983**, *13*, 193.
15. *Jaikin Neu*; Basotherm Company: Biberach-Riss, Germany, 1985.
16. Ravaine, D.; Seminel, A.; Charbouillot, Y.; Vincens, M. *J. Non-Cryst. Solids* **1986**, *82*, 210.
17. Schmidt, H.; Popall, M.; Rousseau, F.; Poinsignon, C.; Armand, M. In *Proc. 2nd Intl. Symp. on Polymer Electrolytes*; Stosati, B., Ed.; Elsevier: London; p 325.
18. Schmitt, H. W.; Walker, J. E. *FEBS Lett.* **1977**, *81*, 403.
19. Schmidt, H.; Strutz, J.; Gerritsen, H.-G.; Mühlmann, H. *German patent 35 18 738*, 1986.

50 Evaporation and Surface Tension Effects in Dip Coating

Alan J. Hurd
Sandia National Laboratories, Ceramic Processing Science Department

CONTENTS

Evaporation sets an important time scale for the formation of structure in sol–gel films during dip coating, and surface tension is the dominant driving force influencing that structure. The action and interplay of these two phenomena were evaluated by experiments with pure and binary solvents. From the optically measured thickness of the steady-state film profile, accelerated evaporation near the drying line that sets stringent constraints on the time available for network formation was found. In binary solvents, there is evidence for strong flows driven by surface tension gradients; this flow gives rise to capillary instabilities. Aided by these flows, differential evaporation leads to regions rich in the nonvolatile component near the drying line.

Dip coating is the deposition of a solid film on a substrate by immersion in a sol or solution, withdrawal, and drying. The simplicity of dip coating, a cousin of painting, belies the fact that films of very high quality can be applied. Indeed, optical-quality films of controlled index and thickness are readily obtainable with simple, inexpensive apparati. Complex shapes can be coated in one step; this simplicity is not always possible with evaporative or sputtering techniques. For bulky objects, dip coating is far easier to scale up than vacuum techniques. Finally, the admirable purity of solution chemistry, such as the popular sol–gel route, can be exploited.

According to a review by Schroeder [1], the technology of spin coating or dip coating inorganic sols to make stable films was pioneered in Germany and became widely known after World War II. The physics involved in spin and dip coating has been reviewed [2] and remains under intense study [3,4]. The flexibility of sol–gel chemistry is demonstrated by the wide range of oxides and mixed oxides that have been used to form coatings [5,6].

The term "controlled index" means that the refractive index can be made *smaller* than that of the bulk precursor by controlling the microstructure via the porosity. When silica is deposited, for example, the film index n can be varied over a wide range [7] from $n = 1.1$ to 1.5. This process control makes sol–gel coatings interesting for many optical, electronic, and sensor applications, but the evolution of the microstructure during film formation is not well understood, in spite of efforts to survey the variables [8,9]. This chapter reviews the important factors determining the microstructure of dip-coated films and explores at length two of them, evaporation and surface tension.

Although easy to accomplish, the process of dip coating is complex because it proceeds through overlapping stages: When the substrate is withdrawn slowly from a sol, a film of liquid, several micrometers thick at the bottom, becomes hydrodynamically entrained on the surface. If the solvent wets the substrate, the film thins through gravitational draining, capillary-driven flows, and evaporation. When the recession speed of the drying film (relative to the substrate) matches the withdrawal rate, steady-state conditions prevail, and the entrained film terminates in a well-defined drying line that is stationary with respect to the reservoir surface. As described later, the presence of this edge in the evaporating film leads to dramatic effects. Meanwhile, the precursors in the entrained liquid experience a rapidly concentrating environment; they tend to gel or jam through chemical or physical interactions. Most likely, a transient chemical or physical gel network occurs fleetingly in the thin liquid film under these conditions, and it is my view that the porosity of the deposited film is a remnant of this network.

Often the reservoir sol is unstable with respect to aggregation and gelation. However, the process of film formation forces reactions at a much accelerated rate: Although the bulk reservoir might require several hours or days to gel, the transit time from entrainment to drying line is of the order 10 sec. Here is the first competition of time scales for film deposition. Clearly, network formation through the usual diffusion-limited and reaction-limited schemes can be frustrated by the accelerating effects of evaporation. In this context, the evaporation can be viewed as a strong force field coupling to the suspended particles — analogous to a centrifugal or electrophoretic field — and forcing them to crowd together. (The centrifugal acceleration causing an equivalent rate of crowding is as much as 10^6g.) When the crowding is rapid, particles do not have time to find low-energy configurations, so porous microstructures result. Thus, sol–gel films can differ greatly in structure from bulk xerogels or aerogels [10] if desired.

Just before he died, Ralph K. Iler (personal communication, 1985) was working on a device for electrophoretically depositing particles with a controlled degree of order. In addition to the direct current field for deposition, he imposed an alternating current component. At low frequencies, the ac component tended to unsnarl packing defects, whereas at high frequencies it apparently created a dipolar interaction between particles.

Although the evaporation creates varying physicochemical states of the entrained sol with height from the reservoir, it is easy to show that in most situations the sol is essentially homogeneous across its thickness. {If it were not homogeneous, the particles might collect near the air interface of the entrained film as a sort of skin that would impede evaporation [11,12]. In most situations the particles' transport by diffusion is fast enough to keep the concentration constant through the thickness: For a liquid film of thickness $\Delta r \approx 1$ μm and a diffusivity D_0 of 10^{-6} cm^2/sec (appropriate for a 10-Å moiety), it takes only a time $\Delta t \sim \Delta r^2/D_o = 10^{-2}$ s to relax thickness concentration gradients that might build up. For most positions on the film, this time scale can be considered short compared to other processes. Only very near the drying line itself would it be possible for concentration gradients normal to the substrate to "lock in." Some evidence suggests that such inhomogeneities exist (Fabes, B., personal communication and poster presentation at the Spring Meeting of the Materials Research Society, San Francisco, CA, 1990) through the thickness, but there does not appear to be a deep enough data base to conjecture about their origin.

Gravitational draining creates hydrodynamic shear throughout the entrained film. Unlike particle concentration, the shear rate is not the same throughout the thickness. In fact it must be zero at the air interface (because the gas cannot exert a shear force on the film, assuming, for the moment, no surface tension gradients exist) and nonzero at the substrate, which provides the force of lifting from the reservoir. Thus the shear rate must be maximal at the substrate: If z is the distance normal to the substrate and h is the thickness of the entrained liquid film, then the velocity (u) satisfying the Navier–Stokes equation $d^2u/dz^2 = 0$ is parabolic:

$$u = -u_0\left(1 + \frac{h(x)z}{\lambda^2} - \frac{z^2}{\lambda^2}\right)$$

Here λ is a characteristic length of order 10 μm given by $\lambda^2 = \rho g/\eta u_0$, where ρ is liquid density, g is gravitational acceleration, η is shear viscosity, x is height and u_0 is substrate withdrawal velocity. The shear rate $\dot{\gamma}$ is due solely to gravitational back flow in the absence of surface-driven forces:

$$\dot{\gamma} = \frac{du}{dz} = -\frac{u_0}{\lambda^2}(h - z)$$

Thus, the shear is greatest near the reservoir, where the film is thickest, and next to the substrate wall z is 0.

It is interesting to consider whether shear-induced particle encounters are slow compared to diffusional transport. The relevant parameter is the Peclet number

$$P_e = \frac{6\pi a^3}{kT}\frac{du}{dz}$$

(for spheres of radius a, where k is the Boltzmann constant and T is the temperature), which is the ratio of these time scales. When P_e exceeds 1, it is well known that the structure of the dispersion is constantly forced by the shear to nonrandom states, typically ordered sheets or strings. For $P_e \ll 1$, diffusion randomizes the structure. Gravitationally driven shear is generally not strong enough to create "shear-induced ordering," even near the substrate at the base of the entrained film, where P_e is only 10^{-4}. However, shear driven by surface tension gradients can be quite large, so shear-induced order might be exploited to affect film microstructure [13] in coatings derived from colloidal suspensions [14].

Indeed, surface tension is arguably the dominant force in dip coating, at least at the point of entrainment and at the drying line where interfacial curvatures are significant. At the point of withdrawal from the reservoir, a meniscus forms to balance the pressure imposed by the curved surface against that of the gravitational "head." The relatively large volume of liquid pulled into the gravitational meniscus (radius of curvature about 1 mm) is indicative of the large surface tension of dip-coating liquids, and the effect [15] of surface tension is large enough to change the dependence of the entrained thickness $h(0)$ (proportional to deposited mass) on withdrawal speed u_0 from $u_0^{1/2}$ to $u_0^{2/3}$. The deposited *thickness* need not scale in this way with withdrawal speed because coating porosity can vary with other factors [16,17]. Between the meniscus and the drying line, the radius of curvature is too large (~10 km!) for surface tension to have any effect, but, at the drying line, capillary pressures are again significant.

Probably the most important surface tension effect is that of "capillary collapse" of the transient networks as they are invaded by the gas phase during the final phase of evaporation [18]. This process is identical to the collapse of a sponge upon drying. The network can resist the invading menisci up to the point when their radii are small enough that the capillary stresses exceed the yield stress of the network [19]. The network then compresses uniaxially until its modulus becomes high enough again to resist the capillary stress while the solvent is completely removed. (Uniaxial compression is not the only possible mode of collapse. Large, stable pores could form, for example, by the lateral retraction of material.) By now the structure is denser and, possibly, completely different from that of the transient network. Nevertheless, the extent of capillary collapse can be controlled to some extent [7] by physical and chemical means to achieve a desired porosity.

Surprisingly large capillary pressures are theoretically possible in drying films. (In order of magnitude, there is little decrease in surface tension at a rapidly evaporating interface, or even a boiling one.) Because the final pore size can be smaller than 1 nm, the pressure in the liquid during the final stages of drying could exceed −100 atm (−10,000 kPa)! The negative sign indicates that the liquid is under tension. Although most liquids can support large tensions in small pores if no gas is present [20,21] owing to suppression of nucleation, it is not clear how the liquid in open pores is similarly prevented or delayed from boiling away [22a]. Nevertheless, strong evidence for suppressed vaporization is the hysteresis in adsorption isothermus of microporous solids [20]: Liquid is reluctant to leave microsize cavities once it has filled them, because, presumably, the pores are smaller than a critical vapor nucleus. Moreover, recent experiments in deflection of sol–gel-coated beams [22b] indicates drying stresses of enormous negative pressure (2000 atm, or 21,000 kPa) generated by the capillary pressure.

The choice of solvent mixtures may be the route to controlling capillary forces. The main subject of the remainder of this chapter is the variety of effects that occur in mixed solvent systems during dip coating. These effects arise when differential volatility gives rise to concentration gradients, thence to surface tension gradients, which have surprisingly large effects on flow. A few of these effects are discussed. First a description is given of the evaporation constraints, because these determine, to first order, the local composition of the film.

EXPERIMENTS ON EVAPORATING THIN FILMS

Clean substrates of silicon were used to entrain liquid films of various compositions in a dip-coating geometry, as in Figure 50.1 with $\alpha = 0$. Interference images were obtained in reflection with monochromatic light. The liquid index of refraction and the angle of incidence were known, so the film thickness profiles were obtained from the position of the interference fringes.

The process of evaporation during dip coating was studied only recently [2,22a], although the essential physics of evaporation has been well known for over a century. According to Fuchs [23], James Clerk Maxwell wrote an article on diffusion for the *Encyclopedia Britannica* in which he considered the stationary evaporation of a spherical droplet in an infinite medium. (Maxwell was interested in wet bulb thermometry.) Not only did he realize that the rate of mass loss by the droplet is limited by vapor diffusion away from the surface, he correctly assumed that the vapor concentration at the surface of the drop is equal to its equilibrium saturation concentration (true when the vapor mean free path is small compared to the dimensions of the droplet). All that remains is to solve the steady-state diffusion equation for

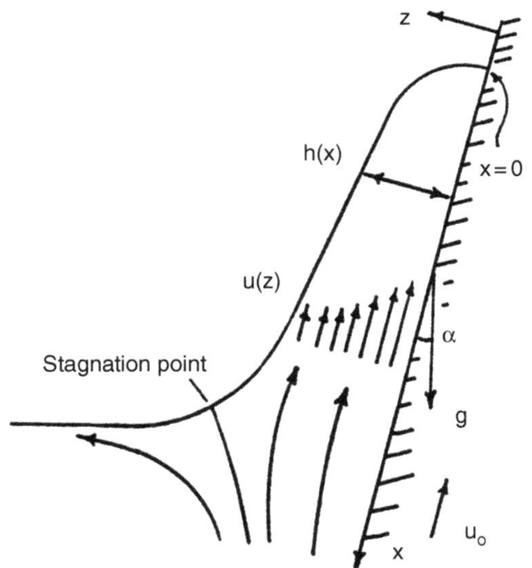

FIGURE 50.1 Geometry of dip coating. The angle of the substrate α can be adjusted to vary the effects of gravity g. A stagnation point in the velocity field u occurs in the gravitational meniscus. The thickness $h(x)$ is in the range 0 to 10 μm, and the height x is of order 1 cm. (Reproduced with permission from reference 22a. Copyright 1990.)

the vapor concentration c,

$$\nabla^2 c = 0 \tag{50.1}$$

on a sphere of radius a, with an additional boundary condition at infinity. The flux from the surface, which can be defined as a local evaporation rate (E), is governed through Fick's law by the vapor concentration gradients there:

$$E = \text{mass loss per unit area per unit time}$$
$$= -D\frac{\delta c}{\delta r}|r = a \tag{50.2}$$

Although the diffusion constant D does not appear in the steady-state diffusion equation, it does appear in kinetic factors such as the time it takes for a droplet of a given initial mass to evaporate. The solution to Equation (50.1) is a concentration c that decreases as r^{-1} from the droplet so that the total evaporation rate $4\pi a^2 E$ from Equation (50.2) is proportional to the product aD. Thus, it is not the surface area that controls the rate of mass loss, but the radius.

Other geometries can readily be worked out. As a useful analogy, the concentration c in Equation (50.1) can be viewed as the electrostatic potential around a conductor of potential c_0. The analog to the local evaporation rate is the electric field, evaluated at the surface of the conductor. By this analogy a fresh set of intuitive ideas can be brought to bear on evaporation problems. For example, it is not surprising that the vapor density around an infinite cylindrical source drops logarithmically with radial distance, and that the evaporation rate varies inversely with the radius of the cylinder. Similarly, the vapor concentration above an infinite sea drops linearly with distance, whereas the evaporation rate is constant everywhere on the surface.

Dip-coating geometries are not perfect one-, two-, or three-dimensional structures, however. Usually a sharp boundary is involved, such as the edge formed by the drying line. When coating on flat substrates, the film can be considered a thin, finite sheet. A dip-coating film can be approximated by a semi-infinite sheet, mathematically formed from a wedge in the limit that the wedge angle approaches zero. The edge carries a field singularity [24] of the form $r^{-1/2}$. In fact, for an arbitrarily shaped finite sheet, the field singularity remains $r^{-1/2}$ as long as the edge is locally straight on the scale of the sheet's thickness. This fact can be seen from the exact solution for the field above a thin disk, which can be readily shown to be encircled by a field singularity of the form $r^{-1/2}$, where r is the distance to the edge [25].

Such evaporation singularities can be very easily observed because the thickness profile of a thin liquid film locally reflects the rate of solvent removal. For very thin films, especially near the drying line, back flow due to gravitational draining can be neglected. The mass flux $h(x + dx)u_0$ carried into a fluid element dx, as shown in Figure 50.1, is balanced by the flux carried out by the substrate $h(x)u_0$ and the mass lost through evaporation $E(x)dx$; this observation leads to the continuity equation

$$\frac{dh}{dx} = \frac{E(x)}{u_0} \tag{50.3}$$

which can be integrated for $E \sim x^{-\eta}$ to give

$$h(x) \sim x^{1-\eta} \quad (\eta < 1) \tag{50.4}$$

The thickness profile $h(x)$ of an evaporating ethanol film during dip coating (Figure 50.2) follows $h(x) \sim x^{1/2}$, giving $\eta = 1/2$ as expected for the edge of a sheet. Profiles of this type were derived from optical interferometry, including wedge fringes (Figure 50.3) and imaging ellipsometry [26]. Optical profiles of evaporating sessile drops (Figure 50.4) also approach the substrate parabolically near the edge, as expected [25].

An interesting case is $\eta = 1$, because it arises in fiber formation [27] and fiber coating. Ideally, according to Equation (50.3), $h \sim \ln(x/x_0)$, where x_0 is the position of the drying line. In reality, the logarithmic profile is difficult to observe because small fibers do not entrain thick liquid layers owing to the large capillary pressures developed by the small curvature around the fiber itself. Nevertheless, it is possible to see the integrated singularity change with cylinders of decreasing radii (Figure 50.5). The needlelike singularity $\eta = 1$ should pertain as long as the radius of

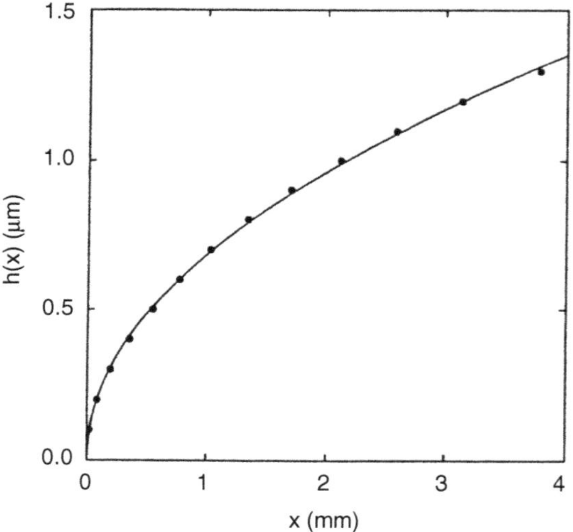

FIGURE 50.2 Thickness profile of dip-coated ethanol film. The profile is fit quite well by the form $h \sim x^{\eta}$ with $\eta = 1/2$. (Reproduced with permission from reference 22a. Copyright 1990.)

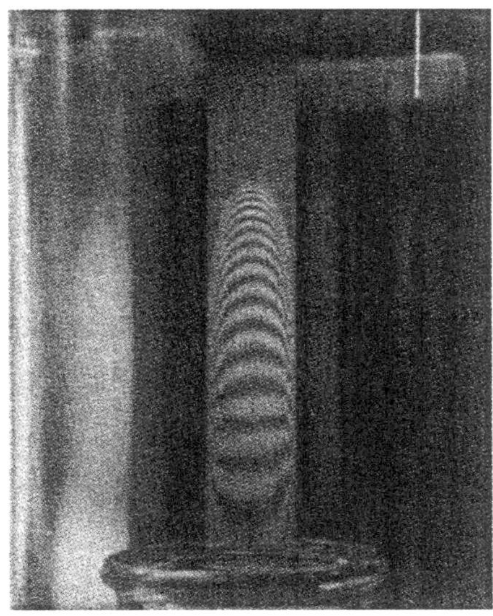

FIGURE 50.3 Optical interferogram of steady-state ethanol film. The drying line can be clearly seen.

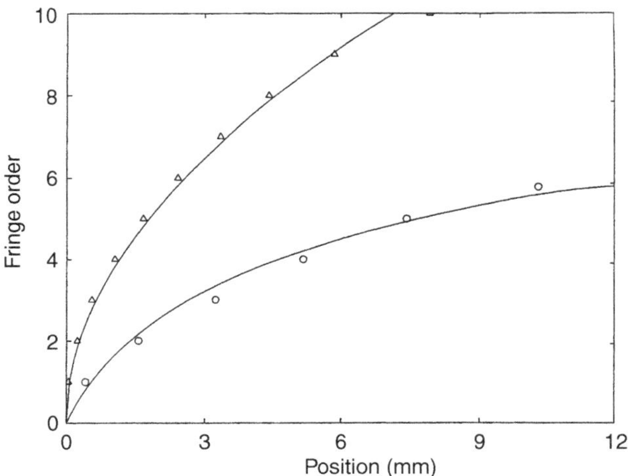

FIGURE 50.5 Thickness profiles of steady-state films on (triangles) a 1-cm strip and (circles) a 1.78-mm-diameter cylinder. The strip is well fit by a square-root function $h \sim x^{1/2}$, whereas the cylinder, being near the needlelike geometric limit, is better fit by $h \sim \ln(x - x_0)$.

the cylinder is small compared to the height of the drying line.

MULTICOMPONENT SOLVENTS AND SURFACE TENSION GRADIENTS

A comparison of Figure 50.6, showing a film from a binary mixture of ethanol and water, with the pure ethanol film in Figure 50.3 shows a striking new feature at the drying line. The binary solvent has an additional "foot" extending from a false drying line, as shown in Figure 50.7. Doubling the amount of water in the reservoir (18% H_2O in Figure 50.6) doubles the extent of the foot. Clearly, then, the foot is a water-rich phase that outlasts the alcohol because of a lower volatility. The well-behaved wetting of the foot suggests that its composition is not pure water, which is very difficult to use to wet a substrate symmetrically without extensive hydrogen peroxide cleaning. This forgiving wetting nature, like

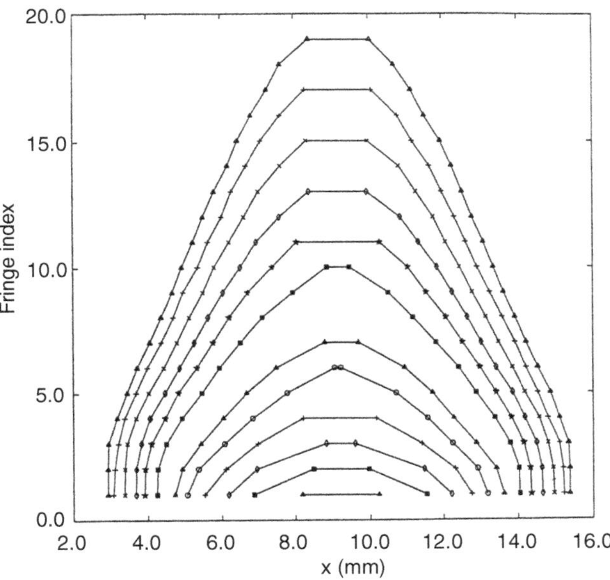

FIGURE 50.4 Thickness profiles of sessile methanol drop. A curve was drawn every 0.5 sec. At the edge of the drop a parabolic section can be seen, as predicted by theory. (Each fringe index, or order, represents roughly 100 nm in thickness.)

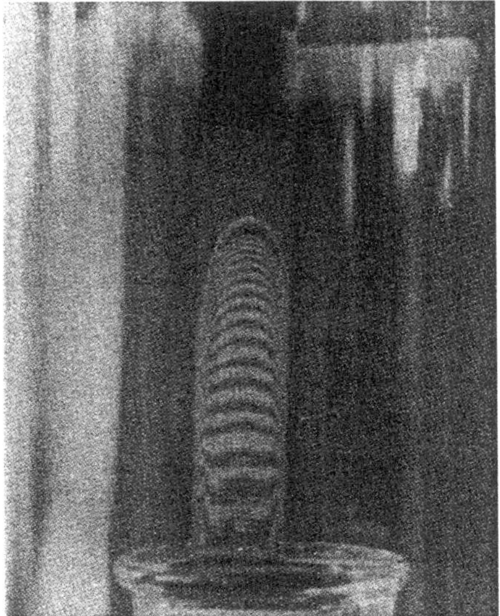

FIGURE 50.6 Optical interferogram of 82:18 (vol) ethanol–water film. A water-rich feature appears at the drying line.

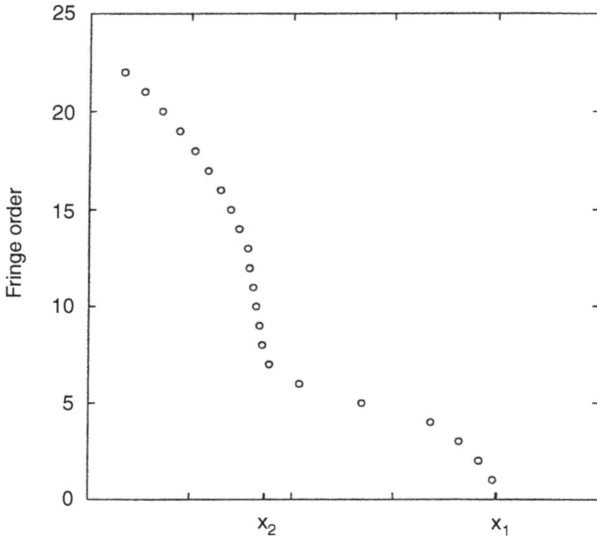

FIGURE 50.7 Thickness profile of 50:50 (vol) propanol–water film. The double-chin "phase separation" is due to differential volatilities and surface-driven flows. (Reproduced with permission from reference 13. Copyright 1991.)

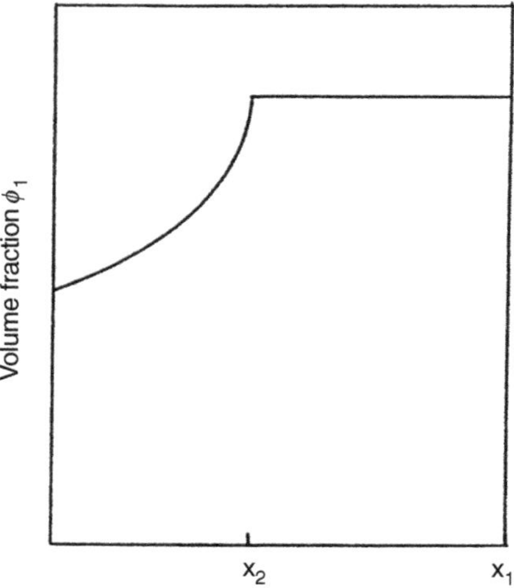

FIGURE 50.8 Schematic plot of volume fraction of water-rich phase ϕ_1 for propanol–water film, based on Figure 52.7.

that of pure alcohol, implies that the water-rich foot approaches a composition that contains at least some alcohol acting as a surfactant. Clearly the evaporating mixture is far from the boiling azeotrope, which for water and ethanol is 95% alcohol.

The binary mixture profile consists of two parabolas; thus, the evaporation of each component is unaffected by the vapor concentration of the other. That is, each component has an independent evaporation singularity. If the water-rich phase is denoted phase 1, and the ethanol phase 2, then the independent profiles are additive:

$$h_1 = a_1 x^{1/2} \qquad x > 0 \tag{50.5a}$$

$$h_2 = a_2 (x - x_2)^{1/2} \qquad x > x_2,\ h = h_1 + h_2 \tag{50.5b}$$

$$h_2 = 0 \qquad x < x_2 \tag{50.5c}$$

where h is the total thickness and x_2 is the position of the false drying line (Figure 50.7). The volume fractions $\phi_1 = h_1/h$ and $\phi_2 = h_2/h$ are plotted schematically in Figure 50.8.

Differential volatility is not the only physics responsible for the "foot" feature of binary solvents. Flows due to surface tension gradients are clearly present. A simple linear mixing law (known to underestimate the gradients potentially present) for the total surface tension σ in terms of its constituents serves to estimate the magnitude of the flow,

$$\sigma = \phi_1 \sigma_1 + \phi_2 \sigma_2 \tag{50.6}$$

and, therefore, the surface tension gradients can be calculated,

$$\frac{d\sigma}{dx} = (\sigma_1 - \sigma_2)\frac{d\phi_1}{dx} - (\sigma_1 - \sigma_2)(x - x_2)^{-1/2} \quad (x > x_2) \tag{50.7a}$$

$$\frac{d\sigma}{dx} = 0 \quad (0 < x < x_2) \tag{50.7b}$$

assuming the no-flow profiles of Equation (50.4) hold as adopted for Equations (50.5*a*)–(50.5*c*). Not surprisingly, the abrupt disappearance of phase 2 at the false drying line x_2 leads to singular gradients.

But the simple no-flow picture of Equation (50.4) can no longer hold in view of Equation (53.7*a*) and Equation (50.7*b*). At the liquid–vapor boundary, the viscous shear force must balance the force imposed by surface tension gradients, $\eta du/dz = d\sigma/dx$ $(z = h)$. This boundary condition leads to a linear flow profile toward the drying line,

$$u = \frac{1}{\eta}\frac{d\sigma}{dx} z - u_0 \tag{50.8}$$

so the profiles h_1 and h_2 have to be recalculated. Profiles for static menisci of binary solvents have been calculated [28].

However, far from the singularity at the false drying line x_2, Equation (50.8) should describe the physics accurately enough so that the strength of the surface-driven flows can be appreciated. Figure 50.9 shows the thickness profile of a binary mixture of toluene and methanol during film formation. The flows are strong enough to distort greatly the foot profile, creating, in fact, a thickened "toe" of toluene near the drying line. A crude estimation on

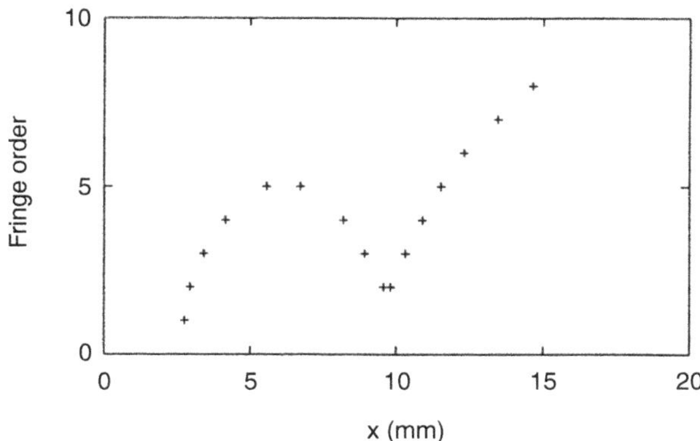

FIGURE 50.9 Thickness profile of 50:50 (vol) methanol–toluene film. A steady-state "bubble" of toluene forms because of strong surface-driven flows. (Reproduced with permission from reference 13. Copyright 1991.)

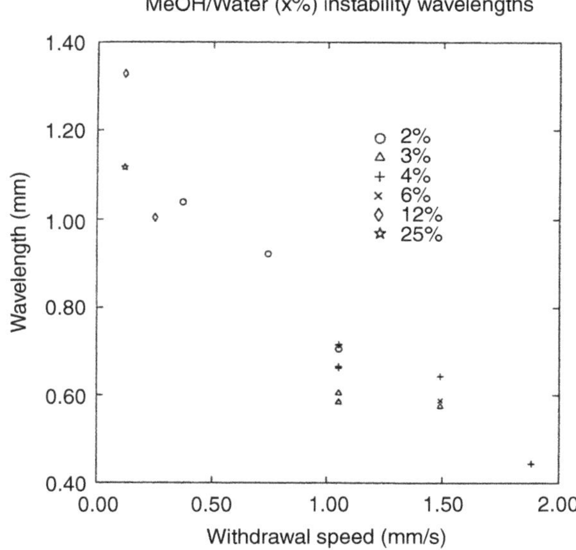

FIGURE 50.11 Wavelength of rib instability as function of withdrawal speed for several concentrations of methanol–water.

the shear rate in the thin region is $\Delta\sigma/\Delta x \approx (10\ \text{dyne/cm})/(10^{-1}\ \text{cm})$, hence $du/dz \sim 10^4\ \text{s}^{-1}$. For this sort of flow, P_e exceeds 1 for particles over 30 nm or so; thus shear-induced ordering would be expected.

Temperature gradients can often be considered small in dip coating; this assumption is not generally true in thin-film evaporation problems [28]. However, because the substrate moves relatively rapidly past the drying line, it constantly supplies energy for evaporation and renews the thermal field.

An interesting manifestation of surface flows in dip coating is a "rib instability" often observed near the reservoir in binary solvent systems and in some pure solvents (ethanol). Figure 50.10 shows a set of wavy interference fringes, indicative of a 20-nm thickness undulation that has a wave vector that runs perpendicular to the withdrawal direction. Typically, this undulation is a standing wave, but under some conditions (not well defined at this time) the ribs have been observed to fluctuate in position considerably. A study of the instability in water and methanol reveals that its wavelength is inversely dependent on the withdrawal velocity u_0 and that it is insensitive to solvent composition (Figure 50.11). An increase in u_0 fattens the entrained film [18]; this result would be expected to increase the wavelength, not decrease it. Furthermore, although the instability has been observed rarely in pure solvents, it is more definite in binary mixtures; hence it is probably aided by surface-driven flows, but, as the data show, the wavelength is relatively insensitive to the mixture.

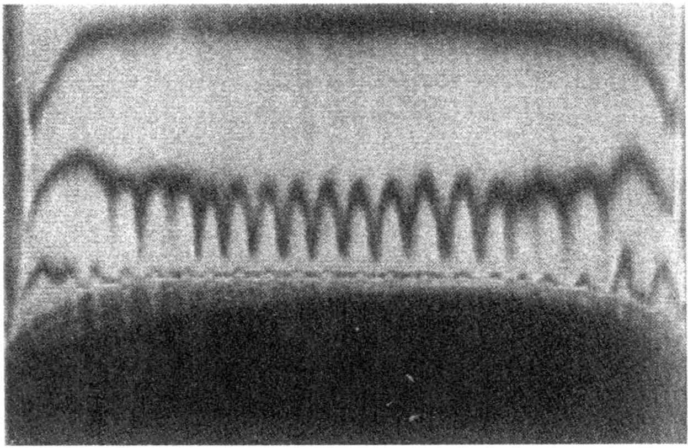

FIGURE 50.10 Imaging ellipsometry of rib instability near the reservoir meniscus (dark region).

These observations suggest that the ribs are related to the "Plateau–Rayleigh instability" of cylindrical surfaces of liquids [29], such as a jet of water, the cylindrical surface for dip coating being the concavity in the gravitational meniscus. First treated in the 19th century by Plateau and Lord Rayleigh, this instability results from the fact that a cylindrical liquid surface can decrease its area by undulating longitudinally; from dynamic considerations, one particular wavelength emerges as the fastest-growing unstable mode (although not necessarily the wavelength that fully develops beyond the linear-response regime). Unstable conditions can exist for concave cylindrical surfaces, but the complex dynamics determining the wavelength represent an unsolved problem. Because higher withdrawal speeds decrease the wavelength and, eventually, smooth out the ribs altogether, it probably takes some time for the instability to organize. The balance point for this competition of time scales is generally within the processing window of typical dip-coating operations.

SUMMARY: TIME SCALES

The limitation on structure formation in dip coating is best appreciated by considering the mean separation between two reactant molecules. By the conservation of nonvolatile mass, the concentration at x is inversely related to the entrained film thickness [22], and because the mean separation $\langle \Delta s \rangle$ is the inverse cube root of the concentration,

$$\Delta s \sim x^{1/6} \quad (50.9)$$

From this precipitous function, shown in Figure 50.12, it is possible to understand the dominant factor for a molecule or particle entrained in the film. In the transit from reservoir to drying line, the first 98% of the trip is relatively bland: Δs decreases by only 50% during this phase. In only the last 2% of the transit time between entrainment and the encounter with the drying line, representing a few tens of milliseconds, the remaining interparticle distance is covered. Dense gel structures of the type developed in quiescent reaction-limited (low sticking probability) samples are unlikely to assemble under these conditions, because the precursors have less time to explore tight-fitting configurations. However, it would seem equally unlikely that delicate diffusion-limited (high sticking probability) structures would arise given the near-ballistic transport. Further, concentration fluctuations normal to the substrate might not have time to smooth out.

In any case, subsequent events usually overwhelm the situation. The capillary pressures during the final drying stage are so large that even the least tenuous structures will collapse partially. The reduction of surface tension at the drying line, perhaps by surfactants or by critical point methods, would appear to be a promising way to prevent complete collapse.

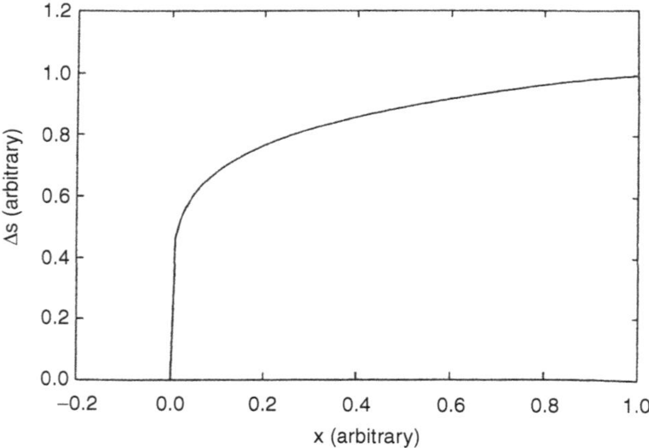

FIGURE 50.12 Interparticle distance near drying line. The square-root singularity in the evaporation rate gives rise to a rapid collapse of network structures following $x^{1/6}$.

ACKNOWLEDGMENTS

I thank Jeff Brinker for a long, statisfying collaboration on film formation, Randy Schunk for a critical reading of the manuscript, George Scherer for pointing out J.H.L. Voncken's work, and Don Stuart for tireless experimental work. This work was supported by Sandia National Laboratories under DOE Contract DE–AC04–76–DP00789.

REFERENCES

1. Schroeder, H. In *Physics of Thin Films*; Haas, B.; Thun, R. E., Eds.; *Advances in Research and Development*; Academic Press: New York, 1969; Vol. 5, pp 87–141.
2. Scriven, L. E. In *Better Ceramics Through Chemistry III*; Brinker, C. J.; Clark, D. E.; Ulrich, D. R., Eds.; Materials Research Society: Pittsburgh, PA, 1988; Vol. 121, pp 717–729.
3. Schmidt, H.; Rinn, G.; Nass, R.; Sporn, D. In *Better Ceramics Through Chemistry III*; Brinker, C. J.; Clark, D. E.; Ulrich, D. R., Eds.; Materials Research Society: Pittsburgh, PA, 1988; Vol. 121, pp 743–754.
4. Brinker, C. J. *NATO ASI Ser. Ser. E* **1988**, *141*, 261–278.
5. Puyané, R.; González-Oliver, C. J. R. *Proc. SPIE Int. Soc. Opt. Eng.* **1983**, *401*, 307–311.
6. LaCourse, W. C.; Kim, S. *Ceram. Eng. Sci. Proc.* **1987**, *8*, 1128–1135.
7. Brinker, C. J. *Ceram. Eng. Sci. Proc.* **1988**, *9*, 1103–1109.
8. Bräutigam, U.; Bürger, H.; Vogel, W. *J. Non-Cryst. Solids* **1989**, *110*, 163–169.
9. Melpolder, S. M.; Coltrain, B. K. In *Better Ceramics Through Chemistry III*; Brinker, C. J.; Clark, D. E.; Ulrich, D. R., Eds.; Materials Research Society: Pittsburgh, PA, 1988; Vol. 121, pp 811–822.
10. Brinker, C. J.; Scherer, G. W. *Sol–Gel Science*, 1st ed.; Academic Press: Boston, MA, 1990; Chapter 13.
11. Brown, G. L. *J. Poly. Sci.* **1956**, *22*, 423–426.
12. de la Court, F. H. *Proc. Xth FATIPEC (Federations d'Associations de Techniciens des Industries de Peintures, Vernis, Emaux, et Encre d'Imprimerie de l'Europe) Congress* **1970**, pp 293–297.
13. Brinker, C. J.; Hurd, A. J.; Frye, G. C.; Schunk, R. P.; Ashley, C. S. *J. Ceram. Soc. Japan* **1991**, *99*, 862–877.
14. Floch, H. G.; Priotton, J. J. *Am. Ceram. Soc. Bull.* **1990**, *69*, 1141–1143.
15. Landau, L.; Levich, B. *Acta Physicochim. (URSS)* **1942**, *17*, 42–54.
16. Guglielmi, M. *Proc. SPIE Int. Soc. Opt. Eng.* **1989**, *1128*, 55–62.
17. Nisnevich, Ya. D. *Zh. Prikl. Khim.* **1986**, *59*, 1406–1409.
18. Poehlein, G. W.; Vanderhoff, J. W.; Witmeyer, R. J. *Polym. Prepr. Am. Chem. Soc. Div. Polym. Chem.* **1975**, *16*, 268–272.
19. Scherer, G. W. *J. Am. Ceram. Soc.* **1990**, *73*, 3–14.

20. Burgess, C. G. V.; Everett, D. H. *J. Colloid Interface Sci.* **1970**, *33*, 611–614.
21. Zheng, Q.; Durben, D. J.; Wolfe, G. H.; Angell, C. A. *Science (Washington, DC)* **1991**, *254*, 829–832.
22. (a) Hurd, A. J.; Brinker, C. J. In *Better Ceramics Through Chemistry IV*; Zelinski, B. J. J.; Brinker, C. J.; Clark, D. E.; Ulrich, D. R., Eds.; Materials Research Society: Pittsburgh, PA, 1990; Vol. 180, pp 575–581. (b) Voncken, J. H. L.; Lijzenga, C.; Kumar, K. P.; Keizer, K.; Burggraff, A. J.; Bonekamp, B. C. *J. Mater Sci. (UK)* **1992**, *27*, 472–478.
23. Fuchs, N. A. *Evaporation and Droplet Growth in Gaseous Media*, 1st ed.; Pergamon Press: London, 1959.
24. Jackson, J. D. *Classical Electrodynamics*, 2nd ed.; Wiley: New York, 1975; Section 2.11.
25. Tranter, C. J. *Integral Transforms in Mathematical Physics*, 1st ed.; Chapman Hall: London, 1951; pp 50 and 99.
26. Hurd, A. J. In *Better Ceramics Through Chemistry III*; Brinker, C. J.; Clark, D. E.; Ulrich, D. R., Eds.; Materials Research Society: Pittsburgh, PA, 1988; Vol. 121, pp 731–742.
27. Sakka, S.; Kamiya, K.; Yoko, Y. In *Inorganic and Organometallic Polymers*; Zeldin, M.; Wynne, K. J.; Allcock, H. R., Eds.; American Chemical Society: Washington, DC, 1988; pp 345–353.
28. Parks, C. J.; Wayner, P. C. *AIChE J.* **1987**, *33*, 1–10.
29. Quéré, D.; di Meglio, J.-M.; Brochard-Wyart, F. *Science (Washington, DC)* **1990**, *249*, 1256–1260.

Part 7

Silica Coatings

Michael R. Baloga
DuPont Company

In 1956, Ralph K. Iler filed a patent application for a coating process that deposited dense shells of hydrated amorphous silica on cores of other solid materials. On May 5, 1959, patent 2,888,366 was awarded to Iler and his assignor, DuPont.

Dense silica coating technology was first exploited by the pigments industry. Premium pigments such as titanium oxide were among the earliest examples of the application of Iler's technology. In 1966, DuPont introduced the first commercial silica-encapsulated pigment under the pigment-grade designation, Ti-Pure R-960. This pigment was rapidly accepted by the manufacturers of automotive paints to create coatings that achieved new levels of durability and color stability upon exposure to sunlight, oxygen, and water. Finally, automobiles would have paint finishes that lasted the lifetime of the vehicle.

The colored pigments industry was also quick to follow this lead, and a second line of DuPont pigments, the Krolar products, soon became available as color-stable pigments that did not fade in sunlight. The automotive and industrial paint industries were major consumers of these products.

In the 1970s, after the expiration of Iler's dense silica patent, major pigment manufacturers used the silica encapsulation technology and added R-960 counterparts to their product lines.

The economic impact of Iler's silica coating process will continue to play a major role in the coatings industry, from the perspective of the pigment manufacturers as well as the coatings producer. A conservative estimate of $300 million is placed on the value of dense silica coated titanium dioxide pigments that were sold in 1990 in the world pigment marketplace. The total value of manufactured coatings products is estimated to be in excess of $4 billion.

Iler began the revolution in surface treatment technology. Its specific contribution was to advance surface coating chemistry from state of the art to state of the science. The pigment surface chemist of the 1990s has become an architect of pigment surface structures.

51 Nanostructuring Metals and Semiconductors with Silica from Monolayers to Crystals

Luis M. Liz-Marzan
University of Vigo, Department of Chemistry

Paul Mulvaney
University of Melbourne, School of Chemistry

CONTENTS

In this chapter we discuss a novel approach to the concept of nanostructuring, which is based on coating metal or semiconductor nanoparticles with silica, so that the final morphology involves a silica sphere of the desired size containing a core placed precisely at its center. Although such concepts have been proposed before, it has only recently been possible to synthesize such coated materials in a reproducible manner. These composite spheres can then be used as the building blocks of the nanostructured material. The interest of these systems is due to the unique optical and electronic properties of nanosized metal and semiconductor particles.

The outline of the paper is therefore as follows: In section 1 we introduce the main interaction forces acting on colloidal particles, as well as the concept of nanostructured materials, in the form of 2D and 3D assemblies. We discuss the main stabilization techniques employed in the synthesis of nanoparticles in solution. Then we outline in section 2 the procedures involved in silica coating, and discuss its advantages as a general stabilization technique. Section 3 deals with the special properties of both metal and semiconductor nanoparticles, summarizing their treatment by Mie theory. In the subsequent section, we illustrate how Mie theory can be applied to the linear optical properties of silica coated metal particles. In section 5 we present the first results on 2D and 3D assemblies of core-shell particles, and finally we demonstrate in section 6 the effect of silica shells on chemical reactions performed on the cores. Such reactions can also be used for the preparation of complex structures.

INTRODUCTION

Nanostructured materials are assemblies of nanosized units which display characteristic properties at a macroscopic scale. The size range of such units lies within the colloidal range, so that the properties of the assemblies can be tuned by varying their colloidal properties, mainly particle size, surface properties, interparticle interactions, and interparticle distance.

The main interaction forces acting on colloidal systems are van der Waals attractive forces, and electrostatic and steric repulsive forces [1]. We shall introduce briefly these forces, which should help us to understand the stability matters dealt with in this paper.

van der Waals Forces

The van der Waals forces affecting colloidal particles are of the same nature as those between atoms, molecules and ions, but since particles contain many molecules, the forces are larger, and often have a longer range over which they are felt. The van der Waals energy between a pair of atoms or molecules is inversely proportional to the sixth power of their distance R.

$$V = -\frac{\lambda_{1,2}}{R^6} \quad (51.1)$$

where $\lambda_{1,2}$ is the London constant. We assume that the pair energies are additive, so that the attraction energy between two particles with volumes V_1 and V_2 is given by,

$$V_{att} = -\int_{V_1}\int_{V_2} \frac{N_1 N_2 \lambda_{1,2}}{R^6} dV_1 dV_2 \quad (51.2)$$

where N_1 and N_2 are the numbers of molecules per unit volume in particles 1 and 2 respectively.

This equation has been integrated for a number of geometries [2]. For the case of two parallel infinite plates of the same material at a distance H, the attraction energy per unit area is,

$$V_{att} = -\frac{A}{12\pi H^2} \quad (51.3)$$

where A is the Hamaker constant $A = \pi^2 N^2 \lambda$.

For two identical spheres with radius a and distance between centers R, the energy can be simplified, when the distance between the sphere surfaces ($H = R - 2a$) is much smaller than a, to

$$V_{att} \approx -\frac{A}{12}\left(\frac{L}{H} + 2\ln\frac{H}{L}\right) \approx -\frac{Aa}{12H} \quad (51.4)$$

with $L = a + 3/4\ H$, though the last approximation is not very good. For two particles of material 1, embedded in medium 2, the relative Hamaker constant can be calculated as a linear combination,

$$A = A_{1,1} + A_{2,2} - 2A_{1,2} \quad (51.5)$$

Hamaker constants (or their combinations) are always positive for interactions between identical particles [3], and values for metals in water can be of the order of 15–75 kT, while they are of just 1–12 kT for oxides and halides. This means that attractions between metal particles will be markedly higher and stabilization (by any means) will be harder to achieve.

Electrostatic Forces

In polar media particles are usually charged. Since the colloid as a whole must remain neutral, small ions with opposite charge accumulate around the particles forming a more or less diffuse double layer. Such a double layer promotes repulsion when a second particle (with its corresponding double layer) approaches, thus acting as a shield preventing coagulation.

The electric field in solution is given by the Poisson–Boltzmann relation,

$$\nabla^2\Psi = \frac{-1}{\varepsilon\varepsilon_0}\sum_i z_i e n_{i,\infty} e^{-z_i e\Psi/kT} \quad (51.6)$$

where ε_0 and ε are respectively the permittivity of vacuum and the relative permittivity of the medium, $n_{i,\infty}$ and z_i are the concentration at infinite distance and the valence of the ion considered, and e the charge of the electron. When $z_i e\Psi/kT < 1$, (low surface potential) Equation (51.7) can be written,

$$\nabla^2\Psi = \kappa^2\Psi \quad (51.7)$$

where the Debye–Hückel length, $1/\kappa$, also known as double layer thickness, is defined as,

$$\frac{1}{\kappa} = \left(\frac{\varepsilon\varepsilon_0 RT}{2F^2 I}\right)^{1/2} \quad (51.8)$$

with F Faraday constant and I the ionic strength, $I = 1/2\ \Sigma c_i z_i^2$.

Assuming constant surface potential, the repulsion potential between two spherical particles is given by,

$$V_{rep} = 2\pi\varepsilon\varepsilon_0 a\Psi^2 \ln\left(1 + e^{-\kappa H}\right) \quad (51.9)$$

when $\kappa a \gg 1$, and,

$$V_{rep} = 2\pi\varepsilon\varepsilon_0 a\Psi^2 e^{-\kappa H} \quad (51.10)$$

when $\kappa a \ll 1$.

The total interaction potential for charged colloids will be thus given by a combination of van der Waals attraction and electrostatic repulsion, which is known as DLVO interaction theory [4,5]. Figure 51.1 shows the total interaction energy as a function of the interparticle distance for different ionic strengths. It can be observed that attraction always wins out at short distances and at large distances, while repulsion may win at intermediate distances ($H \approx 1/\kappa$), which is represented as a maximum, if the repulsion is strong enough.

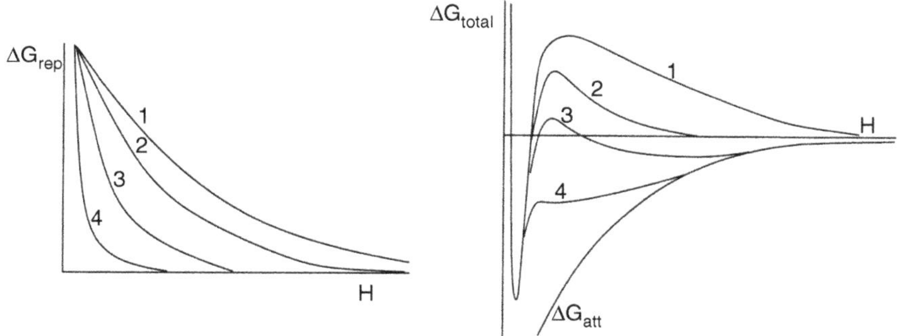

FIGURE 51.1 Double layer repulsion and total free energy of two particles as a function of distance H at constant surface potential, for several values of ionic strength (1 lowest, 4 highest).

Steric Repulsion

The approach of particles to each others can be prevented by the build up of a physical barrier constituted by an adsorbed layer of (usually nonionic) macromolecules. The thicker this layer, the longer the separation between particles and accordingly the more stable the dispersion. This kind of stabilization process is called steric stabilization and is mostly used in nonpolar media, where electrostatic stabilization is difficult to achieve [5,6].

Stabilizing molecules can be mainly of two types. The first type includes polymers of high molecular weight, with low but finite adsorption energy per monomer, the adsorption taking place through a configuration based on branches oriented toward the solution and tails at the ends of the molecule which can attach to two separate points of the surface. The second type is constituted by macromolecules with relatively low molecular weight, an anchor group easily adsorbing and a tail compatible with the solvent (surfactants).

The protecting action itself shows two different elements. When the tails of molecules bonded to two particles approach each other, the increase in local concentration leads to an increase in free energy (repulsion), which is an *osmotic effect.* At the same time, some tails may not fit in the space between the particles, decreasing the number of conformational possibilities, which promotes a loss of entropy. This is the *volume restriction effect.*

Steric repulsion can be modified or even eliminated either by modifying the solubility of the protecting chains (i.e., changing the solvent), or by desorption, which is usually more difficult to promote.

Ordered Colloidal Arrays

In nature, ordered colloidal systems are unusual for two reasons. First, large scale lattices require monodispersity of the individual particle units within the lattice. Second, both diffusion limited aggregation and reaction limited aggregation in solution produce fractal structures rather than ordered colloidal arrays [7,8]. One exception to this is opal, which exhibits very strong iridescence. This mineral is composed of densely packed, ordered microcrystals of colloidal silica, as first shown by electron microscopy in the 1960s by Sanders [9]. Artificial systems that display ordered structures and optical diffraction include colloidal latex [10], colloidal gold [11] and colloidal silica [12]. More recently, ordered colloidal arrays have been proposed as a means to create optical rejection filters [13], tunable filters, fast optical switches and as high density memory devices for computers [14]. We now consider the basic approaches to the synthesis and characterization of 2D and 3D crystals.

3D Structures

The preparation of ordered arrays of colloidal particles in three dimensions has been studied in the past primarily as a model for phase transition processes [15]. Colloid dispersions have concentrations of $10^{13}\,cm^{-3}$, and elastic constants of 10 dyn cm^{-2}, whereas in a typical atomic system, the density is around $10^{22}\,cm^{-3}$ and the elastic modulus 10^{10} dyn cm^{-3}. Since the elasticity scales almost exactly with particle concentration, the particle interaction energies in both types of system must be of a similar magnitude. Thus, the transitions of materials between their gas, liquid, glass and crystalline states can be observed in colloid suspensions. The use of colloid particles offers several experimental advantages over the direct study of atomic phase transitions. Since the diffusion constants of colloid particles are at least 3 orders of magnitudes lower than those of atomic particles in the same fluid (scaling as $1/R$), the crystallization processes are slowed down by at least 3 orders of magnitude in time, enabling accurate kinetic measurements to be made. Furthermore, with particles a few tenths of a micron in size, time resolved (static and dynamic) light scattering measurements can be made with monochromatic laser light rather than just static x-ray or neutron scattering measurements of the colloid structure [16].

The sensitivity of current CCD detectors and their submicrosecond risetimes enables direct kinetic profiles of the time evolving ordering to be obtained, whereas neutron scattering often requires averaging of signals over several minutes. A further advantage of larger particles is that video-enhanced microscopy allows real time images of the motion of individual particles to be made. Photon correlation spectroscopy reveals the variation in diffusion coefficient of individual particles as the volume fraction is increased. The diffusion coefficient measured depends on the timescale over which measurements are made. For short correlation delay times, results show that for particles of equal size, there is short-time diffusion within the 3D dispersion up to volume fractions close to 0.50, though the long-time diffusion coefficient decreases steadily with volume fraction, appearing to be zero at volume fractions of 0.50. The theoretical freezing point for hard-sphere systems is predicted to occur at $\phi = 0.495$. Thus, it seems that long time scale values, which reflect particle motion through the lattice, cease as expected near the freezing point. However, smaller diffusional excursions, where the particle essentially does not sample the lattice, continue [15]. This means that slow conversion of colloid glasses into crystals can occur even at high volume fractions. These phase transitions have important implications for nanoparticle arrays, since diffusion coefficients are higher for smaller particles, so that recrystallization out of glass states may be more effective. Since the optical properties, particularly for metal colloid 3D crystals, are strongly dependent on the radial distribution function, slow changes in the absorption coefficient can be expected. These will be discussed in a later section.

A third important advantage of colloidal systems over atomic ones is that the interparticle forces can be varied readily via the electrolyte concentration and surface charge density on the particles. In general, the interparticle potential used in ordering studies is not the DLVO potential, because the separation between particles is significantly larger than the range of van der Waals forces, and this term is usually dropped. Instead a screened Coulomb potential is used, usually referred to as the Yukawa potential,

$$V(R) = \frac{4\pi\varepsilon\varepsilon_0 a^2 \Psi^2}{r} e^{-\kappa(R-2a)} \quad (51.11)$$

where a is the particle radius and R the interparticle distance. It is important to notice that the incorporation of the van der Waals term produces not just a primary minimum at small separations, but depending on surface charge and electrolyte concentrations will also introduce a weaker *secondary minimum* at separations of several κ^{-1}. More recently, another interparticle potential has been introduced that predicts attractive forces between identical particles at much larger particle separations [17]. The Sogami potential has been used to explain a number of observations that cannot be explained by DLVO potentials, such as the appearance of macroscopic, long-lived voids in concentrated colloids [18]. There is considerable debate over the validity of the assumptions underlying the derivation of this potential [19,20], that are outlined in detail in reference 16.

The thermodynamic behavior of colloidal systems can be treated in the same way as that of atomic systems by considering the collection of colloid particles as a supramolecular fluid dispersed in a continuous medium, as was first shown by Onsager [21]. However, the introduction of a fluid between the particles does have important implications for the dynamics of the particle interactions. Since the exchange of energy and momentum with the solvent bath is faster than the typical timescales involved in diffusional jumps in solution, colloid particles display Brownian motion, and therefore obey the Smoluchowski equation, rather than Newtonian mechanics, as is the case in atomic systems. Pusey has reviewed many of the experimental aspects of 3D colloid crystallization [22]. The phase behavior of hard-sphere systems was first studied by Pusey and Van Megen [23] with sterically stabilized polymethyl methacrylate, and later by Smits et al. [24] with octadecyl coated silica spheres. In every case, increasing volume fractions leads to a fluid-solid phase transition (at $\phi > 0.494$), where the solid phase is crystalline. At larger concentrations ($\phi > 0.58$) a glass phase is encountered. When the particles are charged, they become soft, and then the phase transitions take place at lower volume fractions.

Concentrated colloidal dispersions in this size regime show a variety of unusual effects such as shear induced crystallization and Bragg diffraction, which can be directly observed with the naked eye [25]. Likewise, the use of video enhanced microscopy offers visual verification of colloid dynamics for particles with sizes accessible to confocal optics. The creation of ordered arrays of nanometre sized particles opens up the possibility to tailor the optical properties not just through the interparticle spacing, but by modulation of the size-dependent properties of the particles themselves. The fundamental difficulty to be overcome is that as the particles become smaller, their crystallization kinetics will again be accelerated, and characterization will no longer be possible through light scattering. Instead electron diffraction, SANS and SAXS techniques will be important, as well as newer scanning probe microscopy methods such as STM and AFM for crystal imaging.

To date there has been little work on the synthesis of 3D nanoparticle crystals due to the difficulty in the synthesis of large amounts of sufficiently monodisperse particles. One system which has great promise is alkanethiol stabilized colloidal gold and silver [26,27]. Landman has reported the creation of millimeter sized crystals based on core colloid particles of just 5–10 nm in diameter [28]. Macroscopic amounts can generally be prepared by evaporation of the surfactant stabilized materials without coalescence, and the color of the powder clearly is consistent with the initial quantized semiconductor or nanosized metal core material [29,30]. 3D crystallization is observed for polynuclear CdS clusters, as found by Dance and coworkers [31].

2D Crystals

The creation of 2D crystals of both micron sized and nanometre sized particles remains a somewhat empirical process due to the ill-defined role of the substrate or surface on which nucleation takes place. Perrin first observed diffusion and ordering of micron sized gamboge 2D crystals in 1909 under an optical microscope [32]. Several techniques have been proposed for the formation of 2D arrays at either solid-liquid surfaces or at the air-water interface. Pieranski [33], Murray and van Winkle [34] and later Micheletto et al. [14] have simply evaporated latex dispersions. Dimitrov and coworkers used a dip-coating procedure, which can produce continuous 2D arrays [35,36]. The method involves the adsorption of particles from the bulk solution at the tri-contact phase line. Evaporation of the thin water film leads to an attractive surface capillary force which aids condensation into an ordered structure. By withdrawing the film at the same rate as deposition is occurring, a continuous film of monolayered particles is created. Since the rate of deposition is measured with a CCD camera, it is not possible to use nanometer sized particles with this method, unless a nonoptical monitor for the deposition process can be found.

The incorporation of nanoparticles into 2D crystals is quite difficult. As the particle size becomes smaller, capillary forces at the air-water interface and increased thermal (Brownian) motion make crystallization processes more difficult. The most promising methods for creating macroscopic 2D nanoparticle arrays are through modified LB film techniques. This procedure has been pioneered by Fendler, Kotov and colleagues [37]. They have nucleated the particles themselves directly beneath surfactant monolayers and then compressed them into ordered structures. This method has been applied to a wide range of metal and semiconductor materials (Ag, Au, CdS, PbS). Direct observation of particle morphology is possible using scanning tunnelling microscopy [38,39]. The resultant particulate films can be transferred to a solid substrate. Matsumoto has reported a new iris diaphragm based LB trough using circular rather than linear compression of the particle monolayers [40]. Matsumoto and coworkrs have also prepared 2D protein crystals on mercury surfaces. Their technique allows high resolution images of proteins to be obtained. Because of the regular arrays formed, spectral analysis can be used to enhance features and make more accurate determinations of lattice parameters [41]. This appears to produce the biggest single domain 2D arrays to date. The improvement is linked to the decrease in surface stress that occurs due to local particle concentration gradients, and which forces the particles to nucleate as 2D polycrystalline arrays during compression. Alternatively, the particles can be synthesized in solution, rendered hydrophobic using physisorption or chemisorption of surfactants, and these derivatized particles compressed. Capped or derivatized particles can also contain terminal functionalities which enable them to covalently bond to other surfaces. Alivisatos has bonded CdSe nanocrystals to both gold and aluminum surfaces [42]. Bawendi et al. used a modified LB procedure to fabricate submonolayers of capped CdSe nanocrystals bound to substrates [43].

An alternative concept is to drive the particles to the surface either magnetically or electrically. Polymer coated magnetite particles will form ordered arrays at the air-water interface, in which the particle spacing is tunable not just through the Debye screening parameter but by an external magnetic field. Bentzon et al. reported magnetic superlattice formation from surfactant stabilized iron-carbon alloy particles just 6.9 nm in diameter. These oxidized to antiferromagnetic hematite in air, but retained the ordered structure [44]. Electrophoretic deposition provides a generic method for deposition of nanoparticles onto a conducting substrate. 15 nm citrate stabilized gold particles were deposited onto a carbon coated TEM grid [26]. This permitted direct observation of the 2D arrays in the electron microscope. In Figure 51.2, an electron micrograph of such ordered gold particles is shown [26]. Fourier transformation confirms hexagonal close packing. Due to surface nucleation occurring simultaneously over the grid, single domains do not form, but rather polycrystalline arrays. The interparticle spacing can be used to estimate the thickness of surface stabilizers. Large crystalline arrays of sub-10 nm particles were observed by this technique for the first time. The interparticle spacing was controlled by alkanethiols of various chain lengths, and by tuning the particle size. Electron diffraction could be used to monitor the 2D crystal structure. In all cases the structures were hexagonal, though small regimes of cubic phases were also observed. However deposition onto other surfaces requires an alternative technique such as AFM to monitor the coating process.

Pileni et al. have recently found that alkanethiol capped particles will spontaneously form 2D arrays on surfaces dipped into the colloidal solution [45]. In this case, hydrophobicity of the immersed surface is a critical parameter determining the transfer process. Increased

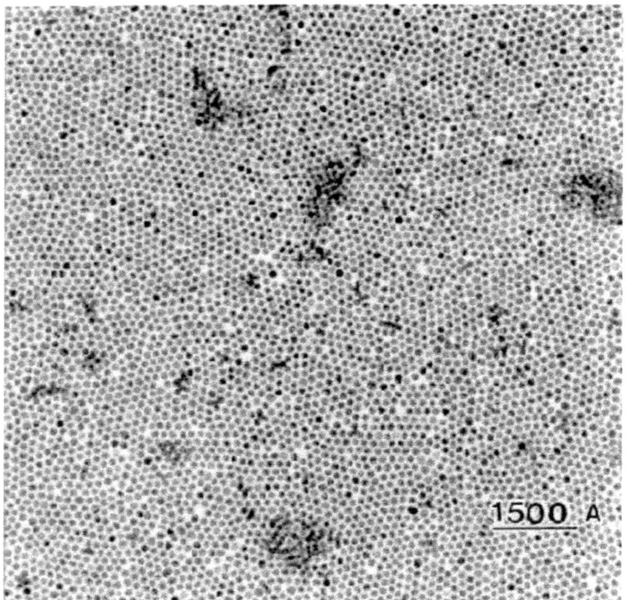

FIGURE 51.2 Electrophoretic deposition of negatively charged gold colloid particles onto positively charged TEM grids produces large regions of ordered particle domains. Adapted from reference [26]. Copyright 1993. American Chemical Society. With permission.

particle sizes should aid adsorption since the decrease in surface energy per particle adsorbed will increase. Consistent with this, Pileni et al. observed better crystals for larger particle sizes. (They also suggested that this could be attributed to an increase in the van der Waals interaction between larger particles.) Böhmer has recently employed video microscopy to monitor in real time adsorption of colloidal latex particles to a substrate which is rotated [46]. The rotation induces a convection current of particles towards the surface. The flow field in this geometry is known, and the probability of adhesion to the substrate could be quantified as a function of tangential and vertical fluid velocity, particle size and substrate hydrophobicity.

That the optical properties of such 2D/3D colloid arrays will depend on particle packing densities has been demonstrated by Dusemund et al., who measured the reflectivity of gold colloid films as a function of colloid volume fraction [47]. Although the films were not ordered they showed a clear shift in surface plasmon position with increasing particle volume fraction. The only compromise was that the volume fraction could not be directly determined on the samples from which spectra were taken, but had to be measured on separately prepared TEM grids. Several groups have found that clustered quantum dots exhibit red-shifted fluorescence, a question recently reviewed by Weller [48].

The first systematic reports on the electrical properties of 2D nanoparticle arrays appeared in 1995. Schon used phosphane stabilized Au_{55} particles, and employing AC impedance measurements, demonstrated that the interparticle conductivity was due to electron hopping [49,50]. Terrill et al. in 1995 presented comprehensive results on the conductivity of alkanethiol derivatized gold particle arrays [51,52]. They used interdigitated array electrodes (IDAEs) to measure both the AC and DC conductivity. The temperature dependence of the conductivity followed the Arrhenius equation. Since chain melting of the alkane thiol stabilizers occurred simultaneously, long equilibration times were crucial to obtaining accurate plots. The data could also be well fitted to cermet models of $\ln(\sigma)$ vs $T^{-1/2}$.

These nanoparticle arrays offer substantial improvements in rigidity and chemical stability over purely organic thin films. In purely organic thin films, polymers or surfactants with functionalized headgroups are deposited in a layer-by-layer method on an appropriate substrate by using Langmuir–Blodgett monolayers. Such films can attain a high degree of order due to close packing of the hydrocarbon chains of the surfactants. Several obstacles have been identified on the way to broader application of organic LB films:

- low stability against ultraviolet radiation;
- spontaneous flip-flop motion of surfactant molecules leading to slow disorganization of the LB assembly;
- fixed thickness of a single monolayer, which results in the necessity of multiple deposition cycles.

Importantly, physical and chemical properties of nanoparticulate films are markedly different from those of the bulk materials. For example, magnetic nanoparticles can be prepared where only one magnetic domain is present, so that the rotation or alignment of the whole particle implies the rotation or alignment of the magnetic moment [53]. Semiconductor nanoparticles possess strong nonlinear optical properties due to increased oscillator strength within excitonic transitions [54–56]. Electrooptical shifts can be induced in metal particles because the surface plasmon band position depends on the free electron concentration; electron injection can be used to modulate the peak position [57].

In Figure 51.3, the spectrum of colloidal silver before and after electron injection from a gold mesh electrode shows that cathodic polarization leads to a blue shift of the surface plasmon band. The particles initially had a redox potential of +0.15 V vs Ag/AgCl, whereas afterwards it was at −0.6 V. The charge transferred corresponded to a double layer capacitance of 80 microfarads cm^{-2}. Femtosecond heating of the conduction electrons in small particles results in shifts in the surface plasmon band. Initial relaxation due to electron-lattice collisions takes

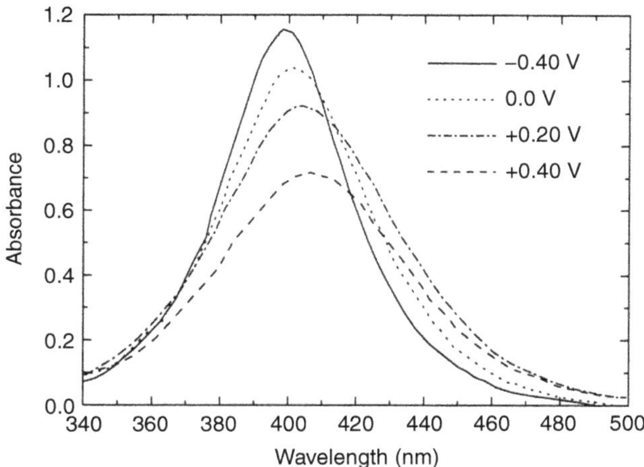

FIGURE 51.3 Spectra of colloidal silver at two different potentials showing the blue shift in the position of the surface plasmon band due to the increase in electron concentration. Adapted from reference [58]. Copyright 1997. American Chemical Society. With permission.

place on a picosecond timescale, then slower nanosecond shifts occur as heat from the particle lattice is transferred to the solvent bath [58].

USE OF SILICA COATING AS A GENERAL STABILIZATION TECHNIQUE

Silica particles show an enhanced stability which has been the subject of intense study [59,60]. It has been suggested that water structuring occurs at the particle surface due to the presence of oligomeric silica "hairs," which consist of —Si(OH)$_2$—O—Si(OH)$_2$—OH groups which can lose protons to become silicic acid [59]. However it has been shown that better agreement with experiment is found if a water-swollen silica-gel layer is considered to be present at the surface [60]. This provides silica surfaces with short-range steric stabilization, while long-range interactions are dominated by DLVO forces, which are, however, in turn affected by these gel layers, effectively enhancing the double layer repulsion. A consequence of the mentioned hairs is the rough surface morphology of silica particles made in water.

An elegant method proposed by Stöber et al. [61] for the preparation of monodisperse spherical silica particles with controlled size comprises the hydrolysis and condensation of alkoxysilanes in mixtures of ammonia, water, and a lower alcohol. The reactions involved can be represented by:

$$\text{—Si—OR} + \text{HOR}' \Longleftrightarrow \text{—Si—OR}' + \text{HOR}$$
$$\text{—Si—OR} + \text{HO—Si—} \Longleftrightarrow \text{—Si—O—Si—} + \text{HOR}$$

where R and R′ stand for a hydrogen atom or an alkoxy group.

Furthermore, the surface properties of silica particles can be modified by classical silylation agents which yield stable colloids in different environments [62–65]. This makes these systems much more suitable as probes for experimental applications [66,67].

All these properties of silica particles and surfaces have been long used for the *protection* of colloids of different sorts through their coating with silica. Additionally, silica is chemically inert, and does not affect redox reactions at the core surface, except through physical blocking of surface sites. We shall see later that the silica shell is optically transparent so that chemical reactions can be monitored spectroscopically. Finally, and most obviously, the shell prevents coagulation during chemical reactions so that particle coalescence does not occur, and concentrated dispersions of nanosized semiconducting, magnetic, or metallic materials can be created.

Iler [68] already in the late 1950s described a method for the coating in water of particles with "at least one dimension which is less than about 5 microns," and which nature is typically of metal oxides and silicates. This method comprises the deposition of silica from a sodium silicate solution with a pH between 8 and 11, which is termed *active silica*, and it is mainly aimed for thin shells.

In the nineties, Ohmori and Matijevic used Stöber synthesis [61] for the homogeneous coating of hematite (α-Fe_2O_3) particles [69], adapting the method subsequently for the preparation of silica coated iron [70]. This procedure was later improved by Thies–Wessie et al. [71].

A combination of the previous two synthetic methods was used for the coating of magnetite and boehmite particles by Philipse et al. [72]. The surface properties of these materials made it necessary to device such a two-step coating procedure.

A basically different and innovative way was devised by Chang et al. for silica coating CdS colloids [73] by performing successive reactions in microemulsions.

All these coating procedures involve surfaces with a significant chemical or electrostatic affinity for silica. In the case of (noble) metal particles, the coating process is further complicated by the very low affinity for silica. Patil et al. [74] proposed the formation of Au—Si or Ag—Si particles by reaction in the gas phase, and further oxidation of the Si shell to SiO_2. Such a process is much more difficult to perform in solution, where the *vitreophobicity* is enhanced because noble metals do not form a passivating oxide film in solution. Furthermore, there are usually adsorbed carboxylic acids or other organic anions present on the surface to stabilize the particles against coagulation, which also render the particle surface *vitreophobic*. A first attempt to overcome this vitreophobic character involved the hetero-coagulation of small gold and silica colloids dispersed in water, followed by silica growth in ethanol [75]. This resulted in a mixture of labelled and unlabelled silica particles, with a rather low concentration of the labelled ones.

A much more efficient approach was later developed for gold [76,77] and silver [78], involving the modification of the particle surface to make it *vitreophilic*. The process used for doing this is sketched in Figure 51.4. Silane coupling agents [79] are used as surface primers to form a hydrated silica monolayer bonded to the metal substrate, which then achieves chemical affinity for silica and can be coated by means of the procedures previously described. The large complexation constant for gold and silver amines [80] is a strong driving force for the adsorption of (3-aminopropyl) trimethoxysilane (APS) molecules at the particle surface, even displacing the previously adsorbed stabilizing moieties. In the case of semiconductor particles like CdS, silane coupling agents with mercapto groups can be used [81]. Hydrolysis of the surface-bonded siloxane moieties to form silane triols occurs within minutes, while condensation is much slower, especially at low concentrations [79]. At pH 7, there is ionization of the silane triols (their isoelectric point is pH 2–3), and this ensures that there is adequate negative surface charge on the gold sol during stabilizer exchange to maintain sol stability. In a second step, active silica is added to the dispersion, which promotes the formation of a thin, dense and relatively homogeneous silica layer around the particles [68], using the silanol groups as anchor points. At this stage, the particles can be transferred into ethanol and the silica layer thickness can be increased in a controlled way.

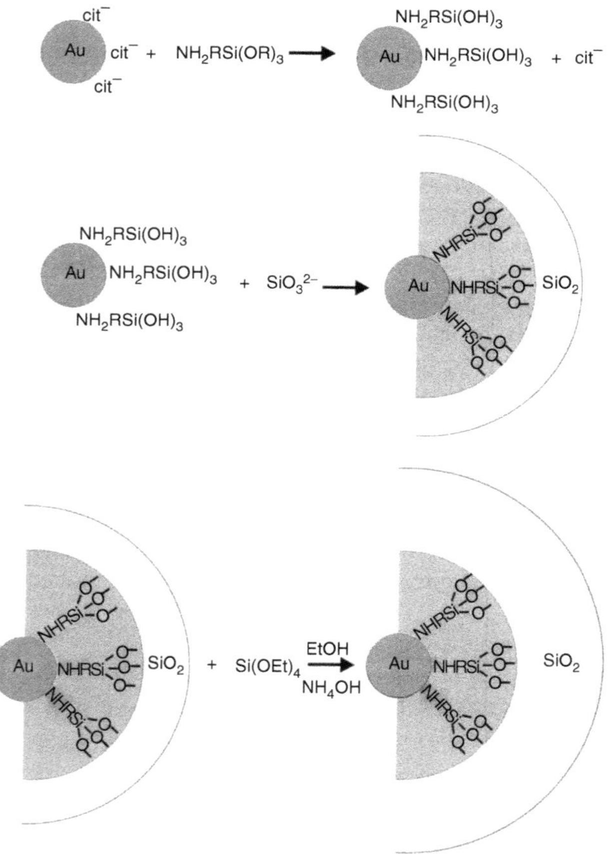

FIGURE 51.4 Sketch of the surface reactions involved in the formation of a thin silica shell on citrate stabilized gold particles.

Influence of APS. Both gold and silver particles can be stabilized in water by citrate anions [82], which can be displaced by amines from the metal surface [83]. An APS concentration was chosen so that the number of molecules per metal particle was close to one monolayer (but still slightly below it), which has been observed to be sufficient for the preparation of homogeneous coatings [77]. When more than one monolayer is added, bridging flocculation occurs over minutes to days depending on the APS concentration. The structure of the APS monolayer at the particle surface is thought to be uniform, with the amino groups complexed to gold surface atoms and the silane groups facing the solvent (see Figure 51.4). A pH of 5 was found to be optimal, so that APS has a net negative charge, with at least two of the siloxy groups ionized, to ensure the sol is not destabilized by the adsorption of APS and the subsequent displacement of citrate groups [78]. At this pH, the APS amine group is still protonated, so that it is preferentially oriented with the amino group towards the surface during adsorption because of the alignment induced by the diffuse double layer around the silver colloid particles, and this may assist in the formation of a more cohesive, ordered monolayer.

Role of active silica. Two parameters are important for the first silica coating: pH and silicate concentration. As indicated by Iler [67] the pH should be kept between 8 and 10, so that the solubility of the silicate species present in the solution is reduced, and polymerization/precipitation must occur at a sufficient rate to homogeneously coat the particles, but still slow enough to avoid the formation of silica nuclei. At the same time, the silicate concentration also plays a role in determining the coating rate. It was found [76–78] that a large excess of silicate ($[SiO_2]/[M] \approx 10$) is necessary to achieve a visible silica layer in a rather short period of time. Figure 51.5 shows a high resolution micrograph of an Ag—SiO_2 particle where the deposition process was allowed to proceed for 48 h. It can be observed that the amorphous silica shell is homogeneously deposited onto the crystalline Ag core (lattice distance = 2.3 Å). Longer periods lead to a further increase in silica layer thickness, but small silica particles also nucleate out of the solution. For gold, a centrifugation/redispersion process was found

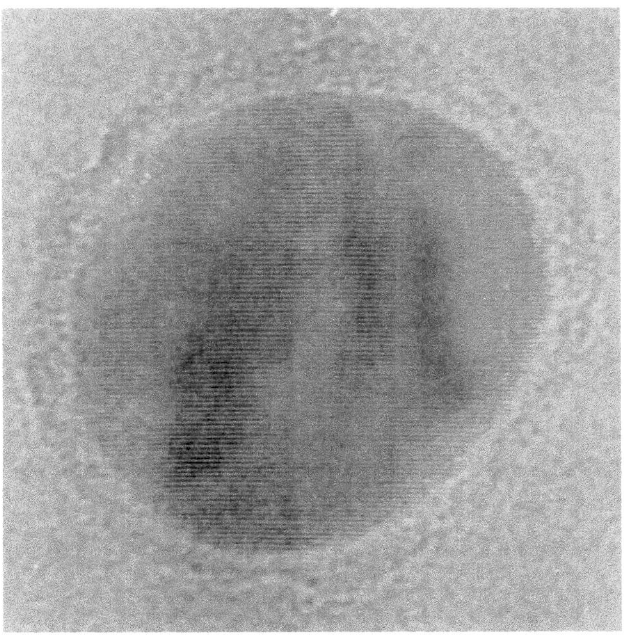

FIGURE 51.5 High resolution transmission electron micrograph of a 9 nm silver particle coated with a thin silica layer by silicate deposition for 48 h.

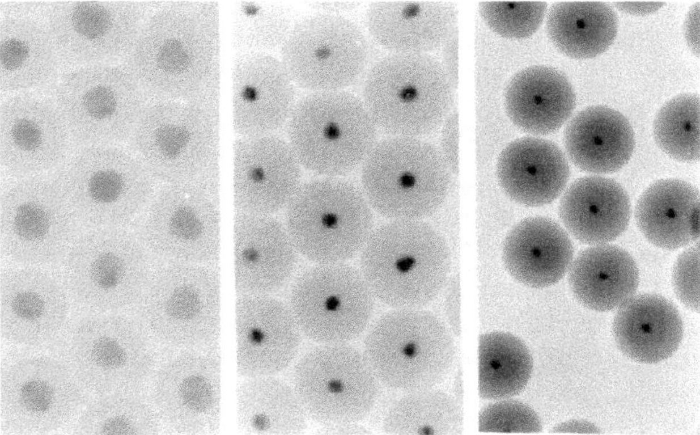

FIGURE 51.6 Transmission electron micrographs of Au—SiO_2 particles produced during the extensive growth of the silica shell around 15 nm Au particles with TES. Silica shell thicknesses are, from left to right: 8, 31, and 78 nm.

to provide a dispersion rather clean of silica nuclei, which can be readily transferred into ethanol.

Transfer into ethanol. The control of the silica layer thickness is much easier to achieve if the particles are dispersed in ethanol, or in ethanol/water mixtures. In such mixtures, the addition of tetraethoxysilane (TES) provides the system with monomeric silica, which leads to slow homogeneous particle growth [61]. However, prior to the preparation of stable dispersions in ethanol, the silica layer formed in the aqueous dispersion must be sufficiently thick to shield the strong van der Waals interactions between the metal particles [84]. When the particles are transferred from water into a solvent with a lower polarity, such as ethanol, there is a reduction in surface charge. If the silica layer around the particles is too thin, the van der Waals forces are still strong and partial flocculation occurs, even if TES and ammonia are added immediately after the transfer. This problem is partially offset by the fact that when the aqueous dispersion of precoated particles is diluted with ethanol, the solubility of silicate is drastically lowered, and therefore polymerization occurs [85]. Consequently the transfer into ethanol causes (most of) the silicate ions still remaining in solution to condense onto the metal–silica nuclei already present in the dispersion. The extent of the polymerization depends mainly on the ethanol/water ratio. An ethanol/water ratio of 4:1 was found to provide a homogeneous growth while minimizing the concentration of core-free silica particles formed [78].

Extensive growth. Further growth of the particles is achieved in the ethanol/water mixtures using Stöber method [61]. It is important to notice that, if the first silica coating in water is skipped, significant aggregation takes place in ethanol before silica from TES hydrolysis can grow onto the particles. This is mainly due to a reduction in electrostatic stabilization of gold particles in ethanol, when stabilized only with APS/citrate. By this method, the formation of alcosols of silica particles of any selected size with the gold particles placed precisely at their centers is possible. Figure 51.6 shows three selected steps in the seeded growth of a Au—SiO_2 sample. When the particle size increases, the system becomes more monodisperse because of a reduction in the relative size distribution. This extensive growth also leads to a smoothing of the particle surface because the growth takes place through monomer addition.

SPECIAL OPTICAL AND ELECTRONIC PROPERTIES OF METAL AND SEMICONDUCTOR NANOPARTICLES

The synthesis of metallic colloids in the nanometer size range was already performed by Faraday [86], who studied for the first time gold sols in a systematic way. The great interest which this sort of systems is still raising mainly arises from the very small dimensions that can be achieved, which give properties to the particles that are very different from those of bulk metals. Several reviews have been published about the special electronic properties of metallic [87–90] and semiconductor [87,91–93] nanoparticles, which are due to a decrease in the density of states within the valence band and the conduction band when the particle size is decreased. Particles which exhibit such *quantum size effects* represent a transition state between the quasicontinuous density of states of the bulk material and the discrete energy level structure characteristic of atoms and molecules. The preparation of ultrafine particles can be facilitated greatly by careful choice of the ligands or stabilizers used to prevent particle coalescence. For example, in aqueous solution, polymeric stabilizers are very efficacious dispersants [94–98], whereas, in organic media, long chain surfactants or chemically specific ligands are most commonly used [99–107]. Alternatively, stabilization can be achieved through compartmentalization in micelles and microemulsions [108–113] or under surfactant monolayers [37,114–117], while immobilization in glasses [118,119] or sol-gels [120–122] is the preferred technique when redox reactions of the particles with the matrix need to be avoided. The surface chemistry of metallic particles largely depends on the environment where they are dispersed. More specifically, the adsorption of some ions [123–125] or other metals [126–134], the surface charge [57,135] or the optical and electronic properties of the dispersing medium [136] have a strong influence on the optical properties of the dispersions. Particles with nonspherical geometries can also be prepared [137–141], which display different optical and electronic properties as well.

The optical properties of dispersions of spherical particles can be predicted by Mie theory. This theory provides expressions for the extinction cross section of spherical particles with a frequency dependent dielectric function $\varepsilon = \varepsilon' + i\varepsilon''$, embedded in a medium of dielectric function ε_m, as [142–144]

$$C_{ext} = \frac{2\pi}{k^2}\sum(2n+1)\mathrm{Re}(a_n + b_n) \qquad (51.12)$$

where $k = 2\pi\sqrt{\varepsilon_m}/\lambda$, and a_n and b_n are the scattering coefficients, which are functions of the radius R and the wavelength λ in terms of Ricatti–Bessel functions. The extinction cross section of a particle can be normalized to give the extinction cross section per unit area,

$$Q_{ext} = \frac{C_{ext}}{\pi R^2} \qquad (51.13)$$

If we want to relate this to the extinction coefficient in units of M^{-1} cm^{-1}, we should use the following relationship:

$$\varepsilon(M^{-1}cm^{-1}) = (3 \times 10^{-3})\frac{V_m Q_{ext}}{4(2.303)R} \qquad (51.14)$$

where V_m (cm^3 mol^{-1}) is the molar volume of the material. For very small particles where $kR \ll 1$, only the first, electric dipole term in Equation (51.12) is significant, and

$$C_{ext} = \frac{24\pi^2 R^3 \varepsilon_m^{3/2}}{\pi R^2}\frac{\varepsilon''}{(\varepsilon' + 2\varepsilon_m)^2 + \varepsilon''^2} \qquad (51.15)$$

In the case of many metals, the region of absorption up to the bulk plasma frequency (in the UV) is dominated by the free electron behaviour, and the dielectric response is well described by the simple Drude model. According to this theory [145], the real and imaginary parts of the dielectric function may be written as,

$$\varepsilon' = \varepsilon^\infty - \omega_p^2/(\omega^2 + \omega_d^2) \qquad (51.16)$$

$$\varepsilon'' = \omega_p^2\omega_d/\omega(\omega^2 + \omega_d^2) \qquad (51.17)$$

where ε^∞ is the high frequency dielectric constant due to interband and core transitions and ω_p is the bulk plasma frequency,

$$\omega_p^2 = Ne^2/m\varepsilon_0 \qquad (51.18)$$

in terms of N, the concentration of free electrons in the metal, and m, the effective mass of the electron. ω_d is the relaxation or damping frequency, which is related to the mean free path of the conduction electrons, R_{bulk}, and the velocity of electrons at the Fermi energy, v_f, by

$$\omega_d = V_f/R_{bulk} \qquad (51.19)$$

When the particle radius, R, is smaller than the mean free path in the bulk metal, conduction electrons are additionally scattered by the surface, and the mean free path, R_{eff}, becomes size dependent as

$$\frac{1}{R_{eff}} = \frac{1}{R} + \frac{1}{R_{bulk}} \qquad (51.20)$$

which was verified by Kreibig for gold and silver particles [146,147]. The advantage of the Drude model is that it allows changes in the absorption spectrum to be interpreted directly in terms of the material properties of the metal. The origin of the strong color changes displayed by small particles lies in the denominator of Equation (51.15), which predicts the existence of an absorption peak when

$$\varepsilon' = -2\varepsilon_m \quad (\text{if } \varepsilon'' \text{ small}) \qquad (51.21)$$

From Equation (51.16) it can be seen that over the whole frequency regime below the bulk plasma frequency of the metal, ε' is negative which is due to the electrons oscillating out of phase with the electric field vector of the light wave. This is why metal particles display absorption spectra which are strong functions of the size parameter kR. In a small metal particle the dipole created by the electric field of the light wave sets up a surface polarization charge, which effectively acts as a restoring force for the free electrons. The net result is that, when condition (51.21) is fulfilled, the long wavelength absorption by the bulk metal is condensed into a single surface plasmon band.

In the case of semiconductor nanocrystallites, the free electron concentration is orders of magnitude smaller, even in degenerately doped materials (i.e. ω_p is smaller), and as a result surface plasmon absorption occurs in the IR, rather than in the visible part of the spectrum. Semiconductor crystallites therefore do not change colour significantly when the particle size is decreased below the wavelength of visible light, although the IR spectrum can be affected. It should be noted that the strong color changes observed when semiconductor crystallites are in the quantum size regime ($R < 50$ Å) are due to the changing electronic band structure of the crystal, which causes the dielectric function of the material itself to change.

INFLUENCE OF SILICA COATING ON THE OPTICAL PROPERTIES

Equqtion (51.15) provides the extinction cross section for spherical particles in a dielectric medium. When the particles are coated by a surface layer, the optical properties of both the core and shell materials must be considered. The extinction cross section of a concentric core-shell sphere is given by [144],

$$C_{ext} = 4\pi R^2 k^* \times \mathrm{Im}\left\{ \frac{(\varepsilon_{\mathrm{shell}} - \varepsilon_{\mathrm{m}})(\varepsilon_{\mathrm{core}} - 2\varepsilon_{\mathrm{shell}}) + (1-g)(\varepsilon_{\mathrm{core}} - \varepsilon_{\mathrm{shell}})(\varepsilon_m + 2\varepsilon_{\mathrm{shell}})}{(\varepsilon_{\mathrm{shell}} + 2\varepsilon_{\mathrm{m}})(\varepsilon_{\mathrm{core}} + 2\varepsilon_{\mathrm{shell}}) + (1-g)(\varepsilon_{\mathrm{shell}} - 2\varepsilon_m)(\varepsilon_{\mathrm{core}} - \varepsilon_{\mathrm{shell}})} \right\} \quad (51.22)$$

where $\varepsilon_{\mathrm{core}}$ is the complex dielectric function of the core material, $\varepsilon_{\mathrm{shell}}$ is that of the shell, ε_{m} is the real dielectric function of the surrounding medium, g is the volume fraction of the shell layer, and R is the radius of the coated particle. As expected, when $g = 0$, Equation (51.22) reduces to Equation (51.15) for an uncoated sphere, and for $g = 1$, Equation (51.22) yields the extinction cross section for a sphere of the shell material.

The optical properties of core-shell metal-metal particles have been studied by Henglein and coworkers [129,130,148,149] on colloids prepared by underpotential deposition of metal ions onto previously formed metallic cores. Some deposition experiments were performed with metals which do not show pronounced surface plasmon absorption because of damping by the d-d interband transitions (Pt, Pd, Pb, ...). In those cases, the measured UV-visible spectra were quite accurately reproduced by calculations using Equation (51.22). However, in the case of metals with plasmon absorption bands, like gold and silver, electronic interactions between core and shell distort the spectrum, so that agreement is not achieved between experiment and theory [131,150].

For the case of silica as a shell material, there is no risk of interactions, since silica is electronically inert (it does not exchange charge with the gold particles). However, its refractive index is different from that of gold, and also from water and ethanol. This renders $M-SiO_2$ particles model systems for the study of optical properties.

The extinction coefficients of $Au-SiO_2$ particles have been calculated systematically and compared with experimental spectra using Equation (51.22) [77]. The optical constants for gold were taken from values for evaporated gold films [151]. Values at intermediate wavelengths were calculated by interpolation, and were corrected for the effect of the small particle size on the dielectric constants for gold using Equation (51.20). These data have been shown to give good agreement for a variety of solvents [136], but there can be a slight mismatch because the position of the surface plasmon band maximum depends weakly on the particle size, and peak positions for aqueous gold sols have been reported to lie variously between 514 and 521 nm [82,108,134]. The gold sols used in reference [77] had a maximum at 518 nm, whereas calculated spectra obtained using the dielectric data of Johnson and Christy had peaks at 521.5 nm, a difference of 3.5 nm. This difference was probably due to the limited accuracy with which the dielectric function of the bulk metal can be determined. It has been shown previously that if the particles have a surface plasmon band at 520 nm initially, the shift in the peak position with solvent refractive index is accurately predicted using the dielectric data of Johnson and Christy [151]. The main effect of the discrepancy is that the peak positions were systematically observed to be at shorter wavelengths than predicted using Mie theory. For silica, a dispersionless dielectric constant was used, taken as the square of the refractive index (1.456), while for ethanol the value at 589 nm of 1.362 was adopted, since experimental values were obtained at this wavelength and this value is close to the plasmon absorption band.

The influence of the silica layer on the optical properties of the suspension is shown in Figure 51.7 for Au—SiO_2 particles dispersed in ethanol. Initially, as the shell thickness is increased, there is an increase in the intensity of the plasmon absorption band, as well as a red shift in the position of the absorption maximum. This is due to the increase in the local refractive index around the particles. However, when the silica shell is sufficiently large, scattering becomes significant, resulting in a strong increase in the absorbance at shorter wavelengths. This effect promotes a blue shift of the surface plasmon band, and a weakening in the apparent intensity of the plasmon band. Eventually at shell thicknesses above 80 nm, the scattering almost completely masks the surface plasmon band. The final colloid is very turbid

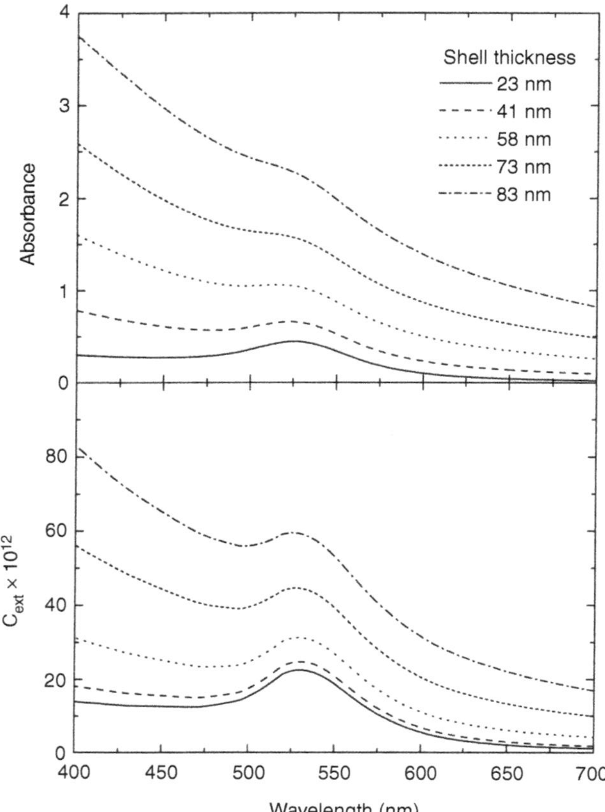

FIGURE 51.7 Influence of thick silica shells on the UV-visible spectra of ethanolic Au—SiO_2 colloids. Top: experimental; bottom: calculated by Mie theory. Adapted from reference [77]. Copyright 1996. American Chemical Society. With permission.

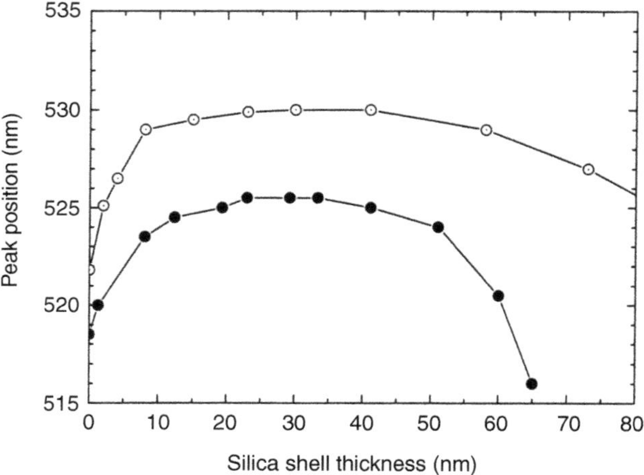

FIGURE 51.8 Variation in the position of the plasmon band of 15 nm gold particles with silica shell thickness. Solid circles are experimental data, open circles are calculated positions. Adapted from reference [77]. Copyright 1996. American Chemical Society. With permission.

and slightly pink in appearance. All these effects are accounted for by Mie theory for core-shell particles described above, as shown in Figure 51.7.

In Figure 51.8, the observed position of the maximum of the plasmon band is plotted vs shell thickness and compared to the calculated spectra. The same trends are present in the experimental and the calculated spectra, though the precise peak position is not the same. The larger scattering contribution observed experimentally is due partly to the presence of gold-free silica particles, which scatter but do not absorb.

Strong changes to the absorption spectra are also observed when the particles (after surface modification) are dispersed in ethanol/toluene mixtures of different concentrations. These mixtures cover a wide range of refractive indices (between 1.36 and 1.49), which allows optical index matching of the silica shell with the solvent. As the refractive index of the solvent approaches that of silica, the scattering decreases dramatically, and the plasmon band of gold cores becomes more pronounced. In Figure 51.9 the peak position is compared for identical

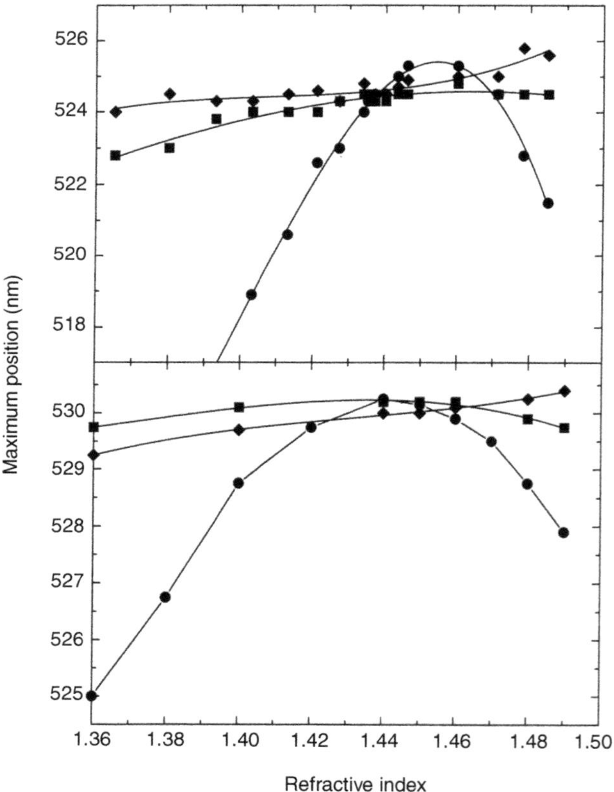

FIGURE 51.9 Variation of the plasmon band position with solvent refractive index for gold particles coated with thin (diamonds, 10 nm), intermediate (squares, 45 nm) and thick (circles, 83 nm) silica shells. Top: experimental; bottom: calculated. Adapted from reference [77]. Copyright 1996. American Chemical Society. With permission.

gold particles coated with thin (10 nm), intermediate (45 nm), and thick (83 nm) silica shells in solvents with different refractive indices. As expected, the effect is larger for particles with thicker shells, and the peak position at the match point (equal refractive index of silica and solvent) basically coincides in all three cases. The slow increase in peak position for the thin silica shell particles arises from the fact that the red shift due to the increase in solvent refractive index is larger than the blue-shift due to scattering. Predictions using Mie theory are shown in the same figure. There is again a systematic error in the predicted position of the peak, but the qualitative trend is exactly the same as for the measured values.

Silica coating allows an investigation [152] of the effect of temperature on the plasmon band of (silica coated) gold colloids in both water and ethanol, and how it directly relates to the refractive index change of the solvent. The average core diameter was 16 nm, and the average silica shell thickness was 5 nm. The silica shell was then able to confer a high stability without surface modification, even in a less polar solvent like ethanol. This means that centrifugation could be performed without loss of stability. In fact, no change was observed in the particles after the solvent exchange process, as confirmed by the slight variation in the spectrum (measured at 20°C), which almost quantitatively coincided with the calculated variation due to the increase in solvent refractive index from 1.333 to 1.362.

For the calculations, two different sets of optical constants were used for gold. The first set was taken from values for evaporated thin films [151], while the second corresponds to a single crystal [153]. The (uncoated) gold sols used had a maximum at 518.5 nm (in water), whereas the data of Johnson and Christy lead to a predicted peak at 521.5 nm, and those of Weaver at 527, a difference of 3.5 and 9.5 nm respectively.

For the coated particles used here, the maxima of the experimental spectra corresponded to 523.5 nm in water and 524.5 in ethanol, while the calculations yielded 527.5 and 528.5 respectively using data for thin films, and 531.5 and 532 using single crystal dielectric constants. We can see that the shift between measured and calculated peak positions is the same (within experimental error) for uncoated and coated particles.

A first attempt was made to measure the effect of temperature on the absorption spectra of citrate stabilized gold sols. However, at temperatures above about 40°C the stability of the dispersions was hindered, as observed by oscillations in the intensity of the measured spectra. This effect was not observed when silica coated particles were used.

Figure 51.10 shows the change in the spectra of aqueous dispersions of Au—SiO_2 particles, when increasing temperature from 14 to 70°C. It can clearly be observed that only the part of the spectra around the plasmon band (500–550 nm) varies with temperature, while no significant changes occurred at higher or lower wavelengths. The intensity around the maximum steadily decreases with increasing temperature (i.e. with decreasing refraction index). No significant variation in the position of the maximum is observed due to the very small refractive index variation (of the order of 10^{-4} °C^{-1}). Extinction coefficients calculated using the two mentioned sets of dielectric data show the same tendency, as shown in the lower part of Figure 51.10, where the intensity at the maximum is plotted versus temperature. For both sets, the experimental variation is dramatically higher than the calculated one, by a factor of ≈6.

Precisely the same trend was observed when the temperature was changed from 16 to 50°C in an alcosol of *the same particles* (after solvent exchange). However, as shown in Figure 51.11, the decrease in intensity with

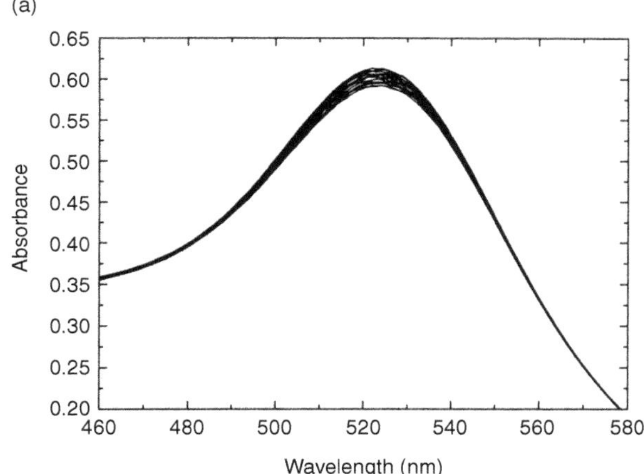

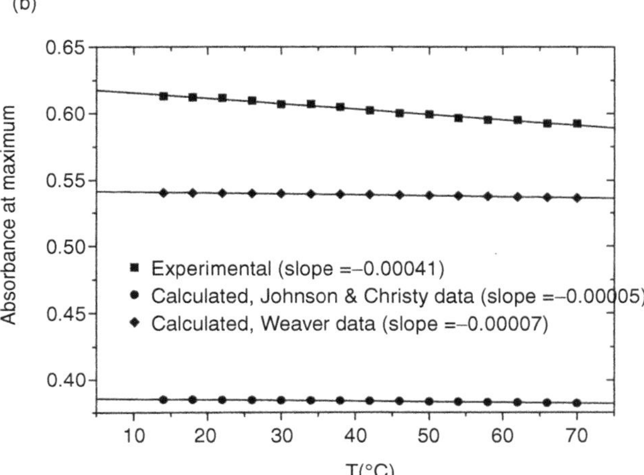

FIGURE 51.10 Influence of temperature on the UV–visible spectra of gold-silica hydrosols. (a) experimental spectra; (b) Intensity variation at maximum position for the experimental and calculated, spectra.

(a)

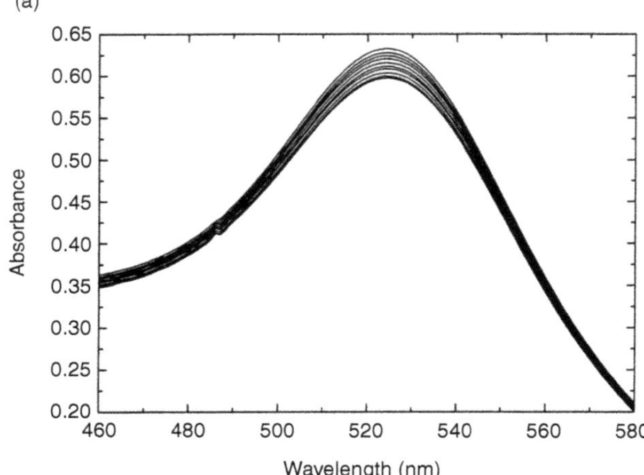

(b)

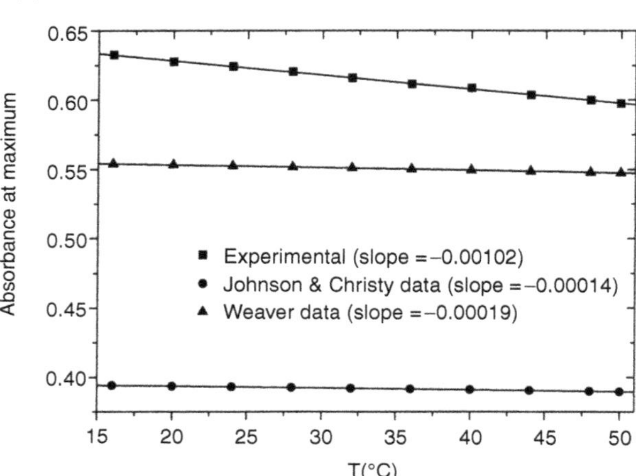

FIGURE 51.11 Influence of temperature on the UV–visible spectra of gold-silica alcosols. (a) experimental spectra; (b) Intensity variation at maximum position for the experimental and calculated spectra.

increasing temperature is noticeably faster, both in the experimental and in the calculated data. Again, the experimental spectra decrease ca. 6 times faster than the calculated ones. The decrease is about three times larger than that found with aqueous dispersions in each case, which agrees with the fact that the change in refractive index in the alcohol is also about three times larger over this temeperature range.

The influence of temperature on the optical spectra of Au nanoparticles dispersed in glass was previously measured by Doremus [118]. This author also found a decreased intensity at the peak position with increasing temperature, in an extent comparable to that shown in Figure 51.10 and Figure 51.11. Doremus attributes this shift to the change in the surface electron concentration on the particles due to thermal expansion of gold.

A different reasoning for the observed enhanced variation in the absorption spectra with respect to those predicted by refractive index variation is the mechanism of thermal line broadening due to shape vibrations of the particles [154], which predicts a temperature dependence of the width proportional to $T^{1/2}$.

It should be noted however, that the calculations did not include changes to the dielectric constants of the metal with temperature. Kreibig has previously investigated [155] the change in peak intensity for silver clusters over a much wider temperature range and found that $\varepsilon_2(T = 300\ K) - \varepsilon_2(T = 1.6\ K) = 0.07/a$ (nm) at the peak. This is a gradient of $d\varepsilon_2/dT = 2.1 \times 10^{-4}\ K^{-1}$ for $a = 10$ nm. For gold, the shift is much higher, $\varepsilon_2(300) - \varepsilon_2(1.6\ K) = 0.4$ for our colloid particles or about 0.07 over 50°C. Such an increase might be enough to explain our data. It is difficult to separate out the effects on the plasmon resonance from the effect of temperature on the interband transitions. Accurate calculations should thus take this effect into account.

2D AND 3D ASSEMBLIES OF CORE-SHELL PARTICLES

As shown in the introduction, one of the most interesting applications of metal and semiconductor nanoparticles is the formation of nanostructured materials. Colloidal silica is almost unique in its ability to naturally form both 2D and 3D arrays. This stems from short range water structuring or gel layers that form at the silica water interface. The existence of this repulsive steric interaction is sensitive to the cations present in the colloid solution during concentration of the particles. Tetraalkyl ammonium cations for example afford almost complete protection against coalescence into a van der Waals primary minimum [59]. This suggests that silica coated nanoparticles would be more readily structured than surfactant stabilized materials in which limited coating thickness, chain melting and chemical oxidation all conspire to limit flexibility and application. Furthermore, silica coated particles open broader possibilities over the ones stabilized by organic moieties for controlling the structure of ordered 2D and 3D systems, due to the possibility of finer surface modification and variation of the thickness of the silica shell. Particulate films from silica coated nanoparticles can in principle be prepared by many different ways including self-assembly, Langmuir–Blodget (LB) technology, electrodeposition and spin coating. Silica coated nanoparticles have the advantage that they can be modified with appropriate sylylating agents [62–65] to make the particles marginally hydrophobic. Upon spreading on water surfaces, such particles do not sink down into the aqueous subphase but stay on the surface. It is essential to attain a proper hydrophilic/hydrophobic balance of the particle surfaces because too high a hydrophobicity can cause surface agglomeration and formation of disordered multilayers. After compression on the water surface, the nanoparticulate film can be transferred onto a solid substrate by means of LB deposition.

The LB technique provides the highest degree of order among methods of film preparation. Nevertheless, other techniques, particularly self-assembly and spin-coating, can be used as well due to their experimental simplicity and technological importance.

Results have been reported on the preparation of monoparticulate films composed of $M-SiO_2$ particles. Self-assembly tested on silicon wafers (see Figure 51.12) yielded much better packed films in the presence of a cationic polymer, such as polydimethyldiallyl-ammonium bromide [156] while in any case the films obtained are rather disordered. It is important however that packed films can be obtained from particles with varying particle sizes (shell thicknesses). This shows that intercorrespacing can indeed be tailored through the control of silica shell thickness. The acquisition of optical measurements as a function of shell thickness has been reported [156].

On the other hand, in Figure 51.13 we show a comparison of AFM and TEM images of ordered 2D nanoparticle clusters of $Au—SiO_2$ particles. The 15 nm gold particles were coated with about 80 nm silica. They were deposited electrophoretically onto TEM grids. As can be seen, the two images are remarkably similar, with the prime differences being that (a) there are edge effects due to the finite tip angle [157] and the defects such as missing particles and channels disappear due to swelling of the particle images in the AFM.

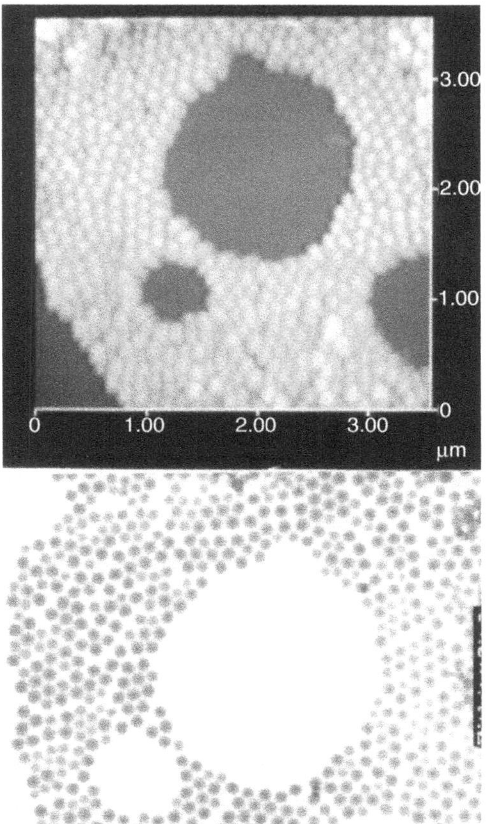

FIGURE 51.13 AFM image (left) and TEM image (right) of $Au-SiO_2$ clusters on a TEM grid. Fourier spectra of the cluster give the identical lattice constant for both images. Thus, tip artifacts in AFM, which broaden the particle image, do not destroy the information on the particle-particle pair distribution function.

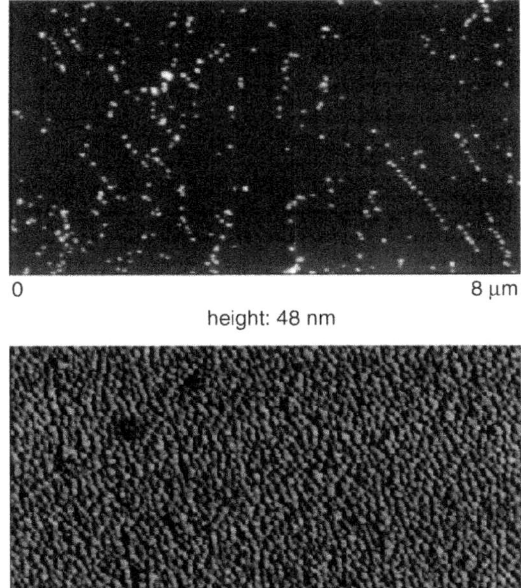

FIGURE 51.12 AFM images of the surface of self-assembled thin films of $Au-SiO_2$ particles (core size 15 nm, total size 35 nm). Bottom: adsorbed to a bare silicon substrate; Top: onto silicon with polydimethyldiallylammonium bromide.

The preparation of 3D (ordered) assemblies of the same sort of particles can also be accomplished by playing around with surface properties, which determine the interparticle interactions, and (see Introduction) the colloid phase diagram.

Concentrated dispersions have been prepared [158] of samples with relatively thick silica shells around the gold cores (core diameter = 15 nm, total diameter = 184 nm), and with the outer surface coated with 3-(trimethoxysilyl) propylmethacrylate (TPM). The TPM coated particles (which we denote $Au—SiO_2$-TPM) are slightly charged due to the presence of residual hydroxide ions on the silica surface [63], and the sols are stable in solvents with a low polarity, such as ethanol and toluene, as well as in ethanol-toluene mixtures. For TPM coated silica spheres, [159] the fluid-solid phase transition has been observed at a particle volume fraction of $\phi = 0.194$, and the crystal-glass transition at $\phi = 0.224$.

Concentrated dispersions of the Au–SiO_2-TPM particles were prepared by centrifugation in a 1 mm path length glass cuvette at low speed (3000 rpm), to prevent irreversible particle aggregation.

With similar experiments performed on gold-free TPM-silica spheres, van Duijneveldt et al. [160] demonstrated using light diffraction experiments that a concentration gradient is obtained within the sample due to slow upward diffusion of particles within the initial sediment. It was shown that the top of the sample can rearrange to a crystalline structure on a time scale of one week, while the rest of the sample remains in the amorphous state for months.

Since the gold cores are too small to influence the particle density or to affect the van der Waals interactions between the silica coated particles, a similar phase behaviour is expected for these core-shell particles. Note however that for silica shells of less than 10 nm thickness, agglomeration would be expected due to dispersion forces between the gold cores.

The changing optical properties can be used to monitor changes in the phase behavior of our colloids. To this end, a *color phase diagram* can be constructed. This was achieved by repeatedly centrifuging the sample in the same 1 mm path length cuvette, removing a small fraction of the supernatant, and redispersing by means of a vortex mixer. The optical properties of these samples have been characterized through reflectance spectra, since their absorbance is too high for a standard spectrophotometer. A summary of the results for volume fractions ranging from 0.10 to 0.40 is shown in Figure 51.14. While the fluid phase is turbid reddish-brown, as indicated by a broad band at high wavelengths, the crystalline phase is bright red, showing Bragg diffraction in the form of bright spots (iridescence) all over the sample, which is observed in the reflectance spectrum as intense, narrow bands centered between 580 and 600 nm depending on concentration (16–19%). At higher concentrations, the enhanced absorption at longer wavelengths adds a browner tinge to the sol colour in transmitted light. When the cuvette is tilted, a bright green colour is observed over the whole sample. This colour is complementary to the red-brown in transmitted light and is attributed to coherent reflection of light by the densely packed gold particles. Glasses impregnated with concentrated colloidal gold also show this sharp change as they are viewed alternatively with transmitted or reflected light [161]. In the amorphous part of the phase diagram, the sample displays different colors ranging from dark yellow through green to blue, as indicated by the bands

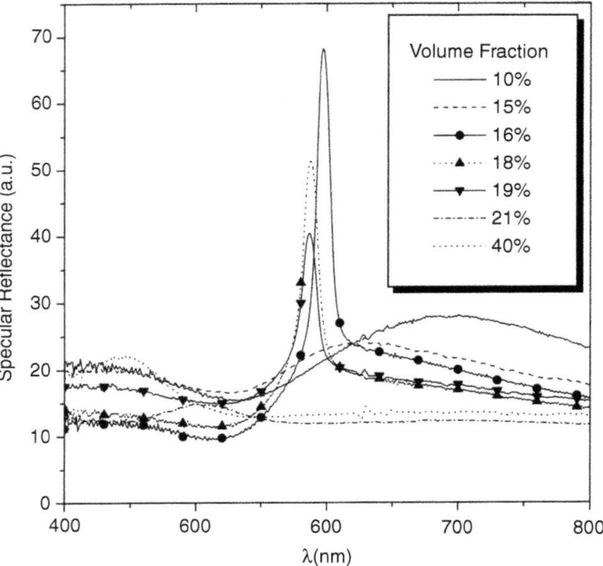

FIGURE 51.14 Specular reflectance spectra of a concentrated Au–SiO_2 dispersion with increasing volume fractions from 0.15 up to 0.40.

at lower wavelengths. These colors result from direct reflection of incident white light, while the sample invariably looks dark red when observed with transmitted light.

In principle, two mechanisms can be proposed to explain this optical behavior. On the one hand, Maxwell–Garnett theory states that if the volume fraction of particles is high, then the particles experience an average dielectric constant given by,

$$\varepsilon(\text{eff}) = \phi\varepsilon\ (\text{medium}) + (1 - \phi)\varepsilon\ (\text{particle}) \quad (51.23)$$

where ϕ is the volume fraction. So in a concentrated dispersion, the plasmon band shifts to longer wavelengths. At a critical point, [47] the sample reflects light like a bulk sample, because of interactions between the scattered waves from the individual dipoles (or particles).

On the other hand, the light scattering for dilute sols is given accurately by the Rayleigh formula, and shows just a monotonic rise into the ultraviolet. However, if the sol crystallizes, there will be enhanced light scattering at particular wavelengths (Bragg diffraction), and this can drastically alter the perceived colloid colour.

In order to establish whether the gold core or the silica shell is primarily responsible for the observed colour changes, the spectra were measured at three different volume fractions. The first sample was a dispersion of Au—SiO_2—TPM particles in ethanol, the second was a dispersion of the same particles in a mixture of ethanol and toluene (40/60 v/v; refractive index very close to that of silica), and the third sample was a dispersion of TPM-silica particles (average radius = 90 nm) in ethanol. The UV-visible spectra of the index matched sample shows simply a progressive increase in absorbance due to the increasing particle concentration, but hardly any variation in the band position or shape. For such large silica shell thicknesses, there is little plasmon coupling perceived even at high volume fractions. In the case of the pure silica sample, at particle concentrations larger than 20%, structuring is observed due to the liquid-solid phase transition. The spectra of the Au—SiO_2—TPM sample are simply a combination of the other two, confirming the fact that the scattering by the silica shell and the core gold metal absorption are independent processes which both contribute to the observed sample color. This must be a consequence of the structuring of the gold cores within the sample.

The nanosized gold core acts as an optical label for the silica sols allowing, through surface plasmon modulation, particle interactions to be monitored spectroscopically. In particular spectral shifts during crystallization provide an interesting method for observing phase separation and colloid crystal nucleation.

Finally, it has been observed that when these concentrated dispersions are allowed to dry in air, a bright red solid product is formed, which easily racks. SEM and AFM images of such dried colloids reveal extremely regular, large 3D Au–SiO_2 crystals (Figure 51.15 and Figure 51.16). In this case, the spheres are in contact, and the interparticle spacing is about 140 nm. (Notice that when silica particles are dried they shrink due to their large porosity, so that EM sizes are consistently smaller than sizes in solution, which have been measured by dynamic light scattering.) There is no evidence for optical diffraction effects from this sample, since the spacing is below visible wavelength, and the colour is similar to that of the particles in a concentrated dispersion.

FIGURE 51.15 SEM micrograph of a 3D array of Au–SiO_2 particles prepared by slowly drying a concentrated alcosol.

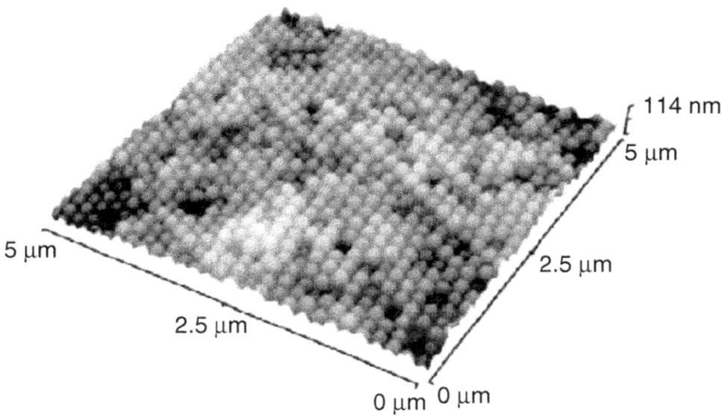

FIGURE 51.16 AFM image of the top surface of the same ordered 3D array of Au–SiO_2 particles shown in Figure 51.15.

DAMPING OF CHEMICAL REACTIONS THROUGH SILICA COATING

We have introduced earlier that the use of silica shells around nanoparticles as a general stabilization technique against flocculation in several media. Apart from this, the silica shells can also be envisaged as *chemical stabilizers*, meaning that they could slow down or even prevent chemical reactions from occurring at the cores.

The question which arises is whether the cores can be accessed by reactants in the surrounding solution or whether they are rendered chemically inert by silica deposition. Extensive studies by Lecloux et al. [162] and by van Blaaderen et al. [163,164] on the properties of silica particles grown using TES have clearly demonstrated that the particles are significantly less dense than bulk silica [162], with micropores ranging in size from 2 nm up to 50 nm. According to this, pore size could easily be the determining factor for the accessibility of reactant moieties to the cores.

Giersig et al. have examined [165] the effect of molecular iodine, a strong oxidant, on the absorption spectrum of Ag–SiO_2 particles. Molecular iodine oxidizes silver according to

$$2Ag + I_2 \rightarrow 2AgI \qquad (51.24)$$

In Figure 51.17, we show the spectra before and after the addition of molecular I_2 to a dispersion of Ag–SiO_2 particles with a shell of thickness 10 nm, and initial Ag cores of 9 nm diameter. As can be seen, the surface plasmon band of the silver core located at 390 nm disappears, and after 30 min the spectrum reveals the typical 420 nm peak of colloidal β-AgI. Calculated spectra of

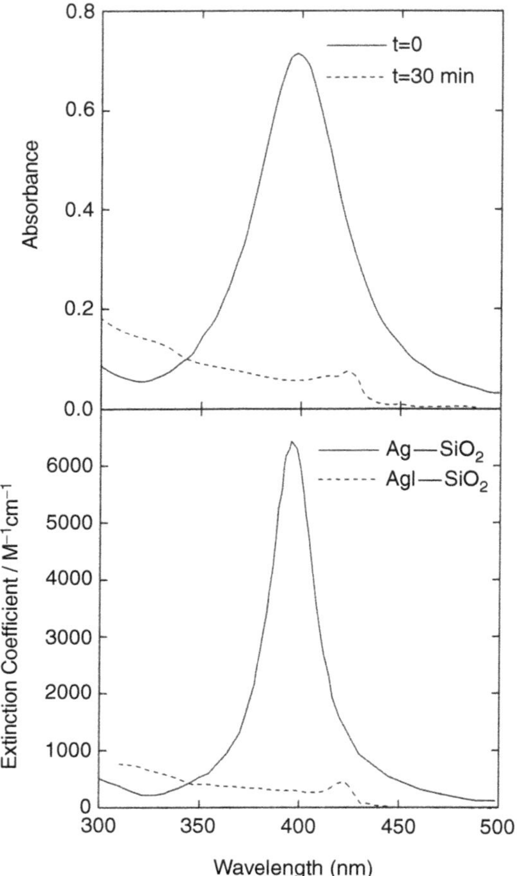

FIGURE 51.17 Top: Absorbance spectrum of Ag—SiO_2 (particles 9 nm core with 10 nm silica shells) before and after admission of 2 mM I2. (b, bottom) The calculated Mie spectra of 8 nm Ag and AgI spheres embedded in 4 nm thick silica shells in water (n = 1.33). Adapted from reference [165]. Copyright 1996, VCH. With permission.

Ag—SiO_2 and AgI—SiO_2 in water (core diameter 8 nm, total diameter 16 nm) are shown as well. The excellent agreement confirms that the observed absorption changes are due to metallic silver oxidation and formation of colloidal AgI, and demonstrates that molecular iodine can diffuse through the 10 nm silica shell and completely oxidize the silver cores. The fundamental mechanistic problem is that the molar volume of the hexagonal β-AgI phase is 41 $cm^3\ mol^{-1}$, whilst that of face-centerd cubic silver is just 10.26 $cm^3\ mol^{-1}$. Thus the core volume must quadruple in size in order to accommodate the nascent AgI nucleus. There appear to be only two ways for such a drastic volume increase to be realized. Either the silica shell has an enormous porosity and can accommodate the silver iodide as a myriad nanoporous filaments, or more catastrophically, the silica shell must undergo violent rupture when iodine is added to the solution.

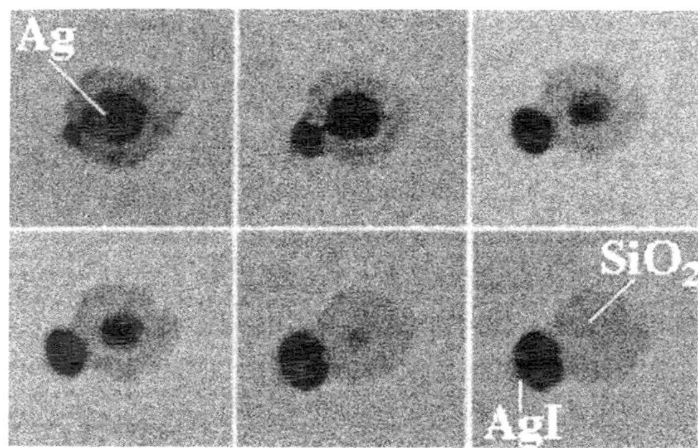

FIGURE 51.18 Electron micrographs of a single Ag–SiO_2 particle as a function of the time after exposure to I_2. Silica shell thickness 10.0 nm. Core diameter 9.0 nm. Reprinted from ref. 165. Copyright 1996, VCH.

In order to capture the dynamics of this nanoscale physical transformation, the kinetics of core corrosion was monitored directly by following the reaction of *one* Ag—SiO_2 particle in the TEM [165]. The dramatic results are shown in Figure 51.18 through a series of images resulting from iodine corrosion of a single Ag–SiO_2 particle. In the absence of iodine, we resolve the characteristic [111] lattice planes of fcc silver metal with $a = 2.36$ Å, and the amorphous silica shell (see Figure 51.5). After contact with I_2, a small nucleus erupts onto the surface of the silica shell and then expands. At the same time, the silver core is seen to shrink, and eventually disappears altogether. Note that only a single nucleus is observed on the silica surface. By the completion of the reaction, a single particle can be seen adsorbed to the exterior of a silica shell. Spectroscopic monitoring of the reaction in solution at 390 nm showed that the reaction lasted 20–25 minutes, slightly longer than the time found by direct TEM analysis. Conversely, the reaction of uncoated silver particles with iodine was completed within several seconds. In the final image, the silica core is seen to have collapsed. Extensive electron beam irradiation of hollow nanosilica shells was frequently observed to cause condensation of the core. The whole silica particle is apparently involved in the implosion, since the diameter of the silica shell afterwards is always clearly smaller than before the beam induced compression. The final product particle exhibits an increased lattice spacing of 3.4 Å, and a change in crystal symmetry from cubic to hexagonal. Based on this, we can unambiguously assign the observed lattice lines to the [110] plane of hexagonal AgI. Electron irradiated Ag—SiO_2 was quite stable in the absence of iodine. From the TEM images, it is clear that large volume changes occur during the reaction. The absorbed AgI particle has an apparent diameter of 17.5 nm. This compares with a silver core of just 9.0 nm diameter. If we assume that the AgI is roughly hemispherical, then its volume is indeed about 4 times that of the core as predicted from the respective molar volumes.

Based on TEM observations of several hundred such corroded particles, we find that about 90% of Ag—SiO_2 particles appear to react by core-to-surface transport and subsequent surface nucleation to form AgI% SiO_2 (the % is used to indicate exterior surface growth, rather than inclusion within, SiO_2). In about 10% of cases, AgI—SiO_2 was formed. However in some few cases, the entire silica shell was found to have disintegrated and totally uncoated AgI particles resulted. Clearly, the chemical reactivity of the silver core toward I_2 is sufficient to drive metal oxidation despite the necessity for large pressure increases within the core.

Whilst surface nucleation of AgI clearly occurs, the mechanism itself is less obvious. We see from Figure 51.18 that the initial phase of the reaction results in rapid formation of a thick filament of AgI extending from the core to the surface. By the completion of the core corrosion process, there is no longer evidence for a bridging filament of AgI between the surface and the core. AgI fibers with thicknesses >10 Å would have been detectable with our instrument. Instead we suggest the following pore transport model, based on the known semiconducting properties of AgI. In the first stage, molecular iodine diffuses through the pores to the silver surface where it begins to corrode the metal surface. Since the volume of the corroded core is larger than the untarnished Ag, the pore rapidly begins to fill with AgI. The expansion exerts strong pressure on the pore, causing it to dilate, or even to rupture, which in turn facilitates the access of molecular iodine to the silver particle surface. The initial AgI nucleation must occur at the silver particle surface. Once AgI has formed, the system becomes a three phase one, with iodine on the outer surface and a reservoir of silver metal on the inner surface. Though AgI is primarily an ionic conductor [166], there is a measurable contribution from electronic charge carriers. Thus electrons can be transported through the AgI layer, and the AgI will not act as a passivation layer on the silver surface. In fact, it is well known that silver metal deposited onto silver halide crystals spontaneously diffuse into the crystal [167,168]. The silver particle can thus act as an electron source for the growing AgI nucleus. As AgI fills the small access pores, both electrons and silver ions (or silver atoms) diffuse through the AgI lattice to the AgI/SiO_2 pore interface thereby facilitating reaction. As soon as one pore fills up completely, continued corrosion on the external silica surface is possible and subsequent reaction then occurs preferentially at this point. Therefore surface AgI formation is inevitable if silver metal can migrate through the nascent AgI forming in the pores and access the external solution. According to this simple model, the AgI actually forms at the surface and not in the core at all. The AgI phase that emerges onto the external silica surface grows by spherical diffusion in the bulk solution, where the I_2 concentration is much higher, and diffusion is unimpeded. This may explain the hemispherical particle shape adopted by many of the AgI particles. It is apparent that such complex morphology changes could not have been predicted from the spectroscopic data shown in Figure 51.17 alone.

The physiochemical effects in the case of negative core volume changes have also been studied using cyanide ion to drive the aerial oxidation of gold [165] and silver [78]:

$$4M + 8CN^- + O_2 + 8H^+ \longrightarrow 4M(CN)_2^- + H_2O \quad (51.25)$$

where M stands for Au or Ag. Note that the dissolution of the core requires the transport through the silica shell of both molecular oxygen and cyanide ions, and for complete dissolution outward diffusion of metal cyanide anions.

As an example, the decrease in absorption of light is shown in Figure 51.19 as the Au—SiO_2 dispersion changes from red to colorless. The inset of the same figure shows the full-width at half-maximum (FWHM) of the gold plasmon band. The band width increases during the first stage of dissolution, remaining nearly constant afterwards. The surface plasmon band width increases with decreasing particle diameter, so this broadening is consistent with a gradual dissolution of the particles.

The actual kinetics of colloidal metal oxidation depends strongly on the preparation conditions used for the silica deposition. In Figure 51.20, we show the

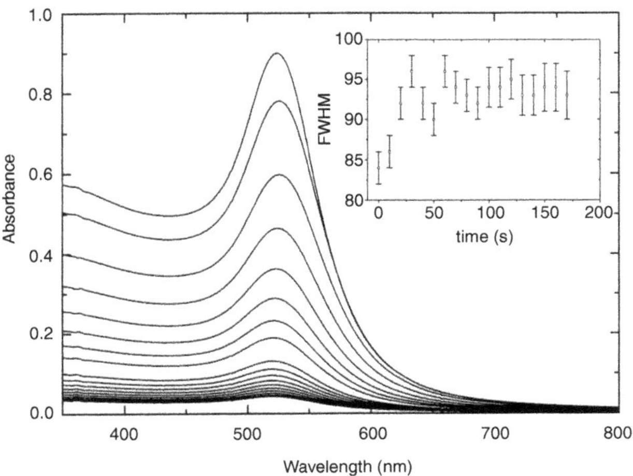

FIGURE 51.19 Evolution of the UV–visible spectrumn of an Au—SiO_2 0.5 mM sol after addition of 1 mM KCN. The time interval between consecutive spectra is 10 sec. The inset is a plot of the half width at middle height vs. time.

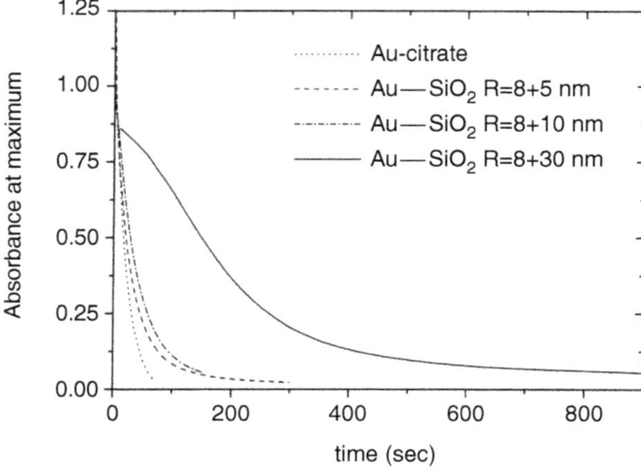

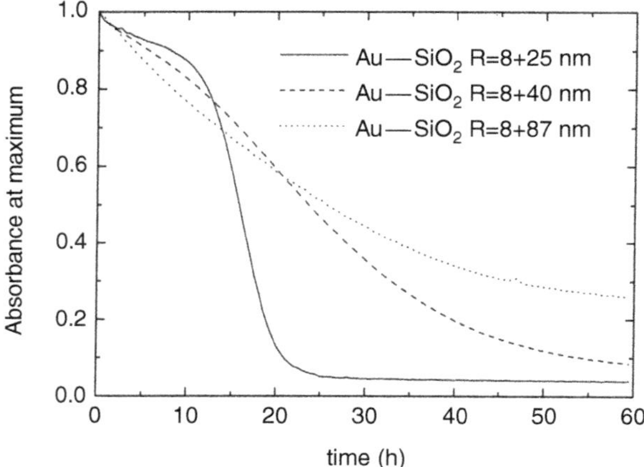

FIGURE 51.20 Time traces of the absorbance at 525 nm of Au—SiO_2 sols with increasing shell thickness. Upper plot: shells formed from silicate deposition; lower plot: shells formed by growth with TES.

effects of various silica deposition routes on the eventual rate of dissolution. The rate of fastest dissolution is observed for uncoated, citrate stabilized sols, which are found to be oxidized very quickly by cyanide ion. Silica shells grown by silicate deposition in water promote a thickness dependent retardation of the reaction, while the effect of those grown by TES hydrolysis is much more dramatic. This can easily be related to the pore size of the different silicas. Sodium silicate forms oligomers in solution, which condense to form silica, yielding relatively large pores, whilst TES silica is formed from the condensation of monomers arising from the slow hydrolysis of TES. It has been observed that boiling of the M—SiO_2 sol also leads to a slowing down of the reaction, presumably due to compression of the silica shell through pore shrinkage.

In Figure 51.21, we show an electron micrograph of hollow silica particles obtained after exposure of the coated particles to 1 mM cyanide ion at pH 10.5.

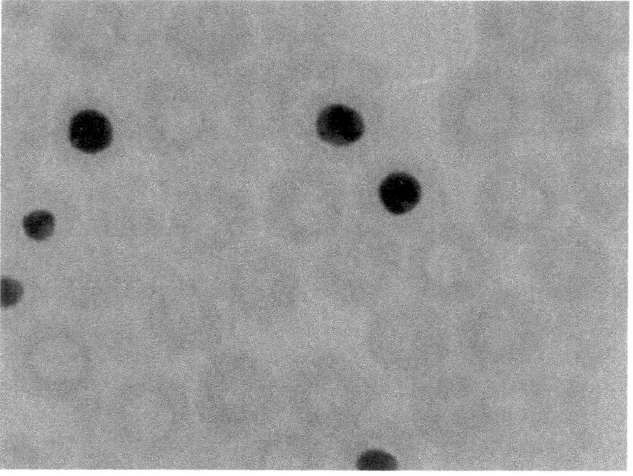

FIGURE 51.21 Electron micrograph of Au—SiO_2 (15 nm core, 10 nm shell) after exposure to 1 mM KCN at pH 10.5 in air. Hollow silica shells are obtained through complete core dissolution. Some non-dissolved cores are shown as well.

Because the initial particles were quite homogeneous in size, the cores dissolved at similar rates, indicating relatively similar pore structuring in the individual gold particles [165]. After 10 minutes, the gold cores were completely dissolved, and hollow nanosized silica shells remained.

We have also examined [78] the conversion of silver cores into the more noble metal gold via oxidation with $AuCl_4^-$. This process was expected to be quite facile given the rapid kinetics for cyanide oxidation. However the deposition of gold within the core via:

$$AuCl_4^- + 3Ag \longrightarrow Au + 3AgCl + Cl^- \qquad (51.26)$$

was found to be very sluggish. It seems likely that the AgCl formed precipitates out within the core volume and passivates the silver surface, or that the large size of the tetrachloroaurate ion reduces its diffusion through the pore structure of the silica shell. With uncoated particles, complete oxidation is observed after some 48 hours with the colloid changing from yellow to red, and the surface plasmon band of the colloidal silver shifting from 395 nm to 530 nm. When the silver is silica coated, the kinetics are much slower. After 4 days a weak emergent surface plasmon resonance is observable at 495 nm, but the silver band is not completely absent.

Other reactions, such as the oxidation of Ag by sulfide ions to form Ag_2S, or its dissolution by ammonia to form a soluble $[Ag(NH_3)_2]^+$ complex, leading to hollow shells have been studied as well [78], which again support the contention that the silica shells are porous, though the pore size can be decreased by prolonged boiling, thus practically sealing the cores.

The inhibition of one last reaction was studied, which has practical applications with respect to the stabilization of CdS semiconductor particles by means of silica coating [81,169].

CdS can be degraded under the influence of light in the presence of dissolved oxygen. The action of oxygen consists of an oxidation of sulfide radicals arising through the formation of hole-anion pairs at the particle surface [170]:

$$S_S^{\bullet -} + O_2 \longrightarrow S_S + O_2^{\bullet -} \tag{51.27}$$

where the subscript S indicates a surface atom.

This process can be inhibited through the addition of excess sulfide ions, though an alternative and cleaner way is to carry out the reaction under nitrogen atmosphere. When the synthesis is performed in the absence of oxygen, the colloid is stable for weeks, allowing for modifications to be performed on it without further precautions.

The influence of silica coating on the stability of CdS colloids with respect to photodegradation was monitored through the temporal decay of the UV-visible spectrum (see Figure 51.22).

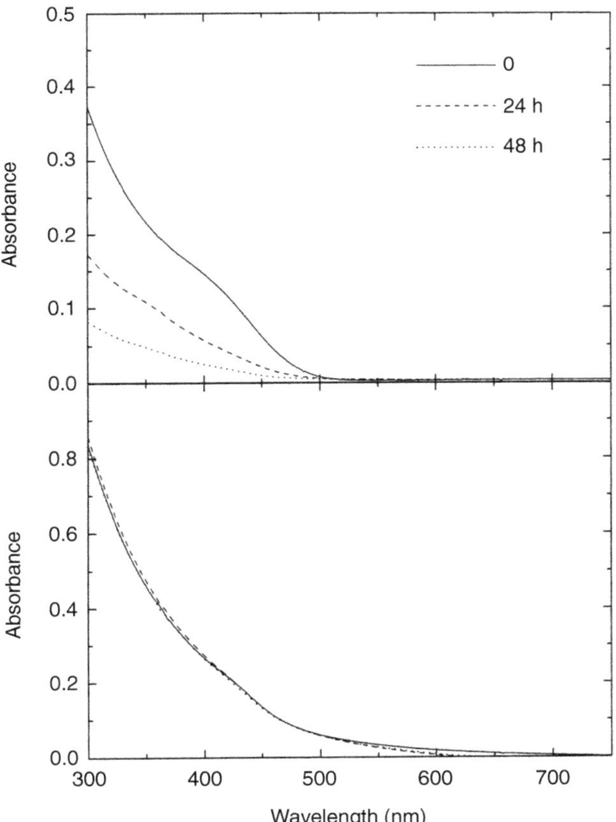

FIGURE 51.22 Time evolution of the absorbance spectra of CdS sols. Upper plot: citrate stabilized; lower plot: coated with a 20 nm thick silica shell.

Silica coated sols proved to be far more stable against photodegradation than the uncoated ones, so that samples could be stored in light for several months with negligible variation of their absorption spectra. This can be related to the very small pore size of silica prepared by means of Stöber synthesis, which implies growth through monomer addition. This makes it very difficult for O_2 molecules to reach the particle surface, thus strongly reducing the oxidation process. Such a protection will be of great importance in the preparation of nanostructured materials with uniformly distributed semiconductor cores.

SUMMARY

This paper demonstrates the main advantages of the modification of metal and semiconductor nanoparticles by means of silica coating.

First of all, silica coating provides very effective protection against flocculation, which is a fundamental requirement when concentrated dispersions are to be prepared. The characteristic properties of silica surfaces provide enhanced stability in aqueous dispersions, and allow for more elaborate surface modifications which can ultimately lead to the preparation of colloids readily dispersed in a wide range of organic solvents.

Through the silica coating procedure described, the optical properties of both dilute and concentrated dispersions of metal nanoparticles can be modulated. The influence of the silica shell on the optical properties has been described in detail, and shown to promote noticeable shifts of the plasmon band position in dilute systems, while coupling absorption and scattering in concentrated systems, which gives rise to rich, concentration-dependent optical behavior.

The assembly of such core-shell particles, both in two and three dimensions, further enables tailored complex structures to be created with new properties. The formation of even more complex structures can be accomplished by making use of the semipermeable nature of the silica shell. The slow diffusion of reagents through the shell allows selective core etching and chemical conversion to be effected. The kinetics of such reactions are controlled by the porosity and the thickness of the shell, so that the cores can be made resistant to almost any chemical reagent.

ACKNOWLEDGMENTS

The following persons are acknowledged for their contributions to the work presented here. T. Ung (Ag—SiO_2), M.A. Correa-Duarte (CdS—SiO_2), N.A. Kotov (2D arrays), M. Giersig (HRTEM), C. Serra (AFM, SEM).

REFERENCES

1. Overbeek, J.Th.G. 1982. *Adv. Colloid Interface Sci.* 16,17.
2. Israelachvili, J.N. 1985. *Intermolecular and Surface Forces*, Academic Press, London.
3. Visser, J. 1972. *Adv. Colloid Interface Sci.* 3, 311.
4. Verwey, E.J.W. & Overbeek, J.Th.G. 1948. Theory of the Stability of Lyophobic Colloids. Elsevier, Amsterdam.
5. Derjaguin, B.V. & Landau, L. 1941. *Acta Physicochim. URSS* 14, 633.
6. Sato, T. & Ruch, R. 1980. *Stabilization of Colloidal Dispersions by Polymer Adsorption*. Marcel Dekker inc., New York.
7. Mandelbrot, B.B. 1983. *The Fractal Geometry of Nature*. Freeman, New York.
8. Weitz, D.A.; Lin, M.Y.; Huang, J.S.; Witten, T.A.; Sinha, S.K. & Gethner, J.S. 1981. In Pynn, R. & Skjeltorp, A., (Editors), *Scaling Phenomena in Disordered Systems*. Pleum Press, New York, p. 171.
9. Sanders, J.V. 1964. *Nature* 204, 1151.
10. Hiltner, P.A. & Krieger, I.M. 1969. *J. Phys. Chem.* 73, 2386.
11. Okamoto, S. & Hachisu, S. 1977. *J. Colloid Interface Sci.* 62, 172.
12. Philipse, A.P. 1989. *J. Mater. Sci. Lett.* 8, 1371.
13. Tse, A.S.; Wu, Z. & Asher, S.A. 1995. *Macromolecules* 28, 6533–6538.
14. Micheletto, R.; Fukuda, H.; Ohtsu, M. 1995. *Langmuir* 11, 3333.
15. Ottewill, R.H. 1989. *Langmuir* 5, 4.
16. Arora, A.K. & Tata, B.V.R. (Editors). 1996. *Ordering and Phase Transitions in charged Colloids*. VCH, New York.
17. Sogami, I. & Ise, N. 1984. *J. Chem. Phys.* 81, 6320.
18. Ise, N.; Okubo, T.; Sugimura, T.; Ito, K. & Nolte, H.J. 1983. *J. Chem. Phys.* 78, 536.
19. Overbeek, J.Th.G. 1987. *J. Chem. Phys.* 87, 4406.
20. Woodward, C.E. 1988. *J. Chem. Phys.* 89, 5140.
21. Onsager, L. 1933. *Chem. Rev.* 13, 73.
22. Pusey, P.N. 1991. In Hansen, J.P.; Levesque, D. & Zinn-Justin, J. (Editors). *Liquids, Freezing, and Glass Transitions*. Elsevier, Amsterdam, pp. 763–942.
23. Pusey, P.N. & Van Megen, W. 1986. Nature 320, 340.
24. Smits, C.; Briels, W.J.; Dhont, J.K.G. & Lekkerkerker, H.N.W. 1989. *Progr. Colloid Polym. Sci.* 79, 287.
25. Imhof, A.; van Blaaderen, A. & Dhont, J.K.G. 1994. *Langmuir* 10, 3477.
26. Giersig, M. & Mulvaney, P. 1993. Langmuir, 9, 3408.
27. Brust, M.; Walker, M.; Bethell, D.; Schiffrin, D.J. & Whyman, R. 1994. *J. Chem. Soc., Chem. Commun.* 801.
28. Whetten, R.L.; Khory, J.T.; Alvarez, M.M.; Murthy, S.; Vezmar, I.; Wang, Z.L.; Stephens, P.W.; Cleveland, C.L.; Luedtke, W.D. & Landman, U. 1996. *Adv. Mater* 8, 428.
29. Fojtik, A.; Weller, H.; Koch, U. & Henglein, A. 1984. *Ber. Bunsenges. Phys. Chem.* 88, 969.
30. Resch, U.; Eychmüller, A.; Haase, M. & Weller, H. 1992. *Langmuir*, 8, 2215.
31. Lee, G.S.H.; Craig, D.C.; Ma, I.; Scudder, M.; Bailey, T.D. & Dance, I.G. 1988. *J. Am Chem. Soc.* 110, 4863.
32. Perrin, J. 1909. *Ann. Chim. Phys.* 18, 5.
33. Pieranski, P., Strzelecki, L. & Pansu, B. 1983. *Phys. Rev. Lett.* 50, 900.
34. Murray, C.A. & van Winkle, D.H. 1987, 58, 1200.
35. Dimitrov, A.S. & Nagayama, K. 1996. *Langmuir*, 12, 1303.
36. Dimitrov, A.S.; Dushkin, C.D.; Yoshimura, H. & Nagayama, K. 1994. *Langmuir*, 10, 432.
37. Kotov, N.A.; Meldrum, F.C.; Wu, C. & Fendler, J.H. 1994. *J. Phys. Chem.* 98, 2735.
38. Meldrum, F.C.; Kotov, N.A. & Fendler, J.H. 1994. *J. Phys. Chem.* 98, 4506.
39. Kotov, N.A.; Meldrum, F.C.; Wu, C. & Fendler, J.H. 1994. *J. Phys. Chem.* 98, 8827.
40. Matsumoto, M.; Tsujii, Y.; K. Nakamura, K. & Yoshimoto, T. 1996. *Thin Solid Films* 280, 238.
41. Yoshimura, H.; Matsumoto, M.; Endo, S. & Nagayama, K. 1990. *Ultramicroscopy* 32, 265.
42. Colvin, V.L.; Goldstein, A.N. & Alivisatos, A. P. 1992. *J. Am. Chem. Soc.* 114, 5221.
43. Dabbousi, B.O.; Murray, C.B.; Rubner, M.F. & Bawendi, M.G. 1994. *Chem. Mater.* 6, 216.
44. Bentzon, M.D.; v. Wonterghem, J. & Thölén, A. 1987. Proc. 45th Ann. Meeting of Elec. Micros. Soc. of America, p. 360.
45. Motte, L.; Billoudet, F. & Pileni, M.P. 1995. *J. Phys. Chem.* 99, 16425.
46. Böhmer, M., 1997. *Langmuir* 12, 5747.
47. Dusemund, B.; Hoffmann, A.; Salzmann, T.; Kreibig, U. & Schmid, G. 1991. *Z. Phys. D* 20, 305.
48. Weller, H. 1996. Angew. Chem. Int. Ed. Engl. 35, 1079.
49. Schon, G. & Simon, U. 1995. *Colloid Polym. Sci.* 273, 101.
50. Schon, G. & Simon, U. 1995. *Colloid Polym. Sci.* 273, 202.
51. Terrill, R.H.; Postlethwaite, T.A.; Chen, C.; Poon, C.-D.; Terzis, A.; Chen, A.; Hutchison, J.E.; Clark, M.R.; Wignall, G.; Londono, J.D.; Superfine, R.; Falvo, M.; Johnson, C.S. Jr.; Samulski, E.T. & Murray, R.W. 1995. *J. Am. Chem. Soc.* 117, 12537.
52. Surridge, N.A.; Zvanut, M.E.; Keene, R.F.; Sosnoff, C.S.; Silver, M. & Murray, R.W. 1992, *J. Phys. Chem.* 96, 962.
53. Kotov, N.A.; Dekany, I. & Fendler, J.H. 1996. *Adv. Mater.* 8, 637.
54. Jain, R.K. & Lind, R.C., 1983. *J. Opt. Soc. Am.* 73, 647.
55. Flytzanis, C.; Hache, F.; Klein, M.C.; Ricard, D. & Rousdignol, P. 1991. *Prog. Opt.* 29, 323.
56. Haus, J.W.; Zhou, H.S.; Takami, S.; Hirasawa, M.; Honma, I. & Komiyayma, H.; 1993. *J. Appl. Phys.* 73, 1043.
57. Mulvaney, P. 1991. In Texter, J.D. (Editor). *Electrochemistry of Colloids and Dispersions.* VCH, New York, p. 345.
58. Ung, T.; Dunstan, D.; Giersig, M. & Mulvaney, P. 1997. *Langmuir*, 13, 1773.

59. Iler, R.K. 1979. *The Chemistry of Silica,* Wiley, New York.
60. Vigil, G.; Xu, Z.; Steinberg, S. & Israelachvili, J. 1994. *J. Colloid Interface Sci.* 165, 367.
61. Stöber, W.; Fink, A. & Bohn, E. 1968. *J. Colloid Interface Sci.* 26, 62.
62. Van Helden, A.K.; Jansen, J.W. & Vrij, A. 1981. *J. Colloid Interface Sci.* 81, 354.
63. Philipse, A.P. & Vrij, A. 1989. *J. Colloid Interface Sci.* 128, 121.
64. Badley, R.D.; Ford, W.T.; McEnroe, F.J. & Assink, R.A. 1990. *Langmuir* 6, 792.
65. Brandiss, S. & Margel, S. 1993. *Langmuir* 9, 1232.
66. Philipse, A.P. & Vrij, A. 1988. *J. Chem. Phys.* 88, 6459.
67. Van Blaaderen, A. 1993. *Adv. Mater.* 5, 52.
68. Iler, R.K. 1959. U.S. Patent No. 2,885,366.
69. Ohmori, M. & Matijevic, E. 1992. *J. Colloid Interface Sci.* 150, 594.
70. Ohmori, M. & Matijevic, E. 1993. *J. Colloid Interface Sci.* 160, 288.
71. Thies-Wessie, D.M.E.; Philipse, A.P. & Kluijtmans, S.G.J.M. 1995. *J. Colloid Interface Sci.* 174, 211.
72. (a) Philipse, A.P.; van Bruggen, M.P.B. & Pathmamanoharan, C. 1994. Langmuir 10, 92; (b) Philipse, A.P.; Nechifor, A.M. & Pathmamanoharan, C. 1994. *Langmuir* 10, 4451.
73. Chang, S.; Liu, L. & Asher, S. A. 1994. *J. Am. Chem. Soc.* 116, 6739.
74. Patil, A.N.; Andres, R.P. & Otsuka, N. 1994. *J. Phys. Chem.* 98, 9247.
75. Liz-Marzán, L.M. & Philipse, A.P. 1995. *J. Colloid Interface Sci.* 176, 459.
76. Liz-Marzán, L.M.; Giersig, M. & Mulvaney, P. 1996. *J. Chem. Soc., Chem. Commun.*, 731.
77. Liz-Marzán, L.M.; Giersig, M. & Mulvaney, P. 1996. *Langmuir* 12, 4329.
78. Ung, T.; Liz-Marzán, L.M. & Mulvaney, P. 1998. *Langmuir*, 14, 3740.
79. Plueddermann, E.P. 1991. *Silane Coupling Agents*, Plenum Press, New York, second edition.
80. Puddephatt, R.J. 1978. *The Chemistry of Gold*, Elsevier, Amsterdam.
81. Correa-Duarte, M.A.; Giersig, M. & Liz-Marzán, L.M. 1998. *Chem. Phys. Lett.*, 286, 497.
82. (a) Turkevich, J.; Garton, G. & Stevenson, P.C. 1954. J. Colloid Sci., Suppl. 1 26 (b) Heard, S.M.; Grieser, F.; Barraclough, C.G. & Sanders, J.V. 1983. *J. Colloid Interface Sci.* 93, 545.
83. Larson, I.; Chan, D.Y.C.; Drummond, C.J. & Grieser, F. 1997. *Langmuir* 13, 2429.
84. Biggs, S. & Mulvaney, P. 1994. *J. Chem. Phys.* 100, 8501.
85. Buining, P.A.; Liz-Marzán, L.M. & Philipse, A.P. 1996. *J. Colloid Interface Sci.* 179, 318.
86. Faraday, M. 1857. *Philos. Trans. R. Soc. London* 147, 145.
87. Henglein, A. 1993. *J. Phys. Chem.* 97, 5457; 1989. *Chem. Rev.* 89, 1861.
88. Belloni, J. 1996. *Chem. Opinion* 2, 184.
89. Mulvaney, P. 1996. *Langmuir* 12, 788.
90. Schmid, G. 1992. *Chem. Rev.* 92, 1709.
91. Weller, H. 1993. *Adv. Mater.* 5, 88; 1993. *Angew. Chem. Int. Ed. Engl.* 32, 41.
92. Wang, Y. & Herron, N. 1991. *J. Phys. Chem.* 95, 525.
93. Alivisatos, A.P. 1996. *J. Phys. Chem.* 100, 13226.
94. Kiwi, J. & Grätzel, M. 1979. *J. Am. Chem. Soc.* 101, 7214
95. Lee, P.C. & Meisel, D. 1982. *J. Phys. Chem.* 86, 3391.
96. Furlong, D.N.; Laukonis, A.; Sasse, W.H.F. & Sanders, J.V. 1984. *J. Chem. Soc., Faraday Trans.* 1 80, 571.
97. Lin, S.-T.; Franklin, M.T. & Klabunde, K.J. 1986. *Langmuir* 2, 259.
98. Weller, H.; Fojtik, A. & Henglein, A. 1985. *Chem. Phys. Lett.* 154, 473.
99. Hirai, H.; Aizawa, H. & Shiozaki, H. 1992. *Chem. Lett.* 1527.
100. Deshpande, V.M.; Singh, P. & Narasimhan, C.S. 1990. *J. Chem. Soc., Chem. Commun.* 1181.
101. Bradley, J.S.; Hill, E.W.; Behal, S.; Klein, C.; Chaudret, B. & Duteil, A. 1992. *Chem. Mater.* 4, 1234.
102. Liz-Marzán, L.M. & Lado-Touriño, I. 1996. *Langmuir* 12, 3585.
103. Brust, M.; Fink, J.; Bethell, D.; Schiffrin, D.J. & Kiely, C. 1995. *J. Chem. Soc., Chem. Commun.* 1655.
104. Esumi, K.; Tano, T.; Torigoe, K. & Meguro, K. 1990. *Chem. Mater.* 2, 564.
105. Duteil, A.; Schmid, G. & Meyer-Zaika, W.J. 1995. *Chem. Soc., Chem. Commun.* 31.
106. Andrews, M.P. & Ozin, G.A. 1986. *J. Phys. Chem.* 90, 2929.
107. Murray, C.B.; Norris, D.J. & Bawendi, M.G. 1993. *J. Am. Chem. Soc.* 115, 8706.
108. Wilcoxon, J.P.; Williamson, R.L. & Baughman, R.J. 1993. *J. Chem. Phys.* 98, 9933.
109. Petit, C.; Lixon, P. & Pileni, M.P. 1993. *J. Phys. Chem.* 97, 12974.
110. Barnickel, P.; Wokaun, A.; Sager, W. & Eicke, H.F. 1992. *J. Colloid Interface Sci.* 148, 80.
111. Towey, T.F.; Khan-Lodhi, A. & Robinson, B.H. 1990. *J. Chem. Soc. Faraday Trans.* 86, 3757.
112. Motte, L.; Petit, C.; Boulanger, L.; Lizon, P. & Pileni, M.P. 1992. *Langmuir* 8, 1049.
113. Lianos, P. & Thomas, J.K. 1986. *Chem. Phys. Lett.* 125, 299.
114. Fendler, J.H. & Meldrum, F.C. 1995. *Adv. Mater.* 7, 607.
115. Meldrum, F.C.; Kotov, N.A. & Fendler, J.H. 1994. *Langmuir* 10, 2035.
116. Yi, K.C.; Sánchez Mendieta, V.; López Castañares, R.; Meldrum, F.C.; Wu, C. & Fendler, J.H. 1995. *J. Phys. Chem.* 99, 9868.
117. Kotov, N.A.; Darbello Zaniquelli, M.E.; Meldrum, F.C. & Fendler, J.H. 1993. *Langmuir* 9, 3710.
118. Doremus, R.H. 1964. *J. Chem. Phys.* 40, 2389.
119. Doremus, R.H. 1965. *J. Chem. Phys.* 42, 414.
120. Akbarian, F.; Dunn, B.S. & Zink, J.I. 1995. *J. Phys. Chem.* 99, 3892.
121. Weaver, S.; Taylor, D.; Gale, W. & Mills, G. 1996. *Langmuir* 12, 4618.

122. Liz-Marzán, L. & Philipse, A.P. 1994. *Colloid Surf. A* 90, 95.
123. Linnert, T.; Mulvaney, P. & Henglein, A. 1991. *Ber. Bunsenges. Phys. Chem.* 95, 838.
124. Henglein, A.; Mulvaney, P. & Linnert, T. 1991. *Faraday Discuss. Chem. Soc.* 92, 31.
125. Strelow, F.; Fojtik, A. & Henglein, A. 1994. *J. Phys. Chem.* 98, 3032.
126. Papavassiliou, G.C. 1976. *J. Phys. F* 6, L103.
127. Marignier, J.; Belloni, J.; Delcourt, M. & Chevalier, J. 1985. *Nature* 317, 344.
128. Sermon, P.A.; Thomas, J.M.; Keryou, K. & Millward, G.R. 1987. *Angew. Chem. Int. Ed. Engl.* 26, 918.
129. Henglein, A.; Mulvaney, P.; Holzwarth, A.; Sosebee, T.E. & Fojtik, A. 1992. *Ber. Bunsenges. Phys. Chem.* 96, 754.
130. Mulvaney, P.; Giersig, M. & Henglein, A. 1992. *J. Phys. Chem.* 96, 10419.
131. Mulvaney, P.; Giersig, M. & Henglein, A. 1993. *J. Phys. Chem.* 97, 7061.
132. Toshima, N.; Harada, M.; Yamazaki, Y. & Kiyotaka, A. 1992. *J. Phys. Chem.* 96, 9927.
133. Bradley, J.S.; Hill, E.W.; Chaudret, B. & Duteil, A. 1995. *Langmuir* 11, 693.
134. Liz-Marzán, L.M. & Philipse, A.P. 1995. *J. Phys. Chem.* 99, 15120.
135. Rostalski, J. & Quinten, M. 1996. *Colloid Polym. Sci.* 274, 648.
136. Underwood, S. & Mulvaney, P. 1994. *Langmuir* 10, 3427.
137. Esumi, K.; Matsuhisa, K. & Torigoe, K. 1995. *Langmuir* 11, 3285.
138. Tanori, J. & Pileni, M.P. 1995. *Adv. Mater.* 7, 862–864
139. Lisiecki, I.; Billoudet, F. & Pileni, M.P. 1996. *J. Phys. Chem.* 100, 4160.
140. van der Zande, B.I.; Böhmer, M.R.; Fokkink, L.G.J. & Schönenberger, C. 1997. *J. Phys. Chem. B* 101, 852.
141. Routkevitch, D.; Bigioni, T.; Moskovits, M. & Xu, J.M. 1996. *J. Phys. Chem.* 100, 14037.
142. van der Hulst, H.C. 1957. *Light Scattering by Small Particles*, Wiley, New York.
143. Kerker., M. 1969. *The Scattering of Light and Other Electromagnetic Radiation*, Academic Press, New York.
144. Bohren, C.F. & Huffman, D.F. 1983. *Absorption and Scattering of Light by Small Particles*, Wiley, New York.
145. Kittel, C. 1956. *Introduction to Solid State Physics*, Wiley, New York, second edition.
146. Kreibig, U. 1974. *J. Phys. F: Met. Phys.* 4, 999.
147. Kreibig, U. 1977. *J. Phys.* (Paris) 38, C2–97.
148. Henglein, A.; Mulvaney, P.; Linnert, A. & Holzwarth, A. 1992. *J. Phys. Chem.* 96, 2411.
149. Henglein, F.; Mulvaney, P. & Henglein, A. 1994. *Ber. Bunsenges. Phys. Chem.* 98, 180.
150. Morris, R.H. & Collins, L.F. 1964. *J. Chem. Phys.* 41, 3357.
151. Johnson, P.B. & Christy, R.W. 1972. *Phys. Rev. B* 6, 4370.
152. Liz-Marzán, L.M. & Mulvaney, P. 1997. *New J. Chem.*, submitted.
153. Weaver, J.H.; Krafka, C.; Lynch, D.W. & Koch, E.E. (Editors). 1981. *Optical Properties of Metals* Vol 2. Physics Data Series No. 18-2; Fachinformationszentrum, Karlsruhe.
154. Brack, M. 1993. *Rev. Mod. Phys.* 65, 701.
155. U. Kreibig, M. Vollmer. 1995. *Optical Properties of Metal Clusters*. Springer Series in Materials Science 25, Heidelberg, p. 319.
156. Ung, T.; Liz-Marzán, L.K. & Mulvaney, P. 2001. *J. Phys. Chem. B* 105, 3441.
157. Giersig, M. & Mulvaney, P. 1996. *J. Chem. Soc. Faraday Trans.* 92, 3137.
158. Liz-Marzán, L.M. & Mulvaney, P. 1997. Unpublished results.
159. Dhont, J.K.G.; Smits, C. & Lekkerkerker, H.N.W. 1992. *J. Colloid Interface Sci.* 152, 386.
160. van Duijneveldt, J.S.; Dhont, J.K.G. & Lekkerkerker, H.N.W. 1993. *J. Chem. Phys* 99, 6941.
161. Savage, G. 1975. *Glass and Glassware*. Octopus Books, London.
162. Le Cloux, A. et al. 1985. *Surf. Sci.* 156, 256.
163. van Blaaderen, A. & Vrij, A. 1993. *J. Colloid Interface Sci.* 156, 1.
164. van Blaaderen, A. & Vrij, A. 1992. *J. Non-Cryst. Solids* 149, 161.
165. Giersig, M.; Ung, T.; Liz-Marzán, L.M. & Mulvaney, P. 1997. *Adv. Mater.* 9, 570.
166. Hoshino, H. & Shimoji, M. 1974. *J. Phys. Chem. Solids* 35, 321.
167. Malinowiski, J. 1974. Photograph Sci. Eng. 18, 363; James, T.H. 1986. *Adv. Photochem.* 13, 329.
168. Jaenicke, W.; Tischer, R.P. & Gerischer, H. 1995. *Z. Elektrochem.* 59, 448.
169. Liz-Marzán, L.M. et al. 1997. Australian Patent Application.
170. Henglein, A. 1982. *Ber. Bunsenges. Phys. Chem.* 86, 301.

52 Surface Chemistry of Silica Coatings of Titania

D. Neil Furlong
Commonwealth Scientific and Industrial Research Organization, Division of Chemicals and Polymers

CONTENTS

The dense silica (DS) process involves the exposure of titania particles to aqueous silica solutions of increasing silica concentration. The process is examined in this chapter by relating silica adsorption on titania surfaces to solution pH and concentration and to the various monomeric, multimeric, and polymeric species present in aqueous "solutions" of silica. Microelectrophoresis and gas adsorption studies reveal that adsorption of monomeric silica occurs via hydrated cation sites that constitute only approximately 40% of titania surfaces. These "anchoring" sites provide a base for complete surface coverage and buildup of silica multilayers (coatings), a buildup that occurs when the silica concentration is increased sufficiently at the chosen pH (around 10 in the DS process) to induce polymerization.

The coating of silica from aqueous solution onto titania — an area of science and technology that has been underpinned by the Iler patent of 1959 [1] describing the composition and process for production of the so-called "dense silica coatings" (DS coatings) — is discussed in this chapter. The patent, though broad in its specifications with respect to the types of substrates that could be coated, was clearly targeted at the titania pigment business. Because of its technological impact, the patent spawned a period of research activity, reported in the open scientific literature mainly to the early 1980s, aimed at probing the nature of the DS deposition process and of the coatings themselves. Of course, many of these findings may already have been recorded in DuPont files. Nevertheless, the open research has proven very useful to the fundamental understanding of oxide–oxide interactions, as well possibly to industry. In recent years relatively little new work on the details of deposition have appeared, although the properties and performance of various DS coatings remain very much of interest.

This chapter is a look black at a "parcel" of surface chemistry research that grew out of an extraordinarily useful piece of technology, and it focuses mainly on the specific chemical interactions of deposition. The characterization of the physicochemical properties and processing variables of DS coatings, particularly with regard to their use in pigments, is the focus of Chapter 53.

WHY COAT TITANIA WITH SILICA?

This question is easily answered if pigment performance is the main concern; the answer is improved dispersibility and paint film durability. Dispersibility relates mainly to the surface chemistry of the silica coating–paint film [2] interface and less to the specifics of titania–silica interactions and is not further considered here. The durability of paint films, on the other hand, relates more to the photoredox properties of titania and the consequent "damage" to organic resins in paint films. DS coatings were designed to provide a barrier to such photoredox reactions. Titania particles absorb ultraviolet light of wavelength below around 380 nm; these absorptions give rise to charge centers (valence band holes and conduction band electrons) with redox potentials appropriate for oxidation–reduction of organic media. The strong oxidizing potential of photogenerated surface holes at ca. 2.6 V (with respect to the normal hydrogen electrode) was particularly threatening to alkyd resins [3–5], for example, which formed the basis for paint films of the past.

The ability of surface photogenerated redox centers to participate in electron-transfer reactions with, for example, organic resins depends on titania particle size (relating to internal charge recombination rates that can also be influenced by metal ion dopants) and other parameters such as resin–surface affinity and availability of oxygen [6]. However, paint film "chalking" (i.e., pigment–resin phase separation on exposure of the film to atmospheric ultraviolet rays) was a familiar sight when uncoated titania pigments were used. Uniform and coherent silica coatings of thickness around 2 nm or more presented a dielectric barrier to surface redox reactions and hence effectively retarded paint film chalking. Iler's DS coatings were such coatings. Although in the patent they were depicted rather stylistically (Figure 52.1), electron microscopy shows their real coherence and uniformity (Figure 52.2), and their retardation of resin oxidation has often been demonstrated [3–7].

Should a researcher not interested in pigment chemistry be interested in silica coatings on titania? It is interesting to digress at this point and consider briefly the question of coatings from the opposite viewpoint — useful photocatalysis. The photocatalytic power of titania particles has been investigated in a number of potentially useful organic oxidation reactions [8–11], the attraction of titania being its inherent chemical (i.e., dark) stability. In particular, titania has been central to the quest for efficient systems for the photolysis–photoreduction of water [12]. Much that has been learned from the photocatalytic behavior of pigmentary titania has found its place in this very current sphere of research activity. For example, the efficient photooxidation of water requires the combination of the photocatalytic power of titania with "hole-storing catalysts," such as ruthenium dioxide and vanadia. The main reason for this combination of catalysts is the need to harness the single electron-hole pair generated per absorbed photon with the desired multielectron reactions at the titania–medium interface. Hence, the laying down of an oxide catalyst onto colloidal particles of titania has become an important catalyst preparation procedure.

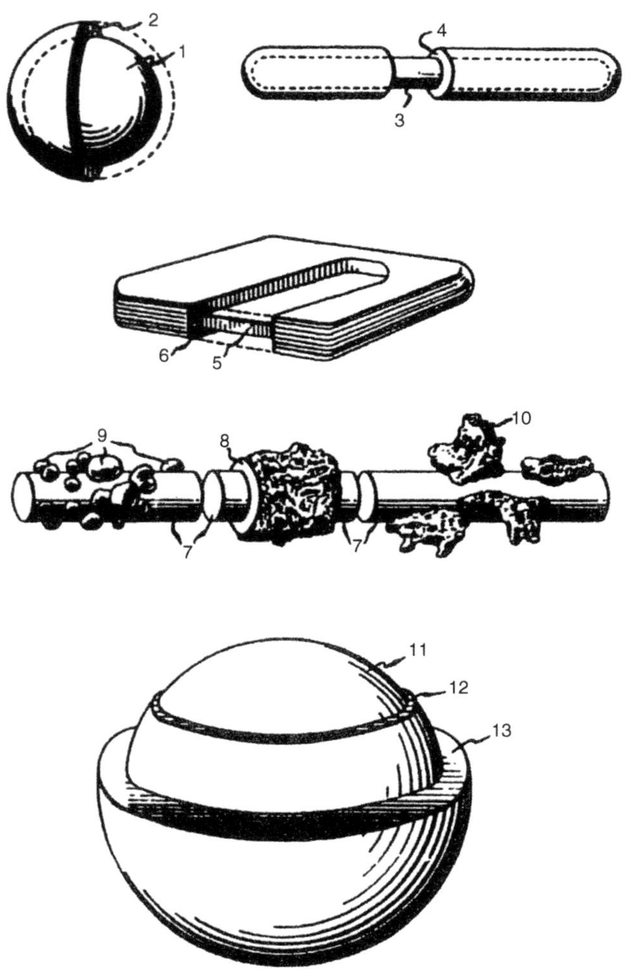

FIGURE 52.1 Schematic representations [1] of dense silica coatings formed by the Iler process.

Of course silica coatings, being dielectric, will not be useful direct participants in water photolysis–photoreduction. This fact is demonstrated by the data of Tada et al. [13], reproduced in Figure 52.3, in which silica coatings ca. 1 nm in thickness (deposited by the so-called liquid-phase deposition or LPD process [14]) inhibit the photoreduction of aqueous silver ions. Researchers of aqueous photoredox chemistry have, however, used another aspect of silica surface chemistry to advantage — that is, the negative surface charge of silica when dispersed in aqueous solutions. Silica is negatively charged above ca. pH 2 [15], in contrast to titania, which has an isoelectric point around 5–6 [16]. The surface charge of both silica and titania can be controlled by adjustment of the dispersion pH, and hence the adsorption of potentially photoreactive (ionic) solutes can be encouraged or retarded. When surface photochemical reactions involve a change of solute charge, the electric field at the silica– or titania–solution interface can enhance the separation of

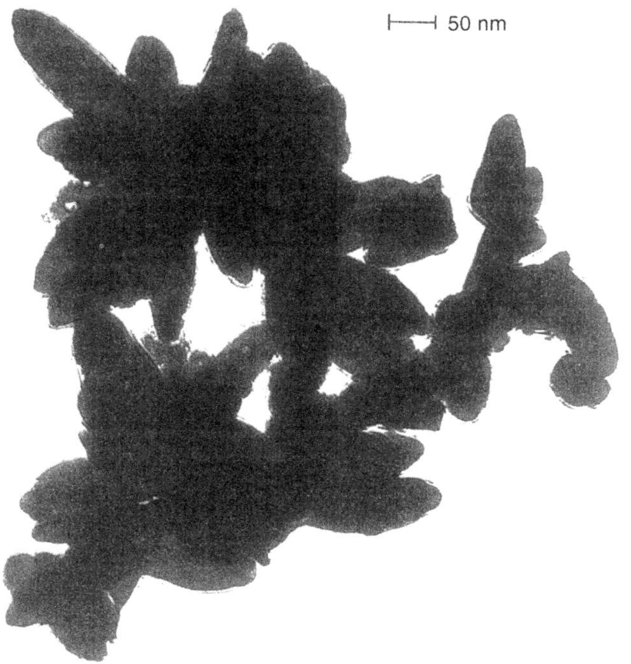

FIGURE 52.2 Transmission electron micrograph of dense-silica-coated titania particles: magnification, 200,000; 5.0 wt% silica loading. (Reproduced with permission from reference 23. Copyright 1979.)

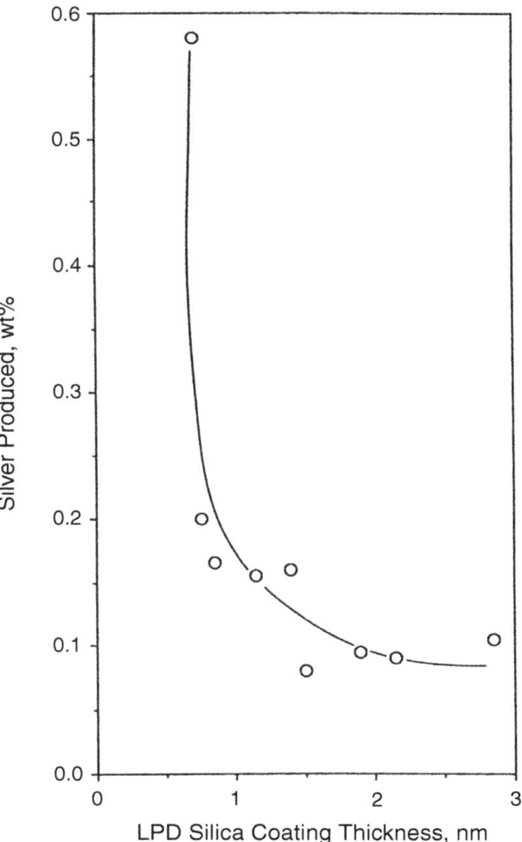

FIGURE 52.3 Dependence of the amount of Ag photodeposited on silica-coated titania particles on the thickness of silica coating. (Reproduced with permission from reference 13. Copyright 1989.)

photochemical products; thus, back-reactions are inhibited and the yields of useful products are improved. To date, such effects have been demonstrated only for either silica or titania alone [17]. However, the potential for using partially silica-coated titania to give subtle control of useful interfacial photochemistry seems attractive. Therefore, the surface chemistry of silica-coated titania is useful in terms of the inhibition of surface photochemistry, as with pigments, and has potential for control of "preparative" surface photochemistry.

HOW TO COAT TITANIA WITH SILICA

The various "dry" processes such as vacuum evaporation or chemical vapor deposition [18] are not discussed here. These processes are central, for example, to semiconductor technology, but are not so readily used with titania powders. With titania (and other oxides) dispersed in aqueous solution, effective silica coating can be achieved (without really trying) merely by allowing an aqueous dispersion of titania to equilibrate in a Pyrex vessel [19]. "Aqueous" silica that is leached from the Pyrex vessel will deposit onto particles. The leaching rate is greater at alkaline pH, and hence so is the coating rate. It is possible to completely mask titania surfaces in this manner, although even at high pH values the process is rather slow (on the scale of days) and somewhat irreproducible. Deposition by leaching is more often than not an undesirable process and not the basis of preparative coating technology. A number of other processes are used to prepare coatings and these include the Iler DS process [1] and the LPD process [13] of the Nippon Sheet Glass Company (Figure 52.4).

Iler approached DS silica coatings via a detailed and unparalleled understanding of the pH–concentration control of the aqueous chemistry of aqueous sodium silicate solutions [20]. The preparation sequence shown in Figure 52.4a is one of a number in the Iler 1959 patent [1] and was used to prepare coated titania powders described later in this chapter. Variants on the original patent, some including coating additives other than silica [21], continue to appear in the patent literature. The "successful" binding of aqueous silica species to the surface of titania relates to the interactions between titania surfaces and the various silica–silicate species present in solution, and in particular to the progressive exposure of the titania surface to an increasing "aqueous silica" concentration. The concentration may be increased sufficiently during the coating procedure to cause precipitation of amorphous silica.

Proponents of LPD coatings have shown that this method produces effective coatings, but the number of characterization studies on both the process and the coatings have been far fewer than those on the Iler process. The silver reduction data [13] in Figure 52.3 suggest that these coatings are either not effectively uniform or are partially permeable to aqueous ions. The use of fluoride solutions would intuitively render the LPD process less attractive from a processing point of view. A key difference between the LPD and DS processes is that the former exposes titania to supersaturated silica solutions at all stages of coating. The speciation chemistry of silica-supersaturated H_2SiF_6 solutions is much less well documented than is that of the aqueous silica solutions of the DS process.

Because of the wide industrial application of the Iler DS process, this chapter focuses on the silica–titantia interactions in that process.

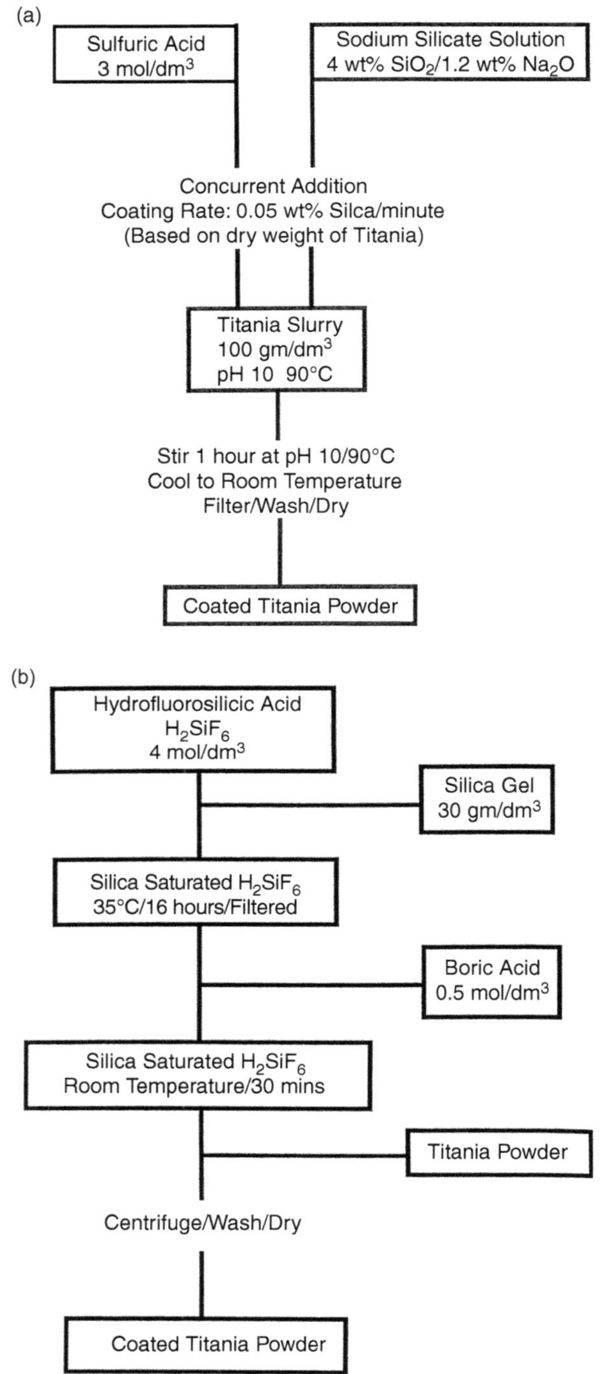

FIGURE 52.4 Sequences for preparation of (a) Iler DS coatings [1] on titania particles. Sequences for preparation of (b) LPD silica coatings [18] on titania particles.

ILER DS DEPOSITION PROCESS

Solution Chemistry of Aqueous Sodium Silicate

The Iler DS process is based on the exposure of the substrate particles to an increasing concentration of "aqueous" silica at 90–100°C and a pH of ca. 9–10. The term "aqueous silica" hides much, in that it can include monomeric species (silicic acid and its conjugate bases), multimers, and fine particulates [20]. Although the direct experimental evidence describing the form of the multimers is somewhat inconclusive, it is generally felt that $Si_4O_6(OH)_6^{2-}$ is dominant. To gain an understanding of the types of silica species interacting with titania during DS deposition, a description of the pH–concentration domains of the various silica species is required. Figure 52.5 describes such a distribution at 25°C and was

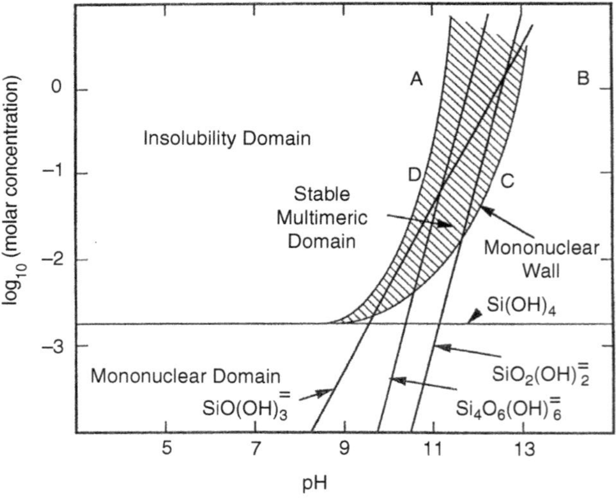

FIGURE 52.5 Calculated solubility diagram for amorphous silica at 25°C. The monomolecular wall represents the lower limit for the stability of multinuclear silica species. Points A, B, C, and D are from experimental light-scattering data [20,42] and provide confirmation of the region of existence of stable multimers. (Reproduced with permission from reference 23. Copyright 1979.)

compiled by Stumm et al. [22] from experimental equilibrium constants. "Indicators" (A, B, C, and D in Figure 52.5) have been added [23] to the Stumm diagram. These indicators arise from published light-scattering experiments and help to define the pH–concentration domain in which multimers are present as precursors to bulk precipitation of silica.

Figure 52.5 shows that the solubility of amorphous silica is independent of pH between 4 and 9; above pH 9 the solubility increases because of the formation of monosilicate, disilicate, and multimer ions. DS coatings are deposited at 90°C — equilibrium constants are not available at this temperature. Silica solubility data are available [24], and hence it is possible to represent the insolubility–solubility domains of silica at this temperature (if not the details of individual speciation), as in Figure 52.6. The predominant reason behind Iler proposing that coatings be deposited at ca. 90°C would be the increase by about a factor of 2.5 available in solubility compared to 25°C, as well as possible "dehydration" and porosity aspects of resultant films [25]. Included in Figure 52.6 are dashed lines that describe the "trajectory" of successive aliquots of aqueous silicate solution (pH ca. 13.5, point S) added to the titania slurry (at pH ca. 10). The steadily increasing silica concentration in the slurry is depicted by the solid vertical arrow in Figure 52.6.

In summary, in the DS process titania particles are exposed to an increasing concentration of monomeric silica species, and, depending on the coating level required (i.e., the total amount of silica added) this concentration may be increased to the extent where precipitation of silica will occur. Multimeric species also may be present in part of the process.

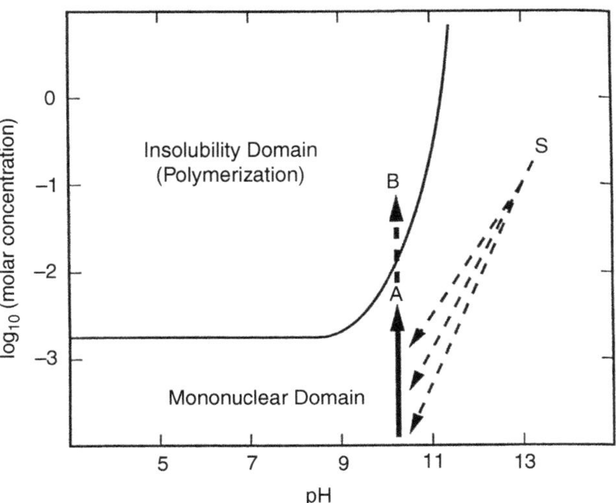

FIGURE 52.6 The course of a DS coating procedure (at 90°C) related to the solubility domains for amorphous silica. S marks the concentration of sodium silicate solution (at pH ca. 13.5) added to the titania slurry. The vertical arrow represents the change in "aqueous silica" concentration during a coating procedure (at pH 10). Point A represents coating procedures that stop short of the precipitation edge, and point B represents those during which the edge is traversed. (Reproduced with permission from reference 23. Copyright 1979.)

Adsorption of Aqueous Silica onto Titania — pH and Concentration

A number of detailed experimental and modeling studies [26–32] have reported the adsorption of aqueous silicate onto oxides such as alumina and goethite and various other (often soil constituent) particulates such as fluorides. Studies of anion adsorption onto titania are not so common [33]. This situation is somewhat strange given the technical importance of titania. The solubility of titania in water between pH 3 and 12 is only approximately 10^{-6} mol/dm^3 [24]. Therefore, the formation of complexes between aqueous Ti(IV) species and aqueous silica will be insignificant in DS slurries. Such complex formation has often complicated the analysis of silicate adsorption on more soluble substrates [29] and makes it somewhat difficult to formulate generalized adsorption models [32].

In the adsorption study with titania [33], the levels of silica used were such that dispersions were always above the solubility limit for amorphous silica. Although the final concentration of "aqueous silica" in a typical DS preparation procedure might be between 10^{-2} and 10^{-1} mol/dm^3 and hence close to or above the solubility limit, the DS procedure certainly involves exposure of titania surfaces to soluble silica species prior to precipitation. Therefore, the value of the previous adsorption study [33] for understanding the details of DS deposition is limited. The final silica concentration reached depends on the levels of titania used and the degree of coating desired. Such "final" concentrations refer to added silica — in fact bulk precipitation may never occur even if the solubility edge of Figure 52.5 or Figure 52.6 is traversed [27].

Figure 52.7–Figure 52.9 describe changes in aqueous silica concentration, as the pH is lowered from ca. 13 to ca. 6, at various total silica concentrations (up to and including 10^{-2} mol/dm^3). Runs were performed in the absence and presence of titania. This series of experiments was performed at 25°C [34], but nevertheless provides a framework for the understanding of DS deposition.

Concentration of 10^{-4} mol/dm^3 Aqueous Silica

This concentration is well below the "solubility edge" at all pH values — the blank result (Figure 52.7) confirms that no silica was precipitated when the pH was reduced from 13 to 7. In the presence of titania, adsorption occurred at all pH values, with a clear adsorption peak at pH 9.5. A similarly positioned adsorption maximum has been reported on goethite [26,27,30] and gibbsite [26]

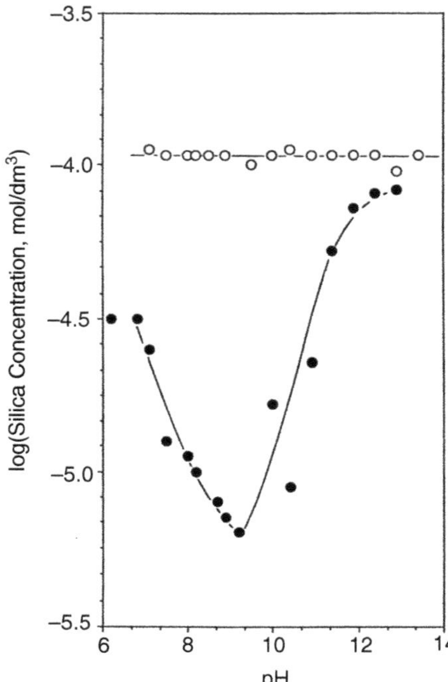

FIGURE 52.7 Aqueous silica concentration vs. pH for 10^{-4} mol/dm^3 total silica with pH decreased from an initial value of 13.5: ○, sodium silicate solution (blank); •, 20 m^2/dm^3 titania added (at pH 13.5).

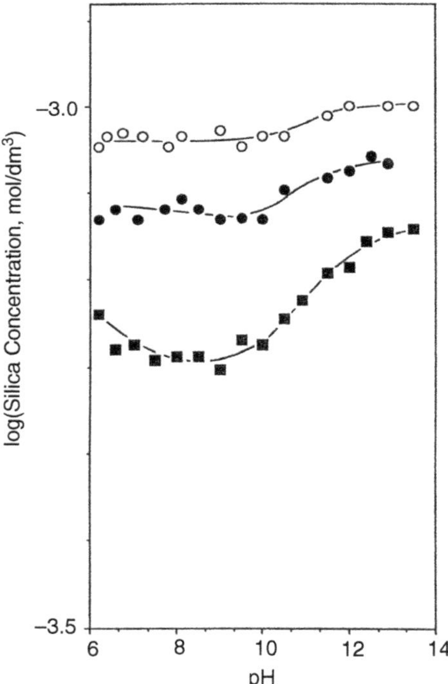

FIGURE 52.8 Aqueous silica concentration vs. pH for 10^{-3} mol/dm^3 total silica with pH decreased from an initial value of 13.5: ○, sodium silicate solution (blank); •, 20 m^2/dm^3 titania added; and ■, 60 m^2/dm^3 titania added (at pH 13.5).

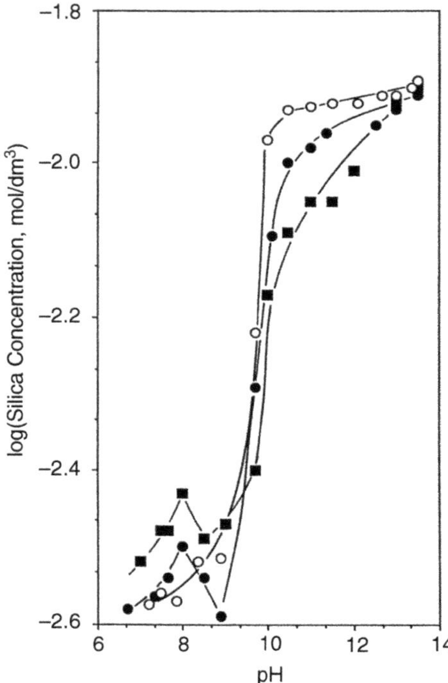

FIGURE 52.9 Aqueous silica concentration vs. pH for 1.25×10^{-2} mol/dm^3 total silica with pH decreased from an initial value of 13.5: ○, sodium silicate solution (blank); •, 20 m^2/dm^3 titania added; and ■, 50 m^2/dm^3 titania added (at pH 13.5).

and interestingly also on some but not all fluoride minerals [29]. By contrast, Howard and Parfitt [33] saw no such maximum on titania. However, as stated earlier, these workers were estimating adsorption in the presence of precipitated silica, a somewhat different experimental system than that represented by Figure 52.7. Interestingly, Howard and Parfitt did observe desorption of silica when the pH was subsequently decreased from 10 — results in accord with the data in Figure 52.7. At all the pH values in Figure 52.7 the titania particles will be negatively charged, hence adsorption is clearly not in response to electrostatic interactions, but rather in spite of them.

The adsorption maximum at pH 9–10 has been seen by others [26] as reflecting the first pK_a of silicic acid (p$K_a^1 = 9.45$ at 25°C), although it has been argued by different workers that this maximum in turn was an indication that $SiO(OH)_3^-$ adsorbed alone [29] or that $Si(OH)_4$ and $SiO(OH)_3^-$ adsorbed concurrently [26]. Early modeling studies [26,28] related silica adsorption to exchange with substrate surface hydroxyl groups, although clearly this situation would not be the case for adsorption on fluoride minerals. Davis and Leckie [30] and others [28] alluded to the possibility of more than one type of surface adsorption site, although some workers [28] conceded that experimental verification may be difficult. Marinakis and Shergold [29] proposed the direct binding of aqueous $SiO(OH)_3^-$ to surface cation sites on calcium fluoride. Barrow and Bowden, longtime contributors toward the understanding of anion adsorption, have extended [32] modeling work to specifically include both surface hydroxyls and surface cations (with coordinated water) on oxide surfaces. In so doing they concluded that the predominant adsorbing species is the divalent $SiO_2(OH)_2^-$ and rationalize this finding with the observed maximum in adsorption at around pH 9–12. They also state that the pK_a values for silicic acid (9.45 and 12.56) are close enough to make it difficult to clearly distinguish between the predominant adsorbing species at various pH values. In the experimental results represented in Figure 52.7 (total silica concentration of 10^{-4} mol/dm^3), calculations show that $SiO(OH)_3^-$ is the greatly dominant species at pH 9.5. This fact leads to the conclusion, contrary to that of Barrow and Bowden [32], that $SiO(OH)_3^-$ is the predominant adsorbing species giving the observed adsorption maximum.

Concentration of 10^{-3} mol/dm^3 Aqueous Silica

This silica blank (Figure 52.8) showed a small decrease in concentration as the pH was decreased from ca. 12 to ca. 9–10; at lower pH values the concentration was constant. The aqueous silica diagram (Figure 52.5) shows an equilibrium solubility of ca. 2.5×10^{-3} mol/dm^3; hence the decrease evident in the blank seems unlikely to indicate precipitation. The likelihood is that silicon polyanions were removed from solution. In the presence of titania, some adsorption is apparent at all pH values. The adsorption peak is still evident at around pH 9, although much less so than in Figure 52.7. Hence it appears that monomer adsorption and multimer "precipitation" occurred concurrently in this slurry. The level of adsorption increased linearly with the increase of titania loading (Figure 52.8); this indicates that the adsorption density was independent of the equilibrium silica concentration in the range of 5.5 to 7.8×10^{-4} mol/dm^3 (see the adsorption isotherm in Figure 52.10). At a total silica concentration of 10^{-3} mol/dm^3 and pH 9–10, the dominant silica species is $SiO(OH)^{3-}$; $SiO_2(OH)^{2-}$ will be present as a very minor species. It seems then that the former is "controlling" when it comes to adsorption.

Concentration of 1.25×10^{-2} mol/dm^3 Aqueous Silica

At this concentration the precipitation edge for silica is at pH 10.0; this result is confirmed by the blank run in Figure 52.9, as was the limiting solubility of precipitated silica of ca. 2.5×10^{-3} mol/dm^3 at pH values less than 9. In the presence of titania, precipitation losses still dominate, although some extraction of silica occurs above the

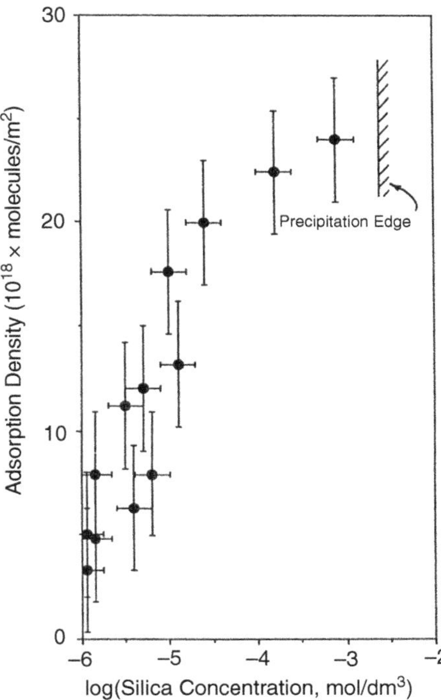

FIGURE 52.10 Adsorption isotherm for aqueous silica on titania at pH 9.5.

pH of precipitation, and the "maximum" in adsorption at pH ca. 9 is still evident.

Adsorption Isotherm at pH 9.5

The data in Figure 52.7–Figure 52.9 demonstrate how silica can adsorb onto titania as both monomers and polymers, the nature of the species depending also on the total concentration of "aqueous silica." Figure 52.10, showing the adsorption at pH 9.5, comes from many such uptake–pH experiments in which the equilibrium silica concentration did not exceed 10^{-3} mol/dm^3. Hence Figure 52.10 describes only monomer adsorption and indicates a limiting adsorption density of ca. 2.5×10^{18} molecules per square meter. This density corresponds to ca. 0.4 nm^2 per adsorbed silica species. The density of hydroxyl groups on titania surfaces corresponds to ca. 0.16 nm^2 per hydroxyl. Therefore, there is not a simple correlation between silica uptake and titania surface hydroxyls and adsorption does not proceed to complete surface coverage.

Howard and Parfitt [33] also observed that adsorption on titania did not proceed to complete surface coverage, stopping at a little less than half coverage. Similar incomplete coverage was also noted on goethite [27] and on some fluoride minerals [29]. The adsorption density may be limited by the size of the adsorbing silicate species, although an area of ca. 0.4 nm^2 for the cross-sectional area of monomeric silica seems unreasonably large. There seems little other reason why all surface hydroxyls would not react equally with aqueous silica. As discussed, other surface sites may participate in adsorption [32]. With the value of hindsight it can be said that silicate adsorption onto titania surfaces in fact occurs preferentially on hydrated surface cation sites (known to be at a density of ca. 3×10^{18} per square meter); the following discussion presents experimental results that lead to this conclusion.

Interaction of Aqueous Silica with Titania

Microelectrophoresis

This method is useful for monitoring the surface charge characteristics of the "coated" titania either after or during deposition. There is a difference of ca. 3 pH units between the isoelectric point of titania (ca. pH 5) and that of silica (ca. pH 2), and the progressive shift in the isoelectric point as coating proceeds indicates the extent of silica coating. Figure 52.11 [23] shows such a progression for 0.6–5 wt% DS coatings on rutile particles ca. 0.1×0.4 µm in size (specific surface area 20 m^2/g). Such titania particles are not typical of commercial pigments (which are usually ca. 0.2 µm in diameter) but were useful for this "diagnostic" study in that they were very pure and without the normal pigmentary additives. Each coated titania was prepared according to the DS scheme (Figure 52.4a), and the level of silica incorporation was increased by increasing the amount of aqueous silicate solution added. Each coated titania powder was dried in air after preparation. The "final" silica concentration corresponded to point A on Figure 52.6 for the 0.62 wt% DS sample and increased to point B for the 4.99 wt% DS sample. Discrete coatings were only perceptible by electron microscopy for the 2.55 and 4.99 wt% DS samples (Figure 52.2). There was no evidence of "bulk" silica

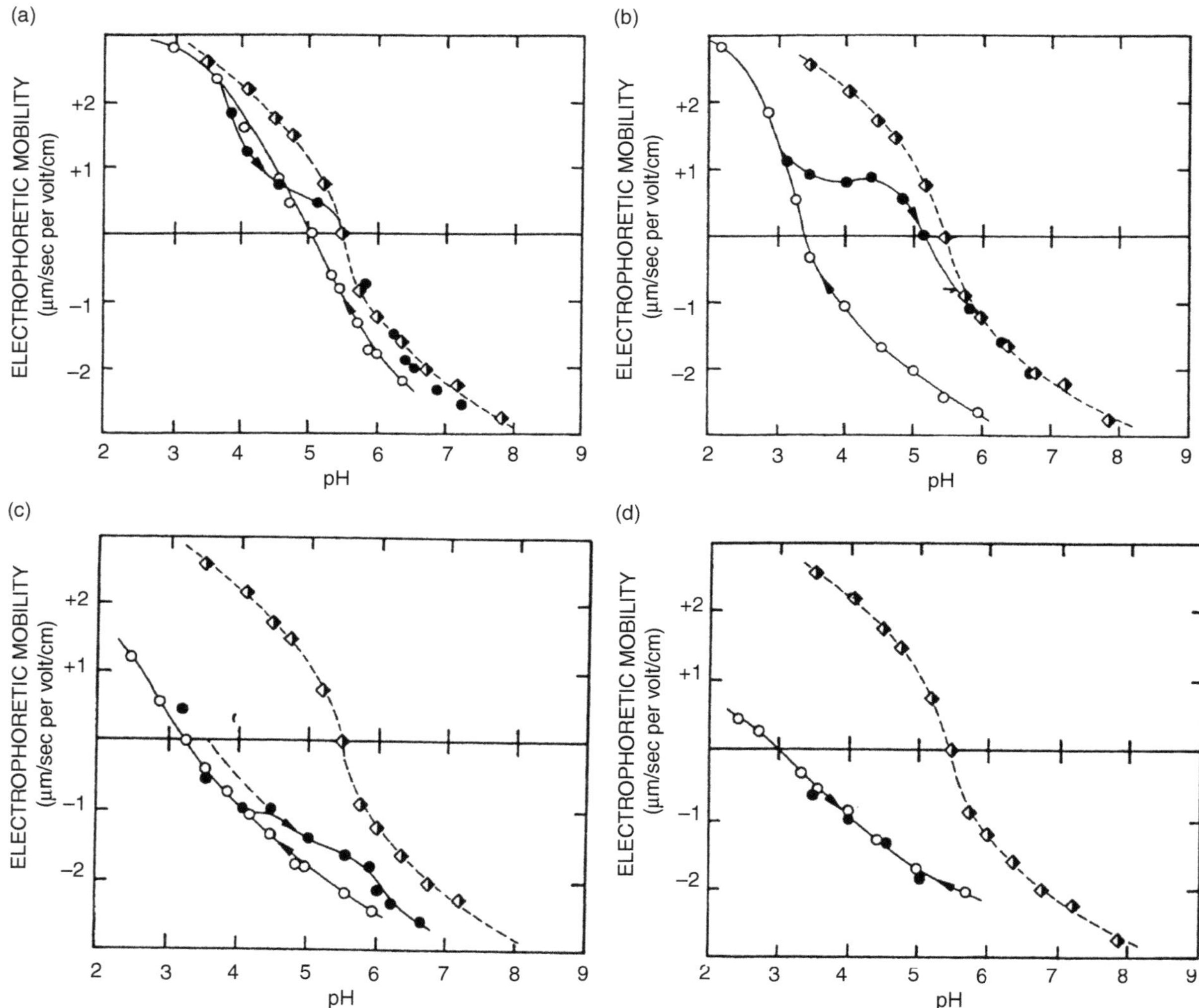

FIGURE 52.11 Electrophoretic mobility–pH curves for DS-coated titanias: ○, decreasing pH; •, increasing pH; and ◆, uncoated titania. Silica loading was (a) 0.62 and (b) 1.35 wt%. (Reproduced with permission from reference 23. Copyright 1979.) Silica loading was (c) 2.55 and (d) 4.99 wt%.

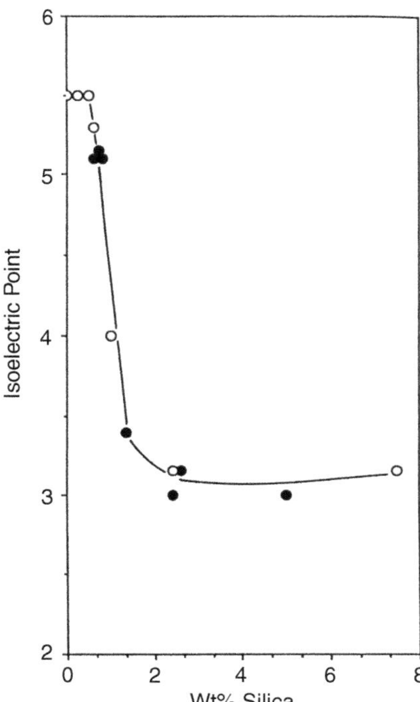

FIGURE 52.12 Isoelectric points of DS-coated titanias: ○, titania dispersed in aqueous silica; •, dried, coated titania powders. The base titania was 20 m^2/g.

precipitation in any of the samples. Types of isoelectric point shift similar to those in Figure 52.11 are seen [23] when the electrophoretic mobility of the titania particles is determined, as a function of pH, in the presence of aqueous sodium silicate. Figure 52.12 is a compilation of isoelectric point data from both DS-coated titania and titania–aqueous silicate experiments.

Figure 52.11 and Figure 52.12 show the following:

- The isoelectric point was apparently unaffected at low silica coverages — below ca. 0.5 wt %. This fact indicates that silica adsorbs preferentially onto surface sites in a manner that does not affect average surface charge–pH behavior. A similar suggestion was made by Kononov et al. [35] in describing silica adsorption onto fluorite.
- The addition of between 0.5 and 1.4 wt% silica resulted in a shift of isoelectric point proportional to the silica loading. This finding is consistent with the "rule of thumb" of electrokinetics that the isoelectric point of a mixed surface is the surface-area-weighted average of the isoelectric point values of the components [15]. Hence it appears that this surface concentration regime corresponds to the completion of silica monolayer surface coverage of titania particles. This relationship was confirmed by electron microscopy [36]. Adsorbed silica in this concentration regime is relatively "fragile," in that it is apparently removed from titania particles during the "increasing pH" stage of the electrophoresis experiment (Figure 52.11).
- The particles exhibited a silica-like surface at greater than 1.5 wt% silica, and the silica coating was resistant to subsequent increases in pH.

If it is assumed that silica deposition results from 1:1 titania–silicate interactions, it can be shown [23] that the loading range of 0.5 to 1.4 wt% corresponds to ca. 40–100% monolayer coverage of titania particles; this result gives these interesting conclusions:

1. The first ca. 40% of surface coverage does not result in changes to surface-pH-dependent electrochemistry. As shown previously [37] with IR spectroscopy, ca. 40% of cations on the surfaces of rutile particles carry coordinated water molecules and the remainder are hydroxylated. The first type of surface site would not contribute to the pH dependence of surface charge — it seems from the electrokinetic data of silica-coated titanias that it is just these sites that preferentially adsorb aqueous silica. Such a finding is consistent with the observed maximum silica adsorption coverage (Figure 52.10). This specific silica–titania interaction is further discussed in the following "Gas Adsorption" section.
2. The electrokinetics of titania surfaces are completely converted to those of a silica surface with only one monolayer of silica. The indications are that silica binding proceeds beyond the level associated with the specific binding discussed earlier, although the adsorption isotherm (Figure 52.10) does not indicate such. Detailed assessment [23] of a wide range of coating experiments shows that complete surface coverage requires the concentration of aqueous silica to be increased to a level likely to induce polymerization, although the quantitative determination of the necessary concentration in a titania slurry at 90°C and pH 10 (as in the DS coating procedure) is not possible. However, if the polymerization domain is avoided, adsorption of monomer inevitably leads only to partial surface coverage. It seems likely then that complete monolayer (and indeed subsequent multilayer) coverage occurs via the specific binding sites discussed earlier and not by interactions between aqueous silica species and surface hydroxyls on titania surfaces. If such anchoring sites are formed by exposure to monomeric silica, polymerization proceeds laterally

across surfaces to effectively mask titania surface functionality at equivalent to monolayer coverage.

Gas Adsorption

Gas-phase infrared spectroscopy studies [37–39] of titania surfaces have revealed that surface cation sites on titania are dehydrated on outgassing between 150 and ca. 250°C. Surface dehydroxylation really only begins at above 250°C and is essentially complete by 400°C. Surface dehydration exposes very polar surface sites, and hence its evolution can be followed by measuring the energetics of interaction, for example, by using microcalorimetry with a polar probe gas (e.g., nitrogen). Representative microcalorimetry data [40,41] for titania and DS-coated titanias are presented in Figure 52.13. The differential energy of nitrogen adsorption is plotted as a function of surface coverage by nitrogen (surface coverage deduced from the adsorption isotherm determined concurrently). For the present context the trends are clear. Curves A, for uncoated titania, demonstrate the emergence of "high-energy surface sites" when the outgassing temperature is increased from 150 to 250°C and show that these sites occupy ca. 40% of titania surfaces (assuming, as is the norm, that nitrogen adsorption will proceed sequentially from "high" to "low" energy sites). Not surprisingly, for titania particles with a ca. 2-nm uniform DS coating (curves C), such active sites (not present on silica surfaces) are not uncovered on outgassing. The general form of the calorimetry curves for this sample over the outgassing range of 150–400°C is very silicalike [40,41]. Interestingly, the titania particles with only around half-monolayer coverage of silica also show general silica-like behavior, and in particular there is no indication of active titania cation surface sites (curves B). Hence the gas adsorption microcalorimetry confirms the conclusions from microelectrophoresis; namely, silica deposition by the DS process proceeds via a specific solute–surface interaction as a precursor to "multilayer deposition." Undoubtedly, just such a sequence ensures coating adhesion and uniformity, two trademarks of the DS process.

The surface hydration–hydroxylation structure of titania, proved previously mainly by IR studies using dry titania powders, also seems to hold when these powders are dispersed in water. An interesting approach, therefore, is to probe directly the uptake of water from the gas phase by DS-coated rutile surfaces [42]. Water adsorption isotherms are presented in Figure 52.14. The dual nature of titania surface sites, a property not seen with other common oxides such as silica and alumina, leads to an unusual type of water adsorption isotherm for titania. The isotherm shows two distinct "knees" (Figure 52.14) connected by a region where adsorption increases linearly with the partial vapor pressure of water. The explanation for this adsorption behavior is rather complex [42] and beyond the scope of this chapter. This behavior is believed to be due to the presence of hydrated surface cation sites.

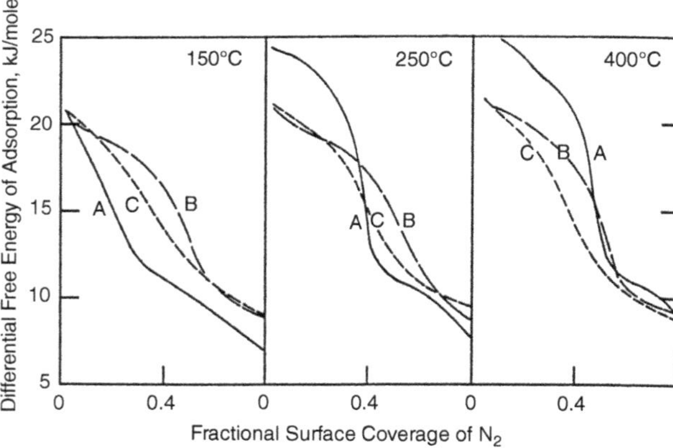

FIGURE 52.13 Differential energies of adsorption of nitrogen at 77 K plotted against nitrogen surface coverage. The outgassing temperature is indicated for each set of curves. Adsorbents were titania (curve A), titania with 0.6 wt% DS silica coating (curve B), and titania with 2.6 wt% DS silica coating (curve C). (Reproduced with permission from reference 41. Copyright 1980.)

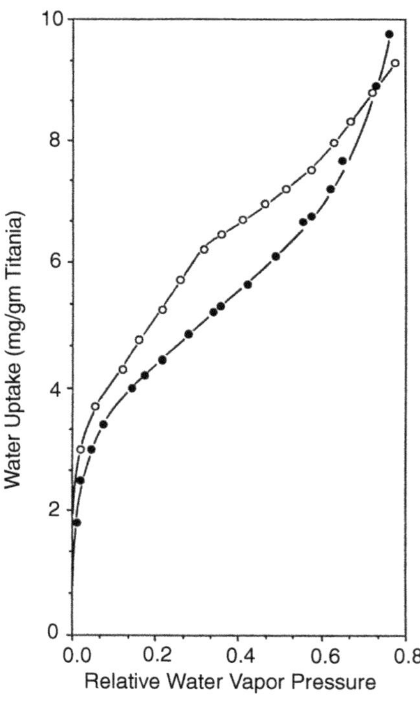

FIGURE 52.14 Adsorption isotherms of water vapor (at 25°C) on titania (○) and 0.6 wt% DS-coated titania (•). The outgassing temperature was 150°C. (Reproduced with permission from reference 42. Copyright 1986.)

What is noteworthy is that this rather specific behavior is not exhibited when ca. 40% of the titania surface is covered with DS silica.

SUMMARY

Iler DS coatings have been a major success in the pigment world, and the Iler process may well be of value in the world of "useful" photocatalysis. Studies of silica adsorption on titania, via uptake and electrophoresis measurements and gas adsorption characterization procedures, have enabled an explanation of the nature of specific interactions between aqueous silica and titania. Binding is proposed to occur preferentially via hydrated cation sites on titania, and the occurrence of such binding is concluded to provide the basis for the subsequent surface polymerization necessary for the buildup of coherent multilayer silica.

ACKNOWLEDGMENTS

As a young, raw researcher I was privileged to receive very positive encouragement from Ralph Iler. I have also benefited greatly by working with Tom Healy, Ken Sing, and Geoff Parfitt (deceased). Silicate adsorption data presented in this paper were the results of an honors program by Janice Grant of the University of Melbourne.

REFERENCES

1. Iler, R. K. U.S. Patent 2,885,366, 1959.
2. Parfitt, G. D. *Croatica Chem. Acta*, **1980**, *53*, 333.
3. Simpson, L. A. *J. Oil Colour Chem. Assoc.* **1981**, *64*, 490.
4. Simpson, L. A. *Polymers Paint Colour J.* **1986**, *176*, 408.
5. Cutrone, L.; Moulton, D. V.; Simpson, L. A. *Pigment Resin Technol.* **1989**, *18*, 16.
6. Simpson, L. A. *Australian OCCA Proc. News* **1983** (May), 6.
7. Egerton, T. A.; King, C. J. *J. Oil Colour Chem. Assoc.* **1979**, *62*, 386.
8. Cundall, R. B.; Hulme, B.; Rudham, R.; Salim, M. S. *J. Oil Colour Chem. Assoc.* **1978**, *61*, 351.
9. Furlong, D. N.; Wells, D.; Sasse W. H. F. *Aust. J. Chem.* **1986**, *39*, 757.
10. Day, R. E.; Egerton, T. A. *Colloids Surf.* **1987**, *23*, 137.
11. Blatt, E.; Furlong, D. N.; Mau, A. W.-H.; Sasse, W. H. F.; Wells, D. *Aust. J. Chem.* **1989**, *42*, 1351.
12. *Photoinduced Electron Transfer*; Fox, M. A.; Chanon, M., Eds.; Elsevier: New York, 1988; Parts A through D.
13. Tada, H.; Saitoh, Y.; Miyata, K.; Kawahara, H. *J. Jpn. Soc. Colour Mater.* **1989**, *62*, 399.
14. Nagayama, H. Japanese Patent 58–161944, 1983.
15. Parks, G. A. *Chem. Rev.* **1965**, *65*, 177.
16. Furlong, D. N.; Parfitt, G. D. *J. Colloid Interface Sci.* **1979**, *69*, 409.
17. Furlong, D. N.; Johansen, O.; Launikonis, A.; Loder, J. W.; Mau, A. W.-H.; Sasse, W. H. F. *Aust. J. Chem.* **1985**, *38*, 363.
18. Nagayama, H.; Honda, H.; Kawahara, H. *J. Electrochem. Soc. Solid State Sci. Technol.* **1988**, *135*, 2013.
19. Furlong, D. N.; Freeman, P. A.; Lau, A. C. M. *J. Colloid Interface Sci.* **1981**, *80*, 20.
20. Iler, R. K. *The Chemistry of Silica*; Wiley: New York, 1979.
21. Jacobsen, H. W. Eur. Pat. Appl. EP 245,984, 1987.
22. Stumm, W.; Huper, H.; Champlin, R. L. *Environ. Sci. Technol.* **1967**, *1*, 221.
23. Furlong, D. N.; Sing, K. S. W.; Parfitt, G. D. *J. Colloid Interface Sci.* **1979**, *69*, 409.
24. Garrels, R. M.; Christ, C. L. *Solutions, Minerals and Equilibria*; Harper and Row: New York, 1965.
25. Iler, R. K., personal communication, 1984.
26. Hingston, F. J.; Pasner, A. M.; Quirk, J. P. *J. Soil Sci.* **1972**, *23*, 177.
27. Yokoyama, T.; Nakazato, T.; Tarutani, T. *Bull. Chem. Soc. Jpn.* **1980**, *53*, 850.
28. Sigg, L.; Stumm, W. *Colloids Surf.* **1980**, *2*, 101.
29. Marinakis, K. I.; Shergold, H. L. *Int. J. Mineral. Proc.* **1985**, *14*, 177.
30. Davis, J. A.; Leckie, J. O. *J. Colloid Interface Sci.* **1980**, *74*, 32.
31. Goldberg, S. *Soil Sci. Soc. Am. J.* **1985**, *49*, 851.
32. Barrow, N. J.; Bowden, J. W. *J. Colloid Interface Sci.* **1987**, *119*, 236.
33. Howard, P. B.; Parfitt, G. D. *Croatica Chem. Acta* **1977**, *50*, 15.
34. Grant, J. B. Sc. Thesis, University of Melbourne, 1980.
35. Kononov, O. V.; Barskii, L. A.; Ratmirova, L. P. *Russ. J. Phys. Chem.* **1973**, *47*, 1651.
36. Furlong, D. N., unpublished data.
37. Jones, P.; Hockey, J. A. *Trans. Faraday Soc.* **1971**, *67*, 2679.
38. Parkyns, N. D.; Sing, K. S. W. In *Colloid Science*; Everett, D. H., Ed.; The Chemical Society: London, 1975; Vol. 2, Chapter 1.
39. Parfitt, G. D. In *Progress in Surface and Membrane Science*; Danielli, J. F.; Rosenberg, M. D.; Cadenhead, D. A., Eds.; Academic: New York, 1976; Vol. 11, p 181.
40. Furlong, D. N.; Rouquerol, F.; Rouquerol, J.; Sing, K. S. W. *J. Chem. Soc. Faraday Trans. I* **1980**, *76*, 774.
41. Furlong, D. N.; Rouquerol, F.; Rouquerol, J.; Sing, K. S. W. *J. Colloid Interface Sci.* **1980**, *75*, 68.
42. Furlong, D. N.; Sing, K. S. W.; Parfitt, G. D. *Adsorption Sci. Technol.* **1986**, *3*, 25.

53 Dense Silica Coatings on Micro- and Nanoparticles by Deposition of Monosilicic Acid

Horacio E. Bergna, Lawrence E. Firment, and Dennis G. Swartzfager
DuPont Company

CONTENTS

The upper limit to the thickness of dense silica coatings on hydroxylated surfaces that can be obtained with conventional coating techniques was found to be extended significantly by coating with monosilicic acid. An example with submicrometer α-alumina particles as a substrate showed dense, uniform coatings up to at least 800 Å thick. The silica coatings and coating mechanism were characterized by chemical analysis, electrokinetic potential, nitrogen surface area (Brunauer–Emmett–Teller), particle size, x-ray photoelectron spectroscopy, diffuse reflectance Fourier transform infrared spectroscopy, secondary ion mass spectroscopy, and transmission electron microscopy.

Silica coatings are applied to particulate materials to modify surface characteristics that interfere with the exploitation of desired bulk properties, as with titania pigments that may photocatalyze the degradation of their vehicle, surfaces of selective zeolite catalysts that may promote undesired reactions, and fillers for plastics that may not disperse in their matrix.

An upper limit to the thickness of dense silica coatings on hydroxylated surfaces can be obtained with some of the conventional coating techniques [1]. We have found that this limit can be extended significantly by coating with monosilicic acid (MSA).

The deposition of silica from water was discussed by Iler [2]. The mechanism of deposition of monomeric silica is different from that of the deposition of colloidal particles. Monomeric silica is deposited from supersaturated solution in two known ways: as a deposit of $Si(OH)_4$ on a solid surface or as colloidal particles forming in the supersaturated solution.

Silanol groups SiOH condense with OH groups of MOH surfaces, where M is a metal that will form a silicate at the pH and temperature involved. The reaction is represented as follows:

$$\begin{array}{c} | \\ -\mathrm{M}-\mathrm{OH} \\ | \\ \mathrm{O} \\ | \\ -\mathrm{M}-\mathrm{OH} \\ | \end{array} + \begin{array}{c} \mathrm{OH} \quad \mathrm{OH} \\ \diagdown \ \diagup \\ \mathrm{Si} \\ \diagup \ \diagdown \\ \mathrm{OH} \quad \mathrm{OH} \end{array} \longrightarrow \begin{array}{c} | \\ -\mathrm{M}-\mathrm{O} \\ | \\ \mathrm{O} \\ | \\ -\mathrm{M}-\mathrm{O} \\ | \end{array} \begin{array}{c} \diagdown \quad \diagup \mathrm{OH} \\ \mathrm{Si} \\ \diagup \quad \diagdown \mathrm{OH} \end{array} + 2H_2O \tag{53.1}$$

Further deposition of $Si(OH)_4$ is on silica, thus a layer of silica is built up.

The second known way of deposition of monomeric silica first involves polymerization in the solution. Iler described this deposition as follows: "If an insufficient area of a receptive solid surface is available to accept silica rapidly, and if the concentration of $Si(OH)_4$ is greater than 200–300 ppm (depending on pH), polymerization occurs with formation first of low polymers such as the cyclic tetramer; then these further condense to form

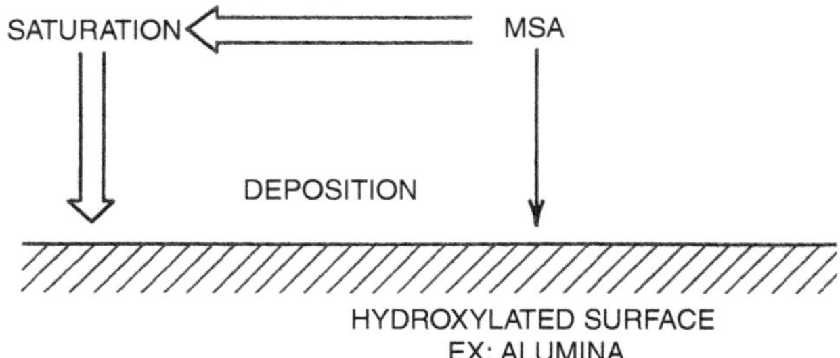

FIGURE 53.1 Silica deposition on hydroxylated surface; first step: before supersaturation.

small three-dimensional polymers which are colloidal particles" [2].

Monomeric silica is also deposited in a third way by living organisms as biogenic amorphous silica through still-unknown mechanisms. The underlying principles of nature's mechanism of deposition are being investigated as a first step to biomimetic processing of ceramics [3].

With the knowledge that monomeric silica is deposited from supersaturated solution as a deposit of $Si(OH)_4$ on a solid surface or as colloidal particles forming in the supersaturated solution, a general picture of the deposition of MSA on a hydroxylated surface may be visualized as occurring in three steps. In the first step, even before supersaturation it is assumed that at least some MSA may react with the hydroxylated surface (Figure 53.1). After supersaturation a number of mechanisms involving homonucleation and heteronucleation of MSA take place (Figure 53.2). Part of the MSA may react directly with the hydroxylated surface, and another part may nucleate to form hydrated silica polymers. The hydrated silica polymers may take several different paths depending on the conditions of the system. In one case, the hydrous silica polymer forms porous deposits on the hydroxylated surface. In another, the hydrous silica polymerizes further, either gelling or forming discrete particles of colloidal silica. Both the gel or the aquasol may attach to the hydroxylated surface, forming either "flaky" or porous deposits. Meanwhile, the MSA in solution may deposit in the interstices of the gel or in the interstices of the flaky or porous deposits formed on the surface by either the gel or the aquasol.

What path the MSA follows is determined by factors such as composition of the substrate surface, addition rate of the MSA if it is being added to the system, slurry concentration if the system involves a dispersion of particles as substrates, ionic strength of the electrolyte solution in which the particles are dispersed, MSA concentration at any given time, and especially pH and temperature of the system.

If all these factors are properly controlled, MSA can be deposited on the hydroxylated surface, forming dense silica layers. Once a layer that completely covers the original hydroxylated surface is formed, any further deposition is on silica, thus an increasingly thicker dense silica layer is built up on the substrate surface (Figure 53.3).

In this chapter processes for coating α-alumina particles with MSA and for characterizing the coated particles to show the MSA is deposited as monomeric units to form a dense silica coating are described. The same procedures can be applied to coating titania and ρ-zeolite particles [4].

EXPERIMENTAL DETAILS

Materials

MSA was prepared by the following procedure [5,6]: a solution of sodium metasilicate was prepared by dissolving

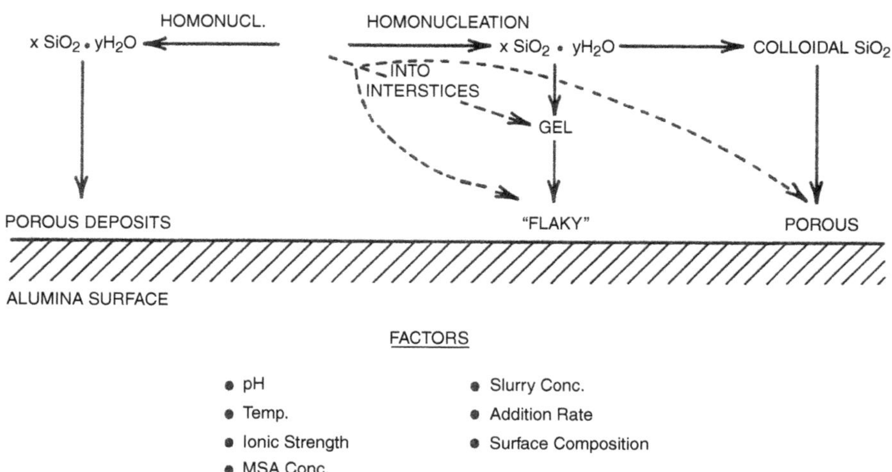

FIGURE 53.2 Silica deposition on hydroxylated surface; second step: after supersaturation.

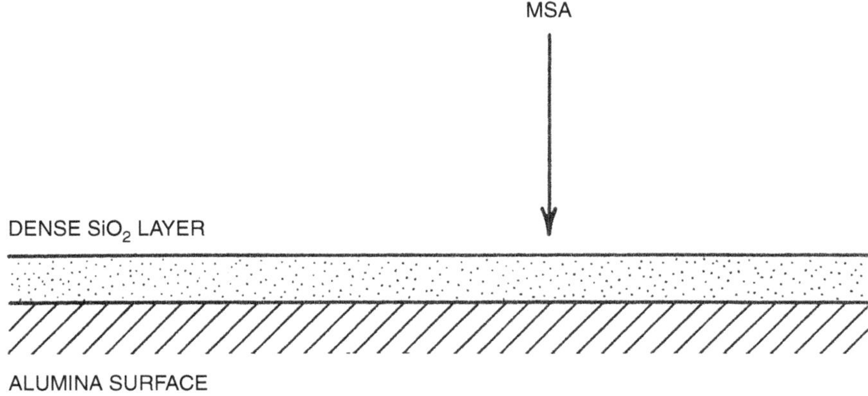

FIGURE 53.3 Silica deposition on hydroxylated surface; third step: after complete surface coverage.

30 g of pulverized, Fisher reagent grade $Na_2SiO_3 \cdot 9H_2O$ in 100 ml of 0.1 *N* NaOH. The silica content of this reagent, designated solution A, was 2.28%. A separate solution B was prepared, consisting of 0.025 *N* H_2SO_4, and was cooled to 0–5°C. Meanwhile, a quantity of sulfonic acid cation-exchange resin (Dowex HCR-W2-H) was washed with distilled water until washings were colorless. A 1.5-g sample of the resulting washed resin was added to 100 ml of solution B in a beaker that was stirred and cooled in an ice bath at about 5°C. At this point, 5 ml of solution A was added by intermittent jets of about 0.3 ml each, delivered by a 1-ml syringe and fine-tipped hypodermic needle. The pH of the resulting mixture was continuously maintained below 2.5 by delaying additions of solution A until the pH of the stirred, cooled mixture dropped below 2. After 5 ml of solution A had been added in this fashion, the pH of the mixture was about 2.15. The resulting clear solution of silicic acid was stored temporarily at 0–5°C in an ice bath to prevent premature polymerization. The calculated concentration of this solution of silicic acid was 3 mg of SiO_2 per ml. The same procedure was used to prepare silicic acid solution with a concentration of 2 mg of SiO_2 per ml.

The α-alumina was superground Alcoa A16 alumina classified by sedimentation in water and kept as a 10% aqueous slurry of pH 3.5–4.5. The Brunauer–Emmett–Teller (BET) specific surface area of the alumina product was 10–13.5 m^2/g.

Equipment

All coating experiments were performed in an automated reactor facility of a glass flask equipped with agitator, inlets for liquids, condenser, pH and temperature sensors, and controlling and recording instrumentation. The agitator consisted of a stir motor (Cole–Palmer model 4370-00) coupled to a glass stir rod with a Teflon blade positioned near the bottom of a glass flask. The flask was placed in a heating mantle, and the MSA was kept between 2 and 3°C by a chiller with a circulating pump.

The aqueous alumina slurry was placed in the flask. The temperature of the slurry and MSA feed, the pH, the feeding rates of MSA and acid (concentrated HCl) or base (concentrated NH_4OH), and the stirring rates were controlled, measured, and recorded automatically (Kaye III Digitstrip recorder). The information obtained was monitored on the color screen of a Digital VT125 computer.

Procedure

Temperature of the alumina slurry was adjusted to 60°C, and pH was maintained at 8.5 with automatic additions of dilute acid or base as needed. At this temperature and pH the MSA solution was added at a rate of ca. 2 g of SiO_2 per 1000 m^2 per hour to prevent formation of colloidal silica [2,3].

At the end of the MSA treatment, the slurry was allowed to cool to room temperature and centrifuged to separate the solid residue, made of coated alumina. The residue was washed by redispersing it in distilled water and centrifuging. The cake was dried in vacuum at 110°C for 12 h.

Characterization Techniques

The dry samples of silica-coated alumina were analyzed by inductively coupled plasma–atomic emission spectroscopy (ICP–AES). Specific surface area was determined by nitrogen adsorption (BET). Particle-size determination was made by low-angle forward scattering of light from a laser beam (Leeds and Northrup's Microtrac particle sizer) and by monitoring sedimentation with a finely collimated beam of low-energy x-rays and a detector (Micromeritics' Sedigraph 5100).

The samples were characterized by x-ray diffraction (XRD); transmission and high-resolution electron microscopy (TEM and HREM); microelectrophoresis; x-ray photoelectron spectroscopy (XPS), also known as electron spectroscopy for chemical analysis (ESCA); secondary ion mass spectrometry (SIMS); and diffuse reflectance Fourier transform (DRIFT) infrared absorption spectroscopy.

The transmission electron microscopy was done with a 100-kV accelerating potential (Hitachi 600). Powder samples were dispersed onto a carbon film on a Cu grid for TEM examination. The surface analysis techniques used, XPS and SIMS, were described earlier [7]. X-ray photoelectron spectroscopy was done with a DuPont 650 instrument and Mg $K_{\propto}$ radiation (10 kV and 30 mA). The samples were held in a cup for XPS analysis. Secondary ion mass spectrometry and depth profiling was done with a modified 3M instrument that was equipped with an Extranuclear quadrupole mass spectrometer and used 2-kV Ne ions at a current density of 0.5 $\mu A/cm^2$. A low-energy electron flood gun was employed for charge compensation on these insulating samples. The secondary ions were detected at 90° from the primary ion direction. The powder was pressed into In foil for the SIMS work.

Electrokinetic potential of the alumina samples redispersed in 0.001 *N* KNO_3 was measured over a pH range between 2 and 9.5 with a Pen Chem System 300 instrument. The coated alumina powder was ultrasonically dispersed at 0.001 wt% into 300 ml of 0.001 *N* KNO_3 for ca. 15 min and immediately placed under N_2 atmosphere. A 50-ml portion was placed onto a titration stirrer under N_2 atmosphere that was fitted with a pH probe, a mechanical stirrer, and a port for addition of titrant. A portion of sample was pumped into the S3000 cell, which was fitted into a constant temperature bath set at 25°C. This sample portion was used to rinse the cell of the previous sample. A second portion of sample was pumped into the cell. The pH was recorded and the zeta potential was measured. Two measurements were taken, one at the front stationary layer and the second at the back stationary layer. The histograms were then combined and averaged. The pH was adjusted with either 0.01 *N* KOH or 0.01 *N* HNO_3, and measurements were repeated for each desired pH value. Once all the desired pH versus zeta potential data were obtained, the data were transferred to an IBM PC and plotted. Estimates of isoelectric points (IEPs) were made from the plots obtained.

Infrared analysis of the coated and uncoated alumina particles was done by DRIFT with a 180° backscattering configuration and referenced to powdered (<30-μm particle size) KCl. A Nicolet 7199 interferometer operating at 4-cm^{-1} resolution was used to average 250 interferograms for an improved signal-to-noise ratio. The evacuated cell was capable of pressures of 10^{-6} torr (10^{-3} Pa) and temperatures to 400°C.

RESULTS

The original, uncoated α-alumina had a specific surface area of 10–13.5 m^2/g. Solid spherical particles of α-alumina 150 nm in diameter would have a surface area of 10 m^2/g. TEMs of the particles show that they are irregularly shaped and have a broad distribution of particle size between 0.1 and 0.8 μm (Figure 53.4).

The silica content by chemical analysis, the specific surface area, and the particle-size analysis of the original alumina and the silica-coated products are included in Table 53.1. For these alumina particles, theoretical monolayer coverage corresponds to 1 wt% silica. Within the error of the measurements, the specific surface area appears to remain fairly constant throughout the coating process. The constancy of surface area with silica level suggests that dense coatings are formed. Available particle-size measurements obtained by two different techniques suggest that the silica coating may cause some aggregation of the alumina particles at relatively low levels of silica and may become significant at higher coverages, starting at 6–9-wt% SiO_2.

Figure 53.5 shows plots of electrokinetic potential versus pH for the alumina samples at various levels of silica coverage and the changing character of the interface of the particles in aqueous solution with increasing coverage with silica, from pure alumina (IEP ca. 9.0) to pure silica (IEP ca. 2.0). Most of the change in electrokinetic potential has occurred by 1-wt% silica loading; thus, most of the alumina surface has been covered by silica.

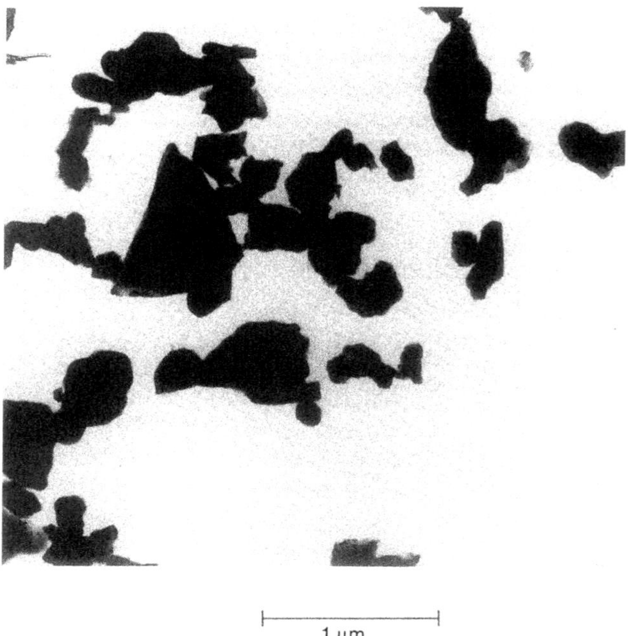

FIGURE 53.4 Transmission electron micrograph of superground, dispersed, and peptized Alcoa A16 α-alumina. (Reprinted with permission from reference 7. Copyright 1989.)

TABLE 53.1
MSA-Coated Submicrometer α-Alumina Particle-Size Analysis

Bulk chemical analysis (wt% SiO_2)	BET surface area (m^2/g)	D_{50} (μm)	
		Microtrac	Sedigraph
<0.05	10–13.4	0.5	0.3
0.12	10.8	0.5	0.3
0.32	12.3		0.13
0.55	12.5		0.12
1.19	14.4		
1.275	13.0		
2.03	14.0		0.26
2.55		0.55	0.08
2.69	9.8		0.42
3.72	13.4	0.44	0.42
4.89	13.5		
5.70	13.9	0.59–0.76	
6.60	13.7	0.62	0.54
8.14	11.7	0.89	0.71
8.41	13.5		0.60
23.38	9.9		1.35
27.0	10.5		

Note: No entry indicates that data were not determined. D_{50} is the 50th percentile of the size distribution.

Figure 53.6 shows the infrared spectra of the surface of α-alumina particles before coating and after coating with 1-wt% of silica. The peak at 3743 cm^{-1} is characteristic of isolated surface Si–OH groups. The strong Si–OH absorption at the low 1% silica level shows that the silica is present largely at the particle surfaces as a coating.

Surface analysis results from XPS (ESCA) and SIMS as well as IEPS of samples with a coverage corresponding

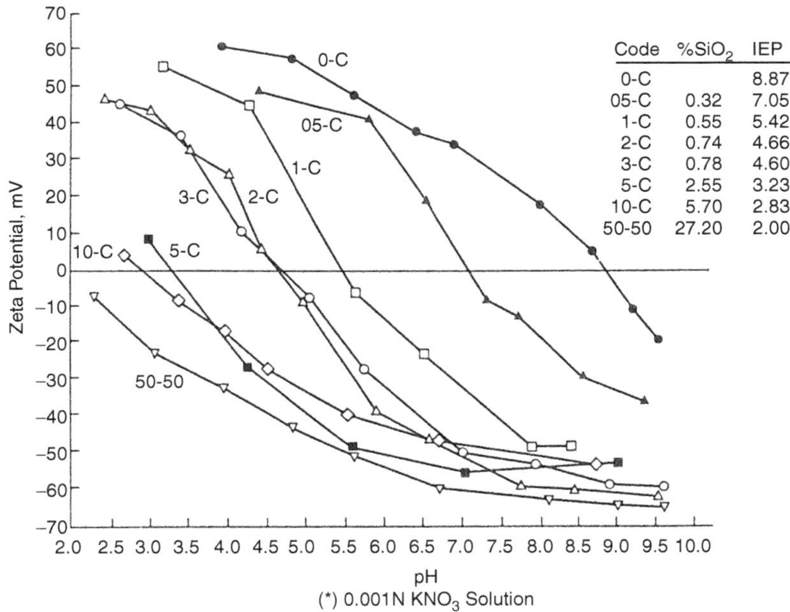

FIGURE 53.5 Electrokinetic (zeta) potential versus pH of alumina and silica-coated alumina particles dispersed in 0.001 *N* KNO_3 solution.

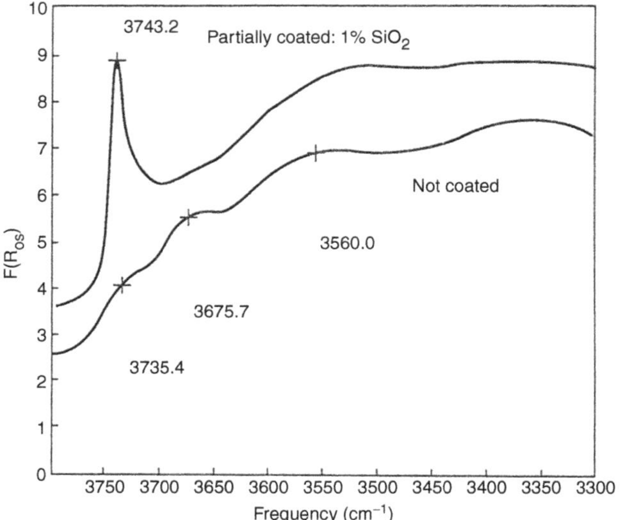

FIGURE 53.6 DRIFT spectra of the surface of α-alumina particles before and after coating with 1-wt% silica. $F(R_{os})$ is the Kubelka–Munk function of reflectance.

FIGURE 53.7 High-resolution electron micrograph of the coated edge of an alumina particle: total silica content, 14.5 wt% SiO_2; silica layer thickness, ca. 130 Å. (Reprinted with permission from reference 7. Copyright 1989.)

to 0.3–5.9-wt% SiO_2 are shown in Table 53.2. At higher silica levels, the surface of the alumina particles is completely covered by a multilayer of silica and the thickness can be measured by TEM. Figure 53.7 shows, as an example, a micrograph of the coated edge of an alumina particle. These measurements are plotted in Figure 53.8 and included in Table 53.3 with BET results.

Figure 53.8 includes a plot of the calculated thickness of silica uniformly covering the surface of spherical alumina particles 150 nm in diameter (i.e., spherical particles having approximately the specific surface area of the original α-alumina particles) as a function of

TABLE 53.2
MSA Coating of Submicrometer α-Alumina Particles, First Stage: Hydrous Silica on Hydroxylated Alumina Surface

W/O SiO_2	XPS			SIMS		
	Si:Al	Calculated thickness (Å)	Coverage	Si:Al + Si	Approximate average thickness (Å)	IEP (pH)
0.12	0.00	0.00	none			8.87
0.30	0.05	1.46	spotty	15	<5	
0.32	0.005	0.15	spotty			7.05
0.55	0.09	2.58	spotty	20	<5	5.42
0.74	0.09	2.58	spotty	30	5	4.66
0.785	0.10	2.86	spotty			4.60
1.19	0.09	2.58				
1.27	0.16	4.45		32	6	
2.03	0.31	8.10		57	6	
2.23	0.37	9.44	layer ca. 10 Å	60	7	3.0
2.55	0.44	10.94	layer ca. 11 Å	70	7–8	3.23
2.69	0.50	12.16		78	10	
4.89	0.95	20.00	layer 20 Å	85	25	
5.70						2.83

Note: No entry indicates that data were not determined.

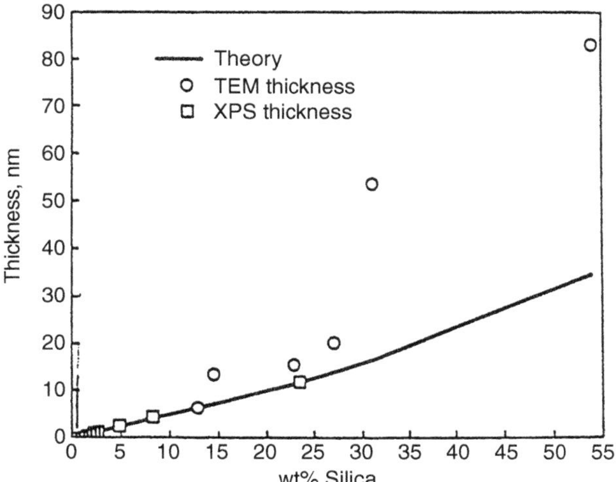

FIGURE 53.8 Silica layer thickness versus silica loading determined by TEM and XPS (ESCA) methods for silica-coated α-alumina particles. The theoretical curve was calculated for uniform silica coating of dense spherical α-alumina particles with equivalent 150-nm diameters. (Reprinted with permission from reference 7. Copyright 1989.)

silica weight percent. The formula for obtaining this plot is

$$t = r\left\{\left[\frac{\rho_{\mathrm{Al}}}{\left(\frac{1}{X_{\mathrm{Si}}}-1\right)\rho_{\mathrm{Si}}}+1\right]^{1/3}-1\right\} \tag{53.2}$$

where t is the coating thickness in nanometers; r is the particle radius, taken as 75 nm; ρ_{Al} is the density of α-alumina, 3.97 $\mathrm{g/cm^3}$; ρ_{Si} is the density of the silica coating, taken as 2.2 $\mathrm{g/cm^3}$; and X_{Si} is the weight fraction of silica in the coated particle.

TABLE 53.3
MSA Coating of Submicrometer α-Alumina Particles, Second Stage: Hydrous Silica on First Silica Layer

Bulk chemical analysis (wt% SiO_2)	HREM thickness (Å)	BET surface area (m^2/g)
Uncoated particles	—	10–13.4
5.70	35	13.9
12.90	60	
14.50	130	
22.50	150	
27.00	200	10.5
31.00	335	
	535	
54.00	830	3.0

Note: No entry indicates that data were not determined.

A comparison of the results obtained by TEM, XPS, and SIMS was reported previously [6] and is included here for reference. In the 15–30-wt% range, there is good agreement between the coating thicknesses measured by TEM and the thickness calculated for uniform coating on spheres. Above 30 wt%, the thicknesses measured by TEM are much higher than the calculation. In this silica loading range, the measured surface area of the particles falls from about 12 (10–13.5) to 3 $\mathrm{m^2/g}$. Apparently, in preparations at this high loading, the particles are agglomerated and the parameters of the calculation are no longer appropriate. Increases in average particle size were measured at lower silica levels, 6–9 wt%. Because there are fewer particles to coat, the coatings are thicker. We were unable to image without ambiguity coatings less than 35 Å thick.

Also plotted in Figure 53.8 is an estimate of thickness obtained from ratios of the Si and Al signals in the XPS spectra. This estimate results from the relationship between the measured Si:Al ratio and the thickness t of a uniform silica coating [7]:

$$t = \lambda \ln\left(n_{\mathrm{Si}}/n_{\mathrm{Al}} + 1\right) \tag{53.3}$$

where $n_{\mathrm{Si}}/n_{\mathrm{Al}}$ is the ratio of photoelectron signals from Si and Al multiplied by the appropriate sensitivity factors and λ is the electron mean free path.

There is quite good agreement between the TEM results, XPS results, and calculated thickness for the 10–20-wt% silica preparations. The low-silica region is plotted on an expanded scale in Figure 53.9; both the $n_{\mathrm{Si}}/n_{\mathrm{Al}}$ ratios and the derived thickness are plotted. Over

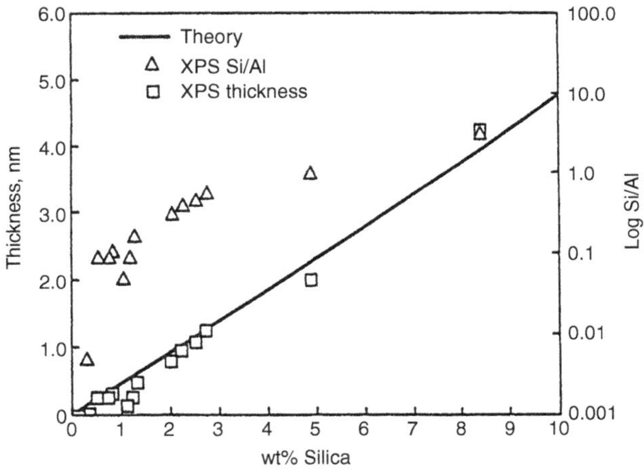

FIGURE 53.9 Low-silica region of an expanded scale plot of logarithm of Si:Al atomic ratio and derived XPS thickness versus silica loading. The theoretical curve was calculated for uniform silica coating of dense spherical α-alumina particles with equivalent 150-nm diameters. (Reprinted with permission from reference 7. Copyright 1989.)

more than 2 orders of magnitude of n_{Si}/n_{Al}, there is good agreement between the XPS thickness and that calculated for uniform coating. Any deviations from a uniform coating would decrease the n_{Si}/n_{Al} ratios and bring the XPS thicknesses below the calculated curve. The agreement is good even for coatings less than 1 nm thick, which approach atomic dimensions. These data show that the silica deposits as a molecular coating over a wide range of thicknesses.

Figure 53.10 shows SIMS depth profiles of surface composition versus time as the surface atoms are removed by 2-keV neon ion bombardment. There can be problems with quantitation of SIMS data associated with dependence of sputtering and ion yields on surface composition, but in this case of similar oxide coatings of substrates, these problems are minimized. The sputtering yields of SiO_2 and Al_2O_3 have been found to be similar [8]. The abscissa is really in units of ion dose, but it was calibrated for depth by profiling through a 28-wt% coating thickness of 20 nm as measured by TEM. The calibration of ion dose to depth is then approximately 1×10^4 nm cm^2 C^{-1}. This calibration curve (compressed by a factor of 8) is shown, along with some representative depth profiles of samples of lower silica content. The ordinate is in units of fractional coverage of the surface by silica ($n_{Si}/(n_{Si} + n_{Al})$). The coverage was calibrated at 100% silica with the signal from a thickly coated sample and at 0% with an uncoated sample; linear interpolation was used in between.

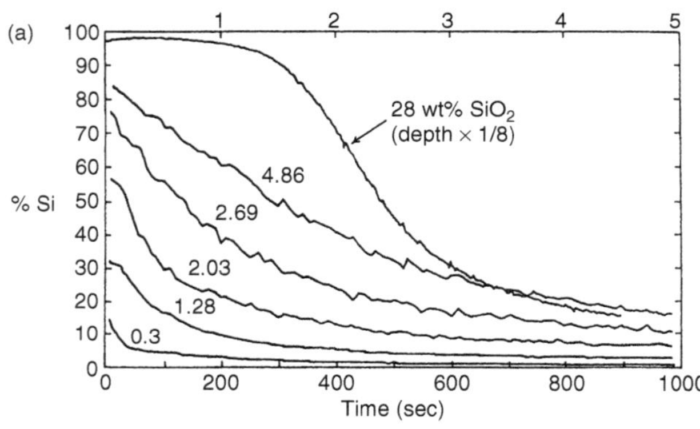

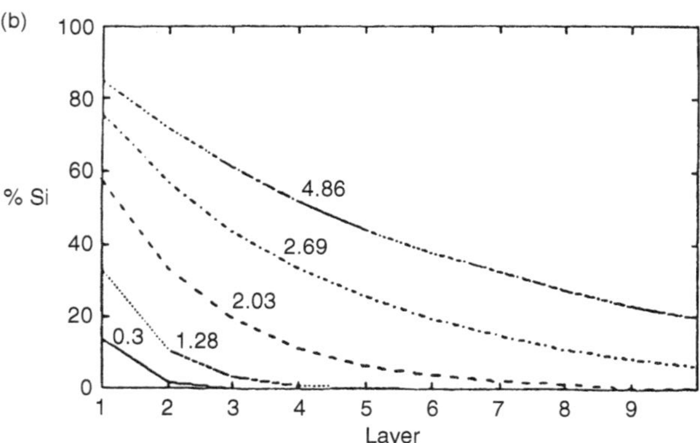

FIGURE 53.10 Depth profiles (a) of silica-coated alumina particles at different silica loadings obtained with SIMS. The calibration curve (28 wt% SiO_2) was compressed by a factor of 8. Model depth profiles (b) of silica-coated particles in the growth of the silica coating by random attachment of the silica units. The silica layer depth was converted to an approximate number of silica layers at 0.6 nm per layer. (Reprinted with permission from reference 7. Copyright 1989.)

DISCUSSION

In this work we tried to create conditions to deposit MSA on the hydroxylated surface of α-alumina in dense silica layers. The surface-sensitive electrokinetic measurements, DRIFT, SIMS, and XPS, show that the coating grows by deposition of molecular units on the surface of the alumina, whereas TEM, XPS, and surface area measurements show that thick, nonporous silica coatings can be grown. Characterization of particles coated with submonolayer or thicker MSA give insights into the nature of the coatings and the deposition mechanism.

At less than 5-wt% silica, there is some uncoated alumina surface. Just 1-wt% silica should be enough to coat the entire surface if it were deposited in a uniform monolayer (3.5 Å thick). The coatings become thicker even before complete coverage is obtained. Evidently, the silica is not deposited in complete layers on the alumina.

Although the silica is clearly not deposited complete layer by complete layer, the IEP, DRIFT, XPS, TEM, and SIMS data (Figures 53.5, 53.6, 53.9, and 53.10) show that the added silica is in the form of a coating on the alumina particles and is not present as an aggregated silica phase of silica particles. TEM data show this directly for coatings of large thickness. The other characterization methods show that the composition and chemistry of the particle surfaces are transformed from those of alumina to those of silica much more rapidly than the bulk composition is altered. For example, the SIMS data show that the particle surfaces are more than 80% silica when the bulk silica composition is 4.86 wt% (Figure 53.10). Also, if the silica was present as particles similar in size to the alumina particles and uniformly distributed through the alumina, the XPS-derived ratio n_{Si}/n_{Al} would approximate the bulk ratio of silica to alumina instead of the ratios 2 orders of magnitude higher that were measured.

IEP and SIMS characterizations can be used to compare the particle surface compositions dry and in aqueous solution. In Figure 53.11, the initial Si surface coverages and the IEP data are plotted together versus the amount of silica added. Schwarz et al. [9] showed a linear relationship between zero point of charge and silica content in a set of aluminosilicates. A more complex relationship is suggested by the curves in

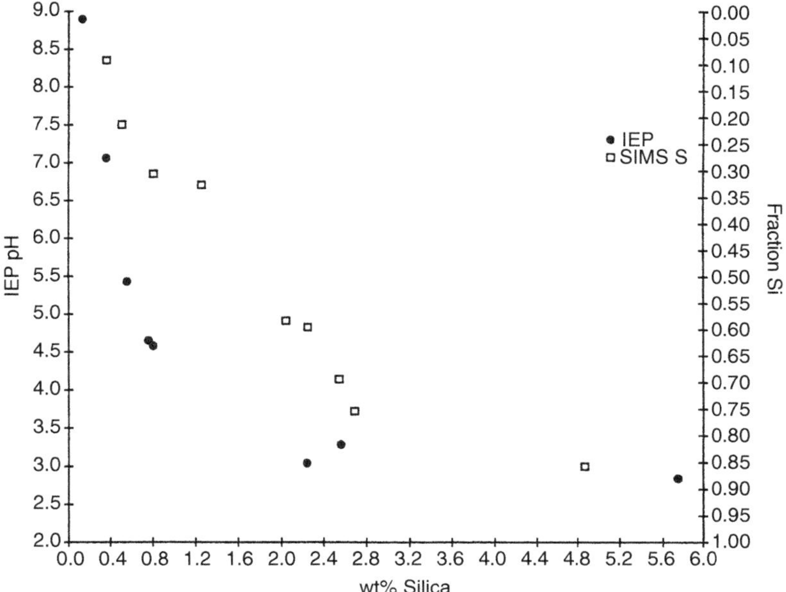

FIGURE 53.11 Relationship between isoelectric point, SIMS Si surface coverage, and total weight percent silica on the alumina surface. (Reprinted from reference 7. Copyright 1989.)

Figure 53.11 for these coated alumina particles, but the data suggest similar surface compositions in both dry and aqueous environments.

Plotted in Figure 53.8 and Figure 53.9 is Equation (53.2), the thickness calculated for uniform layers of silica on spherical alumina particles. The XPS- and TEM-derived coating thicknesses are in fair agreement with the assumption of complete coatings of constant thickness. The TEMs are consistent with the notion that the thick coatings are uniform. A comparison of model calculations and experiment shows that XPS is not sensitive to the details of the distribution of coating thicknesses [7].

The SIMS depth profile of the 28-wt% coating shows a plateau at nearly 100% Si for 10 nm before Al is detected (Figure 53.10). All of the alumina surface is covered at high loadings.

One mechanism of coating growth by random attachment of silica units to the particle surface with equal preference for bare alumina or previously deposited silica can be modeled. In this case, the rate of loss of alumina during growth is proportional to the fraction of alumina exposed. This mechanism predicts that the amount of bare alumina, n_{Al}, measured with SIMS should decay exponentially with added silica. A plot of $-\ln[n_{Al}]$ versus silica weight percent should yield a straight line. Such a plot is shown in Figure 53.12. The fraction of alumina exposed is merely 1 minus the silica fraction measured at the start of the depth profiles, as in Figure 53.10. The measurement is in good agreement with this model.

Growth of the silica coating by random attachment of the silica units also predicts that the depth profiles will have a certain form. The amount of silica two layers thick will simply be the square of the fraction that is one layer thick. In general, the fraction of surface coated with N layers of silica would be

$$f(N) = f(1)^N \tag{53.4}$$

where $f(N)$ is the fraction of the surface covered N layers thick. Figure 53.10b compares model depth profiles calculated for the fractional coverages measured at the start of the depth profiles to the complete depth profiles. The only adjustable parameter was the conversion of silica

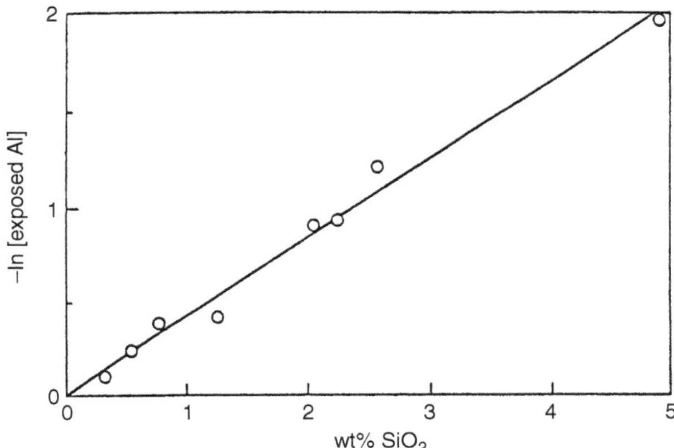

FIGURE 53.12 Plot of $-\ln[n_{Al}]$ as measured by SIMS versus weight percent silica. The linear plot indicates a random growth mechanism. (Reprinted from reference 7. Copyright 1989.)

layer number to depth, about 0.6 nm per layer. The size of a SiO_4 tetrahedron can be estimated to be 0.35 nm. The agreement between the calculated and observed profiles is quite good and gives good support to the random growth mechanism.

Random fixing of the silica units on the particle surface means either that the approaching units stick where they first contact the surface or that the adsorbed silica units are mobile but have no preference for where they become permanently fixed.

After complete coverage of the alumina surface is achieved at high silica loadings, the distribution of thicknesses will be Gaussian with a standard deviation of the square root of the average thickness. With the random growth mechanism established, the coating thicknesses and thickness distributions of silica coatings prepared in this way can be easily predicted.

ACKNOWLEDGMENTS

W.B. Hambleton, Jr., designed and automated synthesis facility and prepared the coated alumina samples; E.E. Carroll, Jr., made the surface area measurements; M.L. and M.J. Van Kavelaar performed the TEM measurements; B.F. Burgess and the Galbraith Laboratories of Knoxville, TN, made the ICP–AES determinations; R.E. Johnson, Jr., and J.V. Hughes, Jr., made the electrokinetic potential measurements; D.B. Chase developed the DRIFT spectra; and P.E. Bierstedt performed the XPS tests. We are grateful to D.P. Button for encouragement and support to do his work. We dedicate this chapter to the late Ralph K. Iler, who was an invaluable source of inspiration at the early stages of our work with MSA and zeolites.

REFERENCES

1. Button, D.P., private communication, 1986.
2. Iler, R.K. *The Chemistry of Silica*; Wiley: New York, 1979; p. 87.
3. Ulrich, D.R. *Chem. Eng. News* **1990**, *68*, 40.
4. Bergna, H.E.; Corbin, D.R.; Sonnidesen, J.C. U.S. Patent 4,683,334.
5. Alexander, G.B. *J. Am. Chem. Soc.* **1953**, *75*, 2887.
6. Bergna, H.E.; Corbin, D.R.; Sonnidesen, J.C. U.S. Patent 4,752,596.
7. Firment, L.E.; Bergna, H.E.; Swartzfager, D.G.; Bierstedt, P.E.; Van Kavelaar, M.L. *Surf. Interface Anal.* **1989**, *14*, 46.
8. Betz, G.; Wehner, G.K. *Top. Appl. Phys.* **1983**, *52*, 11.
9. Schwarz, J.A.; Driscoll, C.T.; Bhanot, A.K. *J. Colloid Interface Sci.* **1984**, *97*, 55.

Part 8

Uses of Colloidal Silicas

Robert E. Patterson
The PQ Corporation
James S. Falcone, Jr.
West Chester University

This section strives to bring the reader up to date on the practical applications of silica. We include four chapters that were previously published in *The Colloid Chemistry of Silica* and supplement them with diverse, yet fundamental discussions on the more current uses of silica. Excluded from our coverage of the uses of silica are commercially important, *naturally* occurring siliceous materials such as sand, quartz, opal, diatomaceous earth, and chert.

Three of the chapters from *The Colloid Chemistry of Silica* summarize many diverse uses for the various synthetic forms of silica. In Chapter 54, Payne surveys the manufacture and applications of silica sols. Then, in Chapter 55, Falcone discusses the general chemistry of soluble silicas, primarily as an introduction to the extensive literature on these chemically complex and industrially important inorganic polymers. In Chapter 58, Birchall explores the uses of silica by nature. Silicon has long been regarded as an essential nutrient, yet its precise place in biochemistry remains elusive. Birchall presents strong evidence that silicon is not bioessential per se, but rather that aluminum can be biotoxic unless inhibited by biologically available silica. In Chapter 60, Patterson covers silica gels and precipitates and discusses a less well-known but likely important use of silica as an essential sequestrant in biological systems.

We bring several new chapters to print. In Chapter 59, Falcone extends the discussion of silica chemistry and focuses it upon biological applications. In Chapter 57, Otterstedt and Greenwood describe in detail some specific applications of colloidal silica to such areas as the production of cement, lead-acid batteries, paper, and industrial coatings and polishing agents. The remaining five chapters in this section (Chapters 56, 61–64), by Bergna, describe other industrial uses of silica and include their methods of preparation.

54 Applications of Colloidal Silica: Past, Present, and Future

Charles C. Payne
Nalco Chemical Company

CONTENTS

Early uses of colloidal silica for catalysis, ceramics, paper, and textile applications, strength enhancement in rubber, tobacco treatment, and medicine are discussed. A historical view of the development of applications is highlighted, and future uses are discussed.

When Thomas Graham prepared silica colloids in the 1860s [1,2], he couldn't have envisioned its many applications today. The list of applications in Iler's 1979 edition of *The Chemistry of Silica* [3] is long and varied. Sometimes silica is used to promote adhesion and sometimes to prevent adhesion. These opposing properties from the same material indicate that applications involving silica sols can be quite complex. Bungenburg de Jong [4], in reviewing the origins of colloid science, pointed out that Graham introduced the term "colloids" for substances that "in solution" showed only a very slow diffusion velocity compared with other substances such as sugar and salts. Most of the early literature [5] made no distinction between silicates, polysilicates, and what ultimately has come to be known as colloidal silica. Colloidal silica has discrete particles that are generally somewhat spherical and amorphous.

PREPARATION PROCEDURES

EARLY

Thomas Graham was not the first person to attempt to prepare a colloidal silica. As early as 1747, Pott made a "semisolution of silica," and as early as 1820, a reference is made to the preparation of a sol of "hydrated silica" [6]. In 1853, a French researcher named Fremy [7] prepared a dilute colloidal silica from silicon sulfide. By 1864, silica colloids were being prepared by the dialysis of gels and by the hydrolysis of silicate esters [8]. All products were very dilute.

Colloidal silica technology proceeded slowly. The early products were not suitable for most applications because (1) only low silica concentrations were available, (2) the materials were not stable with time, or (3) the products did not have reproducible properties. By 1915, only one patent for preparation of a commercial colloidal silica had been issued [9]. Using electrodialysis (Figure 54.1) to make a product, Schwerin marketed a 2.4% silica sol as a stable material. Schwerin suggested the use of this product for medicinal purposes but did not specify how or where it should be used.

1941 TO 1963

The year 1941 was the turning point toward commercial silica sol production methods. Bird [10], at Nalco, found that low-molecular-weight silicic acids could be prepared by ion exchange. If a small amount of sodium hydroxide was added to these materials, particles were formed. These materials could then be concentrated with conventional evaporator techniques to about 27.5% silica. Nalco

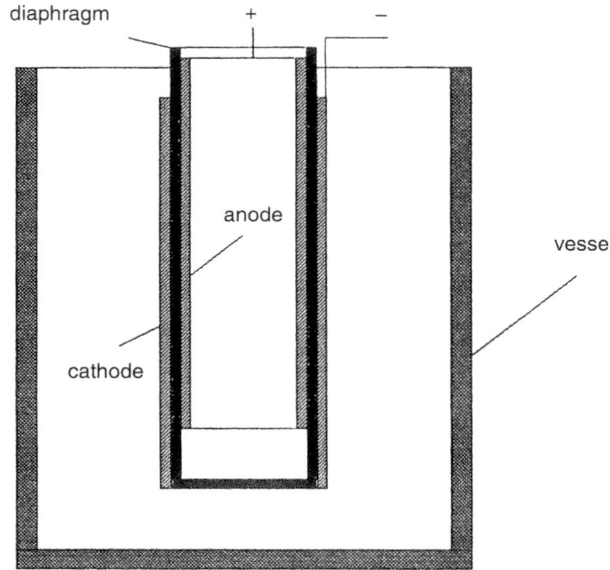

FIGURE 54.1 Electrodialysis cell of Schwerin [10].

licensed the ion-exchange procedure to DuPont, which continued to work on preparation throughout the 1940s. Monsanto [11–14], during the years 1941 to 1951, continued to develop products by peptizing silica gels and thus prepared materials with a broad particle-size distribution, and by 1963 products containing 50% silica were prepared by gel peptization techniques.

The year 1951 is significant for the preparation of colloidal silicas other than by the conventional gel peptization technique. Using Bird's idea of preparing low-molecular-weight silicic acids by ion exchange, Bechtold and Snyder [15], two DuPont scientists under Iler's direction, developed the ingenious method of growing the particles to any given size while concentrating the products to up to 35% silica. This research, coupled with the work by Joseph Rule [16], another DuPont scientist, and his coworkers, determined the parameters needed to keep the products stable at high silica concentrations; thus, the Bechtold and Snyder process became a practical method that turned colloidal silica into a stable product with predictable properties. By 1959, commercial products were available with up to 50% silica concentrations. Rule's work was important in making stable products, but it was also important for using these materials in different applications.

APPLICATIONS

To 1933

Griessbach [17] in 1933 summarized the various methods for preparing and using colloidal silicas up to that time. The methods of preparation of sols included dialysis, electrodialysis, gel peptization, and hydrolysis of the silicate esters or silicon tetrachloride. Silica sols were used in numerous applications at that time. Silica sols were a primary binder in the synthesis of catalysts used in the production of sulfuric acid or the dehydration of alcohols. In ceramics, they were used in glazes and other coatings. Cements and such materials as plaster of Paris could be coated with colloidal silica to improve their resistance to water and acidic substances. Colloidal silica was incorporated into paper and sprayed onto textiles or wood to either strengthen or protect the substrate. It was added to metal solutions such as silver or gold to improve stability so that these materials could be used in medicines. Silica in combination with surface-active agents also showed tendencies to emulsify, especially when the sol became destabilized and the viscosity of the system increased. Rubber latex emulsions were coagulated with colloidal silica to strengthen the final rubber product. A novel use for its time was the addition of colloidal silica to tobacco to help with the fermentation process and to aid in the adsorption of nicotine. Apparently, the fact that silica adsorbed and retained water helped to keep the tobacco fresh. The use of silica sols in medicine appears to be an area of great speculation in 1933. Colloidal silica was claimed to be useful for treating subcutaneous wounds, tuberculosis, and many circulation problems such as hardening of the arteries. Griessbach [17] also suggested the immobilization of such enzymes as those involved in the conversion of amylopectin into simple starches.

Many of these applications are practiced today. Other applications exist only for specialized situations. For example, silica sols are a desirable source of silica for catalyst substrates. For catalyst preparations, however, colloidal silica is generally used only for specialized applications.

Griessbach [17] showed that five companies produced silica sols in 1933. Most products were very dilute, and only one could be considered concentrated. The most concentrated sol available at that time was a product called Kieselsol I.G. (made by I.G. Farbenindustries), which contained 10% silica and was stabilized with ammonium hydroxide. The characteristics of most of the sols were undetermined. Moreover, the difficulty of making a reproducible product that would perform in a predictable manner for specific uses was technically impossible at that time. This difficulty, most likely, was the major reason that colloidal silica applications did not increase rapidly. Iler [18] pointed out that colloidal silica was not accepted for wide commercial use until methods were discovered for producing sols with high concentrations that would not gel or settle with time. The first steps to achieving that goal occurred in 1941 (discussed in the preceding section on preparation procedures).

1933 to 1955

Iler's 1955 edition of *Colloidal Chemistry of Silica and Silicates* [5] devotes seven pages to the uses of silica sols at that time. The areas of use included floor waxes, textiles, organic polymers, water treatment, and miscellaneous areas such as in cements and as a binder for luminescent materials used in television picture tubes.

Two areas of major interest in 1955 were paper antiskid and investment castings. Wilson [19] used colloidal silica to increase the friction between paper surfaces. In this application (Figure 54.2), the silica sol is typically diluted to 1–7% as silica solids and then applied to multiwall bags, corrugated boxes, and linerboards. Various methods of applying the silica sol were used. Typical applicators were sprays, segmented applicator rollers, knife blades, full roll coaters, sponges, and brushes.

Testing to ensure that the colloidal silica has been applied correctly consisted of determining either static or dynamic coefficients of friction with a slide angle tester (Figure 54.3). The coefficient of friction is defined as the numerical value for the tangent of the angle needed to start sliding (static) or the value needed to maintain sliding (kinetic).

Collins [20] used colloidal silica as a binder for investment casting applications. Investment casting techniques are useful for the production of aircraft parts, various industrial castings, dental and jewelry parts, and sporting equipment such as golf club irons and propellers for outboard motors. In a typical procedure, colloidal silica is mixed with a refractory grain such as quartz or alumina, with or without a gelling agent, formed into a ceramic mold, dried, and fired. Other investment castings with expendable patterns included ceramic shell molds; precoats for solid molds; backups for solid molds, "tamp and pack" methods, and solid molds with no precoats.

Colloidal silica was used in general foundry applications such as gunning mixes, ceramic mold facings for core boxes and the like, mold washes, and semi-permanent molds with renewable facings. It was also used in the production of wallboard (Figure 54.4).

1955 to 1962

In 1962, Monsanto [21] reviewed the patent literature and listed the promising applications for colloidal silica at that time. They grouped the areas of application into the categories cellulosics, ceramics, electricity, floor waxes, flotation, insulating coatings, refractory molds, paints, photography, printing, paper, rubber, textiles, and miscellaneous; "miscellaneous" tends to involve some type of coating or inorganic–organic composition.

An area of interest at that time was the incorporation of colloidal silica into emulsion systems. Typical emulsions included floor waxes or rubber latices. Conventional wax compositions ordinarily included certain extenders or modifiers in the wax dispersions. These may comprise wax-soluble or water-dispersible resins (natural or synthetic). These formulations can produce coatings with a pleasing appearance; however, many lack slip resistance. The incorporation of colloidal silica into the formulation produces coatings that prevent slipping.

Iler [22,23] described typical formulations with Carnuba wax, which is dispersed in water containing colloidal silica, a fatty acid like oleic acid, triethanolamine, and potassium hydroxide. This composition can then be applied, for example, as a thin coating to linoleum floor.

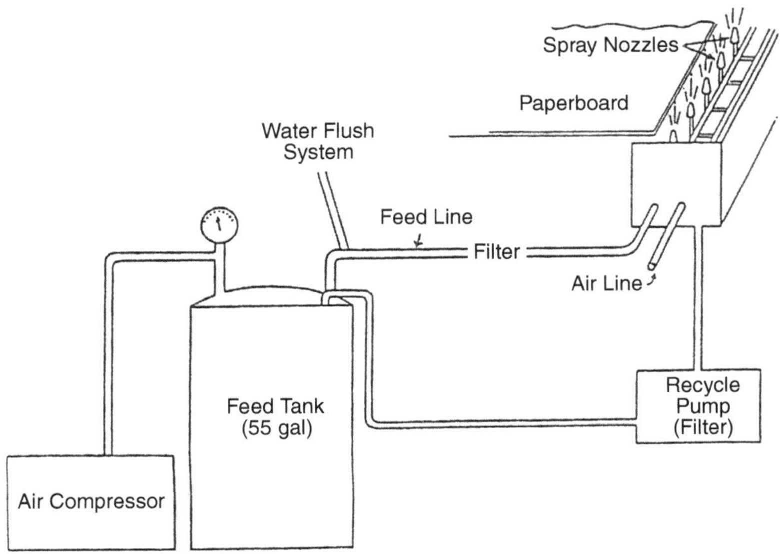

FIGURE 54.2 Spray unit for paper antiskid application.

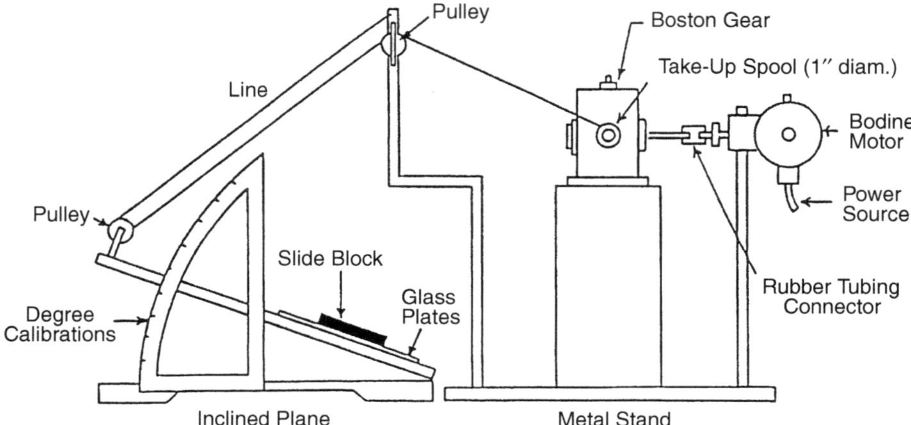

FIGURE 54.3 Testing equipment for skid resistance.

The product forms a lustrous coating without rubbing while providing a slip resistance.

Mixtures of rubber latices or elastomer foams were modified with colloidal silica to give improved properties. Typical processes involved drying, gelling, or coagulating the colloidal silica within the elastomer system. Silica sols were used with phenolic, formaldehyde-based, melamine, polyester, acrylic, vinyl or styrene polymer–copolymer, polyamide, and styrene–butadiene rubber systems to provide strength to films and coatings.

A typical rubber formulation, for example, that demonstrates the surface area properties of silica sol can be seen in the Talalay process for making foam rubber [24]; the elastomer foam is in a latex form, to which is added an accelerator, an antioxidant, a vulcanizing agent, and carbon black. The mixture is then foamed, either mechanically or chemically, gelled, vulcanized to make the rubber, washed, and dried. The dried material is then postdipped in colloidal silica to reinforce the walls and prevent the crumbling effect seen on poorly made foam rubber pillows and so forth. Originally, a 20-nm silica sol was used. If a colloidal silica with a smaller particle size is used, the corresponding higher surface area allows less silica to be used in the system; thus the cost of the whole process is reduced.

1962 TO 1979

When Iler's book, *The Chemistry of Silica*, came out in 1979 [3], the section on the applications for colloidal silica had increased to 21 pages, compared to seven pages in the 1955 book [5]. The increase in the number of applications was largely due to the efforts of Iler. As technical manager of DuPont's colloidal silica area, Iler

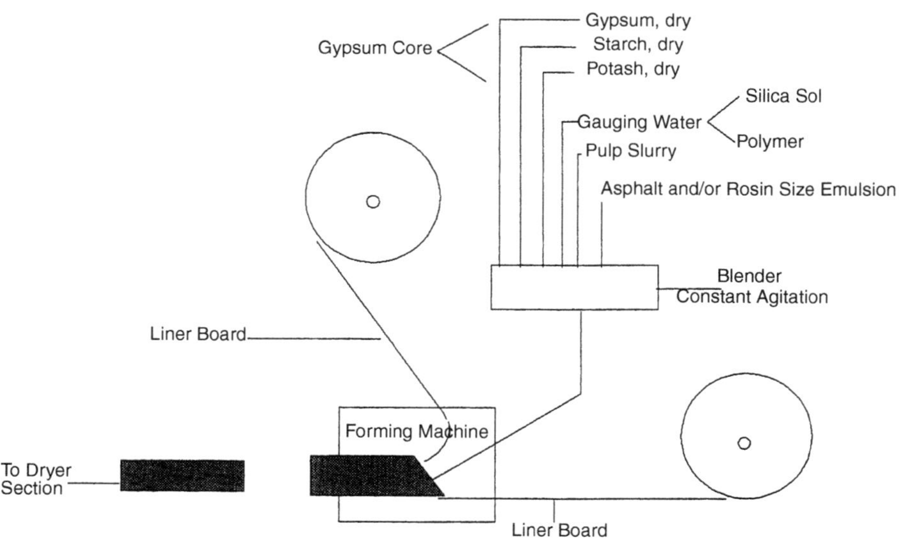

FIGURE 54.4 Typical flow diagram of wallboard production.

enlisted the help of some excellent researchers whose job it was to develop new application areas. As a result, numerous application patents were issued during this time period. The areas of application were many and varied during the period of 1962 to 1979. They can broadly be subdivided into binding and nonbinding systems. For example, silica sols are used as a binder for thermal insulation (Figure 54.5) or for catalyst manufacture and as a polishing agent for silicon wafers in a nonbinding application.

Stiles, McClellan, and Sowards [25–28] showed that colloidal silica is a good source of raw material for making the base catalyst. Silica sols have the advantage of a uniform distribution and known particle size. Thus, the catalyst manufacturer can predict the pore size and volume of the final product. Colloidal silicas also offer the advantage of low levels of sodium (compared to sodium silicate), a known catalyst poison, and therefore require less washing and processing to remove the unwanted cation. In addition, colloidal silicas, because they are liquid, can be mixed with metal salts such as bismuth and molybdenum and spray-dried. Uniform dispersion of the silica with the expensive "active" metals gives a highly desirable catalyst with no "hot" spots. Consequently, high yields of expected products result. Such catalysts can be used in the conversion of propylene and ammonia to acrylonitrile [29,30].

One application area between 1962 and 1979 that has an impact on today's problems is the manufacture of wallboard. Wallboard was prepared by sandwiching a gypsum mixture between heavy papers. A typical core consists of gypsum, starch, potash, a pulp slurry, an asphalt or rosin-size emulsion, and gauging water (i.e., a colloidal silica plus polymer binder system). The mixture is blended together and placed between liners made on linerboard machines.

One way this application can be used to solve today's problems was presented early in 1990 by Air Products and Chemicals Inc. A power plant in Indiana was going to use a high-sulfur coal to generate its electricity. The acidic gas emissions, which consist of sulfur dioxide, was scrubbed with a limestone plus water mixture. The resulting calcium sulfite is oxidized to form calcium sulfate, which is then washed, centrifuged, rinsed, and dried. The final product is gypsum that contains less than 10% water. Because the average modern house is estimated to use 8000 square feet of wallboard, the process solves the gas emission problem while producing a low-cost construction material.

Colloidal silicas can also be used as a fine polishing abrasive for silicon wafers. Typical operations [31,32] involve feeding a dilute silica "slurry" (i.e., silica sol) that has been adjusted with a caustic agent onto a revolving polishing wheel containing a polishing pad. The silicon wafer is fixed to a polishing head and placed in contact with the polishing pad. As the system rotates, the high pH and the abrasive property of the colloidal silica removes silicon from the surface of the water. The resulting wafer has a mirrorlike surface onto which an electrical circuit can be placed.

Other applications of this time period can be found in Iler's discussion and references [3].

1979 to Present

The literature from 1979 to the present shows that the major application areas for colloidal silica involve coatings and "inorganic–organic compositions." Most of these compositions use the silica sol in a binder application. Japanese research far exceeds that of all other countries.

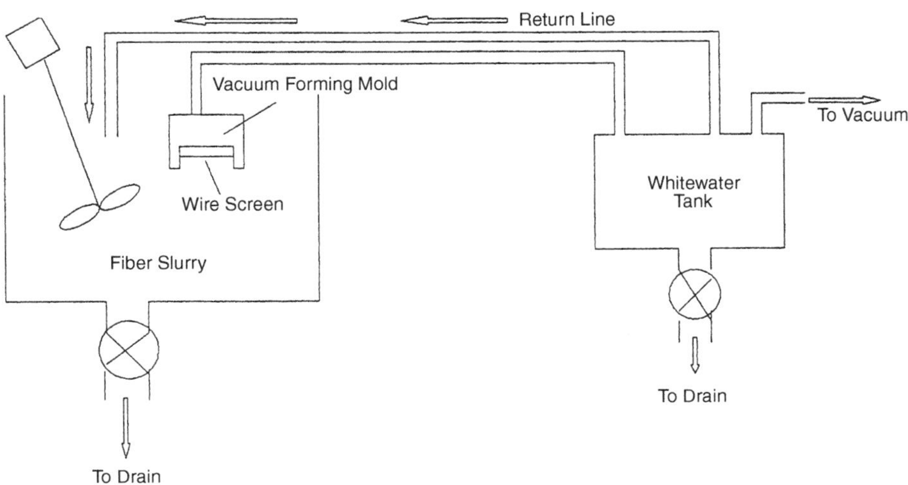

FIGURE 54.5 Vacuum forming process for preparing thermal insulation.

Coatings is a diversified application area involving a substrate and some type of surface covering. The type of substrate material to be coated can be plastics such as polycarbonate or polypropylene, metals, silicone rubber, ceramic- or refractory-like materials, fabrics like nylon and rayon, photographic films, paper materials, walls (painted or not), eyeglass lenses, inorganic fibers such as fiberglass, stone for buildings, investment casting molds, leather or leather substitutes, and concrete surfaces. Typical coating materials tend to be polymeric in nature, although some application areas use the colloidal silica without organic polymers. For example, Nippon Steel Corporation [33] disclosed the use of a colloidal silica in combination with phosphates and chromic oxide or chromate to make a nondirectional magnetic film for steel sheets. Toshiba Silicone Company, Ltd. [34], on the other hand, uses a silane–colloidal silica mixture to coat glass lenses.

Inorganic–organic composition uses are similar to coatings uses except that no substrate is involved. A variety of mixtures can be used in which the colloidal silica is added to impart strength. An example of this type of system was disclosed by Asahi Glass Company, Ltd. [35]; colloidal silica was added to isophorone diisocyanate and a pentaerythritol triacetate–tetraacetate mixture to produce a material for a polycarbonate plate.

Colloidal silica continues to be used in the conventional application areas such as catalysis, paper antiskid, refractory insulation, and photography. Its use, however, is dwarfed by that of coatings and inorganic–organic compositions.

The Future

Two types of silica sol products are needed in the 1990s: specialty products and organosols. Specialty products are used in high-technology areas. Price is generally unimportant if they work. Typical examples of specialty products are monodisperse sols [i.e., one particle size, low sodium, and low metal (aluminum, iron, etc.) concentrations, and no aggregation].

Monodisperse silica sols are important in the previously mentioned Talalay process for making foam rubber pillows. If a 4-nm silica sol is normally used, then a substitution of a 3.5-nm particle at the same silica dosage would theoretically increase the strength of the rubber by 14.3%. If, however, the particle size is changed from 4.5 to 5.0 nm, the strength of the rubber would decrease by 11 to 20% at the same silica dosage. Under these conditions, the crumbling effect of the foam rubber would return.

The second application area of interest is organosols. Organosols involve the use of nonaqueous systems, for example, in making magnetic colloids and recording media, high-technology ceramic composites, and catalyst supports [36]. The high solid loadings, the lack of aggregation, and the improved uniformity of the colloid and the final product can only be obtained from colloidal systems. Silica organosols are dispersions of silica colloids in an organic solvent. Silica organosols can be used as a low-temperature binder [37], as an adhesion promoter (Schmidt, K. E., personal communication), or as a silica source for magnesia refractories. For example, magnesia refractories use silica as a binder to make high-temperature materials through the formation of a magnesium silicate called Fosterite. If a silica sol in water is used, the magnesia grain reacts to form magnesium hydroxide. The reaction with silica then proceeds by a different phase diagram. The firing temperature must be raised to almost 2000°F (1100°C) before the cold crush strength increases with increasing firing temperature. When an organosol is used, the cold crush strength increases with increasing firing temperature immediately, as expected.

Currently, there are now greater demands on those industries in which colloidal silica has been used in the past to develop higher quality products. Thus, a cement to which colloidal silica was added to improve its strength properties years ago must now be UV-resistant and also have better than 90% reflectivity [38]. One application of current interest is in the area of papermaking [39–43]. Colloidal silica is mixed with starch and slurried paper fibers. It is then formed into paper on the wire of a paper machine. The addition of the silica sol to the system gives improved dewatering of the flocced slurry as well as a high retention of the paper fines and fillers. The result is lower heating costs for drying the paper and better mat formation. Considerable savings can be realized for the papermaker using this system. Greater retention properties mean that more recycled, and therefore cheaper, paper can be added to the paper furnish (formula) without sacrificing strength, because recycled paper contains more fines, and paper strength depends on fiber–fiber interactions.

CONCLUSIONS

Where will future applications come from? One source is universities, and a second is extension of existing applications. Some applications of colloidal silica originating in universities include the following:

- grinding aid for pharmaceutical formulations (Modena University, Modena, Italy, 1988) [44]
- stabilizing agent for organic compounds subject to temperature–humidity degradation conditions (Hamburg University, Hamburg, Germany, 1989) [45]
- mixtures (silica sol + montmorillonite clay) to give pillared clays with super galleries (Michigan State University, East Lansing, MI, 1988) [46]

- catalytic agent for hydrolysis of silane-coupling agents (Nihon University, Tokyo, Japan, 1988 [47]

Silica applications are, indeed, a global endeavor.

Extending ideas of other applications involves the technique known as an "association of ideas." For example, in U.S. Patent 4,637,867 [48], colloidal silica is used as a dispersing agent (and probably a crystal modifier) for preparing fine crystal particles of maleic anhydride. This same idea can be used for making crumb rubbers in elastomer systems [49] into fine-grain emulsions in photographic systems [50].

One might argue that silica sols can easily be used as a dispersing agent, but that this kind of argument does not apply to other systems. The use of colloidal silica as a frictionizing agent, however, shows that the technique works for other systems. Historically, silica sols have been used in paper antiskid applications, floor waxes, hot pressing railroad engine drive wheels, and polishing applications. One Japanese researcher used the frictionizing properties of silica for improving nonwoven fiber black-board erasers. Extending this same idea, why not use silica sols for cleaning cloths and floor mops or for rubber or nonrubber pencil erasers?

Historically, colloidal silica has been tried in many applications. In many cases, the reason that it is not used is not that it didn't work, but the cost of the products is more than the customer wants to pay for improvements. Colloidal silica has been used in the making of opals, in the balancing of large-diameter industrial saws, as an anticaking agent for explosives, and as a frictionizing agent for baseball bats to improve the hitting performance of minor league players. Colloidal silica will continue to be a versatile product with an applicability limited only by the imagination of the researcher.

REFERENCES

1. Graham, T. *J. Chem. Soc.* **1862**, *15*, 216.
2. Graham, T. *J. Chem. Soc.* **1864**, *17*, 318.
3. Iler, R. *The Chemistry of Silica*; Wiley-Interscience: New York, 1979; pp 415–438.
4. Bungenburg de Jong, H. G. In *Colloid Science*; H. R. Kruyt, Ed.; Elsevier: New York, 1949; Vol. II, pp 1–5.
5. Iler, R. *The Colloidal Chemistry of Silica and Silicates*; Cornell University Press: New York, 1955; p 87.
6. Hauser, E. A. *Silicic Science*; Van Nostrand: Princeton, NJ, 1955; p 54.
7. Fremy, E. *Ann. Chem. Phys.* **1853** (3), *Bd. 38*, S312–344.
8. Graham, T. *Ann. Chem.* **1862**, *Bd 123*, S860–861.
9. Schwerin, B. U.S. Patent 1,132,394, 1915.
10. Bird, P. U.S. Patent 2,244,325, 1941.
11. White, J. F. U.S. Patent 2,285,477, 1942.
12. White, J. F. U.S. Patent 2,375,738, 1945.
13. Trail, H. S. U.S. Patent 2,572,578, 1951.
14. Trail, H. S. U.S. Patent 2,573,743, 1951.
15. Bechtold, M. F.; Snyder, O. E. U.S. Patent 2,574,902, 1951.
16. Rule, J. M. U.S. Patent 2,577,485, 1951.
17. Griessbach, R. *Chem. Ztg.* **1933**, *57*, Nr 26, S253–260, 274–276.
18. Iler, R. *The Colloidal Chemistry of Silica and Silicates*; Cornell University Press: New York, 1955; p 89.
19. Wilson, I. V. U.S. Patent 2,643,048, 1953.
20. Collins, P. F. U.S. Patent 2,380,945, 1945.
21. Monsanto Technical Bulletin I-237, *Monsanto Silicas for Industry*, 1962.
22. Iler, R. U.S. Patent 2,597,871, 1947.
23. Iler, R. U.S. Patent 2,726,961, 1955.
24. Talalay, A. et al. U.S. Patent 2,926,390, 1960.
25. McClellan, W. R. U.S. Patent 3,415,886, 1968.
26. Stiles, A. B.; McClellan, W. R. U.S. Patent 3,497,461, 1970.
27. Sowards, D. M.; Stiles, A. B. U.S. Patent 3,518,206, 1970.
28. McClellan W. R.; Stiles, A. B. U.S. Patent 3,678,139, 1972.
29. Callahan, J. L. U.S. Patent 2,974,110, 1961.
30. Callahan, J. L.; Szabo, J. J.; Gertisser, B. U.S. Patent 3,322,847, 1967.
31. Walsh, R. J.; Herzog, A. H. U.S. Patent 3,170,273, 1965.
32. Sears, G. W. U.S. Patent 3,922,393, 1975.
33. Nippon Steel Corp., Japanese Patent 57/192,222, 1982.
34. Toshiba Silicon Company, Ltd. Japanese Patent 60/166,355, 1985.
35. Asahi Glass Company, Ltd. Japanese Patent 60/137,939, 1985.
36. Smith, T. W. U.S. Patent, 4,252,671, 1979.
37. Bikadi, Z.; Guder, H. European Patent EP 390276, 1989.
38. Page, C. H.; Thanavala, D. N.; Thombare, C. H.; Kamat, R. D.; Bapat, V. S. Indian Patent 163,979, 1988.
39. Batelson, P. G. Canadian Patent 1,154,564, 1983.
40. Andersson, K.; Andersson, K.; Sandstrom, A.; Stroem, K.; Basla, P. *Nord. Pulp Pap. Res. J.* **1986**, *1* (2), 26–30.
41. Johnson, K. A. U.S. Patent 4,643,801, 1987.
42. Sofia, S. C.; Johnson, K. A.; Crill, M. S.; Roop, M. J.; Gotberg, S. R.; Nigrella, A. S.; Hutchinson, L. S. U.S. Patent 4,795,531, 1989.
43. Rushmere, J. D. U.S. Patent 4,798,653, 1989.
44. Forni, F.; Coppi, G.; Iannuccelli, V.; Vandelli, M. A.; Cameroni, R. *Acta Pharm Suec* **1988**, *25* (3), 173–180.
45. Krahn, F. U.; Mielck, J. B. *Int. J. Pharm.* **1989**, *53* (1), 25–34.
46. Moini, A.; Pinnavaia, T. J. *Solid State Ionics* **1988**, *26* (2), 119–123.
47. Nishiyama, N.; Asakura, T.; Houe, K. *J. Colloid Interface Sci.* **1988**, *124* (1), 14–21.
48. Herbst, R. U.S. Patent 4,637,867, 1987.
49. Payne, C. C., unpublished data.
50. Saleck, W.; Himmelmann; Huckstadt, H.; Meyer, R. Belgian Patent 766095, 1971.

55 The Uses of Soluble Silica

James S. Falcone, Jr.
West Chester University, Department of Chemistry

CONTENTS

The uses of soluble silica are reviewed in a concise manner. Ideas originally summarized by Ralph K. Iler are expanded and discussed with up-to-date references. Key technological factors associated with the application of soluble silica in the broad categories are described. The categories are: Adhesives, binders, and deflocculants; Cleaners and detergents; and Raw materials. These technological factors are related to current understanding of the chemical properties of these complex inorganic polymer solutions.

Ralph K. Iler [1], in *The Chemistry of Silica*, classified the uses of the soluble silicates into three categories:

- Adhesives, binders, and deflocculants' function is dependent primarily on the presence of polysilicate ions. The soluble silicate used has a ratio range of 2.5 to 4.0, where the ratio is defined as one-half of the ratio of moles of Si to moles of cation (e.g., Na^+ or tetramethylammonium ion, TMA^+).
- Cleaners and detergents' function is primarily due to controlled alkalinity using silicates with ratios generally lower than 2.5.
- Raw materials for the production of precipitated forms of silica, sols and gels from solutions with ratios equal to 3.3 or greater.

An analysis of the trends in the U.S. production statistics for sodium silicates from the early 20th century to date is shown in Figure 55.1. It suggests a macro growth pattern in volume of roughly 10 million kg/yr when the silicate production is calculated as 40° Bé 3.2 ratio equivalent (Bé refers to the Baumé density scale). Growth plateaus as seen in the 1930s and the 1960s reflect, in a way, the three categories in the order described. In the first third of the century, silicate uses were predominantly in the first class. In mid-century there was a rise in the use of silicates in synthetic detergents. The overall volume of silicates grew after World War II in spite of great volume losses in applications of adhesives to both natural and synthetic organic polymers. New growth in production the final third of the century is due to increased use of soluble silicates as intermediates in making many classes of silica-based performance materials.

The diversity of uses for the alkali silicates is a result of both their structural complexity [2,3] and their complex reactivity. One might view them [4] as silica dissolved and/or dispersed in an hydroxide ion-rich aqueous system. In the chemical processing industry (CPI), they are valued as a reactive source of $(SiO_2)_n$ structural units. Soluble silica as an intermediate can be reacted with acids and bases to form a wide range of final products ranging from seemingly simple condensed forms of relatively pure noncrystalline silica, precipitates, gels, and sols, to highly complex crystalline metallosilicates like those found in the broad class of aluminosilicates, zeolites. The key properties of an intermediate are likely to be silica concentration, ratio, supporting cation, type and level of impurities, and consistency in these factors. These factors will be particularly important in the manufacture

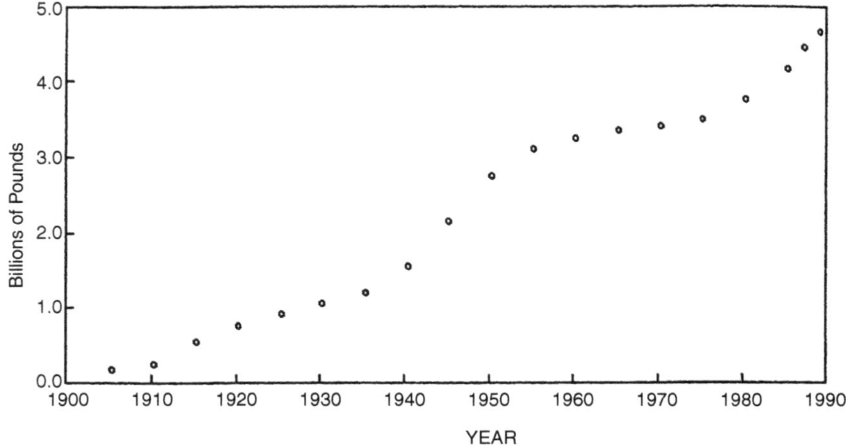

FIGURE 55.1 Sodium silicate production since 1900 as 40° Bé 3.3 ratio solution (data smoothed to show trends). CGR since 1925 is 2.6%.

of catalysts, highly selective sorbents, and other high-performance materials where trace metals could effect final product performance.

When alkali silicates are used as components, rather than reactants, in systems where partial contributions to overall performance are the dominant role, the utility factors are generally not as easy to identify. This is because these systems usually depend on the surface and solution chemical properties of the wide range of highly hydrophilic polymeric silicate ions deliverable from soluble silicate products or their proprietary modifications. In most cases, however, one or two of the many possible influences of these complex anions clearly express themselves in final product performance at a level sufficient to justify their use.

These major values may be seen as resulting from the following broad functions: hydrogen ion buffering, metal ion complexation, and specific adsorption. The sodium silicates are salts of a weak acid, silicic acid ($pK_a \sim 9.8$), and a strong base, sodium hydroxide. Their chemistry is complicated by the fact that silicic acid is by no means a well-defined substance. In fact, it appears that silicic acid might be best viewed as a complex hydrous polymer which varies in Si, O, and H composition and connectivity between Si–O–Si or number of shared SiO_4 tetrahedra corners. These variations influence silanol acidity and number of bonded and non-bonded oxygen atoms. The possible structures are difficult to study directly and model quantitatively; however, they are indirectly observable from the patterns of behavior in systems containing them. Evidence from studies of the interactions of soluble silica, metal ions, and oxide surfaces strongly suggests the presence of a wide range of metastable soluble silica species influencing the balance of other interaction in the system. For example:

- Silica adsorbs on gamma–Al_2O_3 in a broad pH range [5,6], which might be explained by assuming that the silica species in solution have a similarly broad range of pK_a values (between 6 and 10). The silica adsorption does not exhibit the sharp maximum value normally seen for weak acids at the pH value equal to the acid's pK_a value. Silica adsorption was reduced by the presence of divalent metal ions, possibly as a result of reduced silicate species activity.
- The addition of solutions of soluble silica to oxide mineral suspensions increases the magnitude of the negative surface charge on the mineral particles with higher ratio silicates (increased oligomers and higher polymers) being more active. Soluble silica species also attenuate the influence of multivalent cations on the surface charge [7,8].
- Highly polymerized silicate anions appear to interact with metal ions in solution in a manner analogous to silica gel, and the interaction decreases as the degree of silicate polymerization decreases [9]. This behavior is consistent with Iler's generalization [10] that silica suspended in solutions of polyvalent metal ions begins to adsorb these ions when the pH value is raised to within 1 to 2 pH units of the hydroxide activity at which the corresponding metal hydroxide would precipitate. It is likely that much of the behavior of silica hydrogel and the larger soluble silicate polymers towards metal ions can be attributed to metal ion adsorption in the interfacial region of the larger silica oligomers/polymers. This results in localized concentration near the siliceous surface in excess of the metal ion hydroxide solubility product [11].
- It also is well known that the addition of acids and bases shifts the polymer equilibrium in

solutions of silicates as a result of the following generalized scheme:

$$\gt Si{-}OH + {}^{-}O{-}Si\lt \longrightarrow \gt Si{-}O{-}Si\lt + OH^-$$

With these generalizations in mind, one can interpret the performance enhancements that are seen when soluble silicates are added to many industrial processes.

Controlled Buffer

By the appropriate choice of ratio of silica to NaOH, one can effectively buffer hydrogen ion activity in the important industrial range of 9 to 11. When systems performance is dependent on anionic surfactants and sequestrants, this property is valued greatly.

Corrosion Inhibition

The presence of soluble silica in water exposed to various metals leads to the formation of a surface less susceptible to corrosion. A likely explanation is the formation of 'metallosilicate complexes' at the metal water interface after an initial disruption of the metal oxide layer and formation of an active site. This modified surface is expected to be more resistant to subsequent corrosive action via lowered surface activity and/or reduced diffusion.

Red Water Control

Dissolved metal ions like iron and manganese play havoc with the aesthetics of ceramics systems that come into contact with waters containing them (e.g., bathtubs, wash basins, and stucco sprayed by lawn sprinklers). The addition of high-ratio silicates will effectively eliminate this problem through the formation of 'metallosilicate complexes,' which remain suspended in the water [12].

Bleach Stabilization

When added to hydrogen peroxide based bleaching systems, soluble silicates are known to significantly enhance bleach performance. Many hypotheses have been put forth to explain this process, including buffering, peroxysilicate complex formation, and modification of peroxide equilibrium. However, the most recent and plausible explanation is that the silicate inactivates iron and manganese species which catalyze peroxide decomposition [13].

Controlled Gelation

Solutions of soluble silica can be added to permeable matrices and jelled in place either by post-addition of acid, the action of acids in place, or the controlled reaction of a gelation agent added directly with the silicate. For example, CO_2 is used in foundry applications where silicate jelled by CO_2 gas binds sand to make metal molds. Hydrolyzable esters, like diacetin or triacetin, can be mixed with silicate solutions, yielding time-controlled setting systems. Recently, mixtures of portland cement and silicates have been used to gel liquid waste systems. Various proprietary systems involving soluble silica are used to accomplish objectives which involve stabilizing soils and blocking fluid flows. The strengths and set times of these systems are generally a function of the concentration of the silica and the pH value of the solution.

Coagulation–Dispersion

Soluble silica, particularly in the form called activated silica, is used in water treatment as a coagulant aide, where it improves the settling properties of alum-induced flocs, apparently via densification. In addition, silicate added by itself to a mineral suspension can cause beneficial conditioning, coagulation, or dispersion, depending on the treatment levels and system. Judicious use of silicate can lead to an enhancement of mineral flotation processes which employ hydrophobic collectors, generally by dispersion improvement and suppression of the flotation of unwanted oxides. This is done by maintaining negative surface charges and/or hydrophilic surfaces on unwanted components. However, silicates if overused can lead to total suppression of a mineral slurry due to the adsorption of silicate on all minerals, which causes all surfaces to become hydrophilic, and thus non-floating.

Antiredeposition–Sacrificial Agent

The universal suppression described, while it is a negative in mineral flotation, becomes a positive value in such applications as de-inking and detergency, where the maintainance of particulate surfaces in a negative charge state aids in the intended separation action (e.g., soil from cloth and ink from pulp fiber). In a related manner, the presence of anionic polysilicates will often improve a system whose performance is dependent on the effectiveness of anionic surfactant or polyelectrolytes. This improvement is due to a sacrificial effect whereby the silicate preferentially "sorbs" on active sites in the system and helps to maintain high activity for the costly active ingredients which in the absence of silicate would be adsorbed.

Deflocculation

Soluble silicates suppress the formation of ordered structures within clay slurries, thus increasing the solids which can be incorporated into a clay water system. This interesting surface phenomenon finds practical expression in the manufacture of bricks and cement.

These examples attest to the diversity of values to be extracted from these important industrial materials. This brief review only highlights the applied chemistry in this system. The interested reader is directed to the general references that follow for further insight.

REFERENCES

1. Iler, R. K. *The Chemistry of Silica*; Wiley: New York, 1979; p 667.
2. Dent Glasser, L. S.; Lachowski, E. E. *J. Chem. Soc. Dalton Trans.* **1980**, *393*, 390.
3. Falcone, Jr., J. S. In *Cements Research Progress — 1988*; Brown, P. W., Ed.; American Ceramic Society: Westerville, OH, 1989; 277.
4. McLaughlin, J. R. "The Properties, Applications and Markets for Alkaline Solutions of Soluble Silicate", paper presented to CMRA in New York, May, 1976.
5. Hingston, F. J.; Atkinson, R. J.; Posner, A. M.; Quirk, J. P. *Nature (London)* **1967**, *215*, 1459.
6. Huang, C. P. *Earth Planet. Sci. Lett.* **1975**, *27*, 265.
7. Hazel, F. J. *J. Phys. Chem.* **1945**, *49*, 520.
8. Tsai, F.; Falcone, J. S. paper presented at ACS/CSJ Chem. Congress, Honolulu, HI, Apr. 6, 1979.
9. Falcone, Jr., J. S. In *Soluble Silicates*, ACS Symposium Series 194; Falcone, Jr., J. S., Ed.; American Chemical Society: Washington, DC, 1982; p 133.
10. Iler, R. K. *The Chemistry of Silica*; Wiley: New York; p 667.
11. Ananthapadman, K. P.; Sumasundaran, P. *Colloid and Surf.* **1985**, *13*, 151.
12. Browman, M. G.; Robinson, R. B.; Reed, G. D. *Environ. Sci. Technol.* **1989**, *23*, 566.
13. Colodette, J. L.; Rothenberg, S.; Dence, C. W. *J. Pulp and Paper Sci.* **1989**, *15*, J3.

ADDITIONAL READING

Falcone, Jr., J. S.; Boyce, S. D. In *Encyclopedia of Polymer Science and Engineering*, 2nd ed.; Kroschwitz, J. I., Ed.; Wiley: New York, 1989; Vol. 15, pp 178–204.

Liebau, F. *Structural Chemistry of Silicates*; Springer-Verlag: Berlin, Germany, 1985.

Dent Glasser, L. S. *Chemistry in Britain* **1982**, Jan, 33.

Ingri, N. In *Biochemistry of Silica and Related Compounds*; Bendtz, G.; Lindquist, I., Eds.; Plenum: New York, 1978; p 3.

Barby, D.; et al. In *The Modern Inorganic Chemical Industry*; Thomson, R., Ed.; Chemical Society: London, 1977; p 320.

Wills, J. H. In *Encyclopedia of Chemical Technology*, 2nd ed.; Wiley: New York, 1969; Vol. 18, pp 134–166.

Vail, J. G., *Soluble Silicates*; ACS Monograph 116; Van Nostrand Reinhold Co., Inc.: New York, 1952; Vol. 1 and Vol. 2.

56 Attrition Resistant Catalysts, Catalyst Precursors and Catalyst Supports and Process for Preparing Same

Horacio E. Bergna
DuPont Experimental Station

CONTENTS

The use of silica as a support for catalysts or as a binder for catalyst particles is well known. The silica provides strength and attrition resistance and it acts to disperse the catalyst particles.

U.S. Pat. No. 2,904,580 discloses a process for producing acrylonitrile from propylene in a fluid bed reactor using a catalyst consisting essentially of bismuth phosphomolybdate supported on silica. The catalyst precursors in solution were added to an aqueous solution of an aqueous colloidal silica sol containing 30 wt% silica. U.S. Pat. No. 3,425,958 discloses a process for improving the flow properties of particulate silica (average particle size about 1–100 μm) by bringing the particulate silica into contact with a solution of silicic acid and removing the solvent from the silica. When a catalyst is to be deposited onto the particulate silica, the catalyst precursor and any catalyst additive required are added to the silicic acid solution along with the particulate silica and the solvent is then removed. Heating is disclosed as a convenient way to remove the solvent. U.S. Pat. No. 3,772,212 discloses a fluidized bed catalyst for production of aromatic nitriles, the catalyst comprises a vanadium oxide, a chromium oxide, and a boron oxide as catalyst components and silica as a carrier and is prepared by spray-drying a silica sol containing vanadium, chromium and boron compounds. An aqueous 30 wt% colloidal silica sol was used. U.S. Pat. No. 3,746,657 discloses a process for making a fluidized bed catalyst comprising an oxide of molybdenum and a supporting material derived from a colloidal sol of an oxide of silicon, aluminum, titanium or zirconium, said process comprising preparing a slurry of the catalyst components and spray drying the slurry. The silica sol employed is preferably a low alkali aqueous silica sol containing 30–50 wt% SiO_2. U.S. Pat. No. 3,044,965 discloses a process for making a fluidized bed catalyst consisting of bismuth silico-molybdate or bismuth silico-phospho-molybdate comprising forming a slurry of the appropriate metal compounds and silica and spray drying. A low alkali aqueous silica sol containing 30 wt% SiO_2 was used as the silica source. U.S. Pat. No. 4,014,927 discloses a process for the production of unsaturated acids by catalytic oxidation of the corresponding unsaturated aldehydes in the presence of molybdenum-vanadium-iron based catalysts prepared by forming a solution or slurry of source compounds of the above metals and a source of silicon such as a silicate, water glass, or colloidal silica, and then drying by either evaporation or spray drying. U.S. Pat. No. 4,092,354 discloses a process for producing acrylic acid by the gas phase oxidation of acrolein comprising contacting acrolein and molecular oxygen over a metal oxide catalyst containing Mo, V, Cu, and at least one of Fe, Co, Ni, and Mg. The catalyst, carrier is chosen from a group of materials, including silica sol and silica gel. Catalyst precursors in solution are added to a silica so containing 20 wt% SiO_2. The mixture is evaporated and then calcined.

U.S. Pat. No. 3,313,737 discloses a process for preparing an improved silica acid sol for use as a binder. The silicic acid sol is prepared from an alkyl silicate which is hydrolyzed by water in the presence of a mutual solvent by the catalytic action of a strong acid. These sols contain 10–30 wt% SiO_2. U.S. Pat. No. 3,920,578 discloses rapid-gelling binder vehicles produced by admixing a water-soluble, alkaline ionic silicate with a colloidal amorphous silica aquasol having a median particle diameter of 50 Å to 0.5 μm. U.S. Pat. No. 3,894,964 discloses production of shaped bodies of zeolites of improved mechanical resistance (i.e., compression strength) using a silicic acid gel as a binder.

The unstable silicic acid sol that is used has a silicic acid content greater than 10 wt%, usually 25–35 wt%, and a silicic acid surface area greater than 150 m^2/g. U.S. Pat. No. 3,296,151 discloses a process for the production of substantially spherical, silica bonded, zeolitic molecular sieve granules comprising forming a suspension by adding powdery molecular sieve zeolite to an aqueous silica sol, said silica sol having a surface area of 150–400 m^2/g on drying and said silica sol used in a concentration of 10–40 wt% SiO_2; forming a suspension of finely divided magnesium oxide; admixing the two suspensions to produce a product having 0.1–3 wt% MgO; introducing the mixed suspensions dropwise into a liquid which is immiscible with water, whereby spherical granules are formed by sol–gel conversion; and separating the granules from the liquid and drying the granules. U.S. Pat. No. 3,356,450 discloses a process for preparing substantially pure zeolite granules starting with zeolite particles bound with silicic acid. The silica sols used to make the starting particles have surface areas of 150–400 m^2/g and at least 10 wt% SiO_2, and usually of the order of about 25 to about 35 wt% SiO_2. British Patent Specification No. 974,644 discloses a process for the production of molecular sieve pellets bonded with silicic acid, which comprises forming a plastic composition from a molecular sieve zeolite and an aqueous silica sol having a specific surface area between 150 and 400 m^2/g and a SiO_2 concentration between 10 and 40% by weight. Preferably, the silica sol is produced by ion exchange of sodium silicate and subsequent thermal treatment at a pH of 9 to 10. The patent further discloses that it has also been proposed to use, as binding agents, silicic acid esters which are hydrolyzed to silica gel by the water which is added to the mixture. The gels formed by the hydrolysis of the esters have a specific surface area of about 800 m^2/g and consequently consist of extremely fine particles. The use of the silicic acid esters is said to be too expensive for practical use, but experiments to replace them by normal commercial stable aqueous silica sols with a specific surface area between 100 and 200 m^2/g failed because the resulting granules had insufficient bonding strength and disintegrated.

U.S. Pat. No. 4,112,032 discloses a process for making porous silica-containing articles having pore diameters ranging between about 100 Å and 1 μm by combining a silicate solution containing at least 20% SiO_2 and a colloidal silica solution containing 40 wt% SiO_2 and then adding an organic gelation agent. Particulate matter less than about 74 μm in diameter and selected from the group consisting essentially of alumina, titania, silica, zirconia, carbon, silicon carbide, silicon nitride, iron oxides, and catalytically active transition metal oxides can be added before the gelation agent to produce porous silica articles containing a powder phase dispersed therein. U.S. Pat. No. 3,629,148 discloses a process for making an attriton resistant iron-containing catalyst from a bismuth phosphomolybdate catalyst by forming a mixture of the prescribed ingredients in a silica dispersion and spray drying. The silica dispersion contained 30 wt% SiO_2. U.S. Pat. No. 4,453,006 discloses a two-step process for preparing attrition resistant supported solid oxidation catalysts containing any known elements, preferably those containing molybdenum and used for the vapor phase oxidation of propylene or isobutylene to prepare unsaturated aldehydes and acids. The process comprises adding fumed silica to a mixture containing one or more active ingredients of the catalyst, drying said mixture, adding to this mixture in solution a silica or silica-containing compound other than fumed silica, and drying and calcining the mixture. The amount of fused silica can be 5–95% of the total silica used, with 15–65% being preferred. The silica used in the second addition can be silica sol, silica gel, diatomaceous earth or any precursor to silica, such as silicate, that preferably has a surface area of 50 m^2/g or more. The silica sols used in the examples have silica contents of 40 wt%. U.S. Pat No. 4,400,306 discloses a process for preparing supported attrition-resistant catalysts for fluid bed reactors by impregnating a performed support, for example, silica, but also alumina, alumina-silica, zirconia, and niobia, with a metal alkoxide of at least one metal selected from vanadium, molybdenum, antimony, copper, niobium, tantalum, zinc, zirconium, boron and mixtures thereof, and contacting the impregnated support with a solution of at least one additional catalyst component in situ, and drying the catalyst-containing support.

The preparation of mixed oxide compositions of vanadium and phosphorus and the use of these as catalysts for the oxidation of hydrocarbons such as n-butane to maleic anhydride is known in the art. In U.S. Pat. No. 4,111,963 the importance of reducing the vanadium used in a vanadium/phosphorus oxide (V/P/O) catalyst to the four oxidation state is described. Preferred is the use of concentrated hydrochloric acid as the reaction medium to bring about this reduction and preferred catalysts have a phosphorus to vanadium atom ratio of 1:2 to 2:1 and a porosity of at least 35%. In U.S. Pat. No. 3,864,280 the reduction of the vanadium in such a catalyst system to an average valence state of 3.9–4.6 is emphasized; the atomic ratio of phosphorus to vanadium is 0.9–1.8:1. Isobutyl alcohol is used as a solvent for the catalyst preparation, with the indication that an increase in catalyst surface area, over that obtained from use of an aqueous system, is achieved. The addition of promoters to the V/P/O catalyst compositions used for the oxidation of hydrocarbons to maleic anhydride is also disclosed in the art. Thus, in U.S. Pat. Nos. 4,062,873 and 4,064,070 are disclosed vanadium/phosphorus/silicon oxide catalyst compositions made in an organic medium. In U.S. Pat. Nos. 4,132,670 and 4,187,235 are disclosed processes for preparing high surface area V/P/O catalyst. Anhydrous alcohols of 1–10 carbon atoms and 1–3 hydroxyl groups are used to reduce the vanadium to a valence of 4.0–4.6. Also disclosed, as in U.S. Pat. Nos. 4,371,702 and 4,442,226, are V/P/O catalysts containing the promoter comprising silicon and at least one of indium, antimony and tantalum, the Si/V atom ratio being in the range 0.02–3.0:1.0, the (In + Sb + Ta)/V atom ratio being in the range 0.005–0.2:1.0 and the P/v atom ratio being in the range 0.9–1.3:1.0, said catalyst being prepared in an aqueous of organic liquid medium by the procedure wherein the appropriate vanadium species substantially of valence +4 is contacted with the promoter or promoter precursors and thereafter with the appropriate phosphorus species.

The attrition resistance of the V/P/O catalyst is particularly important when the oxidation process is carried out in a fluid bed or recirculating solids reactor. U.S. Pat. Nos. 4,317,778, 4,351,773, and 4,374,043 disclose processes for preparing fluid bed V/P/O catalysts in which an aqueous slurry of comminuted catalyst precursor is spray dried. Preferably, the catalyst precursor is uncalcined when it is made into a slurry. Examples are given in which an aqueous slurry of the catalyst precursor and a silica sol is spray dried to provide the catalysts 80 wt% V/P/O-20 wt% SiO_2 and 70 wt% V/P/O-30 wt% SiO_2. The products are described as uniform, microspheroidal catalyst particles. U.S. Pat. No. 4,127,591 discloses a process for preparing fluid bed V/P/O catalysts containing potassium and iron in which an aqueous slurry of the catalyst precursor is spray dried. Examples are given in which an aqueous slurry of the catalyst precursors in solution and a silica sol, 20 wt% silica, is spray dried to provide the catalyst 65 wt% V/P/K/Fe/O-35 wt% SiO_2. Silica content of the catalyst is to be between 25 and 70 wt%. British Patent Specification No. 1,285,075 discloses a process for preparing attrition-resistant V/P/O catalysts for fluid bed reactors by spray drying a mixture of a vanadium compound, a phosphorus compound, and an aqueous silica sol. The silica sols used in the examples contained 30–35 wt% SiO_2. British Patent Specification No. 2,118,060 discloses a process for preparing a catalyst comprising oxides of vanadium and phosphorus by mixing two crystalline oxides, each containing vanadium and phosphorus and each with a specified x-ray diffraction pattern, with a silica sol, spray drying the resultant slurry, and calcining the particles obtained. The silica sols used were 20–40% silica sol solutions and 40% colloidal silica solutions. U.S. Pat. Nos. 4,062,873 and 4,064,070 disclose processes for preparing a catalyst comprising oxides of vanadium, phosphorus, and silicon by coprecipitating vanadium oxide and silica or a silica precursor. Phosphorus can be coprecipitated with the vanadium oxide and silica or silica precursor or added later to form the catalyst precursor, which is then calcined to give the silica-containing catalyst. The catalysts of the examples contain 0.7–5.3 wt% silica which is distributed uniformly throughout the pellet. Russian Patent No. 215,882 discloses a method for preparing a V/P/O catalyst which is said to have increased activity and increased mechanical strength. Industrial large-pore silica gel is impregnated with a heated solution of oxalic acid, phosphoric acid, and vanadium pentoxide, dried, and activated. U.S. Pat. No. 4,388,221 discloses a process for preparing V/P/O Sn-containing catalysts comprising mixing the catalyst precursor, a binder, solvent and mordenite to form an impregnated mordenite which is then calcined. Silica is one of the suggested binders and the binder is said to comprise 0–10 wt% of the finished composite catalyst.

The objective of this invention is to provide a method for making attrition resistant catalysts, catalyst precursors and catalyst supports.

SUMMARY OF THE PROCESS

This process provides attrition resistant catalyst, catalyst precursor and catalyst support particles and the process of their preparation comprising:

1. Forming a slurry comprised of catalyst, catalyst precursor or catalyst support particles dispersed in a solution of a solute which consists essentially of an oxide precursor of particle size no greater than 5 nm, the relative amounts of the particles and oxide precursor chosen so that the weight of the oxide formed in steps (1) and (2) is about 3–15% of the total weight of the particles and the oxide.
2. Spray drying the slurry to form porous microspheres of attrition resistant catalyst, catalyst precursor or catalyst support particles.
3. Calcining the spray dried microspheres at an elevated temperature which is below the temperature which is substantially deleterious to the catalyst or catalyst support, to produce attrition resistant catalyst or catalyst support particles.

The preferred solvent is water, the preferred oxide is SiO_2, and the preferred solute is silicic acid, a precursor to anhydrous silica. The process of such preferred embodiments comprises:

1. Forming a slurry comprised of catalyst, catalyst precursor or catalyst support particles dispersed in an aqueous silicic acid solution equivalent to a weight of SiO_2 not exceeding about 6 wt%, the relative amounts of the particles and silicic acid chosen so that the weight of the SiO_2

formed is about 3–15% of the total weight of the particles and the SiO_2.
2. Spray drying the slurry to form porous microspheres of attrition resistant catalyst, catalyst precursor or catalyst support particles.
3. Calcining the spray dried microspheres at an elevated temperature which is below the temperature which is substantially deleterious to the catalyst or catalyst support, to produce attrition resistant SiO_2-containing catalyst or catalyst support particles.

Preferably, the silicic acid is polysilicic acid (PSA) having an equivalent concentration of SiO_2 not exceeding about 5 wt%; the catalyst, catalyst precursor or catalyst support particles are less than about 10 μm in diameter; the microporous spheroidal particles produced by spray drying have diameters of from about 10 μm to about 300 μm; and the relative amounts of particles to be spray dried and SiO_2 are chosen so that the weight of the SiO_2 is about 5–12% of the total weight of the particles and the SiO_2.

When this process is used for preparing attrition resistant V/P/O catalysts for the oxidation of hydrocarbons to maleic anhydride, using conventional procedures such as disclosed in the back-ground section of this specification, the particles used to form the slurry of step (1) are V/P/O catalyst precursor particles with a particle size preferably from about 0.5 μm to about 10 μm and more preferably from about 0.5 μm to about 3 μm. As indicated above, the relative amounts of precursor particles to be spray dried and SiO_2 are preferably chosen so that the weight of the SiO_2 is about 5–12% of the total weight of the particles and the SiO_2. More preferably, the relative amounts of precursor particles to be spray dried and SiO_2 are chosen so that the weight of the SiO_2 is about 10% of the total weight of the particles and the SiO_2. Typically, the spray dried catalyst precursor particles are calcined in stagnant air or in a low flow of air at about 375–400°C for about 1–6 h and then activated.

This process also provides a process for preparing attrition resistant, SiO_2-containing V/P/O catalyst precursor particles comprising:

1. Forming a slurry comprised of V/P/O catalyst precursor particles dispersed in an aqueous silicic acid solution equivalent to a weight of SiO_2 not exceeding about 6 wt%, the relative amounts of the particles and silicic acid chosen so that the weight of the SiO_2 formed is about 10–15% of the total weight of the particles and the SiO_2 Figure 56.2.
2. Spray drying the slurry to form porous microspheres of attrition resistant, SiO_2-containing V/P/O catalyst precursor particles.

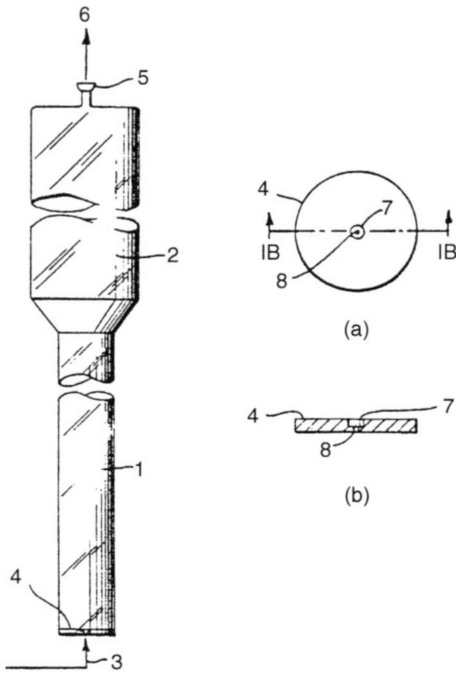

FIGURE 56.1 Attrition mill used for the determination of attrition. (a) and (b) are views of a disc held in place at the bottom of the mill to retain the catalyst.

The process also provides uses for the aforesaid attrition resistant catalyst, catalyst precursor and catalyst support particles.

DETAILED DESCRIPTION OF THE PROCESS

For further comprehension of the process, and of the objects and advantages thereof, reference may be made to the following description and to the appended claims in which the various novel features of the invention are more particularly set forth.

Supplementary to the aforesaid summary, the process of this process comprises forming a sufficiently stable slurry comprised of catalyst, catalyst precursor or catalyst support particles dispersed in a solution of a solute which consists essentially of an oxide precursor, spray drying the slurry to form porous microspheres, and calcining the spray dried microspheres. This process results in the formation of an oxide-rich layer at the periphery of each calcined microsphere. This oxide-rich surface layer is typically 5–10 μm thick and contains substantially all the oxide provided by the oxide precursor solute. Since substantially all of the oxide is in the peripheral layer, good attrition resistance is attained with a small amount of oxide, that is, the weight of the oxide is about 3–15%, preferably about 5–12% (except as noted above

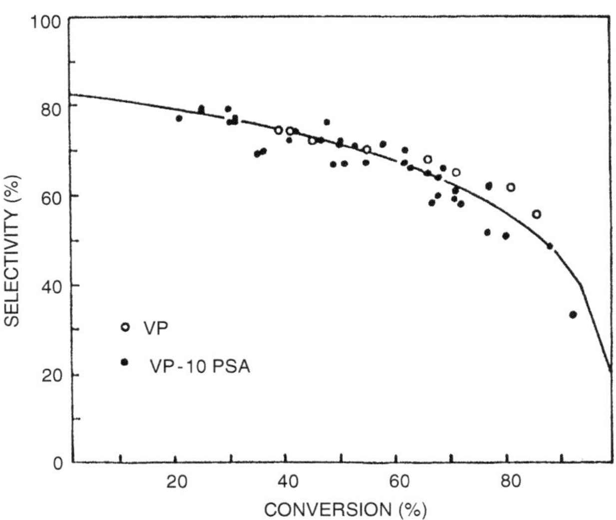

FIGURE 56.2 A plot of selectivity versus conversion for V/P/O catalyst—10 wt% SiO_2(PSA).

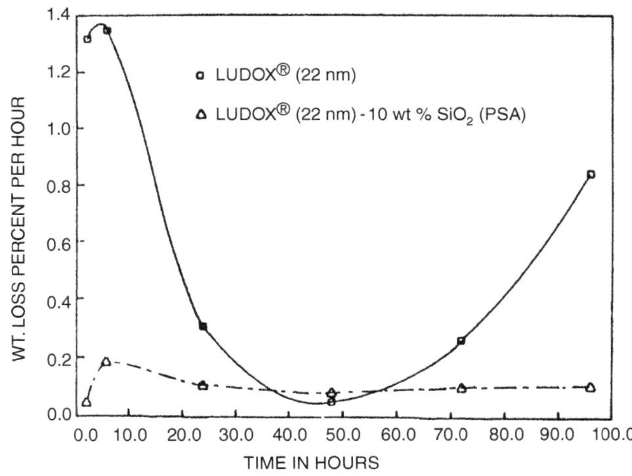

FIGURE 56.4 Results of attrition mill tests for Ludox® (22 nm) catalyst support and for Ludox® (22 nm)—5 wt% SiO_2(PSA) catalyst support. Weight loss percent per hour is plotted versus time in hours.

for the SiO_2-containing V/P/O catalyst), of the total weight of the spray dried particle. The oxide must be chosen so that it has no deleterious effect on the catalytic performance of the particular catalyst being used. The oxide can be inert or show catalytic activity for the particular process being run. When an attrition resistant catalyst is prepared, it is found that this peripheral layer does not affect the microstructure and phase development of the catalytically active phase during the calcination and activation steps and the morphology of the microspheres is such that it allows the reactants access to the catalytically active phase. This oxide-rich layer has a deleterious effect on catalyst performance.

The slurry which is spray dried in the process of this process comprises a catalyst, catalyst precursor or catalyst support particles dispersed in a solution of a solute which consists essentially of an oxide precursor. The catalyst, catalyst precursor or catalyst support particles used may be obtained by synthesis or by comminuting larger particles of crystalline, polycrystalline or mixed amorphous and crystalline phases. Typically these particles are of the order of about 0.5 to about 10 μm in size. It is preferred to use particles that are of the order of about 0.5 to about 3 μm. Examples of types of comminuted particles are those of the catalyst precursor of the V/P/O catalyst for known maleic

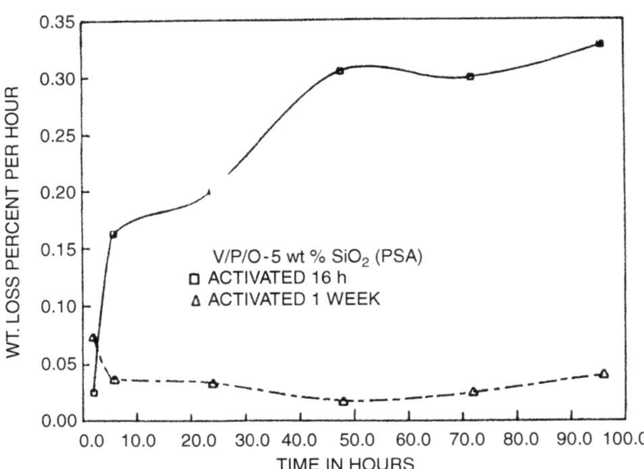

FIGURE 56.3 Results of attrition mill tests for V/P/O catalyst—5 wt% SiO_2(PSA). Weight loss percent per hour is plotted versus time in hours.

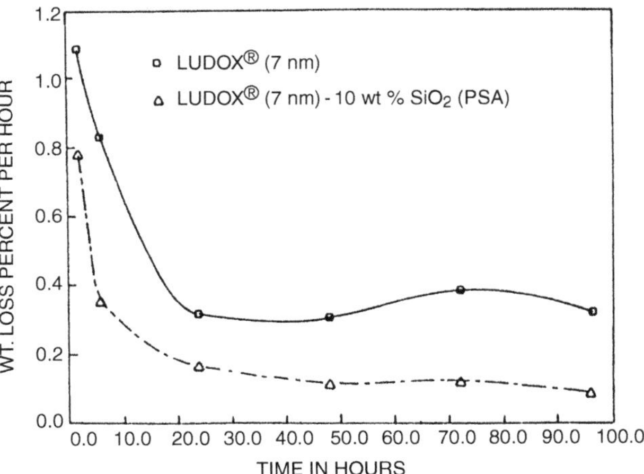

FIGURE 56.5 Results of attrition mill tests for Ludox® (7 nm) catalyst support and for Ludox® (7 nm)—5 wt% SiO_2(PSA) catalyst support. Weight loss percent per hour is plotted versus time in hours.

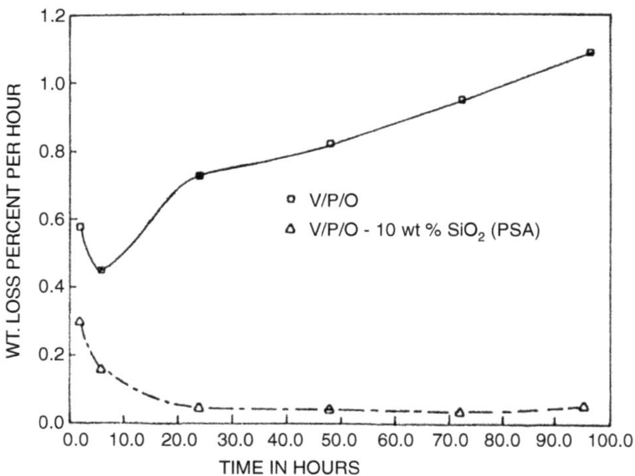

FIGURE 56.6 Results of attrition mill tests for V/P/O catalyst—10 wt% SiO_2(PSA) and for V/P/O catalyst. Weight loss percent per hour is plotted versus time in hours.

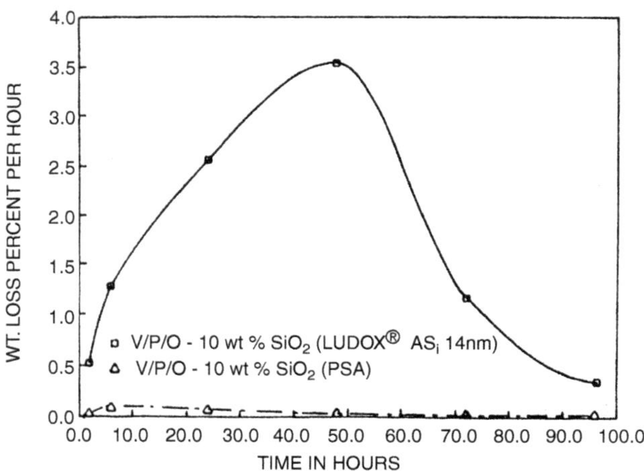

FIGURE 56.8 Results of attrition mill tests for V/P/O catalyst—10 wt% SiO_2(14 nm particle size Ludox® AS) and V/P/O catalyst—10 wt% SiO_2(PSA). Weight loss percent per hour is plotted versus time in hours.

anhydride processes, those of a multicomponent molybdate catalyst for known acrylonitrile processes, and those of the catalyst support alpha alumina. Other such particles include those of fused silica, kaolin, amorphous aluminosilicates, zeolites, zirconia, and titania. The particles used may also be fine particles such as the amorphous particles to 7–200 nm in diameter found in colloidal silica which can be used to form catalyst support particles. Aggregated amorphous silica powders can also be used for this purpose.

The solvent used in the slurry is a solvent for the oxide precursor. Water is preferred. The solute consists essentially of an oxide precursor of subcolloidal particle size. "Subcolloidal particles" (size) are defined herein as particles for which the largest dimension is no greater than 5 nm. The solute particles must not aggregate, precipitate or gel during or following the formation of the solution or in contact with the catalyst, catalyst precursor or catalyst support particles. The solute particles must provide a sufficiently stable solution and slurry to permit spray drying. Because the solute particles with the above properties are much smaller than the voids or spaces between the catalyst, catalyst precursor or catalyst support particles, when the slurry is spray dried, the

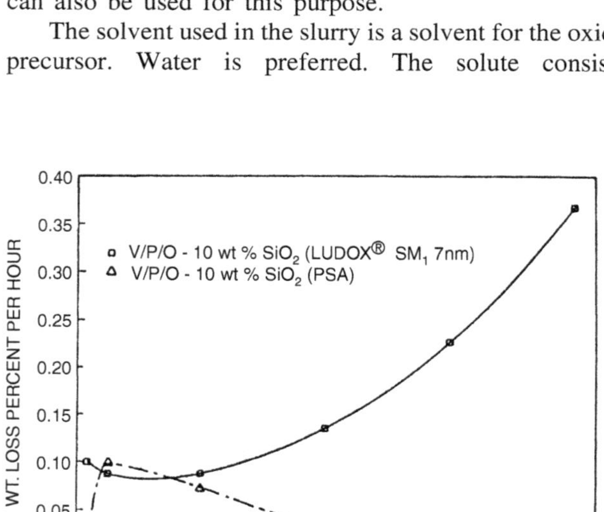

FIGURE 56.7 Results of attrition mill tests for V/P/O catalyst—10 wt% SiO_2 (7 nm particle size Ludox® SM) and V/P/O catalyst—10 wt% SiO_2(PSA). Weight loss percent per hour is plotted versus time in hours.

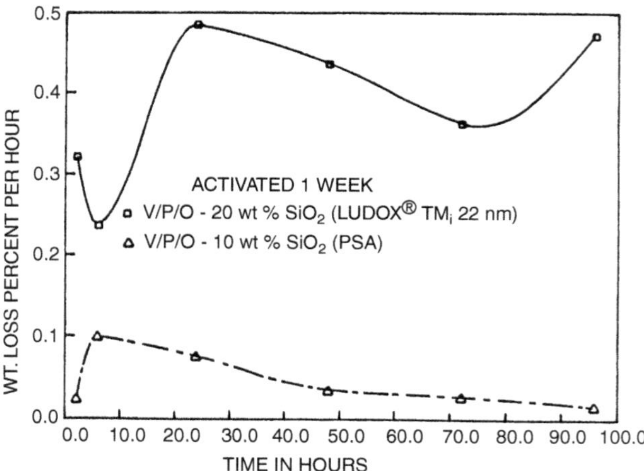

FIGURE 56.9 Results of attrition mill tests for V/P/O catalyst—20 wt% SiO_2 (22 nm particle size Ludox® and V/P/O catalyst—10 wt% SiO_2(PSA). Weight loss percent per hour is plotted versus time in hours.

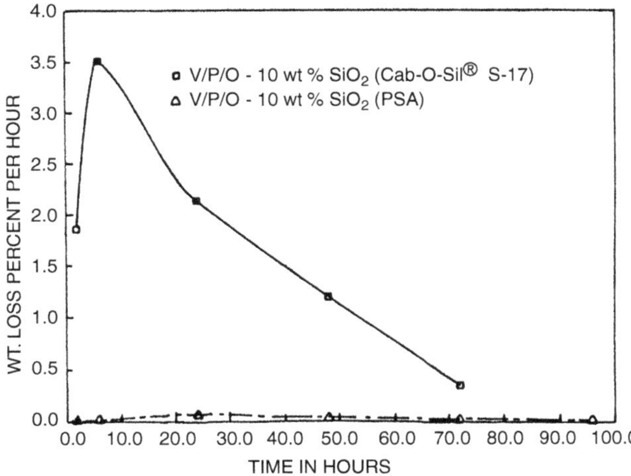

FIGURE 56.10 Results of attrition mill tests for V/P/O catalyst—10 wt% SiO_2(Cab-O-Sil® S-17) and V/P/O catalyst—10 wt% SiO_2(PSA). Weight loss percent per hour is plotted versus time in hours.

solute particles can flow with the solvent from the interior to the peripheral region of the porous microsphere formed by evaporation of the solvent in a droplet of the spray. These solute particles then remain in this peripheral region as the drying is completed and form a hard peripheral composite shell of catalyst, catalyst precursor or catalyst support particles and oxide. The oxide can be chosen from the group comprising SiO_2, Al_2O_3, P_2O_5, TiO_2, ZrO_2, MgO, Cr_2O_3, and rare earth oxides. Examples of solutes

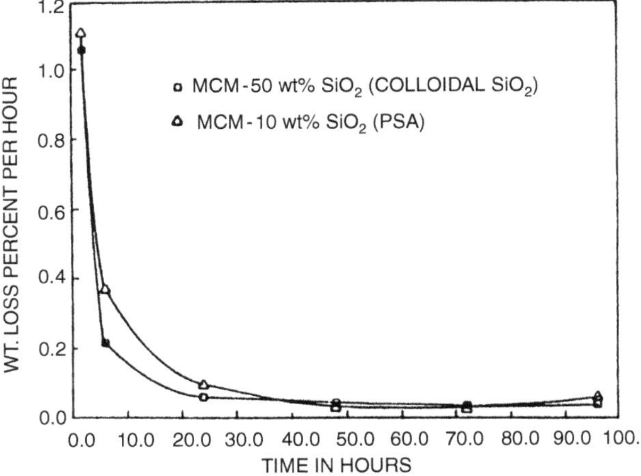

FIGURE 56.11 Results of attrition mill tests for a multicomponent molybdate catalyst—10 wt% SiO_2(PSA) and a multicomponent molybdate catalyst containing about 50 wt% SiO_2 (colloidal silica), said SiO_2 dispersed substantially uniformly throughout the composite particles. Weight loss percent per hour is plotted versus time in hours.

for these oxides are silicic acid, basic aluminum chloride, phosphoric acid, titanyl oxychloride, hydrolyzed zirconyl nitrate, magnesium acetate, hydrolyzed basic chromic chloride ($Cr(OH)_2Cl_4$) and hydrolyzed basic nitrates of rare earths. The preferred oxide is SiO_2 and the preferred solute or oxide precursor is silicic acid, especially polysilicic acid.

The method of removing the liquid from the slurry is critical. Spray drying determines the distribution of the oxide in the dry particles of oxide and catalyst, catalyst precursor or catalyst support to provide the product of this invention. The spray drying may be carried using conventional spray drying techniques and equipment. The chamber product from the chamber of the spray dryer is typically made up of porous spheroidal particles with diameters of about 30 to about 300 μm. The cyclone product collected from the cyclone of the spray dryer is made up of porous spheroidal particles with somewhat smaller diameters. These spray dried particles may be sieved to obtain a fraction of particles with a narrower size distribution. The spray dried spheroidal particles are referred to herein as microspheres. As is well-known in the spray drying art, many of the porous microspheres produced have a void in the center with one or two openings to the outside. Such particles are referred to in the art as Amphora I-type and Amphora II-type particles, respectively.

The spray dried porous microspheres are then calcined. Regardless of whether the spray dried particles are comprised of oxide and catalyst, catalyst precursor or catalyst support, sintering is almost always necessary in order to achieve high attrition resistance. When the spray dried particles are comprised of oxide and catalyst precursor, calcining not only results in sintering but also generates the catalyst. Calcination conditions, such as temperature, time, and type of atmosphere, depend on the composition of the catalyst or catalyst support and the amount and nature of the oxide used. The calcination temperature must be sufficiently high to result in sintering and, when catalyst precursor is present, in catalyst generation, but it must be below temperatures deleterious to the catalyst or catalyst support. Some catalysts are subjected to an activation process before use. Activation can be carried out as part of the calcination process or subsequent to it.

The aqueous silicic acid solution that is useful in this invention contains silica of the proper particle size, that is, no greater than 5 nm, and provides a solution of sufficient stability to allow the formation of the slurry and subsequent spray drying. The silicic acid can be in the form of a monomer or in the form of low molecular weight polymeric units. For a review of the characteristics of silicic acid, see R. K. Iler, *The Chemistry of Silica*, John Wiley and Sons, NY., 1979. Monomeric silicic acid $Si(OH)_4$ has never been isolated. It is a very weak acid and exists only in dilute aqueous solutions. At a concentration greater than about 100–200 ppm as SiO_2, the monomer polymerizes by condensation to form dimer and higher molecular weight species of silicic acid. The preferred form of silicic acid is polysilicic acid. For the purposes of this invention polysilicic acids are defined (following Iler, op. cit., p. 287) as those silicic acids that have been formed and partially polymerized in the pH range 1–4 and consist of ultimate silica particles generally 7 smaller than 3–4 nm diameter. Polysilicic acid, that is, oligomers of monosilicic acid, is comprised of polymers with molecular weights (as SiO_2) up to about 100,000, whether consisting of highly hydrated silica or dense spherical particles less than about 5 nm in diameter and generally smaller than 3–4 nm diameter. Polysilicic acid has sometimes been referred to in the literature as "active" silica.

The term polysilicic acid is justified, particularly in view of the very high specific surface area and the high proportion of SiOH groups. For particles with dimensions less than 5 nm, less than half of all the silicon atoms are present at SiO_2, that is, as silica, whereas more than half are each associated with at least one hydroxyl group. These silanol groups form silicon-oxygen-metal atom bonds with polybasic metal cations, as in the case of monosilicic acid. However, polysilicic acids differ from the monomer in that they form addition complexes with certain classes of polar organic molecules through hydrogen bond formation. Also, they can be isolated and esterified not only with alcohols under dehydrating conditions, but also with trimethylsilanol, even in aqueous solution.

Polysilicic acid made of 1.5–4 nm diameter discrete particles and having a pH of about 2.5–3.0 is not stable and gels at a relatively fast rate depending on concentration and temperature. Once a polysilicic acid solution has been exposed to alkaline conditions, it is rapidly converted to colloidal silica particles larger than 4–5 nm diameter. Thereafter, silica assumes different characteristics and can be stabilized as colloidal silica sols in the pH range 8–10. Thus, colloidal silica as opposed to polysilicic acid is made of highly polymerized species or particles generally larger than about 5 nm.

Aqueous solutions of polysilicic acid can be prepared by adding a thin stream of sodium silicate solution with an SiO_2:Na_2O ratio of 3.25:1.0 into the vortex of a violently stirred solution of H_2SO_4 kept at 0–5°C, stopping the addition when the pH rises to about 1.7. Polysilicic acid solutions can also be made continuously by bringing together solutions of sodium silicate and acid in a zone of intense turbulence and in such proportions that the mixture has a pH about 1.5–2.0. Residual electrolytes increase the ionic strength of the solution and result in destabilization followed by premature gelling of the polysilicic acid. Therefore, polysilicic acid formed by a method in which electrolyte byproducts are produced should be separated promptly from the electrolyte by-products.

The preferred method for the preparation of polysilicic acid is by deionization of a sodium silicate solution with an ion exchange resin at room temperature. In this way the polysilicic acid solution is substantially free of electrolytes and, therefore, is more stable.

Solutions free from the sodium salt can also be obtained by hydrolyzing methyl or ethyl silicate in water at pH 2 with a strong acid as a catalyst for hydrolysis and temporary stabilizer for the silicic acid.

The aqueous solutions of polysilicic acid used in the process of this invention have a concentration of SiO_2 not exceeding about 6 wt%. Very low concentrations of SiO_2 provide even greater assurance of having the desired small particle size and stability; however, very large volumes of solution are required to supply the total amount of SiO_2 required by the process. It is preferred that the aqueous solutions of polysilicic acid used in the process of this invention have a concentration of SiO_2 not exceeding about 5 wt%.

The slurry which is spray dried is prepared by gradually adding catalyst, catalyst precursor or catalyst support particles to an aqueous silicic acid solution. The slurry is stirred until a uniform dispersion is obtained. The relative amounts of silicic acid solution and particles are chosen so that the weight of the SiO_2 represents 3–15% of the total weight of the particles and the SiO_2.

The small particle size of the silica is important not only in enabling the silica to flow to the peripheral region of the porous microsphere but also in forming the hard peripheral oxide-rich shell. Particles of silica 2–3 nm in diameter sinter together to some extent even under the temperature conditions encountered in a conventional spray drying process, whereas particles 10–100 nm do not sinter below 700–1000°C. As a result, attrition resistance of the catalyst, catalyst precursor or catalyst support particle is a function of the particle size and degree of aggreggation of the silica formed by dehydration.

Immediately following is a discussion of results realized by carrying out experiments involving the spray drying of V/P/O catalyst precursor particles and various sources of silica, to demonstrate the importance of the size of the silica particles used. It is to be understood, in this specification, that "(PSA)" following "SiO_2" indicates the source of the SiO_2 as polysilicic acid.

Discrete particles of silica 2–3 nm in diameter, such as those present in the polysilicic acid described above, form hard shells on the resulting porous microspheres under conventional drying conditions. The "green" attrition resistance, that is, the attrition resistance before calcinations, of the porous microspheres of, for example, a

V/P/O catalyst precursor—10 wt% SiO_2 prepared using the PSA of this invention, is as high as the attrition resistance of these microspheres after calcining at 400°C for 1 h. Calcination of these microspheres of V/P/O catalyst precursor—10 wt% SiO_2(PSA) is necessary to convert the precursor to the catalyst. However, since the "green" attrition resistance is so high, calcination can be carried out when convenient, for example when the microspheres are in the reactor.

The "green" attrition resistance of the porous microspheres of V/P/O catalyst precursor—10 wt% SiO_2 is significantly lower when a colloidal sol of particles of 5 nm diameter is used as the source of the silica instead of PSA. When a colloidal sol of particles of 14 nm diameter is used as the source of the silica, the "green" attrition resistance is even lower. When a colloidal sol of particles of 22 nm diameter is used as the source of the silica, significant "green" attrition resistance is not realized even when the amount of silica is increased to 20 wt%. Furthermore, calcination of the porous microspheres of V/P/O catalyst precursor—10 wt% SiO_2 does not result in adequate attrition resistance when the silica particle diameter in the silica source exceeds 5 nm. The attrition test results for microspheres made using a silica source with silica particles of a nominal 5 nm diameter may vary from sample to sample, probably because of the variation of particle size distribution common in such silica sols.

A solution of Al_2O_3 precursor that is useful in this invention is a basic aluminum chloride aqueous solution which consists of high molecular weight units or hexagonal ultimate particles about 1–2 nm in diameter. When dried and calcined, basic aluminum chloride yields aluminum oxide. The "green" attrition resistance of the porous microspheres of V/P/O catalyst precursor—10 wt% Al_2O_3(basic aluminum chloride) is poor is contrast to that of the V/P/O catalyst precursor—10 wt% SiO_2(PSA) described above. However, calcination at 400°C. for 1 h converts the chloride into amorphous aluminum oxide, and the peripheral alumina-rich shell confers high attrition resistance to the microspheres of V/P/O catalyst. Alumina degrades the catalyst performance and, therefore, is not the oxide of preferred choice for V/P/O catalysts.

Attrition resistance was measured using an apparatus (Figure 56.1) in which the conditions are similar to but more severe than those experienced by the catalyst or catalyst support in actual operation. The apparatus is comprised of a tube 1 to contain the catalyst and, connected to this tube, a larger diameter upper Section 2 which serves as an elutriator-dust collector. Means (not shown) vary the pressure and flow rate of air fed through air supply line 3 to a disc 4 containing a 0.0160 in. (0.406 mm) diameter orifice 8 opening up into a 1/16 in. (1.6 mm) hole 7.

The principle of operation involves transfer of energy from a high velocity gas jet passing through a precisely-sized orifice to catalyst particles which in turn collide with other particles. Fine particles (particles with diameters less than about 16 μm) produced from these impacts are entrained in the upward gas flow and exit the mill in exit gas flow 6. The top of the mill shown includes a 35/25 spherical joint 5 to which is attached means (not shown) for collecting the fine particles. Particles with diameters equal to or greater than about 16 μm fall back and concentrate on the outer walls of upper Section 2. They agglomerate there and drop back into the tube 1 due to the action of an electromechanical vibrator (not shown). At elapsed times of 2, 6, 24, 48, 72, and 96 h, the flask and filter are removed, dried at 80°C in a vacuum oven, and weighed. The hourly rate of solids carry over is calculated, recorded and plotted. The hole in the perforated plate should be drilled to close tolerances because the attrition depends markedly on the diameter, that is, on gas velocity. Some erosion of the holes occurs during use, and the plate must be replaced when the pressure drop through the plate deviates significantly from that obtained with a newly prepared plate under the same flow conditions. The apparatus of Figure 58.1 is substantially geometrically equivalent to that described by W. L. Forsythe, Jr. and W. R. Hertwig, *Ind. and Eng. Chem.* 41, 1200 (1949).

The following examples are intended to demonstrate, but not limit, various embodiments of the invention. All temperatures are in degrees Celsius.

EXPERIMENTAL

This is an example of the preparation and testing of a V/P/O—10% SiO_2(PSA) catalyst of the invention.

The preparation comprises the following steps: synthesis of the V/P/O catalyst precursor, preparation of the 5 wt% SiO_2 solution of PSA, preparation and spray drying of the V/P/O precursor-PSA slurry, calcination and activation.

A V/P/O catalyst precursor containing a promoter comprised of 2 wt% SiO_2 and 2.5 atom% it. In was prepared following Example 1 of U.S. Pat. No. 4,371,702. A 15 gallon (56.8 l) crystallizer type kettle was charged with 3600 g of comminuted V_2O_5, 36 l of isobutyl alcohol and 3.6 l of benzyl alcohol. The liquids were stirred while the V_2O_5 was added. The mixture was heated at reflux for 14 h. 1152 g of $Si(OEt)_4$ was then added and the mixture was heated at reflux for 4 h. 114 g of In metal (corresponding to 2.5 atom % In) dissolved in acetic acid and isobutyl alcohol was then added and heating at reflux was continued for 2 h. At this time an additional 3240 g of $Si(OEt)_4$ was added and the mixture was heated at reflux for 14 h. 5400 g of 85% H_3PO_4

were added over a 2-h period at the rate of 45 ml/10 min. The mixture was heated at reflux for 20 h. The slurry was filtered and the filtrate was recycled until it was clear. The solid was dried in air at 110° for 2 days. The solid was comminuted to a powder with particles of size between 1 and 3 μm. 7068 g of V/P/O catalyst precursor was thus produced.

A 5 wt% SiO_2 polysilicic acid solution was prepared by diluting 1014 g of JM grade sodium silicate solution (300 g SiO_2) with 4985 g of distilled water in an 8-l stainless steel beaker. The solution was stirred for a few minutes and then filtered through folded filter paper to give a clear water-like filtrate. This clear filtrate with pH of 11.3 was stirred vigorously while Dowex® HCR-W2-H resin, a strongly acidic nuclear sulfonic acid cation exchanger supplied by Dow Chemical Company, was added to reduce the pH. When the pH was about 6.8, excess resin was added to reduce the pH below 5.5 rapidly, thus avoiding microgel formation. When the pH reached 3.0 ± 0.1, the resin was filtered off and the clear filtrate was used within an hour to prepare the V/P/O precursor-PSA slurry for spray drying.

To a mixing bowl, four thousand four hundred and forty four gram of the 5 wt% SiO_2 PSA solution was added and with the mixer on low speed, 2000 g of V/P/O comminuted catalyst precursor was added in small portions over a period of 30–45 min. The resulting slurry, containing 34.48% solids (90% V/P/O catalyst precursor and 10% SiO_2), had a pH of 2.5 ± 0.1.

The slurry was sprayed dried at the rate of 150 mL/min with atomizer area pressure set at 8 psi (55.2 kPa) and a chamber temperature of 245° ± 5°.

Tyler standard sieves were stacked in the order of No. 40, 60, 100, 200, and 325 along with a lid and a receiving pan and placed on a sieve shaker. 200 g of chamber product was placed on the No. 40 sieve and the sieve shaker was run for 15 min. Each sieve fraction was bottled and weighed separately. The amounts obtained by screening 724 g of chamber product are shown in Table 56.1.

Each of three 50 g samples of the −100 325 fraction was spread out in a thin layer on a fine mesh stainless steel screen on top of a quartz boat to allow air to flow under and over the sample. The three boats were placed end to end on a belt of a 3-zone, 7 foot (2.1 m)-long belt furnace. The first zone was maintained at 285°, the second at 385°, and the third at 374°. The belt speed was 0.25 in. (6.35 mm) per minute so that 4.5 h were required for the boat to travel the length of the furnace. The combined weight of the three calcined samples was 133 g; the product was V/P/O catalyst—10% SiO_2(PSA).

Twenty grams of the V/P/O catalyst—10% SiO_2(PSA) was tested for attrition resistance using the attrition mill and procedure described above. The results are shown in Table 56.2.

The weight of the column residue at the end of the test was 19.19 g, 96% of the catalyst weight at the start.

In 1.5% n-butane/16% oxygen/82.5% nitrogen 60 g of the calcined sample was activated for 16 h at 460°. The production of maleic anhydride (using a feed of 1.5% n-butane in air) was measured at temperatures ranging from 440° to 330° in a fluid bed reactor, and the results obtained are plotted in Figure 56.2 on the basis of

TABLE 56.1

Screen sizes	Screen opening (mm)	Amount collected (g)	Wt%
−40 + 60	0.42–0.250	8	1
−60 + 100	0.250–0.149	45	6
−100 + 325	−0.149–0.044	596	82
−325	0.044	74	10

TABLE 56.2

Cumulative time in mill (hours)	%Weight loss/hour
2	0.0751
6	0.0752
24	0.0476
48	0.0318
72	0.0385
96	0.0475

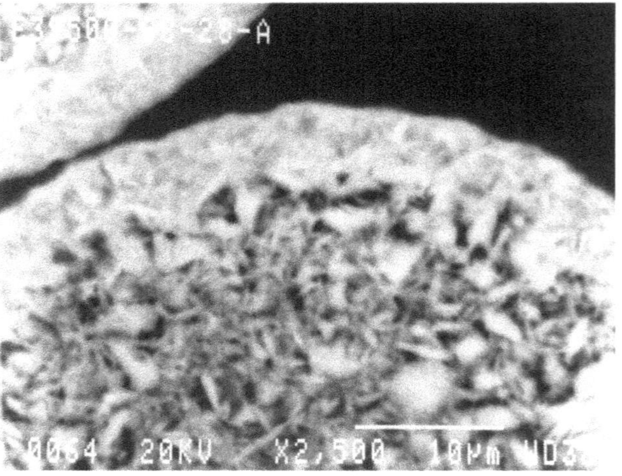

FIGURE 56.12 Cross-section of attrition resistant catalyst showing silica rich surface.

butane conversion versus selectivity for maleic anhydride, along with the results for eight other such samples (a total of nine samples, with four data points per sample). Also plotted in Figure 58.2 are the results obtained after activating and testing three samples (four data points per sample) made with no addition of PSA. The same reactor and reaction conditions were used throughout. The results show that the addition of PSA has no significant effect on the butane conversion and the selectivity for maleic anhydride.

57 Some Important, Fairly New Uses of Colloidal Silica/Silica Sol

Jan-Erik Otterstedt
Chalmers University of Technology

Peter Greenwood
Eka Chemicals (Akzo Nobel)

CONTENTS

Dilute silica sols were prepared and studied over 70 years ago. Their uses as binders in catalyst preparation, as glazes on ceramics, as coatings on concrete and plaster of Paris, as agents for treating paper and textiles, and several other applications were investigated [1]. These early silica sols contained less than 10% by weight of silica, were fairly unstable and did not have reproducible properties. Iler [2] predicted that colloidal silica would not be accepted for wide commercial use before these shortcomings were remedied.

Product development work in several industrial laboratories resulted in the production of concentrated silica sols of high stability and very reproducible properties. Iler [3] describes numerous applications of such sols.

Otterstedt and Brandreth [4] discusses functions that can be achieved by using colloidal silica in various applications. Most of these functions depend on the presence of a high specific surface area of a special chemical nature.

Here we will focus our attention on the use of colloidal silica to make high quality concrete, as retention aid in paper making, as polishing agents for silicon wafers, to provide solid electrolytes in lead-acid batteries and as components in high quality coatings because these applications are very large, relatively new and/or fast growing.

COLLOIDAL SILICA IN CEMENT AND CONCRETE

Cement

The following outline of cement is an adapted extract from Rodney Cotterill's fascinating odyssey into the material world [5].

What Is Cement and Concrete?

Already in ancient times it was known that the reaction between calcium oxide, also called lime or quicklime, and water could yield a binder in building construction. The Etruscans, for instance, added water to lime to form calcium hydroxide, or slaked lime, which they mixed with sand and stone into what today would be called a primitive concrete. The Romans discovered a way to improve cement making by burning a mixture of volcanic ash, which essentially consisted of silica, and lime. The many impressive constructions that have lasted to our days testify to the durability of their cement.

However, the Roman cement technology fell into oblivion and high quality cement became available first in 1824 when an Englishman Joseph Aspdin invented Portland cement, or modern cement. Modern cement is made by grinding a mixture of limestone and clays, with a weight ratio of about 80 to 20, and several other minor components with water to a slurry. This slurry is passed down a rotating kiln and first loses water and then carbon dioxide as the temperature gradually increases downward the kiln. In the last temperature zone, where the temperature is 1200–1500°C, the material sinters and melts to clinker. After cooling, the clinker is ground, together with a small amount of gypsum, which controls the reactivity of the cement with water, into a fine powder. The specific surface area of the particles, which is inversely proportional to particle size, determines the rate of reaction when water is added to the powder. The different grades of commercial cement powder are usually given designations that indicate how rapidly the cement paste becomes rigid and gains strength. Table 57.1 shows the composition and specific surface area of three common grades of Swedish cement.

TABLE 57.1
Composition and Specific Surface Area of Swedish Cements

Compound	OPC Sl	OPC Sk	SRCP D
Cao (%)	62.2	64.3	54.6
SiO_2 (%)	20.0	19.8	21.6
Al_2O_3 (%)	4.53	5.21	3.46
Fe_2O_3 (%)	2.23	3.04	4.75
K_2O (%)	1.42	1.30	0.75
MgO (%)	3.37	1.45	1.02
Na_2O (%)		0.11	0.06
Blaine (m^2/kg)	363	400	323

Manufacturer: Sl = Slite, Sk = Skövde, D = Degerhamn.
Note: Courtesy Euroc Research AB.

The metals oxides in Table 57.1 are not present as such in cement powder but instead as four different major compounds, alite, $3CaO{*}SiO_2$, 40–65% by weight, belite, $2CaO{*}SiO_2$, 10–25% by weight, aluminate, $3CaO{*}Al_2O_3$, up to 10% by weight, and ferrite, $4CaO{*}Al_2O_3$, also up to 10% by weight. There are also present small amounts, a fraction of a percent usually, of free lime, magnesium oxide, sodium sulfate and potassium sulfate. These trace compounds can influence the final properties of the material, for example, concrete, to a much higher degree, and sometimes in a negative way, than their abundance in the cement powder might suggest. Gypsum, which is added when the clinker is ground to a powder, is present in amounts between 2 and 5% by weight.

Modern cement, for example, Portland cement, contains more components and is a much better binder than primitive cement. Another important difference between Portland cement and primitive cement is that the former will set and harden under water. Cement paste, that is, a slurry of cement powder and water is usually mixed with sand or stone when it is used in building construction. The term sand refers to particles smaller than 2 mm and the term stone refers to particles larger than 2 mm. A mixture of inorganic materials, which may include sand and stone, and having a particle size distribution in the range from about 0.01 to 100 mm, is called an aggregate [6]. Mortar is a mixture of cement paste and sand. If aggregate is added to cement paste, the mixture is called concrete. The weight ratio of cement paste to aggregate in concrete is usually in the range up to 1:6. Concrete may contain additives such as setting and hardening additives, usually called accelerators, or workability additives, usually called superplasticizers. The worldwide production of cement amounts to about 1.5 billion metric tons.

Hydration of Cement

What happens when water reacts with the different components of the cement powder is a central question in cement science. In order to answer this complicated question scientists have studied the rates of reaction and heat liberation when water has been added to the different compounds separately.

Alite reacts with water to form calcium silicate hydrate and calcium hydroxide, which is also known as portlandite. The hardened paste has high strength when the reaction is completed, and because alite is the most abundant compound in cement, it also makes the dominant contribution to the mechanical properties of the final product. The hydration reaction proceeds at an appreciable rate a few hours after the addition of water and lasts up to about 20 days. The reaction of alite with water is accelerated by aluminate and gypsum.

Belite reacts with water at a slower rate than alite but the end product is the same. It takes about 2 days for the hardening process to get started and about a year to be completed. The mechanical strength of fully hydrated belite is similar to that of hydrated alite.

The hydration of the aluminate phase is very fast and it is essentially over within the first few hours. The contribution of the final product to the mechanical strength of the hardened cement paste is fairly low. It is also susceptible to attack by sulfate ions, which leads to expansion and weakening of the final product.

The final product of the reaction of ferrite with water is not known but its contribution to the ultimate strength of the paste is modest. Like alite and belite it is not attacked by dissolved sulfates.

The main features of the hydration reaction of the main components in cement are shown in Table 57.2.

In the hydration reaction alite absorbs about 40% by weight of water, of which 24% is chemically bound, and releases 500 J/g. For belite, 21% by weight of water is chemically absorbed, only 250 J of heat per gram are released, and less than half the amount of slaked lime is formed compared with the reaction of alite with water. Hydration of the aluminate phase is the reaction, which consumes most water, up to twice its own weight of water can absorbed in the final product, and releases most heat, 900 J/gram.

When water is added to the mixture of cement powder and aggregate the reaction of the main components, as well as some others, which will be discussed subsequently, will get under way. Although the hydration process is not understood down to the finest detail, much insight and understanding has been gained by the advent of modern analytical equipment, for example, the scanning electron microscope. The initial hydration, lasting a few minutes, involves the alite-water and aluminate-water reactions and rapidly leads to formation of a hydrous gel of colloidal silica and alumina particles at the interface between the water and the cement grains. The gel envelops the cement particles and the doubly charged calcium ions diffuse rapidly out of the gel and into the surrounding water, where the calcium ion concentration is controlled by the precipitation of crystals of calcium hydroxide. Removing calcium ions from the originally homogeneous gel, resulting in a gel, which essentially is a silica gel, leads to a build up of osmotic pressure, which periodically causes rupture of the water–gel interface. When this happens, the calcium hydroxide and silica components are brought together, and a precipitate of calcium silicate is formed in a shape similar to a volcano crater. Repeated rupturing of the water–gel interface at the craters leads to the formation of hollow needle-shaped projections, known as fibrils, sticking out from the cement grains like the burs of a burdock. The fibrils can lock together by a Velcro-like mechanism, forming a strong bond between the cement particles.

Gypsum is an important extra factor in the hydration process. Although a minor component of the cement powder, present in an amount corresponding to between 2 and 5% of the total weight, it effectively regulates the activity of the aluminate phase. A few minutes after the addition of water, needle-shaped crystals of ettringite, $3CaO{*}Al_2O_3{*}3CaSO_4{*}3H_2O$, appear at the surface of the aluminate particles, slowing down the aluminate-water reaction. The alite-water reaction proceeds at its slower

TABLE 57.2
Hydration Reactions

	Chemically bound water weight %	Heat of hydration joules/g[a]	$Ca(OH)_2$ formed weight %
$2(3CaO{*}SiO_2) + 6H_2O = 3CaO{*}2SiO_2 + 3Ca(OH)_2$ Alite / aquagel of Ca-silicate	24	500	48.7
$2(2CaO{*}SiO_2) + 4H_2O = 3CaO{*}2SiO_2 + Ca(OH)_2$ Belite / aquagel of Ca-silicate	21	250	21.5
$3CaO{*}Al_2O_3 + 6H_2O = 3CaO{*}Al_2O_3{*}6H_2O$ Aluminate / aquagel of Ca-aluminate	40	900	—
Typical values for fully hydrated Portland cement	25	400	15–25

[a]Heats of hydration after 28 days.

rate, as does the belite-water reaction with its still slower rate, producing calcium-silicate hydrate fibrils. Eventually, the ferrite-water reaction gets started, producing final products, the structures of which are still unknown. In addition to the crystals of ettringite, plate-shaped crystals of monosulfate are also formed. After about 5 hr the cement paste is set into an open three-dimensional network, filled with colloidal particles. At this point the strength is quite low but the paste is mechanically stable. The hardening process now begins and lasts up to about a month. The hydration products increase in amount and the fibrils, which are either amorphous or fine-grain crystalline, increase in length. The number and size of the calcium sulfate crystals increase. As these reactions proceed, more and more of the available volume becomes filled, and ultimately there is a considerable amount of interaction and bonding between the individual structures, i.e., the various hydration products and the aggregate particles.

Flaws in Cement

Cement and concrete are ceramics and are therefore brittle, which is a consequence of their hardness, but not weak materials. On the contrary, they are very strong materials for example modern concrete, which in this case means a concrete containing superplasticizers and having low water to cement ratio, for example, 0.3, has a compressive strength of up to 100 MPa per square meter. However, defects, for example, in the form of holes or gaps between particles, will seriously weaken the material. Applied stresses will concentrate at the tip of a flaw and will wedge and propagate it right through the ceramic material. In order to estimate the effect of flaws on the mechanical properties of materials [7] studied the distribution of stress in a large plate with a defect in the form of an elliptical hole of length, L, and radius of curvature, r, at the narrow end. He calculated that the stress, which was uniformly applied to the material, far from the hole, was increased by a factor of $2(L/r)^{0.5}$ near the narrow end of the hole. The compressive strength of hardened cement made from a paste containing too much water – for example, a water-cement ratio of 0.7 instead of a more suitable value of about 0.4 – may well drop to about 10 MPa per square meter, due to a cement structure containing many pores. A reduction of the compressive strength by a factor of 10, because stress at the narrow end of the pores has increased by the same factor, would for instance correspond to elliptical holes of length 5 μm and radius of curvature of 0.2 μm. Holes of that size and roughly that shape can be seen in incorrectly made concrete [8].

A necessary condition for concrete to obtain its ultimate strength is thus that it is able to form the densest possible structure, that is, a structure as free from pores as possible, during the hardening process. Now, there are several reasons why the structure of hardened cement may deviate from this ideal structure. Too much water in the cement paste is one. A water-cement ratio of about 0.4 corresponds to the minimum amount of water required to react with the individual components of the paste and keep the paste workable, which corresponds to a paste containing conventional flow additives, plasticizers or water reducers. With so called superplasticizers, the water-cement-ratio can be reduced to about 0.3. However, more water in the paste makes it easier to handle and some builders may be tempted to add extra water so as to make their job easier, but this extra water is not used up in the hydration reactions, and this leads to a rather porous solid with a strength below the ultimate. Reactions that are accompanied by an increase in volume are detrimental. Any expansion that occurs when the solidification processes of the other components are underway can open small cracks, which can seriously weaken the hardened cement. The aluminate phase reacts vigorously with water under strong heat evolution and expansion. Gypsum moderates the activity of the aluminate phase and is a critical component in modern cement. Moreover, the aluminate phase is susceptible to attack by sulfates, which interact with it and causes expansion. The minor components, see Table 57.1, do not always exert an advantageous influence, and modern cement standards specify a maximum content of these compounds. The reactions of the free oxides of calcium and magnesium with water to hydroxides are accompanied by an increase in volume. The other two minor components in cement, sodium sulfate and potassium sulfate, accelerate the hydration reaction and promote rapid setting of the paste. Early congealing is inconsistent with high ultimate strength.

SILICA IN CEMENT

For reasons of utilizing waste materials and decreasing overall energy consumption certain inorganic materials, called mineral additives, such as fly ash and ground granulated blast furnace slag are added to the cement paste. Mineral additives take part in the hydration reaction and thereby make a substantial contribution to the hydration product. For reasons of obtaining durability and strength above the normal range, silica in the form of silica fume or colloidal silica is being used. Cement containing mineral additives is often called composite cement.

Silica Fume

Silica fume is a by-product, in the form of a very finely particulate powder, of the production of silica or silica alloys in an electric furnace. High-quality silica fume consists of spherical particles, which have a density of 2200 kg m^{-3} and a BET specific surface area of 15–25 m^2 g^{-1}, corresponding to an average particle size

TABLE 57.3
Chemical Composition of Silica Fume

Compound	% Weight	Compound	% Weight
SiO_2	94–98	K_2O	0.2–0.7
Al_2O_3	0.1–0.4	Na_2O	0.1–0.4
Fe_2O_3	0.02–0.15	C	0.2–1.3
MgO	0.3–0.9	S	0.1–0.3
CaO	0.08–0.3		

Source: Adapted from reference [9].

from about 100 to 200 nm [9]. The chemical composition of silica fume is shown in Table 57.3.

Silica fume, like other mineral additives, has pozzolanic activity, that is, it reacts with $Ca(OH)_2$, formed during the hydration of alite and belite (see the first two equations, Table 57.2), and produces more calcium-silicate aquagel, the actual binder material in cement. However, being made in high heat, the surface of the silica fume particles contains very few hydroxyl groups, or silanol groups, which are necessary for reaction with water and calcium hydroxide. It will therefore take some time before the particle surface has become rehydroxylated in the warm, highly alkaline environment of the cement paste and the pozzolanic activity of silica fume typically reaches a high value first in the period 7–14 days after mixing.

The fine particles of silica fume fill spaces between clinker grains, producing a denser paste. It also densifies the interfacial transition zone between cement paste and aggregate, which increases the strength and lowers the permeability. Papadakis [10] investigated the effect of adding between 5 and 15% by weight of silica fume to concrete and found that the compressive strength increased by 10% at 5% addition and by 20% at 15% addition.

Colloidal Silica

Iler [11] defines colloidal silica as stable dispersions or sols of discrete particles of amorphous silica in water, called aquasols or hydrosols, or in an organic solvent, then called organosols. Commercial silica sols are fluid, the viscosity is less than 35 mPas, and stable toward gelling and settling in the pH range between 8 and 10. They have been stabilized, or brought into this pH range, by adding an alkali, for example NaOH, KOH, LiOH, or NH_4OH, to the sol. The silica particles are negatively charged and charge neutrality is brought about by the presence of positively charged counter ions, for example, Na^+, K^+, Li^+, and NH_4^+. There are also available commercial silica sols consisting of positively charged particles, which have been stabilized at pH of about 2 by adsorption of polycations of for instance aluminium onto the surface of the particles. Most commercial silica sols are quite monodisperse and consist of dense, discrete spheres with a range of diameters between about 5 and 100 nm. The maximum concentration depends on particle size and is 15% by weight for 5 nm particles, 30% by weight for 8 nm particles and at least 50% for 100 nm particles. There are also commercial sols that have deliberately been made polydisperse or where the particles are not discrete spheres but instead chains of linked spheres. The appearance of silica sols depends on particle size, particle size distribution and concentration. They look milky if the particle size is large and the concentration is high, opalescent if the size is intermediate or clear and almost colorless when the diameter of the particles is in the smallest size range.

In contrast to silica fume, the surface of the particles of colloidal silica is fully hydroxylated and contains 4.6 OH silanol groups per nm^2 [12]. This fact, together with the much higher specific surface area, makes the pozzolanic activity of colloidal silica much higher than that of silica fume. Wagner and Hauk [13] mixed 15 nm colloidal silica with cement paste and noted a 36% increase, compared with a reference paste without colloidal silica, of the early strength, that is the early strength development during the first 1–7 days. In fact, Skarp and Sarkar [14] pointed out that ultrafine silica particles will harden the cement paste very fast because most of the available water is consumed in the early stage of gel formation, due to the very high pozzolanic activity of colloidal silica. The resulting high early strength, however, is gained at the expense of low final strength, caused by the pore structure created during the very rapid early gel formation. On the other hand, they claim that this problem, caused by excessively high pozzolanic activity of colloidal silica, has been solved and they report that small amounts of colloidal silica added to the concrete mixture, 0.15–0.20% silica based on the weight of the concrete mixture, significantly increased the final strength, reduced the chloride ion permeability and increased the sulfate resistance of the concrete. The properties of the colloidal silicas used are shown in Table 57.4.

Had the sols contained spherical particles of uniform size the particle sizes of sols A and B would have

TABLE 57.4
Physical Properties of Colloidal Silica

Product	Sp. Surface area (m^2/g)	Solids (%)	Particle size (nm)
A	400	24	35
B	80	50	45

Source: From reference [14].

TABLE 57.5
Effect of Colloidal Silica on the Compressive Strength (psi) of Concrete

Type silica	SiO_2, %	1 Day	% Increase	7 Days	28 Days	% Increase
-(Sample1)	—	4.300	—	6.840	8.680	—
A(Sample2)	0.15	5.300	23	8.010	9.680	12
B(Sample3)	0.15	5.580	30	8.030	9.840	13
B(Sample4)	0.20	4.470	4	8.280	9.970	15

Source: Adapted from Skarp and Sarkar [14].

been 7 and 34 nm, respectively. Instead the average particle sizes are considerably higher, more so for A than for B, indicating that the sols are polydisperse, sol A being the most polydisperse and also containing the smallest particles.

The cement pastes, with or without colloidal silica, had a water to cement ratio of 0.35 and contained sulfonated naftalene formaldehyde resin (NSF) superplasticizer. The colloidal silica was added to the concrete mixture after the superplasticizer so as to minimize premature gelling. The compressive strength of the concrete samples are shown in Table 57.5.

Addition of colloidal silica increases the 1-day strength by up to 30% and the 28-day strength by up to 15%. Colloidal silica of type B may be somewhat more effective than type A, although the increase of the 1 day strength is only 4% at 0.20% of type B, as compared to 30% at only 0.15% of the same type of silica sol. Obviously, judicious choice of the average particle size and the particle size of silica sols makes it possible to fine-tune the pozzolanic activity of the silica so that significant increases of both the early stage strength and the final strength of the concrete can be accomplished.

Tables 57.6 and 57.7 show that addition of colloidal silica to a concrete mixture will substantially reduce chloride ion permeability and enhance sulfate resistance.

Greenwood et al. [15] showed that the smaller particles in the sol provided most of the sulfate resistance whereas the larger particles provided the chloride resistance, but the two particle size regions appeared to interact and gave rise to significant synergism.

The availability of modern workability additives, e.g., superplasticizers such as polycarboxylates, has made it possible to develop highly fluid concrete, HFC, which does not bleed or segregate in use. Self-compacting concrete, SCC, (in the U.S.: self-consolidating concrete) is a particular type of HFC, which achieves significant benefits and advantages in many types of constructions. Thus, by using SCC it is possible to fill the mould completely and uniformly, even moulds of difficult and complicated shapes. There is no need to vibrate the material so as to eliminate voids and holes formed when conventional, often sluggish concrete is poured into the form. Moreover, the quality of the concrete surface is often very good, minimizing the need for expensive and time-consuming after-treatment.

Skarp et al. [14], however, pointed out that poor stability, that is bleeding or segregation, and loss of workability are two main concerns when working with SCC. A concrete mixture is said to be workable if it can be maintained in fluid form until the casting moment. The term workability time is defined as the time the concrete mixture remains workable. They attribute the instability to deficiencies in mix design and the loss of workability to incompatability between the cement and the superplasticizer.

Greenwood et al. [16] showed that small amounts, 0.2% by weight of SiO_2, of colloidal silica of small particle size, corresponding to a specific surface area of 900 $m^2\ g^{-1}$, significantly increased the workability time

TABLE 57.6
Effect of Colloidal Silica on the Chloride Ion Permeability of Concrete

Silica, %	0	0.1	0.15	0.20
Chloride ion permeability, coulomb	3600	3200	2400	1700

Source: Adapted from reference [14].

TABLE 57.7
Effect of Colloidal Silica on the Sulfate Resistance of Concrete

Weeks	Control % expansion	Sol A, 0.13% SiO_2, % expansion	Sol B, 0.13% SiO_2 % expansion
4	0.01	<0.01	<0.01
8	0.021	<0.01	<0.01
12	0.036	0.015	0.014
16	0.050	0.017	0.016

Source: Adapted from reference [14].

of concrete mixtures containing polycarboxylates as superplasticizers. It required much larger amounts of colloidal silica of larger particle size, corresponding to a specific surface area of 80 $m^2 g^{-1}$, 1.25% by weight of SiO_2, or of fumed silica, 10% by weight of SiO_2, to achieve the same results. In contrast to the control, containing no silica, the concrete mixtures containing either colloidal or fumed silica showed no bleeding.

Aluminum-modified sols, compared with unmodified sols of the same specific surface area, as additives in concrete mixture containing polycarboxylates as superplasticizers achieved not only increased workability time but also improved strength, Greenwood et al. [17].

COLLOIDAL SILICA AS RETENTION AID IN PAPER MAKING

Colloidal silica was introduced as retention aid in paper making less than 25 years ago. The home page of Compozil® states that in year 2000, 347 paper machines all over the world with a combined production of 26.3 million metric tons of paper and paper board used colloidal silica as retention aid, which probably makes it the largest application of colloidal silica today. Otterstedt and Brandreth [18] described the use of colloidal silica as retention aid in paper making and compared it with other types of retention aids. This section is a substantially abbreviated version of their work, but supplemented with the most recent developments of colloidal silica as retention aid in paper making.

The word paper is derived from papyrus, a sheet made in ancient times by pressing together very thin strips of an Egyptian reed, cyperus papyrus. The modern material, paper) consists of sheet materials that are comprised of bonded, flexible, cellulose fibers which, while very short) 0.5–4 mm, are about 100 times as long as they are wide. Small particle fillers or pigments, in the form of clays or other inorganic materials are used to improve the properties of paper, that is, opacity, brightness and printability, or to improve the economics of the papermaking process. In this chapter we will focus on the use of small particles as process aids to improve retention and dewatering on paper machines.

Fillers

Mineral fillers in the form of small particles are used in paper for various reasons. There has always been the economic incentive to substitute low-cost fillers and extenders for some high-cost fibers in paper, but there is also the incentive to improve several of the properties of paper. The use of fillers increases opacity and brightness of the paper and also improves printability by making printing ink absorption more uniform, gives higher gloss after calendering and leads to better "feel" and dimensional stability.

The disadvantages of using fillers in paper are reduced mechanical strength, caused by the filler particles interfering with the hydrogen bonding between the cellulose fibers, heavier paper, greater wear on the wire of the paper machine, and higher content of fine material in the circulating water system.

Pigments, which are also small particles of inorganic materials, are used to improve the optical properties of paper and are usually more expensive than cellulose fibers. Pigments are also often made synthetically, whereas most fillers are ground minerals. The most common types of fillers are kaolin or clay, the most important filler, ($Al_2O_3{*}2\ SiO_2{*}2\ H_2O$), talc ($3MgO{*}H_2O$), calcium carbonate ($CaCO_3$), gypsum ($CaSO_4{*}2H_2O$) and mica ($3Al_2O_3{*}K_2O{*}6\ SiO_2{*}2\ H_2O$).

Paper-Making

Different paper machines have various configurations at the wet end of the machine, but Figure 57.1 shows schematically a representative setup. In the mixing chest fibers and paper chemicals are mixed to an aqueous slurry, the furnish, containing about 0.5–2% fiber. Some of the chemicals may be added at a later stage, for example, to the machine chest or before, or into a pump. From the head box, the furnish is filtered on a wire screen, where the fibers adhere weakly to one another. When more water is removed from the mat formed on the screen by suction, the sheet becomes stronger, but is still relatively weak. When the sheet is dried it becomes still stronger, and becomes the material known as paper. Modern paper machines produce an endless paper sheet, up to 10 m wide, at a speed of over 20 m/sec, that is, one hectare (more than two acres) every 50 sec. The machine is

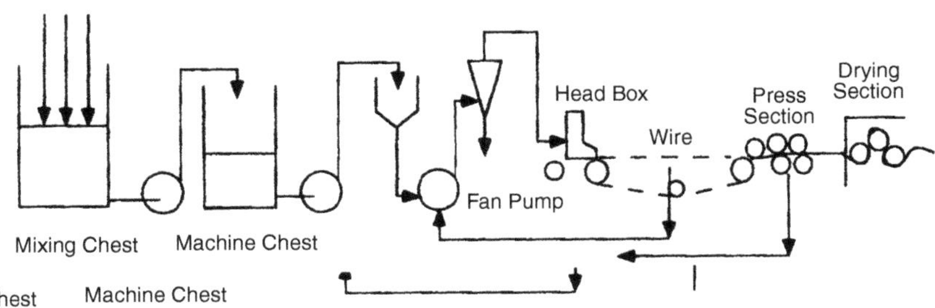

FIGURE 57.1 The wet end of a paper machine. From Otterstedt and Brandreth, [18]. Courtesy Plenum Press.

more than 100 m long and produces about 250,000 metric tons per year.

Environmental and economic pressures have reduced water usage in paper production in the last 30–40 yr from 80 to 90 m^3 per metric ton to less than 10 m^3/ton. During the last decade many efforts have been made to reduce the use of water even more with the ultimate objective of achieving a paper mill that is 100% closed.

The Problem

The achievement of a closed or nearly closed paper mill with respect to water usage is intimately related to the retention of fiber fines and chemicals and other additives in the furnish on the wire. Poor retention will cause the fines and other small particles to go through the wire with the water and make reuse of the back water difficult or impossible; see Figure 57.1. The nature of the problem is further illustrated by Figure 57.2, showing the dimensions in the wet end, and Figure 57.3, which compares the size of the holes in the wire with the sizes of the cellulose fibers, fines, filler particles and the various chemical additives present in the furnish.

The difficulty in retention is further aggravated by the fact that all the particles of the furnish are negatively charged and therefore have no bonding to each other to form aggregates large enough not to pass through the holes of the wire.

The obvious solution to the problem is therefore to put into the system particles or additives of opposite charge to cause agglomeration of the paper components to larger clumps that cannot go thorough the wire. This is accomplished by so-called retention aids.

Retention, Retention Mechanisms and Retention Aids

The term retention refers to the holding back of the components of the stock during dewatering. The fibers are retained on the wire whereas fillers, fines, and additives of colloidal size may be washed through the mat formed

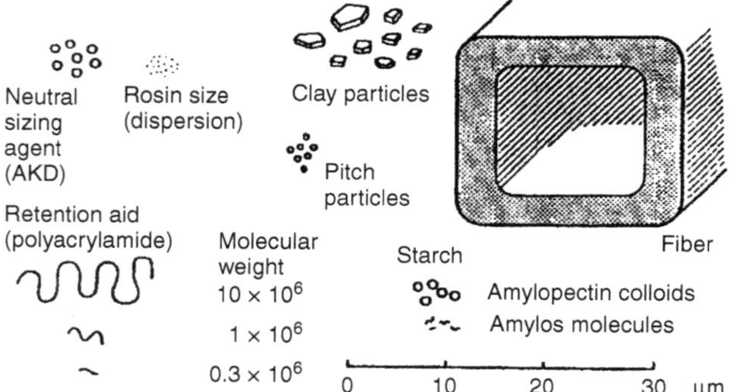

FIGURE 57.2 Dimensions in the wet end of the papermaking process. From Otterstedt and Brandreth, [18]. Courtesy Plenum Press.

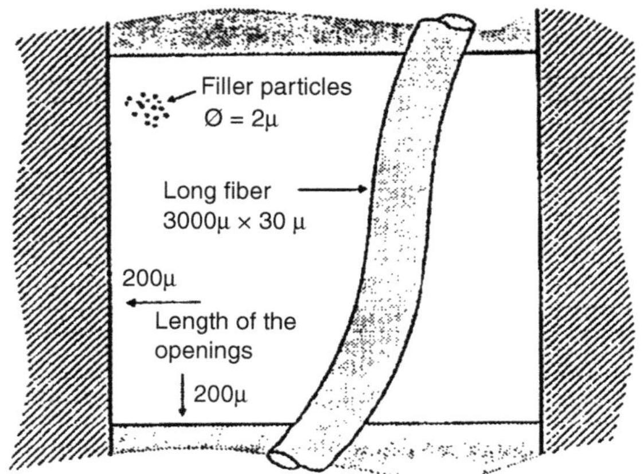

FIGURE 57.3 Small particles on the wire in the papermaking process. From Otterstedt and Brandreth [18]. Courtesy Plenum Press.

on the screen. Retention is accomplished by a combination of mechanical means, i.e., filtration, and the physicochemical mechanism of agglomeration or flocculation.

Mechanical retention during sheet formation on the wire may be considered a filtration process. The fibers in the stock, which are 500–4000 µm long and 20–100 µm thick, are captured on the wire and form a three-dimensional network consisting of 2–100 layers of fibers on the wire. As the layers form, they capture progressively smaller fibers and other colloidal particles in the stock suspension, making the pore structure gradually finer with the largest pores on the wire side and the finest on the top side. Mechanical retention is least efficient in the beginning of the sheet formation and, although it becomes more effective as more layers from, it cannot retain a satisfactorily high proportion of the finest components of the stock. The losses for newsprint are typically about 50%.

By adding special chemicals, retention aids, to the stock, the fines and other colloidal components can be made to flocculate or aggregate into agglomerates too large to go through the wire. Retention aids may consist of either one component or two components. They can act by changing the electrostatic repulsion forces between colloidal particles or affect the stability of colloids by adsorbing on two or more particles causing them to form larger aggregates.

Although good retention is most likely attained by the joint action of more than one mechanism and a given retention aid may act by several mechanisms, it is still useful to distinguish between some principal types of aggregation mechanisms.

There are no sharp distinctions between the terms of coagulation and flocculation, but here coagulation denotes aggregation by the action of low molecular weight electrolytes whereas flocculation means aggregation brought about by polymers, which can be natural or synthetic.

In high-speed modern paper machines the floc is subjected to high shear, which may tear the floc apart. The trend toward reduced water usage in the production of paper increases the amounts of soluble anionic wood polymers and electrolytes in the stock, which will also affect the retention on the wire.

The principal types of retention mechanisms are:

1. *Charge neutralization*: Electrolytes are the simplest kind of coagulants that can be used to improve retention. They act fully in accord with the DLVO theory by screening the charges and compressing the electrolytic double layer of the negatively charged particles of the stock, thus allowing attractive forces to come into play and aggregate the particles. Aggregation by charge neutralization is a fairly slow process which has lost some of its importance as the speed of paper machines has become ever faster.
2. *Hetero-coagulation*: This mechanism involves adsorption of oppositely charged particles, e.g., complexes of resin acids and aluminum sulfate, on the surfaces of fibers and filler particles. Hetero-coagulation is sensitive to soluble anionic wood polymers and electrolytes, with which cationic sizing particles, preferentially interact.
3. *Patch flocculation-Patching*: Patching resembles charge neutralization, but is different. In this mechanism cationic polymers are strongly adsorbed in a flat configuration on the negative surfaces of the particles, on which they form cationic patches. Adsorption leads to partial charge neutralization and electrostatic attraction between oppositely charged patches on different particles leads to flocculation. If the cationic polymer is small and the patches are smaller in size than the thickness of the electrolytic double layer-which depends on the concentration of electrolytes-aggregation will take place by the mechanism of charge neutralization.

   A characteristic difference between charge neutralization and patching is that the rate of coagulation for the former mechanism increases with electrolyte concentration. Once an optimal electrolyte concentration has been attained, however, the rate of flocculation by patching will decrease with electrolyte content due to the fact that the electrolyte cations will force the polymer from the particle surface.

   Relatively short-chained cationic polymers of average molar mass and high charge density are suitable for patch flocculation. Modified polyethylene imines, polyamines, and polyamide-amine-epichlorohydrin resins are in this category.
4. *Bridging*: In this mechanism flocculation is accomplished by long-chain, i.e., high molar mass, polymers forming binding bridges between particles. For effective bridging to occur, it is very important that the polymers adsorbing on the surface of the particles form loops and tails that protrude into the solution. To what extent this happens depends on the type of polymers, contact time, and properties of the surface of the particles to be flocculated. Suitable polymers are weakly charged or nonionic, that is, high molecular-mass polyacrylamide and polyethylene oxide. Flocs made by bridging are large but fairly easily broken by shearing, which may tear the bridging polymers and retard the process of re-flocculation.
5. *Complex flocculation*: Flocculation by any of the four mechanisms described above can be accomplished by only one flocculant or retention aid. Much more effective flocculation and retention can be achieved by using combinations of retention aids. The most common combinations are between oppositely charged retention aids, which can form complexes of varying strength with each other. It is, however, also possible to use combinations of nonionic retention aids that can form complexes by hydrogen bonding.

Some important types of retention aids are the following natural or synthetic polymers:

Polyethylene imines, PEI, are strongly cationic and strongly branched polymers with a molar mass between 100,000 and 1,000,00 (g/mole)

Polyethylene amines, contain secondary amine groups and are linear, strongly cationic polymers with a molar mass of about 100,000.

Polyacrylamides, PAM, are nonionic polymers with a molar mass of about 1,000,000.

Cationic polyacrylamides, CPAM, contain teriary amine groups which can be quaternized. Molar mass is about 1,000,000.

Anionic polyacrylamides, A-PAM, can be synthesized by co-polymerizing acrylamide with acrylic acid. It contains anionic carboxyl groups and has a molar mass of about 1,000,000.

Polyethylene oxide, PEO, is nonionic and has a molar mass of about 1,000,000.

Cationic starch is modified natural polymer. Molar mass about 100,000,000.

Dual Retention Aid Systems

Cationic natural and synthetic polymers have long been used to improve retention of fines and fillers on the wire of paper machines. Such polymers, that is, cationic starch or cationic polyacrylamide, produce a high degree of flocculation in the furnish. This floc, however, is not very strong and is easily broken and redispersed by hydraulic shear. Furthermore, when long-chain polymers are used, chain rupture and rearrangement of the polymer fragments on the particle surfaces may occur. Nevertheless, single-component retention aids improve the first-pass retention, though not to the same degree as dual retention aid systems.

Such systems have been used in the paper industry for many years. Component one, a cationic polymer, usually of the patching type, is first added to the furnish, followed by the addition of the second component, an anionic polymer of the bridging type. Figure 57.4 schematically compares single component and dual retention aid systems. The application of retention aids has been optimized in the sense that the retention maximum in the figure corresponds to a zeta potential of value zero, i.e., the charges on the positive components in the system exactly balance the charges on the negative component, which may be difficult to accomplish in an actual situation. When an optimal amount of cationic component, in this case cationic starch, in the single-component system, is added, the furnish system has no charge and flocculation and retention are maximized. In the dual system cationic starch has to be present in the furnish so as to reach zero-potential after the given amount of the second component, an anionic polymer, has been added. Thus, the maximum in flocculation and retention is not only higher than for the single component system, but it also occurs at larger dosages of cationic starch, which is beneficial since starch is not only a retention aid, but is also an additive that increases the dry strength of paper.

In the last 10–15 years a special kind of dual retention aid, a micro particle — containing flocculant system, often referred to as microparticulate retention/dewatering aid, was developed. In one of the commonly used commercial systems, the so-called Compozil®system, comprise colloidal silica in combination with cationic starch or cationic synthetic polymers. Andersson and Larson [19] and Andersson and Lindgren [20]. In this system, the cationic polymer is added first and the extensive flocs then formed are broken down and partially redispersed by high-shear forces. The anionic microparticles are added just before the paper is formed and cause final flocculation of the furnish. A dual retention system, having colloidal silica as the anionic component has the following characteristics [21]:

- strong, reversible flocculation,
- more effective dewatering in the wire and press sections,
- formation on the wire yields sheets of higher porosity and permeability.

The Compozil® system was recently studied by Andersson and Lindgren [20]. They used a Britt Dynamic Drainage Jar to investigate the retention effects of combinations of various types of anionic colloidal silica, ACS, with either cationic starch or polyacrylamides of different charge density. The furnish consisted of a 60/40 mixture of fully bleached birch and pine sulfate pulps with 30% (based on total solids) chalk as the filler. The solids content and pH of the furnish were 0.5% and 8.1, respectively. The polyacrylamides had charge densities between 2 and 25% cationicity, corresponding to between 0.25 and 3.0 meq/g, and a molecular weight of 5×10^6. The cationic starch had a degree of cationic substitution of 0.4, corresponding to 0.25 meq/g.

The anionic colloidal silica used in this study was either monodisperse colloidal silica with a particle size of about 4 nm or so-called structured colloidal silica, consisting of linear aggregates of about 4 nm particles. Structured colloidal silica is, like monodisperse colloidal silica, characterized by its specific surface area and charge density, which decreases with pH but can be maintained high even at pH as low as 3 and 4 by aluminizing the

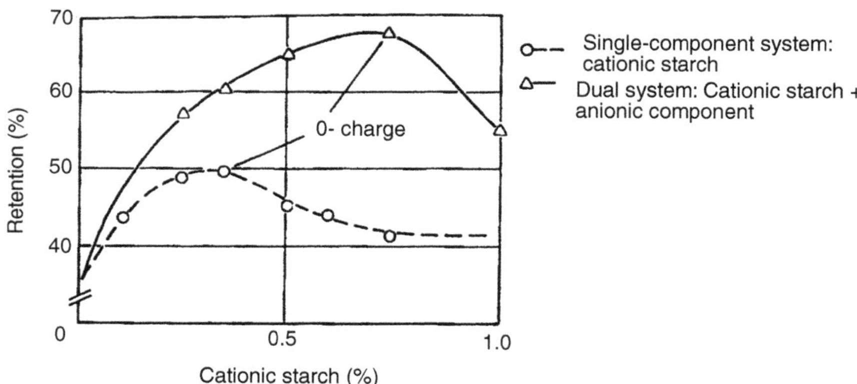

FIGURE 57.4 Single-component and dual retention aid systems. From Andersson and Larsson [19]. Courtesy Arbor Publications.

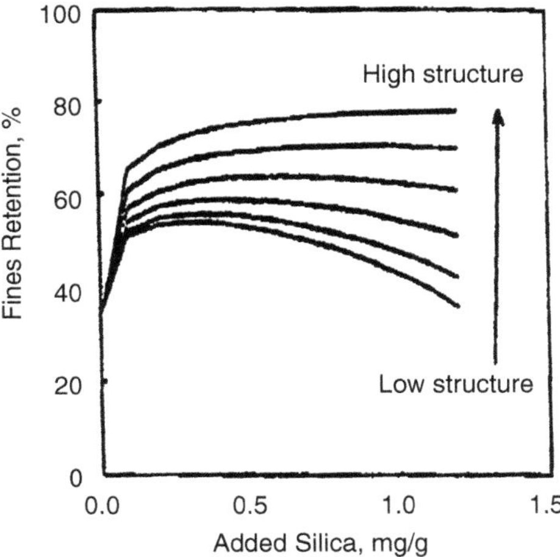

FIGURE 57.5 Retention model for the CPAM-ACS system. From Andersson and Lindgren [20]. Courtesy Arbor Publication.

colloidal silica with sodium aluminate, but also by some other properties. One is the S-value, which is defined as the percentage of silica in the dispersed phase and can be obtained from viscosity measurements [22]. A high S-value indicates well-dispersed, non-aggregated colloidal particles, whereas a low value suggests that the primary particles have formed microaggregates, perhaps linear structures containing up to 7–8 primary particles. Another one is the average size of the microaggregates, A, as determined by dynamic light scattering, DLS. Other ones are the length, *L*, and the width, *W*, of the microaggregates as determined by DLS and viscosity measurements.

From their results, Andersson and Lindgren concluded that for both the cationic starch-ACS and CPAM-ACS systems the main flocculation mechanisms were electrostatic interactions, for example, charge neutralization and bridging.

They also used their data to construct a model for the system CPAM-ACS, shown in Figure 57.5 for a constant dosage of CPAM. Each curve shows the predicted retention for an ACS with a constant S-value. As expected, maximum retention increases with increasing structure, or degree of microaggregation, of the ACS.

Greenwood et al. [23] showed that chemical modification of ACS, comprising stabilization of structured colloidal silica by amines instead of by NaOH, which is the most common method, not only improved the retention but also the dewatering on the wire of the paper machine, see Figure 57.1. They found that quaternary amines gave the best results, followed by tertiary and secondary amines.

COLLOIDAL SILICA IN LEAD-ACID BATTERIES

Lead-acid batteries are one of the most common type of batteries. Most lead-acid batteries are flooded, that is, they have a liquid electrolyte as in standard car batteries, but a significant and growing number have a solid electrolyte. Some of the advantages of a solid electrolyte in lead-acid batteries are little or no spill or splashing of highly corrosive sulphuric acid in case of accidents, no leakage if the battery is placed sideways or even upside-down and longer life time, because no accumulation of precipitated lead at the bottom of the battery, which may cause discharge, can occur.

There are two different methods of immobilising the electrolyte, a solution of lead sulfate, that is, making a solid electrolyte, in VRLA, that is, *v*alve-*r*egulated *l*ead-*a*cid batteries. In AGM-VRLA batteries, the electrolyte is immobilized by being absorbed in *a*bsorptive *g*lass-fiber *m*ats, placed between the electrodes. This type of batteries can produce high starting currents and be rapidly recharged, and are used in for example, uninterrupted power supply systems(UPS). In GEL-VRLA batteries, the electrolyte is immobilized by being absorbed within the very fine pores of a silica gel, which can be made from different silica materials. GEL-VRLA batteries have long life span, good cycling characteristics and relatively low current, and they are used in applications such as telecommunication and solar energy devices and motive power applications, for example, golf cars, wheel chairs and loading trucks [25]. GEL-VRLA batteries are predicted to be used as a second battery in cars to supply steady, nonsurging power to the increasing number of electronic components in modern cars.

Silica gels can be made from different starting materials, for example, sodium silicate solutions, fumed silica or silica sols. Judging from the patent literature, fumed silica appears to be the most common starting material for making silica gels for GEL-VRLA batteries, but recently, in the last several years, some very promising work has been reported on making silica gels for batteries from colloidal silica in the form of silica sols.

Gelling of Silica Sols

At the ACS National Meeting, in Washington, D.C., 1990, Paul Yates gave a talk on the "Kinetics of Gel Formation of Silica Sols" [24]. He described that the gellation of silica sols is kinetically quite different from that of soluble silicates although the same factors are important, that is silica concentration, pH, salt content, temperature and particle size of the sols. Expressions were derived for the quantitative prediction of the gel times of colloidal silica dispersions over a wide range of these variables. The following exposé is a summary of Yates presentation.

Types of Gels

There are three types of silica gels, of which the first results from neutralizing dilute aqueous solutions of a silicate and is formed by the polymerization of silicic acid.

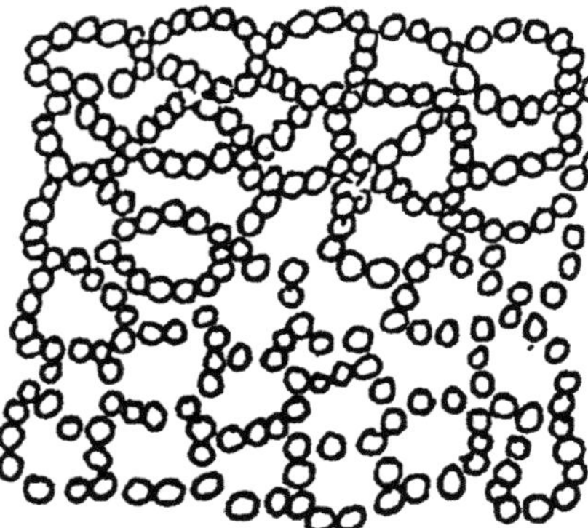

FIGURE 57.6 Structure of collision gel. From Yates [24].

The polysilicate ions extend as a three-dimensional cross-linked network throughout the solution.

The second type results from the collision of preformed colloidal silica particles to form three-dimensional chains of such particles, bonded at their junction points with siloxane bonds; see Figure 57.6.

The third type is a hybrid of the first two with polysilicic acid chains joining preformed colloidal silica particles. Hybrid gels are formed by neutralization of mixtures of silicates and colloidal silica sols.

Common Features in Gel Formation

Although the quantitative kinetic expressions for each of the three types of gel are different, they respond qualitatively in a similar way to most important variables. For all types or gels, gel times become shorter at higher temperatures, at higher concentrations, and in the presence of increasing concentrations of neutral salts.

The effect of pH is complex. Starting with strongly acid (low pH) systems, gel times initially decrease rapidly as the OH^- concentration is increased, then pass through a minimum and finally increase rapidly as the pH continues to increase. For all types of gels, the effect salts in decreasing gel times is much more pronounced on the basic side than on the acid side, and the location of the minimum in the gel time versus pH curve is a very sensitive function of the neutral salt content and even of the specific salt employed.

The Central Mechanism in Gel Formation

The similarities described exist because the central polymerization mechanism is essentially the same for all types of gels. This mechanism also shows the key role played by the hydroxyl ion.

The First Role of Hydroxyl Ions — As a Catalyst

A silicon atom in silicic acid or at a surface normally has a coordination number of 4. The coordination number can be momentarily expanded by adosorption of a hydroxyl ion simultaneously with adsorption of a sixth group such as a silanol group belonging to another silicic acid molecule or colloidal particle. This transition complex is unstable and water condenses out between the two silanol groups to form a permanent siloxane bond. The coordination number drops back to four and the hydroxyl ion is desorbed, regenerating it to continue its catalytic role elsewhere in the solution.

The Second Role of Hydroxyl Ions — Charge Repulsion

If the only role played by hydroxyl ions were a catalytic one, gel times would continuously decrease as the pH increased. The observed minimums in gel times and their rapid increase at high pH shows that the hydroxyl ion plays a dual role in the mechanism of formation of silica gels.

Silicic acid and silanol groups on the surface of colloidal silica particles are weak acids, and lose protons in basic solutions, thereby acquiring a negative electrostatic charge. This charge repels other negatively charged groups which attempt to approach. The charge increases rapidly as the pH increases.

For polysilicic acid, charge repulsion interferes with polymerization when the charged groups on the polysilicic acid polymer repel the negatively charged hydroxyl ion catalyst, which must be adsorbed on the already negatively charged polymer to perform its function.

In the collision of colloidal silica to form gels, the dominant charge repulsion is between the two negatively charged silica particles, which must collide before siloxane bonds can be formed at their surfaces.

The charge repulsion term will not be mathematically same for these different types of polymerization, but the effect on all three types of gels is that charge repulsion opposes the favorable catalytic effect of the hydroxyl ions, and at a sufficiently high pH, counterbalances it to lead to a minimum in the gel time.

The Role of Neutral Salts — Screening of Charge Repulsion

The reason why gel times are always decreased by the addition of a neutral salt is that when negatively charged groups of polysilicic acid, or surface groups on the colloidal particles, can be screened by a swarm of positively charged cations from neutral salts, their repulsion of approaching negatively charged species is substantially decreased. Higher the concentration of neutral salts, higher will be the probability of a number positive ions being located in screening positions around a negatively charged group.

Specific Salt Effects — Effect of Cations

The specific effects of salts come from differences in their screening ability. The larger size of the hydrated positively charged cation in the neutral salt, the more efficiently it can screen and the more rapid the gel time. The effect is particularly pronounced with large organic cations such as guanidine, tetramethylammonium, and tetraethanolammonium ions.

Effect of Anions

Even the anion of a neutral salt shows some specific salt effects. It probably does this through an indirect mechanism.

If the cations of a neutral salt are to be preferentially adsorbed as a double layer in the vicinity of the negatively charged groups of colloidal silica particles or polysilicic acid molecules, they will have to be recruited from the otherwise homogeneous mixture of anions and cations of the salt in the solution. This can only take place by removing them from the charge field of their own anions. The higher this charge field is, the less likely are they to concentrate around the polymer or colloid, and the less effective the salt will be in screening.

Neutral salts containing highly charged anions such as sulfate are not as effective screening agents as those containing monovalent anions such as chloride. Acetate salts are better screening agents than sodium chloride.

The Salt Effect Is Proportional to the pH

From the explanation just given, it is obvious that the salt effect will not be very strong on the acid side, since silicic acid polymers or colloidal silica surfaces do not have a high charge at low pH. It becomes more and more pronounced as the pH increases, since as the charge increases, the importance of screening will also increase.

The Effect of Temperature

The effect of temperature was not specifically studied this work, but previous work in the literature shows that the activation energy for the central polymerization mechanism is about 80,000 J/mole. This means that the temperature coefficient will be about a factor of two in the gel time for every 10°C. change in the temperature. Activation energies will probably be different on the acid and the basic side, since the temperature coefficient for the charge repulsion effect will enter into the total activation energy in basic solution.

Differences between the Polymerization and Collision Mechanisms of Gel Formation

Although many of the features are common to both mechanisms they will not be mathematically the same. In the case of the formation of gels by means of collision of colloidal silica particles, the important charge repulsion is that between two colloidal silica particles, whereas in silicic acid polymerization the charge repulsion is between the charged groups on the silicic acid polymer and the hydroxyl ion. Screening effects of neutral salts will also be quantitatively different.

Collision gels are extremely sensitive to the surface chemistry and surface composition of the colloidal particles. Small amounts of aluminum, for example, can greatly change the rates of collision. This is because such changes in surface chemistry strongly influence the charge on the colloidal silica particles, and thus the repulsion to be expected between two of them at any particular pH.

The basic structural units, which are doing the polymerization, are different in the two cases also, and the relationship of the concentration of silica to the polymerization rate will not be the same.

Kinetics of Collision Gels

Gels formed by the collision of colloidal silica particles in solutions containing only traces of silicate ions have entirely different kinetics than the other two types of gels. Under comparable conditions, gel times are 100 to several thousand times as long. The quantitative response to variables such concentration, salt content, pH, and the surface area silica is also quite different.

The equation derived by Yates [24] for gels prepared from deionized "Ludox HS" mixtures of varying concentrations, pH values, and salt contents is given below as Equation (57.1).

$$\begin{aligned}\log t = {} & 5.85\text{-pH-}\log(\Phi/(1-2.58\Phi)) \\ & + (1.333 - 1.482\Phi)(0.032 - 0.1183 \log c) \\ & \times (\text{pH} - 2.34)^2. \end{aligned} \tag{57.1}$$

Where t = gel time in minutes; Φ = volume fraction of silica; c = salt concentration.

This equation reproduced the gel times of 39 gels from solutions containing 10, 20, 30 and 40 wt.% SiO_2, at pH values of 3.5, 5.0, 6.0, 7.0, and 8.5, and salt concentrations of .01, .03, 0.1, and 0.3 N, with an average error of 0.11 units in the log gel time value. This is within the probable experimental reproducibility of this data.

Figure 57.7 shows the ability of this equation to reproduce gel times at a constant (0.1 N) salt concentration over a range of pH and silica concentrations. The solid lines were calculated from Equation (57.1).

It might be appropriate to review the physical meaning of each term in Equation (57.1). This will be done to indicate qualitatively how each of the variables affect this type of gel formation.

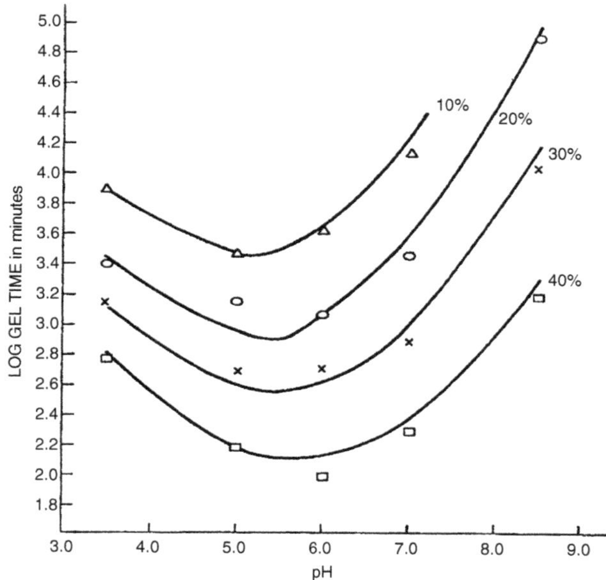

FIGURE 57.7 Effect of silica concentration on the gel time of collision gels at 0.1 normal salt concentration. From Yates [24].

The first term is a constant which includes the specific reaction rate constant for siloxane bond formation, the ionization product constant for water, the viscosity of water, and other miscellaneous numerical constants.

The second term, (−pH) expresses the catalytic effect of the hydroxyl ion in decreasing gel time.

The third term, which involves Φ (the volume fraction of silica), expresses the effect of increasing the silica concentration in decreasing the gel time. The expression $\Phi/(1 - K\Phi)$ is used instead of Φ itself as a concentration variable, since the particles will physically touch one another long before the silica volume fraction becomes one (corresponding to a 100% concentration). The constant in the denominator of this expression, which has the value of 2.58 for the particular sample of deionized "Ludox" used in these gelling experiments, is identical to the constant, which appears in the Einstein–Mooney equation for the viscosity of spherical colloidal particles. This will vary with the degree of hydration and aggregation, or the % solids in the dispersed phase, of the silica particles.

The last and most complicated term in Equation (57.1) is the electrical repulsion term. As can be seen, for a fixed surface area (or particle size) of the colloidal silica, the charge repulsion is a function of the pH, the volume fraction, and the concentration of neutral salt.

The term $(\text{pH-2.34})^2$ occurs because the electrical potential of the silica surface is proportional to the amount of (OH^-) adsorbed, and therefore is inversely proportional to the (H^+) ion concentration. The number 2.34 represents the zero point of charge for this particular colloidal silica, or the pH at which substantially no hydroxyl ions are adsorbed. If some other sol were used, which had a different concentration of aluminum or other impurity atoms at the surface, this zero point of charge might occur at a different pH. This is the case, for example, with aluminum-modified "Ludox", which probably has a zero point of charge somewhere around a pH of 1.00 or lower. This term in the pH is squared because we are dealing with a charge repulsion between charged particles which are essentially identical to each other.

The term (0.032−0.1183 log c) accounts for the screening effect at the salt in decreasing charge repulsion. It will be noted that screening increases only with the logarithm of salt concentration, indicating much less sensitivity to variation in salt content than for the previous type of gels where the slope of the electrical repulsion term was proportional to the reciprocal of the salt concentration.

The constant .032 in the expression in proportional to the slope the charge repulsion term would have in a one molar salt solution when log c would equal 0.

The number 0.1183 in front of log c determines how rapidly the slope of the repulsion term changes for a given change in salt concentration. This will probably vary with both the chemical nature and the valence type of the neutral salt as was observed for the other types of gels.

The last term (1.333−1.482Φ) expresses how the charge repulsion or the electrical work of bringing the particles together will vary with the volume fraction of particles. It is perhaps not immediately obvious why the work of repulsion should depend on the volume fraction. This can be explained by reference to Figure 57.8, which is a schematic plot of the charge repulsion as a function of the distance between the particles. The difference between the bottom of the curve in Figure 57.8 and the top shows the electrical work which must be performed to bring widely separated particles into the gel configuration where the maximum repulsion will exist when the particles are in actual physical contact at tangent points between them. The difference between the top and other marks corresponding to silica sol concentrations represents the electrical work of repulsion involved in bringing the sol particles, each from their equilibrium distances apart, into a gel structure. It is obvious that the electrical work is less if a more concentrated sol is used. To put it simply, a significant part of the work of electrical repulsion to bring the particles into a gelling configuration has already been done in the process of concentrating the sol.

The constant 1.333 in this expression represents the maximum electrical work which might be found if sols were used which were so dilute that their electrical fields did not overlap. The slope 1.482 is proportional to the rate at which the electrical work changes per unit change in the volume fraction of particles.

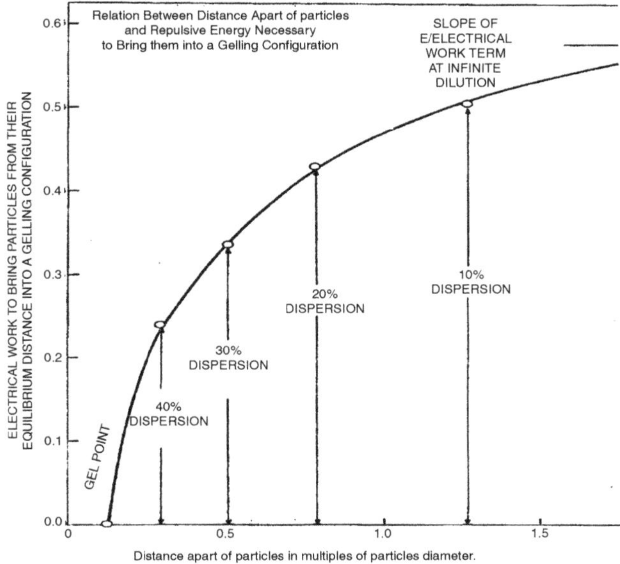

FIGURE 57.8 From Yates [24].

The slope of the repulsive energy term will also be a function of the particle size or surface area of the sol, since the repulsion between particles depends on size as well as on distance apart, charge, and salt content. This did not appear explicitly in Equation (57.1), since all of the data for evaluation of this equation came from sols of the same surface area.

Equation (57.1), and analogous equations for polymerisation and hybrid gels but not shown here, is very useful for predicting the gelling behavior of silica sols above the point of zero charge of silica. Silica gels for batteries, however, are made at pH well below 0. Thus, Lambert et al. [25] reported that they prepared a gel of this kind by mixing a silica sol, with a particle diameter of 12 nm and a silica concentration of 40% by weight, with concentrated sulfuric acid, 98% by weight. The concentration of sulfuric acid was more than 10 molar and the pH <-1. Using a pH value of -1 and electrolyte, SO_4^{-2}, concentration of 1 m, Equation (57.1) predicts that the gelling time would be less than a minute (the higher the anion concentration the shorter the gelling time), when it in an actual experiment was >12 h. Obviously, Equation (57.1) is at best only heuristically useful at very acidic condition. At these conditions, the various terms in the equation will be the same, but the actual values of the constants would have to be recalculated in order to make the equation fit experimental data more closely.

Wang [26] claimed that a solid electrolyte, in the form of a gel containing about 20% by weight of silica and made by mixing a de-ionized silica sol, adjusted to pH between 8 and 14, with sulphuric acid, compared with a conventional fluid electrolyte consisting primarily of sulfuric acid, increased the capacity of lead-acid storage batteries by more than 30%.

Sielemann et al. [27] compares solid electrolytes, consisting of collision gels made from either silica sols with a specific surface area between 100 to 500 m^2g^{-1} or from fumed silica, in lead-acid batteries. The solid electrolyte made from the silica sol was prepared directly in the battery container, whereas the one made from fumed silica had to be made in separate step. The fumed silica, in the form of very light, fluffy powder, was mixed with the sulfuric acid and the other components of the solid electrolyte in a special vessel. The slow-gelling mixture was then poured into the battery container where it eventually solidified. The performances of lead-acid batteries containing the two different types of solid electrolyte were very similar, perhaps with a slight edge for the silica sol battery since it had a somewhat higher $20*1_0$ discharge current.

Lambert et al. [25] have made a comparative study of solid electrolytes made from silica sol and fumed silica and claim the following advantages for electrolytes made with silica sol:

- Simplified handling and mixing of the electrolyte
- No liquid separates from the gel after solidification
- High silica concentration in gelling additive
- Increased residual gel strength
- Controlled gel time
- Less impurities (e.g. iron and chloride)
- Lower cost (~USD 2.60–3.50/kg pure silica from silica sol versus ~USD 9.00–18.00/kg fumed silica)

COLLOIDAL SILICA IN COATINGS

The coating process modifies the surface of a material, providing a gradual difference in composition or property between the surface and the bulk.

Iler [3] gives 36 references to how colloidal silica can be used in coatings on various substrates. Organic coating compositions with improved adhesion, hardness and durability can be obtained by adding colloidal silica in the form of silica sols to organic polymer dispersions. Inorganic coatings may use silica as the main component or as a binder in the coating composition.

Here we will highlight some very recent applications of colloidal silica in coating compositions.

Shop Primers for Steel Substrates

Steel used in the shipbuilding industry and for other large-scale structures such as oil production platforms is often exposed to the weather during storage before construction and during construction, and it is generally protected against corrosion by a coating called "shop primer" or "preconstruction coating". Although the main purpose of the shop coating is to provide temporary corrosion protection during construction it is preferred that the primer can remain on the steel surface during and after fabrication. Thus, steel coated with shop primer must be weldable and compatible with and have good adhesion to the different types of anticorrosive coatings used on ships and other steel constructions.

Many modern shop primers are solvent borne and based on prehydrolyzed tetraethyl orthosilicate binders and zinc powders. They contain substantial amounts of volatile organic solvents, typically about 650 g per liter, to stabilize the paint binder and allow the primer to be applied as a thin film, about 20 μm thick. Obviously, the use of such primers will give rise to environmental concerns.

Much effort has therefore been expended to develop water-based shop primers with the advantages but without the disadvantages of organic solvent-based primers. In one approach, alkali silicate binders, e.g. aqueous solution of 3.3 ratio sodium silicate, were used. However, such binders contain relative large amounts of alkali metal cations, which remain in the coating after it has dried and after subsequent coatings have been applied to the steel surface. Exposure to water, for instance seawater, may cause blistering, that is, local delamination, due to the presence of too much alkali metal ions in the primer.

In a series of patent applications [28–31] Davies et al. show that the problems caused by high amounts of alkali metal ions could be overcome by using silica sols as the water-based binder instead of solutions of alkali silicates. The preferred sols had a particle size between 5 and 10 nm and a SiO_2/Na_2O mole ratio of about 50:1. They also found that they could increase the pot life of the shop primer by using aluminum-modified silica sols.

Inorganic Paints

Silicate coatings are used in construction to provide protective and decorative coatings on concrete and mortal. Such coatings may be completely inorganic in nature but may also contain up to 5% by weight of organic material, for example, in the form of a polymer latex. The binder, often a solution 3.3 molar ration sodium silicate, is mixed with pigments and fillers of different kinds. In the case of completely inorganic paints there is a problem with stability toward gelling and such paints are sold as two-component systems, one component consisting of the silicate binder solution and the other consisting of a dry powder of the other components of the paint. The two components are mixed prior to use and the resulting paint has a pot life of a few days.

An important property of inorganic paints is their resistance to water. A silicate based paint based on e.g. 3.3 molar ratio sodium silicate will have poor water resistance immediately and some time after the application of the paint, although it will improve with time, as alkali gradually becomes neutralized by the carbon dioxide in the atmosphere, that is, the SiO_2/Na_2O molar ratio of the silicate binder slowly increases. The silicates in a silicate-based coating can also react with Ca^{2+} in a cementitious substrate and form insoluble calcium silicates, which also contributes to improved water resistance. In theory, the molar ratio of a silicate solution may be increased by mixing it with a silica sol. However, if an 3.3 ratio sodium silicate solution is mixed with an alkali-stabilized silica sol coagulation and gelling occurs, but Iler [32] reports that this does not happen if potassium silicate is used. Stable mixtures of colloidal silica and potassium silicate can be prepared with a silica concentration of 15–30 wt.% and with SiO_2/K_2O molar ratios of 11:1 to 24:1. Greenwood and Otterstedt [33] studied such mixtures of silica sols and solutions of potassium silicate as binders in silicate paints and found that the pot life of the paint could be increased to about two months, but that it gelled after that time.

Heiberger and Schläffer [34] claim an inorganic paint with excellent water resistance and a pot life of at least 6 months. In a typical preparation they mix a potassium silicate solution, with a solids content of 30 wt.%, a silica sol, having a particle size of 9 nm and containing 20 wt.% SiO_2, in such proportions that the SiO_2/K_2O molar ratio is 10:1, and an aqueous solution of N,N′-Di(2-hydroxypropyl)-N,N′-tetramethylhexylenediamin. Next, they stir into this mixture pigment, filler and a butylacrylate-methylmetacrylate copolymer. The high water resistance of this paint is most likely due to the high SiO_2/K_2O molar ratio of the silicate binder.

Hard, Scratch-Resistant Thin Coatings

Many substrates, for example, many wood products, are provided with protective and decorative coatings of organic polymers, for example, radiation- or heat-curable polyacrylates, polyacrylate copolymers, polycarbonates or terephthalic resins, so as to improve gloss, dryness (no tack), and abrasion resistance and scratch resistance.

In many applications, the inherent properties of coating polymers are not adequate to meet very demanding specifications on for example, surface hardness, transparancy and scratch resistance. Much work has therefore been expended to develop high-quality surface coatings an improvements of surface properties have been achieved by incorporating colloidal silica and silanes in the coating formulation.

In one particularly interesting development Jacquinot and Eranian [35] and Wilhelm et al. [36] prepare an organosol of silica particles with a particle size between 10 and 50 nm in a reactive organic solvent also containing a photo-initiator and a vinylsilane, about .01 to 0.1 millimole of silane per m^2 of silica surface, as a coupling agent. The reactive solvent may be an acrylic monomer such as tripropylene glycol diacrylate or ethoxylated trimethylolpropane triacrylate. The preparation of the organosol is for instance described in patent example 1 [36]. 122 g of a silica sol, containing 40 wt.% SiO_2 with an average particle size of 50 nm, which has been de-cationized to pH of 2 with an ion exchange resin in the acid form, is mixed with 396.4 g isopropanol, 26 g vinyltrimethoxysilane (corresponding to 0.53 g silane per m^2 of silica surface) and 125 g ethoxylated trimethylolpropane triacrylate. The mixture is vacuum distilled at a pressure between 50 and 110 mbar and a temperature of 35°C for about 4 hours. After filtration the organosol is a slightly yellow, transparent liquid containing 30 wt.% SiO_2 and 0.3 wt.% water and having a viscosity of 304 mPa.s at 20°C.

In patent example 5 [36] it is described how a coating formulation is prepared by mixing 50 parts by weight of the organosol of patent example 1 with 50 parts by weight of urethane-acrylate oligomer and 5 parts by weight of a photo-initiator. A polycarbonate support is coated with 50 g m^2 of the formulation and the coating is cured by exposure to ultraviolet radiation. Abrasion and scratch resistance were measured and compared with those of a reference sample of polycarbonate coated with the formulation from which the silica organosol had been excluded. Optical transmission was 100% after abrasion in a Taber abrasion test as compared to only 70% for the reference. Scratch resistance was measured according to the pencil hardness test and was found to be 7 H versus 4 H for the reference.

Colloidal Silica in Polymer Latices

In DuPont's brochure on Ludox® Colloidal Silica it is described that incorporation of silica particles in the form of a silica sol in polymer lateces will improve abrasion resistance and modulus of polymer coatings. Greenwood and Otterstedt [37] studied the effect of mixing silica sols of different particle sizes with various water-based wood lacquers. The modulus of films made from lateces of copolymers of urethane-acrylates containing up to 50 wt.% SiO_2, based on the polymer weight, could be increased by several hundred percent. As expected, the reinforcing effect increased with decreasing particle size and increasing concentration of silica. However, the stability of the latex-silica sol mixture toward gelling decreases with decreasing particle size. Moreover, the stability towards gelling is very sensitive to the particular surfactant system used to make the latex and there are systems that do not form stable mixtures with silica sols of any particle size. From the point of view of stability and also of ensuring a uniform distribution of silica particles in the coating, it would obviously be advantageous if the silica sol was present during the polymerisation of the latex particles, each one of which would then be a nano-composite consisting of nano-sized silica particles distributed in a polymer matrix.

Actually, one of the first nanocomposites of this kind, although the purpose was not to improve the mechanical properties of polymers, was prepared by Kirkland, and Iler and McQueston, [38] when they synthesized polymer-silica composites with average diameters ranging from 500 nm to 20 μm by copolymerisation of either melamine or urea with formaldehyde in the presence of a silica sol. Calcination yielded microporous silica spheres, which were used as chromatographic column packing under the trade name Zorbex®.

Percy and coworkers [39,40] synthesized colloidal dispersions of polymer-silica nanocomposite particles by homopolymerizing 4-vinylpyridine or copolymerizing 4-vinylpyridine with either methyl methacrylate, styrene, n-butyl acrylate or n-butyl methacrylate in the presence of fine-particle silica sols using a free-radical in aqueous media at 60°C. No surfactants were used and a strong acid-based interaction was assumed to be a prerequisite for nanocomposite formation. The nanocomposite particles had comparatively narrow size distributions with mean particle diameters of 150–250 nm and silica contents between 8 and 54 wt.%. The colloidal dispersions were stable at solids contents above 20 wt.%.

Percy and Armes [41] showed that poly(methyl methacrylate)-silica nanocomposite particles can be readily prepared in aqueous alcoholic media at around ambient temperature without using either auxiliary comonomers such as 4-vinylpyridine or surfactants. In this work the silica sol was an organosol of 20 nm silica particles in isopropyl alcohol with a solids content of 30 wt.%.

Colloidal Silica and Ink Jet

Imaging devices such as ink jet printers are well known methods for printing various information on different substrates or receptors, which can be transparent (e.g. polymer films) or opaque (e.g. sheets of paper and paper board). Imaging with either ink jet printers or pen plotters involves depositing ink on the surface of the receptors. Many types of ink consist of an organic dye dissolved in a mixture of water with a water miscible organic liquid having a boiling point of at least 150°C. There are also inks available, in which pigments instead of dyes are used as the coloring agent. Imaging devices normally utilize inks that can be exposed to air for long time without drying. It is therefore desirable that the surface of the receptors be dry and non-tacky to the touch, even after absorption of significant amounts of ink, soon after imaging. Receptors must thus have a rapid absorption rate of ink to give uniform

TABLE 57.8
Colloidal Silica in Coatings for Ink Jet Applications

Reference	Coating system	Silica sol	Comments
[42]	Silica sol + polymeric binder + silane coupling agent + high boiling solvent	Elongated(structured) or spherical particles. Average particle size: up to 200 nm	Polymeric binder: e.g., PU-dispersion Advantage: quick drying of ink, low haze, reduced image bleeding, improved shelf-life
[43]	Silica sol + PVA 100 parts SiO_2 to 25 parts PVA on dry bases	Agglomerated particles. Agglomerated surface area: 100–400 m^2/g	Highly transparent porous coating
[44]	Fumed silica + silica sol + PVA	Particle size: 22, 35 and 50 nm SiO2 from sol (on dry bases): about 20 wt.%	Glossy coating, high liquid absorption capacity, crack resistant, non-brittle surface
[45]	Silica sol + water-soluble resin, e.g.PVA. 15 parts PVA to 100 parts SiO_2 (on dry bases)	Particle size: <200 nm Particle charge: anionic and cationic	High gloss, high ink absorption, excellent water resistance, "photo quality coating"
[46]	Silica sol + polymeric binder + silane coupling agent = image recording Layer	Particle size: 22 nm Ammonium stabilized	Porous solvent absorbing polyolefin layer + image recording layer: Improved color retention Higher optical density
[47]	2 layers of "chain"-silica sol + ink-receiving layer	Primary particle size: 3–40 nm "chain" length: 40–200 nm	Excellent ink absorption High color density Good color reproducibility

print, free from coalescence and banding. This can be accomplished, if the receptor is nonporous or not porous enough, by providing the surface of the receptor with a thin, porous coating.

In the last 20 years, there has been an intensive development of coating formulations for ink jet receptors, and colloidal silica plays a very significant role in this development work. Thus, a patent search, covering the period from 1971 to 2002 and using the search words "Ink Jet and Colloidal Silica," gave 1382 hits.

We have selected six patents or patent applications between 2000 and 2002, which demonstrate the use of colloidal silica in coatings for ink jet applications, and compiled the relevant information in Table 57.8, and one patent application describing the use of silica sols in pigmented inks.

Aluminum-modified silica sol with a particle size of 12 nm in a pigmented ink-jet ink formulation gave a print with improved optical density and superior rub resistance compared with a reference ink [48].

COLLOIDAL SILICA AS POLISHING AGENT FOR ELECTRONIC PRODUCTS

The advancement of high technology products, including computers, has been remarkable in recent years, and parts to be used in such products have been developed for ever higher integration and speed. Paralleling this progress, the design rules for semiconductor devices have been increasingly refined. The depth of focus in a process tends to be shallow and the requirements for the planarization of the pattern-forming surface are becoming increasingly severe.

Electronic devices made on silicon chips must be connected to each other by means of interconnecting metallic tracks to constitute the desired electronic circuit. Interconnected metallic levels are electrically insulated from each other by being encapsulated in a dielectric layer. The interconnecting metallic tracks are often made by a metal reactive ionic etching procedure. Sputtering is used to deposit an aluminium or aluminum alloy film, approximately 10–12 μm thick. The design is transferred onto the film by photolithography and the metallic tracks are created using reactive ionic etching. The tracks are next covered with a dielectric layer of silica, about 2 μm, obtained by decomposition of tetraethylorthosilicate in the vapor phase. Chemical-mechanical planarization is used to planarize the dielectric layer.

Chemical-mechanical planarization, "CMP", processes are widely used to remove material from the surface of a substrate in the production of a wide variety of microelectronics. In a typical CMP process, the

TABLE 57.9
Colloidal Silica as Polishing Agent for Microelectronic Products

Reference	Substrate	Silica sol	Slurry	Comments
[49]	Display grid Al,Cu,W	Ammonium stabilized Average particle size: 25 nm Size range: 12–50 nm Solids: 30% pH: 9–11	pH: 9–11	Well controlled polishing rates over large surfaces
[50]	Cu	Average particle size: 35 nm Size range: 10–100 nm Solids: 30% pH: 2–3.5	pH: 2–3.5	Homogeneous and regular polishing No dishing
[51]	Ni–P plated substrate of Al alloy (Hard disc)	Average particle size: 35 nm pH: alkaline	pH:2–5% SiO_2: 7	Buffer solution is important High stock removal rate No surface defects
[52]	Polymer with low dielectric constant	Cationic sol Average particle size: 25 or 50 nm	pH: 3.5	Good polishing speed Uniformity of polishing No scratching of polished surface
[53]	Cu, Ta	Silica sol + fumed silica Average particle size: <20 nm	pH: 2–5% SiO_2: 5	Improved polishing selectivity for Cu and Ta Suppressed polishing of insulating layer

surface to be polished is pressed against a polishing pad in the presence of a slurry under controlled conditions of chemistry, pressure temperature and velocity. The slurry generally contains small particles that abrade the surface, and chemicals that etch and/or oxidize the newly formed surface. The polishing pad is usually a planar pad made of a polymeric material, for example, polyurethane. As the pad and substrate move relative to each other, material is removed from the surface mechanically by the abrasive particles and chemically by the etchants and/or oxidants in the slurry.

Just like in the case of coating formulations for jet ink receptors, intensive efforts have been made in the last 15 to 20 years to develop polishing formulations for microelectronic products containing colloidal particles of different compositions. The final step in most polishing processes usually employs colloidal particles of silica. We have summarized information on colloidal silica as polishing agent from an number of recent patents in a table, Table 57.9.

REFERENCES

1. Griessbach, R. *Chem. Ztg.* **1933**, 57, No.26, pp 253–260, 274–276.
2. Iler, R.K. *The Colloidal chemistry of Silica and Silicates*; Cornell University Press: New York, **1955**, p 89.
3. Iler, R.K. *The Chemistry of Silica*; Wiley-Interscience: New York, **1979**, pp 415–438.
4. Otterstedt, J-E. and Brandreth, D.A. *Small Particles Technology*; Plenum Press: New York, **1998**, pp 432–433.
5. Cotterill, R. *Material World*; Cambridge University Press: Cambridge, **1985**, pp 124–131.
6. Bergqvist, H.; Chandra, S. U.S. Patent 5,932,000, **1999**
7. Inglis, G.E. *Trans. Inst. Nav. Archit.* **1913**, 55, p 219
8. Eka Chemicals; *Technical Bulletin on Cembinder products* No.2, **2000**
9. Hara, N.; Inoue, N. *Cem. Concr. Res.* **1980**, 10, p 677
10. Papadakis, V. G. *Chem. Concr. Res.* **1999**, 29, pp 79–86
11. Iler, R.K. *The Chemistry of Silica*; Wiley-Interscience: New York, **1979**, p 312
12. Bergna, H.E. *The Colloid Chemistry of Silica*; Advances in Chemistry Series 234; American Chemical Society: Washington D.C., **1994**, p 30
13. Wagner and Hauk *Hochsch. Archit. Bauwes-Weimar*, **1994**, 40, 5/6/7, pp 183–187
14. Skarp, U.; Sarkar, S. *Proceedings from the World of Concrete*, Las Vegas, **2001**, February 27–March 2.
15. Greenwood, P.; Bergqvist, H.; Skarp, U. U.S. Patent Application 2002/0011191, **2002**
16. Greenwood, P.; Bergqvist, H.; Skarp, U. U.S. Patent 6,387,173, **2002**
17. Greenwood, P.; Bergqvist, H.; Skarp, U. PCT WO 01/98227, **2001**
18. Otterstedt, J-E. and Brandreth, D.A. *Small Particles Technology*; Plenum Press: New York, **1998**, pp 407–429
19. Andersson, K.; Larsson, H. *Nordisk Cellulosa* **1984**, No.1, p 57
20. Andersson, K.; Lindgen, E. *Nordic Pulp and Paper Research Journal* **1996**, Nr.1, p 15

21. Lindström, T.; Hallgren, H.; Hedborg, F. *Nordic Pulp and Paper Research Journal* **1989**, 4:2, p 99
22. Dalton, R. L.; Iler, R. K. *J. Phys. Chem.* **1956**, 60, p 955
23. Greenwood, P.; Linsten, M. O.; Johansson-Vestin, H. E. U.S. Patent 6,379,500, **2002**
24. Yates, P. C.; *"Kinetics of Gel Formation of Colloidal Silica Sols", Ralph K. Iler Memorial Symposium on the Colloidal Chemistry of Silica*, ACS National Meeting, Washington D.C., **1990**
25. Lambert, W. H.; Greenwood, P.; Reed, M. C. *Journal of Power Sources*, **2002**, 107, pp 173–179
26. Wang, W. U.S. Patent 6,218,052, **2001**
27. Sielemann, O.; Niepraschk, H.; Nemec-Losert, P. U.S. Patent 5,664,321, **1997**
28. Davies, G. H.; Jackson, P. A.; McCormack, P. PCT WO 00/55260, **2000**
29. Davies, G. H.; Jackson, P. A. PCT WO 02/22746, **2002**
30. Davies, G. H.; Jackson, P. A.; McCormack, P.; Banim, F. PCT WO 00/55261, **2000**
31. Davies, G. H.; Greenwood, P.; Jackson, P. A. PCT WO 02/22745, **2002**
32. Iler, R.K. *The Chemistry of Silica*; Wiley-Interscience: New York, **1979**, p 145
33. Greenwood, P.; Otterstedt, J-E. Unpublished results, **1995**
34. Heiberger, F.; Schläffer, H. PCT WO 01/53419, **2001**
35. Jacquinot, E.; Eranian, A. U.S. Patent 6,136,912, **2000**
36. Wilhelm, D.; Eranian, A.; Vu, N.C. U.S. 2001/0027223, **2001**
37. Greenwood, P.; Otterstedt, J-E. Unpublished results, **1996**
38. Kirkland, J.J. U.S. Patent 3,782,075, **1974;** Iler, R.K.; McQueston, H.J. U.S. Patent 4,010,242, **1977**
39. Percy, M.J.; Barthet, C.; Lobb, J.C.; Khan, M.A.; Lascelles, S.F.; Vamvakaki, M.; Armes, S.P. *Langmuir* **2000**, 16, pp 6913–6920
40. Amalvy, J.I.; Percy, M.J.; Armes, S.P. *Langmuir* **2001**, 17, pp 4770–4778
41. Percy, M.J.; Armes, S.P. *Langmuir* **2002**, 18, pp 4562–4565
42. Paff, A.; Miller, A. PCT WO 00/71360, **2000**
43. Noguchi, T.; Tajiri, K. EP 1 118 584 A1, **2001**
44. Darsillo, M.S.; Fluck, D.J.; Laufhutte, R. U.S. Patent 6, 284,819, **2001**
45. Liu, B.; Nemoto, H.; Ikezawa, H. U.S. 2002/0034613, **2002**
46. Chu, L.; Romano, C.E.; Chen, C.C. EP 0 983 867, **1999**
47. Otani, T.; Ono, A.; Ilmori, Y.; Chatani, A.; Kondo, N.; Ueno, T.; Kuruyama, Y. EP 1 016 546, **1999**
48. Martin, T.W.; Bugner, D. EP 0 976 797, **1999**
49. Alwan, J.J.; Carpenter, C.M. U.S. Patent 6,271,139, **2001**
50. Jacquinot, E.; Letourneau, P.; Rivoire, M. U.S. Patent 6,302,765, **2001**
51. Shemo, D.S.; Rader, W.S.; Owaki, T. U.S. Patent 6,332,831, **2001**
52. Jacquinot, E.; Letourneau, P.; Rivoire, M. U.S. Patent 6,362,108, **2002**
53. Ina, K.; Rader, W.S.; Shemo, D.S.; Hori, T. U.S. Patent 6, 355,075, **2002**

58 Silicon–Aluminum Interactions and Biology

J. D. Birchall
Keele University, Department of Chemistry

CONTENTS

Silicon is listed as an essential element. Its removal from the diet of experimental animals has been shown to result in reduced growth rate (reversed on silicon supplementation) and changes to bone formation and the synthesis of collagenous connective tissue. However, in spite of much effort, no organic binding (e.g., to proteins) of silicon has been convincingly demonstrated under physiological conditions in which silicon exists as silicic acid, $Si(OH)_4$, and no biochemical rationale has been proposed to account for the effects of silicon deficiency. However, recent research indicates that a major role for silicon (as silicic acid) is to reduce the bioavailability of aluminum, which is toxic when it gains entry into biological systems, but which is normally largely excluded. The formation of subcolloidal hydroxyaluminosilicate species is shown to prevent the absorption of aluminum in fish via gill epithelia. The generality of this effect is discussed. The symptoms of silicon deficiency in experimental animals seem likely to result from aluminum toxicity, so that the environmental balance for the two elements may be critical. This chapter reviews the present position.

In Ralph K. Iler's magnum opus, *The Chemistry of Silica* [1], the final chapter is a review of silica in biological systems. Iler was fascinated by this least understood and most challenging aspect of silica chemistry, perhaps because of his instinctive feel for its fundamental importance. The headings within that last chapter reveal its sweep — from the "Origin of Life" to the "Essential Role of Silica in Mammals." He reviewed the concept of silicates as substrates for the formation of complex molecules from simple compounds at the earliest stages in the formation of living organisms, including algae, fungi, insects, plants, mammals, and humans. He discussed the formation of biogenic silica in plants (phytoliths) and in the frustules or exoskeletons of diatoms and the fibrogenic effect of silica and silicates in the lung, the most important manifestation of toxicity. Silica being ubiquitous throughout biological systems, Iler was intrigued by the possibility that it plays a vital role. Silica is used as structural material in diatoms and some sponges, and its deposition in plants can be advantageous in strengthening, stiffening, and hardening leaf surfaces, stalks, or the barbs of nettles. As important as such a "mechanical" role is, it is not a biochemically fundamental one.

The element silicon is the second most abundant element in the earth's crust after oxygen. It is described in modern texts of bioinorganic chemistry as an essential trace element largely as a result of the classical experiments of Carlisle [2] and Schwarz and Milne [3]. These workers, in independent experiments, maintained rats and chicks on a synthetic diet deficient in silicon. Using a similar technique, Schwarz had previously shown the essentiality of selenium and other elements. Silicon deficiency in the experimental animals produced a significant reduction in growth rate, reversed on silicon (as silicate) supplementation, with profound changes to the formation of bone and collagenous connective tissue such as cartilage. Since these experiments were reported [1972], there has been a search for the mechanism of action of silicon. The first hypothesis in explanation was based on the claim that silicon was bound in the biopolymers of connective tissue — collagen and polysaccharides — and acted as a labile cross-link, so influencing tertiary structure and integrity [4]. However, the reported silicon content of isolated biopolymers declined as isolation and analytical techniques improved, so that this role became increasingly unlikely and the hypothesis untenable [5]. Essential trace elements such as Fe, Cu, Se, and Zn have defined binding sites, usually within a protein structure, and the metal–protein complex is the functional entity, for example, an enzyme. However, in spite of much searching, no specific binding site for silicon has been found. No evidence exists for the formation of Si–C bonds in biological systems, and such chemistry as exists is that of silicic acid, $Si(OH)_4$. Organic complexes of silicic acid are few and are unstable at physiological pH (e.g., complexes with 1,2-dihydroxyphenols such as catechol). Much speculation [5] has occurred as to the possible binding of silicic acid to appropriately spaced hydroxyl groups on minority sugars in polysaccharides, but no convincing experimental evidence for such binding has been presented. (Reference 6 is a review of this topic.) By what mechanism, then, does silicon (as silicic acid) deficiency produce the reported pathological changes?

THE SEARCH FOR MECHANISM

A key issue in the mechanism underlying the essentiality of an element is its location at tissue, cellular, and molecular levels, and as noted no *molecular* binding of silicon has been observed. Plasma contains silicic acid in the concentration range 5–10 μM/l, so that all tissue is exposed to it. Noting that osteogenesis was impaired in silicon-deficient experimental animals, Carlisle conducted a microprobe scan across a bone section and found silicon to be concentrated (0.5%) locally at the mineralization front [7]. A major effect of silicon deficiency was on the synthesis of the preosseous, collagenous matrix that is mineralized to form bone proper. Carlisle noted that the activity of prolylhydroxylase (a key enzyme in collagen synthesis) in tissue cultures of cartilage from silicon-deficient chicks was low but was increased when silicon (silicate) was added to the culture [8]. The conclusion drawn was that silicon was a cofactor in the proper functioning of this enzyme. The known cofactors for prolylhydroxylase are iron, oxygen, ascorbate, and 2-oxyglutarate, and it is extremely difficult to see how silicic acid can engage in any chemistry with any of these cofactors, or with the enzymic protein. However, the observed increase in activity prompted by silicic acid addition requires explanation.

The first clue to a possible explanation came from an entirely different direction. Hexokinase, with adenosine triphosphate (ATP)-Mg^{2+}, is involved in the first step in the metabolism of glucose, the formation of glucose-6-phosphate. The activity of hexokinase is low and can be raised by the addition of citrate [9]. The low activity is due to contamination with aluminum, which binds to ATP 10^7 times more strongly than the required Mg^{2+} and blocks phosphate transfer. The role of citrate, a strong complexing agent for aluminum, is to remove aluminum from ATP and so allow Mg^{2+} to bind. Aluminum is almost as ubiquitous as silicon, and the two elements, in Iler's words, "have a unique affinity" [1]. There appears to be no *organic* chemistry of silicon in biological systems, so the question becomes, Could the effects of its deficiency be related to inorganic chemistry, with aluminum (now known to be toxic) being the interacting metal?

A TEST OF THE CONCEPT: PROLYLHYDROXYLASE ACTIVITY

Prolylhydroxylase, the collagen synthesis enzyme that has low activity in silicon-deficient tissue, requires iron that cycles between Fe^{3+} and Fe^{2+}. The apoenzyme will bind aluminum (less strongly than iron), and the enzyme is then, of course, inactive. An experiment was conducted [6] in which the apoenzyme was presented with iron first and then aluminum, all other essential cofactors being present. Activity, as measured by hydroxyproline production, was reduced by 20% of the control level. When the apoenzyme was presented with aluminum first and then iron, activity was reduced by 55% of the control levels. Silicic acid alone had no effect on the activity, which remained at the control level (Table 58.1).

TABLE 58.1
Effect of Silicic Acid on Inhibition of Prolylhydroxylase by Aluminum

Addition (100 μM)	% Inhibition of control[a]
$Si(OH)_4$	0
Fe then Al	20
Al then Fe	55
Al + 600 μM ($Si(OH)_4$)	0

[a]Assays were done by measuring the release of ^3H as ^3H_2O from ^3H-prolylprotocollagen, which accompanies the hydroxylation of proline at position 4.

However, when the "Al first" experiment was repeated in the presence of a sixfold excess of silicic acid (600 μM) over aluminum, the inhibiting effect of aluminum was completely suppressed. Clearly, in the presence of silicic acid, aluminum is removed from competition with iron for binding in prolylhydroxylase.

This experiment strongly supported the developing hypothesis that conditions in which silicon levels are (artificially) low allow the manifestation of aluminum toxicity. However, as will be shown later, aluminum challenges Ca^{2+} and Mg^{2+} rather than iron in vivo.

ALUMINUM IN BIOLOGICAL SYSTEMS

Aluminum has long been regarded as innocuous. It is not used in biological systems and, strangely, for so ubiquitous an element, is largely excluded; iron, with a very similar charge-to-radius ratio, is essential and actively sought. The perception of aluminum as innocuous has changed over the past 2 decades. Aluminum is indisputably the agent responsible for the disorders observed in patients undergoing hemodialysis for renal insufficiency when aluminum-containing dialysate is used. Plasma aluminum can then rise from <1 μM to >5 μM with three potential consequences [10]:

1. a microcytic anemia not responding to iron therapy, but responding to reduced plasma aluminum levels
2. a progressive deterioration in cognitive function with eventual dementia (dialysis encephalopathy).
3. a disorder of bone (dialysis osteomalacia) with inactive osteoblasts, bone pain, and spontaneous fracture. In this, aluminum is found at the growth front (Figure 58.1). (Silicon has been found in the same location).

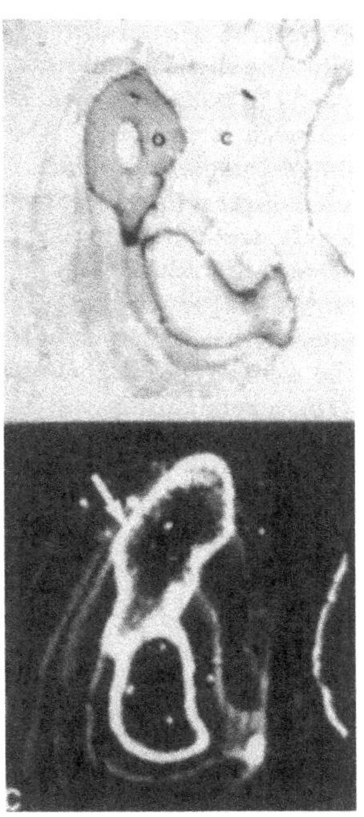

FIGURE 58.1 Aluminum (arrowed) at the growth front in bone from patient with dialysis osteomalacia. (Reproduced with permission from reference 35. Copyright 1990.)

Now that aluminum is recognized as the cause of these disorders, levels are carefully monitored, and aluminum is removed from dialysis fluid by reverse osmosis.

Aluminum is the major toxic entity for aquatic life in waters affected by "acid rain" and is the major cause of poor crop yield in acidic soils in which root growth is stunted [11,12].

The mechanisms underlying these effects are not well-understood, but there is a consensus that, once within the biological milieu, aluminum displaces Mg^{2+} from key sites where this metal is an essential cofactor and is involved in the disturbance of Ca^{2+} manipulation (*see* later).

In dialysis with aluminum-containing water, the metal bypasses the normal exclusion mechanisms, and in acidic waters high in aluminum (5–10 μM), bioavailable aluminum is adsorbed at fish gill epithelia and then absorbed systemically. The dietary intake of aluminum in humans is of the order of 10–20 mg/day [13], but only a small fraction is absorbed in the gut. The exclusion mechanism is unknown. Absorption is increased by citrate and other aluminum-complexing agents.

The facts that aluminum ingress in dialysis gives rise to an encephalopathy, that aluminum levels are raised in the brains of patients with Alzheimer's disease [14], and that aluminum, colocalized with silicon, is found at the core of the senile plaques characteristic of Alzheimer's disease [15] have prompted concern that dietary aluminum can, in susceptible individuals, provoke or be a cofactor in Alzheimer's disease. In one recent epidemiological study [16], a relationship was found between the aluminum content of drinking water and the incidence of Alzheimer's disease. Two criticisms were made of this study: (1) a lack of dose–response correlation with near-maximum increase in Alzheimer's disease incidence at the lowest levels of aluminum concentration, and (2) even at the highest level of aluminum (>111 ppb) included in this study, the daily aluminum intake from water would be <0.5 mg, whereas the intake from food would be 10–20 mg. (In 1982, about 4×10^6 pounds of aluminum compounds were used as food additives in the U.S. [17]). This apparent paradox is resolved if the hypothesis that silicon limits aluminum bioavailability is adopted. Aluminum and silicon concentrations in potable water are inversely related. High aluminum and low silicon levels are found in soft waters from high, well-weathered, acidic catchment areas requiring aluminum coagulation treatment for clarification. Conversely, hard, mineralized waters from still-weathering geology are high in silicon, often 10-fold or more than the level in soft water. It has been proposed [18] that this epidemiological study in fact revealed a relationship between the silicon content of water and its role in suppressing the absorption of the aluminum contained in food.

A TEST OF THE Al–Si BALANCE HYPOTHESIS

Fish in acidic waters containing aluminum have high mortality due to gill damage and loss of osmo- and ionoregulatory function. In one experiment [19], Atlantic salmon fry were exposed to acidic water (pH 5) containing a toxic level of aluminum (ca. 7 μM) with low (0.60 μM) and high (93 μM) levels of silicon as silicic acid. The "pure," control water (at pH 5) contained low levels of both elements, 0.85 μM aluminum and 0.66 μM silicon. Survival curves are shown in Figure 58.2. In the 7 μM Al, low-silicon conditions, 50% of fish were dead within 26 h, and all fish were dead within 48 h. Gill damage was obvious, and the fish contained 2 μM Al per gram of dry mass. With the high silicon level, no fish died in the duration of the experiment, gill structure remained normal, and the fish contained only 0.40 μM Al per gram of dry mass, 10% less than that absorbed from the control water.

Clearly, in the presence of the high level of silicon, aluminum was prevented from binding at gill epithelial surfaces and systemic absorption. This exclusion occurred at the interface between creature and the external environment, and a fundamental question is, Is this a general effect, not only at the fish gill, but also at plant root membranes and in the gastrointestinal tract of mammals and humans?

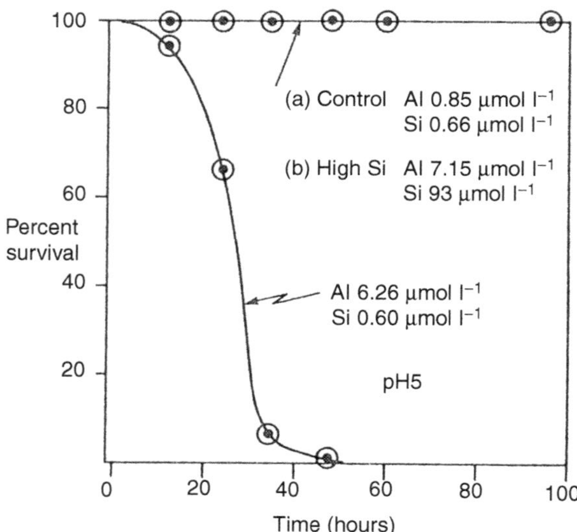

FIGURE 58.2 Survival curves for Atlantic salmon fry exposed to control (low Al) water and water containing ca. 7 μM Al with high and low silicic acid levels. The top curve shows both control and high-Si results. (Reproduced with permission from reference 35. Copyright 1990.)

THE EXCLUSION MECHANISM

Iler has remarked (p 193, reference 1), that "there is a peculiar affinity between the oxides of aluminum and silicon." This affinity results from the isostructural nature of $(SiO_4)^{4-}$ and $(AlO_4)^{5-}$, which is responsible for the vast range of natural aluminosilicates and synthetic zeolites. Synthetic zeolites are synthesized by the reaction of aluminate and silicate anions at high temperature and pH. However, interactions occur between silicic acid and hydroxyaluminum ions in dilute ($<10^{-4}$ *M*) solution at near-neutral pH, and these interactions are of biological and environmental significance in reducing aluminum bioavailability. Hydroxyaluminum cations were shown [20] to react with silicic acid in solutions of pH 4 upwards to form clear solutions containing nondialyzable hydroxyaluminosilicate species with a limiting Si:Al ratio of about 0.5. A concentration of at least 100 μM $Si(OH)_4$ is required. When such solutions are heated, the poorly crystalline mineral imogolite is precipitated. This unidimensional, tubular structure has the ideal composition $(HO)_3Al_2O_3$ SiOH and can be considered as a single gibbsite sheet with the inner surface hydroxyls replaced by silicic acid. The stable, clear, unheated solutions appear to contain fragments of this structure. The species have been detected by the infrared examination of solids recovered by freeze-drying solutions [20] and by ion-exchange experiments [21,22]. Although the apparent solubility of aluminum is raised at the pH of normally minimum solubility in the presence of silicic acid (Figure 58.3), the hydroxyaluminosilicate species present limit the bioavailability of aluminum. The interaction of aluminum with various binding groups is reduced when silicic acid is present, as is illustrated by ion-exchange experiments using an iminodiacetate functional resin (Figure 58.4) [22]. The fall in aluminum retention (Figure 58.4a) from about pH 7 is seen to correspond to the formation of hydroxyaluminosilicate species (Figure 58.4b). Similar results [21] were obtained with sulfonate and phosphonate functional resins with the pH of onset of reduced aluminum binding being >5 and >6.6, respectively. Such experiments reflect the stability of the hydroxyaluminosilicate species with respect to ligand–Al binding.

Fish gill epithelia and associated mucus contain such binding groups and, as in ion-exchange resins, aluminum binding is similarly reduced in the presence of silicic acid. This reduction is possibly aided by a pH more alkaline than that of the bulk water within a boundary layer proximate to the gill surface and resulting from NH_3 and CO_2 excretion.

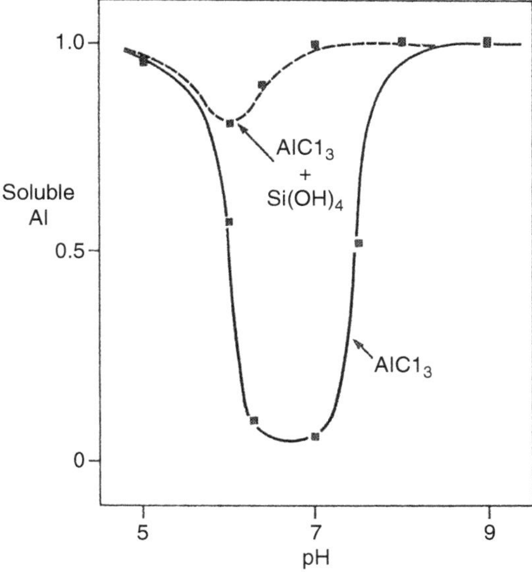

FIGURE 58.3 Fraction of aluminum remaining in 20-h-old solution after filtration through a 0.2-μM membrane as a function of pH. Solutions contained 0.1 mM $AlCl_3$ with and without 0.5 mM $Si(OH)_4$. (Reproduced with permission from reference 35. Copyright 1990.)

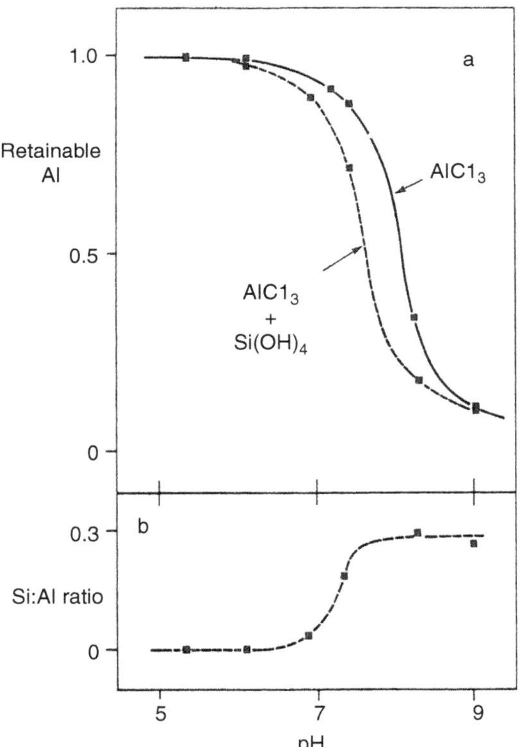

FIGURE 58.4 (a) Fraction of aluminum retained on iminodiacetate functional resin as a function of pH. Solutions contained 0.1 mM $AlCl_3$ with and without 0.5 mM $Si(OH)_2$. (b) The Si:Al ratio of the retained species. (Reproduced with permission from reference 35. Copyright 1990.)

At pH 6.2, the formation of the solid hydroxide phase gibbsite limits the level of dissolved aluminum to about 10^{-7} *M*. The formation of imogolite reduces this level to 10^{-11} *M*, so that the formation of hydroxyaluminosilicate phases reduces the concentration of biologically available aluminum to levels well below those producing toxic effects.

The formation of hydroxyaluminosilicate species at near-neutral pH is unique. No interactions occur between silicic acid and Ca^{2+} or Mg^{2+} at less than pH 10, so that the transport and binding of these cations is unhindered. The interactions of Fe^{3+} with silicic acid at near-neutral pH are very different from those of aluminum. Acidic solutions containing Fe^{3+} (10^{-4} M) with a threefold molar excess of silicic acid remain clear on neutralization: no visible precipitate forms. Instead, Fe–O polymers are formed as spherical hydrated ferric oxide particles 10–15 nm in diameter, stabilized against growth, aggregation, and precipitation by an adsorbed layer of silicic acid [23].

Such sols present iron in a readily available form to chlorotic plants, probably because the minute particles are easily reduced to Fe^{2+} and solubilized by root exudates [24]. Thus, silicic acid distinguishes between aluminum and iron as regards biological availability. This ability to distinguish may have been important in primitive biological systems.

THE SIGNIFICANCE OF THE ENVIRONMENTAL Si:Al BALANCE

Biological systems are not tolerant of internalized aluminum, which normally is excluded or, as in some plants, rendered immobile, possibly by binding to phytate. In mammals and humans, once internalized, aluminum is bound and carried in the iron-transport protein transferrin [25], which can bind two M^{3+} ions per molecule. Although aluminum is bound much less strongly than iron, (log *K* ca. 12 for Al^{3+} and log *K* ca. 20 for Fe^{3+}, where *K* is the stability constant) there is normally an abundance of transferrin with empty sites. Aluminum appears to be concentrated in cells and tissue with high transferrin receptor density, a conclusion supported by experiments using ^{67}Ga-loaded transferrin, in which the marker is found in areas of the rat brain with high receptor density (cerebral cortex, hippocampus, septum, and amygdala) [26]. These areas are selectively vulnerable in Alzheimer's disease. Aluminum-loaded transferrin is internalized by cultured neuroblastoma cells [27] and, presumably, by cerebrovascular endothelial cells because aluminum crosses the blood–brain barrier. Aluminum has also been detected within osteocytes at the junction of osteoid and mineralized bone in patients with dialysis osteomalacia [28] and is almost certainly responsible for the low level of cellular activity. A key point is that ferritin is inefficient at loading with aluminum, so that there is no safe "sink" for the element. Events at the intracellular level ultimately responsible for toxicity remain unclear, but aluminum impairs glucose utilization and cholinergic activity in the rat brain [29]. GTP–GDP (guanosine 5′-triphosphate–guanosine diphosphate) nucleotide exchange is inhibited by aluminum [30] and, in vitro, aluminum stimulates phosphatidylinositol (PtdIns) hydrolysis and inhibits PtdIns(4,5)P_2 (where (4,5)P_2 is inositol 4,5-bisphosphate) hydrolysis [31]. In vitro, aluminum inhibits tetrahydrobiopterin synthesis [32]. Significantly, aluminum has been reported to increase the permeability of the blood–brain barrier [33], and indeed, endothelial "leakiness" is a constant theme, as is alteration in the manipulation of Ca^{2+}. Recent experiments [34] suggest that this latter alteration may result from interference in the phosphatidylinositol-derived Ca^{2+} intracellular second messenger system. In rat pancreatic acinar cells, the microinjection of aluminum eliminated the acetylcholine-evoked mobilization of Ca^{2+} from cytoplasmic stores. Early experiments indicated that this inhibitory effect of aluminum is absent in the presence of silicic acid. This Ca^{2+}-mobilizing system is ubiquitous in biological systems, and its alteration may account for many of the toxic effects of aluminum in plants, animals, and humans.

In view of these various effects of internalized aluminum, the exclusion of the element from biological systems is fundamentally important [35]. The low solubility of aluminum at neutral pH is one mechanism, but acidity increases availability. Citrate [36] and other chelators (e.g., maltol [37]) appear to increase absorption in the gastrointestinal tract. An important question is, How general throughout biological systems is the observed effect of silicic acid in excluding aluminum from fish? Is this mechanism operating at plant root membranes and within the gastrointestinal tract?

SILICIC ACID: GEOCHEMISTRY AND HEALTH

The essential trace elements (and the toxic elements) are ultimately obtained from the earth's crust via crops and water, although human activity can modify occurrence and availability. The multidisciplinary subject of environmental geochemistry and health studies the relationship between disease epidemiology and geochemistry. Most success has come in understanding the relationship between trace-element deficiency in farm animals and local geochemistry, for example, the effects of Cu deficiency in cattle. The recognition of the goiter of iodine deficiency and the effects of fluorine on the incidence of dental caries are examples of success, as is the recognition of selenium deficiency as the cause of endemic cardiac myopathy in the Keshan district of China [38].

A negative association between silica concentration and the incidence of ischemic heart disease (IHD) has been reported [39,40] and is associated with the lower incidence of IHD in hard-water areas [41,42]. Silicon is associated negatively with IHD mortality (correlation coefficient $r = 0.66$) and positively ($r = +0.68$) with water hardness [42]. However, no explanation for this observation has been possible. A marked geographical variation occurs in the incidence of ischemic heart disease in the British Isles. A high incidence occurs in the north and west, and a low incidence in the south and east, the relative odds of a major IHD event in males changing from 1.0 in the south of England to 3.03 in Scotland [43]. This difference reflects, inter alia, a difference in geology between the regions with hard water in the south and east and soft, peaty water from upland areas in the north and west. Silca concentration can differ by an order of magnitude (<10 to >200 μM). The water from upland catchment areas is frequently treated with aluminum for clarification, and with very low silicic acid levels, residual aluminum is likely to be biologically available.

However, with the major part of the intake of aluminum being food (and antacid medication), the important question is the effect of *silicic acid* in water on the absorption of the aluminum in food [18]. The extent of exclusion would not then be a linear function of silicic acid concentration, because a minimum level of about 100 μM Si is required for the formation of stable hydroxyaluminosilicates, in which the Si:Al ratio is about 0.5 and which have minimum solubility (Figure 58.5).

Silicic acid is readily absorbed, and plasma levels rise rapidly following intake. In normal subjects excretion is

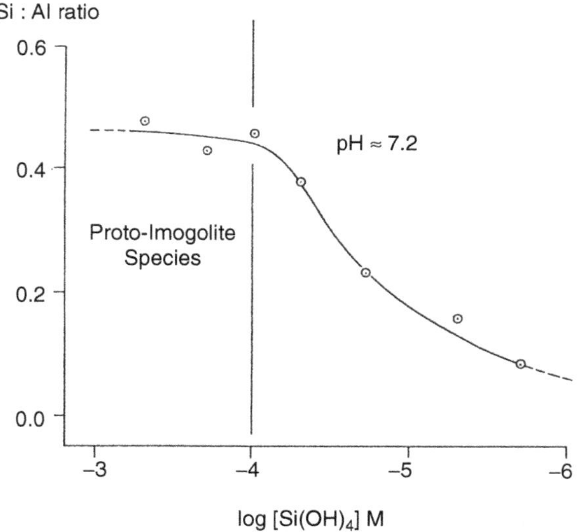

FIGURE 58.5 The Si:Al ratio of hydroxyaluminosilicate species formed in solutions containing 10 μM Al^{3+} at pH 7.2. All solutions were aged 20 h and held at 20°C. (Private communication, J. S. Chappell).

rapid, average plasma levels are 5–10 μM, and all cells and tissues contain silicon, which may be concentrated in some cells or cellular compartments, for example, in the osteoblast. Little is known of the interactions between this silicic acid and internalized aluminum. With aluminum bound strongly to transferrin and the low plasma silicic acid levels, interaction is unlikely except at sites of local concentration. Codeposited aluminum and silicon (as amorphous aluminosilicate) has, so far as is known, been reported only at the core of senile plaques. Separate groups of workers report silicon in artery walls [44] and aluminum [45] in artery walls, but no studies have been made of the association and balance of the two elements in tissue. Such studies will be required if progress is to be made. Some workers have reported an inverse relationship between the level of silicon in arterial tissue and the degree of sclerotic damage and calcification [46]. The link between all these various observations may be the ability of internalized aluminum to increase the permeability of endo- and epithelial membranes and to alter intracellular Ca^{2+} manipulation and the ability of silicic acid to exclude aluminum from entry in biological systems and, possibly, its ability to modify the effects of internalized aluminum.

CONCLUSIONS

The most recent research indicates that in its biological effects, silicon (always as silicic acid) is inextricably linked with aluminum. Good evidence suggests that silicic acid reduces the absorption of manganese in plants and thereby moderates the toxicity associated with excess manganese [47,48]. However, the predominant association is with aluminum because of the prevalence of the two elements and the unique affinity of one for the other. The need in primeval biological systems to "select out" aluminum thus probably involved high silicic acid levels at near-neutral pH, and it will be important to understand the anthropogenic disturbance of the interactions of these two elements that, with oxygen, constitute 80% of the earth's crust. This area is one of the most exciting in bioinorganic chemistry.

REFERENCES

1. Iler, R. K. *The Chemistry of Silica*; Wiley: New York, 1979.
2. Carlisle, E. M. *Science* **1972**, *178*, 619–621.
3. Schwarz, K.; Milne, D. B. *Nature* **1972**, *239*, 333–334.
4. Schwarz, K. *Proc. Nat. Acad. Sci. USA* **1973**, *70*, No 5, 1608–1612.
5. Schwarz, K. In *Biochemistry of Silicon and Related Problems*; Bendz, G.; Lindqvist, I., Eds.; Plenum: New York, 1978; pp 207–230.
6. Birchall, J. D.; Espie, A. W. In *Silicon Biochemistry*; Ciba Foundation Symposium 121; Wiley: Chichester, England 1986; p 140–159.
7. Carlisle, E. M. *Science* **1970**, *167*, 179–280.
8. Carlisle, E. M.; Alpenfels, W. F. *Fed Proc.* **1984**, *43*, 680.
9. Viola, R. E.; Morrison, J. F.; Cleland, W. W. *Biochemistry* **1980**, *19*, 3131–3157.
10. Kerr, D. N. S.; Ward, M. K. In *Metal Ions in Biological Systems, Vol. 24; Aluminum and its Role in Biology*; Sigel, H.; Sigel, A., Eds.; Dekker: New York, 1988; pp 217–258.
11. Driscoll, C. T. *Environ. Health Perspec.* **1985**, *66*, 93–104.
12. Taylor, G. In reference 10, pp 123–163.
13. Jones, K. C.; Bennett, B. G. "Exposure Committment Assessment of Environmental Pollutants," Report No 33, Vol 4. Monitoring and Assessment Research Centre, King's College, University of London, London, 1985.
14. Crapper, D. R.; Krishnan, S. S.; Dalton, A. J. *Science* **1973**, *180*, 511–513.
15. Candy, J. M.; Oakley, A. E.; Klinowski, J.; Perry, R. H.; Fairburn, A.; Carpenter, T.; Atack, J. R.; Blessed, G.; Edwardson, J. E. *Lancet* **1986**, *i*, 354–357.
16. Martyn, C.; Barker, L. J. P.; Osmond, C.; Hams, E. C.; Edwardson, J. A.; Lacy, R. F. *Lancet* **1989**, *i*, 59–62.
17. Committee on Food Additive Survey Data, "Poundage of Uptake of Food Chemicals, 1982"; National Academy Press: Washington, DC, 1984.
18. Birchall, J. D.; Chappell, J. S. *Lancet* **1989**, *i*, 114.
19. Birchall, J. D.; Exley, C.; Chappell, J. S.; Phillips, M. J. *Nature* **1989**, *338*, No. 6211, 146–148.
20. Farmer, V. C.; Frazer, A. R.; Tait, M. *Geochim. Cosmochim. Acta* **1979**, *4*, 1417–1420.
21. Birchall, J. D.; Chappell, J. S. *Clin. Chem.* **1988**, *34/2*, 265–267.
22. Chappell, J. S.; Birchall, J. D. *Inorg. Chim. Acta* **1988**, *153*, 1–4.
23. Birchall, J. D.; Espie, A. W. In reference 6.
24. Demolon, A.; Bastiss, E. *Compt. Rend.* **1944**, *219*, 293–296.
25. Trapp, G. A. *Life Sci.* **1983**, *33*, 311–316.
26. Candy, J. M.; Edwardson, J. A.; Faircloth, R.; Keith, A. B.; Morris, C. N.; Pullen, R. G. L. *J. Physiol.* **1987**, *391*, 34.
27. Morris, C. M.; Candy, J. M.; Court, C. A.; Edwardson, J. A.; Perry, R. M.; Moshtaghie, A. A.; Skillen, A.; Fairbairn, A. *Biochem. Soc. Trans.* **1987**, *15*, 498.
28. Schmidt, P. F.; Zumkley, H.; Barckhaus, R.; Winterberg, B. In *Microbeam Analysis*; Russell, R. E., Ed.; San Francisco Press: San Francisco, CA, 1989; p 50–54.
29. Johnson G. V. W.; Jope, R. S. *Toxicology* **1986**, *40*, 93–102.
30. Miller, J. L.; Hubbard, C. M.; Litman, B. J.; MacDonald, T. L. *J. Biolog. Chem.* **1989**, *264*, No 1, 243–250.
31. McDonald, L. J.; Mamrack, M. D. *Biochem. Biophys. Res. Comm.* **1988**, *155*, No. 1, 203–208.
32. Cowburn, J. D.; Blair, J. A. *Lancet* **1989**, *i*, 99.
33. Banks, W. A.; Kastin, A. J. *Lancet* **1983**, *ii*, 1227–1229.
34. Wakui, M.; Itaya, K.; Birchall, J. D.; Petersen, O. H. *Febs. Lett.* **1990**, *267*, No 2, 301–304.

35. Birchall, J. D. *Chem. Brit.* **1990**, *February*, 141–144.
36. Kruck, T. P. A.; McLachlan, D. R. In reference 10, pp 285–314.
37. Nelson, W. O.; Lutz, T. G.; Orvig, C. In *Environmental Chemistry and Toxicology of Aluminum*; Lewis, T. E., Ed.; Lewis Publishers: Chelsea, Michigan, 1989; pp 271–287.
38. Xu, Guang-Iu; Jiang, Yi-fang. In *Proceedings of the First International Symposium on Geochemistry and Health*; Thornton, I., Ed.; Science Reviews Ltd, 40, The Fairway, Northwood, Middlesex, HA6 3DY, United Kingdom, 1985; pp 192–204.
39. Schroeder, H. A. *J. Am. Med. Assn.* **1966**, *195*, 81–85.
40. Schwarz, K.; Punsar, S.; Ricci, B. A.; Karvonen, M. J. *Lancet* **1977**, *i*, 538–539.
41. Masironi, R. *Phil. Trans R. Soc. Lond. B288*, **1979**, 193–203.
42. Pocock, S. J.; Shaper, A. G.; Powell, P.; Packham, R. F. In reference 38, pp 141–157.
43. Elford, J.; Thomson, A. G.; Phillips, A. N.; Shaper, A. G. *Lancet* **1989**, *i*, 343–346.
44. Carlisle, E. M. In reference 9, pp 123–139.
45. Zinsser, H. H.; Butt, E. M.; Leonard, J. *J. Am. Geriatr. Soc.* **1957**, *5*, 20–26.
46. Loeper, J.; Loeper, J. G.; Lemaire, A. *Press Med.* **1966**, *74*, 865–867.
47. Lewin, J.; Reimann, B. E. F. *Annu. Rev. Plant Physiol.* **1969**, *20*, 289–304.
48. Foy, C. D.; Chancey, R. L. *Annu. Rev. Plant Physiol.* **1978**, *29*, 511–566.

59 Silica in Biology

James S. Falcone, Jr.
West Chester University, Department of Chemistry

CONTENTS

INTRODUCTION

Derek Birchall would have enjoyed the renewed interest in the colloidal chemistry of silica with emphasis on biological application. He was looking forward to a future meeting [1] like the one in memory of Ralph Iler at the 200th ACS Meeting in Washington, DC in 1990 where researchers from around the world could share their views. In 1989, he described the value of colloidal science in understanding the real world, calling it the "midwife of invention." In his words it is the "... region of study of matter too big for atomic and molecular scientists and too small for engineers." [2] He quoted Graham's [3] perceptive observation that "... in nature there are no abrupt transitions, and the distinctions of class are never absolute." Of the well-studied colloidal systems silica is surely the most studied and possibly the least understood. While it is understood that the study of structural biological materials extends to many different inorganic phases, mostly calcium carbonate and hydroxyapatite [4], our discussion today will focus on silica chemistry as pertains to biological systems.

The fascination with the similarity of growing siliceous systems and living organisms and the inferences drawn from this behavior stretches back to at least early last century. In the words of Vail [5]

> Metastable dilute silicate solutions have been shown to yield cell-like structures closely suggestive of forms of animate organisms

and

> Herrera [6]... (following on the work of Moore and Evans [7] and Bastian [8])... concluded that colloidal silicates yield structures most like natural forms when produced from reaction-mixtures of very low concentration, when contact between precipitating agent is made slowly and when the viscosity of the reacting solution is great.

It speculated that life was formed in gelled silicate solutions. So what are these suspected sources of animation, these seminal substances awaiting the midwife? We begin an answer with an attempt at a description.

Solid amorphous [9] and crystalline [10] silica polymorphs and aqueous or soluble silica solutions [11–13] while appearing to be simple are actually *each* highly complex systems in their own right [14–17]. Couple this with the immense complexity extant in biological contexts one can only be humbled by the prospects. Franks' paperback summary of his seven volume treatise on Water [18] is an excellent starting point for any journey of discovery in aqueous systems behavior. A full recognition of the issues associated with understanding the role water plays in natural processes will be helpful when sorting through the greater complexity inherent in systems containing silica in intimate contact with water.

Ralph Iler [19] in The Chemistry of Silica teaches that silica comprises roughly 60% of the earth's crust. Present in nature in a seemingly limitless number of compounds with literally hundreds of different structures, the chemistry of silica is often likened to carbon chemistry. Like carbon, silicon apparently also plays a significant role in the well being of man. Unlike carbon, silicon does not readily bond to itself, bond to carbon or form double bonds. What silicon does is form an extremely strong bond with oxygen yielding tetrahedral structures that readily join together to form polymers. Thus in nature we study the chemistry of silica, not silicon, from the poorly understood and likely rare, if not nonexistent, silicic acid, $Si(OH)_4$, to the well understood crystalline forms, like quartz, SiO_2. *Silica* ends up being a generic name for any of the amorphous polymorphs of a complex hydrous polymer that varies in Si, O and H composition and connectivity between silicon atoms via oxygen atoms, where connectivity is the number of shared SiO_4-tetrahedra corners. As shown in Figure 61.1 the connectivity of each individual silicon in a structure is denoted by a superscripted Q^n, where n is the number of shared tetrahedra and ranges from 0 to 4. For example, Q^0 is a monomeric silicate species and Q^2 would signify silicon in an unbranched chain or ring site. Variations in polymorphs, the number of Si atoms per species (connectivity) and extent of hydrolysis influence electrostatic and the poorly understood nonelectrostatic interactions between silica and its environment. The study of polymorphism has gained a new prominence among those concerned with the performance of materials [20].

Colloidal suspensions of silica, primarily in the form of silicate ions, have been used commercially for over one hundred years [21]. Dent Glasser once called the silicate ions in solution the "Cinderella anions" because their chemistry was so intractable that they had little appeal to classical inorganic chemists [22]. They are generally added to systems to supply aqueous silica species, which through trial and error and experience are known to modify behavior hopefully in some desired manner. They are also used as raw materials to make silicon containing performance materials. Apparently, nature discovered and used the value of these impossibly complex raw materials a few eons earlier without the aid of any scientific understanding [23]. Unfortunately for us, the secrets have been well incorporated into the nature of things evading reductionist attempts at scientific discovery. Tipping [24], in discussing the need to understand processes — biological, geological, physical and chemical — in natural waters, rightly points out that a synthesis of materials characterization, laboratory simulation and field observation is required when attempting to make progress in very complex systems.

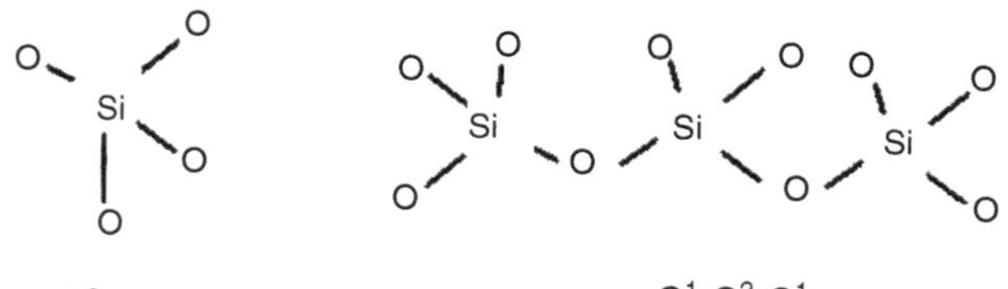

FIGURE 59.1 Representations of Q^n structures in silicates. The oxygen atoms are shown without H atoms or charges. Each silicon center is at the center of a SiO_4 tetrahedron. Tetrahedra only form linkages at vertices to form oligomers and polymers.

Over the years there have been many anecdotes and much controversy over the wide variations in silica properties reported in the literature and the mechanisms of the processes by which silica and silicate species exert their influence. Much of the discussion results from the fact that both the exact nature of the speciation of silica in diluted solution and at the interface between silicas and silicates and water and how this speciation changes in mixtures with other substances is unclear (25,26). Also, from a very practical standpoint silica in some form is already present in many natural systems and in vitro experiments must seek to replicate the speciation to have any chance of learning from them. Complicating the situation, the extensive literature contains many observations drawn from experiments carried out under conditions where soluble silica might have been present in an uncontrolled manner and at concentrations sufficient to exert significant effects or conclusions drawn from insufficient mathematical models. The high potential for artifacts, which can influence the behavior of silica, even in the simplest systems, was well characterized by Hazel (my first college chemistry professor) [27], Debye [28] and Iler [29] who made it clear that one had to be aware of the purity and/or preparation history of the silica.

We attempt to introduce the reader to the extensive scientific literature in this area up to 2002. We have tried to be as comprehensive as possible, but realize only the surface is breached. We have ignored the whole area of chromatographic silica [30] that teaches much about silica–molecule interaction. First, we outline, starting from a solution perspective, an understanding of silica as a complex and generally metastable inorganic polymer and how it interacts with other substances. Then, we discuss the chemistry from a biological perspective and conclude with a brief overview of recent work aimed at learning from nature with the intent of applying the design principles to the synthesis of new performance materials. An extensive, but not exhaustive, bibliography is included to direct the reader to even more extensive studies. We hope that this background will allow the reader to pursue special interests and enhance ability to evaluate the diversity of ideas that one will encounter.

SILICA CHEMISTRY

The chemistry of silica is basically a study of metastability due to the minimal energy variations between the various polymorphs [31]. Many contributors to the field seek to develop a molecular thermodynamic description (see e.g., Dove [32] for a review of recent quantitative treatments) that is valid provided one can identify and characterize a sufficiently long-lived collection of species with determinable stoichiometries. However, mathematical models must be carefully calibrated with the reality of silica behavior in reacting systems likely far from equilibrium and controlled by poorly understood kinetic factors. A solubility diagram for silica is shown in Figure 59.2 after Stumm and Morgan [33]. Iler [34] concluded based on reactions with molybdic acid that solutions of sodium silicates in the so-called "stable multimeric domain," which might be better defined as the fluid metastable domain, were comprised primarily of particles ranging in size from 0.8 to 2.0 nm. Krumrine and Falcone [35], following up on ideas expressed by Dent Glasser et al. (see Figure 59.2) found it of practical value to view solutions in this metastable region as 2 phase systems, very high surface area silica dispersed in an oligomeric silica solution phase. The description of silica speciation chemistry under idealized conditions [36,37] has recently been greatly improved as a result of the use of ^{29}Si-FTNMR.

In Figure 59.2, the values of silica concentration versus pH value or increasing hydroxide concentration are shown along with estimates of regions where silica exists in complex states. The points and estimated linear boundary between particles and "true solution," a solubility line, are based on ultracentrifuge studies [38].

However, factors regulating soluble silica activity in real systems remain poorly understood because several factors propagate their effects to conceal the secrets of silica's amazing diversity of influences. The factors are polymorph metastability (kinetic control of growth and dissolution) modified by poorly understood external influences on growth and dissolution and polymorph specific binding properties and influence on crystallization of other substances.

Our discussion of silica chemistry starts with a model for the nature of amorphous silica grown from such solutions and moves to a discussion of how silica polymorphs in water interact with their surroundings.

HOW SILICA GROWS

Wijnen et al. [39,40] studied the structure of silica gels as they formed using SAXS and NMR at pH values of 3.9 and 7–8. The growth of particles was found to be fractal. The amorphous silica is constructed in a clear and regular pattern. Initially under conditions called diffusion limited cluster–cluster aggregation, silicate monomers and oligomers rearrange or aggregate to form essentially spherical primary particles, A_0, with radii smaller than 1 nm. They appear to be of molecular size and are likely oligomeric silicate ions. The presence of Q^4 containing species is an indicator of these particles whose size appears pH independent in contrast to Iler's model [41] and number appears invariant with time. Cluster aggregation at both high and low pH values appears to occur faster than growth for the particles once formed.

Within about 10 min, at a rate that is a function of $[OH^-]$ and, if present, $[F^-]$ these primary particles

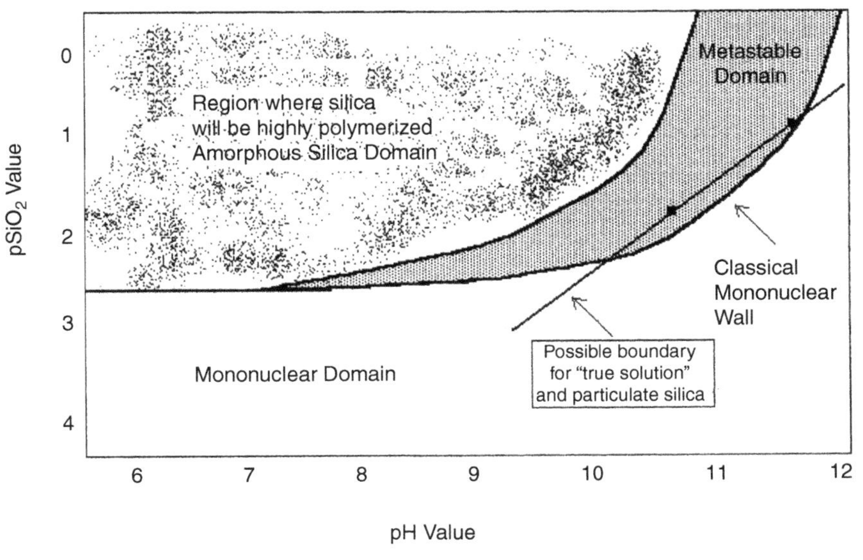

FIGURE 59.2 A solubility diagram for silica.

aggregate to form spherical clusters called fractal aggregates, A_1, with sizes near 2 nm. No evidence is seen for changes in the size of the primary particles within the A_1 particles. Within 40 minutes these fractal aggregates cluster to form primary spherical particles, A_2, with sizes on the order of 8 nm. Over several days these particles do not appear to grow even though they are continuing to form. They simply cluster to form larger aggregates, A_f. They appear to have a finite cut-off size of around 100 nm. These larger fractal aggregates are not confined to spherical geometry. They vary from roughly spherical dense grape-like structures to more open octopus-like aggregations. These ultimate particles are the stuff of silica gels. These structures continue to change over a year, the Q^3/Q^4 content appears to increase and the particles lose their fractal nature. The primary difference between the high and low pH systems seems to be aging time.

Hazel suggested more than 30 yr ago that hydrogen bonded structures are likely to be the basis for polymer formation because the energy of activation is low and entropy considerations are lessened due to fewer geometric restrictions. He also indicated that larger polymers with lower charge densities might depolymerize more readily due to greater susceptibility to OH^- attack [42]. These ideas seem consistent with the "metastable" species observed in the above study.

The key variables in the formation of these larger A_f particles in the absence of organic substances seem to be temperature, pH value, silica concentration and the presence of metal ions like Mg^{2+} or Al^{3+}. It is highly likely that the early stages of particle formation are critical. Once formed these particles exhibit a fairly high surface area, but they seem to be denser at their core. In other words they possess less ordered more reactive surfaces towards the outside of the clusters. Additionally, at RT they are likely suspended in a solution containing oligomers with a total silica concentration on the order of 200 ppm [43]. As suggested by Wijnen, these particles likely age as more active silica (in solution and near surfaces) re-react in the interior of the clusters. The initially formed primary particles appear to grow larger and become denser (Figure 59.3). The net result is a pH dependent reorganization or coarsening of the particles with some overall shrinkage and reduction in overall surface area not unlike the mechanism discussed later for diatom formation.

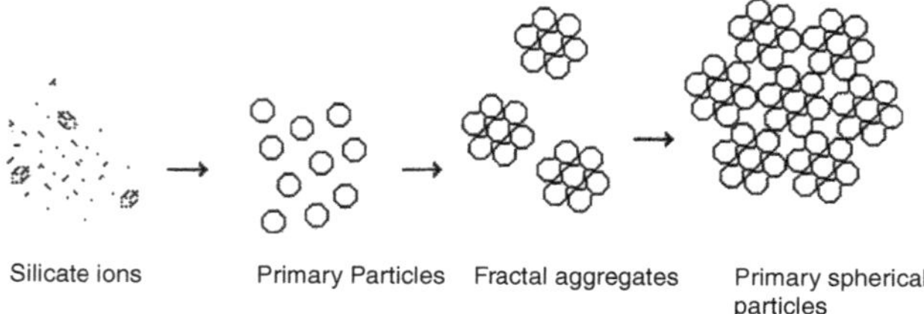

FIGURE 59.3 Representations of growing silica particles after Wijnen et al.

Wijnens et al. [44] suggested that the dissolution of silica gels in aqueous hydroxides was controlled by diffusion through the cation layer surrounding the silica surface. In the series Li^+ to Cs^+, K^+ had the highest rate of dissolution. They also observed that TMA^+ inhibited dissolution, but that the presence of Na^+ increased the rate. Recent studies of the rate of silica depolymerization in dilute solution [45] show that the presence of metal ion species generally decrease the rate of depolymerization of the silica. By analogy, we might infer that rate of depolymerization is a measure of silica site reactivity and trends should be similar to dissolution. Thus, metals present near or on the silica surface could decrease reactivity in both processes. Alkali metal ions decrease depolymerization the least (~10%) followed by alkaline earth ions like Mg^{+2} (~30–40%) and transition metal ions (~50–60%). As mentioned above, it is expected that siliceous species are more mobile at the surface of silica particles and it is likely, in most cases, that trace metals attenuate this mobility, likely by structuring the local environment in some way, thus reducing reactivity. As will be mentioned later one must also be aware of the possibility of ion interaction with specific silanol sites.

The similarity to nanocrystal formation and growth of Zeolite A from colloidal solution is striking. For example, Mintova et al. [46] observed the formation of 3 to 5 nm radius amorphous particles in an initially clear aluminosilicate solution. Upon addition of tetramethylammonium hydroxide, TMAOH, 20–40 nm amorphous aggregates formed in 5 min. The TMA^+ ion is thought to destabilize the primary particles. Over a period of 10 days at RT ordered Zeolite A was seen to emerge within these aggregates. Ultimately, the nanocrystals grew to fully crystalline 20 to 40 nm Zeolite A nanoparticles. Curiously, TMA^+ is known to stabilize forms of the cubic octamer silicate ion, $[Si_8O_{20}]^{8-}$, (also called the double four member ring, D4R ion, or Q_8^3 structure) in solution [47]. In methanol–water mixtures the formation of D4R species is almost quantitative in TMA silicate solutions [48]. The addition of Na^+ is known to destabilize the D4R [49] oligomers.

The colloidal aggregation described above is not surprising in light of studies since Faraday's work with gold sol and more recent studies, such as Lin et al. [50] that point out the apparent universal nature of colloid aggregation. This is due to the thermodynamic instability of the small particles to aggregation. They are maintained in solution by strong repulsive forces, but aggregation is accelerated if repulsion is lessened.

THE SILICA SURFACE–WATER INTERFACE

The silica surface has been the subject of much debate in the literature mostly concerning the fact that silica surface Chemistry defies the predictions of electrostatic based theories (DVLO or double layer models). The properties of pure silica surfaces are variable and are determined largely by the concentration, distribution and nature of silanol groups (≡Si-O-H) on the surface [51–54]. Texter, Klier and Zettlemoyer [55] discussed the changes that silica induces on water at distances greater than molecular dimensions. Their words

> Often the various processes occur simultaneously to such a degree of complexity that it is difficult to resolve them even with the most sophisticated modern experimental means, resulting in controversial interpretations of observations made in water–silica and water–silicate systems

These words are as true today as in 1978. More recently, in studying the stability of interacting silica surfaces Isrealachvili and Pashley [56,57] saw that repulsive hydration forces between silica surfaces decay with an exponential distance dependence of characteristic length ~1–2 nm. These repulsive forces are sufficient to overcome electrostatic forces leading to a breakdown of continuum theory (e.g., DVLO) predictions. The forces are due to the ordered layering of water molecules bound to hydrated surface groups. The energy required to dehydrate these surfaces leads to the large repulsion. Van Oss et al. [58] said these forces, sometimes called "hydrophobic interaction" if attractive or "hydration pressure" if repulsive, that include hydrogen bonding and can result from the interaction of oriented water molecules near surfaces and ions are not merely supplementary refinements, but a drastic correction. They can be one to two orders of magnitude greater than electrostatic and electrodynamic forces and their omission in classic discussions of the behavior of colloids in polar media lead to many of the interpretation anomalies. Etzler [59] suggested that the structure of water is enhanced or entropy is decreased as one approaches a solid surface due to the increase in hydrogen bonding and that this ordering seems independent of temperature in the range 10 to 40°C.

More recently, Honig and Nicholls [60] have also pointed out that nonelectrostatic influences can often dominate interactions in biological systems. The thermodynamic driving force in many cases is a result of strong long-range attractive effects "hydrophobic forces [61]" and close packing. Based on interactions with water, charged and polar groups play a role in generating unique structures as a result of the rearrangements to form ion-pairs, hydrogen bonds and other associations to minimize the total free energy cost of the process [62]. These processes are often called cooperative organization, self-organization or self-assembly. They are processes that result in complex patterns that emerge from the local interactions among many sub-units responding to a few simple rules [63]. The concept of entropy as a force for organization has taken on more of a following in recent years as scientists and engineers awaken to the materials design possibilities inherent in the "gentle force of entropy" [64]. Nonelectrostatic forces thus play a major role in the chemistry of silica surfaces, surfaces that adsorb silica and silicon metal that has oxidized in water. Understanding them and their influence is paramount in understanding variations in silica's behavior in aqueous systems.

SILICA–METAL ION INTERACTIONS

The mechanism of hydrothermal reactions of siliceous species with metal salts at or near room temperature is not well understood. Generally, crystalline phases do not form readily unless the pH value is high, time is long or temperature is high. Studies have shown that maintaining high water activity enhances the reactivity of the silica. The results of many studies over the years are consistent with a model where ion interactions lead to surface precipitation reactions [65–67] and, if incorporated into the silica structure in so-called "structural Lewis sites", lead to increases in the acidity of adjacent silanol sites. Generally, it is thought that at low metal ion concentrations the metal ions adsorb with Langmuir behavior forming a monolayer on the surface of the silica. Increased metal ion concentration leads to the formation of a continuously varying solid solution on the surface of the silica. The activity of the silica, largely associated with surface area (associated with the morphology of the clusters that form the surface), is also an important factor in the formation of the modified particle surface. The composition of this solid solution surface varies between that of the silica and the hydroxide of the "sorbing" metal ion. Ananthapadmanabhan and Somasundaran [68] explained the sorption of Co^{+2} and Ca^{+2} on silica in terms of such a surface precipitation. For example, the precipitation of $Co(OH)_2$ near the silica surface at pH values lower than that required to precipitate Co^{+2} in bulk solution can be simply due to higher than bulk Co^{2+} or OH^- concentrations near the silica surface.

Hazel et al. were the first to study the strong interaction between metal ions, specifically Fe^{+3}, and silicate species [69]. Later, when studying corrosion inhibition of zinc by silica, Hazel et al. [70] measured the interaction of Zn^{+2} ion and silicates and observed slight, but definite, pH value shifts in pH titration inflection points. They suggested that hydroxyl ion was being adsorbed, or occluded, by the amorphous siliceous precipitates. Chemisorption of silica at a zinc hydroxide surface appeared to occur independent of the surface charge or the exact chemical composition of the surface. Falcone [71] measured the activities of Ca^{+2}, Mg^{+2} and Cu^{+2} in the presence of soluble silica. Polymerized silicate anions also shifted the pMe vs. pH curves to lower pH values for these ions. It was suggested that large silicate ions with more acidic surface silanol groups could readily adsorb nucleating metal hydroxides and direct the formation of more ordered, less soluble phases or that intermediate sized silicate species formed complexes with the ions. There was also evidence for the formation of different phases or morphologies of the same phase of metal ion species being formed as the silica differed. Using ^{29}Si-NMR, Harris and Newman [72] suggested that Cr^{3+} binds to the cyclic trimer ion, Q_3^2, whereas Mn^{2+} binds to the linear $Q_1Q_2Q_1$ anion and McCormick, Bell and Radke [73] suggest that Al^{3+} is complexed strongly by the cyclic trimer in aqueous solution.

James and Healy [74] studied the influence of Co^{+2} on the surface charge of quartz as one varied system pH value. Their model suggests a progression of reactivity for "sorbed" species. Metal ions near the surface of silica particles in the Langmuir layer would be influenced strongly by their proximity to the field of the silicon ions in the silica surface. The Si^{+4} is expected to increase the polarization of M–O bonds leading to decreased reactivity. As metal ions "sorb" further from the original silica surface at higher pH values they behaved more like their hydroxide. Falcone and Tsai [75] added solutions of soluble silica to a quartz suspension containing Fe^{+3}, Pb^{+2} and mixtures and measured changes in zeta potential with pH value following James and Healy's work. They found that the systems behaved like quartz at low pH values, but as OH^- activity increased the multivalent ion likely adsorbed causing the charge to reverse and become positive. At high pH values the charge would reverse again as the surface now behaved like the hydroxide of the adsorbing metal ion. Larger soluble silica polymorphs appeared to attenuate the influence of multivalent cations on the surface charge of quartz [76]. This was attributed to either sequestration of the ions by larger silicate anions or adsorption of silicate anions onto the positively charge quartz. This study also suggested that when Ca^{+2} and Mg^{+2} ions were present, silicate anions, due to their relative indifference towards these ions compared to other sequestering anions like phosphate, might be more effective in attenuating the influence of Pb^{+2} and Fe^{+3}.

More recent x-ray absorption studies of Co^{+2} sorption complexes on quartz [77] suggest that for quartz at relatively low density coverage, first-sorbed Co species provide energetically favorable nucleation sites for the subsequent formation of multinuclear hydroxide-like surface complexes. This is also seen for Ni^{2+} adsorbed onto silica [78] where the first species adsorbed are seen as nickel phyllosilicates and not hydroxides. Subsequent layers are either more nickel phyllosilicate or nickel hydroxide depending on the activity of the silica. If the silica is more active, for example due to smaller particles or greater surface area, then more metal-silicate species are formed either at the surface or in solution and adsorb. If the silica is less active (less soluble), due to larger particles or less surface area (generally as a result of aging) then metal hydroxides are more likely on the surface.

The ability to reduce $Co^{+2/+3}$ species on the surface of silica to cobalt metal varies with the way the cobalt–silica material is made. As more cobalt is bound to the silica substrate in the cobalt silicate form compared to hydroxide form the reduction temperature increases. In catalysis this is called a support effect. These interactions are now being studied using analogies to coordination chemistry applied to ligand-surface reactions [79]. Recently, Barth et al. [80] have observed significant variations in reducibility and subsequent catalyst performance with the extent of cobalt coating implying that the first cobalt layer is not easily reducible but subsequent more hydroxide-like layers are more readily reduced.

SILICA COMPLEXATION WITH POLYOLS

It has long been known that silicate anions prefer a four-coordinate or tetrahedral configuration of oxygen atoms around the central silicon atom. In aqueous solution silicon is found in six-coordinate complexes with 1,2-benzenediol, 2-hydroxypyridine-N-oxide and tropolone. Recently, Kinrade et al. [81] observed that addition of high concentrations of aliphatic polyhydroxy molecules to aqueous silicate solutions yielded ^{29}Si-NMR signals suggesting the formation of five- and six-coordinate silicon complexes. These complexes only formed when there were 4 or more adjacent hydroxyl groups in the molecule with two being in the *threo* position. The polyols that yielded higher coordination structures also enhance silica "solubility." Modeling suggested that they wrap around the silicate anion, possibly inhibiting hydrolytic breakdown, and gain extra stability from hydrogen bonding. The authors note the similarity of their model polyols to naturally occurring structures in cells and suggest that such hypervalent silicate complexes may play a vital role in the solution chemistry of silicon. The possibility of forming Si–O–C bonds has generally been discounted [82].

In an attempt to better understand the role played by oxalate ion in increasing the dissolution rate of quartz, Drever et al. [83] found that oxalate had negligible interaction with soluble Si or the siliceous surface. They suggested that increased dissolution could be due to Na^+. They suggested that organic acids could modify dissolution rates of silicate minerals by many mechanisms including pH modification, cation complexation and influence on metal ion speciation. These factors are reviewed in depth. The possibility that solute transport might be rate controlling in natural weathering processes and the implications of that possibility are discussed [84].

DIATOMS

We start our look at the role of silica in biology with the humble diatom. Much of the discussion below is summarized from Kröger and Sumper [85]. Silica precipitation in diatoms produces a wide range of spectacular patterns. The biogenesis of the siliceous components of the diatom occurs within a biochemical reaction vessel or organelle named the silica deposition vesicle (SDV). The silica within diatom cells is amorphous and contains traces of iron and aluminum. It is actively transported to the SDV by Na^+-dependent transporter proteins called silicic acid transporters (SITs) [86]. This transport of silica into the SDV is thought to be an integral component of the silicification puzzle. The species transported appears to be either $Si(OH)_4$ or $SiO(OH)_3^-$ and coupled to Na^+. It is generally assumed that only monomeric forms of soluble silica are present in these systems prior to silica growth. It is believed that colloidal oligomeric silica would be too difficult to transport and too unreliable a raw material source. Transport is blocked by sulfhydril blocking agents. The solutions appear supersaturated with respect to silica ($pSiO_2$ values range from 1 to 2). There is evidence to suggest that silica is drawn into the cell as needed for silica incorporation. Thus, intracellular silica concentration does not appear to drive the process; rather it is regulated by it. It is speculated that silica activity is attenuated by complex formation and would involve some kind of steady state between bound and unbound silica species. However, the exact speciation of the intracellular soluble silica species is unknown. Incorporation of silica reduces unbound species creating an activity gradient that drives the efflux.

THE GROWTH OF SILICA WITHIN DIATOMS

Once inside the SDVs, silica precipitates. There are two theories of silica morphogenesis, silica forms and grows:

- on the surface of species specific templates involving organized assemblies of organic macromolecules with specific spatial distributions
- via diffusion-limited silica particle growth (or self-organization) moderated by non-siliceous components that control the kinetic patterns.

The search for specific macromolecules has identified three families of highly hydrophilic diatom cell wall proteins called frustulins, HEPs and silaffins. The latter two are more strongly bound to the cell walls and can only be removed by anhydrous HF treatment. The frustulins can be removed by ETDA treatment. It is thought that they are bound via Ca^{2+} bridging. The HEPs are thought to be associated with the development of the macrostructure of the diatom. The siliffins are highly basic relatively low molar mass polypeptides containing oligo-*N*-methylpropylamine units attached to a lysine residue. This is the only oligomeric alkyl modified protein described to date. These siliffins have been found to greatly accelerate silica precipitation from metastable silicic acid solutions. They appear to be completely co-precipitated with the silica and the amount of silica precipitated is proportional to the siliffins added as long as it is the limiting reagent. In vitro studies show that the siliffins seem to direct the structure based on their size. The lower MW siliffins produces larger primary aggregates (~600 nm) while a mixture of siliffins yielded spheres on the order of 50 nm. Kröger et al. [87] noticed that siliffins meet both criteria for what Iler [88] called silica flocculation (primary particle destabilization and aggregation above) in that they are both cationic and hydrogen bonding polymers and that they might provide clues for synthesis of siloxane-based materials. Perry [89] observed in vitro that polymerization of silica occurred in stages built from particles ranging from 2–16 nm in size in the absence of structurally directing polysaccharides. Certain substances, like cellulose, were seen to direct very specific particle size distributions (diameter = 4 ± 0.5 nm).

Silica readily participates in the formation of very strong structures due to its ability to reinforce structures with reduced mechanical anisotropy [90]. The diatom biosilica is made up of tightly associated 100 nm spherical bodies. They appear to be non-fractal [91]. Void spaces between these spheres are then filled with silica to yield a smooth appearing surface. It has been suggested [92] that the catalytic effect of the oligo-*N*-methylpropylamine chain modification of the lysine residue on silica precipitation is due to the spacing of adjacent protonated and unprotonated amine groups which can hydrogen bind to silanol groups and facilitate polymerization via a displacement reaction with a silanolate group as nucleophile and water as the leaving group. Williams et al. [93] found that the structural organization of biogenic silica in *Discinid Brachiopods* could be influenced by glycosamineglycans (GAGs) and β-chitin membranous vesicles. The silica "tablets", also no more than 100 nm thick, are separated by GAGs and precipitated apatite.

Recent Hartree–Fock simulations [94] of serine–silicate complexes, identified as likely templating candidates [95] in SDVs, seem to suggest that five-coordinate complexes are not favored in the acidic environment of the SDV; however, four coordinate $serO-Si(OH)_3$ were 10 kcal mol^{-1} more exothermic than $serOH-Si(OH)_4$ interactions. At the basic pH values of seawater, silicic acid was favored over Si–O–C bonds. In a recent Minireview, Perry and Keeling-Tucker [96] discuss research in the area of biosilicification and conclude that biological organisms exert controls not easily mimicked in vitro.

Wetherbee, Crawford and Mulvaney [97] share this view and also provide an in-depth review of this complex area.

SILICA INFLUENCE ON CRYSTALIZATION

Damen and Ten Cate [98] studied calcium phosphate precipitation with and without the presence of silicic acid and showed that polysilicic acid, not monomer, acted as a substrate for hydroxyapatite nucleation, caused a 60% reduction in the induction period in seeded reactions and overcame part of the inhibitory effect of phosprotein. In all cases hydroxyapatite grew in the presence of silica even when the system was seeded with a different polymorph of calcium phosphate. It was concluded that silica promotes dental calculus formation.

Cornell and Giovanoli [99,100] observed that the presence of silicate strongly inhibits the transformation of ferrihydrite to goethite and/or hematite in alkaline media when the concentration exceeds about 10 ppm. When concentrations are low enough for the reactions to proceed, goethite (roughly 0–6 ppm) nucleation is inhibited. The morphology of both forms is changed. Silicate, in 0.1-mM range, seems to behave in a manner similar to hydroxy-carboxylic acids. The concentrations are too low to form monolayer coverage; therefore, the authors suggest that due to the four coordinating sites that silicates can link different particles of ferrihydrite forming an immobilized network. When the silicate was added to ferric nitrate solution prior to ferrihydrite precipitation it was more active than that added after ferrihydrite formation. Interestingly, polynuclear hydroxy-Al ions, but not monomers, also seem to disorder precipitates forming in Fe solutions under acidic conditions [101].

ADSORPTION ON BACTERIAL SURFACES

Silica, and iron and aluminum containing silicates have been observed to be in close association with bacteria. When present they appear to help preserve the shape of cellular surfaces [102]. Electropositive amine groups could bind anionic species at physiological pH values; however, recent studies indicate that most silicate binding to the cell surface occurs on positive sites produced by cation binding possibly to $-COO^-$ or phosphate sites [103]. It is believed that cell surfaces generically promote mineral development by lowering the total free energy required for precipitation. Cells, in general, can play an active role by metabolically modifying the micro-reaction environment or can passively enhance precipitation. The formation of amorphous silica containing species appears to occur via a passive mechanism. Based on studies carried out with living and dead cells the rate of formation appears to be associated with the extent of favorable surface sites.

Ueshima and Tazaki [104] describe mineral formation in the acidic polysaccharides associated with microbial cell surfaces. They find that polysaccharides, associated with extracellular polymeric substances (EPS), direct the preferential formation of nontronite, a sodium-iron (III) phyllosilicate in simulation studies. It is suggested that the chain structure of the polysaccharides affect layer silicate orientation. They observed only Si-bearing amorphous iron hydroxides forming outside of the EPS.

MICROBIAL INFLUENCE ON SILICATE WEATHERING

Silicates dissolve faster in microbially active organic rich waters than predicted on the basis of bulk groundwater geochemical environments [105–107]. Studies of surface corrosion on weathered quartz suggests crystallographically controlled dissolution caused by silica, and possibly alumina, chelating organic acids produced within adhering cells and released at the cell–mineral interface. Observations of aluminosilicate weathering in vitro showed that several strains of bacteria increased the release of Si and Al from biotite (also Fe) and feldspar by up to two orders of magnitude compared to abiotic conditions and that growth seemed to occur preferentially along cleavage steps and edges [108]. Whether biotic weathering is a coincidence of metabolism or a useful strategy in colony formation is unclear; however, recent studies [109] suggest that silicate weathering is sometimes driven by microbial nutrient requirements. Geologists see potential to better understand the rock record and improve geochemical modeling [110,111] based on a better understanding of this possible mineral — microorganism symbiosis.

SILICA TOXICOLOGY

Ralph Iler [112] had a deep interest in silica toxicology. During the early 1980s he was exploring the hypothesis that, in my words, the hydrophilic/hydrophobic profile of the siliceous surface was key to very strong specific enzyme adsorption behavior which made the presence of silica toxic in the lungs. Fubini and Wallace [113] present a general mechanism for the toxicity of silica in the lungs. They discuss the factors that interested Iler, the hydrogen bonding role of surface silanol groups, clearance from the lungs aided by Al(III), inhibition of clearance and other factors. They suggest that the toxicity of pure quartz compared to clays might be due to a lack of surface functionality and interactions that have a prophylactic effect.

Iler also had a deep personal interest in senile dementia of the Alzheimer-type and having read the literature extensively saw several studies, which suggested to him that low silicon environments facilitated disease. Studies that suggested that high aluminum environments (thus low silica activity due to strong complexation) correlated to high incidence of brain or nerve related disease fascinated him. At that time, silica's role was viewed as direct. He expressed great frustration in the mid-1980s with the lack of research combining both a chemical and medical viewpoint and the lack of information on metal chelate chemistry — especially with molecules that are in living systems [114].

Birchall's [115] seminal studies in the late 1980s strongly suggested a different role for silicon's essentiality in biological systems. Silicon in the form of one or more of its polymorphs might serve as an essential sequestrant or masking agent protecting organisms against aluminum poisoning [116]. A 2 level 2-factor study of the survival rates of Atlantic salmon fry involving Al and Si at constant pH value showed this. The influence of silica on alumina described by Birchall [117] could be inferred from the model for the solubility product K_{aas} of an amorphous aluminosilicate (aas) described by Paces [118].

$$K_{\text{aas}} = K_a^{(1-x)} K_s^x$$

Using an empirical expression for the silica molefraction, x, in the aluminosilicates found in natural water as a function of pH value ($x = 1.24 - 0.135*\text{pH}$) in the above expression for K_{aas} for an ideal solid phase (where K_a and K_s are the respective solubility constants for alumina and silica) one gets $\text{p}K_{\text{aas}} = 5.89 - 1.59*\text{pH}$ for the reaction:

$$\text{Al}_{(1-x)}\text{Si}_x + (3-3x)H^+ \Leftrightarrow (1-x)\text{Al}^{3+} + x\text{H}_4\text{SiO}_4 + (3-5x)\text{H}_2\text{O}$$

The pAl vs. pH curves at varying pSi developed from this empirical K value clearly showed the attenuating influence of silica on aluminum activity in acidic solutions. One becomes aware of the fact that nonideal solid solutions control the activity of Al and Si in natural water.

Browne and Driscoll [119] reported that 95% of total inorganic Al in natural water appeared to be associated with soluble Al-Si complexes. They warned that many of the studies of the true solubility of Al-Si minerals might be in error due to undetected soluble complexes. Since complexation experiments with monomeric silica and aluminum imply that such complexes in surficial waters do not exceed 2–3% of the total dissolved aluminum [120], polymers must dominate. More recently, Taylor et al. [121] used solutions of soluble silica made in such a way to maintain a high concentration of larger sized silica polymorphs to study the interaction with Al. They observed that the affinity for Al was at least 10^6 times greater when compared to monomeric silica. They also observed that polymers were stabilized by aluminum ion as expected based on earlier discussions.

Recent studies by Perry and Keeling-Tucker [122] using tris-catecholato silicon (IV) complexes as the source of silica extend the range of influence of silica to include calcium and iron in addition to Al. Some form of silica appears to influence the mineralization of iron oxide and calcium phosphate phases as seen by others. Silica shows an affinity for Al species comparable to organic complexing ligands.

BIOACTIVE SILICA CONTAINING MATERIALS

Hench, with funding from the U.S. Army Medical R&D Command, discovered glass compositions that bonded intimately with living tissues [123]. The intense study of these materials resulted in the development of a fairly elaborate mechanism for the bonding of glass to living bone. In this mechanism, bone formation and growth at the surface of bioactive glass results from a cascade of inorganic and organic reactions. An important intermediate in bioglass bonding is an apatite layer that forms at the glass surface. The apatite only forms on the silica gel, it will not form on quartz [124].

It also appears that specific biological growth genes are activated by soluble silicon species [125] in the highly chem-adsorbing inorganic gel layer. Hench discusses the implications of studies showing a critical concentration of soluble silicon is required for bone mineralization [126,127] and an increase in coronary heart disease is associated with loss of silicon within arterial walls. Finally, in an updated alternative origins of life scenario, bioactive ceramic substrates are advanced as an inorganic template or catalyst for the self-assembly of polypeptides leading first to prebiotic micelles and eventually to primitive forms of RNA/DNA which when transferred to the micelles were able to more efficiently carry out their biosynthetic design [128].

MATERIALS APPLICATIONS

The area of materials application of nature's designs or biomimetic synthesis, the incorporation of molecular biology into inorganic material science [129–131] presents two general opportunities:

- The ability to develop novel rational synthetic routes to organized microarchitectures.

  The headline in the News of the Week section of *C&ENews* is INORGANIC BIOCATALYSIS and reports that diatom peptide forms silica nanosphere for diffraction grating [132,133]. The peptides are embedded into a polyacrylate film in a manner that controls their placement and retains their activity. Exposure of the matrix to silicic acid results in the deposition of silica nanospheres yielding a composite with controlled variations in refractive index.

- The opportunity to revisit and possibly redefine some of the classical rules.

  Bladgen and Davey [134] in discussing organic systems suggest that reaction steps produce by-products that can inadvertently direct the polymorphic outcome of a crystallization process. To quote them, their research is causing them to ask questions such as:

  > "How valid is Ostwald's Rule?
  > What role do solvents play in determining polymorph appearance?
  > What role does preordering (molecular aggregation) in the supersaturated liquid phase determine the subsequent nucleation and growth processes in polymorphic systems?
  > The answer to these questions would make a wise man happy."

  Similar questions have been on the minds of zeolite researchers for about 40 yr.

It is clear that much of what is known about silica and mechanisms by which the apparently infinite variety of forms develop falls under both categories. We must be vigilant to avoid artifacts in the behavior of silica and the models and ideas used to represent their behavior. The systems problems involved in making progress are daunting. Most researchers are aware of the immense challenges, but soldier on since the rewards of better understanding seem limitless [135].

ORGANIC TEMPLATING

Vrieling et al. [136] discuss silicon biomineralization in diatoms as a source of inspiration for novel approaches to silica synthesis. They foresee value in the possibility that unique structures can be synthesized free of the impurities inherent in current large-scale method of synthesis. They relate the assumed structural directing properties of components of the SDV to the templating properties of organic molecules (such as tetramethyl- and tetrapropylamine, see organic templating) in zeolite synthesis. They point out that silica concentration can be elevated to levels approaching 1% by weight. At this concentration colloidal species are highly likely.

Organic templating or the molecular manipulation of microstructure [137–140] has been of great interest to those seeking controlled particle morphology especially for applications where a combination of molecular sieving and chemically active framework sites could be combined to optimize performance as inorganic thin films, catalysts or sorbents. Of particular interest is the ability to form materials with tailored properties at low temperatures. It is not surprising that uncertainties associated with the mechanism of self-assembly are fairly high. The key questions seem to revolve around the manner in which the "template" organizes an emerging phase and whether or not this phase influences the "template."

Biz and Occelli [141] discuss these issues from the perspective of catalysts design and characterization. They note the striking similarity between the needs of the material scientist and the results produced by nature. The goal of the material scientist is to produce thermally and mechanically stable materials with features in what is called the mesoporous range between microporous (pore diameter <2 nm) and macroporous (pore diameter >50 nm). There exists variation among authors in the use of the prefix meso. Generally, it refers to intermediate domains between roughly 1 to 100 nm [142] or what one might historically call the lower end of the colloidal domain.

Pinnavaia et al. [143] found that lamellar liquid crystal phases of electrically neutral gemini surfactants of the type $C_nH_{2n+1}NH(CH_2)_2NH_2$ which normally undergo transition to spongelike phases will, in the presence of tetraethylorthosilicate (TEOS) in the process of hydrolyzing to silica, form vesicular hierarchical structures as a result of hydrogen bonding interactions with the polymerizing silica. TEOS alone generally makes spheres [144] and the cationic form of the surfactants yields conventional morphologies. These mesostructured silicas possess remarkable thermal and hydrothermal stability due to their 3D-pore network and vesiclelike morphology. The silica layers are roughly 3 nm thick. Results from studies using cetyltrimethylammonium ion as surfactant in clear basic silicate solutions suggest that the silica species–surfactant interactions are weak and suggest that the important step in the process is the direction of the formation of siliceous species [145]. Fowler et al. [146] utilized tobacco mosaic virus particles to prepare unusual mesostructured silicas with periodicity on the order of 20 nm.

BIOENCAPSULATION

Our final topic maybe the most important in that it might be the one where the widest and most immediate practical applications result [147]. In this area silica, metal oxides and other polymers offer the potential to immobilize biological materials to create "living ceramics" with applications as electrochemical sensors, catalysts and possibly artificial organs [148]. There are still many hurdles in the development of widely applicable techniques, such as biocompatibility and integrity of the biofilms; however, the promise is great as evidenced by the number of valuable characteristics already observed, including; superior optical properties, controllable porosity, improved biostability of entrapped biologicals and ability to function in diverse environments. Walcarius [149] reviews the field of electroanalytical biosensors based on silica and silica-containing matrices. Systems with high bio-stability and reactivity can be prepared by adding the biomolecules after initiating the gelling process. Additionally fume silica has been found to improve the properties of carbon paste biosensors. He foresees extensive opportunities due to the surface reactivity of silica, its ion exchange capability and the potential for introducing sieving properties.

REFERENCES*

1. Birchall, D. Letter to J. Falcone, September 12, 1990.
2. Birchall, D. *Chemistry & Industry* (1989) 3 July, p. 403.
3. Graham, T. (1886) *Phil. Transactions.*
4. Calvert, P. "Biomimetic Processing" in *Processing of Ceramics, Part 2*, R.J. Brook (ed.) Vol. 17B (VCH Verlagsgesellschaft mbH: Weinheim, FDR, 1996) pp. 51–82.
5. Vail, J.G. "*Soluble Silicates in Industry*", The Chemical Catalog Company, Inc., New York (1928), p. 83.
6. Herrera, *J. Lab. Clin. Med.* (1919) **4** 479.
7. Moore and Evans, *Proc. Royal Soc. (London) Series B* (1915) **89** 17.
8. Bastian, *Nature* (1914) **92** 579.
9. Willey, J.D., *Geochim. Cosmochim. Acta* (1980) **44** 573.
10. Price, G.D. "The Riches of Silica Revealed", (1984) *Nature* **310** (23 Aug.) 631.
11. Dent Glasser, L.S.; Lachowski, E.E. *J. Chem. Soc. Dalton Trans.* (1980) **393**, 390.
12. Falcone, Jr., J.S. In *Cements Research Progress — 1988*; Brown, P.W., Ed.; American Ceramic Society: Westerville, OH, (1989) p 277.
13. Falcone, Jr., J.S.; Blumberg, J.G. "*Anthropogenic Silicates*" in The Handbook of Environmental Chemistry, Vol. 3. O. Hutzinger, Ed. (Springer-Verlag: Berlin, 1992), pp. 367–382.
14. Ingri, N. "Aqueous Silicic Acid, Silicates and Silicate Complexes" in *Biochemistry of Silicon and Related Problems*, G. Bendz and I. Lindqvist, Eds. (Plenum Press: New York, 1978), p. 3.
15. Heaney, P.J.; Prewitt, C.T.; Gibbs, G.V.; Eds. *SILICA: Physical Behavior, Geochemistry and Materials Applications*, MSA: Washington, DC, 1994.
16. Liebau, F. *Structural Chemistry of Silicates*; Springer-Verlag: Berlin, 1985.
17. Bergna, H., Ed. *The Colloid Chemistry of Silica* Adv. in Chemistry Series 234 ACS: Washington, DC, 1994.
18. Franks, F. *Water*, Royal Society of Chemistry: London, 1984.
19. Iler, R.K.; *The Chemistry of Silica*; John Wiley & Sons: New York, NY (1979) p. 667.
20. Blagden, N.; Davey, R. "Polymorphs Take Shape", *Chemistry in Britain* (1999) March, p. 44.
21. Falcone, Jr., J.S. "*Uses of Soluble Silica*" in The Colloid Chemistry of Silica, H. Bergna, Ed., Adv. in Chemistry Series 234 ACS: Washington, DC (1994) p. 595.
22. Dent Glasser, L.S., "Structures and Reactions of Silicates" a lecture presented in London, Scientific Societies' Lecture Theatre, 22 March 1988.
23. Weiner, S.; Addadi, L.; Wagner, H.D. "Materials Design in Biology", *Materials Science and Engineering C* (2000) **11** 1–8.
24. Tipping, E. "Colloids in the Aquatic Environment", *Chemistry and Industry* (1988) 1 August, 485.
25. Roggendorf, H.; Grond, W.; Hurbanic, M. *Glastech. Ber. Glass Sci. Technol.* (1996) **69**(7) 216–231.
26. Falcone, J.S., Ed. *Soluble Silicates* (1982), ACS Books: Washington, DC.
27. Hazel, J.F. In *Principles and Applications of Water Chemistry*, S.D. Faust and J.V. Hunter, Eds. (Wiley & Sons: New York, 1967), p. 301.
28. Nauman, R.V.; Debye, P. *J. Phys. Chem.* (1951) **55** 1; (1961) **65** 5.
29. Iler, R. "Effect of Adsorbed Alumina on the Solubility of Amorphous Silica in Water", *J. Coll. Interface Sci.* (1973) **43**(2) 399.
30. Unger, K.; Kumar, D.; Ehwald, V.; Grossmann, F. "Adsorption on Silica Surfaces from Solution and Its Impact on Chromatographic Separation Techniques" in *Adsorption on Silica Surfaces*, E. Papirer, Ed. (Marcel Dekker, Inc.: New York, 2000), Chap. 17, p. 565.
31. Navrotsky, A. "Thermodynamics of Crystalline and Amorphous Silica" in *SILICA: Physical Behavior, Geochemistry and Materials Applications*, Heaney, P.J.; Prewitt, C.T., Gibbs, G.V., Eds. (MSA: Washington, DC, 1994) pp. 309–329.
32. Dove, P.M. "Kinetic and Thermodynamic Controls on Silica Reactivity in Weathering Environments" in *Chemical Weathering Rates of Silicate Minerals*, A.F. White and S.L. Brantley, Eds. (MSA: Washington, DC, 1995) Chapter 6, pp. 235–290.
33. Stumm, W.; Morgan, J.J. *Aquatic Chemistry*, 3rd Edn. (John Wiley & Sons, Inc.: New York, 1996), p. 368 and earlier editions.
34. Iler, R. "Colloidal Components in Solution of Sodium Silicate" in *Soluble Silicates*, Falcone, Jr., J.S., Ed. (ACS Books: Washington, DC, 1982), p. 95.
35. Krumrine, P.H.; Falcone, Jr., J. S. "Rock Dissolution and Consumption Phenomena in an Alkaline Recovery System", *SPE Reservoir Engineering* (1988) (2) 62.
36. Falcone Jr., J.S. "Silicon Compounds: Synthetic Inorganic Silicates" in *Kirk-Othmer: Encyclopedia of Chemical Technology*, 4th Edn. (John Wiley & Sons, Inc., NY) 1997 **22** 1.
37. Sjöberg, S. "Silica in the Environment", *J. Non-Cryst. Solids* (1996) **196** 51–57.
38. Andersson, K.R.; Dent Glasser, L.S.; Smith, D.N. "Polymerization and Colloid Formation in Silicate Solution" in *Soluble Silicates*, Falcone, Jr., J.S. Ed. (ACS Books: Washington, DC, 1982), p. 115.

*(To Jan. 2002).

39. Wijnen, P.W. J.G. "A Spectroscopic Study of Silica Gel Formation form Aqueous Silicate solution", Ph.D. Thesis, Eindhoven University, Holland (1990).
40. Wijnen, W.P. J.G. et al. *J. Colloid Interface Sci.* (1991) **145**(1) 17.
41. Iler, R. "The Chemistry of Silica" *op. cit.*
42. Hazel, J.F., *op. cit.*
43. Fleming, B.A. *J. Colloid Interface Sci.* (1986) **110** 40.
44. Wijnens, P.W.J.G.; Beelen, T.P.M.; De Haan, J.W.; Van De Ven, L.J.M.; Van Santen, R.A. *Colloids and Surfaces* (1990) **45** 255.
45. Dietzel, M., Usdowski, M. *Colloid Polym. Sci.* (1995) **273** 590–597.
46. Mintova, S.; Olson, N.H.; Valtchev, V.; Bein, T. *Science* (1999) **283** (12 Feb.) 958
47. Knight, C.T. G. *Zeolites* (1989) **9**(9) 448.
48. Hasegawa, I.; Sakka, S.; Kuroda, K.; Kato, C. *J. Molecular Liquids* (1987) **34** 307.
49. McCormick, A.V.; Bell, A.T.; Radke, C.J.; "Effect of Alkali Metal Cations on Silicate Structures in Aqueous Solution," in *Perspectives in Molecular Sieve Science*, W.H. Flank and T.E. Whyte, Jr., Eds. American Chemical Society, ACS Symposium Series 368, (1988); Chapter 14, 222–235.
50. Lin, M.Y. et al. *Nature* (1989) **339** (1 June) 360 and comments by Rarity, J. p. 340.
51. Iler, R.K. *The Chemistry of Silica*, Wiley Interscience: New York, 1979.
52. Kondo, S.; Igarashi, M.; Nakai, K. *Colloids and Surfaces* (1992) **63** 33.
53. Unger, K.K. *The Colloid Chemistry of Silica*, H. Bergna, Ed. American Chemical Society: Washington, DC, 1994.
54. Chuang, I.-S.; Maciel, G. E. *J. Am. Chem. Soc.* (1996) **118** 401.
55. Texter, J.; Klier, K.; Zettlemoyer, A. "Water at Surfaces" *Progress in Surface and Membrane Science* (1978) Vol. 12, Academic Press: New York, p. 342.
56. Israelachvili, J.N.; Pashley, R.M. *Nature* (1982) **300** 341.
57. Israelachvili, J.N.; Pashley, R.M. *Nature* (1983) **306** (Nov 17) 249.
58. van Oss, C.J.; Giese, R.F.; Costanzo, P.M. *Clays and Clay Minerals* (1990) **38**(2) 151.
59. Etzler, F.M. *Langmuir* (1988) **4**(4) 878.
60. Honig, B.; Nicholls, A. *Science* (1995) **268** (26 May), p. 114.
61. Tsao, Y.-H.; Evans, D.F.; Wennerström, H. *Science* (1993) **262** 547.
62. Jeffrey, G.A.; Saenger, W. *Hydrogen Bonding and Biological Structures* (Springer-Verlag: New York, 1991).
63. Pepper, J.W.; Hoelzer, G. *Science* (2001) **294** (16 November) 1466.
64. Kerstenbaum, D. *Science* (1998) **279** (20 March) 1849.
65. Sermon, P.A.; Sivalingam, *J. Colloids and Surfaces* (1992) **63** 59.
66. Farley, K.J.; Dzombak, D.A.; Morel, F.M. M. *J. Colloid Interface Sci.* (1985) **106** 226.
67. Scheidegger, A.M.; Sparks, D.L. *Chemical Geology* (1996) **132** 157.
68. Ananthapadmanabhan, K.P.; Sumasundaran, P. *Colloid and Surf.* (1985) **13** 151.
69. Hazel, F.; Schock, R.; Gordon, J. *JACS* (1949) **71** 2256.
70. Hazel, J.F. *et al. J. Electrochem. Soc.* (1952) **99**(7) 301.
71. Falcone, Jr. J.S. "The Effect of Degree of Polymerization of Silicates on Their Interactions with Cations in Solution", in *Soluble Silicates*, Falcone, J.S., Ed. (1982) ACS Books: Washington, DC.
72. Harris, R.K.; Newman, R.H. *J. Chem. Soc. FT2* (1977) **79**(9) 1204.
73. McCormick, A.V.; Bell, A.T.; Radke, C.J. *Stud. Surf. Sci. Catal.* (1986) **28** 247; *J. Phys. Chem.* (1989) **93** 1741.
74. James, R.O.; Healy, T.W. *J. Colloid Interface Sci.* (1972) **40** 53.
75. Tsai, F.; Falcone, J.S. Paper presented at ACS/CSJ Chem. Congress, Honolulu, HI, Apr. 6, 1979.
76. Hazel, F.J.. *J. Phys. Chem.* (1945) **49** 520.
77. O'Day, P.A.; Chisholm-Brause, C.J.; Towle, S.N.; Parks, G.A.; Brown, G.E. Jr. *Geochim. et Cosmochim. Acta* (1996) **60** 2515
78. Burattin, P.; Che, M.; Louis, C. *J. Phys. Chem. B* (1998) **102** 2772.
79. Dyrek, K.; Che, M. *Chem. Rev.* (1997) **97** 305.
80. Barth, R.; Falcone, Jr., J.S; Vorce, S.; McLennan, J.; Outland, B.; Arnoth, E. "Cobalt–Silica Interaction and the Dehydrogenation of 2-Propanol" submitted to *Catalysis Communications* (2002).
81. Kinrade, S.D.; Del Nin, J.W.; Schach, A.S.; Sloan, T.A.; Wilson, K.L.; Knight, C.T.G. *Science* (1999) **285** (3 September) 1542.
82. Birchall, J.D. *Chem. Soc. Rev.* (1995) **24** 351.
83. Poulson, S; Drever, J.I.; Stillings, L.L. *Chem. Geol.* (1997) **140**(1–2) 1.
84. Drever, J.I.; Stillings, L.L. *Colloids Surf. A*, (1997) **120**(1–3) 167.
85. Kröger, N.; Sumper, M. "The Biochemistry of Silica Formation in Diatoms" in *Biomineralization*, Bäuerlein, E., Ed., Chapter 11, (Wiley-VCH: Weinheim, FRG, 2000) pp. 151–170.
86. Hildenbrand, M. "Silicic Acid Transport and Its Control During Cell Wall Silicification in Diatoms" in *Biomineralization*, Bäuerlein, E., Ed., Chapter 2, (Wiley-VCH: Weinheim, FRG, 2000) pp. 171–188.
87. Kröger, N.; Deutzmann, R.; Sumper, M. *Science* (1999) **286** (5 November) 1129.
88. Iler, R. *The Chemistry of Silica*, *op. cit.*
89. Perry, C.C."Biogenic Silica: A model of Amorphous Structure Control" in *Growth, Dissolution and Pattern formation in Geosystems* B. Jamtviet and P. Meakin Eds. Kluwer Academic Publishers: Netherlands, 1999. Chapter 11, pp. 237–251, for an in-depth review.
90. Weiner, S.; Addadi, L.; Wagner, H.D. "Materials Design in Biology", *Material Sci. and Engineering C11* (2000) 1.

91. Vrieling, E.G.; Beelen, T.P.M.; van Santen, R.A.; Gieskes, W.W.C. *J. Biotechnology* (1999) **70** 39–51.
92. Kröger, N.; Sumper, M. "The Biochemistry of Silica Formation in Diatoms" in *Biomineralization*, Bäuerlein, E., Ed., Chapter 11 (Wiley-VCH: Weinheim, FRG, 2000), pp. 151–170.
93. Williams, A.; Cusack, M.; Buckman, J.D.; Stachel, T. *Science* (1998) **279** (27 March) 2094.
94. Sahai, N.; Tossell, J.A. *Geochim. Cosmochim. Acta* (2001) **65**(13) 2043.
95. Shimizu, K.; Cha, J.; Stucky, G.; Morse, D. *Proc. Natl. Acad. Sci. USA* (1998) **95** 6234.
96. Perry, C.C.; Keeling-Tucker, T. *J. Biol. Inorg. Chem.* (2000) (5) 537.
97. Wetherbee, R.; Crawford, S.; Mulvaney, P. "The Nanostructure and Development of Diatom Silica" in *Biomineralization*, Bäuerlein, E., Ed., Chapter 13 (Wiley-VCH: Weinheim, FRG, 2000), pp. 189–206.
98. Damen, J.J.; Ten Cate, J.M. *J. Dent. Res.* (1989) **68**(9) 1355.
99. Cornell, R.M.; Giovanoli, R. *J. Chem. Soc., Chem Commun.* (1987) 413.
100. Cornell, R.M.; Giovanoli, R.; Schindler, P.W. *Clays and Clay Minerals* (1987) **35**(1) 21.
101. Singh, S.; Kodama, H. *Clays and Clay Minerals* (1994) **42**(5) 606.
102. Fortin, D.; Beveridge, T.J. "Mechanistic Routes to Biomineral Surface Development" in *Biomineralization*, Bäuerlein, E., Ed., Chapter 2 (Wiley-VCH: Weinheim, FRG, 2000), pp. 7–24.
103. Fortin, D.; Ferris, F.G.; Scott, S.D. *Am. Mineral.* (1998) **83**(11–12) 1399.
104. Ueshima, M.; Tazaki, K. *Clays and Clay Minerals* (2001) **49**(4) 292.
105. See *Chemical Weathering Rates of Silicate Minerals* A.F. White and S.L. Brantley, Eds. (MSA: Washington, DC, 1995) for an in-depth review of this important topic.
106. Kalinowski, B.E.; Liermannn, L.J.; Givens, S.; Brantley, S.L. *Chem. Geol.* (2000) **169**(3–4) 357.
107. Hiebert F.K.; Bennett, P.C. *Science* (1992) **258** (9 Oct.) 278.
108. Barker, W.W; Welch, S.A.; Chu, S.; Banfield, J.F. *Am. Mineral.* (1998) **83**(11–12) 1551.
109. Bennett, P.C.; Rogers, J.R.; Choi, W.J. *Geomicrobiol. J.* (2001) **18** 3.
110. Lucas, Y. Luizão, F.J.; Chauvel, A.; Rouiller, J.; Nahon, D. *Science* (1993) **260** (23 April) 521.
111. Berner, R.A. in *Chemical Weathering Rates of Silicate Minerals,* A.F. White and S.L. Brantley, Eds. (MSA: Washington, DC, 1995, Chapter 13, pp. 565–581.
112. Iler, R., various personal communications 1980 to 1983.
113. Fubini, B.; Wallace, W.E. "Modulation of Silica Pathogenicity by Surface Processes" in *Adsorption on Silica Surfaces*, E. Papirer, Ed. (Marcel Dekker, Inc.: New York, 2000), p. 645.
114. Iler, R. Letter to J. Falcone, September 16, 1983.
115. Birchall, J.D. "Silicon–Aluminum Interaction in Biology" in *The Colloid Chemistry of Silica* H. Bergna, Ed., Adv. in Chemistry Series 234 ACS: Washington, DC, 1994, p. 601.
116. Birchall, J.D. "The Role of Silicon in Biology", *Chem. in Britain* (1990) (2) 141.
117. Falcone, Jr., J.S. 1991, unpublished data.
118. Pačes, T. *Geochim. Cosmochim. Acta* (1978) **42** 1487.
119. Browne, B.A.; Driscoll, C.T. *Science* (1992) **256**(19 June) 1667.
120. Pokrovski, G.S.; Schoot, J.; Harrichoury, J.; Sergeyev, A. S. *Geochim. Cosmochim. Acta* (1996) **60**(4) 2495.
121. Taylor, P.D.; Jugdaohsingh, R.; Powell, J.J. *JACS* (1997) **119** 8852.
122. Perry, C.C.; Keeling-Tucker, T. *J. Inorg. Biochem.* (1998) **69** 181; (2000) **78** 331.
123. Hench, L.L. *Sol-Gel Silica*, Noyes Publications: Westwood, NJ, 1998.
124. Ohiura, K. et al. *J. Biomed. Material Res.* (1991) **25** 357.
125. Hench, L.L. *op. cit.*, p. 117.
126. Carlisle, E.M. in *Silicon Biochemistry* (D. Evered and M. O'Connor, Eds.) Wiley, New York, 1986.
127. Schwarz, K., in *Biochemistry of Silicon and Related Problems* (G. Bendz and I. Lindqvist, Eds.) Plenum Press: New York, 1978, pp. 207–230.
128. Hench, L.L., "Bioceramics and the Origins of Life", *J. Biomed. Maters. Res.* (1989) **23** 685.
129. Peppas, N.A.; Langer, R. "New Challenges in Biomaterials" *Science* (1994) **263** (25 March) 1715.
130. Mann, S. et al. *Science* (1993) **261** (3 September) 1286.
131. Perry, C.C. "Biomaterials, Synthetic Synthesis, Fabrication, and Application" in *Encyclopedia of Physical Science and Technology, Third Edition*, Vol. 2 (Academic Press, 2002), pp. 173–191.
132. Stone, M. et al. *Nature* (2001) **413** 291.
133. Henry, C. *C&ENews* (2001) September 24, p. 14.
134. Blagden, N.; Davey, R., *op. cit.*
135. Mark, J.E.; Calvert, P.D. *Materials Sci. and Eng., C1* (1994) 159–173.
136. Vrieling, E.G. et al. *op. cit.*, p. 48.
137. Monnier, A.; Schüth, F.; Huo, Q.; Kumar, D.; Margolese, D.; Maxwell, R.S.; Stucky, G.D.; Krishnamurty, M.; Petroff, P.; Firouzi, A.; Janicke, M.; Chemelka, B.F. *Science* (1993) **261** (3 September) 1299.
138. Aksay, I.A. et al. *Science* (1996) **273** (16 August) 892.
139. Stupp, S.I.; Braun, P.V. *Science* (1997) **277** (29 Aug.) 1242.
140. Weiner, S.; Addadi, L. *J. Mater. Chem.* (1997) **7** 689.
141. Biz, S.; Occelli, M.L. *Catal. Rev.–Sci. Eng.* (1998) **40**(3) 329–407.
142. Dagani, R. *C&Enews* (1995) November 6, p. 7.
143. Kim, S.K.; Zhang, W.; Pinnavaia, T.J. *Science* (1998) **282** (13 Nov.) 1302.
144. Stöber, W.; Fink, A.; Bohn, E.; *J. Colloid Interface Sci.* (1968) **26** 62.

145. Frasch, J.; Lebeau, B.; Soulard, M.; Patarin, J.; Zana, R. *Stud. Surf. Sci. Catal.* (2000) **129** 147.
146. Fowler, C.E.; Shenton, W.; Stubbs, G.; Mann, S. *Adv. Materials* (2001) **13**(16) 1266.
147. Dave, B.; Dunn, B.; Valentine, J.; Zink, J. "Sol–Gel Matrix for Protein Entrapment" in *Immobilized Biomolecules in Analysis*, T. Cass and F.S. Ligler, Eds., Oxford Univ. Press: Oxford, 1998, Chapter 7, pp. 113–134.
148. Gill, I.; Ballesteros, A. *Trends in Biotechnology* (2000) **18**(7) 282.
149. Walcarius, A. "Analytical Applications of Silica-Modified Electrodes — A Comprehensive Review", *Electroanalysis* (1998) **10**(18) 1217.

60 Preparation and Uses of Silica Gels and Precipitated Silicas

Robert E. Patterson
The PQ Corporation

CONTENTS

An overview of the nomenclature, manufacture, and uses of synthetic silica gels and precipitated silicas that are of significant commercial importance is presented. Typical manufacturing processes are reviewed, and differences in the structure of silica gels and precipitated silicas that come about as a result of their methods of production are discussed. Applications covered include uses as a reinforcing agent, carrier, anticaking and free-flow agent, thickener, adsorbent, defoamer, catalyst support, abrasive-polishing agent, antiblock agent, and flatting agent.

Iler devoted one of the seven chapters of *The Chemistry of Silica* [1] entirely to the manufacture, characterization, and uses of "silica gels and powders." In the 160 pages of Chapter 5, Iler covered so many products, processes, and applications that it is difficult for anyone not already in the field to distinguish between what is of commercial importance and what is of theoretical interest.

The aim of this chapter is more limited; the emphasis is to provide an overview of the uses of synthetic silica gels and precipitated silicas that are of significant commercial importance. Excluded from this discussion are naturally occurring silicas, including products such as diatomaceous earth and so-called "amorphous silica" minerals (which are actually microcrystalline). Also excluded are fumed and arc silicas, forms of synthetic silica made at high temperature (in contrast to silica gels and precipitated silicas, which are generally made in aqueous solution and consequently have surface chemical properties quite different from high-temperature silicas) and products that are not properly classified as silicas (such as insoluble metal silicates). Chapter 14, by Ferch, covers fumed and arc silicas, as well as silica gels and precipitated silicas (although from a somewhat different perspective than here).

NOMENCLATURE AND MANUFACTURE

In *The Chemistry of Silica* [1], Iler used the term *silica powders* as a broad category encompassing silica gels, precipitated silicas, and fumed or pyrogenic silicas. These are all forms of synthetic amorphous silicon dioxide, a broad category that also includes another form — silica sols or colloidal silica — that is not a powder, but rather a dispersion of discrete silica particles in a liquid medium,

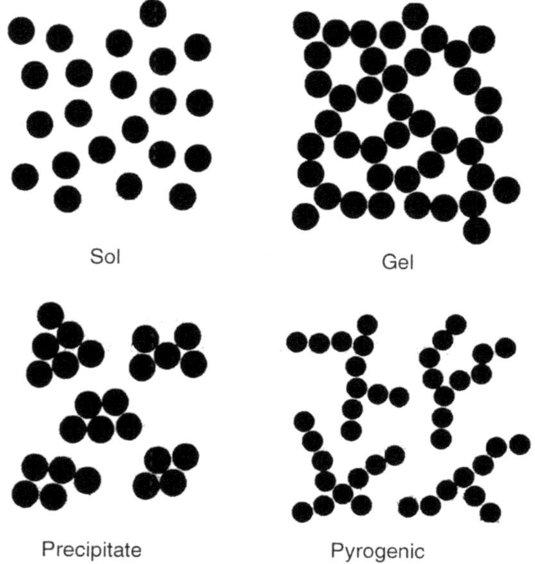

FIGURE 60.1 Modes of aggregation of primary silica particles in commercially important amorphous silicas.

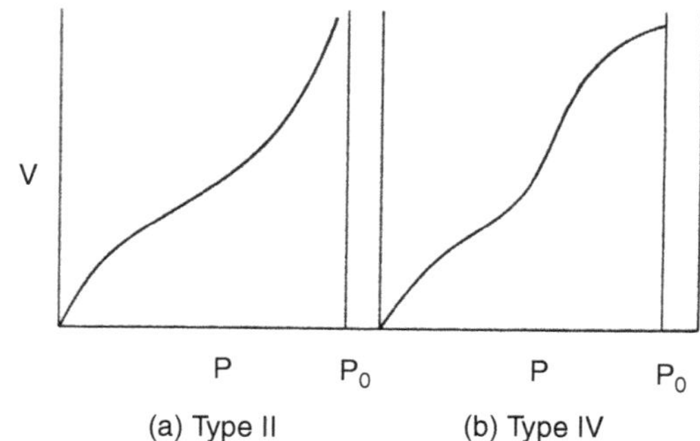

FIGURE 60.2 Typical Brunauer adsorption isotherms for (left) precipitated silica (Type II) and (right) silica gel (Type IV) (V, volume adsorbed; P, pressure; and P_o, saturation pressure).

typically water. Figure 60.1 illustrates the modes of aggregation of primary silica particles for each of these forms of synthetic amorphous silica.

In this chapter, only gels and precipitates are covered. Payne discusses the preparation and uses of colloidal silicas in Chapter 54, and Ferch discusses the preparation and uses of pyrogenic silicas in Chapter 14.

Because silica gels and precipitated silicas are both produced by wet processes, they possess physical and chemical properties that are similar enough to provide for substantial overlap in applications. In contrast, pyrogenic silicas, which are produced by thermal processes, possess unique properties that confer performance characteristics that cannot generally be matched by wet-process silicas in specific applications.

Silica Gels

Silica gels and precipitated silicas can usually be distinguished on the basis of pore structure. Silica gels give a Type IV nitrogen adsorption isotherm, whereas precipitated silicas give a Type II isotherm [2]. The general shapes of these isotherms are shown in Figure 60.2. Mercury intrusion-extrusion isotherms exhibit a hysteresis effect in gels because the stronger pore structure remains relatively intact, whereas in precipitates only the intrusion curve can be measured because the high pressure during intrusion breaks down the pore structure. However, in practice these terms refer to the method by which the powder was manufactured. While many published research papers describe silica gels made by relatively expensive routes (e.g., hydrolysis of tetraethoxysilane), in practice all silica gels and precipitated silicas of major commercial importance are derived from sodium silicate, with the reaction conditions adjusted to yield one type of silica or the other.

Figure 60.3 shows a typical manufacturing process for *acid-set* silica gels. Silica gels are formed by the acidification of sodium silicate solution under conditions that form a three-dimensional network of silica polymers that entirely enclose the liquid phase. The liquid that forms just after the mixing of the acid and silicate is referred to as a *hydrosol*; in time the hydrosol sets into a rigid gel. In the acid-set process shown here, gelation typically occurs over a period of minutes on a moving belt, although some manufacturers use tanks instead. The term *acid-set* refers to the fact that a stoichiometric excess of acid over silicate is used. The resultant large gel mass must be crushed before further processing.

An alternative process (not shown) involves much more rapid gelation, in which case *gel beads* are possible. In practice, beads are made under alkaline conditions, so the resultant gel is called *base-set*.

Silica hydrosols can be formed under a wide range of pH. At constant silica concentration and temperature, the rate of gelation is strongly dependent upon pH. In highly acidic solution, the silicic acid polymerizes to form nuclei that grow to 1–3 nm, which then aggregate into chains. As the pH of formation increases, these nuclei grow to a larger size, with less aggregation because of repulsion of the more negatively charged particles. Thus, the surface area of silica gels becomes lower as the pH of formation is increased.

After gelation is complete, the next step is typically to wash the raw gel to remove the soluble salts. The conditions of washing affect other gel properties as well. If the washed gel is dried only enough to remove excess surface water, the product is called a *hydrogel* (Iler used the much less common term "aquagel"). Remarkably, hydrogels may contain up to 70% water (by weight), yet still remain relatively free-flowing. If the washed gel is

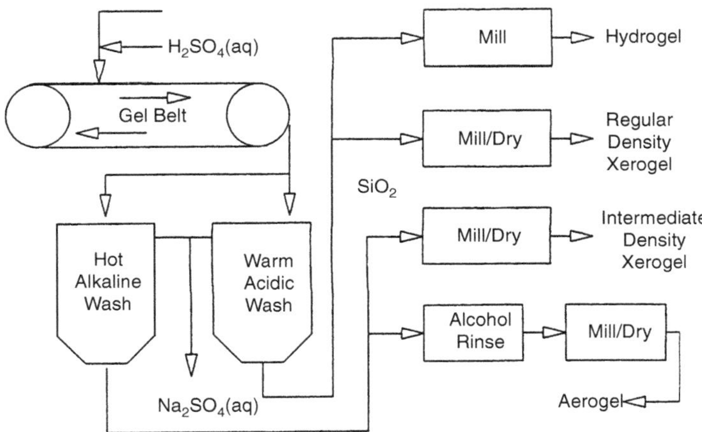

FIGURE 60.3 Typical manufacturing process for silica gels.

substantially dried, the product is called a *xerogel*. Milling may accompany drying to achieve a desired particle-size distribution.

Silica gels are further distinguished by the degree to which their pore structure is reinforced during processing. Acid-set gels washed under acidic conditions retain their high surface area, often over 700 m^2/g. Such products are referred to as *regular-density* gels. When a regular-density gel is dried, surface tension forces the pore structure to collapse to a high degree. This collapse can be reduced by washing the gel under hot alkaline conditions, a process known as *hydrothermal treatment*.

Iler [3] describes a solution redeposition mechanism of hydrothermal treatment in detail. Because the solubility of silica in water increases rapidly above pH 8, reinforcement can occur through the deposition of dissolved silica at the juncture of primary particles owing to the reduction of solubility in areas having a negative radius of curvature. The effect is to cement the primary particles together enough so that they resist collapse upon drying. A silica made in this way is referred to as an *intermediate density* gel because the density of the dried silica structure is intermediate between that of the regular-(high-)density xerogel and the low-density aerogel.

An alternative mechanism of hydrothermal treatment involves a physical reorientation of particles at constant total volume to form clumps that have much larger pores [4]. Electron micrographs of successive changes in morphology as a function of steeping time in hot water are offered in support of this interpretation.

The final category of silica gels of commercial importance is the *aerogel*, which was first reported by Kistler in the 1930s [5,6]. Aerogels, like xerogels, are dry gels, but they are made in such a way as to prevent pore collapse upon drying. One method is to replace the water in the hydrogel with a water-miscible liquid of much lower surface tension, such as alcohol or acetone. The liquid is then removed by heating the gel in an autoclave to above the critical point, then releasing the pressure. In this way a liquid-vapor interface never forms, so surface tension forces never have an opportunity to collapse the pore structure. This method of production, while effective, is sufficiently expensive to have greatly inhibited significant commercial development of silica aerogels.

Only recently, with the development of more economical processes [7,8], have practical applications come within reach. These processes avoid the supercritical drying step by reacting the silica gel in the wet state with hydrophobing agents such as trimethylchlorosilane and hexamethyldisiloxane, which lowers the capillary pressure during drying by increasing the contact angle rather than reducing the surface tension. The next step is to replace expensive silica sources such as tetraethoxysilane with sodium silicate.

Although aerogels can have exceptionally large pore volumes, shrinkage will occur if the aerogel is brought into contact with water and dried again. If stability is necessary, the intermediate density gel with its reinforced pore structure is the better choice. All of these different forms of silica gel are illustrated in Figure 60.4.

The final silica gel may be used as is, or it may be surface treated to impart specific chemical or physical characteristics required for certain applications. For example, the normally hydrophilic surface of the gel may be made hydrophobic, or a catalytically active metal compound may be coated onto the particles.

In summary, the processing parameters that have the greatest impact on gel properties, and consequently on suitability for various end uses, are the following:

- pH and method of gel formation
- pH and temperature of washing
- method of drying or milling
- surface treatment

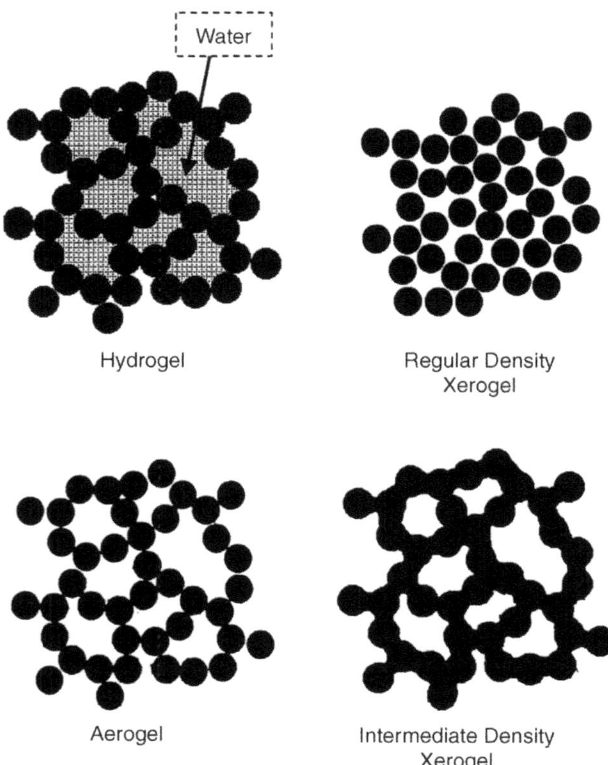

FIGURE 60.4 Subcategories of silica gels.

Precipitated Silicas

Figure 60.5 shows a typical manufacturing process for precipitated silicas. Precipitated silicas are formed by the acidification of sodium silicate solution under conditions that form primary particles that coagulate into clusters. In contrast to the conditions in a gel, the entire phase is not enclosed by the solid silica phase.

Precipitation is carried out under alkaline conditions, generally at lower concentrations than are employed in gel-making. Properties are varied by the choice of agitation, duration of precipitation, rate of addition of reactants, temperature, and concentration. Under typical conditions, the primary silica particles grow to sizes larger than 4–5 nm and are coagulated into aggregates by the sodium ion contributed by the sodium silicate raw material. In fact, this self-coagulation must be avoided if the objective is to produce a stable sol instead of a precipitate. The subject of silica sols is covered in detail in Chapter 54, by Payne; the practical means of achieving the objective is to use an ion-exchange resin rather than a mineral acid as a source of H^+, and at the same time Na^+ will be removed from the solution.

The precipitated silica slurry is next washed to remove soluble salts. The conditions of washing, although important, have less effect on the final product properties than for silica gels. Washing is typically conducted in filter presses.

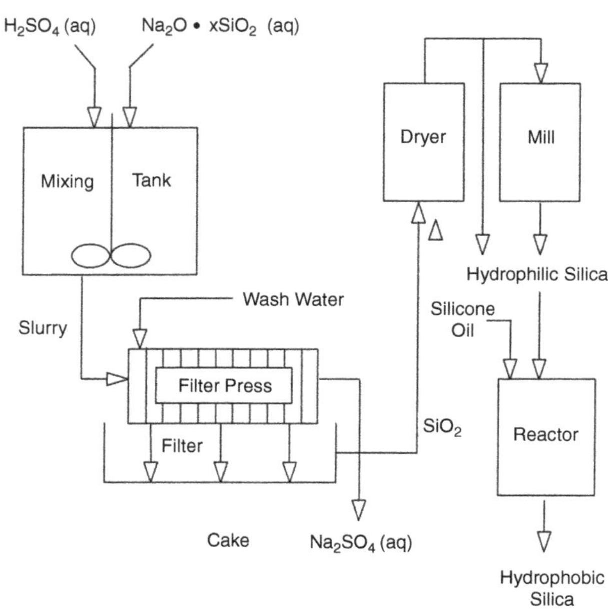

FIGURE 60.5 Typical manufacturing process for precipitated silicas.

For applications requiring exceptionally low levels of soluble salt impurities, the extent of washing is critical.

The resultant filter cake is dried by one of several methods. Most common are spray drying and rotary drying, which give rise to different particle shapes, degrees of agglomeration, and (to a lesser extent) porosity. The dried silica may be subjected to milling and classifying steps to achieve specific particle-size distributions. Figure 60.5 shows an optional step in which the silica is reacted with a chemical, in this case silicone oil (polydimethylsiloxane), to render it hydrophobic.

In summary, the processing parameters that have the greatest impact on precipitated silica properties, and consequently on suitability for various end uses, are the following:

- conditions during precipitation: time, temperature, concentration, agitation, and method of addition of reactants
- degree of washing
- method of drying or milling
- surface treatment

APPLICATIONS

Reinforcing Agent

The use of silica as a reinforcing agent is probably the largest single application. The chief end use is rubber reinforcement, primarily in the production of tires. This is a major market for precipitated silicas [9], but not for silica gels. A reinforcing filler may be defined as one that improves the modulus and failure properties (tensile strength, tear resistance, and abrasion resistance) of the final vulcanizate [10]. Modern so-called "green" tires utilize precipitated silicas to achieve lower rolling resistance, and thus lower fuel consumption.

Historically, carbon black has been the first and foremost reinforcing agent for natural and synthetic rubbers. The degree of reinforcement increases roughly with decreasing particle size [11]. Silica can be used to replace or complement carbon black; an advantage of silica is that it permits the production of durable rubber goods in colors other than black (an aspect important in the manufacture of shoe soles, for example). The degree of reinforcement by silica increases with the quantity of "bound rubber" (rubber that cannot be extracted by solvents), which forms by a free-radical mechanism during milling [12].

Another important reinforcement application is in silicone rubber elastomers. Historically, fumed silicas have played the major role here, but recently precipitated silicas have been developed that possess the characteristics required for this application [13]. Compared to conventional precipitated silicas, a product designed for this end use must have higher purity (to impart acceptable electrical properties, because silicone rubbers are often used as insulating materials) and lower water adsorption (to prevent bubbles from forming during extrusion and to impart resistance against moisture pickup). Good dispersibility is also important.

For example, in heat-cured rubber systems a common loading with synthetic silica is around 30%. For economic reasons, precipitated silicas are used in increasing volumes in silicone elastomers such as high-consistency rubber (HCR), liquid silicone rubber (LSR), and two-component room temperature vulcanized elastomers (RTV2). The reinforcement provided by precipitated silicas is only slightly lower than that provided by fumed silicas. However, fumed silicas are necessary in silicone sealants and other high-end applications because of their desirable properties like low moisture content.

Carrier

The ability to manufacture precipitated silicas and silica gels having large pore volumes gives rise to their use as carriers of numerous liquids; a silica can be made to absorb up to 3 times its weight of many liquids [14].

Because silicas can be manufactured to conform to food-grade regulations, they can be used to absorb essential oils, flavors, and feed supplements such as choline chloride and vitamin E oil. The absorption of pesticides is also a major application. Often the liquid is released when the powder comes into contact with water, owing to the higher affinity of the silica surface for water than for the absorbed liquid. Potential benefits include more convenient dispensing and improved stability when sensitive materials are handled.

In the last few years, new microgranulate carriers based on precipitated silicas have been successfully introduced to the major carrier applications. Based on advanced drying technology, these products offer the same good absorption characteristics but improved flow behavior and dust-free handling.

Silica gels can also be used to encapsulate molecules. For example, biological molecules such as enzymes have been embedded within silica gels, where they retain their bioactivity and remain accessible to external reagents by diffusion through the pore structure [15].

Anticaking and Free-Flow Agent

Many powdered products of commercial importance must be treated to prevent the individual particles from sticking together. Such products include fire extinguisher powders, whey solids, urea used in animal feeds, milk substitutes, cocoa, powdered coffee creamers, spices, citric acid, and many others. The mechanism of interparticle attachment varies with the powder in question, but can include moisture pickup, fusing in materials of low melting point, static charge buildup, or simple agglomeration as a result of surface forces. Both precipitated silicas and silica gels are used as anticaking and free-flow agents. The closely related precipitated silicates are also widely used. These products generally perform the desired function by coating and separating the powder particles and by adsorbing moisture within their pore structures.

Thickener

Silicas are widely employed as thickeners in liquid systems. Fumed silicas are often the cost-effective choice despite their higher price per unit weight. However, precipitated silicas and silica gels also find wide application in this end use. A noteworthy example is the use of wet-process silicas to thicken toothpaste.

The mechanism of thickening depends greatly on the polarity of the liquid and the degree of hydrophilic or hydrophobic character of the silica. Iler [16] provided an excellent summary of the principles involved. The best thickening occurs when silica particles are to interact with each other in such a way as to produce an open network throughout the liquid. This situation is favored when there are portions of the silica surface that have low forces of attraction to the liquid; these areas are free to bond to each other on neighboring particles. On the other hand, if the entire silica surface has low forces of attraction to the liquid, the particles clump together in dense aggregates, reducing the thickening effectiveness. At the other extreme, if the entire surface has strong forces of attraction to the liquid, so that the particles are completely disaggregated, the thickening effect is minimized.

Thus, the characteristics of a silica surface that affect the energy of particle–particle versus particle–liquid contacts, as well as the particle size and degree of dispersion, determine the potential of that silica to thicken. The thickening behavior of a silica can be altered by modification of its surface, such as by the partial attachment of hydrophobic groups.

A growing application is the use of silica gels and precipitated silicas in polyester resins. Fumed silicas still dominate where the highest thickening performance combined with the highest clarity is required. But the lower cost of the wet-process silicas makes them attractive in many cases. Precipitated silicas as well as silica gels cause polyester resins to become thixotropic, or shear thinning. This means that the resin has high viscosity when it is at rest and low viscosity when it is pumped or mixed. The high resting viscosity allows low-cost filler materials like quartz and limestone to be incorporated in the resin as a stable suspensions. When the resin is sheared it can be easily pumped through spray nozzles, injected into cavities, mixed with glass fibers, or hand applied. Unlike organic viscosity modifiers, silicas are stable to changes in temperature and are UV resistant.

Adsorbent

A familiar adsorbent application of silica is its use as a desiccant. The small packets of powder included with most electronic products usually contain a desiccant-grade silica gel. For this purpose, a regular-density acid-set xerogel is most effective, owing to its high surface area on which water can be adsorbed. The small average pore size of such a gel, a result of unimpeded shrinkage upon drying of the regular-density hydrogel, is not a disadvantage because water molecules are small enough to penetrate these pores.

An adsorbent application of major commercial importance is the adsorption of potentially haze-forming proteins from beer. This process is called *chillproofing* because the haze forms when the beer is cooled after a period of aging subsequent to bottling. Silica hydrogels and intermediate-density xerogels are employed, because both have pores of sufficient size (>8 nm average diameter) to admit the proteins in question. Silica hydrogels are preferred by some brewers because they are non-dusty and filter better; silica xerogels are preferred by other brewers because they are more effective on beers that are difficult to chillproof. Modified silica gels in which a metal such as magnesium is incorporated into the surface structure have been shown to provide enhanced protein adsorption [17].

Silica hydrogels are used in the refining of edible oils to adsorb phospholipids, trace metals, and soaps [18]. The adsorption capacity depends on the ease of hydration of the adsorbates, so best performance demands careful control of moisture content in the system [19]. Silica hydrogel in combination with alumina has been found to be useful for purifying used cooking oils in order to extend their life and enhance the quality of fried foods [20].

Silica gels are also used as selective adsorbents in column chromatography. The definitive reference is Unger [21]. An exceptionally narrow particle-size distribution is necessary to prevent excessive back-pressure and to produce sharp peaks; precipitated silicas are not generally suitable because their agglomerates break down much more easily than do the stronger particles found in gels. High purity is also important to prevent contamination of the product being separated. The pore-size distribution must be matched to the separation being attempted. In some cases, totally nonporous silicas are preferred precisely because of their lack of pore diffusion effects [22]. The surface of the silica is often reacted with a silane to create a *reverse-phase* packing, so-called because the elution sequence of solutes is the reverse of that for a normal-phase packing (i.e., an untreated silica). In fact, the presence of residual silanol groups on a reverse-phase packing is generally deleterious to the chromatographic separation [23]. The stability of the bonded phase toward hydrolysis is affected by the properties of the underlying silica, sometimes in ways that are poorly understood: silicas with seemingly similar physical and chemical properties can exhibit different stabilities of an attached alkyl phase [24].

A recent development in adsorbent applications is the use of silica gel for cat litter [25]. The advantage over convention clay litters is much longer life; manufacturers claim that one cat can use one litter box filled with silica gel for up to a month before free liquid or odors become a problem. 1–10 mm beads or granules are used to provide a satisfactory bed for the cat to walk on; because of their irregular shape, granules are said to "track" and scatter less than beads. The pore properties of the silica must be designed to provide high liquid adsorption capacity while avoiding particle breakdown upon wetting.

Defoamer

Silicas are the most widely used active particles in defoamer formulations. Precipitated silicas are used almost exclusively. To be effective, the silica must be reacted with an agent, typically polydimethylsiloxane, to render the surface hydrophobic. The mechanism of bubble breaking is the dewetting of the silica particle by the foam lamella, which creates a defect in the film that leads to its rupture. The criterion for dewetting is a three-phase contact angle of 90° or more (the three phases are the aqueous foam lamella, the silica particle, and the carrier oil in which the particle is dispersed). Patterson has identified the properties of silica that optimize performance in the largest U.S. end use, pulp and paper defoaming, for the two common methods of hydrophobing: dry-roast [26] and in-situ [27]. Other major defoaming applications include paint and coatings, textile dye baths, and (mainly in Europe) laundry detergents.

Catalyst Support

Silica gel is used as a catalyst support because of its resistance to high temperature and the availability of gels of controlled pore and particle size. A catalytically active coating is applied to the gel, usually in the form of a transition metal compound. Silica-supported catalysts are widely used in the production of high-density polyethylene (HDPE), which is used to make bottles, film, pipe, and wire/cable coatings.

The three types of catalysts employed commercially in the manufacture of polyethylene are chromium-silica, Ziegler-Natta, and metallocene compositions. Silica is an integral part of the chromium-silica catalysts, which date back to the 1950s [28], where it is both a support and a component essential for activity. While a support is not required for activity of the Ziegler-Natta and metallocene catalysts, silica supports nevertheless contribute to efficient polymerization in gas and slurry processes. The support provides for an even distribution of active sites and results in better control of the resultant polymer morphology. Granular, spray-dried, and microspherical silicas are all used, and in some systems high pore volume is necessary for increased activity and better polymer morphology [29].

Abrasive-Polishing Agent

Both silica gels and precipitates are found as abrasives (as well as thickeners) in toothpaste. Silicas are especially useful in the production of clear gel-type toothpastes, because it is possible to match the index of refraction of the liquid components of the formulation to that of the silica to provide the desired level of abrasivity while retaining the transparency of the toothpaste. This cannot be done with conventional abrasive agents, such as calcium carbonate. The concept of combining an abrasive silica with a thickening silica has become the basis for many standard formulations and now represents one of the high-volume silica applications. This combination allows the formulator to independently adjust abrasivity level, cleaning power, and viscosity.

Semiconductor silicon wafers, from which modern integrated circuits are made, must be polished to a high level of flatness. This was originally accomplished using alkaline slurries to precipitated silicas. The mechanism of polishing is generally believed to be chemical–mechanical; the high pH of the slurry leads to oxidation of the silicon surface, followed by mechanical removal of the oxidized layer under the action of the silica particles and the polishing pad. The process is now called chemical–mechanical planarization (CMP). In recent years the use of precipitated silicas in this application has largely been supplanted by the use of silica sols and particularly colloidal dispersions of fumed silicas, which reduce the chance for scratching from a poorly dispersed aggregate.

Antiblock Agent

Antiblocking is the prevention of the adhesion of two plastic films in contact with one another. This effect can be achieved by incorporating fine silica particles in the surface of the film. The relatively low refractive index of silica makes the particles less easy to see. Good control of particle size distribution is important.

Flatting Agent

Silica xerogels and precipitated silicas are used extensively as flatting agents in paint and coatings. The silica particles protrude from the surface, and the resulting increase in roughness reduces the gloss of the coating. Particle size and degree of dispersion are obviously of importance. Some commercially available silica flatting agents are coated with wax to improve the redispersibility of silica that settles in the can during storage and to improve the scratch resistance of the dry coating. More recently, silica hydrogels have also been used as flatting agents, in this case in waterborne coatings [30].

Battery Separators

Polyethylene sheets are widely used in automobile batteries to separate the individual electrochemical cells. The polyethylene is highly filled with porous silica particles to provide a path for the migration of conductive ions. This has become a major market for precipitated silicas because the filler loading can be as high as 70%.

Paper

Precipitated silicas (or silicates) are added to pulp used in the manufacture of paper. In high-quality papers the silica can act as a partial, lower-cost substitute for titanium dioxide pigment. In thinner newsprint, which often uses recycled fibers, silica prevents bleeding and strike-through of the printing ink.

A more recent application for silicas is in the specialty papers used for inkjet printing. Silica gels and precipitated silicas offer superior absorption of water-based inks, which allows for high resolution printing. Silica may be also found in high-quality direct thermal papers because of its superior insulation and absorption properties.

Insulation

Aerogels are particularly well suited for insulation applications because of their exceptionally low density, thermal stability, and high transparency. In fact, they can have a thermal conductivity only one-third that of polyurethane or polystyrene foam, and with recent process improvements that reduce the cost of manufacture by an order of magnitude their practical use in certain construction applications is now feasible [31]. The insulating properties can be enhanced through the addition of IR opacifiers [32]. The high transparency of aerogels makes them suitable as insulation in windows or translucent panels.

Other Uses

Numerous other uses of silica gels and precipitated silicas are found in the literature, but in this article the focus has been on applications of significant commercial importance. While is it difficult to predict what applications could become significant in the future, one recent development in particular appears promising.

Silica gel has been shown to reduce the flammability of polymers when incorporated as an additive to the plastic. The silica apparently works by enhancing char formation [33,34]. Silica has the advantage of being an environmentally safe fire retardant compared to halogenated compounds, which have been implicated in the formation of dioxins and furans during incineration. The cost of the added silica relative to other fire-retardant additives and degradation of the polymer's physical properties at high loadings are practical issues that must be resolved for this application to become significant.

ACKNOWLEDGMENTS

The author is grateful to Dr. Holger Glaum of Degussa Corporation for reviewing and contributing to the sections on precipitated silicas.

REFERENCES

1. Iler, R. K., *The Chemistry of Silica*; Wiley: New York, 1979; pp 462–621.
2. Adamson, A. W. *Physical Chemistry of Surfaces*; Wiley: New York, 1976; p 566.
3. Iler, *op. cit.*, pp. 529–531.
4. D. Barby, in G. D. Parfitt and K. S. W. Sing, eds., *Characterization of Powder Surfaces*, Academic Press, Inc., New York, 1976, pp. 353–425.
5. Kistler, S. S., *Nature*, **127**, 741 (1931).
6. Kistler, S. S., *J. Phys. Chem.*, **36**, 52–64 (1932).
7. Deshpande, R.; Smith, D.; and Brinker, C. Jeffrey; U. S. Patent 5,565,142 (1996).
8. Schwertfeger, F.; Frank, D.; and Schmidt, M.; *J. Non-Cryst. Solids*, **225**, 24–29 (1998).
9. *Sipernat Precipitated Silicas and Silicates*, technical bulletin from Degussa Corporation.
10. Boonstra, B. B., in *Rubber Technology*; Morton, M., Ed.; Van Nostrand Reinhold: New York, 1973; p 55.
11. Billmeyer, F. W., *Textbook of Polymer Science*; Wiley: New York, 1971; p 546.
12. Iler, *op. cit.*, pp. 582.
13. *Degussa Silicas for Silicone Rubber*; Technical Bulletin, Pigments No. 12; Degussa AG: Frankfurt, Germany, 1982.
14. *SYLOID Multifunctional Silicas for the Food Industry*; Davison Chemical Division, W. R. Grace: Baltimore, MD.
15. Livage, J.; Coradin, T.; and Roux, C.; *J. Phys.: Condens. Matter*, **13** (33), R673–R691 (2001).
16. Iler, *op. cit.*, pp. 588–590.
17. Berg, K.; Witt, R.; and Derolf, R.; U. S. Patent 5,149,553 (1992).
18. Welsh, W., and Parent, Y.; U. S. Patent 4,629,588 (1987).
19. Nock, A., *Proceedings of the 86th American Oil Chemists' Society Meeting*, San Antonio, Texas (1995).
20. Seybold, J., U. S. Patent 5,391,385 (1995).
21. Unger, K. K., *Porous Silica: Its Properties and Use as Support in Column Liquid Chromatography*; Elsevier: New York, 1979.
22. Hanson, M.; and Unger, K.; *LC-GC*, **15** (4), 364–368 (1997).
23. Cox, G.; *J. Chromatogr.*, **656** (1–2), 353–367 (1993).
24. Sagliano, N.; Hartwick, R.; Patterson, R.; Woods, B.; Bass, J.; and Miller, N.; *J. Chromatogr.*, **458**, 225–240 (1988).
25. Schlueter, D.; Schlueter, R.; and Moberg, R.; U.S. Patent 5,970,915 (1999).
26. Patterson, R. E., in *1988 Nonwoven Conference*; TAPPI Press: Atlanta, GA, 1988; pp 39–48.
27. Patterson, R. E., *Colloids Surf. A*, **74**, 115–126 (1993).
28. Hogan, J., and Banks, R.; U.S. Patent 2,825,721 (1958).
29. Pullukat, T.; Shinomoto, R.; and Gillings, C.; *Plastics, Rubber and Composites Processing and Applications*, **27**, 8–11 (1988).
30. Sestrick, M.; Plichta, B.; and Schneider, H.; *Pitture Vernici Eur.*, **73** (10), 22–25 (1997).
31. "Hoechst finds an economical route to produce aerogels," *Chemical Engineering*, July 1997, pp. 21–22.
32. Fricke, J.; *High Temp.—High Pressures*, **25** (4), 379–390 (1993).
33. Gilman, J.; Ritchie, S.; Kashiwagi, T.; and Lomakin, S.; *Fire Mater.*, **21** (1), 23–32 (1997).
34. Kashiwagi, T.; Gilman, J.; Butler, K.; Harris, R.; Shields, J.; and Asaho, A.; *Fire Mater.*, **24** (6), 277–289 (2000).

ADDITIONAL READING

Alexander, G., *Silica and Me: The Career of an Industrial Chemist;* American Chemical Society: Washington, DC, 1973.

Barby, D., in *Characterization of Powder Surfaces*; Parfitt, G. D.; Sing, K. S. W., Eds.; Academic Press: New York, 1976; pp. 353–425.

Fricke, J., "Aerogels," *Scientific American*, May 1988; pp. 92–97.

Gregg, S. J.; Sing, K. S. W., *Adsorption, Surface Area, and Porosity*; Academic Press: New York, 1967.

Patterson, R. E., *Silica*, in Kirk-Othmer Encyclopedia of Chemical Technology, Fourth Edition, Vol. 21, 977–1005; John Wiley & Sons, New York, 1997.

Soluble Silicates; Falcone, J. S., Jr., Ed.; ACS Symposium Series 194; American Chemical Society: Washington, DC, 1982.

Vail, J. G., *Soluble Silicates in Industry*; American Chemical Society Monograph Series 46; The Chemical Catalog Co.: New York, 1928.

Vail, J. G., *Soluble Silicates: Their Properties and Uses*; American Chemical Society Monograph Series 116; Reinhold: New York, 1952; Vols. 1 and 2.

61 Foundry Mold or Core Compositions and Method

Horacio E. Bergna
DuPont Experimental Station

CONTENTS

Compositions and method for producing foundry sand cores or molds of initial high strength but with essentially no strength after casting metals above 700°C which involves shaping and setting a composition containing foundry sand and a binder comprising sodium, potassium, or lithium silicate and sufficient amorphous silica so that the fraction of the total silica in the binder solution which is present as amorphous silica is from 2 to 75%, the amorphous silica having a particle size in the range from about 2 to 500 nm and the binder having a molar ratio of silica to alkali metal oxide ranging from 3.5:1 to 10.1.

BACKGROUND

In the metal casting industry molten metal is cast into molds containing sand cores made from foundry sand and binders. These sand cores are conventionally bonded with organic resins, which during curing and during casting of the metal, decompose and evolve byproducts which are odoriferous, offensive fumes that are not only skin irritants but in most cases toxic. The molds themselves are made from foundry sand bonded with oils, clays and/or organic resins. Thus, during their use, similar problems can occur.

A great percentage of the sand binders used by the foundry industry are made of phenol- and urea-formaldehyde resins, phenolic- and oil-isocyanate resins, and furan resins. Almost all these binders and their decomposition products such as ketones, aldehydes and ammonia are toxic. The principal effect on man is dermatitis, which occurs not so much from completely polymerized resins, but rather from the excess of free phenol, free formaldehyde, alcohol or hexamethylenetramine used as a catalyst. Formaldehyde has an irritating effect on the eyes, mucous membrane and skin. It has a pungent and suffocating odor and numerous cases of dermatitis have been reported among workers handling it. Phenol is a well-known poison and is not only a skin irritant but is a local anesthetic as well, so that burns may not be felt until serious damage has been done. Besides being capable of causing dermatitis it can do organic damage to the body. Furfuryl alcohol defats the skin and contact with it has to be avoided. Hexamethylenetetramine is a primary skin irritant which can cause dermatitis by direct action on the skin at the site of contact. Urea decomposes to carbon dioxide and ammonia, the latter of which is intolerable in toxic concentrations. In addition to the binders, some processes use flammable gases such as triethylamine as a curing agent. Capturing or destroying gases, smoke and objectionable odors are only temporary, stop-gap expensive solutions. New binders are needed that completely eliminate the sources of offensive odors and toxic gases.

Many of the organic binders are hot setting and therefore require heating to cure. Hot molds not only add hazards and complicate pollution control problems but add economical problems related to increased use of energy and increased equipment, maintenance and operation costs.

An alternative is to use inorganic cold setting binders, such as sodium silicate, which set at room temperature without producing objectionable gases or vapors. The use of silicates, however, results in the silicate bond remaining too strong after casting, so that the core is still coherent, and has to be removed by use of violent mechanical agitation or by dissolving the silicate bond with a strong, hot aqueous alkali. The problem may be lessened to a degree by using sodium silicate solutions admixed with organic materials such as sugar, but even in this case the core is still coherent after casting and requires extreme measures for removal such as violent mechanical agitation.

Thus, there is a need to create a binder for sand in making cores and molds for casting metals such as aluminum, bronze, or iron, that will have satisfactory high strength before the metal is cast, retain sufficient hot strength and dimensional stability during the hot metal pouring, but which will have strength after the metal has been cast and cooled, that the sand can be readily shaken out of the cavities formed by the cores; the binder also should be one that will not evolve unreasonable amounts of objectionable fumes when the sand cores and molds are subjected to molten metal.

SUMMARY

I have discovered that molds and sand cores of initial high strength but with essentially no strength after casting metals above 700°C can be made by bonding foundry sand with an aqueous solution of sodium, potassium, or lithium silicate or their mixtures and amorphous colloidal silica the amounts of silicate and amorphous colloidal silica being such that the overall molar ratio of SiO_2/alkali metal oxide (M_2O) is from 3.5:1 to 10:1, preferably 4:1 to 6:1, the fraction of the total silica present as amorphous colloidal silica is from 2 to 75% by weight, preferably 2 to 50%, and most preferably 10 to 50%, the amorphous colloidal silica having a particle size in the range from about 2 to 500 nm, and the 98 to 25% balance of the total silica being in the form of silicate ions. The amorphous colloidal silica in the binder comprises both the amorphous colloidal silica component of the mixture and the amorphous colloidal silica fraction inherently present in aqueous solution of alkali metal silicates of ratio more than about 2.5.

In alkali metal aqueous solutions containing more than 2.5 mols of SiO_2 per mole of M_2O, it is found by ultrafiltration, according to a procedure referred to herein as the Gore procedure, that at the concentrations used in this invention part of the silica in solution is ionic and part of it is colloidal, the colloidal fraction being retained by the ultrafilter while the ionic silicate passes through. In the case of sodium silicate, for example, concentrated commercial silicate solutions are available having a SiO_2/Na_2O ratio as high as 3.8/1.0 and these concentrated solutions therefore contain a substantial proportion of the silica present in the colloidal state. The colloidal fraction consists of a range of sizes less than 5 nm diameter and down to near 1 nm, with a substantial amount of 2 or 3 nm diameter. These units are so small that solubility equilibrium is rapidly established so that if the solution is diluted with water the units pass into solution forming lower molecular weight ionic species.

The higher the ratio of concentrated aqueous solutions of alkali metal silicates the higher the colloidal silica content, but for each ratio the colloidal silica content decreases with dilution of the solution.

To prepare a binder having SiO_2/M_2O ratio of 3.5 to 3.8 it is therefore not necessary to add any colloidal silica if an alkali metal solution is used already in the ratio range. On the other hand, if an alkali metal silica solution with ratio lower than 3.5 is used, it is necessary to add at least some colloidal silica in the form of a sol to prepare our binder.

Silica aquasols (water dispersions of colloidal amorphous silica) containing only a small amount of alkali as a stabilizer are commercially available and are described in the preferred aspects of this invention.

In summary, binder compositions of our invention comprise (1) aqueous solutions of alkali metal oxide silicates with or without amorphous silica present therein and (2) amorphous colloidal silica, if the silicate does not have any amorphous silica present therein or if the level of amorphous silica in the silicate is not sufficient.

The core and mold compositions of the invention have the additional advantage in that they can be made cold setting, that is, heating to set the binder system is not necessary. Thus, they can be set with CO_2 or a suitable acid releasing curing agent.

Preferred for use in the compositions of the invention are binder wherein the alkali metal silicate is sodium silicate and at least 10% of the amorphous silica is obtained from a silica sol.

In preferred embodiments of the composition of the invention carbonaceous materials and/or film forming resin adhesives are employed. These materials can add desirable properties with respect to shake-out and storage life. The employment of these optional, but preferred, materials is described in greater detail in the following paragraphs.

Thus, I have found sand core or mold compositions of foundry sand and binder wherein the composition consists essentially of 85 to 97 parts by weight of foundry sand and 3 to 15 parts by weight of an aqueous binder comprising an aqueous sodium, potassium or lithium silicate solution or mixtures thereof and amorphous silica, the amorphous silica in the silicate solution determined by the Gore test procedure, the binder characterized by (1) a molar ratio of silica to alkali metal oxide of from 3.5:1 to 10:1; (2) a weight fraction of the total silica present as amorphous silica is from 2 to 75%; and (3) a weight fraction of the total silica present as silicate ions is from 98 to 25% and the amorphous silica has a particle size of from 2 to 500 nm and the sand core of mold possesses a compressive strength sufficiently low to permit easy crushing after said core or mold is used in preparing a metal casting.

Accordingly, the present invention also includes a method for making a sand core or a sand mold useful in the casting of molten metal which comprises mixing 85 to 97 parts by weight of foundry sand with 3 to 15 parts by weight of a binder which comprises am aqueous sodium, potassium or lithium silicate solution or mixtures thereof with amorphous silica having a particle size of from 2 to 500 nm, the amount of silicate and amorphous silica being adjusted to form a binder with (1) a molar ratio of silica to alkali metal oxide ranging from 3.5:1 to 10:1, (2) the weight fraction of total silica present as amorphous silica of from 2 to 75%; and (3) a weight fraction of the total silica present as silicate ions of from 98 to 25%, the amorphous silica present in the silica solution is determined by the Gore test procedure, forming the sand and binder mixtures into the desired shape and setting the formed mixture.

DESCRIPTION

Foundry Sand

The compositions of the invention will contain between 85 and 97 parts by weight of foundry sand, preferably between 90 and 96 parts by weight. The amount of binder used is related to sand type and particle size in that with small sand particles and more angular surfaces, more binder mixture will be necessary.

The type of foundry sand used is not critical and the useful foundry sands include all of the ones conventionally used in the metal casting industry. Thus, these sands can be zircon sands (zirconium silicates), silica sands, for example, quartz, aluminum silicate, chromite, olivine, staurolite and their mixtures.

The particle size of the foundry sand again is not critical and American Foundrymen's Society (AFS) particle sizes of 25 to 275 GFN can be employed. GFN stands for grain fineness number and is approximately the number of meshes per inch of that sieve which would just pass the sample of its grains were of uniform size, that is, the average of the sizes of grains in the sample. It is approximately proportional to the surface area per unit weight of sand exclusive of clay.

The useful sands can be washed sands or they can be unwashed sands and contain small amount of impurities, that is, clay. If recycle sands are used, an adjustment may have to be made to the binder mixture to take into account any silicate present in such sands.

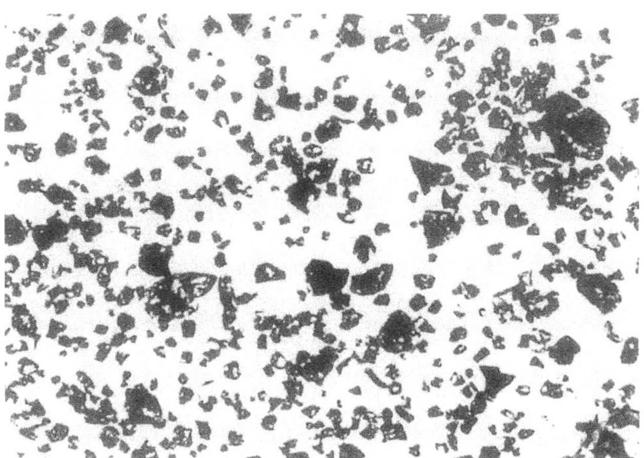

FIGURE 61.1 Chromite sand. × 15

Various minerals can be used as sand additives to optimize mold or core performance. For instance, alumina or clay products can be used to improve the high temperature strength and shake-out characteristics of the sand cores.

Conventional refractory grain alumina powders, kaolin, and Western bentonite can be used. Kaolin is preferred in amounts between 0.5 to 10% by weight of the sand. An example of a kaolin grade useful for this purpose is Freeport Kaolin Co.'s "Nusheen" unpulverized kaolin material which consists of kaolinite particles with a specific surface area of about 16 m^2/g.

Binder System

The compositions of the invention contain 3 to 15 parts, per 100 parts of sand binder mixture by weight, of a binder system comprising a water soluble alkali metal

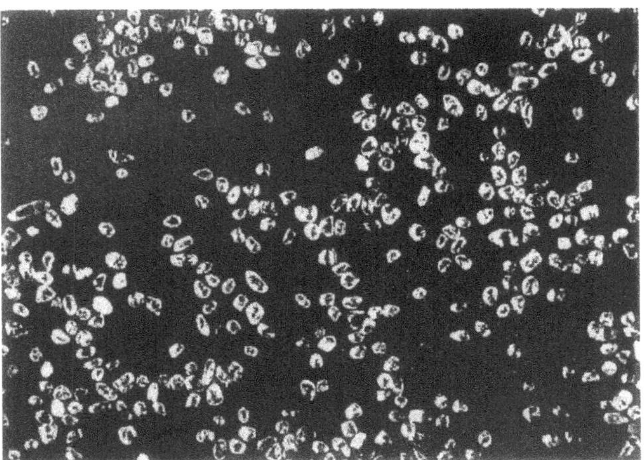

FIGURE 61.2 Zircon sand. × 15

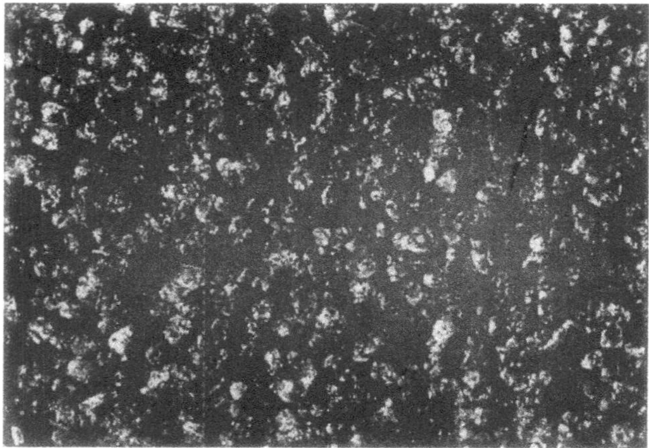

FIGURE 61.3 Olivine sand. × 15

silicate and amorphous colloidal silica. The key is to have very finely divided amorphous silica particles of colloidal size dispersed within the alkali metal silicate bond. It is inherent in the nature of water soluble alkali metal silicates having a molar ratio SiO_2/alkali metal oxide (M_2O) above about 2.5, that colloidal silica is present. In the case of silicates having a ratio higher than 3.5, the colloidal silica content is such that they may be employed without adding more colloidal silica, but in the case of alkali metal silicates of lower/silica/alkali metal oxide ratio there is little or no amorphous colloidal silica present so that amorphous colloidal silica must be added in order to produce the cores and molds of the present invention.

In order for the foundry core or mold to become weak after heating and cooling, it is helpful to have crystalline silica such as cristobalite formed throughout the binder mass by spontaneous nucleation at high temperatures. Such nucleation apparently occurs at the surface of particles of amorphous colloidal silica. Hence, the larger the area of such surface, the weaker the resulting core after heating and cooling. If enough amorphous silica is colloidally subdivided and dispersed within the silicate, then within one gram of such silicate binder there can exist dozens of square meters of amorphous silica surface. The smaller the particles, the more rapid the loss of core strength after heating at 700°C and cooling.

The useful water soluble silicate component of the mixture includes the commercially available sodium, potassium or lithium silicate or their mixtures. Sodium silicate is preferred. These silicates are usually used as solutions; however, their hydrates can be used provided that water is mixed into the binder, either prior to or during application to the sand. The useful sodium silicate aqueous solutions have a weight ratio of silica to sodium oxide ranging from 1.9:1 to 3.75:1 and a concentration of silica and sodium oxide of about 30 to 50% by weight. As stated above, a fraction of the silica in the useful water soluble sodium silicate of SiO_2/M_2O ratio higher than 2.5 is in the form of very small particle size amorphous colloidal silicate. Alkali metal silicates with SiO_2/alkali metal oxide ratio higher than about 3.5:1 are referred to as high ratio alkali silicates or alkali polysilicates although they contain in fact a certain proportion of colloidal silica. In essence high ratio alkali metal silicate aqueous solutions can be conceived as mixtures of alkali metal ions, silicate ions and colloidal silica. High ratio alkali metal silicate solutions contain varying

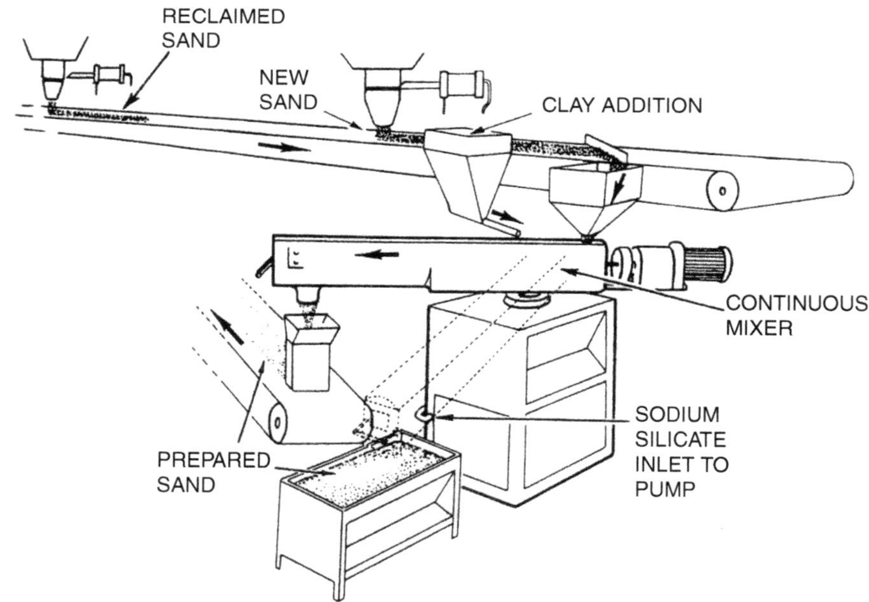

FIGURE 61.4 Sand preparation plant based on a continuous mixer.

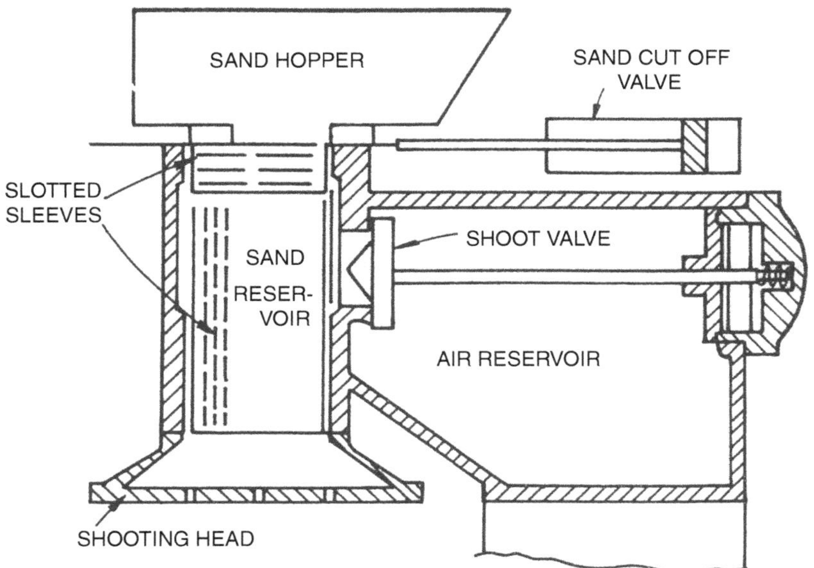

FIGURE 61.5 Section through typical core shooter.

amounts of monomeric silicate ions, polysilicate ions and colloidal silica micelles or particles. The type, size of the ions and micelles or particles, and distribution depend for each alkali metal on ratio and concentration. Aqueous solutions of moderate concentration of the metasilicate ratio, namely SiO_2/alkali metal oxide 1:1, or more contain mainly the monomeric silicate ions. In disilicate aqueous solutions of moderate concentration, with SiO_2/M_2O of 2/1 only the simple metasilicate and disilicate ions are present. Aqueous solution of silicates with greater ratios contain monomeric silicate ions, dimeric silicate ions, and polymeric silicate ions (trimers, tetramers, pentamers, etc.)

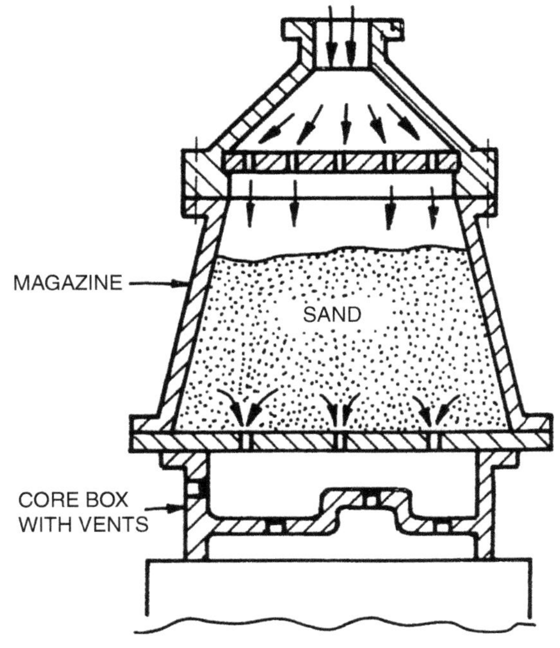

FIGURE 61.6 Section through typical core blower.

The degree of polymerization of the silica is silicate solutions may be expressed as the number of silicate groups formed in the average molecule of silicic or polysilicic acid corresponding to the alkali metal silicate. The degree of polymerization increases with the silicate. Whereas for example a sodium silicate solution of ratio 0.5:1 may have an average silica molecular weight of 60 corresponding to one molecule of SiO_2, sodium silicate solutions of ratio 1, 2, 3.5, and 4.0 are formed to have average molecular weights of about 70, 150, 325, and 400 respectively. This is the reason why as mentioned above high ratio silicates containing a large proportion of polymeric ions are also known as "polysilicates."

Silicate polymer ions with a corresponding silica molecular weight above about 600 are sufficiently large to be considered as very small silica particles and will hereinafter be referred to as colloidal silica or colloidal SiO_2. Colloidal particles are generally defined as particles with a particle size between about 1 nm and 500 to 1000 nm. This particle size range constitutes the colloidal range and is not limited by a sharply defined boundary.

Alkali metal silicates with an "average" silica molecular weight higher than around 200 to 300 have a fraction of their silicate ions present as polysilicate ions in the colloidal range. The higher the average molecular weight the higher the fraction of polysilicate ions in the colloidal range and the higher the molecular weight or particle

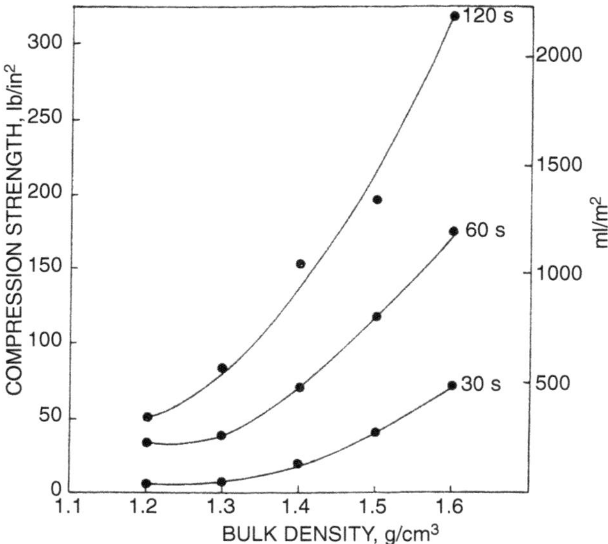

FIGURE 61.7 Influence of core density on as-gassed strength.

size of polymer ions or particles in the colloidal range. For example, a sodium silicate solution ratio 3.25:1 may contain more than 2 and 3 and as much as 15% by weight of the total silicate or silica in the form of colloidal silica. Sodium silicate solutions ratios 3.75:1 and 5:9 may contain more than 8 or 10 and 33% by weight of the total silica respectively in the form of polysilicate ions or colloidal silicate. Higher ratio sodium silicate solutions of various ratios eventually reach a state of equilibrium in which the colloidal silica fraction has a certain particle size distribution. In the case of sodium silicate aqueous solutions ratio 3.25 to 4 at equilibrium the colloidal silica fraction has a particle size smaller than 5 nm.

High ratio sodium silicate solutions may be prepared by simply adding dilute silica aquasols (colloidal dispersions of silica in water) to dilute low ratio sodium silicate solutions. In this case and until equilibrium is reached, average particle size of the colloidal silica fraction will be determined by time and silica particle size distribution of the original sol and the original silicate solution.

Increase in the ratio of alkali metal silicate solutions containing a constant concentration of silica causes an increase is viscosity even to the point of gelling or solidification. For this reason the maximum practical concentrations for alkali metal silicate solutions decrease with increasing ratio. Maximum practical concentration is the maximum concentration of SiO_2 plus Na_2O in solution at which the silicate solution flows like a fluid by gravity and is stable to gelation for long periods of time. The following table illustrates as an example the case of sodium silicate aqueous solutions.

Approximate SiO_2/Na_2O molar ratio	Approximate maximum practical concentration, % wt.
1.95	55
2.40	47
2.90	43
3.25	39
3.75	32
5.0	<20

Above a certain concentration which decreases with increasing silica-soda ratio as explained above, sodium silicate aqueous solutions become very viscous and are stable for only a limited period of time. Stability in this case means resistance to gelling. More stable solutions can be made at lower sodium silicate concentrations but this may become impractical in a foundry binder. The high water content of very high ratio (more than 4 to 5) sodium silicate solutions at practical viscosities prevent their extended use as a foundry binder in the present invention. Excessively high water content in a foundry binder means unacceptable weak sand molds or cores and detrimental quantities of steam evolving when the molten metal is poured into the sand mold-core assembly.

I have discovered ways of using high ratio alkali metal silicates as foundry sand binders without introducing excessive amounts of water into the sand and without employing unstable commodities.

A practical way of using high ratio silicate as binders for foundry sands is to mix concentrated silica aquasols and concentrated sodium silicate aqueous solutions in situ, that is on the surface of the sand grains, thus forming the high ratio silicate on the sand surface.

Concentrated sodium silicate aqueous solutions cannot be mixed with concentrated silica aquasols without almost immediate gelling. It would be very impractical or simply impossible to mix gels formed in this manner with sand using the means available today in common foundry practice.

However, I have discovered that effective mixing and binding effect is obtained with sand if the concentrated silica sol is mixed first with the sand to form a uniform and continuous film on the surface of the sand grains. The concentrated sodium silicate solution is then added to the sand mass in a second, separate step and the sodium silicate then mixed with the colloidal silica film on the surface of the sand, gelling in situ to form an intimately and uniformly mixed binder within the sand mass. The sand mix thus formed in the mixer can be molded by any of the various processes available in foundry practice and hardened to form strong molds or cores.

When sand molds or cores made with low ratio (less than about 3.5) silicate binders get dry either by exposure

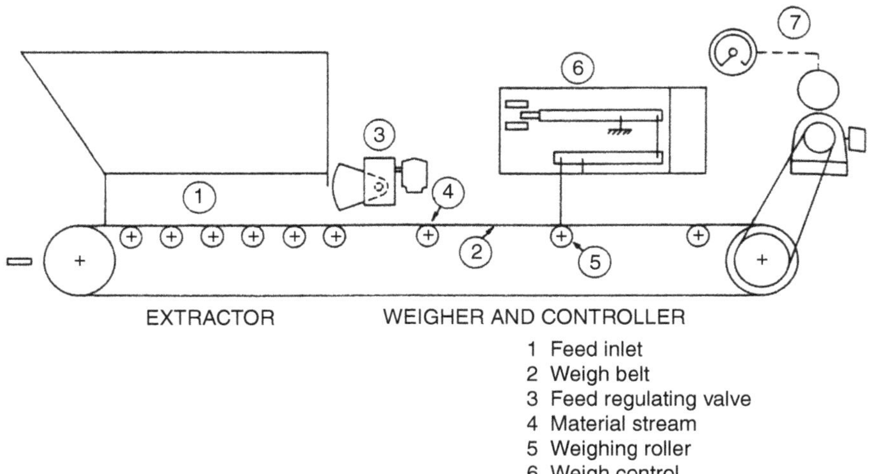

FIGURE 61.8 Operation of a constant weight feeder.

to a dry atmosphere or by heating, they become harder. On the other hand, when sand molds or cores made with very high ratio silicate as binders get dry either by exposure to a dry atmosphere or by heating they tend to become weak and friable. This is because the overall strength of the mold or core is primarily dependent on the mechanical properties of the solid film formed by the silicate adhesive when it sets. The separation of adhesive bonds is rarely the breaking away of the solid-liquid interface but more generally a rupture either within the adhesive film or within the body of the material to which the adhesive was applied. Cracks or other faults within the adhesive film are more likely to account for low bond strength than rupture at the interface.

The formation of crystalline silica within the mass of the binder contributes to weaken the bond between sand grains after heating and cooling the molds and/or cores, therefore, providing easier core shake-out and separation of the metal from the mold. Conventional sodium silicate binders form a glass on the surface of the sand grains when the molds or cores are heated to high temperatures. When the mold or core cools down to room temperature the glass becomes very rigid forming a very strong bond, therefore, hardening the mold or core. For this reason a core made with such a binder is very difficult to break up and remove from the cavity of a cast metal during the foundry operation known as shake-out.

When colloidal silica is embedded in a matrix of sodium silicate it tends to crystallize and form cristobalite at the temperatures the cores reach when metals are cast. Due to the difference in thermal expansion coefficient, the expansions and contractions of the cristobalite crystals embedded in the glass matrix tend to crack the binder film surrounding the sand grains therefore weakening the mold or core. This weakening effect has to be added to the already mentioned weakening effect due to the cracking of high ratio silicate films on dehydration. Due to these weakening mechanisms a core made with the high ratio silicates covered by this invention is very easy to break up and remove or separate from the cast metal during the shake-out operation.

Thus the difference in behavior between low and high ratio silicate binders for sand molds and cores can be understood by observing films formed on silica glass plates by slow evaporation of for example aqueous solutions of sodium silicate of various ratios.

The low silicate/soda ratio (2.0) sodium silicate solution dries in air at room temperature very slowly forming a very viscous, smooth, clear film. At higher ratio (2.4) drying is faster and the silicate film obtained shows some cracks. At very high ratios (3.25 and 4.0) sodium silicate solutions include substantial amounts of very small particle size colloidal silica and drying is even faster: cracking is even more extensive and the film tends to lose integrity. A silica sol of particle size 14 nm and SiO_2/Na_2O ratio 90 does not form a continuous film under the same drying conditions.

Low ratio silicate binders thus form on the sand surface viscous, smooth films which do not form cracks on drying. On the other hand, the films formed on the sand surface by high ratio silicate binders, crack on drying thus weakening the sand core or mold. For these reasons cores made with low ratio silicate binders outside the scope of the present invention become stronger when they are heated at high temperatures by molten metals in the pouring operation of the casting process. On the other hand, cores made with high ratio silicate binders within the present invention are reasonably strong when just made, but become weak and friable during the casting operation.

In the practice of this invention a compromise has to be made when choosing a binder composition by selecting one with a SiO_2/Na_2O ratio not so high that the sand molds or cores will weaken to unacceptable levels by merely drying at room temperature when exposed to the atmosphere, and not so low that the sand molds or cores will form a cohesive, solid glass bond when the core or mold is heated in the casting operation so that the core or mold becomes very strong when cooled down to room temperature and cannot be separated easily from the metal casting. The room temperature, as-made strength of sand molds or cores obtained with high ratio silicate binders of this invention may be upgraded by the addition to the silicate bonded sand mix of a fugitive film-forming resin adhesive in the form of a water solution or water dispersion. In this case, as explained below in more detail, the molds or cores become stronger by drying at room temperature. However, when heated to high temperatures during the casting process the resin adhesive decomposes evolving harmless vapors and the weakened core and mold can be easily separated from the cast metal.

If a preformed sodium polysilicate having a molar ratio of silica to alkali metal oxide in the range of 3.5 to 10 is employed before it gels, the same effects as with the amorphous silica sodium silicate system will be obtained. An aqueous sodium polysilicate containing 10 to 30% by weight silica and sodium oxide and having a silica to sodium oxide weight ratio of 4.2:1 to 6.0:1 can be produced as described in U.S. Pat. No. 3,492,137.

Similarly, the high ratio lithium silicates of Iler U.S. Pat. No. 2,668,149 or the potassium polysilicates of Woltersdorp, application Ser. No. 728,926, filed May 14, 1968, now Defensive Publication 728,926, dated Jan. 7, 1969, can be employed as the binder provided the requirements as to molar ratio, particle size and amount of amorphous silica are followed.

Furthermore, alkali metal polysilicates stabilized by quaternary ammonium compounds or guanidine and its salts can also be employed. Some stabilized polysilicates of this type are described in U.S. Pat. No. 3,625,722. This method, however, has the disadvantage of producing unpleasant odors on casting due to the thermal decomposition of the organic molecule.

Complexed metal ion stabilized alkali metal polysilicates can also be used, such as copper ethylenediamine hydroxide stabilized sodium polysilicate made by mixing copper ethylenediamine with colloidal silica and then the silicate, or the stabilized polysilicates of U.S. Pat. No. 3,715,224.

The useful amorphous silica are those having a particle size in the range from about 2 nanometers to 500 nanometers. In addition to the amorphous silica already present in aqueous solutions of high ratio alkali metal silicates, such silicas can be obtained from silica sol (colloidal dispersions of silica in liquids), colloidal silica powders, or submicron particles of silica. The silica sols and colloidal silica powders, particularly the sols, are preferred in view of the shake-out properties of the binders made from them.

Gore Procedure

The amount of colloidal silica present in an aqueous solution of high ratio alkali metal silicate can be determined for example by ultrafiltration. Ultrafiltration refers to the efficient selective retention of solutes by solvent flow through an anisotropic "skinned" membrane such as the Amicon "Diaflo" ultrafiltration membranes made by the Amicon Corporation of Lexington, Massachusetts. In ultrafiltration solutes, colloids or particles of dimensions larger than the specified membrane "cut-off" are quantitatively retained in solution, while solutes smaller than the uniform minute skin pores pass unhindered with solvent through the supportive membrane substructure.

Amicon "Diaflo" ultrafiltration membranes offer a selection of macrosolute retentions ranging from 500 to 300,000 molecular weight as calibrated with globular macrosolutes. These values correspond to pore sizes between about 1 and 15 nm. Each membrane is characterized by its nominal cut-off, that is, its ability to retain molecules larger than those of a given size.

For effective ultrafiltration, equipment must be optimized to promote the highest transmembrane flow and selectivity. A major problem which must be overcome is concentration polarization, the accumulation of a gradient of retained macrosolute above the membrane. The extent of polarization is determined by the macrosolute concentration and diffusivity, temperature effects on solution viscosity and system geometry. If left undisturbed, concentration polarization restricts solvent and solute transport through the membrane and can even alter membrane selectivity by forming a gel layer on the membrane surface — in effect, a secondary membrane — increasing rejection of normally permeating species.

An effective way of providing polarization control is the use of stirred cells. Magnetic stirring provides high ultrafiltration rates.

A recommended procedure is to use an Amicon ultrafilter Model 202, with a pressure cell of 100 ml capacity and a 62 mm diameter ultrafilter membrane operated at 25°C with magnetic stirring with air pressure at around 50 psi.

In the case of sodium silicate for example, an aqueous solution diluted with water, is placed in the cell. An Amicon PM-10 membrane, 1.8 nm diameter pores, is used. Pressure is applied and filtrate collected. In some cases, water is fed in to replace the volume passing through the filter into the filtrate. The solution in the filter cell is concentrated until the filtration rate is only a few ml per hour.

The filtrate is collected in progressive fractions, and they and the final concentrated solution from the cell are examined: Volumes are noted and SiO_2 and Na_2O concentrations in grams per ml are determined by chemical analysis.

In some cases, the concentrated solution on the filter is further washed by adding water under pressure, as fast as filtrate is removed. In these cases there is further depolymerization or dissolution of the colloid fraction.

The percentage of colloidal silica, based on total silica, is indicated by the amount of residual silica that does not pass through the filter. These represent maximum values for the amount of colloid present, since some ionic soluble silica is still present. In further examples the residual soluble silica is subtracted and the composition of the colloid is calculated.

It is not necessary to isolate the pure colloid, but only to measure the concentration of SiO_2 and Na_2O as ultrafiltration proceeds. Since the concentration of "soluble" sodium silicate in the filtrate is about the same as in the solution in the cell if this colloid is present only at low concentration, the amount and composition of colloid can be calculated by difference.

Allowance should be made in interpreting results obtained with this method for the fact that every time water is added to the system some depolymerization of colloid or polysilicate ions probably occurs.

The colloidal amorphous silicas useful in preparing the compositions of the invention have a specific surface area greater than 5 m^2/g and generally in the range of 50 to 800 m^2/g and preferably in the range of 50 to 250 m^2/g. The specific surface area is determined by nitrogen adsorption according to the BET method. The ultimate particle size of the silica used is in the colloidal range, and is generally in the range of 20 to 500 nm, preferably 12 to 60 nm. Thus, the silica sols of the desired particle size range described by M.F. Bechtold and O.E. Snyder in U.S. Pat. No. 2,574,902; J.M. Rule in U.S. Pat. No. 2,577,484; or G.B. Alexander in U.S. Pat. No. 2,750,345 can be used.

Positive silica sols and alumina modified silica sols wherein the ultimate silica particles have been modified and/or made electrically positive by partially or completely coating the particle surface with aluminum compounds can also be used in the present invention as a source of amorphous silica. Such sols are described for example by G.B. Alexander and G.H. Bolt in U.S. Pat. No. 3,007,878 and by G.B. Alexander and R.K. Iler in U.S. Pat. No. 2,892,797. The advantage of these sols is that in some cases they form more stable mixtures with sodium silicate aqueous solutions than the unmodified silica sols.

Certain very finely divided colloidal silica powders such as those made by the "fume process" by burning a mixture of silicon tetrachloride and methane, have a sufficiently discrete, particulate structure that such powders can be dispersed in water by colloid milling to give a sol useful in this invention. It is also obvious that such a powder can also be colloid milled directly into a solution of silicate.

Very finely divided colloidal silica powders can also be obtained by treating certain silicate minerals such as clay or calcium silicate with acid, followed by suitable heat treatment in an alkaline medium. Similarly, finely divided colloidal silicas can be produced by precipitating silica from a solution of sodium silicate with carbon dioxide. Such precipitated silicas are commonly used as reinforcing fillers, for elastomers because they are extremely finely divided, and the ultimate particles are easily broken apart. Finely divided aerogels of silicas may be employed, such as those described by Kistler in U.S. Pat. Nos. 2,093,454 and 2,249,767.

The finely divided colloidal silica powders useful in the composition of the invention are characterized by having specific surface areas as determined by nitrogen adsorption according to the BET method, of from 5 to 800 m^2/g and preferably 50 to 250 m^2/g, and being further characterized by the fact that the aggregates of ultimate silica particles are generally less than 10 microns in diameter.

The amounts and types of amorphous silica that can be dispersed within the soluble silicate depends to a considerable extent on the amount of grinding or mixing that is done to disintegrate and disperse particles of amorphous silica in the silicate bond. Thus, for example, it is possible to start with fused silica glass and grind it to the point where a substantial amount is present as particles smaller than a micron. The inclusion of a high concentration of this type of material can provide sufficient surface for nucleation of cristobalite or tridymite within the alkali metal silicate glass bond when the sand core or mold reaches high temperature during the metal casting operation. Also, finely divided natural forms of silica such as volcanic glasses which, in the presence of alkali silicates, can be devitrified, may be used, providing they are

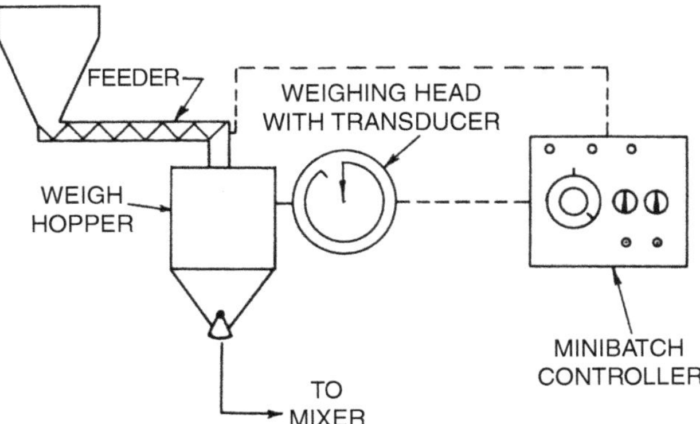

FIGURE 61.9 Schematic diagram of Minibatch automatic weighing system.

sufficiently finely divided and well dispersed in the sodium, potassium or lithium silicate solution used as the binder.

The compositions of the invention will have 2 to 75% of the total silica present in the binder present as amorphous silica, preferably 10 to 50% the balance of the total silica being in the form of silicate ions. As the specific surface area of the amorphous silica increases, lesser amounts of it will be required in the binder mixture.

There is a practical maximum concentration of amorphous silica that can be dispersed in the aqueous silicate solution. It is often desirable to incorporate as high a concentration of amorphous silica as possible, yet still have a workable fluid binder to apply to the sand. If the proportion of amorphous silica to soluble silicate is too low, then the shake-out will be adversely affected. On the other hand, if the ratio of amorphous silica to soluble silicate is too high, the mixture will be too viscous and must be thinned with water. Also, there will not be enough binder to fill the spaces between the amorphous silica particles in the bond, and it will be weak. In generaly, the higher the content of amorphous silica relative to sodium or potassium silicate, the weaker the initial bond as set by carbon dioxide. Conversely, the more silicate in the binder, the higher will be the initial and retained strengths.

The binder system should have a molar ratio of silica to alkali metal oxide which ranges from 3.5 to 10, preferably 3.5 to 7. This ratio is significant because the ratios of soluble potassium, lithium or sodium silicates commercially available as solutions lie within a relatively narrow range. Most of sodium silicates are within the range of SiO_2/Na_2O of about 2:1 to 3.75:1. Thus, overall ratios of binder compositions obtained by admixing colloidal silica, such as ratios 4:1, 5:1, 7:1 are mainly an indication of what proportions of colloidal silica and soluble silicates were mixed since the amount of amorphous silica in the soluble silicate at ratios of 2:1 to 3.75:1 are small.

However, in the ratio range of about 3.5:1 to 4.0:1, compositions of a specified ratio are not necessarily equivalent. Thus, a potassium silicate having an SiO_2/K_2O ratio of 3.9:1, in which there is a distribution of polysilicate ions, but relatively small amount of colloidal silica, differs considerably from a mixture made by mixing a potassium silicate solution of SiO_2/K_2O of 2.0:1 with colloidal silica having a particle size of, for example, 14 nanometers. In the latter case, the colloidal particles will remain as such in solution over a considerable period of time. Such a composition has two advantages over the more homogeneous one in that the low ratio of silicate has a higher binding power giving greater initial strength, while the higher content of colloidal particles results in a major reduction in the strength in the core after casting the metal.

OPTIONAL ADDITIVES

In the casting of some metals, for example, iron or steel, very high casting temperatures are involved, that is, 2500 to 2900°F. If the mass of the core is small relative to the mass of the cast metal during such high temperature casting, there may be some vitrification of the silicate thus creating shake-out problems. To alleviate this situation a carbonaceous material can be added to the core composition. These carbonaceous materials assist the binder of the invention in providing excellent shake-out, particularly after the core has been subjected to very high temperatures.

The useful carbonaceous materials should have the following characteristics:

(a) It should not interfere with the binder system.
(b) It should have a particle size or primary aggregate equivalent diameter sufficiently large to leave discontinuities in the glass formed by the binder at very high temperatures, as it burns off partially or completely. It should also have a particle size which is not large enough to weaken the sand core as fabricated, and specially not larger than the particle size of the sand itself. Thus the particle size or primary aggregate equivalent diameter should range between 0.1 μm and 75 μm, preferably between 5 μm and 50 μm. When the ultimate particle size of the carbonaceous material is smaller than 0.1 μm it is generally coalesced or it tends to coalesce in the sand mix into primary aggregates larger than 0.1 μm.
(c) It should not be too avid for water, otherwise it would subtract from the binder system, drying up the sand and making it impossible or difficult to mold.

Preferred for use are pitch, tar, coal-tar pitch, pitch compounds, asphaltenes, carbon black, and sea coal, and most preferred are pitch and carbon black.

Pitch is a byproduct from coke making and oil refining and is distilled off at around 350°F. It has a melting range from 285 to 315°F, is highly volatile, high in carbon and extremely low in ash. Following is a typical analysis of coal-tar pitch in weight percent:

Volatile	47.37%
Fixed carbon	52.43
Ash	0.2
Sulfur	0.5

Pitch is a material resistant to moisture absorption and is often used as a binder or as an additive for foundry sand cores and molds.

Sea coal is a common name used to describe any ground coal employed as an additive to foundry sands. Sea coal is used in foundry sands primarily to prevent wetting of the sand grains by the molten metal, thus preventing burnon and improving the surface finish of castings. It is also used as a stabilizer and to promote chilling of the metal.

Following is a typical analysis of sea coal given on a dry basis:

	Weight percent
Ash	5.10%
Sulfur	0.51
Volatile carbonaceous material	40.00
Fixed carbon	53.80

Ultimate analysis	Weight percent
Hydrogen	5.20%
Carbon	81.29
Nitrogen	1.50
Oxygen	6.40
Sulfur	0.51
Ash	5.00

Tar is generally defined as a thick, heavy, dark brown or black liquid obtained by the distillation of wood, coal, peat, petroleum and other organic materials. The chemical composition of a tar varies with the temperature at which it is recovered and raw material from which it is obtained.

Carbon blacks are a family of industrial carbons, essentially elemental carbon, produced either by partial combustion or thermal decomposition of liquid or gaseous hydrocarbons. They differ from commercial carbons such as cokes and charcoals by the fact that carbon blacks are particulate and are composed of spherical particles, quasigraphitic in structure and of colloidal dimensions. Many grades and types of carbon black are produced commercially ranging in ultimate particle size from less than 10 to 400 nm. In most grades ultimate particles are coalesced or fused into primary aggregates, which are the smallest dispersible unit of carbon black. The number of ultimate particles making up the primary aggregate gives rise to "structure" — the greater the number of particles per aggregate, the higher the structure of the carbon black.

When mixed with sand fine particle size carbon blacks are coalesced into aggregates in the sand mix, therefore they leave discontinuities in the binder phase when burned off during the high temperature casting operation.

An example of a commercial carbon black is Regal 660, sold by the Cabot Corporation of Boston, Mass., which has the following characteristics:

Nigrometer index	83
Nitrogen surface area	112 m^2/g
Oil (DEP) absorption	62 cc/100 g
Fixed carbon	99%

The carbonaceous material should be present in the core composition in the amount of 0.5 to 4 wt% based on the foundry sand, preferably 1 to 2 wt%.

The amount of carbonaceous material, for example, pitch, needed depends, to some degree, on the refractoriness of the binder used which is in turn a function of the silica/alkali molar ratio, and on the temperature to which the core will be subjected during casting. When a SiO_2/Na_2O ratio of 5:1 sodium polysilicate is used as a binder, no pitch is needed if the core is used for nonferrous metal castings since in these cases the core temperature will not exceed about 1200°C. If the same binder is used for small cores in massive iron castings, 2% of pitch is useful to help break up the silicate glass formed.

In the event it is desirable to make cores and store them for extended periods of time prior to use, I have discovered that the addition of a film-forming resin adhesive in the form of a water solution or water dispersion, drastically extends the storage life of foundry sand cores made with the binder of the invention. Thus the use of these materials enable the formed cores to retain sufficient strength and hardness during storage.

Useful film-forming resin adhesives include polyvinyl esters and ethers and their copolymers and interpolymers with ethylene and vinyl monomers, acrylic resins and their copolymers, polyvinyl alcohol, water dispersions of polyolefin resins, polystyrene copolymers such as polystyrene butadiene, polyamide resins, natural rubber dispersions, and natural and modified carbohydrartes (starch or carboxycellulose). Particularly preferred for use are aqueous dispersions of polyvinyl acetate and vinyl acetate-ethylene copolymers.

The polymer resin should be in a state of subdivision suitable for uniform distribution on the sand grains to form an adhesive film and hold the sand grains strongly together. It is preferred that resin dispersions be between 40 and 60% by weight solids. The higher the concentration of solids, the better, as less water will have to be removed, however, with concentrations above 60% by weight it can be difficult to mix the dispersion into the sand. With resin solutions, for example, solutions of polyvinyl alcohol, concentrations of 4 to 20% solids are preferred.

Useful polyvinyl acetate dispersions are milkwhite, high-solids dispersions of vinyl acetate homopolymer in water. Such dispersions have excellent mechanical and chemical stability. Typical properties of a preferred polyvinyl acetate dispersion are given in Table 61.1. Commercially available dispersions with similar

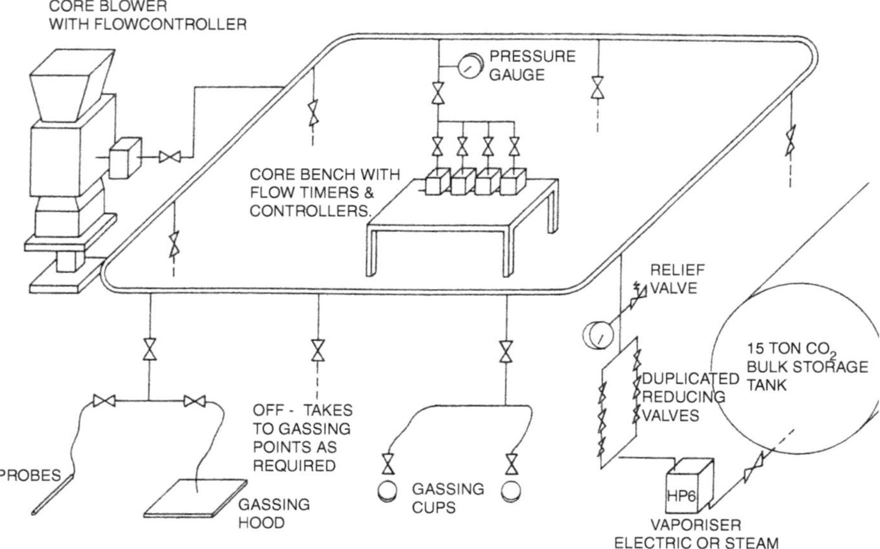

FIGURE 61.10 Diagram of low pressure CO_2 ring main.

characteristics are Monsanto's S-55L, Borden's Polyco 11755, Air Products' Vynac XX-210, and Seydel Wooley's Seycorez C-79.

The useful vinyl acetate-ethylene copolymers are milk-white dispersions of 55 w/o solids in water with a viscosity between 12 and 45 poises. DuPont's Elvace is a commercially available dispersion with these characteristics.

The useful polyvinyl alcohol (PVA) is a water soluble synthetic resin 85 to 99.8% hydrolyzed. DuPont's Elvanol resins and Goshenol GL-05, 85% hydrolyzed, low viscosity PVA are examples of suitable commercially available materials. Elvanol grades give 4% water solutions with a viscosity ranging from 3.5 to 65 Cp at 20°C as measured by the Hoeppler falling ball method. Water solutions of PVA at low concentrations (up to about 10–15 weight percent) or concentrated aqueous colloidal dispersions of the water insoluble polymer resins mix uniformly with sand and provide good adhesion. Very concentrated water solutions of PVA (higher than 20 wt%) are too viscous and do not mix well enough with sand.

To obtain optimum adhesion, the film forming resin dispersion or solution should be added such that it does not gel or coagulate either the silica or the sodium silicate before adding them to the sand. For instance, the polymer resin dispersions can be mixed with the silica before adding to the sand because both are compatible and do

FIGURE 61.11 Combined CO_2 flow rate meter and timer.

FIGURE 61.12 CO_2 mold and core assembly for gear wheel.

FIGURE 61.13 Fully machined 12 ton cast iron roll.

not gel when mixed together. The mixtures can be added to sand and they will form an adhesive film on the surface of the sand grains. After the silica and the polymer resin dispersion have been mixed with the sand, the sodium silicate solution can be added to the sand and although it will thicken in contact with the silica and the polymer resin dispersion, it will do so *in situ*, that is, fairly uniformly distributed on a preformed film of silica and polymer resin.

If before adding to the sand the sodium silicate is mixed with the concentrated polymer dispersion and the silica, it thickens and gels and it cannot subsequently be mixed adequately with the sand. Instead of distributing fairly uniformly on the surface of the sand grains, it would tend to form lumps and distribute unevenly in the sand.

Alcoholic solutions of the polymer resins may be used but are not recommended as additives to the silica-sodium silicate binder because they get very thick in contact with the binder and tend to gel faster than the aqueous dispersions and therefore do not distribute as uniformly on the sand grains. However, dilute alcoholic solutions of polymer resins can be used as such or mixed with commercial zircon core washes to coat the surface of the cores and give improved hardness and storage life to the cores. In this case the gel forms on the surface of the sand core already set, and it air dries fairly fast or it is dried

FIGURE 61.14 Cast iron electric motor stator case.

TABLE 61.1
Typical Properties of a Preferred Polyvinyl Acetate Homopolymer Aqueous Dispersion

Solid, %	55
Brookfield viscosity, P[a]	8.5–10
pH	4–6
Molecular weight (number average)	30,000–60,000 (mostly crosslinked)
Average particle size, microns	1–2 (range from 0.1 to 4)
Density (25°C), approx. lb./gal.	9.2
Surface tension (25°C), approx. dynes/cm.	55
Min. film formation temperature[b]	
°C	17
°F	63
Residual monomer as vinyl acetate, % max.	1.0
Particle charge	essentially nonionic

[a]Brookfield model LVF, No. 2 spindle at 6 rpm or No. 3
[b]ASTM D2354,

almost instantaneously by lighting the alcohol to extinction of the flame, therefore preventing the possible diffusion of the alcohol into the core.

The use of a water solution or water dispersion of a polymer resin produces sand cores with the silica-sodium silicate binder having as gassed mechanical

FIGURE 61.15 Resin-silicate bonded automobile body core, coated ready for use.

strength somewhat lower than that of sand cores made with silica-sodium silicate binder without the polymer resin solution or dispersion. This may be due to the weakening of the sodium silicate bond caused by the dilution produced by the water of the polymer resin solution or dispersion. However, drying of the core on storage, more than overcomes this effect and after very few days the cores show a much higher mechanical strength than the one obtained immediately after gassing with CO_2.

Two mechanisms may contribute to the hardening and strengthening on storage provided by the polymer resin. One is the thickening in situ of the adhesive film of silicapolymer resin-sodium silicate on the sand grains due to the "salting-out" effect caused by electrolyte formation on gassing with CO_2. More important is the thickening and solidification of the film caused initially by the CO_2 blown through the sand grains and specially the subsequent evaporation of the water from the sand core on storage.

Under these conditions the polymer resin macromolecules and/or colloidal particles are expected to coalesce and form an effective adhesive bond between the sand grains and reinforce the sodium polysilicate binder.

In the case of the polyvinyl esters the alkaline hydrolysis caused by the mixing with the sodium silicate will tend to form in the already formed uniform film, polyvinyl alcohol, perhaps an even better adhesive than the ester itself.

The colloidal silica-resin, for example, polyvinyl acetate components of the binder can be used in the form of a stable liquid mixture, the carbonaceous material being optionally present. Thus uniform mixtures containing colloidal silica and polyvinyl acetate within the relative amounts specified in this invention, such as 1.94 parts by weight of 40% aqueous colloidal silica and 2 parts by weight of 55% polyvinyl acetate aqueous dispersion, can be made by mixing the two components in a beaker. The mixture is stable and uniform and can be used within the working day. Overnight the mixture tends to separate in two layers and can be stirred up to make it uniform.

FIGURE 61.16 Cylinder block casting made with resin-silicate bonded core.

One method of providing a stable, pourable mixture of colloidal silica-polyvinyl acetate with or without the carbonaceous material, e.g., pitch, is to make the liquid phase slightly thixotropic but not viscous. In other words, to make it so that it sets to a weak gel structure at once when undisturbed (to maintain all particles in uniform suspension) but when stirred, or even tilted to pour, the yield point is so weak as to permit ready transfer of the material and easy blending with the sand.

Thixotropic suspensions with the characteristics described above can be prepared using a three component suspending agent system disclosed in U.S. Pat. No. 3,852,085, issued Dec. 3, 1974. This system consists of (a) carboxymethyl cellulose and (b) carboxyvinyl polymer in a total amount of about 36 to 65 wt% with the relative amount of (a) to (b) varying from a weight percent ratio of about 1:4 to 4:1 and (c) magnesium montmorillonite clay in a concentration of about 35 to 64 wt%.

The useful compositions will contain between 95 and $99\frac{1}{2}$% by weight of the binder components and between $\frac{1}{2}$ and 5% by weight of the suspending agent system. In a composition containing only the colloidal silica and resin, 15 to 35% of the binder will be silica solids and 15 to 35% of the binder will be resin solids. In a three component binder, 5 to 20% will be silica solids, 5 to 20% resin solids and 5 to 40% will be carbonaceous matter.

This suspension system can be used with dispersions containing a maximum solid content of 55% by weight of polymer resin and colloidal silica or polymer resin, colloidal silica and carbonaceous material such as pitch. The minimum solid content is only limited by the amount of water that is practical to add to the sand mix to obtain practical cores.

For example, to prepare a colloidal silica-polyvinyl acetate-pitch suspension 0.67 parts by weight of Benaqua (magnesium montmorillonite sold by the National Lead Co.) can be dispersed in 235 parts by weight of water with low shear mixing; 0.67 parts by weight of CMC-7H (carboxymethyl cellulose) and 0.67 parts by weight of Carbopol 941 (water soluble carboxyvinyl polymer) can be added and dissolved using low shear mixing; 0.15 parts by weight of a 1% solution of GE-60 (silicone-based emulsion) can be added as an antifoam agent; 194 parts by weight of Ludox HS-40 (aqueous colloidal silica dispersion sold by E.I. DuPont de Nemours & Co.) can be added and mixed with moderate shear mixing; 200 parts by weight of Gelva S-55L (polyvinyl acetate aqueous dispersion sold by the Monsanto Company) can be added and mixed with moderate shear mixing; then 200 parts by weight of "O" Pitch sold by the Ashland Chemical Company can be added and mixed with moderate shear mixing. A fluid suspension containing colloidal silicapolyvinyl acetate and pitch is obtained at a suitable ratio to be used as a component of the silicate binder system of the invention.

Alternatively, 58 parts by weight of water can be used instead of 235 parts by weight of water and in this case a uniform, stable suspension is obtained which is more viscous than the previously described, but still pourable and mixes well with sand.

Alternatively, pitch can be omitted from the preparation, and fluid suspensions containing colloidal silica-polyvinyl acetate are obtained at a suitable ratio to be used as components of the silicate binder system of the invention.

Application of the Binder

The binder mixture of the invention can be applied to the sand in various ways. Thus, if the binder mixture has sufficient shelf life, it can be formulated, stored, and applied to the sand when needed. The silicate and amorphous silica can be stored separately and then mixed together when needed and applied. Furthermore, they can be applied separately to the sand. If this latter procedure is used, it is preferred to first apply the amorphous silica, mix it into the sand, then apply the silicate and mix again. However, the silicate can be applied first.

Uniform sand mixes can be prepared by adding the binder to the sand in conventional foundry mixer, muller, or mix-mixers, or laboratory or kitchen mixers, and mixing for sufficient time to obtain a good admixture of the sand and binder, for example, for several minutes. When added separately, it is desirable to mix each component for less than two minutes to avoid undue drying.

If an alkali metal polysilicate solution is used as a binder, it should be mixed directly with the sand. If on the other hand colloidal silica and sodium silicate solution are added separately to the sand, it is preferable to add the silica sol first and to mix it thoroughly with the sand before adding the sodium silicate. Once the sodium silicate is added, the mix should not be kept too long in the mixer. A period of two minutes stirring is generally optimum for the sodium silicate.

Dry colloidal silicas such as pyrogenic amorphous silica do not mix well with the sand and in addition they tend to absorb water from the sand-binder system. Therefore, dry colloidal silica powders should be added to the sand in the form of a paste made with water or water should be added to the sand to help mix the dry silica powder. The amount of water made to use the peste should be enough to assure good mixing of the silica powder and yet not too much to affect the strength of the core or mold when it is hardened. Generally the amount of water needed in this case is no more than around 3% by weight of sand.

When the film forming resin or pitch are incorporated into the core composition, if the components are added separately to the sand, the resin should be added to the sand before the silicate. The resin can be added to the sand before or after the colloidal silica. The order in which the pitch is added is not critical with respect to either the silica or the silicate.

When materials such as clays or oxides are used as additives besides the binder, they should be mixed thoroughly with the sand in the sand mixer before adding the binder.

In some cases it is found convenient to use a release agent mixed with the sand to prevent the core or mold from sticking to the core box or pattern after setting. In these cases a conventional core or mold release such as kerosene or Mabco Release Agent "G" supplied by the M.A. Bell Company of St. Louis, Mo., should be added to the sand mix in the last 20 sec of the 2 min period of mixing the sodium silicate.

If the sand mix is not going to be used immediately, it should not be allowed to dry or react with atmospheric CO_2. The mix should therefore be stored in a tightly closed container or plastic bag from where the air has been squeezed out before sealing until it is ready to be used. If a slightly hard layer forms on the top surface of the sand due to air left inside the container, the hard layer should be discarded before using the sand to make cores or molds.

A practical way of checking uniformity of the sand mix and observe changes in the sand mix, such as reaction with the atmospheric CO_2, is to add a few grams of an indicator such as phenolphthalein at the beginning of the mixing operation. The phenolphthalein can be added in the form of a fine powder before adding the sodium silicate or dissolved in the sodium silicate or in the silica sol. Usually 160 mg of phenolphthalein per kilogram of sand is sufficient to develop a deep pink color in the sand mix.

Conventional foundry practice can be followed to form and set the sand core or mold. The sand can be compacted by being rammed, squeezed or pressed into the core box either by hand or automatically, or can be blown into the core box with air under pressure.

The formed sand mix can be hardened very fast at room temperature by gassing the sand with CO_2 for a few seconds. Optimum gassing time can be determined either by measuring the hardness or the strength of the core or by observing the change of color of the sand mix when an indicator such as phenolphthalein has been previously added to the sand.

Thermal hardening can be used for cores made with the binder compositions of the invention instead of CO_2 hardening. For instance, high strength cores can be obtained in a very short time by forming the sand mix in a hot box at temperatures between 100 and 300°C. In general, the higher the temperature the shorter the time required to achieve a certain strength level. On the other hand at a fixed temperature in general, the core strength increases with time of heating. However, thermal hardening is not a preferred setting process for the compositions of the invention because cores made in this way do not have as good shake-out characteristics as those made by CO_2 hardening.

Another fast hardening process that can be used is CO_2 gassing in a warm box (about 60 to 80°C) or gassing with heated CO_2.

When fast hardening is not required, cores with the binders of the invention can be set with other common curing agents used for the systems known in the art as silicate no-bakes. These curing agents are organic materials which are latent acids such as ethyl acetate, formamide, and acetins. Most of these agents contain glycerol mono-, di-, or tri-acetates or any other material which can release or decompose into an acid substance which in turn produces hardening of the alkali metal silicate. Furthermore, such a hardening process can produce cores having long shelf life without the need for a film-forming resin adhesive, i.e., polyvinyl acetate.

Conventional water based on alcohol based core washes can be used to treat the surface of the cores. This type of treatment is in some cases to improve the surface of the metal casting or the hardness and shelf life of the core. Shelf life is the period of time after making for which the sand core is useful.

Polyvinyl acetate homopolymers and copolymers can be used as core washes for sand cores as aqueous dispersions, in organic solvent solutions or mixed with zircon or graphite in aqueous or alcoholic suspensions. Polyvinyl alcohol or partially hydrolyzed polyvinyl alcohol can be used in aqueous solutions, organic solvent dispersions or mixed with zircon or graphite.

FIGURE 61.17 Vibrating shake-out unit.

Polyvinyl Alcohol or Hydrolyzed Polyvinyl Acetate

Five percent by weight to 20% by weight in water solutions or 5% by weight to 40% by weight in alcoholic solutions. More concentrated solutions are too thick to obtain uniform coating of the cores, more dilute solutions are too thin to provide satisfactory protective coating on the core surface.

Polymer Resin Aqueous Dispersions and Alcoholic Solutions

Five percent by weight to 40% by weight of polymer resin such as polyvinyl acetate homopolymer or copolymer in water solutions or 5% by weight to 25% by weight of polymer resin such as polyvinyl acetate homopolymer or copolymer in alcoholic solutions.

Polymer Resin-Zircon or Graphite Mixtures

In water based core washes: 15 to 25% by weight of polymer resin such as polyvinyl acetate homopolymer or copolymer and 30 to 50% by weight of zircon (25 to 50% by weight of water).

In alcohol based core washes: 5 to 10% by weight of polymer resin such as polyvinyl acetate homopolymer or copolymer and 30 to 50% by weight of zircon or graphite (40 to 60% alcohol).

The alcohols useful in the above core washes include methanol and ethanol.

Satisfactory polymer resin-zircon core washes are made for example by slurrying 1 part by weight of a commercial zircon core wash (as shipped by the supplier in the form of a wet powder) in 1 part by weight of 55% polyvinyl acetate aqueous dispersion if the core wash is intended to be used shortly after preparation. More dilute slurries are preferred for core wash compositions intended to be stored for some time before using. In this case the 1 part by weight of the zircon wet powder should be slurried in 1 part by weight of water before mixing with 1 part by weight of 55% polyvinyl acetate aqueous dispersion.

Aqueous polyvinyl acetate or zircon-polyvinyl acetate or graphite-polyvinyl acetate core washes are applied on the core surface by common foundry practices such as dipping, spraying, brushing, and so on, and allowing the core to air dry before using.

Sand cores coated with alcohol base polyvinyl acetate or zircon-polyvinyl acetate are lighted immediately after one wash application as in common foundry practice with alcohol base zircon core washes.

Concentration of polyvinyl alcohol aqueous solutions to give satisfactory core washes with adequate viscosity depends on molecular weight of the polymer. Polyvinyl alcohol solutions can also be used as a mixture with zircon or graphite core wash.

FIGURE 61.18 "Tyne God," Newcastle upon Tyne Civic Centre. This sculpture, by David Wynne, was made in 1968 and produced entirely by the CO_2-silicate process.

Casting Metals

Sand molds and cores made with the binder compositions of the invention can be used to cast most metals, such as gray, ductile and malleable iron, steel, aluminum, copper-based alloys such as brass or bronze. Steel is usually cast at around 2900°F, iron at about 2650°F, brass and bronze at around 2100°F and aluminum at about 1300°F.

With the molds or cores of the invention it is desirable that the core have an initial strength such that it can be handled without undue care and that it will stand up during the casting of the molten metal, i.e., will not wash away or distort. In standard American Foundrymen's Society lab tests this means that the core should have a compressive strength of at least 100 psi and preferably over 150 psi.

It is desirable that the hardness of freshly made cores exceed 5, preferably 10. The greater the hardness, the better, particularly at the time of metal pouring when it should exceed 10 and preferably 20.

Scratch hardness of cured cores can be measured with commercial hardness tester No. 674 available from Harry W. Dietert Co., 9330 Roselawn Avenue, Detroit, Michigan. This is a practical, pocket-sized instrument for measuring the surface and sub-surface hardness of baked cores and dry sand molds.

ACKNOWLEDGMENTS

We thank The Castings Development Centre of the United Kingdom for allowing us to use Figures 61.1 to 61.18 from the book *The CO_2-Silicate Process in Foundries* by K.E.L. Nicholas, BSc., ARIC, AIM (Sheffield, UK: British Cast Iron Research Association, 1972).

Foundry figures courtesy of Martin Fallon, British Cast Iron Research association taken from *The CO_2-Silicate Process in Foundries* by K.E.L. Nicholas, BSc., ARIC, AIM (Sheffield, UK: British Cast Iron Research Association, 1972).

62 Silica Supported Catalysts and Method of Preparation

Horacio E. Bergna
Dupont Experimental Station

CONTENTS

This application is a continuation-in-part of our co-pending application S.N. 337,534, filed January 14, 1964 and now abandoned, which is in turn a continuation-in-part of application S.N. 68,338 filed November 10, 1960, also now abandoned.

This invention relates to catalysts. More particularly it is directed to amorphous silica powders modified with a metal-containing constituent, to processes for making such powders by drying silica sols containing said modifiers, to a process for making metal-modified amorphous silica catalysts by molding the metal-modified amorphous silica powders, and to the uniformly porous and uniformly metal-modified catalysts so produced.

More particularly the invention is directed to powders comprising amorphous silica in the form of powdered aggregates of ultimate particles, the aggregates having a coalescence factor of less than 30%, being substantially free of nonvolatile constituents other than silica, and having a bulk density greater than 0.2 g/cm^3, and the ultimate particles being spheroidal and substantially uniform in size as observed by electron microscopy, being non-porous and having an average diameter of from 5 to 200 nm, and a metal-containing constituent of the group consisting of metals, metal oxides and compounds which decompose to metal oxides on heating to 1000°C in air, the metal in said constituent being present in the amount of from .01 to 5% by volume, based on the silica, the metal-containing constituent being dispersed substantially homogeneously in the powder composition.

The invention is further particularly directed to processes for making the above-described powders, the processes comprising the steps of making a silica sol which is substantially free of nonvolatile constituents other than silica and in which the silica is amorphous and in the form of non-porous, substantially uniform-sized, discrete ultimate particles having an average size in the range of 5 to 200 nm, dispersing the metal-containing constituent in said sol, and drying the sol before any substantial silica coalescence occurs therein and at a rate which is substantially instantaneous after the silica concentration in the continuous liquid–sol phase approaches about 50% by weight.

The invention is still further particularly directed to processes for making metal-modified amorphous silica catalysts, said processes comprising preparing a metal-modified silica powder as above described, compacting the powder under pressure to a uniform density of from 1 to 2 g SiO_2 per cm^3 to form a compact containing between the ultimate particles, voids which are of substantially uniform size, and heating the compact at a temperature of up to 1000°C until volatile and pyrolyzable constituents are removed and a uniformly porous, strong amorphous silica compact is obtained.

The processes and compositions of this invention are concerned with amorphous silica, in contradistinction to crystalline silica, hereinafter sometimes referred to as cristobalite. Such terms as fused silica, vitreous silica, and vitrified silica have often been used interchangeably to describe amorphous silica bodies. The presence of crystallinity in a silica body is readily ascertainable by such standard techniques as studying an x-ray diffraction pattern of the silica body.

By this invention metal-modified silica-based catalysts are produced in such a way that the metal-modified, spheroidal, silica particles of uniform size become closely packed, thus having a porosity in the interstices which is of a uniform size and distribution throughout the catalyst body. A further advantage of the catalysts of this invention is that the pores, being between particles which are already closely packed, resist further collapse when the catalysts are subjected to the elevated temperatures required in catalytic reactions.

According to the present invention the surface of the silica is uniformly modified by selected metals. The metal is evenly distributed over the surface of the ultimate silica particles. The manner of modification, however, is highly important in that the metal-containing modifier is introduced into a dispersion of amorphous silica in a manner such that it becomes uniformly and intimately dispersed with respect to said silica, the dispersion converted to a powder without separation of said components, and the powder compressed to form the metal-modified, amorphous silica catalyst.

The metal-modified, amorphous silica powders are novel starting materials for making catalysts. In them, the amorphous silica is present in the form of nonporous, ultimate particles in the size range of 5 to 200 nm these ultimates being jointed together into aggregates which by reason of their methods of preparation are lightly coalesced, and the metal modifier is present in the proportion from 0.01 to 5% by volume, based on the silica, and is uniformly dispersed. The powders lend themselves to compaction to such a degree that catalysts made therefrom are particularly unique because of the porous form they obtain as hereafter described.

The small size and absence of substantial coalescence between ultimate particles are both important requisites for the metal-modified amorphous silica to have for making the catalysts of this invention. To successfully convert metal-modified silica powders to uniformly porous amorphous silica catalysts, the ultimates must be compacted in such a manner as to avoid a substantial number of large pores. If ultimate particles less than 200 nm in diameter are loosely joined, that is, having coalescence factors less than 30%, as with the products of this invention, they can be compressed into catalysts having uniform distribution of pores which are of the same general size as the ultimates themselves. If, on the other hand, coalescence is high, bonds between the ultimates are difficult to break and compacts prepared there-from contain large, non-uniform voids which are impossible to remove unless temperatures in excess of crystallization point are employed.

It is also important that the powders do not contain excessive amounts of non-volatile impurities which can cause crystallization at catalytic reaction temperatures.

The effect which the metals and compounds of metals have on the crystallization rate of the silica is not sufficiently pronounced to preclude their use under the conditions of the present invention. Desirable results were achieved by having the metal modifier present in a highly dispersed state so that upon densification of the powder the modifier is not concentrated into large crystals which would be objectionable. Furthermore, by reason of the highly dispersed condition of the metal modifier, the proportion of it which must be used is confined to a minimum, which is economically advantageous.

MAKING THE METAL-MODIFIED AMORPHOUS SILICA POWDERS

To make a metal-modified, amorphous silica powder according to a process of the present invention, one must introduce the metal modifier in a manner which insures that it is highly dispersed. To do this, one first prepares a suitable silica sol and then dries the sol under certain critical conditions with the metal-modifying agent dispersed therein to obtain catalysts having the desired properties. The sols which are suitable as starting materials and the conditions which are critical in the drying step will now be described in greater detail.

The starting sol must first of all be a dispersion of amorphous silica ultimate particles. Since the silica in the ultimate product is to be amorphous it is highly important that the silica particles in the starting sol be free of crystallinity. Methods for determing the presence of crystallinity by x-ray techniques are discussed in R.B. Sosman's *The Properties of Silica*, Chemical Catalog Company, Inc., New York, 1927, page 207.

The ultimate silica particles in the sol must be discrete and nonporous. The fact of whether or not they are discrete can be readily observed by examination of an electron micrograph. The fact of whether or not they are nonporous can be determined by comparing the surface area of the particles as calculated from the size determined in an electron micrograph with the surface area as determined from nitrogen adsorption measurements. If the latter is not more than 20% greater than the former, the particles can be considered to be dense, whereas a variation substantially greater than this would indicate that the ultimate particles contain pores or surface irregularities too small to be observed on the electron micrographs.

The ultimate silica particles in the sol must be discrete, rather than aggregated. The greater the aggregation of the particles, the greater the chance for development of macroscopic holes in the compacted catalysts, thus leading to heterogeneous porosity. A method for measuring the degree of aggregation of the ultimate particles in the sol is described in U.S. Patent 2,750,345, issued June 12, 1956, to Guy B. Alexander, at column 7, lines 11 to 55. This test provides a method for determining the percent solids in the dispersed phase of a silica sol (percent S). The greater the aggregation, the lower the percent S. The sols used in this invention should have a percent S greater than 55.

Also as observed by electron microscopy the ultimate silica particles in the sol should be uniformly-sized and spheroidal. By "uniformly-sized" is meant that 75% of the total number of particles have a diameter in the range from 0.5D to 2D, where D is the number average particle diameter. The uniform size of the particles is important in obtaining uniform voids in the formed catalysts. The uniformity of the particles can be determined by methods described in the *Journal of Physical Chemistry*, 57 (1953), page 932.

The size of the ultimate silica particles in the sol can also be determined by methods described in the Journal article just mentioned. The average particle size should be in the range of 5 to 200 nm, preferably between 7 and 200 nm and more preferably between 13 and 110 nm.

Sols containing silica particles smaller than 5 nm, such as those obtained by neutralizing sodium silicate with sulfuric acid, for instance, are not satisfactory because the particles tend to aggregate very rapidly and irreversibly. When such sols are freed of stabilizing ions they show rapid increase in viscosity and tend to gel in extremely short times. Accordingly, sols containing ultimate silica particles larger than 5 nm in average diameter are used.

The continuous liquid phase of the silica sol is preferably water — that is, the sols are aquasols. Although organosols can be used, ordinarily any advantages to be gained from the use of an organosol are more than offset by the added cost.

The art is already familiar with methods for making silica aquasols of discrete, dense amorphous silica particles in the size range above specified, and any of these methods may be used to make the starting material employed in processes of the present invention. A particularly preferred type of silica aquasol is described in United States Patent 2,574,902, issued November 13, 1951, to Bechtold and Snyder. This patent describes how the size of the ultimate silica particles is increased into the desired size range by adding quantities of a low molecular weight silica feed sol to a heel sol containing particles which have been grown substantially. To minimize the amount of alkali contained in the silica particles it is preferred to use ammonium hydroxide as the stabilizing alkali used to adjust the ratio as described in the patent.

Regardless of its method of preparation, a silica sol to be useful as a starting material in the processes of this invention should be free of alkali and alkaline earth metals. These, if present, can be removed by deionizing the silica with a mixture of anion and cation exchange resins in accordance with processes described in United States Patent 2,577,485, issued December 4, 1951, to J.M. Rule. The sol is preferred temporarily stabilized against gelling or aggregation of the ultimate spheroidal particles by adjusting the pH to between 8 and 9 with a volatile base such as ammonia.

Alternatively, the sol can be purified by such methods as dialysis and other similar methods known to those skilled in the art.

Generally speaking, the total content of potassium, sodium and calcium oxides, for example, in the powder should not exceed about 0.1%. The exact level depends to some extent upon whether these impurities are on the surface of the particles or are locked within the silica structure. Thus, a few hundredths of a percent of combined sodium within the particles is not as harmful as this amount of combined sodium on the surface of the particle.

The metal-modifying agents used with the amorphous silica in the composition and processes of this invention can be metals, metal oxides, or compounds of metals which decompose to metal oxides upon heating in air to 1000°C. Thus it is seen that the modifying agents can be any of the numerous metals and metal compounds which are used for catalytic purposes. These can include such metals as copper, cadmium, manganese, calcium, magnesium, zinc, chromium, aluminum, molybdenum, tungsten, vanadium, cerium, mercury, barium, tin, lead, bismuth, thallium, iron, cobalt, and nickel; as their salts such as nitrate, formate, acetate, chloride, sulfate and the like. These metals can also be used as their oxides. Also they can include precious metals such as platinum, palladium, rhodium, ruthenium, iridium and osmium in their elemental forms or as their salts.

The proportion of metal added should be from about 0.01 to 5% by volume, based on the silica. It will be understood that this proportion is the total amount of modifying metal present but that more than a single metal can be used to attain this proportion. In such instances there can be found in the final supported catalysts of this invention, mixed oxides of the metals, compounds of two or more of the metals present with each other such as chromites, maganites, molybdates, tungstates, vanadates, cerates, stannates and ferrites.

One of the advantages of starting with a silica sol is that when such colloidal solutions are dried, the surface of the silica particles becomes coated with the metal or metal oxide or their soluble precursors, in molecular dimensions, and little, if any, excess metal modifier will be present in locations other than on said particle surfaces. Thus, the metal modifier is located in such a manner as to exert its maximum influence on catalytic activities and harmful excesses are not concentrated in isolated areas.

The metal modifier is preferably introduced into the silica sol as a dispersion in a liquid medium. It will be understood that the term "dispersion" includes either a true solution or a colloidal suspension. A particularly preferred method is to add to the silica sol an aqueous solution of a compound of the metal, such as a salt, while mixing with sufficient vigor to effect homogeneity substantially instantaneously, and remove the liquid phase of the resulting dispersion under conditions which minimize any segregation of the modifier and silica, as here-inafter more particularly described.

If the modifying agent is in the form of the elemental metal or the metal oxide, hydrous oxide, hydroxide, or other insoluble compound, no problem of coagulation of the silica particles will ordinarily be encountered when it is added to the sol. However, if the metal modifier is in the form of a soluble salt, it it usually desirable to introduce it into the silica sol immediately before the drying step so that the extent to which it can promote coalescence of the silica is minimized. Since the proportion of metal compound is generally quite small the coalescence effect in any event is not very pronounced.

Having prepared a suitable amorphous silica starting sol and added the metal modifier as described earlier, the sol is dried to a powder at a rate which is substantially instantaneous after the silica concentration in the continuous liquid/sol phase approaches about 50%. After the sol has been deionized, and sometimes after salts of the modifying metal have been added, the sol is in a metastable state and has a pronounced tendency to gel. This tendency is marked, in its incipient stages, by an increase in the viscosity of the sol, the increase being due to the individual ultimate silica particles coalescing into aggregates. Coalesced aggregates are not easily broken down when subjected to compaction, and as a result the large voids which are inherent in the compact persist into the catalysts. By quickly drying the sol this tendency is minimized and excessive aggregation is avoided.

Whatever the method used for drying the sol, it should be one which completes the drying very rapidly — that is, practically instantaneously — after the concentration of silica has started to rise. Thus, a batch drying operation wherein the sol is run into trays and water gradually evaporated off is completely unsuitable. When the concentration of silica in the liquid phase approaches about 50%, the rate of aggregation of the ultimate silica particles is very rapid and the coalescence factor of products thus obtained is above the permissible limit.

Spray-drying is a preferred method for drying the sol provided the operation is conducted within the limits above mentioned. Once these limits are recognized, those skilled in the art of spray-drying will have no difficulty in setting up the drying operation to accomplish the desired result.

The most important factors in this spray-drying operation are (a) the fineness of the spray droplets, which in turn will control the size of the silica aggregates in the powder, (b) the concentration of silica in the sol being sprayed, it is generally being desired to keep the silica concentration below the point where the viscosity will interfere the production of very fine spray droplets, (c) the temperature of the drying air, and (d) the removal of physically adsorbed water from the dried aggregate particles at relatively low temperature. Generally speaking, the conditions employed in conventional spray-drying equipment will be found satisfactory, provided the powder product is removed from the heated zone as rapidly as it is dried so as to minimize coalescence of the ultimate spheroidal particles within the aggregates.

Freeze-drying is another method of converting a silica sol to a dry powder with minimum coalescence of the ultimate amorphous silica particles. In this method the sol is cooled to a low temperature and then frozen very fast as a thin film; thereafter the frozen water is sublimed. The process is less preferred than spray-drying because it produces aggregates in the form of relatively large platelets many microns in diameter and several microns in thickness, this peculiar shape being the result of the concentration and aggregation of the ultimate spheroidal particles between the ice crystals at the moment of freezing. Such a powder may be further pulverized, preferably by a procedure which will not compact the silica structure, such as in a micronizer or gas jet compact mill.

Drum-drying the sol is the least preferred procedure but can be used under suitable circumstances. In this method a thin film is applied to the surface of a rotating, heated drum and the dried residue is continuously removed as the drum rotates. This drying process is particularly applicable when the amorphous silica ultimate particles are larger than about 50 nm in average diameter, since these larger particles have a smaller tendency to coalesce and coalescence is the principal problem encountered in drum-drying. Since rapid and complete removal of free water from the silica aggregates at the lowest possible temperature is essential, a vacuum drum-drier which will evaporate the water at a temperature not much higher than ordinary room temperature gives the best results. The particular drying procedure to be employed will be selected principally in view of the size of the ultimate spheroidal silica particles involved. Thus, for particles larger than, say, 50 nm in diameter, various methods of drying can be employed. For drying sols in which the particles are smaller than about 50 nm in diameter, the conditions are more critical and the method must be selected with a view to considerations herein more fully described. Spray-drying can be employed with all sizes of spheroidal colloidal silica particles with which this invention is concerned.

The choice of spray-drying conditions employed becomes very important in the case of ultimate silica-particles smaller than 20 nm in diameter, and are highly critical for particles in the range from 5 to 10 nm in diameter. In the latter case one employs a sol thoroughly purified from non-volatile, ionic contaminants and the pH is in the range of 2 to 4 or 8 to 9, the lower range being obtained by acidification with a volatile acid and the upper range being obtained by adjustment with ammonium hydroxide. Also, the spray droplets must exceedingly fine to permit as nearly instantaneous drying as possible and the resulting powder must be removed from the heated zone as rapidly as possible.

The substantially spheroidal silica powder particles obtained by spray-drying a sol can readily pack together with a minimum of voids. On the other hand, the aggregates obtained by freeze-drying and drum-drying the sol are not easy to pack together because of their irregular shape and thus are not preferred. However, even in this latter instance the irregularly shaped aggregates can be more readily packed together with minimum voids by having an organic lubricant present in the sol when it is dried. Products dried with an organic lubricant present are found to compact more readily than when the lubricant is not used.

Even when an organic lubricant is used in conventional manner, the force required to compact a freeze-dried or drum-dried powder is in the range of about 5 to 10 tons per square inch. The organic lubricant appears to aid in maintaining the particles in coherent shapes of the desired degree of packing after the molding force is released The elasticity of the silica aggregates might disturb the compacted arrangement after the force is released, if the lubricant did not prevent it.

The product obtained by drying the metal-modified silica sols described above, whether spray-dried or dried with an organic lubricant by freeze-drying or drum-drying is a powder consisting of aggregates of loosely coalesed amorphous silica ultimate particles having the metal modifier homogeneously dispersed therethrough, as more fully described.

CHARACTERISTICS OF THE METAL-MODIFIED AMORPHOUS SILICA POWDER

In the metal-modified amorphous silica powders of this invention three kinds of silica particles can be recognized. These are: (1) the smallest units discernible by the electron microscope, herein sometimes called "ultimate particles," consisting of amorphous silica and appearing in electron micrographs substantially as spheres, (2) aggregates made up of a multiplicity of the ultimate particles joined together in chains, rings, or three-dimensional networks by siloxane (Si–O–Si) bridges, and (3) powder particles of microscopic or macroscopic size, which can be made up of a multiplicity of the aggregates loosely bonded by secondary bonds such as hydrogen bridging or van der Waal's forces. The metal-modifying constituent is homogeneously dispersed throughout the powder. Prior to firing it does not appear to penetrate into the ultimate silica particles but is dispersed throughout the aggregates and throughout the powder particles.

The ultimate amorphous silica particles have an average size in the range of 5 to 200 nm — that is, the ultimate particle size of the silica in the original sol persists through the drying step. The particle size can be easily observed in electron micrographs for ultimates larger than, say, 7 nm; for ultimates smaller than this, the size can be calculated from specific surface area measurements.

The aggregates in the novel silica powders are made up of a number of ultimate particles lightly coalesced at their points of contact. In the spray-dried products the aggregates are generally smaller than about 10 μm.

The size of the powder particles in the metal-modified silica powders of this invention is not critical. As a matter of convenience powders which pass through a 50-mesh screen are preferred because they are readily adaptable to compacting operations to form the catalysts of this invention. If for any reason the products obtained by drying sols as described earlier are not sufficiently fine, they can be ground by conventional methods. Such grinding does not, of course, affect the size of the ultimate particles present.

The metal-modified, amorphous silica powders are free of crystalline silica, and the silica aggregates in them have a relatively low degree of coalescence. The latter is an extremely important characteristic in that it renders the powders suitable for the compaction step involved. There is thus a cooperation between the manner of drying the sol and the compacting step.

The degree of coalescence of the aggregates is expressed quantitatively in terms of the "coalescence factor." This coalescence factor is less than 30%. The coalescence factor is an indication of the extent to which the ultimate particles are joined together. In aggregates where the ultimate particles are so firmly joined together that they have substantial Si–O–Si linkages between the ultimates and cannot be compressed to compacts with small, uniform pores, the coalescence factor is considerably higher than 30%. The coalescence factor is well understood in the art and is described, for instance, in United States Patent 2,731,326, issued January 17, 1956, to Guy B. Alexander.

The degree of coalescence, or "coalescence factor," is determined according to the method given in the just-mentioned United States patent, at column 12, lines 25 and the following. The method involves measuring the percentage transmission of light through the silica–water dispersion, measured with light having a wave length of 400 nm. This method is suitable for use with silica powders in which the average particle size of spheroidal amorphous silica particles is less than 50 nm. If the silica particles are larger than this, light having a wave length of 700 nm should be used.

The metal-modified amorphous silica powders have a packed bulk density of at least 0.2 g/cm^3. "Packed" density is measured by placing a weighed quantity of sample in a graduated cylinder, and tapping the cylinder until the volume is essentially constant.

MAKING CATALYSTS FROM THE METAL-MODIFIED AMORPHOUS SILICA POWDERS

Catalysts of metal modified amorphous silica are made from the powders by compacting the powder under pressure. The powder may be evenly flowed into a steel mold or die and compacted with a plunger. Hydrostatic compaction around a mandrel in a rubber mold which is squeezed by hydrostatic pressure can be employed for making cylindrical, hollow objects. Powders can be roll-formed into sheets or bars or can be extruded through a die to form rods.

The molded catalysts can be further shaped by cutting or crushing either after the molding at room temperatures or after calcining at temperatures below 850°C. Usually the mechanical strength of the catalysts is greatly improved by baking at temperatures ranging from 400 to 800°C.

In the compaction step the pressure applied should be sufficient to produce a uniformly dense compact having a density of at least 1 g/cm^3 after removal from the die. The density can be determined from the weight of the sample and its dimensions. For this purpose it is convenient to use a die in the shape of a rectangular parallelopiped.

The pressures involved in the compaction operation are in the order of 1–15 tons per square inch (t.s.i.). At pressures lower than 1 t.s.i., the molding powder does not achieve a high enough formed density. The catalysts formed by the pressing operation have considerable dry strength and can easily be handled without damage.

If, in making the metal-modified silica powder, the metal modifier was added to the silica in the form of a compound which decomposes to a metal oxide on heating to elevated temperatures in air, such compound can be converted to the oxide by heating above the decomposition temperature either before or after the compaction operation.

If an organic lubricant was present in the metal-modified silica powder, it can be removed after the compaction step by volatilization or by burning it out in air. This is done by heating the formed catalyst to a temperature not exceeding the sintering temperature. Satisfactory results can be obtained by heating the formed catalyst slowly from room temperature to the range of 400 to 600°C at a rate no faster than 50°C per hour, preferably no faster than 25°C per hour, and holding at the maximum temperature for 2–6 h. Because the silica in the original powder was colloidal the catalysts at this point will have considerable strength because they have small voids at many points of contact of the silica, and hence can be easily handled.

The heating of the shaped catalyst is usually conducted in air, but it can be carried out in such reducing gases as hydrogen. In the latter case, some of the preferred metal modifiers are reduced to the corresponding metal. For example, if the metal modifier is a nickel compound such as nickel oxide, it is reduced to nickel metal.

THE POROUS, METAL-MODIFIED SILICA CATALYSTS

The porous, metal-modified, shaped silica catalysts obtained as above are novel and unique products. They consist essentially of the amorphous silica modified with the added metal. They have a density of from 1 to 2 g/cm^3. In these instances where the catalyst is in the form of irregular shaped pieces, the density of the individual pieces can be determined by the method of mercury displacement in an accurately calibrated pycnometer as described by M. Burr, *Roczniki Chemie*, 31,293 (1957).

The catalysts are relatively strong and contain uniform-sized voids no larger than 1 μm in largest dimension. Moreover, 95% of the voids, or "pores" are no larger than 300 in largest dimension. If the preferred silica dispersions are used, 95% of the pores are no larger than 200 nm. If a sol containing 15 nm particles is used, 95% of the pores are no larger than 15 nm.

The voids — that is, the open spaces between the ultimate particles — in the porous silica catalysts are substantially uniform in size. The pore volume and pore size distribution can be calculated from an analysis of gas adsorption–desorption isotherms. Small pores are filled at lower pressures and large pores at higher pressures. As the equilibrium pressure is increased in an adsorption experiment, condensation takes place in increasingly larger pores. Structure curves for solids may be drawn by plotting volume adsorption against the pore radius calculated from the relative pressure by means of the Kelvin equation. The pore size distribution curve is obtained by differentiating the structure plot. In catalysts prepared from the powders of this invention, not more than 5% of the total void volume is present in the form of pores having a size greater than twice the average pore size.

Utility — The catalysts of the present invention can be used in the same ways as prior art catalysts containing the same active catalytic materials. Catalysts of the present invention are particularly valuable for the treatment of combustion gases emitted from gas-burning devices such as stoves, furnaces, refrigerators, clothes driers, and the like, and for treatment of the fumes from automobile exhaust and other internal combustion exhausts. The manganese chromite catalysts are particularly to be preferred for such uses.

The catalysts of this invention are also useful for catalyzing hydrogenation-dehydrogenation reactions and catalytic cracking of hydrocarbons in the same manner as prior art catalysts containing the same active catalytic materials are so used.

The products and processes of the present invention will be more readily understood and practiced by reference to the following illustrative examples.

Example 1

A 7-g quantity of manganese metal as its nitrate salt is added slowly to 950 ml of water at 90°C. The mixture is stirred with a magnetic stirrer until all of the solid is dissolved. The solution is filtered while hot and then cooled and made up to a liter. The solution thus formed contains 7 g of manganese per liter of water.

A 320-g quantity of an ammonia-stabilized silica sol (30.9% silica) containing 15 nm silica particles prepared as described in United States Patent 2,574,902, issued November 12, 1951, to Bechtold and Snyder, is deionized with a sulfonic acid-type cationic resin in the hydrogen form and then with a polyamine-type anionic resin in the hydroxyl form. This cationic resin is filtered from the sol before the sol is treated with the anionic resin; the pH of the sol after double deionization is about 3.

The silica sol is charged to a Waring blender and while it is vigorously agitated, 135 ml of the manganese nitrate solution described earlier is added dropwise over a 15-min period. The mixture is agitated for one hour after the addition of the manganese solution is completed.

At this point, 43 g of a polyethylene glycol having an average molecular weight 3000 to 3700, dissolved in 43 ml of water, is added to the silica sol–manganese solution in the blender. The mixture is stirred for 15 min in the blender.

The solution described earlier is transferred to a glass to prevent possible segregation of the manganese and silica. The flask is connected to a Hi-Vac Cenco pump and in this manner the frozen water is sublimed off and solids remaining are freeze dried. A dry, soft powder is obtained by the freeze-drying.

The powder described earlier is formed into bars by placing approximately 3 g of it in a 2-inch by $\frac{1}{4}$ inch steel die and pressing in a hydraulic press to 10 t.s.i pressure in about 3 min, holding the pressure for about 1 min and releasing the pressure slowly. Bars, measuring 2 inch by $\frac{1}{4}$ inch by $\frac{1}{4}$ inch, having a high degree of green strength and which can be handled without damage are obtained.

The pressed bars described earlier are calcined in air at 750°C until all of the manganese nitrate is converted to manganese oxide. They are then cooled and are ready for use as silica-supported catalysts either as is or the bars can be broken up by grinding or crushing into granular pellets, 100% of which pass through an 18 mesh sieve.

Example 2

A portion of the silica sol-manganese nitrate-polyethylene glycol solution described in Example 1 is diluted with 3 volumes of water and spray-dried in a spray dryer using two-fluid atomization. Dryers of this type are described in the Chemical Engineer's Handbook, 3rd Edition by John H. Perry, McGraw-Hill, New York (1950), page 840. The inlet air temperature is 300°C and the outlet air temperature is 110°C. The fine, dry powder that is collected in a cyclone separator is pressed into bars and calcined to form silica-supported catalysts as described in Example 1.

Example 3

A silica sol-manganese nitrate solution is prepared in a Waring blender, as described in Example 1. To this is added 43 g of soluble starch. The mixture is stirred for 15 min after the addition of the soluble starch. The solution is drum-dried by feeding it slowly to the nip of a double drum-drier having 6-in.-diameter rolls. The roll clearance is 2 mils and the roll speed 2 r/min. The surface temperature of the rolls is 263°F. The dried powder is screened through a 50-mesh sieve and pressed into bars and calcined to form silica-supported catalysts as described in Example 1.

Example 4

For the manganese nitrate salts used to form the solution of Examples 1 through 3, there can be substituted stoichiometric equivalent amounts of the following metal salts to obtain similar results:

(1) Nickel formate
(2) Chromium nitrate
(3) Ammonium tungstate
(4) Calcium chloride
(5) Ammonium molybdate
(6) Vanadyl sulfate
(7) Cupric acetate
(8) Cadmium chloride
(9) Zinc nitrate
(10) Mercuric nitrate
(11) Barium acetate
(12) Stannous chloride
(13) Lead nitrate
(14) Bismuth nitrate
(15) Thallium nitrate
(16) Ferric chloride
(17) Cobalt acetate
(18) Cerium sulfate

Example 5

A 7-g quantity of manganese and chromium metals as their nitrate salts, which metals are in the stoichiometric equivalent proportions required to form manganese chromite, is added to 950 ml of water at 90°C. The mixture is stirred with a magnetic stirrer until all of the solids are dissolved. The solution is filtered while hot and then cooled and made up to a liter. The solution thus formed contains 7 grams of manganese-chromium metal per liter of water.

A 21.2% silica sol of 100 nm particles prepared as described in United States Patent 2,574,902 is deionized as described earlier in Example 1. A 135 cm^3 portion of the manganese nitrate-chromium nitrate solution described earlier is added to 467 g of the silica sol just described, in a Waring blender. The mixture is agitated for 1 h, and then 25 g of the solid polyethylene glycol of Example 1 are added to this mixture and stirred for 15 min.

The sol is freeze-dried and the resulting powder is pressed into bars as described in Example 1. These bars are then calcined at 825°C resulting in a silica-supported manganese chromite catalyst.

Example 6

A 6460 g portion of a 15% silica sol of 7 nm average diameter particles as described in United States Patent 2,750,345 is deionized as described earlier in Example 1.

A 20.7 cm^3 portion of each of the metal salt solutions described in Examples 1 and 4 is added a 323 g portion of the just described silica sol in a 20 min period while agitating the sols in a Waring blender. The mixtures are agitated for one hour after the addition of the respective metal solutions. A total of 43 g of solid polyethylene glycol of Example 1 is then added to each of the mixtures and stirring is continued for 15 min. The sols are freeze-dried as described and the resulting powders are pressed into bars as described in Example 1.

Three of the bars resulting from each sample metal-modified silica powder are then calcined at 450, 550, and 800°C respectively to produce silica supported metallic catalysts corresponding to the metal modified silica powder precursor.

Example 7

A 3.8 g quantity of calcium metal as its chloride salt is added slowly to 950 ml of water at 90°C. The mixture is stirred with a magnetic stirrer until all the solid is dissolved. The solution is filtered while hot and then cooled and made up to a liter. The solution thus formed contains 3.8 g of calcium per liter of water.

One hundred and thirty-five milliliters of the above calcium solution is added to a 320 g quantity of the deionized silica sol of Example 1 by the same procedures described in Example 1 including the addition of 43 g of a polyethylene glycol.

The resulting solution is then freeze-dried, pressed into bars, and calcined as described in Example 1 to form a silica-supported calcium catalyst.

Example 8

A 20 g quantity of platinum metal as its chloride salt is added to 900 ml of water at 90°C. The mixture is stirred with a magnetic stirrer until all the solid is dissolved. The solution is filtered while hot and then cooled and made up to 1 l. The solution thus formed contains 20 g of platinum per liter of water.

One hundred and thirty-five milliliters of this platinum solution is added to a 320 g quantity of the deionized silica sol of Example 1 as there described including the addition of a polyethylene glycol.

The resulting solution is then spray-dried as described in Example 2, then pressed into bars and calcined as described in Example 1 to form a silica-supported platinum catalyst.

Example 9

Other precious metal catalysts supported on silica can be made by substituting stoichiometric equivalent amounts of the following metal salts for the platinum chloride of Example 8:

(1) Palladium nitrate
(2) Rhodium chloride
(3) Sodium ruthenate
(4) Iridium chloride
(5) Osmium chloride

Example 10

A 5% by weight silica aquasol containing silica particles of 15 nm average diameter is prepared. The pH of the sol is adjusted to 4.0 by addition of HCl. 600 parts by weight of this sol is rapidly agitated and 65 parts of an aqueous solution containing 3.25 parts of alumina as basic aluminum chloride [$Al_2(OH)_5C1$] is added to the sol over a period of 30 seconds. The sol is then spray dried to provide fine dry powder which is collected in a cyclone separator, pressed into bars, and calcined as described in Example 1 to form a porous body.

63 Molded Amorphous Silica Bodies and Molding Powders for Manufacture of Same

Horacio E. Bergna and Frank A. Simko, Jr.
DuPont Experimental Station

CONTENTS

This chapter relates to strong, anhydrous, amorphous silica bodies and to their manufacture by molding and sintering amorphous colloidal silica powders at a temperature of at least 1000°C but below the devitrification temperature of the molded bodies.

The invention is particularly directed to a process for making amorphous silica bodies having a density from 91 to 100% of the theoretical density of amorphous silica, by compacting aggregates of amorphous silica ultimate particles substantially free of non-volatile constituents other than silica, said aggregates having a particle size in the range from 0.1 to 50 μm, said powder having a tapped bulk density of at least 0.2 g/cm^3, the ultimate amorphous silica particles being spheroidal and substantially uniform in size, as observed by electron microscopy, being dense, having a coalescence factor of less than about 30%, and having an average diameter of from 5 to 300 nm, the powder being compacted under pressure to a uniform density of at least about 1 g/cm^3, and heating the compacted material at a temperature above 1000°C but below the devitrification temperature of the material, until the density of the resulting body is increased to at least 2.0 g/cm^3, and cooling before any cristobalite is formed by devitrification of the amorphous silica.

The invention is more particularly directed to a process for making unusually strong amorphous silica bodies containing interconnecting macropores uniformly distributed throughout the structure, the process being identical to that described earlier, but with the further limitations that the ultimate amorphous silica particles in the aggregates have a coalescence factor of from 10 to 30% the average diameter of the aggregates is at least 10 times the average diameter of the ultimate amorphous silica particles, the compaction of the powder is stopped while spaces still remain between the aggregates, these spaces having an average diameter of at least twice that of the diameter of the ultimate silica particles, and the heating is continued until the density of the sintered body is in the range from 91 to 99% of the theoretical density of amorphous silica.

The invention is also directed to a process for making translucent to transparent anhydrous amorphous silica bodies substantially free from crystalline silica and substantially free from porosity, the density of the bodies ranging from 99 to 100% of the density of amorphous silica, the process being identical to that first described earlier, but with the further limitations that compaction of the powder is continued until there remain essentially no spaces between the aggregates, and the compacted material is heated until a dense, transparent, amorphous silica body having a density greater than 99% of theoretical is obtained. Also in this process it is preferred that the aggregates consist of ultimate silica particles having a coalescence factor of less than 10%.

Adapted from U.S. Patent 3,301,635, filed July 1, 1965, and issued Jan. 31, 1967.

The invention is additionally directed to novel opaque anhydrous amorphous silica bodies substantially free of crystalline silica and nonvolatile constituents other than silica, characterized by having a strength greater than that of 100% dense, transparent amorphous silica, the density of the bodies being in the range from 91 to 99% of the theoretical density of amorphous silica, the major proportion of the porosity being present in the form of interconnecting pores from 20 to 5000 nm in average diameter and preferably between 40 and 1000 nm in average diameter, the pores being of such uniform size that at least 50% of the pores have a diameter between 0.5 and 2.0 times the average pore diameter and essentially no pores are more than 5 times the average diameter, the pores being uniformly distributed throughout the body, as observed by electron microscopy. It the most preferred bodies of the invention, the macropores have an average diameter of between 100 and 500 nm, the density lies between 95 and 98% of theoretical, and the silica has a purity of at least 99.4%.

The invention is also directed to a silica powder preferred for use in the process of this invention, the particles of which consist of aggregates of dense amorphous silica ultimate particles substantially free from nonvolatile constituents other than silica, the aggregates being spheroidal in shape, having an average diameter of 0.5 to 50 μm, the powder consisting of such aggregates, having a tapped bulk density greater than 0.2 g/cm^3, the ultimate particles being spheroidal and substantially uniform in size as observed by electron microscopy, being dense, having a coalescence factor of less than 30%, and having an average diameter of at least 5 but less than 300 nm, preferably less than 200 nm, and even more preferably less than 50 nm.

The processes and compositions of this invention are concerned with amorphous silica, in contradistinction to crystalline silica, hereinafter sometimes referred to as cristobalite. Such terms as fused silica, vitreous silica, vitrified silica, and fused quartz have often been used interchangeably to describe amorphous silica bodies. The presence of crystallinity in a silica body is readily ascertainable by such standard techniques as studying an x-ray diffraction pattern of the silica body.

Amorphous or "fused" silica has long been recognized as a very valuable material of construction, especially for use at elevated temperatures, because it has great strength and a coefficient of expansion so low as to make it resistant to thermal shock. On the other hand, lack of practicable means of fabrication has impeded its widespread use even in view of its outstanding properties.

Conventional methods for fabricating fused silica hitherto have involved heating the silica above its softening point (above 1700°C), casting it into the desired shape, and then quickly chilling it through the temperature range at which conversion of the amorphous form to the crystalline form (devitrification) will occur. Shaping of the silica at the extremely high temperatures above the softening point is very difficult and not well adapted to the preparation of intricate shapes or indeed of any shapes other than rudimentary and simple forms. It will also be apparent that at such high temperatures corrosion and erosion of molds and other shaping means is very rapid.

In the method just described there is also always present the danger that the shaped article will not be cooled sufficiently rapidly to avoid some devitrification. This difficulty is inherent in the fact that the molding temperatures used are so high. Any devitrification, of course, seriously detracts from the strength of the shaped article and decreases its resistance to thermal shock.

The present invention provides novel amorphous silica bodies substantially free of crystallinity and novel methods for their production which avoid the necessity of heating the silica above its minimum devitrification temperature which is around 1200°–1300°C. The processes utilize amorphous silica powders having certain essential characteristics which make them capable of being shaped into desired forms by compaction and being sintered to produce strong bodies without devitrification. In the molding powders used in the process of this invention the amorphous silica is present in the form of aggregates of dense ultimate spheroidal particles in the size range of about 5 to about 300 nm. These ultimates, by reason of their methods of preparation, are so lightly coalesced as to lend themselves to compaction to such a degree that articles made by sintering the shaped compacts can have a density approaching or equal to theoretical.

The size and shape of the aggregates is of importance. Aggregates having an average diameter in the range from 0.1 to 50 μm may be employed. Uniformity of size of the aggregates in the preferred size range becomes more critical with increasing strength of the aggregates, that is, with increasing coalescence of the ultimate particles within the aggregates. It is also more critical in making the unusually strong, opaque bodies than in making the dense, translucent to transparent bodies.

The process of this invention may be employed to produce extremely dense, nonporous amorphous silica bodies having a density greater than 99 and even as high as 100% of theoretical, such bodies being translucent to transparent like the shaped vitreous silica bodies of the prior art. The process may also be employed to produce a novel body of amorphous silica, having a density of from 91 to 99% of theoretical, wherein there are interconnected macropores, the presence of which is accompanied by unusually high strength in the silica body. Such bodies are from 1.5 to 3 times stronger than conventional commercial fused silica, or "fused quartz" when specimens are cut and tested by the same method.

By the term "macropores," we distinguish pores larger than 20 nm from the "micropores" in conventional silica gel which are generally smaller than about 20 nm in average diameter, and more commonly 5 nm or smaller. It is a theory of the invention that these channel-like macropores are presumably acting either as regions for the relief of internal stresses, or as cavities to stop the propagation of cracks when the sample is placed under a breaking load. It is ordinarily supposed that in ceramic bodies or in glass, porosity is a source of mechanical weakness. It is our theory, however, that in the case of pores such as those in the silica bodies of the present invention, which are continuously interconnected in three dimensions, and where these pores make up a network of fine channels, and where they are evenly distributed in three dimensions throughout the silica body, then in this situation, a small degree of porosity ranging from about 1% to about 10% is beneficial. It is theorized that such pores improve the mechanical strength of the body. This hypothesis or theory is supported by the fact that as the bodies of this invention are being sintered at high temperature, those specimens which are cooled and tested while they still contain a degree of porosity ranging from 1 to 9% by volume are considerably stronger than those which have been further fired until the porosity has been essentially reduced to zero.

The particular type of aggregate powder to be employed in the process of this invention is selected with regard to which type of molded silica body is to be produced — either the essentially translucent body with a density greater than 99% of theoretical, similar to fused silica bodies of the prior art, or the novel pore-reinforced, relatively opaque molded silica body of the invention. For the first-mentioned type of dense, translucent silica body, the type of aggregate powder selected is such that when molded under pressure there will remain between the aggregates substantially no pores or macropores larger than the pores between the ultimate particles within the aggregates. On the other hand, to provide a pore-reinforced body, the type of aggregate is selected so that after the powder is compressed there will remain between the aggregates, a certain degree of macroporosity which will remain in the body after it has been heated to eliminate the micropores within the aggregate. This point will be further described in relation to the drawings.

The small size of the ultimate particles is an important requisite for an amorphous silica molding powder if it is to be used to prepare strong opaque bodies having a density greater than 91% of theoretical. The formation even of a very small amount of cristobalite by devitrification of the amorphous silica greatly weakens the body since, when it is cooled to room temperature, the cristobalite introduces severe strains in the surrounding amorphous silica. To avoid the formation of crystalline silica during fabrication it is necessary that compacts of the powder sinter rapidly to dense objects at temperatures below that at which crystalline silica begins to form at an appreciable rate. The larger the pores in the body to be sintered, the higher the temperature required. With amorphous silica powders composed of ultimate particles less than 300 nm in diameter macropores are small and sintering can be readily accomplished. On the other hand, a powder consisting of coarser ultimate particles of amorphous silica of a particle size, for example, larger than a micron, which can be obtained by pulverizing fused silica glass gives a body with very large pores. This cannot be sintered to greater than 91% of theoretical density at a temperature low enough to preclude the formation of cristobalite because of the very large spaces or pores between the large particles, which cannot be eliminated at a temperature less than about 1300°C, above which devitrification becomes rapid.

The absence of strong coalescence between ultimate particles is also important in preparing dense, transparent silica bodies by the process of this invention. Aggregates having a coalescence factor of up to 30% may be employed if the amorphous silica object to be produced is to have a theoretical density of greater than 91%. However, if the product is to be translucent amorphous silica object, the porosity must be very low and the density greater than about 99% or greater than 99.5% of the theoretical value. In the latter case, the coalescence factor should be as low as possible, preferably less than 10%. If the coalescence factor is greater than about 30%, then the aggregate particles tend to be hard and are difficult to press together sufficiently to avoid leaving an excessive amount of porosity and thus result in a fired density of less than 91%.

It is important that the silica powders do not contain nonvolatile impurities which catalyze crystallization. Sodium is one of the commonest impurities. The silica should be of high purity in regard to nonvolatile substances, since most impurities catalyze crystallization of the silica at elevated temperature. Thus the presence of such contaminants effectively reduces the maximum temperature which can be employed for densification and formation of cristobalite is difficult to avoid, even using powders composed of ultimates having a diameter less than 300 nm and a coalescence factor less than 30%.

See Figures 63.1–63.5.

If a powder is used which is similar to that shown in Figure 63.4, except that the ultimate particles within the aggregates have a coalescence factor of less than 10%, the macropores between aggregates in the molded body will be smaller than the voids **9** between aggregates shown in Figure 63.5. The softer aggregates are more easily compressed and more closely packed, so that the voids between the aggregates in the molded body are essentially as small as the voids between the ultimates within the aggregate particle. Upon sintering such a molded body at a temperature between 1000°C and about 1200°C all of the pores — those between aggregates as well as those between the ultimates — are eliminated,

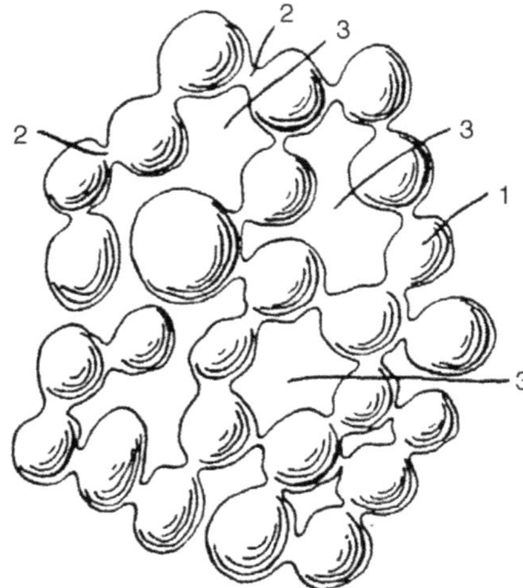

FIGURE 63.1 Diagrammatic view of a section of a silica molding powder showing amorphous silica spheroidal ultimate particles **1**, lightly coalesced (less than 30%) at their contact points **2** to form micropore voids **3**.

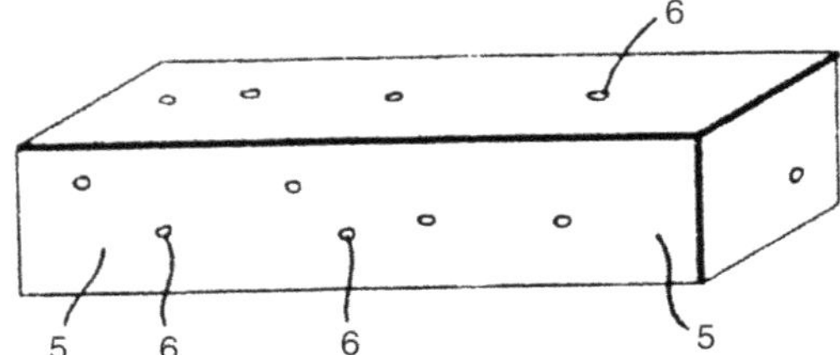

FIGURE 63.3 Solid bar **5** of amorphous silica obtained by sintering a shaped compact of Figure 63.2 at 1000 to 1200°C until the density is between about 92 and 99.5% of the theoretical density of amorphous silica, showing the persistence of macropores as interconnecting channels **6**.

resulting in a transparent body having only occasional very small voids and having a density greater than 99.5% of the theoretical density of amorphous silica.

MAKING THE MOLDING POWDERS

To make an amorphous silica molding powder suitable for use in the process of the present invention one first prepares a suitable silica sol and then dries the sol under certain critical conditions which lead to products having desired properties. The sols which are suitable as starting materials and the conditions which are critical in the drying step will now be described in greater detail.

The starting sol must first of all be a dispersion of amorphous silica ultimate particles. Since the silica in the ultimate product is to be amorphous it is highly important that the silica particles in the starting sol be free of crystallinity. Methods for determining the presence of crystallinity by x-ray techniques are discussed in R.B. Sosman's *The Properties of Silica*, Chemical Catalog Company, Inc., New York, 1927, page 207.

Also, as observed by electron microscopy the ultimate silica particles in the sol should be uniform-sized and spheroidal. By "uniform-sized" is meant that 75% of the total number of particles have a diameter in the range from 5D to 2D, where D is the number average particle diameter. The uniform size of the particles is important in avoiding micropores in the formed articles. The uniformity of the particles can be determined by methods described in the *Journal of Physical Chemistry*, 57 (1953), page 932.

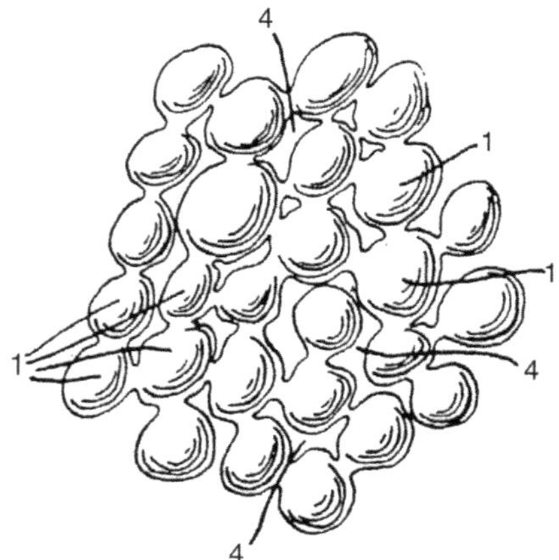

FIGURE 63.2 Product obtained by compaction of a powder of Figure 63.1 to a density of at least 1 g/cm^3, showing the ultimate amorphous silicia particles **1** enclosing substantially uniform micropore voids of reduced volume **4**.

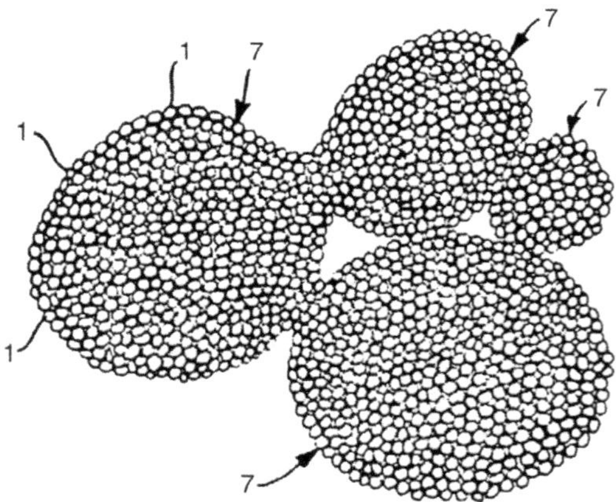

FIGURE 63.4 Multiplicity of spheroidal amorphous silica powder particles **7** such as obtained by spray-drying a silica sol by a process of the invention, the individual powder particles having the structure of the aggregate of Figure 63.2 and being made up of ultimate amorphous silica particles **1**.

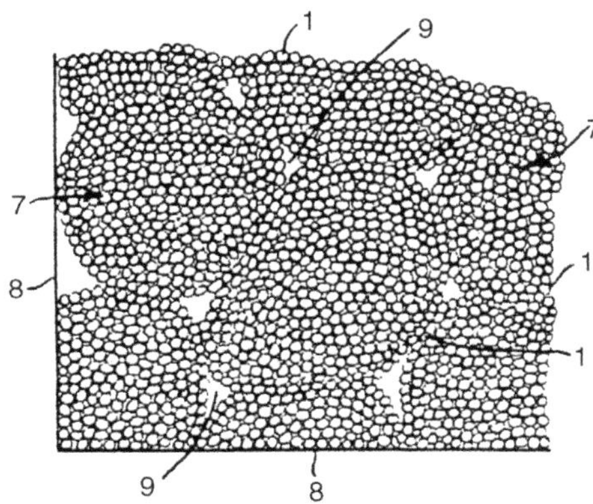

FIGURE 63.5 Cross section of a molded body obtained by compacting the material of Figure 63.4, showing how the aggregate powder particles **7** have been crowded together into a relatively uniform structure with a micropore voids **9**, lines **8** representing a portion of the mold wall.

The size of the ultimate silica particles in the sol can also be determined by methods described in the Journal article just mentioned. For making the novel silica body of this invention, the average ultimate particle size should be in the range of 5 to 300 nm, preferably between 7 and 200 nm and more preferably between 13 and 110 nm, in order to attain a desirable degree of coalescence. Particles larger than about 300 nm when cold pressed or molded give only very weakly coherent bodies which disintegrate during handling. With particles below the preferred limit of 200 nm there is no problem of obtaining coherent molded bodies.

Sols containing silica particles smaller than 5 nm such as those obtained by neutralizing sodium silicate with sulfuric acid, for instance are not satisfactory because the particles tend to aggregate very rapidly and irreversibly. When such sols are freed of stabilizing ions they show rapid increase in viscosity and tend to gel in extremely short times. Accordingly sols containing ultimate silica particles larger than 5 nm in average diameter are used.

The continuous liquid phase of the silica sol is preferably water — that is, the sols are aquasols. Although organosols can be used, ordinarily any advantages to be gained from the use of an organosol are more than offset by the added cost.

The art is already familiar with methods for making silica aquasols of discrete, dense amorphous silica particles in the size range above specified, and any of these methods may be used to make the starting material employed in processes of the present invention. A particularly preferred type of silica aquasol is described in United States Patent No. 2,574,902, issued November 13, 1951, to Bechtold and Snyder. This patent describes how the size of the ultimate silica particles is increased into the desired size range by adding quantities of a low molecular weight silica feed sol to a heel sol containing silica particles which have been grown substantially. To minimize the amount of alkali contained in the silica particles it is preferred to use ammonium hydroxide as the stabilizing alkali used to adjust the ratio as described in the patent.

Regardless of its method of preparation, a silica sol to be useful in the processes of this invention should be purified of all nonvolatile constituents other than the silica. By "nonvolatile" is meant constituents which will not be volatilized off or burned out at temperatures below or equal to the sintering temperature of the silica. Inorganic anions and cations are included among nonvolatile constituents and can be removed, if soluble, by deionizing the silica sol with a mixture of anion and cation exchange resins in accordance with processes described in United States Patent No. 2,577,485, issued December 4, 1951, to J.M. Rule. The sol is preferably temporarily stabilized against gelling or aggregation of the ultimate spheroidal particles by adjusting the pH to between 8 and 9 with a volatile base such as ammonia. Alternatively, the sol may be temporarily stabilized by adjusting the pH to the range 1 to 3 with a volatile strong acid such as hydrochloric or nitric acid.

As a further alternative, the sol can be purified by such methods as dialysis and other similar methods known to those skilled in the art.

The purity of the silica powder obtained by drying the sol is important in that it must not contain substances which promote the devitrification of silica at the sintering temperature. Thus, substantial quantities of impurities such as the alkali and alkaline earth oxides are especially to be avoided and these compounds should be removed as thoroughly as possible from the sol employed in making the silica powder. Generally speaking, the total content of potassium, sodium and calcium oxides, for example, in the powder should not exceed about 0.1% and preferably less than 0.01%. The exact level depends to some extent upon whether these impurities are on the surface of the particles or are locked within the silica structure. Thus, a few hundredths of a percent of combined sodium within the particles is not as harmful as this amount of combined sodium on the surface of the particles.

It should be noted, however, that small traces of alkali metal or alkaline earth metal oxides may act to aid the sintering of the silica and permit it to occur at a lower temperature than otherwise would be possible. While the final sintered, dense, ceramic body will contain at least 99% SiO_2 it will be understood that trace amounts of alkali and alkaline earth metal oxides as discussed earlier are not excluded.

Having prepared a suitable amorphous silica starting sol, as described above, the sol is dried to a powder. The drying method and conditions employed will depend upon the degree of coalescence between the ultimate particles that can be tolerated or may be desired, and will also depend upon the size and shape of the aggregate particles to be produced. To prepare dense transparent silica bodies, powders having a low coalescence, that is, below 30% and preferably below 10% are employed. On the other hand, to prepare the opaque bodies of higher strength which contain interconnected macropores throughout, it is permissible, in fact preferred, to employ a powder in which the aggregates are sufficiently firm and coherent that when they are compacted into a mold there will remain between the aggregates macropores of the desired size. For this purpose, therefore, aggregates in which the ultimate particles have a coalescence factor of between 10 and 30% or higher are preferred.

To prepare a silica powder having a coalescence factor of less than 30%, the sol is dried to a powder before any substantial increase has occurred in its viscosity and at a rate which is substantially instantaneous after the silica concentration in the continuous liquid–sol phase approaches about 50%. After the sol has been deionized, so that stabilizing ions have been removed, it is in a metastable state and has a pronounced tendency to gel. This tendency is marked, in its incipient stages, by an increase in the viscosity of the sol, the increase being due to the individual ultimate silica particles coalescing into aggregates. Coalesced aggregates are not easily broken down when subjected to compaction, and as a result the large voids which are inherent in the compact persist into the final fired object. By quickly drying the sol this tendency is minimized and excessive aggregation may be avoided if a very dense silica body is to be made.

Whatever the method used for drying the sol, it should be one which completes the drying very rapidly — that is, practically instantaneously — after the concentration of silica has started to rise if a powder of low coalescence factor is desired. Thus, a batch drying operation wherein the sol is run into trays and water gradually evaporated off gives a powder with a higher degree of coalescence. When the concentration of silica in the liquid phase approaches about 50% the rate of aggregation of the ultimate silica particles is very rapid and the coalescence factor of products thus obtained is high. The aggregate particles so obtained are of macroscopic size; indeed the dried material is often in the form of lumps of silica gel. These are far too coarse to be used directly in the process of the present invention, although the material may be suitably subdivided by grinding for making the strong opaque bodies of this invention.

Spray-drying is a preferred method for drying the sol provided the operation is conducted within the limits mentioned earlier. Once these limits are recognized, those skilled in the art of spray-drying will have no difficulty in setting up the drying operation to accomplish the desired result.

The most important factors in this spray-drying operation are (a) the fineness of the spray droplets, which in turn will control the size of the silica aggregates in the powder, (b) concentration of silica in the sol being sprayed, it is generally being desired to keep the silica concentration below the point where the viscosity will interfere the production of very fine spray droplets, (c) the temperature of the drying air, and (d) the removal of physically adsorbed water from the dried aggregate particles at relatively low temperature. Generally speaking, the conditions employed in conventional spray drying equipment will be found satisfactory, providing the powder product is removed from the heated zone as rapidly as it is dried so as to minimize coalescence of the ultimate spheroidal particles within the aggregates.

Freeze-drying is another method of converting a silica sol to a dry powder with minimum coalescence of the ultimate amorphous silica particles. In this method the sol is cooled to a low temperature and then frozen very fast as a thin film; thereafter the frozen water is sublimed. For making the strong opaque bodies of this invention containing interconnecting macropores the freeze-drying process is less preferred than spray-drying because it produces aggregates in the form of relatively large platelets many microns in diameter and several microns in thickness, this peculiar shape being the result of the concentration and aggregation of the ultimate spheroidal particles between the ice crystals at the moment of freezing. On the other hand, where the ultimate particles in the starting sol are relatively large, that is, 50–300 nm in size, the freeze-drying process results in a powder having a very low coalescence, for example, 10% or less. Such powders, as noted earlier, are particularly useful in preparing the dense translucent silica bodies. An ideal powder for making dense bodies is that disclosed by G.B. Alexander "Pulverulent Silica Products," U.S. Patent 3,041,140, issued June 26, 1962. This powder, made by freeze-drying a deionized sol containing ultimate particles in the size range 50–300 nm, exhibits a coalescence factor of from 0.1 to 10%. The coalescence between ultimate particles is so low in this powder that inspite of the irregular shape of the aggregates, substantially all macroscopic spaces between aggregates are eliminated upon molding, so that sintering results in a very dense, translucent amorphous silica body.

Drum-drying the sol is the least preferred procedure but can be used under suitable circumstances. In this method a thin film is applied to the surface of a rotating, heated drum and the dried residue is continuously removed as the drum rotates. This drying process is particularly applicable when the amorphous silica ultimate particles are larger than about 50 nm in average diameter since these larger particles have a smaller tendency to coalesce; coalescence tends to be excessive in drum drying. Since rapid and complete removal of free water from the silica aggregates at the lowest possible temperature is essential to minimize coalescence a vacuum drum drier which will evaporate the water at a temperature not much higher than ordinary room temperature gives the best results. The particular drying procedure to be employed will be selected principally in view of the size of the ultimate spheroidal silica particles involved. Thus, for particles larger than, say, 50 nm in diameter, various methods of drying can be employed. For sols in which the particles are smaller than about 50 nm in diameter, it is essential that spray-drying be used to prepare the molding powder in order to avoid excessive coalescence of the ultimates. Spray drying can be employed with all sizes of spheroidal silica particles with which this invention is concerned.

The use of spray-drying becomes very important in making low-coalescence silica powder in the case of ultimate silica particles smaller than 20 nm in diameter, and is especially important for particles in the range from 5 to 10 nm in diameter. In the latter case one employs a sol thoroughly purified from nonvolatile, ionic contaminants and the pH is in the range of 1–3 or 8–10, the lower range being obtained by acidification with a volatile acid such as nitric, and the upper range being obtained by adjustment with ammonium hydroxide. Also, the spray droplets must be exceedingly fine to permit as nearly instantaneous drying as possible and the resulting powder must be removed from the heated zone as rapidly as possible.

As will be seen from Figure 63.4 and Figure 63.5, the substantially spheroidal silica powder particles obtained by spray-drying a sol can readily pack together with uniform continuous macropores remaining between aggregates. Due to the uniform size and spheroidal shape of the aggregates, the pores between aggregates are in the form of interconnected channels and are uniformly distributed throughout the molded compact. Upon sintering such a body the micropores between ultimate particles within the aggregates are eliminated, but the macropores between aggregates remain, resulting in the strong, opaque bodies of this invention.

CHARACTERISTICS OF THE MOLDING POWDER

In the molding powders of this invention two kinds of particles can be recognized. These are: (1) the smallest units discernible by the electron microscope, herein sometimes called "ultimate particles," consisting of dense amorphous silica and appearing in electron micrographs substantially as spheres, (2) aggregates made up of a multiplicity of the ultimate particles joined together in three-dimensional networks by siloxane (Si—O—Si) bridges.

The aggregates constitute the powder particles, and appear as separate particles when the powder is shaken in a liquid medium which deflocculates the particles, such, for example, as normal propyl alcohol. For observation, such a suspension may be diluted out to the point where the powder particles may be spread out on a surface such as a glass slide, and observed individually. In the dry state, or in suspension in non-polar liquids, such as a hydrocarbon oil, the aggregate particles may be loosely held together by van der Waals forces, and in this clumped condition it is not possible to observe the individual aggregates or powder particles. However, if the powder is dispersed in a water-miscible alcohol or in water containing sufficient ammonia to give a pH of about 9, the clumps of aggregates are dispersed and the individual aggregates may then be observed and measured by conventional microscope methods.

The ultimate amorphous silica particles have an average size in the range of 5 to 300 nm — that is, the ultimate particles size of the silica in the original sol persists through the drying step. The particle size can be easily observed in electron micrographs for ultimates larger than, say 7 nm; for ultimates smaller than this, the size can be inferred from specific surface area measurements.

The aggregates in the molding powders are made up of a great many ultimate particles coalesced at their points of contact. The aggregates are generally in the range of about 0.1 to 50 μm, preferably 0.1 to 10 μm in diameter. When prepared by spray-drying a silica sol, the aggregates are from 0.2 to 5 μm in diameter; particles smaller than about 0.2 μm are difficult to collect; aggregates larger than about 10 μm tend to hollow, and as the size of these hollow spheres increases, their strength decreases and they tend to break up during compaction. However, spray-dried spheres up to 50 μm in diameter are useful.

The size of the aggregates is of less critical importance when the aggregate structure is soft and weak due to the ultimate particles having a low coalescence factor. Particularly in the case of ultimate silica particles larger than about 50 nm, when the coalescence factor is less than 10%, the powder is so soft that it is difficult to determine the aggregate size, since the aggregates break apart when subjected even to mild mechanical forces. As stated above such a soft powder is useful when a 100% dense, transparent, pore-free silica body is to be prepared.

On the other hand, if the coalescence factor is in the range from 10 to 30%, the aggregates are sufficiently coherent that the aggregate size can be determined with relative ease. In this case, where the aggregates are sufficiently strong and firm to retain their individual character, the size of the aggregate particles is of importance, since this in turn, together with the degree of pressure applied in molding, determines the size and distribution of the inter-aggregate spaces which retain their identity through the sintering process and appear in the final product as interconnected macropores distributed through the dense amorphous silica matrix. This distribution is determined by the particle size and distribution of particle sizes of the original aggregates.

The aggregate size in the silica molding powder may range from 0.1 to 50 μm. However, it will be apparent that in the case of aggregates having a diameter of 100 nm, the ultimate particles must be much smaller than this — for example, from 5 to 25 nm in diameter. This size range of the aggregate particles will be from about 4 times the diameter of the ultimate particles to 50 μm. Thus in the case of 300-nm particles, the aggregates will range from 1.2 to 50 μm, while with 5-nm particles the range will be from 20 nm to 50 μm.

Generally speaking, when the novel macropore-reinforced silica body of this invention is produced by pressing coherent aggregates, in which the coalescence factor of the ultimate particles is from 10 to 30%, the average diameter of the reinforcing interconnected macropores in the ultimate molded silica body will be less than one-tenth the diameter of the silica aggregates in the powder originally employed. A preferred aspect of this invention is the powder from a spray-dried silica sol consisting of spheroidal aggregates from about 0.2 to 2 μm in diameter with most of the silica present as aggregates of around 1 μm in diameter, made up of ultimate silica particles about 15 nm in diameter. This powder gave a remarkably strong, white, opaque, amorphous silica ceramic body containing about 3% by volume of uniformly distributed interconnected macropores about 0.1 μm in diameter, and the body was from 1.5 to 2.0 times as strong in transverse bending strength as similarly cut bars of transparent, pure, fused silica.

The size of the silica aggregates used for pure silica bodies of maximum density and low porosity is also particularly important in powders of relatively high coalescence factor. The higher the degree of coalescence of the ultimate particles in the powder aggregates, the harder the aggregate particles, the less they are deformed, and therefore the larger the spaces between the compressed but imperfectly consolidated aggregates. Thus in making the silica ceramic with minimum porosity from a powder of high coalescence factor, it is essential that the particle size of the aggregates be as small as possible, and preferably about ten times the diameter of the ultimate particles within the aggregates. Thus, for example, if colloidal silica of 15 nm particle size, originally stabilized with ammonia and otherwise essentially free from sodium or other cationic non-volatile impurities, is dried on a drum dryer to produce macroscopic flakes of silica, it will be necessary to pulverize this gel until the gel fragments or aggregate particles are of the order of 150 nm in diameter, which is ten times the ultimate particle size. Once this is accomplished, then the powder in the form of these very fine aggregates can be compacted under pressure, so that when the aggregate particles are close-packed, the space in the interstices between the individual aggregate particles will be as small as 15 nm, although some larger spaces may be present, because of imperfect packing of the aggregates.

To produce a novel silica body of unique strength, a uniform distribution of interconnected macropores of uniform size within a dense, non-porous silica matrix is the desirable objective. To achieve this, aggregates consisting of ultimate silica particles in the range of 10 to 30 nm in size, in the form of aggregates about 300 to 500 nm in average diameter are preferred. Due to the relatively small ultimate particle size within the aggregates, this portion of the mass is sinterable to a 100% dense, non-porous condition, with great rapidity in the temperature range of 1000 to 1200°C. At the same time, a compact or molded mass made of such aggregate particles with an average particle size of 400 nm, will contain macropores between the compressed aggregate particles, ranging from about 25 to 50 nm in diameter, depending upon the local perfection of packing of the aggregates.

It should be noted that in this situation, the aggregates must possess a certain degree of mechanical strength and the coalescence factor of the ultimate particles should be preferably greater than 10%. When such a molded and compressed body is sintered, the micropores within the aggregates disappear and the macropores between the aggregates remain as continuous channels. If the original aggregate particles are essentially all within the size range of 300 to 500 nm diameter, or an average of 400 nm, then the average distance between macropores after sintering will be of the same order of magnitude, diminished by the percentage of overall linear shrinkage during the final firing. By this means white, opaque, amorphous silica containing a uniform network of interconnecting channels around 50 nm diameter, is obtained with transverse bending strength of the order of 25,000 p.s.i., in comparison with transparent, fused silica with a bending strength of 10,000 to 12,000 p.s.i.

The degree of coalescence of ultimate particles is expressed quantitatively in terms of the "coalescence factor." This coalescence factor is less than 30%. The coalescence factor is an indication of the extent to which the ultimate particles are joined together. In aggregates where the ultimate particles are very firmly joined together, so that they have substantial Si—O—Si linkages between the ultimates and cannot be compressed to compacts with small, uniform pores, the coalescence factor is considerably higher than 30%. The coalescence factor is well understood in the art and described, for instance, in United States Patent No. 2,731,326, issued January 17, 1956, to Guy B. Alexander.

The degree of coalescence, or "coalescence factor," is determined according to the method given in the just-mentioned United States Patent at column 12, line 24 and the following. The method involves measuring the percentage transmission of light through the silica-water dispersion, measured with light having a wave length of 400 nm.

This method is suitable for use with silica powders in which the average particle size of spheroidal amorphous silica particles is less than 50 nm. If the silica particles are larger than this, light having a wave length of 700 nm or larger should be used.

It should be understood that in the case of ultimate silicate particles larger than about 50 nm, the degree of coalescence obtained in powders dried under conditions to minimize the degree of coalescence, such as drying from aqueous solution at a pH of about 2, or by freeze-drying in the absence of strong alkali, or by spray-drying, the degree of coalescence will generally be less than about 30%. In the case of particles larger than about 100 nm, the degree of coalescence in the dry powder is of the order of 10% or less, when, for example, the sol is dried from aqueous solution at a pH of 2 or 3 with a volatile strong acid such as nitric or hydrochloric, and especially if normal propyl alcohol is added to the mixture in the last stages of drying, to minimize the surface tension of the evaporating liquid. It may be assumed that for particles larger than about 50 nm in diameter, the aggregate powders prepared in accordance with the teachings of this invention will have a coalescence factor of less than 30%.

The molding powders have a "tapped" bulk density of at least 0.2 g/cm^3. "Tapped" density is measured by placing a weighed quantity of sample in a graduated cylinder, and tapping the cylinder until the volume is essentially constant. If the bulk density is less than this minimum, it will be found that the powders are extremely difficult to compact uniformly, and will give compacts having internal strains and in which stratification of the solids will be present.

The molding powders are substantially free of nonvolatile impurities. The term "nonvolatile" as here used means that the impurities will not distill out or burn out at the temperature of sintering — that is, in the range of 1000–1200°C.

MAKING USEFUL ARTICLES FROM MOLDING POWDERS

Useful articles of amorphous silica are made from the molding powders by compacting the powder under pressure to form an article of the desired shape and approximately the desired dimensions and then heating the compact at elevated temperatures.

Techniques for forming shaped articles from powders are well known in such arts as powder metallurgy and any of such techniques can be used for making articles from the molding powders of this invention. For instance, the powder may be evenly flowed into a steel mold or die and compacted with a plunger. Hydrostatic compaction around a mandrel or in a rubber mold which is squeezed by hydrostatic pressure can be employed for making cylindrical, hollow objects. Powders can be roll-formed into sheets or bars or can be extruded through a die to form rods.

The molded objects can be further shaped by cutting or machining either after the molding at room temperatures or after prefiring at temperatures below 1000°C. Usually the mechanical strength of the molded body is greatly improved by baking at temperatures ranging from 400 to 800°C. Such bodies can be drilled, turned on a lathe, or otherwise machined. The resulting shaped object will, of course, be considerably larger than the final sintered object, but since the shrinkage during sintering is essentially uniform in all directions, there is produced a miniaturized replica in which the absolute values of the linear dimensions are smaller than the originally machined dimensions, yet the relative dimensions are the same.

In the compaction step the pressure applied should be sufficient to produce a uniformly dense compact having a density of at least 1 g/cm^3 after removal from the die.

If the pressure is released after compaction, this should be done as slowly as is required to avoid the setting up of internal strains within the compact.

The pressures involved in the compaction operation are in the order of from 1 to 20 tons per square inch (t.s.i.). At pressures lower than 1 t.s.i., the molding powder does not achieve a high enough formed density. Pressures higher than 20 t.s.i. have no advantage and have a disadvantage in that the very small amount of air entrapped in the powder and die is compressed so highly at these pressures that when the pressure is released the air expands rapidly, causing local dislocations and flaws. The articles formed by the pressing operation have considerable dry strength and can easily be handled without damage, especially in the case of powders in which the ultimate particles are 100 nm in diameter or less. In the case of powders containing larger ultimate particles, the strength of the pressed but unfired articles becomes progressively weaker with increasing ultimate particle size, and considerable care must be exercised in handling these bodies. Nevertheless, the resistance to devitrification during firing improves as the specific surface area of the powder diminishes, or as the ultimate particle size increases. Consequently the weaker character of the molded bodies made from powders of larger ultimate particle size is compensated for in a practical way by the wider latitude permitted in the temperature at which the bodies must be fired to achieve high density without devitrification.

The molded bodies are placed into a furnace and heated at, for example, 50°C per hour to 600°C. The volatile material will burn out as the samples are being heated to the firing temperature. It is important that the bodies not be heated suddenly or too rapidly to 600°C while the volatile material is being removed and preferably not faster than 100°C per hour up to 800°C. The heating rate may be more rapid once the body has begun to sinter above 800 or 1000°C.

It will be apparent to one skilled in the art that the larger the molded body the more slowly it must be heated in order to permit the evolution of moisture without disrupting the molding. It is a further advantage of bodies molded from silica powders consisting of ultimate particles in the range from 100 to 300 nm in diameter, that the evolved moisture or steam escapes from the molded mass more readily than when much finer ultimate particles are employed. At the same time it must be kept in mind that the molded bodies made with these coarser ultimate particles are weaker and less coherent. Generally speaking, the larger the size of the ultimate particles of which the original molding consists, the more rapidly the molded compact may be heated without danger of disintegration from the internal pressure of the evolved gases.

SINTERING THE COMPACT

The molded compacts are sintered by heating them at a relatively uniform rate until the sintering temperature, in the range between 1000°C and the devitrification temperature of the silica is reached, and maintaining them at this temperature until their density is in the range of about 91% to about 100% of the theoretical density of amorphous silica. The novel character of the compacts makes it possible to sinter them to the desired density in this temperature range without devitrification.

The maximum sintering temperature is determined principally by the size of the ultimate spheroidal, colloidal silica particles in the molding powder. Very large colloidal particles, such as around 300 nm in diameter, require a sintering temperature of around 1200–1300°C. Smaller colloidal particles, around 15 nm in diameter, require a sintering temperature of around 1100°C, while particles from 5 to 10 nm in diameter may be fired as low as 1000 to 1050°C. The exact firing time and temperature can be determined by experiment for each particular molding powder; as explained below the conditions depend among other factors on trace impurities.

On the other hand, large colloidal particles such as around 100 nm or more in diameter, devitrify less readily than smaller particles such as 15 nm in diameter. Thus, while the larger colloidal particles require a somewhat higher sintering temperature, they also possess some resistance to devitrification which makes this higher firing temperature feasible in producing nondevitrified amorphous silica bodies.

Both sintering rate and rate of devitrification increase with increasing amounts of moisture in the furnace atmosphere, and the amount of trace impurities in the silica.

It has been observed that devitrification begins primarily at the surface of colloidal particles and for a given firing temperature, bodies of smaller colloidal particles begin to devitrify sooner than bodies of larger particles. Thus, the rate of devitrification is primarily a surface phenomenon and depends upon the extent of silica surface present in the body and the trace impurities within the particles appear to accelerate the rate of sintering.

For any given molding powder, the time and temperature of sintering is determined by experiment, the temperature and time being that required to sinter the body to a density in the range of 91% to 100% of theoretical. Heating should not be continued after the desired density has been reached since further heating serves no useful purpose and increases the possibility of devitrification. Devitrification is generally observed on the cooled specimens in the form of microscopic, white or opaque, crystalline regions of cristobalite, which tend to crack away from the surrounding amorphous silica. Such regions invariably weaken the final body and thus devitrification must be avoided.

The novel anhydrous amorphous silica bodies of this invention having a density between 91 and 99% of theoretical have been found to possess outstanding strength in comparison with the ordinary fused silica bodies known in the art. The greatest strength is exhibited by bodies having a density in the range of 96 to 99% of theoretical and the optimum strength appears to be obtained at a density of about 97% of the theoretical density of amorphous silica.

Electron microscope inspection of these novel bodies reveals that the interconnecting macropores are evenly distributed throughout the bodies. These pores are generally tetragonal or triangular in shape. The size of a pore is the diameter of a circle of area equal to that of the pore cross-section. The pores are of such uniformity of size that at least 50% of the pores have a diameter of from 0.5 to 2.0 D where D is the average pore diameter. Occasional large pores should be avoided and essentially no pores larger than 5 D should be present. The actual size of the pores will depend upon the size of the silica aggregates in the molding powder used to prepare the body. In general, it can be said that in the final body essentially none of the pores will have a diameter exceeding half the average diameter of the aggregate particles in the molding powder used and in most cases the average maximum dimension of the pores will be around one-tenth the average diameter of the aggregates. Thus, for example, where a molding powder consisting of 200 nm ultimates and 2 μm aggregates is used, the resulting body will contain pores of which no more than about 1% will be larger than 1 μm in diameter and the average dimension of the pores will be about 0.1 μm. It will be understood that in bodies prepared from molding powders consisting of smaller aggregates the pores will be even smaller, down to a minimum of 20 nm, but generally not smaller than 50 nm.

The molded, sintered products are useful as ceramic materials, particularly wherever materials having chemical inertness and a low coefficient of thermal expansion and resistance to thermal shock are needed. Typical of these uses are in nose cones for missiles and in electrical insulators, kiln furniture, cooking ware and covers for electric stove heating elements. The pores may be sealed later if desired by surface fusion or impregnation with colloidal silica or sodium silicate solution and refiring at about 1000°C.

The invention will be better understood by reference to the following illustrative examples:

Example 1

A silica aquasol, prepared by the process of Example 3 of U.S. Patent No. 2,574,902, issued November 13, 1951, to Max F. Bechtold and Omar E. Snyder and containing silica particles having an average diameter of about 17 nm, but using ammonia as the alkali-stabilizing agent, is diluted to 10% SiO_2 and deionized in accordance with the process of U.S. Patent No. 2,577,485, issued December 4, 1951, to Joseph M. Rule. One hundred parts by weight of this diluted sol are stirred for three hours with 10 parts by weight of the hydrogen form of a sulfonated polystyrene ion exchange resin, the resin is drained out, and the sol is then mixed with 10 parts by weight of a cation exchange resin in the hydrogen form and 10 parts by weight of an anion exchange resin in the hydroxyl form, to remove the last traces of anions and cations.

The deionized sol is filtered to remove traces of resin, and is spray-dried in a spray drier using two-fluid atomization. This type of drier is described in *The Chemical Engineer's Handbook*, *3rd Edition*, by John H. Perry, McGraw-Hill, New York, 1950 p. 840. The inlet temperature is 300°C and the outlet temperature is 110°C. The dried powder is separated from the air stream in a cyclone-type separator and dust bag filter. The powder is stored out of contact with moisture.

The dried powder is molded by compaction into a bar by placing approximately 3 g of it in a 2-inch-long, 1/4-inch-wide steel die and pressing it with a corresponding plunger under a pressure of 10 t.s.i. This pressure is applied, held for 1 min, then released slowly over a period of 1 min. A coherent bar 2 inches long by 1/4 inch wide and about 1/4 inch thick having a density of about 1 g/cm^3 is obtained. When removed from the die, the bar is sufficiently strong and coherent to be handled without disintegration and is essentially free from micropores. Care is taken not to damage the corners and edges during handling. Contamination of the bar with NaCl is avoided by handling only with clean gloves.

The bar is dried for twenty-four hours in an air oven at 50°C, to remove any residual water vapor. The dried bar is then placed on a pure silica support in an electrically heated muffle and heated from room temperature to 1200°C at a rate of 50°C per hour, then held at 1200°C for 2 h to sinter the silica particles. The sintered product is removed from the furnace and cooled by dropping it into water.

X-ray analysis indicates that the fired material is amorphous silica. Translucent specimens prepared in this manner have a density of over 99% of the theoretical density of pure fused amorphous silica as calculated from the dimensions and weight of the sample. The modulus of rupture is 5000 p.s.i., as measured by the conventional "transverse bend" technique of applying a measured increasing load at the middle of the bar which is supported across a 1-inch span.

Example 2

A dry, pulverulent silica powder is prepared by deionizing a 30% SiO_2, amorphous silica aquasol in which the silica particles have an average diameter of 15 nm, with a mixture of cation exchange resin in the hydrogen form and anion exchange resin in the hydroxyl form, until its pH is 3.8 and its specific resistance is about 5000 Ω, and heating the deionized sol in an autoclave, the temperature being raised to 325°C over a 4.5-h period and held at 325°C for 5 h, whereby a 12% SiO_2 aquasol having a pH of 9.2 is obtained, deionizing the sol with cation exchange resin in the hydrogen form until the pH is 3.1, freeze-drying this sol, and vacuum-subliming off the water without melting the frozen product, whereby a dry, pulverulent silica having a surface area of 36 m^2/g (corresponding to an average particle diameter of about 83 nm) and a degree of coalescence of 1.5% is obtained, all as described more fully and as claimed in Example 1 of Alexander U.S. Patent 3,041,140.

The powder is pressed into bars, further dried, and heated to the sintering temperature, all as described in Example 1. The specimens are cooled by removing them from the furnace and dropping them into water.

The density of the fired silica ceramic is about 99% of theoretical and the modulus of rupture is 7500 p.s.i.

Example 3

A silica sol having an average particle diameter of 25 nm is prepared in accordance with the process of Example 3 of the above-mentioned U.S. Patent No. 2,574,902. The sol is diluted to a concentration of 10% by weight of SiO_2, and then acidified to a pH of 1.5 with hydrochloric acid. After standing in this acid condition for twenty-four hours at room temperature, the mixture is deionized by adding per 100 parts by weight of sol, 10 parts by weight of the hydrogen form of a cation exchange resin and 10 parts by weight of the hydroxyl form of an anion exchange resin, agitated slowly for two hours, and the mixture is then filtered to remove the resin. The pH of the filtered sol is about 4.

This purified sol is then spray-dried as described in Example 1. The freshly dried powder is pressed into bars and tested, as in Example 1, with the exception that the material is sintered for 10 h at a temperature of 1130°C. The modulus of rupture of the resulting dense, amorphous silica is found to be as strong as clear fused silica.

Example 4

A 17% silica sol prepared in accordance with the process of Example 1 of U.S. Patent 2,750,345, issued June 12, 1956, to Guy B. Alexander, containing colloidal amorphous silica particles having an average diameter of 7 nm, is diluted with water to a concentration of 4% by weight of SiO_2, acidified with nitric acid to a pH of 1.5, permitted to stand for six hours, then deionized with a mixture of the hydrogen and hydroxyl forms of ion exchange resins as described in Example 2, earlier. The resulting purified dilute sol has a pH of 3.7. It is then spray-dried by the method employed in Example 1, but the silica powder is removed continuously from the collecting cyclone and cooled to 0°C.

The freshly prepared powder, within an hour after being dried, is pressed into bars by the mehod described in Example 1, dried overnight at 50°C, and fired to a temperature of 1080°C at the rate of 50°C per hour and held at 1080°C for 6 h. The density of the bars is about 91% of theoretical yet is about as strong as fused amorphous silica.

Example 5

A 50% silica sol consisting of particles averaging 105 nm in diameter is prepared by deionizing a 30% SiO_2, 15 nm-sized particle, amorphous silica aquasol by successive treatments with cation exchange resin in the hydrogen form and anion exchange resin in the hydroxyl form until the pH is 3.2, diluting the sol, heating it for eighty-seven hours at the temperature of the steam bath, and again deionizing with cation exchange resin to a pH of 3.05, autoclaving the resulting sol at 340°C for 6 h to give a sol product containing 10.6% SiO_2, having a pH of 8.0, and concentrating this sol by evaporation to 50% SiO_2 by weight.

This sol is freeze-dried by placing about 150 mL in a 1.5 L spherical glass flask, which is rotated in a freezing mixture of solid carbon dioxide and acetone. The frozen sol forms a solid layer on the walls of the flask. The flask is then connected with a high vacuum in a conventional freeze-drying apparatus, and the water is removed while the mass is still in a frozen state.

The dried powder is placed in a mold and subjected to a gradually increasing pressure of 25 t.s.i. over a period of three minutes and the pressure then gradually reduced over a period of one minute. The coherent, molded bar is further dried in air at 50°C and then placed in an electrically heated muffle and heated to 800°C, while supported on a fused silica plate. Another electric muffle, adjacent to the first, is heated to a temperature of 1050°C. The silica bar, not yet sintered at 800°C is quickly transferred to the muffle at 1050°C and left for 15 min. It is observed that the specimen shrinks perceptibly in dimensions, and becomes more translucent. As soon as this change occurs, which requires 15 min at 1050°C, the specimen is removed from the furnace and cooled by dropping it into water.

The amorphous silica body has a density indistinguishable from that of fused silica and a modulus of rupture of 5500 p.s.i. It is so translucent as to be almost transparent in some regions. However, in areas where the cold, pressed bar is handled with the fingers prior to sintering, there are small, devitrified areas of the surface containing white, opaque, cristobalite crystals. It is apparent that contamination of the surface of the material, probably from sodium chloride absorbed from the fingers, causes some devitrification of the surface.

Example 6

Five liters of an ammonia-stabilized silica sol of the type described in Example 1, containing 5% silica, is deionized as described in Example 1. Concentrated nitric acid is added to the sol until the pH drops to 1.5. The acidified sol is allowed to stand overnight. An anionic exchange resin in the hydroxyl form is then added to the sol until the pH is raised to 2. The resin is filtered from the sol. A 50–50 by volume mixture of cationic and anionic exchange resins is added to the sol until the pH cannot be raised any more. The mixture is stirred for twenty-five minutes. The pH at this point is 3.5. The resins are filtered from the sol. This procedure removes the alkali from the interior of the silica particles. Four cubic centimeters of concentrated nitric acid is added to the purified sol to increase its stability toward gelation.

The sol is diluted to 1% silica with distilled water and spray-dried as described in Example 1. The molding powder resulting is found to contain 5.8% water, 0.035% sodium, 0.33% non-siliceous ash. It has a coalescence factor of 24% and a aggregate size of approximately 2 μm.

The molding powder is compacted as described in Example 1. The density of the compact is 1.1 g/cm^3. The samples are fired by heating to 1200°C at 100°C/hr. The samples are held at this temperature for various lengths of time and quenched from 1200°C into cold water. The following table shows the densities and transverse bend strengths of samples following quenching after various hold times at temperature.

Hold time at 1200°C per hour	Density, percent of theoretical[a]	Transverse bend strength, p.s.i.
1/2	94	5000
1	97	8200

[a]Actual density (g./cc.) as calculated from weight and dimensions divided by 2.2 (theoretical density of amorphous silica) and multiplied by 100.

EXAMPLE 7

An eighteen-kilogram portion of an ammonia-stabilized silica sol of the type described as the starting material in Example 1, containing 1.5% silica, is deionized as in Example 1. The sol is spray-dried in the drier described in Example 1, at the conditions described in that example. The total time for drying the entire sol is 147 minutes, although the drying rate for any given amount being sprayed is practically instantaneous.

The resulting molding powder is found to have the following properties:

2.7% water
0.3% total sodium as Na
0.30% non-siliceous ash
Coalescence factor = 20%
Aggregate size = 1.3 μm

The powder is compacted into bars as described in Example 1. The density after compaction is 1.1 g/cm^3.

The samples are fired by heating to 1200°C at 100°C/h. The samples are held at temperature for various lengths of time and quenched from 1200°C into cold water. The following table shows the densities and transverse bend strengths of the samples after quenching after various hold times at temperature:

Hold time at 1200°C per hour	Density, percent of theoretical	Transverse bend strength, p.s.i.
1/4	91	9300
2	96	9400

EXAMPLE 8

Six liters of an ammonia-stabilized silica sol of the type described in Example 1, containing 10% silica, is deionized as described in Example 1. The sol is then spray-dried in the drier described in Example 1.

The following are the operating data:

Drying air temperatures — 350°C
Drying air rate — 166#/h
Atomizing air temperature — 340°C
Atomizing air rate — 72#/h
Nozzle cooling air rate — 15#/h
Quench air rate — 140#/h

Drying atomizing nozzle cooling and quench air pressure — 75 p.s.i.g.

The total sol is dried in approximately 50 min, the rate of drying of the sol droplets after spraying being substantially instantaneous. The exit air temperature during the drying is 110°C.

The powder is found to contain approximately 2.5% water by Fischer titration, 0.03% sodium, 0.23% non-siliceous ash. It has a coalescence factor of 23%. The aggregate size is approximately 1.5 μm, as observed in the electron microscope.

The molding powder described earlier is compacted as described in Example 1. The density of the sample is 1.1 g/cm^3 after compacting at 10 t.s.i. The compacted samples are fired by heating to 1200°C at 100°C/h. The samples are held at temperature for various lengths of time and quenched from 1200°C into cold water. The following table shows the densities and transverse bend strengths of the samples after quenching after various hold times at the stated temperature.

Hold time at 1200°C per hour	Density, percent of theoretical	Transverse bend strength, p.s.i.
1 . . .	91	9,000
3 . . .	95	11,000

EXAMPLE 9

Fifteen hundred grams of an ammonia-stabilized silica sol of the type described in Example 1 containing 30% silica is deionized, as in Example 1, and the sol is spraydried in the drier of that example. The conditions for this drying are as follows:

Drying air temperature — 354°C.
Drying air rate — 166#/h
Atomizing air temperature — 400°C.
Atomizing air rate — 87#/h
Nozzle cooling air rate — 10#/h
Quench air rate — 110#/h

The sol is dried in 10 min. The exit air temperature is 135°C.

The resulting molding powder is found to contain 2% water, 0.032% sodium, 0.15% non-siliceous ash. It has a coalescence factor of 26%. The aggregate size is 2.5 μm.

The molding powder is compacted as described in Example 1. The density of the compact is 1.04 g/cm^3. The samples are fired by heating to 1200°C at the rate of 100°C/h. They are held at temperature for various lengths of time and quenched from 1200°C into cold water. When held at 1200°C for 3 h the samples are 97% of theoretical density and have a transverse bend strength after quenching of 9000 p.s.i.

Example 10

One liter of an ammonia-stabilized silica sol of the type described in Example 1, containing 30% silica, is allowed to dry in air at room temperature. The gel obtained is pulverized to particles smaller than half a millimeter in diameter and dried in a vacuum oven at 110°C overnight.

One hundred grams of the dry silica powder are placed in a two quart vented steel mill half filled with low-carbon steel balls, one-quarter of an inch in diameter, and containing 500 cc. of water. The powder is ballmilled for five days at 80 r.p.m., water being added as necessary to keep the mixture semi-fluid.

After the milling operation the paste is removed from the mill, separated from the steel balls and treated with warm aqua regia until all metal worn from the balls is dissolved. The silica is then washed with 1% hydrochloric acid until all the iron is separated. This treatment also thoroughly eliminates sodium as well as other metal ions, leaving a pure white silica suspension. The treatments and washing operations are carried out by suspending the silica in the reagents, or the wash water and then permitting it to settle out of the liquid which is then drawn off.

The silica paste is finally washed with distilled water until the wash water is free from chloride ion, and adjusted to pH 9 with ammonia. The aqueous suspension is diluted to a volume of 1 l and allowed to settle in a container filled to a depth of 50 cm for 3 h and the liquid from the top 20 cm is removed. This fraction contains only silica aggregates of about 0.5 μm equivalent diameter and smaller.

The residue is reslurried as before with more water, and again allowed to settle. The liquid from the top 20 cm is again extracted and combined with the fraction form the first operation. This suspension is then placed in a container to a depth of 20 cm and permitted to settle for 3 h and the upper 10-cm layer is discarded and the residue is rediluted and the operation is repeated. These combined fractions are then dried on a steam bath and the obtained residue dried in a vacuum over at 75°C overnight.

The powder is made of aggregates from 0.5 to 0.35 μm in diameter. Several batches of this material are prepared by processing an additional 100 g lot of dried silica.

The powder is compacted as described in Example 1. The density of the sample is about 1 g/cm^3 after being compacted at 10 t.s.i. The compacted samples are sintered by heating to 1100°C at 100°C/h. The samples are held at temperature for 5 h and quenched from 1100°C into water at room temperature. Slight variations from batch to batch of powder require that the exact firing temperature for each batch be determined in a preliminary test to achieve the desired density.

The white, opaque bodies fabricated in this manner have a density of 95–98% of theoretical and contain interconnected channels of uniformly sized pores of 50 nm average diameter, distributed throughout the dense amorphous silica matrix. These bodies are over twice as strong as optical grade fused silica, as measured by transverse bend tests on specimens of equal size cut in the same way by diamond sawing.

Compaction pressure, 25°C (t.s.i.)	Position in the muffle	T, g./cc.[a]	B, g./cc.[b]	P, percent[c]	Transverse bend, p.s.i.	Max. percent closed porosity
10 ...	Back, left hand side (a) ...	2.19	2.13	2.94	8,300	15
10 ...	Back, left hand side (b) ...	2.19	2.13	2.74	4,600	15
10 ...	Back, right hand side (c) ...	2.19	2.04	6.92	10,000	7
10 ...	Back, right hand side (d) ...	2.18	2.10	3.61	9,000	22
20 ...	Center, right hand side (e) ...	2.18	2.17	0.5	5,000	70
10 ...	Center, left hand side (f) ...	2.19	2.09	4.34	16,000	10
15 ...	Front, right hand side (g) ...	2.19	2.13	2.63	20,000	15
10 ...	Front, left hand side (h) ...	2.18	2.03	6.9	12,000	13

[a]The Apparent Specific Gravity, T, of that portion of the test specimen which is impervious to boiling water (ASTM method C20−46).

[b]The bulk density, B, in g/cm^3 of a specimen is the quotient of its dry weight divided by the exterior volume, including pores. (ASTM method C20−46.)

[c]The Apparent Porosity, P, expresses as a percentage the relationship of the volume of the pores of the specimen to its exterior volume.

Example 11

The silica aquasol of Example 1 is diluted, deionized, and spray-dried following the method described in Example 1.

The freshly dried powder has a specific surface area of 175 m^2/g and a coalescence factor of 24.5. Analysis by flame photometry shows a sodium content of 0.03%. Spectrographic analysis gives the following results:

	Parts per million
Iron	150–750
Aluminum	150–750
Nickel	50–250
Zirconium	500–2500
Titanium	200–1000
Copper	3–15
Magnesium	25–250
Boron	20–100
Calcium	20–100

The bulk density of this powder is 0.65 g/cc. Electron microscopy inspection of this powder shows spheroidal aggregates with an average diameter of about 0.6 μm.

Different portions of the powder are molded by compaction at room temperature at 10, 15, and 20 t.s.i. pressure into 2-inch-long, 1/4-inch-wide bars following the procedure indicated in Example 1.

The bars are placed on pure silica supports and distributed along the length of an electrically heated muffle and heated from room temperature to 1100°C at a rate of 50°C per hour, then held at 1100°C for 5 h. The sintered bars are removed from the furnace and cooled by dropping them into water.

The following table shows the densities and transverse bend strengths of samples removed from various parts of the muffle.

It is to be observed that the values in the Table for Apparent Specific Gravity, T, are only approximate. For a fused silica completely devoid of porosity or for a body containing only open pores, the T value should be 2.2, that is, the theoretical density of silica. However, even for such a body the measured T value may be something less than 2.2, say 2.18 or 2.19. Thus, it is not possible to determine from the T values in the table whether the bodies actually contain a small amount of closed porosity. The T values less than 2.20 may be the result of impreciseness of the experimental method.

However, on the assumption that a T value below 2.20 is the result of presence of closed pores, the maximum possible closed porosity as a percent of total porosity is calculated as follows:

$$\text{Percent closed porosity} = \frac{1 - T/2.20}{1 - B/2.20} \times 100$$

The values of maximum percent closed porosity for the bars for this example are given in the table.

It is apparent that in all of the bars, except possibly sample (e), the major proportion of the porosity is present in the form of interconnected pores that form channels throughout the bars.

Electron microscope inspection of a film replica of a polished surface of each of the bars as well as direct transmission electron micrographs of thin sections of each bar, reveals that these channels are evenly distributed throughout the bars. The channels are generally tetragonal or triangular in shape and have a diameter (calculated as the diameter of a circle having the same area as the pore cross section) of about 0.1–0.5 μm. The pores are of such uniformity of size that practically no pores smaller than 0.1 μm or larger than 1 μm are found in the bars.

The invention claimed is:

1. A process for making amorphous silica bodies having a density from 91 to 100% of the theoretical density of amorphous silica which comprises the steps (1) compacting under pressure to a uniform density of at least about 1 g/cm^3 a powder consisting essentially of aggregates of amorphous silica ultimate particles substantially free of non-volatile constituents other than silica, said aggregates having a particle size in the range from 0.1 to 50 μ, said powder having a tapped bulk density of at least 0.2 g/cm^3, the ultimate amorphous silica particles being spheroidal and substantially uniform in size as observed by electron microscopy, being dense, having a coalescence factor of less than about 30% and having an average diameter from 5 to 300 nm; (2) releasing the pressure and then heating the compacted material at a temperature above 1000°C but below the devitrification temperature of the material until the density of the resulting body is increased to at least 2.0 g/cm^3; (3) cooling the body before any cristobalite is formed by devitrification of the amorphous silica.

2. A process as defined in claim **1** wherein the ultimate amorphous silica particles have an average diameter from 5 to 200 nm.

3. A process for making strong amorphous silica bodies containing interconnected macropores uniformly distributed throughout the bodies which comprises the steps (1) compacting under pressure to a uniform density of at least about 1 g/cm^3 a powder consisting essentially of aggregates of amorphous silica ultimate particles substantially free of non-volatile constituents other than silica, said aggregates having a particle size in the range from 0.1 to 50 μm, said powder having a tapped bulk density of at least 0.2 g/cm^3 the ultimate amorphous silica particles being spheroidal and substantially uniform in size as observed by electron microscopy, being dense, having a coalescence factor of 10 to

30%, and having an average diameter from 5 to 300 nm, the average diameter of the aggregates being at least 10 times the average of the ultimate amorphous silica particles, the compaction of the powder being stopped while spaces still remain between the aggregates, these spaces having an average diameter of at least twice that of the diameter of the ultimate silica particles; (2) releasing the pressure and then heating the compacted material at a temperature above 1000°C but below the devitrification temperature of the material until the density of the resulting body is in the range from 91 to 99% of the theoretical density of amorphous silica; (3) cooling before any cristobalite is formed by devitrification of the amorphous silica.

4. A process as defined in claim **3** wherein the ultimate amorphous silica particles have an average diameter from 5 to 200 nm.

5. A a process for making translucent anhydrous amorphous silica bodies substantially free from crystalline silica and substantially free from porosity, the density of the bodies ranging from 99 to 100% of the density of amorphous silica which comprises the steps (1) compacting under pressure a powder consisting essentially of aggregates of amorphous silica ultimate particles substantially free of non-volatile constituents other than silica, said aggregates having a particle size in the range from 0.1 to 50 μm, said powder having a tapped bulk density of at least 0.2 g/cm^3, the ultimate amorphous silica particles being spheroidal and substantially uniform in size as observed by electron microscopy, being dense, having a coalescence factor of less than about 30% and having an average diameter from 5 to 300 nm, the compaction of the powder being continued until there remain essentially no spaces between the aggregates; (2) releasing the pressure and then heating the compacted material at a temperature above 1000°C but below the devitrification temperature of the material until a dense, translucent amorphous silica body having a density greater than 99% of theoretical is obtained; (3) cooling the body before any cristobalite is formed by devitrification of the amorphous silica.

6. A process as defined in claim **5** wherein the ultimate amorphous silica particles have an average diameter from 5 to 200 nm.

7. The process of claim **5** wherein the powder aggregates consist of ultimate silica particles having a coalescence factor of less than 10%.

8. An opaque anhydrous amorphous silica body having a strength greater than that of 100% dense, transparent amorphous silica and being characterized by being substantially free of crystalline silica and non-volatile constituents other than silica, having a density of from 91 to 99% of the theoretical density of amorphous silica, the major proportion of the porosity being in the form of interconnecting macropores from 20 to 5000 nm in average diameter, the pores being of such uniform size that at least 50% have a diameter between 0.5 and 2.0 times the average pore diameter and essentially no pores are more than 5 times the average diameter, the pores being uniformly distributed throughout the body as observed by electron microscopy.

9. An opaque anhydrous amorphous silica body as defined in claim **8** wherein the interconnected macropores have an average diameter between 40 and 1000 nm.

10. An opaque anhydrous amorphous silica body as defined in claim **9** having a density of from 95 to 98% the theoretical density of amorphous silica.

11. An opaque anhydrous amorphous silica body as defined in claim **10** consisting of at least 99.4% pure silica.

12. An opaque anhydrous amorphous silica body as defined in claim **11** in which the interconnected macropores have an average diameter between 100 and 500 nm.

13. A process as defined in claim **3** wherein the powder aggregates are spheroidal in shape.

14. An amorphous silica powder consisting essentially of aggregates of dense amorphous silica ultimate particles substantially free from non-volatile constituents other than silica, the aggregates being spheroidal in shape, substantially uniform in size, and having an average diameter of between 0.5 and 50 μm, the powder having a tapped bulk density greater than 0.2 g/cm^3, the ultimate particles being spheroidal and substantially uniform in size as observed by electron microscopy, being dense, having a coalescence factor of less than 30% and having an average diameter of at least 5 but less than 50 nm.

64 High Ratio Silicate Foundry Sand Binders

Horacio E. Bergna
DuPont Experimental Station

CONTENTS

Binder solution for preparing sand cores of initial high strength but with essentially no strength after casting metals above 700°C, said binder solution comprising an aqueous solution of sodium, potassium or lithium silicate having an overall molar ratio of SiO_2/alkali metal oxide from 3.5:1 to 10:1, and containing sufficient amorphous silica so that the fraction of the total silica in the binder solution which is present as amorphous silica is from 2 to 75%, the amorphous silica having a particle size in the range from ~2 to 500.

BACKGROUND OF THE INVENTION

In the meta casting industry molten metal is cast into molds containing sand cores made from foundry sand and binders. These sand cores are conventionally bonded with organic resins which, during curing and during casting of the metal, decompose and evolve byproducts which are odoriferous, offensive fumes which are not only skin irritants but in most cases toxic. The molds themselves are made from foundry sand bonded with oils, clays and/or organic resins. Thus, during their use, similar problems can occur.

A great percentage of the sand binders used by the foundry industry are made of phenol- and urea-formaldehyde resins, phenolic- and oil-isocyanate resins, and furan resins. Almost all of these binders and their decomposition products such as ketones, aldehydes, and ammonia are toxic. The principal effect on man is dermatitis which occurs not so much from completely polymerized resins, but rather from the excess of free phenol, free formaldehyde, alcohol or hexamethylenetetramine used as a catalyst. Formaldehyde has an irritating effect on the eyes, mucous membrane and skin. It has a pungent and suffocating odor and numerous cases of dermatitis have been reported among workers handling it. Phenol is a well-known poison and is not only a skin irritant but is a local anesthetic as well, so that burns may not be felt until serious damage has been done. Besides being capable of causing dermatitis it can do organic damage to the body. Furfuryl alcohol defats the skin and contact with it has to be avoided. Hexamethylenetetramine is a primary skin irritant which can cause dermatitis by direct action on the skin at the site of contact. Urea decomposes to carbon dioxide and ammonia, the latter of which is intolerable in toxic concentrations. In addition to the binders, some processes use flammable gases such as triethylamine as a curing agent. Capturing or destroying gases, smoke and objectionable odors are only temporary, stop-gas expensive solutions. New binders are needed that completely eliminate the sources of offensive odors and toxic gases.

Many of the organic binders are hot setting and therefore require heating to cure. Hot molds not only add hazards and complicate pollution control problems but add economical problems related to increased use of energy and increased equipment, maintenance and operation costs.

An alternative is to use inorganic cold setting binders, such as sodium silicate, which set at room temperature without producing objectionable gases or vapors. The use of silicates, however, results in the silicate bond remaining too strong after casting, so that the core is still coherent, and has to be removed by use of violent mechanical agitation or by dissolving the silicate bond with a strong, hot aqueous alkali. The problem may be lessened to a degree by using sodium silicate solutions admixed with organic materials such as sugar, but even in this case the core is still coherent after casting and requires extreme measures for removal such as violent mechanical agitation.

Thus, there is a need to create a binder for sand in making cores and molds for casting metals such as aluminum, bronze, or iron that will have satisfactory high strength before the metal is cast, retain sufficient hot strength and dimensional stability during the hot metal pouring, but which will have such strength after the metal has been cast and cooled, that the sand can be readily shaken out of the cavities formed by the cores; the binder also should be one that will not evolve unreasonable amounts of objectional fumes when the sand cores and molds are subjected to molten metal.

SUMMARY OF THE INVENTION

I have discovered that molds and sand cores of initial high strength but with essentially no strength after casting metals above 700°C can be made by bonding foundry sand with an aqueous solution of sodium, potassium, or lithium silicate or their mixtures and amorphous colloidal silica the amounts of silicate and amorphous colloidal silica being such that the overall molar ratio of SiO_2/alkali metal oxide (M_2O) is from 3.5:1 to 10:1, preferably 4:1 to 6:1, the fraction of the total silica present as amorphous colloidal silica is from 2 to 75% by weight, preferably 2 to 50%, and most preferably 10 to 50%, the amorphous colloidal silica having a particle size in the range from about 2 to 500 nm, and the 98 to 25% balance of the total silica being in the form of silicate ions. The percent solids of the aqueous binder solution being 20–55% by weight. The amorphous colloidal silica in the binder comprises both the amorphous colloidal silica component of the mixture and the amorphous colloidal silica fraction inherently present in aqueous solutions of alkali metal silicates of ratio more than about 2.5.

In alkali metal aqueous solutions containing more than 2.5 mols of SiO_2 per mole of M_2O, it is found by ultrafiltration, according to a procedure referred to herein as the Gore Procedure, that the concentrations used in this invention part of the silica in solution is ionic and part of it is colloidal, the colloidal fraction being retained by the ultrafilter while the ionic silicate passes through. In the case of sodium silicate for example, concentrated commercial silicate solutions are available having a SiO_2/Na_2O ratio as high as 3.8/1.0 and these concentrated solutions therefore contain a substantial proportion of the silica present in the colloidal state. The colloidal fraction consists of a range of sizes <5 nm diameter and down to near 1 nm, with a substantial amount of 2 or 3 nm diameter. These units are so small that solubility equilibrium is rapidly established so that if the solution is diluted with water the units pass into solution forming lower molecular weight ionic species.

The higher the ratio of concentrated aqueous solutions of alkali metal silicates the higher the colloidal silica content, but for each ratio the colloidal silica content decreases with dilution of the solution.

To prepare a binder having SiO_2/M_2O ratio of 3.5 to 3.8 it is therefore not necessary to add any colloidal silica if an alkali metal silicate solution is used already in the ratio range. On the other hand, if an alkali metal silica solution with ratio lower than 3.5 is used, it is necessary to add at least some colloidal silica in the form of a sol to prepare our binder.

Silica aquasols (water dispersions of colloidal amorphous silica) containing only a small amount of alkali as a stabilizer are commercially available and are described in the preferred aspects of this invention.

In summary, binder compositions of our invention comprise (1) aqueous solutions of alkali metal oxide silicates with or without amorphous silica present therein and (2) amorphous colloidal silica, if the silicate does not have any amorphous silica present therein or if the level of amorphous silica in the silicate is not sufficient.

The core and mold compositions of the invention have the additional advantage in that they can be made cold-setting, that is, heating to set the binder system is not necessary. Thus, they can be set with CO_2 or a suitable acid releasing curing agent.

Preferred for use in the compositions of the invention are binder wherein the alkali metal silicate is sodium silicate and at least 10% of the amorphous silica is obtained from a silica sol.

In preferred embodiments of the composition of the invention carbonaceous materials and/or film forming resin adhesives are employed. These materials can add desirable properties with respect to shake-out and storage life. The employment of these optional, but preferred, materials is described in greater detail in the following paragraphs.

Thus, I have found sand core or mold compositions of foundry sand and binder wherein the composition consists essentially of 85–97 parts by weight of foundry sand and 3–15 parts by weight of an aqueous binder comprising an aqueous sodium, potassium or lithium silicate solution or mixtures thereof with 20–55% solids content and amorphous silica, the amorphous silica in the silicate solution determined by the Gore test procedure, the binder characterized by (1) a molar ratio of silica to alkali metal oxide of from 3.5:1 to 10:1; (2) a weight fraction of the total silica present as amorphous silica is from 2 to 75%; and (3) a weight fraction of the total silica present as silicate ions is from 98 to 25% and the amorphous silica has a particle size from 2 to 500 nm and the sand core or mold possesses a compressive strength sufficiently low to permit easy crushing after said core or mold is used in preparing a metal casting.

Accordingly, the present invention also includes a method for making a sand core or a sand mold useful in the casting of molten metal which comprises mixing 85 to 97 parts by weight of foundry sand with 3 to 15 parts by weight of a binder which comprises an aqueous sodium, potassium, or lithium silicate solution or mixtures thereof with 20–55% solids content with amorphous silica having a particle size of from 2 to 500 nm, the amount of silicate and amorphous silica being adjusted to form a binder with (1) a molar ratio silica to alkali metal oxide ranging from 3.5:1 to 10:1; (2) the weight fraction of total silica present as amorphous silica from 2–75%; and (3) a weight fraction of the total silica present as silicate ions from 98–25%; the amorphous silica present in the silica solution is determined by the Gore test procedure, forming the sand and binder mixtures into the desired shape and setting the formed mixture.

DESCRIPTION OF THE INVENTION

Foundry Sand

The compositions of the invention will contain between 85 and 97 parts by weight of foundry sand, preferably between 90 and 96 parts by weight. The amount of binder used is related to sand type and particle size in that with small sand particles and more angular surfaces, more binder mixture will be necessary.

The type of foundry sand used is not critical and the useful foundry sands include all of the ones conventionally used in the metal casting industry. Thus, these sands can be zircon sands (zirconium silicates), silica sands, for example, quartz, aluminum silicate, chromite, olivine, staurolite and their mixtures.

The particle size of the foundry sand again is not critical and American Foundrymen's Society (AFS) particle sizes of 25 to 275 GFN can be employed. GFN stands for Grain Fineness Number and is approximately the number of meshes per inch of that sieve which would just pass the sample if its grains were of uniform size, that is, the average of the sizes of grains in the sample. It is approximately proportional to the surface area per unit weight of sand exclusive of clay.

The useful sands can be washed or unwashed sands and contain small amount of impurities, that is, clay. If recycle sands are used, an adjustment may have to be made to the binder mixture to take into account any silicate present in such sands.

Various minerals can be used as sand additives to optimize mold or core performance. For instance, alumina or clay powders can be used to improve the high temperature strength and shake-out characteristics of the sand cores.

Conventional refractory grain alumina powders, kaolin, and Western bentonite can be used. Kaolin is preferred in amounts between 0.5 to 10% by weight of the sand. An example of a kaolin grade useful for this purpose is Freeport Kaolin Co.'s "Nusheen" unpulverized kaolin material which consists of kaolinite particles with a specific surface area of about 16 m^2/g.

Binder System

The compositions of the invention contain 3 to 15 parts, per 100 parts of sand binder mixture by weight, of a binder system comprising a water soluble alkali metal silicate, amorphous colloidal silica and water. The key is to have very finely divided amorphous silica particles of colloidal size dispersed within the alkali metal silicate bond. It is inherent in the nature of water soluble alkali metal silicates having a molar ratio SiO_2/alkali metal oxide (M_2O) above about 2.5, that colloidal silica is present. In the case of silicates having a ratio higher than 3.5, the colloidal silica content is such that they may be employed without adding more colloidal silica, but in the case of alkali metal silicates of lower silica-/alkali metal oxide ratio there is little or no amorphous colloidal silica present so that amorphous colloidal silica must be added in order to produce the cores and molds of the present invention.

In order for the foundry core or mold to become weak after heating and cooling, it is helpful to have crystalline silica such as cristobalite formed throughout the binder mass by spontaneous nucleation at high temperatures. Such nucleation apparently occurs at the surface of particles of amorphous colloidal silica. Hence, the larger the area of such surface, weaker the resulting core after heating and cooling. If enough amorphous silica is colloidally subdivided and dispersed within the silicate, then within one gram of such silicate binder there can exist dozens of square meters of amorphous silica surface. The smaller the particles, the more rapid the loss of core strength after heating at 700°C and cooling.

The useful water soluble silicate component of the mixture includes the commercially available sodium, potassium, or lithium silicate or their mixtures. Sodium silicate is preferred. These silicates are usually used as solutions; however, their hydrates can be used provided that water is mixed into the binder, either prior to or during application to the sand. The useful sodium silicate aqueous solutions have a weight ratio of silica to sodium oxide ranging from 1.9:1 to 3.75:1 and a concentration of silica and sodium oxide about 30–50% by weight. As stated above, a fraction of the silica in the useful water soluble sodium silicate of SiO_2/M_2O ratio higher than 2.5 is in the form of very small particle size amorphous colloidal silicate. Alkali metal silicates with SiO_2/alkali metal oxide ratio higher than about 3.5:1 are referred to as high ratio alkali silicates or alkali polysilicates although they contain in fact a certain proportion of colloidal silica. In essence high ratio alkali metal silicate aqueous solutions can be conceived as mixtures of alkali metal ions, silicate ions, and colloidal silica. High ratio alkali metal silicate solutions contain varying amounts of monomeric silicate ions, polysilicate ions, and colloidal silica micelles as particles. The type, size of the ions and micelles or particles, and distribution depend for each alkali metal on ratio and concentration. Aqueous solutions of moderate concentration of the metasilicate ratio, namely SiO_2/alkali metal oxide 1:1, or more contain mainly the monomeric silicate ions. In disilicate aqueous solutions of moderate concentration, with SiO_2/M_2O of 2/1 only the simple metasilicate and disilicate ions are present. Aqueous solutions of silicates with greater ratios contain monomeric silicate ions, dimeric silicate ions, and polymeric silicate ions (trimers, tetramers, pentamers, etc.).

The degree of polymerization of the silica in silicate solutions may be expressed as the number of silicate groups formed in the average molecule of silicic or polysilicic acid corresponding to the alkali metal silicate. The degree of polymerization increases with the ratio of the silicate. Whereas, for example of sodium silicate solution of ratio 0.5:1 may have an average silica molecular weight of 60 corresponding to one molecule of SiO_2, sodium silicate solutions of ratio 1, 2, 3.5, and 4.0 are formed to have average molecular weights of about 70, 150, 325, and 400 respectively. This is the reason why as mentioned above high ratio silicates containing a large proportion of polymeric ions are also known as "polysilicates."

Silicate polymer ions with a corresponding silica molecular weight above about 600 are sufficiently large to be considered as very small silica particles and will hereinafter be referred to as colloidal silica or colloidal SiO_2. Colloidal particles are generally defined as particles with a particle size between about 1 and 500–1000 nm. This particle size range constitutes the colloidal range and is not limited by a sharply defined boundary.

Alkali metal silicates with an "average" silica molecular weight higher than around 200 to 300 have a fraction of their silicate ions present as polysilicate ions in the colloidal range. Higher the average molecular weight higher the fraction of polysilicate ions in the colloidal range and higher the molecular weight or particle size of polymer ions or particles in the colloidal range. For example, a sodium silicate solution ratio 3.35:1 may contain more than 2 or 3 and as much as 15 percent by weight of the total silicate or silica in the form of colloidal silica. Sodium silicate solutions ratios 3.75:1 and 5:9 may contain more than 8 or 10 and as much as 33% by weight of the total silica respectively in the form of polysilicate ions or colloidal silicate. Higher ratio sodium silicate solutions of various ratios eventually reach a state of equilibrium in which the colloidal silica fraction has a certain particle size distribution. In the case of sodium silicate aqueous solutions ratio 3.25 to 4 at equilibrium the colloidal silica fraction has a particle size smaller than 5 nm.

High ratio sodium silicate solutions may be prepared by simply adding dilute silica aquasols (colloidal dispersions of silica in water) to dilute low ratio sodium silicate solutions. In this case and until equilibrium is reached, average particle size of the colloidal silica fraction will be determined by time and silica particle size distribution of the original sol and the original silicate solution.

Increase in the ratio of alkali metal silicate solutions containing a constant concentration of silica causes an increase is viscosity even to the point of gelling or solidification. For this reason the maximum practical concentrations for alkali metal silicate solutions decrease with increasing ratio. Maximum practical concentration is the maximum concentration of SiO_2 plus Na_2O in solution at which the silicate solution flows like a fluid by gravity and is stable to gelation for long periods of time. Table 64.1 illustrates as an example the case of sodium silicate aqueous solutions. Reducing excessive amounts of water into the sand and without employing unstable commodities.

A practical way of using high ratio silicate as binders for foundry sands is to mix concentrated silica aquasols and concentrated sodium silicate aqueous solutions

TABLE 64.1
Maximum Practical Concentration, (% wt)

Approximate SiO_2/Na_2O molar ratio	Approximate maximum practical concentration, % wt.
1.95	55
2.40	47
2.90	43
3.25	39
3.75	32
5.0	<20

in situ, that is on the surface of the sand grains, thus forming the high ratio silicate on the sand surface.

Concentrated sodium silicate aqueous solutions cannot be mixed with concentrated silica aquasols without almost immediate gelling. It would be very impractical or simply impossible to mix gels formed in this manner with sand using the means available today in common foundry practice.

However, I have discovered that effective mixing and binding effect is obtained with sand if the concentrated silica sol is mixed first with the sand to form a uniform and continuous film on the surface of the sand grains. The concentrated sodium silicate solution is then added to the sand mass in a second, separate step and the sodium silicate then mixed with the colloidal silica film on the surface of the sand, gelling in situ to form an intimately and uniformly mixed binder within the sand mass. The sand mix thus formed in the mixer can be molded by any of the various processes available in foundry practice and hardened to form strong molds or cores.

Above a certain concentration which decreases with increasing silica-soda ratio as explained above, sodium silicate aqueous solutions become very viscous and are stable for only a limited period of time. Stability in this case means resistance to gelling. More stable solutions can be made at lower sodium silicate concentrations but this may become impractical in a foundry binder. The high water content of very high ratio (>4 to 5) sodium silicate solutions at practical viscosities prevent their extended use as a foundry binder in the present invention. Excessively high water content in a foundry binder means unacceptably weak sand molds or cores and detrimental quantities of steam evolving when the molten metal is poured into the sand mold-core assembly.

Thus, the compositions of this invention involve percent solids in the aqueous binder of from 20–55%. Based on 3–15 parts by weight of binder in the composition of this invention this translates to 1.35–12% by weight water in the binder based on the final composition.

I have discovered ways of using high ratio alkali metal silicates as foundry sand binders without introduction.

When sand molds or cores made with low ratio (<3.5) silicate binders get dry either by exposure to a dry atmosphere or by heating, they become harder. On the other hand, when sand molds or cores made with very high ratio silicate as binders get dry either by exposure to a dry atmosphere or by heating they tend to become weak and friable. This is because the overall strength of the mold or core is primarily dependent on the mechanical properties of the solid film formed by the silicate adhesive when it sets. The separation of adhesive bonds is rarely the breaking away of the solid–liquid interface but more generally a rupture either within the adhesive film or within the body of the material to which the adhesive was applied. Cracks or other faults within the adhesive film are more likely to account for low bond strength than rupture at the interface.

The formation of crystalline silica within the mass of the binder contributes to weaken the bond between sand grains after heating and cooling the molds and/or cores, therefore, providing easier core shake-out and separation of the metal from the mold. Conventional sodium silicate binders form a glass on the surface of the sand grains when the molds or cores are heated to high temperatures. When the mold or core cools down to room temperature the glass becomes very rigid forming a very strong bond, therefore, hardening the mold or core. For this reason a core made with such a binder is very difficult to break up and remove from the cavity of a cast metal during the foundry operation known as shake-out.

When colloidal silica is embedded in a matrix of sodium silicate it tends to crystallize and form cristobalite at the temperatures the cores reach when metals are cast. Due to the difference in thermal expansion coefficient, the expansions and contractions of the cristobalite crystals embedded in the glass matrix tend to crack the binder film surrounding the sand grains therefore weakening the mold or core. This weakening effect has to be added to the already mentioned weakening effect due to the cracking of high ratio silicate films on dehydration. Due to these weakening mechanisms a core made with the high ratio silicates covered by this invention is very easy to break up and remove or separate from the cast metal during the shake-out operation.

Thus the difference in behavior between low and high ratio silicate binders for sand molds and cores can be understood by observing films formed on silica glass plates by slow evaporation of for example, aqueous solutions of sodium silicate of various ratios.

The low silicate/soda ratio (2.0) sodium silicate solution dries in air at room temperature very slowly forming a very viscous, smooth, clear film. At higher ratio (2.4) drying is faster and the silicate film obtained shows some cracks. At very high ratios (3.25 and 4.0) sodium silicate solutions include substantial amounts of very small particle size colloidal silica and drying is even faster: cracking is even more extensive and the film tends to lose integrity. A silica sol of particle size 14 nm and SiO_2/Na_2O ratio 90 does not form a continuous film under the same drying conditions.

Low ratio silicate binders thus form on the sand surface viscous, smooth films which do not form cracks on drying. On the other hand, the films formed on the sand surface by high ratio silicate binders, crack on drying thus weakening the sand core or mold. For these reasons cores made with low ratio silicate binders outside the scope of the present invention become stronger when they are heated at high temperatures by molten metals in the pouring operation of the casting process. On the other hand, cores made with high ratio silicate binders within the present invention are reasonably strong when just made, but become weak and friable during the casting operation.

In the practice of this invention a compromise has to be made when choosing a binder composition by selecting one with a SiO_2/Na_2O ratio not so high that the sand molds or cores will weaken to unacceptable levels by merely drying at room temperature when exposed to the atmosphere, and not so low that the sand molds or cores will form a cohesive, solid glass bond when the core or mold is heated in the casting operation so that the core or mold becomes very strong when cooled down to room temperature and cannot be separated easily from the metal casting. The room temperature, as-made strength of sand molds or cores obtained with high ratio silicate binders of this invention may be upgraded by the addition to the silicate bonded sand mix of a fugitive film-forming resin adhesive in the form of a water solution or water dispersion. In this case, as explained below in more detail, the molds or cores become stronger by drying at room temperature. However, when heated to high temperatures during the casting process the resin adhesive decomposes evolving harmless vapors and the weakened core and mold can be easily separated from the cast metal.

If a preformed sodium polysilicate having a molar ratio of silica to alkali metal oxide in the range of 3.5–10 is employed before it gels, the same effects as with the amorphous silica sodium silicate system will be obtained. An aqueous sodium polysilicate containing 10–30% by weight silica and sodium oxide and having a silica to sodium oxide weight ratio of 4.2:1 to 6.0:1 can be produced as described in U.S. Pat. No. 3,492,137.

Similarly, the high ratio lithium silicates of Iler U.S. Pat. No. 2,668,149 or the potassium polysilicates of Woltersdorp, application Ser. No. 728,926, filed May 14, 1968, now Defensive Publication 728,926, dated Jan. 7, 1969, can be employed as the binder provided the requirements as to molar ratio, particle size and amount of amorphous silica are followed.

Furthermore, alkali metal polysilicates stabilized by quaternary ammonium compounds or guanidine and its salts can also be employed. Some stabilized polysilicates of this type are described in U.S. Pat. No. 3,625,722. This method, however, has the disadvantage of producing unpleasant odors on casting due to the thermal decomposition of the organic molecule.

Complexed metal ion stabilized alkali metal polysilicates can also be used, such as copper ethylenediamine hydroxide stabilized sodium polysilicate made by mixing copper ethylenediamine with colloidal silica and then the silicate, or the stabilized polysilicates of U.S. Pat. No. 3,715,224.

The useful amorphous silica are those having a particle size in the range from about 2 to 500 nm. In addition to the amorphous silica already present in aqueous solutions of high ratio alkali metal silicates, such silicas can be obtained from silica sol (colloidal dispersions of silica in liquids), colloidal silica powders, or submicron particles of silica. The silica sols and colloidal silica powders, particularly the sols, are preferred in view of the shake-out properties of the binders made from them.

Gore Procedure

The amount of colloidal silica present in an aqueous solution of high ratio alkali metal silicate can be determined for example by ultrafiltration. Ultrafiltration refers to the efficient selective retention of solutes by solvent flow through an anisotropic "skinned" membrane such as the Amicon "Diaflo" ultrafiltration membranes made by the Amicon Corporation of Lexington, Mass. In ultrafiltration solutes, colloids or particles of dimensions larger than the specified membrane "cut-off" are quantitatively retained in solution, while solutes smaller than the uniform minute skin pores pass unhindered with solvent through the supportive membrane substructure.

Amicon "Diaflo" ultrafiltration membrane offer a selection of macrosolute retentions ranging from 500 to 300,000 molecular weight as calibrated with globular macrosolutes. These values correspond to pore sizes between about 1 and 15 nm. Each membrane is characterized by its nominal cut-off, that is, its ability to retain molecules larger than those of a given size.

For effective ultrafiltration, equipment must be optimized to promote the highest transmembrane flow and selectivity. A major problem which must be overcome is concentration polarization, the accumulation of a gradient of retained macrosolute above the membrane. The extent of polarization is determined by the macrosolute concentration and diffusivity, temperature effects on solution viscosity and system geometry. If left undisturbed, concentration polarization restricts solvent and solute transport through the membrane and can even alter membrane selectivity by forming a gel layer on the membrane surface — in effect, a secondary membrane — increasing rejection of normally permeating species.

An effective way of providing polarization control is the use of stirred cells. Magnetic stirring provides high ultrafiltration rates.

A recommended procedure is to use a Amicon ultrafilter Model 202, with a pressure cell of 200 ml capacity and a 62 mm diameter ultrafilter membrane operated at 25°C. with magnetic stirring with air pressure at around 50 psi.

In the case of sodium silicate for example, an aqueous solution diluted with water, is placed in the cell. An Amicon PM-10 membrane, 1.8 nm diameter pores, is used. Pressure is applied and filtrate collected. In some cases, water is fed in to replace the volume passing through the filter into the filtrate. The solution in the filter cell is concentrated until the filtration rate is only a few milliliter per hour.

The filtrate is collected in progressive fractions, and they and the final concentrated solution from the cell are examined: Volumes are noted and SiO_2 and Na_2O concentrations in grams per milliliter are determined by chemical analysis.

In some cases, the concentrated solution on the filter is further washed by adding water under pressure, as fast as filtrate is removed. In these cases there is further depolymerization or dissolution of the colloid fraction.

The percentage of colloidal silica, based on total silica, is indicated by the amount of residual silica that does not pass through the filter. These represent maximum values for the amount of colloid present, since some ionic soluble silica is still present. In further examples the residual soluble silica is subtracted and the composition of the colloid is calculated.

It is not necessary to isolate the pure colloid, but only to measure the concentration of SiO_2 and Na_2O as ultrafiltration proceeds. Since the concentration of "soluble" sodium silicate in the filtrate is about the same as in the solution in the cell if this colloid is present only at low concentration, the amount and composition of colloid can be calculated by difference.

Allowance should be made in interpreting results obtained with this method for the fact that every time water is added to the system some depolymerization of colloid or polysilicate ions probably occurs.

The colloidal amorphous silicas useful in preparing the compositions of the invention have a specific surface area greater than 5 m^2/g and generally in the range of 50–800 m^2/g and preferably in the range of 50–250 m^2/g. The specific surface area is determined by nitrogen adsorption according to the BET method. The ultimate particle size of the silica used is in the colloidal range, and is generally in the range of 20–500 nm, preferably 12–60 nm. Thus, the silica sols of the desired particle size range described by M.F. Bechtold and O.E. Snyder in U.S. Pat. No. 2,574,902; J.M. Rule in U.S. Pat. No. 2,577,484; or G.B. Alexander in U.S. Pat. No. 2,750,345 can be used.

Positive silica sols and alumina modified silica sols wherein the ultimate silica particles have been modified and/or made electrically positive by partially or completely coating the particle surface with aluminum compounds can also be used in the present invention as a source of amorphous silica. Such sols are described for example by G.B. Alexander and G.H. Bolt in U.S. Pat. No. 3,007,878 and by G.B. Alexander and R.K. Iler in U.S. Pat. No. 2,892,797. The advantage of these sols is that in some cases they form more stable mixtures with sodium silicate aqueous solutions than the unmodified silica sols.

Certain very finely divided colloidal silica powders such as those made by the "fume process" by burning a mixture of silicon tetrachloride and methane, have a sufficiently discrete, particulate structure that such powders can be dispersed in water by colloid milling to give a sol useful in this invention. It is also obvious that such a powder can also be colloid milled directly into a solution of silicate.

Very finely divided colloidal silica powders can also be obtained by treating certain silicate minerals such as clay or calcium silicate with acid, followed by suitable heat treatment in an alkaline medium. Similarly, finely divided colloidal silicas can be produced by precipitating silica from a solution of sodium silicate with carbon dioxide. Such precipitated silicas are commonly used as reinforcing fillers, for elastomers because they are extremely finely divided, and the ultimate particles are easily broken apart. Finely divided aerogels of silicas may be employed, such as those described by Kistler in U.S. Pat. Nos. 2,093,454 and 2,249,767.

The finely divided colloidal silica powders useful in the composition of the invention are characterized by having specific surface areas as determined by nitrogen adsorption according to the BET method, of from 5 to 800 m^2/g and preferably 50 to 250 m^2/g, and being further characterized by the fact that the aggregates of ultimate silica particles are generally <10 μm in diameter.

The amounts and types of amorphous silica that can be dispersed within the soluble silicate depends to a considerable extent on the amount of grinding or mixing that is done to disintegrate and disperse particles of amorphous silica in the silicate bond. Thus, for example, it is possible to start with fused silica glass and grind it to the point where a substantial amount is present as particles smaller than a micron. The inclusion of a high concentration of this type of material can provide sufficient surface for nucleation of cristobalite or tridymite within the alkali metal silicate glass bond when the sand core or mold reaches high temperature during the metal casting operation. Also, finely divided natural forms of silica such as volcanic glasses which, in the presence of alkali silicates, can be devitrified, may be used, providing they are sufficiently finely divided and well dispersed in the sodium, potassium, or lithium silicate solution used as the binder.

The compositions of the invention will have 2 to 75% of the total silica present in the binder present as amorphous silica, preferably 10 to 50% the balance of the total silica being in the form of silicate ions. As the specific surface area of the amorphous silica increases, lesser amounts of it will be required in the binder mixture.

There is a practical maximum concentration of amorphous silica that can be dispersed in the aqueous silicate solution. It is often desirable to incorporate as high a concentration of amorphous silica as possible, yet still have a workable fluid binder to apply to the sand. If the proportion of amorphous silica to soluble silicate is too low, then the shake-out will be adversely affected. On the other hand, if the ratio of amorphous silica to soluble silicate is too high, the mixture will be too viscous and must be thinned with water. Also, there will not be enough binder to fill the spaces between the amorphous silica particles in the bond, and it will be weak. In generaly, the higher the content of amorphous silica relative to sodium or potassium silicate, the weaker the initial bond as set by carbon dioxide. Conversely, the more silicate in the binder, the higher will be the initial and retained strengths.

The binder system should have a molar ratio of silica to alkali metal oxide which ranges from 3.5 to 10, preferably 3.5 to 7. This ratio is significant because the ratios of soluble potassium, lithium, or sodium silicates commercially available as solutions lie within a relatively narrow range. Most of sodium silicates are within the range of SiO_2/Na_2O of about 2:1 to 3.75:1. Thus, over-all ratios of binder compositions obtained by admixing colloidal silica, such as ratios of 4:1, 5:1, 7:1 are mainly an indication of what proportions of colloidal silica and soluble silicates were mixed since the amount of amorphous silica in the soluble silicate at ratios of 2:1 to 3.75:1 are small.

However, in the ratio range of about 3.5:1 to 4.0:1, compositions of a specified ratio are not necessarily equivalent. Thus, a potassium silicate having an SiO_2/K_2O ratio of 3.9:1, in which there is a distribution of polysilicate ions, but relatively small amount of colloidal silica, differs considerably from a mixture made by mixing a potassium silicate solution of SiO_2/K_2O of 2.0:1 with colloidal silica having a particle size of, for example, 14 nm. In the latter case, the colloidal particle will remain as such in solution over a considerable period of time. Such a composition has two advantages over the more homogeneous one in that the low ratio of silicate has a higher binding power giving greater initial strength, while the higher content of colloidal particles results in a major reduction in the strength in the core after casting the metal.

OPTIONAL ADDITIVES

In the casting of some metals, for example, iron or steel, very high casting temperatures are involved, that is, 2500 to 2900°F. If the mass of the core is small relative to the mass of the cast metal during such high temperature casting, there may be some vitrification of the silicate thus creating shake-out problems. To alleviate this situation a carbonaceous material can be added to the core composition. These carbonaceous materials assist the binder of the invention in providing excellent shake-out, particularly after the core has been subjected to very high temperatures.

The useful carbonaceous materials should have the following characteristics:

(a) It should not interfere with the binder system.
(b) It should have a particle size or primary aggregate equivalent diameter sufficiently large to leave discontinuities in the glass formed by the binder at very high temperatures, as it burns off partially or completely. It should also have a particle size which is not large enough to weaken the sand core as fabricated, and specially not larger than the particle size of the sand itself. Thus the particle size or primary aggregate equivalent diameter should range between 0.1 and 75 μm, preferably between 5 and 50 μm. When the ultimate particle size of the carbonaceous material is smaller than 0.1 μm it is generally coalesced or it tends to coalesce in the sand mix into primary aggregates larger than 0.1 μm.
(c) It should not be too avid for water, otherwise it would subtract from the binder system, drying up the sand and making it impossible or difficult to mold.

Preferred for use are pitch, tar, coal-tar pitch, pitch compounds, asphaltenes, carbon black, and sea coal, and most preferred are pitch and carbon black.

Pitch is a by-product from coke making and oil refining and is distilled off at around 350°F. It has a melting range from 285 to 315°F, is highly volatile, high in carbon and extremely low in ash. Following is a typical analysis of coal-tar pitch in weight percent:

Volatile	47.37%
Fixed carbon	52.43
Ash	0.2
Sulfur	0.5

Pitch is a material resistant to moisture absorption and is often used as a binder or as an additive for foundry sand cores and molds.

Sea coal is a common name used to describe any ground coal employed as an additive to foundry sands. Sea coal is used in foundry sands primarily to prevent wetting of the sand grains by the molten metal, thus preventing burn-on and improving the surface finish of castings. It is also used as a stabilizer and to promote chilling of the metal.

Following is a typical analysis of sea coal given on a dry basis:

	Weight percent
Ash	5.10%
Sulfur	0.51
Volatile carbonaceous material	40.00
Fixed Carbon	53.20
Ultimate analysis	
Hydrogen	5.20%
Carbon	81.29
Nitrogen	1.50
Oxygen	6.40
Sulfur	0.51
Ash	5.00

Tar is generally defined as a thick, heavy, dark brown, or black liquid obtained by the distillation of wood, coal, peat, petroleum, and other organic materials. The chemical composition of a tar varies with the temperature at which it is recovered and raw material from which it is obtained.

Carbon blacks are a family of industrial carbons, essentially elemental carbon, produced either by partial combustion or thermal decomposition of liquid or gaseous hydrocarbons. They differ from commercial carbons such as cokes and charcoals by the fact that carbon blacks are particulate and are composed of spherical particles, quasigraphitic in structure and of colloidal dimensions. Many grades and types of carbon black are produced commercially ranging in ultimate particle size from <10 to 400 nm. In most grades ultimate particles are coalesced or fused into primary aggregates, which are the smallest dispersible unit of carbon black. The number of ultimate particles making up the primary aggregate gives rise to "structure" — the greater the number of particles per aggregate, the higher the structure of the carbon black.

When mixed with sand fine particle size carbon blacks are coalesced into aggregates in the sand mix, therefore they leave discontinuities in the binder phase when burned off during the high temperature casting operation.

An example of a commercial carbon black is Regal 660, sold by the Cabot Corporation of Boston, Mass., which has the following characteristics:

Nigrometer index	83
Nitrogen surface area	112 m^2/g
Oil (DBP) absorption	62 cc/100 g
Fixed carbon	99%

The carbonaceous material should be present in the core composition in the amount of 0.5–4 wt% based on the foundry sand, preferably 1–2 wt%.

The amount of carbonaceous material, for example, pitch, needed depends, to some degree, on the refractoriness of the binder used which is in turn a function of the silica/alkali molar ratio, and on the temperature to which the core will be subjected during casting. When a SiO_2/Na_2O ratio of 5:1 sodium polysilicate is used as a binder, no pitch is needed if the core is used for nonferrous metal castings since in these cases the core temperature will not exceed about 1200°C. If the same binder is used for small cores in massive iron castings, 2% of pitch is useful to help break up the silicate glass formed.

In the event it is desirable to make cores and store them for extended periods of time prior to use, I have discovered that the addition of a film-forming resin adhesive in the form of a water solution or water dispersion, drastically extends the storage life of foundry sand cores made with the binder of the invention. Thus the use of these materials enable the formed cores to retain sufficient strength and hardness during storage.

Useful film-forming resin adhesives include polyvinyl-esters and ethers and their copolymers and interpolymers with ethylene and vinyl monomers, acrylic resins and their copolymers, polyvinyl alcohol, water dispersions of polyolefin resins, polystyrene copolymers such as polystyrene butadiene, polyamide resins, natural rubber dispersions, and natural and modified carbohydrates (starch or carboxycellulose). Particularly preferred for use are aqueous dispersions of polyvinyl acetate and vinyl acetate-ethylene copolymers.

The polymer resin should be in a state of subdivision suitable for uniform distribution on the sand grains to form an adhesive film and hold the sand grains strongly together. It is preferred that resin dispersions be between 40 and 60% by weight solids. The higher the concentration of solids, the better, as less water will have to be removed, however, with concentrations above 60% by weight it can be difficult to mix the dispersion into the sand. With resin solutions, for example, solutions of polyvinyl alcohol, concentrations of 4 to 20% solids are preferred.

Useful polyvinyl acetate dispersions are milk-white, high-solids dispersions of vinyl acetate homopolymer in water. Such dispersions have excellent mechanical and chemical stability. Typical properties of a preferred polyvinyl acetate dispersion are given in Table 64.2. Commercially available dispersions with similar characteristics are Monsanto's S-55L, Borden's "Polyco" 11755, Air Products' "Vynac" XX-210, and Seydel Wooley's "Seycorez" C-79.

The useful vinyl acetate–ethylene copolymers are milk-white dispersions of 55 w/o solids in water with a viscosity between 12 and 45 poises. DuPont's "Elvace" is a commercially available dispersion with these characteristics.

The useful polyvinyl alcohol (PVA) is a water soluble synthetic resin 85–99.8% hydrolyzed. Du Pont's "Elvanol" resins and Goshenol GL-05, 85% hydrolyzed, low viscosity PVA are examples of suitable commercially available materials. "Elvanol" grades give 4% water solutions with a viscosity ranging from 3.5 to 65 Cp at 20°C as measured by the Hoeppler falling ball method. Water solutions of PVA at low concentrations (up to about 10–15 wt%) or concentrated aqueous colloidal dispersions of the water insoluble polymer resins mix uniformly with sand and provide good adhesion. Very concentrated water solutions of PVA (higher than 20 wt%) are too viscous and do not mix well enough with sand.

To obtain optimum adhesion, the film forming resin dispersion or solution should be added such that it does not gel or coagulate either the silica or the sodium silicate before adding them to the sand. For instance, the polymer resin dispersions can be mixed with the silica before adding to the sand because both are compatible and do not gel when mixed together. The mixtures can be added to sand and they will form an adhesive film on the surface of the sand grains. After the silica and the polymer resin dispersion have been mixed with the sand, the sodium silicate solution can be added to the sand and although it will thicken in contact with the silica and the polymer resin dispersion, it will do so in situ, that is, fairly uniformly distributed on a preformed film of silica and polymer resin.

If before adding to the sand the sodium silicate is mixed with the concentrated polymer dispersion and the silica, it thickens and gels and it cannot subsequently be mixed adequately with the sand. Instead of distributing fairly uniformly on the surface of the sand grains, it would tend to form lumps and distribute unevenly in the sand.

Alcoholic solutions of the polymer resins may be used but are not recommended as additives to the silica-sodium silicate binder because they get very thick in contact with the binder and tend to gel faster than the aqueous dispersions and therefore do not distribute as uniformly on the sand grains. However, dilute alcoholic solutions of polymer resins can be used as such or mixed with commercial zircon core washes to coat the surface of the cores and give improved hardness and storage life to the cores. In this case the gel forms on the surface of the sand core already set, and it air dries fairly fast or it is dried almost instantaneously by lighting the alcohol to extinction of the flame, therefore preventing the possible diffusion of the alcohol into the core.

The use of a water solution or water dispersion of a polymer resin produces sand cores with the silica-sodium silicate binder having as gassed mechanical strength somewhat lower than that of sand cores made with silica-sodium silicate binder without the polymer resin solution or dispersion. This may be due to the weakening of the sodium silicate bond caused by the dilution produced by the water of the polymer resin solution or dispersion. However, drying of the core on storage, more than overcomes this effect and after very few days the cores show a much higher mechanical strength than the one obtained immediately after gassing with CO_2.

Two mechanisms may contribute to the hardening and strengthening on storage provided by the polymer resin. One is the thickening in situ of the adhesive film of silica–polymer resin–sodium silicate on the sand grains due to the "salting-out" effect caused by electrolyte formation on gassing with CO_2. More important is the thickening and solidification of the film caused initially by the CO_2 blown through the sand grains and specially the subsequent evaporation of the water from the sand core on storage.

TABLE 64.2
Typical Properties of a Preferred Polyvinyl Acetate Homopolymer Aqueous Dispersion

Solids, %	55
Brookfield viscosity, P[a]	8.5–10
pH	4–6
Molecular weight (number average)	30,000–60,000 (mostly crosslinked)
Average particle size, μ	1–2 (range from 0.1 to 4)
Density (25°C), approx. lb/gal	9.2
Surface tension (25°C), approx. dynes/cm	55
Min. film formation temperature[b]	
°C	17
°F	63
Residual monomer as vinyl acetate, % max.	1.0
Particle charge	Essentially nonionic

[a]Brookfield model LVF, No. 2 spindle at 6 rpm or No. 3.

[b]ASTM D2354.

Under these conditions the polymer resin macromolecules and/or colloidal particles are expected to coalesce and form an effective adhesive bond between the sand grains and reinforce the sodium polysilicate binder.

In the case of the polyvinyl esters the alkaline hydrolysis caused by the mixing with the sodium silicate will tend to form in the already formed uniform film, polyvinyl alcohol, perhaps an even better adhesive than the ester itself.

The colloidal silica-resin, for example, polyvinyl acetate components of the binder can be used in the form of a stable liquid mixture, the carbonaceous material being optionally present. Thus uniform mixtures containing colloidal silica and polyvinyl acetate within the relative amounts specified in this invention, such as 1.94 parts by weight of 40% aqueous colloidal silica and 2 parts by weight of 55% polyvinyl acetate aqueous dispersion, can be made by mixing the two components in a beaker. The mixture is stable and uniform and can be used within the working day. Overnight the mixture tends to separate in two layers and can be stirred up to make it uniform.

One method of providing a stable, pourable mixture of colloidal silica–polyvinyl acetate with or without the carbonaceous material, for example, pitch, is to make the liquid phase slightly thixotropic but not viscous. In other words, to make it so that it sets to a weak gel structure at once when undisturbed (to maintain all particles in uniform suspension) but when stirred, or even tilted to pour, the yield point is so weak as to permit ready transfer of the material and easy blending with the sand.

Thixotropic suspensions with the characteristics described above can be prepared using a three component suspending agent system disclosed in U.S. Pat. No. 3,852,085, issued Dec. 3, 1974. This system consists of (a) carboxymethyl cellulose and (b) carboxyvinyl polymer in a total amount of about 36–65 wt% with relative amount of (a) to (b) varying from a weight percent ratio of about 1:4 to 4:1 and (c) magnesium montmorillonite clay in a concentration of about 35–64 wt%.

The useful compositions will contain between 95 and 99.5% by weight of the binder components and between 0.5 and 5% by weight of the suspending agent system. In a composition containing only the colloidal silica and resin, 15–35% of the binder will be silica solids and 15–35% of the binder will be resin solids. In a three component binder, 5–20% will be silica solids, 5–20% resin solids and 5–40% will be carbonaceous matter.

This suspension system can be used with dispersions containing a maximum solid content of 55% by weight of polymer resin and colloidal silica or polymer resin, colloidal silica and carbonaceous material such as pitch. The minimum solid content is only limited by the amount of water that is practical to add to the sand mix to obtain practical cores.

For example, to prepare a colloidal silica–polyvinyl acetate-pitch suspension 0.67 parts by weight of "Benaqua" (magnesium montmorillonite sold by the National Lead Co.) can be dispersed in 235 parts by weight of water with low shear mixing; 0.67 parts by weight of CMC-7H (carboxymethyl cellulose) and 0.67 parts by weight of Carbopol 941 (water soluble carboxyvinyl polymer) can be added and dissolved using low shear mixing; 0.15 parts by weight of a 1% solution of GE-60 (silicone-based emulsion) can be added as an antifoam agent; 194 parts by weight of "Ludox" HS-40 (aqueous colloidal silica dispersion sold by E.I. du Pont de Nemours & Co.) can be added and mixed with moderate shear mixing; 200 parts by weight of Gelva S-55L (polyvinyl acetate aqueous dispersions sold by the Monsanto Company) can be added and mixed with moderate shear mixing; then 200 parts by weight of "O" Pitch sold by the Ashland Chemical Company can be added and mixed with moderate shear mixing. A fluid suspension containing colloidal silica–polyvinyl acetate and pitch is obtained at a suitable ratio to be used as a component of the silicate binder system of the invention.

Alternatively, 58 parts by weight of water can be used instead of 235 parts by weight of water and in this case a uniform, stable suspension is obtained which is more viscous than the previously described, but still pourable and mixes well with sand.

Alternatively, pitch can be omitted from the preparation, and fluid suspensions containing colloidal silica-polyvinyl acetate are obtained at a suitable ratio to be used as components of the silicate binder system of the invention.

Application of the Binder

The binder mixture of the invention can be applied to the sand in various ways. Thus, if the binder mixture has sufficient shelf life, it can be formulated, stored, and applied to the sand when needed. The silicate and amorphous silica can be stored separately and then mixed together when needed and applied. Furthermore, they can be applied separately to the sand. If this latter procedure is used, it is preferred to first apply the amorphous silica, mix it into the sand, then apply the silicate and mix again. However, the silicate can be applied first.

Uniform sand mixes can be prepared by adding the binder to the sand in conventional foundry mixer, muller, or mix-mixers, or laboratory or kitchen mixers, and mixing for sufficient time to obtain a good admixture of the sand and binder, for example, for several minutes. When added separately, it is desirable to mix each component for less than two minutes to avoid undue drying.

If an alkali metal polysilicate solution is used as a binder, it should be mixed directly with the sand. If on the other hand colloidal silica and sodium silicate solution are added separately to the sand, it is preferable to add the silica sol first and to mix it thoroughly with the sand before adding the sodium silicate. Once the sodium silicate is added, the mix should not be kept too long in the mixer. A period of two minutes stirring is generally optimum for the sodium silicate.

Dry colloidal silicas such as pyrogenic amorphous silica do not mix well with the sand and in addition they tend to absorb water from the sand-binder system. Therefore, dry colloidal silica powders should be added to the sand in the form of a paste made with water or water should be added to the sand to help mix the dry silica powder. The amount of water made to use the paste should be enough to assure good mixing of the silica powder and yet not too much to affect the strength of the core or mold when it is hardened. Generally the amount of water needed in this case is no more than around 3% by weight of sand.

When the film forming resin or pitch are incorporated into the core composition, if the components are added separately to the sand, the resin should be added to the sand before the silicate. The resin can be added to the sand before or after the colloidal silica. The order in which the pitch is added is not critical with respect to either the silica or the silicate.

When materials such as clays or oxides are used as additives besides the binder, they should be mixed thoroughly with the sand in the sand mixer before adding the binder.

In some cases it is found convenient to use a release agent mixed with the sand to prevent the core or mold from sticking to the core box or pattern after setting. In these cases a conventional core or mold release such as kerosene or Mabco Release Agent "G" supplied by the M.A. Bell Company of St. Louis, Mo., should be added to the sand mix in the last 20 sec of the 2 min. period of mixing the sodium silicate.

If the sand mix is not going to be used immediately, it should not be allowed to dry or react with atmospheric CO_2. The mix should therefore be stored in a tightly closed container or plastic bag from where the air has been squeezed out before sealing until it is ready to be used. If a slightly hard layer forms on the top surface of the sand due to air left inside the container, the hard layer should be discarded before using the sand to make cores or molds.

A practical way of checking uniformity of the sand mix and observe changes in the sand mix, such as reaction with the atmospheric CO_2, is to add a few grams of an indicator such as phenolphthalein at the beginning of the mixing operation. The phenolphthalein can be added in the form of a fine powder before adding the sodium silicate or dissolved in the sodium silicate or in the silica sol. Usually 160 mg of phenolphthalein per kilogram of sand is sufficient to develop a deep pink color in the sand mix.

Conventional foundry practice can be followed to form and set the sand core or mold. The sand can be compacted by being rammed, squeezed, or pressed into the core box either by hand or automatically, or can be blown into the core box with air under pressure.

The formed sand mix can be hardened very fast at room temperature by gassing the sand with CO_2 for a few seconds. Optimum gassing time can be determined either by measuring the hardness or the strength of the core or by observing the change of color of the sand mix when an indicator such as phenolphthalein has been previously added to the sand.

Thermal hardening can be used for cores made with the binder compositions of the invention instead of CO_2 hardening. For instance, high strength cores can be obtained in a very short time by forming the sand mix in a hot box at temperatures between 100 and 300°C. In general, the higher the temperature the shorter the time required to achieve a certain strength level. On the other hand at a fixed temperature in general, the core strength increases with time of heating. However, thermal hardening is not a preferred setting process for the compositions of the invention because cores made in this way do not have as good shake-out characteristics as those made by CO_2 hardening.

Another fast hardening process that can be used is CO_2 gassing in a warm box (about 60 to 80°C) or gassing with heated CO_2.

When fast hardening is not required, cores with the binders of the invention can be set with other common curing agents used for the systems known in the art as silicate no-bakes. These curing agents are organic materials which are latent acids such as ethyl acetate, formamide, and acetins. Most of these agents contain glycerol mono-, di-, or tri-acetates or any other material which can release or decompose into an acid substance which in turn produces hardening of the alkali metal silicate. Furthermore, such a hardening process can produce cores having long shelf life without the need for a film-forming resin adhesive, that is, polyvinyl acetate.

Conventional water based or alcohol based core washes can be used to treat the surface of the cores. This type of treatment is in some cases to improve the surface of the metal casting or the hardness and shelf life of the core. Shelf life is the period of time after making for which the sand core is useful.

Polyvinyl acetate homopolymers and copolymers can be used as core washes for sand cores as aqueous dispersions, in organic solvent solutions or mixed with zircon or graphite in aqueous or alcoholic suspensions. Polyvinyl alcohol or partially hydrolyzed polyvinyl alcohol can be used in aqueous solutions, organic solvent dispersions or mixed with zircon or graphite.

Polyvinyl alcohol or hydrolyzed polyvinyl acetate: 5 to 20 wt% in water solutions or 5–40 wt% in alcoholic solutions. More concentrated solutions are too thick to obtain uniform coating of the cores, more dilute solutions are too thin to provide satisfactory protective coating on the core surface.

Polymer resin aqueous dispersions and alcoholic solutions: Five percent by weight to 40% by weight of polymer resin such as polyvinyl acetate homopolymer or copolymer in water solutions or 5% by weight to 25% by weight of polymer resin such as polyvinyl acetate homopolymer or copolymer in alcoholic solutions.

Polymer resin-zircon or graphite mixtures: In water based core washes: 15–25% by weight of polymer resin such as polyvinyl acetate homopolymer or copolymer and 30–50% by weight of zircon (25 to 50 percent by weight of water).

In alcohol based core washes: 5–10% by weight of polymer resin such as polyvinyl acetate homopolymer or copolymer and 30–50% by weight of zircon or graphite (40–60% alcohol).

The alcohols useful in the above core washes include methanol and ethanol.

Satisfactory polymer resin-zircon core washes are made for example by slurrying 1 part by weight of a commercial zircon core wash (as shipped by the supplier in the form of a wet powder) in 1 part by weight of 55% polyvinyl acetate aqueous dispersion if the core wash is intended to be used shortly after preparation. More dilute slurries are preferred for core wash compositions intended to be stored for some time before using. In this case the 1 part by weight of the zircon wet powder should be slurried in 1 part by weight of water before mixing with 1 part by weight of 55% polyvinyl acetate aqueous dispersion.

Aqueous polyvinyl acetate or zircon-polyvinyl acetate or graphite-polyvinyl acetate core washes are applied on the core surface by common foundry practices such as dipping, spraying, brushing, and so on, and allowing the core to air dry before using.

Sand cores coated with alcohol base polyvinyl acetate or zircon-polyvinyl acetate are lighted immediately after one wash application as in common foundry practice with alcohol base zircon core washes.

Concentration of polyvinyl alcohol aqueous solutions to give satisfactory core washes with adequate viscosity depends on molecular weight of the polymer. Polyvinyl alcohol solutions can also be used as a mixture with zircon or graphite core wash.

CASTING METALS

Sand molds and cores made with the binder compositions of the invention can be used to cast most metals, such as gray, ductile and malleable iron, steel, aluminum, copper-based alloys such as brass or bronze. Steel is usually cast at around 2900°F, iron at about 2650°F, brass and bronze at around 2100°F and aluminum at about 1300°F.

With the molds or cores of the invention it is desirable that the core have an initial strength such that it can be handled without undue care and that it will stand up during the casting of the molten metal, that is, will not wash away or distort. In standard American Foundrymen's Society lab tests this means that the core should have a compressive strength of at least 100 psi and preferably over 150 psi.

It is desirable that the hardness of freshly made cores exceed 5, preferably 10. The greater the hardness, the better, particularly at the time of metal pouring when it should exceed 10 and preferably 20.

Scratch hardness of cured cores can be measured with commercial hardness tester No. 674 available from Harry W. Dietert Co., 9330 Roselawn Avenue, Detroit, Mich. This is a practical, pocket-sized instrument for measuring the surface and sub-surface hardness of baked cores and dry sand molds.

The tester has three abrading points which are loaded by a calibrated spring which exerts a constant pressure. These abrading points are rotated in a circle $\frac{3}{8}''$ in diameter. To obtain the hardness values, the lower end of the instrument is held against the sand surface and the abrading points are rotated three revolutions. The hardness values are actually obtained by measuring the depth to which the abrading points penetrate. The maximum hardness value indicated by this tester is 100 for zero penetration. When the abrading points move down a distance of 0.250 in., the hardness of the core is zero. Intermediate values are read from the instrument dials.

The core should, after the metal has been cast and cooled, have a retained strength such that it can be shaken out without the use of undue energy. This corresponds to a compressive strength in lab tests of, preferably, <50 psi.

The following examples are offered to illustrate various embodiments of the invention. All parts and percentages are by weight unless otherwise indicated.

EXAMPLE 1

This is an example of the use of guanidine stabilized sodium polysilicate (SiO_2/Na_2O ratio 5:1) prepared according to Example 1 of patent application Ser. No. 287,037, filed Sept. 7, 1972, as a binder for foundry sand cores. These sand cores were used to make aluminum castings in a nonferrous metal foundry.

The binder sample was made with 1890 g of sodium silicate Du Pont Grade No. 20 (SiO_2/Na_2O molar ratio 3.25:1, 28.4% SiO_2, 8.7% Na_2O), 56 g of water, 539 g of 1.3 *M* guanidine hydroxide and 1015 g of Ludox® HS, a commercial colloidal silica sol containing 30% SiO_2 of particle size of about 14 nm.

The sand mix was prepared in the following way: 90 g of kaolin and 2 g of phenolphthalein were added to 10 lbs. of sand while stirring in a 10-lb. capacity Clearfield mixer 0.5 lbs. of binder solution were also added to the sand while stirring and the sand was mixed for a total of two minutes.

The sand used was a mixture of 50 parts of Houston's subangular bank sand AFS No. 40–45 and 50 parts of No. 1 Millcreek, Okla. AFS 99 ground sand. The sand when used was at room temperature (75°F). Humidity of the room was about 80%. The binder mixed readily with the sand showing excellent mixing characteristics. Flowability of the mix was also excellent.

The sand mix was placed in a polyethylene bag and sealed. The sand mix was used the following day to make sand cores. Three to four pound sand cores were made by filling wooden core boxes with the sand mix, compacting the sand by hand and gassing it for about 15–25 sec with CO_2 gas at an estimated pressure of 20–30 lbs.

The color of the sand is deep pink due to the phenolphthalein added. After gassing the cores had the natural color of the original sand. Good release of the core was observed when the core box was opened to remove the core. The cores were immersed in a conventional alcohol zircon core wash and flamed before using. This is common practice with core washes for sodium silicate sand cores.

The cores were assembled in a sand mold and used to make an aluminum casting. Aluminum was poured at a temperature of about 1375°F. When pouring was completed the casting was allowed to cool for about 15 min inside the sand mold assembly. The aluminum casting was removed from the mold when still hot and the sand core was observed before shake-out. Shake-out was very easy; the core broke up and flowed like unbonded sand upon touching. No offensive odors were noticed during the casting and cooling.

The aluminum castings had very good surface finish and were used in normal production.

EXAMPLE 2

This is an example of the use of sand cores made with the binder solution of Example 1, to make gray iron castings.

Two 10 lb sand mix batches were made by adding 0.5 lbs of the binder solution and 0.7 g of phenolphthalein to 10 lbs of Houston subangular bank sand AFS 45–50, while stirring in a 10 lb capacity clearfield mixer and mixing for 2 min. The binder mixed very well with the sand and gave a uniform and mix containing 5% of binder by weight of sand. The sand mix showed excellent flowability. The sand mix was kept in a closed polyethylene bag for 4 h before using.

Two more 10 lb sand mix batches were made by adding 23 g of "Nusheen" kaolin powder furnished by the Freeport Kaolin Co., 0.5 lbs of the binder solution and 0.7 g of phenolphthalein to 10 lbs of the same Houston sand AFS 45–50, while stirring in a 10 lb Clearfield mixer, and mixing for 2 min. The kaolin powder and the binder mixed readily with the sand and a uniform sand mix with excellent flowability containing 5% of binder and 0.5% of kaolin by weight of sand was obtained in this manner. The sand mix was kept in a closed polyethylene bag for about 4 h before using.

Sand cores were made by placing the sand mixes into a half-bottle shaped aluminum core box with no parting agent, placing iron rods longitudinally in the mix, tapping the sand, and gassing the core with CO_2 until the core surface developed enough hardness but the sand still had a light color. The gassing was accomplished by placing a CO_2 probe for 5 to 10 in. in different parts of the sand core until it was uniformly hardened.

Six core halves with the shape of half-bottles were obtained in this manner and all were dried at 450°F for 1 min. No core wash was applied to the surface of the cores. Two half-bottle shaped parts made with sand mix containing no kaolin were assembled and glued together with a conventional silicate core paste furnished by the M.A. Bell Co. of St. Louis, Mo. under the trade name of "Fast-Dry," to form a bottle-shaped sand core.

Two half-bottle shaped parts made with sand mix containing 0.5% of kaolin by weight of sand were also assembled and glued together with the same core paste to form a second bottle-shaped sand core.

A third bottle-shaped core was made by assembling and pasting together one half-bottle shaped core part prepared with sand containing 0.5% by weight of kaolin and one half-bottle shaped core part prepared with sand containing no kaolin.

Three full bottle-shaped sand cores were obtained in this manner and they were assembled inside a sand mold. Gray iron at about 2650°F. was poured into the mold and allowed to cool for about one hour before removing from the mold. Shake-out of all three cores, with and without kaolin, was very easy: The sand core broke up and flowed out when tapped with an iron bar.

EXAMPLE 3

This is an example of the use of a lithium polysilicate solution as a binder for foundry sand cores. The sand cores made with this binder were used to cast brass metal parts.

The lithium polysilicate solution contained 20 wt% of silica and 2.1 wt% of lithium oxide, therefore the SiO_2/Li_2O ratio was 4.8:1. Density of the solution is 9.8 lbs/gal (specific gravity 1.17 g/cc); viscosity 10 cp; pH 11.

The sand mix was prepared by adding 0.1 lb of "Nusheen" kaolin powder, 2 g of phenolphthalein powder, and 1 lb of lithium polysilicate binder solution to 10 lbs of a sand mixture (50 wt% Houston sand AFS 50 and 50 weight percent #1 Millcreek, Okla., sand AFS 90) in a 10 lb. Clearfield sand mixer while stirring. The mix was stirred for one minute and a half and 30 g of a conventional release agent commercially available from the M.A. Bell Co. of St. Louis, Mo., under the trade name of Mabco Release Agent "G," was added while stirring. The mix was stirred for a total time of 2 min.

During the operation it was observed that the binder mixed readily with the sand. The sand mix obtained had very good flowability and it was kept overnight in a closed polyethylene bag before using to make sand cores.

Cores were made by ramming the sand mix with a tamper in a wood core box painted with aluminum paint. CO_2 gassing was applied for 5 to 10 sec from each end of the U shaped cores or through a center hole in the case of cylindrical type cores. When the core boxes were opened, the hard, strong sand cores released without difficulty. The cores were immersed in a conventional zircon-alcohol core wash and flamed before using.

The cores were assembled into sand molds and molten brass was poured at about 2100°F. The metal was allowed to cool to about room temperature. The sand core broke up very easily and flowed from inside the casting without difficulty.

EXAMPLE 4

This is an example of the use of the guanidine stabilized sodium polysilicate (SiO_2/Na_2O ratio 5:1) of Example 1 to make sand cores and test them according to American Foundrymen's Society standard methods.

The sand mix was prepared by adding 30 g of the binder solution and 100 mg of phenolphthalein powder to 570 g Portage 515 sand. Portage 515 is a sand from Portage, Wis., with an AFS (American Foundrymen's Society) Grain Fineness Number as defined in page 5–8 of the seventh edition (1963) of the AFS Foundry Sand Handbook, of 67–71. In this example the AFS number was 68. Phenolphthalein is added only as an indicator for optimum gassing time with CO_2.

The addition of the sodium silicate to the sand was made gradually while the sand was stirred at speed setting 2 in a "Kitchenaid" mixer Hobart K45. The sand was mixed for a total of ten minutes.

AFS standard and specimens for foundry sand mixtures were used for making tests. The specimens are cylindrically shaped and exactly 2 ± 0.001 in. (508 cm) diameter and $2 \pm 1/32$ in. (5.08 cm) height prepared in a standard sand rammer. The standard sand rammer and the standard procedure to make test specimens are described in sections 4–5 and 4–9 respectively of the aforementioned Foundry Sand Handbook. In this example 170 g of the sand mixed were used to fall within AFS specimen height specifications after ramming.

AFS standard specimens prepared in this manner were strong enough to be handled and in this case they had a pink color due to the phenolphthalein indicator added to the alkaline mix.

A Dietert CO_2 gassing fixture set No. 655 supplied by the Harry W. Dietert Co. of Detroit, Mich. was used to harden the sand specimens by making CO_2 gas flow through them at a controlled rate for an optimum period of time. The CO_2 setting equipment consists of a pressure reducer and flow meter, and gassing fixtures for the standard 2 in. diameter precision specimen tube where the sand specimen is rammed.

The flow meter is calibrated in terms of gas flow at atmospheric pressure from 0 to 15 l/min. A constant gas flow of 3 l/min was used and the optimum gassing time of each sand mix was determined by testing a number of cores made at different gassing times. The change of color of the phenolphthalein in the sand during gassing indicated the degree of neutralization reached by the alkaline silicate and could be used as a preliminary guidance to try to estimate the hardening of the sample.

After gassing the compressive strength of the standard sand specimens was measured in a motor driven Dietert No. 400 Universal Sand Strength Machine equipped with a No. 410 high dry strength accessory to increase the range of compression strength to 280 psi.

Evaluation of the shake-out characteristics of the sand cores made with the binder compositions was made with the AFS non-standard Retained Strength test. The standard, hardened-by-gassing, 2″ × 2″ sand specimens were soaked in an electric muffle furnace at 850°C for 12 min in their own atmosphere, then removed from the furnace and allowed to cool to just above room temperature, and tested in the Universal Sand Strength Machine.

Some specimens made with commercial silicates as a comparison sometimes gave strength values higher than 280 psi and were therefore tested in an Instron Machine.

Gassing times and strength values obtained with guanidine stabilized sodium polysilicate bonded AFS 68 Portage 515 sand are give in Table 64.3.

Employing the methods of preparation of the sand mix, forming and hardening the sand core specimen, and testing compression strength given in this Example 4, different binder compositions of the invention were used to make and test sand cores. The binder compositions used are described subsequently. Testing results obtained are included in the Table 64.3.

TABLE 64.3
Compressive strength (psi)

Binder	As gassed	Retained after 850°C — 12 min
Guanidine stabilized sodium polysilicate	160	10
Example A	160	30
Example B	190	30
Example C	185	30
Example D	160	10
Example E	200	25
Example F	180	10
Example G	100	<10

EXAMPLES

A. Kaolin (2% by weight) mixed with the sand before adding the 5% guanidine stabilized sodium polysilicate of this Example 4 and mixing for 2 min.
B. 5% Tetramethylammonium hydroxide (TMAH) stabilized sodium polysilicate made according to teachings of U.S. Pat. No. 3,625,722.
C. Kaolin (0.5% by weight) mixed with the sand before adding the T.M.A.H. stabilized sodium polysilicate of Sample B and mixing for 2 min.
D. 5% Of sodium polysilicate SiO_2/Na_2O molar ratio 3.75:1 made by dissolving fine colloidal silica powder (HiSil 233) in sodium silicate SiO_2/Na_2O molar ratio 3.25:1.
E. 5% Of sodium polysilicate SiO_2/Na_2O molar ratio 6.5:1 stabilized with copper ethylene-diamine hydroxide.
F. 10% Of lithium polysilicate SiO_2/Li_2O molar ratio 4.8:1 made according to the teachings of U.S. Pat. No. 2,668,149.
G. 10% Of potassium polysilicate SiO_2/K_2O molar ratio of 5:1.

EXAMPLE 5

Amorphous silica-sodium silicate binder composition of SiO_2/Na_2O ratio 5:1 can be formed directly on the sand by addition of colloidal silica sol of uniform particle diameter about 14 nm to the sand, mixing, and then adding sodium silicate SiO_2/Na_2O molar ratio 3.25:1 and mixing for 2 min.

14.96 g of W. R. Grace & Co. Ludox® HS − 40 (40 w/o SiO_2) poured into 745 g of Portage 515 sand in a Hobart K-45 mixer while stirring at speed setting 2. Then adding 40 g of Du Pont sodium silicate grade No. 20 (SiO_2/Na_2O molar ratio 3.25:1) and mixing for 2 more min.

Standard AFS 2″ × 2″ samples made by ramming, then gassing for 30 sec with CO_2 at a flow rate of 3 l/min have a compressive strength of 200 psi and a retained compressive strength at room temperature after soaking in a furnace at 850°C for 12 min and cooling, of 20 psi.

EXAMPLE 6

Amorphous silica-sodium silicate binder composition of SiO_2/Na_2O ratio 5:1 formed directly on the sand as in Example 5 but using a colloidal silica sol of uniform particle diameter about 25 nm instead of 14 nm, with the same sodium silicate.

12 g of W. R. Grace & Co. Ludox® TM-50 (50 w/o SiO_2)
40 g of Du Pont sodium silicate No. 20
748 g of Portage 515 sand
CO_2 gassing time = 30 sec
Compressive strength = 230 psi
Retained strength (850°C — 12 min) = 15 psi
Retained strength (1375°C — 12 min) = 35 psi

EXAMPLE 7

Amorphous silica-sodium silicate binder composition of SiO_2/Na_2O ratio 5:1 formed directly on the sand as in Example 5 but using a colloidal silica sol of uniform particle diameter about 25 nm instead of 14 nm, and using sodium silicate SiO_2/Na_2O molar ratio 3.75:1 instead of 3.25:1.

6.76 g of W. R. Grace & Co. Ludox® TM-50 (50 w/o SiO_2)
40 g of Phila. Quartz Co. sodium silicate grade S 35
753.24 g of Portage 515 sand
CO_2 gassing time = 30 sec
Compressive strength = 180 psi
Retained strength (850°C — 12 min) = 15 psi

EXAMPLE 8

Amorphous silica-sodium silicate binder composition made with the same components and using the same forming method directly on the sand as used in Example 5, except that relative amounts of silica sol and sodium silicate are calculated to give a final SiO_2/Na_2O molar ratio 8:1 in the mixture.

32 g of W. R. Grace & Co. Ludox® TM-50
40 g of Du Pont sodium silicate No. 20
728 g of Portage 515 sand
CO_2 gassing time = 30 sec

Compressive strength = 210 psi
Retained strength (850°C — 12 min) = 20 psi

EXAMPLE 9

Amorphous silica-sodium silicate binder compositions made by first mixing the colloidal amorphous silica as a paste with the sand, then adding the sodium silicate and mixing for two minutes.

3.61 g of Cab-O-Sil M-5 pyrogenic silica powder mixed with 14.4 g of water made a thick paste which was mixed with 475 g of Portage 515 sand in a Hobart K-45 mixer. To the uniform sand-silica mixture, 25 g of Du Pont sodium silicate No. 20 added and mixed for two minutes.

Standard AFS 2″ × 2″ samples made by ramming, then gassing for 30 seconds with CO_2 at a flow rate of 3 liters/minute. Compressive strength measured: 210 psi. Retained strength (850°C — 12 min): 15 psi.

EXAMPLE 10

This is an example of the use of an amorphous silica-sodium silicate composition of SiO_2/Na_2O ratio 5:1 as a binder for foundry sand cores, a polyvinyl acetate aqueous dispersion as a co-binder and additive for durability, and pitch as an aid to improve shake-out and casting surface finish (see Table 64.4).

An amorphous silica-sodium silicate binder composition of SiO_2/Na_2O ratio 5:1 is formed directly on the sand by addition of colloidal silica sol of uniform particle diameter about 15 nm to the sand, mixing, and then adding sodium silicate SiO_2/Na_2O molar ratio 3.25:1 and mixing for an additional period of time.

The sand mix is prepared in the following way: 16 grams of "O" Pitch sold by Ashland Chemical Company of Columbus, Ohio are added to 800 grams of Portage 515 sand supplied by Martin Marietta Aggregates of Rukton, Ill., in a "Kitchen-Aid" Hobart K-45 mixer while

TABLE 64.4
Compressive Strength and Core Hardness versus Elapsed Time on Storage at 73 ± 2°F and 50% Relative Humidity of Portage 515 Sand Cores Made with 5% Sodium Silicate — 1.94% Silica Sol — 2% Polyvinyl Acetate Aqueous Dispersion — 2% Pitch Uncoated and Coated with Various Core Washes (Grades of Binder Components in Table 66.5)

Core wash		Core properties as made	Time elapsed on storage				
			1 Day	2 Days	3 Days	One week	One month
No core wash	C.S.[a]	165	180	200	260	260	275
	Hardness[b]	30	40	45	45	45	45
Polyvinyl acetate water based core wash (75% Monsanto Gelva S-55L in water)	Compr. Str.	150	90	280	90	90	90
	Hardness	90		90			
Commercial zircon core wash (50% "Lite-Off" A ethanol dispersion)[c]	C.S.	170		65	>280		335
	Hardness	60					60
Commercial graphite core wash (50% Pyrokote[d] ethanol dispersion)	C.S.	170		75	>280		75
	Hardness	55					
Polyvinyl acetate alcohol based core wash (75% Monsanto Gelva V7-M50 in methanol)	C.S.	170			>280	>335	>100
	Hardness	>100			>100		
Polyvinyl acetate-zircon water based core wash (1 part Gelva S-55L, 1 part Lite-Off A, 1 part water)	C.S.	130				280	
	Hardness	30				100	

[a]C.S. Compressive Strength, psi. American Foundrymen's Society Standard Method for Bulked Cores.

[b]Hardness. Core (Scratch) Hardness.

[c]Lite-Off A is a product of M. A. Bell Co., St. Louis, Mo.

[d]Pyrokote Supreme 114-5X supplied by Penna. Foundry Supply and Sand Co., Philadelphia, Pennsylvania.

TABLE 64.5
Mechanical Properties versus Elapsed Time on Storage at 73 ± 2°F and 50% Relative Humidity of Portage 515 Sand Cores Made with 5% Sodium Silicate[a] — 1.94% Silica sol[b]— 2% Polyvinyl Acetate Aqueous Dispersion[c] — 2% Pitch[d]

Mechanical properties	Core properties as made	Elapsed time since making core				
		1 Day	2 Days	3 Days	One week	One month
Compressive Strength, psi	165	180	200	260	260	275
Core (Scratch) Hardness	30	40	45	45	45	45
Tensile Strength, psi[c]	25	25	40	45	45	45

[a]Du Pont Sodium Silicate No. 9: 29 w/o SiO_2: 8.9 w/o Na_2O.

[b]W. R. Grace & Co. Ludox® HS-40: 40 w/o SiO_2.

[c]Monsanto Gelva S:55L: 55 w/o polyvinyl acetate.

[d]Ashland Chemical "O" Pitch powder.

stirring at speed setting 2 and mixed thoroughly with the sand.

14.70 Grams of "Ludox" HS-40 colloidal silica sold by W. R. Grace & Co., and 16 grams of Gelva S-55L polyvinyl acetate aqueous dispersion, sold by Monsanto Chemical Company, are mixed in a plastic beaker, added to the sand-pitch mix and mixed in the Hobart mixer for 2 min.

Finally, 40 grams of Du Pont sodium silicate grade No. 9 (SiO_2/Na_2O molar ratio 3.25:1) are added and mixed for 2 more min (see Table 64.5).

AFS (American Foundrymen's Society) standard specimens for foundry sand mixtures are made immediately after the mixing is completed, as described in Example 4. The specimens are set by gassing with carbon dioxide using the equipment and procedure of Example 4. Optimum gassing time for the composition of this example is 20 sec.

Gassed cores are separated in two groups: one group of cores is left untreated, the second group of cores is coated by immersion in various core wash compositions given in Table 67.2. Cores coated with water based and methanol based core washes are allowed to air dry, whereas cores coated with ethanol based core washes standard, hardened-by-gassing, 2″ × 2″ sand specimens are soaked in an electric muffle furnace at 850 or 1375°C for 12 min in their own atmosphere, then removed from the furnace and allowed to cool to just above room temperature, and tested in the Universal Sand Strength Machine. For all cores prepared as described in this example, both 850 and 1375°C retained strength values were less than 25 psi.

Some specimens made with commercial silicates as a comparison sometimes give retained strength values higher than 280 psi and are therefore tested on an Instron Machine.

EXAMPLE 11

This example describes the preparation of sand cores bonded with amorphous silica, sodium silicate and polyvinyl acetate and their use in casting 2.5″ grey iron and brass pipe tees. The sand mix is prepared in a Carver "S" mixer by adding to 400 lbs of sand (Whitehead Brothers "E" sand with an AFS number 92.2), a mixture consisting of 10.5 lbs Du Pont "Ludox" HS-40 and 9 lbs. Monsanto Gelva S-55L polyvinyl acetate aqueous dispersion, and 9 lbs of pitch (Ashland Chemicals "O" Grade). After 5 min 27 lbs of Du Pont No. 9 sodium silicate are added and mixing is continued for a further five minutes. The free flowing, uniformly brown mix is then discharged to a storage bin.

The cores are formed by air blowing the mix into a steel pattern comprising twin $2\frac{1}{2}''$ tees and gassing with carbondioxide at 65 psi for 3.5 sec. The cores are immediately removed from the pattern and placed on storage trays. 150 cores are made in 18 minutes, each weighing about $2\frac{1}{4}$ lbs. No fumes or odors are detected during the mixing or core preparation and the cores have adequate strength for normal handling in the foundry. They have a very smooth surface with an AFS hardness number of about 20. The cores are positioned in oil bonded sand molds, enclosed by steel boxes and grey iron is poured at about 2700°F. Ninety cores are used within a few hours of preparation and 58 are stored for three days at relative humidity of about 25% at about 18°C. The cores which are stored for three days are both stronger and harder than when first made.

After pouring the iron the cores are cooled almost to room temperature. No offensive odors are detected during metal pouring or cooling. The cores are then very weak and shake-out readily with excellent surface peel from the iron. After final cleanup by wet drum tumbling and shot blasting, the pipe tees have a much smoother internal surface than those made in normal production using a commercial, proprietary silicate binder. In addition to having a rougher surface some of the tees made using cores with the commercial binder still had sand adhering to the internal surface after cleanup.

Two of the cores prepared as described above are coated by brushing on a slurry consisting of 50% zircon and 20% polyvinyl acetate methanolic dipersion (Monsanto Gelva V7-M50) and 30% methanol. The alcohol is allowed to air dry leaving a hard coating of zircon bonded with polyvinyl acetate. The hardness is measured as 90 AFS and shows no change after storing for three days at about 25% relative humidity and about 18°.

The cores are positioned in molds, and brass is poured at 2120°F. After cooling to room temperature the cores collapse readily and shake-out is easily accomplished with excellent peel from the metal surface. No offensive fumes are detected during metal pouring and cooling. The internal surface of the brass tees is very clean and smooth.

EXAMPLE 12

The procedure of Example 11 is repeated using Houston, subangular bank sand, AFS number 45, and omitting the pitch. Half of the cores are coated by immersing them in an agitated slurry containing 50% graphite (Pyrokote), 10% Monsanto Gelva S-55L polyvinyl acetate, and 40% alcohol, allowing them to drain and igniting the alcohol to burn off completely. The other half are similarly treated with an aqueous slurry containing 75% Monsanto (Gelva S-55L) polyvinyl acetate dispersion, allowing them to drain and air dry.

After storing for two weeks at about 80% relative humidity and 30°C all the cores are strong and hard (AFS hardness number 80–90). The cores are positioned in the molds and brass is poured at 2150°F and allowed to cool to about room temperature. No offensive fumes are detected during metal pouring and cooling. Core breakdown is very easy in all cases and the shake-out sand is granular and free flowing. Surface peel and internal surface finish are excellent in the case of tees made from cores treated with the graphite polyvinyl acetete wash and very good for cores coated with polyvinyl acetate alone. No sand residues are observed on the internal surfaces of tees cast from any of the cores.

EXAMPLE 13

This example describes the preparation of sand cores bonded with colloidal silica powder, sodium silicate and polyvinyl acetate ethylene copolymer and their use in the production of cast iron end plates for boilers.

Two thousand pounds of Portage No. 515 sand, AFS number 68 are charged to a batch muller. Forty pounds of pitch (Ashland Chemical Co. "O" grade) are thoroughly mixed with the sand over a period of three minutes. Twenty pounds of Cab-O-Sil M-5 pyrogenic silica powder, as a thick paste with 80 pounds of water, and 40 pounds of Du Pont's "Elvace" 1873, a 55% aqueous dispersion of polyvinyl acetate/ethylene copolymer (13% ethylene) are then added to the mulled mixture over a period of two minutes. One hundred six pounds of Du Pont No. 20 sodium silicate are then added and the mixing continued for an additional two minutes. Half a

TABLE 64.6
Mechanical Properties versus Elapsed Time on Storage at 73°F and 50% Relative Humidity of Portage 515 Sand Cores Made with 5.3% Sodium Silicate[a] — 1.93% Silica Sol[b] and Either 0.26% Triacetin[c] or 0.26% Ethyl Acetate[d]

Mechanical properties		Elapsed time since making core				
		1 Day	3 Days	5 Days	One week	After heating 850°C for 12 min
Compressive	Triacetin		570	685		90
Strength, psi	Ethyl acetate	260			685	100
Core (Scratch)	Triacetin		80	85		
Hardness	Ethyl acetate	98			90	

[a]Du Pont Sodium Silicate No. 9; 29 w/o SiO_2; 8.9 w/o Na_2O.

[b]W. R. Grace & Co. Ludox® HS-40; 40 w/o SiO_2.

[c]Eastman Kodak glycerol triacetate.

[d]Fisher Scientific Co. ACS grade ethyl acetate.

minute from the end of the mixing period, 1.5 pounds of M. A. Bell's "G" grade flow agent are added. The free flowing mix is discharged into a bin. Cores are made by hand ramming the mix into the two halves of a split core box. The two halves are clamped together and the core is gassed with carbon dioxide at 30 psi for a period of 30 seconds. No fumes or odors are detected during maxing and core preparation. The core is then stripped from the pattern and after storing for several days at about 50% humidity and 25°C it is assembled in the mold. Iron is poured at 2650°F and after cooling to about 1500°F the molds are broken away. Examination of the cores shows them to be quite friable and they collapse immediately on a vibrator table and shake-out as granular lump free sand. The boiler end plates are free from defects, dimensionally accurate and have excellent surface finish.

EXAMPLE 14

This is an example of the use of esters as setting agents for the high ratio silicate binders of this invention (see Table 64.6).

The sand mix is prepared by mixing 14.7 g of "Ludox" HS-40 and 2 grams of Triacetin (glycerol triacetate sold by Eastman Kodak), with 760 g of Portage 515 sand using a "Kitchen-Aid" mixer, Hobart K45. The sand is mixed for a total of 2 min and 40 g of sodium silicate ratio 3.25 (Du Pont No. 9) are then added. After an additional 2 min mixing the free flowing sand mix is used immediately to prepare standard 2″ diameter cylinders as described in Example 4. Cores are similarly made using 2 g of ethyl acetate (ACS grade sold by Fisher Scientific Co.) in place of Triacetin. Cores are stored at 73°F and 50% relative humidity. The compressive strength, hardness, and shake-out characteristics are evaluated as described in Example 4 and the results are tabulated in Table 64.1.

In addition to very good initial strength and hardness, both strength and hardness increase on storage and the loss of strength after heating the cores for 12 min at 850°C is indicative of good shake-out.

A 400 pound sand mix is made in a Carver "S" mixer as described in Example 11 adding 8 pounds of pitch in

TABLE 64.7
Mechanical Properties versus Elapsed Time on Storage at 73 ± 2°F and 50% Relative Humidity of Portage 515 Sand Cores Made with 5% Sodium Silicate — 1.94% Silica Sol — 2% Polyvinyl Acetate Aqueous Dispersion Uncoated and Coated with Various Core Washes

Core wash				Mechanical properties as made	Elapsed time on storage: 1 Day	2 Days	3 Days	One week	Two weeks	One month
	Strength, psi									
	Tensile	30	40	50	55	60				
No core wash	Strength, psi									
	Core	35	50	50	50	50				
	Scratch									
	Hardness									
Polyvinyl acetate-zircon water-based core wash (1 part Monasanto Gelva S-55L; 1 part "Lite-Off" A; 1 part water)	Compressive Strength	150			290		440			
	Hardness	25	95		425	95		95		
Polyvinyl acetate-zircon alcohol-based core wash (1.0 parts Gelva V7-50 diluted with methanol to 20% polyvinyl acetate; 1 part "Lite-Off" A)	C.S.	170			425			450		
	Hardness				100			100		
Polyvinyl acetate water-based core wash (75% Gelva S-55L in H_2O)	C.S.	170			275			315		
	Tensile	25				60		140		
	Hardness	100				100	100			
Polyvinyl acetate alcohol-based core wash (75% Gelva V7-M50 in methanol)	C.S.				320			395		
	Hardness				100			100		

addition to "Ludox" HS-40, sodium silicate No. 9 and Triacetin at the same levels on the sand as described above. Cores for $2\frac{1}{3}''$ pipe tees are made as described in Example 11 except that the cores are not gassed with CO_2. After allowing them to harden in the pattern for 5 minutes the pattern is stripped and the cores are stored for three days before being assembled in the molds. Ductile iron is poured at about 2700°C and the castings are allowed to cool for about two hours inside the mold assembly. After removing the castings from the molds the cores collapse readily in a vibrator and the recovered sand is granular and free from lumps. No odors are produced during the entire operation and the castings have very good interior surface finish.

EXAMPLE 15

This is an example of heat setting the high ratio sodium silicate binder of this invention (see Table 64.7).

A sand mix is prepared in a Hobart K45 mixer by adding 12 g of "Ludox" TM-50, 16 g of polyvinyl acetate dispersion (Monsanto Gelva S-55L) and 40 g of sodium silicate ratio 3.25 Du Pont No. 9) to 750 g of Portage 515 sand. The mixing time is ten minutes and the free flowing mix is used to prepare standard 2″ diameter cylinders as described in Example 4. The cores are carefully removed from the compacting cylinder and heated for 1 hour in an air oven at 100°C. The strength and hardness of the cured cores are as follows:

Compressive strength = 1200 psi
AFS Hardness = 95
Retained strength (850 + C — 12 min) = 150 psi

EXAMPLE 16

This is an example of the use of dextrin with high ratio sodium silicate binder to produce cores which retain excellent strength and hardness when stored for several weeks.

A sand mix is prepared as described in Example 14 by mixing 14.7 g "Ludox" HS-40, 16 g of 50% aqueous solution of dextrin (sold by Industrial Products Chemicals, Pikesville, Md.) previously mixed with 40 grams of sodium silicate ratio 3.25 (Du Pont No. 9), with 760 g of Portage 515 sand. Standard cores are prepared and set by gassing with carbon dioxide as described in Example 4. Compressive strength and hardness

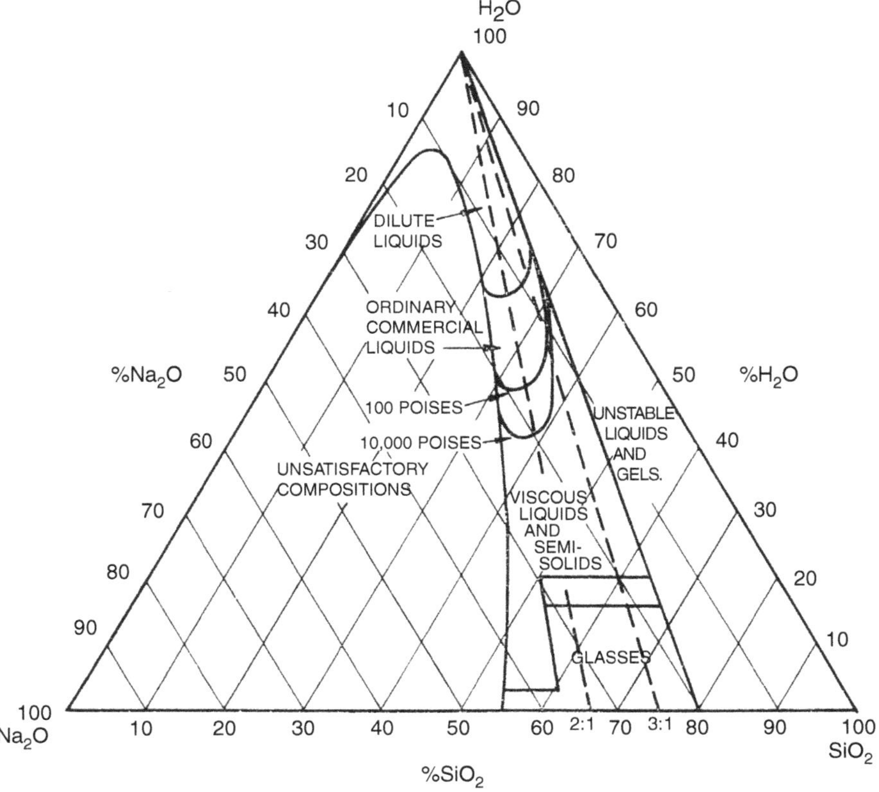

FIGURE 64.1 Compositional diagram of sodium silicates showing properties of commercial grades (courtesy U.K. Centre).

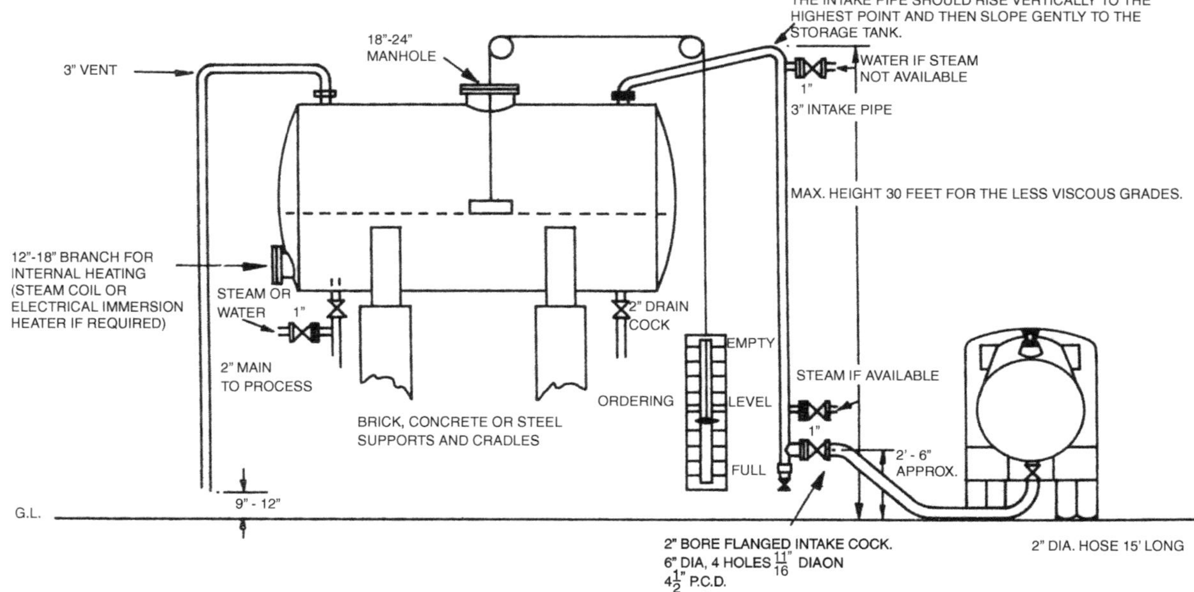

FIGURE 64.2 Typical layout of sodium silicate bulk storage installation (courtesy of U.K. Centre)

measurements when freshly made and after storing for one week at about 50% relative humidity and 23°C show these cores to have excellent storage life. Loss of strength after heating for 12 minutes at 850°C and 1375°C in indicative of good shake-out.

	Initial	After one week
Compressive Strength, psi	135	150
Core (Scratch) Hardness	30	50
Retained Strength (850°C — 12 min), psi	20	
Retained Strength (1375°C — 12 min), psi	50	

EXAMPLE 17

Amorphous silica-sodium silicate-polyvinyl acetate of SiO_2/Na_2O molar ratio 5:1 formed directly on the sand by addition of a uniform, stable mixture of aqueous silica sol of uniform particle diameter about 12 nanometers to the sand, mixing and then adding sodium silicate SiO_2/Na_2O molar ratio 3.25:1 and mixing for 2 min.

14.96 g of W. R. Grace "Ludox" HS-40 (40 w/o SiO_2) are mixed in a beaker with 16 g of Monsanto Gelva S-55L polyvinyl acetate aqueous dispersion (55 w/o polyvinyl acetate) and poured into 740 g of Portage 515 sand in a Hobart "Kitchen-Aid" K45 mixer, stirred at speed setting 2 for two minutes. Then adding 40 g of Du Pont sodium silicate grade No. 9 (29 w/o SiO_2, 8.9 w/o Na_2O) and mixing for 2 more min.

Standard AFS 2″ × 2″ samples made by ramming, then gassing for 20 seconds with CO_2 at a flow rate of 5 l/min, are allowed to age at about 23°C and 50% humidity, others are immersed in polyvinyl acetate or polyvinyl acetate-zircon water-based core washes and allowed to air dry. Samples treated with core washes are allowed to age under the same conditions as the untreated specimens. Compressive strength and core scratch hardness, and in some cases tensile strength is determined the day of making the cores and after several periods of time.

Results obtained are shown on the table.

ACKNOWLEDGMENT

We thank the Castings Development Centre of the United Kingdom for allowing us to use Figure 64.1 and Figure 64.2 from the book "The CO_2-Silicate Process in Foundries" by K.E.L. Nicholas, B.Sc, ARIC, AIM.

Part 9

NMR of Silica Edge

65 On the Silica Edge: An NMR Point of View

A.P. Legrand, H. Hommel and J.B. d'Espinose de la Caillerie
The City of Paris Industrial Physics and Chemistry Higher Educational Institution, Quantum Physics Laboratory

CONTENTS

INTRODUCTION

"The properties of amorphous silicas of high specific surface area, from the smallest colloidal particles to macroscopic gels, depend largely on the chemistry of the surface of the solid phase" [1].

Silicas are useful for numerous applications. The most recent application is certainly their use in the tire manufacturing process (Energy Tires of Michelin Cie). As styrene-butadiene rubber compatibility with an organo phobic silica is questionable contrarily with carbon black, it was necessary to investigate the surface composition to adapt it to such application. To do this, different methods have to be used. Infra-red spectrometry was the first physical method to demonstrate the existence of single and/or associated hydroxyls groups (Figure 65.1). More recently solid state NMR was not only able to identify geminal, the existence of which was sometimes debated, but also single or associated hydroxyls.

USEFUL NMR METHODS

Owing to the chemical composition of silica, two nuclei are usually analyzed. They are ^{1}H, as abundant spin and ^{29}Si as a rare one. Numerous experiments have been done by cross polarization (CP) between proton and silicon nuclei and magic angle spinning (MAS). As it is a classical method analyzed elsewhere [2], it is not described here in details. It is only necessary to notice that the silicon signal is enhanced if the distance to proton is short enough to provide an efficient dipolar interaction. So this means that the method allows to distinguish the different types of hydroxylated silicon species, but also those in the neighbourhood of protons and consequently not chemically bonded to them. Using CP/MAS the determination of the relative geminal/single proportion f_g is done assuming simple Lorentzian or Gaussian lines for each species. This enables to determine the f_g and moreover using the MAS experiment only it is possible to determine another factor f_s:

$$f_g=\frac{\overset{HO\ \ OH}{\text{-Si-}}}{\overset{HO\ \ OH}{\text{-Si-}}+\overset{OH}{\text{-Si-}}}\qquad f_S=\frac{\overset{HO\ \ OH}{\text{-Si-}}+\overset{OH}{\text{-Si-}}}{\overset{HO\ \ OH}{\text{-Si-}}+\overset{OH}{\text{-Si-}}+\text{-Si-}}$$

from these two measurements the relative proportion of geminal Q^2, single Q^3 on the totality of silicon atoms are determined. An example of such analysis is given in Figure 65.2.

Nevertheless in some cases, when amounts of Q^2 and Q^3 is important, MAS only is sufficient, as shown in Figure 65.3.

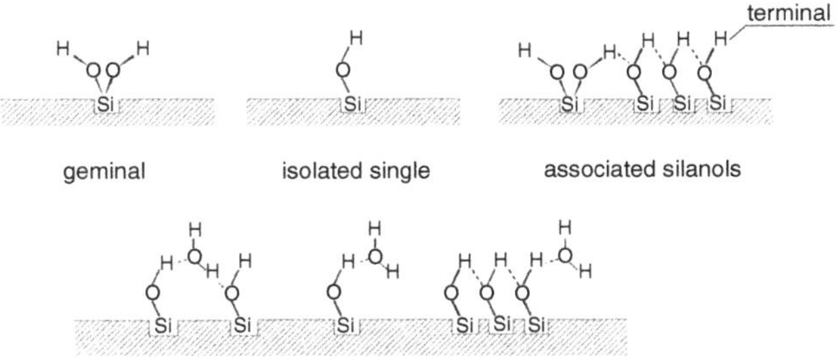

FIGURE 65.1 Different types of hydroxyls and water adsorbed molecules.

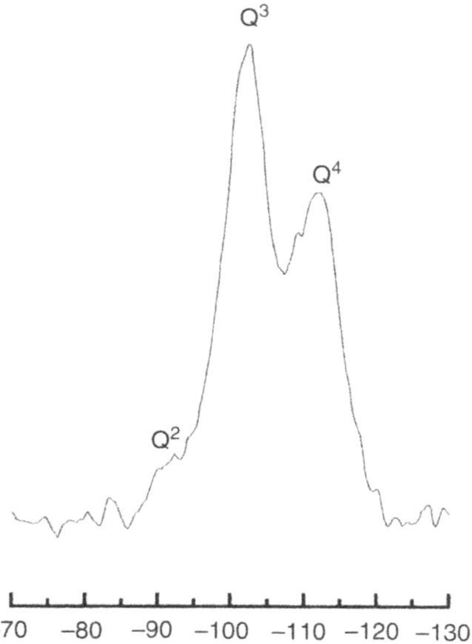

FIGURE 65.2 ^{29}Si CP/MAS spectrum of a precipitated silica and of the corresponding identification into three species of silicon environment namely Q^4, Q^3, and Q^2.

EXPERIMENTAL RESULTS: ^{29}Si ANALYSIS

Different authors using ^{29}Si NMR CP/MAS [3–6] and ^{29}Si NMR MAS only on MCM silicas, have determined the relative proportions of single (Q^3) or geminal (Q^2) silicons. The corresponding site content determinations are plotted in Figure 65.4. A scattered distribution is observed, the origin of which has to be discussed. In particular why Q^2 site content is systematically inferior to that of Q^3.

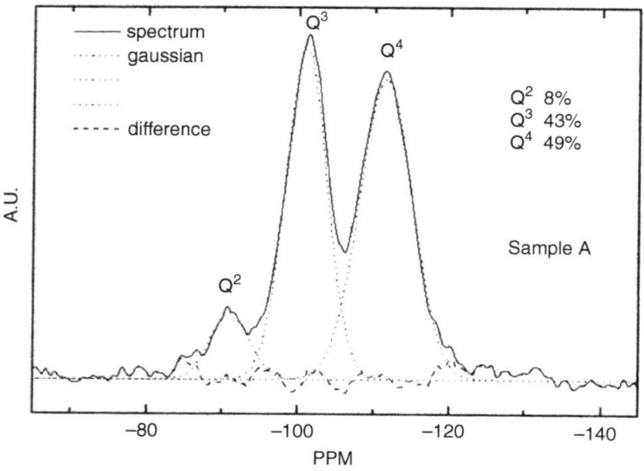

FIGURE 65.3 MAS only spectrum of a MCM silica with its template and decomposition into three site species owing the quantitative determination.

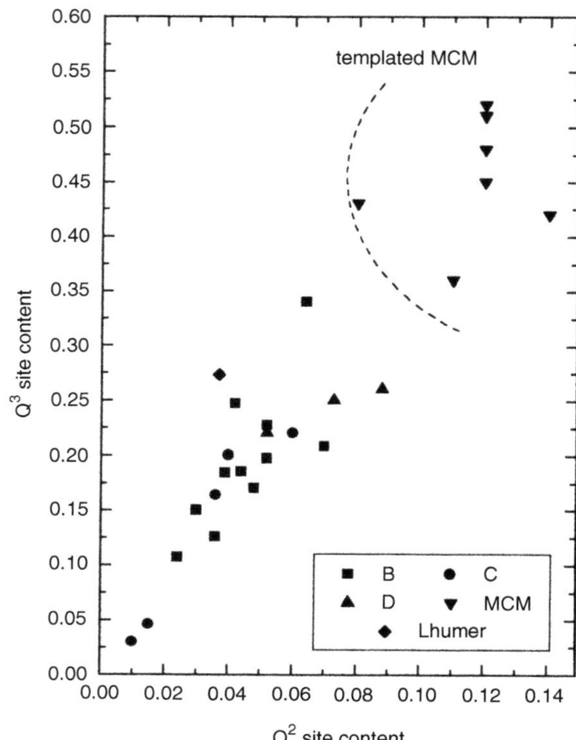

FIGURE 65.4 Q^2 and Q^3 site contents of different silicas. B: from reference 4; C: from reference 3; D: from reference 5; MCM: different MCM silicas with templated, heat treates and MCM [6].

SURFACE MODELS

The silica surface is the border between the solid and a gas, a liquid or another solid. This surface contains different hydroxyls which are responsible for the adsorption properties. Iler [1] suggested that "since the density and the refractive index of amorphous silica is close to that of cristobalite and tridymite, the concentration of surface hydroxyls might be estimated from the crystal structure."

Such consideration has been used to take into account the experimental results to understand why there is a maximum amount of hydroxyls per nm^2 of about 4.9 [7] to 3.9 [8]. Peri and Hensley [9] have studied hydration and dehydration of β-cristobalite (100) face as a model surface justifying such values. Sindorf and Maciel [10,11] using ^{29}Si solid state NMR, clearly demonstrated the presence of geminal $=Si(OH)_2$ and single $\equiv Si(OH)$ on silicas. They considered "that the surface is quite heterogeneous and may contain segments of surface resembling both the (100) (for geminal) and (111) (for single) faces of the cristobalite."

STATISTICAL MODEL

One can assume a square flat array representing the surface silica, where at each intersection is located a silicon, bonded to four oxygen atoms, themselves bonded to silicons. Using a random model of adsorption of water molecules, which are susceptible to break the Si—O—Si bridge when adsorbed, it is possible to estimate the relative proportions of Q^2 and Q^3 through this Monte Carlo method. The different proportions so obtained, constitute cloud into which the main experimental points set [5].

A more formal approach can be done. Single SiO_4 tetrahedron is considered as the basic building blocks of the silica framework, without referring to any subunit formed by several SiO_4 tetrahedra. x_i denotes the fraction of silicon atoms bearing i OH usually denominated Q^{4-i}. If p the probability of formation of an OH group and $(1-p)$ that of disappearance of an OH (e.g., by condensation). To take into account of a large hydroxyl site content observed in MCM silicas, a weighting factor ε is introduced. This has to be understand so that after a breaking of the first Si—O—Si bridge, another breaking at the same Si site has the probability εp instead of p. Such probability signifies and easier ($\varepsilon > 1$) or more difficult ($\varepsilon < 1$) breaking. (Nevertheless one has to take into account that $\varepsilon p \leq 1$.) Then the kinetic equations for the different population are:

$$\frac{dx_0}{dt} = -4px_0 + (1-p)x_1$$
$$\frac{dx_1}{dt} = -4px_0 - (1-p+3\varepsilon p)x_1 + 2(1-\varepsilon p)x_2$$
$$\frac{dx_2}{dt} = 3\varepsilon p x_1 - 2(1-\varepsilon p+\varepsilon p)x_2 + 3(1-\varepsilon p)x_3$$
$$\frac{dx_3}{dt} = 2\varepsilon p x_2 - (3(1-\varepsilon p)+\varepsilon p)x_3 + 4(1-\varepsilon p)x_4$$
$$\frac{dx_4}{dt} = \varepsilon p x_3 - 4(1-\varepsilon p)x_4 \qquad (65.1)$$

A stationary state may be discussed equating each derivative to zero. Moreover this system of five coupled linear equations is constrained by the relation:

$$x_0 + x_1 + x_2 + x_3 + x_4 = 1$$

The solution is:

$$x_0 = \frac{(1-p)(1-\varepsilon p)^3}{\Sigma} \qquad x_1 = \frac{4(1-\varepsilon p)^3 p}{\Sigma}$$
$$x_0 = \frac{(1-p)(1-\varepsilon p)^3}{\Sigma} \qquad x_1 = \frac{4(1-\varepsilon p)^3 p}{\Sigma}$$
$$x_2 = \frac{6\varepsilon(1-\varepsilon p)^2 p^2}{\Sigma} \qquad x_3 = \frac{4\varepsilon^2(1-\varepsilon p)p^3}{\Sigma}$$
$$x_4 = \frac{\varepsilon^3 p^4}{\Sigma}$$

with:

$$\Sigma = (1-p)(1-\varepsilon p)^3 + 4(1-\varepsilon p)^3 p + 6\varepsilon p(1-\varepsilon p)^2 p + 4\varepsilon^2(1-\varepsilon p)p^3 + \varepsilon^3 p^4$$

As shown earlier, only Q^3 and Q^2, as hydroxylated species, are observed by NMR. The corresponding site contents are x_1 and x_2. Parameteric plots as a function of p of such quantities are given in Figure 65.5, taking into account different values of the adaptation factor ε. Only a part of such curves are meaningful relatively to the experimental points (Figure 65.6). The $\varepsilon < 1$ and $\varepsilon > 1$ domains demonstrate the difference in the evolution of the relative proportions of single and geminal species. For example with $\varepsilon < 1$ the mechanism favours relatively the Q^3 site content contrarily to the case $\varepsilon > 1$.

Comparison between experimental determinations and theoretical curves is done in Figure 65.7. As it can be seen the dispersed experimental points are significantly included in the range $2.0 \geq \varepsilon \geq 1.0$, except those which have a low amount of hydroxyl. Such results are in favour of an increase of the probability of formation of a geminal after the formation of a single hydroxyl. Although the simplicity of the model, this underlines the importance of the local silicon structure on the hydration/dehydration process. This have been equally considered by Brinker et al. [12] who have shown the difference for strained and unstrained Si—O—Si bonds in such process.

For a large amount of hydroxyls, as observed on some templated MCM silicas [13], contrarily to other silicas, it seems that a diminution of probability of formation of a geminal after the formation of a single hydroxyl is obtained. This can be observed on Figure 65.7, where the experimental points are scattered in $\varepsilon < 1$ range.

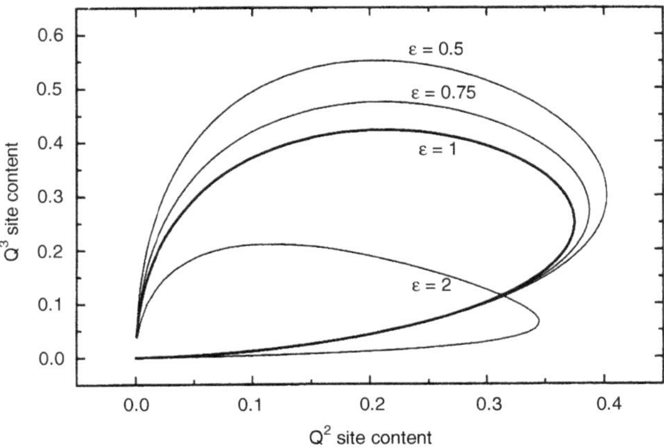

FIGURE 65.5 Theoretical site content distribution assuming Equation (67.1).

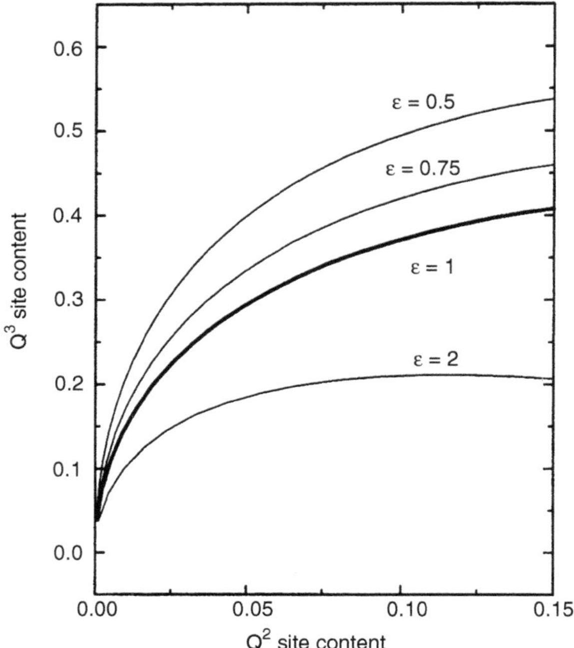

FIGURE 65.6 Enlarged part of Figure 67.5 showing the influence of the values given to the adaptation parameter ε.

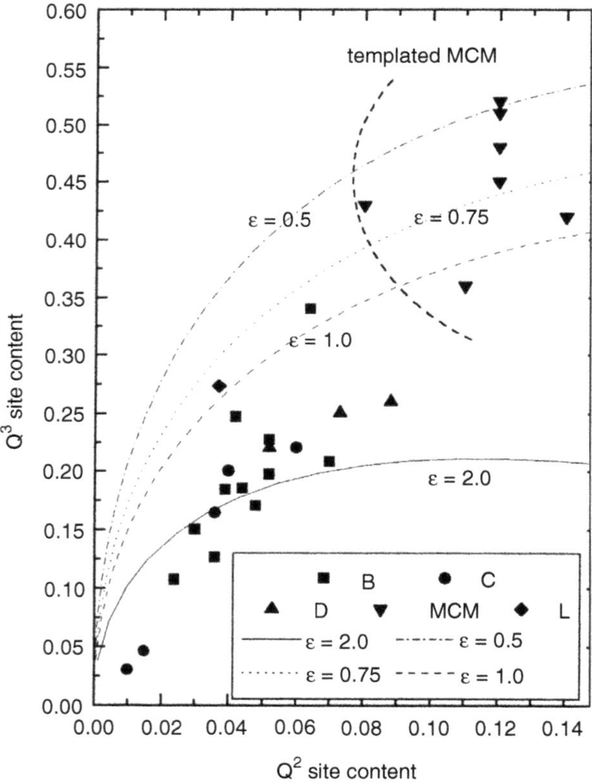

FIGURE 65.7 Comparison between experimental and theoretical equations.

This may be interpreted so that the surface of such silicas is so constrained that the first breaking relaxes the silica structure enough so that a new breaking at the same site is more difficult. Another interpretation could be, as the template extraction is not done, that the silicon are loosely bonded to each other. It should be noted that contrarily to conventional silicas, MCM materials examined here contain large amount of organic cations. It follows that in order to balance the charge, part of siland groups have to be deprotonated, tentatively explaining their peculiar value of the adaptation factor ε. Such open structure, after heat treatment, getting back to the less hydroxylated domain as Luhmer [6] has shown.

EXPERIMENTAL RESULTS: PROTON ANALYSIS

^{1}H Analysis

Infra-red spectroscopy demonstrated the presence of single and associated hydroxyls [14]. Moreover ^{1}H NMR is able to observe such species. Bronniman et al. [15] used CRAMPS to do it on silicas. Although there are some technical difficulties to fluently use this method, another easier is to take advantage of higher speed of rotation now available for the MAS [16]. Figure 65.8 presents two spectra of pyrogenic silicas. Although there are the small amount of hydroxyls at the surface of such materials, the different type of hydroxyls and adserbed

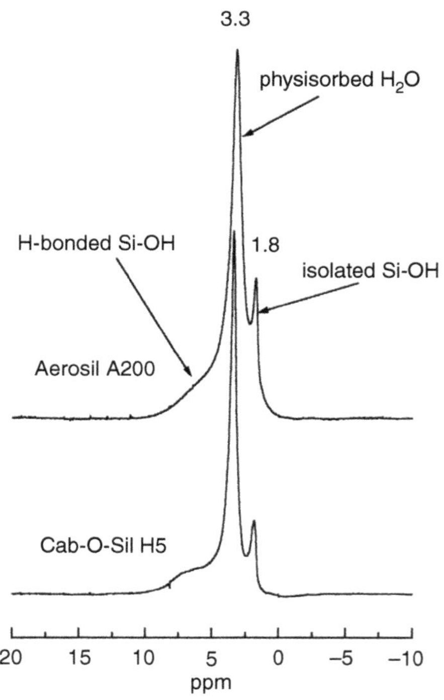

FIGURE 65.8 One-plus ^{1}H MAS only NMR spectra of as-received Cabot Cab—O—Sil H5 and Degussa Aerosil A200 silicas [16].

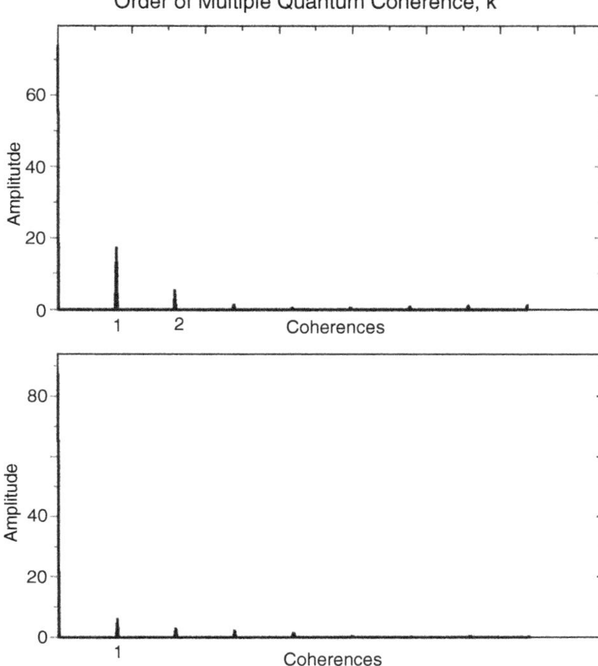

FIGURE 65.9 Multiple quantum spectra $I(k)$ versus k obtained with the single quantum propagator:

$$P = \frac{\tau}{4}, \pi(y), \frac{\tau}{2}, \pi(-y), \frac{\tau}{2}, \pi(-y), \frac{\tau}{2}, \pi(y), \frac{\tau}{4}.$$

The whole sequence is:

$$\left[\frac{\pi}{1}(-x), nP, \frac{\pi}{4}(x)\right]_{\varphi}\left[\frac{\pi}{4}(x), nP, \frac{\pi}{4}(-x)\right]_{\varphi=0}\frac{\pi}{2}(x)$$

acquisition;

where τ is the duration of the π pulse, $n = 4$ and φ is incremented eight times by step of $\pi/8$. (top) evacuated precipitated silica, (bottom) same silica after a heat treatment of 2 h at 450°C under vacuum (courtesy of P. Tougne).

water are observed. This method can be used to follow the evolution of such species in particular during heat treatment.

N-Quantum Coherence or Spin Counting

Gerstein et al. [17] proposed to use N-coherence spectrometry to determine proton size clusters on silicas. This method would be interesting to correlate the experimental observations concerning the grafting ability of different silicas. It is assumed that pyrogenic silicas, owing to their regular porosity and small amount of hydroxyls, these ones are randomly distributed on the surface. Precipitated silicas on the contrary, due to a complex surface and large amount of hydroxyls, are able to present small clusters of hydroxyls. Figure 65.9 shows an example of such phenomenon. Heat treatment contribute to a diminution of the cluster size.

CONCLUSION

Nuclear magnetic resonance spectrometry contributed significantly to the increase in the knowledge on silicas. The main characteristic of this method is its ability to analyze the surface as well as bulk properties of the material. As it is a local analysis of the neighbourhood of different atoms, it is efficient for the study of amorphous materials as silicas [18].

REFERENCES

1. R.K. Iler, The Chemistry of Silica, Wiley, New York, 1979.
2. H. Hommel, A.P. Legrand, C. Dorémieux, J.B. d'Espinose de la Caillerie, in: A.P. Legrand (Ed.), Chapter 3. The Surface Properties of Silicas, Wiley, Chichester, 1998.
3. S. Leonardelli, L. Facchini, C. Frétigny, P. Tougne, A.P. Legrand, J. Am. Chem. Soc. 114 (1992) 6412.
4. L. Jelinek, P. Dong, C. Rojas-Pazos, H. Taïbi, E.sz. Kovatz, Langmuir 8 (1992) 2152.
5. A.P. Legrand, H. Taïbi, H. Hommel, P. Tougne, S. Leonardelli, J. Non-Cryst. Solids 155 (1993) 122.
6. M. Luhmer, J.B. d'Espinose, H. Hommel, A.P. Legrand, Magn. Reson. Imaging 14 (7/8) (1996) 911.
7. L.T. Zhuravlev, Colloids Surf. 74 (1) (1993) 71.
8. H.P. Boehm, Adv. Catal. 16 (1966) 533.
9. J.B. Peri, A.L. Hensley, J. Phys. Chem. 72 (1968) 2926.
10. D.W. Sindorf, G.E. Maciel, J. Phys. Chem. 87 (1983) 5516.
11. D.W. Sindorf, G.E. Maciel, J. Am. Chem. Soc. 105 (1983) 1487.
12. C.J. Brinker, R.J. Kirkpatrick, D.R. Tallant, B.C. Bunker, B. Montez, J. Non-Cryst. Solids. 104 (1988) 139.
13. J.Y. Piquemal, J.M. Manoli, P. Beaunier, A. Ensuque, P. Tougne, A.P. Legrand, J.M. Bregeault, Micro. Meso. Mat. (1999) in press.
14. A. Burneau, J.P. Gallas, in: A.P. Legrand (Ed.), Vibrational Spectroscopies. The Surface Properties of Silicas, Wiley, Chichester, 1998.
15. C.E. Bronnimann, R.C. Zeigler, G.E. Maciel, J. Am. Chem. Soc. 110 (1988) 2023.
16. J.B. d'Espinose de la Caillerie, M.R. Aimeur, Y. El Kortobi, A.P. Legrand, J. Colloid Interface Sci. 194 (1997) 434.
17. B.C. Gerstein, M. Pruski, S.J. Hwang, Anal. Chim. Acta 283 (1993) 1059.
18. A.P. Legrand, H. Hommel, in: M. Nardin & E. Papirer, (Eds.). On the Border of a Magnetic Resonance Point of View. Powders and Fibers: Interfacial Science and Applications. Taylor and Francis Group (to be published).

Part 10

Research in Russia

Horacio E. Bergna
H.E. Bergna Consultants

The Republic of Russia is one of the countries that has contributed more to the development of the science of colloidal silica. The list of Russian researchers on this subject is endless.

L.T. Zhuravlev mentions in historical order Lomonorov, 1723 for his work on properties of natural colloidal solutions, and coagulation and crystallization processes. Lovitz, 1785; Reiss, 1809; Borshchov, 1869; the famous Mendeleev, 1871; Veimarn, 1904 author of one of the first textbooks on colloid chemistry; Grdroitz, 1908, colloidal soils; Zelinsky, proteins.

Zhuravlev mentions that large scale, systematic research in colloid chemistry began in Russia in the 1920s with Dumansky (lyophobic colloids), Peskov, Rebinder, Zhukov, Deryagin, Frumskin, Dubinin, Kiselev, and outside Russia, in the Ukraine, Neimark, Chuiko, Gun-Ko, Pentyuk, V.K. Pogorsly, Lobanov, and Grebenyuk.

Zhuravlev, who attended the Ralph K. Iler Memorial Symposium together with 190 other scientists from Argentina, Asia, North America, Australia and New Zealand is today one of the most distinguished Russian colloid chemist.

The challenge for scientists is to develop an understanding of the stability of colloidal silica since the classic Derjaguin–Landau–Verwey–Overbeek (DLVO) theory and more recent stability theories do not explain.

66 Colloid Chemistry of Silica: Research in the Former Soviet Union

L.T. Zhuravlev
Russian Academy of Sciences, Institute of Physical Chemistry

CONTENTS

Research of Soviet scientists is surveyed. Contributions to various aspects of the colloid chemistry of silica are examined: preparation and stabilization of silica hydrosols; preparation of silica gels; structural characterization of silicas; surface chemistry elucidation; adsorption and ion-exchange property examination; and geometric and chemical modification of silicas, silica coatings, and so forth.

Silica is one of the most widespread substances on earth. The content of SiO_2 in the lithosphere is thought to be 58.3%, and the percentage of SiO_2 in independent rocks (quartz, opal, and chalcedony) is approximately 12%. Earth is apparently the most siliceous part of the universe: the content of SiO_2 in lunar soil is 41% and in rocky meteorites is 21% on average. Silica occurs not only as minerals and dissolved in water, but also is contained, in small quantities, in many organisms, in which it plays an important and not yet completely understood part in the living process. Traditional building and other materials based on silica, such as cement, concrete, firebrick, silicate glasses, rough and fine ceramics, and enamels, occupy a significant place in human life.

The past two decades saw a rapid growth in those fields of science and technology that deal with production and utilization of various colloid and microheterogeneous forms of silica with developed surfaces, such as sols, gels, and powders. The colloid chemistry of silica embraces a wide range of diverse scientific and applied problems. It is an important, independent, and progressive field of colloid chemistry that is closely interwoven with a number of physicochemical and other sciences. Research has led to the creation of production technology for new materials containing silica and having valuable properties.

Various fields use efficient silica adsorbents and selective adsorbents, silica carriers of the active phase in catalysis, silica fillers for polymeric systems, silica thickeners for dispersion mediums, silica binding agents for molding materials, silica adsorbents and carriers for gas chromatography, and so forth. Chemical modification of the surface of dispersion silica has received a large amount of interest; this process allows researchers to change adsorption properties and performance data when synthesizing composite materials.

To understand the mechanisms of adsorption, adhesion, chromatographic separation of mixtures, filling of polymeric systems, and so forth, the nature of interaction of different substances with the surface of silica must be understood. In all such phenomena, the porous structure and the chemistry of the surface silica particles are important.

Colloid chemistry of silica has been researched on a large scale in the former Soviet Union. This chapter discusses not only recent scientific advances there, but also the most important trends in the chemistry of amorphous silica during the last 30–40 years. The most active study in the former Soviet Union of different aspects of colloid chemistry of silica is carried out in numerous scientific organizations in Moscow, Leningrad, and Kiev. Research in this field is also conducted in Minsk, Novosibirsk, Gorki, Kazan, Saratov, Vladivostok, Irkutsk, Tashkent, Vilnius, Baku, Yerevan, Tbilisi, Kishinev, and other cities. Large- and small-capacity industrial production of different kinds of sols, gels, and powders of amorphous silica is realized in a number of chemical plants in the area.

This chapter constitutes a brief review of research carried out in the former Soviet Union on the colloid chemistry of silica. A comprehensive survey of this field would involve an analysis of several thousand publications by Soviet scientists and lies outside the scope of this review. The main trends of research in the former Soviet Union are discussed and illustrated with appropriate examples.

Currently, the definitive book on the chemistry of silica is the monograph by Ralph K. Iler [1]. This book contains 350 references to publications by Soviet researchers. The books by A. V. Kiselev and co-workers [2,3], also published in the United States, are noteworthy for their description of Soviet research on silica surface chemistry and the adsorption properties of oxide adsorbents such as silica. Somewhat more attention is devoted in this chapter to the surface chemistry of amorphous silica, because this field is of special interest to me [4,5].

HISTORICAL SKETCH

The earliest reference to colloidal systems in Russia appeared in 1763, when Lomonosow described the properties of natural colloidal solutions. Even then he distinguished between the coagulation and crystallization processes. In 1785 Lovits investigated the process of adsorption from solutions. The phenomena that later became known as electroosmosis and electrophoresis were discovered in 1809 by Reiss. The concept of the colloidal state as a highly dispersed state of a given phase in a dispersion medium was developed by Borshchov (1869). Mendeleev suggested in 1871 that the general colloidal state of a substance depends on the complexity of its composition and the size of the particle. The works of Shvedov on structure formation in solutions of gelatin and the works of Sabaneev and Lyubavin on the determination of the molecular weight of colloids by the cryoscopic method appeared at the end of the 19th century. Veimarn, the author of one of the first textbooks on colloid chemistry (1904) showed that any substance can be converted into the colloidal state by sharply decreasing its solubility with the right solvent. In 1908 Gedroits studied the colloidal properties of soil, and Zelinsky reported in 1914 the results of investigations into the colloidal structure of proteins.

Large-scale, systematic research in colloid chemistry began in the Soviet Union in the 1920s. Among the prominent chemists in this field are Dumansky (lyophilic colloids), Peskov (stable disperse systems), Rebinder (surface-active substances, physicochemical mechanics), Zhukov (electro-surface phenomena), and Deryagin (surface forces). Fundamental interdisciplinary research was carried out by Frumkin, Balandin, Boreskov, Chmutov, Laskorin, and others. The outstanding schools in the former Soviet Union in the field of adsorption developed under the leadership of Dubinin, A. V. Kiselev, and Neimark.

Dubinin (Institute of Physical Chemistry, the U.S.S.R. Academy of Sciences, Moscow) and colleagues (Zaverina, Radushkevich, Timofeev, Bering, Serpinsky, Zhukovskaya, Nikolaev, Sarakhov, Isirikyan, Zolotarev, Yakubov, Bakaev, Onusaitis, Voloshchuk, and others) conducted a theoretical analysis of the sorption phenomena in porous substances, including silica, and applied these phenomena in practice [6–11].

A. V. Kiselev (Institute of Physical Chemistry, the U.S.S.R. Academy of Sciences, Moscow; Moscow State University) and colleagues (Avgul, Shcherbakova, Dreving, Dzhigit, Lygin, Muttik, Nikitin, Lopatkin, Davydov, Kuznetsov, Belyakova, Berezin, Eltekov, Zhuravlev, and others) studied the dependence of adsorption and the energy of adsorption on the structure and the chemical nature of the adsorbent surface and on the properties of the adsorbed substances. In their investigations, oxide-type adsorbents such as amorphous silica (silica gels, porous glasses, aerosils, and aerosilogels) were widely used [12–17]. A. Kiselev made notable contributions to the study of amorphous silica, quartz, diatomite, silicates, and zeolites. The number of publications in the field of silica chemistry written by A. Kiselev or in collaboration with colleagues exceeds 600 and includes several monographs and textbooks [2,3,17,18], reviews [12,19–25], and patents [26,27].

Neimark (Institute of Physical Chemistry, the Ukrainian S.S.R. Academy of Sciences, Kiev) and colleagues (Slinyakova, Sheinfain, Piontkovskaya, Vysotsky, Chertov, Khatset, Rastenko, Il'in, Chuiko, Tertykh, and others) studied the synthesis and control of porous structures and the modification of the silica surface and other mineral sorbents. The theoretical principles for the formation of a porous structure were worked out [28,29]. Numerous silica gels were prepared that had pores ranging from very wide to very fine; the latter had molecular-sieve properties [28–32].

PREPARATION AND STABILIZATION OF SILICA HYDROSOLS

The preparation and stabilization of silica hydrosols were investigated by Deryagin, Churaev, and co-workers (Institute of Physical Chemistry, the U.S.S.R. Academy of Sciences, Moscow) [33–36]; Frolov, Shabanova, and co-workers (Mendeleev Chemico-Technological Institute, Moscow) [37–41]; Rebinder, Shchukin, Kontorovich, and co-workers (Institute of Physical Chemistry, the U.S.S.R. Academy of Sciences, Moscow) [42–48]; Strazhesko, Strelko, Vysotsky, and co-workers (Institute of Physical Chemistry, the Ukrainian S.S.R. Academy of Sciences, Kiev) [49–55]; Neimark, Sheifain, and co-workers (Institute of Physical Chemistry, the Ukrainian S.S.R. Academy of Sciences, Kiev) [28–32]; Belotserkovsky, Kolosentsev, and co-workers (Technological Institute, Leningrad) [59–64]; Kazantseva et al. (Institute of Geology and Geophysics, Siberian Division of the U.S.S.R. Academy of Sciences, Novosibirsk) [65]; Ryabenko et al. (All-Union Research Institute of Chemical Reagents and High-Purity Substances, Moscow) [66]; and others.

In 1940 Derjaguin [34] studied the subject of coagulation of lyophobic colloids in the presence of an electrolyte. Later Derjaguin and Landau [35] and Verwey and Overbeek [67] developed a theory describing the stabilization of sols (the theory of DLVO). Such stabilization is due to the electrostatic repulsion of diffusion components in the double electrolytic layer formed during the adsorption of the electrolyte ions on the particle surface. The research carried out by Derjaguin into the aggregate stability of colloidal systems [36] showed that solvation shells consisting of solvent molecules hindered the conglomeration of particles because of the elasticity and higher viscosity; cleavage and separation of the particles resulted.

Rebinder [48] investigated the effect of the surface (adsorption) layers on the properties of colloidal systems. When the lyophobic dispersion systems are stable, the structural–mechanical stabilization occurs where the protecting layers of the micelle-forming surface-active agents or high-molecular compounds are formed at the interface boundary.

Research of the problems of the formation of disperse systems and the physicochemical mechanics of disperse structures was conducted by Rebinder, Shchukin, Kontorovich, and co-workers [42–47] particularly of the problems of structure formation in silica hydrosols. The properties of silica sols (their aggregate stability, the nature of the contacts between the individual particles) depend on many factors. Estimates of the diameter of globules in hydrogels of silicic acid were made by using narrow angle X-ray scattering. Within a wide range of time and temperature of aging, concentration of the solid phase, and pH of the sols (i.e., the conditions for the formation of gels), the radius of the globules varied from 15 to 37 Å. Light-scattering methods were used to study the regularities of the solid-phase aggregations. The scattering centers that were formed (of the order of magnitude of several hundred angstroms) belong to loose aggregates made up of primary particles. The rate of formation of solid-phase particles increased with an increase in the concentration of SiO_2 in sols and with an increase in pH. In supersaturated solutions of silica, the strength of contact scattered between the real particles in silica sols is of several orders of magnitude. Thus, the contacts could be subdivided into two main types: contacts of the coagulation type with strength $p \leq 10^{-2}$ dyne being due to the van der Waals interaction, and phase contacts with strength $p \geq 10^{-1}$ dyne being due to the coalescence of particles via the polycondensation mechanism. The effect of temperature on the syneresis kinetics of acid and alkaline hydrogels of polysilicic acids was studied. Above and below the isoelectric point spontaneous shrinkage took place at the same effective activation energy despite the different reaction mechanism of polycondensation within the pH range under consideration. The activation energy in this case was similar to the energy of the hydrogen bonds. The authors concluded that the rate of spontaneous shrinkage of the hydrogels of polysilicic acids is determined mainly by the interaction of particles through the surface forces that lead to the formation of easily mobile coagulative contacts.

Frolov, Shabanova, and co-workers [37–39] studied the transition of a sol into a gel and the aggregate stability of colloidal silica. Their aim was to develop a technology for the production of highly-concentrated silica sols and to use them as binders, catalyst supports, polymer fillers, adsorbents, and so forth. Kinetic studies were made of polycondensation and gel formation in aqueous solutions of silicic acids. At the stage of particle growth, polycondensation proceeds in the diffusion–kinetic region. With changes in pH, temperature, concentration, and the nature of electrolytes, the properties of the surface layers have a decisive effect on heterogeneous polycondensation, gel formation, and aggregate stability. The maximum rate of gel formation at pH 5.5–6.5 corresponds to the transition in the gels' properties from a condensational type to a coagulational thixotropic type capable of undergoing subsequent peptization. The aggregate stability and kinetics of gel formation are limiting processes during the formation of the porous structure in the final product (silica gel). The results were analyzed from the point of view of the DLVO stability theory; the splitting pressure was taken into account.

Kolosentsev and Belotserkovsky [59–61] used the ion-exchange methods proposed by Bird [68] and Bechtold and Snyder [69] in preparing stable aqueous silica sols containing ~50 wt% SiO_2, with particle size ranging from 65 to 380 nm.

Frolov, Khorkin, Shabanova, and co-workers (Mendeleev Chemico-Technological Institute, Moscow; Branch of the Polymer Research Institute, Saratov) carried out the experimental production [41] of highly concentrated silica sol by using the same ion-exchange methods [68,69]. The concentration of SiO_2 sols obtained was $\sim$40% by weight or more.

Lipkind et al. (Experimental Plant of the All-Union Oil Processing Research Institute, Gorky; Research Institute of the Chemical Industry, Kemerovo) set up the industrial production of sols made from silicic acid by the electrodialysis method [70]. The capacity of the experimental pilot plant is 100 L of sol per hour.

Frolov, et al [40] used the ion-exchange method in conjunction with electrodialysis for preparing SiO_2 hydrosols. The process consisted of the following: ion-exchange, electromigration of H^+ and Na^+ ions in an ion-exchange resin that involved the regeneration of the ionite, and electrodialysis of the starting solutions. The cationic resins were in the strongly acidic H+ form. The efficiency of the process was raised when ion exchange was carried out in a direct current electric field. The process makes possible the continuous production of highly concentrated hydrosols.

Ryabenko et al. [66] developed a method for synthesizing highly pure silica by heterogeneous hydrolysis of tetraethoxysilane followed by the concentration of the sol of polysilicic acid and thermal treatment.

Strazhesko, Strelko, Vysotsky, and co-workers [49–51] investigated the polymerization of monosilicic acid. Polysilicic acid is more acidic than monosilicic acid, and the dissociation constant of the acidic centers on the surface of polysilicic acid is 2 to 3 orders of magnitude greater than that of the monomer [49–51]. Strazhesko, Vysotsky, and co-workers [52–55] showed that the isoelectric point represents not only the minimal rate of the gel formation process but also its syneresis. As a result, mechanically strong silica gels were obtained that had a maximum specific surface area.

Strelko and co-workers [71–73] investigated the properties of the siloxane bond and the polymerization mechanism for silicic acid on the basis of the difference in the electronic structure of the Si–O bonds in silanol and the siloxane groups of silica. A kinetic equation was proposed to describe the polymerization of silicic acid throughout the entire pH range. The authors believe that the molecular mechanisms for the formation of globular skeletons in silica gels are based on the polymerization and depolymerization of silicic acids.

Kazantseva et al. [65] synthesized monodisperse silica sols by hydrolysis of a mixture of tetraethoxysilane and $C_2H_5OH-H_2O-NH_3$ at room temperatures. The degree of monodispersion of the silica sols increased with time and corresponded to an increase in the diameter of the spherical particles. When hydrolysis had proceeded for 3.5–4 h, the same uniform size of the particles reached $\sim$300 nm.

Churaev, Nikologorodskaya, and co-workers [33] investigated the Brownian and electrophoretic motion of silica hydrosol particles in aqueous solutions of an electrolyte at different concentrations of poly(ethylene oxide) (PEO) in the disperse medium. The adsorption isotherms of PEO on the surface of silica particles were obtained. The thickness of the adsorption layers of PEO was determined as a function of the electrolyte concentration and the pH of the dispersed medium. The results can be used in an analysis of the flocculation and stabilization conditions for colloidal dispersions of silica (with non-ionogenic water-soluble polymers of the PEO type).

PREPARATION OF SILICA GELS AND POWDERS, GEOMETRIC MODIFICATION OF SILICAS, AND STUDIES OF STRUCTURAL CHARACTERISTICS

Silica gels, porous glasses, and silica powders were prepared by A. Kiselev, Nikitin, and co-workers (Moscow State University, Moscow) [2,3,26,27,74–81]; Dzisko, Fenelonov, and co-workers (Institute of Catalysis, Siberian Division of the U.S.S.R. Academy of Sciences, Novosibirsk) [82,83]; Zhdanov and co-workers (Institute of Silicate Chemistry, the U.S.S.R. Academy of Sciences, Leningrad) [84–87]; Belotserkovsky, Kolosentsev, and co-workers (Technological Institute, Leningrad) [59–61]; Neimark, Sheinfain, and co-workers (Institute of Physical Chemistry, the Ukrainian S.S.R. Academy of Sciences, Kiev) [28,29]; Chuiko (Institute of Surface Chemistry, the Ukrainian S.S.R. Academy of Sciences, Kiev) [88]; and others.

Porous glasses were first prepared and described by Grebenshchikov and co-workers (State Optical Institute, Leningrad) [89,90]. This research was continued by Zhdanov and co-workers [85–87]. Usually, porous glasses are prepared by treating sodium-boron silicate glasses with inorganic acids. The skeleton of porous glasses consists of an acid-resistant sponge network of the silica phase and of the secondary structure formed from highly dispersed hydroxylated silica having a globular structure. The latter precipitated inside the network cavities during the extraction of B_2O_3 and Na_2O. The specific surface area and porosity of such glasses depend on the leaching conditions: the concentration of the acid, the temperature of the solution, and so forth, as well as on the conditions of the thermal treatment of the initial glasses. A very valuable property of porous glasses is that their structure can be regulated within a wide range of porosity. In addition, they are mechanically and chemically stable. Such glasses are made at the Experimental Plant of the All-Union Research Institute of Oil Processing in Gorky [91].

Dobychin and co-workers (State Optical Institute, Leningrad) [92,93] carried out a systematic investigation of the structure and adsorption properties of porous glasses and of the structural changes in them.

Plachenov and co-workers (Technological Institute, Leningrad) [94,95] developed methods of synthesizing active metal oxides, including silicas. A systematic study of the secondary structure of porous substances was carried out [96,97].

Belotserkovsky, Kolosentsev, and co-workers (Technological Institute, Leningrad) (59–64) worked on the synthesis and application of various silica gels. They were able to vary within certain limits the porous structure of the sorbents by compressing them into a powder form. They proposed that silicas could be obtained by ion-exchange, peptization, and other methods.

Sychev and co-workers [98] (Technological Institute, Leningrad) investigated SiO_2–P_2O_5 xerogels, which are of practical interest in the field of optical systems and other areas. The optimal conditions were established for preparing solutions containing the necessary oxides and for converting them into gels. Following thermal treatment at 800°C, the average pore size of the xerogel was ~50 nm.

Komarov, Kuznetsova, and co-workers (Institute of General and Inorganic Chemistry, the Byelorussian S.S.R. Academy of Sciences, Minsk) [99–102] examined the possibility of forming and controlling the porous structure of xerogel during the gelling syneresis and the drying of the gel. The effects of different surface-active agents on the structure formation, the pH dependence of the micellar solutions, the temperature, the presence of electrolytes, and the solubilizing water-soluble organic substances were investigated. The surface-active agents were alkyl-substituted quaternary salts of pyridinium and ammonia (cation- and anion-type surface-active agents), and the solubilizers were decyl alcohol, capric acid, and others. A mechanism describing the gel formation in the presence of surface-active agents and solubilizers was proposed.

Lipkind et al. (Experimental Plant of the All-Union Research Institute of Oil Processing, Gorky) [103] worked out ways of setting up the industrial production of mineral adsorbents, including silica gels.

The most important adsorbents in gas-adsorption chromatography are macroporous silica [26,27,74]. Silica carriers used in chromatography were described by Gavrilova, Bryzgalova, and co-workers (Moscow State University) [104–106].

Khotimchenko, Vasiloi, Demskaya, and co-workers (The Scientific Industrial Association Quartz, Leningrad) [107,108] used sol–gel technology for making quartz glass. A heat-resistant composite was obtained from tetraethoxysilane by using quartz fibers (diameter $d = 5–10\ \mu m$) as the filler. The activation of quartz glass with organic pigments was investigated.

Pryanishnikov, Chepizhnyi, and co-workers (State Research Institute of Glass, Moscow) [109] developed superfine silica fibers (Sivol) by thermochemical treatment of a natural mineral, chrysotile asbestos, in the presence of inorganic acids. The physicochemical properties of these fibers and the starting material were investigated by Pryanishnikov, Osipov, Guy, and co-workers (State Research Institute of Glass, and the Institute of Physical Chemistry, the U.S.S.R. Academy of Sciences; both in Moscow) [110–112]. These superfine fibers have much greater porosity than the starting material. The pores were distributed according to their diameter within a broad range, with a maximum at 5–6 nm. The specific surface areas as measured by Kr and water adsorption was $S_{Kr} \ll S_{H_2O}$. These measurements, as well as the data obtained by low-angle X-ray scattering, show that these superfine fibers contain fine ultramicropores comparable in diameter with the size of water molecules. Electron microscopic investigations show the fibers to be characterized by a multilayered fibril structure with a well-defined internal channel in the center.

Il'in, Turutina, and co-workers (Institute of Physical Chemistry, the Ukrainian S.S.R. Academy of Sciences, Kiev) [113–115] investigated the cation processes for obtaining crystalline porous silicas. The nature of the cation and the composition of the systems M_2O–SiO_2–H_2O (where M is Li^+, Na^+, or K^+) affect the rate of crystallization, the structure, and the adsorption properties of silica sorbents of a new class of microporous hydrated polysilicates (Siolit). These polysilicates are intermediate meta-stable products of the transformation of amorphous silica into a dense crystalline modification. The ion-exchange adsorption of alkali and alkaline earth metals by these polysilicates under acidic conditions increases with an increase in the crystallographic radius and the basicity of the cations; under alkaline conditions, the selectivity has a reverse order. The polysilicates exhibit preferential sorption of alkali cations in the presence of which the hydrothermal synthesis of silica was carried out. This phenomenon is known as the memory effect.

Korneev, Agafonov, and co-workers (Technological Institute and Pigment Science Industrial Association, both in Leningrad) [116,117] synthesized aqueous solutions of systems that have high modulus and low alkalinity and are based on silicates of quaternary ammonia (SQA). The decomposition of SQA took place at ~190°C. The calcined product of SQA has amorphous and crystalline phases. Aqueous solutions of quaternary ammonium silicates can be used as binders; they are not flammable, explosive, or toxic.

In 1961 Stishov and Popova (Moscow State University, Chair of Geochemistry, and Institute of High-Pressure

Physics, the U.S.S.R. Academy of Sciences, respectively; both in Moscow) [118] for the first time synthesized modified crystalline silica having a very dense octahedral structure, which they named Stishovite.

The geometrical modification of the structure of silica adsorbents (by hydrothermal treatment in autoclave, by calcining in air and in an atmosphere of water vapor, by treatment at the hydrogel stage, etc.) was carried out by Nikitin, A. Kiselev, Bebris, Akshinskaya, and co-workers (Moscow State University) [26,27,74–81,119]; Chertov, Neimark, and co-workers (Institute of Physical Chemistry, the Ukrainian S.S.R. Academy of Sciences, Kiev) [56–58]; Zhuravlev, Gorelik, and co-workers (Institute of Physical Chemistry, the U.S.S.R. Academy of Sciences, Moscow) [4,77,80,120–122]; Lipkind and co-workers (Experimental Plant of the All-Union Research Institute of Oil Processing, Gorky) [26,123]; and others.

Nikitin, A. Kiselev, and co-workers [74,124,125] developed the method of geometric modification of the adsorbent structure. This advance led to the industrial production in the Soviet Union of a new class of adsorbents — macroporous silica, in the form of silica gels and silochroms (the latter is the technical name for homogeneous large-pore aerosilogels [26,27]; the diameters of the pores vary from hundreds to thousands of angstroms. These silica gels have found wide application in gas and liquid chromatography, in the immobilization of ferments, and so forth [74,75,119]. Different methods of geometric modification were studied (i.e., predetermined changes in the structure of the pores and the silica skeleton): hydrothermal treatment inside the autoclave; calcining in air and in the atmosphere of water vapor; and calcining in the presence of alkaline additives [74,76–78]. An investigation was carried out of the mechanism of the changes in the structure of silica during hydrothermal treatment: an increase in the diameter of the pores (a decrease in the specific surface area) was due to the dissolution of small silica particles and to the precipitation of this silica on the larger particles, primarily at the point of contact of the particles. In the course of hydrothermal treatment of silica gels ultrapores, in addition to very wide pores (macropores), are produced that are accessible mainly to water molecules but inaccessible to krypton and benzene molecules [76,77]. The thermal treatment of silica gels in air and in water vapor atmosphere prevents the formation at microheterogeneities of the surface of the sample and enhances the geometric uniformity of macroporous silica gels [74,75]. The mechanism of the sintering of silica gels under different conditions of thermal treatment was studied [79–81].

The hydrothermal modification of SiO_2 was investigated by Chertov, Neimark, and co-workers at the hydrogel and xerogel stages [56–58]. The hydrothermal treatment of hydrogels makes it possible to vary the structure of silica gels within a wide range (e.g., specific surface area from 800 to 20 m^2/g, and volume of pores from 0.20 to 2.20 cm^3/g).

Lazarev, Panasyuk, Danchevskaya, and co-workers (Institute of General and Inorganic Chemistry; the U.S.S.R. Academy of Sciences, Moscow State University, Moscow) [126] investigated the phase structural transformations occurring at an elevated temperature in amorphous silica under the influence of an active medium such as water. Such investigations have been carried out to meet the requirements of industry, because fine-grain quartz obtained in such processes serves as a raw material in the production of quartz glass. The starting silica gel was subjected to drastic hydrothermal treatment in an autoclave (300–400°C at pressures of several tens of atmospheres). A thermodynamic analysis was made of the processes, and the enthalpy and energy of the activation were determined. A stepwise mechanism was proposed to describe the water-induced transformation of amorphous silica into a crystalline form (quartz). The high activity of water at elevated temperatures leads to the breaking of the siloxane bonds and the formation of intraskeletal silanol groups. The latter again interact with each other and form new, and thermodynamically more stable, siloxane bonds. This process is accompanied by the removal of interglobular water. Thus, hydrothermal treatment causes crystallization that yields quartz without transforming SiO_2 into a soluble state.

Dubinin, A. Kiselev, Deryagin, Chmutov, Zhdanov, Neimark, Serpinsky, Bering, Radushkevich, Karnaukhov, Plachenov, Nikitin, Dzisko, Keltsev, Torocheshnikov, Dobychin, and others made a significant contribution to elucidating the geometrical structure of dispersed and porous substances, in particular for silica. The structural characteristics of silica were investigated by theoretical analysis, adsorption methods, measurement of apparent density, mercury porosimetry, kinetic methods, narrow-angle X-ray scattering, electron microscopy, and other methods.

The classification of pores according to their size, proposed by Dubinin, has been adopted by IUPAC [127,128]: the classifications are macropores ($d > 2000$–4000 Å), mesopores (30–32 Å $< d < 2000$–4000 Å), supermicropores (12–14 Å $< d < 30$–32 Å), and micropores ($d < 12$–14 Å). This classification system has proved useful and is used in the study of amorphous silica having different structural characteristics [4].

A. Kiselev developed the adsorption–structural method of investigation [129], which made possible a rational classification of adsorbents [130–132]. Dubinin, Radushkevitch, Bering, Serpinsky, and others have developed on the basis of their experimental results a theory on the physical adsorption of gases and vapors in microporous adsorbents that they call the theory of volume filling of micropores. The theory is applicable to almost all the adsorption systems, including microporous silica gels and porous glasses [133,134].

Plachenov [96,97] estimated the limiting dimensions of the equivalent radii of mesopores. The distribution of the volume of the mesopores calculated from their equivalent radii was compared with capillary condensation measurements. The macroporous and the intermediate mesoporous structure were determined, as were the volume of the micropores and the constant in the equation of the isotherm adsorption theory describing the volume filling of micropores [133,134].

Karnaukhov, A. Kiselev, and co-workers [12,14,135–142] investigated the hypothetical geometrical structure describes by a corpuscular model in the form of a system of particles (globules) of uniform size in contact with one another. The diameter of the particles and the number of contacts between particles were taken as the main parameters of the model. An approximate theory of polymolecular adsorption and capillary condensation was developed. The simultaneous process of polymolecular adsorption and capillary condensation was analyzed for porous systems with a globular structure. It was shown that the contribution of both processes determines the form of the adsorption isotherm. The theoretical results obtained were compared with experimental data for different silica adsorbents by using different methods.

Zhuravlev, Gorelik, and co-workers [77,80,120–122,143] carried out kinetic studies of water vapor adsorption and isotopic exchange (D_2O + ≡Si–OH) for different types of amorphous silica. A relationship was established between the nature of the porosity and the shape of the kinetic curve. For example, for bidispersed silica gels containing both wide mesopores and very fine ultramicropores, the kinetic plot consists of two sections: a very short period due to the mass transfer of water vapors through transport mesopores, and a very long period (tens of hours) due to the diffusion of water molecules inside very fine pores that have diameters comparable with that of the water molecule. Thus, such plots provide information on the type of pores present in the silica sample.

With the aid of a kinetic model for a biporous sorbent [144], Zolotarev et al. [143] described the diffusion and kinetics of sorption for water–silica gel systems. On the basis of this theoretical model, an explanation was given for the differences in the experimental kinetic curves for water–vapor sorption on biporous silica gels [143]. The coefficients for the mass transfer of water in the mesopores and ultramicropores were also determined.

SURFACE CHEMISTRY OF AMORPHOUS SILICA

In the 1930s, studies of the condensation processes of silicic acids showed that hydroxyl (silanol) groups, ≡Si–OH, should be present on the surface of silicates and silicas [145–147]. On the basis of measurements of the heat of wetting and a comparison of the adsorption data with the date from chemical analysis and the corresponding results reported in the literature, A. Kiselev [147] suggested [1936] that the water evolved during calcination from silica gel is not physically adsorbed water but is formed from OH groups that are chemically held on the silica surface. This suggestion led to an understanding of the dehydroxylation mechanism [147].

Yaroslavsky and Terenin (Leningrad State University) [148], by using an infrared spectroscopy method, proved for the first time (1949) the existence of hydroxyl groups on the silica surface (porous glass). This fact was soon confirmed by other researchers [149–153].

Zhdanov (Institute of Silicate Chemistry, the U.S.S.R. Academy of Sciences, Leningrad) [154] showed (1949) that the adsorption of water vapor by SiO_2 (porous glasses, silica gels) strongly depends on the temperature of the preliminary thermal treatment of the adsorbent. Calcination of 300–500°C resulted in a sharp decrease in the adsorption of H_2O at low values of pressure over initial pressure p/p_0 (<0.3), and the adsorption isotherms were found to be irreversible. On the other hand, the adsorption isotherms of water on silica subjected to calcination in vacuo at <500°C (but after the sample was kept in contact with water vapor or liquid water at room temperature) again became reversible; that is, the adsorption activity of SiO_2 was restored.

In the 1950s, A. Kiselev, Zhdanov, and co-workers [12,84,155–159] showed that when the adsorption isotherms of water are expressed as absolute isotherms (referred to as the unit surface of the SiO_2 sample), widely different forms of amorphous silica having a completely hydroxylated state adsorb the same amount of water at the same relative pressure ($p/p_0 < 0.3$). Thus the plots of absolute adsorption isotherms for different samples showed that the surfaces of these samples are of a similar nature. The adsorption properties of nonporous silica and silica having large pores (i.e., an absence of micropores) depend above all on the presence of OH groups and on the degree of hydroxylation of the surface.

On the basis of the concept of the "absolute" adsorption properties of SiO_2, an attempt was made by A. Kiselev, Zhdanov, and co-workers [12,158,160] to establish a value for the concentration of OH groups (or the silanol number, α_{OH}) on the surface of amorphous completely hydroxylated silica. The most probable value was $\alpha_{OH} = 6.9$–7.5 OH groups per square nanometer^2. However, the possible presence of structurally bound water inside the silica was not taken into account [161–163], so a somewhat higher calculated value for α_{OH} resulted.

Other workers in this field were critical of the concept put forward by A. Kiselev and Zhdanov. Sidorov (Leningrad State University) [151,153], in infrared spectroscopic studies of water adsorption at a low degree of coverage, had observed that the intensity of the absorption band due to the free surface OH groups ($\nu_{OH} = 3750\,cm^{-1}$) remained unchanged. Similar experiments were carried out by other researchers. It was concluded by Sidorov and by V. Kiselev and co-workers (Moscow State University, Physics Department) [164–166] that the initially free H_2O molecules are adsorbed not on OH group sites but on more active centers that they called centers of the II-type.

The main ideas put forward by V. Kiselev and co-workers [164–166] are as follows:

1. The extent of hydration and hydroxylation of silica surface, and also the adsorption properties (per unit of silica surface), are different for samples having different specific surface area under the same conditions of preliminary treatment. The extent decreases with an increase in the specific surface area.
2. Coordinately bound water firmly held in the form of H_2O molecules exists on nonhydroxyl centers of the II-type; these centers are coordinately unsaturated Si atoms on the surface of the sample. The concentration of the II-type centers is small: $\alpha_{Si} = 0.1–0.3$ Si atoms per square nanometer^2.
3. Coordinately bound water is removed in vacuo at ~400°C. Above this temperature there remains on the surface of silica a hydroxylated layer that has a heterogeneous composition. The NMR data [164–166] point to the presence of siloxane groups; single, isolated OH groups; paired vicinal and paired geminal OH groups; and probably triple $-Si(OH)_3$ groups.

The presence of coordinated water was confirmed by Chuiko and co-workers (Institute of Surface Chemistry, the Ukrainian S.S.R. Academy of Sciences, Kiev) [88,167]. Experiments were carried out on pyrogenic silica (aerosil) with different methods. The main conclusions of these studies are as follows:

1. The topography of SiO_2 is determined by the index of face (1 1 1) and face (1 0 0) in β-cristobalite; the silanol number of the starting aerosil is $\alpha_{OH} \cong 1.7$ OH groups per square nanometer.
2. The adsorption centers of the coordinated water are coordinately unsaturated Si atoms on the silica surface to which OH groups are attached. The adsorbed water molecule (with the coordinate bond) near the adsorption center can be located either over the face of the silicon–oxygen tetrahedron

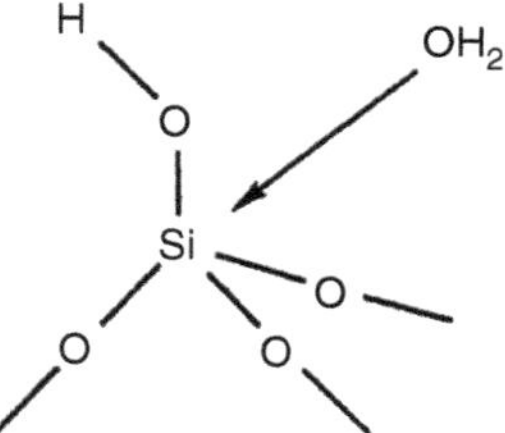

or under the Si atom of the silanol group

3. The concentration of the II-type centers is high, $\alpha_{Si} \cong 1.7$ Si atoms per square nanometer; that is, all the surface Si atoms with attached OH groups belong to these centers.
4. The centers of the II-type ensure a strong retention of coordinated (molecular) water on the SiO_2 surface up to ~650°C, with the energy of the interaction of H_2O molecule with such a center being up to 62 kcal/mol (259 kJ/mol) [167].

On the other hand, some investigations confirm the main conclusions of A. Kiselev and Zhdanov. Galkin, Lygin, and co-workers (Moscow State University) [3,168–170] investigated the changes in absorption band intensity due to valence vibrations of free OH groups ($\nu_{OH} = 3750\,cm^{-1}$) during the adsorption of water under conditions in which the infrared heating of the sample was excluded. When water was gradually introduced to a system originally at a low degree of coverage, the intensity of the absorption band decreased. Thus, the surface OH groups act as the water adsorption centers. Other investigations [171–173] also indicate the important role played by silanol groups as the centers of molecular adsorption of water.

Zhuravlev (Institute of Physical Chemistry, the U.S.S.R. Academy of Sciences, Moscow) [4,5] carried out a systematic investigation of the hydroxylated surface of amorphous silicas by using the deutero-exchange method and the programmed-temperature method. The main conclusions are as follows:

1. The temperature threshold corresponding to the removal of physically adsorbed water from the hydroxylated SiO_2 surface lies at 190°C.

2. The maximum hydroxylated state of the surface of SiO_2 samples (the degree of surface coverage with OH groups, $\theta_{OH} = 1$) is obtained after preliminary treatment of the samples in vacuo at 180–200°C, when for each Si atom there is approximately one OH group. The average value of the silanol number for such a state, $\alpha_{OH} = 4.6$ OH groups per square nanometer, is a physicochemical constant (it is independent of the type of the amorphous silica used, the method of preparing it, and the structural characteristics, that is, the specific surface area, the type of pores, the distribution of the pores according to their diameter, the packing density of the particles, and the skeleton structure of SiO_2).
3. The kinetic order of the thermal desorption of water in the range 200–1100°C is $n = 2$. Thus, the removal of water from the surface (associative desorption) takes place owing to the interaction of pairs of hydroxyl groups that leads to the formation of siloxane bonds. The degree of coverage with silanol groups, θ_{OH}, in this temperature range varies from 1 to ~0.
4. With the degree of coverage of OH groups in the range $1 \geq \theta_{OH} > 0.5$, the activation energy of desorption, E_d, changes from 16.5 to ~25 kcal/mol (~105 kJ/mol). Lateral interaction (hydrogen bonding) occurs between the OH groups. The chemisorption of water (dissociative adsorption) involving the breaking of the siloxane bridges and the formation of new silanol groups is a rapid nonactivated (or weakly activated) process.
5. At the degree of coverage $\theta_{OH} < 0.5$, the energy of activation, E_d, increases substantially [from ~25 to 50 kcal/mol (~105 to 210 kJ/mol) and probably higher] owing to the fact that there exist on the surface of SiO_2 only isolated OH groups, and for the condensation reaction to take place, there must be an activated process involving the migration of protons and the evolution, during the final stage, of H_2O molecules due to the accidental interaction in pairs of OH groups (which draw close to one another to a distance of ~0.3 nm). The chemisorption of water is a slow and strongly activated process.
6. In addition to surface OH groups, there may exist OH groups inside the silica skeleton and inside the very fine ultramicropores [143] that have diameters comparable to that of the water molecule. The relationship between the amount of surface OH groups and the intraskeleton OH groups depends on the method of preparation and subsequent treatment of silica [162].

INVESTIGATIONS OF SILICA SURFACE PROCESSES BY PHYSICAL AND PHYSICOCHEMICAL METHODS

Infrared and ultraviolet spectral methods — electron paramagnetic resonance (EPR) and NMR spectroscopy, luminescent probing, mass spectrometric methods, calorimetric methods, and other procedures — were used in investigating surface processes of SiO_2 by Terenin, Yaroslavsky, Sidorov, and co-workers (Leningrad State University) [148,150–153]; Kurbatov and Neuymin (Naval Medical Academy, Leningrad) [149]; Lygin, Galkin, Davydov, and co-workers (Moscow State University) [168–171,174–183]; Titova, Arutyunyan, Kosheleva, and co-workers (Institute of Physical Chemistry, the U.S.S.R. Academy of Sciences, Moscow) [3,172,173,184]; Vedeneeva and Musatov (Institute of Crystallography, the U.S.S.R. Academy of Sciences, Moscow) [185,186]; Zhdanov et al. (Institute of Silicate Chemistry, the U.S.S.R. Academy of Sciences, Leningrad) [173]; Kvlividze, Egorow, V. Kiselev, Chukin, and co-workers (Moscow State University; All-Union Research Institute of Oil Processing, Moscow) [164–166,187–190]; Bakaev, Pribylov, and co-workers (Institute of Physical Chemistry, the U.S.S.R. Academy of Sciences, Moscow) [191,192]; Butyagin, Radtsig, and co-workers (Institute of Chemical Physics, the U.S.S.R. Academy of Sciences, Moscow) [193–197]; Kotov, Pshezhetskii, and co-workers (Physico-Chemical Institute, Moscow) [198,199]; Chuiko, Sobolev, Tertykh, Eremenko, and co-workers (Institute of Surface Chemistry, the Ukrainian S.S.R. Academy of Sciences, Kiev) [88,200–207]; Zhuravlev, Agzamkhodzhaev, Gorelik, Shengeliya, and co-workers (Institute of Physical Chemistry, the U.S.S.R. Academy of Sciences, Moscow) [77,120–122,176,208–210]; Kazansky, Kustov, and co-workers (Institute of Organic Chemistry, the U.S.S.R. Academy of Sciences, Moscow) [211–214]; Mastikhin et al. (Institute of Catalysis, Siberian Division of the U.S.S.R. Academy of Sciences, Novosibirsk) [215]; and others.

Bondarenko, V. Kiselev, and co-workers [216] carried out a mass spectroscopic analysis of the gaseous products formed during the thermal treatment (300–1000°C) of various crystalline and amorphous silicas. The only product formed was water.

Lygin, Galkin, Davydov, Shchepalin, and co-workers [3,168–171,174,183,217] carried out a series of infrared spectroscopic investigations of silica. The problems they studied include hydroxyl coverage, surface chemical reactions, and the adsorption mechanism of molecules having different structures. They established that high-temperature dehydroxylation of the SiO_2 surface changed the properties of the OH groups. Research has been conducted into the behavior of hydroxyl groups and electron-acceptor centers of silica surface in samples modified with admixtures, Al_2O_3, and B_2O_3. The reaction of surface OH groups with different chemical reagents was examined, the particular, with $ClSi(CH_3)_3$ and $Cl_2Si(CH_3)_2$. The behavior of intraskeletal hydroxyl groups (absorption band with a maximum of $3650\ cm^{-1}$) in relation to the conditions of the preliminary treatment of silica sample was investigated. The surface silanol groups were divided into free, isolated, and perturbed groups that were linked by a hydrogen bond. Spectral and energy evidence of the interaction of OH groups with molecules having different electronic structures in the process of physical adsorption was established. Before complete coverage of the silica surface with a monolayer of water, a number of associated H_2O molecules were formed. On the dehydroxylated sections of the SiO_2 surface, chemisorption of water takes place in addition to molecular adsorption.

Analogous infrared spectroscopic investigations were carried out by Titova, Arutyunyan, Kosheleva, and co-workers [3,172,173,184]. The structure of associates silanol groups on hydroxylated silica surface was examined: in addition to free, isolated OH groups, associations with intramolecular hydrogen bond both by hydrogen atom and by oxygen atom, in the form of mutually perturbed silanol pairs, were found. With a change in the temperature, reversible transitions take place between the free and associated silanol groups. Molecularly adsorbed water on the surface of highly pure silica is held owing to the formation of hydrogen bonds with the surface hydroxyl groups.

Using ^{19}F NMR spectroscopy, Bakaev and Pribylov [191] determined the location of $CFCL_3$ molecules with respect to one another in an adsorbed state on silica gel at 77.4 K. In adsorption of $CFCL_3$, the OH groups being first removed from the surface of the silica gel and under the conditions of a low degree of coverage, there is an increase of the second moment M_2 line with an increase in adsorption. Such a dependence is due to the nonuniform nature of the SiO_2 surface, that is, to the presence of surface adsorption centers that differ in their adsorption activity.

Krasilnikov, V. Kiselev, and co-workers [218] investigated the oxidizing properties of the SiO_2 surface in the process of its dehydroxylation by thermal treatment (300–900°C) in air. The oxidizing capacity was very low: for the same surface the number of oxidizing equivalents was smaller by 3 orders of magnitude than the number of OH groups. The amount of adsorbed oxygen increases with an increase in the calcination temperature. It was concluded that oxygen is adsorbed by the surface radicals produced during the dehydroxylation of the surface. Surface radicals of the $\equiv Si{-}O\cdot$ type are centers of high activity.

Kuznetsov, Dzhigit, Muttik, and co-workers (Moscow State University) [15,217,219–226] carried out calorimetric investigations and obtained the characteristics of adsorbed substances having different geometrical and electronic structures. They also measured the energy of adsorption for hydroxylated and dehydroxylated surfaces of silica. The difference in the heat of adsorption, ΔQ_{OH}, for a fully hydroxylated and strongly dehydroxylated silica surface (degree of coverage $\theta \cong 0.5$) was determined. The difference represents the contribution due to the energy of the specific interaction of the surface hydroxyl groups with the linkages and the bonds of different organic molecules having locally concentrated electron density. The differential heat of adsorption of water vapor as a function of the degree of coverage on a hydroxylated silica surface was measured [15].

The dependence of the differential heat of adsorption of *n*-alkanes, alcohols, and other substances was investigated by Isirikyan and co-workers [227–229], who used the calorimetric method for silica gels possessing different degrees of porosity and hydroxylation. The heat of adsorption was always sharply lower at a low degree of coverage owing to the nonuniform surface of silica. The adsorption isotherms and the differential heats of adsorption for benzene vapor are higher than those of hexane. The greater value for the adsorption energy of benzene is due to the additional specific dispersion interaction of the molecule with the OH groups on silica (due to the local concentration of electron density on the periphery of the benzene ring attributable to π electrons). The heat of adsorption of benzene increases with an increase in the concentration of the silanol groups, whereas the differential heat of adsorption of hexane remains the same for SiO_2 with different degrees of hydroxylation of the surface. For *n*-alkanes in the region of high relative pressures, p/p_0, there appears a sharp maximum in the differential heat of adsorption. This effect is explained by capillary condensation theory [135].

Egorov, Murina, V. Kiselev, and co-workers [230] measured the heat of wetting of silica for water and other liquids in relation to the degree of hydroxylation. The plot of the heat of wetting versus the calcination temperature showed a maximum corresponding to 200–300°C. The heat of wetting decreased at higher calcination temperatures, that is, with a decrease in the degree of hydroxylation of the SiO_2 surface.

Berezin and co-workers (Institute of Physical Chemistry, the U.S.S.R. Academy of Sciences, Moscow) [231–237] measured by the calorimetric method the heat capacity of benzene, water, and other substances adsorbed on silica gel at low temperatures and different degrees of adsorption. In monlayered coverage, the adsorbate does not undergo phase transition. But in polymolecular coverage, there is a range of temperatures (below the normal melting temperature of a given substance) within which the adsorbate undergoes phase transition corresponding to the volume melting of the substance. These results are of practical interest in solving problems related to the freezing of water in different porous substances.

Gusev and co-workers (State University, Kazan) [238,239] investigated the dielectric relaxation of adsorbed water in silica gel, aerosil, and quartz. Characteristic of these systems is the presence of two disperse regions of dielectric permeability, the position of the amplitude of which depends on temperature, humidity, specific surface area, and porosity. The relaxation time of these regions increases with an increase in the diameter of the pores. The distribution parameters of these time periods are much greater than the corresponding relaxation periods for free water and ice. Disperse region I appears only when the humidity corresponds to a monolayer. Disperse region II is due to the reorientation of water molecules caused by the nonuniform surface of the sample. On a silica surface the phase transition temperature of water–ice is lowered.

Zhilenkov, Glazun, and co-workers (Agricultural Institute, Voronezh) [240–243] measured the dielectric permeability ε of water and other substances adsorbed on a silica surface. At a low degree of coverage, the permeability was proportional to the concentration of the adsorbed substance (benzene, chloroform, acetone, and water). In this case, the ε of the adsorbate did not differ from that of liquid. When the degree of coverage with water was high, there was a notable decrease in growth of the dielectric permeability in comparison with that of an ordinary liquid. The decrease in ϵ was explained as being due to electrostatic attraction on the silica surface inside the narrow pores; this attraction decreases the ability of the dipole moment of water molecule to orient itself in the external electrical field, a situation that causes a decrease in the polarization. In the adsorbed structural water at low temperatures, the dielectric relaxation is 5–6 orders of magnitude greater than that of ordinary ice. This observation illustrates the nonuniform nature of the silica surface, with active centers having different adsorption potentials.

PHYSICAL AND CHEMICAL ADSORPTION

A great number of physical adsorption investigations of different substances on surfaces of silica have been conducted. For example, Shcherbakova, Belyakova, Aristov, and co-workers (Moscow State University; Institute of Physical Chemistry, the U.S.S.R. Academy of Sciences, Moscow) [13,244–253] investigated the physical adsorption of different substances on silica gels as a function of the degree of coverage of the silica surface with hydroxyl groups. A quantitative dependence was established for the physical adsorption of nitrogen, hexane, benzene, water, methanol, and other substances in the vapor phase on the concentration of the OH groups on the surface of silica. Adsorption due mainly to the dispersion interaction (e.g., hexane) is only slightly dependent on the concentration of OH groups. Adsorption of benzene vapor, on the other hand, is very sensitive to the presence of silanol groups on the silica surface. Adsorption of methanol vapor on SiO_2 depends on the concentration of OH groups on the surface of the adsorbent, too. A decrease in the degree of the surface hydroxylation of silica gel leads to an increase in the chemisorption of water and methanol.

The physical modification of silica surfaces by various regents was investigated too. For example, Bugnina and Serpinsky [254] investigated the adsorption of N_2 and H_2O vapors at two different temperature intervals on macroporous silica gel that was physically modified with dibutyl phthalate (DBPh). When DBPh in the sorption state on SiO_2 surface loses its lateral mobility (at $-194°C$), it in fact modifies the starting silica gel. It changes the gel's porous structure, specific surface area, and nature of the surface. But when DBPh retains its surface mobility (adsorption at room temperature) during the subsequent adsorption of the second adsorbate (H_2O), it acts as one of the components of the binary mixture; these components are simultaneously adsorbed on SiO_2.

Chemical modification of the silica surface was the subject of investigations carried out by Lisichkin, Kudryavtsev, Staroverov, and co-workers (Moscow State University) [255–257]; A. Kiselev, Nikitin, Shcherbakova, and co-workers (Moscow State University) [258–264]; Chuiko, Tertykh, and coworkers (Institute of Surface Chemistry, the Ukrainian S.S.R. Academy of Sciences, Kiev) [88,200–203,265–269]; Neimark, Sheinfain, and co-workers (Institute of Physical Chemistry, the Ukrainian S.S.R. Academy of Sciences, Kiev) [270–273]; Aleskovsky, Kolstov, and co-workers (Technological Institute, Leningrad; Leningrad State University) [274–281]; Yuffa and co-workers (State University, Tyumen) [279,282]; Trofimchuk and co-workers (State University, Kiev) [283]; Fillipov (Institute of Physical Chemistry, the Ukrainian S.S.R. Academy of Sciences, Kiev) [284]; Reikhsfeld and Skvortsov (Technological Institute, Leningrad) [285]; Ermakov et al. (Institute of Catalysis, Siberian Division of the U.S.S.R. Academy of Sciences, Novosibirsk) [286]; Smit et al. (Institute of Organic Chemistry, the U.S.S.R. Academy of Sciences, Moscow) [287]; and others.

The grafting of organic and inorganic compounds and of metalloorganic complexes onto the surface of amorphous silica is a subject of intense research in the former Soviet Union. As is well known, modified silicas are widely used in sorption, catalysis, and chromatography, as fillers for polymers, and as thickening agents in dispersed media.

The first mention of chemical modification of silica in the former Soviet Union goes back to 1950, when A. Kiselev, Dzhigit, and co-workers [248] observed the irreversible adsorption of methanol vapors on silica gel. The mechanism of the chemisorption of methanol on dehydroxylated SiO_2, which involves the breaking of the siloxane bond, was described by Belyakova et al. [251].

Shcherbakova, Petrova, and co-workers [258–263] studied the chemical modification of silica for applications in gas chromatography (e.g., in the adsorption separation of hydrocarbons by gas chromatographic methods). The adsorption properties are investigated as a function of the degree of surface modification with $ClSi(CH_3)_3$. A number of silica samples were chemically modified so that they would have the desired adsorption properties [17]. Chemical modification is an effective means of changing the shape of adsorption isotherms.

Fundamental research into adsorption and chemical modification has helped to elucidate chromatographic processes. This research has led to new developments in gas adsorption chromatography by A. Kiselev, Yashin (Experimental Design Office, Dzerzhinsk), Poshkus (Institute of Chemistry and Chemical Technology, the Lithuanian Academy of Sciences, Vilnius) and co-workers [2,17,288–290]. Together with theoretical investigations and molecular statistical calculations, this method can be used to determine the structural parameters of the adsorbate molecule (this direction of research is referred to as chromatoscopy).

Lisichkin, Kudryavtsev, Staroverov, and co-workers [255–257] investigated the chemical modification of mineral sorbents, in particular, of amorphous silica. The structure of a grafted organic layer depends on the structure of the modifying agent, the structural characteristics of the carrier (SiO_2), the conditions of modification, and so forth. Modified silicas are classified according to two groups: those containing a monomolecular organic layer on the surface of the sample and those containing high-molecular-weight compounds.

The most widely used method of synthesizing modified silicas of the first type is reaction between the silanol groups of SiO_2 surface and organosilicon compounds. Zhuravlev and co-workers [209] showed that, owing to the steric limitations in the modification reactions, not all surface silanol groups enter into the reaction. In a majority of cases, the distribution of the modifying agent on the silica surface during surface coverage is of an island nature. The same is true of modification with chlorosilanes and of reactions of amines and amino acids and with alkyl halides. In practice, there is a range of pore size, and consequently there may be cases in which the grafted layer has a mixed structure: inside the wide pores the grafted molecular chains will be highly mobile (labile structure), whereas inside the narrow pores, owing to steric hindrance, the grafted layer will have a rigid structure.

The introduction of functional groups into a grafted molecule can affect the structure of the grafted layer because of the possible interaction of the functional groups with the residual silanol groups on the silica surface. In this respect, the most often investigated are silicas modified by γ-aminopropyltriethoxysilane. These systems show the presence of hydrogen bonds between the amino groups and the residual OH groups.

When the surface of silica contains high-molecular-weight compounds, the simplest way to make such samples is by the adsorption of organic polymers from solutions. The best results were obtained by using heterochain nitrogen-containing oligomers derived from polyethyleneimine because they are firmly adsorbed on the hydroxylated surface of silica. In general, such silicas are obtained by polymerizing monomers on the surface of unmodified silica or on the surface of silica-containing, chemically grafted initiators of radical polymerization (peroxides, unsaturated compounds of the vinyltrichlorosilane type, etc.).

Radiation can also be used to initiate the polymerization. The modification in such reactions mainly results in island-type surface coverage. In polymerization reactions, often no covalent bond is formed between the polymer and the silica carrier. As a result, the modifying layer is held on the surface by adsorption forces or by geometric factors. With regard to the state of the macromolecules on the silica surface, two cases should be analyzed: the conformation of the adsorbed polymer and the conformation of the polymer chain firmly held by one end of the silica surface. At present, investigations are being carried out on the adsorption on the surface of silica of various types of polymers, in particular polystyrene, polymethylmethacrylate, and so forth. In the general case, the surface of the polymer-coated silica is less uniform compared with analogous silica modified with grafted compounds [255–257].

Surface chemical reactions were employed by Neimark, Sheinfain, and co-workers [270–272] to synthesize a large number of chemically modified silica gels. Approximately 40 types of organosylyl groups on silica surface were obtained, and their adsorption properties were investigated [273].

Chuiko, Tertykh, and co-workers [88,200–203, 265–269,291] studied chemical reactions on the silica surface. As a result, practical applications were found [88,291] for methylaerosils; butoxyaerosils; aerosils modified with monoethanolamine, ethylene glycol, and α-oxypropionic acid; and other highly dispersed organosilicas. To classify reactions on the silica surfaces, use was made of the electrophilic and nucleophilic mechanisms developed by Ingold [292].

Tarasevich and co-workers [293,294] (Institute of Colloid Chemistry and the Chemistry of Water, the Ukrainian S.S.R. Academy of Sciences, Kiev) investigated the formation of a modified polyorganohydridesiloxane layer on the surface of amorphous silica (aerosil, silochrom, etc.). The modifying agents used were polyalkylhydridesiloxanes $(RSiHO)_n$, where $n = 10-15$ and R is CH_3 and C_2H_5. The degree of polymerization was controlled by thermal treatment. The adsorption isotherms of water vapors, benzene, and *n*-hexane were investigated for the modified and unmodified silicas. The modified samples had a hydrophobic surface with a specific heat of wetting of $\leq 20-30$ mJ/m^2. On the basis of the results of this study, inert, heat-stable (up to $\sim 450^\circ C$) chromatographic carriers were developed.

Ganichenko, V. Kiselev, and co-workers (Moscow State University) [295] investigated the adsorption of vapors of water, benzene, and cyclohexane on the chemically modified (methylated) surface of large-pore silica gels. The partial substitution of the surface hydroxyl groups by methyl groups caused a decrease in the adsorption capacity.

The differential heat of adsorption of different vapors on SiO_2 samples subjected to chemical modification (esterification, etc.) was measured by Kuznetsov and co-workers (Moscow State University) [224,225].

Aleskovsky, Kolstov, Volkova, and co-workers investigated the chemical modification of silica surfaces by inorganic compounds [274–281]. The materials obtained are useful in heterogeneous catalysis, electronics, and so forth. The method for molecular lamination was developed [276–278] by interacting volatile and readily hydrolyzing halides of transitional metals with the surface of hydroxylated silica. This method makes it possible to modify the silica surface with mono- and polymolecular layers of V, Cr, Ti, Fe, Al, B, and so forth.

To prepare highly dispersed metallic adsorbents, S. Kiselev and Sokolova (Institute of Physical Chemistry, the U.S.S.R. Academy of Sciences, Moscow) [296] coated the surface of SiO_2 samples with platinum by reacting platinochloric acid with the amino groups of modified aerosils. Subsequent thermal treatment (250°C) resulted in a reduction to platinum by the hydrogen of the amino groups.

ADSORPTION FROM SOLUTIONS

Eltekov, Pavlova, Shikalova, Bogacheva and co-workers (Moscow State University; Institute of Physical Chemistry, the U.S.S.R. Academy of Sciences, Moscow) [17,290,297–307] investigated the effect of the chemical nature and the dimensions of the pores in silica adsorbents on the adsorption from solutions. The adsorption of benzene from *n*-hexane solutions on silica gels with a hydroxylated and a dehydroxylated surface was studied. Dehydroxylation sharply lowered the heat of adsorption of benzene. A comparison was made of the adsorption isotherms for a series of *n*-hexane solutions of aromatic hydrocarbons on hydroxylated silica gel. The intermolecular interaction of aromatic hydrocarbons with this adsorbent is stronger than the interaction of saturated hydrocarbons because aromatic hydrocarbons form hydrogen bonds with the silanol groups on the SiO_2 surface.

The adsorption from three-component solutions (*n*-hexane, benzene, and dioxane) on the hydroxylated silica surface was studied. The effect of the chemical properties of the SiO_2 surface and the nature of the solvent on the adsorption of polymers was investigated. The adsorption of macromolecules on nonporous and fairly large-pore silica is determined by their conformational transformations. The adsorption equilibrium is often established very slowly: the process may take up to several months.

The adsorption isotherms of polystyrene of different molecular weights were determined with CCl_4, toluene, and other solvents for silica gels with different porosity and different degrees of hydroxylation and surface modification. Adsorption of polystyrene from toluene on the hydroxylated surface of SiO_2 was negative, because the solvent molecules enter into a specific interaction with the surface OH groups. The accessibility of the porous silica gel surface to the macromolecules of polystyrene depends on the size of the pores, the molecular mass distribution in the starting polystyrene, and the temperature. Many different high-molecular-weight compositions were investigated, including oligomers, polymers, proteins, and viruses.

Larionov and co-workers (Institute of Physical Chemistry, the U.S.S.R. Academy of Sciences, Moscow) [308–312] carried out systematic theoretical and experimental investigations of the adsorption from liquid solutions of nonelectrolytes on silica adsorbents. They studied the adsorption of individual substances and binary liquid solutions (benzene/carbon tetrachloride, carbon tetrachloride/isooctane, benzene/isooctane, etc.) on SiO_2 samples with different degrees of porosity but identical surface chemical properties. The experimental results were compared with the theoretical calculations carried out by the Gibbs' method. This procedure made it possible to calculate the dependence of the enthalpy, entropy, and free energy of wetting on the concentration and to obtain expressions describing the chemical potential of the components of adsorption solutions. The considerable difference between the activity coefficients of the volume and the adsorption solutions was explained on the basis of the compression of the adsorption solution in the adsorption field. It was shown that the adsorption equilibrium of solutions can be calculated from the adsorption properties of individual substances. For the adsorption of solutions on hydroxylated, nonporous and large-pore SiO_2 samples, the proposed theoretical method for calculating the adsorption equilibrium of solutions gave results in agreement with experimental data.

V. Kiselev, Ganichenko, and co-workers [313,314] investigated the adsorption of different substances from solutions on the surface of silicas with different degrees of hydroxylation. They studied adsorption of phenol from heptane solutions and the adsorption of aliphatic alcohols from solutions of CCl_4. The extent of the hydroxylation has a marked effect on the magnitude of the maximum absorption. This effect was especially noticeable in the adsorption of methanol. The maximum adsorption decreases with an increase in the calcination temperature of SiO_2.

Nikitin, Khokhlova, Voroshilova and co-workers (Moscow State University) [119,264] studied the use of macroporous silica for adsorption from aqueous solutions, chromatographic retention, and covalent immobilization of biopolymers. The porous structures of SiO_2 were examined, including the surface chemistry of silica and the chemical nature of the biopolymers under the conditions necessary for the adsorption–desorption and gel chromatography of macromolecules. The effect of dehydroxylation and chemical modification of silica with organic and inorganic substances on the adsorption of proteins and some other problems were also studied. By using various proteins (urease, ferritin, hemoglobin, and others), they carried out an investigation to ascertain how the amount of covalently bound proteins depends on the geometric parameters of silica gels, with the hydrodynamic radii of the proteins taken into account.

Zuevsky, Kulaev, and co-workers (Physico-Chemical Medicine Research Institute of RSFSR Public Health Ministry, Moscow) [315,316] carried out an investigation of specific silica hemosorbents for the treatment of blood. Such silica hemosorbents are capable of adsorbing cholesterol from the blood. Macroporous silicas were used for this purpose [26,27]. Their surface is accessible to proteins with a molecular mass up to 70,000 daltons. These adsorbents proved more effective than carbon hemosorbents.

ION EXCHANGE

Strazhesko, Vysotskii, Kirichenko, and co-workers (Kiev State University; Institute of Physical Chemistry, the Ukrainian S.S.R. Academy of Sciences, Kiev) [49–55] investigated the mechanism of ion exchange of Na^+, Cs+, Ca^{2+}, Sr^{2+}, Ba^{2+}, and other ions on silica surfaces. Changes in the magnitude of pK_a on the silica surface were determined as a function of surface neutralization. By using the method of ion-exchange sorption of rubidium ions on silica gels, it was found that the isoelectric point lay at pH $\sim$ 1.5. On the basis of these investigations, methods for synthesizing highly selective sorbents and ion exchangers were developed.

Strelko, Kartel, and co-workers (Institute of General and Inorganic Chemistry, the Ukrainian S.S.R. Academy of Sciences, Kiev) investigated cation exchange on dispersed silicas [71,72]. Such a sorption takes place not only at pH > 9 but also in neutral and even acidic solutions.

Using the numerical experimental method, Ulberg and Churaev (Institute of Physical Chemistry, the U.S.S.R. Academy of Sciences, Moscow) [317] calculated the adsorption energy of ions on the molecularly smooth surface of fused quartz. For ions to be adsorbed, the presence of charged centers is essential. In the case of quartz for a silica–electrolyte system, the ionized OH groups (or the $\equiv SiO^-$ groups) with a negative charge act as such centers. The calculations were based on a model for the localized adsorption of hydrogen ions and electrolyte cations on the same charged centers. The experimental values of the cation and hydrogen ion adsorption potential accorded well with the theoretical values that took into account coulombic and other forces.

Bartenev (Institute of Physical Chemistry, the U.S.S.R. Academy of Sciences, Moscow) [318] carried out a systematic investigation of the relaxation and destruction processes in vitreous quartz. The rupture energy of the Si–O bond is less for alkalisilicate glasses than for vitreous quartz. Also, the strength of the Si–O bond is reduced under the influence of hydroxyl groups in the glass.

ACTION OF IONIZING RADIATION AND MECHANICAL ACTIVATION

The effect of different types of ionizing radiation and mechanical activation on silicas was investigated by Krylova and Dolin (Institute of Electrochemistry, the U.S.S.R. Academy of Sciences, Moscow) [319,320]; Strelko et al. (Institute of Physical Chemistry, the Ukrainian S.S.R. Academy of Sciences, Kiev) [321]; V. Kiselev and co-workers (Moscow State University [322,323]; Garibov, Gezalov, and co-workers (Institute of Chemistry, Azerbaijan S.S.R. Academy of Sciences, Baku) [324,325]; Pshezhetskii, Kotov, and co-workers (Physico-Chemical Institute, Moscow) [198,199]; Kazansky, Pariisky, and co-workers (Institute of Organic Chemistry, the U.S.S.R. Academy of Sciences, Moscow) [326–328]; Ermatov and co-workers (Institute of Nuclear Physics, the Kazakh S.S.R. Academy of Sciences, Alma-Ata) [329–331]; Voevodsky (Institute of Chemical Physics, Siberian Division of the U.S.S.R. Academy of Sciences, Novosibirsk) [332]; Krylov, Butyagin, Radtsig, and co-workers (Institute of Chemical Physics, the U.S.S.R. Academy of Sciences, Moscow) [193–197,323, 333]; Khrustalev, Zhuravlev, and co-workers (Institute of Physical Chemistry, the U.S.S.R. Academy of Sciences, Moscow) [208]; Boldyrev (Institute of Catalysis, Siberian Division of the U.S.S.R Academy of Sciences, Novosibirsk) [334]; Spitsyn, Pirogova, Ryabov, and co-workers (Institute of Physical Chemistry, the U.S.S.R. Academy of Sciences, Moscow) [335–338]; Klevikov, Fenelonov, and co-workers (Institute of Catalysis, Siberian Division of the U.S.S.R. Academy of Sciences, Novosibirsk) [339]; and others.

Krylova and Dolin [319,320] investigated the radiolysis of water adsorbed on silica gel induced by γ-radiation (^{60}Co). The radiolysis products, H_2 and CO_2, formed as a result of the energy transfer from the solid adsorbent silica gel to the molecules of the adsorbed water. The oxidation component of radiolysis is spent on the oxidation of trace amounts of organic compounds present on the sample surface; carbon dioxide results. Radiolysis was investigated in relation to the conditions of the preliminary treatment of SiO_2, the amount of adsorbed water, and the intensity of the radiation. The formation of hydrogen from the surface OH groups of silica gel was also possible. A likely mechanism of radiolysis was described.

Tagieva and V. Kiselev [322] showed that when quartz and silica were exposed to γ-radiation (^{60}Co source), a strong dehydroxylation of the surface occurred. The adsorption of water vapor by SiO_2 following such an exposure decreased sharply.

Butyagin [193] and Radtsig and co-workers [194–197] investigated the tribochemical process leading to the formation of different types of structural defects on the silica surface (quartz glass and aerosil). The homolytic breaking of the siloxane bonds is the primary act of mechanical activation. In addition to the paramagnetic centers $\equiv$Si· and $\equiv$SiO· described earlier [193,340], Ratsig and co-workers identified radicals of the type

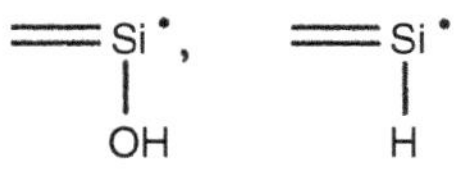

$$=\!\!\!\!=\mathrm{Si}^{\bullet}(\mathrm{OH}),\quad =\!\!\!\!=\mathrm{Si}^{\bullet}(\mathrm{H})$$

(paramagnetic centers), as well as the sililene (=Si:), silanone (=Si=O), and siladioxyrane

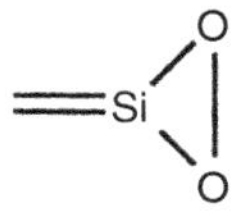

$$=\!\!\!\!=\mathrm{Si}\langle\mathrm{O{-}O}\rangle$$

groups (dimagnetic centers). When the samples were heated in vacuo or in an inert atmosphere, the concentration of the defects decreased and the siloxane bonds were restored. Determinations were made of the conditions for stabilization of the defects and the nature of their transformations. The reactivity of the defects toward different substances in the gas phase was examined for O_2, CO, CO_2, N_2O, H_2, and others. The experimental results (obtained by the EPR spectroscopic method and infrared spectroscopic method) and theoretical calculations made it possible to determine the geometrical and electronic characteristics of different types of silica defects in the ground state and in the excited state and also to make thermochemical calculations for a number of processes.

V. Kiselev, Krasilnikov, and co-workers investigated the physicochemical properties of silica, including tribochemical coercion [164–166,187,341–343]. They found [344] that during the dry grinding of crystalline and amorphous silicas, fine particles became aggregated and the aggregated particles had an ultraporous structure. When grinding was carried out in water, no aggregation took place. The presence of an amorphous layer on the surface of the quartz was established by the differential thermal analysis (DTA) method. The same effects were observed earlier by Khodakov, Rebinder, and co-workers [345–347].

Khurstalev, Zhuravlev, and co-workers [208] investigated the evolution of molecular hydrogen and CO_2 from mechanically activated silica (both crystalline and fused quartz). With the cleaving or scratching of the plate-like samples, there was a reproducible effect of H_2 evolution. As shown earlier by Derjaguin, Krotova, Karasev and co-workers [348,349], from a freshly formed solid surface mechanoelectrons having an energy up to 100 keV, which have a radiochemical effect, are emitted. Experiments were carried out [208] in which samples are subjected to preliminary thermal treatment in vacuo (from room temperature to $\sim$200°C). Thus [4,5], the formation of gaseous H_2 is a result of the dissociative ionization by mechanoelectrons of the residual adsorbed water molecules and silanol groups on the silica surface that involve the breaking of the H–O bond. Such a surface mechanism of hydrogen evolution in the mechanical activation of silica is apparently predominant, as compared to hydrogen evolution from the bulk of a silica sample (from fine pores, inclusions, etc.).

SILICA COATINGS

Pashchenko, Voronkov, Svidersky, and co-workers (Polytechnical Institute, Kiev; Institute of Organic Chemistry, Siberian Division of the U.S.S.R. Academy of Sciences, Irkutsk [350–352] investigated silicon–organic coatings. New types of coatings were developed. The coatings consist of bilayered silicon–organic coatings based on the hydrolysis products of tetraethoxysilane and polyorganohydridesiloxane [352]. The product of hydrolysis of $Si(OC_2H_5)_4$, silicic acid, is strongly adsorbed on the surface of many different materials and interacts with organosilicon compounds containing $\equiv$Si–OH,$\equiv$Si–H, and other groups. The silica surface, which is the substrate, and the hydrophobic layer become chemically bound and form a hydrophobic film. The effect of seawater, organic solvents, and the adhesion of ice on such coatings, as well as thermal and biological stability, were investigated [352].

PROPERTIES OF WATER IN DISPERSED SILICA

Derjaguin, Churaev, Zorin, Muller, Sobolev, and co-workers (Institute of Physical Chemistry, the U.S.S.R. Academy of Sciences, Moscow) [33,353–359] investigated the properties of water in dispersed media and porous bodies, including silica systems. The physicochemical and mechanical properties of dispersed systems depend on the properties of the water present in them. The kinetics of mass-exchange processes is determined by the mobility and energy of the boundary between water and the solid phase. The authors studied the properties of thin layers of water in quartz capillaries [354] and fine-porous glasses [355]. A quantitative investigation was made of the effect of water slipping in micrometer-size quartz capillaries with a smooth surface that was made hydrophobic by methylation.

Within the layers limited by hydrophobic walls, the water molecule dipoles are oriented parallel to the surface. The effect of ordered orientation spreads to a considerable distance; that is, it is of a long-range nature. Such an orientation of water molecules causes a decrease in density near the walls and an increase in the mobility of the molecules in the tangential direction. This situation is interpreted as a decrease in the viscosity of the boundary layers. From a macroscopic point of view, this effect can manifest itself as the slipping of water on the hydrophobic substrate.

In layers bound by the hydrophilic surfaces, the situation changes. Near such a surface, the water dipoles are oriented normal to the surface. This geometry results in an increase in the density of water and a decrease in the tangential mobility of water molecules within layers that are several nanometers thick. From a macroscopic point of view, there should be an increase in the viscosity of the boundary layers of water. With a decrease in the radius of the quartz hydrophilic capillary, the average viscosity of water increases.

Churaev, Zorin, Novikova, and co-workers [356,357] investigated the evaporation of liquids from quartz capillaries ($0.05\ \mu m \le R \le 2\ \mu m$; R is the capillary radius), including the problem of the stability of metastable films on the SiO_2 surface. They also determined the critical pressure at which such films lose their stability, as well as the effect of the hydrophilic nature (the presence of OH groups) of the quartz surface on the film flow. For a hydrophilic surface, there is some increase in the rate of evaporation of water, whereas a hydrophobic surface causes a decrease in the evaporation rate from the capillaries.

Sobolev, Churaev, and co-workers [358] studied the effect of the composition of aqueous electrolyte solutions (KOH, KCl, $BaCl_2$, and others) on the potential and the surface charge of fused quartz. The Shtern potential, ψ, and the charge, σ, of the molecularly smooth virgin surface were determined for solutions of different concentrations and at different pH values.

Gribanova, Chernoberezhsky, Friedrichsberg, and co-workers (Leningrad State University) [360–363] investigated the aggregative stability and electrical conductivity of quartz suspensions in aqueous electrolyte solutions, the dependence of contact angles on quartz and silica glass surfaces on the pH value of solutions, and so forth.

Rebinder (Institute of Physical Chemistry, the U.S.S.R. Academy of Sciences, Moscow) [364] examined how the water layer is bound to the dispersed materials in the course of drying. The isothermic free energy (or characteristic binding strength of water on the surface) of the free water equals zero:

$$-\Delta F = A = RT \ln (p_0/p) \qquad (\text{at } p = p_0)$$

R is the gas constant, and T is the temperature. In the region of physical adsorbed water, the magnitude of A increases continuously. In chemically bound water (OH groups for SiO_2), A will increase abruptly. Such a leapwise increase in A was predicted by Rebinder [364] and experimentally confirmed by Zhuravlev [5].

THEORETICAL INVESTIGATIONS

A. Kiselev, Avgul, Poshkus, and co-workers (Moscow State University) [17,20–22,288,365,366] carried out a theoretical investigation of the thermodynamics of adsorption processes. This work for the first time combined into one thermodynamic analysis the adsorption studies of vapors, gases, pure liquids, and solutions on the surface of solids (including SiO_2) and liquids. A. Kiselev, Poshkus, Lopatkin, and co-workers [17,288,267–370] developed a molecular-statistical theory of adsorption. A. Kiselev, Lopatkin, Bezus, Pham, and co-workers performed molecular-statistical calculations for gas adsorption on silicalite [367,368] and zeolites [369,371,372]. Berezin, A. Kiselev, and co-workers did a thermodynamic analysis of mono- and polymolecular layers on silica surfaces and developed a thermodynamic theory of phase transition of the adsorbate [231–233]. A theory of intermolecular interaction on the surface of solids was developed by A. Kiselev and co-workers [17,20–22,288,365]. Using the quasichemical equilibrium method, A. Kiselev developed a general equation for the adsorption isotherm that takes into account the adsorbate–adsorbent and adsorbate–adsorbate interactions. This equation provides a satisfactory description of all known cases of adsorption on adsorbents with a uniform surface [17,288,365].

Il'in (Moscow State University [373] investigated the nature of the adsorption interactions on the assumption that the adsorbate molecule has a dipole structure and an electrical field is present owing to the charged surface of the adsorbent.

Poshkus (Institute of Physical Chemistry, the U.S.S.R. Academy of Sciences, Moscow) [374] calculated the energy of interaction between the hydroxyl groups of silica and the benzene molecule. The main contribution to the intermolecular interaction between the benzene molecule (locally concentrated electron density due to π-electrons on the periphery of the benzene ring) and the hydroxyl group (two-point model with dipole moment $\mu = 1.6$ D) in the case where the OH group was in a perpendicular direction with respect to the plane of the ring comes from coulomb attraction. Other types of interactions were also considered (induction, dispersion, and repulsion).

Babkin and A. Kiselev (Moscow State University) [375] calculated the dispersion component of the adsorption potential for molecules of benzene and n-hexane for a model of chemically modified ($ClSi(CH_3)_3$) silica. The model consisted of two parts: the silica skeleton and a modifying layer of trimethylsilyl groups. The energy of interaction of the molecules with the adsorbent consists of the energy of interaction with the silica lattice and the energy of interaction with the modifying layer.

Voronkov, Klyuchnikov, and co-workers (Institute of Organic Chemistry, Siberian Division of the U.S.S.R. Academy of Sciences, Irkutsk) [376,377] developed a precise method of combusting organosilicon compounds in a calorimetric bomb to prepare hydrated amorphous silica and measure its standard heat of formation. For samples of SiO_2 that had specific surface areas varied from 140 to 300 m^2/g, the heat of formation was

$$\Delta H^0{}_{f,298.15}SiO_2 = (-0.1681125S - 888.96) \text{ kJ/mol}$$

Lygin and co-workers (Moscow State University) [378–381] used the methods of quantum chemistry in calculating cluster models of the surface structure of silica. The selection of cluster models was based on experimental infrared spectral data. Quantum chemical calculations were made of models describing the defects of the dehydroxylated silica surface that had been thermally treated. Calculations were carried out on models describing the surface structures of water molecules interacting with silanol groups on the silica surface.

Dyatkina, Shchegolev, and co-workers (Institute of General and Inorganic Chemistry, the U.S.S.R. Academy of Sciences, Moscow) [382,383] carried out quantum chemical calculations by using the method of molecular orbits, and they analyzed the peculiarities of the electronic structure of the oxyanion SiO_4^{4-},$Si(OH)_4$, and the fragment [Si–O–Si]. The results of the calculations were used in interpreting the X-ray spectra of amorphous silica, quartz, and stishovite on the basis of the covalent model of their structure. Calculations for the [Si–O–Si] fragment showed the possible angular mobility of this fragment (from 100° to 180°) to be due to the redistribution of electron density. An analysis of the effective charges on Si and O atoms pointed to the covalent nature of the Si–O bond.

Dikov, Dolin, and co-workers (Institute of Geology of Ore Deposits, Petrography, Mineralogy, and Geochemistry, the U.S.S.R. Academy of Sciences, Moscow) [384] carried out an X-ray analysis of the valence bands of the main silicate-forming elements for the most widely distributed types of silicate structures. The experimental data were interpreted on the basis of a structural analysis and with the aid of quantum chemical conceptions within the framework of the theory of molecular orbitals. In silicates, distortions of the silicon–oxygen tetrahedron were present. This distortion was manifested in changes in the Si–O distances and the O–Si–O and Si–O–Si angles.

Grivtsov, Zhuravlev, and co-workers (Institute of Physical Chemistry, the U.S.S.R. Academy of Sciences, Moscow) [385,386] resorted to numerical modeling of molecular dynamics in investigating problems of water adsorption by the hydroxylated surface of the face (0001) in β-tridymite. β-tridymite was chosen as a model form of silica because such a crystalline modification is close in density to that of amorphous silica. The boundary layer in SiO_2 was considered when each surface Si atom held one OH group. The rotational mobility of the hydroxyl groups is an important factor in the adsorption of water.

The development of the theory of surface forces by Derjaguin, Churaev, and Muller [353] made it possible to carry out quantitative estimations of wetting in relation to the properties of the solid support (hydrophilic or hydrophobic surface of SiO_2, in particular) as well as of the effect of the interaction between the wetting substances and the surface.

Rusanov and co-workers (Leningrad State University) [387,388] gave a thermodynamic description of adsorption phenomena for a multicomponent system.

Tovbin (Physico-Chemical Institute, Moscow) [389] gave a theoretical description of the physicochemical processes taking place at the atomic–molecular level on the surface at the gas–solid interface.

Pryanishnikov (State Research Institute of Glass, Moscow) [390] carried out a theoretical analysis of the silica structure and analyzed the experimental data on the structure and properties of various modifications of silica.

Romankov, Lepilin, and co-workers [391,392] investigated adsorption dynamics and developed a general equation describing such a process.

NATURAL MINERAL SORBENTS

Aripov and Agzamkhodzhaev (Institute of Chemistry, the Uzbek S.S.R. Academy of Sciences, Tashkent) [393,394];

Ovcharenko and Tarasevich (Institute of Colloid Chemistry and the Chemistry of Water, the Ukrainian S.S.R. Academy of Sciences, Kiev) [395,396]; Bykov and co-workers (Far-East State University, Vladivostok) [397,398]; Gryazev and co-workers (State University, Saratov) [399,400]; Komarov (Institute of General and Inorganic Chemistry, the Byelorussian S.S.R. Academy of Sciences, Minsk) [401,402]; Tsitsishvili and co-workers (Institute of Physical and Organic Chemistry, the Georgian S.S.R. Academy of Sciences, Tbilisi) [403, 404]; Kerdivarenko, Russu, and co-workers (Chemical Institute, the Moldavian Academy of Sciences, Kishinev) [405,406]; and others investigated natural mineral sorbents that are high in silica content.

REFERENCES

1. Iler, R. K. *The Chemistry of Silica: Solubility, Polymerization, Colloid and Surface Properties and Biochemistry*; Wiley: New York, 1979.
2. Kiselev, A. V.; Yashin Ya. I. *Gas-Adsorption Chromatography* (transl. from Russian; Plenum Press: New York, 1969.
3. Kiselev, A. V.; Lygin, V. I. *Infrared Spectra of Surface Compounds* (transl. from Russian); Wiley: New York, 1975.
4. Zhuravlev, L. T. *Langmuir* **1987**, *3*, 316.
5. Zhuravlev, L. T. *Pure Appl. Chem.* **1989**, *61*, 1969.
6. Dubinin, M. M. In *Main Problems of Physical Adsorption Theory*; Dubinin, M. M.; Serpinskii, V. V., Eds.; Nauka: Moscow, **1970;** p 251.
7. Bering, B. P.; Dubinin, M. M.; Serpinsky, V. V. *J. Colloid Interface Sci.* **1966**, *21*, 378.
8. Voloshchuk, A. M.; Dubinin, M. M.; Zolotarev, P. P. In *Adsorption and Porosity*; Dubinin, M. M.; Serpinskii, V. V., Eds.; Nauka: Moscow, 1976; p 285.
9. Dubinin, M. M.; Polstyanov, E. F. *Zh. Fiz. Khim.* **1966**, *40*, 1169.
10. Dubinin, M. M.; Bering, B. P.; Serpinskii, V. V. In *Recent Progress in Surface Science*; Danielli, J. F.; Pankhurst, K. G. A.; Riddeford, A. C., Eds.; Academic: New York, 1964; p 42.
11. Dubinin, M. M. *Zh. Fiz. Khim.* **1987**, *61*, 1301.
12. Kiselev, A. V. In *Surface Chemical Compounds and Their Role in Adsorption Phenomena*; Kiselev, A. V., Ed.; MGU Press: Moscow, 1957; p 90.
13. Belyakova, L. D.; Kiselev, A. V. *Dokl. Akad. Nauk SSSR* **1958**, *119*, 298.
14. Karnaukhov, A. P.; Kiselev, A. V. *Zh. Fiz. Khim.* **1960**, *34*, 2146.
15. Dzhigit, O. M.; Kiselev, A. V.; Muttik, G. G. *Kolloidn. Zh.* **1961**, *23*, 553.
16. Kiselev, A. V.; Petrova, R. S.; Shcherbakova, K. D. *Kinet. Katal.* **1964**, *5*, 526.
17. Kiselev, A. V. *Intermolecular Interactions in Adsorption and Chromatography*; Vysshaya Shkola: Moscow, 1986.
18. Kiselev, A. V. et al. *Experimental Methods in Adsorption and Molecular Chromatography*; Kiselev, A. V.; Dreving, V. P., Eds.; MGU Press: Moscow, 1973.
19. Kiselev, A. V. *Proc First All-Union Conf. Gas Chromatogr.*; Acad. Sci. Publishers: Moscow, 1960; p 45.
20. Kiselev, A. V. *Zh. Fiz. Khim.* **1961**, *35*, 233.
21. Kiselev, A. V. *Zh. Fiz. Khim.* **1964**, *38*, 2753.
22. Kiselev, A. V. *Discuss Faraday Soc., Intermolecular Forces* **1965**, *40*, 205.
23. Kiselev, A. V. Proc. Third All-Union Conf. Gas Chromatogr.; DF OKBA: Dzerzhinsk, 1966; p 15.
24. Kiselev, A. V. *Zh. Fiz. Khim.* **1967**, *41*, 2470.
25. Kiselev, A. V. *J. Chromatogr.* **1970**, *49*, 84.
26. Kiselev, A. V.; Kustova, G. L.; Lipkind, B. A.; Nikitin, Yu. S. U.S. Patent 3,888,972, 1975.
27. Bebris, N. K.; Kiselev, A. V.; Nikitin, Yu. S.; Yashin, Ya. I. U.S. Patent 3,869,409, 1975.
28. Neimark, I. E. In *Preparation, Structure and Characteristics of Sorbents*; Goskhimizdat: Leningard, 1959; p 112.
29. Neimark, I. E.; Sheinfain, R. Yu. *Silica Gel, Its Preparation, Characteristics and Application*; Naukova Dumka: Kiev, 1973.
30. Sheinfain, R. Yu; Neimark, I. E. *Kinet. Kataliz* **1970**, *8*, 433.
31. Neimark, I. E.; Slinyakova, I. B. *Kolloidn. Zhur.* **1956**, *18*, 219.
32. Neimark, I. E. In *Adsorption and Porosity*; Dubinin, M. M.; Serpinsky, V. V., Eds.; Nauka: Moscow, 1976; p 57.
33. Nikologorskaya, E. A.; Kasaikin, V. A.; Churaev, N. V. *Kolloidn. Zhur.* **1990**, *52*, 489.
34. Deryagin, V. B. *Trans. Faraday Soc.* **1940**, *36*, 203.
35. Deryagin, B. V.; Landau, L. D. *Acta Physicochim. USSR* **1941**, *14*, 633.
36. Deryagin, B. V. *Theory of Colloid Stability and Thin Films*; Nauka: Moscow, 1986.
37. Frolov, Yu. G.; Shabanova, N. A. *Langmuir* **1987**, *3*, 640.
38. Frolov, Yu. G.; Shabanova, N. A.; Savochkina, T. V. *Kolloidn. Zhur.* **1980**, *42*, 1015.
39. Shabanova, N. A.; Frolov, Yu. G. *Izv. Vuzov. Khimiya i Khimich Tekhnologiya* **1985**, *28*, 1.
40. Frolov, Yu. G.; Grodsky, A. S.; Kleschevnikova, S. I.; Pashchenko, L. A.; Rastegina, L. L. In *Preparation and Application of Silica Hydrosols*; Frolov, Yu. G., Ed.; Proceed. of Mendeleev Chemico-Technological Institute: Moscow, 1979; Issue 107, p 31.
41. Frolov, Yu. G.; Khorkin, A. A.; Lebedev, E. N.; Shabanova, N. A.; Kuzmin, M. P.; Reshetnikova, L. V.; Khon, V. N. In *Preparation and Application of Silica Hydrosols*; Frolov, Yu. G., Ed.; Proceed. of Mendeleev Chemico-Technological Institute: Moscow, 1979; Issue 107, p 21.
42. Kontorovich, S. I.; Kononenko, V. G.; Shchukin, E. D. In *Preparation and Application of Silica Hydrosols*; Frolov, Yu. G., Ed.; Moscow, 1979; Issue 107, p 58.
43. Kontorovich, S. I.; Lavrova, K. A.; Plavnik, G. M.; Shchukin, E. D.; Rebinder, P. A. *Dokl. Akad. Nauk USSR* **1971**, *196*, 633.

44. Kontorovich, S. I.; Lavrova, K. A.; Kononenko, V. G.; Shchukin, E. D. *Kolloidn. Zhur.* **1973**, *35*, 1062.
45. Kontorovich, S. I.; Lankin, Ya. I.; Aleshinsky, V. V.; Amelina, E. A.; Shchukin, E. D. *Kolloidn. Zhur.* **1980**, *42*, 639.
46. Yaminsky, V. V.; Pchelin, V. A.; Amelina, E. A.; Shchukin, E. D. *Coagulative Contacts in Dispersed Systems*; Khimiya: Moscow, 1982.
47. Kontorovich, S. I.; Ponomareva, T. P.; Sokolova, L. N.; Sokolova, N. P.; Kochetkova, E. I.; Shchukin, E. D. *Kolloidn. Zhur.* **1988**, *50*, 1100.
48. Rebinder, P. A. *Surface Phenomena in Dispersed Systems: Colloidal Chemistry, Selected Works*; Moscow, 1978.
49. Belyakov, V. N.; Soltievsky, N. M.; Strazhesko, D. N.; Strelko, V. V. *Ukrain. Khimich. Zhur.* **1974**, *40*, 236.
50. Strazhesko, D. N.; Yankovskaya, G. F. *Ukrain. Khimich. Zhur.* **1959**, *25*, 471.
51. Strazhesko, D. N.; Strelko, V. V.; Belyakov, V. N.; Rubanik, S. C. *J. Chromatogr.* **1974**, *102*, 191.
52. Vysotskii, Z. Z.; Strazhesko, D. N. In *Adsorption and Adsorbents*; Strazhesko, D. N., Ed.; Wiley: New York, 1974; p 55.
53. Vysotskii, Z. Z.; Strazhesko, D. N. In *Adsorption and Adsorbents.* Strazhesko, D. N., Eds.; Wiley: New York, 1973; Vol. 1, p 63.
54. Kirichenko, L. F.; Vysotskii, Z. Z. *Dokl. Akad. Nauk USSR* **1967**, *175*, 635.
55. Klimentova, V. P.; Kirichenko, L. F.; Vysotskii, Z. Z. In *Research in Surface Forces*; Deryagin, B. V., Ed.; Consult. Bureau: New York, 1975; Vol. 4, p 77.
56. Chertov, V. M.; Dzhambaeva, D. V.; Plachinda, A. S.; Neimark, I. E. *Dokl. Akad. Nauk USSR* **1965**, *161*, 1149.
57. Chertov, V. M.; Dzhambaeva, D. B.; Neimark, I. E. *Kolloidn. Zhur.* **1965**, *27*, 279.
58. Chertov, V. M.; Neimark, I. E. *Ukrain. Khimich. Zhur.* **1969**, *35*, 499.
59. Kolosentsev, S. D.; Belotserkovsky, G. M. In *Preparation and Application of Silica Hydrosols*; Frolov, Yu. G., Ed.; Proceed. of Mendeleev Chemico-Technological Institute: Moscow, 1979; Issue 107, p 44.
60. Kolosentsev, S. D.; Belotserkovsky, G. M.; Plachenov, T. G. In *Preparation, Structure and Characteristics of Sorbents*; Plachenov, T. G.; Aleksandrov, N. S. Eds.; Proceed. of Leningrad Technological Institute: Leningrad, 1971; Issue 1, p 32.
61. Kolosentsev, S. D.; Belotserkovsky, G. M.; Plachenov, T. G. *Zhur. Prikl. Khim.* **1975**, *48*, 252.
62. Dobruskin, V. Kh.; Belotserkovsky, G. M.; Karelskaya, V. F.; Plachenov, T. G. *Zhur. Prikl. Khim.* **1967**, *40*, 2443.
63. Belotserkovsky, G. M.; Novgorodov, V. N.; Dobruskin, V. Kh.; Plachenov, T. G. *Zhur. Prikl. Khim.* **1969**, *42*, 2749.
64. Belotserkovsky, G. M.; Kolosentsev, S. D. *Preparation of Silica Gel*; Technological Institute: Leningrad, 1975.
65. Kazantseva, L. K.; Kalinin, D. V.; Deniskina, N. D. In *Physico-Chemical Investigations of Sulphuric and Silicate Systems*; Institute of Geology and Geophysics, Siberian Division of the USSR Academy of Sciences: Novosibirsk, 1984; p 59.
66. Ryabenko, E. A.; Kuznetsov, A. I.; Shalumov, B. Z.; Loginov, A. F.; D'yakova, V. V. In *Preparation and Application of Silica Hydrosols*; Frolov, Yu. G., Ed.; Proceed. of Mendeleev Chemico-Technological Institute: Moscow, 1979; Issue 107, p 39.
67. Verwey, E. J. W.; Overbeek, J. Th. G. *Theory of Stability of Lyophobic Colloids*; Elsevier: Amsterdam, 1948.
68. Bird, P. G. U.S. Patent 2,244,325, 1941.
69. Bechtold, M. F.; Snyder, O. E. U.S. Patent 2,574,902, 1951.
70. Lipkind, B. A.; Drozhzhennikov, S. V.; Burylov, V. A.; Tezikov, I. I. In *Preparation and Application of Silica Hydrosols*; Frolov, Yu. G., Ed.; Proceed. of Mendeleev Chemico-Technological Institute: Moscow, 1979; Issue 107, p 26.
71. Laskorin, B. N.; Strelko, V. V.; Strazhesko, D. N.; Denisov, V. I. *Sorbents on Basis of Silica Gels in Radiochemistry*; Atomizdat: Moscow, 1977.
72. Strelko, V. V.; Shvets, D. A.; Kartel, N. T.; Suprunenko, K. A.; Doroshenko, V. I.; Kabakchi, A. M. *Radiation-Chemical Processes in Heterogeneous Systems on Basis of Dispersed Oxides*; Energoizdat: Moscow, 1981.
73. Strelko, V. V. *Adsorption and Adsorbents* **1974**, Kiev: No. 2, p 65.
74. Kiselev, A. V.; Nikitin, Yu. S. In *Modern Problems of Physical Chemistry*; Gerasimov, Ya. I.; Akishin, P. A., Eds.; MGU Press: Moscow, 1968; Vol. 3, p 195.
75. Kiselev, A. V.; Nikitin, Yu. S. In *Chromatography (Itogi nauki i tekhniki)*; VINITI: Moscow, 1978; Vol 2, p 5.
76. Kiselev, A. V.; Nikitin, Yu. S.; Oganesyan, E. B. *Kolloidn. Zhur.* **1969**, *31*, 525.
77. Gorelik, R. L.; Davydov, V. Ya.; Zhuravlev, L. T.; Curthoys, G.; Kiselev, A. V.; Nikitin, Yu. S. *Kolloidn. Zhur.* **1973**, *35*, 456.
78. Bebris, N. K.; Bruk, A. I.; Vyakhirev, D. A.; Kiselev, A. V.; Nikitin, Yu. S. *Kolloidn, Zhur.* **1972**, *34*, 491.
79. Kiselev, A. V.; Nikitin, Yu. S.; Oganesyan, E. B. *Kolloidn. Zhur.* **1967**, *29*, 95.
80. Gorelik, R. L.; Zhuravlev, L. T.; Kiselev, A. V.; Nikitin, Yu. S.; Oganesyan, E. B.; Shengeliya, K. Ya. *Kolloidn. Zhur.* **1971**, *33*, 51.
81. Kiselev, A. V.; Nikitin, Yu. S.; Oganesyan, E. B. *Kolloidn. Zhur.* **1978**, *40*, 37.
82. Simonova, L. G.; Fenelonov, V. B.; Dzisko, V. A.; Kryukova, G. N.; Shmachkova, V. P. *Kinet. Kataliz* **1982**, *23*, 138.
83. Fenelonov, V. B.; Simonova, L. G.; Gavrilov, V. Yu.; Dzisko, V. A. *Kinet. Kataliz* **1982**, *23*, 44.
84. Zhdanov, S. P. In *Surface Chemical Compounds and Their Role in Adsorption Phenomena*; Kiselev, A. V., Eds.; MGU Press: Moscow, 1957; p 129.
85. Zhdanov, S. P. *Summary of Dissertation.* Leningrad, Institute of Silicate Chemistry, the USSR Academy of Sciences, 1952.

86. Zhdanov, S. P.; Koromaldi, E. V. *Izv. Akad. Nauk USSR, Ser. Khim.* **1959**, *4*, 626.
87. Zhdanov, S. P. *Zhur. Vsesoyuzn. Khimich. Obshchestva im. Mendeleeva* **1989**, *34*, 298.
88. Chuiko, A. A. *Teoret. Experim. Khimiya* **1987**, *23*, 597.
89. Grebenshchikov, I. V.; Favorskaya, T. C. *Proceed. of State Optical Institute Leningrad* **1931**, *7*, Issue 72, 1.
90. Grebenshchikov, I. V.; Molchanova, O. S. *Zhur. Obshchey Khim.* **1942**, *12*, 588.
91. Kolikov, V. M.; Mchedlishvili, B. V. In *Chromatography of Biopolymers on Macroporous Silica*; Nauka: Leningrad, 1986; p 189.
92. Ragulin, G. K.; Aleksandrova, N. E.; Dobychin, D. P. *Zhur. Neorganich. Khim.* **1976**, *21*, 2724.
93. Aleksandrova, N. E.; Ragulin, G. K.; Burkat, T. M.; Dobychin, D. P.; Zinyakov, V. M.; Krasii, B. V. In *Investigation of Adsorption Processes and Adsorbents*; Dubinin, M. M., Ed.; Fan: Tashkent, 1979; p 112.
94. Plachenov, T. F.; Filyanskaya, E. D. *Izv. Vuzov USSR* **1958**, *1*, 78.
95. Plachenov, T. G.; Belotserkovsky, G. M.; Karelskaya, V. F. In *Natural Mineral Sorbents*; The Ukrainian SSR Academy of Sci. Publishers: Kiev, 1960; p 43.
96. Plachenov, T. F. In *Main Problems of Physical Adsorption Theory*; Dubinin, M. M.; Serpinskii, V. V., Eds.; Nauka: Moscow, 1970; p 312.
97. Plachenov, T. G. In *Adsorption and Porosity*; Dubinin, M. M.; Serpinskii, V. V., Eds.; Nauka: Moscow, 1976; p 191.
98. Barvinok, G. M.; Sychev, M. M.; Kondratenko, N. E. *Zhur. Prikl. Khim.* **1989**, *4*, 721.
99. Komarov, V. S.; Kuznetsova, T. F. *Vestnik Akad. Nauk Byelorussiun SSR, Ser. Khim.* **1986**, *2*, 26.
100. Kuznetsova, T. F.; Komarov, V. S. *Vestnik Akad. Nauk Byelorussiun SSR, Ser. Khim.* **1986**, *3*, 18.
101. Kuznetsova, T. F.; Komarov, V. S.; Barkatina, E. N. *Vestnik Akad. Nauk Byelorussiun SSR, Ser. Khim.* **1986**, *5*, 24.
102. Kuznetsova, T. F.; Kryukova, E. P.; Barkatina, E. N.; Goryaeva, L. E. *Zhur. Prikl. Khim.* **1990**, *1*, 71.
103. Lipkind, B. A.; Kapatsinsky, S. V.; Kustova, G. L.; Maslova, A. A. In *Preparation, Structure and Characteristics of Sorbents*; Goskhimizdat: Leningrad, 1959; p 156.
104. Bryzgalova, N. I.; Gavrilova, T. B.; Kiselev, A. V.; Khokhlova, T. D. *Neftekhimiya* **1968**, *8*, 915.
105. Bryzgalova, N. I.; Wu Van Tyeu; Gavrilova, T. B.; Kiselev, A. V. *Neftekhimiya* **1969**, *9*, 463.
106. Bakunets, V. V.; Bryzgalova, N. I.; Gavrilova, T. B.; Kiselev, A. V.; Ter-Oganesyan, G. T. In *Analytical Applications of Chromatographic Processes.* NIFKHI Publishers: Moscow, 1976; Issue 26, p 9.
107. Vasiloy, Yu. V.; Demskaya, A. L.; Sokolova, A. P.; Khotimchenko V. S. *Chemistry and Practical Application of Silicon-Organic Compounds.* Theses of papers. Leningrad, 1989, 132.
108. Vasiloy, Yu. V.; Demskaya, A. L.; Kozlova, M. A.; Khotimchenko, V. S.; Khudobina, I. V.; Shkonda, P. A. *Chemistry and Practical Application of Silicon-Organic Compounds.* Theses of papers. Leningrad, 1989, 156.
109. Pryanishnikov, V. P.; Gusynin, V. F.; Sorokin, N. F.; Chepizhny, K. I. *Author. Certificate, Inventions and Discoveries, USSR.* No. 579246; Published in *Invent. Bulletin*, USSR, 1977, No. 41.
110. Guy, A. P.; Zhuravlev, L. T.; Osipov, A. N.; Pryanishnikov, V. P. *Second All-Union Conference on Composition Polymeric Materials.* Theses of papers. Tashkent, 1983 Part I, 90.
111. Belyakova, L. D.; Voloshchuk, A. M.; Guy, A. P.; Zhuravlev, L. T.; Pryanishnikov, V. P.; Shevchenko, T. I. *Izv. Akad. Nauk USSR, Ser. Khim.* **1988**, *4*, 731.
112. Plavnik, G. M.; Khrustaleva, G. N.; Troshkin, G. N.; Guy, A. P. *Izv. Akad. Nauk USSR, Ser. Khim.* **1990**, *5*, 978.
113. Belyakova, L. A.; Il'in, V. G. *Teoret. Experim. Khimiya* **1976**, *12*, 420.
114. Turutina, N. V.; Il'in, V. G.; Kurilenko, M. S. *Teoret. Experim. Khimiya* **1977**, *14*, 656.
115. Il'in, V. G.; Voloshinets, V. G.; Turutina, N. V.; Bobonich, F. M. *Dokl. Akad. Ukrain. Nauk Ukrain. SSR, Ser. B. Geol., Khim., Biolog. nauki*, **1988**, *2*, 43.
116. Danilov, V. V.; Blen, E. V.; Korneev, V. I.; Agafonov, G. I. *Zhur. Prikl. Khim.* **1987**, *7*, 1508.
117. Blen, E. V.; Korneev, V. I.; Danilov, V. V.; Lykov, A. D.; Agafonov, G. I. *Zhur. Prikl. Khim.* **1989**, *7*, 1471.
118. Stishov, S. M.; Popova, S. V. *Geokhimiya* **1961**, *10*, 837.
119. Khokhlova, T. D.; Nikitin, Yu. S.; Voroshilova, O. I. *Zhur. Vsesouzn, Khimich. Obshchestva im. Mendeleeva* **1989**, *34*, 363.
120. Gorelik, R. L.; Zhuravlev, L. T.; Kiselev, A. V. *Kinet. Kataliz* **1971**, *12*, 447.
121. Gorelik, R. L.; Zhdanov, S. P.; Zhuravlev, L. T.; Kiselev, A. V.; Luk'yanovich, V. M.; Nikitin, Yu. S. *Kolloidn. Zhur.* **1972**, *34*, 677.
122. Gorelik, R. L.; Zhdanov, S. P.; Zhuravlev, L. T.; Kiselev, A. V.; Luk'yanovich, V. M.; Malikova, I. Ya.; Nikitin, Yu. S.; Sheshenina, Z. E. *Kolloidn. Zhur.* **1973**, *35*, 911.
123. Burylov, V. A.; Dobruskin, V. Kh.; Belotserkovsky, G. M.; Drozhzhenikov, S. V.; Zolotov, V. T.; Kostina, N. D.; Lipkind, B. A.; Monetov, A. G.; Pishchaev, P. M.; Slepneva, A. T. *Khimich. Promyshlen. Ukrainy* **1969**, *6*, 19.
124. Nikitin, Yu. S. In *Main Problems of Physical Adsorption Theory*; Dubinin, M. M.; Serpinskii, V. V., Eds.; Nauka: Moscow, 1970; p 303.
125. Nikitin, Yu. S. *Summary of Dissertation*; Moscow State University: Moscow, 1975.
126. Lazarev, V. B.; Panasyuk, G. P.; Danchevskaya, M. N.; Budova, G. P. In *Advances in Inorganic Chemistry*; Spitsyn, V. I., Ed.; MIR Publishers: Moscow, 1983; p 196.
127. Dubinin, M. M. *J. Colloid Interface Sci.* **1967**, *23*, 487.
128. Dubinin, M. M. *J. Colloid Interface Sci.* **1974**, *46*, 351.
129. Kiselev, A. V. In *Problems of Kinetics and Catalysis*; Acad. Sci. Publishers: Moscow, 1948; Vol. 5, p 230.
130. Kiselev, A. V. *Zh. Fiz. Khim.* **1949**, *23*, 452.
131. Kiselev, A. V. *Vestn. MGU, Ser. Khim.* **1949**, *11*, 111.

132. Kiselev, A. V. In *Methods of Structure Investigation of High-Dispersed and Porous Solids*; Acad. Sci. Publishers: Moscow, 1953; p 86.
133. Dubinin, M. M.; Yakubov, T. S. *Izv. Akad. Nauk USSR, Ser. Khim.* **1977**, *10*, 2428.
134. Dubinin, M. M. *Prog. Surf. Membr. Sci.* **1975**, *9*, 1.
135. Karnaukhov, A. P.; Kiselev, A. V. *Zh. Fiz. Khim.* **1957**, *31*, 2635.
136. Karnaukhov, A. P. *Kinet. Kataliz* **1971**, *12*, 1235.
137. Karnaukhov, A. P. *Kinet. Kataliz* **1982**, *23*, 1439.
138. Karnaukhov, A. P. In *Adsorption and Porosity*; Dubinin, M. M.; Serpinsky, V. V., Eds.; Nauka: Moscow, 1976; p 7.
139. Dzisko, V. A.; Karnaukhov, A. P.; Tarasova, D. V. *Physico-Chemical Principles of Synthesis of Oxide Catalysts*; Nauka: Novosibirsk, 1978.
140. Gavrilov, V. Yu.; Karnaukhov, A. P.; Fenelonov, V. B. *Kinet. Kataliz* **1978**, *19*, 1549.
141. Aristov, B. G.; Karnaukhov, A. P.; Kiselev, A. V. *Zh. Fiz. Khim.* **1962**, *36*, 2153.
142. Aristov, B. G.; Davydov, V. Ya.; Karnaukhov, A. P.; Kiselev, A. V. *Zh. Fiz. Khim.* **1962**, *36*, 2758.
143. Zolotarev, P. P.; Zhuravlev, L. T.; Ugrozov, V. V. *Kolloidn. Z.* **1984**, *46*, 247.
144. Zolotarev, P. P.; Dubinin, M. M. *Dokl. Akad. Nauk SSSR* **1973**, *210*, 136.
145. Hofmann, U.; Endell, K.; Wilm, D. *Ang. Chem.* **1934**, *47*, 539.
146. Rideal, E. K. *Trans. Faraday Soc.* **1936**, *32*, 4.
147. Kiselev, A. V. *Kolloidn. Zhur.* **1936**, *2*, 17.
148. Yaroslavsky, N. G.; Terenin, A. N. *Dokl. Akad. Nauk USSR* **1949**, *66*, 885.
149. Kurbatov, L. N.; Neuymin, G. G. *Dokl. Akad. Nauk USSR* **1949**, *68*, 34.
150. Yaroslavsky, N. G. *Zhur. Fiz. Khim.* **1950**, *24*, 68.
151. Sidorov, A. N. *Dokl. Akad. Nauk USSR* **1954**, *95*, 1235.
152. Terenin, A. N. In *Surface Chemical Compounds and Their Role in Adsorption Phenomena*; Kiselev, A. V., Ed.; MGU Press: Moscow, 1957; p 206.
153. Nikitin, V. A.; Sidorov, A. N.; Karyakin, A. V. *Zh. Fiz. Khim.* **1956**, *30*, 117.
154. Zhdanov, S. P. *Dokl. Akad. Nauk USSR* **1949**, *68*, 99.
155. Zhdanov, S. P. In *Preparation, Structure and Characteristics of Sorbents*; Goskhimizdat: Leningrad, 1959; p 166.
156. Belyakova, L. D.; Kiselev, A. V. In *Preparation, Structure and Characteristics of Sorbents*; Goskhimizdat: Leningrad, 1959; p 180.
157. Muttik, G. G. In *Preparation, Structure and Characteristics of Sorbents*; Goskhimizdat: Leningrad, 1959; p 193.
158. Belyakova, L. D.; Dzhigit, O. M.; Kiselev, A. V.; Muttik, G. G.; Shcherbakova, K. D. *Zh. Fiz. Khim.* **1959**, *33*, 2624.
159. Zhdanov, S. P. *Zh. Fiz. Khim.* **1962**, *36*, 2098.
160. Zhdanov, S. P.; Kiselev, A. V. *Zh. Fiz. Khim.* **1957**, *31*, 2213.
161. Zhuravlev, L. T.; Kiselev, A. V. *Kolloidn. Zhur.* **1962**, *24*, 22.
162. Zhuravlev, L. T. In *Main Problems of Physical Adsorption Theory*; Dubinin, M. M.; Serpinskii, V. V., Eds.; Nauka: Moscow, 1970; p 309.
163. Zhuravlev, L. T.; Kiselev, A. V. In *Surface Area Determination*; Everett, D. H., Ed.; Butterworth: London, 1970; p 155.
164. Egorow, M. M.; Kvlividze, W. I.; Kiselev, V. F.; Krassilnikow, K. G.; Kolloid, Z. Z. *Polym.* **1966**, *B212*, 126.
165. Ignat'eva, L. A.; Kiselev, V. F.; Chukin, G. D. *Dokl. Akad. Nauk USSR* **1968**, *181*, 914.
166. Ignat'eva, L. A.; Kvlividze, V. I.; Kiselev, V. F. In *Bound Water in Dispersed Systems*; Kiselev, V. F., Kvlividze, V. I., Eds.; MGU Press: Moscow, 1970; Issue 1, p 56.
167. Gorlov, Yu. I.; Golovatyi, V. G.; Konoplya, M. M., Chuiko, A. A. *Teoret. Experim. Khimiya* **1980**, *16*, 202.
168. Galkin, G. A.; Kiselev, A. V.; Lygin, V. I. *Zhur. Fiz. Khim.* **1968**, *42*, 1470.
169. Galkin, G. A.; Kiselev, A. V.; Lygin, V. I. *Zhur. Fiz. Khim.* **1969**, *43*, 1992.
170. Galkin, G. A. *Zhur. Prikl. Spektroskop.* **1975**, *23*, 104.
171. Davydov, V. Ya.; Kiselev, A. V.; Lokutsievsky, V. A.; Lygin, V. I. *Zhur. Fiz. Khim.* **1973**, *47*, 809.
172. Arutyunyan, B. S.; Kiselev, A. V.; Titova, T. I. *Dokl. Akad. Nauk USSR* **1980**, *251*, 1148.
173. Zhdanov, S. P.; Kosheleva, L. A.; Titova, T. I. *Langmuir* **1987**, *3*, 960.
174. Kiselev, A. V.; Lygin, V. I. *Uspekhi Khimii* **1962**, *31*, 351.
175. Galkin, G. A. *Zhur. Prikl. Spektroskop.* **1976**, *24*, 53.
176. Agzamkhodzhaev, A. A.; Galkin, G. A.; Zhuravlev, L. T. In *Main Problems of Physical Adsorption Theory*; Dubinin, M. M., Serpinsky, V. V., Eds.; Nauka: Moscow, 1970; p 168.
177. Davydov, V. Ya.; Kiselev, A. V.; Lygin, V. I. *Kolloidn. Zhur.* **1963**, *25*, 152.
178. Davydov, V. Ya.; Kiselev, A. V.; Lygin, V. I. *Zh. Fiz. Khim.* **1963**, *37*, 469.
179. Davydov, V. Ya.; Kiselev, A. V. *Zh. Fiz. Khim.* **1963**, *37*, 2593.
180. Davydov, V. Ya.; Zhuravlev, L. T.; Kiselev, A. V. *Zh. Fiz. Khim.* **1964**, *38*, 2047.
181. Davydov, V. Ya.; Kiselev, A. V.; Zhuravlev, L. T. *Trans. Faraday Soc.* **1964**, *60*, 2254.
182. Volkov, A. V.; Kiselev, A. V.; Lygin, V. I.; Titova, T. I.; Shchepalin, K. L. *Kolloidn. Zhur.* **1976**, *38*, 32.
183. Kiselev, A. V.; Lygin, V. I.; Shchepalin, K. L. *Kolloidn. Zhur.* **1976**, *38*, 163.
184. Arutyunyan, B. S.; Volodin, V. Ya.; Kiselev, A. V.; Tarasov, N. N.; Titova, T. I. *Kolloidn. Zhur.* **1980**, *42*, 430.
185. Vedeneeva, N. E. In *Surface Chemical Compounds and Their Role in Adsorption Phenomena*; Kiselev, A. V., Ed.; MGU Press: Moscow, 1975; p 243.
186. Musatov, I. K. *Summary of Dissertation*; Institute of Crystallography, the USSR Academy of Sciences: Moscow, 1955.
187. Kiselev, V. F. *Surface Phenomena on Semiconductors and Dielectrics*; Nauka: Moscow, 1970.

188. Golovanova, G. F.; Ivanova, N. N.; Kvlividze, V. I.; Neimark, I. E.; Khrustaleva, S. V.; Chukin, G. D.; Sheinfain, R. Yu. *Teoret. Experim. Khimiya* **1973**, *9*, 383.
189. Chukin, G. D.; Malavich, V. I. *Zhur. Prikl. Spektroskop.* **1976**, *24*, 536.
190. Chukin, G. D.; Malavich, V. I. *Zhur. Struktur. Khim.* **1977**, *18*, 97.
191. Bakaev, V. A.; Pribylov, A. A. *Izv. Akad. Nauk USSR, Ser. Khim.* **1986**, *8*, 1756.
192. Pribylov, A. A.; Bakaev, V. A.; Thamm, H. *Zeolites* **1988**, *8*, 302.
193. Butyagin, P. Yu. *Uspekhi Khimii* **1984**, *53*, 1769.
194. Radtsig, V. A.; Bobyshev, A. A. *Phys. Stat. Sol. (B)* **1986**, *133*, 621.
195. Bobyshev, A. A.; Radtsig, V. A. *Kinet. Kataliz.* **1988**, *29*, 638.
196. Bobyshev, A. A.; Radtsig, V. A. *Khimich. Fizika* **1988**, *7*, 950.
197. Radtsig, V. A.; Senchenya, I. N.; Bobyshev, A. A.; Kazansky, V. B. *Kinet. Kataliz* **1989**, *30*, 1334.
198. Pshezhetskii, S. Ya.; Kotov, A. G.; Milinchuk, V. K.; Roginsky, V. A.; Tupikov, V. I. *EPR of Free Radicals in Radiochemistry*; Khimiya: Moscow, 1972.
199. Shamonina, N. F.; Kotov, A. G.; Pshezhetskii, S. Ya. *Khimiya Vysokikh Energ.* **1971**, *5*, 63.
200. Tertykh, V. A.; Belyakova, L. A. *Zhur. Vsesoyuzn. Khimich. Obshchestva im. Mendeleeva* **1989**, *34*, 395.
201. Tertykh, V. A.; Pavlov, V. V. *Adsorption and Adsorbents* **1978**, *6*, 67.
202. Tertykh, V. A.; Ogenko, V. M. *Teoret. Experim. Khimiya* **1975**, *11*, 827.
203. Yanishpolsky, V. V.; Tertykh, V. A.; Lyubinsky, G. V. *Ukrain. Biokhimich. Zhur.* **1979**, *51*, 324.
204. Sobolev, V. A.; Tertykh, V. A.; Chuiko, A. A. *Zhur. Prikl. Spectroskop.* **1970**, *13*, 646.
205. Sobolev, V. A.; Khoma, M. I.; Furman, V. I.; Ivanov, V. S.; Chebotarev, E. V.; Vatamanyuk, B. I. *Ukr. Khimich. Zhur.* **1976**, *42*, 142.
206. Eremenko, A. M.; Smirnova, N. P.; Tropinov, A. G. *Ukr. Fizich. Zhur.* **1982**, *27*, 1510.
207. Eremenko, A. M. In *Spectroscopy of Molecules and Crystals*; Naukova Dumka: Kiev 1981; Part II, p 137.
208. Guy, A. P.; Glazunov, M. P.; Zhuravlev, L. T.; Khrustalev, Yu. A.; Shengeliya K. Ya. *Dokl. Akad. Nauk USSR* **1984**, *277*, 388.
209. Belyakova, L. D.; Gerasimova, G. A.; Zhuravlev, L. T.; Kudryavtsev, G. V.; Lisichkin, G. V.; Ovchinnikova, N. S.; Platonova, N. P.; Shevchenko, T. I. *Izv. Akad. Nauk USSR, Ser. Khim.* **1989**, *5*, 983.
210. Agzamkhodzhaev, A. A.; Zhuravlev, L. T.; Kiselev, A. F.; Shengeliya, K. Ya. *Izv. Akad. Nauk USSR, Ser. Khim.* **1969**, *10*, 211.
211. Kazansky, V. B. *Kinet. Kataliz.* **1980**, *21*, 159.
212. Kustov, L. M.; Borovkov, V. Yu.; Kazansky, V. B. *J. Catalysis* **1981**, *72*, 149.
213. Kustov, L. M.; Alekseev, A. A.; Borovkov, V. Yu.; Kazansky, V. B. *Dokl. Akad. Nauk USSR* **1981**, *261*, 1374.
214. Kustov, L. M.; Borovkov, V. Yu.; Kazansky, V. B. *Zhur. Fiz. Khim.* **1985**, *59*, 2213.
215. Mastikhin, V. M.; Mudrakovskii, I. L.; Kotsarenko, N. S.; Karakchiev, L. G.; Pelmenshchikov, A. G.; Zamaraev, K. I. *React. Kinet. Catal. Lett.* **1985**, *27*, 447.
216. Bondarenko, A. V.; Kiselev, V. F.; Krasilnikov, K. G. *Kinet. Kataliz.* **1961**, *2*, 590.
217. Akshinskaya, N. V.; Davydov, V. Ya.; Zhuravlev, L. T.; Curthoys, G.; Kiselev, A. V.; Kuznetsov, B. V.; Nikitin, Yu. S.; Rybina, N. V. *Kolloidn. Zhur.* **1964**, *26*, 529.
218. Krasilnikov, K. G.; Kiselev, V. F.; Sysoev, E. A. *Dokl. Akad. Nauk USSR* **1957**, *116*, 990.
219. Dzhigit, O. M.; Kiselev, A. V.; Muttik, G. G. *Kolloidn. Zhur.* **1962**, *24*, 15.
220. Avgul, N. N.; Belyakova, L. D.; Vorob'eva, L. D.; Kiselev, A. V.; Muttik, G. G.; Chistozvonova, O. S.; Checherina, N. Yu. *Kolloidn. Zhur.* **1974**, *36*, 928.
221. Avgul, N. N.; Dzhigit, O. M.; Kiselev, A. V. In *Methods of Catalysts and Catalytic Reactions.* Acad. Sci. Publishers: Novosibirsk, 1965; Vol II, 21.
222. Davydov, V. Ya.; Kiselev, A. V.; Kuznetsov, B. V. *Zhur. Fiz. Khim.* **1965**, *39*, 2058.
223. Ekabson, Ya. Ya.; Kiselev, A. V.; Kuznetsov, B. V.; Nikitin, Yu. S.; *Kolloidn. Zhur.* **1970**, *32*, 41.
224. Kiselev, A. V.; Kuznetsov, B. V.; Lanin, S. N. *Kolloidn. Zhur.* **1976**, *38*, 158.
225. Kiselev, A. V.; Kuznetsov, B. V.; Lanin, S. N. *J. Colloid Interface Sci.* **1979**, *69*, 148.
226. Babkin, I. Yu; Kiselev, A. V. *Zh. Fiz. Khim.* **1963**, *37*, 228.
227. Isirikyan, A. A.; Kiselev, A. V. *Dokl. Akad. Nauk SSSR* **1957**, *115*, 343.
228. Isirikyan, A. A.; Kiselev, A. V.; Frolov, B. A. *Zh. Fiz. Khim.* **1959**, *33*, 389.
229. Nguen, Thi Min Hien; Isirikyan, A. A.; Serpinsky, V. V. *Izv. Akad. Nauk USSR, Ser. Khim.* **1986**, *6*, 1419.
230. Egorov, M. M.; Kiselev, V. F.; Krasilnikov, K. G.; Murina, V. V. *Zh. Fiz. Khim.* **1959**, *33*, 65.
231. Berezin, G. I.; Kiselev, A. V.; Kozlov, A. A.; Kuznetsova, L. V.; Firsova, A. A. *Zh. Fiz. Khim.* **1970**, *44*, 541.
232. Berezin, G. I.; Kiselev, A. V. *J. Colloid Interface Sci.* **1972**, *38*, 227.
233. Berezin, G. I.; Kiselev, A. V.; Sagatelyan, R. T.; Sinitsyn, V. A. *J. Colloid Interface Sci.* **1972**, *38*, 338.
234. Berezin, G. I.; Kiselev, A. V. *J. Chem. Soc. Faraday Trans. 1* **1962**, *78*, 1345.
235. Berezin, G. I.; Kozlov, A. A.; Kuznetsova, L. V. In *Main Problems of Physical Adsorption Theory*; Dubinin, M. M., Serpinsky, V. V., Eds.; Nauka: Moscow: 1970; p 425.
236. Berezin, G. I.; Kiselev, A. V.; Kozlov, A. A.; Kuznetsova, L. V. *Zh. Fiz. Khim.* **1970**, *44*, 1569.
237. Berezin, G. I. *Zh. Fiz. Khim.* **1968**, *42*, 563.
238. Gusev, A. A.; Borovkova, M. A.; Gusev, Yu. A. In *Collection of Post-Graduate Works: Precise Sci., Physics*; State University Press: Kazan, 1977; p 95.
239. Gusev, A. A. *Summary of Dissertation*; State University Press: Kazan, 1979.

240. Zhilenkov, I. V. *Izv. Akad. Nauk USSR, Ser. Khim.* **1957**, 232.
241. Zhilenkov, I. V. In *Main Problems of Physical Adsorption Theory*; Dubinin, M. M.; Serpinsky, V. V., Eds.; Nauka: Moscow, 1970; p 235.
242. Saushkin, V. V.; Zhilenkov, I. V. In *Adsorption in Micropores*; Dubinin, M. M.; Serpinsky, V. V., Eds; Nauka: Moscow, 1983; p 100.
243. Glazun, B. A.; Zhilenkov, I. V.; Rakityanskaya, M. F. *Zhur. Fiz. Khim.* **1969**, *43*, 2397.
244. Kiselev, A. V.; Mikos, N. N.; Romanchuk, M. A.; Shcherbakova, K. D. *Zh. Fiz. Khim.* **1947**, *21*, 1223.
245. Avgul, N. N.; Dzhigit, O. M.; Dreving, V. P.; Gur'ev, M. V.; Kiselev, A. V.; Likhacheva, O. A. *Dokl. Akad. Nauk USSR*, **1951**, *77*, 77.
246. Avgul, N. N.; Dzhigit, O. M.; Isirikyan, A. A.; Kiselev, A. V.; Shcherbakova, K. D. *Dokl. Akad. Nauk USSR* **1951**, *77*, 625.
247. Kiselev, A. V.; Krasilnikov, K. G.; Pokrovsky, N. L.; Avgul, N. N.; Dzhigit, O. M.; Shcherbakova, K. D. *Zh. Fiz. Khim.* **1952**, *26*, 986.
248. Dzhigit, O. M.; Kiselev, A. V.; Mikos-Avgul, N. N.; Shcherbakova, K. D. *Dokl. Akad. Nauk USSR* **1950**, *70*, 441.
249. Avgul, N. N.; Dzhigit, O. M.; Kiselev, A. V.; Shcherbakova, K. D. *Zh. Fiz. Khim.* **1952**, *26*, 977.
250. Belyakova, L. D.; Dzhigit, O. M.; Kiselev, A. V. *Zh. Fiz. Khim.* **1957**, *31*, 1577.
251. Belyakova, L. D.; Kiselev, A. V. *Zh. Fiz. Khim.* **1959**, *33*, 1534.
252. Aristov, B. G.; Kiselev, A. V. *Zh. Fiz. Khim.* **1963**, *37*, 2520.
253. Aristov, B. G.; Kiselev, A. V. *Zh. Fiz. Khim.* **1964**, *38*, 1984.
254. Bugnina, G. A.; Serpinskii, V. V. *Zh. Fiz. Khim.* **1966**, *40*, 887.
255. Lisichkin, G. V.; Kudryavtsev, G. V.; Serdan, A. A.; Staroverov, S. M.; Yuffa, A. Ya. *Modified Silicas in Sorption, Catalysis and Chromatography*; Khimiya: Moscow, 1986.
256. Lisichkin, G. V. *Zhur. Vsesoyuzn. Khimich. Obshchestva im. Mendeleeva* **1989**, *34*, 291.
257. Kudryavtsev, G. V.; Staroverov, S. M. *Zhur. Vsesoyuzn. Khimich. Obshchestva im. Mendeleeva* **1989**, *34*, 308.
258. Kiselev, A. V.; Kovaleva, N. V.; Korolev, A. Ya.; Shcherbakova, K. D. *Dokl. Akad. Nauk USSR* **1959**, *124*, 617.
259. Babkin, I. Yu.; Vasil'eva, V. S.; Drogaleva, I. V.; Kiselev, A. V.; Korolev, A. Ya.; Shcherbakova, K. D. *Dokl. Akad. Nauk USSR* **1959**, *129*, 131.
260. Kiselev, A. V.; Korolev, A. Ya.; Petrova, R. S.; Shcherbakova, K. D. *Kolloidn. Zhur.* **1960**, *22*, 671.
261. Vasil'eva, V. S.; Drogaleva, I. V.; Kiselev, A. V.; Korolev, A. Ya.; Shcherbakova, K. D. *Dokl. Akad. Nauk SSSR* **1961**, *136*, 852.
262. Akshinskaya, N. V.; Kiselev, A. V.; Nikitin, Yu. S.; Petrova, R. S.; Chuikina, V. K.; Shcherbakova, K. D. *Zh. Fiz. Khim.* **1962**, *36*, 1121.
263. Kiselev, A. V.; Petrova, R. S.; Shcherbakova, K. D. *Kinet. Kataliz.* **1964**, *5*, 526.
264. Khokhlova, T. D.; Skoraya, L. A.; Nikitin, Yu. S.; Zinov'eva, M. V. In *Chromatography in Biology and Medicine*; Moscow, 1986; p 186.
265. Brey, V. V.; Gorlov, Yu. I.; Chuiko, A. A. *Teoret. Experim. Khimiya* **1986**, *22*, 378.
266. Tertykh, V. A.; Pavlov, V. V.; Tkachenko, K. I.; Chuiko, A. A. *Teoret. Experim. Khimiya* **1975**, *11*, 174.
267. Tertykh, V. A.; Chuiko, A. A.; Neimark, I. E. *Teoret. Experim. Khimiya* **1965**, *1*, 400.
268. Chuiko, A. A., Guba, G. Ya.; Pavlov, V. V.; Voronin, E. F. *Ukrain. Khimich. Zhur.* **1986**, *52*, 605.
269. Burushkina, T. N.; Chuiko, A. A.; Khaber, N. V.; Manchenko, L. V. *Teoret. Experim. Khimiya* **1968**, *4*, 570.
270. Neimark, I. E.; Sheinfain, R. Yu.; Svintsova, L. G. *Dokl. Akad. Nauk USSR* **1956**, *108*, 87.
271. Neimark, I. E. *Neftekhimiya* **1963**, *3*, 149.
272. Neimark, I. E. In *Adsorption and Porosity*; Dubinin, M. M.; Serpinsky, V. V., Eds.; Nauka: Moscow, 1976; p 27.
273. Neimark, I. E. *Synthetic Mineral Adsorbents and Carriers of Catalysts*; Naukova Dumka: Kiev, 1982.
274. Kolstov, S. I.; Volkova, A. N.; Aleskovsky, V. B. *Zhur. Prikl. Khim.* **1969**, *42*, 73.
275. Kolstov, S. I.; Volkova, A. N.; Aleskovsky, V. B. *Zhur. Prikl. Khim.* **1969**, *42*, 2028.
276. Aleskovsky, V. B. *Zhur. Prikl. Khim.* **1974**, *47*, 2145.
277. Aleskovsky, V. B. *Vestnik Akad. Nauk USSR* **1975**, *6*, 45.
278. Aleskovsky, V. B. *Chemistry of Solids*; Vysshaya Shkola: Moscow, 1978.
279. Aleskovsky, V. B.; Yuffa, A. Ya.; *Zhur. Vsesoyuzn. Khimich. Obshchestva im. Mendeleeva* **1989**, *34*, 317.
280. Postnova, A. M.; Postnov, V. N.; Kolstov, S. I. *Zhur. Prikl. Khim.* **1984**, *57*, 1456.
281. Kovalkov, V. I.; Malygin, A. A.; Kolstov, S. I.; Aleskovsky, V. B. *Zhur. Prikl. Khimii* **1976**, *49*, 2355.
282. Yuffa, A. Ya. In *Chemistry of Heterogenized Compounds*; TGU Press: Tyumen, 1985; p 176.
283. Skopenko, V. V.; Trofimchuk, A. K.; Zaitsev, V. N. *Zhur. Neorganich, Khim.* **1982**, *27*, 2579.
284. Fillipov, A. P. *Teoret. Experim. Khimiya*, **1983**, *4*, 463.
285. Reikhsfeld, V. O.; Skvortsov, N. K. In *Chemistry of Heterogenized Compounds*; TGU Press: Tyumen, 1985; p 111.
286. Ermakov, Yu. I.; Zakharov, V. A.; Kuznetsov, B. N. *Grafted Complexes on Oxide Carriers in Catalysis*; Nauka: Novosibirsk, 1980.
287. Smit, V. A.; Simonyan, S. O.; Tarosov, V. A.; Shashkov, A. S.; Mamyan, S. S.; Gybin, A. S.; Ibragimov, I. I. *Izv. Akad. Nauk USSR, Ser. Khim.* **1988**, *12*, 2796.
288. Kiselev, A. V.; Poshkus, D. P.; Yashin, Ya. I. *Molecular Principles of Adsorption Chromatography*; Khimiya: Moscow, 1986.
289. Kiselev, A. F.; Iogansen, A. V.; Sakodynsky, K. I.; Sakharov, V. M.; Yashin, Ya. I.; Karnaukhov, A. P.; Buyanova, N. E.; Kurkchi, G. A. *Physico-Chemical*

Application of Gas Chromatography; Khimiya: Moscow, 1973.
290. Kiselev, A. V.; Yashin, Ya. I. *Gas- and Liquid-Adsorption Chromatography*; Khimiya: Moscow, 1975.
291. Pavlov, V. V.; Pavlik, G. E.; Khaber, N. V. *Chemistry and Practical Use of Chemical Modified Aerosils*; Znanie: Kiev, 1979.
292. Ingold, C. K. *Structure and Mechanism in Organic Chemistry*, 2nd ed.; Cornell University Press: Ithaca, NY, 1969.
293. Nazarenko, A. V.; Tarasevich, Yu. I.; Bondarenko, S. V., Lantukh, G. V. *Teoret. Experim. Khimiya* **1989**, *6*, 753.
294. Bondarenko, S. V.; Nazarenko, A. V.; Tarasevich, Yu. I. *Zhur. Prikl. Khim.* **1989**, *6*, 1252.
295. Ganichenko, L. G.; Dubinin, M. M.; Zaverina, E. D.; Kiselev, V. F.; Krasilnikov, K. G. *Izv. Akad. Nauk USSR, Ser. Khim.* **1960**, *9*, 1535.
296. Kiselev, S. A.; Sokolova, N. P. *Kolloidn. Zhur.* **1981**, *43*, 165.
297. Kiselev, A. V.; Khopina, V. V.; Eltekov, Yu. A. *Izv. Akad. Nauk SSSR, Ser. Khim.* **1958**, *6*, 664.
298. Kiselev, A. V.; Pavlova, L. F. *Izv. Akad. Nauk SSSR, Ser. Khim.* **1962**, *12*, 2121.
299. Kiselev, A. V.; Pavlova, L. F. *Neftekhimiya* **1962**, *11*, 861.
300. Kiselev, A. V.; Shikalova, I. V. *Kolloidn. Zhur.* **1962**, *24*, 687.
301. Bogacheva, E. K.; Kiselev, A. V.; Nikitin, Yu. S.; Eltekov, Yu. A. *Zh. Fiz. Khim.* **1965**, *39*, 1777.
302. Bogacheva, E. K.; Kiselev, A. V.; Nikitin, Yu. S.; Eltekov, Yu. A. *Vysokomolekular. Soedin.* **1968**, *10A*, 574.
303. Kiselev, A. V.; Shikalova, I. V. *Kolloidn. Zhur.* **1970**, *32*, 702.
304. Zhdanov, S. P.; Kiselev, A. V.; Koromaldi, E. V.; Nazansky, A. S.; Eltekov, Yu. A. *Kolloidn. Zhur.* **1977**, *39*, 354.
305. Eltekov, Yu. A.; Kiselev, A. V. *J. Polymer. Sci., Polymer Symposium* **1977**, *61*, 431.
306. Davydov, V. Ya.; Kiselev, A. V.; Sapozhnikov, Yu. M. *Kolloidn. Zhur.* **1979**, *41*, 333.
307. Chuduk, N. A.; Eltekov, Yu. A.; Kiselev, A. V. *J. Colloid Interface Sci.* **1981**, *84*, 149.
308. Larionov, O. G.; Kurbanbekov, E. In *Physical Adsorption from Multi-Component Phases*; Dubinin, M. M., Serpinsky, V. V., Eds.; Nauka: Moscow, 1972; p 85.
309. Larionov, O. G.; Popov, E. A.; Chmutov, K. V. *Zh. Fiz. Khim.* **1974**, *48*, 2348.
310. Larionov, O. G. In *Physical Adsorption: Proc. Second Czechoslovak Conference on Physical Adsorption, Liblice* **1975**, 43.
311. Kazaryan, S. A.; Larionov, O. G.; Chmutov, K. V.; Yudilevich, M. D. *Zh. Fiz. Khim.* **1973**, *47*, 1619.
312. Kazaryan, S. A.; Larionov, O. G.; Chmutov, K. V.; Yudilevich, M.D. *Zh. Fiz. Khim.* **1973**, *47*, 2170.
313. Kiselev, V. F.; Krasilnikov, K. G. *Zh. Fiz. Khim.* **1958**, *32*, 1435.
314. Ganichenko, L. G.; Kiselev, V. F.; Krasilnikov, K. G. *Dokl. Akad. Nauk USSR* **1959**, *125*, 1277.
315. Zuevsky, V. V.; Kulaev, D. V.; Rabovsky, A. B.; Shmatkov, B. A. *Zhur. Vsesoyuzn, Khimich. Obshchestva im. Mendeleeva* **1989**, *34*, 325.
316. Zuevsky, V. V.; Shmatkov, B. A.; Turaev, A. N. In *Chromatography in Biology and Medicine*; Proceed. of International Symposium: Moscow, 1986; p 107.
317. Ulberg, D. E.; Churaev, N. V. *Kolloidn. Zhur.* **1988**, *50*, 1158.
318. Bartenev, G. M. *Fizika Khimiya Stekla* **1984**, *10*, 41.
319. Krylova, Z. L.; Dolin, P. I. *Kinet. Kataliz.* **1966**, *7*, 977.
320. Krylova, Z. L.; Dolin, P. I. *Khimaya Vysokikh Energii* **1969**, *3*, 152.
321. Strelko, V. V.; Suprunenko, K. A. *Khimiya Vysokikh Energ.* **1968**, *2*, 258.
322. Tagieva, M. M.; Kiselev, V. F. *Zh. Fiz. Khim.* **1961**, *35*, 1381.
323. Kiselev, V. F.; Krylov, O. V. *Electronic Phenomena on Semiconductors and Dielectrics in Adsorption and Catalysis*; Nauka: Moscow, 1979.
324. Garibov, A. A.; Bakirov, M. Ya.; Velibekaro, G. Z.; Elchiev, Ya. M. *Khimiya Vysokikh Energ.* **1984**, *18*, 506.
325. Garibov, A. A.; Gezalov, Kh. B.; Velibekova, G. Z.; Khudiev, A. T.; Ramazanova, M. Kh.; Kasumov, R. D.; Agaev, T. N.; Gasanov, A. M. *Khimiya Vysokikh Energ.* **1987**, *21*, 505.
326. Pariiskii, G. B.; Kazanskii, V. B. *Kinet. Kataliz* **1964**, *5*, 96.
327. Pariiskii, G. B.; Mishchenko, Yu. A; Kazanskii, V. B. *Kinet. Kataliz* **1965**, *6*, 625.
328. Brotikovsky, O. I.; Zhidomirov, G. M.; Kazanskii, V. B.; Mashchenko, A. I.; Shelimov, B. N. *Kinet. Kataliz* **1971**, *12*, 700.
329. Ermatov, S. E. *Izv. Akad. Nauk Kazakh. USSR Ser. fiz.-mat. nauk* **1971**, *2*, 60.
330. Ermatov, S. E.; Vakhabov, M.; Tuseev, T. In *Applied Nuclear Physics, Part I*; Fan: Tashkent, 1973; p 145.
331. Ermatov, S. E. *Radiation Stimulated Adsorption*; Nauka: Alma-Ata, 1973.
332. Voevodsky, V. V. *Physics and Chemistry of Elementary Chemical Processes*; Nauka: Moscow, 1969.
333. Krylov, O. V. *Khimiya Vysokikh Energ.* **1971**, *5*, 179.
334. Boldyrev, V. V. *Experimental Methods in Mechanochemistry of Inorganic Substances*; Nauka: Novosibirsk, 1983.
335. Spitsyn, V. I.; Pirogova, G. N.; Ryabov, A. I.; Kritskaya, V. E.; Naselsky, S. P.; Klimashina, E. V. *Dokl. Akad. Nauk USSR* **1986**, *289*, 1434.
336. Spitsyn, V. I.; Pirogova, G. N.; Ryabov, A. I.; Kritskaya, V. E.; Glazunov, P. Ya.; Naselsky, S. P. *Dokl. Akad. Nauk USSR* **1987**, *293*, 1148.
337. Spitsyn, V. I.; Pirogova, G. N.; Kritskaya, V. E.; Ryabov, A. I. *Izv. Akad. Nauk USSR, Ser. Neorganich Mater.* **1982**, *18*, 74.
338. Pirogova, G. N.; Ryabov, A. I.; Kritskaya, V. E. *Khimiya Vysokikh Energ.* **1988**, *22*, 322.

339. Klevikov, D. P.; Fenelonov, V. B.; Gavrilov, V. Yu; Zolotarevsky, B. V.; Goldenberg, G. I.; Dovbii, Z. A. *Kolloidn. Zhur.* **1989**, *51*, 278.
340. Hochstrasser, G.; Antonini, I. E. *Surface Sci.* **1972**, *32*, 644.
341. Kiselev, V. F. *Zh. Fiz. Khim.* **1960**, *34*, 698.
342. Egorov, M. M.; Kiselev, V. F. *Zh. Fiz. Khim.* **1962**, *36*, 318.
343. K.vlividze, V. I.; Kiselev, V. F. In *Problems of Kinetics and Catalysis*; Nauka: Moscow, 1967; p 302.
344. Egorov, M. M.; Kiselev, V. F.; Krasilnikov, K. G. *Zh. Fiz. Khim.* **1961**, *35*, 2031.
345. Khodakov, G. S.; Rebinder, P. A. *Dokl. Akad. Nauk USSR* **1959**, *127*, 1070.
346. Khodakov, G. S. *Physics of Grinding*; Nauka: Moscow, 1972.
347. Kiselev, V. F.; Krasilnikov, K. G.; Khodakov, G. S. *Dokl. Akad. Nauk USSR* **1960**, *130*, 1273.
348. Krotova, N. A.; Karasev, V. V. *Dokl. Akad. Nauk USSR* **1953**, *92*, 607.
349. Deryagin, V. B.; Krotova, N. A.; Smilga, V. P. *Adhesion of Solids*; Nauka: Moscow, 1973.
350. Paschenko, A. A.; Voronkov, M. G. *Silicon-Organic Protective Coatings*; Tekhnika: Kiev, 1969.
351. Pashchenko, A. A.; Voronkov, M. G.; Mikhailenko, L. A. *Hydrophobic Coatings*; Naukova Dumka: Kiev, 1973.
352. Pashchenko, A. A. et al. *Polyfunctional Element-Organic Coatings*; Pashchenko, A. A., Ed.; Vyshcha Sckola: Kiev, 1987.
353. Deryagin, B. V.; Churaev, N. V.; Muller, V. M. *Surface Forces*; Nauka: Moscow, 1985.
354. Churaev, N. V.; Sobolev, V. D.; Somov, A. N. *J. Colloid Interface Sci.* **1984**, *97*, 574.
355. Deryagin, B. V.; Zheleznyi, B. V.; Zorin, Z. M. In *Surface Forces in Thin Films and Stability of Colloids*; Nauka: Moscow, 1974; p 90.
356. Zorin, Z. M.; Novikova, A. V.; Churaev, N. V. In *Questions of Form Production Physics and Phase Transitions*; Tula, 1973; p 42.
357. Zorin, Z. M.; Novikova, A. V.; Petrov, A. K.; Churaev, N. V. In *Surface Forces in Thin Films and Stability of Colloids*; Nauka: Moscow, 1974; p 94.
358. Nosenko, N. V.; Sergeeva, I. P.; Sobolev, V. D.; Churaev, N. V. *Kolloidn. Zhur.* **1989**, *51*, 786.
359. Deryagin, B. V.; Churaev, N. V. *Wetting Films.* Nauka: Moscow, 1984.
360. Chardymskaya, E. Yu.; Sidorova, M. P.; Friedrichsberg, D. A.; Kulepova, E. V. *Kolloidn. Zhur.* **1986**, *48*, 589.
361. Golikova, E. V.; Gimanova, I. M.; Chernoberezhsky, Yu. M. *Vestnik LGU, Ser. 4* **1986**, *1*, 69.
362. Gribanova, E. V.; Cherkashina, L. M. *Kolloidn. Zhur.* **1989**, *51*, 854.
363. Gribanova, E. V.; Cherkashina, L. M. *Kolloidn. Zhur.* **1989**, *51*, 1069.
364. Rebinder, P. A. In *All-Union Scientific Conference on the Intensification of the Drying Processes and Improving the Quality of Materials in Drying Processes*; Profizdat: Moscow, 1958; p 20.
365. Gerasimov, Ya. I.; Dreving, V. P.; Eremin, E. N.; Kiselev, A. V.; Lebedev, V. P.; Panchenkov, G. M.; Shlygin, A. I. *Course of Physical Chemistry*, 3rd ed.; Gerasimov, Ya. I, Ed.; Khimiya: Moscow, 1973; Vol. 1, Chapters 16–19.
366. Avgul, N. N.; Kiselev, A. V.; Poshkus, D. P. *Adsorption of Gases and Vapours on Homogeneous Surfaces.* Khimiya: Moscow, 1975.
367. Kiselev, A. V.; Lopatkin, A. A.; Shulga, A. A. *Dokl. Akad. Nauk USSR* **1984**, *275*, 916.
368. Kiselev, A. V.; Lopatkin, A. A.; Shulga, A. A. *Zeolites* **1985**, *5*, 261.
369. Bezus, A. G.; Kiselev, A. V.; Lopatkin, A. A.; Quang Du, Pham *J. Chem. Soc. Faraday Trans. 2* **1978**, *74*, 367.
370. Lopatkin, A. A. *Theoretical Principles of Physical Adsorption*; MGU Press: Moscow, 1983.
371. Kiselev, A. V.; Quang Du, Pham. *J. Chem. Soc. Faraday Trans. 2* **1981**, *77*, 1.
372. Kiselev, A. V.; Quang Du, Pham. *J. Chem. Soc. Faraday Trans. 2* **1981**, *77*, 17.
373. Il'in, B. V. *Nature of Adsorption Strengths*; Gostekhizdat: Moscow, 1952.
374. Poshkus, D. P. In *Preparation, Structure and Characteristics of Sorbents*; Goskhimizdat: Leningrad, 1959; p 270.
375. Babkin, I. Yu; Kiselev, A. V. *Dokl. Akad. Nauk USSR* **1959**, *129*, 357.
376. Klyuchnikov, V. A. *Summary of Dissertation*; State University Press: Kalinin, 1979.
377. Klyuchnikov, V. A.; Voronkov, M. G.; Pepekin, V. I.; Popov, V. T.; Balykova, I. A.; Kuz'mina, E. S. *Dokl. Akad. Nauk USSR* **1988**, *298*, 398.
378. Dunken, H.; Lygin, V. I. *Quantenchemie der Adsorption an Festkörperoberflachen*; Verlag Chemie, Weinheim: New York, 1978.
379. Lygin, V. I.; Magomedbekov, G. Kh.; Lygina, I. A. *Zhur. Struktur. Khim.* **1981**, *22*, 156.
380. Lygin, V. I.; Serazetdinov, A. D.; Khlopova, Z. G.; Shchepalin, K. L. In *Catalysis: Fundamental and Applied Researches*; MGU Press: Moscow, 1987; p 223.
381. Lygin, V. I.; Serazetdinov, A. D.; Ryabenko, E. A.; Chertikhina, O. I. *Phys. Stat. Sol. (B)* **1989**, *156*, 205.
382. Shchegolev, B. F.; Dyatkina, M. E. *Zhur. Struk. Khimii* **1974**, *15*, 323.
383. Shchegolev, B. F.; Dyatkina, M. D. *Zhur. Struk. Khimii* **1974**, *15*, 325.
384. Dikov, Yu. P.; Brytov, I. A.; Romashenko, Yu. N.; Dolin, S. P. *Peculiarity of Electronic Structure of Silicates*; Nauka: Moscow, 1979.
385. Grivtsov, A. G.; Zhuravlev, L. T.; Gerasimova, G. A.; Bulatova, I. V.; Khazin, L. G. *Preprint of the Institute of Applied Mathematics, USSR Academy of Sciences,* **1983**, *142.*
386. Grivtsov, A. G.; Zhuravlev, L. T.; Gerasimova, G. A.; Khazin, L. G. *J. Colloid Interface Sci.* **1988**, *126*, 397.
387. Rusanov, A. I. *Phase Equilibrium and Surface Phenomena*; Khimiya: Leningrad, 1967.

388. Rusanov, A. I.; Levichev, S. A.; Zharov, V. T. *Surface Division of Substances*; Khimiya: Leningrad, 1981.
389. Tovbin, Yu. K. *Theory of Physico-Chemical Processes on Boundary of Gas-Solid*; Nauka: Moscow, 1990.
390. Pryanishnikov, V. P. *System of Silica.* Stroyizdat: Leningrad, 1971.
391. Romankov, P. G.; Lepilin, V. N. *Uninterrupted Adsorption of Gases and Vapours*; Leningrad, 1968.
392. Serpionova, E. N. *Industrial Adsorption of Gases and Vapours*; Vysshaya Shkola: Moscow, 1969.
393. Aripov, E. A. *Natural Mineral Sorbents, Their Activation and Modification*; Fan: Tashkent, 1970.
394. Agzamkhodzhaev, A. A. In *Investigation of Adsorption Processes and Adsorbents*; Dubinin, M. M., Ed.; Fan: Tashkent, 1979; p 218.
395. Tarasevich, Yu. I.; Ovcharenko, F. D. *Adsorption on Clayey Minerals*; Naukova Dumka: Kiev, 1975.
396. Tarasevich, Yu. I. *Structure and Surface Chemistry of Schistose Silicates*; Naukova Dumka: Kiev, 1988.
397. Bykov, V. P.; Gerasimova, V. G.; Gritsyuk, A. A.; Iovenko, L. M.; Shcherbatyuk, N. E. In *Natural Sorbents*; Bykov, V. T., Ed.; Nauka: Moscow, 1967; p 104.
398. Bykov, V. T. *Sorbtion Characteristics and Structure of Bleach Soils*; Acad. Sci. Publishers: Vladivostok, 1953.
399. Gryazev, N. N. In *Surface Chemical Compounds and Their Role in Adsorption Phenomena*; Kiselev, A. V., Ed.; MGU Press: Moscow, 1957; p 196.
400. Kiselev, A. V.; Gryazev, N. N. In *Natural Mineral Sorbents*; The Ukrainian SSR Academy of Sci. Publishers: Kiev, 1960; p 24.
401. Komarov, V. S. *Adsorption–Structural, Physico-Chemical and Catalytic Characteristics of Clays of Byelorussia*; Nauka i Tekhnika: Minsk, 1970.
402. Komarov, V. S. In *Investigation of Adsorption Processes and Adsorbents*; Dubinin, M. M., Ed.; Fan: Tashkent, 1979; p 186.
403. Tsitsishvili, G. V.; Shuakrishvili, M. S.; Barnabishvili, D. N. In *Natural Sorbents*; Bykov, V. T., Ed.; Nauka: Moscow, 1967; p 45.
404. Tsitsishvili, G. V. In *Natural Zeolites*; Metsniereba: Tbilisi, 1978.
405. Kerdivarenko, M. A. *Moldavian Natural Adsorbents and Technology of Their Application*; Shteentsa: Kishinev, 1975.
406. Russu, V. I.; Okopnaya, N. T.; Stratulat, G. V.; Ropot, V. M. In *Investigation of Adsorption Processes and Adsorbents*; Dubinin, M. M., Ed.; Fan: Tashkent, 1979; p 257.

Part 11

Analytical Methods

Horacio E. Bergna
DuPont Experimental Station

High quality, cost effective analytical services are essential for the characterization of research samples. Analytical methods include Optical Microscopy, Scanning Electron Microscopy, Transmission Electron Microscopy, Electron Probe X-Ray Microanalysis, Electron Spectroscopy for Chemical Analysis, X-Ray Fluorescence, X-Ray Diffraction, Thermal Analysis (Differential Scanning Calorimetry, DSC, Thermogravimetric Analysis, TGA), and Micro-Fourier Transform Infrared Spectroscopy.

The chapter of this section includes description, in some cases graphical, of Sample Preparation and all the analytical methods mentioned above.

67 Integrated Analytical Methods

Horacio E. Bergna
DuPont Experimental Station

CONTENTS

The means for obtaining high quality, cost effective analytical services are described. These services are essential for accurate and economical analysis with rapid turn around time. Good laboratory practices, including computerized sample tracking, instrument calibration, maintenance of equipment and written documentation of the analytical methods assure high quality of the analytical data.

Analytical services include: optical microscopy, scanning electron microscopy, transmission electron microscopy, electron probe microanalysis, scanning auger microanalysis, electron spectroscopy for chemical analysis, x-ray fluorescence, x-ray diffraction, thermal analysis (DSC, DTA, TGA, TMA) and Micro-Fourier transform infrared spectroscopy.

Sample preparation: One of the most critical steps in any analysis is the proper preparation of the sample for the analysis. Sample preparation methods include: sectioning, grinding, polishing, etching (chemical, plasma and ion), vacuum and glow discharge metallizing, critical point drying, electrochemical drying, electrochemical thinning, ultramicrotomy and provisions for filtering various products.

All phases of the sample preparation should be documented to facilitate reproducibility of future analysis of similar samples.

Precautions should be taken to insure proper identification of samples from the initial receipt to the final return to the originator.

All samples submitted must be accompanied by a material safety data sheet (MSDS) to insure safe handling of the material during the sample preparation.

Materials for Analysis may include: ceramics, metals, plastic, paper, paint, pharmaceuticals, electronic components, thick and thin film circuits and biological materials.

SAMPLE PREPARATION

One of the most critical steps in any analysis is the proper preparation of the sample before analysis.

Sample preparation facilities should include: sectioning, grinding, polishing, etching (chemical plasma & ion), vacuum & glow discharge metallizing, critical point drying, electrochemical thinning, ultramicrotomy and provisions for filtering various materials.

Materials for analysis may include ceramics, plastic, paper, paint, pharmaceuticals, electronic componets, thick and thin film circuits and biological materials.

All phases of the sample preparation should be documented to facilitate reproducibility of future analysis of similar samples

Precautions should be taken to insure proper identification of samples from the initial receipt to the final return of the originator.

OPTICAL MICROSCOPY

Optical microscopy is an efficient and inexpensive means for characterizing the morphological features of a material over a wide range of magnifications.

Bright field, dark field, polarized and phase contrast methods are utilized to depict specific features or optical properties of the material.

The optical investigation provides valuable microstructure information and in addition can be utilized for selecting specific features requiring more detailed analysis.

Video Taping can be utilized for documenting dynamic experiments and to improve the cost effectiveness of documenting extensive microstructure investigations.

In many investigations the optical evaluation is sufficient to resolve the specific problem under investigation, thus eliminating more costly methods of analysis.

Quantitative image analysis: A computer based quantitative image analysis system interfaced to a high quality optical microscope provides a means for counting and sizing features in the optical image. Measurements for Length, Diameter, Area, Longest Dimension and a host of other parameters can be automatically and accurately assessed.

QUANTITATIVE IMAGE ANALYSIS

Quantitative image analysis (QIA) provides a means for automatically and accurately measuring the physical dimensions of specific features in a material.

Measurements for length, diameter, area, longest dimension, perimeter, aspect ratio, sphericity, percent area and a host of other parameters can be obtained provided there is sufficient contrast between the features of interest and the surround.

Analysis can be performed directly with the optical microscope which is an integral part of the computer based system. Images obtained from other imaging systems (scanning electron microscope, transmission electron microscope, electron probe x-ray microanalyzer, etc.) can also be processed with the image analysis system.

A second image analysis system interfaced to a scanning electron microscope/energy dispersive x-ray system allows direct assessment of features based on the features morphology and or chemistry.

TRANSMISSION ELECTRON MICROSCOPY AND SELECTED AREA ELECTRON DIFFRACTION

Transmission electron microscopy (TEM) can be utilized to resolve even finer detail (<10 Å) that can be obtained by optical microscopy or scanning electron microscopy.

Image formation requires the electron beam to penetrate completely through the material being investigated and therefore imposes stringent requirements on the sample preparation methods. Micron has the facilities and expertise to meet these stringent requirements. Ultramicrotomy, Thin Foil and Replication techniques are routinely employed by Micron personnel.

The ability to perform selected area electron diffraction (S.A.E.D.) extends the usefulness of the TEM in that information concerning the crystallographic structure of a material can be obtained and utilized to identify crystalline compounds.

ELECTRON PROBE X-RAY MICROANALYSIS

Electron probe x-ray microanalysis (EPA) provides a means for obtaining quanlitative and quantitative elemental microanalysis within regions as small as a few cubic micrometers.

All elements in the periodic table except hydrogen, helium and lithium can be detected. Detection limits for elements in the mid range of the periodic table $\sim 500 - 1000$ ppm. Detection limits are somewhat poorer for elements at either end of the periodic table.

The electron probe x-ray microanalyzer at Micron Inc. is computer controlled and equipped with both energy dispersive and wave length dispersive x-ray spectrometers thus assuring clients of obtaining cost effective analysis.

Modes of Analysis

Survey scan: Qualitative identification of all elements within the area of interest.

Elemental distribution images: Photographic images depicting variations in elemental concentration as variations in grey level of the image, brightest areas corresponding to areas of highest concentration.

Elemental concentration profiles: Graphical presentation depicting variations in elemental concentration along a given line of the sample.

Quantitative elemental analysis: Raw data corrected for background, absorption, fluorescence and atomic number effects to produce quantitative results with accuracy of 2–3% relative to the amount present.

ESCA

Electron spectroscopy for chemical analysis (ESCA) is an analytical method for characterizing the chemistry of thin (20–30 Å) surface films.

Range of detectable elements: All except hydrogen and helium
Detection limits: Few tenths of one percent
Analyzed region: 1/4 inch diameter area
Depth of analysis: 20–30 Å

Modes of Analysis

Survey scan — Qualitative identification of all elements except hydrogen and helium within the outermost 20–30 Å.

High resolution segmented scans — Analysis for specific elements allows precise measure of peak binding energy (related to chemical state) and integrated peak area (related to atomic concentration). Raw data is corrected for sample charging effects and photoelectric cross-section yield.

Concentration depth profiles — The ability to perform in situ argon ion etching provides a means for characterizing the chemical composition as a function of depth in the sample. Ion etching rate is approximately 20–30 Å/min.

SCANNING AUGER MICROANALYSIS

Scanning auger microanalysis (SAM) is an analytical method for characterizing the surface (20–30 Å) chemistry of materials.

Range of detectable elements — All except hydrogen and helium

Detection limits — Few tenths of one percent

Area sampled — Submicrometer to several millimeters

Depth of analysis — 20–30 Å

Modes of Analysis

Survey scan — Qualitative identification of elements present within the excited volume.

Elemental mapping — Photographic images depicting variations in elemental concentration as variations in grey level of the image.

Concentration depth profiles – The ability to perform in situ argon ion etching provides a means for characterizing the elemental composition as a function of depth in the sample. Sensitivity factors are applied to the raw data to yield atomic concentrations.

X-RAY DIFFRACTION

X-ray diffraction (XRD) is an analytical method for characterizing the crystallographic structure of a material.

Computer matching of the x-ray diffraction pattern with the JCPDS files allows rapid identification of unknown crystalline substances.

Quantitative analysis for specific crystalline phases or compounds can also be obtained.

Computerized x-ray diffractometer methods are routinely employed: However if the amount of sample is limited, x-ray powder camera methods can be employed which requires a very small amount of material.

X-RAY FLUORESCENCE

X-ray fluorescence (XRF) is an analytical method for obtaining qualitative and quantitative elemental analysis of bulk samples (liquids or solids).

The excellent sensitivity of XRF makes it an ideal method for analyzing low concentrations of specific elements in addition to the major and minor constituents.

Computer data reduction software minimizes the need for multiple standards and assures clientele of accurate and cost effective quantitative elemental analysis.

Range of detectable elements — Fluorine through the periodic table.

Detection limits — Low parts per million.

Area sampled — 1/4 × 1/2 in.

Depth of analysis — Several micrometers.

THERMAL ANALYSIS (DSC, TGA)

Differential scanning calorimetry (DSC) measures the rate and degree of heat change as a function of time or temperature.

Thermogravimetric analysis (TGA) monitors the change in mass of a substance as a function of temperature or time as the sample is subjected to a controlled temperature program.

Thermal Properties Which Can Be Measured

Temperatures and heats of fusion, vaporization and crystallization

Heat of reaction, including polymerization, oxidation and combustion

Temperature and heat of decomposition (e.g., dehydration)

Heat of solution, adsorption, or desorption

Specific heat, activation energy, polymer crystallinity, curing time

Material purity, thermal and oxidative stability

Reactivity and reaction rates

Temperature of glass transition, softening, linear coefficient of expansion.

THERMO NICOLET NEXUS/CONTINUUM FTIR SYSTEM

The new FTIR system provides the highest performance available with a wide variety of sampling options.

The Omnic software controls the data collection, data reduction and validation software, ensuring clientele of high quality analytical data.

The ability to simultaneously view the sample and collect the spectrum is a major improvement over previous methods.

The large selection of reference spectra available in the commercial data bases and our internal data base improve the success of identifying unknown components.

The validation software ensures that the instrument is operating within the required specifications.

SCANNING ELECTRON MICROSCOPY AND ENERGY DISPERSIVE X-RAY ANALYSIS

Scanning electron microscopy (SEM) is a means of obtaining high resolution three dimensional like images of solid samples.

Variations in the surface topography of a material are depicted as variations in grey level of the image. Detail as small as 70 Å can be resolved on most samples. Dynamic experiments can be documented with the aid of a conventional video recorder.

Energy dispersive x-ray analysis (EDS) extends the usefulness of the SEM in that elemental analysis can be performed within regions as small as a few cubic micrometers. All elements from fluorine ($Z = 9$) through the periodic table can be detected with sensitivities of a few tenths of one percent.

Modes of Analysis

EDS spectra: Graphical plot of peaks identifying elements detected within the area analyzed. The area analyzed can be adjusted to encompass submicrometer or several millimeters. The intensity of peaks (peak height) is related to the elemental concentration.

BSE & ISC imaging: Sensitive to variations in the average atomic weight of the material. Brightest areas in the image correspond to the highest atomic weight.

Elemental mapping: Photographic images depicting the distribution of the elements of interest. Variations in grey level of the image are related to variations in the elemental concentration. Images can be obtained for all elements of interest.

Concentration profiles: Graphical plot of the elemental concentration as a function of distance along a given line on the sample.

Chemical typing and/or SIZING: Classification of features (phases or particles) according to elemental composition, size or shape.

Standardless semiquantitative analysis: Quick semiquantitative estimate of elemental concentration.

Quantitative analysis: Raw data is corrected for background, absorption,

SYNOPSIS ON ELECTRON BEAM/SOLID INTERACTIONS TO CHARACTERIZE SAMPLES

Abbreviations

Electron Probe x-Ray Microanalysis (EPA)
Scanning Electron Microscopy (SEM)
Energy Dispersive x-Ray Spectroscopy (EDS)
Wavelength Dispersive x-Ray Spectroscopy (WDS)

SEM and EPA

The area of a sample to be examined is irradiated with a finely focused Electron Beam which may be static or swept in a raster across the surface of the sample. The type of signals produced when the electron beam impinges on a sample surface include Auger electrons, secondary electrons (SE), backscattered electrons (BE), characteristic x-rays, and photons of various energies. They are obtained from specific emission volumes within the sample and these approximate depths:

Auger electrons	10A
Secondary electrons	50–500A
Backscattered electrons	2000A
Characteristic x-rays	2–5 μm

The primary signals of interest in the SEM and EPA are:

(1) SE — variation in secondary electron emission that takes place as the electron beam is rastered across the surface of a sample due to differences in surface topography.
(2) BE — variation in backscattered electron emission that takes place as the electron beam is rastered across the surface of a sample due to differences in sample composition (average atomic number).
(3) EDS/WDS — characteristic x-rays which are emitted as a result of a static or rastered electron beam impinging on a sample. The analysis of this radiation by EDS (simultaneous analysis for fluorine to uranium) or WDS (sequential analysis for beryllium to uranium) yields compositional, element specific, information of

both qualitative and quantitative nature. The EDS detector possesses enough sensitivity to provide x-ray spectral data gathering at low electron beam currents but lacks the resolution of a WDS detector leading to possible peak overlaps.

Sample Preparation (General)

Conductive non-volatile samples can be observed directly while nonconductive samples are either carbon or gold/palladium coated, to avoid sample charging, prior to analysis in the SEM or EPA. Sample preparation techniques commonly used, if needed, include:

(1) Surface Analysis
- Critical point drying to retain sample structure
- Plasma etching to reveal subsurface defects/particles
- Chemical etching to show subsurface morphology
- Dispersion for particle size/chemical typing analysis

(2) Cross Sectional Analysis
- Liquid Nitrogen fracturing for fracture surface morphology
- Embedding/polishing for phase identification, coating thickness determinations, elemental spatial distributions mapping, elemental diffusion gradient/penetration determinations, quantitative analysis, and so on.
- Razor sectioning for coarse fiber count
- Microtoming for fiber diameter, aspect ratio, and so on, determinations.

In general, sample preparation is governed by information needed for problem solving.

RESULTS

Quantitative elemental line profiles, showing a vanadium to phosphorus ratio of 1.6, suggest the following compound of the pellets:

VPO_4 and/or $(VO)_2P_2O_7$

with a theoretical composition:

~31.5%V
~19.1%P
~49.4%O

having a vanadium to phosphorus ratio of 1.65. The approximately 80% recovery vs theoretical values are due to the sample porosity.

The silica concentration (sample-200/particle 1) of the total sample particle is approximately 5% with the highest concentration of silica approximately 4 μm from sample OD surface (~20%) and gradually decreasing to zero in approximately 24 μm in the interior of this ~70 μm particle. The core (~22 μm) is practically void of silica (trace).

The line profile spatial resolution is estimated to be between 2–3 μm and the elemental detectability approximately a tenth of a percent for the investigated elements.

ACKNOWLEDGMENT

We want to thank Mr. James Ficca, Jr., President of Micron Inc. of Wilmington, Delaware, for the use of his text.

Index

B

C

D

E

G

H

I

K

L

M

N

Q

R

T

U

Milton Keynes UK
Ingram Content Group UK Ltd.
UKHW050130071024
449327UK00029B/2533

9 780367 577933